Commercial Renovation Costs with RSMeans data

Thomas Lane, Senior Editor

2020
41st annual edition

Chief Data Officer
Noam Reininger

Vice President, Data
Tim Duggan

Principal Engineer
Bob Mewis (1, 4)

Contributing Editors
Brian Adams (21, 22)
Paul Cowan
Christopher Babbitt
Sam Babbitt
Stephen Bell
Michelle Curran
Antonio D'Aulerio (26, 27, 28, 48)

Matthew Doheny (8, 9, 10)
John Gomes (13, 41)
Derrick Hale, PE (2, 31, 32, 33, 34, 35, 44, 46)
Barry Hutchinson
Joseph Kelble (14, 23, 25)
Scott Keller (3, 5)
Charles Kibbee
Gerard Lafond, PE
Thomas Lane (6, 7)
Thomas Lyons
Jake MacDonald (11, 12)
John Melin, P.E.
Elisa Mello
Matthew Sorrentino

Kevin Souza
David Yazbek

Production Manager
Debbie Panarelli

Production
Jonathan Forgit
Sharon Larsen
Sheryl Rose
Janice Thalin

Data Quality Manager
Joseph Ingargiola

Innovation
Ray Diwakar
Kedar Gaikwad
Srini Narla

Cover Design
Blaire Collins

Data Analytics
David Byars
Ellen D'amico
Thomas Hauger
Cameron Jagoe
Matthew Kelliher-Gibson
Renee Rudicil

Numbers in italics are the divisional responsibilities for each editor. Please contact the designated editor directly with any questions.

RSMeans data from Gordian
Construction Publishers & Consultants
1099 Hingham Street, Suite 201
Rockland, MA 02370
United States of America
1.800.448.8182
RSMeans.com

Copyright 2019 by The Gordian Group Inc.
All rights reserved.
Cover photo © iStock.com/Pulvret80

Printed in the United States of America
ISSN 2161-4628
ISBN 978-1-950656-02-8

0040 $319.99 per copy (in United States)
Price is subject to change without prior notice.

Related Data and Services

Our engineers recommend the following products and services to complement *Commercial Renovation Costs with RSMeans data:*

Annual Cost Data Books

2020 Facilities Construction Costs with RSMeans data
2020 Facilities Maintenance & Repair Costs with RSMeans data

Reference Books

Estimating Building Costs
RSMeans Estimating Handbook
Green Building: Project Planning & Estimating
How to Estimate with RSMeans data
Plan Reading & Material Takeoff
Project Scheduling & Management for Construction
Universal Design for Style, Comfort & Safety

Seminars and In-House Training

Facilities Construction Estimating
Training for our online estimating solution
Practical Project Management for Construction Professionals
Scheduling with MSProject for Construction Professionals
Mechanical & Electrical Estimating

RSMeans data

For access to the latest cost data, an intuitive search, and an easy-to-use estimate builder, take advantage of the time savings available from our online application.

To learn more visit: **RSMeans.com/online**

Enterprise Solutions

Building owners, facility managers, building product manufacturers, and attorneys across the public and private sectors engage with RSMeans data Enterprise to solve unique challenges where trusted construction cost data is critical.

To learn more visit: **RSMeans.com/Enterprise**

Custom Built Data Sets

Building and Space Models: Quickly plan construction costs across multiple locations based on geography, project size, building system component, product options, and other variables for precise budgeting and cost control.

Predictive Analytics: Accurately plan future builds with custom graphical interactive dashboards, negotiate future costs of tenant build-outs, and identify and compare national account pricing.

Consulting

Building Product Manufacturing Analytics: Validate your claims and assist with new product launches.

Third-Party Legal Resources: Used in cases of construction cost or estimate disputes, construction product failure vs. installation failure, eminent domain, class action construction product liability, and more.

API

For resellers or internal application integration, RSMeans data is offered via API. Deliver Unit, Assembly, and Square Foot Model data within your interface. To learn more about how you can provide your customers with the latest in localized construction cost data visit: **RSMeans.com/API**

Table of Contents

Foreword

The Value of RSMeans data from Gordian

Since 1942, RSMeans data has been the industry-standard materials, labor, and equipment cost information database for contractors, facility owners and managers, architects, engineers, and anyone else that requires the latest localized construction cost information. More than 75 years later, the objective remains the same: to provide facility and construction professionals with the most current and comprehensive construction cost database possible.

With the constant influx of new construction methods and materials, in addition to ever-changing labor and material costs, last year's cost data is not reliable for today's designs, estimates, or budgets. Gordian's cost engineers apply real-world construction experience to identify and quantify new building products and methodologies, adjust productivity rates, and adjust costs to local market conditions across the nation. This adds up to more than 22,000 hours in cost research annually. This unparalleled construction cost expertise is why so many facility and construction professionals rely on RSMeans data year over year.

About Gordian

Gordian originated in the spirit of innovation and a strong commitment to helping clients reach and exceed their construction goals. In 1982, Gordian's chairman and founder, Harry H. Mellon, created Job Order Contracting while serving as chief engineer at the Supreme Headquarters Allied Powers Europe. Job Order Contracting is a unique indefinite delivery/indefinite quantity (IDIQ) process, which enables facility owners to complete a substantial number of repair, maintenance, and construction projects with a single, competitively awarded contract. Realizing facility and infrastructure owners across various industries could greatly benefit from the time and cost saving advantages of this innovative construction procurement solution, he established Gordian in 1990.

Continuing the commitment to providing the most relevant and accurate facility and construction data, software, and expertise in the industry, Gordian enhanced the fortitude of its data with the acquisition of RSMeans in 2014. And in an effort to expand its facility management capabilities, Gordian acquired Sightlines, the leading provider of facilities benchmarking data and analysis, in 2015.

Our Offerings

Gordian is the leader in facility and construction cost data, software, and expertise for all phases of the building life cycle. From planning to design, procurement, construction, and operations, Gordian's solutions help clients maximize efficiency, optimize cost savings, and increase building quality with its highly specialized data engineers, software, and unique proprietary data sets.

Our Commitment

At Gordian, we do more than talk about the quality of our data and the usefulness of its application. We stand behind all of our RSMeans data— from historical cost indexes to construction materials and techniques—to craft current costs and predict future trends. If you have any questions about our products or services, please call us toll-free at 800.448.8182 or visit our website at gordian.com.

MasterFormat® 2016/ MasterFormat® 2018 Comparison Table

This table compares the 2016 edition of the Construction Specifications Institute's MasterFormat® to the expanded 2018 edition. For your convenience, all revised 2016 numbers and titles are listed along with the corresponding 2018 numbers and titles.

CSI 2016 MF ID	CSI 2016 MF Description	CSI 2018 MF ID	CSI 2018 MF Description
015632	Temporary Security	015733	Temporary Security
019308	Facility Maintenance Equipment	019308	Facilities Maintenance, Equipment
024200	Removal and Salvage of Construction Materials	024200	Removal and Diversion of Construction Materials
040130	Unit Masonry Cleaning	04012052	Cleaning Masonry
068010	Composite Decking	067300	Composite Decking
072127	Reflective Insulation	072153	Reflective Insulation
072610	Above-Grade Vapor Retarders	072613	Above-Grade Vapor Retarders
074473	Metal Faced Panels	074433	Metal Faced Panels
075430	Ketone Ethylene Ester Roofing	075416	Ketone Ethylene Ester Roofing
077280	Vents	077280	Vent Options
087125	Weatherstripping	087125	Door Weatherstripping
087530	Weatherstripping	087530	Window Weatherstripping
096223	Bamboo Flooring	096436	Bamboo Flooring
099103	Paint Restoration	090190	Maintenance of Painting and Coating
102833	Laundry Accessories	102823	Laundry Accessories
117610	Operating Room Equipment	117610	Equipment for Operating Rooms
117710	Radiology Equipment	117710	Equipment for Radiology
122310	Wood Interior Shutters	122313	Wood Interior Shutters
123580	Commercial Kitchen Casework	123539	Commercial Kitchen Casework
124636	Desk Accessories	124113	Desk Accessories
141210	Dumbwaiters	141000	Dumbwaiters
211113	Facility Water Distribution Piping	211113	Facility Fire Suppression Piping
233715	Louvers	233715	Air Outlets and Inlets, HVAC Louvers
260580	Wiring Connections	260583	Wiring Connections
270110	Operation and Maintenance of Communications Systems	270110	Operation and Maintenance of Communication Systems
272123	Data Communications Switches and Hubs	272129	Data Communications Switches and Hubs
283149	Carbon-Monoxide Detection Sensors	284611.21	Carbon-Monoxide Detection Sensors
284621	Fire Alarm	284620	Fire Alarm
316233	Drilled Micropiles	316333	Drilled Micropiles
323420	Fabricated Pedestrian Bridges	323413	Fabricated Pedestrian Bridges
337543	Shunt Reactors	337253	Shunt Reactors
350100	Operation and Maint. of Waterway & Marine Construction	350100	Operation and Maintenance of Waterway and Marine Construction

How the Cost Data Is Built: An Overview

Unit Prices*

All cost data have been divided into 50 divisions according to the MasterFormat® system of classification and numbering.

Assemblies*

The cost data in this section have been organized in an "Assemblies" format. These assemblies are the functional elements of a building and are arranged according to the 7 elements of the UNIFORMAT II classification system. For a complete explanation of a typical "Assembly", see "RSMeans data: Assemblies—How They Work."

Residential Models*

Model buildings for four classes of construction—economy, average, custom, and luxury—are developed and shown with complete costs per square foot.

Commercial/Industrial/Institutional Models*

This section contains complete costs for 77 typical model buildings expressed as costs per square foot.

Green Commercial/Industrial/Institutional Models*

This section contains complete costs for 25 green model buildings expressed as costs per square foot.

References*

This section includes information on Equipment Rental Costs, Crew Listings, Historical Cost Indexes, City Cost Indexes, Location Factors, Reference Tables, and Change Orders, as well as a listing of abbreviations.

- **Equipment Rental Costs:** Included are the average costs to rent and operate hundreds of pieces of construction equipment.
- **Crew Listings:** This section lists all the crews referenced in the cost data. A crew is composed of more than one trade classification and/or the addition of power equipment to any trade classification. Power equipment is included in the cost of the crew. Costs are shown both with bare labor rates and with the installing contractor's overhead and profit added. For each, the total crew cost per eight-hour day and the composite cost per labor-hour are listed.

Unit Cost data

Assembly Cost data

Square Foot Models

- **Historical Cost Indexes**: These indexes provide you with data to adjust construction costs over time.
- **City Cost Indexes**: All costs in this data set are U.S. national averages. Costs vary by region. You can adjust for this by CSI Division to over 730 cities in 900+ 3-digit zip codes throughout the U.S. and Canada by using this data.
- **Location Factors**: You can adjust total project costs to over 730 cities in 900+ 3-digit zip codes throughout the U.S. and Canada by using the weighted number, which applies across all divisions.
- **Reference Tables**: At the beginning of selected major classifications in the Unit Prices are reference numbers indicators. These numbers refer you to related information in the Reference Section. In this section, you'll find reference tables, explanations, and estimating information that support how we develop the unit price data, technical data, and estimating procedures.
- **Change Orders**: This section includes information on the factors that influence the pricing of change orders.

- **Abbreviations**: A listing of abbreviations used throughout this information, along with the terms they represent, is included.

Index (printed versions only)

A comprehensive listing of all terms and subjects will help you quickly find what you need when you are not sure where it occurs in MasterFormat®.

Conclusion

This information is designed to be as comprehensive and easy to use as possible.

The Construction Specifications Institute (CSI) and Construction Specifications Canada (CSC) have produced the 2018 edition of MasterFormat®, a system of titles and numbers used extensively to organize construction information.

All unit prices in the RSMeans cost data are now arranged in the 50-division MasterFormat® 2018 system.

---

* Not all information is available in all data sets

Note: The material prices in RSMeans cost data are "contractor's prices." They are the prices that contractors can expect to pay at the lumberyards, suppliers'/distributors' warehouses, etc. Small orders of specialty items would be higher than the costs shown, while very large orders, such as truckload lots, would be less. The variation would depend on the size, timing, and negotiating power of the contractor. The labor costs are primarily for new construction or major renovation rather than repairs or minor alterations. With reasonable exercise of judgment, the figures can be used for any building work.

Estimating with RSMeans data: Unit Prices

Following these steps will allow you to complete an accurate estimate using RSMeans data: Unit Prices.

1. Scope Out the Project

- Think through the project and identify the CSI divisions needed in your estimate.
- Identify the individual work tasks that will need to be covered in your estimate.
- The Unit Price data have been divided into 50 divisions according to CSI MasterFormat® 2018.
- In printed versions, the Unit Price Section Table of Contents on page 1 may also be helpful when scoping out your project.
- Experienced estimators find it helpful to begin with Division 2 and continue through completion. Division 1 can be estimated after the full project scope is known.

2. Quantify

- Determine the number of units required for each work task that you identified.
- Experienced estimators include an allowance for waste in their quantities. (Waste is not included in our Unit Price line items unless otherwise stated.)

3. Price the Quantities

- Use the search tools available to locate individual Unit Price line items for your estimate.
- Reference Numbers indicated within a Unit Price section refer to additional information that you may find useful.
- The crew indicates who is performing the work for that task. Crew codes are expanded in the Crew Listings in the Reference Section to include all trades and equipment that comprise the crew.
- The Daily Output is the amount of work the crew is expected to complete in one day.
- The Labor-Hours value is the amount of time it will take for the crew to install one unit of work.
- The abbreviated Unit designation indicates the unit of measure upon which the crew, productivity, and prices are based.
- Bare Costs are shown for materials, labor, and equipment needed to complete the Unit Price line item. Bare costs do not include waste, project overhead, payroll insurance, payroll taxes, main office overhead, or profit.
- The Total Incl O&P cost is the billing rate or invoice amount of the installing contractor or subcontractor who performs the work for the Unit Price line item.

4. Multiply

- Multiply the total number of units needed for your project by the Total Incl O&P cost for each Unit Price line item.
- Be careful that your take off unit of measure matches the unit of measure in the Unit column.
- The price you calculate is an estimate for a completed item of work.
- Keep scoping individual tasks, determining the number of units required for those tasks, matching each task with individual Unit Price line items, and multiplying quantities by Total Incl O&P costs.
- An estimate completed in this manner is priced as if a subcontractor, or set of subcontractors, is performing the work. The estimate does not yet include Project Overhead or Estimate Summary components such as general contractor markups on subcontracted work, general contractor office overhead and profit, contingency, and location factors.

5. Project Overhead

- Include project overhead items from Division 1-General Requirements.
- These items are needed to make the job run. They are typically, but not always, provided by the general contractor. Items include, but are not limited to, field personnel, insurance, performance bond, permits, testing, temporary utilities, field office and storage facilities, temporary scaffolding and platforms, equipment mobilization and demobilization, temporary roads and sidewalks, winter protection, temporary barricades and fencing, temporary security, temporary signs, field engineering and layout, final cleaning, and commissioning.
- Each item should be quantified and matched to individual Unit Price line items in Division 1, then priced and added to your estimate.
- An alternate method of estimating project overhead costs is to apply a percentage of the total project cost—usually 5% to 15% with an average of 10% (see General Conditions).
- Include other project related expenses in your estimate such as:
  - Rented equipment not itemized in the Crew Listings
  - Rubbish handling throughout the project (see section 02 41 19.19)

6. Estimate Summary

- Include sales tax as required by laws of your state or county.
- Include the general contractor's markup on self-performed work, usually 5% to 15% with an average of 10%.
- Include the general contractor's markup on subcontracted work, usually 5% to 15% with an average of 10%.
- Include the general contractor's main office overhead and profit:
  - RSMeans data provides general guidelines on the general contractor's main office overhead (see section 01 31 13.60 and Reference Number R013113-50).
  - Markups will depend on the size of the general contractor's operations, projected annual revenue, the level of risk, and the level of competition in the local area and for this project in particular.
- Include a contingency, usually 3% to 5%, if appropriate.
- Adjust your estimate to the project's location by using the City Cost Indexes or the Location Factors in the Reference Section:
  - Look at the rules in "How to Use the City Cost Indexes" to see how to apply the Indexes for your location.
  - When the proper Index or Factor has been identified for the project's location, convert it to a multiplier by dividing it by 100, then multiply that multiplier by your estimated total cost. The original estimated total cost will now be adjusted up or down from the national average to a total that is appropriate for your location.

Editors' Note:
We urge you to spend time reading and understanding the supporting material. An accurate estimate requires experience, knowledge, and careful calculation. The more you know about how we at RSMeans developed the data, the more accurate your estimate will be. In addition, it is important to take into consideration the reference material such as Equipment Listings, Crew Listings, City Cost Indexes, Location Factors, and Reference Tables.

How to Use the Cost Data: The Details

What's Behind the Numbers? The Development of Cost Data

RSMeans data engineers continually monitor developments in the construction industry in order to ensure reliable, thorough, and up-to-date cost information. While overall construction costs may vary relative to general economic conditions, price fluctuations within the industry are dependent upon many factors. Individual price variations may, in fact, be opposite to overall economic trends. Therefore, costs are constantly tracked and complete updates are performed yearly. Also, new items are frequently added in response to changes in materials and methods.

Costs in U.S. Dollars

All costs represent U.S. national averages and are given in U.S. dollars. The City Cost Index (CCI) with RSMeans data can be used to adjust costs to a particular location. The CCI for Canada can be used to adjust U.S. national averages to local costs in Canadian dollars. No exchange rate conversion is necessary because it has already been factored in.

G The processes or products identified by the green symbol in our publications have been determined to be environmentally responsible and/or resource-efficient solely by RSMeans data engineering staff. The inclusion of the green symbol does not represent compliance with any specific industry association or standard.

Material Costs

RSMeans data engineers contact manufacturers, dealers, distributors, and contractors all across the U.S. and Canada to determine national average material costs. If you have access to current material costs for your specific location, you may wish to make adjustments to reflect differences from the national average. Included within material costs are fasteners for a normal installation. RSMeans data engineers use manufacturers' recommendations, written specifications, and/or standard construction practices for the sizing and spacing of fasteners. Adjustments to material costs may be required for your specific application or location. The manufacturer's warranty is assumed. Extended warranties are not included in the material costs. **Material costs do not include sales tax.**

Labor Costs

Labor costs are based upon a mathematical average of trade-specific wages in 30 major U.S. cities. The type of wage (union, open shop, or residential) is identified on the inside back cover of printed publications or selected by the estimator when using the electronic products. Markups for the wages can also be found on the inside back cover of printed publications and/or under the labor references found in the electronic products.

- If wage rates in your area vary from those used, or if rate increases are expected within a given year, labor costs should be adjusted accordingly.

Labor costs reflect productivity based on actual working conditions. In addition to actual installation, these figures include time spent during a normal weekday on tasks, such as material receiving and handling, mobilization at the site, site movement, breaks, and cleanup.

Productivity data is developed over an extended period so as not to be influenced by abnormal variations and reflects a typical average.

Equipment Costs

Equipment costs include not only rental but also operating costs for equipment under normal use. The operating costs include parts and labor for routine servicing, such as the repair and replacement of pumps, filters, and worn lines. Normal operating expendables, such as fuel, lubricants, tires, and electricity (where applicable), are also included. Extraordinary operating expendables with highly variable wear patterns, such as diamond bits and blades, are excluded. These costs are included under materials. Equipment rental rates are obtained from industry sources throughout North America—contractors, suppliers, dealers, manufacturers, and distributors.

Rental rates can also be treated as reimbursement costs for contractor-owned equipment. Owned equipment costs include depreciation, loan payments, interest, taxes, insurance, storage, and major repairs.

Equipment costs do not include operators' wages.

Equipment Cost/Day—The cost of equipment required for each crew is included in the Crew Listings in the Reference Section (small tools that are considered essential everyday tools are not listed out separately). The Crew Listings itemize specialized tools and heavy equipment along with labor trades. The daily cost of itemized equipment included in a crew is based on dividing the weekly bare rental rate by 5 (number of working days per week), then adding the hourly operating cost times 8 (the number of hours per day). This Equipment Cost/Day is shown in the last column of the Equipment Rental Costs in the Reference Section.

Mobilization, Demobilization—The cost to move construction equipment from an equipment yard or rental company to the job site and back again is not included in equipment costs. Mobilization (to the site) and demobilization (from the site) costs can be found in the Unit Price Section. If a piece of equipment is already at the job site, it is not appropriate to utilize mobilization or demobilization costs again in an estimate.

Overhead and Profit

Total Cost including O&P for the installing contractor is shown in the last column of the Unit Price and/or Assemblies. This figure is the sum of the bare material cost plus 10% for profit, the bare labor cost plus total overhead and profit, and the bare equipment cost plus 10% for profit. Details for the calculation of overhead and profit on labor are shown on the inside back cover of the printed product and in the Reference Section of the electronic product.

General Conditions

Cost data in this data set are presented in two ways: Bare Costs and Total Cost including O&P (Overhead and Profit). General Conditions, or General Requirements, of the contract should also be added to the Total Cost including O&P when applicable. Costs for General Conditions are listed in Division 1 of the Unit Price Section and in the Reference Section.

General Conditions for the installing contractor may range from 0% to 10% of the Total Cost including O&P. For the general or prime contractor, costs for General Conditions may range from 5% to 15% of the Total Cost including O&P, with a figure of 10% as the most typical allowance. If applicable, the Assemblies and Models sections use costs that include the installing contractor's overhead and profit (O&P).

Factors Affecting Costs

Costs can vary depending upon a number of variables. Here's a listing of some factors that affect costs and points to consider.

Quality—The prices for materials and the workmanship upon which productivity is based represent sound construction work. They are also in line with industry standard and manufacturer specifications and are frequently used by federal, state, and local governments.

Overtime—We have made no allowance for overtime. If you anticipate premium time or work beyond normal working hours, be sure to make an appropriate adjustment to your labor costs.

Productivity—The productivity, daily output, and labor-hour figures for each line item are based on an eight-hour work day in daylight hours in moderate temperatures and up to a 14' working height unless otherwise indicated. For work that extends beyond normal work hours or is performed under adverse conditions, productivity may decrease.

Size of Project—The size, scope of work, and type of construction project will have a significant impact on cost. Economies of scale can reduce costs for large projects. Unit costs can often run higher for small projects.

Location—Material prices are for metropolitan areas. However, in dense urban areas, traffic and site storage limitations may increase costs. Beyond a 20-mile radius of metropolitan areas, extra trucking or transportation charges may also increase the material costs slightly. On the other hand, lower wage rates may be in effect. Be sure to consider both of these factors when preparing an estimate, particularly if the job site is located in a central city or remote rural location. In addition, highly specialized subcontract items may require travel and per-diem expenses for mechanics.

Other Factors—

- season of year
- contractor management
- weather conditions
- local union restrictions
- building code requirements
- availability of:
  - adequate energy
  - skilled labor
  - building materials
- owner's special requirements/restrictions
- safety requirements
- environmental considerations
- access

Unpredictable Factors—General business conditions influence "in-place" costs of all items. Substitute materials and construction methods may have to be employed. These may affect the installed cost and/or life cycle costs. Such factors may be difficult to evaluate and cannot necessarily be predicted on the basis of the job's location in a particular section of the country. Thus, where these factors apply, you may find significant but unavoidable cost variations for which you will have to apply a measure of judgment to your estimate.

Rounding of Costs

In printed publications only, all unit prices in excess of $5.00 have been rounded to make them easier to use and still maintain adequate precision of the results.

How Subcontracted Items Affect Costs

A considerable portion of all large construction jobs is usually subcontracted. In fact, the percentage done by subcontractors is constantly increasing and may run over 90%. Since the workers employed by these companies do nothing else but install their particular products, they soon become experts in that line. As a result, installation by these firms is accomplished so efficiently that the total in-place cost, even with the general contractor's overhead and profit, is no more, and often less, than if the principal contractor had handled the installation. Companies that deal with construction specialties are anxious to have their products perform well and, consequently, the installation will be the best possible.

Contingencies

The allowance for contingencies generally provides for unforeseen construction difficulties. On alterations or repair jobs, 20% is not too much. If drawings are final and only field contingencies are being considered, 2% or 3% is probably sufficient and often nothing needs to be added. Contractually, changes in plans will be covered by extras. The contractor should consider inflationary price trends and possible material shortages during the course of the job. These escalation factors are dependent upon both economic conditions and the anticipated time between the estimate and actual construction. If drawings are not complete or approved, or a budget cost is wanted, it is wise to add 5% to 10%. Contingencies, then, are a matter of judgment.

Important Estimating Considerations

The productivity, or daily output, of each craftsman or crew assumes a well-managed job where tradesmen with the proper tools and equipment, along with the appropriate construction materials, are present. Included are daily set-up and cleanup time, break time, and plan layout time. Unless otherwise indicated, time for material movement on site (for items that can be transported by hand) of up to 200' into the building and to the first or second floor is also included. If material has to be transported by other means, over greater distances, or to higher floors, an additional allowance should be considered by the estimator.

While horizontal movement is typically a sole function of distances, vertical transport introduces other variables that can significantly impact productivity. In an occupied building, the use of elevators (assuming access, size, and required protective measures are acceptable) must be understood at the time of the estimate. For new construction, hoist wait and cycle times can easily be 15 minutes and may result in scheduled access extending beyond the normal work day. Finally, all vertical transport will impose strict weight limits likely to preclude the use of any motorized material handling.

The productivity, or daily output, also assumes installation that meets manufacturer/designer/ standard specifications. A time allowance for quality control checks, minor adjustments, and any task required to ensure proper function or operation is also included. For items that require connections to services, time is included for positioning, leveling, securing the unit, and making all the necessary connections (and start up where applicable) to ensure a complete installation. Estimating of the services themselves (electrical, plumbing, water, steam, hydraulics, dust collection, etc.) is separate.

In some cases, the estimator must consider the use of a crane and an appropriate crew for the installation of large or heavy items. For those situations where a crane is not included in the assigned crew and as part of the line item cost,

then equipment rental costs, mobilization and demobilization costs, and operator and support personnel costs must be considered.

Labor-Hours

The labor-hours expressed in this publication are derived by dividing the total daily labor-hours for the crew by the daily output. Based on average installation time and the assumptions listed above, the labor-hours include: direct labor, indirect labor, and nonproductive time. A typical day for a craftsman might include but is not limited to:

- Direct Work
  - ☐ Measuring and layout
  - ☐ Preparing materials
  - ☐ Actual installation
  - ☐ Quality assurance/quality control
- Indirect Work
  - ☐ Reading plans or specifications
  - ☐ Preparing space
  - ☐ Receiving materials
  - ☐ Material movement
  - ☐ Giving or receiving instruction
  - ☐ Miscellaneous
- Non-Work
  - ☐ Chatting
  - ☐ Personal issues
  - ☐ Breaks
  - ☐ Interruptions (i.e., sickness, weather, material or equipment shortages, etc.)

If any of the items for a typical day do not apply to the particular work or project situation, the estimator should make any necessary adjustments.

Final Checklist

Estimating can be a straightforward process provided you remember the basics. Here's a checklist of some of the steps you should remember to complete before finalizing your estimate.

Did you remember to:

- factor in the City Cost Index for your locale?
- take into consideration which items have been marked up and by how much?
- mark up the entire estimate sufficiently for your purposes?
- read the background information on techniques and technical matters that could impact your project time span and cost?
- include all components of your project in the final estimate?
- double check your figures for accuracy?
- call RSMeans data engineers if you have any questions about your estimate or the data you've used? Remember, Gordian stands behind all of our products, including our extensive RSMeans data solutions. If you have any questions about your estimate, about the costs you've used from our data, or even about the technical aspects of the job that may affect your estimate, feel free to call the Gordian RSMeans editors at 1.800.448.8182.

Using Minimum Labor/Equipment Charges for Small Quantities

Estimating small construction or repair tasks often creates situations in which the quantity of work to be performed is very small. When this occurs, the labor and/or equipment costs to perform the work may be too low to allow for the crew to get to the job, receive instructions, find materials, get set up, perform the work, clean up, and get to the next job. In these situations, the estimator should compare the developed labor and/or equipment costs for performing the work (e.g., quantity × labor and/or equipment costs) with the *minimum labor/equipment charge* within the Unit Price section of the data set. (These minimum labor/equipment charge line items appear only in *Facilities Construction Costs with RSMeans data* and *Commercial Renovation Costs with RSMeans data*.)

If the labor and/or equipment costs developed by the estimator are LOWER THAN the *minimum labor/equipment charge* listed at the bottom of specific sections of Unit Price costs, the estimator should adjust the developed costs upward to the *minimum labor/equipment charge*. The proper use of a *minimum labor/equipment charge* results in having enough money in the estimate to cover the contractor's higher cost of performing a very small amount of work during a partial workday.

A *minimum labor/equipment charge* should be used only when the task being estimated is the only task the crew will perform at the job site that day. **If, however, the crew will be able to perform other tasks at the job site that day, the use of a *minimum labor/equipment charge* is not appropriate.**

08 52 Wood Windows

08 52 10 – Plain Wood Windows

08 52 10.40 Casement Window		Crew	Daily Output	Labor-Hours	Unit	Material	2020 Bare Costs Labor	Equipment	Total	Total Incl O&P
0010	**CASEMENT WINDOW**, including frame, screen and grilles									
0100	2'-0" x 3'-0" H, double insulated glass [G]	1 Carp	10	.800	Ea.	275	42.50		317.50	370
0150	Low E glass [G]		10	.800		285	42.50		327.50	385
0200	2'-0" x 4'-6" high, double insulated glass [G]		9	.889		390	47		437	505
0250	Low E glass [G]		9	.889		440	47		487	560
9000	Minimum labor/equipment charge	1 Carp	3	2.667	Job		142		142	227

Example:

Establish the bid price to install two casement windows. Assume installation of 2' × 4'-6" metal clad windows with insulating glass [Unit Price line number 08 52 10.40 0250], and that this is the only task this crew will perform at the job site that day.

Solution:

Step One — Develop the Bare Labor Cost for this task:
Bare Labor Cost = 2 windows @ $47 each = $94.00

Step Two — Evaluate the *minimum labor/equipment charge* for this Unit Price section against the developed Bare Labor Cost for this task:
Minimum labor/equipment charge = $142.00 (compare with $94.00)

Step Three — Choose to adjust the developed labor cost upward to the *minimum labor/equipment charge*.

Step Four — Develop the bid price for this task (including O&P):
Add together the marked-up Bare Material Cost for this task and the marked-up *minimum labor/equipment charge* for this Unit Price section.

2 × ($440.00 + 10%) + ($142.00 + 60.3%)

= 2 × ($440.00 + $44.00) + ($142.00 + $85.63)

= 2 × ($484.00) + $227.63

= $968.00 + $227.63

= $1,195.63

ANSWER: $1,195.63 is the correct bid price to use. This sum takes into consideration the material cost (with 10% for profit) for these two windows, plus the *minimum labor/equipment charge* (with O&P included) for this section of the Unit Price.

Unit Price Section

Table of Contents

Table of Contents (cont.)

Table of Contents (cont.)

RSMeans data: Unit Prices— How They Work

All RSMeans data: Unit Prices are organized in the same way.

03 30 Cast-In-Place Concrete

03 30 53 – Miscellaneous Cast-In-Place Concrete

03 30 53.40 Concrete In Place		Crew	Daily Output	Labor-Hours	Unit	Material	2020 Bare Costs Labor	Equipment	Total	Total Incl O&P
0010 **CONCRETE IN PLACE**	R033105-10									
0020 Including forms (4 uses), Grade 60 rebar, concrete (Portland cement	R033105-20									
0050 Type I), placement and finishing unless otherwise indicated	R033105-70									
0300 Beams (3500 psi), 5 kip/L.F., 10' span	R033105-85	C-14A	15.62	12.804	C.Y.	370	675	28	1,073	1,500
0350 25' span		"	18.55	10.782		390	570	23.50	983.50	1,375
0500 Chimney foundations (5000 psi), over 5 C.Y.		C-14C	32.22	3.476		178	175	.83	353.83	475
0510 (3500 psi), under 5 C.Y.		"	23.71	4.724		208	238	1.13	447.13	610
0700 Columns, square (4000 psi), 12" x 12", up to 1% reinforcing by area		C-14A	11.96	16.722		410	885	36.50	1,331.50	1,900
3540 Equipment pad (3000 psi), 3' x 3' x 6" thick		C-14H	45	1.067	Ea.	50.50	55	.60	106.10	144
3550 4' x 4' x 6" thick			30	1.600		78	82.50	.90	161.40	218
3560 5' x 5' x 8" thick			18	2.667		138	138	1.49	277.49	375
3570 6' x 6' x 8" thick			14	3.429		190	177	1.92	368.92	495
3580 8' x 8' x 10" thick			8	6		395	310	3.36	708.36	935
3590 10' x 10' x 12" thick			5	9.600		695	495	5.40	1,195.40	1,550

It is important to understand the structure of RSMeans data: Unit Prices so that you can find information easily and use it correctly.

---

① Line Numbers

Line Numbers consist of 12 characters, which identify a unique location in the database for each task. The first 6 or 8 digits conform to the Construction Specifications Institute MasterFormat® 2018. The remainder of the digits are a further breakdown in order to arrange items in understandable groups of similar tasks. Line numbers are consistent across all of our publications, so a line number in any of our products will always refer to the same item of work.

② Descriptions

Descriptions are shown in a hierarchical structure to make them readable. In order to read a complete description, read up through the indents to the top of the section. Include everything that is above and to the left that is not contradicted by information below. For instance, the complete description for line 03 30 53.40 3550 is "Concrete in place, including forms (4 uses), Grade 60 rebar, concrete (Portland cement Type 1), placement and finishing unless otherwise indicated; Equipment pad (3000 psi), 4' × 4' × 6" thick."

③ RSMeans data

When using **RSMeans data**, it is important to read through an entire section to ensure that you use the data that most closely matches your work. Note that sometimes there is additional information shown in the section that may improve your price. There are frequently lines that further describe, add to, or adjust data for specific situations.

④ Reference Information

Gordian's RSMeans engineers have created **reference** information to assist you in your estimate. **If** there is information that applies to a section, it will be indicated at the start of the section. The Reference Section is located in the back of the data set.

⑤ Crews

Crews include labor and/or equipment necessary to accomplish each task. In this case, Crew C-14H is used. Gordian's RSMeans staff selects a crew to represent the workers and equipment that are typically used for that task. In this case, Crew C-14H consists of one carpenter foreman (outside), two carpenters, one rodman, one laborer, one cement finisher, and one gas engine vibrator. Details of all crews can be found in the Reference Section.

Crews - Renovation

Crew No.	Bare Costs		Incl. Subs O & P		Cost Per Labor-Hour	
Crew C-14H	Hr.	Daily	Hr.	Daily	Bare Costs	Incl. O&P
1 Carpenter Foreman (outside)	$55.15	$441.20	$88.40	$707.20	$51.65	$82.24
2 Carpenters	53.15	850.40	85.15	1362.40		
1 Rodman (reinf.)	56.40	451.20	88.85	710.80		
1 Laborer	42.10	336.80	67.45	539.60		
1 Cement Finisher	49.95	399.60	78.45	627.60		
1 Gas Engine Vibrator		26.85		29.54	.56	.62
48 L.H., Daily Totals		$2506.05		$3977.14	$52.21	$82.86

6 Daily Output

The **Daily Output** is the amount of work that the crew can do in a normal 8-hour workday, including mobilization, layout, movement of materials, and cleanup. In this case, crew C-14H can install thirty 4' × 4' × 6" thick concrete pads in a day. Daily output is variable and based on many factors, including the size of the job, location, and environmental conditions. RSMeans data represents work done in daylight (or adequate lighting) and temperate conditions.

7 Labor-Hours

The figure in the **Labor-Hours** column is the amount of labor required to perform one unit of work—in this case the amount of labor required to construct one 4' × 4' equipment pad. This figure is calculated by dividing the number of hours of labor in the crew by the daily output (48 labor-hours divided by 30 pads = 1.6 hours of labor per pad). Multiply 1.6 times 60 to see the value in minutes: 60 × 1.6 = 96

minutes. Note: the labor-hour figure is not dependent on the crew size. A change in crew size will result in a corresponding change in daily output, but the labor-hours per unit of work will not change.

8 Unit of Measure

All RSMeans data: Unit Prices include the typical **Unit of Measure** used for estimating that item. For concrete-in-place the typical unit is cubic yards (C.Y.) or each (Ea.). For installing broadloom carpet it is square yard and for gypsum board it is square foot. The estimator needs to take special care that the unit in the data matches the unit in the take-off. Unit conversions may be found in the Reference Section.

9 Bare Costs

Bare Costs are the costs of materials, labor, and equipment that the installing contractor pays. They represent the cost, in U.S. dollars, for one unit of work. They do not include any markups for profit or labor burden.

10 Bare Total

The **Total column** represents the total bare cost for the installing contractor in U.S. dollars. In this case, the sum of $78 for material + $82.50 for labor + $.90 for equipment is $161.40.

11 Total Incl O&P

The **Total Incl O&P column** is the total cost, including overhead and profit, that the installing contractor will charge the customer. This represents the cost of materials plus 10% profit, the cost of labor plus labor burden and 10% profit, and the cost of equipment plus 10% profit. It does not include the general contractor's overhead and profit. Note: See the inside back cover of the printed product or the Reference Section of the electronic product for details on how the labor burden is calculated.

National Average

The RSMeans data in our print publications represent a "national average" cost. This data should be modified to the project location using the **City Cost Indexes** *or* **Location Factors** *tables found in the Reference Section. Use the Location Factors to adjust estimate totals if the project covers multiple trades. Use the City Cost Indexes (CCI) for single trade*

projects or projects where a more detailed analysis is required. All figures in the two tables are derived from the same research. The last row of data in the CCI—the weighted average—is the same as the numbers reported for each location in the location factor table.

RSMeans data: Unit Prices— How They Work (Continued)

Project Name: Pre-Engineered Steel Building				Architect: As Shown				
Location:	Anywhere, USA						01/01/20	R+R
Line Number	Description	Qty	Unit	Material	Labor	Equipment	SubContract	Estimate Total
03 30 53.40 3940	Strip footing, 12" x 24", reinforced	15	C.Y.	$2,565.00	$1,770.00	$8.40	$0.00	
03 30 53.40 3950	Strip footing, 12" x 36", reinforced	34	C.Y.	$5,610.00	$3,196.00	$15.30	$0.00	
03 11 13.65 3000	Concrete slab edge forms	500	L.F.	$165.00	$1,345.00	$0.00	$0.00	
03 22 11.10 0200	Welded wire fabric reinforcing	150	C.S.F.	$2,872.50	$4,350.00	$0.00	$0.00	
03 31 13.35 0300	Ready mix concrete, 4000 psi for slab on grade	278	C.Y.	$35,306.00	$0.00	$0.00	$0.00	
03 31 13.70 4300	Place, strike off & consolidate concrete slab	278	C.Y.	$0.00	$5,309.80	$136.22	$0.00	
03 35 13.30 0250	Machine float & trowel concrete slab	15,000	S.F.	$0.00	$9,900.00	$750.00	$0.00	
03 15 16.20 0140	Cut control joints in concrete slab	950	L.F.	$47.50	$418.00	$57.00	$0.00	
03 39 23.13 0300	Sprayed concrete curing membrane	150	C.S.F.	$1,867.50	$1,065.00	$0.00	$0.00	
Division 03	**Subtotal**			**$48,433.50**	**$27,353.80**	**$966.92**	**$0.00**	**$76,754.22**
08 36 13.10 2650	Manual 10' x 10' steel sectional overhead door	8	Ea.	$11,000.00	$3,760.00	$0.00	$0.00	
08 36 13.10 2860	Insulation and steel back panel for OH door	800	S.F.	$4,000.00	$0.00	$0.00	$0.00	
Division 08	**Subtotal**			**$15,000.00**	**$3,760.00**	**$0.00**	**$0.00**	**$18,760.00**
13 34 19.50 1100	Pre-Engineered Steel Building, 100' x 150' x 24'	15,000	SF Flr.	$0.00	$0.00	$0.00	$375,000.00	
13 34 19.50 6050	Framing for PESB door opening, 3' x 7'	4	Opng.	$0.00	$0.00	$0.00	$2,400.00	
13 34 19.50 6100	Framing for PESB door opening, 10' x 10'	8	Opng.	$0.00	$0.00	$0.00	$9,800.00	
13 34 19.50 6200	Framing for PESB window opening, 4' x 3'	6	Opng.	$0.00	$0.00	$0.00	$3,510.00	
13 34 19.50 5750	PESB door, 3' x 7', single leaf	4	Opng.	$2,920.00	$736.00	$0.00	$0.00	
13 34 19.50 7750	PESB sliding window, 4' x 3' with screen	6	Opng.	$2,940.00	$630.00	$67.80	$0.00	
13 34 19.50 6550	PESB gutter, eave type, 26 ga., painted	300	L.F.	$2,415.00	$864.00	$0.00	$0.00	
13 34 19.50 8650	PESB roof vent, 12" wide x 10' long	15	Ea.	$570.00	$3,465.00	$0.00	$0.00	
13 34 19.50 6900	PESB insulation, vinyl faced, 4" thick	27,400	S.F.	$11,782.00	$10,138.00	$0.00	$0.00	
Division 13	**Subtotal**			**$20,627.00**	**$15,833.00**	**$67.80**	**$390,710.00**	**$427,237.80**
	Subtotal			$84,060.50	$46,946.80	$1,034.72	$390,710.00	$522,752.02
Division 01	**General Requirements @ 7%**			5,884.24	3,286.28	72.43	27,349.70	
	Estimate Subtotal			$89,944.74	$50,233.08	$1,107.15	$418,059.70	$522,752.02
	Sales Tax @ 5%			4,497.24		55.36	10,451.49	
	Subtotal A			94,441.97	50,233.08	1,162.51	428,511.19	
	GC O & P			9,444.20	29,788.21	116.25	42,851.12	
	Subtotal B			103,886.17	80,021.29	1,278.76	471,362.31	$656,548.53
	Contingency @ 5%							32,827.43
	Subtotal C							$689,375.96
	Bond @ $12/1000 +10% O&P							9,099.76
	Subtotal D							$698,475.72
	Location Adjustment Factor				115.50			108,263.74
	Grand Total							$806,739.45

This estimate is based on an interactive spreadsheet. You are free to download it and adjust it to your methodology.
A copy of this spreadsheet is available at **RSMeans.com/2020books.**

Sample Estimate

This sample demonstrates the elements of an estimate, including a tally of the RSMeans data lines and a summary of the markups on a contractor's work to arrive at a total cost to the owner. The Location Factor with RSMeans data is added at the bottom of the estimate to adjust the cost of the work to a specific location.

---

① Work Performed

The body of the estimate shows the RSMeans data selected, including the line number, a brief description of each item, its take-off unit and quantity, and the bare costs of materials, labor, and equipment. This estimate also includes a column titled "SubContract." This data is taken from the column "Total Incl O&P" and represents the total that a subcontractor would charge a general contractor for the work, including the sub's markup for overhead and profit.

② Division 1, General Requirements

This is the first division numerically but the last division estimated. Division 1 includes project-wide needs provided by the general contractor. These requirements vary by project but may include temporary facilities and utilities, security, testing, project cleanup, etc. For small projects a percentage can be used—typically between 5% and 15% of project cost. For large projects the costs may be itemized and priced individually.

③ Sales Tax

If the work is subject to state or local sales taxes, the amount must be added to the estimate. Sales tax may be added to material costs, equipment costs, and subcontracted work. In this case, sales tax was added in all three categories. It was assumed that approximately half the subcontracted work would be material cost, so the tax was applied to 50% of the subcontract total.

④ GC O&P

This entry represents the general contractor's markup on material, labor, equipment, and subcontractor costs. Our standard markup on materials, equipment, and subcontracted work is 10%. In this estimate, the markup on the labor performed by the GC's workers uses "Skilled Workers Average" shown in Column F on the table "Installing Contractor's Overhead & Profit," which can be found on the inside back cover of the printed product or in the Reference Section of the electronic product.

⑤ Contingency

A factor for contingency may be added to any estimate to represent the cost of unknowns that may occur between the time that the estimate is performed and the time the project is constructed. The amount of the allowance will depend on the stage of design at which the estimate is done and the contractor's assessment of the risk involved. Refer to section 01 21 16.50 for contingency allowances.

⑥ Bonds

Bond costs should be added to the estimate. The figures here represent a typical performance bond, ensuring the owner that if the general contractor does not complete the obligations in the construction contract the bonding company will pay the cost for completion of the work.

⑦ Location Adjustment

Published prices are based on national average costs. If necessary, adjust the total cost of the project using a location factor from the "Location Factor" table or the "City Cost Index" table. Use location factors if the work is general, covering multiple trades. If the work is by a single trade (e.g., masonry) use the more specific data found in the "City Cost Indexes."

Estimating Tips
01 20 00 Price and Payment Procedures

- Allowances that should be added to estimates to cover contingencies and job conditions that are not included in the national average material and labor costs are shown in Section 01 21.

- When estimating historic preservation projects (depending on the condition of the existing structure and the owner's requirements), a 15–20% contingency or allowance is recommended, regardless of the stage of the drawings.

01 30 00 Administrative Requirements

- Before determining a final cost estimate, it is good practice to review all the items listed in Subdivisions 01 31 and 01 32 to make final adjustments for items that may need customizing to specific job conditions.

- Requirements for initial and periodic submittals can represent a significant cost to the General Requirements of a job. Thoroughly check the submittal specifications when estimating a project to determine any costs that should be included.

01 40 00 Quality Requirements

- All projects will require some degree of quality control. This cost is not included in the unit cost of construction listed in each division. Depending upon the terms of the contract, the various costs of inspection and testing can be the responsibility of either the owner or the contractor. Be sure to include the required costs in your estimate.

01 50 00 Temporary Facilities and Controls

- Barricades, access roads, safety nets, scaffolding, security, and many more requirements for the execution of a safe project are elements of direct cost. These costs can easily be overlooked when preparing an estimate. When looking through the major classifications of this subdivision, determine which items apply to each division in your estimate.

- Construction equipment rental costs can be found in the Reference Section in Section 01 54 33. Operators' wages are not included in equipment rental costs.

- Equipment mobilization and demobilization costs are not included in equipment rental costs and must be considered separately.

- The cost of small tools provided by the installing contractor for his workers is covered in the "Overhead" column on the "Installing Contractor's Overhead and Profit" table that lists labor trades, base rates, and markups. Therefore, it is included in the "Total Incl. O&P" cost of any unit price line item.

01 70 00 Execution and Closeout Requirements

- When preparing an estimate, thoroughly read the specifications to determine the requirements for Contract Closeout. Final cleaning, record documentation, operation and maintenance data, warranties and bonds, and spare parts and maintenance materials can all be elements of cost for the completion of a contract. Do not overlook these in your estimate.

Reference Numbers

Reference numbers are shown at the beginning of some major classifications. These numbers refer to related items in the Reference Section. The reference information may be an estimating procedure, an alternate pricing method, or technical information.

Note: Not all subdivisions listed here necessarily appear. ■

Same Data. Simplified.

Enjoy the convenience and efficiency of accessing your costs anywhere:

- **Skip the multiplier** by setting your location
- **Quickly search,** edit, favorite and share costs
- **Stay on top of price changes** with automatic updates

Discover more at rsmeans.com/online

01 11 Summary of Work

01 11 31 – Professional Consultants

01 11 31.10 Architectural Fees

		Crew	Daily Output	Labor-Hours	Unit	Material	2020 Bare Costs Labor	Equipment	Total	Total Incl O&P
0010	**ARCHITECTURAL FEES** R011110-10									
0020	For new construction									
0060	Minimum				Project				4.90%	4.90%
0090	Maximum								16%	16%
0100	For alteration work, to $500,000, add to new construction fee								50%	50%
0150	Over $500,000, add to new construction fee				↓				25%	25%

01 11 31.20 Construction Management Fees

		Crew	Daily Output	Labor-Hours	Unit	Material	2020 Bare Costs Labor	Equipment	Total	Total Incl O&P
0010	**CONSTRUCTION MANAGEMENT FEES**									
0060	For work to $100,000				Project				10%	10%
0070	To $250,000								9%	9%
0090	To $1,000,000								6%	6%
0100	To $5,000,000								5%	5%
0110	To $10,000,000				↓				4%	4%

01 11 31.30 Engineering Fees

		Crew	Daily Output	Labor-Hours	Unit	Material	2020 Bare Costs Labor	Equipment	Total	Total Incl O&P
0010	**ENGINEERING FEES** R011110-30									
0020	Educational planning consultant, minimum				Project				.50%	.50%
0100	Maximum				"				2.50%	2.50%
0400	Elevator & conveying systems, minimum				Contrct				2.50%	2.50%
0500	Maximum								5%	5%
1000	Mechanical (plumbing & HVAC), minimum								4.10%	4.10%
1100	Maximum				↓				10.10%	10.10%
1200	Structural, minimum				Project				1%	1%
1300	Maximum				"				2.50%	2.50%

01 11 31.75 Renderings

		Crew	Daily Output	Labor-Hours	Unit	Material	2020 Bare Costs Labor	Equipment	Total	Total Incl O&P
0010	**RENDERINGS** Color, matted, 20" x 30", eye level,									
0020	1 building, minimum				Ea.	2,250			2,250	2,475
0050	Average					3,200			3,200	3,525
0100	Maximum					5,150			5,150	5,675
1000	5 buildings, minimum					4,250			4,250	4,675
1100	Maximum				↓	8,525			8,525	9,375

01 21 Allowances

01 21 16 – Contingency Allowances

01 21 16.50 Contingencies

		Crew	Daily Output	Labor-Hours	Unit	Material	2020 Bare Costs Labor	Equipment	Total	Total Incl O&P
0010	**CONTINGENCIES**, Add to estimate									
0020	Conceptual stage				Project				20%	20%
0050	Schematic stage								15%	15%
0100	Preliminary working drawing stage (Design Dev.)								10%	10%
0150	Final working drawing stage				↓				3%	3%

01 21 53 – Factors Allowance

01 21 53.50 Factors

		Crew	Daily Output	Labor-Hours	Unit	Material	2020 Bare Costs Labor	Equipment	Total	Total Incl O&P
0010	**FACTORS** Cost adjustments R012153-10									
0100	Add to construction costs for particular job requirements									
0500	Cut & patch to match existing construction, add, minimum				Costs	2%	3%			
0550	Maximum					5%	9%			
0800	Dust protection, add, minimum					1%	2%			
0850	Maximum					4%	11%			
1100	Equipment usage curtailment, add, minimum					1%	1%			
1150	Maximum					3%	10%			
1400	Material handling & storage limitation, add, minimum				↓	1%	1%			

For customer support on your Commercial Renovation Costs with RSMeans data, call 800.448.8182.

01 21 53 – Factors Allowance

01 21 53.50 Factors		Crew	Daily Output	Labor-Hours	Unit	Material	2020 Bare Costs Labor	Equipment	Total	Total Incl O&P
1450	Maximum				Costs	6%	7%			
1700	Protection of existing work, add, minimum					2%	2%			
1750	Maximum					5%	7%			
2000	Shift work requirements, add, minimum						5%			
2050	Maximum						30%			
2300	Temporary shoring and bracing, add, minimum					2%	5%			
2350	Maximum				▼	5%	12%			

01 21 53.60 Security Factors

		Crew	Daily Output	Labor-Hours	Unit	Material	2020 Bare Costs Labor	Equipment	Total	Total Incl O&P
0010	**SECURITY FACTORS** R012153-60									
0100	Additional costs due to security requirements									
0110	Daily search of personnel, supplies, equipment and vehicles									
0120	Physical search, inventory and doc of assets, at entry				Costs		30%			
0130	At entry and exit						50%			
0140	Physical search, at entry						6.25%			
0150	At entry and exit						12.50%			
0160	Electronic scan search, at entry						2%			
0170	At entry and exit						4%			
0180	Visual inspection only, at entry						.25%			
0190	At entry and exit						.50%			
0200	ID card or display sticker only, at entry						.12%			
0210	At entry and exit				▼		.25%			
0220	Day 1 as described below, then visual only for up to 5 day job duration									
0230	Physical search, inventory and doc of assets, at entry				Costs		5%			
0240	At entry and exit						10%			
0250	Physical search, at entry						1.25%			
0260	At entry and exit						2.50%			
0270	Electronic scan search, at entry						.42%			
0280	At entry and exit				▼		.83%			
0290	Day 1 as described below, then visual only for 6-10 day job duration									
0300	Physical search, inventory and doc of assets, at entry				Costs		2.50%			
0310	At entry and exit						5%			
0320	Physical search, at entry						.63%			
0330	At entry and exit						1.25%			
0340	Electronic scan search, at entry						.21%			
0350	At entry and exit				▼		.42%			
0360	Day 1 as described below, then visual only for 11-20 day job duration									
0370	Physical search, inventory and doc of assets, at entry				Costs		1.25%			
0380	At entry and exit						2.50%			
0390	Physical search, at entry						.31%			
0400	At entry and exit						.63%			
0410	Electronic scan search, at entry						.10%			
0420	At entry and exit				▼		.21%			
0430	Beyond 20 days, costs are negligible									
0440	Escort required to be with tradesperson during work effort				Costs		6.25%			

01 21 55 – Job Conditions Allowance

01 21 55.50 Job Conditions

		Crew	Daily Output	Labor-Hours	Unit	Material	2020 Bare Costs Labor	Equipment	Total	Total Incl O&P
0010	**JOB CONDITIONS** Modifications to applicable									
0020	cost summaries									
0100	Economic conditions, favorable, deduct				Project				2%	2%
0200	Unfavorable, add								5%	5%
0300	Hoisting conditions, favorable, deduct								2%	2%
0400	Unfavorable, add				▼				5%	5%

For customer support on your Commercial Renovation Costs with RSMeans data, call 800.448.8182.

11

01 21 Allowances

01 21 55 – Job Conditions Allowance

01 21 55.50 Job Conditions	Crew	Daily Output	Labor-Hours	Unit	Material	2020 Bare Costs Labor	Equipment	Total	Total Incl O&P
0500	General contractor management, experienced, deduct				Project				2%	2%
0600	Inexperienced, add								10%	10%
0700	Labor availability, surplus, deduct								1%	1%
0800	Shortage, add								10%	10%
0900	Material storage area, available, deduct								1%	1%
1000	Not available, add								2%	2%
1100	Subcontractor availability, surplus, deduct								5%	5%
1200	Shortage, add								12%	12%
1300	Work space, available, deduct								2%	2%
1400	Not available, add								5%	5%

01 21 63 – Taxes

01 21 63.10 Taxes

	01 21 63.10 Taxes					Unit	Material	Labor	Equipment	Total	Total Incl O&P
0010	**TAXES**	R012909-80									
0020	Sales tax, State, average					%	5.08%				
0050	Maximum	R012909-85					7.50%				
0200	Social Security, on first $118,500 of wages							7.65%			
0300	Unemployment, combined Federal and State, minimum	R012909-86						.60%			
0350	Average							9.60%			
0400	Maximum							12%			

01 31 Project Management and Coordination

01 31 13 – Project Coordination

01 31 13.20 Field Personnel

	01 31 13.20 Field Personnel	Crew	Daily Output	Labor-Hours	Unit	Material	Labor	Equipment	Total	Total Incl O&P
0010	**FIELD PERSONNEL**									
0020	Clerk, average				Week		495		495	790
0160	General purpose laborer, average	1 Clab	.20	40			1,675		1,675	2,700
0180	Project manager, minimum						2,175		2,175	3,475
0200	Average						2,500		2,500	4,000
0220	Maximum						2,850		2,850	4,575
0240	Superintendent, minimum						2,125		2,125	3,400
0260	Average						2,325		2,325	3,725
0280	Maximum						2,650		2,650	4,225

01 31 13.30 Insurance

	01 31 13.30 Insurance				Unit	Material	Labor	Equipment	Total	Total Incl O&P
0010	**INSURANCE**	R013113-40								
0020	Builders risk, standard, minimum				Job				.24%	.24%
0050	Maximum	R013113-50							.64%	.64%
0200	All-risk type, minimum								.25%	.25%
0250	Maximum	R013113-60							.62%	.62%
0400	Contractor's equipment floater, minimum				Value				.50%	.50%
0450	Maximum				"				1.50%	1.50%
0600	Public liability, average				Job				2.02%	2.02%
0810	Workers' compensation & employer's liability									
2000	Range of 35 trades in 50 states, excl. wrecking & clerical, min.				Payroll		1.37%			
2100	Average						10.60%			
2200	Maximum						120.29%			

01 31 13.40 Main Office Expense

	01 31 13.40 Main Office Expense				Unit	Material	Labor	Equipment	Total	Total Incl O&P
0010	**MAIN OFFICE EXPENSE** Average for General Contractors									
0020	As a percentage of their annual volume									
0030	Annual volume to $300,000, minimum				% Vol.				20%	
0040	Maximum								30%	

01 31 Project Management and Coordination

01 31 13 – Project Coordination

01 31 13.40 Main Office Expense

		Crew	Daily Output	Labor-Hours	Unit	Material	2020 Bare Costs Labor	Equipment	Total	Total Incl O&P
0060	To $500,000, minimum				% Vol.				17%	
0070	Maximum								22%	
0080	To $1,000,000, minimum								16%	
0090	Maximum								19%	
0110	To $3,000,000, minimum								14%	
0120	Maximum								16%	
0130	To $5,000,000, minimum								8%	
0140	Maximum								10%	

01 31 13.50 General Contractor's Mark-Up

		Crew	Daily Output	Labor-Hours	Unit	Material	Labor	Equipment	Total	Total Incl O&P
0010	**GENERAL CONTRACTOR'S MARK-UP** on Change Orders									
0200	Extra work, by subcontractors, add				%				10%	10%
0250	By General Contractor, add								15%	15%
0400	Omitted work, by subcontractors, deduct all but								5%	5%
0450	By General Contractor, deduct all but								7.50%	7.50%
0600	Overtime work, by subcontractors, add								15%	15%
0650	By General Contractor, add								10%	10%

01 31 13.80 Overhead and Profit

		Crew	Daily Output	Labor-Hours	Unit	Material	Labor	Equipment	Total	Total Incl O&P
0010	**OVERHEAD & PROFIT** Allowance to add to items in this R013113-50									
0020	book that do not include Subs O&P, average				%				25%	
0100	Allowance to add to items in this book that R013113-55									
0110	do include Subs O&P, minimum				%				5%	5%
0150	Average								10%	10%
0200	Maximum								15%	15%
0290	Typical, by size of project, under $50,000								40%	
0310	$50,000 to $100,000								35%	
0320	$100,000 to $500,000								25%	
0330	$500,000 to $1,000,000								20%	

01 31 13.90 Performance Bond

		Crew	Daily Output	Labor-Hours	Unit	Material	Labor	Equipment	Total	Total Incl O&P
0010	**PERFORMANCE BOND**									
0020	For buildings, minimum				Job				.60%	.60%
0100	Maximum				"				2.50%	2.50%

01 31 14 – Facilities Services Coordination

01 31 14.20 Lock Out/Tag Out

		Crew	Daily Output	Labor-Hours	Unit	Material	Labor	Equipment	Total	Total Incl O&P
0010	**LOCK OUT / TAG OUT**									
0020	Miniature circuit breaker lock out device	1 Elec	220	.036	Ea.	24	2.23		26.23	30
0030	Miniature pin circuit breaker lock out device		220	.036		19.90	2.23		22.13	25.50
0040	Single circuit breaker lock out device		220	.036		21	2.23		23.23	26.50
0050	Multi-pole circuit breaker lock out device (15 to 225 Amp)		210	.038		19.90	2.34		22.24	25.50
0060	Large 3 pole circuit breaker lock out device (over 225 Amp)		210	.038		21	2.34		23.34	27
0080	Square D I-Line circuit breaker lock out device		210	.038		33	2.34		35.34	39.50
0090	Lock out disconnect switch, 30 to 100 Amp		330	.024		11	1.49		12.49	14.40
0100	100 to 400 Amp		330	.024		11	1.49		12.49	14.40
0110	Over 400 Amp		330	.024		11	1.49		12.49	14.40
0120	Lock out hasp for multiple lockout tags		200	.040		5.60	2.45		8.05	9.95
0130	Electrical cord plug lock out device		220	.036		10.50	2.23		12.73	15
0140	Electrical plug prong lock out device (3-wire grounding plug)		220	.036		6.95	2.23		9.18	11.10
0150	Wall switch lock out		200	.040		25	2.45		27.45	32
0160	Fire alarm pull station lock out	1 Stpi	200	.040		18.75	2.62		21.37	24.50
0170	Sprinkler valve tamper and flow switch lock out device	1 Skwk	220	.036		16.60	1.99		18.59	21.50
0180	Lock out sign		330	.024		18.90	1.33		20.23	23
0190	Lock out tag		440	.018		5	1		6	7.10

01 41 Regulatory Requirements

01 41 26 – Permit Requirements

01 41 26.50 Permits	Crew	Daily Output	Labor-Hours	Unit	Material	2020 Bare Costs Labor	Equipment	Total	Total Incl O&P
0010 **PERMITS**									
0020 Rule of thumb, most cities, minimum				Job				.50%	.50%
0100 Maximum				"				2%	2%

01 45 Quality Control

01 45 23 – Testing and Inspecting Services

01 45 23.50 Testing	Crew	Daily Output	Labor-Hours	Unit	Material	2020 Bare Costs Labor	Equipment	Total	Total Incl O&P
0010 **TESTING** and Inspecting Services									
1800 Compressive test, cylinder, delivered to lab, ASTM C 39				Ea.				36	40
1900 Picked up by lab, 30 minute round trip, add to above	1 Skwk	16	.500			27.50		27.50	43.50
1950 1 hour round trip, add to above		8	1			55		55	87.50
2000 2 hour round trip, add to above	↓	4	2			110		110	175
2200 Compressive strength, cores (not incl. drilling), ASTM C 42								95	105
2300 Patching core holes, to 12" diameter	1 Cefi	22	.364	↓	36.50	18.15		54.65	68.50
4730 Soil testing									
4750 Moisture content, ASTM D 2216				Ea.				18	20
4780 Permeability test, double ring infiltrometer								500	550
4800 Permeability, var. or constant head, undist., ASTM D 2434								264	290
4850 Recompacted								250	275
4900 Proctor compaction, 4" standard mold, ASTM D 698								230	253
4950 6" modified mold								110	120
5100 Shear tests, triaxial, minimum								410	450
5150 Maximum								545	600
5550 Technician for inspection, per day, earthwork								556	612
5650 Bolting								556	612
5750 Roofing								556	612
5790 Welding				↓				556	612
5820 Non-destructive metal testing, dye penetrant				Day				556	612
5840 Magnetic particle								556	612
5860 Radiography								556	612
5880 Ultrasonic				↓				556	612
6000 Welding certification, minimum				Ea.				91	100
6100 Maximum				"				364	400
7000 Underground storage tank									
7520 Volumetric tightness, <= 12,000 gal.				Ea.	470			470	520
7530 12,000 – 29,999 gal.					610			610	675
7540 >= 30,000 gal.				↓	665			665	730
7600 Vadose zone (soil gas) sampling, 10-40 samples, min.				Day				1,375	1,500
7610 Maximum				"				2,275	2,500
7700 Ground water monitoring incl. drilling 3 wells, min.				Total				4,550	5,000
7710 Maximum				"				6,375	7,000
8000 X-ray concrete slabs				Ea.				182	200

01 51 Temporary Utilities

01 51 13 – Temporary Electricity

01 51 13.80 Temporary Utilities	Crew	Daily Output	Labor-Hours	Unit	Material	2020 Bare Costs Labor	Equipment	Total	Total Incl O&P
0010 **TEMPORARY UTILITIES**									
0350 Lighting, lamps, wiring, outlets, 40,000 S.F. building, 8 strings	1 Elec	34	.235	CSF Flr	5.85	14.45		20.30	28.50
0360 16 strings	"	17	.471		11.75	29		40.75	57.50
0400 Power for temp lighting only, 6.6 KWH, per month								.92	1.01
0430 11.8 KWH, per month								1.65	1.82
0450 23.6 KWH, per month								3.30	3.63
0600 Power for job duration incl. elevator, etc., minimum								53	58
0650 Maximum				↓				110	121
0675 Temporary cooling				Ea.	1,025			1,025	1,125

01 52 Construction Facilities

01 52 13 – Field Offices and Sheds

01 52 13.20 Office and Storage Space

	Crew	Daily Output	Labor-Hours	Unit	Material	2020 Bare Costs Labor	Equipment	Total	Total Incl O&P
0010 **OFFICE AND STORAGE SPACE**									
0020 Office trailer, furnished, no hookups, 20' x 8', buy	2 Skwk	1	16	Ea.	9,525	880		10,405	11,900
0250 Rent per month					195			195	214
0300 32' x 8', buy	2 Skwk	.70	22.857		15,300	1,250		16,550	18,800
0350 Rent per month					245			245	269
0400 50' x 10', buy	2 Skwk	.60	26.667		30,300	1,475		31,775	35,600
0450 Rent per month					355			355	390
0500 50' x 12', buy	2 Skwk	.50	32		25,800	1,750		27,550	31,200
0550 Rent per month					460			460	505
0700 For air conditioning, rent per month, add					51.50			51.50	56.50
0800 For delivery, add per mile				Mile	12.20			12.20	13.40
1200 Storage boxes, 20' x 8', buy	2 Skwk	1.80	8.889	Ea.	3,150	490		3,640	4,250
1250 Rent per month					86			86	94.50
1300 40' x 8', buy	2 Skwk	1.40	11.429		3,800	625		4,425	5,175
1350 Rent per month				↓	126			126	139

01 54 Construction Aids

01 54 16 – Temporary Hoists

01 54 16.50 Weekly Forklift Crew

	Crew	Daily Output	Labor-Hours	Unit	Material	2020 Bare Costs Labor	Equipment	Total	Total Incl O&P
0010 **WEEKLY FORKLIFT CREW**									
0100 All-terrain forklift, 45' lift, 35' reach, 9000 lb. capacity	A-3P	.20	40	Week		2,125	1,875	4,000	5,375

01 54 19 – Temporary Cranes

01 54 19.50 Daily Crane Crews

	Crew	Daily Output	Labor-Hours	Unit	Material	2020 Bare Costs Labor	Equipment	Total	Total Incl O&P
0010 **DAILY CRANE CREWS** for small jobs, portal to portal									
0100 12-ton truck-mounted hydraulic crane	A-3H	1	8	Day		475	725	1,200	1,525
0200 25-ton	A-3I	1	8			475	800	1,275	1,625
0300 40-ton	A-3J	1	8			475	1,275	1,750	2,150
0400 55-ton	A-3K	1	16			880	1,500	2,380	3,025
0500 80-ton	A-3L	1	16	↓		880	2,250	3,130	3,850
0900 If crane is needed on a Saturday, Sunday or Holiday									
0910 At time-and-a-half, add				Day		50%			
0920 At double time, add				"		100%			

01 54 19.60 Monthly Tower Crane Crew

	Crew	Daily Output	Labor-Hours	Unit	Material	2020 Bare Costs Labor	Equipment	Total	Total Incl O&P
0010 **MONTHLY TOWER CRANE CREW**, excludes concrete footing									
0100 Static tower crane, 130' high, 106' jib, 6200 lb. capacity	A-3N	.05	176	Month		10,400	37,300	47,700	57,500

For customer support on your Commercial Renovation Costs with RSMeans data, call 800.448.8182.

15

01 54 Construction Aids

01 54 23 – Temporary Scaffolding and Platforms

01 54 23.60 Pump Staging

	01 54 23.60 Pump Staging	Crew	Daily Output	Labor-Hours	Unit	Material	2020 Bare Costs Labor	2020 Bare Costs Equipment	Total	Total Incl O&P
0010	**PUMP STAGING**, Aluminum R015423-20									
0200	24' long pole section, buy				Ea.	380			380	420
0300	18' long pole section, buy					310			310	345
0400	12' long pole section, buy					199			199	219
0500	6' long pole section, buy					125			125	138
0600	6' long splice joint section, buy					78			78	86
0700	Pump jack, buy					150			150	165
0900	Foldable brace, buy					56			56	61.50
1000	Workbench/back safety rail support, buy					76.50			76.50	84
1100	Scaffolding planks/workbench, 14" wide x 24' long, buy					675			675	745
1200	Plank end safety rail, buy					360			360	395
1250	Safety net, 22' long, buy					355			355	395
1300	System in place, 50' working height, per use based on 50 uses	2 Carp	84.80	.189	C.S.F.	6.25	10.05		16.30	23
1400	100 uses		84.80	.189		3.13	10.05		13.18	19.50
1500	150 uses		84.80	.189		2.09	10.05		12.14	18.35

01 54 23.70 Scaffolding

	01 54 23.70 Scaffolding	Crew	Daily Output	Labor-Hours	Unit	Material	2020 Bare Costs Labor	2020 Bare Costs Equipment	Total	Total Incl O&P
0010	**SCAFFOLDING** R015423-10									
0015	Steel tube, regular, no plank, labor only to erect & dismantle									
0090	Building exterior, wall face, 1 to 5 stories, 6'-4" x 5' frames	3 Carp	8	3	C.S.F.		159		159	255
0200	6 to 12 stories	4 Carp	8	4			213		213	340
0301	13 to 20 stories	5 Clab	8	5			211		211	335
0460	Building interior, wall face area, up to 16' high	3 Carp	12	2			106		106	170
0560	16' to 40' high		10	2.400			128		128	204
0800	Building interior floor area, up to 30' high		150	.160	C.C.F.		8.50		8.50	13.60
0900	Over 30' high	4 Carp	160	.200	"		10.65		10.65	17.05
0906	Complete system for face of walls, no plank, material only rent/mo				C.S.F.	34			34	37.50
0908	Interior spaces, no plank, material only rent/mo				C.C.F.	3.88			3.88	4.27
0910	Steel tubular, heavy duty shoring, buy									
0920	Frames 5' high 2' wide				Ea.	89.50			89.50	98.50
0925	5' high 4' wide					106			106	116
0930	6' high 2' wide					107			107	118
0935	6' high 4' wide					115			115	126
0940	Accessories									
0945	Cross braces				Ea.	17.05			17.05	18.80
0950	U-head, 8" x 8"					20.50			20.50	22.50
0955	J-head, 4" x 8"					14.90			14.90	16.40
0960	Base plate, 8" x 8"					16.05			16.05	17.65
0965	Leveling jack					35			35	38
1000	Steel tubular, regular, buy									
1100	Frames 3' high 5' wide				Ea.	92.50			92.50	102
1150	5' high 5' wide					108			108	119
1200	6'-4" high 5' wide					90.50			90.50	99.50
1350	7'-6" high 6' wide					151			151	166
1500	Accessories, cross braces					18.85			18.85	20.50
1550	Guardrail post					23.50			23.50	25.50
1600	Guardrail 7' section					8.20			8.20	9.05
1650	Screw jacks & plates					26			26	28.50
1700	Sidearm brackets					23.50			23.50	26
1750	8" casters					37.50			37.50	41
1800	Plank 2" x 10" x 16'-0"					64.50			64.50	71
1900	Stairway section					292			292	320
1910	Stairway starter bar					32.50			32.50	35.50

01 54 23 – Temporary Scaffolding and Platforms

01 54 23.70 Scaffolding		Crew	Daily Output	Labor-Hours	Unit	Material	2020 Bare Costs Labor	Equipment	Total	Total Incl O&P
1920	Stairway inside handrail				Ea.	54.50			54.50	60
1930	Stairway outside handrail					87.50			87.50	96
1940	Walk-thru frame guardrail				↓	42			42	46.50
2000	Steel tubular, regular, rent/mo.									
2100	Frames 3' high 5' wide				Ea.	4.50			4.50	4.95
2150	5' high 5' wide					4.50			4.50	4.95
2200	6'-4" high 5' wide					5.70			5.70	6.30
2250	7'-6" high 6' wide					9.90			9.90	10.90
2500	Accessories, cross braces					.90			.90	.99
2550	Guardrail post					.90			.90	.99
2600	Guardrail 7' section					.90			.90	.99
2650	Screw jacks & plates					1.80			1.80	1.98
2700	Sidearm brackets					1.80			1.80	1.98
2750	8" casters					7.20			7.20	7.90
2800	Outrigger for rolling tower					2.70			2.70	2.97
2850	Plank 2" x 10" x 16'-0"					9.95			9.95	10.90
2900	Stairway section					31.50			31.50	35
2940	Walk-thru frame guardrail				↓	2.25			2.25	2.48
3000	Steel tubular, heavy duty shoring, rent/mo.									
3250	5' high 2' & 4' wide				Ea.	8.45			8.45	9.30
3300	6' high 2' & 4' wide					8.45			8.45	9.30
3500	Accessories, cross braces					.90			.90	.99
3600	U-head, 8" x 8"					2.50			2.50	2.75
3650	J-head, 4" x 8"					2.50			2.50	2.75
3700	Base plate, 8" x 8"					.90			.90	.99
3750	Leveling jack					2.47			2.47	2.72
5700	Planks, 2" x 10" x 16'-0", labor only to erect & remove to 50' H	3 Carp	72	.333			17.70		17.70	28.50
5800	Over 50' high	4 Carp	80	.400	↓		21.50		21.50	34

01 54 23.75 Scaffolding Specialties

		Crew	Daily Output	Labor-Hours	Unit	Material	Labor	Equipment	Total	Total Incl O&P
0010	**SCAFFOLDING SPECIALTIES**									
1200	Sidewalk bridge, heavy duty steel posts & beams, including									
1210	parapet protection & waterproofing (material cost is rent/month)									
1220	8' to 10' wide, 2 posts	3 Carp	15	1.600	L.F.	46	85		131	187
1230	3 posts	"	10	2.400	"	71	128		199	282
1500	Sidewalk bridge using tubular steel scaffold frames including									
1510	planking (material cost is rent/month)	3 Carp	45	.533	L.F.	8.40	28.50		36.90	55
1600	For 2 uses per month, deduct from all above					50%				
1700	For 1 use every 2 months, add to all above					100%				
1900	Catwalks, 20" wide, no guardrails, 7' span, buy				Ea.	153			153	168
2000	10' span, buy					213			213	234
3720	Putlog, standard, 8' span, with hangers, buy					80			80	88
3730	Rent per month					16.30			16.30	17.95
3750	12' span, buy					101			101	111
3755	Rent per month					20.50			20.50	22.50
3760	Trussed type, 16' span, buy					262			262	288
3770	Rent per month					24.50			24.50	27
3790	22' span, buy					284			284	315
3795	Rent per month					32.50			32.50	35.50
4000	7 step					1,125			1,125	1,250
4100	Rolling towers, buy, 5' wide, 7' long, 10' high					1,375			1,375	1,500
4200	For additional 5' high sections, to buy				↓	255			255	280
4300	Complete incl. wheels, railings, outriggers,									

01 54 23 – Temporary Scaffolding and Platforms

01 54 23.75 Scaffolding Specialties	Crew	Daily Output	Labor-Hours	Unit	Material	2020 Bare Costs Labor	Equipment	Total	Total Incl O&P
4350	21' high, to buy				Ea.	2,325			2,325	2,550
4400	Rent/month = 5% of purchase cost				"	116			116	128
5000	Motorized work platform, mast climber									
5050	Base unit, 50' W, less than 100' tall, rent/mo				Ea.	3,625			3,625	3,975
5100	Less than 200' tall, rent/mo					3,700			3,700	4,075
5150	Less than 300' tall, rent/mo					4,475			4,475	4,925
5200	Less than 400' tall, rent/mo					5,225			5,225	5,750
5250	Set up and demob, per unit, less than 100' tall	B-68F	16.60	1.446	C.S.F.		80.50	16.85	97.35	147
5300	Less than 200' tall		25	.960			53.50	11.20	64.70	97.50
5350	Less than 300' tall		30	.800			44.50	9.35	53.85	81.50
5400	Less than 400' tall		33	.727			40.50	8.50	49	74
5500	Mobilization (price includes freight in and out) per unit	B-34N	1	16	Ea.		845	680	1,525	2,075

01 54 23.80 Staging Aids

01 54 23.80 Staging Aids	Crew	Daily Output	Labor-Hours	Unit	Material	2020 Bare Costs Labor	Equipment	Total	Total Incl O&P
0010	**STAGING AIDS** and fall protection equipment									
0100	Sidewall staging bracket, tubular, buy				Ea.	61.50			61.50	67.50
0110	Cost each per day, based on 250 days use				Day	.25			.25	.27
0200	Guard post, buy				Ea.	53.50			53.50	59
0210	Cost each per day, based on 250 days use				Day	.21			.21	.24
0300	End guard chains, buy per pair				Pair	47			47	51.50
0310	Cost per set per day, based on 250 days use				Day	.25			.25	.27
1010	Cost each per day, based on 250 days use				"	.04			.04	.05
1100	Wood bracket, buy				Ea.	28			28	31
1110	Cost each per day, based on 250 days use				Day	.11			.11	.12
2010	Cost per pair per day, based on 250 days use					.56			.56	.61
3010	Cost each per day, based on 250 days use					.23			.23	.25
3100	Aluminum scaffolding plank, 20" wide x 24' long, buy				Ea.	635			635	695
3110	Cost each per day, based on 250 days use				Day	2.53			2.53	2.79
4000	Nylon full body harness, lanyard and rope grab				Ea.	186			186	205
4010	Cost each per day, based on 250 days use				Day	.74			.74	.82
4100	Rope for safety line, 5/8" x 100' nylon, buy				Ea.	61.50			61.50	67.50
4110	Cost each per day, based on 250 days use				Day	.25			.25	.27
4200	Permanent U-Bolt roof anchor, buy				Ea.	31			31	34
4300	Temporary (one use) roof ridge anchor, buy				"	6.40			6.40	7
5000	Installation (setup and removal) of staging aids									
5010	Sidewall staging bracket	2 Carp	64	.250	Ea.		13.30		13.30	21.50
5020	Guard post with 2 wood rails	"	64	.250			13.30		13.30	21.50
5030	End guard chains, set	1 Carp	64	.125			6.65		6.65	10.65
5100	Roof shingling bracket		96	.083			4.43		4.43	7.10
5200	Ladder jack		64	.125			6.65		6.65	10.65
5300	Wood plank, 2" x 10" x 16'	2 Carp	80	.200			10.65		10.65	17.05
5310	Aluminum scaffold plank, 20" x 24'	"	40	.400			21.50		21.50	34
5410	Safety rope	1 Carp	40	.200			10.65		10.65	17.05
5420	Permanent U-Bolt roof anchor (install only)	2 Carp	40	.400			21.50		21.50	34
5430	Temporary roof ridge anchor (install only)	1 Carp	64	.125			6.65		6.65	10.65

01 54 26 – Temporary Swing Staging

01 54 26.50 Swing Staging

01 54 26.50 Swing Staging	Crew	Daily Output	Labor-Hours	Unit	Material	2020 Bare Costs Labor	Equipment	Total	Total Incl O&P
0010	**SWING STAGING**, 500 lb. cap., 2' wide to 24' long, hand operated									
0020	Steel cable type, with 60' cables, buy				Ea.	6,325			6,325	6,950
0030	Rent per month				"	630			630	695
0600	Lightweight (not for masons) 24' long for 150' height,									
0610	manual type, buy				Ea.	12,000			12,000	13,200
0620	Rent per month					1,200			1,200	1,325

01 54 Construction Aids

01 54 26 – Temporary Swing Staging

01 54 26.50 Swing Staging	Crew	Daily Output	Labor-Hours	Unit	Material	2020 Bare Costs Labor	Equipment	Total	Total Incl O&P
0700	Powered, electric or air, to 150' high, buy				Ea.	32,100			32,100	35,300
0710	Rent per month					2,250			2,250	2,475
0780	To 300' high, buy					32,600			32,600	35,900
0800	Rent per month					2,275			2,275	2,525
1000	Bosun's chair or work basket 3' x 3.5', to 300' high, electric, buy					13,400			13,400	14,700
1010	Rent per month					935			935	1,025
2200	Move swing staging (setup and remove)	E-4	2	16	Move		930	73.50	1,003.50	1,600

01 54 36 – Equipment Mobilization

01 54 36.50 Mobilization

		Crew	Daily Output	Labor-Hours	Unit	Material	2020 Bare Costs Labor	Equipment	Total	Total Incl O&P
0010	MOBILIZATION (Use line item again for demobilization) R015436-50									
0015	Up to 25 mi. haul dist. (50 mi. RT for mob/demob crew)									
1200	Small equipment, placed in rear of, or towed by pickup truck	A-3A	4	2	Ea.		106	44	150	215
1300	Equipment hauled on 3-ton capacity towed trailer	A-3Q	2.67	3			159	92.50	251.50	350
1400	20-ton capacity	B-34U	2	8			410	221	631	890
1500	40-ton capacity	B-34N	2	8			425	340	765	1,050
1600	50-ton capacity	B-34V	1	24			1,300	985	2,285	3,100
1700	Crane, truck-mounted, up to 75 ton (driver only)	1 Eqhv	4	2			118		118	185
1800	Over 75 ton (with chase vehicle)	A-3E	2.50	6.400			345	70.50	415.50	625
2400	Crane, large lattice boom, requiring assembly	B-34W	.50	144			7,425	7,025	14,450	19,500
2500	For each additional 5 miles haul distance, add						10%	10%		
3000	For large pieces of equipment, allow for assembly/knockdown									
3100	For mob/demob of micro-tunneling equip, see Section 33 05 07.36									
3200	For mob/demob of pile driving equip, see Section 31 62 19.10									

01 55 Vehicular Access and Parking

01 55 23 – Temporary Roads

01 55 23.50 Roads and Sidewalks

		Crew	Daily Output	Labor-Hours	Unit	Material	2020 Bare Costs Labor	Equipment	Total	Total Incl O&P
0010	ROADS AND SIDEWALKS Temporary									
0050	Roads, gravel fill, no surfacing, 4" gravel depth	B-14	715	.067	S.Y.	3.38	2.97	.30	6.65	8.80
0100	8" gravel depth	"	615	.078	"	6.75	3.45	.35	10.55	13.35

01 56 Temporary Barriers and Enclosures

01 56 13 – Temporary Air Barriers

01 56 13.60 Tarpaulins

		Crew	Daily Output	Labor-Hours	Unit	Material	2020 Bare Costs Labor	Equipment	Total	Total Incl O&P
0010	TARPAULINS									
0020	Cotton duck, 10-13.13 oz./S.Y., 6' x 8'				S.F.	.75			.75	.83
0050	30' x 30'					.63			.63	.69
0100	Polyvinyl coated nylon, 14-18 oz., minimum					1.49			1.49	1.64
0150	Maximum					1.49			1.49	1.64
0200	Reinforced polyethylene 3 mils thick, white					.05			.05	.06
0300	4 mils thick, white, clear or black					.16			.16	.18
0400	5.5 mils thick, clear					.19			.19	.21
0500	White, fire retardant					.60			.60	.66
0600	12 mils, oil resistant, fire retardant					.50			.50	.55
0700	8.5 mils, black					.26			.26	.29
0720	Steel reinforced polyethylene, 20 mils thick					.54			.54	.59
0730	Polyester reinforced w/integral fastening system, 11 mils thick					.20			.20	.22
0740	Polyethylene, reflective, 23 mils thick					1.34			1.34	1.47

For customer support on your Commercial Renovation Costs with RSMeans data, call 800.448.8182.

19

01 56 Temporary Barriers and Enclosures

01 56 13 – Temporary Air Barriers

01 56 13.90 Winter Protection	Crew	Daily Output	Labor-Hours	Unit	Material	2020 Bare Costs Labor	Equipment	Total	Total Incl O&P
0010 **WINTER PROTECTION**									
0100 Framing to close openings	2 Clab	500	.032	S.F.	.51	1.35		1.86	2.72
0200 Tarpaulins hung over scaffolding, 8 uses, not incl. scaffolding		1500	.011		.25	.45		.70	1
0250 Tarpaulin polyester reinf. w/integral fastening system, 11 mils thick		1600	.010		.22	.42		.64	.91
0300 Prefab fiberglass panels, steel frame, 8 uses	↓	1200	.013	↓	2.78	.56		3.34	3.96

01 56 16 – Temporary Dust Barriers

01 56 16.10 Dust Barriers, Temporary	Crew	Daily Output	Labor-Hours	Unit	Material	2020 Bare Costs Labor	Equipment	Total	Total Incl O&P
0010 **DUST BARRIERS, TEMPORARY**, erect and dismantle									
0020 Spring loaded telescoping pole & head, to 12', erect and dismantle	1 Clab	240	.033	Ea.		1.40		1.40	2.25
0025 Cost per day (based upon 250 days)				Day	.22			.22	.24
0030 To 21', erect and dismantle	1 Clab	240	.033	Ea.		1.40		1.40	2.25
0035 Cost per day (based upon 250 days)				Day	.72			.72	.79
0040 Accessories, caution tape reel, erect and dismantle	1 Clab	480	.017	Ea.		.70		.70	1.12
0045 Cost per day (based upon 250 days)				Day	.33			.33	.37
0060 Foam rail and connector, erect and dismantle	1 Clab	240	.033	Ea.		1.40		1.40	2.25
0065 Cost per day (based upon 250 days)				Day	.10			.10	.11
0070 Caution tape	1 Clab	384	.021	C.L.F.	2.49	.88		3.37	4.15
0080 Zipper, standard duty		60	.133	Ea.	10.55	5.60		16.15	20.50
0090 Heavy duty		48	.167	"	10.65	7		17.65	23
0100 Polyethylene sheet, 4 mil		37	.216	Sq.	2.62	9.10		11.72	17.50
0110 6 mil	↓	37	.216	"	4	9.10		13.10	19
1000 Dust partition, 6 mil polyethylene, 1" x 3" frame	2 Carp	2000	.008	S.F.	.33	.43		.76	1.04
1080 2" x 4" frame	"	2000	.008	"	.39	.43		.82	1.11
1085 Negative air machine, 1800 CFM				Ea.	890			890	975
1090 Adhesive strip application, 2" width	1 Clab	192	.042	C.L.F.	6.75	1.75		8.50	10.25
4000 Dust & infectious control partition, adj. to 10' high, obscured, 4' panel	2 Carp	90	.178	Ea.	580	9.45		589.45	655
4010 3' panel		90	.178		550	9.45		559.45	620
4020 2' panel	↓	90	.178		430	9.45		439.45	490
4030 1' panel	1 Carp	90	.089		285	4.72		289.72	325
4040 6" panel	"	90	.089		265	4.72		269.72	300
4050 2' panel with HEPA filtered discharge port	2 Carp	90	.178		500	9.45		509.45	565
4060 3' panel with 32" door		90	.178		895	9.45		904.45	1,000
4070 4' panel with 36" door		90	.178		995	9.45		1,004.45	1,125
4080 4'-6" panel with 44" door		90	.178		1,200	9.45		1,209.45	1,350
4090 Hinged corner		80	.200		185	10.65		195.65	221
4100 Outside corner		80	.200		150	10.65		160.65	182
4110 T post		80	.200		150	10.65		160.65	182
4120 Accessories, ceiling grid clip		360	.044		7.45	2.36		9.81	12
4130 Panel locking clip		360	.044		5.15	2.36		7.51	9.45
4140 Panel joint closure strip		360	.044		8	2.36		10.36	12.60
4150 Screw jack	↓	360	.044		6.65	2.36		9.01	11.10
4160 Digital pressure difference guage					275			275	305
4180 Combination lockset	1 Carp	13	.615		200	32.50		232.50	273
4185 Sealant tape, 2" wide	1 Clab	192	.042	C.L.F.	6.75	1.75		8.50	10.25
4190 System in place, including door and accessories									
4200 Based upon 25 uses	2 Carp	51	.314	L.F.	9.85	16.65		26.50	37.50
4210 Based upon 50 uses		51	.314		4.94	16.65		21.59	32
4230 Based upon 100 uses	↓	51	.314	↓	2.47	16.65		19.12	29

01 56 Temporary Barriers and Enclosures

01 56 23 – Temporary Barricades

01 56 23.10 Barricades

		Crew	Daily Output	Labor-Hours	Unit	Material	2020 Bare Costs Labor	Equipment	Total	Total Incl O&P
0010	**BARRICADES**									
0020	5' high, 3 rail @ 2" x 8", fixed	2 Carp	20	.800	L.F.	6.75	42.50		49.25	75.50
0150	Movable	"	30	.533	"	5.60	28.50		34.10	51.50
0300	Stock units, 58' high, 8' wide, reflective, buy				Ea.	211			211	232
0350	With reflective tape, buy				"	425			425	465
0400	Break-a-way 3" PVC pipe barricade									
0410	with 3 ea. 1' x 4' reflectorized panels, buy				Ea.	153			153	168
0500	Barricades, plastic, 8" x 24" wide, foldable					61			61	67
0800	Traffic cones, PVC, 18" high					10.95			10.95	12.05
0850	28" high					23.50			23.50	25.50
0900	Barrels, 55 gal., with flasher	1 Clab	96	.083		112	3.51		115.51	129
1000	Guardrail, wooden, 3' high, 1" x 6" on 2" x 4" posts	2 Carp	200	.080	L.F.	1.44	4.25		5.69	8.40
1100	2" x 6" on 4" x 4" posts	"	165	.097		2.80	5.15		7.95	11.35
1200	Portable metal with base pads, buy					14.45			14.45	15.90
1250	Typical installation, assume 10 reuses	2 Carp	600	.027		2.31	1.42		3.73	4.81
1300	Barricade tape, polyethylene, 7 mil, 3" wide x 300' long roll	1 Clab	128	.063	Ea.	7.50	2.63		10.13	12.45
3000	Detour signs, set up and remove									
3010	Reflective aluminum, MUTCD, 24" x 24", post mounted	1 Clab	20	.400	Ea.	2.80	16.85		19.65	30
4000	Roof edge portable barrier stands and warning flags, 50 uses	1 Rohe	9100	.001	L.F.	.07	.03		.10	.12
4010	100 uses	"	9100	.001	"	.03	.03		.06	.09

01 56 26 – Temporary Fencing

01 56 26.50 Temporary Fencing

		Crew	Daily Output	Labor-Hours	Unit	Material	2020 Bare Costs Labor	Equipment	Total	Total Incl O&P
0010	**TEMPORARY FENCING**									
0020	Chain link, 11 ga., 4' high	2 Clab	400	.040	L.F.	2	1.68		3.68	4.90
0100	6' high		300	.053		4.96	2.25		7.21	9.05
0200	Rented chain link, 6' high, to 1000' (up to 12 mo.)		400	.040		3.37	1.68		5.05	6.40
0250	Over 1000' (up to 12 mo.)		300	.053		4.14	2.25		6.39	8.15
0350	Plywood, painted, 2" x 4" frame, 4' high	A-4	135	.178		6.90	8.95		15.85	22
0400	4" x 4" frame, 8' high	"	110	.218		13.45	10.95		24.40	32.50
0500	Wire mesh on 4" x 4" posts, 4' high	2 Carp	100	.160		11.15	8.50		19.65	26
0550	8' high	"	80	.200		16.60	10.65		27.25	35.50

01 56 29 – Temporary Protective Walkways

01 56 29.50 Protection

		Crew	Daily Output	Labor-Hours	Unit	Material	2020 Bare Costs Labor	Equipment	Total	Total Incl O&P
0010	**PROTECTION**									
0020	Stair tread, 2" x 12" planks, 1 use	1 Carp	75	.107	Tread	5.15	5.65		10.80	14.80
0100	Exterior plywood, 1/2" thick, 1 use		65	.123		2.05	6.55		8.60	12.75
0200	3/4" thick, 1 use		60	.133		2.84	7.10		9.94	14.45
2200	Sidewalks, 2" x 12" planks, 2 uses		350	.023	S.F.	.86	1.22		2.08	2.90
2300	Exterior plywood, 2 uses, 1/2" thick		750	.011		.34	.57		.91	1.29
2400	5/8" thick		650	.012		.38	.65		1.03	1.47
2500	3/4" thick		600	.013		.47	.71		1.18	1.66

For customer support on your Commercial Renovation Costs with RSMeans data, call 800.448.8182.

21

01 57 Temporary Controls

01 57 33 – Temporary Security

01 57 33.50 Watchman	Crew	Daily Output	Labor-Hours	Unit	Material	2020 Bare Costs Labor	Equipment	Total	Total Incl O&P
0010 **WATCHMAN**									
0020 Service, monthly basis, uniformed person, minimum				Hr.				27.65	30.40
0100 Maximum								56	61.62
0200 Person and command dog, minimum								31	34
0300 Maximum								60	65

01 58 Project Identification

01 58 13 – Temporary Project Signage

01 58 13.50 Signs	Crew	Daily Output	Labor-Hours	Unit	Material	2020 Bare Costs Labor	Equipment	Total	Total Incl O&P
0010 **SIGNS**									
0020 High intensity reflectorized, no posts, buy				Ea.	27			27	29.50

01 66 Product Storage and Handling Requirements

01 66 19 – Material Handling

01 66 19.10 Material Handling	Crew	Daily Output	Labor-Hours	Unit	Material	2020 Bare Costs Labor	Equipment	Total	Total Incl O&P
0010 **MATERIAL HANDLING**									
0020 Above 2nd story, via stairs, per C.Y. of material per floor	2 Clab	145	.110	C.Y.		4.65		4.65	7.45
0030 Via elevator, per C.Y. of material		240	.067			2.81		2.81	4.50
0050 Distances greater than 200', per C.Y. of material per each addl 200'		300	.053			2.25		2.25	3.60

01 71 Examination and Preparation

01 71 23 – Field Engineering

01 71 23.13 Construction Layout	Crew	Daily Output	Labor-Hours	Unit	Material	2020 Bare Costs Labor	Equipment	Total	Total Incl O&P
0010 **CONSTRUCTION LAYOUT**									
1100 Crew for layout of building, trenching or pipe laying, 2 person crew	A-6	1	16	Day		845	31	876	1,375
1200 3 person crew	A-7	1	24			1,375	30.50	1,405.50	2,200
1400 Crew for roadway layout, 4 person crew	A-8	1	32			1,775	30.50	1,805.50	2,850

01 74 Cleaning and Waste Management

01 74 13 – Progress Cleaning

01 74 13.20 Cleaning Up	Crew	Daily Output	Labor-Hours	Unit	Material	2020 Bare Costs Labor	Equipment	Total	Total Incl O&P
0010 **CLEANING UP**									
0020 After job completion, allow, minimum				Job				.30%	.30%
0040 Maximum				"				1%	1%
0052 Cleanup of floor area, continuous, per day, during const.	A-5	16	1.125	M.S.F.	2.47	48	3.06	53.53	83
0100 Final by GC at end of job	"	11.50	1.565	"	2.62	67	4.26	73.88	115

01 76 Protecting Installed Construction

01 76 13 – Temporary Protection of Installed Construction

01 76 13.20 Temporary Protection	Crew	Daily Output	Labor-Hours	Unit	Material	2020 Bare Costs Labor	Equipment	Total	Total Incl O&P
0010 **TEMPORARY PROTECTION**									
0020 Flooring, 1/8" tempered hardboard, taped seams	2 Carp	1500	.011	S.F.	.50	.57		1.07	1.46
0030 Peel away carpet protection	1 Clab	3200	.003	"	.13	.11		.24	.31

01 91 Commissioning

01 91 13 – General Commissioning Requirements

01 91 13.50 Building Commissioning

	Crew	Daily Output	Labor-Hours	Unit	Material	2020 Bare Costs Labor	Equipment	Total	Total Incl O&P
0010 **BUILDING COMMISSIONING**									
0100 Systems operation and verification during turnover				%				.25%	.25%
0150 Including all systems subcontractors								.50%	.50%
0200 Systems design assistance, operation, verification and training								.50%	.50%
0250 Including all systems subcontractors								1%	1%

01 93 Facility Maintenance

01 93 09 – Facility Equipment

01 93 09.50 Moving Equipment

	Crew	Daily Output	Labor-Hours	Unit	Material	2020 Bare Costs Labor	Equipment	Total	Total Incl O&P
0010 **MOVING EQUIPMENT**, Remove and reset, 100' distance,									
0020 No obstructions, no assembly or leveling unless noted									
0100 Annealing furnace, 24' overall	B-67	4	4	Ea.		218	67.50	285.50	415
0200 Annealing oven, small		14	1.143			62	19.30	81.30	118
0240 Very large		1	16			870	270	1,140	1,650
0400 Band saw, small		12	1.333			72.50	22.50	95	138
0440 Large		8	2			109	33.50	142.50	207
0500 Blue print copy machine		7	2.286			124	38.50	162.50	237
0600 Bonding mill, 6"		7	2.286			124	38.50	162.50	237
0620 12"		6	2.667			145	45	190	276
0640 18"		4	4			218	67.50	285.50	415
0660 24"		2	8			435	135	570	830
0700 Boring machine (jig)	B-68	7	3.429			188	38.50	226.50	335
0800 Bridgeport mill, standard	B-67	14	1.143			62	19.30	81.30	118
1000 Calibrator, 6 unit	"	14	1.143			62	19.30	81.30	118
1100 Comparitor, bench top	2 Clab	14	1.143			48		48	77
1140 Floor mounted	B-67	7	2.286			124	38.50	162.50	237
1200 Computer, desk top	2 Clab	25	.640			27		27	43
1300 Copy machine	"	25	.640			27		27	43
1500 Deflasher	B-67	14	1.143			62	19.30	81.30	118
1600 Degreaser, small		14	1.143			62	19.30	81.30	118
1640 Large 24' overall		1	16			870	270	1,140	1,650
1700 Desk with chair	2 Clab	25	.640			27		27	43
1725 Desk only	"	22	.727			30.50		30.50	49
1750 Chair only	1 Clab	300	.027			1.12		1.12	1.80
1800 Dial press	B-67	7	2.286			124	38.50	162.50	237
1900 Drafting table	2 Clab	14	1.143			48		48	77
2000 Drill press, bench top	"	14	1.143			48		48	77
2040 Floor mounted	B-67	14	1.143			62	19.30	81.30	118
2080 Industrial radial	"	7	2.286			124	38.50	162.50	237
2100 Dust collector, portable	2 Clab	25	.640			27		27	43
2140 Stationary, small	B-67	7	2.286			124	38.50	162.50	237
2180 Stationary, large	"	2	8			435	135	570	830

For customer support on your Commercial Renovation Costs with RSMeans data, call 800.448.8182.

23

01 93 09 – Facility Equipment

01 93 09.50 Moving Equipment	Crew	Daily Output	Labor-Hours	Unit	Material	2020 Bare Costs Labor	Equipment	Total	Total Incl O&P
2300	Electric discharge machine	B-68	7	3.429	Ea.		188	38.50	226.50	335
2400	Environmental chamber walls, including assembly	4 Clab	18	1.778	L.F.		75		75	120
2600	File cabinet	2 Clab	25	.640	Ea.		27		27	43
2800	Grinder/sander, pedestal mount	B-67	14	1.143			62	19.30	81.30	118
3000	Hack saw, power	2 Clab	24	.667			28		28	45
3100	Hydraulic press	B-67	14	1.143			62	19.30	81.30	118
3500	Laminar flow tables	"	14	1.143			62	19.30	81.30	118
3600	Lathe, bench	2 Clab	14	1.143			48		48	77
3640	6"	B-67	14	1.143			62	19.30	81.30	118
3680	10"		13	1.231			67	21	88	127
3720	12"		12	1.333			72.50	22.50	95	138
4000	Milling machine		8	2			109	33.50	142.50	207
4100	Molding press, 25 ton		5	3.200			174	54	228	330
4140	60 ton		4	4			218	67.50	285.50	415
4180	100 ton		2	8			435	135	570	830
4220	150 ton		1.50	10.667			580	180	760	1,100
4260	200 ton		1	16			870	270	1,140	1,650
4300	300 ton		.75	21.333			1,150	360	1,510	2,200
4700	Oil pot stand		14	1.143			62	19.30	81.30	118
5000	Press, 10 ton		14	1.143			62	19.30	81.30	118
5040	15 ton		12	1.333			72.50	22.50	95	138
5080	20 ton		10	1.600			87	27	114	166
5120	30 ton		8	2			109	33.50	142.50	207
5160	45 ton		6	2.667			145	45	190	276
5200	60 ton		4	4			218	67.50	285.50	415
5240	75 ton		2.50	6.400			350	108	458	665
5280	100 ton		2	8			435	135	570	830
5500	Raised floor, including assembly	2 Carp	250	.064	S.F.		3.40		3.40	5.45
5600	Rolling mill, 6"	B-67	7	2.286	Ea.		124	38.50	162.50	237
5640	9"		6	2.667			145	45	190	276
5680	12"		4	4			218	67.50	285.50	415
5720	13"		3.50	4.571			249	77	326	475
5760	18"		2	8			435	135	570	830
5800	25"		1	16			870	270	1,140	1,650
6000	Sander, floor stand		14	1.143			62	19.30	81.30	118
6100	Screw machine		7	2.286			124	38.50	162.50	237
6200	Shaper, 16"		14	1.143			62	19.30	81.30	118
6300	Shear, power assist		4	4			218	67.50	285.50	415
6400	Slitter, 6"		14	1.143			62	19.30	81.30	118
6440	8"		13	1.231			67	21	88	127
6480	10"		12	1.333			72.50	22.50	95	138
6520	12"		11	1.455			79	24.50	103.50	150
6560	16"		10	1.600			87	27	114	166
6600	20"		8	2			109	33.50	142.50	207
6640	24"		6	2.667			145	45	190	276
6800	Snag and tap machine		7	2.286			124	38.50	162.50	237
6900	Solder machine (auto)		7	2.286			124	38.50	162.50	237
7000	Storage cabinet metal, small	2 Clab	36	.444			18.70		18.70	30
7040	Large		25	.640			27		27	43
7100	Storage rack open, small		14	1.143			48		48	77
7140	Large		7	2.286			96		96	154
7200	Surface bench, small	B-67	14	1.143			62	19.30	81.30	118
7240	Large	"	5	3.200			174	54	228	330

For customer support on your Commercial Renovation Costs with RSMeans data, call 800.448.8182.

01 93 Facility Maintenance

01 93 09 – Facility Equipment

01 93 09.50 Moving Equipment	Crew	Daily Output	Labor-Hours	Unit	Material	2020 Bare Costs Labor	2020 Bare Costs Equipment	Total	Total Incl O&P
7300	Surface grinder, large wet	B-68	5	4.800	Ea.		264	54	318	470
7500	Time check machine	2 Clab	14	1.143			48		48	77
8000	Welder, 30 kVA (bench)		14	1.143			48		48	77
8100	Work bench with chair		25	.640			27		27	43
8900	For moving distance to 200', add						25%			

For customer support on your Commercial Renovation Costs with RSMeans data, call 800.448.8182.

Division Notes

	CREW	DAILY OUTPUT	LABOR-HOURS	UNIT	BARE COSTS				TOTAL INCL O&P
					MAT.	LABOR	EQUIP.	TOTAL	

Estimating Tips

02 30 00 Subsurface Investigation

In preparing estimates on structures involving earthwork or foundations, all information concerning soil characteristics should be obtained. Look particularly for hazardous waste, evidence of prior dumping of debris, and previous stream beds.

02 40 00 Demolition and Structure Moving

The costs shown for selective demolition do not include rubbish handling or disposal. These items should be estimated separately using RSMeans data or other sources.

- Historic preservation often requires that the contractor remove materials from the existing structure, rehab them, and replace them. The estimator must be aware of any related measures and precautions that must be taken when doing selective demolition and cutting and patching. Requirements may include special handling and storage, as well as security.

- In addition to Subdivision 02 41 00, you can find selective demolition items in each division. Example: Roofing demolition is in Division 7.
- Absent of any other specific reference, an approximate demolish-in-place cost can be obtained by halving the new-install labor cost. To remove for reuse, allow the entire new-install labor figure.

02 40 00 Building Deconstruction

This section provides costs for the careful dismantling and recycling of most low-rise building materials.

02 50 00 Containment of Hazardous Waste

This section addresses on-site hazardous waste disposal costs.

02 80 00 Hazardous Material Disposal/Remediation

This subdivision includes information on hazardous waste handling, asbestos remediation, lead remediation, and mold remediation. See reference numbers

RO28213-20 and RO28319-60 for further guidance in using these unit price lines.

02 90 00 Monitoring Chemical Sampling, Testing Analysis

This section provides costs for on-site sampling and testing hazardous waste.

Reference Numbers

Reference numbers are shown at the beginning of some major classifications. These numbers refer to related items in the Reference Section. The reference information may be an estimating procedure, an alternate pricing method, or technical information.

Note: Not all subdivisions listed here necessarily appear. ■

Same Data. Simplified.

Enjoy the convenience and efficiency of accessing your costs anywhere:

- **Skip the multiplier** by setting your location
- **Quickly search,** edit, favorite and share costs
- **Stay on top of price changes** with automatic updates

Discover more at rsmeans.com/online

02 21 Surveys

02 21 13 – Site Surveys

02 21 13.09 Topographical Surveys	Crew	Daily Output	Labor-Hours	Unit	Material	2020 Bare Costs Labor	Equipment	Total	Total Incl O&P
0010 **TOPOGRAPHICAL SURVEYS**									
0020 Topographical surveying, conventional, minimum	A-7	3.30	7.273	Acre	23.50	415	9.30	447.80	690
0100 Maximum	A-8	.60	53.333	"	63	2,975	51	3,089	4,825

02 21 13.13 Boundary and Survey Markers

	Crew	Daily Output	Labor-Hours	Unit	Material	2020 Bare Costs Labor	Equipment	Total	Total Incl O&P
0010 **BOUNDARY AND SURVEY MARKERS**									
0300 Lot location and lines, large quantities, minimum	A-7	2	12	Acre	36	690	15.35	741.35	1,125
0320 Average	"	1.25	19.200		63.50	1,100	24.50	1,188	1,825
0400 Small quantities, maximum	A-8	1	32		76.50	1,775	30.50	1,882	2,950

02 32 Geotechnical Investigations

02 32 13 – Subsurface Drilling and Sampling

02 32 13.10 Boring and Exploratory Drilling

	Crew	Daily Output	Labor-Hours	Unit	Material	2020 Bare Costs Labor	Equipment	Total	Total Incl O&P
0010 **BORING AND EXPLORATORY DRILLING**									
0020 Borings, initial field stake out & determination of elevations	A-6	1	16	Day		845	31	876	1,375
0100 Drawings showing boring details				Total		335		335	425
0200 Report and recommendations from P.E.						775		775	970
0300 Mobilization and demobilization	B-55	4	6			263	305	568	755
0350 For over 100 miles, per added mile		450	.053	Mile		2.34	2.69	5.03	6.70
0600 Auger holes in earth, no samples, 2-1/2" diameter		78.60	.305	L.F.		13.40	15.40	28.80	38.50
0650 4" diameter		67.50	.356			15.60	17.95	33.55	45
0800 Cased borings in earth, with samples, 2-1/2" diameter		55.50	.432		20.50	18.95	22	61.45	77
0850 4" diameter		32.60	.736		29	32.50	37	98.50	125
1000 Drilling in rock, "BX" core, no sampling	B-56	34.90	.458			22	43.50	65.50	82.50
1050 With casing & sampling		31.70	.505		20.50	24	48	92.50	113
1200 "NX" core, no sampling		25.92	.617			29.50	58.50	88	111
1250 With casing and sampling		25	.640		21	30.50	60.50	112	138
1400 Borings, earth, drill rig and crew with truck mounted auger	B-55	1	24	Day		1,050	1,200	2,250	3,000
1450 Rock using crawler type drill	B-56	1	16	"		760	1,525	2,285	2,875
1500 For inner city borings, add, minimum								10%	10%
1510 Maximum								20%	20%

02 41 Demolition

02 41 13 – Selective Site Demolition

02 41 13.15 Hydrodemolition

	Crew	Daily Output	Labor-Hours	Unit	Material	2020 Bare Costs Labor	Equipment	Total	Total Incl O&P
0010 **HYDRODEMOLITION**									
0015 Hydrodemolition, concrete pavement									
0120 Includes removal but exlcudes disposal of concrete									
0130 2" depth	B-5E	1000	.064	S.F.		3	2.75	5.75	7.80
0410 4" depth		800	.080			3.75	3.44	7.19	9.75
0420 6" depth		600	.107			5	4.59	9.59	13
0520 8" depth		300	.213			10	9.20	19.20	26

02 41 13.16 Hydroexcavation

	Crew	Daily Output	Labor-Hours	Unit	Material	2020 Bare Costs Labor	Equipment	Total	Total Incl O&P
0010 **HYDROEXCAVATION**									
0015 Hydroexcavation									
0110 Mobilization or Demobilization	B-6D	8	2.500	Ea.		121	158	279	365
0130 Normal Conditions		48	.417	B.C.Y.		20	26.50	46.50	61
0140 Adverse Conditions		30	.667	"		32	42	74	97.50
0160 Minimum labor/equipment charge		4	5	Ea.		242	315	557	730

02 41 13.17 Demolish, Remove Pavement and Curb

		Crew	Daily Output	Labor-Hours	Unit	Material	2020 Bare Costs Labor	2020 Bare Costs Equipment	Total	Total Incl O&P
0010	**DEMOLISH, REMOVE PAVEMENT AND CURB** R024119-10									
5010	Pavement removal, bituminous roads, up to 3" thick	B-38	690	.058	S.Y.		2.76	1.80	4.56	6.35
5050	4"-6" thick		420	.095			4.53	2.96	7.49	10.45
5100	Bituminous driveways		640	.063			2.98	1.94	4.92	6.85
5200	Concrete to 6" thick, hydraulic hammer, mesh reinforced		255	.157			7.45	4.87	12.32	17.20
5300	Rod reinforced		200	.200			9.50	6.20	15.70	22
5400	Concrete, 7"-24" thick, plain		33	1.212	C.Y.		57.50	37.50	95	133
5500	Reinforced		24	1.667	"		79.50	51.50	131	183
5590	Minimum labor/equipment charge		6	6.667	Job		315	207	522	735
5600	With hand held air equipment, bituminous, to 6" thick	B-39	1900	.025	S.F.		1.12	.19	1.31	1.99
5700	Concrete to 6" thick, no reinforcing		1600	.030			1.33	.22	1.55	2.37
5800	Mesh reinforced		1400	.034			1.52	.26	1.78	2.70
5900	Rod reinforced		765	.063			2.78	.47	3.25	4.94
5990	Minimum labor/equipment charge	B-38	6	6.667	Job		315	207	522	735
6000	Curbs, concrete, plain	B-6	360	.067	L.F.		3.05	.59	3.64	5.50
6100	Reinforced		275	.087			3.99	.78	4.77	7.20
6200	Granite		360	.067			3.05	.59	3.64	5.50
6300	Bituminous		528	.045			2.08	.41	2.49	3.75
6390	Minimum labor/equipment charge		6	4	Job		183	35.50	218.50	330

02 41 13.23 Utility Line Removal

		Crew	Daily Output	Labor-Hours	Unit	Material	Labor	Equipment	Total	Total Incl O&P
0010	**UTILITY LINE REMOVAL**									
0015	No hauling, abandon catch basin or manhole	B-6	7	3.429	Ea.		157	30.50	187.50	283
0020	Remove existing catch basin or manhole, masonry		4	6			274	53.50	327.50	495
0030	Catch basin or manhole frames and covers, stored		13	1.846			84.50	16.45	100.95	152
0040	Remove and reset		7	3.429			157	30.50	187.50	283
0900	Hydrants, fire, remove only	B-21A	5	8			420	85.50	505.50	750
0950	Remove and reset	"	2	20			1,050	214	1,264	1,875
0990	Minimum labor/equipment charge	2 Plum	2	8	Job		515		515	795
2900	Pipe removal, sewer/water, no excavation, 12" diameter	B-6	175	.137	L.F.		6.25	1.22	7.47	11.30
2930	15"-18" diameter	B-12Z	150	.160			7.65	9.75	17.40	23
2960	21"-24" diameter		120	.200			9.55	12.20	21.75	28.50
3000	27"-36" diameter		90	.267			12.75	16.30	29.05	38
3200	Steel, welded connections, 4" diameter	B-6	160	.150			6.85	1.34	8.19	12.35
3300	10" diameter		80	.300			13.70	2.67	16.37	25
3390	Minimum labor/equipment charge		3	8	Job		365	71.50	436.50	660

02 41 13.30 Minor Site Demolition

		Crew	Daily Output	Labor-Hours	Unit	Material	Labor	Equipment	Total	Total Incl O&P
0010	**MINOR SITE DEMOLITION** R024119-10									
0100	Roadside delineators, remove only	B-80	175	.183	Ea.		8.50	5.85	14.35	19.95
0110	Remove and reset	"	100	.320	"		14.90	10.20	25.10	35
0400	Minimum labor/equipment charge	B-6	4	6	Job		274	53.50	327.50	495
0800	Guiderail, corrugated steel, remove only	B-80A	100	.240	L.F.		10.10	8.20	18.30	25
0850	Remove and reset	"	40	.600	"		25.50	20.50	46	63
0860	Guide posts, remove only	B-80B	120	.267	Ea.		11.95	1.97	13.92	21
0870	Remove and reset	B-55	50	.480	"		21	24	45	60
0890	Minimum labor/equipment charge	2 Clab	4	4	Job		168		168	270
1000	Masonry walls, block, solid	B-5	1800	.031	C.F.		1.45	.85	2.30	3.23
1200	Brick, solid		900	.062			2.90	1.70	4.60	6.45
1400	Stone, with mortar		900	.062			2.90	1.70	4.60	6.45
1500	Dry set		1500	.037			1.74	1.02	2.76	3.88
1600	Median barrier, precast concrete, remove and store	B-3	430	.112	L.F.		5.25	5.40	10.65	14.30
1610	Remove and reset	"	390	.123	"		5.80	5.95	11.75	15.80
1650	Minimum labor/equipment charge	A-1	4	2	Job		84	28	112	166

For customer support on your Commercial Renovation Costs with RSMeans data, call 800.448.8182.

29

02 41 Demolition

02 41 13 – Selective Site Demolition

02 41 13.30 Minor Site Demolition

		Crew	Daily Output	Labor-Hours	Unit	Material	2020 Bare Costs Labor	Equipment	Total	Total Incl O&P
4000	Sidewalk removal, bituminous, 2" thick	B-6	350	.069	S.Y.		3.14	.61	3.75	5.65
4010	2-1/2" thick		325	.074			3.38	.66	4.04	6.05
4050	Brick, set in mortar		185	.130			5.95	1.16	7.11	10.65
4100	Concrete, plain, 4"		160	.150			6.85	1.34	8.19	12.35
4110	Plain, 5"		140	.171			7.85	1.53	9.38	14.15
4120	Plain, 6"		120	.200			9.15	1.78	10.93	16.45
4200	Mesh reinforced, concrete, 4"		150	.160			7.30	1.43	8.73	13.15
4210	5" thick		131	.183			8.40	1.63	10.03	15.10
4220	6" thick	↓	112	.214	↓		9.80	1.91	11.71	17.65
4290	Minimum labor/equipment charge	B-39	12	4	Job		177	30	207	315

02 41 13.33 Railtrack Removal

		Crew	Daily Output	Labor-Hours	Unit	Material	2020 Bare Costs Labor	Equipment	Total	Total Incl O&P
0010	**RAILTRACK REMOVAL**									
3500	Railroad track removal, ties and track	B-13	330	.170	L.F.		7.80	1.76	9.56	14.35
3600	Ballast	B-14	500	.096	C.Y.		4.25	.43	4.68	7.20
3700	Remove and re-install, ties & track using new bolts & spikes		50	.960	L.F.		42.50	4.27	46.77	72.50
3800	Turnouts using new bolts and spikes		1	48	Ea.		2,125	214	2,339	3,600
3890	Minimum labor/equipment charge	↓	5	9.600	Job		425	42.50	467.50	720

02 41 13.60 Selective Demolition Fencing

		Crew	Daily Output	Labor-Hours	Unit	Material	2020 Bare Costs Labor	Equipment	Total	Total Incl O&P
0010	**SELECTIVE DEMOLITION FENCING** R024119-10									
1600	Fencing, barbed wire, 3 strand	2 Clab	430	.037	L.F.		1.57		1.57	2.51
1650	5 strand	"	280	.057			2.41		2.41	3.85
1700	Chain link, posts & fabric, 8'-10' high, remove only	B-6	445	.054	↓		2.47	.48	2.95	4.45

02 41 16 – Structure Demolition

02 41 16.13 Building Demolition

		Crew	Daily Output	Labor-Hours	Unit	Material	2020 Bare Costs Labor	Equipment	Total	Total Incl O&P
0010	**BUILDING DEMOLITION** Large urban projects, incl. 20 mi. haul R024119-10									
0011	No foundation or dump fees, C.F. is vol. of building standing									
0020	Steel	B-8	21500	.003	C.F.		.15	.13	.28	.38
0050	Concrete		15300	.004			.20	.19	.39	.53
0080	Masonry		20100	.003			.16	.14	.30	.41
0100	Mixture of types	↓	20100	.003			.16	.14	.30	.41
0500	Small bldgs, or single bldgs, no salvage included, steel	B-3	14800	.003			.15	.16	.31	.41
0600	Concrete		11300	.004			.20	.21	.41	.55
0650	Masonry		14800	.003			.15	.16	.31	.41
0700	Wood	↓	14800	.003			.15	.16	.31	.41
0750	For buildings with no interior walls, deduct				↓				30%	30%
1000	Demolition single family house, one story, wood 1600 S.F.	B-3	1	48	Ea.		2,275	2,325	4,600	6,150
1020	3200 S.F.		.50	96			4,525	4,625	9,150	12,300
1200	Demolition two family house, two story, wood 2400 S.F.		.67	71.964			3,400	3,475	6,875	9,225
1220	4200 S.F.		.38	128			6,025	6,175	12,200	16,400
1300	Demolition three family house, three story, wood 3200 S.F.		.50	96			4,525	4,625	9,150	12,300
1320	5400 S.F.	↓	.30	160			7,550	7,725	15,275	20,500
5000	For buildings with no interior walls, deduct				↓				30%	30%

02 41 16.17 Building Demolition Footings and Foundations

		Crew	Daily Output	Labor-Hours	Unit	Material	2020 Bare Costs Labor	Equipment	Total	Total Incl O&P
0010	**BUILDING DEMOLITION FOOTINGS AND FOUNDATIONS** R024119-10									
0200	Floors, concrete slab on grade,									
0240	4" thick, plain concrete	B-13L	5000	.003	S.F.		.19	.42	.61	.76
0280	Reinforced, wire mesh		4000	.004			.24	.52	.76	.94
0300	Rods		4500	.004			.21	.46	.67	.84
0400	6" thick, plain concrete		4000	.004			.24	.52	.76	.94
0420	Reinforced, wire mesh		3200	.005			.30	.65	.95	1.17
0440	Rods	↓	3600	.004	↓		.26	.58	.84	1.04

02 41 Demolition

02 41 16 – Structure Demolition

02 41 16.17 Building Demolition Footings and Foundations

		Crew	Daily Output	Labor-Hours	Unit	Material	2020 Bare Costs Labor	2020 Bare Costs Equipment	Total	Total Incl O&P
1000	Footings, concrete, 1' thick, 2' wide	B-13L	300	.053	L.F.		3.16	6.95	10.11	12.55
1080	1'-6" thick, 2' wide		250	.064			3.79	8.30	12.09	15.10
1120	3' wide		200	.080			4.74	10.40	15.14	18.85
1140	2' thick, 3' wide		175	.091			5.40	11.90	17.30	21.50
1200	Average reinforcing, add								10%	10%
1220	Heavy reinforcing, add								20%	20%
2000	Walls, block, 4" thick	B-13L	8000	.002	S.F.		.12	.26	.38	.48
2040	6" thick		6000	.003			.16	.35	.51	.63
2080	8" thick		4000	.004			.24	.52	.76	.94
2100	12" thick		3000	.005			.32	.69	1.01	1.25
2200	For horizontal reinforcing, add								10%	10%
2220	For vertical reinforcing, add								20%	20%
2400	Concrete, plain concrete, 6" thick	B-13L	4000	.004			.24	.52	.76	.94
2420	8" thick		3500	.005			.27	.59	.86	1.07
2440	10" thick		3000	.005			.32	.69	1.01	1.25
2500	12" thick		2500	.006			.38	.83	1.21	1.50
2600	For average reinforcing, add								10%	10%
2620	For heavy reinforcing, add								20%	20%
9000	Minimum labor/equipment charge	A-1	2	4	Job		168	56	224	330

02 41 16.19 Minor Building Deconstruction

			Crew	Daily Output	Labor-Hours	Unit	Material	2020 Bare Costs Labor	2020 Bare Costs Equipment	Total	Total Incl O&P
0010	**MINOR BUILDING DECONSTRUCTION**	R024119-10									
0011	For salvage, avg house										
0225	2 story, pre 1970 house, 1400 S.F., labor for all salv mat, min	G	6 Clab	25	1.920	SF Flr.		81		81	130
0230	Maximum	G	"	15	3.200	"		135		135	216
0235	Salvage of carpet, tackless	G	2 Clab	2400	.007	S.F.		.28		.28	.45
0240	Wood floors, incl. denailing and packaging	G	3 Clab	270	.089	"		3.74		3.74	6
0260	Wood doors and trim, standard	G	1 Clab	16	.500	Ea.		21		21	33.50
0280	Base or cove mouldings, incl. denailing and packaging	G		500	.016	L.F.		.67		.67	1.08
0320	Closet shelving and trim, incl. denailing and packaging	G		18	.444	Set		18.70		18.70	30
0340	Kitchen cabinets, uppers and lowers, prefab type	G	2 Clab	24	.667	L.F.		28		28	45
0360	Kitchen cabinets, uppers and lowers, built-in	G		12	1.333	"		56		56	90
0370	Bath fixt, incl. toilet, tub, vanity, and med cabinet	G		3	5.333	Set		225		225	360

02 41 19 – Selective Demolition

02 41 19.13 Selective Building Demolition

0010	**SELECTIVE BUILDING DEMOLITION**
0020	Costs related to selective demolition of specific building components
0025	are included under Common Work Results (XX 05)
0030	in the component's appropriate division.

02 41 19.16 Selective Demolition, Cutout

			Crew	Daily Output	Labor-Hours	Unit	Material	2020 Bare Costs Labor	2020 Bare Costs Equipment	Total	Total Incl O&P
0010	**SELECTIVE DEMOLITION, CUTOUT**	R024119-10									
0020	Concrete, elev. slab, light reinforcement, under 6 C.F.		B-9	65	.615	C.F.		26	5.50	31.50	48
0050	Light reinforcing, over 6 C.F.			75	.533	"		22.50	4.76	27.26	42
0200	Slab on grade to 6" thick, not reinforced, under 8 S.F.			85	.471	S.F.		20	4.20	24.20	36.50
0250	8-16 S.F.			175	.229	"		9.70	2.04	11.74	17.80
0255	For over 16 S.F. see Line 02 41 16.17 0400										
0600	Walls, not reinforced, under 6 C.F.		B-9	60	.667	C.F.		28.50	5.95	34.45	52
0650	6-12 C.F.		"	80	.500	"		21.50	4.47	25.97	39
0655	For over 12 C.F. see Line 02 41 16.17 2500										
1000	Concrete, elevated slab, bar reinforced, under 6 C.F.		B-9	45	.889	C.F.		38	7.95	45.95	69.50
1050	Bar reinforced, over 6 C.F.			50	.800	"		34	7.15	41.15	62.50
1200	Slab on grade to 6" thick, bar reinforced, under 8 S.F.			75	.533	S.F.		22.50	4.76	27.26	42
1250	8-16 S.F.			150	.267	"		11.35	2.38	13.73	21

02 41 Demolition

02 41 19 – Selective Demolition

02 41 19.16 Selective Demolition, Cutout

		Crew	Daily Output	Labor-Hours	Unit	Material	2020 Bare Costs Labor	Equipment	Total	Total Incl O&P
1255	For over 16 S.F. see Line 02 41 16.17 0440									
1400	Walls, bar reinforced, under 6 C.F.	B-9	50	.800	C.F.		34	7.15	41.15	62.50
1450	6-12 C.F.	"	70	.571	"		24.50	5.10	29.60	44.50
1455	For over 12 C.F. see Lines 02 41 16.17 2500 and 2600									
2000	Brick, to 4 S.F. opening, not including toothing									
2040	4" thick	B-9	30	1.333	Ea.		56.50	11.90	68.40	104
2060	8" thick		18	2.222			94.50	19.85	114.35	173
2080	12" thick		10	4			170	35.50	205.50	310
2400	Concrete block, to 4 S.F. opening, 2" thick		35	1.143			48.50	10.20	58.70	89
2420	4" thick		30	1.333			56.50	11.90	68.40	104
2440	8" thick		27	1.481			63	13.25	76.25	116
2460	12" thick		24	1.667			71	14.90	85.90	129
2600	Gypsum block, to 4 S.F. opening, 2" thick		80	.500			21.50	4.47	25.97	39
2620	4" thick		70	.571			24.50	5.10	29.60	44.50
2640	8" thick		55	.727			31	6.50	37.50	56.50
2800	Terra cotta, to 4 S.F. opening, 4" thick		70	.571			24.50	5.10	29.60	44.50
2840	8" thick		65	.615			26	5.50	31.50	48
2880	12" thick	▼	50	.800	▼		34	7.15	41.15	62.50
4000	For toothing masonry, see Section 04 01 20.50									
6000	Walls, interior, not including re-framing,									
6010	openings to 5 S.F.									
6100	Drywall to 5/8" thick	1 Clab	24	.333	Ea.		14.05		14.05	22.50
6200	Paneling to 3/4" thick		20	.400			16.85		16.85	27
6300	Plaster, on gypsum lath		20	.400			16.85		16.85	27
6340	On wire lath	▼	14	.571	▼		24		24	38.50
7000	Wood frame, not including re-framing, openings to 5 S.F.									
7200	Floors, sheathing and flooring to 2" thick	1 Clab	5	1.600	Ea.		67.50		67.50	108
7310	Roofs, sheathing to 1" thick, not including roofing		6	1.333			56		56	90
7410	Walls, sheathing to 1" thick, not including siding		7	1.143	▼		48		48	77
8500	Minimum labor/equipment charge	▼	4	2	Job		84		84	135

02 41 19.18 Selective Demolition, Disposal Only

		Crew	Daily Output	Labor-Hours	Unit	Material	2020 Bare Costs Labor	Equipment	Total	Total Incl O&P
0010	**SELECTIVE DEMOLITION, DISPOSAL ONLY** R024119-10									
0015	Urban bldg w/salvage value allowed									
0020	Including loading and 5 mile haul to dump									
0200	Steel frame	B-3	430	.112	C.Y.		5.25	5.40	10.65	14.30
0300	Concrete frame		365	.132			6.20	6.35	12.55	16.90
0400	Masonry construction		445	.108			5.10	5.20	10.30	13.80
0500	Wood frame	▼	247	.194	▼		9.15	9.35	18.50	25

02 41 19.19 Selective Demolition

		Crew	Daily Output	Labor-Hours	Unit	Material	2020 Bare Costs Labor	Equipment	Total	Total Incl O&P
0010	**SELECTIVE DEMOLITION**, Rubbish Handling R024119-10									
0020	The following are to be added to the demolition prices									
0050	The following are components for a complete chute system									
0100	Top chute circular steel, 4' long, 18" diameter	B-1C	15	1.600	Ea.	285	68.50	20.50	374	450
0102	23" diameter		15	1.600		310	68.50	20.50	399	475
0104	27" diameter		15	1.600		330	68.50	20.50	419	500
0106	30" diameter		15	1.600		355	68.50	20.50	444	525
0108	33" diameter		15	1.600		380	68.50	20.50	469	555
0110	36" diameter		15	1.600		405	68.50	20.50	494	580
0112	Regular chute, 18" diameter		15	1.600		214	68.50	20.50	303	370
0114	23" diameter		15	1.600		238	68.50	20.50	327	395
0116	27" diameter		15	1.600		262	68.50	20.50	351	420
0118	30" diameter	▼	15	1.600	▼	268	68.50	20.50	357	430

02 41 19.19 Selective Demolition	Crew	Daily Output	Labor-Hours	Unit	Material	2020 Bare Costs Labor	Equipment	Total	Total Incl O&P
0120	33" diameter	B-1C	15	1.600	Ea.	310	68.50	20.50	399	475
0122	36" diameter		15	1.600		330	68.50	20.50	419	500
0124	Control door chute, 18" diameter		15	1.600		405	68.50	20.50	494	580
0126	23" diameter		15	1.600		430	68.50	20.50	519	605
0128	27" diameter		15	1.600		450	68.50	20.50	539	630
0130	30" diameter		15	1.600		475	68.50	20.50	564	660
0132	33" diameter		15	1.600		500	68.50	20.50	589	685
0134	36" diameter		15	1.600		520	68.50	20.50	609	710
0136	Chute liners, 14 ga., 18"-30" diameter		15	1.600		228	68.50	20.50	317	385
0138	33"-36" diameter		15	1.600		284	68.50	20.50	373	450
0140	17% thinner chute, 30" diameter		15	1.600		233	68.50	20.50	322	390
0142	33% thinner chute, 30" diameter		15	1.600		176	68.50	20.50	265	325
0144	Top chute cover	1 Clab	24	.333		155	14.05		169.05	194
0146	Door chute cover	"	24	.333		155	14.05		169.05	194
0148	Top chute trough	2 Clab	12	1.333		530	56		586	670
0150	Bolt down frame & counter weights, 250 lb.	B-1	4	6		4,325	257		4,582	5,150
0152	500 lb.		4	6		6,375	257		6,632	7,400
0154	750 lb.		4	6		11,100	257		11,357	12,600
0156	1000 lb.		2.67	8.989		12,000	385		12,385	13,800
0158	1500 lb.		2.67	8.989		14,000	385		14,385	16,000
0160	Chute warning light system, 5 stories	B-1C	4	6		8,525	257	77.50	8,859.50	9,875
0162	10 stories	"	2	12		14,800	515	155	15,470	17,300
0164	Dust control device for dumpsters	1 Clab	8	1		158	42		200	242
0166	Install or replace breakaway cord		8	1		28	42		70	98
0168	Install or replace warning sign		16	.500		11.10	21		32.10	45.50
0600	Dumpster, weekly rental, 1 dump/week, 6 C.Y. capacity (2 tons)				Week	415			415	455
0700	10 C.Y. capacity (3 tons)					480			480	530
0725	20 C.Y. capacity (5 tons) R024119-20					565			565	625
0800	30 C.Y. capacity (7 tons)					730			730	800
0840	40 C.Y. capacity (10 tons)					775			775	850
0900	Alternate pricing for dumpsters									
0910	Delivery, average for all sizes				Ea.	75			75	82.50
0920	Haul, average for all sizes					235			235	259
0930	Rent per day, average for all sizes					20			20	22
0940	Rent per month, average for all sizes					80			80	88
0950	Disposal fee per ton, average for all sizes				Ton	88			88	97
2000	Load, haul, dump and return, 0'-50' haul, hand carried	2 Clab	24	.667	C.Y.		28		28	45
2005	Wheeled		37	.432			18.20		18.20	29
2040	0'-100' haul, hand carried		16.50	.970			41		41	65.50
2045	Wheeled		25	.640			27		27	43
2050	Forklift	A-3R	25	.320			16.95	10.25	27.20	38
2080	Haul and return, add per each extra 100' haul, hand carried	2 Clab	35.50	.451			18.95		18.95	30.50
2085	Wheeled		54	.296			12.45		12.45	20
2120	For travel in elevators, up to 10 floors, add		140	.114			4.81		4.81	7.70
2130	0'-50' haul, incl. up to 5 riser stairs, hand carried		23	.696			29.50		29.50	47
2135	Wheeled		35	.457			19.25		19.25	31
2140	6-10 riser stairs, hand carried		22	.727			30.50		30.50	49
2145	Wheeled		34	.471			19.80		19.80	31.50
2150	11-20 riser stairs, hand carried		20	.800			33.50		33.50	54
2155	Wheeled		31	.516			21.50		21.50	35
2160	21-40 riser stairs, hand carried		16	1			42		42	67.50
2165	Wheeled		24	.667			28		28	45
2170	0'-100' haul, incl. 5 riser stairs, hand carried		15	1.067			45		45	72

02 41 19 – Selective Demolition

02 41 19.19 Selective Demolition	Crew	Daily Output	Labor-Hours	Unit	Material	2020 Bare Costs Labor	2020 Bare Costs Equipment	Total	Total Incl O&P
2175	Wheeled	2 Clab	23	.696	C.Y.		29.50		29.50	47
2180	6-10 riser stairs, hand carried		14	1.143			48		48	77
2185	Wheeled		21	.762			32		32	51.50
2190	11-20 riser stairs, hand carried		12	1.333			56		56	90
2195	Wheeled		18	.889			37.50		37.50	60
2200	21-40 riser stairs, hand carried		8	2			84		84	135
2205	Wheeled		12	1.333			56		56	90
2210	Haul and return, add per each extra 100' haul, hand carried		35.50	.451			18.95		18.95	30.50
2215	Wheeled		54	.296			12.45		12.45	20
2220	For each additional flight of stairs, up to 5 risers, add		550	.029	Flight		1.22		1.22	1.96
2225	6-10 risers, add		275	.058			2.45		2.45	3.92
2230	11-20 risers, add		138	.116			4.88		4.88	7.80
2235	21-40 risers, add		69	.232			9.75		9.75	15.65
3000	Loading & trucking, including 2 mile haul, chute loaded	B-16	45	.711	C.Y.		31.50	12.75	44.25	64.50
3040	Hand loading truck, 50' haul	"	48	.667			29.50	11.95	41.45	60.50
3080	Machine loading truck	B-17	120	.267			12.40	5.35	17.75	25.50
3120	Wheeled 50' and ramp dump loaded	2 Clab	24	.667			28		28	45
5000	Haul, per mile, up to 8 C.Y. truck	B-34B	1165	.007			.34	.49	.83	1.08
5100	Over 8 C.Y. truck	"	1550	.005			.25	.37	.62	.81

02 41 19.20 Selective Demolition, Dump Charges

		Crew	Daily Output	Labor-Hours	Unit	Material	Labor	Equipment	Total	Total Incl O&P
0010	**SELECTIVE DEMOLITION, DUMP CHARGES**	R024119-10								
0020	Dump charges, typical urban city, tipping fees only									
0100	Building construction materials				Ton	74			74	81
0200	Trees, brush, lumber					63			63	69.50
0300	Rubbish only					63			63	69.50
0500	Reclamation station, usual charge					74			74	81

02 41 19.21 Selective Demolition, Gutting

		Crew	Daily Output	Labor-Hours	Unit	Material	Labor	Equipment	Total	Total Incl O&P
0010	**SELECTIVE DEMOLITION, GUTTING**	R024119-10								
0020	Building interior, including disposal, dumpster fees not included									
0500	Residential building									
0560	Minimum	B-16	400	.080	SF Flr.		3.54	1.43	4.97	7.30
0580	Maximum	"	360	.089	"		3.94	1.59	5.53	8.05
0900	Commercial building									
1000	Minimum	B-16	350	.091	SF Flr.		4.05	1.64	5.69	8.30
1020	Maximum		250	.128	"		5.65	2.29	7.94	11.60
3000	Minimum labor/equipment charge		4	8	Job		355	143	498	730

02 41 19.25 Selective Demolition, Saw Cutting

		Crew	Daily Output	Labor-Hours	Unit	Material	Labor	Equipment	Total	Total Incl O&P
0010	**SELECTIVE DEMOLITION, SAW CUTTING**	R024119-10								
0015	Asphalt, up to 3" deep	B-89	1050	.015	L.F.	.20	.76	.98	1.94	2.51
0020	Each additional inch of depth	"	1800	.009		.07	.45	.57	1.09	1.40
1200	Masonry walls, hydraulic saw, brick, per inch of depth	B-89B	300	.053		.07	2.67	5	7.74	9.80
1220	Block walls, solid, per inch of depth	"	250	.064		.07	3.21	6	9.28	11.70
2000	Brick or masonry w/hand held saw, per inch of depth	A-1	125	.064		.05	2.69	.89	3.63	5.35
5000	Wood sheathing to 1" thick, on walls	1 Carp	200	.040			2.13		2.13	3.41
5020	On roof	"	250	.032			1.70		1.70	2.72
9000	Minimum labor/equipment charge	A-1	2	4	Job		168	56	224	330

02 41 19.27 Selective Demolition, Torch Cutting

		Crew	Daily Output	Labor-Hours	Unit	Material	Labor	Equipment	Total	Total Incl O&P
0010	**SELECTIVE DEMOLITION, TORCH CUTTING**	R024119-10								
0020	Steel, 1" thick plate	E-25	333	.024	L.F.	.90	1.43	.04	2.37	3.37
0040	1" diameter bar	"	600	.013	Ea.	.15	.79	.02	.96	1.49
1000	Oxygen lance cutting, reinforced concrete walls									
1040	12"-16" thick walls	1 Clab	10	.800	L.F.		33.50		33.50	54

02 41 Demolition

02 41 19 – Selective Demolition

02 41 19.27 Selective Demolition, Torch Cutting		Crew	Daily Output	Labor-Hours	Unit	Material	2020 Bare Costs Labor	Equipment	Total	Total Incl O&P
1080	24" thick walls	1 Clab	6	1.333	L.F.		56		56	90
1090	Minimum labor/equipment charge	E-25	2	4	Job		238	6.40	244.40	395
1100	See Section 05 05 21.10									

02 42 Removal and Diversion of Construction Materials

02 42 10 – Building Deconstruction

02 42 10.10 Estimated Salvage Value or Savings

02 42 10.10			Crew	Daily Output	Labor-Hours	Unit	Material	2020 Bare Costs Labor	Equipment	Total	Total Incl O&P
0010	**ESTIMATED SALVAGE VALUE OR SAVINGS**										
0015	Excludes material handling, packaging, container costs and										
0020	transportation for salvage or disposal										
0050	All items in Section 02 42 10.10 are credit deducts and not costs										
0100	Copper wire salvage value	G				Lb.	1.70			1.70	1.87
0200	Brass salvage value	G					1.25			1.25	1.38
0250	Bronze						1.45			1.45	1.60
0300	Steel	G					.05			.05	.06
0304	304 Stainless Steel						.30			.30	.33
0316	316 Stainless Steel						.45			.45	.50
0400	Cast Iron	G					.06			.06	.07
0600	Aluminum siding	G					.33			.33	.36
0610	Aluminum wire						.47			.47	.52

02 42 10.20 Deconstruction of Building Components

02 42 10.20			Crew	Daily Output	Labor-Hours	Unit	Material	2020 Bare Costs Labor	Equipment	Total	Total Incl O&P
0010	**DECONSTRUCTION OF BUILDING COMPONENTS**										
0012	Buildings one or two stories only										
0015	Excludes material handling, packaging, container costs and										
0020	transportation for salvage or disposal										
0050	Deconstruction of plumbing fixtures										
0100	Wall hung or countertop lavatory	G	2 Clab	16	1	Ea.		42		42	67.50
0110	Single or double compartment kitchen sink	G		14	1.143			48		48	77
0120	Wall hung urinal	G		14	1.143			48		48	77
0130	Floor mounted	G		8	2			84		84	135
0140	Floor mounted water closet	G		16	1			42		42	67.50
0150	Wall hung	G		14	1.143			48		48	77
0160	Water fountain, free standing	G		16	1			42		42	67.50
0170	Wall hung or deck mounted	G		12	1.333			56		56	90
0180	Bathtub, steel or fiberglass	G		10	1.600			67.50		67.50	108
0190	Cast iron	G		8	2			84		84	135
0200	Shower, single	G		6	2.667			112		112	180
0210	Group	G		7	2.286			96		96	154
0300	Deconstruction of electrical fixtures										
0310	Surface mounted incandescent fixtures	G	2 Clab	48	.333	Ea.		14.05		14.05	22.50
0320	Fluorescent, 2 lamp	G		32	.500			21		21	33.50
0330	4 lamp	G		24	.667			28		28	45
0340	Strip fluorescent, 1 lamp	G		40	.400			16.85		16.85	27
0350	2 lamp	G		32	.500			21		21	33.50
0400	Recessed drop-in fluorescent fixture, 2 lamp	G		27	.593			25		25	40
0410	4 lamp	G		18	.889			37.50		37.50	60
0500	Deconstruction of appliances										
0510	Cooking stoves	G	2 Clab	26	.615	Ea.		26		26	41.50
0520	Dishwashers	G	"	26	.615	"		26		26	41.50
0600	Deconstruction of millwork and trim										
0610	Cabinets, wood	G	2 Carp	40	.400	L.F.		21.50		21.50	34

For customer support on your Commercial Renovation Costs with RSMeans data, call 800.448.8182.

35

02 42 10 – Building Deconstruction

02 42 10.20 Deconstruction of Building Components		Crew	Daily Output	Labor-Hours	Unit	Material	2020 Bare Costs Labor	2020 Bare Costs Equipment	Total	Total Incl O&P
0620	Countertops	G 2 Carp	100	.160	L.F.		8.50		8.50	13.60
0630	Wall paneling, 1" thick	G	500	.032	S.F.		1.70		1.70	2.72
0640	Ceiling trim	G	500	.032	L.F.		1.70		1.70	2.72
0650	Wainscoting	G	500	.032	S.F.		1.70		1.70	2.72
0660	Base, 3/4"-1" thick	G	600	.027	L.F.		1.42		1.42	2.27
0700	Deconstruction of doors and windows									
0710	Doors, wrap, interior, wood, single, no closers	G 2 Carp	21	.762	Ea.	4.75	40.50		45.25	70.50
0720	Double	G	13	1.231		9.50	65.50		75	115
0730	Solid core, single, exterior or interior	G	10	1.600		4.75	85		89.75	141
0740	Double	G	8	2		9.50	106		115.50	180
0810	Windows, wrap, wood, single									
0812	with no casement or cladding	G 2 Carp	21	.762	Ea.	4.75	40.50		45.25	70.50
0820	with casement and/or cladding	G "	18	.889	"	4.75	47		51.75	81
0900	Deconstruction of interior finishes									
0910	Drywall for recycling	G 2 Clab	1775	.009	S.F.		.38		.38	.61
0920	Plaster wall, first floor	G	1775	.009			.38		.38	.61
0930	Second floor	G	1330	.012			.51		.51	.81
1000	Deconstruction of roofing and accessories									
1010	Built-up roofs	G 2 Clab	570	.028	S.F.		1.18		1.18	1.89
1020	Gutters, fascia and rakes	G "	1140	.014	L.F.		.59		.59	.95
2000	Deconstruction of wood components									
2010	Roof sheeting	G 2 Clab	570	.028	S.F.		1.18		1.18	1.89
2020	Main roof framing	G	760	.021	L.F.		.89		.89	1.42
2030	Porch roof framing	G	445	.036			1.51		1.51	2.43
2040	Beams 4" x 8"	G B-1	375	.064			2.74		2.74	4.39
2050	4" x 10"	G	300	.080			3.42		3.42	5.50
2055	4" x 12"	G	250	.096			4.11		4.11	6.60
2060	6" x 8"	G	250	.096			4.11		4.11	6.60
2065	6" x 10"	G	200	.120			5.15		5.15	8.20
2070	6" x 12"	G	170	.141			6.05		6.05	9.65
2075	8" x 12"	G	126	.190			8.15		8.15	13.05
2080	10" x 12"	G	100	.240			10.25		10.25	16.45
2100	Ceiling joists	G 2 Clab	800	.020			.84		.84	1.35
2150	Wall framing, interior	G	1230	.013			.55		.55	.88
2160	Sub-floor	G	2000	.008	S.F.		.34		.34	.54
2170	Floor joists	G	2000	.008	L.F.		.34		.34	.54
2200	Wood siding (no lead or asbestos)	G	1300	.012	S.F.		.52		.52	.83
2300	Wall framing, exterior	G	1600	.010	L.F.		.42		.42	.67
2400	Stair risers	G	53	.302	Ea.		12.70		12.70	20.50
2500	Posts	G	800	.020	L.F.		.84		.84	1.35
3000	Deconstruction of exterior brick walls									
3010	Exterior brick walls, first floor	G 2 Clab	200	.080	S.F.		3.37		3.37	5.40
3020	Second floor	G	64	.250	"		10.55		10.55	16.85
3030	Brick chimney	G	100	.160	C.F.		6.75		6.75	10.80
4000	Deconstruction of concrete									
4010	Slab on grade, 4" thick, plain concrete	G B-9	500	.080	S.F.		3.40	.71	4.11	6.25
4020	Wire mesh reinforced	G	470	.085			3.62	.76	4.38	6.65
4030	Rod reinforced	G	400	.100			4.25	.89	5.14	7.80
4110	Foundation wall, 6" thick, plain concrete	G	160	.250			10.65	2.23	12.88	19.45
4120	8" thick	G	140	.286			12.15	2.55	14.70	22.50
4130	10" thick	G	120	.333			14.15	2.98	17.13	26
9000	Deconstruction process, support equipment as needed									
9010	Daily use, portal to portal, 12-ton truck-mounted hydraulic crane crew	G A-3H	1	8	Day		475	725	1,200	1,525

02 42 Removal and Diversion of Construction Materials

02 42 10 – Building Deconstruction

02 42 10.20 Deconstruction of Building Components

		Crew	Daily Output	Labor-Hours	Unit	Material	Labor	Equipment	Total	Total Incl O&P
9020	Daily use, skid steer and operator	G A-3C	1	8	Day		425	400	825	1,100
9030	Daily use, backhoe 48 HP, operator and labor	G "	1	8	▼		425	400	825	1,100

02 42 10.30 Deconstruction Material Handling

		Crew	Daily Output	Labor-Hours	Unit	Material	Labor	Equipment	Total	Total Incl O&P
0010	**DECONSTRUCTION MATERIAL HANDLING**									
0012	Buildings one or two stories only									
0100	Clean and stack brick on pallet	G 2 Clab	1200	.013	Ea.		.56		.56	.90
0200	Haul 50' and load rough lumber up to 2" x 8" size	G	2000	.008	"		.34		.34	.54
0210	Lumber larger than 2" x 8"	G	3200	.005	B.F.		.21		.21	.34
0300	Finish wood for recycling stack and wrap per pallet	G	8	2	Ea.	38	84		122	177
0350	Light fixtures		6	2.667		68.50	112		180.50	255
0375	Windows		6	2.667		64.50	112		176.50	251
0400	Miscellaneous materials	▼	8	2	▼	19	84		103	156
1000	See Section 02 41 19.19 for bulk material handling									

02 43 Structure Moving

02 43 13 – Structure Relocation

02 43 13.13 Building Relocation

		Crew	Daily Output	Labor-Hours	Unit	Material	Labor	Equipment	Total	Total Incl O&P
0010	**BUILDING RELOCATION**									
0011	One day move, up to 24' wide									
0020	Reset on existing foundation				Total				11,500	11,500
0040	Wood or steel frame bldg., based on ground floor area	G B-4	185	.259	S.F.		11.30	2.67	13.97	21
0060	Masonry bldg., based on ground floor area	G "	137	.350			15.25	3.60	18.85	28.50
0200	For 24'-42' wide, add				▼				15%	15%

02 58 Snow Control

02 58 13 – Snow Fencing

02 58 13.10 Snow Fencing System

		Crew	Daily Output	Labor-Hours	Unit	Material	Labor	Equipment	Total	Total Incl O&P
0010	**SNOW FENCING SYSTEM**									
7001	Snow fence on steel posts 10' OC, 4' high	B-1	500	.048	L.F.	.39	2.05		2.44	3.72

02 65 Underground Storage Tank Removal

02 65 10 – Underground Tank and Contaminated Soil Removal

02 65 10.30 Removal of Underground Storage Tanks

		Crew	Daily Output	Labor-Hours	Unit	Material	Labor	Equipment	Total	Total Incl O&P
0010	**REMOVAL OF UNDERGROUND STORAGE TANKS** R026510-20									
0011	Petroleum storage tanks, non-leaking									
0100	Excavate & load onto trailer									
0110	3,000 gal. to 5,000 gal. tank	G B-14	4	12	Ea.		530	53.50	583.50	905
0120	6,000 gal. to 8,000 gal. tank	G B-3A	3	13.333			600	229	829	1,200
0130	9,000 gal. to 12,000 gal. tank	G "	2	20	▼		900	345	1,245	1,800
0190	Known leaking tank, add				%				100%	100%
0200	Remove sludge, water and remaining product from bottom									
0201	of tank with vacuum truck									
0300	3,000 gal. to 5,000 gal. tank	G A-13	5	1.600	Ea.		85	148	233	295
0310	6,000 gal. to 8,000 gal. tank	G	4	2			106	184	290	370
0320	9,000 gal. to 12,000 gal. tank	G ▼	3	2.667	▼		141	246	387	490
0390	Dispose of sludge off-site, average				Gal.				6.25	6.80
0400	Insert inert solid CO_2 "dry ice" into tank									

02 65 Underground Storage Tank Removal

02 65 10 – Underground Tank and Contaminated Soil Removal

02 65 10.30 Removal of Underground Storage Tanks		Crew	Daily Output	Labor-Hours	Unit	Material	2020 Bare Costs Labor	Equipment	Total	Total Incl O&P
0401	For cleaning/transporting tanks (1.5 lb./100 gal. cap)	G	1 Clab	500	.016	Lb.	1.22	.67		1.89	2.42
0503	Disconnect and remove piping	G	1 Plum	160	.050	L.F.		3.22		3.22	4.98
0603	Transfer liquids, 10% of volume	G	"	1600	.005	Gal.		.32		.32	.50
0703	Cut accessway into underground storage tank	G	1 Clab	5.33	1.501	Ea.		63		63	101
0813	Remove sludge, wash and wipe tank, 500 gal.	G	1 Plum	8	1			64.50		64.50	99.50
0823	3,000 gal.	G		6.67	1.199			77.50		77.50	120
0833	5,000 gal.	G		6.15	1.301			84		84	130
0843	8,000 gal.	G		5.33	1.501			96.50		96.50	150
0853	10,000 gal.	G		4.57	1.751			113		113	174
0863	12,000 gal.	G	↓	4.21	1.900	↓		122		122	189
1020	Haul tank to certified salvage dump, 100 miles round trip										
1023	3,000 gal. to 5,000 gal. tank					Ea.				760	830
1026	6,000 gal. to 8,000 gal. tank									880	960
1029	9,000 gal. to 12,000 gal. tank					↓				1,050	1,150
1100	Disposal of contaminated soil to landfill										
1110	Minimum					C.Y.				145	160
1111	Maximum					"				400	440
1120	Disposal of contaminated soil to										
1121	bituminous concrete batch plant										
1130	Minimum					C.Y.				80	88
1131	Maximum					"				115	125
1203	Excavate, pull, & load tank, backfill hole, 8,000 gal. +	G	B-12C	.50	32	Ea.		1,625	1,900	3,525	4,650
1213	Haul tank to certified dump, 100 miles rt, 8,000 gal. +	G	B-34K	1	8			390	855	1,245	1,575
1223	Excavate, pull, & load tank, backfill hole, 500 gal.	G	B-11C	1	16			790	214	1,004	1,475
1233	Excavate, pull, & load tank, backfill hole, 3,000-5,000 gal.	G	B-11M	.50	32			1,575	465	2,040	3,000
1243	Haul tank to certified dump, 100 miles rt, 500 gal.	G	B-34L	1	8			425	196	621	880
1253	Haul tank to certified dump, 100 miles rt, 3,000-5,000 gal.	G	B-34M	1	8	↓		425	820	1,245	1,575
2010	Decontamination of soil on site incl poly tarp on top/bottom										
2011	Soil containment berm and chemical treatment										
2020	Minimum	G	B-11C	100	.160	C.Y.	7.65	7.90	2.14	17.69	23.50
2021	Maximum	G	"	100	.160		9.90	7.90	2.14	19.94	25.50
2050	Disposal of decontaminated soil, minimum									135	150
2055	Maximum					↓				400	440

02 81 Transportation and Disposal of Hazardous Materials

02 81 20 – Hazardous Waste Handling

02 81 20.10 Hazardous Waste Cleanup/Pickup/Disposal

		Crew	Daily Output	Labor-Hours	Unit	Material	2020 Bare Costs Labor	Equipment	Total	Total Incl O&P
0010	**HAZARDOUS WASTE CLEANUP/PICKUP/DISPOSAL**										
0100	For contractor rental equipment, i.e., dozer,										
0110	Front end loader, dump truck, etc., see 01 54 33 Reference Section										
1000	Solid pickup										
1100	55 gal. drums					Ea.				240	265
1120	Bulk material, minimum					Ton				190	210
1130	Maximum					"				595	655
1200	Transportation to disposal site										
1220	Truckload = 80 drums or 25 C.Y. or 18 tons										
1260	Minimum					Mile				3.95	4.45
1270	Maximum					"				7.25	7.98
3000	Liquid pickup, vacuum truck, stainless steel tank										
3100	Minimum charge, 4 hours										
3110	1 compartment, 2200 gallon					Hr.				140	155

02 81 20 – Hazardous Waste Handling

02 81 20.10 Hazardous Waste Cleanup/Pickup/Disposal	Crew	Daily Output	Labor-Hours	Unit	Material	2020 Bare Costs Labor	Equipment	Total	Total Incl O&P
3120	2 compartment, 5000 gallon				Hr.				200	225
3400	Transportation in 6900 gallon bulk truck				Mile				7.95	8.75
3410	In teflon lined truck				"				10.20	11.25
5000	Heavy sludge or dry vacuumable material				Hr.				140	155
6000	Dumpsite disposal charge, minimum				Ton				140	155
6020	Maximum				"				415	455

02 82 Asbestos Remediation

02 82 13 – Asbestos Abatement

02 82 13.39 Asbestos Remediation Plans and Methods

		Crew	Daily Output	Labor-Hours	Unit	Material	Labor	Equipment	Total	Total Incl O&P
0010	**ASBESTOS REMEDIATION PLANS AND METHODS**									
0100	Building Survey-Commercial Building				Ea.				2,200	2,400
0200	Asbestos Abatement Remediation Plan				"				1,350	1,475

02 82 13.41 Asbestos Abatement Equipment

		Crew	Daily Output	Labor-Hours	Unit	Material	Labor	Equipment	Total	Total Incl O&P
0010	**ASBESTOS ABATEMENT EQUIPMENT** R028213-20									
0011	Equipment and supplies, buy									
0200	Air filtration device, 2000 CFM				Ea.	950			950	1,050
0250	Large volume air sampling pump, minimum					315			315	350
0260	Maximum					345			345	380
0300	Airless sprayer unit, 2 gun					2,625			2,625	2,875
0350	Light stand, 500 watt					41			41	45
0400	Personal respirators									
0410	Negative pressure, 1/2 face, dual operation, minimum				Ea.	30			30	33
0420	Maximum					34			34	37.50
0450	P.A.P.R., full face, minimum					780			780	855
0460	Maximum					1,225			1,225	1,325
0470	Supplied air, full face, including air line, minimum					289			289	320
0480	Maximum					500			500	550
0500	Personnel sampling pump					237			237	261
1500	Power panel, 20 unit, including GFI					475			475	525
1600	Shower unit, including pump and filters					1,125			1,125	1,250
1700	Supplied air system (type C)					3,700			3,700	4,075
1750	Vacuum cleaner, HEPA, 16 gal., stainless steel, wet/dry					1,275			1,275	1,400
1760	55 gallon					1,475			1,475	1,600
1800	Vacuum loader, 9-18 ton/hr.					99,000			99,000	109,000
1900	Water atomizer unit, including 55 gal. drum					305			305	340
2000	Worker protection, whole body, foot, head cover & gloves, plastic					7.45			7.45	8.20
2500	Respirator, single use					30			30	33
2550	Cartridge for respirator					5.15			5.15	5.65
2570	Glove bag, 7 mil, 50" x 64"					8.75			8.75	9.60
2580	10 mil, 44" x 60"					7.90			7.90	8.65
6000	Disposable polyethylene bags, 6 mil, 3 C.F.					.77			.77	.85
6300	Disposable fiber drums, 3 C.F.					18.90			18.90	21
6400	Pressure sensitive caution labels, 3" x 5"					3.08			3.08	3.39
6450	11" x 17"					7.75			7.75	8.55
6500	Negative air machine, 1800 CFM					890			890	975

02 82 13.42 Preparation of Asbestos Containment Area

		Crew	Daily Output	Labor-Hours	Unit	Material	Labor	Equipment	Total	Total Incl O&P
0010	**PREPARATION OF ASBESTOS CONTAINMENT AREA**									
0100	Pre-cleaning, HEPA vacuum and wet wipe, flat surfaces	A-9	12000	.005	S.F.	.01	.31		.32	.51
0200	Protect carpeted area, 2 layers 6 mil poly on 3/4" plywood	"	1000	.064		2.10	3.76		5.86	8.25

02 82 Asbestos Remediation

02 82 13 – Asbestos Abatement

02 82 13.42 Preparation of Asbestos Containment Area

		Crew	Daily Output	Labor-Hours	Unit	Material	2020 Bare Costs Labor	Equipment	Total	Total Incl O&P
0300	Separation barrier, 2" x 4" @ 16", 1/2" plywood ea. side, 8' high	2 Carp	400	.040	S.F.	3.08	2.13		5.21	6.80
0310	12' high		320	.050		3.02	2.66		5.68	7.60
0320	16' high		200	.080		2.99	4.25		7.24	10.10
0400	Personnel decontam. chamber, 2" x 4" @ 16", 3/4" ply ea. side		280	.057		3.55	3.04		6.59	8.75
0450	Waste decontam. chamber, 2" x 4" studs @ 16", 3/4" ply ea. side		360	.044		3.55	2.36		5.91	7.70
0500	Cover surfaces with polyethylene sheeting									
0501	Including glue and tape									
0550	Floors, each layer, 6 mil	A-9	8000	.008	S.F.	.04	.47		.51	.79
0551	4 mil		9000	.007		.03	.42		.45	.69
0560	Walls, each layer, 6 mil		6000	.011		.04	.63		.67	1.04
0561	4 mil		7000	.009		.03	.54		.57	.88
0570	For heights above 14', add						20%			
0575	For heights above 20', add						30%			
0580	For fire retardant poly, add					100%				
0590	For large open areas, deduct					10%	20%			
0600	Seal floor penetrations with foam firestop to 36 sq. in.	2 Carp	200	.080	Ea.	13.90	4.25		18.15	22
0610	36 sq. in. to 72 sq. in.		125	.128		28	6.80		34.80	41.50
0615	72 sq. in. to 144 sq. in.		80	.200		55.50	10.65		66.15	78
0620	Wall penetrations, to 36 sq. in.		180	.089		13.90	4.72		18.62	23
0630	36 sq. in. to 72 sq. in.		100	.160		28	8.50		36.50	44
0640	72 sq. in. to 144 sq. in.		60	.267		55.50	14.15		69.65	83.50
0800	Caulk seams with latex	1 Carp	230	.035	L.F.	.17	1.85		2.02	3.15
0900	Set up neg. air machine, 1-2k CFM/25 M.C.F. volume	1 Asbe	4.30	1.860	Ea.		109		109	173
0950	Set up and remove portable shower unit	2 Asbe	4	4	"		235		235	370

02 82 13.43 Bulk Asbestos Removal

		Crew	Daily Output	Labor-Hours	Unit	Material	2020 Bare Costs Labor	Equipment	Total	Total Incl O&P
0010	**BULK ASBESTOS REMOVAL**									
0020	Includes disposable tools and 2 suits and 1 respirator filter/day/worker									
0100	Beams, W 10 x 19	A-9	235	.272	L.F.	.68	16		16.68	26.50
0110	W 12 x 22		210	.305		.76	17.90		18.66	29.50
0120	W 14 x 26		180	.356		.89	21		21.89	34
0130	W 16 x 31		160	.400		1	23.50		24.50	38
0140	W 18 x 40		140	.457		1.15	27		28.15	44
0150	W 24 x 55		110	.582		1.46	34		35.46	55.50
0160	W 30 x 108		85	.753		1.89	44		45.89	72
0170	W 36 x 150		72	.889		2.23	52		54.23	85.50
0200	Boiler insulation		480	.133	S.F.	.40	7.85		8.25	12.85
0210	With metal lath, add				%				50%	50%
0300	Boiler breeching or flue insulation	A-9	520	.123	S.F.	.31	7.25		7.56	11.80
0310	For active boiler, add				%				100%	100%
0400	Duct or AHU insulation	A-10B	440	.073	S.F.	.18	4.28		4.46	7
0500	Duct vibration isolation joints, up to 24 sq. in. duct	A-9	56	1.143	Ea.	2.87	67		69.87	109
0520	25 sq. in. to 48 sq. in. duct		48	1.333		3.35	78.50		81.85	128
0530	49 sq. in. to 76 sq. in. duct		40	1.600		4.01	94		98.01	153
0600	Pipe insulation, air cell type, up to 4" diameter pipe		900	.071	L.F.	.18	4.17		4.35	6.80
0610	4" to 8" diameter pipe		800	.080		.20	4.70		4.90	7.65
0620	10" to 12" diameter pipe		700	.091		.23	5.35		5.58	8.75
0630	14" to 16" diameter pipe		550	.116		.29	6.85		7.14	11.15
0650	Over 16" diameter pipe		650	.098	S.F.	.25	5.80		6.05	9.40
0700	With glove bag up to 3" diameter pipe		200	.320	L.F.	9.40	18.80		28.20	40.50
1000	Pipe fitting insulation up to 4" diameter pipe		320	.200	Ea.	.50	11.75		12.25	19.15
1100	6" to 8" diameter pipe		304	.211		.53	12.35		12.88	20
1110	10" to 12" diameter pipe		192	.333		.84	19.55		20.39	32

For customer support on your Commercial Renovation Costs with RSMeans data, call 800.448.8182.

02 82 Asbestos Remediation

02 82 13 – Asbestos Abatement

02 82 13.43 Bulk Asbestos Removal	Crew	Daily Output	Labor-Hours	Unit	Material	2020 Bare Costs Labor	Equipment	Total	Total Incl O&P
1120	14" to 16" diameter pipe	A-9	128	.500	Ea.	1.25	29.50		30.75	48
1130	Over 16" diameter pipe		176	.364	S.F.	.91	21.50		22.41	35
1200	With glove bag, up to 8" diameter pipe		75	.853	L.F.	6.55	50		56.55	86.50
2000	Scrape foam fireproofing from flat surface		2400	.027	S.F.	.07	1.57		1.64	2.55
2100	Irregular surfaces		1200	.053		.13	3.13		3.26	5.10
3000	Remove cementitious material from flat surface		1800	.036		.09	2.09		2.18	3.41
3100	Irregular surface		1000	.064		.11	3.76		3.87	6.10
4000	Scrape acoustical coating/fireproofing, from ceiling		3200	.020		.05	1.17		1.22	1.92
5000	Remove VAT and mastic from floor by hand		2400	.027		.07	1.57		1.64	2.55
5100	By machine	A-11	4800	.013		.03	.78	.01	.82	1.29
5150	For 2 layers, add				%				50%	50%
6000	Remove contaminated soil from crawl space by hand	A-9	400	.160	C.F.	.40	9.40		9.80	15.35
6100	With large production vacuum loader	A-12	700	.091	"	.23	5.35	1.05	6.63	9.90
7000	Radiator backing, not including radiator removal	A-9	1200	.053	S.F.	.13	3.13		3.26	5.10
8000	Cement-asbestos transite board and cement wall board	2 Asbe	1000	.016		.16	.94		1.10	1.67
8100	Transite shingle siding	A-10B	750	.043		.21	2.51		2.72	4.21
8200	Shingle roofing	"	2000	.016		.08	.94		1.02	1.58
8250	Built-up, no gravel, non-friable	B-2	1400	.029		.08	1.21		1.29	2.04
8260	Bituminous flashing	1 Rofc	300	.027		.08	1.23		1.31	2.25
8300	Asbestos millboard, flat board and VAT contaminated plywood	2 Asbe	1000	.016		.08	.94		1.02	1.58
9000	For type B (supplied air) respirator equipment, add				%				10%	10%

02 82 13.44 Demolition In Asbestos Contaminated Area

		Crew	Daily Output	Labor-Hours	Unit	Material	Labor	Equipment	Total	Total Incl O&P
0010	**DEMOLITION IN ASBESTOS CONTAMINATED AREA**									
0200	Ceiling, including suspension system, plaster and lath	A-9	2100	.030	S.F.	.08	1.79		1.87	2.92
0210	Finished plaster, leaving wire lath		585	.109		.27	6.40		6.67	10.50
0220	Suspended acoustical tile		3500	.018		.05	1.07		1.12	1.75
0230	Concealed tile grid system		3000	.021		.05	1.25		1.30	2.05
0240	Metal pan grid system		1500	.043		.11	2.51		2.62	4.09
0250	Gypsum board		2500	.026		.06	1.50		1.56	2.45
0260	Lighting fixtures up to 2' x 4'		72	.889	Ea.	2.23	52		54.23	85.50
0400	Partitions, non load bearing									
0410	Plaster, lath, and studs	A-9	690	.093	S.F.	.90	5.45		6.35	9.65
0450	Gypsum board and studs	"	1390	.046	"	.12	2.70		2.82	4.42
9000	For type B (supplied air) respirator equipment, add				%				10%	10%

02 82 13.45 OSHA Testing

		Crew	Daily Output	Labor-Hours	Unit	Material	Labor	Equipment	Total	Total Incl O&P
0010	**OSHA TESTING**									
0100	Certified technician, minimum				Day		340		340	340
0110	Maximum						670		670	670
0121	Industrial hygienist, minimum						385		385	385
0130	Maximum						755		755	755
0200	Asbestos sampling and PCM analysis, NIOSH 7400, minimum	1 Asbe	8	1	Ea.	15.90	58.50		74.40	111
0210	Maximum		4	2		27	117		144	216
1000	Cleaned area samples		8	1		203	58.50		261.50	315
1100	PCM air sample analysis, NIOSH 7400, minimum		8	1		17.70	58.50		76.20	113
1110	Maximum		4	2		2.30	117		119.30	189
1200	TEM air sample analysis, NIOSH 7402, minimum								80	106
1210	Maximum								360	450

02 82 13.46 Decontamination of Asbestos Containment Area

		Crew	Daily Output	Labor-Hours	Unit	Material	Labor	Equipment	Total	Total Incl O&P
0010	**DECONTAMINATION OF ASBESTOS CONTAINMENT AREA**									
0100	Spray exposed substrate with surfactant (bridging)									
0200	Flat surfaces	A-9	6000	.011	S.F.	.41	.63		1.04	1.44
0250	Irregular surfaces		4000	.016	"	.45	.94		1.39	1.99

For customer support on your Commercial Renovation Costs with RSMeans data, call 800.448.8182.

41

02 82 Asbestos Remediation

02 82 13 – Asbestos Abatement

02 82 13.46 Decontamination of Asbestos Containment Area	Crew	Daily Output	Labor-Hours	Unit	Material	2020 Bare Costs Labor	Equipment	Total	Total Incl O&P
0300 Pipes, beams, and columns	A-9	2000	.032	L.F.	.60	1.88		2.48	3.64
1000 Spray encapsulate polyethylene sheeting		8000	.008	S.F.	.41	.47		.88	1.19
1100 Roll down polyethylene sheeting		8000	.008	"		.47		.47	.74
1500 Bag polyethylene sheeting		400	.160	Ea.	1.01	9.40		10.41	16
2000 Fine clean exposed substrate, with nylon brush		2400	.027	S.F.		1.57		1.57	2.48
2500 Wet wipe substrate		4800	.013			.78		.78	1.24
2600 Vacuum surfaces, fine brush		6400	.010			.59		.59	.93
3000 Structural demolition									
3100 Wood stud walls	A-9	2800	.023	S.F.		1.34		1.34	2.13
3500 Window manifolds, not incl. window replacement		4200	.015			.89		.89	1.42
3600 Plywood carpet protection		2000	.032			1.88		1.88	2.98
4000 Remove custom decontamination facility	A-10A	8	3	Ea.	15.15	176		191.15	297
4100 Remove portable decontamination facility	3 Asbe	12	2	"	14.05	117		131.05	201
5000 HEPA vacuum, shampoo carpeting	A-9	4800	.013	S.F.	.11	.78		.89	1.36
9000 Final cleaning of protected surfaces	A-10A	8000	.003	"		.18		.18	.28

02 82 13.47 Asbestos Waste Pkg., Handling, and Disp.

	Crew	Daily Output	Labor-Hours	Unit	Material	Labor	Equipment	Total	Total Incl O&P
0010 **ASBESTOS WASTE PACKAGING, HANDLING, AND DISPOSAL**									
0100 Collect and bag bulk material, 3 C.F. bags, by hand	A-9	400	.160	Ea.	.77	9.40		10.17	15.75
0200 Large production vacuum loader	A-12	880	.073		.99	4.27	.84	6.10	8.75
1000 Double bag and decontaminate	A-9	960	.067		.77	3.91		4.68	7.05
2000 Containerize bagged material in drums, per 3 C.F. drum	"	800	.080		18.90	4.70		23.60	28.50
3000 Cart bags 50' to dumpster	2 Asbe	400	.040			2.35		2.35	3.72
5000 Disposal charges, not including haul, minimum				C.Y.				61	67
5020 Maximum				"				355	395
9000 For type B (supplied air) respirator equipment, add				%				10%	10%

02 82 13.48 Asbestos Encapsulation With Sealants

	Crew	Daily Output	Labor-Hours	Unit	Material	Labor	Equipment	Total	Total Incl O&P
0010 **ASBESTOS ENCAPSULATION WITH SEALANTS**									
0100 Ceilings and walls, minimum	A-9	21000	.003	S.F.	.42	.18		.60	.74
0110 Maximum		10600	.006		.52	.35		.87	1.13
0200 Columns and beams, minimum		13300	.005		.42	.28		.70	.91
0210 Maximum		5325	.012		.63	.71		1.34	1.81
0300 Pipes to 12" diameter including minor repairs, minimum		800	.080	L.F.	.57	4.70		5.27	8.10
0310 Maximum		400	.160	"	1.27	9.40		10.67	16.30

02 83 Lead Remediation

02 83 19 – Lead-Based Paint Remediation

02 83 19.21 Lead Paint Remediation Plans and Methods

	Crew	Daily Output	Labor-Hours	Unit	Material	Labor	Equipment	Total	Total Incl O&P
0010 **LEAD PAINT REMEDIATION PLANS AND METHODS**									
0100 Building Survey-Commercial Building				Ea.				2,050	2,250
0200 Lead Abatement Remediation Plan								1,225	1,350
0300 Lead Paint Testing, AAS Analysis								51	56
0400 Lead Paint Testing, X-Ray Fluorescence								51	56

02 83 19.22 Preparation of Lead Containment Area

	Crew	Daily Output	Labor-Hours	Unit	Material	Labor	Equipment	Total	Total Incl O&P
0010 **PREPARATION OF LEAD CONTAINMENT AREA**									
0020 Lead abatement work area, test kit, per swab	1 Skwk	16	.500	Ea.	3.40	27.50		30.90	47
0025 For dust barriers see Section 01 56 16.10									
0050 Caution sign	1 Skwk	48	.167	Ea.	9	9.15		18.15	24.50
0100 Pre-cleaning, HEPA vacuum and wet wipe, floor and wall surfaces	3 Skwk	5000	.005	S.F.	.01	.26		.27	.43
0105 Ceiling, 6'-11' high		4100	.006		.07	.32		.39	.59
0108 12'-15' high		3550	.007		.07	.37		.44	.67

02 83 Lead Remediation

02 83 19 – Lead-Based Paint Remediation

	02 83 19.22 Preparation of Lead Containment Area	Crew	Daily Output	Labor-Hours	Unit	Material	2020 Bare Costs Labor	Equipment	Total	Total Incl O&P
0115	Over 15' high	3 Skwk	3000	.008	S.F.	.09	.44		.53	.80
0500	Cover surfaces with polyethylene sheeting									
0550	Floors, each layer, 6 mil	3 Skwk	8000	.003	S.F.	.04	.16		.20	.31
0560	Walls, each layer, 6 mil	"	6000	.004	"	.04	.22		.26	.40
0570	For heights above 14', add						20%			
2400	Post abatement cleaning of protective sheeting, HEPA vacuum & wet wipe	3 Skwk	5000	.005	S.F.	.01	.26		.27	.43
2450	Doff, bag and seal protective sheeting		12000	.002		.01	.11		.12	.18
2500	Post abatement cleaning, HEPA vacuum & wet wipe	↓	5000	.005	↓	.01	.26		.27	.43

02 83 19.23 Encapsulation of Lead-Based Paint

		Crew	Daily Output	Labor-Hours	Unit	Material	2020 Bare Costs Labor	Equipment	Total	Total Incl O&P
0010	**ENCAPSULATION OF LEAD-BASED PAINT**									
0020	Interior, brushwork, trim, under 6"	1 Pord	240	.033	L.F.	2.45	1.48		3.93	5.05
0030	6"-12" wide		180	.044		3.19	1.97		5.16	6.65
0040	Balustrades		300	.027		2.07	1.18		3.25	4.16
0050	Pipe to 4" diameter		500	.016		1.27	.71		1.98	2.53
0060	To 8" diameter		375	.021		1.56	.95		2.51	3.22
0070	To 12" diameter		250	.032		2.36	1.42		3.78	4.86
0080	To 16" diameter		170	.047	↓	3.50	2.09		5.59	7.15
0090	Cabinets, ornate design		200	.040	S.F.	3.19	1.78		4.97	6.35
0100	Simple design	↓	250	.032	"	2.42	1.42		3.84	4.92
0110	Doors, 3' x 7', both sides, incl. frame & trim									
0120	Flush	1 Pord	6	1.333	Ea.	30	59		89	127
0130	French, 10-15 lite		3	2.667		6.45	118		124.45	195
0140	Panel		4	2		36.50	89		125.50	181
0150	Louvered	↓	2.75	2.909	↓	34.50	129		163.50	243
0160	Windows, per interior side, per 15 S.F.									
0170	1-6 lite	1 Pord	14	.571	Ea.	21.50	25.50		47	64
0180	7-10 lite		7.50	1.067		23.50	47.50		71	101
0190	12 lite		5.75	1.391		32	62		94	133
0200	Radiators		8	1	↓	74.50	44.50		119	153
0210	Grilles, vents		275	.029	S.F.	2.11	1.29		3.40	4.37
0220	Walls, roller, drywall or plaster		1000	.008		.63	.36		.99	1.25
0230	With spunbonded reinforcing fabric		720	.011		.73	.49		1.22	1.58
0240	Wood		800	.010		.72	.44		1.16	1.50
0250	Ceilings, roller, drywall or plaster		900	.009		.73	.39		1.12	1.43
0260	Wood		700	.011	↓	.87	.51		1.38	1.77
0270	Exterior, brushwork, gutters and downspouts		300	.027	L.F.	2.04	1.18		3.22	4.12
0280	Columns		400	.020	S.F.	1.48	.89		2.37	3.04
0290	Spray, siding	↓	600	.013	"	1.06	.59		1.65	2.11
0300	Miscellaneous									
0310	Electrical conduit, brushwork, to 2" diameter	1 Pord	500	.016	L.F.	1.27	.71		1.98	2.53
0320	Brick, block or concrete, spray		500	.016	S.F.	1.27	.71		1.98	2.53
0330	Steel, flat surfaces and tanks to 12"		500	.016		1.27	.71		1.98	2.53
0340	Beams, brushwork		400	.020		1.48	.89		2.37	3.04
0350	Trusses	↓	400	.020	↓	1.48	.89		2.37	3.04

02 83 19.26 Removal of Lead-Based Paint

		Crew	Daily Output	Labor-Hours	Unit	Material	2020 Bare Costs Labor	Equipment	Total	Total Incl O&P
0010	**REMOVAL OF LEAD-BASED PAINT** R028319-60									
0011	By chemicals, per application									
0050	Baseboard, to 6" wide	1 Pord	64	.125	L.F.	.84	5.55		6.39	9.70
0070	To 12" wide		32	.250	"	1.49	11.10		12.59	19.30
0200	Balustrades, one side		28	.286	S.F.	1.49	12.70		14.19	21.50
1400	Cabinets, simple design		32	.250		1.39	11.10		12.49	19.20
1420	Ornate design	↓	25	.320	↓	1.55	14.20		15.75	24

For customer support on your Commercial Renovation Costs with RSMeans data, call 800.448.8182.

43

02 83 Lead Remediation

02 83 19 – Lead-Based Paint Remediation

02 83 19.26 Removal of Lead-Based Paint

		Crew	Daily Output	Labor-Hours	Unit	Material	2020 Bare Costs Labor	Equipment	Total	Total Incl O&P
1600	Cornice, simple design	1 Pord	60	.133	S.F.	1.59	5.90		7.49	11.15
1620	Ornate design		20	.400		5.50	17.75		23.25	34
2800	Doors, one side, flush		84	.095		1.88	4.23		6.11	8.75
2820	Two panel		80	.100		1.39	4.44		5.83	8.60
2840	Four panel		45	.178		1.40	7.90		9.30	14.10
2880	For trim, one side, add		64	.125	L.F.	.80	5.55		6.35	9.70
3000	Fence, picket, one side		30	.267	S.F.	1.39	11.85		13.24	20.50
3200	Grilles, one side, simple design		30	.267		1.40	11.85		13.25	20.50
3220	Ornate design		25	.320		1.46	14.20		15.66	24
3240	Handrails		90	.089	L.F.	1.40	3.95		5.35	7.80
4400	Pipes, to 4" diameter		90	.089		1.88	3.95		5.83	8.30
4420	To 8" diameter		50	.160		3.76	7.10		10.86	15.45
4440	To 12" diameter		36	.222		5.65	9.85		15.50	22
4460	To 16" diameter		20	.400		7.55	17.75		25.30	36.50
4500	For hangers, add		40	.200	Ea.	2.54	8.90		11.44	16.90
4800	Siding		90	.089	S.F.	1.36	3.95		5.31	7.75
5000	Trusses, open		55	.145	SF Face	2.05	6.45		8.50	12.50
6200	Windows, one side only, double-hung, 1/1 light, 24" x 48" high		4	2	Ea.	25.50	89		114.50	169
6220	30" x 60" high		3	2.667		33.50	118		151.50	225
6240	36" x 72" high		2.50	3.200		41	142		183	271
6280	40" x 80" high		2	4		51	178		229	340
6400	Colonial window, 6/6 light, 24" x 48" high		2	4		51	178		229	340
6420	30" x 60" high		1.50	5.333		67	237		304	450
6440	36" x 72" high		1	8		101	355		456	675
6480	40" x 80" high		1	8		101	355		456	675
6600	8/8 light, 24" x 48" high		2	4		51	178		229	340
6620	40" x 80" high		1	8		101	355		456	675
6800	12/12 light, 24" x 48" high		1	8		101	355		456	675
6820	40" x 80" high		.75	10.667		136	475		611	900
6840	Window frame & trim items, included in pricing above									
7000	Hand scraping and HEPA vacuum, less than 4 S.F.	1 Pord	8	1	Ea.	.77	44.50		45.27	71.50
8000	Collect and bag bulk material, 3 C.F. bags, by hand		30	.267	"	.77	11.85		12.62	19.65
9000	Minimum labor/equipment charge		3	2.667	Job		118		118	188

02 87 Biohazard Remediation

02 87 13 – Mold Remediation

02 87 13.16 Mold Remediation Preparation and Containment

		Crew	Daily Output	Labor-Hours	Unit	Material	2020 Bare Costs Labor	Equipment	Total	Total Incl O&P
0010	**MOLD REMEDIATION PREPARATION AND CONTAINMENT**									
0015	Mold remediation plans and methods									
0020	Initial inspection, areas to 2,500 S.F.				Total				268	295
0030	Areas to 5,000 S.F.								440	485
0032	Areas to 10,000 S.F.								440	485
0040	Testing, air sample each								126	139
0050	Swab sample								115	127
0060	Tape sample								125	140
0070	Post remediation air test								126	139
0080	Mold abatement plan, area to 2,500 S.F.								1,225	1,325
0090	Areas to 5,000 S.F.								1,625	1,775
0095	Areas to 10,000 S.F.								2,550	2,800
0100	Packup & removal of contents, average 3 bedroom home, excl. storage								8,150	8,975
0110	Average 5 bedroom home, excl. storage								15,300	16,800

02 87 Biohazard Remediation

02 87 13 – Mold Remediation

02 87 13.16 Mold Remediation Preparation and Containment	Crew	Daily Output	Labor-Hours	Unit	Material	2020 Bare Costs Labor	2020 Bare Costs Equipment	Total	Total Incl O&P
0600 For demolition in mold contaminated areas, see Section 02 87 13.33									
0610 For personal protection equipment, see Section 02 82 13.41									
6010 Preparation of mold containment area									
6100 Pre-cleaning, HEPA vacuum and wet wipe, flat surfaces	A-9	12000	.005	S.F.	.01	.31		.32	.51
6300 Separation barrier, 2" x 4" @ 16", 1/2" plywood ea. side, 8' high	2 Carp	400	.040		3.70	2.13		5.83	7.50
6310 12' high		320	.050		3.52	2.66		6.18	8.15
6320 16' high		200	.080		2.40	4.25		6.65	9.45
6400 Personnel decontam. chamber, 2" x 4" @ 16", 3/4" ply ea. side		280	.057		4.47	3.04		7.51	9.80
6450 Waste decontam. chamber, 2" x 4" studs @ 16", 3/4" ply each side	↓	360	.044	↓	4.54	2.36		6.90	8.75
6500 Cover surfaces with polyethylene sheeting									
6501 Including glue and tape									
6550 Floors, each layer, 6 mil	A-9	8000	.008	S.F.	.04	.47		.51	.79
6551 4 mil		9000	.007		.03	.42		.45	.69
6560 Walls, each layer, 6 mil		6000	.011		.04	.63		.67	1.04
6561 4 mil	↓	7000	.009	↓	.03	.54		.57	.88
6570 For heights above 14', add						20%			
6575 For heights above 20', add						30%			
6580 For fire retardant poly, add					100%				
6590 For large open areas, deduct					10%	20%			
6600 Seal floor penetrations with foam firestop to 36 sq. in.	2 Carp	200	.080	Ea.	13.90	4.25		18.15	22
6610 36 sq. in. to 72 sq. in.		125	.128		28	6.80		34.80	41.50
6615 72 sq. in. to 144 sq. in.		80	.200		55.50	10.65		66.15	78
6620 Wall penetrations, to 36 sq. in.		180	.089		13.90	4.72		18.62	23
6630 36 sq. in. to 72 sq. in.		100	.160		28	8.50		36.50	44
6640 72 sq. in. to 144 sq. in.	↓	60	.267	↓	55.50	14.15		69.65	83.50
6800 Caulk seams with latex caulk	1 Carp	230	.035	L.F.	.17	1.85		2.02	3.15
6900 Set up neg. air machine, 1-2k CFM/25 M.C.F. volume	1 Asbe	4.30	1.860	Ea.		109		109	173

02 87 13.33 Removal and Disposal of Materials With Mold

	Crew	Daily Output	Labor-Hours	Unit	Material	Labor	Equipment	Total	Total Incl O&P
0010 **REMOVAL AND DISPOSAL OF MATERIALS WITH MOLD**									
0015 Demolition in mold contaminated area									
0200 Ceiling, including suspension system, plaster and lath	A-9	2100	.030	S.F.	.08	1.79		1.87	2.92
0210 Finished plaster, leaving wire lath		585	.109		.27	6.40		6.67	10.50
0220 Suspended acoustical tile		3500	.018		.05	1.07		1.12	1.75
0230 Concealed tile grid system		3000	.021		.05	1.25		1.30	2.05
0240 Metal pan grid system		1500	.043		.11	2.51		2.62	4.09
0250 Gypsum board		2500	.026		.06	1.50		1.56	2.45
0255 Plywood		2500	.026	↓	.06	1.50		1.56	2.45
0260 Lighting fixtures up to 2' x 4'	↓	72	.889	Ea.	2.23	52		54.23	85.50
0400 Partitions, non load bearing									
0410 Plaster, lath, and studs	A-9	690	.093	S.F.	.90	5.45		6.35	9.65
0450 Gypsum board and studs		1390	.046		.12	2.70		2.82	4.42
0465 Carpet & pad		1390	.046	↓	.12	2.70		2.82	4.42
0600 Pipe insulation, air cell type, up to 4" diameter pipe		900	.071	L.F.	.18	4.17		4.35	6.80
0610 4" to 8" diameter pipe		800	.080		.20	4.70		4.90	7.65
0620 10" to 12" diameter pipe		700	.091		.23	5.35		5.58	8.75
0630 14" to 16" diameter pipe	↓	550	.116	↓	.29	6.85		7.14	11.15
0650 Over 16" diameter pipe	↓	650	.098	S.F.	.25	5.80		6.05	9.40
9000 For type B (supplied air) respirator equipment, add				%				10%	10%

Division Notes

	CREW	DAILY OUTPUT	LABOR-HOURS	UNIT	BARE COSTS				TOTAL INCL O&P
					MAT.	LABOR	EQUIP.	TOTAL	

Estimating Tips
General

- Carefully check all the plans and specifications. Concrete often appears on drawings other than structural drawings, including mechanical and electrical drawings for equipment pads. The cost of cutting and patching is often difficult to estimate. See Subdivision 03 81 for Concrete Cutting, Subdivision 02 41 19.16 for Cutout Demolition, Subdivision 03 05 05.10 for Concrete Demolition, and Subdivision 02 41 19.19 for Rubbish Handling (handling, loading, and hauling of debris).

- Always obtain concrete prices from suppliers near the job site. A volume discount can often be negotiated, depending upon competition in the area. Remember to add for waste, particularly for slabs and footings on grade.

03 10 00 Concrete Forming and Accessories

- A primary cost for concrete construction is forming. Most jobs today are constructed with prefabricated forms. The selection of the forms best suited for the job and the total square feet of forms required for efficient concrete forming and placing are key elements in estimating concrete construction. Enough forms must be available for erection to make efficient use of the concrete placing equipment and crew.

- Concrete accessories for forming and placing depend upon the systems used. Study the plans and specifications to ensure that all special accessory requirements have been included in the cost estimate, such as anchor bolts, inserts, and hangers.

- Included within costs for forms-in-place are all necessary bracing and shoring.

03 20 00 Concrete Reinforcing

- Ascertain that the reinforcing steel supplier has included all accessories, cutting, bending, and an allowance for lapping, splicing, and waste. A good rule of thumb is 10% for lapping, splicing, and waste. Also, 10% waste should be allowed for welded wire fabric.

- The unit price items in the subdivisions for Reinforcing In Place, Glass Fiber Reinforcing, and Welded Wire Fabric include the labor to install accessories such as beam and slab bolsters, high chairs, and bar ties and tie wire. The material cost for these accessories is not included; they may be obtained from the Accessories Subdivisions.

03 30 00 Cast-In-Place Concrete

- When estimating structural concrete, pay particular attention to requirements for concrete additives, curing methods, and surface treatments. Special consideration for climate, hot or cold, must be included in your estimate. Be sure to include requirements for concrete placing equipment and concrete finishing.

- For accurate concrete estimating, the estimator must consider each of the following major components individually: forms, reinforcing steel, ready-mix concrete, placement of the concrete, and finishing of the top surface. For faster estimating, Subdivision 03 30 53.40 for Concrete-In-Place can be used; here, various items of concrete work are presented that include the costs of all five major components (unless specifically stated otherwise).

03 40 00 Precast Concrete
03 50 00 Cast Decks and Underlayment

- The cost of hauling precast concrete structural members is often an important factor. For this reason, it is important to get a quote from the nearest supplier. It may become economically feasible to set up precasting beds on the site if the hauling costs are prohibitive.

Reference Numbers

Reference numbers are shown at the beginning of some major classifications. These numbers refer to related items in the Reference Section. The reference information may be an estimating procedure, an alternate pricing method, or technical information.

Note: Not all subdivisions listed here necessarily appear. ■

Same Data. Simplified.

Enjoy the convenience and efficiency of accessing your costs anywhere:

- **Skip the multiplier** by setting your location
- **Quickly search,** edit, favorite and share costs
- **Stay on top of price changes** with automatic updates

Discover more at rsmeans.com/online

03 01 Maintenance of Concrete

03 01 30 – Maintenance of Cast-In-Place Concrete

03 01 30.62 Concrete Patching

		Crew	Daily Output	Labor-Hours	Unit	Material	2020 Bare Costs Labor	2020 Bare Costs Equipment	Total	Total Incl O&P
0010	**CONCRETE PATCHING**									
0100	Floors, 1/4" thick, small areas, regular grout	1 Cefi	170	.047	S.F.	1.61	2.35		3.96	5.45
0150	Epoxy grout	"	100	.080	"	9.90	4		13.90	17.20
2000	Walls, including chipping, cleaning and epoxy grout									
2100	1/4" deep	1 Cefi	65	.123	S.F.	7.80	6.15		13.95	18.25
2150	1/2" deep		50	.160		15.60	8		23.60	29.50
2200	3/4" deep		40	.200		23.50	10		33.50	41
9000	Minimum labor/equipment charge		4.50	1.778	Job		89		89	139

03 01 30.71 Concrete Crack Repair

		Crew	Daily Output	Labor-Hours	Unit	Material	2020 Bare Costs Labor	2020 Bare Costs Equipment	Total	Total Incl O&P
0010	**CONCRETE CRACK REPAIR**									
1000	Structural repair of concrete cracks by epoxy injection (ACI RAP-1)									
1001	suitable for horizontal, vertical and overhead repairs									
1010	Clean/grind concrete surface(s) free of contaminants	1 Cefi	400	.020	L.F.		1		1	1.57
1015	Rout crack with v-notch crack chaser, if needed	C-32	600	.027		.03	1.23	.25	1.51	2.26
1020	Blow out crack with oil-free dry compressed air (1 pass)	C-28	3000	.003			.13	.01	.14	.22
1030	Install surface-mounted entry ports (spacing = concrete depth)	1 Cefi	400	.020	Ea.	1.71	1		2.71	3.45
1040	Cap crack at surface with epoxy gel (per side/face)		400	.020	L.F.	.31	1		1.31	1.92
1050	Snap off ports, grind off epoxy cap residue after injection		200	.040	"		2		2	3.14
1100	Manual injection with 2-part epoxy cartridge, excludes prep									
1110	Up to 1/32" (0.03125") wide x 4" deep	1 Cefi	160	.050	L.F.	.16	2.50		2.66	4.09
1120	6" deep		120	.067		.24	3.33		3.57	5.50
1130	8" deep		107	.075		.31	3.73		4.04	6.20
1140	10" deep		100	.080		.39	4		4.39	6.75
1150	12" deep		96	.083		.47	4.16		4.63	7.05
1210	Up to 1/16" (0.0625") wide x 4" deep		160	.050		.31	2.50		2.81	4.27
1220	6" deep		120	.067		.47	3.33		3.80	5.75
1230	8" deep		107	.075		.63	3.73		4.36	6.55
1240	10" deep		100	.080		.78	4		4.78	7.15
1250	12" deep		96	.083		.94	4.16		5.10	7.60
1310	Up to 3/32" (0.09375") wide x 4" deep		160	.050		.47	2.50		2.97	4.44
1320	6" deep		120	.067		.71	3.33		4.04	6.05
1330	8" deep		107	.075		.94	3.73		4.67	6.90
1340	10" deep		100	.080		1.18	4		5.18	7.60
1350	12" deep		96	.083		1.41	4.16		5.57	8.10
1410	Up to 1/8" (0.125") wide x 4" deep		160	.050		.63	2.50		3.13	4.61
1420	6" deep		120	.067		.94	3.33		4.27	6.30
1430	8" deep		107	.075		1.26	3.73		4.99	7.25
1440	10" deep		100	.080		1.57	4		5.57	8.05
1450	12" deep		96	.083		1.88	4.16		6.04	8.60
1500	Pneumatic injection with 2-part bulk epoxy, excludes prep									
1510	Up to 5/32" (0.15625") wide x 4" deep	C-31	240	.033	L.F.	.27	1.66	1.32	3.25	4.36
1520	6" deep		180	.044		.40	2.22	1.77	4.39	5.85
1530	8" deep		160	.050		.54	2.50	1.99	5.03	6.70
1540	10" deep		150	.053		.67	2.66	2.12	5.45	7.25
1550	12" deep		144	.056		.80	2.78	2.21	5.79	7.65
1610	Up to 3/16" (0.1875") wide x 4" deep		240	.033		.32	1.66	1.32	3.30	4.42
1620	6" deep		180	.044		.48	2.22	1.77	4.47	5.95
1630	8" deep		160	.050		.64	2.50	1.99	5.13	6.80
1640	10" deep		150	.053		.80	2.66	2.12	5.58	7.40
1650	12" deep		144	.056		.96	2.78	2.21	5.95	7.85
1710	Up to 1/4" (0.25") wide x 4" deep		240	.033		.43	1.66	1.32	3.41	4.54
1720	6" deep		180	.044		.64	2.22	1.77	4.63	6.15

03 01 30 – Maintenance of Cast-In-Place Concrete

03 01 30.71 Concrete Crack Repair

		Crew	Daily Output	Labor-Hours	Unit	Material	2020 Bare Costs Labor	2020 Bare Costs Equipment	Total	Total Incl O&P
1730	8" deep	C-31	160	.050	L.F.	.86	2.50	1.99	5.35	7.05
1740	10" deep		150	.053		1.07	2.66	2.12	5.85	7.70
1750	12" deep	↓	144	.056	↓	1.29	2.78	2.21	6.28	8.20
2000	Non-structural filling of concrete cracks by gravity-fed resin (ACI RAP-2)									
2001	suitable for individual cracks in stable horizontal surfaces only									
2010	Clean/grind concrete surface(s) free of contaminants	1 Cefi	400	.020	L.F.		1		1	1.57
2020	Rout crack with v-notch crack chaser, if needed	C-32	600	.027		.03	1.23	.25	1.51	2.26
2030	Blow out crack with oil-free dry compressed air (1 pass)	C-28	3000	.003			.13	.01	.14	.22
2040	Cap crack with epoxy gel at underside of elevated slabs, if needed	1 Cefi	400	.020		.31	1		1.31	1.92
2050	Insert backer rod into crack, if needed		400	.020		.03	1		1.03	1.60
2060	Partially fill crack with fine dry sand, if needed		800	.010		.02	.50		.52	.80
2070	Apply two beads of sealant alongside crack to form reservoir, if needed	↓	400	.020	↓	.28	1		1.28	1.88
2100	Manual filling with squeeze bottle of 2-part epoxy resin									
2110	Full depth crack up to 1/16" (0.0625") wide x 4" deep	1 Cefi	300	.027	L.F.	.10	1.33		1.43	2.20
2120	6" deep		240	.033		.15	1.66		1.81	2.77
2130	8" deep		200	.040		.19	2		2.19	3.35
2140	10" deep		185	.043		.24	2.16		2.40	3.66
2150	12" deep		175	.046		.29	2.28		2.57	3.91
2210	Full depth crack up to 1/8" (0.125") wide x 4" deep		270	.030		.19	1.48		1.67	2.53
2220	6" deep		215	.037		.29	1.86		2.15	3.24
2230	8" deep		180	.044		.39	2.22		2.61	3.92
2240	10" deep		165	.048		.49	2.42		2.91	4.33
2250	12" deep		155	.052		.58	2.58		3.16	4.69
2310	Partial depth crack up to 3/16" (0.1875") wide x 1" deep		300	.027		.07	1.33		1.40	2.17
2410	Up to 1/4" (0.250") wide x 1" deep		270	.030		.10	1.48		1.58	2.43
2510	Up to 5/16" (0.3125") wide x 1" deep		250	.032		.12	1.60		1.72	2.64
2610	Up to 3/8" (0.375") wide x 1" deep	↓	240	.033	↓	.15	1.66		1.81	2.77

03 01 30.72 Concrete Surface Repairs

		Crew	Daily Output	Labor-Hours	Unit	Material	2020 Bare Costs Labor	2020 Bare Costs Equipment	Total	Total Incl O&P
0010	**CONCRETE SURFACE REPAIRS**									
1000	Spall repairs using low-pressure spraying (ACI RAP-3)									
1001	Suitable for vertical and overhead repairs up to 3" thick									
1010	Sound the concrete surface to locate delaminated areas	1 Cefi	2000	.004	S.F.		.20		.20	.31
1020	Remove concrete in repair areas to fully expose reinforcing bars	B-9E	160	.100			4.60	.20	4.80	7.50
1030	Mark the perimeter of each repair area	1 Cefi	2000	.004	↓		.20		.20	.31
1040	Saw cut the perimeter of each repair area down to reinforcing bars	B-89C	160	.050	L.F.	.04	2.50	.38	2.92	4.38
1050	If corroded rebar is exposed, remove concrete to 3/4" below corroded rebars									
1051	Single layer of #4 rebar	B-9E	64	.250	S.F.		11.50	.51	12.01	18.80
1052	#5 rebar		53.30	.300			13.80	.61	14.41	22.50
1053	#6 rebar		45.70	.350			16.10	.71	16.81	26.50
1054	#7 rebar		40	.400			18.40	.81	19.21	30
1055	#8 rebar		35.60	.449			20.50	.91	21.41	34
1056	Double layer of #4 rebar		35.60	.449			20.50	.91	21.41	34
1057	#5 rebar		32	.500			23	1.02	24.02	37.50
1058	#6 rebar		29.10	.550			25.50	1.12	26.62	41
1059	#7 rebar		26.70	.599			27.50	1.22	28.72	45
1060	#8 rebar	↓	24.60	.650	↓		30	1.32	31.32	49
1100	See section 03 21 11.60 for placement of supplemental reinforcement bars									
1104	See section 02 41 19.19 for handling & removal of debris									
1106	See section 02 41 19.20 for disposal of debris									
1110	Final cleaning by sandblasting	E-11	1100	.029	S.F.	.51	1.36	.27	2.14	3.11
1120	By high pressure water	C-29	500	.016			.67	.19	.86	1.29
1130	Blow off dust and debris with oil-free dry compressed air	C-28	16000	.001	↓		.03		.03	.04

03 01 30.72 Concrete Surface Repairs	Crew	Daily Output	Labor-Hours	Unit	Material	2020 Bare Costs Labor	Equipment	Total	Total Incl O&P
1140	Mix up repair material with concrete mixer	C-30	200	.040	C.F.	7.65	1.68	.70	10.03	11.85
1150	Place repair material by low-pressure spray in lifts up to 3" thick	C-9	200	.160	"		7.50	2.32	9.82	14.40
1160	Screed and initial float of repair material by hand	C-10	2000	.012	S.F.		.57		.57	.90
1170	Float and steel trowel finish repair material by hand	"	1265	.019			.90		.90	1.42
1180	Cure with sprayed membrane curing compound	2 Cefi	6200	.003	↓	.12	.13		.25	.34
2000	Surface repairs using form-and-pour techniques (ACI RAP-4)									
2001	Suitable for walls, columns, beam sides/bottoms, and slab soffits									
2010	Sound the concrete surface to locate delaminated areas	1 Cefi	2000	.004	S.F.		.20		.20	.31
2020	Remove concrete in repair areas to fully expose reinforcing bars	B-9E	160	.100			4.60	.20	4.80	7.50
2030	Mark the perimeter of each repair area	1 Cefi	2000	.004	↓		.20		.20	.31
2040	Saw cut the perimeter of each repair area down to reinforcing bars	B-89C	160	.050	L.F.	.04	2.50	.38	2.92	4.38
2050	If corroded rebar is exposed, remove concrete to 3/4" below corroded rebars									
2051	Single layer of #4 rebar	B-9E	64	.250	S.F.		11.50	.51	12.01	18.80
2052	#5 rebar		53.30	.300			13.80	.61	14.41	22.50
2053	#6 rebar		45.70	.350			16.10	.71	16.81	26.50
2054	#7 rebar		40	.400			18.40	.81	19.21	30
2055	#8 rebar		35.60	.449			20.50	.91	21.41	34
2056	Double layer of #4 rebar		35.60	.449			20.50	.91	21.41	34
2057	#5 rebar		32	.500			23	1.02	24.02	37.50
2058	#6 rebar		29.10	.550			25.50	1.12	26.62	41
2059	#7 rebar		26.70	.599			27.50	1.22	28.72	45
2060	#8 rebar	↓	24.60	.650	↓		30	1.32	31.32	49
2100	See section 03 21 11.60 for placement of supplemental reinforcement bars									
2105	See section 02 41 19.19 for handling & removal of debris									
2110	See section 02 41 19.20 for disposal of debris									
2120	For slab soffit repairs, drill 2" diam. holes to 6" deep for conc placement	B-89A	16.50	.970	Ea.	.26	47	6.90	54.16	83
2130	For each additional inch of slab thickness in same hole, add	"	1080	.015		.04	.72	.11	.87	1.32
2140	Drill holes for expansion shields, 3/4" diameter x 4" deep	1 Carp	45	.178	↓		9.45		9.45	15.15
2150	Final cleaning by sandblasting	E-11	1100	.029	S.F.	.51	1.36	.27	2.14	3.11
2160	By high pressure water	C-29	500	.016			.67	.19	.86	1.29
2170	Blow off dust and debris with oil-free dry compressed air	C-28	16000	.001	↓		.03		.03	.04
2180	Build wood forms including chutes where needed for concrete placement	C-2	960	.050	SFCA	3.19	2.58		5.77	7.65
2190	Insert expansion shield, coil rod, coil tie and plastic cone in each hole	1 Carp	32	.250	Ea.	4.88	13.30		18.18	27
2200	Install wood forms and fasten with coil rod and coil nut	C-2	960	.050	SFCA	.48	2.58		3.06	4.67
2210	Mix up bagged repair material by hand with a mixing paddle	1 Cefi	30	.267	C.F.	7.65	13.30		20.95	29.50
2220	By concrete mixer	C-30	200	.040		7.65	1.68	.70	10.03	11.85
2230	Place repair material by hand and consolidate	C-6A	200	.080	↓		4	.24	4.24	6.55
2240	Remove wood forms	C-2	1600	.030	SFCA		1.55		1.55	2.48
2250	Chip/grind off extra concrete at fill chutes	B-9E	16	1	Ea.	.12	46	2.03	48.15	75.50
2260	Break ties and patch voids	1 Cefi	270	.030	"	.09	1.48		1.57	2.42
2270	Cure with sprayed membrane curing compound	2 Cefi	6200	.003	S.F.	.12	.13		.25	.34
3000	Surface repairs using form-and-pump techniques (ACI RAP-5)									
3001	Suitable for walls, columns, beam sides/bottoms, and slab soffits									
3010	Sound the concrete surface to locate delaminated areas	1 Cefi	2000	.004	S.F.		.20		.20	.31
3020	Remove concrete in repair areas to fully expose reinforcing bars	B-9E	160	.100			4.60	.20	4.80	7.50
3030	Mark the perimeter of each repair area	1 Cefi	2000	.004	↓		.20		.20	.31
3040	Saw cut the perimeter of each repair area down to reinforcing bars	B-89C	160	.050	L.F.	.04	2.50	.38	2.92	4.38
3050	If corroded rebar is exposed, remove concrete to 3/4" below corroded rebars									
3051	Single layer of #4 rebar	B-9E	64	.250	S.F.		11.50	.51	12.01	18.80
3052	#5 rebar		53.30	.300			13.80	.61	14.41	22.50
3053	#6 rebar		45.70	.350			16.10	.71	16.81	26.50
3054	#7 rebar		40	.400			18.40	.81	19.21	30
3055	#8 rebar	↓	35.60	.449			20.50	.91	21.41	34

03 01 30.72 Concrete Surface Repairs

		Crew	Daily Output	Labor-Hours	Unit	Material	2020 Bare Costs Labor	Equipment	Total	Total Incl O&P
3056	Double layer of #4 rebar	B-9E	35.60	.449	S.F.		20.50	.91	21.41	34
3057	#5 rebar		32	.500			23	1.02	24.02	37.50
3058	#6 rebar		29.10	.550			25.50	1.12	26.62	41
3059	#7 rebar		26.70	.599			27.50	1.22	28.72	45
3060	#8 rebar		24.60	.650			30	1.32	31.32	49
3100	See section 03 21 11.60 for placement of supplemental reinforcement bars									
3104	See section 02 41 19.19 for handling & removal of debris									
3106	See section 02 41 19.20 for disposal of debris									
3110	Drill holes for expansion shields, 3/4" diameter x 4" deep	1 Carp	45	.178	Ea.		9.45		9.45	15.15
3120	Final cleaning by sandblasting	E-11	1100	.029	S.F.	.51	1.36	.27	2.14	3.11
3130	By high pressure water	C-29	500	.016			.67	.19	.86	1.29
3140	Blow off dust and debris with oil-free dry compressed air	C-28	16000	.001			.03		.03	.04
3150	Build wood forms and drill holes for injection ports	C-2	960	.050	SFCA	3.19	2.58		5.77	7.65
3160	Assemble PVC injection ports with ball valves and install in forms	1 Plum	8	1	Ea.	91.50	64.50		156	201
3170	Insert expansion shield, coil rod, coil tie and plastic cone in each hole	1 Carp	32	.250	"	4.88	13.30		18.18	27
3180	Install wood forms and fasten with coil rod and coil nut	C-2	960	.050	SFCA	.48	2.58		3.06	4.67
3190	Mix up self-consolidating repair material with concrete mixer	C-30	200	.040	C.F.	7.85	1.68	.70	10.23	12.10
3200	Place repair material by pump and pressurize	C-9	200	.160	"		7.50	2.32	9.82	14.40
3210	Remove wood forms	C-2	1600	.030	SFCA		1.55		1.55	2.48
3220	Chip/grind off extra concrete at injection ports	B-9E	64	.250	Ea.	.12	11.50	.51	12.13	18.95
3230	Break ties and patch voids	1 Cefi	270	.030	"	.09	1.48		1.57	2.42
3240	Cure with sprayed membrane curing compound	2 Cefi	6200	.003	S.F.	.12	.13		.25	.34
4000	Vertical & overhead spall repair by hand application (ACI RAP-6)									
4001	Generally recommended for thin repairs that are cosmetic in nature									
4010	Remove concrete in repair areas to fully expose reinforcing bars	B-9E	160	.100	S.F.		4.60	.20	4.80	7.50
4020	Saw cut the perimeter of each repair area to 1/2" depth	B-89C	160	.050	L.F.	.02	2.50	.38	2.90	4.36
4030	If corroded rebar is exposed, remove concrete to 3/4" below corroded rebars									
4031	Single layer of #4 rebar	B-9E	64	.250	S.F.		11.50	.51	12.01	18.80
4032	#5 rebar		53.30	.300			13.80	.61	14.41	22.50
4033	#6 rebar		45.70	.350			16.10	.71	16.81	26.50
4034	#7 rebar		40	.400			18.40	.81	19.21	30
4035	#8 rebar		35.60	.449			20.50	.91	21.41	34
4036	Double layer of #4 rebar		35.60	.449			20.50	.91	21.41	34
4037	#5 rebar		32	.500			23	1.02	24.02	37.50
4038	#6 rebar		29.10	.550			25.50	1.12	26.62	41
4039	#7 rebar		26.70	.599			27.50	1.22	28.72	45
4040	#8 rebar		24.60	.650			30	1.32	31.32	49
4080	See section 03 21 11.60 for placement of supplemental reinforcement bars									
4090	See section 02 41 19.19 for handling & removal of debris									
4095	See section 02 41 19.20 for disposal of debris									
4100	Final cleaning by sandblasting	E-11	1100	.029	S.F.	.51	1.36	.27	2.14	3.11
4110	By high pressure water	C-29	500	.016			.67	.19	.86	1.29
4120	Blow off dust and debris with oil-free dry compressed air	C-28	16000	.001			.03		.03	.04
4130	Brush on bonding material	1 Cefi	320	.025		.11	1.25		1.36	2.08
4140	Mix up bagged repair material by hand with a mixing paddle	"	30	.267	C.F.	77.50	13.30		90.80	106
4150	By concrete mixer	C-30	200	.040		77.50	1.68	.70	79.88	88.50
4160	Place repair material by hand in lifts up to 2" deep for vertical repairs	C-10	48	.500			23.50		23.50	37.50
4170	Up to 1" deep for overhead repairs		36	.667			31.50		31.50	50
4180	Screed and initial float of repair material by hand		2000	.012	S.F.		.57		.57	.90
4190	Float and steel trowel finish repair material by hand		1265	.019			.90		.90	1.42
4200	Cure with sprayed membrane curing compound	2 Cefi	6200	.003		.12	.13		.25	.34
5000	Spall repair of horizontal concrete surfaces (ACI RAP-7)									
5001	Suitable for structural slabs, slabs-on-grade, balconies, interior floors									

For customer support on your Commercial Renovation Costs with RSMeans data, call 800.448.8182.

51

03 01 Maintenance of Concrete

03 01 30 – Maintenance of Cast-In-Place Concrete

03 01 30.72 Concrete Surface Repairs	Crew	Daily Output	Labor-Hours	Unit	Material	2020 Bare Costs		Total	Total Incl O&P
						Labor	Equipment		
5010	Sound the concrete surface to locate delaminated areas	1 Cefi	2000	.004	SF Flr.		.20		.20	.31
5020	Mark the perimeter of each repair area	"	2000	.004	"		.20		.20	.31
5030	Saw cut the perimeter of each repair area down to reinforcing bars	B-89	1060	.015	L.F.	.04	.76	.98	1.78	2.31
5040	Remove concrete in repair areas to fully expose reinforcing bars	B-9E	160	.100	S.F.		4.60	.20	4.80	7.50
5050	If corroded rebar is exposed, remove concrete to 3/4" below corroded rebars									
5051	Single layer of #4 rebar	B-9E	64	.250	S.F.		11.50	.51	12.01	18.80
5052	#5 rebar		53.30	.300			13.80	.61	14.41	22.50
5053	#6 rebar		45.70	.350			16.10	.71	16.81	26.50
5054	#7 rebar		40	.400			18.40	.81	19.21	30
5055	#8 rebar		35.60	.449			20.50	.91	21.41	34
5056	Double layer of #4 rebar		35.60	.449			20.50	.91	21.41	34
5057	#5 rebar		32	.500			23	1.02	24.02	37.50
5058	#6 rebar		29.10	.550			25.50	1.12	26.62	41
5059	#7 rebar		26.70	.599			27.50	1.22	28.72	45
5060	#8 rebar		24.60	.650			30	1.32	31.32	49
5100	See section 03 21 11.60 for placement of supplemental reinforcement bars									
5110	See section 02 41 19.19 for handling & removal of debris									
5115	See section 02 41 19.20 for disposal of debris									
5120	Final cleaning by sandblasting	E-11	1100	.029	S.F.	.51	1.36	.27	2.14	3.11
5130	By high pressure water	C-29	500	.016			.67	.19	.86	1.29
5140	Blow off dust and debris with oil-free dry compressed air	C-28	16000	.001			.03		.03	.04
5160	Brush on bonding material	1 Cefi	320	.025		.11	1.25		1.36	2.08
5170	Mix up bagged repair material by hand with a mixing paddle	"	30	.267	C.F.	77.50	13.30		90.80	106
5180	By concrete mixer	C-30	200	.040		77.50	1.68	.70	79.88	88.50
5190	Place repair material by hand, consolidate and screed	C-6A	200	.080			4	.24	4.24	6.55
5200	Bull float, manual float & broom finish	C-10	1850	.013	S.F.		.61		.61	.97
5210	Bull float, manual float & trowel finish	"	1265	.019			.90		.90	1.42
5220	Cure with sprayed membrane curing compound	2 Cefi	6200	.003		.12	.13		.25	.34
6000	Spall repairs by the Preplaced Aggregate method (ACI RAP-9)									
6001	Suitable for vertical, overhead, and underwater repairs									
6010	Sound the concrete surface to locate delaminated areas	1 Cefi	2000	.004	S.F.		.20		.20	.31
6020	Remove concrete in repair areas to fully expose reinforcing bars	B-9E	160	.100			4.60	.20	4.80	7.50
6030	Mark the perimeter of each repair area	1 Cefi	2000	.004			.20		.20	.31
6040	Saw cut the perimeter of each repair area down to reinforcing bars	B-89C	160	.050	L.F.	.04	2.50	.38	2.92	4.38
6050	If corroded rebar is exposed, remove concrete to 3/4" below corroded rebars									
6051	Single layer of #4 rebar	B-9E	64	.250	S.F.		11.50	.51	12.01	18.80
6052	#5 rebar		53.30	.300			13.80	.61	14.41	22.50
6053	#6 rebar		45.70	.350			16.10	.71	16.81	26.50
6054	#7 rebar		40	.400			18.40	.81	19.21	30
6055	#8 rebar		35.60	.449			20.50	.91	21.41	34
6056	Double layer of #4 rebar		35.60	.449			20.50	.91	21.41	34
6057	#5 rebar		32	.500			23	1.02	24.02	37.50
6058	#6 rebar		29.10	.550			25.50	1.12	26.62	41
6059	#7 rebar		26.70	.599			27.50	1.22	28.72	45
6060	#8 rebar		24.60	.650			30	1.32	31.32	49
6100	See section 03 21 11.60 for placement of supplemental reinforcement bars									
6104	See section 02 41 19.19 for handling & removal of debris									
6106	See section 02 41 19.20 for disposal of debris									
6110	Drill holes for expansion shields, 3/4" diameter x 4" deep	1 Carp	45	.178	Ea.		9.45		9.45	15.15
6120	Final cleaning by sandblasting	E-11	1100	.029	S.F.	.51	1.36	.27	2.14	3.11
6130	By high pressure water	C-29	500	.016			.67	.19	.86	1.29
6140	Blow off dust and debris with oil-free dry compressed air	C-28	16000	.001			.03		.03	.04
6150	Build wood forms and drill holes for injection ports	C-2	960	.050	SFCA	3.19	2.58		5.77	7.65

52

03 01 30 – Maintenance of Cast-In-Place Concrete

03 01 30.72 Concrete Surface Repairs

		Crew	Daily Output	Labor-Hours	Unit	Material	2020 Bare Costs Labor	2020 Bare Costs Equipment	Total	Total Incl O&P
6160	Assemble PVC injection ports with ball valves and install in forms	1 Plum	8	1	Ea.	91.50	64.50		156	201
6170	Insert expansion shield, coil rod, coil tie and plastic cone in each hole	1 Carp	32	.250	"	4.88	13.30		18.18	27
6180	Install wood forms and fasten with coil rod and coil nut	C-2	960	.050	SFCA	.48	2.58		3.06	4.67
6185	Install pre-washed gap-graded aggregate as forms are installed	B-24	160	.150	C.F.	1.20	7.25		8.45	12.85
6190	Mix up pre-bagged grout material with concrete mixer	C-30	200	.040		16.25	1.68	.70	18.63	21.50
6200	Place grout by pump from bottom up	C-9	200	.160	↓		7.50	2.32	9.82	14.40
6210	Remove wood forms	C-2	1600	.030	SFCA		1.55		1.55	2.48
6220	Chip/grind off extra concrete at injection ports	B-9E	64	.250	Ea.	.12	11.50	.51	12.13	18.95
6230	Break ties and patch voids	1 Cefi	270	.030	"	.09	1.48		1.57	2.42
6240	Cure with sprayed membrane curing compound	2 Cefi	6200	.003	S.F.	.12	.13		.25	.34
7000	Leveling & reprofiling of vertical & overhead surfaces (ACI RAP-10)									
7001	Used to provide an acceptable surface for aesthetic or protective coatings									
7010	Clean the surface by sandblasting, leaving an open profile	E-11	1100	.029	S.F.	.51	1.36	.27	2.14	3.11
7020	By high pressure water	C-29	500	.016			.67	.19	.86	1.29
7030	Blow off dust and debris with oil-free dry compressed air	C-28	16000	.001	↓		.03		.03	.04
7040	Mix up bagged repair material by hand with a mixing paddle	1 Cefi	30	.267	C.F.	77.50	13.30		90.80	106
7050	Place repair material by hand with steel trowel to fill pores & honeycombs	C-10	1265	.019	S.F.		.90		.90	1.42
7060	Cure with sprayed membrane curing compound	2 Cefi	6200	.003	"	.12	.13		.25	.34
8000	Concrete repair by shotcrete application (ACI RAP-12)									
8001	Suitable for vertical and overhead surfaces									
8010	Sound the concrete surface to locate delaminated areas	1 Cefi	2000	.004	S.F.		.20		.20	.31
8020	Remove concrete in repair areas to fully expose reinforcing bars	B-9E	160	.100			4.60	.20	4.80	7.50
8030	Mark the perimeter of each repair area	1 Cefi	2000	.004	↓		.20		.20	.31
8040	Saw cut the perimeter of each repair area down to reinforcing bars	B-89C	160	.050	L.F.	.04	2.50	.38	2.92	4.38
8050	If corroded rebar is exposed, remove concrete to 3/4" below corroded rebars									
8051	Single layer of #4 rebar	B-9E	64	.250	S.F.		11.50	.51	12.01	18.80
8052	#5 rebar		53.30	.300			13.80	.61	14.41	22.50
8053	#6 rebar		45.70	.350			16.10	.71	16.81	26.50
8054	#7 rebar		40	.400			18.40	.81	19.21	30
8055	#8 rebar		35.60	.449			20.50	.91	21.41	34
8056	Double layer of #4 rebar		35.60	.449			20.50	.91	21.41	34
8057	#5 rebar		32	.500			23	1.02	24.02	37.50
8058	#6 rebar		29.10	.550			25.50	1.12	26.62	41
8059	#7 rebar		26.70	.599			27.50	1.22	28.72	45
8060	#8 rebar	↓	24.60	.650	↓		30	1.32	31.32	49
8100	See section 03 21 11.60 for placement of supplemental reinforcement bars									
8110	Final cleaning by sandblasting	E-11	1100	.029	S.F.	.51	1.36	.27	2.14	3.11
8120	By high pressure water	C-29	500	.016			.67	.19	.86	1.29
8130	Blow off dust and debris with oil-free dry compressed air	C-28	16000	.001	↓		.03		.03	.04
8140	Mix up repair material with concrete mixer	C-30	200	.040	C.F.	7.65	1.68	.70	10.03	11.85
8150	Place repair material by wet shotcrete process to full depth of repair	C-8C	200	.240	"		11.10	2.98	14.08	21
8160	Carefully trim excess material and steel trowel finish without overworking	C-10	1265	.019	S.F.		.90		.90	1.42
8170	Cure with sprayed membrane curing compound	2 Cefi	6200	.003	"	.12	.13		.25	.34
9000	Surface repair by Methacrylate flood coat (ACI RAP-13)									
9010	Suitable for healing and sealing horizontal surfaces only									
9020	Large cracks must previously have been repaired or filled									
9100	Shotblast entire surface to remove contaminants	A-1A	4000	.002	S.F.		.11	.05	.16	.23
9200	Blow off dust and debris with oil-free dry compressed air	C-28	16000	.001			.03		.03	.04
9300	Flood coat surface w/Methacrylate, distribute w/broom/squeegee, no prep.	3 Cefi	8000	.003		1.03	.15		1.18	1.37
9400	Lightly broadcast even coat of dry silica sand while sealer coat is tacky	1 Cefi	8000	.001	↓	.02	.05		.07	.10

For customer support on your Commercial Renovation Costs with RSMeans data, call 800.448.8182.

53

03 05 Common Work Results for Concrete

03 05 05 – Selective Demolition for Concrete

03 05 05.10 Selective Demolition, Concrete

		Crew	Daily Output	Labor-Hours	Unit	Material	2020 Bare Costs Labor	2020 Bare Costs Equipment	Total	Total Incl O&P
0010	**SELECTIVE DEMOLITION, CONCRETE** R024119-10									
0012	Excludes saw cutting, torch cutting, loading or hauling									
0050	Break into small pieces, reinf. less than 1% of cross-sectional area	B-9	24	1.667	C.Y.		71	14.90	85.90	129
0060	Reinforcing 1% to 2% of cross-sectional area		16	2.500			106	22.50	128.50	195
0070	Reinforcing more than 2% of cross-sectional area	↓	8	5	↓		213	44.50	257.50	390
0150	Remove whole pieces, up to 2 tons per piece	E-18	36	1.111	Ea.		64.50	41.50	106	150
0160	2-5 tons per piece		30	1.333			77	49.50	126.50	180
0170	5-10 tons per piece		24	1.667			96.50	62	158.50	225
0180	10-15 tons per piece	↓	18	2.222			129	82.50	211.50	300
0250	Precast unit embedded in masonry, up to 1 C.F.	D-1	16	1			47		47	76
0260	1-2 C.F.		12	1.333			62.50		62.50	101
0270	2-5 C.F.		10	1.600			75		75	121
0280	5-10 C.F.	↓	8	2	↓		94		94	152
0990	For hydrodemolition see Section 02 41 13.15									
1910	Minimum labor/equipment charge	B-9	2	20	Job		850	179	1,029	1,550

03 05 13 – Basic Concrete Materials

03 05 13.20 Concrete Admixtures and Surface Treatments

		Crew	Daily Output	Labor-Hours	Unit	Material	2020 Bare Costs Labor	2020 Bare Costs Equipment	Total	Total Incl O&P
0010	**CONCRETE ADMIXTURES AND SURFACE TREATMENTS**									
0040	Abrasives, aluminum oxide, over 20 tons				Lb.	2.51			2.51	2.76
0050	1 to 20 tons					2.61			2.61	2.87
0070	Under 1 ton					2.71			2.71	2.98
0100	Silicon carbide, black, over 20 tons					4.36			4.36	4.80
0110	1 to 20 tons					4.53			4.53	4.99
0120	Under 1 ton				↓	4.71			4.71	5.20
0200	Air entraining agent, .7 to 1.5 oz. per bag, 55 gallon drum				Gal.	24.50			24.50	27
0220	5 gallon pail					36.50			36.50	40
0300	Bonding agent, acrylic latex, 250 S.F./gallon, 5 gallon pail					28			28	31
0320	Epoxy resin, 80 S.F./gallon, 4 gallon case				↓	69			69	76
0400	Calcium chloride, 50 lb. bags, T.L. lots				Ton	1,525			1,525	1,675
0420	Less than truckload lots				Bag	24.50			24.50	27
0500	Carbon black, liquid, 2 to 8 lb. per bag of cement				Lb.	17			17	18.70
0600	Concrete admixture, integral colors, dry pigment, 5 lb. bag				Ea.	31			31	34.50
0610	10 lb. bag					38.50			38.50	42
0620	25 lb. bag				↓	78.50			78.50	86
0920	Dustproofing compound, 250 S.F./gal., 5 gallon pail				Gal.	7.70			7.70	8.50
1010	Epoxy based, 125 S.F./gal., 5 gallon pail				"	63.50			63.50	69.50
1100	Hardeners, metallic, 55 lb. bags, natural (grey)				Lb.	.92			.92	1.01
1200	Colors					1.43			1.43	1.57
1300	Non-metallic, 55 lb. bags, natural grey					.74			.74	.82
1320	Colors				↓	.91			.91	1
1550	Release agent, for tilt slabs, 5 gallon pail				Gal.	18.05			18.05	19.90
1570	For forms, 5 gallon pail					13.35			13.35	14.65
1590	Concrete release agent for forms, 100% biodegradable, zero VOC, 5 gal. pail G					23.50			23.50	26
1595	55 gallon drum G					19.30			19.30	21
1600	Sealer, hardener and dustproofer, epoxy-based, 125 S.F./gal., 5 gallon unit					63.50			63.50	69.50
1620	3 gallon unit					69.50			69.50	76.50
1630	Sealer, solvent-based, 250 S.F./gal., 55 gallon drum					23			23	25.50
1640	5 gallon pail					33			33	36
1650	Sealer, water based, 350 S.F., 55 gallon drum					23.50			23.50	26
1660	5 gallon pail					26			26	29
1900	Set retarder, 100 S.F./gal., 1 gallon pail				↓	6.95			6.95	7.65
2000	Waterproofing, integral 1 lb. per bag of cement				Lb.	4.02			4.02	4.42

03 05 Common Work Results for Concrete

03 05 13 – Basic Concrete Materials

03 05 13.20 Concrete Admixtures and Surface Treatments	Crew	Daily Output	Labor-Hours	Unit	Material	2020 Bare Costs Labor	Equipment	Total	Total Incl O&P
2100	Powdered metallic, 40 lbs. per 100 S.F., standard colors				Lb.	4.77			4.77	5.25
2120	Premium colors				↓	6.70			6.70	7.35
3000	For colored ready-mix concrete, add to prices in section 03 31 13.35									
3100	Subtle shades, 5 lb. dry pigment per C.Y., add				C.Y.	31			31	34.50
3400	Medium shades, 10 lb. dry pigment per C.Y., add					38.50			38.50	42
3700	Deep shades, 25 lb. dry pigment per C.Y., add				↓	78.50			78.50	86

03 05 13.85 Winter Protection

		Crew	Daily Output	Labor-Hours	Unit	Material	Labor	Equipment	Total	Total Incl O&P
0010	**WINTER PROTECTION**									
0012	For heated ready mix, add				C.Y.	5.35			5.35	5.90
0100	Temporary heat to protect concrete, 24 hours	2 Clab	50	.320	M.S.F.	203	13.45		216.45	245
0200	Temporary shelter for slab on grade, wood frame/polyethylene sheeting									
0201	Build or remove, light framing for short spans	2 Carp	10	1.600	M.S.F.	345	85		430	515
0210	Large framing for long spans	"	3	5.333	"	460	283		743	965
0710	Electrically heated pads, 15 watts/S.F., 20 uses				S.F.	.58			.58	.64

03 11 Concrete Forming

03 11 13 – Structural Cast-In-Place Concrete Forming

03 11 13.20 Forms In Place, Beams and Girders

		Crew	Daily Output	Labor-Hours	Unit	Material	Labor	Equipment	Total	Total Incl O&P
0010	**FORMS IN PLACE, BEAMS AND GIRDERS** R031113-40									
0500	Exterior spandrel, job-built plywood, 12" wide, 1 use R031113-60	C-2	225	.213	SFCA	3.30	11		14.30	21.50
0650	4 use		310	.155		1.07	8		9.07	14
1000	18" wide, 1 use		250	.192		2.91	9.90		12.81	19.10
1150	4 use		315	.152		.95	7.85		8.80	13.65
1500	24" wide, 1 use		265	.181		2.65	9.35		12	17.90
1650	4 use		325	.148		.86	7.65		8.51	13.15
2000	Interior beam, job-built plywood, 12" wide, 1 use		300	.160		4.02	8.25		12.27	17.65
2150	4 use		377	.127		1.31	6.55		7.86	12
2500	24" wide, 1 use		320	.150		2.71	7.75		10.46	15.40
2650	4 use		395	.122		.87	6.30		7.17	11
3000	Encasing steel beam, hung, job-built plywood, 1 use		325	.148		3.24	7.65		10.89	15.75
3150	4 use	↓	430	.112	↓	1.05	5.75		6.80	10.40
9000	Minimum labor/equipment charge	2 Carp	2	8	Job		425		425	680

03 11 13.25 Forms In Place, Columns

		Crew	Daily Output	Labor-Hours	Unit	Material	Labor	Equipment	Total	Total Incl O&P
0010	**FORMS IN PLACE, COLUMNS** R031113-40									
0500	Round fiberglass, 4 use per mo., rent, 12" diameter	C-1	160	.200	L.F.	9.95	10.10		20.05	27
0550	16" diameter R031113-60		150	.213		11.85	10.75		22.60	30
0600	18" diameter		140	.229		13.25	11.50		24.75	33
0650	24" diameter		135	.237		16.50	11.95		28.45	37.50
0700	28" diameter		130	.246		18.45	12.40		30.85	40.50
0800	30" diameter		125	.256		19.25	12.90		32.15	41.50
0850	36" diameter		120	.267		25.50	13.45		38.95	49.50
1500	Round fiber tube, recycled paper, 1 use, 8" diameter [G]		155	.206		2.96	10.40		13.36	19.90
1550	10" diameter [G]		155	.206		3.91	10.40		14.31	21
1600	12" diameter [G]		150	.213		4.54	10.75		15.29	22
1650	14" diameter [G]		145	.221		4.54	11.10		15.64	23
1700	16" diameter [G]		140	.229		5.65	11.50		17.15	24.50
1720	18" diameter [G]		140	.229		6.60	11.50		18.10	25.50
1750	20" diameter [G]		135	.237		7.90	11.95		19.85	28
1800	24" diameter [G]		130	.246		11.40	12.40		23.80	32.50
1850	30" diameter [G]		125	.256		14.80	12.90		27.70	37

For customer support on your Commercial Renovation Costs with RSMeans data, call 800.448.8182.

55

03 11 Concrete Forming

03 11 13 – Structural Cast-In-Place Concrete Forming

03 11 13.25 Forms In Place, Columns

		Crew	Daily Output	Labor-Hours	Unit	Material	2020 Bare Costs Labor	Equipment	Total	Total Incl O&P
1900	36" diameter	C-1	115	.278	L.F.	20.50	14		34.50	45
1950	42" diameter		100	.320		44.50	16.10		60.60	75
2000	48" diameter		85	.376		55.50	18.95		74.45	91.50
2200	For seamless type, add					15%				
3000	Round, steel, 4 use per mo., rent, regular duty, 14" diameter	C-1	145	.221	L.F.	18.40	11.10		29.50	38
3050	16" diameter		125	.256		18.75	12.90		31.65	41
3100	Heavy duty, 20" diameter		105	.305		20.50	15.35		35.85	47
3150	24" diameter		85	.376		22	18.95		40.95	55
3200	30" diameter		70	.457		25.50	23		48.50	65
3250	36" diameter		60	.533		27.50	27		54.50	73
3300	48" diameter		50	.640		39.50	32.50		72	95
3350	60" diameter		45	.711		50	36		86	113
4500	For second and succeeding months, deduct					50%				
5000	Job-built plywood, 8" x 8" columns, 1 use	C-1	165	.194	SFCA	2.88	9.75		12.63	18.80
5050	2 use		195	.164		1.65	8.25		9.90	15.05
5100	3 use		210	.152		1.15	7.70		8.85	13.55
5150	4 use		215	.149		.95	7.50		8.45	13.05
5500	12" x 12" columns, 1 use		180	.178		2.80	8.95		11.75	17.45
5550	2 use		210	.152		1.54	7.70		9.24	14
5600	3 use		220	.145		1.12	7.35		8.47	13
5650	4 use		225	.142		.91	7.15		8.06	12.50
6000	16" x 16" columns, 1 use		185	.173		2.81	8.70		11.51	17.05
6050	2 use		215	.149		1.50	7.50		9	13.65
6100	3 use		230	.139		1.13	7		8.13	12.50
6150	4 use		235	.136		.92	6.85		7.77	12
6500	24" x 24" columns, 1 use		190	.168		3.19	8.50		11.69	17.10
6550	2 use		216	.148		1.75	7.45		9.20	13.90
6600	3 use		230	.139		1.27	7		8.27	12.65
6650	4 use		238	.134		1.04	6.75		7.79	12
7000	36" x 36" columns, 1 use		200	.160		2.33	8.05		10.38	15.45
7050	2 use		230	.139		1.32	7		8.32	12.70
7100	3 use		245	.131		.93	6.60		7.53	11.55
7150	4 use		250	.128		.76	6.45		7.21	11.20
7400	Steel framed plywood, based on 50 uses of purchased									
7420	forms, and 4 uses of bracing lumber									
7500	8" x 8" column	C-1	340	.094	SFCA	2.22	4.74		6.96	10.05
7550	10" x 10"		350	.091		1.98	4.61		6.59	9.55
7600	12" x 12"		370	.086		1.68	4.36		6.04	8.85
7650	16" x 16"		400	.080		1.30	4.03		5.33	7.90
7700	20" x 20"		420	.076		1.16	3.84		5	7.40
7750	24" x 24"		440	.073		.81	3.66		4.47	6.75
7755	30" x 30"		440	.073		1.04	3.66		4.70	7
7760	36" x 36"		460	.070		.92	3.51		4.43	6.60
9000	Minimum labor/equipment charge	2 Carp	2	8	Job		425		425	680

03 11 13.35 Forms In Place, Elevated Slabs

		Crew	Daily Output	Labor-Hours	Unit	Material	2020 Bare Costs Labor	Equipment	Total	Total Incl O&P
0010	**FORMS IN PLACE, ELEVATED SLABS** R031113-40									
1000	Flat plate, job-built plywood, to 15' high, 1 use	C-2	470	.102	S.F.	4.09	5.25		9.34	12.95
1150	4 use R031113-60		560	.086		1.33	4.43		5.76	8.55
1500	15' to 20' high ceilings, 4 use		495	.097		1.32	5		6.32	9.45
2000	Flat slab, drop panels, job-built plywood, to 15' high, 1 use		449	.107		4.85	5.50		10.35	14.20
2150	4 use		544	.088		1.58	4.56		6.14	9.05
2250	15' to 20' high ceilings, 4 use		480	.100		2.70	5.15		7.85	11.20

03 11 Concrete Forming

03 11 13 – Structural Cast-In-Place Concrete Forming

03 11 13.35 Forms In Place, Elevated Slabs

		Crew	Daily Output	Labor-Hours	Unit	Material	2020 Bare Costs Labor	Equipment	Total	Total Incl O&P
3500	Floor slab, with 1-way joist pans, 1 use	C-2	415	.116	S.F.	8.30	5.95		14.25	18.70
3650	4 use		500	.096		5	4.96		9.96	13.45
4500	With 2-way waffle domes, 1 use		405	.119		8.90	6.10		15	19.60
4550	4 use		470	.102		5.60	5.25		10.85	14.60
5000	Box out for slab openings, over 16" deep, 1 use		190	.253	SFCA	3.75	13.05		16.80	25
5050	2 use		240	.200	"	2.06	10.35		12.41	18.80
5500	Shallow slab box outs, to 10 S.F.		42	1.143	Ea.	12.60	59		71.60	108
5550	Over 10 S.F. (use perimeter)		600	.080	L.F.	1.69	4.13		5.82	8.45
6000	Bulkhead forms for slab, with keyway, 1 use, 2 piece		500	.096		1.80	4.96		6.76	9.95
6100	3 piece (see also edge forms)		460	.104		2.06	5.40		7.46	10.90
6200	Slab bulkhead form, 4-1/2" high, exp metal, w/keyway & stakes	G C-1	1200	.027		.94	1.34		2.28	3.18
6210	5-1/2" high	G	1100	.029		1.19	1.47		2.66	3.66
6215	7-1/2" high	G	960	.033		1.40	1.68		3.08	4.23
6220	9-1/2" high	G	840	.038		1.56	1.92		3.48	4.80
6500	Curb forms, wood, 6" to 12" high, on elevated slabs, 1 use		180	.178	SFCA	1.80	8.95		10.75	16.35
6550	2 use		205	.156		.99	7.85		8.84	13.70
6600	3 use		220	.145		.72	7.35		8.07	12.55
6650	4 use		225	.142		.58	7.15		7.73	12.15
7000	Edge forms to 6" high, on elevated slab, 4 use		500	.064	L.F.	.22	3.23		3.45	5.40
7070	7" to 12" high, 1 use		162	.198	SFCA	1.33	9.95		11.28	17.40
7080	2 use		198	.162		.73	8.15		8.88	13.85
7090	3 use		222	.144		.53	7.25		7.78	12.25
7101	4 use		350	.091		.22	4.61		4.83	7.65
7500	Depressed area forms to 12" high, 4 use		300	.107	L.F.	.83	5.40		6.23	9.50
7550	12" to 24" high, 4 use		175	.183		1.13	9.20		10.33	16
8000	Perimeter deck and rail for elevated slabs, straight		90	.356		12.70	17.90		30.60	42.50
8050	Curved		65	.492		17.45	25		42.45	58.50
8500	Void forms, round plastic, 8" high x 3" diameter	G	450	.071	Ea.	2.05	3.58		5.63	8
8550	4" diameter	G	425	.075		2.32	3.79		6.11	8.65
8600	6" diameter	G	400	.080		4.05	4.03		8.08	10.90
8650	8" diameter	G	375	.085		7	4.30		11.30	14.60
9000	Minimum labor/equipment charge	2 Carp	2	8	Job		425		425	680

03 11 13.40 Forms In Place, Equipment Foundations

		Crew	Daily Output	Labor-Hours	Unit	Material	2020 Bare Costs Labor	Equipment	Total	Total Incl O&P
0010	**FORMS IN PLACE, EQUIPMENT FOUNDATIONS** R031113-40									
0020	1 use	C-2	160	.300	SFCA	2.95	15.50		18.45	28.50
0050	2 use R031113-60		190	.253		1.62	13.05		14.67	23
0100	3 use		200	.240		1.18	12.40		13.58	21
0150	4 use		205	.234		.96	12.10		13.06	20.50
9000	Minimum labor/equipment charge	1 Carp	3	2.667	Job		142		142	227

03 11 13.45 Forms In Place, Footings

		Crew	Daily Output	Labor-Hours	Unit	Material	2020 Bare Costs Labor	Equipment	Total	Total Incl O&P
0010	**FORMS IN PLACE, FOOTINGS** R031113-40									
0020	Continuous wall, plywood, 1 use	C-1	375	.085	SFCA	6.85	4.30		11.15	14.45
0150	4 use R031113-60	"	485	.066	"	2.23	3.32		5.55	7.80
1500	Keyway, 4 use, tapered wood, 2" x 4"	1 Carp	530	.015	L.F.	.25	.80		1.05	1.56
1550	2" x 6"	"	500	.016	"	.34	.85		1.19	1.73
3000	Pile cap, square or rectangular, job-built plywood, 1 use	C-1	290	.110	SFCA	3.16	5.55		8.71	12.40
3150	4 use		383	.084		1.03	4.21		5.24	7.90
5000	Spread footings, job-built lumber, 1 use		305	.105		2.28	5.30		7.58	10.95
5150	4 use		414	.077		.74	3.89		4.63	7.05
9000	Minimum labor/equipment charge	1 Carp	3	2.667	Job		142		142	227

For customer support on your Commercial Renovation Costs with RSMeans data, call 800.448.8182.

57

03 11 13 – Structural Cast-In-Place Concrete Forming

03 11 13.47 Forms In Place, Gas Station Forms		Crew	Daily Output	Labor-Hours	Unit	Material	2020 Bare Costs Labor	Equipment	Total	Total Incl O&P
0010	**FORMS IN PLACE, GAS STATION FORMS**									
0050	Curb fascia, with template, 12 ga. steel, left in place, 9" high	G	1 Carp	50	.160	L.F.	15.05	8.50		23.55	30
1000	Sign or light bases, 18" diameter, 9" high	G		9	.889	Ea.	95	47		142	180
1050	30" diameter, 13" high	G		8	1	"	151	53		204	251
1990	Minimum labor/equipment charge			2	4	Job		213		213	340
2000	Island forms, 10' long, 9" high, 3'-6" wide	G	C-1	10	3.200	Ea.	420	161		581	725
2050	4' wide	G		9	3.556		435	179		614	765
2500	20' long, 9" high, 4' wide	G		6	5.333		700	269		969	1,200
2550	5' wide	G		5	6.400		725	325		1,050	1,325
9000	Minimum labor/equipment charge		1 Carp	3	2.667	Job		142		142	227

03 11 13.50 Forms In Place, Grade Beam

			Crew	Daily Output	Labor-Hours	Unit	Material	2020 Bare Costs Labor	Equipment	Total	Total Incl O&P
0010	**FORMS IN PLACE, GRADE BEAM**	R031113-40									
0020	Job-built plywood, 1 use		C-2	530	.091	SFCA	3.15	4.68		7.83	10.95
0150	4 use	R031113-60	"	605	.079	"	1.02	4.10		5.12	7.70
9000	Minimum labor/equipment charge		2 Carp	2	8	Job		425		425	680

03 11 13.65 Forms In Place, Slab On Grade

			Crew	Daily Output	Labor-Hours	Unit	Material	2020 Bare Costs Labor	Equipment	Total	Total Incl O&P
0010	**FORMS IN PLACE, SLAB ON GRADE**	R031113-40									
1000	Bulkhead forms w/keyway, wood, 6" high, 1 use	R031113-60	C-1	510	.063	L.F.	1.10	3.16		4.26	6.25
1400	Bulkhead form for slab, 4-1/2" high, exp metal, incl. keyway & stakes	G		1200	.027		.94	1.34		2.28	3.18
1410	5-1/2" high	G		1100	.029		1.19	1.47		2.66	3.66
1420	7-1/2" high	G		960	.033		1.40	1.68		3.08	4.23
1430	9-1/2" high	G		840	.038		1.56	1.92		3.48	4.80
2000	Curb forms, wood, 6" to 12" high, on grade, 1 use			215	.149	SFCA	2.19	7.50		9.69	14.40
2150	4 use			275	.116	"	.71	5.85		6.56	10.20
3000	Edge forms, wood, 4 use, on grade, to 6" high			600	.053	L.F.	.33	2.69		3.02	4.67
3050	7" to 12" high			435	.074	SFCA	.75	3.71		4.46	6.80
3060	Over 12"			350	.091	"	.97	4.61		5.58	8.45
3500	For depressed slabs, 4 use, to 12" high			300	.107	L.F.	.81	5.40		6.21	9.50
3550	To 24" high			175	.183		1.04	9.20		10.24	15.90
4000	For slab blockouts, to 12" high, 1 use			200	.160		.82	8.05		8.87	13.80
4050	To 24" high, 1 use			120	.267		1.04	13.45		14.49	22.50
4100	Plastic (extruded), to 6" high, multiple use, on grade			800	.040		6.50	2.02		8.52	10.40
5000	Screed, 24 ga. metal key joint, see Section 03 15 16.30										
5020	Wood, incl. wood stakes, 1" x 3"		C-1	900	.036	L.F.	.87	1.79		2.66	3.83
5050	2" x 4"			900	.036	"	.90	1.79		2.69	3.86
6000	Trench forms in floor, wood, 1 use			160	.200	SFCA	2.05	10.10		12.15	18.40
6150	4 use			185	.173	"	.67	8.70		9.37	14.70
8760	Void form, corrugated fiberboard, 4" x 12", 4' long	G		3000	.011	S.F.	3.64	.54		4.18	4.86
8770	6" x 12", 4' long			3000	.011		4.47	.54		5.01	5.80
8780	1/4" thick hardboard protective cover for void form		2 Carp	1500	.011		.72	.57		1.29	1.70
9000	Minimum labor/equipment charge		1 Carp	2	4	Job		213		213	340

03 11 13.85 Forms In Place, Walls

			Crew	Daily Output	Labor-Hours	Unit	Material	2020 Bare Costs Labor	Equipment	Total	Total Incl O&P
0010	**FORMS IN PLACE, WALLS**	R031113-10									
0100	Box out for wall openings, to 16" thick, to 10 S.F.		C-2	24	2	Ea.	28	103		131	196
0150	Over 10 S.F. (use perimeter)	R031113-40	"	280	.171	L.F.	2.45	8.85		11.30	16.90
0250	Brick shelf, 4" w, add to wall forms, use wall area above shelf										
0260	1 use	R031113-60	C-2	240	.200	SFCA	2.70	10.35		13.05	19.50
0350	4 use			300	.160	"	1.08	8.25		9.33	14.45
0500	Bulkhead, wood with keyway, 1 use, 2 piece			265	.181	L.F.	2.19	9.35		11.54	17.40
0600	Bulkhead forms with keyway, 1 piece expanded metal, 8" wall	G	C-1	1000	.032		1.40	1.61		3.01	4.12
0610	10" wall	G		800	.040		1.56	2.02		3.58	4.95

03 11 Concrete Forming

03 11 13 – Structural Cast-In-Place Concrete Forming

03 11 13.85 Forms In Place, Walls

		Crew	Daily Output	Labor-Hours	Unit	Material	2020 Bare Costs Labor	Equipment	Total	Total Incl O&P
0620	12" wall [G]	C-1	525	.061	L.F.	1.87	3.07		4.94	7
0700	Buttress, to 8' high, 1 use	C-2	350	.137	SFCA	4.85	7.10		11.95	16.70
0850	4 use		480	.100	"	1.60	5.15		6.75	10
1000	Corbel or haunch, to 12" wide, add to wall forms, 1 use		150	.320	L.F.	2.59	16.50		19.09	29.50
1150	4 use		180	.267	"	.84	13.75		14.59	23
2000	Wall, job-built plywood, to 8' high, 1 use		370	.130	SFCA	2.99	6.70		9.69	14.05
2050	2 use		435	.110		1.91	5.70		7.61	11.25
2100	3 use		495	.097		1.39	5		6.39	9.55
2150	4 use		505	.095		1.13	4.91		6.04	9.10
2400	Over 8' to 16' high, 1 use		280	.171		3.32	8.85		12.17	17.85
2450	2 use		345	.139		1.44	7.20		8.64	13.10
2500	3 use		375	.128		1.03	6.60		7.63	11.75
2550	4 use		395	.122		.84	6.30		7.14	11
2700	Over 16' high, 1 use		235	.204		2.94	10.55		13.49	20
2750	2 use		290	.166		1.62	8.55		10.17	15.50
2800	3 use		315	.152		1.18	7.85		9.03	13.90
2850	4 use		330	.145		.96	7.50		8.46	13.10
4000	Radial, smooth curved, job-built plywood, 1 use		245	.196		2.65	10.10		12.75	19.10
4150	4 use		335	.143		.86	7.40		8.26	12.80
4200	Below grade, job-built plywood, 1 use		225	.213		2.86	11		13.86	21
4210	2 use		225	.213		1.58	11		12.58	19.40
4220	3 use		225	.213		1.31	11		12.31	19.10
4230	4 use		225	.213		.93	11		11.93	18.70
4600	Retaining wall, battered, job-built plyw'd, to 8' high, 1 use		300	.160		2.18	8.25		10.43	15.65
4750	4 use		390	.123		.71	6.35		7.06	11
4900	Over 8' to 16' high, 1 use		240	.200		2.39	10.35		12.74	19.15
5050	4 use		320	.150		.78	7.75		8.53	13.25
7500	Lintel or sill forms, 1 use	1 Carp	30	.267		3.51	14.15		17.66	26.50
7560	4 use	"	37	.216		1.14	11.50		12.64	19.65
7800	Modular prefabricated plywood, based on 20 uses of purchased									
7820	forms, and 4 uses of bracing lumber									
7860	To 8' high	C-2	800	.060	SFCA	1.15	3.10		4.25	6.20
8060	Over 8' to 16' high		600	.080		1.20	4.13		5.33	7.90
8600	Pilasters, 1 use		270	.178		3.49	9.20		12.69	18.55
8660	4 use		385	.125		1.13	6.45		7.58	11.55
9010	Steel framed plywood, based on 50 uses of purchased									
9020	forms, and 4 uses of bracing lumber									
9060	To 8' high	C-2	600	.080	SFCA	.79	4.13		4.92	7.45
9260	Over 8' to 16' high		450	.107		.79	5.50		6.29	9.70
9460	Over 16' to 20' high		400	.120		.79	6.20		6.99	10.80
9475	For elevated walls, add						10%			
9480	For battered walls, 1 side battered, add					10%	10%			
9485	For battered walls, 2 sides battered, add					15%	15%			
9900	Minimum labor/equipment charge	2 Carp	2	8	Job		425		425	680

03 11 16 – Architectural Cast-in-Place Concrete Forming

03 11 16.13 Concrete Form Liners

		Crew	Daily Output	Labor-Hours	Unit	Material	2020 Bare Costs Labor	Equipment	Total	Total Incl O&P
0010	**CONCRETE FORM LINERS**									
5750	Liners for forms (add to wall forms), ABS plastic									
5800	Aged wood, 4" wide, 1 use	1 Carp	256	.031	SFCA	3.34	1.66		5	6.35
5820	2 use		256	.031		1.84	1.66		3.50	4.68
5830	3 use		256	.031		1.34	1.66		3	4.13
5840	4 use		256	.031		1.09	1.66		2.75	3.85

03 11 16.13 Concrete Form Liners

		Crew	Daily Output	Labor-Hours	Unit	Material	2020 Bare Costs Labor	2020 Bare Costs Equipment	Total	Total Incl O&P
5900	Fractured rope rib, 1 use	1 Carp	192	.042	SFCA	4.91	2.21		7.12	8.95
5925	2 use		192	.042		2.70	2.21		4.91	6.50
5950	3 use		192	.042		1.96	2.21		4.17	5.70
6000	4 use		192	.042		1.60	2.21		3.81	5.30
6100	Ribbed, 3/4" deep x 1-1/2" OC, 1 use		224	.036		5.05	1.90		6.95	8.65
6125	2 use		224	.036		2.79	1.90		4.69	6.10
6150	3 use		224	.036		2.03	1.90		3.93	5.25
6200	4 use		224	.036		1.65	1.90		3.55	4.85
6300	Rustic brick pattern, 1 use		224	.036		3.47	1.90		5.37	6.85
6325	2 use		224	.036		1.91	1.90		3.81	5.15
6350	3 use		224	.036		1.39	1.90		3.29	4.57
6400	4 use		224	.036		1.13	1.90		3.03	4.28
6500	3/8" striated, random, 1 use		224	.036		3.54	1.90		5.44	6.95
6525	2 use		224	.036		1.95	1.90		3.85	5.20
6550	3 use		224	.036		1.42	1.90		3.32	4.60
6600	4 use		224	.036		1.15	1.90		3.05	4.31

03 15 05.70 Shores

			Crew	Daily Output	Labor-Hours	Unit	Material	2020 Bare Costs Labor	2020 Bare Costs Equipment	Total	Total Incl O&P
0010	**SHORES**										
0020	Erect and strip, by hand, horizontal members										
0500	Aluminum joists and stringers	G	2 Carp	60	.267	Ea.		14.15		14.15	22.50
0600	Steel, adjustable beams	G		45	.356			18.90		18.90	30.50
0700	Wood joists			50	.320			17		17	27.50
0800	Wood stringers			30	.533			28.50		28.50	45.50
1000	Vertical members to 10' high	G		55	.291			15.45		15.45	25
1050	To 13' high	G		50	.320			17		17	27.50
1100	To 16' high	G		45	.356			18.90		18.90	30.50
1500	Reshoring	G		1400	.011	S.F.	.63	.61		1.24	1.66
1600	Flying truss system	G	C-17D	9600	.009	SFCA		.49	.08	.57	.86
1760	Horizontal, aluminum joists, 6-1/4" high x 5' to 21' span, buy	G				L.F.	16.40			16.40	18.05
1770	Beams, 7-1/4" high x 4' to 30' span	G				"	19.70			19.70	21.50
1810	Horizontal, steel beam, W8x10, 7' span, buy	G				Ea.	64.50			64.50	71
1830	10' span	G					75			75	82.50
1920	15' span	G					129			129	142
1940	20' span	G					181			181	199
1970	Steel stringer, W8x10, 4' to 16' span, buy	G				L.F.	7.45			7.45	8.20
3000	Rent for job duration, aluminum joist @ 2' OC, per mo.	G				SF Flr.	.41			.41	.45
3050	Steel W8x10	G					.19			.19	.21
3060	Steel adjustable	G					.19			.19	.21
3500	#1 post shore, steel, 5'-7" to 9'-6" high, 10,000# cap., buy	G				Ea.	161			161	178
3550	#2 post shore, 7'-3" to 12'-10" high, 7800# capacity	G					186			186	205
3600	#3 post shore, 8'-10" to 16'-1" high, 3800# capacity	G					204			204	224
5010	Frame shoring systems, steel, 12,000#/leg, buy										
5040	Frame, 2' wide x 6' high	G				Ea.	107			107	118
5250	X-brace	G					17.05			17.05	18.80
5550	Base plate	G					16.05			16.05	17.65
5600	Screw jack	G					35			35	38
5650	U-head, 8" x 8"	G					20.50			20.50	22.50

03 15 Concrete Accessories

03 15 05 – Concrete Forming Accessories

03 15 05.85 Stair Tread Inserts

		Crew	Daily Output	Labor-Hours	Unit	Material	2020 Bare Costs Labor	Equipment	Total	Total Incl O&P
0010	**STAIR TREAD INSERTS**									
0105	Cast nosing insert, abrasive surface, pre-drilled, includes screws									
0110	Aluminum, 3" wide x 3' long	1 Cefi	32	.250	Ea.	58	12.50		70.50	83
0120	4' long		31	.258		75	12.90		87.90	103
0130	5' long	↓	30	.267	↓	90	13.30		103.30	120
0135	Extruded nosing insert, black abrasive strips, continuous anchor									
0140	Aluminum, 3" wide x 3' long	1 Cefi	64	.125	Ea.	35.50	6.25		41.75	49
0150	4' long		60	.133		48	6.65		54.65	63.50
0160	5' long	↓	56	.143		62.50	7.15		69.65	80
0165	Extruded nosing insert, black abrasive strips, pre-drilled, incl. screws									
0170	Aluminum, 3" wide x 3' long	1 Cefi	32	.250	Ea.	45	12.50		57.50	69
0180	4' long		31	.258		59.50	12.90		72.40	85.50
0190	5' long	↓	30	.267	↓	83.50	13.30		96.80	113
9000	Minimum labor/equipment charge	1 Carp	4	2	Job		106		106	170

03 15 13 – Waterstops

03 15 13.50 Waterstops

		Crew	Daily Output	Labor-Hours	Unit	Material	2020 Bare Costs Labor	Equipment	Total	Total Incl O&P
0010	**WATERSTOPS**, PVC and Rubber									
0020	PVC, ribbed 3/16" thick, 4" wide	1 Carp	155	.052	L.F.	1.55	2.74		4.29	6.10
0050	6" wide		145	.055		2.78	2.93		5.71	7.75
0500	With center bulb, 6" wide, 3/16" thick		135	.059		2.68	3.15		5.83	8
0550	3/8" thick		130	.062		4.85	3.27		8.12	10.60
0600	9" wide x 3/8" thick	↓	125	.064	↓	8.15	3.40		11.55	14.40

03 15 16 – Concrete Construction Joints

03 15 16.20 Control Joints, Saw Cut

		Crew	Daily Output	Labor-Hours	Unit	Material	2020 Bare Costs Labor	Equipment	Total	Total Incl O&P
0010	**CONTROL JOINTS, SAW CUT**									
0100	Sawcut control joints in green concrete									
0120	1" depth	C-27	2000	.008	L.F.	.03	.40	.06	.49	.73
0140	1-1/2" depth		1800	.009		.05	.44	.06	.55	.83
0160	2" depth	↓	1600	.010	↓	.07	.50	.07	.64	.93
0180	Sawcut joint reservoir in cured concrete									
0182	3/8" wide x 3/4" deep, with single saw blade	C-27	1000	.016	L.F.	.05	.80	.11	.96	1.44
0184	1/2" wide x 1" deep, with double saw blades		900	.018		.10	.89	.12	1.11	1.64
0186	3/4" wide x 1-1/2" deep, with double saw blades	↓	800	.020		.20	1	.14	1.34	1.94
0190	Water blast joint to wash away laitance, 2 passes	C-29	2500	.003			.13	.04	.17	.26
0200	Air blast joint to blow out debris and air dry, 2 passes	C-28	2000	.004	↓		.20	.02	.22	.33
0300	For backer rod, see Section 07 91 23.10									
0340	For joint sealant, see Section 03 15 16.30 or 07 92 13.20									
0900	For replacement of joint sealant, see Section 07 01 90.81									

03 15 16.30 Expansion Joints

			Crew	Daily Output	Labor-Hours	Unit	Material	2020 Bare Costs Labor	Equipment	Total	Total Incl O&P
0010	**EXPANSION JOINTS**										
0020	Keyed, cold, 24 ga., incl. stakes, 3-1/2" high	G	1 Carp	200	.040	L.F.	.89	2.13		3.02	4.39
0050	4-1/2" high	G		200	.040		.94	2.13		3.07	4.44
0100	5-1/2" high	G		195	.041		1.19	2.18		3.37	4.80
0150	7-1/2" high	G		190	.042		1.40	2.24		3.64	5.15
0160	9-1/2" high	G	↓	185	.043		1.56	2.30		3.86	5.40
0300	Poured asphalt, plain, 1/2" x 1"		1 Clab	450	.018		.24	.75		.99	1.46
0350	1" x 2"			400	.020		.97	.84		1.81	2.42
0500	Neoprene, liquid, cold applied, 1/2" x 1"			450	.018		2.53	.75		3.28	3.98
0550	1" x 2"			400	.020		9.95	.84		10.79	12.30
0700	Polyurethane, poured, 2 part, 1/2" x 1"			400	.020		1.40	.84		2.24	2.89
0750	1" x 2"			350	.023		5.60	.96		6.56	7.75

For customer support on your Commercial Renovation Costs with RSMeans data, call 800.448.8182.

61

03 15 16 – Concrete Construction Joints

03 15 16.30 Expansion Joints		Crew	Daily Output	Labor-Hours	Unit	Material	2020 Bare Costs Labor	Equipment	Total	Total Incl O&P
0900	Rubberized asphalt, hot or cold applied, 1/2" x 1"	1 Clab	450	.018	L.F.	.32	.75		1.07	1.55
0950	1" x 2"		400	.020		1.27	.84		2.11	2.75
1100	Hot applied, fuel resistant, 1/2" x 1"		450	.018		.48	.75		1.23	1.73
1150	1" x 2"		400	.020		1.91	.84		2.75	3.45
2000	Premolded, bituminous fiber, 1/2" x 6"	1 Carp	375	.021		.44	1.13		1.57	2.30
2050	1" x 12"		300	.027		2.03	1.42		3.45	4.50
2140	Concrete expansion joint, recycled paper and fiber, 1/2" x 6" G		390	.021		.43	1.09		1.52	2.22
2150	1/2" x 12" G		360	.022		.86	1.18		2.04	2.84
2250	Cork with resin binder, 1/2" x 6"		375	.021		1.36	1.13		2.49	3.32
2300	1" x 12"		300	.027		3.50	1.42		4.92	6.10
2500	Neoprene sponge, closed cell, 1/2" x 6"		375	.021		2.43	1.13		3.56	4.49
2550	1" x 12"		300	.027		8.90	1.42		10.32	12
2750	Polyethylene foam, 1/2" x 6"		375	.021		.59	1.13		1.72	2.47
2800	1" x 12"		300	.027		3.11	1.42		4.53	5.70
3000	Polyethylene backer rod, 3/8" diameter		460	.017		.03	.92		.95	1.52
3050	3/4" diameter		460	.017		.07	.92		.99	1.55
3100	1" diameter		460	.017		.15	.92		1.07	1.64
3500	Polyurethane foam, with polybutylene, 1/2" x 1/2"		475	.017		1.25	.90		2.15	2.81
3550	1" x 1"		450	.018		3.19	.95		4.14	5
3750	Polyurethane foam, regular, closed cell, 1/2" x 6"		375	.021		.93	1.13		2.06	2.84
3800	1" x 12"		300	.027		3.32	1.42		4.74	5.90
4000	Polyvinyl chloride foam, closed cell, 1/2" x 6"		375	.021		2.53	1.13		3.66	4.60
4050	1" x 12"		300	.027		8.70	1.42		10.12	11.80
4250	Rubber, gray sponge, 1/2" x 6"		375	.021		2.12	1.13		3.25	4.15
4300	1" x 12"		300	.027		7.65	1.42		9.07	10.65
4400	Redwood heartwood, 1" x 4"		400	.020		1.18	1.06		2.24	2.99
4450	1" x 6"		375	.021		1.79	1.13		2.92	3.79
5000	For installation in walls, add						75%			
5250	For installation in boxouts, add						25%			

03 15 19 – Cast-In Concrete Anchors

03 15 19.05 Anchor Bolt Accessories

		Crew	Daily Output	Labor-Hours	Unit	Material	2020 Bare Costs Labor	Equipment	Total	Total Incl O&P
0010	**ANCHOR BOLT ACCESSORIES**									
0015	For anchor bolts set in fresh concrete, see Section 03 15 19.10									
8150	Anchor bolt sleeve, plastic, 1" diameter bolts	1 Carp	60	.133	Ea.	11.60	7.10		18.70	24
8500	1-1/2" diameter		28	.286		21.50	15.20		36.70	48
8600	2" diameter		24	.333		19.10	17.70		36.80	49.50
8650	3" diameter		20	.400		35	21.50		56.50	72.50

03 15 19.10 Anchor Bolts

		Crew	Daily Output	Labor-Hours	Unit	Material	2020 Bare Costs Labor	Equipment	Total	Total Incl O&P
0010	**ANCHOR BOLTS**									
0015	Made from recycled materials									
0025	Single bolts installed in fresh concrete, no templates									
0030	Hooked w/nut and washer, 1/2" diameter, 8" long G	1 Carp	132	.061	Ea.	1.62	3.22		4.84	6.95
0040	12" long G		131	.061		1.80	3.25		5.05	7.20
0050	5/8" diameter, 8" long G		129	.062		3.39	3.30		6.69	9.05
0060	12" long G		127	.063		4.18	3.35		7.53	9.95
0070	3/4" diameter, 8" long G		127	.063		4.18	3.35		7.53	9.95
0080	12" long G		125	.064		5.20	3.40		8.60	11.20
0090	2-bolt pattern, including job-built 2-hole template, per set									
0100	J-type, incl. hex nut & washer, 1/2" diameter x 6" long G	1 Carp	21	.381	Set	5.60	20.50		26.10	38.50
0110	12" long G		21	.381		6.35	20.50		26.85	39.50
0120	18" long G		21	.381		7.40	20.50		27.90	40.50
0130	3/4" diameter x 8" long G		20	.400		11.10	21.50		32.60	46

03 15 19 – Cast-In Concrete Anchors

03 15 19.10 Anchor Bolts		Crew	Daily Output	Labor- Hours	Unit	Material	2020 Bare Costs Labor	Equipment	Total	Total Incl O&P
0140	12" long	G 1 Carp	20	.400	Set	13.15	21.50		34.65	48.50
0150	18" long	G	20	.400		16.30	21.50		37.80	52
0160	1" diameter x 12" long	G	19	.421		26.50	22.50		49	65
0170	18" long	G	19	.421		31	22.50		53.50	70.50
0180	24" long	G	19	.421		37.50	22.50		60	77
0190	36" long	G	18	.444		50.50	23.50		74	93.50
0200	1-1/2" diameter x 18" long	G	17	.471		47.50	25		72.50	92
0210	24" long	G	16	.500		56	26.50		82.50	104
0300	L-type, incl. hex nut & washer, 3/4" diameter x 12" long	G	20	.400		18.20	21.50		39.70	54
0310	18" long	G	20	.400		22.50	21.50		44	59
0320	24" long	G	20	.400		27	21.50		48.50	63.50
0330	30" long	G	20	.400		33.50	21.50		55	71
0340	36" long	G	20	.400		38	21.50		59.50	75.50
0350	1" diameter x 12" long	G	19	.421		24.50	22.50		47	63
0360	18" long	G	19	.421		29.50	22.50		52	68.50
0370	24" long	G	19	.421		36	22.50		58.50	75.50
0380	30" long	G	19	.421		42	22.50		64.50	82.50
0390	36" long	G	18	.444		47.50	23.50		71	90.50
0400	42" long	G	18	.444		57.50	23.50		81	101
0410	48" long	G	18	.444		64	23.50		87.50	109
0420	1-1/4" diameter x 18" long	G	18	.444		43.50	23.50		67	86
0430	24" long	G	18	.444		51.50	23.50		75	94.50
0440	30" long	G	17	.471		59	25		84	105
0450	36" long	G	17	.471		66.50	25		91.50	114
1000	4-bolt pattern, including job-built 4-hole template, per set									
1100	J-type, incl. hex nut & washer, 1/2" diameter x 6" long	G 1 Carp	19	.421	Set	8.50	22.50		31	45.50
1110	12" long	G	19	.421		9.95	22.50		32.45	47
1120	18" long	G	18	.444		12.10	23.50		35.60	51.50
1130	3/4" diameter x 8" long	G	17	.471		19.45	25		44.45	61.50
1140	12" long	G	17	.471		23.50	25		48.50	66
1150	18" long	G	17	.471		30	25		55	73
1160	1" diameter x 12" long	G	16	.500		50	26.50		76.50	97.50
1170	18" long	G	15	.533		59.50	28.50		88	111
1180	24" long	G	15	.533		72	28.50		100.50	125
1190	36" long	G	15	.533		98	28.50		126.50	154
1200	1-1/2" diameter x 18" long	G	13	.615		92	32.50		124.50	154
1210	24" long	G	12	.667		109	35.50		144.50	177
1300	L-type, incl. hex nut & washer, 3/4" diameter x 12" long	G	17	.471		33.50	25		58.50	77
1310	18" long	G	17	.471		42.50	25		67.50	86.50
1320	24" long	G	17	.471		51	25		76	96
1330	30" long	G	16	.500		64	26.50		90.50	113
1340	36" long	G	16	.500		73	26.50		99.50	123
1350	1" diameter x 12" long	G	16	.500		46	26.50		72.50	93
1360	18" long	G	15	.533		56.50	28.50		85	108
1370	24" long	G	15	.533		69.50	28.50		98	122
1380	30" long	G	15	.533		81.50	28.50		110	135
1390	36" long	G	15	.533		92.50	28.50		121	148
1400	42" long	G	14	.571		112	30.50		142.50	172
1410	48" long	G	14	.571		125	30.50		155.50	187
1420	1-1/4" diameter x 18" long	G	14	.571		84.50	30.50		115	142
1430	24" long	G	14	.571		100	30.50		130.50	159
1440	30" long	G	13	.615		115	32.50		147.50	180
1450	36" long	G	13	.615		131	32.50		163.50	197

For customer support on your Commercial Renovation Costs with RSMeans data, call 800.448.8182.

63

03 15 Concrete Accessories

03 15 19 – Cast-In Concrete Anchors

03 15 19.45 Machinery Anchors		Crew	Daily Output	Labor-Hours	Unit	Material	2020 Bare Costs Labor	Equipment	Total	Total Incl O&P
0010	**MACHINERY ANCHORS**, heavy duty, incl. sleeve, floating base nut,									
0020	lower stud & coupling nut, fiber plug, connecting stud, washer & nut.									
0030	For flush mounted embedment in poured concrete heavy equip. pads.									
0200	Stud & bolt, 1/2" diameter	G	E-16	40	.400	Ea.	61.50	23.50	3.68	88.68	111
0300	5/8" diameter	G		35	.457		67	26.50	4.20	97.70	122
0500	3/4" diameter	G		30	.533		79.50	31	4.90	115.40	144
0600	7/8" diameter	G		25	.640		93.50	37.50	5.90	136.90	171
0800	1" diameter	G		20	.800		102	47	7.35	156.35	197
0900	1-1/4" diameter	G		15	1.067		133	62.50	9.80	205.30	260

03 21 Reinforcement Bars

03 21 11 – Plain Steel Reinforcement Bars

03 21 11.60 Reinforcing In Place

			Crew	Daily Output	Labor-Hours	Unit	Material	2020 Bare Costs Labor	Equipment	Total	Total Incl O&P
0010	**REINFORCING IN PLACE**, 50-60 ton lots, A615 Grade 60		R032110-10								
0020	Includes labor, but not material cost, to install accessories										
0030	Made from recycled materials										
0102	Beams & Girders, #3 to #7	G	4 Rodm	3200	.010	Lb.	.56	.56		1.12	1.51
0152	#8 to #18	G		5400	.006		.56	.33		.89	1.15
0202	Columns, #3 to #7	G		3000	.011		.56	.60		1.16	1.57
0252	#8 to #18	G		4600	.007		.56	.39		.95	1.24
0402	Elevated slabs, #4 to #7	G		5800	.006		.56	.31		.87	1.11
0502	Footings, #4 to #7	G		4200	.008		.56	.43		.99	1.30
0552	#8 to #18	G		7200	.004		.56	.25		.81	1.01
0602	Slab on grade, #3 to #7	G		4200	.008		.56	.43		.99	1.30
0702	Walls, #3 to #7	G		6000	.005		.56	.30		.86	1.09
0752	#8 to #18	G		8000	.004		.56	.23		.79	.98
0900	For other than 50-60 ton lots										
1000	Under 10 ton job, #3 to #7, add						25%	10%			
1010	#8 to #18, add						20%	10%			
1050	10-50 ton job, #3 to #7, add						10%				
1060	#8 to #18, add						5%				
1100	60-100 ton job, #3 to #7, deduct						5%				
1110	#8 to #18, deduct						10%				
1150	Over 100 ton job, #3 to #7, deduct						10%				
1160	#8 to #18, deduct						15%				
2000	Unloading & sorting, add to above		C-5	100	.560	Ton		31.50	5.80	37.30	56
2200	Crane cost for handling, 90 picks/day, up to 1.5 tons/bundle, add to above			135	.415			23.50	4.30	27.80	41
2210	1.0 ton/bundle			92	.609			34	6.30	40.30	61
2220	0.5 ton/bundle			35	1.600			90	16.60	106.60	160
2400	Dowels, 2 feet long, deformed, #3	G	2 Rodm	520	.031	Ea.	.47	1.74		2.21	3.24
2410	#4	G		480	.033		.83	1.88		2.71	3.87
2420	#5	G		435	.037		1.30	2.07		3.37	4.70
2430	#6	G		360	.044		1.87	2.51		4.38	6
2450	Longer and heavier dowels, add	G		725	.022	Lb.	.62	1.24		1.86	2.64
2500	Smooth dowels, 12" long, 1/4" or 3/8" diameter	G		140	.114	Ea.	.87	6.45		7.32	11.10
2520	5/8" diameter	G		125	.128		1.53	7.20		8.73	13.05
2530	3/4" diameter	G		110	.145		1.89	8.20		10.09	15
2600	Dowel sleeves for CIP concrete, 2-part system										
2610	Sleeve base, plastic, for 5/8" smooth dowel sleeve, fasten to edge form		1 Rodm	200	.040	Ea.	.54	2.26		2.80	4.14
2615	Sleeve, plastic, 12" long, for 5/8" smooth dowel, snap onto base			400	.020		1.54	1.13		2.67	3.47

03 21 Reinforcement Bars

03 21 11 – Plain Steel Reinforcement Bars

03 21 11.60 Reinforcing In Place

		Crew	Daily Output	Labor-Hours	Unit	Material	2020 Bare Costs Labor	Equipment	Total	Total Incl O&P
2620	Sleeve base, for 3/4" smooth dowel sleeve	1 Rodm	175	.046	Ea.	.54	2.58		3.12	4.65
2625	Sleeve, 12" long, for 3/4" smooth dowel		350	.023		1.22	1.29		2.51	3.37
2630	Sleeve base, for 1" smooth dowel sleeve		150	.053		.71	3.01		3.72	5.50
2635	Sleeve, 12" long, for 1" smooth dowel	▼	300	.027		1.67	1.50		3.17	4.21
2700	Dowel caps, visual warning only, plastic, #3 to #8	2 Rodm	800	.020		.37	1.13		1.50	2.19
2720	#8 to #18		750	.021		.89	1.20		2.09	2.88
2750	Impalement protective, plastic, #4 to #9	▼	800	.020	▼	1.18	1.13		2.31	3.08
9000	Minimum labor/equipment charge	1 Rodm	4	2	Job		113		113	178

03 21 21 – Composite Reinforcement Bars

03 21 21.11 Glass Fiber-Reinforced Polymer Reinf. Bars

		Crew	Daily Output	Labor-Hours	Unit	Material	2020 Bare Costs Labor	Equipment	Total	Total Incl O&P
0010	**GLASS FIBER-REINFORCED POLYMER REINFORCEMENT BARS**									
0020	Includes labor, but not material cost, to install accessories									
0050	#2 bar, .043 lb./L.F.	4 Rodm	9500	.003	L.F.	.41	.19		.60	.75
0100	#3 bar, .092 lb./L.F.		9300	.003		.65	.19		.84	1.03
0150	#4 bar, .160 lb./L.F.		9100	.004		.94	.20		1.14	1.34
0200	#5 bar, .258 lb./L.F.		8700	.004		1.14	.21		1.35	1.58
0250	#6 bar, .372 lb./L.F.		8300	.004		1.92	.22		2.14	2.45
0300	#7 bar, .497 lb./L.F.		7900	.004		2.59	.23		2.82	3.21
0350	#8 bar, .620 lb./L.F.		7400	.004		2.45	.24		2.69	3.08
0400	#9 bar, .800 lb./L.F.		6800	.005		4.22	.27		4.49	5.05
0450	#10 bar, 1.08 lb./L.F.	▼	5800	.006	▼	3.83	.31		4.14	4.70
0500	For bends, add per bend				Ea.	1.64			1.64	1.80

03 22 Fabric and Grid Reinforcing

03 22 11 – Plain Welded Wire Fabric Reinforcing

03 22 11.10 Plain Welded Wire Fabric

			Crew	Daily Output	Labor-Hours	Unit	Material	2020 Bare Costs Labor	Equipment	Total	Total Incl O&P
0010	**PLAIN WELDED WIRE FABRIC** ASTM A185 R032205-30										
0020	Includes labor, but not material cost, to install accessories										
0030	Made from recycled materials										
0050	Sheets										
0100	6 x 6 - W1.4 x W1.4 (10 x 10) 21 lb./C.S.F.	G	2 Rodm	35	.457	C.S.F.	16.25	26		42.25	58.50
0200	6 x 6 - W2.1 x W2.1 (8 x 8) 30 lb./C.S.F.	G		31	.516		19.15	29		48.15	67
0300	6 x 6 - W2.9 x W2.9 (6 x 6) 42 lb./C.S.F.	G		29	.552		27.50	31		58.50	79.50
0400	6 x 6 - W4 x W4 (4 x 4) 58 lb./C.S.F.	G		27	.593		38.50	33.50		72	95
0500	4 x 4 - W1.4 x W1.4 (10 x 10) 31 lb./C.S.F.	G		31	.516		24.50	29		53.50	72.50
0600	4 x 4 - W2.1 x W2.1 (8 x 8) 44 lb./C.S.F.	G		29	.552		30.50	31		61.50	82.50
0650	4 x 4 - W2.9 x W2.9 (6 x 6) 61 lb./C.S.F.	G		27	.593		48.50	33.50		82	106
0700	4 x 4 - W4 x W4 (4 x 4) 85 lb./C.S.F.	G	▼	25	.640	▼	60.50	36		96.50	124
0750	Rolls										
0800	2 x 2 - #14 galv., 21 lb./C.S.F., beam & column wrap	G	2 Rodm	6.50	2.462	C.S.F.	48.50	139		187.50	273
0900	2 x 2 - #12 galv. for gunite reinforcing	G	"	6.50	2.462	"	73	139		212	300
9000	Minimum labor/equipment charge		1 Rodm	4	2	Job		113		113	178

03 22 13 – Galvanized Welded Wire Fabric Reinforcing

03 22 13.10 Galvanized Welded Wire Fabric

			Crew	Daily Output	Labor-Hours	Unit	Material	2020 Bare Costs Labor	Equipment	Total	Total Incl O&P
0010	**GALVANIZED WELDED WIRE FABRIC**										
0100	Add to plain welded wire pricing for galvanized welded wire						Lb.	.26			.26	.28

For customer support on your Commercial Renovation Costs with RSMeans data, call 800.448.8182.

65

03 22 Fabric and Grid Reinforcing

03 22 16 – Epoxy-Coated Welded Wire Fabric Reinforcing

03 22 16.10 Epoxy-Coated Welded Wire Fabric	Crew	Daily Output	Labor-Hours	Unit	Material	2020 Bare Costs Labor	Equipment	Total	Total Incl O&P
0010 **EPOXY-COATED WELDED WIRE FABRIC**									
0100 Add to plain welded wire pricing for epoxy-coated welded wire				Lb.	.54			.54	.60

03 24 Fibrous Reinforcing

03 24 05 – Reinforcing Fibers

03 24 05.30 Synthetic Fibers

	Crew	Daily Output	Labor-Hours	Unit	Material	Labor	Equipment	Total	Total Incl O&P
0010 **SYNTHETIC FIBERS**									
0100 Synthetic fibers, add to concrete				Lb.	5.05			5.05	5.55
0110 1-1/2 lb./C.Y.				C.Y.	7.80			7.80	8.55

03 24 05.70 Steel Fibers

		Crew	Daily Output	Labor-Hours	Unit	Material	Labor	Equipment	Total	Total Incl O&P
0010 **STEEL FIBERS**										
0140 ASTM A850, Type V, continuously deformed, 1-1/2" long x 0.045" diam.										
0150 Add to price of ready mix concrete	G				Lb.	1.23			1.23	1.35
0205 Alternate pricing, dosing at 5 lb./C.Y., add to price of RMC	G				C.Y.	6.15			6.15	6.75
0210 10 lb./C.Y.	G					12.30			12.30	13.55
0215 15 lb./C.Y.	G					18.45			18.45	20.50
0220 20 lb./C.Y.	G					24.50			24.50	27
0225 25 lb./C.Y.	G					31			31	34
0230 30 lb./C.Y.	G					37			37	40.50
0235 35 lb./C.Y.	G					43			43	47.50
0240 40 lb./C.Y.	G					49			49	54
0250 50 lb./C.Y.	G					61.50			61.50	67.50
0275 75 lb./C.Y.	G					92.50			92.50	101
0300 100 lb./C.Y.	G			▼		123			123	135

03 30 Cast-In-Place Concrete

03 30 53 – Miscellaneous Cast-In-Place Concrete

03 30 53.40 Concrete In Place

		Crew	Daily Output	Labor-Hours	Unit	Material	Labor	Equipment	Total	Total Incl O&P
0010 **CONCRETE IN PLACE**	R033105-10									
0020 Including forms (4 uses), Grade 60 rebar, concrete (Portland cement	R033105-20									
0050 Type I), placement and finishing unless otherwise indicated	R033105-70									
0300 Beams (3500 psi), 5 kip/L.F., 10' span	R033105-85	C-14A	15.62	12.804	C.Y.	370	675	28	1,073	1,500
0350 25' span		"	18.55	10.782		390	570	23.50	983.50	1,375
0500 Chimney foundations (5000 psi), over 5 C.Y.		C-14C	32.22	3.476		178	175	.83	353.83	475
0510 (3500 psi), under 5 C.Y.		"	23.71	4.724		208	238	1.13	447.13	610
0700 Columns, square (4000 psi), 12" x 12", up to 1% reinforcing by area		C-14A	11.96	16.722		410	885	36.50	1,331.50	1,900
0720 Up to 2% reinforcing by area			10.13	19.743		645	1,050	43	1,738	2,425
0740 Up to 3% reinforcing by area			9.03	22.148		965	1,175	48.50	2,188.50	2,975
0800 16" x 16", up to 1% reinforcing by area			16.22	12.330		325	650	27	1,002	1,450
0820 Up to 2% reinforcing by area			12.57	15.911		550	840	35	1,425	2,000
0840 Up to 3% reinforcing by area			10.25	19.512		850	1,025	42.50	1,917.50	2,625
0900 24" x 24", up to 1% reinforcing by area			23.66	8.453		278	445	18.50	741.50	1,050
0920 Up to 2% reinforcing by area			17.71	11.293		490	595	24.50	1,109.50	1,525
0940 Up to 3% reinforcing by area			14.15	14.134		780	745	31	1,556	2,100
1000 36" x 36", up to 1% reinforcing by area			33.69	5.936		247	315	13	575	785
1020 Up to 2% reinforcing by area			23.32	8.576		435	455	18.80	908.80	1,225
1040 Up to 3% reinforcing by area			17.82	11.223		725	595	24.50	1,344.50	1,775
1100 Columns, round (4000 psi), tied, 12" diameter, up to 1% reinforcing by area			20.97	9.537		415	505	21	941	1,275

03 30 Cast-In-Place Concrete

03 30 53 – Miscellaneous Cast-In-Place Concrete

03 30 53.40 Concrete In Place		Crew	Daily Output	Labor-Hours	Unit	Material	2020 Bare Costs Labor	Equipment	Total	Total Incl O&P
1120	Up to 2% reinforcing by area	C-14A	15.27	13.098	C.Y.	645	695	28.50	1,368.50	1,850
1140	Up to 3% reinforcing by area		12.11	16.515		955	875	36	1,866	2,500
1200	16" diameter, up to 1% reinforcing by area		31.49	6.351		340	335	13.90	688.90	920
1220	Up to 2% reinforcing by area		19.12	10.460		575	555	23	1,153	1,525
1240	Up to 3% reinforcing by area		13.77	14.524		855	770	32	1,657	2,200
1300	20" diameter, up to 1% reinforcing by area		41.04	4.873		325	258	10.65	593.65	780
1320	Up to 2% reinforcing by area		24.05	8.316		540	440	18.20	998.20	1,325
1340	Up to 3% reinforcing by area		17.01	11.758		845	620	26	1,491	1,950
1400	24" diameter, up to 1% reinforcing by area		51.85	3.857		315	204	8.45	527.45	685
1420	Up to 2% reinforcing by area		27.06	7.391		555	390	16.20	961.20	1,250
1440	Up to 3% reinforcing by area		18.29	10.935		840	580	24	1,444	1,875
1500	36" diameter, up to 1% reinforcing by area		75.04	2.665		315	141	5.85	461.85	575
1520	Up to 2% reinforcing by area		37.49	5.335		525	282	11.70	818.70	1,050
1540	Up to 3% reinforcing by area	▼	22.84	8.757		815	465	19.20	1,299.20	1,650
1900	Elevated slab (4000 psi), flat slab with drops, 125 psf Sup. Load, 20' span	C-14B	38.45	5.410		300	285	11.35	596.35	800
1950	30' span		50.99	4.079		320	215	8.55	543.55	710
2100	Flat plate, 125 psf Sup. Load, 15' span		30.24	6.878		274	365	14.45	653.45	895
2150	25' span		49.60	4.194		287	221	8.80	516.80	680
2300	Waffle const., 30" domes, 125 psf Sup. Load, 20' span		37.07	5.611		310	296	11.80	617.80	830
2350	30' span		44.07	4.720		290	249	9.90	548.90	725
2500	One way joists, 30" pans, 125 psf Sup. Load, 15' span		27.38	7.597		405	400	15.95	820.95	1,100
2550	25' span		31.15	6.677		375	350	14	739	990
2700	One way beam & slab, 125 psf Sup. Load, 15' span		20.59	10.102		295	535	21	851	1,200
2750	25' span		28.36	7.334		281	385	15.40	681.40	940
2900	Two way beam & slab, 125 psf Sup. Load, 15' span		24.04	8.652		285	455	18.15	758.15	1,075
2950	25' span	▼	35.87	5.799	▼	249	305	12.20	566.20	775
3100	Elevated slabs, flat plate, including finish, not									
3110	including forms or reinforcing									
3150	Regular concrete (4000 psi), 4" slab	C-8	2613	.021	S.F.	1.69	1	.16	2.85	3.62
3200	6" slab		2585	.022		2.48	1.01	.16	3.65	4.50
3250	2-1/2" thick floor fill		2685	.021		1.10	.97	.15	2.22	2.94
3300	Lightweight, 110 #/C.F., 2-1/2" thick floor fill		2585	.022		1.31	1.01	.16	2.48	3.23
3400	Cellular concrete, 1-5/8" fill, under 5000 S.F.		2000	.028		.90	1.31	.21	2.42	3.29
3450	Over 10,000 S.F.		2200	.025		.86	1.19	.19	2.24	3.04
3500	Add per floor for 3 to 6 stories high		31800	.002			.08	.01	.09	.14
3520	For 7 to 20 stories high	▼	21200	.003	▼		.12	.02	.14	.22
3540	Equipment pad (3000 psi), 3' x 3' x 6" thick	C-14H	45	1.067	Ea.	50.50	55	.60	106.10	144
3550	4' x 4' x 6" thick		30	1.600		78	82.50	.90	161.40	218
3560	5' x 5' x 8" thick		18	2.667		138	138	1.49	277.49	375
3570	6' x 6' x 8" thick		14	3.429		190	177	1.92	368.92	495
3580	8' x 8' x 10" thick		8	6		395	310	3.36	708.36	935
3590	10' x 10' x 12" thick	▼	5	9.600	▼	695	495	5.40	1,195.40	1,550
3800	Footings (3000 psi), spread under 1 C.Y.	C-14C	28	4	C.Y.	203	201	.96	404.96	545
3825	1 C.Y. to 5 C.Y.		43	2.605		240	131	.63	371.63	475
3850	Over 5 C.Y.		75	1.493		220	75	.36	295.36	360
3900	Footings, strip (3000 psi), 18" x 9", unreinforced	C-14L	40	2.400		154	118	.67	272.67	360
3920	18" x 9", reinforced	C-14C	35	3.200		181	161	.77	342.77	455
3925	20" x 10", unreinforced	C-14L	45	2.133		150	105	.60	255.60	335
3930	20" x 10", reinforced	C-14C	40	2.800		172	141	.67	313.67	415
3935	24" x 12", unreinforced	C-14L	55	1.745		148	86	.49	234.49	300
3940	24" x 12", reinforced	C-14C	48	2.333		171	118	.56	289.56	375
3945	36" x 12", unreinforced	C-14L	70	1.371		144	67.50	.38	211.88	267
3950	36" x 12", reinforced	C-14C	60	1.867		165	94	.45	259.45	330

For customer support on your Commercial Renovation Costs with RSMeans data, call 800.448.8182.

67

03 30 53.40 Concrete In Place	Crew	Daily Output	Labor-Hours	Unit	Material	2020 Bare Costs Labor	Equipment	Total	Total Incl O&P
4000	Foundation mat (3000 psi), under 10 C.Y.	C-14C	38.67	2.896	C.Y.	246	146	.70	392.70	505
4050	Over 20 C.Y.	↓	56.40	1.986		217	100	.48	317.48	400
4200	Wall, free-standing (3000 psi), 8" thick, 8' high	C-14D	45.83	4.364		197	230	9.55	436.55	590
4250	14' high		27.26	7.337		232	385	16.05	633.05	890
4260	12" thick, 8' high		64.32	3.109		179	164	6.80	349.80	465
4270	14' high		40.01	4.999		190	263	10.95	463.95	640
4300	15" thick, 8' high		80.02	2.499		173	132	5.45	310.45	405
4350	12' high		51.26	3.902		173	205	8.55	386.55	530
4500	18' high	↓	48.85	4.094		194	215	8.95	417.95	570
4520	Handicap access ramp (4000 psi), railing both sides, 3' wide	C-14H	14.58	3.292	L.F.	390	170	1.84	561.84	705
4525	5' wide		12.22	3.928		400	203	2.20	605.20	765
4530	With 6" curb and rails both sides, 3' wide	↓	8.55	5.614	↓	400	290	3.14	693.14	905
4535	5' wide	↓	7.31	6.566	↓	405	340	3.68	748.68	995
4650	Slab on grade (3500 psi), not including finish, 4" thick	C-14E	60.75	1.449	C.Y.	147	74	.45	221.45	280
4700	6" thick	"	92	.957	"	141	49	.30	190.30	234
4701	Thickened slab edge (3500 psi), for slab on grade poured									
4702	monolithically with slab; depth is in addition to slab thickness;									
4703	formed vertical outside edge, earthen bottom and inside slope									
4705	8" deep x 8" wide bottom, unreinforced	C-14L	2190	.044	L.F.	4.07	2.16	.01	6.24	7.95
4710	8" x 8", reinforced	C-14C	1670	.067		6.70	3.38	.02	10.10	12.80
4715	12" deep x 12" wide bottom, unreinforced	C-14L	1800	.053		8.30	2.63	.01	10.94	13.40
4720	12" x 12", reinforced	C-14C	1310	.086		13.15	4.31	.02	17.48	21.50
4725	16" deep x 16" wide bottom, unreinforced	C-14L	1440	.067		14	3.29	.02	17.31	20.50
4730	16" x 16", reinforced	C-14C	1120	.100		19.75	5.05	.02	24.82	30
4735	20" deep x 20" wide bottom, unreinforced	C-14L	1150	.083		21	4.12	.02	25.14	30
4740	20" x 20", reinforced	C-14C	920	.122		28.50	6.15	.03	34.68	41.50
4745	24" deep x 24" wide bottom, unreinforced	C-14L	930	.103		30	5.10	.03	35.13	41
4750	24" x 24", reinforced	C-14C	740	.151	↓	39.50	7.60	.04	47.14	55.50
4751	Slab on grade (3500 psi), incl. troweled finish, not incl. forms									
4760	or reinforcing, over 10,000 S.F., 4" thick	C-14F	3425	.021	S.F.	1.62	1	.01	2.63	3.37
4820	6" thick		3350	.021		2.36	1.02	.01	3.39	4.22
4840	8" thick		3184	.023		3.24	1.08	.01	4.33	5.25
4900	12" thick		2734	.026		4.85	1.25	.01	6.11	7.35
4950	15" thick	↓	2505	.029		6.10	1.37	.01	7.48	8.85
5000	Slab on grade (3000 psi), incl. broom finish, not incl. forms									
5001	or reinforcing, 4" thick	C-14G	2873	.019	S.F.	1.67	.91	.01	2.59	3.30
5010	6" thick		2590	.022		2.62	1.01	.01	3.64	4.49
5020	8" thick	↓	2320	.024	↓	3.41	1.13	.01	4.55	5.55
5200	Lift slab in place above the foundation, incl. forms, reinforcing,									
5210	concrete (4000 psi) and columns, over 20,000 S.F./floor	C-14B	2113	.098	S.F.	7.75	5.20	.21	13.16	17.05
5250	10,000 S.F. to 20,000 S.F./floor		1650	.126		8.45	6.65	.26	15.36	20
5300	Under 10,000 S.F./floor	↓	1500	.139	↓	9.15	7.30	.29	16.74	22
5500	Lightweight, ready mix, including screed finish only,									
5510	not including forms or reinforcing									
5550	1:4 (2500 psi) for structural roof decks	C-14B	260	.800	C.Y.	125	42	1.68	168.68	206
5600	1:6 (3000 psi) for ground slab with radiant heat	C-14F	92	.783		134	37	.29	171.29	207
5650	1:3:2 (2000 psi) with sand aggregate, roof deck	C-14B	260	.800		114	42	1.68	157.68	194
5700	Ground slab (2000 psi)	C-14F	107	.673		114	32	.25	146.25	176
5900	Pile caps (3000 psi), incl. forms and reinf., sq. or rect., under 10 C.Y.	C-14C	54.14	2.069		200	104	.50	304.50	385
5950	Over 10 C.Y.		75	1.493		189	75	.36	264.36	330
6000	Triangular or hexagonal, under 10 C.Y.		53	2.113		153	106	.51	259.51	340
6050	Over 10 C.Y.	↓	85	1.318		171	66.50	.32	237.82	294
6200	Retaining walls (3000 psi), gravity, 4' high see Section 32 32	C-14D	66.20	3.021	↓	173	159	6.60	338.60	450

03 30 Cast-In-Place Concrete

03 30 53 – Miscellaneous Cast-In-Place Concrete

03 30 53.40 Concrete In Place

		Crew	Daily Output	Labor-Hours	Unit	Material	2020 Bare Costs Labor	Equipment	Total	Total Incl O&P
6250	10' high	C-14D	125	1.600	C.Y.	166	84	3.50	253.50	320
6300	Cantilever, level backfill loading, 8' high		70	2.857		184	150	6.25	340.25	450
6350	16' high		91	2.198		179	116	4.81	299.81	385
6800	Stairs (3500 psi), not including safety treads, free standing, 3'-6" wide	C-14H	83	.578	LF Nose	6.40	30	.32	36.72	55
6850	Cast on ground		125	.384	"	5.40	19.85	.22	25.47	37.50
7000	Stair landings, free standing		200	.240	S.F.	5.10	12.40	.13	17.63	25.50
7050	Cast on ground		475	.101	"	4.15	5.20	.06	9.41	12.95
9000	Minimum labor/equipment charge	2 Carp	1	16	Job		850		850	1,350

03 31 Structural Concrete

03 31 13 – Heavyweight Structural Concrete

03 31 13.25 Concrete, Hand Mix

		Crew	Daily Output	Labor-Hours	Unit	Material	2020 Bare Costs Labor	Equipment	Total	Total Incl O&P
0010	**CONCRETE, HAND MIX** for small quantities or remote areas									
0050	Includes bulk local aggregate, bulk sand, bagged Portland									
0060	cement (Type I) and water, using gas powered cement mixer									
0125	2500 psi	C-30	135	.059	C.F.	4.33	2.49	1.04	7.86	9.90
0130	3000 psi		135	.059		4.68	2.49	1.04	8.21	10.30
0135	3500 psi		135	.059		4.88	2.49	1.04	8.41	10.50
0140	4000 psi		135	.059		5.10	2.49	1.04	8.63	10.80
0145	4500 psi		135	.059		5.40	2.49	1.04	8.93	11.10
0150	5000 psi		135	.059		5.80	2.49	1.04	9.33	11.50
0300	Using pre-bagged dry mix and wheelbarrow (80-lb. bag = 0.6 C.F.)									
0340	4000 psi	1 Clab	48	.167	C.F.	8.50	7		15.50	20.50

03 31 13.30 Concrete, Volumetric Site-Mixed

		Crew	Daily Output	Labor-Hours	Unit	Material	2020 Bare Costs Labor	Equipment	Total	Total Incl O&P
0010	**CONCRETE, VOLUMETRIC SITE-MIXED**									
0015	Mixed on-site in volumetric truck									
0020	Includes local aggregate, sand, Portland cement (Type I) and water									
0025	Excludes all additives and treatments									
0100	3000 psi, 1 C.Y. mixed and discharged				C.Y.	213			213	234
0110	2 C.Y.					164			164	180
0120	3 C.Y.					143			143	157
0130	4 C.Y.					125			125	138
0140	5 C.Y.					113			113	125
0200	For truck holding/waiting time past first 2 on-site hours, add				Hr.	91.50			91.50	101
0210	For trip charge beyond first 20 miles, each way, add				Mile	3.65			3.65	4.02
0220	For each additional increase of 500 psi, add				Ea.	4.64			4.64	5.10

03 31 13.35 Heavyweight Concrete, Ready Mix

		Crew	Daily Output	Labor-Hours	Unit	Material	2020 Bare Costs Labor	Equipment	Total	Total Incl O&P
0010	**HEAVYWEIGHT CONCRETE, READY MIX**, delivered									
0012	Includes local aggregate, sand, Portland cement (Type I) and water									
0015	Excludes all additives and treatments R033105-20									
0020	2000 psi				C.Y.	107			107	117
0100	2500 psi					110			110	121
0150	3000 psi					129			129	142
0200	3500 psi					124			124	137
0300	4000 psi					127			127	140
0350	4500 psi					131			131	144
0400	5000 psi					135			135	149
0411	6000 psi					139			139	153
0412	8000 psi					146			146	161
0413	10,000 psi					154			154	169

For customer support on your Commercial Renovation Costs with RSMeans data, call 800.448.8182.

69

03 31 Structural Concrete

03 31 13 – Heavyweight Structural Concrete

03 31 13.35 Heavyweight Concrete, Ready Mix

		Crew	Daily Output	Labor-Hours	Unit	Material	2020 Bare Costs Labor	Equipment	Total	Total Incl O&P
0414	12,000 psi				C.Y.	161			161	177
1000	For high early strength (Portland cement Type III), add					10%				
1300	For winter concrete (hot water), add					5.35			5.35	5.90
1410	For mid-range water reducer, add					4.18			4.18	4.60
1420	For high-range water reducer/superplasticizer, add					6.40			6.40	7
1430	For retarder, add					3.45			3.45	3.80
1440	For non-Chloride accelerator, add					6.80			6.80	7.45
1450	For Chloride accelerator, per 1%, add					4.10			4.10	4.51
1460	For fiber reinforcing, synthetic (1 lb./C.Y.), add					8.10			8.10	8.90
1500	For Saturday delivery, add					9.20			9.20	10.10
1510	For truck holding/waiting time past 1st hour per load, add				Hr.	109			109	120
1520	For short load (less than 4 C.Y.), add per load				Ea.	85			85	93.50
2000	For all lightweight aggregate, add				C.Y.	45%				

03 31 13.70 Placing Concrete

		Crew	Daily Output	Labor-Hours	Unit	Material	2020 Bare Costs Labor	Equipment	Total	Total Incl O&P
0010	**PLACING CONCRETE** R033105-70									
0020	Includes labor and equipment to place, level (strike off) and consolidate									
0050	Beams, elevated, small beams, pumped	C-20	60	1.067	C.Y.		48	7.75	55.75	85
0100	With crane and bucket	C-7	45	1.600			73	24	97	143
0200	Large beams, pumped	C-20	90	.711			32	5.15	37.15	56.50
0250	With crane and bucket	C-7	65	1.108			50.50	16.60	67.10	99
0400	Columns, square or round, 12" thick, pumped	C-20	60	1.067			48	7.75	55.75	85
0450	With crane and bucket	C-7	40	1.800			82.50	27	109.50	161
0600	18" thick, pumped	C-20	90	.711			32	5.15	37.15	56.50
0650	With crane and bucket	C-7	55	1.309			60	19.65	79.65	117
0800	24" thick, pumped	C-20	92	.696			31.50	5.05	36.55	55.50
0850	With crane and bucket	C-7	70	1.029			47	15.45	62.45	92
1000	36" thick, pumped	C-20	140	.457			20.50	3.31	23.81	36.50
1050	With crane and bucket	C-7	100	.720			33	10.80	43.80	64.50
1200	Duct bank, direct chute	C-6	155	.310			13.55	.35	13.90	22
1400	Elevated slabs, less than 6" thick, pumped	C-20	140	.457			20.50	3.31	23.81	36.50
1450	With crane and bucket	C-7	95	.758			34.50	11.35	45.85	67.50
1500	6" to 10" thick, pumped	C-20	160	.400			18.05	2.90	20.95	32
1550	With crane and bucket	C-7	110	.655			30	9.80	39.80	58.50
1600	Slabs over 10" thick, pumped	C-20	180	.356			16.05	2.58	18.63	28.50
1650	With crane and bucket	C-7	130	.554			25.50	8.30	33.80	49.50
1900	Footings, continuous, shallow, direct chute	C-6	120	.400			17.50	.45	17.95	28.50
1950	Pumped	C-20	150	.427			19.25	3.09	22.34	34
2000	With crane and bucket	C-7	90	.800			36.50	12	48.50	71
2100	Footings, continuous, deep, direct chute	C-6	140	.343			15	.38	15.38	24.50
2150	Pumped	C-20	160	.400			18.05	2.90	20.95	32
2200	With crane and bucket	C-7	110	.655			30	9.80	39.80	58.50
2400	Footings, spread, under 1 C.Y., direct chute	C-6	55	.873			38	.98	38.98	62
2450	Pumped	C-20	65	.985			44.50	7.15	51.65	79
2500	With crane and bucket	C-7	45	1.600			73	24	97	143
2600	Over 5 C.Y., direct chute	C-6	120	.400			17.50	.45	17.95	28.50
2650	Pumped	C-20	150	.427			19.25	3.09	22.34	34
2700	With crane and bucket	C-7	100	.720			33	10.80	43.80	64.50
2900	Foundation mats, over 20 C.Y., direct chute	C-6	350	.137			6	.15	6.15	9.75
2950	Pumped	C-20	400	.160			7.25	1.16	8.41	12.80
3000	With crane and bucket	C-7	300	.240			11	3.60	14.60	21.50
3200	Grade beams, direct chute	C-6	150	.320			14	.36	14.36	23
3250	Pumped	C-20	180	.356			16.05	2.58	18.63	28.50

03 31 13.70 Placing Concrete	Crew	Daily Output	Labor-Hours	Unit	Material	2020 Bare Costs Labor	2020 Bare Costs Equipment	Total	Total Incl O&P
3300	With crane and bucket	C-7	120	.600	C.Y.		27.50	9	36.50	53.50
3500	High rise, for more than 5 stories, pumped, add per story	C-20	2100	.030			1.38	.22	1.60	2.43
3510	With crane and bucket, add per story	C-7	2100	.034			1.57	.51	2.08	3.06
3700	Pile caps, under 5 C.Y., direct chute	C-6	90	.533			23.50	.60	24.10	37.50
3750	Pumped	C-20	110	.582			26.50	4.22	30.72	46.50
3800	With crane and bucket	C-7	80	.900			41	13.50	54.50	80.50
3850	Pile cap, 5 C.Y. to 10 C.Y., direct chute	C-6	175	.274			12	.31	12.31	19.50
3900	Pumped	C-20	200	.320			14.45	2.32	16.77	25.50
3950	With crane and bucket	C-7	150	.480			22	7.20	29.20	43
4000	Over 10 C.Y., direct chute	C-6	215	.223			9.75	.25	10	15.85
4050	Pumped	C-20	240	.267			12.05	1.93	13.98	21.50
4100	With crane and bucket	C-7	185	.389			17.80	5.85	23.65	35
4300	Slab on grade, up to 6" thick, direct chute	C-6	110	.436			19.10	.49	19.59	31
4350	Pumped	C-20	130	.492			22	3.57	25.57	39.50
4400	With crane and bucket	C-7	110	.655			30	9.80	39.80	58.50
4600	Over 6" thick, direct chute	C-6	165	.291			12.70	.33	13.03	21
4650	Pumped	C-20	185	.346			15.60	2.51	18.11	28
4700	With crane and bucket	C-7	145	.497			22.50	7.45	29.95	44
4900	Walls, 8" thick, direct chute	C-6	90	.533			23.50	.60	24.10	37.50
4950	Pumped	C-20	100	.640			29	4.64	33.64	51
5000	With crane and bucket	C-7	80	.900			41	13.50	54.50	80.50
5050	12" thick, direct chute	C-6	100	.480			21	.54	21.54	34
5100	Pumped	C-20	110	.582			26.50	4.22	30.72	46.50
5200	With crane and bucket	C-7	90	.800			36.50	12	48.50	71
5300	15" thick, direct chute	C-6	105	.457			20	.51	20.51	32.50
5350	Pumped	C-20	120	.533			24	3.87	27.87	43
5400	With crane and bucket	C-7	95	.758	▼		34.50	11.35	45.85	67.50
5600	Wheeled concrete dumping, add to placing costs above									
5610	Walking cart, 50' haul, add	C-18	32	.281	C.Y.		11.90	4	15.90	23.50
5620	150' haul, add		24	.375			15.85	5.35	21.20	31.50
5700	250' haul, add	▼	18	.500			21	7.10	28.10	42
5800	Riding cart, 50' haul, add	C-19	80	.113			4.76	1.92	6.68	9.75
5810	150' haul, add		60	.150			6.35	2.56	8.91	12.95
5900	250' haul, add	▼	45	.200	▼		8.45	3.41	11.86	17.30
6000	Concrete in-fill for pan-type metal stairs and landings. Manual placement									
6010	includes up to 50' horizontal haul from point of concrete discharge.									
6100	Stair pan treads, 2" deep									
6110	Flights in 1st floor level up/down from discharge point	C-8A	3200	.015	S.F.		.68		.68	1.07
6120	2nd floor level		2500	.019			.87		.87	1.38
6130	3rd floor level		2000	.024			1.08		1.08	1.72
6140	4th floor level	▼	1800	.027	▼		1.20		1.20	1.91
6200	Intermediate stair landings, pan-type 4" deep									
6210	Flights in 1st floor level up/down from discharge point	C-8A	2000	.024	S.F.		1.08		1.08	1.72
6220	2nd floor level		1500	.032			1.44		1.44	2.29
6230	3rd floor level		1200	.040			1.80		1.80	2.87
6240	4th floor level	▼	1000	.048	▼		2.16		2.16	3.44
9000	Minimum labor/equipment charge	C-6	2	24	Job		1,050	27	1,077	1,700

For customer support on your Commercial Renovation Costs with RSMeans data, call 800.448.8182.

71

03 35 Concrete Finishing

03 35 13 – High-Tolerance Concrete Floor Finishing

03 35 13.30 Finishing Floors, High Tolerance	Crew	Daily Output	Labor-Hours	Unit	Material	2020 Bare Costs Labor	Equipment	Total	Total Incl O&P
0010 **FINISHING FLOORS, HIGH TOLERANCE**									
0012 Finishing of fresh concrete flatwork requires that concrete									
0013 first be placed, struck off & consolidated									
0015 Basic finishing for various unspecified flatwork									
0100 Bull float only	C-10	4000	.006	S.F.		.28		.28	.45
0125 Bull float & manual float		2000	.012			.57		.57	.90
0150 Bull float, manual float & broom finish, w/edging & joints		1850	.013			.61		.61	.97
0200 Bull float, manual float & manual steel trowel		1265	.019			.90		.90	1.42
0210 For specified Random Access Floors in ACI Classes 1, 2, 3 and 4 to achieve									
0215 Composite Overall Floor Flatness and Levelness values up to FF35/FL25									
0250 Bull float, machine float & machine trowel (walk-behind)	C-10C	1715	.014	S.F.		.66	.05	.71	1.11
0300 Power screed, bull float, machine float & trowel (walk-behind)	C-10D	2400	.010			.47	.07	.54	.82
0350 Power screed, bull float, machine float & trowel (ride-on)	C-10E	4000	.006			.28	.06	.34	.52
0352 For specified Random Access Floors in ACI Classes 5, 6, 7 and 8 to achieve									
0354 Composite Overall Floor Flatness and Levelness values up to FF50/FL50									
0356 Add for two-dimensional restraightening after power float	C-10	6000	.004	S.F.		.19		.19	.30
0358 For specified Random or Defined Access Floors in ACI Class 9 to achieve									
0360 Composite Overall Floor Flatness and Levelness values up to FF100/FL100									
0362 Add for two-dimensional restraightening after bull float & power float	C-10	3000	.008	S.F.		.38		.38	.60
0364 For specified Superflat Defined Access Floors in ACI Class 9 to achieve									
0366 Minimum Floor Flatness and Levelness values of FF100/FL100									
0368 Add for 2-dim'l restraightening after bull float, power float, power trowel	C-10	2000	.012	S.F.		.57		.57	.90
9100 Minimum labor/equipment charge	"	2	12	Job		570		570	895

03 35 29 – Tooled Concrete Finishing

03 35 29.60 Finishing Walls

	Crew	Daily Output	Labor-Hours	Unit	Material	2020 Bare Costs Labor	Equipment	Total	Total Incl O&P
0010 **FINISHING WALLS**									
0020 Break ties and patch voids	1 Cefi	540	.015	S.F.	.04	.74		.78	1.21
0050 Burlap rub with grout	"	450	.018		.04	.89		.93	1.44
0300 Bush hammer, green concrete	B-39	1000	.048			2.12	.36	2.48	3.78
0350 Cured concrete	"	650	.074			3.27	.55	3.82	5.80
0500 Acid etch	1 Cefi	575	.014		.14	.69		.83	1.24
0700 Sandblast, light penetration	E-11	1100	.029		.51	1.36	.27	2.14	3.11
0750 Heavy penetration	"	375	.085		1.03	3.98	.80	5.81	8.60
0850 Grind form fins flush	1 Clab	700	.011	L.F.		.48		.48	.77
9000 Minimum labor/equipment charge	C-10	2	12	Job		570		570	895

03 35 33 – Stamped Concrete Finishing

03 35 33.50 Slab Texture Stamping

	Crew	Daily Output	Labor-Hours	Unit	Material	2020 Bare Costs Labor	Equipment	Total	Total Incl O&P
0010 **SLAB TEXTURE STAMPING**									
0050 Stamping requires that concrete first be placed, struck off, consolidated,									
0060 bull floated and free of bleed water. Decorative stamping tasks include:									
0100 Step 1 - first application of dry shake colored hardener	1 Cefi	6400	.001	S.F.	.43	.06		.49	.57
0110 Step 2 - bull float		6400	.001			.06		.06	.10
0130 Step 3 - second application of dry shake colored hardener		6400	.001		.21	.06		.27	.33
0140 Step 4 - bull float, manual float & steel trowel	3 Cefi	1280	.019			.94		.94	1.47
0150 Step 5 - application of dry shake colored release agent	1 Cefi	6400	.001		.10	.06		.16	.21
0160 Step 6 - place, tamp & remove mats	3 Cefi	2400	.010		.87	.50		1.37	1.73
0170 Step 7 - touch up edges, mat joints & simulated grout lines	1 Cefi	1280	.006			.31		.31	.49
0300 Alternate stamping estimating method includes all tasks above	4 Cefi	800	.040		1.60	2		3.60	4.90
0400 Step 8 - pressure wash @ 3000 psi after 24 hours	1 Cefi	1600	.005			.25		.25	.39
0500 Step 9 - roll 2 coats cure/seal compound when dry	"	800	.010		.69	.50		1.19	1.54

03 35 Concrete Finishing

03 35 43 – Polished Concrete Finishing

03 35 43.10 Polished Concrete Floors	Crew	Daily Output	Labor-Hours	Unit	Material	2020 Bare Costs Labor	Equipment	Total	Total Incl O&P
0010 **POLISHED CONCRETE FLOORS**									
0015 Processing of cured concrete to include grinding, honing,									
0020 and polishing of interior floors with 22" segmented diamond									
0025 planetary floor grinder (2 passes in different directions per grit)									
0100 Removal of pre-existing coatings, dry, with carbide discs using									
0105 dry vacuum pick-up system, final hand sweeping									
0110 Glue, adhesive or tar	J-4	1.60	15	M.S.F.	22	710	139	871	1,300
0120 Paint, epoxy, 1 coat		3.60	6.667		22	315	62	399	595
0130 2 coats		1.80	13.333		22	630	124	776	1,150
0200 Grinding and edging, wet, including wet vac pick-up and auto									
0205 scrubbing between grit changes									
0210 40-grit diamond/metal matrix	J-4A	1.60	20	M.S.F.	41.50	920	251	1,212.50	1,775
0220 80-grit diamond/metal matrix		2	16		41.50	735	200	976.50	1,450
0230 120-grit diamond/metal matrix		2.40	13.333		41.50	615	167	823.50	1,200
0240 200-grit diamond/metal matrix		2.80	11.429		41.50	525	143	709.50	1,050
0300 Spray on dye or stain (1 coat)	1 Cefi	16	.500		209	25		234	269
0400 Spray on densifier/hardener (2 coats)	"	8	1		315	50		365	425
0410 Auto scrubbing after 2nd coat, when dry	J-4B	16	.500			21	11.15	32.15	46
0500 Honing and edging, wet, including wet vac pick-up and auto									
0505 scrubbing between grit changes									
0510 100-grit diamond/resin matrix	J-4A	2.80	11.429	M.S.F.	41.50	525	143	709.50	1,050
0520 200-grit diamond/resin matrix	"	2.80	11.429	"	41.50	525	143	709.50	1,050
0530 Dry, including dry vacuum pick-up system, final hand sweeping									
0540 400-grit diamond/resin matrix	J-4A	2.80	11.429	M.S.F.	41.50	525	143	709.50	1,050
0600 Polishing and edging, dry, including dry vac pick-up and hand									
0605 sweeping between grit changes									
0610 800-grit diamond/resin matrix	J-4A	2.80	11.429	M.S.F.	41.50	525	143	709.50	1,050
0620 1500-grit diamond/resin matrix		2.80	11.429		41.50	525	143	709.50	1,050
0630 3000-grit diamond/resin matrix		2.80	11.429		41.50	525	143	709.50	1,050
0700 Auto scrubbing after final polishing step	J-4B	16	.500			21	11.15	32.15	46

03 37 Specialty Placed Concrete

03 37 13 – Shotcrete

03 37 13.30 Gunite (Dry-Mix)

	Crew	Daily Output	Labor-Hours	Unit	Material	2020 Bare Costs Labor	Equipment	Total	Total Incl O&P
0010 **GUNITE (DRY-MIX)**									
0020 Typical in place, 1" layers, no mesh included	C-16	2000	.028	S.F.	.42	1.31	.20	1.93	2.75
0300 Typical in place, including mesh, 2" thick, flat surfaces		1000	.056		1.57	2.62	.39	4.58	6.30
0350 Curved surfaces		500	.112		1.57	5.25	.78	7.60	10.90
0500 4" thick, flat surfaces		750	.075		2.41	3.49	.52	6.42	8.75
0550 Curved surfaces		350	.160		2.41	7.50	1.12	11.03	15.75
0900 Prepare old walls, no scaffolding, good condition	C-10	1000	.024			1.14		1.14	1.79
0950 Poor condition	"	275	.087			4.13		4.13	6.55
1100 For high finish requirement or close tolerance, add						50%			
1150 Very high						110%			
9000 Minimum labor/equipment charge	C-10	1	24	Job		1,125		1,125	1,800

03 37 13.60 Shotcrete (Wet-Mix)

	Crew	Daily Output	Labor-Hours	Unit	Material	2020 Bare Costs Labor	Equipment	Total	Total Incl O&P
0010 **SHOTCRETE (WET-MIX)**									
0020 Wet mix, placed @ up to 12 C.Y./hour, 3000 psi	C-8C	80	.600	C.Y.	136	27.50	7.45	170.95	201
0100 Up to 35 C.Y./hour	C-8E	240	.200	"	122	9.10	2.70	133.80	151
1010 Fiber reinforced, 1" thick	C-8C	1740	.028	S.F.	1	1.27	.34	2.61	3.49

For customer support on your Commercial Renovation Costs with RSMeans data, call 800.448.8182.

73

03 37 Specialty Placed Concrete

03 37 13 – Shotcrete

03 37 13.60 Shotcrete (Wet-Mix)	Crew	Daily Output	Labor-Hours	Unit	Material	2020 Bare Costs Labor	Equipment	Total	Total Incl O&P
1020	2" thick	C-8C	900	.053	S.F.	1.99	2.46	.66	5.11	6.85
1030	3" thick	↓	825	.058		2.99	2.69	.72	6.40	8.35
1040	4" thick	↓	750	.064	↓	3.98	2.96	.79	7.73	9.95

03 39 Concrete Curing

03 39 13 – Water Concrete Curing

03 39 13.50 Water Curing

		Crew	Daily Output	Labor-Hours	Unit	Material	2020 Bare Costs Labor	Equipment	Total	Total Incl O&P
0010	WATER CURING									
0015	With burlap, 4 uses assumed, 7.5 oz.	2 Clab	55	.291	C.S.F.	15.65	12.25		27.90	37
0100	10 oz.	"	55	.291	"	28	12.25		40.25	50.50
0400	Curing blankets, 1" to 2" thick, buy				S.F.	.26			.26	.29
9000	Minimum labor/equipment charge	1 Clab	5	1.600	Job		67.50		67.50	108

03 39 23 – Membrane Concrete Curing

03 39 23.13 Chemical Compound Membrane Concrete Curing

		Crew	Daily Output	Labor-Hours	Unit	Material	2020 Bare Costs Labor	Equipment	Total	Total Incl O&P
0010	CHEMICAL COMPOUND MEMBRANE CONCRETE CURING									
0300	Sprayed membrane curing compound	2 Clab	95	.168	C.S.F.	12.45	7.10		19.55	25
0700	Curing compound, solvent based, 400 S.F./gal., 55 gallon lots				Gal.	23			23	25.50
0720	5 gallon lots					33			33	36
0800	Curing compound, water based, 250 S.F./gal., 55 gallon lots					23.50			23.50	26
0820	5 gallon lots			↓	↓	26			26	29

03 39 23.23 Sheet Membrane Concrete Curing

		Crew	Daily Output	Labor-Hours	Unit	Material	2020 Bare Costs Labor	Equipment	Total	Total Incl O&P
0010	SHEET MEMBRANE CONCRETE CURING									
0200	Curing blanket, burlap/poly, 2-ply	2 Clab	70	.229	C.S.F.	20.50	9.60		30.10	38

03 41 Precast Structural Concrete

03 41 13 – Precast Concrete Hollow Core Planks

03 41 13.50 Precast Slab Planks

		Crew	Daily Output	Labor-Hours	Unit	Material	2020 Bare Costs Labor	Equipment	Total	Total Incl O&P
0010	PRECAST SLAB PLANKS	R034105-30								
0020	Prestressed roof/floor members, grouted, solid, 4" thick	C-11	2400	.030	S.F.	8.55	1.72	.95	11.22	13.30
0050	6" thick		2800	.026		8.55	1.47	.81	10.83	12.70
0100	Hollow, 8" thick		3200	.023		11.90	1.29	.71	13.90	15.90
0150	10" thick		3600	.020		9.95	1.15	.63	11.73	13.45
0200	12" thick	↓	4000	.018	↓	11.60	1.03	.57	13.20	15.10

03 41 23 – Precast Concrete Stairs

03 41 23.50 Precast Stairs

		Crew	Daily Output	Labor-Hours	Unit	Material	2020 Bare Costs Labor	Equipment	Total	Total Incl O&P
0010	PRECAST STAIRS									
0020	Precast concrete treads on steel stringers, 3' wide	C-12	75	.640	Riser	149	33.50	6.30	188.80	225
0300	Front entrance, 5' wide with 48" platform, 2 risers		16	3	Flight	675	158	29.50	862.50	1,025
0350	5 risers		12	4		1,125	211	39	1,375	1,600
0500	6' wide, 2 risers		15	3.200		750	168	31.50	949.50	1,125
0550	5 risers	↓	11	4.364		1,250	230	43	1,523	1,750
1200	Basement entrance stairwell, 6 steps, incl. steel bulkhead door	B-51	22	2.182		1,875	94.50	8.90	1,978.40	2,200
1250	14 steps	"	11	4.364	↓	3,350	189	17.85	3,556.85	4,000

03 41 Precast Structural Concrete

03 41 33 – Precast Structural Pretensioned Concrete

03 41 33.10 Precast Beams

		Crew	Daily Output	Labor-Hours	Unit	Material	2020 Bare Costs Labor	2020 Bare Costs Equipment	Total	Total Incl O&P
0010	**PRECAST BEAMS** R034105-30									
0011	L-shaped, 20' span, 12" x 20"	C-11	32	2.250	Ea.	4,500	129	71	4,700	5,225
0060	18" x 36"		24	3		6,350	172	94.50	6,616.50	7,375
0100	24" x 44"		22	3.273		7,450	187	103	7,740	8,625
0150	30' span, 12" x 36"		24	3		8,800	172	94.50	9,066.50	10,100
0200	18" x 44"		20	3.600		10,500	206	113	10,819	12,000
0250	24" x 52"		16	4.500		12,800	258	142	13,200	14,700
0400	40' span, 12" x 52"		20	3.600		13,300	206	113	13,619	15,100
0450	18" x 52"		16	4.500		14,900	258	142	15,300	17,000
0500	24" x 52"		12	6		17,100	345	189	17,634	19,600
1200	Rectangular, 20' span, 12" x 20"		32	2.250		3,925	129	71	4,125	4,575
1250	18" x 36"		24	3		5,800	172	94.50	6,066.50	6,750
1300	24" x 44"		22	3.273		6,850	187	103	7,140	7,975
1400	30' span, 12" x 36"		24	3		6,475	172	94.50	6,741.50	7,500
1450	18" x 44"		20	3.600		9,000	206	113	9,319	10,400
1500	24" x 52"		16	4.500		11,200	258	142	11,600	12,900
1600	40' span, 12" x 52"		20	3.600		10,800	206	113	11,119	12,300
1650	18" x 52"		16	4.500		13,000	258	142	13,400	14,900
1700	24" x 52"		12	6		14,900	345	189	15,434	17,200
2000	"T" shaped, 20' span, 12" x 20"		32	2.250		5,575	129	71	5,775	6,400
2050	18" x 36"		24	3		7,175	172	94.50	7,441.50	8,250
2100	24" x 44"		22	3.273		8,075	187	103	8,365	9,300
2200	30' span, 12" x 36"		24	3		9,950	172	94.50	10,216.50	11,300
2250	18" x 44"		20	3.600		12,500	206	113	12,819	14,300
2300	24" x 52"		16	4.500		13,300	258	142	13,700	15,300
2500	40' span, 12" x 52"		20	3.600		16,200	206	113	16,519	18,300
2550	18" x 52"		16	4.500		16,700	258	142	17,100	18,900
2600	24" x 52"		12	6		17,800	345	189	18,334	20,400

03 41 33.15 Precast Columns

		Crew	Daily Output	Labor-Hours	Unit	Material	2020 Bare Costs Labor	2020 Bare Costs Equipment	Total	Total Incl O&P
0010	**PRECAST COLUMNS** R034105-30									
0020	Rectangular to 12' high, 16" x 16"	C-11	120	.600	L.F.	278	34.50	18.90	331.40	380
0050	24" x 24"	"	96	.750	"	370	43	23.50	436.50	500

03 45 Precast Architectural Concrete

03 45 13 – Faced Architectural Precast Concrete

03 45 13.50 Precast Wall Panels

		Crew	Daily Output	Labor-Hours	Unit	Material	2020 Bare Costs Labor	2020 Bare Costs Equipment	Total	Total Incl O&P
0010	**PRECAST WALL PANELS**									
2200	Fiberglass reinforced cement with urethane core									
2210	R20, 8' x 8', 5" plain finish	E-2	750	.075	S.F.	29.50	4.27	2.26	36.03	42
2220	Exposed aggregate or brick finish	"	600	.093	"	48	5.35	2.83	56.18	64.50

03 48 Precast Concrete Specialties

03 48 43 – Precast Concrete Trim

03 48 43.40 Precast Lintels

		Crew	Daily Output	Labor-Hours	Unit	Material	2020 Bare Costs Labor	Equipment	Total	Total Incl O&P
0010	**PRECAST LINTELS**, smooth gray, prestressed, stock units only									
0800	4" wide x 8" high x 4' long	D-10	28	1.143	Ea.	35.50	59	15.30	109.80	150
0850	8' long		24	1.333		81	69	17.85	167.85	220
1000	6" wide x 8" high x 4' long		26	1.231		58.50	63.50	16.45	138.45	185
1050	10' long		22	1.455		149	75.50	19.45	243.95	305
1200	8" wide x 8" high x 4' long		24	1.333		61	69	17.85	147.85	198
1250	12' long		20	1.600		203	83	21.50	307.50	380
1275	For custom sizes, types, colors, or finishes of precast lintels, add					150%				

03 48 43.90 Precast Window Sills

		Crew	Daily Output	Labor-Hours	Unit	Material	2020 Bare Costs Labor	Equipment	Total	Total Incl O&P
0010	**PRECAST WINDOW SILLS**									
0600	Precast concrete, 4" tapers to 3", 9" wide	D-1	70	.229	L.F.	23	10.70		33.70	42.50
0650	11" wide	"	60	.267	"	37	12.50		49.50	61

03 51 Cast Roof Decks

03 51 13 – Cementitious Wood Fiber Decks

03 51 13.50 Cementitious/Wood Fiber Planks

		Crew	Daily Output	Labor-Hours	Unit	Material	2020 Bare Costs Labor	Equipment	Total	Total Incl O&P
0010	**CEMENTITIOUS/WOOD FIBER PLANKS**									
0050	Plank, beveled edge, 1" thick	2 Carp	1000	.016	S.F.	3.68	.85		4.53	5.40
0100	1-1/2" thick		975	.016		5.60	.87		6.47	7.55
0150	T&G, 2" thick		950	.017		3.33	.90		4.23	5.10
0200	2-1/2" thick		925	.017		3.71	.92		4.63	5.55
1000	Bulb tee, sub-purlin and grout, 6' span, add	E-1	5000	.005		2.22	.27	.03	2.52	2.91
1100	8' span	"	4200	.006		2.22	.32	.04	2.58	3

03 52 Lightweight Concrete Roof Insulation

03 52 16 – Lightweight Insulating Concrete

03 52 16.13 Lightweight Cellular Insulating Concrete

		Crew	Daily Output	Labor-Hours	Unit	Material	2020 Bare Costs Labor	Equipment	Total	Total Incl O&P
0010	**LIGHTWEIGHT CELLULAR INSULATING CONCRETE** R035216-10									
0020	Portland cement and foaming agent **G**	C-8	50	1.120	C.Y.	136	52.50	8.20	196.70	241

03 52 16.16 Lightweight Aggregate Insulating Concrete

		Crew	Daily Output	Labor-Hours	Unit	Material	2020 Bare Costs Labor	Equipment	Total	Total Incl O&P
0010	**LIGHTWEIGHT AGGREGATE INSULATING CONCRETE** R035216-10									
0100	Poured vermiculite or perlite, field mix,									
0110	1:6 field mix **G**	C-8	50	1.120	C.Y.	264	52.50	8.20	324.70	380
0200	Ready mix, 1:6 mix, roof fill, 2" thick **G**		10000	.006	S.F.	1.47	.26	.04	1.77	2.08
0250	3" thick **G**		7700	.007	"	2.20	.34	.05	2.59	3.02

03 53 Concrete Topping

03 53 16 – Iron-Aggregate Concrete Topping

03 53 16.50 Floor Topping

		Crew	Daily Output	Labor-Hours	Unit	Material	2020 Bare Costs Labor	Equipment	Total	Total Incl O&P
0010	**FLOOR TOPPING**									
0400	Integral topping/finish, on fresh concrete, using 1:1:2 mix, 3/16" thick	C-10B	1000	.040	S.F.	.14	1.81	.31	2.26	3.37
0450	1/2" thick		950	.042		.37	1.91	.33	2.61	3.80
0500	3/4" thick		850	.047		.56	2.13	.37	3.06	4.40
0600	1" thick		750	.053		.74	2.41	.42	3.57	5.10
0800	Granolithic topping, on fresh or cured concrete, 1:1:1-1/2 mix, 1/2" thick		590	.068		.41	3.07	.53	4.01	5.90
0820	3/4" thick		580	.069		.62	3.12	.54	4.28	6.25
0850	1" thick		575	.070		.82	3.15	.55	4.52	6.50

03 53 Concrete Topping

03 53 16 – Iron-Aggregate Concrete Topping

03 53 16.50 Floor Topping

		Crew	Daily Output	Labor-Hours	Unit	Material	2020 Bare Costs Labor	Equipment	Total	Total Incl O&P
0950	2" thick	C-10B	500	.080	S.F.	1.65	3.62	.63	5.90	8.25

03 54 Cast Underlayment

03 54 13 – Gypsum Cement Underlayment

03 54 13.50 Poured Gypsum Underlayment

		Crew	Daily Output	Labor-Hours	Unit	Material	Labor	Equipment	Total	Total Incl O&P
0010	**POURED GYPSUM UNDERLAYMENT**									
0400	Underlayment, gypsum based, self-leveling 2500 psi, pumped, 1/2" thick	C-8	24000	.002	S.F.	.44	.11	.02	.57	.68
0500	3/4" thick		20000	.003		.66	.13	.02	.81	.96
0600	1" thick		16000	.004		.88	.16	.03	1.07	1.26
1400	Hand placed, 1/2" thick	C-18	450	.020		.44	.85	.28	1.57	2.16
1500	3/4" thick	"	300	.030		.66	1.27	.43	2.36	3.23

03 54 16 – Hydraulic Cement Underlayment

03 54 16.50 Cement Underlayment

		Crew	Daily Output	Labor-Hours	Unit	Material	Labor	Equipment	Total	Total Incl O&P
0010	**CEMENT UNDERLAYMENT**									
2510	Underlayment, P.C. based, self-leveling, 4100 psi, pumped, 1/4" thick	C-8	20000	.003	S.F.	1.75	.13	.02	1.90	2.16
2520	1/2" thick		19000	.003		3.50	.14	.02	3.66	4.10
2530	3/4" thick		18000	.003		5.25	.15	.02	5.42	6.05
2540	1" thick		17000	.003		7	.15	.02	7.17	7.95
2550	1-1/2" thick		15000	.004		10.50	.17	.03	10.70	11.85
2560	Hand placed, 1/2" thick	C-18	450	.020		3.50	.85	.28	4.63	5.55
2610	Topping, P.C. based, self-leveling, 6100 psi, pumped, 1/4" thick	C-8	20000	.003		2.09	.13	.02	2.24	2.53
2620	1/2" thick		19000	.003		4.18	.14	.02	4.34	4.84
2630	3/4" thick		18000	.003		6.25	.15	.02	6.42	7.15
2660	1" thick		17000	.003		8.35	.15	.02	8.52	9.45
2670	1-1/2" thick		15000	.004		12.55	.17	.03	12.75	14.10
2680	Hand placed, 1/2" thick	C-18	450	.020		4.18	.85	.28	5.31	6.25

03 63 Epoxy Grouting

03 63 05 – Grouting of Dowels and Fasteners

03 63 05.10 Epoxy Only

		Crew	Daily Output	Labor-Hours	Unit	Material	Labor	Equipment	Total	Total Incl O&P
0010	**EPOXY ONLY**									
1500	Chemical anchoring, epoxy cartridge, excludes layout, drilling, fastener									
1530	For fastener 3/4" diam. x 6" embedment	2 Skwk	72	.222	Ea.	4.78	12.20		16.98	24.50
1535	1" diam. x 8" embedment		66	.242		7.15	13.30		20.45	29
1540	1-1/4" diam. x 10" embedment		60	.267		14.35	14.65		29	39.50
1545	1-3/4" diam. x 12" embedment		54	.296		24	16.25		40.25	52.50
1550	14" embedment		48	.333		28.50	18.30		46.80	60.50
1555	2" diam. x 12" embedment		42	.381		38	21		59	75.50
1560	18" embedment		32	.500		48	27.50		75.50	96

03 81 Concrete Cutting

03 81 13 – Flat Concrete Sawing

03 81 13.50 Concrete Floor/Slab Cutting

03 81 13.50 Concrete Floor/Slab Cutting	Crew	Daily Output	Labor-Hours	Unit	Material	2020 Bare Costs Labor	Equipment	Total	Total Incl O&P
0010 **CONCRETE FLOOR/SLAB CUTTING**									
0050 Includes blade cost, layout and set-up time									
0300 Saw cut concrete slabs, plain, up to 3" deep	B-89	1060	.015	L.F.	.12	.76	.98	1.86	2.40
0320 Each additional inch of depth		3180	.005		.04	.25	.33	.62	.80
0400 Mesh reinforced, up to 3" deep		980	.016		.13	.82	1.06	2.01	2.60
0420 Each additional inch of depth		2940	.005		.04	.27	.35	.66	.87
0500 Rod reinforced, up to 3" deep		800	.020		.17	1	1.29	2.46	3.19
0520 Each additional inch of depth		2400	.007		.06	.33	.43	.82	1.06
0590 Minimum labor/equipment charge		2	8	Job		400	515	915	1,200

03 81 13.75 Concrete Saw Blades

03 81 13.75 Concrete Saw Blades	Crew	Daily Output	Labor-Hours	Unit	Material	2020 Bare Costs Labor	Equipment	Total	Total Incl O&P
0010 **CONCRETE SAW BLADES**									
3000 Blades for saw cutting, included in cutting line items									
3020 Diamond, 12" diameter				Ea.	249			249	274
3040 18" diameter					405			405	445
3080 24" diameter					585			585	645
3120 30" diameter					865			865	950
3160 36" diameter					1,150			1,150	1,250
3200 42" diameter					2,450			2,450	2,700

03 81 16 – Track Mounted Concrete Wall Sawing

03 81 16.50 Concrete Wall Cutting

03 81 16.50 Concrete Wall Cutting	Crew	Daily Output	Labor-Hours	Unit	Material	2020 Bare Costs Labor	Equipment	Total	Total Incl O&P
0010 **CONCRETE WALL CUTTING**									
0750 Includes blade cost, layout and set-up time									
0800 Concrete walls, hydraulic saw, plain, per inch of depth	B-89B	250	.064	L.F.	.04	3.21	6	9.25	11.70
0820 Rod reinforcing, per inch of depth		150	.107	"	.06	5.35	10.05	15.46	19.55
0890 Minimum labor/equipment charge		2	8	Job		400	750	1,150	1,450

03 82 Concrete Boring

03 82 13 – Concrete Core Drilling

03 82 13.10 Core Drilling

03 82 13.10 Core Drilling	Crew	Daily Output	Labor-Hours	Unit	Material	2020 Bare Costs Labor	Equipment	Total	Total Incl O&P
0010 **CORE DRILLING**									
0015 Includes bit cost, layout and set-up time									
0020 Reinforced concrete slab, up to 6" thick									
0100 1" diameter core	B-89A	17	.941	Ea.	.20	45.50	6.70	52.40	80.50
0150 For each additional inch of slab thickness in same hole, add		1440	.011		.03	.54	.08	.65	.99
0200 2" diameter core		16.50	.970		.26	47	6.90	54.16	83
0250 For each additional inch of slab thickness in same hole, add		1080	.015		.04	.72	.11	.87	1.32
0300 3" diameter core		16	1		.40	48.50	7.15	56.05	86
0350 For each additional inch of slab thickness in same hole, add		720	.022		.07	1.08	.16	1.31	1.96
0500 4" diameter core		15	1.067		.45	51.50	7.60	59.55	91.50
0550 For each additional inch of slab thickness in same hole, add		480	.033		.07	1.62	.24	1.93	2.92
0700 6" diameter core		14	1.143		.65	55.50	8.15	64.30	98
0750 For each additional inch of slab thickness in same hole, add		360	.044		.11	2.15	.32	2.58	3.91
0900 8" diameter core		13	1.231		1.07	59.50	8.80	69.37	106
0950 For each additional inch of slab thickness in same hole, add		288	.056		.18	2.69	.40	3.27	4.94
1100 10" diameter core		12	1.333		1.20	64.50	9.50	75.20	115
1150 For each additional inch of slab thickness in same hole, add		240	.067		.20	3.23	.48	3.91	5.90
1300 12" diameter core		11	1.455		1.84	70.50	10.40	82.74	126
1350 For each additional inch of slab thickness in same hole, add		206	.078		.31	3.77	.55	4.63	6.95
1500 14" diameter core		10	1.600		1.84	77.50	11.40	90.74	139
1550 For each additional inch of slab thickness in same hole, add		180	.089		.31	4.31	.63	5.25	7.95

03 82 Concrete Boring

03 82 13 – Concrete Core Drilling

03 82 13.10 Core Drilling

		Crew	Daily Output	Labor-Hours	Unit	Material	2020 Bare Costs Labor	Equipment	Total	Total Incl O&P
1700	18" diameter core	B-89A	9	1.778	Ea.	3.06	86	12.70	101.76	155
1750	For each additional inch of slab thickness in same hole, add		144	.111		.51	5.40	.79	6.70	10.05
1754	24" diameter core		8	2		4.15	97	14.30	115.45	175
1756	For each additional inch of slab thickness in same hole, add		120	.133		.69	6.45	.95	8.09	12.10
1760	For horizontal holes, add to above						20%	20%		
1770	Prestressed hollow core plank, 8" thick									
1780	1" diameter core	B-89A	17.50	.914	Ea.	.27	44.50	6.55	51.32	78.50
1790	For each additional inch of plank thickness in same hole, add		3840	.004		.03	.20	.03	.26	.39
1794	2" diameter core		17.25	.928		.35	45	6.60	51.95	79.50
1796	For each additional inch of plank thickness in same hole, add		2880	.006		.04	.27	.04	.35	.52
1800	3" diameter core		17	.941		.53	45.50	6.70	52.73	81
1810	For each additional inch of plank thickness in same hole, add		1920	.008		.07	.40	.06	.53	.79
1820	4" diameter core		16.50	.970		.60	47	6.90	54.50	83.50
1830	For each additional inch of plank thickness in same hole, add		1280	.013		.07	.61	.09	.77	1.15
1840	6" diameter core		15.50	1.032		.87	50	7.35	58.22	89
1850	For each additional inch of plank thickness in same hole, add		960	.017		.11	.81	.12	1.04	1.54
1860	8" diameter core		15	1.067		1.43	51.50	7.60	60.53	92.50
1870	For each additional inch of plank thickness in same hole, add		768	.021		.18	1.01	.15	1.34	1.97
1880	10" diameter core		14	1.143		1.60	55.50	8.15	65.25	99.50
1890	For each additional inch of plank thickness in same hole, add		640	.025		.20	1.21	.18	1.59	2.36
1900	12" diameter core		13.50	1.185		2.45	57.50	8.45	68.40	104
1910	For each additional inch of plank thickness in same hole, add		548	.029		.31	1.42	.21	1.94	2.83
1999	Drilling, core, minimum labor/equipment charge		2	8	Job		390	57	447	685
3000	Bits for core drilling, included in drilling line items									
3010	Diamond, premium, 1" diameter				Ea.	80.50			80.50	89
3020	2" diameter					104			104	114
3030	3" diameter					159			159	175
3040	4" diameter					180			180	198
3060	6" diameter					262			262	288
3080	8" diameter					430			430	470
3110	10" diameter					480			480	530
3120	12" diameter					735			735	810
3140	14" diameter					735			735	810
3180	18" diameter					1,225			1,225	1,350
3240	24" diameter					1,650			1,650	1,825

03 82 16 – Concrete Drilling

03 82 16.10 Concrete Impact Drilling

		Crew	Daily Output	Labor-Hours	Unit	Material	2020 Bare Costs Labor	Equipment	Total	Total Incl O&P
0010	**CONCRETE IMPACT DRILLING**									
0020	Includes bit cost, layout and set-up time, no anchors									
0050	Up to 4" deep in concrete/brick floors/walls									
0100	Holes, 1/4" diameter	1 Carp	75	.107	Ea.	.07	5.65		5.72	9.20
0150	For each additional inch of depth in same hole, add		430	.019		.02	.99		1.01	1.60
0200	3/8" diameter		63	.127		.05	6.75		6.80	10.85
0250	For each additional inch of depth in same hole, add		340	.024		.01	1.25		1.26	2.01
0300	1/2" diameter		50	.160		.05	8.50		8.55	13.65
0350	For each additional inch of depth in same hole, add		250	.032		.01	1.70		1.71	2.73
0400	5/8" diameter		48	.167		.10	8.85		8.95	14.30
0450	For each additional inch of depth in same hole, add		240	.033		.02	1.77		1.79	2.87
0500	3/4" diameter		45	.178		.11	9.45		9.56	15.25
0550	For each additional inch of depth in same hole, add		220	.036		.03	1.93		1.96	3.13
0600	7/8" diameter		43	.186		.17	9.90		10.07	16.05
0650	For each additional inch of depth in same hole, add		210	.038		.04	2.03		2.07	3.29

03 82 Concrete Boring

03 82 16 – Concrete Drilling

03 82 16.10 Concrete Impact Drilling		Crew	Daily Output	Labor-Hours	Unit	Material	2020 Bare Costs		Total	Total Incl O&P
							Labor	Equipment		
0700	1" diameter	1 Carp	40	.200	Ea.	.18	10.65		10.83	17.25
0750	For each additional inch of depth in same hole, add		190	.042		.04	2.24		2.28	3.64
0800	1-1/4" diameter		38	.211		.33	11.20		11.53	18.30
0850	For each additional inch of depth in same hole, add		180	.044		.08	2.36		2.44	3.87
0900	1-1/2" diameter		35	.229		.43	12.15		12.58	19.90
0950	For each additional inch of depth in same hole, add	▼	165	.048	▼	.11	2.58		2.69	4.25
1000	For ceiling installations, add						40%			

Estimating Tips
04 05 00 Common Work Results for Masonry

- The terms mortar and grout are often used interchangeably—and incorrectly. Mortar is used to bed masonry units, seal the entry of air and moisture, provide architectural appearance, and allow for size variations in the units. Grout is used primarily in reinforcing masonry construction and to bond the masonry to the reinforcing steel. Common mortar types are M (2500 psi), S (1800 psi), N (750 psi), and O (350 psi), and they conform to ASTM C270. Grout is either fine or coarse and conforms to ASTM C476, and in-place strengths generally exceed 2500 psi. Mortar and grout are different components of masonry construction and are placed by entirely different methods. An estimator should be aware of their unique uses and costs.

- Mortar is included in all assembled masonry line items. The mortar cost, part of the assembled masonry material cost, includes all ingredients, all labor, and all equipment required. Please see reference number R040513-10.

- Waste, specifically the loss/droppings of mortar and the breakage of brick and block, is included in all unit cost lines that include mortar and masonry units in this division. A factor of 25% is added for mortar and 3% for brick and concrete masonry units.

- Scaffolding or staging is not included in any of the Division 4 costs. Refer to Subdivision 01 54 23 for scaffolding and staging costs.

04 20 00 Unit Masonry

- The most common types of unit masonry are brick and concrete masonry. The major classifications of brick are building brick (ASTM C62), facing brick (ASTM C216), glazed brick, fire brick, and pavers. Many varieties of texture and appearance can exist within these classifications, and the estimator would be wise to check local custom and availability within the project area. For repair and remodeling jobs, matching the existing brick may be the most important criteria.

- Brick and concrete block are priced by the piece and then converted into a price per square foot of wall. Openings less than two square feet are generally ignored by the estimator because any savings in units used are offset by the cutting and trimming required.

- It is often difficult and expensive to find and purchase small lots of historic brick. Costs can vary widely. Many design issues affect costs, selection of mortar mix, and repairs or replacement of masonry materials. Cleaning techniques must be reflected in the estimate.

- All masonry walls, whether interior or exterior, require bracing. The cost of bracing walls during construction should be included by the estimator, and this bracing must remain in place until permanent bracing is complete. Permanent bracing of masonry walls is accomplished by masonry itself, in the form of pilasters or abutting wall corners, or by anchoring the walls to the structural frame. Accessories in the form of anchors, anchor slots, and ties are used, but their supply and installation can be by different trades. For instance, anchor slots on spandrel beams and columns are supplied and welded in place by the steel fabricator, but the ties from the slots into the masonry are installed by the bricklayer. Regardless of the installation method, the estimator must be certain that these accessories are accounted for in pricing.

Reference Numbers
Reference numbers are shown at the beginning of some major classifications. These numbers refer to related items in the Reference Section. The reference information may be an estimating procedure, an alternate pricing method, or technical information.

Note: Not all subdivisions listed here necessarily appear. ■

Same Data. Simplified.

Enjoy the convenience and efficiency of accessing your costs anywhere:

- **Skip the multiplier** by setting your location
- **Quickly search,** edit, favorite and share costs
- **Stay on top of price changes** with automatic updates

Discover more at rsmeans.com/online

04 01 Maintenance of Masonry

04 01 20 – Maintenance of Unit Masonry

04 01 20.20 Pointing Masonry

		Crew	Daily Output	Labor-Hours	Unit	Material	2020 Bare Costs Labor	2020 Bare Costs Equipment	Total	Total Incl O&P
0010	**POINTING MASONRY**									
0300	Cut and repoint brick, hard mortar, running bond	1 Bric	80	.100	S.F.	.57	5.20		5.77	9.05
0320	Common bond		77	.104		.57	5.40		5.97	9.40
0360	Flemish bond		70	.114		.60	5.95		6.55	10.25
0400	English bond		65	.123		.60	6.40		7	11
0600	Soft old mortar, running bond		100	.080		.57	4.16		4.73	7.40
0620	Common bond		96	.083		.57	4.34		4.91	7.65
0640	Flemish bond		90	.089		.60	4.63		5.23	8.15
0680	English bond		82	.098		.60	5.10		5.70	8.85
0700	Stonework, hard mortar		140	.057	L.F.	.76	2.97		3.73	5.65
0720	Soft old mortar		160	.050	"	.76	2.60		3.36	5.05
1000	Repoint, mask and grout method, running bond		95	.084	S.F.	.76	4.38		5.14	7.95
1020	Common bond		90	.089		.76	4.63		5.39	8.35
1040	Flemish bond		86	.093		.80	4.84		5.64	8.70
1060	English bond		77	.104		.80	5.40		6.20	9.65
2000	Scrub coat, sand grout on walls, thin mix, brushed		120	.067		3.59	3.47		7.06	9.55
2020	Troweled		98	.082		4.99	4.25		9.24	12.35
9000	Minimum labor/equipment charge		3	2.667	Job		139		139	224

04 01 20.30 Pointing CMU

		Crew	Daily Output	Labor-Hours	Unit	Material	2020 Bare Costs Labor	2020 Bare Costs Equipment	Total	Total Incl O&P
0010	**POINTING CMU**									
0300	Cut and repoint block, hard mortar, running bond	1 Bric	190	.042	S.F.	.23	2.19		2.42	3.80
0310	Stacked bond		200	.040		.23	2.08		2.31	3.62
0600	Soft old mortar, running bond		230	.035		.23	1.81		2.04	3.19
0610	Stacked bond		245	.033		.23	1.70		1.93	3.01

04 01 20.41 Unit Masonry Stabilization

		Crew	Daily Output	Labor-Hours	Unit	Material	2020 Bare Costs Labor	2020 Bare Costs Equipment	Total	Total Incl O&P
0010	**UNIT MASONRY STABILIZATION**									
0100	Structural repointing method									
0110	Cut/grind mortar joint	1 Bric	240	.033	L.F.		1.73		1.73	2.80
0120	Clean and mask joint		2500	.003		.13	.17		.30	.41
0130	Epoxy paste and 1/4" FRP rod		240	.033		2.11	1.73		3.84	5.10
0132	3/8" FRP rod		160	.050		2.98	2.60		5.58	7.50
0140	Remove masking		14400	.001			.03		.03	.05
0300	Structural fabric method									
0310	Primer	1 Bric	600	.013	S.F.	1.08	.69		1.77	2.30
0320	Apply filling/leveling paste		720	.011		.89	.58		1.47	1.91
0330	Epoxy, glass fiber fabric		720	.011		9.85	.58		10.43	11.80
0340	Carbon fiber fabric		720	.011		23.50	.58		24.08	27

04 01 20.50 Toothing Masonry

		Crew	Daily Output	Labor-Hours	Unit	Material	2020 Bare Costs Labor	2020 Bare Costs Equipment	Total	Total Incl O&P
0010	**TOOTHING MASONRY**									
0500	Brickwork, soft old mortar	1 Clab	40	.200	V.L.F.		8.40		8.40	13.50
0520	Hard mortar		30	.267			11.25		11.25	18
0700	Blockwork, soft old mortar		70	.114			4.81		4.81	7.70
0720	Hard mortar		50	.160			6.75		6.75	10.80
9000	Minimum labor/equipment charge		4	2	Job		84		84	135

04 01 20.52 Cleaning Masonry

		Crew	Daily Output	Labor-Hours	Unit	Material	2020 Bare Costs Labor	2020 Bare Costs Equipment	Total	Total Incl O&P
0010	**CLEANING MASONRY**									
0200	By chemical, brush and rinse, new work, light construction dust	D-1	1000	.016	S.F.	.06	.75		.81	1.27
0220	Medium construction dust		800	.020		.08	.94		1.02	1.61
0240	Heavy construction dust, drips or stains		600	.027		.11	1.25		1.36	2.14
0260	Low pressure wash and rinse, light restoration, light soil		800	.020		.14	.94		1.08	1.68
0270	Average soil, biological staining		400	.040		.22	1.88		2.10	3.27
0280	Heavy soil, biological and mineral staining, paint		330	.048		.29	2.27		2.56	3.99

04 01 20.52 Cleaning Masonry

		Crew	Daily Output	Labor-Hours	Unit	Material	2020 Bare Costs Labor	Equipment	Total	Total Incl O&P
0300	High pressure wash and rinse, heavy restoration, light soil	D-1	600	.027	S.F.	.10	1.25		1.35	2.13
0310	Average soil, biological staining		400	.040		.15	1.88		2.03	3.19
0320	Heavy soil, biological and mineral staining, paint		250	.064		.20	3		3.20	5.05
0400	High pressure wash, water only, light soil	C-29	500	.016			.67	.19	.86	1.29
0420	Average soil, biological staining		375	.021			.90	.26	1.16	1.72
0440	Heavy soil, biological and mineral staining, paint		250	.032			1.35	.39	1.74	2.59
0800	High pressure water and chemical, light soil		450	.018		.17	.75	.22	1.14	1.63
0820	Average soil, biological staining		300	.027		.26	1.12	.32	1.70	2.45
0840	Heavy soil, biological and mineral staining, paint		200	.040		.35	1.68	.48	2.51	3.61
1200	Sandblast, wet system, light soil	J-6	1750	.018		.34	.84	.17	1.35	1.90
1220	Average soil, biological staining		1100	.029		.51	1.34	.27	2.12	2.99
1240	Heavy soil, biological and mineral staining, paint		700	.046		.69	2.10	.43	3.22	4.55
1400	Dry system, light soil		2500	.013		.34	.59	.12	1.05	1.44
1420	Average soil, biological staining		1750	.018		.51	.84	.17	1.52	2.09
1440	Heavy soil, biological and mineral staining, paint		1000	.032		.69	1.47	.30	2.46	3.41
1800	For walnut shells, add					.92			.92	1.01
1820	For corn chips, add					.96			.96	1.06
2000	Steam cleaning, light soil	A-1H	750	.011			.45	.10	.55	.83
2020	Average soil, biological staining		625	.013			.54	.12	.66	.99
2040	Heavy soil, biological and mineral staining		375	.021			.90	.20	1.10	1.66
4000	Add for masking doors and windows	1 Clab	800	.010		.07	.42		.49	.75
4200	Add for pedestrian protection				Job				10%	10%
9000	Minimum labor/equipment charge	D-4	2	16	"		755	94	849	1,300

04 01 20.70 Brick Washing

		Crew	Daily Output	Labor-Hours	Unit	Material	2020 Bare Costs Labor	Equipment	Total	Total Incl O&P
0010	**BRICK WASHING**									
0012	Acid cleanser, smooth brick surface	1 Bric	560	.014	S.F.	.05	.74		.79	1.26
0050	Rough brick		400	.020		.07	1.04		1.11	1.76
0060	Stone, acid wash		600	.013		.08	.69		.77	1.21
1000	Muriatic acid, price per gallon in 5 gallon lots				Gal.	10.55			10.55	11.60

04 05 05.10 Selective Demolition

		Crew	Daily Output	Labor-Hours	Unit	Material	2020 Bare Costs Labor	Equipment	Total	Total Incl O&P
0010	**SELECTIVE DEMOLITION** R024119-10									
0200	Bond beams, 8" block with #4 bar	2 Clab	32	.500	L.F.		21		21	33.50
0300	Concrete block walls, unreinforced, 2" thick		1200	.013	S.F.		.56		.56	.90
0310	4" thick		1150	.014			.59		.59	.94
0320	6" thick		1100	.015			.61		.61	.98
0330	8" thick		1050	.015			.64		.64	1.03
0340	10" thick		1000	.016			.67		.67	1.08
0360	12" thick		950	.017			.71		.71	1.14
0380	Reinforced alternate courses, 2" thick		1130	.014			.60		.60	.96
0390	4" thick		1080	.015			.62		.62	1
0400	6" thick		1035	.015			.65		.65	1.04
0410	8" thick		990	.016			.68		.68	1.09
0420	10" thick		940	.017			.72		.72	1.15
0430	12" thick		890	.018			.76		.76	1.21
0440	Reinforced alternate courses & vertically 48" OC, 4" thick		900	.018			.75		.75	1.20
0450	6" thick		850	.019			.79		.79	1.27
0460	8" thick		800	.020			.84		.84	1.35
0480	10" thick		750	.021			.90		.90	1.44

04 05 05.10 Selective Demolition	Crew	Daily Output	Labor-Hours	Unit	Material	2020 Bare Costs Labor	2020 Bare Costs Equipment	Total	Total Incl O&P
0490	12" thick	2 Clab	700	.023	S.F.		.96		.96	1.54
1000	Chimney, 16" x 16", soft old mortar	1 Clab	55	.145	C.F.		6.10		6.10	9.80
1020	Hard mortar		40	.200			8.40		8.40	13.50
1030	16" x 20", soft old mortar		55	.145			6.10		6.10	9.80
1040	Hard mortar		40	.200			8.40		8.40	13.50
1050	16" x 24", soft old mortar		55	.145			6.10		6.10	9.80
1060	Hard mortar		40	.200			8.40		8.40	13.50
1080	20" x 20", soft old mortar		55	.145			6.10		6.10	9.80
1100	Hard mortar		40	.200			8.40		8.40	13.50
1110	20" x 24", soft old mortar		55	.145			6.10		6.10	9.80
1120	Hard mortar		40	.200			8.40		8.40	13.50
1140	20" x 32", soft old mortar		55	.145			6.10		6.10	9.80
1160	Hard mortar		40	.200			8.40		8.40	13.50
1200	48" x 48", soft old mortar		55	.145			6.10		6.10	9.80
1220	Hard mortar	▼	40	.200	▼		8.40		8.40	13.50
1250	Metal, high temp steel jacket, 24" diameter	E-2	130	.431	V.L.F.		24.50	13.05	37.55	54.50
1260	60" diameter	"	60	.933			53.50	28.50	82	118
1280	Flue lining, up to 12" x 12"	1 Clab	200	.040			1.68		1.68	2.70
1282	Up to 24" x 24"		150	.053			2.25		2.25	3.60
2000	Columns, 8" x 8", soft old mortar		48	.167			7		7	11.25
2020	Hard mortar		40	.200			8.40		8.40	13.50
2060	16" x 16", soft old mortar		16	.500			21		21	33.50
2100	Hard mortar		14	.571			24		24	38.50
2140	24" x 24", soft old mortar		8	1			42		42	67.50
2160	Hard mortar		6	1.333			56		56	90
2200	36" x 36", soft old mortar		4	2			84		84	135
2220	Hard mortar		3	2.667	▼		112		112	180
2230	Alternate pricing method, soft old mortar		30	.267	C.F.		11.25		11.25	18
2240	Hard mortar	▼	23	.348	"		14.65		14.65	23.50
3000	Copings, precast or masonry, to 8" wide									
3020	Soft old mortar	1 Clab	180	.044	L.F.		1.87		1.87	3
3040	Hard mortar	"	160	.050	"		2.11		2.11	3.37
3100	To 12" wide									
3120	Soft old mortar	1 Clab	160	.050	L.F.		2.11		2.11	3.37
3140	Hard mortar	"	140	.057	"		2.41		2.41	3.85
4000	Fireplace, brick, 30" x 24" opening									
4020	Soft old mortar	1 Clab	2	4	Ea.		168		168	270
4040	Hard mortar		1.25	6.400			269		269	430
4100	Stone, soft old mortar		1.50	5.333			225		225	360
4120	Hard mortar		1	8	▼		335		335	540
5000	Veneers, brick, soft old mortar		140	.057	S.F.		2.41		2.41	3.85
5020	Hard mortar		125	.064			2.69		2.69	4.32
5050	Glass block, up to 4" thick		500	.016			.67		.67	1.08
5100	Granite and marble, 2" thick		180	.044			1.87		1.87	3
5120	4" thick		170	.047			1.98		1.98	3.17
5140	Stone, 4" thick		180	.044			1.87		1.87	3
5160	8" thick		175	.046	▼		1.92		1.92	3.08
5400	Alternate pricing method, stone, 4" thick		60	.133	C.F.		5.60		5.60	9
5420	8" thick		85	.094	"		3.96		3.96	6.35
9000	Minimum labor/equipment charge	▼	2	4	Job		168		168	270

04 05 13.10 Cement	Crew	Daily Output	Labor-Hours	Unit	Material	2020 Bare Costs Labor	Equipment	Total	Total Incl O&P
0010 **CEMENT**									
0100 Masonry, 70 lb. bag, T.L. lots				Bag	11.40			11.40	12.50
0150 L.T.L. lots					12.05			12.05	13.25
0200 White, 70 lb. bag, T.L. lots					18.65			18.65	20.50
0250 L.T.L. lots				↓	22.50			22.50	24.50

04 05 13.20 Lime

	Crew	Daily Output	Labor-Hours	Unit	Material	Labor	Equipment	Total	Total Incl O&P
0010 **LIME**									
0020 Masons, hydrated, 50 lb. bag, T.L. lots				Bag	11.25			11.25	12.40
0050 L.T.L. lots					12.40			12.40	13.65
0200 Finish, double hydrated, 50 lb. bag, T.L. lots					11.90			11.90	13.10
0250 L.T.L. lots				↓	13.10			13.10	14.40

04 05 13.23 Surface Bonding Masonry Mortaring

	Crew	Daily Output	Labor-Hours	Unit	Material	Labor	Equipment	Total	Total Incl O&P
0010 **SURFACE BONDING MASONRY MORTARING**									
0020 Gray or white colors, not incl. block work	1 Bric	540	.015	S.F.	.20	.77		.97	1.47

04 05 13.30 Mortar

	Crew	Daily Output	Labor-Hours	Unit	Material	Labor	Equipment	Total	Total Incl O&P
0010 **MORTAR**									
0020 With masonry cement									
0100 Type M, 1:1:6 mix	1 Brhe	143	.056	C.F.	6.15	2.34		8.49	10.50
0200 Type N, 1:3 mix		143	.056		5.35	2.34		7.69	9.65
0300 Type O, 1:3 mix		143	.056		5.20	2.34		7.54	9.45
0400 Type PM, 1:1:6 mix, 2500 psi		143	.056		6.20	2.34		8.54	10.55
0500 Type S, 1/2:1:4 mix	↓	143	.056	↓	6.55	2.34		8.89	10.95
2000 With Portland cement and lime									
2100 Type M, 1:1/4:3 mix	1 Brhe	143	.056	C.F.	9.10	2.34		11.44	13.75
2200 Type N, 1:1:6 mix, 750 psi		143	.056		7.30	2.34		9.64	11.80
2300 Type O, 1:2:9 mix (Pointing Mortar)		143	.056		8.75	2.34		11.09	13.40
2400 Type PL, 1:1/2:4 mix, 2500 psi		143	.056		6.35	2.34		8.69	10.75
2600 Type S, 1:1/2:4 mix, 1800 psi	↓	143	.056		8.55	2.34		10.89	13.20
2650 Pre-mixed, type S or N					6.25			6.25	6.90
2700 Mortar for glass block	1 Brhe	143	.056	↓	13.25	2.34		15.59	18.30
2900 Mortar for fire brick, dry mix, 10 lb. pail				Ea.	25			25	28

04 05 13.91 Masonry Restoration Mortaring

	Crew	Daily Output	Labor-Hours	Unit	Material	Labor	Equipment	Total	Total Incl O&P
0010 **MASONRY RESTORATION MORTARING**									
0020 Masonry restoration mix				Lb.	.85			.85	.94
0050 White				"	1.40			1.40	1.54

04 05 13.93 Mortar Pigments

	Crew	Daily Output	Labor-Hours	Unit	Material	Labor	Equipment	Total	Total Incl O&P
0010 **MORTAR PIGMENTS**, 50 lb. bags (2 bags per M bricks)									
0020 Color admixture, range 2 to 10 lb. per bag of cement, light colors				Lb.	6.30			6.30	6.95
0050 Medium colors					7.05			7.05	7.75
0100 Dark colors				↓	15.95			15.95	17.55

04 05 13.95 Sand

	Crew	Daily Output	Labor-Hours	Unit	Material	Labor	Equipment	Total	Total Incl O&P
0010 **SAND**, screened and washed at pit									
0020 For mortar, per ton				Ton	21.50			21.50	24
0050 With 10 mile haul					42			42	46
0100 With 30 mile haul				↓	67			67	74
0200 Screened and washed, at the pit				C.Y.	30			30	33
0250 With 10 mile haul					58.50			58.50	64
0300 With 30 mile haul				↓	93.50			93.50	103

For customer support on your Commercial Renovation Costs with RSMeans data, call 800.448.8182.

85

04 05 Common Work Results for Masonry

04 05 13 – Masonry Mortaring

04 05 13.98 Mortar Admixtures	Crew	Daily Output	Labor-Hours	Unit	Material	2020 Bare Costs Labor	2020 Bare Costs Equipment	Total	Total Incl O&P
0010 **MORTAR ADMIXTURES**									
0020 Waterproofing admixture, per quart (1 qt. to 2 bags of masonry cement)				Qt.	4.10			4.10	4.51

04 05 16 – Masonry Grouting

04 05 16.30 Grouting

	Crew	Daily Output	Labor-Hours	Unit	Material	Labor	Equipment	Total	Total Incl O&P
0010 **GROUTING**									
0011 Bond beams & lintels, 8" deep, 6" thick, 0.15 C.F./L.F.	D-4	1480	.022	L.F.	.80	1.02	.13	1.95	2.65
0020 8" thick, 0.2 C.F./L.F.		1400	.023		1.29	1.08	.13	2.50	3.30
0050 10" thick, 0.25 C.F./L.F.		1200	.027		1.34	1.26	.16	2.76	3.65
0060 12" thick, 0.3 C.F./L.F.		1040	.031		1.61	1.45	.18	3.24	4.29
0200 Concrete block cores, solid, 4" thk., by hand, 0.067 C.F./S.F. of wall	D-8	1100	.036	S.F.	.36	1.74		2.10	3.21
0210 6" thick, pumped, 0.175 C.F./S.F.	D-4	720	.044		.94	2.09	.26	3.29	4.67
0250 8" thick, pumped, 0.258 C.F./S.F.		680	.047		1.38	2.22	.28	3.88	5.35
0300 10" thick, pumped, 0.340 C.F./S.F.		660	.048		1.82	2.29	.29	4.40	5.95
0350 12" thick, pumped, 0.422 C.F./S.F.		640	.050		2.26	2.36	.29	4.91	6.60
0500 Cavity walls, 2" space, pumped, 0.167 C.F./S.F. of wall		1700	.019		.89	.89	.11	1.89	2.52
0550 3" space, 0.250 C.F./S.F.		1200	.027		1.34	1.26	.16	2.76	3.65
0600 4" space, 0.333 C.F./S.F.		1150	.028		1.78	1.31	.16	3.25	4.24
0700 6" space, 0.500 C.F./S.F.		800	.040		2.68	1.89	.24	4.81	6.25
0800 Door frames, 3' x 7' opening, 2.5 C.F. per opening		60	.533	Opng.	13.40	25	3.14	41.54	58.50
0850 6' x 7' opening, 3.5 C.F. per opening		45	.711	"	18.75	33.50	4.19	56.44	78.50
2000 Grout, C476, for bond beams, lintels and CMU cores		350	.091	C.F.	5.35	4.31	.54	10.20	13.40
9000 Minimum labor/equipment charge	1 Bric	2	4	Job		208		208	335

04 05 19 – Masonry Anchorage and Reinforcing

04 05 19.05 Anchor Bolts

	Crew	Daily Output	Labor-Hours	Unit	Material	Labor	Equipment	Total	Total Incl O&P
0010 **ANCHOR BOLTS**									
0015 Installed in fresh grout in CMU bond beams or filled cores, no templates									
0020 Hooked, with nut and washer, 1/2" diameter, 8" long	1 Bric	132	.061	Ea.	1.62	3.15		4.77	6.90
0030 12" long		131	.061		1.80	3.18		4.98	7.15
0040 5/8" diameter, 8" long		129	.062		3.39	3.23		6.62	8.95
0050 12" long		127	.063		4.18	3.28		7.46	9.90
0060 3/4" diameter, 8" long		127	.063		4.18	3.28		7.46	9.90
0070 12" long		125	.064		5.20	3.33		8.53	11.15

04 05 19.16 Masonry Anchors

	Crew	Daily Output	Labor-Hours	Unit	Material	Labor	Equipment	Total	Total Incl O&P
0010 **MASONRY ANCHORS**									
0020 For brick veneer, galv., corrugated, 7/8" x 7", 22 ga.	1 Bric	10.50	.762	C	16.70	39.50		56.20	82.50
0100 24 ga.		10.50	.762		10.35	39.50		49.85	75.50
0150 16 ga.		10.50	.762		31.50	39.50		71	99
0200 Buck anchors, galv., corrugated, 16 ga., 2" bend, 8" x 2"		10.50	.762		65	39.50		104.50	136
0250 8" x 3"		10.50	.762		69.50	39.50		109	141
0300 Adjustable, rectangular, 4-1/8" wide									
0350 Anchor and tie, 3/16" wire, mill galv.									
0400 2-3/4" eye, 3-1/4" tie	1 Bric	1.05	7.619	M	475	395		870	1,175
0500 4-3/4" tie		1.05	7.619		515	395		910	1,200
0520 5-1/2" tie		1.05	7.619		555	395		950	1,250
0550 4-3/4" eye, 3-1/4" tie		1.05	7.619		520	395		915	1,200
0570 4-3/4" tie		1.05	7.619		550	395		945	1,250
0580 5-1/2" tie		1.05	7.619		610	395		1,005	1,325
0660 Cavity wall, Z-type, galvanized, 6" long, 1/8" diam.		10.50	.762	C	25	39.50		64.50	91.50
0670 3/16" diameter		10.50	.762		31.50	39.50		71	98.50
0680 1/4" diameter		10.50	.762		40.50	39.50		80	109
0850 8" long, 3/16" diameter		10.50	.762		27	39.50		66.50	94

04 05 19.16 Masonry Anchors	Crew	Daily Output	Labor-Hours	Unit	Material	2020 Bare Costs Labor	Equipment	Total	Total Incl O&P
0855	1/4" diameter	1 Bric	10.50	.762	C	49	39.50		88.50	118
1000	Rectangular type, galvanized, 1/4" diameter, 2" x 6"		10.50	.762		80	39.50		119.50	152
1050	4" x 6"		10.50	.762		91.50	39.50		131	164
1100	3/16" diameter, 2" x 6"		10.50	.762		54.50	39.50		94	124
1150	4" x 6"		10.50	.762		55	39.50		94.50	125
1200	Mesh wall tie, 1/2" mesh, hot dip galvanized									
1400	16 ga., 12" long, 3" wide	1 Bric	9	.889	C	92.50	46.50		139	177
1420	6" wide		9	.889		135	46.50		181.50	224
1440	12" wide		8.50	.941		214	49		263	315
1500	Rigid partition anchors, plain, 8" long, 1" x 1/8"		10.50	.762		241	39.50		280.50	330
1550	1" x 1/4"		10.50	.762		325	39.50		364.50	420
1580	1-1/2" x 1/8"		10.50	.762		261	39.50		300.50	350
1600	1-1/2" x 1/4"		10.50	.762		340	39.50		379.50	440
1650	2" x 1/8"		10.50	.762		330	39.50		369.50	430
1700	2" x 1/4"		10.50	.762		445	39.50		484.50	555
2000	Column flange ties, wire, galvanized									
2300	3/16" diameter, up to 3" wide	1 Bric	10.50	.762	C	88	39.50		127.50	161
2350	To 5" wide		10.50	.762		96	39.50		135.50	170
2400	To 7" wide		10.50	.762		104	39.50		143.50	178
2600	To 9" wide		10.50	.762		110	39.50		149.50	185
2650	1/4" diameter, up to 3" wide		10.50	.762		112	39.50		151.50	187
2700	To 5" wide		10.50	.762		133	39.50		172.50	211
2800	To 7" wide		10.50	.762		150	39.50		189.50	229
2850	To 9" wide		10.50	.762		162	39.50		201.50	243
2900	For hot dip galvanized, add					35%				
4000	Channel slots, 1-3/8" x 1/2" x 8"									
4100	12 ga., plain	1 Bric	10.50	.762	C	244	39.50		283.50	335
4150	16 ga., galvanized	"	10.50	.762	"	155	39.50		194.50	234
4200	Channel slot anchors									
4300	16 ga., galvanized, 1-1/4" x 3-1/2"				C	61.50			61.50	67.50
4350	1-1/4" x 5-1/2"					70.50			70.50	77.50
4400	1-1/4" x 7-1/2"					80			80	88
4500	1/8" plain, 1-1/4" x 3-1/2"					160			160	176
4550	1-1/4" x 5-1/2"					171			171	188
4600	1-1/4" x 7-1/2"					186			186	205
4700	For corrugation, add					86			86	94.50
4750	For hot dip galvanized, add					35%				
5000	Dowels									
5100	Plain, 1/4" diameter, 3" long				C	54.50			54.50	60
5150	4" long					60.50			60.50	66.50
5200	6" long					74			74	81
5300	3/8" diameter, 3" long					72			72	79
5350	4" long					89.50			89.50	98.50
5400	6" long					103			103	113
5500	1/2" diameter, 3" long					105			105	115
5550	4" long					124			124	137
5600	6" long					161			161	177
5700	5/8" diameter, 3" long					144			144	158
5750	4" long					177			177	195
5800	6" long					243			243	267
6000	3/4" diameter, 3" long					179			179	197
6100	4" long					227			227	249
6150	6" long					325			325	355

04 05 19 – Masonry Anchorage and Reinforcing

04 05 19.16 Masonry Anchors	Crew	Daily Output	Labor-Hours	Unit	Material	2020 Bare Costs Labor	Equipment	Total	Total Incl O&P
6300	For hot dip galvanized, add				C	35%				

04 05 19.26 Masonry Reinforcing Bars

04 05 19.26 Masonry Reinforcing Bars		Crew	Daily Output	Labor-Hours	Unit	Material	2020 Bare Costs Labor	Equipment	Total	Total Incl O&P
0010	**MASONRY REINFORCING BARS**									
0015	Steel bars A615, placed horiz., #3 & #4 bars	1 Bric	450	.018	Lb.	.56	.93		1.49	2.12
0020	#5 & #6 bars		800	.010		.56	.52		1.08	1.46
0050	Placed vertical, #3 & #4 bars		350	.023		.56	1.19		1.75	2.54
0060	#5 & #6 bars		650	.012		.56	.64		1.20	1.66
0200	Joint reinforcing, regular truss, to 6" wide, mill std galvanized		30	.267	C.L.F.	23	13.90		36.90	48
0250	12" wide		20	.400		29.50	21		50.50	66
0400	Cavity truss with drip section, to 6" wide		30	.267		23	13.90		36.90	48
0450	12" wide		20	.400		26.50	21		47.50	62.50
0500	Joint reinforcing, ladder type, mill std galvanized									
0600	9 ga. sides, 9 ga. ties, 4" wall	1 Bric	30	.267	C.L.F.	24	13.90		37.90	48.50
0650	6" wall		30	.267		22	13.90		35.90	46.50
0700	8" wall		25	.320		26.50	16.65		43.15	56
0750	10" wall		20	.400		24.50	21		45.50	60.50
0800	12" wall		20	.400		26	21		47	62
1000	Truss type									
1100	9 ga. sides, 9 ga. ties, 4" wall	1 Bric	30	.267	C.L.F.	24.50	13.90		38.40	49.50
1150	6" wall		30	.267		26	13.90		39.90	51
1200	8" wall		25	.320		28	16.65		44.65	58
1250	10" wall		20	.400		24.50	21		45.50	60
1300	12" wall		20	.400		25	21		46	61
1500	3/16" sides, 9 ga. ties, 4" wall		30	.267		28	13.90		41.90	53.50
1550	6" wall		30	.267		36	13.90		49.90	62
1600	8" wall		25	.320		32.50	16.65		49.15	62.50
1650	10" wall		20	.400		39	21		60	76.50
1700	12" wall		20	.400		38.50	21		59.50	75.50
2000	3/16" sides, 3/16" ties, 4" wall		30	.267		40	13.90		53.90	66.50
2050	6" wall		30	.267		41.50	13.90		55.40	68
2100	8" wall		25	.320		40	16.65		56.65	71
2150	10" wall		20	.400		45	21		66	83
2200	12" wall		20	.400		46.50	21		67.50	85
2500	Cavity truss type, galvanized									
2600	9 ga. sides, 9 ga. ties, 4" wall	1 Bric	25	.320	C.L.F.	37	16.65		53.65	68
2650	6" wall		25	.320		36	16.65		52.65	66.50
2700	8" wall		20	.400		51	21		72	89.50
2750	10" wall		15	.533		43.50	28		71.50	93
2800	12" wall		15	.533		49	28		77	99
3000	3/16" sides, 9 ga. ties, 4" wall		25	.320		36.50	16.65		53.15	67
3050	6" wall		25	.320		36.50	16.65		53.15	67.50
3100	8" wall		20	.400		56	21		77	95
3150	10" wall		15	.533		40	28		68	89
3200	12" wall		15	.533		41.50	28		69.50	90.50
3500	For hot dip galvanizing, add				Ton	510			510	560

04 05 23 – Masonry Accessories

04 05 23.13 Masonry Control and Expansion Joints

04 05 23.13 Masonry Control and Expansion Joints		Crew	Daily Output	Labor-Hours	Unit	Material	2020 Bare Costs Labor	Equipment	Total	Total Incl O&P
0010	**MASONRY CONTROL AND EXPANSION JOINTS**									
0020	Rubber, for double wythe 8" minimum wall (Brick/CMU)	1 Bric	400	.020	L.F.	2.46	1.04		3.50	4.39
0025	"T" shaped		320	.025		1.22	1.30		2.52	3.44
0030	Cross-shaped for CMU units		280	.029		1.66	1.49		3.15	4.23
0050	PVC, for double wythe 8" minimum wall (Brick/CMU)		400	.020		1.70	1.04		2.74	3.55

04 05 Common Work Results for Masonry

04 05 23 – Masonry Accessories

04 05 23.13 Masonry Control and Expansion Joints

04 05 23.13 Masonry Control and Expansion Joints	Crew	Daily Output	Labor-Hours	Unit	Material	2020 Bare Costs Labor	2020 Bare Costs Equipment	Total	Total Incl O&P
0120	"T" shaped	1 Bric	320	.025	L.F.	.92	1.30		2.22	3.11
0160	Cross-shaped for CMU units	↓	280	.029	↓	1.14	1.49		2.63	3.65

04 05 23.19 Masonry Cavity Drainage, Weepholes, and Vents

		Crew	Daily Output	Labor-Hours	Unit	Material	2020 Bare Costs Labor	2020 Bare Costs Equipment	Total	Total Incl O&P
0010	**MASONRY CAVITY DRAINAGE, WEEPHOLES, AND VENTS**									
0020	Extruded aluminum, 4" deep, 2-3/8" x 8-1/8"	1 Bric	30	.267	Ea.	41	13.90		54.90	67.50
0050	5" x 8-1/8"		25	.320		53	16.65		69.65	85.50
0100	2-1/4" x 25"		25	.320		89	16.65		105.65	125
0150	5" x 16-1/2"		22	.364		61.50	18.95		80.45	98.50
0175	5" x 24"		22	.364		93	18.95		111.95	133
0200	6" x 16-1/2"		22	.364		99.50	18.95		118.45	140
0250	7-3/4" x 16-1/2"	↓	20	.400		92.50	21		113.50	136
0400	For baked enamel finish, add					35%				
0500	For cast aluminum, painted, add					60%				
1000	Stainless steel ventilators, 6" x 6"	1 Bric	25	.320		230	16.65		246.65	280
1050	8" x 8"		24	.333		257	17.35		274.35	310
1100	12" x 12"		23	.348		289	18.10		307.10	350
1150	12" x 6"		24	.333		279	17.35		296.35	335
1200	Foundation block vent, galv., 1-1/4" thk, 8" high, 16" long, no damper	↓	30	.267		17.45	13.90		31.35	41.50
1250	For damper, add				↓	3.34			3.34	3.67
1450	Drainage and ventilation fabric	2 Bric	1450	.011	S.F.	.82	.57		1.39	1.83

04 05 23.95 Wall Plugs

		Crew	Daily Output	Labor-Hours	Unit	Material	2020 Bare Costs Labor	2020 Bare Costs Equipment	Total	Total Incl O&P
0010	**WALL PLUGS** (for nailing to brickwork)									
0020	25 ga., galvanized, plain	1 Bric	10.50	.762	C	26.50	39.50		66	93
0050	Wood filled	"	10.50	.762	"	73	39.50		112.50	145

04 21 Clay Unit Masonry

04 21 13 – Brick Masonry

04 21 13.13 Brick Veneer Masonry

		Crew	Daily Output	Labor-Hours	Unit	Material	2020 Bare Costs Labor	2020 Bare Costs Equipment	Total	Total Incl O&P
0010	**BRICK VENEER MASONRY**, T.L. lots, excl. scaff., grout & reinforcing R042110-20									
0015	Material costs incl. 3% brick and 25% mortar waste									
2000	Standard, sel. common, 4" x 2-2/3" x 8" (6.75/S.F.) R042110-50	D-8	230	.174	S.F.	4.52	8.35		12.87	18.45
2020	Red, 4" x 2-2/3" x 8", running bond		220	.182		4.86	8.70		13.56	19.45
2050	Full header every 6th course (7.88/S.F.)		185	.216		5.65	10.35		16	23
2100	English, full header every 2nd course (10.13/S.F.)		140	.286		7.25	13.70		20.95	30
2150	Flemish, alternate header every course (9.00/S.F.)		150	.267		6.45	12.80		19.25	27.50
2200	Flemish, alt. header every 6th course (7.13/S.F.)		205	.195		5.15	9.35		14.50	21
2250	Full headers throughout (13.50/S.F.)		105	.381		9.65	18.25		27.90	40
2300	Rowlock course (13.50/S.F.)		100	.400		9.65	19.15		28.80	41.50
2350	Rowlock stretcher (4.50/S.F.)		310	.129		3.27	6.20		9.47	13.60
2400	Soldier course (6.75/S.F.)		200	.200		4.86	9.60		14.46	21
2450	Sailor course (4.50/S.F.)		290	.138		3.27	6.60		9.87	14.30
2600	Buff or gray face, running bond (6.75/S.F.)		220	.182		5.15	8.70		13.85	19.75
2700	Glazed face brick, running bond		210	.190		14	9.15		23.15	30
2750	Full header every 6th course (7.88/S.F.)		170	.235		16.30	11.30		27.60	36
3000	Jumbo, 6" x 4" x 12" running bond (3.00/S.F.)		435	.092		5.40	4.41		9.81	13
3050	Norman, 4" x 2-2/3" x 12" running bond (4.5/S.F.)		320	.125		6.65	6		12.65	17.05
3100	Norwegian, 4" x 3-1/5" x 12" (3.75/S.F.)		375	.107		6.05	5.10		11.15	14.90
3150	Economy, 4" x 4" x 8" (4.50/S.F.)		310	.129		4.63	6.20		10.83	15.10
3200	Engineer, 4" x 3-1/5" x 8" (5.63/S.F.)		260	.154		3.89	7.35		11.24	16.20
3250	Roman, 4" x 2" x 12" (6.00/S.F.)	↓	250	.160		7.65	7.65		15.30	21

For customer support on your Commercial Renovation Costs with RSMeans data, call 800.448.8182.

89

04 21 13.13 Brick Veneer Masonry

		Crew	Daily Output	Labor-Hours	Unit	Material	2020 Bare Costs Labor	Equipment	Total	Total Incl O&P
3300	S.C.R., 6" x 2-2/3" x 12" (4.50/S.F.)	D-8	310	.129	S.F.	6.30	6.20		12.50	16.95
3350	Utility, 4" x 4" x 12" (3.00/S.F.)	↓	360	.111		5.40	5.35		10.75	14.55
3360	For less than truck load lots, add					.10%				
3400	For cavity wall construction, add						15%			
3450	For stacked bond, add						10%			
3500	For interior veneer construction, add						15%			
3550	For curved walls, add				↓		30%			
9000	Minimum labor/equipment charge	D-1	2	8	Job		375		375	605

04 21 13.14 Thin Brick Veneer

		Crew	Daily Output	Labor-Hours	Unit	Material	2020 Bare Costs Labor	Equipment	Total	Total Incl O&P
0010	**THIN BRICK VENEER**									
0015	Material costs incl. 3% brick and 25% mortar waste									
0020	On & incl. metal panel support sys, modular, 2-2/3" x 5/8" x 8", red	D-7	92	.174	S.F.	9.70	7.70		17.40	22.50
0100	Closure, 4" x 5/8" x 8"		110	.145		9.40	6.40		15.80	20.50
0110	Norman, 2-2/3" x 5/8" x 12"		110	.145		9.65	6.40		16.05	20.50
0120	Utility, 4" x 5/8" x 12"		125	.128		9.10	5.65		14.75	18.85
0130	Emperor, 4" x 3/4" x 16"		175	.091		10.30	4.04		14.34	17.65
0140	Super emperor, 8" x 3/4" x 16"	↓	195	.082	↓	11.15	3.62		14.77	17.95
0150	For L shaped corners with 4" return, add				L.F.	9.25			9.25	10.20
0200	On masonry/plaster back-up, modular, 2-2/3" x 5/8" x 8", red	D-7	137	.117	S.F.	4.66	5.15		9.81	13.25
0210	Closure, 4" x 5/8" x 8"		165	.097		4.37	4.28		8.65	11.50
0220	Norman, 2-2/3" x 5/8" x 12"		165	.097		4.61	4.28		8.89	11.75
0230	Utility, 4" x 5/8" x 12"		185	.086		4.09	3.82		7.91	10.50
0240	Emperor, 4" x 3/4" x 16"		260	.062		5.25	2.72		7.97	10.05
0250	Super emperor, 8" x 3/4" x 16"	↓	285	.056	↓	6.15	2.48		8.63	10.65
0260	For L shaped corners with 4" return, add				L.F.	10			10	11
0270	For embedment into pre-cast concrete panels, add				S.F.	14.40			14.40	15.85

04 21 13.15 Chimney

		Crew	Daily Output	Labor-Hours	Unit	Material	2020 Bare Costs Labor	Equipment	Total	Total Incl O&P
0010	**CHIMNEY**, excludes foundation, scaffolding, grout and reinforcing									
0100	Brick, 16" x 16", 8" flue	D-1	18.20	.879	V.L.F.	26.50	41		67.50	95.50
0150	16" x 20" with one 8" x 12" flue		16	1		42.50	47		89.50	123
0200	16" x 24" with two 8" x 8" flues		14	1.143		63	53.50		116.50	156
0250	20" x 20" with one 12" x 12" flue		13.70	1.168		49.50	55		104.50	143
0300	20" x 24" with two 8" x 12" flues		12	1.333		71	62.50		133.50	179
0350	20" x 32" with two 12" x 12" flues	↓	10	1.600	↓	87	75		162	217

04 21 13.18 Columns

		Crew	Daily Output	Labor-Hours	Unit	Material	2020 Bare Costs Labor	Equipment	Total	Total Incl O&P
0010	**COLUMNS**, solid, excludes scaffolding, grout and reinforcing									
0050	Brick, 8" x 8", 9 brick/V.L.F.	D-1	56	.286	V.L.F.	6.30	13.40		19.70	28.50
0100	12" x 8", 13.5 brick/V.L.F.		37	.432		9.45	20.50		29.95	43.50
0200	12" x 12", 20 brick/V.L.F.		25	.640		14	30		44	64
0300	16" x 12", 27 brick/V.L.F.		19	.842		18.90	39.50		58.40	85
0400	16" x 16", 36 brick/V.L.F.		14	1.143		25	53.50		78.50	114
0500	20" x 16", 45 brick/V.L.F.		11	1.455		31.50	68		99.50	145
0600	20" x 20", 56 brick/V.L.F.		9	1.778		39	83.50		122.50	178
0700	24" x 20", 68 brick/V.L.F.		7	2.286		47.50	107		154.50	226
0800	24" x 24", 81 brick/V.L.F.		6	2.667		56.50	125		181.50	265
1000	36" x 36", 182 brick/V.L.F.		3	5.333	↓	127	250		377	545
9000	Minimum labor/equipment charge	↓	2	8	Job		375		375	605

04 21 13.30 Oversized Brick

		Crew	Daily Output	Labor-Hours	Unit	Material	2020 Bare Costs Labor	Equipment	Total	Total Incl O&P
0010	**OVERSIZED BRICK**, excludes scaffolding, grout and reinforcing									
0100	Veneer, 4" x 2.25" x 16"	D-8	387	.103	S.F.	5.05	4.95		10	13.55
0102	8" x 2.25" x 16", multicell		265	.151		17.55	7.25		24.80	31
0105	4" x 2.75" x 16"	↓	412	.097	↓	5.75	4.65		10.40	13.85

04 21 Clay Unit Masonry

04 21 13 – Brick Masonry

04 21 13.30 Oversized Brick

		Crew	Daily Output	Labor- Hours	Unit	Material	2020 Bare Costs Labor	Equipment	Total	Total Incl O&P
0107	8" x 2.75" x 16", multicell	D-8	295	.136	S.F.	17.50	6.50		24	30
0110	4" x 4" x 16"		460	.087		3.80	4.17		7.97	10.95
0120	4" x 8" x 16"		533	.075		5.10	3.60		8.70	11.45
0122	4" x 8" x 16" multicell		327	.122		15.70	5.85		21.55	27
0125	Loadbearing, 6" x 4" x 16", grouted and reinforced		387	.103		11.60	4.95		16.55	21
0130	8" x 4" x 16", grouted and reinforced		327	.122		12.05	5.85		17.90	22.50
0132	10" x 4" x 16", grouted and reinforced		327	.122		27.50	5.85		33.35	40
0135	6" x 8" x 16", grouted and reinforced		440	.091		15.05	4.36		19.41	23.50
0140	8" x 8" x 16", grouted and reinforced		400	.100		16.05	4.79		20.84	25.50
0145	Curtainwall/reinforced veneer, 6" x 4" x 16"		387	.103		16.85	4.95		21.80	26.50
0150	8" x 4" x 16"		327	.122		19.95	5.85		25.80	31.50
0152	10" x 4" x 16"		327	.122		30.50	5.85		36.35	43
0155	6" x 8" x 16"		440	.091		21	4.36		25.36	30
0160	8" x 8" x 16"		400	.100		28.50	4.79		33.29	39.50
0200	For 1 to 3 slots in face, add					15%				
0210	For 4 to 7 slots in face, add					25%				
0220	For bond beams, add					20%				
0230	For bullnose shapes, add					20%				
0240	For open end knockout, add					10%				
0250	For white or gray color group, add					10%				
0260	For 135 degree corner, add					250%				

04 21 13.35 Common Building Brick

		Crew	Daily Output	Labor- Hours	Unit	Material	2020 Bare Costs Labor	Equipment	Total	Total Incl O&P
0010	**COMMON BUILDING BRICK**, C62, T.L. lots, material only R042110-20									
0020	Standard				M	590			590	650
0050	Select				"	555			555	615

04 21 13.40 Structural Brick

		Crew	Daily Output	Labor- Hours	Unit	Material	2020 Bare Costs Labor	Equipment	Total	Total Incl O&P
0010	**STRUCTURAL BRICK** C652, Grade SW, incl. mortar, scaffolding not incl.									
0100	Standard unit, 4-5/8" x 2-3/4" x 9-5/8"	D-8	245	.163	S.F.	5.25	7.85		13.10	18.45
0120	Bond beam		225	.178		4.44	8.50		12.94	18.65
0140	V cut bond beam		225	.178		4.34	8.50		12.84	18.50
0160	Stretcher quoin, 5-5/8" x 2-3/4" x 9-5/8"		245	.163		8.30	7.85		16.15	22
0180	Corner quoin		245	.163		8.25	7.85		16.10	21.50
0200	Corner, 45 degree, 4-5/8" x 2-3/4" x 10-7/16"		235	.170		8.25	8.15		16.40	22.50

04 21 13.45 Face Brick

		Crew	Daily Output	Labor- Hours	Unit	Material	2020 Bare Costs Labor	Equipment	Total	Total Incl O&P
0010	**FACE BRICK** Material Only, C216, T.L. lots R042110-20									
0300	Standard modular, 4" x 2-2/3" x 8"				M	605			605	665
0450	Economy, 4" x 4" x 8"					900			900	990
0510	Economy, 4" x 4" x 12"					1,425			1,425	1,550
0550	Jumbo, 6" x 4" x 12"					1,625			1,625	1,800
0610	Jumbo, 8" x 4" x 12"					1,625			1,625	1,800
0650	Norwegian, 4" x 3-1/5" x 12"					1,475			1,475	1,600
0710	Norwegian, 6" x 3-1/5" x 12"					1,725			1,725	1,900
0850	Standard glazed, plain colors, 4" x 2-2/3" x 8"					1,925			1,925	2,100
1000	Deep trim shades, 4" x 2-2/3" x 8"					2,350			2,350	2,600
1080	Jumbo utility, 4" x 4" x 12"					1,650			1,650	1,800
1120	4" x 8" x 8"					1,975			1,975	2,175
1140	4" x 8" x 16"					6,250			6,250	6,875
1260	Engineer, 4" x 3-1/5" x 8"					575			575	635
1350	King, 4" x 2-3/4" x 10"					600			600	660
1770	Standard modular, double glazed, 4" x 2-2/3" x 8"					2,950			2,950	3,250
1850	Jumbo, colored glazed ceramic, 6" x 4" x 12"					3,050			3,050	3,350
2050	Jumbo utility, glazed, 4" x 4" x 12"					6,125			6,125	6,725

For customer support on your Commercial Renovation Costs with RSMeans data, call 800.448.8182.

91

04 21 Clay Unit Masonry

04 21 13 – Brick Masonry

04 21 13.45 Face Brick	Crew	Daily Output	Labor-Hours	Unit	Material	2020 Bare Costs Labor	Equipment	Total	Total Incl O&P
2100	4" x 8" x 8"				M	7,200			7,200	7,925
2150	4" x 16" x 8"					7,875			7,875	8,675
3050	Used brick					480			480	530
3150	Add for brick to match existing work, minimum					5%			
3200	Maximum					50%			

04 21 26 – Glazed Structural Clay Tile Masonry

04 21 26.10 Structural Facing Tile

		Crew	Daily Output	Labor-Hours	Unit	Material	2020 Bare Costs Labor	Equipment	Total	Total Incl O&P
0010	**STRUCTURAL FACING TILE**, std. colors, excl. scaffolding, grout, reinforcing									
0020	6T series, 5-1/3" x 12", 2.3 pieces per S.F., glazed 1 side, 2" thick	D-8	225	.178	S.F.	9	8.50		17.50	23.50
0100	4" thick		220	.182		13.50	8.70		22.20	29
0150	Glazed 2 sides		195	.205		16.40	9.85		26.25	34
0250	6" thick		210	.190		17.25	9.15		26.40	34
0300	Glazed 2 sides		185	.216		21	10.35		31.35	40
0400	8" thick		180	.222		23	10.65		33.65	42.50
0500	Special shapes, group 1		400	.100	Ea.	7.80	4.79		12.59	16.30
0550	Group 2		375	.107		12.75	5.10		17.85	22.50
0600	Group 3		350	.114		17.50	5.50		23	28
0650	Group 4		325	.123		36	5.90		41.90	49.50
0700	Group 5		300	.133		43.50	6.40		49.90	58
0750	Group 6		275	.145		59	6.95		65.95	76.50
1000	Fire rated, 4" thick, 1 hr. rating		210	.190	S.F.	19	9.15		28.15	36
1050	2 hr. rating		170	.235		29	11.30		40.30	50
1300	Acoustic, 4" thick		210	.190		40	9.15		49.15	59
1310	6" thick		170	.235		41.50	11.30		52.80	63.50
1320	8" thick		145	.276		43.50	13.20		56.70	69.50
1350	8T acoustic, 4" thick		265	.151		20	7.25		27.25	33.50
1360	6" thick		256	.156		29	7.50		36.50	44
1370	8" thick		208	.192		31.50	9.20		40.70	49.50
2000	8W series, 8" x 16", 1.125 pieces per S.F.									
2050	2" thick, glazed 1 side	D-8	360	.111	S.F.	9.95	5.35		15.30	19.55
2100	4" thick, glazed 1 side		345	.116		14.35	5.55		19.90	25
2150	Glazed 2 sides		325	.123		15.30	5.90		21.20	26.50
2200	6" thick, glazed 1 side		330	.121		28	5.80		33.80	40.50
2210	Glazed 2 sides		250	.160		30.50	7.65		38.15	46
2250	8" thick, glazed 1 side		310	.129		30.50	6.20		36.70	43.50
2260	Glazed 2 sides		220	.182		35	8.70		43.70	52.50
2500	Special shapes, group 1		300	.133	Ea.	18.35	6.40		24.75	30.50
2550	Group 2		280	.143		24.50	6.85		31.35	38
2600	Group 3		260	.154		26	7.35		33.35	40.50
2650	Group 4		250	.160		54	7.65		61.65	71.50
2700	Group 5		240	.167		47.50	8		55.50	65
2750	Group 6		230	.174		104	8.35		112.35	127
2900	Fire rated, 2" thick, glazed 1 side		360	.111	S.F.	13.55	5.35		18.90	23.50
3000	4" thick, glazed 1 side		345	.116		16.75	5.55		22.30	27.50
3020	Glazed 2 sides		345	.116		23	5.55		28.55	34.50
3040	6" thick, glazed 1 side		330	.121		26.50	5.80		32.30	39
3045	Glazed 2 sides		250	.160		24	7.65		31.65	39
3047	8" thick, glazed 1 side		310	.129		30	6.20		36.20	43
3048	Glazed 2 sides		220	.182		32	8.70		40.70	49
3050	Special shapes, group 1, fire rated		300	.133		20.50	6.40		26.90	33.50
3060	Group 2		280	.143		29.50	6.85		36.35	43.50
3070	Group 3		260	.154		28.50	7.35		35.85	43.50

For customer support on your Commercial Renovation Costs with RSMeans data, call 800.448.8182.

04 21 Clay Unit Masonry

04 21 26 – Glazed Structural Clay Tile Masonry

	04 21 26.10 Structural Facing Tile	Crew	Daily Output	Labor-Hours	Unit	Material	2020 Bare Costs Labor	Equipment	Total	Total Incl O&P
3080	Group 4	D-8	250	.160	S.F.	49.50	7.65		57.15	67
3090	Group 5		240	.167		55	8		63	73.50
3100	Acoustic, 4" thick	↓	345	.116	↓	22.50	5.55		28.05	33.50
3120	4W series, 8" x 8", 2.25 pieces per S.F.									
3125	2" thick, glazed 1 side	D-8	360	.111	S.F.	9.75	5.35		15.10	19.35
3130	4" thick, glazed 1 side		345	.116		12.80	5.55		18.35	23
3135	Glazed 2 sides		325	.123		17.10	5.90		23	28.50
3140	6" thick, glazed 1 side		330	.121		17.35	5.80		23.15	28.50
3145	Glazed 2 sides		250	.160		22	7.65		29.65	36.50
3150	8" thick, glazed 1 side		310	.129	↓	26.50	6.20		32.70	39
3155	Special shapes, group 1		300	.133	Ea.	8.25	6.40		14.65	19.45
3160	Group 2	↓	280	.143	"	9.65	6.85		16.50	21.50
3200	For designer colors, add					25%				
3300	For epoxy mortar joints, add				S.F.	1.82			1.82	2
9000	Minimum labor/equipment charge	D-1	2	8	Job		375		375	605

04 21 29 – Terra Cotta Masonry

04 21 29.10 Terra Cotta Masonry Components

		Crew	Daily Output	Labor-Hours	Unit	Material	2020 Bare Costs Labor	Equipment	Total	Total Incl O&P
0010	**TERRA COTTA MASONRY COMPONENTS**									
0020	Coping, split type, not glazed, 9" wide	D-1	90	.178	L.F.	10.65	8.35		19	25
0100	13" wide		80	.200		16.30	9.40		25.70	33
0200	Coping, split type, glazed, 9" wide		90	.178		20.50	8.35		28.85	36
0250	13" wide	↓	80	.200	↓	24.50	9.40		33.90	42
0500	Partition or back-up blocks, scored, in C.L. lots									
0700	Non-load bearing 12" x 12", 3" thick, special order	D-8	550	.073	S.F.	19.40	3.49		22.89	27
0750	4" thick, standard		500	.080		6.80	3.83		10.63	13.65
0800	6" thick		450	.089		9.05	4.26		13.31	16.85
0850	8" thick		400	.100		11.25	4.79		16.04	20
1000	Load bearing, 12" x 12", 4" thick, in walls		500	.080		6.30	3.83		10.13	13.10
1050	In floors		750	.053		6.30	2.56		8.86	11.05
1200	6" thick, in walls		450	.089		9.75	4.26		14.01	17.65
1250	In floors		675	.059		9.75	2.84		12.59	15.35
1400	8" thick, in walls		400	.100		11.90	4.79		16.69	21
1450	In floors		575	.070		11.90	3.33		15.23	18.50
1600	10" thick, in walls, special order		350	.114		28	5.50		33.50	40
1650	In floors, special order		500	.080		28	3.83		31.83	37
1800	12" thick, in walls, special order		300	.133		28.50	6.40		34.90	42
1850	In floors, special order	↓	450	.089		28.50	4.26		32.76	38.50
2000	For reinforcing with steel rods, add to above				↓	15%	5%			
2025	Placed vertical, #3 & #4 bars	1 Bric	350	.023	Lb.	.56	1.19		1.75	2.54
2100	For smooth tile instead of scored, add				S.F.	2.94			2.94	3.23
2200	For L.C.L. quantities, add				"	10%	10%			
9000	Minimum labor/equipment charge	D-1	2	8	Job		375		375	605

04 21 29.20 Terra Cotta Tile

		Crew	Daily Output	Labor-Hours	Unit	Material	2020 Bare Costs Labor	Equipment	Total	Total Incl O&P
0010	**TERRA COTTA TILE**, on walls, dry set, 1/2" thick									
0100	Square, hexagonal or lattice shapes, unglazed	1 Tilf	135	.059	S.F.	5.20	2.93		8.13	10.30
0300	Glazed, plain colors		130	.062		7.90	3.05		10.95	13.45
0400	Intense colors	↓	125	.064	↓	9.25	3.17		12.42	15.10

For customer support on your Commercial Renovation Costs with RSMeans data, call 800.448.8182.

93

04 22 10 – Concrete Masonry Units

04 22 10.11 Autoclave Aerated Concrete Block		Crew	Daily Output	Labor-Hours	Unit	Material	2020 Bare Costs Labor	Equipment	Total	Total Incl O&P
0010	**AUTOCLAVE AERATED CONCRETE BLOCK**, excl. scaffolding, grout & reinforcing									
0050	Solid, 4" x 8" x 24", incl. mortar	G	D-8	600	.067	S.F.	1.55	3.20		4.75	6.85
0060	6" x 8" x 24"	G		600	.067		2.50	3.20		5.70	7.90
0070	8" x 8" x 24"	G		575	.070		3.22	3.33		6.55	8.95
0080	10" x 8" x 24"	G		575	.070		4	3.33		7.33	9.80
0090	12" x 8" x 24"	G		550	.073		4.83	3.49		8.32	10.95

04 22 10.14 Concrete Block, Back-Up

		Crew	Daily Output	Labor-Hours	Unit	Material	Labor	Equipment	Total	Total Incl O&P
0010	**CONCRETE BLOCK, BACK-UP**, C90, 2000 psi									
0020	Normal weight, 8" x 16" units, tooled joint 1 side									
0050	Not-reinforced, 2000 psi, 2" thick	D-8	475	.084	S.F.	1.68	4.04		5.72	8.35
0200	4" thick		460	.087		2.16	4.17		6.33	9.15
0300	6" thick		440	.091		2.72	4.36		7.08	10.05
0350	8" thick		400	.100		2.73	4.79		7.52	10.75
0400	10" thick		330	.121		3.25	5.80		9.05	12.95
0450	12" thick	D-9	310	.155		4.72	7.25		11.97	16.95
1000	Reinforced, alternate courses, 4" thick	D-8	450	.089		2.34	4.26		6.60	9.50
1100	6" thick		430	.093		2.91	4.46		7.37	10.40
1150	8" thick		395	.101		2.94	4.85		7.79	11.10
1200	10" thick		320	.125		3.43	6		9.43	13.45
1250	12" thick	D-9	300	.160		4.90	7.50		12.40	17.50
9000	Minimum labor/equipment charge	D-1	2	8	Job		375		375	605

04 22 10.16 Concrete Block, Bond Beam

		Crew	Daily Output	Labor-Hours	Unit	Material	Labor	Equipment	Total	Total Incl O&P
0010	**CONCRETE BLOCK, BOND BEAM**, C90, 2000 psi									
0020	Not including grout or reinforcing									
0125	Regular block, 6" thick	D-8	584	.068	L.F.	2.77	3.28		6.05	8.35
0130	8" high, 8" thick	"	565	.071		2.90	3.39		6.29	8.70
0150	12" thick	D-9	510	.094		4.09	4.41		8.50	11.65
0525	Lightweight, 6" thick	D-8	592	.068		2.90	3.24		6.14	8.45
0530	8" high, 8" thick	"	575	.070		3.41	3.33		6.74	9.15
0550	12" thick	D-9	520	.092		4.63	4.33		8.96	12.10
2000	Including grout and 2 #5 bars									
2100	Regular block, 8" high, 8" thick	D-8	300	.133	L.F.	5.35	6.40		11.75	16.25
2150	12" thick	D-9	250	.192		7.25	9		16.25	22.50
2500	Lightweight, 8" high, 8" thick	D-8	305	.131		5.90	6.30		12.20	16.60
2550	12" thick	D-9	255	.188		7.80	8.85		16.65	23
9000	Minimum labor/equipment charge	D-1	2	8	Job		375		375	605

04 22 10.19 Concrete Block, Insulation Inserts

		Crew	Daily Output	Labor-Hours	Unit	Material	Labor	Equipment	Total	Total Incl O&P
0010	**CONCRETE BLOCK, INSULATION INSERTS**									
0100	Styrofoam, plant installed, add to block prices									
0200	8" x 16" units, 6" thick				S.F.	1.25			1.25	1.38
0250	8" thick					1.57			1.57	1.73
0300	10" thick					1.47			1.47	1.62
0350	12" thick					1.85			1.85	2.04
0500	8" x 8" units, 8" thick					1.40			1.40	1.54
0550	12" thick					1.47			1.47	1.62

04 22 10.23 Concrete Block, Decorative

		Crew	Daily Output	Labor-Hours	Unit	Material	Labor	Equipment	Total	Total Incl O&P
0010	**CONCRETE BLOCK, DECORATIVE**, C90, 2000 psi									
0020	Embossed, simulated brick face									
0100	8" x 16" units, 4" thick	D-8	400	.100	S.F.	3.23	4.79		8.02	11.30
0200	8" thick		340	.118		3.25	5.65		8.90	12.70
0250	12" thick		300	.133		5.60	6.40		12	16.50
0400	Embossed both sides									

04 22 Concrete Unit Masonry

04 22 10 – Concrete Masonry Units

04 22 10.23 Concrete Block, Decorative

		Crew	Daily Output	Labor-Hours	Unit	Material	2020 Bare Costs Labor	Equipment	Total	Total Incl O&P
0500	8" thick	D-8	300	.133	S.F.	4.88	6.40		11.28	15.70
0550	12" thick	"	275	.145	"	6.10	6.95		13.05	17.95
1000	Fluted high strength									
1100	8" x 16" x 4" thick, flutes 1 side,	D-8	345	.116	S.F.	4.11	5.55		9.66	13.50
1150	Flutes 2 sides		335	.119		5.40	5.70		11.10	15.20
1200	8" thick		300	.133		5.85	6.40		12.25	16.80
1250	For special colors, add					.67			.67	.73
1400	Deep grooved, smooth face									
1450	8" x 16" x 4" thick	D-8	345	.116	S.F.	2.97	5.55		8.52	12.25
1500	8" thick	"	300	.133	"	4.48	6.40		10.88	15.30
2000	Formblock, incl. inserts & reinforcing									
2100	8" x 16" x 8" thick	D-8	345	.116	S.F.	4.22	5.55		9.77	13.65
2150	12" thick	"	310	.129	"	5.50	6.20		11.70	16.05
2500	Ground face									
2600	8" x 16" x 4" thick	D-8	345	.116	S.F.	4.86	5.55		10.41	14.35
2650	6" thick		325	.123		4.90	5.90		10.80	14.95
2700	8" thick		300	.133		5.25	6.40		11.65	16.10
2725	10" thick		280	.143		6.80	6.85		13.65	18.50
2750	12" thick	D-9	265	.181		6.15	8.50		14.65	20.50
2900	For special colors, add, minimum					15%				
2950	For special colors, add, maximum					45%				
4000	Slump block									
4100	4" face height x 16" x 4" thick	D-1	165	.097	S.F.	4.76	4.55		9.31	12.60
4150	6" thick		160	.100		6.25	4.69		10.94	14.50
4200	8" thick		155	.103		8	4.84		12.84	16.60
4250	10" thick		140	.114		14.10	5.35		19.45	24
4300	12" thick		130	.123		12.55	5.75		18.30	23
4400	6" face height x 16" x 6" thick		155	.103		5.80	4.84		10.64	14.20
4450	8" thick		150	.107		8.75	5		13.75	17.75
4500	10" thick		130	.123		14.90	5.75		20.65	26
4550	12" thick		120	.133		14.50	6.25		20.75	26
5000	Split rib profile units, 1" deep ribs, 8 ribs									
5100	8" x 16" x 4" thick	D-8	345	.116	S.F.	4.61	5.55		10.16	14.10
5150	6" thick		325	.123		5.25	5.90		11.15	15.35
5200	8" thick		300	.133		6.55	6.40		12.95	17.60
5225	10" thick		300	.133		7.05	6.40		13.45	18.10
5250	12" thick	D-9	275	.175		7.50	8.20		15.70	21.50
5400	For special deeper colors, 4" thick, add					1.32			1.32	1.45
5450	12" thick, add					1.41			1.41	1.55
5600	For white, 4" thick, add					1.32			1.32	1.45
5650	6" thick, add					1.39			1.39	1.53
5700	8" thick, add					1.48			1.48	1.63
5750	12" thick, add					1.47			1.47	1.62
6000	Split face									
6100	8" x 16" x 4" thick	D-8	350	.114	S.F.	3.90	5.50		9.40	13.15
6150	6" thick		325	.123		4.40	5.90		10.30	14.40
6200	8" thick		300	.133		5.60	6.40		12	16.55
6225	10" thick		275	.145		5.85	6.95		12.80	17.70
6250	12" thick	D-9	270	.178		6.25	8.35		14.60	20.50
6300	For scored, add					.39			.39	.43
6400	For special deeper colors, 4" thick, add					.63			.63	.70
6450	6" thick, add					.77			.77	.85
6500	8" thick, add					.81			.81	.89

For customer support on your Commercial Renovation Costs with RSMeans data, call 800.448.8182.

95

04 22 10.23 Concrete Block, Decorative

		Crew	Daily Output	Labor-Hours	Unit	Material	2020 Bare Costs Labor	Equipment	Total	Total Incl O&P
6550	12" thick, add				S.F.	.86			.86	.94
6650	For white, 4" thick, add					1.32			1.32	1.45
6700	6" thick, add					1.29			1.29	1.42
6750	8" thick, add					1.41			1.41	1.55
6800	12" thick, add					1.41			1.41	1.55
7000	Scored ground face, 2 to 5 scores									
7100	8" x 16" x 4" thick	D-8	340	.118	S.F.	9	5.65		14.65	19.05
7150	6" thick		310	.129		10.10	6.20		16.30	21
7200	8" thick		290	.138		11.45	6.60		18.05	23.50
7250	12" thick	D-9	265	.181		15.65	8.50		24.15	31
8000	Hexagonal face profile units, 8" x 16" units									
8100	4" thick, hollow	D-8	340	.118	S.F.	4.06	5.65		9.71	13.55
8200	Solid		340	.118		5.15	5.65		10.80	14.80
8300	6" thick, hollow		310	.129		4.37	6.20		10.57	14.80
8350	8" thick, hollow		290	.138		4.74	6.60		11.34	15.90
8500	For stacked bond, add						26%			
8550	For high rise construction, add per story	D-8	67.80	.590	M.S.F.		28.50		28.50	45.50
8600	For scored block, add					10%				
8650	For honed or ground face, per face, add				Ea.	1.23			1.23	1.35
8700	For honed or ground end, per end, add				"	1.36			1.36	1.49
8750	For bullnose block, add					10%				
8800	For special color, add					13%				
9000	Minimum labor/equipment charge	D-1	2	8	Job		375		375	605

04 22 10.24 Concrete Block, Exterior

		Crew	Daily Output	Labor-Hours	Unit	Material	2020 Bare Costs Labor	Equipment	Total	Total Incl O&P
0010	**CONCRETE BLOCK, EXTERIOR**, C90, 2000 psi									
0020	Reinforced alt courses, tooled joints 2 sides									
0100	Normal weight, 8" x 16" x 6" thick	D-8	395	.101	S.F.	2.56	4.85		7.41	10.65
0200	8" thick		360	.111		4.16	5.35		9.51	13.20
0250	10" thick		290	.138		4.50	6.60		11.10	15.65
0300	12" thick	D-9	250	.192		5.30	9		14.30	20.50
9000	Minimum labor/equipment charge	D-1	2	8	Job		375		375	605

04 22 10.26 Concrete Block Foundation Wall

		Crew	Daily Output	Labor-Hours	Unit	Material	2020 Bare Costs Labor	Equipment	Total	Total Incl O&P
0010	**CONCRETE BLOCK FOUNDATION WALL**, C90/C145									
0050	Normal-weight, cut joints, horiz joint reinf, no vert reinf.									
0200	Hollow, 8" x 16" x 6" thick	D-8	455	.088	S.F.	3.20	4.21		7.41	10.30
0250	8" thick		425	.094		3.23	4.51		7.74	10.85
0300	10" thick		350	.114		3.73	5.50		9.23	12.95
0350	12" thick	D-9	300	.160		5.20	7.50		12.70	17.85
0500	Solid, 8" x 16" block, 6" thick	D-8	440	.091		3.63	4.36		7.99	11.05
0550	8" thick	"	415	.096		4.38	4.62		9	12.25
0600	12" thick	D-9	350	.137		6.45	6.45		12.90	17.50
1000	Reinforced, #4 vert @ 48"									
1100	Hollow, 8" x 16" block, 4" thick	D-8	455	.088	S.F.	3.29	4.21		7.50	10.40
1125	6" thick		445	.090		4.25	4.31		8.56	11.65
1150	8" thick		415	.096		4.72	4.62		9.34	12.65
1200	10" thick		340	.118		5.65	5.65		11.30	15.35
1250	12" thick	D-9	290	.166		7.60	7.75		15.35	21
1500	Solid, 8" x 16" block, 6" thick	D-8	430	.093		3.64	4.46		8.10	11.20
1600	8" thick	"	405	.099		4.38	4.73		9.11	12.45
1650	12" thick	D-9	340	.141		6.45	6.60		13.05	17.80
9000	Minimum labor/equipment charge	D-1	2	8	Job		375		375	605

04 22 Concrete Unit Masonry

04 22 10 – Concrete Masonry Units

04 22 10.30 Concrete Block, Interlocking		Crew	Daily Output	Labor-Hours	Unit	Material	2020 Bare Costs Labor	Equipment	Total	Total Incl O&P
0010	**CONCRETE BLOCK, INTERLOCKING**									
0100	Not including grout or reinforcing									
0200	8" x 16" units, 2,000 psi, 8" thick	D-1	245	.065	S.F.	3.08	3.06		6.14	8.35
0300	12" thick	"	220	.073		4.57	3.41		7.98	10.55
0400	Including grout & reinforcing, 8" thick	D-4	245	.131		8.90	6.15	.77	15.82	20.50
0450	12" thick		220	.145		10.65	6.85	.86	18.36	23.50
0500	16" thick		185	.173		13	8.15	1.02	22.17	28.50
9000	Minimum labor/equipment charge	D-1	2	8	Job		375		375	605

04 22 10.32 Concrete Block, Lintels

04 22 10.32 Concrete Block, Lintels		Crew	Daily Output	Labor-Hours	Unit	Material	2020 Bare Costs Labor	Equipment	Total	Total Incl O&P
0010	**CONCRETE BLOCK, LINTELS**, C90, normal weight									
0100	Including grout and horizontal reinforcing									
0200	8" x 8" x 8", 1 #4 bar	D-4	300	.107	L.F.	4.54	5.05	.63	10.22	13.75
0250	2 #4 bars		295	.108		4.79	5.10	.64	10.53	14.15
0400	8" x 16" x 8", 1 #4 bar		275	.116		4.24	5.50	.69	10.43	14.20
0450	2 #4 bars		270	.119		4.49	5.60	.70	10.79	14.65
1000	12" x 8" x 8", 1 #4 bar		275	.116		6.05	5.50	.69	12.24	16.20
1100	2 #4 bars		270	.119		6.30	5.60	.70	12.60	16.60
1150	2 #5 bars		270	.119		6.55	5.60	.70	12.85	16.95
1200	2 #6 bars		265	.121		6.90	5.70	.71	13.31	17.50
1500	12" x 16" x 8", 1 #4 bar		250	.128		7.05	6.05	.75	13.85	18.25
1600	2 #3 bars		245	.131		7.10	6.15	.77	14.02	18.50
1650	2 #4 bars		245	.131		7.30	6.15	.77	14.22	18.75
1700	2 #5 bars		240	.133		7.60	6.30	.79	14.69	19.25
9000	Minimum labor/equipment charge	D-1	2	8	Job		375		375	605

04 22 10.33 Lintel Block

04 22 10.33 Lintel Block		Crew	Daily Output	Labor-Hours	Unit	Material	2020 Bare Costs Labor	Equipment	Total	Total Incl O&P
0010	**LINTEL BLOCK**									
3481	Lintel block 6" x 8" x 8"	D-1	300	.053	Ea.	1.37	2.50		3.87	5.55
3501	6" x 16" x 8"		275	.058		2.05	2.73		4.78	6.65
3521	8" x 8" x 8"		275	.058		1.29	2.73		4.02	5.85
3561	8" x 16" x 8"		250	.064		2.01	3		5.01	7.05

04 22 10.34 Concrete Block, Partitions

04 22 10.34 Concrete Block, Partitions		Crew	Daily Output	Labor-Hours	Unit	Material	2020 Bare Costs Labor	Equipment	Total	Total Incl O&P
0010	**CONCRETE BLOCK, PARTITIONS**, excludes scaffolding									
1000	Lightweight block, tooled joints, 2 sides, hollow									
1100	Not reinforced, 8" x 16" x 4" thick	D-8	440	.091	S.F.	2	4.36		6.36	9.25
1150	6" thick		410	.098		2.77	4.68		7.45	10.60
1200	8" thick		385	.104		3.75	4.98		8.73	12.15
1250	10" thick		370	.108		4.18	5.20		9.38	12.95
1300	12" thick	D-9	350	.137		4.28	6.45		10.73	15.10
1500	Reinforced alternate courses, 4" thick	D-8	435	.092		2.18	4.41		6.59	9.50
1600	6" thick		405	.099		2.93	4.73		7.66	10.85
1650	8" thick		380	.105		3.95	5.05		9	12.50
1700	10" thick		365	.110		4.36	5.25		9.61	13.30
1750	12" thick	D-9	345	.139		4.48	6.55		11.03	15.45
4000	Regular block, tooled joints, 2 sides, hollow									
4100	Not reinforced, 8" x 16" x 4" thick	D-8	430	.093	S.F.	2.06	4.46		6.52	9.45
4150	6" thick		400	.100		2.62	4.79		7.41	10.65
4200	8" thick		375	.107		2.63	5.10		7.73	11.15
4250	10" thick		360	.111		3.15	5.35		8.50	12.05
4300	12" thick	D-9	340	.141		4.62	6.60		11.22	15.80
4500	Reinforced alternate courses, 8" x 16" x 4" thick	D-8	425	.094		2.25	4.51		6.76	9.75
4550	6" thick		395	.101		2.81	4.85		7.66	10.95

For customer support on your Commercial Renovation Costs with RSMeans data, call 800.448.8182.

97

04 22 Concrete Unit Masonry

04 22 10 – Concrete Masonry Units

04 22 10.34 Concrete Block, Partitions	Crew	Daily Output	Labor-Hours	Unit	Material	2020 Bare Costs Labor	Equipment	Total	Total Incl O&P
4600	8" thick	D-8	370	.108	S.F.	2.84	5.20		8.04	11.45
4650	10" thick	↓	355	.113		4.67	5.40		10.07	13.90
4700	12" thick	D-9	335	.143		4.81	6.70		11.51	16.15
9000	Minimum labor/equipment charge	D-1	2	8	Job		375		375	605

04 22 10.38 Concrete Brick

0010	**CONCRETE BRICK**, C55, grade N, type 1									
0100	Regular, 4" x 2-1/4" x 8"	D-8	660	.061	Ea.	.60	2.91		3.51	5.35
0125	Rusticated, 4" x 2-1/4" x 8"		660	.061		.67	2.91		3.58	5.45
0150	Frog, 4" x 2-1/4" x 8"		660	.061		.65	2.91		3.56	5.40
0200	Double, 4" x 4-7/8" x 8"	↓	535	.075	↓	1.05	3.58		4.63	6.95

04 22 10.44 Glazed Concrete Block

0010	**GLAZED CONCRETE BLOCK** C744									
0100	Single face, 8" x 16" units, 2" thick	D-8	360	.111	S.F.	12.25	5.35		17.60	22
0200	4" thick		345	.116		12.55	5.55		18.10	23
0250	6" thick		330	.121		12.80	5.80		18.60	23.50
0300	8" thick		310	.129		15.35	6.20		21.55	27
0350	10" thick	↓	295	.136		16.90	6.50		23.40	29
0400	12" thick	D-9	280	.171		18.15	8.05		26.20	33
0700	Double face, 8" x 16" units, 4" thick	D-8	340	.118		16.80	5.65		22.45	27.50
0750	6" thick		320	.125		21.50	6		27.50	33
0800	8" thick		300	.133	↓	23	6.40		29.40	36
1000	Jambs, bullnose or square, single face, 8" x 16", 2" thick		315	.127	Ea.	20	6.10		26.10	32
1050	4" thick		285	.140	"	21	6.75		27.75	34
1200	Caps, bullnose or square, 8" x 16", 2" thick		420	.095	L.F.	19.90	4.56		24.46	29.50
1250	4" thick		380	.105	"	23	5.05		28.05	33
1256	Corner, bullnose or square, 2" thick		280	.143	Ea.	23.50	6.85		30.35	36.50
1258	4" thick		270	.148		26.50	7.10		33.60	41
1260	6" thick		260	.154		29.50	7.35		36.85	44.50
1270	8" thick		250	.160		39.50	7.65		47.15	56
1280	10" thick		240	.167		35.50	8		43.50	52
1290	12" thick		230	.174	↓	37.50	8.35		45.85	55
1500	Cove base, 8" x 16", 2" thick		315	.127	L.F.	11.25	6.10		17.35	22.50
1550	4" thick		285	.140		10.75	6.75		17.50	22.50
1600	6" thick		265	.151		11.45	7.25		18.70	24.50
1650	8" thick	↓	245	.163	↓	12	7.85		19.85	26
9000	Minimum labor/equipment charge	D-1	2	8	Job		375		375	605

04 23 Glass Unit Masonry

04 23 13 – Vertical Glass Unit Masonry

04 23 13.10 Glass Block

0010	**GLASS BLOCK**									
0100	Plain, 4" thick, under 1,000 S.F., 6" x 6"	D-8	115	.348	S.F.	23	16.65		39.65	52.50
0150	8" x 8"		160	.250		14.25	12		26.25	35
0160	end block		160	.250		64.50	12		76.50	90.50
0170	90 degree corner		160	.250		65	12		77	91
0180	45 degree corner		160	.250		67.50	12		79.50	93.50
0200	12" x 12"		175	.229		24	10.95		34.95	44
0210	4" x 8"		160	.250		36.50	12		48.50	59.50
0220	6" x 8"		160	.250		23	12		35	45
0300	1,000 to 5,000 S.F., 6" x 6"	↓	135	.296	↓	22.50	14.20		36.70	48

04 23 Glass Unit Masonry

04 23 13 – Vertical Glass Unit Masonry

04 23 13.10 Glass Block

		Crew	Daily Output	Labor-Hours	Unit	Material	2020 Bare Costs Labor	Equipment	Total	Total Incl O&P
0350	8" x 8"	D-8	190	.211	S.F.	14	10.10		24.10	31.50
0400	12" x 12"		215	.186		23.50	8.90		32.40	40.50
0410	4" x 8"		215	.186		35.50	8.90		44.40	53.50
0420	6" x 8"		215	.186		22.50	8.90		31.40	39.50
0500	Over 5,000 S.F., 6" x 6"		145	.276		22	13.20		35.20	45.50
0700	For solar reflective blocks, add					100%				
1000	Thinline, plain, 3-1/8" thick, under 1,000 S.F., 6" x 6"	D-8	115	.348	S.F.	25	16.65		41.65	54
1050	8" x 8"		160	.250		15.35	12		27.35	36.50
1250	8" x 8"		215	.186		15.30	8.90		24.20	31
1400	For cleaning block after installation (both sides), add		1000	.040		.16	1.92		2.08	3.28
9000	Minimum labor/equipment charge	D-1	2	8	Job		375		375	605

04 27 Multiple-Wythe Unit Masonry

04 27 10 – Multiple-Wythe Masonry

04 27 10.10 Cornices

		Crew	Daily Output	Labor-Hours	Unit	Material	2020 Bare Costs Labor	Equipment	Total	Total Incl O&P
0010	**CORNICES**									
0110	Face bricks, 12 brick/S.F.	D-1	30	.533	SF Face	5.75	25		30.75	47
0150	15 brick/S.F.		23	.696	"	6.90	32.50		39.40	60
9000	Minimum labor/equipment charge		1.50	10.667	Job		500		500	810

04 27 10.30 Brick Walls

		Crew	Daily Output	Labor-Hours	Unit	Material	2020 Bare Costs Labor	Equipment	Total	Total Incl O&P
0010	**BRICK WALLS**, including mortar, excludes scaffolding									
0020	Estimating by number of brick									
0140	Face brick, 4" thick wall, 6.75 brick/S.F.	D-8	1.45	27.586	M	710	1,325		2,035	2,900
0150	Common brick, 4" thick wall, 6.75 brick/S.F.		1.60	25		690	1,200		1,890	2,675
0204	8" thick, 13.50 brick/S.F.		1.80	22.222		710	1,075		1,785	2,500
0250	12" thick, 20.25 brick/S.F.		1.90	21.053		715	1,000		1,715	2,400
0304	16" thick, 27.00 brick/S.F.		2	20		720	960		1,680	2,350
0500	Reinforced, face brick, 4" thick wall, 6.75 brick/S.F.		1.40	28.571		735	1,375		2,110	3,025
0520	Common brick, 4" thick wall, 6.75 brick/S.F.		1.55	25.806		720	1,225		1,945	2,800
0550	8" thick, 13.50 brick/S.F.		1.75	22.857		740	1,100		1,840	2,575
0600	12" thick, 20.25 brick/S.F.		1.85	21.622		740	1,025		1,765	2,500
0650	16" thick, 27.00 brick/S.F.		1.95	20.513		750	985		1,735	2,425
0790	Alternate method of figuring by square foot									
0800	Face brick, 4" thick wall, 6.75 brick/S.F. R042110-50	D-8	215	.186	S.F.	4.78	8.90		13.68	19.65
0850	Common brick, 4" thick wall, 6.75 brick/S.F.		240	.167		4.67	8		12.67	18.05
0900	8" thick, 13.50 brick/S.F.		135	.296		9.60	14.20		23.80	33.50
1000	12" thick, 20.25 brick/S.F.		95	.421		14.45	20		34.45	48.50
1050	16" thick, 27.00 brick/S.F.		75	.533		19.50	25.50		45	63
1200	Reinforced, face brick, 4" thick wall, 6.75 brick/S.F.		210	.190		4.96	9.15		14.11	20
1220	Common brick, 4" thick wall, 6.75 brick/S.F.		235	.170		4.86	8.15		13.01	18.55
1250	8" thick, 13.50 brick/S.F.		130	.308		9.95	14.75		24.70	35
1260	8" thick, 2.25 brick/S.F.		130	.308		1.79	14.75		16.54	26
1300	12" thick, 20.25 brick/S.F.		90	.444		15	21.50		36.50	51
1350	16" thick, 27.00 brick/S.F.		70	.571		20	27.50		47.50	66.50
9000	Minimum labor/equipment charge	D-1	2	8	Job		375		375	605

04 27 10.40 Steps

		Crew	Daily Output	Labor-Hours	Unit	Material	2020 Bare Costs Labor	Equipment	Total	Total Incl O&P
0010	**STEPS**									
0012	Entry steps, select common brick	D-1	.30	53.333	M	555	2,500		3,055	4,675

04 41 Dry-Placed Stone

04 41 10 – Dry Placed Stone

04 41 10.10 Rough Stone Wall

		Crew	Daily Output	Labor-Hours	Unit	Material	2020 Bare Costs Labor	2020 Bare Costs Equipment	Total	Total Incl O&P
0011	**ROUGH STONE WALL**, Dry									
0012	Dry laid (no mortar), under 18" thick [G]	D-1	60	.267	C.F.	14.15	12.50		26.65	35.50
0100	Random fieldstone, under 18" thick [G]	D-12	60	.533		14.15	25.50		39.65	56.50
0150	Over 18" thick [G]	"	63	.508		17	24		41	57.50
0500	Field stone veneer [G]	D-8	120	.333	S.F.	12.95	16		28.95	40.50
0510	Valley stone veneer [G]		120	.333		12.95	16		28.95	40.50
0520	River stone veneer [G]		120	.333		12.95	16		28.95	40.50
0600	Rubble stone walls, in mortar bed, up to 18" thick [G]	D-11	75	.320	C.F.	17.10	15.75		32.85	44.50
9000	Minimum labor/equipment charge	D-1	2	8	Job		375		375	605

04 43 Stone Masonry

04 43 10 – Masonry with Natural and Processed Stone

04 43 10.05 Ashlar Veneer

		Crew	Daily Output	Labor-Hours	Unit	Material	2020 Bare Costs Labor	2020 Bare Costs Equipment	Total	Total Incl O&P
0011	**ASHLAR VENEER** +/- 4" thk, random or random rectangular									
0150	Sawn face, split joints, low priced stone	D-8	140	.286	S.F.	12.15	13.70		25.85	35.50
0200	Medium priced stone		130	.308		15.40	14.75		30.15	41
0300	High priced stone		120	.333		18.45	16		34.45	46.50
0600	Seam face, split joints, medium price stone		125	.320		18.20	15.35		33.55	45
0700	High price stone		120	.333		18.10	16		34.10	46
1000	Split or rock face, split joints, medium price stone		125	.320		10.95	15.35		26.30	37
1100	High price stone		120	.333		17.25	16		33.25	45

04 43 10.10 Bluestone

		Crew	Daily Output	Labor-Hours	Unit	Material	2020 Bare Costs Labor	2020 Bare Costs Equipment	Total	Total Incl O&P
0010	**BLUESTONE**, cut to size									
0500	Sills, natural cleft, 10" wide to 6' long, 1-1/2" thick	D-11	70	.343	L.F.	13.50	16.90		30.40	42.50
0550	2" thick		63	.381		14.80	18.75		33.55	47
0600	Smooth finish, 1-1/2" thick		70	.343		14.45	16.90		31.35	43.50
0650	2" thick		63	.381		18.10	18.75		36.85	50.50
0800	Thermal finish, 1-1/2" thick		70	.343		15.35	16.90		32.25	44.50
0850	2" thick		63	.381		12.20	18.75		30.95	44
1000	Stair treads, natural cleft, 12" wide, 6' long, 1-1/2" thick	D-10	115	.278		14.05	14.40	3.72	32.17	42.50
1050	2" thick		105	.305		15.30	15.75	4.08	35.13	47
1100	Smooth finish, 1-1/2" thick		115	.278		13.45	14.40	3.72	31.57	42
1150	2" thick		105	.305		14.55	15.75	4.08	34.38	46
1300	Thermal finish, 1-1/2" thick		115	.278		13.80	14.40	3.72	31.92	42.50
1350	2" thick		105	.305		13.50	15.75	4.08	33.33	45
9000	Minimum labor/equipment charge	D-1	2.50	6.400	Job		300		300	485

04 43 10.45 Granite

		Crew	Daily Output	Labor-Hours	Unit	Material	2020 Bare Costs Labor	2020 Bare Costs Equipment	Total	Total Incl O&P
0010	**GRANITE**, cut to size									
0050	Veneer, polished face, 3/4" to 1-1/2" thick									
0150	Low price, gray, light gray, etc.	D-10	130	.246	S.F.	27	12.75	3.29	43.04	53.50
0180	Medium price, pink, brown, etc.		130	.246		30.50	12.75	3.29	46.54	57.50
0220	High price, red, black, etc.		130	.246		42.50	12.75	3.29	58.54	71
0300	1-1/2" to 2-1/2" thick, veneer									
0350	Low price, gray, light gray, etc.	D-10	130	.246	S.F.	29	12.75	3.29	45.04	56
0500	Medium price, pink, brown, etc.	"	130	.246	"	34	12.75	3.29	50.04	61
2500	Steps, copings, etc., finished on more than one surface									
2550	Low price, gray, light gray, etc.	D-10	50	.640	C.F.	92	33	8.55	133.55	163
2575	Medium price, pink, brown, etc.		50	.640		120	33	8.55	161.55	194
2600	High price, red, black, etc.		50	.640		147	33	8.55	188.55	224
9000	Minimum labor/equipment charge	D-1	2	8	Job		375		375	605

04 43 Stone Masonry

04 43 10 – Masonry with Natural and Processed Stone

04 43 10.55 Limestone

		Crew	Daily Output	Labor-Hours	Unit	Material	2020 Bare Costs Labor	Equipment	Total	Total Incl O&P
0010	**LIMESTONE**, cut to size									
0020	Veneer facing panels									
0500	Texture finish, light stick, 4-1/2" thick, 5' x 12'	D-4	300	.107	S.F.	22.50	5.05	.63	28.18	33.50
0750	5" thick, 5' x 14' panels	D-10	275	.116		24	6	1.56	31.56	38
1000	Sugarcube finish, 2" thick, 3' x 5' panels		275	.116		29.50	6	1.56	37.06	44
1050	3" thick, 4' x 9' panels		275	.116		29	6	1.56	36.56	43
1200	4" thick, 5' x 11' panels		275	.116		32.50	6	1.56	40.06	47
1400	Sugarcube, textured finish, 4-1/2" thick, 5' x 12'		275	.116		35.50	6	1.56	43.06	50.50
1450	5" thick, 5' x 14' panels		275	.116	↓	35.50	6	1.56	43.06	50.50
2000	Coping, sugarcube finish, top & 2 sides		30	1.067	C.F.	66.50	55	14.25	135.75	178
2100	Sills, lintels, jambs, trim, stops, sugarcube finish, simple		20	1.600		66.50	83	21.50	171	230
2150	Detailed		20	1.600	↓	66.50	83	21.50	171	230
2300	Steps, extra hard, 14" wide, 6" rise	↓	50	.640	L.F.	29	33	8.55	70.55	94.50
3000	Quoins, plain finish, 6" x 12" x 12"	D-12	25	1.280	Ea.	44.50	60.50		105	147
3050	6" x 16" x 24"	"	25	1.280	"	59.50	60.50		120	164
9000	Minimum labor/equipment charge	D-1	2	8	Job		375		375	605

04 43 10.60 Marble

		Crew	Daily Output	Labor-Hours	Unit	Material	2020 Bare Costs Labor	Equipment	Total	Total Incl O&P
0011	**MARBLE**, ashlar, split face, +/- 4" thick, random									
0040	Lengths 1' to 4' & heights 2" to 7-1/2", average	D-8	175	.229	S.F.	17.60	10.95		28.55	37
0100	Base, polished, 3/4" or 7/8" thick, polished, 6" high	D-10	65	.492	L.F.	11.45	25.50	6.60	43.55	61
0300	Carvings or bas-relief, from templates, simple design		80	.400	S.F.	157	20.50	5.35	182.85	212
0350	Intricate design	↓	80	.400	"	355	20.50	5.35	380.85	430
1000	Facing, polished finish, cut to size, 3/4" to 7/8" thick									
1050	Carrara or equal	D-10	130	.246	S.F.	20.50	12.75	3.29	36.54	47
1100	Arabescato or equal		130	.246		38	12.75	3.29	54.04	65.50
1300	1-1/4" thick, Botticino Classico or equal		125	.256		24.50	13.25	3.43	41.18	52
1350	Statuarietto or equal		125	.256		39	13.25	3.43	55.68	68
1500	2" thick, Crema Marfil or equal		120	.267		48.50	13.80	3.57	65.87	79.50
1550	Cafe Pinta or equal	↓	120	.267	↓	65	13.80	3.57	82.37	97.50
1700	Rubbed finish, cut to size, 4" thick									
1740	Average	D-10	100	.320	S.F.	42.50	16.55	4.28	63.33	77.50
1780	Maximum	"	100	.320	"	71	16.55	4.28	91.83	110
2200	Window sills, 6" x 3/4" thick	D-1	85	.188	L.F.	13.10	8.85		21.95	28.50
2500	Flooring, polished tiles, 12" x 12" x 3/8" thick									
2510	Thin set, Giallo Solare or equal	D-11	90	.267	S.F.	15.65	13.15		28.80	38.50
2600	Sky Blue or equal		90	.267		15.05	13.15		28.20	37.50
2700	Mortar bed, Giallo Solare or equal		65	.369		15.75	18.20		33.95	47
2740	Sky Blue or equal	↓	65	.369		15.05	18.20		33.25	46
2780	Travertine, 3/8" thick, Sierra or equal	D-10	130	.246		10.80	12.75	3.29	26.84	36
2790	Silver or equal	"	130	.246		29	12.75	3.29	45.04	56
2800	Patio tile, non-slip, 1/2" thick, flame finish	D-11	75	.320	↓	12.40	15.75		28.15	39
2900	Shower or toilet partitions, 7/8" thick partitions									
3050	3/4" or 1-1/4" thick stiles, polished 2 sides, average	D-11	75	.320	S.F.	47.50	15.75		63.25	77.50
3201	Soffits, add to above prices				"	20%	100%			
3210	Stairs, risers, 7/8" thick x 6" high	D-10	115	.278	L.F.	15.15	14.40	3.72	33.27	44
3360	Treads, 12" wide x 1-1/4" thick	"	115	.278	"	50	14.40	3.72	68.12	82
3500	Thresholds, 3' long, 7/8" thick, 4" to 5" wide, plain	D-12	24	1.333	Ea.	36	63		99	142
3550	Beveled		24	1.333	"	76	63		139	186
3700	Window stools, polished, 7/8" thick, 5" wide	↓	85	.376	L.F.	24.50	17.85		42.35	56
9000	Minimum labor/equipment charge	D-1	2	8	Job		375		375	605

04 43 Stone Masonry

04 43 10 – Masonry with Natural and Processed Stone

04 43 10.75 Sandstone or Brownstone

	04 43 10.75 Sandstone or Brownstone	Crew	Daily Output	Labor-Hours	Unit	Material	2020 Bare Costs Labor	Equipment	Total	Total Incl O&P
0011	**SANDSTONE OR BROWNSTONE**									
0100	Sawed face veneer, 2-1/2" thick, to 2' x 4' panels	D-10	130	.246	S.F.	22	12.75	3.29	38.04	48
0150	4" thick, to 3'-6" x 8' panels		100	.320		22	16.55	4.28	42.83	55
0300	Split face, random sizes	↓	100	.320	↓	14.50	16.55	4.28	35.33	47
9000	Minimum labor/equipment charge	D-1	2.50	6.400	Job		300		300	485

04 43 10.80 Slate

	04 43 10.80 Slate	Crew	Daily Output	Labor-Hours	Unit	Material	2020 Bare Costs Labor	Equipment	Total	Total Incl O&P
0010	**SLATE**									
0040	Pennsylvania - blue gray to black									
0050	Vermont - unfading green, mottled green & purple, gray & purple									
0100	Virginia - blue black									
3100	Stair landings, 1" thick, black, clear	D-1	65	.246	S.F.	21	11.55		32.55	41.50
3200	Ribbon	"	65	.246	"	23	11.55		34.55	43.50
3500	Stair treads, sand finish, 1" thick x 12" wide									
3550	Under 3 L.F.	D-10	85	.376	L.F.	23	19.50	5.05	47.55	61.50
3600	3 L.F. to 6 L.F.	"	120	.267	"	25	13.80	3.57	42.37	53
3700	Ribbon, sand finish, 1" thick x 12" wide									
3750	To 6 L.F.	D-10	120	.267	L.F.	21	13.80	3.57	38.37	49
4000	Stools or sills, sand finish, 1" thick, 6" wide	D-12	160	.200		12	9.50		21.50	28.50
4100	Honed finish		160	.200		11.45	9.50		20.95	28
4200	10" wide		90	.356		18.50	16.85		35.35	47.50
4250	Honed finish		90	.356		17.20	16.85		34.05	46
4400	2" thick, 6" wide		140	.229		19.30	10.85		30.15	38.50
4450	Honed finish		140	.229		18.20	10.85		29.05	37.50
4600	10" wide		90	.356		30	16.85		46.85	60
4650	Honed finish	↓	90	.356		28.50	16.85		45.35	58.50
4800	For lengths over 3', add				↓	25%				
9000	Minimum labor/equipment charge	D-1	2.50	6.400	Job		300		300	485

04 43 10.85 Window Sill

	04 43 10.85 Window Sill	Crew	Daily Output	Labor-Hours	Unit	Material	2020 Bare Costs Labor	Equipment	Total	Total Incl O&P
0010	**WINDOW SILL**									
0020	Bluestone, thermal top, 10" wide, 1-1/2" thick	D-1	85	.188	S.F.	10.30	8.85		19.15	25.50
0050	2" thick		75	.213	"	10.95	10		20.95	28
0100	Cut stone, 5" x 8" plain		48	.333	L.F.	12.95	15.65		28.60	39.50
0200	Face brick on edge, brick, 8" wide		80	.200		3.42	9.40		12.82	18.90
0400	Marble, 9" wide, 1" thick		85	.188		9.25	8.85		18.10	24.50
0900	Slate, colored, unfading, honed, 12" wide, 1" thick		85	.188		10.45	8.85		19.30	26
0950	2" thick	↓	70	.229	↓	10.05	10.70		20.75	28.50
9000	Minimum labor/equipment charge	1 Bric	2	4	Job		208		208	335

04 54 Refractory Brick Masonry

04 54 10 – Refractory Brick Work

04 54 10.10 Fire Brick

	04 54 10.10 Fire Brick	Crew	Daily Output	Labor-Hours	Unit	Material	2020 Bare Costs Labor	Equipment	Total	Total Incl O&P
0010	**FIRE BRICK**									
0012	Low duty, 2000°F, 9" x 2-1/2" x 4-1/2"	D-1	.60	26.667	M	1,675	1,250		2,925	3,875
0050	High duty, 3000°F	"	.60	26.667	"	2,800	1,250		4,050	5,125

04 54 10.20 Fire Clay

	04 54 10.20 Fire Clay	Crew	Daily Output	Labor-Hours	Unit	Material	2020 Bare Costs Labor	Equipment	Total	Total Incl O&P
0010	**FIRE CLAY**									
0020	Gray, high duty, 100 lb. bag				Bag	37			37	40.50
0050	100 lb. drum, premixed (400 brick per drum)				Drum	48.50			48.50	53.50

04 57 Masonry Fireplaces

04 57 10 – Brick or Stone Fireplaces

04 57 10.10 Fireplace

04 57 10.10 Fireplace	Crew	Daily Output	Labor-Hours	Unit	Material	2020 Bare Costs Labor	Equipment	Total	Total Incl O&P
0010 **FIREPLACE**									
0100 Brick fireplace, not incl. foundations or chimneys									
0110 30" x 29" opening, incl. chamber, plain brickwork	D-1	.40	40	Ea.	625	1,875		2,500	3,700
0200 Fireplace box only (110 brick)	"	2	8	"	172	375		547	795
0300 For elaborate brickwork and details, add					35%	35%			
0400 For hearth, brick & stone, add	D-1	2	8	Ea.	215	375		590	840
0410 For steel, damper, cleanouts, add		4	4		18.25	188		206.25	325
0600 Plain brickwork, incl. metal circulator		.50	32		995	1,500		2,495	3,525
0800 Face brick only, standard size, 8" x 2-2/3" x 4"		.30	53.333	M	605	2,500		3,105	4,725
0900 Stone fireplace, fieldstone, add				SF Face	10			10	11
1000 Cut stone, add				"	8.50			8.50	9.35
9000 Minimum labor/equipment charge	D-1	2	8	Job		375		375	605

04 71 Manufactured Brick Masonry

04 71 10 – Simulated or Manufactured Brick

04 71 10.10 Simulated Brick

04 71 10.10 Simulated Brick	Crew	Daily Output	Labor-Hours	Unit	Material	2020 Bare Costs Labor	Equipment	Total	Total Incl O&P
0010 **SIMULATED BRICK**									
0020 Aluminum, baked on colors	1 Carp	200	.040	S.F.	4.53	2.13		6.66	8.40
0050 Fiberglass panels		200	.040		9.90	2.13		12.03	14.30
0100 Urethane pieces cemented in mastic		150	.053		9	2.83		11.83	14.45
0150 Vinyl siding panels		200	.040		11.50	2.13		13.63	16.05

04 72 Cast Stone Masonry

04 72 10 – Cast Stone Masonry Features

04 72 10.10 Coping

04 72 10.10 Coping	Crew	Daily Output	Labor-Hours	Unit	Material	2020 Bare Costs Labor	Equipment	Total	Total Incl O&P
0010 **COPING**, stock units									
0050 Precast concrete, 10" wide, 4" tapers to 3-1/2", 8" wall	D-1	75	.213	L.F.	25.50	10		35.50	44
0100 12" wide, 3-1/2" tapers to 3", 10" wall		70	.229		27.50	10.70		38.20	47.50
0110 14" wide, 4" tapers to 3-1/2", 12" wall		65	.246		31.50	11.55		43.05	53
0150 16" wide, 4" tapers to 3-1/2", 14" wall		60	.267		34	12.50		46.50	57
0300 Limestone for 12" wall, 4" thick		90	.178		16.80	8.35		25.15	32
0350 6" thick		80	.200		24	9.40		33.40	41
0500 Marble, to 4" thick, no wash, 9" wide		90	.178		12.30	8.35		20.65	27
0550 12" wide		80	.200		19.90	9.40		29.30	37
0700 Terra cotta, 9" wide		90	.178		8.15	8.35		16.50	22.50
0750 12" wide		80	.200		8.60	9.40		18	24.50
0800 Aluminum, for 12" wall		80	.200		9.45	9.40		18.85	25.50
9000 Minimum labor/equipment charge		2	8	Job		375		375	605

04 72 20 – Cultured Stone Veneer

04 72 20.10 Cultured Stone Veneer Components

04 72 20.10 Cultured Stone Veneer Components	Crew	Daily Output	Labor-Hours	Unit	Material	2020 Bare Costs Labor	Equipment	Total	Total Incl O&P
0010 **CULTURED STONE VENEER COMPONENTS**									
0110 On wood frame and sheathing substrate, random sized cobbles, corner stones	D-8	70	.571	V.L.F.	10.55	27.50		38.05	56
0120 Field stones		140	.286	S.F.	7.45	13.70		21.15	30
0130 Random sized flats, corner stones		70	.571	V.L.F.	10.55	27.50		38.05	56
0140 Field stones		140	.286	S.F.	8.85	13.70		22.55	31.50
0150 Horizontal lined ledgestones, corner stones		75	.533	V.L.F.	10.35	25.50		35.85	53
0160 Field stones		150	.267	S.F.	7.80	12.80		20.60	29
0170 Random shaped flats, corner stones		65	.615	V.L.F.	10.35	29.50		39.85	59
0180 Field stones		150	.267	S.F.	7.75	12.80		20.55	29

04 72 Cast Stone Masonry

04 72 20 – Cultured Stone Veneer

04 72 20.10 Cultured Stone Veneer Components		Crew	Daily Output	Labor-Hours	Unit	Material	2020 Bare Costs Labor	Equipment	Total	Total Incl O&P
0190	Random shaped/textured face, corner stones	D-8	65	.615	V.L.F.	10.55	29.50		40.05	59
0200	Field stones		130	.308	S.F.	7.70	14.75		22.45	32.50
0210	Random shaped river rock, corner stones		65	.615	V.L.F.	10.55	29.50		40.05	59
0220	Field stones		130	.308	S.F.	7.70	14.75		22.45	32.50
0240	On concrete or CMU substrate, random sized cobbles, corner stones		70	.571	V.L.F.	9.70	27.50		37.20	55
0250	Field stones		140	.286	S.F.	7.05	13.70		20.75	30
0260	Random sized flats, corner stones		70	.571	V.L.F.	9.70	27.50		37.20	55
0270	Field stones		140	.286	S.F.	8.40	13.70		22.10	31.50
0280	Horizontal lined ledgestones, corner stones		75	.533	V.L.F.	9.50	25.50		35	52
0290	Field stones		150	.267	S.F.	7.35	12.80		20.15	28.50
0300	Random shaped flats, corner stones		70	.571	V.L.F.	9.50	27.50		37	55
0310	Field stones		140	.286	S.F.	7.35	13.70		21.05	30
0320	Random shaped/textured face, corner stones		65	.615	V.L.F.	9.70	29.50		39.20	58
0330	Field stones		130	.308	S.F.	7.30	14.75		22.05	32
0340	Random shaped river rock, corner stones		65	.615	V.L.F.	9.70	29.50		39.20	58
0350	Field stones	▽	130	.308	S.F.	7.30	14.75		22.05	32
0360	Cultured stone veneer, #15 felt weather resistant barrier	1 Clab	3700	.002	Sq.	5.25	.09		5.34	5.95
0370	Expanded metal lath, diamond, 2.5 lb./S.Y., galvanized	1 Lath	85	.094	S.Y.	3.80	4.90		8.70	11.85
0390	Water table or window sill, 18" long	1 Bric	80	.100	Ea.	9.95	5.20		15.15	19.35

04 73 Manufactured Stone Masonry

04 73 20 – Simulated or Manufactured Stone

04 73 20.10 Simulated Stone		Crew	Daily Output	Labor-Hours	Unit	Material	2020 Bare Costs Labor	Equipment	Total	Total Incl O&P
0010	**SIMULATED STONE**									
0100	Insulated fiberglass panels, 5/8" ply backer	L-4	200	.120	S.F.	10.85	6		16.85	21.50

Estimating Tips

05 05 00 Common Work Results for Metals

- Nuts, bolts, washers, connection angles, and plates can add a significant amount to both the tonnage of a structural steel job and the estimated cost. As a rule of thumb, add 10% to the total weight to account for these accessories.

- Type 2 steel construction, commonly referred to as "simple construction," consists generally of field-bolted connections with lateral bracing supplied by other elements of the building, such as masonry walls or x-bracing. The estimator should be aware, however, that shop connections may be accomplished by welding or bolting. The method may be particular to the fabrication shop and may have an impact on the estimated cost.

05 10 00 Structural Steel

- Steel items can be obtained from two sources: a fabrication shop or a metals service center. Fabrication shops can fabricate items under more controlled conditions than crews in the field can. They are also more efficient and can produce items more economically. Metal service centers serve as a source of long mill shapes to both fabrication shops and contractors.

- Most line items in this structural steel subdivision, and most items in 05 50 00 Metal Fabrications, are indicated as being shop fabricated. The bare material cost for these shop fabricated items is the "Invoice Cost" from the shop and includes the mill base price of steel plus mill extras, transportation to the shop, shop drawings and detailing where warranted, shop fabrication and handling, sandblasting and a shop coat of primer paint, all necessary structural bolts, and delivery to the job site. The bare labor cost and bare equipment cost for these shop fabricated items are for field installation or erection.

- Line items in Subdivision 05 12 23.40 Lightweight Framing, and other items scattered in Division 5, are indicated as being field fabricated. The bare material cost for these field fabricated items is the "Invoice Cost" from the metals service center and includes the mill base price of steel plus mill extras, transportation to the metals service center, material handling, and delivery of long lengths of mill shapes to the job site. Material costs for structural bolts and welding rods should be added to the estimate. The bare labor cost and bare equipment cost for these items are for both field fabrication and field installation or erection, and include time for cutting, welding, and drilling in the fabricated metal items. Drilling into concrete and fasteners to fasten field fabricated items to other work is not included and should be added to the estimate.

05 20 00 Steel Joist Framing

- In any given project the total weight of open web steel joists is determined by the loads to be supported and the design. However, economies can be realized in minimizing the amount of labor used to place the joists. This is done by maximizing the joist spacing and therefore minimizing the number of joists required to be installed on the job. Certain spacings and locations may be required by the design, but in other cases maximizing the spacing and keeping it as uniform as possible will keep the costs down.

05 30 00 Steel Decking

- The takeoff and estimating of a metal deck involve more than the area of the floor or roof and the type of deck specified or shown on the drawings. Many different sizes and types of openings may exist. Small openings for individual pipes or conduits may be drilled after the floor/roof is installed, but larger openings may require special deck lengths as well as reinforcing or structural support. The estimator should determine who will be supplying this reinforcing. Additionally, some deck terminations are part of the deck package, such as screed angles and pour stops, and others will be part of the steel contract, such as angles attached to structural members and cast-in-place angles and plates. The estimator must ensure that all pieces are accounted for in the complete estimate.

05 50 00 Metal Fabrications

- The most economical steel stairs are those that use common materials, standard details, and most importantly, a uniform and relatively simple method of field assembly. Commonly available A36/A992 channels and plates are very good choices for the main stringers of the stairs, as are angles and tees for the carrier members. Risers and treads are usually made by specialty shops, and it is most economical to use a typical detail in as many places as possible. The stairs should be pre-assembled and shipped directly to the site. The field connections should be simple and straightforward enough to be accomplished efficiently, and with minimum equipment and labor.

Reference Numbers

Reference numbers are shown at the beginning of some major classifications. These numbers refer to related items in the Reference Section. The reference information may be an estimating procedure, an alternate pricing method, or technical information.

Note: Not all subdivisions listed here necessarily appear. ■

Same Data. Simplified.

Enjoy the convenience and efficiency of accessing your costs anywhere:

- **Skip the multiplier** by setting your location
- **Quickly search,** edit, favorite and share costs
- **Stay on top of price changes** with automatic updates

Discover more at rsmeans.com/online

05 05 Common Work Results for Metals

05 05 05 – Selective Demolition for Metals

05 05 05.10 Selective Demolition, Metals

		Crew	Daily Output	Labor-Hours	Unit	Material	2020 Bare Costs Labor	Equipment	Total	Total Incl O&P
0010	**SELECTIVE DEMOLITION, METALS** R024119-10									
0015	Excludes shores, bracing, cutting, loading, hauling, dumping									
0020	Remove nuts only up to 3/4" diameter	1 Sswk	480	.017	Ea.		.96		.96	1.57
0030	7/8" to 1-1/4" diameter		240	.033			1.92		1.92	3.15
0040	1-3/8" to 2" diameter		160	.050			2.88		2.88	4.72
0060	Unbolt and remove structural bolts up to 3/4" diameter		240	.033			1.92		1.92	3.15
0070	7/8" to 2" diameter		160	.050			2.88		2.88	4.72
0140	Light weight framing members, remove whole or cut up, up to 20 lb.		240	.033			1.92		1.92	3.15
0150	21-40 lb.	2 Sswk	210	.076			4.39		4.39	7.20
0160	41-80 lb.	3 Sswk	180	.133			7.70		7.70	12.60
0170	81-120 lb.	4 Sswk	150	.213			12.30		12.30	20
0230	Structural members, remove whole or cut up, up to 500 lb.	E-19	48	.500			28.50	31	59.50	80
0240	1/4-2 tons	E-18	36	1.111			64.50	41.50	106	150
0250	2-5 tons	E-24	30	1.067			61.50	19.35	80.85	121
0260	5-10 tons	E-20	24	2.667			153	84.50	237.50	340
0270	10-15 tons	E-2	18	3.111			178	94	272	390
0340	Fabricated item, remove whole or cut up, up to 20 lb.	1 Sswk	96	.083			4.80		4.80	7.85
0350	21-40 lb.	2 Sswk	84	.190			11		11	18
0360	41-80 lb.	3 Sswk	72	.333			19.20		19.20	31.50
0370	81-120 lb.	4 Sswk	60	.533			31		31	50.50
0380	121-500 lb.	E-19	48	.500			28.50	31	59.50	80
0390	501-1000 lb.	"	36	.667			38	41.50	79.50	107
0500	Steel roof decking, uncovered, bare	B-2	5000	.008	S.F.		.34		.34	.54
2950	Minimum labor/equipment charge	E-19	2	12	Job		680	745	1,425	1,925

05 05 19 – Post-Installed Concrete Anchors

05 05 19.10 Chemical Anchors

		Crew	Daily Output	Labor-Hours	Unit	Material	2020 Bare Costs Labor	Equipment	Total	Total Incl O&P
0010	**CHEMICAL ANCHORS**									
0020	Includes layout & drilling									
1430	Chemical anchor, w/rod & epoxy cartridge, 3/4" diameter x 9-1/2" long	B-89A	27	.593	Ea.	10.15	28.50	4.23	42.88	62
1435	1" diameter x 11-3/4" long		24	.667		20.50	32.50	4.76	57.76	79.50
1440	1-1/4" diameter x 14" long		21	.762		39	37	5.45	81.45	108
1445	1-3/4" diameter x 15" long		20	.800		68.50	39	5.70	113.20	144
1450	18" long		17	.941		82.50	45.50	6.70	134.70	171
1455	2" diameter x 18" long		16	1		113	48.50	7.15	168.65	210
1460	24" long		15	1.067		148	51.50	7.60	207.10	254

05 05 19.20 Expansion Anchors

		Crew	Daily Output	Labor-Hours	Unit	Material	2020 Bare Costs Labor	Equipment	Total	Total Incl O&P
0010	**EXPANSION ANCHORS**									
0100	Anchors for concrete, brick or stone, no layout and drilling									
0200	Expansion shields, zinc, 1/4" diameter, 1-5/16" long, single [G]	1 Carp	90	.089	Ea.	.46	4.72		5.18	8.05
0300	1-3/8" long, double [G]		85	.094		.66	5		5.66	8.75
0400	3/8" diameter, 1-1/2" long, single [G]		85	.094		.71	5		5.71	8.80
0500	2" long, double [G]		80	.100		1.20	5.30		6.50	9.80
0600	1/2" diameter, 2-1/16" long, single [G]		80	.100		1.28	5.30		6.58	9.90
0700	2-1/2" long, double [G]		75	.107		2.17	5.65		7.82	11.50
0800	5/8" diameter, 2-5/8" long, single [G]		75	.107		2.22	5.65		7.87	11.55
0900	2-3/4" long, double [G]		70	.114		3.64	6.05		9.69	13.75
1000	3/4" diameter, 2-3/4" long, single [G]		70	.114		3.66	6.05		9.71	13.80
1100	3-15/16" long, double [G]		65	.123		5.65	6.55		12.20	16.75
2100	Hollow wall anchors for gypsum wall board, plaster or tile									
2300	1/8" diameter, short [G]	1 Carp	160	.050	Ea.	.31	2.66		2.97	4.60
2400	Long [G]		150	.053		.32	2.83		3.15	4.89
2500	3/16" diameter, short [G]		150	.053		.58	2.83		3.41	5.20

05 05 19 – Post-Installed Concrete Anchors

	05 05 19.20 Expansion Anchors		Crew	Daily Output	Labor-Hours	Unit	Material	2020 Bare Costs Labor	Equipment	Total	Total Incl O&P
2600	Long	G	1 Carp	140	.057	Ea.	.75	3.04		3.79	5.70
2700	1/4" diameter, short	G		140	.057		.65	3.04		3.69	5.60
2800	Long	G		130	.062		1.03	3.27		4.30	6.40
3000	Toggle bolts, bright steel, 1/8" diameter, 2" long	G		85	.094		.21	5		5.21	8.25
3100	4" long	G		80	.100		.26	5.30		5.56	8.80
3200	3/16" diameter, 3" long	G		80	.100		.30	5.30		5.60	8.85
3300	6" long	G		75	.107		.49	5.65		6.14	9.65
3400	1/4" diameter, 3" long	G		75	.107		.44	5.65		6.09	9.60
3500	6" long	G		70	.114		.57	6.05		6.62	10.40
3600	3/8" diameter, 3" long	G		70	.114		.88	6.05		6.93	10.70
3700	6" long	G		60	.133		1.58	7.10		8.68	13.10
3800	1/2" diameter, 4" long	G		60	.133		1.86	7.10		8.96	13.40
3900	6" long	G	↓	50	.160	↓	2.37	8.50		10.87	16.20
4000	Nailing anchors										
4100	Nylon nailing anchor, 1/4" diameter, 1" long		1 Carp	3.20	2.500	C	19.75	133		152.75	235
4200	1-1/2" long			2.80	2.857		22	152		174	267
4300	2" long			2.40	3.333		24	177		201	310
4400	Metal nailing anchor, 1/4" diameter, 1" long	G		3.20	2.500		20	133		153	235
4500	1-1/2" long	G		2.80	2.857		25	152		177	271
4600	2" long	G		2.40	3.333	↓	43	177		220	330
8000	Wedge anchors, not including layout or drilling										
8050	Carbon steel, 1/4" diameter, 1-3/4" long	G	1 Carp	150	.053	Ea.	.62	2.83		3.45	5.20
8100	3-1/4" long	G		140	.057		.82	3.04		3.86	5.75
8150	3/8" diameter, 2-1/4" long	G		145	.055		.44	2.93		3.37	5.20
8200	5" long	G		140	.057		.78	3.04		3.82	5.75
8250	1/2" diameter, 2-3/4" long	G		140	.057		.99	3.04		4.03	5.95
8300	7" long	G		125	.064		1.70	3.40		5.10	7.30
8350	5/8" diameter, 3-1/2" long	G		130	.062		2.05	3.27		5.32	7.50
8400	8-1/2" long	G		115	.070		4.36	3.70		8.06	10.70
8450	3/4" diameter, 4-1/4" long	G		115	.070		2.97	3.70		6.67	9.15
8500	10" long	G		95	.084		6.75	4.48		11.23	14.60
8550	1" diameter, 6" long	G		100	.080		8.80	4.25		13.05	16.45
8575	9" long	G		85	.094		11.40	5		16.40	20.50
8600	12" long	G		75	.107		12.30	5.65		17.95	22.50
8650	1-1/4" diameter, 9" long	G		70	.114		23.50	6.05		29.55	35.50
8700	12" long	G	↓	60	.133	↓	30	7.10		37.10	44.50
8750	For type 303 stainless steel, add						350%				
8800	For type 316 stainless steel, add						450%				
8950	Self-drilling concrete screw, hex washer head, 3/16" diam. x 1-3/4" long	G	1 Carp	300	.027	Ea.	.18	1.42		1.60	2.47
8960	2-1/4" long	G		250	.032		.24	1.70		1.94	2.98
8970	Phillips flat head, 3/16" diam. x 1-3/4" long	G		300	.027		.19	1.42		1.61	2.48
8980	2-1/4" long	G		250	.032	↓	.23	1.70		1.93	2.97
9000	Minimum labor/equipment charge			4	2	Job		106		106	170

05 05 21 – Fastening Methods for Metal

05 05 21.10 Cutting Steel

			Crew	Daily Output	Labor-Hours	Unit	Material	2020 Bare Costs Labor	Equipment	Total	Total Incl O&P
0010	**CUTTING STEEL**										
0020	Hand burning, incl. preparation, torch cutting & grinding, no staging										
0050	Steel to 1/4" thick		E-25	400	.020	L.F.	.21	1.19	.03	1.43	2.22
0100	1/2" thick			320	.025		.39	1.49	.04	1.92	2.91
0150	3/4" thick			260	.031		.63	1.83	.05	2.51	3.74
0200	1" thick			200	.040	↓	.90	2.38	.06	3.34	4.96
9000	Minimum labor/equipment charge		↓	2	4	Job		238	6.40	244.40	395

05 05 21 – Fastening Methods for Metal

05 05 21.15 Drilling Steel	Crew	Daily Output	Labor-Hours	Unit	Material	2020 Bare Costs Labor	Equipment	Total	Total Incl O&P
0010 **DRILLING STEEL**									
1910 Drilling & layout for steel, up to 1/4" deep, no anchor									
1920 Holes, 1/4" diameter	1 Sswk	112	.071	Ea.	.10	4.12		4.22	6.85
1925 For each additional 1/4" depth, add		336	.024		.10	1.37		1.47	2.36
1930 3/8" diameter		104	.077		.08	4.43		4.51	7.35
1935 For each additional 1/4" depth, add		312	.026		.08	1.48		1.56	2.51
1940 1/2" diameter		96	.083		.08	4.80		4.88	7.95
1945 For each additional 1/4" depth, add		288	.028		.08	1.60		1.68	2.71
1950 5/8" diameter		88	.091		.16	5.25		5.41	8.75
1955 For each additional 1/4" depth, add		264	.030		.16	1.75		1.91	3.03
1960 3/4" diameter		80	.100		.17	5.75		5.92	9.65
1965 For each additional 1/4" depth, add		240	.033		.17	1.92		2.09	3.33
1970 7/8" diameter		72	.111		.24	6.40		6.64	10.75
1975 For each additional 1/4" depth, add		216	.037		.24	2.14		2.38	3.77
1980 1" diameter		64	.125		.26	7.20		7.46	12.10
1985 For each additional 1/4" depth, add		192	.042		.26	2.40		2.66	4.22
1990 For drilling up, add						40%			
2000 Minimum labor/equipment charge	1 Carp	4	2	Job		106		106	170

05 05 21.90 Welding Steel

	Crew	Daily Output	Labor-Hours	Unit	Material	2020 Bare Costs Labor	Equipment	Total	Total Incl O&P
0010 **WELDING STEEL**, Structural R050521-20									
0020 Field welding, 1/8" E6011, cost per welder, no operating engineer	E-14	8	1	Hr.	5.55	59.50	18.40	83.45	124
0200 With 1/2 operating engineer	E-13	8	1.500		5.55	86	18.40	109.95	165
0300 With 1 operating engineer	E-12	8	2		5.55	112	18.40	135.95	206
0500 With no operating engineer, 2# weld rod per ton	E-14	8	1	Ton	5.55	59.50	18.40	83.45	124
0600 8# E6011 per ton	"	2	4		22	238	73.50	333.50	495
0800 With one operating engineer per welder, 2# E6011 per ton	E-12	8	2		5.55	112	18.40	135.95	206
0900 8# E6011 per ton	"	2	8		22	450	73.50	545.50	825
1200 Continuous fillet, down welding									
1300 Single pass, 1/8" thick, 0.1#/L.F.	E-14	150	.053	L.F.	.28	3.17	.98	4.43	6.60
1400 3/16" thick, 0.2#/L.F.		75	.107		.55	6.35	1.96	8.86	13.15
1500 1/4" thick, 0.3#/L.F.		50	.160		.83	9.50	2.94	13.27	19.75
1610 5/16" thick, 0.4#/L.F.		38	.211		1.11	12.50	3.87	17.48	26
1800 3 passes, 3/8" thick, 0.5#/L.F.		30	.267		1.38	15.85	4.90	22.13	33
2010 4 passes, 1/2" thick, 0.7#/L.F.		22	.364		1.93	21.50	6.70	30.13	45
2200 5 to 6 passes, 3/4" thick, 1.3#/L.F.		12	.667		3.59	39.50	12.25	55.34	82.50
2400 8 to 11 passes, 1" thick, 2.4#/L.F.		6	1.333		6.65	79.50	24.50	110.65	164
2600 For vertical joint welding, add						20%			
2700 Overhead joint welding, add						300%			
2900 For semi-automatic welding, obstructed joints, deduct						5%			
3000 Exposed joints, deduct						15%			
4000 Cleaning and welding plates, bars, or rods									
4010 to existing beams, columns, or trusses	E-14	12	.667	L.F.	1.38	39.50	12.25	53.13	80
9000 Minimum labor/equipment charge	"	4	2	Job		119	37	156	236

05 05 23 – Metal Fastenings

05 05 23.30 Lag Screws

	Crew	Daily Output	Labor-Hours	Unit	Material	2020 Bare Costs Labor	Equipment	Total	Total Incl O&P
0010 **LAG SCREWS**									
0020 Steel, 1/4" diameter, 2" long [G]	1 Carp	200	.040	Ea.	.10	2.13		2.23	3.52
0200 1/2" diameter, 3" long [G]		130	.062		.72	3.27		3.99	6.05
0300 5/8" diameter, 3" long [G]		120	.067		1.26	3.54		4.80	7.10

05 05 Common Work Results for Metals

05 05 23 – Metal Fastenings

05 05 23.50 Powder Actuated Tools and Fasteners

		Crew	Daily Output	Labor-Hours	Unit	Material	2020 Bare Costs Labor	Equipment	Total	Total Incl O&P
0010	**POWDER ACTUATED TOOLS & FASTENERS**									
0020	Stud driver, .22 caliber, single shot				Ea.	159			159	175
0100	.27 caliber, semi automatic, strip				"	460			460	505
0300	Powder load, single shot, .22 cal, power level 2, brown				C	6.25			6.25	6.85
0400	Strip, .27 cal, power level 4, red					10.30			10.30	11.35
0600	Drive pin, .300 x 3/4" long	G	1 Carp	4.80	1.667		4.48	88.50		92.98	147
0700	.300 x 3" long with washer	G	"	4	2		11.95	106		117.95	183

05 05 23.70 Structural Blind Bolts

		Crew	Daily Output	Labor-Hours	Unit	Material	Labor	Equipment	Total	Total Incl O&P
0010	**STRUCTURAL BLIND BOLTS**									
0100	1/4" diameter x 1/4" grip	G	1 Sswk	240	.033	Ea.	1.84	1.92		3.76	5.15
0150	1/2" grip	G		216	.037		1.89	2.14		4.03	5.60
0200	3/8" diameter x 1/2" grip	G		232	.034		3.50	1.99		5.49	7.10
0250	3/4" grip	G		208	.038		3.14	2.22		5.36	7.10
0300	1/2" diameter x 1/2" grip	G		224	.036		6.25	2.06		8.31	10.25
0350	3/4" grip	G		200	.040		9.10	2.31		11.41	13.80
0400	5/8" diameter x 3/4" grip	G		216	.037		11.25	2.14		13.39	15.90
0450	1" grip	G		192	.042		15.70	2.40		18.10	21

05 05 23.87 Weld Studs

		Crew	Daily Output	Labor-Hours	Unit	Material	Labor	Equipment	Total	Total Incl O&P
0010	**WELD STUDS**									
0020	1/4" diameter, 2-11/16" long	G	E-10	1120	.014	Ea.	.37	.84	.86	2.07	2.73
0100	4-1/8" long	G		1080	.015		.35	.87	.90	2.12	2.80
0200	3/8" diameter, 4-1/8" long	G		1080	.015		.40	.87	.90	2.17	2.85
0300	6-1/8" long	G		1040	.015		.52	.90	.93	2.35	3.06
9000	Minimum labor/equipment charge		1 Sswk	2	4	Job		231		231	380

05 12 Structural Steel Framing

05 12 23 – Structural Steel for Buildings

05 12 23.05 Canopy Framing

		Crew	Daily Output	Labor-Hours	Unit	Material	Labor	Equipment	Total	Total Incl O&P
0010	**CANOPY FRAMING**									
0020	6" and 8" members, shop fabricated	G	E-4	3000	.011	Lb.	1.72	.62	.05	2.39	2.97
9000	Minimum labor/equipment charge		1 Sswk	1	8	Job		460		460	755

05 12 23.10 Ceiling Supports

		Crew	Daily Output	Labor-Hours	Unit	Material	Labor	Equipment	Total	Total Incl O&P
0010	**CEILING SUPPORTS**									
1000	Entrance door/folding partition supports, shop fabricated	G	E-4	60	.533	L.F.	29	31	2.45	62.45	85
1100	Linear accelerator door supports	G		14	2.286		131	133	10.50	274.50	375
1200	Lintels or shelf angles, hung, exterior hot dipped galv.	G		267	.120		19.60	6.95	.55	27.10	33.50
1250	Two coats primer paint instead of galv.	G		267	.120		17	6.95	.55	24.50	30.50
1400	Monitor support, ceiling hung, expansion bolted	G		4	8	Ea.	455	465	36.50	956.50	1,300
1450	Hung from pre-set inserts	G		6	5.333		490	310	24.50	824.50	1,075
1600	Motor supports for overhead doors	G		4	8		232	465	36.50	733.50	1,050
1700	Partition support for heavy folding partitions, without pocket	G		24	1.333	L.F.	65.50	77.50	6.10	149.10	206
1750	Supports at pocket only	G		12	2.667		131	155	12.25	298.25	410
2000	Rolling grilles & fire door supports	G		34	.941		56	54.50	4.32	114.82	156
2100	Spider-leg light supports, expansion bolted to ceiling slab	G		8	4	Ea.	187	233	18.35	438.35	605
2150	Hung from pre-set inserts	G		12	2.667	"	201	155	12.25	368.25	490
2400	Toilet partition support	G		36	.889	L.F.	65.50	51.50	4.08	121.08	161
2500	X-ray travel gantry support	G		12	2.667	"	224	155	12.25	391.25	515

For customer support on your Commercial Renovation Costs with RSMeans data, call 800.448.8182.

109

05 12 Structural Steel Framing

05 12 23 – Structural Steel for Buildings

05 12 23.17 Columns, Structural

		Crew	Daily Output	Labor-Hours	Unit	Material	2020 Bare Costs Labor	Equipment	Total	Total Incl O&P
0010	**COLUMNS, STRUCTURAL** R051223-10									
0015	Made from recycled materials									
0020	Shop fab'd for 100-ton, 1-2 story project, bolted connections									
0800	Steel, concrete filled, extra strong pipe, 3-1/2" diameter	E-2	660	.085	L.F.	47.50	4.85	2.57	54.92	62.50
0830	4" diameter		780	.072		53	4.10	2.17	59.27	67
0890	5" diameter		1020	.055		63	3.14	1.66	67.80	76
0930	6" diameter		1200	.047		83.50	2.67	1.41	87.58	98
0940	8" diameter		1100	.051		83.50	2.91	1.54	87.95	98.50
1500	Steel pipe, extra strong, no concrete, 3" to 5" diameter G		16000	.004	Lb.	1.44	.20	.11	1.75	2.02
3300	Structural tubing, square, A500GrB, 4" to 6" square, light section G		11270	.005	"	1.44	.28	.15	1.87	2.21
8090	For projects 75 to 99 tons, add				%	10%				
8092	50 to 74 tons, add					20%				
8094	25 to 49 tons, add					30%	10%			
8096	10 to 24 tons, add					50%	25%			
8098	2 to 9 tons, add					75%	50%			
8099	Less than 2 tons, add					100%	100%			
9000	Minimum labor/equipment charge	1 Sswk	1	8	Job		460		460	755

05 12 23.18 Corner Guards

		Crew	Daily Output	Labor-Hours	Unit	Material	2020 Bare Costs Labor	Equipment	Total	Total Incl O&P
0010	**CORNER GUARDS**									
0020	Steel angle w/anchors, 1" x 1" x 1/4", 1.5#/L.F.	2 Carp	160	.100	L.F.	7.95	5.30		13.25	17.20
0100	2" x 2" x 1/4" angles, 3.2#/L.F.	"	150	.107	"	10.60	5.65		16.25	21
9000	Minimum labor/equipment charge	1 Carp	2	4	Job		213		213	340

05 12 23.20 Curb Edging

		Crew	Daily Output	Labor-Hours	Unit	Material	2020 Bare Costs Labor	Equipment	Total	Total Incl O&P
0010	**CURB EDGING**									
0020	Steel angle w/anchors, shop fabricated, on forms, 1" x 1", 0.8#/L.F. G	E-4	350	.091	L.F.	1.74	5.30	.42	7.46	11.05
0300	4" x 4" angles, 8.2#/L.F. G		275	.116	"	15.30	6.75	.53	22.58	28.50
9000	Minimum labor/equipment charge		4	8	Job		465	36.50	501.50	800

05 12 23.40 Lightweight Framing

		Crew	Daily Output	Labor-Hours	Unit	Material	2020 Bare Costs Labor	Equipment	Total	Total Incl O&P
0010	**LIGHTWEIGHT FRAMING**									
0015	Made from recycled materials									
0200	For load-bearing steel studs see Section 05 41 13.30									
0400	Angle framing, field fabricated, 4" and larger G	E-3	440	.055	Lb.	.83	3.18	.33	4.34	6.50
0450	Less than 4" angles G		265	.091		.86	5.30	.56	6.72	10.20
0600	Channel framing, field fabricated, 8" and larger G		500	.048		.86	2.80	.29	3.95	5.85
0650	Less than 8" channels G		335	.072		.86	4.17	.44	5.47	8.30
1000	Continuous slotted channel framing system, shop fab, simple framing G	2 Sswk	2400	.007		4.45	.38		4.83	5.55
1200	Complex framing G	"	1600	.010		5.05	.58		5.63	6.50
1250	Plate & bar stock for reinforcing beams and trusses G					1.58			1.58	1.74
1300	Cross bracing, rods, shop fabricated, 3/4" diameter G	E-3	700	.034		1.72	2	.21	3.93	5.40
1310	7/8" diameter G		850	.028		1.72	1.65	.17	3.54	4.79
1320	1" diameter G		1000	.024		1.72	1.40	.15	3.27	4.35
1330	Angle, 5" x 5" x 3/8" G		2800	.009		1.72	.50	.05	2.27	2.78
1350	Hanging lintels, shop fabricated G		850	.028		1.72	1.65	.17	3.54	4.79
1380	Roof frames, shop fabricated, 3'-0" square, 5' span G	E-2	4200	.013		1.72	.76	.40	2.88	3.57
1400	Tie rod, not upset, 1-1/2" to 4" diameter, with turnbuckle G	2 Sswk	800	.020		1.87	1.15		3.02	3.94
1420	No turnbuckle G		700	.023		1.80	1.32		3.12	4.14
1500	Upset, 1-3/4" to 4" diameter, with turnbuckle G		800	.020		1.87	1.15		3.02	3.94
1520	No turnbuckle G		700	.023		1.80	1.32		3.12	4.14
9000	Minimum labor/equipment charge		2	8	Job		460		460	755

05 12 Structural Steel Framing

05 12 23 – Structural Steel for Buildings

05 12 23.45 Lintels		Crew	Daily Output	Labor- Hours	Unit	Material	2020 Bare Costs Labor	Equipment	Total	Total Incl O&P
0010	**LINTELS**									
0015	Made from recycled materials									
0020	Plain steel angles, shop fabricated, under 500 lb.	G	1 Bric	550	.015	Lb.	1.11	.76		1.87	2.44
0100	500 to 1,000 lb.	G		640	.013		1.08	.65		1.73	2.24
0200	1,000 to 2,000 lb.	G		640	.013		1.05	.65		1.70	2.20
0300	2,000 to 4,000 lb.	G		640	.013		1.02	.65		1.67	2.17
0500	For built-up angles and plates, add to above	G					1.44			1.44	1.58
0700	For engineering, add to above						.14			.14	.16
0900	For galvanizing, add to above, under 500 lb.						.32			.32	.35
0950	500 to 2,000 lb.						.29			.29	.32
1000	Over 2,000 lb.						.29			.29	.31
2000	Steel angles, 3-1/2" x 3", 1/4" thick, 2'-6" long	G	1 Bric	47	.170	Ea.	15.50	8.85		24.35	31.50
2100	4'-6" long	G		26	.308		28	16		44	56.50
2500	3-1/2" x 3-1/2" x 5/16", 5'-0" long	G		18	.444		41.50	23		64.50	83
2600	4" x 3-1/2", 1/4" thick, 5'-0" long	G		21	.381		35.50	19.85		55.35	71
2700	9'-0" long	G		12	.667		64	34.50		98.50	127
2800	4" x 3-1/2" x 5/16", 7'-0" long	G		12	.667		62	34.50		96.50	124
2900	5" x 3-1/2" x 5/16", 10'-0" long	G		8	1		100	52		152	194
9000	Minimum labor/equipment charge			4	2	Job		104		104	168

05 12 23.60 Pipe Support Framing

0010	**PIPE SUPPORT FRAMING**									
0020	Under 10#/L.F., shop fabricated	G	E-4	3900	.008	Lb.	1.93	.48	.04	2.45	2.94
0200	10.1 to 15#/L.F.	G		4300	.007		1.90	.43	.03	2.36	2.84
0400	15.1 to 20#/L.F.	G		4800	.007		1.87	.39	.03	2.29	2.72
0600	Over 20#/L.F.	G		5400	.006		1.84	.34	.03	2.21	2.62

05 12 23.65 Plates

0010	**PLATES**									
0015	Made from recycled materials									
0020	For connections & stiffener plates, shop fabricated									
0050	1/8" thick (5.1 lb./S.F.)	G				S.F.	7.35			7.35	8.05
0100	1/4" thick (10.2 lb./S.F.)	G					14.65			14.65	16.10
0300	3/8" thick (15.3 lb./S.F.)	G					22			22	24
0400	1/2" thick (20.4 lb./S.F.)	G					29.50			29.50	32
0450	3/4" thick (30.6 lb./S.F.)	G					44			44	48.50
0500	1" thick (40.8 lb./S.F.)	G					58.50			58.50	64.50
2000	Steel plate, warehouse prices, no shop fabrication									
2100	1/4" thick (10.2 lb./S.F.)	G				S.F.	7.20			7.20	7.90

05 12 23.75 Structural Steel Members

0010	**STRUCTURAL STEEL MEMBERS** R051223-10									
0015	Made from recycled materials									
0020	Shop fab'd for 100-ton, 1-2 story project, bolted connections									
0102	Beam or girder, W 6 x 9 R051223-15	G	E-2	600	.093	L.F.	14.25	5.35	2.83	22.43	27.50
0302	W 8 x 10	G		600	.093		15.80	5.35	2.83	23.98	29
0502	x 31	G		550	.102		49	5.80	3.08	57.88	67
0702	W 10 x 22	G		600	.093		35	5.35	2.83	43.18	50.50
0902	x 49	G		550	.102		77.50	5.80	3.08	86.38	98
1102	W 12 x 16	G		880	.064		25.50	3.64	1.93	31.07	36
1302	x 22	G		880	.064		35	3.64	1.93	40.57	46.50
1502	x 26	G		880	.064		41	3.64	1.93	46.57	53
1902	W 14 x 26	G		990	.057		41	3.23	1.71	45.94	52
2102	x 30	G		900	.062		47.50	3.56	1.88	52.94	60

05 12 Structural Steel Framing

05 12 23 – Structural Steel for Buildings

05 12 23.75 Structural Steel Members

		Crew	Daily Output	Labor-Hours	Unit	Material	2020 Bare Costs Labor	Equipment	Total	Total Incl O&P
2702	W 16 x 26 **G**	E-2	1000	.056	L.F.	41	3.20	1.70	45.90	52
2902	x 31 **G**	↓	900	.062		49	3.56	1.88	54.44	62
8490	For projects 75 to 99 tons, add					10%				
8492	50 to 74 tons, add					20%				
8494	25 to 49 tons, add					30%	10%			
8496	10 to 24 tons, add					50%	25%			
8498	2 to 9 tons, add					75%	50%			
8499	Less than 2 tons, add				↓	100%	100%			
9000	Minimum labor/equipment charge	E-2	2	28	Job		1,600	850	2,450	3,525

05 12 23.77 Structural Steel Projects

				Crew	Daily Output	Labor-Hours	Unit	Material	2020 Bare Costs Labor	Equipment	Total	Total Incl O&P
0010	**STRUCTURAL STEEL PROJECTS**											
0015	Made from recycled materials											
0020	Shop fab'd for 100-ton, 1-2 story project, bolted connections											
0700	Offices, hospitals, etc., steel bearing, 1 to 2 stories	R050521-20	**G**	E-5	10.30	7.767	Ton	2,875	445	179	3,499	4,075
1300	Industrial bldgs., 1 story, beams & girders, steel bearing		**G**		12.90	6.202		2,875	355	143	3,373	3,875
1400	Masonry bearing	R051223-10	**G**		10	8		2,875	460	184	3,519	4,100
1600	1 story with roof trusses, steel bearing		**G**		10.60	7.547		3,400	435	174	4,009	4,625
1700	Masonry bearing	R051223-15	**G**	↓	8.30	9.639		3,400	555	222	4,177	4,875
5390	For projects 75 to 99 tons, add							10%				
5392	50 to 74 tons, add							20%				
5394	25 to 49 tons, add							30%	10%			
5396	10 to 24 tons, add							50%	25%			
5398	2 to 9 tons, add							75%	50%			
5399	Less than 2 tons, add						↓	100%	100%			

05 12 23.78 Structural Steel Secondary Members

		Crew	Daily Output	Labor-Hours	Unit	Material	2020 Bare Costs Labor	Equipment	Total	Total Incl O&P
0010	**STRUCTURAL STEEL SECONDARY MEMBERS**									
0015	Made from recycled materials									
0020	Shop fabricated for 20-ton girt/purlin framing package, materials only									
0100	Girts/purlins, C/Z-shapes, includes clips and bolts									
0110	6" x 2-1/2" x 2-1/2", 16 ga., 3.0 lb./L.F.				L.F.	3.88			3.88	4.27
0115	14 ga., 3.5 lb./L.F.					4.53			4.53	4.98
0120	8" x 2-3/4" x 2-3/4", 16 ga., 3.4 lb./L.F.					4.40			4.40	4.84
0125	14 ga., 4.1 lb./L.F.					5.30			5.30	5.85
0130	12 ga., 5.6 lb./L.F.					7.25			7.25	7.95
0135	10" x 3-1/2" x 3-1/2", 14 ga., 4.7 lb./L.F.					6.10			6.10	6.70
0140	12 ga., 6.7 lb./L.F.					8.65			8.65	9.55
0145	12" x 3-1/2" x 3-1/2", 14 ga., 5.3 lb./L.F.					6.85			6.85	7.55
0150	12 ga., 7.4 lb./L.F.				↓	9.55			9.55	10.55
0200	Eave struts, C-shape, includes clips and bolts									
0210	6" x 4" x 3", 16 ga., 3.1 lb./L.F.				L.F.	4.01			4.01	4.41
0215	14 ga., 3.9 lb./L.F.					5.05			5.05	5.55
0220	8" x 4" x 3", 16 ga., 3.5 lb./L.F.					4.53			4.53	4.98
0225	14 ga., 4.4 lb./L.F.					5.70			5.70	6.25
0230	12 ga., 6.2 lb./L.F.					8			8	8.80
0235	10" x 5" x 3", 14 ga., 5.2 lb./L.F.					6.70			6.70	7.40
0240	12 ga., 7.3 lb./L.F.					9.45			9.45	10.40
0245	12" x 5" x 4", 14 ga., 6.0 lb./L.F.					7.75			7.75	8.55
0250	12 ga., 8.4 lb./L.F.				↓	10.85			10.85	11.95
0300	Rake/base angle, excludes concrete drilling and expansion anchors									
0310	2" x 2", 14 ga., 1.0 lb./L.F.	2 Sswk	640	.025	L.F.	1.29	1.44		2.73	3.78
0315	3" x 2", 14 ga., 1.3 lb./L.F.		535	.030		1.68	1.72		3.40	4.68
0320	3" x 3", 14 ga., 1.6 lb./L.F.	↓	500	.032	↓	2.07	1.84		3.91	5.30

05 12 Structural Steel Framing

05 12 23 – Structural Steel for Buildings

05 12 23.78 Structural Steel Secondary Members		Crew	Daily Output	Labor-Hours	Unit	Material	2020 Bare Costs Labor	Equipment	Total	Total Incl O&P
0325	4" x 3", 14 ga., 1.8 lb./L.F.	2 Sswk	480	.033	L.F.	2.33	1.92		4.25	5.70
0600	Installation of secondary members, erection only									
0610	Girts, purlins, eave struts, 16 ga., 6" deep	E-18	100	.400	Ea.		23	14.90	37.90	54
0615	8" deep		80	.500			29	18.60	47.60	67.50
0620	14 ga., 6" deep		80	.500			29	18.60	47.60	67.50
0625	8" deep		65	.615			35.50	23	58.50	83
0630	10" deep		55	.727			42	27	69	98.50
0635	12" deep		50	.800			46.50	30	76.50	108
0640	12 ga., 8" deep		50	.800			46.50	30	76.50	108
0645	10" deep		45	.889			51.50	33	84.50	120
0650	12" deep	↓	40	1	↓		58	37	95	135
0900	For less than 20-ton job lots									
0905	For 15 to 19 tons, add				%	10%				
0910	For 10 to 14 tons, add					25%				
0915	For 5 to 9 tons, add					50%	50%	50%		
0920	For 1 to 4 tons, add					75%	75%	75%		
0925	For less than 1 ton, add				↓	100%	100%	100%		
0990	Minimum labor/equipment charge	E-18	1	40	Job		2,325	1,500	3,825	5,375

05 15 Wire Rope Assemblies

05 15 16 – Steel Wire Rope Assemblies

05 15 16.70 Temporary Cable Safety Railing

0010	**TEMPORARY CABLE SAFETY RAILING**, Each 100' strand incl.									
0020	2 eyebolts, 1 turnbuckle, 100' cable, 2 thimbles, 6 clips									
0025	Made from recycled materials									
0100	One strand using 1/4" cable & accessories	G	2 Sswk	4	4	C.L.F.	158	231		389	555
0200	1/2" cable & accessories	G	"	2	8	"	345	460		805	1,125

05 21 Steel Joist Framing

05 21 13 – Deep Longspan Steel Joist Framing

05 21 13.50 Deep Longspan Joists

0010	**DEEP LONGSPAN JOISTS**									
3010	DLH series, 40-ton job lots, bolted cross bridging, shop primer									
3015	Made from recycled materials									
3040	Spans to 144' (shipped in 2 pieces)	G	E-7	13	6.154	Ton	2,275	355	153	2,783	3,250
3500	For less than 40-ton job lots									
3502	For 30 to 39 tons, add					%	10%			
3504	20 to 29 tons, add						20%			
3506	10 to 19 tons, add						30%			
3507	5 to 9 tons, add						50%	25%		
3508	1 to 4 tons, add						75%	50%		
3509	Less than 1 ton, add					↓	100%	100%		
4010	SLH series, 40-ton job lots, bolted cross bridging, shop primer									
4040	Spans to 200' (shipped in 3 pieces)	G	E-7	13	6.154	Ton	2,350	355	153	2,858	3,350
6100	For less than 40-ton job lots									
6102	For 30 to 39 tons, add					%	10%			
6104	20 to 29 tons, add						20%			
6106	10 to 19 tons, add						30%			
6107	5 to 9 tons, add						50%	25%		

For customer support on your Commercial Renovation Costs with RSMeans data, call 800.448.8182.

113

05 21 Steel Joist Framing

05 21 13 – Deep Longspan Steel Joist Framing

05 21 13.50 Deep Longspan Joists		Crew	Daily Output	Labor-Hours	Unit	Material	2020 Bare Costs Labor	Equipment	Total	Total Incl O&P
6108	1 to 4 tons, add				%	75%	50%			
6109	Less than 1 ton, add					100%	100%			

05 21 16 – Longspan Steel Joist Framing

05 21 16.50 Longspan Joists

			Crew	Daily Output	Labor-Hours	Unit	Material	2020 Bare Costs Labor	Equipment	Total	Total Incl O&P
0010	**LONGSPAN JOISTS**										
2000	LH series, 40-ton job lots, bolted cross bridging, shop primer										
2015	Made from recycled materials										
2040	Longspan joists, LH series, up to 96'	G	E-7	13	6.154	Ton	2,175	355	153	2,683	3,150
2600	For less than 40-ton job lots										
2602	For 30 to 39 tons, add					%	10%				
2604	20 to 29 tons, add						20%				
2606	10 to 19 tons, add						30%				
2607	5 to 9 tons, add						50%	25%			
2608	1 to 4 tons, add						75%	50%			
2609	Less than 1 ton, add						100%	100%			

05 21 19 – Open Web Steel Joist Framing

05 21 19.10 Open Web Joists

			Crew	Daily Output	Labor-Hours	Unit	Material	2020 Bare Costs Labor	Equipment	Total	Total Incl O&P
0010	**OPEN WEB JOISTS**										
0015	Made from recycled materials										
0050	K series, 40-ton lots, horiz. bridging, spans to 30', shop primer	G	E-7	12	6.667	Ton	1,925	385	166	2,476	2,925
0440	K series, 30' to 50' spans	G	"	17	4.706	"	1,950	270	117	2,337	2,725
0800	For less than 40-ton job lots										
0802	For 30 to 39 tons, add					%	10%				
0804	20 to 29 tons, add						20%				
0806	10 to 19 tons, add						30%				
0807	5 to 9 tons, add						50%	25%			
0808	1 to 4 tons, add						75%	50%			
0809	Less than 1 ton, add						100%	100%			
1010	CS series, 40-ton job lots, horizontal bridging, shop primer										
1040	Spans to 30'	G	E-7	12	6.667	Ton	2,000	385	166	2,551	3,000
1500	For less than 40-ton job lots										
1502	For 30 to 39 tons, add					%	10%				
1504	20 to 29 tons, add						20%				
1506	10 to 19 tons, add						30%				
1507	5 to 9 tons, add						50%	25%			
1508	1 to 4 tons, add						75%	50%			
1509	Less than 1 ton, add						100%	100%			
6400	Individual steel bearing plate, 6" x 6" x 1/4" with J-hook	G	1 Bric	160	.050	Ea.	8.60	2.60		11.20	13.70
9000	Minimum labor and equipment, bar joists		F-6	1	40	Job		2,000	470	2,470	3,700

05 31 Steel Decking

05 31 13 – Steel Floor Decking

05 31 13.50 Floor Decking

		Crew	Daily Output	Labor-Hours	Unit	Material	2020 Bare Costs Labor	Equipment	Total	Total Incl O&P
0010	**FLOOR DECKING** R053100-10									
0015	Made from recycled materials									
5100	Non-cellular composite decking, galvanized, 1-1/2" deep, 16 ga. [G]	E-4	3500	.009	S.F.	4.82	.53	.04	5.39	6.20
5120	18 ga. [G]		3650	.009		3	.51	.04	3.55	4.17
5140	20 ga. [G]		3800	.008		3.52	.49	.04	4.05	4.71
5200	2" deep, 22 ga. [G]		3860	.008		2.38	.48	.04	2.90	3.45
5300	20 ga. [G]		3600	.009		3.38	.52	.04	3.94	4.61
5400	18 ga. [G]		3380	.009		3.12	.55	.04	3.71	4.38
5500	16 ga. [G]		3200	.010		4.77	.58	.05	5.40	6.25
5700	3" deep, 22 ga. [G]		3200	.010		2.59	.58	.05	3.22	3.85
5800	20 ga. [G]		3000	.011		3.72	.62	.05	4.39	5.15
5900	18 ga. [G]		2850	.011		3.12	.65	.05	3.82	4.56
6000	16 ga. [G]		2700	.012		5.40	.69	.05	6.14	7.15
9000	Minimum labor/equipment charge	1 Sswk	1	8	Job		460		460	755

05 31 23 – Steel Roof Decking

05 31 23.50 Roof Decking

		Crew	Daily Output	Labor-Hours	Unit	Material	2020 Bare Costs Labor	Equipment	Total	Total Incl O&P
0010	**ROOF DECKING**									
0015	Made from recycled materials									
2100	Open type, 1-1/2" deep, Type B, wide rib, galv., 22 ga., under 50 sq. [G]	E-4	4500	.007	S.F.	2.68	.41	.03	3.12	3.67
2400	Over 500 squares [G]	"	5100	.006	"	1.93	.36	.03	2.32	2.75

05 31 33 – Steel Form Decking

05 31 33.50 Form Decking

		Crew	Daily Output	Labor-Hours	Unit	Material	2020 Bare Costs Labor	Equipment	Total	Total Incl O&P
0010	**FORM DECKING**									
0015	Made from recycled materials									
6100	Slab form, steel, 28 ga., 9/16" deep, Type UFS, uncoated [G]	E-4	4000	.008	S.F.	1.90	.47	.04	2.41	2.89
6200	Galvanized [G]	"	4000	.008	"	1.68	.47	.04	2.19	2.65

05 35 Raceway Decking Assemblies

05 35 13 – Steel Cellular Decking

05 35 13.50 Cellular Decking

		Crew	Daily Output	Labor-Hours	Unit	Material	2020 Bare Costs Labor	Equipment	Total	Total Incl O&P
0010	**CELLULAR DECKING**									
0015	Made from recycled materials									
0200	Cellular units, galv, 1-1/2" deep, Type BC, 20-20 ga., over 15 squares [G]	E-4	1460	.022	S.F.	11.55	1.27	.10	12.92	14.90
0400	3" deep, Type NC, galvanized, 20-20 ga. [G]		1375	.023		12.70	1.35	.11	14.16	16.35
1000	4-1/2" deep, Type JC, galvanized, 18-20 ga. [G]		1100	.029		13.85	1.69	.13	15.67	18.15
1500	For acoustical deck, add					15%				
1900	For multi-story or congested site, add						50%			

05 41 Structural Metal Stud Framing

05 41 13 – Load-Bearing Metal Stud Framing

05 41 13.05 Bracing

		Crew	Daily Output	Labor-Hours	Unit	Material	2020 Bare Costs Labor	Equipment	Total	Total Incl O&P
0010	**BRACING**, shear wall X-bracing, per 10' x 10' bay, one face									
0015	Made of recycled materials									
0120	Metal strap, 20 ga. x 4" wide [G]	2 Carp	18	.889	Ea.	19.35	47		66.35	97
0130	6" wide [G]		18	.889		32	47		79	111
0160	18 ga. x 4" wide [G]		16	1		32	53		85	120
0170	6" wide [G]		16	1		47	53		100	137
0410	Continuous strap bracing, per horizontal row on both faces									

For customer support on your Commercial Renovation Costs with RSMeans data, call 800.448.8182.

115

05 41 Structural Metal Stud Framing

05 41 13 – Load-Bearing Metal Stud Framing

05 41 13.05 Bracing		Crew	Daily Output	Labor-Hours	Unit	Material	2020 Bare Costs Labor	Equipment	Total	Total Incl O&P
0420	Metal strap, 20 ga. x 2" wide, studs 12" OC	G	1 Carp	7	1.143	C.L.F.	57	60.50		117.50	161
0430	16" OC	G		8	1		57	53		110	148
0440	24" OC	G		10	.800		57	42.50		99.50	131
0450	18 ga. x 2" wide, studs 12" OC	G		6	1.333		77	71		148	199
0460	16" OC	G		7	1.143		77	60.50		137.50	182
0470	24" OC	G		8	1		77	53		130	170

05 41 13.10 Bridging		Crew	Daily Output	Labor-Hours	Unit	Material	2020 Bare Costs Labor	Equipment	Total	Total Incl O&P
0010	**BRIDGING**, solid between studs w/1-1/4" leg track, per stud bay										
0015	Made from recycled materials										
0200	Studs 12" OC, 18 ga. x 2-1/2" wide	G	1 Carp	125	.064	Ea.	.93	3.40		4.33	6.45
0210	3-5/8" wide	G		120	.067		1.11	3.54		4.65	6.90
0220	4" wide	G		120	.067		1.11	3.54		4.65	6.90
0230	6" wide	G		115	.070		1.53	3.70		5.23	7.60
0240	8" wide	G		110	.073		1.88	3.87		5.75	8.25
0300	16 ga. x 2-1/2" wide	G		115	.070		1.15	3.70		4.85	7.15
0310	3-5/8" wide	G		110	.073		1.39	3.87		5.26	7.75
0320	4" wide	G		110	.073		1.54	3.87		5.41	7.90
0330	6" wide	G		105	.076		1.94	4.05		5.99	8.65
0340	8" wide	G		100	.080		2.39	4.25		6.64	9.45
1200	Studs 16" OC, 18 ga. x 2-1/2" wide	G		125	.064		1.19	3.40		4.59	6.75
1210	3-5/8" wide	G		120	.067		1.42	3.54		4.96	7.25
1220	4" wide	G		120	.067		1.42	3.54		4.96	7.25
1230	6" wide	G		115	.070		1.96	3.70		5.66	8.05
1240	8" wide	G		110	.073		2.41	3.87		6.28	8.85
1300	16 ga. x 2-1/2" wide	G		115	.070		1.47	3.70		5.17	7.50
1310	3-5/8" wide	G		110	.073		1.78	3.87		5.65	8.15
1320	4" wide	G		110	.073		1.98	3.87		5.85	8.40
1330	6" wide	G		105	.076		2.49	4.05		6.54	9.25
1340	8" wide	G		100	.080		3.07	4.25		7.32	10.20
2200	Studs 24" OC, 18 ga. x 2-1/2" wide	G		125	.064		1.72	3.40		5.12	7.35
2210	3-5/8" wide	G		120	.067		2.05	3.54		5.59	7.95
2220	4" wide	G		120	.067		2.05	3.54		5.59	7.95
2230	6" wide	G		115	.070		2.84	3.70		6.54	9
2240	8" wide	G		110	.073		3.49	3.87		7.36	10.05
2300	16 ga. x 2-1/2" wide	G		115	.070		2.12	3.70		5.82	8.25
2310	3-5/8" wide	G		110	.073		2.58	3.87		6.45	9.05
2320	4" wide	G		110	.073		2.86	3.87		6.73	9.35
2330	6" wide	G		105	.076		3.60	4.05		7.65	10.45
2340	8" wide	G		100	.080		4.44	4.25		8.69	11.70
3000	Continuous bridging, per row										
3100	16 ga. x 1-1/2" channel thru studs 12" OC	G	1 Carp	6	1.333	C.L.F.	51.50	71		122.50	171
3110	16" OC	G		7	1.143		51.50	60.50		112	155
3120	24" OC	G		8.80	.909		51.50	48.50		100	135
4100	2" x 2" angle x 18 ga., studs 12" OC	G		7	1.143		78	60.50		138.50	184
4110	16" OC	G		9	.889		78	47		125	162
4120	24" OC	G		12	.667		78	35.50		113.50	143
4200	16 ga., studs 12" OC	G		5	1.600		98	85		183	244
4210	16" OC	G		7	1.143		98	60.50		158.50	206
4220	24" OC	G		10	.800		98	42.50		140.50	176

05 41 Structural Metal Stud Framing

05 41 13 – Load-Bearing Metal Stud Framing

05 41 13.25 Framing, Boxed Headers/Beams

		Crew	Daily Output	Labor-Hours	Unit	Material	2020 Bare Costs Labor	Equipment	Total	Total Incl O&P
0010	**FRAMING, BOXED HEADERS/BEAMS**									
0015	Made from recycled materials									
0200	Double, 18 ga. x 6" deep	G 2 Carp	220	.073	L.F.	5.55	3.87		9.42	12.30
0210	8" deep	G	210	.076		5.80	4.05		9.85	12.85
0220	10" deep	G	200	.080		7.30	4.25		11.55	14.85
0230	12" deep	G	190	.084		7.95	4.48		12.43	15.90
0300	16 ga. x 8" deep	G	180	.089		6.70	4.72		11.42	14.90
0310	10" deep	G	170	.094		8.30	5		13.30	17.15
0320	12" deep	G	160	.100		9	5.30		14.30	18.40
0400	14 ga. x 10" deep	G	140	.114		9.60	6.05		15.65	20.50
0410	12" deep	G	130	.123		10.50	6.55		17.05	22
1210	Triple, 18 ga. x 8" deep	G	170	.094		8.55	5		13.55	17.40
1220	10" deep	G	165	.097		10.40	5.15		15.55	19.70
1230	12" deep	G	160	.100		11.40	5.30		16.70	21
1300	16 ga. x 8" deep	G	145	.110		9.95	5.85		15.80	20.50
1310	10" deep	G	140	.114		11.90	6.05		17.95	23
1320	12" deep	G	135	.119		13	6.30		19.30	24.50
1400	14 ga. x 10" deep	G	115	.139		13.10	7.40		20.50	26.50
1410	12" deep	G	110	.145		14.45	7.75		22.20	28.50

05 41 13.30 Framing, Stud Walls

		Crew	Daily Output	Labor-Hours	Unit	Material	2020 Bare Costs Labor	Equipment	Total	Total Incl O&P
0010	**FRAMING, STUD WALLS** w/top & bottom track, no openings,									
0020	Headers, beams, bridging or bracing									
0025	Made from recycled materials									
4100	8' high walls, 18 ga. x 2-1/2" wide, studs 12" OC	G 2 Carp	54	.296	L.F.	8.40	15.75		24.15	34.50
4110	16" OC	G	77	.208		6.75	11.05		17.80	25
4120	24" OC	G	107	.150		5.10	7.95		13.05	18.35
4130	3-5/8" wide, studs 12" OC	G	53	.302		9.95	16.05		26	36.50
4140	16" OC	G	76	.211		8	11.20		19.20	27
4150	24" OC	G	105	.152		6.05	8.10		14.15	19.65
4160	4" wide, studs 12" OC	G	52	.308		10.25	16.35		26.60	37.50
4170	16" OC	G	74	.216		8.25	11.50		19.75	27.50
4180	24" OC	G	103	.155		6.20	8.25		14.45	20
4190	6" wide, studs 12" OC	G	51	.314		13.25	16.65		29.90	41
4200	16" OC	G	73	.219		10.65	11.65		22.30	30.50
4210	24" OC	G	101	.158		8.10	8.40		16.50	22.50
4220	8" wide, studs 12" OC	G	50	.320		16.95	17		33.95	46
4230	16" OC	G	72	.222		13.60	11.80		25.40	34
4240	24" OC	G	100	.160		10.25	8.50		18.75	25
4300	16 ga. x 2-1/2" wide, studs 12" OC	G	47	.340		10.50	18.10		28.60	40.50
4310	16" OC	G	68	.235		8.30	12.50		20.80	29
4320	24" OC	G	94	.170		6.10	9.05		15.15	21.50
4330	3-5/8" wide, studs 12" OC	G	46	.348		12.35	18.50		30.85	43
4340	16" OC	G	66	.242		9.80	12.90		22.70	31.50
4350	24" OC	G	92	.174		7.25	9.25		16.50	23
4360	4" wide, studs 12" OC	G	45	.356		13.05	18.90		31.95	45
4370	16" OC	G	65	.246		10.35	13.10		23.45	32.50
4380	24" OC	G	90	.178		7.60	9.45		17.05	23.50
4390	6" wide, studs 12" OC	G	44	.364		16.30	19.35		35.65	49
4400	16" OC	G	64	.250		12.95	13.30		26.25	36
4410	24" OC	G	88	.182		9.60	9.65		19.25	26
4420	8" wide, studs 12" OC	G	43	.372		20	19.80		39.80	53.50
4430	16" OC	G	63	.254		16	13.50		29.50	39

For customer support on your Commercial Renovation Costs with RSMeans data, call 800.448.8182.

117

05 41 13 – Load-Bearing Metal Stud Framing

05 41 13.30 Framing, Stud Walls		Crew	Daily Output	Labor-Hours	Unit	Material	2020 Bare Costs Labor	Equipment	Total	Total Incl O&P
4440	24" OC	G	2 Carp	86	.186	L.F.	11.85	9.90		21.75	29
5100	10' high walls, 18 ga. x 2-1/2" wide, studs 12" OC	G		54	.296		10.05	15.75		25.80	36
5110	16" OC	G		77	.208		8	11.05		19.05	26.50
5120	24" OC	G		107	.150		5.90	7.95		13.85	19.25
5130	3-5/8" wide, studs 12" OC	G		53	.302		11.90	16.05		27.95	38.50
5140	16" OC	G		76	.211		9.45	11.20		20.65	28.50
5150	24" OC	G		105	.152		7	8.10		15.10	20.50
5160	4" wide, studs 12" OC	G		52	.308		12.30	16.35		28.65	39.50
5170	16" OC	G		74	.216		9.75	11.50		21.25	29
5180	24" OC	G		103	.155		7.20	8.25		15.45	21
5190	6" wide, studs 12" OC	G		51	.314		15.80	16.65		32.45	44
5200	16" OC	G		73	.219		12.60	11.65		24.25	32.50
5210	24" OC	G		101	.158		9.35	8.40		17.75	24
5220	8" wide, studs 12" OC	G		50	.320		20.50	17		37.50	50
5230	16" OC	G		72	.222		16.10	11.80		27.90	36.50
5240	24" OC	G		100	.160		11.95	8.50		20.45	26.50
5300	16 ga. x 2-1/2" wide, studs 12" OC	G		47	.340		12.65	18.10		30.75	43
5310	16" OC	G		68	.235		9.95	12.50		22.45	31
5320	24" OC	G		94	.170		7.20	9.05		16.25	22.50
5330	3-5/8" wide, studs 12" OC	G		46	.348		14.90	18.50		33.40	46
5340	16" OC	G		66	.242		11.70	12.90		24.60	33.50
5350	24" OC	G		92	.174		8.50	9.25		17.75	24
5360	4" wide, studs 12" OC	G		45	.356		15.80	18.90		34.70	48
5370	16" OC	G		65	.246		12.40	13.10		25.50	34.50
5380	24" OC	G		90	.178		8.95	9.45		18.40	25
5390	6" wide, studs 12" OC	G		44	.364		19.60	19.35		38.95	52.50
5400	16" OC	G		64	.250		15.45	13.30		28.75	38.50
5410	24" OC	G		88	.182		11.25	9.65		20.90	28
5420	8" wide, studs 12" OC	G		43	.372		24.50	19.80		44.30	58
5430	16" OC	G		63	.254		19.10	13.50		32.60	42.50
5440	24" OC	G		86	.186		13.95	9.90		23.85	31
6190	12' high walls, 18 ga. x 6" wide, studs 12" OC	G		41	.390		18.40	20.50		38.90	53
6200	16" OC	G		58	.276		14.55	14.65		29.20	39.50
6210	24" OC	G		81	.198		10.65	10.50		21.15	28.50
6220	8" wide, studs 12" OC	G		40	.400		23.50	21.50		45	60
6230	16" OC	G		57	.281		18.60	14.90		33.50	44.50
6240	24" OC	G		80	.200		13.60	10.65		24.25	32
6390	16 ga. x 6" wide, studs 12" OC	G		35	.457		23	24.50		47.50	64.50
6400	16" OC	G		51	.314		17.95	16.65		34.60	46.50
6410	24" OC	G		70	.229		12.95	12.15		25.10	33.50
6420	8" wide, studs 12" OC	G		34	.471		28.50	25		53.50	71.50
6430	16" OC	G		50	.320		22	17		39	52
6440	24" OC	G		69	.232		16	12.30		28.30	37.50
6530	14 ga. x 3-5/8" wide, studs 12" OC	G		34	.471		22	25		47	64.50
6540	16" OC	G		48	.333		17.30	17.70		35	47.50
6550	24" OC	G		65	.246		12.45	13.10		25.55	34.50
6560	4" wide, studs 12" OC	G		33	.485		23.50	26		49.50	67.50
6570	16" OC	G		47	.340		18.35	18.10		36.45	49
6580	24" OC	G		64	.250		13.20	13.30		26.50	36
6730	12 ga. x 3-5/8" wide, studs 12" OC	G		31	.516		30.50	27.50		58	78
6740	16" OC	G		43	.372		23.50	19.80		43.30	57.50
6750	24" OC	G		59	.271		16.70	14.40		31.10	41.50
6760	4" wide, studs 12" OC	G		30	.533		33	28.50		61.50	82

05 41 Structural Metal Stud Framing

05 41 13 – Load-Bearing Metal Stud Framing

05 41 13.30 Framing, Stud Walls

		Crew	Daily Output	Labor-Hours	Unit	Material	2020 Bare Costs Labor	Equipment	Total	Total Incl O&P
6770	16" OC	G	2 Carp	42	.381	L.F.	25.50	20.50		46	60.50
6780	24" OC	G		58	.276		18	14.65		32.65	43.50
7390	16' high walls, 16 ga. x 6" wide, studs 12" OC	G		33	.485		29.50	26		55.50	74
7400	16" OC	G		48	.333		23	17.70		40.70	54
7410	24" OC	G		67	.239		16.30	12.70		29	38.50
7420	8" wide, studs 12" OC	G		32	.500		36.50	26.50		63	83
7430	16" OC	G		47	.340		28.50	18.10		46.60	60.50
7440	24" OC	G		66	.242		20	12.90		32.90	42.50
7560	14 ga. x 4" wide, studs 12" OC	G		31	.516		30.50	27.50		58	77.50
7570	16" OC	G		45	.356		23.50	18.90		42.40	56.50
7580	24" OC	G		61	.262		16.60	13.95		30.55	41
7590	6" wide, studs 12" OC	G		30	.533		38.50	28.50		67	87.50
7600	16" OC	G		44	.364		29.50	19.35		48.85	63.50
7610	24" OC	G		60	.267		21	14.15		35.15	45.50
7760	12 ga. x 4" wide, studs 12" OC	G		29	.552		43	29.50		72.50	94.50
7770	16" OC	G		40	.400		33	21.50		54.50	70.50
7780	24" OC	G		55	.291		23	15.45		38.45	50.50
7790	6" wide, studs 12" OC	G		28	.571		54	30.50		84.50	108
7800	16" OC	G		39	.410		41.50	22		63.50	80.50
7810	24" OC	G		54	.296		29	15.75		44.75	57
8590	20' high walls, 14 ga. x 6" wide, studs 12" OC	G		29	.552		47	29.50		76.50	98.50
8600	16" OC	G		42	.381		36	20.50		56.50	72
8610	24" OC	G		57	.281		25.50	14.90		40.40	52
8620	8" wide, studs 12" OC	G		28	.571		53	30.50		83.50	107
8630	16" OC	G		41	.390		40.50	20.50		61	78
8640	24" OC	G		56	.286		28.50	15.20		43.70	56
8790	12 ga. x 6" wide, studs 12" OC	G		27	.593		66.50	31.50		98	124
8800	16" OC	G		37	.432		51	23		74	93
8810	24" OC	G		51	.314		35	16.65		51.65	65
8820	8" wide, studs 12" OC	G		26	.615		81	32.50		113.50	142
8830	16" OC	G		36	.444		62	23.50		85.50	106
8840	24" OC	G		50	.320		43	17		60	74.50
9000	Minimum labor/equipment charge			4	4	Job		213		213	340

05 42 Cold-Formed Metal Joist Framing

05 42 13 – Cold-Formed Metal Floor Joist Framing

05 42 13.05 Bracing

		Crew	Daily Output	Labor-Hours	Unit	Material	2020 Bare Costs Labor	Equipment	Total	Total Incl O&P
0010	**BRACING**, continuous, per row, top & bottom										
0015	Made from recycled materials										
0120	Flat strap, 20 ga. x 2" wide, joists at 12" OC	G	1 Carp	4.67	1.713	C.L.F.	60	91		151	212
0130	16" OC	G		5.33	1.501		57.50	80		137.50	192
0140	24" OC	G		6.66	1.201		55.50	64		119.50	163
0150	18 ga. x 2" wide, joists at 12" OC	G		4	2		76.50	106		182.50	254
0160	16" OC	G		4.67	1.713		75	91		166	229
0170	24" OC	G		5.33	1.501		73.50	80		153.50	209

05 42 13.10 Bridging

		Crew	Daily Output	Labor-Hours	Unit	Material	2020 Bare Costs Labor	Equipment	Total	Total Incl O&P
0010	**BRIDGING**, solid between joists w/1-1/4" leg track, per joist bay										
0015	Made from recycled materials										
0230	Joists 12" OC, 18 ga. track x 6" wide	G	1 Carp	80	.100	Ea.	1.53	5.30		6.83	10.20
0240	8" wide	G		75	.107		1.88	5.65		7.53	11.15
0250	10" wide	G		70	.114		2.39	6.05		8.44	12.40

For customer support on your Commercial Renovation Costs with RSMeans data, call 800.448.8182.

05 42 13.10 Bridging

		Crew	Daily Output	Labor-Hours	Unit	Material	2020 Bare Costs Labor	Equipment	Total	Total Incl O&P
0260	12" wide G	1 Carp	65	.123	Ea.	2.74	6.55		9.29	13.50
0330	16 ga. track x 6" wide G		70	.114		1.94	6.05		7.99	11.90
0340	8" wide G		65	.123		2.39	6.55		8.94	13.15
0350	10" wide G		60	.133		3.06	7.10		10.16	14.70
0360	12" wide G		55	.145		3.53	7.75		11.28	16.30
0440	14 ga. track x 8" wide G		60	.133		3.05	7.10		10.15	14.70
0450	10" wide G		55	.145		3.78	7.75		11.53	16.55
0460	12" wide G		50	.160		4.38	8.50		12.88	18.40
0550	12 ga. track x 10" wide G		45	.178		5.55	9.45		15	21.50
0560	12" wide G		40	.200		5.90	10.65		16.55	23.50
1230	16" OC, 18 ga. track x 6" wide G		80	.100		1.96	5.30		7.26	10.65
1240	8" wide G		75	.107		2.41	5.65		8.06	11.75
1250	10" wide G		70	.114		3.07	6.05		9.12	13.15
1260	12" wide G		65	.123		3.51	6.55		10.06	14.35
1330	16 ga. track x 6" wide G		70	.114		2.49	6.05		8.54	12.50
1340	8" wide G		65	.123		3.07	6.55		9.62	13.90
1350	10" wide G		60	.133		3.93	7.10		11.03	15.65
1360	12" wide G		55	.145		4.52	7.75		12.27	17.35
1440	14 ga. track x 8" wide G		60	.133		3.91	7.10		11.01	15.65
1450	10" wide G		55	.145		4.85	7.75		12.60	17.75
1460	12" wide G		50	.160		5.60	8.50		14.10	19.75
1550	12 ga. track x 10" wide G		45	.178		7.10	9.45		16.55	23
1560	12" wide G		40	.200		7.55	10.65		18.20	25.50
2230	24" OC, 18 ga. track x 6" wide G		80	.100		2.84	5.30		8.14	11.60
2240	8" wide G		75	.107		3.49	5.65		9.14	12.95
2250	10" wide G		70	.114		4.44	6.05		10.49	14.65
2260	12" wide G		65	.123		5.10	6.55		11.65	16.10
2330	16 ga. track x 6" wide G		70	.114		3.60	6.05		9.65	13.70
2340	8" wide G		65	.123		4.44	6.55		10.99	15.40
2350	10" wide G		60	.133		5.70	7.10		12.80	17.60
2360	12" wide G		55	.145		6.55	7.75		14.30	19.60
2440	14 ga. track x 8" wide G		60	.133		5.65	7.10		12.75	17.55
2450	10" wide G		55	.145		7	7.75		14.75	20
2460	12" wide G		50	.160		8.10	8.50		16.60	22.50
2550	12 ga. track x 10" wide G		45	.178		10.30	9.45		19.75	26.50
2560	12" wide G		40	.200		10.95	10.65		21.60	29

05 42 13.25 Framing, Band Joist

		Crew	Daily Output	Labor-Hours	Unit	Material	2020 Bare Costs Labor	Equipment	Total	Total Incl O&P
0010	**FRAMING, BAND JOIST** (track) fastened to bearing wall									
0015	Made from recycled materials									
0220	18 ga. track x 6" deep G	2 Carp	1000	.016	L.F.	1.25	.85		2.10	2.73
0230	8" deep G		920	.017		1.53	.92		2.45	3.17
0240	10" deep G		860	.019		1.95	.99		2.94	3.73
0320	16 ga. track x 6" deep G		900	.018		1.59	.95		2.54	3.25
0330	8" deep G		840	.019		1.95	1.01		2.96	3.77
0340	10" deep G		780	.021		2.50	1.09		3.59	4.50
0350	12" deep G		740	.022		2.88	1.15		4.03	5
0430	14 ga. track x 8" deep G		750	.021		2.49	1.13		3.62	4.56
0440	10" deep G		720	.022		3.09	1.18		4.27	5.30
0450	12" deep G		700	.023		3.57	1.22		4.79	5.90
0540	12 ga. track x 10" deep G		670	.024		4.53	1.27		5.80	7
0550	12" deep G		650	.025		4.81	1.31		6.12	7.40

05 42 Cold-Formed Metal Joist Framing

05 42 13 – Cold-Formed Metal Floor Joist Framing

05 42 13.30 Framing, Boxed Headers/Beams

		Crew	Daily Output	Labor-Hours	Unit	Material	2020 Bare Costs Labor	Equipment	Total	Total Incl O&P
0010	**FRAMING, BOXED HEADERS/BEAMS**									
0015	Made from recycled materials									
0200	Double, 18 ga. x 6" deep	G 2 Carp	220	.073	L.F.	5.55	3.87		9.42	12.30
0210	8" deep	G	210	.076		5.80	4.05		9.85	12.85
0220	10" deep	G	200	.080		7.30	4.25		11.55	14.85
0230	12" deep	G	190	.084		7.95	4.48		12.43	15.90
0300	16 ga. x 8" deep	G	180	.089		6.70	4.72		11.42	14.90
0310	10" deep	G	170	.094		8.30	5		13.30	17.15
0320	12" deep	G	160	.100		9	5.30		14.30	18.40
0400	14 ga. x 10" deep	G	140	.114		9.60	6.05		15.65	20.50
0410	12" deep	G	130	.123		10.50	6.55		17.05	22
0500	12 ga. x 10" deep	G	110	.145		12.65	7.75		20.40	26.50
0510	12" deep	G	100	.160		14	8.50		22.50	29
1210	Triple, 18 ga. x 8" deep	G	170	.094		8.55	5		13.55	17.40
1220	10" deep	G	165	.097		10.40	5.15		15.55	19.70
1230	12" deep	G	160	.100		11.40	5.30		16.70	21
1300	16 ga. x 8" deep	G	145	.110		9.95	5.85		15.80	20.50
1310	10" deep	G	140	.114		11.90	6.05		17.95	23
1320	12" deep	G	135	.119		13	6.30		19.30	24.50
1400	14 ga. x 10" deep	G	115	.139		13.85	7.40		21.25	27
1410	12" deep	G	110	.145		15.25	7.75		23	29
1500	12 ga. x 10" deep	G	90	.178		18.45	9.45		27.90	35.50
1510	12" deep	G	85	.188		20.50	10		30.50	38.50

05 42 13.40 Framing, Joists

		Crew	Daily Output	Labor-Hours	Unit	Material	2020 Bare Costs Labor	Equipment	Total	Total Incl O&P
0010	**FRAMING, JOISTS**, no band joists (track), web stiffeners, headers,									
0020	Beams, bridging or bracing									
0025	Made from recycled materials									
0030	Joists (2" flange) and fasteners, materials only									
0220	18 ga. x 6" deep	G			L.F.	1.84			1.84	2.02
0230	8" deep	G				1.95			1.95	2.15
0240	10" deep	G				2.32			2.32	2.55
0320	16 ga. x 6" deep	G				2.04			2.04	2.24
0330	8" deep	G				2.44			2.44	2.68
0340	10" deep	G				2.85			2.85	3.13
0350	12" deep	G				3.22			3.22	3.55
0430	14 ga. x 8" deep	G				3.07			3.07	3.37
0440	10" deep	G				3.53			3.53	3.88
0450	12" deep	G				4.01			4.01	4.41
0540	12 ga. x 10" deep	G				5.15			5.15	5.65
0550	12" deep	G				5.85			5.85	6.40
1010	Installation of joists to band joists, beams & headers, labor only									
1220	18 ga. x 6" deep	2 Carp	110	.145	Ea.		7.75		7.75	12.40
1230	8" deep		90	.178			9.45		9.45	15.15
1240	10" deep		80	.200			10.65		10.65	17.05
1320	16 ga. x 6" deep		95	.168			8.95		8.95	14.35
1330	8" deep		70	.229			12.15		12.15	19.45
1340	10" deep		60	.267			14.15		14.15	22.50
1350	12" deep		55	.291			15.45		15.45	25
1430	14 ga. x 8" deep		65	.246			13.10		13.10	21
1440	10" deep		45	.356			18.90		18.90	30.50
1450	12" deep		35	.457			24.50		24.50	39
1540	12 ga. x 10" deep		40	.400			21.50		21.50	34

For customer support on your Commercial Renovation Costs with RSMeans data, call 800.448.8182.

121

05 42 Cold-Formed Metal Joist Framing

05 42 13 – Cold-Formed Metal Floor Joist Framing

05 42 13.40 Framing, Joists		Crew	Daily Output	Labor-Hours	Unit	Material	2020 Bare Costs Labor	Equipment	Total	Total Incl O&P
1550	12" deep	2 Carp	30	.533	Ea.		28.50		28.50	45.50
9000	Minimum labor/equipment charge	↓	4	4	Job		213		213	340

05 42 13.45 Framing, Web Stiffeners

	05 42 13.45 Framing, Web Stiffeners		Crew	Daily Output	Labor-Hours	Unit	Material	2020 Bare Costs Labor	Equipment	Total	Total Incl O&P
0010	**FRAMING, WEB STIFFENERS** at joist bearing, fabricated from										
0020	Stud piece (1-5/8" flange) to stiffen joist (2" flange)										
0025	Made from recycled materials										
2120	For 6" deep joist, with 18 ga. x 2-1/2" stud	G	1 Carp	120	.067	Ea.	.83	3.54		4.37	6.60
2130	3-5/8" stud	G		110	.073		.98	3.87		4.85	7.30
2140	4" stud	G		105	.076		1.02	4.05		5.07	7.60
2150	6" stud	G		100	.080		1.29	4.25		5.54	8.20
2160	8" stud	G		95	.084		1.67	4.48		6.15	9
2220	8" deep joist, with 2-1/2" stud	G		120	.067		1.11	3.54		4.65	6.90
2230	3-5/8" stud	G		110	.073		1.31	3.87		5.18	7.65
2240	4" stud	G		105	.076		1.37	4.05		5.42	8
2250	6" stud	G		100	.080		1.73	4.25		5.98	8.70
2260	8" stud	G		95	.084		2.24	4.48		6.72	9.60
2320	10" deep joist, with 2-1/2" stud	G		110	.073		1.38	3.87		5.25	7.70
2330	3-5/8" stud	G		100	.080		1.63	4.25		5.88	8.60
2340	4" stud	G		95	.084		1.69	4.48		6.17	9
2350	6" stud	G		90	.089		2.14	4.72		6.86	9.90
2360	8" stud	G		85	.094		2.77	5		7.77	11.05
2420	12" deep joist, with 2-1/2" stud	G		110	.073		1.66	3.87		5.53	8.05
2430	3-5/8" stud	G		100	.080		1.96	4.25		6.21	8.95
2440	4" stud	G		95	.084		2.04	4.48		6.52	9.40
2450	6" stud	G		90	.089		2.58	4.72		7.30	10.40
2460	8" stud	G		85	.094		3.34	5		8.34	11.65
3130	For 6" deep joist, with 16 ga. x 3-5/8" stud	G		100	.080		1.28	4.25		5.53	8.20
3140	4" stud	G		95	.084		1.37	4.48		5.85	8.65
3150	6" stud	G		90	.089		1.67	4.72		6.39	9.40
3160	8" stud	G		85	.094		2.07	5		7.07	10.30
3230	8" deep joist, with 3-5/8" stud	G		100	.080		1.72	4.25		5.97	8.70
3240	4" stud	G		95	.084		1.84	4.48		6.32	9.15
3250	6" stud	G		90	.089		2.24	4.72		6.96	10
3260	8" stud	G		85	.094		2.77	5		7.77	11.05
3330	10" deep joist, with 3-5/8" stud	G		85	.094		2.12	5		7.12	10.35
3340	4" stud	G		80	.100		2.27	5.30		7.57	11
3350	6" stud	G		75	.107		2.77	5.65		8.42	12.15
3360	8" stud	G		70	.114		3.44	6.05		9.49	13.55
3430	12" deep joist, with 3-5/8" stud	G		85	.094		2.56	5		7.56	10.80
3440	4" stud	G		80	.100		2.74	5.30		8.04	11.50
3450	6" stud	G		75	.107		3.34	5.65		8.99	12.75
3460	8" stud	G		70	.114		4.14	6.05		10.19	14.30
4230	For 8" deep joist, with 14 ga. x 3-5/8" stud	G		90	.089		2.18	4.72		6.90	9.95
4240	4" stud	G		85	.094		2.29	5		7.29	10.50
4250	6" stud	G		80	.100		2.89	5.30		8.19	11.70
4260	8" stud	G		75	.107		3.23	5.65		8.88	12.65
4330	10" deep joist, with 3-5/8" stud	G		75	.107		2.71	5.65		8.36	12.10
4340	4" stud	G		70	.114		2.84	6.05		8.89	12.85
4350	6" stud	G		65	.123		3.59	6.55		10.14	14.45
4360	8" stud	G		60	.133		4	7.10		11.10	15.75
4430	12" deep joist, with 3-5/8" stud	G		75	.107		3.26	5.65		8.91	12.70
4440	4" stud	G		70	.114		3.42	6.05		9.47	13.50

05 42 Cold-Formed Metal Joist Framing

05 42 13 – Cold-Formed Metal Floor Joist Framing

05 42 13.45 Framing, Web Stiffeners

		Crew	Daily Output	Labor-Hours	Unit	Material	2020 Bare Costs Labor	Equipment	Total	Total Incl O&P
4450	6" stud	G	1 Carp	65	.123	Ea.	4.32	6.55		10.87	15.25
4460	8" stud	G		60	.133		4.82	7.10		11.92	16.65
5330	For 10" deep joist, with 12 ga. x 3-5/8" stud	G		65	.123		3.88	6.55		10.43	14.75
5340	4" stud	G		60	.133		4.17	7.10		11.27	15.95
5350	6" stud	G		55	.145		5.25	7.75		13	18.15
5360	8" stud	G		50	.160		6.35	8.50		14.85	20.50
5430	12" deep joist, with 3-5/8" stud	G		65	.123		4.68	6.55		11.23	15.65
5440	4" stud	G		60	.133		5	7.10		12.10	16.85
5450	6" stud	G		55	.145		6.30	7.75		14.05	19.35
5460	8" stud	G		50	.160		7.65	8.50		16.15	22

05 42 23 – Cold-Formed Metal Roof Joist Framing

05 42 23.05 Framing, Bracing

		Crew	Daily Output	Labor-Hours	Unit	Material	2020 Bare Costs Labor	Equipment	Total	Total Incl O&P
0010	**FRAMING, BRACING**										
0015	Made from recycled materials										
0020	Continuous bracing, per row										
0100	16 ga. x 1-1/2" channel thru rafters/trusses @ 16" OC	G	1 Carp	4.50	1.778	C.L.F.	51.50	94.50		146	208
0120	24" OC	G		6	1.333		51.50	71		122.50	171
0300	2" x 2" angle x 18 ga., rafters/trusses @ 16" OC	G		6	1.333		78	71		149	200
0320	24" OC	G		8	1		78	53		131	171
0400	16 ga., rafters/trusses @ 16" OC	G		4.50	1.778		98	94.50		192.50	259
0420	24" OC	G		6.50	1.231		98	65.50		163.50	213

05 42 23.10 Framing, Bridging

		Crew	Daily Output	Labor-Hours	Unit	Material	2020 Bare Costs Labor	Equipment	Total	Total Incl O&P
0010	**FRAMING, BRIDGING**										
0015	Made from recycled materials										
0020	Solid, between rafters w/1-1/4" leg track, per rafter bay										
1200	Rafters 16" OC, 18 ga. x 4" deep	G	1 Carp	60	.133	Ea.	1.42	7.10		8.52	12.90
1210	6" deep	G		57	.140		1.96	7.45		9.41	14.10
1220	8" deep	G		55	.145		2.41	7.75		10.16	15.05
1230	10" deep	G		52	.154		3.07	8.20		11.27	16.50
1240	12" deep	G		50	.160		3.51	8.50		12.01	17.45
2200	24" OC, 18 ga. x 4" deep	G		60	.133		2.05	7.10		9.15	13.60
2210	6" deep	G		57	.140		2.84	7.45		10.29	15.05
2220	8" deep	G		55	.145		3.49	7.75		11.24	16.25
2230	10" deep	G		52	.154		4.44	8.20		12.64	18
2240	12" deep	G		50	.160		5.10	8.50		13.60	19.20

05 42 23.50 Framing, Parapets

		Crew	Daily Output	Labor-Hours	Unit	Material	2020 Bare Costs Labor	Equipment	Total	Total Incl O&P
0010	**FRAMING, PARAPETS**										
0015	Made from recycled materials										
0100	3' high installed on 1st story, 18 ga. x 4" wide studs, 12" OC	G	2 Carp	100	.160	L.F.	5.15	8.50		13.65	19.30
0110	16" OC	G		150	.107		4.40	5.65		10.05	13.95
0120	24" OC	G		200	.080		3.64	4.25		7.89	10.80
0200	6" wide studs, 12" OC	G		100	.160		6.80	8.50		15.30	21
0210	16" OC	G		150	.107		5.80	5.65		11.45	15.50
0220	24" OC	G		200	.080		4.85	4.25		9.10	12.15
1100	Installed on 2nd story, 18 ga. x 4" wide studs, 12" OC	G		95	.168		5.15	8.95		14.10	20
1110	16" OC	G		145	.110		4.40	5.85		10.25	14.25
1120	24" OC	G		190	.084		3.64	4.48		8.12	11.15
1200	6" wide studs, 12" OC	G		95	.168		6.80	8.95		15.75	22
1210	16" OC	G		145	.110		5.80	5.85		11.65	15.80
1220	24" OC	G		190	.084		4.85	4.48		9.33	12.50
2100	Installed on gable, 18 ga. x 4" wide studs, 12" OC	G		85	.188		5.15	10		15.15	22
2110	16" OC	G		130	.123		4.40	6.55		10.95	15.35

For customer support on your Commercial Renovation Costs with RSMeans data, call 800.448.8182.

123

05 42 Cold-Formed Metal Joist Framing

05 42 23 – Cold-Formed Metal Roof Joist Framing

05 42 23.50 Framing, Parapets

		Crew	Daily Output	Labor-Hours	Unit	Material	2020 Bare Costs Labor	Equipment	Total	Total Incl O&P
2120	24" OC	G 2 Carp	170	.094	L.F.	3.64	5		8.64	12
2200	6" wide studs, 12" OC	G	85	.188		6.80	10		16.80	23.50
2210	16" OC	G	130	.123		5.80	6.55		12.35	16.90
2220	24" OC	G	170	.094		4.85	5		9.85	13.35

05 42 23.60 Framing, Roof Rafters

		Crew	Daily Output	Labor-Hours	Unit	Material	2020 Bare Costs Labor	Equipment	Total	Total Incl O&P
0010	**FRAMING, ROOF RAFTERS**									
0015	Made from recycled materials									
0100	Boxed ridge beam, double, 18 ga. x 6" deep	G 2 Carp	160	.100	L.F.	5.55	5.30		10.85	14.60
0110	8" deep	G	150	.107		5.80	5.65		11.45	15.45
0120	10" deep	G	140	.114		7.30	6.05		13.35	17.80
0130	12" deep	G	130	.123		7.95	6.55		14.50	19.25
0200	16 ga. x 6" deep	G	150	.107		5.95	5.65		11.60	15.65
0210	8" deep	G	140	.114		6.70	6.05		12.75	17.10
0220	10" deep	G	130	.123		8.30	6.55		14.85	19.65
0230	12" deep	G	120	.133		9	7.10		16.10	21.50
1100	Rafters, 2" flange, material only, 18 ga. x 6" deep	G				1.84			1.84	2.02
1110	8" deep	G				1.95			1.95	2.15
1120	10" deep	G				2.32			2.32	2.55
1130	12" deep	G				2.67			2.67	2.93
1200	16 ga. x 6" deep	G				2.04			2.04	2.24
1210	8" deep	G				2.44			2.44	2.68
1220	10" deep	G				2.85			2.85	3.13
1230	12" deep	G				3.22			3.22	3.55
2100	Installation only, ordinary rafter to 4:12 pitch, 18 ga. x 6" deep	2 Carp	35	.457	Ea.		24.50		24.50	39
2110	8" deep		30	.533			28.50		28.50	45.50
2120	10" deep		25	.640			34		34	54.50
2130	12" deep		20	.800			42.50		42.50	68
2200	16 ga. x 6" deep		30	.533			28.50		28.50	45.50
2210	8" deep		25	.640			34		34	54.50
2220	10" deep		20	.800			42.50		42.50	68
2230	12" deep		15	1.067			56.50		56.50	91
8100	Add to labor, ordinary rafters on steep roofs						25%			
8110	Dormers & complex roofs						50%			
8200	Hip & valley rafters to 4:12 pitch						25%			
8210	Steep roofs						50%			
8220	Dormers & complex roofs						75%			
8300	Hip & valley jack rafters to 4:12 pitch						50%			
8310	Steep roofs						75%			
8320	Dormers & complex roofs						100%			
9000	Minimum labor/equipment charge	2 Carp	4	4	Job		213		213	340

05 42 23.70 Framing, Soffits and Canopies

		Crew	Daily Output	Labor-Hours	Unit	Material	2020 Bare Costs Labor	Equipment	Total	Total Incl O&P
0010	**FRAMING, SOFFITS & CANOPIES**									
0015	Made from recycled materials									
0130	Continuous ledger track @ wall, studs @ 16" OC, 18 ga. x 4" wide	G 2 Carp	535	.030	L.F.	.95	1.59		2.54	3.59
0140	6" wide	G	500	.032		1.31	1.70		3.01	4.16
0150	8" wide	G	465	.034		1.61	1.83		3.44	4.70
0160	10" wide	G	430	.037		2.05	1.98		4.03	5.40
0230	Studs @ 24" OC, 18 ga. x 4" wide	G	800	.020		.90	1.06		1.96	2.69
0240	6" wide	G	750	.021		1.25	1.13		2.38	3.19
0250	8" wide	G	700	.023		1.53	1.22		2.75	3.64
0260	10" wide	G	650	.025		1.95	1.31		3.26	4.25
1000	Horizontal soffit and canopy members, material only									

05 42 Cold-Formed Metal Joist Framing

05 42 23 – Cold-Formed Metal Roof Joist Framing

05 42 23.70 Framing, Soffits and Canopies		Crew	Daily Output	Labor-Hours	Unit	Material	2020 Bare Costs Labor	Equipment	Total	Total Incl O&P
1030	1-5/8" flange studs, 18 ga. x 4" deep	G			L.F.	1.22			1.22	1.35
1040	6" deep	G				1.55			1.55	1.70
1050	8" deep	G				2			2	2.20
1140	2" flange joists, 18 ga. x 6" deep	G				2.10			2.10	2.31
1150	8" deep	G				2.23			2.23	2.46
1160	10" deep	G				2.65			2.65	2.92
4030	Installation only, 18 ga., 1-5/8" flange x 4" deep		2 Carp	130	.123	Ea.		6.55		6.55	10.50
4040	6" deep			110	.145			7.75		7.75	12.40
4050	8" deep			90	.178			9.45		9.45	15.15
4140	2" flange, 18 ga. x 6" deep			110	.145			7.75		7.75	12.40
4150	8" deep			90	.178			9.45		9.45	15.15
4160	10" deep			80	.200			10.65		10.65	17.05
6010	Clips to attach fascia to rafter tails, 2" x 2" x 18 ga. angle	G	1 Carp	120	.067		.92	3.54		4.46	6.70
6020	16 ga. angle	G	"	100	.080		1.16	4.25		5.41	8.05
9000	Minimum labor/equipment charge		2 Carp	4	4	Job		213		213	340

05 44 Cold-Formed Metal Trusses

05 44 13 – Cold-Formed Metal Roof Trusses

05 44 13.60 Framing, Roof Trusses		Crew	Daily Output	Labor-Hours	Unit	Material	2020 Bare Costs Labor	Equipment	Total	Total Incl O&P
0010	**FRAMING, ROOF TRUSSES**									
0015	Made from recycled materials									
0020	Fabrication of trusses on ground, Fink (W) or King Post, to 4:12 pitch									
0120	18 ga. x 4" chords, 16' span	G	2 Carp	12	1.333	Ea.	57	71		128	177
0130	20' span	G		11	1.455		71.50	77.50		149	203
0140	24' span	G		11	1.455		85.50	77.50		163	219
0150	28' span	G		10	1.600		100	85		185	246
0160	32' span	G		10	1.600		114	85		199	262
0250	6" chords, 28' span	G		9	1.778		126	94.50		220.50	290
0260	32' span	G		9	1.778		144	94.50		238.50	310
0270	36' span	G		8	2		163	106		269	350
0280	40' span	G		8	2		181	106		287	370
1120	5:12 to 8:12 pitch, 18 ga. x 4" chords, 16' span	G		10	1.600		65.50	85		150.50	208
1130	20' span	G		9	1.778		81.50	94.50		176	241
1140	24' span	G		9	1.778		98	94.50		192.50	259
1150	28' span	G		8	2		114	106		220	296
1160	32' span	G		8	2		131	106		237	315
1250	6" chords, 28' span	G		7	2.286		144	121		265	355
1260	32' span	G		7	2.286		165	121		286	375
1270	36' span	G		6	2.667		186	142		328	430
1280	40' span	G		6	2.667		206	142		348	455
2120	9:12 to 12:12 pitch, 18 ga. x 4" chords, 16' span	G		8	2		81.50	106		187.50	260
2130	20' span	G		7	2.286		102	121		223	305
2140	24' span	G		7	2.286		122	121		243	330
2150	28' span	G		6	2.667		143	142		285	385
2160	32' span	G		6	2.667		163	142		305	405
2250	6" chords, 28' span	G		5	3.200		181	170		351	470
2260	32' span	G		5	3.200		206	170		376	500
2270	36' span	G		4	4		232	213		445	595
2280	40' span	G		4	4		258	213		471	625
4900	Minimum labor/equipment charge			4	4	Job		213		213	340
5120	Erection only of roof trusses, to 4:12 pitch, 16' span		F-6	48	.833	Ea.		41.50	9.80	51.30	77.50

05 44 Cold-Formed Metal Trusses

05 44 13 – Cold-Formed Metal Roof Trusses

05 44 13.60 Framing, Roof Trusses		Crew	Daily Output	Labor-Hours	Unit	Material	2020 Bare Costs Labor	Equipment	Total	Total Incl O&P
5130	20' span	F-6	46	.870	Ea.		43.50	10.25	53.75	80.50
5140	24' span		44	.909			45.50	10.70	56.20	84.50
5150	28' span		42	.952			47.50	11.20	58.70	88.50
5160	32' span		40	1			50	11.80	61.80	92.50
5170	36' span		38	1.053			52.50	12.40	64.90	97.50
5180	40' span		36	1.111			55.50	13.10	68.60	103
5220	5:12 to 8:12 pitch, 16' span		42	.952			47.50	11.20	58.70	88.50
5230	20' span		40	1			50	11.80	61.80	92.50
5240	24' span		38	1.053			52.50	12.40	64.90	97.50
5250	28' span		36	1.111			55.50	13.10	68.60	103
5260	32' span		34	1.176			59	13.85	72.85	109
5270	36' span		32	1.250			62.50	14.75	77.25	116
5280	40' span		30	1.333			66.50	15.70	82.20	123
5320	9:12 to 12:12 pitch, 16' span		36	1.111			55.50	13.10	68.60	103
5330	20' span		34	1.176			59	13.85	72.85	109
5340	24' span		32	1.250			62.50	14.75	77.25	116
5350	28' span		30	1.333			66.50	15.70	82.20	123
5360	32' span		28	1.429			71.50	16.85	88.35	133
5370	36' span		26	1.538			77	18.10	95.10	142
5380	40' span		24	1.667			83	19.65	102.65	155
9000	Minimum labor/equipment charge		2	20	Job		1,000	236	1,236	1,850

05 51 Metal Stairs

05 51 13 – Metal Pan Stairs

05 51 13.50 Pan Stairs

			Crew	Daily Output	Labor-Hours	Unit	Material	Labor	Equipment	Total	Total Incl O&P
0010	**PAN STAIRS**, shop fabricated, steel stringers										
0015	Made from recycled materials										
0200	Metal pan tread for concrete in-fill, picket rail, 3'-6" wide	G	E-4	35	.914	Riser	600	53	4.20	657.20	750
0300	4'-0" wide	G		30	1.067		590	62	4.90	656.90	755
0350	Wall rail, both sides, 3'-6" wide	G		53	.604		460	35	2.77	497.77	565
1500	Landing, steel pan, conventional	G		160	.200	S.F.	79.50	11.65	.92	92.07	108

05 51 16 – Metal Floor Plate Stairs

05 51 16.50 Floor Plate Stairs

			Crew	Daily Output	Labor-Hours	Unit	Material	Labor	Equipment	Total	Total Incl O&P
0010	**FLOOR PLATE STAIRS**, shop fabricated, steel stringers										
0015	Made from recycled materials										
0400	Cast iron tread and pipe rail, 3'-6" wide	G	E-4	35	.914	Riser	570	53	4.20	627.20	715
0450	4'-0" wide	G		30	1.067		595	62	4.90	661.90	760
0475	5'-0" wide	G		28	1.143		655	66.50	5.25	726.75	835
0500	Checkered plate tread, industrial, 3'-6" wide	G		28	1.143		385	66.50	5.25	456.75	540
0550	Circular, for tanks, 3'-0" wide	G		33	.970		435	56.50	4.45	495.95	575
0600	For isolated stairs, add							100%			
0800	Custom steel stairs, 3'-6" wide, economy	G	E-4	35	.914		585	53	4.20	642.20	735
0810	Medium priced	G		30	1.067		775	62	4.90	841.90	960
0900	Deluxe	G		20	1.600		970	93	7.35	1,070.35	1,225
1100	For 4' wide stairs, add							5%	5%		
1300	For 5' wide stairs, add							10%	10%		

05 51 Metal Stairs

05 51 19 – Metal Grating Stairs

05 51 19.50 Grating Stairs

	05 51 19.50 Grating Stairs		Crew	Daily Output	Labor- Hours	Unit	Material	2020 Bare Costs Labor	Equipment	Total	Total Incl O&P
0010	**GRATING STAIRS**, shop fabricated, steel stringers, safety nosing on treads										
0015	Made from recycled materials										
0020	Grating tread and pipe railing, 3'-6" wide	G	E-4	35	.914	Riser	390	53	4.20	447.20	520
0100	4'-0" wide	G		30	1.067	"	455	62	4.90	521.90	605
9000	Minimum labor/equipment charge			2	16	Job		930	73.50	1,003.50	1,600
0001	Grating tread and pipe railing, 3'-6" wide			35	.914	Flight	3,025	53	4.20	3,082.20	3,425

05 51 23 – Metal Fire Escapes

05 51 23.25 Fire Escapes

	05 51 23.25 Fire Escapes		Crew	Daily Output	Labor- Hours	Unit	Material	Labor	Equipment	Total	Total Incl O&P
0010	**FIRE ESCAPES**, shop fabricated										
0200	2' wide balcony, 1" x 1/4" bars 1-1/2" OC, with railing	G	2 Sswk	10	1.600	L.F.	63.50	92		155.50	221
0400	1st story cantilevered stair, standard, with railing	G		.50	32	Ea.	2,550	1,850		4,400	5,825
0500	Cable counterweighted, with railing	G		.40	40	"	2,425	2,300		4,725	6,450
0700	36" x 40" platform & fixed stair, with railing	G		.40	40	Flight	1,125	2,300		3,425	5,025
0900	For 3'-6" wide escapes, add to above						100%	150%			

05 51 23.50 Fire Escape Stairs

	05 51 23.50 Fire Escape Stairs		Crew	Daily Output	Labor- Hours	Unit	Material	Labor	Equipment	Total	Total Incl O&P
0010	**FIRE ESCAPE STAIRS**, portable										
0100	Portable ladder					Ea.	125			125	137

05 51 33 – Metal Ladders

05 51 33.13 Vertical Metal Ladders

	05 51 33.13 Vertical Metal Ladders		Crew	Daily Output	Labor- Hours	Unit	Material	Labor	Equipment	Total	Total Incl O&P
0010	**VERTICAL METAL LADDERS**, shop fabricated										
0015	Made from recycled materials										
0020	Steel, 20" wide, bolted to concrete, with cage	G	E-4	50	.640	V.L.F.	92	37	2.94	131.94	165
0100	Without cage	G		85	.376		42	22	1.73	65.73	84.50
0300	Aluminum, bolted to concrete, with cage	G		50	.640		130	37	2.94	169.94	207
0400	Without cage	G		85	.376		51	22	1.73	74.73	94
9000	Minimum labor/equipment charge			2	16	Job		930	73.50	1,003.50	1,600

05 51 33.16 Inclined Metal Ladders

	05 51 33.16 Inclined Metal Ladders		Crew	Daily Output	Labor- Hours	Unit	Material	Labor	Equipment	Total	Total Incl O&P
0010	**INCLINED METAL LADDERS**, shop fabricated										
0015	Made from recycled materials										
3900	Industrial ships ladder, steel, 24" W, grating treads, 2 line pipe rail	G	E-4	30	1.067	Riser	205	62	4.90	271.90	330
4000	Aluminum	G	"	30	1.067	"	300	62	4.90	366.90	435

05 51 33.23 Alternating Tread Ladders

	05 51 33.23 Alternating Tread Ladders		Crew	Daily Output	Labor- Hours	Unit	Material	Labor	Equipment	Total	Total Incl O&P
0010	**ALTERNATING TREAD LADDERS**, shop fabricated										
0015	Made from recycled materials										
0800	Alternating tread ladders, 68-degree angle of incline										
0810	8' vertical rise, steel, 149 lb., standard paint color		B-68G	3	5.333	Ea.	1,950	305	93.50	2,348.50	2,725
0820	Non-standard paint color			3	5.333		2,400	305	93.50	2,798.50	3,250
0830	Galvanized			3	5.333		2,425	305	93.50	2,823.50	3,250
0840	Stainless			3	5.333		3,700	305	93.50	4,098.50	4,650
0850	Aluminum, 87 lb.			3	5.333		2,700	305	93.50	3,098.50	3,575
1010	10' vertical rise, steel, 181 lb., standard paint color			2.75	5.818		2,775	335	102	3,212	3,725
1020	Non-standard paint color			2.75	5.818		3,025	335	102	3,462	3,975
1030	Galvanized			2.75	5.818		3,075	335	102	3,512	4,025
1040	Stainless			2.75	5.818		4,450	335	102	4,887	5,550
1050	Aluminum, 103 lb.			2.75	5.818		3,250	335	102	3,687	4,225
1210	12' vertical rise, steel, 245 lb., standard paint color			2.50	6.400		3,100	370	112	3,582	4,125
1220	Non-standard paint color			2.50	6.400		3,525	370	112	4,007	4,600
1230	Galvanized			2.50	6.400		3,600	370	112	4,082	4,700
1240	Stainless			2.50	6.400		5,200	370	112	5,682	6,450
1250	Aluminum, 103 lb.			2.50	6.400		3,775	370	112	4,257	4,875

05 51 Metal Stairs

05 51 33 – Metal Ladders

05 51 33.23 Alternating Tread Ladders		Crew	Daily Output	Labor-Hours	Unit	Material	2020 Bare Costs Labor	Equipment	Total	Total Incl O&P
1410	14' vertical rise, steel, 281 lb., standard paint color	B-68G	2.25	7.111	Ea.	3,550	410	124	4,084	4,700
1420	Non-standard paint color		2.25	7.111		4,025	410	124	4,559	5,225
1430	Galvanized		2.25	7.111		4,150	410	124	4,684	5,350
1440	Stainless		2.25	7.111		5,950	410	124	6,484	7,350
1450	Aluminum, 136 lb.		2.25	7.111		4,375	410	124	4,909	5,600
1610	16' vertical rise, steel, 317 lb., standard paint color		2	8		4,025	460	140	4,625	5,325
1620	Non-standard paint color		2	8		4,525	460	140	5,125	5,875
1630	Galvanized		2	8		4,675	460	140	5,275	6,050
1640	Stainless		2	8		6,700	460	140	7,300	8,275
1650	Aluminum, 153 lb.		2	8		4,675	460	140	5,275	6,025

05 52 Metal Railings

05 52 13 – Pipe and Tube Railings

05 52 13.50 Railings, Pipe

			Crew	Daily Output	Labor-Hours	Unit	Material	2020 Bare Costs Labor	Equipment	Total	Total Incl O&P
0010	**RAILINGS, PIPE**, shop fab'd, 3'-6" high, posts @ 5' OC										
0015	Made from recycled materials										
0020	Aluminum, 2 rail, satin finish, 1-1/4" diameter	G	E-4	160	.200	L.F.	51.50	11.65	.92	64.07	76.50
0030	Clear anodized	G		160	.200		61.50	11.65	.92	74.07	87.50
0040	Dark anodized	G		160	.200		68	11.65	.92	80.57	95
0080	1-1/2" diameter, satin finish	G		160	.200		58.50	11.65	.92	71.07	84.50
0090	Clear anodized	G		160	.200		66	11.65	.92	78.57	92.50
0100	Dark anodized	G		160	.200		72.50	11.65	.92	85.07	100
0140	Aluminum, 3 rail, 1-1/4" diam., satin finish	G		137	.234		67	13.60	1.07	81.67	97.50
0150	Clear anodized	G		137	.234		83.50	13.60	1.07	98.17	116
0160	Dark anodized	G		137	.234		92	13.60	1.07	106.67	125
0200	1-1/2" diameter, satin finish	G		137	.234		79.50	13.60	1.07	94.17	111
0210	Clear anodized	G		137	.234		90.50	13.60	1.07	105.17	123
0220	Dark anodized	G		137	.234		99.50	13.60	1.07	114.17	133
0500	Steel, 2 rail, on stairs, primed, 1-1/4" diameter	G		160	.200		32	11.65	.92	44.57	55.50
0520	1-1/2" diameter	G		160	.200		31	11.65	.92	43.57	54
0540	Galvanized, 1-1/4" diameter	G		160	.200		41.50	11.65	.92	54.07	66
0560	1-1/2" diameter	G		160	.200		49	11.65	.92	61.57	74
0580	Steel, 3 rail, primed, 1-1/4" diameter	G		137	.234		46	13.60	1.07	60.67	74
0600	1-1/2" diameter	G		137	.234		50.50	13.60	1.07	65.17	79
0620	Galvanized, 1-1/4" diameter	G		137	.234		64.50	13.60	1.07	79.17	94.50
0640	1-1/2" diameter	G		137	.234		77.50	13.60	1.07	92.17	109
0700	Stainless steel, 2 rail, 1-1/4" diam., #4 finish	G		137	.234		141	13.60	1.07	155.67	179
0720	High polish	G		137	.234		228	13.60	1.07	242.67	275
0740	Mirror polish	G		137	.234		285	13.60	1.07	299.67	340
0760	Stainless steel, 3 rail, 1-1/2" diam., #4 finish	G		120	.267		212	15.50	1.22	228.72	260
0770	High polish	G		120	.267		350	15.50	1.22	366.72	410
0780	Mirror finish	G		120	.267		430	15.50	1.22	446.72	495
0900	Wall rail, alum. pipe, 1-1/4" diam., satin finish	G		213	.150		25	8.75	.69	34.44	42.50
0905	Clear anodized	G		213	.150		31	8.75	.69	40.44	49
0910	Dark anodized	G		213	.150		36.50	8.75	.69	45.94	55.50
0915	1-1/2" diameter, satin finish	G		213	.150		27.50	8.75	.69	36.94	45.50
0920	Clear anodized	G		213	.150		34.50	8.75	.69	43.94	53
0925	Dark anodized	G		213	.150		43.50	8.75	.69	52.94	63
0930	Steel pipe, 1-1/4" diameter, primed	G		213	.150		18.50	8.75	.69	27.94	35.50
0935	Galvanized	G		213	.150		27	8.75	.69	36.44	44.50
0940	1-1/2" diameter	G		176	.182		17.75	10.55	.83	29.13	37.50

05 52 Metal Railings

05 52 13 – Pipe and Tube Railings

05 52 13.50 Railings, Pipe

		Crew	Daily Output	Labor-Hours	Unit	Material	2020 Bare Costs Labor	Equipment	Total	Total Incl O&P
0945	Galvanized	G E-4	213	.150	L.F.	27	8.75	.69	36.44	44.50
0955	Stainless steel pipe, 1-1/2" diam., #4 finish	G	107	.299		113	17.40	1.37	131.77	154
0960	High polish	G	107	.299		229	17.40	1.37	247.77	282
0965	Mirror polish	G	107	.299		271	17.40	1.37	289.77	330
2000	2-line pipe rail (1-1/2" T&B) with 1/2" pickets @ 4-1/2" OC,									
2005	attached handrail on brackets									
2010	42" high aluminum, satin finish, straight & level	G E-4	120	.267	L.F.	355	15.50	1.22	371.72	415
2050	42" high steel, primed, straight & level	"	120	.267		149	15.50	1.22	165.72	191
4000	For curved and level rails, add						10%	10%		
4100	For sloped rails for stairs, add						30%	30%		
9000	Minimum labor/equipment charge	1 Sswk	2	4	Job		231		231	380

05 52 16 – Industrial Railings

05 52 16.50 Railings, Industrial

		Crew	Daily Output	Labor-Hours	Unit	Material	2020 Bare Costs Labor	Equipment	Total	Total Incl O&P
0010	**RAILINGS, INDUSTRIAL,** shop fab'd, 3'-6" high, posts @ 5' OC									
0020	2 rail, 3'-6" high, 1-1/2" pipe	G E-4	255	.125	L.F.	39	7.30	.58	46.88	55.50
0200	For 4" high kick plate, 10 ga., add	G				6.20			6.20	6.80
0500	For curved level rails, add					10%	10%			
0550	For sloped rails for stairs, add					30%	30%			
9000	Minimum labor/equipment charge	1 Sswk	2	4	Job		231		231	380

05 55 Metal Stair Treads and Nosings

05 55 13 – Metal Stair Treads

05 55 13.50 Stair Treads

		Crew	Daily Output	Labor-Hours	Unit	Material	2020 Bare Costs Labor	Equipment	Total	Total Incl O&P
0010	**STAIR TREADS,** stringers and bolts not included									
3000	Diamond plate treads, steel, 1/8" thick									
3005	Open riser, black enamel									
3010	9" deep x 36" long	G 2 Sswk	48	.333	Ea.	112	19.20		131.20	155
3020	42" long	G	48	.333		118	19.20		137.20	162
3030	48" long	G	48	.333		123	19.20		142.20	167
3040	11" deep x 36" long	G	44	.364		118	21		139	165
3050	42" long	G	44	.364		125	21		146	172
3060	48" long	G	44	.364		133	21		154	181
3110	Galvanized, 9" deep x 36" long	G	48	.333		177	19.20		196.20	226
3120	42" long	G	48	.333		188	19.20		207.20	239
3130	48" long	G	48	.333		199	19.20		218.20	251
3140	11" deep x 36" long	G	44	.364		183	21		204	236
3150	42" long	G	44	.364		199	21		220	254
3160	48" long	G	44	.364		207	21		228	263
3200	Closed riser, black enamel									
3210	12" deep x 36" long	G 2 Sswk	40	.400	Ea.	135	23		158	187
3220	42" long	G	40	.400		147	23		170	199
3230	48" long	G	40	.400		155	23		178	208
3240	Galvanized, 12" deep x 36" long	G	40	.400		213	23		236	273
3250	42" long	G	40	.400		233	23		256	294
3260	48" long	G	40	.400		244	23		267	305
4000	Bar grating treads									
4005	Steel, 1-1/4" x 3/16" bars, anti-skid nosing, black enamel									
4010	8-5/8" deep x 30" long	G 2 Sswk	48	.333	Ea.	56	19.20		75.20	93
4020	36" long	G	48	.333		65.50	19.20		84.70	104
4030	48" long	G	48	.333		101	19.20		120.20	143

05 55 13 – Metal Stair Treads

05 55 13.50 Stair Treads

		Crew	Daily Output	Labor-Hours	Unit	Material	2020 Bare Costs Labor	Equipment	Total	Total Incl O&P
4040	10-15/16" deep x 36" long	G	2 Sswk	44	.364	Ea.	72.50	21		93.50	115
4050	48" long	G		44	.364		104	21		125	149
4060	Galvanized, 8-5/8" deep x 30" long	G		48	.333		64	19.20		83.20	102
4070	36" long	G		48	.333		76	19.20		95.20	115
4080	48" long	G		48	.333		112	19.20		131.20	155
4090	10-15/16" deep x 36" long	G		44	.364		89	21		110	133
4100	48" long	G		44	.364		115	21		136	162
4200	Aluminum, 1-1/4" x 3/16" bars, serrated, with nosing										
4210	7-5/8" deep x 18" long	G	2 Sswk	52	.308	Ea.	53	17.75		70.75	87.50
4220	24" long	G		52	.308		63	17.75		80.75	98
4230	30" long	G		52	.308		73	17.75		90.75	109
4240	36" long	G		52	.308		188	17.75		205.75	236
4250	8-13/16" deep x 18" long	G		48	.333		71.50	19.20		90.70	110
4260	24" long	G		48	.333		106	19.20		125.20	148
4270	30" long	G		48	.333		122	19.20		141.20	166
4280	36" long	G		48	.333		207	19.20		226.20	260
4290	10" deep x 18" long	G		44	.364		139	21		160	187
4300	30" long	G		44	.364		184	21		205	238
4310	36" long	G		44	.364		224	21		245	281
5000	Channel grating treads										
5005	Steel, 14 ga., 2-1/2" thick, galvanized										
5010	9" deep x 36" long	G	2 Sswk	48	.333	Ea.	106	19.20		125.20	149
5020	48" long	G		48	.333	"	143	19.20		162.20	190
9000	Minimum labor/equipment charge			4	4	Job		231		231	380

05 55 19 – Metal Stair Tread Covers

05 55 19.50 Stair Tread Covers for Renovation

		Crew	Daily Output	Labor-Hours	Unit	Material	2020 Bare Costs Labor	Equipment	Total	Total Incl O&P
0010	**STAIR TREAD COVERS FOR RENOVATION**										
0205	Extruded tread cover with nosing, pre-drilled, includes screws										
0210	Aluminum with black abrasive strips, 9" wide x 3' long	1 Carp	24	.333	Ea.	104	17.70		121.70	144
0220	4' long		22	.364		136	19.35		155.35	181
0230	5' long		20	.400		186	21.50		207.50	238
0240	11" wide x 3' long		24	.333		147	17.70		164.70	190
0250	4' long		22	.364		199	19.35		218.35	249
0260	5' long		20	.400		238	21.50		259.50	295
0305	Black abrasive strips with yellow front strips									
0310	Aluminum, 9" wide x 3' long	1 Carp	24	.333	Ea.	124	17.70		141.70	166
0320	4' long		22	.364		166	19.35		185.35	213
0330	5' long		20	.400		213	21.50		234.50	268
0340	11" wide x 3' long		24	.333		161	17.70		178.70	206
0350	4' long		22	.364		204	19.35		223.35	256
0360	5' long		20	.400		267	21.50		288.50	325
0405	Black abrasive strips with photoluminescent front strips									
0410	Aluminum, 9" wide x 3' long	1 Carp	24	.333	Ea.	157	17.70		174.70	202
0420	4' long		22	.364		178	19.35		197.35	227
0430	5' long		20	.400		223	21.50		244.50	279
0440	11" wide x 3' long		24	.333		162	17.70		179.70	207
0450	4' long		22	.364		216	19.35		235.35	268
0460	5' long		20	.400		270	21.50		291.50	330
9000	Minimum labor/equipment charge		4	2	Job		106		106	170

130

For customer support on your Commercial Renovation Costs with RSMeans data, call 800.448.8182.

05 56 Metal Castings

05 56 13 – Metal Construction Castings

	05 56 13.50 Construction Castings		Crew	Daily Output	Labor-Hours	Unit	Material	2020 Bare Costs Labor	Equipment	Total	Total Incl O&P
0010	**CONSTRUCTION CASTINGS**										
0020	Manhole covers and frames, see Section 33 44 13.13										
0100	Column bases, cast iron, 16" x 16", approx. 65 lb.	G	E-4	46	.696	Ea.	140	40.50	3.19	183.69	224
0200	32" x 32", approx. 256 lb.	G		23	1.391	"	525	81	6.40	612.40	715
0600	Miscellaneous C.I. castings, light sections, less than 150 lb.	G		3200	.010	Lb.	9.10	.58	.05	9.73	11.05
1300	Special low volume items	G		3200	.010	"	11.05	.58	.05	11.68	13.15

05 58 Formed Metal Fabrications

05 58 13 – Column Covers

05 58 13.05 Column Covers

			Crew	Daily Output	Labor-Hours	Unit	Material	2020 Bare Costs Labor	Equipment	Total	Total Incl O&P
0010	**COLUMN COVERS**										
0015	Made from recycled materials										
0020	Excludes structural steel, light ga. metal framing, misc. metals, sealants										
0100	Round covers, 2 halves with 2 vertical joints for backer rod and sealant										
0110	Up to 12' high, no horizontal joints										
0120	12" diameter, 0.125" aluminum, anodized/painted finish	G	2 Sswk	32	.500	V.L.F.	33	29		62	83
0130	Type 304 stainless steel, 16 gauge, #4 brushed finish	G		32	.500		47	29		76	99
0140	Type 316 stainless steel, 16 gauge, #4 brushed finish	G		32	.500		57.50	29		86.50	111
0150	18" diameter, aluminum	G		32	.500		49	29		78	101
0160	Type 304 stainless steel	G		32	.500		70.50	29		99.50	125
0170	Type 316 stainless steel	G		32	.500		86.50	29		115.50	142
0180	24" diameter, aluminum	G		32	.500		65.50	29		94.50	119
0190	Type 304 stainless steel	G		32	.500		94.50	29		123.50	151
0200	Type 316 stainless steel	G		32	.500		115	29		144	174
0210	30" diameter, aluminum	G		30	.533		82	31		113	141
0220	Type 304 stainless steel	G		30	.533		118	31		149	181
0230	Type 316 stainless steel	G		30	.533		144	31		175	210
0240	36" diameter, aluminum	G		30	.533		98.50	31		129.50	159
0250	Type 304 stainless steel	G		30	.533		141	31		172	207
0260	Type 316 stainless steel	G		30	.533		173	31		204	241
0400	Up to 24' high, 2 stacked sections with 1 horizontal joint										
0410	18" diameter, aluminum	G	2 Sswk	28	.571	V.L.F.	51.50	33		84.50	111
0450	Type 304 stainless steel	G		28	.571		74.50	33		107.50	136
0460	Type 316 stainless steel	G		28	.571		91	33		124	154
0470	24" diameter, aluminum	G		28	.571		69	33		102	130
0480	Type 304 stainless steel	G		28	.571		99	33		132	163
0490	Type 316 stainless steel	G		28	.571		121	33		154	187
0500	30" diameter, aluminum	G		24	.667		86	38.50		124.50	158
0510	Type 304 stainless steel	G		24	.667		124	38.50		162.50	199
0520	Type 316 stainless steel	G		24	.667		151	38.50		189.50	230
0530	36" diameter, aluminum	G		24	.667		103	38.50		141.50	177
0540	Type 304 stainless steel	G		24	.667		149	38.50		187.50	226
0550	Type 316 stainless steel	G		24	.667		182	38.50		220.50	263

05 58 23 – Formed Metal Guards

05 58 23.90 Window Guards

			Crew	Daily Output	Labor-Hours	Unit	Material	2020 Bare Costs Labor	Equipment	Total	Total Incl O&P
0010	**WINDOW GUARDS**, shop fabricated										
0015	Expanded metal, steel angle frame, permanent	G	E-4	350	.091	S.F.	25	5.30	.42	30.72	36.50
0025	Steel bars, 1/2" x 1/2", spaced 5" OC	G	"	290	.110	"	17.30	6.40	.51	24.21	30
0030	Hinge mounted, add	G				Opng.	50			50	55
0040	Removable type, add	G				"	31.50			31.50	35

05 58 Formed Metal Fabrications

05 58 23 – Formed Metal Guards

05 58 23.90 Window Guards		Crew	Daily Output	Labor-Hours	Unit	Material	2020 Bare Costs Labor	Equipment	Total	Total Incl O&P
0050	For galvanized guards, add				S.F.	35%				
0070	For pivoted or projected type, add					105%	40%			
0100	Mild steel, stock units, economy	G	E-4	405	.079		6.80	4.59	.36	11.75	15.40
0200	Deluxe	G		405	.079	↓	13.85	4.59	.36	18.80	23
0400	Woven wire, stock units, 3/8" channel frame, 3' x 5' opening	G		40	.800	Opng.	184	46.50	3.67	234.17	282
0500	4' x 6' opening	G	↓	38	.842		293	49	3.87	345.87	410
0800	Basket guards for above, add	G					263			263	290
1000	Swinging guards for above, add	G				↓	90			90	99
9000	Minimum labor/equipment charge		1 Sswk	2	4	Job		231		231	380

05 58 25 – Formed Lamp Posts

05 58 25.40 Lamp Posts

0010	**LAMP POSTS**									
0020	Aluminum, 7' high, stock units, post only	G	1 Carp	16	.500	Ea.	84	26.50		110.50	135
0100	Mild steel, plain	G		16	.500	"	86	26.50		112.50	137
9000	Minimum labor/equipment charge		↓	4	2	Job		106		106	170

05 71 Decorative Metal Stairs

05 71 13 – Fabricated Metal Spiral Stairs

05 71 13.50 Spiral Stairs

			Crew	Daily Output	Labor-Hours	Unit	Material	Labor	Equipment	Total	Total Incl O&P
0010	**SPIRAL STAIRS**										
1805	Shop fabricated, custom ordered										
1810	Aluminum, 5'-0" diameter, plain units	G	E-4	45	.711	Riser	720	41.50	3.26	764.76	860
1820	Fancy units	G		45	.711		1,200	41.50	3.26	1,244.76	1,400
1900	Cast iron, 4'-0" diameter, plain units	G		45	.711		715	41.50	3.26	759.76	855
1920	Fancy units	G		25	1.280		1,400	74.50	5.90	1,480.40	1,675
2000	Steel, industrial checkered plate, 4' diameter	G		45	.711		700	41.50	3.26	744.76	835
2200	6' diameter	G	↓	40	.800	↓	715	46.50	3.67	765.17	865
3100	Spiral stair kits, 12 stacking risers to fit exact floor height										
3110	Steel, flat metal treads, primed, 3'-6" diameter	G	2 Carp	1.60	10	Flight	1,300	530		1,830	2,275
3120	4'-0" diameter	G		1.45	11.034		1,500	585		2,085	2,600
3130	4'-6" diameter	G		1.35	11.852		1,650	630		2,280	2,825
3140	5'-0" diameter	G		1.25	12.800		1,850	680		2,530	3,125
3210	Galvanized, 3'-6" diameter	G		1.60	10		1,875	530		2,405	2,900
3220	4'-0" diameter	G		1.45	11.034		2,150	585		2,735	3,300
3230	4'-6" diameter	G		1.35	11.852		2,325	630		2,955	3,550
3240	5'-0" diameter	G		1.25	12.800		2,575	680		3,255	3,925
3310	Checkered plate tread, primed, 3'-6" diameter	G		1.45	11.034		1,575	585		2,160	2,700
3320	4'-0" diameter	G		1.35	11.852		1,725	630		2,355	2,900
3330	4'-6" diameter	G		1.25	12.800		1,875	680		2,555	3,150
3340	5'-0" diameter	G		1.15	13.913		1,950	740		2,690	3,325
3410	Galvanized, 3'-6" diameter	G		1.45	11.034		2,150	585		2,735	3,325
3420	4'-0" diameter	G		1.35	11.852		2,425	630		3,055	3,650
3430	4'-6" diameter	G		1.25	12.800		2,625	680		3,305	3,975
3440	5'-0" diameter	G		1.15	13.913		2,925	740		3,665	4,375
3510	Red oak covers on flat metal treads, 3'-6" diameter			1.35	11.852		2,375	630		3,005	3,625
3520	4'-0" diameter			1.25	12.800		2,850	680		3,530	4,250
3530	4'-6" diameter			1.15	13.913		3,075	740		3,815	4,575
3540	5'-0" diameter		↓	1.05	15.238	↓	3,300	810		4,110	4,925

05 73 Decorative Metal Railings

05 73 16 – Wire Rope Decorative Metal Railings

05 73 16.10 Cable Railings		Crew	Daily Output	Labor-Hours	Unit	Material	2020 Bare Costs Labor	Equipment	Total	Total Incl O&P
0010	**CABLE RAILINGS**, with 316 stainless steel 1 x 19 cable, 3/16" diameter									
0015	Made from recycled materials									
0100	1-3/4" diameter stainless steel posts x 42" high, cables 4" OC	[G] 2 Sswk	25	.640	L.F.	43.50	37		80.50	109

05 73 23 – Ornamental Railings

05 73 23.50 Railings, Ornamental		Crew	Daily Output	Labor-Hours	Unit	Material	2020 Bare Costs Labor	Equipment	Total	Total Incl O&P
0010	**RAILINGS, ORNAMENTAL**, 3'-6" high, posts @ 6' OC									
0020	Bronze or stainless, hand forged, plain	[G] 2 Sswk	24	.667	L.F.	265	38.50		303.50	355
0100	Fancy	[G]	18	.889		525	51		576	665
0200	Aluminum, panelized, plain	[G]	24	.667		13.35	38.50		51.85	77.50
0300	Fancy	[G]	18	.889		25.50	51		76.50	112
0400	Wrought iron, hand forged, plain	[G]	24	.667		96.50	38.50		135	169
0500	Fancy	[G]	18	.889		231	51		282	340
0550	Steel, panelized, plain	[G]	24	.667		21	38.50		59.50	86.50
0560	Fancy	[G]	18	.889		33	51		84	121
0600	Composite metal/wood/glass, plain		18	.889		145	51		196	243
0700	Fancy		12	1.333		290	77		367	445
9000	Minimum labor/equipment charge	1 Sswk	2	4	Job		231		231	380

05 75 Decorative Formed Metal

05 75 13 – Columns

05 75 13.10 Aluminum Columns		Crew	Daily Output	Labor-Hours	Unit	Material	2020 Bare Costs Labor	Equipment	Total	Total Incl O&P
0010	**ALUMINUM COLUMNS**									
0015	Made from recycled materials									
0020	Aluminum, extruded, stock units, no cap or base, 6" diameter	[G] E-4	240	.133	L.F.	16.85	7.75	.61	25.21	32
0100	8" diameter	[G] "	170	.188	"	18.60	10.95	.86	30.41	39.50
0460	Caps, ornamental, plain	[G]			Set	315			315	345
0470	Fancy	[G]			"	1,750			1,750	1,925
0500	For square columns, add to column prices above				L.F.	50%				

For customer support on your Commercial Renovation Costs with RSMeans data, call 800.448.8182.

133

Division Notes

	CREW	DAILY OUTPUT	LABOR-HOURS	UNIT	BARE COSTS				TOTAL INCL O&P
					MAT.	LABOR	EQUIP.	TOTAL	

Estimating Tips
06 05 00 Common Work Results for Wood, Plastics, and Composites

- Common to any wood-framed structure are the accessory connector items such as screws, nails, adhesives, hangers, connector plates, straps, angles, and hold-downs. For typical wood-framed buildings, such as residential projects, the aggregate total for these items can be significant, especially in areas where seismic loading is a concern. For floor and wall framing, the material cost is based on 10 to 25 lbs. of accessory connectors per MBF. Hold-downs, hangers, and other connectors should be taken off by the piece.

  Included with material costs are fasteners for a normal installation. Gordian's RSMeans engineers use manufacturers' recommendations, written specifications, and/or standard construction practice for the sizing and spacing of fasteners. Prices for various fasteners are shown for informational purposes only. Adjustments should be made if unusual fastening conditions exist.

06 10 00 Carpentry

- Lumber is a traded commodity and therefore sensitive to supply and demand in the marketplace. Even with "budgetary" estimating of wood-framed projects, it is advisable to call local suppliers for the latest market pricing.

- The common quantity unit for wood-framed projects is "thousand board feet" (MBF). A board foot is a volume of wood—1" x 1' x 1' or 144 cubic inches. Board-foot quantities are generally calculated using nominal material dimensions—dressed sizes are ignored. Board foot per lineal foot of any stick of lumber can be calculated by dividing the nominal cross-sectional area by 12. As an example, 2,000 lineal feet of 2 x 12 equates to 4 MBF by dividing the nominal area, 2 x 12, by 12, which equals 2, and multiplying that by 2,000 to give 4,000 board feet. This simple rule applies to all nominal dimensioned lumber.

- Waste is an issue of concern at the quantity takeoff for any area of construction. Framing lumber is sold in even foot lengths, i.e., 8', 10', 12', 14', 16', and depending on spans, wall heights, and the grade of lumber, waste is inevitable. A rule of thumb for lumber waste is 5–10% depending on material quality and the complexity of the framing.

- Wood in various forms and shapes is used in many projects, even where the main structural framing is steel, concrete, or masonry. Plywood as a back-up partition material and 2x boards used as blocking and cant strips around roof edges are two common examples. The estimator should ensure that the costs of all wood materials are included in the final estimate.

06 20 00 Finish Carpentry

- It is necessary to consider the grade of workmanship when estimating labor costs for erecting millwork and an interior finish. In practice, there are three grades: premium, custom, and economy. The RSMeans daily output for base and case moldings is in the range of 200 to 250 L.F. per carpenter per day. This is appropriate for most average custom-grade projects. For premium projects, an adjustment to productivity of 25–50% should be made, depending on the complexity of the job.

Reference Numbers
Reference numbers are shown at the beginning of some major classifications. These numbers refer to related items in the Reference Section. The reference information may be an estimating procedure, an alternate pricing method, or technical information.

Note: Not all subdivisions listed here necessarily appear. ■

Same Data. Simplified.

Enjoy the convenience and efficiency of accessing your costs anywhere:

- **Skip the multiplier** by setting your location
- **Quickly search,** edit, favorite and share costs
- **Stay on top of price changes** with automatic updates

Discover more at rsmeans.com/online

06 05 05.10 Selective Demolition Wood Framing	Crew	Daily Output	Labor-Hours	Unit	Material	2020 Bare Costs Labor	2020 Bare Costs Equipment	Total	Total Incl O&P
0010 **SELECTIVE DEMOLITION WOOD FRAMING** R024119-10									
0100 Timber connector, nailed, small	1 Clab	96	.083	Ea.		3.51		3.51	5.60
0110 Medium		60	.133			5.60		5.60	9
0120 Large		48	.167			7		7	11.25
0130 Bolted, small		48	.167			7		7	11.25
0140 Medium		32	.250			10.55		10.55	16.85
0150 Large		24	.333			14.05		14.05	22.50
2958 Beams, 2" x 6"	2 Clab	1100	.015	L.F.		.61		.61	.98
2960 2" x 8"		825	.019			.82		.82	1.31
2965 2" x 10"		665	.024			1.01		1.01	1.62
2970 2" x 12"		550	.029			1.22		1.22	1.96
2972 2" x 14"		470	.034			1.43		1.43	2.30
2975 4" x 8"	B-1	413	.058			2.49		2.49	3.98
2980 4" x 10"		330	.073			3.11		3.11	4.98
2985 4" x 12"		275	.087			3.73		3.73	6
3000 6" x 8"		275	.087			3.73		3.73	6
3040 6" x 10"		220	.109			4.67		4.67	7.45
3080 6" x 12"		185	.130			5.55		5.55	8.90
3120 8" x 12"		140	.171			7.35		7.35	11.75
3160 10" x 12"		110	.218			9.35		9.35	14.95
3162 Alternate pricing method		1.10	21.818	M.B.F.		935		935	1,500
3170 Blocking, in 16" OC wall framing, 2" x 4"	1 Clab	600	.013	L.F.		.56		.56	.90
3172 2" x 6"		400	.020			.84		.84	1.35
3174 In 24" OC wall framing, 2" x 4"		600	.013			.56		.56	.90
3176 2" x 6"		400	.020			.84		.84	1.35
3178 Alt method, wood blocking removal from wood framing		.40	20	M.B.F.		840		840	1,350
3179 Wood blocking removal from steel framing		.36	22.222	"		935		935	1,500
3180 Bracing, let in, 1" x 3", studs 16" OC		1050	.008	L.F.		.32		.32	.51
3181 Studs 24" OC		1080	.007			.31		.31	.50
3182 1" x 4", studs 16" OC		1050	.008			.32		.32	.51
3183 Studs 24" OC		1080	.007			.31		.31	.50
3184 1" x 6", studs 16" OC		1050	.008			.32		.32	.51
3185 Studs 24" OC		1080	.007			.31		.31	.50
3186 2" x 3", studs 16" OC		800	.010			.42		.42	.67
3187 Studs 24" OC		830	.010			.41		.41	.65
3188 2" x 4", studs 16" OC		800	.010			.42		.42	.67
3189 Studs 24" OC		830	.010			.41		.41	.65
3190 2" x 6", studs 16" OC		800	.010			.42		.42	.67
3191 Studs 24" OC		830	.010			.41		.41	.65
3192 2" x 8", studs 16" OC		800	.010			.42		.42	.67
3193 Studs 24" OC		830	.010			.41		.41	.65
3194 "T" shaped metal bracing, studs at 16" OC		1060	.008			.32		.32	.51
3195 Studs at 24" OC		1200	.007			.28		.28	.45
3196 Metal straps, studs at 16" OC		1200	.007			.28		.28	.45
3197 Studs at 24" OC		1240	.006			.27		.27	.44
3200 Columns, round, 8' to 14' tall		40	.200	Ea.		8.40		8.40	13.50
3202 Dimensional lumber sizes	2 Clab	1.10	14.545	M.B.F.		610		610	980
3250 Blocking, between joists	1 Clab	320	.025	Ea.		1.05		1.05	1.69
3252 Bridging, metal strap, between joists		320	.025	Pr.		1.05		1.05	1.69
3254 Wood, between joists		320	.025	"		1.05		1.05	1.69
3260 Door buck, studs, header & access., 8' high 2" x 4" wall, 3' wide		32	.250	Ea.		10.55		10.55	16.85
3261 4' wide		32	.250			10.55		10.55	16.85
3262 5' wide		32	.250			10.55		10.55	16.85

06 05 05.10 Selective Demolition Wood Framing		Crew	Daily Output	Labor-Hours	Unit	Material	2020 Bare Costs Labor	Equipment	Total	Total Incl O&P
3263	6' wide	1 Clab	32	.250	Ea.		10.55		10.55	16.85
3264	8' wide		30	.267			11.25		11.25	18
3265	10' wide		30	.267			11.25		11.25	18
3266	12' wide		30	.267			11.25		11.25	18
3267	2" x 6" wall, 3' wide		32	.250			10.55		10.55	16.85
3268	4' wide		32	.250			10.55		10.55	16.85
3269	5' wide		32	.250			10.55		10.55	16.85
3270	6' wide		32	.250			10.55		10.55	16.85
3271	8' wide		30	.267			11.25		11.25	18
3272	10' wide		30	.267			11.25		11.25	18
3273	12' wide		30	.267			11.25		11.25	18
3274	Window buck, studs, header & access., 8' high 2" x 4" wall, 2' wide		24	.333			14.05		14.05	22.50
3275	3' wide		24	.333			14.05		14.05	22.50
3276	4' wide		24	.333			14.05		14.05	22.50
3277	5' wide		24	.333			14.05		14.05	22.50
3278	6' wide		24	.333			14.05		14.05	22.50
3279	7' wide		24	.333			14.05		14.05	22.50
3280	8' wide		22	.364			15.30		15.30	24.50
3281	10' wide		22	.364			15.30		15.30	24.50
3282	12' wide		22	.364			15.30		15.30	24.50
3283	2" x 6" wall, 2' wide		24	.333			14.05		14.05	22.50
3284	3' wide		24	.333			14.05		14.05	22.50
3285	4' wide		24	.333			14.05		14.05	22.50
3286	5' wide		24	.333			14.05		14.05	22.50
3287	6' wide		24	.333			14.05		14.05	22.50
3288	7' wide		24	.333			14.05		14.05	22.50
3289	8' wide		22	.364			15.30		15.30	24.50
3290	10' wide		22	.364			15.30		15.30	24.50
3291	12' wide		22	.364			15.30		15.30	24.50
3360	Deck or porch decking		825	.010	L.F.		.41		.41	.65
3400	Fascia boards, 1" x 6"		500	.016			.67		.67	1.08
3440	1" x 8"		450	.018			.75		.75	1.20
3480	1" x 10"		400	.020			.84		.84	1.35
3490	2" x 6"		450	.018			.75		.75	1.20
3500	2" x 8"		400	.020			.84		.84	1.35
3510	2" x 10"		350	.023			.96		.96	1.54
3610	Furring, on wood walls or ceiling		4000	.002	S.F.		.08		.08	.13
3620	On masonry or concrete walls or ceiling		1200	.007	"		.28		.28	.45
3800	Headers over openings, 2 @ 2" x 6"		110	.073	L.F.		3.06		3.06	4.91
3840	2 @ 2" x 8"		100	.080			3.37		3.37	5.40
3880	2 @ 2" x 10"		90	.089			3.74		3.74	6
3885	Alternate pricing method		.26	30.651	M.B.F.		1,300		1,300	2,075
3920	Joists, 1" x 4"		1250	.006	L.F.		.27		.27	.43
3930	1" x 6"		1135	.007			.30		.30	.48
3940	1" x 8"		1000	.008			.34		.34	.54
3950	1" x 10"		895	.009			.38		.38	.60
3960	1" x 12"		765	.010			.44		.44	.71
4200	2" x 4"	2 Clab	1000	.016			.67		.67	1.08
4230	2" x 6"		970	.016			.69		.69	1.11
4240	2" x 8"		940	.017			.72		.72	1.15
4250	2" x 10"		910	.018			.74		.74	1.19
4280	2" x 12"		880	.018			.77		.77	1.23
4281	2" x 14"		850	.019			.79		.79	1.27

06 05 05 – Selective Demolition for Wood, Plastics, and Composites

06 05 05.10 Selective Demolition Wood Framing		Crew	Daily Output	Labor-Hours	Unit	Material	2020 Bare Costs Labor	Equipment	Total	Total Incl O&P
4282	Composite joists, 9-1/2"	2 Clab	960	.017	L.F.		.70		.70	1.12
4283	11-7/8"		930	.017			.72		.72	1.16
4284	14"		897	.018			.75		.75	1.20
4285	16"		865	.019			.78		.78	1.25
4290	Wood joists, alternate pricing method		1.50	10.667	M.B.F.		450		450	720
4500	Open web joist, 12" deep		500	.032	L.F.		1.35		1.35	2.16
4505	14" deep		475	.034			1.42		1.42	2.27
4510	16" deep		450	.036			1.50		1.50	2.40
4520	18" deep		425	.038			1.59		1.59	2.54
4530	24" deep		400	.040			1.68		1.68	2.70
4550	Ledger strips, 1" x 2"	1 Clab	1200	.007			.28		.28	.45
4560	1" x 3"		1200	.007			.28		.28	.45
4570	1" x 4"		1200	.007			.28		.28	.45
4580	2" x 2"		1100	.007			.31		.31	.49
4590	2" x 4"		1000	.008			.34		.34	.54
4600	2" x 6"		1000	.008			.34		.34	.54
4601	2" x 8 or 2" x 10"		800	.010			.42		.42	.67
4602	4" x 6"		600	.013			.56		.56	.90
4604	4" x 8"		450	.018			.75		.75	1.20
5400	Posts, 4" x 4"	2 Clab	800	.020			.84		.84	1.35
5405	4" x 6"		550	.029			1.22		1.22	1.96
5410	4" x 8"		440	.036			1.53		1.53	2.45
5425	4" x 10"		390	.041			1.73		1.73	2.77
5430	4" x 12"		350	.046			1.92		1.92	3.08
5440	6" x 6"		400	.040			1.68		1.68	2.70
5445	6" x 8"		350	.046			1.92		1.92	3.08
5450	6" x 10"		320	.050			2.11		2.11	3.37
5455	6" x 12"		290	.055			2.32		2.32	3.72
5480	8" x 8"		300	.053			2.25		2.25	3.60
5500	10" x 10"		240	.067			2.81		2.81	4.50
5660	T&G floor planks		2	8	M.B.F.		335		335	540
5682	Rafters, ordinary, 16" OC, 2" x 4"		880	.018	S.F.		.77		.77	1.23
5683	2" x 6"		840	.019			.80		.80	1.28
5684	2" x 8"		820	.020			.82		.82	1.32
5685	2" x 10"		820	.020			.82		.82	1.32
5686	2" x 12"		810	.020			.83		.83	1.33
5687	24" OC, 2" x 4"		1170	.014			.58		.58	.92
5688	2" x 6"		1117	.014			.60		.60	.97
5689	2" x 8"		1091	.015			.62		.62	.99
5690	2" x 10"		1091	.015			.62		.62	.99
5691	2" x 12"		1077	.015			.63		.63	1
5795	Rafters, ordinary, 2" x 4" (alternate method)		862	.019	L.F.		.78		.78	1.25
5800	2" x 6" (alternate method)		850	.019			.79		.79	1.27
5840	2" x 8" (alternate method)		837	.019			.81		.81	1.29
5855	2" x 10" (alternate method)		825	.019			.82		.82	1.31
5865	2" x 12" (alternate method)		812	.020			.83		.83	1.33
5870	Sill plate, 2" x 4"	1 Clab	1170	.007			.29		.29	.46
5871	2" x 6"		780	.010			.43		.43	.69
5872	2" x 8"		586	.014			.57		.57	.92
5873	Alternate pricing method		.78	10.256	M.B.F.		430		430	690
5885	Ridge board, 1" x 4"	2 Clab	900	.018	L.F.		.75		.75	1.20
5886	1" x 6"		875	.018			.77		.77	1.23
5887	1" x 8"		850	.019			.79		.79	1.27

06 05 05.10 Selective Demolition Wood Framing	Crew	Daily Output	Labor-Hours	Unit	Material	2020 Bare Costs Labor	Equipment	Total	Total Incl O&P
5888	1" x 10"	2 Clab	825	.019	L.F.		.82		.82	1.31
5889	1" x 12"		800	.020			.84		.84	1.35
5890	2" x 4"		900	.018			.75		.75	1.20
5892	2" x 6"		875	.018			.77		.77	1.23
5894	2" x 8"		850	.019			.79		.79	1.27
5896	2" x 10"		825	.019			.82		.82	1.31
5898	2" x 12"		800	.020			.84		.84	1.35
5900	Hip & valley rafters, 2" x 6"		500	.032			1.35		1.35	2.16
5940	2" x 8"		420	.038			1.60		1.60	2.57
6050	Rafter tie, 1" x 4"		1250	.013			.54		.54	.86
6052	1" x 6"		1135	.014			.59		.59	.95
6054	2" x 4"		1000	.016			.67		.67	1.08
6056	2" x 6"		970	.016			.69		.69	1.11
6070	Sleepers, on concrete, 1" x 2"	1 Clab	4700	.002			.07		.07	.11
6075	1" x 3"		4000	.002			.08		.08	.13
6080	2" x 4"		3000	.003			.11		.11	.18
6085	2" x 6"		2600	.003			.13		.13	.21
6086	Sheathing from roof, 5/16"	2 Clab	1600	.010	S.F.		.42		.42	.67
6088	3/8"		1525	.010			.44		.44	.71
6090	1/2"		1400	.011			.48		.48	.77
6092	5/8"		1300	.012			.52		.52	.83
6094	3/4"		1200	.013			.56		.56	.90
6096	Board sheathing from roof		1400	.011			.48		.48	.77
6100	Sheathing, from walls, 1/4"		1200	.013			.56		.56	.90
6110	5/16"		1175	.014			.57		.57	.92
6120	3/8"		1150	.014			.59		.59	.94
6130	1/2"		1125	.014			.60		.60	.96
6140	5/8"		1100	.015			.61		.61	.98
6150	3/4"		1075	.015			.63		.63	1
6152	Board sheathing from walls		1500	.011			.45		.45	.72
6158	Subfloor/roof deck, with boards		2200	.007			.31		.31	.49
6159	Subfloor/roof deck, with tongue & groove boards		2000	.008			.34		.34	.54
6160	Plywood, 1/2" thick		768	.021			.88		.88	1.41
6162	5/8" thick		760	.021			.89		.89	1.42
6164	3/4" thick		750	.021			.90		.90	1.44
6165	1-1/8" thick		720	.022			.94		.94	1.50
6166	Underlayment, particle board, 3/8" thick	1 Clab	780	.010			.43		.43	.69
6168	1/2" thick		768	.010			.44		.44	.70
6170	5/8" thick		760	.011			.44		.44	.71
6172	3/4" thick		750	.011	Riser		.45		.45	.72
6200	Stairs and stringers, straight run	2 Clab	40	.400	Riser		16.85		16.85	27
6240	With platforms, winders or curves	"	26	.615	"		26		26	41.50
6300	Components, tread	1 Clab	110	.073	Ea.		3.06		3.06	4.91
6320	Riser		80	.100	"		4.21		4.21	6.75
6390	Stringer, 2" x 10"		260	.031	L.F.		1.30		1.30	2.08
6400	2" x 12"		260	.031			1.30		1.30	2.08
6410	3" x 10"		250	.032			1.35		1.35	2.16
6420	3" x 12"		250	.032			1.35		1.35	2.16
6590	Wood studs, 2" x 3"	2 Clab	3076	.005			.22		.22	.35
6600	2" x 4"		2000	.008			.34		.34	.54
6640	2" x 6"		1600	.010			.42		.42	.67
6720	Wall framing, including studs, plates and blocking, 2" x 4"	1 Clab	600	.013	S.F.		.56		.56	.90
6740	2" x 6"		480	.017	"		.70		.70	1.12

06 05 05 – Selective Demolition for Wood, Plastics, and Composites

06 05 05.10 Selective Demolition Wood Framing		Crew	Daily Output	Labor-Hours	Unit	Material	2020 Bare Costs Labor	Equipment	Total	Total Incl O&P
6750	Headers, 2" x 4"	1 Clab	1125	.007	L.F.		.30		.30	.48
6755	2" x 6"		1125	.007			.30		.30	.48
6760	2" x 8"		1050	.008			.32		.32	.51
6765	2" x 10"		1050	.008			.32		.32	.51
6770	2" x 12"		1000	.008			.34		.34	.54
6780	4" x 10"		525	.015			.64		.64	1.03
6785	4" x 12"		500	.016			.67		.67	1.08
6790	6" x 8"		560	.014			.60		.60	.96
6795	6" x 10"		525	.015			.64		.64	1.03
6797	6" x 12"		500	.016			.67		.67	1.08
7000	Trusses									
7050	12' span	2 Clab	74	.216	Ea.		9.10		9.10	14.60
7150	24' span	F-3	66	.606			33	7.15	40.15	60.50
7200	26' span		64	.625			34	7.35	41.35	62
7250	28' span		62	.645			35	7.60	42.60	64.50
7300	30' span		58	.690			37.50	8.10	45.60	69
7350	32' span		56	.714			39	8.40	47.40	71.50
7400	34' span		54	.741			40.50	8.75	49.25	73.50
7450	36' span		52	.769			42	9.05	51.05	76.50
8000	Soffit, T&G wood	1 Clab	520	.015	S.F.		.65		.65	1.04
8010	Hardboard, vinyl or aluminum	"	640	.013			.53		.53	.84
8030	Plywood	2 Carp	315	.051			2.70		2.70	4.32
9000	Minimum labor/equipment charge	1 Clab	4	2	Job		84		84	135
9500	See Section 02 41 19.19 for rubbish handling									

06 05 05.20 Selective Demolition Millwork and Trim

0010	SELECTIVE DEMOLITION MILLWORK AND TRIM	R024119-10								
1000	Cabinets, wood, base cabinets, per L.F.	2 Clab	80	.200	L.F.		8.40		8.40	13.50
1020	Wall cabinets, per L.F.	"	80	.200	"		8.40		8.40	13.50
1060	Remove and reset, base cabinets	2 Carp	18	.889	Ea.		47		47	75.50
1070	Wall cabinets		20	.800			42.50		42.50	68
1072	Oven cabinet, 7' high		11	1.455			77.50		77.50	124
1074	Cabinet door, up to 2' high	1 Clab	66	.121			5.10		5.10	8.20
1076	2' - 4' high	"	46	.174			7.30		7.30	11.75
1100	Steel, painted, base cabinets	2 Clab	60	.267	L.F.		11.25		11.25	18
1120	Wall cabinets		60	.267	"		11.25		11.25	18
1200	Casework, large area		320	.050	S.F.		2.11		2.11	3.37
1220	Selective		200	.080	"		3.37		3.37	5.40
1500	Counter top, straight runs		200	.080	L.F.		3.37		3.37	5.40
1510	L, U or C shapes		120	.133			5.60		5.60	9
1550	Remove and reset, straight runs	2 Carp	50	.320			17		17	27.50
1560	L, U or C shape	"	40	.400			21.50		21.50	34
2000	Paneling, 4' x 8' sheets	2 Clab	2000	.008	S.F.		.34		.34	.54
2100	Boards, 1" x 4"		700	.023			.96		.96	1.54
2120	1" x 6"		750	.021			.90		.90	1.44
2140	1" x 8"		800	.020			.84		.84	1.35
3000	Trim, baseboard, to 6" wide		1200	.013	L.F.		.56		.56	.90
3040	Greater than 6" and up to 12" wide		1000	.016			.67		.67	1.08
3080	Remove and reset, minimum	2 Carp	400	.040			2.13		2.13	3.41
3090	Maximum	"	300	.053			2.83		2.83	4.54
3100	Ceiling trim	2 Clab	1000	.016			.67		.67	1.08
3120	Chair rail		1200	.013			.56		.56	.90
3140	Railings with balusters		240	.067			2.81		2.81	4.50

140

For customer support on your Commercial Renovation Costs with RSMeans data, call 800.448.8182.

06 05 Common Work Results for Wood, Plastics, and Composites

06 05 05 – Selective Demolition for Wood, Plastics, and Composites

	06 05 05.20 Selective Demolition Millwork and Trim	Crew	Daily Output	Labor-Hours	Unit	Material	2020 Bare Costs Labor	Equipment	Total	Total Incl O&P
3160	Wainscoting	2 Clab	700	.023	S.F.		.96		.96	1.54
9000	Minimum labor/equipment charge	1 Clab	4	2	Job		84		84	135

06 05 23 – Wood, Plastic, and Composite Fastenings

06 05 23.10 Nails

		Crew	Daily Output	Labor-Hours	Unit	Material	2020 Bare Costs Labor	Equipment	Total	Total Incl O&P
0010	**NAILS**, material only, based upon 50# box purchase									
0020	Copper nails, plain				Lb.	11.65			11.65	12.85
0400	Stainless steel, plain					10.10			10.10	11.10
0500	Box, 3d to 20d, bright					1.52			1.52	1.67
0520	Galvanized					2.55			2.55	2.81
0600	Common, 3d to 60d, plain					1.13			1.13	1.24
0700	Galvanized					2.48			2.48	2.73
0800	Aluminum					11.05			11.05	12.15
1000	Annular or spiral thread, 4d to 60d, plain					2.85			2.85	3.14
1200	Galvanized					3.49			3.49	3.84
1400	Drywall nails, plain					2.01			2.01	2.21
1600	Galvanized					1.89			1.89	2.08
1800	Finish nails, 4d to 10d, plain					1.57			1.57	1.73
2000	Galvanized					1.93			1.93	2.12
2100	Aluminum					8.75			8.75	9.65
2300	Flooring nails, hardened steel, 2d to 10d, plain					3.88			3.88	4.27
2400	Galvanized					4.09			4.09	4.50
2500	Gypsum lath nails, 1-1/8", 13 ga. flathead, blued					3.64			3.64	4
2600	Masonry nails, hardened steel, 3/4" to 3" long, plain					2.34			2.34	2.57
2700	Galvanized					4.14			4.14	4.55
2900	Roofing nails, threaded, galvanized					2.93			2.93	3.22
3100	Aluminum					7.60			7.60	8.40
3300	Compressed lead head, threaded, galvanized					3.06			3.06	3.37
3600	Siding nails, plain shank, galvanized					2.59			2.59	2.85
3800	Aluminum					5.95			5.95	6.55
5000	Add to prices above for cement coating					.16			.16	.18
5200	Zinc or tin plating					.27			.27	.30
5500	Vinyl coated sinkers, 8d to 16d					2.51			2.51	2.76

06 05 23.50 Wood Screws

		Crew	Daily Output	Labor-Hours	Unit	Material	2020 Bare Costs Labor	Equipment	Total	Total Incl O&P
0010	**WOOD SCREWS**									
0020	#8, 1" long, steel				C	4.03			4.03	4.43
0100	Brass					12.30			12.30	13.50
0200	#8, 2" long, steel					6.90			6.90	7.60
0300	Brass					27.50			27.50	30
0400	#10, 1" long, steel					3.55			3.55	3.91
0500	Brass					15.65			15.65	17.25
0600	#10, 2" long, steel					5.15			5.15	5.65
0700	Brass					26			26	28.50
0800	#10, 3" long, steel					9.20			9.20	10.15
1000	#12, 2" long, steel					7.70			7.70	8.45
1100	Brass					36.50			36.50	40
1500	#12, 3" long, steel					11.55			11.55	12.70
2000	#12, 4" long, steel					26.50			26.50	29
2050	Composite trim screws									
2100	Stainless steel, trim head, #7 x 1-5/8" long				C	27.50			27.50	30
2200	#7 x 2" long					27			27	30
2300	#7 x 2-1/2" long					30.50			30.50	33.50
2400	#7 x 3" long					35.50			35.50	39

For customer support on your Commercial Renovation Costs with RSMeans data, call 800.448.8182.

141

06 05 23.50 Wood Screws		Crew	Daily Output	Labor-Hours	Unit	Material	2020 Bare Costs Labor	Equipment	Total	Total Incl O&P
2500	#9 x 3-1/2" long				C	76.50			76.50	84.50
2600	#9 x 4" long					90			90	99
3000	Stainless steel, flat head, #6 x 1-5/8" long					25.50			25.50	28
3100	#8 x 2" long					18.05			18.05	19.85
3200	#8 x 2-1/2" long					21.50			21.50	23.50
3300	#8 x 3" long					25			25	27.50
3400	#10 x 2" long					20.50			20.50	22.50
3500	#10 x 3-1/2" long					41			41	45.50
3600	#12 x 3" long					47.50			47.50	52
3700	#12 x 4" long					72.50			72.50	80
3800	#12 x 6" long					124			124	137

06 05 23.60 Timber Connectors		Crew	Daily Output	Labor-Hours	Unit	Material	2020 Bare Costs Labor	Equipment	Total	Total Incl O&P
0010	**TIMBER CONNECTORS**									
0020	Add up cost of each part for total cost of connection									
0100	Connector plates, steel, with bolts, straight	2 Carp	75	.213	Ea.	27.50	11.35		38.85	48.50
0110	Tee, 7 ga.		50	.320		41.50	17		58.50	73
0120	T- Strap, 14 ga., 12" x 8" x 2"		50	.320		41.50	17		58.50	73
0150	Anchor plates, 7 ga., 9" x 7"		75	.213		27.50	11.35		38.85	48.50
0200	Bolts, machine, sq. hd. with nut & washer, 1/2" diameter, 4" long	1 Carp	140	.057		.77	3.04		3.81	5.70
0300	7-1/2" long		130	.062		1.41	3.27		4.68	6.80
0500	3/4" diameter, 7-1/2" long		130	.062		3.51	3.27		6.78	9.10
0610	Machine bolts, w/nut, washer, 3/4" diameter, 15" L, HD's & beam hangers		95	.084		6.60	4.48		11.08	14.40
0800	Drilling bolt holes in timber, 1/2" diameter		450	.018	Inch		.95		.95	1.51
0900	1" diameter		350	.023	"		1.22		1.22	1.95
1100	Framing anchor, angle, 3" x 3" x 1-1/2", 12 ga.		175	.046	Ea.	2.53	2.43		4.96	6.65
1150	Framing anchors, 18 ga., 4-1/2" x 2-3/4"		175	.046		2.53	2.43		4.96	6.65
1160	Framing anchors, 18 ga., 4-1/2" x 3"		175	.046		2.53	2.43		4.96	6.65
1170	Clip anchors plates, 18 ga., 12" x 1-1/8"		175	.046		2.53	2.43		4.96	6.65
1250	Holdowns, 3 ga. base, 10 ga. body		8	1		46.50	53		99.50	136
1260	Holdowns, 7 ga. 11-1/16" x 3-1/4"		8	1		46.50	53		99.50	136
1270	Holdowns, 7 ga. 14-3/8" x 3-1/8"		8	1		46.50	53		99.50	136
1275	Holdowns, 12 ga. 8" x 2-1/2"		8	1		46.50	53		99.50	136
1300	Joist and beam hangers, 18 ga. galv., for 2" x 4" joist		175	.046		.85	2.43		3.28	4.83
1400	2" x 6" to 2" x 10" joist		165	.048		1.56	2.58		4.14	5.85
1600	16 ga. galv., 3" x 6" to 3" x 10" joist		160	.050		3.03	2.66		5.69	7.60
1700	3" x 10" to 3" x 14" joist		160	.050		4.64	2.66		7.30	9.35
1800	4" x 6" to 4" x 10" joist		155	.052		3.59	2.74		6.33	8.35
1900	4" x 10" to 4" x 14" joist		155	.052		5	2.74		7.74	9.90
2000	Two-2" x 6" to two-2" x 10" joists		150	.053		3.98	2.83		6.81	8.90
2100	Two-2" x 10" to two-2" x 14" joists		150	.053		4.55	2.83		7.38	9.55
2300	3/16" thick, 6" x 8" joist		145	.055		75	2.93		77.93	87
2400	6" x 10" joist		140	.057		77.50	3.04		80.54	90.50
2500	6" x 12" joist		135	.059		79.50	3.15		82.65	92.50
2700	1/4" thick, 6" x 14" joist		130	.062		82	3.27		85.27	95.50
2900	Plywood clips, extruded aluminum H clip, for 3/4" panels					.24			.24	.26
3000	Galvanized 18 ga. back-up clip					.19			.19	.21
3200	Post framing, 16 ga. galv. for 4" x 4" base, 2 piece	1 Carp	130	.062		15.25	3.27		18.52	22
3300	Cap		130	.062		25	3.27		28.27	33
3500	Rafter anchors, 18 ga. galv., 1-1/2" wide, 5-1/4" long		145	.055		.41	2.93		3.34	5.15
3600	10-3/4" long		145	.055		1.15	2.93		4.08	5.95
3800	Shear plates, 2-5/8" diameter		120	.067		3.20	3.54		6.74	9.20
3900	4" diameter		115	.070		2.69	3.70		6.39	8.85

06 05 23 – Wood, Plastic, and Composite Fastenings

06 05 23.60 Timber Connectors

		Crew	Daily Output	Labor-Hours	Unit	Material	2020 Bare Costs Labor	Equipment	Total	Total Incl O&P
4000	Sill anchors, embedded in concrete or block, 25-1/2" long	1 Carp	115	.070	Ea.	16.10	3.70		19.80	23.50
4100	Spike grids, 3" x 6"		120	.067		1.34	3.54		4.88	7.15
4400	Split rings, 2-1/2" diameter		120	.067		3.31	3.54		6.85	9.35
4500	4" diameter		110	.073		3.51	3.87		7.38	10.05
4550	Tie plate, 20 ga., 7" x 3-1/8"		110	.073		3.51	3.87		7.38	10.05
4560	5" x 4-1/8"		110	.073		3.51	3.87		7.38	10.05
4575	Twist straps, 18 ga., 12" x 1-1/4"		110	.073		3.51	3.87		7.38	10.05
4580	16" x 1-1/4"		110	.073		3.51	3.87		7.38	10.05
4600	Strap ties, 20 ga., 2-1/16" wide, 12-13/16" long		180	.044		1.01	2.36		3.37	4.89
4700	16 ga., 1-3/8" wide, 12" long		180	.044		1.01	2.36		3.37	4.89
4800	1-1/4" wide, 21-5/8" long		160	.050		2.81	2.66		5.47	7.35
5000	Toothed rings, 2-5/8" or 4" diameter		90	.089		2.33	4.72		7.05	10.10
5200	Truss plates, nailed, 20 ga., up to 32' span	▼	17	.471	Truss	15.15	25		40.15	56.50
5400	Washers, 2" x 2" x 1/8"				Ea.	.48			.48	.53
5500	3" x 3" x 3/16"				"	1.26			1.26	1.39
9000	Minimum labor/equipment charge	1 Carp	4	2	Job		106		106	170

06 05 23.80 Metal Bracing

		Crew	Daily Output	Labor-Hours	Unit	Material	2020 Bare Costs Labor	Equipment	Total	Total Incl O&P
0010	**METAL BRACING**									
0302	Let-in, "T" shaped, 22 ga. galv. steel, studs at 16" OC	1 Carp	580	.014	L.F.	.85	.73		1.58	2.11
0402	Studs at 24" OC		600	.013		.85	.71		1.56	2.08
0502	Steel straps, 16 ga. galv. steel, studs at 16" OC		600	.013		1.10	.71		1.81	2.35
0602	Studs at 24" OC	▼	620	.013	▼	1.10	.69		1.79	2.31

06 11 Wood Framing

06 11 10 – Framing with Dimensional, Engineered or Composite Lumber

06 11 10.01 Forest Stewardship Council Certification

		Crew	Daily Output	Labor-Hours	Unit	Material	2020 Bare Costs Labor	Equipment	Total	Total Incl O&P
0010	**FOREST STEWARDSHIP COUNCIL CERTIFICATION**									
0020	For Forest Stewardship Council (FSC) cert dimension lumber, add G					65%				

06 11 10.02 Blocking

		Crew	Daily Output	Labor-Hours	Unit	Material	2020 Bare Costs Labor	Equipment	Total	Total Incl O&P
0010	**BLOCKING**									
2600	Miscellaneous, to wood construction									
2620	2" x 4"	1 Carp	.17	47.059	M.B.F.	710	2,500		3,210	4,775
2625	Pneumatic nailed		.21	38.095		720	2,025		2,745	4,050
2660	2" x 8"		.27	29.630		750	1,575		2,325	3,350
2665	Pneumatic nailed	▼	.33	24.242	▼	760	1,300		2,060	2,900
2720	To steel construction									
2740	2" x 4"	1 Carp	.14	57.143	M.B.F.	710	3,025		3,735	5,650
2780	2" x 8"		.21	38.095	"	750	2,025		2,775	4,075
9000	Minimum labor/equipment charge	▼	4	2	Job		106		106	170

06 11 10.04 Wood Bracing

		Crew	Daily Output	Labor-Hours	Unit	Material	2020 Bare Costs Labor	Equipment	Total	Total Incl O&P
0010	**WOOD BRACING**									
0012	Let-in, with 1" x 6" boards, studs @ 16" OC	1 Carp	150	.053	L.F.	.83	2.83		3.66	5.45
0202	Studs @ 24" OC	"	230	.035	"	.83	1.85		2.68	3.87

06 11 10.06 Bridging

		Crew	Daily Output	Labor-Hours	Unit	Material	2020 Bare Costs Labor	Equipment	Total	Total Incl O&P
0010	**BRIDGING**									
0012	Wood, for joists 16" OC, 1" x 3"	1 Carp	130	.062	Pr.	.70	3.27		3.97	6
0017	Pneumatic nailed		170	.047		.79	2.50		3.29	4.88
0102	2" x 3" bridging		130	.062		.72	3.27		3.99	6.05
0107	Pneumatic nailed		170	.047		.77	2.50		3.27	4.85
0302	Steel, galvanized, 18 ga., for 2" x 10" joists at 12" OC	▼	130	.062	▼	1.71	3.27		4.98	7.15

06 11 Wood Framing

06 11 10 – Framing with Dimensional, Engineered or Composite Lumber

06 11 10.06 Bridging

		Crew	Daily Output	Labor-Hours	Unit	Material	2020 Bare Costs Labor	Equipment	Total	Total Incl O&P
0352	16" OC	1 Carp	135	.059	Pr.	1.72	3.15		4.87	6.95
0402	24" OC		140	.057		2.63	3.04		5.67	7.75
0602	For 2" x 14" joists at 16" OC		130	.062		1.98	3.27		5.25	7.45
0902	Compression type, 16" OC, 2" x 8" joists		200	.040		1.43	2.13		3.56	4.98
1002	2" x 12" joists	↓	200	.040	↓	1.43	2.13		3.56	4.98

06 11 10.10 Beam and Girder Framing

		Crew	Daily Output	Labor-Hours	Unit	Material	2020 Bare Costs Labor	Equipment	Total	Total Incl O&P
0010	**BEAM AND GIRDER FRAMING** R061110-30									
1000	Single, 2" x 6"	2 Carp	700	.023	L.F.	.69	1.22		1.91	2.70
1005	Pneumatic nailed		812	.020		.69	1.05		1.74	2.44
1020	2" x 8"		650	.025		1	1.31		2.31	3.20
1025	Pneumatic nailed		754	.021		1.01	1.13		2.14	2.92
1040	2" x 10"		600	.027		1.33	1.42		2.75	3.73
1045	Pneumatic nailed		696	.023		1.34	1.22		2.56	3.44
1060	2" x 12"		550	.029		1.74	1.55		3.29	4.39
1065	Pneumatic nailed		638	.025		1.76	1.33		3.09	4.07
1080	2" x 14"		500	.032		2.38	1.70		4.08	5.35
1085	Pneumatic nailed		580	.028		2.40	1.47		3.87	4.99
1100	3" x 8"		550	.029		2.95	1.55		4.50	5.70
1120	3" x 10"		500	.032		3.57	1.70		5.27	6.65
1140	3" x 12"		450	.036		4.83	1.89		6.72	8.35
1160	3" x 14"	↓	400	.040		5.30	2.13		7.43	9.25
1170	4" x 6"	F-3	1100	.036		3.06	1.98	.43	5.47	7
1180	4" x 8"		1000	.040		4.33	2.17	.47	6.97	8.75
1200	4" x 10"		950	.042		5.45	2.29	.50	8.24	10.20
1220	4" x 12"		900	.044		5.80	2.42	.52	8.74	10.85
1240	4" x 14"		850	.047		6.70	2.56	.55	9.81	12.05
1250	6" x 8"		525	.076		8.35	4.14	.90	13.39	16.75
1260	6" x 10"		500	.080		7.10	4.35	.94	12.39	15.80
1290	8" x 12"	↓	300	.133		17.45	7.25	1.57	26.27	32.50
2000	Double, 2" x 6"	2 Carp	625	.026		1.37	1.36		2.73	3.69
2005	Pneumatic nailed		725	.022		1.39	1.17		2.56	3.41
2020	2" x 8"		575	.028		2	1.48		3.48	4.57
2025	Pneumatic nailed		667	.024		2.02	1.28		3.30	4.26
2040	2" x 10"		550	.029		2.65	1.55		4.20	5.40
2045	Pneumatic nailed		638	.025		2.68	1.33		4.01	5.10
2060	2" x 12"		525	.030		3.48	1.62		5.10	6.45
2065	Pneumatic nailed		610	.026		3.51	1.39		4.90	6.10
2080	2" x 14"		475	.034		4.76	1.79		6.55	8.10
2085	Pneumatic nailed		551	.029		4.80	1.54		6.34	7.75
3000	Triple, 2" x 6"		550	.029		2.06	1.55		3.61	4.74
3005	Pneumatic nailed		638	.025		2.08	1.33		3.41	4.43
3020	2" x 8"		525	.030		3	1.62		4.62	5.90
3025	Pneumatic nailed		609	.026		3.03	1.40		4.43	5.55
3040	2" x 10"		500	.032		3.98	1.70		5.68	7.10
3045	Pneumatic nailed		580	.028		4.03	1.47		5.50	6.80
3060	2" x 12"		475	.034		5.20	1.79		6.99	8.60
3065	Pneumatic nailed		551	.029		5.25	1.54		6.79	8.25
3080	2" x 14"		450	.036		7.70	1.89		9.59	11.55
3085	Pneumatic nailed		522	.031	↓	7.20	1.63		8.83	10.55
9000	Minimum labor/equipment charge	1 Carp	2	4	Job		213		213	340

06 11 Wood Framing

06 11 10 – Framing with Dimensional, Engineered or Composite Lumber

06 11 10.12 Ceiling Framing		Crew	Daily Output	Labor-Hours	Unit	Material	2020 Bare Costs Labor	Equipment	Total	Total Incl O&P
0010	**CEILING FRAMING**									
6000	Suspended, 2" x 3"	2 Carp	1000	.016	L.F.	.42	.85		1.27	1.83
6050	2" x 4"		900	.018		.47	.95		1.42	2.03
6100	2" x 6"		800	.020		.69	1.06		1.75	2.45
6150	2" x 8"		650	.025		1	1.31		2.31	3.20
9000	Minimum labor/equipment charge	1 Carp	4	2	Job		106		106	170

06 11 10.14 Posts and Columns

		Crew	Daily Output	Labor-Hours	Unit	Material	2020 Bare Costs Labor	Equipment	Total	Total Incl O&P
0010	**POSTS AND COLUMNS**									
0100	4" x 4"	2 Carp	390	.041	L.F.	2.04	2.18		4.22	5.75
0150	4" x 6"		275	.058		3.06	3.09		6.15	8.30
0200	4" x 8"		220	.073		4.33	3.87		8.20	10.95
0250	6" x 6"		215	.074		5.35	3.96		9.31	12.25
0300	6" x 8"		175	.091		8.35	4.86		13.21	16.95
0350	6" x 10"		150	.107		7.10	5.65		12.75	16.90
9000	Minimum labor/equipment charge	1 Carp	2	4	Job		213		213	340

06 11 10.18 Joist Framing

		Crew	Daily Output	Labor-Hours	Unit	Material	2020 Bare Costs Labor	Equipment	Total	Total Incl O&P
0010	**JOIST FRAMING** R061110-30									
2000	Joists, 2" x 4"	2 Carp	1250	.013	L.F.	.47	.68		1.15	1.61
2005	Pneumatic nailed		1438	.011		.48	.59		1.07	1.48
2100	2" x 6"		1250	.013		.69	.68		1.37	1.84
2105	Pneumatic nailed		1438	.011		.69	.59		1.28	1.71
2150	2" x 8"		1100	.015		1	.77		1.77	2.34
2155	Pneumatic nailed		1265	.013		1.01	.67		1.68	2.19
2200	2" x 10"		900	.018		1.33	.95		2.28	2.97
2205	Pneumatic nailed		1035	.015		1.34	.82		2.16	2.80
2250	2" x 12"		875	.018		1.74	.97		2.71	3.47
2255	Pneumatic nailed		1006	.016		1.76	.85		2.61	3.28
2300	2" x 14"		770	.021		2.38	1.10		3.48	4.39
2305	Pneumatic nailed		886	.018		2.40	.96		3.36	4.18
2350	3" x 6"		925	.017		2.21	.92		3.13	3.90
2400	3" x 10"		780	.021		3.57	1.09		4.66	5.70
2450	3" x 12"		600	.027		4.83	1.42		6.25	7.55
2500	4" x 6"		800	.020		3.06	1.06		4.12	5.05
2550	4" x 10"		600	.027		5.45	1.42		6.87	8.25
2600	4" x 12"		450	.036		5.80	1.89		7.69	9.45
2605	Sister joist, 2" x 6"		800	.020		.69	1.06		1.75	2.45
2606	Pneumatic nailed		960	.017		.69	.89		1.58	2.18
2610	2" x 8"		640	.025		1	1.33		2.33	3.23
2611	Pneumatic nailed		768	.021		1.01	1.11		2.12	2.88
2615	2" x 10"		535	.030		1.33	1.59		2.92	4.01
2616	Pneumatic nailed		642	.025		1.34	1.32		2.66	3.60
2620	2" x 12"		455	.035		1.74	1.87		3.61	4.90
2625	Pneumatic nailed		546	.029		1.76	1.56		3.32	4.42
3000	Composite wood joist 9-1/2" deep		.90	17.778	M.L.F.	1,600	945		2,545	3,275
3010	11-1/2" deep		.88	18.182		2,000	965		2,965	3,750
3020	14" deep		.82	19.512		2,375	1,025		3,400	4,250
3030	16" deep		.78	20.513		4,225	1,100		5,325	6,375
4000	Open web joist 12" deep		.88	18.182		3,850	965		4,815	5,800
4002	Per linear foot		880	.018	L.F.	3.86	.97		4.83	5.80
4004	Treated, per linear foot		880	.018	"	4.85	.97		5.82	6.90
4010	14" deep		.82	19.512	M.L.F.	3,900	1,025		4,925	5,950
4012	Per linear foot		820	.020	L.F.	3.91	1.04		4.95	5.95

06 11 10.18 Joist Framing

		Crew	Daily Output	Labor-Hours	Unit	Material	2020 Bare Costs Labor	2020 Bare Costs Equipment	Total	Total Incl O&P
4014	Treated, per linear foot	2 Carp	820	.020	L.F.	5.05	1.04		6.09	7.25
4020	16" deep		.78	20.513	M.L.F.	4,200	1,100		5,300	6,375
4022	Per linear foot		780	.021	L.F.	4.21	1.09		5.30	6.40
4024	Treated, per linear foot		780	.021	"	5.55	1.09		6.64	7.85
4030	18" deep		.74	21.622	M.L.F.	4,400	1,150		5,550	6,675
4032	Per linear foot		740	.022	L.F.	4.39	1.15		5.54	6.65
4034	Treated, per linear foot		740	.022	"	5.90	1.15		7.05	8.30
6000	Composite rim joist, 1-1/4" x 9-1/2"		.90	17.778	M.L.F.	2,075	945		3,020	3,825
6010	1-1/4" x 11-1/2"		.88	18.182		2,250	965		3,215	4,025
6020	1-1/4" x 14-1/2"		.82	19.512		3,200	1,025		4,225	5,175
6030	1-1/4" x 16-1/2"		.78	20.513		2,850	1,100		3,950	4,900
9000	Minimum labor/equipment charge	1 Carp	4	2	Job		106		106	170

06 11 10.24 Miscellaneous Framing

		Crew	Daily Output	Labor-Hours	Unit	Material	2020 Bare Costs Labor	2020 Bare Costs Equipment	Total	Total Incl O&P
0010	**MISCELLANEOUS FRAMING**									
2000	Firestops, 2" x 4"	2 Carp	780	.021	L.F.	.47	1.09		1.56	2.27
2005	Pneumatic nailed		952	.017		.48	.89		1.37	1.96
2100	2" x 6"		600	.027		.69	1.42		2.11	3.02
2105	Pneumatic nailed		732	.022		.69	1.16		1.85	2.62
5000	Nailers, treated, wood construction, 2" x 4"		800	.020		.57	1.06		1.63	2.33
5005	Pneumatic nailed		960	.017		.58	.89		1.47	2.06
5100	2" x 6"		750	.021		.76	1.13		1.89	2.66
5105	Pneumatic nailed		900	.018		.77	.95		1.72	2.36
5120	2" x 8"		700	.023		1.23	1.22		2.45	3.30
5125	Pneumatic nailed		840	.019		1.24	1.01		2.25	2.99
5200	Steel construction, 2" x 4"		750	.021		.57	1.13		1.70	2.45
5220	2" x 6"		700	.023		.76	1.22		1.98	2.79
5240	2" x 8"		650	.025		1.23	1.31		2.54	3.46
7000	Rough bucks, treated, for doors or windows, 2" x 6"		400	.040		.76	2.13		2.89	4.25
7005	Pneumatic nailed		480	.033		.77	1.77		2.54	3.69
7100	2" x 8"		380	.042		1.23	2.24		3.47	4.94
7105	Pneumatic nailed		456	.035		1.24	1.87		3.11	4.36
8000	Stair stringers, 2" x 10"		130	.123		1.33	6.55		7.88	11.95
8100	2" x 12"		130	.123		1.74	6.55		8.29	12.40
8150	3" x 10"		125	.128		3.57	6.80		10.37	14.85
8200	3" x 12"		125	.128		4.83	6.80		11.63	16.20
8870	Laminated structural lumber, 1-1/4" x 11-1/2"		130	.123		2.24	6.55		8.79	12.95
8880	1-1/4" x 14-1/2"		130	.123		3.18	6.55		9.73	14
9000	Minimum labor/equipment charge	1 Carp	4	2	Job		106		106	170

06 11 10.26 Partitions

		Crew	Daily Output	Labor-Hours	Unit	Material	2020 Bare Costs Labor	2020 Bare Costs Equipment	Total	Total Incl O&P
0010	**PARTITIONS**									
0020	Single bottom and double top plate, no waste, std. & better lumber									
0180	2" x 4" studs, 8' high, studs 12" OC	2 Carp	80	.200	L.F.	5.20	10.65		15.85	23
0185	12" OC, pneumatic nailed		96	.167		5.25	8.85		14.10	20
0200	16" OC		100	.160		4.26	8.50		12.76	18.30
0205	16" OC, pneumatic nailed		120	.133		4.31	7.10		11.41	16.10
0300	24" OC		125	.128		3.31	6.80		10.11	14.55
0305	24" OC, pneumatic nailed		150	.107		3.35	5.65		9	12.80
0380	10' high, studs 12" OC		80	.200		6.15	10.65		16.80	24
0385	12" OC, pneumatic nailed		96	.167		6.25	8.85		15.10	21
0400	16" OC		100	.160		4.97	8.50		13.47	19.05
0405	16" OC, pneumatic nailed		120	.133		5.05	7.10		12.15	16.90
0500	24" OC		125	.128		3.79	6.80		10.59	15.05

146

For customer support on your Commercial Renovation Costs with RSMeans data, call 800.448.8182.

06 11 Wood Framing

06 11 10 – Framing with Dimensional, Engineered or Composite Lumber

06 11 10.26 Partitions

		Crew	Daily Output	Labor-Hours	Unit	Material	2020 Bare Costs Labor	Equipment	Total	Total Incl O&P
0505	24" OC, pneumatic nailed	2 Carp	150	.107	L.F.	3.83	5.65		9.48	13.30
0580	12' high, studs 12" OC		65	.246		7.10	13.10		20.20	29
0585	12" OC, pneumatic nailed		78	.205		7.20	10.90		18.10	25.50
0600	16" OC		80	.200		5.70	10.65		16.35	23.50
0605	16" OC, pneumatic nailed		96	.167		5.75	8.85		14.60	20.50
0700	24" OC		100	.160		4.26	8.50		12.76	18.30
0705	24" OC, pneumatic nailed		120	.133		4.31	7.10		11.41	16.10
0780	2" x 6" studs, 8' high, studs 12" OC		70	.229		7.55	12.15		19.70	28
0785	12" OC, pneumatic nailed		84	.190		7.65	10.10		17.75	24.50
0800	16" OC		90	.178		6.15	9.45		15.60	22
0805	16" OC, pneumatic nailed		108	.148		6.25	7.85		14.10	19.45
0900	24" OC		115	.139		4.80	7.40		12.20	17.15
0905	24" OC, pneumatic nailed		138	.116		4.86	6.15		11.01	15.20
0980	10' high, studs 12" OC		70	.229		8.90	12.15		21.05	29.50
0985	12" OC, pneumatic nailed		84	.190		9	10.10		19.10	26
1000	16" OC		90	.178		7.20	9.45		16.65	23
1005	16" OC, pneumatic nailed		108	.148		7.30	7.85		15.15	20.50
1100	24" OC		115	.139		5.50	7.40		12.90	17.90
1105	24" OC, pneumatic nailed		138	.116		5.55	6.15		11.70	15.95
1180	12' high, studs 12" OC		55	.291		10.30	15.45		25.75	36.50
1185	12" OC, pneumatic nailed		66	.242		10.40	12.90		23.30	32
1200	16" OC		70	.229		8.25	12.15		20.40	28.50
1205	16" OC, pneumatic nailed		84	.190		8.35	10.10		18.45	25.50
1300	24" OC		90	.178		6.15	9.45		15.60	22
1305	24" OC, pneumatic nailed		108	.148		6.25	7.85		14.10	19.45
1400	For horizontal blocking, 2" x 4", add		600	.027		.47	1.42		1.89	2.79
1500	2" x 6", add		600	.027		.69	1.42		2.11	3.02
1600	For openings, add	▼	250	.064	▼		3.40		3.40	5.45
1702	Headers for above openings, material only, add				B.F.	.83			.83	.91
9000	Minimum labor/equipment charge	1 Carp	4	2	Job		106		106	170

06 11 10.28 Porch or Deck Framing

		Crew	Daily Output	Labor-Hours	Unit	Material	2020 Bare Costs Labor	Equipment	Total	Total Incl O&P
0010	**PORCH OR DECK FRAMING**									
0100	Treated lumber, posts or columns, 4" x 4"	2 Carp	390	.041	L.F.	1.41	2.18		3.59	5.05
0110	4" x 6"		275	.058		2.24	3.09		5.33	7.40
0120	4" x 8"		220	.073		4.28	3.87		8.15	10.90
0130	Girder, single, 4" x 4"		675	.024		1.41	1.26		2.67	3.57
0140	4" x 6"		600	.027		2.24	1.42		3.66	4.73
0150	4" x 8"		525	.030		4.28	1.62		5.90	7.30
0160	Double, 2" x 4"		625	.026		1.18	1.36		2.54	3.48
0170	2" x 6"		600	.027		1.58	1.42		3	4.01
0180	2" x 8"		575	.028		2.54	1.48		4.02	5.15
0190	2" x 10"		550	.029		2.94	1.55		4.49	5.70
0200	2" x 12"		525	.030		4.75	1.62		6.37	7.85
0210	Triple, 2" x 4"		575	.028		1.77	1.48		3.25	4.32
0220	2" x 6"		550	.029		2.37	1.55		3.92	5.10
0230	2" x 8"		525	.030		3.81	1.62		5.43	6.80
0240	2" x 10"		500	.032		4.41	1.70		6.11	7.55
0250	2" x 12"		475	.034		7.15	1.79		8.94	10.70
0260	Ledger, bolted 4' OC, 2" x 4"		400	.040		.75	2.13		2.88	4.23
0270	2" x 6"		395	.041		.93	2.15		3.08	4.48
0280	2" x 8"		390	.041		1.40	2.18		3.58	5.05
0290	2" x 10"	▼	385	.042	▼	1.59	2.21		3.80	5.30

For customer support on your Commercial Renovation Costs with RSMeans data, call 800.448.8182.

147

	06 11 10.28 Porch or Deck Framing	Crew	Daily Output	Labor-Hours	Unit	Material	2020 Bare Costs Labor	Equipment	Total	Total Incl O&P
0300	2" x 12"	2 Carp	380	.042	L.F.	2.48	2.24		4.72	6.30
0310	Joists, 2" x 4"		1250	.013		.59	.68		1.27	1.74
0320	2" x 6"		1250	.013		.79	.68		1.47	1.96
0330	2" x 8"		1100	.015		1.27	.77		2.04	2.64
0340	2" x 10"		900	.018		1.47	.95		2.42	3.13
0350	2" x 12"		875	.018		1.77	.97		2.74	3.50
0360	Railings and trim, 1" x 4"	1 Carp	300	.027		.54	1.42		1.96	2.86
0370	2" x 2"		300	.027		.48	1.42		1.90	2.79
0380	2" x 4"		300	.027		.58	1.42		2	2.91
0390	2" x 6"		300	.027		.77	1.42		2.19	3.12
0400	Decking, 1" x 4"		275	.029	S.F.	3.09	1.55		4.64	5.90
0410	2" x 4"		300	.027		1.95	1.42		3.37	4.42
0420	2" x 6"		320	.025		1.66	1.33		2.99	3.96
0430	5/4" x 6"		320	.025		2.17	1.33		3.50	4.51
0440	Balusters, square, 2" x 2"	2 Carp	660	.024	L.F.	.48	1.29		1.77	2.59
0450	Turned, 2" x 2"		420	.038		.64	2.03		2.67	3.94
0460	Stair stringer, 2" x 10"		130	.123		1.47	6.55		8.02	12.10
0470	2" x 12"		130	.123		1.77	6.55		8.32	12.45
0480	Stair treads, 1" x 4"		140	.114		3.09	6.05		9.14	13.15
0490	2" x 4"		140	.114		.59	6.05		6.64	10.40
0500	2" x 6"		160	.100		.89	5.30		6.19	9.50
0510	5/4" x 6"		160	.100		1.01	5.30		6.31	9.60
0520	Turned handrail post, 4" x 4"		64	.250	Ea.	36	13.30		49.30	61.50
0530	Lattice panel, 4' x 8', 1/2"		1600	.010	S.F.	.68	.53		1.21	1.60
0535	3/4"		1600	.010	"	1.01	.53		1.54	1.96
0540	Cedar, posts or columns, 4" x 4"		390	.041	L.F.	4.14	2.18		6.32	8.05
0550	4" x 6"		275	.058		6.65	3.09		9.74	12.25
0560	4" x 8"		220	.073		12.50	3.87		16.37	19.95
0800	Decking, 1" x 4"		550	.029		2.85	1.55		4.40	5.60
0810	2" x 4"		600	.027		5.45	1.42		6.87	8.25
0820	2" x 6"		640	.025		9.90	1.33		11.23	13
0830	5/4" x 6"		640	.025		7.55	1.33		8.88	10.45
0840	Railings and trim, 1" x 4"		600	.027		2.85	1.42		4.27	5.40
0860	2" x 4"		600	.027		5.45	1.42		6.87	8.25
0870	2" x 6"		600	.027		9.90	1.42		11.32	13.10
0920	Stair treads, 1" x 4"		140	.114		2.85	6.05		8.90	12.90
0930	2" x 4"		140	.114		5.45	6.05		11.50	15.75
0940	2" x 6"		160	.100		9.90	5.30		15.20	19.35
0950	5/4" x 6"		160	.100		7.55	5.30		12.85	16.80
0980	Redwood, posts or columns, 4" x 4"		390	.041		6.60	2.18		8.78	10.80
0990	4" x 6"		275	.058		12.85	3.09		15.94	19.10
1000	4" x 8"		220	.073		24	3.87		27.87	32.50
1240	Decking, 1" x 4"	1 Carp	275	.029	S.F.	4.03	1.55		5.58	6.90
1260	2" x 6"		340	.024		6.95	1.25		8.20	9.60
1270	5/4" x 6"		320	.025		4.89	1.33		6.22	7.55
1280	Railings and trim, 1" x 4"	2 Carp	600	.027	L.F.	1.19	1.42		2.61	3.58
1310	2" x 6"		600	.027		6.95	1.42		8.37	9.90
1420	Alternative decking, wood/plastic composite, 5/4" x 6" [G]		640	.025		3.44	1.33		4.77	5.90
1440	1" x 4" square edge fir		550	.029		3.10	1.55		4.65	5.90
1450	1" x 4" tongue and groove fir		450	.036		1.53	1.89		3.42	4.71
1460	1" x 4" mahogany		550	.029		2.13	1.55		3.68	4.82
1462	5/4" x 6" PVC		550	.029		3.45	1.55		5	6.25
1465	Framing, porch or deck, alt deck fastening, screws, add	1 Carp	240	.033	S.F.		1.77		1.77	2.84

06 11 Wood Framing

06 11 10 – Framing with Dimensional, Engineered or Composite Lumber

06 11 10.28 Porch or Deck Framing

		Crew	Daily Output	Labor-Hours	Unit	Material	2020 Bare Costs Labor	2020 Bare Costs Equipment	Total	Total Incl O&P
1470	Accessories, joist hangers, 2" x 4"	1 Carp	160	.050	Ea.	.85	2.66		3.51	5.20
1480	2" x 6" through 2" x 12"	▼	150	.053		1.56	2.83		4.39	6.25
1530	Post footing, incl excav, backfill, tube form & concrete, 4' deep, 8" diam.	F-7	12	2.667		18.30	127		145.30	223
1540	10" diameter		11	2.909		26	139		165	251
1550	12" diameter	▼	10	3.200	▼	33.50	152		185.50	281

06 11 10.30 Roof Framing

		Crew	Daily Output	Labor-Hours	Unit	Material	2020 Bare Costs Labor	2020 Bare Costs Equipment	Total	Total Incl O&P
0010	**ROOF FRAMING** R061110-30									
1900	Rough fascia, 2" x 6"	2 Carp	250	.064	L.F.	.69	3.40		4.09	6.20
2000	2" x 8"		225	.071		1	3.78		4.78	7.15
2100	2" x 10"		180	.089		1.33	4.72		6.05	9
2200	2" x 12"		180	.089		1.74	4.72		6.46	9.45
5000	Rafters, to 4 in 12 pitch, 2" x 6", ordinary		1000	.016		.69	.85		1.54	2.11
5060	2" x 8", ordinary		950	.017		1	.90		1.90	2.53
5120	2" x 10", ordinary		630	.025		1.33	1.35		2.68	3.62
5180	2" x 12", ordinary		575	.028		1.74	1.48		3.22	4.28
5250	Composite rafter, 9-1/2" deep		575	.028		1.60	1.48		3.08	4.13
5260	11-1/2" deep		575	.028		2	1.48		3.48	4.57
5300	Hip and valley rafters, 2" x 6", ordinary		760	.021		.69	1.12		1.81	2.54
5360	2" x 8", ordinary		720	.022		1	1.18		2.18	2.99
5420	2" x 10", ordinary		570	.028		1.33	1.49		2.82	3.85
5480	2" x 12", ordinary		525	.030		1.74	1.62		3.36	4.51
5540	Hip and valley jacks, 2" x 6", ordinary		600	.027		.69	1.42		2.11	3.02
5600	2" x 8", ordinary		490	.033		1	1.74		2.74	3.88
5660	2" x 10", ordinary		450	.036		1.33	1.89		3.22	4.49
5720	2" x 12", ordinary	▼	375	.043	▼	1.74	2.27		4.01	5.55
5761	For slopes steeper than 4 in 12, add						30%			
5780	Rafter tie, 1" x 4", #3	2 Carp	800	.020	L.F.	.53	1.06		1.59	2.29
5790	2" x 4", #3		800	.020		.47	1.06		1.53	2.22
5800	Ridge board, #2 or better, 1" x 6"		600	.027		.83	1.42		2.25	3.18
5820	1" x 8"		550	.029		1.29	1.55		2.84	3.90
5840	1" x 10"		500	.032		1.98	1.70		3.68	4.90
5860	2" x 6"		500	.032		.69	1.70		2.39	3.47
5880	2" x 8"		450	.036		1	1.89		2.89	4.13
5900	2" x 10"		400	.040		1.33	2.13		3.46	4.87
5920	Roof cants, split, 4" x 4"		650	.025		2.04	1.31		3.35	4.34
5940	6" x 6"		600	.027		5.35	1.42		6.77	8.15
5960	Roof curbs, untreated, 2" x 6"		520	.031		.69	1.64		2.33	3.37
5980	2" x 12"		400	.040		1.74	2.13		3.87	5.30
6000	Sister rafters, 2" x 6"		800	.020		.69	1.06		1.75	2.45
6020	2" x 8"		640	.025		1	1.33		2.33	3.23
6040	2" x 10"		535	.030		1.33	1.59		2.92	4.01
6060	2" x 12"	▼	455	.035	▼	1.74	1.87		3.61	4.90
9000	Minimum labor/equipment charge	1 Carp	4	2	Job		106		106	170

06 11 10.32 Sill and Ledger Framing

		Crew	Daily Output	Labor-Hours	Unit	Material	2020 Bare Costs Labor	2020 Bare Costs Equipment	Total	Total Incl O&P
0010	**SILL AND LEDGER FRAMING**									
0020	Extruded polystyrene sill sealer, 5-1/2" wide	1 Carp	1600	.005	L.F.	.14	.27		.41	.58
2002	Ledgers, nailed, 2" x 4"	2 Carp	755	.021		.47	1.13		1.60	2.32
2052	2" x 6"		600	.027		.69	1.42		2.11	3.02
2102	Bolted, not including bolts, 3" x 6"		325	.049		2.19	2.62		4.81	6.60
2152	3" x 12"		233	.069		4.81	3.65		8.46	11.15
2602	Mud sills, redwood, construction grade, 2" x 4"		895	.018		2.31	.95		3.26	4.06
2622	2" x 6"	▼	780	.021	▼	3.14	1.09		4.23	5.20

For customer support on your Commercial Renovation Costs with RSMeans data, call 800.448.8182.

149

06 11 10.32 Sill and Ledger Framing

		Crew	Daily Output	Labor-Hours	Unit	Material	2020 Bare Costs Labor	2020 Bare Costs Equipment	Total	Total Incl O&P
4002	Sills, 2" x 4"	2 Carp	600	.027	L.F.	.47	1.42		1.89	2.78
4052	2" x 6"		550	.029		.68	1.55		2.23	3.22
4082	2" x 8"		500	.032		.99	1.70		2.69	3.81
4101	2" x 10"		450	.036		1.31	1.89		3.20	4.47
4121	2" x 12"		400	.040		1.72	2.13		3.85	5.30
4202	Treated, 2" x 4"		550	.029		.57	1.55		2.12	3.10
4222	2" x 6"		500	.032		.75	1.70		2.45	3.55
4242	2" x 8"		450	.036		1.22	1.89		3.11	4.37
4261	2" x 10"		400	.040		1.41	2.13		3.54	4.96
4281	2" x 12"		350	.046		2.30	2.43		4.73	6.40
4402	4" x 4"		450	.036		1.36	1.89		3.25	4.52
4422	4" x 6"		350	.046		2.16	2.43		4.59	6.25
4462	4" x 8"		300	.053		4.18	2.83		7.01	9.15
4481	4" x 10"		260	.062		5.75	3.27		9.02	11.55
9000	Minimum labor/equipment charge	1 Carp	4	2	Job		106		106	170

06 11 10.34 Sleepers

		Crew	Daily Output	Labor-Hours	Unit	Material	2020 Bare Costs Labor	2020 Bare Costs Equipment	Total	Total Incl O&P
0010	**SLEEPERS**									
0100	On concrete, treated, 1" x 2"	2 Carp	2350	.007	L.F.	.31	.36		.67	.92
0150	1" x 3"		2000	.008		.51	.43		.94	1.24
0200	2" x 4"		1500	.011		.61	.57		1.18	1.58
0250	2" x 6"		1300	.012		.84	.65		1.49	1.98
9000	Minimum labor/equipment charge	1 Carp	4	2	Job		106		106	170

06 11 10.36 Soffit and Canopy Framing

		Crew	Daily Output	Labor-Hours	Unit	Material	2020 Bare Costs Labor	2020 Bare Costs Equipment	Total	Total Incl O&P
0010	**SOFFIT AND CANOPY FRAMING**									
1002	Canopy or soffit framing, 1" x 4"	2 Carp	900	.018	L.F.	.53	.95		1.48	2.10
1021	1" x 6"		850	.019		.83	1		1.83	2.51
1042	1" x 8"		750	.021		1.29	1.13		2.42	3.24
1102	2" x 4"		620	.026		.47	1.37		1.84	2.72
1121	2" x 6"		560	.029		.69	1.52		2.21	3.18
1142	2" x 8"		500	.032		1	1.70		2.70	3.82
1202	3" x 4"		500	.032		1.38	1.70		3.08	4.23
1221	3" x 6"		400	.040		2.21	2.13		4.34	5.85
1242	3" x 10"		300	.053		3.57	2.83		6.40	8.45
9000	Minimum labor/equipment charge	1 Carp	4	2	Job		106		106	170

06 11 10.38 Treated Lumber Framing Material

		Crew	Daily Output	Labor-Hours	Unit	Material	2020 Bare Costs Labor	2020 Bare Costs Equipment	Total	Total Incl O&P
0010	**TREATED LUMBER FRAMING MATERIAL**									
0100	2" x 4"				M.B.F.	850			850	935
0110	2" x 6"					755			755	830
0120	2" x 8"					915			915	1,000
0130	2" x 10"					845			845	930
0140	2" x 12"					1,150			1,150	1,275
0200	4" x 4"					1,025			1,025	1,125
0210	4" x 6"					1,075			1,075	1,200
0220	4" x 8"					1,575			1,575	1,725

06 11 10.40 Wall Framing

		Crew	Daily Output	Labor-Hours	Unit	Material	2020 Bare Costs Labor	2020 Bare Costs Equipment	Total	Total Incl O&P
0010	**WALL FRAMING** R061110-30									
0100	Door buck, studs, header, access, 8' high, 2" x 4" wall, 3' wide	1 Carp	32	.250	Ea.	19.60	13.30		32.90	43
0110	4' wide		32	.250		21	13.30		34.30	44.50
0120	5' wide		32	.250		25.50	13.30		38.80	49.50
0130	6' wide		32	.250		27.50	13.30		40.80	52
0140	8' wide		30	.267		37	14.15		51.15	63
0150	10' wide		30	.267		50.50	14.15		64.65	78

For customer support on your Commercial Renovation Costs with RSMeans data, call 800.448.8182.

06 11 10 – Framing with Dimensional, Engineered or Composite Lumber

06 11 10.40 Wall Framing		Crew	Daily Output	Labor-Hours	Unit	Material	2020 Bare Costs Labor	Equipment	Total	Total Incl O&P
0160	12' wide	1 Carp	30	.267	Ea.	73	14.15		87.15	103
0170	2" x 6" wall, 3' wide		32	.250		26.50	13.30		39.80	50.50
0180	4' wide		32	.250		28	13.30		41.30	52
0190	5' wide		32	.250		32.50	13.30		45.80	57
0200	6' wide		32	.250		34.50	13.30		47.80	59.50
0210	8' wide		30	.267		43.50	14.15		57.65	70.50
0220	10' wide		30	.267		57.50	14.15		71.65	85.50
0230	12' wide		30	.267		80	14.15		94.15	111
0240	Window buck, studs, header & access, 8' high 2" x 4" wall, 2' wide		24	.333		21	17.70		38.70	51.50
0250	3' wide		24	.333		24.50	17.70		42.20	55.50
0260	4' wide		24	.333		26.50	17.70		44.20	58
0270	5' wide		24	.333		31.50	17.70		49.20	63
0280	6' wide		24	.333		34.50	17.70		52.20	66.50
0290	7' wide		24	.333		42.50	17.70		60.20	75
0300	8' wide		22	.364		47	19.35		66.35	82.50
0310	10' wide		22	.364		62	19.35		81.35	99
0320	12' wide		22	.364		87.50	19.35		106.85	127
0330	2" x 6" wall, 2' wide		24	.333		29.50	17.70		47.20	61
0340	3' wide		24	.333		33	17.70		50.70	64.50
0350	4' wide		24	.333		35.50	17.70		53.20	67.50
0360	5' wide		24	.333		40.50	17.70		58.20	73
0370	6' wide		24	.333		44.50	17.70		62.20	77.50
0380	7' wide		24	.333		53.50	17.70		71.20	87.50
0390	8' wide		22	.364		58	19.35		77.35	95
0400	10' wide		22	.364		74	19.35		93.35	113
0410	12' wide	▼	22	.364	▼	101	19.35		120.35	142
2002	Headers over openings, 2" x 6"	2 Carp	360	.044	L.F.	.69	2.36		3.05	4.53
2007	2" x 6", pneumatic nailed		432	.037		.69	1.97		2.66	3.91
2052	2" x 8"		340	.047		1	2.50		3.50	5.10
2057	2" x 8", pneumatic nailed		408	.039		1.01	2.08		3.09	4.45
2101	2" x 10"		320	.050		1.33	2.66		3.99	5.70
2106	2" x 10", pneumatic nailed		384	.042		1.34	2.21		3.55	5.05
2152	2" x 12"		300	.053		1.74	2.83		4.57	6.45
2157	2" x 12", pneumatic nailed		360	.044		1.76	2.36		4.12	5.70
2180	4" x 8"		260	.062		4.33	3.27		7.60	10
2185	4" x 8", pneumatic nailed		312	.051		4.35	2.73		7.08	9.15
2191	4" x 10"		240	.067		5.45	3.54		8.99	11.70
2196	4" x 10", pneumatic nailed		288	.056		5.45	2.95		8.40	10.75
2202	4" x 12"		190	.084		5.80	4.48		10.28	13.55
2207	4" x 12", pneumatic nailed		228	.070		5.85	3.73		9.58	12.45
2241	6" x 10"		165	.097		7.05	5.15		12.20	16.05
2246	6" x 10", pneumatic nailed		198	.081		7.10	4.30		11.40	14.75
2251	6" x 12"		140	.114		9	6.05		15.05	19.65
2256	6" x 12", pneumatic nailed		168	.095		9.05	5.05		14.10	18.05
5002	Plates, untreated, 2" x 3"		850	.019		.42	1		1.42	2.07
5007	2" x 3", pneumatic nailed		1020	.016		.43	.83		1.26	1.81
5022	2" x 4"		800	.020		.47	1.06		1.53	2.22
5027	2" x 4", pneumatic nailed		960	.017		.48	.89		1.37	1.95
5041	2" x 6"		750	.021		.69	1.13		1.82	2.57
5046	2" x 6", pneumatic nailed		900	.018		.69	.95		1.64	2.27
5122	Studs, 8' high wall, 2" x 3"		1200	.013		.42	.71		1.13	1.61
5127	2" x 3", pneumatic nailed		1440	.011		.43	.59		1.02	1.42
5142	2" x 4"		1100	.015		.47	.77		1.24	1.75

06 11 Wood Framing

06 11 10 – Framing with Dimensional, Engineered or Composite Lumber

06 11 10.40 Wall Framing

		Crew	Daily Output	Labor-Hours	Unit	Material	2020 Bare Costs Labor	Equipment	Total	Total Incl O&P
5147	2" x 4", pneumatic nailed	2 Carp	1320	.012	L.F.	.48	.64		1.12	1.56
5162	2" x 6"		1000	.016		.69	.85		1.54	2.11
5167	2" x 6", pneumatic nailed		1200	.013		.69	.71		1.40	1.90
5182	3" x 4"		800	.020		1.38	1.06		2.44	3.21
5187	3" x 4", pneumatic nailed		960	.017		1.38	.89		2.27	2.94
8200	For 12' high walls, deduct						5%			
8220	For stub wall, 6' high, add						20%			
8240	3' high, add						40%			
8250	For second story & above, add						5%			
8300	For dormer & gable, add						15%			
9000	Minimum labor/equipment charge	1 Carp	4	2	Job		106		106	170

06 11 10.42 Furring

		Crew	Daily Output	Labor-Hours	Unit	Material	2020 Bare Costs Labor	Equipment	Total	Total Incl O&P
0010	**FURRING**									
0012	Wood strips, 1" x 2", on walls, on wood	1 Carp	550	.015	L.F.	.29	.77		1.06	1.56
0015	On wood, pneumatic nailed		710	.011		.29	.60		.89	1.28
0300	On masonry		495	.016		.31	.86		1.17	1.73
0400	On concrete		260	.031		.31	1.64		1.95	2.97
0600	1" x 3", on walls, on wood		550	.015		.44	.77		1.21	1.72
0605	On wood, pneumatic nailed		710	.011		.44	.60		1.04	1.44
0700	On masonry		495	.016		.47	.86		1.33	1.90
0800	On concrete		260	.031		.47	1.64		2.11	3.14
0850	On ceilings, on wood		350	.023		.44	1.22		1.66	2.43
0855	On wood, pneumatic nailed		450	.018		.44	.95		1.39	1.99
0900	On masonry		320	.025		.47	1.33		1.80	2.65
0950	On concrete		210	.038		.47	2.03		2.50	3.76
9000	Minimum labor/equipment charge		4	2	Job		106		106	170

06 11 10.44 Grounds

		Crew	Daily Output	Labor-Hours	Unit	Material	2020 Bare Costs Labor	Equipment	Total	Total Incl O&P
0010	**GROUNDS**									
0020	For casework, 1" x 2" wood strips, on wood	1 Carp	330	.024	L.F.	.29	1.29		1.58	2.38
0100	On masonry		285	.028		.31	1.49		1.80	2.74
0200	On concrete		250	.032		.31	1.70		2.01	3.07
0400	For plaster, 3/4" deep, on wood		450	.018		.29	.95		1.24	1.83
0500	On masonry		225	.036		.31	1.89		2.20	3.38
0600	On concrete		175	.046		.31	2.43		2.74	4.24
0700	On metal lath		200	.040		.31	2.13		2.44	3.76
9000	Minimum labor/equipment charge		4	2	Job		106		106	170

06 12 Structural Panels

06 12 10 – Structural Insulated Panels

06 12 10.10 OSB Faced Panels

			Crew	Daily Output	Labor-Hours	Unit	Material	2020 Bare Costs Labor	Equipment	Total	Total Incl O&P
0010	**OSB FACED PANELS**										
0100	Structural insul. panels, 7/16" OSB both faces, EPS insul., 3-5/8" T	G	F-3	2075	.019	S.F.	4.18	1.05	.23	5.46	6.50
0110	5-5/8" thick	G		1725	.023		4.41	1.26	.27	5.94	7.15
0120	7-3/8" thick	G		1425	.028		5.50	1.53	.33	7.36	8.85
0130	9-3/8" thick	G		1125	.036		5.95	1.93	.42	8.30	10.10
0140	7/16" OSB one face, EPS insul., 3-5/8" thick	G		2175	.018		4.27	1	.22	5.49	6.55
0150	5-5/8" thick	G		1825	.022		4.62	1.19	.26	6.07	7.30
0160	7-3/8" thick	G		1525	.026		5.40	1.43	.31	7.14	8.50
0170	9-3/8" thick	G		1225	.033		5.85	1.77	.38	8	9.70
0190	7/16" OSB - 1/2" GWB faces, EPS insul., 3-5/8" T	G		2075	.019		3.61	1.05	.23	4.89	5.90

152

06 12 Structural Panels

06 12 10 – Structural Insulated Panels

06 12 10.10 OSB Faced Panels

		Crew	Daily Output	Labor-Hours	Unit	Material	2020 Bare Costs Labor	Equipment	Total	Total Incl O&P
0200	5-5/8" thick	G F-3	1725	.023	S.F.	4.74	1.26	.27	6.27	7.50
0210	7-3/8" thick	G	1425	.028		5	1.53	.33	6.86	8.30
0220	9-3/8" thick	G	1125	.036		5.60	1.93	.42	7.95	9.70
0240	7/16" OSB - 1/2" MRGWB faces, EPS insul., 3-5/8" T	G	2075	.019		4.45	1.05	.23	5.73	6.80
0250	5-5/8" thick	G	1725	.023		4.56	1.26	.27	6.09	7.30
0260	7-3/8" thick	G	1425	.028		5.25	1.53	.33	7.11	8.60
0270	9-3/8" thick	G	1125	.036		5.85	1.93	.42	8.20	9.95
0300	For 1/2" GWB added to OSB skin, add	G				1.52			1.52	1.67
0310	For 1/2" MRGWB added to OSB skin, add	G				1.54			1.54	1.69
0320	For one T1-11 skin, add to OSB-OSB	G				2.22			2.22	2.44
0330	For one 19/32" CDX skin, add to OSB-OSB	G				1.63			1.63	1.79
0500	Structural insulated panel, 7/16" OSB both sides, straw core									
0510	4-3/8" T, walls (w/sill, splines, plates)	G F-6	2400	.017	S.F.	7.30	.83	.20	8.33	9.55
0520	Floors (w/splines)	G	2400	.017		7.30	.83	.20	8.33	9.55
0530	Roof (w/splines)	G	2400	.017		7.30	.83	.20	8.33	9.55
0550	7-7/8" T, walls (w/sill, splines, plates)	G	2400	.017		11.45	.83	.20	12.48	14.15
0560	Floors (w/splines)	G	2400	.017		11.45	.83	.20	12.48	14.15
0570	Roof (w/splines)	G	2400	.017		11.45	.83	.20	12.48	14.15

06 12 19 – Composite Shearwall Panels

06 12 19.10 Steel and Wood Composite Shearwall Panels

		Crew	Daily Output	Labor-Hours	Unit	Material	2020 Bare Costs Labor	Equipment	Total	Total Incl O&P
0010	**STEEL & WOOD COMPOSITE SHEARWALL PANELS**									
0020	Anchor bolts, 36" long (must be placed in wet concrete)	1 Carp	150	.053	Ea.	43.50	2.83		46.33	52.50
0030	On concrete, 2" x 4" & 2" x 6" walls, 7'-10' high, 360 lb. shear, 12" wide	2 Carp	8	2		495	106		601	715
0040	715 lb. shear, 15" wide		8	2		650	106		756	885
0050	1860 lb. shear, 18" wide		8	2		700	106		806	940
0060	2780 lb. shear, 21" wide		8	2		830	106		936	1,075
0070	3790 lb. shear, 24" wide		8	2		780	106		886	1,025
0080	2" x 6" walls, 11'-13' high, 1180 lb. shear, 18" wide		6	2.667		735	142		877	1,025
0090	1555 lb. shear, 21" wide		6	2.667		755	142		897	1,050
0100	2280 lb. shear, 24" wide		6	2.667		895	142		1,037	1,200
0110	For installing above on wood floor frame, add									
0120	Coupler nuts, threaded rods, bolts, shear transfer plate kit	1 Carp	16	.500	Ea.	63	26.50		89.50	112
0130	Framing anchors, angle (2 required)	"	96	.083	"	2.53	4.43		6.96	9.90
0140	For blocking see Section 06 11 10.02									
0150	For installing above, first floor to second floor, wood floor frame, add									
0160	Add stack option to first floor wall panel				Ea.	69.50			69.50	76.50
0170	Threaded rods, bolts, shear transfer plate kit	1 Carp	16	.500		78	26.50		104.50	128
0180	Framing anchors, angle (2 required)	"	96	.083		2.53	4.43		6.96	9.90
0190	For blocking see section 06 11 10.02									
0200	For installing stacked panels, balloon framing									
0210	Add stack option to first floor wall panel				Ea.	69.50			69.50	76.50
0220	Threaded rods, bolts kit	1 Carp	16	.500	"	44.50	26.50		71	91.50

For customer support on your Commercial Renovation Costs with RSMeans data, call 800.448.8182.

153

06 13 Heavy Timber Construction

06 13 23 – Heavy Timber Framing

06 13 23.10 Heavy Framing	Crew	Daily Output	Labor-Hours	Unit	Material	2020 Bare Costs Labor	Equipment	Total	Total Incl O&P
0010 **HEAVY FRAMING**									
0020 Beams, single 6" x 10"	2 Carp	1.10	14.545	M.B.F.	1,575	775		2,350	2,975
0100 Single 8" x 16"		1.20	13.333	"	2,000	710		2,710	3,325
0202 Built from 2" lumber, multiple 2" x 14"		900	.018	B.F.	1.01	.95		1.96	2.62
0212 Built from 3" lumber, multiple 3" x 6"		700	.023		1.46	1.22		2.68	3.56
0222 Multiple 3" x 8"		800	.020		1.46	1.06		2.52	3.31
0232 Multiple 3" x 10"		900	.018		1.42	.95		2.37	3.07
0242 Multiple 3" x 12"		1000	.016		1.60	.85		2.45	3.12
0252 Built from 4" lumber, multiple 4" x 6"		800	.020		1.52	1.06		2.58	3.37
0262 Multiple 4" x 8"		900	.018		1.61	.95		2.56	3.28
0272 Multiple 4" x 10"		1000	.016		1.62	.85		2.47	3.15
0282 Multiple 4" x 12"		1100	.015		1.45	.77		2.22	2.83
0292 Columns, structural grade, 1500Fb, 4" x 4"		450	.036	L.F.	2.32	1.89		4.21	5.60
0302 6" x 6"		225	.071		4.07	3.78		7.85	10.55
0402 8" x 8"		240	.067		8.25	3.54		11.79	14.75
0502 10" x 10"		90	.178		14.60	9.45		24.05	31
0602 12" x 12"		70	.229		18.30	12.15		30.45	39.50
0802 Floor planks, 2" thick, T&G, 2" x 6"		1050	.015	B.F.	1.45	.81		2.26	2.89
0902 2" x 10"		1100	.015		1.46	.77		2.23	2.85
1102 3" thick, 3" x 6"		1050	.015		1.62	.81		2.43	3.08
1202 3" x 10"		1100	.015		1.65	.77		2.42	3.05
1402 Girders, structural grade, 12" x 12"		800	.020		1.52	1.06		2.58	3.38
1502 10" x 16"		1000	.016		2.56	.85		3.41	4.18
2302 Roof purlins, 4" thick, structural grade		1050	.015		1.61	.81		2.42	3.07
9000 Minimum labor/equipment charge	1 Carp	2	4	Job		213		213	340

06 15 Wood Decking

06 15 16 – Wood Roof Decking

06 15 16.10 Solid Wood Roof Decking

	Crew	Daily Output	Labor-Hours	Unit	Material	2020 Bare Costs Labor	Equipment	Total	Total Incl O&P
0010 **SOLID WOOD ROOF DECKING**									
0350 Cedar planks, 2" thick	2 Carp	350	.046	S.F.	7.05	2.43		9.48	11.65
0400 3" thick		320	.050		10.60	2.66		13.26	15.90
0500 4" thick		250	.064		14.15	3.40		17.55	21
0550 6" thick		200	.080		21	4.25		25.25	30.50
0650 Douglas fir, 2" thick		350	.046		3.10	2.43		5.53	7.30
0700 3" thick		320	.050		4.65	2.66		7.31	9.35
0800 4" thick		250	.064		6.20	3.40		9.60	12.30
0850 6" thick		200	.080		9.30	4.25		13.55	17.05
0950 Hemlock, 2" thick		350	.046		3.16	2.43		5.59	7.35
1000 3" thick		320	.050		4.74	2.66		7.40	9.45
1100 4" thick		250	.064		6.30	3.40		9.70	12.40
1150 6" thick		200	.080		9.50	4.25		13.75	17.20
1250 Western white spruce, 2" thick		350	.046		2.02	2.43		4.45	6.10
1300 3" thick		320	.050		3.03	2.66		5.69	7.60
1400 4" thick		250	.064		4.03	3.40		7.43	9.90
1450 6" thick		200	.080		6.05	4.25		10.30	13.45
9000 Minimum labor/equipment charge	1 Carp	2	4	Job		213		213	340

06 15 Wood Decking

06 15 23 – Laminated Wood Decking

06 15 23.10 Laminated Roof Deck

06 15 23.10 Laminated Roof Deck		Crew	Daily Output	Labor-Hours	Unit	Material	2020 Bare Costs Labor	Equipment	Total	Total Incl O&P
0010	**LAMINATED ROOF DECK**									
0020	Pine or hemlock, 3" thick	2 Carp	425	.038	S.F.	5.70	2		7.70	9.45
0100	4" thick		325	.049		7.40	2.62		10.02	12.35
0300	Cedar, 3" thick		425	.038		8.20	2		10.20	12.20
0400	4" thick		325	.049		11.05	2.62		13.67	16.35
0600	Fir, 3" thick		425	.038		6.60	2		8.60	10.45
0700	4" thick		325	.049		8.40	2.62		11.02	13.45
9000	Minimum labor/equipment charge	1 Carp	3	2.667	Job		142		142	227

06 16 Sheathing

06 16 13 – Insulating Sheathing

06 16 13.10 Insulating Sheathing

06 16 13.10 Insulating Sheathing			Crew	Daily Output	Labor-Hours	Unit	Material	2020 Bare Costs Labor	Equipment	Total	Total Incl O&P
0010	**INSULATING SHEATHING**										
0020	Expanded polystyrene, 1#/C.F. density, 3/4" thick, R2.89	G	2 Carp	1400	.011	S.F.	.37	.61		.98	1.38
0030	1" thick, R3.85	G		1300	.012		.45	.65		1.10	1.54
0040	2" thick, R7.69	G		1200	.013		.74	.71		1.45	1.95
0050	Extruded polystyrene, 15 psi compressive strength, 1" thick, R5	G		1300	.012		.68	.65		1.33	1.80
0060	2" thick, R10	G		1200	.013		.83	.71		1.54	2.06
0070	Polyisocyanurate, 2#/C.F. density, 3/4" thick	G		1400	.011		.68	.61		1.29	1.72
0080	1" thick	G		1300	.012		.61	.65		1.26	1.72
0090	1-1/2" thick	G		1250	.013		.76	.68		1.44	1.92
0100	2" thick	G		1200	.013		1.08	.71		1.79	2.33

06 16 23 – Subflooring

06 16 23.10 Subfloor

06 16 23.10 Subfloor			Crew	Daily Output	Labor-Hours	Unit	Material	2020 Bare Costs Labor	Equipment	Total	Total Incl O&P
0010	**SUBFLOOR**	R061636-20									
0011	Plywood, CDX, 1/2" thick		2 Carp	1500	.011	SF Flr.	.68	.57		1.25	1.66
0015	Pneumatic nailed			1860	.009		.68	.46		1.14	1.48
0102	5/8" thick			1350	.012		.77	.63		1.40	1.86
0107	Pneumatic nailed			1674	.010		.77	.51		1.28	1.66
0202	3/4" thick			1250	.013		.95	.68		1.63	2.13
0207	Pneumatic nailed			1550	.010		.95	.55		1.50	1.92
0302	1-1/8" thick, 2-4-1 including underlayment			1050	.015		2.12	.81		2.93	3.63
0440	With boards, 1" x 6", S4S, laid regular			900	.018		1.79	.95		2.74	3.48
0452	1" x 8", laid regular			1000	.016		2.06	.85		2.91	3.62
0462	Laid diagonal			850	.019		2.06	1		3.06	3.86
0502	1" x 10", laid regular			1100	.015		2.49	.77		3.26	3.98
0602	Laid diagonal			900	.018		2.49	.95		3.44	4.25
8990	Subfloor adhesive, 3/8" bead		1 Carp	2300	.003	L.F.	.10	.19		.29	.41
9000	Minimum labor/equipment charge		"	4	2	Job		106		106	170

06 16 26 – Underlayment

06 16 26.10 Wood Product Underlayment

06 16 26.10 Wood Product Underlayment			Crew	Daily Output	Labor-Hours	Unit	Material	2020 Bare Costs Labor	Equipment	Total	Total Incl O&P
0010	**WOOD PRODUCT UNDERLAYMENT**	R061636-20									
0015	Plywood, underlayment grade, 1/4" thick		2 Carp	1500	.011	S.F.	.95	.57		1.52	1.95
0018	Pneumatic nailed			1860	.009		.95	.46		1.41	1.77
0030	3/8" thick			1500	.011		1.05	.57		1.62	2.07
0070	Pneumatic nailed			1860	.009		1.05	.46		1.51	1.89
0102	1/2" thick			1450	.011		1.18	.59		1.77	2.24
0107	Pneumatic nailed			1798	.009		1.18	.47		1.65	2.06
0202	5/8" thick			1400	.011		1.54	.61		2.15	2.66

06 16 26 – Underlayment

06 16 26.10 Wood Product Underlayment		Crew	Daily Output	Labor-Hours	Unit	Material	2020 Bare Costs Labor	Equipment	Total	Total Incl O&P
0207	Pneumatic nailed	2 Carp	1736	.009	S.F.	1.54	.49		2.03	2.48
0302	3/4" thick		1300	.012		1.49	.65		2.14	2.69
0306	Pneumatic nailed		1612	.010		1.49	.53		2.02	2.49
0502	Particle board, 3/8" thick		1500	.011		.39	.57		.96	1.34
0507	Pneumatic nailed		1860	.009		.39	.46		.85	1.16
0602	1/2" thick		1450	.011		.41	.59		1	1.39
0607	Pneumatic nailed		1798	.009		.41	.47		.88	1.21
0802	5/8" thick		1400	.011		.61	.61		1.22	1.64
0807	Pneumatic nailed		1736	.009		.61	.49		1.10	1.46
0902	3/4" thick		1300	.012		.74	.65		1.39	1.86
0907	Pneumatic nailed		1612	.010		.74	.53		1.27	1.66
1100	Hardboard, underlayment grade, 4' x 4', .215" thick G		1500	.011		.72	.57		1.29	1.70
9000	Minimum labor/equipment charge	1 Carp	4	2	Job		106		106	170

06 16 33 – Wood Board Sheathing

06 16 33.10 Board Sheathing

		Crew	Daily Output	Labor-Hours	Unit	Material	Labor	Equipment	Total	Total Incl O&P
0009	**BOARD SHEATHING**									
0010	Roof, 1" x 6" boards, laid horizontal	2 Carp	725	.022	S.F.	1.79	1.17		2.96	3.85
0020	On steep roof		520	.031		1.79	1.64		3.43	4.59
0040	On dormers, hips, & valleys		480	.033		1.79	1.77		3.56	4.81
0050	Laid diagonal		650	.025		1.79	1.31		3.10	4.07
0070	1" x 8" boards, laid horizontal		875	.018		2.06	.97		3.03	3.82
0080	On steep roof		635	.025		2.06	1.34		3.40	4.41
0090	On dormers, hips, & valleys		580	.028		2.10	1.47		3.57	4.66
0100	Laid diagonal		725	.022		2.06	1.17		3.23	4.14
0110	Skip sheathing, 1" x 4", 7" OC	1 Carp	1200	.007		.66	.35		1.01	1.30
0120	1" x 6", 9" OC		1450	.006		.82	.29		1.11	1.37
0180	T&G sheathing/decking, 1" x 6"		1000	.008		1.78	.43		2.21	2.64
0190	2" x 6"		1000	.008		3.95	.43		4.38	5.05
0200	Walls, 1" x 6" boards, laid regular	2 Carp	650	.025		1.79	1.31		3.10	4.07
0210	Laid diagonal		585	.027		1.79	1.45		3.24	4.30
0220	1" x 8" boards, laid regular		765	.021		2.06	1.11		3.17	4.04
0230	Laid diagonal		650	.025		2.06	1.31		3.37	4.36

06 16 36 – Wood Panel Product Sheathing

06 16 36.10 Sheathing

		Crew	Daily Output	Labor-Hours	Unit	Material	Labor	Equipment	Total	Total Incl O&P
0010	**SHEATHING** R061110-30									
0012	Plywood on roofs, CDX									
0032	5/16" thick	2 Carp	1600	.010	S.F.	.60	.53		1.13	1.50
0037	Pneumatic nailed R061636-20		1952	.008		.60	.44		1.04	1.35
0052	3/8" thick		1525	.010		.53	.56		1.09	1.47
0057	Pneumatic nailed		1860	.009		.53	.46		.99	1.31
0102	1/2" thick		1400	.011		.68	.61		1.29	1.72
0103	Pneumatic nailed		1708	.009		.68	.50		1.18	1.55
0202	5/8" thick		1300	.012		.77	.65		1.42	1.90
0207	Pneumatic nailed		1586	.010		.77	.54		1.31	1.71
0302	3/4" thick		1200	.013		.95	.71		1.66	2.18
0307	Pneumatic nailed		1464	.011		.95	.58		1.53	1.97
0502	Plywood on walls, with exterior CDX, 3/8" thick		1200	.013		.53	.71		1.24	1.72
0507	Pneumatic nailed		1488	.011		.53	.57		1.10	1.50
0602	1/2" thick		1125	.014		.68	.76		1.44	1.96
0607	Pneumatic nailed		1395	.011		.68	.61		1.29	1.73
0702	5/8" thick		1050	.015		.77	.81		1.58	2.15
0707	Pneumatic nailed		1302	.012		.77	.65		1.42	1.90

06 16 Sheathing

06 16 36 – Wood Panel Product Sheathing

	06 16 36.10 Sheathing	Crew	Daily Output	Labor-Hours	Unit	Material	2020 Bare Costs Labor	Equipment	Total	Total Incl O&P
0802	3/4" thick	2 Carp	975	.016	S.F.	.95	.87		1.82	2.44
0807	Pneumatic nailed		1209	.013		.95	.70		1.65	2.17
1000	For shear wall construction, add						20%			
1200	For structural 1 exterior plywood, add				S.F.	10%				
3000	Wood fiber, regular, no vapor barrier, 1/2" thick	2 Carp	1200	.013		.64	.71		1.35	1.84
3100	5/8" thick		1200	.013		.70	.71		1.41	1.91
3300	No vapor barrier, in colors, 1/2" thick		1200	.013		.82	.71		1.53	2.04
3400	5/8" thick		1200	.013		.86	.71		1.57	2.09
3600	With vapor barrier one side, white, 1/2" thick		1200	.013		.63	.71		1.34	1.83
3700	Vapor barrier 2 sides, 1/2" thick		1200	.013		.86	.71		1.57	2.09
3800	Asphalt impregnated, 25/32" thick		1200	.013		.32	.71		1.03	1.49
3850	Intermediate, 1/2" thick		1200	.013		.34	.71		1.05	1.51
4500	Oriented strand board, on roof, 7/16" thick [G]		1460	.011		.48	.58		1.06	1.46
4505	Pneumatic nailed [G]		1780	.009		.48	.48		.96	1.30
4550	1/2" thick [G]		1400	.011		.48	.61		1.09	1.50
4555	Pneumatic nailed [G]		1736	.009		.48	.49		.97	1.32
4600	5/8" thick [G]		1300	.012		.66	.65		1.31	1.78
4605	Pneumatic nailed [G]		1586	.010		.66	.54		1.20	1.59
4610	On walls, 7/16" thick [G]		1200	.013		.48	.71		1.19	1.67
4615	Pneumatic nailed [G]		1488	.011		.48	.57		1.05	1.45
4620	1/2" thick [G]		1195	.013		.48	.71		1.19	1.67
4625	Pneumatic nailed [G]		1325	.012		.48	.64		1.12	1.56
4630	5/8" thick [G]		1050	.015		.66	.81		1.47	2.03
4635	Pneumatic nailed [G]		1302	.012		.66	.65		1.31	1.78
4700	Oriented strand board, factory laminated W.R. barrier, on roof, 1/2" thick [G]		1400	.011		.79	.61		1.40	1.84
4705	Pneumatic nailed [G]		1736	.009		.79	.49		1.28	1.66
4720	5/8" thick [G]		1300	.012		.98	.65		1.63	2.13
4725	Pneumatic nailed [G]		1586	.010		.98	.54		1.52	1.94
4730	5/8" thick, T&G [G]		1150	.014		1.06	.74		1.80	2.35
4735	Pneumatic nailed, T&G [G]		1400	.011		1.06	.61		1.67	2.14
4740	On walls, 7/16" thick [G]		1200	.013		.73	.71		1.44	1.94
4745	Pneumatic nailed [G]		1488	.011		.73	.57		1.30	1.72
4750	1/2" thick [G]		1195	.013		.79	.71		1.50	2.01
4755	Pneumatic nailed [G]		1325	.012		.79	.64		1.43	1.90
4800	Joint sealant tape, 3-1/2"		7600	.002	L.F.	.36	.11		.47	.58
4810	Joint sealant tape, 6"		7600	.002	"	.41	.11		.52	.63
9000	Minimum labor/equipment charge	1 Carp	2	4	Job		213		213	340

06 16 43 – Gypsum Sheathing

06 16 43.10 Gypsum Sheathing

		Crew	Daily Output	Labor-Hours	Unit	Material	2020 Bare Costs Labor	Equipment	Total	Total Incl O&P
0010	**GYPSUM SHEATHING**									
0020	Gypsum, weatherproof, 1/2" thick	2 Carp	1125	.014	S.F.	.53	.76		1.29	1.79
0040	With embedded glass mats	"	1100	.015	"	.71	.77		1.48	2.02

For customer support on your Commercial Renovation Costs with RSMeans data, call 800.448.8182.

157

06 17 33 – Wood I-Joists

06 17 33.10 Wood and Composite I-Joists	Crew	Daily Output	Labor-Hours	Unit	Material	2020 Bare Costs Labor	Equipment	Total	Total Incl O&P
0010 **WOOD AND COMPOSITE I-JOISTS**									
0100 Plywood webs, incl. bridging & blocking, panels 24" OC									
1200 15' to 24' span, 50 psf live load	F-5	2400	.013	SF Flr.	1.81	.72		2.53	3.14
1300 55 psf live load		2250	.014		2.26	.76		3.02	3.71
1400 24' to 30' span, 45 psf live load		2600	.012		2.68	.66		3.34	4
1500 55 psf live load		2400	.013		4.77	.72		5.49	6.40

06 17 53 – Shop-Fabricated Wood Trusses

06 17 53.10 Roof Trusses

	Crew	Daily Output	Labor-Hours	Unit	Material	2020 Bare Costs Labor	Equipment	Total	Total Incl O&P
0010 **ROOF TRUSSES**									
0100 Fink (W) or King post type, 2'-0" OC									
0200 Metal plate connected, 4 in 12 slope									
0210 24' to 29' span	F-3	3000	.013	SF Flr.	1.82	.72	.16	2.70	3.33
0300 30' to 43' span		3000	.013		2.40	.72	.16	3.28	3.97
0400 44' to 60' span		3000	.013		2.41	.72	.16	3.29	3.98
0700 Glued and nailed, add					50%				

06 18 Glued-Laminated Construction

06 18 13 – Glued-Laminated Beams

06 18 13.20 Laminated Framing

	Crew	Daily Output	Labor-Hours	Unit	Material	2020 Bare Costs Labor	Equipment	Total	Total Incl O&P
0010 **LAMINATED FRAMING**									
0020 30 lb., short term live load, 15 lb. dead load									
0200 Straight roof beams, 20' clear span, beams 8' OC	F-3	2560	.016	SF Flr.	2.27	.85	.18	3.30	4.05
0300 Beams 16' OC		3200	.013		1.66	.68	.15	2.49	3.07
0500 40' clear span, beams 8' OC		3200	.013		4.31	.68	.15	5.14	6
0600 Beams 16' OC		3840	.010		3.57	.57	.12	4.26	4.96
0800 60' clear span, beams 8' OC	F-4	2880	.017		7.40	.90	.34	8.64	9.90
0900 Beams 16' OC	"	3840	.013		5.55	.67	.26	6.48	7.45
1100 Tudor arches, 30' to 40' clear span, frames 8' OC	F-3	1680	.024		9.60	1.29	.28	11.17	12.95
1200 Frames 16' OC	"	2240	.018		7.50	.97	.21	8.68	10.05
1400 50' to 60' clear span, frames 8' OC	F-4	2200	.022		10.35	1.17	.45	11.97	13.75
1500 Frames 16' OC		2640	.018		8.85	.98	.37	10.20	11.65
1700 Radial arches, 60' clear span, frames 8' OC		1920	.025		9.70	1.34	.51	11.55	13.35
1800 Frames 16' OC		2880	.017		7.75	.90	.34	8.99	10.30
2000 100' clear span, frames 8' OC		1600	.030		10	1.61	.61	12.22	14.25
2100 Frames 16' OC		2400	.020		8.80	1.07	.41	10.28	11.85
2300 120' clear span, frames 8' OC		1440	.033		13.25	1.79	.68	15.72	18.20
2400 Frames 16' OC		1920	.025		12.10	1.34	.51	13.95	16.05
2600 Bowstring trusses, 20' OC, 40' clear span	F-3	2400	.017		6	.91	.20	7.11	8.25
2700 60' clear span	F-4	3600	.013		5.40	.72	.27	6.39	7.40
2800 100' clear span		4000	.012		7.65	.64	.25	8.54	9.70
2900 120' clear span		3600	.013		8.10	.72	.27	9.09	10.40
3000 For less than 1000 B.F., add					20%				
3100 For premium appearance, add to S.F. prices					5%				
3300 For industrial type, deduct					15%				
3500 For stain and varnish, add					5%				
3900 For 3/4" laminations, add to straight					25%				
4100 Add to curved					15%				
4300 Alternate pricing method: (use nominal footage of									
4310 components). Straight beams, camber less than 6"	F-3	3.50	11.429	M.B.F.	3,075	620	135	3,830	4,525
4400 Columns, including hardware		2	20		3,300	1,075	236	4,611	5,600

06 18 Glued-Laminated Construction

06 18 13 – Glued-Laminated Beams

06 18 13.20 Laminated Framing

		Crew	Daily Output	Labor-Hours	Unit	Material	2020 Bare Costs Labor	Equipment	Total	Total Incl O&P
4600	Curved members, radius over 32'	F-3	2.50	16	M.B.F.	3,375	870	188	4,433	5,275
4700	Radius 10' to 32'	↓	3	13.333		3,350	725	157	4,232	5,000
4900	For complicated shapes, add maximum					100%				
5100	For pressure treating, add to straight					35%				
5200	Add to curved				↓	45%				
6000	Laminated veneer members, southern pine or western species									
6050	1-3/4" wide x 5-1/2" deep	2 Carp	480	.033	L.F.	3.36	1.77		5.13	6.55
6100	9-1/2" deep		480	.033		4.56	1.77		6.33	7.85
6150	14" deep		450	.036		7.35	1.89		9.24	11.10
6200	18" deep	↓	450	.036	↓	10.15	1.89		12.04	14.25
6300	Parallel strand members, southern pine or western species									
6350	1-3/4" wide x 9-1/4" deep	2 Carp	480	.033	L.F.	5.05	1.77		6.82	8.40
6400	11-1/4" deep		450	.036		5.65	1.89		7.54	9.30
6450	14" deep		400	.040		8.35	2.13		10.48	12.60
6500	3-1/2" wide x 9-1/4" deep		480	.033		18.10	1.77		19.87	22.50
6550	11-1/4" deep		450	.036		19.70	1.89		21.59	24.50
6600	14" deep		400	.040		22.50	2.13		24.63	28.50
6650	7" wide x 9-1/4" deep		450	.036		33	1.89		34.89	39.50
6700	11-1/4" deep		420	.038		42	2.03		44.03	49
6750	14" deep	↓	400	.040	↓	50	2.13		52.13	58.50
9000	Minimum labor/equipment charge	F-3	2.50	16	Job		870	188	1,058	1,575

06 22 Millwork

06 22 13 – Standard Pattern Wood Trim

06 22 13.15 Moldings, Base

		Crew	Daily Output	Labor-Hours	Unit	Material	2020 Bare Costs Labor	Equipment	Total	Total Incl O&P
0010	**MOLDINGS, BASE**									
5100	Classic profile, 5/8" x 5-1/2", finger jointed and primed	1 Carp	250	.032	L.F.	1.76	1.70		3.46	4.66
5105	Poplar		240	.033		2.18	1.77		3.95	5.25
5110	Red oak		220	.036		2.54	1.93		4.47	5.90
5115	Maple		220	.036		4.15	1.93		6.08	7.65
5120	Cherry		220	.036		4.55	1.93		6.48	8.10
5125	3/4" x 7-1/2", finger jointed and primed		250	.032		2.05	1.70		3.75	4.98
5130	Poplar		240	.033		2.66	1.77		4.43	5.75
5135	Red oak		220	.036		3.70	1.93		5.63	7.15
5140	Maple		220	.036		5.15	1.93		7.08	8.75
5145	Cherry		220	.036		6.10	1.93		8.03	9.85
5150	Modern profile, 5/8" x 3-1/2", finger jointed and primed		250	.032		.95	1.70		2.65	3.77
5155	Poplar		240	.033		1.06	1.77		2.83	4.01
5160	Red oak		220	.036		1.74	1.93		3.67	5
5165	Maple		220	.036		2.70	1.93		4.63	6.05
5170	Cherry		220	.036		2.94	1.93		4.87	6.35
5175	Ogee profile, 7/16" x 3", finger jointed and primed		250	.032		.68	1.70		2.38	3.47
5180	Poplar		240	.033		.76	1.77		2.53	3.68
5185	Red oak		220	.036		.99	1.93		2.92	4.19
5200	9/16" x 3-1/2", finger jointed and primed		250	.032		.68	1.70		2.38	3.47
5205	Pine		240	.033		1.28	1.77		3.05	4.25
5210	Red oak		220	.036		2.92	1.93		4.85	6.30
5215	9/16" x 4-1/2", red oak		220	.036		4.38	1.93		6.31	7.90
5220	5/8" x 3-1/2", finger jointed and primed		250	.032		1.04	1.70		2.74	3.86
5225	Poplar		240	.033		1.06	1.77		2.83	4.01
5230	Red oak	↓	220	.036	↓	1.74	1.93		3.67	5

06 22 13 – Standard Pattern Wood Trim

06 22 13.15 Moldings, Base

		Crew	Daily Output	Labor-Hours	Unit	Material	2020 Bare Costs Labor	Equipment	Total	Total Incl O&P
5235	Maple	1 Carp	220	.036	L.F.	2.70	1.93		4.63	6.05
5240	Cherry		220	.036		2.92	1.93		4.85	6.30
5245	5/8" x 4", finger jointed and primed		250	.032		1.28	1.70		2.98	4.13
5250	Poplar		240	.033		1.37	1.77		3.14	4.35
5255	Red oak		220	.036		1.96	1.93		3.89	5.25
5260	Maple		220	.036		3.02	1.93		4.95	6.40
5265	Cherry		220	.036		3.17	1.93		5.10	6.60
5270	Rectangular profile, oak, 3/8" x 1-1/4"		260	.031		1.58	1.64		3.22	4.35
5275	1/2" x 2-1/2"		255	.031		2.27	1.67		3.94	5.15
5280	1/2" x 3-1/2"		250	.032		2.85	1.70		4.55	5.85
5285	1" x 6"		240	.033		4.06	1.77		5.83	7.30
5290	1" x 8"		240	.033		5.05	1.77		6.82	8.45
5295	Pine, 3/8" x 1-3/4"		260	.031		.50	1.64		2.14	3.17
5300	7/16" x 2-1/2"		255	.031		.82	1.67		2.49	3.57
5305	1" x 6"		240	.033		.84	1.77		2.61	3.76
5310	1" x 8"		240	.033		.95	1.77		2.72	3.89
5315	Shoe, 1/2" x 3/4", primed		260	.031		.59	1.64		2.23	3.27
5320	Pine		240	.033		.43	1.77		2.20	3.31
5325	Poplar		240	.033		.43	1.77		2.20	3.31
5330	Red oak		220	.036		.57	1.93		2.50	3.73
5335	Maple		220	.036		.82	1.93		2.75	4
5340	Cherry		220	.036		.84	1.93		2.77	4.02
5345	11/16" x 1-1/2", pine		240	.033		.87	1.77		2.64	3.80
5350	Caps, 11/16" x 1-3/8", pine		240	.033		.69	1.77		2.46	3.60
5355	3/4" x 1-3/4", finger jointed and primed		260	.031		.93	1.64		2.57	3.64
5360	Poplar		240	.033		1.07	1.77		2.84	4.02
5365	Red oak		220	.036		1.33	1.93		3.26	4.56
5370	Maple		220	.036		1.68	1.93		3.61	4.95
5375	Cherry		220	.036		3.69	1.93		5.62	7.15
5380	Combination base & shoe, 9/16" x 3-1/2" & 1/2" x 3/4", pine		125	.064		1.71	3.40		5.11	7.35
5385	Three piece oak, 6" high		80	.100		5.95	5.30		11.25	15.05
5390	Including 3/4" x 1" base shoe		70	.114		6.60	6.05		12.65	17
5395	Flooring cant strip, 3/4" x 3/4", pre-finished pine		260	.031		.48	1.64		2.12	3.15
5400	For pre-finished, stain and clear coat, add					.59			.59	.65
5405	Clear coat only, add					.45			.45	.50
9000	Minimum labor/equipment charge	1 Carp	4	2	Job		106		106	170

06 22 13.30 Moldings, Casings

		Crew	Daily Output	Labor-Hours	Unit	Material	2020 Bare Costs Labor	Equipment	Total	Total Incl O&P
0010	**MOLDINGS, CASINGS**									
0085	Apron, 9/16" x 2-1/2", pine	1 Carp	250	.032	L.F.	1.51	1.70		3.21	4.38
0090	5/8" x 2-1/2", pine		250	.032		1.81	1.70		3.51	4.71
0110	5/8" x 3-1/2", pine		220	.036		2.27	1.93		4.20	5.60
0300	Band, 11/16" x 1-1/8", pine		270	.030		.77	1.57		2.34	3.37
0310	11/16" x 1-1/2", finger jointed and primed		270	.030		.76	1.57		2.33	3.36
0320	Pine		270	.030		1.02	1.57		2.59	3.64
0330	11/16" x 1-3/4", finger jointed and primed		270	.030		.96	1.57		2.53	3.58
0350	Pine		270	.030		1.09	1.57		2.66	3.72
0355	Beaded, 3/4" x 3-1/2", finger jointed and primed		220	.036		1.03	1.93		2.96	4.23
0360	Poplar		220	.036		1.20	1.93		3.13	4.42
0365	Red oak		220	.036		1.74	1.93		3.67	5
0370	Maple		220	.036		2.69	1.93		4.62	6.05
0375	Cherry		220	.036		3.25	1.93		5.18	6.70
0380	3/4" x 4", finger jointed and primed		220	.036		1.27	1.93		3.20	4.50

06 22 Millwork

06 22 13 – Standard Pattern Wood Trim

06 22 13.30 Moldings, Casings	Crew	Daily Output	Labor-Hours	Unit	Material	2020 Bare Costs Labor	Equipment	Total	Total Incl O&P
0385	Poplar	1 Carp	220	.036	L.F.	1.84	1.93		3.77	5.10
0390	Red oak		220	.036		2.47	1.93		4.40	5.80
0395	Maple		220	.036		2.71	1.93		4.64	6.10
0400	Cherry		220	.036		4.03	1.93		5.96	7.55
0405	3/4" x 5-1/2", finger jointed and primed		200	.040		1.59	2.13		3.72	5.15
0410	Poplar		200	.040		2.04	2.13		4.17	5.65
0415	Red oak		200	.040		2.99	2.13		5.12	6.70
0420	Maple		200	.040		3.95	2.13		6.08	7.75
0425	Cherry		200	.040		4.53	2.13		6.66	8.40
0430	Classic profile, 3/4" x 2-3/4", finger jointed and primed		250	.032		.87	1.70		2.57	3.68
0435	Poplar		250	.032		1.05	1.70		2.75	3.88
0440	Red oak		250	.032		1.57	1.70		3.27	4.45
0445	Maple		250	.032		2.21	1.70		3.91	5.15
0450	Cherry		250	.032		2.54	1.70		4.24	5.50
0455	Fluted, 3/4" x 3-1/2", poplar		220	.036		1.20	1.93		3.13	4.42
0460	Red oak		220	.036		1.74	1.93		3.67	5
0465	Maple		220	.036		3.04	1.93		4.97	6.45
0470	Cherry		220	.036		3.25	1.93		5.18	6.70
0475	3/4" x 4", poplar		220	.036		1.47	1.93		3.40	4.72
0480	Red oak		220	.036		2.31	1.93		4.24	5.65
0485	Maple		220	.036		2.71	1.93		4.64	6.10
0490	Cherry		220	.036		4.03	1.93		5.96	7.55
0495	3/4" x 5-1/2", poplar		200	.040		2.04	2.13		4.17	5.65
0500	Red oak		200	.040		3.08	2.13		5.21	6.80
0505	Maple		200	.040		3.95	2.13		6.08	7.75
0510	Cherry		200	.040		4.52	2.13		6.65	8.40
0515	3/4" x 7-1/2", poplar		190	.042		1.59	2.24		3.83	5.35
0520	Red oak		190	.042		4	2.24		6.24	8
0525	Maple		190	.042		7.20	2.24		9.44	11.50
0530	Cherry		190	.042		8.60	2.24		10.84	13.05
0535	3/4" x 9-1/2", poplar		180	.044		4.21	2.36		6.57	8.40
0540	Red oak		180	.044		6.60	2.36		8.96	11.10
0545	Maple		180	.044		11.45	2.36		13.81	16.40
0550	Cherry		180	.044		12.50	2.36		14.86	17.55
0555	Modern profile, 9/16" x 2-1/4", poplar		250	.032		.88	1.70		2.58	3.69
0560	Red oak		250	.032		.99	1.70		2.69	3.81
0565	11/16" x 2-1/2", finger jointed & primed		250	.032		.77	1.70		2.47	3.57
0570	Pine		250	.032		1.29	1.70		2.99	4.14
0575	3/4" x 2-1/2", poplar		250	.032		.94	1.70		2.64	3.75
0580	Red oak		250	.032		1.26	1.70		2.96	4.11
0585	Maple		250	.032		1.94	1.70		3.64	4.85
0590	Cherry		250	.032		2.75	1.70		4.45	5.75
0595	Mullion, 5/16" x 2", pine		270	.030		.98	1.57		2.55	3.60
0600	9/16" x 2-1/2", finger jointed and primed		250	.032		.98	1.70		2.68	3.80
0605	Pine		250	.032		1.35	1.70		3.05	4.21
0610	Red oak		250	.032		3.40	1.70		5.10	6.45
0615	1-1/16" x 3-3/4", red oak		220	.036		7.05	1.93		8.98	10.85
0620	Ogee, 7/16" x 2-1/2", poplar		250	.032		.79	1.70		2.49	3.59
0625	Red oak		250	.032		.93	1.70		2.63	3.74
0630	9/16" x 2-1/4", finger jointed and primed		250	.032		.61	1.70		2.31	3.39
0635	Poplar		250	.032		.62	1.70		2.32	3.40
0640	Red oak		250	.032		.85	1.70		2.55	3.66
0645	11/16" x 2-1/2", finger jointed and primed		250	.032		.73	1.70		2.43	3.52

06 22 13 – Standard Pattern Wood Trim

06 22 13.30 Moldings, Casings

		Crew	Daily Output	Labor-Hours	Unit	Material	2020 Bare Costs Labor	Equipment	Total	Total Incl O&P
0700	Pine	1 Carp	250	.032	L.F.	1.41	1.70		3.11	4.27
0701	Red oak		250	.032		3.29	1.70		4.99	6.35
0730	11/16" x 3-1/2", finger jointed and primed		220	.036		1.36	1.93		3.29	4.60
0750	Pine		220	.036		1.95	1.93		3.88	5.25
0755	3/4" x 2-1/2", finger jointed and primed		250	.032		.77	1.70		2.47	3.57
0760	Poplar		250	.032		.96	1.70		2.66	3.78
0765	Red oak		250	.032		1.32	1.70		3.02	4.17
0770	Maple		250	.032		1.91	1.70		3.61	4.82
0775	Cherry		250	.032		2.40	1.70		4.10	5.35
0780	3/4" x 3-1/2", finger jointed and primed		220	.036		.86	1.93		2.79	4.05
0785	Poplar		220	.036		1.20	1.93		3.13	4.42
0790	Red oak		220	.036		1.74	1.93		3.67	5
0795	Maple		220	.036		2.68	1.93		4.61	6.05
0800	Cherry		220	.036		3.30	1.93		5.23	6.75
4700	Square profile, 1" x 1", teak		215	.037		2.23	1.98		4.21	5.60
4800	Rectangular profile, 1" x 3", teak		200	.040		6.60	2.13		8.73	10.65
9000	Minimum labor/equipment charge		4	2	Job		106		106	170

06 22 13.35 Moldings, Ceilings

		Crew	Daily Output	Labor-Hours	Unit	Material	2020 Bare Costs Labor	Equipment	Total	Total Incl O&P
0010	**MOLDINGS, CEILINGS**									
0600	Bed, 9/16" x 1-3/4", pine	1 Carp	270	.030	L.F.	1.13	1.57		2.70	3.76
0650	9/16" x 2", pine		270	.030		1.30	1.57		2.87	3.95
0710	9/16" x 1-3/4", oak		270	.030		2.44	1.57		4.01	5.20
1200	Cornice, 9/16" x 1-3/4", pine		270	.030		1.02	1.57		2.59	3.64
1300	9/16" x 2-1/4", pine		265	.030		1.37	1.60		2.97	4.08
1350	Cove, 1/2" x 2-1/4", poplar		265	.030		1.26	1.60		2.86	3.96
1360	Red oak		265	.030		1.73	1.60		3.33	4.47
1370	Hard maple		265	.030		1.71	1.60		3.31	4.45
1380	Cherry		265	.030		2.31	1.60		3.91	5.10
2400	9/16" x 1-3/4", pine		270	.030		1.10	1.57		2.67	3.73
2500	11/16" x 2-3/4", pine		265	.030		2.21	1.60		3.81	5
2510	Crown, 5/8" x 5/8", poplar		300	.027		.55	1.42		1.97	2.88
2520	Red oak		300	.027		.56	1.42		1.98	2.89
2530	Hard maple		300	.027		.76	1.42		2.18	3.11
2540	Cherry		300	.027		.80	1.42		2.22	3.15
2600	9/16" x 3-5/8", pine		250	.032		1.83	1.70		3.53	4.73
2700	11/16" x 4-1/4", pine		250	.032		3.05	1.70		4.75	6.10
2705	Oak		250	.032		6.50	1.70		8.20	9.85
2710	3/4" x 1-3/4", poplar		270	.030		.76	1.57		2.33	3.36
2720	Red oak		270	.030		1.12	1.57		2.69	3.75
2730	Hard maple		270	.030		1.47	1.57		3.04	4.14
2740	Cherry		270	.030		1.81	1.57		3.38	4.51
2750	3/4" x 2", poplar		270	.030		1	1.57		2.57	3.62
2760	Red oak		270	.030		1.50	1.57		3.07	4.17
2770	Hard maple		270	.030		2.05	1.57		3.62	4.78
2780	Cherry		270	.030		2.28	1.57		3.85	5.05
2790	3/4" x 2-3/4", poplar		265	.030		1.08	1.60		2.68	3.76
2800	Red oak		265	.030		1.67	1.60		3.27	4.41
2810	Hard maple		265	.030		2.28	1.60		3.88	5.10
2820	Cherry		265	.030		2.59	1.60		4.19	5.40
2830	3/4" x 3-1/2", poplar		250	.032		1.40	1.70		3.10	4.26
2840	Red oak		250	.032		2.08	1.70		3.78	5
2850	Hard maple		250	.032		2.95	1.70		4.65	5.95

06 22 13 – Standard Pattern Wood Trim

06 22 13.35 Moldings, Ceilings		Crew	Daily Output	Labor-Hours	Unit	Material	2020 Bare Costs Labor	Equipment	Total	Total Incl O&P
2860	Cherry	1 Carp	250	.032	L.F.	3.16	1.70		4.86	6.20
2870	FJP poplar		250	.032		1.06	1.70		2.76	3.89
2880	3/4" x 5", poplar		245	.033		2.04	1.74		3.78	5.05
2890	Red oak		245	.033		3.02	1.74		4.76	6.10
2900	Hard maple		245	.033		3.99	1.74		5.73	7.15
2910	Cherry		245	.033		4.76	1.74		6.50	8.05
2920	FJP poplar		245	.033		1.48	1.74		3.22	4.41
2930	3/4" x 6-1/4", poplar		240	.033		2.45	1.77		4.22	5.55
2940	Red oak		240	.033		3.69	1.77		5.46	6.90
2950	Hard maple		240	.033		4.96	1.77		6.73	8.30
2960	Cherry		240	.033		5.90	1.77		7.67	9.35
2970	7/8" x 8-3/4", poplar		220	.036		4.97	1.93		6.90	8.55
2980	Red oak		220	.036		5.20	1.93		7.13	8.80
2990	Hard maple		220	.036		6.85	1.93		8.78	10.65
3000	Cherry		220	.036		7.85	1.93		9.78	11.75
3010	1" x 7-1/4", poplar		220	.036		4.45	1.93		6.38	8
3020	Red oak		220	.036		6.25	1.93		8.18	9.95
3030	Hard maple		220	.036		9.05	1.93		10.98	13.05
3040	Cherry		220	.036		10.30	1.93		12.23	14.40
3050	1-1/16" x 4-1/4", poplar		250	.032		2.54	1.70		4.24	5.50
3060	Red oak		250	.032		2.86	1.70		4.56	5.85
3070	Hard maple		250	.032		4.20	1.70		5.90	7.35
3080	Cherry		250	.032		5.45	1.70		7.15	8.65
3090	Dentil crown, 3/4" x 5", poplar		250	.032		2.04	1.70		3.74	4.97
3100	Red oak		250	.032		3.02	1.70		4.72	6.05
3110	Hard maple		250	.032		3.98	1.70		5.68	7.10
3120	Cherry		250	.032		4.59	1.70		6.29	7.75
3130	Dentil piece for above, 1/2" x 1/2", poplar		300	.027		3.22	1.42		4.64	5.80
3140	Red oak		300	.027		3.33	1.42		4.75	5.95
3150	Hard maple		300	.027		4.17	1.42		5.59	6.85
3160	Cherry		300	.027		4.14	1.42		5.56	6.80
9000	Minimum labor/equipment charge		4	2	Job		106		106	170

06 22 13.40 Moldings, Exterior

		Crew	Daily Output	Labor-Hours	Unit	Material	2020 Bare Costs Labor	Equipment	Total	Total Incl O&P
0010	**MOLDINGS, EXTERIOR**									
0100	Band board, cedar, rough sawn, 1" x 2"	1 Carp	300	.027	L.F.	.72	1.42		2.14	3.06
0110	1" x 3"		300	.027		1.07	1.42		2.49	3.45
0120	1" x 4"		250	.032		1.43	1.70		3.13	4.29
0130	1" x 6"		250	.032		2.15	1.70		3.85	5.10
0140	1" x 8"		225	.036		2.86	1.89		4.75	6.20
0150	1" x 10"		225	.036		3.56	1.89		5.45	6.95
0160	1" x 12"		200	.040		4.27	2.13		6.40	8.10
0240	STK, 1" x 2"		300	.027		.44	1.42		1.86	2.75
0250	1" x 3"		300	.027		.48	1.42		1.90	2.80
0260	1" x 4"		250	.032		.85	1.70		2.55	3.66
0270	1" x 6"		250	.032		1.39	1.70		3.09	4.25
0280	1" x 8"		225	.036		2.35	1.89		4.24	5.60
0290	1" x 10"		225	.036		2.86	1.89		4.75	6.20
0300	1" x 12"		200	.040		4.44	2.13		6.57	8.30
0310	Pine, #2, 1" x 2"		300	.027		.31	1.42		1.73	2.61
0320	1" x 3"		300	.027		.45	1.42		1.87	2.77
0330	1" x 4"		250	.032		.57	1.70		2.27	3.35
0340	1" x 6"		250	.032		.86	1.70		2.56	3.67

06 22 13 – Standard Pattern Wood Trim

06 22 13.40 Moldings, Exterior		Crew	Daily Output	Labor-Hours	Unit	Material	2020 Bare Costs Labor	Equipment	Total	Total Incl O&P
0350	1" x 8"	1 Carp	225	.036	L.F.	1.33	1.89		3.22	4.49
0360	1" x 10"		225	.036		2.04	1.89		3.93	5.25
0370	1" x 12"		200	.040		2.66	2.13		4.79	6.35
0380	D & better, 1" x 2"		300	.027		.51	1.42		1.93	2.83
0390	1" x 3"		300	.027		.73	1.42		2.15	3.07
0400	1" x 4"		250	.032		.95	1.70		2.65	3.77
0410	1" x 6"		250	.032		1.10	1.70		2.80	3.93
0420	1" x 8"		225	.036		1.64	1.89		3.53	4.83
0430	1" x 10"		225	.036		2.16	1.89		4.05	5.40
0440	1" x 12"		200	.040		2.73	2.13		4.86	6.40
0450	Redwood, clear all heart, 1" x 2"		300	.027		.66	1.42		2.08	3
0460	1" x 3"		300	.027		.98	1.42		2.40	3.35
0470	1" x 4"		250	.032		1.22	1.70		2.92	4.06
0480	1" x 6"		252	.032		1.83	1.69		3.52	4.72
0490	1" x 8"		225	.036		2.43	1.89		4.32	5.70
0500	1" x 10"		225	.036		4.03	1.89		5.92	7.45
0510	1" x 12"		200	.040		4.40	2.13		6.53	8.25
0530	Corner board, cedar, rough sawn, 1" x 2"		225	.036		.72	1.89		2.61	3.82
0540	1" x 3"		225	.036		1.07	1.89		2.96	4.21
0550	1" x 4"		200	.040		1.43	2.13		3.56	4.98
0560	1" x 6"		200	.040		2.15	2.13		4.28	5.75
0570	1" x 8"		200	.040		2.86	2.13		4.99	6.55
0580	1" x 10"		175	.046		3.56	2.43		5.99	7.80
0590	1" x 12"		175	.046		4.27	2.43		6.70	8.60
0670	STK, 1" x 2"		225	.036		.44	1.89		2.33	3.51
0680	1" x 3"		225	.036		.48	1.89		2.37	3.56
0690	1" x 4"		200	.040		.83	2.13		2.96	4.32
0700	1" x 6"		200	.040		1.39	2.13		3.52	4.94
0710	1" x 8"		200	.040		2.35	2.13		4.48	6
0720	1" x 10"		175	.046		2.86	2.43		5.29	7.05
0730	1" x 12"		175	.046		4.44	2.43		6.87	8.80
0740	Pine, #2, 1" x 2"		225	.036		.31	1.89		2.20	3.37
0750	1" x 3"		225	.036		.45	1.89		2.34	3.53
0760	1" x 4"		200	.040		.57	2.13		2.70	4.04
0770	1" x 6"		200	.040		.86	2.13		2.99	4.36
0780	1" x 8"		200	.040		1.33	2.13		3.46	4.87
0790	1" x 10"		175	.046		2.04	2.43		4.47	6.15
0800	1" x 12"		175	.046		2.66	2.43		5.09	6.80
0810	D & better, 1" x 2"		225	.036		.51	1.89		2.40	3.59
0820	1" x 3"		225	.036		.73	1.89		2.62	3.83
0830	1" x 4"		200	.040		.95	2.13		3.08	4.46
0840	1" x 6"		200	.040		1.10	2.13		3.23	4.62
0850	1" x 8"		200	.040		1.64	2.13		3.77	5.20
0860	1" x 10"		175	.046		2.16	2.43		4.59	6.25
0870	1" x 12"		175	.046		2.73	2.43		5.16	6.90
0880	Redwood, clear all heart, 1" x 2"		225	.036		.66	1.89		2.55	3.76
0890	1" x 3"		225	.036		.98	1.89		2.87	4.11
0900	1" x 4"		200	.040		1.22	2.13		3.35	4.75
0910	1" x 6"		200	.040		1.83	2.13		3.96	5.45
0920	1" x 8"		200	.040		2.43	2.13		4.56	6.10
0930	1" x 10"		175	.046		4.03	2.43		6.46	8.35
0940	1" x 12"		175	.046		4.40	2.43		6.83	8.75
0950	Cornice board, cedar, rough sawn, 1" x 2"		330	.024		.72	1.29		2.01	2.85

For customer support on your Commercial Renovation Costs with RSMeans data, call 800.448.8182.

06 22 13.40 Moldings, Exterior		Crew	Daily Output	Labor-Hours	Unit	Material	2020 Bare Costs Labor	Equipment	Total	Total Incl O&P
0960	1" x 3"	1 Carp	290	.028	L.F.	1.07	1.47		2.54	3.53
0970	1" x 4"		250	.032		1.43	1.70		3.13	4.29
0980	1" x 6"		250	.032		2.15	1.70		3.85	5.10
0990	1" x 8"		200	.040		2.86	2.13		4.99	6.55
1000	1" x 10"		180	.044		3.56	2.36		5.92	7.70
1010	1" x 12"		180	.044		4.27	2.36		6.63	8.50
1020	STK, 1" x 2"		330	.024		.44	1.29		1.73	2.54
1030	1" x 3"		290	.028		.48	1.47		1.95	2.88
1040	1" x 4"		250	.032		.85	1.70		2.55	3.66
1050	1" x 6"		250	.032		1.39	1.70		3.09	4.25
1060	1" x 8"		200	.040		2.35	2.13		4.48	6
1070	1" x 10"		180	.044		2.86	2.36		5.22	6.95
1080	1" x 12"		180	.044		4.44	2.36		6.80	8.65
1500	Pine, #2, 1" x 2"		330	.024		.31	1.29		1.60	2.40
1510	1" x 3"		290	.028		.31	1.47		1.78	2.69
1600	1" x 4"		250	.032		.57	1.70		2.27	3.35
1700	1" x 6"		250	.032		.86	1.70		2.56	3.67
1800	1" x 8"		200	.040		1.33	2.13		3.46	4.87
1900	1" x 10"		180	.044		2.04	2.36		4.40	6
2000	1" x 12"		180	.044		2.66	2.36		5.02	6.70
2020	D & better, 1" x 2"		330	.024		.51	1.29		1.80	2.62
2030	1" x 3"		290	.028		.73	1.47		2.20	3.15
2040	1" x 4"		250	.032		.95	1.70		2.65	3.77
2050	1" x 6"		250	.032		1.10	1.70		2.80	3.93
2060	1" x 8"		200	.040		1.64	2.13		3.77	5.20
2070	1" x 10"		180	.044		2.16	2.36		4.52	6.15
2080	1" x 12"		180	.044		2.73	2.36		5.09	6.80
2090	Redwood, clear all heart, 1" x 2"		330	.024		.66	1.29		1.95	2.79
2100	1" x 3"		290	.028		.98	1.47		2.45	3.43
2110	1" x 4"		250	.032		1.22	1.70		2.92	4.06
2120	1" x 6"		250	.032		1.83	1.70		3.53	4.74
2130	1" x 8"		200	.040		2.43	2.13		4.56	6.10
2140	1" x 10"		180	.044		4.03	2.36		6.39	8.20
2150	1" x 12"		180	.044		4.40	2.36		6.76	8.60
2160	3 piece, 1" x 2", 1" x 4", 1" x 6", rough sawn cedar		80	.100		4.31	5.30		9.61	13.25
2180	STK cedar		80	.100		2.68	5.30		7.98	11.45
2200	#2 pine		80	.100		1.75	5.30		7.05	10.40
2210	D & better pine		80	.100		2.56	5.30		7.86	11.30
2220	Clear all heart redwood		80	.100		3.71	5.30		9.01	12.60
2230	1" x 8", 1" x 10", 1" x 12", rough sawn cedar		65	.123		10.65	6.55		17.20	22
2240	STK cedar		65	.123		9.60	6.55		16.15	21
2300	#2 pine		65	.123		6	6.55		12.55	17.10
2320	D & better pine		65	.123		6.50	6.55		13.05	17.65
2330	Clear all heart redwood		65	.123		10.80	6.55		17.35	22.50
2340	Door/window casing, cedar, rough sawn, 1" x 2"		275	.029		.72	1.55		2.27	3.27
2350	1" x 3"		275	.029		1.07	1.55		2.62	3.66
2360	1" x 4"		250	.032		1.43	1.70		3.13	4.29
2370	1" x 6"		250	.032		2.15	1.70		3.85	5.10
2380	1" x 8"		230	.035		2.86	1.85		4.71	6.10
2390	1" x 10"		230	.035		3.56	1.85		5.41	6.85
2395	1" x 12"		210	.038		4.27	2.03		6.30	7.95
2410	STK, 1" x 2"		275	.029		.44	1.55		1.99	2.96
2420	1" x 3"		275	.029		.48	1.55		2.03	3.01

06 22 13.40 Moldings, Exterior		Crew	Daily Output	Labor-Hours	Unit	Material	2020 Bare Costs Labor	Equipment	Total	Total Incl O&P
2430	1" x 4"	1 Carp	250	.032	L.F.	.85	1.70		2.55	3.66
2440	1" x 6"		250	.032		1.39	1.70		3.09	4.25
2450	1" x 8"		230	.035		2.35	1.85		4.20	5.55
2460	1" x 10"		230	.035		2.86	1.85		4.71	6.10
2470	1" x 12"		210	.038		4.44	2.03		6.47	8.15
2550	Pine, #2, 1" x 2"		275	.029		.31	1.55		1.86	2.82
2560	1" x 3"		275	.029		.45	1.55		2	2.98
2570	1" x 4"		250	.032		.57	1.70		2.27	3.35
2580	1" x 6"		250	.032		.86	1.70		2.56	3.67
2590	1" x 8"		230	.035		1.33	1.85		3.18	4.42
2600	1" x 10"		230	.035		2.04	1.85		3.89	5.20
2610	1" x 12"		210	.038		2.66	2.03		4.69	6.15
2620	Pine, D & better, 1" x 2"		275	.029		.51	1.55		2.06	3.04
2630	1" x 3"		275	.029		.73	1.55		2.28	3.28
2640	1" x 4"		250	.032		.95	1.70		2.65	3.77
2650	1" x 6"		250	.032		1.10	1.70		2.80	3.93
2660	1" x 8"		230	.035		1.64	1.85		3.49	4.76
2670	1" x 10"		230	.035		2.16	1.85		4.01	5.35
2680	1" x 12"		210	.038		2.73	2.03		4.76	6.25
2690	Redwood, clear all heart, 1" x 2"		275	.029		.66	1.55		2.21	3.21
2695	1" x 3"		275	.029		.98	1.55		2.53	3.56
2710	1" x 4"		250	.032		1.22	1.70		2.92	4.06
2715	1" x 6"		250	.032		1.83	1.70		3.53	4.74
2730	1" x 8"		230	.035		2.43	1.85		4.28	5.65
2740	1" x 10"		230	.035		4.03	1.85		5.88	7.40
2750	1" x 12"		210	.038		4.40	2.03		6.43	8.10
3500	Bellyband, pine, 11/16" x 4-1/4"		250	.032		2.82	1.70		4.52	5.80
3610	Brickmold, pine, 1-1/4" x 2"		200	.040		2.13	2.13		4.26	5.75
3620	FJP, 1-1/4" x 2"		200	.040		1.03	2.13		3.16	4.54
5100	Fascia, cedar, rough sawn, 1" x 2"		275	.029		.72	1.55		2.27	3.27
5110	1" x 3"		275	.029		1.07	1.55		2.62	3.66
5120	1" x 4"		250	.032		1.43	1.70		3.13	4.29
5200	1" x 6"		250	.032		2.15	1.70		3.85	5.10
5300	1" x 8"		230	.035		2.86	1.85		4.71	6.10
5310	1" x 10"		230	.035		3.56	1.85		5.41	6.85
5320	1" x 12"		210	.038		4.27	2.03		6.30	7.95
5400	2" x 4"		220	.036		1.26	1.93		3.19	4.48
5500	2" x 6"		220	.036		1.88	1.93		3.81	5.15
5600	2" x 8"		200	.040		2.51	2.13		4.64	6.15
5700	2" x 10"		180	.044		3.12	2.36		5.48	7.20
5800	2" x 12"		170	.047		7.45	2.50		9.95	12.15
6120	STK, 1" x 2"		275	.029		.44	1.55		1.99	2.96
6130	1" x 3"		275	.029		.48	1.55		2.03	3.01
6140	1" x 4"		250	.032		.85	1.70		2.55	3.66
6150	1" x 6"		250	.032		1.39	1.70		3.09	4.25
6160	1" x 8"		230	.035		2.35	1.85		4.20	5.55
6170	1" x 10"		230	.035		2.86	1.85		4.71	6.10
6180	1" x 12"		210	.038		4.44	2.03		6.47	8.15
6185	2" x 2"		260	.031		.76	1.64		2.40	3.46
6190	Pine, #2, 1" x 2"		275	.029		.31	1.55		1.86	2.82
6200	1" x 3"		275	.029		.45	1.55		2	2.98
6210	1" x 4"		250	.032		.57	1.70		2.27	3.35
6220	1" x 6"		250	.032		.86	1.70		2.56	3.67

06 22 13 – Standard Pattern Wood Trim

		Crew	Daily Output	Labor-Hours	Unit	Material	2020 Bare Costs Labor	Equipment	Total	Total Incl O&P
06 22 13.40 Moldings, Exterior										
6230	1" x 8"	1 Carp	230	.035	L.F.	1.33	1.85		3.18	4.42
6240	1" x 10"		230	.035		2.04	1.85		3.89	5.20
6250	1" x 12"		210	.038		2.66	2.03		4.69	6.15
6260	D & better, 1" x 2"		275	.029		.51	1.55		2.06	3.04
6270	1" x 3"		275	.029		.73	1.55		2.28	3.28
6280	1" x 4"		250	.032		.95	1.70		2.65	3.77
6290	1" x 6"		250	.032		1.10	1.70		2.80	3.93
6300	1" x 8"		230	.035		1.64	1.85		3.49	4.76
6310	1" x 10"		230	.035		2.16	1.85		4.01	5.35
6312	1" x 12"		210	.038		2.73	2.03		4.76	6.25
6330	Southern yellow, 1-1/4" x 5"		240	.033		3.12	1.77		4.89	6.25
6340	1-1/4" x 6"		240	.033		2.55	1.77		4.32	5.65
6350	1-1/4" x 8"		215	.037		3.83	1.98		5.81	7.40
6360	1-1/4" x 12"		190	.042		6.15	2.24		8.39	10.40
6370	Redwood, clear all heart, 1" x 2"		275	.029		.66	1.55		2.21	3.21
6380	1" x 3"		275	.029		1.22	1.55		2.77	3.82
6390	1" x 4"		250	.032		1.22	1.70		2.92	4.06
6400	1" x 6"		250	.032		1.83	1.70		3.53	4.74
6410	1" x 8"		230	.035		2.43	1.85		4.28	5.65
6420	1" x 10"		230	.035		4.03	1.85		5.88	7.40
6430	1" x 12"		210	.038		4.40	2.03		6.43	8.10
6440	1-1/4" x 5"		240	.033		1.92	1.77		3.69	4.95
6450	1-1/4" x 6"		240	.033		2.29	1.77		4.06	5.35
6460	1-1/4" x 8"		215	.037		3.63	1.98		5.61	7.15
6470	1-1/4" x 12"		190	.042		6.60	2.24		8.84	10.85
6580	Frieze, cedar, rough sawn, 1" x 2"		275	.029		.72	1.55		2.27	3.27
6590	1" x 3"		275	.029		1.07	1.55		2.62	3.66
6600	1" x 4"		250	.032		1.43	1.70		3.13	4.29
6610	1" x 6"		250	.032		2.15	1.70		3.85	5.10
6620	1" x 8"		250	.032		2.86	1.70		4.56	5.85
6630	1" x 10"		225	.036		3.56	1.89		5.45	6.95
6640	1" x 12"		200	.040		4.23	2.13		6.36	8.05
6650	STK, 1" x 2"		275	.029		.44	1.55		1.99	2.96
6660	1" x 3"		275	.029		.48	1.55		2.03	3.01
6670	1" x 4"		250	.032		.85	1.70		2.55	3.66
6680	1" x 6"		250	.032		1.39	1.70		3.09	4.25
6690	1" x 8"		250	.032		2.35	1.70		4.05	5.30
6700	1" x 10"		225	.036		2.86	1.89		4.75	6.20
6710	1" x 12"		200	.040		4.44	2.13		6.57	8.30
6790	Pine, #2, 1" x 2"		275	.029		.31	1.55		1.86	2.82
6800	1" x 3"		275	.029		.45	1.55		2	2.98
6810	1" x 4"		250	.032		.57	1.70		2.27	3.35
6820	1" x 6"		250	.032		.86	1.70		2.56	3.67
6830	1" x 8"		250	.032		1.33	1.70		3.03	4.18
6840	1" x 10"		225	.036		2.04	1.89		3.93	5.25
6850	1" x 12"		200	.040		2.66	2.13		4.79	6.35
6860	D & better, 1" x 2"		275	.029		.51	1.55		2.06	3.04
6870	1" x 3"		275	.029		.73	1.55		2.28	3.28
6880	1" x 4"		250	.032		.95	1.70		2.65	3.77
6890	1" x 6"		250	.032		1.10	1.70		2.80	3.93
6900	1" x 8"		250	.032		1.64	1.70		3.34	4.52
6910	1" x 10"		225	.036		2.16	1.89		4.05	5.40
6920	1" x 12"		200	.040		2.73	2.13		4.86	6.40

06 22 13.40 Moldings, Exterior	Crew	Daily Output	Labor-Hours	Unit	Material	2020 Bare Costs Labor	Equipment	Total	Total Incl O&P
6930	Redwood, clear all heart, 1" x 2"	1 Carp	275	.029	L.F.	.66	1.55		2.21	3.21
6940	1" x 3"		275	.029		.98	1.55		2.53	3.56
6950	1" x 4"		250	.032		1.22	1.70		2.92	4.06
6960	1" x 6"		250	.032		1.83	1.70		3.53	4.74
6970	1" x 8"		250	.032		2.43	1.70		4.13	5.40
6980	1" x 10"		225	.036		4.03	1.89		5.92	7.45
6990	1" x 12"		200	.040		4.40	2.13		6.53	8.25
7000	Grounds, 1" x 1", cedar, rough sawn		300	.027		.37	1.42		1.79	2.68
7010	STK		300	.027		.27	1.42		1.69	2.57
7020	Pine, #2		300	.027		.20	1.42		1.62	2.48
7030	D & better		300	.027		.31	1.42		1.73	2.62
7050	Redwood		300	.027		.40	1.42		1.82	2.71
7060	Rake/verge board, cedar, rough sawn, 1" x 2"		225	.036		.72	1.89		2.61	3.82
7070	1" x 3"		225	.036		1.07	1.89		2.96	4.21
7080	1" x 4"		200	.040		1.43	2.13		3.56	4.98
7090	1" x 6"		200	.040		2.15	2.13		4.28	5.75
7100	1" x 8"		190	.042		2.86	2.24		5.10	6.75
7110	1" x 10"		190	.042		3.56	2.24		5.80	7.50
7120	1" x 12"		180	.044		4.27	2.36		6.63	8.50
7130	STK, 1" x 2"		225	.036		.44	1.89		2.33	3.51
7140	1" x 3"		225	.036		.48	1.89		2.37	3.56
7150	1" x 4"		200	.040		.85	2.13		2.98	4.35
7160	1" x 6"		200	.040		1.39	2.13		3.52	4.94
7170	1" x 8"		190	.042		2.35	2.24		4.59	6.20
7180	1" x 10"		190	.042		2.86	2.24		5.10	6.75
7190	1" x 12"		180	.044		4.44	2.36		6.80	8.65
7200	Pine, #2, 1" x 2"		225	.036		.31	1.89		2.20	3.37
7210	1" x 3"		225	.036		.45	1.89		2.34	3.53
7220	1" x 4"		200	.040		.57	2.13		2.70	4.04
7230	1" x 6"		200	.040		.86	2.13		2.99	4.36
7240	1" x 8"		190	.042		1.33	2.24		3.57	5.05
7250	1" x 10"		190	.042		2.04	2.24		4.28	5.85
7260	1" x 12"		180	.044		2.66	2.36		5.02	6.70
7340	D & better, 1" x 2"		225	.036		.51	1.89		2.40	3.59
7350	1" x 3"		225	.036		.73	1.89		2.62	3.83
7360	1" x 4"		200	.040		.95	2.13		3.08	4.46
7370	1" x 6"		200	.040		1.10	2.13		3.23	4.62
7380	1" x 8"		190	.042		1.64	2.24		3.88	5.40
7390	1" x 10"		190	.042		2.16	2.24		4.40	5.95
7400	1" x 12"		180	.044		2.73	2.36		5.09	6.80
7410	Redwood, clear all heart, 1" x 2"		225	.036		.66	1.89		2.55	3.76
7420	1" x 3"		225	.036		.98	1.89		2.87	4.11
7430	1" x 4"		200	.040		1.22	2.13		3.35	4.75
7440	1" x 6"		200	.040		1.83	2.13		3.96	5.45
7450	1" x 8"		190	.042		2.43	2.24		4.67	6.25
7460	1" x 10"		190	.042		4.03	2.24		6.27	8.05
7470	1" x 12"		180	.044		4.40	2.36		6.76	8.60
7480	2" x 4"		200	.040		2.35	2.13		4.48	6
7490	2" x 6"		182	.044		3.19	2.34		5.53	7.25
7500	2" x 8"		165	.048		5.10	2.58		7.68	9.80
7630	Soffit, cedar, rough sawn, 1" x 2"	2 Carp	440	.036		.72	1.93		2.65	3.89
7640	1" x 3"		440	.036		1.07	1.93		3	4.28
7650	1" x 4"		420	.038		1.43	2.03		3.46	4.81

06 22 13 – Standard Pattern Wood Trim

06 22 13.40 Moldings, Exterior

		Crew	Daily Output	Labor-Hours	Unit	Material	2020 Bare Costs Labor	Equipment	Total	Total Incl O&P
7660	1" x 6"	2 Carp	420	.038	L.F.	2.15	2.03		4.18	5.60
7670	1" x 8"		420	.038		2.86	2.03		4.89	6.40
7680	1" x 10"		400	.040		3.56	2.13		5.69	7.30
7690	1" x 12"		400	.040		4.27	2.13		6.40	8.10
7700	STK, 1" x 2"		440	.036		.44	1.93		2.37	3.58
7710	1" x 3"		440	.036		.48	1.93		2.41	3.63
7720	1" x 4"		420	.038		.85	2.03		2.88	4.18
7730	1" x 6"		420	.038		1.39	2.03		3.42	4.77
7740	1" x 8"		420	.038		2.35	2.03		4.38	5.85
7750	1" x 10"		400	.040		2.86	2.13		4.99	6.55
7760	1" x 12"		400	.040		4.44	2.13		6.57	8.30
7770	Pine, #2, 1" x 2"		440	.036		.31	1.93		2.24	3.44
7780	1" x 3"		440	.036		.45	1.93		2.38	3.60
7790	1" x 4"		420	.038		.57	2.03		2.60	3.87
7800	1" x 6"		420	.038		.86	2.03		2.89	4.19
7810	1" x 8"		420	.038		1.33	2.03		3.36	4.70
7820	1" x 10"		400	.040		2.04	2.13		4.17	5.65
7830	1" x 12"		400	.040		2.66	2.13		4.79	6.35
7840	D & better, 1" x 2"		440	.036		.51	1.93		2.44	3.66
7850	1" x 3"		440	.036		.73	1.93		2.66	3.90
7860	1" x 4"		420	.038		.95	2.03		2.98	4.29
7870	1" x 6"		420	.038		1.10	2.03		3.13	4.45
7880	1" x 8"		420	.038		1.64	2.03		3.67	5.05
7890	1" x 10"		400	.040		2.16	2.13		4.29	5.80
7900	1" x 12"		400	.040		2.73	2.13		4.86	6.40
7910	Redwood, clear all heart, 1" x 2"		440	.036		.66	1.93		2.59	3.83
7920	1" x 3"		440	.036		.98	1.93		2.91	4.18
7930	1" x 4"		420	.038		1.22	2.03		3.25	4.58
7940	1" x 6"		420	.038		1.83	2.03		3.86	5.25
7950	1" x 8"		420	.038		2.43	2.03		4.46	5.90
7960	1" x 10"		400	.040		4.03	2.13		6.16	7.85
7970	1" x 12"		400	.040		4.40	2.13		6.53	8.25
8050	Trim, crown molding, pine, 11/16" x 4-1/4"	1 Carp	250	.032		4.10	1.70		5.80	7.25
8060	Back band, 11/16" x 1-1/16"		250	.032		.92	1.70		2.62	3.73
8070	Insect screen frame stock, 1-1/16" x 1-3/4"		395	.020		2.24	1.08		3.32	4.18
8080	Dentils, 2-1/2" x 2-1/2" x 4", 6" OC		30	.267		1.27	14.15		15.42	24
8100	Fluted, 5-1/2"		165	.048		5.15	2.58		7.73	9.80
8110	Stucco bead, 1-3/8" x 1-5/8"		250	.032		2.53	1.70		4.23	5.50
9000	Minimum labor/equipment charge		4	2	Job		106		106	170

06 22 13.45 Moldings, Trim

0010	**MOLDINGS, TRIM**									
0200	Astragal, stock pine, 11/16" x 1-3/4"	1 Carp	255	.031	L.F.	1.26	1.67		2.93	4.06
0250	1-5/16" x 2-3/16"		240	.033		2.65	1.77		4.42	5.75
0800	Chair rail, stock pine, 5/8" x 2-1/2"		270	.030		1.54	1.57		3.11	4.21
0900	5/8" x 3-1/2"		240	.033		2.36	1.77		4.13	5.45
1000	Closet pole, stock pine, 1-1/8" diameter		200	.040		1.19	2.13		3.32	4.72
1100	Fir, 1-5/8" diameter		200	.040		2.28	2.13		4.41	5.90
3300	Half round, stock pine, 1/4" x 1/2"		270	.030		.25	1.57		1.82	2.80
3350	1/2" x 1"		255	.031		.64	1.67		2.31	3.37
3400	Handrail, fir, single piece, stock, hardware not included									
3450	1-1/2" x 1-3/4"	1 Carp	80	.100	L.F.	2.24	5.30		7.54	10.95
3470	Pine, 1-1/2" x 1-3/4"		80	.100		2.07	5.30		7.37	10.80

06 22 13 – Standard Pattern Wood Trim

06 22 13.45 Moldings, Trim

		Crew	Daily Output	Labor-Hours	Unit	Material	2020 Bare Costs Labor	Equipment	Total	Total Incl O&P
3500	1-1/2" x 2-1/2"	1 Carp	76	.105	L.F.	2.59	5.60		8.19	11.80
3600	Lattice, stock pine, 1/4" x 1-1/8"		270	.030		.40	1.57		1.97	2.96
3700	1/4" x 1-3/4"		250	.032		.72	1.70		2.42	3.51
3800	Miscellaneous, custom, pine, 1" x 1"		270	.030		.42	1.57		1.99	2.98
3850	1" x 2"		265	.030		.84	1.60		2.44	3.49
3900	1" x 3"		240	.033		1.26	1.77		3.03	4.22
4100	Birch or oak, nominal 1" x 1"		240	.033		.40	1.77		2.17	3.28
4200	Nominal 1" x 3"		215	.037		1.20	1.98		3.18	4.49
4400	Walnut, nominal 1" x 1"		215	.037		.65	1.98		2.63	3.89
4500	Nominal 1" x 3"		200	.040		1.96	2.13		4.09	5.55
4700	Teak, nominal 1" x 1"		215	.037		2.91	1.98		4.89	6.35
4800	Nominal 1" x 3"		200	.040		8.75	2.13		10.88	13
4900	Quarter round, stock pine, 1/4" x 1/4"		275	.029		.29	1.55		1.84	2.80
4950	3/4" x 3/4"		255	.031		.54	1.67		2.21	3.26
5600	Wainscot moldings, 1-1/8" x 9/16", 2' high, minimum		76	.105	S.F.	12.70	5.60		18.30	23
5700	Maximum		65	.123	"	16.80	6.55		23.35	29
9000	Minimum labor/equipment charge		4	2	Job		106		106	170

06 22 13.50 Moldings, Window and Door

		Crew	Daily Output	Labor-Hours	Unit	Material	2020 Bare Costs Labor	Equipment	Total	Total Incl O&P
0010	**MOLDINGS, WINDOW AND DOOR**									
2800	Door moldings, stock, decorative, 1-1/8" wide, plain	1 Carp	17	.471	Set	47.50	25		72.50	92.50
2900	Detailed		17	.471	"	112	25		137	163
2960	Clear pine door jamb, no stops, 11/16" x 4-9/16"		240	.033	L.F.	5.45	1.77		7.22	8.85
3150	Door trim set, 1 head and 2 sides, pine, 2-1/2" wide		12	.667	Opng.	24	35.50		59.50	83
3170	3-1/2" wide		11	.727	"	33	38.50		71.50	98.50
3250	Glass beads, stock pine, 3/8" x 1/2"		275	.029	L.F.	.35	1.55		1.90	2.87
3270	3/8" x 7/8"		270	.030		.44	1.57		2.01	3
4850	Parting bead, stock pine, 3/8" x 3/4"		275	.029		.47	1.55		2.02	3
4870	1/2" x 3/4"		255	.031		.46	1.67		2.13	3.18
5000	Stool caps, stock pine, 11/16" x 3-1/2"		200	.040		2.56	2.13		4.69	6.25
5100	1-1/16" x 3-1/4"		150	.053		3.80	2.83		6.63	8.70
5300	Threshold, oak, 3' long, inside, 5/8" x 3-5/8"		32	.250	Ea.	12.70	13.30		26	35.50
5400	Outside, 1-1/2" x 7-5/8"		16	.500	"	48	26.50		74.50	95.50
5900	Window trim sets, including casings, header, stops,									
5910	stool and apron, 2-1/2" wide, FJP	1 Carp	13	.615	Opng.	33	32.50		65.50	89
5950	Pine		10	.800		39	42.50		81.50	111
6000	Oak		6	1.333		79.50	71		150.50	202
9000	Minimum labor/equipment charge		4	2	Job		106		106	170

06 22 13.60 Moldings, Soffits

		Crew	Daily Output	Labor-Hours	Unit	Material	2020 Bare Costs Labor	Equipment	Total	Total Incl O&P
0010	**MOLDINGS, SOFFITS** R061110-30									
0200	Soffits, pine, 1" x 4"	2 Carp	420	.038	L.F.	.53	2.03		2.56	3.82
0210	1" x 6"		420	.038		.82	2.03		2.85	4.14
0220	1" x 8"		420	.038		1.29	2.03		3.32	4.65
0230	1" x 10"		400	.040		1.97	2.13		4.10	5.60
0240	1" x 12"		400	.040		2.60	2.13		4.73	6.25
0250	STK cedar, 1" x 4"		420	.038		.81	2.03		2.84	4.13
0260	1" x 6"		420	.038		1.35	2.03		3.38	4.73
0270	1" x 8"		420	.038		2.31	2.03		4.34	5.80
0280	1" x 10"		400	.040		2.80	2.13		4.93	6.50
0290	1" x 12"		400	.040		4.38	2.13		6.51	8.25
1000	Exterior AC plywood, 1/4" thick		400	.040	S.F.	.99	2.13		3.12	4.50
1050	3/8" thick		400	.040		1.05	2.13		3.18	4.57
1100	1/2" thick		400	.040		1.18	2.13		3.31	4.71

06 22 Millwork

06 22 13 – Standard Pattern Wood Trim

06 22 13.60 Moldings, Soffits		Crew	Daily Output	Labor-Hours	Unit	Material	2020 Bare Costs Labor	Equipment	Total	Total Incl O&P
1150	Polyvinyl chloride, white, solid	1 Carp	230	.035	S.F.	2.24	1.85		4.09	5.40
1160	Perforated	"	230	.035	↓	2.24	1.85		4.09	5.40
1170	Accessories, "J" channel 5/8"	2 Carp	700	.023	L.F.	.56	1.22		1.78	2.57
9000	Minimum labor/equipment charge	"	5	3.200	Job		170		170	272

06 25 Prefinished Paneling

06 25 13 – Prefinished Hardboard Paneling

06 25 13.10 Paneling, Hardboard

0010	PANELING, HARDBOARD		Crew	Daily Output	Labor-Hours	Unit	Material	2020 Bare Costs Labor	Equipment	Total	Total Incl O&P
0050	Not incl. furring or trim, hardboard, tempered, 1/8" thick	G	2 Carp	500	.032	S.F.	.50	1.70		2.20	3.27
0100	1/4" thick	G		500	.032		.58	1.70		2.28	3.36
0300	Tempered pegboard, 1/8" thick	G		500	.032		.43	1.70		2.13	3.19
0400	1/4" thick	G		500	.032		.76	1.70		2.46	3.56
0600	Untempered hardboard, natural finish, 1/8" thick	G		500	.032		.44	1.70		2.14	3.20
0700	1/4" thick	G		500	.032		.53	1.70		2.23	3.30
0900	Untempered pegboard, 1/8" thick	G		500	.032		.49	1.70		2.19	3.26
1000	1/4" thick	G		500	.032		.55	1.70		2.25	3.33
1200	Plastic faced hardboard, 1/8" thick	G		500	.032		.63	1.70		2.33	3.41
1300	1/4" thick	G		500	.032		.87	1.70		2.57	3.68
1500	Plastic faced pegboard, 1/8" thick	G		500	.032		.68	1.70		2.38	3.47
1600	1/4" thick	G		500	.032		.83	1.70		2.53	3.63
1800	Wood grained, plain or grooved, 1/8" thick	G		500	.032		.68	1.70		2.38	3.47
1900	1/4" thick	G		425	.038	↓	1.43	2		3.43	4.78
2100	Moldings, wood grained MDF			500	.032	L.F.	.41	1.70		2.11	3.17
2200	Pine	↓		425	.038	"	1.40	2		3.40	4.75
9000	Minimum labor/equipment charge		1 Carp	2	4	Job		213		213	340

06 25 16 – Prefinished Plywood Paneling

06 25 16.10 Paneling, Plywood

0010	PANELING, PLYWOOD R061636-20	Crew	Daily Output	Labor-Hours	Unit	Material	2020 Bare Costs Labor	Equipment	Total	Total Incl O&P
2400	Plywood, prefinished, 1/4" thick, 4' x 8' sheets									
2410	with vertical grooves. Birch faced, economy	2 Carp	500	.032	S.F.	1.69	1.70		3.39	4.58
2420	Average		420	.038		1.24	2.03		3.27	4.60
2430	Custom		350	.046		1.08	2.43		3.51	5.10
2600	Mahogany, African		400	.040		2.43	2.13		4.56	6.10
2700	Philippine (Lauan)		500	.032		.57	1.70		2.27	3.35
2900	Oak		500	.032		1.36	1.70		3.06	4.22
3000	Cherry		400	.040		2.03	2.13		4.16	5.65
3200	Rosewood		320	.050		3.08	2.66		5.74	7.65
3400	Teak		400	.040		3.26	2.13		5.39	7
3600	Chestnut		375	.043		5.45	2.27		7.72	9.65
3800	Pecan		400	.040		2.57	2.13		4.70	6.25
3900	Walnut, average		500	.032		2.64	1.70		4.34	5.60
3950	Custom		400	.040		2.38	2.13		4.51	6.05
4000	Plywood, prefinished, 3/4" thick, stock grades, economy		320	.050		1.38	2.66		4.04	5.80
4100	Average		224	.071		5.20	3.80		9	11.80
4300	Architectural grade, custom		224	.071		5.30	3.80		9.10	11.95
4400	Luxury		160	.100		5.55	5.30		10.85	14.65
4600	Plywood, "A" face, birch, VC, 1/2" thick, natural		450	.036		1.36	1.89		3.25	4.53
4700	Select		450	.036		1.95	1.89		3.84	5.20
4900	Veneer core, 3/4" thick, natural		320	.050		2.29	2.66		4.95	6.80

For customer support on your Commercial Renovation Costs with RSMeans data, call 800.448.8182.

171

06 25 Prefinished Paneling

06 25 16 – Prefinished Plywood Paneling

06 25 16.10 Paneling, Plywood	Crew	Daily Output	Labor-Hours	Unit	Material	2020 Bare Costs Labor	Equipment	Total	Total Incl O&P
5000	Select	2 Carp	320	.050	S.F.	2.68	2.66		5.34	7.20
5200	Lumber core, 3/4" thick, natural		320	.050		3.05	2.66		5.71	7.60
5500	Plywood, knotty pine, 1/4" thick, A2 grade		450	.036		1.62	1.89		3.51	4.81
5600	A3 grade		450	.036		2.25	1.89		4.14	5.50
5800	3/4" thick, veneer core, A2 grade		320	.050		2.42	2.66		5.08	6.90
5900	A3 grade		320	.050		2.49	2.66		5.15	7
6100	Aromatic cedar, 1/4" thick, plywood		400	.040		2.26	2.13		4.39	5.90
6200	1/4" thick, particle board		400	.040		1.16	2.13		3.29	4.69
9000	Minimum labor/equipment charge	1 Carp	2	4	Job		213		213	340

06 25 26 – Panel System

06 25 26.10 Panel Systems

		Crew	Daily Output	Labor-Hours	Unit	Material	2020 Bare Costs Labor	Equipment	Total	Total Incl O&P
0010	**PANEL SYSTEMS**									
0100	Raised panel, eng. wood core w/wood veneer, std., paint grade	2 Carp	300	.053	S.F.	13.10	2.83		15.93	19
0110	Oak veneer		300	.053		26.50	2.83		29.33	33.50
0120	Maple veneer		300	.053		30	2.83		32.83	37.50
0130	Cherry veneer		300	.053		40	2.83		42.83	49
0300	Class I fire rated, paint grade		300	.053		14	2.83		16.83	19.95
0310	Oak veneer		300	.053		30.50	2.83		33.33	38.50
0320	Maple veneer		300	.053		34.50	2.83		37.33	42.50
0330	Cherry veneer		300	.053		50	2.83		52.83	59.50
0510	Beadboard, 5/8" MDF, standard, primed		300	.053		9.70	2.83		12.53	15.20
0520	Oak veneer, unfinished		300	.053		16	2.83		18.83	22
0530	Maple veneer, unfinished		300	.053		16.50	2.83		19.33	22.50
0610	Rustic paneling, 5/8" MDF, standard, maple veneer, unfinished		300	.053		18.20	2.83		21.03	24.50

06 26 Board Paneling

06 26 13 – Profile Board Paneling

06 26 13.10 Paneling, Boards

		Crew	Daily Output	Labor-Hours	Unit	Material	2020 Bare Costs Labor	Equipment	Total	Total Incl O&P
0010	**PANELING, BOARDS**									
6400	Wood board paneling, 3/4" thick, knotty pine	2 Carp	300	.053	S.F.	2.07	2.83		4.90	6.80
6500	Rough sawn cedar		300	.053		3.48	2.83		6.31	8.35
6700	Redwood, clear, 1" x 4" boards		300	.053		5.05	2.83		7.88	10.10
6900	Aromatic cedar, closet lining, boards		275	.058		2.28	3.09		5.37	7.45
9000	Minimum labor/equipment charge	1 Carp	2	4	Job		213		213	340

06 43 Wood Stairs and Railings

06 43 13 – Wood Stairs

06 43 13.20 Prefabricated Wood Stairs

		Crew	Daily Output	Labor-Hours	Unit	Material	2020 Bare Costs Labor	Equipment	Total	Total Incl O&P
0010	**PREFABRICATED WOOD STAIRS**									
0100	Box stairs, prefabricated, 3'-0" wide									
0110	Oak treads, up to 14 risers	2 Carp	39	.410	Riser	98	22		120	143
0600	With pine treads for carpet, up to 14 risers	"	39	.410	"	63.50	22		85.50	105
1100	For 4' wide stairs, add				Flight	25%				
1550	Stairs, prefabricated stair handrail with balusters	1 Carp	30	.267	L.F.	83	14.15		97.15	114
1700	Basement stairs, prefabricated, pine treads									
1710	Pine risers, 3' wide, up to 14 risers	2 Carp	52	.308	Riser	63.50	16.35		79.85	96
4000	Residential, wood, oak treads, prefabricated		1.50	10.667	Flight	1,275	565		1,840	2,300
4200	Built in place		.44	36.364	"	2,250	1,925		4,175	5,575
4400	Spiral, oak, 4'-6" diameter, unfinished, prefabricated,									

For customer support on your Commercial Renovation Costs with RSMeans data, call 800.448.8182.

06 43 Wood Stairs and Railings

06 43 13 – Wood Stairs

06 43 13.20 Prefabricated Wood Stairs	Crew	Daily Output	Labor-Hours	Unit	Material	2020 Bare Costs Labor	Equipment	Total	Total Incl O&P
4500	incl. railing, 9' high	2 Carp	1.50	10.667	Flight	2,625	565		3,190	3,775
9000	Minimum labor/equipment charge	"	3	5.333	Job		283		283	455

06 43 13.40 Wood Stair Parts

		Crew	Daily Output	Labor-Hours	Unit	Material	2020 Bare Costs Labor	Equipment	Total	Total Incl O&P
0010	**WOOD STAIR PARTS**									
0020	Pin top balusters, 1-1/4", oak, 34"	1 Carp	96	.083	Ea.	5.65	4.43		10.08	13.30
0030	38"		96	.083		5.75	4.43		10.18	13.45
0040	42"		96	.083		6.85	4.43		11.28	14.60
0050	Poplar, 34"		96	.083		3.34	4.43		7.77	10.75
0060	38"		96	.083		5.60	4.43		10.03	13.25
0070	42"		96	.083		5.95	4.43		10.38	13.65
0080	Maple, 34"		96	.083		6	4.43		10.43	13.70
0090	38"		96	.083		5.55	4.43		9.98	13.20
0100	42"		96	.083		7.85	4.43		12.28	15.75
0130	Primed, 34"		96	.083		3.67	4.43		8.10	11.15
0140	38"		96	.083		4.20	4.43		8.63	11.70
0150	42"		96	.083		4.84	4.43		9.27	12.40
0180	Box top balusters, 1-1/4", oak, 34"		60	.133		7.45	7.10		14.55	19.55
0190	38"		60	.133		11.75	7.10		18.85	24.50
0200	42"		60	.133		12.85	7.10		19.95	25.50
0210	Poplar, 34"		60	.133		7.65	7.10		14.75	19.80
0220	38"		60	.133		8.25	7.10		15.35	20.50
0230	42"		60	.133		9.10	7.10		16.20	21.50
0240	Maple, 34"		60	.133		9.75	7.10		16.85	22
0250	38"		60	.133		10.65	7.10		17.75	23
0260	42"		60	.133		11.75	7.10		18.85	24.50
0290	Primed, 34"		60	.133		8.20	7.10		15.30	20.50
0300	38"		60	.133		12.05	7.10		19.15	24.50
0310	42"		60	.133		11.30	7.10		18.40	24
0340	Square balusters, cut from lineal stock, pine, 1-1/16" x 1-1/16"		180	.044	L.F.	1.29	2.36		3.65	5.20
0350	1-5/16" x 1-5/16"		180	.044		1.99	2.36		4.35	5.95
0360	1-5/8" x 1-5/8"		180	.044		2.48	2.36		4.84	6.50
0370	Turned newel, oak, 3-1/2" square, 48" high		8	1	Ea.	94	53		147	188
0380	62" high		8	1		99.50	53		152.50	194
0390	Poplar, 3-1/2" square, 48" high		8	1		46	53		99	136
0400	62" high		8	1		56.50	53		109.50	148
0410	Maple, 3-1/2" square, 48" high		8	1		79	53		132	172
0420	62" high		8	1		79	53		132	172
0430	Square newel, oak, 3-1/2" square, 48" high		8	1		56.50	53		109.50	148
0440	58" high		8	1		65	53		118	157
0450	Poplar, 3-1/2" square, 48" high		8	1		42	53		95	131
0460	58" high		8	1		62	53		115	153
0470	Maple, 3" square, 48" high		8	1		53	53		106	144
0480	58" high		8	1		67	53		120	159
0490	Railings, oak, economy		96	.083	L.F.	9	4.43		13.43	17
0500	Average		96	.083		15.10	4.43		19.53	24
0510	Custom		96	.083		18.75	4.43		23.18	27.50
0520	Maple, economy		96	.083		11.70	4.43		16.13	20
0530	Average		96	.083		13.25	4.43		17.68	21.50
0540	Custom		96	.083		22	4.43		26.43	31.50
0550	Oak, for bending rail, economy		48	.167		26.50	8.85		35.35	43
0560	Average		48	.167		29.50	8.85		38.35	46.50
0570	Custom		48	.167		33	8.85		41.85	50

For customer support on your Commercial Renovation Costs with RSMeans data, call 800.448.8182.

173

06 43 13.40 Wood Stair Parts	Crew	Daily Output	Labor-Hours	Unit	Material	2020 Bare Costs Labor	Equipment	Total	Total Incl O&P
0580 Maple, for bending rail, economy	1 Carp	48	.167	L.F.	29	8.85		37.85	46
0590 Average		48	.167		31.50	8.85		40.35	48.50
0600 Custom		48	.167		36.50	8.85		45.35	54
0610 Risers, oak, 3/4" x 8", 36" long		80	.100	Ea.	12.85	5.30		18.15	22.50
0620 42" long		70	.114		15	6.05		21.05	26.50
0630 48" long		63	.127		17.10	6.75		23.85	29.50
0640 54" long		56	.143		19.25	7.60		26.85	33
0650 60" long		50	.160		21.50	8.50		30	37
0660 72" long		42	.190		25.50	10.10		35.60	44.50
0670 Poplar, 3/4" x 8", 36" long		80	.100		13.10	5.30		18.40	23
0680 42" long		71	.113		15.25	6		21.25	26.50
0690 48" long		63	.127		17.45	6.75		24.20	30
0700 54" long		56	.143		19.60	7.60		27.20	33.50
0710 60" long		50	.160		22	8.50		30.50	37.50
0720 72" long		42	.190		26	10.10		36.10	45
0730 Pine, 1" x 8", 36" long		80	.100		3.86	5.30		9.16	12.75
0740 42" long		70	.114		4.50	6.05		10.55	14.70
0750 48" long		63	.127		5.15	6.75		11.90	16.45
0760 54" long		56	.143		5.80	7.60		13.40	18.50
0770 60" long		50	.160		6.45	8.50		14.95	20.50
0780 72" long		42	.190		7.70	10.10		17.80	24.50
0790 Treads, oak, no returns, 1-1/32" x 11-1/2" x 36" long		32	.250		31.50	13.30		44.80	56
0800 42" long		32	.250		36.50	13.30		49.80	62
0810 48" long		32	.250		42	13.30		55.30	67.50
0820 54" long		32	.250		47	13.30		60.30	73.50
0830 60" long		32	.250		52.50	13.30		65.80	79
0840 72" long		32	.250		63	13.30		76.30	90.50
0850 Mitred return one end, 1-1/32" x 11-1/2" x 36" long		24	.333		38.50	17.70		56.20	70.50
0860 42" long		24	.333		45	17.70		62.70	77.50
0870 48" long		24	.333		51	17.70		68.70	85
0880 54" long		24	.333		57.50	17.70		75.20	92
0890 60" long		24	.333		64	17.70		81.70	99
0900 72" long		24	.333		76.50	17.70		94.20	113
0910 Mitred return two ends, 1-1/32" x 11-1/2" x 36" long		12	.667		45.50	35.50		81	107
0920 42" long		12	.667		53	35.50		88.50	116
0930 48" long		12	.667		61	35.50		96.50	124
0940 54" long		12	.667		68.50	35.50		104	133
0950 60" long		12	.667		76	35.50		111.50	141
0960 72" long		12	.667		91.50	35.50		127	157
0970 Starting step, oak, 48", bullnose		8	1		176	53		229	279
0980 Double end bullnose		8	1		268	53		321	380
1030 Skirt board, pine, 1" x 10"		55	.145	L.F.	1.97	7.75		9.72	14.55
1040 1" x 12"		52	.154	"	2.60	8.20		10.80	15.95
1050 Oak landing tread, 1-1/16" thick		54	.148	S.F.	7.90	7.85		15.75	21.50
1060 Oak cove molding		96	.083	L.F.	.95	4.43		5.38	8.15
1070 Oak stringer molding		96	.083	"	3.68	4.43		8.11	11.15
1090 Rail bolt, 5/16" x 3-1/2"		48	.167	Ea.	3.55	8.85		12.40	18.10
1100 5/16" x 4-1/2"		48	.167		3.04	8.85		11.89	17.55
1120 Newel post anchor		16	.500		12.85	26.50		39.35	56.50
1130 Tapered plug, 1/2"		240	.033		.95	1.77		2.72	3.89
1140 1"		240	.033		.98	1.77		2.75	3.92
9000 Minimum labor/equipment charge		3	2.667	Job		142		142	227

06 43 Wood Stairs and Railings

06 43 16 – Wood Railings

06 43 16.10 Wood Handrails and Railings	Crew	Daily Output	Labor-Hours	Unit	Material	2020 Bare Costs Labor	Equipment	Total	Total Incl O&P
0010 **WOOD HANDRAILS AND RAILINGS**									
0020 Custom design, architectural grade, hardwood, plain	1 Carp	38	.211	L.F.	12.60	11.20		23.80	32
0100 Shaped		30	.267		56	14.15		70.15	84
0300 Stock interior railing with spindles 4" OC, 4' long		40	.200		46.50	10.65		57.15	68
0400 8' long		48	.167	↓	46.50	8.85		55.35	65
9000 Minimum labor/equipment charge	↓	3	2.667	Job		142		142	227

06 44 Ornamental Woodwork

06 44 19 – Wood Grilles

06 44 19.10 Grilles

	Crew	Daily Output	Labor-Hours	Unit	Material	Labor	Equipment	Total	Total Incl O&P
0010 **GRILLES** and panels, hardwood, sanded									
0020 2' x 4' to 4' x 8', custom designs, unfinished, economy	1 Carp	38	.211	S.F.	61	11.20		72.20	85
0050 Average		30	.267		73.50	14.15		87.65	104
0100 Custom		19	.421	↓	70.50	22.50		93	114
9000 Minimum labor/equipment charge	↓	2	4	Job		213		213	340

06 44 33 – Wood Mantels

06 44 33.10 Fireplace Mantels

	Crew	Daily Output	Labor-Hours	Unit	Material	Labor	Equipment	Total	Total Incl O&P
0010 **FIREPLACE MANTELS**									
0015 6" molding, 6' x 3'-6" opening, plain, paint grade	1 Carp	5	1.600	Opng.	490	85		575	675
0100 Ornate, oak		5	1.600		640	85		725	840
0300 Prefabricated pine, colonial type, stock, deluxe		2	4		1,900	213		2,113	2,450
0400 Economy		3	2.667	↓	815	142		957	1,125
9000 Minimum labor/equipment charge	↓	3	2.667	Job		142		142	227

06 44 33.20 Fireplace Mantel Beam

	Crew	Daily Output	Labor-Hours	Unit	Material	Labor	Equipment	Total	Total Incl O&P
0010 **FIREPLACE MANTEL BEAM**									
0020 Rough texture wood, 4" x 8"	1 Carp	36	.222	L.F.	8.80	11.80		20.60	28.50
0100 4" x 10"		35	.229	"	12.45	12.15		24.60	33
0300 Laminated hardwood, 2-1/4" x 10-1/2" wide, 6' long		5	1.600	Ea.	109	85		194	256
0400 8' long		5	1.600	"	162	85		247	315
0600 Brackets for above, rough sawn		12	.667	Pr.	11.50	35.50		47	69.50
0700 Laminated		12	.667	"	22.50	35.50		58	81.50
9000 Minimum labor/equipment charge	↓	4	2	Job		106		106	170

06 44 39 – Wood Posts and Columns

06 44 39.10 Decorative Beams

	Crew	Daily Output	Labor-Hours	Unit	Material	Labor	Equipment	Total	Total Incl O&P
0010 **DECORATIVE BEAMS**									
0020 Rough sawn cedar, non-load bearing, 4" x 4"	2 Carp	180	.089	L.F.	1.61	4.72		6.33	9.30
0100 4" x 6"		170	.094		1.84	5		6.84	10
0200 4" x 8"		160	.100		2.44	5.30		7.74	11.20
0300 4" x 10"		150	.107		4.66	5.65		10.31	14.25
0400 4" x 12"		140	.114		5.50	6.05		11.55	15.80
0500 8" x 8"	↓	130	.123	↓	5	6.55		11.55	16
9000 Minimum labor/equipment charge	1 Carp	3	2.667	Job		142		142	227

06 44 39.20 Columns

	Crew	Daily Output	Labor-Hours	Unit	Material	Labor	Equipment	Total	Total Incl O&P
0010 **COLUMNS**									
0050 Aluminum, round colonial, 6" diameter	2 Carp	80	.200	V.L.F.	21.50	10.65		32.15	41
0100 8" diameter		62.25	.257		20.50	13.65		34.15	44.50
0200 10" diameter		55	.291		24.50	15.45		39.95	52
0250 Fir, stock units, hollow round, 6" diameter		80	.200		31	10.65		41.65	51
0300 8" diameter	↓	80	.200	↓	36.50	10.65		47.15	57.50

06 44 Ornamental Woodwork

06 44 39 – Wood Posts and Columns

06 44 39.20 Columns		Crew	Daily Output	Labor-Hours	Unit	Material	2020 Bare Costs Labor	Equipment	Total	Total Incl O&P
0350	10" diameter	2 Carp	70	.229	V.L.F.	45.50	12.15		57.65	69.50
0360	12" diameter		65	.246		65	13.10		78.10	92.50
0400	Solid turned, to 8' high, 3-1/2" diameter		80	.200		11.30	10.65		21.95	29.50
0500	4-1/2" diameter		75	.213		14.30	11.35		25.65	34
0600	5-1/2" diameter		70	.229		19.75	12.15		31.90	41
0800	Square columns, built-up, 5" x 5"		65	.246		34.50	13.10		47.60	59
0900	Solid, 3-1/2" x 3-1/2"		130	.123		11.95	6.55		18.50	23.50
1600	Hemlock, tapered, T&G, 12" diam., 10' high		100	.160		48	8.50		56.50	66
1700	16' high		65	.246		82.50	13.10		95.60	112
1900	14" diameter, 10' high		100	.160		102	8.50		110.50	126
2000	18' high		65	.246		102	13.10		115.10	133
2200	18" diameter, 12' high		65	.246		204	13.10		217.10	246
2300	20' high		50	.320		125	17		142	166
2500	20" diameter, 14' high		40	.400		201	21.50		222.50	255
2600	20' high		35	.457		188	24.50		212.50	245
2800	For flat pilasters, deduct					33%				
3000	For splitting into halves, add				Ea.	130			130	143
4000	Rough sawn cedar posts, 4" x 4"	2 Carp	250	.064	V.L.F.	4.61	3.40		8.01	10.50
4100	4" x 6"		235	.068		8.55	3.62		12.17	15.20
4200	6" x 6"		220	.073		13.10	3.87		16.97	20.50
4300	8" x 8"		200	.080		21	4.25		25.25	30
9000	Minimum labor/equipment charge	1 Carp	3	2.667	Job		142		142	227

06 48 Wood Frames

06 48 13 – Exterior Wood Door Frames

06 48 13.10 Exterior Wood Door Frames and Accessories

06 48 13.10		Crew	Daily Output	Labor-Hours	Unit	Material	2020 Bare Costs Labor	Equipment	Total	Total Incl O&P
0010	**EXTERIOR WOOD DOOR FRAMES AND ACCESSORIES**									
0400	Exterior frame, incl. ext. trim, pine, 5/4 x 4-9/16" deep	2 Carp	375	.043	L.F.	6.90	2.27		9.17	11.25
0420	5-3/16" deep		375	.043		8.05	2.27		10.32	12.50
0440	6-9/16" deep		375	.043		10	2.27		12.27	14.65
0600	Oak, 5/4 x 4-9/16" deep		350	.046		12.30	2.43		14.73	17.40
0620	5-3/16" deep		350	.046		13.50	2.43		15.93	18.75
0640	6-9/16" deep		350	.046		21	2.43		23.43	27
1000	Sills, 8/4 x 8" deep, oak, no horns		100	.160		8.35	8.50		16.85	23
1020	2" horns		100	.160		21.50	8.50		30	37
1040	3" horns		100	.160		21.50	8.50		30	37
1100	8/4 x 10" deep, oak, no horns		90	.178		6.85	9.45		16.30	22.50
1120	2" horns		90	.178		27.50	9.45		36.95	45.50
1140	3" horns		90	.178		27.50	9.45		36.95	45.50
2000	Wood frame & trim, ext., colonial, 3' opng., fluted pilasters, flat head		22	.727	Ea.	510	38.50		548.50	620
2010	Dentil head		21	.762		620	40.50		660.50	750
2020	Ram's head		20	.800		735	42.50		777.50	875
2100	5'-4" opening, in-swing, fluted pilasters, flat head		17	.941		485	50		535	615
2120	Ram's head		15	1.067		1,325	56.50		1,381.50	1,575
2140	Out-swing, fluted pilasters, flat head		17	.941		480	50		530	610
2160	Ram's head		15	1.067		1,550	56.50		1,606.50	1,800
2400	6'-0" opening, in-swing, fluted pilasters, flat head		16	1		490	53		543	620
2420	Ram's head		10	1.600		1,525	85		1,610	1,800
2460	Out-swing, fluted pilasters, flat head		16	1		485	53		538	620
2480	Ram's head		10	1.600		1,525	85		1,610	1,800
2600	For two sidelights, flat head, add		30	.533	Opng.	330	28.50		358.50	410

06 48 Wood Frames

06 48 13 – Exterior Wood Door Frames

06 48 13.10 Exterior Wood Door Frames and Accessories	Crew	Daily Output	Labor-Hours	Unit	Material	2020 Bare Costs Labor	Equipment	Total	Total Incl O&P
2620	Ram's head, add	2 Carp	20	.800	Opng.	865	42.50		907.50	1,025
2700	Custom birch frame, 3'-0" opening		16	1	Ea.	246	53		299	355
2750	6'-0" opng.		16	1		395	53		448	520
2900	Exterior, modern, plain trim, 3' opng., in-swing, FJP		26	.615		55	32.50		87.50	113
2920	Fir		24	.667		59	35.50		94.50	122
2940	Oak		22	.727		68.50	38.50		107	138

06 48 16 – Interior Wood Door Frames

06 48 16.10 Interior Wood Door Jamb and Frames

		Crew	Daily Output	Labor-Hours	Unit	Material	2020 Bare Costs Labor	Equipment	Total	Total Incl O&P
0010	INTERIOR WOOD DOOR JAMB AND FRAMES									
3000	Interior frame, pine, 11/16" x 3-5/8" deep	2 Carp	375	.043	L.F.	4.27	2.27		6.54	8.35
3020	4-9/16" deep		375	.043		5.40	2.27		7.67	9.60
3200	Oak, 11/16" x 3-5/8" deep		350	.046		2.53	2.43		4.96	6.65
3220	4-9/16" deep		350	.046		10.45	2.43		12.88	15.35
3240	5-3/16" deep		350	.046		18.60	2.43		21.03	24.50
3400	Walnut, 11/16" x 3-5/8" deep		350	.046		9.75	2.43		12.18	14.60
3420	4-9/16" deep		350	.046		12.10	2.43		14.53	17.20
3440	5-3/16" deep		350	.046		9.95	2.43		12.38	14.85
3600	Pocket door frame		16	1	Ea.	113	53		166	209
3800	Threshold, oak, 5/8" x 3-5/8" deep		200	.080	L.F.	3.73	4.25		7.98	10.90
3820	4-5/8" deep		190	.084		4.25	4.48		8.73	11.85
3840	5-5/8" deep		180	.089		6.80	4.72		11.52	15.05
9000	Minimum labor/equipment charge	1 Carp	4	2	Job		106		106	170

06 49 Wood Screens and Exterior Wood Shutters

06 49 19 – Exterior Wood Shutters

06 49 19.10 Shutters, Exterior

		Crew	Daily Output	Labor-Hours	Unit	Material	2020 Bare Costs Labor	Equipment	Total	Total Incl O&P
0010	SHUTTERS, EXTERIOR									
0012	Aluminum, louvered, 1'-4" wide, 3'-0" long	1 Carp	10	.800	Pr.	219	42.50		261.50	310
0200	4'-0" long		10	.800		265	42.50		307.50	360
0300	5'-4" long		10	.800		315	42.50		357.50	415
0400	6'-8" long		9	.889		390	47		437	505
1000	Pine, louvered, primed, each 1'-2" wide, 3'-3" long		10	.800		263	42.50		305.50	355
1100	4'-7" long		10	.800		305	42.50		347.50	405
1250	Each 1'-4" wide, 3'-0" long		10	.800		297	42.50		339.50	395
1350	5'-3" long		10	.800		410	42.50		452.50	520
1500	Each 1'-6" wide, 3'-3" long		10	.800		294	42.50		336.50	395
1600	4'-7" long		10	.800		405	42.50		447.50	515
1620	Cedar, louvered, 1'-2" wide, 5'-7" long		10	.800		360	42.50		402.50	465
1630	Each 1'-4" wide, 2'-2" long		10	.800		195	42.50		237.50	282
1640	3'-0" long		10	.800		257	42.50		299.50	350
1650	3'-3" long		10	.800		270	42.50		312.50	365
1660	3'-11" long		10	.800		310	42.50		352.50	410
1670	4'-3" long		10	.800		345	42.50		387.50	450
1680	5'-3" long		10	.800		410	42.50		452.50	525
1690	5'-11" long		10	.800		455	42.50		497.50	570
1700	Door blinds, 6'-9" long, each 1'-3" wide		9	.889		500	47		547	625
1710	1'-6" wide		9	.889		455	47		502	580
1720	Cedar, solid raised panel, each 1'-4" wide, 3'-3" long		10	.800		375	42.50		417.50	480
1730	3'-11" long		10	.800		375	42.50		417.50	485
1740	4'-3" long		10	.800		370	42.50		412.50	475

06 49 Wood Screens and Exterior Wood Shutters

06 49 19 – Exterior Wood Shutters

06 49 19.10 Shutters, Exterior	Crew	Daily Output	Labor-Hours	Unit	Material	2020 Bare Costs Labor	Equipment	Total	Total Incl O&P
1750	4'-7" long	1 Carp	10	.800	Pr.	385	42.50		427.50	495
1760	4'-11" long		10	.800		420	42.50		462.50	530
1770	5'-11" long		10	.800		590	42.50		632.50	720
1800	Door blinds, 6'-9" long, each 1'-3" wide		9	.889		515	47		562	640
1900	1'-6" wide		9	.889		770	47		817	925
2500	Polystyrene, solid raised panel, each 1'-4" wide, 3'-3" long		10	.800		89	42.50		131.50	166
2600	3'-11" long		10	.800		93.50	42.50		136	171
2700	4'-7" long		10	.800		106	42.50		148.50	185
2800	5'-3" long		10	.800		120	42.50		162.50	200
2900	6'-8" long		9	.889		153	47		200	245
4500	Polystyrene, louvered, each 1'-2" wide, 3'-3" long		10	.800		37.50	42.50		80	109
4600	4'-7" long		10	.800		50	42.50		92.50	123
4750	5'-3" long		10	.800		58	42.50		100.50	132
4850	6'-8" long		9	.889		72.50	47		119.50	156
6000	Vinyl, louvered, each 1'-2" x 4'-7" long		10	.800		67	42.50		109.50	142
6200	Each 1'-4" x 6'-8" long		9	.889	↓	89	47		136	174
9000	Minimum labor/equipment charge	↓	4	2	Job		106		106	170

06 51 Structural Plastic Shapes and Plates

06 51 13 – Plastic Lumber

06 51 13.10 Recycled Plastic Lumber

			Crew	Daily Output	Labor-Hours	Unit	Material	2020 Bare Costs Labor	Equipment	Total	Total Incl O&P
0010	**RECYCLED PLASTIC LUMBER**										
4000	Sheeting, recycled plastic, black or white, 4' x 8' x 1/8"	G	2 Carp	1100	.015	S.F.	1.05	.77		1.82	2.40
4010	4' x 8' x 3/16"	G		1100	.015		1.46	.77		2.23	2.85
4020	4' x 8' x 1/4"	G		950	.017		1.62	.90		2.52	3.21
4030	4' x 8' x 3/8"	G		950	.017		2.80	.90		3.70	4.51
4040	4' x 8' x 1/2"	G		900	.018		3.74	.95		4.69	5.60
4050	4' x 8' x 5/8"	G		900	.018		6.95	.95		7.90	9.15
4060	4' x 8' x 3/4"	G	↓	850	.019	↓	8.10	1		9.10	10.50
4070	Add for colors	G				Ea.	5%				
8500	100% recycled plastic, var colors, NLB, 2" x 2"	G				L.F.	1.89			1.89	2.08
8510	2" x 4"	G					3.96			3.96	4.36
8520	2" x 6"	G					6.20			6.20	6.85
8530	2" x 8"	G					8.80			8.80	9.65
8540	2" x 10"	G					12.45			12.45	13.70
8550	5/4" x 4"	G					4.32			4.32	4.75
8560	5/4" x 6"	G					5.90			5.90	6.50
8570	1" x 6"	G					2.90			2.90	3.19
8580	1/2" x 8"	G					3.41			3.41	3.75
8590	2" x 10" T&G	G					12.60			12.60	13.85
8600	3" x 10" T&G	G					19.80			19.80	22
8610	Add for premium colors	G				↓	20%				

06 51 13.12 Structural Plastic Lumber

			Crew	Daily Output	Labor-Hours	Unit	Material	2020 Bare Costs Labor	Equipment	Total	Total Incl O&P
0010	**STRUCTURAL PLASTIC LUMBER**										
1320	Plastic lumber, posts or columns, 4" x 4"		2 Carp	390	.041	L.F.	10.95	2.18		13.13	15.55
1325	4" x 6"			275	.058		14.45	3.09		17.54	21
1330	4" x 8"			220	.073		21	3.87		24.87	29.50
1340	Girder, single, 4" x 4"			675	.024		10.95	1.26		12.21	14.05
1345	4" x 6"			600	.027		14.45	1.42		15.87	18.15
1350	4" x 8"			525	.030		21	1.62	·	22.62	26
1352	Double, 2" x 4"			625	.026		9.55	1.36		10.91	12.75

06 51 Structural Plastic Shapes and Plates

06 51 13 – Plastic Lumber

06 51 13.12 Structural Plastic Lumber		Crew	Daily Output	Labor-Hours	Unit	Material	2020 Bare Costs Labor	Equipment	Total	Total Incl O&P
1354	2" x 6"	2 Carp	600	.027	L.F.	10.45	1.42		11.87	13.75
1356	2" x 8"		575	.028		18.30	1.48		19.78	22.50
1358	2" x 10"		550	.029		22.50	1.55		24.05	27.50
1360	2" x 12"		525	.030		26	1.62		27.62	31
1362	Triple, 2" x 4"		575	.028		14.35	1.48		15.83	18.15
1364	2" x 6"		550	.029		15.70	1.55		17.25	19.75
1366	2" x 8"		525	.030		27.50	1.62		29.12	32.50
1368	2" x 10"		500	.032		34	1.70		35.70	40
1370	2" x 12"		475	.034		38.50	1.79		40.29	45.50
1372	Ledger, bolted 4' OC, 2" x 4"		400	.040		4.94	2.13		7.07	8.85
1374	2" x 6"		550	.029		5.30	1.55		6.85	8.35
1376	2" x 8"		390	.041		9.30	2.18		11.48	13.70
1378	2" x 10"		385	.042		11.40	2.21		13.61	16.10
1380	2" x 12"		380	.042		13	2.24		15.24	17.90
1382	Joists, 2" x 4"		1250	.013		4.79	.68		5.47	6.35
1384	2" x 6"		1250	.013		5.25	.68		5.93	6.85
1386	2" x 8"		1100	.015		9.15	.77		9.92	11.30
1388	2" x 10"		500	.032		11.45	1.70		13.15	15.25
1390	2" x 12"		875	.018		12.90	.97		13.87	15.75
1392	Railings and trim, 5/4" x 4"	1 Carp	300	.027		4.90	1.42		6.32	7.65
1394	2" x 2"		300	.027		2.20	1.42		3.62	4.69
1396	2" x 4"		300	.027		4.76	1.42		6.18	7.50
1398	2" x 6"		300	.027		5.20	1.42		6.62	7.95

06 52 Plastic Structural Assemblies

06 52 10 – Fiberglass Structural Assemblies

06 52 10.30 Fiberglass Grating

		Crew	Daily Output	Labor-Hours	Unit	Material	2020 Bare Costs Labor	Equipment	Total	Total Incl O&P
0010	**FIBERGLASS GRATING**									
0100	Molded, green (for mod. corrosive environment)									
0140	1" x 4" mesh, 1" thick	2 Sswk	400	.040	S.F.	7.30	2.31		9.61	11.85
0180	1-1/2" square mesh, 1" thick		400	.040		16.40	2.31		18.71	22
0220	1-1/4" thick		400	.040		9.10	2.31		11.41	13.85
0260	1-1/2" thick		400	.040		24	2.31		26.31	30.50
0300	2" square mesh, 2" thick		320	.050		27	2.88		29.88	34.50
1000	Orange (for highly corrosive environment)									
1040	1" x 4" mesh, 1" thick	2 Sswk	400	.040	S.F.	17.85	2.31		20.16	23.50
1080	1-1/2" square mesh, 1" thick		400	.040		19.70	2.31		22.01	25.50
1120	1-1/4" thick		400	.040		18.55	2.31		20.86	24.50
1160	1-1/2" thick		400	.040		21	2.31		23.31	27
1200	2" square mesh, 2" thick		320	.050		21	2.88		23.88	27.50
3000	Pultruded, green (for mod. corrosive environment)									
3040	1" OC bar spacing, 1" thick	2 Sswk	400	.040	S.F.	14.35	2.31		16.66	19.55
3080	1-1/2" thick		320	.050		16.10	2.88		18.98	22.50
3120	1-1/2" OC bar spacing, 1" thick		400	.040		14.95	2.31		17.26	20
3160	1-1/2" thick		400	.040		21.50	2.31		23.81	27.50
4000	Grating support legs, fixed height, no base				Ea.	45.50			45.50	50
4040	With base					38.50			38.50	42.50
4080	Adjustable to 60"					58			58	63.50

179

06 52 Plastic Structural Assemblies

06 52 10 – Fiberglass Structural Assemblies

06 52 10.40 Fiberglass Floor Grating	Crew	Daily Output	Labor-Hours	Unit	Material	2020 Bare Costs Labor	Equipment	Total	Total Incl O&P
0010 **FIBERGLASS FLOOR GRATING**									
0100 Reinforced polyester, fire retardant, 1" x 4" grid, 1" thick	E-4	510	.063	S.F.	14.55	3.65	.29	18.49	22.50
0200 1-1/2" x 6" mesh, 1-1/2" thick		500	.064		16.80	3.72	.29	20.81	25
0300 With grit surface, 1-1/2" x 6" grid, 1-1/2" thick		500	.064		17.15	3.72	.29	21.16	25.50

06 63 Plastic Railings

06 63 10 – Plastic (PVC) Railings

06 63 10.10 Plastic Railings

		Crew	Daily Output	Labor-Hours	Unit	Material	2020 Bare Costs Labor	Equipment	Total	Total Incl O&P
0010	**PLASTIC RAILINGS**									
0100	Horizontal PVC handrail with balusters, 3-1/2" wide, 36" high	1 Carp	96	.083	L.F.	30	4.43		34.43	40
0150	42" high		96	.083		29	4.43		33.43	38.50
0200	Angled PVC handrail with balusters, 3-1/2" wide, 36" high		72	.111		20.50	5.90		26.40	32
0250	42" high		72	.111		30.50	5.90		36.40	43
0300	Post sleeve for 4 x 4 post		96	.083		16.45	4.43		20.88	25
0400	Post cap for 4 x 4 post, flat profile		48	.167	Ea.	15.55	8.85		24.40	31.50
0450	Newel post style profile		48	.167		28.50	8.85		37.35	45.50
0500	Raised corbeled profile		48	.167		35	8.85		43.85	53
0550	Post base trim for 4 x 4 post		96	.083		20.50	4.43		24.93	29.50

06 65 Plastic Trim

06 65 10 – PVC Trim

06 65 10.10 PVC Trim, Exterior

		Crew	Daily Output	Labor-Hours	Unit	Material	2020 Bare Costs Labor	Equipment	Total	Total Incl O&P
0010	**PVC TRIM, EXTERIOR**									
0100	Cornerboards, 5/4" x 6" x 6"	1 Carp	240	.033	L.F.	8.50	1.77		10.27	12.20
0110	Door/window casing, 1" x 4"		200	.040		2.09	2.13		4.22	5.70
0120	1" x 6"		200	.040		2.60	2.13		4.73	6.25
0130	1" x 8"		195	.041		3.43	2.18		5.61	7.25
0140	1" x 10"		195	.041		4.50	2.18		6.68	8.45
0150	1" x 12"		190	.042		5.10	2.24		7.34	9.20
0160	5/4" x 4"		195	.041		2.48	2.18		4.66	6.20
0170	5/4" x 6"		195	.041		3.74	2.18		5.92	7.60
0180	5/4" x 8"		190	.042		4.75	2.24		6.99	8.80
0190	5/4" x 10"		190	.042		5.95	2.24		8.19	10.15
0200	5/4" x 12"		185	.043		6.75	2.30		9.05	11.10
0210	Fascia, 1" x 4"		250	.032		2.09	1.70		3.79	5
0220	1" x 6"		250	.032		2.60	1.70		4.30	5.60
0230	1" x 8"		225	.036		3.43	1.89		5.32	6.80
0240	1" x 10"		225	.036		4.50	1.89		6.39	8
0250	1" x 12"		200	.040		5.10	2.13		7.23	9
0260	5/4" x 4"		240	.033		2.48	1.77		4.25	5.55
0270	5/4" x 6"		240	.033		3.74	1.77		5.51	6.95
0280	5/4" x 8"		215	.037		4.75	1.98		6.73	8.35
0290	5/4" x 10"		215	.037		5.95	1.98		7.93	9.70
0300	5/4" x 12"		190	.042		6.75	2.24		8.99	11
0310	Frieze, 1" x 4"		250	.032		2.09	1.70		3.79	5
0320	1" x 6"		250	.032		2.60	1.70		4.30	5.60
0330	1" x 8"		225	.036		3.43	1.89		5.32	6.80
0340	1" x 10"		225	.036		4.50	1.89		6.39	8
0350	1" x 12"		200	.040		5.10	2.13		7.23	9

06 65 Plastic Trim

06 65 10 – PVC Trim

06 65 10.10 PVC Trim, Exterior

		Crew	Daily Output	Labor-Hours	Unit	Material	2020 Bare Costs Labor	Equipment	Total	Total Incl O&P
0360	5/4" x 4"	1 Carp	240	.033	L.F.	2.48	1.77		4.25	5.55
0370	5/4" x 6"		240	.033		3.74	1.77		5.51	6.95
0380	5/4" x 8"		215	.037		4.75	1.98		6.73	8.35
0390	5/4" x 10"		215	.037		5.95	1.98		7.93	9.70
0400	5/4" x 12"		190	.042		6.75	2.24		8.99	11
0410	Rake, 1" x 4"		200	.040		2.09	2.13		4.22	5.70
0420	1" x 6"		200	.040		2.60	2.13		4.73	6.25
0430	1" x 8"		190	.042		3.43	2.24		5.67	7.35
0440	1" x 10"		190	.042		4.50	2.24		6.74	8.55
0450	1" x 12"		180	.044		5.10	2.36		7.46	9.40
0460	5/4" x 4"		195	.041		2.48	2.18		4.66	6.20
0470	5/4" x 6"		195	.041		3.74	2.18		5.92	7.60
0480	5/4" x 8"		185	.043		4.75	2.30		7.05	8.90
0490	5/4" x 10"		185	.043		5.95	2.30		8.25	10.25
0500	5/4" x 12"		175	.046		6.75	2.43		9.18	11.30
0510	Rake trim, 1" x 4"		225	.036		2.09	1.89		3.98	5.35
0520	1" x 6"		225	.036		2.60	1.89		4.49	5.90
0560	5/4" x 4"		220	.036		2.48	1.93		4.41	5.85
0570	5/4" x 6"		220	.036		3.74	1.93		5.67	7.20
0610	Soffit, 1" x 4"	2 Carp	420	.038		2.09	2.03		4.12	5.55
0620	1" x 6"		420	.038		2.60	2.03		4.63	6.10
0630	1" x 8"		420	.038		3.43	2.03		5.46	7
0640	1" x 10"		400	.040		4.50	2.13		6.63	8.35
0650	1" x 12"		400	.040		5.10	2.13		7.23	9
0660	5/4" x 4"		410	.039		2.48	2.07		4.55	6.05
0670	5/4" x 6"		410	.039		3.74	2.07		5.81	7.45
0680	5/4" x 8"		410	.039		4.75	2.07		6.82	8.50
0690	5/4" x 10"		390	.041		5.95	2.18		8.13	10.05
0700	5/4" x 12"		390	.041		6.75	2.18		8.93	10.90

06 73 Composite Decking

06 73 13 – Composite Decking

06 73 13.10 Woodgrained Composite Decking

		Crew	Daily Output	Labor-Hours	Unit	Material	2020 Bare Costs Labor	Equipment	Total	Total Incl O&P
0010	**WOODGRAINED COMPOSITE DECKING**									
0100	Woodgrained composite decking, 1" x 6"	2 Carp	640	.025	L.F.	4.20	1.33		5.53	6.75
0110	Grooved edge		660	.024		4.39	1.29		5.68	6.90
0120	2" x 6"		640	.025		4.02	1.33		5.35	6.55
0130	Encased, 1" x 6"		640	.025		4.71	1.33		6.04	7.35
0140	Grooved edge		660	.024		4.66	1.29		5.95	7.15
0150	2" x 6"		640	.025		5.45	1.33		6.78	8.15

06 81 Composite Railings

06 81 10 – Encased Railings

06 81 10.10 Encased Composite Railings	Crew	Daily Output	Labor-Hours	Unit	Material	2020 Bare Costs Labor	Equipment	Total	Total Incl O&P
0010 **ENCASED COMPOSITE RAILINGS**									
0100 Encased composite railing, 6' long, 36" high, incl. balusters	1 Carp	16	.500	Ea.	146	26.50		172.50	203
0110 42" high, incl. balusters		16	.500		225	26.50		251.50	290
0120 8' long, 36" high, incl. balusters		12	.667		170	35.50		205.50	243
0130 42" high, incl. balusters		12	.667		178	35.50		213.50	253
0140 Accessories, post sleeve, 4" x 4", 39" long		32	.250		30	13.30		43.30	54.50
0150 96" long		24	.333		76	17.70		93.70	113
0160 6" x 6", 39" long		32	.250		53.50	13.30		66.80	80
0170 96" long		24	.333		159	17.70		176.70	204
0180 Accessories, post skirt, 4" x 4"		96	.083		5.75	4.43		10.18	13.40
0190 6" x 6"		96	.083		8.05	4.43		12.48	15.95
0200 Post cap, 4" x 4", flat		48	.167		9.15	8.85		18	24.50
0210 Pyramid		48	.167		7.95	8.85		16.80	23
0220 Post cap, 6" x 6", flat		48	.167		14.25	8.85		23.10	30
0230 Pyramid		48	.167		10.85	8.85		19.70	26

Estimating Tips
07 10 00 Dampproofing and Waterproofing
- Be sure of the job specifications before pricing this subdivision. The difference in cost between waterproofing and dampproofing can be great. Waterproofing will hold back standing water. Dampproofing prevents the transmission of water vapor. Also included in this section are vapor retarding membranes.

07 20 00 Thermal Protection
- Insulation and fireproofing products are measured by area, thickness, volume, or R-value. Specifications may give only what the specific R-value should be in a certain situation. The estimator may need to choose the type of insulation to meet that R-value.

07 30 00 Steep Slope Roofing
07 40 00 Roofing and Siding Panels
- Many roofing and siding products are bought and sold by the square. One square is equal to an area that measures 100 square feet.

  This simple change in unit of measure could create a large error if the estimator is not observant. Accessories necessary for a complete installation must be figured into any calculations for both material and labor.

07 50 00 Membrane Roofing
07 60 00 Flashing and Sheet Metal
07 70 00 Roofing and Wall Specialties and Accessories
- The items in these subdivisions compose a roofing system. No one component completes the installation, and all must be estimated. Built-up or single-ply membrane roofing systems are made up of many products and installation trades. Wood blocking at roof perimeters or penetrations, parapet coverings, reglets, roof drains, gutters, downspouts, sheet metal flashing, skylights, smoke vents, and roof hatches all need to be considered along with the roofing material. Several different installation trades will need to work together on the roofing system. Inherent difficulties in the scheduling and coordination of various trades must be accounted for when estimating labor costs.

07 90 00 Joint Protection
- To complete the weather-tight shell, the sealants and caulkings must be estimated. Where different materials meet—at expansion joints, at flashing penetrations, and at hundreds of other locations throughout a construction project—caulking and sealants provide another line of defense against water penetration. Often, an entire system is based on the proper location and placement of caulking or sealants. The detailed drawings that are included as part of a set of architectural plans show typical locations for these materials. When caulking or sealants are shown at typical locations, this means the estimator must include them for all the locations where this detail is applicable. Be careful to keep different types of sealants separate, and remember to consider backer rods and primers if necessary.

Reference Numbers
Reference numbers are shown at the beginning of some major classifications. These numbers refer to related items in the Reference Section. The reference information may be an estimating procedure, an alternate pricing method, or technical information.

Note: Not all subdivisions listed here necessarily appear. ∎

Same Data. Simplified.
Enjoy the convenience and efficiency of accessing your costs anywhere:
- **Skip the multiplier** by setting your location
- **Quickly search,** edit, favorite and share costs
- **Stay on top of price changes** with automatic updates

Discover more at rsmeans.com/online

07 01 50 – Maintenance of Membrane Roofing

07 01 50.10 Roof Coatings

		Crew	Daily Output	Labor-Hours	Unit	Material	2020 Bare Costs Labor	2020 Bare Costs Equipment	Total	Total Incl O&P
0010	**ROOF COATINGS**									
0012	Asphalt, brush grade, material only				Gal.	9.50			9.50	10.45
0200	Asphalt base, fibered aluminum coating G					9.40			9.40	10.35
0300	Asphalt primer, 5 gal.				↓	7.15			7.15	7.85
0600	Coal tar pitch, 200 lb. barrels				Ton	1,225			1,225	1,350
0700	Tar roof cement, 5 gal. lots				Gal.	16.15			16.15	17.75
0800	Glass fibered roof & patching cement, 5 gal.				"	9.40			9.40	10.35
0900	Reinforcing glass membrane, 450 S.F./roll				Ea.	61			61	67
1000	Neoprene roof coating, 5 gal., 2 gal./sq.				Gal.	32.50			32.50	35.50
1100	Roof patch & flashing cement, 5 gal.					8.90			8.90	9.75
1200	Roof resaturant, glass fibered, 3 gal./sq.					9.35			9.35	10.25
1600	Reflective roof coating, white, elastomeric, approx. 50 S.F./gal. G				↓	17.35			17.35	19.10

07 01 90 – Maintenance of Joint Protection

07 01 90.81 Joint Sealant Replacement

		Crew	Daily Output	Labor-Hours	Unit	Material	2020 Bare Costs Labor	2020 Bare Costs Equipment	Total	Total Incl O&P
0010	**JOINT SEALANT REPLACEMENT**									
0050	Control joints in concrete floors/slabs									
0100	Option 1 for joints with hard dry sealant									
0110	Step 1: Sawcut to remove 95% of old sealant									
0112	1/4" wide x 1/2" deep, with single saw blade	C-27	4800	.003	L.F.	.02	.17	.02	.21	.31
0114	3/8" wide x 3/4" deep, with single saw blade		4000	.004		.03	.20	.03	.26	.37
0116	1/2" wide x 1" deep, with double saw blades		3600	.004		.06	.22	.03	.31	.44
0118	3/4" wide x 1-1/2" deep, with double saw blades	↓	3200	.005		.11	.25	.03	.39	.55
0120	Step 2: Water blast joint faces and edges	C-29	2500	.003			.13	.04	.17	.26
0130	Step 3: Air blast joint faces and edges	C-28	2000	.004			.20	.02	.22	.33
0140	Step 4: Sand blast joint faces and edges	E-11	2000	.016			.75	.15	.90	1.40
0150	Step 5: Air blast joint faces and edges	C-28	2000	.004	↓		.20	.02	.22	.33
0200	Option 2 for joints with soft pliable sealant									
0210	Step 1: Plow joint with rectangular blade	B-62	2600	.009	L.F.		.42	.07	.49	.75
0220	Step 2: Sawcut to re-face joint faces									
0222	1/4" wide x 1/2" deep, with single saw blade	C-27	2400	.007	L.F.	.02	.33	.05	.40	.59
0224	3/8" wide x 3/4" deep, with single saw blade		2000	.008		.04	.40	.06	.50	.73
0226	1/2" wide x 1" deep, with double saw blades		1800	.009		.07	.44	.06	.57	.85
0228	3/4" wide x 1-1/2" deep, with double saw blades	↓	1600	.010		.15	.50	.07	.72	1.03
0230	Step 3: Water blast joint faces and edges	C-29	2500	.003			.13	.04	.17	.26
0240	Step 4: Air blast joint faces and edges	C-28	2000	.004			.20	.02	.22	.33
0250	Step 5: Sand blast joint faces and edges	E-11	2000	.016			.75	.15	.90	1.40
0260	Step 6: Air blast joint faces and edges	C-28	2000	.004	↓		.20	.02	.22	.33
0290	For saw cutting new control joints, see Section 03 15 16.20									
8910	For backer rod, see Section 07 91 23.10									
8920	For joint sealant, see Section 03 15 16.30 or 07 92 13.20									

184

For customer support on your Commercial Renovation Costs with RSMeans data, call 800.448.8182.

07 05 05 – Selective Demolition for Thermal and Moisture Protection

07 05 05.10 Selective Demo., Thermal and Moist. Protection	Crew	Daily Output	Labor-Hours	Unit	Material	2020 Bare Costs Labor	2020 Bare Costs Equipment	Total	Total Incl O&P
0010 **SELECTIVE DEMO., THERMAL AND MOISTURE PROTECTION**									
0020 Caulking/sealant, to 1" x 1" joint R024119-10	1 Clab	600	.013	L.F.		.56		.56	.90
0120 Downspouts, including hangers		350	.023	"		.96		.96	1.54
0220 Flashing, sheet metal		290	.028	S.F.		1.16		1.16	1.86
0420 Gutters, aluminum or wood, edge hung		240	.033	L.F.		1.40		1.40	2.25
0520 Built-in		100	.080	"		3.37		3.37	5.40
0620 Insulation, air/vapor barrier		3500	.002	S.F.		.10		.10	.15
0670 Batts or blankets		1400	.006	C.F.		.24		.24	.39
0720 Foamed or sprayed in place	2 Clab	1000	.016	B.F.		.67		.67	1.08
0770 Loose fitting	1 Clab	3000	.003	C.F.		.11		.11	.18
0870 Rigid board		3450	.002	B.F.		.10		.10	.16
1120 Roll roofing, cold adhesive		12	.667	Sq.		28		28	45
1170 Roof accessories, adjustable metal chimney flashing		9	.889	Ea.		37.50		37.50	60
1325 Plumbing vent flashing		32	.250	"		10.55		10.55	16.85
1375 Ridge vent strip, aluminum		310	.026	L.F.		1.09		1.09	1.74
1620 Skylight to 10 S.F.		8	1	Ea.		42		42	67.50
2120 Roof edge, aluminum soffit and fascia		570	.014	L.F.		.59		.59	.95
2170 Concrete coping, up to 12" wide	2 Clab	160	.100			4.21		4.21	6.75
2220 Drip edge	1 Clab	1000	.008			.34		.34	.54
2270 Gravel stop		950	.008			.35		.35	.57
2370 Sheet metal coping, up to 12" wide		240	.033			1.40		1.40	2.25
2470 Roof insulation board, over 2" thick	B-2	7800	.005	B.F.		.22		.22	.35
2520 Up to 2" thick	"	3900	.010	S.F.		.44		.44	.70
2620 Roof ventilation, louvered gable vent	1 Clab	16	.500	Ea.		21		21	33.50
2670 Remove, roof hatch	G-3	15	2.133			111		111	176
2675 Rafter vents	1 Clab	960	.008			.35		.35	.56
2720 Soffit vent and/or fascia vent		575	.014	L.F.		.59		.59	.94
2775 Soffit vent strip, aluminum, 3" to 4" wide		160	.050			2.11		2.11	3.37
2820 Roofing accessories, shingle moulding, to 1" x 4"		1600	.005			.21		.21	.34
2870 Cant strip	B-2	2000	.020			.85		.85	1.36
2920 Concrete block walkway	1 Clab	230	.035			1.46		1.46	2.35
3070 Roofing, felt paper, #15		70	.114	Sq.		4.81		4.81	7.70
3125 #30 felt		30	.267	"		11.25		11.25	18
3170 Asphalt shingles, 1 layer	B-2	3500	.011	S.F.		.49		.49	.78
3180 2 layers		1750	.023	"		.97		.97	1.56
3370 Modified bitumen		26	1.538	Sq.		65.50		65.50	105
3420 Built-up, no gravel, 3 ply		25	1.600			68		68	109
3470 4 ply		21	1.905			81		81	130
3620 5 ply		1600	.025	S.F.		1.06		1.06	1.70
3720 5 ply, with gravel		890	.045			1.91		1.91	3.06
3725 Loose gravel removal		5000	.008			.34		.34	.54
3730 Embedded gravel removal		2000	.020			.85		.85	1.36
3870 Fiberglass sheet		1200	.033			1.42		1.42	2.27
4120 Slate shingles		1900	.021			.89		.89	1.43
4170 Ridge shingles, clay or slate		2000	.020	L.F.		.85		.85	1.36
4320 Single ply membrane, attached at seams		52	.769	Sq.		32.50		32.50	52.50
4370 Ballasted		75	.533			22.50		22.50	36.50
4420 Fully adhered		39	1.026			43.50		43.50	70
4550 Roof hatch, 2'-6" x 3'-0"	1 Clab	10	.800	Ea.		33.50		33.50	54
4670 Wood shingles	B-2	2200	.018	S.F.		.77		.77	1.24
4820 Sheet metal roofing	"	2150	.019			.79		.79	1.27
4970 Siding, horizontal wood clapboards	1 Clab	380	.021			.89		.89	1.42
5025 Exterior insulation finish system	"	120	.067			2.81		2.81	4.50

07 05 Common Work Results for Thermal and Moisture Protection

07 05 05 – Selective Demolition for Thermal and Moisture Protection

07 05 05.10 Selective Demo., Thermal and Moist. Protection	Crew	Daily Output	Labor-Hours	Unit	Material	2020 Bare Costs Labor	Equipment	Total	Total Incl O&P
5070	Tempered hardboard, remove and reset	1 Carp	380	.021	S.F.		1.12		1.12	1.79
5120	Tempered hardboard sheet siding	"	375	.021	↓		1.13		1.13	1.82
5170	Metal, corner strips	1 Clab	850	.009	L.F.		.40		.40	.63
5225	Horizontal strips		444	.018	S.F.		.76		.76	1.22
5320	Vertical strips		400	.020			.84		.84	1.35
5520	Wood shingles		350	.023			.96		.96	1.54
5620	Stucco siding		360	.022			.94		.94	1.50
5670	Textured plywood		725	.011			.46		.46	.74
5720	Vinyl siding		510	.016	↓		.66		.66	1.06
5770	Corner strips		900	.009	L.F.		.37		.37	.60
5870	Wood, boards, vertical		400	.020	S.F.		.84		.84	1.35
5880	Steel siding, corrugated/ribbed	↓	402.50	.020	"		.84		.84	1.34
5920	Waterproofing, protection/drain board	2 Clab	3900	.004	B.F.		.17		.17	.28
5970	Over 1/2" thick		1750	.009	S.F.		.38		.38	.62
6020	To 1/2" thick	↓	2000	.008	"		.34		.34	.54
9000	Minimum labor/equipment charge	1 Clab	2	4	Job		168		168	270

07 11 Dampproofing

07 11 13 – Bituminous Dampproofing

07 11 13.10 Bituminous Asphalt Coating

		Crew	Daily Output	Labor-Hours	Unit	Material	Labor	Equipment	Total	Total Incl O&P
0010	**BITUMINOUS ASPHALT COATING**									
0030	Brushed on, below grade, 1 coat	1 Rofc	665	.012	S.F.	.24	.56		.80	1.23
0100	2 coat		500	.016		.48	.74		1.22	1.81
0300	Sprayed on, below grade, 1 coat		830	.010		.24	.45		.69	1.04
0400	2 coat	↓	500	.016	↓	.46	.74		1.20	1.80
0500	Asphalt coating, with fibers				Gal.	9.40			9.40	10.35
0600	Troweled on, asphalt with fibers, 1/16" thick	1 Rofc	500	.016	S.F.	.41	.74		1.15	1.74
0700	1/8" thick		400	.020		.72	.92		1.64	2.41
1000	1/2" thick		350	.023	↓	2.35	1.06		3.41	4.43
9000	Minimum labor/equipment charge		3	2.667	Job		123		123	216

07 11 16 – Cementitious Dampproofing

07 11 16.20 Cementitious Parging

		Crew	Daily Output	Labor-Hours	Unit	Material	Labor	Equipment	Total	Total Incl O&P
0010	**CEMENTITIOUS PARGING**									
0020	Portland cement, 2 coats, 1/2" thick	D-1	250	.064	S.F.	.42	3		3.42	5.30
0100	Waterproofed Portland cement, 1/2" thick, 2 coats	"	250	.064	"	7.25	3		10.25	12.85

07 12 Built-up Bituminous Waterproofing

07 12 13 – Built-Up Asphalt Waterproofing

07 12 13.20 Membrane Waterproofing

		Crew	Daily Output	Labor-Hours	Unit	Material	Labor	Equipment	Total	Total Incl O&P
0010	**MEMBRANE WATERPROOFING**									
0012	On slabs, 1 ply, felt, mopped	G-1	3000	.019	S.F.	.40	.81	.19	1.40	2.06
0015	On walls, 1 ply, felt, mopped		3000	.019		.40	.81	.19	1.40	2.06
0100	On slabs, 1 ply, glass fiber fabric, mopped		2100	.027		.45	1.15	.27	1.87	2.80
0105	On walls, 1 ply, glass fiber fabric, mopped		2100	.027		.45	1.15	.27	1.87	2.80
0300	On slabs, 2 ply, felt, mopped		2500	.022		.80	.97	.23	2	2.82
0305	On walls, 2 ply, felt, mopped		2500	.022		.80	.97	.23	2	2.82
0400	On slabs, 2 ply, glass fiber fabric, mopped		1650	.034		.99	1.47	.34	2.80	4.03
0405	On walls, 2 ply, glass fiber fabric, mopped		1650	.034		.99	1.47	.34	2.80	4.03
0600	On slabs, 3 ply, felt, mopped		2100	.027	↓	1.20	1.15	.27	2.62	3.63

07 12 Built-up Bituminous Waterproofing

07 12 13 – Built-Up Asphalt Waterproofing

07 12 13.20 Membrane Waterproofing	Crew	Daily Output	Labor-Hours	Unit	Material	2020 Bare Costs Labor	Equipment	Total	Total Incl O&P
0605	On walls, 3 ply, felt, mopped	G-1	2100	.027	S.F.	1.20	1.15	.27	2.62	3.63
0700	On slabs, 3 ply, glass fiber fabric, mopped		1550	.036		1.35	1.56	.37	3.28	4.61
0705	On walls, 3 ply, glass fiber fabric, mopped	↓	1550	.036		1.35	1.56	.37	3.28	4.61
0710	Asphaltic hardboard protection board, 1/8" thick	2 Rofc	500	.032		.76	1.48		2.24	3.43
0715	1/4" thick		450	.036		1.51	1.64		3.15	4.54
1000	EPS membrane protection board, 1/4" thick		3500	.005		.33	.21		.54	.73
1050	3/8" thick		3500	.005		.36	.21		.57	.77
1060	1/2" thick		3500	.005	↓	.40	.21		.61	.81
1070	Fiberglass fabric, black, 20/10 mesh		116	.138	Sq.	15.10	6.35		21.45	28
9000	Minimum labor/equipment charge	↓	2	8	Job		370		370	645

07 13 Sheet Waterproofing

07 13 53 – Elastomeric Sheet Waterproofing

07 13 53.10 Elastomeric Sheet Waterproofing and Access.

		Crew	Daily Output	Labor-Hours	Unit	Material	2020 Bare Costs Labor	Equipment	Total	Total Incl O&P
0010	**ELASTOMERIC SHEET WATERPROOFING AND ACCESS.**									
0090	EPDM, plain, 45 mils thick	2 Rofc	580	.028	S.F.	1.50	1.27		2.77	3.88
0100	60 mils thick		570	.028		1.72	1.30		3.02	4.16
0300	Nylon reinforced sheets, 45 mils thick		580	.028		1.62	1.27		2.89	4.01
0400	60 mils thick	↓	570	.028	↓	1.69	1.30		2.99	4.13
0600	Vulcanizing splicing tape for above, 2" wide				C.L.F.	61.50			61.50	67.50
0700	4" wide				"	124			124	136
0900	Adhesive, bonding, 60 S.F./gal.				Gal.	28			28	31
1000	Splicing, 75 S.F./gal.				"	37.50			37.50	41.50
1200	Neoprene sheets, plain, 45 mils thick	2 Rofc	580	.028	S.F.	2.17	1.27		3.44	4.62
1300	60 mils thick		570	.028		2.26	1.30		3.56	4.76
1500	Nylon reinforced, 45 mils thick		580	.028		2.40	1.27		3.67	4.87
1600	60 mils thick		570	.028		2.56	1.30		3.86	5.10
1800	120 mils thick	↓	500	.032	↓	5.25	1.48		6.73	8.40
1900	Adhesive, splicing, 150 S.F./gal. per coat				Gal.	37.50			37.50	41.50
2100	Fiberglass reinforced, fluid applied, 1/8" thick	2 Rofc	500	.032	S.F.	1.72	1.48		3.20	4.48
2200	Polyethylene and rubberized asphalt sheets, 60 mils thick		550	.029		.96	1.34		2.30	3.41
2400	Polyvinyl chloride sheets, plain, 10 mils thick		580	.028		.15	1.27		1.42	2.40
2500	20 mils thick		570	.028		.22	1.30		1.52	2.51
2700	30 mils thick	↓	560	.029	↓	.24	1.32		1.56	2.57
3000	Adhesives, trowel grade, 40-100 S.F./gal.				Gal.	23.50			23.50	26
3100	Brush grade, 100-250 S.F./gal.				"	21			21	23.50
3300	Bitumen modified polyurethane, fluid applied, 55 mils thick	2 Rofc	665	.024	S.F.	1.12	1.11		2.23	3.18
9000	Minimum labor/equipment charge	"	2	8	Job		370		370	645

07 16 Cementitious and Reactive Waterproofing

07 16 16 – Crystalline Waterproofing

07 16 16.20 Cementitious Waterproofing

		Crew	Daily Output	Labor-Hours	Unit	Material	2020 Bare Costs Labor	Equipment	Total	Total Incl O&P
0010	**CEMENTITIOUS WATERPROOFING**									
0020	1/8" application, sprayed on	G-2A	1000	.024	S.F.	.71	.98	.63	2.32	3.15
0050	4 coat cementitious metallic slurry	1 Cefi	1.20	6.667	C.S.F.	75	335		410	610

For customer support on your Commercial Renovation Costs with RSMeans data, call 800.448.8182.

187

07 19 Water Repellents

07 19 19 – Silicone Water Repellents

07 19 19.10 Silicone Based Water Repellents		Crew	Daily Output	Labor-Hours	Unit	Material	2020 Bare Costs Labor	Equipment	Total	Total Incl O&P
0010	**SILICONE BASED WATER REPELLENTS**									
0020	Water base liquid, roller applied	2 Rofc	7000	.002	S.F.	.43	.11		.54	.66
0200	Silicone or stearate, sprayed on CMU, 1 coat	1 Rofc	4000	.002		.40	.09		.49	.60
0300	2 coats		3000	.003		.79	.12		.91	1.09
9000	Minimum labor/equipment charge		3	2.667	Job		123		123	216

07 21 Thermal Insulation

07 21 13 – Board Insulation

07 21 13.10 Rigid Insulation

07 21 13.10 Rigid Insulation			Crew	Daily Output	Labor-Hours	Unit	Material	2020 Bare Costs Labor	Equipment	Total	Total Incl O&P
0010	**RIGID INSULATION**, for walls										
0040	Fiberglass, 1.5#/C.F., unfaced, 1" thick, R4.1	G	1 Carp	1000	.008	S.F.	.33	.43		.76	1.04
0060	1-1/2" thick, R6.2	G		1000	.008		.43	.43		.86	1.15
0080	2" thick, R8.3	G		1000	.008		.49	.43		.92	1.22
0120	3" thick, R12.4	G		800	.010		.56	.53		1.09	1.47
0370	3#/C.F., unfaced, 1" thick, R4.3	G		1000	.008		.57	.43		1	1.31
0390	1-1/2" thick, R6.5	G		1000	.008		.78	.43		1.21	1.54
0400	2" thick, R8.7	G		890	.009		1.09	.48		1.57	1.97
0420	2-1/2" thick, R10.9	G		800	.010		1.12	.53		1.65	2.08
0440	3" thick, R13	G		800	.010		1.65	.53		2.18	2.67
0520	Foil faced, 1" thick, R4.3	G		1000	.008		.85	.43		1.28	1.62
0540	1-1/2" thick, R6.5	G		1000	.008		1.28	.43		1.71	2.09
0560	2" thick, R8.7	G		890	.009		1.62	.48		2.10	2.55
0580	2-1/2" thick, R10.9	G		800	.010		1.91	.53		2.44	2.95
0600	3" thick, R13	G		800	.010		2.01	.53		2.54	3.06
1600	Isocyanurate, 4' x 8' sheet, foil faced, both sides										
1610	1/2" thick	G	1 Carp	800	.010	S.F.	.31	.53		.84	1.19
1620	5/8" thick	G		800	.010		.62	.53		1.15	1.53
1630	3/4" thick	G		800	.010		.43	.53		.96	1.32
1640	1" thick	G		800	.010		.62	.53		1.15	1.53
1650	1-1/2" thick	G		730	.011		.76	.58		1.34	1.77
1660	2" thick	G		730	.011		.74	.58		1.32	1.74
1670	3" thick	G		730	.011		3.11	.58		3.69	4.35
1680	4" thick	G		730	.011		2.58	.58		3.16	3.77
1700	Perlite, 1" thick, R2.77	G		800	.010		.47	.53		1	1.37
1750	2" thick, R5.55	G		730	.011		.86	.58		1.44	1.88
1900	Extruded polystyrene, 25 psi compressive strength, 1" thick, R5	G		800	.010		.62	.53		1.15	1.53
1940	2" thick, R10	G		730	.011		1.22	.58		1.80	2.27
1960	3" thick, R15	G		730	.011		1.67	.58		2.25	2.77
2100	Expanded polystyrene, 1" thick, R3.85	G		800	.010		.29	.53		.82	1.17
2120	2" thick, R7.69	G		730	.011		.58	.58		1.16	1.57
2140	3" thick, R11.49	G		730	.011		.87	.58		1.45	1.89
9000	Minimum labor/equipment charge			4	2	Job		106		106	170

07 21 13.13 Foam Board Insulation

07 21 13.13 Foam Board Insulation			Crew	Daily Output	Labor-Hours	Unit	Material	2020 Bare Costs Labor	Equipment	Total	Total Incl O&P
0010	**FOAM BOARD INSULATION**										
0600	Polystyrene, expanded, 1" thick, R4	G	1 Carp	680	.012	S.F.	.29	.63		.92	1.32
0700	2" thick, R8	G		675	.012	"	.58	.63		1.21	1.65
9000	Minimum labor/equipment charge			4	2	Job		106		106	170

For customer support on your Commercial Renovation Costs with RSMeans data, call 800.448.8182.

07 21 Thermal Insulation

07 21 16 – Blanket Insulation

07 21 16.10 Blanket Insulation for Floors/Ceilings

07 21 16.10 Blanket Insulation for Floors/Ceilings		Crew	Daily Output	Labor-Hours	Unit	Material	2020 Bare Costs Labor	Equipment	Total	Total Incl O&P
0010	**BLANKET INSULATION FOR FLOORS/CEILINGS**									
0020	Including spring type wire fasteners									
2000	Fiberglass, blankets or batts, paper or foil backing									
2100	3-1/2" thick, R13	G	1 Carp	700	.011	S.F.	.41	.61		1.02	1.42
2150	6-1/4" thick, R19	G		600	.013		.51	.71		1.22	1.70
2210	9-1/2" thick, R30	G		500	.016		.86	.85		1.71	2.31
2220	12" thick, R38	G		475	.017		1.04	.90		1.94	2.57
3000	Unfaced, 3-1/2" thick, R13	G		600	.013		.33	.71		1.04	1.50
3010	6-1/4" thick, R19	G		500	.016		.39	.85		1.24	1.79
3020	9-1/2" thick, R30	G		450	.018		.68	.95		1.63	2.26
3030	12" thick, R38	G		425	.019		.77	1		1.77	2.45
9000	Minimum labor/equipment charge			4	2	Job		106		106	170

07 21 16.20 Blanket Insulation for Walls

07 21 16.20 Blanket Insulation for Walls		Crew	Daily Output	Labor-Hours	Unit	Material	2020 Bare Costs Labor	Equipment	Total	Total Incl O&P
0010	**BLANKET INSULATION FOR WALLS**									
0020	Kraft faced fiberglass, 3-1/2" thick, R11, 15" wide	G	1 Carp	1350	.006	S.F.	.30	.32		.62	.83
0030	23" wide	G		1600	.005		.30	.27		.57	.76
0060	R13, 11" wide	G		1150	.007		.34	.37		.71	.96
0080	15" wide	G		1350	.006		.34	.32		.66	.87
0100	23" wide	G		1600	.005		.34	.27		.61	.80
0110	R15, 11" wide	G		1150	.007		.55	.37		.92	1.20
0120	15" wide	G		1350	.006		.55	.32		.87	1.11
0130	23" wide	G		1600	.005		.55	.27		.82	1.04
0140	6" thick, R19, 11" wide	G		1150	.007		.44	.37		.81	1.07
0160	15" wide	G		1350	.006		.44	.32		.76	.98
0180	23" wide	G		1600	.005		.44	.27		.71	.91
0182	R21, 11" wide	G		1150	.007		.70	.37		1.07	1.36
0184	15" wide	G		1350	.006		.70	.32		1.02	1.27
0186	23" wide	G		1600	.005		.70	.27		.97	1.20
0188	9" thick, R30, 11" wide	G		985	.008		.86	.43		1.29	1.64
0200	15" wide	G		1150	.007		.86	.37		1.23	1.54
0220	23" wide	G		1350	.006		.86	.32		1.18	1.45
0230	12" thick, R38, 11" wide	G		985	.008		1.04	.43		1.47	1.83
0240	15" wide	G		1150	.007		1.04	.37		1.41	1.73
0260	23" wide	G		1350	.006		1.04	.32		1.36	1.64
0410	Foil faced fiberglass, 3-1/2" thick, R13, 11" wide	G		1150	.007		.47	.37		.84	1.11
0420	15" wide	G		1350	.006		.47	.32		.79	1.02
0440	23" wide	G		1600	.005		.47	.27		.74	.95
0442	R15, 11" wide	G		1150	.007		.46	.37		.83	1.10
0444	15" wide	G		1350	.006		.46	.32		.78	1.01
0446	23" wide	G		1600	.005		.46	.27		.73	.94
0448	6" thick, R19, 11" wide	G		1150	.007		.63	.37		1	1.28
0460	15" wide	G		1350	.006		.63	.32		.95	1.19
0480	23" wide	G		1600	.005		.63	.27		.90	1.12
0482	R21, 11" wide	G		1150	.007		.65	.37		1.02	1.31
0484	15" wide	G		1350	.006		.65	.32		.97	1.22
0486	23" wide	G		1600	.005		.65	.27		.92	1.15
0488	9" thick, R30, 11" wide	G		985	.008		1.05	.43		1.48	1.85
0500	15" wide	G		1150	.007		1.05	.37		1.42	1.75
0550	23" wide	G		1350	.006		1.05	.32		1.37	1.66
0560	12" thick, R38, 11" wide	G		985	.008		1.14	.43		1.57	1.94
0570	15" wide	G		1150	.007		1.14	.37		1.51	1.84
0580	23" wide	G		1350	.006		1.14	.32		1.46	1.75

07 21 Thermal Insulation

07 21 16 – Blanket Insulation

07 21 16.20 Blanket Insulation for Walls

			Crew	Daily Output	Labor-Hours	Unit	Material	2020 Bare Costs Labor	Equipment	Total	Total Incl O&P
0620	Unfaced fiberglass, 3-1/2" thick, R13, 11" wide	G	1 Carp	1150	.007	S.F.	.33	.37		.70	.95
0820	15" wide	G		1350	.006		.33	.32		.65	.86
0830	23" wide	G		1600	.005		.33	.27		.60	.79
0832	R15, 11" wide	G		1150	.007		.46	.37		.83	1.10
0836	23" wide	G		1600	.005		.46	.27		.73	.94
0838	6" thick, R19, 11" wide	G		1150	.007		.39	.37		.76	1.02
0860	15" wide	G		1150	.007		.39	.37		.76	1.02
0880	23" wide	G		1350	.006		.39	.32		.71	.93
0882	R21, 11" wide	G		1150	.007		.68	.37		1.05	1.34
0886	15" wide	G		1350	.006		.68	.32		1	1.25
0888	23" wide	G		1600	.005		.68	.27		.95	1.18
0890	9" thick, R30, 11" wide	G		985	.008		.68	.43		1.11	1.44
0900	15" wide	G		1150	.007		.68	.37		1.05	1.34
0920	23" wide	G		1350	.006		.68	.32		1	1.25
0930	12" thick, R38, 11" wide	G		985	.008		.77	.43		1.20	1.54
0940	15" wide	G		1000	.008		.77	.43		1.20	1.53
0960	23" wide	G		1150	.007		.77	.37		1.14	1.44
1300	Wall or ceiling insulation, mineral wool batts										
1320	3-1/2" thick, R15	G	1 Carp	1600	.005	S.F.	.82	.27		1.09	1.33
1340	5-1/2" thick, R23	G		1600	.005		1.29	.27		1.56	1.85
1380	7-1/4" thick, R30	G		1350	.006		1.70	.32		2.02	2.37
1700	Non-rigid insul., recycled blue cotton fiber, unfaced batts, R13, 16" wide	G		1600	.005		1.18	.27		1.45	1.73
1710	R19, 16" wide	G		1600	.005		1.45	.27		1.72	2.03
1850	Friction fit wire insulation supports, 16" OC			960	.008	Ea.	.07	.44		.51	.79
9000	Minimum labor/equipment charge			4	2	Job		106		106	170

07 21 19 – Foamed In Place Insulation

07 21 19.10 Masonry Foamed In Place Insulation

			Crew	Daily Output	Labor-Hours	Unit	Material	2020 Bare Costs Labor	Equipment	Total	Total Incl O&P
0010	**MASONRY FOAMED IN PLACE INSULATION**										
0100	Amino-plast foam, injected into block core, 6" block	G	G-2A	6000	.004	Ea.	.17	.16	.11	.44	.59
0110	8" block	G		5000	.005		.20	.20	.13	.53	.69
0120	10" block	G		4000	.006		.26	.25	.16	.67	.87
0130	12" block	G		3000	.008		.34	.33	.21	.88	1.17
0140	Injected into cavity wall	G		13000	.002	B.F.	.06	.08	.05	.19	.25
0150	Preparation, drill holes into mortar joint every 4 V.L.F., 5/8" diameter		1 Clab	960	.008	Ea.		.35		.35	.56
0160	7/8" diameter			680	.012			.50		.50	.79
0170	Patch drilled holes, 5/8" diameter			1800	.004		.04	.19		.23	.34
0180	7/8" diameter			1200	.007		.05	.28		.33	.51
9000	Minimum labor/equipment charge		G-2A	2	12	Job		490	315	805	1,175
9010	Minimum labor/equipment charge		1 Clab	4	2	"		84		84	135

07 21 23 – Loose-Fill Insulation

07 21 23.10 Poured Loose-Fill Insulation

			Crew	Daily Output	Labor-Hours	Unit	Material	2020 Bare Costs Labor	Equipment	Total	Total Incl O&P
0010	**POURED LOOSE-FILL INSULATION**										
0020	Cellulose fiber, R3.8 per inch	G	1 Carp	200	.040	C.F.	.70	2.13		2.83	4.18
0021	4" thick	G		1000	.008	S.F.	.17	.43		.60	.86
0022	6" thick	G		800	.010	"	.28	.53		.81	1.16
0080	Fiberglass wool, R4 per inch	G		200	.040	C.F.	.70	2.13		2.83	4.18
0081	4" thick	G		600	.013	S.F.	.24	.71		.95	1.40
0082	6" thick	G		400	.020	"	.34	1.06		1.40	2.07
0100	Mineral wool, R3 per inch	G		200	.040	C.F.	.56	2.13		2.69	4.03
0101	4" thick	G		600	.013	S.F.	.18	.71		.89	1.34
0102	6" thick	G		400	.020	"	.28	1.06		1.34	2.01
0300	Polystyrene, R4 per inch	G		200	.040	C.F.	1.56	2.13		3.69	5.15

07 21 Thermal Insulation

07 21 23 – Loose-Fill Insulation

	07 21 23.10 Poured Loose-Fill Insulation		Crew	Daily Output	Labor-Hours	Unit	Material	2020 Bare Costs Labor	Equipment	Total	Total Incl O&P
0301	4" thick	G	1 Carp	600	.013	S.F.	.51	.71		1.22	1.71
0302	6" thick	G		400	.020	"	.78	1.06		1.84	2.56
0400	Perlite, R2.78 per inch	G		200	.040	C.F.	5.30	2.13		7.43	9.20
0401	4" thick	G		1000	.008	S.F.	1.76	.43		2.19	2.62
0402	6" thick	G		800	.010	"	2.65	.53		3.18	3.76
9000	Minimum labor/equipment charge			4	2	Job		106		106	170

07 21 23.20 Masonry Loose-Fill Insulation

			Crew	Daily Output	Labor-Hours	Unit	Material	Labor	Equipment	Total	Total Incl O&P
0010	**MASONRY LOOSE-FILL INSULATION**, vermiculite or perlite										
0100	In cores of concrete block, 4" thick wall, .115 C.F./S.F.	G	D-1	4800	.003	S.F.	.61	.16		.77	.92
0200	6" thick wall, .175 C.F./S.F.	G		3000	.005		.93	.25		1.18	1.42
0300	8" thick wall, .258 C.F./S.F.	G		2400	.007		1.36	.31		1.67	2.01
0400	10" thick wall, .340 C.F./S.F.	G		1850	.009		1.80	.41		2.21	2.64
0500	12" thick wall, .422 C.F./S.F.	G		1200	.013		2.23	.63		2.86	3.47
0600	Poured cavity wall, vermiculite or perlite, water repellent	G		250	.064	C.F.	5.30	3		8.30	10.65
0700	Foamed in place, urethane in 2-5/8" cavity	G	G-2A	1035	.023	S.F.	1.39	.95	.61	2.95	3.81
0800	For each 1" added thickness, add	G	"	2372	.010	"	.53	.41	.27	1.21	1.57

07 21 26 – Blown Insulation

07 21 26.10 Blown Insulation

			Crew	Daily Output	Labor-Hours	Unit	Material	Labor	Equipment	Total	Total Incl O&P
0010	**BLOWN INSULATION** Ceilings, with open access										
0020	Cellulose, 3-1/2" thick, R13	G	G-4	5000	.005	S.F.	.24	.21	.08	.53	.68
0030	5-3/16" thick, R19	G		3800	.006		.35	.27	.11	.73	.94
0050	6-1/2" thick, R22	G		3000	.008		.45	.34	.13	.92	1.20
0100	8-11/16" thick, R30	G		2600	.009		.61	.39	.15	1.15	1.47
0120	10-7/8" thick, R38	G		1800	.013		.78	.57	.22	1.57	2.01
1000	Fiberglass, 5.5" thick, R11	G		3800	.006		.24	.27	.11	.62	.81
1050	6" thick, R12	G		3000	.008		.34	.34	.13	.81	1.07
1100	8.8" thick, R19	G		2200	.011		.42	.47	.18	1.07	1.41
1200	10" thick, R22	G		1800	.013		.49	.57	.22	1.28	1.69
1300	11.5" thick, R26	G		1500	.016		.58	.68	.27	1.53	2.03
1350	13" thick, R30	G		1400	.017		.68	.73	.28	1.69	2.22
1450	16" thick, R38	G		1145	.021		.86	.90	.35	2.11	2.77
1500	20" thick, R49	G		920	.026		1.14	1.12	.43	2.69	3.53
9000	Minimum labor/equipment charge			4	6	Job		257	99.50	356.50	520

07 21 29 – Sprayed Insulation

07 21 29.10 Sprayed-On Insulation

			Crew	Daily Output	Labor-Hours	Unit	Material	Labor	Equipment	Total	Total Incl O&P
0010	**SPRAYED-ON INSULATION**										
0020	Fibrous/cementitious, finished wall, 1" thick, R3.7	G	G-2	2050	.012	S.F.	.36	.52	.09	.97	1.33
0100	Attic, 5.2" thick, R19	G		1550	.015	"	.43	.69	.12	1.24	1.70
0200	Fiberglass, R4 per inch, vertical	G		1600	.015	B.F.	.20	.67	.12	.99	1.41
0210	Horizontal	G		1200	.020	"	.20	.89	.16	1.25	1.81
0300	Closed cell, spray polyurethane foam, 2 lb./C.F. density										
0310	1" thick	G	G-2A	6000	.004	S.F.	.53	.16	.11	.80	.98
0320	2" thick	G		3000	.008		1.06	.33	.21	1.60	1.96
0330	3" thick	G		2000	.012		1.59	.49	.32	2.40	2.94
0335	3-1/2" thick	G		1715	.014		1.86	.57	.37	2.80	3.42
0340	4" thick	G		1500	.016		2.12	.66	.42	3.20	3.92
0350	5" thick	G		1200	.020		2.65	.82	.53	4	4.89
0355	5-1/2" thick	G		1090	.022		2.92	.90	.58	4.40	5.40
0360	6" thick	G		1000	.024		3.18	.98	.63	4.79	5.85
9000	Minimum labor/equipment charge		G-2	2	12	Job		530	94	624	955

For customer support on your Commercial Renovation Costs with RSMeans data, call 800.448.8182.

191

07 21 Thermal Insulation

07 21 53 – Reflective Insulation

07 21 53.10 Reflective Insulation Options		Crew	Daily Output	Labor-Hours	Unit	Material	2020 Bare Costs Labor	Equipment	Total	Total Incl O&P
0010	**REFLECTIVE INSULATION OPTIONS**									
0020	Aluminum foil on reinforced scrim	G	1 Carp	19	.421	C.S.F.	15	22.50		37.50	52.50
0100	Reinforced with woven polyolefin	G		19	.421		22.50	22.50		45	61
0500	With single bubble air space, R8.8	G		15	.533		26.50	28.50		55	74.50
0600	With double bubble air space, R9.8	G		15	.533		32.50	28.50		61	81.50
9000	Minimum labor/equipment charge			4	2	Job		106		106	170

07 22 Roof and Deck Insulation

07 22 16 – Roof Board Insulation

07 22 16.10 Roof Deck Insulation

			Crew	Daily Output	Labor-Hours	Unit	Material	2020 Bare Costs Labor	Equipment	Total	Total Incl O&P
0010	**ROOF DECK INSULATION**, fastening excluded										
0016	Asphaltic cover board, fiberglass lined, 1/8" thick		1 Rofc	1400	.006	S.F.	.49	.26		.75	1
0018	1/4" thick			1400	.006		.99	.26		1.25	1.55
0020	Fiberboard low density, 1/2" thick, R1.39	G		1300	.006		.36	.28		.64	.90
0030	1" thick, R2.78	G		1040	.008		.69	.36		1.05	1.38
0080	1-1/2" thick, R4.17	G		1040	.008		1.04	.36		1.40	1.76
0100	2" thick, R5.56	G		1040	.008		1.19	.36		1.55	1.93
0110	Fiberboard high density, 1/2" thick, R1.3	G		1300	.006		.28	.28		.56	.81
0120	1" thick, R2.5	G		1040	.008		.64	.36		1	1.32
0130	1-1/2" thick, R3.8	G		1040	.008		1.24	.36		1.60	1.98
0200	Fiberglass, 3/4" thick, R2.78	G		1300	.006		.63	.28		.91	1.19
0400	15/16" thick, R3.70	G		1300	.006		.84	.28		1.12	1.42
0460	1-1/16" thick, R4.17	G		1300	.006		1.17	.28		1.45	1.79
0600	1-5/16" thick, R5.26	G		1300	.006		1.45	.28		1.73	2.10
0650	2-1/16" thick, R8.33	G		1040	.008		1.63	.36		1.99	2.41
0700	2-7/16" thick, R10	G		1040	.008		1.75	.36		2.11	2.55
0800	Gypsum cover board, fiberglass mat facer, 1/4" thick			1400	.006		.51	.26		.77	1.02
0810	1/2" thick			1300	.006		.67	.28		.95	1.24
0820	5/8" thick			1200	.007		.66	.31		.97	1.27
0830	Primed fiberglass mat facer, 1/4" thick			1400	.006		.50	.26		.76	1.01
0840	1/2" thick			1300	.006		.58	.28		.86	1.14
0850	5/8" thick			1200	.007		.61	.31		.92	1.21
1650	Perlite, 1/2" thick, R1.32	G		1365	.006		.29	.27		.56	.79
1655	3/4" thick, R2.08	G		1040	.008		.39	.36		.75	1.05
1660	1" thick, R2.78	G		1040	.008		.57	.36		.93	1.25
1670	1-1/2" thick, R4.17	G		1040	.008		.83	.36		1.19	1.53
1680	2" thick, R5.56	G		910	.009		1.09	.41		1.50	1.91
1685	2-1/2" thick, R6.67	G		910	.009		1.42	.41		1.83	2.27
1690	Tapered for drainage	G		1040	.008	B.F.	1.09	.36		1.45	1.82
1700	Polyisocyanurate, 2#/C.F. density, 3/4" thick	G		1950	.004	S.F.	.52	.19		.71	.90
1705	1" thick	G		1820	.004		.45	.20		.65	.86
1715	1-1/2" thick	G		1625	.005		.60	.23		.83	1.06
1725	2" thick	G		1430	.006		.92	.26		1.18	1.46
1735	2-1/2" thick	G		1365	.006		1	.27		1.27	1.57
1745	3" thick	G		1300	.006		1.13	.28		1.41	1.74
1755	3-1/2" thick	G		1300	.006		1.67	.28		1.95	2.34
1765	Tapered for drainage	G		1820	.004	B.F.	.53	.20		.73	.94
1900	Extruded polystyrene										
1910	15 psi compressive strength, 1" thick, R5	G	1 Rofc	1950	.004	S.F.	.52	.19		.71	.90
1920	2" thick, R10	G		1625	.005		.68	.23		.91	1.14

07 22 16 – Roof Board Insulation

07 22 16.10 Roof Deck Insulation		Crew	Daily Output	Labor-Hours	Unit	Material	2020 Bare Costs Labor	2020 Bare Costs Equipment	Total	Total Incl O&P
1930	3" thick, R15	G	1 Rofc	1300	.006	S.F.	1.35	.28		1.63	1.99
1932	4" thick, R20	G		1300	.006	↓	1.82	.28		2.10	2.50
1934	Tapered for drainage	G		1950	.004	B.F.	.51	.19		.70	.89
1940	25 psi compressive strength, 1" thick, R5	G		1950	.004	S.F.	1.06	.19		1.25	1.50
1942	2" thick, R10	G		1625	.005		2.01	.23		2.24	2.62
1944	3" thick, R15	G		1300	.006		3.07	.28		3.35	3.88
1946	4" thick, R20	G		1300	.006	↓	4.24	.28		4.52	5.15
1948	Tapered for drainage	G		1950	.004	B.F.	.57	.19		.76	.96
1950	40 psi compressive strength, 1" thick, R5	G		1950	.004	S.F.	.87	.19		1.06	1.29
1952	2" thick, R10	G		1625	.005		1.65	.23		1.88	2.22
1954	3" thick, R15	G		1300	.006		2.39	.28		2.67	3.13
1956	4" thick, R20	G		1300	.006	↓	3.13	.28		3.41	3.95
1958	Tapered for drainage	G		1820	.004	B.F.	.88	.20		1.08	1.33
1960	60 psi compressive strength, 1" thick, R5	G		1885	.004	S.F.	1.09	.20		1.29	1.54
1962	2" thick, R10	G		1560	.005		2.07	.24		2.31	2.69
1964	3" thick, R15	G		1270	.006		3.38	.29		3.67	4.23
1966	4" thick, R20	G		1235	.006	↓	4.20	.30		4.50	5.15
1968	Tapered for drainage	G		1820	.004	B.F.	.99	.20		1.19	1.45
2010	Expanded polystyrene, 1#/C.F. density, 3/4" thick, R2.89	G		1950	.004	S.F.	.22	.19		.41	.57
2020	1" thick, R3.85	G		1950	.004		.29	.19		.48	.65
2100	2" thick, R7.69	G		1625	.005		.58	.23		.81	1.04
2110	3" thick, R11.49	G		1625	.005		.87	.23		1.10	1.36
2120	4" thick, R15.38	G		1625	.005		1.16	.23		1.39	1.68
2130	5" thick, R19.23	G		1495	.005		1.45	.25		1.70	2.03
2140	6" thick, R23.26	G		1495	.005	↓	1.74	.25		1.99	2.34
2150	Tapered for drainage	G	↓	1950	.004	B.F.	.51	.19		.70	.89
2400	Composites with 2" EPS										
2410	1" fiberboard	G	1 Rofc	1325	.006	S.F.	1.48	.28		1.76	2.12
2420	7/16" oriented strand board	G		1040	.008		1.25	.36		1.61	2
2430	1/2" plywood	G		1040	.008		1.47	.36		1.83	2.24
2440	1" perlite	G	↓	1040	.008	↓	1.16	.36		1.52	1.90
2450	Composites with 1-1/2" polyisocyanurate										
2460	1" fiberboard	G	1 Rofc	1040	.008	S.F.	1.21	.36		1.57	1.95
2470	1" perlite	G		1105	.007		.98	.33		1.31	1.67
2480	7/16" oriented strand board	G		1040	.008	↓	.91	.36		1.27	1.62
3000	Fastening alternatives, coated screws, 2" long			3744	.002	Ea.	.06	.10		.16	.24
3010	4" long			3120	.003		.11	.12		.23	.33
3020	6" long			2675	.003		.20	.14		.34	.46
3030	8" long			2340	.003		.28	.16		.44	.59
3040	10" long			1872	.004		.52	.20		.72	.92
3050	Pre-drill and drive wedge spike, 2-1/2"			1248	.006		.40	.30		.70	.96
3060	3-1/2"			1101	.007		.59	.34		.93	1.24
3070	4-1/2"			936	.009		.65	.40		1.05	1.41
3075	3" galvanized deck plates		↓	7488	.001	↓	.10	.05		.15	.20
3080	Spot mop asphalt		G-1	295	.190	Sq.	5.30	8.20	1.92	15.42	22.50
3090	Full mop asphalt		"	192	.292		10.65	12.60	2.95	26.20	37
3110	Low-rise polyurethane adhesive, 12" OC beads		1 Rofc	45	.178		41.50	8.20		49.70	60
3120	6" OC beads			32	.250		83	11.55		94.55	111
3130	4" OC beads			30	.267	↓	125	12.30		137.30	160
9000	Minimum labor/equipment charge		↓	3.25	2.462	Job		114		114	199

For customer support on your Commercial Renovation Costs with RSMeans data, call 800.448.8182.

193

07 24 Exterior Insulation and Finish Systems

07 24 13 – Polymer-Based Exterior Insulation and Finish System

07 24 13.10 Exterior Insulation and Finish Systems

	07 24 13.10 Exterior Insulation and Finish Systems		Crew	Daily Output	Labor-Hours	Unit	Material	2020 Bare Costs Labor	Equipment	Total	Total Incl O&P
0010	**EXTERIOR INSULATION AND FINISH SYSTEMS**										
0095	Field applied, 1" EPS insulation	G	J-1	390	.103	S.F.	1.80	4.73	.30	6.83	9.85
0100	With 1/2" cement board sheathing	G		268	.149		2.58	6.90	.44	9.92	14.25
0105	2" EPS insulation	G		390	.103		2.09	4.73	.30	7.12	10.20
0110	With 1/2" cement board sheathing	G		268	.149		2.87	6.90	.44	10.21	14.60
0115	3" EPS insulation	G		390	.103		2.38	4.73	.30	7.41	10.50
0120	With 1/2" cement board sheathing	G		268	.149		3.16	6.90	.44	10.50	14.90
0125	4" EPS insulation	G		390	.103		2.67	4.73	.30	7.70	10.80
0130	With 1/2" cement board sheathing	G		268	.149		4.22	6.90	.44	11.56	16.10
0140	Premium finish add			1265	.032		.45	1.46	.09	2	2.92
0145	Drainage and ventilation cavity, add		2 Plas	1450	.011		.82	.54		1.36	1.75
0150	Heavy duty reinforcement add		J-1	914	.044		.59	2.02	.13	2.74	4
0160	2.5#/S.Y. metal lath substrate add		1 Lath	75	.107	S.Y.	3.80	5.55		9.35	12.90
0170	3.4#/S.Y. metal lath substrate add		"	75	.107	"	4.72	5.55		10.27	13.90
0180	Color or texture change		J-1	1265	.032	S.F.	.74	1.46	.09	2.29	3.23
0190	With substrate leveling base coat		1 Plas	530	.015		.84	.73		1.57	2.09
0210	With substrate sealing base coat		1 Pord	1224	.007		.14	.29		.43	.61
0370	V groove shape in panel face					L.F.	.70			.70	.77
0380	U groove shape in panel face					"	.86			.86	.95
0433	Crack repair, acrylic rubber, fluid applied, 20 mils thick		1 Plas	350	.023	S.F.	1.55	1.11		2.66	3.48
0437	50 mils thick, reinforced		"	200	.040	"	2.77	1.94		4.71	6.15
0440	For higher than one story, add							25%			

07 25 Weather Barriers

07 25 10 – Weather Barriers or Wraps

07 25 10.10 Weather Barriers

		Crew	Daily Output	Labor-Hours	Unit	Material	2020 Bare Costs Labor	Equipment	Total	Total Incl O&P
0010	**WEATHER BARRIERS**									
0400	Asphalt felt paper, #15	1 Carp	37	.216	Sq.	5.25	11.50		16.75	24
0401	Per square foot	"	3700	.002	S.F.	.05	.11		.16	.24
0450	Housewrap, exterior, spun bonded polypropylene									
0470	Small roll	1 Carp	3800	.002	S.F.	.15	.11		.26	.35
0480	Large roll	"	4000	.002	"	.16	.11		.27	.35
2100	Asphalt felt roof deck vapor barrier, class 1 metal decks	1 Rofc	37	.216	Sq.	21	10		31	41
2200	For all other decks	"	37	.216		15.90	10		25.90	35
2800	Asphalt felt, 50% recycled content, 15 lb., 4 sq./roll	1 Carp	36	.222		6.10	11.80		17.90	25.50
2810	30 lb., 2 sq./roll	"	36	.222		10.45	11.80		22.25	30.50
3000	Building wrap, spun bonded polyethylene	2 Carp	8000	.002	S.F.	.14	.11		.25	.32
9960	Minimum labor/equipment charge	1 Carp	2	4	Job		213		213	340

07 26 Vapor Retarders

07 26 13 – Above-Grade Vapor Retarders

07 26 13.10 Vapor Retarders

			Crew	Daily Output	Labor-Hours	Unit	Material	2020 Bare Costs Labor	Equipment	Total	Total Incl O&P
0010	**VAPOR RETARDERS**										
0020	Aluminum and kraft laminated, foil 1 side	G	1 Carp	37	.216	Sq.	16.05	11.50		27.55	36
0100	Foil 2 sides	G		37	.216		15.50	11.50		27	35.50
0600	Polyethylene vapor barrier, standard, 2 mil	G		37	.216		1.63	11.50		13.13	20
0700	4 mil	G		37	.216		2.62	11.50		14.12	21.50
0900	6 mil	G		37	.216		4	11.50		15.50	23
1200	10 mil	G		37	.216		9.70	11.50		21.20	29

07 26 Vapor Retarders

07 26 13 – Above-Grade Vapor Retarders

07 26 13.10 Vapor Retarders		Crew	Daily Output	Labor-Hours	Unit	Material	2020 Bare Costs Labor	Equipment	Total	Total Incl O&P
1300	Clear reinforced, fire retardant, 8 mil	G	1 Carp	37	.216	Sq.	28	11.50		39.50	49.50
1350	Cross laminated type, 3 mil	G		37	.216		12.90	11.50		24.40	32.50
1400	4 mil	G		37	.216		14.15	11.50		25.65	34
1800	Reinf. waterproof, 2 mil polyethylene backing, 1 side			37	.216		11.35	11.50		22.85	31
1900	2 sides			37	.216		14.35	11.50		25.85	34
2400	Waterproofed kraft with sisal or fiberglass fibers			37	.216		21	11.50		32.50	42
9950	Minimum labor/equipment charge			4	2	Job		106		106	170

07 27 Air Barriers

07 27 13 – Modified Bituminous Sheet Air Barriers

07 27 13.10 Modified Bituminous Sheet Air Barrier

		Crew	Daily Output	Labor-Hours	Unit	Material	2020 Bare Costs Labor	Equipment	Total	Total Incl O&P
0010	**MODIFIED BITUMINOUS SHEET AIR BARRIER**									
0100	SBS modified sheet laminated to polyethylene sheet, 40 mils, 4" wide	1 Carp	1200	.007	L.F.	.32	.35		.67	.93
0120	6" wide		1100	.007		.44	.39		.83	1.11
0140	9" wide		1000	.008		.62	.43		1.05	1.36
0160	12" wide		900	.009		.79	.47		1.26	1.63
0180	18" wide	2 Carp	1700	.009	S.F.	.74	.50		1.24	1.61
0200	36" wide	"	1800	.009		.72	.47		1.19	1.55
0220	Adhesive for above	1 Carp	1400	.006		.32	.30		.62	.84

07 27 26 – Fluid-Applied Membrane Air Barriers

07 27 26.10 Fluid Applied Membrane Air Barrier

		Crew	Daily Output	Labor-Hours	Unit	Material	2020 Bare Costs Labor	Equipment	Total	Total Incl O&P
0010	**FLUID APPLIED MEMBRANE AIR BARRIER**									
0100	Spray applied vapor barrier, 25 S.F./gallon	1 Pord	1375	.006	S.F.	.02	.26		.28	.43

07 31 Shingles and Shakes

07 31 13 – Asphalt Shingles

07 31 13.10 Asphalt Roof Shingles

		Crew	Daily Output	Labor-Hours	Unit	Material	2020 Bare Costs Labor	Equipment	Total	Total Incl O&P
0010	**ASPHALT ROOF SHINGLES**									
0100	Standard strip shingles									
0150	Inorganic, class A, 25 year	1 Rofc	5.50	1.455	Sq.	75	67		142	201
0155	Pneumatic nailed		7	1.143		75	53		128	175
0200	30 year		5	1.600		111	74		185	251
0205	Pneumatic nailed		6.25	1.280		111	59		170	225
0250	Standard laminated multi-layered shingles									
0300	Class A, 240-260 lb./square	1 Rofc	4.50	1.778	Sq.	103	82		185	257
0305	Pneumatic nailed		5.63	1.422		103	65.50		168.50	228
0350	Class A, 250-270 lb./square		4	2		103	92.50		195.50	275
0355	Pneumatic nailed		5	1.600		103	74		177	242
0400	Premium, laminated multi-layered shingles									
0450	Class A, 260-300 lb./square	1 Rofc	3.50	2.286	Sq.	148	106		254	350
0455	Pneumatic nailed		4.37	1.831		148	84.50		232.50	310
0500	Class A, 300-385 lb./square		3	2.667		310	123		433	555
0505	Pneumatic nailed		3.75	2.133		310	98.50		408.50	510
0800	#15 felt underlayment		64	.125		5.25	5.80		11.05	15.90
0825	#30 felt underlayment		58	.138		10.20	6.35		16.55	22.50
0850	Self adhering polyethylene and rubberized asphalt underlayment		22	.364		76	16.80		92.80	113
0900	Ridge shingles		330	.024	L.F.	2.31	1.12		3.43	4.50
0905	Pneumatic nailed		412.50	.019	"	2.31	.90		3.21	4.11
1000	For steep roofs (7 to 12 pitch or greater), add						50%			

07 31 Shingles and Shakes

07 31 13 – Asphalt Shingles

07 31 13.10 Asphalt Roof Shingles

		Crew	Daily Output	Labor-Hours	Unit	Material	2020 Bare Costs Labor	2020 Bare Costs Equipment	Total	Total Incl O&P
9000	Minimum labor/equipment charge	1 Rofc	3	2.667	Job		123		123	216

07 31 16 – Metal Shingles

07 31 16.10 Aluminum Shingles

		Crew	Daily Output	Labor-Hours	Unit	Material	2020 Bare Costs Labor	2020 Bare Costs Equipment	Total	Total Incl O&P
0010	**ALUMINUM SHINGLES**									
0020	Mill finish, .019" thick	1 Carp	5	1.600	Sq.	232	85		317	390
0100	.020" thick	"	5	1.600		256	85		341	420
0300	For colors, add					21.50			21.50	23.50
0600	Ridge cap, .024" thick	1 Carp	170	.047	L.F.	4.02	2.50		6.52	8.45
0700	End wall flashing, .024" thick		170	.047		2.48	2.50		4.98	6.75
0900	Valley section, .024" thick		170	.047		4.08	2.50		6.58	8.50
1000	Starter strip, .024" thick		400	.020		2.01	1.06		3.07	3.91
1200	Side wall flashing, .024" thick		170	.047		2.41	2.50		4.91	6.65
1500	Gable flashing, .024" thick		400	.020		1.89	1.06		2.95	3.78
9000	Minimum labor/equipment charge		3	2.667	Job		142		142	227

07 31 16.20 Steel Shingles

		Crew	Daily Output	Labor-Hours	Unit	Material	2020 Bare Costs Labor	2020 Bare Costs Equipment	Total	Total Incl O&P
0010	**STEEL SHINGLES**									
0012	Galvanized, 26 ga.	1 Rots	2.20	3.636	Sq.	390	168		558	725
0200	24 ga.	"	2.20	3.636		380	168		548	715
0300	For colored galvanized shingles, add					59.50			59.50	65.50
9000	Minimum labor/equipment charge	1 Rots	3	2.667	Job		123		123	216

07 31 26 – Slate Shingles

07 31 26.10 Slate Roof Shingles

			Crew	Daily Output	Labor-Hours	Unit	Material	2020 Bare Costs Labor	2020 Bare Costs Equipment	Total	Total Incl O&P
0010	**SLATE ROOF SHINGLES**										
0100	Buckingham Virginia black, 3/16" - 1/4" thick	G	1 Rots	1.75	4.571	Sq.	560	211		771	985
0200	1/4" thick	G		1.75	4.571		560	211		771	985
0900	Pennsylvania black, Bangor, #1 clear	G		1.75	4.571		500	211		711	920
1200	Vermont, unfading, green, mottled green	G		1.75	4.571		505	211		716	925
1300	Semi-weathering green & gray	G		1.75	4.571		405	211		616	815
1400	Purple	G		1.75	4.571		440	211		651	855
1500	Black or gray	G		1.75	4.571		485	211		696	905
2500	Slate roof repair, extensive replacement			1	8		605	370		975	1,325
2600	Repair individual pieces, scattered			19	.421	Ea.	7.15	19.45		26.60	42
2700	Ridge shingles, slate			200	.040	L.F.	10.20	1.85		12.05	14.45
9000	Minimum labor/equipment charge			3	2.667	Job		123		123	216

07 31 29 – Wood Shingles and Shakes

07 31 29.13 Wood Shingles

			Crew	Daily Output	Labor-Hours	Unit	Material	2020 Bare Costs Labor	2020 Bare Costs Equipment	Total	Total Incl O&P
0010	**WOOD SHINGLES**	R061110-30									
0012	16" No. 1 red cedar shingles, 5" exposure, on roof		1 Carp	2.50	3.200	Sq.	320	170		490	620
0015	Pneumatic nailed			3.25	2.462		320	131		451	560
0200	7-1/2" exposure, on walls			2.05	3.902		213	207		420	565
0205	Pneumatic nailed			2.67	2.996		213	159		372	490
0300	18" No. 1 red cedar perfections, 5-1/2" exposure, on roof			2.75	2.909		286	155		441	565
0305	Pneumatic nailed			3.57	2.241		286	119		405	505
0500	7-1/2" exposure, on walls			2.25	3.556		210	189		399	535
0505	Pneumatic nailed			2.92	2.740		210	146		356	465
0600	Resquared and rebutted, 5-1/2" exposure, on roof			3	2.667		280	142		422	535
0605	Pneumatic nailed			3.90	2.051		280	109		389	485
0900	7-1/2" exposure, on walls			2.45	3.265		206	174		380	505
0905	Pneumatic nailed			3.18	2.516		206	134		340	440
1000	Add to above for fire retardant shingles						60			60	66
1060	Preformed ridge shingles		1 Carp	400	.020	L.F.	4.06	1.06		5.12	6.15

196

07 31 Shingles and Shakes

07 31 29 – Wood Shingles and Shakes

07 31 29.13 Wood Shingles

		Crew	Daily Output	Labor-Hours	Unit	Material	2020 Bare Costs Labor	2020 Bare Costs Equipment	Total	Total Incl O&P
2000	White cedar shingles, 16" long, extras, 5" exposure, on roof	1 Carp	2.40	3.333	Sq.	194	177		371	495
2005	Pneumatic nailed		3.12	2.564		194	136		330	430
2050	5" exposure on walls		2	4		194	213		407	555
2055	Pneumatic nailed		2.60	3.077		194	164		358	475
2100	7-1/2" exposure, on walls		2	4		138	213		351	490
2105	Pneumatic nailed		2.60	3.077		138	164		302	415
2150	"B" grade, 5" exposure on walls		2	4		168	213		381	525
2155	Pneumatic nailed		2.60	3.077		168	164		332	445
2300	For #15 organic felt underlayment on roof, 1 layer, add		64	.125		5.25	6.65		11.90	16.45
2400	2 layers, add	▼	32	.250		10.50	13.30		23.80	33
2600	For steep roofs (7/12 pitch or greater), add to above				▼		50%			
2700	Panelized systems, No.1 cedar shingles on 5/16" CDX plywood									
2800	On walls, 8' strips, 7" or 14" exposure	2 Carp	700	.023	S.F.	6.85	1.22		8.07	9.50
3500	On roofs, 8' strips, 7" or 14" exposure	1 Carp	3	2.667	Sq.	665	142		807	955
3505	Pneumatic nailed		4	2	"	665	106		771	900
9000	Minimum labor/equipment charge	▼	3	2.667	Job		142		142	227

07 31 29.16 Wood Shakes

		Crew	Daily Output	Labor-Hours	Unit	Material	2020 Bare Costs Labor	2020 Bare Costs Equipment	Total	Total Incl O&P
0010	**WOOD SHAKES**									
1100	Hand-split red cedar shakes, 1/2" thick x 24" long, 10" exp. on roof	1 Carp	2.50	3.200	Sq.	340	170		510	645
1105	Pneumatic nailed		3.25	2.462		340	131		471	585
1110	3/4" thick x 24" long, 10" exp. on roof		2.25	3.556		340	189		529	680
1115	Pneumatic nailed		2.92	2.740		340	146		486	610
1200	1/2" thick, 18" long, 8-1/2" exp. on roof		2	4		294	213		507	665
1205	Pneumatic nailed		2.60	3.077		294	164		458	585
1210	3/4" thick x 18" long, 8-1/2" exp. on roof		1.80	4.444		294	236		530	705
1215	Pneumatic nailed		2.34	3.419		294	182		476	615
1255	10" exposure on walls		2	4		283	213		496	650
1260	10" exposure on walls, pneumatic nailed	▼	2.60	3.077		283	164		447	570
1700	Add to above for fire retardant shakes, 24" long					60			60	66
1800	18" long				▼	60			60	66
1810	Ridge shakes	1 Carp	350	.023	L.F.	5.75	1.22		6.97	8.30

07 32 Roof Tiles

07 32 13 – Clay Roof Tiles

07 32 13.10 Clay Tiles

		Crew	Daily Output	Labor-Hours	Unit	Material	2020 Bare Costs Labor	2020 Bare Costs Equipment	Total	Total Incl O&P
0010	**CLAY TILES**, including accessories									
0300	Flat shingle, interlocking, 15", 166 pcs./sq., fireflashed blend	3 Rots	6	4	Sq.	535	185		720	915
0500	Terra cotta red		6	4		545	185		730	925
0600	Roman pan and top, 18", 102 pcs./sq., fireflashed blend	▼	5.50	4.364		495	202		697	900
0640	Terra cotta red	1 Rots	2.40	3.333		585	154		739	915
1100	Barrel mission tile, 18", 166 pcs./sq., fireflashed blend	3 Rots	5.50	4.364		460	202		662	860
1140	Terra cotta red		5.50	4.364		465	202		667	865
1700	Scalloped edge flat shingle, 14", 145 pcs./sq., fireflashed blend		6	4		1,125	185		1,310	1,550
1800	Terra cotta red	▼	6	4		1,050	185		1,235	1,475
3010	#15 felt underlayment	1 Rofc	64	.125		5.25	5.80		11.05	15.90
3020	#30 felt underlayment		58	.138		10.20	6.35		16.55	22.50
3040	Polyethylene and rubberized asph. underlayment	▼	22	.364	▼	76	16.80		92.80	113
9010	Minimum labor/equipment charge	1 Rots	2	4	Job		185		185	325

For customer support on your Commercial Renovation Costs with RSMeans data, call 800.448.8182.

197

07 32 Roof Tiles

07 32 16 – Concrete Roof Tiles

07 32 16.10 Concrete Tiles

		Crew	Daily Output	Labor-Hours	Unit	Material	2020 Bare Costs Labor	Equipment	Total	Total Incl O&P
0010	**CONCRETE TILES**									
0020	Corrugated, 13" x 16-1/2", 90 per sq., 950 lb./sq.									
0050	Earthtone colors, nailed to wood deck	1 Rots	1.35	5.926	Sq.	107	274		381	595
0150	Blues		1.35	5.926		106	274		380	595
0200	Greens		1.35	5.926		117	274		391	610
0250	Premium colors		1.35	5.926		117	274		391	610
0500	Shakes, 13" x 16-1/2", 90 per sq., 950 lb./sq.									
0600	All colors, nailed to wood deck	1 Rots	1.50	5.333	Sq.	153	247		400	600
1500	Accessory pieces, ridge & hip, 10" x 16-1/2", 8 lb. each	"	120	.067	Ea.	3.72	3.08		6.80	9.50
1700	Rake, 6-1/2" x 16-3/4", 9 lb. each					3.72			3.72	4.09
1800	Mansard hip, 10" x 16-1/2", 9.2 lb. each					3.72			3.72	4.09
1900	Hip starter, 10" x 16-1/2", 10.5 lb. each					10.65			10.65	11.75
2000	3 or 4 way apex, 10" each side, 11.5 lb. each					12.20			12.20	13.40
9000	Minimum labor/equipment charge	1 Rots	3	2.667	Job		123		123	216

07 32 19 – Metal Roof Tiles

07 32 19.10 Metal Roof Tiles

		Crew	Daily Output	Labor-Hours	Unit	Material	2020 Bare Costs Labor	Equipment	Total	Total Incl O&P
0010	**METAL ROOF TILES**									
0020	Accessories included, .032" thick aluminum, mission tile	1 Carp	2.50	3.200	Sq.	830	170		1,000	1,175
0200	Spanish tiles		3	2.667	"	550	142		692	830
9000	Minimum labor/equipment charge		3	2.667	Job		142		142	227

07 33 Natural Roof Coverings

07 33 63 – Vegetated Roofing

07 33 63.10 Green Roof Systems

			Crew	Daily Output	Labor-Hours	Unit	Material	2020 Bare Costs Labor	Equipment	Total	Total Incl O&P
0010	**GREEN ROOF SYSTEMS**										
0020	Soil mixture for green roof 30% sand, 55% gravel, 15% soil										
0100	Hoist and spread soil mixture 4" depth up to 5 stories tall roof	G	B-13B	4000	.014	S.F.	.23	.64	.25	1.12	1.55
0150	6" depth	G		2667	.021		.35	.97	.37	1.69	2.33
0200	8" depth	G		2000	.028		.47	1.29	.49	2.25	3.11
0250	10" depth	G		1600	.035		.59	1.61	.61	2.81	3.87
0300	12" depth	G		1335	.042		.70	1.93	.73	3.36	4.65
0310	Alt. man-made soil mix, hoist & spread, 4" deep up to 5 stories tall roof	G		4000	.014		1.92	.64	.25	2.81	3.40
0350	Mobilization 55 ton crane to site	G	1 Eqhv	3.60	2.222	Ea.		132		132	206
0355	Hoisting cost to 5 stories per day (Avg. 28 picks per day)	G	B-13B	1	56	Day		2,575	980	3,555	5,175
0360	Mobilization or demobilization, 100 ton crane to site driver & escort	G	A-3E	2.50	6.400	Ea.		345	70.50	415.50	625
0365	Hoisting cost 6-10 stories per day (Avg. 21 picks per day)	G	B-13C	1	56	Day		2,575	2,275	4,850	6,625
0370	Hoist and spread soil mixture 4" depth 6-10 stories tall roof	G		4000	.014	S.F.	.23	.64	.57	1.44	1.91
0375	6" depth	G		2667	.021		.35	.97	.86	2.18	2.87
0380	8" depth	G		2000	.028		.47	1.29	1.14	2.90	3.83
0385	10" depth	G		1600	.035		.59	1.61	1.43	3.63	4.77
0390	12" depth	G		1335	.042		.70	1.93	1.71	4.34	5.70
0400	Green roof edging treated lumber 4" x 4", no hoisting included	G	2 Carp	400	.040	L.F.	1.41	2.13		3.54	4.96
0410	4" x 6"	G		400	.040		2.24	2.13		4.37	5.85
0420	4" x 8"	G		360	.044		4.28	2.36		6.64	8.50
0430	4" x 6" double stacked	G		300	.053		4.48	2.83		7.31	9.45
0500	Green roof edging redwood lumber 4" x 4", no hoisting included	G		400	.040		6.60	2.13		8.73	10.70
0510	4" x 6"	G		400	.040		12.85	2.13		14.98	17.55
0520	4" x 8"	G		360	.044		24	2.36		26.36	30.50
0530	4" x 6" double stacked	G		300	.053		25.50	2.83		28.33	33
0550	Components, not including membrane or insulation:										

07 33 Natural Roof Coverings

07 33 63 – Vegetated Roofing

07 33 63.10 Green Roof Systems		Crew	Daily Output	Labor-Hours	Unit	Material	2020 Bare Costs Labor	Equipment	Total	Total Incl O&P
0560	Fluid applied rubber membrane, reinforced, 215 mil thick **G**	G-5	350	.114	S.F.	.29	4.80	.55	5.64	9.35
0570	Root barrier **G**	2 Rofc	775	.021		.70	.95		1.65	2.44
0580	Moisture retention barrier and reservoir **G**	"	900	.018		2.66	.82		3.48	4.37
0600	Planting sedum, light soil, potted, 2-1/4" diameter, 2 per S.F. **G**	1 Clab	420	.019		6.35	.80		7.15	8.30
0610	1 per S.F. **G**	"	840	.010		3.18	.40		3.58	4.14
0630	Planting sedum mat per S.F. including shipping (4000 S.F. min) **G**	4 Clab	4000	.008		7.65	.34		7.99	8.95
0640	Installation sedum mat system (no soil required) per S.F. (4000 S.F. min) **G**	"	4000	.008		10.60	.34		10.94	12.25
0645	Note: pricing of sedum mats shipped in full truck loads (4000-5000 S.F.)									

07 41 Roof Panels

07 41 13 – Metal Roof Panels

07 41 13.10 Aluminum Roof Panels

		Crew	Daily Output	Labor-Hours	Unit	Material	2020 Bare Costs Labor	Equipment	Total	Total Incl O&P
0010	**ALUMINUM ROOF PANELS**									
0020	Corrugated or ribbed, .0155" thick, natural	G-3	1200	.027	S.F.	1.02	1.39		2.41	3.32
0300	Painted		1200	.027		1.50	1.39		2.89	3.85
0400	Corrugated, .018" thick, on steel frame, natural finish		1200	.027		1.25	1.39		2.64	3.58
0600	Painted		1200	.027		1.57	1.39		2.96	3.93
0700	Corrugated, on steel frame, natural, .024" thick		1200	.027		1.82	1.39		3.21	4.20
0800	Painted		1200	.027		2.20	1.39		3.59	4.62
0900	.032" thick, natural		1200	.027		3.15	1.39		4.54	5.65
1200	Painted		1200	.027		3.41	1.39		4.80	5.95
9000	Minimum labor/equipment charge	1 Rofc	3	2.667	Job		123		123	216

07 41 13.20 Steel Roofing Panels

		Crew	Daily Output	Labor-Hours	Unit	Material	2020 Bare Costs Labor	Equipment	Total	Total Incl O&P
0010	**STEEL ROOFING PANELS**									
0012	Corrugated or ribbed, on steel framing, 30 ga. galv	G-3	1100	.029	S.F.	1.80	1.52		3.32	4.38
0100	28 ga.		1050	.030		1.63	1.59		3.22	4.31
0300	26 ga.		1000	.032		2	1.67		3.67	4.84
0400	24 ga.		950	.034		3.06	1.76		4.82	6.15
0600	Colored, 28 ga.		1050	.030		1.90	1.59		3.49	4.61
0700	26 ga.		1000	.032		2.15	1.67		3.82	5
0710	Flat profile, 1-3/4" standing seams, 10" wide, standard finish, 26 ga.		1000	.032		4.16	1.67		5.83	7.20
0715	24 ga.		950	.034		4.85	1.76		6.61	8.15
0720	22 ga.		900	.036		6	1.86		7.86	9.55
0725	Zinc aluminum alloy finish, 26 ga.		1000	.032		3.35	1.67		5.02	6.35
0730	24 ga.		950	.034		3.92	1.76		5.68	7.10
0735	22 ga.		900	.036		4.47	1.86		6.33	7.85
0740	12" wide, standard finish, 26 ga.		1000	.032		4.19	1.67		5.86	7.25
0745	24 ga.		950	.034		5.45	1.76		7.21	8.80
0750	Zinc aluminum alloy finish, 26 ga.		1000	.032		4.71	1.67		6.38	7.85
0755	24 ga.		950	.034		3.92	1.76		5.68	7.10
0840	Flat profile, 1" x 3/8" batten, 12" wide, standard finish, 26 ga.		1000	.032		3.66	1.67		5.33	6.65
0845	24 ga.		950	.034		4.31	1.76		6.07	7.50
0850	22 ga.		900	.036		5.15	1.86		7.01	8.65
0855	Zinc aluminum alloy finish, 26 ga.		1000	.032		3.53	1.67		5.20	6.50
0860	24 ga.		950	.034		4.02	1.76		5.78	7.20
0865	22 ga.		900	.036		4.56	1.86		6.42	7.95
0870	16-1/2" wide, standard finish, 24 ga.		950	.034		4.25	1.76		6.01	7.45
0875	22 ga.		900	.036		4.78	1.86		6.64	8.20
0880	Zinc aluminum alloy finish, 24 ga.		950	.034		3.71	1.76		5.47	6.85
0885	22 ga.		900	.036		4.20	1.86		6.06	7.55
0890	Flat profile, 2" x 2" batten, 12" wide, standard finish, 26 ga.		1000	.032		4.24	1.67		5.91	7.30

07 41 Roof Panels

07 41 13 – Metal Roof Panels

07 41 13.20 Steel Roofing Panels

		Crew	Daily Output	Labor-Hours	Unit	Material	2020 Bare Costs Labor	Equipment	Total	Total Incl O&P
0895	24 ga.	G-3	950	.034	S.F.	5.10	1.76		6.86	8.40
0900	22 ga.		900	.036		6.20	1.86		8.06	9.80
0905	Zinc aluminum alloy finish, 26 ga.		1000	.032		4.02	1.67		5.69	7.05
0910	24 ga.		950	.034		4.51	1.76		6.27	7.75
0915	22 ga.		900	.036		5.25	1.86		7.11	8.70
0920	16-1/2" wide, standard finish, 24 ga.		950	.034		4.67	1.76		6.43	7.95
0925	22 ga.		900	.036		5.40	1.86		7.26	8.90
0930	Zinc aluminum alloy finish, 24 ga.		950	.034		4.20	1.76		5.96	7.40
0935	22 ga.		900	.036		4.80	1.86		6.66	8.25
9000	Minimum labor/equipment charge	1 Rofc	2	4	Job		185		185	325

07 41 33 – Plastic Roof Panels

07 41 33.10 Fiberglass Panels

		Crew	Daily Output	Labor-Hours	Unit	Material	2020 Bare Costs Labor	Equipment	Total	Total Incl O&P
0010	**FIBERGLASS PANELS**									
0012	Corrugated panels, roofing, 8 oz./S.F.	G-3	1000	.032	S.F.	2.55	1.67		4.22	5.45
0100	12 oz./S.F.		1000	.032		4.68	1.67		6.35	7.80
0300	Corrugated siding, 6 oz./S.F.		880	.036		2.02	1.90		3.92	5.20
0400	8 oz./S.F.		880	.036		2.55	1.90		4.45	5.80
0500	Fire retardant		880	.036		4.02	1.90		5.92	7.40
0600	12 oz. siding, textured		880	.036		3.78	1.90		5.68	7.15
0700	Fire retardant		880	.036		4.74	1.90		6.64	8.20
0900	Flat panels, 6 oz./S.F., clear or colors		880	.036		2.78	1.90		4.68	6.05
1100	Fire retardant, class A		880	.036		3.64	1.90		5.54	7
1300	8 oz./S.F., clear or colors		880	.036		2.41	1.90		4.31	5.65
9000	Minimum labor/equipment charge	1 Rofc	2	4	Job		185		185	325

07 42 Wall Panels

07 42 13 – Metal Wall Panels

07 42 13.10 Mansard Panels

		Crew	Daily Output	Labor-Hours	Unit	Material	2020 Bare Costs Labor	Equipment	Total	Total Incl O&P
0010	**MANSARD PANELS**									
0600	Aluminum, stock units, straight surfaces	1 Shee	115	.070	S.F.	4.48	4.33		8.81	11.75
0700	Concave or convex surfaces, add		75	.107	"	2.16	6.65		8.81	12.80
0800	For framing, to 5' high, add		115	.070	L.F.	3.91	4.33		8.24	11.10
0900	Soffits, to 1' wide		125	.064	S.F.	2.94	3.99		6.93	9.50
9000	Minimum labor/equipment charge		2.50	3.200	Job		199		199	310

07 42 13.20 Aluminum Siding Panels

		Crew	Daily Output	Labor-Hours	Unit	Material	2020 Bare Costs Labor	Equipment	Total	Total Incl O&P
0010	**ALUMINUM SIDING PANELS**									
0012	Corrugated, on steel framing, .019" thick, natural finish	G-3	775	.041	S.F.	1.72	2.16		3.88	5.30
0100	Painted		775	.041		1.86	2.16		4.02	5.45
0400	Farm type, .021" thick on steel frame, natural		775	.041		1.76	2.16		3.92	5.35
0600	Painted		775	.041		1.86	2.16		4.02	5.45
0700	Industrial type, corrugated, on steel, .024" thick, mill		775	.041		2.42	2.16		4.58	6.05
0900	Painted		775	.041		2.57	2.16		4.73	6.25
1000	.032" thick, mill		775	.041		2.65	2.16		4.81	6.35
1200	Painted		775	.041		3.28	2.16		5.44	7
1300	V-Beam, on steel frame, .032" thick, mill		775	.041		2.94	2.16		5.10	6.65
1500	Painted		775	.041		3.28	2.16		5.44	7
1600	.040" thick, mill		775	.041		3.71	2.16		5.87	7.50
1800	Painted		775	.041		4.31	2.16		6.47	8.15
3800	Horizontal, colored clapboard, 8" wide, plain	2 Carp	515	.031		2.72	1.65		4.37	5.65
3810	Insulated		515	.031		3.17	1.65		4.82	6.15

07 42 Wall Panels

07 42 13 – Metal Wall Panels

07 42 13.20 Aluminum Siding Panels

		Crew	Daily Output	Labor-Hours	Unit	Material	2020 Bare Costs Labor	2020 Bare Costs Equipment	Total	Total Incl O&P
3830	8" embossed, painted	2 Carp	515	.031	S.F.	3.14	1.65		4.79	6.10
3840	Insulated		515	.031		3.22	1.65		4.87	6.20
3860	12" painted, smooth		600	.027		2.94	1.42		4.36	5.50
3870	Insulated		600	.027		3.16	1.42		4.58	5.75
3890	12" embossed, painted		600	.027		3.19	1.42		4.61	5.80
3900	Insulated		515	.031		2.86	1.65		4.51	5.80
4000	Vertical board & batten, colored, non-insulated		515	.031		2.36	1.65		4.01	5.25
4200	For simulated wood design, add					.16			.16	.18
4300	Corners for above, outside	2 Carp	515	.031	V.L.F.	3.92	1.65		5.57	6.95
4500	Inside corners	"	515	.031	"	1.81	1.65		3.46	4.64
9000	Minimum labor/equipment charge	1 Carp	3	2.667	Job		142		142	227

07 42 13.30 Steel Siding

		Crew	Daily Output	Labor-Hours	Unit	Material	2020 Bare Costs Labor	2020 Bare Costs Equipment	Total	Total Incl O&P
0010	**STEEL SIDING**									
0020	Beveled, vinyl coated, 8" wide	1 Carp	265	.030	S.F.	1.87	1.60		3.47	4.63
0050	10" wide	"	275	.029		1.95	1.55		3.50	4.63
0080	Galv, corrugated or ribbed, on steel frame, 30 ga.	G-3	800	.040		1.20	2.09		3.29	4.62
0100	28 ga.		795	.040		1.30	2.10		3.40	4.75
0300	26 ga.		790	.041		1.69	2.11		3.80	5.20
0400	24 ga.		785	.041		2.22	2.13		4.35	5.80
0600	22 ga.		770	.042		2.31	2.17		4.48	5.95
0700	Colored, corrugated/ribbed, on steel frame, 10 yr. finish, 28 ga.		800	.040		2.33	2.09		4.42	5.85
0900	26 ga.		795	.040		2.01	2.10		4.11	5.55
1000	24 ga.		790	.041		2.24	2.11		4.35	5.80
1020	20 ga.		785	.041		2.85	2.13		4.98	6.50
9000	Minimum labor/equipment charge	1 Carp	3	2.667	Job		142		142	227

07 46 Siding

07 46 23 – Wood Siding

07 46 23.10 Wood Board Siding

		Crew	Daily Output	Labor-Hours	Unit	Material	2020 Bare Costs Labor	2020 Bare Costs Equipment	Total	Total Incl O&P
0010	**WOOD BOARD SIDING**									
3200	Wood, cedar bevel, A grade, 1/2" x 6"	1 Carp	295	.027	S.F.	4.46	1.44		5.90	7.20
3300	1/2" x 8"		330	.024		8.50	1.29		9.79	11.40
3500	3/4" x 10", clear grade		375	.021		8.45	1.13		9.58	11.10
3600	"B" grade		375	.021		4.16	1.13		5.29	6.40
3800	Cedar, rough sawn, 1" x 4", A grade, natural		220	.036		7.90	1.93		9.83	11.80
3900	Stained		220	.036		8	1.93		9.93	11.90
4100	1" x 12", board & batten, #3 & Btr., natural		420	.019		4.77	1.01		5.78	6.85
4200	Stained		420	.019		5.20	1.01		6.21	7.30
4400	1" x 8" channel siding, #3 & Btr., natural		330	.024		5.15	1.29		6.44	7.70
4500	Stained		330	.024		5.05	1.29		6.34	7.60
4700	Redwood, clear, beveled, vertical grain, 1/2" x 4"		220	.036		5	1.93		6.93	8.60
4750	1/2" x 6"		295	.027		5.35	1.44		6.79	8.15
4800	1/2" x 8"		330	.024		5.45	1.29		6.74	8
5000	3/4" x 10"		375	.021		5.35	1.13		6.48	7.65
5200	Channel siding, 1" x 10", B grade		375	.021		4.87	1.13		6	7.15
5250	Redwood, T&G boards, B grade, 1" x 4"		220	.036		7.10	1.93		9.03	10.90
5270	1" x 8"		330	.024		8.35	1.29		9.64	11.20
5400	White pine, rough sawn, 1" x 8", natural		330	.024		2.55	1.29		3.84	4.87
5500	Stained		330	.024		2.41	1.29		3.70	4.71
9000	Minimum labor/equipment charge		2	4	Job		213		213	340

For customer support on your Commercial Renovation Costs with RSMeans data, call 800.448.8182.

201

07 46 29 – Plywood Siding

07 46 29.10 Plywood Siding Options	Crew	Daily Output	Labor-Hours	Unit	Material	2020 Bare Costs Labor	Equipment	Total	Total Incl O&P
0010 **PLYWOOD SIDING OPTIONS**									
0900 Plywood, medium density overlaid, 3/8" thick	2 Carp	750	.021	S.F.	1.35	1.13		2.48	3.31
1000 1/2" thick		700	.023		1.58	1.22		2.80	3.69
1100 3/4" thick		650	.025		2.29	1.31		3.60	4.62
1600 Texture 1-11, cedar, 5/8" thick, natural		675	.024		2.59	1.26		3.85	4.87
1700 Factory stained		675	.024		2.93	1.26		4.19	5.25
1900 Texture 1-11, fir, 5/8" thick, natural		675	.024		1.21	1.26		2.47	3.35
2000 Factory stained		675	.024		2.05	1.26		3.31	4.28
2050 Texture 1-11, S.Y.P., 5/8" thick, natural		675	.024		1.44	1.26		2.70	3.60
2100 Factory stained		675	.024		1.54	1.26		2.80	3.71
2200 Rough sawn cedar, 3/8" thick, natural		675	.024		1.28	1.26		2.54	3.43
2300 Factory stained		675	.024		1.59	1.26		2.85	3.77
2500 Rough sawn fir, 3/8" thick, natural		675	.024		.97	1.26		2.23	3.09
2600 Factory stained		675	.024		1.10	1.26		2.36	3.23
2800 Redwood, textured siding, 5/8" thick		675	.024		2.05	1.26		3.31	4.28
3000 Polyvinyl chloride coated, 3/8" thick		750	.021		1.18	1.13		2.31	3.12
9000 Minimum labor/equipment charge	1 Carp	2	4	Job		213		213	340

07 46 33 – Plastic Siding

07 46 33.10 Vinyl Siding

07 46 33.10 Vinyl Siding	Crew	Daily Output	Labor-Hours	Unit	Material	2020 Bare Costs Labor	Equipment	Total	Total Incl O&P
0010 **VINYL SIDING**									
3995 Clapboard profile, woodgrain texture, .048 thick, double 4	2 Carp	495	.032	S.F.	1.09	1.72		2.81	3.95
4000 Double 5		550	.029		1.09	1.55		2.64	3.68
4005 Single 8		495	.032		1.65	1.72		3.37	4.57
4010 Single 10		550	.029		1.98	1.55		3.53	4.66
4015 .044 thick, double 4		495	.032		1.08	1.72		2.80	3.94
4020 Double 5		550	.029		1.10	1.55		2.65	3.70
4025 .042 thick, double 4		495	.032		1.08	1.72		2.80	3.94
4030 Double 5		550	.029		1.08	1.55		2.63	3.67
4035 Cross sawn texture, .040 thick, double 4		495	.032		.72	1.72		2.44	3.55
4040 Double 5		550	.029		.65	1.55		2.20	3.20
4045 Smooth texture, .042 thick, double 4		495	.032		.91	1.72		2.63	3.76
4050 Double 5		550	.029		.80	1.55		2.35	3.37
4055 Single 8		495	.032		.80	1.72		2.52	3.64
4060 Cedar texture, .044 thick, double 4		495	.032		1.18	1.72		2.90	4.05
4065 Double 6		600	.027		1.31	1.42		2.73	3.72
4070 Dutch lap profile, woodgrain texture, .048 thick, double 5		550	.029		1.15	1.55		2.70	3.75
4075 .044 thick, double 4.5		525	.030		1.09	1.62		2.71	3.80
4080 .042 thick, double 4.5		525	.030		.92	1.62		2.54	3.62
4085 .040 thick, double 4.5		525	.030		.72	1.62		2.34	3.40
4100 Shake profile, 10" wide		400	.040		4.04	2.13		6.17	7.85
4105 Vertical pattern, .046 thick, double 5		550	.029		1.77	1.55		3.32	4.43
4110 .044 thick, triple 3		550	.029		1.77	1.55		3.32	4.43
4115 .040 thick, triple 4		550	.029		1.72	1.55		3.27	4.38
4120 .040 thick, triple 2.66		550	.029		1.88	1.55		3.43	4.55
4125 Insulation, fan folded extruded polystyrene, 1/4"		2000	.008		.29	.43		.72	1
4130 3/8"		2000	.008		.32	.43		.75	1.04
4135 Accessories, J channel, 5/8" pocket		700	.023	L.F.	.57	1.22		1.79	2.57
4140 3/4" pocket		695	.023		.63	1.22		1.85	2.65
4145 1-1/4" pocket		680	.024		1.15	1.25		2.40	3.26
4150 Flexible, 3/4" pocket		600	.027		2.85	1.42		4.27	5.40
4155 Under sill finish trim		500	.032		.59	1.70		2.29	3.37
4160 Vinyl starter strip		700	.023		.73	1.22		1.95	2.75

07 46 33 – Plastic Siding

07 46 33.10 Vinyl Siding

		Crew	Daily Output	Labor-Hours	Unit	Material	2020 Bare Costs Labor	2020 Bare Costs Equipment	Total	Total Incl O&P
4165	Aluminum starter strip	2 Carp	700	.023	L.F.	.31	1.22		1.53	2.29
4170	Window casing, 2-1/2" wide, 3/4" pocket		510	.031		1.78	1.67		3.45	4.63
4175	Outside corner, woodgrain finish, 4" face, 3/4" pocket		700	.023		2.35	1.22		3.57	4.54
4180	5/8" pocket		700	.023		2.29	1.22		3.51	4.47
4185	Smooth finish, 4" face, 3/4" pocket		700	.023		2.36	1.22		3.58	4.55
4190	7/8" pocket		690	.023		1.91	1.23		3.14	4.08
4195	1-1/4" pocket		700	.023		1.35	1.22		2.57	3.44
4200	Soffit and fascia, 1' overhang, solid		120	.133		5.15	7.10		12.25	17
4205	Vented		120	.133		5.15	7.10		12.25	17
4207	18" overhang, solid		110	.145		6.05	7.75		13.80	19.05
4208	Vented		110	.145		6.05	7.75		13.80	19.05
4210	2' overhang, solid		100	.160		6.90	8.50		15.40	21
4215	Vented		100	.160		6.90	8.50		15.40	21
4217	3' overhang, solid		100	.160		8.70	8.50		17.20	23
4218	Vented		100	.160		8.70	8.50		17.20	23
4220	Colors for siding and soffits, add				S.F.	.15			.15	.17
4225	Colors for accessories and trim, add				L.F.	.31			.31	.34
9000	Minimum labor/equipment charge	1 Carp	3	2.667	Job		142		142	227

07 46 33.20 Polypropylene Siding

		Crew	Daily Output	Labor-Hours	Unit	Material	2020 Bare Costs Labor	2020 Bare Costs Equipment	Total	Total Incl O&P
0010	**POLYPROPYLENE SIDING**									
4090	Shingle profile, random grooves, double 7	2 Carp	400	.040	S.F.	3.51	2.13		5.64	7.30
4092	Cornerpost for above	1 Carp	365	.022	L.F.	12.85	1.17		14.02	16
4095	Triple 5	2 Carp	400	.040	S.F.	3.72	2.13		5.85	7.50
4097	Cornerpost for above	1 Carp	365	.022	L.F.	12.40	1.17		13.57	15.50
5000	Staggered butt, double 7"	2 Carp	400	.040	S.F.	3.64	2.13		5.77	7.40
5002	Cornerpost for above	1 Carp	365	.022	L.F.	13.35	1.17		14.52	16.55
5010	Half round, double 6-1/4"	2 Carp	360	.044	S.F.	4.27	2.36		6.63	8.50
5020	Shake profile, staggered butt, double 9"	"	510	.031	"	3.80	1.67		5.47	6.85
5022	Cornerpost for above	1 Carp	365	.022	L.F.	10	1.17		11.17	12.85
5030	Straight butt, double 7"	2 Carp	400	.040	S.F.	4.27	2.13		6.40	8.10
5032	Cornerpost for above	1 Carp	365	.022	L.F.	14.55	1.17		15.72	17.85
6000	Accessories, J channel, 5/8" pocket	2 Carp	700	.023		.57	1.22		1.79	2.57
6010	3/4" pocket		695	.023		.63	1.22		1.85	2.65
6020	1-1/4" pocket		680	.024		1.15	1.25		2.40	3.26
6030	Aluminum starter strip		700	.023		.31	1.22		1.53	2.29

07 46 46 – Fiber Cement Siding

07 46 46.10 Fiber Cement Siding

		Crew	Daily Output	Labor-Hours	Unit	Material	2020 Bare Costs Labor	2020 Bare Costs Equipment	Total	Total Incl O&P
0010	**FIBER CEMENT SIDING**									
0020	Lap siding, 5/16" thick, 6" wide, 4-3/4" exposure, smooth texture	2 Carp	415	.039	S.F.	1.41	2.05		3.46	4.83
0025	Woodgrain texture		415	.039		1.41	2.05		3.46	4.83
0030	7-1/2" wide, 6-1/4" exposure, smooth texture		425	.038		2	2		4	5.40
0035	Woodgrain texture		425	.038		2	2		4	5.40
0040	8" wide, 6-3/4" exposure, smooth texture		425	.038		1.31	2		3.31	4.66
0045	Rough sawn texture		425	.038		1.31	2		3.31	4.66
0050	9-1/2" wide, 8-1/4" exposure, smooth texture		440	.036		1.19	1.93		3.12	4.41
0055	Woodgrain texture		440	.036		1.19	1.93		3.12	4.41
0060	12" wide, 10-3/8" exposure, smooth texture		455	.035		2.18	1.87		4.05	5.40
0065	Woodgrain texture		455	.035		2.18	1.87		4.05	5.40
0070	Panel siding, 5/16" thick, smooth texture		750	.021		1.22	1.13		2.35	3.16
0075	Stucco texture		750	.021		1.22	1.13		2.35	3.16
0080	Grooved woodgrain texture		750	.021		1.22	1.13		2.35	3.16
0088	Shingle siding, 48" x 15-1/4" panels, 7" exposure		700	.023		4.31	1.22		5.53	6.70

07 46 Siding

07 46 46 – Fiber Cement Siding

07 46 46.10 Fiber Cement Siding	Crew	Daily Output	Labor-Hours	Unit	Material	2020 Bare Costs Labor	2020 Bare Costs Equipment	Total	Total Incl O&P
0090	Wood starter strip	2 Carp	400	.040	L.F.	.47	2.13		2.60	3.93

07 46 73 – Soffit

07 46 73.10 Soffit Options

	07 46 73.10 Soffit Options	Crew	Daily Output	Labor-Hours	Unit	Material	Labor	Equipment	Total	Total Incl O&P
0010	**SOFFIT OPTIONS**									
0012	Aluminum, residential, .020" thick	1 Carp	210	.038	S.F.	2.25	2.03		4.28	5.70
0100	Baked enamel on steel, 16 or 18 ga.		105	.076		6.45	4.05		10.50	13.60
0300	Polyvinyl chloride, white, solid		230	.035		2.24	1.85		4.09	5.40
0400	Perforated		230	.035		2.24	1.85		4.09	5.40
0500	For colors, add					.15			.15	.17
9000	Minimum labor/equipment charge	1 Carp	3	2.667	Job		142		142	227

07 51 Built-Up Bituminous Roofing

07 51 13 – Built-Up Asphalt Roofing

07 51 13.10 Built-Up Roofing Components

	07 51 13.10 Built-Up Roofing Components	Crew	Daily Output	Labor-Hours	Unit	Material	Labor	Equipment	Total	Total Incl O&P
0010	**BUILT-UP ROOFING COMPONENTS**									
0012	Asphalt saturated felt, #30, 2 sq./roll	1 Rofc	58	.138	Sq.	10.20	6.35		16.55	22.50
0200	#15, 4 sq./roll, plain or perforated, not mopped		58	.138		5.25	6.35		11.60	16.95
0300	Roll roofing, smooth, #65		15	.533		10.30	24.50		34.80	54.50
0500	#90		12	.667		37.50	31		68.50	95
0520	Mineralized		12	.667		36	31		67	93.50
0540	D.C. (double coverage), 19" selvage edge		10	.800		48	37		85	117
0580	Adhesive (lap cement)				Gal.	8.50			8.50	9.35
0800	Steep, flat or dead level asphalt, 10 ton lots, packaged				Ton	885			885	975
9000	Minimum labor/equipment charge	1 Rofc	4	2	Job		92.50		92.50	162

07 51 13.13 Cold-Applied Built-Up Asphalt Roofing

	07 51 13.13 Cold-Applied Built-Up Asphalt Roofing	Crew	Daily Output	Labor-Hours	Unit	Material	Labor	Equipment	Total	Total Incl O&P
0010	**COLD-APPLIED BUILT-UP ASPHALT ROOFING**									
0020	3 ply system, installation only (components listed below)	G-5	50	.800	Sq.		33.50	3.86	37.36	62.50
0100	Spunbond poly. fabric, 1.35 oz./S.Y., 36"W, 10.8 sq./roll				Ea.	133			133	146
0500	Base & finish coat, 3 gal./sq., 5 gal./can				Gal.	8.10			8.10	8.95
0600	Coating, ceramic granules, 1/2 sq./bag				Ea.	23			23	25.50
0700	Aluminum, 2 gal./sq.				Gal.	12.45			12.45	13.70
0800	Emulsion, fibered or non-fibered, 4 gal./sq.				"	6.65			6.65	7.30

07 51 13.20 Built-Up Roofing Systems

	07 51 13.20 Built-Up Roofing Systems	Crew	Daily Output	Labor-Hours	Unit	Material	Labor	Equipment	Total	Total Incl O&P
0010	**BUILT-UP ROOFING SYSTEMS**									
0120	Asphalt flood coat with gravel/slag surfacing, not including									
0140	Insulation, flashing or wood nailers									
0200	Asphalt base sheet, 3 plies #15 asphalt felt, mopped	G-1	22	2.545	Sq.	108	110	26	244	340
0350	On nailable decks		21	2.667		100	115	27	242	340
0500	4 plies #15 asphalt felt, mopped		20	2.800		133	121	28.50	282.50	390
0550	On nailable decks		19	2.947		118	127	30	275	385
0700	Coated glass base sheet, 2 plies glass (type IV), mopped		22	2.545		108	110	26	244	340
0850	3 plies glass, mopped		20	2.800		130	121	28.50	279.50	385
0950	On nailable decks		19	2.947		123	127	30	280	390
1100	4 plies glass fiber felt (type IV), mopped		20	2.800		161	121	28.50	310.50	420
1150	On nailable decks		19	2.947		146	127	30	303	415
1200	Coated & saturated base sheet, 3 plies #15 asph. felt, mopped		20	2.800		109	121	28.50	258.50	365
1250	On nailable decks		19	2.947		102	127	30	259	370
1300	4 plies #15 asphalt felt, mopped		22	2.545		127	110	26	263	360
2000	Asphalt flood coat, smooth surface									
2200	Asphalt base sheet & 3 plies #15 asphalt felt, mopped	G-1	24	2.333	Sq.	102	101	23.50	226.50	315

07 51 Built-Up Bituminous Roofing

07 51 13 – Built-Up Asphalt Roofing

07 51 13.20 Built-Up Roofing Systems

		Crew	Daily Output	Labor-Hours	Unit	Material	2020 Bare Costs Labor	Equipment	Total	Total Incl O&P
2400	On nailable decks	G-1	23	2.435	Sq.	95	105	24.50	224.50	315
2600	4 plies #15 asphalt felt, mopped		24	2.333		120	101	23.50	244.50	335
2700	On nailable decks	↓	23	2.435	↓	113	105	24.50	242.50	335
2900	Coated glass fiber base sheet, mopped, and 2 plies of									
2910	glass fiber felt (type IV)	G-1	25	2.240	Sq.	102	96.50	22.50	221	305
3100	On nailable decks		24	2.333		97	101	23.50	221.50	310
3200	3 plies, mopped		23	2.435		125	105	24.50	254.50	350
3300	On nailable decks		22	2.545		118	110	26	254	350
3800	4 plies glass fiber felt (type IV), mopped		23	2.435		148	105	24.50	277.50	375
3900	On nailable decks		22	2.545		140	110	26	276	375
4000	Coated & saturated base sheet, 3 plies #15 asph. felt, mopped		24	2.333		104	101	23.50	228.50	315
4200	On nailable decks		23	2.435		96	105	24.50	225.50	315
4300	4 plies #15 organic felt, mopped	↓	22	2.545	↓	121	110	26	257	355
4500	Coal tar pitch with gravel/slag surfacing									
4600	4 plies #15 tarred felt, mopped	G-1	21	2.667	Sq.	204	115	27	346	455
4800	3 plies glass fiber felt (type IV), mopped	"	19	2.947	"	168	127	30	325	440
5000	Coated glass fiber base sheet, and 2 plies of									
5010	glass fiber felt (type IV), mopped	G-1	19	2.947	Sq.	172	127	30	329	445
5300	On nailable decks		18	3.111		150	134	31.50	315.50	435
5600	4 plies glass fiber felt (type IV), mopped		21	2.667		233	115	27	375	485
5800	On nailable decks	↓	20	2.800	↓	212	121	28.50	361.50	475

07 51 13.30 Cants

		Crew	Daily Output	Labor-Hours	Unit	Material	2020 Bare Costs Labor	Equipment	Total	Total Incl O&P
0010	**CANTS**									
0012	Lumber, treated, 4" x 4" cut diagonally	1 Rofc	325	.025	L.F.	1.89	1.14		3.03	4.07
0300	Mineral or fiber, trapezoidal, 1" x 4" x 48"		325	.025		.31	1.14		1.45	2.33
0400	1-1/2" x 5-5/8" x 48"		325	.025	↓	.48	1.14		1.62	2.52
9000	Minimum labor/equipment charge	↓	4	2	Job		92.50		92.50	162

07 51 13.40 Felts

		Crew	Daily Output	Labor-Hours	Unit	Material	2020 Bare Costs Labor	Equipment	Total	Total Incl O&P
0010	**FELTS**									
0012	Glass fibered roofing felt, #15, not mopped	1 Rofc	58	.138	Sq.	10.20	6.35		16.55	22.50
0300	Base sheet, #80, channel vented		58	.138		48.50	6.35		54.85	64.50
0400	#70, coated		58	.138		18.70	6.35		25.05	31.50
0500	Cap, #87, mineral surfaced		58	.138		84.50	6.35		90.85	104
0600	Flashing membrane, #65		16	.500		10.30	23		33.30	52
0800	Coal tar fibered, #15, no mopping		58	.138		17.55	6.35		23.90	30.50
0900	Asphalt felt, #15, 4 sq./roll, no mopping		58	.138		5.25	6.35		11.60	16.95
1100	#30, 2 sq./roll		58	.138		10.20	6.35		16.55	22.50
1200	Double coated, #33		58	.138		11.90	6.35		18.25	24.50
1400	#40, base sheet		58	.138		10.95	6.35		17.30	23
1450	Coated and saturated		58	.138		12.20	6.35		18.55	24.50
1500	Tarred felt, organic, #15, 4 sq. rolls		58	.138		14.70	6.35		21.05	27.50
1550	#30, 2 sq. roll	↓	58	.138		25	6.35		31.35	38.50
1700	Add for mopping above felts, per ply, asphalt, 24 lb./sq.	G-1	192	.292		10.65	12.60	2.95	26.20	37
1800	Coal tar mopping, 30 lb./sq.		186	.301		18.50	13	3.05	34.55	47
1900	Flood coat, with asphalt, 60 lb./sq.		60	.933		26.50	40.50	9.45	76.45	110
2000	With coal tar, 75 lb./sq.	↓	56	1	↓	46	43	10.10	99.10	138
9000	Minimum labor/equipment charge	1 Rofc	4	2	Job		92.50		92.50	162

07 51 13.50 Walkways for Built-Up Roofs

		Crew	Daily Output	Labor-Hours	Unit	Material	2020 Bare Costs Labor	Equipment	Total	Total Incl O&P
0010	**WALKWAYS FOR BUILT-UP ROOFS**									
0020	Asphalt impregnated, 3' x 6' x 1/2" thick	1 Rofc	400	.020	S.F.	1.98	.92		2.90	3.80
0100	3' x 3' x 3/4" thick		400	.020	"	5.70	.92		6.62	7.90
0600	100% recycled rubber, 3' x 4' x 3/8" [G]	↓	400	.020	L.F.	7.35	.92		8.27	9.70

For customer support on your Commercial Renovation Costs with RSMeans data, call 800.448.8182.

205

07 51 Built-Up Bituminous Roofing

07 51 13 – Built-Up Asphalt Roofing

07 51 13.50 Walkways for Built-Up Roofs		Crew	Daily Output	Labor-Hours	Unit	Material	2020 Bare Costs Labor	Equipment	Total	Total Incl O&P
0610	3' x 4' x 1/2"	[G] 1 Rofc	400	.020	L.F.	8.20	.92		9.12	10.65
0620	3' x 4' x 3/4"	[G]	400	.020		8.60	.92		9.52	11.05
9000	Minimum labor/equipment charge		2.75	2.909	Job		134		134	235

07 52 Modified Bituminous Membrane Roofing

07 52 13 – Atactic-Polypropylene-Modified Bituminous Membrane Roofing

07 52 13.10 APP Modified Bituminous Membrane

		Crew	Daily Output	Labor-Hours	Unit	Material	2020 Bare Costs Labor	Equipment	Total	Total Incl O&P
0010	**APP MODIFIED BITUMINOUS MEMBRANE** R075213-30									
0020	Base sheet, #15 glass fiber felt, nailed to deck	1 Rofc	58	.138	Sq.	11.85	6.35		18.20	24
0030	Spot mopped to deck	G-1	295	.190		15.55	8.20	1.92	25.67	33.50
0040	Fully mopped to deck	"	192	.292		21	12.60	2.95	36.55	48.50
0050	#15 organic felt, nailed to deck	1 Rofc	58	.138		6.90	6.35		13.25	18.75
0060	Spot mopped to deck	G-1	295	.190		10.60	8.20	1.92	20.72	28
0070	Fully mopped to deck	"	192	.292		15.90	12.60	2.95	31.45	43
2100	APP mod., smooth surf. cap sheet, poly. reinf., torched, 160 mils	G-5	2100	.019	S.F.	.84	.80	.09	1.73	2.42
2150	170 mils		2100	.019		.76	.80	.09	1.65	2.34
2200	Granule surface cap sheet, poly. reinf., torched, 180 mils		2000	.020		.96	.84	.10	1.90	2.64
2250	Smooth surface flashing, torched, 160 mils		1260	.032		.84	1.33	.15	2.32	3.42
2300	170 mils		1260	.032		.76	1.33	.15	2.24	3.34
2350	Granule surface flashing, torched, 180 mils		1260	.032		.96	1.33	.15	2.44	3.56
2400	Fibrated aluminum coating	1 Rofc	3800	.002		.09	.10		.19	.27
2450	Seam heat welding	"	205	.039	L.F.	.09	1.80		1.89	3.25

07 52 16 – Styrene-Butadiene-Styrene Modified Bituminous Membrane Roofing

07 52 16.10 SBS Modified Bituminous Membrane

		Crew	Daily Output	Labor-Hours	Unit	Material	2020 Bare Costs Labor	Equipment	Total	Total Incl O&P
0010	**SBS MODIFIED BITUMINOUS MEMBRANE**									
0080	Mod. bit. rfng., SBS mod, gran surf. cap sheet, poly. reinf.									
0650	120 to 149 mils thick	G-1	2000	.028	S.F.	1.36	1.21	.28	2.85	3.93
0750	150 to 160 mils		2000	.028		1.80	1.21	.28	3.29	4.41
1600	Smooth surface cap sheet, mopped, 145 mils		2100	.027		.83	1.15	.27	2.25	3.22
1620	Lightweight base sheet, fiberglass reinforced, 35 to 47 mil		2100	.027		.29	1.15	.27	1.71	2.63
1625	Heavyweight base/ply sheet, reinforced, 87 to 120 mil thick		2100	.027		.93	1.15	.27	2.35	3.33
1650	Granulated walkpad, 180 to 220 mils	1 Rofc	400	.020		1.86	.92		2.78	3.67
1700	Smooth surface flashing, 145 mils	G-1	1260	.044		.83	1.92	.45	3.20	4.77
1800	150 mils		1260	.044		.51	1.92	.45	2.88	4.42
1900	Granular surface flashing, 150 mils		1260	.044		.69	1.92	.45	3.06	4.62
2000	160 mils		1260	.044		.78	1.92	.45	3.15	4.72
2010	Elastomeric asphalt primer	1 Rofc	2600	.003		.17	.14		.31	.44
2015	Roofing asphalt, 30 lb./square	G-1	19000	.003		.13	.13	.03	.29	.40
2020	Cold process adhesive, 20 to 30 mils thick	1 Rofc	750	.011		.26	.49		.75	1.15
2025	Self adhering vapor retarder, 30 to 45 mils thick	G-5	2150	.019		1.07	.78	.09	1.94	2.65
2050	Seam heat welding	1 Rofc	205	.039	L.F.	.09	1.80		1.89	3.25

07 53 Elastomeric Membrane Roofing

07 53 16 – Chlorosulfonate-Polyethylene Roofing

07 53 16.10 Chlorosulfonated Polyethylene Roofing	Crew	Daily Output	Labor-Hours	Unit	Material	2020 Bare Costs Labor	Equipment	Total	Total Incl O&P
0010 **CHLOROSULFONATED POLYETHYLENE ROOFING**									
0800 Chlorosulfonated polyethylene (CSPE)									
0900 45 mils, heat welded seams, plate attachment	G-5	35	1.143	Sq.	249	48	5.50	302.50	365
1100 Heat welded seams, plate attachment and ballasted		26	1.538		260	64.50	7.40	331.90	405
1200 60 mils, heat welded seams, plate attachment		35	1.143		335	48	5.50	388.50	455
1300 Heat welded seams, plate attachment and ballasted		26	1.538		345	64.50	7.40	416.90	500

07 53 23 – Ethylene-Propylene-Diene-Monomer Roofing

07 53 23.20 Ethylene-Propylene-Diene-Monomer Roofing

	Crew	Daily Output	Labor-Hours	Unit	Material	Labor	Equipment	Total	Total Incl O&P
0010 **ETHYLENE-PROPYLENE-DIENE-MONOMER ROOFING (EPDM)**									
3500 Ethylene-propylene-diene-monomer (EPDM), 45 mils, 0.28 psf									
3600 Loose-laid & ballasted with stone (10 psf)	G-5	51	.784	Sq.	92	33	3.78	128.78	163
3700 Mechanically attached		35	1.143		85.50	48	5.50	139	184
3800 Fully adhered with adhesive		26	1.538		120	64.50	7.40	191.90	253
4500 60 mils, 0.40 psf									
4600 Loose-laid & ballasted with stone (10 psf)	G-5	51	.784	Sq.	112	33	3.78	148.78	186
4700 Mechanically attached		35	1.143		105	48	5.50	158.50	205
4800 Fully adhered with adhesive		26	1.538		139	64.50	7.40	210.90	274
4810 45 mil, 0.28 psf, membrane only					55			55	60.50
4820 60 mil, 0.40 psf, membrane only					72			72	79.50
4850 Seam tape for membrane, 3" x 100' roll				Ea.	49.50			49.50	54.50
4900 Batten strips, 10' sections					4.05			4.05	4.46
4910 Cover tape for batten strips, 6" x 100' roll					181			181	199
4930 Plate anchors				M	82.50			82.50	91
4970 Adhesive for fully adhered systems, 60 S.F./gal.				Gal.	21			21	23

07 53 29 – Polyisobutylene Roofing

07 53 29.10 Polyisobutylene Roofing

	Crew	Daily Output	Labor-Hours	Unit	Material	Labor	Equipment	Total	Total Incl O&P
0010 **POLYISOBUTYLENE ROOFING**									
7500 Polyisobutylene (PIB), 100 mils, 0.57 psf									
7600 Loose-laid & ballasted with stone/gravel (10 psf)	G-5	51	.784	Sq.	215	33	3.78	251.78	298
7700 Partially adhered with adhesive		35	1.143		253	48	5.50	306.50	370
7800 Hot asphalt attachment		35	1.143		241	48	5.50	294.50	355
7900 Fully adhered with contact cement		26	1.538		266	64.50	7.40	337.90	415

07 54 Thermoplastic Membrane Roofing

07 54 16 – Ketone Ethylene Ester Roofing

07 54 16.10 Ketone Ethylene Ester Roofing

	Crew	Daily Output	Labor-Hours	Unit	Material	Labor	Equipment	Total	Total Incl O&P
0010 **KETONE ETHYLENE ESTER ROOFING**									
0100 Ketone ethylene ester roofing, 50 mil, fully adhered	G-5	26	1.538	Sq.	216	64.50	7.40	287.90	360
0120 Mechanically attached		35	1.143		140	48	5.50	193.50	244
0140 Ballasted with stone		51	.784		148	33	3.78	184.78	225
0160 50 mil, fleece backed, adhered w/hot asphalt	G-1	26	2.154		160	93	22	275	360
0180 Accessories, pipe boot	1 Rofc	32	.250	Ea.	26	11.55		37.55	48.50
0200 Pre-formed corners		32	.250	"	9.75	11.55		21.30	31
0220 Ketone clad metal, including up to 4 bends		330	.024	S.F.	4.02	1.12		5.14	6.40
0240 Walkway pad	2 Rofc	800	.020	"	4.27	.92		5.19	6.30
0260 Stripping material	1 Rofc	310	.026	L.F.	.99	1.19		2.18	3.18

For customer support on your Commercial Renovation Costs with RSMeans data, call 800.448.8182.

207

07 54 Thermoplastic Membrane Roofing

07 54 19 – Polyvinyl-Chloride Roofing

07 54 19.10 Polyvinyl-Chloride Roofing (PVC)	Crew	Daily Output	Labor-Hours	Unit	Material	2020 Bare Costs Labor	Equipment	Total	Total Incl O&P
0010 **POLYVINYL-CHLORIDE ROOFING (PVC)**									
8200 Heat welded seams									
8700 Reinforced, 48 mils, 0.33 psf									
8750 Loose-laid & ballasted with stone/gravel (12 psf)	G-5	51	.784	Sq.	127	33	3.78	163.78	202
8800 Mechanically attached		35	1.143		120	48	5.50	173.50	222
8850 Fully adhered with adhesive		26	1.538		166	64.50	7.40	237.90	305
8860 Reinforced, 60 mils, 0.40 psf									
8870 Loose-laid & ballasted with stone/gravel (12 psf)	G-5	51	.784	Sq.	129	33	3.78	165.78	204
8880 Mechanically attached		35	1.143		121	48	5.50	174.50	224
8890 Fully adhered with adhesive		26	1.538		167	64.50	7.40	238.90	305

07 54 23 – Thermoplastic-Polyolefin Roofing

07 54 23.10 Thermoplastic Polyolefin Roofing (T.P.O.)	Crew	Daily Output	Labor-Hours	Unit	Material	2020 Bare Costs Labor	Equipment	Total	Total Incl O&P
0010 **THERMOPLASTIC POLYOLEFIN ROOFING (T.P.O.)**									
0100 45 mil, loose laid & ballasted with stone (1/2 ton/sq.)	G-5	51	.784	Sq.	86.50	33	3.78	123.28	157
0120 Fully adhered		25	1.600		76	67	7.70	150.70	209
0140 Mechanically attached		34	1.176		75.50	49.50	5.65	130.65	176
0160 Self adhered		35	1.143		79	48	5.50	132.50	177
0180 60 mil membrane, heat welded seams, ballasted		50	.800		100	33.50	3.86	137.36	173
0200 Fully adhered		25	1.600		89	67	7.70	163.70	224
0220 Mechanically attached		34	1.176		93	49.50	5.65	148.15	195
0240 Self adhered		35	1.143		106	48	5.50	159.50	207

07 55 Protected Membrane Roofing

07 55 10 – Protected Membrane Roofing Components

07 55 10.10 Protected Membrane Roofing Components	Crew	Daily Output	Labor-Hours	Unit	Material	2020 Bare Costs Labor	Equipment	Total	Total Incl O&P
0010 **PROTECTED MEMBRANE ROOFING COMPONENTS**									
0100 Choose roofing membrane from 07 50									
0120 Then choose roof deck insulation from 07 22									
0130 Filter fabric	2 Rofc	10000	.002	S.F.	.10	.07		.17	.24
0140 Ballast, 3/8" - 1/2" in place	G-1	36	1.556	Ton	22	67	15.75	104.75	159
0150 3/4" - 1-1/2" in place	"	36	1.556	"	22	67	15.75	104.75	159
0200 2" concrete blocks, natural	1 Clab	115	.070	S.F.	3.58	2.93		6.51	8.65
0210 Colors	"	115	.070	"	3.71	2.93		6.64	8.75

07 56 Fluid-Applied Roofing

07 56 10 – Fluid-Applied Roofing Elastomers

07 56 10.10 Elastomeric Roofing	Crew	Daily Output	Labor-Hours	Unit	Material	2020 Bare Costs Labor	Equipment	Total	Total Incl O&P
0010 **ELASTOMERIC ROOFING**									
0020 Acrylic, 44% solids, 2 coats, on corrugated metal	2 Rofc	2400	.007	S.F.	.56	.31		.87	1.16
0025 On smooth metal		3000	.005		.45	.25		.70	.92
0030 On foam or modified bitumen		1500	.011		.90	.49		1.39	1.84
0035 On concrete		1500	.011		.90	.49		1.39	1.84
0040 On tar and gravel		1500	.011		.90	.49		1.39	1.84
0045 36% solids, 2 coats, on corrugated metal		2400	.007		.51	.31		.82	1.10
0050 On smooth metal		3000	.005		.41	.25		.66	.88
0055 On foam or modified bitumen		1500	.011		.82	.49		1.31	1.76
0060 On concrete		1500	.011		.82	.49		1.31	1.76
0065 On tar and gravel		1500	.011		.82	.49		1.31	1.76

For customer support on your Commercial Renovation Costs with RSMeans data, call 800.448.8182.

07 56 Fluid-Applied Roofing

07 56 10 – Fluid-Applied Roofing Elastomers

07 56 10.10 Elastomeric Roofing	Crew	Daily Output	Labor-Hours	Unit	Material	2020 Bare Costs Labor	Equipment	Total	Total Incl O&P
0070	Primer if required, 2 coats on corrugated metal	2 Rofc	2400	.007	S.F.	.51	.31		.82	1.10
0075	On smooth metal		3000	.005		.51	.25		.76	.99
0080	On foam or modified bitumen		1500	.011		.69	.49		1.18	1.62
0085	On concrete		1500	.011		.52	.49		1.01	1.43
0090	On tar & gravel/rolled roof		1500	.011		.69	.49		1.18	1.62
0110	Acrylic rubber, fluid applied, 20 mils thick	G-5	2000	.020		2.20	.84	.10	3.14	4
0120	50 mils, reinforced		1200	.033		3.23	1.40	.16	4.79	6.20
0130	For walking surface, add		900	.044		1.34	1.86	.21	3.41	4.97
0300	Neoprene, fluid applied, 20 mil thick, not reinforced	G-1	1135	.049		1.31	2.13	.50	3.94	5.70
0600	Non-woven polyester, reinforced		960	.058		1.43	2.52	.59	4.54	6.65
0700	5 coat neoprene deck, 60 mil thick, under 10,000 S.F.		325	.172		4.37	7.45	1.74	13.56	19.70
0900	Over 10,000 S.F.		625	.090		4.37	3.87	.91	9.15	12.55
9000	Minimum labor/equipment charge	1 Rofc	2	4	Job		185		185	325

07 57 Coated Foamed Roofing

07 57 13 – Sprayed Polyurethane Foam Roofing

07 57 13.10 Sprayed Polyurethane Foam Roofing (S.P.F.)	Crew	Daily Output	Labor-Hours	Unit	Material	2020 Bare Costs Labor	Equipment	Total	Total Incl O&P
0010	**SPRAYED POLYURETHANE FOAM ROOFING (S.P.F.)**									
0100	Primer for metal substrate (when required)	G-2A	3000	.008	S.F.	.51	.33	.21	1.05	1.35
0200	Primer for non-metal substrate (when required)		3000	.008		.19	.33	.21	.73	.99
0300	Closed cell spray, polyurethane foam, 3 lb./C.F. density, 1", R6.7		15000	.002		.64	.07	.04	.75	.87
0400	2", R13.4		13125	.002		1.29	.08	.05	1.42	1.60
0500	3", R18.6		11485	.002		1.93	.09	.06	2.08	2.34
0550	4", R24.8		10080	.002		2.58	.10	.06	2.74	3.07
0700	Spray-on silicone coating		2500	.010		1.28	.39	.25	1.92	2.36

07 58 Roll Roofing

07 58 10 – Asphalt Roll Roofing

07 58 10.10 Roll Roofing	Crew	Daily Output	Labor-Hours	Unit	Material	2020 Bare Costs Labor	Equipment	Total	Total Incl O&P
0010	**ROLL ROOFING**									
0100	Asphalt, mineral surface									
0200	1 ply #15 organic felt, 1 ply mineral surfaced									
0300	Selvage roofing, lap 19", nailed & mopped	G-1	27	2.074	Sq.	69	89.50	21	179.50	256
0400	3 plies glass fiber felt (type IV), 1 ply mineral surfaced									
0500	Selvage roofing, lapped 19", mopped	G-1	25	2.240	Sq.	121	96.50	22.50	240	325
0600	Coated glass fiber base sheet, 2 plies of glass fiber									
0700	Felt (type IV), 1 ply mineral surfaced selvage									
0800	Roofing, lapped 19", mopped	G-1	25	2.240	Sq.	130	96.50	22.50	249	335
0900	On nailable decks	"	24	2.333	"	119	101	23.50	243.50	335
1000	3 plies glass fiber felt (type III), 1 ply mineral surfaced									
1100	Selvage roofing, lapped 19", mopped	G-1	25	2.240	Sq.	121	96.50	22.50	240	325

07 61 Sheet Metal Roofing

07 61 13 – Standing Seam Sheet Metal Roofing

07 61 13.10 Standing Seam Sheet Metal Roofing, Field Fab.	Crew	Daily Output	Labor-Hours	Unit	Material	2020 Bare Costs Labor	2020 Bare Costs Equipment	Total	Total Incl O&P
0010 **STANDING SEAM SHEET METAL ROOFING, FIELD FABRICATED**									
0400 Copper standing seam roofing, over 10 squares, 16 oz., 125 lb./sq.	1 Shee	1.30	6.154	Sq.	1,025	385		1,410	1,725
0600 18 oz., 140 lb./sq.		1.20	6.667		1,150	415		1,565	1,900
0700 20 oz., 150 lb./sq.		1.10	7.273		1,300	455		1,755	2,150
1200 For abnormal conditions or small areas, add					25%	100%			
1300 For lead-coated copper, add					25%				
1400 Lead standing seam roofing, 5 lb./S.F.	1 Shee	1.30	6.154		1,625	385		2,010	2,400
1500 Zinc standing seam roofing, .020" thick		1.20	6.667		1,475	415		1,890	2,275
1510 .027" thick		1.15	6.957		1,825	435		2,260	2,675
1520 .032" thick		1.10	7.273		1,950	455		2,405	2,850
1530 .040" thick		1.05	7.619		2,575	475		3,050	3,575
9000 Minimum labor/equipment charge		2	4	Job		249		249	390

07 61 16 – Batten Seam Sheet Metal Roofing

07 61 16.10 Batten Seam Sheet Metal Roofing, Field Fab.

	Crew	Daily Output	Labor-Hours	Unit	Material	2020 Bare Costs Labor	2020 Bare Costs Equipment	Total	Total Incl O&P
0010 **BATTEN SEAM SHEET METAL ROOFING, FIELD FABRICATED**									
0012 Copper batten seam roofing, over 10 sq., 16 oz., 130 lb./sq.	1 Shee	1.10	7.273	Sq.	1,300	455		1,755	2,125
0020 Lead batten seam roofing, 5 lb./S.F.		1.20	6.667		1,700	415		2,115	2,525
0100 Zinc/copper alloy batten seam roofing, .020" thick		1.20	6.667		1,550	415		1,965	2,350
0200 Copper roofing, batten seam, over 10 sq., 18 oz., 145 lb./sq.		1	8		1,450	500		1,950	2,375
0300 20 oz., 160 lb./sq.		1	8		1,650	500		2,150	2,600
0500 Stainless steel batten seam roofing, type 304, 28 ga.		1.20	6.667		680	415		1,095	1,400
0600 26 ga.		1.15	6.957		770	435		1,205	1,525
0800 Zinc, copper alloy roofing, batten seam, .027" thick		1.15	6.957		1,900	435		2,335	2,775
0900 .032" thick		1.10	7.273		2,025	455		2,480	2,925
1000 .040" thick		1.05	7.619		2,675	475		3,150	3,700
9000 Minimum labor/equipment charge		2	4	Job		249		249	390

07 61 19 – Flat Seam Sheet Metal Roofing

07 61 19.10 Flat Seam Sheet Metal Roofing, Field Fabricated

	Crew	Daily Output	Labor-Hours	Unit	Material	2020 Bare Costs Labor	2020 Bare Costs Equipment	Total	Total Incl O&P
0010 **FLAT SEAM SHEET METAL ROOFING, FIELD FABRICATED**									
0900 Copper flat seam roofing, over 10 squares, 16 oz., 115 lb./sq.	1 Shee	1.20	6.667	Sq.	945	415		1,360	1,700
0950 18 oz., 130 lb./sq.		1.15	6.957		1,050	435		1,485	1,850
1000 20 oz., 145 lb./sq.		1.10	7.273		1,225	455		1,680	2,025
1008 Zinc flat seam roofing, .020" thick		1.20	6.667		1,325	415		1,740	2,100
1010 .027" thick		1.15	6.957		1,625	435		2,060	2,450
1020 .032" thick		1.12	7.143		1,725	445		2,170	2,600
1030 .040" thick		1.05	7.619		2,300	475		2,775	3,275
1100 Lead flat seam roofing, 5 lb./S.F.		1.30	6.154		1,450	385		1,835	2,200
9000 Minimum labor/equipment charge		2.75	2.909	Job		181		181	284

07 62 Sheet Metal Flashing and Trim

07 62 10 – Sheet Metal Trim

07 62 10.10 Sheet Metal Cladding

	Crew	Daily Output	Labor-Hours	Unit	Material	2020 Bare Costs Labor	2020 Bare Costs Equipment	Total	Total Incl O&P
0010 **SHEET METAL CLADDING**									
0100 Aluminum, up to 6 bends, .032" thick, window casing	1 Carp	180	.044	S.F.	1.95	2.36		4.31	5.95
0200 Window sill		72	.111	L.F.	1.95	5.90		7.85	11.60
0300 Door casing		180	.044	S.F.	1.95	2.36		4.31	5.95
0400 Fascia		250	.032		1.95	1.70		3.65	4.87
0500 Rake trim		225	.036		1.95	1.89		3.84	5.20
0700 .024" thick, window casing		180	.044		1.47	2.36		3.83	5.40
0800 Window sill		72	.111	L.F.	1.47	5.90		7.37	11.05

07 62 Sheet Metal Flashing and Trim

07 62 10 – Sheet Metal Trim

07 62 10.10 Sheet Metal Cladding

07 62 10.10 Sheet Metal Cladding		Crew	Daily Output	Labor-Hours	Unit	Material	2020 Bare Costs Labor	Equipment	Total	Total Incl O&P
0900	Door casing	1 Carp	180	.044	S.F.	1.47	2.36		3.83	5.40
1000	Fascia		250	.032		1.47	1.70		3.17	4.34
1100	Rake trim		225	.036		1.47	1.89		3.36	4.65
1200	Vinyl coated aluminum, up to 6 bends, window casing		180	.044		1.94	2.36		4.30	5.90
1300	Window sill		72	.111	L.F.	1.94	5.90		7.84	11.60
1400	Door casing		180	.044	S.F.	1.94	2.36		4.30	5.90
1500	Fascia		250	.032		1.94	1.70		3.64	4.85
1600	Rake trim		225	.036		1.94	1.89		3.83	5.15

07 65 Flexible Flashing

07 65 10 – Sheet Metal Flashing

07 65 10.10 Sheet Metal Flashing and Counter Flashing

07 65 10.10 Sheet Metal Flashing and Counter Flashing		Crew	Daily Output	Labor-Hours	Unit	Material	2020 Bare Costs Labor	Equipment	Total	Total Incl O&P
0010	**SHEET METAL FLASHING AND COUNTER FLASHING**									
0011	Including up to 4 bends									
0020	Aluminum, mill finish, .013" thick	1 Rofc	145	.055	S.F.	1.01	2.55		3.56	5.55
0030	.016" thick		145	.055		1.15	2.55		3.70	5.75
0060	.019" thick		145	.055		1.50	2.55		4.05	6.10
0100	.032" thick		145	.055		1.38	2.55		3.93	6
0200	.040" thick		145	.055		2.33	2.55		4.88	7
0300	.050" thick		145	.055		3.22	2.55		5.77	8
0325	Mill finish 5" x 7" step flashing, .016" thick		1920	.004	Ea.	.15	.19		.34	.51
0350	Mill finish 12" x 12" step flashing, .016" thick		1600	.005	"	.60	.23		.83	1.06
0400	Painted finish, add				S.F.	.34			.34	.37
1000	Mastic-coated 2 sides, .005" thick	1 Rofc	330	.024		1.89	1.12		3.01	4.04
1100	.016" thick		330	.024		2.09	1.12		3.21	4.26
1600	Copper, 16 oz. sheets, under 1000 lb.		115	.070		8.15	3.21		11.36	14.60
1700	Over 4000 lb.		155	.052		8.15	2.38		10.53	13.15
1900	20 oz. sheets, under 1000 lb.		110	.073		10.45	3.36		13.81	17.40
2000	Over 4000 lb.		145	.055		9.95	2.55		12.50	15.40
2200	24 oz. sheets, under 1000 lb.		105	.076		15.30	3.52		18.82	23
2300	Over 4000 lb.		135	.059		14.55	2.74		17.29	21
2500	32 oz. sheets, under 1000 lb.		100	.080		21	3.70		24.70	29.50
2600	Over 4000 lb.		130	.062		19.75	2.84		22.59	26.50
2700	W shape for valleys, 16 oz., 24" wide		100	.080	L.F.	16.65	3.70		20.35	25
5800	Lead, 2.5 lb./S.F., up to 12" wide		135	.059	S.F.	6.10	2.74		8.84	11.50
5900	Over 12" wide		135	.059		4.05	2.74		6.79	9.25
8900	Stainless steel sheets, 32 ga.		155	.052		3.51	2.38		5.89	8.05
9000	28 ga.		155	.052		5	2.38		7.38	9.65
9100	26 ga.		155	.052		5.10	2.38		7.48	9.75
9200	24 ga.		155	.052		5.85	2.38		8.23	10.55
9290	For mechanically keyed flashing, add					40%				
9320	Steel sheets, galvanized, 20 ga.	1 Rofc	130	.062	S.F.	1.22	2.84		4.06	6.30
9322	22 ga.		135	.059		1.27	2.74		4.01	6.20
9324	24 ga.		140	.057		.97	2.64		3.61	5.70
9326	26 ga.		148	.054		.85	2.50		3.35	5.30
9328	28 ga.		155	.052		.73	2.38		3.11	4.98
9340	30 ga.		160	.050		.62	2.31		2.93	4.72
9400	Terne coated stainless steel, .015" thick, 28 ga.		155	.052		8.20	2.38		10.58	13.15
9500	.018" thick, 26 ga.		155	.052		9.15	2.38		11.53	14.20
9600	Zinc and copper alloy (brass), .020" thick		155	.052		10.10	2.38		12.48	15.25
9700	.027" thick		155	.052		12.25	2.38		14.63	17.65

07 65 Flexible Flashing

07 65 10 – Sheet Metal Flashing

07 65 10.10 Sheet Metal Flashing and Counter Flashing

	07 65 10.10 Sheet Metal Flashing and Counter Flashing	Crew	Daily Output	Labor-Hours	Unit	Material	2020 Bare Costs Labor	2020 Bare Costs Equipment	Total	Total Incl O&P
9800	.032" thick	1 Rofc	155	.052	S.F.	15.45	2.38		17.83	21
9900	.040" thick		155	.052		20.50	2.38		22.88	26.50
9950	Minimum labor/equipment charge		3	2.667	Job		123		123	216

07 65 12 – Fabric and Mastic Flashings

07 65 12.10 Fabric and Mastic Flashing and Counter Flashing

	07 65 12.10 Fabric and Mastic Flashing and Counter Flashing	Crew	Daily Output	Labor-Hours	Unit	Material	Labor	Equipment	Total	Total Incl O&P
0010	**FABRIC AND MASTIC FLASHING AND COUNTER FLASHING**									
1300	Asphalt flashing cement, 5 gallon				Gal.	11.35			11.35	12.50
4900	Fabric, asphalt-saturated cotton, specification grade	1 Rofc	35	.229	S.Y.	3.26	10.55		13.81	22
5000	Utility grade		35	.229		1.55	10.55		12.10	20
5300	Close-mesh fabric, saturated, 17 oz./S.Y.		35	.229		2.21	10.55		12.76	21
5500	Fiberglass, resin-coated		35	.229		1.24	10.55		11.79	19.85
8500	Shower pan, bituminous membrane, 7 oz.		155	.052	S.F.	1.95	2.38		4.33	6.30

07 65 13 – Laminated Sheet Flashing

07 65 13.10 Laminated Sheet Flashing

	07 65 13.10 Laminated Sheet Flashing	Crew	Daily Output	Labor-Hours	Unit	Material	Labor	Equipment	Total	Total Incl O&P
0010	**LAMINATED SHEET FLASHING**, Including up to 4 bends									
0500	Aluminum, fabric-backed 2 sides, mill finish, .004" thick	1 Rofc	330	.024	S.F.	1.58	1.12		2.70	3.70
0700	.005" thick		330	.024		1.88	1.12		3	4.03
0750	Mastic-backed, self adhesive		460	.017		3.41	.80		4.21	5.15
0800	Mastic-coated 2 sides, .004" thick		330	.024		1.58	1.12		2.70	3.70
2800	Copper, paperbacked 1 side, 2 oz.		330	.024		2.16	1.12		3.28	4.34
2900	3 oz.		330	.024		3.15	1.12		4.27	5.45
3100	Paperbacked 2 sides, 2 oz.		330	.024		2.39	1.12		3.51	4.59
3150	3 oz.		330	.024		2.24	1.12		3.36	4.42
3200	5 oz.		330	.024		3.65	1.12		4.77	6
3250	7 oz.		330	.024		6.85	1.12		7.97	9.45
3400	Mastic-backed 2 sides, copper, 2 oz.		330	.024		2.15	1.12		3.27	4.33
3500	3 oz.		330	.024		2.57	1.12		3.69	4.79
3700	5 oz.		330	.024		3.90	1.12		5.02	6.25
3800	Fabric-backed 2 sides, copper, 2 oz.		330	.024		2.05	1.12		3.17	4.22
4000	3 oz.		330	.024		2.88	1.12		4	5.15
4100	5 oz.		330	.024		4.03	1.12		5.15	6.40
4300	Copper-clad stainless steel, .015" thick, under 500 lb.		115	.070		7.05	3.21		10.26	13.35
4400	Over 2000 lb.		155	.052		7.05	2.38		9.43	11.90
4600	.018" thick, under 500 lb.		100	.080		7.95	3.70		11.65	15.20
4700	Over 2000 lb.		145	.055		8.10	2.55		10.65	13.35
8550	Shower pan, 3 ply copper and fabric, 3 oz.		155	.052		4.12	2.38		6.50	8.70
8600	7 oz.		155	.052		5.45	2.38		7.83	10.10
9300	Stainless steel, paperbacked 2 sides, .005" thick		330	.024		4.04	1.12		5.16	6.40

07 65 19 – Plastic Sheet Flashing

07 65 19.10 Plastic Sheet Flashing and Counter Flashing

	07 65 19.10 Plastic Sheet Flashing and Counter Flashing	Crew	Daily Output	Labor-Hours	Unit	Material	Labor	Equipment	Total	Total Incl O&P
0010	**PLASTIC SHEET FLASHING AND COUNTER FLASHING**									
7300	Polyvinyl chloride, black, 10 mil	1 Rofc	285	.028	S.F.	.27	1.30		1.57	2.57
7400	20 mil		285	.028		.26	1.30		1.56	2.56
7600	30 mil		285	.028		.33	1.30		1.63	2.63
7700	60 mil		285	.028		.88	1.30		2.18	3.24
7900	Black or white for exposed roofs, 60 mil		285	.028		1.11	1.30		2.41	3.49
8060	PVC tape, 5" x 45 mils, for joint covers, 100 L.F./roll				Ea.	183			183	201
8850	Polyvinyl chloride, 30 mil	1 Rofc	160	.050	S.F.	1.47	2.31		3.78	5.65

07 65 Flexible Flashing

07 65 23 – Rubber Sheet Flashing

07 65 23.10 Rubber Sheet Flashing and Counter Flashing	Crew	Daily Output	Labor-Hours	Unit	Material	2020 Bare Costs Labor	2020 Bare Costs Equipment	Total	Total Incl O&P
0010 **RUBBER SHEET FLASHING AND COUNTER FLASHING**									
4810 EPDM 90 mils, 1" diameter pipe flashing	1 Rofc	32	.250	Ea.	21	11.55		32.55	43
4820 2" diameter		30	.267		21	12.30		33.30	44.50
4830 3" diameter		28	.286		24	13.20		37.20	49.50
4840 4" diameter		24	.333		29.50	15.40		44.90	59.50
4850 6" diameter		22	.364		29.50	16.80		46.30	61.50
8100 Rubber, butyl, 1/32" thick		285	.028	S.F.	2.56	1.30		3.86	5.10
8200 1/16" thick		285	.028		3.83	1.30		5.13	6.50
8300 Neoprene, cured, 1/16" thick		285	.028		2.77	1.30		4.07	5.30
8400 1/8" thick		285	.028		6.15	1.30		7.45	9.05

07 65 26 – Self-Adhering Sheet Flashing

07 65 26.10 Self-Adhering Sheet or Roll Flashing

	Crew	Daily Output	Labor-Hours	Unit	Material	Labor	Equipment	Total	Total Incl O&P
0010 **SELF-ADHERING SHEET OR ROLL FLASHING**									
0020 Self-adhered flashing, 25 mil cross laminated HDPE, 4" wide	1 Rofc	960	.008	L.F.	.20	.38		.58	.89
0040 6" wide		896	.009		.31	.41		.72	1.06
0060 9" wide		832	.010		.46	.44		.90	1.28
0080 12" wide		768	.010		.61	.48		1.09	1.51

07 71 Roof Specialties

07 71 16 – Manufactured Counterflashing Systems

07 71 16.10 Roof Drain Boot

	Crew	Daily Output	Labor-Hours	Unit	Material	Labor	Equipment	Total	Total Incl O&P
0010 **ROOF DRAIN BOOT**									
0100 Cast iron, 4" diameter	1 Shee	125	.064	L.F.	46	3.99		49.99	57
0300 4" x 3"		125	.064		72.50	3.99		76.49	86.50
0400 5" x 4"		125	.064		135	3.99		138.99	155

07 71 16.20 Pitch Pockets, Variable Sizes

	Crew	Daily Output	Labor-Hours	Unit	Material	Labor	Equipment	Total	Total Incl O&P
0010 **PITCH POCKETS, VARIABLE SIZES**									
0100 Adjustable, 4" to 7", welded corners, 4" deep	1 Rofc	48	.167	Ea.	21.50	7.70		29.20	37.50
0200 Side extenders, 6"	"	240	.033	"	3.34	1.54		4.88	6.35

07 71 19 – Manufactured Gravel Stops and Fasciae

07 71 19.10 Gravel Stop

	Crew	Daily Output	Labor-Hours	Unit	Material	Labor	Equipment	Total	Total Incl O&P
0010 **GRAVEL STOP**									
0020 Aluminum, .050" thick, 4" face height, mill finish	1 Shee	145	.055	L.F.	6.95	3.44		10.39	13.05
0080 Duranodic finish		145	.055		8.25	3.44		11.69	14.45
0100 Painted		145	.055		7.60	3.44		11.04	13.75
0300 6" face height		135	.059		7.45	3.69		11.14	13.95
0350 Duranodic finish		135	.059		8.25	3.69		11.94	14.85
0400 Painted		135	.059		8.85	3.69		12.54	15.55
0600 8" face height		125	.064		7.95	3.99		11.94	14.95
0650 Duranodic finish		125	.064		9.40	3.99		13.39	16.60
0700 Painted		125	.064		9.60	3.99		13.59	16.80
0900 12" face height, .080" thick, 2 piece		100	.080		11	4.98		15.98	19.90
0950 Duranodic finish		100	.080		10.95	4.98		15.93	19.85
1000 Painted		100	.080		14.45	4.98		19.43	23.50
1200 Copper, 16 oz., 3" face height		145	.055		28.50	3.44		31.94	37
1300 6" face height		135	.059		37	3.69		40.69	46.50
1350 Galv steel, 24 ga., 4" leg, plain, with continuous cleat, 4" face		145	.055		6.60	3.44		10.04	12.70
1360 6" face height		145	.055		6.95	3.44		10.39	13.05
1500 Polyvinyl chloride, 6" face height		135	.059		5.85	3.69		9.54	12.20

07 71 19 – Manufactured Gravel Stops and Fasciae

07 71 19.10 Gravel Stop

		Crew	Daily Output	Labor-Hours	Unit	Material	2020 Bare Costs Labor	Equipment	Total	Total Incl O&P
1600	9" face height	1 Shee	125	.064	L.F.	7.05	3.99		11.04	14
1800	Stainless steel, 24 ga., 6" face height		135	.059		17	3.69		20.69	24.50
1900	12" face height		100	.080		22.50	4.98		27.48	33
2100	20 ga., 6" face height		135	.059		22.50	3.69		26.19	30.50
2200	12" face height		100	.080		27	4.98		31.98	37.50
9000	Minimum labor/equipment charge		3.50	2.286	Job		142		142	223

07 71 19.30 Fascia

		Crew	Daily Output	Labor-Hours	Unit	Material	2020 Bare Costs Labor	Equipment	Total	Total Incl O&P
0010	**FASCIA**									
0100	Aluminum, reverse board and batten, .032" thick, colored, no furring incl.	1 Shee	145	.055	S.F.	7.20	3.44		10.64	13.35
0200	Residential type, aluminum	1 Carp	200	.040	L.F.	2.15	2.13		4.28	5.80
0220	Vinyl	"	200	.040	"	2.43	2.13		4.56	6.10
0300	Steel, galv and enameled, stock, no furring, long panels	1 Shee	145	.055	S.F.	5.65	3.44		9.09	11.60
0600	Short panels		115	.070	"	5.35	4.33		9.68	12.70
9000	Minimum labor/equipment charge		4	2	Job		125		125	195

07 71 23 – Manufactured Gutters and Downspouts

07 71 23.10 Downspouts

		Crew	Daily Output	Labor-Hours	Unit	Material	2020 Bare Costs Labor	Equipment	Total	Total Incl O&P
0010	**DOWNSPOUTS**									
0020	Aluminum, embossed, .020" thick, 2" x 3"	1 Shee	190	.042	L.F.	.95	2.62		3.57	5.15
0100	Enameled		190	.042		1.37	2.62		3.99	5.60
0300	.024" thick, 2" x 3"		180	.044		2.17	2.77		4.94	6.75
0400	3" x 4"		140	.057		2.29	3.56		5.85	8.10
0600	Round, corrugated aluminum, 3" diameter, .020" thick		190	.042		2.13	2.62		4.75	6.45
0700	4" diameter, .025" thick		140	.057		3.21	3.56		6.77	9.15
0900	Wire strainer, round, 2" diameter		155	.052	Ea.	2.05	3.22		5.27	7.30
1000	4" diameter		155	.052		2.66	3.22		5.88	8
1200	Rectangular, perforated, 2" x 3"		145	.055		2.36	3.44		5.80	8
1300	3" x 4"		145	.055		3.42	3.44		6.86	9.15
1500	Copper, round, 16 oz., stock, 2" diameter		190	.042	L.F.	8	2.62		10.62	12.90
1600	3" diameter		190	.042		9	2.62		11.62	14
1800	4" diameter		145	.055		10.65	3.44		14.09	17.15
1900	5" diameter		130	.062		14.85	3.83		18.68	22.50
2100	Rectangular, corrugated copper, stock, 2" x 3"		190	.042		9.20	2.62		11.82	14.25
2200	3" x 4"		145	.055		9.40	3.44		12.84	15.70
2400	Rectangular, plain copper, stock, 2" x 3"		190	.042		13.25	2.62		15.87	18.70
2500	3" x 4"		145	.055		13.85	3.44		17.29	20.50
2700	Wire strainers, rectangular, 2" x 3"		145	.055	Ea.	17.15	3.44		20.59	24.50
2800	3" x 4"		145	.055		20	3.44		23.44	27.50
3000	Round, 2" diameter		145	.055		7.65	3.44		11.09	13.85
3100	3" diameter		145	.055		10.40	3.44		13.84	16.85
3300	4" diameter		145	.055		12	3.44		15.44	18.60
3400	5" diameter		115	.070		23	4.33		27.33	32
3600	Lead-coated copper, round, stock, 2" diameter		190	.042	L.F.	25.50	2.62		28.12	32
3700	3" diameter		190	.042		24.50	2.62		27.12	31
3900	4" diameter		145	.055		27.50	3.44		30.94	36
4000	5" diameter, corrugated		130	.062		26	3.83		29.83	34.50
4200	6" diameter, corrugated		105	.076		32	4.75		36.75	42.50
4300	Rectangular, corrugated, stock, 2" x 3"		190	.042		15.25	2.62		17.87	21
4500	Plain, stock, 2" x 3"		190	.042		27.50	2.62		30.12	34.50
4600	3" x 4"		145	.055		40.50	3.44		43.94	50
4800	Steel, galvanized, round, corrugated, 2" or 3" diameter, 28 ga.		190	.042		2.30	2.62		4.92	6.65
4900	4" diameter, 28 ga.		145	.055		2.19	3.44		5.63	7.80
5100	5" diameter, 26 ga.		130	.062		3.86	3.83		7.69	10.25

For customer support on your Commercial Renovation Costs with RSMeans data, call 800.448.8182.

07 71 23 – Manufactured Gutters and Downspouts

07 71 23.10 Downspouts

	Crew	Daily Output	Labor-Hours	Unit	Material	2020 Bare Costs Labor	Equipment	Total	Total Incl O&P
5400	6" diameter, 28 ga.	1 Shee	105	.076	L.F.	3.72	4.75		8.47	11.55
5500	26 ga.		105	.076		4.08	4.75		8.83	11.95
5700	Rectangular, corrugated, 28 ga., 2" x 3"		190	.042		2.38	2.62		5	6.75
5800	3" x 4"		145	.055		1.96	3.44		5.40	7.55
6000	Rectangular, plain, 28 ga., galvanized, 2" x 3"		190	.042		3.98	2.62		6.60	8.50
6100	3" x 4"		145	.055		4.83	3.44		8.27	10.70
6300	Epoxy painted, 24 ga., corrugated, 2" x 3"		190	.042		2.53	2.62		5.15	6.90
6400	3" x 4"		145	.055		2.99	3.44		6.43	8.70
6600	Wire strainers, rectangular, 2" x 3"		145	.055	Ea.	20.50	3.44		23.94	28
6700	3" x 4"		145	.055		19.30	3.44		22.74	26.50
6900	Round strainers, 2" or 3" diameter		145	.055		4.50	3.44		7.94	10.35
7000	4" diameter		145	.055		6.55	3.44		9.99	12.60
9000	Minimum labor/equipment charge		4	2	Job		125		125	195

07 71 23.20 Downspout Elbows

	Crew	Daily Output	Labor-Hours	Unit	Material	2020 Bare Costs Labor	Equipment	Total	Total Incl O&P
0010	**DOWNSPOUT ELBOWS**									
0020	Aluminum, embossed, 2" x 3", .020" thick	1 Shee	100	.080	Ea.	1	4.98		5.98	8.90
0100	Enameled		100	.080		2.45	4.98		7.43	10.50
0200	Embossed, 3" x 4", .025" thick		100	.080		2.96	4.98		7.94	11.05
0300	Enameled		100	.080		3.81	4.98		8.79	12
0400	Embossed, corrugated, 3" diameter, .020" thick		100	.080		3.15	4.98		8.13	11.25
0500	4" diameter, .025" thick		100	.080		7	4.98		11.98	15.50
0600	Copper, 16 oz., 2" diameter		100	.080		10.90	4.98		15.88	19.80
0700	3" diameter		100	.080		10.30	4.98		15.28	19.15
0800	4" diameter		100	.080		12.95	4.98		17.93	22
1000	Rectangular, 2" x 3" corrugated		100	.080		10.20	4.98		15.18	19.05
1100	3" x 4" corrugated		100	.080		15.15	4.98		20.13	24.50
1300	Vinyl, 2-1/2" diameter, 45 or 75 degree bend		100	.080		4.04	4.98		9.02	12.25
1400	Tee Y junction		75	.107		13.10	6.65		19.75	25
9000	Minimum labor/equipment charge		4	2	Job		125		125	195

07 71 23.30 Gutters

	Crew	Daily Output	Labor-Hours	Unit	Material	2020 Bare Costs Labor	Equipment	Total	Total Incl O&P
0010	**GUTTERS**									
0012	Aluminum, stock units, 5" K type, .027" thick, plain	1 Shee	125	.064	L.F.	2.85	3.99		6.84	9.40
0100	Enameled		125	.064		2.94	3.99		6.93	9.50
0300	5" K type, .032" thick, plain		125	.064		3.79	3.99		7.78	10.40
0400	Enameled		125	.064		3.58	3.99		7.57	10.20
0700	Copper, half round, 16 oz., stock units, 4" wide		125	.064		8.60	3.99		12.59	15.75
0900	5" wide		125	.064		7.65	3.99		11.64	14.70
1000	6" wide		118	.068		10.90	4.22		15.12	18.55
1200	K type, 16 oz., stock, 5" wide		125	.064		8.75	3.99		12.74	15.85
1300	6" wide		125	.064		8.45	3.99		12.44	15.55
1500	Lead coated copper, 16 oz., half round, stock, 4" wide		125	.064		17.55	3.99		21.54	25.50
1600	6" wide		118	.068		20	4.22		24.22	28.50
1800	K type, stock, 5" wide		125	.064		18.55	3.99		22.54	27
1900	6" wide		125	.064		21	3.99		24.99	29.50
2100	Copper clad stainless steel, K type, 5" wide		125	.064		7.60	3.99		11.59	14.60
2200	6" wide		125	.064		10.30	3.99		14.29	17.60
2400	Steel, galv, half round or box, 28 ga., 5" wide, plain		125	.064		2.40	3.99		6.39	8.90
2500	Enameled		125	.064		2.24	3.99		6.23	8.70
2700	26 ga., stock, 5" wide		125	.064		2.66	3.99		6.65	9.20
2800	6" wide		125	.064		2.64	3.99		6.63	9.15
3000	Vinyl, O.G., 4" wide	1 Carp	115	.070		1.39	3.70		5.09	7.45
3100	5" wide		115	.070		1.51	3.70		5.21	7.55

07 71 23.30 Gutters		Crew	Daily Output	Labor-Hours	Unit	Material	2020 Bare Costs Labor	Equipment	Total	Total Incl O&P
3200	4" half round, stock units	1 Carp	115	.070	L.F.	1.42	3.70		5.12	7.45
3250	Joint connectors				Ea.	3.10			3.10	3.41
3300	Wood, clear treated cedar, fir or hemlock, 3" x 4"	1 Carp	100	.080	L.F.	12.70	4.25		16.95	21
3400	4" x 5"	"	100	.080	"	24.50	4.25		28.75	34
5000	Accessories, end cap, K type, aluminum 5"	1 Shee	625	.013	Ea.	.74	.80		1.54	2.06
5010	6"		625	.013		1.49	.80		2.29	2.89
5020	Copper, 5"		625	.013		3.41	.80		4.21	5
5030	6"		625	.013		3.59	.80		4.39	5.20
5040	Lead coated copper, 5"		625	.013		13.20	.80		14	15.75
5050	6"		625	.013		14.10	.80		14.90	16.75
5060	Copper clad stainless steel, 5"		625	.013		3.58	.80		4.38	5.20
5070	6"		625	.013		3.58	.80		4.38	5.20
5080	Galvanized steel, 5"		625	.013		1.57	.80		2.37	2.98
5090	6"		625	.013		2.72	.80		3.52	4.24
5100	Vinyl, 4"	1 Carp	625	.013		6.50	.68		7.18	8.25
5110	5"	"	625	.013		6.85	.68		7.53	8.65
5120	Half round, copper, 4"	1 Shee	625	.013		4.65	.80		5.45	6.35
5130	5"		625	.013		5.25	.80		6.05	7
5140	6"		625	.013		8.50	.80		9.30	10.60
5150	Lead coated copper, 5"		625	.013		14.85	.80		15.65	17.60
5160	6"		625	.013		22	.80		22.80	26
5170	Copper clad stainless steel, 5"		625	.013		5	.80		5.80	6.75
5180	6"		625	.013		4.62	.80		5.42	6.35
5190	Galvanized steel, 5"		625	.013		2.80	.80		3.60	4.33
5200	6"		625	.013		3.47	.80		4.27	5.05
5210	Outlet, aluminum, 2" x 3"		420	.019		.72	1.19		1.91	2.65
5220	3" x 4"		420	.019		1.18	1.19		2.37	3.16
5230	2-3/8" round		420	.019		.66	1.19		1.85	2.59
5240	Copper, 2" x 3"		420	.019		7.80	1.19		8.99	10.45
5250	3" x 4"		420	.019		8.80	1.19		9.99	11.55
5260	2-3/8" round		420	.019		4.87	1.19		6.06	7.20
5270	Lead coated copper, 2" x 3"		420	.019		26	1.19		27.19	31
5280	3" x 4"		420	.019		30	1.19		31.19	35
5290	2-3/8" round		420	.019		26.50	1.19		27.69	31
5300	Copper clad stainless steel, 2" x 3"		420	.019		7.30	1.19		8.49	9.90
5310	3" x 4"		420	.019		8.35	1.19		9.54	11
5320	2-3/8" round		420	.019		4.87	1.19		6.06	7.20
5330	Galvanized steel, 2" x 3"		420	.019		3.81	1.19		5	6.05
5340	3" x 4"		420	.019		6.10	1.19		7.29	8.55
5350	2-3/8" round		420	.019		4.95	1.19		6.14	7.30
5360	K type mitres, aluminum		65	.123		4.94	7.65		12.59	17.45
5370	Copper		65	.123		17.45	7.65		25.10	31
5380	Lead coated copper		65	.123		56	7.65		63.65	73.50
5390	Copper clad stainless steel		65	.123		28	7.65		35.65	42.50
5400	Galvanized steel		65	.123		28	7.65		35.65	43
5420	Half round mitres, copper		65	.123		70.50	7.65		78.15	89.50
5430	Lead coated copper		65	.123		91.50	7.65		99.15	113
5440	Copper clad stainless steel		65	.123		57.50	7.65		65.15	75
5450	Galvanized steel		65	.123		33.50	7.65		41.15	49
5460	Vinyl mitres and outlets		65	.123		11	7.65		18.65	24
5470	Sealant		940	.009	L.F.	.01	.53		.54	.84
5480	Soldering		96	.083	"	.29	5.20		5.49	8.45
9000	Minimum labor/equipment charge		3.75	2.133	Job		133		133	208

For customer support on your Commercial Renovation Costs with RSMeans data, call 800.448.8182.

07 71 Roof Specialties

07 71 23 – Manufactured Gutters and Downspouts

07 71 23.35 Gutter Guard	Crew	Daily Output	Labor-Hours	Unit	Material	2020 Bare Costs Labor	Equipment	Total	Total Incl O&P
0010 **GUTTER GUARD**									
0020 6" wide strip, aluminum mesh	1 Carp	500	.016	L.F.	2.46	.85		3.31	4.07
0100 Vinyl mesh		500	.016	"	2.89	.85		3.74	4.54
9000 Minimum labor/equipment charge		4	2	Job		106		106	170

07 71 26 – Reglets

07 71 26.10 Reglets and Accessories

07 71 26.10 Reglets and Accessories	Crew	Daily Output	Labor-Hours	Unit	Material	2020 Bare Costs Labor	Equipment	Total	Total Incl O&P
0010 **REGLETS AND ACCESSORIES**									
0020 Reglet, aluminum, .025" thick, in parapet	1 Carp	225	.036	L.F.	1.84	1.89		3.73	5.05
0300 16 oz. copper		225	.036		6.55	1.89		8.44	10.25
0400 Galvanized steel, 24 ga.		225	.036		1.37	1.89		3.26	4.54
0600 Stainless steel, .020" thick		225	.036		4.21	1.89		6.10	7.65
0900 Counter flashing for above, 12" wide, .032" aluminum	1 Shee	150	.053		2.09	3.32		5.41	7.50
1200 16 oz. copper		150	.053		6.35	3.32		9.67	12.20
1300 Galvanized steel, 26 ga.		150	.053		1.33	3.32		4.65	6.65
1500 Stainless steel, .020" thick		150	.053		6.80	3.32		10.12	12.70
9000 Minimum labor/equipment charge	1 Carp	3	2.667	Job		142		142	227

07 71 29 – Manufactured Roof Expansion Joints

07 71 29.10 Expansion Joints

07 71 29.10 Expansion Joints	Crew	Daily Output	Labor-Hours	Unit	Material	2020 Bare Costs Labor	Equipment	Total	Total Incl O&P
0010 **EXPANSION JOINTS**									
0300 Butyl or neoprene center with foam insulation, metal flanges									
0400 Aluminum, .032" thick for openings to 2-1/2"	1 Rofc	165	.048	L.F.	12.85	2.24		15.09	18.05
0600 For joint openings to 3-1/2"		165	.048		12.40	2.24		14.64	17.55
0610 For joint openings to 5"		165	.048		14.75	2.24		16.99	20
0620 For joint openings to 8"		165	.048		16.70	2.24		18.94	22.50
0700 Copper, 16 oz. for openings to 2-1/2"		165	.048		21.50	2.24		23.74	28
0900 For joint openings to 3-1/2"		165	.048		19.70	2.24		21.94	25.50
0910 For joint openings to 5"		165	.048		22.50	2.24		24.74	29
0920 For joint openings to 8"		165	.048		25	2.24		27.24	31
1000 Galvanized steel, 26 ga. for openings to 2-1/2"		165	.048		10.80	2.24		13.04	15.80
1200 For joint openings to 3-1/2"		165	.048		10.55	2.24		12.79	15.50
1210 For joint openings to 5"		165	.048		14.10	2.24		16.34	19.45
1220 For joint openings to 8"		165	.048		15.20	2.24		17.44	20.50
1300 Lead-coated copper, 16 oz. for openings to 2-1/2"		165	.048		35.50	2.24		37.74	43
1500 For joint openings to 3-1/2"		165	.048		37.50	2.24		39.74	45
1600 Stainless steel, .018", for openings to 2-1/2"		165	.048		13.35	2.24		15.59	18.55
1800 For joint openings to 3-1/2"		165	.048		14.15	2.24		16.39	19.45
1810 For joint openings to 5"		165	.048		15.25	2.24		17.49	20.50
1820 For joint openings to 8"		165	.048		23.50	2.24		25.74	29.50
1900 Neoprene, double-seal type with thick center, 4-1/2" wide		125	.064		14.95	2.96		17.91	21.50
1950 Polyethylene bellows, with galv steel flat flanges		100	.080		7.20	3.70		10.90	14.35
1960 With galvanized angle flanges		100	.080		7.75	3.70		11.45	15
2000 Roof joint with extruded aluminum cover, 2"	1 Shee	115	.070		31.50	4.33		35.83	41.50
2100 Expansion, stainless flange, foam center, standard	1 Rofc	100	.080		13.80	3.70		17.50	21.50
2200 Large	"	100	.080		18.35	3.70		22.05	26.50
2500 Roof to wall joint with extruded aluminum cover	1 Shee	115	.070		30.50	4.33		34.83	40.50
2700 Wall joint, closed cell foam on PVC cover, 9" wide	1 Rofc	125	.064		6.10	2.96		9.06	11.90
2800 12" wide	"	115	.070		7	3.21		10.21	13.30
9000 Minimum labor/equipment charge	1 Shee	3	2.667	Job		166		166	260

07 71 Roof Specialties

07 71 43 – Drip Edge

07 71 43.10 Drip Edge, Rake Edge, Ice Belts	Crew	Daily Output	Labor-Hours	Unit	Material	2020 Bare Costs Labor	Equipment	Total	Total Incl O&P
0010 **DRIP EDGE, RAKE EDGE, ICE BELTS**									
0020 Aluminum, .016" thick, 5" wide, mill finish	1 Carp	400	.020	L.F.	.59	1.06		1.65	2.35
0100 White finish		400	.020		.65	1.06		1.71	2.42
0200 8" wide, mill finish		400	.020		1.48	1.06		2.54	3.33
0300 Ice belt, 28" wide, mill finish		100	.080		8	4.25		12.25	15.60
0310 Vented, mill finish		400	.020		2.29	1.06		3.35	4.22
0320 Painted finish		400	.020		2.54	1.06		3.60	4.49
0400 Galvanized, 5" wide		400	.020		.62	1.06		1.68	2.38
0500 8" wide, mill finish		400	.020		.84	1.06		1.90	2.62
0510 Rake edge, aluminum, 1-1/2" x 1-1/2"		400	.020		.35	1.06		1.41	2.09
0520 3-1/2" x 1-1/2"		400	.020		.51	1.06		1.57	2.26
9000 Minimum labor/equipment charge		4	2	Job		106		106	170

07 72 Roof Accessories

07 72 23 – Relief Vents

07 72 23.10 Roof Vents

	Crew	Daily Output	Labor-Hours	Unit	Material	2020 Bare Costs Labor	Equipment	Total	Total Incl O&P
0010 **ROOF VENTS**									
0020 Mushroom shape, for built-up roofs, aluminum	1 Rofc	30	.267	Ea.	74.50	12.30		86.80	104
0100 PVC, 6" high		30	.267	"	26	12.30		38.30	50
9000 Minimum labor/equipment charge		2.75	2.909	Job		134		134	235

07 72 23.20 Vents

	Crew	Daily Output	Labor-Hours	Unit	Material	2020 Bare Costs Labor	Equipment	Total	Total Incl O&P
0010 **VENTS**									
0100 Soffit or eave, aluminum, mill finish, strips, 2-1/2" wide	1 Carp	200	.040	L.F.	.49	2.13		2.62	3.95
0200 3" wide		200	.040		.47	2.13		2.60	3.93
0300 Enamel finish, 3" wide		200	.040		.54	2.13		2.67	4
0400 Mill finish, rectangular, 4" x 16"		72	.111	Ea.	1.53	5.90		7.43	11.15
0500 8" x 16"		72	.111		2.60	5.90		8.50	12.30
2420 Roof ventilator	Q-9	16	1		54	56		110	148
2500 Vent, roof vent	1 Rofc	24	.333		27	15.40		42.40	56.50

07 72 26 – Ridge Vents

07 72 26.10 Ridge Vents and Accessories

	Crew	Daily Output	Labor-Hours	Unit	Material	2020 Bare Costs Labor	Equipment	Total	Total Incl O&P
0010 **RIDGE VENTS AND ACCESSORIES**									
0100 Aluminum strips, mill finish	1 Rofc	160	.050	L.F.	2.67	2.31		4.98	7
0150 Painted finish		160	.050	"	4.23	2.31		6.54	8.70
0200 Connectors		48	.167	Ea.	5.20	7.70		12.90	19.20
0300 End caps		48	.167	"	2.49	7.70		10.19	16.25
0400 Galvanized strips		160	.050	L.F.	3.81	2.31		6.12	8.25
0430 Molded polyethylene, shingles not included		160	.050	"	2.85	2.31		5.16	7.20
0440 End plugs		48	.167	Ea.	2.49	7.70		10.19	16.25
0450 Flexible roll, shingles not included		160	.050	L.F.	2.54	2.31		4.85	6.85
2300 Ridge vent strip, mill finish	1 Shee	155	.052	"	3.95	3.22		7.17	9.40

07 72 33 – Roof Hatches

07 72 33.10 Roof Hatch Options

	Crew	Daily Output	Labor-Hours	Unit	Material	2020 Bare Costs Labor	Equipment	Total	Total Incl O&P
0010 **ROOF HATCH OPTIONS**									
0500 2'-6" x 3', aluminum curb and cover	G-3	10	3.200	Ea.	900	167		1,067	1,250
0520 Galvanized steel curb and aluminum cover		10	3.200		850	167		1,017	1,200
0540 Galvanized steel curb and cover		10	3.200		610	167		777	935
0600 2'-6" x 4'-6", aluminum curb and cover		9	3.556		1,150	186		1,336	1,575
0800 Galvanized steel curb and aluminum cover		9	3.556		1,025	186		1,211	1,425

07 72 Roof Accessories

07 72 33 - Roof Hatches

07 72 33.10 Roof Hatch Options

		Crew	Daily Output	Labor-Hours	Unit	Material	2020 Bare Costs Labor	Equipment	Total	Total Incl O&P
0900	Galvanized steel curb and cover	G-3	9	3.556	Ea.	995	186		1,181	1,400
1100	4' x 4' aluminum curb and cover		8	4		1,700	209		1,909	2,200
1120	Galvanized steel curb and aluminum cover		8	4		1,800	209		2,009	2,300
1140	Galvanized steel curb and cover		8	4		1,150	209		1,359	1,575
1200	2'-6" x 8'-0", aluminum curb and cover		6.60	4.848		2,175	253		2,428	2,800
1400	Galvanized steel curb and aluminum cover		6.60	4.848		1,900	253		2,153	2,500
1500	Galvanized steel curb and cover	▼	6.60	4.848		1,200	253		1,453	1,725
1800	For plexiglass panels, 2'-6" x 3'-0", add to above				▼	485			485	535
9000	Minimum labor/equipment charge	2 Carp	2	8	Job		425		425	680

07 72 36 - Smoke Vents

07 72 36.10 Smoke Hatches

		Crew	Daily Output	Labor-Hours	Unit	Material	2020 Bare Costs Labor	Equipment	Total	Total Incl O&P
0010	**SMOKE HATCHES**									
0200	For 3'-0" long, add to roof hatches from Section 07 72 33.10				Ea.	25%	5%			
0250	For 4'-0" long, add to roof hatches from Section 07 72 33.10					20%	5%			
0300	For 8'-0" long, add to roof hatches from Section 07 72 33.10				▼	10%	5%			

07 72 36.20 Smoke Vent Options

		Crew	Daily Output	Labor-Hours	Unit	Material	2020 Bare Costs Labor	Equipment	Total	Total Incl O&P
0010	**SMOKE VENT OPTIONS**									
0100	4' x 4' aluminum cover and frame	G-3	13	2.462	Ea.	2,200	128		2,328	2,625
0200	Galvanized steel cover and frame		13	2.462		1,725	128		1,853	2,100
0300	4' x 8' aluminum cover and frame		8	4		3,275	209		3,484	3,925
0400	Galvanized steel cover and frame	▼	8	4	▼	2,750	209		2,959	3,350
9000	Minimum labor/equipment charge	2 Carp	2	8	Job		425		425	680

07 72 53 - Snow Guards

07 72 53.10 Snow Guard Options

		Crew	Daily Output	Labor-Hours	Unit	Material	2020 Bare Costs Labor	Equipment	Total	Total Incl O&P
0010	**SNOW GUARD OPTIONS**									
0100	Slate & asphalt shingle roofs, fastened with nails	1 Rofc	160	.050	Ea.	12.45	2.31		14.76	17.75
0200	Standing seam metal roofs, fastened with set screws		48	.167		17.75	7.70		25.45	33
0300	Surface mount for metal roofs, fastened with solder		48	.167	▼	7.75	7.70		15.45	22
0400	Double rail pipe type, including pipe	▼	130	.062	L.F.	35	2.84		37.84	43.50

07 72 80 - Vent Options

07 72 80.30 Vent Options

		Crew	Daily Output	Labor-Hours	Unit	Material	2020 Bare Costs Labor	Equipment	Total	Total Incl O&P
0010	**VENT OPTIONS**									
0020	Plastic, for insulated decks, 1 per M.S.F.	1 Rofc	40	.200	Ea.	24	9.25		33.25	42.50
0100	Heavy duty		20	.400		56.50	18.50		75	95
0300	Aluminum	▼	30	.267		21.50	12.30		33.80	45
0800	Polystyrene baffles, 12" wide for 16" OC rafter spacing	1 Carp	90	.089		.53	4.72		5.25	8.15
0900	For 24" OC rafter spacing		110	.073	▼	.88	3.87		4.75	7.15
9000	Minimum labor/equipment charge	▼	3	2.667	Job		142		142	227

07 76 Roof Pavers

07 76 16 – Roof Decking Pavers

07 76 16.10 Roof Pavers and Supports	Crew	Daily Output	Labor-Hours	Unit	Material	2020 Bare Costs Labor	Equipment	Total	Total Incl O&P
0010 **ROOF PAVERS AND SUPPORTS**									
1000 Roof decking pavers, concrete blocks, 2" thick, natural	1 Clab	115	.070	S.F.	3.58	2.93		6.51	8.65
1100 Colors		115	.070	"	3.71	2.93		6.64	8.75
1200 Support pedestal, bottom cap		960	.008	Ea.	2.66	.35		3.01	3.49
1300 Top cap		960	.008		4.88	.35		5.23	5.90
1400 Leveling shims, 1/16"		1920	.004		1.22	.18		1.40	1.62
1500 1/8"		1920	.004		1.22	.18		1.40	1.62
1600 Buffer pad		960	.008		2.54	.35		2.89	3.35
1700 PVC legs (4" SDR 35)		2880	.003	Inch	.14	.12		.26	.34
2000 Alternate pricing method, system in place		101	.079	S.F.	7.20	3.33		10.53	13.25

07 81 Applied Fireproofing

07 81 16 – Cementitious Fireproofing

07 81 16.10 Sprayed Cementitious Fireproofing

	Crew	Daily Output	Labor-Hours	Unit	Material	2020 Bare Costs Labor	Equipment	Total	Total Incl O&P
0010 **SPRAYED CEMENTITIOUS FIREPROOFING**									
0050 Not including canvas protection, normal density									
0100 Per 1" thick, on flat plate steel	G-2	3000	.008	S.F.	.59	.36	.06	1.01	1.29
0200 Flat decking		2400	.010		.59	.44	.08	1.11	1.45
0400 Beams		1500	.016		.59	.71	.13	1.43	1.92
0500 Corrugated or fluted decks		1250	.019		.88	.85	.15	1.88	2.50
0700 Columns, 1-1/8" thick		1100	.022		.66	.97	.17	1.80	2.46
0800 2-3/16" thick		700	.034		1.53	1.52	.27	3.32	4.41
0900 For canvas protection, add		5000	.005		.12	.21	.04	.37	.51
1000 Not including canvas protection, high density									
1100 Per 1" thick, on flat plate steel	G-2	3000	.008	S.F.	2.28	.36	.06	2.70	3.15
1110 On flat decking		2400	.010		2.28	.44	.08	2.80	3.31
1120 On beams		1500	.016		2.28	.71	.13	3.12	3.78
1130 Corrugated or fluted decks		1250	.019		2.28	.85	.15	3.28	4.04
1140 Columns, 1-1/8" thick		1100	.022		2.56	.97	.17	3.70	4.55
1150 2-3/16" thick		1100	.022		5.15	.97	.17	6.29	7.40
1170 For canvas protection, add		5000	.005		.12	.21	.04	.37	.51
1200 Not including canvas protection, retrofitting									
1210 Per 1" thick, on flat plate steel	G-2	1500	.016	S.F.	.52	.71	.13	1.36	1.84
1220 On flat decking		1200	.020		.52	.89	.16	1.57	2.16
1230 On beams		750	.032		.52	1.42	.25	2.19	3.12
1240 Corrugated or fluted decks		625	.038		.77	1.70	.30	2.77	3.90
1250 Columns, 1-1/8" thick		550	.044		.58	1.94	.34	2.86	4.11
1260 2-3/16" thick		500	.048		1.16	2.13	.38	3.67	5.10
1400 Accessories, preliminary spattered texture coat		4500	.005		.05	.24	.04	.33	.48
1410 Bonding agent	1 Plas	1000	.008		.10	.39		.49	.73
9000 Minimum labor/equipment charge	G-2	3	8	Job		355	63	418	635

07 84 13 – Penetration Firestopping

07 84 13.10 Firestopping		Crew	Daily Output	Labor-Hours	Unit	Material	2020 Bare Costs Labor	Equipment	Total	Total Incl O&P
0010	**FIRESTOPPING**	R078413-30								
0100	Metallic piping, non insulated									
0110	Through walls, 2" diameter	1 Carp	16	.500	Ea.	3.67	26.50		30.17	46.50
0120	4" diameter		14	.571		8.20	30.50		38.70	57.50
0130	6" diameter		12	.667		12.60	35.50		48.10	71
0140	12" diameter		10	.800		25	42.50		67.50	95.50
0150	Through floors, 2" diameter		32	.250		1.86	13.30		15.16	23.50
0160	4" diameter		28	.286		3.28	15.20		18.48	28
0170	6" diameter		24	.333		6.45	17.70		24.15	35.50
0180	12" diameter		20	.400		12.80	21.50		34.30	48
0190	Metallic piping, insulated									
0200	Through walls, 2" diameter	1 Carp	16	.500	Ea.	8.60	26.50		35.10	52
0210	4" diameter		14	.571		12.55	30.50		43.05	62.50
0220	6" diameter		12	.667		16.75	35.50		52.25	75.50
0230	12" diameter		10	.800		28.50	42.50		71	99.50
0240	Through floors, 2" diameter		32	.250		3.27	13.30		16.57	25
0250	4" diameter		28	.286		6.30	15.20		21.50	31.50
0260	6" diameter		24	.333		8.70	17.70		26.40	38
0270	12" diameter		20	.400		12.95	21.50		34.45	48
0280	Non metallic piping, non insulated									
0290	Through walls, 2" diameter	1 Carp	12	.667	Ea.	52.50	35.50		88	115
0300	4" diameter		10	.800		101	42.50		143.50	179
0310	6" diameter		8	1		186	53		239	290
0330	Through floors, 2" diameter		16	.500		36.50	26.50		63	82.50
0340	4" diameter		6	1.333		62	71		133	183
0350	6" diameter		6	1.333		135	71		206	263
0370	Ductwork, insulated & non insulated, round									
0380	Through walls, 6" diameter	1 Carp	12	.667	Ea.	16.55	35.50		52.05	75
0390	12" diameter		10	.800		28.50	42.50		71	99.50
0400	18" diameter		8	1		33.50	53		86.50	122
0410	Through floors, 6" diameter		16	.500		8.50	26.50		35	52
0420	12" diameter		14	.571		14.60	30.50		45.10	64.50
0430	18" diameter		12	.667		17.60	35.50		53.10	76.50
0440	Ductwork, insulated & non insulated, rectangular									
0450	With stiffener/closure angle, through walls, 6" x 12"	1 Carp	8	1	Ea.	21	53		74	108
0460	12" x 24"		6	1.333		36	71		107	154
0470	24" x 48"		4	2		84	106		190	262
0480	With stiffener/closure angle, through floors, 6" x 12"		10	.800		10.40	42.50		52.90	79.50
0490	12" x 24"		8	1		20	53		73	107
0500	24" x 48"		6	1.333		42	71		113	160
0510	Multi trade openings									
0520	Through walls, 6" x 12"	1 Carp	2	4	Ea.	52	213		265	400
0530	12" x 24"	"	1	8		189	425		614	890
0540	24" x 48"	2 Carp	1	16		670	850		1,520	2,100
0550	48" x 96"	"	.75	21.333		2,550	1,125		3,675	4,625
0560	Through floors, 6" x 12"	1 Carp	2	4		44.50	213		257.50	390
0570	12" x 24"	"	1	8		69.50	425		494.50	755
0580	24" x 48"	2 Carp	.75	21.333		154	1,125		1,279	2,000
0590	48" x 96"	"	.50	32		365	1,700		2,065	3,125
0600	Structural penetrations, through walls									
0610	Steel beams, W8 x 10	1 Carp	8	1	Ea.	75	53		128	168
0620	W12 x 14		6	1.333		109	71		180	233
0630	W21 x 44		5	1.600		154	85		239	305

For customer support on your Commercial Renovation Costs with RSMeans data, call 800.448.8182.

221

07 84 Firestopping

07 84 13 – Penetration Firestopping

07 84 13.10 Firestopping	Crew	Daily Output	Labor-Hours	Unit	Material	2020 Bare Costs Labor	Equipment	Total	Total Incl O&P
0640 W36 x 135	1 Carp	3	2.667	Ea.	229	142		371	480
0650 Bar joists, 18" deep		6	1.333		42	71		113	161
0660 24" deep		6	1.333		59	71		130	179
0670 36" deep		5	1.600		88.50	85		173.50	234
0680 48" deep		4	2		120	106		226	300
0690 Construction joints, floor slab at exterior wall									
0700 Precast, brick, block or drywall exterior									
0710 2" wide joint	1 Carp	125	.064	L.F.	10.30	3.40		13.70	16.80
0720 4" wide joint	"	75	.107	"	16.60	5.65		22.25	27.50
0730 Metal panel, glass or curtain wall exterior									
0740 2" wide joint	1 Carp	40	.200	L.F.	12.45	10.65		23.10	31
0750 4" wide joint	"	25	.320	"	18.65	17		35.65	48
0760 Floor slab to drywall partition									
0770 Flat joint	1 Carp	100	.080	L.F.	14.55	4.25		18.80	23
0780 Fluted joint		50	.160		16.35	8.50		24.85	31.50
0790 Etched fluted joint		75	.107		20.50	5.65		26.15	31.50
0800 Floor slab to concrete/masonry partition									
0810 Flat joint	1 Carp	75	.107	L.F.	15.45	5.65		21.10	26
0820 Fluted joint	"	50	.160	"	17.65	8.50		26.15	33
0830 Concrete/CMU wall joints									
0840 1" wide	1 Carp	100	.080	L.F.	20.50	4.25		24.75	29.50
0850 2" wide		75	.107		30	5.65		35.65	42
0860 4" wide		50	.160		36	8.50		44.50	53
0870 Concrete/CMU floor joints									
0880 1" wide	1 Carp	200	.040	L.F.	20.50	2.13		22.63	26.50
0890 2" wide		150	.053		25	2.83		27.83	32
0900 4" wide		100	.080		36.50	4.25		40.75	47

07 91 Preformed Joint Seals

07 91 13 – Compression Seals

07 91 13.10 Compression Seals

		Crew	Daily Output	Labor-Hours	Unit	Material	2020 Bare Costs Labor	Equipment	Total	Total Incl O&P
0010	**COMPRESSION SEALS**									
4900	O-ring type cord, 1/4"	1 Bric	472	.017	L.F.	.40	.88		1.28	1.87
4910	1/2"		440	.018		1.01	.95		1.96	2.64
4920	3/4"		424	.019		2	.98		2.98	3.79
4930	1"		408	.020		3.93	1.02		4.95	5.95
4940	1-1/4"		384	.021		9.40	1.08		10.48	12.10
4950	1-1/2"		368	.022		11.50	1.13		12.63	14.50
4960	1-3/4"		352	.023		13.50	1.18		14.68	16.75
4970	2"		344	.023		22.50	1.21		23.71	27

07 91 16 – Joint Gaskets

07 91 16.10 Joint Gaskets

		Crew	Daily Output	Labor-Hours	Unit	Material	2020 Bare Costs Labor	Equipment	Total	Total Incl O&P
0010	**JOINT GASKETS**									
4400	Joint gaskets, neoprene, closed cell w/adh, 1/8" x 3/8"	1 Bric	240	.033	L.F.	.36	1.73		2.09	3.20
4500	1/4" x 3/4"		215	.037		.66	1.94		2.60	3.86
4700	1/2" x 1"		200	.040		1.71	2.08		3.79	5.25
4800	3/4" x 1-1/2"		165	.048		1.85	2.52		4.37	6.10

07 91 Preformed Joint Seals

07 91 23 – Backer Rods

07 91 23.10 Backer Rods		Crew	Daily Output	Labor-Hours	Unit	Material	2020 Bare Costs Labor	2020 Bare Costs Equipment	Total	Total Incl O&P
0010	**BACKER RODS**									
0032	Backer rod, polyethylene, 1/4" diameter	1 Bric	460	.017	L.F.	.03	.91		.94	1.49
0052	1/2" diameter		460	.017		.04	.91		.95	1.51
0072	3/4" diameter		460	.017		.07	.91		.98	1.53
0092	1" diameter	↓	460	.017	↓	.15	.91		1.06	1.62

07 91 26 – Joint Fillers

07 91 26.10 Joint Fillers

		Crew	Daily Output	Labor-Hours	Unit	Material	Labor	Equipment	Total	Total Incl O&P
0010	**JOINT FILLERS**									
4360	Butyl rubber filler, 1/4" x 1/4"	1 Bric	290	.028	L.F.	.23	1.44		1.67	2.57
4365	1/2" x 1/2"		250	.032		.92	1.67		2.59	3.70
4370	1/2" x 3/4"		210	.038		1.38	1.98		3.36	4.71
4375	3/4" x 3/4"		230	.035		2.06	1.81		3.87	5.20
4380	1" x 1"	↓	180	.044	↓	2.75	2.31		5.06	6.75
4390	For coloring, add					12%				
4980	Polyethylene joint backing, 1/4" x 2"	1 Bric	2.08	3.846	C.L.F.	14.25	200		214.25	340
4990	1/4" x 6"		1.28	6.250	"	32.50	325		357.50	560
5600	Silicone, room temp vulcanizing foam seal, 1/4" x 1/2"		1312	.006	L.F.	.46	.32		.78	1.01
5610	1/2" x 1/2"		656	.012		.92	.64		1.56	2.04
5620	1/2" x 3/4"		442	.018		1.37	.94		2.31	3.03
5630	3/4" x 3/4"		328	.024		2.06	1.27		3.33	4.32
5640	1/8" x 1"		1312	.006		.46	.32		.78	1.01
5650	1/8" x 3"		442	.018		1.37	.94		2.31	3.03
5670	1/4" x 3"		295	.027		2.75	1.41		4.16	5.30
5680	1/4" x 6"		148	.054		5.50	2.81		8.31	10.60
5690	1/2" x 6"		82	.098		11	5.10		16.10	20.50
5700	1/2" x 9"		52.50	.152		16.50	7.95		24.45	31
5710	1/2" x 12"	↓	33	.242	↓	22	12.60		34.60	44.50

07 92 Joint Sealants

07 92 13 – Elastomeric Joint Sealants

07 92 13.10 Masonry Joint Sealants

		Crew	Daily Output	Labor-Hours	Unit	Material	Labor	Equipment	Total	Total Incl O&P
0010	**MASONRY JOINT SEALANTS**, 1/2" x 1/2" joint									
0050	Re-caulk only, oil base	1 Bric	225	.036	L.F.	.58	1.85		2.43	3.62
0100	Acrylic latex		205	.039		.24	2.03		2.27	3.54
0200	Polyurethane		200	.040		.72	2.08		2.80	4.15
0300	Silicone		195	.041		.69	2.14		2.83	4.21
1000	Cut out and re-caulk, oil base		145	.055		.58	2.87		3.45	5.25
1050	Acrylic latex		130	.062		.24	3.20		3.44	5.45
1100	Polyurethane		125	.064		.72	3.33		4.05	6.20
1150	Silicone		120	.067	↓	.69	3.47		4.16	6.35
9000	Minimum labor/equipment charge	↓	4	2	Job		104		104	168

07 92 13.20 Caulking and Sealant Options

		Crew	Daily Output	Labor-Hours	Unit	Material	Labor	Equipment	Total	Total Incl O&P
0010	**CAULKING AND SEALANT OPTIONS**									
0050	Latex acrylic based, bulk				Gal.	31			31	34
0055	Bulk in place 1/4" x 1/4" bead	1 Bric	300	.027	L.F.	.10	1.39		1.49	2.35
0060	1/4" x 3/8"		294	.027		.16	1.42		1.58	2.47
0065	1/4" x 1/2"		288	.028		.22	1.45		1.67	2.58
0075	3/8" x 3/8"		284	.028		.24	1.47		1.71	2.64
0080	3/8" x 1/2"		280	.029		.32	1.49		1.81	2.76
0085	3/8" x 5/8"	↓	276	.029	↓	.40	1.51		1.91	2.89

07 92 Joint Sealants

07 92 13 – Elastomeric Joint Sealants

07 92 13.20 Caulking and Sealant Options		Crew	Daily Output	Labor-Hours	Unit	Material	2020 Bare Costs Labor	Equipment	Total	Total Incl O&P
0095	3/8" x 3/4"	1 Bric	272	.029	L.F.	.48	1.53		2.01	3
0100	1/2" x 1/2"		275	.029		.43	1.51		1.94	2.93
0105	1/2" x 5/8"		269	.030		.54	1.55		2.09	3.09
0110	1/2" x 3/4"		263	.030		.65	1.58		2.23	3.27
0115	1/2" x 7/8"		256	.031		.76	1.63		2.39	3.46
0120	1/2" x 1"		250	.032		.87	1.67		2.54	3.64
0125	3/4" x 3/4"		244	.033		.97	1.71		2.68	3.83
0130	3/4" x 1"		225	.036		1.30	1.85		3.15	4.42
0135	1" x 1"		200	.040		1.73	2.08		3.81	5.25
0190	Cartridges				Gal.	33.50			33.50	37
0200	11 fl. oz. cartridge				Ea.	2.89			2.89	3.18
0500	1/4" x 1/2"	1 Bric	288	.028	L.F.	.24	1.45		1.69	2.60
0600	1/2" x 1/2"		275	.029		.47	1.51		1.98	2.97
0800	3/4" x 3/4"		244	.033		1.06	1.71		2.77	3.93
0900	3/4" x 1"		225	.036		1.42	1.85		3.27	4.55
1000	1" x 1"		200	.040		1.77	2.08		3.85	5.30
1400	Butyl based, bulk				Gal.	43.50			43.50	48
1500	Cartridges				"	43.50			43.50	48
1700	1/4" x 1/2", 154 L.F./gal.	1 Bric	288	.028	L.F.	.28	1.45		1.73	2.65
1800	1/2" x 1/2", 77 L.F./gal.	"	275	.029	"	.57	1.51		2.08	3.07
2300	Polysulfide compounds, 1 component, bulk				Gal.	88.50			88.50	97.50
2400	Cartridges				"	145			145	159
2600	1 or 2 component, in place, 1/4" x 1/4", 308 L.F./gal.	1 Bric	300	.027	L.F.	.29	1.39		1.68	2.56
2700	1/2" x 1/4", 154 L.F./gal.		288	.028		.57	1.45		2.02	2.97
2900	3/4" x 3/8", 68 L.F./gal.		272	.029		1.30	1.53		2.83	3.90
3000	1" x 1/2", 38 L.F./gal.		250	.032		2.33	1.67		4	5.25
3200	Polyurethane, 1 or 2 component				Gal.	54			54	59.50
3300	Cartridges				"	67.50			67.50	74.50
3500	Bulk, in place, 1/4" x 1/4"	1 Bric	300	.027	L.F.	.18	1.39		1.57	2.43
3655	1/2" x 1/4"		288	.028		.35	1.45		1.80	2.73
3800	3/4" x 3/8"		272	.029		.80	1.53		2.33	3.34
3900	1" x 1/2"		250	.032		1.40	1.67		3.07	4.23
4100	Silicone rubber, bulk				Gal.	68.50			68.50	75.50
4200	Cartridges				"	51.50			51.50	56.50
4300	Bulk in place, 1/4" x 1/2", 154 L.F./gal.	1 Bric	235	.034	L.F.	.45	1.77		2.22	3.35

07 92 16 – Rigid Joint Sealants

07 92 16.10 Rigid Joint Sealants

		Crew	Daily Output	Labor-Hours	Unit	Material	2020 Bare Costs Labor	Equipment	Total	Total Incl O&P
0010	**RIGID JOINT SEALANTS**									
5802	Tapes, sealant, PVC foam adhesive, 1/16" x 1/4"				L.F.	.09			.09	.10
5902	1/16" x 1/2"					.09			.09	.10
5952	1/16" x 1"					.16			.16	.18
6002	1/8" x 1/2"					.09			.09	.10

07 92 19 – Acoustical Joint Sealants

07 92 19.10 Acoustical Sealant

		Crew	Daily Output	Labor-Hours	Unit	Material	2020 Bare Costs Labor	Equipment	Total	Total Incl O&P
0010	**ACOUSTICAL SEALANT**									
0020	Acoustical sealant, elastomeric, cartridges				Ea.	8.65			8.65	9.55
0025	In place, 1/4" x 1/4"	1 Bric	300	.027	L.F.	.35	1.39		1.74	2.63
0030	1/4" x 1/2"		288	.028		.71	1.45		2.16	3.12
0035	1/2" x 1/2"		275	.029		1.41	1.51		2.92	4.01
0040	1/2" x 3/4"		263	.030		2.12	1.58		3.70	4.89
0045	3/4" x 3/4"		244	.033		3.18	1.71		4.89	6.25
0050	1" x 1"		200	.040		5.65	2.08		7.73	9.60

07 95 13.50 Expansion Joint Assemblies	Crew	Daily Output	Labor-Hours	Unit	Material	2020 Bare Costs Labor	Equipment	Total	Total Incl O&P
0010 **EXPANSION JOINT ASSEMBLIES**									
0200 Floor cover assemblies, 1" space, aluminum	1 Sswk	38	.211	L.F.	26	12.15		38.15	48.50
0300 Bronze		38	.211		52.50	12.15		64.65	78
0500 2" space, aluminum		38	.211		18.55	12.15		30.70	40.50
0600 Bronze		38	.211		61	12.15		73.15	87
0800 Wall and ceiling assemblies, 1" space, aluminum		38	.211		17.20	12.15		29.35	39
0900 Bronze		38	.211		54	12.15		66.15	79.50
1100 2" space, aluminum		38	.211		22	12.15		34.15	44
1200 Bronze		38	.211		52.50	12.15		64.65	77.50
1400 Floor to wall assemblies, 1" space, aluminum		38	.211		20	12.15		32.15	42
1500 Bronze or stainless		38	.211		65	12.15		77.15	91.50
1700 Gym floor angle covers, aluminum, 3" x 3" angle		46	.174		19.70	10.05		29.75	38
1800 3" x 4" angle		46	.174		22	10.05		32.05	40.50
2000 Roof closures, aluminum, flat roof, low profile, 1" space		57	.140		24	8.10		32.10	40
2100 High profile		57	.140		34	8.10		42.10	51
2300 Roof to wall, low profile, 1" space		57	.140		23.50	8.10		31.60	39.50
2400 High profile		57	.140		34	8.10		42.10	51
9000 Minimum labor/equipment charge		2	4	Job		231		231	380

Division Notes

		CREW	DAILY OUTPUT	LABOR-HOURS	UNIT	BARE COSTS				TOTAL INCL O&P
						MAT.	LABOR	EQUIP.	TOTAL	

Estimating Tips

08 10 00 Doors and Frames

All exterior doors should be addressed for their energy conservation (insulation and seals).

- Most metal doors and frames look alike, but there may be significant differences among them. When estimating these items, be sure to choose the line item that most closely compares to the specification or door schedule requirements regarding:
  - □ type of metal
  - □ metal gauge
  - □ door core material
  - □ fire rating
  - □ finish
- Wood and plastic doors vary considerably in price. The primary determinant is the veneer material. Lauan, birch, and oak are the most common veneers. Other variables include the following:
  - □ hollow or solid core
  - □ fire rating
  - □ flush or raised panel
  - □ finish
- Door pricing includes bore for cylindrical locksets and mortise for hinges.

08 30 00 Specialty Doors and Frames

- There are many varieties of special doors, and they are usually priced per each. Add frames, hardware, or operators required for a complete installation.

08 40 00 Entrances, Storefronts, and Curtain Walls

- Glazed curtain walls consist of the metal tube framing and the glazing material. The cost data in this subdivision is presented for the metal tube framing alone or the composite wall. If your estimate requires a detailed takeoff of the framing, be sure to add the glazing cost and any tints.

08 50 00 Windows

- Steel windows are unglazed and aluminum can be glazed or unglazed. Some metal windows are priced without glass. Refer to 08 80 00 Glazing for glass pricing. The grade C indicates commercial grade windows, usually ASTM C-35.
- All wood windows and vinyl are priced preglazed. The glazing is insulating glass. Add the cost of screens and grills if required and not already included.

08 70 00 Hardware

- Hardware costs add considerably to the cost of a door. The most efficient method to determine the hardware requirements for a project is to review the door and hardware schedule together. One type of door may have different hardware, depending on the door usage.
- Door hinges are priced by the pair, with most doors requiring 1-1/2 pairs per door. The hinge prices do not include installation labor because it is included in door installation.

Hinges are classified according to the frequency of use, base material, and finish.

08 80 00 Glazing

- Different openings require different types of glass. The most common types are:
  - □ float
  - □ tempered
  - □ insulating
  - □ impact-resistant
  - □ ballistic-resistant
- Most exterior windows are glazed with insulating glass. Entrance doors and window walls, where the glass is less than 18" from the floor, are generally glazed with tempered glass. Interior windows and some residential windows are glazed with float glass.
- Coastal communities require the use of impact-resistant glass, dependent on wind speed.
- The insulation or 'u' value is a strong consideration, along with solar heat gain, to determine total energy efficiency.

Reference Numbers

Reference numbers are shown at the beginning of some major classifications. These numbers refer to related items in the Reference Section. The reference information may be an estimating procedure, an alternate pricing method, or technical information.

Note: Not all subdivisions listed here necessarily appear. ■

Same Data. Simplified.

Enjoy the convenience and efficiency of accessing your costs anywhere:

- **Skip the multiplier** by setting your location
- **Quickly search,** edit, favorite and share costs
- **Stay on top of price changes** with automatic updates

Discover more at rsmeans.com/online

---

08 01 Operation and Maintenance of Openings

08 01 11 – Operation and Maintenance of Metal Doors and Frames

08 01 11.10 Door and Window Maintenance		Crew	Daily Output	Labor-Hours	Unit	Material	2020 Bare Costs Labor	2020 Bare Costs Equipment	Total	Total Incl O&P
0010	**DOOR & WINDOW MAINTENANCE**									
1130	Install door	2 Carp	13.07	1.224	Ea.		65		65	104
1140	Remove deadbolt	1 Carp	13.85	.578			30.50		30.50	49
1150	Remove panic bar		7.69	1.040			55.50		55.50	88.50
1420	Remove door, plane to fit, rail, reinstall door		9.60	.833			44.50		44.50	71

08 01 14 – Operation and Maintenance of Wood Doors

08 01 14.15 Door and Window Maintenance

		Crew	Daily Output	Labor-Hours	Unit	Material	2020 Bare Costs Labor	Equipment	Total	Total Incl O&P
0010	**DOOR & WINDOW MAINTENANCE**									
0050	Rehang single door	1 Carp	32	.250	Ea.		13.30		13.30	21.50
0100	Rehang double door		16	.500	Pr.		26.50		26.50	42.50
0350	Install window sash cord		16	.500	Ea.	5	26.50		31.50	48

08 01 53 – Operation and Maintenance of Plastic Windows

08 01 53.81 Solid Vinyl Replacement Windows

			Crew	Daily Output	Labor-Hours	Unit	Material	2020 Bare Costs Labor	Equipment	Total	Total Incl O&P
0010	**SOLID VINYL REPLACEMENT WINDOWS** R085313-20										
0020	Double-hung, insulated glass, up to 83 united inches	G	2 Carp	8	2	Ea.	183	106		289	370
0040	84 to 93	G		8	2		195	106		301	385
0060	94 to 101	G		6	2.667		195	142		337	440
0080	102 to 111	G		6	2.667		207	142		349	455
0100	112 to 120	G		6	2.667		223	142		365	470
0120	For each united inch over 120, add	G		800	.020	Inch	2.92	1.06		3.98	4.91
0140	Casement windows, one operating sash, 42 to 60 united inches	G		8	2	Ea.	282	106		388	480
0160	61 to 70	G		8	2		305	106		411	510
0180	71 to 80	G		8	2		340	106		446	540
0200	81 to 96	G		8	2		370	106		476	575
0220	Two operating sash, 58 to 78 united inches	G		8	2		655	106		761	890
0240	79 to 88	G		8	2		755	106		861	1,000
0260	89 to 98	G		8	2		950	106		1,056	1,225
0280	99 to 108	G		6	2.667		780	142		922	1,075
0300	109 to 121	G		6	2.667		855	142		997	1,175
0320	Two operating, one fixed sash, 73 to 108 united inches	G		8	2		620	106		726	850
0340	109 to 118	G		8	2		675	106		781	910
0360	119 to 128	G		6	2.667		770	142		912	1,075
0380	129 to 138	G		6	2.667		800	142		942	1,100
0400	139 to 156	G		6	2.667		990	142		1,132	1,300
0420	Four operating sash, 98 to 118 united inches	G		8	2		1,325	106		1,431	1,625
0440	119 to 128	G		8	2		1,375	106		1,481	1,700
0460	129 to 138	G		6	2.667		1,425	142		1,567	1,775
0480	139 to 148	G		6	2.667		1,475	142		1,617	1,850
0500	149 to 168	G		6	2.667		1,575	142		1,717	1,975
0520	169 to 178	G		6	2.667		1,625	142		1,767	2,025
0560	Fixed picture window, up to 63 united inches	G		8	2		181	106		287	370
0580	64 to 83	G		8	2		219	106		325	410
0600	84 to 101	G		8	2		273	106		379	470
0620	For each united inch over 101, add	G		900	.018	Inch	3.39	.95		4.34	5.25
0800	Cellulose fiber insulation, poured into sash balance cavity	G	1 Carp	36	.222	C.F.	.70	11.80		12.50	19.65
0820	Silicone caulking at perimeter	G	"	800	.010	L.F.	.17	.53		.70	1.03
2000	Impact resistant replacement windows										
2005	Laminated glass, 120 MPH rating, measure in united inches										
2010	Installation labor does not cover any rework of the window opening										
2020	Double-hung, insulated glass, up to 101 united inches		2 Carp	8	2	Ea.	555	106		661	780
2025	For each united inch over 101, add			80	.200	Inch	3.48	10.65		14.13	21
2100	Casement windows, impact resistant, up to 60 united inches			8	2	Ea.	515	106		621	735

08 01 Operation and Maintenance of Openings

08 01 53 – Operation and Maintenance of Plastic Windows

08 01 53.81 Solid Vinyl Replacement Windows	Crew	Daily Output	Labor-Hours	Unit	Material	2020 Bare Costs Labor	2020 Bare Costs Equipment	Total	Total Incl O&P
2120	61 to 70	2 Carp	8	2	Ea.	545	106		651	765
2130	71 to 80		8	2		570	106		676	800
2140	81 to 100		8	2	▼	590	106		696	815
2150	For each united inch over 100, add		80	.200	Inch	5.30	10.65		15.95	23
2200	Awning windows, impact resistant, up to 60 united inches		8	2	Ea.	525	106		631	750
2220	61 to 70		8	2		550	106		656	775
2230	71 to 80		8	2	▼	560	106		666	790
2240	For each united inch over 80, add		80	.200	Inch	4.99	10.65		15.64	22.50
2300	Picture windows, impact resistant, up to 63 united inches		8	2	Ea.	380	106		486	590
2320	63 to 83		8	2		410	106		516	620
2330	84 to 101		8	2	▼	450	106		556	665
2340	For each united inch over 101, add	▼	80	.200	Inch	3.35	10.65		14	20.50

08 05 Common Work Results for Openings

08 05 05 – Selective Demolition for Openings

08 05 05.10 Selective Demolition Doors

		Crew	Daily Output	Labor-Hours	Unit	Material	2020 Bare Costs Labor	2020 Bare Costs Equipment	Total	Total Incl O&P
0010	**SELECTIVE DEMOLITION DOORS** R024119-10									
0200	Doors, exterior, 1-3/4" thick, single, 3' x 7' high	1 Clab	16	.500	Ea.		21		21	33.50
0202	Doors, exterior, 3' - 6' wide x 7' high		12	.667			28		28	45
0210	3' x 8' high		10	.800			33.50		33.50	54
0215	Double, 3' x 8' high		6	1.333			56		56	90
0220	Double, 6' x 7' high		12	.667			28		28	45
0500	Interior, 1-3/8" thick, single, 3' x 7' high		20	.400			16.85		16.85	27
0520	Double, 6' x 7' high		16	.500			21		21	33.50
0700	Bi-folding, 3' x 6'-8" high		20	.400			16.85		16.85	27
0720	6' x 6'-8" high		18	.444			18.70		18.70	30
0900	Bi-passing, 3' x 6'-8" high		16	.500			21		21	33.50
0940	6' x 6'-8" high		14	.571			24		24	38.50
0960	Interior metal door 1-3/4" thick, 3'-0" x 6'-8"		18	.444			18.70		18.70	30
0980	Interior metal door 1-3/4" thick, 3'-0" x 7'-0"		18	.444			18.70		18.70	30
1000	Interior wood door 1-3/4" thick, 3'-0" x 6'-8"		20	.400			16.85		16.85	27
1020	Interior wood door 1-3/4" thick, 3'-0" x 7'-0"	▼	20	.400			16.85		16.85	27
1100	Door demo, floor door	2 Sswk	5	3.200			184		184	300
1500	Remove and reset, hollow core	1 Carp	8	1			53		53	85
1520	Solid		6	1.333			71		71	114
2000	Frames, including trim, metal	▼	8	1			53		53	85
2200	Wood	2 Carp	32	.500	▼		26.50		26.50	42.50
2201	Alternate pricing method	1 Carp	200	.040	L.F.		2.13		2.13	3.41
2205	Remove door hardware	"	45.70	.175	Ea.		9.30		9.30	14.90
3000	Special doors, counter doors	2 Carp	6	2.667			142		142	227
3100	Double acting		10	1.600			85		85	136
3200	Floor door (trap type), or access type		8	2			106		106	170
3300	Glass, sliding, including frames		12	1.333			71		71	114
3400	Overhead, commercial, 12' x 12' high		4	4			213		213	340
3440	up to 20' x 16' high		3	5.333			283		283	455
3445	up to 35' x 30' high		1	16			850		850	1,350
3500	Residential, 9' x 7' high		8	2			106		106	170
3540	16' x 7' high		7	2.286			121		121	195
3600	Remove and reset, small		4	4			213		213	340
3620	Large		2.50	6.400			340		340	545
3700	Roll-up grille		5	3.200			170		170	272

08 05 05.10 Selective Demolition Doors

		Crew	Daily Output	Labor-Hours	Unit	Material	2020 Bare Costs Labor	Equipment	Total	Total Incl O&P
3800	Revolving door, no wire or connections	2 Carp	2	8	Ea.		425		425	680
3900	Storefront swing door	↓	3	5.333			283		283	455
3902	Cafe/bar swing door	2 Clab	8	2			84		84	135
4000	Residential lockset, exterior	1 Carp	28	.286			15.20		15.20	24.50
4224	Pocket door, no frame		8	1			53		53	85
4300	Remove and reset deadbolt	↓	8	1			53		53	85
5500	Remove door closer	1 Clab	10	.800			33.50		33.50	54
5520	Remove push/pull plate		20	.400			16.85		16.85	27
5540	Remove door bolt		30	.267			11.25		11.25	18
5560	Remove kick/mop plate		20	.400			16.85		16.85	27
5585	Remove panic device		10	.800			33.50		33.50	54
5590	Remove mail slot		45	.178			7.50		7.50	12
5595	Remove peep hole	↓	45	.178			7.50		7.50	12
5604	Remove 9' x 7' garage door	2 Carp	9.50	1.684			89.50		89.50	143
5624	16' x 7' garage door		8	2			106		106	170
5626	9' x 7' swing up garage door		7	2.286			121		121	195
5628	16' x 7' swing up garage door	↓	7	2.286			121		121	195
5630	Garage door demolition, remove garage door track	1 Carp	6	1.333			71		71	114
5644	Remove overhead door opener	1 Clab	5	1.600			67.50		67.50	108
5774	Remove heavy gage sectional door, 20 ga., 8' x 8'	2 Carp	7	2.286			121		121	195
5784	10' x 10'		6	2.667			142		142	227
5794	12' x 12'		5	3.200			170		170	272
5805	14' x 14'	↓	4	4			213		213	340
6384	Remove French door unit	1 Clab	8	1			42		42	67.50
7100	Remove double swing pneumatic doors, openers and sensors	2 Skwk	.50	32	Opng.		1,750		1,750	2,800
7570	Remove shock absorbing door	2 Sswk	1.90	8.421	"		485		485	795
9000	Minimum labor/equipment charge	1 Carp	4	2	Job		106		106	170

08 05 05.20 Selective Demolition of Windows

		Crew	Daily Output	Labor-Hours	Unit	Material	2020 Bare Costs Labor	Equipment	Total	Total Incl O&P
0010	**SELECTIVE DEMOLITION OF WINDOWS** [R024119-10]									
0200	Aluminum, including trim, to 12 S.F.	1 Clab	16	.500	Ea.		21		21	33.50
0240	To 25 S.F.		11	.727			30.50		30.50	49
0280	To 50 S.F.		5	1.600			67.50		67.50	108
0320	Storm windows/screens, to 12 S.F.		27	.296			12.45		12.45	20
0360	To 25 S.F.		21	.381			16.05		16.05	25.50
0400	To 50 S.F.		16	.500	↓		21		21	33.50
0600	Glass, up to 10 S.F./window		200	.040	S.F.		1.68		1.68	2.70
0620	Over 10 S.F./window		150	.053	"		2.25		2.25	3.60
1000	Steel, including trim, to 12 S.F.		13	.615	Ea.		26		26	41.50
1020	To 25 S.F.		9	.889			37.50		37.50	60
1040	To 50 S.F.		4	2			84		84	135
2000	Wood, including trim, to 12 S.F.		22	.364			15.30		15.30	24.50
2020	To 25 S.F.		18	.444			18.70		18.70	30
2060	To 50 S.F.		13	.615			26		26	41.50
2065	To 180 S.F.	↓	8	1	↓		42		42	67.50
4400	Remove skylight, prefabricated glass block with metal frame	G-3	180	.178	S.F.		9.30		9.30	14.65
4410	Remove skylight, plstc domes, flush/curb mtd	2 Carp	395	.041	"		2.15		2.15	3.45
4420	Remove skylight, plstc/glass up to 2' x 3'	1 Carp	15	.533	Ea.		28.50		28.50	45.50
4440	Remove skylight, plstc/glass up to 4' x 6'	2 Carp	10	1.600			85		85	136
4480	Remove roof window up to 3' x 4'	1 Carp	8	1			53		53	85
4500	Remove roof window up to 4' x 6'	2 Carp	6	2.667			142		142	227
5020	Remove and reset window, up to a 2' x 2' window	1 Carp	6	1.333			71		71	114
5040	Up to a 3' x 3' window	↓	4	2	↓		106		106	170

08 05 Common Work Results for Openings

08 05 05 – Selective Demolition for Openings

08 05 05.20 Selective Demolition of Windows	Crew	Daily Output	Labor-Hours	Unit	Material	2020 Bare Costs Labor	Equipment	Total	Total Incl O&P
5080	Up to a 4' x 5' window	1 Carp	2	4	Ea.		213		213	340
6000	Screening only	1 Clab	4000	.002	S.F.		.08		.08	.13
9000	Minimum labor/equipment charge	"	4	2	Job		84		84	135

08 11 Metal Doors and Frames

08 11 16 – Aluminum Doors and Frames

08 11 16.10 Entrance Doors

	08 11 16.10 Entrance Doors	Crew	Daily Output	Labor-Hours	Unit	Material	2020 Bare Costs Labor	Equipment	Total	Total Incl O&P
0010	**ENTRANCE DOORS** and frame, aluminum, narrow stile									
0011	Including standard hardware, clear finish, no glass									
0012	Top and bottom offset pivots, 1/4" beveled glass stops, threshold									
0013	Dead bolt lock with inside thumb screw, standard push pull									
0020	3'-0" x 7'-0" opening	2 Sswk	2	8	Ea.	1,075	460		1,535	1,925
0025	Anodizing aluminum entr. door & frame, add					115			115	126
0030	3'-6" x 7'-0" opening	2 Sswk	2	8		865	460		1,325	1,700
0100	3'-0" x 10'-0" opening, 3' high transom		1.80	8.889		1,325	510		1,835	2,325
0200	3'-6" x 10'-0" opening, 3' high transom		1.80	8.889		1,400	510		1,910	2,400
0280	5'-0" x 7'-0" opening		2	8		1,575	460		2,035	2,475
0300	6'-0" x 7'-0" opening		1.30	12.308		1,050	710		1,760	2,325
0301	6'-0" x 7'-0" opening		1.30	12.308	Pr.	1,050	710		1,760	2,325
0400	6'-0" x 10'-0" opening, 3' high transom		1.10	14.545	"	1,500	840		2,340	3,025
0520	3'-0" x 7'-0" opening, wide stile		2	8	Ea.	905	460		1,365	1,750
0540	3'-6" x 7'-0" opening		2	8		990	460		1,450	1,850
0560	5'-0" x 7'-0" opening		2	8		1,500	460		1,960	2,400
0580	6'-0" x 7'-0" opening		1.30	12.308	Pr.	1,600	710		2,310	2,925
0600	7'-0" x 7'-0" opening		1	16	"	2,275	920		3,195	4,000
1200	For non-standard size, add				Leaf	80%				
1250	For installation of non-standard size, add						20%			
1300	Light bronze finish, add				Leaf	36%				
1400	Dark bronze finish, add					25%				
1500	For black finish, add					40%				
1600	Concealed panic device, add					1,075			1,075	1,175
1700	Electric striker release, add				Opng.	325			325	360
1800	Floor check, add				Leaf	705			705	780
1900	Concealed closer, add				"	575			575	630
9000	Minimum labor/equipment charge	2 Carp	4	4	Job		213		213	340

08 11 63 – Metal Screen and Storm Doors and Frames

08 11 63.23 Aluminum Screen and Storm Doors and Frames

	08 11 63.23 Aluminum Screen and Storm Doors and Frames	Crew	Daily Output	Labor-Hours	Unit	Material	2020 Bare Costs Labor	Equipment	Total	Total Incl O&P
0010	**ALUMINUM SCREEN AND STORM DOORS AND FRAMES**									
0020	Combination storm and screen									
0420	Clear anodic coating, 2'-8" wide	2 Carp	14	1.143	Ea.	238	60.50		298.50	360
0440	3'-0" wide	"	14	1.143	"	192	60.50		252.50	310
0500	For 7'-0" door height, add					8%				
1020	Mill finish, 2'-8" wide	2 Carp	14	1.143	Ea.	249	60.50		309.50	370
1040	3'-0" wide	"	14	1.143		268	60.50		328.50	395
1100	For 7'-0" door, add					8%				
1520	White painted, 2'-8" wide	2 Carp	14	1.143		269	60.50		329.50	395
1540	3'-0" wide	"	14	1.143		315	60.50		375.50	445
1600	For 7'-0" door, add					8%				
2000	Wood door & screen, see Section 08 14 33.20									
9000	Minimum labor/equipment charge	1 Carp	4	2	Job		106		106	170

08 12 13 – Hollow Metal Frames

08 12 13.13 Standard Hollow Metal Frames		Crew	Daily Output	Labor-Hours	Unit	Material	2020 Bare Costs Labor	Equipment	Total	Total Incl O&P
0010	**STANDARD HOLLOW METAL FRAMES**									
0020	16 ga., up to 5-3/4" jamb depth									
0025	3'-0" x 6'-8" single	G	2 Carp	16	1	Ea.	217	53		270	325
0028	3'-6" wide, single	G		16	1		270	53		323	380
0030	4'-0" wide, single	G		16	1		305	53		358	420
0040	6'-0" wide, double	G		14	1.143		256	60.50		316.50	380
0045	8'-0" wide, double	G		14	1.143		248	60.50		308.50	370
0100	3'-0" x 7'-0" single	G		16	1		214	53		267	320
0110	3'-6" wide, single			16	1		221	53		274	330
0112	4'-0" wide, single			16	1		252	53		305	360
0140	6'-0" wide, double			14	1.143		268	60.50		328.50	395
0145	8'-0" wide, double			14	1.143		236	60.50		296.50	360
1000	16 ga., up to 4-7/8" deep, 3'-0" x 7'-0" single	G		16	1		202	53		255	305
1140	6'-0" wide, double	G		14	1.143		215	60.50		275.50	335
1200	16 ga., 8-3/4" deep, 3'-0" x 7'-0" single	G		16	1		206	53		259	310
1240	6'-0" wide, double	G		14	1.143		310	60.50		370.50	440
2800	14 ga., up to 3-7/8" deep, 3'-0" x 7'-0" single	G		16	1		251	53		304	360
2840	6'-0" wide, double	G		14	1.143		360	60.50		420.50	495
3000	14 ga., up to 5-3/4" deep, 3'-0" x 6'-8" single	G		16	1		159	53		212	260
3002	3'-6" wide, single	G		16	1		267	53		320	380
3005	4'-0" wide, single	G		16	1		158	53		211	259
3600	up to 5-3/4" jamb depth, 4'-0" x 7'-0" single			15	1.067		256	56.50		312.50	375
3620	6'-0" wide, double			12	1.333		201	71		272	335
3640	8'-0" wide, double			12	1.333		300	71		371	445
3700	8'-0" high, 4'-0" wide, single			15	1.067		305	56.50		361.50	425
3740	8'-0" wide, double			12	1.333		345	71		416	495
4000	6-3/4" deep, 4'-0" x 7'-0" single	G		15	1.067		310	56.50		366.50	430
4020	6'-0" wide, double	G		12	1.333		360	71		431	510
4040	8'-0" wide, double	G		12	1.333		258	71		329	400
4100	8'-0" high, 4'-0" wide, single	G		15	1.067		188	56.50		244.50	297
4140	8'-0" wide, double	G		12	1.333		450	71		521	610
4400	8-3/4" deep, 4'-0" x 7'-0", single			15	1.067		430	56.50		486.50	565
4440	8'-0" wide, double			12	1.333		455	71		526	615
4500	4'-0" x 8'-0", single			15	1.067		480	56.50		536.50	615
4540	8'-0" wide, double	▼		12	1.333		490	71		561	650
4900	For welded frames, add						59.50			59.50	65.50
5380	Steel frames, KD, 14 ga., "B" label, to 5-3/4" throat, to 3'-0" x 7'-0"		2 Carp	15	1.067		248	56.50		304.50	365
5400	14 ga., "B" label, up to 5-3/4" deep, 4'-0" x 7'-0" single	G		15	1.067		190	56.50		246.50	300
5440	8'-0" wide, double	G		12	1.333		243	71		314	380
5800	6-3/4" deep, 7'-0" high, 4'-0" wide, single	G		15	1.067		169	56.50		225.50	277
5840	8'-0" wide, double	G		12	1.333		335	71		406	485
6200	8-3/4" deep, 4'-0" x 7'-0" single	G		15	1.067		269	56.50		325.50	385
6240	8'-0" wide, double	G		12	1.333	▼	370	71		441	525
6300	For "A" label use same price as "B" label										
6400	For baked enamel finish, add						30%	15%			
6500	For galvanizing, add						20%				
6600	For hospital stop, add					Ea.	251			251	276
6620	For hospital stop, stainless steel, add					"	118			118	130
7900	Transom lite frames, fixed, add		2 Carp	155	.103	S.F.	49	5.50		54.50	63
8000	Movable, add		"	130	.123	"	64	6.55		70.55	81
9000	Minimum labor/equipment charge		1 Carp	4	2	Job		106		106	170

08 12 Metal Frames

08 12 13 – Hollow Metal Frames

08 12 13.15 Borrowed Lites	Crew	Daily Output	Labor-Hours	Unit	Material	2020 Bare Costs Labor	Equipment	Total	Total Incl O&P
0010 **BORROWED LITES**									
0100 Hollow metal section 20 ga., 3-1/2" with glass stop	2 Carp	100	.160	L.F.	19.90	8.50		28.40	35.50
0110 3-3/4" with glass stop		100	.160		9.15	8.50		17.65	23.50
0120 4" with glass stop		100	.160		9.30	8.50		17.80	24
0130 4-5/8" with glass stop		100	.160		9.60	8.50		18.10	24
0140 4-7/8" with glass stop		100	.160		9.80	8.50		18.30	24.50
0150 5" with glass stop		100	.160		10.65	8.50		19.15	25.50
0160 5-3/8" with glass stop		100	.160		12.55	8.50		21.05	27.50
0300 Hollow metal section 18 ga., 3-1/2" with glass stop		80	.200		9.90	10.65		20.55	28
0310 3-3/4" with glass stop		80	.200		10.10	10.65		20.75	28
0320 4" with glass stop		80	.200		14	10.65		24.65	32.50
0330 4-5/8" with glass stop		80	.200		10.75	10.65		21.40	29
0340 4-7/8" with glass stop		80	.200		11	10.65		21.65	29
0350 5" with glass stop		80	.200		11.10	10.65		21.75	29.50
0360 5-3/8" with glass stop		80	.200		11.10	10.65		21.75	29.50
0370 6-5/8" with glass stop		80	.200		12.35	10.65		23	30.50
0380 6-7/8" with glass stop		80	.200		12.55	10.65		23.20	31
0390 7-1/4" with glass stop		80	.200		13.10	10.65		23.75	31.50
0500 Mullion section 18 ga., 3-1/2" with double stop		50	.320		23	17		40	53
0510 3-3/4" with double stop		50	.320		23.50	17		40.50	53.50
0530 4-5/8" with double stop		50	.320		25	17		42	55
0540 4-7/8" with double stop		50	.320		25.50	17		42.50	55.50
0550 5" with double stop		50	.320		26	17		43	56
0560 5-3/8" with double stop		50	.320		26.50	17		43.50	56.50
0570 6-5/8" with double stop		50	.320		29	17		46	59
0580 6-7/8" with double stop		50	.320		29.50	17		46.50	60
0590 7-1/4" with double stop		50	.320		30	17		47	60.50
1000 Assembled frame 2'-0" x 2'-0" 16 ga., 3-3/4" with glass stop		10	1.600	Ea.	217	85		302	375
1010 3'-0" x 2'-0"		10	1.600		217	85		302	375
1020 4'-0" x 2'-0"		10	1.600		218	85		303	375
1030 5'-0" x 2'-0"		10	1.600		224	85		309	385
1040 6'-0" x 2'-0"		8	2		325	106		431	530
1050 7'-0" x 2'-0"		8	2		325	106		431	530
1060 8'-0" x 2'-0"		8	2		325	106		431	530
1100 2'-0" x 7'-0"		8	2		325	106		431	530
1110 3'-0" x 7'-0"		8	2		325	106		431	530
1120 4'-0" x 7'-0"		8	2		325	106		431	530
1130 5'-0" x 7'-0"		8	2		325	106		431	530
1140 6'-0" x 7'-0"		6	2.667		350	142		492	610
1150 7'-0" x 7'-0"		6	2.667		350	142		492	610
1160 8'-0" x 7'-0"		6	2.667		350	142		492	610
1200 Assembled frame 2'-0" x 2'-0" 16 ga., 4-7/8" with glass stop		10	1.600		218	85		303	375
1210 3'-0" x 2'-0"		10	1.600		218	85		303	375
1220 4'-0" x 2'-0"		8	2		218	106		324	410
1230 5'-0" x 2'-0"		8	2		325	106		431	530
1240 6'-0" x 2'-0"		8	2		325	106		431	530
1250 7'-0" x 2'-0"		8	2		325	106		431	530
1260 8'-0" x 2'-0"		8	2		325	106		431	530
1300 2'-0" x 7'-0"		8	2		325	106		431	530
1310 3'-0" x 7'-0"		8	2		325	106		431	530
1320 4'-0" x 7'-0"		8	2		325	106		431	530
1330 5'-0" x 7'-0"		6	2.667		325	142		467	585

For customer support on your Commercial Renovation Costs with RSMeans data, call 800.448.8182.

233

08 12 Metal Frames

08 12 13 – Hollow Metal Frames

08 12 13.15 Borrowed Lites

		Crew	Daily Output	Labor-Hours	Unit	Material	2020 Bare Costs Labor	Equipment	Total	Total Incl O&P
1340	6'-0" x 7'-0"	2 Carp	6	2.667	Ea.	350	142		492	610
1350	7'-0" x 7'-0"		6	2.667		360	142		502	625
1360	8'-0" x 7'-0"		6	2.667		360	142		502	625
1400	Assembled frame 2'-0" x 3'-0" 16 ga., 6-1/8" with glass stop		6	2.667		219	142		361	465
1410	3'-0" x 3'-0"		8	2		218	106		324	410
1420	4'-0" x 3'-0"		8	2		218	106		324	410
1430	5'-0" x 3'-0"		6	2.667		325	142		467	585
1440	6'-0" x 3'-0"		6	2.667		300	142		442	555
1450	7'-0" x 3'-0"		6	2.667		325	142		467	585
1460	8'-0" x 3'-0"		10	1.600		325	85		410	495

08 12 13.25 Channel Metal Frames

			Crew	Daily Output	Labor-Hours	Unit	Material	2020 Bare Costs Labor	Equipment	Total	Total Incl O&P
0010	**CHANNEL METAL FRAMES**										
0020	Steel channels with anchors and bar stops										
0100	6" channel @ 8.2#/L.F., 3' x 7' door, weighs 150#	G	E-4	13	2.462	Ea.	259	143	11.30	413.30	530
0200	8" channel @ 11.5#/L.F., 6' x 8' door, weighs 275#	G		9	3.556		475	207	16.30	698.30	880
0300	8' x 12' door, weighs 400#	G		6.50	4.923		690	286	22.50	998.50	1,250
0400	10" channel @ 15.3#/L.F., 10' x 10' door, weighs 500#	G		6	5.333		860	310	24.50	1,194.50	1,475
0500	12' x 12' door, weighs 600#	G		5.50	5.818		1,025	340	26.50	1,391.50	1,725
0600	12" channel @ 20.7#/L.F., 12' x 12' door, weighs 825#	G		4.50	7.111		1,425	415	32.50	1,872.50	2,300
0700	12' x 16' door, weighs 1000#	G		4	8		1,725	465	36.50	2,226.50	2,700
0800	For frames without bar stops, light sections, deduct						15%				
0900	Heavy sections, deduct						10%				
9000	Minimum labor/equipment charge		E-4	4	8	Job		465	36.50	501.50	800

08 13 Metal Doors

08 13 13 – Hollow Metal Doors

08 13 13.13 Standard Hollow Metal Doors

			Crew	Daily Output	Labor-Hours	Unit	Material	2020 Bare Costs Labor	Equipment	Total	Total Incl O&P
0010	**STANDARD HOLLOW METAL DOORS**	R081313-20									
0015	Flush, full panel, hollow core										
0017	When noted doors are prepared but do not include glass or louvers										
0020	1-3/8" thick, 20 ga., 2'-0" x 6'-8"	G	2 Carp	20	.800	Ea.	345	42.50		387.50	450
0040	2'-8" x 6'-8"	G		18	.889		355	47		402	470
0060	3'-0" x 6'-8"	G		17	.941		360	50		410	475
0100	3'-0" x 7'-0"			17	.941		375	50		425	490
0120	For vision lite, add						111			111	122
0140	For narrow lite, add						115			115	126
0320	Half glass, 20 ga., 2'-0" x 6'-8"	G	2 Carp	20	.800		695	42.50		737.50	835
0340	2'-8" x 6'-8"	G		18	.889		610	47		657	745
0360	3'-0" x 6'-8"	G		17	.941		605	50		655	745
0400	3'-0" x 7'-0"	G		17	.941		475	50		525	605
0410	1-3/8" thick, 18 ga., 2'-0" x 6'-8"	G		20	.800		465	42.50		507.50	585
0420	3'-0" x 6'-8"	G		17	.941		470	50		520	595
0425	3'-0" x 7'-0"	G		17	.941		460	50		510	585
0450	For vision lite, add						111			111	122
0452	For narrow lite, add						115			115	126
0460	Half glass, 18 ga., 2'-0" x 6'-8"	G	2 Carp	20	.800		635	42.50		677.50	770
0465	2'-8" x 6'-8"	G		18	.889		660	47		707	800
0470	3'-0" x 6'-8"	G		17	.941		645	50		695	790
0475	3'-0" x 7'-0"	G		17	.941		605	50		655	745
0500	Hollow core, 1-3/4" thick, full panel, 20 ga., 2'-8" x 6'-8"	G		18	.889		495	47		542	620
0520	3'-0" x 6'-8"	G		17	.941		495	50		545	625

For customer support on your Commercial Renovation Costs with RSMeans data, call 800.448.8182.

08 13 13.13 Standard Hollow Metal Doors	Crew	Daily Output	Labor-Hours	Unit	Material	2020 Bare Costs Labor	2020 Bare Costs Equipment	Total	Total Incl O&P
0640	3'-0" x 7'-0" G	2 Carp	17	.941	Ea.	515	50		565	645
0680	4'-0" x 7'-0" G		15	1.067		650	56.50		706.50	805
0700	4'-0" x 8'-0" G		13	1.231		755	65.50		820.50	935
1000	18 ga., 2'-8" x 6'-8" G		17	.941		525	50		575	655
1020	3'-0" x 6'-8" G		16	1		510	53		563	645
1120	3'-0" x 7'-0" G		17	.941		505	50		555	635
1150	3'-6" x 7'-0"	1 Carp	14	.571		555	30.50		585.50	660
1180	4'-0" x 7'-0" G	2 Carp	14	1.143		740	60.50		800.50	915
1200	4'-0" x 8'-0" G	"	17	.941		850	50		900	1,025
1212	For vision lite, add					111			111	122
1214	For narrow lite, add					115			115	126
1230	Half glass, 20 ga., 2'-8" x 6'-8" G	2 Carp	20	.800		575	42.50		617.50	700
1240	3'-0" x 6'-8" G		18	.889		570	47		617	700
1260	3'-0" x 7'-0" G		18	.889		600	47		647	735
1280	Embossed panel, 1-3/4" thick, poly core, 20 ga., 3'-0" x 7'-0" G		18	.889		425	47		472	545
1290	Half glass, 1-3/4" thick, poly core, 20 ga., 3'-0" x 7'-0" G		18	.889		625	47		672	760
1320	18 ga., 2'-8" x 6'-8" G		18	.889		645	47		692	785
1340	3'-0" x 6'-8" G		17	.941		640	50		690	785
1360	3'-0" x 7'-0" G		17	.941		655	50		705	800
1380	4'-0" x 7'-0" G		15	1.067		785	56.50		841.50	955
1400	4'-0" x 8'-0" G		14	1.143		880	60.50		940.50	1,075
1500	Flush full panel, 16 ga., steel hollow core									
1520	2'-0" x 6'-8" G	2 Carp	20	.800	Ea.	580	42.50		622.50	705
1530	2'-8" x 6'-8" G		20	.800		580	42.50		622.50	705
1540	3'-0" x 6'-8" G		20	.800		495	42.50		537.50	615
1560	2'-8" x 7'-0" G		18	.889		590	47		637	725
1570	3'-0" x 7'-0" G		18	.889		550	47		597	680
1580	3'-6" x 7'-0" G		18	.889		675	47		722	820
1590	4'-0" x 7'-0" G		18	.889		745	47		792	890
1600	2'-8" x 8'-0" G		18	.889		725	47		772	875
1620	3'-0" x 8'-0" G		18	.889		740	47		787	890
1630	3'-6" x 8'-0" G		18	.889		820	47		867	980
1640	4'-0" x 8'-0" G		18	.889		865	47		912	1,025
1650	1-13/16", 14 ga., 2'-8" x 7'-0" G		10	1.600		1,150	85		1,235	1,400
1670	3'-0" x 7'-0" G		10	1.600		1,125	85		1,210	1,350
1690	3'-6" x 7'-0" G		10	1.600		1,225	85		1,310	1,475
1700	4'-0" x 7'-0" G		10	1.600		1,300	85		1,385	1,550
1720	Insulated, 1-3/4" thick, full panel, 18 ga., 3'-0" x 6'-8" G		15	1.067		500	56.50		556.50	645
1740	2'-8" x 7'-0" G		16	1		505	53		558	640
1760	3'-0" x 7'-0" G		15	1.067		520	56.50		576.50	665
1800	4'-0" x 8'-0" G		13	1.231		810	65.50		875.50	1,000
1820	Half glass, 18 ga., 3'-0" x 6'-8" G		16	1		630	53		683	780
1840	2'-8" x 7'-0" G		17	.941		660	50		710	805
1860	3'-0" x 7'-0" G		16	1		690	53		743	845
1900	4'-0" x 8'-0" G		14	1.143		590	60.50		650.50	750
2000	For vision lite, add					111			111	122
2010	For narrow lite, add					115			115	126
8100	For bottom louver, add					243			243	267
8110	For baked enamel finish, add					30%	15%			
8120	For galvanizing, add					20%				
8190	Commercial door, flush steel, fire door see section 08 13 13 15									
8270	For dutch door shelf, add to standard door				Ea.	300			300	330
9000	Minimum labor/equipment charge	1 Carp	4	2	Job		106		106	170

08 13 13.15 Metal Fire Doors

		Crew	Daily Output	Labor-Hours	Unit	Material	2020 Bare Costs Labor	Equipment	Total	Total Incl O&P
0010	**METAL FIRE DOORS** R081313-20									
0015	Steel, flush, "B" label, 90 minutes									
0020	Full panel, 20 ga., 2'-0" x 6'-8"	2 Carp	20	.800	Ea.	420	42.50		462.50	530
0040	2'-8" x 6'-8"		18	.889		435	47		482	555
0060	3'-0" x 6'-8"		17	.941		435	50		485	560
0080	3'-0" x 7'-0"		17	.941		450	50		500	575
0140	18 ga., 3'-0" x 6'-8"		16	1		530	53		583	670
0160	2'-8" x 7'-0"		17	.941		560	50		610	700
0180	3'-0" x 7'-0"		16	1		510	53		563	645
0200	4'-0" x 7'-0"		15	1.067		645	56.50		701.50	800
0220	For "A" label, 3 hour, 18 ga., use same price as "B" label									
0240	For vision lite, add				Ea.	183			183	201
0300	Full panel, 16 ga., 2'-0" x 6'-8"	2 Carp	20	.800		320	42.50		362.50	420
0310	2'-8" x 6'-8"		18	.889		560	47		607	690
0320	3'-0" x 6'-8"		17	.941		450	50		500	575
0350	2'-8" x 7'-0"		17	.941		580	50		630	715
0360	3'-0" x 7'-0"		16	1		560	53		613	705
0370	4'-0" x 7'-0"		15	1.067		725	56.50		781.50	890
0520	Flush, "B" label, 90 minutes, egress core, 20 ga., 2'-0" x 6'-8"		18	.889		680	47		727	825
0540	2'-8" x 6'-8"		17	.941		685	50		735	830
0560	3'-0" x 6'-8"		16	1		685	53		738	840
0580	3'-0" x 7'-0"		16	1		715	53		768	870
0640	Flush, "A" label, 3 hour, egress core, 18 ga., 3'-0" x 6'-8"		15	1.067		740	56.50		796.50	905
0660	2'-8" x 7'-0"		16	1		780	53		833	940
0680	3'-0" x 7'-0"		15	1.067		410	56.50		466.50	545
0700	4'-0" x 7'-0"		14	1.143		945	60.50		1,005.50	1,150
9000	Minimum labor/equipment charge	1 Carp	4	2	Job		106		106	170

08 13 13.20 Residential Steel Doors

			Crew	Daily Output	Labor-Hours	Unit	Material	2020 Bare Costs Labor	Equipment	Total	Total Incl O&P
0010	**RESIDENTIAL STEEL DOORS**										
0020	Prehung, insulated, exterior										
0030	Embossed, full panel, 2'-8" x 6'-8"	G	2 Carp	17	.941	Ea.	410	50		460	530
0040	3'-0" x 6'-8"	G		15	1.067		300	56.50		356.50	420
0060	3'-0" x 7'-0"	G		15	1.067		355	56.50		411.50	480
0070	5'-4" x 6'-8", double	G		8	2		590	106		696	820
0220	Half glass, 2'-8" x 6'-8"	G		17	.941		330	50		380	445
0240	3'-0" x 6'-8"	G		16	1		330	53		383	450
0260	3'-0" x 7'-0"	G		16	1		400	53		453	525
0270	5'-4" x 6'-8", double	G		8	2		985	106		1,091	1,250
1320	Flush face, full panel, 2'-8" x 6'-8"	G		16	1		335	53		388	455
1340	3'-0" x 6'-8"	G		15	1.067		335	56.50		391.50	460
1360	3'-0" x 7'-0"	G		15	1.067		315	56.50		371.50	435
1380	5'-4" x 6'-8", double	G		8	2		810	106		916	1,075
1420	Half glass, 2'-8" x 6'-8"	G		17	.941		350	50		400	465
1440	3'-0" x 6'-8"	G		16	1		350	53		403	470
1460	3'-0" x 7'-0"	G		16	1		455	53		508	590
1480	5'-4" x 6'-8", double	G		8	2		680	106		786	915
1500	Sidelight, full lite, 1'-0" x 6'-8" with grille	G					395			395	435
1510	1'-0" x 6'-8", low E	G					410			410	450
1520	1'-0" x 6'-8", half lite	G					299			299	330
1530	1'-0" x 6'-8", half lite, low E	G					300			300	330
2300	Interior, residential, closet, bi-fold, 2'-0" x 6'-8"	G	2 Carp	16	1		251	53		304	360
2330	3'-0" wide	G		16	1		284	53		337	400

08 13 Metal Doors

08 13 13 – Hollow Metal Doors

08 13 13.20 Residential Steel Doors		Crew	Daily Output	Labor-Hours	Unit	Material	2020 Bare Costs Labor	Equipment	Total	Total Incl O&P
2360	4'-0" wide	G	2 Carp	15	1.067	Ea.	293	56.50		349.50	410
2400	5'-0" wide	G		14	1.143		370	60.50		430.50	505
2420	6'-0" wide	G		13	1.231		300	65.50		365.50	435
9000	Minimum labor/equipment charge		1 Carp	4	2	Job		106		106	170

08 13 13.25 Doors Hollow Metal

0010	**DOORS HOLLOW METAL**										
0500	Exterior, commercial, flush, 20 ga., 1-3/4" x 7'-0" x 2'-6" wide	G	2 Carp	15	1.067	Ea.	287	56.50		343.50	405
0530	2'-8" wide	G		15	1.067		345	56.50		401.50	470
0560	3'-0" wide	G		14	1.143		345	60.50		405.50	480
0590	3'-6" wide	G		14	1.143		375	60.50		435.50	515
1000	18 ga., 1-3/4" x 7'-0" x 2'-6" wide	G		15	1.067		300	56.50		356.50	420
1030	2'-8" wide	G		15	1.067		305	56.50		361.50	425
1060	3'-0" wide	G		14	1.143		296	60.50		356.50	425
1500	16 ga., 1-3/4" x 7'-0" x 2'-6" wide	G		15	1.067		385	56.50		441.50	515
1530	2'-8" wide	G		15	1.067		385	56.50		441.50	510
1560	3'-0" wide	G		14	1.143		455	60.50		515.50	600
1590	3'-6" wide	G		14	1.143		735	60.50		795.50	910
2900	Fire door, "A" label, 18 gauge, 1-3/4" x 2'-6" x 7'-0"	G		15	1.067		275	56.50		331.50	395
2930	2'-8" wide	G		15	1.067		275	56.50		331.50	395
2960	3'-0" wide	G		14	1.143		275	60.50		335.50	405
2990	3'-6" wide	G		14	1.143		495	60.50		555.50	645
3100	"B" label, 2'-6" wide	G		15	1.067		655	56.50		711.50	810
3130	2'-8" wide	G		15	1.067		675	56.50		731.50	830
3160	3'-0" wide	G		14	1.143		670	60.50		730.50	835

08 13 16 – Aluminum Doors

08 13 16.10 Commercial Aluminum Doors

0010	**COMMERCIAL ALUMINUM DOORS**, flush, no glazing										
5000	Flush panel doors, pair of 2'-6" x 7'-0"		2 Sswk	2	8	Pr.	1,250	460		1,710	2,100
5050	3'-0" x 7'-0", single			2.50	6.400	Ea.	730	370		1,100	1,400
5100	Pair of 3'-0" x 7'-0"			2	8	Pr.	1,500	460		1,960	2,400
5150	3'-6" x 7'-0", single			2.50	6.400	Ea.	955	370		1,325	1,650

08 14 Wood Doors

08 14 13 – Carved Wood Doors

08 14 13.10 Types of Wood Doors, Carved

0010	**TYPES OF WOOD DOORS, CARVED**										
3000	Solid wood, 1-3/4" thick stile and rail										
3020	Mahogany, 3'-0" x 7'-0", six panel		2 Carp	14	1.143	Ea.	1,125	60.50		1,185.50	1,350
3030	With two lites			10	1.600		3,475	85		3,560	3,950
3040	3'-6" x 8'-0", six panel			10	1.600		1,800	85		1,885	2,100
3050	With two lites			8	2		3,050	106		3,156	3,550
3100	Pine, 3'-0" x 7'-0", six panel			14	1.143		655	60.50		715.50	820
3110	With two lites			10	1.600		880	85		965	1,100
3120	3'-6" x 8'-0", six panel			10	1.600		1,125	85		1,210	1,375
3130	With two lites			8	2		2,000	106		2,106	2,400
3200	Red oak, 3'-0" x 7'-0", six panel			14	1.143		1,325	60.50		1,385.50	1,575
3210	With two lites			10	1.600		2,675	85		2,760	3,075
3220	3'-6" x 8'-0", six panel			10	1.600		2,825	85		2,910	3,225
3230	With two lites			8	2		3,575	106		3,681	4,125
4000	Hand carved door, mahogany										

For customer support on your Commercial Renovation Costs with RSMeans data, call 800.448.8182.

237

08 14 Wood Doors

08 14 13 – Carved Wood Doors

08 14 13.10 Types of Wood Doors, Carved		Crew	Daily Output	Labor-Hours	Unit	Material	2020 Bare Costs Labor	Equipment	Total	Total Incl O&P
4020	3'-0" x 7'-0", simple design	2 Carp	14	1.143	Ea.	1,875	60.50		1,935.50	2,175
4030	Intricate design		11	1.455		3,825	77.50		3,902.50	4,325
4040	3'-6" x 8'-0", simple design		10	1.600		2,875	85		2,960	3,300
4050	Intricate design		8	2		3,875	106		3,981	4,450
4400	For custom finish, add					640			640	705
4600	Side light, mahogany, 7'-0" x 1'-6" wide, 4 lites	2 Carp	18	.889		1,200	47		1,247	1,375
4610	6 lites		14	1.143		2,800	60.50		2,860.50	3,175
4620	8'-0" x 1'-6" wide, 4 lites		14	1.143		2,225	60.50		2,285.50	2,550
4630	6 lites		10	1.600		2,375	85		2,460	2,750
4640	Side light, oak, 7'-0" x 1'-6" wide, 4 lites		18	.889		1,425	47		1,472	1,625
4650	6 lites		14	1.143		2,475	60.50		2,535.50	2,825
4660	8'-0" x 1'-6" wide, 4 lites		14	1.143		1,325	60.50		1,385.50	1,575
4670	6 lites		10	1.600		2,475	85		2,560	2,850

08 14 16 – Flush Wood Doors

08 14 16.09 Smooth Wood Doors

		Crew	Daily Output	Labor-Hours	Unit	Material	2020 Bare Costs Labor	Equipment	Total	Total Incl O&P
0010	**SMOOTH WOOD DOORS**									
0015	Flush, interior, hollow core									
0025	Lauan face, 1-3/8", 3'-0" x 6'-8"	2 Carp	17	.941	Ea.	53.50	50		103.50	139
0030	4'-0" x 6'-8"		16	1		134	53		187	232
0080	1-3/4", 2'-0" x 6'-8"		17	.941		60	50		110	146
0085	3'-0" x 6'-8"	1 Carp	16	.500		53.50	26.50		80	101
0108	3'-0" x 7'-0"	2 Carp	16	1		226	53		279	335
0112	Pair of 3'-0" x 7'-0"		9	1.778	Pr.	153	94.50		247.50	320
0140	Birch face, 1-3/8", 2'-6" x 6'-8"		17	.941	Ea.	122	50		172	214
0180	3'-0" x 6'-8"		17	.941		101	50		151	191
0200	4'-0" x 6'-8"		16	1		167	53		220	269
0202	1-3/4", 2'-0" x 6'-8"		17	.941		69	50		119	156
0204	2'-4" x 7'-0"		16	1		133	53		186	232
0206	2'-6" x 7'-0"		16	1		137	53		190	236
0210	3'-0" x 7'-0"		16	1		174	53		227	277
0214	Pair of 3'-0" x 7'-0"		9	1.778	Pr.	274	94.50		368.50	450
0220	Oak face, 1-3/8", 2'-0" x 6'-8"		17	.941	Ea.	117	50		167	208
0280	3'-0" x 6'-8"		17	.941		112	50		162	203
0300	4'-0" x 6'-8"		16	1		148	53		201	248
0305	1-3/4", 2'-6" x 6'-8"		17	.941		137	50		187	230
0310	3'-0" x 7'-0"		16	1		272	53		325	385
0320	Walnut face, 1-3/8", 2'-0" x 6'-8"		17	.941		206	50		256	305
0340	2'-6" x 6'-8"		17	.941		201	50		251	300
0380	3'-0" x 6'-8"		17	.941		210	50		260	310
0400	4'-0" x 6'-8"		16	1		231	53		284	340
0430	For 7'-0" high, add					31.50			31.50	35
0440	For 8'-0" high, add					46			46	50.50
0480	For prefinishing, clear, add					57			57	63
0500	For prefinishing, stain, add					67.50			67.50	74
1320	M.D. overlay on hardboard, 1-3/8", 2'-0" x 6'-8"	2 Carp	17	.941		134	50		184	227
1340	2'-6" x 6'-8"		17	.941		135	50		185	229
1380	3'-0" x 6'-8"		17	.941		136	50		186	230
1400	4'-0" x 6'-8"		16	1		231	53		284	340
1420	For 7'-0" high, add					18.95			18.95	21
1440	For 8'-0" high, add					42			42	46
1720	H.P. plastic laminate, 1-3/8", 2'-0" x 6'-8"	2 Carp	16	1		286	53		339	400
1740	2'-6" x 6'-8"		16	1		282	53		335	395

08 14 16.09 Smooth Wood Doors	Crew	Daily Output	Labor-Hours	Unit	Material	2020 Bare Costs Labor	Equipment	Total	Total Incl O&P
1780	3'-0" x 6'-8"	2 Carp	15	1.067	Ea.	305	56.50		361.50	425
1800	4'-0" x 6'-8"	↓	14	1.143		405	60.50		465.50	545
1820	For 7'-0" high, add					22.50			22.50	24.50
1840	For 8'-0" high, add					39			39	43
2020	Particle core, lauan face, 1-3/8", 2'-6" x 6'-8"	2 Carp	15	1.067		105	56.50		161.50	207
2040	3'-0" x 6'-8"		14	1.143		108	60.50		168.50	217
2080	3'-0" x 7'-0"		13	1.231		128	65.50		193.50	246
2085	4'-0" x 7'-0"		12	1.333		140	71		211	268
2110	1-3/4", 3'-0" x 7'-0"		13	1.231		226	65.50		291.50	355
2120	Birch face, 1-3/8", 2'-6" x 6'-8"		15	1.067		120	56.50		176.50	223
2140	3'-0" x 6'-8"		14	1.143		130	60.50		190.50	241
2180	3'-0" x 7'-0"		13	1.231		143	65.50		208.50	262
2200	4'-0" x 7'-0"		12	1.333		156	71		227	286
2205	1-3/4", 3'-0" x 7'-0"		13	1.231		211	65.50		276.50	335
2220	Oak face, 1-3/8", 2'-6" x 6'-8"		15	1.067		132	56.50		188.50	237
2240	3'-0" x 6'-8"		14	1.143		145	60.50		205.50	258
2280	3'-0" x 7'-0"		13	1.231		152	65.50		217.50	272
2300	4'-0" x 7'-0"		12	1.333		176	71		247	310
2305	1-3/4", 3'-0" x 7'-0"		13	1.231		256	65.50		321.50	385
2320	Walnut face, 1-3/8", 2'-0" x 6'-8"		15	1.067		138	56.50		194.50	243
2340	2'-6" x 6'-8"		14	1.143		155	60.50		215.50	268
2380	3'-0" x 6'-8"		13	1.231		175	65.50		240.50	297
2400	4'-0" x 6'-8"	↓	12	1.333		229	71		300	365
2440	For 8'-0" high, add					46.50			46.50	51.50
2460	For 8'-0" high walnut, add					42.50			42.50	46.50
2720	For prefinishing, clear, add					41.50			41.50	45.50
2740	For prefinishing, stain, add					59.50			59.50	65.50
3320	M.D. overlay on hardboard, 1-3/8", 2'-6" x 6'-8"	2 Carp	14	1.143		189	60.50		249.50	305
3340	3'-0" x 6'-8"		13	1.231		213	65.50		278.50	340
3380	3'-0" x 7'-0"		12	1.333		216	71		287	350
3400	4'-0" x 7'-0"	↓	10	1.600		298	85		383	465
3440	For 8'-0" height, add					42			42	46
3460	For solid wood core, add					75.50			75.50	83
3720	H.P. plastic laminate, 1-3/8", 2'-6" x 6'-8"	2 Carp	13	1.231		172	65.50		237.50	294
3740	3'-0" x 6'-8"		12	1.333		194	71		265	330
3780	3'-0" x 7'-0"		11	1.455		208	77.50		285.50	350
3800	4'-0" x 7'-0"	↓	8	2		415	106		521	630
3840	For 8'-0" height, add					42			42	46
3860	For solid wood core, add					44			44	48.50
4000	Exterior, flush, solid core, birch, 1-3/4" x 2'-6" x 7'-0"	2 Carp	15	1.067		165	56.50		221.50	273
4020	2'-8" wide		15	1.067		239	56.50		295.50	355
4040	3'-0" wide	↓	14	1.143		182	60.50		242.50	298
4045	3'-0" x 8'-0"	1 Carp	8	1		530	53		583	665
4100	Oak faced 1-3/4" x 2'-6" x 7'-0"	2 Carp	15	1.067		238	56.50		294.50	355
4120	2'-8" wide		15	1.067		264	56.50		320.50	380
4140	3'-0" wide	↓	14	1.143		255	60.50		315.50	380
4160	Walnut faced, 1-3/4" x 3'-0" x 6'-8"	1 Carp	17	.471		470	25		495	555
4180	3'-6" wide	"	17	.471		640	25		665	745
4200	Walnut faced, 1-3/4" x 2'-6" x 7'-0"	2 Carp	15	1.067		340	56.50		396.50	465
4220	2'-8" wide		15	1.067		375	56.50		431.50	500
4240	3'-0" wide	↓	14	1.143		315	60.50		375.50	445
4300	For 6'-8" high door, deduct from 7'-0" door					18.70			18.70	20.50
5000	Wood doors, for vision lite, add					111			111	122

08 14 Wood Doors

08 14 16 – Flush Wood Doors

08 14 16.09 Smooth Wood Doors	Crew	Daily Output	Labor-Hours	Unit	Material	2020 Bare Costs Labor	2020 Bare Costs Equipment	Total	Total Incl O&P
5010	Wood doors, for narrow lite, add				Ea.	115			115	126
5015	Wood doors, for bottom (or top) louver, add					243			243	267
9000	Minimum labor/equipment charge	1 Carp	4	2	Job		106		106	170

08 14 16.10 Wood Doors Decorator

		Crew	Daily Output	Labor-Hours	Unit	Material	Labor	Equipment	Total	Total Incl O&P
0010	**WOOD DOORS DECORATOR**									
1800	Exterior, flush, solid wood core, birch 1-3/4" x 2'-6" x 7'-0"	2 Carp	15	1.067	Ea.	370	56.50		426.50	495
1820	2'-8" wide		15	1.067		360	56.50		416.50	485
1840	3'-0" wide		14	1.143		360	60.50		420.50	500
1900	Oak faced, 1-3/4" x 2'-6" x 7'-0"		15	1.067		189	56.50		245.50	299
1920	2'-8" wide		15	1.067		495	56.50		551.50	635
1940	3'-0" wide		14	1.143		465	60.50		525.50	610
2100	Walnut faced, 1-3/4" x 2'-6" x 7'-0"		15	1.067		475	56.50		531.50	610
2120	2'-8" wide		15	1.067		510	56.50		566.50	650
2140	3'-0" wide		14	1.143		475	60.50		535.50	625

08 14 16.20 Wood Fire Doors

		Crew	Daily Output	Labor-Hours	Unit	Material	Labor	Equipment	Total	Total Incl O&P
0010	**WOOD FIRE DOORS**									
0020	Particle core, 7 face plys, "B" label,									
0040	1 hour, birch face, 1-3/4" x 2'-6" x 6'-8"	2 Carp	14	1.143	Ea.	465	60.50		525.50	615
0080	3'-0" x 6'-8"		13	1.231		540	65.50		605.50	700
0090	3'-0" x 7'-0"		12	1.333		500	71		571	665
0100	4'-0" x 7'-0"		12	1.333		730	71		801	920
0140	Oak face, 2'-6" x 6'-8"		14	1.143		585	60.50		645.50	745
0180	3'-0" x 6'-8"		13	1.231		525	65.50		590.50	680
0190	3'-0" x 7'-0"		12	1.333		585	71		656	755
0200	4'-0" x 7'-0"		12	1.333		685	71		756	870
0240	Walnut face, 2'-6" x 6'-8"		14	1.143		545	60.50		605.50	700
0280	3'-0" x 6'-8"		13	1.231		555	65.50		620.50	715
0290	3'-0" x 7'-0"		12	1.333		620	71		691	800
0300	4'-0" x 7'-0"		12	1.333		865	71		936	1,075
0440	M.D. overlay on hardboard, 2'-6" x 6'-8"		15	1.067		370	56.50		426.50	500
0480	3'-0" x 6'-8"		14	1.143		460	60.50		520.50	605
0490	3'-0" x 7'-0"		13	1.231		490	65.50		555.50	640
0500	4'-0" x 7'-0"		12	1.333		475	71		546	635
0740	90 minutes, birch face, 1-3/4" x 2'-6" x 6'-8"		14	1.143		345	60.50		405.50	475
0780	3'-0" x 6'-8"		13	1.231		410	65.50		475.50	555
0790	3'-0" x 7'-0"		12	1.333		460	71		531	620
0800	4'-0" x 7'-0"		12	1.333		555	71		626	725
0840	Oak face, 2'-6" x 6'-8"		14	1.143		470	60.50		530.50	615
0880	3'-0" x 6'-8"		13	1.231		520	65.50		585.50	680
0890	3'-0" x 7'-0"		12	1.333		555	71		626	725
0900	4'-0" x 7'-0"		12	1.333		680	71		751	865
0940	Walnut face, 2'-6" x 6'-8"		14	1.143		545	60.50		605.50	700
0980	3'-0" x 6'-8"		13	1.231		465	65.50		530.50	615
0990	3'-0" x 7'-0"		12	1.333		570	71		641	740
1000	4'-0" x 7'-0"		12	1.333		695	71		766	880
1140	M.D. overlay on hardboard, 2'-6" x 6'-8"		15	1.067		455	56.50		511.50	590
1180	3'-0" x 6'-8"		14	1.143		435	60.50		495.50	580
1190	3'-0" x 7'-0"		13	1.231		515	65.50		580.50	675
1200	4'-0" x 7'-0"		12	1.333		595	71		666	765
1240	For 8'-0" height, add					86			86	94.50
1260	For 8'-0" height walnut, add					105			105	115
2200	Custom architectural "B" label, flush, 1-3/4" thick, birch,									

240

08 14 Wood Doors

08 14 16 – Flush Wood Doors

08 14 16.20 Wood Fire Doors

		Crew	Daily Output	Labor-Hours	Unit	Material	2020 Bare Costs Labor	Equipment	Total	Total Incl O&P
2210	Solid core									
2220	2'-6" x 7'-0"	2 Carp	15	1.067	Ea.	425	56.50		481.50	560
2260	3'-0" x 7'-0"		14	1.143		370	60.50		430.50	505
2300	4'-0" x 7'-0"		13	1.231		565	65.50		630.50	725
2420	4'-0" x 8'-0"		11	1.455		565	77.50		642.50	750
2480	For oak veneer, add					50%				
2500	For walnut veneer, add					75%				
9000	Minimum labor/equipment charge	1 Carp	4	2	Job		106		106	170

08 14 33 – Stile and Rail Wood Doors

08 14 33.10 Wood Doors Paneled

		Crew	Daily Output	Labor-Hours	Unit	Material	2020 Bare Costs Labor	Equipment	Total	Total Incl O&P
0010	**WOOD DOORS PANELED**									
0020	Interior, six panel, hollow core, 1-3/8" thick									
0040	Molded hardboard, 2'-0" x 6'-8"	2 Carp	17	.941	Ea.	65.50	50		115.50	153
0060	2'-6" x 6'-8"		17	.941		69	50		119	156
0070	2'-8" x 6'-8"		17	.941		72	50		122	159
0080	3'-0" x 6'-8"		17	.941		89.50	50		139.50	179
0140	Embossed print, molded hardboard, 2'-0" x 6'-8"		17	.941		69	50		119	156
0160	2'-6" x 6'-8"		17	.941		69	50		119	156
0180	3'-0" x 6'-8"		17	.941		89.50	50		139.50	179
0540	Six panel, solid, 1-3/8" thick, pine, 2'-0" x 6'-8"		15	1.067		166	56.50		222.50	274
0560	2'-6" x 6'-8"		14	1.143		164	60.50		224.50	279
0580	3'-0" x 6'-8"		13	1.231		150	65.50		215.50	270
1020	Two panel, bored rail, solid, 1-3/8" thick, pine, 1'-6" x 6'-8"		16	1		257	53		310	370
1040	2'-0" x 6'-8"		15	1.067		335	56.50		391.50	460
1060	2'-6" x 6'-8"		14	1.143		380	60.50		440.50	520
1340	Two panel, solid, 1-3/8" thick, fir, 2'-0" x 6'-8"		15	1.067		173	56.50		229.50	281
1360	2'-6" x 6'-8"		14	1.143		230	60.50		290.50	350
1380	3'-0" x 6'-8"		13	1.231		420	65.50		485.50	565
1740	Five panel, solid, 1-3/8" thick, fir, 2'-0" x 6'-8"		15	1.067		310	56.50		366.50	435
1760	2'-6" x 6'-8"		14	1.143		400	60.50		460.50	540
1780	3'-0" x 6'-8"		13	1.231		400	65.50		465.50	545
4190	Exterior, Knotty pine, paneled, 1-3/4", 3'-0" x 6'-8"		16	1		880	53		933	1,050
4195	Double 1-3/4", 3'-0" x 6'-8"		16	1		1,750	53		1,803	2,000
4200	Ash, paneled, 1-3/4", 3'-0" x 6'-8"		16	1		1,000	53		1,053	1,175
4205	Double 1-3/4", 3'-0" x 6'-8"		16	1		2,000	53		2,053	2,275
4210	Cherry, paneled, 1-3/4", 3'-0" x 6'-8"		16	1		1,150	53		1,203	1,350
4215	Double 1-3/4", 3'-0" x 6'-8"		16	1		2,300	53		2,353	2,600
4230	Ash, paneled, 1-3/4", 3'-0" x 8'-0"		16	1		1,350	53		1,403	1,550
4235	Double 1-3/4", 3'-0" x 8'-0"		16	1		2,675	53		2,728	3,025
4240	Hard maple, paneled, 1-3/4", 3'-0" x 8'-0"		16	1		1,350	53		1,403	1,550
4245	Double 1-3/4", 3'-0" x 8'-0"		16	1		2,675	53		2,728	3,025
4250	Cherry, paneled, 1-3/4", 3'-0" x 8'-0"		16	1		1,350	53		1,403	1,550
4255	Double 1-3/4", 3'-0" x 8'-0"		16	1		2,675	53		2,728	3,025
9000	Minimum labor/equipment charge	1 Carp	4	2	Job		106		106	170

08 14 33.20 Wood Doors Residential

		Crew	Daily Output	Labor-Hours	Unit	Material	2020 Bare Costs Labor	Equipment	Total	Total Incl O&P
0010	**WOOD DOORS RESIDENTIAL**									
0200	Exterior, combination storm & screen, pine									
0260	2'-8" wide	2 Carp	10	1.600	Ea.	330	85		415	495
0280	3'-0" wide		9	1.778		345	94.50		439.50	530
0300	7'-1" x 3'-0" wide		9	1.778		365	94.50		459.50	550
0400	Full lite, 6'-9" x 2'-6" wide		11	1.455		355	77.50		432.50	515
0420	2'-8" wide		10	1.600		345	85		430	515

For customer support on your Commercial Renovation Costs with RSMeans data, call 800.448.8182.

241

08 14 33 - Stile and Rail Wood Doors

08 14 33.20 Wood Doors Residential	Crew	Daily Output	Labor-Hours	Unit	Material	2020 Bare Costs Labor	Equipment	Total	Total Incl O&P
0440	3'-0" wide	2 Carp	9	1.778	Ea.	355	94.50		449.50	540
0500	7'-1" x 3'-0" wide		9	1.778		385	94.50		479.50	570
0700	Dutch door, pine, 1-3/4" x 2'-8" x 6'-8", 6 panel		12	1.333		740	71		811	930
0720	Half glass		10	1.600		1,025	85		1,110	1,250
0800	3'-0" wide, 6 panel		12	1.333		660	71		731	845
0820	Half glass		10	1.600		1,050	85		1,135	1,275
1000	Entrance door, colonial, 1-3/4" x 6'-8" x 2'-8" wide		16	1		625	53		678	775
1020	6 panel pine, 3'-0" wide		15	1.067		610	56.50		666.50	760
1100	8 panel pine, 2'-8" wide		16	1		735	53		788	895
1120	3'-0" wide		15	1.067		670	56.50		726.50	830
1200	For tempered safety glass lites (min. of 2), add					92.50			92.50	102
1300	Flush, birch, solid core, 1-3/4" x 6'-8" x 2'-8" wide	2 Carp	16	1		169	53		222	271
1320	3'-0" wide		15	1.067		153	56.50		209.50	260
1350	7'-0" x 2'-8" wide		16	1		141	53		194	240
1360	3'-0" wide		15	1.067		149	56.50		205.50	255
1380	For tempered safety glass lites, add					120			120	132
1420	6'-8" x 3'-0" wide, fir	2 Carp	16	1		540	53		593	680
1720	Carved mahogany 3'-0" x 6'-8"		15	1.067		1,600	56.50		1,656.50	1,850
1800	Lauan, solid core, 1-3/4" x 7'-0" x 2'-4" wide		16	1		143	53		196	243
1810	2'-6" wide		15	1.067		150	56.50		206.50	256
1820	2'-8" wide		9	1.778		187	94.50		281.50	355
1830	3'-0" wide		16	1		204	53		257	310
1840	3'-4" wide		16	1		276	53		329	390
1850	Pair of 3'-0" wide		15	1.067	Pr.	405	56.50		461.50	540
2700	Interior, closet, bi-fold, w/hardware, no frame or trim incl.									
2720	Flush, birch, 2'-6" x 6'-8"	2 Carp	13	1.231	Ea.	80.50	65.50		146	194
2740	3'-0" wide		13	1.231		78	65.50		143.50	191
2760	4'-0" wide		12	1.333		123	71		194	249
2780	5'-0" wide		11	1.455		121	77.50		198.50	257
2800	6'-0" wide		10	1.600		137	85		222	287
2820	Flush, hardboard, primed, 6'-8" x 2'-6" wide		13	1.231		77.50	65.50		143	191
2840	3'-0" wide		13	1.231		95.50	65.50		161	210
2860	4'-0" wide		12	1.333		156	71		227	286
2880	5'-0" wide		11	1.455		206	77.50		283.50	350
2900	6'-0" wide		10	1.600		193	85		278	350
3000	Raised panel pine, 6'-6" or 6'-8" x 2'-6" wide		13	1.231		207	65.50		272.50	335
3020	3'-0" wide		13	1.231		305	65.50		370.50	440
3040	4'-0" wide		12	1.333		330	71		401	480
3060	5'-0" wide		11	1.455		425	77.50		502.50	595
3080	6'-0" wide		10	1.600		470	85		555	650
3200	Louvered, pine, 6'-6" or 6'-8" x 2'-6" wide		13	1.231		170	65.50		235.50	292
3220	3'-0" wide		13	1.231		284	65.50		349.50	420
3240	4'-0" wide		12	1.333		256	71		327	395
3260	5'-0" wide		11	1.455		276	77.50		353.50	430
3280	6'-0" wide		10	1.600		335	85		420	500
4400	Bi-passing closet, incl. hardware and frame, no trim incl.									
4420	Flush, lauan, 6'-8" x 4'-0" wide	2 Carp	12	1.333	Opng.	178	71		249	310
4440	5'-0" wide		11	1.455		196	77.50		273.50	340
4460	6'-0" wide		10	1.600		177	85		262	330
4600	Flush, birch, 6'-8" x 4'-0" wide		12	1.333		270	71		341	410
4620	5'-0" wide		11	1.455		227	77.50		304.50	375
4640	6'-0" wide		10	1.600		360	85		445	530
4800	Louvered, pine, 6'-8" x 4'-0" wide		12	1.333		525	71		596	690

08 14 Wood Doors

08 14 33 – Stile and Rail Wood Doors

	08 14 33.20 Wood Doors Residential	Crew	Daily Output	Labor-Hours	Unit	Material	2020 Bare Costs Labor	Equipment	Total	Total Incl O&P
4820	5'-0" wide	2 Carp	11	1.455	Opng.	740	77.50		817.50	940
4840	6'-0" wide		10	1.600		770	85		855	980
5000	Paneled, pine, 6'-8" x 4'-0" wide		12	1.333		505	71		576	670
5020	5'-0" wide		11	1.455		760	77.50		837.50	960
5040	6'-0" wide		10	1.600		900	85		985	1,125
5042	8'-0" wide	▼	12	1.333	▼	1,000	71		1,071	1,225
6100	Folding accordion, closet, including track and frame									
6120	Vinyl, 2 layer, stock	2 Carp	10	1.600	Ea.	81	85		166	225
6200	Rigid PVC	"	10	1.600	"	124	85		209	273
7310	Passage doors, flush, no frame included									
7320	Hardboard, hollow core, 1-3/8" x 6'-8" x 1'-6" wide	2 Carp	18	.889	Ea.	43	47		90	123
7330	2'-0" wide		18	.889		48	47		95	128
7340	2'-6" wide		18	.889		55	47		102	136
7350	2'-8" wide		18	.889		57	47		104	139
7360	3'-0" wide		17	.941		58	50		108	144
7420	Lauan, hollow core, 1-3/8" x 6'-8" x 1'-6" wide		18	.889		59.50	47		106.50	141
7440	2'-0" wide		18	.889		63	47		110	145
7450	2'-4" wide		18	.889		71.50	47		118.50	155
7460	2'-6" wide		18	.889		71.50	47		118.50	155
7480	2'-8" wide		18	.889		74	47		121	157
7500	3'-0" wide		17	.941		90.50	50		140.50	180
7700	Birch, hollow core, 1-3/8" x 6'-8" x 1'-6" wide		18	.889		78	47		125	161
7720	2'-0" wide		18	.889		86	47		133	170
7740	2'-6" wide		18	.889		87.50	47		134.50	172
7760	2'-8" wide		18	.889		91	47		138	176
7780	3'-0" wide		17	.941		94.50	50		144.50	184
8000	Pine louvered, 1-3/8" x 6'-8" x 1'-6" wide		19	.842		160	45		205	248
8020	2'-0" wide		18	.889		177	47		224	270
8040	2'-6" wide		18	.889		200	47		247	296
8060	2'-8" wide		18	.889		220	47		267	320
8080	3'-0" wide		17	.941		233	50		283	335
8300	Pine paneled, 1-3/8" x 6'-8" x 1'-6" wide		19	.842		201	45		246	293
8320	2'-0" wide		18	.889		251	47		298	350
8330	2'-4" wide		18	.889		252	47		299	355
8340	2'-6" wide		18	.889		265	47		312	365
8360	2'-8" wide		18	.889		271	47		318	375
8380	3'-0" wide		17	.941		291	50		341	400
8804	Pocket door, 6 panel pine, 2'-6" x 6'-8" with frame		10.50	1.524		395	81		476	565
8814	2'-8" x 6'-8"		10.50	1.524		390	81		471	555
8824	3'-0" x 6'-8"		10.50	1.524		400	81		481	570
9000	Passage doors, flush, no frame, birch, solid core, 1-3/8" x 2'-4" x 7'-0"		16	1		133	53		186	231
9020	2'-8" wide		16	1		135	53		188	233
9040	3'-0" wide		16	1		167	53		220	269
9060	3'-4" wide		15	1.067	▼	320	56.50		376.50	440
9080	Pair of 3'-0" wide		9	1.778	Pr.	296	94.50		390.50	475
9100	Lauan, solid core, 1-3/8" x 7'-0" x 2'-4" wide		16	1	Ea.	171	53		224	273
9120	2'-8" wide		16	1		152	53		205	253
9140	3'-0" wide		16	1		201	53		254	305
9160	3'-4" wide		15	1.067	▼	218	56.50		274.50	330
9180	Pair of 3'-0" wide		9	1.778	Pr.	425	94.50		519.50	615
9200	Hardboard, solid core, 1-3/8" x 7'-0" x 2'-4" wide		16	1	Ea.	177	53		230	279
9220	2'-8" wide		16	1		180	53		233	283
9240	3'-0" wide		16	1		186	53		239	290

08 14 Wood Doors

08 14 33 – Stile and Rail Wood Doors

08 14 33.20 Wood Doors Residential	Crew	Daily Output	Labor-Hours	Unit	Material	2020 Bare Costs Labor	Equipment	Total	Total Incl O&P
9260	3'-4" wide	2 Carp	15	1.067	Ea.	375	56.50		431.50	500
9900	Minimum labor/equipment charge	1 Carp	4	2	Job		106		106	170

08 14 35 – Torrified Doors

08 14 35.10 Torrified Exterior Doors

	08 14 35.10 Torrified Exterior Doors	Crew	Daily Output	Labor-Hours	Unit	Material	2020 Bare Costs Labor	Equipment	Total	Total Incl O&P
0010	**TORRIFIED EXTERIOR DOORS**									
0020	Wood doors made from torrified wood, exterior									
0030	All doors require a finish be applied, all glass is insulated									
0040	All doors require pilot holes for all fasteners									
0100	6 panel, paint grade poplar, 1-3/4" x 3'-0" x 6'-8"	2 Carp	12	1.333	Ea.	1,425	71		1,496	1,700
0120	Half glass 3'-0" x 6'-8"	"	12	1.333		1,550	71		1,621	1,825
0200	Side lite, full glass, 1-3/4" x 1'-2" x 6'-8"					1,025			1,025	1,125
0220	Side lite, half glass, 1-3/4" x 1'-2" x 6'-8"					995			995	1,100
0300	Raised face, 2 panel, paint grade poplar, 1-3/4" x 3'-0" x 7'-0"	2 Carp	12	1.333		1,425	71		1,496	1,700
0320	Side lite, raised face, half glass, 1-3/4" x 1'-2" x 7'-0"					1,150			1,150	1,275
0500	6 panel, Fir, 1-3/4" x 3'-0" x 6'-8"	2 Carp	12	1.333		2,050	71		2,121	2,375
0520	Half glass 3'-0" x 6'-8"	"	12	1.333		2,025	71		2,096	2,350
0600	Side lite, full glass, 1-3/4" x 1'-2" x 6'-8"					1,025			1,025	1,125
0620	Side lite, half glass, 1-3/4" x 1'-2" x 6'-8"					1,050			1,050	1,175
0700	6 panel, Mahogany, 1-3/4" x 3'-0" x 6'-8"	2 Carp	12	1.333		2,525	71		2,596	2,900
0800	Side lite, full glass, 1-3/4" x 1'-2" x 6'-8"					1,150			1,150	1,250
0820	Side lite, half glass, 1-3/4" x 1'-2" x 6'-8"					1,150			1,150	1,275

08 14 40 – Interior Cafe Doors

08 14 40.10 Cafe Style Doors

	08 14 40.10 Cafe Style Doors	Crew	Daily Output	Labor-Hours	Unit	Material	2020 Bare Costs Labor	Equipment	Total	Total Incl O&P
0010	**CAFE STYLE DOORS**									
6520	Interior cafe doors, 2'-6" opening, stock, panel pine	2 Carp	16	1	Ea.	430	53		483	560
6540	3'-0" opening	"	16	1	"	455	53		508	585
6550	Louvered pine									
6560	2'-6" opening	2 Carp	16	1	Ea.	345	53		398	460
8000	3'-0" opening		16	1		370	53		423	495
8010	2'-6" opening, hardwood		16	1		375	53		428	500
8020	3'-0" opening		16	1		425	53		478	550
9000	Minimum labor/equipment charge	1 Carp	4	2	Job		106		106	170

08 16 Composite Doors

08 16 13 – Fiberglass Doors

08 16 13.10 Entrance Doors, Fibrous Glass

	08 16 13.10 Entrance Doors, Fibrous Glass		Crew	Daily Output	Labor-Hours	Unit	Material	2020 Bare Costs Labor	Equipment	Total	Total Incl O&P
0010	**ENTRANCE DOORS, FIBROUS GLASS**										
0020	Exterior, fiberglass, door, 2'-8" wide x 6'-8" high	G	2 Carp	15	1.067	Ea.	276	56.50		332.50	395
0040	3'-0" wide x 6'-8" high	G		15	1.067		279	56.50		335.50	395
0060	3'-0" wide x 7'-0" high	G		15	1.067		520	56.50		576.50	660
0080	3'-0" wide x 6'-8" high, with two lites	G		15	1.067		345	56.50		401.50	470
0100	3'-0" wide x 8'-0" high, with two lites	G		15	1.067		535	56.50		591.50	680
0110	Half glass, 3'-0" wide x 6'-8" high	G		15	1.067		485	56.50		541.50	625
0120	3'-0" wide x 6'-8" high, low E	G		15	1.067		525	56.50		581.50	670
0130	3'-0" wide x 8'-0" high	G		15	1.067		605	56.50		661.50	755
0140	3'-0" wide x 8'-0" high, low E	G		15	1.067		685	56.50		741.50	845
0150	Side lights, 1'-0" wide x 6'-8" high	G					284			284	310
0160	1'-0" wide x 6'-8" high, low E	G					305			305	335
0180	1'-0" wide x 6'-8" high, full glass	G					340			340	375
0190	1'-0" wide x 6'-8" high, low E	G					385			385	420

08 16 Composite Doors

08 16 14 – French Doors

08 16 14.10 Exterior Doors With Glass Lites		Crew	Daily Output	Labor-Hours	Unit	Material	2020 Bare Costs Labor	Equipment	Total	Total Incl O&P
0010	**EXTERIOR DOORS WITH GLASS LITES**									
0020	French, Fir, 1-3/4", 3'-0" wide x 6'-8" high	2 Carp	12	1.333	Ea.	650	71		721	830
0025	Double		12	1.333		1,300	71		1,371	1,550
0030	Maple, 1-3/4", 3'-0" wide x 6'-8" high		12	1.333		725	71		796	915
0035	Double		12	1.333		1,450	71		1,521	1,725
0040	Cherry, 1-3/4", 3'-0" wide x 6'-8" high		12	1.333		850	71		921	1,050
0045	Double		12	1.333		1,700	71		1,771	2,000
0100	Mahogany, 1-3/4", 3'-0" wide x 8'-0" high		10	1.600		875	85		960	1,100
0105	Double		10	1.600		1,750	85		1,835	2,050
0110	Fir, 1-3/4", 3'-0" wide x 8'-0" high		10	1.600		1,300	85		1,385	1,550
0115	Double		10	1.600		2,625	85		2,710	3,000
0120	Oak, 1-3/4", 3'-0" wide x 8'-0" high		10	1.600		1,975	85		2,060	2,275
0125	Double	▼	10	1.600	▼	3,925	85		4,010	4,450

08 17 Integrated Door Opening Assemblies

08 17 13 – Integrated Metal Door Opening Assemblies

08 17 13.10 Hollow Metal Doors and Frames

08 17 13.10 Hollow Metal Doors and Frames		Crew	Daily Output	Labor-Hours	Unit	Material	2020 Bare Costs Labor	Equipment	Total	Total Incl O&P
0010	**HOLLOW METAL DOORS AND FRAMES**									
0100	Prehung, flush, 18 ga., 1-3/4" x 6'-8" x 2'-8" wide	1 Carp	16	.500	Ea.	640	26.50		666.50	750
0120	3'-0" wide		15	.533		640	28.50		668.50	750
0130	3'-6" wide		13	.615		740	32.50		772.50	865
0140	4'-0" wide		10	.800		740	42.50		782.50	885
0300	Double, 1-3/4" x 3'-0" x 6'-8"		5	1.600		1,175	85		1,260	1,425
0350	3'-0" x 8'-0"	▼	4	2	▼	1,425	106		1,531	1,750

08 17 13.20 Stainless Steel Doors and Frames

08 17 13.20 Stainless Steel Doors and Frames			Crew	Daily Output	Labor-Hours	Unit	Material	2020 Bare Costs Labor	Equipment	Total	Total Incl O&P
0010	**STAINLESS STEEL DOORS AND FRAMES**										
0020	Stainless steel (304) prehung 24 ga. 2'-6" x 6'-8" door w/16 g frame	G	2 Carp	6	2.667	Ea.	1,900	142		2,042	2,300
0025	2'-8" x 6'-8"	G		6	2.667		1,950	142		2,092	2,375
0030	3'-0" x 6'-8"	G		6	2.667		1,925	142		2,067	2,350
0040	3'-0" x 7'-0"	G		6	2.667		2,000	142		2,142	2,425
0050	4'-0" x 7'-0"	G		5	3.200		2,200	170		2,370	2,675
0100	Stainless steel (316) prehung 24 ga. 2'-6" x 6'-8" door w/16 g frame	G		6	2.667		2,500	142		2,642	2,975
0110	2'-8" x 6'-8"	G		6	2.667		2,300	142		2,442	2,750
0120	3'-0" x 6'-8"	G		6	2.667		2,400	142		2,542	2,875
0150	3'-0" x 7'-0"	G		6	2.667		2,550	142		2,692	3,025
0160	4'-0" x 7'-0"	G		6	2.667		3,175	142		3,317	3,700
0300	Stainless steel (304) prehung 18 ga. 2'-6" x 6'-8" door w/16 g frame	G		6	2.667		2,850	142		2,992	3,350
0310	2'-8" x 6'-8"	G		6	2.667		2,900	142		3,042	3,400
0320	3'-0" x 6'-8"	G		6	2.667		2,950	142		3,092	3,450
0350	3'-0" x 7'-0"	G		6	2.667		2,975	142		3,117	3,500
0360	4'-0" x 7'-0"	G		5	3.200		3,125	170		3,295	3,700
0500	Stainless steel, prehung door, foam core, 14 ga., 3'-0" x 7'-0"	G		5	3.200		3,125	170		3,295	3,700
0600	Stainless steel, prehung double door, foam core, 14 ga., 3'-0" x 7'-0"	G	▼	4	4	▼	6,150	213		6,363	7,125

08 17 23 – Integrated Wood Door Opening Assemblies

08 17 23.10 Pre-Hung Doors

08 17 23.10 Pre-Hung Doors		Crew	Daily Output	Labor-Hours	Unit	Material	2020 Bare Costs Labor	Equipment	Total	Total Incl O&P
0010	**PRE-HUNG DOORS**									
0300	Exterior, wood, comb. storm & screen, 6'-9" x 2'-6" wide	2 Carp	15	1.067	Ea.	262	56.50		318.50	380
0320	2'-8" wide		15	1.067		350	56.50		406.50	475
0340	3'-0" wide	▼	15	1.067		330	56.50		386.50	455
0360	For 7'-0" high door, add				▼	45.50			45.50	50

For customer support on your Commercial Renovation Costs with RSMeans data, call 800.448.8182.

245

08 17 23 – Integrated Wood Door Opening Assemblies

08 17 23.10 Pre-Hung Doors		Crew	Daily Output	Labor-Hours	Unit	Material	2020 Bare Costs Labor	Equipment	Total	Total Incl O&P
1600	Entrance door, flush, birch, solid core									
1620	4-5/8" solid jamb, 1-3/4" x 6'-8" x 2'-8" wide	2 Carp	16	1	Ea.	299	53		352	415
1640	3'-0" wide		16	1		405	53		458	535
1642	5-5/8" jamb		16	1		350	53		403	470
1680	For 7'-0" high door, add					25.50			25.50	28
2000	Entrance door, colonial, 6 panel pine									
2020	4-5/8" solid jamb, 1-3/4" x 6'-8" x 2'-8" wide	2 Carp	16	1	Ea.	670	53		723	825
2040	3'-0" wide	"	16	1		705	53		758	860
2060	For 7'-0" high door, add					57			57	63
2200	For 5-5/8" solid jamb, add					44			44	48.50
2230	French style, exterior, 1 lite, 1-3/4" x 3'-0" x 6'-8"	1 Carp	14	.571		755	30.50		785.50	880
2235	9 lites	"	14	.571		755	30.50		785.50	880
2245	15 lites	2 Carp	14	1.143		815	60.50		875.50	995
2250	Double, 15 lites, 2'-0" x 6'-8", 4'-0" opening		7	2.286	Pr.	1,300	121		1,421	1,625
2260	2'-6" x 6'-8", 5'-0" opening		7	2.286		1,425	121		1,546	1,775
2280	3'-0" x 6'-8", 6'-0" opening		7	2.286		1,700	121		1,821	2,075
2430	3'-0" x 7'-0", 15 lites		14	1.143	Ea.	1,000	60.50		1,060.50	1,200
2432	Two 3'-0" x 7'-0"		7	2.286	Pr.	2,075	121		2,196	2,475
2435	3'-0" x 8'-0"		14	1.143	Ea.	1,075	60.50		1,135.50	1,275
2437	Two, 3'-0" x 8'-0"		7	2.286	Pr.	2,200	121		2,321	2,625
4000	Interior, passage door, 4-5/8" solid jamb									
4350	Paneled, primed, hollow core, 2'-8" wide	2 Carp	17	.941	Ea.	140	50		190	234
4360	3'-0" wide		17	.941		191	50		241	291
4400	Lauan, flush, solid core, 1-3/8" x 6'-8" x 2'-6" wide		17	.941		194	50		244	294
4420	2'-8" wide		17	.941		194	50		244	294
4440	3'-0" wide		16	1		212	53		265	320
4600	Hollow core, 1-3/8" x 6'-8" x 2'-6" wide		17	.941		136	50		186	230
4620	2'-8" wide		17	.941		136	50		186	229
4640	3'-0" wide		16	1		145	53		198	245
4700	For 7'-0" high door, add					41			41	45
5000	Birch, flush, solid core, 1-3/8" x 6'-8" x 2'-6" wide	2 Carp	17	.941		315	50		365	425
5020	2'-8" wide		17	.941		210	50		260	310
5040	3'-0" wide		16	1		315	53		368	430
5200	Hollow core, 1-3/8" x 6'-8" x 2'-6" wide		17	.941		260	50		310	365
5220	2'-8" wide		17	.941		273	50		323	380
5240	3'-0" wide		16	1		277	53		330	390
5280	For 7'-0" high door, add					35			35	38.50
5500	Hardboard paneled, 1-3/8" x 6'-8" x 2'-6" wide	2 Carp	17	.941		156	50		206	252
5520	2'-8" wide		17	.941		163	50		213	259
5540	3'-0" wide		16	1		165	53		218	266
6000	Pine paneled, 1-3/8" x 6'-8" x 2'-6" wide		17	.941		278	50		328	385
6020	2'-8" wide		17	.941		294	50		344	405
6040	3'-0" wide		16	1		305	53		358	420
8200	Birch, flush, solid core, 1-3/4" x 6'-8" x 2'-4" wide	1 Carp	17	.471		237	25		262	300
8220	2'-6" wide		17	.471		219	25		244	281
8240	2'-8" wide		17	.471		231	25		256	294
8260	3'-0" wide		16	.500		235	26.50		261.50	300
8280	3'-6" wide		15	.533		370	28.50		398.50	455
8500	Pocket door frame with lauan, flush, hollow core, 1-3/8" x 3'-0" x 6'-8"		17	.471		320	25		345	395
9000	Minimum labor/equipment charge		4	2	Job		106		106	170

246

For customer support on your Commercial Renovation Costs with RSMeans data, call 800.448.8182.

08 31 Access Doors and Panels

08 31 13 – Access Doors and Frames

08 31 13.10 Types of Framed Access Doors

08 31 13.10 Types of Framed Access Doors		Crew	Daily Output	Labor-Hours	Unit	Material	2020 Bare Costs Labor	Equipment	Total	Total Incl O&P
0010	**TYPES OF FRAMED ACCESS DOORS**									
1000	Fire rated door with lock									
1100	Metal, 12" x 12"	1 Carp	10	.800	Ea.	164	42.50		206.50	248
1150	18" x 18"		9	.889		221	47		268	320
1200	24" x 24"		9	.889		345	47		392	455
1250	24" x 36"		8	1		350	53		403	470
1300	24" x 48"		8	1		415	53		468	540
1350	36" x 36"		7.50	1.067		515	56.50		571.50	655
1400	48" x 48"		7.50	1.067		670	56.50		726.50	825
1600	Stainless steel, 12" x 12"		10	.800		340	42.50		382.50	445
1650	18" x 18"		9	.889		375	47		422	490
1700	24" x 24"		9	.889		550	47		597	680
1750	24" x 36"	↓	8	1	↓	655	53		708	810
2000	Flush door for finishing									
2100	Metal 8" x 8"	1 Carp	10	.800	Ea.	37.50	42.50		80	109
2150	12" x 12"	"	10	.800	"	37.50	42.50		80	110
3000	Recessed door for acoustic tile									
3100	Metal, 12" x 12"	1 Carp	4.50	1.778	Ea.	57	94.50		151.50	214
3150	12" x 24"		4.50	1.778		111	94.50		205.50	273
3200	24" x 24"		4	2		107	106		213	288
3250	24" x 36"	↓	4	2	↓	165	106		271	350
4000	Recessed door for drywall									
4100	Metal 12" x 12"	1 Carp	6	1.333	Ea.	76.50	71		147.50	198
4150	12" x 24"		5.50	1.455		119	77.50		196.50	254
4200	24" x 36"	↓	5	1.600	↓	177	85		262	330
6000	Standard door									
6100	Metal, 8" x 8"	1 Carp	10	.800	Ea.	30	42.50		72.50	101
6150	12" x 12"		10	.800		30.50	42.50		73	102
6200	18" x 18"		9	.889		35.50	47		82.50	115
6250	24" x 24"		9	.889		50	47		97	131
6300	24" x 36"		8	1		88	53		141	182
6350	36" x 36"		8	1		103	53		156	198
6500	Stainless steel, 8" x 8"		10	.800		80.50	42.50		123	157
6550	12" x 12"		10	.800		97	42.50		139.50	174
6600	18" x 18"		9	.889		177	47		224	271
6650	24" x 24"		9	.889	↓	276	47		323	380
9000	Minimum labor/equipment charge	↓	4	2	Job		106		106	170

08 31 13.20 Bulkhead/Cellar Doors

08 31 13.20 Bulkhead/Cellar Doors		Crew	Daily Output	Labor-Hours	Unit	Material	2020 Bare Costs Labor	Equipment	Total	Total Incl O&P
0010	**BULKHEAD/CELLAR DOORS**									
0020	Steel, not incl. sides, 44" x 62"	1 Carp	5.50	1.455	Ea.	705	77.50		782.50	900
0100	52" x 73"		5.10	1.569		865	83.50		948.50	1,075
0500	With sides and foundation plates, 57" x 45" x 24"		4.70	1.702		895	90.50		985.50	1,125
0600	42" x 49" x 51"		4.30	1.860	↓	600	99		699	820
9000	Minimum labor/equipment charge	↓	2	4	Job		213		213	340

08 31 13.30 Commercial Floor Doors

08 31 13.30 Commercial Floor Doors		Crew	Daily Output	Labor-Hours	Unit	Material	2020 Bare Costs Labor	Equipment	Total	Total Incl O&P
0010	**COMMERCIAL FLOOR DOORS**									
0020	Aluminum tile, steel frame, one leaf, 2' x 2' opng.	2 Sswk	3.50	4.571	Opng.	725	264		989	1,225
0050	3'-6" x 3'-6" opening		3.50	4.571		1,325	264		1,589	1,900
0500	Double leaf, 4' x 4' opening		3	5.333		1,650	305		1,955	2,300
0550	5' x 5' opening		3	5.333	↓	2,700	305		3,005	3,475
9000	Minimum labor/equipment charge	↓	2	8	Job		460		460	755

08 31 Access Doors and Panels

08 31 13 – Access Doors and Frames

08 31 13.35 Industrial Floor Doors		Crew	Daily Output	Labor-Hours	Unit	Material	2020 Bare Costs Labor	Equipment	Total	Total Incl O&P
0010	**INDUSTRIAL FLOOR DOORS**									
0020	Steel 300 psf L.L., single leaf, 2' x 2', 175#	2 Sswk	6	2.667	Opng.	810	154		964	1,150
0050	3' x 3' opening, 300#		5.50	2.909		1,300	168		1,468	1,700
0300	Double leaf, 4' x 4' opening, 455#		5	3.200		2,450	184		2,634	3,000
0350	5' x 5' opening, 645#		4.50	3.556		2,925	205		3,130	3,525
1000	Aluminum, 300 psf L.L., single leaf, 2' x 2', 60#		6	2.667		720	154		874	1,050
1050	3' x 3' opening, 100#		5.50	2.909		995	168		1,163	1,375
1500	Double leaf, 4' x 4' opening, 160#		5	3.200		2,825	184		3,009	3,400
1550	5' x 5' opening, 235#		4.50	3.556		3,250	205		3,455	3,900
9000	Minimum labor/equipment charge		2	8	Job		460		460	755

08 32 Sliding Glass Doors

08 32 13 – Sliding Aluminum-Framed Glass Doors

08 32 13.10 Sliding Aluminum Doors

08 32 13.10 Sliding Aluminum Doors		Crew	Daily Output	Labor-Hours	Unit	Material	2020 Bare Costs Labor	Equipment	Total	Total Incl O&P
0010	**SLIDING ALUMINUM DOORS**									
0350	Aluminum, 5/8" tempered insulated glass, 6' wide									
0400	Premium	2 Carp	4	4	Ea.	1,600	213		1,813	2,125
0450	Economy		4	4		890	213		1,103	1,325
0500	8' wide, premium		3	5.333		1,775	283		2,058	2,400
0550	Economy		3	5.333		1,575	283		1,858	2,175
0600	12' wide, premium		2.50	6.400		3,100	340		3,440	3,975
0650	Economy		2.50	6.400		1,675	340		2,015	2,375
4000	Aluminum, baked on enamel, temp glass, 6'-8" x 10'-0" wide		4	4		1,200	213		1,413	1,650
4020	Insulating glass, 6'-8" x 6'-0" wide		4	4		975	213		1,188	1,425
4040	8'-0" wide		3	5.333		1,175	283		1,458	1,750
4060	10'-0" wide		2	8		1,450	425		1,875	2,275
4080	Anodized, temp glass, 6'-8" x 6'-0" wide		4	4		495	213		708	885
4100	8'-0" wide		3	5.333		605	283		888	1,125
4120	10'-0" wide		2	8		710	425		1,135	1,450
5000	Aluminum sliding glass door system									
5010	Sliding door 4' wide opening single side	2 Carp	2	8	Ea.	6,475	425		6,900	7,775
5015	8' wide opening single side		2	8		9,700	425		10,125	11,400
5020	Telescoping glass door system, 4' wide opening biparting		2	8		5,525	425		5,950	6,750
5025	8' wide opening biparting		2	8		6,475	425		6,900	7,775
5030	Folding glass door, 4' wide opening biparting		2	8		8,625	425		9,050	10,200
5035	8' wide opening biparting		2	8		10,800	425		11,225	12,600
5040	ICU-CCU sliding telescoping glass door, 4' x 7', single side opening		2	8		3,300	425		3,725	4,300
5045	8' x 7', single side opening		2	8		5,000	425		5,425	6,175
7000	Electric swing door operator and control, single door w/sensors	1 Carp	4	2		2,950	106		3,056	3,400
7005	Double door w/sensors		2	4		5,925	213		6,138	6,850
7010	Electric folding door operator and control, single door w/sensors		4	2		6,950	106		7,056	7,825
7015	Bi-folding door		4	2		8,175	106		8,281	9,175
7020	Electric swing door operator and control, single door		4	2		2,950	106		3,056	3,425

08 32 19 – Sliding Wood-Framed Glass Doors

08 32 19.10 Sliding Wood Doors

08 32 19.10 Sliding Wood Doors		Crew	Daily Output	Labor-Hours	Unit	Material	2020 Bare Costs Labor	Equipment	Total	Total Incl O&P
0010	**SLIDING WOOD DOORS**									
0020	Wood, tempered insul. glass, 6' wide, premium	2 Carp	4	4	Ea.	1,525	213		1,738	2,025
0100	Economy		4	4		1,250	213		1,463	1,725
0150	8' wide, wood, premium		3	5.333		1,950	283		2,233	2,600
0200	Economy		3	5.333		1,600	283		1,883	2,200

08 32 Sliding Glass Doors

08 32 19 – Sliding Wood-Framed Glass Doors

08 32 19.10 Sliding Wood Doors

		Crew	Daily Output	Labor-Hours	Unit	Material	2020 Bare Costs Labor	Equipment	Total	Total Incl O&P
0235	10' wide, wood, premium	2 Carp	2.50	6.400	Ea.	2,875	340		3,215	3,700
0240	Economy		2.50	6.400		2,450	340		2,790	3,225
0250	12' wide, wood, premium		2.50	6.400		3,275	340		3,615	4,150
0300	Economy	↓	2.50	6.400	↓	2,675	340		3,015	3,475
9000	Minimum labor/equipment charge	1 Carp	2	4	Job		213		213	340

08 32 19.15 Sliding Glass Vinyl-Clad Wood Doors

		Crew	Daily Output	Labor-Hours	Unit	Material	2020 Bare Costs Labor	Equipment	Total	Total Incl O&P
0010	**SLIDING GLASS VINYL-CLAD WOOD DOORS**									
0020	Glass, sliding vinyl-clad, insul. glass, 6'-0" x 6'-8" G	2 Carp	4	4	Opng.	1,575	213		1,788	2,100
0025	6'-0" x 6'-10" high G		4	4		1,775	213		1,988	2,300
0030	6'-0" x 8'-0" high G		4	4		2,175	213		2,388	2,750
0050	5'-0" x 6'-8" high G		4	4		1,625	213		1,838	2,150
0100	8'-0" x 6'-10" high G		4	4		2,125	213		2,338	2,675
0150	8'-0" x 8'-0" high G		4	4		2,475	213		2,688	3,075
0500	4 leaf, 9'-0" x 6'-10" high G		3	5.333		3,675	283		3,958	4,500
0550	9'-0" x 8'-0" high G		3	5.333		4,050	283		4,333	4,900
0600	12'-0" x 6'-10" high G		3	5.333		4,350	283		4,633	5,225
0650	12'-0" x 8'-0" high G	↓	3	5.333	↓	4,350	283		4,633	5,225
9000	Minimum labor/equipment charge	1 Carp	4	2	Job		106		106	170

08 33 Coiling Doors and Grilles

08 33 13 – Coiling Counter Doors

08 33 13.10 Counter Doors, Coiling Type

		Crew	Daily Output	Labor-Hours	Unit	Material	2020 Bare Costs Labor	Equipment	Total	Total Incl O&P
0010	**COUNTER DOORS, COILING TYPE**									
0020	Manual, incl. frame and hardware, galv. stl., 4' roll-up, 6' long	2 Carp	2	8	Opng.	1,325	425		1,750	2,125
0300	Galvanized steel, UL label		1.80	8.889		1,325	470		1,795	2,200
0600	Stainless steel, 4' high roll-up, 6' long		2	8		2,200	425		2,625	3,100
0700	10' long		1.80	8.889		2,625	470		3,095	3,650
2000	Aluminum, 4' high, 4' long		2.20	7.273		1,725	385		2,110	2,500
2020	6' long		2	8		2,000	425		2,425	2,875
2040	8' long		1.90	8.421		2,300	450		2,750	3,250
2060	10' long		1.80	8.889		2,300	470		2,770	3,275
2080	14' long		1.40	11.429		2,975	605		3,580	4,250
2100	6' high, 4' long		2	8		2,025	425		2,450	2,900
2120	6' long		1.60	10		1,825	530		2,355	2,850
2140	10' long	↓	1.40	11.429	↓	2,400	605		3,005	3,625
9000	Minimum labor/equipment charge	1 Carp	2	4	Job		213		213	340

08 33 16 – Coiling Counter Grilles

08 33 16.10 Coiling Grilles

		Crew	Daily Output	Labor-Hours	Unit	Material	2020 Bare Costs Labor	Equipment	Total	Total Incl O&P
0010	**COILING GRILLES**									
2020	Aluminum, manual operated, mill finish	2 Sswk	82	.195	S.F.	30	11.25		41.25	51.50
2040	Bronze anodized		82	.195	"	46	11.25		57.25	69
2060	Steel, manual operated, 10' x 10' high		1	16	Opng.	2,650	920		3,570	4,425
2080	15' x 8' high	↓	.80	20	"	2,975	1,150		4,125	5,150
3000	For safety edge bottom bar, electric, add				L.F.	49.50			49.50	54.50
8000	For motor operation, add	2 Sswk	5	3.200	Opng.	1,300	184		1,484	1,725
9000	Minimum labor/equipment charge	"	1	16	Job		920		920	1,500

For customer support on your Commercial Renovation Costs with RSMeans data, call 800.448.8182.

249

08 33 23 – Overhead Coiling Doors

08 33 23.10 Coiling Service Doors	Crew	Daily Output	Labor-Hours	Unit	Material	2020 Bare Costs Labor	Equipment	Total	Total Incl O&P
0010 **COILING SERVICE DOORS** Steel, manual, 20 ga., incl. hardware									
0050 8' x 8' high	2 Sswk	1.60	10	Ea.	710	575		1,285	1,725
0130 12' x 12' high, standard		1.20	13.333		2,175	770		2,945	3,650
0160 10' x 20' high, standard		.50	32		2,600	1,850		4,450	5,900
2000 Class A fire doors, manual, 20 ga., 8' x 8' high		1.40	11.429		1,575	660		2,235	2,825
2100 10' x 10' high		1.10	14.545		2,150	840		2,990	3,750
2200 20' x 10' high		.80	20		4,400	1,150		5,550	6,750
2300 12' x 12' high		1	16		3,400	920		4,320	5,250
2400 20' x 12' high		.80	20		4,775	1,150		5,925	7,150
2500 14' x 14' high		.60	26.667		3,700	1,525		5,225	6,600
2600 20' x 16' high		.50	32		5,950	1,850		7,800	9,550
2700 10' x 20' high		.40	40		4,725	2,300		7,025	8,975
3000 For 18 ga. doors, add				S.F.	1.52			1.52	1.67
3300 For enamel finish, add				"	2.84			2.84	3.12
3600 For safety edge bottom bar, pneumatic, add				L.F.	24.50			24.50	27
4000 For weatherstripping, extruded rubber, jambs, add					19.15			19.15	21
4100 Hood, add					8.65			8.65	9.50
4200 Sill, add					8.85			8.85	9.75
4500 Motor operators, to 14' x 14' opening	2 Sswk	5	3.200	Ea.	1,200	184		1,384	1,625
4700 For fire door, additional fusible link, add				"	77.50			77.50	85.50
9000 Minimum labor/equipment charge	2 Sswk	1	16	Job		920		920	1,500

08 34 Special Function Doors

08 34 13 – Cold Storage Doors

08 34 13.10 Doors for Cold Area Storage

	Crew	Daily Output	Labor-Hours	Unit	Material	2020 Bare Costs Labor	Equipment	Total	Total Incl O&P
0010 **DOORS FOR COLD AREA STORAGE**									
0020 Single, 20 ga. galvanized steel									
0300 Horizontal sliding, 5' x 7', manual operation, 3.5" thick	2 Carp	2	8	Ea.	3,150	425		3,575	4,125
0400 4" thick		2	8		3,650	425		4,075	4,675
0500 6" thick		2	8		3,400	425		3,825	4,425
0800 5' x 7', power operation, 2" thick		1.90	8.421		5,675	450		6,125	6,975
0900 4" thick		1.90	8.421		5,675	450		6,125	6,975
1000 6" thick		1.90	8.421		6,475	450		6,925	7,825
1300 9' x 10', manual operation, 2" insulation		1.70	9.412		4,450	500		4,950	5,700
1400 4" insulation		1.70	9.412		4,825	500		5,325	6,100
1500 6" insulation		1.70	9.412		5,675	500		6,175	7,050
1800 Power operation, 2" insulation		1.60	10		7,825	530		8,355	9,450
1900 4" insulation		1.60	10		7,975	530		8,505	9,625
2000 6" insulation		1.70	9.412		8,225	500		8,725	9,825
2300 For stainless steel face, add					25%				
3000 Hinged, lightweight, 3' x 7'-0", 2" thick	2 Carp	2	8	Ea.	1,300	425		1,725	2,100
3050 4" thick		1.90	8.421		1,625	450		2,075	2,500
3300 Polymer doors, 3' x 7'-0"		1.90	8.421		1,350	450		1,800	2,200
3350 6" thick		1.40	11.429		2,375	605		2,980	3,600
3600 Stainless steel, 3' x 7'-0", 4" thick		1.90	8.421		1,725	450		2,175	2,625
3650 6" thick		1.40	11.429		2,875	605		3,480	4,125
3900 Painted, 3' x 7'-0", 4" thick		1.90	8.421		1,350	450		1,800	2,200
3950 6" thick		1.40	11.429		2,325	605		2,930	3,550
5000 Bi-parting, electric operated									
5010 6' x 8' opening, galv. faces, 4" thick for cooler	2 Carp	.80	20	Opng.	7,350	1,075		8,425	9,800

250

For customer support on your Commercial Renovation Costs with RSMeans data, call 800.448.8182.

08 34 Special Function Doors

08 34 13 – Cold Storage Doors

08 34 13.10 Doors for Cold Area Storage

	08 34 13.10 Doors for Cold Area Storage	Crew	Daily Output	Labor-Hours	Unit	Material	2020 Bare Costs Labor	Equipment	Total	Total Incl O&P
5050	For freezer, 4" thick	2 Carp	.80	20	Opng.	8,100	1,075		9,175	10,600
5300	For door buck framing and door protection, add		2.50	6.400		680	340		1,020	1,300
6000	Galvanized batten door, galvanized hinges, 4' x 7'		2	8		1,875	425		2,300	2,725
6050	6' x 8'		1.80	8.889		2,475	470		2,945	3,475
6500	Fire door, 3 hr., 6' x 8', single slide		.80	20		8,725	1,075		9,800	11,300
6550	Double, bi-parting	↓	.70	22.857	↓	11,300	1,225		12,525	14,500
9000	Minimum labor/equipment charge	1 Carp	2	4	Job		213		213	340

08 34 36 – Darkroom Doors

08 34 36.10 Various Types of Darkroom Doors

	08 34 36.10 Various Types of Darkroom Doors	Crew	Daily Output	Labor-Hours	Unit	Material	2020 Bare Costs Labor	Equipment	Total	Total Incl O&P
0010	**VARIOUS TYPES OF DARKROOM DOORS**									
0015	Revolving, standard, 2 way, 36" diameter	2 Carp	3.10	5.161	Opng.	3,350	274		3,624	4,150
0020	41" diameter		3.10	5.161		3,850	274		4,124	4,700
0050	3 way, 51" diameter		1.40	11.429		4,125	605		4,730	5,500
1000	4 way, 49" diameter		1.40	11.429		4,075	605		4,680	5,450
2000	Hinged safety, 2 way, 41" diameter		2.30	6.957		4,050	370		4,420	5,075
2500	3 way, 51" diameter		1.40	11.429		5,025	605		5,630	6,500
3000	Pop out safety, 2 way, 41" diameter		3.10	5.161		4,875	274		5,149	5,800
4000	3 way, 51" diameter		1.40	11.429		5,375	605		5,980	6,900
5000	Wheelchair-type, pop out, 51" diameter		1.40	11.429		5,925	605		6,530	7,500
5020	72" diameter	↓	.90	17.778	↓	10,400	945		11,345	13,000

08 34 53 – Security Doors and Frames

08 34 53.20 Steel Door

	08 34 53.20 Steel Door	Crew	Daily Output	Labor-Hours	Unit	Material	2020 Bare Costs Labor	Equipment	Total	Total Incl O&P
0010	**STEEL DOOR** flush with ballistic core and welded frame both 14 ga.									
0120	UL 752 Level 3, 1-3/4", 3'-0" x 7'-0"	2 Carp	1.50	10.667	Opng.	2,700	565		3,265	3,850
0125	1-3/4", 3'-6" x 7'-0"		1.50	10.667		2,950	565		3,515	4,150
0130	1-3/4", 4'-0" x 7'-0"		1.20	13.333		3,175	710		3,885	4,600
0150	UL 752 Level 8, 1-3/4", 3'-0" x 7'-0"		1.50	10.667		14,100	565		14,665	16,400
0155	1-3/4", 3'-6" x 7'-0"		1.50	10.667		15,400	565		15,965	17,800
0160	1-3/4", 4'-0" x 7'-0"		1.20	13.333		15,300	710		16,010	18,000
1000	Safe room sliding door and hardware, 1-3/4", 3'-0" x 7'-0" UL 752 Level 3		.50	32		24,300	1,700		26,000	29,400
1050	Safe room swinging door and hardware, 1-3/4", 3'-0" x 7'-0" UL 752 Level 3	↓	.50	32	↓	29,100	1,700		30,800	34,700

08 34 53.30 Wood Ballistic Doors

	08 34 53.30 Wood Ballistic Doors	Crew	Daily Output	Labor-Hours	Unit	Material	2020 Bare Costs Labor	Equipment	Total	Total Incl O&P
0010	**WOOD BALLISTIC DOORS** with frames and hardware									
0050	Wood, 1-3/4", 3'-0" x 7'-0" UL 752 Level 3	2 Carp	1.50	10.667	Opng.	2,700	565		3,265	3,850

08 34 56 – Security Gates

08 34 56.10 Gates

	08 34 56.10 Gates	Crew	Daily Output	Labor-Hours	Unit	Material	2020 Bare Costs Labor	Equipment	Total	Total Incl O&P
0010	**GATES**									
0015	Driveway gates include mounting hardware									
0500	Wood, security gate, driveway, dual, 10' wide	H-4	.80	25	Opng.	3,825	1,250		5,075	6,200
0505	12' wide		.80	25		3,700	1,250		4,950	6,025
0510	15' wide		.80	25		4,400	1,250		5,650	6,800
0600	Steel, security gate, driveway, single, 10' wide		.80	25		1,825	1,250		3,075	3,975
0605	12' wide		.80	25		2,150	1,250		3,400	4,350
0620	Steel, security gate, driveway, dual, 12' wide		.80	25		2,200	1,250		3,450	4,375
0625	14' wide		.80	25		2,525	1,250		3,775	4,750
0630	16' wide		.80	25		2,500	1,250		3,750	4,725
0700	Aluminum, security gate, driveway, dual, 10' wide		.80	25		3,275	1,250		4,525	5,575
0705	12' wide		.80	25		4,050	1,250		5,300	6,450
0710	16' wide	↓	.80	25	↓	4,625	1,250		5,875	7,050
1000	Security gate, driveway, opener 12 VDC				Ea.	840			840	925
1010	Wireless				↓	1,950			1,950	2,150

For customer support on your Commercial Renovation Costs with RSMeans data, call 800.448.8182.

251

08 34 Special Function Doors

08 34 56 – Security Gates

08 34 56.10 Gates		Crew	Daily Output	Labor-Hours	Unit	Material	2020 Bare Costs Labor	2020 Bare Costs Equipment	Total	Total Incl O&P
1020	Security gate, driveway, opener 24 VDC				Ea.	1,500			1,500	1,625
1030	Wireless					1,650			1,650	1,825
1040	Security gate, driveway, opener 12 VDC, solar panel 10 watt	1 Elec	2	4		495	245		740	925
1050	20 watt	"	2	4		635	245		880	1,075

08 34 73 – Sound Control Door Assemblies

08 34 73.10 Acoustical Doors

		Crew	Daily Output	Labor-Hours	Unit	Material	2020 Bare Costs Labor	2020 Bare Costs Equipment	Total	Total Incl O&P
0010	**ACOUSTICAL DOORS**									
0020	Including framed seals, 3' x 7', wood, 40 STC rating	2 Carp	1.50	10.667	Ea.	3,875	565		4,440	5,175
0100	Steel, 41 STC rating		1.50	10.667		3,850	565		4,415	5,125
0200	45 STC rating		1.50	10.667		4,175	565		4,740	5,500
0300	48 STC rating		1.50	10.667		4,925	565		5,490	6,300
0400	52 STC rating		1.50	10.667		5,400	565		5,965	6,850
9000	Minimum labor/equipment charge	1 Carp	4	2	Job		106		106	170

08 36 Panel Doors

08 36 13 – Sectional Doors

08 36 13.10 Overhead Commercial Doors

		Crew	Daily Output	Labor-Hours	Unit	Material	2020 Bare Costs Labor	2020 Bare Costs Equipment	Total	Total Incl O&P
0010	**OVERHEAD COMMERCIAL DOORS**									
1000	Stock, sectional, heavy duty, wood, 1-3/4" thick, 8' x 8' high	2 Carp	2	8	Ea.	1,250	425		1,675	2,050
1200	12' x 12' high		1.50	10.667		2,775	565		3,340	3,950
1300	Chain hoist, 14' x 14' high		1.30	12.308		3,825	655		4,480	5,250
1600	20' x 16' high		.65	24.615		5,975	1,300		7,275	8,675
2100	For medium duty custom door, deduct					5%	5%			
2150	For medium duty stock doors, deduct					10%	5%			
2300	Fiberglass and aluminum, heavy duty, sectional, 12' x 12' high	2 Carp	1.50	10.667	Ea.	3,425	565		3,990	4,650
2450	Chain hoist, 20' x 20' high	"	.50	32		7,775	1,700		9,475	11,300
2900	For electric trolley operator, 1/3 HP, to 12' x 12', add	1 Carp	2	4		1,125	213		1,338	1,575
2950	Over 12' x 12', 1/2 HP, add		1	8		1,250	425		1,675	2,050
2980	Overhead, for row of clear lites, add		1	8		169	425		594	865
9000	Minimum labor/equipment charge	2 Carp	1.50	10.667	Job		565		565	910

08 36 13.20 Residential Garage Doors

		Crew	Daily Output	Labor-Hours	Unit	Material	2020 Bare Costs Labor	2020 Bare Costs Equipment	Total	Total Incl O&P
0010	**RESIDENTIAL GARAGE DOORS**									
0050	Hinged, wood, custom, double door, 9' x 7'	2 Carp	4	4	Ea.	935	213		1,148	1,375
0070	16' x 7'		3	5.333		1,300	283		1,583	1,875
0200	Overhead, sectional, incl. hardware, fiberglass, 9' x 7', standard		5	3.200		975	170		1,145	1,350
0220	Deluxe		5	3.200		1,200	170		1,370	1,600
0300	16' x 7', standard		6	2.667		1,950	142		2,092	2,375
0320	Deluxe		6	2.667		2,325	142		2,467	2,800
0500	Hardboard, 9' x 7', standard		8	2		750	106		856	995
0520	Deluxe		8	2		905	106		1,011	1,175
0600	16' x 7', standard		6	2.667		1,325	142		1,467	1,675
0620	Deluxe		6	2.667		1,600	142		1,742	2,000
0700	Metal, 9' x 7', standard		8	2		1,000	106		1,106	1,275
0720	Deluxe		6	2.667		1,025	142		1,167	1,350
0800	16' x 7', standard		6	2.667		1,200	142		1,342	1,550
0820	Deluxe		5	3.200		1,450	170		1,620	1,875
0900	Wood, 9' x 7', standard		8	2		1,075	106		1,181	1,350
0920	Deluxe		8	2		2,325	106		2,431	2,725
1000	16' x 7', standard		6	2.667		1,750	142		1,892	2,150
1020	Deluxe		6	2.667		3,150	142		3,292	3,700

08 36 Panel Doors

08 36 13 – Sectional Doors

08 36 13.20 Residential Garage Doors		Crew	Daily Output	Labor-Hours	Unit	Material	2020 Bare Costs Labor	Equipment	Total	Total Incl O&P
1800	Door hardware, sectional	1 Carp	4	2	Ea.	375	106		481	585
1810	Door tracks only		4	2		173	106		279	360
1820	One side only		7	1.143		129	60.50		189.50	239
4000	For electric operator, economy, add		8	1		460	53		513	590
4100	Deluxe, including remote control	▼	8	1	▼	640	53		693	785
4500	For transmitter/receiver control, add to operator				Total	119			119	131
4600	Transmitters, additional				"	67			67	74
6000	Replace section, on sectional door, fiberglass, 9' x 7'	1 Carp	4	2	Ea.	740	106		846	985
6020	16' x 7'		3.50	2.286		765	121		886	1,050
6200	Hardboard, 9' x 7'		4	2		176	106		282	365
6220	16' x 7'		3.50	2.286		248	121		369	470
6300	Metal, 9' x 7'		4	2		228	106		334	420
6320	16' x 7'		3.50	2.286		335	121		456	565
6500	Wood, 9' x 7'		4	2		125	106		231	310
6520	16' x 7'	▼	3.50	2.286		249	121		370	470
7010	Garage doors, row of lites				▼	133			133	146
9000	Minimum labor/equipment charge	1 Carp	2.50	3.200	Job		170		170	272

08 36 19 – Multi-Leaf Vertical Lift Doors

08 36 19.10 Sectional Vertical Lift Doors

		Crew	Daily Output	Labor-Hours	Unit	Material	2020 Bare Costs Labor	Equipment	Total	Total Incl O&P
0010	**SECTIONAL VERTICAL LIFT DOORS**									
0020	Motorized, 14 ga. steel, incl. frame and control panel									
0050	16' x 16' high	L-10	.50	48	Ea.	22,600	2,825	940	26,365	30,500
0100	10' x 20' high		1.30	18.462		42,800	1,075	360	44,235	49,300
0120	15' x 20' high		1.30	18.462		51,500	1,075	360	52,935	59,000
0140	20' x 20' high		1	24		61,500	1,400	470	63,370	70,500
0160	25' x 20' high		1	24		59,000	1,400	470	60,870	67,500
0170	32' x 24' high		.75	32		54,500	1,875	630	57,005	63,500
0180	20' x 25' high		1	24		60,500	1,400	470	62,370	69,500
0200	25' x 25' high		.70	34.286		78,000	2,025	675	80,700	90,000
0220	25' x 30' high		.70	34.286		83,500	2,025	675	86,200	95,500
0240	30' x 30' high		.70	34.286		99,000	2,025	675	101,700	113,000
0260	35' x 30' high	▼	.70	34.286	▼	98,500	2,025	675	101,200	112,500

08 38 Traffic Doors

08 38 19 – Rigid Traffic Doors

08 38 19.20 Double Acting Swing Doors

		Crew	Daily Output	Labor-Hours	Unit	Material	2020 Bare Costs Labor	Equipment	Total	Total Incl O&P
0010	**DOUBLE ACTING SWING DOORS**									
0020	Including frame, closer, hardware and vision panel									
1000	Polymer, 7'-0" high, 4'-0" wide	2 Carp	4.20	3.810	Pr.	2,100	202		2,302	2,625
1025	6'-0" wide		4	4		2,225	213		2,438	2,800
1050	6'-8" wide	▼	4	4	▼	2,425	213		2,638	3,025
2000	3/4" thick, stainless steel									
2010	Stainless steel, 7' high opening, 4' wide	2 Carp	4	4	Pr.	2,625	213		2,838	3,225
2050	7' wide		3.80	4.211	"	2,800	224		3,024	3,425
9000	Minimum labor/equipment charge	▼	2	8	Job		425		425	680

08 38 19.30 Shock Absorbing Doors

		Crew	Daily Output	Labor-Hours	Unit	Material	2020 Bare Costs Labor	Equipment	Total	Total Incl O&P
0010	**SHOCK ABSORBING DOORS**									
0020	Rigid, no frame, 1-1/2" thick, 5' x 7'	2 Sswk	1.90	8.421	Opng.	1,550	485		2,035	2,500
0100	8' x 8'		1.80	8.889		2,025	510		2,535	3,075
0500	Flexible, no frame, insulated, .16" thick, economy, 5' x 7'		2	8		1,775	460		2,235	2,725

For customer support on your Commercial Renovation Costs with RSMeans data, call 800.448.8182.

253

08 38 Traffic Doors

08 38 19 – Rigid Traffic Doors

08 38 19.30 Shock Absorbing Doors	Crew	Daily Output	Labor-Hours	Unit	Material	2020 Bare Costs Labor	Equipment	Total	Total Incl O&P
0600	Deluxe	2 Sswk	1.90	8.421	Opng.	2,675	485		3,160	3,750
1000	8' x 8' opening, economy		2	8		2,825	460		3,285	3,850
1100	Deluxe		1.90	8.421		3,575	485		4,060	4,725
9000	Minimum labor/equipment charge		2	8	Job		460		460	755

08 41 Entrances and Storefronts

08 41 13 – Aluminum-Framed Entrances and Storefronts

08 41 13.20 Tube Framing

		Crew	Daily Output	Labor-Hours	Unit	Material	2020 Bare Costs Labor	Equipment	Total	Total Incl O&P
0010	**TUBE FRAMING**, For window walls and storefronts, aluminum stock									
0050	Plain tube frame, mill finish, 1-3/4" x 1-3/4"	2 Glaz	103	.155	L.F.	11.15	7.90		19.05	25
0150	1-3/4" x 4"		98	.163		14.95	8.35		23.30	30
0200	1-3/4" x 4-1/2"		95	.168		17	8.60		25.60	32.50
0250	2" x 6"		89	.180		26	9.15		35.15	43
0350	4" x 4"		87	.184		27.50	9.40		36.90	45.50
0400	4-1/2" x 4-1/2"		85	.188		31	9.60		40.60	49.50
0450	Glass bead		240	.067		3.36	3.40		6.76	9.15
1000	Flush tube frame, mill finish, 1/4" glass, 1-3/4" x 4", open header		80	.200		14.90	10.20		25.10	32.50
1050	Open sill		82	.195		12	9.95		21.95	29
1100	Closed back header		83	.193		21	9.85		30.85	39
1150	Closed back sill		85	.188		20	9.60		29.60	37.50
1160	Tube fmg., spandrel cover both sides, alum 1" wide	1 Sswk	85	.094	S.F.	98	5.45		103.45	117
1170	Tube fmg., spandrel cover both sides, alum 2" wide	"	85	.094	"	40	5.45		45.45	53
1200	Vertical mullion, one piece	2 Glaz	75	.213	L.F.	21.50	10.90		32.40	41
1250	Two piece		73	.219		23	11.20		34.20	43
1300	90° or 180° vertical corner post		75	.213		34.50	10.90		45.40	55.50
1400	1-3/4" x 4-1/2", open header		80	.200		17.45	10.20		27.65	35.50
1450	Open sill		82	.195		14.40	9.95		24.35	32
1500	Closed back header		83	.193		21	9.85		30.85	38.50
1550	Closed back sill		85	.188		20	9.60		29.60	37.50
1600	Vertical mullion, one piece		75	.213		22.50	10.90		33.40	42.50
1650	Two piece		73	.219		24	11.20		35.20	44.50
1700	90° or 180° vertical corner post		75	.213		24.50	10.90		35.40	44.50
2000	Flush tube frame, mil fin.,ins. glass w/thml brk, 2" x 4-1/2", open header		75	.213		15.95	10.90		26.85	35
2050	Open sill		77	.208		13.45	10.60		24.05	31.50
2100	Closed back header		78	.205		14.85	10.45		25.30	33
2150	Closed back sill		80	.200		15.75	10.20		25.95	33.50
2200	Vertical mullion, one piece		70	.229		16.95	11.65		28.60	37.50
2250	Two piece		68	.235		18.30	12		30.30	39
2300	90° or 180° vertical corner post		70	.229		19.05	11.65		30.70	39.50
5000	Flush tube frame, mill fin., thermal brk., 2-1/4" x 4-1/2", open header		74	.216		16.50	11.05		27.55	36
5050	Open sill		75	.213		14.25	10.90		25.15	33
5100	Vertical mullion, one piece		69	.232		19.35	11.85		31.20	40.50
5150	Two piece		67	.239		21	12.20		33.20	42.50
5200	90° or 180° vertical corner post		69	.232		18.80	11.85		30.65	39.50
5295	Flush tube frame, mill finish, 4-1/2" x 4-1/2"		85	.188		31	9.60		40.60	49.50
5300	Plain tube frame, mill finish, 2" x 3" jamb		115	.139		12.15	7.10		19.25	24.50
5310	2" x 4"		115	.139		16.45	7.10		23.55	29.50
5320	3" x 5-1/2"		107	.150		29	7.65		36.65	44
5330	Head, 2" x 3"		200	.080		21	4.08		25.08	29.50
5340	Mullion, 2" x 3"		200	.080		21.50	4.08		25.58	30
5350	2" x 4"		115	.139		27.50	7.10		34.60	41.50

08 41 13 – Aluminum-Framed Entrances and Storefronts

08 41 13.20 Tube Framing

		Crew	Daily Output	Labor-Hours	Unit	Material	2020 Bare Costs Labor	Equipment	Total	Total Incl O&P
5360	3" x 5-1/2"	2 Glaz	107	.150	L.F.	36	7.65		43.65	51.50
5370	4" corner mullion		90	.178		30.50	9.05		39.55	48
5380	Horizontal, 2" x 3"		115	.139		25	7.10		32.10	39
5390	3" x 5-1/2"		107	.150		29	7.65		36.65	43.50
5430	Sill section, 1/8" x 6"		133	.120		35.50	6.15		41.65	49.50
5440	1/8" x 7"		133	.120		43	6.15		49.15	57
5450	1/8" x 8-1/2"		133	.120		50	6.15		56.15	65
5460	Column covers, aluminum, 1/8" x 26"		57	.281		69	14.30		83.30	99
5470	1/8" x 34"		53	.302		87	15.40		102.40	120
5480	1/8" x 38"		53	.302		99.50	15.40		114.90	134
5700	Vertical mullions, clear finish, 1/4" thick glass		528	.030		15.25	1.55		16.80	19.25
5720	3/8" thick		445	.036		13.70	1.83		15.53	18.05
5740	1/2" thick		400	.040		14.25	2.04		16.29	18.95
5760	3/4" thick		344	.047		16.30	2.37		18.67	21.50
5780	1" thick		304	.053		15.10	2.68		17.78	21
6980	Door stop (snap in)	↓	380	.042	↓	3.61	2.15		5.76	7.40
7000	For joints, 90°, clip type, add				Ea.	26.50			26.50	29
7050	Screw spline joint, add					24.50			24.50	27
7100	For joint other than 90°, add				↓	51			51	56
8000	For bronze anodized aluminum, add					15%				
8020	For black finish, add					30%				
8050	For stainless steel materials, add					350%				
8100	For monumental grade, add					53%				
8150	For steel stiffener, add	2 Glaz	200	.080	L.F.	11.55	4.08		15.63	19.20
8200	For 2 to 5 stories, add per story				Story		8%			
9000	Minimum labor/equipment charge	2 Glaz	2	8	Job		410		410	650

08 41 19 – Stainless-Steel-Framed Entrances and Storefronts

08 41 19.10 Stainless-Steel and Glass Entrance Unit

		Crew	Daily Output	Labor-Hours	Unit	Material	2020 Bare Costs Labor	Equipment	Total	Total Incl O&P
0010	**STAINLESS-STEEL AND GLASS ENTRANCE UNIT**, narrow stiles									
0020	3' x 7' opening, including hardware, minimum	2 Sswk	1.60	10	Opng.	7,650	575		8,225	9,375
0050	Average		1.40	11.429		8,075	660		8,735	9,950
0100	Maximum	↓	1.20	13.333		8,700	770		9,470	10,800
1000	For solid bronze entrance units, statuary finish, add					64%				
1100	Without statuary finish, add				↓	45%				
2000	Balanced doors, 3' x 7', economy	2 Sswk	.90	17.778	Ea.	10,300	1,025		11,325	13,100
2100	Premium		.70	22.857	"	16,800	1,325		18,125	20,700
9000	Minimum labor/equipment charge	↓	2	8	Job		460		460	755

08 41 26 – All-Glass Entrances and Storefronts

08 41 26.10 Window Walls Aluminum, Stock

		Crew	Daily Output	Labor-Hours	Unit	Material	2020 Bare Costs Labor	Equipment	Total	Total Incl O&P
0010	**WINDOW WALLS ALUMINUM, STOCK**, including glazing									
0020	Minimum	H-2	160	.150	S.F.	51.50	7.20		58.70	68
0050	Average		140	.171		71.50	8.25		79.75	91.50
0100	Maximum	↓	110	.218	↓	188	10.50		198.50	224
0500	For translucent sandwich wall systems, see Section 07 41 33.10									
0850	Cost of the above walls depends on material,									
0860	finish, repetition, and size of units.									
0870	The larger the opening, the lower the S.F. cost									

For customer support on your Commercial Renovation Costs with RSMeans data, call 800.448.8182.

255

08 42 Entrances

08 42 26 – All-Glass Entrances

08 42 26.10 Swinging Glass Doors

		Crew	Daily Output	Labor-Hours	Unit	Material	2020 Bare Costs Labor	Equipment	Total	Total Incl O&P
0010	**SWINGING GLASS DOORS**									
0020	Including hardware, 1/2" thick, tempered, 3' x 7' opening	2 Glaz	2	8	Opng.	2,400	410		2,810	3,275
0100	6' x 7' opening		1.40	11.429	"	4,650	585		5,235	6,050
9000	Minimum labor/equipment charge		2	8	Job		410		410	650

08 42 36 – Balanced Door Entrances

08 42 36.10 Balanced Entrance Doors

		Crew	Daily Output	Labor-Hours	Unit	Material	2020 Bare Costs Labor	Equipment	Total	Total Incl O&P
0010	**BALANCED ENTRANCE DOORS**									
0020	Hardware & frame, alum. & glass, 3' x 7', econ.	2 Sswk	.90	17.778	Ea.	7,200	1,025		8,225	9,600
0150	Premium		.70	22.857	"	8,725	1,325		10,050	11,800
9000	Minimum labor/equipment charge		1	16	Job		920		920	1,500

08 43 Storefronts

08 43 13 – Aluminum-Framed Storefronts

08 43 13.10 Aluminum-Framed Entrance Doors and Frames

		Crew	Daily Output	Labor-Hours	Unit	Material	2020 Bare Costs Labor	Equipment	Total	Total Incl O&P
0010	**ALUMINUM-FRAMED ENTRANCE DOORS AND FRAMES**									
0015	Standard hardware and glass stops but no glass									
0020	Entrance door, 3' x 7' opening, clear anodized finish	2 Sswk	7	2.286	Opng.	680	132		812	965
0040	Bronze finish		7	2.286		835	132		967	1,125
0060	Black finish		7	2.286		815	132		947	1,125
0200	3'-6" x 7'-0", mill finish		7	2.286		835	132		967	1,125
0220	Bronze finish		7	2.286		960	132		1,092	1,275
0240	Black finish		7	2.286		985	132		1,117	1,300
0500	6' x 7' opening, clear finish		6	2.667		1,075	154		1,229	1,425
0520	Bronze finish		6	2.667		1,175	154		1,329	1,550
0600	Door frame for above doors 3'-0" x 7'-0", mill finish		6	2.667		560	154		714	865
0620	Bronze finish		6	2.667		595	154		749	905
0640	Black finish		6	2.667		670	154		824	990
0700	3'-6" x 7'-0", mill finish		6	2.667		410	154		564	700
0720	Bronze finish		6	2.667		410	154		564	700
0740	Black finish		6	2.667		410	154		564	700
0800	6'-0" x 7'-0", mill finish		6	2.667		410	154		564	700
0820	Bronze finish		6	2.667		415	154		569	705
0840	Black finish		6	2.667		445	154		599	740
1000	With 3' high transom above, 3' x 7' opening, clear finish		5.50	2.909		635	168		803	975
1050	Bronze finish		5.50	2.909		630	168		798	970
1100	Black finish		5.50	2.909		695	168		863	1,025
1300	3'-6" x 7'-0" opening, clear finish		5.50	2.909		420	168		588	735
1320	Bronze finish		5.50	2.909		440	168		608	760
1340	Black finish		5.50	2.909		450	168		618	770
1500	6' x 7' opening, clear finish		5.50	2.909		695	168		863	1,050
1550	Bronze finish		5.50	2.909		785	168		953	1,150
1600	Black finish		5.50	2.909		870	168		1,038	1,225
8000	For 8' high doors, add				Ea.	174			174	174
9000	Minimum labor/equipment charge	2 Sswk	4	4	Job		231		231	380

08 43 13.20 Storefront Systems

		Crew	Daily Output	Labor-Hours	Unit	Material	2020 Bare Costs Labor	Equipment	Total	Total Incl O&P
0010	**STOREFRONT SYSTEMS**, aluminum frame clear 3/8" plate glass									
0020	incl. 3' x 7' door with hardware (400 sq. ft. max. wall)									
0500	Wall height to 12' high, commercial grade	2 Glaz	150	.107	S.F.	27.50	5.45		32.95	38.50
0600	Institutional grade		130	.123		32	6.30		38.30	45
0700	Monumental grade		115	.139		84.50	7.10		91.60	104

For customer support on your Commercial Renovation Costs with RSMeans data, call 800.448.8182.

08 43 Storefronts

08 43 13 – Aluminum-Framed Storefronts

08 43 13.20 Storefront Systems		Crew	Daily Output	Labor-Hours	Unit	Material	2020 Bare Costs Labor	Equipment	Total	Total Incl O&P
1000	6' x 7' door with hardware, commercial grade	2 Glaz	135	.119	S.F.	76	6.05		82.05	93
1100	Institutional grade		115	.139		50	7.10		57.10	66.50
1200	Monumental grade	↓	100	.160		117	8.15		125.15	141
1500	For bronze anodized finish, add					15%				
1600	For black anodized finish, add					36%				
1700	For stainless steel framing, add to monumental				↓	78%				
9000	Minimum labor/equipment charge	2 Glaz	1	16	Job		815		815	1,300

08 43 29 – Sliding Storefronts

08 43 29.10 Sliding Panels		Crew	Daily Output	Labor-Hours	Unit	Material	2020 Bare Costs Labor	Equipment	Total	Total Incl O&P
0010	**SLIDING PANELS**									
0020	Mall fronts, aluminum & glass, 15' x 9' high	2 Glaz	1.30	12.308	Opng.	4,300	630		4,930	5,750
0100	24' x 9' high		.70	22.857		6,125	1,175		7,300	8,625
0200	48' x 9' high, with fixed panels	↓	.90	17.778		11,000	905		11,905	13,600
0500	For bronze finish, add				↓	17%				
9000	Minimum labor/equipment charge	2 Glaz	1	16	Job		815		815	1,300

08 45 Translucent Wall and Roof Assemblies

08 45 10 – Translucent Roof Assemblies

08 45 10.10 Skyroofs		Crew	Daily Output	Labor-Hours	Unit	Material	2020 Bare Costs Labor	Equipment	Total	Total Incl O&P
0010	**SKYROOFS**									
1200	Skylights, circular, clear, double glazed acrylic									
1230	30" diameter	2 Carp	3	5.333	Ea.	3,075	283		3,358	3,825
1250	60" diameter		3	5.333		4,075	283		4,358	4,950
1290	96" diameter	↓	2	8	↓	5,100	425		5,525	6,300
1300	Skylight Barrel Vault, clear, double glazed, acrylic									
1330	3'-0" x 12'-0"	G-3	3	10.667	Ea.	5,100	555		5,655	6,500
1350	4'-0" x 12'-0"		3	10.667		5,625	555		6,180	7,050
1390	5'-0" x 12'-0"	↓	2	16		6,125	835		6,960	8,075
1400	Skylight Pyramid, aluminum frame, clear low E laminated glass									
1410	The glass is installed in the frame except where noted									
1430	Square, 3' x 3'	G-3	3	10.667	Ea.	6,125	555		6,680	7,625
1440	4' x 4'		3	10.667		7,150	555		7,705	8,725
1450	5' x 5', glass must be field installed		3	10.667		8,175	555		8,730	9,850
1460	6' x 6', glass must be field installed	↓	2	16	↓	10,200	835		11,035	12,500
1550	Install pre-cut laminated glass in aluminum frame on a flat roof	2 Glaz	55	.291	SF Surf		14.85		14.85	23.50
1560	Install pre-cut laminated glass in aluminum frame on a sloped roof	"	40	.400	"		20.50		20.50	32.50
9000	Minimum labor/equipment charge	2 Carp	8	2	Job		106		106	170

08 51 Metal Windows

08 51 13 – Aluminum Windows

08 51 13.10 Aluminum Sash		Crew	Daily Output	Labor-Hours	Unit	Material	2020 Bare Costs Labor	Equipment	Total	Total Incl O&P
0010	**ALUMINUM SASH**									
0020	Stock, grade C, glaze & trim not incl., casement	2 Sswk	200	.080	S.F.	43	4.61		47.61	54.50
0050	Double-hung		200	.080		42	4.61		46.61	54
0100	Fixed casement		200	.080		18.45	4.61		23.06	28
0150	Picture window		200	.080		19.75	4.61		24.36	29
0200	Projected window		200	.080		38.50	4.61		43.11	50
0250	Single-hung		200	.080		17.40	4.61		22.01	26.50
0300	Sliding	↓	200	.080	↓	23	4.61		27.61	33

08 51 Metal Windows

08 51 13 – Aluminum Windows

08 51 13.10 Aluminum Sash

		Crew	Daily Output	Labor-Hours	Unit	Material	2020 Bare Costs Labor	Equipment	Total	Total Incl O&P
1000	Mullions for above, tubular	2 Sswk	240	.067	L.F.	6.55	3.84		10.39	13.50
9000	Minimum labor/equipment charge	1 Sswk	2	4	Job		231		231	380

08 51 13.20 Aluminum Windows

		Crew	Daily Output	Labor-Hours	Unit	Material	2020 Bare Costs Labor	Equipment	Total	Total Incl O&P
0010	**ALUMINUM WINDOWS**, incl. frame and glazing, commercial grade									
1000	Stock units, casement, 3'-1" x 3'-2" opening	2 Sswk	10	1.600	Ea.	395	92		487	580
1050	Add for storms					126			126	138
1600	Projected, with screen, 3'-1" x 3'-2" opening	2 Sswk	10	1.600		370	92		462	560
1700	Add for storms					123			123	135
2000	4'-5" x 5'-3" opening	2 Sswk	8	2		410	115		525	645
2100	Add for storms					132			132	145
2500	Enamel finish windows, 3'-1" x 3'-2"	2 Sswk	10	1.600		380	92		472	565
2600	4'-5" x 5'-3"		8	2		425	115		540	660
3000	Single-hung, 2' x 3' opening, enameled, standard glazed		10	1.600		219	92		311	390
3100	Insulating glass		10	1.600		265	92		357	445
3300	2'-8" x 6'-8" opening, standard glazed		8	2		375	115		490	600
3400	Insulating glass		8	2		480	115		595	720
3700	3'-4" x 5'-0" opening, standard glazed		9	1.778		315	102		417	520
3800	Insulating glass		9	1.778		345	102		447	545
4000	Sliding aluminum, 3' x 2' opening, standard glazed		10	1.600		218	92		310	390
4100	Insulating glass		10	1.600		241	92		333	415
4300	5' x 3' opening, standard glazed		9	1.778		345	102		447	550
4400	Insulating glass		9	1.778		385	102		487	595
4600	8' x 4' opening, standard glazed		6	2.667		365	154		519	650
4700	Insulating glass		6	2.667		590	154		744	900
5000	9' x 5' opening, standard glazed		4	4		550	231		781	985
5100	Insulating glass		4	4		850	231		1,081	1,325
5500	Sliding, with thermal barrier and screen, 6' x 4', 2 track		8	2		725	115		840	990
5700	4 track		8	2		915	115		1,030	1,200
6000	For above units with bronze finish, add					15%				
6200	For installation in concrete openings, add					8%				
7000	Double-hung, insulating glass, 2'-0" x 2'-0"	2 Sswk	11	1.455		118	84		202	267
7020	2'-0" x 2'-6"		11	1.455		134	84		218	284
7040	3'-0" x 1'-6"		10	1.600		271	92		363	450
7060	3'-0" x 2'-0"		10	1.600		315	92		407	500
7080	3'-0" x 2'-6"		10	1.600		385	92		477	570
7100	3'-0" x 3'-0"		10	1.600		425	92		517	620
7120	3'-0" x 3'-6"		10	1.600		450	92		542	645
7140	3'-0" x 4'-0"		10	1.600		485	92		577	685
7160	3'-0" x 5'-0"		10	1.600		530	92		622	735
7180	3'-0" x 6'-0"		9	1.778		565	102		667	790

08 51 13.30 Impact Resistant Aluminum Windows

		Crew	Daily Output	Labor-Hours	Unit	Material	2020 Bare Costs Labor	Equipment	Total	Total Incl O&P
0010	**IMPACT RESISTANT ALUMINUM WINDOWS**, incl. frame and glazing									
0100	Single-hung, impact resistant, 2'-8" x 5'-0"	2 Carp	9	1.778	Ea.	1,250	94.50		1,344.50	1,525
0120	3'-0" x 5'-0"		9	1.778		1,350	94.50		1,444.50	1,650
0130	4'-0" x 5'-0"		9	1.778		1,450	94.50		1,544.50	1,750
0250	Horizontal slider, impact resistant, 5'-5" x 5'-2"		9	1.778		1,625	94.50		1,719.50	1,925

08 51 23 – Steel Windows

08 51 23.10 Steel Sash

		Crew	Daily Output	Labor-Hours	Unit	Material	2020 Bare Costs Labor	Equipment	Total	Total Incl O&P
0010	**STEEL SASH** Custom units, glazing and trim not included									
0100	Casement, 100% vented	2 Sswk	200	.080	S.F.	69.50	4.61		74.11	83.50
0200	50% vented		200	.080		56	4.61		60.61	69
0300	Fixed		200	.080		29	4.61		33.61	39

08 51 Metal Windows

08 51 23 – Steel Windows

08 51 23.10 Steel Sash

		Crew	Daily Output	Labor-Hours	Unit	Material	2020 Bare Costs Labor	Equipment	Total	Total Incl O&P
1000	Projected, commercial, 40% vented	2 Sswk	200	.080	S.F.	53	4.61		57.61	66
1100	Intermediate, 50% vented		200	.080		62.50	4.61		67.11	76
1500	Industrial, horizontally pivoted		200	.080		56	4.61		60.61	69.50
1600	Fixed		200	.080		32.50	4.61		37.11	43
2000	Industrial security sash, 50% vented		200	.080		61.50	4.61		66.11	75
2100	Fixed		200	.080		49	4.61		53.61	61.50
2500	Picture window		200	.080		32	4.61		36.61	42.50
3000	Double-hung		200	.080		62	4.61		66.61	76
5000	Mullions for above, open interior face		240	.067	L.F.	10.95	3.84		14.79	18.35
5100	With interior cover		240	.067	"	18.10	3.84		21.94	26
5200	Single glazing for above, add	2 Glaz	200	.080	S.F.	7.20	4.08		11.28	14.45
6000	Double glazing for above, add		200	.080		13.40	4.08		17.48	21
6100	Triple glazing for above, add		85	.188		12.65	9.60		22.25	29.50
9000	Minimum labor/equipment charge	1 Sswk	2	4	Job		231		231	380

08 51 23.20 Steel Windows

		Crew	Daily Output	Labor-Hours	Unit	Material	2020 Bare Costs Labor	Equipment	Total	Total Incl O&P
0010	**STEEL WINDOWS** Stock, including frame, trim and insul. glass R085123-10									
1000	Custom units, double-hung, 2'-8" x 4'-6" opening	2 Sswk	12	1.333	Ea.	730	77		807	925
1100	2'-4" x 3'-9" opening		12	1.333		600	77		677	785
1500	Commercial projected, 3'-9" x 5'-5" opening		10	1.600		1,275	92		1,367	1,550
1600	6'-9" x 4'-1" opening		7	2.286		1,675	132		1,807	2,075
2000	Intermediate projected, 2'-9" x 4'-1" opening		12	1.333		710	77		787	905
2100	4'-1" x 5'-5" opening		10	1.600		1,450	92		1,542	1,750
9000	Minimum labor/equipment charge	1 Sswk	3	2.667	Job		154		154	252

08 51 66 – Metal Window Screens

08 51 66.10 Screens

		Crew	Daily Output	Labor-Hours	Unit	Material	2020 Bare Costs Labor	Equipment	Total	Total Incl O&P
0010	**SCREENS**									
0020	For metal sash, aluminum or bronze mesh, flat screen	2 Sswk	1200	.013	S.F.	4.63	.77		5.40	6.35
0500	Wicket screen, inside window		1000	.016		6.95	.92		7.87	9.15
0800	Security screen, aluminum frame with stainless steel cloth		1200	.013		24.50	.77		25.27	28.50
0900	Steel grate, painted, on steel frame		1600	.010		13.95	.58		14.53	16.25
1000	Screens for solar louvers		160	.100		25.50	5.75		31.25	37.50
4000	See Section 05 58 23.90 for window guards									

08 52 Wood Windows

08 52 10 – Plain Wood Windows

08 52 10.20 Awning Window

		Crew	Daily Output	Labor-Hours	Unit	Material	2020 Bare Costs Labor	Equipment	Total	Total Incl O&P
0010	**AWNING WINDOW**, Including frame, screens and grilles									
0100	34" x 22", insulated glass	1 Carp	10	.800	Ea.	300	42.50		342.50	400
0200	Low E glass		10	.800		330	42.50		372.50	430
0300	40" x 28", insulated glass		9	.889		315	47		362	425
0400	Low E glass		9	.889		350	47		397	460
0500	48" x 36", insulated glass		8	1		480	53		533	610
0600	Low E glass		8	1		505	53		558	640
4000	Impact windows, minimum, add					60%				
4010	Impact windows, maximum, add					160%				
9000	Minimum labor/equipment charge	1 Carp	4	2	Job		106		106	170

08 52 10.30 Wood Windows

		Crew	Daily Output	Labor-Hours	Unit	Material	2020 Bare Costs Labor	Equipment	Total	Total Incl O&P
0010	**WOOD WINDOWS**, double-hung									
0020	Including frame, double insulated glass, screens and grilles									
0040	Double-hung, 2'-2" x 3'-4" high	2 Carp	15	1.067	Ea.	221	56.50		277.50	335

08 52 10.30 Wood Windows

		Crew	Daily Output	Labor-Hours	Unit	Material	2020 Bare Costs Labor	2020 Bare Costs Equipment	Total	Total Incl O&P
0060	2'-2" x 4'-4"	2 Carp	14	1.143	Ea.	236	60.50		296.50	360
0080	2'-6" x 3'-4"		13	1.231		233	65.50		298.50	360
0100	2'-6" x 4'-0"		12	1.333		238	71		309	375
0120	2'-6" x 4'-8"		12	1.333		254	71		325	395
0140	2'-10" x 3'-4"		10	1.600		229	85		314	390
0160	2'-10" x 4'-0"		10	1.600		262	85		347	425
0180	3'-7" x 3'-4"		9	1.778		262	94.50		356.50	440
0200	3'-7" x 5'-4"		9	1.778		300	94.50		394.50	480
0220	3'-10" x 5'-4"		8	2		525	106		631	745
3800	Triple glazing for above, add					25%				

08 52 10.40 Casement Window

			Crew	Daily Output	Labor-Hours	Unit	Material	2020 Bare Costs Labor	2020 Bare Costs Equipment	Total	Total Incl O&P
0010	**CASEMENT WINDOW**, including frame, screen and grilles										
0100	2'-0" x 3'-0" H, double insulated glass	G	1 Carp	10	.800	Ea.	275	42.50		317.50	370
0150	Low E glass	G		10	.800		285	42.50		327.50	385
0200	2'-0" x 4'-6" high, double insulated glass	G		9	.889		390	47		437	505
0250	Low E glass	G		9	.889		440	47		487	560
0260	Casement 4'-2" x 4'-2" double insulated glass	G		11	.727		940	38.50		978.50	1,075
0270	4'-0" x 4'-0" Low E glass	G		11	.727		575	38.50		613.50	690
0290	6'-4" x 5'-7" Low E glass	G		9	.889		1,200	47		1,247	1,400
0300	2'-4" x 6'-0" high, double insulated glass	G		8	1		470	53		523	605
0350	Low E glass	G		8	1		450	53		503	580
0522	Vinyl-clad, premium, double insulated glass, 2'-0" x 3'-0"	G		10	.800		281	42.50		323.50	380
0524	2'-0" x 4'-0"	G		9	.889		335	47		382	440
0525	2'-0" x 5'-0"	G		8	1		365	53		418	485
0528	2'-0" x 6'-0"	G		8	1		410	53		463	535
0600	3'-0" x 5'-0"	G		8	1		685	53		738	840
0700	4'-0" x 3'-0"	G		8	1		770	53		823	930
0710	4'-0" x 4'-0"	G		8	1		645	53		698	795
0720	4'-8" x 4'-0"	G		8	1		710	53		763	865
0730	4'-8" x 5'-0"	G		6	1.333		830	71		901	1,025
0740	4'-8" x 6'-0"	G		6	1.333		920	71		991	1,125
0750	6'-0" x 4'-0"	G		6	1.333		835	71		906	1,025
0800	6'-0" x 5'-0"	G		6	1.333		1,025	71		1,096	1,250
0900	5'-6" x 5'-6"	G	2 Carp	15	1.067		1,500	56.50		1,556.50	1,750
8190	For installation, add per leaf							15%			
8200	For multiple leaf units, deduct for stationary sash										
8220	2' high					Ea.	24.50			24.50	27
8240	4'-6" high						28			28	30.50
8260	6' high						37			37	41
8300	Impact windows, minimum, add						60%				
8310	Impact windows, maximum, add						160%				
9000	Minimum labor/equipment charge		1 Carp	3	2.667	Job		142		142	227

08 52 10.50 Double-Hung

			Crew	Daily Output	Labor-Hours	Unit	Material	2020 Bare Costs Labor	2020 Bare Costs Equipment	Total	Total Incl O&P
0010	**DOUBLE-HUNG**, Including frame, screens and grilles										
0100	2'-0" x 3'-0" high, low E insul. glass	G	1 Carp	10	.800	Ea.	193	42.50		235.50	280
0200	3'-0" x 4'-0" high, double insulated glass	G		9	.889		290	47		337	395
0300	4'-0" x 4'-6" high, low E insulated glass	G		8	1		325	53		378	445
8000	Impact windows, minimum, add						60%				
8010	Impact windows, maximum, add						160%				
9000	Minimum labor/equipment charge		1 Carp	3	2.667	Job		142		142	227

08 52 Wood Windows

08 52 10 – Plain Wood Windows

08 52 10.55 Picture Window

	08 52 10.55 Picture Window	Crew	Daily Output	Labor-Hours	Unit	Material	2020 Bare Costs Labor	Equipment	Total	Total Incl O&P
0010	**PICTURE WINDOW**, Including frame and grilles									
0100	3'-6" x 4'-0" high, double insulated glass	2 Carp	12	1.333	Ea.	440	71		511	600
0150	Low E glass		12	1.333		450	71		521	610
0200	4'-0" x 4'-6" high, double insulated glass		11	1.455		545	77.50		622.50	725
0250	Low E glass		11	1.455		560	77.50		637.50	740
0300	5'-0" x 4'-0" high, double insulated glass		11	1.455		610	77.50		687.50	800
0350	Low E glass		11	1.455		620	77.50		697.50	805
0400	6'-0" x 4'-6" high, double insulated glass		10	1.600		655	85		740	855
0450	Low E glass	▼	10	1.600	▼	645	85		730	845

08 52 10.65 Wood Sash

	08 52 10.65 Wood Sash	Crew	Daily Output	Labor-Hours	Unit	Material	2020 Bare Costs Labor	Equipment	Total	Total Incl O&P
0010	**WOOD SASH**, Including glazing but not trim									
0050	Custom, 5'-0" x 4'-0", 1" double glazed, 3/16" thick lites	2 Carp	3.20	5	Ea.	244	266		510	695
0100	1/4" thick lites		5	3.200		293	170		463	590
0200	1" thick, triple glazed		5	3.200		430	170		600	745
0300	7'-0" x 4'-6" high, 1" double glazed, 3/16" thick lites		4.30	3.721		430	198		628	790
0400	1/4" thick lites		4.30	3.721		495	198		693	855
0500	1" thick, triple glazed		4.30	3.721		560	198		758	935
0600	8'-6" x 5'-0" high, 1" double glazed, 3/16" thick lites		3.50	4.571		585	243		828	1,025
0700	1/4" thick lites		3.50	4.571		640	243		883	1,100
0800	1" thick, triple glazed	▼	3.50	4.571	▼	725	243		968	1,175
0900	Window frames only, based on perimeter length				L.F.	4.10			4.10	4.51
1200	Window sill, stock, per lineal foot					9.05			9.05	9.95
1250	Casing, stock				▼	3.36			3.36	3.70

08 52 10.70 Sliding Windows

	08 52 10.70 Sliding Windows		Crew	Daily Output	Labor-Hours	Unit	Material	2020 Bare Costs Labor	Equipment	Total	Total Incl O&P
0010	**SLIDING WINDOWS**										
0100	3'-0" x 3'-0" high, double insulated	G	1 Carp	10	.800	Ea.	286	42.50		328.50	385
0120	Low E glass	G		10	.800		310	42.50		352.50	415
0200	4'-0" x 3'-6" high, double insulated	G		9	.889		425	47		472	540
0220	Low E glass	G		9	.889		420	47		467	540
0300	6'-0" x 5'-0" high, double insulated	G		8	1		495	53		548	625
0320	Low E glass	G		8	1	▼	605	53		658	750
9000	Minimum labor/equipment charge		▼	3	2.667	Job		142		142	227

08 52 13 – Metal-Clad Wood Windows

08 52 13.10 Awning Windows, Metal-Clad

	08 52 13.10 Awning Windows, Metal-Clad	Crew	Daily Output	Labor-Hours	Unit	Material	2020 Bare Costs Labor	Equipment	Total	Total Incl O&P
0010	**AWNING WINDOWS, METAL-CLAD**									
2000	Metal-clad, awning deluxe, double insulated glass, 34" x 22"	1 Carp	9	.889	Ea.	251	47		298	355
2050	36" x 25"		9	.889		280	47		327	385
2100	40" x 22"		9	.889		297	47		344	400
2150	40" x 30"		9	.889		345	47		392	455
2200	48" x 28"		8	1		355	53		408	475
2250	60" x 36"	▼	8	1	▼	380	53		433	505

08 52 13.20 Casement Windows, Metal-Clad

	08 52 13.20 Casement Windows, Metal-Clad		Crew	Daily Output	Labor-Hours	Unit	Material	2020 Bare Costs Labor	Equipment	Total	Total Incl O&P
0010	**CASEMENT WINDOWS, METAL-CLAD**										
0100	Metal-clad, deluxe, dbl. insul. glass, 2'-0" x 3'-0" high	G	1 Carp	10	.800	Ea.	300	42.50		342.50	400
0120	2'-0" x 4'-0" high	G		9	.889		325	47		372	435
0130	2'-0" x 5'-0" high	G		8	1		360	53		413	480
0140	2'-0" x 6'-0" high	G	▼	8	1		390	53		443	515
0300	Metal-clad, casement, bldrs mdl, 6'-0" x 4'-0", dbl. insul. glass, 3 panels		2 Carp	10	1.600		1,250	85		1,335	1,525
0310	9'-0" x 4'-0", 4 panels			8	2		1,600	106		1,706	1,925
0320	10'-0" x 5'-0", 5 panels			7	2.286		2,175	121		2,296	2,600
0330	12'-0" x 6'-0", 6 panels		▼	6	2.667	▼	2,750	142		2,892	3,250

For customer support on your Commercial Renovation Costs with RSMeans data, call 800.448.8182.

261

08 52 Wood Windows

08 52 13 – Metal-Clad Wood Windows

08 52 13.30 Double-Hung Windows, Metal-Clad

	08 52 13.30 Double-Hung Windows, Metal-Clad		Crew	Daily Output	Labor-Hours	Unit	Material	2020 Bare Costs Labor	Equipment	Total	Total Incl O&P
0010	**DOUBLE-HUNG WINDOWS, METAL-CLAD**										
0100	Metal-clad, deluxe, dbl. insul. glass, 2'-6" x 3'-0" high	G	1 Carp	10	.800	Ea.	294	42.50		336.50	395
0120	3'-0" x 3'-6" high	G		10	.800		330	42.50		372.50	435
0140	3'-0" x 4'-0" high	G		9	.889		340	47		387	450
0160	3'-0" x 4'-6" high	G		9	.889		365	47		412	475
0180	3'-0" x 5'-0" high	G		8	1		395	53		448	520
0200	3'-6" x 6'-0" high	G	▼	8	1	▼	475	53		528	610

08 52 13.35 Picture and Sliding Windows Metal-Clad

	08 52 13.35 Picture and Sliding Windows Metal-Clad		Crew	Daily Output	Labor-Hours	Unit	Material	Labor	Equipment	Total	Total Incl O&P
0010	**PICTURE AND SLIDING WINDOWS METAL-CLAD**										
2000	Metal-clad, dlx picture, dbl. insul. glass, 4'-0" x 4'-0" high		2 Carp	12	1.333	Ea.	385	71		456	540
2100	4'-0" x 6'-0" high			11	1.455		575	77.50		652.50	755
2200	5'-0" x 6'-0" high			10	1.600		635	85		720	835
2300	6'-0" x 6'-0" high		▼	10	1.600		725	85		810	935
2400	Metal-clad, dlx sliding, dbl. insul. glass, 3'-0" x 3'-0" high	G	1 Carp	10	.800		330	42.50		372.50	430
2420	4'-0" x 3'-6" high	G		9	.889		400	47		447	515
2440	5'-0" x 4'-0" high	G		9	.889		480	47		527	605
2460	6'-0" x 5'-0" high	G	▼	8	1	▼	770	53		823	930
9000	Minimum labor/equipment charge		2 Carp	2.75	5.818	Job		310		310	495

08 52 13.40 Bow and Bay Windows, Metal-Clad

	08 52 13.40 Bow and Bay Windows, Metal-Clad	Crew	Daily Output	Labor-Hours	Unit	Material	Labor	Equipment	Total	Total Incl O&P
0010	**BOW AND BAY WINDOWS, METAL-CLAD**									
0100	Metal-clad, deluxe, dbl. insul. glass, 8'-0" x 5'-0" high, 4 panels	2 Carp	10	1.600	Ea.	1,725	85		1,810	2,025
0120	10'-0" x 5'-0" high, 5 panels		8	2		1,875	106		1,981	2,225
0140	10'-0" x 6'-0" high, 5 panels		7	2.286		2,200	121		2,321	2,625
0160	12'-0" x 6'-0" high, 6 panels		6	2.667		2,950	142		3,092	3,475
0400	Double-hung, bldrs. model, bay, 8' x 4' high, dbl. insul. glass		10	1.600		1,375	85		1,460	1,650
0440	Low E glass		10	1.600		1,475	85		1,560	1,750
0460	9'-0" x 5'-0" high, dbl. insul. glass		6	2.667		1,475	142		1,617	1,850
0480	Low E glass		6	2.667		1,550	142		1,692	1,950
0500	Metal-clad, deluxe, dbl. insul. glass, 7'-0" x 4'-0" high		10	1.600		1,325	85		1,410	1,575
0520	8'-0" x 4'-0" high		8	2		1,375	106		1,481	1,675
0540	8'-0" x 5'-0" high		7	2.286		1,400	121		1,521	1,750
0560	9'-0" x 5'-0" high	▼	6	2.667	▼	1,500	142		1,642	1,875

08 52 16 – Plastic-Clad Wood Windows

08 52 16.10 Bow Window

	08 52 16.10 Bow Window	Crew	Daily Output	Labor-Hours	Unit	Material	Labor	Equipment	Total	Total Incl O&P
0010	**BOW WINDOW** including frames, screens, and grilles									
0020	End panels operable									
1000	Bow type, casement, wood, bldrs. mdl., 8' x 5' dbl. insul. glass, 4 panel	2 Carp	10	1.600	Ea.	1,600	85		1,685	1,875
1050	Low E glass		10	1.600		1,325	85		1,410	1,575
1100	10'-0" x 5'-0", dbl. insul. glass, 6 panels		6	2.667		1,375	142		1,517	1,750
1200	Low E glass, 6 panels		6	2.667		1,475	142		1,617	1,850
1300	Vinyl-clad, bldrs. model, dbl. insul. glass, 6'-0" x 4'-0", 3 panel		10	1.600		1,050	85		1,135	1,275
1340	9'-0" x 4'-0", 4 panel		8	2		1,450	106		1,556	1,775
1380	10'-0" x 6'-0", 5 panels		7	2.286		2,375	121		2,496	2,800
1420	12'-0" x 6'-0", 6 panels		6	2.667		3,100	142		3,242	3,650
2000	Bay window, 8' x 5', dbl. insul. glass		10	1.600		1,975	85		2,060	2,275
2050	Low E glass		10	1.600		2,375	85		2,460	2,725
2100	12'-0" x 6'-0", dbl. insul. glass, 6 panels		6	2.667		2,450	142		2,592	2,925
2200	Low E glass		6	2.667		3,175	142		3,317	3,725
2300	Vinyl-clad, premium, dbl. insul. glass, 8'-0" x 5'-0"		10	1.600		1,825	85		1,910	2,125
2340	10'-0" x 5'-0"		8	2		2,500	106		2,606	2,925
2380	10'-0" x 6'-0"		7	2.286		3,075	121		3,196	3,575

08 52 Wood Windows

08 52 16 – Plastic-Clad Wood Windows

	08 52 16.10 **Bow Window**	Crew	Daily Output	Labor-Hours	Unit	Material	2020 Bare Costs Labor	Equipment	Total	Total Incl O&P
2420	12'-0" x 6'-0"	2 Carp	6	2.667	Ea.	3,325	142		3,467	3,875
3300	Vinyl-clad, premium, dbl. insul. glass, 7'-0" x 4'-6"		10	1.600		1,400	85		1,485	1,650
3340	8'-0" x 4'-6"		8	2		1,450	106		1,556	1,775
3380	8'-0" x 5'-0"		7	2.286		1,525	121		1,646	1,875
3420	9'-0" x 5'-0"		6	2.667		1,550	142		1,692	1,950
9000	Minimum labor/equipment charge	▼	2.50	6.400	Job		340		340	545

08 52 16.15 Awning Window Vinyl-Clad

		Crew	Daily Output	Labor-Hours	Unit	Material	2020 Bare Costs Labor	Equipment	Total	Total Incl O&P
0010	**AWNING WINDOW VINYL-CLAD** including frames, screens, and grilles									
0240	Vinyl-clad, 34" x 22"	1 Carp	10	.800	Ea.	272	42.50		314.50	365
0280	36" x 28"		9	.889		310	47		357	415
0300	36" x 36"		9	.889		350	47		397	460
0340	40" x 22"		10	.800		298	42.50		340.50	395
0360	48" x 28"		8	1		375	53		428	495
0380	60" x 36"	▼	8	1	▼	515	53		568	650

08 52 16.20 Half Round, Vinyl-Clad

		Crew	Daily Output	Labor-Hours	Unit	Material	2020 Bare Costs Labor	Equipment	Total	Total Incl O&P
0010	**HALF ROUND, VINYL-CLAD**, double insulated glass, incl. grille									
0800	14" height x 24" base	2 Carp	9	1.778	Ea.	415	94.50		509.50	605
1040	15" height x 25" base		8	2		385	106		491	595
1060	16" height x 28" base		7	2.286		430	121		551	670
1080	17" height x 29" base	▼	7	2.286		445	121		566	685
2000	19" height x 33" base	1 Carp	6	1.333		465	71		536	625
2100	20" height x 35" base		6	1.333		465	71		536	630
2200	21" height x 37" base	▼	6	1.333		485	71		556	650
2250	23" height x 41" base	2 Carp	6	2.667		535	142		677	815
2300	26" height x 48" base		6	2.667		560	142		702	840
2350	30" height x 56" base	▼	6	2.667		650	142		792	940
3000	36" height x 67" base	1 Carp	4	2		1,150	106		1,256	1,450
3040	38" height x 71" base	2 Carp	5	3.200		1,025	170		1,195	1,400
3050	40" height x 75" base	"	5	3.200		1,325	170		1,495	1,725
5000	Elliptical, 71" x 16"	1 Carp	11	.727		1,000	38.50		1,038.50	1,150
5100	95" x 21"	"	10	.800	▼	1,425	42.50		1,467.50	1,650

08 52 16.35 Double-Hung Window

			Crew	Daily Output	Labor-Hours	Unit	Material	2020 Bare Costs Labor	Equipment	Total	Total Incl O&P
0010	**DOUBLE-HUNG WINDOW** including frames, screens, and grilles										
0300	Vinyl-clad, premium, double insulated glass, 2'-6" x 3'-0"	G	1 Carp	10	.800	Ea.	365	42.50		407.50	470
0305	2'-6" x 4'-0"	G		10	.800		385	42.50		427.50	490
0400	3'-0" x 3'-6"	G		10	.800		355	42.50		397.50	460
0500	3'-0" x 4'-0"	G		9	.889		405	47		452	520
0600	3'-0" x 4'-6"	G		9	.889		445	47		492	565
0700	3'-0" x 5'-0"	G		8	1		465	53		518	600
0790	3'-4" x 5'-0"	G		8	1		470	53		523	605
0800	3'-6" x 6'-0"	G		8	1		525	53		578	665
0820	4'-0" x 5'-0"	G		7	1.143		570	60.50		630.50	730
0830	4'-0" x 6'-0"	G	▼	7	1.143	▼	715	60.50		775.50	885

08 52 16.40 Transom Windows

		Crew	Daily Output	Labor-Hours	Unit	Material	2020 Bare Costs Labor	Equipment	Total	Total Incl O&P
0010	**TRANSOM WINDOWS**									
0050	Vinyl-clad, premium, dbl. insul. glass, 32" x 8"	1 Carp	16	.500	Ea.	189	26.50		215.50	251
0100	36" x 8"		16	.500		209	26.50		235.50	273
0110	36" x 12"		16	.500		225	26.50		251.50	290
2000	Custom sizes, up to 350 sq. in.		12	.667		288	35.50		323.50	370
2100	351 to 750 sq. in.		12	.667		325	35.50		360.50	410
2200	751 to 1150 sq. in.		11	.727		430	38.50		468.50	530
2300	1151 to 1450 sq. in.	▼	11	.727	▼	500	38.50		538.50	610

08 52 Wood Windows

08 52 16 – Plastic-Clad Wood Windows

08 52 16.40 Transom Windows

		Crew	Daily Output	Labor-Hours	Unit	Material	2020 Bare Costs Labor	2020 Bare Costs Equipment	Total	Total Incl O&P
2400	1451 to 1850 sq. in.	2 Carp	12	1.333	Ea.	530	71		601	700
2500	1851 to 2250 sq. in.		12	1.333		655	71		726	835
2600	2251 to 2650 sq. in.		11	1.455		700	77.50		777.50	895
2700	2651 to 3050 sq. in.		11	1.455		750	77.50		827.50	950
2800	3051 to 3450 sq. in.		11	1.455		810	77.50		887.50	1,025
2900	3451 to 3850 sq. in.		10	1.600		880	85		965	1,100
3000	3851 to 4250 sq. in.		10	1.600		870	85		955	1,100
3100	4251 to 4650 sq. in.		10	1.600		885	85		970	1,100
3200	4651 to 5050 sq. in.		10	1.600		965	85		1,050	1,175
3300	5051 to 5450 sq. in.		9	1.778		1,125	94.50		1,219.50	1,400
3400	5451 to 5850 sq. in.		9	1.778		1,000	94.50		1,094.50	1,250
3600	6251 to 6650 sq. in.		8	2		1,250	106		1,356	1,550
3700	6651 to 7050 sq. in.		8	2		1,275	106		1,381	1,575

08 52 16.45 Trapezoid Windows

		Crew	Daily Output	Labor-Hours	Unit	Material	2020 Bare Costs Labor	2020 Bare Costs Equipment	Total	Total Incl O&P
0010	**TRAPEZOID WINDOWS**									
0900	20" base x 44" leg x 53" leg	2 Carp	13	1.231	Ea.	410	65.50		475.50	555
1000	24" base x 90" leg x 102" leg		8	2		730	106		836	975
3000	36" base x 40" leg x 22" leg		12	1.333		455	71		526	620
3010	36" base x 44" leg x 25" leg		13	1.231		480	65.50		545.50	635
3050	36" base x 26" leg x 48" leg		9	1.778		485	94.50		579.50	680
3100	36" base x 42" legs, 50" peak		9	1.778		570	94.50		664.50	775
3200	36" base x 60" leg x 81" leg		11	1.455		745	77.50		822.50	945
4320	44" base x 23" leg x 56" leg		11	1.455		580	77.50		657.50	760
4350	44" base x 59" leg x 92" leg		10	1.600		925	85		1,010	1,150
4500	46" base x 15" leg x 46" leg		8	2		460	106		566	675
4550	46" base x 16" leg x 48" leg		8	2		485	106		591	700
4600	46" base x 50" leg x 80" leg		7	2.286		725	121		846	995
6600	66" base x 12" leg x 42" leg		8	2		625	106		731	860
6650	66" base x 12" legs, 28" peak		9	1.778		490	94.50		584.50	690
6700	68" base x 23" legs, 31" peak		8	2		625	106		731	860

08 52 16.70 Vinyl-Clad, Premium, DBL. Insul. Glass

			Crew	Daily Output	Labor-Hours	Unit	Material	2020 Bare Costs Labor	2020 Bare Costs Equipment	Total	Total Incl O&P
0010	**VINYL-CLAD, PREMIUM, DBL. INSUL. GLASS**										
1000	Sliding, 3'-0" x 3'-0"	G	1 Carp	10	.800	Ea.	645	42.50		687.50	780
1050	4'-0" x 3'-6"	G		9	.889		695	47		742	840
1100	5'-0" x 4'-0"	G		9	.889		940	47		987	1,100
1150	6'-0" x 5'-0"	G		8	1		1,200	53		1,253	1,400

08 52 50 – Window Accessories

08 52 50.10 Window Grille or Muntin

		Crew	Daily Output	Labor-Hours	Unit	Material	2020 Bare Costs Labor	2020 Bare Costs Equipment	Total	Total Incl O&P
0010	**WINDOW GRILLE OR MUNTIN**, snap in type									
0020	Standard pattern interior grilles									
2000	Wood, awning window, glass size, 28" x 16" high	1 Carp	30	.267	Ea.	30.50	14.15		44.65	56
2060	44" x 24" high		32	.250		44	13.30		57.30	70
2100	Casement, glass size, 20" x 36" high		30	.267		34.50	14.15		48.65	60.50
2180	20" x 56" high		32	.250		46	13.30		59.30	72.50
2200	Double-hung, glass size, 16" x 24" high		24	.333	Set	61	17.70		78.70	95.50
2280	32" x 32" high		34	.235	"	141	12.50		153.50	175
2500	Picture, glass size, 48" x 48" high		30	.267	Ea.	128	14.15		142.15	164
2580	60" x 68" high		28	.286	"	206	15.20		221.20	251
2600	Sliding, glass size, 14" x 36" high		24	.333	Set	41	17.70		58.70	73.50
2680	36" x 36" high		22	.364	"	47	19.35		66.35	83
9000	Minimum labor/equipment charge		5	1.600	Job		85		85	136

08 52 Wood Windows

08 52 66 – Wood Window Screens

08 52 66.10 Wood Screens		Crew	Daily Output	Labor-Hours	Unit	Material	2020 Bare Costs Labor	Equipment	Total	Total Incl O&P
0010	**WOOD SCREENS**									
0020	Over 3 S.F., 3/4" frames	2 Carp	375	.043	S.F.	4.79	2.27		7.06	8.90
0100	1-1/8" frames		375	.043		8.45	2.27		10.72	12.95
0200	Rescreen wood frame	▼	500	.032	▼	1.70	1.70		3.40	4.59
9000	Minimum labor/equipment charge	1 Carp	4	2	Job		106		106	170

08 52 69 – Wood Storm Windows

08 52 69.10 Storm Windows

08 52 69.10 Storm Windows			Crew	Daily Output	Labor-Hours	Unit	Material	2020 Bare Costs Labor	Equipment	Total	Total Incl O&P
0010	**STORM WINDOWS**, aluminum residential										
0300	Basement, mill finish, incl. fiberglass screen										
0320	1'-10" x 1'-0" high	G	2 Carp	30	.533	Ea.	37.50	28.50		66	87
0340	2'-9" x 1'-6" high	G		30	.533		41.50	28.50		70	91
0360	3'-4" x 2'-0" high	G	▼	30	.533	▼	44	28.50		72.50	94
1600	Double-hung, combination, storm & screen										
1700	Custom, clear anodic coating, 2'-0" x 3'-5" high		2 Carp	30	.533	Ea.	97.50	28.50		126	153
1720	2'-6" x 5'-0" high			28	.571		133	30.50		163.50	196
1740	4'-0" x 6'-0" high			25	.640		244	34		278	325
1800	White painted, 2'-0" x 3'-5" high			30	.533		128	28.50		156.50	187
1820	2'-6" x 5'-0" high			28	.571		189	30.50		219.50	257
1840	4'-0" x 6'-0" high			25	.640		297	34		331	380
2000	Clear anodic coating, 2'-0" x 3'-5" high	G		30	.533		98	28.50		126.50	154
2020	2'-6" x 5'-0" high	G		28	.571		126	30.50		156.50	187
2040	4'-0" x 6'-0" high	G		25	.640		133	34		167	201
2400	White painted, 2'-0" x 3'-5" high	G		30	.533		104	28.50		132.50	160
2420	2'-6" x 5'-0" high	G		28	.571		102	30.50		132.50	161
2440	4'-0" x 6'-0" high	G		25	.640		120	34		154	187
2600	Mill finish, 2'-0" x 3'-5" high	G		30	.533		86.50	28.50		115	141
2620	2'-6" x 5'-0" high	G		28	.571		103	30.50		133.50	162
2640	4'-0" x 6'-8" high	G	▼	25	.640	▼	130	34		164	199
4000	Picture window, storm, 1 lite, white or bronze finish										
4020	4'-6" x 4'-6" high		2 Carp	25	.640	Ea.	147	34		181	217
4040	5'-8" x 4'-6" high			20	.800		154	42.50		196.50	237
4400	Mill finish, 4'-6" x 4'-6" high			25	.640		141	34		175	210
4420	5'-8" x 4'-6" high		▼	20	.800	▼	173	42.50		215.50	258
4600	3 lite, white or bronze finish										
4620	4'-6" x 4'-6" high		2 Carp	25	.640	Ea.	176	34		210	249
4640	5'-8" x 4'-6" high			20	.800		178	42.50		220.50	263
4800	Mill finish, 4'-6" x 4'-6" high			25	.640		155	34		189	226
4820	5'-8" x 4'-6" high		▼	20	.800	▼	176	42.50		218.50	262
6000	Sliding window, storm, 2 lite, white or bronze finish										
6020	3'-4" x 2'-7" high		2 Carp	28	.571	Ea.	150	30.50		180.50	214
6040	4'-4" x 3'-3" high			25	.640		161	34		195	233
6060	5'-4" x 6'-0" high		▼	20	.800	▼	233	42.50		275.50	325
8000	PVC framed										
8100	Double-hung, combination, storm & screen										
8120	2'-6" x 3'-5"		2 Carp	15	1.067	Ea.	135	56.50		191.50	239
8140	4' x 6'		"	12.50	1.280	"	176	68		244	305
8600	Single lite picture storm										
8620	4'-6" x 4'-6"		2 Carp	12.50	1.280	Ea.	195	68		263	325
8640	5'-8" x 4'-6"		"	10	1.600	"	221	85		306	380
9000	Interior storm window										
9100	Storm window interior glass		1 Glaz	107	.075	S.F.	6.60	3.81		10.41	13.35
9410	Minimum labor/equipment charge		1 Carp	4	2	Job		106		106	170

08 53 13 – Vinyl Windows

08 53 13.20 Vinyl Single-Hung Windows		Crew	Daily Output	Labor-Hours	Unit	Material	2020 Bare Costs Labor	Equipment	Total	Total Incl O&P
0010	**VINYL SINGLE-HUNG WINDOWS**, insulated glass									
0020	Grids, low E, J fin, extension jambs									
0130	25" x 41"	G	2 Carp	20	.800	Ea.	201	42.50		243.50	289
0140	25" x 49"	G		18	.889		208	47		255	305
0150	25" x 57"	G		17	.941		225	50		275	325
0160	25" x 65"	G		16	1		239	53		292	350
0170	29" x 41"	G		18	.889		202	47		249	298
0180	29" x 53"	G		18	.889		217	47		264	315
0190	29" x 57"	G		17	.941		235	50		285	340
0200	29" x 65"	G		16	1		279	53		332	390
0210	33" x 41"	G		20	.800		216	42.50		258.50	305
0220	33" x 53"	G		18	.889		231	47		278	330
0230	33" x 57"	G		17	.941		279	50		329	385
0240	33" x 65"	G		16	1		284	53		337	400
0250	37" x 41"	G		20	.800		236	42.50		278.50	330
0260	37" x 53"	G		18	.889		271	47		318	375
0270	37" x 57"	G		17	.941		282	50		332	390
0280	37" x 65"	G		16	1		320	53		373	440

08 53 13.30 Vinyl Double-Hung Windows		Crew	Daily Output	Labor-Hours	Unit	Material	2020 Bare Costs Labor	Equipment	Total	Total Incl O&P
0010	**VINYL DOUBLE-HUNG WINDOWS**, insulated glass									
0100	Grids, low E, J fin, ext. jambs, 21" x 53"	G	2 Carp	18	.889	Ea.	213	47		260	310
0102	21" x 37"	G		18	.889		249	47		296	350
0104	21" x 41"	G		18	.889		259	47		306	360
0106	21" x 49"	G		18	.889		263	47		310	365
0110	21" x 57"	G		17	.941		283	50		333	390
0120	21" x 65"	G		16	1		300	53		353	415
0128	25" x 37"	G		20	.800		249	42.50		291.50	340
0130	25" x 41"	G		20	.800		258	42.50		300.50	350
0140	25" x 49"	G		18	.889		260	47		307	360
0145	25" x 53"	G		18	.889		285	47		332	390
0150	25" x 57"	G		17	.941		293	50		343	400
0160	25" x 65"	G		16	1		269	53		322	380
0162	25" x 69"	G		16	1		315	53		368	435
0164	25" x 77"	G		16	1		360	53		413	480
0168	29" x 37"	G		18	.889		258	47		305	360
0170	29" x 41"	G		18	.889		278	47		325	380
0172	29" x 49"	G		18	.889		283	47		330	385
0180	29" x 53"	G		18	.889		292	47		339	395
0190	29" x 57"	G		17	.941		305	50		355	415
0200	29" x 65"	G		16	1		320	53		373	440
0202	29" x 69"	G		16	1		345	53		398	465
0205	29" x 77"	G		16	1		340	53		393	460
0208	33" x 37"	G		20	.800		273	42.50		315.50	370
0210	33" x 41"	G		20	.800		278	42.50		320.50	375
0215	33" x 49"	G		20	.800		295	42.50		337.50	395
0220	33" x 53"	G		18	.889		310	47		357	415
0230	33" x 57"	G		17	.941		287	50		337	395
0240	33" x 65"	G		16	1		330	53		383	450
0242	33" x 69"	G		16	1		360	53		413	480
0246	33" x 77"	G		16	1		340	53		393	460
0250	37" x 41"	G		20	.800		305	42.50		347.50	405
0255	37" x 49"	G		20	.800		320	42.50		362.50	420

08 53 13 – Vinyl Windows

08 53 13.30 Vinyl Double-Hung Windows

			Crew	Daily Output	Labor-Hours	Unit	Material	2020 Bare Costs Labor	Equipment	Total	Total Incl O&P
0260	37" x 53"	G	2 Carp	18	.889	Ea.	330	47		377	435
0270	37" x 57"	G		17	.941		350	50		400	465
0280	37" x 65"	G		16	1		380	53		433	505
0282	37" x 69"	G		16	1		395	53		448	520
0286	37" x 77"	G		16	1		390	53		443	515
0300	Solid vinyl, average quality, double insulated glass, 2'-0" x 3'-0"	G	1 Carp	10	.800		295	42.50		337.50	395
0310	3'-0" x 4'-0"	G		9	.889		247	47		294	345
0330	Premium, double insulated glass, 2'-6" x 3'-0"	G		10	.800		305	42.50		347.50	405
0340	3'-0" x 3'-6"	G		9	.889		335	47		382	440
0350	3'-0" x 4'-0"	G		9	.889		345	47		392	455
0360	3'-0" x 4'-6"	G		9	.889		360	47		407	470
0370	3'-0" x 5'-0"	G		8	1		400	53		453	525
0380	3'-6" x 6'-0"	G		8	1		420	53		473	545

08 53 13.40 Vinyl Casement Windows

			Crew	Daily Output	Labor-Hours	Unit	Material	2020 Bare Costs Labor	Equipment	Total	Total Incl O&P
0010	**VINYL CASEMENT WINDOWS**, insulated glass										
0015	Grids, low E, J fin, extension jambs, screens										
0100	One lite, 21" x 41"	G	2 Carp	20	.800	Ea.	350	42.50		392.50	455
0110	21" x 47"	G		20	.800		360	42.50		402.50	465
0120	21" x 53"	G		20	.800		390	42.50		432.50	500
0128	24" x 35"	G		19	.842		305	45		350	405
0130	24" x 41"	G		19	.842		355	45		400	460
0140	24" x 47"	G		19	.842		380	45		425	485
0150	24" x 53"	G		19	.842		400	45		445	510
0158	28" x 35"	G		19	.842		350	45		395	455
0160	28" x 41"	G		19	.842		310	45		355	415
0170	28" x 47"	G		19	.842		400	45		445	510
0180	28" x 53"	G		19	.842		390	45		435	500
0184	28" x 59"	G		19	.842		440	45		485	550
0188	Two lites, 33" x 35"	G		18	.889		525	47		572	650
0190	33" x 41"	G		18	.889		510	47		557	635
0200	33" x 47"	G		18	.889		540	47		587	665
0210	33" x 53"	G		18	.889		550	47		597	680
0212	33" x 59"	G		18	.889		605	47		652	745
0215	33" x 72"	G		18	.889		620	47		667	760
0220	41" x 41"	G		18	.889		580	47		627	710
0230	41" x 47"	G		18	.889		580	47		627	715
0240	41" x 53"	G		17	.941		625	50		675	765
0242	41" x 59"	G		17	.941		660	50		710	805
0246	41" x 72"	G		17	.941		685	50		735	835
0250	47" x 41"	G		17	.941		560	50		610	695
0260	47" x 47"	G		17	.941		585	50		635	720
0270	47" x 53"	G		17	.941		620	50		670	765
0272	47" x 59"	G		17	.941		695	50		745	845
0280	56" x 41"	G		15	1.067		600	56.50		656.50	750
0290	56" x 47"	G		15	1.067		620	56.50		676.50	775
0300	56" x 53"	G		15	1.067		695	56.50		751.50	855
0302	56" x 59"	G		15	1.067		720	56.50		776.50	880
0310	56" x 72"	G		15	1.067		880	56.50		936.50	1,050
0340	Solid vinyl, premium, double insulated glass, 2'-0" x 3'-0" high	G	1 Carp	10	.800		276	42.50		318.50	375
0360	2'-0" x 4'-0" high	G		9	.889		300	47		347	405
0380	2'-0" x 5'-0" high	G		8	1		345	53		398	460

For customer support on your Commercial Renovation Costs with RSMeans data, call 800.448.8182.

267

08 53 13 – Vinyl Windows

08 53 13.50 Vinyl Picture Windows		Crew	Daily Output	Labor-Hours	Unit	Material	2020 Bare Costs Labor	Equipment	Total	Total Incl O&P
0010	**VINYL PICTURE WINDOWS**, insulated glass									
0120	Grids, low E, J fin, ext. jambs, 47" x 35"	2 Carp	12	1.333	Ea.	290	71		361	435
0130	47" x 41"		12	1.333		430	71		501	585
0140	47" x 47"		12	1.333		355	71		426	505
0150	47" x 53"		11	1.455		410	77.50		487.50	580
0160	71" x 35"		11	1.455		420	77.50		497.50	585
0170	71" x 41"		11	1.455		440	77.50		517.50	610
0180	71" x 47"		11	1.455		470	77.50		547.50	645

08 53 13.60 Vinyl Half Round Windows		Crew	Daily Output	Labor-Hours	Unit	Material	Labor	Equipment	Total	Total Incl O&P
0010	**VINYL HALF ROUND WINDOWS**, Including grille, J fin, low E, ext. jambs									
0100	10" height x 20" base	2 Carp	8	2	Ea.	345	106		451	550
0110	15" height x 30" base		8	2		355	106		461	560
0120	17" height x 34" base		7	2.286		395	121		516	630
0130	19" height x 38" base		7	2.286		655	121		776	915
0140	19" height x 33" base		7	2.286		550	121		671	800
0150	24" height x 48" base	1 Carp	6	1.333		585	71		656	760
0160	25" height x 50" base	"	6	1.333		560	71		631	730
0170	30" height x 60" base	2 Carp	6	2.667		620	142		762	910

08 54 Composite Windows

08 54 13 – Fiberglass Windows

08 54 13.10 Fiberglass Single-Hung Windows

			Crew	Daily Output	Labor-Hours	Unit	Material	Labor	Equipment	Total	Total Incl O&P
0010	**FIBERGLASS SINGLE-HUNG WINDOWS**										
0100	Grids, low E, 18" x 24"	G	2 Carp	18	.889	Ea.	390	47		437	505
0110	18" x 40"	G		17	.941		400	50		450	520
0130	24" x 40"	G		20	.800		400	42.50		442.50	510
0230	36" x 36"	G		17	.941		420	50		470	545
0250	36" x 48"	G		20	.800		455	42.50		497.50	570
0260	36" x 60"	G		18	.889		505	47		552	630
0280	36" x 72"	G		16	1		550	53		603	690
0290	48" x 40"	G		16	1		550	53		603	690

08 54 13.30 Fiberglass Slider Windows

			Crew	Daily Output	Labor-Hours	Unit	Material	Labor	Equipment	Total	Total Incl O&P
0010	**FIBERGLASS SLIDER WINDOWS**										
0100	Grids, low E, 36" x 24"	G	2 Carp	20	.800	Ea.	390	42.50		432.50	500
0110	36" x 36"	G	"	20	.800	"	395	42.50		437.50	505

08 54 13.50 Fiberglass Bay Windows

			Crew	Daily Output	Labor-Hours	Unit	Material	Labor	Equipment	Total	Total Incl O&P
0010	**FIBERGLASS BAY WINDOWS**										
0150	48" x 36"	G	2 Carp	11	1.455	Ea.	1,125	77.50		1,202.50	1,375

08 56 Special Function Windows

08 56 46 – Radio-Frequency-Interference Shielding Windows

08 56 46.10 Radio-Frequency-Interference Mesh	Crew	Daily Output	Labor-Hours	Unit	Material	2020 Bare Costs Labor	Equipment	Total	Total Incl O&P
0010 **RADIO-FREQUENCY-INTERFERENCE MESH**									
0100 16 mesh copper 0.011" wire				S.F.	4.88			4.88	5.35
0150 22 mesh copper 0.015" wire					6			6	6.60
0200 100 mesh copper 0.022" wire					8.60			8.60	9.50
0250 100 mesh stainless steel 0.0012" wire					14.90			14.90	16.40

08 61 Roof Windows

08 61 13 – Metal Roof Windows

08 61 13.10 Roof Windows

08 61 13.10 Roof Windows	Crew	Daily Output	Labor-Hours	Unit	Material	2020 Bare Costs Labor	Equipment	Total	Total Incl O&P
0010 **ROOF WINDOWS**, fixed high perf tmpd glazing, metallic framed									
0020 46" x 21-1/2", Flashed for shingled roof	1 Carp	8	1	Ea.	298	53		351	415
0100 46" x 28"		8	1		335	53		388	450
0125 57" x 44"		6	1.333		390	71		461	545
0130 72" x 28"		7	1.143		390	60.50		450.50	530
0150 Fixed, laminated tempered glazing, 46" x 21-1/2"		8	1		480	53		533	615
0175 46" x 28"		8	1		535	53		588	670
0200 57" x 44"		6	1.333		495	71		566	660
0500 Vented flashing set for shingled roof, 46" x 21-1/2"		7	1.143		540	60.50		600.50	695
0525 46" x 28"		6	1.333		600	71		671	770
0550 57" x 44"		5	1.600		750	85		835	960
0560 72" x 28"		5	1.600		750	85		835	960
0575 Flashing set for low pitched roof, 46" x 21-1/2"		7	1.143		580	60.50		640.50	735
0600 46" x 28"		7	1.143		635	60.50		695.50	800
0625 57" x 44"		5	1.600		790	85		875	1,000
0650 Flashing set for curb, 46" x 21-1/2"		7	1.143		680	60.50		740.50	850
0675 46" x 28"		7	1.143		745	60.50		805.50	915
0700 57" x 44"		5	1.600		920	85		1,005	1,125

08 62 Unit Skylights

08 62 13 – Domed Unit Skylights

08 62 13.10 Domed Skylights

08 62 13.10 Domed Skylights		Crew	Daily Output	Labor-Hours	Unit	Material	2020 Bare Costs Labor	Equipment	Total	Total Incl O&P
0010 **DOMED SKYLIGHTS**										
0020 Skylight, fixed dome type, 22" x 22"	G	G-3	12	2.667	Ea.	214	139		353	455
0030 22" x 46"	G		10	3.200		273	167		440	565
0040 30" x 30"	G		12	2.667		288	139		427	535
0050 30" x 46"	G		10	3.200		375	167		542	680
0110 Fixed, double glazed, 22" x 27"	G		12	2.667		274	139		413	520
0120 22" x 46"	G		10	3.200		320	167		487	615
0130 44" x 46"	G		10	3.200		430	167		597	735
0210 Operable, double glazed, 22" x 27"	G		12	2.667		410	139		549	670
0220 22" x 46"	G		10	3.200		485	167		652	800
0230 44" x 46"	G		10	3.200		965	167		1,132	1,325
9000 Minimum labor/equipment charge			2	16	Job		835		835	1,325

08 62 13.20 Skylights

08 62 13.20 Skylights		Crew	Daily Output	Labor-Hours	Unit	Material	2020 Bare Costs Labor	Equipment	Total	Total Incl O&P
0010 **SKYLIGHTS**, flush or curb mounted										
2120 Ventilating insulated plexiglass dome with										
2130 curb mounting, 36" x 36"	G	G-3	12	2.667	Ea.	490	139		629	755
2150 52" x 52"	G		12	2.667		665	139		804	955
2160 28" x 52"	G		10	3.200		525	167		692	840

08 62 Unit Skylights

08 62 13 – Domed Unit Skylights

08 62 13.20 Skylights

		Crew	Daily Output	Labor-Hours	Unit	Material	2020 Bare Costs Labor	Equipment	Total	Total Incl O&P
2170	36" x 52" G	G-3	10	3.200	Ea.	630	167		797	955
2180	For electric opening system, add G					335			335	370
2210	Operating skylight, with thermopane glass, 24" x 48" G	G-3	10	3.200		595	167		762	920
2220	32" x 48" G		9	3.556		620	186		806	980
2300	Insulated safety glass with aluminum frame G		160	.200	S.F.	95	10.45		105.45	122
4000	Skylight, solar tube kit, incl. dome, flashing, diffuser, 1 pipe, 10" diam. G	1 Carp	2	4	Ea.	300	213		513	675
4010	14" diam. G		2	4		365	213		578	745
4020	21" diam. G		2	4		440	213		653	825
4030	Accessories for, 1' long x 9" diam. pipe G		24	.333		53	17.70		70.70	86.50
4040	2' long x 9" diam. pipe G		24	.333		41.50	17.70		59.20	74
4050	4' long x 9" diam. pipe G		20	.400		73.50	21.50		95	115
4060	1' long x 13" diam. pipe G		24	.333		73.50	17.70		91.20	110
4070	2' long x 13" diam. pipe G		24	.333		54	17.70		71.70	87.50
4080	4' long x 13" diam. pipe G		20	.400		101	21.50		122.50	145
4090	6.5" turret ext. for 21" diam. pipe G		16	.500		105	26.50		131.50	158
4100	12' long x 21" diam. flexible pipe G		12	.667		91.50	35.50		127	158
4110	45 degree elbow, 10" G		16	.500		168	26.50		194.50	227
4120	14" G		16	.500		83.50	26.50		110	135
4130	Interior decorative ring, 9" G		20	.400		48.50	21.50		70	87.50
4140	13" G		20	.400		70	21.50		91.50	111

08 63 Metal-Framed Skylights

08 63 13 – Domed Metal-Framed Skylights

08 63 13.20 Skylight Rigid Metal-Framed

		Crew	Daily Output	Labor-Hours	Unit	Material	2020 Bare Costs Labor	Equipment	Total	Total Incl O&P
0010	**SKYLIGHT RIGID METAL-FRAMED** skylight framing is aluminum									
0050	Fixed acrylic double domes, curb mount, 25-1/2" x 25-1/2"	G-3	10	3.200	Ea.	222	167		389	510
0060	25-1/2" x 33-1/2"		10	3.200		250	167		417	540
0070	25-1/2" x 49-1/2"		6	5.333		294	278		572	765
0080	33-1/2" x 33-1/2"		8	4		320	209		529	680
0090	37-1/2" x 25-1/2"		8	4		258	209		467	615
0100	37-1/2" x 37-1/2"		8	4		253	209		462	610
0110	37-1/2" x 49-1/2"		6	5.333		435	278		713	915
0120	49-1/2" x 33-1/2"		6	5.333		385	278		663	865
0130	49-1/2" x 49-1/2"		6	5.333		495	278		773	980
1000	Fixed tempered glass, curb mount, 17-1/2" x 33-1/2"		10	3.200		159	167		326	440
1020	17-1/2" x 49-1/2"		6	5.333		180	278		458	640
1030	25-1/2" x 25-1/2"		10	3.200		159	167		326	440
1040	25-1/2" x 33-1/2"		10	3.200		189	167		356	470
1050	25-1/2" x 37-1/2"		8	4		201	209		410	550
1060	25-1/2" x 49-1/2"		6	5.333		229	278		507	690
1070	25-1/2" x 73-1/2"		6	5.333		365	278		643	845
1080	33-1/2" x 33-1/2"		8	4		220	209		429	570
2000	Manual vent tempered glass & screen, curb, 25-1/2" x 25-1/2"		10	3.200		420	167		587	725
2020	25-1/2" x 37-1/2"		10	3.200		480	167		647	795
2030	25-1/2" x 49-1/2"		8	4		520	209		729	900
2040	33-1/2" x 33-1/2"		8	4		555	209		764	940
2050	33-1/2" x 49-1/2"		6	5.333		700	278		978	1,200
2060	37-1/2" x 37-1/2"		6	5.333		635	278		913	1,150
2070	49-1/2" x 49-1/2"		6	5.333		820	278		1,098	1,350
3000	Electric vent tempered glass, curb mount, 25-1/2" x 25-1/2"		10	3.200		1,075	167		1,242	1,450
3020	25-1/2" x 37-1/2"		10	3.200		1,150	167		1,317	1,550

08 63 Metal-Framed Skylights

08 63 13 – Domed Metal-Framed Skylights

	08 63 13.20 Skylight Rigid Metal-Framed	Crew	Daily Output	Labor-Hours	Unit	Material	2020 Bare Costs Labor	Equipment	Total	Total Incl O&P
3030	25-1/2" x 49-1/2"	G-3	8	4	Ea.	1,225	209		1,434	1,675
3040	33-1/2" x 33-1/2"		8	4		1,250	209		1,459	1,700
3050	33-1/2" x 49-1/2"		6	5.333		1,325	278		1,603	1,900
3060	37-1/2" x 37-1/2"		6	5.333		1,300	278		1,578	1,875
3070	49-1/2" x 49-1/2"		6	5.333		1,425	278		1,703	2,000

08 71 Door Hardware

08 71 13 – Automatic Door Operators

08 71 13.10 Automatic Openers Commercial

		Crew	Daily Output	Labor-Hours	Unit	Material	2020 Bare Costs Labor	Equipment	Total	Total Incl O&P
0010	**AUTOMATIC OPENERS COMMERCIAL**									
0020	Pneumatic door opener incl. motion sens, control box, tubing, compressor									
0050	For single swing door, per opening	2 Skwk	.80	20	Ea.	4,900	1,100		6,000	7,125
0100	Pair, per opening		.50	32	Opng.	7,875	1,750		9,625	11,500
1000	For single sliding door, per opening		.60	26.667		5,275	1,475		6,750	8,125
1300	Bi-parting pair		.50	32		7,800	1,750		9,550	11,400
1420	Electronic door opener incl. motion sens, 12 V control box, motor									
1450	For single swing door, per opening	2 Skwk	.80	20	Opng.	3,925	1,100		5,025	6,075
1500	Pair, per opening		.50	32		7,825	1,750		9,575	11,400
1600	For single sliding door, per opening		.60	26.667		5,150	1,475		6,625	7,975
1700	Bi-parting pair		.50	32		5,725	1,750		7,475	9,100
1750	Handicap actuator buttons, 2, incl. 12 V DC wiring, add	1 Carp	1.50	5.333	Pr.	495	283		778	1,000
2000	Electric panic button for door	2 Skwk	.50	32	Opng.	232	1,750		1,982	3,050

08 71 20.15 Hardware

		Crew	Daily Output	Labor-Hours	Unit	Material	2020 Bare Costs Labor	Equipment	Total	Total Incl O&P
0010	**HARDWARE**									
0020	Average hardware cost									
1000	Door hardware, apartment, interior	1 Carp	4	2	Door	570	106		676	795
1300	Average, door hardware, motel/hotel interior, with access card		4	2		700	106		806	940
1500	Hospital bedroom, average quality		4	2		720	106		826	960
2000	High quality		3	2.667		725	142		867	1,025
2250	School, single exterior, incl. lever, incl. panic device		3	2.667		1,600	142		1,742	2,000
2500	Single interior, regular use, lever included		3	2.667		585	142		727	870
2600	Average, door hdwe. set, school, classroom, ANSI F88, incl. lever		3	2.667		915	142		1,057	1,225
2850	Stairway, single interior		3	2.667		645	142		787	935
3100	Double exterior, with panic device		2	4	Pr.	3,025	213		3,238	3,700
6020	Add for fire alarm door holder, electro-magnetic	1 Elec	4	2	Ea.	92	123		215	290

08 71 20.30 Door Closers

		Crew	Daily Output	Labor-Hours	Unit	Material	2020 Bare Costs Labor	Equipment	Total	Total Incl O&P
0010	**DOOR CLOSERS** adjustable backcheck, multiple mounting									
0015	and rack and pinion									
0020	Standard Regular Arm	1 Carp	6	1.333	Ea.	211	71		282	345
0040	Hold open arm		6	1.333		99.50	71		170.50	223
0100	Fusible link		6.50	1.231		205	65.50		270.50	330
1500	Concealed closers, normal use, head, pivot hung, interior		5.50	1.455		390	77.50		467.50	555
1510	Exterior		5.50	1.455		460	77.50		537.50	630
1520	Overhead concealed, all sizes, regular arm		5.50	1.455		246	77.50		323.50	395
1525	Concealed arm		5	1.600		365	85		450	535
1530	Concealed in door, all sizes, regular arm		5.50	1.455		385	77.50		462.50	545
1535	Concealed arm		5	1.600		272	85		357	435
1560	Floor concealed, all sizes, single acting		2.20	3.636		570	193		763	935
1565	Double acting		2.20	3.636		675	193		868	1,050
1570	Interior, floor, offset pivot, single acting		3.50	2.286		955	121		1,076	1,250

08 71 20.30 Door Closers

		Crew	Daily Output	Labor-Hours	Unit	Material	2020 Bare Costs Labor	Equipment	Total	Total Incl O&P
1590	Exterior	1 Carp	3.50	2.286	Ea.	1,050	121		1,171	1,350
1610	Hold open arm		6	1.333		455	71		526	615
1620	Double acting, standard arm		6	1.333		905	71		976	1,100
1630	Hold open arm		6	1.333		935	71		1,006	1,150
1640	Floor, center hung, single acting, bottom arm		6	1.333		480	71		551	645
1650	Double acting		6	1.333		540	71		611	710
1660	Offset hung, single acting, bottom arm		6	1.333		645	71		716	825
2000	Backcheck and adjustable power, hinge face mount									
5000	For cast aluminum cylinder, deduct				Ea.	41.50			41.50	45.50
5040	For delayed action, add					55			55	60.50
5080	For fusible link arm, add					54			54	59.50
5120	For shock absorbing arm, add					55			55	60.50
5160	For spring power adjustment, add					41.50			41.50	46
6000	Closer-holder, hinge face mount, all sizes, exposed arm	1 Carp	6.50	1.231		235	65.50		300.50	365
6500	Electro magnetic closer/holder									
6510	Single point, no detector	1 Carp	4	2	Ea.	595	106		701	825
6515	Including detector		4	2		795	106		901	1,050
6520	Multi-point, no detector		4	2		995	106		1,101	1,275
6524	Including detector		4	2		1,550	106		1,656	1,875
6550	Electric automatic operators									
6555	Operator	1 Carp	4	2	Ea.	2,150	106		2,256	2,550
6570	Wall plate actuator		4	2		269	106		375	465
7000	Electronic closer-holder, hinge facemount, concealed arm		5	1.600		490	85		575	670
7400	With built-in detector		5	1.600		620	85		705	815
9000	Minimum labor/equipment charge		4	2	Job		106		106	170

08 71 20.36 Panic Devices

		Crew	Daily Output	Labor-Hours	Unit	Material	2020 Bare Costs Labor	Equipment	Total	Total Incl O&P
0010	**PANIC DEVICES** R087110-10									
0015	Touch bars various styles									
0040	Single door exit only, rim device, wide stile									
0050	Economy US28	1 Carp	5	1.600	Ea.	287	85		372	450
0060	Standard duty US28		5	1.600		820	85		905	1,050
0065	US26D		5	1.600		810	85		895	1,025
0070	US10		5	1.600		880	85		965	1,100
0075	US3		5	1.600		1,000	85		1,085	1,225
0080	Night latch, economy US28		5	1.600		365	85		450	535
0085	Standard duty, night latch US28		5	1.600		910	85		995	1,125
0090	US26D		5	1.600		990	85		1,075	1,225
0095	US10		5	1.600		1,050	85		1,135	1,275
0100	US3		5	1.600		1,225	85		1,310	1,475
0500	Single door exit only, rim device, narrow stile									
0540	Economy US28	1 Carp	5	1.600	Ea.	470	85		555	650
0550	Standard duty US28		5	1.600		895	85		980	1,125
0565	US26D		5	1.600		815	85		900	1,025
0570	US10		5	1.600		900	85		985	1,125
0575	US3		5	1.600		1,125	85		1,210	1,350
0580	Economy with night latch US28		5	1.600		355	85		440	525
0585	Standard duty US28		5	1.600		985	85		1,070	1,200
0590	US26D		5	1.600		1,100	85		1,185	1,325
0595	US10		5	1.600		990	85		1,075	1,225
0600	US32D		5	1.600		1,275	85		1,360	1,525
1040	Single door exit only, surface vertical rod									
1050	Surface vertical rod, economy US28	1 Carp	4	2	Ea.	540	106		646	760

08 71 Door Hardware

08 71 13 – Automatic Door Operators

08 71 20.36 Panic Devices

		Crew	Daily Output	Labor-Hours	Unit	Material	2020 Bare Costs Labor	Equipment	Total	Total Incl O&P
1060	Surface vertical rod, standard duty US28	1 Carp	4	2	Ea.	1,075	106		1,181	1,350
1065	US26D		4	2		1,150	106		1,256	1,450
1070	US10		4	2		1,175	106		1,281	1,475
1075	US3		4	2		1,200	106		1,306	1,475
1080	US32D		4	2		1,425	106		1,531	1,750
1500	Single door exit, rim, narrow stile, economy, surface vertical rod									
1540	Surface vertical rod, narrow, economy US28	1 Carp	4	2	Ea.	540	106		646	765
2040	Single door exit only, concealed vertical rod									
2042	These devices will require separate and extra door preparation									
2060	Concealed rod, standard duty US28	1 Carp	4	2	Ea.	1,125	106		1,231	1,400
2065	US26D		4	2		1,225	106		1,331	1,525
2070	US10		4	2		1,350	106		1,456	1,650
2560	Concealed rod, narrow, standard duty US28		4	2		1,275	106		1,381	1,575
2565	US26D		4	2		1,500	106		1,606	1,825
2567	US26		4	2		1,575	106		1,681	1,900
2570	US10		4	2		1,325	106		1,431	1,650
2575	US3		4	2		1,550	106		1,656	1,875
3040	Single door exit only, mortise device, wide stile									
3042	These devices must be combined with a mortise lock									
3060	Mortise, standard duty US28	1 Carp	4	2	Ea.	1,175	106		1,281	1,450
3065	US26D		4	2		1,150	106		1,256	1,450
3067	US26		4	2		1,400	106		1,506	1,725
3070	US10		4	2		1,275	106		1,381	1,575
3075	US3		4	2		1,175	106		1,281	1,475
3080	US32D		4	2		1,225	106		1,331	1,525
3500	Single door exit only, mortise device, narrow stile									
3510	These devices must be combined with a mortise lock									
3540	Mortise, economy US28	1 Carp	4	2	Ea.	425	106		531	635

08 71 20.40 Lockset

		Crew	Daily Output	Labor-Hours	Unit	Material	2020 Bare Costs Labor	Equipment	Total	Total Incl O&P
0010	**LOCKSET**, Standard duty									
0020	Non-keyed, passage, w/sect. trim	1 Carp	12	.667	Ea.	81	35.50		116.50	146
0100	Privacy		12	.667		77.50	35.50		113	143
0400	Keyed, single cylinder function		10	.800		163	42.50		205.50	247
0420	Hotel (see also Section 08 71 20.15)		8	1		210	53		263	315
0500	Lever handled, keyed, single cylinder function		10	.800		159	42.50		201.50	243
1000	Heavy duty with sectional trim, non-keyed, passages		12	.667		129	35.50		164.50	199
1100	Privacy		12	.667		160	35.50		195.50	232
1400	Keyed, single cylinder function		10	.800		201	42.50		243.50	289
1420	Hotel		8	1		580	53		633	725
1600	Communicating		10	.800		325	42.50		367.50	425
1690	For re-core cylinder, add					62.50			62.50	68.50
3800	Cipher lockset w/key pad (security item)	1 Carp	13	.615		1,050	32.50		1,082.50	1,225
3900	Cipher lockset with dial for swinging doors (security item)		13	.615		1,950	32.50		1,982.50	2,200
3920	with dial for swinging doors & drill resistant plate (security item)		12	.667		2,375	35.50		2,410.50	2,650
3950	Cipher lockset with dial for safe/vault door (security item)		12	.667		1,675	35.50		1,710.50	1,875
3980	Keyless, pushbutton type									
4000	Residential/light commercial, deadbolt, standard	1 Carp	9	.889	Ea.	156	47		203	247
4010	Heavy duty		9	.889		258	47		305	360
4020	Industrial, heavy duty, with deadbolt		9	.889		475	47		522	600
4030	Key override		9	.889		470	47		517	595
4040	Lever activated handle		9	.889		480	47		527	605
4050	Key override		9	.889		495	47		542	615

08 71 Door Hardware

08 71 13 – Automatic Door Operators

08 71 20.40 Lockset

		Crew	Daily Output	Labor-Hours	Unit	Material	2020 Bare Costs Labor	2020 Bare Costs Equipment	Total	Total Incl O&P
4060	Double sided pushbutton type	1 Carp	8	1	Ea.	960	53		1,013	1,125
4070	Key override		8	1	↓	875	53		928	1,050
9000	Minimum labor/equipment charge	↓	6	1.333	Job		71		71	114

08 71 20.41 Dead Locks

		Crew	Daily Output	Labor-Hours	Unit	Material	Labor	Equipment	Total	Total Incl O&P
0010	**DEAD LOCKS**									
0011	Mortise heavy duty outside key (security item)	1 Carp	9	.889	Ea.	193	47		240	289
0020	Double cylinder		9	.889		193	47		240	288
0100	Medium duty, outside key		10	.800		117	42.50		159.50	196
0110	Double cylinder		10	.800		143	42.50		185.50	225
1000	Tubular, standard duty, outside key		10	.800		54	42.50		96.50	128
1010	Double cylinder		10	.800		71.50	42.50		114	147
1200	Night latch, outside key		10	.800		58	42.50		100.50	132
1203	Deadlock night latch		7.70	1.039		58	55		113	152
1420	Deadbolt lock, single cylinder		10	.800		48.50	42.50		91	122
1440	Double cylinder	↓	10	.800	↓	67.50	42.50		110	142

08 71 20.42 Mortise Locksets

		Crew	Daily Output	Labor-Hours	Unit	Material	Labor	Equipment	Total	Total Incl O&P
0010	**MORTISE LOCKSETS**, Comm., wrought knobs & full escutcheon trim									
0015	Assumes mortise is cut									
0020	Non-keyed, passage, Grade 3	1 Carp	9	.889	Ea.	164	47		211	256
0030	Grade 1		8	1		440	53		493	570
0040	Privacy set, Grade 3		9	.889		176	47		223	269
0050	Grade 1		8	1		485	53		538	620
0100	Keyed, office/entrance/apartment, Grade 2		8	1		205	53		258	310
0110	Grade 1		7	1.143		510	60.50		570.50	660
0120	Single cylinder, typical, Grade 3		8	1		199	53		252	305
0130	Grade 1		7	1.143		540	60.50		600.50	695
0200	Hotel, room, Grade 3		7	1.143		234	60.50		294.50	355
0210	Grade 1 (see also Section 08 71 20.15)		6	1.333		570	71		641	740
0300	Double cylinder, Grade 3		8	1		241	53		294	350
0310	Grade 1		7	1.143		560	60.50		620.50	715
1000	Wrought knobs and sectional trim, non-keyed, passage, Grade 3		10	.800		138	42.50		180.50	220
1010	Grade 1		9	.889		440	47		487	560
1040	Privacy, Grade 3		10	.800		161	42.50		203.50	245
1050	Grade 1		9	.889		495	47		542	620
1100	Keyed, entrance, office/apartment, Grade 3		9	.889		239	47		286	340
1103	Install lockset		6.92	1.156		239	61.50		300.50	360
1110	Grade 1		8	1		565	53		618	705
1120	Single cylinder, Grade 3		9	.889		236	47		283	335
1130	Grade 1	↓	8	1	↓	520	53		573	660
2000	Cast knobs and full escutcheon trim									
2010	Non-keyed, passage, Grade 3	1 Carp	9	.889	Ea.	287	47		334	390
2020	Grade 1		8	1		425	53		478	555
2040	Privacy, Grade 3		9	.889		340	47		387	445
2050	Grade 1		8	1		450	53		503	580
2120	Keyed, single cylinder, Grade 3		8	1		345	53		398	465
2123	Mortise lock		6.15	1.301		345	69		414	490
2130	Grade 1		7	1.143		550	60.50		610.50	700
3000	Cast knob and sectional trim, non-keyed, passage, Grade 3		10	.800		221	42.50		263.50	310
3010	Grade 1		10	.800		395	42.50		437.50	505
3040	Privacy, Grade 3		10	.800		231	42.50		273.50	320
3050	Grade 1		10	.800		480	42.50		522.50	595
3100	Keyed, office/entrance/apartment, Grade 3		9	.889		261	47		308	365

08 71 20.42 Mortise Locksets

		Crew	Daily Output	Labor-Hours	Unit	Material	2020 Bare Costs Labor	Equipment	Total	Total Incl O&P
3110	Grade 1	1 Carp	9	.889	Ea.	605	47		652	740
3120	Single cylinder, Grade 3		9	.889		273	47		320	375
3130	Grade 1		9	.889		550	47		597	685
3190	For re-core cylinder, add					90.50			90.50	99.50

08 71 20.44 Anti-Ligature Locksets Grade 1

		Crew	Daily Output	Labor-Hours	Unit	Material	2020 Bare Costs Labor	Equipment	Total	Total Incl O&P
0010	**ANTI-LIGATURE LOCKSETS GRADE 1**									
0100	Anti-ligature cylindrical locksets									
0500	Arch handle passage set, US32D	1 Carp	8	1	Ea.	535	53		588	675
0510	privacy set		8	1		575	53		628	715
0520	classroom set		8	1		590	53		643	735
0530	storeroom set		8	1		590	53		643	735
0550	Lever handle passage set US32D		8	1		405	53		458	530
0560	Hospital/privacy set		8	1		420	53		473	545
0570	Office set		8	1		425	53		478	550
0580	Classroom set		8	1		435	53		488	565
0590	Storeroom set		8	1		420	53		473	550
0600	Exit set		8	1		515	53		568	650
0610	Entry set		8	1		445	53		498	575
0620	Asylum set		8	1		485	53		538	620
0630	Back to back dummy set		8	1		310	53		363	425
0690	Anti-ligature mortise locksets									
0700	Knob handle privacy set, US32D	1 Carp	8	1	Ea.	605	53		658	750
0710	Office set		8	1		590	53		643	730
0720	Classroom set		8	1		590	53		643	730
0730	Hotel set		8	1		605	53		658	750
0740	Apartment set		8	1		605	53		658	750
0750	Institutional privacy set		8	1		650	53		703	800
0760	Lever handle office set		8	1		585	53		638	730
0770	Classroom set sectional trim		8	1		585	53		638	730
0780	Hotel set		8	1		600	53		653	745
0790	Apartment set		8	1		600	53		653	745
0800	Institutional privacy set		8	1		650	53		703	800
0810	Office set US32D		8	1		630	53		683	775
0820	Classroom set escutcheon trim		8	1		745	53		798	905
0830	Hotel set escutcheon trim		8	1		765	53		818	925
0840	Apartment set escutcheon trim		8	1		765	53		818	925
0850	Institutional privacy set escutcheon trim		8	1		765	53		818	925

08 71 20.45 Peepholes

		Crew	Daily Output	Labor-Hours	Unit	Material	2020 Bare Costs Labor	Equipment	Total	Total Incl O&P
0010	**PEEPHOLES**									
2010	Peephole	1 Carp	32	.250	Ea.	9.45	13.30		22.75	32
2020	Peephole, wide view	"	32	.250	"	11.25	13.30		24.55	34

08 71 20.50 Door Stops

		Crew	Daily Output	Labor-Hours	Unit	Material	2020 Bare Costs Labor	Equipment	Total	Total Incl O&P
0010	**DOOR STOPS**									
0020	Holder & bumper, floor or wall	1 Carp	32	.250	Ea.	36.50	13.30		49.80	61.50
1300	Wall bumper, 4" diameter, with rubber pad, aluminum		32	.250		13.55	13.30		26.85	36.50
1600	Door bumper, floor type, aluminum		32	.250		3.94	13.30		17.24	26
1620	Brass		32	.250		12.95	13.30		26.25	36
1630	Bronze		32	.250		22	13.30		35.30	45.50
1900	Plunger type, door mounted		32	.250		32	13.30		45.30	57
2500	Holder, floor type, aluminum		32	.250		21.50	13.30		34.80	45.50
2520	Wall type, aluminum		32	.250		35.50	13.30		48.80	60.50
2530	Overhead type, bronze		32	.250		116	13.30		129.30	150

08 71 Door Hardware

08 71 13 – Automatic Door Operators

08 71 20.50 Door Stops

		Crew	Daily Output	Labor-Hours	Unit	Material	2020 Bare Costs Labor	Equipment	Total	Total Incl O&P
2540	Plunger type, aluminum	1 Carp	32	.250	Ea.	32	13.30		45.30	57
2560	Brass		32	.250		47	13.30		60.30	73
3000	Electromagnetic, wall mounted, US3		3	2.667		298	142		440	555
3020	Floor mounted, US3		3	2.667		345	142		487	605
4000	Doorstop, ceiling mounted		3	2.667		19.35	142		161.35	249
4030	Doorstop, header		3	2.667	▼	45.50	142		187.50	277
9000	Minimum labor/equipment charge	▼	6	1.333	Job		71		71	114

08 71 20.55 Push-Pull Plates

		Crew	Daily Output	Labor-Hours	Unit	Material	2020 Bare Costs Labor	Equipment	Total	Total Incl O&P
0010	**PUSH-PULL PLATES**									
0090	Push plate, 0.050 thick, 3" x 12", aluminum	1 Carp	12	.667	Ea.	6.20	35.50		41.70	64
0100	4" x 16"		12	.667		15.50	35.50		51	74
0110	6" x 16"		12	.667		8.30	35.50		43.80	66
0120	8" x 16"		12	.667		9.75	35.50		45.25	68
0200	Push plate, 0.050 thick, 3" x 12", brass		12	.667		14.20	35.50		49.70	72.50
0210	4" x 16"		12	.667		17.75	35.50		53.25	76.50
0220	6" x 16"		12	.667		27.50	35.50		63	87
0230	8" x 16"		12	.667		35.50	35.50		71	96.50
0250	Push plate, 0.050 thick, 3" x 12", satin brass		12	.667		14.45	35.50		49.95	73
0260	4" x 16"		12	.667		17.95	35.50		53.45	77
0270	6" x 16"		12	.667		27.50	35.50		63	87
0280	8" x 16"		12	.667		36	35.50		71.50	96.50
0490	Push plate, 0.050 thick, 3" x 12", bronze		12	.667		17.15	35.50		52.65	76
0500	4" x 16"		12	.667		27	35.50		62.50	86.50
0510	6" x 16"		12	.667		32.50	35.50		68	93
0520	8" x 16"		12	.667		39.50	35.50		75	101
0600	Push plate, antimicrobial copper alloy finish, 3.5" x 15"		13	.615		54	32.50		86.50	112
0610	4" x 16"		13	.615		60.50	32.50		93	119
0620	6" x 16"		13	.615		23	32.50		55.50	78
0630	6" x 20"		13	.615		42	32.50		74.50	99
0740	Push plate, 0.050 thick, 3" x 12", stainless steel		12	.667		12.10	35.50		47.60	70.50
0750	4" x 16"		12	.667		30	35.50		65.50	90
0760	6" x 16"		12	.667		19	35.50		54.50	78
0780	8" x 16"		12	.667		24	35.50		59.50	83.50
0790	Push plate, 0.050 thick, 3" x 12", satin stainless steel		12	.667		6.85	35.50		42.35	64.50
0810	4" x 16"		12	.667		8.40	35.50		43.90	66.50
0820	6" x 16"		12	.667		11.95	35.50		47.45	70
0830	8" x 16"		12	.667		16.15	35.50		51.65	75
0980	Pull plate, 0.050 thick, 3" x 12", aluminum		12	.667		26.50	35.50		62	86
1000	4" x 16"		12	.667		29.50	35.50		65	89.50
1050	Pull plate, 0.050 thick, 3" x 12", brass		12	.667		41	35.50		76.50	102
1060	4" x 16"		12	.667		36	35.50		71.50	96.50
1080	Pull plate, 0.050 thick, 3" x 12", bronze		12	.667		51.50	35.50		87	114
1100	4" x 16"		12	.667		59.50	35.50		95	123
1180	Pull plate, 0.050 thick, 3" x 12", stainless steel		12	.667		47	35.50		82.50	109
1200	4" x 16"		12	.667		62	35.50		97.50	125
1250	Pull plate, 0.050 thick, 3" x 12", chrome		12	.667		47	35.50		82.50	109
1270	4" x 16"		12	.667		49	35.50		84.50	111
1500	Pull handle and push bar, aluminum		11	.727		125	38.50		163.50	199
2000	Bronze		10	.800	▼	169	42.50		211.50	254
9800	Minimum labor/equipment charge	▼	5	1.600	Job		85		85	136

08 71 20.60 Entrance Locks	Crew	Daily Output	Labor-Hours	Unit	Material	2020 Bare Costs Labor	Equipment	Total	Total Incl O&P
0010 **ENTRANCE LOCKS**									
0015 Cylinder, grip handle deadlocking latch	1 Carp	9	.889	Ea.	190	47		237	285
0020 Deadbolt		8	1		195	53		248	299
0100 Push and pull plate, dead bolt		8	1		241	53		294	350
0900 For handicapped lever, add					158			158	173

08 71 20.65 Thresholds	Crew	Daily Output	Labor-Hours	Unit	Material	Labor	Equipment	Total	Total Incl O&P
0010 **THRESHOLDS**									
0011 Threshold 3' long saddles aluminum	1 Carp	48	.167	L.F.	11.10	8.85		19.95	26.50
0100 Aluminum, 8" wide, 1/2" thick		12	.667	Ea.	56.50	35.50		92	119
0500 Bronze		60	.133	L.F.	40.50	7.10		47.60	56.50
0600 Bronze, panic threshold, 5" wide, 1/2" thick		12	.667	Ea.	165	35.50		200.50	239
0700 Rubber, 1/2" thick, 5-1/2" wide		20	.400		43	21.50		64.50	81.50
0800 2-3/4" wide		20	.400		54.50	21.50		76	93.50
1950 ADA compliant thresholds									
2000 Threshold, wood oak 3-1/2" wide x 24" long	1 Carp	12	.667	Ea.	12.55	35.50		48.05	71
2010 3-1/2" wide x 36" long		12	.667		16.90	35.50		52.40	75.50
2020 3-1/2" wide x 48" long		12	.667		22.50	35.50		58	82
2030 4-1/2" wide x 24" long		12	.667		14.55	35.50		50.05	73
2040 4-1/2" wide x 36" long		12	.667		21.50	35.50		57	80.50
2050 4-1/2" wide x 48" long		12	.667		28	35.50		63.50	88
2060 6-1/2" wide x 24" long		12	.667		20	35.50		55.50	79
2070 6-1/2" wide x 36" long		12	.667		30.50	35.50		66	90.50
2080 6-1/2" wide x 48" long		12	.667		40.50	35.50		76	102
2090 Threshold, wood cherry 3-1/2" wide x 24" long		12	.667		15.05	35.50		50.55	73.50
2100 3-1/2" wide x 36" long		12	.667		36	35.50		71.50	96.50
2110 3-1/2" wide x 48" long		12	.667		40	35.50		75.50	101
2120 4-1/2" wide x 24" long		12	.667		19.20	35.50		54.70	78
2130 4-1/2" wide x 36" long		12	.667		38.50	35.50		74	99.50
2140 4-1/2" wide x 48" long		12	.667		47	35.50		82.50	109
2150 6-1/2" wide x 24" long		12	.667		29	35.50		64.50	89
2160 6-1/2" wide x 36" long		12	.667		52.50	35.50		88	115
2170 6-1/2" wide x 48" long		12	.667		62.50	35.50		98	126
2180 Threshold, wood walnut 3-1/2" wide x 24" long		12	.667		18.35	35.50		53.85	77
2190 3-1/2" wide x 36" long		12	.667		36	35.50		71.50	96.50
2200 3-1/2" wide x 48" long		12	.667		41.50	35.50		77	103
2210 4-1/2" wide x 24" long		12	.667		29.50	35.50		65	89
2220 4-1/2" wide x 36" long		12	.667		46.50	35.50		82	108
2230 4-1/2" wide x 48" long		12	.667		59	35.50		94.50	122
2240 6-1/2" wide x 24" long		12	.667		45	35.50		80.50	107
2250 6-1/2" wide x 36" long		12	.667		72.50	35.50		108	137
2260 6-1/2" wide x 48" long		12	.667		94.50	35.50		130	161
2300 Threshold, aluminum 4" wide x 36" long		12	.667		26	35.50		61.50	86
2310 4" wide x 48" long		12	.667		31.50	35.50		67	91.50
2320 4" wide x 72" long		12	.667		47	35.50		82.50	109
2330 5" wide x 36" long		12	.667		34.50	35.50		70	95
2340 5" wide x 48" long		12	.667		41	35.50		76.50	102
2350 5" wide x 72" long		12	.667		76	35.50		111.50	141
2360 6" wide x 36" long		12	.667		47	35.50		82.50	109
2370 6" wide x 48" long		12	.667		50.50	35.50		86	113
2380 6" wide x 72" long		12	.667		73.50	35.50		109	138
2390 7" wide x 36" long		12	.667		51.50	35.50		87	114
2400 7" wide x 48" long		12	.667		69	35.50		104.50	133

For customer support on your Commercial Renovation Costs with RSMeans data, call 800.448.8182.

277

08 71 Door Hardware

08 71 20.65 Thresholds

		Crew	Daily Output	Labor-Hours	Unit	Material	2020 Bare Costs Labor	Equipment	Total	Total Incl O&P
2410	7" wide x 72" long	1 Carp	12	.667	Ea.	98.50	35.50		134	165
2500	Threshold, ramp, aluminum or rubber 24" x 24"		12	.667		173	35.50		208.50	247
9000	Minimum labor/equipment charge		4	2	Job		106		106	170

08 71 20.75 Door Hardware Accessories

		Crew	Daily Output	Labor-Hours	Unit	Material	2020 Bare Costs Labor	Equipment	Total	Total Incl O&P
0010	**DOOR HARDWARE ACCESSORIES**									
0050	Door closing coordinator, 36" (for paired openings up to 56")	1 Carp	8	1	Ea.	131	53		184	229
0060	48" (for paired openings up to 84")		8	1		139	53		192	238
0070	56" (for paired openings up to 96")		8	1		148	53		201	248
1000	Knockers, brass, standard		16	.500		65.50	26.50		92	115
1100	Deluxe		10	.800		97.50	42.50		140	175
2000	Torsion springs for overhead doors									
2050	1-3/4" diam., 32" long, 0.243" spring	1 Carp	8	1	Ea.	60	53		113	151
2060	1-3/4" diam., 32" long, 0.250" spring		8	1		43	53		96	132
2070	2" diam., 32" long, 0.243" spring		8	1		44.50	53		97.50	134
2080	2" diam., 32" long, 0.253" spring		8	1		53	53		106	144
4500	Rubber door silencers		540	.015		.41	.79		1.20	1.71
9000	Minimum labor/equipment charge		6	1.333	Job		71		71	114

08 71 20.80 Hasps

		Crew	Daily Output	Labor-Hours	Unit	Material	2020 Bare Costs Labor	Equipment	Total	Total Incl O&P
0010	**HASPS**, steel assembly									
0015	3"	1 Carp	26	.308	Ea.	5.95	16.35		22.30	32.50
0020	4-1/2"		13	.615		7.10	32.50		39.60	60.50
0040	6"		12.50	.640		10.45	34		44.45	66

08 71 20.90 Hinges

		Crew	Daily Output	Labor-Hours	Unit	Material	2020 Bare Costs Labor	Equipment	Total	Total Incl O&P
0010	**HINGES**									
0012	Full mortise, avg. freq., steel base, USP, 4-1/2" x 4-1/2"				Pr.	47.50			47.50	52.50
0100	5" x 5", USP					68.50			68.50	75
0200	6" x 6", USP					144			144	158
0400	Brass base, 4-1/2" x 4-1/2", US10					73.50			73.50	81
0500	5" x 5", US10					78			78	85.50
0600	6" x 6", US10					171			171	188
0800	Stainless steel base, 4-1/2" x 4-1/2", US32					93			93	102
0900	For non removable pin, add (security item)				Ea.	5.80			5.80	6.40
0910	For floating pin, driven tips, add					3.29			3.29	3.62
0930	For hospital type tip on pin, add					14.85			14.85	16.35
0940	For steeple type tip on pin, add					20.50			20.50	23
0950	Full mortise, high frequency, steel base, 3-1/2" x 3-1/2", US26D				Pr.	31.50			31.50	34.50
1000	4-1/2" x 4-1/2", USP					67.50			67.50	74
1100	5" x 5", USP					54			54	59.50
1200	6" x 6", USP					143			143	157
1400	Brass base, 3-1/2" x 3-1/2", US4					53.50			53.50	58.50
1430	4-1/2" x 4-1/2", US10					74.50			74.50	82
1500	5" x 5", US10					139			139	152
1600	6" x 6", US10					175			175	192
1800	Stainless steel base, 4-1/2" x 4-1/2", US32					107			107	117
1810	5" x 4-1/2", US32					141			141	155
1930	For hospital type tip on pin, add				Ea.	14.65			14.65	16.15
1950	Full mortise, low frequency, steel base, 3-1/2" x 3-1/2", US26D				Pr.	33			33	36.50
2000	4-1/2" x 4-1/2", USP					25			25	27.50
2100	5" x 5", USP					54			54	59.50
2200	6" x 6", USP					97			97	107
2300	4-1/2" x 4-1/2", US3					18.75			18.75	20.50
2310	5" x 5", US3					43			43	47.50

278

08 71 13 – Automatic Door Operators

08 71 20.90 Hinges		Crew	Daily Output	Labor-Hours	Unit	Material	2020 Bare Costs Labor	Equipment	Total	Total Incl O&P
2400	Brass bass, 4-1/2" x 4-1/2", US10				Pr.	58.50			58.50	64.50
2500	5" x 5", US10					93			93	102
2800	Stainless steel base, 4-1/2" x 4-1/2", US32					80			80	88
8000	Install hinge	1 Carp	34	.235	▼		12.50		12.50	20
08 71 20.91 Special Hinges										
0010	**SPECIAL HINGES**									
0015	Paumelle, high frequency									
0020	Steel base, 6" x 4-1/2", US10				Pr.	138			138	152
0100	Brass base, 5" x 4-1/2", US10				Ea.	226			226	248
0200	Paumelle, average frequency, steel base, 4-1/2" x 3-1/2", US10				Pr.	93.50			93.50	103
0400	Olive knuckle, low frequency, brass base, 6" x 4-1/2", US10				Ea.	151			151	167
1000	Electric hinge with concealed conductor, average frequency									
1010	Steel base, 4-1/2" x 4-1/2", US26D				Pr.	345			345	380
1100	Bronze base, 4-1/2" x 4-1/2", US26D				"	350			350	385
1200	Electric hinge with concealed conductor, high frequency									
1210	Steel base, 4-1/2" x 4-1/2", US26D				Pr.	261			261	287
1600	Double weight, 800 lb., steel base, removable pin, 5" x 6", USP					535			535	590
1700	Steel base-welded pin, 5" x 6", USP					172			172	190
1800	Triple weight, 2000 lb., steel base, welded pin, 5" x 6", USP					700			700	770
2000	Pivot reinf., high frequency, steel base, 7-3/4" door plate, USP					182			182	200
2200	Bronze base, 7-3/4" door plate, US10				▼	248			248	273
3000	Swing clear, full mortise, full or half surface, high frequency,									
3010	Steel base, 5" high, USP				Pr.	150			150	165
3200	Swing clear, full mortise, average frequency									
3210	Steel base, 4-1/2" high, USP				Pr.	133			133	146
4000	Wide throw, average frequency, steel base, 4-1/2" x 6", USP					94.50			94.50	104
4200	High frequency, steel base, 4-1/2" x 6", USP					121			121	133
4440	US26D				▼	93			93	102
4600	Spring hinge, single acting, 6" flange, steel				Ea.	49.50			49.50	54.50
4700	Brass					96			96	106
4900	Double acting, 6" flange, steel					89			89	98
4950	Brass				▼	133			133	146
5000	T-strap, galvanized, 4"				Pr.	22			22	24.50
5010	6"					31			31	34
5020	8"				▼	60.50			60.50	66.50
8000	Continuous hinges									
8010	Steel, piano, 2" x 72"	1 Carp	20	.400	Ea.	24	21.50		45.50	60.50
8020	Brass, piano, 1-1/16" x 30"		30	.267		9.45	14.15		23.60	33
8030	Acrylic, piano, 1-3/4" x 12"		40	.200		15.65	10.65		26.30	34.50
8040	Aluminum, door, standard duty, 7'		3	2.667		120	142		262	360
8050	Heavy duty, 7'		3	2.667		152	142		294	395
8060	8'		3	2.667		183	142		325	430
8070	Steel, door, heavy duty, 7'		3	2.667		210	142		352	460
8080	8'		3	2.667		263	142		405	515
8090	Stainless steel, door, heavy duty, 7'		3	2.667		268	142		410	520
8100	8'		3	2.667		300	142		442	555
9000	Continuous hinge, steel, full mortise, heavy duty, 96"		2	4		725	213		938	1,150
9200	Continuous geared hinge, aluminum, full mortise, standard duty, 83"		3	2.667		157	142		299	400
9250	Continuous geared hinge, aluminum, full mortise, heavy duty, 83"	▼	3	2.667	▼	225	142		367	475

08 71 Door Hardware

08 71 13 – Automatic Door Operators

08 71 20.95 Kick Plates		Crew	Daily Output	Labor-Hours	Unit	Material	2020 Bare Costs Labor	Equipment	Total	Total Incl O&P
0010	**KICK PLATES**									
0020	Stainless steel, .050", 16 ga., 8" x 28", US32	1 Carp	15	.533	Ea.	43	28.50		71.50	92.50
0030	8" x 30"		15	.533		46.50	28.50		75	96.50
0040	8" x 34"		15	.533		52.50	28.50		81	103
0050	10" x 28"		15	.533		86	28.50		114.50	140
0060	10" x 30"		15	.533		93	28.50		121.50	148
0070	10" x 34"		15	.533		104	28.50		132.50	161
0080	Mop/Kick, 4" x 28"		15	.533		38.50	28.50		67	88
0090	4" x 30"		15	.533		41	28.50		69.50	90.50
0100	4" x 34"		15	.533		46.50	28.50		75	96.50
0110	6" x 28"		15	.533		44	28.50		72.50	94
0120	6" x 30"		15	.533		53.50	28.50		82	104
0130	6" x 34"		15	.533		60	28.50		88.50	112
0500	Bronze, .050", 8" x 28"		15	.533		77.50	28.50		106	131
0510	8" x 30"		15	.533		73	28.50		101.50	126
0520	8" x 34"		15	.533		75.50	28.50		104	129
0530	10" x 28"		15	.533		79	28.50		107.50	133
0540	10" x 30"		15	.533		86	28.50		114.50	140
0550	10" x 34"		15	.533		96	28.50		124.50	152
0560	Mop/Kick, 4" x 28"		15	.533		37	28.50		65.50	86
0570	4" x 30"		15	.533		41	28.50		69.50	90.50
0580	4" x 34"		15	.533		41.50	28.50		70	91
0590	6" x 28"		15	.533		51.50	28.50		80	102
0600	6" x 30"		15	.533		52	28.50		80.50	103
0610	6" x 34"		15	.533		64.50	28.50		93	117
1000	Acrylic, .125", 8" x 26"		15	.533		28	28.50		56.50	76.50
1010	8" x 36"		15	.533		39.50	28.50		68	89
1020	8" x 42"		15	.533		45.50	28.50		74	95.50
1030	10" x 26"		15	.533		35.50	28.50		64	84.50
1040	10" x 36"		15	.533		49.50	28.50		78	100
1050	10" x 42"		15	.533		57.50	28.50		86	109
1060	Mop/Kick, 4" x 26"		15	.533		16.70	28.50		45.20	64
1070	4" x 36"		15	.533		23.50	28.50		52	71.50
1080	4" x 42"		15	.533		27	28.50		55.50	75.50
1090	6" x 26"		15	.533		25	28.50		53.50	73
1100	6" x 36"		15	.533		34.50	28.50		63	83.50
1110	6" x 42"		15	.533		40.50	28.50		69	90
1220	Brass, .050", 8" x 26"		15	.533		72.50	28.50		101	126
1230	8" x 36"		15	.533		72	28.50		100.50	125
1240	8" x 42"		15	.533		88.50	28.50		117	143
1250	10" x 26"		15	.533		73.50	28.50		102	126
1260	10" x 36"		15	.533		94	28.50		122.50	150
1270	10" x 42"		15	.533		108	28.50		136.50	165
1320	Mop/Kick, 4" x 26"		15	.533		27	28.50		55.50	75
1330	4" x 36"		15	.533		37.50	28.50		66	86.50
1340	4" x 42"		15	.533		43.50	28.50		72	93.50
1350	6" x 26"		15	.533		40.50	28.50		69	90
1360	6" x 36"		15	.533		48.50	28.50		77	99
1370	6" x 42"		15	.533		57.50	28.50		86	109
1800	Aluminum, .050", 8" x 26"		15	.533		40.50	28.50		69	90
1810	8" x 36"		15	.533		37.50	28.50		66	87
1820	8" x 42"		15	.533		44.50	28.50		73	94.50

08 71 Door Hardware

08 71 13 - Automatic Door Operators

08 71 20.95 Kick Plates		Crew	Daily Output	Labor-Hours	Unit	Material	2020 Bare Costs Labor	Equipment	Total	Total Incl O&P
1830	10" x 26"	1 Carp	15	.533	Ea.	34	28.50		62.50	83
1840	10" x 36"		15	.533		47	28.50		75.50	97
1850	10" x 42"		15	.533		65.50	28.50		94	118
1860	Mop/Kick, 4" x 26"		15	.533		16.80	28.50		45.30	64
1870	4" x 36"		15	.533		22	28.50		50.50	70
1880	4" x 42"		15	.533		27	28.50		55.50	75
1890	6" x 26"		15	.533		24.50	28.50		53	72.50
1900	6" x 36"		15	.533		33	28.50		61.50	82
1910	6" x 42"		15	.533		38.50	28.50		67	88
9000	Minimum labor/equipment charge		6	1.333	Job		71		71	114

08 71 21 - Astragals

08 71 21.10 Exterior Mouldings, Astragals

		Crew	Daily Output	Labor-Hours	Unit	Material	2020 Bare Costs Labor	Equipment	Total	Total Incl O&P
0010	**EXTERIOR MOULDINGS, ASTRAGALS**									
0400	One piece, overlapping cadmium plated steel, flat, 3/16" x 2"	1 Carp	90	.089	L.F.	4.12	4.72		8.84	12.10
0600	Prime coated steel, flat, 1/8" x 3"		90	.089		6.25	4.72		10.97	14.40
0800	Stainless steel, flat, 3/32" x 1-5/8"		90	.089		18.10	4.72		22.82	27.50
1000	Aluminum, flat, 1/8" x 2"		90	.089		4.16	4.72		8.88	12.15
1200	Nail on, "T" extrusion		120	.067		2.21	3.54		5.75	8.15
1300	Vinyl bulb insert		105	.076		2.78	4.05		6.83	9.55
1600	Screw on, "T" extrusion		90	.089		4.16	4.72		8.88	12.15
1700	Vinyl insert		75	.107		4.65	5.65		10.30	14.20
2000	"L" extrusion, neoprene bulbs		75	.107		4.51	5.65		10.16	14.05
2100	Neoprene sponge insert		75	.107		6.75	5.65		12.40	16.55
2200	Magnetic		75	.107		10.50	5.65		16.15	20.50
2400	Spring hinged security seal, with cam		75	.107		6.75	5.65		12.40	16.50
2600	Spring loaded locking bolt, vinyl insert		45	.178		9.15	9.45		18.60	25
2800	Neoprene sponge strip, "Z" shaped, aluminum		60	.133		8.25	7.10		15.35	20.50
2900	Solid neoprene strip, nail on aluminum strip		90	.089		3.82	4.72		8.54	11.75
3000	One piece stile protection									
3020	Neoprene fabric loop, nail on aluminum strips	1 Carp	60	.133	L.F.	1.25	7.10		8.35	12.75
3110	Flush mounted aluminum extrusion, 1/2" x 1-1/4"		60	.133		6.85	7.10		13.95	18.85
3140	3/4" x 1-3/8"		60	.133		4.13	7.10		11.23	15.90
3160	1-1/8" x 1-3/4"		60	.133		5	7.10		12.10	16.85
3300	Mortise, 9/16" x 3/4"		60	.133		4.13	7.10		11.23	15.90
3320	13/16" x 1-3/8"		60	.133		4.20	7.10		11.30	15.95
3600	Spring bronze strip, nail on type		105	.076		2.16	4.05		6.21	8.90
3620	Screw on, with retainer		75	.107		2.69	5.65		8.34	12.05
3800	Flexible stainless steel housing, pile insert, 1/2" door		105	.076		7.25	4.05		11.30	14.50
3820	3/4" door		105	.076		8.05	4.05		12.10	15.35
4000	Extruded aluminum retainer, flush mount, pile insert		105	.076		2.49	4.05		6.54	9.25
4080	Mortise, felt insert		90	.089		4.50	4.72		9.22	12.50
4160	Mortise with spring, pile insert		90	.089		3.62	4.72		8.34	11.55
4400	Rigid vinyl retainer, mortise, pile insert		105	.076		2.85	4.05		6.90	9.65
4600	Wool pile filler strip, aluminum backing		105	.076		2.94	4.05		6.99	9.75
5000	Two piece overlapping astragal, extruded aluminum retainer									
5010	Pile insert	1 Carp	60	.133	L.F.	3.26	7.10		10.36	14.95
5020	Vinyl bulb insert		60	.133		1.91	7.10		9.01	13.45
5040	Vinyl flap insert		60	.133		3.97	7.10		11.07	15.70
5060	Solid neoprene flap insert		60	.133		6.50	7.10		13.60	18.50
5080	Hypalon rubber flap insert		60	.133		6.65	7.10		13.75	18.65
5090	Snap on cover, pile insert		60	.133		9.75	7.10		16.85	22
5400	Magnetic aluminum, surface mounted		60	.133		25.50	7.10		32.60	39.50

08 71 Door Hardware

08 71 21 – Astragals

08 71 21.10 Exterior Mouldings, Astragals

	08 71 21.10 Exterior Mouldings, Astragals	Crew	Daily Output	Labor-Hours	Unit	Material	2020 Bare Costs Labor	Equipment	Total	Total Incl O&P
5500	Interlocking aluminum, 5/8" x 1" neoprene bulb insert	1 Carp	45	.178	L.F.	5.30	9.45		14.75	21
5600	Adjustable aluminum, 9/16" x 21/32", pile insert		45	.178		17.65	9.45		27.10	34.50
5800	Magnetic, adjustable, 9/16" x 21/32"		45	.178		23	9.45		32.45	40.50
6000	Two piece stile protection									
6010	Cloth backed rubber loop, 1" gap, nail on aluminum strips	1 Carp	45	.178	L.F.	4.33	9.45		13.78	19.90
6040	Screw on aluminum strips		45	.178		6.50	9.45		15.95	22.50
6100	1-1/2" gap, screw on aluminum extrusion		45	.178		5.80	9.45		15.25	21.50
6240	Vinyl fabric loop, slotted aluminum extrusion, 1" gap		45	.178		2.28	9.45		11.73	17.65
6300	1-1/4" gap		45	.178		6.15	9.45		15.60	22

08 71 25 – Door Weatherstripping

08 71 25.10 Mechanical Seals, Weatherstripping

	08 71 25.10 Mechanical Seals, Weatherstripping	Crew	Daily Output	Labor-Hours	Unit	Material	2020 Bare Costs Labor	Equipment	Total	Total Incl O&P
0010	**MECHANICAL SEALS, WEATHERSTRIPPING**									
1000	Doors, wood frame, interlocking, for 3' x 7' door, zinc	1 Carp	3	2.667	Opng.	58.50	142		200.50	292
1100	Bronze		3	2.667		71	142		213	305
1300	6' x 7' opening, zinc		2	4		60.50	213		273.50	405
1400	Bronze		2	4		71	213		284	420
1700	Wood frame, spring type, bronze									
1800	3' x 7' door	1 Carp	7.60	1.053	Opng.	24.50	56		80.50	116
1900	6' x 7' door	"	7	1.143	"	32	60.50		92.50	133
2200	Metal frame, spring type, bronze									
2300	3' x 7' door	1 Carp	3	2.667	Opng.	52	142		194	284
2400	6' x 7' door	"	2.50	3.200	"	53.50	170		223.50	330
2500	For stainless steel, spring type, add					133%				
2700	Metal frame, extruded sections, 3' x 7' door, aluminum	1 Carp	3	2.667	Opng.	31	142		173	261
2800	Bronze		3	2.667		90	142		232	325
3100	6' x 7' door, aluminum		1.50	5.333		38	283		321	495
3200	Bronze		1.50	5.333		148	283		431	615
3500	Threshold weatherstripping									
3650	Door sweep, flush mounted, aluminum	1 Carp	25	.320	Ea.	22	17		39	51.50
3700	Vinyl		25	.320		19.90	17		36.90	49.50
5000	Garage door bottom weatherstrip, 12' aluminum, clear		14	.571		26	30.50		56.50	77
5010	Bronze		14	.571		94.50	30.50		125	153
5050	Bottom protection, rubber		14	.571		54.50	30.50		85	109
5100	Threshold		14	.571		70.50	30.50		101	126
9000	Minimum labor/equipment charge		3	2.667	Job		142		142	227

08 75 Window Hardware

08 75 10 – Window Handles and Latches

08 75 10.10 Handles and Latches

	08 75 10.10 Handles and Latches	Crew	Daily Output	Labor-Hours	Unit	Material	2020 Bare Costs Labor	Equipment	Total	Total Incl O&P
0010	**HANDLES AND LATCHES**									
1000	Handles, surface mounted, aluminum	1 Carp	24	.333	Ea.	6.15	17.70		23.85	35.50
1020	Brass		24	.333		5.35	17.70		23.05	34.50
1040	Chrome		24	.333		8.65	17.70		26.35	38
1200	Window handles window crank ADA		24	.333		17.45	17.70		35.15	47.50
1500	Recessed, aluminum		12	.667		3.90	35.50		39.40	61.50
1520	Brass		12	.667		5.60	35.50		41.10	63
1540	Chrome		12	.667		4.63	35.50		40.13	62
2000	Latches, aluminum		20	.400		4.41	21.50		25.91	39
2020	Brass		20	.400		6.70	21.50		28.20	41.50
2040	Chrome		20	.400		4.39	21.50		25.89	39

08 75 Window Hardware

08 75 10 – Window Handles and Latches

08 75 10.10 Handles and Latches

	Crew	Daily Output	Labor-Hours	Unit	Material	2020 Bare Costs Labor	Equipment	Total	Total Incl O&P
9000 Minimum labor/equipment charge	1 Carp	6	1.333	Job		71		71	114

08 75 10.15 Window Opening Control

	Crew	Daily Output	Labor-Hours	Unit	Material	2020 Bare Costs Labor	Equipment	Total	Total Incl O&P
0010 **WINDOW OPENING CONTROL**									
0015 Window stops									
0020 For double-hung window	1 Carp	35	.229	Ea.	40.50	12.15		52.65	64
0030 Cam action for sliding window		35	.229		4.55	12.15		16.70	24.50
0040 Thumb screw for sliding window		35	.229		4.91	12.15		17.06	25
0100 Window guards, child safety bars for single or double-hung									
0110 14" to 17" wide max vert opening 26"	1 Carp	16	.500	Ea.	68.50	26.50		95	118
0120 17" to 23" wide max vert opening 26"		16	.500		78	26.50		104.50	129
0130 23" to 36" wide max vert opening 26"		16	.500		89	26.50		115.50	141
0140 35" to 58" wide max vert opening 26"		16	.500		103	26.50		129.50	157
0150 58" to 90" wide max vert opening 26"		16	.500		139	26.50		165.50	196
0160 73" to 120" wide max vert opening 26"		16	.500		243	26.50		269.50	310

08 75 30 – Window Weatherstripping

08 75 30.10 Mechanical Weather Seals

	Crew	Daily Output	Labor-Hours	Unit	Material	2020 Bare Costs Labor	Equipment	Total	Total Incl O&P
0010 **MECHANICAL WEATHER SEALS**, Window, double-hung, 3' x 5'									
0020 Zinc	1 Carp	7.20	1.111	Opng.	22.50	59		81.50	120
0100 Bronze		7.20	1.111		42.50	59		101.50	142
0200 Vinyl V strip		7	1.143		11.60	60.50		72.10	110
0500 As above but heavy duty, zinc		4.60	1.739		21	92.50		113.50	171
0600 Bronze		4.60	1.739		79	92.50		171.50	235
9000 Minimum labor/equipment charge	1 Clab	4.60	1.739	Job		73		73	117

08 79 Hardware Accessories

08 79 20 – Door Accessories

08 79 20.10 Door Hardware Accessories

	Crew	Daily Output	Labor-Hours	Unit	Material	2020 Bare Costs Labor	Equipment	Total	Total Incl O&P
0010 **DOOR HARDWARE ACCESSORIES**									
0140 Door bolt, surface, 4"	1 Carp	32	.250	Ea.	18.80	13.30		32.10	42
0160 Door latch	"	12	.667	"	8.80	35.50		44.30	66.50
0200 Sliding closet door									
0220 Track and hanger, single	1 Carp	10	.800	Ea.	67.50	42.50		110	143
0240 Double		8	1		85	53		138	179
0260 Door guide, single		48	.167		32.50	8.85		41.35	49.50
0280 Double		48	.167		41.50	8.85		50.35	59.50
0600 Deadbolt and lock cover plate, brass or stainless steel		30	.267		33	14.15		47.15	58.50
0620 Hole cover plate, brass or chrome		35	.229		8.70	12.15		20.85	29
2240 Mortise lockset, passage, lever handle		9	.889		173	47		220	266
4000 Security chain, standard		18	.444		15	23.50		38.50	54.50
4100 Deluxe		18	.444		55	23.50		78.50	98.50

For customer support on your Commercial Renovation Costs with RSMeans data, call 800.448.8182.

283

08 81 10 – Float Glass

08 81 10.10 Various Types and Thickness of Float Glass		Crew	Daily Output	Labor-Hours	Unit	Material	2020 Bare Costs Labor	Equipment	Total	Total Incl O&P
0010	**VARIOUS TYPES AND THICKNESS OF FLOAT GLASS**									
0020	3/16" plain	2 Glaz	130	.123	S.F.	7.10	6.30		13.40	17.90
0200	Tempered, clear		130	.123		7.50	6.30		13.80	18.30
0300	Tinted		130	.123		7.60	6.30		13.90	18.40
0600	1/4" thick, clear, plain		120	.133		8.40	6.80		15.20	20
0700	Tinted		120	.133		9.75	6.80		16.55	21.50
0800	Tempered, clear		120	.133		7.95	6.80		14.75	19.60
0900	Tinted		120	.133		12.15	6.80		18.95	24.50
1600	3/8" thick, clear, plain		75	.213		11.70	10.90		22.60	30.50
1700	Tinted		75	.213		16.85	10.90		27.75	36
1800	Tempered, clear		75	.213		18.45	10.90		29.35	38
1900	Tinted		75	.213		21	10.90		31.90	41
2200	1/2" thick, clear, plain		55	.291		18.95	14.85		33.80	44.50
2300	Tinted		55	.291		29.50	14.85		44.35	56
2400	Tempered, clear		55	.291		27.50	14.85		42.35	54
2500	Tinted		55	.291		28	14.85		42.85	54
2800	5/8" thick, clear, plain		45	.356		29.50	18.15		47.65	61.50
2900	Tempered, clear		45	.356		33.50	18.15		51.65	66
3200	3/4" thick, clear, plain		35	.457		38	23.50		61.50	79.50
3300	Tempered, clear		35	.457		45	23.50		68.50	86.50
3600	1" thick, clear, plain		30	.533		63	27		90	113
8900	For low emissivity coating for 3/16" & 1/4" only, add to above					18%				
9000	Minimum labor/equipment charge	1 Glaz	2	4	Job		204		204	325

08 81 17 – Fire Glass

08 81 17.10 Fire Resistant Glass

		Crew	Daily Output	Labor-Hours	Unit	Material	2020 Bare Costs Labor	Equipment	Total	Total Incl O&P
0010	**FIRE RESISTANT GLASS**									
0020	Fire glass minimum	2 Glaz	40	.400	S.F.	42.50	20.50		63	79
0030	Mid range		40	.400		85.50	20.50		106	127
0050	High end		40	.400		375	20.50		395.50	445

08 81 20 – Vision Panels

08 81 20.10 Full Vision

		Crew	Daily Output	Labor-Hours	Unit	Material	2020 Bare Costs Labor	Equipment	Total	Total Incl O&P
0010	**FULL VISION**, window system with 3/4" glass mullions									
0020	Up to 10' high	H-2	130	.185	S.F.	71	8.85		79.85	92
0100	10' to 20' high, minimum		110	.218		75.50	10.50		86	100
0150	Average		100	.240		81	11.55		92.55	107
0200	Maximum		80	.300		91	14.40		105.40	123
9000	Minimum labor/equipment charge	1 Glaz	2	4	Job		204		204	325

08 81 25 – Glazing Variables

08 81 25.10 Applications of Glazing

		Crew	Daily Output	Labor-Hours	Unit	Material	2020 Bare Costs Labor	Equipment	Total	Total Incl O&P
0010	**APPLICATIONS OF GLAZING** R088110-10									
0600	For glass replacement, add				S.F.		100%			
0700	For gasket settings, add				L.F.	6.40			6.40	7
0900	For sloped glazing, add				S.F.		26%			
2000	Fabrication, polished edges, 1/4" thick				Inch	.61			.61	.67
2100	1/2" thick					1.44			1.44	1.58
2500	Mitered edges, 1/4" thick					1.44			1.44	1.58
2600	1/2" thick					2.38			2.38	2.62

08 81 Glass Glazing

08 81 30 – Insulating Glass

08 81 30.10 Reduce Heat Transfer Glass		Crew	Daily Output	Labor-Hours	Unit	Material	2020 Bare Costs Labor	Equipment	Total	Total Incl O&P
0010	**REDUCE HEAT TRANSFER GLASS**									
0015	2 lites 1/8" float, 1/2" thk under 15 S.F.									
0020	Clear R088110-10 [G]	2 Glaz	95	.168	S.F.	10.60	8.60		19.20	25.50
0100	Tinted [G]		95	.168		14.75	8.60		23.35	30
0200	2 lites 3/16" float, for 5/8" thk unit, 15 to 30 S.F., clear [G]		90	.178		14.25	9.05		23.30	30
0400	1" thk, dbl. glazed, 1/4" float, 30 to 70 S.F., clear [G]		75	.213		17.10	10.90		28	36
0500	Tinted [G]		75	.213		24	10.90		34.90	44
2000	Both lites, light & heat reflective [G]		85	.188		33	9.60		42.60	51.50
2500	Heat reflective, film inside, 1" thick unit, clear [G]		85	.188		28.50	9.60		38.10	47
2600	Tinted [G]		85	.188		31	9.60		40.60	49.50
3000	Film on weatherside, clear, 1/2" thick unit [G]		95	.168		20	8.60		28.60	36
3100	5/8" thick unit [G]		90	.178		21	9.05		30.05	37.50
3200	1" thick unit [G]	↓	85	.188		28.50	9.60		38.10	47
3350	Clear heat reflective film on inside [G]	1 Glaz	50	.160		13.65	8.15		21.80	28
3360	Heat reflective film on inside, tinted [G]		25	.320		14.30	16.30		30.60	41.50
3370	Heat reflective film on the inside, metalized [G]	↓	20	.400		19.40	20.50		39.90	54
5000	Spectrally selective film, on ext., blocks solar gain/allows 70% of light [G]	2 Glaz	95	.168	↓	15.80	8.60		24.40	31
9000	Minimum labor/equipment charge	1 Glaz	2	4	Job		204		204	325

08 81 40 – Plate Glass

08 81 40.10 Plate Glass		Crew	Daily Output	Labor-Hours	Unit	Material	2020 Bare Costs Labor	Equipment	Total	Total Incl O&P
0010	**PLATE GLASS** Twin ground, polished,									
0020	3/16" thick	2 Glaz	100	.160	S.F.	6	8.15		14.15	19.65
0100	1/4" thick		94	.170		8.20	8.70		16.90	23
0200	3/8" thick		60	.267		13.90	13.60		27.50	37
0300	1/2" thick	↓	40	.400	↓	27	20.50		47.50	62.50

08 81 55 – Window Glass

08 81 55.10 Sheet Glass		Crew	Daily Output	Labor-Hours	Unit	Material	2020 Bare Costs Labor	Equipment	Total	Total Incl O&P
0010	**SHEET GLASS** (window), clear float, stops, putty bed									
0015	1/8" thick, clear float	2 Glaz	480	.033	S.F.	4.27	1.70		5.97	7.40
0500	3/16" thick, clear		480	.033		6.65	1.70		8.35	10
0600	Tinted		480	.033		8.15	1.70		9.85	11.65
0700	Tempered		480	.033	↓	10	1.70		11.70	13.70
9000	Minimum labor/equipment charge	↓	5	3.200	Job		163		163	261

08 81 65 – Wire Glass

08 81 65.10 Glass Reinforced With Wire		Crew	Daily Output	Labor-Hours	Unit	Material	2020 Bare Costs Labor	Equipment	Total	Total Incl O&P
0010	**GLASS REINFORCED WITH WIRE**									
0012	1/4" thick rough obscure	2 Glaz	135	.119	S.F.	27.50	6.05		33.55	39.50
1000	Polished wire, 1/4" thick, diamond, clear		135	.119		33	6.05		39.05	45.50
1500	Pinstripe, obscure	↓	135	.119	↓	55	6.05		61.05	70

For customer support on your Commercial Renovation Costs with RSMeans data, call 800.448.8182.

285

08 83 Mirrors

08 83 13 – Mirrored Glass Glazing

08 83 13.10 Mirrors		Crew	Daily Output	Labor-Hours	Unit	Material	2020 Bare Costs Labor	Equipment	Total	Total Incl O&P
0010	**MIRRORS**, No frames, wall type, 1/4" plate glass, polished edge									
0100	Up to 5 S.F.	2 Glaz	125	.128	S.F.	9.85	6.55		16.40	21.50
0200	Over 5 S.F.		160	.100		9.45	5.10		14.55	18.55
0500	Door type, 1/4" plate glass, up to 12 S.F.		160	.100		9.55	5.10		14.65	18.65
1000	Float glass, up to 10 S.F., 1/8" thick		160	.100		6.75	5.10		11.85	15.55
1100	3/16" thick		150	.107		7.85	5.45		13.30	17.35
1500	12" x 12" wall tiles, square edge, clear		195	.082		2.80	4.18		6.98	9.80
1600	Veined		195	.082		6.60	4.18		10.78	13.95
2000	1/4" thick, stock sizes, one way transparent		125	.128		21	6.55		27.55	34
2010	Bathroom, unframed, laminated		160	.100		16	5.10		21.10	26
2500	Tempered		160	.100		18.75	5.10		23.85	28.50

08 83 13.15 Reflective Glass

0010	**REFLECTIVE GLASS**									
0100	1/4" float with fused metallic oxide fixed	G	2 Glaz	115	.139	S.F.	18.30	7.10		25.40	31.50
0500	1/4" float glass with reflective applied coating	G	"	115	.139	"	14.95	7.10		22.05	28

08 84 Plastic Glazing

08 84 10 – Plexiglass Glazing

08 84 10.10 Plexiglass Acrylic

		Crew	Daily Output	Labor-Hours	Unit	Material	Labor	Equipment	Total	Total Incl O&P
0010	**PLEXIGLASS ACRYLIC**, clear, masked,									
0020	1/8" thick, cut sheets	2 Glaz	170	.094	S.F.	13.90	4.80		18.70	23
0200	Full sheets		195	.082		6.20	4.18		10.38	13.55
0500	1/4" thick, cut sheets		165	.097		14.65	4.95		19.60	24
0600	Full sheets		185	.086		10.10	4.41		14.51	18.15
0900	3/8" thick, cut sheets		155	.103		23	5.25		28.25	33.50
1000	Full sheets		180	.089		17	4.53		21.53	26
1300	1/2" thick, cut sheets		135	.119		34	6.05		40.05	47
1400	Full sheets		150	.107		22	5.45		27.45	33
1700	3/4" thick, cut sheets		115	.139		32	7.10		39.10	46.50
1800	Full sheets		130	.123		18.35	6.30		24.65	30
2100	1" thick, cut sheets		105	.152		36	7.75		43.75	52
2200	Full sheets		125	.128		22	6.55		28.55	35
3000	Colored, 1/8" thick, cut sheets		170	.094		20	4.80		24.80	29.50
3200	Full sheets		195	.082		11.95	4.18		16.13	19.85
3500	1/4" thick, cut sheets		165	.097		22.50	4.95		27.45	32.50
3600	Full sheets		185	.086		16.35	4.41		20.76	25
4000	Mirrors, untinted, cut sheets, 1/8" thick		185	.086		16.15	4.41		20.56	25
4200	1/4" thick		180	.089		18.70	4.53		23.23	28

08 84 20 – Polycarbonate

08 84 20.10 Thermoplastic

		Crew	Daily Output	Labor-Hours	Unit	Material	Labor	Equipment	Total	Total Incl O&P
0010	**THERMOPLASTIC**, clear, masked, cut sheets									
0020	1/8" thick	2 Glaz	170	.094	S.F.	16	4.80		20.80	25.50
0500	3/16" thick		165	.097		20	4.95		24.95	30
1000	1/4" thick		155	.103		19.20	5.25		24.45	29.50
1500	3/8" thick		150	.107		30	5.45		35.45	41.50
9000	Minimum labor/equipment charge	1 Glaz	2	4	Job		204		204	325

08 85 Glazing Accessories

08 85 10 – Miscellaneous Glazing Accessories

08 85 10.10 Glazing Gaskets	Crew	Daily Output	Labor-Hours	Unit	Material	2020 Bare Costs Labor	Equipment	Total	Total Incl O&P
0010	**GLAZING GASKETS**, Neoprene for glass tongued mullion									
0015	1/4" glass	2 Glaz	200	.080	L.F.	.98	4.08		5.06	7.60
0020	3/8"		200	.080		1.18	4.08		5.26	7.80
0040	1/2"		200	.080		1.29	4.08		5.37	7.90
0060	3/4"		180	.089		1.98	4.53		6.51	9.45
0080	1"		180	.089		1.64	4.53		6.17	9.05
1000	Glazing compound, wood	1 Glaz	58	.138		1.31	7.05		8.36	12.70
1005	Glazing compound, per window, up to 30 L.F.					1.31			1.31	1.45
1006	Glazing compound, per window, up to 30 L.F.				Ea.	39.50			39.50	43.50
1020	Metal	1 Glaz	58	.138	L.F.	1.31	7.05		8.36	12.70

08 85 20 – Structural Glazing Sealants

08 85 20.10 Structural Glazing Adhesives

		Crew	Daily Output	Labor-Hours	Unit	Material	Labor	Equipment	Total	Total Incl O&P
0010	**STRUCTURAL GLAZING ADHESIVES**									
0050	Structural glazing adhesive, 1/4" x 1/4" joint seal	1 Glaz	200	.040	L.F.	14.65	2.04		16.69	19.35

08 87 Glazing Surface Films

08 87 13 – Solar Control Films

08 87 13.10 Solar Films On Glass

		Crew	Daily Output	Labor-Hours	Unit	Material	Labor	Equipment	Total	Total Incl O&P
0010	**SOLAR FILMS ON GLASS** (glass not included)									
1000	Bronze, 20% VLT	H-2	180	.133	S.F.	1.67	6.40		8.07	12.10
1020	50% VLT		180	.133		1.87	6.40		8.27	12.30
1050	Neutral, 20% VLT		180	.133		1.87	6.40		8.27	12.30
1100	Silver, 15% VLT		180	.133		1.08	6.40		7.48	11.45
1120	35% VLT		180	.133		3.14	6.40		9.54	13.70
1150	68% VLT		180	.133		.42	6.40		6.82	10.70
3000	One way mirror with night vision 5% in and 15% out VLT		180	.133		3.15	6.40		9.55	13.70

08 87 16 – Glass Safety Films

08 87 16.10 Safety Films On Glass

		Crew	Daily Output	Labor-Hours	Unit	Material	Labor	Equipment	Total	Total Incl O&P
0010	**SAFETY FILMS ON GLASS** (glass not included)									
0015	Safety film helps hold glass together when broken									
0050	Safety film, clear, 2 mil	H-2	180	.133	S.F.	1.26	6.40		7.66	11.65
0100	Safety film, clear, 4 mil		180	.133		1.88	6.40		8.28	12.30
0110	Safety film, tinted, 4 mil		180	.133		3.60	6.40		10	14.20
0150	Safety film, black tint, 8 mil		180	.133		2.32	6.40		8.72	12.80

08 87 26 – Bird Control Film

08 87 26.10 Bird Control Film

		Crew	Daily Output	Labor-Hours	Unit	Material	Labor	Equipment	Total	Total Incl O&P
0010	**BIRD CONTROL FILM**									
0050	Patterned, adhered to glass	2 Glaz	180	.089	S.F.	7	4.53		11.53	14.95
0200	Decals small, adhered to glass	1 Glaz	50	.160	Ea.	2.57	8.15		10.72	15.90
0250	Decals large, adhered to glass		50	.160	"	8.20	8.15		16.35	22
0300	Decals set of 4, adhered to glass		25	.320	Set	7.55	16.30		23.85	34.50
0400	Bird control film tape		10	.800	Roll	20.50	41		61.50	88

08 87 33 – Electrically Tinted Window Film

08 87 33.20 Window Film

		Crew	Daily Output	Labor-Hours	Unit	Material	Labor	Equipment	Total	Total Incl O&P
0010	**WINDOW FILM** adhered on glass (glass not included)									
0015	Film is pre-wired and can be trimmed									
0100	Window film 4 S.F. adhered on glass	1 Glaz	200	.040	S.F.	224	2.04		226.04	250
0120	6 S.F.		180	.044		335	2.27		337.27	375
0160	9 S.F.		170	.047		505	2.40		507.40	560

08 87 Glazing Surface Films

08 87 33 – Electrically Tinted Window Film

08 87 33.20 Window Film	Crew	Daily Output	Labor-Hours	Unit	Material	2020 Bare Costs Labor	Equipment	Total	Total Incl O&P
0180	10 S.F.	1 Glaz	150	.053	S.F.	585	2.72		587.72	645
0200	12 S.F.		144	.056		670	2.83		672.83	745
0220	14 S.F.		140	.057		785	2.91		787.91	865
0240	16 S.F.		128	.063		895	3.19		898.19	990
0260	18 S.F.		126	.063		1,000	3.24		1,003.24	1,100
0280	20 S.F.		120	.067		1,125	3.40		1,128.40	1,225

08 87 53 – Security Films On Glass

08 87 53.10 Security Film Adhered On Glass

		Crew	Daily Output	Labor-Hours	Unit	Material	Labor	Equipment	Total	Total Incl O&P
0010	**SECURITY FILM ADHERED ON GLASS** (glass not included)									
0020	Security film, clear, 7 mil	1 Glaz	200	.040	S.F.	1.13	2.04		3.17	4.50
0030	8 mil		200	.040		2.04	2.04		4.08	5.50
0040	9 mil		200	.040		1.12	2.04		3.16	4.49
0050	10 mil		200	.040		1.93	2.04		3.97	5.40
0060	12 mil		200	.040		1.91	2.04		3.95	5.35
0075	15 mil		200	.040		2.41	2.04		4.45	5.90
0100	Security film, sealed with structural adhesive, 7 mil		180	.044		5.75	2.27		8.02	9.95
0110	8 mil		180	.044		6.70	2.27		8.97	10.95
0140	14 mil		180	.044		7.35	2.27		9.62	11.65

08 88 Special Function Glazing

08 88 52 – Prefabricated Glass Block Windows

08 88 52.10 Prefabricated Glass Block Windows

		Crew	Daily Output	Labor-Hours	Unit	Material	Labor	Equipment	Total	Total Incl O&P
0010	**PREFABRICATED GLASS BLOCK WINDOWS**									
0015	Includes frame and silicone seal									
0020	Glass Block Window 16" x 8"	2 Glaz	6	2.667	Ea.	168	136		304	400
0025	16" x 16"		6	2.667		192	136		328	430
0030	16" x 24"		6	2.667		223	136		359	460
0050	16" x 48"		4	4		360	204		564	720
0100	Glass Block Window 24" x 8"		6	2.667		181	136		317	415
0110	24" x 16"		6	2.667		232	136		368	470
0120	24" x 24"		6	2.667		292	136		428	535
0150	24" x 48"		6	2.667		430	136		566	690
0200	Glass Block Window 32" x 8"		6	2.667		182	136		318	415
0210	32" x 16"		6	2.667		266	136		402	510
0220	32" x 24"		4	4		330	204		534	690
0250	32" x 48"		4	4		525	204		729	905
0300	Glass Block Window 40" x 8"		6	2.667		229	136		365	470
0310	40" x 16"		6	2.667		305	136		441	550
0320	40" x 24"		6	2.667		375	136		511	625
0350	40" x 48"		6	2.667		615	136		751	895
0400	Glass Block Window 48" x 8"		6	2.667		274	136		410	515
0410	48" x 16"		6	2.667		360	136		496	610
0420	48" x 24"		6	2.667		430	136		566	690
0450	48" x 48"		6	2.667		700	136		836	985
0500	Glass Block Window 56" x 8"		6	2.667		272	136		408	515
0510	56" x 16"		4	4		405	204		609	770
0520	56" x 24"		4	4		500	204		704	875
0550	56" x 48"		4	4		785	204		989	1,200

08 88 52 – Prefabricated Glass Block Windows

08 88 52.20 Prefab. Glass Block Windows Hurricane Resistant	Crew	Daily Output	Labor-Hours	Unit	Material	2020 Bare Costs Labor	Equipment	Total	Total Incl O&P
0010 **PREFABRICATED GLASS BLOCK WINDOWS HURRICANE RESISTANT**									
0015 Includes frame and silicone seal									
3020 Glass Block Window 16" x 16"	2 Glaz	6	2.667	Ea.	430	136		566	690
3030 16" x 24"		6	2.667		515	136		651	785
3040 16" x 32"		4	4		590	204		794	975
3050 16" x 40"		4	4		645	204		849	1,025
3060 16" x 48"		4	4		680	204		884	1,075
3070 16" x 56"		4	4		845	204		1,049	1,250
3080 Glass Block Window 24" x 24"		6	2.667		600	136		736	875
3090 24" x 32"		4	4		680	204		884	1,075
3100 24" x 40"		4	4		835	204		1,039	1,250
3110 24" x 48"		4	4		890	204		1,094	1,300
3120 24" x 56"		4	4		920	204		1,124	1,325
3130 Glass Block Window 32" x 32"		4	4		760	204		964	1,175
3140 32" x 40"		4	4		565	204		769	945
3150 32" x 48"		4	4		1,000	204		1,204	1,425
3160 32" x 56"		4	4		865	204		1,069	1,275
3200 Glass Block Window 40" x 40"		4	4		1,000	204		1,204	1,425
3210 40" x 48"		3	5.333		1,225	272		1,497	1,750
3220 40" x 56"		3	5.333		1,350	272		1,622	1,925
3260 Glass Block Window 48" x 48"		3	5.333		1,425	272		1,697	1,975
3270 48" x 56"		3	5.333		1,575	272		1,847	2,150
3310 Glass Block Window 56" x 56"		3	5.333		1,750	272		2,022	2,350

08 88 52.30 Prefabricated Acrylic Block Casement Windows

	Crew	Daily Output	Labor-Hours	Unit	Material	2020 Bare Costs Labor	Equipment	Total	Total Incl O&P
0010 **PREFABRICATED ACRYLIC BLOCK CASEMENT WINDOWS**									
0015 Windows include frame and nail fin.									
3050 6" x 6" x 1-3/4" Block Casement Window 2 x 2 Pattern	2 Glaz	6	2.667	Ea.	450	136		586	710
3055 2 x 3		6	2.667		465	136		601	730
3060 2 x 4		6	2.667		510	136		646	775
3065 2 x 5		6	2.667		540	136		676	810
3070 2 x 6		6	2.667		595	136		731	870
3075 2 x 7		4	4		710	204		914	1,100
3080 2 x 8		4	4		740	204		944	1,150
3085 2 x 9		4	4		810	204		1,014	1,225
3090 3 x 3		4	4		490	204		694	865
3095 3 x 4		4	4		540	204		744	920
3100 3 x 5		4	4		610	204		814	995
3110 3 x 6		4	4		645	204		849	1,025
3115 3 x 7		4	4		740	204		944	1,150
3120 3 x 8		4	4		815	204		1,019	1,225
3125 3 x 9		3	5.333		880	272		1,152	1,400
3130 4 x 4		3	5.333		635	272		907	1,125
3135 4 x 5		3	5.333		675	272		947	1,175
3140 4 x 6		3	5.333		725	272		997	1,225
3145 4 x 7		3	5.333		765	272		1,037	1,275
3150 4 x 8		3	5.333		880	272		1,152	1,400
3500 8" x 8" x 1-3/4" Block Casement Window 2 x 2 Pattern		6	2.667		415	136		551	670
3510 2 x 3		6	2.667		460	136		596	720
3515 2 x 4		6	2.667		515	136		651	780
3520 2 x 5		6	2.667		595	136		731	870
3525 2 x 6		4	4		660	204		864	1,050
3530 2 x 7		4	4		695	204		899	1,100

For customer support on your Commercial Renovation Costs with RSMeans data, call 800.448.8182.

289

08 88 Special Function Glazing

08 88 52 – Prefabricated Glass Block Windows

08 88 52.30 Prefabricated Acrylic Block Casement Windows	Crew	Daily Output	Labor-Hours	Unit	Material	2020 Bare Costs Labor	Equipment	Total	Total Incl O&P
3535	2 x 8	2 Glaz	3	5.333	Ea.	755	272		1,027	1,275
3540	2 x 9		3	5.333		795	272		1,067	1,300
3545	2 x 10		3	5.333		800	272		1,072	1,325
3550	2 x 11		3	5.333		815	272		1,087	1,325
3555	3 x 3		4	4		515	204		719	890
3560	3 x 4		4	4		570	204		774	950
3565	3 x 5		4	4		685	204		889	1,075
3570	3 x 6		4	4		750	204		954	1,150
3575	3 x 7		4	4		855	204		1,059	1,275
3580	3 x 8		3	5.333		950	272		1,222	1,475
3585	3 x 9		3	5.333		875	272		1,147	1,400
3590	4 x 4		3	5.333		700	272		972	1,200
3595	4 x 5		3	5.333		855	272		1,127	1,375
3600	4 x 6		3	5.333		920	272		1,192	1,450

08 88 56 – Ballistics-Resistant Glazing

08 88 56.10 Laminated Glass

		Crew	Daily Output	Labor-Hours	Unit	Material	2020 Bare Costs Labor	Equipment	Total	Total Incl O&P
0010	**LAMINATED GLASS**									
0020	Clear float .03" vinyl 1/4" thick	2 Glaz	90	.178	S.F.	13.30	9.05		22.35	29
0100	3/8" thick		78	.205		31	10.45		41.45	50.50
0200	.06" vinyl, 1/2" thick		65	.246		28	12.55		40.55	51
1000	5/8" thick		90	.178		32.50	9.05		41.55	50.50
2000	Bullet-resisting, 1-3/16" thick, to 15 S.F.		16	1		126	51		177	220
2100	Over 15 S.F.		16	1		157	51		208	255
2200	2" thick, to 15 S.F.		10	1.600		152	81.50		233.50	297
2300	Over 15 S.F.		10	1.600		133	81.50		214.50	276
2500	2-1/4" thick, to 15 S.F.		12	1.333		216	68		284	345
2600	Over 15 S.F.		12	1.333		205	68		273	335
2700	Level 2 (.357 magnum), NIJ and UL		12	1.333		56	68		124	171
2750	Level 3A (.44 magnum) NIJ, UL 3		12	1.333		60	68		128	175
2800	Level 4 (AK-47) NIJ, UL 7 & 8		12	1.333		82.50	68		150.50	200
2850	Level 5 (M-16) UL		12	1.333		81.50	68		149.50	199
2900	Level 3 (7.62 Armor Piercing) NIJ, UL 4 & 5		12	1.333		96	68		164	215

08 91 Louvers

08 91 16 – Operable Wall Louvers

08 91 16.10 Movable Blade Louvers

		Crew	Daily Output	Labor-Hours	Unit	Material	2020 Bare Costs Labor	Equipment	Total	Total Incl O&P
0010	**MOVABLE BLADE LOUVERS**									
0100	PVC, commercial grade, 12" x 12"	1 Shee	20	.400	Ea.	82.50	25		107.50	130
0110	16" x 16"		14	.571		96	35.50		131.50	162
0120	18" x 18"		14	.571		98.50	35.50		134	165
0130	24" x 24"		14	.571		141	35.50		176.50	211
0140	30" x 30"		12	.667		170	41.50		211.50	251
0150	36" x 36"		12	.667		200	41.50		241.50	285
0300	Stainless steel, commercial grade, 12" x 12"		20	.400		230	25		255	292
0310	16" x 16"		14	.571		310	35.50		345.50	395
0320	18" x 18"		14	.571		310	35.50		345.50	395
0330	20" x 20"		14	.571		360	35.50		395.50	450
0340	24" x 24"		14	.571		425	35.50		460.50	525
0350	30" x 30"		12	.667		475	41.50		516.50	590
0360	36" x 36"		12	.667		580	41.50		621.50	700

08 91 Louvers

08 91 19 – Fixed Louvers

08 91 19.10 Aluminum Louvers

		Crew	Daily Output	Labor-Hours	Unit	Material	2020 Bare Costs Labor	Equipment	Total	Total Incl O&P
0010	**ALUMINUM LOUVERS**									
0020	Aluminum with screen, residential, 8" x 8"	1 Carp	38	.211	Ea.	21.50	11.20		32.70	41.50
0100	12" x 12"		38	.211		17.85	11.20		29.05	37.50
0200	12" x 18"		35	.229		22.50	12.15		34.65	44
0250	14" x 24"		30	.267		32.50	14.15		46.65	58
0300	18" x 24"		27	.296		33.50	15.75		49.25	62
0500	24" x 30"		24	.333		70.50	17.70		88.20	106
0700	Triangle, adjustable, small		20	.400		73.50	21.50		95	115
0800	Large		15	.533		94	28.50		122.50	149
2100	Midget, aluminum, 3/4" deep, 1" diameter		85	.094		.98	5		5.98	9.10
2150	3" diameter		60	.133		3.22	7.10		10.32	14.90
2200	4" diameter		50	.160		6	8.50		14.50	20
2250	6" diameter	▼	30	.267		3.87	14.15		18.02	27
3000	PVC, commercial grade, 12" x 12"	1 Shee	20	.400		97	25		122	146
3010	12" x 18"		20	.400		109	25		134	159
3020	12" x 24"		20	.400		147	25		172	200
3030	14" x 24"		20	.400		153	25		178	208
3100	Aluminum, commercial grade, 12" x 12"		20	.400		170	25		195	226
3110	24" x 24"		16	.500		234	31		265	305
3120	24" x 36"		16	.500		335	31		366	415
3130	36" x 24"		16	.500		345	31		376	430
3140	36" x 36"		14	.571		430	35.50		465.50	530
3150	36" x 48"		10	.800		465	50		515	590
3160	48" x 36"		10	.800		500	50		550	630
3170	48" x 48"		10	.800		455	50		505	580
3180	60" x 48"		8	1		515	62.50		577.50	670
3190	60" x 60"	▼	8	1	▼	750	62.50		812.50	925

08 91 19.20 Steel Louvers

		Crew	Daily Output	Labor-Hours	Unit	Material	2020 Bare Costs Labor	Equipment	Total	Total Incl O&P
0010	**STEEL LOUVERS**									
3300	Galvanized steel, fixed blades, commercial grade, 18" x 18"	1 Shee	20	.400	Ea.	163	25		188	218
3310	24" x 24"		20	.400		218	25		243	279
3320	24" x 36"		16	.500		284	31		315	360
3330	36" x 24"		16	.500		291	31		322	370
3340	36" x 36"		14	.571		355	35.50		390.50	450
3350	36" x 48"		10	.800		420	50		470	545
3360	48" x 36"		10	.800		410	50		460	530
3370	48" x 48"		10	.800		510	50		560	640
3380	60" x 48"		10	.800		560	50		610	695
3390	60" x 60"	▼	10	.800	▼	670	50		720	820

08 91 26 – Door Louvers

08 91 26.10 Steel Louvers, 18 Gauge, Fixed Blade

		Crew	Daily Output	Labor-Hours	Unit	Material	2020 Bare Costs Labor	Equipment	Total	Total Incl O&P
0010	**STEEL LOUVERS, 18 GAUGE, FIXED BLADE**									
0050	12" x 12", with powder coat	1 Carp	20	.400	Ea.	101	21.50		122.50	145
0055	18" x 12"		20	.400		106	21.50		127.50	151
0060	18" x 18"		20	.400		109	21.50		130.50	154
0065	24" x 12"		20	.400		123	21.50		144.50	169
0070	24" x 18"		20	.400		130	21.50		151.50	177
0075	24" x 24"		20	.400		164	21.50		185.50	215
0100	12" x 12", galvanized		20	.400		82	21.50		103.50	124
0105	18" x 12"		20	.400		94	21.50		115.50	137
0115	24" x 12"		20	.400		110	21.50		131.50	155

08 91 Louvers

08 91 26 – Door Louvers

08 91 26.10 Steel Louvers, 18 Gauge, Fixed Blade		Crew	Daily Output	Labor-Hours	Unit	Material	2020 Bare Costs Labor	Equipment	Total	Total Incl O&P
0125	24" x 24"	1 Carp	20	.400	Ea.	150	21.50		171.50	199
0300	6" x 12", painted		20	.400		45.50	21.50		67	84
0320	10" x 16"		20	.400		71	21.50		92.50	113
0340	12" x 22"		20	.400		108	21.50		129.50	152
0360	14" x 14"		20	.400		109	21.50		130.50	153
0400	18" x 18"		20	.400		85	21.50		106.50	128
0420	20" x 26"		20	.400		147	21.50		168.50	196
0440	22" x 22"		20	.400		139	21.50		160.50	187
0460	26" x 26"		20	.400		226	21.50		247.50	282
3000	Fire rated									
3010	12" x 12"	1 Carp	10	.800	Ea.	183	42.50		225.50	269
3050	18" x 18"		10	.800		268	42.50		310.50	365
3100	24" x 12"		10	.800		274	42.50		316.50	370
3120	24" x 18"		10	.800		300	42.50		342.50	400
3130	24" x 24"		10	.800		300	42.50		342.50	400

08 91 26.20 Aluminum Louvers

		Crew	Daily Output	Labor-Hours	Unit	Material	Labor	Equipment	Total	Total Incl O&P
0010	**ALUMINUM LOUVERS**, 18 ga., fixed blade, clear anodized									
0050	6" x 12"	1 Carp	20	.400	Ea.	65.50	21.50		87	106
0060	10" x 10"		20	.400		61.50	21.50		83	102
0065	10" x 14"		20	.400		70	21.50		91.50	111
0080	12" x 22"		20	.400		98	21.50		119.50	142
0090	14" x 14"		20	.400		88	21.50		109.50	131
0100	18" x 10"		20	.400		73	21.50		94.50	114
0110	18" x 18"		20	.400		131	21.50		152.50	178
0120	22" x 22"		20	.400		159	21.50		180.50	209
0130	26" x 26"		20	.400		203	21.50		224.50	258

08 95 Vents

08 95 13 – Soffit Vents

08 95 13.10 Wall Louvers

		Crew	Daily Output	Labor-Hours	Unit	Material	Labor	Equipment	Total	Total Incl O&P
0010	**WALL LOUVERS**									
2340	Baked enamel finish	1 Carp	200	.040	L.F.	4.81	2.13		6.94	8.70
2400	Under eaves vent, aluminum, mill finish, 16" x 4"		48	.167	Ea.	2.01	8.85		10.86	16.40
2500	16" x 8"		48	.167	"	2.85	8.85		11.70	17.35

08 95 16 – Wall Vents

08 95 16.10 Louvers

		Crew	Daily Output	Labor-Hours	Unit	Material	Labor	Equipment	Total	Total Incl O&P
0010	**LOUVERS**									
0020	Redwood, 2'-0" diameter, full circle	1 Carp	16	.500	Ea.	310	26.50		336.50	390
0100	Half circle		16	.500		210	26.50		236.50	274
0200	Octagonal		16	.500		214	26.50		240.50	278
0300	Triangular, 5/12 pitch, 5'-0" at base		16	.500		655	26.50		681.50	765
1000	Rectangular, 1'-4" x 1'-3"		16	.500		28.50	26.50		55	74
1100	Rectangular, 1'-4" x 1'-8"		16	.500		43	26.50		69.50	90
1200	1'-4" x 2'-2"		15	.533		50.50	28.50		79	102
1300	1'-9" x 2'-2"		15	.533		62	28.50		90.50	114
1400	2'-3" x 2'-2"		14	.571		75.50	30.50		106	132
1700	2'-4" x 2'-11"		13	.615		75.50	32.50		108	136
2000	Aluminum, 12" x 16"		25	.320		24.50	17		41.50	54.50
2010	16" x 20"		25	.320		35	17		52	66
2020	24" x 30"		25	.320		69	17		86	104

08 95 16.10 Louvers		Crew	Daily Output	Labor-Hours	Unit	Material	2020 Bare Costs Labor	Equipment	Total	Total Incl O&P
2100	6' triangle	1 Carp	12	.667	Ea.	193	35.50		228.50	270
3100	Round, 2'-2" diameter		16	.500		200	26.50		226.50	263
7000	Vinyl gable vent, 8" x 8"		38	.211		16.55	11.20		27.75	36
7020	12" x 12"		38	.211		32	11.20		43.20	53
7080	12" x 18"		35	.229		41	12.15		53.15	65
7200	18" x 24"		30	.267		58.50	14.15		72.65	87
9000	Minimum labor/equipment charge		3.50	2.286	Job		121		121	195

Division Notes

		CREW	DAILY OUTPUT	LABOR-HOURS	UNIT	BARE COSTS				TOTAL INCL O&P
						MAT.	LABOR	EQUIP.	TOTAL	

Estimating Tips

General

- Room Finish Schedule: A complete set of plans should contain a room finish schedule. If one is not available, it would be well worth the time and effort to obtain one.

09 20 00 Plaster and Gypsum Board

- Lath is estimated by the square yard plus a 5% allowance for waste. Furring, channels, and accessories are measured by the linear foot. An extra foot should be allowed for each accessory miter or stop.
- Plaster is also estimated by the square yard. Deductions for openings vary by preference, from zero deduction to 50% of all openings over 2 feet in width. The estimator should allow one extra square foot for each linear foot of horizontal interior or exterior angle located below the ceiling level. Also, double the areas of small radius work.
- Drywall accessories, studs, track, and acoustical caulking are all measured by the linear foot. Drywall taping is figured by the square foot. Gypsum wallboard is estimated by the square foot. No material deductions should be made for door or window openings under 32 S.F.

09 60 00 Flooring

- Tile and terrazzo areas are taken off on a square foot basis. Trim and base materials are measured by the linear foot. Accent tiles are listed per each. Two basic methods of installation are used. Mud set is approximately 30% more expensive than thin set.

The cost of grout is included with tile unit price lines unless otherwise noted. In terrazzo work, be sure to include the linear footage of embedded decorative strips, grounds, machine rubbing, and power cleanup.

- Wood flooring is available in strip, parquet, or block configuration. The latter two types are set in adhesives with quantities estimated by the square foot. The laying pattern will influence labor costs and material waste. In addition to the material and labor for laying wood floors, the estimator must make allowances for sanding and finishing these areas, unless the flooring is prefinished.
- Sheet flooring is measured by the square yard. Roll widths vary, so consideration should be given to use the most economical width, as waste must be figured into the total quantity. Consider also the installation methods available—direct glue down or stretched. Direct glue-down installation is assumed with sheet carpet unit price lines unless otherwise noted.

09 70 00 Wall Finishes

- Wall coverings are estimated by the square foot. The area to be covered is measured—length by height of the wall above the baseboards—to calculate the square footage of each wall. This figure is divided by the number of square feet in the single roll which is being used. Deduct, in full, the areas of openings such as doors and windows. Where a pattern match is required allow 25–30% waste.

09 80 00 Acoustic Treatment

- Acoustical systems fall into several categories. The takeoff of these materials should be by the square foot of area with a 5% allowance for waste. Do not forget about scaffolding, if applicable, when estimating these systems.

09 90 00 Painting and Coating

- New line items created for cut-ins with reference diagram.
- A major portion of the work in painting involves surface preparation. Be sure to include cleaning, sanding, filling, and masking costs in the estimate.
- Protection of adjacent surfaces is not included in painting costs. When considering the method of paint application, an important factor is the amount of protection and masking required. These must be estimated separately and may be the determining factor in choosing the method of application.

Reference Numbers

Reference numbers are shown at the beginning of some major classifications. These numbers refer to related items in the Reference Section. The reference information may be an estimating procedure, an alternate pricing method, or technical information.

Note: Not all subdivisions listed here necessarily appear. ■

Same Data. Simplified.

Enjoy the convenience and efficiency of accessing your costs anywhere:

- **Skip the multiplier** by setting your location
- **Quickly search**, edit, favorite and share costs
- **Stay on top of price changes** with automatic updates

Discover more at rsmeans.com/online

09 01 70 – Maintenance of Wall Finishes

09 01 70.10 Gypsum Wallboard Repairs		Crew	Daily Output	Labor-Hours	Unit	Material	2020 Bare Costs Labor	Equipment	Total	Total Incl O&P
0010	**GYPSUM WALLBOARD REPAIRS**									
0100	Fill and sand, pin/nail holes	1 Carp	960	.008	Ea.		.44		.44	.71
0110	Screw head pops		480	.017			.89		.89	1.42
0120	Dents, up to 2" square		48	.167		.01	8.85		8.86	14.20
0130	2" to 4" square		24	.333		.04	17.70		17.74	28.50
0140	Cut square, patch, sand and finish, holes, up to 2" square		12	.667		.04	35.50		35.54	57
0150	2" to 4" square		11	.727		.09	38.50		38.59	62
0160	4" to 8" square		10	.800		.24	42.50		42.74	68.50
0170	8" to 12" square		8	1		.47	53		53.47	85.50
0180	12" to 32" square		6	1.333		1.58	71		72.58	116
0210	16" by 48"		5	1.600		2.68	85		87.68	139
0220	32" by 48"		4	2		4.45	106		110.45	175
0230	48" square		3.50	2.286		6.30	121		127.30	202
0240	60" square		3.20	2.500		9.55	133		142.55	224
0500	Skim coat surface with joint compound		1600	.005	S.F.	.04	.27		.31	.47
0510	Prepare, retape and refinish joints		60	.133	L.F.	.70	7.10		7.80	12.10
9000	Minimum labor/equipment charge		2	4	Job		213		213	340

09 01 90 – Maintenance of Painting and Coating

09 01 90.92 Sanding

09 01 90.92 Sanding		Crew	Daily Output	Labor-Hours	Unit	Material	2020 Bare Costs Labor	Equipment	Total	Total Incl O&P
0010	**SANDING** and puttying interior trim, compared to									
0100	Painting 1 coat, on quality work				L.F.		100%			
0300	Medium work						50%			
0400	Industrial grade						25%			
0500	Surface protection, placement and removal									
0510	Surface protection, placement and removal, basic drop cloths	1 Pord	6400	.001	S.F.		.06		.06	.09
0520	Masking with paper		800	.010		.07	.44		.51	.79
0530	Volume cover up (using plastic sheathing or building paper)		16000	.001			.02		.02	.04

09 01 90.93 Exterior Surface Preparation

09 01 90.93 Exterior Surface Preparation		Crew	Daily Output	Labor-Hours	Unit	Material	2020 Bare Costs Labor	Equipment	Total	Total Incl O&P
0010	**EXTERIOR SURFACE PREPARATION**									
0015	Doors, per side, not incl. frames or trim									
0020	Scrape & sand									
0030	Wood, flush	1 Pord	616	.013	S.F.		.58		.58	.92
0040	Wood, detail		496	.016			.72		.72	1.14
0050	Wood, louvered		280	.029			1.27		1.27	2.01
0060	Wood, overhead		616	.013			.58		.58	.92
0070	Wire brush									
0080	Metal, flush	1 Pord	640	.013	S.F.		.56		.56	.88
0090	Metal, detail		520	.015			.68		.68	1.08
0100	Metal, louvered		360	.022			.99		.99	1.57
0110	Metal or fibr., overhead		640	.013			.56		.56	.88
0120	Metal, roll up		560	.014			.63		.63	1.01
0130	Metal, bulkhead		640	.013			.56		.56	.88
0140	Power wash, based on 2500 lb. operating pressure									
0150	Metal, flush	A-1H	2240	.004	S.F.		.15	.03	.18	.28
0160	Metal, detail		2120	.004			.16	.04	.20	.29
0170	Metal, louvered		2000	.004			.17	.04	.21	.31
0180	Metal or fibr., overhead		2400	.003			.14	.03	.17	.25
0190	Metal, roll up		2400	.003			.14	.03	.17	.25
0200	Metal, bulkhead		2200	.004			.15	.03	.18	.29
0400	Windows, per side, not incl. trim									
0410	Scrape & sand									
0420	Wood, 1-2 lite	1 Pord	320	.025	S.F.		1.11		1.11	1.76

09 01 90.93 Exterior Surface Preparation

	Crew	Daily Output	Labor-Hours	Unit	Material	2020 Bare Costs Labor	Equipment	Total	Total Incl O&P
0430 Wood, 3-6 lite	1 Pord	280	.029	S.F.		1.27		1.27	2.01
0440 Wood, 7-10 lite		240	.033			1.48		1.48	2.35
0450 Wood, 12 lite		200	.040			1.78		1.78	2.82
0460 Wood, Bay/Bow	▼	320	.025	▼		1.11		1.11	1.76
0470 Wire brush									
0480 Metal, 1-2 lite	1 Pord	480	.017	S.F.		.74		.74	1.18
0490 Metal, 3-6 lite		400	.020			.89		.89	1.41
0500 Metal, Bay/Bow	▼	480	.017	▼		.74		.74	1.18
0510 Power wash, based on 2500 lb. operating pressure									
0520 1-2 lite	A-1H	4400	.002	S.F.		.08	.02	.10	.14
0530 3-6 lite		4320	.002			.08	.02	.10	.14
0540 7-10 lite		4240	.002			.08	.02	.10	.15
0550 12 lite		4160	.002			.08	.02	.10	.15
0560 Bay/Bow	▼	4400	.002	▼		.08	.02	.10	.14
0600 Siding, scrape and sand, light=10-30%, med.=30-70%									
0610 Heavy=70-100% of surface to sand									
0650 Texture 1-11, light	1 Pord	480	.017	S.F.		.74		.74	1.18
0660 Med.		440	.018			.81		.81	1.28
0670 Heavy		360	.022			.99		.99	1.57
0680 Wood shingles, shakes, light		440	.018			.81		.81	1.28
0690 Med.		360	.022			.99		.99	1.57
0700 Heavy		280	.029			1.27		1.27	2.01
0710 Clapboard, light		520	.015			.68		.68	1.08
0720 Med.		480	.017			.74		.74	1.18
0730 Heavy	▼	400	.020	▼		.89		.89	1.41
0740 Wire brush									
0750 Aluminum, light	1 Pord	600	.013	S.F.		.59		.59	.94
0760 Med.		520	.015			.68		.68	1.08
0770 Heavy	▼	440	.018	▼		.81		.81	1.28
0780 Pressure wash, based on 2500 lb. operating pressure									
0790 Stucco	A-1H	3080	.003	S.F.		.11	.02	.13	.21
0800 Aluminum or vinyl		3200	.003			.11	.02	.13	.20
0810 Siding, masonry, brick & block	▼	2400	.003	▼		.14	.03	.17	.25
1300 Miscellaneous, wire brush									
1310 Metal, pedestrian gate	1 Pord	100	.080	S.F.		3.55		3.55	5.65
1320 Aluminum chain link, both sides		250	.032			1.42		1.42	2.26
1400 Existing galvanized surface, clean and prime, prep for painting	▼	380	.021	▼	.13	.93		1.06	1.62
8000 For chemical washing, see Section 04 01 30									
8010 For steam cleaning, see Section 04 01 30.20									

09 01 90.94 Interior Surface Preparation

	Crew	Daily Output	Labor-Hours	Unit	Material	2020 Bare Costs Labor	Equipment	Total	Total Incl O&P
0010 **INTERIOR SURFACE PREPARATION**									
0020 Doors, per side, not incl. frames or trim									
0030 Scrape & sand									
0040 Wood, flush	1 Pord	616	.013	S.F.		.58		.58	.92
0050 Wood, detail		496	.016			.72		.72	1.14
0060 Wood, louvered	▼	280	.029	▼		1.27		1.27	2.01
0070 Wire brush									
0080 Metal, flush	1 Pord	640	.013	S.F.		.56		.56	.88
0090 Metal, detail		520	.015			.68		.68	1.08
0100 Metal, louvered	▼	360	.022	▼		.99		.99	1.57
0110 Hand wash									
0120 Wood, flush	1 Pord	2160	.004	S.F.		.16		.16	.26

For customer support on your Commercial Renovation Costs with RSMeans data, call 800.448.8182.

297

09 01 Maintenance of Finishes

09 01 90 – Maintenance of Painting and Coating

09 01 90.94 Interior Surface Preparation

		Crew	Daily Output	Labor-Hours	Unit	Material	2020 Bare Costs Labor	Equipment	Total	Total Incl O&P
0130	Wood, detail	1 Pord	2000	.004	S.F.		.18		.18	.28
0140	Wood, louvered		1360	.006			.26		.26	.41
0150	Metal, flush		2160	.004			.16		.16	.26
0160	Metal, detail		2000	.004			.18		.18	.28
0170	Metal, louvered	↓	1360	.006	↓		.26		.26	.41
0400	Windows, per side, not incl. trim									
0410	Scrape & sand									
0420	Wood, 1-2 lite	1 Pord	360	.022	S.F.		.99		.99	1.57
0430	Wood, 3-6 lite		320	.025			1.11		1.11	1.76
0440	Wood, 7-10 lite		280	.029			1.27		1.27	2.01
0450	Wood, 12 lite		240	.033			1.48		1.48	2.35
0460	Wood, Bay/Bow	↓	360	.022	↓		.99		.99	1.57
0470	Wire brush									
0480	Metal, 1-2 lite	1 Pord	520	.015	S.F.		.68		.68	1.08
0490	Metal, 3-6 lite		440	.018			.81		.81	1.28
0500	Metal, Bay/Bow	↓	520	.015	↓		.68		.68	1.08
0600	Walls, sanding, light=10-30%, medium=30-70%,									
0610	heavy=70-100% of surface to sand									
0650	Walls, sand									
0660	Gypsum board or plaster, light	1 Pord	3077	.003	S.F.		.12		.12	.18
0670	Gypsum board or plaster, medium		2160	.004			.16		.16	.26
0680	Gypsum board or plaster, heavy		923	.009			.38		.38	.61
0690	Wood, T&G, light		2400	.003			.15		.15	.23
0700	Wood, T&G, medium		1600	.005			.22		.22	.35
0710	Wood, T&G, heavy	↓	800	.010	↓		.44		.44	.71
0720	Walls, wash									
0730	Gypsum board or plaster	1 Pord	3200	.003	S.F.		.11		.11	.18
0740	Wood, T&G		3200	.003			.11		.11	.18
0750	Masonry, brick & block, smooth		2800	.003			.13		.13	.20
0760	Masonry, brick & block, coarse	↓	2000	.004	↓		.18		.18	.28
8000	For chemical washing, see Section 04 01 30									
8010	For steam cleaning, see Section 04 01 30.20									
9010	Minimum labor/equipment charge	1 Pord	3	2.667	Job		118		118	188

09 01 90.95 Scrape After Fire Damage

		Crew	Daily Output	Labor-Hours	Unit	Material	2020 Bare Costs Labor	Equipment	Total	Total Incl O&P
0010	**SCRAPE AFTER FIRE DAMAGE**									
0050	Boards, 1" x 4"	1 Pord	336	.024	L.F.		1.06		1.06	1.68
0060	1" x 6"		260	.031			1.37		1.37	2.17
0070	1" x 8"		207	.039			1.72		1.72	2.72
0080	1" x 10"		174	.046			2.04		2.04	3.24
0500	Framing, 2" x 4"		265	.030			1.34		1.34	2.13
0510	2" x 6"		221	.036			1.61		1.61	2.55
0520	2" x 8"		190	.042			1.87		1.87	2.97
0530	2" x 10"		165	.048			2.15		2.15	3.42
0540	2" x 12"		144	.056			2.47		2.47	3.92
1000	Heavy framing, 3" x 4"		226	.035			1.57		1.57	2.50
1010	4" x 4"		210	.038			1.69		1.69	2.69
1020	4" x 6"		191	.042			1.86		1.86	2.95
1030	4" x 8"		165	.048			2.15		2.15	3.42
1040	4" x 10"		144	.056			2.47		2.47	3.92
1060	4" x 12"		131	.061	↓		2.71		2.71	4.31
2900	For sealing, light damage		825	.010	S.F.	.15	.43		.58	.85
2920	Heavy damage	↓	460	.017	"	.35	.77		1.12	1.62

09 01 Maintenance of Finishes

09 01 90 – Maintenance of Painting and Coating

09 01 90.95 Scrape After Fire Damage	Crew	Daily Output	Labor-Hours	Unit	Material	2020 Bare Costs Labor	2020 Bare Costs Equipment	Total	Total Incl O&P
3000	For sandblasting, see Section 04 01 30.20									
9000	Minimum labor/equipment charge	1 Pord	3	2.667	Job		118		118	188

09 05 Common Work Results for Finishes

09 05 05 – Selective Demolition for Finishes

09 05 05.10 Selective Demolition, Ceilings

		Crew	Daily Output	Labor-Hours	Unit	Material	2020 Bare Costs Labor	2020 Bare Costs Equipment	Total	Total Incl O&P
0010	**SELECTIVE DEMOLITION, CEILINGS** R024119-10									
0200	Ceiling, gypsum wall board, furred and nailed or screwed	2 Clab	800	.020	S.F.		.84		.84	1.35
0220	On metal frame		760	.021			.89		.89	1.42
0240	On suspension system, including system		720	.022			.94		.94	1.50
1000	Plaster, lime and horse hair, on wood lath, incl. lath		700	.023			.96		.96	1.54
1020	On metal lath		570	.028			1.18		1.18	1.89
1100	Gypsum, on gypsum lath		720	.022			.94		.94	1.50
1120	On metal lath		500	.032			1.35		1.35	2.16
1200	Suspended ceiling, mineral fiber, 2' x 2' or 2' x 4'		1500	.011			.45		.45	.72
1250	On suspension system, incl. system		1200	.013			.56		.56	.90
1500	Tile, wood fiber, 12" x 12", glued		900	.018			.75		.75	1.20
1540	Stapled		1500	.011			.45		.45	.72
1580	On suspension system, incl. system		760	.021			.89		.89	1.42
2000	Wood, tongue and groove, 1" x 4"		1000	.016			.67		.67	1.08
2040	1" x 8"		1100	.015			.61		.61	.98
2400	Plywood or wood fiberboard, 4' x 8' sheets		1200	.013			.56		.56	.90
2500	Remove & refinish textured ceiling	1 Plas	222	.036		.04	1.75		1.79	2.83
9000	Minimum labor/equipment charge	1 Clab	2	4	Job		168		168	270

09 05 05.20 Selective Demolition, Flooring

		Crew	Daily Output	Labor-Hours	Unit	Material	2020 Bare Costs Labor	2020 Bare Costs Equipment	Total	Total Incl O&P
0010	**SELECTIVE DEMOLITION, FLOORING** R024119-10									
0200	Brick with mortar	2 Clab	475	.034	S.F.		1.42		1.42	2.27
0400	Carpet, bonded, including surface scraping		2000	.008			.34		.34	.54
0480	Tackless		9000	.002			.07		.07	.12
0550	Carpet tile, releasable adhesive		5000	.003			.13		.13	.22
0560	Permanent adhesive		1850	.009			.36		.36	.58
0600	Composition, acrylic or epoxy		400	.040			1.68		1.68	2.70
0700	Concrete, scarify skin	A-1A	225	.036			1.95	.92	2.87	4.12
0800	Resilient, sheet goods	2 Clab	1400	.011			.48		.48	.77
0820	For gym floors	"	900	.018			.75		.75	1.20
0850	Vinyl or rubber cove base	1 Clab	1000	.008	L.F.		.34		.34	.54
0860	Vinyl or rubber cove base, molded corner	"	1000	.008	Ea.		.34		.34	.54
0870	For glued and caulked installation, add to labor						50%			
0900	Vinyl composition tile, 12" x 12"	2 Clab	1000	.016	S.F.		.67		.67	1.08
2000	Tile, ceramic, thin set		675	.024			1		1	1.60
2020	Mud set		625	.026			1.08		1.08	1.73
2200	Marble, slate, thin set		675	.024			1		1	1.60
2220	Mud set		625	.026			1.08		1.08	1.73
2600	Terrazzo, thin set		450	.036			1.50		1.50	2.40
2620	Mud set		425	.038			1.59		1.59	2.54
2640	Terrazzo, cast in place		300	.053			2.25		2.25	3.60
3000	Wood, block, on end	1 Carp	400	.020			1.06		1.06	1.70
3200	Parquet		450	.018			.95		.95	1.51
3400	Strip flooring, interior, 2-1/4" x 25/32" thick		325	.025			1.31		1.31	2.10
3500	Exterior, porch flooring, 1" x 4"		220	.036			1.93		1.93	3.10
3800	Subfloor, tongue and groove, 1" x 6"		325	.025			1.31		1.31	2.10

09 05 05 – Selective Demolition for Finishes

09 05 05.20 Selective Demolition, Flooring

		Crew	Daily Output	Labor-Hours	Unit	Material	2020 Bare Costs Labor	Equipment	Total	Total Incl O&P
3820	1" x 8"	1 Carp	430	.019	S.F.		.99		.99	1.58
3840	1" x 10"		520	.015			.82		.82	1.31
4000	Plywood, nailed		600	.013			.71		.71	1.14
4100	Glued and nailed		400	.020			1.06		1.06	1.70
4200	Hardboard, 1/4" thick		760	.011			.56		.56	.90
8000	Remove flooring, bead blast, simple floor plan	A-1A	1000	.008		.05	.44	.21	.70	.99
8100	Complex floor plan		400	.020		.05	1.10	.52	1.67	2.38
8150	Mastic only		1500	.005		.05	.29	.14	.48	.68
9000	Minimum labor/equipment charge	1 Clab	4	2	Job		84		84	135
9050	For grinding concrete floors, see Section 03 35 43.10									

09 05 05.30 Selective Demolition, Walls and Partitions

		Crew	Daily Output	Labor-Hours	Unit	Material	2020 Bare Costs Labor	Equipment	Total	Total Incl O&P
0010	SELECTIVE DEMOLITION, WALLS AND PARTITIONS R024119-10									
0020	Walls, concrete, reinforced	B-39	120	.400	C.F.		17.70	2.98	20.68	31.50
0025	Plain	"	160	.300			13.30	2.23	15.53	23.50
0100	Brick, 4" to 12" thick	B-9	220	.182			7.75	1.62	9.37	14.20
0200	Concrete block, 4" thick		1150	.035	S.F.		1.48	.31	1.79	2.71
0280	8" thick		1050	.038			1.62	.34	1.96	2.96
0300	Exterior stucco 1" thick over mesh		3200	.013			.53	.11	.64	.97
1000	Gypsum wallboard, nailed or screwed	1 Clab	1000	.008			.34		.34	.54
1010	2 layers		400	.020			.84		.84	1.35
1020	Glued and nailed		900	.009			.37		.37	.60
1500	Fiberboard, nailed		900	.009			.37		.37	.60
1520	Glued and nailed		800	.010			.42		.42	.67
1568	Plenum barrier, sheet lead		300	.027			1.12		1.12	1.80
1600	Glass block		65	.123			5.20		5.20	8.30
2000	Movable walls, metal, 5' high		300	.027			1.12		1.12	1.80
2020	8' high		400	.020			.84		.84	1.35
2200	Metal or wood studs, finish 2 sides, fiberboard	B-1	520	.046			1.97		1.97	3.16
2250	Lath and plaster		260	.092			3.95		3.95	6.35
2300	Gypsum wallboard		520	.046			1.97		1.97	3.16
2350	Plywood		450	.053			2.28		2.28	3.65
2800	Paneling, 4' x 8' sheets	1 Clab	475	.017			.71		.71	1.14
3000	Plaster, lime and horsehair, on wood lath		400	.020			.84		.84	1.35
3020	On metal lath		335	.024			1.01		1.01	1.61
3400	Gypsum or perlite, on gypsum lath		410	.020			.82		.82	1.32
3420	On metal lath		300	.027			1.12		1.12	1.80
3450	Plaster, interior gypsum, acoustic, or cement		60	.133	S.Y.		5.60		5.60	9
3500	Stucco, on masonry		145	.055			2.32		2.32	3.72
3510	Commercial 3-coat		80	.100			4.21		4.21	6.75
3520	Interior stucco		25	.320			13.45		13.45	21.50
3753	Remove damaged glass block	B-1	134.62	.178	S.F.		7.65		7.65	12.20
3760	Tile, ceramic, on walls, thin set	1 Clab	300	.027			1.12		1.12	1.80
3765	Mud set		250	.032			1.35		1.35	2.16
3800	Toilet partitions, slate or marble		5	1.600	Ea.		67.50		67.50	108
3820	Metal or plastic		8	1	"		42		42	67.50
5000	Wallcovering, vinyl	1 Pape	700	.011	S.F.		.51		.51	.81
5010	With release agent		1500	.005			.24		.24	.38
5025	Wallpaper, 2 layers or less, by hand		250	.032			1.43		1.43	2.26
5035	3 layers or more		165	.048			2.16		2.16	3.43
9000	Minimum labor/equipment charge	1 Clab	4	2	Job		84		84	135

09 05 Common Work Results for Finishes

09 05 71 – Acoustic Underlayment

09 05 71.10 Acoustical Underlayment	Crew	Daily Output	Labor-Hours	Unit	Material	2020 Bare Costs Labor	Equipment	Total	Total Incl O&P
0010 **ACOUSTICAL UNDERLAYMENT**									
0100 Rubber underlayment, 5/64"	1 Tilf	275	.029	S.F.	.32	1.44		1.76	2.61
0110 1/8" thk.		275	.029		.52	1.44		1.96	2.83
0120 13/64" thk.		270	.030		.71	1.47		2.18	3.08
0130 15/64" thk.		270	.030		.94	1.47		2.41	3.33
0140 23/64" thk.		265	.030		1.29	1.49		2.78	3.76
0150 15/32" thk.		265	.030		1.50	1.49		2.99	3.99
0400 Cork underlayment, 3/32"		275	.029		.40	1.44		1.84	2.70
0410 1/8" thk.		275	.029		.41	1.44		1.85	2.71
0420 5/32" thk.		275	.029		.48	1.44		1.92	2.79
0430 15/64" thk.		270	.030		.62	1.47		2.09	2.98
0440 15/32" thk.		265	.030		1.30	1.49		2.79	3.77
0600 Rubber cork underlayment, 3/64"		275	.029		.31	1.44		1.75	2.60
0610 13/64" thk.		270	.030		1.60	1.47		3.07	4.06
0620 3/8" thk.		270	.030		2.70	1.47		4.17	5.25
0630 25/64" thk.		270	.030		2.79	1.47		4.26	5.35
0800 Foam underlayment, 5/64"		275	.029		.55	1.44		1.99	2.87
0810 1/8" thk.		275	.029		.67	1.44		2.11	3
0820 15/32" thk.	▼	265	.030	▼	.54	1.49		2.03	2.93
4000 Nylon matting 0.4" thick, with carbon black spinerette									
4010 plus polyester fabric, on floor	D-7	1600	.010	S.F.	1.44	.44		1.88	2.27

09 21 Plaster and Gypsum Board Assemblies

09 21 16 – Gypsum Board Assemblies

09 21 16.23 Gypsum Board Shaft Wall Assemblies

	Crew	Daily Output	Labor-Hours	Unit	Material	2020 Bare Costs Labor	Equipment	Total	Total Incl O&P
0010 **GYPSUM BOARD SHAFT WALL ASSEMBLIES**									
0020 Cavity type on 25 ga. J-track & C-H studs, 24" OC									
0030 1" thick coreboard wall liner on shaft side									
0040 2-hour assembly with double layer									
0060 5/8" f.r. gyp bd on rm side, 2-1/2" J-track & C-H studs	2 Carp	220	.073	S.F.	2.99	3.87		6.86	9.50
0065 4" J-track & C-H studs		220	.073		3.15	3.87		7.02	9.65
0070 6" J-track & C-H studs	▼	220	.073	▼	3.36	3.87		7.23	9.90
0100 3-hour assembly with triple layer									
0300 5/8" f.r. gyp bd on rm side, 2-1/2" J-track & C-H studs	2 Carp	180	.089	S.F.	2.66	4.72		7.38	10.45
0305 4" J-track & C-H studs		180	.089		2.82	4.72		7.54	10.65
0310 6" J-track & C-H studs	▼	180	.089	▼	3.03	4.72		7.75	10.90
0400 4-hour assembly, 1" coreboard, 5/8" fire rated gypsum board									
0600 and 3/4" galv. metal furring channels, 24" OC, with									
0700 Dbl. layer 5/8" f.r. gyp bd on rm side, 2-1/2" trk. & C-H studs	2 Carp	110	.145	S.F.	3.36	7.75		11.11	16.10
0705 4" J-track & C-H studs		110	.145		3.15	7.75		10.90	15.85
0710 6" J-track & C-H studs	▼	110	.145		3.36	7.75		11.11	16.10
0900 For taping & finishing, add per side	1 Carp	1050	.008	▼	.05	.41		.46	.71
1000 For insulation, see Section 07 21									
5200 For work over 8' high, add	2 Carp	3060	.005	S.F.		.28		.28	.45
5300 For distribution cost 3 stories and above, add per story	"	6100	.003	"		.14		.14	.22

09 21 16.33 Partition Wall

	Crew	Daily Output	Labor-Hours	Unit	Material	2020 Bare Costs Labor	Equipment	Total	Total Incl O&P
0010 **PARTITION WALL** Stud wall, 8' to 12' high									
0050 1/2", interior, gypsum board, std, tape & finish 2 sides									
0500 Installed on and incl., 2" x 4" wood studs, 16" OC	2 Carp	310	.052	S.F.	1.29	2.74		4.03	5.80
1000 Metal studs, NLB, 25 ga., 16" OC, 3-5/8" wide	▼	350	.046	▼	1.17	2.43		3.60	5.20

For customer support on your Commercial Renovation Costs with RSMeans data, call 800.448.8182.

301

09 21 16 – Gypsum Board Assemblies

09 21 16.33 Partition Wall	Crew	Daily Output	Labor-Hours	Unit	Material	2020 Bare Costs Labor	Equipment	Total	Total Incl O&P
1200	6" wide	2 Carp	330	.048	S.F.	1.30	2.58		3.88	5.55
1400	Water resistant, on 2" x 4" wood studs, 16" OC		310	.052		1.45	2.74		4.19	6
1600	Metal studs, NLB, 25 ga., 16" OC, 3-5/8" wide		350	.046		1.33	2.43		3.76	5.35
1800	6" wide		330	.048		1.46	2.58		4.04	5.75
2000	Fire res., 2 layers, 1-1/2 hr., on 2" x 4" wood studs, 16" OC		210	.076		2.15	4.05		6.20	8.85
2200	Metal studs, NLB, 25 ga., 16" OC, 3-5/8" wide		250	.064		2.03	3.40		5.43	7.70
2400	6" wide		230	.070		2.16	3.70		5.86	8.30
2600	Fire & water res., 2 layers, 1-1/2 hr., 2" x 4" studs, 16" OC		210	.076		2.15	4.05		6.20	8.85
2800	Metal studs, NLB, 25 ga., 16" OC, 3-5/8" wide		250	.064		2.03	3.40		5.43	7.70
3000	6" wide	↓	230	.070	↓	2.16	3.70		5.86	8.30
3200	5/8", interior, gypsum board, standard, tape & finish 2 sides									
3400	Installed on and including 2" x 4" wood studs, 16" OC	2 Carp	300	.053	S.F.	1.29	2.83		4.12	5.95
3600	24" OC		330	.048		1.17	2.58		3.75	5.40
3800	Metal studs, NLB, 25 ga., 16" OC, 3-5/8" wide		340	.047		1.17	2.50		3.67	5.30
4000	6" wide		320	.050		1.30	2.66		3.96	5.70
4200	24" OC, 3-5/8" wide		360	.044		1.06	2.36		3.42	4.95
4400	6" wide		340	.047		1.16	2.50		3.66	5.30
4800	Water resistant, on 2" x 4" wood studs, 16" OC		300	.053		1.51	2.83		4.34	6.20
5000	24" OC		330	.048		1.39	2.58		3.97	5.65
5200	Metal studs, NLB, 25 ga. 16" OC, 3-5/8" wide		340	.047		1.39	2.50		3.89	5.55
5400	6" wide		320	.050		1.52	2.66		4.18	5.95
5600	24" OC, 3-5/8" wide		360	.044		1.28	2.36		3.64	5.20
5800	6" wide		340	.047		1.38	2.50		3.88	5.55
6000	Fire resistant, 2 layers, 2 hr., on 2" x 4" wood studs, 16" OC		205	.078		2.15	4.15		6.30	9
6200	24" OC		235	.068		2.03	3.62		5.65	8.05
6400	Metal studs, NLB, 25 ga., 16" OC, 3-5/8" wide		245	.065		2.06	3.47		5.53	7.80
6600	6" wide		225	.071		2.16	3.78		5.94	8.45
6800	24" OC, 3-5/8" wide		265	.060		1.92	3.21		5.13	7.25
7000	6" wide		245	.065		2.02	3.47		5.49	7.75
7200	Fire & water resistant, 2 layers, 2 hr., 2" x 4" studs, 16" OC		205	.078		2.15	4.15		6.30	9
7400	24" OC		235	.068		2.03	3.62		5.65	8.05
7600	Metal studs, NLB, 25 ga., 16" OC, 3-5/8" wide		245	.065		2.03	3.47		5.50	7.80
7800	6" wide		225	.071		2.16	3.78		5.94	8.45
8000	24" OC, 3-5/8" wide		265	.060		1.92	3.21		5.13	7.25
8200	6" wide	↓	245	.065	↓	2.02	3.47		5.49	7.75
8600	1/2" blueboard, mesh tape both sides									
8620	Installed on and including 2" x 4" wood studs, 16" OC	2 Carp	300	.053	S.F.	1.45	2.83		4.28	6.15
8640	Metal studs, NLB, 25 ga., 16" OC, 3-5/8" wide		340	.047		1.33	2.50		3.83	5.50
8660	6" wide	↓	320	.050	↓	1.46	2.66		4.12	5.85
8800	Hospital security partition, 5/8" fiber reinf. high abuse gyp. bd.									
8810	Mtl. studs, NLB, 20 ga., 16" OC, 3-5/8" wide, w/sec. mesh, gyp. bd.	2 Carp	208	.077	S.F.	4.67	4.09		8.76	11.70
9000	Exterior, 1/2" gypsum sheathing, 1/2" gypsum finished, interior,									
9100	including foil faced insulation, metal studs, 20 ga.									
9200	16" OC, 3-5/8" wide	2 Carp	290	.055	S.F.	1.90	2.93		4.83	6.80
9400	6" wide		270	.059		2.15	3.15		5.30	7.40
9600	Partitions, for work over 8' high, add	↓	1530	.010	↓		.56		.56	.89

09 22 Supports for Plaster and Gypsum Board

09 22 03 – Fastening Methods for Finishes

09 22 03.20 Drilling Plaster/Drywall	Crew	Daily Output	Labor-Hours	Unit	Material	2020 Bare Costs Labor	Equipment	Total	Total Incl O&P
0010 **DRILLING PLASTER/DRYWALL**									
1100 Drilling & layout for drywall/plaster walls, up to 1" deep, no anchor									
1200 Holes, 1/4" diameter	1 Carp	150	.053	Ea.	.01	2.83		2.84	4.55
1300 3/8" diameter		140	.057		.01	3.04		3.05	4.88
1400 1/2" diameter		130	.062		.01	3.27		3.28	5.25
1500 3/4" diameter		120	.067		.01	3.54		3.55	5.70
1600 1" diameter		110	.073		.02	3.87		3.89	6.20
1700 1-1/4" diameter		100	.080		.04	4.25		4.29	6.85
1800 1-1/2" diameter		90	.089		.05	4.72		4.77	7.60
1900 For ceiling installations, add						40%			

09 22 13 – Metal Furring

09 22 13.13 Metal Channel Furring

09 22 13.13 Metal Channel Furring	Crew	Daily Output	Labor-Hours	Unit	Material	2020 Bare Costs Labor	Equipment	Total	Total Incl O&P
0010 **METAL CHANNEL FURRING**									
0030 Beams and columns, 7/8" hat channels, galvanized, 12" OC	1 Lath	155	.052	S.F.	.46	2.69		3.15	4.71
0050 16" OC		170	.047		.37	2.45		2.82	4.25
0070 24" OC		185	.043		.25	2.25		2.50	3.79
0100 Ceilings, on steel, 7/8" hat channels, galvanized, 12" OC		210	.038		.42	1.98		2.40	3.57
0300 16" OC		290	.028		.37	1.44		1.81	2.66
0400 24" OC		420	.019		.25	.99		1.24	1.82
0600 1-5/8" hat channels, galvanized, 12" OC		190	.042		.56	2.19		2.75	4.05
0700 16" OC		260	.031		.50	1.60		2.10	3.06
0900 24" OC		390	.021		.34	1.07		1.41	2.04
0930 7/8" hat channels with sound isolation clips, 12" OC		120	.067		1.73	3.47		5.20	7.35
0940 16" OC		100	.080		1.30	4.16		5.46	7.95
0950 24" OC		165	.048		.87	2.52		3.39	4.90
0960 1-5/8" hat channels, galvanized, 12" OC		110	.073		1.88	3.79		5.67	8
0970 16" OC		100	.080		1.41	4.16		5.57	8.05
0980 24" OC		155	.052		.94	2.69		3.63	5.25
1000 Walls, 7/8" hat channels, galvanized, 12" OC		235	.034		.42	1.77		2.19	3.23
1200 16" OC		265	.030		.37	1.57		1.94	2.87
1300 24" OC		350	.023		.25	1.19		1.44	2.13
1500 1-5/8" hat channels, galvanized, 12" OC		210	.038		.56	1.98		2.54	3.73
1600 16" OC		240	.033		.50	1.73		2.23	3.27
1800 24" OC		305	.026		.34	1.37		1.71	2.51
1920 7/8" hat channels with sound isolation clips, 12" OC		125	.064		1.73	3.33		5.06	7.10
1940 16" OC		100	.080		1.30	4.16		5.46	7.95
1950 24" OC		150	.053		.87	2.78		3.65	5.30
1960 1-5/8" hat channels, galvanized, 12" OC		115	.070		1.88	3.62		5.50	7.70
1970 16" OC		95	.084		1.41	4.38		5.79	8.40
1980 24" OC		140	.057		.94	2.97		3.91	5.70
3000 Z Furring, walls, 1" deep, 25 ga., 24" OC		350	.023		1.37	1.19		2.56	3.36
3010 48" OC		700	.011		.68	.59		1.27	1.68
3020 1-1/2" deep, 24" OC		345	.023		1.59	1.21		2.80	3.64
3030 48" OC		695	.012		.80	.60		1.40	1.81
3040 2" deep, 24" OC		340	.024		1.91	1.22		3.13	4.02
3050 48" OC		690	.012		.96	.60		1.56	1.99
3060 1" deep, 20 ga., 24" OC		350	.023		2.26	1.19		3.45	4.34
3070 48" OC		700	.011		1.13	.59		1.72	2.17
3080 1-1/2" deep, 24" OC		345	.023		2.60	1.21		3.81	4.74
3090 48" OC		695	.012		1.30	.60		1.90	2.37
4000 2" deep, 24" OC		340	.024		3.19	1.22		4.41	5.40
4010 48" OC		690	.012		1.59	.60		2.19	2.69

For customer support on your Commercial Renovation Costs with RSMeans data, call 800.448.8182.

303

09 22 Supports for Plaster and Gypsum Board

09 22 13 – Metal Furring

09 22 13.13 Metal Channel Furring	Crew	Daily Output	Labor-Hours	Unit	Material	2020 Bare Costs Labor	2020 Bare Costs Equipment	Total	Total Incl O&P
9000	Minimum labor/equipment charge	1 Lath	4	2	Job		104		104	163

09 22 16 – Non-Structural Metal Framing

09 22 16.13 Non-Structural Metal Stud Framing

		Crew	Daily Output	Labor-Hours	Unit	Material	2020 Bare Costs Labor	2020 Bare Costs Equipment	Total	Total Incl O&P
0010	**NON-STRUCTURAL METAL STUD FRAMING**									
1600	Non-load bearing, galv., 8' high, 25 ga. 1-5/8" wide, 16" OC	1 Carp	619	.013	S.F.	.28	.69		.97	1.41
1610	24" OC		950	.008		.21	.45		.66	.95
1620	2-1/2" wide, 16" OC		613	.013		.36	.69		1.05	1.51
1630	24" OC		938	.009		.27	.45		.72	1.03
1640	3-5/8" wide, 16" OC		600	.013		.43	.71		1.14	1.61
1650	24" OC		925	.009		.32	.46		.78	1.09
1660	4" wide, 16" OC		594	.013		.48	.72		1.20	1.68
1670	24" OC		925	.009		.36	.46		.82	1.14
1680	6" wide, 16" OC		588	.014		.57	.72		1.29	1.79
1690	24" OC		906	.009		.43	.47		.90	1.22
1700	20 ga. studs, 1-5/8" wide, 16" OC		494	.016		.38	.86		1.24	1.80
1710	24" OC		763	.010		.28	.56		.84	1.20
1720	2-1/2" wide, 16" OC		488	.016		.48	.87		1.35	1.92
1730	24" OC		750	.011		.36	.57		.93	1.31
1740	3-5/8" wide, 16" OC		481	.017		.55	.88		1.43	2.02
1750	24" OC		738	.011		.41	.58		.99	1.37
1760	4" wide, 16" OC		475	.017		.67	.90		1.57	2.17
1770	24" OC		738	.011		.50	.58		1.08	1.47
1780	6" wide, 16" OC		469	.017		.81	.91		1.72	2.34
1790	24" OC		725	.011		.61	.59		1.20	1.61
2000	Non-load bearing, galv., 10' high, 25 ga. 1-5/8" wide, 16" OC		495	.016		.26	.86		1.12	1.67
2100	24" OC		760	.011		.19	.56		.75	1.11
2200	2-1/2" wide, 16" OC		490	.016		.35	.87		1.22	1.77
2250	24" OC		750	.011		.25	.57		.82	1.19
2300	3-5/8" wide, 16" OC		480	.017		.41	.89		1.30	1.87
2350	24" OC		740	.011		.30	.57		.87	1.25
2400	4" wide, 16" OC		475	.017		.46	.90		1.36	1.93
2450	24" OC		740	.011		.34	.57		.91	1.29
2500	6" wide, 16" OC		470	.017		.54	.90		1.44	2.05
2550	24" OC		725	.011		.40	.59		.99	1.38
2600	20 ga. studs, 1-5/8" wide, 16" OC		395	.020		.36	1.08		1.44	2.11
2650	24" OC		610	.013		.26	.70		.96	1.41
2700	2-1/2" wide, 16" OC		390	.021		.45	1.09		1.54	2.25
2750	24" OC		600	.013		.33	.71		1.04	1.50
2800	3-5/8" wide, 16" OC		385	.021		.52	1.10		1.62	2.34
2850	24" OC		590	.014		.38	.72		1.10	1.57
2900	4" wide, 16" OC		380	.021		.63	1.12		1.75	2.49
2950	24" OC		590	.014		.47	.72		1.19	1.66
3000	6" wide, 16" OC		375	.021		.77	1.13		1.90	2.66
3050	24" OC		580	.014		.56	.73		1.29	1.79
3060	Non-load bearing, galv., 12' high, 25 ga. 1-5/8" wide, 16" OC		413	.019		.25	1.03		1.28	1.92
3070	24" OC		633	.013		.18	.67		.85	1.28
3080	2-1/2" wide, 16" OC		408	.020		.33	1.04		1.37	2.03
3090	24" OC		625	.013		.24	.68		.92	1.35
3100	3-5/8" wide, 16" OC		400	.020		.39	1.06		1.45	2.13
3110	24" OC		617	.013		.28	.69		.97	1.41
3120	4" wide, 16" OC		396	.020		.44	1.07		1.51	2.20
3130	24" OC		617	.013		.32	.69		1.01	1.45

09 22 16 – Non-Structural Metal Framing

	09 22 16.13 Non-Structural Metal Stud Framing	Crew	Daily Output	Labor-Hours	Unit	Material	2020 Bare Costs Labor	Equipment	Total	Total Incl O&P
3140	6" wide, 16" OC	1 Carp	392	.020	S.F.	.52	1.08		1.60	2.31
3150	24" OC		604	.013		.38	.70		1.08	1.54
3160	20 ga. studs, 1-5/8" wide, 16" OC		329	.024		.34	1.29		1.63	2.45
3170	24" OC		508	.016		.25	.84		1.09	1.61
3180	2-1/2" wide, 16" OC		325	.025		.43	1.31		1.74	2.57
3190	24" OC		500	.016		.31	.85		1.16	1.71
3200	3-5/8" wide, 16" OC		321	.025		.50	1.32		1.82	2.67
3210	24" OC		492	.016		.36	.86		1.22	1.78
3220	4" wide, 16" OC		317	.025		.61	1.34		1.95	2.82
3230	24" OC		492	.016		.44	.86		1.30	1.87
3240	6" wide, 16" OC		313	.026		.73	1.36		2.09	2.99
3250	24" OC		483	.017		.53	.88		1.41	2
3260	Non-load bearing, galv., 16' high, 25 ga. 4" wide, 12" OC		195	.041		.56	2.18		2.74	4.11
3270	16" OC		275	.029		.44	1.55		1.99	2.97
3280	24" OC		400	.020		.32	1.06		1.38	2.05
3290	6" wide, 12" OC		190	.042		.67	2.24		2.91	4.32
3300	16" OC		280	.029		.52	1.52		2.04	3.01
3310	24" OC		400	.020		.38	1.06		1.44	2.12
3320	20 ga. studs, 2-1/2" wide, 12" OC		180	.044		.55	2.36		2.91	4.39
3330	16" OC		254	.032		.43	1.67		2.10	3.16
3340	24" OC		390	.021		.32	1.09		1.41	2.10
3350	3-5/8" wide, 12" OC		170	.047		.65	2.50		3.15	4.72
3360	16" OC		251	.032		.51	1.69		2.20	3.27
3370	24" OC		384	.021		.37	1.11		1.48	2.18
3380	4" wide, 12" OC		170	.047		.78	2.50		3.28	4.86
3390	16" OC		247	.032		.61	1.72		2.33	3.43
3400	24" OC		384	.021		.45	1.11		1.56	2.26
3410	6" wide, 12" OC		175	.046		.94	2.43		3.37	4.92
3420	16" OC		245	.033		.74	1.74		2.48	3.59
3430	24" OC		400	.020		.54	1.06		1.60	2.29
3440	Non-load bearing, galv., 20' high, 25 ga. 6" wide, 12" OC		125	.064		.65	3.40		4.05	6.15
3450	16" OC		220	.036		.51	1.93		2.44	3.66
3460	24" OC		360	.022		.37	1.18		1.55	2.30
3470	20 ga. studs, 4" wide, 12" OC		120	.067		.76	3.54		4.30	6.55
3480	16" OC		215	.037		.61	1.98		2.59	3.84
3490	6" wide, 12" OC		115	.070		.92	3.70		4.62	6.90
3500	16" OC		215	.037		.72	1.98		2.70	3.96
3510	24" OC	↓	331	.024	↓	.52	1.28		1.80	2.63
5000	For load bearing studs, see Section 05 41 13.30									
9000	Minimum labor/equipment charge	1 Carp	4	2	Job		106		106	170

09 22 26 – Suspension Systems

09 22 26.13 Ceiling Suspension Systems

		Crew	Daily Output	Labor-Hours	Unit	Material	2020 Bare Costs Labor	Equipment	Total	Total Incl O&P
0010	**CEILING SUSPENSION SYSTEMS** for gypsum board or plaster									
8000	Suspended ceilings, including carriers									
8200	1-1/2" carriers, 24" OC with:									
8300	7/8" channels, 16" OC	1 Lath	275	.029	S.F.	.59	1.51		2.10	3.02
8320	24" OC		310	.026		.47	1.34		1.81	2.62
8400	1-5/8" channels, 16" OC		205	.039		.72	2.03		2.75	3.98
8420	24" OC	↓	250	.032	↓	.56	1.67		2.23	3.22
8600	2" carriers, 24" OC with:									
8700	7/8" channels, 16" OC	1 Lath	250	.032	S.F.	.69	1.67		2.36	3.37
8720	24" OC	↓	285	.028	↓	.57	1.46		2.03	2.92

09 22 26 – Suspension Systems

09 22 26.13 Ceiling Suspension Systems	Crew	Daily Output	Labor-Hours	Unit	Material	2020 Bare Costs Labor	Equipment	Total	Total Incl O&P
8800	1-5/8" channels, 16" OC	1 Lath	190	.042	S.F.	.82	2.19		3.01	4.34
8820	24" OC		225	.036		.66	1.85		2.51	3.62

09 22 36 – Lath

09 22 36.13 Gypsum Lath

	09 22 36.13 Gypsum Lath	Crew	Daily Output	Labor-Hours	Unit	Material	Labor	Equipment	Total	Total Incl O&P
0010	**GYPSUM LATH** R092000-50									
0020	Plain or perforated, nailed, 3/8" thick	1 Lath	85	.094	S.Y.	3.51	4.90		8.41	11.50
0100	1/2" thick		80	.100		2.70	5.20		7.90	11.10
0300	Clipped to steel studs, 3/8" thick		75	.107		3.51	5.55		9.06	12.55
0400	1/2" thick		70	.114		2.70	5.95		8.65	12.25
1500	For ceiling installations, add		216	.037			1.93		1.93	3.02
1600	For columns and beams, add		170	.047			2.45		2.45	3.84
9000	Minimum labor/equipment charge		4.25	1.882	Job		98		98	153

09 22 36.23 Metal Lath

	09 22 36.23 Metal Lath	Crew	Daily Output	Labor-Hours	Unit	Material	Labor	Equipment	Total	Total Incl O&P
0010	**METAL LATH** R092000-50									
0020	Diamond, expanded, 2.5 lb./S.Y., painted				S.Y.	4.04			4.04	4.44
0100	Galvanized					3.80			3.80	4.18
0300	3.4 lb./S.Y., painted					4.23			4.23	4.65
0400	Galvanized					4.72			4.72	5.20
0600	For #15 asphalt sheathing paper, add					.47			.47	.52
0900	Flat rib, 1/8" high, 2.75 lb., painted					3.51			3.51	3.86
1000	Foil backed					3.69			3.69	4.06
1200	3.4 lb./S.Y., painted					4.29			4.29	4.72
1300	Galvanized					4.60			4.60	5.05
1500	For #15 asphalt sheathing paper, add					.47			.47	.52
1800	High rib, 3/8" high, 3.4 lb./S.Y., painted					4.83			4.83	5.30
1900	Galvanized					4.01			4.01	4.41
2400	3/4" high, painted, .60 lb./S.F.				S.F.	.77			.77	.85
2500	.75 lb./S.F.				"	1.61			1.61	1.77
2800	Stucco mesh, painted, 3.6 lb.				S.Y.	4.17			4.17	4.59
3000	K-lath, perforated, absorbent paper, regular					4.36			4.36	4.80
3100	Heavy duty					5.15			5.15	5.65
3300	Waterproof, heavy duty, grade B backing					5.05			5.05	5.55
3400	Fire resistant backing					5.55			5.55	6.15
3600	2.5 lb. diamond painted, on wood framing, on walls	1 Lath	85	.094		4.04	4.90		8.94	12.10
3700	On ceilings		75	.107		4.04	5.55		9.59	13.15
3900	3.4 lb. diamond painted, on wood framing, on walls		80	.100		4.29	5.20		9.49	12.85
4000	On ceilings		70	.114		4.29	5.95		10.24	14
4200	3.4 lb. diamond painted, wired to steel framing		75	.107		4.29	5.55		9.84	13.40
4300	On ceilings		60	.133		4.29	6.95		11.24	15.55
4600	Cornices, wired to steel		35	.229		4.29	11.90		16.19	23.50
4800	Screwed to steel studs, 2.5 lb.		80	.100		4.04	5.20		9.24	12.60
4900	3.4 lb.		75	.107		4.23	5.55		9.78	13.35
5100	Rib lath, painted, wired to steel, on walls, 2.5 lb.		75	.107		3.51	5.55		9.06	12.55
5200	3.4 lb.		70	.114		4.83	5.95		10.78	14.60
5400	4.0 lb.		65	.123		5.70	6.40		12.10	16.30
5500	For self-furring lath, add					.12			.12	.13
5700	Suspended ceiling system, incl. 3.4 lb. diamond lath, painted	1 Lath	15	.533		4.30	28		32.30	48
5800	Galvanized	"	15	.533		4.64	28		32.64	48.50
6000	Hollow metal stud partitions, 3.4 lb. painted lath both sides									
6010	Non-load bearing, 25 ga., w/rib lath, 2-1/2" studs, 12" OC	1 Lath	20.30	.394	S.Y.	13.60	20.50		34.10	47
6300	16" OC		21.10	.379		12.75	19.75		32.50	45
6350	24" OC		22.70	.352		11.95	18.35		30.30	41.50

09 22 Supports for Plaster and Gypsum Board

09 22 36 – Lath

09 22 36.23 Metal Lath

		Crew	Daily Output	Labor-Hours	Unit	Material	2020 Bare Costs Labor	Equipment	Total	Total Incl O&P
6400	3-5/8" studs, 16" OC	1 Lath	19.50	.410	S.Y.	13.35	21.50		34.85	48
6600	24" OC		20.40	.392		12.35	20.50		32.85	45.50
6700	4" studs, 16" OC		20.40	.392		13.80	20.50		34.30	47
6900	24" OC		21.60	.370		12.70	19.30		32	44
7000	6" studs, 16" OC		19.50	.410		14.55	21.50		36.05	49.50
7100	24" OC		21.10	.379		13.25	19.75		33	45.50
7200	L.B. partitions, 16 ga., w/rib lath, 2-1/2" studs, 16" OC		20	.400		13.05	21		34.05	47
7300	3-5/8" studs, 16 ga.		19.70	.406		14.70	21		35.70	49
7500	4" studs, 16 ga.		19.50	.410		15.50	21.50		37	50.50
7600	6" studs, 16 ga.		18.70	.428		18.10	22.50		40.60	55
9000	Minimum labor/equipment charge	▼	4.25	1.882	Job		98		98	153

09 23 Gypsum Plastering

09 23 13 – Acoustical Gypsum Plastering

09 23 13.10 Perlite or Vermiculite Plaster

		Crew	Daily Output	Labor-Hours	Unit	Material	2020 Bare Costs Labor	Equipment	Total	Total Incl O&P
0010	PERLITE OR VERMICULITE PLASTER	R092000-50								
0020	In 100 lb. bags, under 200 bags				Bag	19.10			19.10	21
0300	2 coats, no lath included, on walls	J-1	92	.435	S.Y.	5.95	20	1.27	27.22	40
0400	On ceilings		79	.506		5.95	23.50	1.48	30.93	45
0900	3 coats, no lath included, on walls		74	.541		6.65	25	1.58	33.23	48.50
1000	On ceilings	▼	63	.635		6.65	29.50	1.86	38.01	56
1700	For irregular or curved surfaces, add to above						30%			
1800	For columns and beams, add to above						50%			
1900	For soffits, add to ceiling prices				▼		40%			
9000	Minimum labor/equipment charge	1 Plas	1	8	Job		390		390	620

09 23 20 – Gypsum Plaster

09 23 20.10 Gypsum Plaster On Walls and Ceilings

		Crew	Daily Output	Labor-Hours	Unit	Material	2020 Bare Costs Labor	Equipment	Total	Total Incl O&P
0010	GYPSUM PLASTER ON WALLS AND CEILINGS	R092000-50								
0020	80# bag, less than 1 ton				Bag	15.90			15.90	17.50
0300	2 coats, no lath included, on walls	J-1	105	.381	S.Y.	3.90	17.55	1.12	22.57	33.50
0400	On ceilings		92	.435		3.90	20	1.27	25.17	37.50
0900	3 coats, no lath included, on walls		87	.460		5.55	21	1.35	27.90	41.50
1000	On ceilings	▼	78	.513	▼	5.55	23.50	1.50	30.55	45.50
1600	For irregular or curved surfaces, add						30%			
1800	For columns & beams, add						50%			
9000	Minimum labor/equipment charge	1 Plas	1	8	Job		390		390	620

09 24 Cement Plastering

09 24 23 – Cement Stucco

09 24 23.40 Stucco

		Crew	Daily Output	Labor-Hours	Unit	Material	2020 Bare Costs Labor	Equipment	Total	Total Incl O&P
0010	STUCCO	R092000-50								
0015	3 coats 7/8" thick, float finish, with mesh, on wood frame	J-2	63	.762	S.Y.	7.65	36	1.86	45.51	67.50
0100	On masonry construction, no mesh incl.	J-1	67	.597		3.02	27.50	1.75	32.27	49.50
0300	For trowel finish, add	1 Plas	170	.047			2.29		2.29	3.64
0600	For coloring, add	J-1	685	.058		.43	2.69	.17	3.29	4.95
0700	For special texture, add	"	200	.200		1.50	9.20	.59	11.29	17
0900	For soffits, add	J-2	155	.310		2.29	14.60	.76	17.65	26.50
1000	Stucco, with bonding agent, 3 coats, on walls, no mesh incl.	J-1	200	.200		4.23	9.20	.59	14.02	20
1200	Ceilings	▼	180	.222		3.83	10.25	.65	14.73	21

09 24 Cement Plastering

09 24 23 – Cement Stucco

09 24 23.40 Stucco

		Crew	Daily Output	Labor-Hours	Unit	Material	2020 Bare Costs Labor	Equipment	Total	Total Incl O&P
1300	Beams	J-1	80	.500	S.Y.	3.83	23	1.47	28.30	42.50
1500	Columns	↓	100	.400	↓	3.83	18.45	1.17	23.45	35
1550	Minimum labor/equipment charge	1 Plas	1	8	Job		390		390	620
1600	Mesh, galvanized, nailed to wood, 1.8 lb.	1 Lath	60	.133	S.Y.	7.50	6.95		14.45	19.10
1800	3.6 lb.		55	.145		4.17	7.55		11.72	16.45
1900	Wired to steel, galvanized, 1.8 lb.		53	.151		7.50	7.85		15.35	20.50
2100	3.6 lb.		50	.160	↓	4.17	8.35		12.52	17.65
9000	Minimum labor/equipment charge	↓	4	2	Job		104		104	163

09 25 Other Plastering

09 25 23 – Lime Based Plastering

09 25 23.10 Venetian Plaster

		Crew	Daily Output	Labor-Hours	Unit	Material	2020 Bare Costs Labor	Equipment	Total	Total Incl O&P
0010	**VENETIAN PLASTER**									
0100	Walls, 1 coat primer, roller applied	1 Plas	950	.008	S.F.	.17	.41		.58	.84
0200	Plaster, 3 coats, incl. sanding	2 Plas	700	.023		.41	1.11		1.52	2.22
0210	For pigment, light colors add per S.F. plaster					.02			.02	.02
0220	For pigment, dark colors add per S.F. plaster					.04			.04	.04
0300	For sealer/wax coat incl. burnishing, add	1 Plas	300	.027	↓	.38	1.30		1.68	2.48

09 26 Veneer Plastering

09 26 13 – Gypsum Veneer Plastering

09 26 13.20 Blueboard

		Crew	Daily Output	Labor-Hours	Unit	Material	2020 Bare Costs Labor	Equipment	Total	Total Incl O&P
0010	**BLUEBOARD** For use with thin coat									
0100	plaster application see Section 09 26 13.80									
1000	3/8" thick, on walls or ceilings, standard, no finish included	2 Carp	1900	.008	S.F.	.33	.45		.78	1.08
1100	With thin coat plaster finish		875	.018		.43	.97		1.40	2.04
1400	On beams, columns, or soffits, standard, no finish included		675	.024		.38	1.26		1.64	2.44
1450	With thin coat plaster finish		475	.034		.48	1.79		2.27	3.40
3000	1/2" thick, on walls or ceilings, standard, no finish included		1900	.008		.33	.45		.78	1.08
3100	With thin coat plaster finish		875	.018		.43	.97		1.40	2.04
3300	Fire resistant, no finish included		1900	.008		.33	.45		.78	1.08
3400	With thin coat plaster finish		875	.018		.43	.97		1.40	2.04
3450	On beams, columns, or soffits, standard, no finish included		675	.024		.38	1.26		1.64	2.44
3500	With thin coat plaster finish		475	.034		.48	1.79		2.27	3.40
3700	Fire resistant, no finish included		675	.024		.38	1.26		1.64	2.44
3800	With thin coat plaster finish		475	.034		.48	1.79		2.27	3.40
5000	5/8" thick, on walls or ceilings, fire resistant, no finish included		1900	.008		.35	.45		.80	1.11
5100	With thin coat plaster finish		875	.018		.45	.97		1.42	2.06
5500	On beams, columns, or soffits, no finish included		675	.024		.40	1.26		1.66	2.46
5600	With thin coat plaster finish		475	.034		.51	1.79		2.30	3.43
6000	For high ceilings, over 8' high, add		3060	.005			.28		.28	.45
6500	For distribution costs 3 stories and above, add per story	↓	6100	.003	↓		.14		.14	.22
9000	Minimum labor/equipment charge	1 Carp	2	4	Job		213		213	340

09 26 13.80 Thin Coat Plaster

		Crew	Daily Output	Labor-Hours	Unit	Material	2020 Bare Costs Labor	Equipment	Total	Total Incl O&P
0010	**THIN COAT PLASTER** R092000-50									
0012	1 coat veneer, not incl. lath	J-1	3600	.011	S.F.	.10	.51	.03	.64	.97
1000	In 50 lb. bags				Bag	14			14	15.40

09 28 Backing Boards and Underlayments

09 28 13 – Cementitious Backing Boards

09 28 13.10 Cementitious Backerboard	Crew	Daily Output	Labor-Hours	Unit	Material	2020 Bare Costs Labor	Equipment	Total	Total Incl O&P
0010 **CEMENTITIOUS BACKERBOARD**									
0070 Cementitious backerboard, on floor, 3' x 4' x 1/2" sheets	2 Carp	525	.030	S.F.	.92	1.62		2.54	3.61
0080 3' x 5' x 1/2" sheets		525	.030		.80	1.62		2.42	3.47
0090 3' x 6' x 1/2" sheets		525	.030		.78	1.62		2.40	3.46
0100 3' x 4' x 5/8" sheets		525	.030		.98	1.62		2.60	3.68
0110 3' x 5' x 5/8" sheets		525	.030		1	1.62		2.62	3.70
0120 3' x 6' x 5/8" sheets		525	.030		.98	1.62		2.60	3.68
0150 On wall, 3' x 4' x 1/2" sheets		350	.046		.92	2.43		3.35	4.90
0160 3' x 5' x 1/2" sheets		350	.046		.80	2.43		3.23	4.76
0170 3' x 6' x 1/2" sheets		350	.046		.78	2.43		3.21	4.75
0180 3' x 4' x 5/8" sheets		350	.046		.98	2.43		3.41	4.97
0190 3' x 5' x 5/8" sheets		350	.046		1	2.43		3.43	4.99
0200 3' x 6' x 5/8" sheets		350	.046		.98	2.43		3.41	4.97
0250 On counter, 3' x 4' x 1/2" sheets		180	.089		.92	4.72		5.64	8.55
0260 3' x 5' x 1/2" sheets		180	.089		.80	4.72		5.52	8.40
0270 3' x 6' x 1/2" sheets		180	.089		.78	4.72		5.50	8.40
0300 3' x 4' x 5/8" sheets		180	.089		.98	4.72		5.70	8.65
0310 3' x 5' x 5/8" sheets		180	.089		1	4.72		5.72	8.65
0320 3' x 6' x 5/8" sheets		180	.089		.98	4.72		5.70	8.65

09 29 Gypsum Board

09 29 10 – Gypsum Board Panels

09 29 10.30 Gypsum Board	Crew	Daily Output	Labor-Hours	Unit	Material	2020 Bare Costs Labor	Equipment	Total	Total Incl O&P
0010 **GYPSUM BOARD** on walls & ceilings R092910-10									
0100 Nailed or screwed to studs unless otherwise noted									
0110 1/4" thick, on walls or ceilings, standard, no finish included	2 Carp	1330	.012	S.F.	.38	.64		1.02	1.44
0115 1/4" thick, on walls or ceilings, flexible, no finish included		1050	.015		.54	.81		1.35	1.89
0117 1/4" thick, on columns or soffits, flexible, no finish included		1050	.015		.54	.81		1.35	1.89
0130 1/4" thick, standard, no finish included, less than 800 S.F.		510	.031		.38	1.67		2.05	3.09
0150 3/8" thick, on walls, standard, no finish included		2000	.008		.35	.43		.78	1.07
0200 On ceilings, standard, no finish included		1800	.009		.35	.47		.82	1.15
0250 On beams, columns, or soffits, no finish included		675	.024		.35	1.26		1.61	2.41
0300 1/2" thick, on walls, standard, no finish included		2000	.008		.33	.43		.76	1.04
0350 Taped and finished (level 4 finish)		965	.017		.38	.88		1.26	1.83
0390 With compound skim coat (level 5 finish)		775	.021		.44	1.10		1.54	2.24
0400 Fire resistant, no finish included		2000	.008		.38	.43		.81	1.10
0450 Taped and finished (level 4 finish)		965	.017		.43	.88		1.31	1.88
0490 With compound skim coat (level 5 finish)		775	.021		.49	1.10		1.59	2.30
0500 Water resistant, no finish included		2000	.008		.41	.43		.84	1.13
0550 Taped and finished (level 4 finish)		965	.017		.46	.88		1.34	1.92
0590 With compound skim coat (level 5 finish)		775	.021		.52	1.10		1.62	2.33
0600 Prefinished, vinyl, clipped to studs		900	.018		.52	.95		1.47	2.08
0700 Mold resistant, no finish included		2000	.008		.44	.43		.87	1.16
0710 Taped and finished (level 4 finish)		965	.017		.49	.88		1.37	1.95
0720 With compound skim coat (level 5 finish)		775	.021		.55	1.10		1.65	2.36
1000 On ceilings, standard, no finish included		1800	.009		.33	.47		.80	1.12
1050 Taped and finished (level 4 finish)		765	.021		.38	1.11		1.49	2.20
1090 With compound skim coat (level 5 finish)		610	.026		.44	1.39		1.83	2.71
1100 Fire resistant, no finish included		1800	.009		.38	.47		.85	1.18
1150 Taped and finished (level 4 finish)		765	.021		.43	1.11		1.54	2.25
1195 With compound skim coat (level 5 finish)		610	.026		.49	1.39		1.88	2.77

09 29 10 – Gypsum Board Panels

09 29 10.30 Gypsum Board		Crew	Daily Output	Labor-Hours	Unit	Material	2020 Bare Costs Labor	Equipment	Total	Total Incl O&P
1200	Water resistant, no finish included	2 Carp	1800	.009	S.F.	.41	.47		.88	1.21
1250	Taped and finished (level 4 finish)		765	.021		.46	1.11		1.57	2.29
1290	With compound skim coat (level 5 finish)		610	.026		.52	1.39		1.91	2.80
1310	Mold resistant, no finish included		1800	.009		.44	.47		.91	1.24
1320	Taped and finished (level 4 finish)		765	.021		.49	1.11		1.60	2.32
1330	With compound skim coat (level 5 finish)		610	.026		.55	1.39		1.94	2.83
1350	Sag resistant, no finish included		1600	.010		.33	.53		.86	1.21
1360	Taped and finished (level 4 finish)		765	.021		.38	1.11		1.49	2.20
1370	With compound skim coat (level 5 finish)		610	.026		.44	1.39		1.83	2.71
1500	On beams, columns, or soffits, standard, no finish included		675	.024		.38	1.26		1.64	2.44
1550	Taped and finished (level 4 finish)		540	.030		.38	1.57		1.95	2.94
1590	With compound skim coat (level 5 finish)		475	.034		.44	1.79		2.23	3.35
1600	Fire resistant, no finish included		675	.024		.38	1.26		1.64	2.44
1650	Taped and finished (level 4 finish)		540	.030		.43	1.57		2	2.99
1690	With compound skim coat (level 5 finish)		475	.034		.49	1.79		2.28	3.41
1700	Water resistant, no finish included		675	.024		.47	1.26		1.73	2.54
1750	Taped and finished (level 4 finish)		540	.030		.46	1.57		2.03	3.03
1790	With compound skim coat (level 5 finish)		475	.034		.52	1.79		2.31	3.44
1800	Mold resistant, no finish included		675	.024		.51	1.26		1.77	2.58
1810	Taped and finished (level 4 finish)		540	.030		.49	1.57		2.06	3.06
1820	With compound skim coat (level 5 finish)		475	.034		.55	1.79		2.34	3.47
1850	Sag resistant, no finish included		675	.024		.38	1.26		1.64	2.44
1860	Taped and finished (level 4 finish)		540	.030		.38	1.57		1.95	2.94
1870	With compound skim coat (level 5 finish)		475	.034		.44	1.79		2.23	3.35
2000	5/8" thick, on walls, standard, no finish included		2000	.008		.33	.43		.76	1.04
2050	Taped and finished (level 4 finish)		965	.017		.38	.88		1.26	1.83
2090	With compound skim coat (level 5 finish)		775	.021		.44	1.10		1.54	2.24
2100	Fire resistant, no finish included		2000	.008		.38	.43		.81	1.10
2150	Taped and finished (level 4 finish)		965	.017		.43	.88		1.31	1.88
2195	With compound skim coat (level 5 finish)		775	.021		.49	1.10		1.59	2.30
2200	Water resistant, no finish included		2000	.008		.44	.43		.87	1.16
2250	Taped and finished (level 4 finish)		965	.017		.49	.88		1.37	1.95
2290	With compound skim coat (level 5 finish)		775	.021		.55	1.10		1.65	2.36
2300	Prefinished, vinyl, clipped to studs		900	.018		.95	.95		1.90	2.56
2510	Mold resistant, no finish included		2000	.008		.54	.43		.97	1.27
2520	Taped and finished (level 4 finish)		965	.017		.59	.88		1.47	2.06
2530	With compound skim coat (level 5 finish)		775	.021		.65	1.10		1.75	2.47
3000	On ceilings, standard, no finish included		1800	.009		.33	.47		.80	1.12
3050	Taped and finished (level 4 finish)		765	.021		.38	1.11		1.49	2.20
3090	With compound skim coat (level 5 finish)		615	.026		.44	1.38		1.82	2.70
3100	Fire resistant, no finish included		1800	.009		.38	.47		.85	1.18
3150	Taped and finished (level 4 finish)		765	.021		.43	1.11		1.54	2.25
3190	With compound skim coat (level 5 finish)		615	.026		.49	1.38		1.87	2.76
3200	Water resistant, no finish included		1800	.009		.44	.47		.91	1.24
3250	Taped and finished (level 4 finish)		765	.021		.49	1.11		1.60	2.32
3290	With compound skim coat (level 5 finish)		615	.026		.55	1.38		1.93	2.82
3300	Mold resistant, no finish included		1800	.009		.54	.47		1.01	1.35
3310	Taped and finished (level 4 finish)		765	.021		.59	1.11		1.70	2.43
3320	With compound skim coat (level 5 finish)		615	.026		.65	1.38		2.03	2.93
3500	On beams, columns, or soffits, no finish included		675	.024		.38	1.26		1.64	2.44
3550	Taped and finished (level 4 finish)		475	.034		.44	1.79		2.23	3.35
3590	With compound skim coat (level 5 finish)		380	.042		.51	2.24		2.75	4.15
3600	Fire resistant, no finish included		675	.024		.44	1.26		1.70	2.50

09 29 10 – Gypsum Board Panels

	09 29 10.30 Gypsum Board	Crew	Daily Output	Labor-Hours	Unit	Material	2020 Bare Costs Labor	2020 Bare Costs Equipment	Total	Total Incl O&P
3650	Taped and finished (level 4 finish)	2 Carp	475	.034	S.F.	.50	1.79		2.29	3.42
3690	With compound skim coat (level 5 finish)		380	.042		.49	2.24		2.73	4.13
3700	Water resistant, no finish included		675	.024		.51	1.26		1.77	2.58
3750	Taped and finished (level 4 finish)		475	.034		.55	1.79		2.34	3.47
3790	With compound skim coat (level 5 finish)		380	.042		.56	2.24		2.80	4.21
3800	Mold resistant, no finish included		675	.024		.62	1.26		1.88	2.70
3810	Taped and finished (level 4 finish)		475	.034		.65	1.79		2.44	3.58
3820	With compound skim coat (level 5 finish)		380	.042		.68	2.24		2.92	4.34
4000	Fireproofing, beams or columns, 2 layers, 1/2" thick, incl finish		330	.048		.86	2.58		3.44	5.10
4010	Mold resistant		330	.048		.98	2.58		3.56	5.20
4050	5/8" thick		300	.053		.86	2.83		3.69	5.50
4060	Mold resistant		300	.053		1.18	2.83		4.01	5.85
4100	3 layers, 1/2" thick		225	.071		1.29	3.78		5.07	7.45
4110	Mold resistant		225	.071		1.47	3.78		5.25	7.65
4150	5/8" thick		210	.076		1.29	4.05		5.34	7.90
4160	Mold resistant		210	.076		1.77	4.05		5.82	8.45
5050	For 1" thick coreboard on columns		480	.033		.89	1.77		2.66	3.82
5100	For foil-backed board, add					.21			.21	.23
5200	For work over 8' high, add	2 Carp	3060	.005			.28		.28	.45
5270	For textured spray, add	2 Lath	1600	.010		.04	.52		.56	.86
5300	For distribution cost 3 stories and above, add per story	2 Carp	6100	.003			.14		.14	.22
5350	For finishing inner corners, add		950	.017	L.F.	.11	.90		1.01	1.55
5355	For finishing outer corners, add		1250	.013		.24	.68		.92	1.36
5500	For acoustical sealant, add per bead	1 Carp	500	.016		.04	.85		.89	1.41
5550	Sealant, 1 quart tube				Ea.	7.15			7.15	7.90
6000	Gypsum sound dampening panels									
6010	1/2" thick on walls, multi-layer, lightweight, no finish included	2 Carp	1500	.011	S.F.	2.25	.57		2.82	3.39
6015	Taped and finished (level 4 finish)		725	.022		2.30	1.17		3.47	4.41
6020	With compound skim coat (level 5 finish)		580	.028		2.36	1.47		3.83	4.95
6025	5/8" thick on walls, for wood studs, no finish included		1500	.011		2.46	.57		3.03	3.62
6030	Taped and finished (level 4 finish)		725	.022		2.51	1.17		3.68	4.64
6035	With compound skim coat (level 5 finish)		580	.028		2.57	1.47		4.04	5.20
6040	For metal stud, no finish included		1500	.011		2.22	.57		2.79	3.35
6045	Taped and finished (level 4 finish)		725	.022		2.27	1.17		3.44	4.38
6050	With compound skim coat (level 5 finish)		580	.028		2.33	1.47		3.80	4.91
6055	Abuse resist, no finish included		1500	.011		4.23	.57		4.80	5.55
6060	Taped and finished (level 4 finish)		725	.022		4.28	1.17		5.45	6.60
6065	With compound skim coat (level 5 finish)		580	.028		4.34	1.47		5.81	7.10
6070	Shear rated, no finish included		1500	.011		5.60	.57		6.17	7.10
6075	Taped and finished (level 4 finish)		725	.022		5.65	1.17		6.82	8.15
6080	With compound skim coat (level 5 finish)		580	.028		5.75	1.47		7.22	8.65
6085	For SCIF applications, no finish included		1500	.011		5.30	.57		5.87	6.75
6090	Taped and finished (level 4 finish)		725	.022		5.35	1.17		6.52	7.80
6095	With compound skim coat (level 5 finish)		580	.028		5.40	1.47		6.87	8.30
6100	1-3/8" thick on walls, THX certified, no finish included		1500	.011		9.35	.57		9.92	11.20
6105	Taped and finished (level 4 finish)		725	.022		9.40	1.17		10.57	12.25
6110	With compound skim coat (level 5 finish)		580	.028		9.50	1.47		10.97	12.80
6115	5/8" thick on walls, score & snap installation, no finish included		2000	.008		1.99	.43		2.42	2.87
6120	Taped and finished (level 4 finish)		965	.017		2.04	.88		2.92	3.66
6125	With compound skim coat (level 5 finish)		775	.021		2.10	1.10		3.20	4.07
7020	5/8" thick on ceilings, for wood joists, no finish included		1200	.013		2.46	.71		3.17	3.85
7025	Taped and finished (level 4 finish)		510	.031		2.51	1.67		4.18	5.45
7030	With compound skim coat (level 5 finish)		410	.039		2.57	2.07		4.64	6.15

09 29 Gypsum Board

09 29 10 – Gypsum Board Panels

09 29 10.30 Gypsum Board

		Crew	Daily Output	Labor-Hours	Unit	Material	2020 Bare Costs Labor	2020 Bare Costs Equipment	Total	Total Incl O&P
7035	For metal joists, no finish included	2 Carp	1200	.013	S.F.	2.22	.71		2.93	3.58
7040	Taped and finished (level 4 finish)		510	.031		2.27	1.67		3.94	5.15
7045	With compound skim coat (level 5 finish)		410	.039		2.33	2.07		4.40	5.90
7050	Abuse resist, no finish included		1200	.013		4.23	.71		4.94	5.80
7055	Taped and finished (level 4 finish)		510	.031		4.28	1.67		5.95	7.40
7060	With compound skim coat (level 5 finish)		410	.039		4.34	2.07		6.41	8.10
7065	Shear rated, no finish included		1200	.013		5.60	.71		6.31	7.35
7070	Taped and finished (level 4 finish)		510	.031		5.65	1.67		7.32	8.90
7075	With compound skim coat (level 5 finish)		410	.039		5.75	2.07		7.82	9.60
7080	For SCIF applications, no finish included		1200	.013		5.30	.71		6.01	7
7085	Taped and finished (level 4 finish)		510	.031		5.35	1.67		7.02	8.55
7090	With compound skim coat (level 5 finish)		410	.039		5.40	2.07		7.47	9.25
8010	5/8" thick on ceilings, score & snap installation, no finish included		1600	.010		1.99	.53		2.52	3.04
8015	Taped and finished (level 4 finish)		680	.024		2.04	1.25		3.29	4.25
8020	With compound skim coat (level 5 finish)		545	.029		2.10	1.56		3.66	4.81
9000	Minimum labor/equipment charge	1 Carp	2	4	Job		213		213	340

09 29 10.50 High Abuse Gypsum Board

		Crew	Daily Output	Labor-Hours	Unit	Material	2020 Bare Costs Labor	2020 Bare Costs Equipment	Total	Total Incl O&P
0010	**HIGH ABUSE GYPSUM BOARD**, fiber reinforced, nailed or									
0100	screwed to studs unless otherwise noted									
0110	1/2" thick, on walls, no finish included	2 Carp	1800	.009	S.F.	.82	.47		1.29	1.66
0120	Taped and finished (level 4 finish)		870	.018		.87	.98		1.85	2.53
0130	With compound skim coat (level 5 finish)		700	.023		.93	1.22		2.15	2.97
0150	On ceilings, no finish included		1620	.010		.82	.53		1.35	1.74
0160	Taped and finished (level 4 finish)		690	.023		.87	1.23		2.10	2.93
0170	With compound skim coat (level 5 finish)		550	.029		.93	1.55		2.48	3.50
0210	5/8" thick, on walls, no finish included		1800	.009		.87	.47		1.34	1.72
0220	Taped and finished (level 4 finish)		870	.018		.92	.98		1.90	2.58
0230	With compound skim coat (level 5 finish)		700	.023		.98	1.22		2.20	3.03
0250	On ceilings, no finish included		1620	.010		.87	.53		1.40	1.80
0260	Taped and finished (level 4 finish)		690	.023		.92	1.23		2.15	2.98
0270	With compound skim coat (level 5 finish)		550	.029		.98	1.55		2.53	3.56
0272	5/8" thick, on roof sheathing for 1-hour rating		1620	.010		.87	.53		1.40	1.80
0310	5/8" thick, on walls, very high impact, no finish included		1800	.009		.95	.47		1.42	1.81
0320	Taped and finished (level 4 finish)		870	.018		1	.98		1.98	2.67
0330	With compound skim coat (level 5 finish)		700	.023		1.06	1.22		2.28	3.12
0350	On ceilings, no finish included		1620	.010		.95	.53		1.48	1.89
0360	Taped and finished (level 4 finish)		690	.023		1	1.23		2.23	3.07
0370	With compound skim coat (level 5 finish)		550	.029		1.06	1.55		2.61	3.65
0400	High abuse, gypsum core, paper face									
0410	1/2" thick, on walls, no finish included	2 Carp	1800	.009	S.F.	.74	.47		1.21	1.57
0420	Taped and finished (level 4 finish)		870	.018		.79	.98		1.77	2.44
0430	With compound skim coat (level 5 finish)		700	.023		.85	1.22		2.07	2.88
0450	On ceilings, no finish included		1620	.010		.74	.53		1.27	1.65
0460	Taped and finished (level 4 finish)		690	.023		.79	1.23		2.02	2.84
0470	With compound skim coat (level 5 finish)		550	.029		.85	1.55		2.40	3.41
0510	5/8" thick, on walls, no finish included		1800	.009		.82	.47		1.29	1.66
0520	Taped and finished (level 4 finish)		870	.018		.87	.98		1.85	2.53
0530	With compound skim coat (level 5 finish)		700	.023		.93	1.22		2.15	2.97
0550	On ceilings, no finish included		1620	.010		.82	.53		1.35	1.74
0560	Taped and finished (level 4 finish)		690	.023		.87	1.23		2.10	2.93
0570	With compound skim coat (level 5 finish)		550	.029		.93	1.55		2.48	3.50
1000	For high ceilings, over 8' high, add		2750	.006			.31		.31	.50

09 29 Gypsum Board

09 29 10 – Gypsum Board Panels

09 29 10.50 High Abuse Gypsum Board		Crew	Daily Output	Labor-Hours	Unit	Material	2020 Bare Costs Labor	Equipment	Total	Total Incl O&P
1010	For distribution cost 3 stories and above, add per story	2 Carp	5500	.003	S.F.		.15		.15	.25

09 29 15 – Gypsum Board Accessories

09 29 15.10 Accessories, Gypsum Board

		Crew	Daily Output	Labor-Hours	Unit	Material	Labor	Equipment	Total	Total Incl O&P
0010	**ACCESSORIES, GYPSUM BOARD**									
0020	Casing bead, galvanized steel	1 Carp	2.90	2.759	C.L.F.	24.50	147		171.50	262
0100	Vinyl		3	2.667		23.50	142		165.50	253
0300	Corner bead, galvanized steel, 1" x 1"		4	2		15.90	106		121.90	188
0400	1-1/4" x 1-1/4"		3.50	2.286		17.35	121		138.35	214
0600	Vinyl		4	2		19.95	106		125.95	192
0900	Furring channel, galv. steel, 7/8" deep, standard		2.60	3.077		35	164		199	300
1000	Resilient		2.55	3.137		27	167		194	297
1100	J trim, galvanized steel, 1/2" wide		3	2.667		23	142		165	253
1120	5/8" wide		2.95	2.712		31	144		175	265
1500	Z stud, galvanized steel, 1-1/2" wide		2.60	3.077		43.50	164		207.50	310
9000	Minimum labor/equipment charge		3	2.667	Job		142		142	227

09 30 Tiling

09 30 13 – Ceramic Tiling

09 30 13.20 Ceramic Tile Repairs

		Crew	Daily Output	Labor-Hours	Unit	Material	Labor	Equipment	Total	Total Incl O&P
0010	**CERAMIC TILE REPAIRS**									
1000	Grout removal, carbide tipped, rotary grinder	1 Clab	240	.033	L.F.		1.40		1.40	2.25

09 30 13.45 Ceramic Tile Accessories

		Crew	Daily Output	Labor-Hours	Unit	Material	Labor	Equipment	Total	Total Incl O&P
0010	**CERAMIC TILE ACCESSORIES**									
0100	Spacers, 1/8"				C	2.27			2.27	2.50
1310	Sealer for natural stone tile, installed	1 Tilf	650	.012	S.F.	.05	.61		.66	1.01

09 30 29 – Metal Tiling

09 30 29.10 Metal Tile

		Crew	Daily Output	Labor-Hours	Unit	Material	Labor	Equipment	Total	Total Incl O&P
0010	**METAL TILE** 4' x 4' sheet, 24 ga., tile pattern, nailed									
0200	Stainless steel	2 Carp	512	.031	S.F.	28	1.66		29.66	33.50
0400	Aluminized steel	"	512	.031	"	20.50	1.66		22.16	25
9000	Minimum labor/equipment charge	1 Carp	4	2	Job		106		106	170

09 30 95 – Tile & Stone Setting Materials and Specialties

09 30 95.10 Moisture Resistant, Anti-Fracture Membrane

		Crew	Daily Output	Labor-Hours	Unit	Material	Labor	Equipment	Total	Total Incl O&P
0010	**MOISTURE RESISTANT, ANTI-FRACTURE MEMBRANE**									
0200	Elastomeric membrane, 1/16" thick	D-7	275	.058	S.F.	1.14	2.57		3.71	5.25

09 31 Thin-Set Tiling

09 31 13 – Thin-Set Ceramic Tiling

09 31 13.10 Thin-Set Ceramic Tile

		Crew	Daily Output	Labor-Hours	Unit	Material	Labor	Equipment	Total	Total Incl O&P
0010	**THIN-SET CERAMIC TILE**									
0700	Cove base, 4-1/4" x 4-1/4"	D-7	128	.125	L.F.	4.11	5.50		9.61	13.15
1000	6" x 4-1/4" high		137	.117		5	5.15		10.15	13.60
1300	Sanitary cove base, 6" x 4-1/4" high		124	.129		4.84	5.70		10.54	14.25
1600	6" x 6" high		117	.137		5	6.05		11.05	14.95
2500	Bullnose trim, 4-1/4" x 4-1/4"		128	.125		4.35	5.50		9.85	13.45
2800	2" x 6"		124	.129		4.75	5.70		10.45	14.20
3300	Ceramic tile, porcelain type, 1 color, color group 2, 1" x 1"		183	.087	S.F.	6.25	3.86		10.11	12.90

For customer support on your Commercial Renovation Costs with RSMeans data, call 800.448.8182.

313

09 31 Thin-Set Tiling

09 31 13 – Thin-Set Ceramic Tiling

09 31 13.10 Thin-Set Ceramic Tile

		Crew	Daily Output	Labor-Hours	Unit	Material	2020 Bare Costs Labor	Equipment	Total	Total Incl O&P
3310	2" x 2" or 2" x 1"	D-7	190	.084	S.F.	6.35	3.72		10.07	12.85
3350	For random blend, 2 colors, add					1.02			1.02	1.12
3360	4 colors, add					1.43			1.43	1.57
4300	Specialty tile, 4-1/4" x 4-1/4" x 1/2", decorator finish	D-7	183	.087		12.85	3.86		16.71	20
4500	Add for epoxy grout, 1/16" joint, 1" x 1" tile		800	.020		.67	.88		1.55	2.12
4600	2" x 2" tile		820	.020		.63	.86		1.49	2.04
4610	Add for epoxy grout, 1/8" joint, 8" x 8" x 3/8" tile, add		900	.018		1.43	.79		2.22	2.80
4800	Pregrouted sheets, walls, 4-1/4" x 4-1/4", 6" x 4-1/4"									
4810	and 8-1/2" x 4-1/4", 4 S.F. sheets, silicone grout	D-7	240	.067	S.F.	5.70	2.94		8.64	10.90
5100	Floors, unglazed, 2 S.F. sheets,									
5110	urethane adhesive	D-7	180	.089	S.F.	2.17	3.92		6.09	8.55
5400	Walls, interior, 4-1/4" x 4-1/4" tile		190	.084		3.10	3.72		6.82	9.25
5500	6" x 4-1/4" tile		190	.084		3.25	3.72		6.97	9.40
5700	8-1/2" x 4-1/4" tile		190	.084		5.95	3.72		9.67	12.40
5800	6" x 6" tile		175	.091		4.08	4.04		8.12	10.85
5810	8" x 8" tile		170	.094		5.20	4.16		9.36	12.25
5820	12" x 12" tile		160	.100		5	4.42		9.42	12.40
5830	16" x 16" tile		150	.107		5.45	4.71		10.16	13.40
6000	Decorated wall tile, 4-1/4" x 4-1/4", color group 1		270	.059		3.52	2.62		6.14	7.95
6100	Color group 4		180	.089		54.50	3.92		58.42	66
9300	Ceramic tiles, recycled glass, standard colors, 2" x 2" thru 6" x 6" [G]		190	.084		23	3.72		26.72	31
9310	6" x 6" [G]		175	.091		23	4.04		27.04	31.50
9320	8" x 8" [G]		170	.094		23.50	4.16		27.66	32.50
9330	12" x 12" [G]		160	.100		23.50	4.42		27.92	33
9340	Earthtones, 2" x 2" to 4" x 8" [G]		190	.084		27	3.72		30.72	35.50
9350	6" x 6" [G]		175	.091		27	4.04		31.04	36
9360	8" x 8" [G]		170	.094		27	4.16		31.16	36.50
9370	12" x 12" [G]		160	.100		27	4.42		31.42	37
9380	Deep colors, 2" x 2" to 4" x 8" [G]		190	.084		30.50	3.72		34.22	40
9390	6" x 6" [G]		175	.091		30.50	4.04		34.54	40
9400	8" x 8" [G]		170	.094		32	4.16		36.16	42
9410	12" x 12" [G]		160	.100		32	4.42		36.42	42.50

09 31 33 – Thin-Set Stone Tiling

09 31 33.10 Tiling, Thin-Set Stone

		Crew	Daily Output	Labor-Hours	Unit	Material	2020 Bare Costs Labor	Equipment	Total	Total Incl O&P
0010	**TILING, THIN-SET STONE**									
3000	Floors, natural clay, random or uniform, color group 1	D-7	183	.087	S.F.	4.55	3.86		8.41	11.05
3100	Color group 2	"	183	.087	"	5.95	3.86		9.81	12.55

09 32 Mortar-Bed Tiling

09 32 13 – Mortar-Bed Ceramic Tiling

09 32 13.10 Ceramic Tile

		Crew	Daily Output	Labor-Hours	Unit	Material	2020 Bare Costs Labor	Equipment	Total	Total Incl O&P
0010	**CERAMIC TILE**									
0600	Cove base, 4-1/4" x 4-1/4" high	D-7	91	.176	L.F.	4.22	7.75		11.97	16.80
0900	6" x 4-1/4" high		100	.160		5.10	7.05		12.15	16.70
1200	Sanitary cove base, 6" x 4-1/4" high		93	.172		4.95	7.60		12.55	17.35
1500	6" x 6" high		84	.190		5.10	8.40		13.50	18.85
2400	Bullnose trim, 4-1/4" x 4-1/4"		82	.195		4.43	8.60		13.03	18.35
2700	2" x 6" bullnose trim		84	.190		4.82	8.40		13.22	18.50
6210	Wall tile, 4-1/4" x 4-1/4", better grade	1 Tilf	50	.160	S.F.	9.55	7.90		17.45	23
6240	2" x 2"		50	.160		7.75	7.90		15.65	21

09 32 Mortar-Bed Tiling

09 32 13 – Mortar-Bed Ceramic Tiling

09 32 13.10 Ceramic Tile

		Crew	Daily Output	Labor-Hours	Unit	Material	2020 Bare Costs Labor	Equipment	Total	Total Incl O&P
6250	6" x 6"	1 Tilf	55	.145	S.F.	9.75	7.20		16.95	22
6260	8" x 8"	↓	60	.133		9.75	6.60		16.35	21
6600	Crystalline glazed, 4-1/4" x 4-1/4", plain	D-7	100	.160		4.77	7.05		11.82	16.30
6700	4-1/4" x 4-1/4", scored tile		100	.160		6.60	7.05		13.65	18.30
6900	6" x 6" plain		93	.172		5.70	7.60		13.30	18.15
7000	For epoxy grout, 1/16" joints, 4-1/4" tile, add		800	.020		.42	.88		1.30	1.84
7200	For tile set in dry mortar, add		1735	.009			.41		.41	.64
7300	For tile set in Portland cement mortar, add		290	.055	↓	.17	2.44		2.61	4.01
9500	Minimum labor/equipment charge	↓	3.25	4.923	Job		217		217	340

09 32 16 – Mortar-Bed Quarry Tiling

09 32 16.10 Quarry Tile

		Crew	Daily Output	Labor-Hours	Unit	Material	2020 Bare Costs Labor	Equipment	Total	Total Incl O&P
0010	**QUARRY TILE**									
0100	Base, cove or sanitary, to 5" high, 1/2" thick	D-7	110	.145	L.F.	6.35	6.40		12.75	17
0300	Bullnose trim, red, 6" x 6" x 1/2" thick		120	.133		5.10	5.90		11	14.80
0400	4" x 4" x 1/2" thick		110	.145		5.15	6.40		11.55	15.70
0600	4" x 8" x 1/2" thick, using 8" as edge		130	.123	↓	5.40	5.45		10.85	14.45
0700	Floors, 1,000 S.F. lots, red, 4" x 4" x 1/2" thick		120	.133	S.F.	8.55	5.90		14.45	18.60
0900	6" x 6" x 1/2" thick		140	.114		8.70	5.05		13.75	17.50
1000	4" x 8" x 1/2" thick	↓	130	.123		6.65	5.45		12.10	15.80
1300	For waxed coating, add					.77			.77	.85
1500	For non-standard colors, add					.49			.49	.54
1600	For abrasive surface, add					.53			.53	.58
1800	Brown tile, imported, 6" x 6" x 3/4"	D-7	120	.133		7.20	5.90		13.10	17.10
1900	8" x 8" x 1"		110	.145		9.60	6.40		16	20.50
2100	For thin set mortar application, deduct		700	.023			1.01		1.01	1.58
2700	Stair tread, 6" x 6" x 3/4", plain		50	.320		7.80	14.15		21.95	30.50
2800	Abrasive		47	.340		9.10	15.05		24.15	33.50
3000	Wainscot, 6" x 6" x 1/2", thin set, red		105	.152		6.55	6.75		13.30	17.75
3100	Non-standard colors		105	.152	↓	6.55	6.75		13.30	17.75
3300	Window sill, 6" wide, 3/4" thick		90	.178	L.F.	8.85	7.85		16.70	22
3400	Corners		80	.200	Ea.	6.10	8.85		14.95	20.50
9000	Minimum labor/equipment charge	↓	3.25	4.923	Job		217		217	340

09 32 23 – Mortar-Bed Glass Mosaic Tiling

09 32 23.10 Glass Mosaics

		Crew	Daily Output	Labor-Hours	Unit	Material	2020 Bare Costs Labor	Equipment	Total	Total Incl O&P
0010	**GLASS MOSAICS** 3/4" tile on 12" sheets, standard grout									
1020	1" tile on 12" sheets, opalescent finish	D-7	73	.219	S.F.	18.40	9.70		28.10	35
1040	1" x 2" tile on 12" sheet, blend		73	.219		21	9.70		30.70	38
1060	2" tile on 12" sheet, blend		73	.219		17.15	9.70		26.85	34
1080	5/8" x random tile, linear, on 12" sheet, blend		73	.219		25	9.70		34.70	42.50
1600	Dots on 12" sheet		73	.219		25	9.70		34.70	42.50
1700	For glass mosaic tiles set in dry mortar, add		290	.055		.45	2.44		2.89	4.32
1720	For glass mosaic tiles set in Portland cement mortar, add		290	.055	↓	.01	2.44		2.45	3.83
1730	For polyblend sanded tile grout	↓	96.15	.166	Lb.	2.19	7.35		9.54	13.90

For customer support on your Commercial Renovation Costs with RSMeans data, call 800.448.8182.

315

09 34 Waterproofing-Membrane Tiling

09 34 13 – Waterproofing-Membrane Ceramic Tiling

09 34 13.10 Ceramic Tile Waterproofing Membrane	Crew	Daily Output	Labor-Hours	Unit	Material	2020 Bare Costs Labor	Equipment	Total	Total Incl O&P
0010 **CERAMIC TILE WATERPROOFING MEMBRANE**									
0020 On floors, including thinset									
0030 Fleece laminated polyethylene grid, 1/8" thick	D-7	250	.064	S.F.	2.13	2.83		4.96	6.75
0040 5/16" thick	"	250	.064	"	2.61	2.83		5.44	7.30
0050 On walls, including thinset									
0060 Fleece laminated polyethylene sheet, 8 mil thick	D-7	480	.033	S.F.	2.30	1.47		3.77	4.84
0070 Accessories, including thinset									
0080 Joint and corner sheet, 4 mils thick, 5" wide	1 Tilf	240	.033	L.F.	1.35	1.65		3	4.06
0090 7-1/4" wide		180	.044		1.72	2.20		3.92	5.35
0100 10" wide		120	.067	↓	2.09	3.30		5.39	7.45
0110 Pre-formed corners, inside		32	.250	Ea.	7.85	12.40		20.25	28
0120 Outside		32	.250		7.70	12.40		20.10	28
0130 2" flanged floor drain with 6" stainless steel grate		16	.500	↓	390	25		415	470
0140 EPS, sloped shower floor		480	.017	S.F.	5.55	.83		6.38	7.40
0150 Curb	↓	32	.250	L.F.	14.10	12.40		26.50	35

09 35 Chemical-Resistant Tiling

09 35 13 – Chemical-Resistant Ceramic Tiling

09 35 13.10 Chemical-Resistant Ceramic Tiling

		Crew	Daily Output	Labor-Hours	Unit	Material	Labor	Equipment	Total	Total Incl O&P
0010	**CHEMICAL-RESISTANT CERAMIC TILING**									
0100	4-1/4" x 4-1/4" x 1/4", 1/8" joint	D-7	130	.123	S.F.	12.25	5.45		17.70	22
0200	6" x 6" x 1/2" thick		120	.133		9.95	5.90		15.85	20
0300	8" x 8" x 1/2" thick		110	.145		11.35	6.40		17.75	22.50
0400	4-1/4" x 4-1/4" x 1/4", 1/4" joint		130	.123		13.05	5.45		18.50	23
0500	6" x 6" x 1/2" thick		120	.133		11.15	5.90		17.05	21.50
0600	8" x 8" x 1/2" thick		110	.145		12	6.40		18.40	23.50
0700	4-1/4" x 4-1/4" x 1/4", 3/8" joint		130	.123		13.80	5.45		19.25	23.50
0800	6" x 6" x 1/2" thick		120	.133		12.15	5.90		18.05	22.50
0900	8" x 8" x 1/2" thick	↓	110	.145	↓	13.25	6.40		19.65	24.50

09 35 16 – Chemical-Resistant Quarry Tiling

09 35 16.10 Chemical-Resistant Quarry Tiling

		Crew	Daily Output	Labor-Hours	Unit	Material	Labor	Equipment	Total	Total Incl O&P
0010	**CHEMICAL-RESISTANT QUARRY TILING**									
0100	4" x 8" x 1/2" thick, 1/8" joint	D-7	130	.123	S.F.	11.45	5.45		16.90	21
0200	6" x 6" x 1/2" thick		120	.133		11.50	5.90		17.40	22
0300	8" x 8" x 1/2" thick		110	.145		10.55	6.40		16.95	21.50
0400	4" x 8" x 1/2" thick, 1/4" joint		130	.123		12.75	5.45		18.20	22.50
0500	6" x 6" x 1/2" thick		120	.133		12.65	5.90		18.55	23
0600	8" x 8" x 1/2" thick		110	.145		11.25	6.40		17.65	22.50
0700	4" x 8" x 1/2" thick, 3/8" joint		130	.123		13.90	5.45		19.35	24
0800	6" x 6" x 1/2" thick		120	.133		13.65	5.90		19.55	24.50
0900	8" x 8" x 1/2" thick	↓	110	.145	↓	12.50	6.40		18.90	24

09 51 Acoustical Ceilings

09 51 13 – Acoustical Panel Ceilings

09 51 13.10 Ceiling, Acoustical Panel	Crew	Daily Output	Labor-Hours	Unit	Material	2020 Bare Costs Labor	Equipment	Total	Total Incl O&P
0010 **CEILING, ACOUSTICAL PANEL**									
0100 Fiberglass boards, film faced, 2' x 2' or 2' x 4', 5/8" thick	1 Carp	625	.013	S.F.	1.34	.68		2.02	2.56
0120 3/4" thick		600	.013		3.09	.71		3.80	4.54
0130 3" thick, thermal, R11		450	.018		3.87	.95		4.82	5.75

09 51 14 – Acoustical Fabric-Faced Panel Ceilings

09 51 14.10 Ceiling, Acoustical Fabric-Faced Panel

	Crew	Daily Output	Labor-Hours	Unit	Material	Labor	Equipment	Total	Total Incl O&P
0010 **CEILING, ACOUSTICAL FABRIC-FACED PANEL**									
0100 Glass cloth faced fiberglass, 3/4" thick	1 Carp	500	.016	S.F.	3.57	.85		4.42	5.30
0120 1" thick		485	.016		3.48	.88		4.36	5.25
0130 1-1/2" thick, nubby face		475	.017		2.80	.90		3.70	4.51

09 51 23 – Acoustical Tile Ceilings

09 51 23.10 Suspended Acoustic Ceiling Tiles

	Crew	Daily Output	Labor-Hours	Unit	Material	Labor	Equipment	Total	Total Incl O&P
0010 **SUSPENDED ACOUSTIC CEILING TILES**, not including									
0100 suspension system									
1110 Mineral fiber tile, lay-in, 2' x 2' or 2' x 4', 5/8" thick, fine texture	1 Carp	625	.013	S.F.	.85	.68		1.53	2.03
1115 Rough textured		625	.013		.64	.68		1.32	1.79
1125 3/4" thick, fine textured		600	.013		2.16	.71		2.87	3.52
1130 Rough textured		600	.013		1.99	.71		2.70	3.33
1135 Fissured		600	.013		2.16	.71		2.87	3.52
1150 Tegular, 5/8" thick, fine textured		470	.017		1.28	.90		2.18	2.86
1155 Rough textured		470	.017		1.40	.90		2.30	2.99
1165 3/4" thick, fine textured		450	.018		2.38	.95		3.33	4.13
1170 Rough textured		450	.018		1.51	.95		2.46	3.17
1175 Fissured		450	.018		2.83	.95		3.78	4.62
1185 For plastic film face, add					.86			.86	.95
1190 For fire rating, add					.54			.54	.59
3720 Mineral fiber, 24" x 24" or 48", reveal edge, painted, 5/8" thick	1 Carp	600	.013		1.36	.71		2.07	2.64
3740 3/4" thick		575	.014		1.72	.74		2.46	3.07
5020 66-78% recycled content, 3/4" thick **G**		600	.013		2.08	.71		2.79	3.43
5040 Mylar, 42% recycled content, 3/4" thick **G**		600	.013		4.66	.71		5.37	6.30
6000 Remove & install new ceiling tiles, min fiber, 2' x 2' or 2' x 4', 5/8"thk.		335	.024		.85	1.27		2.12	2.97
9000 Minimum labor/equipment charge		4	2	Job		106		106	170

09 51 23.30 Suspended Ceilings, Complete

	Crew	Daily Output	Labor-Hours	Unit	Material	Labor	Equipment	Total	Total Incl O&P
0010 **SUSPENDED CEILINGS, COMPLETE**, incl. standard									
0100 suspension system but not incl. 1-1/2" carrier channels									
0600 Fiberglass ceiling board, 2' x 4' x 5/8", plain faced	1 Carp	500	.016	S.F.	2.27	.85		3.12	3.86
0700 Offices, 2' x 4' x 3/4"		380	.021		4.02	1.12		5.14	6.20
0800 Mineral fiber, on 15/16" T bar susp. 2' x 2' x 3/4" lay-in board		345	.023		3.32	1.23		4.55	5.65
0810 2' x 4' x 5/8" tile		380	.021		2.44	1.12		3.56	4.47
0820 Tegular, 2' x 2' x 5/8" tile on 9/16" grid		250	.032		2.92	1.70		4.62	5.95
0830 2' x 4' x 3/4" tile		275	.029		3	1.55		4.55	5.80
0900 Luminous panels, prismatic, acrylic		255	.031		4.08	1.67		5.75	7.15
1200 Metal pan with acoustic pad, steel		75	.107		5.15	5.65		10.80	14.75
1300 Painted aluminum		75	.107		3.83	5.65		9.48	13.30
1500 Aluminum, degreased finish		75	.107		5.10	5.65		10.75	14.70
1600 Stainless steel		75	.107		9.65	5.65		15.30	19.75
1800 Tile, Z bar suspension, 5/8" mineral fiber tile		150	.053		2.15	2.83		4.98	6.90
1900 3/4" mineral fiber tile		150	.053		2.44	2.83		5.27	7.20
2402 For strip lighting, see Section 26 51 13.50									
2500 For rooms under 500 S.F., add				S.F.		25%			
9000 Minimum labor/equipment charge	1 Carp	2	4	Job		213		213	340

09 51 Acoustical Ceilings

09 51 33 – Acoustical Metal Pan Ceilings

09 51 33.10 Ceiling, Acoustical Metal Pan

		Crew	Daily Output	Labor-Hours	Unit	Material	2020 Bare Costs Labor	Equipment	Total	Total Incl O&P
0010	**CEILING, ACOUSTICAL METAL PAN**									
0100	Metal panel, lay-in, 2' x 2', sq. edge	1 Carp	500	.016	S.F.	11.10	.85		11.95	13.55
0110	Tegular edge		500	.016		14.20	.85		15.05	16.95
0120	2' x 4', sq. edge		500	.016		13.95	.85		14.80	16.70
0130	Tegular edge		500	.016		14.05	.85		14.90	16.80
0140	Perforated alum. clip-in, 2' x 2'		500	.016		14.20	.85		15.05	16.95
0150	2' x 4'		500	.016		11.60	.85		12.45	14.10

09 51 53 – Direct-Applied Acoustical Ceilings

09 51 53.10 Ceiling Tile

		Crew	Daily Output	Labor-Hours	Unit	Material	2020 Bare Costs Labor	Equipment	Total	Total Incl O&P
0010	**CEILING TILE**, stapled or cemented									
0100	12" x 12" or 12" x 24", not including furring									
0600	Mineral fiber, vinyl coated, 5/8" thick	1 Carp	300	.027	S.F.	2.23	1.42		3.65	4.72
0700	3/4" thick		300	.027		3.15	1.42		4.57	5.75
0900	Fire rated, 3/4" thick, plain faced		300	.027		1.47	1.42		2.89	3.89
1000	Plastic coated face		300	.027		2.26	1.42		3.68	4.76
1200	Aluminum faced, 5/8" thick, plain		300	.027		1.92	1.42		3.34	4.38
3000	Wood fiber tile, 1/2" thick		400	.020		1.23	1.06		2.29	3.05
3100	3/4" thick		400	.020		1.21	1.06		2.27	3.03
3300	For flameproofing, add					.08			.08	.09
3400	For sculptured 3 dimensional, add					.33			.33	.36
3900	For ceiling primer, add					.12			.12	.13
4000	For ceiling cement, add					.42			.42	.46
9000	Minimum labor/equipment charge	1 Carp	4	2	Job		106		106	170

09 53 Acoustical Ceiling Suspension Assemblies

09 53 23 – Metal Acoustical Ceiling Suspension Assemblies

09 53 23.30 Ceiling Suspension Systems

		Crew	Daily Output	Labor-Hours	Unit	Material	2020 Bare Costs Labor	Equipment	Total	Total Incl O&P
0010	**CEILING SUSPENSION SYSTEMS** for boards and tile									
0050	Class A suspension system, 15/16" T bar, 2' x 4' grid	1 Carp	800	.010	S.F.	.93	.53		1.46	1.88
0300	2' x 2' grid		650	.012		1.16	.65		1.81	2.33
0310	25% recycled steel, 2' x 4' grid ☐G		800	.010		.91	.53		1.44	1.85
0320	2' x 2' grid ☐G		650	.012		1.13	.65		1.78	2.29
0350	For 9/16" grid, add					.16			.16	.18
0360	For fire rated grid, add					.09			.09	.10
0370	For colored grid, add					.22			.22	.24
0400	Concealed Z bar suspension system, 12" module	1 Carp	520	.015		.86	.82		1.68	2.26
0600	1-1/2" carrier channels, 4' OC, add	"	470	.017		.12	.90		1.02	1.58
0700	Carrier channels for ceilings with									
0900	recessed lighting fixtures, add	1 Carp	460	.017	S.F.	.22	.92		1.14	1.72
3000	Seismic ceiling bracing, IBC Site Class D, Occupancy Category II									
3050	For ceilings less than 2500 S.F.									
3060	Seismic clips at attached walls	1 Carp	180	.044	Ea.	.98	2.36		3.34	4.86
3100	For ceilings greater than 2500 S.F., add									
3120	Seismic clips, joints at cross tees	1 Carp	120	.067	Ea.	2	3.54		5.54	7.90
3140	At cross tees and mains, mains field cut	"	60	.133	"	2	7.10		9.10	13.55
3200	Compression posts, telescopic, attached to structure above									
3210	To 30" high	1 Carp	26	.308	Ea.	40	16.35		56.35	70
3220	30" to 48" high		25.50	.314		42	16.65		58.65	72.50
3230	48" to 84" high		25	.320		50.50	17		67.50	83
3240	84" to 102" high		24.50	.327		57.50	17.35		74.85	91

09 53 23 – Metal Acoustical Ceiling Suspension Assemblies

09 53 23.30 Ceiling Suspension Systems	Crew	Daily Output	Labor-Hours	Unit	Material	2020 Bare Costs Labor	Equipment	Total	Total Incl O&P
3250	102" to 120" high	1 Carp	24	.333	Ea.	83	17.70		100.70	120
3260	120" to 144" high		24	.333		92	17.70		109.70	130
3300	Stabilizer bars									
3310	12" long	1 Carp	240	.033	Ea.	1.11	1.77		2.88	4.06
3320	24" long		235	.034		1.04	1.81		2.85	4.04
3330	36" long		230	.035		1.02	1.85		2.87	4.08
3340	48" long		220	.036		.82	1.93		2.75	4.01
3400	Wire support for light fixtures, per L.F. height to structure above									
3410	Less than 10 lb.	1 Carp	400	.020	L.F.	.30	1.06		1.36	2.03
3420	10 lb. to 56 lb.	"	240	.033	"	.59	1.77		2.36	3.49
3500	Retrofit existing suspended ceiling grid to current code									
3510	Less than 2500 S.F. (using clips @ perimeter)	1 Carp	1455	.006	S.F.	.06	.29		.35	.54
3520	Greater than 2500 S.F. (using compression posts and clips)		550	.015		.42	.77		1.19	1.70
4000	Remove and replace ceiling grid, class A, 15/16" T bar, 2' x 2' grid		435	.018		1.16	.98		2.14	2.85
4010	2' x 4' grid		535	.015		.93	.79		1.72	2.30

09 54 Specialty Ceilings

09 54 16 – Luminous Ceilings

09 54 16.10 Ceiling, Luminous

		Crew	Daily Output	Labor-Hours	Unit	Material	2020 Bare Costs Labor	Equipment	Total	Total Incl O&P
0010	**CEILING, LUMINOUS**									
0020	Translucent lay-in panels, 2' x 2'	1 Carp	500	.016	S.F.	23.50	.85		24.35	27
0030	2' x 6'	"	500	.016	"	18.05	.85		18.90	21

09 54 23 – Linear Metal Ceilings

09 54 23.10 Metal Ceilings

		Crew	Daily Output	Labor-Hours	Unit	Material	2020 Bare Costs Labor	Equipment	Total	Total Incl O&P
0010	**METAL CEILINGS**									
0015	Solid alum. planks, 3-1/4" x 12', open reveal	1 Carp	500	.016	S.F.	2.38	.85		3.23	3.98
0020	Closed reveal		500	.016		3.05	.85		3.90	4.72
0030	7-1/4" x 12', open reveal		500	.016		4.07	.85		4.92	5.85
0040	Closed reveal		500	.016		5.20	.85		6.05	7.05
0050	Metal, open cell, 2' x 2', 6" cell		500	.016		8.80	.85		9.65	11
0060	8" cell		500	.016		9.70	.85		10.55	12.05
0070	2' x 4', 6" cell		500	.016		5.80	.85		6.65	7.70
0080	8" cell		500	.016		5.80	.85		6.65	7.70

09 54 26 – Suspended Wood Ceilings

09 54 26.10 Wood Ceilings

		Crew	Daily Output	Labor-Hours	Unit	Material	2020 Bare Costs Labor	Equipment	Total	Total Incl O&P
0010	**WOOD CEILINGS**									
1000	4"-6" wood slats on heavy duty 15/16" T-bar grid	2 Carp	250	.064	S.F.	26	3.40		29.40	34

09 54 33 – Decorative Panel Ceilings

09 54 33.20 Metal Panel Ceilings

		Crew	Daily Output	Labor-Hours	Unit	Material	2020 Bare Costs Labor	Equipment	Total	Total Incl O&P
0010	**METAL PANEL CEILINGS**									
0020	Lay-in or screwed to furring, not including grid									
0100	Tin ceilings, 2' x 2' or 2' x 4', bare steel finish	2 Carp	300	.053	S.F.	2.22	2.83		5.05	7
0120	Painted white finish		300	.053	"	3.53	2.83		6.36	8.40
0140	Copper, chrome or brass finish		300	.053	L.F.	6.30	2.83		9.13	11.50
0200	Cornice molding, 2-1/2" to 3-1/2" wide, 4' long, bare steel finish		200	.080	S.F.	2.17	4.25		6.42	9.20
0220	Painted white finish		200	.080		2.87	4.25		7.12	9.95
0240	Copper, chrome or brass finish		200	.080		3.96	4.25		8.21	11.15
0320	5" to 6-1/2" wide, 4' long, bare steel finish		150	.107		3.37	5.65		9.02	12.80
0340	Painted white finish		150	.107		3.89	5.65		9.54	13.40

09 54 Specialty Ceilings

09 54 33 – Decorative Panel Ceilings

09 54 33.20 Metal Panel Ceilings		Crew	Daily Output	Labor-Hours	Unit	Material	2020 Bare Costs Labor	Equipment	Total	Total Incl O&P
0360	Copper, chrome or brass finish	2 Carp	150	.107	S.F.	5.70	5.65		11.35	15.35
0420	Flat molding, 3-1/2" to 5" wide, 4' long, bare steel finish		250	.064		3.89	3.40		7.29	9.70
0440	Painted white finish		250	.064		3.94	3.40		7.34	9.80
0460	Copper, chrome or brass finish		250	.064		7.15	3.40		10.55	13.30

09 61 Flooring Treatment

09 61 19 – Concrete Floor Staining

09 61 19.40 Floors, Interior		Crew	Daily Output	Labor-Hours	Unit	Material	2020 Bare Costs Labor	Equipment	Total	Total Incl O&P
0010	**FLOORS, INTERIOR**									
0300	Acid stain and sealer									
0310	Stain, one coat	1 Pord	650	.012	S.F.	.15	.55		.70	1.03
0320	Two coats		570	.014		.29	.62		.91	1.31
0330	Acrylic sealer, one coat		2600	.003		.25	.14		.39	.50
0340	Two coats		1400	.006		.51	.25		.76	.96

09 62 Specialty Flooring

09 62 19 – Laminate Flooring

09 62 19.10 Floating Floor		Crew	Daily Output	Labor-Hours	Unit	Material	2020 Bare Costs Labor	Equipment	Total	Total Incl O&P
0010	**FLOATING FLOOR**									
8300	Floating floor, laminate, wood pattern strip, complete	1 Clab	133	.060	S.F.	4.95	2.53		7.48	9.50
8310	Components, T&G wood composite strips					4.36			4.36	4.79
8320	Film					.21			.21	.23
8330	Foam					.30			.30	.33
8340	Adhesive					.47			.47	.52
8350	Installation kit					.24			.24	.26
8360	Trim, 2" wide x 3' long				L.F.	5.20			5.20	5.70
8370	Reducer moulding				"	6.15			6.15	6.80

09 62 29 – Cork Flooring

09 62 29.10 Cork Tile Flooring			Crew	Daily Output	Labor-Hours	Unit	Material	2020 Bare Costs Labor	Equipment	Total	Total Incl O&P
0010	**CORK TILE FLOORING**										
2200	Cork tile, standard finish, 1/8" thick	G	1 Tilf	315	.025	S.F.	5.25	1.26		6.51	7.70
2250	3/16" thick	G		315	.025		6.20	1.26		7.46	8.75
2300	5/16" thick	G		315	.025		6.65	1.26		7.91	9.30
2350	1/2" thick	G		315	.025		6.95	1.26		8.21	9.60
2500	Urethane finish, 1/8" thick	G		315	.025		5.10	1.26		6.36	7.55
2550	3/16" thick	G		315	.025		8.20	1.26		9.46	10.95
2600	5/16" thick	G		315	.025		7.50	1.26		8.76	10.20
2650	1/2" thick	G		315	.025		7.80	1.26		9.06	10.50

For customer support on your Commercial Renovation Costs with RSMeans data, call 800.448.8182.

09 63 Masonry Flooring

09 63 13 – Brick Flooring

09 63 13.10 Miscellaneous Brick Flooring

		Crew	Daily Output	Labor-Hours	Unit	Material	2020 Bare Costs Labor	2020 Bare Costs Equipment	Total	Total Incl O&P
0010	**MISCELLANEOUS BRICK FLOORING**									
0020	Acid-proof shales, red, 8" x 3-3/4" x 1-1/4" thick	D-7	.43	37.209	M	725	1,650		2,375	3,375
0050	2-1/4" thick	D-1	.40	40		1,075	1,875		2,950	4,225
0200	Acid-proof clay brick, 8" x 3-3/4" x 2-1/4" thick G		.40	40		1,025	1,875		2,900	4,150
0250	9" x 4-1/2" x 3" G		95	.168	S.F.	4.42	7.90		12.32	17.60
0260	Cast ceramic, pressed, 4" x 8" x 1/2", unglazed	D-7	100	.160		7.55	7.05		14.60	19.40
0270	Glazed		100	.160		10.10	7.05		17.15	22
0280	Hand molded flooring, 4" x 8" x 3/4", unglazed		95	.168		10	7.45		17.45	22.50
0290	Glazed		95	.168		11.40	7.45		18.85	24
0300	8" hexagonal, 3/4" thick, unglazed		85	.188		9.95	8.30		18.25	24
0310	Glazed		85	.188		18	8.30		26.30	33
0400	Heavy duty industrial, cement mortar bed, 2" thick, not incl. brick	D-1	80	.200		1.18	9.40		10.58	16.45
0450	Acid-proof joints, 1/4" wide	"	65	.246		1.74	11.55		13.29	20.50
0500	Pavers, 8" x 4", 1" to 1-1/4" thick, red	D-7	95	.168		4.39	7.45		11.84	16.50
0510	Ironspot	"	95	.168		5.65	7.45		13.10	17.85
0540	1-3/8" to 1-3/4" thick, red	D-1	95	.168		4.24	7.90		12.14	17.40
0560	Ironspot		95	.168		5.60	7.90		13.50	18.90
0580	2-1/4" thick, red		90	.178		4.32	8.35		12.67	18.20
0590	Ironspot		90	.178		6.70	8.35		15.05	21
0700	Paver, adobe brick, 6" x 12", 1/2" joint G		42	.381		1.47	17.85		19.32	30.50
0710	Mexican red, 12" x 12" G	1 Tilf	48	.167		2.08	8.25		10.33	15.25
0720	Saltillo, 12" x 12" G	"	48	.167		1.66	8.25		9.91	14.80
0800	For sidewalks and patios with pavers, see Section 32 14 16.10									
0870	For epoxy joints, add	D-1	600	.027	S.F.	3.15	1.25		4.40	5.50
0880	For Furan underlayment, add	"	600	.027		2.60	1.25		3.85	4.88
0890	For waxed surface, steam cleaned, add	A-1H	1000	.008		.21	.34	.08	.63	.85
9000	Minimum labor/equipment charge	1 Bric	2	4	Job		208		208	335

09 63 40 – Stone Flooring

09 63 40.10 Marble

		Crew	Daily Output	Labor-Hours	Unit	Material	2020 Bare Costs Labor	2020 Bare Costs Equipment	Total	Total Incl O&P
0010	**MARBLE**									
0020	Thin gauge tile, 12" x 6", 3/8", white Carara	D-7	60	.267	S.F.	17.05	11.75		28.80	37.50
0100	Travertine		60	.267		9.50	11.75		21.25	29
0200	12" x 12" x 3/8", thin set, floors		60	.267		11.65	11.75		23.40	31.50
0300	On walls		52	.308		10	13.60		23.60	32.50
1000	Marble threshold, 4" wide x 36" long x 5/8" thick, white		60	.267	Ea.	11.65	11.75		23.40	31.50
9000	Minimum labor/equipment charge		3	5.333	Job		235		235	370

09 63 40.20 Slate Tile

		Crew	Daily Output	Labor-Hours	Unit	Material	2020 Bare Costs Labor	2020 Bare Costs Equipment	Total	Total Incl O&P
0010	**SLATE TILE**									
0020	Vermont, 6" x 6" x 1/4" thick, thin set	D-7	180	.089	S.F.	7.85	3.92		11.77	14.75
9000	Minimum labor/equipment charge	"	3	5.333	Job		235		235	370

09 64 Wood Flooring

09 64 16 – Wood Block Flooring

09 64 16.10 End Grain Block Flooring

		Crew	Daily Output	Labor-Hours	Unit	Material	2020 Bare Costs Labor	Equipment	Total	Total Incl O&P
0010	**END GRAIN BLOCK FLOORING**									
0020	End grain flooring, coated, 2" thick	1 Carp	295	.027	S.F.	3.78	1.44		5.22	6.45
0400	Natural finish, 1" thick, fir		125	.064		3.91	3.40		7.31	9.75
0600	1-1/2" thick, pine		125	.064		3.83	3.40		7.23	9.65
0700	2" thick, pine		125	.064		5.70	3.40		9.10	11.70
9000	Minimum labor/equipment charge		2	4	Job		213		213	340

09 64 23 – Wood Parquet Flooring

09 64 23.10 Wood Parquet

		Crew	Daily Output	Labor-Hours	Unit	Material	2020 Bare Costs Labor	Equipment	Total	Total Incl O&P
0010	**WOOD PARQUET** flooring									
5200	Parquetry, 5/16" thk, no finish, oak, plain pattern	1 Carp	160	.050	S.F.	5.50	2.66		8.16	10.30
5300	Intricate pattern		100	.080		10.55	4.25		14.80	18.40
5500	Teak, plain pattern		160	.050		6.05	2.66		8.71	10.95
5600	Intricate pattern		100	.080		10.35	4.25		14.60	18.20
5650	13/16" thick, select grade oak, plain pattern		160	.050		11.35	2.66		14.01	16.75
5700	Intricate pattern		100	.080		17.05	4.25		21.30	25.50
5800	Custom parquetry, including finish, plain pattern		100	.080		17.75	4.25		22	26.50
5900	Intricate pattern		50	.160		24.50	8.50		33	40.50
6700	Parquetry, prefinished white oak, 5/16" thick, plain pattern		160	.050		8.80	2.66		11.46	13.95
6800	Intricate pattern		100	.080		8.60	4.25		12.85	16.30
7000	Walnut or teak, parquetry, plain pattern		160	.050		9.20	2.66		11.86	14.40
7100	Intricate pattern		100	.080		16.40	4.25		20.65	25
7200	Acrylic wood parquet blocks, 12" x 12" x 5/16",									
7210	Irradiated, set in epoxy	1 Carp	160	.050	S.F.	10.90	2.66		13.56	16.25

09 64 29 – Wood Strip and Plank Flooring

09 64 29.10 Wood

		Crew	Daily Output	Labor-Hours	Unit	Material	2020 Bare Costs Labor	Equipment	Total	Total Incl O&P
0010	**WOOD**	R061110-30								
0020	Fir, vertical grain, 1" x 4", not incl. finish, grade B & better	1 Carp	255	.031	S.F.	3.65	1.67		5.32	6.70
0100	Grade C & better		255	.031		3.43	1.67		5.10	6.45
0300	Flat grain, 1" x 4", not incl. finish, grade B & better		255	.031		4.15	1.67		5.82	7.25
0400	Grade C & better		255	.031		4	1.67		5.67	7.05
4000	Maple, strip, 25/32" x 2-1/4", not incl. finish, select		170	.047		5.10	2.50		7.60	9.60
4100	#2 & better		170	.047		5.15	2.50		7.65	9.65
4300	33/32" x 3-1/4", not incl. finish, #1 grade		170	.047		6.05	2.50		8.55	10.65
4400	#2 & better		170	.047		5.25	2.50		7.75	9.75
4600	Oak, white or red, 25/32" x 2-1/4", not incl. finish									
4700	#1 common	1 Carp	170	.047	S.F.	3.62	2.50		6.12	8
4900	Select quartered, 2-1/4" wide		170	.047		4.48	2.50		6.98	8.95
5000	Clear		170	.047		4.43	2.50		6.93	8.90
6100	Prefinished, white oak, prime grade, 2-1/4" wide		170	.047		6.60	2.50		9.10	11.25
6200	3-1/4" wide		185	.043		5.85	2.30		8.15	10.10
6400	Ranch plank		145	.055		6.90	2.93		9.83	12.30
6500	Hardwood blocks, 9" x 9", 25/32" thick		160	.050		7.90	2.66		10.56	12.95
7400	Yellow pine, 3/4" x 3-1/8", T&G, C & better, not incl. finish		200	.040		1.93	2.13		4.06	5.55
7500	Refinish wood floor, sand, 2 coats poly, wax, soft wood	1 Clab	400	.020		.23	.84		1.07	1.60
7600	Hardwood		130	.062		.23	2.59		2.82	4.40
7800	Sanding and finishing, 2 coats polyurethane		295	.027		.23	1.14		1.37	2.08
7900	Subfloor and underlayment, see Section 06 16									
8015	Transition molding, 2-1/4" wide, 5' long	1 Carp	19.20	.417	Ea.	21	22		43	59
9000	Minimum labor/equipment charge	"	2	4	Job		213		213	340

09 64 Wood Flooring

09 64 36 - Bamboo Flooring

09 64 36.10 Flooring, Bamboo

		Crew	Daily Output	Labor-Hours	Unit	Material	2020 Bare Costs Labor	Equipment	Total	Total Incl O&P
0010	**FLOORING, BAMBOO**									
8600	Flooring, wood, bamboo strips, unfinished, 5/8" x 4" x 3'	G 1 Carp	255	.031	S.F.	6	1.67		7.67	9.25
8610	5/8" x 4" x 4'	G	275	.029		6.25	1.55		7.80	9.35
8620	5/8" x 4" x 6'	G	295	.027		6.85	1.44		8.29	9.85
8630	Finished, 5/8" x 4" x 3'	G	255	.031		6.60	1.67		8.27	9.90
8640	5/8" x 4" x 4'	G	275	.029		6.95	1.55		8.50	10.15
8650	5/8" x 4" x 6'	G	295	.027		5.10	1.44		6.54	7.95
8660	Stair treads, unfinished, 1-1/16" x 11-1/2" x 4'	G	18	.444	Ea.	57.50	23.50		81	102
8670	Finished, 1-1/16" x 11-1/2" x 4'	G	18	.444		91	23.50		114.50	138
8680	Stair risers, unfinished, 5/8" x 7-1/2" x 4'	G	18	.444		21.50	23.50		45	61.50
8690	Finished, 5/8" x 7-1/2" x 4'	G	18	.444		40.50	23.50		64	82.50
8700	Stair nosing, unfinished, 6' long	G	16	.500		47	26.50		73.50	94.50
8710	Finished, 6' long	G	16	.500		45	26.50		71.50	92

09 65 Resilient Flooring

09 65 10 - Resilient Tile Underlayment

09 65 10.10 Latex Underlayment

		Crew	Daily Output	Labor-Hours	Unit	Material	Labor	Equipment	Total	Total Incl O&P
0010	**LATEX UNDERLAYMENT**									
3600	Latex underlayment, 1/8" thk., cementitious for resilient flooring	1 Tilf	160	.050	S.F.	1.29	2.48		3.77	5.30
4000	Liquid, fortified				Gal.	30.50			30.50	33.50

09 65 13 - Resilient Base and Accessories

09 65 13.13 Resilient Base

		Crew	Daily Output	Labor-Hours	Unit	Material	Labor	Equipment	Total	Total Incl O&P
0010	**RESILIENT BASE**									
0690	1/8" vinyl base, 2-1/2" H, straight or cove, standard colors	1 Tilf	315	.025	L.F.	.78	1.26		2.04	2.83
0700	4" high		315	.025		1.17	1.26		2.43	3.26
0710	6" high		315	.025		1.50	1.26		2.76	3.62
0720	Corners, 2-1/2" high		315	.025	Ea.	2.27	1.26		3.53	4.47
0730	4" high		315	.025		3.08	1.26		4.34	5.35
0740	6" high		315	.025		2.91	1.26		4.17	5.15
0800	1/8" rubber base, 2-1/2" H, straight or cove, standard colors		315	.025	L.F.	1.20	1.26		2.46	3.29
1100	4" high		315	.025		1.43	1.26		2.69	3.54
1110	6" high		315	.025		2.08	1.26		3.34	4.26
1150	Corners, 2-1/2" high		315	.025	Ea.	2.52	1.26		3.78	4.74
1153	4" high		315	.025		2.60	1.26		3.86	4.83
1155	6" high		315	.025		3.58	1.26		4.84	5.90
1450	For premium color/finish add					50%				
1500	Millwork profile	1 Tilf	315	.025	L.F.	6.45	1.26		7.71	9

09 65 13.23 Resilient Stair Treads and Risers

		Crew	Daily Output	Labor-Hours	Unit	Material	Labor	Equipment	Total	Total Incl O&P
0010	**RESILIENT STAIR TREADS AND RISERS**									
0300	Rubber, molded tread, 12" wide, 5/16" thick, black	1 Tilf	115	.070	L.F.	14.35	3.44		17.79	21
0400	Colors		115	.070		15	3.44		18.44	22
0600	1/4" thick, black		115	.070		13.45	3.44		16.89	20
0700	Colors		115	.070		16.25	3.44		19.69	23.50
0900	Grip strip safety tread, colors, 5/16" thick		115	.070		20.50	3.44		23.94	28
1000	3/16" thick		120	.067		16.70	3.30		20	23.50
1200	Landings, smooth sheet rubber, 1/8" thick		120	.067	S.F.	8.60	3.30		11.90	14.60
1300	3/16" thick		120	.067	"	9.35	3.30		12.65	15.40
1500	Nosings, 3" wide, 3/16" thick, black		140	.057	L.F.	4.79	2.83		7.62	9.70
1600	Colors		140	.057		6.50	2.83		9.33	11.60
1800	Risers, 7" high, 1/8" thick, flat		250	.032		8.55	1.58		10.13	11.90

For customer support on your Commercial Renovation Costs with RSMeans data, call 800.448.8182.

323

09 65 Resilient Flooring

09 65 13 − Resilient Base and Accessories

09 65 13.23 Resilient Stair Treads and Risers

		Crew	Daily Output	Labor-Hours	Unit	Material	2020 Bare Costs Labor	2020 Bare Costs Equipment	Total	Total Incl O&P
1900	Coved	1 Tilf	250	.032	L.F.	9.55	1.58		11.13	13
2100	Vinyl, molded tread, 12" wide, colors, 1/8" thick		115	.070		6.05	3.44		9.49	12.05
2200	1/4" thick		115	.070		8.15	3.44		11.59	14.35
2300	Landing material, 1/8" thick		200	.040	S.F.	6.35	1.98		8.33	10.10
2400	Riser, 7" high, 1/8" thick, coved		175	.046	L.F.	3.39	2.26		5.65	7.25
2500	Tread and riser combined, 1/8" thick		80	.100	"	9.90	4.95		14.85	18.65
9000	Minimum labor/equipment charge		3	2.667	Job		132		132	207

09 65 13.37 Vinyl Transition Strips

		Crew	Daily Output	Labor-Hours	Unit	Material	2020 Bare Costs Labor	2020 Bare Costs Equipment	Total	Total Incl O&P
0010	**VINYL TRANSITION STRIPS**									
0100	Various mats. to various mats., adhesive applied, 1/4" to 1/8"	1 Tilf	315	.025	L.F.	1.51	1.26		2.77	3.63
0105	0.08" to 1/8"		315	.025		1.36	1.26		2.62	3.47
0110	0.08" to 1/4"		315	.025		1.49	1.26		2.75	3.61
0115	1/4" to 3/8"		315	.025		1.38	1.26		2.64	3.49
0120	1/4" to 1/2"		315	.025		1.38	1.26		2.64	3.49
0125	1/4" to 0.08"		315	.025		1.49	1.26		2.75	3.61
0200	Vinyl wheeled trans. strips, carpet to var. mats., 1/4" to 1/8" x 2-1/2"		315	.025		5.15	1.26		6.41	7.60
0205	1/4" to 1/8" x 4"		315	.025		6.25	1.26		7.51	8.85
0210	Various mats. to various mats. 1/4" to 0.08" x 2-1/2"		315	.025		5.15	1.26		6.41	7.65
0215	Carpet to various materials, 1/4" to flush x 2-1/2"		315	.025		4.29	1.26		5.55	6.70
0220	1/4" to flush x 4"		315	.025		6.60	1.26		7.86	9.20
0225	Various materials to resilient, 3/8" to 1/8" x 2-1/2"		315	.025		4.29	1.26		5.55	6.70
0230	Carpet to various materials, 3/8" to 1/4" x 2-1/2"		315	.025		5.90	1.26		7.16	8.45
0235	1/4" to 1/4" x 2-1/2"		315	.025		6.60	1.26		7.86	9.20
0240	Various materials to resilient, 1/8" to 1/8" x 2-1/2"		315	.025		5	1.26		6.26	7.45
0245	Various materials to var. mats., 1/8" to flush x 2-1/2"		315	.025		3.35	1.26		4.61	5.65
0250	3/8" to flush x 4"		315	.025		6.75	1.26		8.01	9.35
0255	1/2" to flush x 4"		315	.025		9.05	1.26		10.31	11.90
0260	Various materials to resilient, 1/8" to 0.08" x 2-1/2"		315	.025		3.88	1.26		5.14	6.25
0265	0.08" to 0.08" x 2-1/2"		315	.025		3.68	1.26		4.94	6
0270	3/8" to 0.08" x 2-1/2"		315	.025		3.68	1.26		4.94	6

09 65 16 − Resilient Sheet Flooring

09 65 16.10 Rubber and Vinyl Sheet Flooring

			Crew	Daily Output	Labor-Hours	Unit	Material	2020 Bare Costs Labor	2020 Bare Costs Equipment	Total	Total Incl O&P
0010	**RUBBER AND VINYL SHEET FLOORING**										
5500	Linoleum, sheet goods	G	1 Tilf	360	.022	S.F.	3.63	1.10		4.73	5.70
5900	Rubber, sheet goods, 36" wide, 1/8" thick			120	.067		9.25	3.30		12.55	15.30
5950	3/16" thick			100	.080		10	3.96		13.96	17.20
6000	1/4" thick			90	.089		11.75	4.40		16.15	19.85
8000	Vinyl sheet goods, backed, .065" thick, plain pattern/colors			250	.032		4.52	1.58		6.10	7.45
8050	Intricate pattern/colors			200	.040		3.82	1.98		5.80	7.30
8100	.080" thick, plain pattern/colors			230	.035		4.15	1.72		5.87	7.25
8150	Intricate pattern/colors			200	.040		6.40	1.98		8.38	10.15
8200	.125" thick, plain pattern/colors			230	.035		4.19	1.72		5.91	7.30
8250	Intricate pattern/colors			200	.040		7.35	1.98		9.33	11.15
8400	For welding seams, add			100	.080	L.F.	.24	3.96		4.20	6.45
8450	For integral cove base, add			175	.046	"	.82	2.26		3.08	4.44
8700	Adhesive cement, 1 gallon per 200 to 300 S.F.					Gal.	33			33	36
8800	Asphalt primer, 1 gallon per 300 S.F.						15.10			15.10	16.60
8900	Emulsion, 1 gallon per 140 S.F.						19.95			19.95	22

For customer support on your Commercial Renovation Costs with RSMeans data, call 800.448.8182.

09 65 Resilient Flooring

09 65 19 – Resilient Tile Flooring

09 65 19.19 Vinyl Composition Tile Flooring

		Daily Output	Labor-Hours	Unit	Material	2020 Bare Costs Labor	2020 Bare Costs Equipment	Total	Total Incl O&P
		Crew							
0010	**VINYL COMPOSITION TILE FLOORING**								
7000	Vinyl composition tile, 12" x 12", 1/16" thick	1 Tilf	500	.016	S.F.	1.23	.79		2.02	2.59
7050	Embossed		500	.016		2.72	.79		3.51	4.23
7100	Marbleized		500	.016		2.72	.79		3.51	4.23
7150	Solid		500	.016		3.51	.79		4.30	5.10
7200	3/32" thick, embossed		500	.016		1.53	.79		2.32	2.92
7250	Marbleized		500	.016		3.13	.79		3.92	4.68
7300	Solid		500	.016		2.89	.79		3.68	4.42
7350	1/8" thick, marbleized		500	.016		2.47	.79		3.26	3.96
7400	Solid		500	.016		1.75	.79		2.54	3.17
7450	Conductive		500	.016		5.90	.79		6.69	7.70

09 65 19.23 Vinyl Tile Flooring

		Crew	Daily Output	Labor-Hours	Unit	Material	Labor	Equipment	Total	Total Incl O&P
0010	**VINYL TILE FLOORING**									
7500	Vinyl tile, 12" x 12", 3/32" thick, standard colors/patterns	1 Tilf	500	.016	S.F.	3.58	.79		4.37	5.20
7550	1/8" thick, standard colors/patterns		500	.016		5.25	.79		6.04	7
7600	1/8" thick, premium colors/patterns		500	.016		6.90	.79		7.69	8.85
7650	Solid colors		500	.016		3.09	.79		3.88	4.64
7700	Marbleized or Travertine pattern		500	.016		6.35	.79		7.14	8.20
7750	Florentine pattern		500	.016		7.20	.79		7.99	9.15
7800	Premium colors/patterns		500	.016		7.25	.79		8.04	9.20
9500	Minimum labor/equipment charge		4	2	Job		99		99	155

09 65 19.33 Rubber Tile Flooring

		Crew	Daily Output	Labor-Hours	Unit	Material	Labor	Equipment	Total	Total Incl O&P
0010	**RUBBER TILE FLOORING**									
6050	Rubber tile, marbleized colors, 12" x 12", 1/8" thick	1 Tilf	400	.020	S.F.	6	.99		6.99	8.15
6100	3/16" thick		400	.020		8.40	.99		9.39	10.80
6300	Special tile, plain colors, 1/8" thick		400	.020		8.30	.99		9.29	10.65
6350	3/16" thick		400	.020		11.15	.99		12.14	13.80

09 65 33 – Conductive Resilient Flooring

09 65 33.10 Conductive Rubber and Vinyl Flooring

		Crew	Daily Output	Labor-Hours	Unit	Material	Labor	Equipment	Total	Total Incl O&P
0010	**CONDUCTIVE RUBBER AND VINYL FLOORING**									
1700	Conductive flooring, rubber tile, 1/8" thick	1 Tilf	315	.025	S.F.	7.35	1.26		8.61	10.05
1800	Homogeneous vinyl tile, 1/8" thick	"	315	.025	"	7.35	1.26		8.61	10.05

09 65 66 – Resilient Athletic Flooring

09 65 66.10 Resilient Athletic Flooring

		Crew	Daily Output	Labor-Hours	Unit	Material	Labor	Equipment	Total	Total Incl O&P
0010	**RESILIENT ATHLETIC FLOORING**									
1000	Recycled rubber rolled goods, for weight rooms, 3/8" thk.	1 Tilf	315	.025	S.F.	2.86	1.26		4.12	5.10
1050	Interlocking 2' x 2' squares, rubber, 1/4" thk.		310	.026		1.93	1.28		3.21	4.12
1055	5/16" thk.		310	.026		2.59	1.28		3.87	4.85
1060	3/8" thk.		300	.027		2.94	1.32		4.26	5.30
1065	1/2" thk.		310	.026		3.73	1.28		5.01	6.10
2000	Vinyl sheet flooring, 1/4" thk.		315	.025		4.93	1.26		6.19	7.35

For customer support on your Commercial Renovation Costs with RSMeans data, call 800.448.8182.

325

09 66 13 – Portland Cement Terrazzo Flooring

09 66 13.10 Portland Cement Terrazzo		Crew	Daily Output	Labor-Hours	Unit	Material	2020 Bare Costs Labor	Equipment	Total	Total Incl O&P
0010	**PORTLAND CEMENT TERRAZZO,** cast-in-place									
0020	Cove base, 6" high, 16 ga. zinc	1 Mstz	20	.400	L.F.	3.75	19.75		23.50	35
0100	Curb, 6" high and 6" wide		6	1.333		6.60	66		72.60	110
0300	Divider strip for floors, 14 ga., 1-1/4" deep, zinc		375	.021		1.43	1.05		2.48	3.22
0400	Brass		375	.021		2.83	1.05		3.88	4.76
0600	Heavy top strip 1/4" thick, 1-1/4" deep, zinc		300	.027		2.26	1.32		3.58	4.55
1200	For thin set floors, 16 ga., 1/2" x 1/2", zinc		350	.023		1.40	1.13		2.53	3.31
1500	Floor, bonded to concrete, 1-3/4" thick, gray cement	J-3	75	.213	S.F.	3.84	9.60	3.54	16.98	23
1600	White cement, mud set		75	.213		4.53	9.60	3.54	17.67	24
1800	Not bonded, 3" total thickness, gray cement		70	.229		4.73	10.25	3.80	18.78	25.50
1900	White cement, mud set		70	.229		5.55	10.25	3.80	19.60	26.50
9000	Minimum labor/equipment charge	1 Mstz	1	8	Job		395		395	620

09 66 16 – Terrazzo Floor Tile

09 66 16.10 Tile or Terrazzo Base

		Crew	Daily Output	Labor-Hours	Unit	Material	2020 Bare Costs Labor	Equipment	Total	Total Incl O&P
0010	**TILE OR TERRAZZO BASE**									
0020	Scratch coat only	1 Mstz	150	.053	S.F.	.58	2.63		3.21	4.76
0500	Scratch and brown coat only		75	.107	"	1.15	5.25		6.40	9.50
9000	Minimum labor/equipment charge		1	8	Job		395		395	620

09 66 16.13 Portland Cement Terrazzo Floor Tile

		Crew	Daily Output	Labor-Hours	Unit	Material	2020 Bare Costs Labor	Equipment	Total	Total Incl O&P
0010	**PORTLAND CEMENT TERRAZZO FLOOR TILE**									
1200	Floor tiles, non-slip, 1" thick, 12" x 12"	D-1	60	.267	S.F.	25.50	12.50		38	48
1300	1-1/4" thick, 12" x 12"		60	.267		26	12.50		38.50	49
1500	16" x 16"		50	.320		28	15		43	55.50
1600	1-1/2" thick, 16" x 16"		45	.356		26	16.70		42.70	55.50
1800	For Venetian terrazzo, add					7.75			7.75	8.55
1900	For white cement, add					.73			.73	.80

09 66 16.30 Terrazzo, Precast

		Crew	Daily Output	Labor-Hours	Unit	Material	2020 Bare Costs Labor	Equipment	Total	Total Incl O&P
0010	**TERRAZZO, PRECAST**									
0020	Base, 6" high, straight	1 Mstz	70	.114	L.F.	12.65	5.65		18.30	23
0100	Cove		60	.133		16.95	6.60		23.55	29
0300	8" high, straight		60	.133		16.45	6.60		23.05	28.50
0400	Cove		50	.160		24.50	7.90		32.40	39.50
0600	For white cement, add					.60			.60	.66
0700	For 16 ga. zinc toe strip, add					2.33			2.33	2.56
0900	Curbs, 4" x 4" high	1 Mstz	40	.200		43.50	9.90		53.40	63.50
1000	8" x 8" high	"	30	.267		47	13.15		60.15	72
2400	Stair treads, 1-1/2" thick, non-slip, three line pattern	2 Mstz	70	.229		54	11.30		65.30	77
2500	Nosing and two lines		70	.229		54	11.30		65.30	77
2700	2" thick treads, straight		60	.267		64.50	13.15		77.65	91.50
2800	Curved		50	.320		84.50	15.80		100.30	118
3000	Stair risers, 1" thick, to 6" high, straight sections		60	.267		15.30	13.15		28.45	37.50
3100	Cove		50	.320		19.50	15.80		35.30	46.50
3300	Curved, 1" thick, to 6" high, vertical		48	.333		27	16.45		43.45	56
3400	Cove		38	.421		43	21		64	80
3600	Stair tread and riser, single piece, straight, smooth surface		60	.267		66.50	13.15		79.65	94
3700	Non skid surface		40	.400		86	19.75		105.75	126
3900	Curved tread and riser, smooth surface		40	.400		96	19.75		115.75	136
4000	Non skid surface		32	.500		118	24.50		142.50	168
4200	Stair stringers, notched, 1" thick		25	.640		38	31.50		69.50	91.50
4300	2" thick		22	.727		45	36		81	106
4500	Stair landings, structural, non-slip, 1-1/2" thick		85	.188	S.F.	42.50	9.30		51.80	61.50
4600	3" thick		75	.213		59	10.55		69.55	81.50

09 66 Terrazzo Flooring

09 66 16 – Terrazzo Floor Tile

	09 66 16.30 Terrazzo, Precast	Crew	Daily Output	Labor-Hours	Unit	Material	2020 Bare Costs Labor	Equipment	Total	Total Incl O&P
4800	Wainscot, 12" x 12" x 1" tiles	1 Mstz	12	.667	S.F.	9.75	33		42.75	62.50
4900	16" x 16" x 1-1/2" tiles	"	8	1	↓	18.75	49.50		68.25	98
9500	Minimum labor/equipment charge	1 Tilf	2	4	Job		198		198	310

09 66 23 – Resinous Matrix Terrazzo Flooring

09 66 23.16 Epoxy-Resin Terrazzo Flooring

		Crew	Daily Output	Labor-Hours	Unit	Material	2020 Bare Costs Labor	Equipment	Total	Total Incl O&P
0010	**EPOXY-RESIN TERRAZZO FLOORING**									
1800	Epoxy terrazzo, 1/4" thick, chemical resistant, granite chips	J-3	200	.080	S.F.	6.60	3.60	1.33	11.53	14.35
1900	Recycled porcelain	"	150	.107	"	10.40	4.79	1.77	16.96	21

09 67 Fluid-Applied Flooring

09 67 13 – Elastomeric Liquid Flooring

09 67 13.13 Elastomeric Liquid Flooring

		Crew	Daily Output	Labor-Hours	Unit	Material	2020 Bare Costs Labor	Equipment	Total	Total Incl O&P
0010	**ELASTOMERIC LIQUID FLOORING**									
0020	Cementitious acrylic, 1/4" thick	C-6	520	.092	S.F.	1.81	4.04	.10	5.95	8.55
0200	Methyl methacrylate, 1/4" thick	C-8A	3000	.016	"	7.20	.72		7.92	9.10

09 67 23 – Resinous Flooring

09 67 23.23 Resinous Flooring

		Crew	Daily Output	Labor-Hours	Unit	Material	2020 Bare Costs Labor	Equipment	Total	Total Incl O&P
0010	**RESINOUS FLOORING**									
1200	Heavy duty epoxy topping, 1/4" thick,									
1300	500 to 1,000 S.F.	C-6	420	.114	S.F.	6.20	5	.13	11.33	14.95
1500	1,000 to 2,000 S.F.		450	.107		5.25	4.67	.12	10.04	13.40
1600	Over 10,000 S.F.	↓	480	.100	↓	5	4.37	.11	9.48	12.60

09 67 26 – Quartz Flooring

09 67 26.26 Quartz Flooring

		Crew	Daily Output	Labor-Hours	Unit	Material	2020 Bare Costs Labor	Equipment	Total	Total Incl O&P
0010	**QUARTZ FLOORING**									
0600	Epoxy, with colored quartz chips, broadcast, 3/8" thick	C-6	675	.071	S.F.	3.22	3.11	.08	6.41	8.60
0700	1/2" thick		490	.098		4.52	4.28	.11	8.91	11.95
0900	Troweled, minimum		560	.086		3.64	3.75	.10	7.49	10.10
1000	Maximum	↓	480	.100	↓	6.15	4.37	.11	10.63	13.85

09 68 Carpeting

09 68 05 – Carpet Accessories

09 68 05.11 Flooring Transition Strip

		Crew	Daily Output	Labor-Hours	Unit	Material	2020 Bare Costs Labor	Equipment	Total	Total Incl O&P
0010	**FLOORING TRANSITION STRIP**									
0107	Clamp down brass divider, 12' strip, vinyl to carpet	1 Tilf	31.25	.256	Ea.	15	12.65		27.65	36.50
0117	Vinyl to hard surface	"	31.25	.256	"	15	12.65		27.65	36.50

09 68 10 – Carpet Pad

09 68 10.10 Commercial Grade Carpet Pad

		Crew	Daily Output	Labor-Hours	Unit	Material	2020 Bare Costs Labor	Equipment	Total	Total Incl O&P
0010	**COMMERCIAL GRADE CARPET PAD**									
9000	Sponge rubber pad, 20 oz./sq. yd.	1 Tilf	150	.053	S.Y.	4.85	2.64		7.49	9.50
9100	40 to 62 oz./sq. yd.		150	.053		8.85	2.64		11.49	13.85
9200	Felt pad, 20 oz./sq. yd.		150	.053		6.40	2.64		9.04	11.20
9300	32 to 56 oz./sq. yd.		150	.053		12.10	2.64		14.74	17.50
9400	Bonded urethane pad, 2.7 density		150	.053		5.95	2.64		8.59	10.70
9500	13.0 density		150	.053		8.05	2.64		10.69	13
9600	Prime urethane pad, 2.7 density		150	.053		3.69	2.64		6.33	8.20
9700	13.0 density	↓	150	.053	↓	7.95	2.64		10.59	12.90

09 68 Carpeting

09 68 13 – Tile Carpeting

09 68 13.10 Carpet Tile	Crew	Daily Output	Labor-Hours	Unit	Material	2020 Bare Costs Labor	Equipment	Total	Total Incl O&P
0010 **CARPET TILE**									
0100 Tufted nylon, 18" x 18", hard back, 20 oz.	1 Tilf	80	.100	S.Y.	28	4.95		32.95	39
0110 26 oz.		80	.100		26.50	4.95		31.45	37
0200 Cushion back, 20 oz.		80	.100		26	4.95		30.95	36.50
0210 26 oz.		80	.100		31	4.95		35.95	42.50
1100 Tufted, 24" x 24", hard back, 24 oz. nylon		80	.100		32	4.95		36.95	43
1180 35 oz.		80	.100		36.50	4.95		41.45	48
5060 42 oz.		80	.100		46.50	4.95		51.45	59
6000 Electrostatic dissapative carpet tile, 24" x 24", 24 oz.		80	.100		38.50	4.95		43.45	50.50
6100 Electrostatic dissapative carpet tile for access floors, 24" x 24", 24 oz.	▼	80	.100		48	4.95		52.95	61

09 68 16 – Sheet Carpeting

09 68 16.10 Sheet Carpet

09 68 16.10 Sheet Carpet	Crew	Daily Output	Labor-Hours	Unit	Material	2020 Bare Costs Labor	Equipment	Total	Total Incl O&P
0010 **SHEET CARPET**									
0700 Nylon, level loop, 26 oz., light to medium traffic	1 Tilf	75	.107	S.Y.	24.50	5.30		29.80	35.50
0720 28 oz., light to medium traffic		75	.107		32.50	5.30		37.80	44.50
0900 32 oz., medium traffic		75	.107		41	5.30		46.30	53.50
1100 40 oz., medium to heavy traffic		75	.107		48	5.30		53.30	61
2920 Nylon plush, 30 oz., medium traffic		75	.107		28	5.30		33.30	39
3000 36 oz., medium traffic		75	.107		37.50	5.30		42.80	49.50
3100 42 oz., medium to heavy traffic		70	.114		46.50	5.65		52.15	60
3200 46 oz., medium to heavy traffic		70	.114		53.50	5.65		59.15	68
3300 54 oz., heavy traffic		70	.114		62	5.65		67.65	77
4110 Wool, level loop, 40 oz., medium traffic		70	.114		115	5.65		120.65	136
4500 50 oz., medium to heavy traffic		70	.114		111	5.65		116.65	131
4700 Patterned, 32 oz., medium to heavy traffic		70	.114		102	5.65		107.65	121
4900 48 oz., heavy traffic	▼	70	.114	▼	112	5.65		117.65	132
5000 For less than full roll (approx. 1500 S.F.), add					25%				
5100 For small rooms, less than 12' wide, add						25%			
5200 For large open areas (no cuts), deduct						25%			
5600 For bound carpet baseboard, add	1 Tilf	300	.027	L.F.	1.79	1.32		3.11	4.04
5610 For stairs, not incl. price of carpet, add	"	30	.267	Riser		13.20		13.20	20.50
5620 For borders and patterns, add to labor						18%			
8950 For tackless, stretched installation, add padding from 09 68 10.10 to above									
9850 For brand-named specific fiber, add				S.Y.	25%				
9900 Carpet cleaning machine, rent				Day				38	42
9910 Minimum labor/equipment charge	1 Tilf	3	2.667	Job		132		132	207

09 68 20 – Athletic Carpet

09 68 20.10 Indoor Athletic Carpet

09 68 20.10 Indoor Athletic Carpet	Crew	Daily Output	Labor-Hours	Unit	Material	2020 Bare Costs Labor	Equipment	Total	Total Incl O&P
0010 **INDOOR ATHLETIC CARPET**									
3700 Polyethylene, in rolls, no base incl., landscape surfaces	1 Tilf	275	.029	S.F.	4.47	1.44		5.91	7.20
3800 Nylon action surface, 1/8" thick		275	.029		4.17	1.44		5.61	6.85
3900 1/4" thick		275	.029		6	1.44		7.44	8.85
4000 3/8" thick		275	.029		7.55	1.44		8.99	10.55
5500 Polyvinyl chloride, sheet goods for gyms, 1/4" thick		80	.100		7.25	4.95		12.20	15.75
5600 3/8" thick	▼	60	.133	▼	9.95	6.60		16.55	21.50

09 69 Access Flooring

09 69 13 – Rigid-Grid Access Flooring

09 69 13.10 Access Floors	Crew	Daily Output	Labor-Hours	Unit	Material	2020 Bare Costs Labor	Equipment	Total	Total Incl O&P
0010	**ACCESS FLOORS**								
0015	Access floor pkg. including panel, pedestal, & stringers 1500 lbs. load								
0100	Package pricing, conc. fill panels, no fin., 6" ht.	4 Carp	750	.043	S.F.	18.45	2.27		20.72	24
0105	12" ht.		750	.043		18.55	2.27		20.82	24
0110	18" ht.		750	.043		19.20	2.27		21.47	24.50
0115	24" ht.		750	.043		19.05	2.27		21.32	24.50
0120	Package pricing, steel panels, no fin., 6" ht.		750	.043		15.15	2.27		17.42	20.50
0125	12" ht.		750	.043		20	2.27		22.27	25.50
0130	18" ht.		750	.043		20.50	2.27		22.77	26
0135	24" ht.		750	.043		20.50	2.27		22.77	26
0140	Package pricing, wood core panels, no fin., 6" ht.		750	.043		13.65	2.27		15.92	18.65
0145	12" ht.		750	.043		13.90	2.27		16.17	18.90
0150	18" ht.		750	.043		13.90	2.27		16.17	18.90
0155	24" ht.		750	.043		14.15	2.27		16.42	19.20
0160	Pkg. pricing, conc. fill pnls, no stringers, no fin., 6" ht, 1250 lbs. load		700	.046		11.05	2.43		13.48	16.05
0165	12" ht.		700	.046		10.25	2.43		12.68	15.15
0170	18" ht.		700	.046		11.20	2.43		13.63	16.20
0175	24" ht.		700	.046		12.15	2.43		14.58	17.30
0250	Panels, 2' x 2' conc. fill, no fin.	2 Carp	500	.032		23	1.70		24.70	27.50
0255	With 1/8" high pressure laminate		500	.032		57.50	1.70		59.20	66
0260	Metal panels with 1/8" high pressure laminate		500	.032		68	1.70		69.70	77.50
0265	Wood core panels with 1/8" high pressure laminate		500	.032		44.50	1.70		46.20	51.50
0400	Aluminum panels, no fin.		500	.032		35	1.70		36.70	41
0600	For carpet covering, add					9.25			9.25	10.15
0700	For vinyl floor covering, add					9.90			9.90	10.90
0900	For high pressure laminate covering, add					8.15			8.15	8.95
0910	For snap on stringer system, add	2 Carp	1000	.016		1.60	.85		2.45	3.12
1000	Machine cutouts after initial installation	1 Carp	50	.160	Ea.	20.50	8.50		29	36
1050	Pedestals, 6" to 12"	2 Carp	85	.188		9.10	10		19.10	26
1100	Air conditioning grilles, 4" x 12"	1 Carp	17	.471		72.50	25		97.50	120
1150	4" x 18"	"	14	.571		99	30.50		129.50	158
1200	Approach ramps, steel	2 Carp	60	.267	S.F.	28	14.15		42.15	53.50
1300	Aluminum	"	40	.400	"	34.50	21.50		56	72
1500	Handrail, 2 rail, aluminum	1 Carp	15	.533	L.F.	126	28.50		154.50	184

09 72 Wall Coverings

09 72 13 – Cork Wall Coverings

09 72 13.10 Covering, Cork Wall

		Crew	Daily Output	Labor-Hours	Unit	Material	2020 Bare Costs Labor	Equipment	Total	Total Incl O&P
0010	**COVERING, CORK WALL**									
0600	Cork tiles, light or dark, 12" x 12" x 3/16"	1 Pape	240	.033	S.F.	4.26	1.48		5.74	7.05
0700	5/16" thick		235	.034		3.40	1.52		4.92	6.15
0900	1/4" basket weave		240	.033		3.29	1.48		4.77	6
1000	1/2" natural, non-directional pattern		240	.033		6.80	1.48		8.28	9.85
1100	3/4" natural, non-directional pattern		240	.033		11.85	1.48		13.33	15.35
1200	Granular surface, 12" x 36", 1/2" thick		385	.021		1.38	.93		2.31	2.99
1300	1" thick		370	.022		1.72	.96		2.68	3.42
1500	Polyurethane coated, 12" x 12" x 3/16" thick		240	.033		4.17	1.48		5.65	6.95
1600	5/16" thick		235	.034		6	1.52		7.52	9
1800	Cork wallpaper, paperbacked, natural		480	.017		1.68	.74		2.42	3.03
1900	Colors		480	.017		2.89	.74		3.63	4.36

09 72 Wall Coverings

09 72 16 – Vinyl-Coated Fabric Wall Coverings

09 72 16.13 Flexible Vinyl Wall Coverings

		Crew	Daily Output	Labor-Hours	Unit	Material	2020 Bare Costs Labor	Equipment	Total	Total Incl O&P
0010	**FLEXIBLE VINYL WALL COVERINGS**									
3000	Vinyl wall covering, fabric-backed, lightweight, type 1 (12-15 oz./S.Y.)	1 Pape	640	.013	S.F.	1.45	.56		2.01	2.48
3300	Medium weight, type 2 (20-24 oz./S.Y.)		480	.017		1.06	.74		1.80	2.35
3400	Heavy weight, type 3 (28 oz./S.Y.)		435	.018		1.59	.82		2.41	3.05
3600	Adhesive, 5 gal. lots (18 S.Y./gal.)				Gal.	12.60			12.60	13.85

09 72 16.16 Rigid-Sheet Vinyl Wall Coverings

		Crew	Daily Output	Labor-Hours	Unit	Material	2020 Bare Costs Labor	Equipment	Total	Total Incl O&P
0010	**RIGID-SHEET VINYL WALL COVERINGS**									
0100	Acrylic, modified, semi-rigid PVC, .028" thick	2 Carp	330	.048	S.F.	1.32	2.58		3.90	5.60
0110	.040" thick	"	320	.050	"	2.05	2.66		4.71	6.50

09 72 19 – Textile Wall Coverings

09 72 19.10 Textile Wall Covering

		Crew	Daily Output	Labor-Hours	Unit	Material	2020 Bare Costs Labor	Equipment	Total	Total Incl O&P
0010	**TEXTILE WALL COVERING**, including sizing; add 10-30% waste @ takeoff									
0020	Silk	1 Pape	640	.013	S.F.	4.53	.56		5.09	5.85
0030	Cotton		640	.013		6.90	.56		7.46	8.50
0040	Linen		640	.013		1.88	.56		2.44	2.95
0050	Blend		640	.013		3.23	.56		3.79	4.43
0090	Grass cloth, natural fabric [G]		400	.020		1.94	.89		2.83	3.55
0100	Grass cloths with lining paper [G]		400	.020		1.30	.89		2.19	2.85
0110	Premium texture/color [G]		350	.023		3.29	1.02		4.31	5.25

09 72 20 – Natural Fiber Wall Covering

09 72 20.10 Natural Fiber Wall Covering

		Crew	Daily Output	Labor-Hours	Unit	Material	2020 Bare Costs Labor	Equipment	Total	Total Incl O&P
0010	**NATURAL FIBER WALL COVERING**, including sizing; add 10-30% waste @ takeoff									
0015	Bamboo	1 Pape	640	.013	S.F.	2.20	.56		2.76	3.30
0030	Burlap		640	.013		1.90	.56		2.46	2.97
0045	Jute		640	.013		1.31	.56		1.87	2.32
0060	Sisal		640	.013		1.72	.56		2.28	2.77

09 72 23 – Wallpapering

09 72 23.10 Wallpaper

		Crew	Daily Output	Labor-Hours	Unit	Material	2020 Bare Costs Labor	Equipment	Total	Total Incl O&P
0010	**WALLPAPER** including sizing; add 10-30% waste @ takeoff									
0050	Aluminum foil	1 Pape	275	.029	S.F.	1.05	1.30		2.35	3.22
0100	Copper sheets, .025" thick, vinyl backing		240	.033		5.60	1.48		7.08	8.50
0300	Phenolic backing		240	.033		7.25	1.48		8.73	10.30
2400	Gypsum-based, fabric-backed, fire resistant									
2500	for masonry walls, 21 oz./S.Y.	1 Pape	800	.010	S.F.	.85	.45		1.30	1.65
2600	Average		720	.011		1.34	.50		1.84	2.26
2700	Small quantities		640	.013		.85	.56		1.41	1.82
3700	Wallpaper, average workmanship, solid pattern, low cost paper		640	.013		.63	.56		1.19	1.57
3900	Basic patterns (matching required), avg. cost paper		535	.015		1.28	.67		1.95	2.47
4000	Paper at $85 per double roll, quality workmanship		435	.018		2.20	.82		3.02	3.72
5990	Wallpaper removal, 1 layer		800	.010		.02	.45		.47	.73
6000	Wallpaper removal, 3 layer		400	.020		.07	.89		.96	1.49
9000	Minimum labor/equipment charge		2	4	Job		178		178	283

09 74 Flexible Wood Sheets

09 74 16 – Flexible Wood Veneers

09 74 16.10 Veneer, Flexible Wood	Crew	Daily Output	Labor-Hours	Unit	Material	2020 Bare Costs Labor	2020 Bare Costs Equipment	Total	Total Incl O&P
0010 **VENEER, FLEXIBLE WOOD**									
0100 Flexible wood veneer, 1/32" thick, plain woods	1 Pape	100	.080	S.F.	2.50	3.56		6.06	8.40
0110 Exotic woods	"	95	.084	"	3.77	3.75		7.52	10.10

09 77 Special Wall Surfacing

09 77 30 – Fiberglass Reinforced Panels

09 77 30.10 Fiberglass Reinforced Plastic Panels

	Crew	Daily Output	Labor-Hours	Unit	Material	2020 Bare Costs Labor	2020 Bare Costs Equipment	Total	Total Incl O&P
0010 **FIBERGLASS REINFORCED PLASTIC PANELS**, .090" thick									
0020 On walls, adhesive mounted, embossed surface	2 Carp	640	.025	S.F.	1.19	1.33		2.52	3.44
0030 Smooth surface		640	.025		1.47	1.33		2.80	3.75
0040 Fire rated, embossed surface		640	.025		2.12	1.33		3.45	4.46
0050 Nylon rivet mounted, on drywall, embossed surface		480	.033		1.18	1.77		2.95	4.14
0060 Smooth surface		480	.033		1.47	1.77		3.24	4.46
0070 Fire rated, embossed surface		480	.033		2.12	1.77		3.89	5.15
0080 On masonry, embossed surface		320	.050		1.18	2.66		3.84	5.55
0090 Smooth surface		320	.050		1.47	2.66		4.13	5.90
0100 Fire rated, embossed surface		320	.050		2.12	2.66		4.78	6.60
0110 Nylon rivet and adhesive mounted, on drywall, embossed surface		240	.067		1.35	3.54		4.89	7.20
0120 Smooth surface		240	.067		1.43	3.54		4.97	7.25
0130 Fire rated, embossed surface		240	.067		2.34	3.54		5.88	8.25
0140 On masonry, embossed surface		190	.084		1.35	4.48		5.83	8.65
0150 Smooth surface		190	.084		1.43	4.48		5.91	8.70
0160 Fire rated, embossed surface		190	.084		2.34	4.48		6.82	9.70
0170 For moldings, add	1 Carp	250	.032	L.F.	.29	1.70		1.99	3.04
0180 On ceilings, for lay in grid system, embossed surface		400	.020	S.F.	1.19	1.06		2.25	3.01
0190 Smooth surface		400	.020		1.47	1.06		2.53	3.32
0200 Fire rated, embossed surface		400	.020		2.12	1.06		3.18	4.03

09 77 43 – Panel Systems

09 77 43.20 Slatwall Panels and Accessories

	Crew	Daily Output	Labor-Hours	Unit	Material	2020 Bare Costs Labor	2020 Bare Costs Equipment	Total	Total Incl O&P
0010 **SLATWALL PANELS AND ACCESSORIES**									
0100 Slatwall panel, 4' x 8' x 3/4" T, MDF, paint grade	1 Carp	500	.016	S.F.	1.58	.85		2.43	3.10
0110 Melamine finish		500	.016		2.13	.85		2.98	3.70
0120 High pressure plastic laminate finish		500	.016		3.38	.85		4.23	5.10
0125 Wood veneer		500	.016		4.23	.85		5.08	6
0130 Aluminum channel inserts, add					2.66			2.66	2.93
0200 Accessories, corner forms, 8' L				L.F.	5.70			5.70	6.30
0210 T-connector, 8' L					7.45			7.45	8.20
0220 J-mold, 8' L					1.25			1.25	1.38
0230 Edge cap, 8' L					1.74			1.74	1.91
0240 Finish end cap, 8' L					3.39			3.39	3.73
0300 Display hook, metal, 4" L				Ea.	.47			.47	.52
0310 6" L					.52			.52	.57
0320 8" L					.56			.56	.62
0330 10" L					.54			.54	.59
0340 12" L					.51			.51	.56
0350 Acrylic, 4" L					1.27			1.27	1.40
0360 6" L					.97			.97	1.07
0400 Waterfall hanger, metal, 12"-16"					3.92			3.92	4.31
0410 Acrylic					11.35			11.35	12.50
0500 Shelf bracket, metal, 8"					2.06			2.06	2.27

09 77 Special Wall Surfacing

09 77 43 – Panel Systems

09 77 43.20 Slatwall Panels and Accessories	Crew	Daily Output	Labor-Hours	Unit	Material	2020 Bare Costs Labor	Equipment	Total	Total Incl O&P
0510	10"				Ea.	2.04			2.04	2.24
0520	12"					2.26			2.26	2.49
0530	14"					2.49			2.49	2.74
0540	16"					3.07			3.07	3.38
0550	Acrylic, 8"					3.94			3.94	4.33
0560	10"					4.10			4.10	4.51
0570	12"					4.64			4.64	5.10
0580	14"					6			6	6.60
0600	Shelf, acrylic, 12" x 16" x 1/4"					23			23	25
0610	12" x 24" x 1/4"					25			25	27.50

09 81 Acoustic Insulation

09 81 13 – Acoustic Board Insulation

09 81 13.10 Acoustic Board Insulation

		Crew	Daily Output	Labor-Hours	Unit	Material	Labor	Equipment	Total	Total Incl O&P
0010	**ACOUSTIC BOARD INSULATION**									
0020	Cellulose fiber board, 1/2" thk.	1 Carp	800	.010	S.F.	.91	.53		1.44	1.85

09 81 16 – Acoustic Blanket Insulation

09 81 16.10 Sound Attenuation Blanket

		Crew	Daily Output	Labor-Hours	Unit	Material	Labor	Equipment	Total	Total Incl O&P
0010	**SOUND ATTENUATION BLANKET**									
0020	Blanket, 1" thick	1 Carp	925	.009	S.F.	.28	.46		.74	1.05
0500	1-1/2" thick		920	.009		.33	.46		.79	1.10
1000	2" thick		915	.009		.40	.46		.86	1.18
1500	3" thick		910	.009		.59	.47		1.06	1.40
2000	Wall hung, STC 18-21, 1" thick, 4' x 20'	2 Carp	22	.727	Ea.	550	38.50		588.50	665
2010	10' x 20'	"	19	.842		1,375	45		1,420	1,600
2020	Wall hung, STC 27-28, 3" thick, 4' x 20'	3 Carp	12	2		590	106		696	820
2030	10' x 20'	"	9	2.667		1,475	142		1,617	1,850
3400	Urethane plastic foam, open cell, on wall, 2" thick	2 Carp	2050	.008	S.F.	3.29	.41		3.70	4.28
3500	3" thick		1550	.010		4.37	.55		4.92	5.70
3600	4" thick		1050	.015		6.15	.81		6.96	8.05
3700	On ceiling, 2" thick		1700	.009		3.28	.50		3.78	4.41
3800	3" thick		1300	.012		4.37	.65		5.02	5.85
3900	4" thick		900	.018		6.15	.95		7.10	8.25
9000	Minimum labor/equipment charge	1 Carp	5	1.600	Job		85		85	136

09 84 Acoustic Room Components

09 84 13 – Fixed Sound-Absorptive Panels

09 84 13.10 Fixed Panels

		Crew	Daily Output	Labor-Hours	Unit	Material	Labor	Equipment	Total	Total Incl O&P
0010	**FIXED PANELS** Perforated steel facing, painted with									
0100	Fiberglass or mineral filler, no backs, 2-1/4" thick, modular									
0200	space units, ceiling or wall hung, white or colored	1 Carp	100	.080	S.F.	9.95	4.25		14.20	17.70
0300	Fiberboard sound deadening panels, 1/2" thick	"	600	.013	"	.33	.71		1.04	1.50
0500	Fiberglass panels, 4' x 8' x 1" thick, with									
0600	glass cloth face for walls, cemented	1 Carp	155	.052	S.F.	9.15	2.74		11.89	14.50
0700	1-1/2" thick, dacron covered, inner aluminum frame,									
0710	wall mounted	1 Carp	300	.027	S.F.	8.85	1.42		10.27	11.95
0900	Mineral fiberboard panels, fabric covered, 30" x 108",									
1000	3/4" thick, concealed spline, wall mounted	1 Carp	150	.053	S.F.	6.35	2.83		9.18	11.55
9000	Minimum labor/equipment charge	"	4	2	Job		106		106	170

For customer support on your Commercial Renovation Costs with RSMeans data, call 800.448.8182.

09 91 Painting

09 91 13 – Exterior Painting

09 91 13.30 Fences

		Crew	Daily Output	Labor-Hours	Unit	Material	2020 Bare Costs Labor	Equipment	Total	Total Incl O&P
0010	**FENCES**									
0100	Chain link or wire metal, one side, water base									
0110	Roll & brush, first coat	1 Pord	960	.008	S.F.	.08	.37		.45	.68
0120	Second coat		1280	.006		.08	.28		.36	.52
0130	Spray, first coat		2275	.004		.08	.16		.24	.34
0140	Second coat	↓	2600	.003	↓	.08	.14		.22	.31
0150	Picket, water base									
0160	Roll & brush, first coat	1 Pord	865	.009	S.F.	.09	.41		.50	.75
0170	Second coat		1050	.008		.09	.34		.43	.64
0180	Spray, first coat		2275	.004		.09	.16		.25	.35
0190	Second coat	↓	2600	.003	↓	.09	.14		.23	.32
0200	Stockade, water base									
0210	Roll & brush, first coat	1 Pord	1040	.008	S.F.	.09	.34		.43	.64
0220	Second coat		1200	.007		.09	.30		.39	.57
0230	Spray, first coat		2275	.004		.09	.16		.25	.35
0240	Second coat		2600	.003	↓	.09	.14		.23	.32
9000	Minimum labor/equipment charge	↓	2	4	Job		178		178	282

09 91 13.42 Miscellaneous, Exterior

		Crew	Daily Output	Labor-Hours	Unit	Material	2020 Bare Costs Labor	Equipment	Total	Total Incl O&P
0010	**MISCELLANEOUS, EXTERIOR**									
0100	Railing, ext., decorative wood, incl. cap & baluster									
0110	Newels & spindles @ 12" OC									
0120	Brushwork, stain, sand, seal & varnish									
0130	First coat	1 Pord	90	.089	L.F.	.88	3.95		4.83	7.20
0140	Second coat	"	120	.067	"	.88	2.96		3.84	5.65
0150	Rough sawn wood, 42" high, 2" x 2" verticals, 6" OC									
0160	Brushwork, stain, each coat	1 Pord	90	.089	L.F.	.29	3.95		4.24	6.55
0170	Wrought iron, 1" rail, 1/2" sq. verticals									
0180	Brushwork, zinc chromate, 60" high, bars 6" OC									
0190	Primer	1 Pord	130	.062	L.F.	.85	2.73		3.58	5.30
0200	Finish coat		130	.062		1.13	2.73		3.86	5.60
0210	Additional coat	↓	190	.042	↓	1.31	1.87		3.18	4.42
0220	Shutters or blinds, single panel, 2' x 4', paint all sides									
0230	Brushwork, primer	1 Pord	20	.400	Ea.	.72	17.75		18.47	29
0240	Finish coat, exterior latex		20	.400		.65	17.75		18.40	28.50
0250	Primer & 1 coat, exterior latex		13	.615		1.21	27.50		28.71	45
0260	Spray, primer		35	.229		1.04	10.15		11.19	17.25
0270	Finish coat, exterior latex		35	.229		1.38	10.15		11.53	17.60
0280	Primer & 1 coat, exterior latex	↓	20	.400	↓	1.13	17.75		18.88	29
0290	For louvered shutters, add				S.F.	10%				
0300	Stair stringers, exterior, metal									
0310	Roll & brush, zinc chromate, to 14", each coat	1 Pord	320	.025	L.F.	.38	1.11		1.49	2.17
0320	Rough sawn wood, 4" x 12"									
0330	Roll & brush, exterior latex, each coat	1 Pord	215	.037	L.F.	.10	1.65		1.75	2.73
0340	Trellis/lattice, 2" x 2" @ 3" OC with 2" x 8" supports									
0350	Spray, latex, per side, each coat	1 Pord	475	.017	S.F.	.10	.75		.85	1.30
0450	Decking, ext., sealer, alkyd, brushwork, sealer coat		1140	.007		.10	.31		.41	.60
0460	1st coat		1140	.007		.11	.31		.42	.61
0470	2nd coat		1300	.006		.08	.27		.35	.52
0500	Paint, alkyd, brushwork, primer coat		1140	.007		.11	.31		.42	.61
0510	1st coat		1140	.007		.14	.31		.45	.64
0520	2nd coat		1300	.006		.10	.27		.37	.54
0600	Sand paint, alkyd, brushwork, 1 coat	↓	150	.053	↓	.14	2.37		2.51	3.92

For customer support on your Commercial Renovation Costs with RSMeans data, call 800.448.8182.

333

09 91 Painting

09 91 13 – Exterior Painting

09 91 13.42 Miscellaneous, Exterior	Crew	Daily Output	Labor-Hours	Unit	Material	2020 Bare Costs Labor	2020 Bare Costs Equipment	Total	Total Incl O&P
9000 Minimum labor/equipment charge	1 Pord	2	4	Job		178		178	282

09 91 13.60 Siding Exterior

	Crew	Daily Output	Labor-Hours	Unit	Material	2020 Bare Costs Labor	2020 Bare Costs Equipment	Total	Total Incl O&P
0010 **SIDING EXTERIOR**, Alkyd (oil base)									
0450 Steel siding, oil base, paint 1 coat, brushwork	2 Pord	2015	.008	S.F.	.11	.35		.46	.68
0500 Spray		4550	.004		.17	.16		.33	.43
0800 Paint 2 coats, brushwork		1300	.012		.22	.55		.77	1.11
1000 Spray		2750	.006		.15	.26		.41	.58
1200 Stucco, rough, oil base, paint 2 coats, brushwork		1300	.012		.22	.55		.77	1.11
1400 Roller		1625	.010		.23	.44		.67	.94
1600 Spray		2925	.005		.24	.24		.48	.66
1800 Texture 1-11 or clapboard, oil base, primer coat, brushwork		1300	.012		.14	.55		.69	1.02
2000 Spray		4550	.004		.14	.16		.30	.40
2400 Paint 2 coats, brushwork		810	.020		.32	.88		1.20	1.74
2600 Spray		2600	.006		.36	.27		.63	.82
3400 Stain 2 coats, brushwork		950	.017		.20	.75		.95	1.41
4000 Spray		3050	.005		.22	.23		.45	.61
4200 Wood shingles, oil base primer coat, brushwork		1300	.012		.13	.55		.68	1.01
4400 Spray		3900	.004		.12	.18		.30	.42
5000 Paint 2 coats, brushwork		810	.020		.27	.88		1.15	1.68
5200 Spray		2275	.007		.25	.31		.56	.78
6500 Stain 2 coats, brushwork		950	.017		.20	.75		.95	1.41
7000 Spray		2660	.006		.27	.27		.54	.72
8000 For latex paint, deduct					10%				
8100 For work over 12' H, from pipe scaffolding, add						15%			
8200 For work over 12' H, from extension ladder, add						25%			
8300 For work over 12' H, from swing staging, add						35%			
9000 Minimum labor/equipment charge	1 Pord	2	4	Job		178		178	282

09 91 13.62 Siding, Misc.

	Crew	Daily Output	Labor-Hours	Unit	Material	2020 Bare Costs Labor	2020 Bare Costs Equipment	Total	Total Incl O&P
0010 **SIDING, MISC.**, latex paint									
0100 Aluminum siding									
0110 Brushwork, primer	2 Pord	2275	.007	S.F.	.07	.31		.38	.57
0120 Finish coat, exterior latex		2275	.007		.06	.31		.37	.57
0130 Primer & 1 coat exterior latex		1300	.012		.14	.55		.69	1.02
0140 Primer & 2 coats exterior latex		975	.016		.20	.73		.93	1.38
0150 Mineral fiber shingles									
0160 Brushwork, primer	2 Pord	1495	.011	S.F.	.14	.48		.62	.90
0170 Finish coat, industrial enamel		1495	.011		.20	.48		.68	.97
0180 Primer & 1 coat enamel		810	.020		.34	.88		1.22	1.76
0190 Primer & 2 coats enamel		540	.030		.54	1.32		1.86	2.68
0200 Roll, primer		1625	.010		.16	.44		.60	.86
0210 Finish coat, industrial enamel		1625	.010		.22	.44		.66	.93
0220 Primer & 1 coat enamel		975	.016		.37	.73		1.10	1.57
0230 Primer & 2 coats enamel		650	.025		.59	1.09		1.68	2.38
0240 Spray, primer		3900	.004		.12	.18		.30	.42
0250 Finish coat, industrial enamel		3900	.004		.18	.18		.36	.49
0260 Primer & 1 coat enamel		2275	.007		.30	.31		.61	.83
0270 Primer & 2 coats enamel		1625	.010		.48	.44		.92	1.22
0280 Waterproof sealer, first coat		4485	.004		.12	.16		.28	.39
0290 Second coat		5235	.003		.12	.14		.26	.35
0300 Rough wood incl. shingles, shakes or rough sawn siding									
0310 Brushwork, primer	2 Pord	1280	.013	S.F.	.14	.56		.70	1.04
0320 Finish coat, exterior latex		1280	.013		.11	.56		.67	1

09 91 13 – Exterior Painting

09 91 13.62 Siding, Misc.

		Crew	Daily Output	Labor-Hours	Unit	Material	2020 Bare Costs Labor	Equipment	Total	Total Incl O&P
0330	Primer & 1 coat exterior latex	2 Pord	960	.017	S.F.	.25	.74		.99	1.46
0340	Primer & 2 coats exterior latex		700	.023		.36	1.02		1.38	2
0350	Roll, primer		2925	.005		.19	.24		.43	.60
0360	Finish coat, exterior latex		2925	.005		.13	.24		.37	.53
0370	Primer & 1 coat exterior latex		1790	.009		.32	.40		.72	.98
0380	Primer & 2 coats exterior latex		1300	.012		.45	.55		1	1.36
0390	Spray, primer		3900	.004		.16	.18		.34	.47
0400	Finish coat, exterior latex		3900	.004		.10	.18		.28	.40
0410	Primer & 1 coat exterior latex		2600	.006		.26	.27		.53	.72
0420	Primer & 2 coats exterior latex		2080	.008		.36	.34		.70	.94
0430	Waterproof sealer, first coat		4485	.004		.22	.16		.38	.50
0440	Second coat	▼	4485	.004	▼	.12	.16		.28	.39
0450	Smooth wood incl. butt, T&G, beveled, drop or B&B siding									
0460	Brushwork, primer	2 Pord	2325	.007	S.F.	.10	.31		.41	.60
0470	Finish coat, exterior latex		1280	.013		.11	.56		.67	1
0480	Primer & 1 coat exterior latex		800	.020		.21	.89		1.10	1.64
0490	Primer & 2 coats exterior latex		630	.025		.32	1.13		1.45	2.14
0500	Roll, primer		2275	.007		.12	.31		.43	.63
0510	Finish coat, exterior latex		2275	.007		.11	.31		.42	.63
0520	Primer & 1 coat exterior latex		1300	.012		.23	.55		.78	1.12
0530	Primer & 2 coats exterior latex		975	.016		.35	.73		1.08	1.54
0540	Spray, primer		4550	.004		.09	.16		.25	.35
0550	Finish coat, exterior latex		4550	.004		.10	.16		.26	.36
0560	Primer & 1 coat exterior latex		2600	.006		.19	.27		.46	.64
0570	Primer & 2 coats exterior latex		1950	.008		.29	.36		.65	.90
0580	Waterproof sealer, first coat		5230	.003		.12	.14		.26	.36
0590	Second coat	▼	5980	.003	▼	.12	.12		.24	.33
0600	For oil base paint, add					10%				
9000	Minimum labor/equipment charge	1 Pord	2	4	Job		178		178	282

09 91 13.70 Doors and Windows, Exterior

		Crew	Daily Output	Labor-Hours	Unit	Material	2020 Bare Costs Labor	Equipment	Total	Total Incl O&P
0010	**DOORS AND WINDOWS, EXTERIOR**									
0100	Door frames & trim, only									
0110	Brushwork, primer	1 Pord	512	.016	L.F.	.07	.69		.76	1.17
0120	Finish coat, exterior latex		512	.016		.08	.69		.77	1.19
0130	Primer & 1 coat, exterior latex		300	.027	▼	.15	1.18		1.33	2.04
0135	2 coats, exterior latex, both sides		15	.533	Ea.	7.25	23.50		30.75	45.50
0140	Primer & 2 coats, exterior latex	▼	265	.030	L.F.	.23	1.34		1.57	2.38
0150	Doors, flush, both sides, incl. frame & trim									
0160	Roll & brush, primer	1 Pord	10	.800	Ea.	4.96	35.50		40.46	62
0170	Finish coat, exterior latex		10	.800		6.20	35.50		41.70	63.50
0180	Primer & 1 coat, exterior latex		7	1.143		11.15	50.50		61.65	93
0190	Primer & 2 coats, exterior latex		5	1.600		17.30	71		88.30	132
0200	Brushwork, stain, sealer & 2 coats polyurethane	▼	4	2	▼	31	89		120	175
0210	Doors, French, both sides, 10-15 lite, incl. frame & trim									
0220	Brushwork, primer	1 Pord	6	1.333	Ea.	2.48	59		61.48	96.50
0230	Finish coat, exterior latex		6	1.333		3.09	59		62.09	97.50
0240	Primer & 1 coat, exterior latex		3	2.667		5.55	118		123.55	194
0250	Primer & 2 coats, exterior latex		2	4		8.50	178		186.50	291
0260	Brushwork, stain, sealer & 2 coats polyurethane	▼	2.50	3.200	▼	11.15	142		153.15	238
0270	Doors, louvered, both sides, incl. frame & trim									
0280	Brushwork, primer	1 Pord	7	1.143	Ea.	4.96	50.50		55.46	86
0290	Finish coat, exterior latex	▼	7	1.143	▼	6.20	50.50		56.70	87.50

For customer support on your Commercial Renovation Costs with RSMeans data, call 800.448.8182.

335

09 91 13.70 Doors and Windows, Exterior

		Crew	Daily Output	Labor-Hours	Unit	Material	2020 Bare Costs Labor	2020 Bare Costs Equipment	Total	Total Incl O&P
0300	Primer & 1 coat, exterior latex	1 Pord	4	2	Ea.	11.15	89		100.15	153
0310	Primer & 2 coats, exterior latex		3	2.667		16.95	118		134.95	207
0320	Brushwork, stain, sealer & 2 coats polyurethane	↓	4.50	1.778	↓	31	79		110	159
0330	Doors, panel, both sides, incl. frame & trim									
0340	Roll & brush, primer	1 Pord	6	1.333	Ea.	4.96	59		63.96	99.50
0350	Finish coat, exterior latex		6	1.333		6.20	59		65.20	101
0360	Primer & 1 coat, exterior latex		3	2.667		11.15	118		129.15	200
0370	Primer & 2 coats, exterior latex		2.50	3.200		16.95	142		158.95	245
0380	Brushwork, stain, sealer & 2 coats polyurethane	↓	3	2.667	↓	31	118		149	222
0400	Windows, per ext. side, based on 15 S.F.									
0410	1 to 6 lite									
0420	Brushwork, primer	1 Pord	13	.615	Ea.	.98	27.50		28.48	44.50
0430	Finish coat, exterior latex		13	.615		1.22	27.50		28.72	45
0440	Primer & 1 coat, exterior latex		8	1		2.20	44.50		46.70	73
0450	Primer & 2 coats, exterior latex		6	1.333		3.35	59		62.35	97.50
0460	Stain, sealer & 1 coat varnish	↓	7	1.143	↓	4.40	50.50		54.90	85.50
0470	7 to 10 lite									
0480	Brushwork, primer	1 Pord	11	.727	Ea.	.98	32.50		33.48	52.50
0490	Finish coat, exterior latex		11	.727		1.22	32.50		33.72	53
0500	Primer & 1 coat, exterior latex		7	1.143		2.20	50.50		52.70	83
0510	Primer & 2 coats, exterior latex		5	1.600		3.35	71		74.35	117
0520	Stain, sealer & 1 coat varnish	↓	6	1.333	↓	4.40	59		63.40	99
0530	12 lite									
0540	Brushwork, primer	1 Pord	10	.800	Ea.	.98	35.50		36.48	57.50
0550	Finish coat, exterior latex		10	.800		1.22	35.50		36.72	58
0560	Primer & 1 coat, exterior latex		6	1.333		2.20	59		61.20	96.50
0570	Primer & 2 coats, exterior latex		5	1.600		3.35	71		74.35	117
0580	Stain, sealer & 1 coat varnish	↓	6	1.333		4.45	59		63.45	99
0590	For oil base paint, add				↓	10%				
9000	Minimum labor/equipment charge	1 Pord	2	4	Job		178		178	282

09 91 13.80 Trim, Exterior

		Crew	Daily Output	Labor-Hours	Unit	Material	2020 Bare Costs Labor	2020 Bare Costs Equipment	Total	Total Incl O&P
0010	**TRIM, EXTERIOR**									
0100	Door frames & trim (see Doors, interior or exterior)									
0110	Fascia, latex paint, one coat coverage									
0120	1" x 4", brushwork	1 Pord	640	.013	L.F.	.02	.56		.58	.91
0130	Roll		1280	.006		.03	.28		.31	.47
0140	Spray		2080	.004		.02	.17		.19	.29
0150	1" x 6" to 1" x 10", brushwork		640	.013		.09	.56		.65	.98
0160	Roll		1230	.007		.09	.29		.38	.56
0170	Spray		2100	.004		.07	.17		.24	.35
0180	1" x 12", brushwork		640	.013		.09	.56		.65	.98
0190	Roll		1050	.008		.09	.34		.43	.64
0200	Spray		2200	.004	↓	.07	.16		.23	.34
0210	Gutters & downspouts, metal, zinc chromate paint									
0220	Brushwork, gutters, 5", first coat	1 Pord	640	.013	L.F.	.40	.56		.96	1.31
0230	Second coat		960	.008		.38	.37		.75	1
0240	Third coat		1280	.006		.30	.28		.58	.77
0250	Downspouts, 4", first coat		640	.013		.40	.56		.96	1.31
0260	Second coat		960	.008		.38	.37		.75	1
0270	Third coat		1280	.006	↓	.30	.28		.58	.77
0280	Gutters & downspouts, wood									
0290	Brushwork, gutters, 5", primer	1 Pord	640	.013	L.F.	.07	.56		.63	.95

09 91 Painting

09 91 13 – Exterior Painting

09 91 13.80 Trim, Exterior

		Crew	Daily Output	Labor-Hours	Unit	Material	2020 Bare Costs Labor	Equipment	Total	Total Incl O&P
0300	Finish coat, exterior latex	1 Pord	640	.013	L.F.	.07	.56		.63	.96
0310	Primer & 1 coat exterior latex		400	.020		.15	.89		1.04	1.57
0320	Primer & 2 coats exterior latex		325	.025		.23	1.09		1.32	1.99
0330	Downspouts, 4", primer		640	.013		.07	.56		.63	.95
0340	Finish coat, exterior latex		640	.013		.07	.56		.63	.96
0350	Primer & 1 coat exterior latex		400	.020		.15	.89		1.04	1.57
0360	Primer & 2 coats exterior latex		325	.025		.11	1.09		1.20	1.87
0370	Molding, exterior, up to 14" wide									
0380	Brushwork, primer	1 Pord	640	.013	L.F.	.08	.56		.64	.97
0390	Finish coat, exterior latex		640	.013		.09	.56		.65	.98
0400	Primer & 1 coat exterior latex		400	.020		.18	.89		1.07	1.60
0410	Primer & 2 coats exterior latex		315	.025		.18	1.13		1.31	1.98
0420	Stain & fill		1050	.008		.12	.34		.46	.67
0430	Shellac		1850	.004		.16	.19		.35	.47
0440	Varnish		1275	.006		.12	.28		.40	.57
9000	Minimum labor/equipment charge		2	4	Job		178		178	282

09 91 13.90 Walls, Masonry (CMU), Exterior

		Crew	Daily Output	Labor-Hours	Unit	Material	2020 Bare Costs Labor	Equipment	Total	Total Incl O&P
0010	**WALLS, MASONRY (CMU), EXTERIOR**									
0360	Concrete masonry units (CMU), smooth surface									
0370	Brushwork, latex, first coat	1 Pord	640	.013	S.F.	.06	.56		.62	.95
0380	Second coat		960	.008		.05	.37		.42	.64
0390	Waterproof sealer, first coat		736	.011		.28	.48		.76	1.08
0400	Second coat		1104	.007		.28	.32		.60	.82
0410	Roll, latex, paint, first coat		1465	.005		.07	.24		.31	.46
0420	Second coat		1790	.004		.06	.20		.26	.38
0430	Waterproof sealer, first coat		1680	.005		.28	.21		.49	.65
0440	Second coat		2060	.004		.28	.17		.45	.58
0450	Spray, latex, paint, first coat		1950	.004		.06	.18		.24	.35
0460	Second coat		2600	.003		.05	.14		.19	.27
0470	Waterproof sealer, first coat		2245	.004		.28	.16		.44	.56
0480	Second coat		2990	.003		.28	.12		.40	.50
0490	Concrete masonry unit (CMU), porous									
0500	Brushwork, latex, first coat	1 Pord	640	.013	S.F.	.12	.56		.68	1.01
0510	Second coat		960	.008		.06	.37		.43	.66
0520	Waterproof sealer, first coat		736	.011		.28	.48		.76	1.08
0530	Second coat		1104	.007		.28	.32		.60	.82
0540	Roll latex, first coat		1465	.005		.09	.24		.33	.48
0550	Second coat		1790	.004		.06	.20		.26	.38
0560	Waterproof sealer, first coat		1680	.005		.28	.21		.49	.65
0570	Second coat		2060	.004		.28	.17		.45	.58
0580	Spray latex, first coat		1950	.004		.07	.18		.25	.36
0590	Second coat		2600	.003		.05	.14		.19	.27
0600	Waterproof sealer, first coat		2245	.004		.28	.16		.44	.56
0610	Second coat		2990	.003		.28	.12		.40	.50
9000	Minimum labor/equipment charge		2	4	Job		178		178	282

09 91 23 – Interior Painting

09 91 23.20 Cabinets and Casework

		Crew	Daily Output	Labor-Hours	Unit	Material	2020 Bare Costs Labor	Equipment	Total	Total Incl O&P
0010	**CABINETS AND CASEWORK**									
1000	Primer coat, oil base, brushwork	1 Pord	650	.012	S.F.	.07	.55		.62	.95
2000	Paint, oil base, brushwork, 1 coat		650	.012		.12	.55		.67	1
3000	Stain, brushwork, wipe off		650	.012		.10	.55		.65	.98
4000	Shellac, 1 coat, brushwork		650	.012		.13	.55		.68	1.02

09 91 23.20 Cabinets and Casework

		Crew	Daily Output	Labor-Hours	Unit	Material	2020 Bare Costs Labor	2020 Bare Costs Equipment	Total	Total Incl O&P
4500	Varnish, 3 coats, brushwork, sand after 1st coat	1 Pord	325	.025	S.F.	.29	1.09		1.38	2.06
5000	For latex paint, deduct				↓	10%				
6300	Strip, prep and refinish wood furniture									
6310	Remove paint using chemicals, wood furniture	1 Pord	28	.286	S.F.	1.55	12.70		14.25	21.50
6320	Prep for painting, sanding		75	.107		.25	4.74		4.99	7.75
6350	Stain and wipe, brushwork		600	.013		.10	.59		.69	1.05
6355	Spray applied		900	.009		.09	.39		.48	.73
6360	Sealer or varnish, brushwork		1080	.007		.10	.33		.43	.63
6365	Spray applied		2100	.004		.10	.17		.27	.38
6370	Paint, primer, brushwork		720	.011		.07	.49		.56	.86
6375	Spray applied		2100	.004		.07	.17		.24	.34
6380	Finish coat, brushwork		810	.010		.12	.44		.56	.83
6385	Spray applied		2100	.004	↓	.11	.17		.28	.40
9010	Minimum labor/equipment charge	↓	2	4	Job		178		178	282

09 91 23.33 Doors and Windows, Interior Alkyd (Oil Base)

		Crew	Daily Output	Labor-Hours	Unit	Material	2020 Bare Costs Labor	2020 Bare Costs Equipment	Total	Total Incl O&P
0010	**DOORS AND WINDOWS, INTERIOR ALKYD (OIL BASE)**									
0500	Flush door & frame, 3' x 7', oil, primer, brushwork	1 Pord	10	.800	Ea.	4.21	35.50		39.71	61
1000	Paint, 1 coat		10	.800		4.82	35.50		40.32	62
1400	Stain, brushwork, wipe off		18	.444		2.06	19.75		21.81	34
1600	Shellac, 1 coat, brushwork		25	.320		2.77	14.20		16.97	25.50
1800	Varnish, 3 coats, brushwork, sand after 1st coat		9	.889		6.05	39.50		45.55	69
2000	Panel door & frame, 3' x 7', oil, primer, brushwork		6	1.333		2.64	59		61.64	97
2200	Paint, 1 coat		6	1.333		4.82	59		63.82	99.50
2600	Stain, brushwork, panel door, 3' x 7', not incl. frame		16	.500		2.06	22		24.06	38
2800	Shellac, 1 coat, brushwork		22	.364		2.77	16.15		18.92	28.50
3000	Varnish, 3 coats, brushwork, sand after 1st coat	↓	7.50	1.067	↓	6.05	47.50		53.55	81.50
4400	Windows, including frame and trim, per side									
4600	Colonial type, 6/6 lites, 2' x 3', oil, primer, brushwork	1 Pord	14	.571	Ea.	.42	25.50		25.92	41
5800	Paint, 1 coat		14	.571		.76	25.50		26.26	41.50
6200	3' x 5' opening, 6/6 lites, primer coat, brushwork		12	.667		1.04	29.50		30.54	48
6400	Paint, 1 coat		12	.667		1.90	29.50		31.40	49
6800	4' x 8' opening, 6/6 lites, primer coat, brushwork		8	1		2.23	44.50		46.73	73
7000	Paint, 1 coat		8	1		4.06	44.50		48.56	75
8000	Single lite type, 2' x 3', oil base, primer coat, brushwork		33	.242		.42	10.75		11.17	17.55
8200	Paint, 1 coat		33	.242		.76	10.75		11.51	17.95
8600	3' x 5' opening, primer coat, brushwork		20	.400		1.04	17.75		18.79	29
8800	Paint, 1 coat		20	.400		1.90	17.75		19.65	30
9200	4' x 8' opening, primer coat, brushwork		14	.571		2.23	25.50		27.73	43
9400	Paint, 1 coat		14	.571	↓	4.06	25.50		29.56	45
9900	Minimum labor/equipment charge	↓	2	4	Job		178		178	282

09 91 23.35 Doors and Windows, Interior Latex

		Crew	Daily Output	Labor-Hours	Unit	Material	2020 Bare Costs Labor	2020 Bare Costs Equipment	Total	Total Incl O&P
0010	**DOORS & WINDOWS, INTERIOR LATEX**									
0100	Doors, flush, both sides, incl. frame & trim									
0110	Roll & brush, primer	1 Pord	10	.800	Ea.	4.17	35.50		39.67	61
0120	Finish coat, latex		10	.800		5.95	35.50		41.45	63
0130	Primer & 1 coat latex		7	1.143		10.10	50.50		60.60	91.50
0140	Primer & 2 coats latex		5	1.600		15.70	71		86.70	130
0160	Spray, both sides, primer		20	.400		4.39	17.75		22.14	33
0170	Finish coat, latex		20	.400		6.20	17.75		23.95	35
0180	Primer & 1 coat latex		11	.727		10.70	32.50		43.20	63.50
0190	Primer & 2 coats latex	↓	8	1	↓	16.60	44.50		61.10	89
0200	Doors, French, both sides, 10-15 lite, incl. frame & trim									

338

For customer support on your Commercial Renovation Costs with RSMeans data, call 800.448.8182.

09 91 23 – Interior Painting

09 91 23.35 Doors and Windows, Interior Latex

	Crew	Daily Output	Labor-Hours	Unit	Material	2020 Bare Costs Labor	2020 Bare Costs Equipment	Total	Total Incl O&P
0210 Roll & brush, primer	1 Pord	6	1.333	Ea.	2.08	59		61.08	96.50
0220 Finish coat, latex		6	1.333		2.96	59		61.96	97.50
0230 Primer & 1 coat latex		3	2.667		5.05	118		123.05	194
0240 Primer & 2 coats latex		2	4		7.85	178		185.85	291
0260 Doors, louvered, both sides, incl. frame & trim									
0270 Roll & brush, primer	1 Pord	7	1.143	Ea.	4.17	50.50		54.67	85
0280 Finish coat, latex		7	1.143		5.95	50.50		56.45	87
0290 Primer & 1 coat, latex		4	2		9.85	89		98.85	152
0300 Primer & 2 coats, latex		3	2.667		16.05	118		134.05	206
0320 Spray, both sides, primer		20	.400		4.39	17.75		22.14	33
0330 Finish coat, latex		20	.400		6.20	17.75		23.95	35
0340 Primer & 1 coat, latex		11	.727		10.70	32.50		43.20	63.50
0350 Primer & 2 coats, latex		8	1		17	44.50		61.50	89
0360 Doors, panel, both sides, incl. frame & trim									
0370 Roll & brush, primer	1 Pord	6	1.333	Ea.	4.39	59		63.39	99
0380 Finish coat, latex		6	1.333		5.95	59		64.95	101
0390 Primer & 1 coat, latex		3	2.667		10.10	118		128.10	199
0400 Primer & 2 coats, latex		2.50	3.200		16.05	142		158.05	244
0420 Spray, both sides, primer		10	.800		4.39	35.50		39.89	61.50
0430 Finish coat, latex		10	.800		6.20	35.50		41.70	63.50
0440 Primer & 1 coat, latex		5	1.600		10.70	71		81.70	125
0450 Primer & 2 coats, latex		4	2		17	89		106	160
0460 Windows, per interior side, based on 15 S.F.									
0470 1 to 6 lite									
0480 Brushwork, primer	1 Pord	13	.615	Ea.	.82	27.50		28.32	44.50
0490 Finish coat, enamel		13	.615		1.17	27.50		28.67	45
0500 Primer & 1 coat enamel		8	1		1.99	44.50		46.49	72.50
0510 Primer & 2 coats enamel		6	1.333		3.16	59		62.16	97.50
0530 7 to 10 lite									
0540 Brushwork, primer	1 Pord	11	.727	Ea.	.82	32.50		33.32	52.50
0550 Finish coat, enamel		11	.727		1.17	32.50		33.67	53
0560 Primer & 1 coat enamel		7	1.143		1.99	50.50		52.49	82.50
0570 Primer & 2 coats enamel		5	1.600		3.16	71		74.16	116
0590 12 lite									
0600 Brushwork, primer	1 Pord	10	.800	Ea.	.82	35.50		36.32	57.50
0610 Finish coat, enamel		10	.800		1.17	35.50		36.67	58
0620 Primer & 1 coat enamel		6	1.333		1.99	59		60.99	96
0630 Primer & 2 coats enamel		5	1.600		3.16	71		74.16	116
0650 For oil base paint, add					10%				
9000 Minimum labor/equipment charge	1 Pord	2	4	Job		178		178	282

09 91 23.39 Doors and Windows, Interior Latex, Zero Voc

		Crew	Daily Output	Labor-Hours	Unit	Material	2020 Bare Costs Labor	2020 Bare Costs Equipment	Total	Total Incl O&P
0010 **DOORS & WINDOWS, INTERIOR LATEX, ZERO VOC**										
0100 Doors flush, both sides, incl. frame & trim										
0110 Roll & brush, primer	G	1 Pord	10	.800	Ea.	5.70	35.50		41.20	63
0120 Finish coat, latex	G		10	.800		6.60	35.50		42.10	64
0130 Primer & 1 coat latex	G		7	1.143		12.25	50.50		62.75	94
0140 Primer & 2 coats latex	G		5	1.600		18.50	71		89.50	134
0160 Spray, both sides, primer	G		20	.400		6	17.75		23.75	34.50
0170 Finish coat, latex	G		20	.400		6.90	17.75		24.65	35.50
0180 Primer & 1 coat latex	G		11	.727		13	32.50		45.50	66
0190 Primer & 2 coats latex	G		8	1		19.55	44.50		64.05	92
0200 Doors, French, both sides, 10-15 lite, incl. frame & trim										

For customer support on your Commercial Renovation Costs with RSMeans data, call 800.448.8182.

339

09 91 Painting

09 91 23 – Interior Painting

09 91 23.39 Doors and Windows, Interior Latex, Zero Voc

			Crew	Daily Output	Labor-Hours	Unit	Material	2020 Bare Costs Labor	Equipment	Total	Total Incl O&P
0210	Roll & brush, primer	G	1 Pord	6	1.333	Ea.	2.84	59		61.84	97
0220	Finish coat, latex	G		6	1.333		3.29	59		62.29	97.50
0230	Primer & 1 coat latex	G		3	2.667		6.15	118		124.15	195
0240	Primer & 2 coats latex	G		2	4		9.25	178		187.25	292
0360	Doors, panel, both sides, incl. frame & trim										
0370	Roll & brush, primer	G	1 Pord	6	1.333	Ea.	6	59		65	101
0380	Finish coat, latex	G		6	1.333		6.60	59		65.60	101
0390	Primer & 1 coat, latex	G		3	2.667		12.25	118		130.25	202
0400	Primer & 2 coats, latex	G		2.50	3.200		18.85	142		160.85	247
0420	Spray, both sides, primer	G		10	.800		6	35.50		41.50	63
0430	Finish coat, latex	G		10	.800		6.90	35.50		42.40	64
0440	Primer & 1 coat, latex	G		5	1.600		13	71		84	127
0450	Primer & 2 coats, latex	G		4	2		20	89		109	163
0460	Windows, per interior side, based on 15 S.F.										
0470	1 to 6 lite										
0480	Brushwork, primer	G	1 Pord	13	.615	Ea.	1.12	27.50		28.62	44.50
0490	Finish coat, enamel	G		13	.615		1.30	27.50		28.80	45
0500	Primer & 1 coat enamel	G		8	1		2.42	44.50		46.92	73
0510	Primer & 2 coats enamel	G		6	1.333		3.72	59		62.72	98
9000	Minimum labor/equipment charge			2	4	Job		178		178	282

09 91 23.40 Floors, Interior

		Crew	Daily Output	Labor-Hours	Unit	Material	2020 Bare Costs Labor	Equipment	Total	Total Incl O&P
0010	**FLOORS, INTERIOR**									
0100	Concrete paint, latex									
0110	Brushwork									
0120	1st coat	1 Pord	975	.008	S.F.	.15	.36		.51	.75
0130	2nd coat		1150	.007		.10	.31		.41	.60
0140	3rd coat		1300	.006		.08	.27		.35	.52
0150	Roll									
0160	1st coat	1 Pord	2600	.003	S.F.	.20	.14		.34	.44
0170	2nd coat		3250	.002		.12	.11		.23	.30
0180	3rd coat		3900	.002		.09	.09		.18	.24
0190	Spray									
0200	1st coat	1 Pord	2600	.003	S.F.	.17	.14		.31	.41
0210	2nd coat		3250	.002		.10	.11		.21	.27
0220	3rd coat		3900	.002		.08	.09		.17	.22

09 91 23.44 Anti-Slip Floor Treatments

		Crew	Daily Output	Labor-Hours	Unit	Material	2020 Bare Costs Labor	Equipment	Total	Total Incl O&P
0010	**ANTI-SLIP FLOOR TREATMENTS**									
1000	Walking surface treatment, ADA compliant, mop on and rinse									
1100	For tile, terrazzo, stone or smooth concrete	1 Pord	4000	.002	S.F.	.34	.09		.43	.52
1110	For marble		4000	.002		.25	.09		.34	.41
1120	For wood		4000	.002		.23	.09		.32	.39
1130	For baths and showers		500	.016		.39	.71		1.10	1.56
2000	Granular additive for paint or sealer, add to paint cost					.02			.02	.02

09 91 23.52 Miscellaneous, Interior

		Crew	Daily Output	Labor-Hours	Unit	Material	2020 Bare Costs Labor	Equipment	Total	Total Incl O&P
0010	**MISCELLANEOUS, INTERIOR**									
2400	Floors, conc./wood, oil base, primer/sealer coat, brushwork	2 Pord	1950	.008	S.F.	.09	.36		.45	.68
2450	Roller		5200	.003		.09	.14		.23	.32
2600	Spray		6000	.003		.09	.12		.21	.29
2650	Paint 1 coat, brushwork		1950	.008		.10	.36		.46	.69
2800	Roller		5200	.003		.10	.14		.24	.33
2850	Spray		6000	.003		.11	.12		.23	.31
3000	Stain, wood floor, brushwork, 1 coat		4550	.004		.10	.16		.26	.36

09 91 Painting

09 91 23 – Interior Painting

09 91 23.52 Miscellaneous, Interior

		Crew	Daily Output	Labor-Hours	Unit	Material	2020 Bare Costs Labor	Equipment	Total	Total Incl O&P
3200	Roller	2 Pord	5200	.003	S.F.	.10	.14		.24	.33
3250	Spray		6000	.003		.10	.12		.22	.30
3400	Varnish, wood floor, brushwork		4550	.004		.10	.16		.26	.36
3450	Roller		5200	.003		.10	.14		.24	.33
3600	Spray	▼	6000	.003		.11	.12		.23	.31
3800	Grilles, per side, oil base, primer coat, brushwork	1 Pord	520	.015		.14	.68		.82	1.23
3850	Spray		1140	.007		.15	.31		.46	.65
3920	Paint 2 coats, brushwork		325	.025		.49	1.09		1.58	2.28
3940	Spray	▼	650	.012	▼	.56	.55		1.11	1.49
4600	Miscellaneous surfaces, metallic paint, spray applied									
4610	Water based, non-tintable, warm silver	1 Pord	1140	.007	S.F.	.59	.31		.90	1.14
4620	Rusted iron		1140	.007		1.04	.31		1.35	1.63
4630	Low VOC, tintable	▼	1140	.007	▼	.40	.31		.71	.93
5000	Pipe, 1"-4" diameter, primer or sealer coat, oil base, brushwork	2 Pord	1250	.013	L.F.	.09	.57		.66	1
5100	Spray		2165	.007		.09	.33		.42	.62
5350	Paint 2 coats, brushwork		775	.021		.23	.92		1.15	1.71
5400	Spray		1240	.013		.25	.57		.82	1.19
6300	13"-16" diameter, primer or sealer coat, brushwork		310	.052		.37	2.29		2.66	4.04
6450	Spray		540	.030		.52	1.32		1.84	2.66
6500	Paint 2 coats, brushwork		195	.082		.91	3.64		4.55	6.80
6550	Spray	▼	310	.052	▼	1.01	2.29		3.30	4.75
7000	Trim, wood, incl. puttying, under 6" wide									
7200	Primer coat, oil base, brushwork	1 Pord	650	.012	L.F.	.03	.55		.58	.91
7250	Paint, 1 coat, brushwork		650	.012		.06	.55		.61	.94
7450	3 coats		325	.025		.18	1.09		1.27	1.94
7500	Over 6" wide, primer coat, brushwork		650	.012		.07	.55		.62	.95
7550	Paint, 1 coat, brushwork		650	.012		.13	.55		.68	1.01
7650	3 coats		325	.025	▼	.37	1.09		1.46	2.14
8000	Cornice, simple design, primer coat, oil base, brushwork		650	.012	S.F.	.07	.55		.62	.95
8250	Paint, 1 coat		650	.012		.13	.55		.68	1.01
8350	Ornate design, primer coat		350	.023		.07	1.02		1.09	1.69
8400	Paint, 1 coat		350	.023		.13	1.02		1.15	1.75
8600	Balustrades, primer coat, oil base, brushwork		520	.015		.07	.68		.75	1.16
8900	Trusses and wood frames, primer coat, oil base, brushwork		800	.010		.07	.44		.51	.79
8950	Spray		1200	.007		.07	.30		.37	.55
9220	Paint 2 coats, brushwork		500	.016		.25	.71		.96	1.40
9240	Spray		600	.013		.27	.59		.86	1.24
9260	Stain, brushwork, wipe off		600	.013		.10	.59		.69	1.05
9280	Varnish, 3 coats, brushwork	▼	275	.029	▼	.29	1.29		1.58	2.37
9350	For latex paint, deduct					10%				
9900	Minimum labor/equipment charge	1 Pord	2	4	Job		178		178	282

09 91 23.62 Electrostatic Painting

		Crew	Daily Output	Labor-Hours	Unit	Material	2020 Bare Costs Labor	Equipment	Total	Total Incl O&P
0010	**ELECTROSTATIC PAINTING**									
0100	In shop									
0200	Flat surfaces (lockers, casework, elevator doors, etc.)									
0300	One coat	1 Pord	200	.040	S.F.	.65	1.78		2.43	3.53
0400	Two coats	"	120	.067	"	.94	2.96		3.90	5.75
0500	Irregular surfaces (furniture, door frames, etc.)									
0600	One coat	1 Pord	150	.053	S.F.	.65	2.37		3.02	4.47
0700	Two coats	"	100	.080	"	.94	3.55		4.49	6.70
0800	On site									
0900	Flat surfaces (lockers, casework, elevator doors, etc.)									

09 91 Painting

09 91 23 – Interior Painting

09 91 23.62 Electrostatic Painting

		Crew	Daily Output	Labor-Hours	Unit	Material	2020 Bare Costs Labor	Equipment	Total	Total Incl O&P
1000	One coat	1 Pord	150	.053	S.F.	.65	2.37		3.02	4.47
1100	Two coats	"	100	.080	"	.94	3.55		4.49	6.70
1200	Irregular surfaces (furniture, door frames, etc.)									
1300	One coat	1 Pord	115	.070	S.F.	.65	3.09		3.74	5.60
1400	Two coats	↓	70	.114	↓	.94	5.05		5.99	9.10
2000	Anti-microbial coating, hospital application	↓	150	.053	↓	.06	2.37		2.43	3.82

09 91 23.72 Walls and Ceilings, Interior

		Crew	Daily Output	Labor-Hours	Unit	Material	2020 Bare Costs Labor	Equipment	Total	Total Incl O&P
0010	**WALLS AND CEILINGS, INTERIOR**									
0100	Concrete, drywall or plaster, latex, primer or sealer coat									
0150	Smooth finish, cut-in by brush	1 Pord	1150	.007	L.F.	.02	.31		.33	.51
0200	Brushwork		1150	.007	S.F.	.06	.31		.37	.56
0240	Roller		1350	.006		.06	.26		.32	.49
0280	Spray		2750	.003	↓	.05	.13		.18	.27
0290	Sand finish, cut-in by brush		975	.008	L.F.	.02	.36		.38	.60
0300	Brushwork		975	.008	S.F.	.06	.36		.42	.65
0340	Roller		1150	.007		.06	.31		.37	.56
0380	Spray		2275	.004		.05	.16		.21	.31
0590	Paint 2 coats, smooth finish, cut-in by brush		680	.012		.04	.52		.56	.87
0800	Brushwork		680	.012		.15	.52		.67	1
0840	Roller		800	.010		.15	.44		.59	.88
0880	Spray		1625	.005	↓	.14	.22		.36	.51
0890	Sand finish, cut-in by brush		605	.013	L.F.	.04	.59		.63	.97
0900	Brushwork		605	.013	S.F.	.15	.59		.74	1.10
0940	Roller		1020	.008		.15	.35		.50	.72
0980	Spray		1700	.005		.14	.21		.35	.49
1600	Glaze coating, 2 coats, spray, clear		1200	.007		.56	.30		.86	1.09
1640	Multicolor	↓	1200	.007	↓	.87	.30		1.17	1.43
1660	Painting walls, complete, including surface prep, primer &									
1670	2 coats finish, on drywall or plaster, with roller	1 Pord	325	.025	S.F.	.22	1.09		1.31	1.99
1700	For oil base paint, add					10%				
1800	For ceiling installations, add				↓		25%			
2000	Masonry or concrete block, primer/sealer, latex paint									
2090	Primer, smooth finish, cut-in by brush	1 Pord	1000	.008	L.F.	.04	.36		.40	.61
2100	Brushwork		1000	.008	S.F.	.17	.36		.53	.75
2110	Roller		1150	.007		.11	.31		.42	.61
2180	Spray		2400	.003	↓	.10	.15		.25	.34
2190	Sand finish, cut-in by brush		850	.009	L.F.	.03	.42		.45	.69
2200	Brushwork		850	.009	S.F.	.11	.42		.53	.78
2210	Roller		975	.008		.11	.36		.47	.70
2280	Spray		2050	.004	↓	.10	.17		.27	.39
2590	Primer plus one finish coat, smooth cut-in by brush		525	.015	L.F.	.08	.68		.76	1.15
2800	Brushwork		525	.015	S.F.	.30	.68		.98	1.40
2810	Roller		615	.013		.19	.58		.77	1.13
2880	Spray		1200	.007	↓	.17	.30		.47	.66
2890	Sand finish, cut-in by brush		450	.018	L.F.	.05	.79		.84	1.30
2900	Brushwork		450	.018	S.F.	.19	.79		.98	1.46
2910	Roller		515	.016		.19	.69		.88	1.30
2980	Spray		1025	.008		.17	.35		.52	.74
3600	Glaze coating, 3 coats, spray, clear		900	.009		.80	.39		1.19	1.51
3620	Multicolor		900	.009		1.08	.39		1.47	1.82
4000	Block filler, 1 coat, brushwork		425	.019		.13	.84		.97	1.48
4100	Silicone, water repellent, 2 coats, spray	↓	2000	.004	↓	.48	.18		.66	.80

09 91 Painting

09 91 23 – Interior Painting

09 91 23.72 Walls and Ceilings, Interior

		Crew	Daily Output	Labor-Hours	Unit	Material	2020 Bare Costs Labor	Equipment	Total	Total Incl O&P
4120	For oil base paint, add				S.F.	10%				
8200	For work 8'-15' H, add						10%			
8300	For work over 15' H, add						20%			
8400	For light textured surfaces, add						10%			
8410	Heavy textured, add						25%			
9900	Minimum labor/equipment charge	1 Pord	2	4	Job		178		178	282

09 91 23.74 Walls and Ceilings, Interior, Zero VOC Latex

		Crew	Daily Output	Labor-Hours	Unit	Material	2020 Bare Costs Labor	Equipment	Total	Total Incl O&P
0010	**WALLS AND CEILINGS, INTERIOR, ZERO VOC LATEX**									
0100	Concrete, dry wall or plaster, latex, primer or sealer coat									
0190	Smooth finish, cut-in by brush	1 Pord	1150	.007	L.F.	.02	.31		.33	.51
0200	Brushwork [G]		1150	.007	S.F.	.07	.31		.38	.57
0240	Roller [G]		1350	.006		.07	.26		.33	.50
0280	Spray [G]		2750	.003		.05	.13		.18	.27
0290	Sand finish, cut-in by brush		975	.008	L.F.	.02	.36		.38	.60
0300	Brushwork [G]		975	.008	S.F.	.07	.36		.43	.66
0340	Roller [G]		1150	.007		.08	.31		.39	.58
0380	Spray [G]		2275	.004		.06	.16		.22	.32
0790	Paint 2 coats, smooth finish, cut-in by brush		680	.012	L.F.	.04	.52		.56	.87
0800	Brushwork [G]		680	.012	S.F.	.04	.52		.56	.87
0840	Roller [G]		800	.010		.17	.44		.61	.89
0880	Spray [G]		1625	.005		.14	.22		.36	.51
0890	Sand finish, cut-in by brush		605	.013	L.F.	.04	.59		.63	.97
0900	Brushwork [G]		605	.013	S.F.	.16	.59		.75	1.10
0940	Roller [G]		1020	.008		.17	.35		.52	.73
0980	Spray [G]		1700	.005		.14	.21		.35	.49
1800	For ceiling installations, add [G]						25%			
8200	For work 8' - 15' H, add						10%			
8300	For work over 15' H, add						20%			
9900	Minimum labor/equipment charge	1 Pord	2	4	Job		178		178	282

09 91 23.75 Dry Fall Painting

		Crew	Daily Output	Labor-Hours	Unit	Material	2020 Bare Costs Labor	Equipment	Total	Total Incl O&P
0010	**DRY FALL PAINTING**									
0100	Sprayed on walls, gypsum board or plaster									
0220	One coat	1 Pord	2600	.003	S.F.	.08	.14		.22	.31
0250	Two coats		1560	.005		.16	.23		.39	.54
0280	Concrete or textured plaster, one coat		1560	.005		.08	.23		.31	.45
0310	Two coats		1300	.006		.16	.27		.43	.61
0340	Concrete block, one coat		1560	.005		.08	.23		.31	.45
0370	Two coats		1300	.006		.16	.27		.43	.61
0400	Wood, one coat		877	.009		.08	.40		.48	.73
0430	Two coats		650	.012		.16	.55		.71	1.05
0440	On ceilings, gypsum board or plaster									
0470	One coat	1 Pord	1560	.005	S.F.	.08	.23		.31	.45
0500	Two coats		1300	.006		.16	.27		.43	.61
0530	Concrete or textured plaster, one coat		1560	.005		.08	.23		.31	.45
0560	Two coats		1300	.006		.16	.27		.43	.61
0570	Structural steel, bar joists or metal deck, one coat		1560	.005		.08	.23		.31	.45
0580	Two coats		1040	.008		.16	.34		.50	.72
9900	Minimum labor/equipment charge		2	4	Job		178		178	282

For customer support on your Commercial Renovation Costs with RSMeans data, call 800.448.8182.

343

09 93 Staining and Transparent Finishing

09 93 23 – Interior Staining and Finishing

09 93 23.10 Varnish

	09 93 23.10 Varnish	Crew	Daily Output	Labor-Hours	Unit	Material	2020 Bare Costs Labor	Equipment	Total	Total Incl O&P
0010	**VARNISH**									
0012	1 coat + sealer, on wood trim, brush, no sanding included	1 Pord	400	.020	S.F.	.07	.89		.96	1.49
0020	1 coat + sealer, on wood trim, brush, no sanding included, no VOC		400	.020		.22	.89		1.11	1.65
0100	Hardwood floors, 2 coats, no sanding included, roller		1890	.004	↓	.15	.19		.34	.47
9000	Minimum labor/equipment charge	↓	4	2	Job		89		89	141

09 96 High-Performance Coatings

09 96 23 – Graffiti-Resistant Coatings

09 96 23.10 Graffiti-Resistant Treatments

	09 96 23.10 Graffiti-Resistant Treatments	Crew	Daily Output	Labor-Hours	Unit	Material	2020 Bare Costs Labor	Equipment	Total	Total Incl O&P
0010	**GRAFFITI-RESISTANT TREATMENTS**, sprayed on walls									
0100	Non-sacrificial, permanent non-stick coating, clear, on metals	1 Pord	2000	.004	S.F.	2.05	.18		2.23	2.54
0200	Concrete		2000	.004		2.33	.18		2.51	2.84
0300	Concrete block		2000	.004		3.02	.18		3.20	3.60
0400	Brick		2000	.004		3.42	.18		3.60	4.04
0500	Stone		2000	.004		3.42	.18		3.60	4.04
0600	Unpainted wood		2000	.004		3.94	.18		4.12	4.62
2000	Semi-permanent cross linking polymer primer, on metals		2000	.004		.62	.18		.80	.96
2100	Concrete		2000	.004		.74	.18		.92	1.09
2200	Concrete block		2000	.004		.92	.18		1.10	1.30
2300	Brick		2000	.004		.74	.18		.92	1.09
2400	Stone		2000	.004		.74	.18		.92	1.09
2500	Unpainted wood		2000	.004		1.03	.18		1.21	1.41
3000	Top coat, on metals		2000	.004		.56	.18		.74	.89
3100	Concrete		2000	.004		.64	.18		.82	.98
3200	Concrete block		2000	.004		.89	.18		1.07	1.26
3300	Brick		2000	.004		.74	.18		.92	1.10
3400	Stone		2000	.004		.74	.18		.92	1.10
3500	Unpainted wood		2000	.004		.89	.18		1.07	1.26
5000	Sacrificial, water based, on metal		2000	.004		.34	.18		.52	.65
5100	Concrete		2000	.004		.34	.18		.52	.65
5200	Concrete block		2000	.004		.34	.18		.52	.65
5300	Brick		2000	.004		.34	.18		.52	.65
5400	Stone		2000	.004		.34	.18		.52	.65
5500	Unpainted wood	↓	2000	.004	↓	.34	.18		.52	.65
8000	Cleaner for use after treatment									
8100	Towels or wipes, per package of 30				Ea.	.72			.72	.79
8200	Aerosol spray, 24 oz. can				"	19			19	21

09 96 46 – Intumescent Painting

09 96 46.10 Coatings, Intumescent

	09 96 46.10 Coatings, Intumescent	Crew	Daily Output	Labor-Hours	Unit	Material	2020 Bare Costs Labor	Equipment	Total	Total Incl O&P
0010	**COATINGS, INTUMESCENT**, spray applied									
0100	On exterior structural steel, 0.25" d.f.t.	1 Pord	475	.017	S.F.	.50	.75		1.25	1.74
0150	0.51" d.f.t.		350	.023		.50	1.02		1.52	2.16
0200	0.98" d.f.t.		280	.029		.50	1.27		1.77	2.56
0300	On interior structural steel, 0.108" d.f.t.		300	.027		.49	1.18		1.67	2.42
0350	0.310" d.f.t.		150	.053		.49	2.37		2.86	4.30
0400	0.670" d.f.t.	↓	100	.080	↓	.49	3.55		4.04	6.20

09 96 High-Performance Coatings

09 96 53 – Elastomeric Coatings

09 96 53.10 Coatings, Elastomeric	Crew	Daily Output	Labor-Hours	Unit	Material	2020 Bare Costs Labor	Equipment	Total	Total Incl O&P
0010 **COATINGS, ELASTOMERIC**									
0020 High build, water proof, one coat system									
0100 Concrete, brush	1 Pord	650	.012	S.F.	.29	.55		.84	1.19

09 96 56 – Epoxy Coatings

09 96 56.20 Wall Coatings

	Crew	Daily Output	Labor-Hours	Unit	Material	2020 Bare Costs Labor	Equipment	Total	Total Incl O&P
0010 **WALL COATINGS**									
0100 Acrylic glazed coatings, matte	1 Pord	525	.015	S.F.	.38	.68		1.06	1.49
0200 Gloss		305	.026		.78	1.16		1.94	2.71
0300 Epoxy coatings, solvent based		525	.015		.47	.68		1.15	1.59
0400 Water based		170	.047		.38	2.09		2.47	3.74
2400 Sprayed perlite or vermiculite, 1/16" thick, solvent based		2935	.003		.28	.12		.40	.50
2500 Water based		640	.013		.87	.56		1.43	1.84
2700 Vinyl plastic wall coating, solvent based		735	.011		.41	.48		.89	1.22
2800 Water based		240	.033		1	1.48		2.48	3.45
3000 Urethane on smooth surface, 2 coats, solvent based		1135	.007		.35	.31		.66	.89
3100 Water based		665	.012		.62	.53		1.15	1.53

09 97 Special Coatings

09 97 35 – Dry Erase Coatings

09 97 35.10 Dry Erase Coatings

	Crew	Daily Output	Labor-Hours	Unit	Material	2020 Bare Costs Labor	Equipment	Total	Total Incl O&P
0010 **DRY ERASE COATINGS**									
0020 Dry erase coatings, clear, roller applied	1 Pord	1325	.006	S.F.	2.14	.27		2.41	2.78

Division Notes

	CREW	DAILY OUTPUT	LABOR-HOURS	UNIT	BARE COSTS				TOTAL INCL O&P
					MAT.	LABOR	EQUIP.	TOTAL	

Estimating Tips
General

- The items in this division are usually priced per square foot or each.

- Many items in Division 10 require some type of support system or special anchors that are not usually furnished with the item. The required anchors must be added to the estimate in the appropriate division.

- Some items in Division 10, such as lockers, may require assembly before installation. Verify the amount of assembly required. Assembly can often exceed installation time.

10 20 00 Interior Specialties

- Support angles and blocking are not included in the installation of toilet compartments, shower/dressing compartments, or cubicles. Appropriate line items from Division 5 or 6 may need to be added to support the installations.

- Toilet partitions are priced by the stall. A stall consists of a side wall, pilaster, and door with hardware. Toilet tissue holders and grab bars are extra.

- The required acoustical rating of a folding partition can have a significant impact on costs. Verify the sound transmission coefficient rating of the panel priced against the specification requirements.

- Grab bar installation does not include supplemental blocking or backing to support the required load. When grab bars are installed at an existing facility, provisions must be made to attach the grab bars to a solid structure.

Reference Numbers

Reference numbers are shown at the beginning of some major classifications. These numbers refer to related items in the Reference Section. The reference information may be an estimating procedure, an alternate pricing method, or technical information.

Note: Not all subdivisions listed here necessarily appear. ■

Same Data. Simplified.

Enjoy the convenience and efficiency of accessing your costs anywhere:

- **Skip the multiplier** by setting your location
- **Quickly search,** edit, favorite and share costs
- **Stay on top of price changes** with automatic updates

Discover more at rsmeans.com/online

10 11 13 – Chalkboards

10 11 13.13 Fixed Chalkboards

		Crew	Daily Output	Labor-Hours	Unit	Material	2020 Bare Costs Labor	Equipment	Total	Total Incl O&P
0010	**FIXED CHALKBOARDS** Porcelain enamel steel									
3900	Wall hung									
4000	Aluminum frame and chalktrough									
4300	3' x 5'	2 Carp	15	1.067	Ea.	315	56.50		371.50	440
4600	4' x 12'	"	13	1.231	"	565	65.50		630.50	725
4700	Wood frame and chalktrough									
4800	3' x 4'	2 Carp	16	1	Ea.	194	53		247	298
5300	4' x 8'	"	13	1.231	"	375	65.50		440.50	515
5400	Liquid chalk, white porcelain enamel, wall hung									
5420	Deluxe units, aluminum trim and chalktrough									
5450	4' x 4'	2 Carp	16	1	Ea.	315	53		368	430
5550	4' x 12'	"	12	1.333	"	635	71		706	810
5700	Wood trim and chalktrough									
5900	4' x 4'	2 Carp	16	1	Ea.	765	53		818	925
6200	4' x 8'		14	1.143	"	1,050	60.50		1,110.50	1,250
9000	Minimum labor/equipment charge		3	5.333	Job		283		283	455

10 11 13.43 Portable Chalkboards

		Crew	Daily Output	Labor-Hours	Unit	Material	Labor	Equipment	Total	Total Incl O&P
0010	**PORTABLE CHALKBOARDS**									
0100	Freestanding, reversible									
0120	Economy, wood frame, 4' x 6'									
0140	Chalkboard both sides				Ea.	630			630	695
0160	Chalkboard one side, cork other side				"	625			625	685
0200	Standard, lightweight satin finished aluminum, 4' x 6'									
0220	Chalkboard both sides				Ea.	565			565	625
0240	Chalkboard one side, cork other side				"	610			610	675
0300	Deluxe, heavy duty extruded aluminum, 4' x 6'									
0320	Chalkboard both sides				Ea.	1,150			1,150	1,250
0340	Chalkboard one side, cork other side				"	1,175			1,175	1,300

10 11 16 – Markerboards

10 11 16.53 Electronic Markerboards

		Crew	Daily Output	Labor-Hours	Unit	Material	Labor	Equipment	Total	Total Incl O&P
0010	**ELECTRONIC MARKERBOARDS**									
0100	Wall hung or free standing, 3' x 4' to 4' x 6'	2 Carp	8	2	S.F.	86.50	106		192.50	265
0150	5' x 6' to 4' x 8'		8	2	"	62.50	106		168.50	239
0500	Interactive projection module for existing whiteboards		8	2	Ea.	1,300	106		1,406	1,625

10 11 23 – Tackboards

10 11 23.10 Fixed Tackboards

		Crew	Daily Output	Labor-Hours	Unit	Material	Labor	Equipment	Total	Total Incl O&P
0010	**FIXED TACKBOARDS**									
2120	Prefabricated, 1/4" cork, 3' x 5' with aluminum frame	2 Carp	16	1	Ea.	146	53		199	246
2140	Wood frame	"	16	1	"	168	53		221	269
2300	Glass enclosed cabinets, alum., cork panel, hinged doors									
2600	4' x 7', 3 door	2 Carp	10	1.600	Ea.	2,125	85		2,210	2,450
9000	Minimum labor/equipment charge	"	4	4	Job		213		213	340

10 11 23.20 Control Boards

		Crew	Daily Output	Labor-Hours	Unit	Material	Labor	Equipment	Total	Total Incl O&P
0010	**CONTROL BOARDS**									
1000	Hospital patient display board, 4-color custom design									
1010	Porcelain steel dry erase board, 36" x 24"	2 Carp	7.50	2.133	Ea.	238	113		351	445

10 13 Directories

10 13 10 – Building Directories

10 13 10.10 Directory Boards	Crew	Daily Output	Labor-Hours	Unit	Material	2020 Bare Costs Labor	Equipment	Total	Total Incl O&P
0010 **DIRECTORY BOARDS**									
0050 Plastic, glass covered, 30" x 20"	2 Carp	3	5.333	Ea.	179	283		462	650
0100 36" x 48"		2	8		880	425		1,305	1,650
0900 Outdoor, weatherproof, black plastic, 36" x 24"		2	8		700	425		1,125	1,450
1000 36" x 36"		1.50	10.667		815	565		1,380	1,800
9000 Minimum labor/equipment charge	1 Carp	1	8	Job		425		425	680

10 14 Signage

10 14 19 – Dimensional Letter Signage

10 14 19.10 Exterior Signs	Crew	Daily Output	Labor-Hours	Unit	Material	2020 Bare Costs Labor	Equipment	Total	Total Incl O&P
0010 **EXTERIOR SIGNS**									
3900 Plaques, custom, 20" x 30", for up to 450 letters, cast aluminum	2 Carp	4	4	Ea.	1,875	213		2,088	2,400
4000 Cast bronze		4	4		1,950	213		2,163	2,500
4200 30" x 36", up to 900 letters, cast aluminum		3	5.333		2,925	283		3,208	3,675
4300 Cast bronze		3	5.333		4,200	283		4,483	5,075
5100 Exit signs, 24 ga. alum., 14" x 12" surface mounted	1 Carp	30	.267		52	14.15		66.15	79.50
5200 10" x 7"	"	20	.400		32.50	21.50		54	70
6400 Replacement sign faces, 6" or 8"	1 Clab	50	.160		66.50	6.75		73.25	84.50
9000 Minimum labor/equipment charge	1 Carp	4	2	Job		106		106	170

10 14 23 – Panel Signage

10 14 23.13 Engraved Panel Signage	Crew	Daily Output	Labor-Hours	Unit	Material	2020 Bare Costs Labor	Equipment	Total	Total Incl O&P
0010 **ENGRAVED PANEL SIGNAGE**, interior									
1010 Flexible door sign, adhesive back, w/Braille, 5/8" letters, 4" x 4"	1 Clab	32	.250	Ea.	35	10.55		45.55	55
1050 6" x 6"		32	.250		50	10.55		60.55	72
1100 8" x 2"		32	.250		35.50	10.55		46.05	56
1150 8" x 4"		32	.250		44.50	10.55		55.05	66
1200 8" x 8"		32	.250		74	10.55		84.55	98
1250 12" x 2"		32	.250		36	10.55		46.55	56.50
1300 12" x 6"		32	.250		42.50	10.55		53.05	64
1350 12" x 12"		32	.250		158	10.55		168.55	191
1500 Graphic symbols, 2" x 2"		32	.250		12.05	10.55		22.60	30
1550 6" x 6"		32	.250		31	10.55		41.55	51
1600 8" x 8"		32	.250		39	10.55		49.55	59.50
2010 Corridor, stock acrylic, 2-sided, with mounting bracket, 2" x 8"	1 Carp	24	.333		29.50	17.70		47.20	61
2020 2" x 10"		24	.333		34	17.70		51.70	66
2050 3" x 8"		24	.333		27	17.70		44.70	58
2060 3" x 10"		24	.333		39	17.70		56.70	71.50
2070 3" x 12"		24	.333		46.50	17.70		64.20	79.50
2100 4" x 8"		24	.333		23.50	17.70		41.20	54.50
2110 4" x 10"		24	.333		40	17.70		57.70	72.50
2120 4" x 12"		24	.333		49.50	17.70		67.20	83

10 14 26 – Post and Panel/Pylon Signage

10 14 26.10 Post and Panel Signage	Crew	Daily Output	Labor-Hours	Unit	Material	2020 Bare Costs Labor	Equipment	Total	Total Incl O&P
0010 **POST AND PANEL SIGNAGE**, Alum., incl. 2 posts,									
0011 2-sided panel (blank), concrete footings per post									
0025 24" W x 18" H	B-1	25	.960	Ea.	905	41		946	1,075
0050 36" W x 24" H		25	.960		950	41		991	1,125
0100 57" W x 36" H		25	.960		2,150	41		2,191	2,425
0150 81" W x 46" H		25	.960		2,225	41		2,266	2,525

10 14 Signage

10 14 43 – Photoluminescent Signage

10 14 43.10 Photoluminescent Signage	Crew	Daily Output	Labor-Hours	Unit	Material	2020 Bare Costs Labor	Equipment	Total	Total Incl O&P
0010 **PHOTOLUMINESCENT SIGNAGE**									
0011 Photoluminescent exit sign, plastic, 7"x10"	1 Carp	32	.250	Ea.	24	13.30		37.30	48
0022 Polyester		32	.250		15.20	13.30		28.50	38
0033 Aluminum		32	.250		53.50	13.30		66.80	80
0044 Plastic, 10"x14"		32	.250		59.50	13.30		72.80	87
0055 Polyester		32	.250		35.50	13.30		48.80	60.50
0066 Aluminum		32	.250		70.50	13.30		83.80	99
0111 Directional exit sign, plastic, 10"x14"		32	.250		58.50	13.30		71.80	85.50
0122 Polyester		32	.250		46	13.30		59.30	72
0133 Aluminum		32	.250		68.50	13.30		81.80	96.50
0153 Kick plate exit sign, 10"x34"		28	.286		278	15.20		293.20	330
0173 Photoluminescent tape, 1"x60'		300	.027	L.F.	.50	1.42		1.92	2.82
0183 2"x60'		290	.028	"	1.01	1.47		2.48	3.46
0211 Photoluminescent directional arrows		42	.190	Ea.	1.30	10.10		11.40	17.65

10 14 53 – Traffic Signage

10 14 53.20 Traffic Signs

	Crew	Daily Output	Labor-Hours	Unit	Material	2020 Bare Costs Labor	Equipment	Total	Total Incl O&P
0010 **TRAFFIC SIGNS**									
0012 Stock, 24" x 24", no posts, .080" alum. reflectorized	B-80	70	.457	Ea.	93.50	21.50	14.60	129.60	153
0100 High intensity		70	.457		93.50	21.50	14.60	129.60	153
0300 30" x 30", reflectorized		70	.457		65	21.50	14.60	101.10	122
0400 High intensity		70	.457		172	21.50	14.60	208.10	239
0600 Guide and directional signs, 12" x 18", reflectorized		70	.457		34	21.50	14.60	70.10	87
0700 High intensity		70	.457		65.50	21.50	14.60	101.60	122
0900 18" x 24", stock signs, reflectorized		70	.457		48.50	21.50	14.60	84.60	103
1000 High intensity		70	.457		53.50	21.50	14.60	89.60	109
1200 24" x 24", stock signs, reflectorized		70	.457		59	21.50	14.60	95.10	115
1300 High intensity		70	.457		63.50	21.50	14.60	99.60	120
1500 Add to above for steel posts, galvanized, 10'-0" upright, bolted		200	.160		44	7.45	5.10	56.55	66
1600 12'-0" upright, bolted		140	.229		38.50	10.65	7.30	56.45	67.50
1800 Highway road signs, aluminum, over 20 S.F., reflectorized		350	.091	S.F.	16.45	4.26	2.92	23.63	28
2000 High intensity		350	.091		16.45	4.26	2.92	23.63	28
2200 Highway, suspended over road, 80 S.F. min., reflectorized		165	.194		19.35	9.05	6.20	34.60	42.50
2300 High intensity		165	.194		19.20	9.05	6.20	34.45	42
9000 Replace directional sign		6	5.333	Ea.	172	249	170	591	770

10 17 Telephone Specialties

10 17 16 – Telephone Enclosures

10 17 16.10 Commercial Telephone Enclosures

	Crew	Daily Output	Labor-Hours	Unit	Material	2020 Bare Costs Labor	Equipment	Total	Total Incl O&P
0010 **COMMERCIAL TELEPHONE ENCLOSURES**									
0300 Shelf type, wall hung, recessed	2 Carp	5	3.200	Ea.	885	170		1,055	1,250
0400 Surface mount		5	3.200	"	1,700	170		1,870	2,150
9000 Minimum labor/equipment charge		4	4	Job		213		213	340

For customer support on your Commercial Renovation Costs with RSMeans data, call 800.448.8182.

10 21 Compartments and Cubicles

10 21 13 – Toilet Compartments

10 21 13.13 Metal Toilet Compartments		Crew	Daily Output	Labor-Hours	Unit	Material	2020 Bare Costs Labor	2020 Bare Costs Equipment	Total	Total Incl O&P
0010	**METAL TOILET COMPARTMENTS**									
0110	Cubicles, ceiling hung									
0200	Powder coated steel	2 Carp	4	4	Ea.	575	213		788	975
0500	Stainless steel	"	4	4		1,100	213		1,313	1,550
0600	For handicap units, add					465			465	510
0900	Floor and ceiling anchored									
1000	Powder coated steel	2 Carp	5	3.200	Ea.	580	170		750	910
1300	Stainless steel	"	5	3.200		1,225	170		1,395	1,625
1400	For handicap units, add					370			370	405
1610	Floor anchored									
1700	Powder coated steel	2 Carp	7	2.286	Ea.	615	121		736	870
2000	Stainless steel	"	7	2.286		1,350	121		1,471	1,675
2100	For handicap units, add					360			360	395
2200	For juvenile units, deduct					44.50			44.50	49
2450	Floor anchored, headrail braced									
2500	Powder coated steel	2 Carp	6	2.667	Ea.	385	142		527	650
2804	Stainless steel	"	6	2.667		1,050	142		1,192	1,400
2900	For handicap units, add					325			325	360
3000	Wall hung partitions, powder coated steel	2 Carp	7	2.286		675	121		796	940
3300	Stainless steel	"	7	2.286		1,775	121		1,896	2,150
3400	For handicap units, add					325			325	360
4000	Screens, entrance, floor mounted, 58" high, 48" wide									
4200	Powder coated steel	2 Carp	15	1.067	Ea.	240	56.50		296.50	355
4500	Stainless steel	"	15	1.067	"	970	56.50		1,026.50	1,175
4650	Urinal screen, 18" wide									
4704	Powder coated steel	2 Carp	6.15	2.602	Ea.	226	138		364	470
5004	Stainless steel	"	6.15	2.602	"	535	138		673	805
5100	Floor mounted, headrail braced									
5300	Powder coated steel	2 Carp	8	2	Ea.	234	106		340	425
5600	Stainless steel	"	8	2	"	580	106		686	810
5750	Pilaster, flush									
5800	Powder coated steel	2 Carp	10	1.600	Ea.	275	85		360	435
6100	Stainless steel		10	1.600		610	85		695	810
6300	Post braced, powder coated steel		10	1.600		167	85		252	320
6800	Powder coated steel		10	1.600		176	85		261	330
7800	Wedge type, powder coated steel		10	1.600		138	85		223	288
8100	Stainless steel		10	1.600		615	85		700	810
9000	Minimum labor/equipment charge	1 Carp	2.50	3.200	Job		170		170	272

10 21 13.16 Plastic-Laminate-Clad Toilet Compartments

10 21 13.16 Plastic-Laminate-Clad Toilet Compartments		Crew	Daily Output	Labor-Hours	Unit	Material	2020 Bare Costs Labor	2020 Bare Costs Equipment	Total	Total Incl O&P
0010	**PLASTIC-LAMINATE-CLAD TOILET COMPARTMENTS**									
0110	Cubicles, ceiling hung									
0300	Plastic laminate on particle board	2 Carp	4	4	Ea.	550	213		763	945
0600	For handicap units, add				"	465			465	510
0900	Floor and ceiling anchored									
1100	Plastic laminate on particle board	2 Carp	5	3.200	Ea.	800	170		970	1,150
1400	For handicap units, add				"	370			370	405
1610	Floor mounted									
1800	Plastic laminate on particle board	2 Carp	7	2.286	Ea.	540	121		661	790
2450	Floor mounted, headrail braced									
2600	Plastic laminate on particle board	2 Carp	6	2.667	Ea.	810	142		952	1,125
3400	For handicap units, add					325			325	360
4300	Entrance screen, floor mtd., plas. lam., 58" high, 48" wide	2 Carp	15	1.067		520	56.50		576.50	660

351

10 21 Compartments and Cubicles

10 21 13 – Toilet Compartments

10 21 13.16 Plastic-Laminate-Clad Toilet Compartments	Crew	Daily Output	Labor-Hours	Unit	Material	2020 Bare Costs Labor	Equipment	Total	Total Incl O&P
4800	Urinal screen, 18" wide, ceiling braced, plastic laminate	2 Carp	8	2	Ea.	210	106		316	400
5400	Floor mounted, headrail braced		8	2		208	106		314	400
5900	Pilaster, flush, plastic laminate		10	1.600		430	85		515	610
6400	Post braced, plastic laminate		10	1.600		237	85		322	395
6700	Wall hung, bracket supported									
6900	Plastic laminate on particle board	2 Carp	10	1.600	Ea.	97	85		182	242
7450	Flange supported									
7500	Plastic laminate on particle board	2 Carp	10	1.600	Ea.	251	85		336	410
9000	Minimum labor/equipment charge	1 Carp	2.50	3.200	Job		170		170	272

10 21 13.19 Plastic Toilet Compartments

10 21 13.19 Plastic Toilet Compartments	Crew	Daily Output	Labor-Hours	Unit	Material	2020 Bare Costs Labor	Equipment	Total	Total Incl O&P
0010	**PLASTIC TOILET COMPARTMENTS**									
0110	Cubicles, ceiling hung									
0250	Phenolic	2 Carp	4	4	Ea.	885	213		1,098	1,325
0260	Polymer plastic	"	4	4		1,025	213		1,238	1,475
0600	For handicap units, add					465			465	510
0900	Floor and ceiling anchored									
1050	Phenolic	2 Carp	5	3.200	Ea.	940	170		1,110	1,300
1060	Polymer plastic	"	5	3.200		1,200	170		1,370	1,600
1400	For handicap units, add					370			370	405
1610	Floor mounted									
1750	Phenolic	2 Carp	7	2.286	Ea.	880	121		1,001	1,175
1760	Polymer plastic	"	7	2.286		855	121		976	1,125
2100	For handicap units, add					360			360	395
2200	For juvenile units, deduct					44.50			44.50	49
2450	Floor mounted, headrail braced									
2550	Phenolic	2 Carp	6	2.667	Ea.	740	142		882	1,050
3600	Polymer plastic		6	2.667		870	142		1,012	1,175
3810	Entrance screen, polymer plastic, flr. mtd., 48"x58"		6	2.667		495	142		637	770
3820	Entrance screen, polymer plastic, flr. to clg pilaster, 48"x58"		6	2.667		590	142		732	875
6110	Urinal screen, polymer plastic, pilaster flush, 18" w		6	2.667		430	142		572	700
7110	Wall hung		6	2.667		167	142		309	410
7710	Flange mounted		6	2.667		920	142		1,062	1,250
9000	Minimum labor/equipment charge	1 Carp	2.50	3.200	Job		170		170	272

10 21 13.40 Stone Toilet Compartments

10 21 13.40 Stone Toilet Compartments	Crew	Daily Output	Labor-Hours	Unit	Material	2020 Bare Costs Labor	Equipment	Total	Total Incl O&P
0010	**STONE TOILET COMPARTMENTS**									
0100	Cubicles, ceiling hung, marble	2 Marb	2	8	Ea.	1,650	405		2,055	2,475
0600	For handicap units, add					465			465	510
0800	Floor & ceiling anchored, marble	2 Marb	2.50	6.400		1,775	325		2,100	2,475
1400	For handicap units, add					370			370	405
1600	Floor mounted, marble	2 Marb	3	5.333		1,050	271		1,321	1,600
2400	Floor mounted, headrail braced, marble	"	3	5.333		1,225	271		1,496	1,800
2900	For handicap units, add					325			325	360
4100	Entrance screen, floor mounted marble, 58" high, 48" wide	2 Marb	9	1.778		735	90.50		825.50	955
4600	Urinal screen, 18" wide, ceiling braced, marble	D-1	6	2.667		770	125		895	1,050
5100	Floor mounted, headrail braced									
5200	Marble	D-1	6	2.667	Ea.	660	125		785	925
5700	Pilaster, flush, marble		9	1.778		860	83.50		943.50	1,075
6200	Post braced, marble		9	1.778		845	83.50		928.50	1,075
9000	Minimum labor/equipment charge	1 Carp	2.50	3.200	Job		170		170	272

For customer support on your Commercial Renovation Costs with RSMeans data, call 800.448.8182.

10 21 Compartments and Cubicles

10 21 14 – Toilet Compartment Components

10 21 14.13 Metal Toilet Compartment Components

		Crew	Daily Output	Labor-Hours	Unit	Material	2020 Bare Costs Labor	2020 Bare Costs Equipment	Total	Total Incl O&P
0010	**METAL TOILET COMPARTMENT COMPONENTS**									
0100	Pilasters									
0110	Overhead braced, powder coated steel, 7" wide x 82" high	2 Carp	22.20	.721	Ea.	72.50	38.50		111	142
0120	Stainless steel		22.20	.721		132	38.50		170.50	207
0130	Floor braced, powder coated steel, 7" wide x 70" high		23.30	.687		106	36.50		142.50	176
0140	Stainless steel		23.30	.687		227	36.50		263.50	310
0150	Ceiling hung, powder coated steel, 7" wide x 83" high		13.30	1.203		130	64		194	245
0160	Stainless steel		13.30	1.203		320	64		384	450
0170	Wall hung, powder coated steel, 3" wide x 58" high		18.90	.847		142	45		187	228
0180	Stainless steel	↓	18.90	.847	↓	234	45		279	330
0200	Panels									
0210	Powder coated steel, 31" wide x 58" high	2 Carp	18.90	.847	Ea.	139	45		184	225
0220	Stainless steel		18.90	.847		340	45		385	445
0230	Powder coated steel, 53" wide x 58" high		18.90	.847		177	45		222	266
0240	Stainless steel		18.90	.847		405	45		450	515
0250	Powder coated steel, 63" wide x 58" high		18.90	.847		176	45		221	265
0260	Stainless steel	↓	18.90	.847	↓	485	45		530	600
0300	Doors									
0310	Powder coated steel, 24" wide x 58" high	2 Carp	14.10	1.135	Ea.	125	60.50		185.50	234
0320	Stainless steel		14.10	1.135		325	60.50		385.50	455
0330	Powder coated steel, 26" wide x 58" high		14.10	1.135		161	60.50		221.50	275
0340	Stainless steel		14.10	1.135		335	60.50		395.50	465
0350	Powder coated steel, 28" wide x 58" high		14.10	1.135		178	60.50		238.50	292
0360	Stainless steel		14.10	1.135		330	60.50		390.50	460
0370	Powder coated steel, 36" wide x 58" high		14.10	1.135		182	60.50		242.50	297
0380	Stainless steel	↓	14.10	1.135	↓	395	60.50		455.50	530
0400	Headrails									
0410	For powder coated steel, 62" long	2 Carp	65	.246	Ea.	20	13.10		33.10	43
0420	Stainless steel		65	.246		20.50	13.10		33.60	43.50
0430	For powder coated steel, 84" long		50	.320		30.50	17		47.50	61
0440	Stainless steel		50	.320		30.50	17		47.50	61
0450	For powder coated steel, 120" long		30	.533		41	28.50		69.50	90.50
0460	Stainless steel	↓	30	.533	↓	41	28.50		69.50	90.50
9000	Minimum labor/equipment charge	1 Carp	4	2	Job		106		106	170

10 21 14.16 Plastic-Laminate Clad Toilet Compartment Components

		Crew	Daily Output	Labor-Hours	Unit	Material	2020 Bare Costs Labor	2020 Bare Costs Equipment	Total	Total Incl O&P
0010	**PLASTIC-LAMINATE CLAD TOILET COMPARTMENT COMPONENTS**									
0100	Pilasters									
0110	Overhead braced, 7" wide x 82" high	2 Carp	22.20	.721	Ea.	102	38.50		140.50	175
0130	Floor anchored, 7" wide x 70" high		23.30	.687		108	36.50		144.50	177
0150	Ceiling hung, 7" wide x 83" high		13.30	1.203		130	64		194	245
0180	Wall hung, 3" wide x 58" high	↓	18.90	.847	↓	98	45		143	180
0200	Panels									
0210	31" wide x 58" high	2 Carp	18.90	.847	Ea.	169	45		214	258
0230	51" wide x 58" high		18.90	.847		208	45		253	300
0250	63" wide x 58" high	↓	18.90	.847	↓	243	45		288	340
0300	Doors									
0310	24" wide x 58" high	2 Carp	14.10	1.135	Ea.	154	60.50		214.50	266
0330	26" wide x 58" high		14.10	1.135		159	60.50		219.50	271
0350	28" wide x 58" high		14.10	1.135		164	60.50		224.50	277
0370	36" wide x 58" high	↓	14.10	1.135	↓	174	60.50		234.50	289
0400	Headrails									
0410	62" long	2 Carp	65	.246	Ea.	26	13.10		39.10	49.50

For customer support on your Commercial Renovation Costs with RSMeans data, call 800.448.8182.

353

10 21 Compartments and Cubicles

10 21 14 – Toilet Compartment Components

10 21 14.16 Plastic-Laminate Clad Toilet Compartment Components

		Crew	Daily Output	Labor-Hours	Unit	Material	2020 Bare Costs Labor	Equipment	Total	Total Incl O&P
0430	84" long	2 Carp	60	.267	Ea.	31.50	14.15		45.65	57
0450	120" long		30	.533		42	28.50		70.50	92
9000	Minimum labor/equipment charge	1 Carp	4	2	Job		106		106	170

10 21 14.19 Plastic Toilet Compartment Components

		Crew	Daily Output	Labor-Hours	Unit	Material	2020 Bare Costs Labor	Equipment	Total	Total Incl O&P
0010	**PLASTIC TOILET COMPARTMENT COMPONENTS**									
0100	Pilasters									
0110	Overhead braced, polymer plastic, 7" wide x 82" high	2 Carp	22.20	.721	Ea.	116	38.50		154.50	190
0120	Phenolic		22.20	.721		140	38.50		178.50	216
0130	Floor braced, polymer plastic, 7" wide x 70" high		23.30	.687		185	36.50		221.50	262
0140	Phenolic		23.30	.687		135	36.50		171.50	207
0150	Ceiling hung, polymer plastic, 7" wide x 83" high		13.30	1.203		179	64		243	299
0160	Phenolic		13.30	1.203		168	64		232	286
0180	Wall hung, phenolic, 3" wide x 58" high		18.90	.847		105	45		150	188
0200	Panels									
0203	Polymer plastic, 18" wide x 55" high	2 Carp	18.90	.847	Ea.	310	45		355	415
0206	Phenolic, 18" wide x 58" high		18.90	.847		256	45		301	355
0210	Polymer plastic, 31" wide x 55" high		18.90	.847		340	45		385	445
0220	Phenolic, 31" wide x 58" high		18.90	.847		274	45		319	370
0223	Polymer plastic, 48" wide x 55" high		18.90	.847		520	45		565	645
0226	Phenolic, 48" wide x 58" high		18.90	.847		460	45		505	580
0230	Polymer plastic, 51" wide x 55" high		18.90	.847		490	45		535	610
0240	Phenolic, 51" wide x 58" high		18.90	.847		435	45		480	550
0250	Polymer plastic, 63" wide x 55" high		18.90	.847		690	45		735	830
0260	Phenolic, 63" wide x 58" high		18.90	.847		475	45		520	595
0300	Doors									
0310	Polymer plastic, 24" wide x 55" high	2 Carp	14.10	1.135	Ea.	252	60.50		312.50	375
0320	Phenolic, 24" wide x 58" high		14.10	1.135		340	60.50		400.50	465
0330	Polymer plastic, 26" wide x 55" high		14.10	1.135		262	60.50		322.50	385
0340	Phenolic, 26" wide x 58" high		14.10	1.135		360	60.50		420.50	490
0350	Polymer plastic, 28" wide x 55" high		14.10	1.135		284	60.50		344.50	405
0360	Phenolic, 28" wide x 58" high		14.10	1.135		375	60.50		435.50	510
0370	Polymer plastic, 36" wide x 55" high		14.10	1.135		335	60.50		395.50	465
0380	Phenolic, 36" wide x 58" high		14.10	1.135		520	60.50		580.50	665
0400	Headrails									
0410	For polymer plastic, 62" long	2 Carp	65	.246	Ea.	22.50	13.10		35.60	46
0420	Phenolic		65	.246		22.50	13.10		35.60	45.50
0430	For polymer plastic, 84" long		50	.320		31	17		48	61.50
0440	Phenolic		50	.320		39.50	17		56.50	71
0450	For polymer plastic, 120" long		30	.533		44	28.50		72.50	93.50
0460	Phenolic		30	.533		42	28.50		70.50	91.50
9000	Minimum labor/equipment charge	1 Carp	4	2	Job		106		106	170

10 21 23 – Cubicle Curtains and Track

10 21 23.16 Cubicle Track and Hardware

		Crew	Daily Output	Labor-Hours	Unit	Material	2020 Bare Costs Labor	Equipment	Total	Total Incl O&P
0010	**CUBICLE TRACK AND HARDWARE**									
0020	Curtain track, box channel, ceiling mounted	1 Carp	135	.059	L.F.	6.75	3.15		9.90	12.45
0100	Suspended		100	.080		8.35	4.25		12.60	16
0550	Polyester, antimicrobial, 9' ceiling height		425	.019		19.40	1		20.40	23
0560	8' ceiling height		425	.019		17.30	1		18.30	20.50

10 22 16 – Folding Gates

10 22 16.10 Security Gates

	Crew	Daily Output	Labor-Hours	Unit	Material	2020 Bare Costs Labor	Equipment	Total	Total Incl O&P
0010 **SECURITY GATES**									
0015 For roll up type, see Section 08 33 13.10									
0300 Scissors type folding gate, ptd. steel, single, 6-1/2' high, 5-1/2'wide	2 Sswk	4	4	Opng.	235	231		466	640
0350 6-1/2' wide		4	4		240	231		471	645
0400 7-1/2' wide		4	4		223	231		454	625
0600 Double gate, 8' high, 8' wide		2.50	6.400		385	370		755	1,025
0650 10' wide		2.50	6.400		455	370		825	1,100
0700 12' wide		2	8		605	460		1,065	1,425
0750 14' wide		2	8		625	460		1,085	1,450
0900 Door gate, folding steel, 4' wide, 61" high		4	4		144	231		375	540
1000 71" high		4	4		179	231		410	575
1200 81" high		4	4		196	231		427	595
1300 Window gates, 2' to 4' wide, 31" high		4	4		90	231		321	480
1500 55" high		3.75	4.267		125	246		371	545
1600 79" high		3.50	4.571		148	264		412	595

10 22 19 – Demountable Partitions

10 22 19.43 Demountable Composite Partitions

	Crew	Daily Output	Labor-Hours	Unit	Material	2020 Bare Costs Labor	Equipment	Total	Total Incl O&P
0010 **DEMOUNTABLE COMPOSITE PARTITIONS**, add for doors									
0100 Do not deduct door openings from total L.F.									
0900 Demountable gypsum system on 2" to 2-1/2"									
1000 Steel studs, 9' high, 3" to 3-3/4" thick									
1200 Vinyl-clad gypsum	2 Carp	48	.333	L.F.	61	17.70		78.70	95.50
1300 Fabric clad gypsum		44	.364		152	19.35		171.35	198
1500 Steel clad gypsum		40	.400		170	21.50		191.50	221
1600 1.75 system, aluminum framing, vinyl-clad hardboard,									
1800 Paper honeycomb core panel, 1-3/4" to 2-1/2" thick									
1900 9' high	2 Carp	48	.333	L.F.	102	17.70		119.70	141
2100 7' high		60	.267		91.50	14.15		105.65	124
2200 5' high		80	.200		77.50	10.65		88.15	103
2250 Unitized gypsum system									
2300 Unitized panel, 9' high, 2" to 2-1/2" thick									
2350 Vinyl-clad gypsum	2 Carp	48	.333	L.F.	132	17.70		149.70	174
2400 Fabric clad gypsum	"	44	.364	"	217	19.35		236.35	269
2500 Unitized mineral fiber system									
2510 Unitized panel, 9' high, 2-1/4" thick, aluminum frame									
2550 Vinyl-clad mineral fiber	2 Carp	48	.333	L.F.	131	17.70		148.70	173
2600 Fabric clad mineral fiber	"	44	.364	"	195	19.35		214.35	246
2800 Movable steel walls, modular system									
2900 Unitized panels, 9' high, 48" wide									
3100 Baked enamel, pre-finished	2 Carp	60	.267	L.F.	148	14.15		162.15	186
3200 Fabric clad steel	"	56	.286	"	214	15.20		229.20	261
5500 For acoustical partitions, add, unsealed				S.F.	2.38			2.38	2.62
5550 Sealed				"	11.10			11.10	12.20
5700 For doors, see Section 08 16									
5800 For door hardware, see Section 08 71									
6100 In-plant modular office system, w/prehung hollow core door									
6200 3" thick polystyrene core panels									
6250 12' x 12', 2 wall	2 Clab	3.80	4.211	Ea.	4,975	177		5,152	5,750
6300 4 wall		1.90	8.421		7,125	355		7,480	8,425
6350 16' x 16', 2 wall		3.60	4.444		8,175	187		8,362	9,300
6400 4 wall		1.80	8.889		11,200	375		11,575	12,900
9000 Minimum labor/equipment charge	2 Carp	3	5.333	Job		283		283	455

10 22 Partitions

10 22 23 – Portable Partitions, Screens, and Panels

10 22 23.13 Wall Screens

	10 22 23.13 Wall Screens	Crew	Daily Output	Labor-Hours	Unit	Material	2020 Bare Costs Labor	Equipment	Total	Total Incl O&P
0010	**WALL SCREENS**, divider panels, free standing, fiber core									
0020	Fabric face straight									
0100	3'-0" long, 4'-0" high	2 Carp	100	.160	L.F.	136	8.50		144.50	164
0200	5'-0" high		90	.178		126	9.45		135.45	154
0500	6'-0" high		75	.213		127	11.35		138.35	157
0900	5'-0" long, 4'-0" high		175	.091		61.50	4.86		66.36	76
1000	5'-0" high		150	.107		84.50	5.65		90.15	102
1500	6'-0" high		125	.128		113	6.80		119.80	135
1600	6'-0" long, 5'-0" high		162	.099		113	5.25		118.25	132
3200	Economical panels, fabric face, 4'-0" long, 5'-0" high		132	.121		59.50	6.45		65.95	75.50
3250	6'-0" high		112	.143		61.50	7.60		69.10	79.50
3300	5'-0" long, 5'-0" high		150	.107		34.50	5.65		40.15	47
3350	6'-0" high		125	.128		52.50	6.80		59.30	68.50
3450	Acoustical panels, 60 to 90 NRC, 3'-0" long, 5'-0" high		90	.178		87	9.45		96.45	111
3550	6'-0" high		75	.213		88	11.35		99.35	115
3600	5'-0" long, 5'-0" high		150	.107		61	5.65		66.65	76
3650	6'-0" high		125	.128		59	6.80		65.80	76
3700	6'-0" long, 5'-0" high		162	.099		53	5.25		58.25	67
3750	6'-0" high		138	.116		83.50	6.15		89.65	102
3800	Economy acoustical panels, 40 NRC, 4'-0" long, 5'-0" high		132	.121		59.50	6.45		65.95	75.50
3850	6'-0" high		112	.143		61.50	7.60		69.10	79.50
3900	5'-0" long, 6'-0" high		125	.128		52.50	6.80		59.30	68.50
3950	6'-0" long, 5'-0" high		162	.099		46	5.25		51.25	59
9000	Minimum labor/equipment charge		3	5.333	Job		283		283	455

10 22 33 – Accordion Folding Partitions

10 22 33.10 Partitions, Accordion Folding

	10 22 33.10 Partitions, Accordion Folding	Crew	Daily Output	Labor-Hours	Unit	Material	2020 Bare Costs Labor	Equipment	Total	Total Incl O&P
0010	**PARTITIONS, ACCORDION FOLDING**									
0100	Vinyl covered, over 150 S.F., frame not included									
0300	Residential, 1.25 lb./S.F., 8' maximum height	2 Carp	300	.053	S.F.	22	2.83		24.83	29
0400	Commercial, 1.75 lb./S.F., 8' maximum height		225	.071		25.50	3.78		29.28	34
0900	Acoustical, 3 lb./S.F., 17' maximum height		100	.160		27.50	8.50		36	44
1200	5 lb./S.F., 20' maximum height		95	.168		44.50	8.95		53.45	63
1500	Vinyl-clad wood or steel, electric operation, 5.0 psf		160	.100		55	5.30		60.30	69.50
1900	Wood, non-acoustic, birch or mahogany, to 10' high		300	.053		30	2.83		32.83	37
9000	Minimum labor/equipment charge		4	4	Job		213		213	340

10 22 39 – Folding Panel Partitions

10 22 39.10 Partitions, Folding Panel

	10 22 39.10 Partitions, Folding Panel	Crew	Daily Output	Labor-Hours	Unit	Material	2020 Bare Costs Labor	Equipment	Total	Total Incl O&P
0010	**PARTITIONS, FOLDING PANEL**, acoustic, wood									
0100	Vinyl faced, to 18' high, 6 psf, economy trim	2 Carp	60	.267	S.F.	70.50	14.15		84.65	100
0150	Standard trim		45	.356		84	18.90		102.90	123
0200	Premium trim		30	.533		108	28.50		136.50	165
0400	Plastic laminate or hardwood finish, standard trim		60	.267		72.50	14.15		86.65	102
0500	Premium trim		30	.533		77	28.50		105.50	131
0600	Wood, low acoustical type, 4.5 psf, to 14' high		50	.320		52.50	17		69.50	85.50
9000	Minimum labor/equipment charge		4	4	Job		213		213	340

10 26 Wall and Door Protection

10 26 23 – Protective Wall Covering

10 26 23.10 Wall Covering, Protective

	10 26 23.10 Wall Covering, Protective	Crew	Daily Output	Labor-Hours	Unit	Material	2020 Bare Costs Labor	Equipment	Total	Total Incl O&P
0010	**WALL COVERING, PROTECTIVE**									
1775	Wall end guard, stainless steel, 16 ga., 36" tall, screwed to studs	1 Skwk	30	.267	Ea.	26	14.65		40.65	52

10 26 23.13 Impact Resistant Wall Protection

		Crew	Daily Output	Labor-Hours	Unit	Material	Labor	Equipment	Total	Total Incl O&P
0010	**IMPACT RESISTANT WALL PROTECTION**									
0100	Vinyl wall protection, complete instl. incl. panels, trim, adhesive									
0110	.040" thk., std. colors	2 Carp	320	.050	S.F.	15.60	2.66		18.26	21.50
0120	.040" thk., element patterns		300	.053		17.35	2.83		20.18	23.50
0130	.060" thk., std. colors		300	.053		17.90	2.83		20.73	24
0140	.060" thk., element patterns		280	.057		18.90	3.04		21.94	26
1775	Wall protection, stainless steel, 16 ga., 48" x 36" tall, screwed to studs	2 Skwk	500	.032		8	1.76		9.76	11.60

10 26 33 – Door and Frame Protection

10 26 33.10 Protection, Door and Frame

		Crew	Daily Output	Labor-Hours	Unit	Material	Labor	Equipment	Total	Total Incl O&P
0010	**PROTECTION, DOOR AND FRAME**									
0100	Door frame guard, vinyl, 3" x 3" x 4'	1 Carp	30	.267	Ea.	88.50	14.15		102.65	120
0110	Door frame guard, vinyl, 3" x 3" x 8'		26	.308		134	16.35		150.35	173
0120	Door frame guard, stainless steel, 3" x 3" x 4'		30	.267		440	14.15		454.15	510
0500	Steel door track/wheel guards, 4'-0" high	E-4	22	1.455		116	84.50	6.70	207.20	274

10 28 Toilet, Bath, and Laundry Accessories

10 28 13 – Toilet Accessories

10 28 13.13 Commercial Toilet Accessories

		Crew	Daily Output	Labor-Hours	Unit	Material	Labor	Equipment	Total	Total Incl O&P
0010	**COMMERCIAL TOILET ACCESSORIES**									
0200	Curtain rod, stainless steel, 5' long, 1" diameter	1 Carp	13	.615	Ea.	26.50	32.50		59	82
0300	1-1/4" diameter		13	.615		29	32.50		61.50	84.50
0350	Chrome, 1" diameter		13	.615		26	32.50		58.50	81
0360	For vinyl curtain, add		1950	.004	S.F.	.97	.22		1.19	1.42
0400	Diaper changing station, horizontal, wall mounted, plastic		10	.800	Ea.	246	42.50		288.50	340
0420	Vertical		10	.800		221	42.50		263.50	310
0430	Oval shaped		10	.800		305	42.50		347.50	405
0440	Recessed, with stainless steel flange		6	1.333		515	71		586	680
0500	Dispenser units, combined soap & towel dispensers,									
0510	Mirror and shelf, flush mounted	1 Carp	10	.800	Ea.	345	42.50		387.50	450
0600	Towel dispenser and waste receptacle,									
0610	18 gallon capacity	1 Carp	10	.800	Ea.	365	42.50		407.50	475
0800	Grab bar, straight, 1-1/4" diameter, stainless steel, 18" long		24	.333		30	17.70		47.70	61
0900	24" long		23	.348		29	18.50		47.50	61.50
1000	30" long		22	.364		26.50	19.35		45.85	60
1100	36" long		20	.400		34.50	21.50		56	72
1105	42" long		20	.400		40	21.50		61.50	78
1120	Corner, 36" long		20	.400		95	21.50		116.50	138
1200	1-1/2" diameter, 24" long		23	.348		33.50	18.50		52	66.50
1300	36" long		20	.400		37	21.50		58.50	74.50
1310	42" long		18	.444		33.50	23.50		57	75
1500	Tub bar, 1-1/4" diameter, 24" x 36"		14	.571		90	30.50		120.50	148
1600	Plus vertical arm		12	.667		100	35.50		135.50	167
1900	End tub bar, 1" diameter, 90° angle, 16" x 32"		12	.667		111	35.50		146.50	179
2010	Tub/shower/toilet, 2-wall, 36" x 24"		12	.667		90.50	35.50		126	157
2300	Hand dryer, surface mounted, electric, 115 volt, 20 amp		4	2		425	106		531	635
2400	230 volt, 10 amp		4	2		645	106		751	880
2450	Hand dryer, touch free, 1400 watt, 81,000 rpm		4	2		1,350	106		1,456	1,650

10 28 13 – Toilet Accessories

10 28 13.13 Commercial Toilet Accessories	Crew	Daily Output	Labor-Hours	Unit	Material	2020 Bare Costs Labor	Equipment	Total	Total Incl O&P
2600	Hat and coat strip, stainless steel, 4 hook, 36" long	1 Carp	24	.333	Ea.	68	17.70		85.70	104
2700	6 hook, 60" long		20	.400		118	21.50		139.50	164
3000	Mirror, with stainless steel 3/4" square frame, 18" x 24"		20	.400		51.50	21.50		73	90.50
3100	36" x 24"		15	.533		104	28.50		132.50	160
3200	48" x 24"		10	.800		186	42.50		228.50	272
3300	72" x 24"		6	1.333		305	71		376	450
3500	With 5" stainless steel shelf, 18" x 24"		20	.400		185	21.50		206.50	237
3600	36" x 24"		15	.533		236	28.50		264.50	305
3700	48" x 24"		10	.800		246	42.50		288.50	340
3800	72" x 24"		6	1.333		277	71		348	420
4100	Mop holder strip, stainless steel, 5 holders, 48" long		20	.400		76.50	21.50		98	119
4200	Napkin/tampon dispenser, recessed		15	.533		560	28.50		588.50	660
4220	Semi-recessed		6.50	1.231		370	65.50		435.50	510
4250	Napkin receptacle, recessed		6.50	1.231		171	65.50		236.50	293
4300	Robe hook, single, regular		96	.083		20.50	4.43		24.93	29.50
4400	Heavy duty, concealed mounting		56	.143		25.50	7.60		33.10	40
4600	Soap dispenser, chrome, surface mounted, liquid		20	.400		58.50	21.50		80	98
5000	Recessed stainless steel, liquid		10	.800		163	42.50		205.50	247
5600	Shelf, stainless steel, 5" wide, 18 ga., 24" long		24	.333		82	17.70		99.70	119
5700	48" long		16	.500		168	26.50		194.50	228
5800	8" wide shelf, 18 ga., 24" long		22	.364		63	19.35		82.35	101
5900	48" long		14	.571		145	30.50		175.50	208
6000	Toilet seat cover dispenser, stainless steel, recessed		20	.400		181	21.50		202.50	233
6050	Surface mounted		15	.533		32.50	28.50		61	81.50
6100	Toilet tissue dispenser, surface mounted, SS, single roll		30	.267		16	14.15		30.15	40
6200	Double roll		24	.333		23	17.70		40.70	53.50
6240	Plastic, twin/jumbo dbl. roll		24	.333		30.50	17.70		48.20	62
6290	Toilet seat	1 Plum	40	.200		26.50	12.90		39.40	49
6400	Towel bar, stainless steel, 18" long	1 Carp	23	.348		40	18.50		58.50	73.50
6500	30" long		21	.381		48.50	20.50		69	85.50
6700	Towel dispenser, stainless steel, surface mounted		16	.500		41.50	26.50		68	88.50
6800	Flush mounted, recessed		10	.800		67.50	42.50		110	143
6900	Plastic, touchless, battery operated		16	.500		107	26.50		133.50	161
7000	Towel holder, hotel type, 2 guest size		20	.400		58	21.50		79.50	97.50
7200	Towel shelf, stainless steel, 24" long, 8" wide		20	.400		65	21.50		86.50	106
7400	Tumbler holder, for tumbler only		30	.267		37	14.15		51.15	63.50
7410	Tumbler holder, recessed		20	.400		9.40	21.50		30.90	44.50
7500	Soap, tumbler & toothbrush		30	.267		20	14.15		34.15	44.50
7510	Tumbler & toothbrush holder		20	.400		13.45	21.50		34.95	49
8000	Waste receptacles, stainless steel, with top, 13 gallon		10	.800		335	42.50		377.50	435
8100	36 gallon		8	1		460	53		513	590
9000	Minimum labor/equipment charge		5	1.600	Job		85		85	136
9996	Bathroom access., grab bar, straight, 1-1/2" diam., SS, 42" L install only		18	.444	Ea.		23.50		23.50	38

10 28 16 – Bath Accessories

10 28 16.20 Medicine Cabinets

		Crew	Daily Output	Labor-Hours	Unit	Material	2020 Bare Costs Labor	Equipment	Total	Total Incl O&P
0010	**MEDICINE CABINETS**									
0020	With mirror, sst frame, 16" x 22", unlighted	1 Carp	14	.571	Ea.	98.50	30.50		129	157
0100	Wood frame		14	.571		128	30.50		158.50	190
0300	Sliding mirror doors, 20" x 16" x 4-3/4", unlighted		7	1.143		130	60.50		190.50	241
0400	24" x 19" x 8-1/2", lighted		5	1.600		219	85		304	375
0600	Triple door, 30" x 32", unlighted, plywood body		7	1.143		355	60.50		415.50	490
0700	Steel body		7	1.143		355	60.50		415.50	490

10 28 Toilet, Bath, and Laundry Accessories

10 28 16 – Bath Accessories

10 28 16.20 Medicine Cabinets

		Crew	Daily Output	Labor-Hours	Unit	Material	2020 Bare Costs Labor	Equipment	Total	Total Incl O&P
0900	Oak door, wood body, beveled mirror, single door	1 Carp	7	1.143	Ea.	170	60.50		230.50	284
1000	Double door		6	1.333		370	71		441	520
1200	Hotel cabinets, stainless, with lower shelf, unlighted		10	.800		218	42.50		260.50	310
1300	Lighted		5	1.600		330	85		415	495
9000	Minimum labor/equipment charge		4	2	Job		106		106	170

10 28 19 – Tub and Shower Enclosures

10 28 19.10 Partitions, Shower

		Crew	Daily Output	Labor-Hours	Unit	Material	2020 Bare Costs Labor	Equipment	Total	Total Incl O&P
0010	**PARTITIONS, SHOWER** floor mounted, no plumbing									
0400	Cabinet, one piece, fiberglass, 32" x 32"	2 Carp	5	3.200	Ea.	625	170		795	960
0420	36" x 36"		5	3.200		550	170		720	875
0440	36" x 48"		5	3.200		1,375	170		1,545	1,800
0460	Acrylic, 32" x 32"		5	3.200		335	170		505	640
0480	36" x 36"		5	3.200		1,050	170		1,220	1,425
0500	36" x 48"		5	3.200		1,300	170		1,470	1,700
0520	Shower door for above, clear plastic, 24" wide	1 Carp	8	1		185	53		238	289
0540	28" wide		8	1		241	53		294	350
0560	Tempered glass, 24" wide		8	1		260	53		313	370
0580	28" wide		8	1		293	53		346	405
4100	Shower doors, economy plastic, 24" wide	1 Shee	9	.889		134	55.50		189.50	234
4200	Tempered glass door, economy		8	1		279	62.50		341.50	405
4700	Deluxe, tempered glass, chrome on brass frame, 42" to 44"		8	1		485	62.50		547.50	635
4800	39" to 48" wide		1	8		700	500		1,200	1,550
9990	Minimum labor/equipment charge		2.50	3.200	Job		199		199	310

10 31 Manufactured Fireplaces

10 31 13 – Manufactured Fireplace Chimneys

10 31 13.10 Fireplace Chimneys

		Crew	Daily Output	Labor-Hours	Unit	Material	2020 Bare Costs Labor	Equipment	Total	Total Incl O&P
0010	**FIREPLACE CHIMNEYS**									
0500	Chimney dbl. wall, all stainless, over 8'-6", 7" diam., add to fireplace	1 Carp	33	.242	V.L.F.	94	12.90		106.90	124
0600	10" diameter, add to fireplace		32	.250		103	13.30		116.30	135
0700	12" diameter, add to fireplace		31	.258		181	13.70		194.70	221
0800	14" diameter, add to fireplace		30	.267		216	14.15		230.15	261
1000	Simulated brick chimney top, 4' high, 16" x 16"		10	.800	Ea.	515	42.50		557.50	635
1100	24" x 24"		7	1.143	"	575	60.50		635.50	730

10 31 13.20 Chimney Accessories

		Crew	Daily Output	Labor-Hours	Unit	Material	2020 Bare Costs Labor	Equipment	Total	Total Incl O&P
0010	**CHIMNEY ACCESSORIES**									
0020	Chimney screens, galv., 13" x 13" flue	1 Bric	8	1	Ea.	56.50	52		108.50	146
0050	24" x 24" flue		5	1.600		128	83.50		211.50	275
0200	Stainless steel, 13" x 13" flue		8	1		101	52		153	196
0250	20" x 20" flue		5	1.600		147	83.50		230.50	297
2400	Squirrel and bird screens, galvanized, 8" x 8" flue		16	.500		56	26		82	104
2450	13" x 13" flue		12	.667		66	34.50		100.50	129
9000	Minimum labor/equipment charge		3.50	2.286	Job		119		119	192

10 31 16 – Manufactured Fireplace Forms

10 31 16.10 Fireplace Forms

		Crew	Daily Output	Labor-Hours	Unit	Material	2020 Bare Costs Labor	Equipment	Total	Total Incl O&P
0010	**FIREPLACE FORMS**									
1800	Fireplace forms, no accessories, 32" opening	1 Bric	3	2.667	Ea.	735	139		874	1,025
1900	36" opening		2.50	3.200		940	167		1,107	1,300
2000	40" opening		2	4		1,250	208		1,458	1,700
2100	78" opening		1.50	5.333		1,825	278		2,103	2,450

10 31 Manufactured Fireplaces

10 31 23 - Prefabricated Fireplaces

10 31 23.10 Fireplace, Prefabricated	Crew	Daily Output	Labor-Hours	Unit	Material	2020 Bare Costs Labor	Equipment	Total	Total Incl O&P
0010 **FIREPLACE, PREFABRICATED**, free standing or wall hung									
0100 With hood & screen, painted	1 Carp	1.30	6.154	Ea.	1,650	325		1,975	2,325
0150 Average		1	8		1,775	425		2,200	2,625
0200 Stainless steel		.90	8.889		3,300	470		3,770	4,375
1500 Simulated logs, gas fired, 40,000 BTU, 2' long, manual safety pilot		7	1.143	Set	560	60.50		620.50	715
1600 Adjustable flame remote pilot		6	1.333		1,325	71		1,396	1,575
1700 Electric, 1,500 BTU, 1'-6" long, incandescent flame		7	1.143		265	60.50		325.50	390
1800 1,500 BTU, LED flame		6	1.333		345	71		416	490
2000 Fireplace, built-in, 36" hearth, radiant		1.30	6.154	Ea.	740	325		1,065	1,350
2100 Recirculating, small fan		1	8		905	425		1,330	1,675
2150 Large fan		.90	8.889		2,250	470		2,720	3,225
2200 42" hearth, radiant		1.20	6.667		1,100	355		1,455	1,775
2300 Recirculating, small fan		.90	8.889		1,300	470		1,770	2,175
2350 Large fan		.80	10		1,475	530		2,005	2,475
2400 48" hearth, radiant		1.10	7.273		2,500	385		2,885	3,375
2500 Recirculating, small fan		.80	10		2,825	530		3,355	3,950
2550 Large fan		.70	11.429		2,675	605		3,280	3,900
3000 See through, including doors		.80	10		2,325	530		2,855	3,425
3200 Corner (2 wall)		1	8		3,750	425		4,175	4,775

10 32 Fireplace Specialties

10 32 13 - Fireplace Dampers

10 32 13.10 Dampers

10 32 13.10 Dampers	Crew	Daily Output	Labor-Hours	Unit	Material	2020 Bare Costs Labor	Equipment	Total	Total Incl O&P
0010 **DAMPERS**									
0800 Damper, rotary control, steel, 30" opening	1 Bric	6	1.333	Ea.	123	69.50		192.50	248
0850 Cast iron, 30" opening		6	1.333		124	69.50		193.50	248
0880 36" opening		6	1.333		126	69.50		195.50	250
0900 48" opening		6	1.333		175	69.50		244.50	305
0920 60" opening		6	1.333		355	69.50		424.50	505
1000 84" opening, special order		5	1.600		920	83.50		1,003.50	1,150
1050 96" opening, special order		4	2		935	104		1,039	1,200
1200 Steel plate, poker control, 60" opening		8	1		325	52		377	445
1250 84" opening, special order		5	1.600		595	83.50		678.50	790
1400 "Universal" type, chain operated, 32" x 20" opening		8	1		254	52		306	365
1450 48" x 24" opening		5	1.600		380	83.50		463.50	550

10 32 23 - Fireplace Doors

10 32 23.10 Doors

10 32 23.10 Doors	Crew	Daily Output	Labor-Hours	Unit	Material	2020 Bare Costs Labor	Equipment	Total	Total Incl O&P
0010 **DOORS**									
0400 Cleanout doors and frames, cast iron, 8" x 8"	1 Bric	12	.667	Ea.	57	34.50		91.50	119
0450 12" x 12"		10	.800		90.50	41.50		132	167
0500 18" x 24"		8	1		152	52		204	251
0550 Cast iron frame, steel door, 24" x 30"		5	1.600		325	83.50		408.50	495
1600 Dutch oven door and frame, cast iron, 12" x 15" opening		13	.615		133	32		165	198
1650 Copper plated, 12" x 15" opening		13	.615		260	32		292	340

10 35 Stoves

10 35 13 - Heating Stoves

10 35 13.10 Wood Burning Stoves		Crew	Daily Output	Labor-Hours	Unit	Material	2020 Bare Costs Labor	Equipment	Total	Total Incl O&P
0010	**WOOD BURNING STOVES**									
0015	Cast iron, less than 1,500 S.F.	2 Carp	1.30	12.308	Ea.	1,500	655		2,155	2,700
0020	1,500 to 2,000 S.F.		1	16		2,550	850		3,400	4,150
0030	greater than 2,000 S.F.		.80	20		2,925	1,075		4,000	4,925
0050	For gas log lighter, add					49.50			49.50	54.50

10 43 Emergency Aid Specialties

10 43 13 - Defibrillator Cabinets

10 43 13.05 Defibrillator Cabinets		Crew	Daily Output	Labor-Hours	Unit	Material	2020 Bare Costs Labor	Equipment	Total	Total Incl O&P
0010	**DEFIBRILLATOR CABINETS**, not equipped, stainless steel									
0050	Defibrillator cabinet, stainless steel with strobe & alarm 12" x 27"	1 Carp	10	.800	Ea.	430	42.50		472.50	540
0100	Automatic External Defibrillator	"	30	.267	"	1,400	14.15		1,414.15	1,550

10 44 Fire Protection Specialties

10 44 13 - Fire Protection Cabinets

10 44 13.53 Fire Equipment Cabinets		Crew	Daily Output	Labor-Hours	Unit	Material	2020 Bare Costs Labor	Equipment	Total	Total Incl O&P
0010	**FIRE EQUIPMENT CABINETS**, not equipped, 20 ga. steel box									
0040	Recessed, D.S. glass in door, box size given									
1000	Portable extinguisher, single, 8" x 12" x 27", alum. door & frame	Q-12	8	2	Ea.	142	114		256	330
1100	Steel door and frame	"	8	2	"	164	114		278	355
3000	Hose rack assy., 1-1/2" valve & 100' hose, 24" x 40" x 5-1/2"									
3200	Steel door and frame	Q-12	6	2.667	Ea.	259	152		411	520
4000	Hose rack assy., 2-1/2" x 1-1/2" valve, 100' hose, 24" x 40" x 8"									
4200	Steel door and frame	Q-12	6	2.667	Ea.	385	152		537	660
5000	Hose rack assy., 2-1/2" x 1-1/2" valve, 100' hose									
5010	and extinguisher, 30" x 40" x 8"									
5200	Steel door and frame	Q-12	5	3.200	Ea.	299	182		481	610

10 44 16 - Fire Extinguishers

10 44 16.13 Portable Fire Extinguishers		Crew	Daily Output	Labor-Hours	Unit	Material	2020 Bare Costs Labor	Equipment	Total	Total Incl O&P
0010	**PORTABLE FIRE EXTINGUISHERS**									
0140	CO_2, with hose and "H" horn, 10 lb.				Ea.	300			300	330
1000	Dry chemical, pressurized									
1040	Standard type, portable, painted, 2-1/2 lb.				Ea.	44			44	48.50
1080	10 lb.					96			96	106
1100	20 lb.					138			138	152
1120	30 lb.					430			430	470
2000	ABC all purpose type, portable, 2-1/2 lb.					23.50			23.50	26
2080	9-1/2 lb.					52			52	57
3500	Halotron 1, 2-1/2 lb.					133			133	146
3600	5 lb.					207			207	228
3700	60 lb.					425			425	470

10 51 13 – Metal Lockers

10 51 13.10 Lockers	Crew	Daily Output	Labor-Hours	Unit	Material	2020 Bare Costs Labor	Equipment	Total	Total Incl O&P
0011 **LOCKERS** steel, baked enamel, knock down construction									
0012 Body- 24ga, door- 16 or 18ga									
0110 1-tier locker, 12" x 15" x 72"	1 Shee	20	.400	Ea.	271	25		296	335
0120 18" x 15" x 72"		20	.400		335	25		360	410
0130 12" x 18" x 72"		20	.400		230	25		255	292
0140 18" x 18" x 72"		20	.400		299	25		324	370
0410 2- tier, 12" x 15" x 36"		30	.267		290	16.60		306.60	345
0420 18" x 15" x 36"		30	.267		239	16.60		255.60	289
0430 12" x 18" x 36"		30	.267		320	16.60		336.60	380
0440 18" x 18" x 36"		30	.267		255	16.60		271.60	305
0450 3- tier 12" x 15" x 24"		30	.267		293	16.60		309.60	345
0455 12" x 18" x 24"		30	.267		300	16.60		316.60	355
0480 4- tier 12" x 15" x 18"		30	.267		375	16.60		391.60	435
0485 12" x 18" x 18"		30	.267		390	16.60		406.60	455
0500 Two person, 18" x 15" x 72"		20	.400		325	25		350	400
0510 18" x 18" x 72"		20	.400		305	25		330	375
0520 Duplex, 15" x 15" x 72"		20	.400		345	25		370	420
0530 15" x 21" x 72"		20	.400		385	25		410	465
1098 All welded, ventilated athletic lockers, body- 16 ga., door- 14 ga.									
1100 1-tier, 12" x 18" x 72"	1 Shee	7.50	1.067	Ea.	350	66.50		416.50	490
1110 18" x 15" x 72"		7.50	1.067		555	66.50		621.50	720
1115 18" x 18" x 72"		7.50	1.067		570	66.50		636.50	730
1130 2-tier, 12" x 15" x 36"		7.50	1.067		580	66.50		646.50	745
1135 12" x 18" x 36"		7.50	1.067		620	66.50		686.50	790
1140 18" x 15" x 36"		7.50	1.067		675	66.50		741.50	850
1145 18" x 18" x 36"		7.50	1.067		690	66.50		756.50	860
1160 3-tier, 12" x 18" x 24"		7.50	1.067		795	66.50		861.50	980
1180 4-tier, 12" x 18" x 18"		7.50	1.067		605	66.50		671.50	770
1190 5H box, 12" x 18" x 14.4"		7.50	1.067		620	66.50		686.50	790
1195 18" x 18" x 14.4"		7.50	1.067		795	66.50		861.50	980
1196 6H box, 12" x 15" x 12"		7.50	1.067		535	66.50		601.50	690
1197 12" x 18" x 12"		7.50	1.067		605	66.50		671.50	770
1198 18" x 18" x 12"		7.50	1.067		755	66.50		821.50	935
2399 Standard duty lockers, body- 24 ga., doors- 16-18 ga.									
2400 16-person locker unit with clothing rack									
2500 72" wide x 15" deep x 72" high	1 Shee	15	.533	Ea.	610	33		643	725
2550 18" deep		15	.533	"	590	33		623	700
3250 Rack w/24 wire mesh baskets		1.50	5.333	Set	460	330		790	1,025
3260 30 baskets		1.25	6.400		420	400		820	1,075
3270 36 baskets		.95	8.421		450	525		975	1,325
3280 42 baskets		.80	10		465	625		1,090	1,475
3600 For hanger rods, add				Ea.	2.53			2.53	2.78
3650 For number plate kit, 100 plates #1 - #100, add	1 Shee	4	2		85.50	125		210.50	289
3700 For locker base, closed front panel		90	.089		5.40	5.55		10.95	14.65
3710 End panel, bolted		36	.222		8.35	13.85		22.20	30.50
3800 For sloping top, 12" wide		24	.333		31.50	21		52.50	67
3810 15" wide		24	.333		32	21		53	67.50
3820 18" wide		24	.333		33	21		54	69
3850 Sloping top end panel, 12" deep		72	.111		15.95	6.90		22.85	28.50
3860 15" deep		72	.111		17.25	6.90		24.15	30
3870 18" deep		72	.111		19.75	6.90		26.65	32.50
3900 For finish end panels, steel, 60" high, 15" deep		12	.667		30.50	41.50		72	98.50
3910 72" high, 12" deep		12	.667		27.50	41.50		69	95

10 51 Lockers

10 51 13 – Metal Lockers

	10 51 13.10 Lockers	Crew	Daily Output	Labor-Hours	Unit	Material	2020 Bare Costs Labor	Equipment	Total	Total Incl O&P
3920	18" deep	1 Shee	12	.667	Ea.	44	41.50		85.50	113
5000	For "ready to assemble" lockers,									
5010	Add to labor						75%			
5020	Deduct from material					20%				
6000	Heavy duty for detention facility, tamper proof, 14 ga. welded steel, solid									
6100	24" W x 24" D x 74" H, single tier	1 Shee	18	.444	Ea.	610	27.50		637.50	715
6110	Double tier		18	.444		560	27.50		587.50	660
6120	Triple tier		18	.444		610	27.50		637.50	715
9000	Minimum labor/equipment charge		2.50	3.200	Job		199		199	310

10 55 Postal Specialties

10 55 23 – Mail Boxes

10 55 23.10 Commercial Mail Boxes

		Crew	Daily Output	Labor-Hours	Unit	Material	2020 Bare Costs Labor	Equipment	Total	Total Incl O&P
0010	**COMMERCIAL MAIL BOXES**									
0020	Horiz., key lock, 5"H x 6"W x 15"D, alum., rear load	1 Carp	34	.235	Ea.	40.50	12.50		53	64.50
0100	Front loading		34	.235		40.50	12.50		53	64.50
0200	Double, 5"H x 12"W x 15"D, rear loading		26	.308		70.50	16.35		86.85	104
0300	Front loading		26	.308		77.50	16.35		93.85	112
0500	Quadruple, 10"H x 12"W x 15"D, rear loading		20	.400		145	21.50		166.50	193
0600	Front loading		20	.400		94.50	21.50		116	138
1600	Vault type, horizontal, for apartments, 4" x 5"		34	.235		27.50	12.50		40	50.50
1700	Alphabetical directories, 120 names		10	.800		103	42.50		145.50	182
1900	Letter slot, residential		20	.400		80	21.50		101.50	122
2000	Post office type		8	1		33.50	53		86.50	122
9000	Minimum labor/equipment charge		5	1.600	Job		85		85	136

10 56 Storage Assemblies

10 56 13 – Metal Storage Shelving

10 56 13.10 Shelving

		Crew	Daily Output	Labor-Hours	Unit	Material	2020 Bare Costs Labor	Equipment	Total	Total Incl O&P
0010	**SHELVING**									
0020	Metal, industrial, cross-braced, 3' W, 12" D	1 Sswk	175	.046	SF Shlf	7.60	2.64		10.24	12.65
0100	24" D		330	.024		5.65	1.40		7.05	8.55
2200	Wide span, 1600 lb. capacity per shelf, 6' W, 24" D		380	.021		7.25	1.21		8.46	10
2400	36" D		440	.018		6.55	1.05		7.60	8.95
9000	Minimum labor/equipment charge	1 Carp	4	2	Job		106		106	170

10 57 Wardrobe and Closet Specialties

10 57 23 – Closet and Utility Shelving

10 57 23.19 Wood Closet and Utility Shelving

		Crew	Daily Output	Labor-Hours	Unit	Material	2020 Bare Costs Labor	Equipment	Total	Total Incl O&P
0010	**WOOD CLOSET AND UTILITY SHELVING**									
0020	Pine, clear grade, no edge band, 1" x 8"	1 Carp	115	.070	L.F.	3.35	3.70		7.05	9.60
0100	1" x 10"		110	.073		4.17	3.87		8.04	10.80
0200	1" x 12"		105	.076		5.05	4.05		9.10	12.05
0600	Plywood, 3/4" thick with lumber edge, 12" wide		75	.107		1.91	5.65		7.56	11.20
0700	24" wide		70	.114		3.40	6.05		9.45	13.50
0900	Bookcase, clear grade pine, shelves 12" OC, 8" deep, per S.F. shelf		70	.114	S.F.	10.90	6.05		16.95	22
1000	12" deep shelves		65	.123	"	16.35	6.55		22.90	28.50

10 57 Wardrobe and Closet Specialties

10 57 23 – Closet and Utility Shelving

10 57 23.19 Wood Closet and Utility Shelving	Crew	Daily Output	Labor-Hours	Unit	Material	2020 Bare Costs Labor	Equipment	Total	Total Incl O&P
1200 Adjustable closet rod and shelf, 12" wide, 3' long	1 Carp	20	.400	Ea.	13.80	21.50		35.30	49
1300 8' long		15	.533	"	25.50	28.50		54	73.50
1500 Prefinished shelves with supports, stock, 8" wide		75	.107	L.F.	5.85	5.65		11.50	15.55
1600 10" wide		70	.114	"	5.65	6.05		11.70	16
9000 Minimum labor/equipment charge	▼	4	2	Job		106		106	170

10 73 Protective Covers

10 73 13 – Awnings

10 73 13.10 Awnings, Fabric

		Crew	Daily Output	Labor-Hours	Unit	Material	2020 Bare Costs Labor	Equipment	Total	Total Incl O&P
0010	**AWNINGS, FABRIC**									
0020	Including acrylic canvas and frame, standard design									
0100	Door and window, slope, 3' high, 4' wide	1 Carp	4.50	1.778	Ea.	755	94.50		849.50	980
0110	6' wide		3.50	2.286		970	121		1,091	1,275
0120	8' wide		3	2.667		1,175	142		1,317	1,525
0200	Quarter round convex, 4' wide		3	2.667		1,175	142		1,317	1,525
0210	6' wide		2.25	3.556		1,525	189		1,714	1,975
0220	8' wide		1.80	4.444		1,875	236		2,111	2,425
0300	Dome, 4' wide		7.50	1.067		455	56.50		511.50	590
0310	6' wide		3.50	2.286		1,025	121		1,146	1,325
0320	8' wide		2	4		1,800	213		2,013	2,350
0350	Elongated dome, 4' wide		1.33	6.015		1,700	320		2,020	2,375
0360	6' wide		1.11	7.207		2,025	385		2,410	2,850
0370	8' wide	▼	1	8		2,375	425		2,800	3,275
1000	Entry or walkway, peak, 12' long, 4' wide	2 Carp	.90	17.778		5,375	945		6,320	7,450
1010	6' wide		.60	26.667		8,300	1,425		9,725	11,400
1020	8' wide		.40	40		11,500	2,125		13,625	16,000
1100	Radius with dome end, 4' wide		1.10	14.545		4,075	775		4,850	5,750
1110	6' wide		.70	22.857		6,575	1,225		7,800	9,175
1120	8' wide	▼	.50	32	▼	9,325	1,700		11,025	13,000
2000	Retractable lateral arm awning, manual									
2010	To 12' wide, 8'-6" projection	2 Carp	1.70	9.412	Ea.	1,225	500		1,725	2,150
2020	To 14' wide, 8'-6" projection		1.10	14.545		1,425	775		2,200	2,825
2030	To 19' wide, 8'-6" projection		.85	18.824		1,925	1,000		2,925	3,725
2040	To 24' wide, 8'-6" projection	▼	.67	23.881		2,425	1,275		3,700	4,700
2050	Motor for above, add	1 Carp	2.67	3	▼	1,050	159		1,209	1,400
3000	Patio/deck canopy with frame									
3010	12' wide, 12' projection	2 Carp	2	8	Ea.	1,725	425		2,150	2,575
3020	16' wide, 14' projection	"	1.20	13.333		2,675	710		3,385	4,075
9000	For fire retardant canvas, add					7%				
9010	For lettering or graphics, add					35%				
9020	For painted or coated acrylic canvas, deduct					8%				
9030	For translucent or opaque vinyl canvas, add					10%				
9040	For 6 or more units, deduct				▼	20%	15%			

10 73 16 – Canopies

10 73 16.10 Canopies, Residential

		Crew	Daily Output	Labor-Hours	Unit	Material	2020 Bare Costs Labor	Equipment	Total	Total Incl O&P
0010	**CANOPIES, RESIDENTIAL** Prefabricated									
0500	Carport, free standing, baked enamel, alum., .032", 40 psf									
0520	16' x 8', 4 posts	2 Carp	3	5.333	Ea.	4,800	283		5,083	5,725
0600	20' x 10', 6 posts		2	8		4,950	425		5,375	6,125
0605	30' x 10', 8 posts	▼	2	8	▼	7,425	425		7,850	8,850

10 73 16 – Canopies

10 73 16.10 Canopies, Residential

		Crew	Daily Output	Labor-Hours	Unit	Material	2020 Bare Costs Labor	Equipment	Total	Total Incl O&P
1000	Door canopies, extruded alum., .032", 42" projection, 4' wide	1 Carp	8	1	Ea.	212	53		265	320
1020	6' wide	"	6	1.333		305	71		376	455
1040	8' wide	2 Carp	9	1.778		445	94.50		539.50	640
1060	10' wide		7	2.286		560	121		681	815
1080	12' wide		5	3.200		620	170		790	950
1200	54" projection, 4' wide	1 Carp	8	1		254	53		307	365
1220	6' wide	"	6	1.333		320	71		391	470
1240	8' wide	2 Carp	9	1.778		425	94.50		519.50	620
1260	10' wide		7	2.286		870	121		991	1,150
1280	12' wide		5	3.200		940	170		1,110	1,300
1300	Painted, add					20%				
1310	Bronze anodized, add					50%				
3000	Window awnings, aluminum, window 3' high, 4' wide	1 Carp	10	.800		152	42.50		194.50	235
3020	6' wide	"	8	1		227	53		280	335
3040	9' wide	2 Carp	9	1.778		400	94.50		494.50	590
3060	12' wide	"	5	3.200		515	170		685	840
3100	Window, 4' high, 4' wide	1 Carp	10	.800		284	42.50		326.50	380
3120	6' wide	"	8	1		335	53		388	455
3140	9' wide	2 Carp	9	1.778		600	94.50		694.50	810
3160	12' wide	"	5	3.200		760	170		930	1,100
3200	Window, 6' high, 4' wide	1 Carp	10	.800		370	42.50		412.50	475
3220	6' wide	"	8	1		385	53		438	510
3240	9' wide	2 Carp	9	1.778		960	94.50		1,054.50	1,200
3260	12' wide	"	5	3.200		1,450	170		1,620	1,875
3400	Roll-up aluminum, 2'-6" wide	1 Carp	14	.571		208	30.50		238.50	278
3420	3' wide		12	.667		213	35.50		248.50	292
3440	4' wide		10	.800		273	42.50		315.50	370
3460	6' wide		8	1		285	53		338	400
3480	9' wide	2 Carp	9	1.778		400	94.50		494.50	590
3500	12' wide	"	5	3.200		510	170		680	835
3600	Window awnings, canvas, 24" drop, 3' wide	1 Carp	30	.267	L.F.	61	14.15		75.15	89.50
3620	4' wide		40	.200		46	10.65		56.65	67.50
3700	30" drop, 3' wide		30	.267		61.50	14.15		75.65	90
3720	4' wide		40	.200		50	10.65		60.65	72
3740	5' wide		45	.178		47	9.45		56.45	66.50
3760	6' wide		48	.167		43.50	8.85		52.35	62
3780	8' wide		48	.167		40	8.85		48.85	58
3800	10' wide		50	.160		41.50	8.50		50	59
3900	Repair canvas, minimum		8	1	Ea.		53		53	85
3920	Maximum		4	2	"		106		106	170
9000	Minimum labor/equipment charge		3	2.667	Job		142		142	227

10 73 16.20 Metal Canopies

		Crew	Daily Output	Labor-Hours	Unit	Material	2020 Bare Costs Labor	Equipment	Total	Total Incl O&P
0010	**METAL CANOPIES**									
0020	Wall hung, .032", aluminum, prefinished, 8' x 10'	K-2	1.30	18.462	Ea.	2,500	1,025	630	4,155	5,100
0300	8' x 20'		1.10	21.818		4,200	1,200	745	6,145	7,375
1000	12' x 20'		1	24		5,825	1,325	820	7,970	9,450
2300	Aluminum entrance canopies, flat soffit, .032"									
2500	3'-6" x 4'-0", clear anodized	2 Carp	4	4	Ea.	1,050	213		1,263	1,500
9000	Minimum labor/equipment charge	"	2	8	Job		425		425	680

10 74 Manufactured Exterior Specialties

10 74 23 – Cupolas

10 74 23.10 Wood Cupolas

10 74 23.10 Wood Cupolas	Crew	Daily Output	Labor-Hours	Unit	Material	2020 Bare Costs Labor	Equipment	Total	Total Incl O&P
0010 **WOOD CUPOLAS**									
0020 Stock units, pine, painted, 18" sq., 28" high, alum. roof	1 Carp	4.10	1.951	Ea.	320	104		424	515
0100 Copper roof		3.80	2.105		298	112		410	510
0300 23" square, 33" high, aluminum roof		3.70	2.162		455	115		570	685
0400 Copper roof		3.30	2.424		605	129		734	870
0600 30" square, 37" high, aluminum roof		3.70	2.162		660	115		775	910
0700 Copper roof		3.30	2.424		785	129		914	1,075
0900 Hexagonal, 31" wide, 46" high, copper roof		4	2		995	106		1,101	1,275
1000 36" wide, 50" high, copper roof		3.50	2.286		1,750	121		1,871	2,125
1200 For deluxe stock units, add to above					25%				
1400 For custom built units, add to above					50%	50%			
9000 Minimum labor/equipment charge	1 Carp	2.75	2.909	Job		155		155	248

10 74 29 – Steeples

10 74 29.10 Prefabricated Steeples

10 74 29.10 Prefabricated Steeples	Crew	Daily Output	Labor-Hours	Unit	Material	2020 Bare Costs Labor	Equipment	Total	Total Incl O&P
0010 **PREFABRICATED STEEPLES**									
4000 Steeples, translucent fiberglass, 30" square, 15' high	F-3	2	20	Ea.	9,925	1,075	236	11,236	12,900
4150 25' high	"	1.80	22.222		11,500	1,200	262	12,962	14,800
4600 Aluminum, baked finish, 16" square, 14' high					6,250			6,250	6,850
4640 35' high, 8' base					38,700			38,700	42,600
4680 152' high, custom					660,500			660,500	727,000

10 74 33 – Weathervanes

10 74 33.10 Residential Weathervanes

10 74 33.10 Residential Weathervanes	Crew	Daily Output	Labor-Hours	Unit	Material	2020 Bare Costs Labor	Equipment	Total	Total Incl O&P
0010 **RESIDENTIAL WEATHERVANES**									
0020 Residential types, 18" to 24"	1 Carp	8	1	Ea.	138	53		191	237
0100 24" to 48"		2	4	"	1,900	213		2,113	2,425
9000 Minimum labor/equipment charge		4	2	Job		106		106	170

10 74 46 – Window Wells

10 74 46.10 Area Window Wells

10 74 46.10 Area Window Wells	Crew	Daily Output	Labor-Hours	Unit	Material	2020 Bare Costs Labor	Equipment	Total	Total Incl O&P
0010 **AREA WINDOW WELLS**, Galvanized steel									
0020 20 ga., 3'-2" wide, 1' deep	1 Sswk	29	.276	Ea.	18.70	15.90		34.60	46.50
0100 2' deep		23	.348		30	20		50	66
0300 16 ga., 3'-2" wide, 1' deep		29	.276		25.50	15.90		41.40	54
0400 3' deep		23	.348		48.50	20		68.50	86.50
0600 Welded grating for above, 15 lb., painted		45	.178		103	10.25		113.25	130
0700 Galvanized		45	.178		126	10.25		136.25	155
0900 Translucent plastic cap for above		60	.133		21.50	7.70		29.20	36

10 86 Security Mirrors and Domes

10 86 10 – Security Mirrors

10 86 10.10 Exterior Traffic Control Mirrors

10 86 10.10 Exterior Traffic Control Mirrors	Crew	Daily Output	Labor-Hours	Unit	Material	2020 Bare Costs Labor	Equipment	Total	Total Incl O&P
0010 **EXTERIOR TRAFFIC CONTROL MIRRORS**									
0100 Convex, stainless steel, 20 ga., 26" diameter	1 Carp	12	.667	Ea.	188	35.50		223.50	264

10 86 20 – Security Domes

10 86 20.10 Domes

10 86 20.10 Domes	Crew	Daily Output	Labor-Hours	Unit	Material	2020 Bare Costs Labor	Equipment	Total	Total Incl O&P
0010 **DOMES** for security cameras (CCTV)									
0100 Ceiling mounted, 10" diameter	1 Carp	30	.267	Ea.	12.30	14.15		26.45	36
0110 12" diameter	"	30	.267	"	8.85	14.15		23	32

Estimating Tips
General

- The items in this division are usually priced per square foot or each. Many of these items are purchased by the owner for installation by the contractor. Check the specifications for responsibilities and include time for receiving, storage, installation, and mechanical and electrical hookups in the appropriate divisions.

- Many items in Division 11 require some type of support system that is not usually furnished with the item. Examples of these systems include blocking for the attachment of casework and support angles for ceiling-hung projection screens. The required blocking or supports must be added to the estimate in the appropriate division.

- Some items in Division 11 may require assembly or electrical hookups. Verify the amount of assembly required or the need for a hard electrical connection and add the appropriate costs.

Reference Numbers

Reference numbers are shown at the beginning of some major classifications. These numbers refer to related items in the Reference Section. The reference information may be an estimating procedure, an alternate pricing method, or technical information.

Same Data. Simplified.

Enjoy the convenience and efficiency of accessing your costs anywhere:

- **Skip the multiplier** by setting your location
- **Quickly search,** edit, favorite and share costs
- **Stay on top of price changes** with automatic updates

Discover more at rsmeans.com/online

---

11 11 Vehicle Service Equipment

11 11 13 – Compressed-Air Vehicle Service Equipment

11 11 13.10 Compressed Air Equipment

		Crew	Daily Output	Labor-Hours	Unit	Material	2020 Bare Costs Labor	2020 Bare Costs Equipment	Total	Total Incl O&P
0010	**COMPRESSED AIR EQUIPMENT**									
0030	Compressors, electric, 1-1/2 HP, standard controls	L-4	1.50	16	Ea.	505	800		1,305	1,825
0550	Dual controls		1.50	16		1,050	800		1,850	2,425
0600	5 HP, 115/230 volt, standard controls		1	24		1,725	1,200		2,925	3,825
0650	Dual controls	↓	1	24	↓	3,225	1,200		4,425	5,475

11 11 19 – Vehicle Lubrication Equipment

11 11 19.10 Lubrication Equipment

		Crew	Daily Output	Labor-Hours	Unit	Material	Labor	Equipment	Total	Total Incl O&P
0010	**LUBRICATION EQUIPMENT**									
3000	Lube equipment, 3 reel type, with pumps, not including piping	L-4	.50	48	Set	10,400	2,400		12,800	15,300
3100	Hose reel, including hose, oil/lube, 1000 psi	2 Sswk	2	8	Ea.	480	460		940	1,275
3200	Grease, 5000 psi		2	8		585	460		1,045	1,400
3300	Air, 50', 160 psi		2	8		605	460		1,065	1,425
3350	25', 160 psi	↓	2	8	↓	365	460		825	1,150

11 11 33 – Vehicle Spray Painting Equipment

11 11 33.10 Spray Painting Equipment

		Crew	Daily Output	Labor-Hours	Unit	Material	Labor	Equipment	Total	Total Incl O&P
0010	**SPRAY PAINTING EQUIPMENT**									
4000	Spray painting booth, 26' long, complete	L-4	.40	60	Ea.	11,600	3,000		14,600	17,500

11 12 Parking Control Equipment

11 12 13 – Parking Key and Card Control Units

11 12 13.10 Parking Control Units

		Crew	Daily Output	Labor-Hours	Unit	Material	Labor	Equipment	Total	Total Incl O&P
0010	**PARKING CONTROL UNITS**									
5100	Card reader	1 Elec	2	4	Ea.	1,475	245		1,720	2,000
5120	Proximity with customer display	2 Elec	1	16		6,025	980		7,005	8,125
6000	Parking control software, basic functionality	1 Elec	.50	16		24,000	980		24,980	27,900
6020	Multi-function	"	.20	40	↓	96,000	2,450		98,450	109,500

11 12 16 – Parking Ticket Dispensers

11 12 16.10 Ticket Dispensers

		Crew	Daily Output	Labor-Hours	Unit	Material	Labor	Equipment	Total	Total Incl O&P
0010	**TICKET DISPENSERS**									
5900	Ticket spitter with time/date stamp, standard	2 Elec	2	8	Ea.	6,625	490		7,115	8,025
5920	Mag stripe encoding	"	2	8	"	18,900	490		19,390	21,600

11 12 26 – Parking Fee Collection Equipment

11 12 26.13 Parking Fee Coin Collection Equipment

		Crew	Daily Output	Labor-Hours	Unit	Material	Labor	Equipment	Total	Total Incl O&P
0010	**PARKING FEE COIN COLLECTION EQUIPMENT**									
5200	Cashier booth, average	B-22	1	30	Ea.	10,600	1,475	284	12,359	14,400
5300	Collector station, pay on foot	2 Elec	.20	80		111,500	4,900		116,400	130,500
5320	Credit card only	"	.50	32	↓	20,600	1,975		22,575	25,700

11 12 26.23 Fee Equipment

		Crew	Daily Output	Labor-Hours	Unit	Material	Labor	Equipment	Total	Total Incl O&P
0010	**FEE EQUIPMENT**									
5600	Fee computer	1 Elec	1.50	5.333	Ea.	13,000	325		13,325	14,800

11 12 33 – Parking Gates

11 12 33.13 Lift Arm Parking Gates

		Crew	Daily Output	Labor-Hours	Unit	Material	Labor	Equipment	Total	Total Incl O&P
0010	**LIFT ARM PARKING GATES**									
5000	Barrier gate with programmable controller	2 Elec	3	5.333	Ea.	3,300	325		3,625	4,125
5020	Industrial		3	5.333		5,225	325		5,550	6,250
5050	Non-programmable, with reader and 12' arm		3	5.333		1,700	325		2,025	2,375
5500	Exit verifier	↓	1	16		19,600	980		20,580	23,100
5700	Full sign, 4" letters	1 Elec	2	4	↓	1,250	245		1,495	1,725

11 12 Parking Control Equipment

11 12 33 - Parking Gates

11 12 33.13 Lift Arm Parking Gates

		Crew	Daily Output	Labor-Hours	Unit	Material	2020 Bare Costs Labor	Equipment	Total	Total Incl O&P
5800	Inductive loop	2 Elec	4	4	Ea.	209	245		454	610
5950	Vehicle detector, microprocessor based	1 Elec	3	2.667		470	164		634	765
7100	Traffic spike unit, flush mount, spring loaded, 72" L	B-89	4	4		1,275	201	258	1,734	2,000
7200	Surface mount, 72" L	2 Skwk	10	1.600	↓	2,425	88		2,513	2,800

11 13 Loading Dock Equipment

11 13 13 - Loading Dock Bumpers

11 13 13.10 Dock Bumpers

		Crew	Daily Output	Labor-Hours	Unit	Material	2020 Bare Costs Labor	Equipment	Total	Total Incl O&P
0010	**DOCK BUMPERS** Bolts not included									
0012	2" x 6" to 4" x 8", average	1 Carp	300	.027	B.F.	.74	1.42		2.16	3.08
0050	Bumpers, lam. rubber blocks 4-1/2" thick, 10" high, 14" long		26	.308	Ea.	60.50	16.35		76.85	92.50
0200	24" long		22	.364		103	19.35		122.35	145
0300	36" long		17	.471		340	25		365	415
0500	12" high, 14" long	↓	25	.320	↓	81	17		98	117

11 13 16 - Loading Dock Seals and Shelters

11 13 16.10 Dock Seals and Shelters

		Crew	Daily Output	Labor-Hours	Unit	Material	2020 Bare Costs Labor	Equipment	Total	Total Incl O&P
0010	**DOCK SEALS AND SHELTERS**									
6200	Shelters, fabric, for truck or train, scissor arms, minimum	1 Carp	1	8	Ea.	2,250	425		2,675	3,150
6300	Maximum	"	.50	16	"	2,550	850		3,400	4,150

11 13 19 - Stationary Loading Dock Equipment

11 13 19.10 Dock Equipment

		Crew	Daily Output	Labor-Hours	Unit	Material	2020 Bare Costs Labor	Equipment	Total	Total Incl O&P
0010	**DOCK EQUIPMENT**									
2200	Dock boards, heavy duty, 60" x 60", aluminum, 5,000 lb. capacity				Ea.	1,375			1,375	1,525
4200	Platform lifter, 6' x 6', portable, 3,000 lb. capacity				"	9,800			9,800	10,800
6000	Dock leveler, 15 ton capacity									
6100	I-beam construction, mechanical, 6' x 8'	E-16	.50	32	Ea.	5,425	1,875	294	7,594	9,375
6150	Hydraulic		.50	32		6,375	1,875	294	8,544	10,400
6200	Formed beam deck construction, mechanical, 6' x 8'		.50	32		4,175	1,875	294	6,344	8,000
6250	Hydraulic		.50	32		5,725	1,875	294	7,894	9,700
6300	Edge of dock leveler, mechanical, 15 ton capacity		2	8		1,250	470	73.50	1,793.50	2,225
6301	Edge of dock leveler, mechanical, 10 ton capacity	↓	2	8	↓	1,250	470	73.50	1,793.50	2,225
7000	22.5 ton capacity									
7100	Vertical storing dock leveler, hydraulic, 6' x 6'	E-16	.40	40	Ea.	8,300	2,350	370	11,020	13,400

11 21 Retail and Service Equipment

11 21 13 - Cash Registers and Checking Equipment

11 21 13.10 Checkout Counter

		Crew	Daily Output	Labor-Hours	Unit	Material	2020 Bare Costs Labor	Equipment	Total	Total Incl O&P
0010	**CHECKOUT COUNTER**									
0020	Supermarket conveyor, single belt	2 Clab	10	1.600	Ea.	3,775	67.50		3,842.50	4,275
0100	Double belt, power take-away		9	1.778		5,175	75		5,250	5,825
0400	Double belt, power take-away, incl. side scanning		7	2.286		5,975	96		6,071	6,725
0800	Warehouse or bulk type	↓	6	2.667	↓	6,950	112		7,062	7,800
1000	Scanning system, 2 lanes, w/registers, scan gun & memory				System	18,800			18,800	20,700
1100	10 lanes, single processor, full scan, with scales				"	178,500			178,500	196,500
2000	Register, restaurant, minimum				Ea.	800			800	880
2100	Maximum					3,575			3,575	3,925
2150	Store, minimum					800			800	880
2200	Maximum				↓	3,575			3,575	3,925

For customer support on your Commercial Renovation Costs with RSMeans data, call 800.448.8182.

369

11 21 Retail and Service Equipment

11 21 33 – Checkroom Equipment

11 21 33.10 Clothes Check Equipment

		Crew	Daily Output	Labor-Hours	Unit	Material	2020 Bare Costs Labor	2020 Bare Costs Equipment	Total	Total Incl O&P
0010	**CLOTHES CHECK EQUIPMENT**									
0030	Clothes check rack, free standing, st. stl., 2-tier, 90 bag capacity				Ea.	1,575			1,575	1,725
0050	Wall mounted, 45 bag capacity	L-2	8	2		905	93		998	1,150
0100	Garment checking bag, green mesh fabric, 21" H x 17" W with 4.5" hook			↓		28			28	31

11 21 53 – Barber and Beauty Shop Equipment

11 21 53.10 Barber Equipment

		Crew	Daily Output	Labor-Hours	Unit	Material	2020 Bare Costs Labor	2020 Bare Costs Equipment	Total	Total Incl O&P
0010	**BARBER EQUIPMENT**									
0020	Chair, hydraulic, movable, minimum	1 Carp	24	.333	Ea.	640	17.70		657.70	735
0050	Maximum	"	16	.500		4,150	26.50		4,176.50	4,625
0200	Wall hung styling station with mirrors, minimum	L-2	8	2		600	93		693	810
0300	Maximum	"	4	4		2,775	186		2,961	3,350
0500	Sink, hair washing basin, rough plumbing not incl.	1 Plum	8	1		530	64.50		594.50	685
1000	Sterilizer, liquid solution for tools					171			171	189
1100	Total equipment, rule of thumb, per chair, minimum	L-8	1	20		2,225	1,100		3,325	4,200
1150	Maximum	"	1	20		6,175	1,100		7,275	8,525

11 21 73 – Commercial Laundry and Dry Cleaning Equipment

11 21 73.13 Dry Cleaning Equipment

		Crew	Daily Output	Labor-Hours	Unit	Material	2020 Bare Costs Labor	2020 Bare Costs Equipment	Total	Total Incl O&P
0010	**DRY CLEANING EQUIPMENT**									
2000	Dry cleaners, electric, 20 lb. capacity, not incl. rough-in	L-1	.20	80	Ea.	35,100	5,025		40,125	46,400
2050	25 lb. capacity		.17	94.118		52,000	5,925		57,925	66,000
2100	30 lb. capacity		.15	107		55,000	6,700		61,700	71,000
2150	60 lb. capacity	↓	.09	178	↓	79,000	11,200		90,200	104,500

11 21 73.16 Drying and Conditioning Equipment

		Crew	Daily Output	Labor-Hours	Unit	Material	2020 Bare Costs Labor	2020 Bare Costs Equipment	Total	Total Incl O&P
0010	**DRYING AND CONDITIONING EQUIPMENT**									
0100	Dryers, not including rough-in									
1500	Industrial, 30 lb. capacity	1 Plum	2	4	Ea.	3,350	258		3,608	4,075
1600	50 lb. capacity	"	1.70	4.706		4,550	305		4,855	5,500
4700	Lint collector, ductwork not included, 8,000 to 10,000 CFM	Q-10	.30	80	↓	9,900	4,650		14,550	18,200

11 21 73.19 Finishing Equipment

		Crew	Daily Output	Labor-Hours	Unit	Material	2020 Bare Costs Labor	2020 Bare Costs Equipment	Total	Total Incl O&P
0010	**FINISHING EQUIPMENT**									
3500	Folders, blankets & sheets, minimum	1 Elec	.17	47.059	Ea.	35,600	2,875		38,475	43,600
3700	King size with automatic stacker		.10	80		64,500	4,900		69,400	78,000
3800	For conveyor delivery, add	↓	.45	17.778	↓	16,600	1,100		17,700	20,000

11 21 73.23 Commercial Ironing Equipment

		Crew	Daily Output	Labor-Hours	Unit	Material	2020 Bare Costs Labor	2020 Bare Costs Equipment	Total	Total Incl O&P
0010	**COMMERCIAL IRONING EQUIPMENT**									
4500	Ironers, institutional, 110", single roll	1 Elec	.20	40	Ea.	34,400	2,450		36,850	41,700

11 21 73.26 Commercial Washers and Extractors

		Crew	Daily Output	Labor-Hours	Unit	Material	2020 Bare Costs Labor	2020 Bare Costs Equipment	Total	Total Incl O&P
0010	**COMMERCIAL WASHERS AND EXTRACTORS**, not including rough-in									
6000	Combination washer/extractor, 20 lb. capacity	L-6	1.50	8	Ea.	6,725	505		7,230	8,175
6100	30 lb. capacity		.80	15		10,300	950		11,250	12,800
6200	50 lb. capacity		.68	17.647		12,900	1,125		14,025	15,900
6300	75 lb. capacity		.30	40		19,200	2,525		21,725	25,000
6350	125 lb. capacity	↓	.16	75	↓	29,900	4,750		34,650	40,300

11 21 73.33 Coin-Operated Laundry Equipment

		Crew	Daily Output	Labor-Hours	Unit	Material	2020 Bare Costs Labor	2020 Bare Costs Equipment	Total	Total Incl O&P
0010	**COIN-OPERATED LAUNDRY EQUIPMENT**									
0990	Dryer, gas fired									
1000	Commercial, 30 lb. capacity, coin operated, single	1 Plum	3	2.667	Ea.	3,475	172		3,647	4,100
1100	Double stacked	"	2	4	"	8,275	258		8,533	9,500
5290	Clothes washer									
5300	Commercial, coin operated, average	1 Plum	3	2.667	Ea.	1,350	172		1,522	1,775

11 21 Retail and Service Equipment

11 21 83 – Photo Processing Equipment

11 21 83.13 Darkroom Equipment

		Crew	Daily Output	Labor-Hours	Unit	Material	2020 Bare Costs Labor	Equipment	Total	Total Incl O&P
0010	**DARKROOM EQUIPMENT**									
0020	Developing sink, 5" deep, 24" x 48"	Q-1	2	8	Ea.	500	465		965	1,275
0050	48" x 52"		1.70	9.412		1,250	545		1,795	2,225
0200	10" deep, 24" x 48"		1.70	9.412		1,625	545		2,170	2,650
0250	24" x 108"		1.50	10.667		3,875	620		4,495	5,200
0500	Dryers, dehumidified filtered air, 36" x 25" x 68" high	L-7	6	4.667		2,175	239		2,414	2,775
3000	Viewing lights, 20" x 24"		6	4.667		276	239		515	685
3100	20" x 24" with color correction		6	4.667		410	239		649	830

11 22 Banking Equipment

11 22 13 – Vault Equipment

11 22 13.16 Safes

		Crew	Daily Output	Labor-Hours	Unit	Material	2020 Bare Costs Labor	Equipment	Total	Total Incl O&P
0010	**SAFES**									
0800	Money, "B" label, 9" x 14" x 14"				Ea.	605			605	670
0900	Tool resistive, 24" x 24" x 20"					4,175			4,175	4,600
1050	Tool and torch resistive, 24" x 24" x 20"					9,525			9,525	10,500

11 22 16 – Teller and Service Equipment

11 22 16.13 Teller Equipment Systems

		Crew	Daily Output	Labor-Hours	Unit	Material	2020 Bare Costs Labor	Equipment	Total	Total Incl O&P
0010	**TELLER EQUIPMENT SYSTEMS**									
0020	Alarm system, police	2 Elec	1.60	10	Ea.	5,075	615		5,690	6,525
0400	Bullet resistant teller window, 44" x 60"	1 Glaz	.60	13.333		4,625	680		5,305	6,150
0500	48" x 60"	"	.60	13.333		6,475	680		7,155	8,200
3000	Counters for banks, frontal only	2 Carp	1	16	Station	1,925	850		2,775	3,450
3100	Complete with steel undercounter		.50	32	"	3,700	1,700		5,400	6,800
5400	Partitions, bullet-resistant, 1-3/16" glass, 8' high		10	1.600	L.F.	205	85		290	360
5600	Pass thru, bullet-res. window, painted steel, 24" x 36"	2 Sswk	1.60	10	Ea.	3,025	575		3,600	4,275

11 30 Residential Equipment

11 30 13 – Residential Appliances

11 30 13.15 Cooking Equipment

		Crew	Daily Output	Labor-Hours	Unit	Material	2020 Bare Costs Labor	Equipment	Total	Total Incl O&P
0010	**COOKING EQUIPMENT**									
0020	Cooking range, 30" free standing, 1 oven, minimum	2 Clab	10	1.600	Ea.	465	67.50		532.50	620
0050	Maximum	"	4	4		2,200	168		2,368	2,700
0350	Built-in, 30" wide, 1 oven, minimum	1 Elec	6	1.333		870	82		952	1,075
0400	Maximum	2 Carp	2	8		2,000	425		2,425	2,875
0900	Countertop cooktops, 4 burner, standard, minimum	1 Elec	6	1.333		340	82		422	495
0950	Maximum		3	2.667		2,025	164		2,189	2,475
1250	Microwave oven, minimum		4	2		98.50	123		221.50	297
1300	Maximum		2	4		470	245		715	900
5380	Oven, built-in, standard		4	2		960	123		1,083	1,250
5390	Deluxe		2	4		2,975	245		3,220	3,650

11 30 13.16 Refrigeration Equipment

		Crew	Daily Output	Labor-Hours	Unit	Material	2020 Bare Costs Labor	Equipment	Total	Total Incl O&P
0010	**REFRIGERATION EQUIPMENT**									
5450	Refrigerator, no frost, 6 C.F.	2 Clab	15	1.067	Ea.	365	45		410	470
5500	Refrigerator, no frost, 10 C.F. to 12 C.F., minimum		10	1.600		455	67.50		522.50	610
5600	Maximum		6	2.667		560	112		672	800
6790	Energy-star qualified, 18 C.F., minimum [G]	2 Carp	4	4		510	213		723	905
6795	Maximum [G]		2	8		1,950	425		2,375	2,800

For customer support on your Commercial Renovation Costs with RSMeans data, call 800.448.8182.

371

11 30 Residential Equipment

11 30 13 – Residential Appliances

11 30 13.16 Refrigeration Equipment

		Crew	Daily Output	Labor-Hours	Unit	Material	2020 Bare Costs Labor	2020 Bare Costs Equipment	Total	Total Incl O&P
6797	21.7 C.F., minimum	G 2 Carp	4	4	Ea.	1,025	213		1,238	1,475
6799	Maximum	G	4	4		1,800	213		2,013	2,325

11 30 13.17 Kitchen Cleaning Equipment

		Crew	Daily Output	Labor-Hours	Unit	Material	2020 Bare Costs Labor	2020 Bare Costs Equipment	Total	Total Incl O&P
0010	**KITCHEN CLEANING EQUIPMENT**									
2750	Dishwasher, built-in, 2 cycles, minimum	L-1	4	4	Ea.	310	252		562	730
2800	Maximum		2	8		445	505		950	1,275
3100	Energy-star qualified, minimum	G	4	4		405	252		657	840
3110	Maximum	G	2	8		1,975	505		2,480	2,950

11 30 13.18 Waste Disposal Equipment

		Crew	Daily Output	Labor-Hours	Unit	Material	2020 Bare Costs Labor	2020 Bare Costs Equipment	Total	Total Incl O&P
0010	**WASTE DISPOSAL EQUIPMENT**									
1750	Compactor, residential size, 4 to 1 compaction, minimum	1 Carp	5	1.600	Ea.	705	85		790	910
1800	Maximum	"	3	2.667		1,150	142		1,292	1,500
3300	Garbage disposal, sink type, minimum	L-1	10	1.600		111	101		212	277
3350	Maximum	"	10	1.600		216	101		317	390

11 30 13.19 Kitchen Ventilation Equipment

		Crew	Daily Output	Labor-Hours	Unit	Material	2020 Bare Costs Labor	2020 Bare Costs Equipment	Total	Total Incl O&P
0010	**KITCHEN VENTILATION EQUIPMENT**									
4150	Hood for range, 2 speed, vented, 30" wide, minimum	L-3	5	3.200	Ea.	107	184		291	410
4200	Maximum		3	5.333		1,000	305		1,305	1,575
4300	42" wide, minimum		5	3.200		151	184		335	455
4330	Custom		5	3.200		1,800	184		1,984	2,275
4350	Maximum		3	5.333		2,175	305		2,480	2,875
4400	Ventless hood, 2 speed, 30" wide, minimum	1 Elec	8	1		115	61.50		176.50	221
4450	Maximum		7	1.143		630	70		700	805
4510	36" wide, minimum		7	1.143		78.50	70		148.50	194
4550	Maximum		6	1.333		785	82		867	990
4580	42" wide, minimum		6	1.333		171	82		253	315
4600	Maximum		5	1.600		630	98		728	840
4650	For vented 1 speed, deduct from maximum					74			74	81.50

11 30 13.24 Washers

		Crew	Daily Output	Labor-Hours	Unit	Material	2020 Bare Costs Labor	2020 Bare Costs Equipment	Total	Total Incl O&P
0010	**WASHERS**									
5000	Residential, 4 cycle, average	1 Plum	3	2.667	Ea.	990	172		1,162	1,375
6750	Energy star qualified, front loading, minimum	G	3	2.667		880	172		1,052	1,225
6760	Maximum	G	1	8		1,925	515		2,440	2,900
6764	Top loading, minimum	G	3	2.667		735	172		907	1,075
6766	Maximum	G	3	2.667		1,125	172		1,297	1,525

11 30 13.25 Dryers

		Crew	Daily Output	Labor-Hours	Unit	Material	2020 Bare Costs Labor	2020 Bare Costs Equipment	Total	Total Incl O&P
0010	**DRYERS**									
0500	Gas fired residential, 16 lb. capacity, average	1 Plum	3	2.667	Ea.	720	172		892	1,050
6770	Electric, front loading, energy-star qualified, minimum	G L-2	3	5.333		480	248		728	925
6780	Maximum	G "	2	8		1,500	370		1,870	2,225

11 30 15 – Miscellaneous Residential Appliances

11 30 15.13 Sump Pumps

		Crew	Daily Output	Labor-Hours	Unit	Material	2020 Bare Costs Labor	2020 Bare Costs Equipment	Total	Total Incl O&P
0010	**SUMP PUMPS**									
6400	Cellar drainer, pedestal, 1/3 HP, molded PVC base	1 Plum	3	2.667	Ea.	141	172		313	420
6450	Solid brass	"	2	4	"	243	258		501	665
6460	Sump pump, see also Section 22 14 29.16									

11 30 Residential Equipment

11 30 33 – Retractable Stairs

11 30 33.10 Disappearing Stairway

		Crew	Daily Output	Labor-Hours	Unit	Material	2020 Bare Costs Labor	Equipment	Total	Total Incl O&P
0010	**DISAPPEARING STAIRWAY** No trim included									
0020	One piece, yellow pine, 8'-0" ceiling	2 Carp	4	4	Ea.	278	213		491	645
0030	9'-0" ceiling		4	4		305	213		518	675
0040	10'-0" ceiling		3	5.333		250	283		533	730
0050	11'-0" ceiling		3	5.333		360	283		643	850
0060	12'-0" ceiling		3	5.333		355	283		638	845
0100	Custom grade, pine, 8'-6" ceiling, minimum	1 Carp	4	2		129	106		235	310
0150	Average		3.50	2.286		292	121		413	515
0200	Maximum		3	2.667		279	142		421	530
0500	Heavy duty, pivoted, from 7'-7" to 12'-10" floor to floor		3	2.667		1,375	142		1,517	1,725
0600	16'-0" ceiling		2	4		1,600	213		1,813	2,100
0800	Economy folding, pine, 8'-6" ceiling		4	2		178	106		284	365
0900	9'-6" ceiling		4	2		232	106		338	425
1100	Automatic electric, aluminum, floor to floor height, 8' to 9'	2 Carp	1	16		9,200	850		10,050	11,500
1400	11' to 12'		.90	17.778		9,725	945		10,670	12,200
1700	14' to 15'		.70	22.857		10,400	1,225		11,625	13,400
9000	Minimum labor/equipment charge	1 Carp	2	4	Job		213		213	340

11 32 Unit Kitchens

11 32 13 – Metal Unit Kitchens

11 32 13.10 Commercial Unit Kitchens

		Crew	Daily Output	Labor-Hours	Unit	Material	2020 Bare Costs Labor	Equipment	Total	Total Incl O&P
0010	**COMMERCIAL UNIT KITCHENS**									
1500	Combination range, refrigerator and sink, 30" wide, minimum	L-1	2	8	Ea.	710	505		1,215	1,550
1550	Maximum	"	1	16	"	1,100	1,000		2,100	2,750
1640	Combination range, refrigerator, sink, microwave									
1660	Oven and ice maker	L-1	.80	20	Ea.	4,900	1,250		6,150	7,325

11 41 Foodservice Storage Equipment

11 41 13 – Refrigerated Food Storage Cases

11 41 13.10 Refrigerated Food Cases

		Crew	Daily Output	Labor-Hours	Unit	Material	2020 Bare Costs Labor	Equipment	Total	Total Incl O&P
0010	**REFRIGERATED FOOD CASES**									
0030	Dairy, multi-deck, 12' long	Q-5	3	5.333	Ea.	11,500	315		11,815	13,100
0100	For rear sliding doors, add					2,050			2,050	2,250
0200	Delicatessen case, service deli, 12' long, single deck	Q-5	3.90	4.103		8,750	242		8,992	10,000
0300	Multi-deck, 18 S.F. shelf display		3	5.333		8,375	315		8,690	9,700
0400	Freezer, self-contained, chest-type, 30 C.F.		3.90	4.103		4,800	242		5,042	5,650
0500	Glass door, upright, 78 C.F.		3.30	4.848		9,375	286		9,661	10,700
0600	Frozen food, chest type, 12' long		3.30	4.848		7,575	286		7,861	8,775
0700	Glass door, reach-in, 5 door		3	5.333		10,400	315		10,715	11,900
0800	Island case, 12' long, single deck		3.30	4.848		7,900	286		8,186	9,125
0900	Multi-deck		3	5.333		10,100	315		10,415	11,600
1000	Meat case, 12' long, single deck		3.30	4.848		7,875	286		8,161	9,100
1050	Multi-deck		3.10	5.161		10,200	305		10,505	11,700
1100	Produce, 12' long, single deck		3.30	4.848		6,475	286		6,761	7,575
1200	Multi-deck		3.10	5.161		8,550	305		8,855	9,875

11 41 13 – Refrigerated Food Storage Cases

11 41 13.20 Refrigerated Food Storage Equipment		Crew	Daily Output	Labor-Hours	Unit	Material	2020 Bare Costs Labor	Equipment	Total	Total Incl O&P
0010	**REFRIGERATED FOOD STORAGE EQUIPMENT**									
2350	Cooler, reach-in, beverage, 6' long	Q-1	6	2.667	Ea.	3,125	155		3,280	3,675
4300	Freezers, reach-in, 44 C.F.		4	4		4,425	232		4,657	5,225
4500	68 C.F.		3	5.333		5,575	310		5,885	6,600
4600	Freezer, pre-fab, 8' x 8' w/refrigeration	2 Carp	.45	35.556		9,200	1,900		11,100	13,100
4620	8' x 12'		.35	45.714		10,900	2,425		13,325	15,900
4640	8' x 16'		.25	64		13,200	3,400		16,600	20,000
4660	8' x 20'		.17	94.118		20,500	5,000		25,500	30,500
4680	Reach-in, 1 compartment	Q-1	4	4		3,350	232		3,582	4,025
4685	Energy star rated [G]	R-18	7.80	3.333		3,575	180		3,755	4,200
4700	2 compartment	Q-1	3	5.333		5,000	310		5,310	5,975
4705	Energy star rated [G]	R-18	6.20	4.194		3,750	226		3,976	4,475
4710	3 compartment	Q-1	3	5.333		6,225	310		6,535	7,325
4715	Energy star rated [G]	R-18	5.60	4.643		5,250	250		5,500	6,150
8320	Refrigerator, reach-in, 1 compartment		7.80	3.333		3,600	180		3,780	4,225
8325	Energy star rated [G]		7.80	3.333		3,575	180		3,755	4,200
8330	2 compartment		6.20	4.194		3,425	226		3,651	4,125
8335	Energy star rated [G]		6.20	4.194		3,750	226		3,976	4,475
8340	3 compartment		5.60	4.643		4,825	250		5,075	5,675
8345	Energy star rated [G]		5.60	4.643		5,250	250		5,500	6,150
8350	Pre-fab, with refrigeration, 8' x 8'	2 Carp	.45	35.556		6,950	1,900		8,850	10,700
8360	8' x 12'		.35	45.714		8,900	2,425		11,325	13,700
8370	8' x 16'		.25	64		11,500	3,400		14,900	18,200
8380	8' x 20'		.17	94.118		14,600	5,000		19,600	24,100
8390	Pass-thru/roll-in, 1 compartment	R-18	7.80	3.333		4,100	180		4,280	4,800
8400	2 compartment		6.24	4.167		5,375	224		5,599	6,275
8410	3 compartment		5.60	4.643		9,625	250		9,875	11,000
8420	Walk-in, alum, door & floor only, no refrig, 6' x 6' x 7'-6"	2 Carp	1.40	11.429		6,800	605		7,405	8,450
8430	10' x 6' x 7'-6"		.55	29.091		9,475	1,550		11,025	12,900
8440	12' x 14' x 7'-6"		.25	64		13,200	3,400		16,600	20,000
8450	12' x 20' x 7'-6"		.17	94.118		13,100	5,000		18,100	22,400
8451	12' x 20' x 7'-6"		.17	94.118		6,525	5,000		11,525	15,200
8460	Refrigerated cabinets, mobile					6,350			6,350	6,975
8470	Refrigerator/freezer, reach-in, 1 compartment	R-18	5.60	4.643		6,950	250		7,200	8,000
8480	2 compartment	"	4.80	5.417		7,575	292		7,867	8,800

11 41 13.30 Wine Cellar

		Crew	Daily Output	Labor-Hours	Unit	Material	Labor	Equipment	Total	Total Incl O&P
0010	**WINE CELLAR**, refrigerated, Redwood interior, carpeted, walk-in type									
0020	6'-8" high, including racks									
0200	80" W x 48" D for 900 bottles	2 Carp	1.50	10.667	Ea.	4,200	565		4,765	5,525
0250	80" W x 72" D for 1300 bottles		1.33	12.030		5,725	640		6,365	7,325
0300	80" W x 94" D for 1900 bottles		1.17	13.675		6,725	725		7,450	8,575
0400	80" W x 124" D for 2500 bottles		1	16		7,800	850		8,650	9,950

11 41 33 – Foodservice Shelving

11 41 33.20 Metal Food Storage Shelving

		Crew	Daily Output	Labor-Hours	Unit	Material	Labor	Equipment	Total	Total Incl O&P
0010	**METAL FOOD STORAGE SHELVING**									
8600	Stainless steel shelving, louvered 4-tier, 20" x 3'	1 Clab	6	1.333	Ea.	1,500	56		1,556	1,750
8605	20" x 4'		6	1.333		1,650	56		1,706	1,925
8610	20" x 6'		6	1.333		1,750	56		1,806	2,025
8615	24" x 3'		6	1.333		2,050	56		2,106	2,350
8620	24" x 4'		6	1.333		2,525	56		2,581	2,875
8625	24" x 6'		6	1.333		3,450	56		3,506	3,900

11 41 Foodservice Storage Equipment

11 41 33 – Foodservice Shelving

11 41 33.20 Metal Food Storage Shelving

		Crew	Daily Output	Labor-Hours	Unit	Material	2020 Bare Costs Labor	Equipment	Total	Total Incl O&P
8630	Flat 4-tier, 20" x 3'	1 Clab	6	1.333	Ea.	1,300	56		1,356	1,525
8635	20" x 4'		6	1.333		1,400	56		1,456	1,650
8640	20" x 5'		6	1.333		1,675	56		1,731	1,950
8645	24" x 3'		6	1.333		885	56		941	1,050
8650	24" x 4'		6	1.333		2,300	56		2,356	2,650
8655	24" x 6'		6	1.333		2,825	56		2,881	3,200
8700	Galvanized shelving, louvered 4-tier, 20" x 3'		6	1.333		810	56		866	980
8705	20" x 4'		6	1.333		900	56		956	1,075
8710	20" x 6'		6	1.333		940	56		996	1,125
8715	24" x 3'		6	1.333		790	56		846	960
8720	24" x 4'		6	1.333		945	56		1,001	1,125
8725	24" x 6'		6	1.333		1,350	56		1,406	1,600
8730	Flat 4-tier, 20" x 3'		6	1.333		750	56		806	915
8735	20" x 4'		6	1.333		685	56		741	840
8740	20" x 6'		6	1.333		985	56		1,041	1,175
8745	24" x 3'		6	1.333		770	56		826	935
8750	24" x 4'		6	1.333		570	56		626	715
8755	24" x 6'		6	1.333		955	56		1,011	1,150
8760	Stainless steel dunnage rack, 24" x 3'		8	1		325	42		367	430
8765	24" x 4'		8	1		400	42		442	510
8770	Galvanized dunnage rack, 24" x 3'		8	1		146	42		188	228
8775	24" x 4'	▼	8	1	▼	191	42		233	279

11 42 Food Preparation Equipment

11 42 10 – Commercial Food Preparation Equipment

11 42 10.10 Choppers, Mixers and Misc. Equipment

		Crew	Daily Output	Labor-Hours	Unit	Material	2020 Bare Costs Labor	Equipment	Total	Total Incl O&P
0010	**CHOPPERS, MIXERS AND MISC. EQUIPMENT**									
1700	Choppers, 5 pounds	R-18	7	3.714	Ea.	3,375	200		3,575	4,000
1720	16 pounds		5	5.200		2,550	280		2,830	3,225
1740	35 to 40 pounds	▼	4	6.500		3,675	350		4,025	4,600
1840	Coffee brewer, 5 burners	1 Plum	3	2.667		960	172		1,132	1,325
1850	Coffee urn, twin 6 gallon urns		2	4		2,525	258		2,783	3,175
1860	Single, 3 gallon	▼	3	2.667		1,825	172		1,997	2,275
3000	Fast food equipment, total package, minimum	6 Skwk	.08	600		220,000	32,900		252,900	294,500
3100	Maximum	"	.07	686		286,500	37,600		324,100	375,000
3800	Food mixers, bench type, 20 quarts	L-7	7	4		3,150	205		3,355	3,775
3850	40 quarts		5.40	5.185		8,325	265		8,590	9,575
3900	60 quarts		5	5.600		11,500	287		11,787	13,200
4040	80 quarts		3.90	7.179		17,300	365		17,665	19,600
4100	Floor type, 20 quarts		15	1.867		3,725	95.50		3,820.50	4,250
4120	60 quarts		14	2		11,000	102		11,102	12,300
4140	80 quarts		12	2.333		19,200	119		19,319	21,400
4160	140 quarts	▼	8.60	3.256		27,100	167		27,267	30,100
6700	Peelers, small	R-18	8	3.250		2,100	175		2,275	2,575
6720	Large	"	6	4.333		3,175	233		3,408	3,850
6800	Pulper/extractor, close coupled, 5 HP	1 Plum	1.90	4.211		2,475	271		2,746	3,150
8580	Slicer with table	R-18	9	2.889	▼	4,325	156		4,481	5,025

11 43 Food Delivery Carts and Conveyors

11 43 13 – Food Delivery Carts

11 43 13.10 Mobile Carts, Racks and Trays

	11 43 13.10 Mobile Carts, Racks and Trays	Crew	Daily Output	Labor-Hours	Unit	Material	2020 Bare Costs Labor	2020 Bare Costs Equipment	Total	Total Incl O&P
0010	**MOBILE CARTS, RACKS AND TRAYS**									
1650	Cabinet, heated, 1 compartment, reach-in	R-18	5.60	4.643	Ea.	4,375	250		4,625	5,175
1655	Pass-thru roll-in		5.60	4.643		3,475	250		3,725	4,200
1660	2 compartment, reach-in		4.80	5.417		5,975	292		6,267	7,025
1670	Mobile					3,925			3,925	4,325
2000	Hospital food cart, hot and cold service, 20 tray capacity					15,500			15,500	17,000
6850	Mobile rack w/pan slide					1,525			1,525	1,675
9180	Tray and silver dispenser, mobile	1 Clab	16	.500		675	21		696	780

11 44 Food Cooking Equipment

11 44 13 – Commercial Ranges

11 44 13.10 Cooking Equipment

	11 44 13.10 Cooking Equipment	Crew	Daily Output	Labor-Hours	Unit	Material	2020 Bare Costs Labor	2020 Bare Costs Equipment	Total	Total Incl O&P
0010	**COOKING EQUIPMENT**									
0020	Bake oven, gas, one section	Q-1	8	2	Ea.	5,675	116		5,791	6,425
0300	Two sections		7	2.286		9,375	133		9,508	10,500
0600	Three sections		6	2.667		12,400	155		12,555	13,800
0900	Electric convection, single deck	L-7	4	7		5,575	360		5,935	6,700
1300	Broiler, without oven, standard	Q-1	8	2		3,650	116		3,766	4,200
1550	Infrared	L-7	4	7		7,500	360		7,860	8,825
4750	Fryer, with twin baskets, modular model	Q-1	7	2.286		1,525	133		1,658	1,875
5000	Floor model, on 6" legs	"	5	3.200		2,550	186		2,736	3,075
5100	Extra single basket, large					51			51	56
5170	Energy star rated, 50 lb. capacity G	R-18	4	6.500		5,025	350		5,375	6,075
5175	85 lb. capacity G	"	4	6.500		7,725	350		8,075	9,025
5300	Griddle, SS, 24" plate, w/4" legs, elec, 208 V, 3 phase, 3' long	Q-1	7	2.286		2,300	133		2,433	2,750
5550	4' long	"	6	2.667		2,450	155		2,605	2,925
6200	Iced tea brewer	1 Plum	3.44	2.326		705	150		855	1,000
6350	Kettle, w/steam jacket, tilting, w/positive lock, SS, 20 gallons	L-7	7	4		8,650	205		8,855	9,850
6600	60 gallons	"	6	4.667		17,100	239		17,339	19,200
6900	Range, restaurant type, 6 burners and 1 standard oven, 36" wide	Q-1	7	2.286		2,675	133		2,808	3,150
6950	Convection		7	2.286		4,800	133		4,933	5,500
7150	2 standard ovens, 24" griddle, 60" wide		6	2.667		5,225	155		5,380	6,000
7200	1 standard, 1 convection oven		6	2.667		9,750	155		9,905	10,900
7450	Heavy duty, single 34" standard oven, open top		5	3.200		5,400	186		5,586	6,225
7500	Convection oven		5	3.200		6,300	186		6,486	7,200
7700	Griddle top		6	2.667		2,400	155		2,555	2,900
7750	Convection oven		6	2.667		3,125	155		3,280	3,675
7760	Induction cooker, electric	L-7	7	4		1,850	205		2,055	2,350
8850	Steamer, electric 27 KW		7	4		13,300	205		13,505	14,900
9100	Electric, 10 KW or gas 100,000 BTU		5	5.600		7,625	287		7,912	8,825
9150	Toaster, conveyor type, 16-22 slices/minute					1,200			1,200	1,325
9160	Pop-up, 2 slot					99			99	109
9200	For deluxe models of above equipment, add					75%				
9400	Rule of thumb: Equipment cost based									
9410	on kitchen work area									
9420	Office buildings, minimum	L-7	77	.364	S.F.	97	18.60		115.60	137
9450	Maximum		58	.483		158	24.50		182.50	213
9550	Public eating facilities, minimum		77	.364		127	18.60		145.60	170
9600	Maximum		46	.609		207	31		238	278
9750	Hospitals, minimum		58	.483		126	24.50		150.50	178
9800	Maximum		39	.718		231	36.50		267.50	315

11 46 Food Dispensing Equipment

11 46 16 – Service Line Equipment

11 46 16.10 Commercial Food Dispensing Equipment	Crew	Daily Output	Labor-Hours	Unit	Material	2020 Bare Costs Labor	Equipment	Total	Total Incl O&P
0010 **COMMERCIAL FOOD DISPENSING EQUIPMENT**									
1050 Butter pat dispenser	1 Clab	13	.615	Ea.	865	26		891	990
1100 Bread dispenser, counter top		13	.615		825	26		851	950
1900 Cup and glass dispenser, drop in		4	2		560	84		644	750
1920 Disposable cup, drop in		16	.500		750	21		771	855
2650 Dish dispenser, drop in, 12"		11	.727		2,850	30.50		2,880.50	3,200
2660 Mobile		10	.800		3,275	33.50		3,308.50	3,650
3300 Food warmer, counter, 1.2 KW					765			765	840
3550 1.6 KW					2,200			2,200	2,425
3600 Well, hot food, built-in, rectangular, 12" x 20"	R-30	10	2.600		330	129		459	570
3610 Circular, 7 qt.		10	2.600		400	129		529	645
3620 Refrigerated, 2 compartments		10	2.600		2,825	129		2,954	3,300
3630 3 compartments		9	2.889		3,975	143		4,118	4,600
3640 4 compartments		8	3.250		4,750	161		4,911	5,475
4720 Frost cold plate		9	2.889		22,700	143		22,843	25,100
5700 Hot chocolate dispenser	1 Plum	4	2		1,175	129		1,304	1,500
5750 Ice dispenser 567 pound	Q-1	6	2.667		6,075	155		6,230	6,950
6250 Jet spray dispenser	R-18	4.50	5.778		2,125	310		2,435	2,825
6300 Juice dispenser, concentrate	"	4.50	5.778		1,975	310		2,285	2,650
6690 Milk dispenser, bulk, 2 flavor	R-30	8	3.250		2,100	161		2,261	2,575
6695 3 flavor	"	8	3.250		2,650	161		2,811	3,175
8800 Serving counter, straight	1 Carp	40	.200	L.F.	2,000	10.65		2,010.65	2,225
8820 Curved section	"	30	.267	"	2,225	14.15		2,239.15	2,475
8825 Solid surface, see Section 12 36 61.16									
8860 Sneeze guard with lights, 60" L	1 Clab	16	.500	Ea.	325	21		346	390
8900 Sneeze guard, stainless steel and glass, single sided									
8910 Portable, 48" W				Ea.	395			395	435
8920 Portable, 72" W					445			445	490
8930 Adjustable, 36" W	1 Carp	24	.333		360	17.70		377.70	425
8940 Adjustable, 48" W	"	20	.400		320	21.50		341.50	385
9100 Soft serve ice cream machine, medium	R-18	11	2.364		8,125	127		8,252	9,125
9110 Large	"	9	2.889		22,200	156		22,356	24,600

11 46 83 – Ice Machines

11 46 83.10 Commercial Ice Equipment

	Crew	Daily Output	Labor-Hours	Unit	Material	2020 Bare Costs Labor	Equipment	Total	Total Incl O&P
0010 **COMMERCIAL ICE EQUIPMENT**									
5800 Ice cube maker, 50 lbs./day	Q-1	6	2.667	Ea.	1,800	155		1,955	2,225
5810 65 lbs./day, energy star rated		6	2.667		1,675	155		1,830	2,075
5900 250 lbs./day		1.20	13.333		2,525	775		3,300	3,975
5950 300 lbs./day, remote condensing		1.20	13.333		2,500	775		3,275	3,950
6050 500 lbs./day		4	4		2,600	232		2,832	3,200
6060 With bin		1.20	13.333		3,950	775		4,725	5,550
6070 Modular, with bin and condenser		1.20	13.333		4,125	775		4,900	5,725
6090 1000 lbs./day, with bin		1	16		5,300	930		6,230	7,250
6100 Ice flakers, 300 lbs./day		1.60	10		3,750	580		4,330	5,025
6120 600 lbs./day		.95	16.842		4,300	975		5,275	6,250
6130 1000 lbs./day		.75	21.333		4,925	1,225		6,150	7,325
6140 2000 lbs./day		.65	24.615		22,200	1,425		23,625	26,600
6160 Ice storage bin, 500 pound capacity	Q-5	1	16		1,025	945		1,970	2,575
6180 1000 pound	"	.56	28.571		3,050	1,675		4,725	5,975

For customer support on your Commercial Renovation Costs with RSMeans data, call 800.448.8182.

377

11 48 Foodservice Cleaning and Disposal Equipment

11 48 13 – Commercial Dishwashers

11 48 13.10 Dishwashers

11 48 13.10 Dishwashers	Crew	Daily Output	Labor-Hours	Unit	Material	2020 Bare Costs Labor	2020 Bare Costs Equipment	Total	Total Incl O&P
0010 **DISHWASHERS**									
2700 Dishwasher, commercial, rack type									
2720 10 to 12 racks/hour	Q-1	3.20	5	Ea.	3,450	290		3,740	4,250
2730 Energy star rated, 35 to 40 racks/hour G		1.30	12.308		5,500	715		6,215	7,150
2740 50 to 60 racks/hour G		1.30	12.308		10,600	715		11,315	12,700
2800 Automatic, 190 to 230 racks/hour	L-6	.35	34.286		14,600	2,175		16,775	19,400
2820 235 to 275 racks/hour		.25	48		28,000	3,050		31,050	35,500
2840 8,750 to 12,500 dishes/hour		.10	120		47,500	7,600		55,100	64,500
2950 Dishwasher hood, canopy type	L-3A	10	1.200	L.F.	1,250	69		1,319	1,475
2960 Pant leg type	"	2.50	4.800	Ea.	7,825	276		8,101	9,075
5200 Garbage disposal 1.5 HP, 100 GPH	L-1	4.80	3.333		1,875	210		2,085	2,375
5210 3 HP, 120 GPH		4.60	3.478		2,400	219		2,619	3,000
5220 5 HP, 250 GPH		4.50	3.556		2,975	224		3,199	3,625
6750 Pot sink, 3 compartment	1 Plum	7.25	1.103	L.F.	910	71		981	1,100
6760 Pot washer, low temp wash/rinse		1.60	5	Ea.	5,375	320		5,695	6,425
6770 High pressure wash, high temperature rinse		1.20	6.667		37,200	430		37,630	41,600
9170 Trash compactor, small, up to 125 lb. compacted weight	L-4	4	6		21,500	299		21,799	24,200
9175 Large, up to 175 lb. compacted weight	"	3	8		28,500	400		28,900	31,900

11 52 Audio-Visual Equipment

11 52 13 – Projection Screens

11 52 13.10 Projection Screens, Wall or Ceiling Hung

11 52 13.10	Crew	Daily Output	Labor-Hours	Unit	Material	2020 Bare Costs Labor	2020 Bare Costs Equipment	Total	Total Incl O&P
0010 **PROJECTION SCREENS, WALL OR CEILING HUNG**, matte white									
0100 Manually operated, economy	2 Carp	500	.032	S.F.	6.55	1.70		8.25	9.90
0300 Intermediate		450	.036		7.90	1.89		9.79	11.75
0400 Deluxe		400	.040		10	2.13		12.13	14.40
9000 Minimum labor/equipment charge		3	5.333	Job		283		283	455

11 52 16 – Projectors

11 52 16.10 Movie Equipment

11 52 16.10	Crew	Daily Output	Labor-Hours	Unit	Material	2020 Bare Costs Labor	2020 Bare Costs Equipment	Total	Total Incl O&P
0011 **MOVIE EQUIPMENT**									
0020 Changeover, minimum				Ea.	520			520	570
3000 Projection screens, rigid, in wall, acrylic, 1/4" thick	2 Glaz	195	.082	S.F.	48.50	4.18		52.68	59.50
3100 1/2" thick	"	130	.123	"	55.50	6.30		61.80	71
3700 Sound systems, incl. amplifier, mono, minimum	1 Elec	.90	8.889	Ea.	3,825	545		4,370	5,050
3800 Dolby/Super Sound, maximum		.40	20		20,600	1,225		21,825	24,600
4100 Dual system, 2 channel, front surround, minimum		.70	11.429		5,275	700		5,975	6,875
4200 Dolby/Super Sound, 4 channel, maximum		.40	20		19,700	1,225		20,925	23,600
5700 Seating, painted steel, upholstered, minimum	2 Carp	35	.457		151	24.50		175.50	205
5800 Maximum	"	28	.571		550	30.50		580.50	655

11 53 Laboratory Equipment

11 53 33 – Emergency Safety Appliances

11 53 33.13 Emergency Equipment	Crew	Daily Output	Labor-Hours	Unit	Material	2020 Bare Costs Labor	Equipment	Total	Total Incl O&P
0010 **EMERGENCY EQUIPMENT**									
1400 Safety equipment, eye wash, hand held				Ea.	400			400	440
1450 Deluge shower				"	840			840	920

11 53 43 – Service Fittings and Accessories

11 53 43.13 Fittings

11 53 43.13 Fittings	Crew	Daily Output	Labor-Hours	Unit	Material	2020 Bare Costs Labor	Equipment	Total	Total Incl O&P
0010 **FITTINGS**									
1600 Sink, one piece plastic, flask wash, hose, free standing	1 Plum	1.60	5	Ea.	2,075	320		2,395	2,775
1610 Epoxy resin sink, 25" x 16" x 10"	"	2	4	"	227	258		485	650
1630 Steel table, open underneath				L.F.	575			575	635
1640 Doors underneath				"	940			940	1,025
1900 Utility cabinet, stainless steel				Ea.	1,700			1,700	1,875
1950 Utility table, acid resistant top with drawers	2 Carp	30	.533	L.F.	179	28.50		207.50	243
8000 Alternate pricing method: as percent of lab furniture									
8050 Installation, not incl. plumbing & duct work				% Furn.				22%	22%
8100 Plumbing, final connections, simple system								10%	10%
8110 Moderately complex system								15%	15%
8120 Complex system								20%	20%
8150 Electrical, simple system								10%	10%
8160 Moderately complex system								20%	20%
8170 Complex system								35%	35%

11 53 53 – Biological Safety Cabinets

11 53 53.10 Pharmacy Cabinets

11 53 53.10 Pharmacy Cabinets	Crew	Daily Output	Labor-Hours	Unit	Material	2020 Bare Costs Labor	Equipment	Total	Total Incl O&P
0010 **PHARMACY CABINETS**, vertical flow									
0100 Class II, type B2, 6' L	2 Carp	1.50	10.667	Ea.	15,400	565		15,965	17,800

11 57 Vocational Shop Equipment

11 57 10 – Shop Equipment

11 57 10.10 Vocational School Shop Equipment

11 57 10.10 Vocational School Shop Equipment	Crew	Daily Output	Labor-Hours	Unit	Material	2020 Bare Costs Labor	Equipment	Total	Total Incl O&P
0010 **VOCATIONAL SCHOOL SHOP EQUIPMENT**									
0020 Benches, work, wood, average	2 Carp	5	3.200	Ea.	755	170		925	1,100
0100 Metal, average		5	3.200		440	170		610	755
0400 Combination belt & disc sander, 6"		4	4		1,700	213		1,913	2,225
0700 Drill press, floor mounted, 12", 1/2 HP		4	4		470	213		683	855
0800 Dust collector, not incl. ductwork, 6" diameter	1 Shee	1.10	7.273		5,925	455		6,380	7,225
0810 Dust collector bag, 20" diameter	"	5	1.600		600	99.50		699.50	820
1000 Grinders, double wheel, 1/2 HP	2 Carp	5	3.200		239	170		409	535
1300 Jointer, 4", 3/4 HP		4	4		1,500	213		1,713	2,000
1600 Kilns, 16 C.F., to 2000°		4	4		1,875	213		2,088	2,425
1900 Lathe, woodworking, 10", 1/2 HP		4	4		580	213		793	975
2200 Planer, 13" x 6"		4	4		1,325	213		1,538	1,825
2500 Potter's wheel, motorized		4	4		1,150	213		1,363	1,625
2800 Saws, band, 14", 3/4 HP		4	4		1,150	213		1,363	1,600
3100 Metal cutting band saw, 14"		4	4		2,275	213		2,488	2,850
3400 Radial arm saw, 10", 2 HP		4	4		1,350	213		1,563	1,850
3700 Scroll saw, 24"		4	4		625	213		838	1,025
4000 Table saw, 10", 3 HP		4	4		3,150	213		3,363	3,825
4300 Welder AC arc, 30 amp capacity		4	4		3,500	213		3,713	4,175

11 61 Broadcast, Theater, and Stage Equipment

11 61 23 – Folding and Portable Stages

11 61 23.10 Portable Stages	Crew	Daily Output	Labor-Hours	Unit	Material	2020 Bare Costs Labor	2020 Bare Costs Equipment	Total	Total Incl O&P
0010 **PORTABLE STAGES**									
5000 Stages, portable with steps, folding legs, stock, 8" high				SF Stg.	46.50			46.50	51
5100 16" high					61			61	67
5200 32" high					62			62	68.50
5300 40" high					64.50			64.50	71

11 61 33 – Rigging Systems and Controls

11 61 33.10 Controls

	Crew	Daily Output	Labor-Hours	Unit	Material	Labor	Equipment	Total	Total Incl O&P
0010 **CONTROLS**									
0050 Control boards with dimmers and breakers, minimum	1 Elec	1	8	Ea.	19,700	490		20,190	22,500
0150 Maximum	"	.20	40	"	148,500	2,450		150,950	167,000

11 61 43 – Stage Curtains

11 61 43.10 Curtains

	Crew	Daily Output	Labor-Hours	Unit	Material	Labor	Equipment	Total	Total Incl O&P
0010 **CURTAINS**									
0500 Curtain track, straight, light duty	2 Carp	20	.800	L.F.	31	42.50		73.50	103
0700 Curved sections		12	1.333	"	201	71		272	335
1000 Curtains, velour, medium weight		600	.027	S.F.	8.75	1.42		10.17	11.90

11 66 Athletic Equipment

11 66 13 – Exercise Equipment

11 66 13.10 Physical Training Equipment

	Crew	Daily Output	Labor-Hours	Unit	Material	Labor	Equipment	Total	Total Incl O&P
0010 **PHYSICAL TRAINING EQUIPMENT**									
0020 Abdominal rack, 2 board capacity				Ea.	525			525	575
0050 Abdominal board, upholstered					775			775	850
0200 Bicycle trainer, minimum					610			610	670
0300 Deluxe, electric					4,850			4,850	5,350
0400 Barbell set, chrome plated steel, 25 lb.					305			305	335
0420 100 lb.					620			620	680
0450 200 lb.					740			740	815
0500 Weight plates, cast iron, per lb.				Lb.	3.70			3.70	4.07
0520 Storage rack, 10 station				Ea.	1,225			1,225	1,325
0600 Circuit training apparatus, 12 machines minimum	2 Clab	1.25	12.800	Set	36,800	540		37,340	41,400
0700 Average		1	16		41,700	675		42,375	47,000
0800 Maximum		.75	21.333		50,000	900		50,900	56,500
0820 Dumbbell set, cast iron, with rack and 5 pair					475			475	520
0900 Squat racks	2 Clab	5	3.200	Ea.	825	135		960	1,125
4150 Exercise equipment, bicycle trainer					1,125			1,125	1,250
4180 Chinning bar, adjustable, wall mounted	1 Carp	5	1.600		237	85		322	395
4200 Exercise ladder, 16' x 1'-7", suspended	L-2	3	5.333		1,375	248		1,623	1,925
4210 High bar, floor plate attached	1 Carp	4	2		2,725	106		2,831	3,150
4240 Parallel bars, adjustable		4	2		1,625	106		1,731	1,975
4270 Uneven parallel bars, adjustable		4	2		3,950	106		4,056	4,525
4280 Wall mounted, adjustable	L-2	1.50	10.667	Set	955	495		1,450	1,850
4300 Rope, ceiling mounted, 18' long	1 Carp	3.66	2.186	Ea.	208	116		324	415
4330 Side horse, vaulting		5	1.600		1,600	85		1,685	1,900
4360 Treadmill, motorized, deluxe, training type		5	1.600		3,975	85		4,060	4,475
4390 Weight lifting multi-station, minimum	2 Clab	1	16		305	675		980	1,400

11 66 Athletic Equipment

11 66 23 – Gymnasium Equipment

11 66 23.13 Basketball Equipment

		Crew	Daily Output	Labor-Hours	Unit	Material	2020 Bare Costs Labor	Equipment	Total	Total Incl O&P
0010	**BASKETBALL EQUIPMENT**									
1000	Backstops, wall mtd., 6' extended, fixed, minimum	L-2	1	16	Ea.	1,750	745		2,495	3,125
1100	Maximum		1	16		2,075	745		2,820	3,475
1200	Swing up, minimum		1	16		1,750	745		2,495	3,125
1250	Maximum		1	16		3,300	745		4,045	4,825
1300	Portable, manual, heavy duty, spring operated		1.90	8.421		13,200	390		13,590	15,200
1400	Ceiling suspended, stationary, minimum		.78	20.513		4,350	955		5,305	6,350
1450	Fold up, with accessories, maximum	↓	.40	40		6,325	1,850		8,175	9,950
1600	For electrically operated, add	1 Elec	1	8	↓	2,600	490		3,090	3,600
5800	Wall pads, 1-1/2" thick, standard (not fire rated)	2 Carp	640	.025	S.F.	6.85	1.33		8.18	9.70

11 66 23.19 Boxing Ring

		Crew	Daily Output	Labor-Hours	Unit	Material	2020 Bare Costs Labor	Equipment	Total	Total Incl O&P
0010	**BOXING RING**									
4100	Elevated, 22' x 22'	L-4	.10	240	Ea.	6,150	12,000		18,150	26,000
4110	For cellular plastic foam padding, add		.10	240		1,250	12,000		13,250	20,600
4120	Floor level, including posts and ropes only, 20' x 20'		.80	30		4,825	1,500		6,325	7,700
4130	Canvas, 30' x 30'	↓	5	4.800	↓	1,500	239		1,739	2,025

11 66 23.47 Gym Mats

		Crew	Daily Output	Labor-Hours	Unit	Material	2020 Bare Costs Labor	Equipment	Total	Total Incl O&P
0010	**GYM MATS**									
5500	2" thick, naugahyde covered				S.F.	5.35			5.35	5.90
5600	Vinyl/nylon covered					8.55			8.55	9.45
6000	Wrestling mats, 1" thick, heavy duty				↓	5.05			5.05	5.55

11 66 43 – Interior Scoreboards

11 66 43.10 Scoreboards

		Crew	Daily Output	Labor-Hours	Unit	Material	2020 Bare Costs Labor	Equipment	Total	Total Incl O&P
0010	**SCOREBOARDS**									
7000	Baseball, minimum	R-3	1.30	15.385	Ea.	4,125	940	145	5,210	6,150
7200	Maximum		.05	400		19,800	24,400	3,775	47,975	64,000
7300	Football, minimum		.86	23.256		5,225	1,425	220	6,870	8,200
7400	Maximum		.20	100		20,200	6,100	945	27,245	32,700
7500	Basketball (one side), minimum		2.07	9.662		2,475	590	91.50	3,156.50	3,725
7600	Maximum		.30	66.667		8,900	4,075	630	13,605	16,800
7700	Hockey-basketball (four sides), minimum		.25	80		10,100	4,900	755	15,755	19,500
7800	Maximum	↓	.15	133	↓	16,300	8,150	1,250	25,700	31,900

11 66 53 – Gymnasium Dividers

11 66 53.10 Divider Curtains

		Crew	Daily Output	Labor-Hours	Unit	Material	2020 Bare Costs Labor	Equipment	Total	Total Incl O&P
0010	**DIVIDER CURTAINS**									
4500	Gym divider curtain, mesh top, vinyl bottom, manual	L-4	500	.048	S.F.	9.90	2.39		12.29	14.75
4700	Electric roll up	L-7	400	.070	"	13.30	3.58		16.88	20.50

11 71 Medical Sterilizing Equipment

11 71 10 – Medical Sterilizers & Distillers

11 71 10.10 Sterilizers and Distillers

		Crew	Daily Output	Labor-Hours	Unit	Material	2020 Bare Costs Labor	Equipment	Total	Total Incl O&P
0010	**STERILIZERS AND DISTILLERS**									
3010	Portable, top loading, 105-135 degree C, 3 to 30 psi, 50 L chamber				Ea.	9,775			9,775	10,800
3020	Stainless steel basket, 10.7" diam. x 11.8" H					214			214	236
3025	Stainless steel pail, 10.7" diam. x 10.7" H					320			320	350
3050	85 L chamber					17,800			17,800	19,600
3060	Stainless steel basket, 15.3" diam. x 11.5" H					455			455	505
3065	Stainless steel pail, 15.3" diam. x 11" H				↓	675			675	745

11 76 Operating Room Equipment

11 76 10 – Equipment for Operating Rooms

11 76 10.10 Surgical Equipment	Crew	Daily Output	Labor-Hours	Unit	Material	2020 Bare Costs Labor	Equipment	Total	Total Incl O&P
0010 **SURGICAL EQUIPMENT**									
6550 Major surgery table, minimum	1 Sswk	.50	16	Ea.	25,900	920		26,820	30,000
6570 Maximum		.50	16		39,300	920		40,220	44,800
6600 Hydraulic, hand-held control, general surgery		.60	13.333		41,100	770		41,870	46,500
6650 Stationary, universal		.50	16		53,500	920		54,420	60,000
6800 Surgical lights, major operating room, dual head, minimum	2 Elec	1	16		5,975	980		6,955	8,075
6850 Maximum		1	16		40,100	980		41,080	45,600
6900 Ceiling mount articulation, single arm		1	16		4,875	980		5,855	6,850

11 81 Facility Maintenance Equipment

11 81 19 – Vacuum Cleaning Systems

11 81 19.10 Vacuum Cleaning

	Crew	Daily Output	Labor-Hours	Unit	Material	2020 Bare Costs Labor	Equipment	Total	Total Incl O&P
0010 **VACUUM CLEANING**									
0020 Central, 3 inlet, residential	1 Skwk	.90	8.889	Total	1,325	490		1,815	2,250
0200 Commercial		.70	11.429		1,350	625		1,975	2,500
0400 5 inlet system, residential		.50	16		2,025	880		2,905	3,625
0600 7 inlet system, commercial		.40	20		2,600	1,100		3,700	4,600

11 82 Facility Solid Waste Handling Equipment

11 82 26 – Facility Waste Compactors

11 82 26.10 Compactors

	Crew	Daily Output	Labor-Hours	Unit	Material	2020 Bare Costs Labor	Equipment	Total	Total Incl O&P
0010 **COMPACTORS**									
0020 Compactors, 115 volt, 250 lbs./hr., chute fed	L-4	1	24	Ea.	12,600	1,200		13,800	15,700
0100 Hand fed		2.40	10		17,200	500		17,700	19,700
1000 Heavy duty industrial compactor, 0.5 C.Y. capacity		1	24		11,200	1,200		12,400	14,200
1050 1.0 C.Y. capacity		1	24		17,100	1,200		18,300	20,700
1400 For handling hazardous waste materials, 55 gallon drum packer, std.					21,700			21,700	23,900
1410 55 gallon drum packer w/HEPA filter					26,400			26,400	29,100
1420 55 gallon drum packer w/charcoal & HEPA filter					35,200			35,200	38,800
1430 All of the above made explosion proof, add					1,725			1,725	1,900

11 82 39 – Medical Waste Disposal Systems

11 82 39.10 Off-Site Disposal

	Crew	Daily Output	Labor-Hours	Unit	Material	2020 Bare Costs Labor	Equipment	Total	Total Incl O&P
0010 **OFF-SITE DISPOSAL**									
0100 Medical waste disposal, Red Bag system, pick up & treat, 200 lbs./week				Week	208			208	228
0110 Per month				Month	745			745	820
0150 Red bags, 7-10 gal., 1.2 mil, pkg of 500				Ea.	66			66	72.50
0200 15 gal., package of 250					64			64	70
0250 33 gal., package of 250					73			73	80
0300 45 gal., package of 100					59.50			59.50	65.50

11 82 39.20 Disposal Carts

	Crew	Daily Output	Labor-Hours	Unit	Material	2020 Bare Costs Labor	Equipment	Total	Total Incl O&P
0010 **DISPOSAL CARTS**									
2010 Medical waste disposal cart, HDPE, w/lid, 28 gal. capacity				Ea.	289			289	320
2020 96 gal. capacity					335			335	370
2030 150 gal. capacity, low profile					500			500	555
2040 200 gal. capacity					1,000			1,000	1,100

11 82 Facility Solid Waste Handling Equipment

11 82 39 – Medical Waste Disposal Systems

11 82 39.30 Medical Waste Sanitizers

		Crew	Daily Output	Labor-Hours	Unit	Material	2020 Bare Costs Labor	2020 Bare Costs Equipment	Total	Total Incl O&P
0010	**MEDICAL WASTE SANITIZERS**									
2010	Small, hand loaded, 1.5 C.Y., 225 lb. capacity				Ea.	84,000			84,000	92,500
2020	Medium, cart loaded, 6.25 C.Y., 938 lb. capacity					112,000			112,000	123,500
2030	Large, cart loaded, 15 C.Y., 2250 lb. capacity					134,500			134,500	148,000
3010	Cart, aluminum, 75 lb. capacity					2,225			2,225	2,450
3020	95 lb. capacity					2,425			2,425	2,675
4010	Stainless steel, 173 lb. capacity					3,175			3,175	3,500
4020	232 lb. capacity					3,550			3,550	3,900
4030	Cart lift, hydraulic scissor type					6,300			6,300	6,925
4040	Portable aluminum ramp					1,775			1,775	1,950
4050	Fold-down steel tracks					1,450			1,450	1,600
4060	Pull-out drawer, small					6,775			6,775	7,475
4070	Medium					9,800			9,800	10,800
4080	Large					13,800			13,800	15,200
5000	Medical waste treatment, sanitize, on-site									
5010	Less than 15,000 lbs./month				Lb.	.20			.20	.22
5020	Over 15,000 lbs./month				"	.16			.16	.18

11 91 Religious Equipment

11 91 13 – Baptisteries

11 91 13.10 Baptistry

		Crew	Daily Output	Labor-Hours	Unit	Material	Labor	Equipment	Total	Total Incl O&P
0010	**BAPTISTRY**									
0150	Fiberglass, 3'-6" deep, x 13'-7" long,									
0160	steps at both ends, incl. plumbing, minimum	L-8	1	20	Ea.	6,300	1,100		7,400	8,700
0200	Maximum	"	.70	28.571		10,500	1,575		12,075	14,000
0250	Add for filter, heater and lights					1,850			1,850	2,025

11 91 23 – Sanctuary Equipment

11 91 23.10 Sanctuary Furnishings

		Crew	Daily Output	Labor-Hours	Unit	Material	Labor	Equipment	Total	Total Incl O&P
0010	**SANCTUARY FURNISHINGS**									
0020	Altar, wood, custom design, plain	1 Carp	1.40	5.714	Ea.	2,725	305		3,030	3,475
0050	Deluxe		.20	40	"	13,100	2,125		15,225	17,800
5000	Wall cross, aluminum, extruded, 2" x 2" section		34	.235	L.F.	246	12.50		258.50	291
5150	4" x 4" section		29	.276		355	14.65		369.65	415
5300	Bronze, extruded, 1" x 2" section		31	.258		485	13.70		498.70	555
5350	2-1/2" x 2-1/2" section		34	.235		735	12.50		747.50	830

11 97 Security Equipment

11 97 30 – Security Drawers

11 97 30.10 Pass Through Drawer

		Crew	Daily Output	Labor-Hours	Unit	Material	Labor	Equipment	Total	Total Incl O&P
0010	**PASS THROUGH DRAWER**									
0100	Pass-thru drawer for personal items, 18" x 15" x 24"	1 Skwk	2	4	Ea.	2,725	219		2,944	3,350
0110	Including speakers	"	1.50	5.333	"	3,050	293		3,343	3,850

11 98 Detention Equipment

11 98 21 – Detention Windows

11 98 21.13 Visitor Cubicle Windows	Crew	Daily Output	Labor-Hours	Unit	Material	2020 Bare Costs Labor	Equipment	Total	Total Incl O&P
0010 **VISITOR CUBICLE WINDOWS**									
4000 Visitor cubicle, vision panel, no intercom	E-4	2	16	Ea.	3,550	930	73.50	4,553.50	5,500

11 98 30 – Detention Cell Equipment

11 98 30.10 Cell Equipment

	Crew	Daily Output	Labor-Hours	Unit	Material	2020 Bare Costs Labor	Equipment	Total	Total Incl O&P
0010 **CELL EQUIPMENT**									
3000 Toilet apparatus including wash basin, average	L-8	1.50	13.333	Ea.	3,550	740		4,290	5,075

Estimating Tips
General

- The items in this division are usually priced per square foot or each. Most of these items are purchased by the owner and installed by the contractor. Do not assume the items in Division 12 will be purchased and installed by the contractor. Check the specifications for responsibilities and include receiving, storage, installation, and mechanical and electrical hookups in the appropriate divisions.

- Some items in this division require some type of support system that is not usually furnished with the item. Examples of these systems include blocking for the attachment of casework and heavy drapery rods. The required blocking must be added to the estimate in the appropriate division.

Reference Numbers

Reference numbers are shown at the beginning of some major classifications. These numbers refer to related items in the Reference Section. The reference information may be an estimating procedure, an alternate pricing method, or technical information.

Same Data. Simplified.

Enjoy the convenience and efficiency of accessing your costs anywhere:

- **Skip the multiplier** by setting your location
- **Quickly search,** edit, favorite and share costs
- **Stay on top of price changes** with automatic updates

Discover more at rsmeans.com/online

12 21 Window Blinds

12 21 13 – Horizontal Louver Blinds

12 21 13.13 Metal Horizontal Louver Blinds

		Crew	Daily Output	Labor-Hours	Unit	Material	2020 Bare Costs Labor	Equipment	Total	Total Incl O&P
0010	**METAL HORIZONTAL LOUVER BLINDS**									
0020	Horizontal, 1" aluminum slats, solid color, stock	1 Carp	590	.014	S.F.	5.80	.72		6.52	7.55

12 21 13.33 Vinyl Horizontal Louver Blinds

		Crew	Daily Output	Labor-Hours	Unit	Material	Labor	Equipment	Total	Total Incl O&P
0010	**VINYL HORIZONTAL LOUVER BLINDS**									
0100	2" composite, 48" wide, 48" high	1 Carp	30	.267	Ea.	93	14.15		107.15	125
0120	72" high		29	.276		125	14.65		139.65	162
0140	96" high		28	.286		171	15.20		186.20	213
0200	60" wide, 60" high		27	.296		135	15.75		150.75	174
0220	72" high		25	.320		156	17		173	199
0240	96" high		24	.333		233	17.70		250.70	285
0300	72" wide, 72" high		25	.320		206	17		223	254
0320	96" high		23	.348		284	18.50		302.50	340
0400	96" wide, 96" high		20	.400		505	21.50		526.50	590
1000	2" faux wood, 48" wide, 48" high		30	.267		101	14.15		115.15	134
1020	72" high		29	.276		131	14.65		145.65	168
1040	96" high		28	.286		156	15.20		171.20	196
1300	72" wide, 72" high		25	.320		129	17		146	170
1320	96" high		23	.348		194	18.50		212.50	244
1400	96" wide, 96" high		20	.400		265	21.50		286.50	325

12 23 Interior Shutters

12 23 13 – Wood Interior Shutters

12 23 13.10 Wood Interior Shutters

		Crew	Daily Output	Labor-Hours	Unit	Material	Labor	Equipment	Total	Total Incl O&P
0010	**WOOD INTERIOR SHUTTERS**, louvered									
0200	Two panel, 27" wide, 36" high	1 Carp	5	1.600	Set	173	85		258	325
0300	33" wide, 36" high		5	1.600		223	85		308	380
0500	47" wide, 36" high		5	1.600		299	85		384	465
1000	Four panel, 27" wide, 36" high		5	1.600		161	85		246	315
1100	33" wide, 36" high		5	1.600		206	85		291	360
1300	47" wide, 36" high		5	1.600		275	85		360	440
1400	Plantation shutters, 16" x 48"		5	1.600	Ea.	146	85		231	297
1450	16" x 96"		4	2		240	106		346	435
1460	36" x 96"		3	2.667		530	142		672	805

12 23 13.13 Wood Panels

		Crew	Daily Output	Labor-Hours	Unit	Material	Labor	Equipment	Total	Total Incl O&P
0010	**WOOD PANELS**									
3000	Wood folding panels with movable louvers, 7" x 20" each	1 Carp	17	.471	Pr.	95.50	25		120.50	145
4000	Fixed louver type, stock units, 8" x 20" each		17	.471		107	25		132	158
4450	18" x 40" each		17	.471		153	25		178	208

12 24 Window Shades

12 24 13 – Roller Window Shades

12 24 13.10 Shades

		Crew	Daily Output	Labor-Hours	Unit	Material	2020 Bare Costs Labor	Equipment	Total	Total Incl O&P
0010	**SHADES**									
0020	Basswood, roll-up, stain finish, 3/8" slats	1 Carp	300	.027	S.F.	23	1.42		24.42	27.50
0200	7/8" slats		300	.027		21	1.42		22.42	25.50
0900	Mylar, single layer, non-heat reflective		685	.012		3.14	.62		3.76	4.44
0910	Mylar, single layer, heat reflective		685	.012		2.83	.62		3.45	4.10
1000	Double layered, heat reflective		685	.012		6.65	.62		7.27	8.30
1100	Triple layered, heat reflective		685	.012		8.35	.62		8.97	10.15
2000	Polyester, room darkening, with continuous cord, GEI									
2010	36" x 72"	1 Carp	38	.211	Ea.	122	11.20		133.20	152
2020	48" x 72"		28	.286		166	15.20		181.20	208
2030	60" x 72"		23	.348		188	18.50		206.50	237
2040	72" x 72"		19	.421		213	22.50		235.50	270
5011	Insulative shades		125	.064	S.F.	17.15	3.40		20.55	24.50
5030	R10.7		44	.182	"	20.50	9.65		30.15	38
5050	Magnetic clips, set of 20				Set	38			38	41.50
6011	Solar screening, fiberglass ⬛G	1 Carp	85	.094	S.F.	8.25	5		13.25	17.10

12 32 Manufactured Wood Casework

12 32 23 – Hardwood Casework

12 32 23.10 Manufactured Wood Casework, Stock Units

		Crew	Daily Output	Labor-Hours	Unit	Material	2020 Bare Costs Labor	Equipment	Total	Total Incl O&P
0010	**MANUFACTURED WOOD CASEWORK, STOCK UNITS**									
0300	Built-in drawer units, pine, 18" deep, 32" high, unfinished									
0400	Minimum	2 Carp	53	.302	L.F.	140	16.05		156.05	180
0500	Maximum	"	40	.400	"	166	21.50		187.50	217
0700	Kitchen base cabinets, hardwood, not incl. counter tops,									
0710	24" deep, 35" high, prefinished									
0800	One top drawer, one door below, 12" wide	2 Carp	24.80	.645	Ea.	325	34.50		359.50	410
0820	15" wide		24	.667		385	35.50		420.50	480
0840	18" wide		23.30	.687		299	36.50		335.50	390
0860	21" wide		22.70	.705		315	37.50		352.50	405
0880	24" wide		22.30	.717		435	38		473	540
1000	Four drawers, 12" wide		24.80	.645		330	34.50		364.50	420
1020	15" wide		24	.667		335	35.50		370.50	425
1040	18" wide		23.30	.687		370	36.50		406.50	465
1060	24" wide		22.30	.717		415	38		453	520
1200	Two top drawers, two doors below, 27" wide		22	.727		455	38.50		493.50	560
1220	30" wide		21.40	.748		515	39.50		554.50	635
1240	33" wide		20.90	.766		545	40.50		585.50	665
1260	36" wide		20.30	.788		570	42		612	695
1280	42" wide		19.80	.808		595	43		638	725
1300	48" wide		18.90	.847		695	45		740	835
1500	Range or sink base, two doors below, 30" wide		21.40	.748		460	39.50		499.50	570
1520	33" wide		20.90	.766		485	40.50		525.50	600
1540	36" wide		20.30	.788		495	42		537	610
1560	42" wide		19.80	.808		540	43		583	665
1580	48" wide		18.90	.847		545	45		590	670
1800	For sink front units, deduct					188			188	207
2000	Corner base cabinets, 36" wide, standard	2 Carp	18	.889		790	47		837	945
2100	Lazy Susan with revolving door	"	16.50	.970		1,025	51.50		1,076.50	1,200
4000	Kitchen wall cabinets, hardwood, 12" deep with two doors									
4050	12" high, 30" wide	2 Carp	24.80	.645	Ea.	279	34.50		313.50	360

387

12 32 23.10 Manufactured Wood Casework, Stock Units

		Crew	Daily Output	Labor-Hours	Unit	Material	2020 Bare Costs Labor	Equipment	Total	Total Incl O&P
4400	15" high, 30" wide	2 Carp	24	.667	Ea.	285	35.50		320.50	370
4420	33" wide		23.30	.687		350	36.50		386.50	445
4440	36" wide		22.70	.705		355	37.50		392.50	450
4450	42" wide		22.70	.705		410	37.50		447.50	510
4700	24" high, 30" wide		23.30	.687		415	36.50		451.50	520
4720	36" wide		22.70	.705		485	37.50		522.50	595
4740	42" wide		22.30	.717		296	38		334	385
5000	30" high, one door, 12" wide		22	.727		280	38.50		318.50	370
5020	15" wide		21.40	.748		289	39.50		328.50	385
5040	18" wide		20.90	.766		315	40.50		355.50	415
5060	24" wide		20.30	.788		375	42		417	480
5300	Two doors, 27" wide		19.80	.808		430	43		473	545
5320	30" wide		19.30	.829		485	44		529	605
5340	36" wide		18.80	.851		465	45		510	585
5360	42" wide		18.50	.865		535	46		581	660
5380	48" wide		18.40	.870		625	46		671	760
6000	Corner wall, 30" high, 24" wide		18	.889		435	47		482	555
6050	30" wide		17.20	.930		445	49.50		494.50	570
6100	36" wide		16.50	.970		475	51.50		526.50	610
6500	Revolving Lazy Susan		15.20	1.053		139	56		195	243
7000	Broom cabinet, 84" high, 24" deep, 18" wide		10	1.600		785	85		870	1,000
7500	Oven cabinets, 84" high, 24" deep, 27" wide		8	2	↓	1,075	106		1,181	1,350
7750	Valance board trim		396	.040	L.F.	20.50	2.15		22.65	26.50
7780	Toe kick trim	1 Carp	256	.031	"	3.61	1.66		5.27	6.65
7790	Base cabinet corner filler		16	.500	Ea.	50	26.50		76.50	97.50
7800	Cabinet filler, 3" x 24"		20	.400		19.95	21.50		41.45	56
7810	3" x 30"		20	.400		25	21.50		46.50	61.50
7820	3" x 42"		18	.444		35	23.50		58.50	76.50
7830	3" x 80"		16	.500	↓	66.50	26.50		93	116
7850	Cabinet panel	↓	50	.160	S.F.	11.20	8.50		19.70	26
9000	For deluxe models of all cabinets, add					40%				
9500	For custom built in place, add					25%	10%			
9558	Rule of thumb, kitchen cabinets not including									
9560	appliances & counter top, minimum	2 Carp	30	.533	L.F.	215	28.50		243.50	282
9600	Maximum	"	25	.640	"	460	34		494	560
9700	Minimum labor/equipment charge	1 Carp	3	2.667	Job		142		142	227

12 32 23.30 Manufactured Wood Casework Vanities

		Crew	Daily Output	Labor-Hours	Unit	Material	2020 Bare Costs Labor	Equipment	Total	Total Incl O&P
0010	**MANUFACTURED WOOD CASEWORK VANITIES**									
8000	Vanity bases, 2 doors, 30" high, 21" deep, 24" wide	2 Carp	20	.800	Ea.	420	42.50		462.50	530
8050	30" wide		16	1		460	53		513	590
8100	36" wide		13.33	1.200		400	64		464	540
8150	48" wide	↓	11.43	1.400		610	74.50		684.50	790
9000	For deluxe models of all vanities, add to above					40%				
9500	For custom built in place, add to above				↓	25%	10%			

12 32 23.35 Manufactured Wood Casework Hardware

		Crew	Daily Output	Labor-Hours	Unit	Material	2020 Bare Costs Labor	Equipment	Total	Total Incl O&P
0010	**MANUFACTURED WOOD CASEWORK HARDWARE**									
1000	Catches, minimum	1 Carp	235	.034	Ea.	1.56	1.81		3.37	4.62
1040	Maximum	"	80	.100	"	9.25	5.30		14.55	18.70
2000	Door/drawer pulls, handles									
2200	Handles and pulls, projecting, metal, minimum	1 Carp	48	.167	Ea.	5.75	8.85		14.60	20.50
2240	Maximum		36	.222		11.45	11.80		23.25	31.50
2300	Wood, minimum	↓	48	.167	↓	5.85	8.85		14.70	20.50

12 32 Manufactured Wood Casework

12 32 23 – Hardwood Casework

12 32 23.35 Manufactured Wood Casework Hardware	Crew	Daily Output	Labor-Hours	Unit	Material	2020 Bare Costs Labor	Equipment	Total	Total Incl O&P
2340	Maximum	1 Carp	36	.222	Ea.	10.80	11.80		22.60	31
2400	Drawer pulls, antimicrobial copper alloy finish		50	.160		20.50	8.50		29	36
2600	Flush, metal, minimum		48	.167		5.85	8.85		14.70	20.50
2640	Maximum		36	.222		10.95	11.80		22.75	31
2900	Drawer knobs, antimicrobial copper alloy finish		50	.160		13.45	8.50		21.95	28.50
3000	Drawer tracks/glides, minimum		48	.167	Pr.	9.60	8.85		18.45	25
3040	Maximum		24	.333		28	17.70		45.70	59
4000	Cabinet hinges, minimum		160	.050		3.40	2.66		6.06	8
4040	Maximum		68	.118		14.20	6.25		20.45	25.50
7000	Appliance pulls, antimicrobial copper alloy finish		50	.160	L.F.	73.50	8.50		82	94.50

12 34 Manufactured Plastic Casework

12 34 16 – Manufactured Solid-Plastic Casework

12 34 16.10 Outdoor Casework

		Crew	Daily Output	Labor-Hours	Unit	Material	Labor	Equipment	Total	Total Incl O&P
0010	**OUTDOOR CASEWORK**									
0020	Cabinet, base, sink/range, 36"	2 Carp	20.30	.788	Ea.	1,600	42		1,642	1,825
0100	Base, 36"		20.30	.788		1,825	42		1,867	2,100
0200	Filler strip, 1" x 30"		158	.101		35	5.40		40.40	47
0210	Filler strip, 2" x 30"		158	.101		44	5.40		49.40	57

12 35 Specialty Casework

12 35 53 – Laboratory Casework

12 35 53.13 Metal Laboratory Casework

		Crew	Daily Output	Labor-Hours	Unit	Material	Labor	Equipment	Total	Total Incl O&P
0010	**METAL LABORATORY CASEWORK**									
0020	Cabinets, base, door units, metal	2 Carp	18	.889	L.F.	270	47		317	375
0300	Drawer units		18	.889		595	47		642	730
0700	Tall storage cabinets, open, 7' high		20	.800		575	42.50		617.50	700
0900	With glazed doors		20	.800		1,025	42.50		1,067.50	1,200
1300	Wall cabinets, metal, 12-1/2" deep, open		20	.800		227	42.50		269.50	320
1500	With doors		20	.800		445	42.50		487.50	560

12 35 70 – Healthcare Casework

12 35 70.13 Hospital Casework

		Crew	Daily Output	Labor-Hours	Unit	Material	Labor	Equipment	Total	Total Incl O&P
0010	**HOSPITAL CASEWORK**									
3000	Hospital cabinets, stainless steel with glass door(s), lockable									
3010	One door, 24" W x 18" D x 60" H	2 Clab	18	.889	Ea.	3,175	37.50		3,212.50	3,550
3020	Two doors, 36" W x 18" D x 60" H		15	1.067		4,325	45		4,370	4,825
3030	36" W x 24" D x 67" H		15	1.067		5,800	45		5,845	6,450
3040	48" W x 24" D x 66" H		12	1.333		4,925	56		4,981	5,500
3050	48" W x 24" D x 72" H		12	1.333		6,125	56		6,181	6,850
3060	60" W x 24" D x 72" H		9	1.778		7,075	75		7,150	7,900

12 36 Countertops

12 36 16 – Metal Countertops

12 36 16.10 Stainless Steel Countertops	Crew	Daily Output	Labor-Hours	Unit	Material	2020 Bare Costs Labor	Equipment	Total	Total Incl O&P
0010 **STAINLESS STEEL COUNTERTOPS**									
3200 Stainless steel, custom	1 Carp	24	.333	S.F.	184	17.70		201.70	232

12 36 19 – Wood Countertops

12 36 19.10 Maple Countertops

	Crew	Daily Output	Labor-Hours	Unit	Material	2020 Bare Costs Labor	Equipment	Total	Total Incl O&P
0010 **MAPLE COUNTERTOPS**									
2900 Solid, laminated, 1-1/2" thick, no splash	1 Carp	28	.286	L.F.	92	15.20		107.20	126
3000 With square splash		28	.286	"	109	15.20		124.20	145
3400 Recessed cutting block with trim, 16" x 20" x 1"	↓	8	1	Ea.	113	53		166	209

12 36 23 – Plastic Countertops

12 36 23.13 Plastic-Laminate-Clad Countertops

	Crew	Daily Output	Labor-Hours	Unit	Material	2020 Bare Costs Labor	Equipment	Total	Total Incl O&P
0010 **PLASTIC-LAMINATE-CLAD COUNTERTOPS**									
0020 Stock, 24" wide w/backsplash, minimum	1 Carp	30	.267	L.F.	17.75	14.15		31.90	42
0100 Maximum		25	.320		42.50	17		59.50	74
0300 Custom plastic, 7/8" thick, aluminum molding, no splash		30	.267		35.50	14.15		49.65	61.50
0400 Cove splash		30	.267		47	14.15		61.15	74
0600 1-1/4" thick, no splash		28	.286		39.50	15.20		54.70	68
0700 Square splash		28	.286		44	15.20		59.20	73
0900 Square edge, plastic face, 7/8" thick, no splash		30	.267		34.50	14.15		48.65	60.50
1000 With splash	↓	30	.267		45.50	14.15		59.65	73
1200 For stainless channel edge, 7/8" thick, add					3.82			3.82	4.20
1300 1-1/4" thick, add					4.46			4.46	4.91
1500 For solid color suede finish, add				↓	5.95			5.95	6.55
1700 For end splash, add				Ea.	22			22	24
1900 For cut outs, standard, add, minimum	1 Carp	32	.250		17.85	13.30		31.15	41
2000 Maximum		8	1	↓	7.30	53		60.30	93
2100 Postformed, including backsplash and front edge		30	.267	L.F.	12.55	14.15		26.70	36.50
2110 Mitred, add		12	.667	Ea.		35.50		35.50	57
2200 Built-in place, 25" wide, plastic laminate		25	.320	L.F.	68.50	17		85.50	103
9000 Minimum labor/equipment charge	↓	3.75	2.133	Job		113		113	182

12 36 33 – Tile Countertops

12 36 33.10 Ceramic Tile Countertops

	Crew	Daily Output	Labor-Hours	Unit	Material	2020 Bare Costs Labor	Equipment	Total	Total Incl O&P
0010 **CERAMIC TILE COUNTERTOPS**									
2300 Ceramic tile mosaic	1 Carp	25	.320	L.F.	41	17		58	72.50

12 36 40 – Stone Countertops

12 36 40.10 Natural Stone Countertops

	Crew	Daily Output	Labor-Hours	Unit	Material	2020 Bare Costs Labor	Equipment	Total	Total Incl O&P
0010 **NATURAL STONE COUNTERTOPS**									
2500 Marble, stock, with splash, 1/2" thick, minimum	1 Bric	17	.471	L.F.	52.50	24.50		77	97.50
2700 3/4" thick, maximum	"	13	.615	"	129	32		161	194

12 36 53 – Laboratory Countertops

12 36 53.10 Laboratory Countertops and Sinks

	Crew	Daily Output	Labor-Hours	Unit	Material	2020 Bare Costs Labor	Equipment	Total	Total Incl O&P
0010 **LABORATORY COUNTERTOPS AND SINKS**									
0020 Countertops, epoxy resin, not incl. base cabinets, acid-proof, minimum	2 Carp	82	.195	S.F.	50.50	10.35		60.85	72
0030 Maximum	↓	70	.229	↓	48	12.15		60.15	72.50
0040 Stainless steel	↓	82	.195	↓	218	10.35		228.35	257

12 36 61 – Simulated Stone Countertops

12 36 61.16 Solid Surface Countertops

	Crew	Daily Output	Labor-Hours	Unit	Material	2020 Bare Costs Labor	Equipment	Total	Total Incl O&P
0010 **SOLID SURFACE COUNTERTOPS**, Acrylic polymer									
0020 Pricing for orders of 100 L.F. or greater									
0100 25" wide, solid colors	2 Carp	28	.571	L.F.	52	30.50		82.50	106

12 36 61.16 Solid Surface Countertops

		Crew	Daily Output	Labor-Hours	Unit	Material	2020 Bare Costs Labor	Equipment	Total	Total Incl O&P
0200	Patterned colors	2 Carp	28	.571	L.F.	66	30.50		96.50	122
0300	Premium patterned colors		28	.571		88	30.50		118.50	146
0400	With silicone attached 4" backsplash, solid colors		27	.593		61	31.50		92.50	118
0500	Patterned colors		27	.593		77	31.50		108.50	136
0600	Premium patterned colors		27	.593		96	31.50		127.50	157
0700	With hard seam attached 4" backsplash, solid colors		23	.696		61	37		98	126
0800	Patterned colors		23	.696		77	37		114	144
0900	Premium patterned colors		23	.696		96	37		133	165
1000	Pricing for order of 51-99 L.F.									
1100	25" wide, solid colors	2 Carp	24	.667	L.F.	60	35.50		95.50	123
1200	Patterned colors		24	.667		76	35.50		111.50	141
1300	Premium patterned colors		24	.667		101	35.50		136.50	168
1400	With silicone attached 4" backsplash, solid colors		23	.696		70	37		107	136
1500	Patterned colors		23	.696		89	37		126	157
1600	Premium patterned colors		23	.696		111	37		148	181
1700	With hard seam attached 4" backsplash, solid colors		20	.800		70	42.50		112.50	145
1800	Patterned colors		20	.800		89	42.50		131.50	166
1900	Premium patterned colors		20	.800		111	42.50		153.50	190
2000	Pricing for order of 1-50 L.F.									
2100	25" wide, solid colors	2 Carp	20	.800	L.F.	70	42.50		112.50	146
2200	Patterned colors		20	.800		89.50	42.50		132	167
2300	Premium patterned colors		20	.800		119	42.50		161.50	199
2400	With silicone attached 4" backsplash, solid colors		19	.842		82.50	45		127.50	162
2500	Patterned colors		19	.842		104	45		149	187
2600	Premium patterned colors		19	.842		130	45		175	215
2700	With hard seam attached 4" backsplash, solid colors		15	1.067		82.50	56.50		139	182
2800	Patterned colors		15	1.067		104	56.50		160.50	206
2900	Premium patterned colors		15	1.067		130	56.50		186.50	234
3000	Sinks, pricing for order of 100 or greater units									
3100	Single bowl, hard seamed, solid colors, 13" x 17"	1 Carp	3	2.667	Ea.	375	142		517	640
3200	10" x 15"		7	1.143		174	60.50		234.50	289
3300	Cutouts for sinks		8	1			53		53	85
3400	Sinks, pricing for order of 51-99 units									
3500	Single bowl, hard seamed, solid colors, 13" x 17"	1 Carp	2.55	3.137	Ea.	430	167		597	740
3600	10" x 15"		6	1.333		200	71		271	335
3700	Cutouts for sinks		7	1.143			60.50		60.50	97.50
3800	Sinks, pricing for order of 1-50 units									
3900	Single bowl, hard seamed, solid colors, 13" x 17"	1 Carp	2	4	Ea.	505	213		718	895
4000	10" x 15"		4.55	1.758		234	93.50		327.50	410
4100	Cutouts for sinks		5.25	1.524			81		81	130
4200	Cooktop cutouts, pricing for 100 or greater units		4	2		25.50	106		131.50	198
4300	51-99 units		3.40	2.353		29.50	125		154.50	232
4400	1-50 units		3	2.667		34.50	142		176.50	265

12 36 61.17 Solid Surface Vanity Tops

		Crew	Daily Output	Labor-Hours	Unit	Material	2020 Bare Costs Labor	Equipment	Total	Total Incl O&P
0010	**SOLID SURFACE VANITY TOPS**									
0015	Solid surface, center bowl, 17" x 19"	1 Carp	12	.667	Ea.	171	35.50		206.50	245
0020	19" x 25"		12	.667		216	35.50		251.50	294
0030	19" x 31"		12	.667		243	35.50		278.50	325
0040	19" x 37"		12	.667		283	35.50		318.50	365
0050	22" x 25"		10	.800		335	42.50		377.50	440
0060	22" x 31"		10	.800		395	42.50		437.50	505
0070	22" x 37"		10	.800		460	42.50		502.50	575

12 36 Countertops

12 36 61 – Simulated Stone Countertops

12 36 61.17 Solid Surface Vanity Tops

		Crew	Daily Output	Labor-Hours	Unit	Material	2020 Bare Costs Labor	Equipment	Total	Total Incl O&P
0080	22" x 43"	1 Carp	10	.800	Ea.	525	42.50		567.50	650
0090	22" x 49"		10	.800		575	42.50		617.50	700
0110	22" x 55"		8	1		425	53		478	550
0120	22" x 61"		8	1		480	53		533	615
0220	Double bowl, 22" x 61"		8	1		535	53		588	675
0230	Double bowl, 22" x 73"		8	1		1,075	53		1,128	1,275
0240	For aggregate colors, add					35%				
0250	For faucets and fittings, see Section 22 41 39.10									

12 36 61.19 Quartz Agglomerate Countertops

		Crew	Daily Output	Labor-Hours	Unit	Material	2020 Bare Costs Labor	Equipment	Total	Total Incl O&P
0010	**QUARTZ AGGLOMERATE COUNTERTOPS**									
0100	25" wide, 4" backsplash, color group A, minimum	2 Carp	15	1.067	L.F.	68	56.50		124.50	166
0110	Maximum		15	1.067		95	56.50		151.50	196
0120	Color group B, minimum		15	1.067		77	56.50		133.50	176
0130	Maximum		15	1.067		108	56.50		164.50	210
0140	Color group C, minimum		15	1.067		82	56.50		138.50	181
0150	Maximum		15	1.067		125	56.50		181.50	229
0160	Color group D, minimum		15	1.067		89	56.50		145.50	189
0170	Maximum		15	1.067		124	56.50		180.50	227

12 48 Rugs and Mats

12 48 13 – Entrance Floor Mats and Frames

12 48 13.13 Entrance Floor Mats

			Crew	Daily Output	Labor-Hours	Unit	Material	2020 Bare Costs Labor	Equipment	Total	Total Incl O&P
0010	**ENTRANCE FLOOR MATS**										
2000	Recycled rubber tire tile, 12" x 12" x 3/8" thick	G	1 Clab	125	.064	S.F.	11.50	2.69		14.19	16.95
2510	Natural cocoa fiber, 1/2" thick	G		125	.064		8	2.69		10.69	13.10
2520	3/4" thick	G		125	.064		7.15	2.69		9.84	12.15
2530	1" thick	G		125	.064		13.40	2.69		16.09	19.05
3000	Hospital tacky mats, package of 30 with frame					Ea.	60			60	66
3010	4 packages of 30					"	89			89	98

12 51 Office Furniture

12 51 16 – Case Goods

12 51 16.13 Metal Case Goods

		Crew	Daily Output	Labor-Hours	Unit	Material	2020 Bare Costs Labor	Equipment	Total	Total Incl O&P
0010	**METAL CASE GOODS**									
0020	Desks, 29" high, double pedestal, 30" x 60", metal, minimum				Ea.	590			590	645
0030	Maximum				"	915			915	1,000

12 54 Hospitality Furniture

12 54 13 – Hotel and Motel Furniture

12 54 13.10 Hotel Furniture

		Crew	Daily Output	Labor-Hours	Unit	Material	2020 Bare Costs Labor	Equipment	Total	Total Incl O&P
0010	**HOTEL FURNITURE**									
0020	Standard quality set, minimum				Room	2,650			2,650	2,900
0200	Maximum				"	8,625			8,625	9,475

12 54 Hospitality Furniture

12 54 16 - Restaurant Furniture

12 54 16.20 Furniture, Restaurant	Crew	Daily Output	Labor-Hours	Unit	Material	2020 Bare Costs Labor	Equipment	Total	Total Incl O&P
0010 **FURNITURE, RESTAURANT**									
0020 Bars, built-in, front bar	1 Carp	5	1.600	L.F.	315	85		400	480
0200 Back bar		5	1.600	"	230	85		315	390
0500 Booth unit, molded plastic, stub wall and 2 seats, minimum		2	4	Set	350	213		563	725
0600 Maximum		1.50	5.333	"	1,700	283		1,983	2,325
0800 Booth seat, upholstered, foursome, single (end) minimum		5	1.600	Ea.	890	85		975	1,125
0900 Maximum		4	2		1,200	106		1,306	1,500
1000 Foursome, double, minimum		4	2		1,150	106		1,256	1,450
1100 Maximum		3	2.667		1,950	142		2,092	2,350
1300 Circle booth, upholstered, 1/4 circle, minimum		3	2.667		1,625	142		1,767	2,025
1400 Maximum		2	4		1,775	213		1,988	2,300
1500 3/4 circle, minimum		1.50	5.333		7,200	283		7,483	8,375
1600 Maximum	▼	1	8	▼	7,825	425		8,250	9,300

12 55 Detention Furniture

12 55 13 - Detention Bunks

12 55 13.13 Cots

	Crew	Daily Output	Labor-Hours	Unit	Material	2020 Bare Costs Labor	Equipment	Total	Total Incl O&P
0010 **COTS**									
2500 Bolted, single, painted steel	E-4	20	1.600	Ea.	350	93	7.35	450.35	545
2700 Stainless steel	"	20	1.600	"	1,025	93	7.35	1,125.35	1,275

12 56 Institutional Furniture

12 56 43 - Dormitory Furniture

12 56 43.10 Dormitory Furnishings

	Crew	Daily Output	Labor-Hours	Unit	Material	2020 Bare Costs Labor	Equipment	Total	Total Incl O&P
0010 **DORMITORY FURNISHINGS**									
0300 Bunkable bed, twin, minimum				Ea.	390			390	430
0320 Maximum				"	600			600	660

12 56 51 - Library Furniture

12 56 51.10 Library Furnishings

	Crew	Daily Output	Labor-Hours	Unit	Material	2020 Bare Costs Labor	Equipment	Total	Total Incl O&P
0010 **LIBRARY FURNISHINGS**									
1710 Carrels, hardwood, 36" x 24", minimum	1 Carp	5	1.600	Ea.	580	85		665	775
1720 Maximum		4	2		1,825	106		1,931	2,175
1730 Metal, minimum		5	1.600		278	85		363	440
1740 Maximum		4	2	▼	485	106		591	700
6010 Bookshelf, metal, 90" high, 10" shelf, double face		11.50	.696	L.F.	206	37		243	285
6020 Single face	▼	12	.667	"	117	35.50		152.50	185
6050 For 8" shelving, subtract from above					10%				
6070 For 42" high with countertop, subtract from above					20%				

12 56 70 - Healthcare Furniture

12 56 70.10 Furniture, Hospital

	Crew	Daily Output	Labor-Hours	Unit	Material	2020 Bare Costs Labor	Equipment	Total	Total Incl O&P
0010 **FURNITURE, HOSPITAL**									
0020 Beds, manual, minimum				Ea.	800			800	880
0100 Maximum				"	2,950			2,950	3,250
1100 Patient wall systems, not incl. plumbing, minimum				Room	1,375			1,375	1,500
1200 Maximum				"	2,275			2,275	2,500

For customer support on your Commercial Renovation Costs with RSMeans data, call 800.448.8182.

393

12 63 Stadium and Arena Seating

12 63 13 – Stadium and Arena Bench Seating

12 63 13.13 Bleachers		Crew	Daily Output	Labor-Hours	Unit	Material	2020 Bare Costs Labor	Equipment	Total	Total Incl O&P
0010	**BLEACHERS**									
3000	Telescoping, manual to 15 tier, minimum	F-5	65	.492	Seat	114	26.50		140.50	169
3100	Maximum		60	.533		172	28.50		200.50	235
3300	16 to 20 tier, minimum		60	.533		250	28.50		278.50	320
3400	Maximum		55	.582		340	31		371	425
3600	21 to 30 tier, minimum		50	.640		272	34.50		306.50	355
3700	Maximum		40	.800		435	43		478	550
3900	For integral power operation, add, minimum	2 Elec	300	.053		55.50	3.27		58.77	66.50
4000	Maximum	"	250	.064		91	3.93		94.93	106
5000	Benches, folding, in wall, 14' table, 2 benches	L-4	2	12	Set	880	600		1,480	1,925

12 67 Pews and Benches

12 67 13 – Pews

12 67 13.13 Sanctuary Pews

		Crew	Daily Output	Labor-Hours	Unit	Material	Labor	Equipment	Total	Total Incl O&P
0010	**SANCTUARY PEWS**									
1500	Bench type, hardwood, minimum	1 Carp	20	.400	L.F.	111	21.50		132.50	156
1550	Maximum	"	15	.533		172	28.50		200.50	235
1570	For kneeler, add					22			22	24.50

12 93 Interior Public Space Furnishings

12 93 13 – Bicycle Racks

12 93 13.10 Bicycle Racks

		Crew	Daily Output	Labor-Hours	Unit	Material	Labor	Equipment	Total	Total Incl O&P
0010	**BICYCLE RACKS**									
0020	Single side, grid, 1-5/8" OD stl. pipe, w/1/2" bars, galv, 5 bike cap	2 Clab	10	1.600	Ea.	282	67.50		349.50	420
0025	Powder coat finish		10	1.600		288	67.50		355.50	425
0030	Single side, grid, 1-5/8" OD stl. pipe, w/1/2" bars, galv, 9 bike cap		8	2		400	84		484	575
0035	Powder coat finish		8	2		430	84		514	605
0040	Single side, grid, 1-5/8" OD stl. pipe, w/1/2" bars, galv, 18 bike cap		4	4		520	168		688	845
0045	Powder coat finish		4	4		535	168		703	860
0050	S curve, 1-7/8" OD stl. pipe, 11 ga., galv, 5 bike cap		10	1.600		160	67.50		227.50	284
0055	Powder coat finish		10	1.600		160	67.50		227.50	284
0060	S curve, 1-7/8" OD stl. pipe, 11 ga., galv, 7 bike cap		9	1.778		203	75		278	345
0065	Powder coat finish		9	1.778		219	75		294	360
0070	S curve, 1-7/8" OD stl. pipe, 11 ga., galv, 9 bike cap		8	2		320	84		404	485
0075	Powder coat finish		8	2		310	84		394	475
0080	S curve, 1-7/8" OD stl. pipe, 11 ga., galv, 11 bike cap		6	2.667		425	112		537	645
0085	Powder coat finish		6	2.667		395	112		507	615

12 93 23 – Trash and Litter Receptacles

12 93 23.10 Trash Receptacles

			Crew	Daily Output	Labor-Hours	Unit	Material	Labor	Equipment	Total	Total Incl O&P
0010	**TRASH RECEPTACLES**										
0500	Recycled plastic, var colors, round, 32 gal., 28" x 38" high	G	2 Clab	5	3.200	Ea.	710	135		845	995
0510	32 gal., 31" x 32" high	G		5	3.200		770	135		905	1,050
9110	Plastic, with dome lid, 32 gal. capacity			35	.457		61	19.25		80.25	98
9120	Recycled plastic slats, plastic dome lid, 32 gal. capacity			35	.457		297	19.25		316.25	355

Estimating Tips
General

- The items and systems in this division are usually estimated, purchased, supplied, and installed as a unit by one or more subcontractors. The estimator must ensure that all parties are operating from the same set of specifications and assumptions, and that all necessary items are estimated and will be provided. Many times the complex items and systems are covered, but the more common ones, such as excavation or a crane, are overlooked for the very reason that everyone assumes nobody could miss them. The estimator should be the central focus and be able to ensure that all systems are complete.

- It is important to consider factors such as site conditions, weather, shape and size of building, as well as labor availability as they may impact the overall cost of erecting special structures and systems included in this division.

- Another area where problems can develop in this division is at the interface between systems.

The estimator must ensure, for instance, that anchor bolts, nuts, and washers are estimated and included for the air-supported structures and pre-engineered buildings to be bolted to their foundations. Utility supply is a common area where essential items or pieces of equipment can be missed or overlooked because each subcontractor may feel it is another's responsibility. The estimator should also be aware of certain items which may be supplied as part of a package but installed by others, and ensure that the installing contractor's estimate includes the cost of installation. Conversely, the estimator must also ensure that items are not costed by two different subcontractors, resulting in an inflated overall estimate.

13 30 00 Special Structures

- The foundations and floor slab, as well as rough mechanical and electrical, should be estimated, as this work is required for the assembly and erection of the structure. Generally, as noted in the data set, the pre-engineered building comes

as a shell. Pricing is based on the size and structural design parameters stated in the reference section. Additional features, such as windows and doors with their related structural framing, must also be included by the estimator. Here again, the estimator must have a clear understanding of the scope of each portion of the work and all the necessary interfaces.

Reference Numbers

Reference numbers are shown at the beginning of some major classifications. These numbers refer to related items in the Reference Section. The reference information may be an estimating procedure, an alternate pricing method, or technical information.

Note: Not all subdivisions listed here necessarily appear. ■

Same Data. Simplified.

Enjoy the convenience and efficiency of accessing your costs anywhere:

- **Skip the multiplier** by setting your location
- **Quickly search,** edit, favorite and share costs
- **Stay on top of price changes** with automatic updates

Discover more at rsmeans.com/online

13 05 Common Work Results for Special Construction

13 05 05 – Selective Demolition for Special Construction

13 05 05.10 Selective Demolition, Air Supported Structures	Crew	Daily Output	Labor-Hours	Unit	Material	2020 Bare Costs Labor	Equipment	Total	Total Incl O&P
0010 **SELECTIVE DEMOLITION, AIR SUPPORTED STRUCTURES**									
0020 Tank covers, scrim, dbl. layer, vinyl poly w/hdwe., blower & controls									
0050 Round and rectangular R024119-10	B-2	9000	.004	S.F.		.19		.19	.30
0100 Warehouse structures									
0120 Poly/vinyl fabric, 28 oz., incl. tension cables & inflation system	4 Clab	9000	.004	SF Flr.		.15		.15	.24
0150 Reinforced vinyl, 12 oz., 3,000 S.F.	"	5000	.006			.27		.27	.43
0200 12,000 to 24,000 S.F.	8 Clab	20000	.003			.13		.13	.22
0250 Tedlar vinyl fabric, 28 oz. w/liner, to 3,000 S.F.	4 Clab	5000	.006			.27		.27	.43
0300 12,000 to 24,000 S.F.	8 Clab	20000	.003	▼		.13		.13	.22
0350 Greenhouse/shelter, woven polyethylene with liner									
0400 3,000 S.F.	4 Clab	5000	.006	SF Flr.		.27		.27	.43
0450 12,000 to 24,000 S.F.	8 Clab	20000	.003			.13		.13	.22
0500 Tennis/gymnasium, poly/vinyl fabric, 28 oz., incl. thermal liner	4 Clab	9000	.004			.15		.15	.24
0600 Stadium/convention center, teflon coated fiberglass, incl. thermal liner	9 Clab	40000	.002	▼		.08		.08	.12
0700 Doors, air lock, 15' long, 10' x 10'	2 Carp	1.50	10.667	Ea.		565		565	910
0720 15' x 15'		.80	20			1,075		1,075	1,700
0750 Revolving personnel door, 6' diam. x 6'-6" high		1.50	10.667	▼		565		565	910

13 05 05.20 Selective Demolition, Garden Houses

	Crew	Daily Output	Labor-Hours	Unit	Material	Labor	Equipment	Total	Total Incl O&P
0010 **SELECTIVE DEMOLITION, GARDEN HOUSES**									
0020 Prefab, wood, excl. foundation, average	2 Clab	400	.040	SF Flr.		1.68		1.68	2.70

13 05 05.25 Selective Demolition, Geodesic Domes

	Crew	Daily Output	Labor-Hours	Unit	Material	Labor	Equipment	Total	Total Incl O&P
0010 **SELECTIVE DEMOLITION, GEODESIC DOMES**									
0050 Shell only, interlocking plywood panels, 30' diameter	F-5	3.20	10	Ea.		535		535	860
0060 34' diameter		2.30	13.913			745		745	1,200
0070 39' diameter	▼	2	16			860		860	1,375
0080 45' diameter	F-3	2.20	18.182			990	214	1,204	1,800
0090 55' diameter		2	20			1,075	236	1,311	1,975
0100 60' diameter		2	20			1,075	236	1,311	1,975
0110 65' diameter	▼	1.60	25	▼		1,350	295	1,645	2,500

13 05 05.30 Selective Demolition, Greenhouses

	Crew	Daily Output	Labor-Hours	Unit	Material	Labor	Equipment	Total	Total Incl O&P
0010 **SELECTIVE DEMOLITION, GREENHOUSES**									
0020 Resi-type, free standing, excl. foundations, 9' long x 8' wide	2 Clab	160	.100	SF Flr.		4.21		4.21	6.75
0030 9' long x 11' wide		170	.094			3.96		3.96	6.35
0040 9' long x 14' wide		220	.073			3.06		3.06	4.91
0050 9' long x 17' wide		320	.050			2.11		2.11	3.37
0060 Lean-to type, 4' wide		64	.250			10.55		10.55	16.85
0070 7' wide		120	.133	▼		5.60		5.60	9
0080 Geodesic hemisphere, 1/8" plexiglass glazing, 8' diam.		4	4	Ea.		168		168	270
0090 24' diam.		.80	20			840		840	1,350
0100 48' diam.	▼	.40	40	▼		1,675		1,675	2,700

13 05 05.35 Selective Demolition, Hangars

	Crew	Daily Output	Labor-Hours	Unit	Material	Labor	Equipment	Total	Total Incl O&P
0010 **SELECTIVE DEMOLITION, HANGARS**									
0020 T type hangars, prefab, steel, galv roof & walls, incl doors, excl fndtn	E-2	2550	.022	SF Flr.		1.25	.67	1.92	2.76
0030 Circular type, prefab, steel frame, plastic skin, incl foundation, 80' diam	"	.50	112	Total		6,400	3,400	9,800	14,100

13 05 05.45 Selective Demolition, Lightning Protection

	Crew	Daily Output	Labor-Hours	Unit	Material	Labor	Equipment	Total	Total Incl O&P
0010 **SELECTIVE DEMOLITION, LIGHTNING PROTECTION**									
0020 Air terminal & base, copper, 3/8" diam. x 10", to 75' H	1 Clab	16	.500	Ea.		21		21	33.50
0030 1/2" diam. x 12", over 75' H		16	.500			21		21	33.50
0050 Aluminum, 1/2" diam. x 12", to 75' H		16	.500			21		21	33.50
0060 5/8" diam. x 12", over 75' H		16	.500	▼		21		21	33.50
0070 Cable, copper, 220 lb. per thousand feet, to 75' H	▼	640	.013	L.F.		.53		.53	.84

13 05 05 – Selective Demolition for Special Construction

13 05 05.45 Selective Demolition, Lightning Protection	Crew	Daily Output	Labor-Hours	Unit	Material	2020 Bare Costs Labor	Equipment	Total	Total Incl O&P
0080	375 lb. per thousand feet, over 75' H	1 Clab	460	.017	L.F.		.73		.73	1.17
0090	Aluminum, 101 lb. per thousand feet, to 75' H		560	.014			.60		.60	.96
0100	199 lb. per thousand feet, over 75' H		480	.017			.70		.70	1.12
0110	Arrester, 175 V AC, to ground		16	.500	Ea.		21		21	33.50
0120	650 V AC, to ground		13	.615	"		26		26	41.50

13 05 05.50 Selective Demolition, Pre-Engineered Steel Buildings

		Crew	Daily Output	Labor-Hours	Unit	Material	2020 Bare Costs Labor	Equipment	Total	Total Incl O&P
0010	**SELECTIVE DEMOLITION, PRE-ENGINEERED STEEL BUILDINGS**									
0500	Pre-engd. steel bldgs., rigid frame, clear span & multi post, excl. salvage									
0550	3,500 to 7,500 S.F.	L-10	1000	.024	SF Flr.		1.41	.47	1.88	2.80
0600	7,501 to 12,500 S.F.		1500	.016			.94	.31	1.25	1.87
0650	12,501 S.F. or greater		1650	.015			.86	.29	1.15	1.69
0700	Pre-engd. steel building components									
0710	Entrance canopy, including frame 4' x 4'	E-24	8	4	Ea.		230	72.50	302.50	450
0720	4' x 8'	"	7	4.571			263	83	346	515
0730	HM doors, self framing, single leaf	2 Skwk	8	2			110		110	175
0740	Double leaf		5	3.200			176		176	280
0760	Gutter, eave type		600	.027	L.F.		1.46		1.46	2.33
0770	Sash, single slide, double slide or fixed		24	.667	Ea.		36.50		36.50	58.50
0780	Skylight, fiberglass, to 30 S.F.		16	1			55		55	87.50
0785	Roof vents, circular, 12" to 24" diameter		12	1.333			73		73	117
0790	Continuous, 10' long		8	2			110		110	175
0900	Shelters, aluminum frame									
0910	Acrylic glazing, 3' x 9' x 8' high	2 Skwk	2	8	Ea.		440		440	700
0920	9' x 12' x 8' high	"	1.50	10.667	"		585		585	930

13 05 05.60 Selective Demolition, Silos

		Crew	Daily Output	Labor-Hours	Unit	Material	2020 Bare Costs Labor	Equipment	Total	Total Incl O&P
0010	**SELECTIVE DEMOLITION, SILOS**									
0020	Conc stave, indstrl, conical/sloping bott, excl fndtn, 12' diam., 35' H	E-24	.18	178	Ea.		10,200	3,225	13,425	20,100
0030	16' diam., 45' H		.12	267			15,300	4,850	20,150	30,100
0040	25' diam., 75' H		.08	400			23,000	7,250	30,250	45,200
0050	Steel, factory fabricated, 30,000 gal. cap, painted or epoxy lined	L-5	2	28			1,625	290	1,915	2,975

13 05 05.65 Selective Demolition, Sound Control

		Crew	Daily Output	Labor-Hours	Unit	Material	2020 Bare Costs Labor	Equipment	Total	Total Incl O&P
0010	**SELECTIVE DEMOLITION, SOUND CONTROL**									
0120	Acoustical enclosure, 4" thick walls & ceiling panels, 8 lb./S.F.	3 Carp	144	.167	SF Surf		8.85		8.85	14.20
0130	10.5 lb./S.F.		128	.188			9.95		9.95	15.95
0140	Reverb chamber, parallel walls, 4" thick		120	.200			10.65		10.65	17.05
0150	Skewed walls, parallel roof, 4" thick		110	.218			11.60		11.60	18.60
0160	Skewed walls/roof, 4" layer/air space		96	.250			13.30		13.30	21.50
0170	Sound-absorbing panels, painted metal, 2'-6" x 8', under 1,000 S.F.		430	.056			2.97		2.97	4.75
0180	Over 1,000 S.F.		480	.050			2.66		2.66	4.26
0190	Flexible transparent curtain, clear	3 Shee	430	.056			3.48		3.48	5.45
0192	50% clear, 50% foam		430	.056			3.48		3.48	5.45
0194	25% clear, 75% foam		430	.056			3.48		3.48	5.45
0196	100% foam		430	.056			3.48		3.48	5.45
0200	Audio-masking sys., incl. speakers, amplfr., signal gnrtr.									
0205	Ceiling mounted, 5,000 S.F.	2 Elec	4800	.003	S.F.		.20		.20	.31
0210	10,000 S.F.		5600	.003			.18		.18	.27
0220	Plenum mounted, 5,000 S.F.		7600	.002			.13		.13	.20
0230	10,000 S.F.		8800	.002			.11		.11	.17

13 05 05.70 Selective Demolition, Special Purpose Rooms

		Crew	Daily Output	Labor-Hours	Unit	Material	2020 Bare Costs Labor	Equipment	Total	Total Incl O&P
0010	**SELECTIVE DEMOLITION, SPECIAL PURPOSE ROOMS**									
0100	Audiometric rooms, under 500 S.F. surface	4 Carp	200	.160	SF Surf		8.50		8.50	13.60
0110	Over 500 S.F. surface	"	240	.133	"		7.10		7.10	11.35

13 05 05.70 Selective Demolition, Special Purpose Rooms

		Crew	Daily Output	Labor-Hours	Unit	Material	2020 Bare Costs Labor	Equipment	Total	Total Incl O&P
0200	Clean rooms, 12' x 12' soft wall, class 100	1 Carp	.30	26.667	Ea.		1,425		1,425	2,275
0210	Class 1,000		.30	26.667			1,425		1,425	2,275
0220	Class 10,000		.35	22.857			1,225		1,225	1,950
0230	Class 100,000		.35	22.857			1,225		1,225	1,950
0300	Darkrooms, shell complete, 8' high	2 Carp	220	.073	SF Flr.		3.87		3.87	6.20
0310	12' high		110	.145	"		7.75		7.75	12.40
0350	Darkrooms doors, mini-cylindrical, revolving		4	4	Ea.		213		213	340
0400	Music room, practice modular		140	.114	SF Surf		6.05		6.05	9.75
0500	Refrigeration structures and finishes									
0510	Wall finish, 2 coat Portland cement plaster, 1/2" thick	1 Clab	200	.040	S.F.		1.68		1.68	2.70
0520	Fiberglass panels, 1/8" thick		400	.020			.84		.84	1.35
0530	Ceiling finish, polystyrene plastic, 1" to 2" thick		500	.016			.67		.67	1.08
0540	4" thick		450	.018			.75		.75	1.20
0550	Refrigerator, prefab aluminum walk-in, 7'-6" high, 6' x 6' OD	2 Carp	100	.160	SF Flr.		8.50		8.50	13.60
0560	10' x 10' OD		160	.100			5.30		5.30	8.50
0570	Over 150 S.F.		200	.080			4.25		4.25	6.80
0600	Sauna, prefabricated, including heater & controls, 7' high, to 30 S.F.		120	.133			7.10		7.10	11.35
0610	To 40 S.F.		140	.114			6.05		6.05	9.75
0620	To 60 S.F.		175	.091			4.86		4.86	7.80
0630	To 100 S.F.		220	.073			3.87		3.87	6.20
0640	To 130 S.F.		250	.064			3.40		3.40	5.45
0650	Steam bath, heater, timer, head, single, to 140 C.F.	1 Plum	2.20	3.636	Ea.		234		234	360
0660	To 300 C.F.		2.20	3.636			234		234	360
0670	Steam bath, comm. size, w/blow-down assembly, to 800 C.F.		1.80	4.444			286		286	445
0680	To 2,500 C.F.		1.60	5			320		320	500
0690	Steam bath, comm. size, multiple, for motels, apts, 500 C.F., 2 baths		2	4			258		258	400
0700	1,000 C.F., 4 baths		1.40	5.714			370		370	570

13 05 05.75 Selective Demolition, Storage Tanks

		Crew	Daily Output	Labor-Hours	Unit	Material	2020 Bare Costs Labor	Equipment	Total	Total Incl O&P
0010	**SELECTIVE DEMOLITION, STORAGE TANKS**									
0500	Steel tank, single wall, above ground, not incl. fdn., pumps or piping									
0510	Single wall, 275 gallon	Q-1	3	5.333	Ea.		310		310	480
0520	550 thru 2,000 gallon	B-34P	2	12			680	515	1,195	1,650
0530	5,000 thru 10,000 gallon	B-34Q	2	12			690	570	1,260	1,700
0540	15,000 thru 30,000 gallon	B-34S	2	16			955	1,550	2,505	3,225
0600	Steel tank, double wall, above ground not incl. fdn., pumps & piping									
0620	500 thru 2,000 gallon	B-34P	2	12	Ea.		680	515	1,195	1,650

13 05 05.85 Selective Demolition, Swimming Pool Equip

		Crew	Daily Output	Labor-Hours	Unit	Material	2020 Bare Costs Labor	Equipment	Total	Total Incl O&P
0010	**SELECTIVE DEMOLITION, SWIMMING POOL EQUIP**									
0020	Diving stand, stainless steel, 3 meter	2 Clab	3	5.333	Ea.		225		225	360
0030	1 meter		5	3.200			135		135	216
0040	Diving board, 16' long, aluminum		5.40	2.963			125		125	200
0050	Fiberglass		5.40	2.963			125		125	200
0070	Ladders, heavy duty, stainless steel, 2 tread		14	1.143			48		48	77
0080	4 tread		12	1.333			56		56	90
0090	Lifeguard chair, stainless steel, fixed		5	3.200			135		135	216
0100	Slide, tubular, fiberglass, aluminum handrails & ladder, 5', straight		4	4			168		168	270
0110	8', curved		6	2.667			112		112	180
0120	10', curved		3	5.333			225		225	360
0130	12' straight, with platform		2.50	6.400			269		269	430
0140	Removable access ramp, stainless steel		4	4			168		168	270
0150	Removable stairs, stainless steel, collapsible		4	4			168		168	270

13 05 Common Work Results for Special Construction

13 05 05 – Selective Demolition for Special Construction

13 05 05.90 Selective Demolition, Tension Structures

	13 05 05.90 Selective Demolition, Tension Structures	Crew	Daily Output	Labor-Hours	Unit	Material	2020 Bare Costs Labor	Equipment	Total	Total Incl O&P
0010	**SELECTIVE DEMOLITION, TENSION STRUCTURES**									
0020	Steel/alum. frame, fabric shell, 60' clear span, 6,000 S.F.	B-41	2000	.022	SF Flr.		.96	.15	1.11	1.70
0030	12,000 S.F.		2200	.020			.87	.14	1.01	1.54
0040	80' clear span, 20,800 S.F.	↓	2440	.018			.79	.12	.91	1.40
0050	100' clear span, 10,000 S.F.	L-5	4350	.013			.75	.13	.88	1.37
0060	26,000 S.F.		4600	.012			.71	.13	.84	1.29
0070	36,000 S.F.	↓	5000	.011	↓		.65	.12	.77	1.19

13 05 05.95 Selective Demo, X-Ray/Radio Freq Protection

	13 05 05.95 Selective Demo, X-Ray/Radio Freq Protection	Crew	Daily Output	Labor-Hours	Unit	Material	2020 Bare Costs Labor	Equipment	Total	Total Incl O&P
0010	**SELECTIVE DEMO, X-RAY/RADIO FREQ PROTECTION**									
0020	Shielding lead, lined door frame, excl. hdwe., 1/16" thick	1 Clab	4.80	1.667	Ea.		70		70	112
0030	Lead sheets, 1/16" thick	2 Clab	270	.059	S.F.		2.49		2.49	4
0040	1/8" thick		240	.067			2.81		2.81	4.50
0050	Lead shielding, 1/4" thick		270	.059			2.49		2.49	4
0060	1/2" thick	↓	240	.067	↓		2.81		2.81	4.50
0070	Lead glass, 1/4" thick, 2.0 mm LE, 12" x 16"	2 Glaz	16	1	Ea.		51		51	81.50
0080	24" x 36"		8	2			102		102	163
0090	36" x 60"		4	4			204		204	325
0100	Lead glass window frame, with 1/16" lead & voice passage, 36" x 60"		4	4			204		204	325
0110	Lead glass window frame, 24" x 36"	↓	8	2			102		102	163
0120	Lead gypsum board, 5/8" thick with 1/16" lead	2 Clab	320	.050	S.F.		2.11		2.11	3.37
0130	1/8" lead		280	.057			2.41		2.41	3.85
0140	1/32" lead		400	.040	↓		1.68		1.68	2.70
0150	Butt joints, 1/8" lead or thicker, 2" x 7' long batten strip		480	.033	Ea.		1.40		1.40	2.25
0160	X-ray protection, average radiography room, up to 300 S.F., 1/16" lead, min		.50	32	Total		1,350		1,350	2,150
0170	Maximum		.30	53.333			2,250		2,250	3,600
0180	Deep therapy X-ray room, 250 kV cap, up to 300 S.F., 1/4" lead, min		.20	80			3,375		3,375	5,400
0190	Maximum		.12	133	↓		5,625		5,625	9,000
0880	Radio frequency shielding, prefab or screen-type copper or steel, minimum		360	.044	SF Surf		1.87		1.87	3
0890	Average		310	.052			2.17		2.17	3.48
0895	Maximum	↓	290	.055	↓		2.32		2.32	3.72

13 12 Fountains

13 12 13 – Exterior Fountains

13 12 13.10 Outdoor Fountains

	13 12 13.10 Outdoor Fountains	Crew	Daily Output	Labor-Hours	Unit	Material	2020 Bare Costs Labor	Equipment	Total	Total Incl O&P
0010	**OUTDOOR FOUNTAINS**									
0100	Outdoor fountain, 48" high with bowl and figures	2 Clab	2	8	Ea.	435	335		770	1,025
0200	Commercial, concrete or cast stone, 40-60" H, simple		2	8		955	335		1,290	1,600
0220	Average		2	8		1,850	335		2,185	2,575
0240	Ornate		2	8		4,200	335		4,535	5,175
0260	Metal, 72" high		2	8		1,500	335		1,835	2,200
0280	90" high		2	8		2,250	335		2,585	3,025
0300	120" high		2	8		5,275	335		5,610	6,350
0320	Resin or fiberglass, 40-60" H, wall type		2	8		675	335		1,010	1,275
0340	Waterfall type	↓	2	8	↓	1,150	335		1,485	1,800

13 12 23 – Interior Fountains

13 12 23.10 Indoor Fountains

	13 12 23.10 Indoor Fountains	Crew	Daily Output	Labor-Hours	Unit	Material	2020 Bare Costs Labor	Equipment	Total	Total Incl O&P
0010	**INDOOR FOUNTAINS**									
0100	Commercial, floor type, resin or fiberglass, lighted, cascade type	2 Clab	2	8	Ea.	425	335		760	1,000
0120	Tiered type		2	8		490	335		825	1,075
0140	Waterfall type	↓	2	8	↓	315	335		650	885

13 17 Tubs and Pools

13 17 13 – Hot Tubs

13 17 13.10 Redwood Hot Tub System		Crew	Daily Output	Labor-Hours	Unit	Material	2020 Bare Costs Labor	2020 Bare Costs Equipment	Total	Total Incl O&P
0010	**REDWOOD HOT TUB SYSTEM**									
7050	4' diameter x 4' deep	Q-1	1	16	Ea.	3,250	930		4,180	5,000
7150	6' diameter x 4' deep		.80	20		4,975	1,150		6,125	7,275
7200	8' diameter x 4' deep		.80	20		7,300	1,150		8,450	9,825

13 17 33 – Whirlpool Tubs

13 17 33.10 Whirlpool Bath		Crew	Daily Output	Labor-Hours	Unit	Material	2020 Bare Costs Labor	2020 Bare Costs Equipment	Total	Total Incl O&P
0010	**WHIRLPOOL BATH**									
6000	Whirlpool, bath with vented overflow, molded fiberglass									
6100	66" x 36" x 24"	Q-1	1	16	Ea.	910	930		1,840	2,425
6400	72" x 36" x 21"		1	16		1,450	930		2,380	3,025
6500	60" x 34" x 21"		1	16		1,450	930		2,380	3,025
6600	72" x 42" x 23"		1	16		1,350	930		2,280	2,925
6710	For color, add					10%				
6711	For designer colors and trim, add					25%				

13 21 Controlled Environment Rooms

13 21 26 – Cold Storage Rooms

13 21 26.50 Refrigeration

13 21 26.50 Refrigeration		Crew	Daily Output	Labor-Hours	Unit	Material	2020 Bare Costs Labor	2020 Bare Costs Equipment	Total	Total Incl O&P
0010	**REFRIGERATION**									
0020	Curbs, 12" high, 4" thick, concrete	2 Carp	58	.276	L.F.	3.30	14.65		17.95	27
6300	Rule of thumb for complete units, w/o doors & refrigeration, cooler		146	.110	SF Flr.	166	5.80		171.80	192
6400	Freezer		109.60	.146	"	113	7.75		120.75	136

13 24 Special Activity Rooms

13 24 16 – Saunas

13 24 16.50 Saunas and Heaters

13 24 16.50 Saunas and Heaters		Crew	Daily Output	Labor-Hours	Unit	Material	2020 Bare Costs Labor	2020 Bare Costs Equipment	Total	Total Incl O&P
0010	**SAUNAS AND HEATERS**									
0020	Prefabricated, incl. heater & controls, 7' high, 6' x 4', C/C	L-7	2.20	12.727	Ea.	5,250	650		5,900	6,800
1700	Door only, cedar, 2' x 6', w/ 1' x 4' tempered insulated glass window	2 Carp	3.40	4.706		610	250		860	1,075
1800	Prehung, incl. jambs, pulls & hardware	"	12	1.333		620	71		691	800
2500	Heaters only (incl. above), wall mounted, to 200 C.F.					1,025			1,025	1,125
4480	For additional equipment, see Section 11 66 13.10									

13 24 26 – Steam Baths

13 24 26.50 Steam Baths and Components

13 24 26.50 Steam Baths and Components		Crew	Daily Output	Labor-Hours	Unit	Material	2020 Bare Costs Labor	2020 Bare Costs Equipment	Total	Total Incl O&P
0010	**STEAM BATHS AND COMPONENTS**									
0020	Heater, timer & head, single, to 140 C.F.	1 Plum	1.20	6.667	Ea.	2,350	430		2,780	3,275
0500	To 300 C.F.	"	1.10	7.273		2,700	470		3,170	3,675
2000	Multiple, motels, apts., 2 baths, w/blow-down assm., 500 C.F.	Q-1	1.30	12.308		6,650	715		7,365	8,425
2500	4 baths	"	.70	22.857		10,500	1,325		11,825	13,600

13 28 Athletic and Recreational Special Construction

13 28 33 – Athletic and Recreational Court Walls

13 28 33.50 Sport Court		Crew	Daily Output	Labor-Hours	Unit	Material	2020 Bare Costs Labor	Equipment	Total	Total Incl O&P
0010	**SPORT COURT**									
0020	Sport court floors, no. 2 & better maple, 25/32" thick, see 09 64 66.10				SF Flr.					
0300	Squash, regulation court in existing building, minimum				Court	41,900			41,900	46,000
0400	Maximum				"	46,600			46,600	51,500
0450	Rule of thumb for components:									
0470	Walls	3 Carp	.15	160	Court	12,700	8,500		21,200	27,600
0500	Floor	"	.25	96		10,000	5,100		15,100	19,200
0550	Lighting	2 Elec	.60	26.667		2,400	1,625		4,025	5,150

13 34 Fabricated Engineered Structures

13 34 13 – Glazed Structures

13 34 13.13 Greenhouses

		Crew	Daily Output	Labor-Hours	Unit	Material	2020 Bare Costs Labor	Equipment	Total	Total Incl O&P
0010	**GREENHOUSES**, Shell only, stock units, not incl. 2' stub walls,									
0020	foundation, floors, heat or compartments									
0300	Residential type, free standing, 8'-6" long x 7'-6" wide	2 Carp	59	.271	SF Flr.	24	14.40		38.40	49
0400	10'-6" wide		85	.188		42.50	10		52.50	63
0600	13'-6" wide		108	.148		45	7.85		52.85	62
0700	17'-0" wide		160	.100		43.50	5.30		48.80	56
0900	Lean-to type, 3'-10" wide		34	.471		43.50	25		68.50	87.50
1000	6'-10" wide		58	.276		28	14.65		42.65	54.50
1050	8'-0" wide		60	.267		73	14.15		87.15	103
1100	Wall mounted to existing window, 3' x 3'	1 Carp	4	2	Ea.	2,225	106		2,331	2,625
1120	4' x 5'	"	3	2.667		2,600	142		2,742	3,100
3900	Cooling, 1,200 CFM exhaust fan, add					325			325	355
4000	7,850 CFM					1,025			1,025	1,150
4200	For heaters, 10 MBH, add					246			246	270
4300	60 MBH, add					845			845	925
4500	For benches, 2' x 8', add					188			188	207
4600	4' x 10', add					227			227	249
4800	For ventilation & humidity control w/4 integrated outlets, add				Total	273			273	300
4900	For environmental controls and automation, 8 outputs, 9 stages, add				"	1,200			1,200	1,325
5100	For humidification equipment, add				Ea.	325			325	355
5200	For vinyl shading, add				S.F.	.44			.44	.48

13 34 13.19 Swimming Pool Enclosures

		Crew	Daily Output	Labor-Hours	Unit	Material	2020 Bare Costs Labor	Equipment	Total	Total Incl O&P
0010	**SWIMMING POOL ENCLOSURES** Translucent, free standing									
0020	not including foundations, heat or light									
0200	Economy	2 Carp	200	.080	SF Hor.	46.50	4.25		50.75	58
0600	Deluxe	"	70	.229	"	103	12.15		115.15	132

13 34 23 – Fabricated Structures

13 34 23.16 Fabricated Control Booths

		Crew	Daily Output	Labor-Hours	Unit	Material	2020 Bare Costs Labor	Equipment	Total	Total Incl O&P
0010	**FABRICATED CONTROL BOOTHS**									
0100	Guard House, prefab conc. w/bullet resistant doors & windows, roof & wiring									
0110	8' x 8', Level III	L-10	1	24	Ea.	54,500	1,400	470	56,370	63,000
0120	8' x 8', Level IV	"	1	24	"	72,000	1,400	470	73,870	82,000

13 34 23.45 Kiosks

		Crew	Daily Output	Labor-Hours	Unit	Material	2020 Bare Costs Labor	Equipment	Total	Total Incl O&P
0010	**KIOSKS**									
0020	Round, advertising type, 5' diameter, 7' high, aluminum wall, illuminated				Ea.	23,500			23,500	25,900
0100	Aluminum wall, non-illuminated					22,500			22,500	24,800
0500	Rectangular, 5' x 9', 7'-6" high, aluminum wall, illuminated					25,500			25,500	28,100
0600	Aluminum wall, non-illuminated					23,800			23,800	26,200

For customer support on your Commercial Renovation Costs with RSMeans data, call 800.448.8182.

401

13 34 Fabricated Engineered Structures

13 34 63 – Natural Fiber Construction

13 34 63.50 Straw Bale Construction		Crew	Daily Output	Labor-Hours	Unit	Material	2020 Bare Costs Labor	Equipment	Total	Total Incl O&P
0010	**STRAW BALE CONSTRUCTION**									
2020	Straw bales in walls w/modified post and beam frame	G 2 Carp	320	.050	S.F.	6.85	2.66		9.51	11.80

13 42 Building Modules

13 42 63 – Detention Cell Modules

13 42 63.16 Steel Detention Cell Modules

		Crew	Daily Output	Labor-Hours	Unit	Material	Labor	Equipment	Total	Total Incl O&P
0010	**STEEL DETENTION CELL MODULES**									
2000	Cells, prefab., 5' to 6' wide, 7' to 8' high, 7' to 8' deep,									
2010	bar front, cot, not incl. plumbing	E-4	1.50	21.333	Ea.	9,650	1,250	98	10,998	12,700

13 48 Sound, Vibration, and Seismic Control

13 48 13 – Manufactured Sound and Vibration Control Components

13 48 13.50 Audio Masking

		Crew	Daily Output	Labor-Hours	Unit	Material	Labor	Equipment	Total	Total Incl O&P
0010	**AUDIO MASKING**, acoustical enclosure, 4" thick wall and ceiling									
0020	8 lb./S.F., up to 12' span	3 Carp	72	.333	SF Surf	31.50	17.70		49.20	63
0300	Better quality panels, 10.5 #/S.F		64	.375		36	19.95		55.95	71.50
0400	Reverb-chamber, 4" thick, parallel walls	↓	60	.400	↓	44.50	21.50		66	83

13 49 Radiation Protection

13 49 13 – Integrated X-Ray Shielding Assemblies

13 49 13.50 Lead Sheets

		Crew	Daily Output	Labor-Hours	Unit	Material	Labor	Equipment	Total	Total Incl O&P
0010	**LEAD SHEETS**									
0300	Lead sheets, 1/16" thick	2 Lath	135	.119	S.F.	11.40	6.15		17.55	22
0400	1/8" thick		120	.133		24	6.95		30.95	37.50
0500	Lead shielding, 1/4" thick		135	.119		40	6.15		46.15	53.50
0550	1/2" thick	↓	120	.133	↓	91.50	6.95		98.45	112
1200	X-ray protection, average radiography or fluoroscopy									
1210	room, up to 300 S.F. floor, 1/16" lead, economy	2 Lath	.25	64	Total	12,100	3,325		15,425	18,600
1500	7'-0" walls, deluxe	"	.15	107	"	14,600	5,550		20,150	24,700
1600	Deep therapy X-ray room, 250 kV capacity,									
1800	up to 300 S.F. floor, 1/4" lead, economy	2 Lath	.08	200	Total	33,900	10,400		44,300	53,500
1900	7'-0" walls, deluxe		.06	267	"	41,800	13,900		55,700	67,500
1999	Minimum labor/equipment charge	↓	4.50	3.556	Job		185		185	290

13 49 21 – Lead Glazing

13 49 21.50 Lead Glazing

		Crew	Daily Output	Labor-Hours	Unit	Material	Labor	Equipment	Total	Total Incl O&P
0010	**LEAD GLAZING**									
0600	Lead glass, 1/4" thick, 2.0 mm LE, 12" x 16"	2 Glaz	13	1.231	Ea.	435	63		498	580
0700	24" x 36"	"	8	2	"	1,500	102		1,602	1,825
2000	X-ray viewing panels, clear lead plastic									
2010	7 mm thick, 0.3 mm LE, 2.3 lb./S.F.	H-3	139	.115	S.F.	277	5.25		282.25	315
2020	12 mm thick, 0.5 mm LE, 3.9 lb./S.F.		82	.195		385	8.85		393.85	440
2030	18 mm thick, 0.8 mm LE, 5.9 lb./S.F.		54	.296		470	13.50		483.50	535
2040	22 mm thick, 1.0 mm LE, 7.2 lb./S.F.		44	.364		665	16.55		681.55	755
2050	35 mm thick, 1.5 mm LE, 11.5 lb./S.F.		28	.571		1,000	26		1,026	1,150
2060	46 mm thick, 2.0 mm LE, 15.0 lb./S.F.	↓	21	.762	↓	1,300	34.50		1,334.50	1,475

13 49 23.50 Modular Shielding Partitions	Crew	Daily Output	Labor-Hours	Unit	Material	2020 Bare Costs Labor	Equipment	Total	Total Incl O&P
0010 MODULAR SHIELDING PARTITIONS									
4000 X-ray barriers, modular, panels mounted within framework for									
4002 attaching to floor, wall or ceiling, upper portion is clear lead									
4005 plastic window panels 48"H, lower portion is opaque leaded									
4008 steel panels 36"H, structural supports not incl.									
4010 1-section barrier, 36"W x 84"H overall									
4020 0.5 mm LE panels	H-3	6.40	2.500	Ea.	9,025	114		9,139	10,100
4030 0.8 mm LE panels		6.40	2.500		9,725	114		9,839	10,900
4040 1.0 mm LE panels		5.33	3.002		13,500	137		13,637	15,100
4050 1.5 mm LE panels		5.33	3.002		15,800	137		15,937	17,600
4060 2-section barrier, 72"W x 84"H overall									
4070 0.5 mm LE panels	H-3	4	4	Ea.	14,400	182		14,582	16,100
4080 0.8 mm LE panels		4	4		14,800	182		14,982	16,600
4090 1.0 mm LE panels		3.56	4.494		16,300	204		16,504	18,200
5000 1.5 mm LE panels		3.20	5		27,400	227		27,627	30,600
5010 3-section barrier, 108"W x 84"H overall									
5020 0.5 mm LE panels	H-3	3.20	5	Ea.	21,500	227		21,727	24,100
5030 0.8 mm LE panels		3.20	5		22,200	227		22,427	24,800
5040 1.0 mm LE panels		2.67	5.993		24,400	273		24,673	27,300
5050 1.5 mm LE panels		2.46	6.504		41,500	296		41,796	46,100
7000 X-ray barriers, mobile, mounted within framework w/casters on									
7005 bottom, clear lead plastic window panels on upper portion,									
7010 opaque on lower, 30"W x 75"H overall, incl. framework									
7020 24"H upper w/0.5 mm LE, 48"H lower w/0.8 mm LE	1 Carp	16	.500	Ea.	3,975	26.50		4,001.50	4,425
7030 48"W x 75"H overall, incl. framework									
7040 36"H upper w/0.5 mm LE, 36"H lower w/0.8 mm LE	1 Carp	16	.500	Ea.	7,075	26.50		7,101.50	7,850
7050 36"H upper w/1.0 mm LE, 36"H lower w/1.5 mm LE	"	16	.500	"	8,400	26.50		8,426.50	9,275
7060 72"W x 75"H overall, incl. framework									
7070 36"H upper w/0.5 mm LE, 36"H lower w/0.8 mm LE	1 Carp	16	.500	Ea.	8,400	26.50		8,426.50	9,275
7080 36"H upper w/1.0 mm LE, 36"H lower w/1.5 mm LE	"	16	.500	"	10,500	26.50		10,526.50	11,600

For customer support on your Commercial Renovation Costs with RSMeans data, call 800.448.8182.

403

Division Notes

		CREW	DAILY OUTPUT	LABOR-HOURS	UNIT	BARE COSTS				TOTAL INCL O&P
						MAT.	LABOR	EQUIP.	TOTAL	

Division 14

Conveying Equipment

Estimating Tips
General

- Many products in Division 14 will require some type of support or blocking for installation not included with the item itself. Examples are supports for conveyors or tube systems, attachment points for lifts, and footings for hoists or cranes. Add these supports in the appropriate division.

14 10 00 Dumbwaiters
14 20 00 Elevators

- Dumbwaiters and elevators are estimated and purchased in a method similar to buying a car. The manufacturer has a base unit with standard features. Added to this base unit price will be whatever options the owner or specifications require. Increased load capacity, additional vertical travel, additional stops, higher speed, and cab finish options are items to be considered. When developing an estimate for dumbwaiters and elevators, remember that some items needed by the installers may have to be included as part of the general contract.

Examples are:
- ☐ shaftway
- ☐ rail support brackets
- ☐ machine room
- ☐ electrical supply
- ☐ sill angles
- ☐ electrical connections
- ☐ pits
- ☐ roof penthouses
- ☐ pit ladders

Check the job specifications and drawings before pricing.

- Installation of elevators and handicapped lifts in historic structures can require significant additional costs. The associated structural requirements may involve cutting into and repairing finishes, moldings, flooring, etc. The estimator must account for these special conditions.

14 30 00 Escalators and Moving Walks

- Escalators and moving walks are specialty items installed by specialty contractors. There are numerous options associated with these items. For specific options, contact a manufacturer or contractor. In

a method similar to estimating dumbwaiters and elevators, you should verify the extent of general contract work and add items as necessary.

14 40 00 Lifts
14 90 00 Other Conveying Equipment

- Products such as correspondence lifts, chutes, and pneumatic tube systems, as well as other items specified in this subdivision, may require trained installers. The general contractor might not have any choice as to who will perform the installation or when it will be performed. Long lead times are often required for these products, making early decisions in scheduling necessary.

Reference Numbers

Reference numbers are shown at the beginning of some major classifications. These numbers refer to related items in the Reference Section. The reference information may be an estimating procedure, an alternate pricing method, or technical information.

Note: Not all subdivisions listed here necessarily appear. ■

Same Data. Simplified.

Enjoy the convenience and efficiency of accessing your costs anywhere:
- **Skip the multiplier** by setting your location
- **Quickly search,** edit, favorite and share costs
- **Stay on top of price changes** with automatic updates

Discover more at rsmeans.com/online

14 11 Manual Dumbwaiters

14 11 10 – Hand Operated Dumbwaiters

14 11 10.20 Manual Dumbwaiters		Crew	Daily Output	Labor-Hours	Unit	Material	2020 Bare Costs Labor	Equipment	Total	Total Incl O&P
0010	**MANUAL DUMBWAITERS**									
0020	2 stop, hand powered, up to 75 lb. capacity	2 Elev	.75	21.333	Ea.	3,450	1,825		5,275	6,600
0100	76 lb. capacity and up		.50	32	"	7,075	2,750		9,825	12,000
0300	For each additional stop, add		.75	21.333	Stop	1,125	1,825		2,950	4,050

14 12 Electric Dumbwaiters

14 12 10 – Dumbwaiters

14 12 10.10 Electric Dumbwaiters

		Crew	Daily Output	Labor-Hours	Unit	Material	2020 Bare Costs Labor	Equipment	Total	Total Incl O&P
0010	**ELECTRIC DUMBWAITERS**									
0020	2 stop, up to 75 lb. capacity	2 Elev	.13	123	Ea.	8,950	10,500		19,450	26,100
0100	76 lb. capacity and up		.11	145	"	23,600	12,400		36,000	45,000
0600	For each additional stop, add		.54	29.630	Stop	3,575	2,525		6,100	7,825

14 21 Electric Traction Elevators

14 21 13 – Electric Traction Freight Elevators

14 21 13.10 Electric Traction Freight Elevators and Options

			Crew	Daily Output	Labor-Hours	Unit	Material	2020 Bare Costs Labor	Equipment	Total	Total Incl O&P
0010	**ELECTRIC TRACTION FREIGHT ELEVATORS AND OPTIONS**	R142000-10									
0425	Electric freight, base unit, 4000 lb., 200 fpm, 4 stop, std. fin.	R142000-20	2 Elev	.05	320	Ea.	124,500	27,400		151,900	178,500
0450	For 5000 lb. capacity, add	R142000-30					6,350			6,350	6,975
0500	For 6000 lb. capacity, add	R142000-40					14,900			14,900	16,400
0525	For 7000 lb. capacity, add						20,400			20,400	22,400
0550	For 8000 lb. capacity, add						23,800			23,800	26,200
0575	For 10000 lb. capacity, add						33,200			33,200	36,500
0600	For 12000 lb. capacity, add						40,100			40,100	44,100
0625	For 16000 lb. capacity, add						48,900			48,900	53,500
0650	For 20000 lb. capacity, add						55,500			55,500	61,000
0675	For increased speed, 250 fpm, add						14,800			14,800	16,300
0700	300 fpm, geared electric, add						18,500			18,500	20,300
0725	350 fpm, geared electric, add						22,600			22,600	24,800
0750	400 fpm, geared electric, add						26,600			26,600	29,300
0775	500 fpm, gearless electric, add						36,900			36,900	40,500
0800	600 fpm, gearless electric, add						40,900			40,900	45,000
0825	700 fpm, gearless electric, add						48,400			48,400	53,500
0850	800 fpm, gearless electric, add						54,000			54,000	59,000
0875	For class "B" loading, add						3,000			3,000	3,300
0900	For class "C-1" loading, add						7,375			7,375	8,125
0925	For class "C-2" loading, add						8,675			8,675	9,550
0950	For class "C-3" loading, add						12,100			12,100	13,400
0975	For travel over 40 V.L.F., add		2 Elev	7.25	2.207	V.L.F.	620	189		809	970
1000	For number of stops over 4, add		"	.27	59.259	Stop	4,675	5,075		9,750	12,900

14 21 23 – Electric Traction Passenger Elevators

14 21 23.10 Electric Traction Passenger Elevators and Options

			Crew	Daily Output	Labor-Hours	Unit	Material	2020 Bare Costs Labor	Equipment	Total	Total Incl O&P
0010	**ELECTRIC TRACTION PASSENGER ELEVATORS AND OPTIONS**	R142000-10									
1625	Electric pass., base unit, 2,000 lb., 200 fpm, 4 stop, std. fin.	R142000-20	2 Elev	.05	320	Ea.	104,500	27,400		131,900	157,000
1650	For 2,500 lb. capacity, add	R142000-30					4,475			4,475	4,925
1675	For 3,000 lb. capacity, add	R142000-40					4,900			4,900	5,400
1700	For 3,500 lb. capacity, add						6,225			6,225	6,850
1725	For 4,000 lb. capacity, add						7,325			7,325	8,050
1750	For 4,500 lb. capacity, add						10,100			10,100	11,100

14 21 Electric Traction Elevators

14 21 23 – Electric Traction Passenger Elevators

14 21 23.10 Electric Traction Passenger Elevators and Options	Crew	Daily Output	Labor-Hours	Unit	Material	2020 Bare Costs Labor	Equipment	Total	Total Incl O&P
1775	For 5000 lb. capacity, add				Ea.	13,000			13,000	14,300
1800	For increased speed, 250 fpm, geared electric, add					3,550			3,550	3,900
1825	300 fpm, geared electric, add					6,850			6,850	7,525
1850	350 fpm, geared electric, add					8,800			8,800	9,675
1875	400 fpm, geared electric, add					12,800			12,800	14,100
1900	500 fpm, gearless electric, add					30,900			30,900	34,000
1925	600 fpm, gearless electric, add					50,000			50,000	55,000
1950	700 fpm, gearless electric, add					56,500			56,500	62,000
1975	800 fpm, gearless electric, add					62,500			62,500	69,000
2000	For travel over 40 V.L.F., add	2 Elev	7.25	2.207	V.L.F.	740	189		929	1,100
2025	For number of stops over 4, add		.27	59.259	Stop	3,225	5,075		8,300	11,300
2400	Electric hospital, base unit, 4,000 lb., 200 fpm, 4 stop, std fin.		.05	320	Ea.	97,000	27,400		124,400	148,500
2425	For 4,500 lb. capacity, add					6,200			6,200	6,825
2450	For 5,000 lb. capacity, add					8,125			8,125	8,925
2475	For increased speed, 250 fpm, geared electric, add					3,950			3,950	4,350
2500	300 fpm, geared electric, add					7,575			7,575	8,325
2525	350 fpm, geared electric, add					9,125			9,125	10,000
2550	400 fpm, geared electric, add					13,200			13,200	14,600
2575	500 fpm, gearless electric, add					36,300			36,300	40,000
2600	600 fpm, gearless electric, add					54,500			54,500	60,000
2625	700 fpm, gearless electric, add					60,500			60,500	66,500
2650	800 fpm, gearless electric, add					68,000			68,000	75,000
2675	For travel over 40 V.L.F., add	2 Elev	7.25	2.207	V.L.F.	196	189		385	505
2700	For number of stops over 4, add	"	.27	59.259	Stop	4,775	5,075		9,850	13,000

14 21 33 – Electric Traction Residential Elevators

14 21 33.20 Residential Elevators

		Crew	Daily Output	Labor-Hours	Unit	Material	2020 Bare Costs Labor	Equipment	Total	Total Incl O&P
0010	**RESIDENTIAL ELEVATORS**									
7000	Residential, cab type, 1 floor, 2 stop, economy model	2 Elev	.20	80	Ea.	10,700	6,850		17,550	22,300
7100	Custom model		.10	160		18,100	13,700		31,800	40,900
7200	2 floor, 3 stop, economy model		.12	133		15,900	11,400		27,300	35,000
7300	Custom model		.06	267		26,000	22,800		48,800	63,500

14 24 Hydraulic Elevators

14 24 13 – Hydraulic Freight Elevators

14 24 13.10 Hydraulic Freight Elevators and Options

			Crew	Daily Output	Labor-Hours	Unit	Material	2020 Bare Costs Labor	Equipment	Total	Total Incl O&P
0010	**HYDRAULIC FREIGHT ELEVATORS AND OPTIONS**	R142000-10									
1025	Hydraulic freight, base unit, 2,000 lb., 50 fpm, 2 stop, std. fin.	R142000-20	2 Elev	.10	160	Ea.	83,500	13,700		97,200	113,000
1050	For 2,500 lb. capacity, add	R142000-30					4,225			4,225	4,625
1075	For 3,000 lb. capacity, add	R142000-40					5,575			5,575	6,125
1100	For 3,500 lb. capacity, add						9,100			9,100	10,000
1125	For 4,000 lb. capacity, add						10,400			10,400	11,500
1150	For 4,500 lb. capacity, add						12,100			12,100	13,300
1175	For 5,000 lb. capacity, add						16,400			16,400	18,000
1200	For 6,000 lb. capacity, add						16,700			16,700	18,400
1225	For 7,000 lb. capacity, add						25,700			25,700	28,200
1250	For 8,000 lb. capacity, add						31,100			31,100	34,200
1275	For 10,000 lb. capacity, add						32,900			32,900	36,200
1300	For 12,000 lb. capacity, add						40,000			40,000	44,000
1325	For 16,000 lb. capacity, add						53,500			53,500	59,000
1350	For 20,000 lb. capacity, add						59,500			59,500	65,500

For customer support on your Commercial Renovation Costs with RSMeans data, call 800.448.8182.

407

14 24 Hydraulic Elevators

14 24 13 – Hydraulic Freight Elevators

14 24 13.10 Hydraulic Freight Elevators and Options		Crew	Daily Output	Labor-Hours	Unit	Material	2020 Bare Costs Labor	Equipment	Total	Total Incl O&P
1375	For increased speed, 100 fpm, add				Ea.	1,200			1,200	1,325
1400	125 fpm, add					3,150			3,150	3,450
1425	150 fpm, add					4,950			4,950	5,450
1450	175 fpm, add					7,100			7,100	7,800
1475	For class "B" loading, add					2,825			2,825	3,100
1500	For class "C-1" loading, add					7,125			7,125	7,825
1525	For class "C-2" loading, add					8,550			8,550	9,400
1550	For class "C-3" loading, add					11,700			11,700	12,900
1575	For travel over 20 V.L.F., add	2 Elev	7.25	2.207	V.L.F.	895	189		1,084	1,275
1600	For number of stops over 2, add	"	.27	59.259	Stop	2,250	5,075		7,325	10,300

14 24 23 – Hydraulic Passenger Elevators

14 24 23.10 Hydraulic Passenger Elevators and Options

		Crew	Daily Output	Labor-Hours	Unit	Material	2020 Bare Costs Labor	Equipment	Total	Total Incl O&P
0010	**HYDRAULIC PASSENGER ELEVATORS AND OPTIONS** R142000-10									
2050	Hyd. pass., base unit, 1,500 lb., 100 fpm, 2 stop, std. fin. R142000-20	2 Elev	.10	160	Ea.	40,800	13,700		54,500	66,000
2075	For 2,000 lb. capacity, add R142000-30					845			845	930
2100	For 2,500 lb. capacity, add R142000-40					3,175			3,175	3,500
2125	For 3,000 lb. capacity, add					4,550			4,550	5,025
2150	For 3,500 lb. capacity, add					7,625			7,625	8,375
2175	For 4,000 lb. capacity, add					9,175			9,175	10,100
2200	For 4,500 lb. capacity, add					12,100			12,100	13,300
2225	For 5,000 lb. capacity, add					17,000			17,000	18,700
2250	For increased speed, 125 fpm, add					1,225			1,225	1,350
2275	150 fpm, add					2,750			2,750	3,025
2300	175 fpm, add					5,350			5,350	5,875
2325	200 fpm, add					10,300			10,300	11,400
2350	For travel over 12 V.L.F., add	2 Elev	7.25	2.207	V.L.F.	730	189		919	1,100
2375	For number of stops over 2, add		.27	59.259	Stop	1,000	5,075		6,075	8,900
2725	Hydraulic hospital, base unit, 4,000 lb., 100 fpm, 2 stop, std. fin.		.10	160	Ea.	67,000	13,700		80,700	94,500
2775	For 4,500 lb. capacity, add					7,525			7,525	8,275
2800	For 5,000 lb. capacity, add					10,900			10,900	12,000
2825	For increased speed, 125 fpm, add					2,250			2,250	2,475
2850	150 fpm, add					3,750			3,750	4,125
2875	175 fpm, add					6,325			6,325	6,950
2900	200 fpm, add					9,300			9,300	10,200
2925	For travel over 12 V.L.F., add	2 Elev	7.25	2.207	V.L.F.	410	189		599	740
2950	For number of stops over 2, add	"	.27	59.259	Stop	4,675	5,075		9,750	12,900

14 27 Custom Elevator Cabs and Doors

14 27 13 – Custom Elevator Cab Finishes

14 27 13.10 Cab Finishes

		Crew	Daily Output	Labor-Hours	Unit	Material	2020 Bare Costs Labor	Equipment	Total	Total Incl O&P
0010	**CAB FINISHES**									
3325	Passenger elevator cab finishes (based on 3,500 lb. cab size)									
3350	Acrylic panel ceiling				Ea.	795			795	875
3375	Aluminum eggcrate ceiling					715			715	790
3400	Stainless steel doors					4,150			4,150	4,550
3425	Carpet flooring					635			635	695
3450	Epoxy flooring					480			480	525
3475	Quarry tile flooring					920			920	1,025
3500	Slate flooring					1,650			1,650	1,825
3525	Textured rubber flooring					665			665	735

14 27 Custom Elevator Cabs and Doors

14 27 13 – Custom Elevator Cab Finishes

14 27 13.10 Cab Finishes

		Crew	Daily Output	Labor-Hours	Unit	Material	2020 Bare Costs Labor	Equipment	Total	Total Incl O&P
3550	Stainless steel walls				Ea.	4,150			4,150	4,575
3575	Stainless steel returns at door				▼	1,200			1,200	1,300
4450	Hospital elevator cab finishes (based on 3,500 lb. cab size)									
4475	Aluminum eggcrate ceiling				Ea.	725			725	795
4500	Stainless steel doors					4,125			4,125	4,550
4525	Epoxy flooring					475			475	525
4550	Quarry tile flooring					920			920	1,025
4575	Textured rubber flooring					660			660	730
4600	Stainless steel walls					4,700			4,700	5,175
4625	Stainless steel returns at door				▼	930			930	1,025

14 28 Elevator Equipment and Controls

14 28 10 – Elevator Equipment and Control Options

14 28 10.10 Elevator Controls and Doors

		Crew	Daily Output	Labor-Hours	Unit	Material	2020 Bare Costs Labor	Equipment	Total	Total Incl O&P
0010	**ELEVATOR CONTROLS AND DOORS**									
2975	Passenger elevator options									
3000	2 car group automatic controls	2 Elev	.66	24.242	Ea.	3,625	2,075		5,700	7,150
3025	3 car group automatic controls		.44	36.364		5,525	3,100		8,625	10,900
3050	4 car group automatic controls		.33	48.485		9,650	4,150		13,800	17,000
3075	5 car group automatic controls		.26	61.538		14,500	5,275		19,775	24,100
3100	6 car group automatic controls		.22	72.727		22,100	6,225		28,325	33,900
3125	Intercom service		3	5.333		580	455		1,035	1,350
3150	Duplex car selective collective		.66	24.242		4,175	2,075		6,250	7,775
3175	Center opening 1 speed doors		2	8		2,075	685		2,760	3,350
3200	Center opening 2 speed doors		2	8		2,925	685		3,610	4,275
3225	Rear opening doors (opposite front)		2	8		4,475	685		5,160	5,975
3250	Side opening 2 speed doors		2	8		7,650	685		8,335	9,475
3275	Automatic emergency power switching		.66	24.242		1,275	2,075		3,350	4,575
3300	Manual emergency power switching	▼	8	2		540	171		711	860
3625	Hall finishes, stainless steel doors					1,475			1,475	1,625
3650	Stainless steel frames					1,550			1,550	1,700
3675	12 month maintenance contract								4,425	4,425
3700	Signal devices, hall lanterns	2 Elev	8	2		545	171		716	865
3725	Position indicators, up to 3		9.40	1.702		350	146		496	610
3750	Position indicators, per each over 3	▼	32	.500		94.50	43		137.50	170
3775	High speed heavy duty door opener					2,775			2,775	3,050
3800	Variable voltage, O.H. gearless machine, min.	2 Elev	.16	100		34,600	8,550		43,150	51,000
3815	Maximum		.07	229		77,500	19,600		97,100	115,500
3825	Basement installed geared machine	▼	.33	48.485	▼	13,600	4,150		17,750	21,400
3850	Freight elevator options									
3875	Doors, bi-parting	2 Elev	.66	24.242	Ea.	6,250	2,075		8,325	10,100
3900	Power operated door and gate	"	.66	24.242		25,100	2,075		27,175	30,800
3925	Finishes, steel plate floor					1,125			1,125	1,250
3950	14 ga. 1/4" x 4' steel plate walls					2,075			2,075	2,275
3975	12 month maintenance contract								3,400	3,400
4000	Signal devices, hall lanterns	2 Elev	8	2		530	171		701	850
4025	Position indicators, up to 3		9.40	1.702		375	146		521	640
4050	Position indicators, per each over 3		32	.500		103	43		146	179
4075	Variable voltage basement installed geared machine	▼	.66	24.242	▼	21,000	2,075		23,075	26,300
4100	Hospital elevator options									
4125	2 car group automatic controls	2 Elev	.66	24.242	Ea.	3,625	2,075		5,700	7,150

14 28 Elevator Equipment and Controls

14 28 10 – Elevator Equipment and Control Options

14 28 10.10 Elevator Controls and Doors

	Elevator Controls and Doors	Crew	Daily Output	Labor-Hours	Unit	Material	2020 Bare Costs Labor	Equipment	Total	Total Incl O&P
4150	3 car group automatic controls	2 Elev	.44	36.364	Ea.	5,500	3,100		8,600	10,800
4175	4 car group automatic controls		.33	48.485		13,100	4,150		17,250	20,800
4200	5 car group automatic controls		.26	61.538		13,100	5,275		18,375	22,500
4225	6 car group automatic controls		.22	72.727		20,000	6,225		26,225	31,600
4250	Intercom service		3	5.333		550	455		1,005	1,300
4275	Duplex car selective collective		.66	24.242		4,075	2,075		6,150	7,675
4300	Center opening 1 speed doors		2	8		2,075	685		2,760	3,325
4325	Center opening 2 speed doors		2	8		2,750	685		3,435	4,050
4350	Rear opening doors (opposite front)		2	8		4,475	685		5,160	5,975
4375	Side opening 2 speed doors		2	8		6,650	685		7,335	8,350
4400	Automatic emergency power switching		.66	24.242		1,250	2,075		3,325	4,550
4425	Manual emergency power switching	↓	8	2		525	171		696	840
4675	Hall finishes, stainless steel doors					1,550			1,550	1,700
4700	Stainless steel frames					1,550			1,550	1,725
4725	12 month maintenance contract								6,125	6,125
4750	Signal devices, hall lanterns	2 Elev	8	2		520	171		691	835
4775	Position indicators, up to 3		9.40	1.702		355	146		501	615
4800	Position indicators, per each over 3	↓	32	.500		95	43		138	171
4825	High speed heavy duty door opener					2,800			2,800	3,075
4850	Variable voltage, O.H. gearless machine, min.	2 Elev	.16	100		36,400	8,550		44,950	53,000
4865	Maximum	↓	.07	229		79,000	19,600		98,600	117,000
4875	Basement installed geared machine		.33	48.485	↓	18,200	4,150		22,350	26,500
5000	Drilling for piston, casing included, 18" diameter	B-48	80	.700	V.L.F.	65.50	33.50	13.10	112.10	139

14 42 Wheelchair Lifts

14 42 13 – Inclined Wheelchair Lifts

14 42 13.10 Inclined Wheelchair Lifts and Stairclimbers

	INCLINED WHEELCHAIR LIFTS AND STAIRCLIMBERS	Crew	Daily Output	Labor-Hours	Unit	Material	2020 Bare Costs Labor	Equipment	Total	Total Incl O&P
0010	**INCLINED WHEELCHAIR LIFTS AND STAIRCLIMBERS**									
7700	Stair climber (chair lift), single seat, minimum	2 Elev	1	16	Ea.	4,975	1,375		6,350	7,575
7800	Maximum		.20	80		6,825	6,850		13,675	18,000
8700	Stair lift, minimum		1	16		13,500	1,375		14,875	17,000
8900	Maximum	↓	.20	80	↓	21,400	6,850		28,250	34,000

14 42 16 – Vertical Wheelchair Lifts

14 42 16.10 Wheelchair Lifts

	WHEELCHAIR LIFTS	Crew	Daily Output	Labor-Hours	Unit	Material	2020 Bare Costs Labor	Equipment	Total	Total Incl O&P
0010	**WHEELCHAIR LIFTS**									
8000	Wheelchair lift, minimum	2 Elev	1	16	Ea.	6,825	1,375		8,200	9,600
8500	Maximum	"	.50	32	"	16,100	2,750		18,850	22,000

14 45 Vehicle Lifts

14 45 10 – Hydraulic Vehicle Lifts

14 45 10.10 Hydraulic Lifts

	HYDRAULIC LIFTS	Crew	Daily Output	Labor-Hours	Unit	Material	2020 Bare Costs Labor	Equipment	Total	Total Incl O&P
0010	**HYDRAULIC LIFTS**									
2200	Single post, 8,000 lb. capacity	L-4	.40	60	Ea.	7,100	3,000		10,100	12,600
2810	Double post, 6,000 lb. capacity		2.67	8.989		9,775	450		10,225	11,500
2815	9,000 lb. capacity		2.29	10.480		23,200	525		23,725	26,300
2820	15,000 lb. capacity		2	12		26,200	600		26,800	29,900
2822	Four post, 26,000 lb. capacity		1.80	13.333		15,100	665		15,765	17,700
2825	30,000 lb. capacity		1.60	15		58,000	750		58,750	65,000
2830	Ramp style, 4 post, 25,000 lb. capacity	↓	2	12		22,700	600		23,300	26,000

14 45 Vehicle Lifts

14 45 10 - Hydraulic Vehicle Lifts

14 45 10.10 Hydraulic Lifts

		Crew	Daily Output	Labor-Hours	Unit	Material	2020 Bare Costs Labor	Equipment	Total	Total Incl O&P
2835	35,000 lb. capacity	L-4	1	24	Ea.	106,000	1,200		107,200	118,500
2840	50,000 lb. capacity		1	24		121,000	1,200		122,200	135,500
2845	75,000 lb. capacity		1	24		141,000	1,200		142,200	157,000
2850	For drive thru tracks, add, minimum					1,150			1,150	1,275
2855	Maximum					1,975			1,975	2,175
2860	Ramp extensions, 3' (set of 2)					1,200			1,200	1,325
2865	Rolling jack platform					4,175			4,175	4,575
2870	Electric/hydraulic jacking beam					11,400			11,400	12,600
2880	Scissor lift, portable, 6,000 lb. capacity					10,900			10,900	12,000

14 91 Facility Chutes

14 91 33 - Laundry and Linen Chutes

14 91 33.10 Chutes

		Crew	Daily Output	Labor-Hours	Unit	Material	2020 Bare Costs Labor	Equipment	Total	Total Incl O&P
0011	**CHUTES**, linen, trash or refuse									
0050	Aluminized steel, 16 ga., 18" diameter	2 Shee	3.50	4.571	Floor	1,800	285		2,085	2,450
0100	24" diameter		3.20	5		2,000	310		2,310	2,700
0400	Galvanized steel, 16 ga., 18" diameter		3.50	4.571		1,075	285		1,360	1,650
0500	24" diameter		3.20	5		1,225	310		1,535	1,850
0800	Stainless steel, 18" diameter		3.50	4.571		3,375	285		3,660	4,175
0900	24" diameter		3.20	5		3,250	310		3,560	4,075

14 91 82 - Trash Chutes

14 91 82.10 Trash Chutes and Accessories

		Crew	Daily Output	Labor-Hours	Unit	Material	2020 Bare Costs Labor	Equipment	Total	Total Incl O&P
0010	**TRASH CHUTES AND ACCESSORIES**									
2900	Package chutes, spiral type, minimum	2 Shee	4.50	3.556	Floor	2,575	222		2,797	3,200
3000	Maximum	"	1.50	10.667	"	6,750	665		7,415	8,475
9000	Minimum labor/equipment charge	1 Shee	1	8	Job		500		500	780

14 92 Pneumatic Tube Systems

14 92 10 - Conventional, Automatic and Computer Controlled Pneumatic Tube Systems

14 92 10.10 Pneumatic Tube Systems

		Crew	Daily Output	Labor-Hours	Unit	Material	2020 Bare Costs Labor	Equipment	Total	Total Incl O&P
0010	**PNEUMATIC TUBE SYSTEMS**									
0020	100' long, single tube, 2 stations, stock									
0100	3" diameter	2 Stpi	.12	133	Total	3,675	8,750		12,425	17,600
0300	4" diameter	"	.09	178	"	4,600	11,700		16,300	23,100
0400	Twin tube, two stations or more, conventional system									
0700	3" round	2 Stpi	46	.348	L.F.	40	23		63	79.50
1050	Add for blower		2	8	System	5,200	525		5,725	6,525
1110	Plus for each round station, add		7.50	2.133	Ea.	1,375	140		1,515	1,725
1200	Alternate pricing method: base cost, economy model		.75	21.333	Total	5,850	1,400		7,250	8,575
1300	Custom model		.25	64	"	11,700	4,200		15,900	19,300
1500	Plus total system length, add, for economy model		93.40	.171	L.F.	8.25	11.25		19.50	26.50
1600	For custom model		37.60	.426	"	25	28		53	70.50

Division Notes

	CREW	DAILY OUTPUT	LABOR-HOURS	UNIT	BARE COSTS				TOTAL INCL O&P
					MAT.	LABOR	EQUIP.	TOTAL	

Estimating Tips

Pipe for fire protection and all uses is located in Subdivisions 21 11 13 and 22 11 13.

The labor adjustment factors listed in Subdivision 22 01 02.20 also apply to Division 21.

Many, but not all, areas in the U.S. require backflow protection in the fire system. Insurance underwriters may have specific requirements for the type of materials to be installed or design requirements based on the hazard to be protected. Local jurisdictions may have requirements not covered by code. It is advisable to be aware of any special conditions.

For your reference, the following is a list of the most applicable Fire Codes and Standards, which may be purchased from the NFPA, 1 Batterymarch Park, Quincy, MA 02169-7471.

- NFPA 1: Uniform Fire Code
- NFPA 10: Portable Fire Extinguishers
- NFPA 11: Low-, Medium-, and High-Expansion Foam
- NFPA 12: Carbon Dioxide Extinguishing Systems (Also companion 12A)
- NFPA 13: Installation of Sprinkler Systems (Also companion 13D, 13E, and 13R)
- NFPA 14: Installation of Standpipe and Hose Systems
- NFPA 15: Water Spray Fixed Systems for Fire Protection
- NFPA 16: Installation of Foam-Water Sprinkler and Foam-Water Spray Systems
- NFPA 17: Dry Chemical Extinguishing Systems (Also companion 17A)
- NFPA 18: Wetting Agents
- NFPA 20: Installation of Stationary Pumps for Fire Protection
- NFPA 22: Water Tanks for Private Fire Protection
- NFPA 24: Installation of Private Fire Service Mains and their Appurtenances
- NFPA 25: Inspection, Testing and Maintenance of Water-Based Fire Protection

Reference Numbers

Reference numbers are shown at the beginning of some major classifications. These numbers refer to related items in the Reference Section. The reference information may be an estimating procedure, an alternate pricing method, or technical information.

Same Data. Simplified.

Enjoy the convenience and efficiency of accessing your costs anywhere:

- **Skip the multiplier** by setting your location
- **Quickly search,** edit, favorite and share costs
- **Stay on top of price changes** with automatic updates

Discover more at rsmeans.com/online

21 05 Common Work Results for Fire Suppression

21 05 23 – General-Duty Valves for Water-Based Fire-Suppression Piping

21 05 23.50 General-Duty Valves	Crew	Daily Output	Labor-Hours	Unit	Material	2020 Bare Costs Labor	Equipment	Total	Total Incl O&P
0010 **GENERAL-DUTY VALVES,** for water-based fire suppression									
6500 Check, swing, C.I. body, brass fittings, auto. ball drip									
6520 4" size	Q-12	3	5.333	Ea.	420	305		725	930
9990 Minimum labor/equipment charge	1 Spri	3	2.667	Job		169		169	261

21 11 Facility Fire-Suppression Water-Service Piping

21 11 16 – Facility Fire Hydrants

21 11 16.50 Fire Hydrants for Buildings

	Crew	Daily Output	Labor-Hours	Unit	Material	Labor	Equipment	Total	Total Incl O&P
0010 **FIRE HYDRANTS FOR BUILDINGS**									
3750 Hydrants, wall, w/caps, single, flush, polished brass									
3800 2-1/2" x 2-1/2"	Q-12	5	3.200	Ea.	273	182		455	580
4350 Double, projecting, polished brass									
4400 2-1/2" x 2-1/2" x 4"	Q-12	5	3.200	Ea.	290	182		472	600

21 11 19 – Fire-Department Connections

21 11 19.50 Connections for the Fire-Department

	Crew	Daily Output	Labor-Hours	Unit	Material	Labor	Equipment	Total	Total Incl O&P
0010 **CONNECTIONS FOR THE FIRE-DEPARTMENT**									
7140 Standpipe connections, wall, w/plugs & chains									
7280 Double, flush, polished brass									
7300 2-1/2" x 2-1/2" x 4"	Q-12	5	3.200	Ea.	775	182		957	1,125
9990 Minimum labor/equipment charge	1 Spri	4	2	Job		127		127	196

21 12 Fire-Suppression Standpipes

21 12 13 – Fire-Suppression Hoses and Nozzles

21 12 13.50 Fire Hoses and Nozzles

	Crew	Daily Output	Labor-Hours	Unit	Material	Labor	Equipment	Total	Total Incl O&P
0010 **FIRE HOSES AND NOZZLES**									
0200 Adapters, rough brass, straight hose threads									
0220 One piece, female to male, rocker lugs									
0240 1" x 1"				Ea.	56.50			56.50	62
0320 2" x 2"					42.50			42.50	46.50
0380 2-1/2" x 2-1/2"					19.55			19.55	21.50
2200 Hose, less couplings									
2260 Synthetic jacket, lined, 300 lb. test, 1-1/2" diameter	Q-12	2600	.006	L.F.	3.31	.35		3.66	4.18
2270 2" diameter		2200	.007		2.59	.41		3	3.49
2280 2-1/2" diameter		2200	.007		5.90	.41		6.31	7.15
2290 3" diameter		2200	.007		3.27	.41		3.68	4.24
2360 High strength, 500 lb. test, 1-1/2" diameter		2600	.006		2.59	.35		2.94	3.39
2380 2-1/2" diameter		2200	.007		6.05	.41		6.46	7.30
5600 Nozzles, brass									
5620 Adjustable fog, 3/4" booster line				Ea.	150			150	165
5640 1-1/2" leader line					117			117	128
5660 2-1/2" direct connection					199			199	219
5680 2-1/2" playpipe nozzle					238			238	262
5780 For chrome plated, add					8%				
9900 Minimum labor/equipment charge	1 Plum	2	4	Job		258		258	400

21 12 Fire-Suppression Standpipes

21 12 19 – Fire-Suppression Hose Racks

21 12 19.50 Fire Hose Racks	Crew	Daily Output	Labor-Hours	Unit	Material	2020 Bare Costs Labor	2020 Bare Costs Equipment	Total	Total Incl O&P
0010 **FIRE HOSE RACKS**									
2600　Hose rack, swinging, for 1-1/2" diameter hose,									
2620　　Enameled steel, 50' and 75' lengths of hose	Q-12	20	.800	Ea.	72	45.50		117.50	150

21 12 23 – Fire-Suppression Hose Valves

21 12 23.70 Fire Hose Valves

	Crew	Daily Output	Labor-Hours	Unit	Material	2020 Bare Costs Labor	2020 Bare Costs Equipment	Total	Total Incl O&P
0010 **FIRE HOSE VALVES**									
3000　Gate, hose, wheel handle, N.R.S., rough brass, 1-1/2"	1 Spri	12	.667	Ea.	166	42		208	248
3040　　2-1/2", 300 lb.	"	7	1.143		217	72.50		289.50	350
3080　　For polished brass, add					40%				
3090　　For polished chrome, add				▼	50%				

21 13 Fire-Suppression Sprinkler Systems

21 13 13 – Wet-Pipe Sprinkler Systems

21 13 13.50 Wet-Pipe Sprinkler System Components

	Crew	Daily Output	Labor-Hours	Unit	Material	2020 Bare Costs Labor	2020 Bare Costs Equipment	Total	Total Incl O&P
0010 **WET-PIPE SPRINKLER SYSTEM COMPONENTS**									
1220　Water motor gong	1 Spri	4	2	Ea.	480	127		607	725
1800　Firecycle system, controls, includes panel,									
1900　Flexible sprinkler head connectors									
1910　　Braided stainless steel hose with mounting bracket									
1920　　　1/2" and 3/4" outlet size									
1940　　　　40" length	1 Spri	30	.267	Ea.	73	16.85		89.85	106
1960　　　　60" length	"	22	.364	"	85	23		108	129
1982　　May replace hard-pipe armovers									
1984　　for wet and pre-action systems.									
2600　Sprinkler heads, not including supply piping									
3700　　Standard spray, pendent or upright, brass, 135°F to 286°F									
3740　　　1/2" NPT,K5.6	1 Spri	16	.500	Ea.	11.20	31.50		42.70	61.50
3790　For riser & feeder piping, see Section 22 11 13.00									

21 13 16 – Dry-Pipe Sprinkler Systems

21 13 16.50 Dry-Pipe Sprinkler System Components

	Crew	Daily Output	Labor-Hours	Unit	Material	2020 Bare Costs Labor	2020 Bare Costs Equipment	Total	Total Incl O&P
0010 **DRY-PIPE SPRINKLER SYSTEM COMPONENTS**									
0600　Accelerator	1 Spri	8	1	Ea.	900	63.50		963.50	1,100
0800　Air compressor for dry pipe system, automatic, complete									
0820　　30 gal. system capacity, 3/4 HP	1 Spri	1.30	6.154	Ea.	1,200	390		1,590	1,925
2600　Sprinkler heads, not including supply piping									
2640　　Dry, pendent, 1/2" orifice, 3/4" or 1" NPT									
2660　　　3" to 6" length	1 Spri	14	.571	Ea.	145	36		181	216
2700　　　15-1/4" to 18" length	"	14	.571	"	170	36		206	243
6330　Valves and components									
8200　　Dry pipe valve, incl. trim and gauges, 3" size	Q-12	2	8	Ea.	3,075	455		3,530	4,075
8220　　　4" size	"	1	16	"	3,300	910		4,210	5,025

21 13 39 – Foam-Water Systems

21 13 39.50 Foam-Water System Components

	Crew	Daily Output	Labor-Hours	Unit	Material	2020 Bare Costs Labor	2020 Bare Costs Equipment	Total	Total Incl O&P
0010 **FOAM-WATER SYSTEM COMPONENTS**									
2600　Sprinkler heads, not including supply piping									
3600　　Foam-water, pendent or upright, 1/2" NPT	1 Spri	12	.667	Ea.	237	42		279	325

For customer support on your Commercial Renovation Costs with RSMeans data, call 800.448.8182.

415

Division Notes

	CREW	DAILY OUTPUT	LABOR-HOURS	UNIT	BARE COSTS				TOTAL INCL O&P
					MAT.	LABOR	EQUIP.	TOTAL	

Estimating Tips
22 10 00 Plumbing Piping and Pumps

This subdivision is primarily basic pipe and related materials. The pipe may be used by any of the mechanical disciplines, i.e., plumbing, fire protection, heating, and air conditioning.

Note: CPVC plastic piping approved for fire protection is located in 21 11 13.

- The labor adjustment factors listed in Subdivision 22 01 02.20 apply throughout Divisions 21, 22, and 23. CAUTION: the correct percentage may vary for the same items. For example, the percentage add for the basic pipe installation should be based on the maximum height that the installer must install for that particular section. If the pipe is to be located 14' above the floor but it is suspended on threaded rod from beams, the bottom flange of which is 18' high (4' rods), then the height is actually 18' and the add is 20%. The pipe cover, however, does not have to go above the 14' and so the add should be 10%.

- Most pipe is priced first as straight pipe with a joint (coupling, weld, etc.) every 10' and a hanger usually every 10'. There are exceptions with hanger spacing such as for cast iron pipe (5')

and plastic pipe (3 per 10'). Following each type of pipe there are several lines listing sizes and the amount to be subtracted to delete couplings and hangers. This is for pipe that is to be buried or supported together on trapeze hangers. The reason that the couplings are deleted is that these runs are usually long, and frequently longer lengths of pipe are used. By deleting the couplings, the estimator is expected to look up and add back the correct reduced number of couplings.

- When preparing an estimate, it may be necessary to approximate the fittings. Fittings usually run between 25% and 50% of the cost of the pipe. The lower percentage is for simpler runs, and the higher number is for complex areas, such as mechanical rooms.

- For historic restoration projects, the systems must be as invisible as possible, and pathways must be sought for pipes, conduit, and ductwork. While installations in accessible spaces (such as basements and attics) are relatively straightforward to estimate, labor costs may be more difficult to determine when delivery systems must be concealed.

22 40 00 Plumbing Fixtures

- Plumbing fixture costs usually require two lines: the fixture itself and its "rough-in, supply, and waste."

- In the Assemblies Section (Plumbing D2010) for the desired fixture, the System Components Group at the center of the page shows the fixture on the first line. The rest of the list (fittings, pipe, tubing, etc.) will total up to what we refer to in the Unit Price section as "Rough-in, supply, waste, and vent." Note that for most fixtures we allow a nominal 5' of tubing to reach from the fixture to a main or riser.

- Remember that gas- and oil-fired units need venting.

Reference Numbers

Reference numbers are shown at the beginning of some major classifications. These numbers refer to related items in the Reference Section. The reference information may be an estimating procedure, an alternate pricing method, or technical information.

Note: Not all subdivisions listed here necessarily appear. ∎

Same Data. Simplified.

Enjoy the convenience and efficiency of accessing your costs anywhere:

- **Skip the multiplier** by setting your location
- **Quickly search**, edit, favorite and share costs
- **Stay on top of price changes** with automatic updates

Discover more at rsmeans.com/online

---

22 01 02.20 Labor Adjustment Factors		Crew	Daily Output	Labor-Hours	Unit	Material	2020 Bare Costs Labor	Equipment	Total	Total Incl O&P
0010	**LABOR ADJUSTMENT FACTORS** (For Div. 21, 22 and 23) R220102-20									
0100	Labor factors: The below are reasonable suggestions, but									
0110	each project must be evaluated for its own peculiarities, and									
0120	the adjustments be increased or decreased depending on the									
0130	severity of the special conditions.									
1000	Add to labor for elevated installation (Above floor level)									
1080	10' to 14.5' high						10%			
1100	15' to 19.5' high						20%			
1120	20' to 24.5' high						25%			
1140	25' to 29.5' high						35%			
1160	30' to 34.5' high						40%			
1180	35' to 39.5' high						50%			
1200	40' and higher						55%			
2000	Add to labor for crawl space									
2100	3' high						40%			
2140	4' high						30%			
3000	Add to labor for multi-story building									
3010	For new construction (No elevator available)									
3100	Add for floors 3 thru 10						5%			
3110	Add for floors 11 thru 15						10%			
3120	Add for floors 16 thru 20						15%			
3130	Add for floors 21 thru 30						20%			
3140	Add for floors 31 and up						30%			
3170	For existing structure (Elevator available)									
3180	Add for work on floor 3 and above						2%			
4000	Add to labor for working in existing occupied buildings									
4100	Hospital						35%			
4140	Office building						25%			
4180	School						20%			
4220	Factory or warehouse						15%			
4260	Multi dwelling						15%			
5000	Add to labor, miscellaneous									
5100	Cramped shaft						35%			
5140	Congested area						15%			
5180	Excessive heat or cold						30%			
9000	Labor factors: The above are reasonable suggestions, but									
9010	each project should be evaluated for its own peculiarities.									
9100	Other factors to be considered are:									
9140	Movement of material and equipment through finished areas									
9180	Equipment room									
9220	Attic space									
9260	No service road									
9300	Poor unloading/storage area									
9340	Congested site area/heavy traffic									

22 05 05 – Selective Demolition for Plumbing

22 05 05.10 Plumbing Demolition

22 05 05.10 Plumbing Demolition		Crew	Daily Output	Labor-Hours	Unit	Material	2020 Bare Costs Labor	Equipment	Total	Total Incl O&P
0010	**PLUMBING DEMOLITION** R220105-10									
1020	Fixtures, including 10' piping									
1100	Bathtubs, cast iron R024119-10	1 Plum	4	2	Ea.		129		129	199
1120	Fiberglass		6	1.333			86		86	133
1140	Steel		5	1.600			103		103	159
1200	Lavatory, wall hung		10	.800			51.50		51.50	79.50
1220	Counter top		8	1			64.50		64.50	99.50
1300	Sink, single compartment		8	1			64.50		64.50	99.50
1320	Double compartment		7	1.143			73.50		73.50	114
1400	Water closet, floor mounted		8	1			64.50		64.50	99.50
1420	Wall mounted		7	1.143			73.50		73.50	114
1500	Urinal, floor mounted		4	2			129		129	199
1520	Wall mounted		7	1.143			73.50		73.50	114
1600	Water fountains, free standing		8	1			64.50		64.50	99.50
1620	Wall or deck mounted		6	1.333			86		86	133
2000	Piping, metal, up thru 1-1/2" diameter		200	.040	L.F.		2.58		2.58	3.99
2050	2" thru 3-1/2" diameter		150	.053			3.44		3.44	5.30
2100	4" thru 6" diameter	2 Plum	100	.160			10.30		10.30	15.95
2150	8" thru 14" diameter	"	60	.267			17.20		17.20	26.50
2153	16" thru 20" diameter	Q-18	70	.343			21	1.52	22.52	34
2155	24" thru 26" diameter		55	.436			26.50	1.93	28.43	43.50
2156	30" thru 36" diameter		40	.600			36.50	2.66	39.16	60
2160	Plastic pipe with fittings, up thru 1-1/2" diameter	1 Plum	250	.032			2.06		2.06	3.19
2162	2" thru 3" diameter	"	200	.040			2.58		2.58	3.99
2164	4" thru 6" diameter	Q-1	200	.080			4.64		4.64	7.15
2166	8" thru 14" diameter		150	.107			6.20		6.20	9.55
2168	16" diameter		100	.160			9.30		9.30	14.35
2250	Water heater, 40 gal.	1 Plum	6	1.333	Ea.		86		86	133
6000	Remove and reset fixtures, easy access		6	1.333			86		86	133
6100	Difficult access		4	2			129		129	199
9000	Minimum labor/equipment charge		2	4	Job		258		258	400

22 05 23 – General-Duty Valves for Plumbing Piping

22 05 23.20 Valves, Bronze

22 05 23.20 Valves, Bronze		Crew	Daily Output	Labor-Hours	Unit	Material	2020 Bare Costs Labor	Equipment	Total	Total Incl O&P
0010	**VALVES, BRONZE**									
1020	Angle, 150 lb., rising stem, threaded									
1070	3/4"	1 Plum	20	.400	Ea.	230	26		256	292
1080	1"		19	.421		315	27		342	385
1100	1-1/2"		13	.615		525	39.50		564.50	635
1102	Soldered same price as threaded									
1110	2"	1 Plum	11	.727	Ea.	845	47		892	1,000
1300	Ball									
1398	Threaded, 150 psi									
1460	3/4"	1 Plum	20	.400	Ea.	35.50	26		61.50	79
1522	Solder the same price as threaded									
1750	Check, swing, class 150, regrinding disc, threaded									
1850	1/2"	1 Plum	24	.333	Ea.	77	21.50		98.50	118
1860	3/4"		20	.400		94	26		120	143
1870	1"		19	.421		178	27		205	238
1890	1-1/2"		13	.615		300	39.50		339.50	390
1900	2"		11	.727		380	47		427	495
2850	Gate, N.R.S., soldered, 125 psi									
2920	1/2"	1 Plum	24	.333	Ea.	60.50	21.50		82	99.50

22 05 23.20 Valves, Bronze

		Crew	Daily Output	Labor-Hours	Unit	Material	2020 Bare Costs Labor	2020 Bare Costs Equipment	Total	Total Incl O&P
2940	3/4"	1 Plum	20	.400	Ea.	74	26		100	122
2950	1"		19	.421		78.50	27		105.50	129
2970	1-1/2"		13	.615		184	39.50		223.50	265
2980	2"		11	.727		213	47		260	305
4850	Globe, class 150, rising stem, threaded									
4950	1/2"	1 Plum	24	.333	Ea.	106	21.50		127.50	150
4960	3/4"	"	20	.400	"	172	26		198	229
5600	Relief, pressure & temperature, self-closing, ASME, threaded									
5640	3/4"	1 Plum	28	.286	Ea.	257	18.40		275.40	310
5650	1"		24	.333		425	21.50		446.50	500
5660	1-1/4"		20	.400		805	26		831	925
5670	1-1/2"		18	.444		1,550	28.50		1,578.50	1,750
5680	2"		16	.500		1,675	32		1,707	1,900
6400	Pressure, water, ASME, threaded									
6440	3/4"	1 Plum	28	.286	Ea.	119	18.40		137.40	160
6450	1"		24	.333		340	21.50		361.50	410
6470	1-1/2"		18	.444		750	28.50		778.50	865
6900	Reducing, water pressure									
6920	300 psi to 25-75 psi, threaded or sweat									
6940	1/2"	1 Plum	24	.333	Ea.	470	21.50		491.50	555
6950	3/4"		20	.400		550	26		576	645
6960	1"		19	.421		850	27		877	975
8350	Tempering, water, sweat connections									
8400	1/2"	1 Plum	24	.333	Ea.	118	21.50		139.50	163
8440	3/4"	"	20	.400	"	167	26		193	223
8650	Threaded connections									
8700	1/2"	1 Plum	24	.333	Ea.	162	21.50		183.50	211
8740	3/4"	"	20	.400	"	1,050	26		1,076	1,225
8800	Water heater water & gas safety shut off									
8810	Protection against a leaking water heater									
8814	Shut off valve	1 Plum	16	.500	Ea.	196	32		228	266
8818	Water heater dam		32	.250		32.50	16.10		48.60	60.50
8822	Gas control wiring harness		32	.250		24.50	16.10		40.60	52
8830	Whole house flood safety shut off									
8834	Connections									
8838	3/4" NPT	1 Plum	12	.667	Ea.	1,025	43		1,068	1,200
8842	1" NPT		11	.727		1,100	47		1,147	1,275
8846	1-1/4" NPT		10	.800		1,100	51.50		1,151.50	1,275
9000	Minimum labor/equipment charge		4	2	Job		129		129	199

22 05 23.60 Valves, Plastic

		Crew	Daily Output	Labor-Hours	Unit	Material	2020 Bare Costs Labor	2020 Bare Costs Equipment	Total	Total Incl O&P
0010	**VALVES, PLASTIC**									
1150	Ball, PVC, socket or threaded, true union									
1230	1/2"	1 Plum	26	.308	Ea.	39	19.85		58.85	73.50
1240	3/4"		25	.320		50	20.50		70.50	87
1250	1"		23	.348		55.50	22.50		78	95.50
1260	1-1/4"		21	.381		94	24.50		118.50	141
1270	1-1/2"		20	.400		86	26		112	135
1280	2"		17	.471		136	30.50		166.50	196
1650	CPVC, socket or threaded, single union									
1700	1/2"	1 Plum	26	.308	Ea.	61.50	19.85		81.35	98
1720	3/4"		25	.320		81	20.50		101.50	121
1730	1"		23	.348		92.50	22.50		115	137

420

For customer support on your Commercial Renovation Costs with RSMeans data, call 800.448.8182.

22 05 Common Work Results for Plumbing

22 05 23 – General-Duty Valves for Plumbing Piping

22 05 23.60 Valves, Plastic		Crew	Daily Output	Labor-Hours	Unit	Material	2020 Bare Costs Labor	2020 Bare Costs Equipment	Total	Total Incl O&P
1750	1-1/4"	1 Plum	21	.381	Ea.	144	24.50		168.50	197
1760	1-1/2"	↓	20	.400	↓	146	26		172	201
1770	2"	↓	17	.471	↓	202	30.50		232.50	269
3150	Ball check, PVC, socket or threaded									
3290	2"	1 Plum	17	.471	Ea.	149	30.50		179.50	210
9000	Minimum labor/equipment charge	"	3.50	2.286	Job		147		147	228

22 05 76 – Facility Drainage Piping Cleanouts

22 05 76.10 Cleanouts

		Crew	Daily Output	Labor-Hours	Unit	Material	Labor	Equipment	Total	Total Incl O&P
0010	**CLEANOUTS**									
0060	Floor type									
0080	Round or square, scoriated nickel bronze top									
0100	2" pipe size	1 Plum	10	.800	Ea.	183	51.50		234.50	282
0120	3" pipe size		8	1		264	64.50		328.50	390
0140	4" pipe size	↓	6	1.333	↓	284	86		370	445
0980	Round top, recessed for terrazzo									
1000	2" pipe size	1 Plum	9	.889	Ea.	545	57.50		602.50	690
1080	3" pipe size		6	1.333		620	86		706	815
1100	4" pipe size	↓	4	2		710	129		839	980
1120	5" pipe size	Q-1	6	2.667	↓	1,175	155		1,330	1,550
9000	Minimum labor/equipment charge	1 Plum	3	2.667	Job		172		172	266

22 05 76.20 Cleanout Tees

		Crew	Daily Output	Labor-Hours	Unit	Material	Labor	Equipment	Total	Total Incl O&P
0010	**CLEANOUT TEES**									
0100	Cast iron, B&S, with countersunk plug									
0220	3" pipe size	1 Plum	3.60	2.222	Ea.	181	143		324	420
0240	4" pipe size	"	3.30	2.424	"	275	156		431	540
0500	For round smooth access cover, same price									
4000	Plastic, tees and adapters. Add plugs									
4010	ABS, DWV									
4020	Cleanout tee, 1-1/2" pipe size	1 Plum	15	.533	Ea.	14.15	34.50		48.65	68.50
9000	Minimum labor/equipment charge	"	2.75	2.909	Job		187		187	290

22 07 Plumbing Insulation

22 07 16 – Plumbing Equipment Insulation

22 07 16.10 Insulation for Plumbing Equipment

			Crew	Daily Output	Labor-Hours	Unit	Material	Labor	Equipment	Total	Total Incl O&P
0010	**INSULATION FOR PLUMBING EQUIPMENT**										
2900	Domestic water heater wrap kit										
2920	1-1/2" with vinyl jacket, 20 to 60 gal.	G	1 Plum	8	1	Ea.	16.70	64.50		81.20	118
2925	50 to 80 gal.	G	"	8	1	"	28	64.50		92.50	131

22 07 19 – Plumbing Piping Insulation

22 07 19.10 Piping Insulation

			Crew	Daily Output	Labor-Hours	Unit	Material	Labor	Equipment	Total	Total Incl O&P
0010	**PIPING INSULATION**										
0110	Insulation req'd. is based on the surface size/area to be covered										
0230	Insulated protectors (ADA)										
0235	For exposed piping under sinks or lavatories	♿									
0240	Vinyl coated foam, velcro tabs										
0245	P Trap, 1-1/4" or 1-1/2"		1 Plum	32	.250	Ea.	16.95	16.10		33.05	43.50
0260	Valve and supply cover										
0265	1/2", 3/8", and 7/16" pipe size		1 Plum	32	.250	Ea.	16.40	16.10		32.50	43
0280	Tailpiece offset (wheelchair)	♿									
0285	1-1/4" pipe size		1 Plum	32	.250	Ea.	13.15	16.10		29.25	39.50

22 07 19 – Plumbing Piping Insulation

22 07 19.10 Piping Insulation		Crew	Daily Output	Labor-Hours	Unit	Material	2020 Bare Costs Labor	Equipment	Total	Total Incl O&P
0600	Pipe covering (price copper tube one size less than IPS)									
6120	2" wall, 1/2" iron pipe size	G	Q-14	145	.110	L.F.	6.55	5.80		12.35	16.45
6210	4" iron pipe size	G	"	125	.128	"	11.90	6.75		18.65	24
6600	Fiberglass, with all service jacket									
6840	1" wall, 1/2" iron pipe size	G	Q-14	240	.067	L.F.	.92	3.52		4.44	6.60
6860	3/4" iron pipe size	G		230	.070		.98	3.67		4.65	6.90
6870	1" iron pipe size	G		220	.073		1.08	3.84		4.92	7.30
6880	1-1/4" iron pipe size	G		210	.076		1.16	4.02		5.18	7.70
6890	1-1/2" iron pipe size	G		210	.076		1.25	4.02		5.27	7.80
6900	2" iron pipe size	G		200	.080		1.78	4.22		6	8.65
7080	1-1/2" wall, 1/2" iron pipe size	G		230	.070		2.06	3.67		5.73	8.05
7140	2" iron pipe size	G		190	.084		2.79	4.44		7.23	10.10
7160	3" iron pipe size	G		170	.094		3.12	4.97		8.09	11.35
7879	Rubber tubing, flexible closed cell foam									
8100	1/2" wall, 1/4" iron pipe size	G	1 Asbe	90	.089	L.F.	.99	5.20		6.19	9.35
8120	3/8" iron pipe size	G		90	.089		1.05	5.20		6.25	9.40
8130	1/2" iron pipe size	G		89	.090		1.06	5.25		6.31	9.50
8140	3/4" iron pipe size	G		89	.090		1.22	5.25		6.47	9.70
8150	1" iron pipe size	G		88	.091		.86	5.35		6.21	9.40
8170	1-1/2" iron pipe size	G		87	.092		1.89	5.40		7.29	10.65
8180	2" iron pipe size	G		86	.093		2.38	5.45		7.83	11.25
8200	3" iron pipe size	G		85	.094		2.78	5.50		8.28	11.80
8220	4" iron pipe size	G		80	.100		4.12	5.85		9.97	13.85
8300	3/4" wall, 1/4" iron pipe size	G		90	.089		.94	5.20		6.14	9.30
8330	1/2" iron pipe size	G		89	.090		1.13	5.25		6.38	9.60
8340	3/4" iron pipe size	G		89	.090		2.13	5.25		7.38	10.70
8350	1" iron pipe size	G		88	.091		2.10	5.35		7.45	10.75
8360	1-1/4" iron pipe size	G		87	.092		2.46	5.40		7.86	11.25
8370	1-1/2" iron pipe size	G		87	.092		3.07	5.40		8.47	11.95
8380	2" iron pipe size	G		86	.093		4.27	5.45		9.72	13.35
8444	1" wall, 1/2" iron pipe size	G		86	.093		3.18	5.45		8.63	12.15
8445	3/4" iron pipe size	G		84	.095		3.90	5.60		9.50	13.15
8446	1" iron pipe size	G		84	.095		3.49	5.60		9.09	12.70
8447	1-1/4" iron pipe size	G		82	.098		3.83	5.70		9.53	13.25
8448	1-1/2" iron pipe size	G		82	.098		6.15	5.70		11.85	15.80
8449	2" iron pipe size	G		80	.100		7.55	5.85		13.40	17.60
8450	2-1/2" iron pipe size	G		80	.100		8.45	5.85		14.30	18.60
8456	Rubber insulation tape, 1/8" x 2" x 30'	G				Ea.	23			23	25
9600	Minimum labor/equipment charge		1 Plum	4	2	Job		129		129	199

22 11 Facility Water Distribution

22 11 13 – Facility Water Distribution Piping

22 11 13.14 Pipe, Brass

		Crew	Daily Output	Labor-Hours	Unit	Material	2020 Bare Costs Labor	Equipment	Total	Total Incl O&P
0010	**PIPE, BRASS**, Plain end									
0900	Field threaded, coupling & clevis hanger assembly 10' OC									
0920	Regular weight									
1120	1/2" diameter	1 Plum	48	.167	L.F.	8.30	10.75		19.05	25.50
1140	3/4" diameter		46	.174		10.95	11.20		22.15	29.50
1160	1" diameter		43	.186		7.15	12		19.15	26.50
1180	1-1/4" diameter	Q-1	72	.222		23.50	12.90		36.40	46
1200	1-1/2" diameter		65	.246		28	14.30		42.30	52.50

22 11 13 – Facility Water Distribution Piping

22 11 13.14 Pipe, Brass	Crew	Daily Output	Labor-Hours	Unit	Material	2020 Bare Costs Labor	Equipment	Total	Total Incl O&P
1220	2" diameter	Q-1	53	.302	L.F.	18.85	17.50		36.35	48
9000	Minimum labor/equipment charge	1 Plum	4	2	Job		129		129	199

22 11 13.16 Pipe Fittings, Brass

		Crew	Daily Output	Labor-Hours	Unit	Material	2020 Bare Costs Labor	Equipment	Total	Total Incl O&P
0010	**PIPE FITTINGS, BRASS**, Rough bronze, threaded, lead free.									
1000	Standard wt., 90° elbow									
1100	1/2"	1 Plum	12	.667	Ea.	22.50	43		65.50	91
1120	3/4"		11	.727		30	47		77	106
1140	1"		10	.800		48.50	51.50		100	133
1160	1-1/4"	Q-1	17	.941		78.50	54.50		133	171
1180	1-1/2"	"	16	1		97.50	58		155.50	197
1500	45° elbow, 1/8"	1 Plum	13	.615		27.50	39.50		67	92
1580	1/2"		12	.667		27.50	43		70.50	97
1600	3/4"		11	.727		39	47		86	116
1620	1"		10	.800		66.50	51.50		118	153
1640	1-1/4"	Q-1	17	.941		106	54.50		160.50	201
1660	1-1/2"	"	16	1		133	58		191	237
2000	Tee, 1/8"	1 Plum	9	.889		26.50	57.50		84	118
2080	1/2"		8	1		26.50	64.50		91	129
2100	3/4"		7	1.143		37.50	73.50		111	155
2120	1"		6	1.333		67.50	86		153.50	207
2140	1-1/4"	Q-1	10	1.600		116	93		209	271
2160	1-1/2"	"	9	1.778		131	103		234	305
2500	Coupling, 1/8"	1 Plum	26	.308		18.70	19.85		38.55	51
2580	1/2"		15	.533		18.80	34.50		53.30	73.50
2600	3/4"		14	.571		26.50	37		63.50	86
2620	1"		13	.615		45	39.50		84.50	111
2640	1-1/4"	Q-1	22	.727		75	42		117	148
2660	1-1/2"		20	.800		97.50	46.50		144	179
2680	2"		18	.889		161	51.50		212.50	257
9000	Minimum labor/equipment charge	1 Plum	4	2	Job		129		129	199

22 11 13.23 Pipe/Tube, Copper

		Crew	Daily Output	Labor-Hours	Unit	Material	2020 Bare Costs Labor	Equipment	Total	Total Incl O&P
0010	**PIPE/TUBE, COPPER**, Solder joints									
1000	Type K tubing, couplings & clevis hanger assemblies 10' OC									
1100	1/4" diameter	1 Plum	84	.095	L.F.	4.25	6.15		10.40	14.15
1200	1" diameter		66	.121		12.80	7.80		20.60	26
1260	2" diameter		40	.200		27	12.90		39.90	50
2000	Type L tubing, couplings & clevis hanger assemblies 10' OC									
2100	1/4" diameter	1 Plum	88	.091	L.F.	2.67	5.85		8.52	12
2120	3/8" diameter		84	.095		3.34	6.15		9.49	13.15
2140	1/2" diameter		81	.099		3.68	6.35		10.03	13.90
2160	5/8" diameter		79	.101		5.75	6.55		12.30	16.40
2180	3/4" diameter		76	.105		4.71	6.80		11.51	15.70
2200	1" diameter		68	.118		7.60	7.60		15.20	20
2220	1-1/4" diameter		58	.138		12	8.90		20.90	27
2240	1-1/2" diameter		52	.154		11.60	9.90		21.50	28
2260	2" diameter		42	.190		18.85	12.30		31.15	40
2280	2-1/2" diameter	Q-1	62	.258		30	14.95		44.95	56
2300	3" diameter		56	.286		46.50	16.55		63.05	76.50
2320	3-1/2" diameter		43	.372		60	21.50		81.50	99.50
2340	4" diameter		39	.410		65.50	24		89.50	109
2360	5" diameter		34	.471		148	27.50		175.50	204
2380	6" diameter	Q-2	40	.600		151	36		187	222

For customer support on your Commercial Renovation Costs with RSMeans data, call 800.448.8182.

423

22 11 13.23 Pipe/Tube, Copper

		Crew	Daily Output	Labor-Hours	Unit	Material	2020 Bare Costs Labor	2020 Bare Costs Equipment	Total	Total Incl O&P
2410	For other than full hard temper, add				L.F.	21%				
2590	For silver solder, add						15%			
3000	Type M tubing, couplings & clevis hanger assemblies 10' OC									
3140	1/2" diameter	1 Plum	84	.095	L.F.	3.56	6.15		9.71	13.40
3180	3/4" diameter		78	.103		5.05	6.60		11.65	15.75
3200	1" diameter	↓	70	.114	↓	8.50	7.35		15.85	21
4000	Type DWV tubing, couplings & clevis hanger assemblies 10' OC									
4100	1-1/4" diameter	1 Plum	60	.133	L.F.	12.45	8.60		21.05	27
4120	1-1/2" diameter		54	.148		12.60	9.55		22.15	28.50
4140	2" diameter	↓	44	.182		18.30	11.70		30	38
4160	3" diameter	Q-1	58	.276		29	16		45	56.50
4180	4" diameter		40	.400		60.50	23		83.50	103
4200	5" diameter	↓	36	.444		110	26		136	161
4220	6" diameter	Q-2	42	.571	↓	161	34.50		195.50	230
9000	Minimum labor/equipment charge	1 Plum	4	2	Job		129		129	199

22 11 13.25 Pipe/Tube Fittings, Copper

		Crew	Daily Output	Labor-Hours	Unit	Material	2020 Bare Costs Labor	2020 Bare Costs Equipment	Total	Total Incl O&P
0010	**PIPE/TUBE FITTINGS, COPPER**, Wrought unless otherwise noted									
0020	For silver solder, add						15%			
0040	Solder joints, copper x copper									
0070	90° elbow, 1/4"	1 Plum	22	.364	Ea.	3.88	23.50		27.38	40.50
0100	1/2"		20	.400		1.26	26		27.26	41.50
0120	3/4"		19	.421		2.68	27		29.68	45
0130	1"		16	.500		7.05	32		39.05	58
0250	45° elbow, 1/4"		22	.364		7.75	23.50		31.25	44.50
0270	3/8"		22	.364		6.80	23.50		30.30	43.50
0280	1/2"		20	.400		2.75	26		28.75	43
0290	5/8"		19	.421		11.45	27		38.45	54.50
0300	3/4"		19	.421		4.46	27		31.46	47
0310	1"		16	.500		11.20	32		43.20	62.50
0320	1-1/4"		15	.533		16	34.50		50.50	70.50
0330	1-1/2"		13	.615		18.20	39.50		57.70	81.50
0340	2"	↓	11	.727		30.50	47		77.50	106
0350	2-1/2"	Q-1	13	1.231		49.50	71.50		121	165
0360	3"		13	1.231		88	71.50		159.50	207
0370	3-1/2"		10	1.600		122	93		215	277
0380	4"		9	1.778		173	103		276	350
0390	5"	↓	6	2.667		635	155		790	940
0400	6"	Q-2	9	2.667		995	160		1,155	1,350
0450	Tee, 1/4"	1 Plum	14	.571		8.45	37		45.45	66.50
0470	3/8"		14	.571		6.85	37		43.85	64.50
0480	1/2"		13	.615		2.44	39.50		41.94	64
0490	5/8"		12	.667		16.25	43		59.25	84.50
0500	3/4"		12	.667		6.05	43		49.05	73
0510	1"		10	.800		17.05	51.50		68.55	98.50
0520	1-1/4"		9	.889		24.50	57.50		82	116
0530	1-1/2"		8	1		36.50	64.50		101	140
0540	2"	↓	7	1.143		58	73.50		131.50	178
0550	2-1/2"	Q-1	8	2		111	116		227	300
0560	3"		7	2.286		158	133		291	380
0570	3-1/2"		6	2.667		475	155		630	760
0580	4"		5	3.200		365	186		551	685
0590	5"	↓	4	4	↓	1,075	232		1,307	1,550

22 11 13.25 Pipe/Tube Fittings, Copper		Crew	Daily Output	Labor-Hours	Unit	Material	2020 Bare Costs Labor	Equipment	Total	Total Incl O&P
0600	6"	Q-2	6	4	Ea.	1,475	241		1,716	2,000
0612	Tee, reducing on the outlet, 1/4"	1 Plum	15	.533		17.05	34.50		51.55	72
0613	3/8"		15	.533		15.55	34.50		50.05	70
0614	1/2"		14	.571		15.60	37		52.60	74
0615	5/8"		13	.615		30.50	39.50		70	95
0616	3/4"		12	.667		8	43		51	75.50
0617	1"		11	.727		29	47		76	105
0618	1-1/4"		10	.800		32.50	51.50		84	115
0619	1-1/2"		9	.889		32	57.50		89.50	124
0620	2"		8	1		64.50	64.50		129	171
0621	2-1/2"	Q-1	9	1.778		137	103		240	310
0622	3"		8	2		163	116		279	360
0623	4"		6	2.667		299	155		454	570
0624	5"		5	3.200		1,725	186		1,911	2,175
0625	6"	Q-2	7	3.429		2,150	206		2,356	2,700
0626	8"	"	6	4		10,800	241		11,041	12,300
0630	Tee, reducing on the run, 1/4"	1 Plum	15	.533		23.50	34.50		58	79
0631	3/8"		15	.533		31	34.50		65.50	87
0632	1/2"		14	.571		20.50	37		57.50	79.50
0633	5/8"		13	.615		29	39.50		68.50	93.50
0634	3/4"		12	.667		21	43		64	89.50
0635	1"		11	.727		26	47		73	101
0636	1-1/4"		10	.800		41	51.50		92.50	125
0637	1-1/2"		9	.889		72	57.50		129.50	168
0638	2"		8	1		82.50	64.50		147	191
0639	2-1/2"	Q-1	9	1.778		177	103		280	355
0640	3"		8	2		320	116		436	535
0641	4"		6	2.667		585	155		740	885
0642	5"		5	3.200		1,950	186		2,136	2,425
0643	6"	Q-2	7	3.429		2,950	206		3,156	3,575
0644	8"	"	6	4		10,200	241		10,441	11,600
0650	Coupling, 1/4"	1 Plum	24	.333		1.07	21.50		22.57	34
0670	3/8"		24	.333		1.53	21.50		23.03	34.50
0680	1/2"		22	.364		.96	23.50		24.46	37
0690	5/8"		21	.381		4.23	24.50		28.73	42.50
0700	3/4"		21	.381		2.71	24.50		27.21	41
0710	1"		18	.444		5.30	28.50		33.80	50.50
0715	1-1/4"		17	.471		7.80	30.50		38.30	55.50
0716	1-1/2"		15	.533		10.80	34.50		45.30	65
0718	2"		13	.615		15.30	39.50		54.80	78.50
0721	2-1/2"	Q-1	15	1.067		40.50	62		102.50	140
0722	3"		13	1.231		47	71.50		118.50	162
0724	3-1/2"		8	2		124	116		240	315
0726	4"		7	2.286		154	133		287	375
0728	5"		6	2.667		228	155		383	490
0731	6"	Q-2	8	3		355	180		535	670
2000	DWV, solder joints, copper x copper									
2030	90° elbow, 1-1/4"	1 Plum	13	.615	Ea.	18.50	39.50		58	82
2050	1-1/2"		12	.667		24.50	43		67.50	93.50
2070	2"		10	.800		46	51.50		97.50	130
2090	3"	Q-1	10	1.600		85.50	93		178.50	237
2100	4"	"	9	1.778		545	103		648	760
2250	Tee, sanitary, 1-1/4"	1 Plum	9	.889		28.50	57.50		86	120

For customer support on your Commercial Renovation Costs with RSMeans data, call 800.448.8182.

425

22 11 13 – Facility Water Distribution Piping

22 11 13.25 Pipe/Tube Fittings, Copper

		Crew	Daily Output	Labor-Hours	Unit	Material	2020 Bare Costs Labor	2020 Bare Costs Equipment	Total	Total Incl O&P
2270	1-1/2"	1 Plum	8	1	Ea.	35.50	64.50		100	139
2290	2"	↓	7	1.143		55	73.50		128.50	175
2310	3"	Q-1	7	2.286		213	133		346	440
2330	4"	"	6	2.667		520	155		675	815
2400	Coupling, 1-1/4"	1 Plum	14	.571		7.75	37		44.75	65.50
2420	1-1/2"		13	.615		9.60	39.50		49.10	72
2440	2"	↓	11	.727		13.35	47		60.35	87
2460	3"	Q-1	11	1.455		30.50	84.50		115	164
2480	4"	"	10	1.600	↓	68	93		161	218
9000	Minimum labor/equipment charge	1 Plum	4	2	Job		129		129	199

22 11 13.44 Pipe, Steel

		Crew	Daily Output	Labor-Hours	Unit	Material	2020 Bare Costs Labor	2020 Bare Costs Equipment	Total	Total Incl O&P
0010	**PIPE, STEEL** R221113-50									
0020	All pipe sizes are to Spec. A-53 unless noted otherwise									
0050	Schedule 40, threaded, with couplings, and clevis hanger									
0060	assemblies sized for covering, 10' OC									
0540	Black, 1/4" diameter	1 Plum	66	.121	L.F.	6.15	7.80		13.95	18.90
0560	1/2" diameter		63	.127		4.03	8.20		12.23	17.10
0570	3/4" diameter		61	.131		4.40	8.45		12.85	17.90
0580	1" diameter	↓	53	.151		7.80	9.75		17.55	23.50
0590	1-1/4" diameter	Q-1	89	.180		8.20	10.45		18.65	25
0600	1-1/2" diameter		80	.200		8.75	11.60		20.35	27.50
0610	2" diameter		64	.250		6.90	14.50		21.40	30
0620	2-1/2" diameter		50	.320		11.40	18.55		29.95	41
0630	3" diameter		43	.372		14.35	21.50		35.85	49.50
0640	3-1/2" diameter		40	.400		21	23		44	59.50
0650	4" diameter	↓	36	.444	↓	23	26		49	65.50
1280	All pipe sizes are to Spec. A-53 unless noted otherwise									
1281	Schedule 40, threaded, with couplings and clevis hanger									
1282	assemblies sized for covering, 10' OC									
1290	Galvanized, 1/4" diameter	1 Plum	66	.121	L.F.	8.10	7.80		15.90	21
1310	1/2" diameter		63	.127		4.10	8.20		12.30	17.15
1320	3/4" diameter		61	.131		5	8.45		13.45	18.55
1330	1" diameter	↓	53	.151		7.90	9.75		17.65	24
1350	1-1/2" diameter	Q-1	80	.200		9.45	11.60		21.05	28.50
1360	2" diameter		64	.250		7.90	14.50		22.40	31
1370	2-1/2" diameter		50	.320		13.15	18.55		31.70	43
1380	3" diameter		43	.372		16	21.50		37.50	51
1400	4" diameter	↓	36	.444		25.50	26		51.50	68
2000	Welded, sch. 40, on yoke & roll hanger assy's, sized for covering, 10' OC									
2040	Black, 1" diameter	Q-15	93	.172	L.F.	7.15	10	1.14	18.29	24.50
2070	2" diameter		61	.262		6.60	15.20	1.74	23.54	32.50
2090	3" diameter		43	.372		16.65	21.50	2.47	40.62	54.50
2110	4" diameter		37	.432		24	25	2.88	51.88	68.50
2120	5" diameter	↓	32	.500		37	29	3.33	69.33	89
2130	6" diameter	Q-16	36	.667	↓	45	40	2.95	87.95	115
9990	Minimum labor/equipment charge	1 Plum	3	2.667	Job		172		172	266

22 11 13.45 Pipe Fittings, Steel, Threaded

		Crew	Daily Output	Labor-Hours	Unit	Material	2020 Bare Costs Labor	2020 Bare Costs Equipment	Total	Total Incl O&P
0010	**PIPE FITTINGS, STEEL, THREADED** R221113-50									
0020	Cast iron									
0040	Standard weight, black									
0060	90° elbow, straight									
0070	1/4"	1 Plum	16	.500	Ea.	11.95	32		43.95	63

22 11 13.45 Pipe Fittings, Steel, Threaded		Crew	Daily Output	Labor-Hours	Unit	Material	2020 Bare Costs Labor	Equipment	Total	Total Incl O&P
0080	3/8"	1 Plum	16	.500	Ea.	17.30	32		49.30	69
0090	1/2"		15	.533		7.60	34.50		42.10	61.50
0100	3/4"		14	.571		7.90	37		44.90	65.50
0110	1"		13	.615		9.35	39.50		48.85	72
0130	1-1/2"	Q-1	20	.800		18.35	46.50		64.85	91.50
0140	2"		18	.889		28.50	51.50		80	111
0150	2-1/2"		14	1.143		68.50	66.50		135	178
0160	3"		10	1.600		113	93		206	267
0180	4"		6	2.667		209	155		364	470
0500	Tee, straight									
0510	1/4"	1 Plum	10	.800	Ea.	18.70	51.50		70.20	100
0520	3/8"		10	.800		18.20	51.50		69.70	99.50
0540	3/4"		9	.889		13.75	57.50		71.25	104
0550	1"		8	1		12.25	64.50		76.75	113
0570	1-1/2"	Q-1	13	1.231		29	71.50		100.50	142
0580	2"		11	1.455		40.50	84.50		125	175
0590	2-1/2"		9	1.778		105	103		208	275
0600	3"		6	2.667		161	155		316	415
0620	4"		4	4		315	232		547	705
0700	Standard weight, galvanized cast iron									
0720	90° elbow, straight									
0730	1/4"	1 Plum	16	.500	Ea.	21.50	32		53.50	73.50
0740	3/8"		16	.500		21.50	32		53.50	73.50
0750	1/2"		15	.533		22	34.50		56.50	77
0760	3/4"		14	.571		21	37		58	80
0770	1"		13	.615		28	39.50		67.50	92
0790	1-1/2"	Q-1	20	.800		59.50	46.50		106	137
0800	2"		18	.889		76	51.50		127.50	163
0810	2-1/2"		14	1.143		156	66.50		222.50	273
0820	3"		10	1.600		237	93		330	405
0840	4"		6	2.667		435	155		590	720
1100	Tee, straight									
1110	1/4"	1 Plum	10	.800	Ea.	20	51.50		71.50	102
1120	3/8"		10	.800		22.50	51.50		74	105
1140	3/4"		9	.889		34.50	57.50		92	127
1150	1"		8	1		32.50	64.50		97	136
1170	1-1/2"	Q-1	13	1.231		75.50	71.50		147	193
1180	2"		11	1.455		94	84.50		178.50	233
1190	2-1/2"		9	1.778		191	103		294	370
1200	3"		6	2.667		455	155		610	740
1220	4"		4	4		590	232		822	1,000
5000	Malleable iron, 150 lb.									
5020	Black									
5040	90° elbow, straight									
5060	1/4"	1 Plum	16	.500	Ea.	5.90	32		37.90	56.50
5070	3/8"		16	.500		5.90	32		37.90	56.50
5090	3/4"		14	.571		4.12	37		41.12	61.50
5100	1"		13	.615		8.35	39.50		47.85	70.50
5120	1-1/2"	Q-1	20	.800		14.60	46.50		61.10	87.50
5130	2"		18	.889		25.50	51.50		77	108
5140	2-1/2"		14	1.143		71	66.50		137.50	181
5150	3"		10	1.600		104	93		197	258
5170	4"		6	2.667		196	155		351	455

22 11 13.45 Pipe Fittings, Steel, Threaded

		Crew	Daily Output	Labor-Hours	Unit	Material	2020 Bare Costs Labor	Equipment	Total	Total Incl O&P
5450	Tee, straight									
5470	1/4"	1 Plum	10	.800	Ea.	8.55	51.50		60.05	89
5480	3/8"		10	.800		8.55	51.50		60.05	89
5500	3/4"		9	.889		7.85	57.50		65.35	97
5510	1"		8	1		13.40	64.50		77.90	114
5520	1-1/4"	Q-1	14	1.143		21.50	66.50		88	126
5530	1-1/2"		13	1.231		27	71.50		98.50	140
5540	2"		11	1.455		46	84.50		130.50	181
5550	2-1/2"		9	1.778		99	103		202	268
5560	3"		6	2.667		146	155		301	400
5570	3-1/2"		5	3.200		340	186		526	660
5580	4"		4	4		355	232		587	750
5650	Coupling									
5670	1/4"	1 Plum	19	.421	Ea.	7.30	27		34.30	50
5680	3/8"		19	.421		7.30	27		34.30	50
5690	1/2"		19	.421		5.65	27		32.65	48
5700	3/4"		18	.444		6.60	28.50		35.10	52
5710	1"		15	.533		9.90	34.50		44.40	64
5720	1-1/4"	Q-1	26	.615		11.95	35.50		47.45	68
5730	1-1/2"		24	.667		15.80	38.50		54.30	77.50
5740	2"		21	.762		23.50	44		67.50	94
5750	2-1/2"		18	.889		64.50	51.50		116	151
5760	3"		14	1.143		87.50	66.50		154	198
5780	4"		10	1.600		176	93		269	335
6000	For galvanized elbows, tees, and couplings, add					20%				
9990	Minimum labor/equipment charge	1 Plum	4	2	Job		129		129	199

22 11 13.47 Pipe Fittings, Steel

		Crew	Daily Output	Labor-Hours	Unit	Material	2020 Bare Costs Labor	Equipment	Total	Total Incl O&P
0010	**PIPE FITTINGS, STEEL**, flanged, welded & special									
3000	Weld joint, butt, carbon steel, standard weight									
3040	90° elbow, long radius									
3050	1/2" pipe size	Q-15	16	1	Ea.	44	58	6.65	108.65	145
3060	3/4" pipe size		16	1		44	58	6.65	108.65	145
3070	1" pipe size		16	1		26.50	58	6.65	91.15	126
3080	1-1/4" pipe size		14	1.143		26.50	66.50	7.60	100.60	139
3090	1-1/2" pipe size		13	1.231		26.50	71.50	8.20	106.20	148
3100	2" pipe size		10	1.600		22.50	93	10.65	126.15	179
3110	2-1/2" pipe size		8	2		35	116	13.30	164.30	232
3120	3" pipe size		7	2.286		30	133	15.20	178.20	255
3130	4" pipe size		5	3.200		56.50	186	21.50	264	375
3136	5" pipe size		4	4		112	232	26.50	370.50	515
3140	6" pipe size	Q-16	5	4.800		126	289	21.50	436.50	605
3350	Tee, straight									
3360	1/2" pipe size	Q-15	10	1.600	Ea.	116	93	10.65	219.65	283
3370	3/4" pipe size		10	1.600		116	93	10.65	219.65	283
3380	1" pipe size		10	1.600		57.50	93	10.65	161.15	218
3390	1-1/4" pipe size		9	1.778		72	103	11.80	186.80	251
3400	1-1/2" pipe size		8	2		72	116	13.30	201.30	273
3410	2" pipe size		6	2.667		57.50	155	17.75	230.25	320
3420	2-1/2" pipe size		5	3.200		79.50	186	21.50	287	400
3430	3" pipe size		4	4		88.50	232	26.50	347	485
3440	4" pipe size		3	5.333		124	310	35.50	469.50	655
3446	5" pipe size		2.50	6.400		205	370	42.50	617.50	845

22 11 13 – Facility Water Distribution Piping

22 11 13.47 Pipe Fittings, Steel

		Crew	Daily Output	Labor-Hours	Unit	Material	2020 Bare Costs Labor	2020 Bare Costs Equipment	Total	Total Incl O&P
3450	6" pipe size	Q-16	3	8	Ea.	213	480	35.50	728.50	1,025
9805	2" x 1"	1 Plum	50	.160		101	10.30		111.30	127
9810	2" x 1-1/4"		50	.160		117	10.30		127.30	144
9815	2" x 1-1/2"	▼	50	.160		125	10.30		135.30	154
9820	2-1/2" x 1"	Q-1	80	.200		106	11.60		117.60	134
9825	2-1/2" x 1-1/4"		80	.200		156	11.60		167.60	190
9830	2-1/2" x 1-1/2"		80	.200		156	11.60		167.60	190
9835	2-1/2" x 2"		80	.200		106	11.60		117.60	134
9840	3" x 1"		80	.200		115	11.60		126.60	144
9845	3" x 1-1/4"		80	.200		153	11.60		164.60	187
9850	3" x 1-1/2"		80	.200		153	11.60		164.60	187
9852	3" x 2"		80	.200		184	11.60		195.60	220
9856	4" x 1"		67	.239		142	13.85		155.85	179
9858	4" x 1-1/4"		67	.239		187	13.85		200.85	228
9860	4" x 1-1/2"		67	.239		191	13.85		204.85	232
9862	4" x 2"		67	.239		196	13.85		209.85	237
9864	4" x 2-1/2"		67	.239		203	13.85		216.85	245
9866	4" x 3"	▼	67	.239		207	13.85		220.85	250
9870	6" x 1-1/2"	Q-2	50	.480		239	29		268	310
9872	6" x 2"		50	.480		242	29		271	310
9874	6" x 2-1/2"		50	.480		246	29		275	315
9876	6" x 3"		50	.480		274	29		303	345
9878	6" x 4"		50	.480		305	29		334	380
9880	8" x 2"		42	.571		415	34.50		449.50	510
9882	8" x 2-1/2"		42	.571		415	34.50		449.50	510
9884	8" x 3"		42	.571		435	34.50		469.50	535
9886	8" x 4"	▼	42	.571	▼	890	34.50		924.50	1,025
9990	Minimum labor/equipment charge	Q-15	3	5.333	Job		310	35.50	345.50	520

22 11 13.48 Pipe, Fittings and Valves, Steel, Grooved-Joint

		Crew	Daily Output	Labor-Hours	Unit	Material	2020 Bare Costs Labor	2020 Bare Costs Equipment	Total	Total Incl O&P
0010	**PIPE, FITTINGS AND VALVES, STEEL, GROOVED-JOINT**									
0012	Fittings are ductile iron. Steel fittings noted.									
0020	Pipe includes coupling & clevis type hanger assemblies, 10' OC									
1000	Schedule 40, black									
1040	3/4" diameter	1 Plum	71	.113	L.F.	6.55	7.25		13.80	18.45
1050	1" diameter		63	.127		6.45	8.20		14.65	19.75
1060	1-1/4" diameter		58	.138		7.70	8.90		16.60	22.50
1070	1-1/2" diameter		51	.157		8.40	10.10		18.50	25
1080	2" diameter	▼	40	.200		9.70	12.90		22.60	30.50
1090	2-1/2" diameter	Q-1	57	.281		16.50	16.30		32.80	43
1100	3" diameter		50	.320		22	18.55		40.55	52.50
1110	4" diameter		45	.356		26	20.50		46.50	60.50
1120	5" diameter	▼	37	.432	▼	41	25		66	84
3990	Fittings: coupling material required at joints not incl. in fitting price.									
3994	Add 1 selected coupling, material only, per joint for installed price.									
4000	Elbow, 90° or 45°, painted									
4030	3/4" diameter	1 Plum	50	.160	Ea.	82.50	10.30		92.80	107
4040	1" diameter		50	.160		44.50	10.30		54.80	65
4050	1-1/4" diameter		40	.200		44.50	12.90		57.40	69
4060	1-1/2" diameter		33	.242		44.50	15.60		60.10	73
4070	2" diameter	▼	25	.320		44.50	20.50		65	81
4080	2-1/2" diameter	Q-1	40	.400		44.50	23		67.50	85
4090	3" diameter	▼	33	.485	▼	78	28		106	129

For customer support on your Commercial Renovation Costs with RSMeans data, call 800.448.8182.

429

22 11 13 – Facility Water Distribution Piping

22 11 13.48 Pipe, Fittings and Valves, Steel, Grooved-Joint	Crew	Daily Output	Labor-Hours	Unit	Material	2020 Bare Costs Labor	Equipment	Total	Total Incl O&P
4100	4" diameter	Q-1	25	.640	Ea.	84.50	37		121.50	151
4110	5" diameter		20	.800		201	46.50		247.50	293
4250	For galvanized elbows, add					26%				
4690	Tee, painted									
4700	3/4" diameter	1 Plum	38	.211	Ea.	88.50	13.55		102.05	119
4740	1" diameter		33	.242		68.50	15.60		84.10	99
4750	1-1/4" diameter		27	.296		68.50	19.10		87.60	105
4760	1-1/2" diameter		22	.364		68.50	23.50		92	111
4770	2" diameter		17	.471		68.50	30.50		99	122
4780	2-1/2" diameter	Q-1	27	.593		68.50	34.50		103	128
4790	3" diameter		22	.727		93.50	42		135.50	168
4800	4" diameter		17	.941		142	54.50		196.50	241
4810	5" diameter		13	1.231		330	71.50		401.50	475
4900	For galvanized tees, add					24%				
4933	20" diameter	Q-2	16	1.500		725	90		815	935
4940	Flexible, standard, painted									
4950	3/4" diameter	1 Plum	100	.080	Ea.	25	5.15		30.15	35.50
4960	1" diameter		100	.080		25	5.15		30.15	35.50
4970	1-1/4" diameter		80	.100		32.50	6.45		38.95	45.50
4980	1-1/2" diameter		67	.119		35	7.70		42.70	50.50
4990	2" diameter		50	.160		38	10.30		48.30	57.50
5000	2-1/2" diameter	Q-1	80	.200		44	11.60		55.60	66.50
5010	3" diameter		67	.239		48.50	13.85		62.35	75
5020	3-1/2" diameter		57	.281		69.50	16.30		85.80	102
5030	4" diameter		50	.320		70	18.55		88.55	106
5040	5" diameter		40	.400		105	23		128	152
5050	6" diameter	Q-2	50	.480		125	29		154	182
5200	For galvanized couplings, add					33%				
7790	Valves: coupling material required at joints not incl. in valve price.									
7794	Add 1 selected coupling, material only, per joint for installed price.									
9990	Minimum labor/equipment charge	1 Plum	4	2	Job		129		129	199

22 11 13.74 Pipe, Plastic

		Crew	Daily Output	Labor-Hours	Unit	Material	2020 Bare Costs Labor	Equipment	Total	Total Incl O&P
0010	**PIPE, PLASTIC**									
1800	PVC, couplings 10' OC, clevis hanger assemblies, 3 per 10'									
1820	Schedule 40									
1860	1/2" diameter	1 Plum	54	.148	L.F.	4.90	9.55		14.45	20
1870	3/4" diameter		51	.157		5.40	10.10		15.50	21.50
1880	1" diameter		46	.174		9.45	11.20		20.65	28
1890	1-1/4" diameter		42	.190		9.95	12.30		22.25	30
1900	1-1/2" diameter		36	.222		10.25	14.30		24.55	33.50
1910	2" diameter	Q-1	59	.271		12.15	15.75		27.90	38
1920	2-1/2" diameter		56	.286		20	16.55		36.55	48
1930	3" diameter		53	.302		22	17.50		39.50	51.50
1940	4" diameter		48	.333		14.80	19.35		34.15	46.50
1950	5" diameter		43	.372		23.50	21.50		45	59.50
1960	6" diameter		39	.410		30.50	24		54.50	70.50
4100	DWV type, schedule 40, couplings 10' OC, clevis hanger assy's, 3 per 10'									
4210	ABS, schedule 40, foam core type									
4212	Plain end black									
4214	1-1/2" diameter	1 Plum	39	.205	L.F.	8.60	13.20		21.80	30
4216	2" diameter	Q-1	62	.258		9.25	14.95		24.20	33
4218	3" diameter		56	.286		17.60	16.55		34.15	45

430

For customer support on your Commercial Renovation Costs with RSMeans data, call 800.448.8182.

22 11 13 – Facility Water Distribution Piping

22 11 13.74 Pipe, Plastic

		Crew	Daily Output	Labor-Hours	Unit	Material	2020 Bare Costs Labor	Equipment	Total	Total Incl O&P
4220	4" diameter	Q-1	51	.314	L.F.	9.80	18.20		28	39
4222	6" diameter	↓	42	.381	↓	21.50	22		43.50	57.50
4240	To delete coupling & hangers, subtract									
4244	1-1/2" diam. to 6" diam.					43%	48%			
4400	PVC									
4410	1-1/4" diameter	1 Plum	42	.190	L.F.	9.10	12.30		21.40	29
4460	2" diameter	Q-1	59	.271		9.15	15.75		24.90	34.50
4470	3" diameter		53	.302		17.45	17.50		34.95	46
4480	4" diameter		48	.333		19.20	19.35		38.55	51
4490	6" diameter	↓	39	.410	↓	18.60	24		42.60	57.50
5300	CPVC, socket joint, couplings 10' OC, clevis hanger assemblies, 3 per 10'									
5302	Schedule 40									
5304	1/2" diameter	1 Plum	54	.148	L.F.	5.80	9.55		15.35	21
5305	3/4" diameter		51	.157		6.95	10.10		17.05	23.50
5306	1" diameter		46	.174		11.35	11.20		22.55	30
5307	1-1/4" diameter		42	.190		12.85	12.30		25.15	33
5308	1-1/2" diameter	↓	36	.222		12.45	14.30		26.75	35.50
5309	2" diameter	Q-1	59	.271		15.70	15.75		31.45	42
5310	2-1/2" diameter		56	.286		28	16.55		44.55	56.50
5311	3" diameter	↓	53	.302	↓	32	17.50		49.50	62
5360	CPVC, threaded, couplings 10' OC, clevis hanger assemblies, 3 per 10'									
5380	Schedule 40									
5460	1/2" diameter	1 Plum	54	.148	L.F.	6.65	9.55		16.20	22
5470	3/4" diameter		51	.157		8.45	10.10		18.55	25
5480	1" diameter		46	.174		12.90	11.20		24.10	31.50
5490	1-1/4" diameter		42	.190		14.05	12.30		26.35	34.50
5500	1-1/2" diameter	↓	36	.222		13.45	14.30		27.75	37
5510	2" diameter	Q-1	59	.271		16.90	15.75		32.65	43
5520	2-1/2" diameter		56	.286		29.50	16.55		46.05	58
5530	3" diameter	↓	53	.302	↓	33.50	17.50		51	64
9900	Minimum labor/equipment charge	1 Plum	4	2	Job		129		129	199

22 11 13.76 Pipe Fittings, Plastic

		Crew	Daily Output	Labor-Hours	Unit	Material	2020 Bare Costs Labor	Equipment	Total	Total Incl O&P
0010	**PIPE FITTINGS, PLASTIC**									
2700	PVC (white), schedule 40, socket joints									
2760	90° elbow, 1/2"	1 Plum	33.30	.240	Ea.	.52	15.50		16.02	24.50
2810	2"	Q-1	36.40	.440	"	3.11	25.50		28.61	43
4500	DWV, ABS, non pressure, socket joints									
4540	1/4 bend, 1-1/4"	1 Plum	20.20	.396	Ea.	4.48	25.50		29.98	44.50
4560	1-1/2"	"	18.20	.440		3.44	28.50		31.94	48
4570	2"	Q-1	33.10	.483	↓	5.30	28		33.30	49.50
4650	1/8 bend, same as 1/4 bend									
4800	Tee, sanitary									
4820	1-1/4"	1 Plum	13.50	.593	Ea.	5.95	38		43.95	65.50
4830	1-1/2"	"	12.10	.661		5.15	42.50		47.65	71.50
4840	2"	Q-1	20	.800	↓	7.90	46.50		54.40	80
5000	DWV, PVC, schedule 40, socket joints									
5040	1/4 bend, 1-1/4"	1 Plum	20.20	.396	Ea.	8.80	25.50		34.30	49
5060	1-1/2"	"	18.20	.440		2.48	28.50		30.98	46.50
5070	2"	Q-1	33.10	.483		3.91	28		31.91	48
5080	3"		20.80	.769		11.65	44.50		56.15	82
5090	4"		16.50	.970		23	56		79	112
5100	6"	↓	10.10	1.584	↓	79.50	92		171.50	230

For customer support on your Commercial Renovation Costs with RSMeans data, call 800.448.8182.

431

22 11 13.76 Pipe Fittings, Plastic		Crew	Daily Output	Labor-Hours	Unit	Material	2020 Bare Costs Labor	Equipment	Total	Total Incl O&P
5105	8"	Q-2	9.30	2.581	Ea.	122	155		277	375
5110	1/4 bend, long sweep, 1-1/2"	1 Plum	18.20	.440		5.85	28.50		34.35	50.50
5112	2"	Q-1	33.10	.483		6.50	28		34.50	50.50
5114	3"		20.80	.769		15.05	44.50		59.55	85.50
5116	4"	↓	16.50	.970		28.50	56		84.50	119
5150	1/8 bend, 1-1/4"	1 Plum	20.20	.396		6.05	25.50		31.55	46
5215	8"	Q-2	9.30	2.581		117	155		272	370
5250	Tee, sanitary 1-1/4"	1 Plum	13.50	.593		9.45	38		47.45	69.50
5254	1-1/2"	"	12.10	.661		4.41	42.50		46.91	71
5255	2"	Q-1	20	.800		6.50	46.50		53	78.50
5256	3"		13.90	1.151		17.05	67		84.05	122
5257	4"		11	1.455		31	84.50		115.50	165
5259	6"		6.70	2.388		126	139		265	355
5261	8"	Q-2	6.20	3.871		293	233		526	685
5264	2" x 1-1/2"	Q-1	22	.727		5.75	42		47.75	71.50
5266	3" x 1-1/2"		15.50	1.032		12.45	60		72.45	106
5268	4" x 3"		12.10	1.322		37	76.50		113.50	160
5271	6" x 4"	↓	6.90	2.319		122	134		256	340
5314	Combination Y & 1/8 bend, 1-1/2"	1 Plum	12.10	.661		10.75	42.50		53.25	78
5315	2"	Q-1	20	.800		12.90	46.50		59.40	85.50
5317	3"		13.90	1.151		29	67		96	135
5318	4"	↓	11	1.455	↓	58	84.50		142.50	194
5324	Combination Y & 1/8 bend, reducing									
5325	2" x 2" x 1-1/2"	Q-1	22	.727	Ea.	15.15	42		57.15	81.50
5327	3" x 3" x 1-1/2"		15.50	1.032		27	60		87	122
5328	3" x 3" x 2"		15.30	1.046		19.95	60.50		80.45	116
5329	4" x 4" x 2"	↓	12.20	1.311		30.50	76		106.50	152
5331	Wye, 1-1/4"	1 Plum	13.50	.593		12.10	38		50.10	72.50
5332	1-1/2"	"	12.10	.661		8.20	42.50		50.70	75
5333	2"	Q-1	20	.800		7.90	46.50		54.40	80
5334	3"		13.90	1.151		21.50	67		88.50	127
5335	4"		11	1.455		39	84.50		123.50	173
5336	6"	↓	6.70	2.388		114	139		253	340
5337	8"	Q-2	6.20	3.871		217	233		450	600
5341	2" x 1-1/2"	Q-1	22	.727		9.85	42		51.85	76
5342	3" x 1-1/2"		15.50	1.032		14.60	60		74.60	109
5343	4" x 3"		12.10	1.322		32	76.50		108.50	154
5344	6" x 4"	↓	6.90	2.319		87	134		221	305
5345	8" x 6"	Q-2	6.40	3.750		188	226		414	555
5347	Double wye, 1-1/2"	1 Plum	9.10	.879		18.25	56.50		74.75	108
5348	2"	Q-1	16.60	.964		20.50	56		76.50	109
5349	3"		10.40	1.538		42.50	89		131.50	185
5350	4"	↓	8.25	1.939	↓	85.50	112		197.50	269
5353	Double wye, reducing									
5354	2" x 2" x 1-1/2" x 1-1/2"	Q-1	16.80	.952	Ea.	18.60	55		73.60	106
5355	3" x 3" x 2" x 2"		10.60	1.509		31.50	87.50		119	170
5356	4" x 4" x 3" x 3"		8.45	1.893		68	110		178	245
5357	6" x 6" x 4" x 4"	↓	7.25	2.207		239	128		367	460
5374	Coupling, 1-1/4"	1 Plum	20.20	.396		5.70	25.50		31.20	46
5376	1-1/2"	"	18.20	.440		1.19	28.50		29.69	45.50
5378	2"	Q-1	33.10	.483		1.62	28		29.62	45.50
5380	3"		20.80	.769		5.65	44.50		50.15	75
5390	4"	↓	16.50	.970		9.65	56		65.65	97.50

22 11 13.76 Pipe Fittings, Plastic		Crew	Daily Output	Labor-Hours	Unit	Material	2020 Bare Costs Labor	Equipment	Total	Total Incl O&P
5410	Reducer bushing, 2" x 1-1/4"	Q-1	36.50	.438	Ea.	3.38	25.50		28.88	43
5412	3" x 1-1/2"		27.30	.586		10.10	34		44.10	63.50
5414	4" x 2"		18.20	.879		17.35	51		68.35	98
5416	6" x 4"	▼	11.10	1.441		45	83.50		128.50	179
5418	8" x 6"	Q-2	10.20	2.353	▼	89	142		231	315
5500	CPVC, Schedule 80, threaded joints									
5540	90° elbow, 1/4"	1 Plum	32	.250	Ea.	13	16.10		29.10	39.50
5560	1/2"		30.30	.264		7.55	17		24.55	35
5570	3/4"		26	.308		11.30	19.85		31.15	43
5580	1"		22.70	.352		15.85	22.50		38.35	52.50
5590	1-1/4"		20.20	.396		30.50	25.50		56	73
5600	1-1/2"	▼	18.20	.440		33	28.50		61.50	80
5610	2"	Q-1	33.10	.483		44	28		72	92
5620	2-1/2"		24.20	.661		137	38.50		175.50	210
5630	3"	▼	20.80	.769		147	44.50		191.50	230
5730	Coupling, 1/4"	1 Plum	32	.250		16.65	16.10		32.75	43.50
5732	1/2"		30.30	.264		13.75	17		30.75	41.50
5734	3/4"		26	.308		22	19.85		41.85	55
5736	1"		22.70	.352		25	22.50		47.50	62.50
5738	1-1/4"		20.20	.396		26.50	25.50		52	68.50
5740	1-1/2"	▼	18.20	.440		28.50	28.50		57	75.50
5742	2"	Q-1	33.10	.483		33.50	28		61.50	80.50
5744	2-1/2"		24.20	.661		60.50	38.50		99	126
5746	3"	▼	20.80	.769	▼	70	44.50		114.50	147
5900	CPVC, Schedule 80, socket joints									
5904	90° elbow, 1/4"	1 Plum	32	.250	Ea.	12.45	16.10		28.55	38.50
5906	1/2"		30.30	.264		4.88	17		21.88	32
5908	3/4"		26	.308		6.25	19.85		26.10	37.50
5910	1"		22.70	.352		9.90	22.50		32.40	46
5912	1-1/4"		20.20	.396		21.50	25.50		47	63
5914	1-1/2"	▼	18.20	.440		24	28.50		52.50	70
5916	2"	Q-1	33.10	.483		29	28		57	75
5918	2-1/2"		24.20	.661		66.50	38.50		105	133
5920	3"	▼	20.80	.769		75	44.50		119.50	152
5930	45° elbow, 1/4"	1 Plum	32	.250		18.50	16.10		34.60	45.50
5932	1/2"		30.30	.264		5.95	17		22.95	33
5934	3/4"		26	.308		7.80	19.85		27.65	39
5936	1"		22.70	.352		13.70	22.50		36.20	50
5938	1-1/4"		20.20	.396		27	25.50		52.50	69
5940	1-1/2"	▼	18.20	.440		27.50	28.50		56	74.50
5942	2"	Q-1	33.10	.483		27.50	28		55.50	73.50
5944	2-1/2"		24.20	.661		63.50	38.50		102	129
5946	3"	▼	20.80	.769		81.50	44.50		126	159
5990	Coupling, 1/4"	1 Plum	32	.250		13.25	16.10		29.35	39.50
5992	1/2"		30.30	.264		5.15	17		22.15	32
5994	3/4"		26	.308		7.20	19.85		27.05	38.50
5996	1"		22.70	.352		9.70	22.50		32.20	45.50
5998	1-1/4"		20.20	.396		14.55	25.50		40.05	55.50
6000	1-1/2"		18.20	.440		18.25	28.50		46.75	64
6002	2"	Q-1	33.10	.483		21.50	28		49.50	67
6004	2-1/2"		24.20	.661		47.50	38.50		86	112
6006	3"	▼	20.80	.769	▼	51.50	44.50		96	126
9990	Minimum labor/equipment charge	1 Plum	4	2	Job		129		129	199

22 11 Facility Water Distribution

22 11 19 – Domestic Water Piping Specialties

22 11 19.10 Flexible Connectors

		Crew	Daily Output	Labor-Hours	Unit	Material	2020 Bare Costs Labor	Equipment	Total	Total Incl O&P
0010	**FLEXIBLE CONNECTORS**, Corrugated, 5/8" OD, 3/4" ID									
0050	Gas, seamless brass, steel fittings									
0200	12" long	1 Plum	36	.222	Ea.	6.80	14.30		21.10	29.50
0220	18" long		36	.222		8.55	14.30		22.85	31.50
0240	24" long		34	.235		10	15.15		25.15	34.50
0260	30" long		34	.235		12.45	15.15		27.60	37
9000	Minimum labor/equipment charge		4	2	Job		129		129	199

22 11 19.18 Mixing Valve

		Crew	Daily Output	Labor-Hours	Unit	Material	2020 Bare Costs Labor	Equipment	Total	Total Incl O&P
0010	**MIXING VALVE**, Automatic, water tempering.									
0040	1/2" size	1 Stpi	19	.421	Ea.	625	27.50		652.50	735
0050	3/4" size		18	.444	"	660	29		689	770
9000	Minimum labor/equipment charge		5	1.600	Job		105		105	162

22 11 19.38 Water Supply Meters

		Crew	Daily Output	Labor-Hours	Unit	Material	2020 Bare Costs Labor	Equipment	Total	Total Incl O&P
0010	**WATER SUPPLY METERS**									
2000	Domestic/commercial, bronze									
2020	Threaded									
2060	5/8" diameter, to 20 GPM	1 Plum	16	.500	Ea.	54	32		86	110
2080	3/4" diameter, to 30 GPM		14	.571		98.50	37		135.50	165
2100	1" diameter, to 50 GPM		12	.667		149	43		192	231
2300	Threaded/flanged									
2340	1-1/2" diameter, to 100 GPM	1 Plum	8	1	Ea.	365	64.50		429.50	500
2600	Flanged, compound									
9000	Minimum labor/equipment charge	1 Plum	3.25	2.462	Job		159		159	245

22 11 19.42 Backflow Preventers

		Crew	Daily Output	Labor-Hours	Unit	Material	2020 Bare Costs Labor	Equipment	Total	Total Incl O&P
0010	**BACKFLOW PREVENTERS**, Includes valves									
0020	and four test cocks, corrosion resistant, automatic operation									
1000	Double check principle									
1010	Threaded, with ball valves									
1020	3/4" pipe size	1 Plum	16	.500	Ea.	241	32		273	315
1030	1" pipe size		14	.571		253	37		290	335
1040	1-1/2" pipe size		10	.800		620	51.50		671.50	765
1050	2" pipe size		7	1.143		665	73.50		738.50	850
1080	Threaded, with gate valves									
1100	3/4" pipe size	1 Plum	16	.500	Ea.	1,150	32		1,182	1,300
1120	1" pipe size		14	.571		1,150	37		1,187	1,325
1140	1-1/2" pipe size		10	.800		1,225	51.50		1,276.50	1,425
1160	2" pipe size		7	1.143		1,825	73.50		1,898.50	2,125
4000	Reduced pressure principle									
4100	Threaded, bronze, valves are ball									
4120	3/4" pipe size	1 Plum	16	.500	Ea.	530	32		562	635
4140	1" pipe size		14	.571		560	37		597	670
4150	1-1/4" pipe size		12	.667		1,125	43		1,168	1,300
4160	1-1/2" pipe size		10	.800		1,150	51.50		1,201.50	1,350
4180	2" pipe size		7	1.143		1,325	73.50		1,398.50	1,600
5000	Flanged, bronze, valves are OS&Y									
5060	2-1/2" pipe size	Q-1	5	3.200	Ea.	5,475	186		5,661	6,300
5080	3" pipe size		4.50	3.556		6,600	206		6,806	7,575
5100	4" pipe size		3	5.333		7,725	310		8,035	8,975
5120	6" pipe size	Q-2	3	8		12,000	480		12,480	13,900
5600	Flanged, iron, valves are OS&Y									
5660	2-1/2" pipe size	Q-1	5	3.200	Ea.	3,300	186		3,486	3,900

22 11 Facility Water Distribution

22 11 19 – Domestic Water Piping Specialties

22 11 19.42 Backflow Preventers

		Crew	Daily Output	Labor-Hours	Unit	Material	2020 Bare Costs Labor	2020 Bare Costs Equipment	Total	Total Incl O&P
5680	3" pipe size	Q-1	4.50	3.556	Ea.	3,275	206		3,481	3,925
5700	4" pipe size	↓	3	5.333		4,725	310		5,035	5,675
5720	6" pipe size	Q-2	3	8		5,825	480		6,305	7,175
5800	Rebuild 4" diameter reduced pressure	1 Plum	4	2		385	129		514	620
5810	6" diameter		2.66	3.008		405	194		599	745
5820	8" diameter		2	4		505	258		763	955
5830	10" diameter		1.60	5	↓	585	320		905	1,150
9010	Minimum labor/equipment charge		2	4	Job		258		258	400

22 11 19.50 Vacuum Breakers

		Crew	Daily Output	Labor-Hours	Unit	Material	2020 Bare Costs Labor	2020 Bare Costs Equipment	Total	Total Incl O&P
0010	**VACUUM BREAKERS**									
0013	See also backflow preventers Section 22 11 19.42									
1000	Anti-siphon continuous pressure type									
1010	Max. 150 psi - 210°F									
1020	Bronze body									
1030	1/2" size	1 Stpi	24	.333	Ea.	192	22		214	245
1040	3/4" size		20	.400		192	26		218	252
1050	1" size		19	.421		197	27.50		224.50	259
1060	1-1/4" size		15	.533		390	35		425	485
1070	1-1/2" size		13	.615		475	40.50		515.50	590
1080	2" size	↓	11	.727	↓	490	47.50		537.50	615
1200	Max. 125 psi with atmospheric vent									
1210	Brass, in-line construction									
1220	1/4" size	1 Stpi	24	.333	Ea.	143	22		165	191
1230	3/8" size	"	24	.333		143	22		165	191
1260	For polished chrome finish, add				↓	13%				
2000	Anti-siphon, non-continuous pressure type									
2010	Hot or cold water 125 psi - 210°F									
2020	Bronze body									
2030	1/4" size	1 Stpi	24	.333	Ea.	88.50	22		110.50	131
2040	3/8" size		24	.333		88.50	22		110.50	131
2050	1/2" size		24	.333		99	22		121	143
2060	3/4" size		20	.400		119	26		145	171
2070	1" size		19	.421		183	27.50		210.50	245
2080	1-1/4" size		15	.533		320	35		355	410
2090	1-1/2" size		13	.615		350	40.50		390.50	450
2100	2" size		11	.727		585	47.50		632.50	720
2110	2-1/2" size		8	1		1,675	65.50		1,740.50	1,950
2120	3" size	↓	6	1.333	↓	2,225	87.50		2,312.50	2,575
2150	For polished chrome finish, add					50%				
3000	Air gap fitting									
3020	1/2" NPT size	1 Plum	19	.421	Ea.	61.50	27		88.50	110
3030	1" NPT size	"	15	.533		70.50	34.50		105	131
3040	2" NPT size	Q-1	21	.762		138	44		182	221
3050	3" NPT size		14	1.143		273	66.50		339.50	400
3060	4" NPT size	↓	10	1.600	↓	273	93		366	445

22 11 19.54 Water Hammer Arresters/Shock Absorbers

		Crew	Daily Output	Labor-Hours	Unit	Material	2020 Bare Costs Labor	2020 Bare Costs Equipment	Total	Total Incl O&P
0010	**WATER HAMMER ARRESTERS/SHOCK ABSORBERS**									
0490	Copper									
0500	3/4" male IPS for 1 to 11 fixtures	1 Plum	12	.667	Ea.	31	43		74	101
0600	1" male IPS for 12 to 32 fixtures		8	1		51.50	64.50		116	156
0700	1-1/4" male IPS for 33 to 60 fixtures		8	1		50.50	64.50		115	156
0800	1-1/2" male IPS for 61 to 113 fixtures	↓	8	1		73	64.50		137.50	180

22 11 Facility Water Distribution

22 11 19 – Domestic Water Piping Specialties

22 11 19.54 Water Hammer Arresters/Shock Absorbers

		Crew	Daily Output	Labor-Hours	Unit	Material	2020 Bare Costs Labor	Equipment	Total	Total Incl O&P
0900	2" male IPS for 114 to 154 fixtures	1 Plum	8	1	Ea.	108	64.50		172.50	218
1000	2-1/2" male IPS for 155 to 330 fixtures	↓	4	2	↓	330	129		459	565
9000	Minimum labor/equipment charge	↓	3.50	2.286	Job		147		147	228

22 11 19.64 Hydrants

		Crew	Daily Output	Labor-Hours	Unit	Material	2020 Bare Costs Labor	Equipment	Total	Total Incl O&P
0010	**HYDRANTS**									
0050	Wall type, moderate climate, bronze, encased									
0200	3/4" IPS connection	1 Plum	16	.500	Ea.	1,025	32		1,057	1,175
1000	Non-freeze, bronze, exposed									
1100	3/4" IPS connection, 4" to 9" thick wall	1 Plum	14	.571	Ea.	470	37		507	570
1120	10" to 14" thick wall	"	12	.667		445	43		488	555
1280	For anti-siphon type, add				↓	176			176	194
9000	Minimum labor/equipment charge	1 Plum	3	2.667	Job		172		172	266

22 13 Facility Sanitary Sewerage

22 13 16 – Sanitary Waste and Vent Piping

22 13 16.20 Pipe, Cast Iron

		Crew	Daily Output	Labor-Hours	Unit	Material	2020 Bare Costs Labor	Equipment	Total	Total Incl O&P
0010	**PIPE, CAST IRON**, Soil, on clevis hanger assemblies, 5' OC R221113-50									
0020	Single hub, service wt., lead & oakum joints 10' OC									
2120	2" diameter	Q-1	63	.254	L.F.	17.10	14.75		31.85	42
2140	3" diameter		60	.267		26.50	15.45		41.95	53
2160	4" diameter	↓	55	.291		24.50	16.85		41.35	53
2180	5" diameter	Q-2	76	.316		31	19		50	63.50
2200	6" diameter	"	73	.329		42.50	19.80		62.30	77
2220	8" diameter	Q-3	59	.542	↓	64.50	33.50		98	123
4000	No hub, couplings 10' OC									
4100	1-1/2" diameter	Q-1	71	.225	L.F.	16.85	13.05		29.90	38.50
4120	2" diameter		67	.239		20	13.85		33.85	43.50
4140	3" diameter		64	.250		27	14.50		41.50	52
4160	4" diameter	↓	58	.276		25	16		41	52
4180	5" diameter	Q-2	83	.289	↓	36.50	17.40		53.90	67.50
9000	Minimum labor/equipment charge	1 Plum	4	2	Job		129		129	199

22 13 16.30 Pipe Fittings, Cast Iron

		Crew	Daily Output	Labor-Hours	Unit	Material	2020 Bare Costs Labor	Equipment	Total	Total Incl O&P
0010	**PIPE FITTINGS, CAST IRON**, Soil R221113-50									
0040	Hub and spigot, service weight, lead & oakum joints									
0080	1/4 bend, 2"	Q-1	16	1	Ea.	23.50	58		81.50	115
0120	3"		14	1.143		31	66.50		97.50	136
0140	4"	↓	13	1.231		49	71.50		120.50	164
0160	5"	Q-2	18	1.333		68	80		148	199
0180	6"	"	17	1.412		85	85		170	225
0200	8"	Q-3	11	2.909		256	178		434	560
0340	1/8 bend, 2"	Q-1	16	1		16.65	58		74.65	108
0350	3"		14	1.143		26	66.50		92.50	131
0360	4"	↓	13	1.231		38	71.50		109.50	152
0380	5"	Q-2	18	1.333		54	80		134	184
0400	6"	"	17	1.412		64.50	85		149.50	202
0420	8"	Q-3	11	2.909		193	178		371	490
0500	Sanitary tee, 2"	Q-1	10	1.600		32.50	93		125.50	179
0540	3"		9	1.778		53	103		156	217
0620	4"	↓	8	2		65.50	116		181.50	251
0700	5"	Q-2	12	2		129	120		249	330

22 13 16.30 Pipe Fittings, Cast Iron	Crew	Daily Output	Labor-Hours	Unit	Material	2020 Bare Costs Labor	Equipment	Total	Total Incl O&P
0800	6"	Q-2	11	2.182	Ea.	146	131		277	365
0880	8"	Q-3	7	4.571	▼	385	280		665	860
5990	No hub									
6000	Cplg. & labor required at joints not incl. in fitting									
6010	price. Add 1 coupling per joint for installed price									
6020	1/4 bend, 1-1/2"				Ea.	11.65			11.65	12.80
6060	2"					12.60			12.60	13.85
6080	3"					17.65			17.65	19.40
6120	4"					26			26	28.50
6184	1/4 bend, long sweep, 1-1/2"					29.50			29.50	32.50
6186	2"					28			28	30.50
6188	3"					33.50			33.50	37
6189	4"					53			53	58.50
6190	5"					103			103	114
6191	6"					118			118	129
6192	8"					310			310	345
6193	10"					655			655	725
6200	1/8 bend, 1-1/2"					9.70			9.70	10.70
6210	2"					10.90			10.90	12
6212	3"					14.55			14.55	16
6214	4"					19.15			19.15	21
6380	Sanitary tee, tapped, 1-1/2"					23.50			23.50	26
6382	2" x 1-1/2"					21			21	23
6384	2"					22			22	24
6386	3" x 2"					33.50			33.50	36.50
6388	3"					56			56	62
6390	4" x 1-1/2"					30			30	32.50
6392	4" x 2"					34			34	37
6393	4"					34			34	37
6394	6" x 1-1/2"					76.50			76.50	84
6396	6" x 2"					77.50			77.50	85
6459	Sanitary tee, 1-1/2"					16.25			16.25	17.85
6460	2"					17.40			17.40	19.15
6470	3"					21.50			21.50	23.50
6472	4"				▼	41			41	45
8000	Coupling, standard (by CISPI Mfrs.)									
8020	1-1/2"	Q-1	48	.333	Ea.	16.70	19.35		36.05	48.50
8040	2"		44	.364		17.80	21		38.80	52
8080	3"		38	.421		20	24.50		44.50	60
8120	4"	▼	33	.485	▼	23	28		51	69
8300	Coupling, cast iron clamp & neoprene gasket (by MG)									
8310	1-1/2"	Q-1	48	.333	Ea.	9.30	19.35		28.65	40.50
8320	2"		44	.364		12.90	21		33.90	46.50
8330	3"		38	.421		11.55	24.50		36.05	50.50
8340	4"	▼	33	.485	▼	18.85	28		46.85	64
8600	Coupling, stainless steel, heavy duty									
8620	1-1/2"	Q-1	48	.333	Ea.	5.90	19.35		25.25	36.50
8630	2"		44	.364		6.40	21		27.40	39.50
8640	2" x 1-1/2"		44	.364		11.45	21		32.45	45
8650	3"		38	.421		6.70	24.50		31.20	45.50
8660	4"	▼	33	.485		7.55	28		35.55	52
9000	Minimum labor/equipment charge	1 Plum	4	2	Job		129		129	199

For customer support on your Commercial Renovation Costs with RSMeans data, call 800.448.8182.

437

22 13 16.60 Traps		Crew	Daily Output	Labor-Hours	Unit	Material	2020 Bare Costs Labor	Equipment	Total	Total Incl O&P
0010	**TRAPS**									
0030	Cast iron, service weight									
0050	Running P trap, without vent									
1100	2"	Q-1	16	1	Ea.	167	58		225	274
1150	4"	"	13	1.231		167	71.50		238.50	294
1160	6"	Q-2	17	1.412		770	85		855	980
3000	P trap, B&S, 2" pipe size	Q-1	16	1		41.50	58		99.50	135
3040	3" pipe size	"	14	1.143	↓	61.50	66.50		128	170
4700	Copper, drainage, drum trap									
4840	3" x 6" swivel, 1-1/2" pipe size	1 Plum	16	.500	Ea.	315	32		347	395
5100	P trap, standard pattern									
5200	1-1/4" pipe size	1 Plum	18	.444	Ea.	102	28.50		130.50	158
5240	1-1/2" pipe size		17	.471		111	30.50		141.50	170
5260	2" pipe size		15	.533		184	34.50		218.50	255
5280	3" pipe size		11	.727	↓	525	47		572	655
9000	Minimum labor/equipment charge	↓	3	2.667	Job		172		172	266

22 13 16.80 Vent Flashing and Caps

		Crew	Daily Output	Labor-Hours	Unit	Material	Labor	Equipment	Total	Total Incl O&P
0010	**VENT FLASHING AND CAPS**									
0120	Vent caps									
0140	Cast iron									
0180	2-1/2" to 3-5/8" pipe	1 Plum	21	.381	Ea.	52	24.50		76.50	95.50
0190	4" to 4-1/8" pipe	"	19	.421	"	75	27		102	125
0900	Vent flashing									
1000	Aluminum with lead ring									
1050	3" pipe	1 Plum	17	.471	Ea.	6.25	30.50		36.75	54
1060	4" pipe	"	16	.500	"	7.55	32		39.55	58.50
1350	Copper with neoprene ring									
1440	2" pipe	1 Plum	18	.444	Ea.	74	28.50		102.50	126
1450	3" pipe		17	.471		89.50	30.50		120	146
1460	4" pipe		16	.500	↓	89.50	32		121.50	149
9000	Minimum labor/equipment charge	↓	4	2	Job		129		129	199

22 13 19.13 Sanitary Drains

		Crew	Daily Output	Labor-Hours	Unit	Material	Labor	Equipment	Total	Total Incl O&P
0010	**SANITARY DRAINS**									
0400	Deck, auto park, CI, 13" top									
0440	3", 4", 5", and 6" pipe size	Q-1	8	2	Ea.	1,875	116		1,991	2,225
0480	For galvanized body, add				"	1,125			1,125	1,250
2000	Floor, medium duty, CI, deep flange, 7" diam. top									
2040	2" and 3" pipe size	Q-1	12	1.333	Ea.	227	77.50		304.50	370
2080	For galvanized body, add					140			140	154
2120	With polished bronze top				↓	375			375	410
2500	Heavy duty, cleanout & trap w/bucket, CI, 15" top									
2540	2", 3", and 4" pipe size	Q-1	6	2.667	Ea.	8,125	155		8,280	9,200
2560	For galvanized body, add					2,425			2,425	2,675
2580	With polished bronze top				↓	9,700			9,700	10,700

22 13 19.14 Floor Receptors

		Crew	Daily Output	Labor-Hours	Unit	Material	Labor	Equipment	Total	Total Incl O&P
0010	**FLOOR RECEPTORS,** For connection to 2", 3" & 4" diameter pipe									
0200	12-1/2" square top, 25 sq. in. open area	Q-1	10	1.600	Ea.	980	93		1,073	1,225
0300	For grate with 4" diameter x 3-3/4" high funnel, add					330			330	360
0400	For grate with 6" diameter x 6" high funnel, add					264			264	290
0700	For acid-resisting bucket, add				↓	325			325	355

22 13 19 – Sanitary Waste Piping Specialties

22 13 19.14 Floor Receptors

	Crew	Daily Output	Labor-Hours	Unit	Material	2020 Bare Costs Labor	Equipment	Total	Total Incl O&P
0900	For stainless steel mesh bucket liner, add				Ea.	282			282	310
2000	12-5/8" diameter top, 40 sq. in. open area	Q-1	10	1.600	↓	960	93		1,053	1,200
2100	For options, add same prices as square top									
3000	8" x 4" rectangular top, 7.5 sq. in. open area	Q-1	14	1.143	Ea.	1,050	66.50		1,116.50	1,250
3100	For trap primer connections, add				"	110			110	121
9000	Minimum labor/equipment charge	Q-1	3	5.333	Job		310		310	480

22 13 19.15 Sink Waste Treatment

	Crew	Daily Output	Labor-Hours	Unit	Material	2020 Bare Costs Labor	Equipment	Total	Total Incl O&P
0010	**SINK WASTE TREATMENT**, System for commercial kitchens									
0100	includes clock timer & fittings									
0200	System less chemical, wall mounted cabinet	1 Plum	16	.500	Ea.	505	32		537	605
2000	Chemical, 1 gallon, add					41			41	45
2100	6 gallons, add					185			185	203
2200	15 gallons, add					500			500	550
2300	30 gallons, add					940			940	1,025
2400	55 gallons, add				↓	1,600			1,600	1,750

22 13 23 – Sanitary Waste Interceptors

22 13 23.10 Interceptors

	Crew	Daily Output	Labor-Hours	Unit	Material	2020 Bare Costs Labor	Equipment	Total	Total Incl O&P
0010	**INTERCEPTORS**									
0150	Grease, fabricated steel, 4 GPM, 8 lb. fat capacity	1 Plum	4	2	Ea.	1,575	129		1,704	1,925
0200	7 GPM, 14 lb. fat capacity		4	2		2,050	129		2,179	2,475
1000	10 GPM, 20 lb. fat capacity	↓	4	2		2,575	129		2,704	3,025
1120	50 GPM, 100 lb. fat capacity	Q-1	2	8		8,075	465		8,540	9,600
1160	100 GPM, 200 lb. fat capacity		2	8		20,100	465		20,565	22,900
1240	300 GPM, 600 lb. fat capacity	↓	1	16	↓	40,400	930		41,330	45,900
9000	Minimum labor/equipment charge	1 Plum	3	2.667	Job		172		172	266

22 13 29 – Sanitary Sewerage Pumps

22 13 29.14 Sewage Ejector Pumps

	Crew	Daily Output	Labor-Hours	Unit	Material	2020 Bare Costs Labor	Equipment	Total	Total Incl O&P
0010	**SEWAGE EJECTOR PUMPS**, With operating and level controls									
0100	Simplex system incl. tank, cover, pump 15' head									
0500	37 gal. PE tank, 12 GPM, 1/2 HP, 2" discharge	Q-1	3.20	5	Ea.	505	290		795	1,000
0510	3" discharge		3.10	5.161		550	299		849	1,075
0530	87 GPM, .7 HP, 2" discharge		3.20	5		770	290		1,060	1,300
0540	3" discharge		3.10	5.161		835	299		1,134	1,375
0600	45 gal. coated stl. tank, 12 GPM, 1/2 HP, 2" discharge		3	5.333		900	310		1,210	1,475
0610	3" discharge		2.90	5.517		935	320		1,255	1,525
0630	87 GPM, .7 HP, 2" discharge		3	5.333		1,150	310		1,460	1,750
0640	3" discharge		2.90	5.517		1,225	320		1,545	1,850
0660	134 GPM, 1 HP, 2" discharge		2.80	5.714		1,250	330		1,580	1,875
0680	3" discharge		2.70	5.926		1,325	345		1,670	1,975
0700	70 gal. PE tank, 12 GPM, 1/2 HP, 2" discharge		2.60	6.154		970	355		1,325	1,625
0710	3" discharge		2.40	6.667		1,050	385		1,435	1,750
0730	87 GPM, .7 HP, 2" discharge		2.50	6.400		1,275	370		1,645	2,000
0740	3" discharge		2.30	6.957		1,325	405		1,730	2,100
0760	134 GPM, 1 HP, 2" discharge		2.20	7.273		1,375	420		1,795	2,150
0770	3" discharge		2	8	↓	1,450	465		1,915	2,325
9000	Minimum labor/equipment charge	↓	2.50	6.400	Job		370		370	575

For customer support on your Commercial Renovation Costs with RSMeans data, call 800.448.8182.

439

22 14 Facility Storm Drainage

22 14 26 – Facility Storm Drains

22 14 26.13 Roof Drains	Crew	Daily Output	Labor-Hours	Unit	Material	2020 Bare Costs Labor	Equipment	Total	Total Incl O&P
0010 **ROOF DRAINS**									
0140 Cornice, CI, 45° or 90° outlet									
0200 3" and 4" pipe size	Q-1	12	1.333	Ea.	385	77.50		462.50	545
0260 For galvanized body, add					111			111	122
0280 For polished bronze dome, add				↓	114			114	126
3860 Roof, flat metal deck, CI body, 12" CI dome									
3890 3" pipe size	Q-1	14	1.143	Ea.	410	66.50		476.50	555
3900 4" pipe size	"	13	1.231	"	510	71.50		581.50	670
4620 Main, all aluminum, 12" low profile dome									
4640 2", 3" and 4" pipe size	Q-1	14	1.143	Ea.	590	66.50		656.50	745
9000 Minimum labor/equipment charge	1 Plum	4	2	Job		129		129	199

22 14 29 – Sump Pumps

22 14 29.16 Submersible Sump Pumps

	Crew	Daily Output	Labor-Hours	Unit	Material	2020 Bare Costs Labor	Equipment	Total	Total Incl O&P
0010 **SUBMERSIBLE SUMP PUMPS**									
1000 Elevator sump pumps, automatic									
1010 Complete systems, pump, oil detector, controls and alarm									
1020 1-1/2" discharge, does not include the sump pit/tank									
1040 1/3 HP, 115 V	1 Plum	4.40	1.818	Ea.	1,925	117		2,042	2,275
1050 1/2 HP, 115 V		4	2		2,025	129		2,154	2,425
1060 1/2 HP, 230 V		4	2		2,000	129		2,129	2,400
1070 3/4 HP, 115 V		3.60	2.222		2,100	143		2,243	2,550
1080 3/4 HP, 230 V	↓	3.60	2.222	↓	2,075	143		2,218	2,500
1100 Sump pump only									
1110 1/3 HP, 115 V	1 Plum	6.40	1.250	Ea.	179	80.50		259.50	320
1120 1/2 HP, 115 V		5.80	1.379		254	89		343	415
1130 1/2 HP, 230 V		5.80	1.379		294	89		383	460
1140 3/4 HP, 115 V		5.40	1.481		340	95.50		435.50	525
1150 3/4 HP, 230 V	↓	5.40	1.481	↓	380	95.50		475.50	570
1200 Oil detector, control and alarm only									
1210 115 V	1 Plum	8	1	Ea.	1,775	64.50		1,839.50	2,050
1220 230 V	"	8	1	"	1,775	64.50		1,839.50	2,050
7000 Sump pump, automatic									
7100 Plastic, 1-1/4" discharge, 1/4 HP	1 Plum	6.40	1.250	Ea.	168	80.50		248.50	310
7140 1/3 HP		6	1.333		224	86		310	380
7180 1-1/2" discharge, 1/2 HP		5.20	1.538		325	99		424	510
7500 Cast iron, 1-1/4" discharge, 1/4 HP		6	1.333		225	86		311	380
7540 1/3 HP		6	1.333		264	86		350	425
7560 1/2 HP		5	1.600	↓	320	103		423	510
9000 Minimum labor/equipment charge	↓	4	2	Job		129		129	199

22 15 General Service Compressed-Air Systems

22 15 19 – General Service Packaged Air Compressors and Receivers

22 15 19.10 Air Compressors

	Crew	Daily Output	Labor-Hours	Unit	Material	2020 Bare Costs Labor	Equipment	Total	Total Incl O&P
0010 **AIR COMPRESSORS**									
5250 Air, reciprocating air cooled, splash lubricated, tank mounted									
5300 Single stage, 1 phase, 140 psi									
5303 1/2 HP, 17 gal. tank	1 Stpi	3	2.667	Ea.	2,000	175		2,175	2,475
5305 3/4 HP, 30 gal. tank		2.60	3.077		1,700	202		1,902	2,175
5307 1 HP, 30 gal. tank	↓	2.20	3.636	↓	2,075	238		2,313	2,675

22 31 Domestic Water Softeners

22 31 13 – Residential Domestic Water Softeners

22 31 13.10 Residential Water Softeners	Crew	Daily Output	Labor-Hours	Unit	Material	2020 Bare Costs Labor	Equipment	Total	Total Incl O&P
0010 **RESIDENTIAL WATER SOFTENERS**									
7350 Water softener, automatic, to 30 grains per gallon	2 Plum	5	3.200	Ea.	395	206		601	755
7400 To 100 grains per gallon	"	4	4	"	940	258		1,198	1,425

22 33 Electric Domestic Water Heaters

22 33 13 – Instantaneous Electric Domestic Water Heaters

22 33 13.20 Instantaneous Elec. Point-Of-Use Water Heaters

		Crew	Daily Output	Labor-Hours	Unit	Material	2020 Bare Costs Labor	Equipment	Total	Total Incl O&P
0010	**INSTANTANEOUS ELECTRIC POINT-OF-USE WATER HEATERS**									
8965	Point of use, electric, glass lined									
8969	Energy saver									
8970	2.5 gal. single element	G	1 Plum	2.80	2.857	Ea.	370	184		554	690
8971	4 gal. single element	G		2.80	2.857		375	184		559	695
8974	6 gal. single element	G		2.50	3.200		415	206		621	775
8975	10 gal. single element	G		2.50	3.200		470	206		676	835
8976	15 gal. single element	G		2.40	3.333		545	215		760	930
8977	20 gal. single element	G		2.40	3.333		675	215		890	1,075
8978	30 gal. single element	G		2.30	3.478		700	224		924	1,125
8979	40 gal. single element	G		2.20	3.636		1,250	234		1,484	1,725
8988	Commercial (ASHRAE energy std. 90)									
8989	6 gallon	G	1 Plum	2.50	3.200	Ea.	1,000	206		1,206	1,425
8990	10 gallon	G		2.50	3.200		1,175	206		1,381	1,600
8991	15 gallon	G		2.40	3.333		1,150	215		1,365	1,600
8992	20 gallon	G		2.40	3.333		1,350	215		1,565	1,800
8993	30 gallon	G		2.30	3.478		2,825	224		3,049	3,450
8995	Under the sink, copper, w/bracket									
8996	2.5 gallon	G	1 Plum	4	2	Ea.	405	129		534	645
9000	Minimum labor/equipment charge		"	1.75	4.571	Job		295		295	455

22 33 30 – Residential, Electric Domestic Water Heaters

22 33 30.13 Residential, Small-Capacity Elec. Water Heaters

		Crew	Daily Output	Labor-Hours	Unit	Material	2020 Bare Costs Labor	Equipment	Total	Total Incl O&P
0010	**RESIDENTIAL, SMALL-CAPACITY ELECTRIC DOMESTIC WATER HEATERS**									
1000	Residential, electric, glass lined tank, 5 yr., 10 gal., single element	1 Plum	2.30	3.478	Ea.	470	224		694	860
1060	30 gallon, double element		2.20	3.636		1,075	234		1,309	1,525
1080	40 gallon, double element		2	4		1,250	258		1,508	1,775
1100	52 gallon, double element		2	4		1,425	258		1,683	1,975
1120	66 gallon, double element		1.80	4.444		1,925	286		2,211	2,575
1140	80 gallon, double element		1.60	5		2,175	320		2,495	2,900

22 33 33 – Light-Commercial Electric Domestic Water Heaters

22 33 33.10 Commercial Electric Water Heaters

		Crew	Daily Output	Labor-Hours	Unit	Material	2020 Bare Costs Labor	Equipment	Total	Total Incl O&P
0010	**COMMERCIAL ELECTRIC WATER HEATERS**									
4000	Commercial, 100° rise. NOTE: for each size tank, a range of									
4010	heaters between the ones shown is available									
4020	Electric									
4100	5 gal., 3 kW, 12 GPH, 208 volt	1 Plum	2	4	Ea.	4,600	258		4,858	5,450
4160	50 gal., 36 kW, 148 GPH, 208 volt	"	1.80	4.444		11,100	286		11,386	12,600
4480	400 gal., 210 kW, 860 GPH, 480 volt	Q-1	1	16		80,500	930		81,430	90,000

22 34 Fuel-Fired Domestic Water Heaters

22 34 13 – Instantaneous, Tankless, Gas Domestic Water Heaters

22 34 13.10 Instantaneous, Tankless, Gas Water Heaters

		Crew	Daily Output	Labor-Hours	Unit	Material	2020 Bare Costs Labor	2020 Bare Costs Equipment	Total	Total Incl O&P
0010	**INSTANTANEOUS, TANKLESS, GAS WATER HEATERS**									
9410	Natural gas/propane, 3.2 GPM G	1 Plum	2	4	Ea.	575	258		833	1,025
9420	6.4 GPM G		1.90	4.211		780	271		1,051	1,275
9430	8.4 GPM G		1.80	4.444		880	286		1,166	1,400
9440	9.5 GPM G		1.60	5		1,050	320		1,370	1,650

22 34 30 – Residential Gas Domestic Water Heaters

22 34 30.13 Residential, Atmos, Gas Domestic Wtr Heaters

		Crew	Daily Output	Labor-Hours	Unit	Material	Labor	Equipment	Total	Total Incl O&P
0010	**RESIDENTIAL, ATMOSPHERIC, GAS DOMESTIC WATER HEATERS**									
2000	Gas fired, foam lined tank, 10 yr., vent not incl.									
2040	30 gallon	1 Plum	2	4	Ea.	2,100	258		2,358	2,725
2060	40 gallon		1.90	4.211		1,825	271		2,096	2,425
2100	75 gallon		1.50	5.333		2,650	345		2,995	3,450
3000	Tank leak safety, water & gas shut off see 22 05 23.20 8800									

22 34 36 – Commercial Gas Domestic Water Heaters

22 34 36.13 Commercial, Atmos., Gas Domestic Water Htrs.

		Crew	Daily Output	Labor-Hours	Unit	Material	Labor	Equipment	Total	Total Incl O&P
0010	**COMMERCIAL, ATMOSPHERIC, GAS DOMESTIC WATER HEATERS**									
6000	Gas fired, flush jacket, std. controls, vent not incl.									
6040	75 MBH input, 73 GPH	1 Plum	1.40	5.714	Ea.	3,725	370		4,095	4,675
6060	98 MBH input, 95 GPH		1.40	5.714		8,800	370		9,170	10,200
6180	200 MBH input, 192 GPH		.60	13.333		12,100	860		12,960	14,700
6900	For low water cutoff, add		8	1		385	64.50		449.50	525
6960	For bronze body hot water circulator, add		4	2		2,250	129		2,379	2,675

22 34 46 – Oil-Fired Domestic Water Heaters

22 34 46.10 Residential Oil-Fired Water Heaters

		Crew	Daily Output	Labor-Hours	Unit	Material	Labor	Equipment	Total	Total Incl O&P
0010	**RESIDENTIAL OIL-FIRED WATER HEATERS**									
3000	Oil fired, glass lined tank, 5 yr., vent not included, 30 gallon	1 Plum	2	4	Ea.	1,400	258		1,658	1,950
3040	50 gallon	"	1.80	4.444	"	1,375	286		1,661	1,975

22 34 46.20 Commercial Oil-Fired Water Heaters

		Crew	Daily Output	Labor-Hours	Unit	Material	Labor	Equipment	Total	Total Incl O&P
0010	**COMMERCIAL OIL-FIRED WATER HEATERS**									
8000	Oil fired, glass lined, UL listed, std. controls, vent not incl.									
8060	140 gal., 140 MBH input, 134 GPH	Q-1	2.13	7.512	Ea.	25,200	435		25,635	28,400
8080	140 gal., 199 MBH input, 191 GPH		2	8		26,100	465		26,565	29,400
8160	140 gal., 540 MBH input, 519 GPH		.96	16.667		35,600	965		36,565	40,600
8280	201 gal., 1,250 MBH input, 1,200 GPH	Q-2	1.22	19.672		56,000	1,175		57,175	63,500

22 41 Residential Plumbing Fixtures

22 41 13 – Residential Water Closets, Urinals, and Bidets

22 41 13.13 Water Closets

		Crew	Daily Output	Labor-Hours	Unit	Material	Labor	Equipment	Total	Total Incl O&P
0010	**WATER CLOSETS**									
0032	For automatic flush, see Line 22 42 39.10 0972									
0150	Tank type, vitreous china, incl. seat, supply pipe w/stop, 1.6 gpf or noted									
0200	Wall hung									
0400	Two piece, close coupled	Q-1	5.30	3.019	Ea.	420	175		595	730
0960	For rough-in, supply, waste, vent and carrier	"	2.73	5.861	"	1,275	340		1,615	1,925
0999	Floor mounted									
1100	Two piece, close coupled	Q-1	5.30	3.019	Ea.	219	175		394	510
1102	Economy		5.30	3.019		121	175		296	405
1110	Two piece, close coupled, dual flush		5.30	3.019		280	175		455	580
1140	Two piece, close coupled, 1.28 gpf, ADA G		5.30	3.019		320	175		495	620

22 41 Residential Plumbing Fixtures

22 41 13 – Residential Water Closets, Urinals, and Bidets

22 41 13.13 Water Closets

		Crew	Daily Output	Labor-Hours	Unit	Material	2020 Bare Costs Labor	Equipment	Total	Total Incl O&P
1960	For color, add					30%				
1980	For rough-in, supply, waste and vent	Q-1	3.05	5.246	Ea.	340	305		645	845

22 41 16 – Residential Lavatories and Sinks

22 41 16.13 Lavatories

		Crew	Daily Output	Labor-Hours	Unit	Material	2020 Bare Costs Labor	Equipment	Total	Total Incl O&P
0010	**LAVATORIES**, With trim, white unless noted otherwise									
0500	Vanity top, porcelain enamel on cast iron									
0600	20" x 18"	Q-1	6.40	2.500	Ea.	305	145		450	565
0640	33" x 19" oval	"	6.40	2.500	"	520	145		665	800
0860	For color, add					25%				
1000	Cultured marble, 19" x 17", single bowl	Q-1	6.40	2.500	Ea.	124	145		269	360
1120	25" x 22", single bowl	"	6.40	2.500	"	164	145		309	405
1580	For color, same price									
1900	Stainless steel, self-rimming, 25" x 22", single bowl, ledge	Q-1	6.40	2.500	Ea.	315	145		460	575
1960	17" x 22", single bowl		6.40	2.500		305	145		450	560
2600	Steel, enameled, 20" x 17", single bowl		5.80	2.759		126	160		286	385
2900	Vitreous china, 20" x 16", single bowl		5.40	2.963		211	172		383	500
2960	20" x 17", single bowl		5.40	2.963		121	172		293	400
3580	Rough-in, supply, waste and vent for all above lavatories		2.30	6.957		278	405		683	930
4000	Wall hung									
4040	Porcelain enamel on cast iron, 16" x 14", single bowl	Q-1	8	2	Ea.	435	116		551	655
4180	20" x 18", single bowl	"	8	2	"	246	116		362	450
4580	For color, add					30%				
6000	Vitreous china, 18" x 15", single bowl with backsplash	Q-1	7	2.286	Ea.	170	133		303	390
6960	Rough-in, supply, waste and vent for above lavatories	"	1.66	9.639	"	475	560		1,035	1,375
7000	Pedestal type									
7600	Vitreous china, 27" x 21", white	Q-1	6.60	2.424	Ea.	685	141		826	970
7610	27" x 21", colored		6.60	2.424		875	141		1,016	1,175
7620	27" x 21", premium color		6.60	2.424		995	141		1,136	1,325
7660	26" x 20", white		6.60	2.424		665	141		806	945
7670	26" x 20", colored		6.60	2.424		855	141		996	1,150
7680	26" x 20", premium color		6.60	2.424		1,025	141		1,166	1,350
7700	24" x 20", white		6.60	2.424		465	141		606	725
7710	24" x 20", colored		6.60	2.424		580	141		721	850
7720	24" x 20", premium color		6.60	2.424		640	141		781	920
7760	21" x 18", white		6.60	2.424		258	141		399	500
7770	21" x 18", colored		6.60	2.424		289	141		430	530
7990	Rough-in, supply, waste and vent for pedestal lavatories		1.66	9.639		475	560		1,035	1,375
9000	Minimum labor/equipment charge	1 Plum	3	2.667	Job		172		172	266

22 41 16.16 Sinks

		Crew	Daily Output	Labor-Hours	Unit	Material	2020 Bare Costs Labor	Equipment	Total	Total Incl O&P
0010	**SINKS**, With faucets and drain									
2000	Kitchen, counter top style, PE on CI, 24" x 21" single bowl	Q-1	5.60	2.857	Ea.	310	166		476	595
2100	31" x 22" single bowl		5.60	2.857		860	166		1,026	1,200
2200	32" x 21" double bowl		4.80	3.333		390	193		583	730
3000	Stainless steel, self rimming, 19" x 18" single bowl		5.60	2.857		625	166		791	940
3100	25" x 22" single bowl		5.60	2.857		690	166		856	1,025
3200	33" x 22" double bowl		4.80	3.333		1,000	193		1,193	1,400
3300	43" x 22" double bowl		4.80	3.333		1,150	193		1,343	1,575
4000	Steel, enameled, with ledge, 24" x 21" single bowl		5.60	2.857		535	166		701	845
4100	32" x 21" double bowl		4.80	3.333		520	193		713	870
4960	For color sinks except stainless steel, add					10%				
4980	For rough-in, supply, waste and vent, counter top sinks	Q-1	2.14	7.477		320	435		755	1,025
5790	For rough-in, supply, waste & vent, sinks	"	1.85	8.649		320	500		820	1,125

443

For customer support on your Commercial Renovation Costs with RSMeans data, call 800.448.8182.

22 41 Residential Plumbing Fixtures

22 41 19 – Residential Bathtubs

22 41 19.10 Baths

		Crew	Daily Output	Labor-Hours	Unit	Material	2020 Bare Costs Labor	2020 Bare Costs Equipment	Total	Total Incl O&P
0010	**BATHS**									
0100	Tubs, recessed porcelain enamel on cast iron, with trim									
0180	48" x 42"	Q-1	4	4	Ea.	3,150	232		3,382	3,800
0220	72" x 36"	"	3	5.333	"	2,975	310		3,285	3,750
0300	Mat bottom									
0380	5' long	Q-1	4.40	3.636	Ea.	1,275	211		1,486	1,725
0560	Corner 48" x 44"		4.40	3.636		2,925	211		3,136	3,525
2000	Enameled formed steel, 4'-6" long		5.80	2.759		515	160		675	810
4600	Module tub & showerwall surround, molded fiberglass									
4610	5' long x 34" wide x 76" high	Q-1	4	4	Ea.	790	232		1,022	1,225
4750	ADA compliant with 1-1/2" OD grab bar, antiskid bottom									
4760	60" x 32-3/4" x 72" high	Q-1	4	4	Ea.	595	232		827	1,025
4770	60" x 30" x 71" high with molded seat		3.50	4.571		755	265		1,020	1,250
9600	Rough-in, supply, waste and vent, for all above tubs, add		2.07	7.729		445	450		895	1,175
9900	Minimum labor/equipment charge		3	5.333	Job		310		310	480

22 41 23 – Residential Showers

22 41 23.20 Showers

		Crew	Daily Output	Labor-Hours	Unit	Material	2020 Bare Costs Labor	2020 Bare Costs Equipment	Total	Total Incl O&P
0011	**SHOWERS**, Stall, with drain only									
1520	32" square	Q-1	5	3.200	Ea.	1,175	186		1,361	1,550
1530	36" square		4.80	3.333		2,950	193		3,143	3,525
1540	Terrazzo receptor, 32" square		5	3.200		1,375	186		1,561	1,800
1580	36" corner angle		4.80	3.333		2,050	193		2,243	2,550
3000	Fiberglass, one piece, with 3 walls, 32" x 32" square		5.50	2.909		355	169		524	650
3100	36" x 36" square		5.50	2.909		465	169		634	770
4200	Rough-in, supply, waste and vent for above showers		2.05	7.805		410	455		865	1,150

22 41 23.40 Shower System Components

		Crew	Daily Output	Labor-Hours	Unit	Material	2020 Bare Costs Labor	2020 Bare Costs Equipment	Total	Total Incl O&P
0010	**SHOWER SYSTEM COMPONENTS**									
5500	Head, water economizer, 1.6 GPM [G]	1 Plum	24	.333	Ea.	53	21.50		74.50	91.50

22 41 36 – Residential Laundry Trays

22 41 36.10 Laundry Sinks

		Crew	Daily Output	Labor-Hours	Unit	Material	2020 Bare Costs Labor	2020 Bare Costs Equipment	Total	Total Incl O&P
0010	**LAUNDRY SINKS**, With trim									
0020	Porcelain enamel on cast iron, black iron frame									
0050	24" x 21", single compartment	Q-1	6	2.667	Ea.	615	155		770	915
0100	26" x 21", single compartment	"	6	2.667	"	630	155		785	935
2000	Molded stone, on wall hanger or legs									
2020	22" x 23", single compartment	Q-1	6	2.667	Ea.	176	155		331	435
2100	45" x 21", double compartment	"	5	3.200	"	360	186		546	680
3000	Plastic, on wall hanger or legs									
3020	18" x 23", single compartment	Q-1	6.50	2.462	Ea.	145	143		288	380
3100	20" x 24", single compartment		6.50	2.462		165	143		308	400
3200	36" x 23", double compartment		5.50	2.909		219	169		388	500
3300	40" x 24", double compartment		5.50	2.909		287	169		456	575
5000	Stainless steel, counter top, 22" x 17" single compartment		6	2.667		77.50	155		232.50	325
5200	33" x 22", double compartment		5	3.200		93	186		279	390
9600	Rough-in, supply, waste and vent, for all laundry sinks		2.14	7.477		320	435		755	1,025
9810	Minimum labor/equipment charge	1 Plum	3	2.667	Job		172		172	266

22 41 39 – Residential Faucets, Supplies and Trim

22 41 39.10 Faucets and Fittings

		Crew	Daily Output	Labor-Hours	Unit	Material	2020 Bare Costs Labor	2020 Bare Costs Equipment	Total	Total Incl O&P
0010	**FAUCETS AND FITTINGS**									
0150	Bath, faucets, diverter spout combination, sweat	1 Plum	8	1	Ea.	87	64.50		151.50	195

22 41 Residential Plumbing Fixtures

22 41 39 – Residential Faucets, Supplies and Trim

22 41 39.10 Faucets and Fittings	Crew	Daily Output	Labor-Hours	Unit	Material	2020 Bare Costs Labor	Equipment	Total	Total Incl O&P
0200 For integral stops, IPS unions, add				Ea.	111			111	123
0420 Bath, press-bal mix valve w/diverter, spout, shower head, arm/flange	1 Plum	8	1		185	64.50		249.50	305
0500 Drain, central lift, 1-1/2" IPS male		20	.400		50.50	26		76.50	95.50
0600 Trip lever, 1-1/2" IPS male	↓	20	.400	↓	60.50	26		86.50	107
0810 Bidet									
0812 Fitting, over the rim, swivel spray/pop-up drain	1 Plum	8	1	Ea.	279	64.50		343.50	405
1000 Kitchen sink faucets, top mount, cast spout		10	.800		84	51.50		135.50	172
1100 For spray, add		24	.333		17.25	21.50		38.75	52
1200 Wall type, swing tube spout	↓	10	.800	↓	74.50	51.50		126	162
1300 Single control lever handle									
1310 With pull out spray									
1320 Polished chrome	1 Plum	10	.800	Ea.	196	51.50		247.50	296
2000 Laundry faucets, shelf type, IPS or copper unions		12	.667		62	43		105	135
2100 Lavatory faucet, centerset, without drain		10	.800		67.50	51.50		119	154
2120 With pop-up drain	↓	6.66	1.201	↓	54.50	77.50		132	180
2210 Porcelain cross handles and pop-up drain									
2220 Polished chrome	1 Plum	6.66	1.201	Ea.	222	77.50		299.50	365
2230 Polished brass	"	6.66	1.201	"	293	77.50		370.50	445
2260 Single lever handle and pop-up drain									
2280 Satin nickel	1 Plum	6.66	1.201	Ea.	280	77.50		357.50	430
2290 Polished chrome		6.66	1.201		200	77.50		277.50	340
2800 Self-closing, center set		10	.800		151	51.50		202.50	246
2810 Automatic sensor and operator, with faucet head [G]		6.15	1.301		495	84		579	670
4000 Shower by-pass valve with union		18	.444		57.50	28.50		86	108
4100 Shower arm with flange and head		22	.364		21	23.50		44.50	59.50
4140 Shower, hand held, pin mount, massage action, chrome		22	.364		82	23.50		105.50	127
4142 Polished brass		22	.364		162	23.50		185.50	214
4144 Shower, hand held, wall mtd, adj. spray, 2 wall mounts, chrome		20	.400		116	26		142	168
4146 Polished brass		20	.400		237	26		263	300
4148 Shower, hand held head, bar mounted 24", adj. spray, chrome		20	.400		180	26		206	238
4150 Polished brass		20	.400		370	26		396	445
4200 Shower thermostatic mixing valve, concealed, with shower head trim kit	↓	8	1	↓	385	64.50		449.50	520
4220 Shower pressure balancing mixing valve									
4230 With shower head, arm, flange and diverter tub spout									
4240 Chrome	1 Plum	6.14	1.303	Ea.	415	84		499	585
4250 Satin nickel		6.14	1.303		560	84		644	745
4260 Polished graphite		6.14	1.303		545	84		629	730
5000 Sillcock, compact, brass, IPS or copper to hose	↓	24	.333	↓	12.05	21.50		33.55	46.50
6000 Stop and waste valves, bronze									
6100 Angle, solder end 1/2"	1 Plum	24	.333	Ea.	28	21.50		49.50	64
6110 3/4"		20	.400		36	26		62	79.50
6300 Straightway, solder end 3/8"		24	.333		20.50	21.50		42	55.50
6310 1/2"		24	.333		20.50	21.50		42	55.50
6410 Straightway, threaded 1/2"		24	.333		22	21.50		43.50	57
6420 3/4"		20	.400		24.50	26		50.50	67
6430 1"		19	.421		27	27		54	71.50
7800 Water closet, wax gasket		96	.083		1.67	5.35		7.02	10.15
7820 Gasket toilet tank to bowl		32	.250		3.15	16.10		19.25	28.50
7830 Replacement diaphragm washer assy for ballcock valve		12	.667		3.15	43		46.15	70
7850 Dual flush valve		12	.667		133	43		176	214
8000 Water supply stops, polished chrome plate									
8200 Angle, 3/8"	1 Plum	24	.333	Ea.	9.30	21.50		30.80	43
8300 1/2"	↓	22	.364		10.05	23.50		33.55	47

22 41 Residential Plumbing Fixtures

22 41 39 – Residential Faucets, Supplies and Trim

22 41 39.10 Faucets and Fittings	Crew	Daily Output	Labor-Hours	Unit	Material	2020 Bare Costs Labor	Equipment	Total	Total Incl O&P
8400	Straight, 3/8"	1 Plum	26	.308	Ea.	9.40	19.85		29.25	41
8500	1/2"		24	.333		9.75	21.50		31.25	43.50
8600	Water closet, angle, w/flex riser, 3/8"		24	.333		34	21.50		55.50	70.50
9000	Minimum labor/equipment charge		4	2	Job		129		129	199

22 42 Commercial Plumbing Fixtures

22 42 13 – Commercial Water Closets, Urinals, and Bidets

22 42 13.13 Water Closets

	22 42 13.13 Water Closets	Crew	Daily Output	Labor-Hours	Unit	Material	2020 Bare Costs Labor	Equipment	Total	Total Incl O&P
0010	**WATER CLOSETS**									
3000	Bowl only, with flush valve, seat, 1.6 gpf unless noted									
3100	Wall hung	Q-1	5.80	2.759	Ea.	1,100	160		1,260	1,450
3200	For rough-in, supply, waste and vent, single WC		2.56	6.250		1,300	365		1,665	2,000
3360	With floor outlet, 1.28 gpf G		5.80	2.759		550	160		710	850
3362	With floor outlet, 1.28 gpf, ADA G		5.80	2.759		570	160		730	875
3370	For rough-in, supply, waste and vent, single WC		2.84	5.634		385	325		710	925
3390	Floor mounted children's size, 10-3/4" high									
3392	With automatic flush sensor, 1.6 gpf	Q-1	6.20	2.581	Ea.	660	150		810	955
3396	With automatic flush sensor, 1.28 gpf		6.20	2.581	"	610	150		760	900
9000	Minimum labor/equipment charge		4	4	Job		232		232	360

22 42 13.16 Urinals

	22 42 13.16 Urinals	Crew	Daily Output	Labor-Hours	Unit	Material	2020 Bare Costs Labor	Equipment	Total	Total Incl O&P
0010	**URINALS**									
3000	Wall hung, vitreous china, with self-closing valve									
3100	Siphon jet type	Q-1	3	5.333	Ea.	315	310		625	825
3120	Blowout type		3	5.333		480	310		790	1,000
3140	Water saving .5 gpf G		3	5.333		595	310		905	1,125
3300	Rough-in, supply, waste & vent		2.83	5.654		755	330		1,085	1,350
5000	Stall type, vitreous china, includes valve		2.50	6.400		830	370		1,200	1,475
6980	Rough-in, supply, waste and vent		1.99	8.040		510	465		975	1,275
9000	Minimum labor/equipment charge		4	4	Job		232		232	360

22 42 16 – Commercial Lavatories and Sinks

22 42 16.13 Lavatories

	22 42 16.13 Lavatories	Crew	Daily Output	Labor-Hours	Unit	Material	2020 Bare Costs Labor	Equipment	Total	Total Incl O&P
0010	**LAVATORIES**, With trim, white unless noted otherwise									
0020	Commercial lavatories same as residential. See Section 22 41 16									

22 42 16.40 Service Sinks

	22 42 16.40 Service Sinks	Crew	Daily Output	Labor-Hours	Unit	Material	2020 Bare Costs Labor	Equipment	Total	Total Incl O&P
0010	**SERVICE SINKS**									
6650	Service, floor, corner, PE on CI, 28" x 28"	Q-1	4.40	3.636	Ea.	1,125	211		1,336	1,575
6790	For rough-in, supply, waste & vent, floor service sinks		1.64	9.756	"	1,075	565		1,640	2,050
9000	Minimum labor/equipment charge		4	4	Job		232		232	360

22 42 23 – Commercial Showers

22 42 23.30 Group Showers

	22 42 23.30 Group Showers	Crew	Daily Output	Labor-Hours	Unit	Material	2020 Bare Costs Labor	Equipment	Total	Total Incl O&P
0010	**GROUP SHOWERS**									
6000	Group, w/pressure balancing valve, rough-in and rigging not included									
6800	Column, 6 heads, no receptors, less partitions	Q-1	3	5.333	Ea.	9,675	310		9,985	11,100
6900	With stainless steel partitions		1	16		12,600	930		13,530	15,300
7600	5 heads, no receptors, less partitions		3	5.333		6,650	310		6,960	7,775
7620	4 heads (1 ADA compliant) no receptors, less partitions		3	5.333		6,050	310		6,360	7,125
7700	With stainless steel partitions		1	16		6,600	930		7,530	8,700
9000	Minimum labor/equipment charge		4	4	Job		232		232	360

446

22 42 33 – Wash Fountains

22 42 33.20 Commercial Wash Fountains		Crew	Daily Output	Labor-Hours	Unit	Material	2020 Bare Costs Labor	Equipment	Total	Total Incl O&P
0010	**COMMERCIAL WASH FOUNTAINS**									
1900	Group, foot control									
2000	Precast terrazzo, circular, 36" diam., 5 or 6 persons	Q-2	3	8	Ea.	7,675	480		8,155	9,175
2100	54" diam. for 8 or 10 persons		2.50	9.600		10,600	575		11,175	12,500
2400	Semi-circular, 36" diam. for 3 persons		3	8		6,250	480		6,730	7,625
2500	54" diam. for 4 or 5 persons	↓	2.50	9.600	↓	9,700	575		10,275	11,600
5610	Group, infrared control, barrier free									
5614	Precast terrazzo									
5620	Semi-circular 36" diam. for 3 persons	Q-2	3	8	Ea.	8,475	480		8,955	10,100
5630	46" diam. for 4 persons		2.80	8.571		9,100	515		9,615	10,800
5640	Circular, 54" diam. for 8 persons, button control	↓	2.50	9.600		11,200	575		11,775	13,200
5700	Rough-in, supply, waste and vent for above wash fountains	Q-1	1.82	8.791	↓	505	510		1,015	1,350
9000	Minimum labor/equipment charge	Q-2	3	8	Job		480		480	745

22 42 39 – Commercial Faucets, Supplies, and Trim

22 42 39.10 Faucets and Fittings

22 42 39.10 Faucets and Fittings		Crew	Daily Output	Labor-Hours	Unit	Material	Labor	Equipment	Total	Total Incl O&P
0010	**FAUCETS AND FITTINGS**									
0840	Flush valves, with vacuum breaker									
0850	Water closet									
0860	Exposed, rear spud	1 Plum	8	1	Ea.	146	64.50		210.50	261
0870	Top spud		8	1		197	64.50		261.50	315
0880	Concealed, rear spud		8	1		213	64.50		277.50	335
0890	Top spud		8	1		173	64.50		237.50	290
0900	Wall hung		8	1		199	64.50		263.50	320
0910	Dual flush flushometer		12	.667		245	43		288	335
0912	Flushometer retrofit kit	↓	18	.444	↓	18.15	28.50		46.65	64.50
0920	Urinal									
0930	Exposed, stall	1 Plum	8	1	Ea.	197	64.50		261.50	315
0940	Wall (washout)		8	1		156	64.50		220.50	272
0950	Pedestal, top spud		8	1		137	64.50		201.50	250
0960	Concealed, stall		8	1		170	64.50		234.50	287
0970	Wall (washout)	↓	8	1	↓	183	64.50		247.50	300
0971	Automatic flush sensor and operator for									
0972	urinals or water closets, standard　G	1 Plum	8	1	Ea.	485	64.50		549.50	635
0980	High efficiency water saving									
0984	Water closets, 1.28 gpf　G	1 Plum	8	1	Ea.	425	64.50		489.50	570
0988	Urinals, .5 gpf　G	"	8	1	"	425	64.50		489.50	570
2790	Faucets for lavatories									
2800	Self-closing, center set	1 Plum	10	.800	Ea.	151	51.50		202.50	246
2810	Automatic sensor and operator, with faucet head		6.15	1.301		495	84		579	670
3000	Service sink faucet, cast spout, pail hook, hose end	↓	14	.571	↓	76.50	37		113.50	141

22 42 39.30 Carriers and Supports

22 42 39.30 Carriers and Supports		Crew	Daily Output	Labor-Hours	Unit	Material	Labor	Equipment	Total	Total Incl O&P
0010	**CARRIERS AND SUPPORTS,** For plumbing fixtures									
3000	Lavatory, concealed arm									
3050	Floor mounted, single									
3100	High back fixture	1 Plum	6	1.333	Ea.	655	86		741	855
3200	Flat slab fixture		6	1.333		575	86		661	770
3220	ADA compliant	↓	6	1.333	↓	665	86		751	870
6980	Water closet, siphon jet									
7000	Horizontal, adjustable, caulk									
7040	Single, 4" pipe size	1 Plum	5.33	1.501	Ea.	1,125	96.50		1,221.50	1,375
7060	5" pipe size	↓	5.33	1.501	↓	1,125	96.50		1,221.50	1,375

22 42 Commercial Plumbing Fixtures

22 42 39 – Commercial Faucets, Supplies, and Trim

22 42 39.30 Carriers and Supports	Crew	Daily Output	Labor-Hours	Unit	Material	2020 Bare Costs Labor	Equipment	Total	Total Incl O&P
7100	Double, 4" pipe size	1 Plum	5	1.600	Ea.	1,625	103		1,728	1,925
7120	5" pipe size	↓	5	1.600	↓	1,925	103		2,028	2,275
8200	Water closet, residential									
8220	Vertical centerline, floor mount									
8240	Single, 3" caulk, 2" or 3" vent	1 Plum	6	1.333	Ea.	875	86		961	1,100
8260	4" caulk, 2" or 4" vent		6	1.333	"	1,125	86		1,211	1,375
9990	Minimum labor/equipment charge	↓	3.50	2.286	Job		147		147	228

22 43 Healthcare Plumbing Fixtures

22 43 39 – Healthcare Faucets

22 43 39.10 Faucets and Fittings

		Crew	Daily Output	Labor-Hours	Unit	Material	2020 Bare Costs Labor	Equipment	Total	Total Incl O&P
0010	**FAUCETS AND FITTINGS**									
2850	Medical, bedpan cleanser, with pedal valve,	1 Plum	12	.667	Ea.	815	43		858	960
2860	With screwdriver stop valve		12	.667		420	43		463	530
2870	With self-closing spray valve		12	.667		260	43		303	355
2900	Faucet, gooseneck spout, wrist handles, grid drain		10	.800		202	51.50		253.50	300
2940	Mixing valve, knee action, screwdriver stops	↓	4	2	↓	450	129		579	695

22 45 Emergency Plumbing Fixtures

22 45 13 – Emergency Showers

22 45 13.10 Emergency Showers

		Crew	Daily Output	Labor-Hours	Unit	Material	2020 Bare Costs Labor	Equipment	Total	Total Incl O&P
0010	**EMERGENCY SHOWERS**, Rough-in not included									
5000	Shower, single head, drench, ball valve, pull, freestanding	Q-1	4	4	Ea.	390	232		622	790
5200	Horizontal or vertical supply		4	4		615	232		847	1,025
6000	Multi-nozzle, eye/face wash combination		4	4		760	232		992	1,200
6400	Multi-nozzle, 12 spray, shower only		4	4		2,100	232		2,332	2,675
6600	For freeze-proof, add		6	2.667	↓	505	155		660	795
9000	Minimum labor/equipment charge		3	5.333	Job		310		310	480

22 45 16 – Eyewash Equipment

22 45 16.10 Eyewash Safety Equipment

		Crew	Daily Output	Labor-Hours	Unit	Material	2020 Bare Costs Labor	Equipment	Total	Total Incl O&P
0010	**EYEWASH SAFETY EQUIPMENT**, Rough-in not included									
1000	Eye wash fountain									
1400	Plastic bowl, pedestal mounted	Q-1	4	4	Ea.	335	232		567	725

22 47 Drinking Fountains and Water Coolers

22 47 13 – Drinking Fountains

22 47 13.10 Drinking Water Fountains

		Crew	Daily Output	Labor-Hours	Unit	Material	2020 Bare Costs Labor	Equipment	Total	Total Incl O&P
0010	**DRINKING WATER FOUNTAINS**, For connection to cold water supply									
1000	Wall mounted, non-recessed									
2700	Stainless steel, single bubbler, no back	1 Plum	4	2	Ea.	915	129		1,044	1,200
2740	With back		4	2		1,100	129		1,229	1,400
2780	Dual handle, ADA compliant		4	2		760	129		889	1,025
2820	Dual level, ADA compliant		3.20	2.500		1,600	161		1,761	2,000
3980	For rough-in, supply and waste, add	↓	2.21	3.620	↓	229	233		462	610
4000	Wall mounted, semi-recessed									
4200	Poly-marble, single bubbler	1 Plum	4	2	Ea.	960	129		1,089	1,250
4600	Stainless steel, satin finish, single bubbler	"	4	2	"	1,400	129		1,529	1,750

22 47 Drinking Fountains and Water Coolers

22 47 13 – Drinking Fountains

22 47 13.10 Drinking Water Fountains

		Crew	Daily Output	Labor-Hours	Unit	Material	2020 Bare Costs Labor	Equipment	Total	Total Incl O&P
6000	Wall mounted, fully recessed									
6400	Poly-marble, single bubbler	1 Plum	4	2	Ea.	1,775	129		1,904	2,150
6800	Stainless steel, single bubbler		4	2		1,750	129		1,879	2,125
7580	For rough-in, supply and waste, add	↓	1.83	4.372	↓	229	282		511	685
7600	Floor mounted, pedestal type									
8600	Enameled iron, heavy duty service, 2 bubblers	1 Plum	2	4	Ea.	3,675	258		3,933	4,450
8880	For freeze-proof valve system, add		2	4		700	258		958	1,175
8900	For rough-in, supply and waste, add		1.83	4.372	↓	229	282		511	685
9000	Minimum labor/equipment charge	↓	2	4	Job		258		258	400

22 47 16 – Pressure Water Coolers

22 47 16.10 Electric Water Coolers

		Crew	Daily Output	Labor-Hours	Unit	Material	2020 Bare Costs Labor	Equipment	Total	Total Incl O&P
0010	**ELECTRIC WATER COOLERS**									
0100	Wall mounted, non-recessed									
0140	4 GPH	Q-1	4	4	Ea.	700	232		932	1,125
0160	8 GPH, barrier free, sensor operated		4	4		1,125	232		1,357	1,600
1000	Dual height, 8.2 GPH		3.80	4.211		2,025	244		2,269	2,600
1040	14.3 GPH		3.80	4.211		1,750	244		1,994	2,300
3300	Semi-recessed, 8.1 GPH	↓	4	4	↓	915	232		1,147	1,350
4600	Floor mounted, flush-to-wall									
4640	4 GPH	1 Plum	3	2.667	Ea.	840	172		1,012	1,175
4980	For stainless steel cabinet, add					141			141	155
5000	Dual height, 8.2 GPH	1 Plum	2	4	↓	1,225	258		1,483	1,750
9000	Minimum labor/equipment charge	"	2	4	Job		258		258	400

22 66 Chemical-Waste Systems for Lab. and Healthcare Facilities

22 66 53 – Laboratory Chemical-Waste and Vent Piping

22 66 53.30 Glass Pipe

		Crew	Daily Output	Labor-Hours	Unit	Material	2020 Bare Costs Labor	Equipment	Total	Total Incl O&P
0010	**GLASS PIPE**, Borosilicate, couplings & clevis hanger assemblies, 10' OC									
0020	Drainage									
1100	1-1/2" diameter	Q-1	52	.308	L.F.	15.30	17.85		33.15	44.50
1120	2" diameter		44	.364		19.50	21		40.50	54
1140	3" diameter		39	.410		26	24		50	65.50
1160	4" diameter		30	.533		42.50	31		73.50	95
1180	6" diameter	↓	26	.615	↓	75	35.50		110.50	138
9000	Minimum labor/equipment charge	1 Plum	4	2	Job		129		129	199

22 66 53.40 Pipe Fittings, Glass

		Crew	Daily Output	Labor-Hours	Unit	Material	2020 Bare Costs Labor	Equipment	Total	Total Incl O&P
0010	**PIPE FITTINGS, GLASS**									
0020	Drainage, beaded ends									
0040	Coupling & labor required at joints not incl. in fitting									
0050	price. Add 1 per joint for installed price									
0070	90° bend or sweep, 1-1/2"				Ea.	36.50			36.50	40
0090	2"					46.50			46.50	51.50
0100	3"					76			76	83.50
0110	4"					121			121	134
0120	6" (sweep only)					360			360	395
0350	Tee, single sanitary, 1-1/2"					59			59	65
0370	2"					59			59	65
0380	3"					86			86	94.50
0390	4"					157			157	173
0400	6"				↓	420			420	465

For customer support on your Commercial Renovation Costs with RSMeans data, call 800.448.8182.

449

22 66 53 – Laboratory Chemical-Waste and Vent Piping

22 66 53.40 Pipe Fittings, Glass

		Crew	Daily Output	Labor-Hours	Unit	Material	2020 Bare Costs Labor	Equipment	Total	Total Incl O&P
0500	Coupling, stainless steel, TFE seal ring									
0520	1-1/2"	Q-1	32	.500	Ea.	27.50	29		56.50	75.50
0530	2"		30	.533		35	31		66	86.50
0540	3"		25	.640		33.50	37		70.50	94
0550	4"		23	.696		80.50	40.50		121	151
0560	6"		20	.800		181	46.50		227.50	271
9000	Minimum labor/equipment charge	1 Plum	4	2	Job		129		129	199

22 66 53.60 Corrosion Resistant Pipe

		Crew	Daily Output	Labor-Hours	Unit	Material	2020 Bare Costs Labor	Equipment	Total	Total Incl O&P
0010	**CORROSION RESISTANT PIPE**, No couplings or hangers									
0020	Iron alloy, drain, mechanical joint									
1000	1-1/2" diameter	Q-1	70	.229	L.F.	84.50	13.25		97.75	114
1100	2" diameter		66	.242		87	14.05		101.05	117
1120	3" diameter		60	.267		95	15.45		110.45	129
1140	4" diameter		52	.308		117	17.85		134.85	157
2980	Plastic, epoxy, fiberglass filament wound, B&S joint									
3000	2" diameter	Q-1	62	.258	L.F.	12.35	14.95		27.30	36.50
3100	3" diameter		51	.314		14.40	18.20		32.60	44
3120	4" diameter		45	.356		20.50	20.50		41	54.50
3160	8" diameter	Q-2	38	.632		45	38		83	108
3200	12" diameter	"	28	.857		74.50	51.50		126	162
9800	Minimum labor/equipment charge	1 Plum	4	2	Job		129		129	199

22 66 53.70 Pipe Fittings, Corrosion Resistant

		Crew	Daily Output	Labor-Hours	Unit	Material	2020 Bare Costs Labor	Equipment	Total	Total Incl O&P
0010	**PIPE FITTINGS, CORROSION RESISTANT**									
0030	Iron alloy									
0050	Mechanical joint									
0060	1/4 bend, 1-1/2"	Q-1	12	1.333	Ea.	125	77.50		202.50	257
0080	2"		10	1.600		204	93		297	370
0090	3"		9	1.778		245	103		348	430
0100	4"		8	2		282	116		398	490
0160	Tee and Y, sanitary, straight									
0170	1-1/2"	Q-1	8	2	Ea.	136	116		252	330
0180	2"		7	2.286		182	133		315	405
0190	3"		6	2.667		282	155		437	550
0200	4"		5	3.200		520	186		706	855
0360	Coupling, 1-1/2"		14	1.143		75	66.50		141.50	185
0380	2"		12	1.333		85	77.50		162.50	214
0390	3"		11	1.455		89.50	84.50		174	229
0400	4"		10	1.600		101	93		194	254
3000	Epoxy, filament wound									
3030	Quick-lock joint									
3040	90° elbow, 2"	Q-1	28	.571	Ea.	100	33		133	163
3060	3"		16	1		116	58		174	217
3070	4"		13	1.231		158	71.50		229.50	283
3190	Tee, 2"		19	.842		239	49		288	340
3200	3"		11	1.455		289	84.50		373.50	450
3210	4"		9	1.778		350	103		453	545
9000	Minimum labor/equipment charge	1 Plum	4	2	Job		129		129	199

Estimating Tips

The labor adjustment factors listed in Subdivision 22 01 02.20 also apply to Division 23.

23 10 00 Facility Fuel Systems

- The prices in this subdivision for above- and below-ground storage tanks do not include foundations or hold-down slabs, unless noted. The estimator should refer to Divisions 3 and 31 for foundation system pricing. In addition to the foundations, required tank accessories, such as tank gauges, leak detection devices, and additional manholes and piping, must be added to the tank prices.

23 50 00 Central Heating Equipment

- When estimating the cost of an HVAC system, check to see who is responsible for providing and installing the temperature control system. It is possible to overlook controls, assuming that they would be included in the electrical estimate.
- When looking up a boiler, be careful on specified capacity. Some

manufacturers rate their products on output while others use input.

- Include HVAC insulation for pipe, boiler, and duct (wrap and liner).
- Be careful when looking up mechanical items to get the correct pressure rating and connection type (thread, weld, flange).

23 70 00 Central HVAC Equipment

- Combination heating and cooling units are sized by the air conditioning requirements. (See Reference No. R236000-20 for the preliminary sizing guide.)
- A ton of air conditioning is nominally 400 CFM.
- Rectangular duct is taken off by the linear foot for each size, but its cost is usually estimated by the pound. Remember that SMACNA standards now base duct on internal pressure.
- Prefabricated duct is estimated and purchased like pipe: straight sections and fittings.
- Note that cranes or other lifting equipment are not included on any

lines in Division 23. For example, if a crane is required to lift a heavy piece of pipe into place high above a gym floor, or to put a rooftop unit on the roof of a four-story building, etc., it must be added. Due to the potential for extreme variation—from nothing additional required to a major crane or helicopter—we feel that including a nominal amount for "lifting contingency" would be useless and detract from the accuracy of the estimate. When using equipment rental cost data from RSMeans, do not forget to include the cost of the operator(s).

Reference Numbers

Reference numbers are shown at the beginning of some major classifications. These numbers refer to related items in the Reference Section. The reference information may be an estimating procedure, an alternate pricing method, or technical information.

Note: Not all subdivisions listed here necessarily appear. ■

Same Data. Simplified.

Enjoy the convenience and efficiency of accessing your costs anywhere:

- **Skip the multiplier** by setting your location
- **Quickly search,** edit, favorite and share costs
- **Stay on top of price changes** with automatic updates

Discover more at rsmeans.com/online

Note: "Powered in part by CINX™, based on liscensed proprietary information of Harrison Publishing House, Inc."

Note: Trade Service, in part, has been used as a reference source for some of the material prices used in Division 23.

23 05 Common Work Results for HVAC

23 05 05 – Selective Demolition for HVAC

23 05 05.10 HVAC Demolition

		Crew	Daily Output	Labor-Hours	Unit	Material	2020 Bare Costs Labor	2020 Bare Costs Equipment	Total	Total Incl O&P
0010	**HVAC DEMOLITION**	R220105-10								
0100	Air conditioner, split unit, 3 ton	Q-5	2	8	Ea.		470		470	730
0150	Package unit, 3 ton R024119-10	Q-6	3	8	"		490		490	755
0298	Boilers									
0300	Electric, up thru 148 kW	Q-19	2	12	Ea.		715		715	1,100
0310	150 thru 518 kW	"	1	24			1,425		1,425	2,225
0320	550 thru 2,000 kW	Q-21	.40	80			4,900		4,900	7,575
0330	2,070 kW and up	"	.30	107			6,525		6,525	10,100
0340	Gas and/or oil, up thru 150 MBH	Q-7	2.20	14.545			910		910	1,400
0350	160 thru 2,000 MBH		.80	40			2,500		2,500	3,850
0360	2,100 thru 4,500 MBH		.50	64			4,000		4,000	6,175
0370	4,600 thru 7,000 MBH		.30	107			6,650		6,650	10,300
0380	7,100 thru 12,000 MBH		.16	200			12,500		12,500	19,300
0390	12,200 thru 25,000 MBH		.12	267			16,600		16,600	25,700
1000	Ductwork, 4" high, 8" wide	1 Clab	200	.040	L.F.		1.68		1.68	2.70
1020	10" wide		190	.042			1.77		1.77	2.84
1040	14" wide		180	.044			1.87		1.87	3
1100	6" high, 8" wide		165	.048			2.04		2.04	3.27
1120	12" wide		150	.053			2.25		2.25	3.60
1140	18" wide		135	.059			2.49		2.49	4
1200	10" high, 12" wide		125	.064			2.69		2.69	4.32
1220	18" wide		115	.070			2.93		2.93	4.69
1240	24" wide		110	.073			3.06		3.06	4.91
1300	12"-14" high, 16"-18" wide		85	.094			3.96		3.96	6.35
1320	24" wide		75	.107			4.49		4.49	7.20
1340	48" wide		71	.113			4.74		4.74	7.60
1400	18" high, 24" wide		67	.119			5.05		5.05	8.05
1420	36" wide		63	.127			5.35		5.35	8.55
1440	48" wide		59	.136			5.70		5.70	9.15
1500	30" high, 36" wide		56	.143			6		6	9.65
1520	48" wide		53	.151			6.35		6.35	10.20
1540	72" wide		50	.160			6.75		6.75	10.80
1550	Duct heater, electric strip	1 Elec	8	1	Ea.		61.50		61.50	94.50
1850	Minimum labor/equipment charge	1 Clab	3	2.667	Job		112		112	180
2200	Furnace, electric	Q-20	2	10	Ea.		570		570	890
2300	Gas or oil, under 120 MBH	Q-9	4	4			224		224	350
2340	Over 120 MBH	"	3	5.333			299		299	470
2800	Heat pump, package unit, 3 ton	Q-5	2.40	6.667			395		395	610
2840	Split unit, 3 ton		2	8			470		470	730
3000	Mechanical equipment, light items. Unit is weight, not cooling.		.90	17.778	Ton		1,050		1,050	1,625
3600	Heavy items		1.10	14.545	"		860		860	1,325
5090	Remove refrigerant from system	1 Stpi	40	.200	Lb.		13.10		13.10	20.50
9000	Minimum labor/equipment charge	Q-6	3	8	Job		490		490	755

23 05 23 – General-Duty Valves for HVAC Piping

23 05 23.30 Valves, Iron Body

		Crew	Daily Output	Labor-Hours	Unit	Material	2020 Bare Costs Labor	2020 Bare Costs Equipment	Total	Total Incl O&P
0010	**VALVES, IRON BODY**									
1020	Butterfly, wafer type, gear actuator, 200 lb.									
1030	2"	1 Plum	14	.571	Ea.	118	37		155	187
1060	4"	Q-1	5	3.200	"	139	186		325	440
1650	Gate, 125 lb., N.R.S.									
2150	Flanged									
2240	2-1/2"	Q-1	5	3.200	Ea.	630	186		816	975

For customer support on your Commercial Renovation Costs with RSMeans data, call 800.448.8182.

23 05 Common Work Results for HVAC

23 05 23 – General-Duty Valves for HVAC Piping

23 05 23.30 Valves, Iron Body

		Crew	Daily Output	Labor-Hours	Unit	Material	2020 Bare Costs Labor	Equipment	Total	Total Incl O&P
2260	3"	Q-1	4.50	3.556	Ea.	710	206		916	1,100
2280	4"	↓	3	5.333	↓	1,000	310		1,310	1,575
3550	OS&Y, 125 lb., flanged									
3680	4"	Q-1	3	5.333	Ea.	685	310		995	1,225
3690	5"	Q-2	3.40	7.059		1,075	425		1,500	1,850
3700	6"	"	3	8	↓	1,075	480		1,555	1,950
9000	Minimum labor/equipment charge	1 Plum	3	2.667	Job		172		172	266

23 07 HVAC Insulation

23 07 13 – Duct Insulation

23 07 13.10 Duct Thermal Insulation

			Crew	Daily Output	Labor-Hours	Unit	Material	2020 Bare Costs Labor	Equipment	Total	Total Incl O&P
0010	**DUCT THERMAL INSULATION**										
0110	Insulation req'd. is based on the surface size/area to be covered										
3000	Ductwork										
3020	Blanket type, fiberglass, flexible										
3030	Fire rated for grease and hazardous exhaust ducts										
3060	1-1/2" thick		Q-14	84	.190	S.F.	4.49	10.05		14.54	21
3090	Fire rated for plenums										
3100	1/2" x 24" x 25'		Q-14	1.94	8.247	Roll	177	435		612	885
3110	1/2" x 24" x 25'		↓	98	.163	S.F.	3.54	8.60		12.14	17.55
3120	1/2" x 48" x 25'			1.04	15.385	Roll	350	810		1,160	1,650
3126	1/2" x 48" x 25'		↓	104	.154	S.F.	3.49	8.10		11.59	16.70
3140	FSK vapor barrier wrap, .75 lb. density										
3160	1" thick	G	Q-14	350	.046	S.F.	.22	2.41		2.63	4.07
3170	1-1/2" thick	G	↓	320	.050		.27	2.64		2.91	4.48
3180	2" thick	G		300	.053		.33	2.81		3.14	4.82
3190	3" thick	G		260	.062		.51	3.25		3.76	5.70
3200	4" thick	G	↓	242	.066	↓	.69	3.49		4.18	6.30
3210	Vinyl jacket, same as FSK										
9600	Minimum labor/equipment charge		1 Stpi	4	2	Job		131		131	203

23 07 16 – HVAC Equipment Insulation

23 07 16.10 HVAC Equipment Thermal Insulation

		Crew	Daily Output	Labor-Hours	Unit	Material	2020 Bare Costs Labor	Equipment	Total	Total Incl O&P
0010	**HVAC EQUIPMENT THERMAL INSULATION**									
0110	Insulation req'd. is based on the surface size/area to be covered									
9610	Minimum labor/equipment charge	1 Stpi	4	2	Job		131		131	203

23 09 Instrumentation and Control for HVAC

23 09 53 – Pneumatic and Electric Control System for HVAC

23 09 53.10 Control Components

			Crew	Daily Output	Labor-Hours	Unit	Material	2020 Bare Costs Labor	Equipment	Total	Total Incl O&P
0010	**CONTROL COMPONENTS**										
5000	Thermostats										
5030	Manual		1 Stpi	8	1	Ea.	42.50	65.50		108	148
5040	1 set back, electric, timed	G		8	1		77.50	65.50		143	187
5050	2 set back, electric, timed	G	↓	8	1		257	65.50		322.50	385
5200	24 hour, automatic, clock	G	1 Shee	8	1	↓	204	62.50		266.50	325
6000	Valves, motorized zone										
6100	Sweat connections, 1/2" C x C		1 Stpi	20	.400	Ea.	136	26		162	190
6110	3/4" C x C			20	.400		138	26		164	193
6120	1" C x C		↓	19	.421	↓	185	27.50		212.50	246

453

23 09 Instrumentation and Control for HVAC

23 09 53 – Pneumatic and Electric Control System for HVAC

23 09 53.10 Control Components	Crew	Daily Output	Labor-Hours	Unit	Material	2020 Bare Costs Labor	2020 Bare Costs Equipment	Total	Total Incl O&P
9000 Minimum labor/equipment charge	1 Plum	4	2	Job		129		129	199

23 21 Hydronic Piping and Pumps

23 21 20 – Hydronic HVAC Piping Specialties

23 21 20.34 Dielectric Unions

		Crew	Daily Output	Labor-Hours	Unit	Material	2020 Bare Costs Labor	2020 Bare Costs Equipment	Total	Total Incl O&P
0010	**DIELECTRIC UNIONS**, Standard gaskets for water and air									
0020	250 psi maximum pressure									
0280	Female IPT to sweat, straight									
0340	3/4" pipe size	1 Plum	20	.400	Ea.	9.40	26		35.40	50.50
0780	Female IPT to female IPT, straight									
0800	1/2" pipe size	1 Plum	24	.333	Ea.	15.90	21.50		37.40	50.50
0840	3/4" pipe size		20	.400	"	18.25	26		44.25	60
9000	Minimum labor/equipment charge		4	2	Job		129		129	199

23 21 20.46 Expansion Tanks

		Crew	Daily Output	Labor-Hours	Unit	Material	2020 Bare Costs Labor	2020 Bare Costs Equipment	Total	Total Incl O&P
0010	**EXPANSION TANKS**									
2000	Steel, liquid expansion, ASME, painted, 15 gallon capacity	Q-5	17	.941	Ea.	855	55.50		910.50	1,025
2040	30 gallon capacity		12	1.333		985	78.50		1,063.50	1,200
2080	60 gallon capacity		8	2		1,375	118		1,493	1,700
2120	100 gallon capacity		6	2.667		2,000	157		2,157	2,450
3000	Steel ASME expansion, rubber diaphragm, 19 gal. cap. accept.		12	1.333		3,200	78.50		3,278.50	3,650
3020	31 gallon capacity		8	2		3,575	118		3,693	4,125
3040	61 gallon capacity		6	2.667		5,225	157		5,382	6,000
3080	119 gallon capacity		4	4		5,425	236		5,661	6,325
9000	Minimum labor/equipment charge		4	4	Job		236		236	365

23 21 20.70 Steam Traps

		Crew	Daily Output	Labor-Hours	Unit	Material	2020 Bare Costs Labor	2020 Bare Costs Equipment	Total	Total Incl O&P
0010	**STEAM TRAPS**									
0030	Cast iron body, threaded									
0040	Inverted bucket									
0050	1/2" pipe size	1 Stpi	12	.667	Ea.	170	43.50		213.50	255
0100	1" pipe size		9	.889		430	58.50		488.50	565
0120	1-1/4" pipe size		8	1		650	65.50		715.50	815
1000	Float & thermostatic, 15 psi									
1010	3/4" pipe size	1 Stpi	16	.500	Ea.	168	33		201	235
1020	1" pipe size		15	.533		193	35		228	266
1040	1-1/2" pipe size		9	.889		360	58.50		418.50	485
1060	2" pipe size		6	1.333		1,000	87.50		1,087.50	1,225
9000	Minimum labor/equipment charge		4	2	Job		131		131	203

23 21 20.76 Strainers, Y Type, Bronze Body

		Crew	Daily Output	Labor-Hours	Unit	Material	2020 Bare Costs Labor	2020 Bare Costs Equipment	Total	Total Incl O&P
0010	**STRAINERS, Y TYPE, BRONZE BODY**									
0050	Screwed, 125 lb., 1/4" pipe size	1 Stpi	24	.333	Ea.	44	22		66	82.50
0100	1/2" pipe size		20	.400		45.50	26		71.50	90.50
0120	3/4" pipe size		19	.421		57	27.50		84.50	106
0140	1" pipe size		17	.471		66.50	31		97.50	121
0160	1-1/2" pipe size		14	.571		144	37.50		181.50	216
0180	2" pipe size		13	.615		190	40.50		230.50	272
0182	3" pipe size		12	.667		1,200	43.50		1,243.50	1,400
0200	300 lb., 2-1/2" pipe size	Q-5	17	.941		865	55.50		920.50	1,025
0220	3" pipe size		16	1		1,325	59		1,384	1,575
0240	4" pipe size		15	1.067		2,425	63		2,488	2,775
1000	Flanged, 150 lb., 1-1/2" pipe size	1 Stpi	11	.727		545	47.50		592.50	675

23 21 Hydronic Piping and Pumps

23 21 20 – Hydronic HVAC Piping Specialties

23 21 20.76 Strainers, Y Type, Bronze Body

		Crew	Daily Output	Labor-Hours	Unit	Material	2020 Bare Costs Labor	2020 Bare Costs Equipment	Total	Total Incl O&P
1020	2" pipe size	1 Stpi	8	1	Ea.	630	65.50		695.50	790
1030	2-1/2" pipe size	Q-5	5	3.200		1,125	189		1,314	1,550
1040	3" pipe size		4.50	3.556		1,375	210		1,585	1,850
1060	4" pipe size		3	5.333		2,125	315		2,440	2,825
1100	6" pipe size	Q-6	3	8		4,100	490		4,590	5,275
1106	8" pipe size	"	2.60	9.231		4,450	565		5,015	5,775
1500	For 300 lb. rating, add					40%				
9000	Minimum labor/equipment charge	1 Stpi	3.75	2.133	Job		140		140	216

23 21 23 – Hydronic Pumps

23 21 23.13 In-Line Centrifugal Hydronic Pumps

		Crew	Daily Output	Labor-Hours	Unit	Material	2020 Bare Costs Labor	2020 Bare Costs Equipment	Total	Total Incl O&P
0010	**IN-LINE CENTRIFUGAL HYDRONIC PUMPS**									
0600	Bronze, sweat connections, 1/40 HP, in line									
0640	3/4" size	Q-1	16	1	Ea.	275	58		333	390
1000	Flange connection, 3/4" to 1-1/2" size									
1040	1/12 HP	Q-1	6	2.667	Ea.	710	155		865	1,025
1060	1/8 HP		6	2.667		1,250	155		1,405	1,625
1100	1/3 HP		6	2.667		1,400	155		1,555	1,775
1140	2" size, 1/6 HP		5	3.200		1,750	186		1,936	2,200
1180	2-1/2" size, 1/4 HP		5	3.200		2,200	186		2,386	2,700
1220	3" size, 1/4 HP		4	4		2,400	232		2,632	2,975
9000	Minimum labor/equipment charge		3.25	4.923	Job		286		286	440

23 23 Refrigerant Piping

23 23 16 – Refrigerant Piping Specialties

23 23 16.16 Refrigerant Line Sets

		Crew	Daily Output	Labor-Hours	Unit	Material	2020 Bare Costs Labor	2020 Bare Costs Equipment	Total	Total Incl O&P
0010	**REFRIGERANT LINE SETS**, Standard									
0100	Copper tube									
0110	1/2" insulation, both tubes									
0120	Combination 1/4" and 1/2" tubes									
0130	10' set	Q-5	42	.381	Ea.	48.50	22.50		71	88.50
0135	15' set		42	.381		70.50	22.50		93	113
0140	20' set		40	.400		76	23.50		99.50	120
0150	30' set		37	.432		107	25.50		132.50	158
0160	40' set		35	.457		134	27		161	189
0170	50' set		32	.500		166	29.50		195.50	229
0180	100' set		22	.727		360	43		403	460
0300	Combination 1/4" and 3/4" tubes									
0310	10' set	Q-5	40	.400	Ea.	54.50	23.50		78	96.50
0320	20' set		38	.421		96.50	25		121.50	145
0330	30' set		35	.457		144	27		171	200
0340	40' set		33	.485		188	28.50		216.50	251
0350	50' set		30	.533		237	31.50		268.50	310
0380	100' set		20	.800		540	47		587	670
0500	Combination 3/8" & 3/4" tubes									
0510	10' set	Q-5	28	.571	Ea.	62.50	33.50		96	121
0520	20' set		36	.444		97.50	26		123.50	148
0530	30' set		34	.471		132	28		160	189
0540	40' set		31	.516		174	30.50		204.50	238
0550	50' set		28	.571		204	33.50		237.50	277
0580	100' set		18	.889		605	52.50		657.50	745

For customer support on your Commercial Renovation Costs with RSMeans data, call 800.448.8182.

455

23 23 16.16 Refrigerant Line Sets

		Crew	Daily Output	Labor-Hours	Unit	Material	2020 Bare Costs Labor	2020 Bare Costs Equipment	Total	Total Incl O&P
0700	Combination 3/8" & 1-1/8" tubes									
0710	10' set	Q-5	36	.444	Ea.	109	26		135	161
0720	20' set		33	.485		183	28.50		211.50	245
0730	30' set		31	.516		246	30.50		276.50	315
0740	40' set		28	.571		370	33.50		403.50	455
0750	50' set		26	.615		355	36.50		391.50	450
0900	Combination 1/2" & 3/4" tubes									
0910	10' set	Q-5	37	.432	Ea.	66	25.50		91.50	112
0920	20' set		35	.457		117	27		144	171
0930	30' set		33	.485		176	28.50		204.50	238
0940	40' set		30	.533		232	31.50		263.50	305
0950	50' set		27	.593		292	35		327	375
0980	100' set		17	.941		660	55.50		715.50	810
2100	Combination 1/2" & 1-1/8" tubes									
2110	10' set	Q-5	35	.457	Ea.	113	27		140	166
2120	20' set		31	.516		193	30.50		223.50	259
2130	30' set		29	.552		290	32.50		322.50	370
2140	40' set		25	.640		390	38		428	490
2150	50' set		14	1.143		475	67.50		542.50	630
2300	For 1" thick insulation add					30%	15%			
3000	Refrigerant line sets, min-split, flared									
3100	Combination 1/4" & 3/8" tubes									
3120	15' set	Q-5	41	.390	Ea.	69.50	23		92.50	112
3140	25' set		38.50	.416		98.50	24.50		123	146
3160	35' set		37	.432		124	25.50		149.50	177
3180	50' set		30	.533		169	31.50		200.50	235
3200	Combination 1/4" & 1/2" tubes									
3220	15' set	Q-5	41	.390	Ea.	74.50	23		97.50	118
3240	25' set		38.50	.416		101	24.50		125.50	149
3260	35' set		37	.432		130	25.50		155.50	183
3280	50' set		30	.533		176	31.50		207.50	242

23 31 HVAC Ducts and Casings

23 31 13 – Metal Ducts

23 31 13.13 Rectangular Metal Ducts

		Crew	Daily Output	Labor-Hours	Unit	Material	2020 Bare Costs Labor	2020 Bare Costs Equipment	Total	Total Incl O&P
0010	**RECTANGULAR METAL DUCTS**									
0020	Fabricated rectangular, includes fittings, joints, supports,									
0021	allowance for flexible connections and field sketches.									
0030	Does not include "as-built dwgs." or insulation.									
0031	NOTE: Fabrication and installation are combined									
0040	as LABOR cost. Approx. 25% fittings assumed.									
0042	Fabrication/Inst. is to commercial quality standards									
0043	(SMACNA or equiv.) for structure, sealing, leak testing, etc.									
0100	Aluminum, alloy 3003-H14, under 100 lb.	Q-10	75	.320	Lb.	3.20	18.60		21.80	32.50
0110	100 to 500 lb.		80	.300		1.89	17.45		19.34	29.50
0120	500 to 1,000 lb.		95	.253		1.91	14.70		16.61	25
0160	Over 5,000 lb.		145	.166		1.86	9.65		11.51	17.15
0500	Galvanized steel, under 200 lb.		235	.102		.60	5.95		6.55	9.95
0520	200 to 500 lb.		245	.098		.57	5.70		6.27	9.55
0540	500 to 1,000 lb.		255	.094		.54	5.45		5.99	9.15
0560	1,000 to 2,000 lb.		265	.091		.55	5.25		5.80	8.85

23 31 HVAC Ducts and Casings

23 31 13 – Metal Ducts

23 31 13.13 Rectangular Metal Ducts

		Crew	Daily Output	Labor-Hours	Unit	Material	2020 Bare Costs Labor	Equipment	Total	Total Incl O&P
0580	Over 5,000 lb.	Q-10	285	.084	Lb.	.55	4.90		5.45	8.25
1000	Stainless steel, type 304, under 100 lb.		165	.145		4.78	8.45		13.23	18.50
1020	100 to 500 lb.		175	.137		3.88	7.95		11.83	16.75
1030	500 to 1,000 lb.		190	.126		2.77	7.35		10.12	14.55

23 31 13.16 Round and Flat-Oval Spiral Ducts

		Crew	Daily Output	Labor-Hours	Unit	Material	2020 Bare Costs Labor	Equipment	Total	Total Incl O&P
0010	**ROUND AND FLAT-OVAL SPIRAL DUCTS**									
0020	Fabricated round and flat oval spiral, includes hangers,									
0021	supports and field sketches.									
5400	Spiral preformed, steel, galv., straight lengths, max 10" spwg.									
5410	4" diameter, 26 ga.	Q-9	360	.044	L.F.	1.93	2.49		4.42	6
5416	5" diameter, 26 ga.		320	.050		1.98	2.80		4.78	6.55
5420	6" diameter, 26 ga.		280	.057		2.03	3.20		5.23	7.25
5425	7" diameter, 26 ga.		240	.067		2.27	3.74		6.01	8.35
5430	8" diameter, 26 ga.		200	.080		2.90	4.49		7.39	10.25
5440	10" diameter, 26 ga.		160	.100		3.54	5.60		9.14	12.70
9990	Minimum labor/equipment charge	1 Shee	3	2.667	Job		166		166	260

23 31 16 – Nonmetal Ducts

23 31 16.13 Fibrous-Glass Ducts

		Crew	Daily Output	Labor-Hours	Unit	Material	2020 Bare Costs Labor	Equipment	Total	Total Incl O&P
0010	**FIBROUS-GLASS DUCTS**									
3490	Rigid fiberglass duct board, foil reinf. kraft facing									
3500	Rectangular, 1" thick, alum. faced, (FRK), std. weight	Q-10	350	.069	SF Surf	.89	3.99		4.88	7.25
9990	Minimum labor/equipment charge	1 Shee	3	2.667	Job		166		166	260

23 31 16.19 PVC Ducts

		Crew	Daily Output	Labor-Hours	Unit	Material	2020 Bare Costs Labor	Equipment	Total	Total Incl O&P
0010	**PVC DUCTS**									
4000	Rigid plastic, corrosive fume resistant PVC									
4020	Straight, 6" diameter	Q-9	220	.073	L.F.	14.15	4.08		18.23	22
4040	8" diameter		160	.100		20.50	5.60		26.10	31.50
4060	10" diameter		120	.133		26	7.50		33.50	40
4070	12" diameter		100	.160		30	8.95		38.95	47
4080	14" diameter		70	.229		37.50	12.80		50.30	61.50
4090	16" diameter		62	.258		43.50	14.45		57.95	70.50
4100	18" diameter		58	.276		60	15.45		75.45	90
4110	20" diameter	Q-10	75	.320		54.50	18.60		73.10	89
4130	24" diameter	"	55	.436		74.50	25.50		100	122
4250	Coupling, 6" diameter	Q-9	88	.182	Ea.	29	10.20		39.20	48
4270	8" diameter		78	.205		38	11.50		49.50	60
4290	10" diameter		70	.229		43.50	12.80		56.30	67.50
4300	12" diameter		55	.291		47	16.30		63.30	77.50
4310	14" diameter		44	.364		60.50	20.50		81	98.50
4320	16" diameter		40	.400		64	22.50		86.50	106
4330	18" diameter		38	.421		83	23.50		106.50	129
4340	20" diameter	Q-10	55	.436		98.50	25.50		124	148
4360	24" diameter	"	45	.533		125	31		156	186
4470	Elbow, 90°, 6" diameter	Q-9	44	.364		132	20.50		152.50	178
4490	8" diameter		28	.571		148	32		180	212
4510	10" diameter		18	.889		185	50		235	281
4520	12" diameter		15	1.067		179	60		239	291
4540	16" diameter		10	1.600		315	89.50		404.50	485
4550	18" diameter	Q-10	15	1.600		380	93		473	565
4560	20" diameter		14	1.714		640	99.50		739.50	855
4580	24" diameter		12	2		785	116		901	1,050
4750	Elbow 45°, use 90° and deduct					25%				

For customer support on your Commercial Renovation Costs with RSMeans data, call 800.448.8182.

457

23 31 HVAC Ducts and Casings

23 31 16 – Nonmetal Ducts

23 31 16.19 PVC Ducts		Crew	Daily Output	Labor-Hours	Unit	Material	2020 Bare Costs Labor	Equipment	Total	Total Incl O&P
9990	Minimum labor/equipment charge	1 Shee	3	2.667	Job		166		166	260

23 31 16.22 Duct Board

0010	**DUCT BOARD**									
3800	Rigid, resin bonded fibrous glass, FSK									
3810	Temperature, bacteria and fungi resistant									
3820	1" thick	Q-14	150	.107	S.F.	1.68	5.65		7.33	10.80
3830	1-1/2" thick	↓	130	.123		2.22	6.50		8.72	12.75
3840	2" thick	↓	120	.133	↓	3.82	7.05		10.87	15.35

23 33 Air Duct Accessories

23 33 13 – Dampers

23 33 13.13 Volume-Control Dampers

0010	**VOLUME-CONTROL DAMPERS**	Crew	Daily Output	Labor-Hours	Unit	Material	2020 Bare Costs Labor	Equipment	Total	Total Incl O&P
5990	Multi-blade dampers, opposed blade, 8" x 6"	1 Shee	24	.333	Ea.	32	21		53	67.50
5994	8" x 8"		22	.364		34	22.50		56.50	73
5996	10" x 10"		21	.381		33.50	23.50		57	74
6000	12" x 12"		21	.381		46	23.50		69.50	87.50
6020	12" x 18"		18	.444		61.50	27.50		89	111
6030	14" x 10"		20	.400		45	25		70	88.50
6031	14" x 14"		17	.471		44.50	29.50		74	95
6033	16" x 12"		17	.471		44.50	29.50		74	95
6035	16" x 16"		16	.500		55.50	31		86.50	110
6037	18" x 16"		15	.533		75	33		108	135
6038	18" x 18"		15	.533		80	33		113	140
6070	20" x 16"		14	.571		80	35.50		115.50	144
6072	20" x 20"		13	.615		79	38.50		117.50	147
6074	22" x 18"		14	.571		83	35.50		118.50	148
6076	24" x 16"		11	.727		94.50	45.50		140	175
6078	24" x 20"		8	1		114	62.50		176.50	224
6080	24" x 24"		8	1		121	62.50		183.50	231
6110	26" x 26"	↓	6	1.333		143	83		226	287
6133	30" x 30"	Q-9	6.60	2.424		216	136		352	450
6135	32" x 32"		6.40	2.500		236	140		376	480
6180	48" x 36"	↓	5.60	2.857	↓	400	160		560	690
7500	Variable volume modulating motorized damper, incl. elect. mtr.									
7504	8" x 6"	1 Shee	15	.533	Ea.	177	33		210	246
7506	10" x 6"		14	.571		177	35.50		212.50	250
7510	10" x 10"		13	.615		184	38.50		222.50	263
7520	12" x 12"		12	.667		190	41.50		231.50	274
7522	12" x 16"		11	.727		191	45.50		236.50	281
7524	16" x 10"		12	.667		188	41.50		229.50	271
7526	16" x 14"		10	.800		197	50		247	294
7528	16" x 18"		9	.889		203	55.50		258.50	310
7542	18" x 18"		8	1		206	62.50		268.50	325
7544	20" x 14"		8	1		213	62.50		275.50	335
7546	20" x 18"		7	1.143		220	71		291	355
7560	24" x 12"		8	1		215	62.50		277.50	335
7562	24" x 18"		7	1.143		232	71		303	365
7564	24" x 24"		6	1.333		242	83		325	395
7568	28" x 10"		7	1.143		215	71		286	350
7590	30" x 14"	↓	5	1.600	↓	221	99.50		320.50	400

23 33 Air Duct Accessories

23 33 13 – Dampers

23 33 13.13 Volume-Control Dampers

		Crew	Daily Output	Labor-Hours	Unit	Material	2020 Bare Costs Labor	Equipment	Total	Total Incl O&P
7600	30" x 18"	1 Shee	4	2	Ea.	296	125		421	520
7610	30" x 24"		3.80	2.105		395	131		526	640
7700	For thermostat, add		8	1		48.50	62.50		111	151
8000	Multi-blade dampers, parallel blade									
8100	8" x 8"	1 Shee	24	.333	Ea.	119	21		140	164
8140	16" x 10"		20	.400		152	25		177	207
8160	18" x 12"		18	.444		168	27.50		195.50	228
8220	28" x 16"		10	.800		218	50		268	315

23 33 13.16 Fire Dampers

		Crew	Daily Output	Labor-Hours	Unit	Material	2020 Bare Costs Labor	Equipment	Total	Total Incl O&P
0010	**FIRE DAMPERS**									
3000	Fire damper, curtain type, 1-1/2 hr. rated, vertical, 6" x 6"	1 Shee	24	.333	Ea.	35	21		56	71
3020	8" x 6"		22	.364		35	22.50		57.50	74
3240	16" x 14"		18	.444		59	27.50		86.50	109
3400	24" x 20"		8	1		66	62.50		128.50	170

23 33 19 – Duct Silencers

23 33 19.10 Duct Silencers

		Crew	Daily Output	Labor-Hours	Unit	Material	2020 Bare Costs Labor	Equipment	Total	Total Incl O&P
0010	**DUCT SILENCERS**									
9000	Silencers, noise control for air flow, duct				MCFM	87			87	96

23 33 23 – Turning Vanes

23 33 23.13 Air Turning Vanes

		Crew	Daily Output	Labor-Hours	Unit	Material	2020 Bare Costs Labor	Equipment	Total	Total Incl O&P
0010	**AIR TURNING VANES**									
9400	Turning vane components									
9410	Turning vane rail	1 Shee	160	.050	L.F.	.84	3.12		3.96	5.80
9420	Double thick, factory fab. vane		300	.027		1.26	1.66		2.92	3.99
9428	12" high set		170	.047		1.92	2.93		4.85	6.70
9432	14" high set		160	.050		2.24	3.12		5.36	7.35
9900	Minimum labor/equipment charge		4	2	Job		125		125	195

23 33 46 – Flexible Ducts

23 33 46.10 Flexible Air Ducts

		Crew	Daily Output	Labor-Hours	Unit	Material	2020 Bare Costs Labor	Equipment	Total	Total Incl O&P
0010	**FLEXIBLE AIR DUCTS**									
1280	Add to labor for elevated installation									
1282	of prefabricated (purchased) ductwork									
1283	10' to 15' high						10%			
1284	15' to 20' high						20%			
1285	20' to 25' high						25%			
1286	25' to 30' high						35%			
1287	30' to 35' high						40%			
1288	35' to 40' high						50%			
1289	Over 40' high						55%			
1300	Flexible, coated fiberglass fabric on corr. resist. metal helix									
1400	pressure to 12" (WG) UL-181									
1500	Noninsulated, 3" diameter	Q-9	400	.040	L.F.	1.36	2.24		3.60	5
1540	5" diameter		320	.050		1.57	2.80		4.37	6.10
1560	6" diameter		280	.057		1.85	3.20		5.05	7.05
1580	7" diameter		240	.067		2.05	3.74		5.79	8.10
1600	8" diameter		200	.080		2.32	4.49		6.81	9.60
1640	10" diameter		160	.100		2.97	5.60		8.57	12.05
1660	12" diameter		120	.133		3.50	7.50		11	15.55
1900	Insulated, 1" thick, PE jacket, 3" diameter [G]		380	.042		2.68	2.36		5.04	6.65
1910	4" diameter [G]		340	.047		3.26	2.64		5.90	7.70
1920	5" diameter [G]		300	.053		3.42	2.99		6.41	8.45

For customer support on your Commercial Renovation Costs with RSMeans data, call 800.448.8182.

459

23 33 Air Duct Accessories

23 33 46 – Flexible Ducts

23 33 46.10 Flexible Air Ducts

			Crew	Daily Output	Labor-Hours	Unit	Material	2020 Bare Costs Labor	2020 Bare Costs Equipment	Total	Total Incl O&P
1940	6" diameter	G	Q-9	260	.062	L.F.	3.61	3.45		7.06	9.35
1960	7" diameter	G		220	.073		4.16	4.08		8.24	11
1980	8" diameter	G		180	.089		4.04	4.99		9.03	12.25
2020	10" diameter	G		140	.114		5.15	6.40		11.55	15.75
2040	12" diameter	G		100	.160		6.05	8.95		15	21
2060	14" diameter	G		80	.200		7.40	11.20		18.60	25.50
9990	Minimum labor/equipment charge		1 Shee	3	2.667	Job		166		166	260

23 33 53 – Duct Liners

23 33 53.10 Duct Liner Board

			Crew	Daily Output	Labor-Hours	Unit	Material	2020 Bare Costs Labor	2020 Bare Costs Equipment	Total	Total Incl O&P
0010	**DUCT LINER BOARD**										
3490	Board type, fiberglass liner, 3 lb. density										
3680	No finish										
3700	1" thick	G	Q-14	170	.094	S.F.	.73	4.97		5.70	8.70
3710	1-1/2" thick	G		140	.114		1.10	6.05		7.15	10.75
3720	2" thick	G		130	.123		1.49	6.50		7.99	11.95
3940	Board type, non-fibrous foam										
3950	Temperature, bacteria and fungi resistant										
3960	1" thick	G	Q-14	150	.107	S.F.	2.76	5.65		8.41	12
3970	1-1/2" thick	G		130	.123		4.89	6.50		11.39	15.70
3980	2" thick	G		120	.133		4.13	7.05		11.18	15.70

23 34 HVAC Fans

23 34 13 – Axial HVAC Fans

23 34 13.10 Axial Flow HVAC Fans

0010	**AXIAL FLOW HVAC FANS**										
0020	Air conditioning and process air handling										

23 34 14 – Blower HVAC Fans

23 34 14.10 Blower Type HVAC Fans

			Crew	Daily Output	Labor-Hours	Unit	Material	2020 Bare Costs Labor	2020 Bare Costs Equipment	Total	Total Incl O&P
0010	**BLOWER TYPE HVAC FANS**										
2500	Ceiling fan, right angle, extra quiet, 0.10" S.P.										
2540	210 CFM		Q-20	19	1.053	Ea.	360	60		420	490
2580	885 CFM			16	1.250		900	71.50		971.50	1,100
2620	2,960 CFM			11	1.818		1,675	104		1,779	1,975
2640	For wall or roof cap, add		1 Shee	16	.500		305	31		336	385

23 34 16 – Centrifugal HVAC Fans

23 34 16.10 Centrifugal Type HVAC Fans

			Crew	Daily Output	Labor-Hours	Unit	Material	2020 Bare Costs Labor	2020 Bare Costs Equipment	Total	Total Incl O&P
0010	**CENTRIFUGAL TYPE HVAC FANS**										
0200	In-line centrifugal, supply/exhaust booster										
0220	aluminum wheel/hub, disconnect switch, 1/4" S.P.										
0240	500 CFM, 10" diameter connection		Q-20	3	6.667	Ea.	1,625	380		2,005	2,375
0280	1,520 CFM, 16" diameter connection			2	10		1,700	570		2,270	2,775
0320	3,480 CFM, 20" diameter connection			.80	25		2,175	1,425		3,600	4,625
0326	5,080 CFM, 20" diameter connection			.75	26.667		2,375	1,525		3,900	4,975
5000	Utility set, centrifugal, V belt drive, motor										
5020	1/4" S.P., 1,200 CFM, 1/4 HP		Q-20	6	3.333	Ea.	1,925	190		2,115	2,425
5040	1,520 CFM, 1/3 HP			5	4		2,575	229		2,804	3,175
5060	1,850 CFM, 1/2 HP			4	5		2,550	286		2,836	3,275
5080	2,180 CFM, 3/4 HP			3	6.667		2,950	380		3,330	3,850
7000	Roof exhauster, centrifugal, aluminum housing, 12" galvanized										

23 34 HVAC Fans

23 34 16 – Centrifugal HVAC Fans

23 34 16.10 Centrifugal Type HVAC Fans

		Crew	Daily Output	Labor-Hours	Unit	Material	2020 Bare Costs Labor	Equipment	Total	Total Incl O&P
7020	curb, bird screen, back draft damper, 1/4" S.P.									
7100	Direct drive, 320 CFM, 11" sq. damper	Q-20	7	2.857	Ea.	785	163		948	1,125
7120	600 CFM, 11" sq. damper	"	6	3.333	"	880	190		1,070	1,275
9900	Minimum labor/equipment charge	1 Elec	4	2	Job		123		123	189

23 34 23 – HVAC Power Ventilators

23 34 23.10 HVAC Power Circulators and Ventilators

		Crew	Daily Output	Labor-Hours	Unit	Material	2020 Bare Costs Labor	Equipment	Total	Total Incl O&P
0010	**HVAC POWER CIRCULATORS AND VENTILATORS**									
6650	Residential, bath exhaust, grille, back draft damper									
6660	50 CFM	Q-20	24	.833	Ea.	61	47.50		108.50	142
6670	110 CFM		22	.909		116	52		168	209
6680	Light combination, squirrel cage, 100 watt, 70 CFM	↓	24	.833	↓	123	47.50		170.50	210
6700	Light/heater combination, ceiling mounted									
6710	70 CFM, 1,450 watt	Q-20	24	.833	Ea.	139	47.50		186.50	228
6800	Heater combination, recessed, 70 CFM		24	.833		83.50	47.50		131	167
6820	With 2 infrared bulbs		23	.870		100	49.50		149.50	188
6900	Kitchen exhaust, grille, complete, 160 CFM		22	.909		104	52		156	196
6910	180 CFM		20	1		97	57		154	196
6920	270 CFM	↓	18	1.111	↓	214	63.50		277.50	335
6940	Residential roof jacks and wall caps									
6944	Wall cap with back draft damper									
6946	3" & 4" diam. round duct	1 Shee	11	.727	Ea.	26.50	45.50		72	100
6948	6" diam. round duct	"	11	.727	"	76	45.50		121.50	155
6958	Roof jack with bird screen and back draft damper									
6960	3" & 4" diam. round duct	1 Shee	11	.727	Ea.	18.85	45.50		64.35	92
6962	3-1/4" x 10" rectangular duct	"	10	.800	"	35	50		85	117
6980	Transition									
6982	3-1/4" x 10" to 6" diam. round	1 Shee	20	.400	Ea.	36	25		61	78.50
8020	Attic, roof type									
8030	Aluminum dome, damper & curb									
8080	12" diameter, 1,000 CFM (gravity)	1 Elec	10	.800	Ea.	640	49		689	780
8090	16" diameter, 1,500 CFM (gravity)		9	.889		775	54.50		829.50	935
8100	20" diameter, 2,500 CFM (gravity)		8	1		950	61.50		1,011.50	1,150
8110	26" diameter, 4,000 CFM (gravity)		7	1.143		1,150	70		1,220	1,375
8120	32" diameter, 6,500 CFM (gravity)		6	1.333		1,575	82		1,657	1,850
8130	38" diameter, 8,000 CFM (gravity)		5	1.600		2,350	98		2,448	2,725
8140	50" diameter, 13,000 CFM (gravity)	↓	4	2	↓	3,400	123		3,523	3,925
8160	Plastic, ABS dome									
8180	1,050 CFM	1 Elec	14	.571	Ea.	189	35		224	261
8200	1,600 CFM	"	12	.667	"	282	41		323	375
8240	Attic, wall type, with shutter, one speed									
8250	12" diameter, 1,000 CFM	1 Elec	14	.571	Ea.	430	35		465	530
8260	14" diameter, 1,500 CFM		12	.667		465	41		506	580
8270	16" diameter, 2,000 CFM	↓	9	.889	↓	530	54.50		584.50	665
8290	Whole house, wall type, with shutter, one speed									
8300	30" diameter, 4,800 CFM	1 Elec	7	1.143	Ea.	1,125	70		1,195	1,350
8310	36" diameter, 7,000 CFM		6	1.333		1,225	82		1,307	1,475
8320	42" diameter, 10,000 CFM		5	1.600		1,375	98		1,473	1,675
8330	48" diameter, 16,000 CFM	↓	4	2		1,700	123		1,823	2,075
8340	For two speed, add				↓	103			103	113
8350	Whole house, lay-down type, with shutter, one speed									
8360	30" diameter, 4,500 CFM	1 Elec	8	1	Ea.	1,200	61.50		1,261.50	1,425
8370	36" diameter, 6,500 CFM	↓	7	1.143	↓	1,300	70		1,370	1,525

For customer support on your Commercial Renovation Costs with RSMeans data, call 800.448.8182.

461

23 34 HVAC Fans

23 34 23 – HVAC Power Ventilators

23 34 23.10 HVAC Power Circulators and Ventilators	Crew	Daily Output	Labor-Hours	Unit	Material	2020 Bare Costs Labor	Equipment	Total	Total Incl O&P
8380	42" diameter, 9,000 CFM	1 Elec	6	1.333	Ea.	1,425	82		1,507	1,675
8390	48" diameter, 12,000 CFM	↓	5	1.600		1,600	98		1,698	1,925
8440	For two speed, add					77.50			77.50	85
8450	For 12 hour timer switch, add	1 Elec	32	.250	↓	77.50	15.35		92.85	109

23 36 Air Terminal Units

23 36 13 – Constant-Air-Volume Units

23 36 13.10 Constant Volume Mixing Boxes

		Crew	Daily Output	Labor-Hours	Unit	Material	Labor	Equipment	Total	Total Incl O&P
0010	**CONSTANT VOLUME MIXING BOXES**									
5180	Mixing box, includes electric or pneumatic motor									
5200	Constant volume, 150 to 270 CFM	Q-9	12	1.333	Ea.	1,200	75		1,275	1,450
5210	270 to 600 CFM	"	11	1.455	"	1,150	81.50		1,231.50	1,400

23 36 16 – Variable-Air-Volume Units

23 36 16.10 Variable Volume Mixing Boxes

		Crew	Daily Output	Labor-Hours	Unit	Material	Labor	Equipment	Total	Total Incl O&P
0010	**VARIABLE VOLUME MIXING BOXES**									
5180	Mixing box, includes electric or pneumatic motor									
5500	VAV cool only, pneumatic, pressure independent 300 to 600 CFM	Q-9	11	1.455	Ea.	815	81.50		896.50	1,025
5510	500 to 1,000 CFM		9	1.778		835	99.50		934.50	1,075
5520	800 to 1,600 CFM		9	1.778		865	99.50		964.50	1,100
5530	1,100 to 2,000 CFM		8	2		885	112		997	1,150
5540	1,500 to 3,000 CFM		7	2.286		940	128		1,068	1,225
5550	2,000 to 4,000 CFM	↓	6	2.667	↓	950	150		1,100	1,275

23 37 Air Outlets and Inlets

23 37 13 – Diffusers, Registers, and Grilles

23 37 13.10 Diffusers

		Crew	Daily Output	Labor-Hours	Unit	Material	Labor	Equipment	Total	Total Incl O&P
0010	**DIFFUSERS**, Aluminum, opposed blade damper unless noted									
0100	Ceiling, linear, also for sidewall									
0120	2" wide	1 Shee	32	.250	L.F.	21	15.60		36.60	47.50
0160	4" wide		26	.308		24	19.15		43.15	56.50
0180	6" wide		24	.333		30	21		51	65.50
0200	8" wide	↓	22	.364	↓	40.50	22.50		63	80
0500	Perforated, 24" x 24" lay-in panel size, 6" x 6"		16	.500	Ea.	166	31		197	231
0520	8" x 8"		15	.533		181	33		214	251
0530	9" x 9"		14	.571		177	35.50		212.50	250
0540	10" x 10"		14	.571		177	35.50		212.50	251
0560	12" x 12"		12	.667		184	41.50		225.50	268
0590	16" x 16"		11	.727		207	45.50		252.50	299
0600	18" x 18"		10	.800		222	50		272	320
0610	20" x 20"		10	.800		239	50		289	340
0620	24" x 24"		9	.889		284	55.50		339.50	395
1000	Rectangular, 1 to 4 way blow, 6" x 6"		16	.500		43.50	31		74.50	97
1010	8" x 8"		15	.533		63	33		96	121
1014	9" x 9"		15	.533		53.50	33		86.50	111
1016	10" x 10"		15	.533		83.50	33		116.50	144
1020	12" x 6"		15	.533		78	33		111	138
1040	12" x 9"		14	.571		84.50	35.50		120	149
1060	12" x 12"		12	.667		76	41.50		117.50	149
1070	14" x 6"	↓	13	.615		85	38.50		123.50	154

23 37 Air Outlets and Inlets

23 37 13 – Diffusers, Registers, and Grilles

23 37 13.10 Diffusers

		Crew	Daily Output	Labor-Hours	Unit	Material	2020 Bare Costs Labor	Equipment	Total	Total Incl O&P
1074	14" x 14"	1 Shee	12	.667	Ea.	133	41.50		174.50	212
1150	18" x 18"		9	.889		129	55.50		184.50	228
1170	24" x 12"		10	.800		177	50		227	272
2000	T-bar mounting, 24" x 24" lay-in frame, 6" x 6"		16	.500		73.50	31		104.50	130
2020	8" x 8"		14	.571		73.50	35.50		109	137
2040	12" x 12"		12	.667		90	41.50		131.50	164
2060	16" x 16"		11	.727		125	45.50		170.50	208
2080	18" x 18"		10	.800		123	50		173	213
6000	For steel diffusers instead of aluminum, deduct					10%				
9000	Minimum labor/equipment charge	1 Shee	4	2	Job		125		125	195

23 37 13.30 Grilles

		Crew	Daily Output	Labor-Hours	Unit	Material	2020 Bare Costs Labor	Equipment	Total	Total Incl O&P
0010	**GRILLES**									
0020	Aluminum, unless noted otherwise									
1000	Air return, steel, 6" x 6"	1 Shee	26	.308	Ea.	23.50	19.15		42.65	56
1020	10" x 6"		24	.333		23.50	21		44.50	58.50
1080	16" x 8"		22	.364		35	22.50		57.50	74
1100	12" x 12"		22	.364		33.50	22.50		56	72
1120	24" x 12"		18	.444		43.50	27.50		71	91.50
1300	48" x 24"		12	.667		156	41.50		197.50	237
9000	Minimum labor/equipment charge		4	2	Job		125		125	195

23 37 13.60 Registers

		Crew	Daily Output	Labor-Hours	Unit	Material	2020 Bare Costs Labor	Equipment	Total	Total Incl O&P
0010	**REGISTERS**									
0980	Air supply									
1000	Ceiling/wall, O.B. damper, anodized aluminum									
1010	One or two way deflection, adj. curved face bars									
1120	12" x 12"	1 Shee	18	.444	Ea.	37	27.50		64.50	84
1140	14" x 8"		17	.471		22	29.50		51.50	70.50
1240	20" x 6"		18	.444		30.50	27.50		58	77
1340	24" x 8"		13	.615		61.50	38.50		100	128
1350	24" x 18"		12	.667		113	41.50		154.50	189
2700	Above registers in steel instead of aluminum, deduct					10%				
3000	Baseboard, hand adj. damper, enameled steel									
3012	8" x 6"	1 Shee	26	.308	Ea.	8.10	19.15		27.25	39
3020	10" x 6"		24	.333		8.80	21		29.80	42
3040	12" x 5"		23	.348		7.70	21.50		29.20	42.50
3060	12" x 6"		23	.348		11.45	21.50		32.95	46.50
3080	12" x 8"		22	.364		13.85	22.50		36.35	51
3100	14" x 6"		20	.400		10.40	25		35.40	50.50
9000	Minimum labor/equipment charge		4	2	Job		125		125	195

23 37 23 – HVAC Gravity Ventilators

23 37 23.10 HVAC Gravity Air Ventilators

		Crew	Daily Output	Labor-Hours	Unit	Material	2020 Bare Costs Labor	Equipment	Total	Total Incl O&P
0010	**HVAC GRAVITY AIR VENTILATORS**, Includes base									
1280	Rotary ventilators, wind driven, galvanized									
1300	4" neck diameter	Q-9	20	.800	Ea.	52	45		97	128
1400	12" neck diameter		10	1.600		73	89.50		162.50	221
1500	24" neck diameter		8	2		320	112		432	530
1600	For aluminum, add					30%				
9000	Minimum labor/equipment charge	1 Plum	2	4	Job		258		258	400

For customer support on your Commercial Renovation Costs with RSMeans data, call 800.448.8182.

463

23 38 Ventilation Hoods

23 38 13 – Commercial-Kitchen Hoods

23 38 13.10 Hood and Ventilation Equipment	Crew	Daily Output	Labor-Hours	Unit	Material	2020 Bare Costs Labor	Equipment	Total	Total Incl O&P
0010 **HOOD AND VENTILATION EQUIPMENT**									
2970 Exhaust hood, sst, gutter on all sides, 4' x 4' x 2'	1 Carp	1.80	4.444	Ea.	4,075	236		4,311	4,850
2980 4' x 4' x 7'	"	1.60	5	"	6,750	266		7,016	7,850

23 41 Particulate Air Filtration

23 41 13 – Panel Air Filters

23 41 13.10 Panel Type Air Filters

	Crew	Daily Output	Labor-Hours	Unit	Material	2020 Bare Costs Labor	Equipment	Total	Total Incl O&P
0010 **PANEL TYPE AIR FILTERS**									
2950 Mechanical media filtration units									
3000 High efficiency type, with frame, non-supported G				MCFM	41			41	45
3100 Supported type G				"	52.50			52.50	58
5500 Throwaway glass or paper media type, 12" x 36" x 1"				Ea.	2.48			2.48	2.73

23 41 16 – Renewable-Media Air Filters

23 41 16.10 Disposable Media Air Filters

	Crew	Daily Output	Labor-Hours	Unit	Material	2020 Bare Costs Labor	Equipment	Total	Total Incl O&P
0010 **DISPOSABLE MEDIA AIR FILTERS**									
5000 Renewable disposable roll				C.S.F.	5.25			5.25	5.75

23 41 19 – Washable Air Filters

23 41 19.10 Permanent Air Filters

	Crew	Daily Output	Labor-Hours	Unit	Material	2020 Bare Costs Labor	Equipment	Total	Total Incl O&P
0010 **PERMANENT AIR FILTERS**									
4500 Permanent washable G				MCFM	25			25	27.50

23 41 23 – Extended Surface Filters

23 41 23.10 Expanded Surface Filters

	Crew	Daily Output	Labor-Hours	Unit	Material	2020 Bare Costs Labor	Equipment	Total	Total Incl O&P
0010 **EXPANDED SURFACE FILTERS**									
4000 Medium efficiency, extended surface G				MCFM	6.65			6.65	7.35

23 42 Gas-Phase Air Filtration

23 42 13 – Activated-Carbon Air Filtration

23 42 13.10 Charcoal Type Air Filtration

	Crew	Daily Output	Labor-Hours	Unit	Material	2020 Bare Costs Labor	Equipment	Total	Total Incl O&P
0010 **CHARCOAL TYPE AIR FILTRATION**									
0050 Activated charcoal type, full flow				MCFM	650			650	715
0060 Full flow, impregnated media 12" deep					225			225	248
0070 HEPA filter & frame for field erection					410			410	450
0080 HEPA filter-diffuser, ceiling install.				↓	340			340	375

23 43 Electronic Air Cleaners

23 43 13 – Washable Electronic Air Cleaners

23 43 13.10 Electronic Air Cleaners

	Crew	Daily Output	Labor-Hours	Unit	Material	2020 Bare Costs Labor	Equipment	Total	Total Incl O&P
0010 **ELECTRONIC AIR CLEANERS**									
2000 Electronic air cleaner, duct mounted									
2150 1,000 CFM	1 Shee	4	2	Ea.	425	125		550	665
2200 1,200 CFM		3.80	2.105		570	131		701	830
2250 1,400 CFM	↓	3.60	2.222	↓	600	138		738	875

23 51 Breechings, Chimneys, and Stacks

23 51 13 – Draft Control Devices

23 51 13.16 Vent Dampers

		Crew	Daily Output	Labor-Hours	Unit	Material	2020 Bare Costs Labor	Equipment	Total	Total Incl O&P
0010	**VENT DAMPERS**									
5000	Vent damper, bi-metal, gas, 3" diameter	Q-9	24	.667	Ea.	71	37.50		108.50	137
5010	4" diameter		24	.667		70	37.50		107.50	136
5020	5" diameter		23	.696		71	39		110	139
5030	6" diameter		22	.727		70	41		111	141
5040	7" diameter		21	.762		72.50	42.50		115	147
5050	8" diameter		20	.800		77.50	45		122.50	156
9000	Minimum labor/equipment charge	1 Shee	4	2	Job		125		125	195

23 51 13.19 Barometric Dampers

		Crew	Daily Output	Labor-Hours	Unit	Material	2020 Bare Costs Labor	Equipment	Total	Total Incl O&P
0010	**BAROMETRIC DAMPERS**									
1000	Barometric, gas fired system only, 6" size for 5" and 6" pipes	1 Shee	20	.400	Ea.	105	25		130	155
1040	8" size, for 7" and 8" pipes	"	18	.444	"	148	27.50		175.50	207
2000	All fuel, oil, oil/gas, coal									
2020	10" for 9" and 10" pipes	1 Shee	15	.533	Ea.	255	33		288	335
3260	For thermal switch for above, add	"	24	.333	"	105	21		126	149

23 51 23 – Gas Vents

23 51 23.10 Gas Chimney Vents

		Crew	Daily Output	Labor-Hours	Unit	Material	2020 Bare Costs Labor	Equipment	Total	Total Incl O&P
0010	**GAS CHIMNEY VENTS**, Prefab metal, UL listed									
0020	Gas, double wall, galvanized steel									
0080	3" diameter	Q-9	72	.222	V.L.F.	7.60	12.45		20.05	28
0100	4" diameter		68	.235		9.35	13.20		22.55	31
0120	5" diameter		64	.250		10.60	14		24.60	33.50
0140	6" diameter		60	.267		12.65	14.95		27.60	37.50
0160	7" diameter		56	.286		24.50	16		40.50	52
0180	8" diameter		52	.308		26	17.25		43.25	55.50
0200	10" diameter		48	.333		49.50	18.70		68.20	84
0220	12" diameter		44	.364		58.50	20.50		79	96.50
0260	16" diameter		40	.400		140	22.50		162.50	189
0300	20" diameter	Q-10	36	.667		205	39		244	287
0340	24" diameter	"	32	.750		320	43.50		363.50	420

23 51 26 – All-Fuel Vent Chimneys

23 51 26.30 All-Fuel Vent Chimneys, Double Wall, St. Stl.

		Crew	Daily Output	Labor-Hours	Unit	Material	2020 Bare Costs Labor	Equipment	Total	Total Incl O&P
0010	**ALL-FUEL VENT CHIMNEYS, DOUBLE WALL, STAINLESS STEEL**									
7780	All fuel, pressure tight, double wall, 4" insulation, UL listed, 1,400°F.									
7790	304 stainless steel liner, aluminized steel outer jacket									
7800	6" diameter	Q-9	60	.267	V.L.F.	64.50	14.95		79.45	94.50
7804	8" diameter		52	.308		77	17.25		94.25	112
7806	10" diameter		48	.333		87.50	18.70		106.20	126
7808	12" diameter		44	.364		101	20.50		121.50	143
7810	14" diameter		42	.381		113	21.50		134.50	158
7880	For 316 stainless steel liner, add				L.F.	30%				
8000	All fuel, double wall, stainless steel fittings									
8010	Roof support, 6" diameter	Q-9	30	.533	Ea.	120	30		150	179
8030	8" diameter		26	.615		136	34.50		170.50	203
8040	10" diameter		24	.667		151	37.50		188.50	225
8050	12" diameter		22	.727		157	41		198	236
8060	14" diameter		21	.762		167	42.50		209.50	250
8100	Elbow 45°, 6" diameter		30	.533		252	30		282	325
8140	8" diameter		26	.615		286	34.50		320.50	370
8160	10" diameter		24	.667		330	37.50		367.50	425
8180	12" diameter		22	.727		375	41		416	475

23 51 Breechings, Chimneys, and Stacks

23 51 26 – All-Fuel Vent Chimneys

23 51 26.30 All-Fuel Vent Chimneys, Double Wall, St. Stl.	Crew	Daily Output	Labor-Hours	Unit	Material	2020 Bare Costs Labor	Equipment	Total	Total Incl O&P
8200	14" diameter	Q-9	21	.762	Ea.	410	42.50		452.50	515
8300	Insulated tee, 6" diameter		30	.533		296	30		326	370
8360	8" diameter		26	.615		330	34.50		364.50	420
8380	10" diameter		24	.667		350	37.50		387.50	445
8400	12" diameter		22	.727		420	41		461	525
8420	14" diameter		20	.800		490	45		535	610
8500	Boot tee, 6" diameter		28	.571		610	32		642	720
8520	8" diameter		24	.667		670	37.50		707.50	795
8530	10" diameter		22	.727		770	41		811	910
8540	12" diameter		20	.800		905	45		950	1,075
8550	14" diameter		18	.889		945	50		995	1,125
8600	Rain cap with bird screen, 6" diameter		30	.533		297	30		327	370
8640	8" diameter		26	.615		350	34.50		384.50	440
8660	10" diameter		24	.667		410	37.50		447.50	510
8680	12" diameter		22	.727		470	41		511	585
8700	14" diameter		21	.762		535	42.50		577.50	655
8800	Flat roof flashing, 6" diameter		30	.533		111	30		141	169
8840	8" diameter		26	.615		121	34.50		155.50	187
8860	10" diameter		24	.667		132	37.50		169.50	204
8880	12" diameter		22	.727		145	41		186	224
8900	14" diameter		21	.762		147	42.50		189.50	229

23 52 Heating Boilers

23 52 13 – Electric Boilers

23 52 13.10 Electric Boilers, ASME

		Crew	Daily Output	Labor-Hours	Unit	Material	2020 Bare Costs Labor	Equipment	Total	Total Incl O&P
0010	**ELECTRIC BOILERS, ASME**, Standard controls and trim									
1000	Steam, 6 KW, 20.5 MBH	Q-19	1.20	20	Ea.	4,450	1,200		5,650	6,725
1160	60 KW, 205 MBH		1	24		7,250	1,425		8,675	10,200
1300	296 KW, 1,010 MBH		.45	53.333		25,300	3,200		28,500	32,700
1400	592 KW, 2,020 MBH	Q-21	.34	94.118		36,000	5,775		41,775	48,500
1540	1,110 KW, 3,788 MBH	"	.19	168		50,500	10,300		60,800	71,500
2000	Hot water, 7.5 KW, 25.6 MBH	Q-19	1.30	18.462		5,425	1,100		6,525	7,650
2040	30 KW, 102 MBH		1.20	20		5,850	1,200		7,050	8,275
2060	45 KW, 164 MBH		1.20	20		5,700	1,200		6,900	8,100
2070	60 KW, 205 MBH		1.20	20		5,800	1,200		7,000	8,250
2080	75 KW, 256 MBH		1.10	21.818		6,150	1,300		7,450	8,775
2100	90 KW, 307 MBH		1.10	21.818		6,125	1,300		7,425	8,775
9000	Minimum labor/equipment charge	Q-20	1	20	Job		1,150		1,150	1,775

23 52 23 – Cast-Iron Boilers

23 52 23.20 Gas-Fired Boilers

		Crew	Daily Output	Labor-Hours	Unit	Material	2020 Bare Costs Labor	Equipment	Total	Total Incl O&P
0010	**GAS-FIRED BOILERS**, Natural or propane, standard controls, packaged									
1000	Cast iron, with insulated jacket									
2000	Steam, gross output, 81 MBH	Q-7	1.40	22.857	Ea.	2,750	1,425		4,175	5,225
2060	163 MBH		.90	35.556		3,400	2,225		5,625	7,175
2200	440 MBH		.51	62.500		6,250	3,900		10,150	12,900
2320	1,875 MBH		.30	107		25,200	6,650		31,850	38,100
2400	3,570 MBH		.18	182		35,800	11,300		47,100	57,000
2540	6,970 MBH		.10	320		97,000	20,000		117,000	137,500
3000	Hot water, gross output, 80 MBH		1.46	21.918		1,975	1,375		3,350	4,300
3020	100 MBH		1.35	23.704		2,575	1,475		4,050	5,100

For customer support on your Commercial Renovation Costs with RSMeans data, call 800.448.8182.

23 52 Heating Boilers

23 52 23 – Cast-Iron Boilers

23 52 23.20 Gas-Fired Boilers

23 52 23.20 Gas-Fired Boilers		Crew	Daily Output	Labor-Hours	Unit	Material	2020 Bare Costs Labor	Equipment	Total	Total Incl O&P
3040	122 MBH	Q-7	1.10	29.091	Ea.	2,825	1,825		4,650	5,900
3060	163 MBH		1	32		3,275	2,000		5,275	6,700
3200	440 MBH		.58	54.983		5,550	3,425		8,975	11,400
3320	2,000 MBH		.26	125		23,300	7,800		31,100	37,700
3540	6,970 MBH		.09	360		128,000	22,400		150,400	175,500
7000	For tankless water heater, add					10%				
7050	For additional zone valves up to 312 MBH, add					199			199	219
9900	Minimum labor/equipment charge	Q-6	1	24	Job		1,475		1,475	2,275

23 52 23.30 Gas/Oil Fired Boilers

23 52 23.30 Gas/Oil Fired Boilers		Crew	Daily Output	Labor-Hours	Unit	Material	2020 Bare Costs Labor	Equipment	Total	Total Incl O&P
0010	**GAS/OIL FIRED BOILERS**, Combination with burners and controls, packaged									
1000	Cast iron with insulated jacket									
2000	Steam, gross output, 720 MBH	Q-7	.43	74.074	Ea.	15,200	4,625		19,825	23,900
2140	2,700 MBH		.19	166		26,800	10,300		37,100	45,500
2380	6,970 MBH		.09	372		106,000	23,200		129,200	152,500
2900	Hot water, gross output									
2910	200 MBH	Q-6	.62	39.024	Ea.	11,400	2,400		13,800	16,200
2920	300 MBH		.49	49.080		11,400	3,000		14,400	17,200
2930	400 MBH		.41	57.971		13,300	3,550		16,850	20,200
2940	500 MBH		.36	67.039		14,300	4,100		18,400	22,200
3000	584 MBH	Q-7	.44	72.072		14,100	4,500		18,600	22,600
3060	1,460 MBH		.28	113		33,200	7,050		40,250	47,400
3300	13,500 MBH, 403.3 BHP		.04	727		206,000	45,400		251,400	296,500

23 52 23.40 Oil-Fired Boilers

23 52 23.40 Oil-Fired Boilers		Crew	Daily Output	Labor-Hours	Unit	Material	2020 Bare Costs Labor	Equipment	Total	Total Incl O&P
0010	**OIL-FIRED BOILERS**, Standard controls, flame retention burner, packaged									
1000	Cast iron, with insulated flush jacket									
2000	Steam, gross output, 109 MBH	Q-7	1.20	26.667	Ea.	2,375	1,675		4,050	5,175
2060	207 MBH		.90	35.556		3,225	2,225		5,450	6,950
2180	1,084 MBH		.38	85.106		10,800	5,300		16,100	20,100
2200	1,360 MBH		.33	98.160		13,300	6,125		19,425	24,100
2240	2,175 MBH		.24	134		17,400	8,350		25,750	32,000
2340	4,360 MBH		.15	215		40,800	13,400		54,200	65,500
2460	6,970 MBH		.09	364		102,500	22,700		125,200	148,000
3000	Hot water, same price as steam									
4000	For tankless coil in smaller sizes, add				Ea.	15%				

23 52 26 – Steel Boilers

23 52 26.40 Oil-Fired Boilers

23 52 26.40 Oil-Fired Boilers		Crew	Daily Output	Labor-Hours	Unit	Material	2020 Bare Costs Labor	Equipment	Total	Total Incl O&P
0010	**OIL-FIRED BOILERS**, Standard controls, flame retention burner									
5000	Steel, with insulated flush jacket									
7000	Hot water, gross output, 103 MBH	Q-6	1.60	15	Ea.	2,000	920		2,920	3,625
7020	122 MBH		1.45	16.506		2,150	1,000		3,150	3,925
7040	137 MBH		1.36	17.595		2,325	1,075		3,400	4,225
7060	168 MBH		1.30	18.405		2,275	1,125		3,400	4,275
7080	225 MBH		1.22	19.704		4,000	1,200		5,200	6,275
7100	315 MBH		.96	25.105		5,075	1,525		6,600	7,950
7120	420 MBH		.70	34.483		5,925	2,100		8,025	9,775
9000	Minimum labor/equipment charge		1.75	13.714	Job		840		840	1,300

23 52 88 – Burners

23 52 88.10 Replacement Type Burners

23 52 88.10 Replacement Type Burners		Crew	Daily Output	Labor-Hours	Unit	Material	2020 Bare Costs Labor	Equipment	Total	Total Incl O&P
0010	**REPLACEMENT TYPE BURNERS**									
0990	Residential, conversion, gas fired, LP or natural									
1000	Gun type, atmospheric input 50 to 225 MBH	Q-1	2.50	6.400	Ea.	1,025	370		1,395	1,700

23 52 Heating Boilers

23 52 88 – Burners

23 52 88.10 Replacement Type Burners	Crew	Daily Output	Labor-Hours	Unit	Material	2020 Bare Costs Labor	2020 Bare Costs Equipment	Total	Total Incl O&P
1020	100 to 400 MBH	Q-1	2	8	Ea.	1,725	465		2,190	2,625
1040	300 to 1,000 MBH	↓	1.70	9.412	↓	5,125	545		5,670	6,500
3000	Flame retention oil fired assembly, input									
3020	.50 to 2.25 GPH	Q-1	2.40	6.667	Ea.	360	385		745	995
3040	2.0 to 5.0 GPH		2	8		335	465		800	1,075
3060	3.0 to 7.0 GPH		1.80	8.889		625	515		1,140	1,475
3080	6.0 to 12.0 GPH	↓	1.60	10	↓	1,000	580		1,580	2,000

23 54 Furnaces

23 54 13 – Electric-Resistance Furnaces

23 54 13.10 Electric Furnaces

		Crew	Daily Output	Labor-Hours	Unit	Material	2020 Bare Costs Labor	2020 Bare Costs Equipment	Total	Total Incl O&P
0010	**ELECTRIC FURNACES**, Hot air, blowers, std. controls									
0011	not including gas, oil or flue piping									
1000	Electric, UL listed									
1070	10.2 MBH	Q-20	4.40	4.545	Ea.	350	260		610	790
1080	17.1 MBH		4.60	4.348		430	248		678	860
1100	34.1 MBH	↓	4.40	4.545	↓	535	260		795	995

23 54 16 – Fuel-Fired Furnaces

23 54 16.13 Gas-Fired Furnaces

		Crew	Daily Output	Labor-Hours	Unit	Material	2020 Bare Costs Labor	2020 Bare Costs Equipment	Total	Total Incl O&P
0010	**GAS-FIRED FURNACES**									
3000	Gas, AGA certified, upflow, direct drive models									
3020	45 MBH input	Q-9	4	4	Ea.	700	224		924	1,125
3040	60 MBH input		3.80	4.211		730	236		966	1,175
3060	75 MBH input		3.60	4.444		835	249		1,084	1,300
3100	100 MBH input		3.20	5		825	280		1,105	1,350
3120	125 MBH input		3	5.333		855	299		1,154	1,400
3130	150 MBH input		2.80	5.714		1,750	320		2,070	2,425
3140	200 MBH input	↓	2.60	6.154	↓	3,525	345		3,870	4,425

23 54 16.16 Oil-Fired Furnaces

		Crew	Daily Output	Labor-Hours	Unit	Material	2020 Bare Costs Labor	2020 Bare Costs Equipment	Total	Total Incl O&P
0010	**OIL-FIRED FURNACES**									
6000	Oil, UL listed, atomizing gun type burner									
6020	56 MBH output	Q-9	3.60	4.444	Ea.	3,275	249		3,524	4,000
6040	95 MBH output		3.40	4.706		3,400	264		3,664	4,150
6060	134 MBH output		3.20	5		3,325	280		3,605	4,125
6080	151 MBH output		3	5.333		3,700	299		3,999	4,550
6100	200 MBH input		2.60	6.154	↓	4,175	345		4,520	5,150
9000	Minimum labor/equipment charge	↓	2.75	5.818	Job		325		325	510

23 54 24 – Furnace Components for Cooling

23 54 24.10 Furnace Components and Combinations

		Crew	Daily Output	Labor-Hours	Unit	Material	2020 Bare Costs Labor	2020 Bare Costs Equipment	Total	Total Incl O&P
0010	**FURNACE COMPONENTS AND COMBINATIONS**									
0080	Coils, A.C. evaporator, for gas or oil furnaces									
0090	Add-on, with holding charge									
0100	Upflow									
0120	1-1/2 ton cooling	Q-5	4	4	Ea.	244	236		480	635
0130	2 ton cooling		3.70	4.324		305	255		560	730
0140	3 ton cooling		3.30	4.848		430	286		716	910
0150	4 ton cooling		3	5.333		570	315		885	1,100
0160	5 ton cooling	↓	2.70	5.926	↓	615	350		965	1,225
0300	Downflow									
0330	2-1/2 ton cooling	Q-5	3	5.333	Ea.	365	315		680	885

23 54 Furnaces

23 54 24 – Furnace Components for Cooling

23 54 24.10 Furnace Components and Combinations

		Crew	Daily Output	Labor-Hours	Unit	Material	2020 Bare Costs Labor	Equipment	Total	Total Incl O&P
0340	3-1/2 ton cooling	Q-5	2.60	6.154	Ea.	495	365		860	1,100
0350	5 ton cooling	↓	2.20	7.273	↓	510	430		940	1,225
0600	Horizontal									
0630	2 ton cooling	Q-5	3.90	4.103	Ea.	440	242		682	860
0640	3 ton cooling		3.50	4.571		450	270		720	910
0650	4 ton cooling		3.20	5		540	295		835	1,050
0660	5 ton cooling	↓	2.90	5.517	↓	540	325		865	1,100
2000	Cased evaporator coils for air handlers									
2100	1-1/2 ton cooling	Q-5	4.40	3.636	Ea.	310	215		525	675
2110	2 ton cooling		4.10	3.902		375	230		605	765
2120	2-1/2 ton cooling		3.90	4.103		400	242		642	815
2130	3 ton cooling		3.70	4.324		470	255		725	910
2140	3-1/2 ton cooling		3.50	4.571		600	270		870	1,075
2150	4 ton cooling		3.20	5		630	295		925	1,150
2160	5 ton cooling	↓	2.90	5.517	↓	625	325		950	1,200
3010	Air handler, modular									
3100	With cased evaporator cooling coil									
3120	1-1/2 ton cooling	Q-5	3.80	4.211	Ea.	1,050	248		1,298	1,525
3130	2 ton cooling		3.50	4.571		1,100	270		1,370	1,650
3140	2-1/2 ton cooling		3.30	4.848		1,175	286		1,461	1,750
3150	3 ton cooling		3.10	5.161		1,450	305		1,755	2,075
3160	3-1/2 ton cooling		2.90	5.517		1,700	325		2,025	2,375
3170	4 ton cooling		2.50	6.400		1,650	380		2,030	2,400
3180	5 ton cooling	↓	2.10	7.619	↓	2,125	450		2,575	3,025
3500	With no cooling coil									
3520	1-1/2 ton coil size	Q-5	12	1.333	Ea.	1,025	78.50		1,103.50	1,250
3530	2 ton coil size		10	1.600		1,100	94.50		1,194.50	1,350
3540	2-1/2 ton coil size		10	1.600		1,075	94.50		1,169.50	1,325
3554	3 ton coil size		9	1.778		1,075	105		1,180	1,325
3560	3-1/2 ton coil size		9	1.778		1,150	105		1,255	1,400
3570	4 ton coil size		8.50	1.882		1,275	111		1,386	1,575
3580	5 ton coil size	↓	8	2	↓	1,575	118		1,693	1,925
4000	With heater									
4120	5 kW, 17.1 MBH	Q-5	16	1	Ea.	485	59		544	625
4130	7.5 kW, 25.6 MBH		15.60	1.026		830	60.50		890.50	1,000
4140	10 kW, 34.2 MBH		15.20	1.053		1,225	62		1,287	1,450
4150	12.5 kW, 42.7 MBH	↓	14.80	1.081	↓	1,375	64		1,439	1,625

23 55 Fuel-Fired Heaters

23 55 13 – Fuel-Fired Duct Heaters

23 55 13.16 Gas-Fired Duct Heaters

		Crew	Daily Output	Labor-Hours	Unit	Material	2020 Bare Costs Labor	Equipment	Total	Total Incl O&P
0010	**GAS-FIRED DUCT HEATERS**, Includes burner, controls, stainless steel									
0020	heat exchanger. Gas fired, electric ignition									
1000	Outdoor installation, with power venter									
1120	225 MBH output	Q-5	2.50	6.400	Ea.	5,725	380		6,105	6,875
1160	375 MBH output		1.60	10		6,800	590		7,390	8,375
1180	450 MBH output	↓	1.40	11.429	↓	8,300	675		8,975	10,200

For customer support on your Commercial Renovation Costs with RSMeans data, call 800.448.8182.

469

23 55 Fuel-Fired Heaters

23 55 33 – Fuel-Fired Unit Heaters

23 55 33.16 Gas-Fired Unit Heaters

		Crew	Daily Output	Labor-Hours	Unit	Material	2020 Bare Costs Labor	2020 Bare Costs Equipment	Total	Total Incl O&P
0010	**GAS-FIRED UNIT HEATERS**, Cabinet, grilles, fan, ctrls., burner, no piping									
0022	thermostat, no piping. For flue see Section 23 51 23.10									
1000	Gas fired, floor mounted									
1100	60 MBH output	Q-5	10	1.600	Ea.	765	94.50		859.50	990
1180	180 MBH output		6	2.667		1,200	157		1,357	1,575
2000	Suspension mounted, propeller fan, 20 MBH output		8.50	1.882		1,725	111		1,836	2,075
2040	60 MBH output		7	2.286		1,850	135		1,985	2,225
2100	130 MBH output		5	3.200		2,375	189		2,564	2,900
2240	320 MBH output	▼	2	8		4,400	470		4,870	5,575
2500	For powered venter and adapter, add					400			400	440
5000	Wall furnace, 17.5 MBH output	Q-5	6	2.667		710	157		867	1,025
5020	24 MBH output		5	3.200		805	189		994	1,175
5040	35 MBH output		4	4	▼	1,150	236		1,386	1,625
9000	Minimum labor/equipment charge	▼	3.50	4.571	Job		270		270	415

23 56 Solar Energy Heating Equipment

23 56 16 – Packaged Solar Heating Equipment

23 56 16.40 Solar Heating Systems

			Crew	Daily Output	Labor-Hours	Unit	Material	2020 Bare Costs Labor	2020 Bare Costs Equipment	Total	Total Incl O&P
0010	**SOLAR HEATING SYSTEMS**										
0020	System/package prices, not including connecting										
0030	pipe, insulation, or special heating/plumbing fixtures										
0500	Hot water, standard package, low temperature										
0580	2 collectors, circulator, fittings, 120 gal. tank	G	Q-1	.40	40	Ea.	5,750	2,325		8,075	9,900

23 56 19 – Solar Heating Components

23 56 19.50 Solar Heating Ancillary

			Crew	Daily Output	Labor-Hours	Unit	Material	2020 Bare Costs Labor	2020 Bare Costs Equipment	Total	Total Incl O&P
0010	**SOLAR HEATING ANCILLARY**										
2300	Circulators, air										
2310	Blowers										
2550	Booster fan 6" diameter, 120 CFM	G	Q-9	16	1	Ea.	38	56		94	130
2580	8" diameter, 150 CFM	G		16	1		42.50	56		98.50	135
2660	Shutter/damper	G		12	1.333		58.50	75		133.50	182
2670	Shutter motor	G	▼	16	1		146	56		202	249
2800	Circulators, liquid, 1/25 HP, 5.3 GPM	G	Q-1	14	1.143		205	66.50		271.50	325
2820	1/20 HP, 17 GPM	G	"	12	1.333	▼	284	77.50		361.50	430
3000	Collector panels, air with aluminum absorber plate										
3010	Wall or roof mount										
3040	Flat black, plastic glazing										
3080	4' x 8'	G	Q-9	6	2.667	Ea.	685	150		835	990
3300	Collector panels, liquid with copper absorber plate										
3320	Black chrome, tempered glass glazing										
3330	Alum. frame, 4' x 8', 5/32" single glazing	G	Q-1	9.50	1.684	Ea.	1,075	97.50		1,172.50	1,325
3450	Flat black, alum. frame, 3.5' x 7.5'	G		9	1.778		815	103		918	1,050
3500	4' x 8'	G		5.50	2.909		1,025	169		1,194	1,375
3600	Liquid, full wetted, plastic, alum. frame, 4' x 10'	G		5	3.200		335	186		521	655
3650	Collector panel mounting, flat roof or ground rack	G		7	2.286	▼	253	133		386	485
3670	Roof clamps	G	▼	70	.229	Set	3.27	13.25		16.52	24
3700	Roof strap, teflon	G	1 Plum	205	.039	L.F.	25	2.51		27.51	31.50
3900	Differential controller with two sensors										
3930	Thermostat, hard wired	G	1 Plum	8	1	Ea.	112	64.50		176.50	223
4050	Pool valve system	G		2.50	3.200	▼	182	206		388	520

23 56 Solar Energy Heating Equipment

23 56 19 – Solar Heating Components

23 56 19.50 Solar Heating Ancillary	Crew	Daily Output	Labor-Hours	Unit	Material	2020 Bare Costs Labor	Equipment	Total	Total Incl O&P
4070	With 12 VAC actuator	1 Plum	2	4	Ea.	335	258		593	765
4150	Sensors									
4220	Freeze prevention	1 Plum	32	.250	Ea.	25	16.10		41.10	52.50
4300	Heat exchanger									
4315	includes coil, blower, circulator									
4316	and controller for DHW and space hot air									
4330	Fluid to air coil, up flow, 45 MBH	Q-1	4	4	Ea.	330	232		562	725
4380	70 MBH		3.50	4.571		375	265		640	820
4400	80 MBH		3	5.333		495	310		805	1,025
4580	Fluid to fluid package includes two circulating pumps									
4590	expansion tank, check valve, relief valve									
4600	controller, high temperature cutoff and sensors	Q-1	2.50	6.400	Ea.	810	370		1,180	1,475
4650	Heat transfer fluid									
4700	Propylene glycol, inhibited anti-freeze	1 Plum	28	.286	Gal.	23	18.40		41.40	54
4800	Solar storage tanks, knocked down									
5190	6'-3" high, 7' x 7' = 306 C.F./2,000 gallons	Q-10	1.20	20	Ea.	13,900	1,175		15,075	17,100
5210	7' x 14' = 613 C.F./4,000 gallons		.60	40		22,400	2,325		24,725	28,300
5230	10'-6" x 14' = 919 C.F./6,000 gallons		.40	60		24,200	3,500		27,700	32,200
5250	14' x 17'-6" = 1,531 C.F./10,000 gallons	Q-11	.30	107		31,400	6,325		37,725	44,500
7000	Solar control valves and vents									
7070	Air eliminator, automatic 3/4" size	1 Plum	32	.250	Ea.	31	16.10		47.10	59
7090	Air vent, automatic, 1/8" fitting		32	.250		17.90	16.10		34	44.50
7100	Manual, 1/8" NPT		32	.250		2.69	16.10		18.79	28
7120	Backflow preventer, 1/2" pipe size		16	.500		90.50	32		122.50	150
7130	3/4" pipe size		16	.500		89	32		121	148
7150	Balancing valve, 3/4" pipe size		20	.400		59.50	26		85.50	105
7180	Draindown valve, 1/2" copper tube		9	.889		219	57.50		276.50	330
7200	Flow control valve, 1/2" pipe size		22	.364		140	23.50		163.50	190
7220	Expansion tank, up to 5 gal.		32	.250		70.50	16.10		86.60	103
7400	Pressure gauge, 2" dial		32	.250		25	16.10		41.10	52.50
7450	Relief valve, temp. and pressure 3/4" pipe size		30	.267		25.50	17.20		42.70	54.50
7500	Solenoid valve, normally closed									
7520	Brass, 3/4" NPT, 24V	1 Plum	9	.889	Ea.	137	57.50		194.50	239
7530	1" NPT, 24V		9	.889		224	57.50		281.50	335
7750	Vacuum relief valve, 3/4" pipe size		32	.250		31	16.10		47.10	59
7800	Thermometers									
7820	Digital temperature monitoring, 4 locations	1 Plum	2.50	3.200	Ea.	138	206		344	470
7900	Upright, 1/2" NPT	"	8	1	"	39.50	64.50		104	143
8250	Water storage tank with heat exchanger and electric element									
8270	66 gal. with 2" x 2 lb. density insulation	1 Plum	1.60	5	Ea.	2,150	320		2,470	2,875
8300	80 gal. with 2" x 2 lb. density insulation	"	1.60	5	"	1,950	320		2,270	2,650
8500	Water storage module, plastic									
8600	Tubular, 12" diameter, 4' high	1 Carp	48	.167	Ea.	99	8.85		107.85	123
8610	12" diameter, 8' high	"	40	.200		157	10.65		167.65	189
8650	Cap, 12" diameter					20			20	22
9000	Minimum labor/equipment charge	1 Plum	2	4	Job		258		258	400

23 61 Refrigerant Compressors

23 61 16 – Reciprocating Refrigerant Compressors

23 61 16.10 Reciprocating Compressors	Crew	Daily Output	Labor-Hours	Unit	Material	2020 Bare Costs Labor	Equipment	Total	Total Incl O&P
0010 **RECIPROCATING COMPRESSORS**									
0990 Refrigeration, recip. hermetic, switches & protective devices									
1000 10 ton	Q-5	1	16	Ea.	15,200	945		16,145	18,200
1100 20 ton	Q-6	.72	33.333		23,700	2,050		25,750	29,300
1400 50 ton	"	.20	120		34,300	7,350		41,650	49,200
1600 130 ton	Q-7	.21	152	↓	43,400	9,500		52,900	62,500

23 62 Packaged Compressor and Condenser Units

23 62 13 – Packaged Air-Cooled Refrigerant Compressor and Condenser Units

23 62 13.10 Packaged Air-Cooled Refrig. Condensing Units	Crew	Daily Output	Labor-Hours	Unit	Material	2020 Bare Costs Labor	Equipment	Total	Total Incl O&P
0010 **PACKAGED AIR-COOLED REFRIGERANT CONDENSING UNITS**									
0020 Condensing unit									
0030 Air cooled, compressor, standard controls									
0050 1.5 ton	Q-5	2.50	6.400	Ea.	1,025	380		1,405	1,700
0200 2.5 ton		1.70	9.412		1,225	555		1,780	2,200
0300 3 ton		1.30	12.308		1,250	725		1,975	2,500
0400 4 ton		.90	17.778		1,700	1,050		2,750	3,500
0500 5 ton	↓	.60	26.667	↓	2,025	1,575		3,600	4,650

23 64 Packaged Water Chillers

23 64 19 – Reciprocating Water Chillers

23 64 19.10 Reciprocating Type Water Chillers

	Crew	Daily Output	Labor-Hours	Unit	Material	2020 Bare Costs Labor	Equipment	Total	Total Incl O&P
0010 **RECIPROCATING TYPE WATER CHILLERS**, With standard controls									
0494 Water chillers, integral air cooled condenser									
0980 Water cooled, multiple compressor, semi-hermetic, tower not incl.									
1090 45 ton cooling	Q-7	.29	111	Ea.	27,200	6,925		34,125	40,600
1160 100 ton cooling		.18	180		49,800	11,200		61,000	72,500
1200 145 ton cooling	↓	.16	203	↓	61,500	12,600		74,100	87,000

23 64 23 – Scroll Water Chillers

23 64 23.10 Scroll Water Chillers

	Crew	Daily Output	Labor-Hours	Unit	Material	2020 Bare Costs Labor	Equipment	Total	Total Incl O&P
0010 **SCROLL WATER CHILLERS**, With standard controls									
0480 Packaged w/integral air cooled condenser									
0482 10 ton cooling	Q-7	.34	94.118	Ea.	15,200	5,875		21,075	25,800
0490 15 ton cooling		.37	86.486		18,300	5,400		23,700	28,600
0500 20 ton cooling		.34	94.118		23,300	5,875		29,175	34,800
0520 40 ton cooling	↓	.30	108	↓	34,800	6,750		41,550	48,700
0680 Scroll water cooled, single compressor, hermetic, tower not incl.									
0760 10 ton cooling	Q-6	.36	67.039	Ea.	6,575	4,100		10,675	13,600
0900 Scroll water cooled, multiple compressor, hermetic, tower not incl.									

23 64 26 – Rotary-Screw Water Chillers

23 64 26.10 Rotary-Screw Type Water Chillers

	Crew	Daily Output	Labor-Hours	Unit	Material	2020 Bare Costs Labor	Equipment	Total	Total Incl O&P
0010 **ROTARY-SCREW TYPE WATER CHILLERS**, With standard controls									
0110 Screw, liquid chiller, air cooled, insulated evaporator									
0120 130 ton	Q-7	.14	229	Ea.	98,500	14,300		112,800	130,500
0124 160 ton	"	.13	246	"	125,000	15,400		140,400	161,500

472

23 64 Packaged Water Chillers

23 64 33 – Modular Water Chillers

23 64 33.10 Direct Expansion Type Water Chillers	Crew	Daily Output	Labor-Hours	Unit	Material	2020 Bare Costs Labor	2020 Bare Costs Equipment	Total	Total Incl O&P
0010 **DIRECT EXPANSION TYPE WATER CHILLERS**, With standard controls									
8000 Direct expansion, shell and tube type, for built up systems									
8020 1 ton	Q-5	2	8	Ea.	8,125	470		8,595	9,650
8030 5 ton		1.90	8.421		13,500	495		13,995	15,600
8040 10 ton		1.70	9.412		16,700	555		17,255	19,300
9000 Minimum labor/equipment charge	Q-6	1	24	Job		1,475		1,475	2,275

23 65 Cooling Towers

23 65 13 – Forced-Draft Cooling Towers

23 65 13.10 Forced-Draft Type Cooling Towers

	Crew	Daily Output	Labor-Hours	Unit	Material	2020 Bare Costs Labor	2020 Bare Costs Equipment	Total	Total Incl O&P
0010 **FORCED-DRAFT TYPE COOLING TOWERS**, Packaged units									
0070 Galvanized steel									
0080 Induced draft, crossflow									
0100 Vertical, belt drive, 61 tons	Q-6	90	.267	TonAC	203	16.30		219.30	248
0150 100 ton		100	.240		195	14.70		209.70	238
0200 115 ton		109	.220		170	13.45		183.45	208
0250 131 ton		120	.200		204	12.25		216.25	243
0260 162 ton		132	.182		165	11.10		176.10	198
1000 For higher capacities, use multiples									
1500 Induced air, double flow									
1900 Vertical, gear drive, 167 ton	Q-6	126	.190	TonAC	160	11.65		171.65	194
2000 297 ton		129	.186		98.50	11.40		109.90	127
2100 582 ton		132	.182		55	11.10		66.10	77.50
2200 1,016 ton		150	.160		74.50	9.80		84.30	97
3000 For higher capacities, use multiples									
3500 For pumps and piping, add	Q-6	38	.632	TonAC	101	38.50		139.50	171
4000 For absorption systems, add				"	75%	75%			
4100 Cooling water chemical feeder	Q-5	3	5.333	Ea.	425	315		740	955
5000 Fiberglass tower on galvanized steel support structure									
5010 Draw thru									
5100 100 ton	Q-6	1.40	17.143	Ea.	14,000	1,050		15,050	17,000
5120 120 ton		1.20	20		16,700	1,225		17,925	20,300
5140 140 ton		1	24		18,100	1,475		19,575	22,200
5160 160 ton		.80	30		22,900	1,825		24,725	28,100
5180 180 ton		.65	36.923		26,400	2,250		28,650	32,600
5251 600 ton		.16	150		77,500	9,175		86,675	99,000
5252 1,000 ton		.10	250		129,000	15,300		144,300	165,000
5300 For stainless steel support structure, add					30%				
5360 For higher capacities, use multiples of each size									
6000 Stainless steel									
6010 Induced draft, crossflow, horizontal, belt drive									
6100 57 ton	Q-6	1.50	16	Ea.	29,900	980		30,880	34,400
6120 91 ton		.99	24.242		36,700	1,475		38,175	42,700
6140 111 ton		.43	55.814		49,600	3,425		53,025	60,000
6160 126 ton		.22	109		54,000	6,675		60,675	70,000
6170 Induced draft, crossflow, vertical, gear drive									
6172 167 ton	Q-6	.75	32	Ea.	61,500	1,950		63,450	70,500
6174 297 ton		.43	55.814		68,000	3,425		71,425	80,000
6176 582 ton		.23	104		107,000	6,375		113,375	127,500
6180 1,016 ton		.15	160		185,500	9,800		195,300	219,500

For customer support on your Commercial Renovation Costs with RSMeans data, call 800.448.8182.

473

23 65 Cooling Towers

23 65 13 – Forced-Draft Cooling Towers

23 65 13.10 Forced-Draft Type Cooling Towers	Crew	Daily Output	Labor-Hours	Unit	Material	2020 Bare Costs Labor	Equipment	Total	Total Incl O&P
9000	Minimum labor/equipment charge	Q-6	1	24	Job		1,475		1,475	2,275

23 72 Air-to-Air Energy Recovery Equipment

23 72 13 – Heat-Wheel Air-to-Air Energy-Recovery Equipment

23 72 13.10 Heat-Wheel Air-to-Air Energy Recovery Equip.

		Crew	Daily Output	Labor-Hours	Unit	Material	2020 Bare Costs Labor	Equipment	Total	Total Incl O&P
0010	**HEAT-WHEEL AIR-TO-AIR ENERGY RECOVERY EQUIPMENT**									
0100	Air to air									
9000	Minimum labor/equipment charge	1 Shee	4	2	Job		125		125	195

23 73 Indoor Central-Station Air-Handling Units

23 73 39 – Indoor, Direct Gas-Fired Heating and Ventilating Units

23 73 39.10 Make-Up Air Unit

		Crew	Daily Output	Labor-Hours	Unit	Material	2020 Bare Costs Labor	Equipment	Total	Total Incl O&P
0010	**MAKE-UP AIR UNIT**									
0020	Indoor suspension, natural/LP gas, direct fired,									
0032	standard control. For flue see Section 23 51 23.10									
0040	70°F temperature rise, MBH is input									
0100	75 MBH input	Q-6	3.60	6.667	Ea.	5,225	410		5,635	6,350
0220	225 MBH input		2.40	10		7,075	610		7,685	8,750
0300	400 MBH input	↓	1.60	15	↓	16,500	920		17,420	19,600

23 74 Packaged Outdoor HVAC Equipment

23 74 23 – Packaged, Outdoor, Heating-Only Makeup-Air Units

23 74 23.16 Packaged, Ind.-Fired, In/Outdoor

		Crew	Daily Output	Labor-Hours	Unit	Material	2020 Bare Costs Labor	Equipment	Total	Total Incl O&P
0010	**PACKAGED, IND.-FIRED, IN/OUTDOOR**, Heat-Only Makeup-Air Units									
1000	Rooftop unit, natural gas, gravity vent, S.S. exchanger									
1010	70°F temperature rise, MBH is input									
1020	250 MBH	Q-6	4	6	Ea.	16,000	365		16,365	18,300
1040	400 MBH		3.60	6.667		18,300	410		18,710	20,800
1060	550 MBH	↓	3.30	7.273		21,200	445		21,645	24,100
1600	For cleanable filters, add					5%				
1700	For electric modulating gas control, add				↓	10%				
9000	Minimum labor/equipment charge	Q-6	2.75	8.727	Job		535		535	825

23 74 33 – Dedicated Outdoor-Air Units

23 74 33.10 Rooftop Air Conditioners

		Crew	Daily Output	Labor-Hours	Unit	Material	2020 Bare Costs Labor	Equipment	Total	Total Incl O&P
0010	**ROOFTOP AIR CONDITIONERS**, Standard controls, curb, economizer									
1000	Single zone, electric cool, gas heat									
1100	3 ton cooling, 60 MBH heating	Q-5	.70	22.857	Ea.	2,900	1,350		4,250	5,250
1120	4 ton cooling, 95 MBH heating		.61	26.403		3,750	1,550		5,300	6,525
1150	7.5 ton cooling, 170 MBH heating	↓	.50	32.258		5,725	1,900		7,625	9,250
1156	8.5 ton cooling, 170 MBH heating	↓	.46	34.783		7,200	2,050		9,250	11,100
1160	10 ton cooling, 200 MBH heating	Q-6	.67	35.982		9,675	2,200		11,875	14,000
1220	30 ton cooling, 540 MBH heating	Q-7	.47	68.376		36,300	4,275		40,575	46,500
1240	40 ton cooling, 675 MBH heating	"	.35	91.168	↓	44,400	5,700		50,100	57,500
2000	Multizone, electric cool, gas heat, economizer									
2120	20 ton cooling, 360 MBH heating	Q-7	.53	60.038	Ea.	72,000	3,750		75,750	85,500
2180	30 ton cooling, 540 MBH heating		.37	85.562		110,500	5,350		115,850	130,000
2210	50 ton cooling, 540 MBH heating	↓	.23	142	↓	161,000	8,875		169,875	190,500

23 74 Packaged Outdoor HVAC Equipment

23 74 33 – Dedicated Outdoor-Air Units

23 74 33.10 Rooftop Air Conditioners	Crew	Daily Output	Labor-Hours	Unit	Material	2020 Bare Costs Labor	Equipment	Total	Total Incl O&P
2220	70 ton cooling, 1,500 MBH heating	Q-7	.16	199	Ea.	171,500	12,400		183,900	207,500
2240	80 ton cooling, 1,500 MBH heating		.14	229		195,500	14,300		209,800	237,500
2280	105 ton cooling, 1,500 MBH heating	↓	.11	291		223,000	18,200		241,200	273,500
2400	For hot water heat coil, deduct					5%				
2500	For steam heat coil, deduct					2%				
2600	For electric heat, deduct				↓	3%	5%			
9000	Minimum labor/equipment charge	Q-5	1.50	10.667	Job		630		630	975

23 81 Decentralized Unitary HVAC Equipment

23 81 19 – Self-Contained Air-Conditioners

23 81 19.10 Window Unit Air Conditioners

		Crew	Daily Output	Labor-Hours	Unit	Material	Labor	Equipment	Total	Total Incl O&P
0010	**WINDOW UNIT AIR CONDITIONERS**									
4000	Portable/window, 15 amp, 125 V grounded receptacle required									
4500	10,000 BTUH	1 Carp	6	1.333	Ea.	650	71		721	830
4520	12,000 BTUH	L-2	8	2	"	2,000	93		2,093	2,350
9000	Minimum labor/equipment charge	1 Carp	2	4	Job		213		213	340

23 81 19.20 Self-Contained Single Package

		Crew	Daily Output	Labor-Hours	Unit	Material	Labor	Equipment	Total	Total Incl O&P
0010	**SELF-CONTAINED SINGLE PACKAGE**									
0100	Air cooled, for free blow or duct, not incl. remote condenser									
0110	Constant volume									
0200	3 ton cooling	Q-5	1	16	Ea.	4,050	945		4,995	5,900
0210	4 ton cooling	"	.80	20		4,400	1,175		5,575	6,675
0240	10 ton cooling	Q-7	1	32		8,350	2,000		10,350	12,300
0280	30 ton cooling	"	.80	40	↓	26,400	2,500		28,900	33,000
1000	Water cooled for free blow or duct, not including tower									
1010	Constant volume									
1100	3 ton cooling	Q-6	1	24	Ea.	3,750	1,475		5,225	6,400
1120	5 ton cooling	"	1	24		4,875	1,475		6,350	7,650
1140	10 ton cooling	Q-7	.90	35.556		8,725	2,225		10,950	13,000
1180	30 ton cooling	"	.70	45.714		36,600	2,850		39,450	44,700
1240	60 ton cooling	Q-8	.30	107		58,000	6,650	355	65,005	74,500
1300	For hot water or steam heat coils, add				↓	12%	10%			
9000	Minimum labor/equipment charge	Q-5	1	16	Job		945		945	1,450

23 81 23 – Computer-Room Air-Conditioners

23 81 23.10 Computer Room Units

		Crew	Daily Output	Labor-Hours	Unit	Material	Labor	Equipment	Total	Total Incl O&P
0010	**COMPUTER ROOM UNITS**									
1000	Air cooled, includes remote condenser but not									
1020	interconnecting tubing or refrigerant									
1160	6 ton	Q-5	.30	53.333	Ea.	55,000	3,150		58,150	65,500
1240	10 ton	"	.25	64		57,500	3,775		61,275	69,500
1300	20 ton	Q-6	.26	92.308		76,000	5,650		81,650	92,000
1320	22 ton	"	.24	100	↓	77,000	6,125		83,125	94,000

23 81 26 – Split-System Air-Conditioners

23 81 26.10 Split Ductless Systems

		Crew	Daily Output	Labor-Hours	Unit	Material	Labor	Equipment	Total	Total Incl O&P
0010	**SPLIT DUCTLESS SYSTEMS**									
0100	Cooling only, single zone									
0110	Wall mount									
0120	3/4 ton cooling	Q-5	2	8	Ea.	1,125	470		1,595	1,950
0130	1 ton cooling		1.80	8.889		1,250	525		1,775	2,175
0140	1-1/2 ton cooling	↓	1.60	10	↓	1,975	590		2,565	3,075

For customer support on your Commercial Renovation Costs with RSMeans data, call 800.448.8182.

475

23 81 Decentralized Unitary HVAC Equipment

23 81 26 – Split-System Air-Conditioners

23 81 26.10 Split Ductless Systems	Crew	Daily Output	Labor-Hours	Unit	Material	2020 Bare Costs Labor	Equipment	Total	Total Incl O&P
0150	2 ton cooling	Q-5	1.40	11.429	Ea.	2,275	675		2,950	3,550
0160	3 ton cooling	↓	1.20	13.333	↓	2,275	785		3,060	3,725
1000	Ceiling mount									
1020	2 ton cooling	Q-5	1.40	11.429	Ea.	2,000	675		2,675	3,250
1030	3 ton cooling	"	1.20	13.333	"	2,575	785		3,360	4,075
3000	Multizone									
3010	Wall mount									
3020	2 @ 3/4 ton cooling	Q-5	1.80	8.889	Ea.	3,425	525		3,950	4,575
5000	Cooling/Heating									
5010	Wall mount									
5110	1 ton cooling	Q-5	1.70	9.412	Ea.	1,300	555		1,855	2,275
5120	1-1/2 ton cooling	"	1.50	10.667	"	2,000	630		2,630	3,175
7000	Accessories for all split ductless systems									
7010	Add for ambient frost control	Q-5	8	2	Ea.	129	118		247	325
7020	Add for tube/wiring kit (line sets)									
7030	15' kit	Q-5	32	.500	Ea.	91	29.50		120.50	146
7036	25' kit		28	.571		158	33.50		191.50	225
7040	35' kit		24	.667		205	39.50		244.50	287
7050	50' kit	↓	20	.800	↓	206	47		253	299

23 81 29 – Variable Refrigerant Flow HVAC Systems

23 81 29.10 Heat Pump, Gas Driven

		Crew	Daily Output	Labor-Hours	Unit	Material	Labor	Equipment	Total	Total Incl O&P
0010	**HEAT PUMP, GAS DRIVEN**, Variable refrigerant volume (VRV) type									
0020	Not including interconnecting tubing or multi-zone controls									
1000	For indoor fan VRV type AHU see 23 82 19.40									
1010	Multi-zone split									
1020	Outdoor unit									
1100	8 tons cooling, for up to 17 zones	Q-5	1.30	12.308	Ea.	30,200	725		30,925	34,300
1110	Isolation rails		2.60	6.154	Pair	1,075	365		1,440	1,725
1160	15 tons cooling, for up to 33 zones		1	16	Ea.	38,600	945		39,545	43,900
1170	Isolation rails	↓	2.60	6.154	Pair	1,250	365		1,615	1,925
2000	Packaged unit									
2020	Outdoor unit									
2200	11 tons cooling	Q-5	1.30	12.308	Ea.	38,900	725		39,625	43,900
2210	Roof curb adapter	"	2.60	6.154		675	365		1,040	1,300
2220	Thermostat	1 Stpi	1.30	6.154	↓	405	405		810	1,075

23 81 43 – Air-Source Unitary Heat Pumps

23 81 43.10 Air-Source Heat Pumps

		Crew	Daily Output	Labor-Hours	Unit	Material	Labor	Equipment	Total	Total Incl O&P
0010	**AIR-SOURCE HEAT PUMPS**, Not including interconnecting tubing									
1000	Air to air, split system, not including curbs, pads, fan coil and ductwork									
1012	Outside condensing unit only, for fan coil see Section 23 82 19.10									
1020	2 ton cooling, 8.5 MBH heat @ 0°F	Q-5	2	8	Ea.	1,775	470		2,245	2,675
1040	3 ton cooling, 13 MBH heat @ 0°F		1.20	13.333		2,150	785		2,935	3,575
1080	7.5 ton cooling, 33 MBH heat @ 0°F	↓	.45	35.556		3,975	2,100		6,075	7,625
1120	15 ton cooling, 64 MBH heat @ 0°F	Q-6	.50	48		10,400	2,925		13,325	16,000
1130	20 ton cooling, 85 MBH heat @ 0°F		.35	68.571		18,600	4,200		22,800	27,000
1140	25 ton cooling, 119 MBH heat @ 0°F	↓	.25	96	↓	21,800	5,875		27,675	33,100
1500	Single package, not including curbs, pads, or plenums									
1520	2 ton cooling, 6.5 MBH heat @ 0°F	Q-5	1.50	10.667	Ea.	3,325	630		3,955	4,650
1560	3 ton cooling, 10 MBH heat @ 0°F	"	1.20	13.333	"	3,775	785		4,560	5,375
6000	Air to water, single package, excluding storage tank and ductwork									
6010	Includes circulating water pump, air duct connections, digital temperature									
6020	controller with remote tank temp. probe and sensor for storage tank.									

476

23 81 Decentralized Unitary HVAC Equipment

23 81 43 – Air-Source Unitary Heat Pumps

23 81 43.10 Air-Source Heat Pumps

		Crew	Daily Output	Labor-Hours	Unit	Material	2020 Bare Costs Labor	Equipment	Total	Total Incl O&P
6040	Water heating - air cooling capacity									
6110	35.5 MBH heat water, 2.3 ton cool air	Q-5	1.60	10	Ea.	17,300	590		17,890	20,000
6120	58 MBH heat water, 3.8 ton cool air		1.10	14.545		19,800	860		20,660	23,000
6130	76 MBH heat water, 4.9 ton cool air		.87	18.391		23,600	1,075		24,675	27,600
6140	98 MBH heat water, 6.5 ton cool air		.62	25.806		30,500	1,525		32,025	35,900
6150	113 MBH heat water, 7.4 ton cool air		.59	27.119		33,300	1,600		34,900	39,100
6160	142 MBH heat water, 9.2 ton cool air		.52	30.769		40,500	1,825		42,325	47,400
6170	171 MBH heat water, 11.1 ton cool air	▼	.49	32.653	▼	47,500	1,925		49,425	55,000

23 81 46 – Water-Source Unitary Heat Pumps

23 81 46.10 Water Source Heat Pumps

		Crew	Daily Output	Labor-Hours	Unit	Material	2020 Bare Costs Labor	Equipment	Total	Total Incl O&P
0010	**WATER SOURCE HEAT PUMPS**, Not incl. connecting tubing or water source									
2000	Water source to air, single package									
2100	1 ton cooling, 13 MBH heat @ 75°F	Q-5	2	8	Ea.	2,125	470		2,595	3,075
2140	2 ton cooling, 19 MBH heat @ 75°F		1.70	9.412		2,475	555		3,030	3,575
2220	5 ton cooling, 29 MBH heat @ 75°F		.90	17.778	▼	3,400	1,050		4,450	5,350
9000	Minimum labor/equipment charge	▼	1.75	9.143	Job		540		540	835

23 82 Convection Heating and Cooling Units

23 82 16 – Air Coils

23 82 16.10 Flanged Coils

		Crew	Daily Output	Labor-Hours	Unit	Material	2020 Bare Costs Labor	Equipment	Total	Total Incl O&P
0010	**FLANGED COILS**									
1500	Hot water heating, 1 row, 24" x 48"	Q-5	4	4	Ea.	1,850	236		2,086	2,425
9000	Minimum labor/equipment charge	1 Plum	1	8	Job		515		515	795

23 82 16.20 Duct Heaters

		Crew	Daily Output	Labor-Hours	Unit	Material	2020 Bare Costs Labor	Equipment	Total	Total Incl O&P
0010	**DUCT HEATERS**, Electric, 480 V, 3 Ph.									
0020	Finned tubular insert, 500°F									
0100	8" wide x 6" high, 4.0 kW	Q-20	16	1.250	Ea.	825	71.50		896.50	1,025
0120	12" high, 8.0 kW		15	1.333		1,375	76		1,451	1,625
0140	18" high, 12.0 kW		14	1.429		1,925	81.50		2,006.50	2,225
0160	24" high, 16.0 kW		13	1.538		2,475	88		2,563	2,850
0180	30" high, 20.0 kW		12	1.667		3,025	95		3,120	3,475
0300	12" wide x 6" high, 6.7 kW		15	1.333		875	76		951	1,075
0320	12" high, 13.3 kW	▼	14	1.429		1,425	81.50		1,506.50	1,675
0340	18" high, 20.0 kW		13	1.538	▼	1,975	88		2,063	2,300
8000	To obtain BTU multiply kW by 3413									
9900	Minimum labor/equipment charge	1 Shee	4	2	Job		125		125	195

23 82 19 – Fan Coil Units

23 82 19.10 Fan Coil Air Conditioning

		Crew	Daily Output	Labor-Hours	Unit	Material	2020 Bare Costs Labor	Equipment	Total	Total Incl O&P
0010	**FAN COIL AIR CONDITIONING**									
0030	Fan coil AC, cabinet mounted, filters and controls									
0100	Chilled water, 1/2 ton cooling	Q-5	8	2	Ea.	575	118		693	815
0120	1 ton cooling		6	2.667		820	157		977	1,150
0160	2-1/2 ton cooling		5	3.200		1,800	189		1,989	2,275
0180	3 ton cooling	▼	4	4	▼	1,975	236		2,211	2,550
0262	For hot water coil, add					40%	10%			
0320	1 ton cooling	Q-5	6	2.667	Ea.	1,750	157		1,907	2,175
0940	Direct expansion, for use w/air cooled condensing unit, 1-1/2 ton cooling		5	3.200		685	189		874	1,050
1000	5 ton cooling	▼	3	5.333		1,250	315		1,565	1,850
1040	10 ton cooling	Q-6	2.60	9.231		2,250	565		2,815	3,375
1060	20 ton cooling	"	.70	34.286		4,325	2,100		6,425	8,000

23 82 19.10 Fan Coil Air Conditioning

		Crew	Daily Output	Labor-Hours	Unit	Material	2020 Bare Costs Labor	Equipment	Total	Total Incl O&P
1500	For hot water coil, add				Ea.	40%	10%			
1512	For condensing unit add see Section 23 62									
3000	Chilled water, horizontal unit, housing, 2 pipe, fan control, no valves									
3100	1/2 ton cooling	Q-5	8	2	Ea.	1,425	118		1,543	1,750
3110	1 ton cooling		6	2.667		1,675	157		1,832	2,100
3120	1-1/2 ton cooling		5.50	2.909		1,875	172		2,047	2,350
3130	2 ton cooling		5.25	3.048		2,250	180		2,430	2,750
3140	3 ton cooling		4	4		2,775	236		3,011	3,425
3150	3-1/2 ton cooling		3.80	4.211		3,075	248		3,323	3,750
3160	4 ton cooling	↓	3.80	4.211	↓	3,075	248		3,323	3,750
4000	With electric heat, 2 pipe									
4100	1/2 ton cooling	Q-5	8	2	Ea.	1,675	118		1,793	2,025
4105	3/4 ton cooling		7	2.286		1,875	135		2,010	2,250
4110	1 ton cooling		6	2.667		2,050	157		2,207	2,500
4120	1-1/2 ton cooling		5.50	2.909		2,275	172		2,447	2,775
4130	2 ton cooling		5.25	3.048		2,875	180		3,055	3,425
4135	2-1/2 ton cooling		4.80	3.333		3,975	197		4,172	4,675
4140	3 ton cooling		4	4		5,300	236		5,536	6,200
4150	3-1/2 ton cooling		3.80	4.211		6,250	248		6,498	7,250
4160	4 ton cooling		3.60	4.444	↓	7,200	262		7,462	8,325
9000	Minimum labor/equipment charge	↓	3.75	4.267	Job		252		252	390

23 82 19.40 Fan Coil Air Conditioning

			Crew	Daily Output	Labor-Hours	Unit	Material	2020 Bare Costs Labor	Equipment	Total	Total Incl O&P
0010	**FAN COIL AIR CONDITIONING**, Variable refrigerant volume (VRV) type										
0020	Not including interconnecting tubing or multi-zone controls										
0030	For VRV condensing unit see section 23 81 29.10										
0050	Indoor type, ducted										
0100	Vertical concealed										
0130	1 ton cooling	G	Q-5	2.97	5.387	Ea.	2,650	320		2,970	3,400
0140	1.5 ton cooling	G		2.77	5.776		2,675	340		3,015	3,475
0150	2 ton cooling	G		2.70	5.926		2,900	350		3,250	3,725
0160	2.5 ton cooling	G		2.60	6.154		3,050	365		3,415	3,900
0170	3 ton cooling	G		2.50	6.400		3,100	380		3,480	3,975
0180	3.5 ton cooling	G		2.30	6.957		3,250	410		3,660	4,200
0190	4 ton cooling	G		2.19	7.306		3,275	430		3,705	4,275
0200	4.5 ton cooling	G	↓	2.08	7.692	↓	3,600	455		4,055	4,650
0230	Outside air connection possible										
1100	Ceiling concealed										
1130	0.6 ton cooling	G	Q-5	2.60	6.154	Ea.	1,725	365		2,090	2,450
1140	0.75 ton cooling	G		2.60	6.154		1,775	365		2,140	2,525
1150	1 ton cooling	G	↓	2.60	6.154		1,875	365		2,240	2,625
1152	Screening door	G	1 Stpi	5.20	1.538		104	101		205	271
1160	1.5 ton cooling	G	Q-5	2.60	6.154		1,950	365		2,315	2,700
1162	Screening door	G	1 Stpi	5.20	1.538		121	101		222	289
1170	2 ton cooling	G	Q-5	2.60	6.154		2,300	365		2,665	3,075
1172	Screening door	G	1 Stpi	5.20	1.538		104	101		205	271
1180	2.5 ton cooling	G	Q-5	2.60	6.154		2,650	365		3,015	3,475
1182	Screening door	G	1 Stpi	5.20	1.538		181	101		282	355
1190	3 ton cooling	G	Q-5	2.60	6.154		2,825	365		3,190	3,650
1192	Screening door	G	1 Stpi	5.20	1.538		181	101		282	355
1200	4 ton cooling	G	Q-5	2.30	6.957		2,675	410		3,085	3,575
1202	Screening door	G	1 Stpi	5.20	1.538		181	101		282	355
1220	6 ton cooling	G	Q-5	2.08	7.692	↓	4,900	455		5,355	6,100

478

For customer support on your Commercial Renovation Costs with RSMeans data, call 800.448.8182.

23 82 Convection Heating and Cooling Units

23 82 19 – Fan Coil Units

23 82 19.40 Fan Coil Air Conditioning

			Crew	Daily Output	Labor-Hours	Unit	Material	2020 Bare Costs Labor	Equipment	Total	Total Incl O&P
1230	8 ton cooling	G	Q-5	1.89	8.466	Ea.	5,500	500		6,000	6,825
1260	Outside air connection possible										
4050	Indoor type, duct-free										
4100	Ceiling mounted cassette										
4130	0.75 ton cooling	G	Q-5	3.46	4.624	Ea.	2,050	273		2,323	2,675
4140	1 ton cooling	G		2.97	5.387		2,175	320		2,495	2,900
4150	1.5 ton cooling	G		2.77	5.776		2,250	340		2,590	2,975
4160	2 ton cooling	G		2.60	6.154		2,375	365		2,740	3,175
4170	2.5 ton cooling	G		2.50	6.400		2,450	380		2,830	3,275
4180	3 ton cooling	G		2.30	6.957		2,625	410		3,035	3,500
4190	4 ton cooling	G		2.08	7.692		2,925	455		3,380	3,925
4198	For ceiling cassette decoration panel, add	G	1 Stpi	5.20	1.538		310	101		411	500
4230	Outside air connection possible										
4500	Wall mounted										
4530	0.6 ton cooling	G	Q-5	5.20	3.077	Ea.	1,050	182		1,232	1,425
4540	0.75 ton cooling	G		5.20	3.077		1,100	182		1,282	1,475
4550	1 ton cooling	G		4.16	3.846		1,250	227		1,477	1,725
4560	1.5 ton cooling	G		3.77	4.244		1,375	250		1,625	1,900
4570	2 ton cooling	G		3.46	4.624		1,500	273		1,773	2,075
4590	Condensate pump for the above air handlers	G	1 Stpi	6.46	1.238		225	81		306	375
4700	Floor standing unit										
4730	1 ton cooling	G	Q-5	2.60	6.154	Ea.	1,775	365		2,140	2,500
4740	1.5 ton cooling	G		2.60	6.154		2,000	365		2,365	2,775
4750	2 ton cooling	G		2.60	6.154		2,150	365		2,515	2,925
4800	Floor standing unit, concealed										
4830	1 ton cooling	G	Q-5	2.60	6.154	Ea.	1,700	365		2,065	2,425
4840	1.5 ton cooling	G		2.60	6.154		1,925	365		2,290	2,675
4850	2 ton cooling	G		2.60	6.154		2,075	365		2,440	2,825
4880	Outside air connection possible										
8000	Accessories										
8100	Branch divergence pipe fitting										
8110	Capacity under 76 MBH		1 Stpi	2.50	3.200	Ea.	164	210		374	505
8120	76 MBH to 112 MBH			2.50	3.200		188	210		398	530
8130	112 MBH to 234 MBH			2.50	3.200		530	210		740	905
8200	Header pipe fitting										
8210	Max 4 branches										
8220	Capacity under 76 MBH		1 Stpi	1.70	4.706	Ea.	233	310		543	730
8260	Max 8 branches										
8270	76 MBH to 112 MBH		1 Stpi	1.70	4.706	Ea.	475	310		785	1,000
8280	112 MBH to 234 MBH		"	1.30	6.154	"	755	405		1,160	1,450

23 82 29 – Radiators

23 82 29.10 Hydronic Heating

		Crew	Daily Output	Labor-Hours	Unit	Material	2020 Bare Costs Labor	Equipment	Total	Total Incl O&P
0010	**HYDRONIC HEATING**, Terminal units, not incl. main supply pipe									
1000	Radiation									
1100	Panel, baseboard, C.I., including supports, no covers	Q-5	46	.348	L.F.	49	20.50		69.50	85.50
3000	Radiators, cast iron									
3100	Free standing or wall hung, 6 tube, 25" high	Q-5	96	.167	Section	61	9.85		70.85	82
9000	Minimum labor/equipment charge	"	3.20	5	Job		295		295	455

For customer support on your Commercial Renovation Costs with RSMeans data, call 800.448.8182.

479

23 82 Convection Heating and Cooling Units

23 82 36 – Finned-Tube Radiation Heaters

23 82 36.10 Finned Tube Radiation	Crew	Daily Output	Labor-Hours	Unit	Material	2020 Bare Costs Labor	Equipment	Total	Total Incl O&P
0010 **FINNED TUBE RADIATION**, Terminal units, not incl. main supply pipe									
1150 Fin tube, wall hung, 14" slope top cover, with damper									
1200 1-1/4" copper tube, 4-1/4" alum. fin	Q-5	38	.421	L.F.	52.50	25		77.50	96.50
1250 1-1/4" steel tube, 4-1/4" steel fin		36	.444		43	26		69	88
1255 2" steel tube, 4-1/4" steel fin		32	.500		47	29.50		76.50	97
1310 Baseboard, pkgd, 1/2" copper tube, alum. fin, 7" high		60	.267		11.95	15.75		27.70	37.50
1320 3/4" copper tube, alum. fin, 7" high		58	.276		8.05	16.30		24.35	34
1340 1" copper tube, alum. fin, 8-7/8" high		56	.286		21	16.85		37.85	49
1360 1-1/4" copper tube, alum. fin, 8-7/8" high		54	.296		31	17.50		48.50	61.50
1500 Note: fin tube may also require corners, caps, etc.									

23 82 39 – Unit Heaters

23 82 39.16 Propeller Unit Heaters

23 82 39.16 Propeller Unit Heaters	Crew	Daily Output	Labor-Hours	Unit	Material	2020 Bare Costs Labor	Equipment	Total	Total Incl O&P
0010 **PROPELLER UNIT HEATERS**									
3950 Unit heaters, propeller, 115 V 2 psi steam, 60°F entering air									
4000 Horizontal, 12 MBH	Q-5	12	1.333	Ea.	415	78.50		493.50	580
4060 43.9 MBH		8	2		625	118		743	865
4240 286.9 MBH		2	8		1,825	470		2,295	2,725
4260 364 MBH		1.80	8.889		2,325	525		2,850	3,350
4270 404 MBH		1.60	10		2,375	590		2,965	3,525
4300 Vertical diffuser same price									
4310 Vertical flow, 40 MBH	Q-5	11	1.455	Ea.	640	86		726	840
4326 131 MBH	"	4	4		955	236		1,191	1,425
4354 420 MBH (460 V)	Q-6	1.80	13.333		2,525	815		3,340	4,025
4358 500 MBH (460 V)		1.71	14.035		3,125	860		3,985	4,775
4362 570 MBH (460 V)		1.40	17.143		4,075	1,050		5,125	6,100
4366 620 MBH (460 V)		1.30	18.462		4,725	1,125		5,850	6,950
4370 960 MBH (460 V)		1.10	21.818		8,200	1,325		9,525	11,100

23 83 Radiant Heating Units

23 83 33 – Electric Radiant Heaters

23 83 33.10 Electric Heating

23 83 33.10 Electric Heating	Crew	Daily Output	Labor-Hours	Unit	Material	2020 Bare Costs Labor	Equipment	Total	Total Incl O&P
0010 **ELECTRIC HEATING**, not incl. conduit or feed wiring									
1100 Rule of thumb: Baseboard units, including control	1 Elec	4.40	1.818	kW	116	112		228	300
1300 Baseboard heaters, 2' long, 350 watt		8	1	Ea.	29	61.50		90.50	126
1400 3' long, 750 watt		8	1		35	61.50		96.50	133
1600 4' long, 1,000 watt		6.70	1.194		40.50	73.50		114	158
1800 5' long, 935 watt		5.70	1.404		44	86		130	182
2000 6' long, 1,500 watt		5	1.600		57	98		155	214
2400 8' long, 2,000 watt		4	2		70	123		193	266
2600 9' long, 1,680 watt		3.60	2.222		86.50	136		222.50	305
2800 10' long, 1,875 watt		3.30	2.424		162	149		311	405
2950 Wall heaters with fan, 120 to 277 volt									
3170 1,000 watt	1 Elec	6	1.333	Ea.	94.50	82		176.50	230
3180 1,250 watt		5	1.600		107	98		205	269
3190 1,500 watt		4	2		107	123		230	305
3230 2,500 watt		3.50	2.286		236	140		376	475
3250 4,000 watt		2.70	2.963		370	182		552	685
3600 Thermostats, integral		16	.500		29	30.50		59.50	79
3800 Line voltage, 1 pole		8	1		16.30	61.50		77.80	112
3810 2 pole		8	1		23.50	61.50		85	121

23 83 Radiant Heating Units

23 83 33 – Electric Radiant Heaters

23 83 33.10 Electric Heating	Crew	Daily Output	Labor-Hours	Unit	Material	2020 Bare Costs Labor	Equipment	Total	Total Incl O&P
9990 Minimum labor/equipment charge	1 Elec	4	2	Job		123		123	189

23 84 Humidity Control Equipment

23 84 13 – Humidifiers

23 84 13.10 Humidifier Units

0010 **HUMIDIFIER UNITS**									
0520 Steam, room or duct, filter, regulators, auto. controls, 220 V									
0560 22 lb./hr.	Q-5	5	3.200	Ea.	3,050	189		3,239	3,675
0580 33 lb./hr.		4	4		3,150	236		3,386	3,825
0620 100 lb./hr.		3	5.333		5,175	315		5,490	6,175
9000 Minimum labor/equipment charge		3.50	4.571	Job		270		270	415

23 91 Prefabricated Equipment Supports

23 91 10 – Prefabricated Curbs, Pads and Stands

23 91 10.10 Prefabricated Pads and Stands

0010 **PREFABRICATED PADS AND STANDS**									
6000 Pad, fiberglass reinforced concrete with polystyrene foam core									
6090 24" x 36"	1 Shee	12	.667	Ea.	41.50	41.50		83	111
6280 36" x 36"	Q-9	8	2		59.50	112		171.50	242
6300 36" x 40"		7	2.286		64.50	128		192.50	272
6320 36" x 48"		7	2.286		79	128		207	288

For customer support on your Commercial Renovation Costs with RSMeans data, call 800.448.8182.

481

Division Notes

		CREW	DAILY OUTPUT	LABOR-HOURS	UNIT	BARE COSTS				TOTAL INCL O&P
						MAT.	LABOR	EQUIP.	TOTAL	

Estimating Tips

- When estimating material costs for special systems, it is always prudent to obtain manufacturers' quotations for equipment prices and special installation requirements that may affect the total cost.

- For cost modifications for elevated tray installation, add the percentages to labor according to the height of the installation and only to the quantities exceeding the different height levels, not to the total tray quantities. Refer to subdivision 26 01 02.20 for labor adjustment factors.

- Do not overlook the costs for equipment used in the installation.

Reference Numbers

Reference numbers are shown at the beginning of some major classifications. These numbers refer to related items in the Reference Section. The reference information may be an estimating procedure, an alternate pricing method, or technical information.

Same Data. Simplified.

Enjoy the convenience and efficiency of accessing your costs anywhere:

- **Skip the multiplier** by setting your location
- **Quickly search,** edit, favorite and share costs
- **Stay on top of price changes** with automatic updates

Discover more at rsmeans.com/online

Note: Trade Service, in part, has been used as a reference source for some of the material prices used in Division 25.

25 05 Common Work Results for Integrated Automation

25 05 28 – Pathways for Integrated Automation

25 05 28.39 Surface Raceways for Integrated Automation	Crew	Daily Output	Labor-Hours	Unit	Material	2020 Bare Costs Labor	Equipment	Total	Total Incl O&P
0010 **SURFACE RACEWAYS FOR INTEGRATED AUTOMATION**									
1000 Din rail, G type, 15 mm, steel, galvanized, 3', for termination blocks	1 Elec	48	.167	Ea.	12.45	10.25		22.70	29.50
1010 32 mm, steel, galvanized, 3', for termination blocks		48	.167		16.20	10.25		26.45	33.50
1020 35 mm, steel, galvanized, 3', for termination blocks		48	.167		14.55	10.25		24.80	32
1030 Din rail, rail mounted, 35 mm, aluminum, 3', for termination blocks		48	.167		20.50	10.25		30.75	38.50
1040 35 mm, aluminum, 6', for termination blocks		48	.167		25.50	10.25		35.75	44.50
2000 Terminal block, Din mounted, 35 mm, medium screw, connector end bracket		48	.167		2.19	10.25		12.44	18.15
2010 35 mm, large screw, connector end bracket		48	.167		3.43	10.25		13.68	19.50
2020 Large screw, connector end bracket		48	.167		4.88	10.25		15.13	21
3000 Terminal block, feed through, 35 mm, Din mounted, 600V, 20A		48	.167		1	10.25		11.25	16.85
3010 35 mm, Din mounted, 600V, 30A		48	.167		1.29	10.25		11.54	17.15
3020 Terminal block, fuse holder, LED indicator, Din mounted, 110V, 6A		48	.167		9.50	10.25		19.75	26
3030 Din mounted, 220V, 6A		48	.167		7.80	10.25		18.05	24.50
3040 Din mounted, 300V, 6A		48	.167		7.65	10.25		17.90	24

25 36 Integrated Automation Instrumentation & Terminal Devices

25 36 19 – Integrated Automation Current Sensors

25 36 19.10 Integrated Automation Current Sensors

	Crew	Daily Output	Labor-Hours	Unit	Material	Labor	Equipment	Total	Total Incl O&P
0010 **INTEGRATED AUTOMATION CURRENT SENSORS**									
1000 Current limiter inrush, 100 to 480 VAC, 10A, power supply	1 Elec	48	.167	Ea.	86.50	10.25		96.75	111

Estimating Tips
26 05 00 Common Work Results for Electrical

- Conduit should be taken off in three main categories—power distribution, branch power, and branch lighting—so the estimator can concentrate on systems and components, therefore making it easier to ensure all items have been accounted for.
- For cost modifications for elevated conduit installation, add the percentages to labor according to the height of installation and only to the quantities exceeding the different height levels, not to the total conduit quantities. Refer to subdivision 26 01 02.20 for labor adjustment factors.
- Remember that aluminum wiring of equal ampacity is larger in diameter than copper and may require larger conduit.
- If more than three wires at a time are being pulled, deduct percentages from the labor hours of that grouping of wires.
- When taking off grounding systems, identify separately the type and size of wire, and list each unique type of ground connection.

- The estimator should take the weights of materials into consideration when completing a takeoff. Topics to consider include: How will the materials be supported? What methods of support are available? How high will the support structure have to reach? Will the final support structure be able to withstand the total burden? Is the support material included or separate from the fixture, equipment, and material specified?
- Do not overlook the costs for equipment used in the installation. If scaffolding or highlifts are available in the field, contractors may use them in lieu of the proposed ladders and rolling staging.

26 20 00 Low-Voltage Electrical Transmission

- Supports and concrete pads may be shown on drawings for the larger equipment, or the support system may be only a piece of plywood for the back of a panelboard. In either case, they must be included in the costs.

26 40 00 Electrical and Cathodic Protection

- When taking off cathodic protection systems, identify the type and size of cable, and list each unique type of anode connection.

26 50 00 Lighting

- Fixtures should be taken off room by room using the fixture schedule, specifications, and the ceiling plan. For large concentrations of lighting fixtures in the same area, deduct the percentages from labor hours.

Reference Numbers

Reference numbers are shown at the beginning of some major classifications. These numbers refer to related items in the Reference Section. The reference information may be an estimating procedure, an alternate pricing method, or technical information.

Note: Not all subdivisions listed here necessarily appear. ■

Same Data. Simplified.

Enjoy the convenience and efficiency of accessing your costs anywhere:

- **Skip the multiplier** by setting your location
- **Quickly search,** edit, favorite and share costs
- **Stay on top of price changes** with automatic updates

Discover more at rsmeans.com/online

---

26 01 02 – Labor Adjustment

26 01 02.20 Labor Adjustment Factors	Crew	Daily Output	Labor-Hours	Unit	Material	2020 Bare Costs Labor	Equipment	Total	Total Incl O&P
0010 LABOR ADJUSTMENT FACTORS (For Div. 26, 27 and 28) R260519-90									
0100 Subtract from labor for Economy of Scale for Wire									
0110 4-5 wires						25%			
0120 6-10 wires						30%			
0130 11-15 wires						35%			
0140 over 15 wires						40%			
0150 Labor adjustment factors (For Div. 26, 27, 28 and 48)									
0200 Labor factors: The below are reasonable suggestions, but									
0210 each project must be evaluated for its own peculiarities, and									
0220 the adjustments be increased or decreased depending on the									
0230 severity of the special conditions.									
1000 Add to labor for elevated installation (above floor level)									
1010 10' to 14.5' high						10%			
1020 15' to 19.5' high						20%			
1030 20' to 24.5' high						25%			
1040 25' to 29.5' high						35%			
1050 30' to 34.5' high						40%			
1060 35' to 39.5' high						50%			
1070 40' and higher						55%			
2000 Add to labor for crawl space									
2010 3' high						40%			
2020 4' high						30%			
3000 Add to labor for multi-story building									
3100 For new construction (No elevator available)									
3110 Add for floors 3 thru 10						5%			
3120 Add for floors 11 thru 15						10%			
3130 Add for floors 16 thru 20						15%			
3140 Add for floors 21 thru 30						20%			
3150 Add for floors 31 and up						30%			
3200 For existing structure (Elevator available)									
3210 Add for work on floor 3 and above						2%			
4000 Add to labor for working in existing occupied buildings									
4010 Hospital						35%			
4020 Office building						25%			
4030 School						20%			
4040 Factory or warehouse						15%			
4050 Multi-dwelling						15%			
5000 Add to labor, miscellaneous									
5010 Cramped shaft						35%			
5020 Congested area						15%			
5030 Excessive heat or cold						30%			
6000 Labor factors: the above are reasonable suggestions, but									
6100 each project should be evaluated for its own peculiarities									
6200 Other factors to be considered are:									
6210 Movement of material and equipment through finished areas						10%			
6220 Equipment room min security direct access w/authorization						15%			
6230 Attic space						25%			
6240 No service road						25%			
6250 Poor unloading/storage area, no hydraulic lifts or jacks						20%			
6260 Congested site area/heavy traffic						20%			
7000 Correctional facilities (no compounding division 1 adjustment factors)									
7010 Minimum security w/facilities escort						30%			
7020 Medium security w/facilities and correctional officer escort						40%			

For customer support on your Commercial Renovation Costs with RSMeans data, call 800.448.8182.

26 01 Operation and Maintenance of Electrical Systems

26 01 02 – Labor Adjustment

26 01 02.20 Labor Adjustment Factors	Crew	Daily Output	Labor-Hours	Unit	Material	2020 Bare Costs Labor	Equipment	Total	Total Incl O&P
7030	Max security w/facilities & correctional officer escort (no inmate contact)						50%			

26 01 50 – Operation and Maintenance of Lighting

26 01 50.81 Luminaire Replacement

		Crew	Daily Output	Labor-Hours	Unit	Material	Labor	Equipment	Total	Total Incl O&P
0010	**LUMINAIRE REPLACEMENT**									
0962	Remove and install new replacement lens, 2' x 2' troffer	1 Elec	16	.500	Ea.	18	30.50		48.50	67
0964	2' x 4' troffer		16	.500		14.85	30.50		45.35	63.50
3200	Remove and replace (reinstall), lighting fixture	↓	4	2	↓		123		123	189

26 05 Common Work Results for Electrical

26 05 05 – Selective Demolition for Electrical

26 05 05.10 Electrical Demolition

			Crew	Daily Output	Labor-Hours	Unit	Material	Labor	Equipment	Total	Total Incl O&P
0010	**ELECTRICAL DEMOLITION**	R260105-30									
0020	Electrical demolition, conduit to 10' high, incl. fittings & hangers										
0100	Rigid galvanized steel, 1/2" to 1" diameter	R024119-10	1 Elec	242	.033	L.F.		2.03		2.03	3.12
0120	1-1/4" to 2"		"	200	.040			2.45		2.45	3.78
0140	2-1/2" to 3-1/2"		2 Elec	302	.053			3.25		3.25	5
0200	Electric metallic tubing (EMT), 1/2" to 1"		1 Elec	394	.020			1.25		1.25	1.92
0220	1-1/4" to 1-1/2"			326	.025			1.51		1.51	2.32
0240	2" to 3"		↓	236	.034	↓		2.08		2.08	3.20
0400	Wiremold raceway, including fittings & hangers										
0420	No. 3000		1 Elec	250	.032	L.F.		1.96		1.96	3.02
0440	No. 4000		"	217	.037	"		2.26		2.26	3.48
0500	Channels, steel, including fittings & hangers										
0520	3/4" x 1-1/2"		1 Elec	308	.026	L.F.		1.59		1.59	2.45
0540	1-1/2" x 1-1/2"			269	.030			1.82		1.82	2.81
0560	1-1/2" x 1-7/8"		↓	229	.035	↓		2.14		2.14	3.30
0600	Copper bus duct, indoor, 3 phase										
0610	Including hangers & supports										
0620	225 amp		2 Elec	135	.119	L.F.		7.25		7.25	11.20
0640	400 amp			106	.151			9.25		9.25	14.25
0660	600 amp			86	.186			11.40		11.40	17.55
0680	1,000 amp			60	.267			16.35		16.35	25
0700	1,600 amp			40	.400			24.50		24.50	38
0720	3,000 amp		↓	10	1.600	↓		98		98	151
0800	Plug-in switches, 600 V 3 ph., incl. disconnecting										
0820	wire, conduit terminations, 30 amp		1 Elec	15.50	.516	Ea.		31.50		31.50	49
0840	60 amp			13.90	.576			35.50		35.50	54.50
0850	100 amp			10.40	.769			47		47	72.50
0860	200 amp		↓	6.20	1.290			79		79	122
0880	400 amp		2 Elec	5.40	2.963			182		182	280
0900	600 amp			3.40	4.706			289		289	445
0920	800 amp			2.60	6.154			380		380	580
0940	1,200 amp			2	8			490		490	755
0960	1,600 amp		↓	1.70	9.412	↓		575		575	890
1010	Safety switches, 250 or 600 V, incl. disconnection										
1050	of wire & conduit terminations										
1100	30 amp		1 Elec	12.30	.650	Ea.		40		40	61.50
1120	60 amp			8.80	.909			56		56	86
1140	100 amp			7.30	1.096			67		67	104
1160	200 amp		↓	5	1.600	↓		98		98	151

For customer support on your Commercial Renovation Costs with RSMeans data, call 800.448.8182.

487

26 05 05.10 Electrical Demolition	Crew	Daily Output	Labor-Hours	Unit	Material	2020 Bare Costs Labor	2020 Bare Costs Equipment	Total	Total Incl O&P
1210	Panel boards, incl. removal of all breakers,								
1220	conduit terminations & wire connections								
1230	3 wire, 120/240 V, 100A, to 20 circuits	1 Elec	2.60	3.077	Ea.		189		189	291
1240	200 amps, to 42 circuits	2 Elec	2.60	6.154			380		380	580
1241	225 amps, to 42 circuits		2.40	6.667			410		410	630
1250	400 amps, to 42 circuits	▼	2.20	7.273			445		445	685
1260	4 wire, 120/208 V, 125A, to 20 circuits	1 Elec	2.40	3.333			205		205	315
1270	200 amps, to 42 circuits	2 Elec	2.40	6.667	▼		410		410	630
1300	Transformer, dry type, 1 phase, incl. removal of								
1320	supports, wire & conduit terminations								
1340	1 kVA	1 Elec	7.70	1.039	Ea.		63.50		63.50	98
1360	5 kVA		4.70	1.702			104		104	161
1380	10 kVA	▼	3.60	2.222			136		136	210
1400	37.5 kVA	2 Elec	3	5.333			325		325	505
1420	75 kVA	"	2.50	6.400	▼		395		395	605
1440	3 phase to 600 V, primary								
1460	3 kVA	1 Elec	3.87	2.067	Ea.		127		127	195
1480	15 kVA	2 Elec	3.67	4.360			267		267	410
1490	25 kVA		3.51	4.558			280		280	430
1500	30 kVA	▼	3.42	4.678			287		287	440
1530	112.5 kVA	R-3	2.90	6.897			420	65	485	720
1560	500 kVA		1.40	14.286	▼		875	135	1,010	1,500
1570	750 kVA	▼	1.10	18.182	▼		1,100	172	1,272	1,925
1600	Pull boxes & cabinets, sheet metal, incl. removal								
1620	of supports and conduit terminations								
1640	6" x 6" x 4"	1 Elec	31.10	.257	Ea.		15.80		15.80	24.50
1660	12" x 12" x 4"		23.30	.343			21		21	32.50
1720	Junction boxes, 4" sq. & oct.		80	.100			6.15		6.15	9.45
1740	Handy box		107	.075			4.59		4.59	7.05
1760	Switch box		107	.075			4.59		4.59	7.05
1780	Receptacle & switch plates	▼	257	.031	▼		1.91		1.91	2.94
1800	Wire, THW-THWN-THHN, removed from								
1810	in place conduit, to 10' high								
1830	#14	1 Elec	65	.123	C.L.F.		7.55		7.55	11.60
1840	#12		55	.145			8.90		8.90	13.75
1850	#10	▼	45.50	.176			10.80		10.80	16.60
1880	#4	2 Elec	53	.302			18.50		18.50	28.50
1890	#3		50	.320			19.65		19.65	30
1910	1/0		33.20	.482			29.50		29.50	45.50
1920	2/0		29.20	.548			33.50		33.50	52
1930	3/0		25	.640			39.50		39.50	60.50
1980	400 kcmil		17	.941			57.50		57.50	89
1990	500 kcmil	▼	16.20	.988	▼		60.50		60.50	93.50
2000	Interior fluorescent fixtures, incl. supports								
2010	& whips, to 10' high								
2100	Recessed drop-in 2' x 2', 2 lamp	2 Elec	35	.457	Ea.		28		28	43
2110	2' x 2', 4 lamp		30	.533			32.50		32.50	50.50
2120	2' x 4', 2 lamp		33	.485			30		30	46
2140	2' x 4', 4 lamp		30	.533			32.50		32.50	50.50
2160	4' x 4', 4 lamp	▼	20	.800	▼		49		49	75.50
2180	Surface mount, acrylic lens & hinged frame								
2200	1' x 4', 2 lamp	2 Elec	44	.364	Ea.		22.50		22.50	34.50
2210	6" x 4', 2 lamp		44	.364	▼		22.50		22.50	34.50

26 05 05 − Selective Demolition for Electrical

26 05 05.10 Electrical Demolition	Crew	Daily Output	Labor-Hours	Unit	Material	2020 Bare Costs Labor	Equipment	Total	Total Incl O&P
2220	2' x 2', 2 lamp	2 Elec	44	.364	Ea.		22.50		22.50	34.50
2260	2' x 4', 4 lamp		33	.485			30		30	46
2280	4' x 4', 4 lamp		23	.696			42.50		42.50	65.50
2281	4' x 4', 6 lamp	▼	23	.696	▼		42.50		42.50	65.50
2300	Strip fixtures, surface mount									
2320	4' long, 1 lamp	2 Elec	53	.302	Ea.		18.50		18.50	28.50
2340	4' long, 2 lamp		50	.320			19.65		19.65	30
2360	8' long, 1 lamp		42	.381			23.50		23.50	36
2380	8' long, 2 lamp	▼	40	.400	▼		24.50		24.50	38
2400	Pendant mount, industrial, incl. removal									
2410	of chain or rod hangers, to 10' high									
2420	4' long, 2 lamp	2 Elec	35	.457	Ea.		28		28	43
2421	4' long, 4 lamp		35	.457			28		28	43
2440	8' long, 2 lamp	▼	27	.593	▼		36.50		36.50	56
2460	Interior incandescent, surface, ceiling									
2470	or wall mount, to 10' high									
2480	Metal cylinder type, 75 Watt	2 Elec	62	.258	Ea.		15.85		15.85	24.50
2500	150 Watt	"	62	.258	"		15.85		15.85	24.50
2520	Metal halide, high bay									
2540	400 Watt	2 Elec	15	1.067	Ea.		65.50		65.50	101
2560	1,000 Watt		12	1.333			82		82	126
2580	150 Watt, low bay	▼	20	.800	▼		49		49	75.50
2600	Exterior fixtures, incandescent, wall mount									
2620	100 Watt	2 Elec	50	.320	Ea.		19.65		19.65	30
2640	Quartz, 500 Watt		33	.485			30		30	46
2660	1,500 Watt	▼	27	.593			36.50		36.50	56
2680	Wall pack, mercury vapor									
2700	175 Watt	2 Elec	25	.640	Ea.		39.50		39.50	60.50
2720	250 Watt	"	25	.640	"		39.50		39.50	60.50
9000	Minimum labor/equipment charge	1 Elec	4	2	Job		123		123	189
9900	Add to labor for higher elevated installation									
9905	10' to 14.5' high, add						10%			
9910	15' to 20' high, add						20%			
9920	20' to 25' high, add						25%			
9930	25' to 30' high, add						35%			
9940	30' to 35' high, add						40%			
9950	35' to 40' high, add						50%			
9960	Over 40' high, add						55%			

26 05 19 − Low-Voltage Electrical Power Conductors and Cables

26 05 19.20 Armored Cable

		Crew	Daily Output	Labor-Hours	Unit	Material	2020 Bare Costs Labor	Equipment	Total	Total Incl O&P
0010	**ARMORED CABLE**									
0050	600 volt, copper (BX), #14, 2 conductor, solid	1 Elec	2.40	3.333	C.L.F.	39	205		244	360
0100	3 conductor, solid		2.20	3.636		65.50	223		288.50	415
0150	#12, 2 conductor, solid		2.30	3.478		41	213		254	375
0200	3 conductor, solid		2	4		69	245		314	455
0250	#10, 2 conductor, solid		2	4		81.50	245		326.50	470
0300	3 conductor, solid		1.60	5		109	305		414	590
0340	#8, 2 conductor, stranded		1.50	5.333		257	325		582	790
0350	3 conductor, stranded	▼	1.30	6.154		257	380		637	865
9010	600 volt, copper (MC) steel clad, #14, 2 wire	R-1A	8.16	1.961		41.50	108		149.50	213
9020	3 wire		7.21	2.219		77	123		200	274
9030	4 wire	▼	6.69	2.392	▼	200	132		332	425

26 05 19 – Low-Voltage Electrical Power Conductors and Cables

26 05 19.20 Armored Cable

		Crew	Daily Output	Labor-Hours	Unit	Material	2020 Bare Costs Labor	2020 Bare Costs Equipment	Total	Total Incl O&P
9040	#12, 2 wire	R-1A	7.21	2.219	C.L.F.	57	123		180	252
9050	3 wire		6.69	2.392		67.50	132		199.50	278
9070	#10, 2 wire		6.04	2.649		103	146		249	340
9080	3 wire		5.65	2.832		132	156		288	385
9100	#8, 2 wire, stranded		3.94	4.061		161	224		385	520
9110	3 wire, stranded		3.40	4.706		185	260		445	605
9900	Minimum labor/equipment charge	R-1C	4	8	Job		440	25.50	465.50	710

26 05 19.50 Mineral Insulated Cable

		Crew	Daily Output	Labor-Hours	Unit	Material	2020 Bare Costs Labor	2020 Bare Costs Equipment	Total	Total Incl O&P
0010	**MINERAL INSULATED CABLE** 600 volt									
0100	1 conductor, #12	1 Elec	1.60	5	C.L.F.	355	305		660	860
0200	#10		1.60	5		535	305		840	1,050
0400	#8		1.50	5.333		675	325		1,000	1,250
0500	#6		1.40	5.714		710	350		1,060	1,325
0600	#4	2 Elec	2.40	6.667		1,050	410		1,460	1,800
0800	#2		2.20	7.273		1,450	445		1,895	2,250
0900	#1		2.10	7.619		1,650	465		2,115	2,550
1000	1/0		2	8		1,925	490		2,415	2,875
1100	2/0		1.90	8.421		2,425	515		2,940	3,475
1200	3/0		1.80	8.889		2,950	545		3,495	4,075
1400	4/0		1.60	10		3,400	615		4,015	4,675
1410	250 kcmil	3 Elec	2.40	10		3,825	615		4,440	5,150
1420	350 kcmil		1.95	12.308		4,400	755		5,155	5,975
1430	500 kcmil		1.95	12.308		5,550	755		6,305	7,250
1500	2 conductor, #12	1 Elec	1.40	5.714		925	350		1,275	1,575
1600	#10		1.20	6.667		1,150	410		1,560	1,875
1800	#8		1.10	7.273		1,300	445		1,745	2,100
2000	#6		1.05	7.619		1,800	465		2,265	2,700
2100	#4	2 Elec	2	8		2,550	490		3,040	3,550
2200	3 conductor, #12	1 Elec	1.20	6.667		1,075	410		1,485	1,825
2400	#10		1.10	7.273		1,225	445		1,670	2,025
2600	#8		1.05	7.619		1,650	465		2,115	2,525
2800	#6		1	8		2,175	490		2,665	3,125
3000	#4	2 Elec	1.80	8.889		2,850	545		3,395	3,975
3100	4 conductor, #12	1 Elec	1.20	6.667		1,150	410		1,560	1,875
3200	#10		1.10	7.273		1,375	445		1,820	2,175
3400	#8		1	8		1,925	490		2,415	2,875
3600	#6		.90	8.889		2,600	545		3,145	3,725
3620	7 conductor, #12		1.10	7.273		1,575	445		2,020	2,400
3640	#10		1	8		2,150	490		2,640	3,125
3800	Terminations, 600 volt, 1 conductor, #12		8	1	Ea.	18.30	61.50		79.80	115
5500	2 conductor, #12		6.70	1.194		18.30	73.50		91.80	133
5600	#10		6.40	1.250		27.50	76.50		104	149
5800	#8		6.20	1.290		27.50	79		106.50	153
6000	#6		5.70	1.404		27.50	86		113.50	164
6200	#4		5.30	1.509		61	92.50		153.50	210
6400	3 conductor, #12		5.70	1.404		27.50	86		113.50	164
6500	#10		5.50	1.455		27.50	89		116.50	168
6600	#8		5.20	1.538		27.50	94.50		122	176
6800	#6		4.80	1.667		27.50	102		129.50	188
7200	#4		4.60	1.739		61	107		168	231
7400	4 conductor, #12		4.60	1.739		31	107		138	198
7500	#10		4.40	1.818		31	112		143	206

26 05 19 – Low-Voltage Electrical Power Conductors and Cables

26 05 19.50 Mineral Insulated Cable

		Crew	Daily Output	Labor-Hours	Unit	Material	2020 Bare Costs Labor	2020 Bare Costs Equipment	Total	Total Incl O&P
7600	#8	1 Elec	4.20	1.905	Ea.	31	117		148	214
8400	#6		4	2		64	123		187	260
8500	7 conductor, #12		3.50	2.286		30.50	140		170.50	250
8600	#10		3	2.667		66.50	164		230.50	325
8800	Crimping tool, plier type					77			77	84.50
9000	Stripping tool					291			291	320
9200	Hand vise					86.50			86.50	95.50
9500	Minimum labor/equipment charge	1 Elec	4	2	Job		123		123	189

26 05 19.55 Non-Metallic Sheathed Cable

		Crew	Daily Output	Labor-Hours	Unit	Material	2020 Bare Costs Labor	2020 Bare Costs Equipment	Total	Total Incl O&P
0010	**NON-METALLIC SHEATHED CABLE** 600 volt									
0100	Copper with ground wire (Romex)									
0150	#14, 2 conductor	1 Elec	2.70	2.963	C.L.F.	18.40	182		200.40	300
0200	3 conductor		2.40	3.333		26.50	205		231.50	345
0220	4 conductor		2.20	3.636		42	223		265	390
0250	#12, 2 conductor		2.50	3.200		26.50	196		222.50	330
0300	3 conductor		2.20	3.636		41	223		264	390
0320	4 conductor		2	4		65	245		310	450
0350	#10, 2 conductor		2.20	3.636		45	223		268	395
0400	3 conductor		1.80	4.444		66.50	273		339.50	495
0430	#8, 2 conductor		1.60	5		87.50	305		392.50	565
0450	3 conductor		1.50	5.333		117	325		442	635
0500	#6, 3 conductor		1.40	5.714		179	350		529	735
0550	SE type SER aluminum cable, 3 RHW and									
0600	1 bare neutral, 3 #8 & 1 #8	1 Elec	1.60	5	C.L.F.	63	305		368	540
0650	3 #6 & 1 #6	"	1.40	5.714		70.50	350		420.50	620
0700	3 #4 & 1 #6	2 Elec	2.40	6.667		78.50	410		488.50	715
0750	3 #2 & 1 #4		2.20	7.273		143	445		588	845
0800	3 #1/0 & 1 #2		2	8		167	490		657	940
0850	3 #2/0 & 1 #1		1.80	8.889		199	545		744	1,050
0900	3 #4/0 & 1 #2/0		1.60	10		292	615		907	1,275
6500	Service entrance cap for copper SEU									
6600	100 amp	1 Elec	12	.667	Ea.	7.85	41		48.85	71.50
6700	150 amp		10	.800		11.75	49		60.75	88.50
6800	200 amp		8	1		17.85	61.50		79.35	114
9000	Minimum labor/equipment charge		4	2	Job		123		123	189

26 05 19.90 Wire

			Crew	Daily Output	Labor-Hours	Unit	Material	2020 Bare Costs Labor	2020 Bare Costs Equipment	Total	Total Incl O&P
0010	**WIRE**, normal installation conditions in wireway, conduit, cable tray										
0020	600 volt, copper type THW, solid, #14	R260533-22	1 Elec	13	.615	C.L.F.	6.25	38		44.25	65
0030	#12			11	.727		9.85	44.50		54.35	79.50
0040	#10			10	.800		15.95	49		64.95	93
0050	Stranded, #14			13	.615		7.60	38		45.60	66.50
0100	#12			11	.727		13.75	44.50		58.25	83.50
0120	#10			10	.800		22	49		71	99.50
0140	#8			8	1		34.50	61.50		96	133
0160	#6			6.50	1.231		57.50	75.50		133	180
0180	#4		2 Elec	10.60	1.509		87.50	92.50		180	240
0200	#3			10	1.600		118	98		216	280
0220	#2			9	1.778		136	109		245	320
0240	#1			8	2		174	123		297	380
0260	1/0			6.60	2.424		226	149		375	475
0280	2/0			5.80	2.759		390	169		559	690
0300	3/0			5	3.200		278	196		474	605

491

For customer support on your Commercial Renovation Costs with RSMeans data, call 800.448.8182.

26 05 19 – Low-Voltage Electrical Power Conductors and Cables

26 05 19.90 Wire		Crew	Daily Output	Labor-Hours	Unit	Material	2020 Bare Costs Labor	Equipment	Total	Total Incl O&P
0350	4/0	2 Elec	4.40	3.636	C.L.F.	450	223		673	840
0400	250 kcmil	3 Elec	6	4		490	245		735	915
0420	300 kcmil		5.70	4.211		685	258		943	1,150
0450	350 kcmil		5.40	4.444		670	273		943	1,150
0480	400 kcmil		5.10	4.706		1,200	289		1,489	1,775
0490	500 kcmil		4.80	5		965	305		1,270	1,550
0510	750 kcmil		3.30	7.273		1,150	445		1,595	1,950
0540	600 volt, aluminum type THHN, stranded, #6	1 Elec	8	1		20.50	61.50		82	117
0560	#4	2 Elec	13	1.231		25	75.50		100.50	144
0580	#2		10.60	1.509		48	92.50		140.50	196
0600	#1		9	1.778		57	109		166	231
0620	1/0		8	2		80	123		203	278
0640	2/0		7.20	2.222		82	136		218	300
0680	3/0		6.60	2.424		103	149		252	340
0700	4/0		6.20	2.581		131	158		289	390
0720	250 kcmil	3 Elec	8.70	2.759		137	169		306	410
0740	300 kcmil		8.10	2.963		190	182		372	490
0760	350 kcmil		7.50	3.200		192	196		388	510
0780	400 kcmil		6.90	3.478		219	213		432	570
0800	500 kcmil		6	4		246	245		491	650
0850	600 kcmil		5.70	4.211		315	258		573	745
0880	700 kcmil		5.10	4.706		385	289		674	870
9000	Minimum labor/equipment charge	1 Elec	4	2	Job		123		123	189

26 05 26 – Grounding and Bonding for Electrical Systems

26 05 26.80 Grounding

		Crew	Daily Output	Labor-Hours	Unit	Material	Labor	Equipment	Total	Total Incl O&P
0010	**GROUNDING**									
0030	Rod, copper clad, 8' long, 1/2" diameter	1 Elec	5.50	1.455	Ea.	22.50	89		111.50	162
0040	5/8" diameter		5.50	1.455		22.50	89		111.50	162
0050	3/4" diameter		5.30	1.509		36	92.50		128.50	183
0080	10' long, 1/2" diameter		4.80	1.667		24.50	102		126.50	184
0090	5/8" diameter		4.60	1.739		22.50	107		129.50	189
0100	3/4" diameter		4.40	1.818		53	112		165	231
0130	15' long, 3/4" diameter		4	2		55	123		178	250
0260	Wire ground bare armored, #8-1 conductor		2	4	C.L.F.	74.50	245		319.50	460
0280	#4-1 conductor		1.60	5		146	305		451	630
0390	Bare copper wire, stranded, #8		11	.727		40	44.50		84.50	113
0400	#6		10	.800		42	49		91	122
0600	#2	2 Elec	10	1.600		89	98		187	249
0800	3/0		6.60	2.424		305	149		454	565
1000	4/0		5.70	2.807		420	172		592	730
1200	250 kcmil	3 Elec	7.20	3.333		440	205		645	795
1800	Water pipe ground clamps, heavy duty									
2000	Bronze, 1/2" to 1" diameter	1 Elec	8	1	Ea.	29	61.50		90.50	126
2100	1-1/4" to 2" diameter		8	1		31	61.50		92.50	129
2200	2-1/2" to 3" diameter		6	1.333		60.50	82		142.50	193
2400	Grounding, exothermic welding									
2500	Exothermic welding reusable mold, cable to cable, parallel, vertical	1 Elec	8	1	Ea.	340	61.50		401.50	465
2505	Splice single		8	1		293	61.50		354.50	415
2510	Splice single to double		8	1		340	61.50		401.50	470
2520	Cable to cable, termination, Tee		8	1		330	61.50		391.50	460
2530	Cable to rod, termination, Tee		8	1		286	61.50		347.50	410
2535	Cable to rebar over, termination, Tee		8	1		96.50	61.50		158	201

26 05 26 – Grounding and Bonding for Electrical Systems

26 05 26.80 Grounding	Crew	Daily Output	Labor-Hours	Unit	Material	2020 Bare Costs Labor	Equipment	Total	Total Incl O&P
2540	Cable to rod, termination, 90 Deg	1 Elec	8	1	Ea.	330	61.50		391.50	460
2550	Cable to verticle flat steel		8	1		335	61.50		396.50	465
2555	Cable to 45 Deg		8	1		340	61.50		401.50	470
2730	Exothermic weld, 4/0 wire to 1" ground rod		7	1.143		9.45	70		79.45	118
2800	Brazed connections, #6 wire		12	.667		16.15	41		57.15	81
3000	#2 wire		10	.800		21.50	49		70.50	99.50
3100	3/0 wire		8	1		32	61.50		93.50	130
3200	4/0 wire		7	1.143		36.50	70		106.50	149
3400	250 kcmil wire		5	1.600		43	98		141	198
3600	500 kcmil wire		4	2		52.50	123		175.50	247
7000	Exothermic welding kit, multi vertical		8	1		720	61.50		781.50	885
9000	Minimum labor/equipment charge		4	2	Job		123		123	189

26 05 29 – Hangers and Supports for Electrical Systems

26 05 29.30 Fittings and Channel Support

26 05 29.30 Fittings and Channel Support	Crew	Daily Output	Labor-Hours	Unit	Material	2020 Bare Costs Labor	Equipment	Total	Total Incl O&P
0010	**FITTINGS & CHANNEL SUPPORT**									
0020	Rooftop channel support									
0200	2-7/8" L x 1-5/8" W, 12 ga. pre galv., dbl. base	1 Elec	46	.174	Ea.	3.82	10.65		14.47	20.50
0210	2-7/8" L x 1-5/8" W, 12 ga. hot dip galv., dbl. base		46	.174		2.22	10.65		12.87	18.90
0220	2-7/16" L x 1-5/8" W, 12 ga. pre galv., sngl base		48	.167		1.83	10.25		12.08	17.75
0230	2-7/16" L x 1-5/8" W, 12 ga. hot dip galv., sngl base		48	.167		6.50	10.25		16.75	23
0240	13/16" L x 1-5/8" W, 14 ga. pre galv., sngl base		56	.143		1.56	8.75		10.31	15.20
0250	13/16" L x 1-5/8" W, 14 ga. hot dip galv., sngl base		56	.143		6.35	8.75		15.10	20.50
0260	1-5/8" L x 1-5/8" W, 12 ga. pre galv., sngl base		54	.148		1.69	9.10		10.79	15.85
0270	1-5/8" L x 1-5/8" W, 12 ga. hot dip galv., sngl base		54	.148		1.79	9.10		10.89	15.95
0280	1-5/8" x 28", 12 ga. pre galv., dbl. H block base		42	.190		3.87	11.70		15.57	22.50
0290	1-5/8" x 36", 12 ga. pre galv., dbl. H block base		40	.200		4.09	12.25		16.34	23.50
0300	1-5/8" x 42", 12 ga. pre galv., dbl. H block base		34	.235		4.59	14.45		19.04	27
0310	1-5/8" x 50", 12 ga. pre galv., dbl. H block base		38	.211		4.90	12.90		17.80	25.50
0320	1-5/8" x 60", 12 ga. pre galv., dbl. H block base		36	.222		5.40	13.65		19.05	27
0330	13/16" x 8" H, threaded rod pre galv., H block base		35	.229		2.59	14		16.59	24.50
0340	13/16" x 12" H, threaded rod pre galv., H block base		34.60	.231		3.24	14.20		17.44	25.50
0350	1-5/8" x 16" H, threaded rod dbl. H block base		34	.235		4.85	14.45		19.30	27.50
0360	1-5/8" x 12" H, pre galv., dbl. H block base		34.60	.231		9.25	14.20		23.45	32
0370	1-5/8" x 24"H, pre galv., dbl. H block base		34	.235		9.80	14.45		24.25	33
0380	3-1/4" x 24" H, pre galv., dbl. H block base		32	.250		22	15.35		37.35	48
0390	3-1/4" x 36" H, pre galv., dbl. H block base		30	.267		24.50	16.35		40.85	51.50

26 05 33 – Raceway and Boxes for Electrical Systems

26 05 33.13 Conduit

26 05 33.13 Conduit		Crew	Daily Output	Labor-Hours	Unit	Material	2020 Bare Costs Labor	Equipment	Total	Total Incl O&P
0010	**CONDUIT** To 10' high, includes 2 terminations, 2 elbows,	R260533-22									
0020	11 beam clamps, and 11 couplings per 100 L.F.										
1161	Field bends, 45° to 90°, 1/2" diameter		1 Elec	53	.151	Ea.		9.25		9.25	14.25
1162	3/4" diameter			47	.170			10.45		10.45	16.10
1163	1" diameter			44	.182			11.15		11.15	17.15
1164	1-1/4" diameter			23	.348			21.50		21.50	33
1165	1-1/2" diameter			21	.381			23.50		23.50	36
1166	2" diameter			16	.500			30.50		30.50	47
1691	See note on line 26 05 33.13 9995										
1750	Rigid galvanized steel, 1/2" diameter		1 Elec	90	.089	L.F.	2.77	5.45		8.22	11.45
1770	3/4" diameter			80	.100		5.15	6.15		11.30	15.10
1800	1" diameter			65	.123		7.70	7.55		15.25	20
1830	1-1/4" diameter			60	.133		5.35	8.20		13.55	18.45
1850	1-1/2" diameter			55	.145		9.35	8.90		18.25	24

For customer support on your Commercial Renovation Costs with RSMeans data, call 800.448.8182.

493

26 05 33 – Raceway and Boxes for Electrical Systems

26 05 33.13 Conduit		Crew	Daily Output	Labor-Hours	Unit	Material	2020 Bare Costs Labor	Equipment	Total	Total Incl O&P
1870	2" diameter	1 Elec	45	.178	L.F.	9.80	10.90		20.70	27.50
1991	Field bends, 45° to 90°, 1/2" diameter		44	.182	Ea.		11.15		11.15	17.15
1992	3/4" diameter		40	.200			12.25		12.25	18.90
1993	1" diameter		36	.222			13.65		13.65	21
1994	1-1/4" diameter		19	.421			26		26	40
1995	1-1/2" diameter		18	.444			27.50		27.50	42
1996	2" diameter		13	.615	↓		38		38	58
2500	Steel, intermediate conduit (IMC), 1/2" diameter		100	.080	L.F.	1.91	4.91		6.82	9.65
2530	3/4" diameter		90	.089		2.64	5.45		8.09	11.30
2550	1" diameter		70	.114		3.57	7		10.57	14.70
2570	1-1/4" diameter		65	.123		3.95	7.55		11.50	15.95
2600	1-1/2" diameter		60	.133		5.75	8.20		13.95	18.90
2630	2" diameter		50	.160		6.85	9.80		16.65	22.50
2650	2-1/2" diameter		40	.200		9.05	12.25		21.30	29
2670	3" diameter	2 Elec	60	.267		13.55	16.35		29.90	40
2700	3-1/2" diameter		54	.296		18.85	18.20		37.05	48.50
2730	4" diameter	↓	50	.320	↓	12.05	19.65		31.70	43.50
2731	Field bends, 45° to 90°, 1/2" diameter	1 Elec	44	.182	Ea.		11.15		11.15	17.15
2732	3/4" diameter		40	.200			12.25		12.25	18.90
2733	1" diameter		36	.222			13.65		13.65	21
2734	1-1/4" diameter		19	.421			26		26	40
2735	1-1/2" diameter		18	.444			27.50		27.50	42
2736	2" diameter		13	.615	↓		38		38	58
5000	Electric metallic tubing (EMT), 1/2" diameter		170	.047	L.F.	.76	2.89		3.65	5.25
5020	3/4" diameter		130	.062		1.07	3.78		4.85	7
5040	1" diameter		115	.070		1.80	4.27		6.07	8.55
5060	1-1/4" diameter		100	.080		2.94	4.91		7.85	10.80
5080	1-1/2" diameter		90	.089		3.46	5.45		8.91	12.20
5100	2" diameter		80	.100		4.37	6.15		10.52	14.25
5120	2-1/2" diameter		60	.133		4.85	8.20		13.05	17.95
5140	3" diameter	2 Elec	100	.160		6.15	9.80		15.95	22
5160	3-1/2" diameter		90	.178		7.70	10.90		18.60	25.50
5180	4" diameter	↓	80	.200	↓	13.55	12.25		25.80	34
5200	Field bends, 45° to 90°, 1/2" diameter	1 Elec	89	.090	Ea.		5.50		5.50	8.50
5220	3/4" diameter		80	.100			6.15		6.15	9.45
5240	1" diameter		73	.110			6.70		6.70	10.35
5260	1-1/4" diameter		38	.211			12.90		12.90	19.90
5280	1-1/2" diameter		36	.222			13.65		13.65	21
5300	2" diameter		26	.308			18.90		18.90	29
5320	Offsets, 1/2" diameter		65	.123			7.55		7.55	11.60
5340	3/4" diameter		62	.129			7.90		7.90	12.20
5360	1" diameter		53	.151			9.25		9.25	14.25
5380	1-1/4" diameter		30	.267			16.35		16.35	25
5400	1-1/2" diameter		28	.286			17.55		17.55	27
9990	Minimum labor/equipment charge	↓	4	2	Job		123		123	189

26 05 33.16 Boxes for Electrical Systems

0010	**BOXES FOR ELECTRICAL SYSTEMS**									
0020	Pressed steel, octagon, 4"	1 Elec	20	.400	Ea.	2.96	24.50		27.46	41.50
0060	Covers, blank		64	.125		.87	7.65		8.52	12.75
0100	Extension rings		40	.200		4.59	12.25		16.84	24
0150	Square, 4"		20	.400		6.85	24.50		31.35	45.50
0200	Extension rings	↓	40	.200		4.85	12.25		17.10	24.50

26 05 33.16 Boxes for Electrical Systems

		Crew	Daily Output	Labor-Hours	Unit	Material	2020 Bare Costs Labor	2020 Bare Costs Equipment	Total	Total Incl O&P
0250	Covers, blank	1 Elec	64	.125	Ea.	.77	7.65		8.42	12.65
0300	Plaster rings		64	.125		1.66	7.65		9.31	13.65
0652	Switchbox		24	.333		5.60	20.50		26.10	37.50
1100	Concrete, floor, 1 gang		5.30	1.509		112	92.50		204.50	266
9000	Minimum labor/equipment charge		4	2	Job		123		123	189

26 05 33.17 Outlet Boxes, Plastic

		Crew	Daily Output	Labor-Hours	Unit	Material	2020 Bare Costs Labor	2020 Bare Costs Equipment	Total	Total Incl O&P
0010	**OUTLET BOXES, PLASTIC**									
0050	4" diameter, round with 2 mounting nails	1 Elec	25	.320	Ea.	2.89	19.65		22.54	33
0100	Bar hanger mounted		25	.320		5.05	19.65		24.70	35.50
0200	4", square with 2 mounting nails		25	.320		5.55	19.65		25.20	36
0300	Plaster ring		64	.125		2.10	7.65		9.75	14.10
0400	Switch box with 2 mounting nails, 1 gang		30	.267		3.21	16.35		19.56	28.50
0500	2 gang		25	.320		3.95	19.65		23.60	34.50
0600	3 gang		20	.400		5.50	24.50		30	44
0700	Old work box		30	.267		4.59	16.35		20.94	30
9000	Minimum labor/equipment charge		4	2	Job		123		123	189

26 05 33.18 Pull Boxes

		Crew	Daily Output	Labor-Hours	Unit	Material	2020 Bare Costs Labor	2020 Bare Costs Equipment	Total	Total Incl O&P
0010	**PULL BOXES**									
0100	Steel, pull box, NEMA 1, type SC, 6" W x 6" H x 4" D	1 Elec	8	1	Ea.	9.85	61.50		71.35	105
0200	8" W x 8" H x 4" D		8	1		12.90	61.50		74.40	109
0300	10" W x 12" H x 6" D		5.30	1.509		30.50	92.50		123	177
0400	16" W x 20" H x 8" D		4	2		82.50	123		205.50	280
0500	20" W x 24" H x 8" D		3.20	2.500		110	153		263	355
0600	24" W x 36" H x 8" D		2.70	2.963		153	182		335	450
0650	Pull box, hinged, NEMA 1, 6" W x 6" H x 4" D		8	1		16.95	61.50		78.45	113
0800	12" W x 16" H x 6" D		4.70	1.702		49.50	104		153.50	216
1000	20" W x 20" H x 6" D		3.60	2.222		107	136		243	330
1200	20" W x 20" H x 8" D		3.20	2.500		163	153		316	415
1400	24" W x 36" H x 8" D		2.70	2.963		257	182		439	565
1600	24" W x 42" H x 8" D		2	4		380	245		625	795

26 05 33.23 Wireway

		Crew	Daily Output	Labor-Hours	Unit	Material	2020 Bare Costs Labor	2020 Bare Costs Equipment	Total	Total Incl O&P
0010	**WIREWAY** to 10' high									
0020	For higher elevations, see Section 26 05 36.40									
0100	NEMA 1, screw cover w/fittings and supports, 2-1/2" x 2-1/2"	1 Elec	45	.178	L.F.	12.45	10.90		23.35	30.50
0200	4" x 4"	"	40	.200		13.40	12.25		25.65	33.50
0400	6" x 6"	2 Elec	60	.267		23.50	16.35		39.85	51
0600	8" x 8"	"	40	.400		28	24.50		52.50	69
4475	NEMA 3R, screw cover w/fittings and supports, 4" x 4"	1 Elec	36	.222		16.65	13.65		30.30	39.50
4480	6" x 6"	2 Elec	55	.291		18.50	17.85		36.35	48
4485	8" x 8"		36	.444		32.50	27.50		60	78
4490	12" x 12"		18	.889		60.50	54.50		115	151

26 05 33.95 Cutting and Drilling

		Crew	Daily Output	Labor-Hours	Unit	Material	2020 Bare Costs Labor	2020 Bare Costs Equipment	Total	Total Incl O&P
0010	**CUTTING AND DRILLING**									
0100	Hole drilling to 10' high, concrete wall									
0110	8" thick, 1/2" pipe size	R-31	12	.667	Ea.	.27	41	5.20	46.47	69
0120	3/4" pipe size		12	.667		.27	41	5.20	46.47	69
0130	1" pipe size		9.50	.842		.35	51.50	6.60	58.45	87
0140	1-1/4" pipe size		9.50	.842		.35	51.50	6.60	58.45	87
0150	1-1/2" pipe size		9.50	.842		.35	51.50	6.60	58.45	87
0160	2" pipe size		4.40	1.818		.53	112	14.20	126.73	188
0170	2-1/2" pipe size		4.40	1.818		.53	112	14.20	126.73	188
0180	3" pipe size		4.40	1.818		.53	112	14.20	126.73	188

26 05 33.95 Cutting and Drilling		Crew	Daily Output	Labor-Hours	Unit	Material	2020 Bare Costs Labor	2020 Bare Costs Equipment	Total	Total Incl O&P
0190	3-1/2" pipe size	R-31	3.30	2.424	Ea.	.60	149	18.95	168.55	251
0200	4" pipe size		3.30	2.424		.60	149	18.95	168.55	251
0500	12" thick, 1/2" pipe size		9.40	.851		.40	52	6.65	59.05	88
0520	3/4" pipe size		9.40	.851		.40	52	6.65	59.05	88
0540	1" pipe size		7.30	1.096		.52	67	8.55	76.07	114
0560	1-1/4" pipe size		7.30	1.096		.52	67	8.55	76.07	114
0570	1-1/2" pipe size		7.30	1.096		.52	67	8.55	76.07	114
0580	2" pipe size		3.60	2.222		.80	136	17.35	154.15	230
0590	2-1/2" pipe size		3.60	2.222		.80	136	17.35	154.15	230
0600	3" pipe size		3.60	2.222		.80	136	17.35	154.15	230
0610	3-1/2" pipe size		2.80	2.857		.90	175	22.50	198.40	295
0630	4" pipe size		2.50	3.200		.90	196	25	221.90	330
0650	16" thick, 1/2" pipe size		7.60	1.053		.54	64.50	8.20	73.24	109
0670	3/4" pipe size		7	1.143		.54	70	8.95	79.49	118
0690	1" pipe size		6	1.333		.69	82	10.40	93.09	138
0710	1-1/4" pipe size		5.50	1.455		.69	89	11.35	101.04	150
0730	1-1/2" pipe size		5.50	1.455		.69	89	11.35	101.04	150
0750	2" pipe size		3	2.667		1.06	164	21	186.06	276
0770	2-1/2" pipe size		2.70	2.963		1.06	182	23	206.06	305
0790	3" pipe size		2.50	3.200		1.06	196	25	222.06	330
0810	3-1/2" pipe size		2.30	3.478		1.20	213	27	241.20	360
0830	4" pipe size		2	4		1.20	245	31	277.20	415
0850	20" thick, 1/2" pipe size		6.40	1.250		.67	76.50	9.75	86.92	129
0870	3/4" pipe size		6	1.333		.67	82	10.40	93.07	138
0890	1" pipe size		5	1.600		.86	98	12.50	111.36	166
0910	1-1/4" pipe size		4.80	1.667		.86	102	13	115.86	172
0930	1-1/2" pipe size		4.60	1.739		.86	107	13.60	121.46	180
0950	2" pipe size		2.70	2.963		1.33	182	23	206.33	305
0970	2-1/2" pipe size		2.40	3.333		1.33	205	26	232.33	345
0990	3" pipe size		2.20	3.636		1.33	223	28.50	252.83	375
1010	3-1/2" pipe size		2	4		1.50	245	31	277.50	415
1030	4" pipe size		1.70	4.706		1.50	289	37	327.50	485
1050	24" thick, 1/2" pipe size		5.50	1.455		.81	89	11.35	101.16	150
1070	3/4" pipe size		5.10	1.569		.81	96	12.25	109.06	162
1090	1" pipe size		4.30	1.860		1.04	114	14.55	129.59	193
1110	1-1/4" pipe size		4	2		1.04	123	15.60	139.64	207
1130	1-1/2" pipe size		4	2		1.04	123	15.60	139.64	207
1150	2" pipe size		2.40	3.333		1.59	205	26	232.59	345
1170	2-1/2" pipe size		2.20	3.636		1.59	223	28.50	253.09	380
1190	3" pipe size		2	4		1.59	245	31	277.59	415
1210	3-1/2" pipe size		1.80	4.444		1.80	273	34.50	309.30	460
1230	4" pipe size		1.50	5.333		1.80	325	41.50	368.30	555
1500	Brick wall, 8" thick, 1/2" pipe size		18	.444		.27	27.50	3.47	31.24	46
1520	3/4" pipe size		18	.444		.27	27.50	3.47	31.24	46
1540	1" pipe size		13.30	.602		.35	37	4.70	42.05	62.50
1560	1-1/4" pipe size		13.30	.602		.35	37	4.70	42.05	62.50
1580	1-1/2" pipe size		13.30	.602		.35	37	4.70	42.05	62.50
1600	2" pipe size		5.70	1.404		.53	86	10.95	97.48	146
1620	2-1/2" pipe size		5.70	1.404		.53	86	10.95	97.48	146
1640	3" pipe size		5.70	1.404		.53	86	10.95	97.48	146
1660	3-1/2" pipe size		4.40	1.818		.60	112	14.20	126.80	188
1680	4" pipe size		4	2		.60	123	15.60	139.20	207
1700	12" thick, 1/2" pipe size		14.50	.552		.40	34	4.31	38.71	57

26 05 33.95 Cutting and Drilling		Crew	Daily Output	Labor-Hours	Unit	Material	2020 Bare Costs Labor	Equipment	Total	Total Incl O&P
1720	3/4" pipe size	R-31	14.50	.552	Ea.	.40	34	4.31	38.71	57
1740	1" pipe size		11	.727		.52	44.50	5.70	50.72	75.50
1760	1-1/4" pipe size		11	.727		.52	44.50	5.70	50.72	75.50
1780	1-1/2" pipe size		11	.727		.52	44.50	5.70	50.72	75.50
1800	2" pipe size		5	1.600		.80	98	12.50	111.30	166
1820	2-1/2" pipe size		5	1.600		.80	98	12.50	111.30	166
1840	3" pipe size		5	1.600		.80	98	12.50	111.30	166
1860	3-1/2" pipe size		3.80	2.105		.90	129	16.45	146.35	218
1880	4" pipe size		3.30	2.424		.90	149	18.95	168.85	251
1900	16" thick, 1/2" pipe size		12.30	.650		.54	40	5.10	45.64	67.50
1920	3/4" pipe size		12.30	.650		.54	40	5.10	45.64	67.50
1940	1" pipe size		9.30	.860		.69	53	6.70	60.39	89.50
1960	1-1/4" pipe size		9.30	.860		.69	53	6.70	60.39	89.50
1980	1-1/2" pipe size		9.30	.860		.69	53	6.70	60.39	89.50
2000	2" pipe size		4.40	1.818		1.06	112	14.20	127.26	189
2010	2-1/2" pipe size		4.40	1.818		1.06	112	14.20	127.26	189
2030	3" pipe size		4.40	1.818		1.06	112	14.20	127.26	189
2050	3-1/2" pipe size		3.30	2.424		1.20	149	18.95	169.15	251
2070	4" pipe size		3	2.667		1.20	164	21	186.20	276
2090	20" thick, 1/2" pipe size		10.70	.748		.67	46	5.85	52.52	77.50
2110	3/4" pipe size		10.70	.748		.67	46	5.85	52.52	77.50
2130	1" pipe size		8	1		.86	61.50	7.80	70.16	104
2150	1-1/4" pipe size		8	1		.86	61.50	7.80	70.16	104
2170	1-1/2" pipe size		8	1		.86	61.50	7.80	70.16	104
2190	2" pipe size		4	2		1.33	123	15.60	139.93	208
2210	2-1/2" pipe size		4	2		1.33	123	15.60	139.93	208
2230	3" pipe size		4	2		1.33	123	15.60	139.93	208
2250	3-1/2" pipe size		3	2.667		1.50	164	21	186.50	277
2270	4" pipe size		2.70	2.963		1.50	182	23	206.50	305
2290	24" thick, 1/2" pipe size		9.40	.851		.81	52	6.65	59.46	88.50
2310	3/4" pipe size		9.40	.851		.81	52	6.65	59.46	88.50
2330	1" pipe size		7.10	1.127		1.04	69	8.80	78.84	117
2350	1-1/4" pipe size		7.10	1.127		1.04	69	8.80	78.84	117
2370	1-1/2" pipe size		7.10	1.127		1.04	69	8.80	78.84	117
2390	2" pipe size		3.60	2.222		1.59	136	17.35	154.94	231
2410	2-1/2" pipe size		3.60	2.222		1.59	136	17.35	154.94	231
2430	3" pipe size		3.60	2.222		1.59	136	17.35	154.94	231
2450	3-1/2" pipe size		2.80	2.857		1.80	175	22.50	199.30	296
2470	4" pipe size	▼	2.50	3.200	▼	1.80	196	25	222.80	330
3000	Knockouts to 8' high, metal boxes & enclosures									
3020	With hole saw, 1/2" pipe size	1 Elec	53	.151	Ea.		9.25		9.25	14.25
3040	3/4" pipe size		47	.170			10.45		10.45	16.10
3050	1" pipe size		40	.200			12.25		12.25	18.90
3060	1-1/4" pipe size		36	.222			13.65		13.65	21
3070	1-1/2" pipe size		32	.250			15.35		15.35	23.50
3080	2" pipe size		27	.296			18.20		18.20	28
3090	2-1/2" pipe size		20	.400			24.50		24.50	38
4010	3" pipe size		16	.500			30.50		30.50	47
4030	3-1/2" pipe size		13	.615			38		38	58
4050	4" pipe size	▼	11	.727	▼		44.50		44.50	68.50

26 05 39 – Underfloor Raceways for Electrical Systems

26 05 39.30 Conduit In Concrete Slab

	26 05 39.30 Conduit In Concrete Slab	Crew	Daily Output	Labor-Hours	Unit	Material	2020 Bare Costs Labor	Equipment	Total	Total Incl O&P
0010	**CONDUIT IN CONCRETE SLAB** Including terminations,									
0020	fittings and supports									
3230	PVC, schedule 40, 1/2" diameter	1 Elec	270	.030	L.F.	.55	1.82		2.37	3.40
3250	3/4" diameter		230	.035		.59	2.13		2.72	3.94
3270	1" diameter		200	.040		.78	2.45		3.23	4.63
3300	1-1/4" diameter		170	.047		1.05	2.89		3.94	5.60
3330	1-1/2" diameter		140	.057		1.27	3.51		4.78	6.80
3350	2" diameter		120	.067		1.59	4.09		5.68	8.05
4350	Rigid galvanized steel, 1/2" diameter		200	.040		2.34	2.45		4.79	6.35
4400	3/4" diameter		170	.047		4.45	2.89		7.34	9.35
4450	1" diameter		130	.062		6.90	3.78		10.68	13.40
4500	1-1/4" diameter		110	.073		4.53	4.46		8.99	11.85
4600	1-1/2" diameter		100	.080		8.40	4.91		13.31	16.80
4800	2" diameter		90	.089		8.20	5.45		13.65	17.45
9000	Minimum labor/equipment charge		4	2	Job		123		123	189

26 05 39.40 Conduit In Trench

	26 05 39.40 Conduit In Trench	Crew	Daily Output	Labor-Hours	Unit	Material	2020 Bare Costs Labor	Equipment	Total	Total Incl O&P
0010	**CONDUIT IN TRENCH** Includes terminations and fittings									
0020	Does not include excavation or backfill, see Section 31 23 16									
0200	Rigid galvanized steel, 2" diameter	1 Elec	150	.053	L.F.	7.80	3.27		11.07	13.65
0400	2-1/2" diameter	"	100	.080		13	4.91		17.91	22
0600	3" diameter	2 Elec	160	.100		14.55	6.15		20.70	25.50
0800	3-1/2" diameter		140	.114		19.20	7		26.20	32
1000	4" diameter		100	.160		23.50	9.80		33.30	40.50
1200	5" diameter		80	.200		37.50	12.25		49.75	60
1400	6" diameter		60	.267		57.50	16.35		73.85	88.50
9000	Minimum labor/equipment charge	1 Elec	4	2	Job		123		123	189

26 05 83 – Wiring Connections

26 05 83.10 Motor Connections

		Crew	Daily Output	Labor-Hours	Unit	Material	2020 Bare Costs Labor	Equipment	Total	Total Incl O&P
0010	**MOTOR CONNECTIONS**									
0020	Flexible conduit and fittings, 115 volt, 1 phase, up to 1 HP motor	1 Elec	8	1	Ea.	5.90	61.50		67.40	101
9000	Minimum labor/equipment charge	"	4	2	Job		123		123	189

26 05 90 – Residential Applications

26 05 90.10 Residential Wiring

		Crew	Daily Output	Labor-Hours	Unit	Material	2020 Bare Costs Labor	Equipment	Total	Total Incl O&P
0010	**RESIDENTIAL WIRING**									
0020	20' avg. runs and #14/2 wiring incl. unless otherwise noted									
1000	Service & panel, includes 24' SE-AL cable, service eye, meter,									
1010	Socket, panel board, main bkr., ground rod, 15 or 20 amp									
1020	1-pole circuit breakers, and misc. hardware									
1100	100 amp, with 10 branch breakers	1 Elec	1.19	6.723	Ea.	335	410		745	1,000
1110	With PVC conduit and wire		.92	8.696		370	535		905	1,225
1120	With RGS conduit and wire		.73	10.959		570	670		1,240	1,650
1150	150 amp, with 14 branch breakers		1.03	7.767		805	475		1,280	1,625
1170	With PVC conduit and wire		.82	9.756		875	600		1,475	1,875
1180	With RGS conduit and wire		.67	11.940		1,200	735		1,935	2,450
1200	200 amp, with 18 branch breakers	2 Elec	1.80	8.889		1,025	545		1,570	1,975
1220	With PVC conduit and wire		1.46	10.959		1,100	670		1,770	2,225
1230	With RGS conduit and wire		1.24	12.903		1,500	790		2,290	2,875
1800	Lightning surge suppressor	1 Elec	32	.250		85	15.35		100.35	117
2000	Switch devices									
2100	Single pole, 15 amp, ivory, with a 1-gang box, cover plate,									
2110	Type NM (Romex) cable	1 Elec	17.10	.468	Ea.	15.55	28.50		44.05	61

26 05 Common Work Results for Electrical

26 05 90 – Residential Applications

26 05 90.10 Residential Wiring	Crew	Daily Output	Labor-Hours	Unit	Material	2020 Bare Costs Labor	Equipment	Total	Total Incl O&P
2120 Type MC cable	1 Elec	14.30	.559	Ea.	26	34.50		60.50	82
2130 EMT & wire		5.71	1.401		36	86		122	172
2150 3-way, #14/3, type NM cable		14.55	.550		10.10	33.50		43.60	63
2170 Type MC cable		12.31	.650		24.50	40		64.50	88.50
2180 EMT & wire		5	1.600		30.50	98		128.50	185
2200 4-way, #14/3, type NM cable		14.55	.550		18.20	33.50		51.70	72
2220 Type MC cable		12.31	.650		32.50	40		72.50	97.50
2230 EMT & wire		5	1.600		38.50	98		136.50	194
2250 S.P., 20 amp, #12/2, type NM cable		13.33	.600		12.30	37		49.30	70
2270 Type MC cable		11.43	.700		22	43		65	90
2280 EMT & wire		4.85	1.649		35	101		136	195
2290 S.P. rotary dimmer, 600 W, no wiring		17	.471		32	29		61	79.50
2300 S.P. rotary dimmer, 600 W, type NM cable		14.55	.550		35.50	33.50		69	91
2320 Type MC cable		12.31	.650		46	40		86	113
2330 EMT & wire		5	1.600		57.50	98		155.50	214
2350 3-way rotary dimmer, type NM cable		13.33	.600		24	37		61	83
2370 Type MC cable		11.43	.700		34.50	43		77.50	104
2380 EMT & wire	▼	4.85	1.649	▼	46	101		147	207
2400 Interval timer wall switch, 20 amp, 1-30 min., #12/2									
2410 Type NM cable	1 Elec	14.55	.550	Ea.	59	33.50		92.50	117
2420 Type MC cable		12.31	.650		64.50	40		104.50	133
2430 EMT & wire	▼	5	1.600	▼	81.50	98		179.50	241
2500 Decorator style									
2510 S.P., 15 amp, type NM cable	1 Elec	17.10	.468	Ea.	21	28.50		49.50	67
2520 Type MC cable		14.30	.559		32	34.50		66.50	88
2530 EMT & wire		5.71	1.401		41.50	86		127.50	178
2550 3-way, #14/3, type NM cable		14.55	.550		15.65	33.50		49.15	69.50
2570 Type MC cable		12.31	.650		30	40		70	94.50
2580 EMT & wire		5	1.600		36	98		134	191
2600 4-way, #14/3, type NM cable		14.55	.550		24	33.50		57.50	78
2620 Type MC cable		12.31	.650		38	40		78	104
2630 EMT & wire		5	1.600		44	98		142	200
2650 S.P., 20 amp, #12/2, type NM cable		13.33	.600		17.85	37		54.85	76
2670 Type MC cable		11.43	.700		27.50	43		70.50	96
2680 EMT & wire		4.85	1.649		40.50	101		141.50	201
2700 S.P., slide dimmer, type NM cable		17.10	.468		37.50	28.50		66	85.50
2720 Type MC cable		14.30	.559		48	34.50		82.50	106
2730 EMT & wire		5.71	1.401		59.50	86		145.50	198
2750 S.P., touch dimmer, type NM cable		17.10	.468		53	28.50		81.50	103
2770 Type MC cable		14.30	.559		64	34.50		98.50	123
2780 EMT & wire		5.71	1.401		75	86		161	215
2800 3-way touch dimmer, type NM cable		13.33	.600		49.50	37		86.50	111
2820 Type MC cable		11.43	.700		60.50	43		103.50	133
2830 EMT & wire	▼	4.85	1.649	▼	71.50	101		172.50	235
3000 Combination devices									
3100 S.P. switch/15 amp recpt., ivory, 1-gang box, plate									
3110 Type NM cable	1 Elec	11.43	.700	Ea.	22	43		65	90
3120 Type MC cable		10	.800		32.50	49		81.50	112
3130 EMT & wire		4.40	1.818		44	112		156	220
3150 S.P. switch/pilot light, type NM cable		11.43	.700		23.50	43		66.50	91.50
3170 Type MC cable		10	.800		34	49		83	113
3180 EMT & wire		4.43	1.806		45	111		156	221
3190 2-S.P. switches, 2-#14/2, no wiring	▼	14	.571	▼	13	35		48	68.50

26 05 90.10 Residential Wiring	Crew	Daily Output	Labor-Hours	Unit	Material	2020 Bare Costs Labor	Equipment	Total	Total Incl O&P
3200	2-S.P. switches, 2-#14/2, type NM cables	1 Elec	10	.800	Ea.	24	49		73	102
3220	Type MC cable		8.89	.900		38.50	55		93.50	128
3230	EMT & wire		4.10	1.951		46	120		166	235
3250	3-way switch/15 amp recpt., #14/3, type NM cable		10	.800		30	49		79	109
3270	Type MC cable		8.89	.900		44	55		99	134
3280	EMT & wire		4.10	1.951		50	120		170	239
3300	2-3 way switches, 2-#14/3, type NM cables		8.89	.900		38	55		93	127
3320	Type MC cable		8	1		60	61.50		121.50	161
3330	EMT & wire		4	2		56.50	123		179.50	252
3350	S.P. switch/20 amp recpt., #12/2, type NM cable		10	.800		40	49		89	120
3370	Type MC cable		8.89	.900		45.50	55		100.50	135
3380	EMT & wire	▼	4.10	1.951	▼	62.50	120		182.50	253
3400	Decorator style									
3410	S.P. switch/15 amp recpt., type NM cable	1 Elec	11.43	.700	Ea.	27.50	43		70.50	96
3420	Type MC cable		10	.800		38	49		87	118
3430	EMT & wire		4.40	1.818		49.50	112		161.50	227
3450	S.P. switch/pilot light, type NM cable		11.43	.700		29	43		72	97.50
3470	Type MC cable		10	.800		39.50	49		88.50	119
3480	EMT & wire		4.40	1.818		50.50	112		162.50	228
3500	2-S.P. switches, 2-#14/2, type NM cables		10	.800		29.50	49		78.50	108
3520	Type MC cable		8.89	.900		44.50	55		99.50	134
3530	EMT & wire		4.10	1.951		51.50	120		171.50	241
3550	3-way/15 amp recpt., #14/3, type NM cable		10	.800		35.50	49		84.50	115
3570	Type MC cable		8.89	.900		49.50	55		104.50	140
3580	EMT & wire		4.10	1.951		55.50	120		175.50	245
3650	2-3 way switches, 2-#14/3, type NM cables		8.89	.900		43.50	55		98.50	133
3670	Type MC cable		8	1		65.50	61.50		127	167
3680	EMT & wire		4	2		62	123		185	258
3700	S.P. switch/20 amp recpt., #12/2, type NM cable		10	.800		45.50	49		94.50	126
3720	Type MC cable		8.89	.900		51	55		106	141
3730	EMT & wire	▼	4.10	1.951	▼	68	120		188	259
4000	Receptacle devices									
4010	Duplex outlet, 15 amp recpt., ivory, 1-gang box, plate									
4015	Type NM cable	1 Elec	14.55	.550	Ea.	8.90	33.50		42.40	62
4020	Type MC cable		12.31	.650		19.55	40		59.55	83
4030	EMT & wire		5.33	1.501		29.50	92		121.50	175
4050	With #12/2, type NM cable		12.31	.650		10.50	40		50.50	73
4070	Type MC cable		10.67	.750		19.95	46		65.95	93
4080	EMT & wire		4.71	1.699		33	104		137	196
4100	20 amp recpt., #12/2, type NM cable		12.31	.650		19.50	40		59.50	83
4120	Type MC cable		10.67	.750		29	46		75	103
4130	EMT & wire	▼	4.71	1.699	▼	42	104		146	206
4140	For GFI see Section 26 05 90.10 line 4300 below									
4150	Decorator style, 15 amp recpt., type NM cable	1 Elec	14.55	.550	Ea.	14.45	33.50		47.95	68
4170	Type MC cable		12.31	.650		25	40		65	89
4180	EMT & wire		5.33	1.501		35	92		127	181
4200	With #12/2, type NM cable		12.31	.650		16.05	40		56.05	79
4220	Type MC cable		10.67	.750		25.50	46		71.50	99
4230	EMT & wire		4.71	1.699		38.50	104		142.50	203
4250	20 amp recpt., #12/2, type NM cable		12.31	.650		25	40		65	89
4270	Type MC cable		10.67	.750		34.50	46		80.50	109
4280	EMT & wire		4.71	1.699		47.50	104		151.50	213
4300	GFI, 15 amp recpt., type NM cable		12.31	.650		21	40		61	84.50

26 05 90.10 Residential Wiring		Crew	Daily Output	Labor-Hours	Unit	Material	2020 Bare Costs Labor	Equipment	Total	Total Incl O&P
4320	Type MC cable	1 Elec	10.67	.750	Ea.	31.50	46		77.50	106
4330	EMT & wire		4.71	1.699		41.50	104		145.50	206
4350	GFI with #12/2, type NM cable		10.67	.750		22.50	46		68.50	95.50
4370	Type MC cable		9.20	.870		32	53.50		85.50	117
4380	EMT & wire		4.21	1.900		45	117		162	229
4400	20 amp recpt., #12/2, type NM cable		10.67	.750		54	46		100	131
4420	Type MC cable		9.20	.870		63.50	53.50		117	152
4430	EMT & wire		4.21	1.900		76.50	117		193.50	263
4500	Weather-proof cover for above receptacles, add	▼	32	.250	▼	2.02	15.35		17.37	25.50
4550	Air conditioner outlet, 20 amp-240 volt recpt.									
4560	30' of #12/2, 2 pole circuit breaker									
4570	Type NM cable	1 Elec	10	.800	Ea.	61	49		110	143
4580	Type MC cable		9	.889		72	54.50		126.50	163
4590	EMT & wire		4	2		84	123		207	281
4600	Decorator style, type NM cable		10	.800		66	49		115	148
4620	Type MC cable		9	.889		77	54.50		131.50	169
4630	EMT & wire	▼	4	2	▼	88.50	123		211.50	287
4650	Dryer outlet, 30 amp-240 volt recpt., 20' of #10/3									
4660	2 pole circuit breaker									
4670	Type NM cable	1 Elec	6.41	1.248	Ea.	54.50	76.50		131	178
4680	Type MC cable		5.71	1.401		62.50	86		148.50	201
4690	EMT & wire	▼	3.48	2.299	▼	73	141		214	298
4700	Range outlet, 50 amp-240 volt recpt., 30' of #8/3									
4710	Type NM cable	1 Elec	4.21	1.900	Ea.	82.50	117		199.50	270
4720	Type MC cable		4	2		133	123		256	335
4730	EMT & wire		2.96	2.703		105	166		271	370
4750	Central vacuum outlet, type NM cable		6.40	1.250		58	76.50		134.50	182
4770	Type MC cable		5.71	1.401		70.50	86		156.50	210
4780	EMT & wire	▼	3.48	2.299	▼	88	141		229	315
4800	30 amp-110 volt locking recpt., #10/2 circ. bkr.									
4810	Type NM cable	1 Elec	6.20	1.290	Ea.	67	79		146	196
4820	Type MC cable		5.40	1.481		83	91		174	232
4830	EMT & wire	▼	3.20	2.500	▼	98	153		251	345
4900	Low voltage outlets									
4910	Telephone recpt., 20' of 4/C phone wire	1 Elec	26	.308	Ea.	8.80	18.90		27.70	38.50
4920	TV recpt., 20' of RG59U coax wire, F type connector	"	16	.500	"	17.60	30.50		48.10	66.50
4950	Door bell chime, transformer, 2 buttons, 60' of bellwire									
4970	Economy model	1 Elec	11.50	.696	Ea.	57.50	42.50		100	129
4980	Custom model		11.50	.696		111	42.50		153.50	188
4990	Luxury model, 3 buttons	▼	9.50	.842	▼	188	51.50		239.50	287
6000	Lighting outlets									
6050	Wire only (for fixture), type NM cable	1 Elec	32	.250	Ea.	5.95	15.35		21.30	30
6070	Type MC cable		24	.333		11.30	20.50		31.80	44
6080	EMT & wire		10	.800		20.50	49		69.50	98
6100	Box (4"), and wire (for fixture), type NM cable		25	.320		15.05	19.65		34.70	46.50
6120	Type MC cable		20	.400		20.50	24.50		45	60.50
6130	EMT & wire	▼	11	.727	▼	29.50	44.50		74	101
6200	Fixtures (use with line 6050 or 6100 above)									
6210	Canopy style, economy grade	1 Elec	40	.200	Ea.	22	12.25		34.25	43.50
6220	Custom grade		40	.200		53	12.25		65.25	77
6250	Dining room chandelier, economy grade		19	.421		82	26		108	130
6270	Luxury grade		15	.533		1,300	32.50		1,332.50	1,475
6310	Kitchen fixture (fluorescent), economy grade		30	.267		72.50	16.35		88.85	105

For customer support on your Commercial Renovation Costs with RSMeans data, call 800.448.8182.

501

26 05 90 – Residential Applications

26 05 90.10 Residential Wiring		Crew	Daily Output	Labor-Hours	Unit	Material	2020 Bare Costs Labor	Equipment	Total	Total Incl O&P
6320	Custom grade	1 Elec	25	.320	Ea.	147	19.65		166.65	192
6350	Outdoor, wall mounted, economy grade		30	.267		30.50	16.35		46.85	58.50
6360	Custom grade		30	.267		119	16.35		135.35	155
6370	Luxury grade		25	.320		247	19.65		266.65	300
6410	Outdoor PAR floodlights, 1 lamp, 150 watt		20	.400		28	24.50		52.50	68.50
6420	2 lamp, 150 watt each		20	.400		45	24.50		69.50	87.50
6425	Motion sensing, 2 lamp, 150 watt each		20	.400		109	24.50		133.50	158
6430	For infrared security sensor, add		32	.250		95	15.35		110.35	128
6450	Outdoor, quartz-halogen, 300 watt flood		20	.400		40	24.50		64.50	82.50
6600	Recessed downlight, round, pre-wired, 50 or 75 watt trim		30	.267		68	16.35		84.35	100
6610	With shower light trim		30	.267		93.50	16.35		109.85	128
6620	With wall washer trim		28	.286		94	17.55		111.55	130
6630	With eye-ball trim		28	.286		85	17.55		102.55	121
6700	Porcelain lamp holder		40	.200		2.76	12.25		15.01	22
6710	With pull switch		40	.200		10.70	12.25		22.95	30.50
6750	Fluorescent strip, 2-20 watt tube, wrap around diffuser, 24"		24	.333		45.50	20.50		66	81.50
6760	1-34 watt tube, 48"		24	.333		122	20.50		142.50	166
6770	2-34 watt tubes, 48"		20	.400		160	24.50		184.50	214
6800	Bathroom heat lamp, 1-250 watt		28	.286		33.50	17.55		51.05	64
6810	2-250 watt lamps	▼	28	.286	▼	64.50	17.55		82.05	97.50
6820	For timer switch, see Section 26 05 90.10 line 2400									
6900	Outdoor post lamp, incl. post, fixture, 35' of #14/2									
6910	Type NM cable	1 Elec	3.50	2.286	Ea.	325	140		465	570
6920	Photo-eye, add		27	.296		27	18.20		45.20	58
6950	Clock dial time switch, 24 hr., w/enclosure, type NM cable		11.43	.700		72.50	43		115.50	146
6970	Type MC cable		11	.727		83	44.50		127.50	160
6980	EMT & wire	▼	4.85	1.649	▼	93	101		194	258
7000	Alarm systems									
7050	Smoke detectors, box, #14/3, type NM cable	1 Elec	14.55	.550	Ea.	33.50	33.50		67	89
7070	Type MC cable		12.31	.650		44	40		84	110
7080	EMT & wire	▼	5	1.600		50	98		148	206
7090	For relay output to security system, add				▼	10.25			10.25	11.25
8000	Residential equipment									
8050	Disposal hook-up, incl. switch, outlet box, 3' of flex									
8060	20 amp-1 pole circ. bkr., and 25' of #12/2									
8070	Type NM cable	1 Elec	10	.800	Ea.	29	49		78	108
8080	Type MC cable		8	1		39	61.50		100.50	138
8090	EMT & wire	▼	5	1.600	▼	55	98		153	211
8100	Trash compactor or dishwasher hook-up, incl. outlet box,									
8110	3' of flex, 15 amp-1 pole circ. bkr., and 25' of #14/2									
8120	Type NM cable	1 Elec	10	.800	Ea.	16.05	49		65.05	93
8130	Type MC cable		8	1		28	61.50		89.50	126
8140	EMT & wire	▼	5	1.600	▼	41.50	98		139.50	197
8150	Hot water sink dispenser hook-up, use line 8100									
8200	Vent/exhaust fan hook-up, type NM cable	1 Elec	32	.250	Ea.	5.95	15.35		21.30	30
8220	Type MC cable		24	.333		11.30	20.50		31.80	44
8230	EMT & wire	▼	10	.800	▼	20.50	49		69.50	98
8250	Bathroom vent fan, 50 CFM (use with above hook-up)									
8260	Economy model	1 Elec	15	.533	Ea.	18.80	32.50		51.30	71
8270	Low noise model		15	.533		47.50	32.50		80	103
8280	Custom model	▼	12	.667	▼	117	41		158	192
8300	Bathroom or kitchen vent fan, 110 CFM									
8310	Economy model	1 Elec	15	.533	Ea.	66.50	32.50		99	124

26 05 90.10 Residential Wiring	Crew	Daily Output	Labor-Hours	Unit	Material	2020 Bare Costs Labor	Equipment	Total	Total Incl O&P
8320	Low noise model	1 Elec	15	.533	Ea.	96.50	32.50		129	157
8350	Paddle fan, variable speed (w/o lights)									
8360	Economy model (AC motor)	1 Elec	10	.800	Ea.	136	49		185	225
8362	With light kit		10	.800		176	49		225	269
8370	Custom model (AC motor)		10	.800		345	49		394	455
8372	With light kit		10	.800		385	49		434	500
8380	Luxury model (DC motor)		8	1		315	61.50		376.50	440
8382	With light kit		8	1		355	61.50		416.50	485
8390	Remote speed switch for above, add	↓	12	.667	↓	40.50	41		81.50	108
8500	Whole house exhaust fan, ceiling mount, 36", variable speed									
8510	Remote switch, incl. shutters, 20 amp-1 pole circ. bkr.									
8520	30' of #12/2, type NM cable	1 Elec	4	2	Ea.	1,375	123		1,498	1,725
8530	Type MC cable		3.50	2.286		1,400	140		1,540	1,775
8540	EMT & wire	↓	3	2.667	↓	1,425	164		1,589	1,800
8600	Whirlpool tub hook-up, incl. timer switch, outlet box									
8610	3' of flex, 20 amp-1 pole GFI circ. bkr.									
8620	30' of #12/2, type NM cable	1 Elec	5	1.600	Ea.	129	98		227	293
8630	Type MC cable		4.20	1.905		136	117		253	330
8640	EMT & wire	↓	3.40	2.353	↓	149	144		293	385
8650	Hot water heater hook-up, incl. 1-2 pole circ. bkr., box;									
8660	3' of flex, 20' of #10/2, type NM cable	1 Elec	5	1.600	Ea.	29	98		127	183
8670	Type MC cable		4.20	1.905		42	117		159	226
8680	EMT & wire	↓	3.40	2.353	↓	48	144		192	275
9000	Heating/air conditioning									
9050	Furnace/boiler hook-up, incl. firestat, local on-off switch									
9060	Emergency switch, and 40' of type NM cable	1 Elec	4	2	Ea.	57	123		180	252
9070	Type MC cable		3.50	2.286		72	140		212	295
9080	EMT & wire	↓	1.50	5.333	↓	93.50	325		418.50	610
9100	Air conditioner hook-up, incl. local 60 amp disc. switch									
9110	3' sealtite, 40 amp, 2 pole circuit breaker									
9130	40' of #8/2, type NM cable	1 Elec	3.50	2.286	Ea.	144	140		284	375
9140	Type MC cable		3	2.667		212	164		376	485
9150	EMT & wire	↓	1.30	6.154	↓	185	380		565	785
9200	Heat pump hook-up, 1-40 & 1-100 amp 2 pole circ. bkr.									
9210	Local disconnect switch, 3' sealtite									
9220	40' of #8/2 & 30' of #3/2									
9230	Type NM cable	1 Elec	1.30	6.154	Ea.	520	380		900	1,150
9240	Type MC cable		1.08	7.407		550	455		1,005	1,300
9250	EMT & wire	↓	.94	8.511	↓	535	520		1,055	1,400
9500	Thermostat hook-up, using low voltage wire									
9520	Heating only, 25' of #18-3	1 Elec	24	.333	Ea.	6.75	20.50		27.25	39
9530	Heating/cooling, 25' of #18-4	"	20	.400	"	8.60	24.50		33.10	47.50

26 22 Low-Voltage Transformers

26 22 13 – Low-Voltage Distribution Transformers

26 22 13.10 Transformer, Dry-Type	Crew	Daily Output	Labor-Hours	Unit	Material	2020 Bare Costs Labor	Equipment	Total	Total Incl O&P
0010 **TRANSFORMER, DRY-TYPE**									
0050 Single phase, 240/480 volt primary, 120/240 volt secondary									
0100 1 kVA	1 Elec	2	4	Ea.	390	245		635	805
0300 2 kVA		1.60	5		700	305		1,005	1,250
0500 3 kVA		1.40	5.714		845	350		1,195	1,475
0700 5 kVA		1.20	6.667		1,300	410		1,710	2,050
0900 7.5 kVA	2 Elec	2.20	7.273		1,250	445		1,695	2,050
1100 10 kVA		1.60	10		1,625	615		2,240	2,750
1300 15 kVA		1.20	13.333		2,075	820		2,895	3,525
2300 3 phase, 480 volt primary, 120/208 volt secondary									
2310 Ventilated, 3 kVA	1 Elec	1	8	Ea.	1,225	490		1,715	2,100
2700 6 kVA		.80	10		1,225	615		1,840	2,300
2900 9 kVA		.70	11.429		1,450	700		2,150	2,650
3100 15 kVA	2 Elec	1.10	14.545		1,900	890		2,790	3,450
3300 30 kVA		.90	17.778		1,800	1,100		2,900	3,650
3500 45 kVA		.80	20		1,850	1,225		3,075	3,950
9000 Minimum labor/equipment charge	1 Elec	1	8	Job		490		490	755
9300 Energy efficient transformer 3 ph									
9400 15 kVA, 480VAC delta, 208Y/120VAC	R-3	1.50	13.333	Ea.	1,975	815	126	2,916	3,575
9405 30 kVA, 480VAC delta, 208Y/120VAC		1.50	13.333		2,450	815	126	3,391	4,100
9410 45 kVA, 480VAC delta, 208Y/120VAC		1.50	13.333		3,425	815	126	4,366	5,175
9415 75 kVA, 480VAC delta, 208Y/120VAC		1.25	16		4,500	980	151	5,631	6,625
9420 112 kVA, 480VAC delta, 208Y/120VAC		1.25	16		6,700	980	151	7,831	9,050
9425 150 kVA, 480VAC delta, 208Y/120VAC		1.25	16		15,400	980	151	16,531	18,700
9430 225 kVA, 480VAC delta, 208Y/120VAC		1.25	16		10,700	980	151	11,831	13,500
9435 300 kVA, 480VAC delta, 208Y/120VAC		1.25	16		15,000	980	151	16,131	18,200

26 24 Switchboards and Panelboards

26 24 16 – Panelboards

26 24 16.20 Panelboard and Load Center Circuit Breakers

26 24 16.20	Crew	Daily Output	Labor-Hours	Unit	Material	2020 Bare Costs Labor	Equipment	Total	Total Incl O&P
0010 **PANELBOARD AND LOAD CENTER CIRCUIT BREAKERS**									
0050 Bolt-on, 10,000 amp I.C., 120 volt, 1 pole									
0100 15-50 amp	1 Elec	10	.800	Ea.	18.60	49		67.60	96
0200 60 amp		8	1		20.50	61.50		82	117
0300 70 amp		8	1		28	61.50		89.50	125
0350 240 volt, 2 pole									
0400 15-50 amp	1 Elec	8	1	Ea.	56.50	61.50		118	157
0500 60 amp		7.50	1.067		37.50	65.50		103	142
0600 80-100 amp		5	1.600		111	98		209	273
0700 3 pole, 15-60 amp		6.20	1.290		135	79		214	271
0800 70 amp		5	1.600		172	98		270	340
0900 80-100 amp		3.60	2.222		205	136		341	435
1000 22,000 amp I.C., 240 volt, 2 pole, 70-225 amp		2.70	2.963		555	182		737	890
1100 3 pole, 70-225 amp		2.30	3.478		345	213		558	705
1200 14,000 amp I.C., 277 volts, 1 pole, 15-30 amp		8	1		37	61.50		98.50	136
1300 22,000 amp I.C., 480 volts, 2 pole, 70-225 amp		2.70	2.963		525	182		707	855
1400 3 pole, 70-225 amp		2.30	3.478		1,050	213		1,263	1,475
2060 Plug-in tandem, 120/240 V, 2-15 A, 1 pole		11	.727		30	44.50		74.50	102
2070 1-15 A & 1-20 A		11	.727		18.35	44.50		62.85	88.50
2080 2-20 A		11	.727		18.40	44.50		62.90	88.50
2082 Arc fault circuit interrupter, 120/240 V, 1-15 A & 1-20 A, 1 pole		11	.727		57	44.50		101.50	131

26 24 Switchboards and Panelboards

26 24 16 – Panelboards

26 24 16.20 Panelboard and Load Center Circuit Breakers	Crew	Daily Output	Labor-Hours	Unit	Material	2020 Bare Costs Labor	Equipment	Total	Total Incl O&P
5110 NEMA 1 enclosure only, 600V, 3 p, 14k AIC, 100A	1 Elec	2.20	3.636	Ea.	160	223		383	520
9000 Minimum labor/equipment charge		3	2.667	Job		164		164	252

26 24 16.30 Panelboards Commercial Applications

	Crew	Daily Output	Labor-Hours	Unit	Material	2020 Bare Costs Labor	Equipment	Total	Total Incl O&P
0010 **PANELBOARDS COMMERCIAL APPLICATIONS**									
0050 NQOD, w/20 amp 1 pole bolt-on circuit breakers									
0100 3 wire, 120/240 volts, 100 amp main lugs									
0150 10 circuits	1 Elec	1	8	Ea.	935	490		1,425	1,775
0200 14 circuits		.88	9.091		1,050	560		1,610	2,000
0250 18 circuits		.75	10.667		1,150	655		1,805	2,250
0300 20 circuits		.65	12.308		1,275	755		2,030	2,575
0600 4 wire, 120/208 volts, 100 amp main lugs, 12 circuits		1	8		1,025	490		1,515	1,875
0650 16 circuits		.75	10.667		1,175	655		1,830	2,275
0700 20 circuits		.65	12.308		1,350	755		2,105	2,625
0750 24 circuits		.60	13.333		1,275	820		2,095	2,650
0800 30 circuits		.53	15.094		1,675	925		2,600	3,275
0850 225 amp main lugs, 32 circuits	2 Elec	.90	17.778		1,875	1,100		2,975	3,750
0900 34 circuits		.84	19.048		1,925	1,175		3,100	3,900
0950 36 circuits		.80	20		1,975	1,225		3,200	4,050
1000 42 circuits		.68	23.529		2,200	1,450		3,650	4,650
1010 400 amp main lugs, 42 circs		.68	23.529		2,200	1,450		3,650	4,650
1200 NEHB, w/20 amp, 1 pole bolt-on circuit breakers									
1250 4 wire, 277/480 volts, 100 amp main lugs, 12 circuits	1 Elec	.88	9.091	Ea.	1,475	560		2,035	2,475
1300 20 circuits	"	.60	13.333		2,200	820		3,020	3,675
1350 225 amp main lugs, 24 circuits	2 Elec	.90	17.778		2,500	1,100		3,600	4,425
1400 30 circuits		.80	20		2,975	1,225		4,200	5,175
1448 32 circuits		4.90	3.265		3,475	200		3,675	4,100
1450 36 circuits		.72	22.222		3,475	1,375		4,850	5,900
1600 NQOD panel, w/20 amp, 1 pole, circuit breakers									
2000 4 wire, 120/208 volts with main circuit breaker									
2050 100 amp main, 24 circuits	1 Elec	.47	17.021	Ea.	1,825	1,050		2,875	3,625
2100 30 circuits	"	.40	20		2,125	1,225		3,350	4,250
2200 225 amp main, 32 circuits	2 Elec	.72	22.222		3,550	1,375		4,925	6,025
2250 42 circuits		.56	28.571		4,125	1,750		5,875	7,225
2300 400 amp main, 42 circuits		.48	33.333		5,225	2,050		7,275	8,900
2350 600 amp main, 42 circuits		.40	40		7,725	2,450		10,175	12,300
2400 NEHB, with 20 amp, 1 pole circuit breaker									
2450 4 wire, 277/480 volts with main circuit breaker									
2500 100 amp main, 24 circuits	1 Elec	.42	19.048	Ea.	2,875	1,175		4,050	4,950
2550 30 circuits	"	.38	21.053		3,350	1,300		4,650	5,700
2600 225 amp main, 30 circuits	2 Elec	.72	22.222		4,200	1,375		5,575	6,725
2650 42 circuits	"	.56	28.571		5,175	1,750		6,925	8,400
9000 Minimum labor/equipment charge	1 Elec	1	8	Job		490		490	755

For customer support on your Commercial Renovation Costs with RSMeans data, call 800.448.8182.

505

26 27 Low-Voltage Distribution Equipment

26 27 13 – Electricity Metering

26 27 13.10 Meter Centers and Sockets

		Crew	Daily Output	Labor-Hours	Unit	Material	2020 Bare Costs Labor	Equipment	Total	Total Incl O&P
0010	**METER CENTERS AND SOCKETS**									
0100	Sockets, single position, 4 terminal, 100 amp	1 Elec	3.20	2.500	Ea.	52.50	153		205.50	294
0200	150 amp		2.30	3.478		62.50	213		275.50	400
0300	200 amp		1.90	4.211		110	258		368	520
0400	Transformer rated, 20 amp		3.20	2.500		179	153		332	435
0500	Double position, 4 terminal, 100 amp		2.80	2.857		240	175		415	535
0600	150 amp		2.10	3.810		290	234		524	680
0700	200 amp		1.70	4.706		550	289		839	1,050
1100	Meter centers and sockets, three phase, single pos, 7 terminal, 100 amp		2.80	2.857		136	175		311	420
1200	200 amp		2.10	3.810		239	234		473	625
1400	400 amp		1.70	4.706		825	289		1,114	1,350
9000	Minimum labor/equipment charge		3	2.667	Job		164		164	252

26 27 16 – Electrical Cabinets and Enclosures

26 27 16.10 Cabinets

		Crew	Daily Output	Labor-Hours	Unit	Material	2020 Bare Costs Labor	Equipment	Total	Total Incl O&P
0010	**CABINETS**									
7000	Cabinets, current transformer									
7050	Single door, 24" H x 24" W x 10" D	1 Elec	1.60	5	Ea.	172	305		477	660
7100	30" H x 24" W x 10" D		1.30	6.154		165	380		545	760
7150	36" H x 24" W x 10" D		1.10	7.273		410	445		855	1,125
7200	30" H x 30" W x 10" D		1	8		226	490		716	1,000
7250	36" H x 30" W x 10" D		.90	8.889		299	545		844	1,175
7300	36" H x 36" W x 10" D		.80	10		310	615		925	1,275
7500	Double door, 48" H x 36" W x 10" D		.60	13.333		575	820		1,395	1,875
7550	24" H x 24" W x 12" D		1	8		190	490		680	965
9990	Minimum labor/equipment charge		2	4	Job		245		245	380

26 27 23 – Indoor Service Poles

26 27 23.40 Surface Raceway

		Crew	Daily Output	Labor-Hours	Unit	Material	2020 Bare Costs Labor	Equipment	Total	Total Incl O&P
0010	**SURFACE RACEWAY**									
0090	Metal, straight section									
0100	No. 500	1 Elec	100	.080	L.F.	1.30	4.91		6.21	9
0110	No. 700		100	.080		1.41	4.91		6.32	9.10
0400	No. 1500, small pancake		90	.089		2.44	5.45		7.89	11.10
0600	No. 2000, base & cover, blank		90	.089		2.60	5.45		8.05	11.25
0800	No. 3000, base & cover, blank		75	.107		4.85	6.55		11.40	15.45
1000	No. 4000, base & cover, blank		65	.123		7.90	7.55		15.45	20.50
1200	No. 6000, base & cover, blank		50	.160		13.85	9.80		23.65	30.50
2400	Fittings, elbows, No. 500		40	.200	Ea.	2.34	12.25		14.59	21.50
2800	Elbow cover, No. 2000		40	.200		4.23	12.25		16.48	23.50
2880	Tee, No. 500		42	.190		4.19	11.70		15.89	22.50
2900	No. 2000		27	.296		13.65	18.20		31.85	43
3000	Switch box, No. 500		16	.500		14.85	30.50		45.35	63.50
3400	Telephone outlet, No. 1500		16	.500		17.55	30.50		48.05	66.50
3600	Junction box, No. 1500		16	.500		11.55	30.50		42.05	59.50
3800	Plugmold wired sections, No. 2000									
4000	1 circuit, 6 outlets, 3' long	1 Elec	8	1	Ea.	45	61.50		106.50	144
4100	2 circuits, 8 outlets, 6' long	"	5.30	1.509	"	65.50	92.50		158	215
9300	Non-metallic, straight section									
9310	7/16" x 7/8", base & cover, blank	1 Elec	160	.050	L.F.	2	3.07		5.07	6.90
9320	Base & cover w/adhesive		160	.050		1.71	3.07		4.78	6.60
9340	7/16" x 1-5/16", base & cover, blank		145	.055		2.20	3.38		5.58	7.60
9350	Base & cover w/adhesive		145	.055		2.26	3.38		5.64	7.70
9370	11/16" x 2-1/4", base & cover, blank		130	.062		3.02	3.78		6.80	9.10

For customer support on your Commercial Renovation Costs with RSMeans data, call 800.448.8182.

26 27 23 – Indoor Service Poles

26 27 23.40 Surface Raceway	Crew	Daily Output	Labor-Hours	Unit	Material	2020 Bare Costs Labor	Equipment	Total	Total Incl O&P
9380	Base & cover w/adhesive	1 Elec	130	.062	L.F.	3.24	3.78		7.02	9.35
9385	1-11/16" x 5-1/4", two compartment base & cover w/screws		80	.100	▼	10.40	6.15		16.55	21
9400	Fittings, elbows, 7/16" x 7/8"		50	.160	Ea.	1.97	9.80		11.77	17.25
9410	7/16" x 1-5/16"		45	.178		2.04	10.90		12.94	19.05
9420	11/16" x 2-1/4"		40	.200		2.21	12.25		14.46	21.50
9425	1-11/16" x 5-1/4"		28	.286		12.45	17.55		30	40.50
9430	Tees, 7/16" x 7/8"		35	.229		2.53	14		16.53	24.50
9440	7/16" x 1-5/16"		32	.250		2.59	15.35		17.94	26.50
9450	11/16" x 2-1/4"		30	.267		2.69	16.35		19.04	28
9455	1-11/16" x 5-1/4"		24	.333		19.95	20.50		40.45	53.50
9460	Cover clip, 7/16" x 7/8"		80	.100		.51	6.15		6.66	10
9470	7/16" x 1-5/16"		72	.111		.46	6.80		7.26	11
9480	11/16" x 2-1/4"		64	.125		.84	7.65		8.49	12.70
9484	1-11/16" x 5-1/4"		42	.190		2.94	11.70		14.64	21
9486	Wire clip, 1-11/16" x 5-1/4"		68	.118		.58	7.20		7.78	11.75
9490	Blank end, 7/16" x 7/8"		50	.160		.77	9.80		10.57	15.95
9500	7/16" x 1-5/16"		45	.178		.86	10.90		11.76	17.75
9510	11/16" x 2-1/4"		40	.200		1.28	12.25		13.53	20.50
9515	1-11/16" x 5-1/4"		38	.211		6.70	12.90		19.60	27.50
9520	Round fixture box, 5.5" diam. x 1"		25	.320		11.55	19.65		31.20	42.50
9530	Device box, 1 gang		30	.267		5.25	16.35		21.60	31
9540	2 gang		25	.320	▼	7.65	19.65		27.30	38.50
9990	Minimum labor/equipment charge	▼	5	1.600	Job		98		98	151

26 27 26 – Wiring Devices

26 27 26.20 Wiring Devices Elements

		Crew	Daily Output	Labor-Hours	Unit	Material	Labor	Equipment	Total	Total Incl O&P
0010	**WIRING DEVICES ELEMENTS**									
0200	Toggle switch, quiet type, single pole, 15 amp	1 Elec	40	.200	Ea.	.52	12.25		12.77	19.45
0500	20 amp		27	.296		3.46	18.20		21.66	32
0550	Rocker, 15 amp		40	.200		2.82	12.25		15.07	22
0560	20 amp		27	.296		10.85	18.20		29.05	40
0600	3 way, 15 amp		23	.348		1.83	21.50		23.33	35
0850	Rocker, 15 amp		23	.348		8.25	21.50		29.75	42
0860	20 amp		18	.444		12.80	27.50		40.30	56
0900	4 way, 15 amp		15	.533		10.60	32.50		43.10	62
1030	Rocker, 15 amp		15	.533		10	32.50		42.50	61.50
1040	20 amp		11	.727		19.15	44.50		63.65	89.50
1650	Dimmer switch, 120 volt, incandescent, 600 watt, 1 pole [G]		16	.500		22.50	30.50		53	72
2460	Receptacle, duplex, 120 volt, grounded, 15 amp		40	.200		1.62	12.25		13.87	20.50
2470	20 amp		27	.296		10.65	18.20		28.85	40
2480	Ground fault interrupting, 15 amp		27	.296		13.55	18.20		31.75	43
2482	20 amp		27	.296		45	18.20		63.20	77.50
2490	Dryer, 30 amp		15	.533		4.55	32.50		37.05	55.50
2500	Range, 50 amp		11	.727		11.20	44.50		55.70	81
2600	Wall plates, stainless steel, 1 gang		80	.100		2.57	6.15		8.72	12.30
2800	2 gang		53	.151		4.38	9.25		13.63	19.05
3200	Lampholder, keyless		26	.308		19.35	18.90		38.25	50.50
3400	Pullchain with receptacle		22	.364		23	22.50		45.50	59.50
4870	Receptacle box assembly, cast aluminum, 60A, 4P5W, 3P, 120/208V, NEMA PR4		22	.364		735	22.50		757.50	845
4875	100A, 4P5W, 3P, 347/600V, NEMA PR4		22	.364		810	22.50		832.50	925
4880	200A, 4P5W, 3P, 347/600V, NEMA PR4		22	.364		1,750	22.50		1,772.50	1,950
8100	Pin/sleeve, 20A, 480V, DSN1, male inlet, 3 pole, 3 phase, 4 wire		16	.500		48.50	30.50		79	101
8110	20A, 125V, DSN20, male inlet, 3 pole, 3 phase, 3 wire	▼	16	.500		39.50	30.50		70	90.50

26 27 26 – Wiring Devices

26 27 26.20 Wiring Devices Elements	Crew	Daily Output	Labor-Hours	Unit	Material	2020 Bare Costs Labor	Equipment	Total	Total Incl O&P
8120 60A, 480V, DS60, male inlet, 3 pole, 3 phase, 3 wire	1 Elec	16	.500	Ea.	325	30.50		355.50	405
8130 Pin/sleeve, 20A, 480V, DSN10, female RCPT, 3P, 3 phase, 4 wire		16	.500		92.50	30.50		123	149
8140 30A, 480V, DS30, female RCPT, 3P, 3 phase, 4 wire		16	.500		300	30.50		330.50	380
8500 Wiring device terminal strip, 2 pole, 12-2AWG, 300VAC, CSA, 600VAC, 10A		36	.222		3.44	13.65		17.09	25
8505 4 pole, 12-2AWG, 300VAC, CSA, 600VAC, 10A		36	.222		4.87	13.65		18.52	26.50
8510 6 pole, 12-2AWG, 300VAC, CSA, 600VAC, 10A		36	.222		5.50	13.65		19.15	27
8515 8 pole, 12-2AWG, 300VAC, CSA, 600VAC, 10A		30	.267		6.15	16.35		22.50	32
8520 10 pole, 12-2AWG, 300VAC, CSA, 600VAC, 10A		30	.267		6.80	16.35		23.15	32.50
8525 12 pole, 12-2AWG, 300VAC, CSA, 600VAC, 10A		30	.267		7.50	16.35		23.85	33.50
8530 Wiring device terminal strip jumper, 10A		36	.222		.84	13.65		14.49	22
8535 Wiring device terminal strip, 2 pole, 12-22AWG, 300VAC, CSA, 600VAC, 20A		34	.235		3.74	14.45		18.19	26
8540 4 pole, 12-22AWG, 300VAC, CSA, 600VAC, 20A		34	.235		5.10	14.45		19.55	27.50
8545 6 pole, 12-22AWG, 300VAC, CSA, 600VAC, 20A		34	.235		5.80	14.45		20.25	28.50
8550 8 pole, 12-22AWG, 300VAC, CSA, 600VAC, 20A		30	.267		6.45	16.35		22.80	32
8555 10 pole, 12-22AWG, 300VAC, CSA, 600VAC, 20A		30	.267		7.15	16.35		23.50	33
8560 12 pole, 12-22AWG, 300VAC, CSA, 600VAC, 20A		30	.267		7.85	16.35		24.20	33.50
8565 Wiring device terminal strip jumper, 20A		34	.235		1.03	14.45		15.48	23
8570 Wiring device terminal shorting type no cover, 4 pole, 6-10AWG, 600VAC, 50A		30	.267		19.40	16.35		35.75	46.50
8575 6 pole, 6-10AWG, 600VAC, 50A		30	.267		24.50	16.35		40.85	52
8580 8 pole, 6-10AWG, 600VAC, 50A		30	.267		28.50	16.35		44.85	56.50
8585 12 pole, 6-10AWG, 600VAC, 50A		30	.267		32	16.35		48.35	60
8590 Wiring device terminal 2 shorting pin 1/2 cover, 2 pole, 8-4AWG, 600V, 45A		30	.267		20.50	16.35		36.85	47.50
8595 4 pole, 8-4AWG, 600V, 45A		30	.267		25.50	16.35		41.85	53
8600 6 pole, 8-4AWG, 600V, 45A		30	.267		31.50	16.35		47.85	60
8605 8 pole, 8-4AWG, 600V, 45A		30	.267		45	16.35		61.35	74.50
8610 12 pole, 8-4AWG, 600V, 45A		30	.267		54	16.35		70.35	84.50
8615 Wiring device terminal 2 shorting pin W/cover, 4 pole, 8-4AWG, 600V, 45A		30	.267		34	16.35		50.35	62.50
8620 6 pole, 8-4AWG, 600V, 45A		30	.267		52	16.35		68.35	82
8625 8 pole, 8-4AWG, 600V, 45A		30	.267		54.50	16.35		70.85	85
8630 12 pole, 8-4AWG, 600V, 45A		30	.267		65	16.35		81.35	96.50
9000 Minimum labor/equipment charge		4	2	Job		123		123	189

26 27 33 – Power Distribution Units

26 27 33.10 Power Distribution Unit Cabinet

	Crew	Daily Output	Labor-Hours	Unit	Material	2020 Bare Costs Labor	Equipment	Total	Total Incl O&P
0010 **POWER DISTRIBUTION UNIT CABINET**									
0100 Power distribution unit, single cabinet, 50 kVA output, 480/208 input	1 Elec	2	4	Ea.	9,200	245		9,445	10,500
0110 75 kVA output, 480/208 input		2	4		9,975	245		10,220	11,400
0120 100 kVA output, 480/208 input		2	4		10,900	245		11,145	12,400
0130 125 kVA output, 480/208 input		1.50	5.333		12,000	325		12,325	13,700
0140 150 kVA output, 480/208 input		1.50	5.333		14,500	325		14,825	16,500
0150 200 kVA output, 480/208 input		1	8		16,000	490		16,490	18,400
0160 225 kVA output, 480/208 input		1	8		16,900	490		17,390	19,400

26 27 73 – Door Chimes

26 27 73.10 Doorbell System

	Crew	Daily Output	Labor-Hours	Unit	Material	2020 Bare Costs Labor	Equipment	Total	Total Incl O&P
0010 **DOORBELL SYSTEM**, incl. transformer, button & signal									
0100 6" bell	1 Elec	4	2	Ea.	158	123		281	365
0200 Buzzer		4	2		127	123		250	330
1000 Door chimes, 2 notes		16	.500		32	30.50		62.50	82
1020 with ambient light		12	.667		116	41		157	191
1100 Tube type, 3 tube system		12	.667		233	41		274	320
1180 4 tube system		10	.800		495	49		544	620
1900 For transformer & button, add		5	1.600		16.15	98		114.15	169
3000 For push button only		24	.333		.92	20.50		21.42	32.50

For customer support on your Commercial Renovation Costs with RSMeans data, call 800.448.8182.

26 27 Low-Voltage Distribution Equipment

26 27 73 – Door Chimes

26 27 73.10 Doorbell System	Crew	Daily Output	Labor-Hours	Unit	Material	2020 Bare Costs Labor	Equipment	Total	Total Incl O&P
3200	Bell transformer	1 Elec	16	.500	Ea.	24	30.50		54.50	73.50
9000	Minimum labor/equipment charge	↓	4	2	Job		123		123	189

26 28 Low-Voltage Circuit Protective Devices

26 28 16 – Enclosed Switches and Circuit Breakers

26 28 16.10 Circuit Breakers

		Crew	Daily Output	Labor-Hours	Unit	Material	2020 Bare Costs Labor	Equipment	Total	Total Incl O&P
0010	**CIRCUIT BREAKERS** (in enclosure)									
0100	Enclosed (NEMA 1), 600 volt, 3 pole, 30 amp	1 Elec	3.20	2.500	Ea.	530	153		683	815
0200	60 amp		2.80	2.857		645	175		820	975
0400	100 amp		2.30	3.478		735	213		948	1,150
0500	200 amp		1.50	5.333		1,875	325		2,200	2,575
0600	225 amp	↓	1.50	5.333		1,700	325		2,025	2,375
0700	400 amp	2 Elec	1.60	10	↓	2,900	615		3,515	4,150
9000	Minimum labor/equipment charge	1 Elec	4	2	Job		123		123	189

26 28 16.13 Circuit Breakers

		Crew	Daily Output	Labor-Hours	Unit	Material	2020 Bare Costs Labor	Equipment	Total	Total Incl O&P
0010	**CIRCUIT BREAKERS**									
0100	Circuit breaker current limiter, 225 to 400, ampere, 1 pole	1 Elec	48	.167	Ea.	835	10.25		845.25	930
0200	Circuit breaker current limiter, 200 to 500, ampere, 1 pole, thermal magnet		46	.174		690	10.65		700.65	775
0300	300 to 500, ampere, 1 pole, thermal magnet		46	.174		1,275	10.65		1,285.65	1,450
0400	Circuit breaker current limiter, 600 to 1000, ampere, 1 pole, tri pac	↓	46	.174	↓	1,750	10.65		1,760.65	1,950

26 28 16.20 Safety Switches

		Crew	Daily Output	Labor-Hours	Unit	Material	2020 Bare Costs Labor	Equipment	Total	Total Incl O&P
0010	**SAFETY SWITCHES**									
0100	General duty 240 volt, 3 pole NEMA 1, fusible, 30 amp	1 Elec	3.20	2.500	Ea.	59.50	153		212.50	300
0200	60 amp		2.30	3.478		102	213		315	440
0300	100 amp		1.90	4.211		175	258		433	590
0400	200 amp	↓	1.30	6.154		375	380		755	990
0500	400 amp	2 Elec	1.80	8.889	↓	985	545		1,530	1,925
1100	Heavy duty, 600 volt, 3 pole NEMA 1 nonfused									
1110	30 amp	1 Elec	3.20	2.500	Ea.	86	153		239	330
1500	60 amp		2.30	3.478		154	213		367	500
1700	100 amp		1.90	4.211		243	258		501	665
1900	200 amp	↓	1.30	6.154		365	380		745	985
2100	400 amp	2 Elec	1.80	8.889		890	545		1,435	1,825
2300	600 amp		1.20	13.333		1,675	820		2,495	3,075
2500	800 amp		.94	17.021		3,000	1,050		4,050	4,900
2700	1,200 amp	↓	.80	20		3,575	1,225		4,800	5,825
4350	600 volt, 3 pole, fusible, 30 amp	1 Elec	3.20	2.500		166	153		319	420
4380	60 amp		2.30	3.478		202	213		415	555
4400	100 amp		1.90	4.211		365	258		623	805
4420	200 amp	↓	1.30	6.154		520	380		900	1,150
4440	400 amp	2 Elec	1.80	8.889		1,400	545		1,945	2,375
4450	600 amp		1.20	13.333		2,475	820		3,295	3,975
4460	800 amp		.94	17.021		4,700	1,050		5,750	6,775
4480	1,200 amp	↓	.80	20		6,325	1,225		7,550	8,875
5510	Heavy duty, 600 volt, 3 pole 3 ph. NEMA 3R fusible, 30 amp	1 Elec	3.10	2.581		279	158		437	550
5520	60 amp		2.20	3.636		340	223		563	720
5530	100 amp		1.80	4.444		500	273		773	970
5540	200 amp		1.20	6.667		695	410		1,105	1,400
5550	400 amp	2 Elec	1.60	10	↓	1,700	615		2,315	2,825
5700	600 volt, 3 pole NEMA 3R nonfused									

26 28 Low-Voltage Circuit Protective Devices

26 28 16 – Enclosed Switches and Circuit Breakers

26 28 16.20 Safety Switches

		Crew	Daily Output	Labor-Hours	Unit	Material	2020 Bare Costs Labor	Equipment	Total	Total Incl O&P
5710	30 amp	1 Elec	3.10	2.581	Ea.	152	158		310	410
5900	60 amp		2.20	3.636		266	223		489	635
6100	100 amp		1.80	4.444		375	273		648	835
6300	200 amp		1.20	6.667		450	410		860	1,125
6500	400 amp	2 Elec	1.60	10		1,175	615		1,790	2,250
6700	600 amp	"	1	16		2,450	980		3,430	4,175
6900	600 volt, 6 pole NEMA 3R nonfused, 30 amp	1 Elec	2.70	2.963		1,025	182		1,207	1,400
7100	60 amp		2	4		1,950	245		2,195	2,525
7300	100 amp		1.50	5.333		1,875	325		2,200	2,575
7500	200 amp		1.20	6.667		3,500	410		3,910	4,450
9990	Minimum labor/equipment charge		3	2.667	Job		164		164	252

26 28 16.40 Time Switches

		Crew	Daily Output	Labor-Hours	Unit	Material	2020 Bare Costs Labor	Equipment	Total	Total Incl O&P
0010	**TIME SWITCHES**									
0100	Single pole, single throw, 24 hour dial	1 Elec	4	2	Ea.	131	123		254	335
0200	24 hour dial with reserve power		3.60	2.222		675	136		811	950
0300	Astronomic dial		3.60	2.222		256	136		392	490
0400	Astronomic dial with reserve power		3.30	2.424		940	149		1,089	1,250
0500	7 day calendar dial		3.30	2.424		218	149		367	470
0600	7 day calendar dial with reserve power		3.20	2.500		212	153		365	470
0700	Photo cell 2,000 watt		8	1		28.50	61.50		90	126
1080	Load management device, 4 loads		2	4		1,150	245		1,395	1,625
1100	8 loads		1	8		2,400	490		2,890	3,375
9000	Minimum labor/equipment charge		3.50	2.286	Job		140		140	216

26 28 16.50 Meter Socket Entry Hub

		Crew	Daily Output	Labor-Hours	Unit	Material	2020 Bare Costs Labor	Equipment	Total	Total Incl O&P
0010	**METER SOCKET ENTRY HUB**									
0100	Meter socket entry hub closing cap, 3/4 to 4 inch	1 Elec	40	.200	Ea.	2.43	12.25		14.68	21.50
0110	Meter socket entry hub, 3/4 conduit		40	.200		20	12.25		32.25	41
0120	1 inch conduit		40	.200		21	12.25		33.25	42
0130	1.25 inch conduit		40	.200		19.50	12.25		31.75	40.50
0140	1.50 inch conduit		40	.200		19.60	12.25		31.85	40.50
0150	2 inch conduit		40	.200		20.50	12.25		32.75	41.50
0160	2.50 inch conduit		40	.200		62	12.25		74.25	87
0170	3 inch conduit		40	.200		93	12.25		105.25	121

26 29 Low-Voltage Controllers

26 29 13 – Enclosed Controllers

26 29 13.40 Relays

		Crew	Daily Output	Labor-Hours	Unit	Material	2020 Bare Costs Labor	Equipment	Total	Total Incl O&P
0010	**RELAYS** Enclosed (NEMA 1)									
0050	600 volt AC, 1 pole, 12 amp	1 Elec	5.30	1.509	Ea.	89	92.50		181.50	241
0100	2 pole, 12 amp		5	1.600		89	98		187	249
0200	4 pole, 10 amp		4.50	1.778		119	109		228	299

26 29 23 – Variable-Frequency Motor Controllers

26 29 23.10 Variable Frequency Drives/Adj. Frequency Drives

			Crew	Daily Output	Labor-Hours	Unit	Material	2020 Bare Costs Labor	Equipment	Total	Total Incl O&P
0010	**VARIABLE FREQUENCY DRIVES/ADJ. FREQUENCY DRIVES**										
0100	Enclosed (NEMA 1), 460 volt, for 3 HP motor size	G	1 Elec	.80	10	Ea.	1,825	615		2,440	2,950
0110	5 HP motor size	G		.80	10		2,050	615		2,665	3,200
0120	7.5 HP motor size	G		.67	11.940		2,450	735		3,185	3,800
0130	10 HP motor size	G		.67	11.940		2,775	735		3,510	4,175
0140	15 HP motor size	G	2 Elec	.89	17.978		3,450	1,100		4,550	5,500
0150	20 HP motor size	G		.89	17.978		4,350	1,100		5,450	6,475

26 29 Low-Voltage Controllers

26 29 23 – Variable-Frequency Motor Controllers

26 29 23.10 Variable Frequency Drives/Adj. Frequency Drives		Crew	Daily Output	Labor-Hours	Unit	Material	2020 Bare Costs Labor	Equipment	Total	Total Incl O&P
0160	25 HP motor size	G	2 Elec	.67	23.881	Ea.	5,125	1,475		6,600	7,900
0170	30 HP motor size	G		.67	23.881		6,250	1,475		7,725	9,125
0180	40 HP motor size	G		.67	23.881		7,225	1,475		8,700	10,200
0190	50 HP motor size	G		.53	30.189		9,650	1,850		11,500	13,500
0200	60 HP motor size	G	R-3	.56	35.714		11,700	2,175	340	14,215	16,600
0210	75 HP motor size	G		.56	35.714		13,700	2,175	340	16,215	18,800
0220	100 HP motor size	G		.50	40		16,800	2,450	380	19,630	22,600
0230	125 HP motor size	G		.50	40		18,200	2,450	380	21,030	24,200
0240	150 HP motor size	G		.50	40		23,600	2,450	380	26,430	30,200
0250	200 HP motor size	G		.42	47.619		30,600	2,900	450	33,950	38,700
1100	Custom-engineered, 460 volt, for 3 HP motor size	G	1 Elec	.56	14.286		3,425	875		4,300	5,125
1110	5 HP motor size	G		.56	14.286		3,450	875		4,325	5,150
1120	7.5 HP motor size	G		.47	17.021		3,100	1,050		4,150	5,000
1130	10 HP motor size	G		.47	17.021		3,250	1,050		4,300	5,200
1140	15 HP motor size	G	2 Elec	.62	25.806		4,700	1,575		6,275	7,600
1150	20 HP motor size	G		.62	25.806		4,350	1,575		5,925	7,200
1160	25 HP motor size	G		.47	34.043		5,375	2,100		7,475	9,150
1170	30 HP motor size	G		.47	34.043		6,625	2,100		8,725	10,500
1180	40 HP motor size	G		.47	34.043		7,925	2,100		10,025	12,000
1190	50 HP motor size	G		.37	43.243		10,500	2,650		13,150	15,600
1200	60 HP motor size	G	R-3	.39	51.282		15,800	3,125	485	19,410	22,800
1210	75 HP motor size	G		.39	51.282		14,600	3,125	485	18,210	21,500
1220	100 HP motor size	G		.35	57.143		18,500	3,500	540	22,540	26,300
1230	125 HP motor size	G		.35	57.143		19,800	3,500	540	23,840	27,800
1240	150 HP motor size	G		.35	57.143		22,500	3,500	540	26,540	30,800
1250	200 HP motor size	G		.29	68.966		25,700	4,225	650	30,575	35,400
2000	For complex & special design systems to meet specific										
2010	requirements, obtain quote from vendor.										

26 32 Packaged Generator Assemblies

26 32 13 – Engine Generators

26 32 13.16 Gas-Engine-Driven Generator Sets

		Crew	Daily Output	Labor-Hours	Unit	Material	2020 Bare Costs Labor	Equipment	Total	Total Incl O&P
0010	**GAS-ENGINE-DRIVEN GENERATOR SETS**									
0020	Gas or gasoline operated, includes battery,									
0050	charger, & muffler									
0200	7.5 kW	R-3	.83	24.096	Ea.	8,750	1,475	228	10,453	12,200
0300	11.5 kW		.71	28.169		12,400	1,725	266	14,391	16,500
0400	20 kW		.63	31.746		14,600	1,950	300	16,850	19,400
0500	35 kW		.55	36.364		17,400	2,225	345	19,970	22,900

For customer support on your Commercial Renovation Costs with RSMeans data, call 800.448.8182.

511

26 33 Battery Equipment

26 33 19 – Battery Units

26 33 19.10 Battery Units

26 33 19.10 Battery Units	Crew	Daily Output	Labor-Hours	Unit	Material	2020 Bare Costs Labor	2020 Bare Costs Equipment	Total	Total Incl O&P
0010 **BATTERY UNITS**									
0500 Salt water battery, 2.1 kWA, 24V, 750W, 30A, wired in crate, IP22 rated	1 Elec	16.25	.492	Ea.	955	30		985	1,100
0510 2.2 kWh, 48V, 800W, 17A, wired in crate, IP22 rated		16.25	.492		1,200	30		1,230	1,350
0520 25.9 kWh,48V, 11700W, 240A, wired in crate, IP2X rated		16.25	.492		14,200	30		14,230	15,600
0700 Nickel iron nife battery, deep cycle, 100Ah, 12V, 10 cells		14.25	.561		1,025	34.50		1,059.50	1,175
0710 24V, 10 cells		14.25	.561		2,175	34.50		2,209.50	2,450
0720 48V, 10 cells		14.25	.561		4,100	34.50		4,134.50	4,550
0730 Nickel iron nife battery, deep cycle, 300Ah, 12V, 10 cells		12.25	.653		3,025	40		3,065	3,375
0740 24V, 10 cells		12.25	.653		6,050	40		6,090	6,700
0750 48V, 10 cells		12.25	.653		12,700	40		12,740	14,000

26 51 Interior Lighting

26 51 13 – Interior Lighting Fixtures, Lamps, and Ballasts

26 51 13.50 Interior Lighting Fixtures

	Crew	Daily Output	Labor-Hours	Unit	Material	2020 Bare Costs Labor	2020 Bare Costs Equipment	Total	Total Incl O&P
0010 **INTERIOR LIGHTING FIXTURES** Including lamps, mounting									
0030 hardware and connections									
0100 Fluorescent, C.W. lamps, troffer, recess mounted in grid, RS									
0130 Grid ceiling mount									
0200 Acrylic lens, 1' W x 4' L, two 40 watt	1 Elec	5.70	1.404	Ea.	53.50	86		139.50	192
0300 2' W x 2' L, two U40 watt		5.70	1.404		57.50	86		143.50	196
0600 2' W x 4' L, four 40 watt		4.70	1.702		64.50	104		168.50	232
0910 Acrylic lens, 1' W x 4' L, two 32 watt T8 [G]		5.70	1.404		76	86		162	217
0930 2' W x 2' L, two U32 watt T8 [G]		5.70	1.404		115	86		201	260
0940 2' W x 4' L, two 32 watt T8 [G]		5.30	1.509		87	92.50		179.50	239
0950 2' W x 4' L, three 32 watt T8 [G]		5	1.600		80.50	98		178.50	240
0960 2' W x 4' L, four 32 watt T8 [G]		4.70	1.702		75	104		179	244
1000 Surface mounted, RS									
1030 Acrylic lens with hinged & latched door frame									
1100 1' W x 4' L, two 40 watt	1 Elec	7	1.143	Ea.	68.50	70		138.50	184
1200 2' W x 2' L, two U40 watt		7	1.143		73.50	70		143.50	189
1500 2' W x 4' L, four 40 watt		5.30	1.509		87	92.50		179.50	239
1501 2' W x 4' L, six 40 watt T8		5.20	1.538		87	94.50		181.50	241
2100 Strip fixture									
2130 Surface mounted									
2200 4' long, one 40 watt, RS	1 Elec	8.50	.941	Ea.	31.50	57.50		89	124
2300 4' long, two 40 watt, RS	"	8	1		49	61.50		110.50	149
2600 8' long, one 75 watt, SL	2 Elec	13.40	1.194		57.50	73.50		131	176
2700 8' long, two 75 watt, SL	"	12.40	1.290		71	79		150	200
3000 Strip, pendent mounted, industrial, white porcelain enamel									
3100 4' long, two 40 watt, RS	1 Elec	5.70	1.404	Ea.	53	86		139	191
3200 4' long, two 60 watt, HO	"	5	1.600		82.50	98		180.50	242
3300 8' long, two 75 watt, SL	2 Elec	8.80	1.818		101	112		213	283
4220 Metal halide, integral ballast, ceiling, recess mounted									
4230 prismatic glass lens, floating door									
4240 2' W x 2' L, 250 watt	1 Elec	3.20	2.500	Ea.	300	153		453	565
4250 2' W x 2' L, 400 watt	2 Elec	5.80	2.759		370	169		539	665
4260 Surface mounted, 2' W x 2' L, 250 watt	1 Elec	2.70	2.963		340	182		522	655
4270 400 watt	2 Elec	4.80	3.333		405	205		610	760
4280 High bay, aluminum reflector,									
4290 Single unit, 400 watt	2 Elec	4.60	3.478	Ea.	425	213		638	795
4300 Single unit, 1,000 watt		4	4		610	245		855	1,050

26 51 13 – Interior Lighting Fixtures, Lamps, and Ballasts

26 51 13.50 Interior Lighting Fixtures		Crew	Daily Output	Labor-Hours	Unit	Material	2020 Bare Costs Labor	Equipment	Total	Total Incl O&P
4310	Twin unit, 400 watt	2 Elec	3.20	5	Ea.	820	305		1,125	1,375
4320	Low bay, aluminum reflector, 250W DX lamp	1 Elec	3.20	2.500	↓	360	153		513	635
4340	High pressure sodium integral ballast ceiling, recess mounted									
4350	prismatic glass lens, floating door									
4360	2' W x 2' L, 150 watt lamp	1 Elec	3.20	2.500	Ea.	400	153		553	675
4370	2' W x 2' L, 400 watt lamp	2 Elec	5.80	2.759		475	169		644	785
4380	Surface mounted, 2' W x 2' L, 150 watt lamp	1 Elec	2.70	2.963		485	182		667	815
4390	400 watt lamp	2 Elec	4.80	3.333	↓	545	205		750	915
4400	High bay, aluminum reflector,									
4410	Single unit, 400 watt lamp	2 Elec	4.60	3.478	Ea.	390	213		603	760
4430	Single unit, 1,000 watt lamp	"	4	4		545	245		790	980
4440	Low bay, aluminum reflector, 150 watt lamp	1 Elec	3.20	2.500	↓	340	153		493	605
4450	Incandescent, high hat can, round alzak reflector, prewired									
4470	100 watt	1 Elec	8	1	Ea.	68	61.50		129.50	169
4480	150 watt		8	1		104	61.50		165.50	209
4500	300 watt	↓	6.70	1.194	↓	241	73.50		314.50	380
4600	Square glass lens with metal trim, prewired									
4630	100 watt	1 Elec	6.70	1.194	Ea.	55.50	73.50		129	174
4680	150 watt		6.70	1.194		97.50	73.50		171	220
4700	200 watt		6.70	1.194		97.50	73.50		171	220
6010	Vapor tight, incandescent, ceiling mounted, 200 watt		6.20	1.290		78.50	79		157.50	209
6100	Fluorescent, surface mounted, 2 lamps, 4' L, RS, 40 watt		3.20	2.500		119	153		272	365
6850	Vandalproof, surface mounted, fluorescent, two 32 watt T8 G		3.20	2.500		278	153		431	540
6860	Incandescent, one 150 watt		8	1		94.50	61.50		156	199
6900	Mirror light, fluorescent, RS, acrylic enclosure, two 40 watt		8	1		115	61.50		176.50	222
6910	One 40 watt		8	1		99.50	61.50		161	204
6920	One 20 watt	↓	12	.667	↓	84	41		125	156
7500	Ballast replacement, by weight of ballast, to 15' high									
7520	Indoor fluorescent, less than 2 lb.	1 Elec	10	.800	Ea.	25	49		74	103
7540	Two 40W, watt reducer, 2 to 5 lb.		9.40	.851		71.50	52		123.50	159
7560	Two F96 slimline, over 5 lb.		8	1		111	61.50		172.50	217
7580	Vaportite ballast, less than 2 lb.		9.40	.851		25	52		77	108
7600	2 lb. to 5 lb.		8.90	.899		71.50	55		126.50	164
7620	Over 5 lb.		7.60	1.053		111	64.50		175.50	222
7630	Electronic ballast for two tubes		8	1		41.50	61.50		103	140
7640	Dimmable ballast one-lamp G		8	1		105	61.50		166.50	211
7650	Dimmable ballast two-lamp G		7.60	1.053		108	64.50		172.50	219
7660	Dimmable ballast three-lamp	↓	6.90	1.159	↓	123	71		194	245
9000	Minimum labor/equipment charge		3	2.667	Job		164		164	252

26 51 13.70 Residential Fixtures

		Crew	Daily Output	Labor-Hours	Unit	Material	2020 Bare Costs Labor	Equipment	Total	Total Incl O&P
0010	**RESIDENTIAL FIXTURES**									
0400	Fluorescent, interior, surface, circline, 32 watt & 40 watt	1 Elec	20	.400	Ea.	164	24.50		188.50	218
0500	2' x 2', two U-tube 32 watt T8		8	1		118	61.50		179.50	225
0700	Shallow under cabinet, two 20 watt		16	.500		70.50	30.50		101	125
0900	Wall mounted, 4' L, two 32 watt T8, with baffle		10	.800		165	49		214	258
2000	Incandescent, exterior lantern, wall mounted, 60 watt		16	.500		61.50	30.50		92	115
2100	Post light, 150 W, with 7' post		4	2		291	123		414	510
2500	Lamp holder, weatherproof with 150 W PAR		16	.500		35.50	30.50		66	86
2550	With reflector and guard		12	.667		66	41		107	136
2600	Interior pendent, globe with shade, 150 W		20	.400	↓	190	24.50		214.50	247
9000	Minimum labor/equipment charge		4	2	Job		123		123	189

26 51 Interior Lighting

26 51 19 – LED Interior Lighting

26 51 19.10 LED Interior Lighting	Crew	Daily Output	Labor-Hours	Unit	Material	2020 Bare Costs Labor	Equipment	Total	Total Incl O&P
0010 **LED INTERIOR LIGHTING**									
7000 Interior LED fixts, tape rope kit, 120V, 3' strip, 4W, w/5' cord, 3,000K	1 Elec	24	.333	Ea.	7.25	20.50		27.75	39.50
7010 5,000K		24	.333		10.95	20.50		31.45	43.50
7020 Interior LED fixts, tape rope kit, 120V, 6' strip, 8W, w/5' cord, 3,000K		23	.348		15.35	21.50		36.85	50
7030 5,000K		23	.348		22	21.50		43.50	57.50
7040 Interior LED fixts, tape rope kit, 120V, 13' strip, 15W, w/5' cord, 3,000K		22	.364		26	22.50		48.50	63
7050 5,000K		22	.364		28.50	22.50		51	65.50
7060 Interior LED fixts, tape rope kit, 120V, 20' strip, 22W, w/5' cord, 3,000K		21	.381		41	23.50		64.50	81
7070 5,000K		21	.381		41	23.50		64.50	81
7080 Interior LED fixts, tape rope kit, 120V, 33' strip, 37W, w/5' cord, 3,000K		20	.400		60.50	24.50		85	105
7090 5,000K		20	.400		60.50	24.50		85	105
7100 Interior LED fixts, tape rope kit, 120V, 50' strip, 55W, w/5' cord, 3,000K		19	.421		91.50	26		117.50	141
7110 5,000K		19	.421		131	26		157	184
7120 Interior LED tape light, starstrand, dimmable, 12V, super star, 2", 3,000K		25	.320		11.25	19.65		30.90	42.50
7130 Ultra star, 2", 3,000K		25	.320		14.05	19.65		33.70	45.50
7140 Rainbow star, 2", 3,000K		25	.320		11.25	19.65		30.90	42.50
7150 Interior LED tape light, starstrand, dimmable, 24V, elite star, 4", 35K		25	.320		9.80	19.65		29.45	41
7160 12", 35K		25	.320		21	19.65		40.65	53.50
7170 60", 35K		23	.348		98	21.50		119.50	141
7180 120", 35K		21	.381		123	23.50		146.50	171
7190 240", 35K		18	.444		297	27.50		324.50	365
7200 Interior LED tape lighting channel , 36", AL flex, starstrand, elite star	↓	48	.167	↓	21	10.25		31.25	39

26 52 Safety Lighting

26 52 13 – Emergency and Exit Lighting

26 52 13.10 Emergency Lighting and Battery Units

	Crew	Daily Output	Labor-Hours	Unit	Material	2020 Bare Costs Labor	Equipment	Total	Total Incl O&P
0010 **EMERGENCY LIGHTING AND BATTERY UNITS**									
0300 Emergency light units, battery operated									
0350 Twin sealed beam light, 25 W, 6 V each									
0500 Lead battery operated	1 Elec	4	2	Ea.	140	123		263	345
0700 Nickel cadmium battery operated		4	2	"	330	123		453	555
9000 Minimum labor/equipment charge	↓	4	2	Job		123		123	189

26 52 13.16 Exit Signs

	Crew	Daily Output	Labor-Hours	Unit	Material	2020 Bare Costs Labor	Equipment	Total	Total Incl O&P
0010 **EXIT SIGNS**									
0080 Exit light ceiling or wall mount, incandescent, single face	1 Elec	8	1	Ea.	73	61.50		134.50	175
0100 Double face	"	6.70	1.194		50.50	73.50		124	169
1500 Exit sign, 12 V, 1 face, remote end mounted (Type 602A1)	R-19	18	1.111		73	68.50		141.50	186
1780 With emergency battery, explosion proof	"	7.70	2.597	↓	4,675	160		4,835	5,375
9000 Minimum labor/equipment charge	1 Elec	4	2	Job		123		123	189

26 55 Special Purpose Lighting

26 55 33 – Hazard Warning Lighting

26 55 33.10 Warning Beacons

	26 55 33.10 Warning Beacons	Crew	Daily Output	Labor-Hours	Unit	Material	2020 Bare Costs Labor	Equipment	Total	Total Incl O&P
0010	**WARNING BEACONS**									
0015	Surface mount with colored or clear lens									
0100	Rotating beacon									
0110	120V, 40 watt halogen	1 Elec	3.50	2.286	Ea.	107	140		247	335
0120	24V, 20 watt halogen	"	3.50	2.286	"	254	140		394	495
0200	Steady beacon									
0210	120V, 40 watt halogen	1 Elec	3.50	2.286	Ea.	109	140		249	335
0220	24V, 20 watt		3.50	2.286		113	140		253	340
0230	12V DC, incandescent	↓	3.50	2.286	↓	114	140		254	340
0300	Flashing beacon									
0310	120V, 40 watt halogen	1 Elec	3.50	2.286	Ea.	106	140		246	335
0320	24V, 20 watt halogen		3.50	2.286		107	140		247	335
0410	12V DC with two 6V lantern batteries	↓	7	1.143	↓	108	70		178	227

26 55 59 – Display Lighting

26 55 59.10 Track Lighting

	26 55 59.10 Track Lighting	Crew	Daily Output	Labor-Hours	Unit	Material	Labor	Equipment	Total	Total Incl O&P
0010	**TRACK LIGHTING**									
0080	Track, 1 circuit, 4' section	1 Elec	6.70	1.194	Ea.	43	73.50		116.50	161
0100	8' section	2 Elec	10.60	1.509		76	92.50		168.50	227
0300	3 circuits, 4' section	1 Elec	6.70	1.194		117	73.50		190.50	241
0400	8' section	2 Elec	10.60	1.509	↓	130	92.50		222.50	286
9000	Minimum labor/equipment charge	1 Elec	3	2.667	Job		164		164	252

26 55 63 – Detention Lighting

26 55 63.10 Detention Lighting Fixtures

	26 55 63.10 Detention Lighting Fixtures	Crew	Daily Output	Labor-Hours	Unit	Material	Labor	Equipment	Total	Total Incl O&P
0010	**DETENTION LIGHTING FIXTURES**									
1000	Surface mounted, cold rolled steel, 14 ga., 1' x 4'	1 Elec	5.40	1.481	Ea.	835	91		926	1,050
1010	2' x 2'		5.40	1.481		845	91		936	1,075
1020	2' x 4'		5	1.600		975	98		1,073	1,225
1100	12 ga., 1' x 4'		5.40	1.481		1,050	91		1,141	1,300
1110	2' x 2'		5.40	1.481		845	91		936	1,075
1120	2' x 4'		5	1.600		1,050	98		1,148	1,300
3000	Fluorescent vandal resistant light fixture, high abuse troffer		5.40	1.481		222	91		313	385
3010	50" L x 8" W x 4" H		5.40	1.481		196	91		287	355
3020	32W		5.40	1.481		280	91		371	450
3030	T8/32W		5.40	1.481		294	91		385	465
3040	No lamp		5.60	1.429		335	87.50		422.50	505
3050	Lamp included		5.40	1.481		385	91		476	565
3060	16 ga. cold rolled steel		5.40	1.481		310	91		401	480
3070	16 ga. cold rolled steel, ceiling/wall mount		5.40	1.481		315	91		406	485
3080	w/lamp, white		5.40	1.481		244	91		335	410
3090	wo/lamp, bronze		5.60	1.429		223	87.50		310.50	380
3100	w/lamp 6" L x 4-1/2" W x 8-1/2" H		5.40	1.481		77.50	91		168.50	226
3110	w/lamp 5-7/8" W x 4-1/2" D x 8-1/2" H clear		5.40	1.481		61	91		152	207
3120	w/lamp white		5.40	1.481		495	91		586	680
3130	Tall wallpack, w/lamp, bronze		5.40	1.481		117	91		208	268
3140	Tall wallpack, w/lamp, white	↓	5.40	1.481	↓	117	91		208	269

26 56 Exterior Lighting

26 56 19 - LED Exterior Lighting

26 56 19.55 Roadway LED Luminaire

		Crew	Daily Output	Labor-Hours	Unit	Material	2020 Bare Costs Labor	2020 Bare Costs Equipment	Total	Total Incl O&P
0010	**ROADWAY LED LUMINAIRE**									
2000	Exterior fixtures, LED roadway, type 2, 50W, neutral slipfitter bronze	2 Elec	5.20	3.077	Ea.	660	189		849	1,025
2010	3, 50W, neutral slipfitter bronze		5.20	3.077		635	189		824	990
2020	4, 50W, neutral slipfitter bronze		5.20	3.077		635	189		824	990
2030	Exterior fixtures, LED roadway, type 2, 78W, neutral slipfitter bronze		5.20	3.077		630	189		819	980
2040	3, 78W, neutral slipfitter bronze		5.20	3.077		675	189		864	1,025
2050	4, 78W, neutral slipfitter bronze		5.20	3.077		660	189		849	1,025
2060	Exterior fixtures, LED roadway, type 2, 105W, neutral slipfitter bronze		5.20	3.077		820	189		1,009	1,200
2070	3, 105W, neutral slipfitter bronze		5.20	3.077		785	189		974	1,150
2080	4, 105W, neutral slipfitter bronze		5.20	3.077		730	189		919	1,100
2090	Exterior fixtures, LED roadway, type 2, 125W, neutral slipfitter bronze		5.20	3.077		790	189		979	1,150

26 56 23 - Area Lighting

26 56 23.10 Exterior Fixtures

		Crew	Daily Output	Labor-Hours	Unit	Material	2020 Bare Costs Labor	2020 Bare Costs Equipment	Total	Total Incl O&P
0010	**EXTERIOR FIXTURES** With lamps									
0200	Wall mounted, incandescent, 100 watt	1 Elec	8	1	Ea.	51.50	61.50		113	151
0400	Quartz, 500 watt		5.30	1.509		62	92.50		154.50	212
1100	Wall pack, low pressure sodium, 35 watt		4	2		205	123		328	415
1150	55 watt		4	2		241	123		364	455
1160	High pressure sodium, 70 watt		4	2		197	123		320	405
1170	150 watt		4	2		218	123		341	430
1175	High pressure sodium, 250 watt		4	2		218	123		341	430
1180	Metal halide, 175 watt		4	2		221	123		344	435
1190	250 watt		4	2		251	123		374	465
1195	400 watt		4	2		410	123		533	645
1250	Induction lamp, 40 watt		4	2		495	123		618	735
1260	80 watt		4	2		590	123		713	835
1278	LED, poly lens, 26 watt		4	2		310	123		433	530
1280	110 watt		4	2		830	123		953	1,100
1500	LED, glass lens, 13 watt		4	2		310	123		433	530

26 56 33 - Walkway Lighting

26 56 33.10 Walkway Luminaire

		Crew	Daily Output	Labor-Hours	Unit	Material	2020 Bare Costs Labor	2020 Bare Costs Equipment	Total	Total Incl O&P
0010	**WALKWAY LUMINAIRE**									
9000	Minimum labor/equipment charge	1 Elec	3.75	2.133	Job		131		131	201

26 56 33.55 Walkway LED Luminaire

		Crew	Daily Output	Labor-Hours	Unit	Material	2020 Bare Costs Labor	2020 Bare Costs Equipment	Total	Total Incl O&P
0010	**WALKWAY LED LUMINAIRE**									
0600	LED bollard pole, 6" diam. x 36" H, cast alum, surf. mtd., 8W, 6,000K, 120V	1 Elec	8.50	.941	Ea.	385	57.50		442.50	515
0610	6" diam. x 42" H, cast alum, surf. mtd., 12W, 5,100K, 120V		8.50	.941		565	57.50		622.50	710
0620	6" diam. x 42" H, cast alum, surf. mtd., 18W, 5,100K, 120V		8.50	.941		645	57.50		702.50	800
0630	6" diam. x 42" H, cast alum, surf. mtd., 24W, 5,100K, 120V		8.50	.941		505	57.50		562.50	645
0640	9.5" square x 42" H, cast resin, surf. mtd., A19, 9W, 120V		8.50	.941		610	57.50		667.50	760
0650	9.75" diam. x 42" H, cast resin, side mount, dome, 9W, 120V		8.50	.941		695	57.50		752.50	855
0660	9.75" diam. x 43" H, cast resin, burial base, dome, 9W, 120V		8.50	.941		695	57.50		752.50	855
0670	8" diam. x 36" H, concrete security base, w/rebar, 8W, HID		8	1		1,025	61.50		1,086.50	1,225
0680	10" diam. x 49" H, concrete base, w/Sch 40 Steel pipe, 8W		6.75	1.185		1,175	72.50		1,247.50	1,375
0690	12" diam. x 48" H, concrete base, w/Sch 40 Steel pipe, 8W		6.75	1.185		1,325	72.50		1,397.50	1,575
0700	8" diam. x 36" H, concrete, cast aluminum dome, 90 degree, 4.1W		7	1.143		1,025	70		1,095	1,250
0710	8" diam. x 36" H, concrete, cast aluminum dome, 180 degree, 4.1W		7	1.143		1,275	70		1,345	1,500
0720	8" diam. x 48" H, concrete, cast aluminum dome, 360 degree, 4.1W		7	1.143		1,575	70		1,645	1,825

26 56 Exterior Lighting

26 56 36 – Flood Lighting

26 56 36.20 Floodlights	Crew	Daily Output	Labor-Hours	Unit	Material	2020 Bare Costs Labor	Equipment	Total	Total Incl O&P
0010	**FLOODLIGHTS** with ballast and lamp,									
1290	floor mtd, mount with swivel bracket									
1400	Pole mounted, pole not included									
2250	Low pressure sodium, 55 watt	1 Elec	2.70	2.963	Ea.	460	182		642	785
2270	90 watt	"	2	4	"	620	245		865	1,050

26 61 Lighting Systems and Accessories

26 61 23 – Lamps Applications

26 61 23.10 Lamps

		Crew	Daily Output	Labor-Hours	Unit	Material	2020 Bare Costs Labor	Equipment	Total	Total Incl O&P
0010	**LAMPS**									
0080	Fluorescent, rapid start, cool white, 2' long, 20 watt	1 Elec	1	8	C	330	490		820	1,125
0100	4' long, 40 watt		.90	8.889		276	545		821	1,150
0170	4' long, 34 watt energy saver [G]		.90	8.889		790	545		1,335	1,700
0176	2' long, T8, 17 watt energy saver [G]		1	8		350	490		840	1,150
0178	3' long, T8, 25 watt energy saver [G]		.90	8.889		370	545		915	1,250
0180	4' long, T8, 32 watt energy saver [G]		.90	8.889		215	545		760	1,075
0560	Twin tube compact lamp [G]		.90	8.889		435	545		980	1,325
0570	Double twin tube compact lamp [G]		.80	10		950	615		1,565	2,000
0600	Mercury vapor, mogul base, deluxe white, 100 watt		.30	26.667		5,150	1,625		6,775	8,200
0800	400 watt		.30	26.667		4,575	1,625		6,200	7,575
1000	Metal halide, mogul base, 175 watt		.30	26.667		1,050	1,625		2,675	3,675
1100	250 watt		.30	26.667		1,650	1,625		3,275	4,325
1350	High pressure sodium, 70 watt		.30	26.667		1,700	1,625		3,325	4,400
1370	150 watt		.30	26.667		1,625	1,625		3,250	4,325
1500	Low pressure sodium, 35 watt		.30	26.667		15,900	1,625		17,525	19,900
1600	90 watt		.30	26.667		6,375	1,625		8,000	9,525
1800	Incandescent, interior, A21, 100 watt		1.60	5		2,700	305		3,005	3,450
1900	A21, 150 watt		1.60	5		16,000	305		16,305	18,100
2300	R30, 75 watt		1.30	6.154		695	380		1,075	1,350
2500	Exterior, PAR 38, 75 watt		1.30	6.154		1,825	380		2,205	2,575
2600	PAR 38, 150 watt		1.30	6.154		2,050	380		2,430	2,825
9000	Minimum labor/equipment charge		4	2	Job		123		123	189

For customer support on your Commercial Renovation Costs with RSMeans data, call 800.448.8182.

517

Division Notes

	CREW	DAILY OUTPUT	LABOR-HOURS	UNIT	BARE COSTS				TOTAL INCL O&P
					MAT.	LABOR	EQUIP.	TOTAL	

Estimating Tips
27 20 00 Data Communications
27 30 00 Voice Communications
27 40 00 Audio-Video Communications

- When estimating material costs for special systems, it is always prudent to obtain manufacturers' quotations for equipment prices and special installation requirements that may affect the total cost.

- For cost modifications for elevated tray installation, add the percentages to labor according to the height of the installation and only to the quantities exceeding the different height levels, not to the total tray quantities. Refer to subdivision 26 01 02.20 for labor adjustment factors.

- Do not overlook the costs for equipment used in the installation. If scissor lifts and boom lifts are available in the field, contractors may use them in lieu of the proposed ladders and rolling staging.

Reference Numbers
Reference numbers are shown at the beginning of some major classifications. These numbers refer to related items in the Reference Section. The reference information may be an estimating procedure, an alternate pricing method, or technical information.

Note: Not all subdivisions listed here necessarily appear. ∎

Same Data. Simplified.

Enjoy the convenience and efficiency of accessing your costs anywhere:

- **Skip the multiplier** by setting your location
- **Quickly search,** edit, favorite and share costs
- **Stay on top of price changes** with automatic updates

Discover more at rsmeans.com/online

27 13 23.13 Communications Optical Fiber	Crew	Daily Output	Labor-Hours	Unit	Material	2020 Bare Costs Labor	2020 Bare Costs Equipment	Total	Total Incl O&P
0010 **COMMUNICATIONS OPTICAL FIBER**									
0040 Specialized tools & techniques cause installation costs to vary.									
0070 Fiber optic, cable, bulk simplex, single mode	1 Elec	8	1	C.L.F.	23	61.50		84.50	120
0080 Multi mode		8	1		30	61.50		91.50	128
0090 4 strand, single mode		7.34	1.090		38.50	67		105.50	146
0095 Multi mode		7.34	1.090		50.50	67		117.50	159
0100 12 strand, single mode		6.67	1.199		86	73.50		159.50	208
0105 Multi mode		6.67	1.199		99	73.50		172.50	222
0150 Jumper				Ea.	34.50			34.50	38
0200 Pigtail					36			36	40
0300 Connector	1 Elec	24	.333		26.50	20.50		47	60.50
0350 Finger splice		32	.250		37.50	15.35		52.85	65
0400 Transceiver (low cost bi-directional)		8	1		455	61.50		516.50	600
0450 Rack housing, 4 rack spaces, 12 panels (144 fibers)		2	4		615	245		860	1,050
0500 Patch panel, 12 ports		6	1.333		295	82		377	450
0600 Cable connector panel, 6 fiber		6	1.333		295	82		377	450
1000 Cable, 62.5 microns, direct burial, 4 fiber	R-15	1200	.040	L.F.	.93	2.40	.23	3.56	4.99
1020 Indoor, 2 fiber	R-19	1000	.020		.44	1.23		1.67	2.37
1040 Outdoor, aerial/duct	"	1670	.012		.65	.74		1.39	1.85
1060 50 microns, direct burial, 8 fiber	R-22	4000	.009		1.30	.52		1.82	2.24
1080 12 fiber		4000	.009		2.43	.52		2.95	3.48
1100 Indoor, 12 fiber		759	.049		2.01	2.76		4.77	6.45
1120 Connectors, 62.5 micron cable, transmission	R-19	40	.500	Ea.	14.70	30.50		45.20	63.50
1140 Cable splice		40	.500		17.50	30.50		48	67
1160 125 micron cable, transmission		16	1.250		15.45	77		92.45	135
1180 Receiver, 1.2 mile range		20	1		282	61.50		343.50	405
1200 1.9 mile range		20	1		249	61.50		310.50	370
1220 6.2 mile range		5	4		325	246		571	740
1240 Transmitter, 1.2 mile range		20	1		315	61.50		376.50	440
1260 1.9 mile range		20	1		284	61.50		345.50	405
1280 6.2 mile range		5	4		435	246		681	855
1300 Modem, 1.2 mile range		5	4		183	246		429	580
1320 6.2 mile range		5	4		325	246		571	735
1340 1.9 mile range, 12 channel		5	4		2,075	246		2,321	2,650
1360 Repeater, 1.2 mile range		10	2		370	123		493	595
1380 1.9 mile range		10	2		495	123		618	730
1400 6.2 mile range		5	4		910	246		1,156	1,375
1420 1.2 mile range, digital		5	4		455	246		701	880
1510 1/4", black, non-metallic, flexible, tube, liquid tight		8	2.500	C.L.F.	136	154		290	385
1520 3/8"		8	2.500		110	154		264	360
1530 1/2"		7.34	2.725		127	167		294	395
1540 3/4"		7.34	2.725		228	167		395	510
1550 1"		7.34	2.725		271	167		438	555
1560 1-1/4"		6.67	2.999		435	184		619	765
1570 1-1/2"		6.67	2.999		500	184		684	835
1580 2"		6.35	3.150		975	194		1,169	1,375
2040 Fiber optic cable, 48 strand, single mode, steel armor, material		1.67	11.976	M.L.F.	4,700	735		5,435	6,300

27 15 01 – Communications Horizontal Cabling Applications

27 15 01.23 Audio-Video Communications Horizontal Cabling	Crew	Daily Output	Labor-Hours	Unit	Material	2020 Bare Costs Labor	Equipment	Total	Total Incl O&P
0010	**AUDIO-VIDEO COMMUNICATIONS HORIZONTAL CABLING**									
3000	S-video cable, MD4, 1M Pro interconnects, 3'	1 Elec	38	.211	Ea.	34.50	12.90		47.40	58
3010	6'		38	.211		28.50	12.90		41.40	51.50
3020	9'		38	.211		45.50	12.90		58.40	70
3030	16'		36	.222		49	13.65		62.65	75
3040	22'		36	.222		57.50	13.65		71.15	84
3300	HDMI cable, coaxial, HDMI/HDMI, Non plenum, 3', 24 kt terminations		40	.200		13.20	12.25		25.45	33.50
3310	6', 24 kt terminations		40	.200		15.40	12.25		27.65	36
3320	HDMI cable, coaxial, HDMI/HDMI metal, Non plenum, 3', 24 kt terminations		40	.200		15.05	12.25		27.30	35.50
3330	6', 24 kt terminations		40	.200		19.65	12.25		31.90	40.50
3340	15', 24 kt terminations		38	.211		27.50	12.90		40.40	50
3350	25', 24 kt terminations		38	.211		35.50	12.90		48.40	59.50

27 15 10 – Special Communications Cabling

27 15 10.23 Sound and Video Cables and Fittings

		Crew	Daily Output	Labor-Hours	Unit	Material	Labor	Equipment	Total	Total Incl O&P
0010	**SOUND AND VIDEO CABLES & FITTINGS**									
1250	Nonshielded, #22-2 conductor	1 Elec	10	.800	C.L.F.	11.85	49		60.85	88.50

27 15 13 – Communications Copper Horizontal Cabling

27 15 13.13 Communication Cables and Fittings

		Crew	Daily Output	Labor-Hours	Unit	Material	Labor	Equipment	Total	Total Incl O&P
0010	**COMMUNICATION CABLES AND FITTINGS**									
5000	High performance unshielded twisted pair (UTP)									
7000	Category 5, #24, 4 pair solid, PVC jacket	1 Elec	7	1.143	C.L.F.	9.60	70		79.60	119
7100	4 pair solid, plenum		7	1.143		14.60	70		84.60	124
7200	4 pair stranded, PVC jacket		7	1.143		25	70		95	136
7210	Category 5e, #24, 4 pair solid, PVC jacket		7	1.143		15.60	70		85.60	125
7212	4 pair solid, plenum		7	1.143		21.50	70		91.50	132
7214	4 pair stranded, PVC jacket		7	1.143		22	70		92	132
7240	Category 6, #24, 4 pair solid, PVC jacket		7	1.143		16.65	70		86.65	126
7242	4 pair solid, plenum		7	1.143		14	70		84	123
7244	4 pair stranded, PVC jacket		7	1.143		15.45	70		85.45	125
7300	Connector, RJ45, category 5		80	.100	Ea.	.76	6.15		6.91	10.30
7302	Shielded RJ45, category 5		72	.111		1.73	6.80		8.53	12.40
7310	Jack, UTP RJ45, category 3		72	.111		.44	6.80		7.24	11
7312	Category 5		65	.123		6.25	7.55		13.80	18.50
7314	Category 5e		65	.123		4.98	7.55		12.53	17.10
7316	Category 6		65	.123		3.44	7.55		10.99	15.40
7402	RJ45, green		16.50	.485		5.90	30		35.90	52.50
7408	RJ45, orange		16.50	.485		6.15	30		36.15	53
7420	5M, RJ45, 568B, cord set		16.50	.485		35.50	30		65.50	85
7422	3M, RJ45, 568B, cord set		16.50	.485		32	30		62	81.50
7424	1M, RJ45, 568B, cord set		16.50	.485		28	30		58	77
7450	M12 TO RJ45, 2M ultra-loch D-mode, cord set		15.50	.516		40	31.50		71.50	93

For customer support on your Commercial Renovation Costs with RSMeans data, call 800.448.8182.

521

27 32 Voice Communications Terminal Equipment

27 32 26 – Ring-Down Emergency Telephones

27 32 26.10 Emergency Phone Stations

	27 32 26.10 Emergency Phone Stations	Crew	Daily Output	Labor-Hours	Unit	Material	2020 Bare Costs Labor	Equipment	Total	Total Incl O&P
0010	**EMERGENCY PHONE STATIONS**									
0100	Hands free emergency speaker phone w/LED strobe	1 Elec	7.22	1.108	Ea.	860	68		928	1,050
0110	Wall/pole mounted emergency phone station w/LED strobe	"	4.60	1.739		1,525	107		1,632	1,850
2000	Call station, 30", pole mounted emergency phone, 120V/ VoIP, blue beacon	2 Elec	5	3.200		1,100	196		1,296	1,500
2010	Call station, 36"		5	3.200		1,575	196		1,771	2,025
2020	Call station, 5'-8"		4.80	3.333		2,650	205		2,855	3,250
2030	Call station, 9'		4.80	3.333		2,675	205		2,880	3,275

27 41 Audio-Video Systems

27 41 33 – Master Antenna Television Systems

27 41 33.10 TV Systems

		Crew	Daily Output	Labor-Hours	Unit	Material	2020 Bare Costs Labor	Equipment	Total	Total Incl O&P
0010	**TV SYSTEMS,** not including rough-in wires, cables & conduits									
5000	Antenna, small	1 Elec	6	1.333	Ea.	54.50	82		136.50	186
5100	Large	"	4	2	"	230	123		353	440

27 51 Distributed Audio-Video Communications Systems

27 51 16 – Public Address Systems

27 51 16.10 Public Address System

		Crew	Daily Output	Labor-Hours	Unit	Material	2020 Bare Costs Labor	Equipment	Total	Total Incl O&P
0010	**PUBLIC ADDRESS SYSTEM**									
0100	Conventional, office	1 Elec	5.33	1.501	Speaker	171	92		263	330
0200	Industrial		2.70	2.963	"	330	182		512	640
9000	Minimum labor/equipment charge		3.50	2.286	Job		140		140	216

27 51 19 – Sound Masking Systems

27 51 19.10 Sound System

		Crew	Daily Output	Labor-Hours	Unit	Material	2020 Bare Costs Labor	Equipment	Total	Total Incl O&P
0010	**SOUND SYSTEM,** not including rough-in wires, cables & conduits									
2000	Intercom, 30 station capacity, master station	2 Elec	2	8	Ea.	2,400	490		2,890	3,400
2020	10 station capacity	"	4	4		1,475	245		1,720	2,000
3600	House telephone, talking station	1 Elec	1.60	5		625	305		930	1,150
3800	Press to talk, release to listen	"	5.30	1.509		145	92.50		237.50	300
4000	System-on button					87			87	95.50
4200	Door release	1 Elec	4	2		155	123		278	360
4400	Combination speaker and microphone		8	1		265	61.50		326.50	385
4600	Termination box		3.20	2.500		83.50	153		236.50	330
4800	Amplifier or power supply		5.30	1.509		955	92.50		1,047.50	1,200
5000	Vestibule door unit		16	.500	Name	161	30.50		191.50	224
5200	Strip cabinet		27	.296	Ea.	330	18.20		348.20	395
5400	Directory		16	.500	"	156	30.50		186.50	218
9000	Minimum labor/equipment charge		3.50	2.286	Job		140		140	216

Estimating Tips

- When estimating material costs for electronic safety and security systems, it is always prudent to obtain manufacturers' quotations for equipment prices and special installation requirements that may affect the total cost.

- Fire alarm systems consist of control panels, annunciator panels, batteries with rack, charger, and fire alarm actuating and indicating devices. Some fire alarm systems include speakers, telephone lines, door closer controls, and other components. Be careful not to overlook the costs related to installation for these items. Also be aware of costs for integrated automation instrumentation and terminal devices, control equipment, control wiring, and programming. Insurance underwriters may have specific requirements for the type of materials to be installed or design requirements based on the hazard to be protected. Local jurisdictions may have requirements not covered by code. It is advisable to be aware of any special conditions.

- Security equipment includes items such as CCTV, access control, and other detection and identification systems to perform alert and alarm functions. Be sure to consider the costs related to installation for this security equipment, such as for integrated automation instrumentation and terminal devices, control equipment, control wiring, and programming.

Reference Numbers

Reference numbers are shown at the beginning of some major classifications. These numbers refer to related items in the Reference Section. The reference information may be an estimating procedure, an alternate pricing method, or technical information.

Same Data. Simplified.

Enjoy the convenience and efficiency of accessing your costs anywhere:

- **Skip the multiplier** by setting your location
- **Quickly search,** edit, favorite and share costs
- **Stay on top of price changes** with automatic updates

Discover more at rsmeans.com/online

---

28 15 Access Control Hardware Devices

28 15 11 – Integrated Credential Readers and Field Entry Management

28 15 11.11 Standard Card Readers

28 15 11.11 Standard Card Readers	Crew	Daily Output	Labor-Hours	Unit	Material	2020 Bare Costs Labor	Equipment	Total	Total Incl O&P
0010 **STANDARD CARD READERS**									
0015 Card key access									
0020 Computerized system, processor, proximity reader and cards									
0030 Does not include door hardware, lockset or wiring									
0040 Card key system for 1 door				Ea.	1,525			1,525	1,675
0060 Card key system for 2 doors					2,350			2,350	2,600
0080 Card key system for 4 doors					2,500			2,500	2,750
0100 Processor for card key access system					1,175			1,175	1,300
0160 Magnetic lock for electric access, 600 pound holding force					185			185	203
0170 Magnetic lock for electric access, 1200 pound holding force					185			185	203
0200 Proximity card reader					188			188	206

28 15 11.15 Biometric Identity Devices

28 15 11.15 Biometric Identity Devices	Crew	Daily Output	Labor-Hours	Unit	Material	2020 Bare Costs Labor	Equipment	Total	Total Incl O&P
0010 **BIOMETRIC IDENTITY DEVICES**									
0220 Hand geometry scanner, mem of 512 users, excl. striker/power	1 Elec	3	2.667	Ea.	2,450	164		2,614	2,950
0230 Memory upgrade for, adds 9,700 user profiles		8	1		410	61.50		471.50	545
0240 Adds 32,500 user profiles		8	1		695	61.50		756.50	860
0250 Prison type, memory of 256 users, excl. striker/power		3	2.667		2,775	164		2,939	3,300
0260 Memory upgrade for, adds 3,300 user profiles		8	1		385	61.50		446.50	520
0270 Adds 9,700 user profiles		8	1		565	61.50		626.50	720
0280 Adds 27,900 user profiles		8	1		755	61.50		816.50	925
0290 All weather, mem of 512 users, excl. striker/power		3	2.667		4,375	164		4,539	5,050
0300 Facial & fingerprint scanner, combination unit, excl. striker/power		3	2.667		930	164		1,094	1,275
0310 Access for, for initial setup, excl. striker/power		3	2.667		1,350	164		1,514	1,725

28 18 Security Access Detection Equipment

28 18 11 – Security Access Metal Detectors

28 18 11.13 Security Access Metal Detectors

28 18 11.13 Security Access Metal Detectors	Crew	Daily Output	Labor-Hours	Unit	Material	2020 Bare Costs Labor	Equipment	Total	Total Incl O&P
0010 **SECURITY ACCESS METAL DETECTORS**									
0240 Metal detector, hand-held, wand type, unit only				Ea.	129			129	138
0250 Metal detector, walk through portal type, single zone	1 Elec	2	4		3,750	245		3,995	4,500
0260 Multi-zone	"	2	4		4,875	245		5,120	5,725

28 18 13 – Security Access X-Ray Equipment

28 18 13.16 Security Access X-Ray Equipment

28 18 13.16 Security Access X-Ray Equipment	Crew	Daily Output	Labor-Hours	Unit	Material	2020 Bare Costs Labor	Equipment	Total	Total Incl O&P
0010 **SECURITY ACCESS X-RAY EQUIPMENT**									
0290 X-ray machine, desk top, for mail/small packages/letters	1 Elec	4	2	Ea.	3,425	123		3,548	3,975
0300 Conveyor type, incl. monitor		2	4		15,700	245		15,945	17,700
0310 Includes additional features		2	4		27,500	245		27,745	30,700
0320 X-ray machine, large unit, for airports, incl. monitor	2 Elec	1	16		39,400	980		40,380	44,800
0330 Full console	"	.50	32		67,500	1,975		69,475	77,000

28 18 15 – Security Access Explosive Detection Equipment

28 18 15.23 Security Access Explosive Detection Equipment

28 18 15.23 Security Access Explosive Detection Equipment	Crew	Daily Output	Labor-Hours	Unit	Material	2020 Bare Costs Labor	Equipment	Total	Total Incl O&P
0010 **SECURITY ACCESS EXPLOSIVE DETECTION EQUIPMENT**									
0270 Explosives detector, walk through portal type	1 Elec	2	4	Ea.	48,800	245		49,045	54,000
0280 Hand-held, battery operated				"				25,500	28,100

28 23 Video Management System

28 23 23 – Video Surveillance Systems Infrastructure

28 23 23.50 Video Surveillance Equipments	Crew	Daily Output	Labor-Hours	Unit	Material	2020 Bare Costs Labor	Equipment	Total	Total Incl O&P
0010 **VIDEO SURVEILLANCE EQUIPMENTS**									
0200 Video cameras, wireless, hidden in exit signs, clocks, etc., incl. receiver	1 Elec	3	2.667	Ea.	171	164		335	440
0210 Accessories for video recorder, single camera		3	2.667		195	164		359	465
0220 For multiple cameras		3	2.667		1,675	164		1,839	2,100
0230 Video cameras, wireless, for under vehicle searching, complete	↓	2	4	↓	15,200	245		15,445	17,100

28 31 Intrusion Detection

28 31 16 – Intrusion Detection Systems Infrastructure

28 31 16.50 Intrusion Detection

	Crew	Daily Output	Labor-Hours	Unit	Material	2020 Bare Costs Labor	Equipment	Total	Total Incl O&P
0010 **INTRUSION DETECTION**, not including wires & conduits									
0100 Burglar alarm, battery operated, mechanical trigger	1 Elec	4	2	Ea.	289	123		412	510
0200 Electrical trigger		4	2		345	123		468	570
0400 For outside key control, add		8	1		89.50	61.50		151	193
0600 For remote signaling circuitry, add		8	1		143	61.50		204.50	252
0800 Card reader, flush type, standard		2.70	2.963		725	182		907	1,075
1000 Multi-code		2.70	2.963		1,250	182		1,432	1,650
1200 Door switches, hinge switch		5.30	1.509		64.50	92.50		157	214
1400 Magnetic switch		5.30	1.509		105	92.50		197.50	259
1600 Exit control locks, horn alarm		4	2		218	123		341	430
1800 Flashing light alarm		4	2		247	123		370	460
2000 Indicating panels, 1 channel		2.70	2.963		244	182		426	550
2200 10 channel	2 Elec	3.20	5		1,175	305		1,480	1,775
2400 20 channel		2	8		2,525	490		3,015	3,525
2600 40 channel	↓	1.14	14.035		4,650	860		5,510	6,450
2800 Ultrasonic motion detector, 12 V	1 Elec	2.30	3.478		200	213		413	550
3000 Infrared photoelectric detector		4	2		133	123		256	335
3200 Passive infrared detector		4	2		237	123		360	450
3400 Glass break alarm switch		8	1		85.50	61.50		147	189
3420 Switchmats, 30" x 5'		5.30	1.509		116	92.50		208.50	271
3440 30" x 25'		4	2		212	123		335	420
3460 Police connect panel		4	2		283	123		406	500
3480 Telephone dialer		5.30	1.509		400	92.50		492.50	585
3500 Alarm bell		4	2		106	123		229	305
3520 Siren		4	2		152	123		275	355
3540 Microwave detector, 10' to 200'		2	4		585	245		830	1,025
3560 10' to 350'	↓	2	4	↓	1,775	245		2,020	2,325

28 46 Fire Detection and Alarm

28 46 11 – Fire Sensors and Detectors

28 46 11.21 Carbon-Monoxide Detection Sensors

	Crew	Daily Output	Labor-Hours	Unit	Material	2020 Bare Costs Labor	Equipment	Total	Total Incl O&P
0010 **CARBON-MONOXIDE DETECTION SENSORS**									
8400 Smoke and carbon monoxide alarm battery operated photoelectric low profile	1 Elec	24	.333	Ea.	51.50	20.50		72	88
8410 low profile photoelectric battery powered		24	.333		40	20.50		60.50	76
8420 photoelectric low profile sealed lithium		24	.333		70.50	20.50		91	109
8430 Photoelectric low profile sealed lithium smoke and CO with voice combo		24	.333		50.50	20.50		71	87
8500 Carbon monoxide sensor, wall mount 1Mod 1 relay output smoke & heat		24	.333		490	20.50		510.50	570
8700 Carbon monoxide detector, battery operated, wall mounted		16	.500		43	30.50		73.50	94.50
8710 Hardwired, wall and ceiling mounted		8	1		84.50	61.50		146	188
8720 Duct mounted	↓	8	1		350	61.50		411.50	480

28 46 Fire Detection and Alarm

28 46 11 – Fire Sensors and Detectors

28 46 11.21 Carbon-Monoxide Detection Sensors

		Crew	Daily Output	Labor-Hours	Unit	Material	2020 Bare Costs Labor	Equipment	Total	Total Incl O&P
8730	Continuous air monitoring system, PC based, incl. remote sensors, min.	2 Elec	.25	64	Ea.	31,600	3,925		35,525	40,800
8740	Maximum	"	.10	160	↓	56,500	9,825		66,325	77,000

28 46 11.27 Other Sensors

		Crew	Daily Output	Labor-Hours	Unit	Material	2020 Bare Costs Labor	Equipment	Total	Total Incl O&P
0010	**OTHER SENSORS**									
5200	Smoke detector, ceiling type	1 Elec	6.20	1.290	Ea.	120	79		199	255
5240	Smoke detector, addressable type		6	1.333		224	82		306	370
5400	Duct type		3.20	2.500		295	153		448	560
5420	Duct addressable type		3.20	2.500		520	153		673	810
8300	Smoke alarm with integrated strobe light 120 V, 16DB 60 fpm flash rate		16	.500		119	30.50		149.50	178
8310	Photoelectric smoke detector with strobe 120 V, 90 DB ceiling mount		12	.667		227	41		268	310
8320	120 V, 90 DB wall mount	↓	12	.667	↓	202	41		243	285

28 46 11.50 Fire and Heat Detectors

		Crew	Daily Output	Labor-Hours	Unit	Material	2020 Bare Costs Labor	Equipment	Total	Total Incl O&P
0010	**FIRE & HEAT DETECTORS**									
5000	Detector, rate of rise	1 Elec	8	1	Ea.	52	61.50		113.50	152
5100	Fixed temp fire alarm		7	1.143		42	70		112	155
5200	125 V SRF MT-135F		7.25	1.103		33.50	67.50		101	141
5300	MT-195F	↓	7.25	1.103	↓	34	67.50		101.50	142

28 46 20 – Fire Alarm

28 46 20.50 Alarm Panels and Devices

		Crew	Daily Output	Labor-Hours	Unit	Material	2020 Bare Costs Labor	Equipment	Total	Total Incl O&P
0010	**ALARM PANELS AND DEVICES**, not including wires & conduits									
2200	Intercom remote station	1 Elec	8	1	Ea.	89.50	61.50		151	193
2400	Intercom outlet		8	1		64.50	61.50		126	166
2600	Sound system, intercom handset		8	1		485	61.50		546.50	630
3590	Signal device, beacon		8	1		202	61.50		263.50	315
3594	2 zone	↓	1.35	5.926		310	365		675	900
3600	4 zone	2 Elec	2	8		410	490		900	1,200
3610	3 zone		2	8		640	490		1,130	1,450
3800	8 zone		1	16		875	980		1,855	2,475
3810	5 zone		1.50	10.667		700	655		1,355	1,775
3900	10 zone		1.25	12.800		1,025	785		1,810	2,350
4000	12 zone	↓	.67	23.988		2,350	1,475		3,825	4,850
4020	Alarm device, tamper, flow	1 Elec	8	1		230	61.50		291.50	350
4025	Fire alarm, loop expander card		16	.500		680	30.50		710.50	795
4200	Battery and rack		4	2		460	123		583	700
4400	Automatic charger		8	1		525	61.50		586.50	675
4600	Signal bell		8	1		66.50	61.50		128	168
4610	Fire alarm signal bell 10" red 20-24 V P		8	1		154	61.50		215.50	264
4800	Trouble buzzer or manual station		8	1		88.50	61.50		150	192
5425	Duct smoke and heat detector 2 wire		8	1		125	61.50		186.50	233
5430	Fire alarm duct detector controller		3	2.667		254	164		418	530
5435	Fire alarm duct detector sensor kit		8	1		75	61.50		136.50	177
5440	Remote test station for smoke detector duct type		5.30	1.509		54.50	92.50		147	203
5460	Remote fire alarm indicator light		5.30	1.509		25	92.50		117.50	171
5600	Strobe and horn		5.30	1.509		137	92.50		229.50	294
5610	Strobe and horn (ADA type)		5.30	1.509		165	92.50		257.50	325
5620	Visual alarm (ADA type)		6.70	1.194		119	73.50		192.50	244
5800	electric bell		6.70	1.194		54	73.50		127.50	173
6000	Door holder, electro-magnetic		4	2		92	123		215	290
6200	Combination holder and closer		3.20	2.500		146	153		299	395
6600	Drill switch		8	1		400	61.50		461.50	535
6800	Master box		2.70	2.963		6,675	182		6,857	7,600
7000	Break glass station	↓	8	1	↓	58	61.50		119.50	158

For customer support on your Commercial Renovation Costs with RSMeans data, call 800.448.8182.

28 46 Fire Detection and Alarm

28 46 20 – Fire Alarm

28 46 20.50 Alarm Panels and Devices	Crew	Daily Output	Labor-Hours	Unit	Material	2020 Bare Costs Labor	2020 Bare Costs Equipment	Total	Total Incl O&P
7800	Remote annunciator, 8 zone lamp	1 Elec	1.80	4.444	Ea.	212	273		485	655
8000	12 zone lamp	2 Elec	2.60	6.154		415	380		795	1,025
8200	16 zone lamp	"	2.20	7.273	↓	370	445		815	1,100

28 47 Mass Notification

28 47 12 – Notification Systems

28 47 12.10 Mass Notification System

		Crew	Daily Output	Labor-Hours	Unit	Material	2020 Bare Costs Labor	2020 Bare Costs Equipment	Total	Total Incl O&P
0010	**MASS NOTIFICATION SYSTEM**									
0100	Wireless command center, 10,000 devices	2 Elec	1.33	12.030	Ea.	2,325	740		3,065	3,675
0200	Option, email notification					2,100			2,100	2,325
0210	Remote device supervision & monitor					2,650			2,650	2,900
0300	Antenna VHF or UHF, for medium range	1 Elec	4	2		93	123		216	292
0310	For high-power transmitter		2	4		1,125	245		1,370	1,625
0400	Transmitter, 25 watt		4	2		1,750	123		1,873	2,125
0410	40 watt		2.66	3.008		2,025	185		2,210	2,500
0420	100 watt		1.33	6.015		6,600	370		6,970	7,825
0500	Wireless receiver/control module for speaker		8	1		286	61.50		347.50	410
0600	Desktop paging controller, stand alone		4	2		850	123		973	1,125
0700	Auxiliary level inputs	↓	8.36	.957	↓	103	58.50		161.50	204

Division Notes

		CREW	DAILY OUTPUT	LABOR-HOURS	UNIT	BARE COSTS				TOTAL INCL O&P
						MAT.	LABOR	EQUIP.	TOTAL	

Estimating Tips
31 05 00 Common Work Results for Earthwork

- Estimating the actual cost of performing earthwork requires careful consideration of the variables involved. This includes items such as type of soil, whether water will be encountered, dewatering, whether banks need bracing, disposal of excavated earth, and length of haul to fill or spoil sites, etc. If the project has large quantities of cut or fill, consider raising or lowering the site to reduce costs, while paying close attention to the effect on site drainage and utilities.
- If the project has large quantities of fill, creating a borrow pit on the site can significantly lower the costs.
- It is very important to consider what time of year the project is scheduled for completion. Bad weather can create large cost overruns from dewatering, site repair, and lost productivity from cold weather.

Reference Numbers

Reference numbers are shown at the beginning of some major classifications. These numbers refer to related items in the Reference Section. The reference information may be an estimating procedure, an alternate pricing method, or technical information.

Note: Not all subdivisions listed here necessarily appear. ■

Same Data. Simplified.

Enjoy the convenience and efficiency of accessing your costs anywhere:

- **Skip the multiplier** by setting your location
- **Quickly search,** edit, favorite and share costs
- **Stay on top of price changes** with automatic updates

Discover more at rsmeans.com/online

31 05 Common Work Results for Earthwork

31 05 13 – Soils for Earthwork

31 05 13.10 Borrow

	31 05 13.10 Borrow	Crew	Daily Output	Labor-Hours	Unit	Material	2020 Bare Costs Labor	2020 Bare Costs Equipment	Total	Total Incl O&P
0010	**BORROW**									
0020	Spread, 200 HP dozer, no compaction, 2 mile RT haul									
0200	Common borrow	B-15	600	.047	C.Y.	12.60	2.34	4.42	19.36	22.50
0700	Screened loam		600	.047		29	2.34	4.42	35.76	40.50
0800	Topsoil, weed free	↓	600	.047		26.50	2.34	4.42	33.26	37.50
0900	For 5 mile haul, add	B-34B	200	.040	↓		1.96	2.86	4.82	6.30

31 05 16 – Aggregates for Earthwork

31 05 16.10 Borrow

	31 05 16.10 Borrow	Crew	Daily Output	Labor-Hours	Unit	Material	2020 Bare Costs Labor	2020 Bare Costs Equipment	Total	Total Incl O&P
0010	**BORROW**									
0020	Spread, with 200 HP dozer, no compaction, 2 mile RT haul									
0100	Bank run gravel	B-15	600	.047	L.C.Y.	18.60	2.34	4.42	25.36	29
0300	Crushed stone (1.40 tons per C.Y.), 1-1/2"		600	.047		27.50	2.34	4.42	34.26	39
0320	3/4"		600	.047		27.50	2.34	4.42	34.26	39
0340	1/2"		600	.047		27	2.34	4.42	33.76	38
0360	3/8"		600	.047		32	2.34	4.42	38.76	44
0400	Sand, washed, concrete		600	.047		35.50	2.34	4.42	42.26	47.50
0500	Dead or bank sand		600	.047		18.25	2.34	4.42	25.01	28.50
0600	Select structural fill		600	.047		24.50	2.34	4.42	31.26	35.50
0900	For 5 mile haul, add	B-34B	200	.040	↓		1.96	2.86	4.82	6.30

31 06 Schedules for Earthwork

31 06 60 – Schedules for Special Foundations and Load Bearing Elements

31 06 60.14 Piling Special Costs

	31 06 60.14 Piling Special Costs	Crew	Daily Output	Labor-Hours	Unit	Material	2020 Bare Costs Labor	2020 Bare Costs Equipment	Total	Total Incl O&P
0010	**PILING SPECIAL COSTS**									
0011	Piling special costs, pile caps, see Section 03 30 53.40									
0500	Cutoffs, concrete piles, plain	1 Pile	5.50	1.455	Ea.		79		79	126
0600	With steel thin shell, add		38	.211			11.40		11.40	18.30
0700	Steel pile or "H" piles		19	.421	↓		23		23	36.50
0800	Wood piles	↓	38	.211			11.40		11.40	18.30
1000	Testing, any type piles, test load is twice the design load									
1050	50 ton design load, 100 ton test				Ea.				14,000	15,500
1100	100 ton design load, 200 ton test								20,000	22,000
1200	200 ton design load, 400 ton test				↓				28,000	31,000

31 06 60.15 Mobilization

	31 06 60.15 Mobilization	Crew	Daily Output	Labor-Hours	Unit	Material	2020 Bare Costs Labor	2020 Bare Costs Equipment	Total	Total Incl O&P
0010	**MOBILIZATION**									
0020	Set up & remove, air compressor, 600 CFM	A-5	3.30	5.455	Ea.		233	14.85	247.85	390
0100	1,200 CFM	"	2.20	8.182			350	22.50	372.50	585
0200	Crane, with pile leads and pile hammer, 75 ton	B-19	.60	107			5,900	3,350	9,250	13,000
0300	150 ton	"	.36	178	↓		9,825	5,575	15,400	21,800

31 14 Earth Stripping and Stockpiling

31 14 13 – Soil Stripping and Stockpiling

31 14 13.23 Topsoil Stripping and Stockpiling

		Crew	Daily Output	Labor-Hours	Unit	Material	2020 Bare Costs Labor	2020 Bare Costs Equipment	Total	Total Incl O&P
0010	**TOPSOIL STRIPPING AND STOCKPILING**									
1500	Loam or topsoil, remove/stockpile on site									
1510	By hand, 6" deep, 50' haul, less than 100 S.Y.	B-1	100	.240	S.Y.		10.25		10.25	16.45
1520	By skid steer, 6" deep, 100' haul, 101-500 S.Y.	B-62	500	.048			2.20	.36	2.56	3.88
1530	100' haul, 501-900 S.Y.	"	900	.027			1.22	.20	1.42	2.16
1540	200' haul, 901-1,100 S.Y.	B-63	1000	.040			1.77	.18	1.95	3.02
1550	By dozer, 200' haul, 1,101-4,000 S.Y.	B-10B	4000	.003			.16	.38	.54	.66

31 23 Excavation and Fill

31 23 16 – Excavation

31 23 16.13 Excavating, Trench

		Crew	Daily Output	Labor-Hours	Unit	Material	2020 Bare Costs Labor	2020 Bare Costs Equipment	Total	Total Incl O&P
0010	**EXCAVATING, TRENCH**									
0011	Or continuous footing									
0020	Common earth with no sheeting or dewatering included									
1400	By hand with pick and shovel 2' to 6' deep, light soil	1 Clab	8	1	B.C.Y.		42		42	67.50
1500	Heavy soil	"	4	2	"		84		84	135
1700	For tamping backfilled trenches, air tamp, add	A-1G	100	.080	E.C.Y.		3.37	.54	3.91	6
1900	Vibrating plate, add	B-18	180	.133	"		5.70	.92	6.62	10.15
2100	Trim sides and bottom for concrete pours, common earth		1500	.016	S.F.		.68	.11	.79	1.22
2300	Hardpan		600	.040	"		1.71	.28	1.99	3.04
9000	Minimum labor/equipment charge	1 Clab	4	2	Job		84		84	135

31 23 16.14 Excavating, Utility Trench

		Crew	Daily Output	Labor-Hours	Unit	Material	2020 Bare Costs Labor	2020 Bare Costs Equipment	Total	Total Incl O&P
0010	**EXCAVATING, UTILITY TRENCH**									
0011	Common earth									
0050	Trenching with chain trencher, 12 HP, operator walking									
0100	4" wide trench, 12" deep	B-53	800	.010	L.F.		.53	.20	.73	1.05
0150	18" deep		750	.011			.57	.21	.78	1.12
0200	24" deep		700	.011			.61	.22	.83	1.20
0300	6" wide trench, 12" deep		650	.012			.65	.24	.89	1.29
0350	18" deep		600	.013			.71	.26	.97	1.40
0400	24" deep		550	.015			.77	.29	1.06	1.52
0450	36" deep		450	.018			.94	.35	1.29	1.85
0600	8" wide trench, 12" deep		475	.017			.89	.33	1.22	1.76
0650	18" deep		400	.020			1.06	.39	1.45	2.09
0700	24" deep		350	.023			1.21	.45	1.66	2.39
0750	36" deep		300	.027			1.41	.52	1.93	2.79
0900	Minimum labor/equipment charge		2	4	Job		212	78.50	290.50	415
1000	Backfill by hand including compaction, add									
1050	4" wide trench, 12" deep	A-1G	800	.010	L.F.		.42	.07	.49	.74
1100	18" deep		530	.015			.64	.10	.74	1.13
1150	24" deep		400	.020			.84	.14	.98	1.50
1300	6" wide trench, 12" deep		540	.015			.62	.10	.72	1.11
1350	18" deep		405	.020			.83	.13	.96	1.48
1400	24" deep		270	.030			1.25	.20	1.45	2.22
1450	36" deep		180	.044			1.87	.30	2.17	3.33
1600	8" wide trench, 12" deep		400	.020			.84	.14	.98	1.50
1650	18" deep		265	.030			1.27	.20	1.47	2.26
1700	24" deep		200	.040			1.68	.27	1.95	3
1750	36" deep		135	.059			2.49	.40	2.89	4.44
2000	Chain trencher, 40 HP operator riding									
2050	6" wide trench and backfill, 12" deep	B-54	1200	.007	L.F.		.35	.34	.69	.92

For customer support on your Commercial Renovation Costs with RSMeans data, call 800.448.8182.

531

31 23 16 – Excavation

31 23 16.14 Excavating, Utility Trench

		Crew	Daily Output	Labor-Hours	Unit	Material	2020 Bare Costs Labor	Equipment	Total	Total Incl O&P
2100	18" deep	B-54	1000	.008	L.F.		.42	.40	.82	1.10
2150	24" deep		975	.008			.44	.41	.85	1.13
2200	36" deep		900	.009			.47	.45	.92	1.23
2250	48" deep		750	.011			.57	.54	1.11	1.48
2300	60" deep		650	.012			.65	.62	1.27	1.70
2400	8" wide trench and backfill, 12" deep		1000	.008			.42	.40	.82	1.10
2450	18" deep		950	.008			.45	.42	.87	1.17
2500	24" deep		900	.009			.47	.45	.92	1.23
2550	36" deep		800	.010			.53	.50	1.03	1.38
2600	48" deep		650	.012			.65	.62	1.27	1.70
2700	12" wide trench and backfill, 12" deep		975	.008			.44	.41	.85	1.13
2750	18" deep		860	.009			.49	.47	.96	1.28
2800	24" deep		800	.010			.53	.50	1.03	1.38
2850	36" deep		725	.011			.58	.55	1.13	1.52
3000	16" wide trench and backfill, 12" deep		835	.010			.51	.48	.99	1.32
3050	18" deep		750	.011			.57	.54	1.11	1.48
3100	24" deep	↓	700	.011	↓		.61	.57	1.18	1.58
3200	Compaction with vibratory plate, add								35%	35%
5100	Hand excavate and trim for pipe bells after trench excavation									
5200	8" pipe	1 Clab	155	.052	L.F.		2.17		2.17	3.48
5300	18" pipe	"	130	.062	"		2.59		2.59	4.15
9000	Minimum labor/equipment charge	A-1G	4	2	Job		84	13.50	97.50	150

31 23 16.16 Structural Excavation for Minor Structures

		Crew	Daily Output	Labor-Hours	Unit	Material	2020 Bare Costs Labor	Equipment	Total	Total Incl O&P
0010	**STRUCTURAL EXCAVATION FOR MINOR STRUCTURES**									
0015	Hand, pits to 6' deep, sandy soil	1 Clab	8	1	B.C.Y.		42		42	67.50
0100	Heavy soil or clay		4	2			84		84	135
0300	Pits 6' to 12' deep, sandy soil		5	1.600			67.50		67.50	108
0500	Heavy soil or clay		3	2.667			112		112	180
0700	Pits 12' to 18' deep, sandy soil		4	2			84		84	135
0900	Heavy soil or clay		2	4			168		168	270
1100	Hand loading trucks from stock pile, sandy soil		12	.667			28		28	45
1300	Heavy soil or clay	↓	8	1	↓		42		42	67.50
1500	For wet or muck hand excavation, add to above								50%	50%
6000	Machine excavation, for spread and mat footings, elevator pits,									
6001	and small building foundations									
6030	Common earth, hydraulic backhoe, 1/2 C.Y. bucket	B-12E	55	.291	B.C.Y.		14.75	8.20	22.95	32.50
6035	3/4 C.Y. bucket	B-12F	90	.178			9	7.70	16.70	23
6040	1 C.Y. bucket	B-12A	108	.148			7.50	7.35	14.85	19.95
6050	1-1/2 C.Y. bucket	B-12B	144	.111			5.65	4.77	10.42	14.15
6060	2 C.Y. bucket	B-12C	200	.080			4.05	4.76	8.81	11.65
6070	Sand and gravel, 3/4 C.Y. bucket	B-12F	100	.160			8.10	6.95	15.05	20.50
6080	1 C.Y. bucket	B-12A	120	.133			6.75	6.65	13.40	17.95
6090	1-1/2 C.Y. bucket	B-12B	160	.100			5.05	4.30	9.35	12.75
6100	2 C.Y. bucket	B-12C	220	.073			3.68	4.32	8	10.55
6110	Clay, till, or blasted rock, 3/4 C.Y. bucket	B-12F	80	.200			10.15	8.70	18.85	25.50
6120	1 C.Y. bucket	B-12A	95	.168			8.55	8.35	16.90	22.50
6130	1-1/2 C.Y. bucket	B-12B	130	.123			6.25	5.30	11.55	15.65
6140	2 C.Y. bucket	B-12C	175	.091			4.63	5.45	10.08	13.30
6230	Sandy clay & loam, hydraulic backhoe, 1/2 C.Y. bucket	B-12E	60	.267			13.50	7.55	21.05	30
6235	3/4 C.Y. bucket	B-12F	98	.163			8.25	7.10	15.35	21
6240	1 C.Y. bucket	B-12A	116	.138			7	6.85	13.85	18.60
6250	1-1/2 C.Y. bucket	B-12B	156	.103	↓		5.20	4.41	9.61	13.05

31 23 16 – Excavation

31 23 16.16 Structural Excavation for Minor Structures	Crew	Daily Output	Labor-Hours	Unit	Material	2020 Bare Costs Labor	2020 Bare Costs Equipment	Total	Total Incl O&P
9000	Minimum labor/equipment charge	1 Clab	4	2	Job		84		84	135

31 23 16.42 Excavating, Bulk Bank Measure

		Crew	Daily Output	Labor-Hours	Unit	Material	Labor	Equipment	Total	Total Incl O&P
0010	**EXCAVATING, BULK BANK MEASURE** R312316-40									
0011	Common earth piled									
0020	For loading onto trucks, add								15%	15%
0200	Excavator, hydraulic, crawler mtd., 1 C.Y. cap. = 100 C.Y./hr. R312316-45	B-12A	800	.020	B.C.Y.		1.01	.99	2	2.69
1200	Front end loader, track mtd., 1-1/2 C.Y. cap. = 70 C.Y./hr.	B-10N	560	.021			1.11	1.01	2.12	2.86
1500	Wheel mounted, 3/4 C.Y. cap. = 45 C.Y./hr.	B-10R	360	.033			1.73	.84	2.57	3.64
5000	Excavating, bulk bank measure, sandy clay & loam piled									
5020	For loading onto trucks, add								15%	15%
5100	Excavator, hydraulic, crawler mtd., 1 C.Y. cap. = 120 C.Y./hr.	B-12A	960	.017	B.C.Y.		.84	.83	1.67	2.24
9000	Minimum labor/equipment charge	B-10L	2	6	Job		310	201	511	710

31 23 16.46 Excavating, Bulk, Dozer

		Crew	Daily Output	Labor-Hours	Unit	Material	Labor	Equipment	Total	Total Incl O&P
0010	**EXCAVATING, BULK, DOZER**									
0011	Open site									
2000	80 HP, 50' haul, sand & gravel	B-10L	460	.026	B.C.Y.		1.35	.87	2.22	3.09
2200	150' haul, sand & gravel		230	.052			2.71	1.75	4.46	6.20
2400	300' haul, sand & gravel		120	.100			5.20	3.35	8.55	11.85
3000	105 HP, 50' haul, sand & gravel	B-10W	700	.017			.89	.91	1.80	2.40
3200	150' haul, sand & gravel		310	.039			2.01	2.04	4.05	5.40
3300	300' haul, sand & gravel		140	.086			4.45	4.53	8.98	12
4000	200 HP, 50' haul, sand & gravel	B-10B	1400	.009			.44	1.07	1.51	1.88
4200	150' haul, sand & gravel		595	.020			1.05	2.53	3.58	4.43
4400	300' haul, sand & gravel		310	.039			2.01	4.85	6.86	8.50

31 23 23 – Fill

31 23 23.13 Backfill

		Crew	Daily Output	Labor-Hours	Unit	Material	Labor	Equipment	Total	Total Incl O&P
0010	**BACKFILL** R312323-30									
0015	By hand, no compaction, light soil	1 Clab	14	.571	L.C.Y.		24		24	38.50
0100	Heavy soil		11	.727	"		30.50		30.50	49
0300	Compaction in 6" layers, hand tamp, add to above		20.60	.388	E.C.Y.		16.35		16.35	26
0400	Roller compaction operator walking, add	B-10A	100	.120			6.20	1.66	7.86	11.65
0500	Air tamp, add	B-9D	190	.211			8.95	1.67	10.62	16.20
0600	Vibrating plate, add	A-1D	60	.133			5.60	.53	6.13	9.60
0800	Compaction in 12" layers, hand tamp, add to above	1 Clab	34	.235			9.90		9.90	15.85
0900	Roller compaction operator walking, add	B-10A	150	.080			4.15	1.11	5.26	7.75
1000	Air tamp, add	B-9	285	.140			5.95	1.25	7.20	10.95
1100	Vibrating plate, add	A-1E	90	.089			3.74	1.84	5.58	8
1200	Trench, dozer, no compaction, 60 HP	B-10L	425	.028	L.C.Y.		1.46	.94	2.40	3.35
1300	Dozer backfilling, bulk, up to 300' haul, no compaction	B-10B	1200	.010	"		.52	1.25	1.77	2.20
1400	Air tamped, add	B-11B	80	.200	E.C.Y.		9.50	4.97	14.47	20.50
1900	Dozer backfilling, trench, up to 300' haul, no compaction	B-10B	900	.013	L.C.Y.		.69	1.67	2.36	2.93
2000	Air tamped, add	B-11B	80	.200	E.C.Y.		9.50	4.97	14.47	20.50
2350	Spreading in 8" layers, small dozer	B-10B	1060	.011	L.C.Y.		.59	1.42	2.01	2.48
2450	Compacting with vibrating plate, 8" lifts	A-1D	73	.110	E.C.Y.		4.61	.43	5.04	7.90

31 23 23.14 Backfill, Structural

		Crew	Daily Output	Labor-Hours	Unit	Material	Labor	Equipment	Total	Total Incl O&P
0010	**BACKFILL, STRUCTURAL**									
0011	Dozer or F.E. loader									
0020	From existing stockpile, no compaction									
1000	55 HP wheeled loader, 50' haul, common earth	B-11C	200	.080	L.C.Y.		3.95	1.07	5.02	7.45
2000	80 HP, 50' haul, sand & gravel	B-10L	1100	.011			.57	.36	.93	1.29
2010	Sandy clay & loam		1070	.011			.58	.38	.96	1.33

31 23 23.14 Backfill, Structural

		Crew	Daily Output	Labor-Hours	Unit	Material	2020 Bare Costs Labor	Equipment	Total	Total Incl O&P
2020	Common earth	B-10L	975	.012	L.C.Y.		.64	.41	1.05	1.46
2040	Clay		850	.014			.73	.47	1.20	1.67
2400	300' haul, sand & gravel		370	.032			1.68	1.08	2.76	3.84
2410	Sandy clay & loam		360	.033			1.73	1.11	2.84	3.95
2420	Common earth		330	.036			1.89	1.22	3.11	4.31
2440	Clay		290	.041			2.15	1.38	3.53	4.90
3000	105 HP, 50' haul, sand & gravel	B-10W	1350	.009			.46	.47	.93	1.25
3010	Sandy clay & loam		1325	.009			.47	.48	.95	1.27
3020	Common earth		1225	.010			.51	.52	1.03	1.37
3040	Clay		1100	.011			.57	.58	1.15	1.52
3300	300' haul, sand & gravel		465	.026			1.34	1.36	2.70	3.61
3310	Sandy clay & loam		455	.026			1.37	1.39	2.76	3.68
3320	Common earth		415	.029			1.50	1.53	3.03	4.04
3340	Clay		370	.032			1.68	1.71	3.39	4.53

31 23 23.16 Fill By Borrow and Utility Bedding

		Crew	Daily Output	Labor-Hours	Unit	Material	2020 Bare Costs Labor	Equipment	Total	Total Incl O&P
0010	**FILL BY BORROW AND UTILITY BEDDING**									
0049	Utility bedding, for pipe & conduit, not incl. compaction									
0050	Crushed or screened bank run gravel	B-6	150	.160	L.C.Y.	21	7.30	1.43	29.73	36
0100	Crushed stone 3/4" to 1/2"		150	.160		27.50	7.30	1.43	36.23	43.50
0200	Sand, dead or bank		150	.160		18.25	7.30	1.43	26.98	33
0500	Compacting bedding in trench	A-1D	90	.089	E.C.Y.		3.74	.35	4.09	6.40
0600	If material source exceeds 2 miles, add for extra mileage.									
0610	See Section 31 23 23.20 for hauling mileage add.									

31 23 23.17 General Fill

		Crew	Daily Output	Labor-Hours	Unit	Material	2020 Bare Costs Labor	Equipment	Total	Total Incl O&P
0010	**GENERAL FILL**									
0011	Spread dumped material, no compaction									
0020	By dozer	B-10B	1000	.012	L.C.Y.		.62	1.50	2.12	2.63
0100	By hand	1 Clab	12	.667	"		28		28	45
9000	Minimum labor/equipment charge	"	4	2	Job		84		84	135

31 23 23.20 Hauling

		Crew	Daily Output	Labor-Hours	Unit	Material	2020 Bare Costs Labor	Equipment	Total	Total Incl O&P
0010	**HAULING**									
0011	Excavated or borrow, loose cubic yards									
0012	no loading equipment, including hauling, waiting, loading/dumping									
0013	time per cycle (wait, load, travel, unload or dump & return)									
0014	8 C.Y. truck, 15 MPH avg., cycle 0.5 miles, 10 min. wait/ld./uld.	B-34A	320	.025	L.C.Y.		1.22	1.34	2.56	3.43
0016	cycle 1 mile		272	.029			1.44	1.57	3.01	4.03
0018	cycle 2 miles		208	.038			1.88	2.06	3.94	5.25
0020	cycle 4 miles		144	.056			2.72	2.97	5.69	7.60
0022	cycle 6 miles		112	.071			3.50	3.82	7.32	9.80
0024	cycle 8 miles		88	.091			4.45	4.86	9.31	12.45
0026	20 MPH avg., cycle 0.5 mile		336	.024			1.17	1.27	2.44	3.26
0028	cycle 1 mile		296	.027			1.32	1.45	2.77	3.71
0030	cycle 2 miles		240	.033			1.63	1.78	3.41	4.57
0032	cycle 4 miles		176	.045			2.22	2.43	4.65	6.25
0034	cycle 6 miles		136	.059			2.88	3.15	6.03	8.05
0036	cycle 8 miles		112	.071			3.50	3.82	7.32	9.80
0044	25 MPH avg., cycle 4 miles		192	.042			2.04	2.23	4.27	5.70
0046	cycle 6 miles		160	.050			2.45	2.68	5.13	6.85
0048	cycle 8 miles		128	.063			3.06	3.34	6.40	8.55
0050	30 MPH avg., cycle 4 miles		216	.037			1.81	1.98	3.79	5.10
0052	cycle 6 miles		176	.045			2.22	2.43	4.65	6.25
0054	cycle 8 miles		144	.056			2.72	2.97	5.69	7.60

31 23 23.20 Hauling

	Crew	Daily Output	Labor-Hours	Unit	Material	2020 Bare Costs Labor	2020 Bare Costs Equipment	Total	Total Incl O&P
0114	15 MPH avg., cycle 0.5 mile, 15 min. wait/ld./uld.	B-34A	224	.036	L.C.Y.		1.75	1.91	3.66	4.89
0116	cycle 1 mile		200	.040			1.96	2.14	4.10	5.50
0118	cycle 2 miles		168	.048			2.33	2.55	4.88	6.55
0120	cycle 4 miles		120	.067			3.26	3.57	6.83	9.10
0122	cycle 6 miles		96	.083			4.08	4.46	8.54	11.40
0124	cycle 8 miles		80	.100			4.90	5.35	10.25	13.75
0126	20 MPH avg., cycle 0.5 mile		232	.034			1.69	1.85	3.54	4.73
0128	cycle 1 mile		208	.038			1.88	2.06	3.94	5.25
0130	cycle 2 miles		184	.043			2.13	2.33	4.46	5.95
0132	cycle 4 miles		144	.056			2.72	2.97	5.69	7.60
0134	cycle 6 miles		112	.071			3.50	3.82	7.32	9.80
0136	cycle 8 miles		96	.083			4.08	4.46	8.54	11.40
0144	25 MPH avg., cycle 4 miles		152	.053			2.58	2.82	5.40	7.20
0146	cycle 6 miles		128	.063			3.06	3.34	6.40	8.55
0148	cycle 8 miles		112	.071			3.50	3.82	7.32	9.80
0150	30 MPH avg., cycle 4 miles		168	.048			2.33	2.55	4.88	6.55
0152	cycle 6 miles		144	.056			2.72	2.97	5.69	7.60
0154	cycle 8 miles		120	.067			3.26	3.57	6.83	9.10
0214	15 MPH avg., cycle 0.5 mile, 20 min. wait/ld./uld.		176	.045			2.22	2.43	4.65	6.25
0216	cycle 1 mile		160	.050			2.45	2.68	5.13	6.85
0218	cycle 2 miles		136	.059			2.88	3.15	6.03	8.05
0220	cycle 4 miles		104	.077			3.77	4.12	7.89	10.55
0222	cycle 6 miles		88	.091			4.45	4.86	9.31	12.45
0224	cycle 8 miles		72	.111			5.45	5.95	11.40	15.25
0226	20 MPH avg., cycle 0.5 mile		176	.045			2.22	2.43	4.65	6.25
0228	cycle 1 mile		168	.048			2.33	2.55	4.88	6.55
0230	cycle 2 miles		144	.056			2.72	2.97	5.69	7.60
0232	cycle 4 miles		120	.067			3.26	3.57	6.83	9.10
0234	cycle 6 miles		96	.083			4.08	4.46	8.54	11.40
0236	cycle 8 miles		88	.091			4.45	4.86	9.31	12.45
0244	25 MPH avg., cycle 4 miles		128	.063			3.06	3.34	6.40	8.55
0246	cycle 6 miles		112	.071			3.50	3.82	7.32	9.80
0248	cycle 8 miles		96	.083			4.08	4.46	8.54	11.40
0250	30 MPH avg., cycle 4 miles		136	.059			2.88	3.15	6.03	8.05
0252	cycle 6 miles		120	.067			3.26	3.57	6.83	9.10
0254	cycle 8 miles		104	.077			3.77	4.12	7.89	10.55
0314	15 MPH avg., cycle 0.5 mile, 25 min. wait/ld./uld.		144	.056			2.72	2.97	5.69	7.60
0316	cycle 1 mile		128	.063			3.06	3.34	6.40	8.55
0318	cycle 2 miles		112	.071			3.50	3.82	7.32	9.80
0320	cycle 4 miles		96	.083			4.08	4.46	8.54	11.40
0322	cycle 6 miles		80	.100			4.90	5.35	10.25	13.75
0324	cycle 8 miles		64	.125			6.10	6.70	12.80	17.15
0326	20 MPH avg., cycle 0.5 mile		144	.056			2.72	2.97	5.69	7.60
0328	cycle 1 mile		136	.059			2.88	3.15	6.03	8.05
0330	cycle 2 miles		120	.067			3.26	3.57	6.83	9.10
0332	cycle 4 miles		104	.077			3.77	4.12	7.89	10.55
0334	cycle 6 miles		88	.091			4.45	4.86	9.31	12.45
0336	cycle 8 miles		80	.100			4.90	5.35	10.25	13.75
0344	25 MPH avg., cycle 4 miles		112	.071			3.50	3.82	7.32	9.80
0346	cycle 6 miles		96	.083			4.08	4.46	8.54	11.40
0348	cycle 8 miles		88	.091			4.45	4.86	9.31	12.45
0350	30 MPH avg., cycle 4 miles		112	.071			3.50	3.82	7.32	9.80
0352	cycle 6 miles		104	.077			3.77	4.12	7.89	10.55

31 23 23.20 Hauling	Crew	Daily Output	Labor-Hours	Unit	Material	2020 Bare Costs Labor	Equipment	Total	Total Incl O&P
0354	cycle 8 miles	B-34A	96	.083	L.C.Y.		4.08	4.46	8.54	11.40
0414	15 MPH avg., cycle 0.5 mile, 30 min. wait/ld./uld.		120	.067			3.26	3.57	6.83	9.10
0416	cycle 1 mile		112	.071			3.50	3.82	7.32	9.80
0418	cycle 2 miles		96	.083			4.08	4.46	8.54	11.40
0420	cycle 4 miles		80	.100			4.90	5.35	10.25	13.75
0422	cycle 6 miles		72	.111			5.45	5.95	11.40	15.25
0424	cycle 8 miles		64	.125			6.10	6.70	12.80	17.15
0426	20 MPH avg., cycle 0.5 mile		120	.067			3.26	3.57	6.83	9.10
0428	cycle 1 mile		112	.071			3.50	3.82	7.32	9.80
0430	cycle 2 miles		104	.077			3.77	4.12	7.89	10.55
0432	cycle 4 miles		88	.091			4.45	4.86	9.31	12.45
0434	cycle 6 miles		80	.100			4.90	5.35	10.25	13.75
0436	cycle 8 miles		72	.111			5.45	5.95	11.40	15.25
0444	25 MPH avg., cycle 4 miles		96	.083			4.08	4.46	8.54	11.40
0446	cycle 6 miles		88	.091			4.45	4.86	9.31	12.45
0448	cycle 8 miles		80	.100			4.90	5.35	10.25	13.75
0450	30 MPH avg., cycle 4 miles		96	.083			4.08	4.46	8.54	11.40
0452	cycle 6 miles		88	.091			4.45	4.86	9.31	12.45
0454	cycle 8 miles		80	.100			4.90	5.35	10.25	13.75
0514	15 MPH avg., cycle 0.5 mile, 35 min. wait/ld./uld.		104	.077			3.77	4.12	7.89	10.55
0516	cycle 1 mile		96	.083			4.08	4.46	8.54	11.40
0518	cycle 2 miles		88	.091			4.45	4.86	9.31	12.45
0520	cycle 4 miles		72	.111			5.45	5.95	11.40	15.25
0522	cycle 6 miles		64	.125			6.10	6.70	12.80	17.15
0524	cycle 8 miles		56	.143			7	7.65	14.65	19.60
0526	20 MPH avg., cycle 0.5 mile		104	.077			3.77	4.12	7.89	10.55
0528	cycle 1 mile		96	.083			4.08	4.46	8.54	11.40
0530	cycle 2 miles		96	.083			4.08	4.46	8.54	11.40
0532	cycle 4 miles		80	.100			4.90	5.35	10.25	13.75
0534	cycle 6 miles		72	.111			5.45	5.95	11.40	15.25
0536	cycle 8 miles		64	.125			6.10	6.70	12.80	17.15
0544	25 MPH avg., cycle 4 miles		88	.091			4.45	4.86	9.31	12.45
0546	cycle 6 miles		80	.100			4.90	5.35	10.25	13.75
0548	cycle 8 miles		72	.111			5.45	5.95	11.40	15.25
0550	30 MPH avg., cycle 4 miles		88	.091			4.45	4.86	9.31	12.45
0552	cycle 6 miles		80	.100			4.90	5.35	10.25	13.75
0554	cycle 8 miles		72	.111			5.45	5.95	11.40	15.25
1014	12 C.Y. truck, cycle 0.5 mile, 15 MPH avg., 15 min. wait/ld./uld.	B-34B	336	.024			1.17	1.71	2.88	3.74
1016	cycle 1 mile		300	.027			1.31	1.91	3.22	4.19
1018	cycle 2 miles		252	.032			1.55	2.27	3.82	4.98
1020	cycle 4 miles		180	.044			2.18	3.18	5.36	7
1022	cycle 6 miles		144	.056			2.72	3.98	6.70	8.75
1024	cycle 8 miles		120	.067			3.26	4.77	8.03	10.45
1025	cycle 10 miles		96	.083			4.08	5.95	10.03	13.05
1026	20 MPH avg., cycle 0.5 mile		348	.023			1.13	1.65	2.78	3.61
1028	cycle 1 mile		312	.026			1.26	1.84	3.10	4.03
1030	cycle 2 miles		276	.029			1.42	2.08	3.50	4.55
1032	cycle 4 miles		216	.037			1.81	2.65	4.46	5.80
1034	cycle 6 miles		168	.048			2.33	3.41	5.74	7.50
1036	cycle 8 miles		144	.056			2.72	3.98	6.70	8.75
1038	cycle 10 miles		120	.067			3.26	4.77	8.03	10.45
1040	25 MPH avg., cycle 4 miles		228	.035			1.72	2.51	4.23	5.50
1042	cycle 6 miles		192	.042			2.04	2.98	5.02	6.55

31 23 Excavation and Fill

31 23 23 – Fill

31 23 23.20 Hauling		Crew	Daily Output	Labor-Hours	Unit	Material	2020 Bare Costs Labor	Equipment	Total	Total Incl O&P
1044	cycle 8 miles	B-34B	168	.048	L.C.Y.		2.33	3.41	5.74	7.50
1046	cycle 10 miles		144	.056			2.72	3.98	6.70	8.75
1050	30 MPH avg., cycle 4 miles		252	.032			1.55	2.27	3.82	4.98
1052	cycle 6 miles		216	.037			1.81	2.65	4.46	5.80
1054	cycle 8 miles		180	.044			2.18	3.18	5.36	7
1056	cycle 10 miles		156	.051			2.51	3.67	6.18	8.05
1060	35 MPH avg., cycle 4 miles		264	.030			1.48	2.17	3.65	4.76
1062	cycle 6 miles		228	.035			1.72	2.51	4.23	5.50
1064	cycle 8 miles		204	.039			1.92	2.81	4.73	6.15
1066	cycle 10 miles		180	.044			2.18	3.18	5.36	7
1068	cycle 20 miles		120	.067			3.26	4.77	8.03	10.45
1069	cycle 30 miles		84	.095			4.66	6.80	11.46	14.95
1070	cycle 40 miles		72	.111			5.45	7.95	13.40	17.45
1072	40 MPH avg., cycle 6 miles		240	.033			1.63	2.39	4.02	5.25
1074	cycle 8 miles		216	.037			1.81	2.65	4.46	5.80
1076	cycle 10 miles		192	.042			2.04	2.98	5.02	6.55
1078	cycle 20 miles		120	.067			3.26	4.77	8.03	10.45
1080	cycle 30 miles		96	.083			4.08	5.95	10.03	13.05
1082	cycle 40 miles		72	.111			5.45	7.95	13.40	17.45
1084	cycle 50 miles		60	.133			6.55	9.55	16.10	21
1094	45 MPH avg., cycle 8 miles		216	.037			1.81	2.65	4.46	5.80
1096	cycle 10 miles		204	.039			1.92	2.81	4.73	6.15
1098	cycle 20 miles		132	.061			2.97	4.34	7.31	9.50
1100	cycle 30 miles		108	.074			3.63	5.30	8.93	11.65
1102	cycle 40 miles		84	.095			4.66	6.80	11.46	14.95
1104	cycle 50 miles		72	.111			5.45	7.95	13.40	17.45
1106	50 MPH avg., cycle 10 miles		216	.037			1.81	2.65	4.46	5.80
1108	cycle 20 miles		144	.056			2.72	3.98	6.70	8.75
1110	cycle 30 miles		108	.074			3.63	5.30	8.93	11.65
1112	cycle 40 miles		84	.095			4.66	6.80	11.46	14.95
1114	cycle 50 miles		72	.111			5.45	7.95	13.40	17.45
1214	15 MPH avg., cycle 0.5 mile, 20 min. wait/ld./uld.		264	.030			1.48	2.17	3.65	4.76
1216	cycle 1 mile		240	.033			1.63	2.39	4.02	5.25
1218	cycle 2 miles		204	.039			1.92	2.81	4.73	6.15
1220	cycle 4 miles		156	.051			2.51	3.67	6.18	8.05
1222	cycle 6 miles		132	.061			2.97	4.34	7.31	9.50
1224	cycle 8 miles		108	.074			3.63	5.30	8.93	11.65
1225	cycle 10 miles		96	.083			4.08	5.95	10.03	13.05
1226	20 MPH avg., cycle 0.5 mile		264	.030			1.48	2.17	3.65	4.76
1228	cycle 1 mile		252	.032			1.55	2.27	3.82	4.98
1230	cycle 2 miles		216	.037			1.81	2.65	4.46	5.80
1232	cycle 4 miles		180	.044			2.18	3.18	5.36	7
1234	cycle 6 miles		144	.056			2.72	3.98	6.70	8.75
1236	cycle 8 miles		132	.061			2.97	4.34	7.31	9.50
1238	cycle 10 miles		108	.074			3.63	5.30	8.93	11.65
1240	25 MPH avg., cycle 4 miles		192	.042			2.04	2.98	5.02	6.55
1242	cycle 6 miles		168	.048			2.33	3.41	5.74	7.50
1244	cycle 8 miles		144	.056			2.72	3.98	6.70	8.75
1246	cycle 10 miles		132	.061			2.97	4.34	7.31	9.50
1250	30 MPH avg., cycle 4 miles		204	.039			1.92	2.81	4.73	6.15
1252	cycle 6 miles		180	.044			2.18	3.18	5.36	7
1254	cycle 8 miles		156	.051			2.51	3.67	6.18	8.05
1256	cycle 10 miles		144	.056			2.72	3.98	6.70	8.75

	31 23 23.20 Hauling	Crew	Daily Output	Labor-Hours	Unit	Material	2020 Bare Costs Labor	2020 Bare Costs Equipment	Total	Total Incl O&P
1260	35 MPH avg., cycle 4 miles	B-34B	216	.037	L.C.Y.		1.81	2.65	4.46	5.80
1262	cycle 6 miles		192	.042			2.04	2.98	5.02	6.55
1264	cycle 8 miles		168	.048			2.33	3.41	5.74	7.50
1266	cycle 10 miles		156	.051			2.51	3.67	6.18	8.05
1268	cycle 20 miles		108	.074			3.63	5.30	8.93	11.65
1269	cycle 30 miles		72	.111			5.45	7.95	13.40	17.45
1270	cycle 40 miles		60	.133			6.55	9.55	16.10	21
1272	40 MPH avg., cycle 6 miles		192	.042			2.04	2.98	5.02	6.55
1274	cycle 8 miles		180	.044			2.18	3.18	5.36	7
1276	cycle 10 miles		156	.051			2.51	3.67	6.18	8.05
1278	cycle 20 miles		108	.074			3.63	5.30	8.93	11.65
1280	cycle 30 miles		84	.095			4.66	6.80	11.46	14.95
1282	cycle 40 miles		72	.111			5.45	7.95	13.40	17.45
1284	cycle 50 miles		60	.133			6.55	9.55	16.10	21
1294	45 MPH avg., cycle 8 miles		180	.044			2.18	3.18	5.36	7
1296	cycle 10 miles		168	.048			2.33	3.41	5.74	7.50
1298	cycle 20 miles		120	.067			3.26	4.77	8.03	10.45
1300	cycle 30 miles		96	.083			4.08	5.95	10.03	13.05
1302	cycle 40 miles		72	.111			5.45	7.95	13.40	17.45
1304	cycle 50 miles		60	.133			6.55	9.55	16.10	21
1306	50 MPH avg., cycle 10 miles		180	.044			2.18	3.18	5.36	7
1308	cycle 20 miles		132	.061			2.97	4.34	7.31	9.50
1310	cycle 30 miles		96	.083			4.08	5.95	10.03	13.05
1312	cycle 40 miles		84	.095			4.66	6.80	11.46	14.95
1314	cycle 50 miles		72	.111			5.45	7.95	13.40	17.45
1414	15 MPH avg., cycle 0.5 mile, 25 min. wait/ld./uld.		204	.039			1.92	2.81	4.73	6.15
1416	cycle 1 mile		192	.042			2.04	2.98	5.02	6.55
1418	cycle 2 miles		168	.048			2.33	3.41	5.74	7.50
1420	cycle 4 miles		132	.061			2.97	4.34	7.31	9.50
1422	cycle 6 miles		120	.067			3.26	4.77	8.03	10.45
1424	cycle 8 miles		96	.083			4.08	5.95	10.03	13.05
1425	cycle 10 miles		84	.095			4.66	6.80	11.46	14.95
1426	20 MPH avg., cycle 0.5 mile		216	.037			1.81	2.65	4.46	5.80
1428	cycle 1 mile		204	.039			1.92	2.81	4.73	6.15
1430	cycle 2 miles		180	.044			2.18	3.18	5.36	7
1432	cycle 4 miles		156	.051			2.51	3.67	6.18	8.05
1434	cycle 6 miles		132	.061			2.97	4.34	7.31	9.50
1436	cycle 8 miles		120	.067			3.26	4.77	8.03	10.45
1438	cycle 10 miles		96	.083			4.08	5.95	10.03	13.05
1440	25 MPH avg., cycle 4 miles		168	.048			2.33	3.41	5.74	7.50
1442	cycle 6 miles		144	.056			2.72	3.98	6.70	8.75
1444	cycle 8 miles		132	.061			2.97	4.34	7.31	9.50
1446	cycle 10 miles		108	.074			3.63	5.30	8.93	11.65
1450	30 MPH avg., cycle 4 miles		168	.048			2.33	3.41	5.74	7.50
1452	cycle 6 miles		156	.051			2.51	3.67	6.18	8.05
1454	cycle 8 miles		132	.061			2.97	4.34	7.31	9.50
1456	cycle 10 miles		120	.067			3.26	4.77	8.03	10.45
1460	35 MPH avg., cycle 4 miles		180	.044			2.18	3.18	5.36	7
1462	cycle 6 miles		156	.051			2.51	3.67	6.18	8.05
1464	cycle 8 miles		144	.056			2.72	3.98	6.70	8.75
1466	cycle 10 miles		132	.061			2.97	4.34	7.31	9.50
1468	cycle 20 miles		96	.083			4.08	5.95	10.03	13.05
1469	cycle 30 miles		72	.111			5.45	7.95	13.40	17.45

31 23 23 – Fill

31 23 23.20 Hauling		Crew	Daily Output	Labor-Hours	Unit	Material	2020 Bare Costs		Total	Total Incl O&P
							Labor	Equipment		
1470	cycle 40 miles	B-34B	60	.133	L.C.Y.		6.55	9.55	16.10	21
1472	40 MPH avg., cycle 6 miles		168	.048			2.33	3.41	5.74	7.50
1474	cycle 8 miles		156	.051			2.51	3.67	6.18	8.05
1476	cycle 10 miles		144	.056			2.72	3.98	6.70	8.75
1478	cycle 20 miles		96	.083			4.08	5.95	10.03	13.05
1480	cycle 30 miles		84	.095			4.66	6.80	11.46	14.95
1482	cycle 40 miles		60	.133			6.55	9.55	16.10	21
1484	cycle 50 miles		60	.133			6.55	9.55	16.10	21
1494	45 MPH avg., cycle 8 miles		156	.051			2.51	3.67	6.18	8.05
1496	cycle 10 miles		144	.056			2.72	3.98	6.70	8.75
1498	cycle 20 miles		108	.074			3.63	5.30	8.93	11.65
1500	cycle 30 miles		84	.095			4.66	6.80	11.46	14.95
1502	cycle 40 miles		72	.111			5.45	7.95	13.40	17.45
1504	cycle 50 miles		60	.133			6.55	9.55	16.10	21
1506	50 MPH avg., cycle 10 miles		156	.051			2.51	3.67	6.18	8.05
1508	cycle 20 miles		120	.067			3.26	4.77	8.03	10.45
1510	cycle 30 miles		96	.083			4.08	5.95	10.03	13.05
1512	cycle 40 miles		72	.111			5.45	7.95	13.40	17.45
1514	cycle 50 miles		60	.133			6.55	9.55	16.10	21
1614	15 MPH avg., cycle 0.5 mile, 30 min. wait/ld./uld.		180	.044			2.18	3.18	5.36	7
1616	cycle 1 mile		168	.048			2.33	3.41	5.74	7.50
1618	cycle 2 miles		144	.056			2.72	3.98	6.70	8.75
1620	cycle 4 miles		120	.067			3.26	4.77	8.03	10.45
1622	cycle 6 miles		108	.074			3.63	5.30	8.93	11.65
1624	cycle 8 miles		84	.095			4.66	6.80	11.46	14.95
1625	cycle 10 miles		84	.095			4.66	6.80	11.46	14.95
1626	20 MPH avg., cycle 0.5 mile		180	.044			2.18	3.18	5.36	7
1628	cycle 1 mile		168	.048			2.33	3.41	5.74	7.50
1630	cycle 2 miles		156	.051			2.51	3.67	6.18	8.05
1632	cycle 4 miles		132	.061			2.97	4.34	7.31	9.50
1634	cycle 6 miles		120	.067			3.26	4.77	8.03	10.45
1636	cycle 8 miles		108	.074			3.63	5.30	8.93	11.65
1638	cycle 10 miles		96	.083			4.08	5.95	10.03	13.05
1640	25 MPH avg., cycle 4 miles		144	.056			2.72	3.98	6.70	8.75
1642	cycle 6 miles		132	.061			2.97	4.34	7.31	9.50
1644	cycle 8 miles		108	.074			3.63	5.30	8.93	11.65
1646	cycle 10 miles		108	.074			3.63	5.30	8.93	11.65
1650	30 MPH avg., cycle 4 miles		144	.056			2.72	3.98	6.70	8.75
1652	cycle 6 miles		132	.061			2.97	4.34	7.31	9.50
1654	cycle 8 miles		120	.067			3.26	4.77	8.03	10.45
1656	cycle 10 miles		108	.074			3.63	5.30	8.93	11.65
1660	35 MPH avg., cycle 4 miles		156	.051			2.51	3.67	6.18	8.05
1662	cycle 6 miles		144	.056			2.72	3.98	6.70	8.75
1664	cycle 8 miles		132	.061			2.97	4.34	7.31	9.50
1666	cycle 10 miles		120	.067			3.26	4.77	8.03	10.45
1668	cycle 20 miles		84	.095			4.66	6.80	11.46	14.95
1669	cycle 30 miles		72	.111			5.45	7.95	13.40	17.45
1670	cycle 40 miles		60	.133			6.55	9.55	16.10	21
1672	40 MPH avg., cycle 6 miles		144	.056			2.72	3.98	6.70	8.75
1674	cycle 8 miles		132	.061			2.97	4.34	7.31	9.50
1676	cycle 10 miles		120	.067			3.26	4.77	8.03	10.45
1678	cycle 20 miles		96	.083			4.08	5.95	10.03	13.05
1680	cycle 30 miles		72	.111			5.45	7.95	13.40	17.45

31 23 23.20 Hauling		Crew	Daily Output	Labor-Hours	Unit	Material	2020 Bare Costs Labor	2020 Bare Costs Equipment	Total	Total Incl O&P
1682	cycle 40 miles	B-34B	60	.133	L.C.Y.		6.55	9.55	16.10	21
1684	cycle 50 miles		48	.167			8.15	11.95	20.10	26
1694	45 MPH avg., cycle 8 miles		144	.056			2.72	3.98	6.70	8.75
1696	cycle 10 miles		132	.061			2.97	4.34	7.31	9.50
1698	cycle 20 miles		96	.083			4.08	5.95	10.03	13.05
1700	cycle 30 miles		84	.095			4.66	6.80	11.46	14.95
1702	cycle 40 miles		60	.133			6.55	9.55	16.10	21
1704	cycle 50 miles		60	.133			6.55	9.55	16.10	21
1706	50 MPH avg., cycle 10 miles		132	.061			2.97	4.34	7.31	9.50
1708	cycle 20 miles		108	.074			3.63	5.30	8.93	11.65
1710	cycle 30 miles		84	.095			4.66	6.80	11.46	14.95
1712	cycle 40 miles		72	.111			5.45	7.95	13.40	17.45
1714	cycle 50 miles		60	.133			6.55	9.55	16.10	21
2000	Hauling, 8 C.Y. truck, small project cost per hour	B-34A	8	1	Hr.		49	53.50	102.50	138
2100	12 C.Y. truck	B-34B	8	1			49	71.50	120.50	158
2150	16.5 C.Y. truck	B-34C	8	1			49	78.50	127.50	165
2175	18 C.Y. 8 wheel truck	B-34I	8	1			49	89	138	177
2200	20 C.Y. truck	B-34D	8	1			49	80.50	129.50	167
2300	Grading at dump, or embankment if required, by dozer	B-10B	1000	.012	L.C.Y.		.62	1.50	2.12	2.63
2310	Spotter at fill or cut, if required	1 Clab	8	1	Hr.		42		42	67.50
9014	18 C.Y. truck, 8 wheels,15 min. wait/ld./uld.,15 MPH, cycle 0.5 mi.	B-34I	504	.016	L.C.Y.		.78	1.42	2.20	2.80
9016	cycle 1 mile		450	.018			.87	1.59	2.46	3.14
9018	cycle 2 miles		378	.021			1.04	1.89	2.93	3.74
9020	cycle 4 miles		270	.030			1.45	2.64	4.09	5.25
9022	cycle 6 miles		216	.037			1.81	3.31	5.12	6.55
9024	cycle 8 miles		180	.044			2.18	3.97	6.15	7.85
9025	cycle 10 miles		144	.056			2.72	4.96	7.68	9.80
9026	20 MPH avg., cycle 0.5 mile		522	.015			.75	1.37	2.12	2.70
9028	cycle 1 mile		468	.017			.84	1.52	2.36	3.02
9030	cycle 2 miles		414	.019			.95	1.72	2.67	3.41
9032	cycle 4 miles		324	.025			1.21	2.20	3.41	4.35
9034	cycle 6 miles		252	.032			1.55	2.83	4.38	5.60
9036	cycle 8 miles		216	.037			1.81	3.31	5.12	6.55
9038	cycle 10 miles		180	.044			2.18	3.97	6.15	7.85
9040	25 MPH avg., cycle 4 miles		342	.023			1.14	2.09	3.23	4.13
9042	cycle 6 miles		288	.028			1.36	2.48	3.84	4.90
9044	cycle 8 miles		252	.032			1.55	2.83	4.38	5.60
9046	cycle 10 miles		216	.037			1.81	3.31	5.12	6.55
9050	30 MPH avg., cycle 4 miles		378	.021			1.04	1.89	2.93	3.74
9052	cycle 6 miles		324	.025			1.21	2.20	3.41	4.35
9054	cycle 8 miles		270	.030			1.45	2.64	4.09	5.25
9056	cycle 10 miles		234	.034			1.67	3.05	4.72	6.05
9060	35 MPH avg., cycle 4 miles		396	.020			.99	1.80	2.79	3.56
9062	cycle 6 miles		342	.023			1.14	2.09	3.23	4.13
9064	cycle 8 miles		288	.028			1.36	2.48	3.84	4.90
9066	cycle 10 miles		270	.030			1.45	2.64	4.09	5.25
9068	cycle 20 miles		162	.049			2.42	4.41	6.83	8.70
9070	cycle 30 miles		126	.063			3.11	5.65	8.76	11.20
9072	cycle 40 miles		90	.089			4.35	7.95	12.30	15.65
9074	40 MPH avg., cycle 6 miles		360	.022			1.09	1.98	3.07	3.92
9076	cycle 8 miles		324	.025			1.21	2.20	3.41	4.35
9078	cycle 10 miles		288	.028			1.36	2.48	3.84	4.90
9080	cycle 20 miles		180	.044			2.18	3.97	6.15	7.85

31 23 23.20 Hauling		Crew	Daily Output	Labor-Hours	Unit	Material	2020 Bare Costs Labor	2020 Bare Costs Equipment	Total	Total Incl O&P
9082	cycle 30 miles	B-34I	144	.056	L.C.Y.		2.72	4.96	7.68	9.80
9084	cycle 40 miles		108	.074			3.63	6.60	10.23	13.05
9086	cycle 50 miles		90	.089			4.35	7.95	12.30	15.65
9094	45 MPH avg., cycle 8 miles		324	.025			1.21	2.20	3.41	4.35
9096	cycle 10 miles		306	.026			1.28	2.33	3.61	4.62
9098	cycle 20 miles		198	.040			1.98	3.60	5.58	7.15
9100	cycle 30 miles		144	.056			2.72	4.96	7.68	9.80
9102	cycle 40 miles		126	.063			3.11	5.65	8.76	11.20
9104	cycle 50 miles		108	.074			3.63	6.60	10.23	13.05
9106	50 MPH avg., cycle 10 miles		324	.025			1.21	2.20	3.41	4.35
9108	cycle 20 miles		216	.037			1.81	3.31	5.12	6.55
9110	cycle 30 miles		162	.049			2.42	4.41	6.83	8.70
9112	cycle 40 miles		126	.063			3.11	5.65	8.76	11.20
9114	cycle 50 miles		108	.074			3.63	6.60	10.23	13.05
9214	20 min. wait/ld./uld.,15 MPH, cycle 0.5 mi.		396	.020			.99	1.80	2.79	3.56
9216	cycle 1 mile		360	.022			1.09	1.98	3.07	3.92
9218	cycle 2 miles		306	.026			1.28	2.33	3.61	4.62
9220	cycle 4 miles		234	.034			1.67	3.05	4.72	6.05
9222	cycle 6 miles		198	.040			1.98	3.60	5.58	7.15
9224	cycle 8 miles		162	.049			2.42	4.41	6.83	8.70
9225	cycle 10 miles		144	.056			2.72	4.96	7.68	9.80
9226	20 MPH avg., cycle 0.5 mile		396	.020			.99	1.80	2.79	3.56
9228	cycle 1 mile		378	.021			1.04	1.89	2.93	3.74
9230	cycle 2 miles		324	.025			1.21	2.20	3.41	4.35
9232	cycle 4 miles		270	.030			1.45	2.64	4.09	5.25
9234	cycle 6 miles		216	.037			1.81	3.31	5.12	6.55
9236	cycle 8 miles		198	.040			1.98	3.60	5.58	7.15
9238	cycle 10 miles		162	.049			2.42	4.41	6.83	8.70
9240	25 MPH avg., cycle 4 miles		288	.028			1.36	2.48	3.84	4.90
9242	cycle 6 miles		252	.032			1.55	2.83	4.38	5.60
9244	cycle 8 miles		216	.037			1.81	3.31	5.12	6.55
9246	cycle 10 miles		198	.040			1.98	3.60	5.58	7.15
9250	30 MPH avg., cycle 4 miles		306	.026			1.28	2.33	3.61	4.62
9252	cycle 6 miles		270	.030			1.45	2.64	4.09	5.25
9254	cycle 8 miles		234	.034			1.67	3.05	4.72	6.05
9256	cycle 10 miles		216	.037			1.81	3.31	5.12	6.55
9260	35 MPH avg., cycle 4 miles		324	.025			1.21	2.20	3.41	4.35
9262	cycle 6 miles		288	.028			1.36	2.48	3.84	4.90
9264	cycle 8 miles		252	.032			1.55	2.83	4.38	5.60
9266	cycle 10 miles		234	.034			1.67	3.05	4.72	6.05
9268	cycle 20 miles		162	.049			2.42	4.41	6.83	8.70
9270	cycle 30 miles		108	.074			3.63	6.60	10.23	13.05
9272	cycle 40 miles		90	.089			4.35	7.95	12.30	15.65
9274	40 MPH avg., cycle 6 miles		288	.028			1.36	2.48	3.84	4.90
9276	cycle 8 miles		270	.030			1.45	2.64	4.09	5.25
9278	cycle 10 miles		234	.034			1.67	3.05	4.72	6.05
9280	cycle 20 miles		162	.049			2.42	4.41	6.83	8.70
9282	cycle 30 miles		126	.063			3.11	5.65	8.76	11.20
9284	cycle 40 miles		108	.074			3.63	6.60	10.23	13.05
9286	cycle 50 miles		90	.089			4.35	7.95	12.30	15.65
9294	45 MPH avg., cycle 8 miles		270	.030			1.45	2.64	4.09	5.25
9296	cycle 10 miles		252	.032			1.55	2.83	4.38	5.60
9298	cycle 20 miles		180	.044			2.18	3.97	6.15	7.85

31 23 23.20 Hauling	Crew	Daily Output	Labor-Hours	Unit	Material	2020 Bare Costs Labor	2020 Bare Costs Equipment	Total	Total Incl O&P
9300	cycle 30 miles	B-34I	144	.056	L.C.Y.		2.72	4.96	7.68	9.80
9302	cycle 40 miles		108	.074			3.63	6.60	10.23	13.05
9304	cycle 50 miles		90	.089			4.35	7.95	12.30	15.65
9306	50 MPH avg., cycle 10 miles		270	.030			1.45	2.64	4.09	5.25
9308	cycle 20 miles		198	.040			1.98	3.60	5.58	7.15
9310	cycle 30 miles		144	.056			2.72	4.96	7.68	9.80
9312	cycle 40 miles		126	.063			3.11	5.65	8.76	11.20
9314	cycle 50 miles		108	.074			3.63	6.60	10.23	13.05
9414	25 min. wait/ld./uld.,15 MPH, cycle 0.5 mi.		306	.026			1.28	2.33	3.61	4.62
9416	cycle 1 mile		288	.028			1.36	2.48	3.84	4.90
9418	cycle 2 miles		252	.032			1.55	2.83	4.38	5.60
9420	cycle 4 miles		198	.040			1.98	3.60	5.58	7.15
9422	cycle 6 miles		180	.044			2.18	3.97	6.15	7.85
9424	cycle 8 miles		144	.056			2.72	4.96	7.68	9.80
9425	cycle 10 miles		126	.063			3.11	5.65	8.76	11.20
9426	20 MPH avg., cycle 0.5 mile		324	.025			1.21	2.20	3.41	4.35
9428	cycle 1 mile		306	.026			1.28	2.33	3.61	4.62
9430	cycle 2 miles		270	.030			1.45	2.64	4.09	5.25
9432	cycle 4 miles		234	.034			1.67	3.05	4.72	6.05
9434	cycle 6 miles		198	.040			1.98	3.60	5.58	7.15
9436	cycle 8 miles		180	.044			2.18	3.97	6.15	7.85
9438	cycle 10 miles		144	.056			2.72	4.96	7.68	9.80
9440	25 MPH avg., cycle 4 miles		252	.032			1.55	2.83	4.38	5.60
9442	cycle 6 miles		216	.037			1.81	3.31	5.12	6.55
9444	cycle 8 miles		198	.040			1.98	3.60	5.58	7.15
9446	cycle 10 miles		180	.044			2.18	3.97	6.15	7.85
9450	30 MPH avg., cycle 4 miles		252	.032			1.55	2.83	4.38	5.60
9452	cycle 6 miles		234	.034			1.67	3.05	4.72	6.05
9454	cycle 8 miles		198	.040			1.98	3.60	5.58	7.15
9456	cycle 10 miles		180	.044			2.18	3.97	6.15	7.85
9460	35 MPH avg., cycle 4 miles		270	.030			1.45	2.64	4.09	5.25
9462	cycle 6 miles		234	.034			1.67	3.05	4.72	6.05
9464	cycle 8 miles		216	.037			1.81	3.31	5.12	6.55
9466	cycle 10 miles		198	.040			1.98	3.60	5.58	7.15
9468	cycle 20 miles		144	.056			2.72	4.96	7.68	9.80
9470	cycle 30 miles		108	.074			3.63	6.60	10.23	13.05
9472	cycle 40 miles		90	.089			4.35	7.95	12.30	15.65
9474	40 MPH avg., cycle 6 miles		252	.032			1.55	2.83	4.38	5.60
9476	cycle 8 miles		234	.034			1.67	3.05	4.72	6.05
9478	cycle 10 miles		216	.037			1.81	3.31	5.12	6.55
9480	cycle 20 miles		144	.056			2.72	4.96	7.68	9.80
9482	cycle 30 miles		126	.063			3.11	5.65	8.76	11.20
9484	cycle 40 miles		90	.089			4.35	7.95	12.30	15.65
9486	cycle 50 miles		90	.089			4.35	7.95	12.30	15.65
9494	45 MPH avg., cycle 8 miles		234	.034			1.67	3.05	4.72	6.05
9496	cycle 10 miles		216	.037			1.81	3.31	5.12	6.55
9498	cycle 20 miles		162	.049			2.42	4.41	6.83	8.70
9500	cycle 30 miles		126	.063			3.11	5.65	8.76	11.20
9502	cycle 40 miles		108	.074			3.63	6.60	10.23	13.05
9504	cycle 50 miles		90	.089			4.35	7.95	12.30	15.65
9506	50 MPH avg., cycle 10 miles		234	.034			1.67	3.05	4.72	6.05
9508	cycle 20 miles		180	.044			2.18	3.97	6.15	7.85
9510	cycle 30 miles		144	.056			2.72	4.96	7.68	9.80

31 23 Excavation and Fill

31 23 23 – Fill

	31 23 23.20 Hauling	Crew	Daily Output	Labor-Hours	Unit	Material	2020 Bare Costs Labor	2020 Bare Costs Equipment	Total	Total Incl O&P
9512	cycle 40 miles	B-34I	108	.074	L.C.Y.		3.63	6.60	10.23	13.05
9514	cycle 50 miles		90	.089			4.35	7.95	12.30	15.65
9614	30 min. wait/ld./uld.,15 MPH, cycle 0.5 mi.		270	.030			1.45	2.64	4.09	5.25
9616	cycle 1 mile		252	.032			1.55	2.83	4.38	5.60
9618	cycle 2 miles		216	.037			1.81	3.31	5.12	6.55
9620	cycle 4 miles		180	.044			2.18	3.97	6.15	7.85
9622	cycle 6 miles		162	.049			2.42	4.41	6.83	8.70
9624	cycle 8 miles		126	.063			3.11	5.65	8.76	11.20
9625	cycle 10 miles		126	.063			3.11	5.65	8.76	11.20
9626	20 MPH avg., cycle 0.5 mile		270	.030			1.45	2.64	4.09	5.25
9628	cycle 1 mile		252	.032			1.55	2.83	4.38	5.60
9630	cycle 2 miles		234	.034			1.67	3.05	4.72	6.05
9632	cycle 4 miles		198	.040			1.98	3.60	5.58	7.15
9634	cycle 6 miles		180	.044			2.18	3.97	6.15	7.85
9636	cycle 8 miles		162	.049			2.42	4.41	6.83	8.70
9638	cycle 10 miles		144	.056			2.72	4.96	7.68	9.80
9640	25 MPH avg., cycle 4 miles		216	.037			1.81	3.31	5.12	6.55
9642	cycle 6 miles		198	.040			1.98	3.60	5.58	7.15
9644	cycle 8 miles		180	.044			2.18	3.97	6.15	7.85
9646	cycle 10 miles		162	.049			2.42	4.41	6.83	8.70
9650	30 MPH avg., cycle 4 miles		216	.037			1.81	3.31	5.12	6.55
9652	cycle 6 miles		198	.040			1.98	3.60	5.58	7.15
9654	cycle 8 miles		180	.044			2.18	3.97	6.15	7.85
9656	cycle 10 miles		162	.049			2.42	4.41	6.83	8.70
9660	35 MPH avg., cycle 4 miles		234	.034			1.67	3.05	4.72	6.05
9662	cycle 6 miles		216	.037			1.81	3.31	5.12	6.55
9664	cycle 8 miles		198	.040			1.98	3.60	5.58	7.15
9666	cycle 10 miles		180	.044			2.18	3.97	6.15	7.85
9668	cycle 20 miles		126	.063			3.11	5.65	8.76	11.20
9670	cycle 30 miles		108	.074			3.63	6.60	10.23	13.05
9672	cycle 40 miles		90	.089			4.35	7.95	12.30	15.65
9674	40 MPH avg., cycle 6 miles		216	.037			1.81	3.31	5.12	6.55
9676	cycle 8 miles		198	.040			1.98	3.60	5.58	7.15
9678	cycle 10 miles		180	.044			2.18	3.97	6.15	7.85
9680	cycle 20 miles		144	.056			2.72	4.96	7.68	9.80
9682	cycle 30 miles		108	.074			3.63	6.60	10.23	13.05
9684	cycle 40 miles		90	.089			4.35	7.95	12.30	15.65
9686	cycle 50 miles		72	.111			5.45	9.90	15.35	19.60
9694	45 MPH avg., cycle 8 miles		216	.037			1.81	3.31	5.12	6.55
9696	cycle 10 miles		198	.040			1.98	3.60	5.58	7.15
9698	cycle 20 miles		144	.056			2.72	4.96	7.68	9.80
9700	cycle 30 miles		126	.063			3.11	5.65	8.76	11.20
9702	cycle 40 miles		108	.074			3.63	6.60	10.23	13.05
9704	cycle 50 miles		90	.089			4.35	7.95	12.30	15.65
9706	50 MPH avg., cycle 10 miles		198	.040			1.98	3.60	5.58	7.15
9708	cycle 20 miles		162	.049			2.42	4.41	6.83	8.70
9710	cycle 30 miles		126	.063			3.11	5.65	8.76	11.20
9712	cycle 40 miles		108	.074			3.63	6.60	10.23	13.05
9714	cycle 50 miles		90	.089			4.35	7.95	12.30	15.65

543

31 23 23.23 Compaction

		Crew	Daily Output	Labor-Hours	Unit	Material	2020 Bare Costs Labor	Equipment	Total	Total Incl O&P
0010	**COMPACTION**									
5000	Riding, vibrating roller, 6" lifts, 2 passes	B-10Y	3000	.004	E.C.Y.		.21	.19	.40	.54
5020	3 passes		2300	.005			.27	.25	.52	.71
5040	4 passes		1900	.006			.33	.30	.63	.86
5050	8" lifts, 2 passes		4100	.003			.15	.14	.29	.40
5060	12" lifts, 2 passes		5200	.002			.12	.11	.23	.31
5080	3 passes		3500	.003			.18	.17	.35	.46
5100	4 passes		2600	.005			.24	.22	.46	.63
5600	Sheepsfoot or wobbly wheel roller, 6" lifts, 2 passes	B-10G	2400	.005			.26	.56	.82	1.03
5620	3 passes		1735	.007			.36	.78	1.14	1.43
5640	4 passes		1300	.009			.48	1.04	1.52	1.89
5680	12" lifts, 2 passes		5200	.002			.12	.26	.38	.48
5700	3 passes		3500	.003			.18	.39	.57	.70
5720	4 passes		2600	.005			.24	.52	.76	.95
7000	Walk behind, vibrating plate 18" wide, 6" lifts, 2 passes	A-1D	200	.040			1.68	.16	1.84	2.87
7020	3 passes		185	.043			1.82	.17	1.99	3.11
7040	4 passes		140	.057			2.41	.23	2.64	4.10
7200	12" lifts, 2 passes, 21" wide	A-1E	560	.014			.60	.30	.90	1.29
7220	3 passes		375	.021			.90	.44	1.34	1.93
7240	4 passes		280	.029			1.20	.59	1.79	2.58
7500	Vibrating roller 24" wide, 6" lifts, 2 passes	B-10A	420	.029			1.48	.40	1.88	2.77
7520	3 passes		280	.043			2.22	.59	2.81	4.15
7540	4 passes		210	.057			2.96	.79	3.75	5.55
7600	12" lifts, 2 passes		840	.014			.74	.20	.94	1.39
7620	3 passes		560	.021			1.11	.30	1.41	2.08
7640	4 passes		420	.029			1.48	.40	1.88	2.77
8000	Rammer tamper, 6" to 11", 4" lifts, 2 passes	A-1F	130	.062			2.59	.36	2.95	4.54
8050	3 passes		97	.082			3.47	.48	3.95	6.10
8100	4 passes		65	.123			5.20	.72	5.92	9.10
8200	8" lifts, 2 passes		260	.031			1.30	.18	1.48	2.28
8250	3 passes		195	.041			1.73	.24	1.97	3.03
8300	4 passes		130	.062			2.59	.36	2.95	4.54
8400	13" to 18", 4" lifts, 2 passes	A-1G	390	.021			.86	.14	1	1.53
8450	3 passes		290	.028			1.16	.19	1.35	2.07
8500	4 passes		195	.041			1.73	.28	2.01	3.07
8600	8" lifts, 2 passes		780	.010			.43	.07	.50	.77
8650	3 passes		585	.014			.58	.09	.67	1.02
8700	4 passes		390	.021			.86	.14	1	1.53
9000	Water, 3,000 gal. truck, 3 mile haul	B-45	1888	.008		1.22	.45	.44	2.11	2.53
9010	6 mile haul		1444	.011		1.22	.59	.57	2.38	2.90
9020	12 mile haul		1000	.016		1.22	.85	.82	2.89	3.59
9030	6,000 gal. wagon, 3 mile haul	B-59	2000	.004		1.22	.20	.23	1.65	1.90
9040	6 mile haul	"	1600	.005		1.22	.24	.29	1.75	2.05

31 23 23.24 Compaction, Structural

		Crew	Daily Output	Labor-Hours	Unit	Material	2020 Bare Costs Labor	Equipment	Total	Total Incl O&P
0010	**COMPACTION, STRUCTURAL** R312323-30									
0020	Steel wheel tandem roller, 5 tons	B-10E	8	1.500	Hr.		78	32	110	158
0050	Air tamp, 6" to 8" lifts, common fill	B-9	250	.160	E.C.Y.		6.80	1.43	8.23	12.45
0060	Select fill	"	300	.133			5.65	1.19	6.84	10.40
0600	Vibratory plate, 8" lifts, common fill	A-1D	200	.040			1.68	.16	1.84	2.87
0700	Select fill	"	216	.037			1.56	.15	1.71	2.66
9000	Minimum labor/equipment charge	1 Clab	4	2	Job		84		84	135

31 25 Erosion and Sedimentation Controls

31 25 14 – Stabilization Measures for Erosion and Sedimentation Control

31 25 14.16 Rolled Erosion Control Mats and Blankets

		Crew	Daily Output	Labor-Hours	Unit	Material	2020 Bare Costs Labor	Equipment	Total	Total Incl O&P
0010	**ROLLED EROSION CONTROL MATS AND BLANKETS**									
0020	Jute mesh, 100 S.Y. per roll, 4' wide, stapled	B-80A	2400	.010	S.Y.	1.03	.42	.34	1.79	2.18
0070	Paper biodegradable mesh	B-1	2500	.010		.18	.41		.59	.86
0100	Plastic netting, stapled, 2" x 1" mesh, 20 mil	"	2500	.010		.27	.41		.68	.96
0120	Revegetation mat, webbed	2 Clab	1000	.016		2.96	.67		3.63	4.34
0200	Polypropylene mesh, stapled, 6.5 oz./S.Y.	B-1	2500	.010		1.70	.41		2.11	2.53
0300	Tobacco netting, or jute mesh #2, stapled	"	2500	.010		.31	.41		.72	1
0600	Straw in polymeric netting, biodegradable log	A-2	1000	.024	L.F.	6.45	1.05	.20	7.70	9
0705	Sediment Log, Filter Sock, 9"		1000	.024		3.50	1.05	.20	4.75	5.75
0710	Sediment Log, Filter Sock, 12"		1000	.024		4.95	1.05	.20	6.20	7.35
1000	Silt fence, install and maintain, remove	B-62	1300	.018		.45	.84	.14	1.43	1.99

31 31 Soil Treatment

31 31 16 – Termite Control

31 31 16.13 Chemical Termite Control

		Crew	Daily Output	Labor-Hours	Unit	Material	2020 Bare Costs Labor	Equipment	Total	Total Incl O&P
0010	**CHEMICAL TERMITE CONTROL**									
0020	Slab and walls, residential	1 Skwk	1200	.007	SF Flr.	.33	.37		.70	.94
0030	SS mesh, no chemicals, avg 1,400 S.F. home, min		1000	.008		.33	.44		.77	1.06
0100	Commercial, minimum		2496	.003		.35	.18		.53	.67
0200	Maximum		1645	.005		.53	.27		.80	1
0390	Minimum labor/equipment charge		4	2	Job		110		110	175
0400	Insecticides for termite control, minimum		14.20	.563	Gal.	70.50	31		101.50	127
0500	Maximum		11	.727	"	120	40		160	196

31 41 Shoring

31 41 13 – Timber Shoring

31 41 13.10 Building Shoring

		Crew	Daily Output	Labor-Hours	Unit	Material	2020 Bare Costs Labor	Equipment	Total	Total Incl O&P
0010	**BUILDING SHORING**									
0020	Shoring, existing building, with timber, no salvage allowance	B-51	2.20	21.818	M.B.F.	860	945	89	1,894	2,575
1000	On cribbing with 35 ton screw jacks, per box and jack		3.60	13.333	Jack	65	575	54.50	694.50	1,050
1090	Minimum labor/equipment charge		2	24	Ea.		1,050	98	1,148	1,775
1100	Masonry openings in walls, see Section 02 41 19.16									

31 41 16 – Sheet Piling

31 41 16.10 Sheet Piling Systems

		Crew	Daily Output	Labor-Hours	Unit	Material	2020 Bare Costs Labor	Equipment	Total	Total Incl O&P
0010	**SHEET PILING SYSTEMS**									
0020	Sheet piling, 50,000 psi steel, not incl. wales, 22 psf, left in place	B-40	10.81	5.920	Ton	1,875	325	320	2,520	2,925
0100	Drive, extract & salvage	R314116-40	6	10.667	"	525	590	580	1,695	2,150
1200	15' deep excavation, 22 psf, left in place		983	.065	S.F.	21.50	3.60	3.54	28.64	33.50
1300	Drive, extract & salvage		545	.117	"	5.90	6.50	6.40	18.80	24
2100	Rent steel sheet piling and wales, first month				Ton	335			335	370
2200	Per added month				"	33.50			33.50	37
3900	Wood, solid sheeting, incl. wales, braces and spacers,									
3910	drive, extract & salvage, 8' deep excavation	B-31	330	.121	S.F.	2.06	5.40	.78	8.24	11.80
4520	Left in place, 8' deep, 55 S.F./hr.		440	.091	"	3.70	4.06	.58	8.34	11.20
4990	Minimum labor/equipment charge		2	20	Job		895	129	1,024	1,575
5000	For treated lumber add cost of treatment to lumber									

31 43 Concrete Raising

31 43 13 – Pressure Grouting

31 43 13.13 Concrete Pressure Grouting	Crew	Daily Output	Labor-Hours	Unit	Material	2020 Bare Costs Labor	Equipment	Total	Total Incl O&P
0010 **CONCRETE PRESSURE GROUTING**									
0020 Grouting, pressure, cement & sand, 1:1 mix, minimum	B-61	124	.323	Bag	19.30	14.40	2.49	36.19	47
0100 Maximum		51	.784	"	19.30	35	6.05	60.35	84
0200 Cement and sand, 1:1 mix, minimum		250	.160	C.F.	38.50	7.15	1.23	46.88	55.50
0300 Maximum		100	.400		58	17.85	3.08	78.93	96
0400 Epoxy cement grout, minimum		137	.292		785	13.05	2.25	800.30	890
0500 Maximum		57	.702		785	31.50	5.40	821.90	920
0700 Alternate pricing method: (Add for materials)									
0710 5 person crew and equipment	B-61	1	40	Day		1,775	310	2,085	3,200

31 46 Needle Beams

31 46 13 – Cantilever Needle Beams

31 46 13.10 Needle Beams

		Crew	Daily Output	Labor-Hours	Unit	Material	2020 Bare Costs Labor	Equipment	Total	Total Incl O&P
0010	**NEEDLE BEAMS**									
0011	Incl. wood shoring 10' x 10' opening									
0400	Block, concrete, 8" thick	B-9	7.10	5.634	Ea.	59	239	50.50	348.50	505
0420	12" thick		6.70	5.970		68	254	53.50	375.50	540
0800	Brick, 4" thick with 8" backup block		5.70	7.018		68	298	62.50	428.50	625
1000	Brick, solid, 8" thick		6.20	6.452		59	274	57.50	390.50	570
1040	12" thick		4.90	8.163		68	345	73	486	710
1080	16" thick		4.50	8.889		87	380	79.50	546.50	790
2000	Add for additional floors of shoring	B-1	6	4		59	171		230	340
9000	Minimum labor/equipment charge	"	2	12	Job		515		515	820

31 48 Underpinning

31 48 13 – Underpinning Piers

31 48 13.10 Underpinning Foundations

		Crew	Daily Output	Labor-Hours	Unit	Material	2020 Bare Costs Labor	Equipment	Total	Total Incl O&P
0010	**UNDERPINNING FOUNDATIONS**									
0011	Including excavation,									
0020	forming, reinforcing, concrete and equipment									
0100	5' to 16' below grade, 100 to 500 C.Y.	B-52	2.30	24.348	C.Y.	340	1,175	254	1,769	2,550
0200	Over 500 C.Y.		2.50	22.400		305	1,100	234	1,639	2,350
0400	16' to 25' below grade, 100 to 500 C.Y.		2	28		370	1,375	292	2,037	2,900
0500	Over 500 C.Y.		2.10	26.667		350	1,300	278	1,928	2,775
0700	26' to 40' below grade, 100 to 500 C.Y.		1.60	35		405	1,700	365	2,470	3,575
0800	Over 500 C.Y.		1.80	31.111		370	1,525	325	2,220	3,200
0900	For under 50 C.Y., add					10%	40%			

31 62 Driven Piles

31 62 13 – Concrete Piles

31 62 13.23 Prestressed Concrete Piles

		Crew	Daily Output	Labor-Hours	Unit	Material	2020 Bare Costs Labor	Equipment	Total	Total Incl O&P
0010	**PRESTRESSED CONCRETE PILES**, 200 piles									
0020	Unless specified otherwise, not incl. pile caps or mobilization									
3100	Precast, prestressed, 40' long, 10" thick, square	B-19	700	.091	V.L.F.	14.10	5.05	2.87	22.02	26.50
3200	12" thick, square		680	.094		26	5.20	2.96	34.16	40
3400	14" thick, square	↓	600	.107	↓	29	5.90	3.35	38.25	45

31 62 16 – Steel Piles

31 62 16.13 Steel Piles

		Crew	Daily Output	Labor-Hours	Unit	Material	2020 Bare Costs Labor	Equipment	Total	Total Incl O&P
0010	**STEEL PILES**									
0100	Step tapered, round, concrete filled									
0110	8" tip, 12" butt, 60 ton capacity, 30' depth	B-19	760	.084	V.L.F.	18.85	4.65	2.64	26.14	31
0120	60' depth with extension		740	.086		36.50	4.78	2.72	44	51
0130	80' depth with extensions		700	.091		57	5.05	2.87	64.92	73.50
0250	"H" Sections, 50' long, HP8 x 36	↓	640	.100	↓	18.35	5.50	3.14	26.99	32.50

31 62 19 – Timber Piles

31 62 19.10 Wood Piles

		Crew	Daily Output	Labor-Hours	Unit	Material	2020 Bare Costs Labor	Equipment	Total	Total Incl O&P
0010	**WOOD PILES**									
0011	Friction or end bearing, not including									
0050	mobilization or demobilization									
0100	ACZA Treated piles, 1.0 lb/C.F., up to 30' long, 12" butts, 8" points	B-19	625	.102	V.L.F.	13.70	5.65	3.22	22.57	27.50
0200	30' to 39' long, 12" butts, 8" points	"	700	.091	"	14.45	5.05	2.87	22.37	27
0800	ACZA Treated piles, 1.5 lb./C.F.									
0810	friction or end bearing, ASTM class B									
1000	Up to 30' long, 12" butts, 8" points	B-19	625	.102	V.L.F.	18.20	5.65	3.22	27.07	32.50
1100	30' to 39' long, 12" butts, 8" points		700	.091		18.90	5.05	2.87	26.82	32
2700	Mobilization for 10,000 L.F. pile job, add		3300	.019			1.07	.61	1.68	2.37
2800	25,000 L.F. pile job, add	↓	8500	.008	↓		.42	.24	.66	.92

31 62 23 – Composite Piles

31 62 23.13 Concrete-Filled Steel Piles

		Crew	Daily Output	Labor-Hours	Unit	Material	2020 Bare Costs Labor	Equipment	Total	Total Incl O&P
0010	**CONCRETE-FILLED STEEL PILES** no mobilization or demobilization									
2600	Pipe piles, 50' L, 8" diam., 29 lb./L.F., no concrete	B-19	500	.128	V.L.F.	21.50	7.05	4.02	32.57	39
2700	Concrete filled		460	.139		25.50	7.70	4.37	37.57	45
2900	10" diameter, 34 lb./L.F., no concrete		500	.128		24	7.05	4.02	35.07	42
3000	Concrete filled		450	.142		30	7.85	4.47	42.32	50.50
3200	12" diameter, 44 lb./L.F., no concrete		475	.135		31.50	7.45	4.23	43.18	51
3300	Concrete filled		415	.154		35.50	8.50	4.84	48.84	58
3500	14" diameter, 46 lb./L.F., no concrete		430	.149		33.50	8.20	4.67	46.37	55
3600	Concrete filled	↓	355	.180	↓	41.50	9.95	5.65	57.10	67.50

31 63 Bored Piles

31 63 26 – Drilled Caissons

31 63 26.13 Fixed End Caisson Piles

			Crew	Daily Output	Labor-Hours	Unit	Material	2020 Bare Costs Labor	Equipment	Total	Total Incl O&P
0010	**FIXED END CAISSON PILES**										
0015	Including excavation, concrete, 50 lb. reinforcing										
0020	per C.Y., not incl. mobilization, boulder removal, disposal										
0100	Open style, machine drilled, to 50' deep, in stable ground, no	R316326-60									
0110	casings or ground water, 18" diam., 0.065 C.Y./L.F.		B-43	200	.240	V.L.F.	10.20	11.20	3.85	25.25	33
0200	24" diameter, 0.116 C.Y./L.F.		↓	190	.253	"	18.20	11.80	4.06	34.06	43
0210	4' bell diameter, add			20	2.400	Ea.	278	112	38.50	428.50	525
0500	48" diameter, 0.465 C.Y./L.F.		↓	100	.480	V.L.F.	73	22.50	7.70	103.20	125

31 63 Bored Piles

31 63 26 – Drilled Caissons

31 63 26.13 Fixed End Caisson Piles

		Crew	Daily Output	Labor-Hours	Unit	Material	2020 Bare Costs Labor	Equipment	Total	Total Incl O&P
0510	9' bell diameter, add	B-43	2	24	Ea.	1,575	1,125	385	3,085	3,925
1200	Open style, machine drilled, to 50' deep, in wet ground, pulled									
1220	casing and pumping									
1400	24" diameter, 0.116 C.Y./L.F.	B-48	125	.448	V.L.F.	18.20	21.50	8.40	48.10	63
1410	4' bell diameter, add	"	19.80	2.828	Ea.	278	135	53	466	575
1700	48" diameter, 0.465 C.Y./L.F.	B-49	55	1.600	V.L.F.	73	80	29.50	182.50	240
1710	9' bell diameter, add	"	3.30	26.667	Ea.	1,575	1,325	495	3,395	4,400
2300	Open style, machine drilled, to 50' deep, in soft rocks and									
2320	medium hard shales									
2500	24" diameter, 0.116 C.Y./L.F.	B-49	30	2.933	V.L.F.	18.20	147	54.50	219.70	315
2510	4' bell diameter, add		10.90	8.073	Ea.	278	405	149	832	1,100
2800	48" diameter, 0.465 C.Y./L.F.		10	8.800	V.L.F.	73	440	163	676	960
2810	9' bell diameter, add		1.10	80	Ea.	1,175	4,000	1,475	6,650	9,275
3600	For rock excavation, sockets, add, minimum		120	.733	C.F.		37	13.55	50.55	73
3650	Average		95	.926			46.50	17.15	63.65	92.50
3700	Maximum		48	1.833			92	34	126	184
3900	For 50' to 100' deep, add				V.L.F.				7%	7%
4000	For 100' to 150' deep, add								25%	25%
4100	For 150' to 200' deep, add								30%	30%
4200	For casings left in place, add				Lb.	1.38			1.38	1.52
4300	For other than 50 lb. reinf. per C.Y., add or deduct				"	1.29			1.29	1.42
4400	For steel I-beam cores, add	B-49	8.30	10.602	Ton	2,300	530	196	3,026	3,575
4500	Load and haul excess excavation, 2 miles	B-34B	178	.045	L.C.Y.		2.20	3.22	5.42	7.05
4600	For mobilization, 50 mile radius, rig to 36"	B-43	2	24	Ea.		1,125	385	1,510	2,200
4650	Rig to 84"	B-48	1.75	32			1,525	600	2,125	3,050
4700	For low headroom, add								50%	50%
4750	For difficult access, add								25%	25%

31 63 29 – Drilled Concrete Piers and Shafts

31 63 29.13 Uncased Drilled Concrete Piers

		Crew	Daily Output	Labor-Hours	Unit	Material	2020 Bare Costs Labor	Equipment	Total	Total Incl O&P
0010	**UNCASED DRILLED CONCRETE PIERS**									
0020	Unless specified otherwise, not incl. pile caps or mobilization									
0800	Cast in place friction pile, 50' long, fluted,									
0810	tapered steel, 4,000 psi concrete, no reinforcing									
0900	12" diameter, 7 ga.	B-19	600	.107	V.L.F.	29.50	5.90	3.35	38.75	45.50
1200	18" diameter, 7 ga.	"	480	.133	"	44.50	7.35	4.19	56.04	65
1300	End bearing, fluted, constant diameter,									
1320	4,000 psi concrete, no reinforcing									
1340	12" diameter, 7 ga.	B-19	600	.107	V.L.F.	31	5.90	3.35	40.25	47
1400	18" diameter, 7 ga.	"	480	.133	"	49.50	7.35	4.19	61.04	71

Estimating Tips
32 01 00 Operations and Maintenance of Exterior Improvements

- Recycling of asphalt pavement is becoming very popular and is an alternative to removal and replacement. It can be a good value engineering proposal if removed pavement can be recycled, either at the project site or at another site that is reasonably close to the project site. Sections on repair of flexible and rigid pavement are included.

32 10 00 Bases, Ballasts, and Paving

- When estimating paving, keep in mind the project schedule. Also note that prices for asphalt and concrete are generally higher in the cold seasons. Lines for pavement markings, including tactile warning systems and fence lines, are included.

32 90 00 Planting

- The timing of planting and guarantee specifications often dictate the costs for establishing tree and shrub growth and a stand of grass or ground cover. Establish the work performance schedule to coincide with the local planting season. Maintenance and growth guarantees can add 20–100% to the total landscaping cost and can be contractually cumbersome. The cost to replace trees and shrubs can be as high as 5% of the total cost, depending on the planting zone, soil conditions, and time of year.

Reference Numbers

Reference numbers are shown at the beginning of some major classifications. These numbers refer to related items in the Reference Section. The reference information may be an estimating procedure, an alternate pricing method, or technical information.

Note: Not all subdivisions listed here necessarily appear. ■

Same Data. Simplified.

Enjoy the convenience and efficiency of accessing your costs anywhere:

- **Skip the multiplier** by setting your location
- **Quickly search,** edit, favorite and share costs
- **Stay on top of price changes** with automatic updates

Discover more at rsmeans.com/online

32 01 11 – Paving Cleaning

32 01 11.51 Rubber and Paint Removal From Paving

		Crew	Daily Output	Labor-Hours	Unit	Material	2020 Bare Costs Labor	Equipment	Total	Total Incl O&P
0000	**RUBBER AND PAINT REMOVAL FROM PAVING**									
2000	Minimum labor/equipment charge	1 Clab	2	4	Job		168		168	253

32 01 11.51 Rubber and Paint Removal From Paving

		Crew	Daily Output	Labor-Hours	Unit	Material	2020 Bare Costs Labor	Equipment	Total	Total Incl O&P
1000	Prepare and clean, 25,000 S.Y.	B-91A	7.36	2.174	Mile		103	112	215	288
2000	Minimum labor/equipment charge	1 Clab	2	4	Job		168		168	270

32 01 13 – Flexible Paving Surface Treatment

32 01 13.61 Slurry Seal (Latex Modified)

		Crew	Daily Output	Labor-Hours	Unit	Material	2020 Bare Costs Labor	Equipment	Total	Total Incl O&P
0010	**SLURRY SEAL (LATEX MODIFIED)**									
0011	Chip seal, slurry seal, and microsurfacing, see section 32 12 36									

32 01 13.62 Asphalt Surface Treatment

		Crew	Daily Output	Labor-Hours	Unit	Material	2020 Bare Costs Labor	Equipment	Total	Total Incl O&P
0010	**ASPHALT SURFACE TREATMENT**									
3000	Pavement overlay, polypropylene									
3040	6 oz./S.Y., ideal conditions	B-63	10000	.004	S.Y.	1.09	.18	.02	1.29	1.50
3190	Prime coat, bituminous, 0.28 gallon/S.Y.	B-45	2400	.007	C.S.F.	8.20	.35	.34	8.89	10
3200	Tack coat, emulsion, 0.05 gal./S.Y., 1,000 S.Y.		2500	.006	S.Y.	.35	.34	.33	1.02	1.28
3240	10,000 S.Y.		10000	.002		.28	.08	.08	.44	.53
3275	10,000 S.Y.		10000	.002		.53	.08	.08	.69	.80
3280	0.15 gal./S.Y., 1,000 S.Y.		2500	.006		.96	.34	.33	1.63	1.95
3320	10,000 S.Y.		10000	.002		.77	.08	.08	.93	1.07
3780	Rubberized asphalt (latex) seal		5000	.003		1.39	.17	.16	1.72	1.98
5400	Thermoplastic coal-tar, Type I, small or irregular area	B-90	2400	.027		4.32	1.25	.92	6.49	7.75
5450	Roadway or large area		8000	.008		4.32	.37	.27	4.96	5.65
5500	Type II, small or irregular area		2400	.027		5.50	1.25	.92	7.67	9.05
6000	Gravel surfacing on asphalt, screened and rolled	B-11L	160	.100	C.Y.	23	4.94	6.65	34.59	40.50
7000	For subbase treatment, see Section 31 32 13									

32 01 13.64 Sand Seal

		Crew	Daily Output	Labor-Hours	Unit	Material	2020 Bare Costs Labor	Equipment	Total	Total Incl O&P
0010	**SAND SEAL**									
2080	Sand sealing, sharp sand, asphalt emulsion, small area	B-91	10000	.006	S.Y.	1.48	.32	.23	2.03	2.39
2120	Roadway or large area	"	18000	.004	"	1.27	.18	.13	1.58	1.82
3000	Sealing random cracks, min. 1/2" wide, to 1-1/2", 1,000 L.F.	B-77	2800	.014	L.F.	1.65	.62	.37	2.64	3.22
3040	10,000 L.F.		4000	.010	"	1.14	.44	.26	1.84	2.24
3080	Alternate method, 1,000 L.F.		200	.200	Gal.	45	8.70	5.15	58.85	69
3120	10,000 L.F.		325	.123	"	36	5.35	3.16	44.51	52
3200	Multi-cracks (flooding), 1 coat, small area	B-92	460	.070	S.Y.	2.55	2.96	2.57	8.08	10.40
3320	Large area		1425	.022	"	14.35	.96	.83	16.14	18.25
3360	Alternate method, small area		115	.278	Gal.	15.55	11.85	10.30	37.70	47.50
3400	Large area		715	.045	"	14.60	1.91	1.65	18.16	21

32 01 13.66 Fog Seal

		Crew	Daily Output	Labor-Hours	Unit	Material	2020 Bare Costs Labor	Equipment	Total	Total Incl O&P
0010	**FOG SEAL**									
0012	Sealcoating, 2 coat coal tar pitch emulsion over 10,000 S.Y.	B-45	5000	.003	S.Y.	.84	.17	.16	1.17	1.37
0030	1,000 to 10,000 S.Y.	"	3000	.005		.84	.28	.27	1.39	1.67
0100	Under 1,000 S.Y.	B-1	1050	.023		.84	.98		1.82	2.49
0300	Petroleum resistant, over 10,000 S.Y.	B-45	5000	.003		1.40	.17	.16	1.73	1.99
0320	1,000 to 10,000 S.Y.	"	3000	.005		1.40	.28	.27	1.95	2.29
0400	Under 1,000 S.Y.	B-1	1050	.023		1.40	.98		2.38	3.11
0600	Non-skid pavement renewal, over 10,000 S.Y.	B-45	5000	.003		1.37	.17	.16	1.70	1.96
0620	1,000 to 10,000 S.Y.	"	3000	.005		1.37	.28	.27	1.92	2.26
0700	Under 1,000 S.Y.	B-1	1050	.023		1.37	.98		2.35	3.08
0800	Prepare and clean surface for above	A-2	8545	.003			.12	.02	.14	.23
1000	Hand seal asphalt curbing	B-1	4420	.005	L.F.	.59	.23		.82	1.02
1900	Asphalt surface treatment, single course, small area									

32 01 Operation and Maintenance of Exterior Improvements

32 01 13 – Flexible Paving Surface Treatment

32 01 13.66 Fog Seal

		Crew	Daily Output	Labor-Hours	Unit	Material	2020 Bare Costs Labor	2020 Bare Costs Equipment	Total	Total Incl O&P
1901	0.30 gal./S.Y. asphalt material, 20 lb./S.Y. aggregate	B-91	5000	.013	S.Y.	1.41	.65	.45	2.51	3.07
1910	Roadway or large area		10000	.006		1.30	.32	.23	1.85	2.19
1950	Asphalt surface treatment, dbl. course for small area		3000	.021		2.97	1.08	.75	4.80	5.80
1960	Roadway or large area		6000	.011		2.67	.54	.38	3.59	4.20
1980	Asphalt surface treatment, single course, for shoulders		7500	.009		1.46	.43	.30	2.19	2.62

32 01 16 – Flexible Paving Rehabilitation

32 01 16.71 Cold Milling Asphalt Paving

		Crew	Daily Output	Labor-Hours	Unit	Material	Labor	Equipment	Total	Total Incl O&P
0010	**COLD MILLING ASPHALT PAVING**									
5200	Cold planing & cleaning, 1" to 3" asphalt pavmt., over 25,000 S.Y.	B-71	6000	.009	S.Y.		.45	.76	1.21	1.56
5280	5,000 S.Y. to 10,000 S.Y.		4000	.014	"		.68	1.15	1.83	2.34
5285	5,000 S.Y. to 10,000 S.Y.		36000	.002	S.F.		.08	.13	.21	.26
5300	Asphalt pavement removal from conc. base, no haul									
5320	Rip, load & sweep 1" to 3"	B-70	8000	.007	S.Y.		.34	.29	.63	.86
5330	3" to 6" deep	"	5000	.011			.55	.46	1.01	1.36
5340	Profile grooving, asphalt pavement load & sweep, 1" deep	B-71	12500	.004			.22	.37	.59	.75
5350	3" deep		9000	.006			.30	.51	.81	1.04
5360	6" deep		5000	.011			.55	.92	1.47	1.87

32 01 16.73 In Place Cold Reused Asphalt Paving

		Crew	Daily Output	Labor-Hours	Unit	Material	Labor	Equipment	Total	Total Incl O&P
0010	**IN PLACE COLD REUSED ASPHALT PAVING**									
5000	Reclamation, pulverizing and blending with existing base									
5040	Aggregate base, 4" thick pavement, over 15,000 S.Y.	B-73	2400	.027	S.Y.		1.37	1.87	3.24	4.23
5080	5,000 S.Y. to 15,000 S.Y.		2200	.029			1.50	2.04	3.54	4.60
5120	8" thick pavement, over 15,000 S.Y.		2200	.029			1.50	2.04	3.54	4.60
5160	5,000 S.Y. to 15,000 S.Y.		2000	.032			1.65	2.24	3.89	5.05

32 01 16.74 In Place Hot Reused Asphalt Paving

			Crew	Daily Output	Labor-Hours	Unit	Material	Labor	Equipment	Total	Total Incl O&P
0010	**IN PLACE HOT REUSED ASPHALT PAVING**										
5500	Recycle asphalt pavement at site										
5520	Remove, rejuvenate and spread 4" deep	G	B-72	2500	.026	S.Y.	4.58	1.27	3.29	9.14	10.70
5521	6" deep	G	"	2000	.032	"	6.70	1.59	4.11	12.40	14.45

32 01 17 – Flexible Paving Repair

32 01 17.10 Repair of Asphalt Pavement Holes

		Crew	Daily Output	Labor-Hours	Unit	Material	Labor	Equipment	Total	Total Incl O&P
0010	**REPAIR OF ASPHALT PAVEMENT HOLES** (cold patch)									
0100	Flexible pavement repair holes, roadway, light traffic, 1 C.F. size each	B-37A	24	1	Ea.	9.50	44	14.70	68.20	96.50
0150	Group of two, 1 C.F. size each		16	1.500	Set	19	65.50	22	106.50	150
0200	Group of three, 1 C.F. size each		12	2	"	28.50	87.50	29.50	145.50	204
0300	Medium traffic, 1 C.F. size each	B-37B	24	1.333	Ea.	9.50	58	14.70	82.20	119
0350	Group of two, 1 C.F. size each		16	2	Set	19	87	22	128	184
0400	Group of three, 1 C.F. size each		12	2.667	"	28.50	116	29.50	174	249
0500	Highway/heavy traffic, 1 C.F. size each	B-37C	24	1.333	Ea.	9.50	59.50	23	92	131
0550	Group of two, 1 C.F. size each		16	2	Set	19	89.50	34.50	143	202
0600	Group of three, 1 C.F. size each		12	2.667	"	28.50	119	45.50	193	273
0700	Add police officer and car for traffic control				Hr.	54.50			54.50	60
1000	Flexible pavement repair holes, parking lot, bag material, 1 C.F. size each	B-37D	18	.889	Ea.	59	39.50	6.15	104.65	135
1010	Economy bag material, 1 C.F. size each		18	.889	"	37	39.50	6.15	82.65	111
1100	Group of two, 1 C.F. size each		12	1.333	Set	118	59.50	9.25	186.75	235
1110	Group of two, economy bag, 1 C.F. size each		12	1.333		73.50	59.50	9.25	142.25	187
1200	Group of three, 1 C.F. size each		10	1.600		176	71.50	11.10	258.60	320
1210	Economy bag, group of three, 1 C.F. size each		10	1.600		110	71.50	11.10	192.60	247
1300	Flexible pavement repair holes, parking lot, 1 C.F. single hole	A-3A	4	2	Ea.	59	106	44	209	279
1310	Economy material, 1 C.F. single hole		4	2	"	37	106	44	187	255
1400	Flexible pavement repair holes, parking lot, 1 C.F. four holes		2	4	Set	235	212	88	535	685

32 01 17.10 Repair of Asphalt Pavement Holes

		Crew	Daily Output	Labor-Hours	Unit	Material	2020 Bare Costs Labor	Equipment	Total	Total Incl O&P
1410	Economy material, 1 C.F. four holes	A-3A	2	4	Set	147	212	88	447	590
1500	Flexible pavement repair holes, large parking lot, bulk matl, 1 C.F. size	B-37A	32	.750	Ea.	9.50	33	11	53.50	75

32 01 17.20 Repair of Asphalt Pavement Patches

		Crew	Daily Output	Labor-Hours	Unit	Material	2020 Bare Costs Labor	Equipment	Total	Total Incl O&P
0010	**REPAIR OF ASPHALT PAVEMENT PATCHES**									
0100	Flexible pavement patches, roadway, light traffic, sawcut 10-25 S.F.	B-89	12	1.333	Ea.		67	86	153	201
0150	Sawcut 26-60 S.F.		10	1.600			80	103	183	241
0200	Sawcut 61-100 S.F.		8	2			100	129	229	300
0210	Sawcut groups of small size patches		24	.667			33.50	43	76.50	101
0220	Large size patches	▼	16	1			50	64.50	114.50	151
0300	Flexible pavement patches, roadway, light traffic, digout 10-25 S.F.	B-6	16	1.500			68.50	13.35	81.85	124
0350	Digout 26-60 S.F.		12	2			91.50	17.80	109.30	165
0400	Digout 61-100 S.F.	▼	8	3	▼		137	26.50	163.50	248
0450	Add 8 C.Y. truck, small project debris haulaway	B-34A	8	1	Hr.		49	53.50	102.50	138
0460	Add 12 C.Y. truck, small project debris haulaway	B-34B	8	1			49	71.50	120.50	158
0480	Add flagger for non-intersection medium traffic	1 Clab	8	1			42		42	67.50
0490	Add flasher truck for intersection medium traffic or heavy traffic	A-2B	8	1	▼		47.50	24.50	72	103
0500	Flexible pavement patches, roadway, repave, cold, 15 S.F., 4" D	B-37	8	6	Ea.	45	266	32	343	510
0510	6" depth		8	6		68	266	32	366	535
0520	Repave, cold, 20 S.F., 4" depth		8	6		60	266	32	358	525
0530	6" depth		8	6		90.50	266	32	388.50	560
0540	Repave, cold, 25 S.F., 4" depth		8	6		74.50	266	32	372.50	540
0550	6" depth		8	6		113	266	32	411	585
0600	Repave, cold, 30 S.F., 4" depth		8	6		89.50	266	32	387.50	560
0610	6" depth		8	6		136	266	32	434	610
0640	Repave, cold, 40 S.F., 4" depth		8	6		119	266	32	417	590
0650	6" depth		8	6		181	266	32	479	660
0680	Repave, cold, 50 S.F., 4" depth		8	6		149	266	32	447	625
0690	6" depth		8	6		226	266	32	524	710
0720	Repave, cold, 60 S.F., 4" depth		8	6		179	266	32	477	655
0730	6" depth		8	6		271	266	32	569	760
0800	Repave, cold, 70 S.F., 4" depth		7	6.857		209	305	36.50	550.50	755
0810	6" depth		7	6.857		315	305	36.50	656.50	875
0820	Repave, cold, 75 S.F., 4" depth		7	6.857		222	305	36.50	563.50	770
0830	6" depth		7	6.857		340	305	36.50	681.50	900
0900	Repave, cold, 80 S.F., 4" depth		6	8		239	355	42.50	636.50	875
0910	6" depth		6	8		360	355	42.50	757.50	1,000
0940	Repave, cold, 90 S.F., 4" depth		6	8		268	355	42.50	665.50	905
0950	6" depth		6	8		405	355	42.50	802.50	1,050
0980	Repave, cold, 100 S.F., 4" depth		6	8		298	355	42.50	695.50	940
0990	6" depth	▼	6	8	▼	450	355	42.50	847.50	1,100
1000	Add flasher truck for paving operations in medium or heavy traffic	A-2B	8	1	Hr.		47.50	24.50	72	103
1100	Prime coat for repair 15-40 S.F., 25% overspray	B-37A	48	.500	Ea.	8.75	22	7.35	38.10	53
1150	41-60 S.F.		40	.600		17.50	26.50	8.80	52.80	71
1175	61-80 S.F.		32	.750		23.50	33	11	67.50	90.50
1200	81-100 S.F.		32	.750	▼	29	33	11	73	96.50
1210	Groups of patches w/25% overspray	▼	3600	.007	S.F.	.29	.29	.10	.68	.90
1300	Flexible pavement repair patches, street repave, hot, 60 S.F., 4" D	B-37	8	6	Ea.	97	266	32	395	565
1310	6" depth		8	6		147	266	32	445	620
1320	Repave, hot, 70 S.F., 4" depth		7	6.857		113	305	36.50	454.50	650
1330	6" depth		7	6.857		172	305	36.50	513.50	715
1340	Repave, hot, 75 S.F., 4" depth		7	6.857		121	305	36.50	462.50	660
1350	6" depth	▼	7	6.857		184	305	36.50	525.50	725

32 01 17 – Flexible Paving Repair

32 01 17.20 Repair of Asphalt Pavement Patches

		Crew	Daily Output	Labor-Hours	Unit	Material	2020 Bare Costs Labor	Equipment	Total	Total Incl O&P
1360	Repave, hot, 80 S.F., 4" depth	B-37	6	8	Ea.	130	355	42.50	527.50	755
1370	6" depth		6	8		196	355	42.50	593.50	830
1380	Repave, hot, 90 S.F., 4" depth		6	8		146	355	42.50	543.50	770
1390	6" depth		6	8		221	355	42.50	618.50	855
1400	Repave, hot, 100 S.F., 4" depth		6	8		162	355	42.50	559.50	790
1410	6" depth		6	8		245	355	42.50	642.50	880
1420	Pave hot groups of patches, 4" depth		900	.053	S.F.	1.62	2.36	.28	4.26	5.85
1430	6" depth		900	.053	"	2.46	2.36	.28	5.10	6.75
1500	Add 8 C.Y. truck for hot asphalt paving operations	B-34A	1	8	Day		390	430	820	1,100
1550	Add flasher truck for hot paving operations in med/heavy traffic	A-2B	8	1	Hr.		47.50	24.50	72	103
2000	Add police officer and car for traffic control				"	54.50			54.50	60

32 01 17.61 Sealing Cracks In Asphalt Paving

		Crew	Daily Output	Labor-Hours	Unit	Material	2020 Bare Costs Labor	Equipment	Total	Total Incl O&P
0010	**SEALING CRACKS IN ASPHALT PAVING**									
0100	Sealing cracks in asphalt paving, 1/8" wide x 1/2" depth, slow set	B-37A	2000	.012	L.F.	.17	.53	.18	.88	1.22
0110	1/4" wide x 1/2" depth		2000	.012		.20	.53	.18	.91	1.25
0130	3/8" wide x 1/2" depth		2000	.012		.22	.53	.18	.93	1.28
0140	1/2" wide x 1/2" depth		1800	.013		.25	.58	.20	1.03	1.44
0150	3/4" wide x 1/2" depth		1600	.015		.31	.66	.22	1.19	1.63
0160	1" wide x 1/2" depth		1600	.015		.37	.66	.22	1.25	1.69
0165	1/8" wide x 1" depth, rapid set	B-37F	2000	.016		.18	.69	.26	1.13	1.60
0170	1/4" wide x 1" depth		2000	.016		.36	.69	.26	1.31	1.79
0175	3/8" wide x 1" depth		1800	.018		.53	.77	.29	1.59	2.15
0180	1/2" wide x 1" depth		1800	.018		.71	.77	.29	1.77	2.34
0185	5/8" wide x 1" depth		1800	.018		.89	.77	.29	1.95	2.54
0190	3/4" wide x 1" depth		1600	.020		1.07	.87	.33	2.27	2.93
0195	1" wide x 1" depth		1600	.020		1.42	.87	.33	2.62	3.32
0200	1/8" wide x 2" depth, rapid set		2000	.016		.36	.69	.26	1.31	1.79
0210	1/4" wide x 2" depth		2000	.016		.71	.69	.26	1.66	2.18
0220	3/8" wide x 2" depth		1800	.018		1.07	.77	.29	2.13	2.74
0230	1/2" wide x 2" depth		1800	.018		1.42	.77	.29	2.48	3.13
0240	5/8" wide x 2" depth		1800	.018		1.78	.77	.29	2.84	3.52
0250	3/4" wide x 2" depth		1600	.020		2.14	.87	.33	3.34	4.10
0260	1" wide x 2" depth		1600	.020		2.85	.87	.33	4.05	4.89
0300	Add flagger for non-intersection medium traffic	1 Clab	8	1	Hr.		42		42	67.50
0400	Add flasher truck for intersection medium traffic or heavy traffic	A-2B	8	1	"		47.50	24.50	72	103

32 01 19 – Rigid Paving Surface Treatment

32 01 19.61 Sealing of Joints In Rigid Paving

		Crew	Daily Output	Labor-Hours	Unit	Material	2020 Bare Costs Labor	Equipment	Total	Total Incl O&P
0010	**SEALING OF JOINTS IN RIGID PAVING**									
2000	Waterproofing, membrane, tar and fabric, small area	B-63	233	.172	S.Y.	15.40	7.60	.76	23.76	30
2500	Large area		1435	.028		13.60	1.23	.12	14.95	17.05
3000	Preformed rubberized asphalt, small area		100	.400		19.90	17.70	1.78	39.38	52
3500	Large area		367	.109		18.10	4.83	.48	23.41	28
9000	Minimum labor/equipment charge	1 Clab	2	4	Job		168		168	270

32 01 29 – Rigid Paving Repair

32 01 29.61 Partial Depth Patching of Rigid Pavement

		Crew	Daily Output	Labor-Hours	Unit	Material	2020 Bare Costs Labor	Equipment	Total	Total Incl O&P
0010	**PARTIAL DEPTH PATCHING OF RIGID PAVEMENT**									
0100	Rigid pavement repair, roadway, light traffic, 25% pitting 15 S.F.	B-37F	16	2	Ea.	24	87	32.50	143.50	201
0110	Pitting 20 S.F.		16	2		31.50	87	32.50	151	210
0120	Pitting 25 S.F.		16	2		39.50	87	32.50	159	219
0130	Pitting 30 S.F.		16	2		47.50	87	32.50	167	228
0140	Pitting 40 S.F.		16	2		63.50	87	32.50	183	245

For customer support on your Commercial Renovation Costs with RSMeans data, call 800.448.8182.

553

32 01 29.61 Partial Depth Patching of Rigid Pavement

		Crew	Daily Output	Labor-Hours	Unit	Material	2020 Bare Costs Labor	Equipment	Total	Total Incl O&P
0150	Pitting 50 S.F.	B-37F	12	2.667	Ea.	79	116	43.50	238.50	320
0160	Pitting 60 S.F.		12	2.667		95	116	43.50	254.50	340
0170	Pitting 70 S.F.		12	2.667		111	116	43.50	270.50	355
0180	Pitting 75 S.F.		12	2.667		119	116	43.50	278.50	365
0190	Pitting 80 S.F.		8	4		127	174	65	366	490
0200	Pitting 90 S.F.		8	4		143	174	65	382	505
0210	Pitting 100 S.F.		8	4		158	174	65	397	525
0300	Roadway, light traffic, 50% pitting 15 S.F.		12	2.667		47.50	116	43.50	207	286
0310	Pitting 20 S.F.		12	2.667		63.50	116	43.50	223	305
0320	Pitting 25 S.F.		12	2.667		79	116	43.50	238.50	320
0330	Pitting 30 S.F.		12	2.667		95	116	43.50	254.50	340
0340	Pitting 40 S.F.		12	2.667		127	116	43.50	286.50	370
0350	Pitting 50 S.F.		8	4		158	174	65	397	525
0360	Pitting 60 S.F.		8	4		190	174	65	429	560
0370	Pitting 70 S.F.		8	4		222	174	65	461	595
0380	Pitting 75 S.F.		8	4		238	174	65	477	610
0390	Pitting 80 S.F.		6	5.333		253	231	87	571	745
0400	Pitting 90 S.F.		6	5.333		285	231	87	603	780
0410	Pitting 100 S.F.		6	5.333		315	231	87	633	815
1000	Rigid pavement repair, light traffic, surface patch, 2" deep, 15 S.F.		8	4		95	174	65	334	455
1010	Surface patch, 2" deep, 20 S.F.		8	4		127	174	65	366	490
1020	Surface patch, 2" deep, 25 S.F.		8	4		158	174	65	397	525
1030	Surface patch, 2" deep, 30 S.F.		8	4		190	174	65	429	560
1040	Surface patch, 2" deep, 40 S.F.		8	4		253	174	65	492	630
1050	Surface patch, 2" deep, 50 S.F.		6	5.333		315	231	87	633	815
1060	Surface patch, 2" deep, 60 S.F.		6	5.333		380	231	87	698	885
1070	Surface patch, 2" deep, 70 S.F.		6	5.333		445	231	87	763	955
1080	Surface patch, 2" deep, 75 S.F.		6	5.333		475	231	87	793	990
1090	Surface patch, 2" deep, 80 S.F.		5	6.400		505	278	104	887	1,125
1100	Surface patch, 2" deep, 90 S.F.		5	6.400		570	278	104	952	1,175
1200	Surface patch, 2" deep, 100 S.F.		5	6.400		635	278	104	1,017	1,250
2000	Add flagger for non-intersection medium traffic	1 Clab	8	1	Hr.		42		42	67.50
2100	Add flasher truck for intersection medium traffic or heavy traffic	A-2B	8	1			47.50	24.50	72	103
2200	Add police officer and car for traffic control					54.50			54.50	60

32 01 29.70 Full Depth Patching of Rigid Pavement

		Crew	Daily Output	Labor-Hours	Unit	Material	2020 Bare Costs Labor	Equipment	Total	Total Incl O&P
0010	**FULL DEPTH PATCHING OF RIGID PAVEMENT**									
0015	Pavement preparation includes sawcut, remove pavement and replace									
0020	6" of subbase and one layer of reinforcement									
0030	Trucking and haulaway of debris excluded									
0100	Rigid pavement replace, light traffic, prep, 15 S.F., 6" depth	B-37E	16	3.500	Ea.	23	165	58	246	350
0110	Replacement preparation 20 S.F., 6" depth		16	3.500		30.50	165	58	253.50	360
0115	25 S.F., 6" depth		16	3.500		38	165	58	261	370
0120	30 S.F., 6" depth		14	4		46	189	66	301	425
0125	35 S.F., 6" depth		14	4		53.50	189	66	308.50	430
0130	40 S.F., 6" depth		14	4		61	189	66	316	440
0135	45 S.F., 6" depth		14	4		69	189	66	324	450
0140	50 S.F., 6" depth		12	4.667		76.50	220	77	373.50	520
0145	55 S.F., 6" depth		12	4.667		84	220	77	381	530
0150	60 S.F., 6" depth		10	5.600		92	264	92.50	448.50	625
0155	65 S.F., 6" depth		10	5.600		99.50	264	92.50	456	630
0160	70 S.F., 6" depth		10	5.600		107	264	92.50	463.50	640
0165	75 S.F., 6" depth		10	5.600		115	264	92.50	471.50	650

32 01 29 – Rigid Paving Repair

32 01 29.70 Full Depth Patching of Rigid Pavement	Crew	Daily Output	Labor-Hours	Unit	Material	2020 Bare Costs Labor	Equipment	Total	Total Incl O&P
0170	80 S.F., 6" depth	B-37E	8	7	Ea.	122	330	116	568	785
0175	85 S.F., 6" depth		8	7		130	330	116	576	795
0180	90 S.F., 6" depth		8	7		138	330	116	584	805
0185	95 S.F., 6" depth		8	7		145	330	116	591	810
0190	100 S.F., 6" depth		8	7		153	330	116	599	820
0200	Pavement preparation for 8" rigid paving same as 6"									
0290	Pavement preparation for 9" rigid paving same as 10"									
0300	Rigid pavement replace, light traffic, prep, 15 S.F., 10" depth	B-37E	16	3.500	Ea.	38	165	58	261	370
0302	Pavement preparation 10" includes sawcut, remove pavement and									
0304	replace 6" of subbase and two layers of reinforcement									
0306	Trucking and haulaway of debris excluded									
0310	Replacement preparation 20 S.F., 10" depth	B-37E	16	3.500	Ea.	51	165	58	274	385
0315	25 S.F., 10" depth		16	3.500		63.50	165	58	286.50	395
0320	30 S.F., 10" depth		14	4		76.50	189	66	331.50	455
0325	35 S.F., 10" depth		14	4		58.50	189	66	313.50	435
0330	40 S.F., 10" depth		14	4		102	189	66	357	485
0335	45 S.F., 10" depth		12	4.667		115	220	77	412	560
0340	50 S.F., 10" depth		12	4.667		127	220	77	424	575
0345	55 S.F., 10" depth		12	4.667		140	220	77	437	590
0350	60 S.F., 10" depth		10	5.600		153	264	92.50	509.50	690
0355	65 S.F., 10" depth		10	5.600		166	264	92.50	522.50	705
0360	70 S.F., 10" depth		10	5.600		178	264	92.50	534.50	720
0365	75 S.F., 10" depth		8	7		191	330	116	637	860
0370	80 S.F., 10" depth		8	7		204	330	116	650	875
0375	85 S.F., 10" depth		8	7		216	330	116	662	890
0380	90 S.F., 10" depth		8	7		229	330	116	675	905
0385	95 S.F., 10" depth		8	7		242	330	116	688	920
0390	100 S.F., 10" depth		8	7		255	330	116	701	930
0500	Add 8 C.Y. truck, small project debris haulaway	B-34A	8	1	Hr.		49	53.50	102.50	138
0550	Add 12 C.Y. truck, small project debris haulaway	B-34B	8	1			49	71.50	120.50	158
0600	Add flagger for non-intersection medium traffic	1 Clab	8	1			42		42	67.50
0650	Add flasher truck for intersection medium traffic or heavy traffic	A-2B	8	1			47.50	24.50	72	103
0700	Add police officer and car for traffic control					54.50			54.50	60
0990	Concrete will be replaced using quick set mix with aggregate									
1000	Rigid pavement replace, light traffic, repour, 15 S.F., 6" depth	B-37F	18	1.778	Ea.	174	77	29	280	350
1010	20 S.F., 6" depth		18	1.778		232	77	29	338	410
1015	25 S.F., 6" depth		18	1.778		290	77	29	396	475
1020	30 S.F., 6" depth		16	2		350	87	32.50	469.50	560
1025	35 S.F., 6" depth		16	2		405	87	32.50	524.50	620
1030	40 S.F., 6" depth		16	2		465	87	32.50	584.50	685
1035	45 S.F., 6" depth		16	2		525	87	32.50	644.50	750
1040	50 S.F., 6" depth		16	2		580	87	32.50	699.50	815
1045	55 S.F., 6" depth		12	2.667		640	116	43.50	799.50	940
1050	60 S.F., 6" depth		12	2.667		695	116	43.50	854.50	1,000
1055	65 S.F., 6" depth		12	2.667		755	116	43.50	914.50	1,075
1060	70 S.F., 6" depth		12	2.667		815	116	43.50	974.50	1,125
1065	75 S.F., 6" depth		10	3.200		870	139	52	1,061	1,250
1070	80 S.F., 6" depth		10	3.200		930	139	52	1,121	1,300
1075	85 S.F., 6" depth		8	4		990	174	65	1,229	1,425
1080	90 S.F., 6" depth		8	4		1,050	174	65	1,289	1,500
1085	95 S.F., 6" depth		8	4		1,100	174	65	1,339	1,575
1090	100 S.F., 6" depth		8	4		1,150	174	65	1,389	1,625
1099	Concrete will be replaced using 4,500 psi concrete ready mix									

For customer support on your Commercial Renovation Costs with RSMeans data, call 800.448.8182.

555

32 01 29 – Rigid Paving Repair

32 01 29.70 Full Depth Patching of Rigid Pavement

		Crew	Daily Output	Labor-Hours	Unit	Material	2020 Bare Costs Labor	Equipment	Total	Total Incl O&P
1100	Rigid pavement replace, light traffic, repour, 15 S.F., 6" depth	A-2	12	2	Ea.	345	87.50	16.35	448.85	540
1110	20-50 S.F., 6" depth		12	2		345	87.50	16.35	448.85	540
1120	55-65 S.F., 6" depth		10	2.400		380	105	19.60	504.60	610
1130	70-80 S.F., 6" depth		10	2.400		415	105	19.60	539.60	645
1140	85-100 S.F., 6" depth		8	3		485	131	24.50	640.50	770
1200	Repour 15-40 S.F., 8" depth		12	2		345	87.50	16.35	448.85	540
1210	45-50 S.F., 8" depth		12	2		380	87.50	16.35	483.85	580
1220	55-60 S.F., 8" depth		10	2.400		415	105	19.60	539.60	645
1230	65-80 S.F., 8" depth		10	2.400		485	105	19.60	609.60	725
1240	85-90 S.F., 8" depth		8	3		525	131	24.50	680.50	810
1250	95-100 S.F., 8" depth		8	3		555	131	24.50	710.50	845
1300	Repour 15-30 S.F., 10" depth		12	2		345	87.50	16.35	448.85	540
1310	35-40 S.F., 10" depth		12	2		380	87.50	16.35	483.85	580
1320	45 S.F., 10" depth		12	2		415	87.50	16.35	518.85	615
1330	50-65 S.F., 10" depth		10	2.400		485	105	19.60	609.60	725
1340	70 S.F., 10" depth		10	2.400		525	105	19.60	649.60	765
1350	75-80 S.F., 10" depth		10	2.400		555	105	19.60	679.60	800
1360	85-95 S.F., 10" depth		8	3		625	131	24.50	780.50	925
1370	100 S.F., 10" depth		8	3		665	131	24.50	820.50	965

32 06 Schedules for Exterior Improvements

32 06 10 – Schedules for Bases, Ballasts, and Paving

32 06 10.10 Sidewalks, Driveways and Patios

		Crew	Daily Output	Labor-Hours	Unit	Material	2020 Bare Costs Labor	Equipment	Total	Total Incl O&P
0010	**SIDEWALKS, DRIVEWAYS AND PATIOS** No base									
0020	Asphaltic concrete, 2" thick	B-37	720	.067	S.Y.	7.20	2.95	.36	10.51	13
0100	2-1/2" thick	"	660	.073	"	9.10	3.22	.39	12.71	15.65
0110	Bedding for brick or stone, mortar, 1" thick	D-1	300	.053	S.F.	.82	2.50		3.32	4.94
0120	2" thick	"	200	.080		2.05	3.75		5.80	8.30
0130	Sand, 2" thick	B-18	8000	.003		.34	.13	.02	.49	.61
0140	4" thick	"	4000	.006		.68	.26	.04	.98	1.21
0300	Concrete, 3,000 psi, CIP, 6 x 6 - W1.4 x W1.4 mesh,									
0310	broomed finish, no base, 4" thick	B-24	600	.040	S.F.	2.31	1.94		4.25	5.60
0350	5" thick		545	.044		2.86	2.13		4.99	6.55
0400	6" thick		510	.047		3.34	2.28		5.62	7.30
0450	For bank run gravel base, 4" thick, add	B-18	2500	.010		.43	.41	.07	.91	1.21
0520	8" thick, add	"	1600	.015		.87	.64	.10	1.61	2.10
0550	Exposed aggregate finish, add to above, minimum	B-24	1875	.013		.13	.62		.75	1.13
0600	Maximum	"	455	.053		.43	2.55		2.98	4.53
0950	Concrete tree grate, 5' square	B-6	25	.960	Ea.	440	44	8.55	492.55	565
0955	Tree well & cover, concrete, 3' square		25	.960		245	44	8.55	297.55	350
0960	Cast iron tree grate with frame, 2 piece, round, 5' diameter		25	.960		1,075	44	8.55	1,127.55	1,250
0980	Square, 5' side		25	.960		1,075	44	8.55	1,127.55	1,250
1700	Redwood, prefabricated, 4' x 4' sections	2 Carp	316	.051	S.F.	4.89	2.69		7.58	9.65
1750	Redwood planks, 1" thick, on sleepers	"	240	.067	"	4.89	3.54		8.43	11.05
2250	Stone dust, 4" thick	B-62	900	.027	S.Y.	5.25	1.22	.20	6.67	7.95
9000	Minimum labor/equipment charge	D-1	2	8	Job		375		375	605

32 06 10.20 Steps

		Crew	Daily Output	Labor-Hours	Unit	Material	2020 Bare Costs Labor	Equipment	Total	Total Incl O&P
0010	**STEPS**									
0011	Incl. excav., borrow & concrete base as required									
0100	Brick steps	B-24	35	.686	LF Riser	17.95	33		50.95	73
0200	Railroad ties	2 Clab	25	.640		3.75	27		30.75	47

32 06 Schedules for Exterior Improvements

32 06 10 – Schedules for Bases, Ballasts, and Paving

32 06 10.20 Steps		Crew	Daily Output	Labor-Hours	Unit	Material	2020 Bare Costs Labor	Equipment	Total	Total Incl O&P
0300	Bluestone treads, 12" x 2" or 12" x 1-1/2"	B-24	30	.800	LF Riser	44.50	38.50		83	111
0490	Minimum labor/equipment charge	D-1	2	8	Job		375		375	605
0500	Concrete, cast in place, see Section 03 30 53.40									
0600	Precast concrete, see Section 03 41 23.50									

32 11 Base Courses

32 11 23 – Aggregate Base Courses

32 11 23.23 Base Course Drainage Layers

		Crew	Daily Output	Labor-Hours	Unit	Material	2020 Bare Costs Labor	Equipment	Total	Total Incl O&P
0010	**BASE COURSE DRAINAGE LAYERS**									
0011	For Soil Stabilization, see Section 31 32									
0012	For roadways and large areas									
0050	Crushed 3/4" stone base, compacted, 3" deep	B-36C	5200	.008	S.Y.	2.70	.41	.82	3.93	4.50
0100	6" deep		5000	.008		5.40	.42	.85	6.67	7.55
0200	9" deep		4600	.009		8.10	.46	.92	9.48	10.65
0300	12" deep		4200	.010		10.80	.50	1.01	12.31	13.80
0301	Crushed 1-1/2" stone base, compacted to 4" deep	B-36B	6000	.011		4.87	.54	.82	6.23	7.10
0302	6" deep		5400	.012		7.30	.60	.91	8.81	10
0303	8" deep		4500	.014		9.75	.72	1.09	11.56	13.05
0304	12" deep		3800	.017		14.60	.85	1.29	16.74	18.85
0310	Minimum labor/equipment charge	B-36	1	40	Job		1,925	1,800	3,725	5,050
0350	Bank run gravel, spread and compacted									
0370	6" deep	B-32	6000	.005	S.Y.	3.62	.28	.47	4.37	4.93
0390	9" deep		4900	.007		5.45	.35	.57	6.37	7.15
0400	12" deep		4200	.008		7.25	.40	.67	8.32	9.30
6900	For small and irregular areas, add						50%	50%		
6990	Minimum labor/equipment charge	1 Clab	3	2.667	Job		112		112	180
7000	Prepare and roll sub-base, small areas to 2,500 S.Y.	B-32A	1500	.016	S.Y.		.83	1.15	1.98	2.58
8000	Large areas over 2,500 S.Y.	"	3500	.007			.36	.49	.85	1.10
8050	For roadways	B-32	4000	.008			.42	.70	1.12	1.44
9000	Minimum labor/equipment charge	1 Clab	4	2	Job		84		84	135

32 12 Flexible Paving

32 12 16 – Asphalt Paving

32 12 16.13 Plant-Mix Asphalt Paving

		Crew	Daily Output	Labor-Hours	Unit	Material	2020 Bare Costs Labor	Equipment	Total	Total Incl O&P
0010	**PLANT-MIX ASPHALT PAVING**									
0080	Binder course, 1-1/2" thick	B-25	7725	.011	S.Y.	5.55	.53	.35	6.43	7.30
0120	2" thick		6345	.014		7.40	.64	.43	8.47	9.60
0130	2-1/2" thick		5620	.016		9.25	.72	.48	10.45	11.85
0160	3" thick		4905	.018		11.05	.83	.55	12.43	14.15
0170	3-1/2" thick		4520	.019		12.90	.90	.60	14.40	16.30
0300	Wearing course, 1" thick	B-25B	10575	.009		3.66	.43	.28	4.37	5
0340	1-1/2" thick		7725	.012		6.15	.59	.38	7.12	8.10
0380	2" thick		6345	.015		8.25	.71	.46	9.42	10.75
0420	2-1/2" thick		5480	.018		10.20	.83	.54	11.57	13.10
0460	3" thick		4900	.020		12.15	.92	.60	13.67	15.50
0470	3-1/2" thick		4520	.021		14.25	1	.65	15.90	18
0480	4" thick		4140	.023		16.30	1.09	.71	18.10	20.50
0500	Open graded friction course	B-25C	5000	.010		2.55	.45	.47	3.47	4.04
0800	Alternate method of figuring paving costs									

32 12 Flexible Paving

32 12 16 – Asphalt Paving

32 12 16.13 Plant-Mix Asphalt Paving

		Crew	Daily Output	Labor-Hours	Unit	Material	2020 Bare Costs Labor	2020 Bare Costs Equipment	Total	Total Incl O&P
0810	Binder course, 1-1/2" thick	B-25	630	.140	Ton	68	6.45	4.29	78.74	89.50
0811	2" thick		690	.128		68	5.90	3.91	77.81	88
0812	3" thick		800	.110		68	5.10	3.37	76.47	86.50
0813	4" thick		900	.098		68	4.53	3	75.53	85
0850	Wearing course, 1" thick	B-25B	575	.167		73.50	7.85	5.10	86.45	99
0851	1-1/2" thick		630	.152		73.50	7.20	4.66	85.36	97.50
0852	2" thick		690	.139		73.50	6.55	4.26	84.31	96
0853	2-1/2" thick		765	.125		73.50	5.90	3.84	83.24	94.50
0854	3" thick		800	.120		73.50	5.65	3.67	82.82	94
1000	Pavement replacement over trench, 2" thick	B-37	90	.533	S.Y.	7.60	23.50	2.84	33.94	49
1050	4" thick		70	.686		15.10	30.50	3.65	49.25	69
1080	6" thick		55	.873		24	38.50	4.65	67.15	93
3000	Prime coat, emulsion, 0.30 gal./S.Y., 1000 S.Y.	B-45	2500	.006		2.10	.34	.33	2.77	3.20
3100	Tack coat, emulsion, 0.10 gal./S.Y., 1000 S.Y.	"	2500	.006		.70	.34	.33	1.37	1.66

32 12 16.14 Asphaltic Concrete Paving

		Crew	Daily Output	Labor-Hours	Unit	Material	2020 Bare Costs Labor	2020 Bare Costs Equipment	Total	Total Incl O&P
0011	**ASPHALTIC CONCRETE PAVING**, parking lots & driveways									
0015	No asphalt hauling included									
0020	6" stone base, 2" binder course, 1" topping	B-25C	9000	.005	S.F.	1.92	.25	.26	2.43	2.80
0025	2" binder course, 2" topping		9000	.005		2.36	.25	.26	2.87	3.29
0030	3" binder course, 2" topping		9000	.005		2.78	.25	.26	3.29	3.75
0035	4" binder course, 2" topping		9000	.005		3.19	.25	.26	3.70	4.19
0040	1-1/2" binder course, 1" topping		9000	.005		1.72	.25	.26	2.23	2.58
0042	3" binder course, 1" topping		9000	.005		2.33	.25	.26	2.84	3.26
0045	3" binder course, 3" topping		9000	.005		3.23	.25	.26	3.74	4.24
0050	4" binder course, 3" topping		9000	.005		3.63	.25	.26	4.14	4.69
0055	4" binder course, 4" topping		9000	.005		4.08	.25	.26	4.59	5.15
0300	Binder course, 1-1/2" thick		35000	.001		.62	.06	.07	.75	.85
0400	2" thick		25000	.002		.80	.09	.09	.98	1.12
0500	3" thick		15000	.003		1.23	.15	.16	1.54	1.77
0600	4" thick		10800	.004		1.62	.21	.22	2.05	2.35
0800	Sand finish course, 3/4" thick		41000	.001		.31	.06	.06	.43	.49
0900	1" thick		34000	.001		.38	.07	.07	.52	.61
1000	Fill pot holes, hot mix, 2" thick	B-16	4200	.008		.82	.34	.14	1.30	1.59
1100	4" thick		3500	.009		1.20	.41	.16	1.77	2.15
1120	6" thick		3100	.010		1.61	.46	.18	2.25	2.70
1140	Cold patch, 2" thick	B-51	3000	.016		.87	.69	.07	1.63	2.14
1160	4" thick		2700	.018		1.66	.77	.07	2.50	3.14
1180	6" thick		1900	.025		2.58	1.09	.10	3.77	4.70
3000	Prime coat, emulsion, 0.30 gal./S.Y., 1000 S.Y.	B-45	2500	.006	S.Y.	2.10	.34	.33	2.77	3.20
3100	Tack coat, emulsion, 0.10 gal./S.Y., 1000 S.Y.	"	2500	.006	"	.70	.34	.33	1.37	1.66

32 12 36 – Seal Coats

32 12 36.13 Chip Seal

		Crew	Daily Output	Labor-Hours	Unit	Material	2020 Bare Costs Labor	2020 Bare Costs Equipment	Total	Total Incl O&P
0010	**CHIP SEAL**									
0011	Excludes crack repair and flush coat									
1000	Fine - PMCRS-2h (20lbs/sy, 1/4" (No.10), .30gal/sy app. rate)									
1010	Small, irregular areas	B-91	5000	.013	S.Y.	1.10	.65	.45	2.20	2.73
1020	Parking Lot	B-91D	15000	.007		1.10	.34	.22	1.66	2
1030	Roadway	"	30000	.003		1.10	.17	.11	1.38	1.60
1090	For Each .5% Latex Additive, Add					.23			.23	.25
1100	Medium Fine - PMCRS-2h (25lbs/sy, 5/16" (No.8), .35gal/sy app. rate)									
1110	Small, irregular areas	B-91	4000	.016	S.Y.	1.29	.81	.56	2.66	3.31
1120	Parking Lot	B-91D	12000	.009		1.29	.42	.28	1.99	2.39

32 12 36.13 Chip Seal	Crew	Daily Output	Labor-Hours	Unit	Material	2020 Bare Costs Labor	Equipment	Total	Total Incl O&P
1130 Roadway	B-91D	24000	.004	S.Y.	1.29	.21	.14	1.64	1.90
1190 For Each .5% Latex Additive, Add					.28			.28	.31
1200 Medium - PMCRS-2h (30lbs/sy, 3/8" (No.6), .40gal/sy app. rate)									
1210 Small, irregular areas	B-91	3330	.019	S.Y.	1.41	.97	.68	3.06	3.83
1220 Parking Lot	B-91D	10000	.010		1.41	.51	.34	2.26	2.73
1230 Roadway	"	20000	.005		1.41	.25	.17	1.83	2.13
1290 For Each .5% Latex Additive, Add					.34			.34	.37
1300 Course - PMCRS-2h (30lbs/sy, 1/2" (No.4), .40gal/sy app. rate)									
1310 Small, irregular areas	B-91	2500	.026	S.Y.	1.39	1.29	.90	3.58	4.55
1320 Parking Lot	B-91D	7500	.014		1.39	.68	.45	2.52	3.09
1330 Roadway	"	15000	.007		1.39	.34	.22	1.95	2.31
1390 For Each .5% Latex Additive, Add					.34			.34	.37
1400 Double - PMCRS-2h (Course Base w/ Fine Top)									
1410 Small, irregular areas	B-91	2000	.032	S.Y.	2.36	1.62	1.13	5.11	6.40
1420 Parking Lot	B-91D	6000	.017		2.36	.85	.56	3.77	4.56
1430 Roadway	"	12000	.009		2.36	.42	.28	3.06	3.58
1490 For Each .5% Latex Additive, Add					.56			.56	.62

32 12 36.14 Flush Coat

	Crew	Daily Output	Labor-Hours	Unit	Material	2020 Bare Costs Labor	Equipment	Total	Total Incl O&P
0010 **FLUSH COAT**									
0011 Fog Seal w/ Sand Cover.18 gal/sy, 6lbs/sy									
1010 Small, irregular areas	B-91	2000	.032	S.Y.	.52	1.62	1.13	3.27	4.38
1020 Parking lot		6000	.011		.52	.54	.38	1.44	1.84
1030 Roadway		12000	.005		.52	.27	.19	.98	1.22

32 12 36.33 Slurry Seal

	Crew	Daily Output	Labor-Hours	Unit	Material	2020 Bare Costs Labor	Equipment	Total	Total Incl O&P
0010 **SLURRY SEAL**									
0011 Includes sweeping and cleaning of area									
1000 Type I-PMCQS-1h-EAS (12lbs/sy, 1/8", 20% asphalt emulsion)									
1010 Small, irregular areas	B-90	8000	.008	S.Y.	.98	.37	.27	1.62	1.97
1020 Parking Lot		25000	.003		.98	.12	.09	1.19	1.36
1030 Roadway		50000	.001		.98	.06	.04	1.08	1.22
1090 For Each .5% Latex Additive, Add					.15			.15	.16
1100 Type II-PMCQS-1h-EAS (15lbs/sy, 1/4", 18% asphalt emulsion)									
1110 Small, irregular areas	B-90	6000	.011	S.Y.	1.10	.50	.37	1.97	2.40
1120 Parking lot		20000	.003		1.10	.15	.11	1.36	1.57
1130 Roadway		40000	.002		1.10	.07	.05	1.22	1.39
1200 Type III-PMCQS-1h-EAS (25lbs/sy, 3/8", 15% asphalt emulsion)									
1210 Small, irregular areas	B-90	4000	.016	S.Y.	1.51	.75	.55	2.81	3.45
1220 Parking lot		12000	.005		1.51	.25	.18	1.94	2.26
1230 Roadway		24000	.003		1.51	.12	.09	1.72	1.96
1290 For Each .5% Latex Additive, Add					.29			.29	.32

32 12 36.36 Microsurfacing

	Crew	Daily Output	Labor-Hours	Unit	Material	2020 Bare Costs Labor	Equipment	Total	Total Incl O&P
0010 **MICROSURFACING**									
1100 Type II-MSE (20lbs/sy, 1/4", 18% microsurfacing emulsion)									
1110 Small, irregular areas	B-90	5000	.013	S.Y.	1.65	.60	.44	2.69	3.24
1120 Parking lot		15000	.004		1.65	.20	.15	2	2.29
1130 Roadway		30000	.002		1.65	.10	.07	1.82	2.05
1200 Type IIIa-MSE (32lbs/sy, 3/8", 15% microsurfacing emulsion)									
1210 Small, irregular areas	B-90	3000	.021	S.Y.	2.29	1	.73	4.02	4.92
1220 Parking lot		9000	.007		2.29	.33	.24	2.86	3.32
1230 Roadway		18000	.004		2.29	.17	.12	2.58	2.92

32 13 Rigid Paving

32 13 13 – Concrete Paving

32 13 13.25 Concrete Pavement, Highways

		Crew	Daily Output	Labor-Hours	Unit	Material	2020 Bare Costs Labor	Equipment	Total	Total Incl O&P
0010	**CONCRETE PAVEMENT, HIGHWAYS**									
0015	Including joints, finishing and curing									
0020	Fixed form, 12' pass, unreinforced, 6" thick	B-26	3000	.029	S.Y.	26	1.38	1.18	28.56	32
0100	8" thick		2750	.032		35.50	1.50	1.29	38.29	43
0400	12" thick		1800	.049		50.50	2.30	1.96	54.76	62
0410	Conc. pavement, w/jt., fnsh.& curing, fix form, 24' pass, unreinforced, 6"T		6000	.015		24.50	.69	.59	25.78	28.50
0430	8" thick		5500	.016		33.50	.75	.64	34.89	39
0460	12" thick		3600	.024		49	1.15	.98	51.13	56.50
0520	Welded wire fabric, sheets for rigid paving 2.33 lb./S.Y.	2 Rodm	389	.041		1.46	2.32		3.78	5.25
0530	Reinforcing steel for rigid paving 12 lb./S.Y.		666	.024		7.15	1.35		8.50	10
0540	Reinforcing steel for rigid paving 18 lb./S.Y.		444	.036		10.70	2.03		12.73	14.95
0620	Slip form, 12' pass, unreinforced, 6" thick	B-26A	5600	.016		25	.74	.65	26.39	29.50
0624	8" thick		5300	.017		34.50	.78	.69	35.97	39.50
0630	12" thick		3470	.025		49.50	1.19	1.05	51.74	57.50
0632	15" thick		2890	.030		62.50	1.43	1.26	65.19	72.50
0640	Slip form, 24' pass, unreinforced, 6" thick		11200	.008		24.50	.37	.33	25.20	28
0644	8" thick		10600	.008		33	.39	.34	33.73	37
0650	12" thick		6940	.013		48.50	.60	.53	49.63	54.50
0652	15" thick		5780	.015		60.50	.71	.63	61.84	69
0700	Finishing, broom finish small areas	2 Cefi	120	.133			6.65		6.65	10.45
1000	Curing, with sprayed membrane by hand	2 Clab	1500	.011		1.12	.45		1.57	1.95

32 14 Unit Paving

32 14 13 – Precast Concrete Unit Paving

32 14 13.18 Precast Concrete Plantable Pavers

		Crew	Daily Output	Labor-Hours	Unit	Material	2020 Bare Costs Labor	Equipment	Total	Total Incl O&P
0010	**PRECAST CONCRETE PLANTABLE PAVERS** (50% grass)									
0300	3/4" crushed stone base for plantable pavers, 6" depth	B-62	1000	.024	S.Y.	4.20	1.10	.18	5.48	6.55
0400	8" depth		900	.027		5.60	1.22	.20	7.02	8.30
0500	10" depth		800	.030		7	1.37	.22	8.59	10.10
0600	12" depth		700	.034		8.40	1.57	.25	10.22	12
0700	Hydro seeding plantable pavers	B-81A	20	.800	M.S.F.	12.10	35.50	47	94.60	122
0800	Apply fertilizer and seed to plantable pavers	1 Clab	8	1	"	46	42		88	119

32 14 16 – Brick Unit Paving

32 14 16.10 Brick Paving

		Crew	Daily Output	Labor-Hours	Unit	Material	2020 Bare Costs Labor	Equipment	Total	Total Incl O&P
0010	**BRICK PAVING**									
0012	4" x 8" x 1-1/2", without joints (4.5 bricks/S.F.)	D-1	110	.145	S.F.	2.66	6.80		9.46	13.90
0100	Grouted, 3/8" joint (3.9 bricks/S.F.)		90	.178		2.17	8.35		10.52	15.85
0200	4" x 8" x 2-1/4", without joints (4.5 bricks/S.F.)		110	.145		2.51	6.80		9.31	13.75
0300	Grouted, 3/8" joint (3.9 bricks/S.F.)		90	.178		2.17	8.35		10.52	15.85
0455	Pervious brick paving, 4" x 8" x 3-1/4", without joints (4.5 bricks/S.F.)		110	.145		3.70	6.80		10.50	15.05
0500	Bedding, asphalt, 3/4" thick	B-25	5130	.017		.68	.79	.53	2	2.59
0540	Course washed sand bed, 1" thick	B-18	5000	.005		.38	.21	.03	.62	.79
0580	Mortar, 1" thick	D-1	300	.053		.68	2.50		3.18	4.79
0620	2" thick		200	.080		1.37	3.75		5.12	7.55
1500	Brick on 1" thick sand bed laid flat, 4.5/S.F.		100	.160		3.05	7.50		10.55	15.45
2000	Brick pavers, laid on edge, 7.2/S.F.		70	.229		4.61	10.70		15.31	22.50
2500	For 4" thick concrete bed and joints, add		595	.027		1.31	1.26		2.57	3.49
2800	For steam cleaning, add	A-1H	950	.008		.12	.35	.08	.55	.79
9000	Minimum labor/equipment charge	1 Bric	2	4	Job		208		208	335

32 14 Unit Paving

32 14 23 – Asphalt Unit Paving

32 14 23.10 Asphalt Blocks

		Crew	Daily Output	Labor-Hours	Unit	Material	2020 Bare Costs Labor	Equipment	Total	Total Incl O&P
0010	**ASPHALT BLOCKS**									
0020	Rectangular, 6" x 12" x 1-1/4", w/bed & neopr. adhesive	D-1	135	.119	S.F.	10.20	5.55		15.75	20
0100	3" thick		130	.123		14.30	5.75		20.05	25
0300	Hexagonal tile, 8" wide, 1-1/4" thick		135	.119		10.20	5.55		15.75	20
0400	2" thick		130	.123		14.30	5.75		20.05	25
0500	Square, 8" x 8", 1-1/4" thick		135	.119		10.20	5.55		15.75	20
0600	2" thick		130	.123		14.30	5.75		20.05	25
9000	Minimum labor/equipment charge	1 Bric	2	4	Job		208		208	335

32 14 40 – Stone Paving

32 14 40.10 Stone Pavers

		Crew	Daily Output	Labor-Hours	Unit	Material	2020 Bare Costs Labor	Equipment	Total	Total Incl O&P
0010	**STONE PAVERS**									
1300	Slate, natural cleft, irregular, 3/4" thick	D-1	92	.174	S.F.	9.75	8.15		17.90	24
1350	Random rectangular, gauged, 1/2" thick		105	.152		21	7.15		28.15	34.50
1400	Random rectangular, butt joint, gauged, 1/4" thick		150	.107		22.50	5		27.50	33
1450	For sand rubbed finish, add					9.65			9.65	10.65
1550	Granite blocks, 3-1/2" x 3-1/2" x 3-1/2"	D-1	92	.174		21.50	8.15		29.65	36.50
1600	4" to 12" long, 3" to 5" wide, 3" to 5" thick	"	98	.163		17.95	7.65		25.60	32

32 16 Curbs, Gutters, Sidewalks, and Driveways

32 16 13 – Curbs and Gutters

32 16 13.13 Cast-in-Place Concrete Curbs and Gutters

		Crew	Daily Output	Labor-Hours	Unit	Material	2020 Bare Costs Labor	Equipment	Total	Total Incl O&P
0010	**CAST-IN-PLACE CONCRETE CURBS AND GUTTERS**									
0290	Forms only, no concrete									
0300	Concrete, wood forms, 6" x 18", straight	C-2	500	.096	L.F.	3.16	4.96		8.12	11.40
0400	6" x 18", radius	"	200	.240		3.30	12.40		15.70	23.50
0404	Concrete, wood forms, 6" x 18", straight & concrete	C-2A	500	.096		6.75	4.91		11.66	15.30
0406	6" x 18", radius	"	200	.240		6.90	12.25		19.15	27
0415	Machine formed, 6" x 18", straight	B-69A	2000	.024		5.15	1.11	.61	6.87	8.10
0416	6" x 18", radius	"	900	.053		5.20	2.46	1.35	9.01	11.15

32 16 13.23 Precast Concrete Curbs and Gutters

		Crew	Daily Output	Labor-Hours	Unit	Material	2020 Bare Costs Labor	Equipment	Total	Total Incl O&P
0010	**PRECAST CONCRETE CURBS AND GUTTERS**									
0550	Precast, 6" x 18", straight	B-29	700	.080	L.F.	9.15	3.68	1.21	14.04	17.30
0600	6" x 18", radius	"	325	.172	"	12.20	7.95	2.61	22.76	29

32 16 13.33 Asphalt Curbs

		Crew	Daily Output	Labor-Hours	Unit	Material	2020 Bare Costs Labor	Equipment	Total	Total Incl O&P
0010	**ASPHALT CURBS**									
0012	Curbs, asphaltic, machine formed, 8" wide, 6" high, 40 L.F./ton	B-27	1000	.032	L.F.	1.64	1.36	.27	3.27	4.28
0100	8" wide, 8" high, 30 L.F./ton		900	.036		2.19	1.51	.30	4	5.15
0150	Asphaltic berm, 12" W, 3" to 6" H, 35 L.F./ton, before pavement		700	.046		.04	1.95	.39	2.38	3.60
0200	12" W, 1-1/2" to 4" H, 60 L.F./ton, laid with pavement	B-2	1050	.038		.02	1.62		1.64	2.62

32 16 13.43 Stone Curbs

		Crew	Daily Output	Labor-Hours	Unit	Material	2020 Bare Costs Labor	Equipment	Total	Total Incl O&P
0010	**STONE CURBS**									
1000	Granite, split face, straight, 5" x 16"	D-13	275	.175	L.F.	16.50	8.80	1.56	26.86	34
1100	6" x 18"	"	250	.192		21.50	9.65	1.71	32.86	41.50
1300	Radius curbing, 6" x 18", over 10' radius	B-29	260	.215		26.50	9.90	3.26	39.66	48.50
1400	Corners, 2' radius	"	80	.700	Ea.	89	32	10.60	131.60	161
1600	Edging, 4-1/2" x 12", straight	D-13	300	.160	L.F.	8.25	8.05	1.43	17.73	23.50
1800	Curb inlets (guttermouth) straight	B-29	41	1.366	Ea.	198	63	20.50	281.50	340
2000	Indian granite (Belgian block)									
2100	Jumbo, 10-1/2" x 7-1/2" x 4", grey	D-1	150	.107	L.F.	9.65	5		14.65	18.70

For customer support on your Commercial Renovation Costs with RSMeans data, call 800.448.8182.

561

32 16 Curbs, Gutters, Sidewalks, and Driveways

32 16 13 – Curbs and Gutters

32 16 13.43 Stone Curbs

		Crew	Daily Output	Labor-Hours	Unit	Material	2020 Bare Costs Labor	Equipment	Total	Total Incl O&P
2150	Pink	D-1	150	.107	L.F.	9.55	5		14.55	18.60
2200	Regular, 9" x 4-1/2" x 4-1/2", grey		160	.100		4.64	4.69		9.33	12.70
2250	Pink		160	.100		6.75	4.69		11.44	15.05
2300	Cubes, 4" x 4" x 4", grey		175	.091		3.34	4.29		7.63	10.60
2350	Pink		175	.091		3.99	4.29		8.28	11.35
2400	6" x 6" x 6", pink		155	.103		12.60	4.84		17.44	21.50
2500	Alternate pricing method for Indian granite									
2550	Jumbo, 10-1/2" x 7-1/2" x 4" (30 lb.), grey				Ton	550			550	605
2600	Pink					555			555	610
2650	Regular, 9" x 4-1/2" x 4-1/2" (20 lb.), grey					325			325	360
2700	Pink					475			475	520
2750	Cubes, 4" x 4" x 4" (5 lb.), grey					405			405	445
2800	Pink					515			515	565
2850	6" x 6" x 6" (25 lb.), pink					490			490	540
2900	For pallets, add					22			22	24

32 17 Paving Specialties

32 17 13 – Parking Bumpers

32 17 13.13 Metal Parking Bumpers

		Crew	Daily Output	Labor-Hours	Unit	Material	2020 Bare Costs Labor	Equipment	Total	Total Incl O&P
0010	**METAL PARKING BUMPERS**									
0015	Bumper rails for garages, 12 ga. rail, 6" wide, with steel									
0020	posts 12'-6" OC, minimum	E-4	190	.168	L.F.	22	9.80	.77	32.57	41
0030	Average		165	.194		27.50	11.30	.89	39.69	49.50
0100	Maximum		140	.229		33	13.30	1.05	47.35	59
1300	Pipe bollards, conc. filled/paint, 8' L x 4' D hole, 6" diam.	B-6	20	1.200	Ea.	635	55	10.70	700.70	800
1400	8" diam.		15	1.600		740	73	14.25	827.25	945
1500	12" diam.		12	2		1,025	91.50	17.80	1,134.30	1,300
8900	Security bollards, SS, lighted, hyd., incl. controls, group of 3	L-7	.06	509		43,000	26,000		69,000	89,000
8910	Group of 5	"	.04	683		75,000	34,900		109,900	138,000
9000	Minimum labor/equipment charge	E-4	2	16	Job		930	73.50	1,003.50	1,600

32 17 13.16 Plastic Parking Bumpers

			Crew	Daily Output	Labor-Hours	Unit	Material	2020 Bare Costs Labor	Equipment	Total	Total Incl O&P
0010	**PLASTIC PARKING BUMPERS**										
1600	Bollards, recycled plastic, 5" x 5" x 5' L	G	B-6	20	1.200	Ea.	134	55	10.70	199.70	246
1610	Wheel stops, recycled plastic, yellow, 4" x 6" x 6' L	G		20	1.200		48	55	10.70	113.70	151
1620	Speed bumps, recycled plastic, yellow, 3" H x 10" W x 6' L	G		20	1.200		173	55	10.70	238.70	289
1630	3" H x 10" W x 9' L	G		20	1.200		285	55	10.70	350.70	415

32 17 13.19 Precast Concrete Parking Bumpers

		Crew	Daily Output	Labor-Hours	Unit	Material	2020 Bare Costs Labor	Equipment	Total	Total Incl O&P
0010	**PRECAST CONCRETE PARKING BUMPERS**									
1000	Wheel stops, precast concrete incl. dowels, 6" x 10" x 6'-0"	B-2	120	.333	Ea.	63.50	14.15		77.65	92
1100	8" x 13" x 6'-0"	"	120	.333		74	14.15		88.15	104
1540	Bollards, precast concrete, free standing, round, 15" diam. x 36" H	B-6	14	1.714		244	78.50	15.25	337.75	410
1550	16" diam. x 30" H		14	1.714		294	78.50	15.25	387.75	465
1560	18" diam. x 34" H		14	1.714		425	78.50	15.25	518.75	605
1570	18" diam. x 72" H		14	1.714		680	78.50	15.25	773.75	885
1572	26" diam. x 18" H		12	2		425	91.50	17.80	534.30	635
1574	26" diam. x 30" H		12	2		650	91.50	17.80	759.30	880
1578	26" diam. x 50" H		12	2		960	91.50	17.80	1,069.30	1,225
1580	Square, 22" x 32" H		12	2		480	91.50	17.80	589.30	695
1585	22" x 41" H		12	2		745	91.50	17.80	854.30	985
1590	Hexagon, 26" x 41" H		10	2.400		595	110	21.50	726.50	855

562

32 17 13 – Parking Bumpers

32 17 13.26 Wood Parking Bumpers	Crew	Daily Output	Labor-Hours	Unit	Material	2020 Bare Costs Labor	Equipment	Total	Total Incl O&P
0010 **WOOD PARKING BUMPERS**									
0020 Parking barriers, timber w/saddles, treated type									
0100 4" x 4" for cars	B-2	520	.077	L.F.	3.16	3.27		6.43	8.75
0200 6" x 6" for trucks	"	520	.077	"	6.55	3.27		9.82	12.45

32 17 23 – Pavement Markings

32 17 23.13 Painted Pavement Markings

	Crew	Daily Output	Labor-Hours	Unit	Material	2020 Bare Costs Labor	Equipment	Total	Total Incl O&P
0010 **PAINTED PAVEMENT MARKINGS**									
0020 Acrylic waterborne, white or yellow, 4" wide, less than 3,000 L.F.	B-78	20000	.002	L.F.	.11	.10	.05	.26	.35
0200 6" wide, less than 3,000 L.F.		11000	.004		.16	.19	.10	.45	.59
0500 8" wide, less than 3,000 L.F.		10000	.005		.21	.21	.11	.53	.69
0600 12" wide, less than 3,000 L.F.		4000	.012		.32	.52	.26	1.10	1.47
0620 Arrows or gore lines		2300	.021	S.F.	.22	.90	.46	1.58	2.20
0640 Temporary paint, white or yellow, less than 3,000 L.F.		15000	.003	L.F.	.12	.14	.07	.33	.43
0660 Removal	1 Clab	300	.027			1.12		1.12	1.80
0680 Temporary tape	2 Clab	1500	.011		.53	.45		.98	1.30
0710 Thermoplastic, white or yellow, 4" wide, less than 6,000 L.F.	B-79	15000	.003		.42	.12	.12	.66	.78
0730 6" wide, less than 6,000 L.F.		14000	.003		.63	.12	.12	.87	1.03
0740 8" wide, less than 6,000 L.F.		12000	.003		.84	.15	.14	1.13	1.31
0750 12" wide, less than 6,000 L.F.		6000	.007		1.24	.29	.29	1.82	2.15
0760 Arrows		660	.061	S.F.	.69	2.64	2.63	5.96	7.90
0770 Gore lines		2500	.016		.69	.70	.69	2.08	2.64
0780 Letters		660	.061		.70	2.64	2.63	5.97	7.90

32 17 23.14 Pavement Parking Markings

	Crew	Daily Output	Labor-Hours	Unit	Material	2020 Bare Costs Labor	Equipment	Total	Total Incl O&P
0010 **PAVEMENT PARKING MARKINGS**									
0790 Layout of pavement marking	A-2	25000	.001	L.F.		.04	.01	.05	.08
0800 Lines on pvmt., parking stall, paint, white, 4" wide	B-78B	400	.045	Stall	5.05	1.95	1.01	8.01	9.75
0825 Parking stall, small quantities	2 Pord	80	.200		10.10	8.90		19	25.50
0830 Lines on pvmt., parking stall, thermoplastic, white, 4" wide	B-79	300	.133		20	5.80	5.80	31.60	37.50
1000 Street letters and numbers	B-78B	1600	.011	S.F.	.82	.49	.25	1.56	1.96

32 31 Fences and Gates

32 31 13 – Chain Link Fences and Gates

32 31 13.20 Fence, Chain Link Industrial

	Crew	Daily Output	Labor-Hours	Unit	Material	2020 Bare Costs Labor	Equipment	Total	Total Incl O&P
0010 **FENCE, CHAIN LINK INDUSTRIAL**									
0011 Schedule 40, including concrete									
0020 3 strands barb wire, 2" post @ 10' OC, set in concrete, 6' H									
0200 9 ga. wire, galv. steel, in concrete	B-80C	240	.100	L.F.	19.95	4.38	1.04	25.37	30
0248 Fence, add for vinyl coated fabric				S.F.	.71			.71	.78
0300 Aluminized steel	B-80C	240	.100	L.F.	22.50	4.38	1.04	27.92	33
0301 Fence, wrought iron		240	.100		43	4.38	1.04	48.42	55
0303 Fence, commercial 4' high		240	.100		32.50	4.38	1.04	37.92	43.50
0304 Fence, commercial 6' high		240	.100		58	4.38	1.04	63.42	72
0500 6 ga. wire, galv. steel		240	.100		23.50	4.38	1.04	28.92	34
0600 Aluminized steel		240	.100		31	4.38	1.04	36.42	42
0800 6 ga. wire, 6' high but omit barbed wire, galv. steel		250	.096		21.50	4.21	1	26.71	31.50
0900 Aluminized steel, in concrete		250	.096		28.50	4.21	1	33.71	39
0920 8' H, 6 ga. wire, 2-1/2" line post, galv. steel, in concrete		180	.133		33.50	5.85	1.39	40.74	48
0940 Aluminized steel, in concrete		180	.133		41	5.85	1.39	48.24	56
1100 Add for corner posts, 3" diam., galv. steel, in concrete		40	.600	Ea.	91.50	26.50	6.25	124.25	150
1200 Aluminized steel, in concrete		40	.600		92.50	26.50	6.25	125.25	151

32 31 13.20 Fence, Chain Link Industrial

		Crew	Daily Output	Labor-Hours	Unit	Material	2020 Bare Costs Labor	2020 Bare Costs Equipment	Total	Total Incl O&P
1300	Add for braces, galv. steel	B-80C	80	.300	Ea.	42.50	13.15	3.13	58.78	71.50
1350	Aluminized steel		80	.300		49	13.15	3.13	65.28	78.50
1400	Gate for 6' high fence, 1-5/8" frame, 3' wide, galv. steel		10	2.400		209	105	25	339	425
1500	Aluminized steel, in concrete		10	2.400		218	105	25	348	435
2000	5'-0" high fence, 9 ga., no barbed wire, 2" line post, in concrete									
2010	10' OC, 1-5/8" top rail, in concrete									
2100	Galvanized steel, in concrete	B-80C	300	.080	L.F.	22.50	3.51	.83	26.84	31.50
2200	Aluminized steel, in concrete		300	.080	"	21	3.51	.83	25.34	29.50
2400	Gate, 4' wide, 5' high, 2" frame, galv. steel, in concrete		10	2.400	Ea.	229	105	25	359	450
2500	Aluminized steel, in concrete		10	2.400	"	191	105	25	321	405
3100	Overhead slide gate, chain link, 6' high, to 18' wide, in concrete		38	.632	L.F.	95	27.50	6.60	129.10	157
3110	Cantilever type, in concrete	B-80	48	.667		168	31	21.50	220.50	258
3120	8' high, in concrete		24	1.333		176	62	42.50	280.50	340
3130	10' high, in concrete		18	1.778		218	83	56.50	357.50	435
9000	Minimum labor/equipment charge		2	16	Job		745	510	1,255	1,725

32 31 13.25 Fence, Chain Link Residential

		Crew	Daily Output	Labor-Hours	Unit	Material	2020 Bare Costs Labor	2020 Bare Costs Equipment	Total	Total Incl O&P
0010	**FENCE, CHAIN LINK RESIDENTIAL**									
0011	Schedule 20, 11 ga. wire, 1-5/8" post									
3300	Residential, 11 ga. wire, 1-5/8" line post @ 10' OC									
3310	1-3/8" top rail									
3350	4' high, galvanized steel	B-80C	475	.051	L.F.	12.20	2.21	.53	14.94	17.55
3400	Aluminized		475	.051	"	15.80	2.21	.53	18.54	21.50
3600	Gate, 3' wide, 1-3/8" frame, galv. steel		10	2.400	Ea.	96.50	105	25	226.50	300
3700	Aluminized		10	2.400	"	150	105	25	280	360
3900	3' high, galvanized steel		620	.039	L.F.	127	1.70	.40	129.10	143
4000	Aluminized		620	.039	"	131	1.70	.40	133.10	147
4200	Gate, 3' wide, 1-3/8" frame, galv. steel		12	2	Ea.	100	87.50	21	208.50	273
4300	Aluminized		12	2	"	101	87.50	21	209.50	274
9000	Minimum labor/equipment charge	B-1	2	12	Job		515		515	820

32 31 13.26 Tennis Court Fences and Gates

		Crew	Daily Output	Labor-Hours	Unit	Material	2020 Bare Costs Labor	2020 Bare Costs Equipment	Total	Total Incl O&P
0010	**TENNIS COURT FENCES AND GATES**									
3150	Tennis courts, 11 ga. wire, 1-3/4" mesh, 2-1/2" line posts									
3170	1-5/8" top rail, 3" corner and gate posts									
3190	10' high, galvanized steel	B-80	190	.168	L.F.	30	7.85	5.35	43.20	51.50
3210	Vinyl covered 9 ga. wire	"	190	.168	"	36.50	7.85	5.35	49.70	58.50

32 31 13.40 Fence, Fabric and Accessories

		Crew	Daily Output	Labor-Hours	Unit	Material	2020 Bare Costs Labor	2020 Bare Costs Equipment	Total	Total Incl O&P
0010	**FENCE, FABRIC & ACCESSORIES**									
8000	Components only, fabric, galvanized, 4' high	B-80C	585	.041	L.F.	3.56	1.80	.43	5.79	7.25
8040	6' high		430	.056	"	5.95	2.45	.58	8.98	11.10
8200	Posts, line post, 4' high		34	.706	Ea.	16	31	7.35	54.35	75
8240	6' high		26	.923	"	19.55	40.50	9.60	69.65	97
8400	Top rails, 1-3/8" OD		1000	.024	L.F.	1.77	1.05	.25	3.07	3.91
8440	1-5/8" OD		1000	.024	"	2.89	1.05	.25	4.19	5.15

32 31 13.80 Residential Chain Link Gate

		Crew	Daily Output	Labor-Hours	Unit	Material	2020 Bare Costs Labor	2020 Bare Costs Equipment	Total	Total Incl O&P
0010	**RESIDENTIAL CHAIN LINK GATE**									
0110	Residential 4' gate, single incl. hardware and concrete	B-80C	10	2.400	Ea.	128	105	25	258	335
0120	5'		10	2.400		138	105	25	268	350
0130	6'		10	2.400		149	105	25	279	360
0510	Residential 4' gate, double incl. hardware and concrete		10	2.400		228	105	25	358	445
0520	5'		10	2.400		244	105	25	374	465
0530	6'		10	2.400		284	105	25	414	505

32 31 Fences and Gates

32 31 13 – Chain Link Fences and Gates

32 31 13.82 Internal Chain Link Gate

		Crew	Daily Output	Labor-Hours	Unit	Material	2020 Bare Costs Labor	Equipment	Total	Total Incl O&P
0010	**INTERNAL CHAIN LINK GATE**									
0110	Internal 6' gate, single incl. post flange, hardware and concrete	B-80C	10	2.400	Ea.	279	105	25	409	500
0120	8'		10	2.400		320	105	25	450	545
0130	10'		10	2.400		435	105	25	565	675
0510	Internal 6' gate, double incl. post flange, hardware and concrete		10	2.400		510	105	25	640	755
0520	8'		10	2.400		585	105	25	715	840
0530	10'		10	2.400		740	105	25	870	1,000

32 31 13.84 Industrial Chain Link Gate

		Crew	Daily Output	Labor-Hours	Unit	Material	2020 Bare Costs Labor	Equipment	Total	Total Incl O&P
0010	**INDUSTRIAL CHAIN LINK GATE**									
0110	Industrial 8' gate, single incl. hardware and concrete	B-80C	10	2.400	Ea.	475	105	25	605	720
0120	10'		10	2.400		540	105	25	670	790
0510	Industrial 8' gate, double incl. hardware and concrete		10	2.400		740	105	25	870	1,000
0520	10'		10	2.400		845	105	25	975	1,125

32 31 13.88 Chain Link Transom

		Crew	Daily Output	Labor-Hours	Unit	Material	2020 Bare Costs Labor	Equipment	Total	Total Incl O&P
0010	**CHAIN LINK TRANSOM**									
0110	Add for, single transom, 3' wide, incl. components & hardware	B-80C	10	2.400	Ea.	118	105	25	248	325
0120	Add for, double transom, 6' wide, incl. components & hardware	"	10	2.400	"	127	105	25	257	335

32 31 19 – Decorative Metal Fences and Gates

32 31 19.10 Decorative Fence

		Crew	Daily Output	Labor-Hours	Unit	Material	2020 Bare Costs Labor	Equipment	Total	Total Incl O&P
0010	**DECORATIVE FENCE**									
5300	Tubular picket, steel, 6' sections, 1-9/16" posts, 4' high	B-80C	300	.080	L.F.	37	3.51	.83	41.34	47.50
5400	2" posts, 5' high		240	.100		44	4.38	1.04	49.42	56
5600	2" posts, 6' high		200	.120		53	5.25	1.25	59.50	68.50
5700	Staggered picket, 1-9/16" posts, 4' high		300	.080		39.50	3.51	.83	43.84	50
5800	2" posts, 5' high		240	.100		45	4.38	1.04	50.42	57.50
5900	2" posts, 6' high		200	.120		53	5.25	1.25	59.50	68.50
6200	Gates, 4' high, 3' wide	B-1	10	2.400	Ea.	335	103		438	535
6300	5' high, 3' wide		10	2.400		245	103		348	435
6400	6' high, 3' wide		10	2.400		288	103		391	480
6500	4' wide		10	2.400		345	103		448	545

32 31 23 – Plastic Fences and Gates

32 31 23.20 Fence, Recycled Plastic

			Crew	Daily Output	Labor-Hours	Unit	Material	2020 Bare Costs Labor	Equipment	Total	Total Incl O&P
0010	**FENCE, RECYCLED PLASTIC**										
9015	Fence rail, made from recycled plastic, various colors, 2 rail	G	B-1	150	.160	L.F.	9.45	6.85		16.30	21.50
9018	3 rail	G		150	.160		12.15	6.85		19	24.50
9020	4 rail	G		150	.160		15.85	6.85		22.70	28.50
9030	Fence pole, made from recycled plastic, various colors, 7'	G		96	.250	Ea.	50.50	10.70		61.20	72.50
9040	Stockade fence, made from recycled plastic, various colors, 4' high	G	B-80C	160	.150	L.F.	35.50	6.55	1.56	43.61	51
9050	6' high	G		160	.150	"	36.50	6.55	1.56	44.61	52.50
9060	6' pole	G		96	.250	Ea.	39	10.95	2.61	52.56	63.50
9070	9' pole	G		96	.250	"	49	10.95	2.61	62.56	74
9080	Picket fence, made from recycled plastic, various colors, 3' high	G		160	.150	L.F.	24	6.55	1.56	32.11	38.50
9090	4' high	G		160	.150	"	35	6.55	1.56	43.11	50.50
9100	3' high gate	G		8	3	Ea.	26.50	131	31.50	189	274
9110	4' high gate	G		8	3		34.50	131	31.50	197	282
9120	5' high pole	G		96	.250		36.50	10.95	2.61	50.06	61
9130	6' high pole	G		96	.250		42	10.95	2.61	55.56	67
9140	Pole cap only	G					6.05			6.05	6.65
9150	Keeper pins only	G					.43			.43	.47

32 31 Fences and Gates

32 31 26 – Wire Fences and Gates

32 31 26.10 Fences, Misc. Metal

		Crew	Daily Output	Labor-Hours	Unit	Material	2020 Bare Costs Labor	Equipment	Total	Total Incl O&P
0010	**FENCES, MISC. METAL**									
0012	Chicken wire, posts @ 4', 1" mesh, 4' high	B-80C	410	.059	L.F.	3.53	2.57	.61	6.71	8.65
0100	2" mesh, 6' high		350	.069		4.01	3	.71	7.72	10
0200	Galv. steel, 12 ga., 2" x 4" mesh, posts 5' OC, 3' high		300	.080		2.80	3.51	.83	7.14	9.60
0300	5' high		300	.080		3.37	3.51	.83	7.71	10.25
0400	14 ga., 1" x 2" mesh, 3' high		300	.080		3.48	3.51	.83	7.82	10.35
0500	5' high		300	.080		4.50	3.51	.83	8.84	11.45
1000	Kennel fencing, 1-1/2" mesh, 6' long, 3'-6" wide, 6'-2" high	2 Clab	4	4	Ea.	495	168		663	815
1050	12' long		4	4		700	168		868	1,050
1200	Top covers, 1-1/2" mesh, 6' long		15	1.067		141	45		186	227
1250	12' long		12	1.333		210	56		266	320
4492	Security fence, prison grade, barbed wire, set in concrete, 10' high	B-80	22	1.455	L.F.	64	68	46.50	178.50	230
4494	Security fence, prison grade, razor wire, set in concrete, 10' high		18	1.778		65.50	83	56.50	205	267
4500	Security fence, prison grade, set in concrete, 12' high		25	1.280		60	59.50	41	160.50	206
4600	16' high		20	1.600		79.50	74.50	51	205	263
4990	Security fence, prison grade, set in concrete, 10' high		25	1.280		60	59.50	41	160.50	206

32 31 29 – Wood Fences and Gates

32 31 29.20 Fence, Wood Rail

		Crew	Daily Output	Labor-Hours	Unit	Material	2020 Bare Costs Labor	Equipment	Total	Total Incl O&P
0010	**FENCE, WOOD RAIL**									
0012	Picket, No. 2 cedar, Gothic, 2 rail, 3' high	B-1	160	.150	L.F.	8.60	6.40		15	19.75
0050	Gate, 3'-6" wide	B-80C	9	2.667	Ea.	75	117	28	220	300
0600	Open rail, rustic, No. 1 cedar, 2 rail, 3' high		160	.150	L.F.	6.30	6.55	1.56	14.41	19.15
0650	Gate, 3' wide		9	2.667	Ea.	88.50	117	28	233.50	315
1200	Stockade, No. 2 cedar, treated wood rails, 6' high		160	.150	L.F.	9.55	6.55	1.56	17.66	22.50
1250	Gate, 3' wide		9	2.667	Ea.	109	117	28	254	340
3300	Board, shadow box, 1" x 6", treated pine, 6' high		160	.150	L.F.	12.65	6.55	1.56	20.76	26
3400	No. 1 cedar, 6' high		150	.160		25	7	1.67	33.67	40.50
3900	Basket weave, No. 1 cedar, 6' high		160	.150		36.50	6.55	1.56	44.61	52
3950	Gate, 3'-6" wide	B-1	8	3	Ea.	251	128		379	480
4000	Treated pine, 6' high		150	.160	L.F.	17.95	6.85		24.80	30.50
4200	Gate, 3'-6" wide		9	2.667	Ea.	203	114		317	405
8000	Posts only, 4' high		40	.600		12.85	25.50		38.35	55
8040	6' high		30	.800		17.90	34		51.90	74.50
9000	Minimum labor/equipment charge	1 Clab	2	4	Job		168		168	270

32 32 Retaining Walls

32 32 29 – Timber Retaining Walls

32 32 29.10 Landscape Timber Retaining Walls

		Crew	Daily Output	Labor-Hours	Unit	Material	2020 Bare Costs Labor	Equipment	Total	Total Incl O&P
0010	**LANDSCAPE TIMBER RETAINING WALLS**									
0100	Treated timbers, 6" x 6"	1 Clab	265	.030	L.F.	2.76	1.27		4.03	5.10
0110	6" x 8'	"	200	.040	"	6.65	1.68		8.33	10
0120	Drilling holes in timbers for fastening, 1/2"	1 Carp	450	.018	Inch		.95		.95	1.51
0130	5/8"	"	450	.018	"		.95		.95	1.51
0140	Reinforcing rods for fastening, 1/2"	1 Clab	312	.026	L.F.	.41	1.08		1.49	2.19
0150	5/8"	"	312	.026	"	.65	1.08		1.73	2.44
0160	Reinforcing fabric	2 Clab	2500	.006	S.Y.	2.22	.27		2.49	2.87
0170	Gravel backfill		28	.571	C.Y.	16.90	24		40.90	57
0180	Perforated pipe, 4" diameter with silt sock		1200	.013	L.F.	1.05	.56		1.61	2.06
0190	Galvanized 60d common nails	1 Clab	625	.013	Ea.	.21	.54		.75	1.09
0200	20d common nails	"	3800	.002	"	.05	.09		.14	.19

32 32 Retaining Walls

32 32 53 – Stone Retaining Walls

32 32 53.10 Retaining Walls, Stone

	Crew	Daily Output	Labor-Hours	Unit	Material	2020 Bare Costs Labor	Equipment	Total	Total Incl O&P
0010 **RETAINING WALLS, STONE**									
0015 Including excavation, concrete footing and									
0020 stone 3' below grade. Price is exposed face area.									
0200 Decorative random stone, to 6' high, 1'-6" thick, dry set	D-1	35	.457	S.F.	77	21.50		98.50	120
0300 Mortar set		40	.400		78.50	18.75		97.25	117
0500 Cut stone, to 6' high, 1'-6" thick, dry set		35	.457		80	21.50		101.50	123
0600 Mortar set		40	.400		80.50	18.75		99.25	119
0800 Random stone, 6' to 10' high, 2' thick, dry set		45	.356		80	16.70		96.70	115
0900 Mortar set		50	.320		83	15		98	116
1100 Cut stone, 6' to 10' high, 2' thick, dry set		45	.356		80.50	16.70		97.20	116
1200 Mortar set		50	.320		83.50	15		98.50	116
9000 Minimum labor/equipment charge		2	8	Job		375		375	605

32 33 Site Furnishings

32 33 33 – Site Manufactured Planters

32 33 33.10 Planters

	Crew	Daily Output	Labor-Hours	Unit	Material	2020 Bare Costs Labor	Equipment	Total	Total Incl O&P
0010 **PLANTERS**									
0012 Concrete, sandblasted, precast, 48" diameter, 24" high	2 Clab	15	1.067	Ea.	640	45		685	775
0100 Fluted, precast, 7' diameter, 36" high		10	1.600		1,625	67.50		1,692.50	1,875
0300 Fiberglass, circular, 36" diameter, 24" high		15	1.067		745	45		790	890
0400 60" diameter, 24" high		10	1.600		1,175	67.50		1,242.50	1,375
9000 Minimum labor/equipment charge	1 Clab	2	4	Job		168		168	270

32 33 43 – Site Seating and Tables

32 33 43.13 Site Seating

	Crew	Daily Output	Labor-Hours	Unit	Material	2020 Bare Costs Labor	Equipment	Total	Total Incl O&P
0010 **SITE SEATING**									
0012 Seating, benches, park, precast conc., w/backs, wood rails, 4' long	2 Clab	5	3.200	Ea.	610	135		745	885
0100 8' long		4	4		1,125	168		1,293	1,525
0300 Fiberglass, without back, one piece, 4' long		10	1.600		1,725	67.50		1,792.50	2,000
0400 8' long		7	2.286		2,300	96		2,396	2,675
0500 Steel barstock pedestals w/backs, 2" x 3" wood rails, 4' long		10	1.600		1,375	67.50		1,442.50	1,625
0510 8' long		7	2.286		1,725	96		1,821	2,050
0515 Powder coated steel, 4" x 4" plastic slats, 6' long		8	2		540	84		624	730
0520 3" x 8" wood plank, 4' long		10	1.600		1,525	67.50		1,592.50	1,775
0530 8' long		7	2.286		1,550	96		1,646	1,875
0540 Backless, 4" x 4" wood plank, 4' square		10	1.600		1,275	67.50		1,342.50	1,500
0550 8' long		7	2.286		1,175	96		1,271	1,450
0560 Powder coated steel, with back and 2 anti-vagrant dividers, 6' long		8	2		1,200	84		1,284	1,450
0600 Aluminum pedestals, with backs, aluminum slats, 8' long		8	2		480	84		564	665
0610 15' long		5	3.200		1,025	135		1,160	1,350
0620 Portable, aluminum slats, 8' long		8	2		470	84		554	655
0630 15' long		5	3.200		505	135		640	775
0800 Cast iron pedestals, back & arms, wood slats, 4' long		8	2		470	84		554	655
0820 8' long		5	3.200		1,050	135		1,185	1,375
0840 Backless, wood slats, 4' long		8	2		715	84		799	920
0860 8' long		5	3.200		1,200	135		1,335	1,550
1700 Steel frame, fir seat, 10' long		10	1.600		410	67.50		477.50	560
2000 Benches, park, with back, galv. stl. frame, 4" x 4" plastic slats, 6' long		7	2.286		545	96		641	755
9000 Minimum labor/equipment charge		2	8	Job		335		335	540

For customer support on your Commercial Renovation Costs with RSMeans data, call 800.448.8182.

567

32 84 Planting Irrigation

32 84 23 – Underground Sprinklers

32 84 23.10 Sprinkler Irrigation System	Crew	Daily Output	Labor-Hours	Unit	Material	2020 Bare Costs Labor	Equipment	Total	Total Incl O&P
0010 **SPRINKLER IRRIGATION SYSTEM**									
0011 For lawns									
0800 Residential system, custom, 1" supply	B-20	2000	.012	S.F.	.27	.56		.83	1.20
0900 1-1/2" supply	"	1800	.013	"	.42	.63		1.05	1.46

32 91 Planting Preparation

32 91 13 – Soil Preparation

32 91 13.23 Structural Soil Mixing

	Crew	Daily Output	Labor-Hours	Unit	Material	2020 Bare Costs Labor	Equipment	Total	Total Incl O&P
0010 **STRUCTURAL SOIL MIXING**									
0100 Rake topsoil, site material, harley rock rake, ideal	B-6	33	.727	M.S.F.		33.50	6.50	40	60
0200 Adverse	"	7	3.429			157	30.50	187.50	283
0300 Screened loam, york rake and finish, ideal	B-62	24	1			45.50	7.40	52.90	80.50
0400 Adverse	"	20	1.200			55	8.90	63.90	97
1000 Remove topsoil & stock pile on site, 75 HP dozer, 6" deep, 50' haul	B-10L	30	.400			21	13.40	34.40	47
1050 300' haul		6.10	1.967			102	66	168	234
1100 12" deep, 50' haul		15.50	.774			40	26	66	91.50
1150 300' haul		3.10	3.871			201	129	330	455
1200 200 HP dozer, 6" deep, 50' haul	B-10B	125	.096			4.98	12.05	17.03	21
1250 300' haul		30.70	.391			20.50	49	69.50	86
1300 12" deep, 50' haul		62	.194			10.05	24.50	34.55	42.50
1350 300' haul		15.40	.779			40.50	97.50	138	171
1400 Alternate method, 75 HP dozer, 50' haul	B-10L	860	.014	C.Y.		.72	.47	1.19	1.65
1450 300' haul	"	114	.105			5.45	3.52	8.97	12.45
1500 200 HP dozer, 50' haul	B-10B	2660	.005			.23	.57	.80	.99
1600 300' haul	"	570	.021			1.09	2.64	3.73	4.62
1800 Rolling topsoil, hand push roller	1 Clab	3200	.003	S.F.		.11		.11	.17
1850 Tractor drawn roller	B-66	10666	.001	"		.04	.02	.06	.08
2000 Root raking and loading, residential, no boulders	B-6	53.30	.450	M.S.F.		20.50	4.01	24.51	37
2100 With boulders		32	.750			34.50	6.70	41.20	62
2200 Municipal, no boulders		200	.120			5.50	1.07	6.57	9.90
2300 With boulders		120	.200			9.15	1.78	10.93	16.45
2400 Large commercial, no boulders	B-10B	400	.030			1.56	3.76	5.32	6.60
2500 With boulders	"	240	.050			2.59	6.25	8.84	11
3000 Scarify subsoil, residential, skid steer loader w/scarifiers, 50 HP	B-66	32	.250			13.25	7.55	20.80	29
3050 Municipal, skid steer loader w/scarifiers, 50 HP	"	120	.067			3.53	2.01	5.54	7.75
3100 Large commercial, 75 HP, dozer w/scarifier	B-10L	240	.050			2.59	1.67	4.26	5.90
3500 Screen topsoil from stockpile, vibrating screen, wet material (organic)	B-10P	200	.060	C.Y.		3.11	5.85	8.96	11.35
3550 Dry material	"	300	.040			2.07	3.90	5.97	7.55
3600 Mixing with conditioners, manure and peat	B-10R	550	.022			1.13	.55	1.68	2.38
3650 Mobilization add for 2 days or less operation	B-34K	3	2.667	Job		131	286	417	525
3800 Spread conditioned topsoil, 6" deep, by hand	B-1	360	.067	S.Y.	5.90	2.85		8.75	11.05
3850 300 HP dozer	B-10M	27	.444	M.S.F.	635	23	65.50	723.50	810
4000 Spread soil conditioners, alum. sulfate, 1#/S.Y., hand push spreader	1 Clab	17500	.001	S.Y.	23	.02		23.02	25.50
4050 Tractor spreader	B-66	700	.011	M.S.F.	2,575	.61	.35	2,575.96	2,825
4100 Fertilizer, 0.2#/S.Y., push spreader	1 Clab	17500	.001	S.Y.	.08	.02		.10	.12
4150 Tractor spreader	B-66	700	.011	M.S.F.	9.10	.61	.35	10.06	11.35
4200 Ground limestone, 1#/S.Y., push spreader	1 Clab	17500	.001	S.Y.	.14	.02		.16	.18
4250 Tractor spreader	B-66	700	.011	M.S.F.	15.55	.61	.35	16.51	18.45
4400 Manure, 18#/S.Y., push spreader	1 Clab	2500	.003	S.Y.	8.10	.13		8.23	9.10
4450 Tractor spreader	B-66	280	.029	M.S.F.	900	1.51	.86	902.37	995
4500 Perlite, 1" deep, push spreader	1 Clab	17500	.001	S.Y.	11.15	.02		11.17	12.30

32 91 Planting Preparation

32 91 13 – Soil Preparation

32 91 13.23 Structural Soil Mixing

		Crew	Daily Output	Labor-Hours	Unit	Material	2020 Bare Costs Labor	Equipment	Total	Total Incl O&P
4550	Tractor spreader	B-66	700	.011	M.S.F.	1,225	.61	.35	1,225.96	1,350
4600	Vermiculite, push spreader	1 Clab	17500	.001	S.Y.	8.45	.02		8.47	9.35
4650	Tractor spreader	B-66	700	.011	M.S.F.	940	.61	.35	940.96	1,025
5000	Spread topsoil, skid steer loader and hand dress	B-62	270	.089	C.Y.	26.50	4.06	.66	31.22	36
5100	Articulated loader and hand dress	B-100	320	.038		26.50	1.95	3	31.45	35.50
5200	Articulated loader and 75 HP dozer	B-10M	500	.024		26.50	1.24	3.53	31.27	35
5300	Road grader and hand dress	B-11L	1000	.016		26.50	.79	1.06	28.35	31.50
6000	Tilling topsoil, 20 HP tractor, disk harrow, 2" deep	B-66	450	.018	M.S.F.		.94	.54	1.48	2.06
6050	4" deep		360	.022			1.18	.67	1.85	2.58
6100	6" deep	↓	270	.030	↓		1.57	.89	2.46	3.44
6150	26" rototiller, 2" deep	A-1J	1250	.006	S.Y.		.27	.04	.31	.48
6200	4" deep		1000	.008			.34	.05	.39	.60
6250	6" deep	↓	750	.011	↓		.45	.07	.52	.80

32 91 13.26 Planting Beds

		Crew	Daily Output	Labor-Hours	Unit	Material	2020 Bare Costs Labor	Equipment	Total	Total Incl O&P
0010	**PLANTING BEDS**									
0100	Backfill planting pit, by hand, on site topsoil	2 Clab	18	.889	C.Y.		37.50		37.50	60
0200	Prepared planting mix, by hand	"	24	.667			28		28	45
0300	Skid steer loader, on site topsoil	B-62	340	.071			3.23	.52	3.75	5.70
0400	Prepared planting mix	"	410	.059			2.68	.43	3.11	4.73
1000	Excavate planting pit, by hand, sandy soil	2 Clab	16	1			42		42	67.50
1100	Heavy soil or clay	"	8	2			84		84	135
1200	1/2 C.Y. backhoe, sandy soil	B-11C	150	.107			5.25	1.43	6.68	9.90
1300	Heavy soil or clay	"	115	.139			6.90	1.86	8.76	12.90
2000	Mix planting soil, incl. loam, manure, peat, by hand	2 Clab	60	.267		46.50	11.25		57.75	69
2100	Skid steer loader	B-62	150	.160	↓	46.50	7.30	1.18	54.98	64
3000	Pile sod, skid steer loader	"	2800	.009	S.Y.		.39	.06	.45	.69
3100	By hand	2 Clab	400	.040			1.68		1.68	2.70
4000	Remove sod, F.E. loader	B-10S	2000	.006			.31	.21	.52	.72
4100	Sod cutter	B-12K	3200	.005			.25	.30	.55	.73
4200	By hand	2 Clab	240	.067	↓		2.81		2.81	4.50

32 91 19 – Landscape Grading

32 91 19.13 Topsoil Placement and Grading

		Crew	Daily Output	Labor-Hours	Unit	Material	2020 Bare Costs Labor	Equipment	Total	Total Incl O&P
0010	**TOPSOIL PLACEMENT AND GRADING**									
0700	Furnish and place, truck dumped, screened, 4" deep	B-10S	1300	.009	S.Y.	3.65	.48	.33	4.46	5.15
0800	6" deep		820	.015	"	4.68	.76	.52	5.96	6.90
0810	Minimum labor/equipment charge	↓	2	6	Ea.		310	212	522	725
0900	Fine grading and seeding, incl. lime, fertilizer & seed,									
1000	With equipment	B-14	1000	.048	S.Y.	.49	2.12	.21	2.82	4.17
2000	Minimum labor/equipment charge	1 Clab	4	2	Job		84		84	135

For customer support on your Commercial Renovation Costs with RSMeans data, call 800.448.8182.

569

32 92 19 – Seeding

32 92 19.13 Mechanical Seeding

	32 92 19.13 Mechanical Seeding	Crew	Daily Output	Labor-Hours	Unit	Material	2020 Bare Costs Labor	Equipment	Total	Total Incl O&P
0010	**MECHANICAL SEEDING**									
0020	Mechanical seeding, 215 lb./acre	B-66	1.50	5.333	Acre	595	283	161	1,039	1,275
0100	44 lb./M.S.Y.	"	2500	.003	S.Y.	.23	.17	.10	.50	.63
0101	44 lb./M.S.Y.	1 Clab	13950	.001	S.F.	.03	.02		.05	.07
0300	Fine grading and seeding incl. lime, fertilizer & seed,									
0310	with equipment	B-14	1000	.048	S.Y.	.47	2.12	.21	2.80	4.15
0400	Fertilizer hand push spreader, 35 lb./M.S.F.	1 Clab	200	.040	M.S.F.	10.70	1.68		12.38	14.45
0600	Limestone hand push spreader, 50 lb./M.S.F.		180	.044		5.60	1.87		7.47	9.15
0800	Grass seed hand push spreader, 4.5 lb./M.S.F.	↓	180	.044	↓	25.50	1.87		27.37	31
1000	Hydro or air seeding for large areas, incl. seed and fertilizer	B-81	8900	.003	S.Y.	.64	.13	.07	.84	.98
1100	With wood fiber mulch added	"	8900	.003	"	2.07	.13	.07	2.27	2.56
1300	Seed only, over 100 lb., field seed, minimum				Lb.	2.08			2.08	2.29
1400	Maximum					1.97			1.97	2.17
1500	Lawn seed, minimum					1.42			1.42	1.56
1600	Maximum				↓	2.70			2.70	2.97
1800	Aerial operations, seeding only, field seed	B-58	50	.480	Acre	675	22	46	743	830
1900	Lawn seed		50	.480		460	22	46	528	595
2100	Seed and liquid fertilizer, field seed		50	.480		760	22	46	828	920
2200	Lawn seed	↓	50	.480	↓	545	22	46	613	685
9000	Minimum labor/equipment charge	1 Clab	4	2	Job		84		84	135

32 92 23 – Sodding

32 92 23.10 Sodding Systems

	32 92 23.10 Sodding Systems	Crew	Daily Output	Labor-Hours	Unit	Material	2020 Bare Costs Labor	Equipment	Total	Total Incl O&P
0010	**SODDING SYSTEMS**									
0020	Sodding, 1" deep, bluegrass sod, on level ground, over 8 M.S.F.	B-63	22	1.818	M.S.F.	223	80.50	8.05	311.55	385
0200	4 M.S.F.		17	2.353		286	104	10.45	400.45	495
0300	1,000 S.F.		13.50	2.963		335	131	13.15	479.15	595
0500	Sloped ground, over 8 M.S.F.		6	6.667		223	295	29.50	547.50	750
0600	4 M.S.F.		5	8		286	355	35.50	676.50	920
0700	1,000 S.F.		4	10		335	445	44.50	824.50	1,125
1000	Bent grass sod, on level ground, over 6 M.S.F.		20	2		305	88.50	8.90	402.40	485
1100	3 M.S.F.		18	2.222		320	98.50	9.85	428.35	525
1200	Sodding 1,000 S.F. or less		14	2.857		345	127	12.70	484.70	595
1500	Sloped ground, over 6 M.S.F.		15	2.667		305	118	11.85	434.85	535
1600	3 M.S.F.		13.50	2.963		320	131	13.15	464.15	580
1700	1,000 S.F.	↓	12	3.333	↓	345	148	14.80	507.80	630

32 93 Plants

32 93 13 – Ground Covers

32 93 13.10 Ground Cover Plants

	32 93 13.10 Ground Cover Plants	Crew	Daily Output	Labor-Hours	Unit	Material	2020 Bare Costs Labor	Equipment	Total	Total Incl O&P
0010	**GROUND COVER PLANTS**									
0012	Plants, pachysandra, in prepared beds	B-1	15	1.600	C	88.50	68.50		157	207
0200	Vinca minor, 1 yr., bare root, in prepared beds		12	2	"	81.50	85.50		167	227
0600	Stone chips, in 50 lb. bags, Georgia marble		520	.046	Bag	4.78	1.97		6.75	8.40
0700	Onyx gemstone		260	.092		16.90	3.95		20.85	25
0800	Quartz		260	.092	↓	17.15	3.95		21.10	25
0900	Pea gravel, truckload lots	↓	28	.857	Ton	11.80	36.50		48.30	71.50

For customer support on your Commercial Renovation Costs with RSMeans data, call 800.448.8182.

32 93 Plants

32 93 33 – Shrubs

32 93 33.10 Shrubs and Trees

	Crew	Daily Output	Labor-Hours	Unit	Material	2020 Bare Costs Labor	Equipment	Total	Total Incl O&P
0010	**SHRUBS AND TREES**									
0011	Evergreen, in prepared beds, B&B									
0100	Arborvitae pyramidal, 4'-5'	B-17	30	1.067	Ea.	104	49.50	21.50	175	218
0150	Globe, 12"-15"	B-1	96	.250		24	10.70		34.70	43
0300	Cedar, blue, 8'-10'	B-17	18	1.778		241	82.50	35.50	359	435
0500	Hemlock, Canadian, 2-1/2'-3'	B-1	36	.667		33	28.50		61.50	82
0550	Holly, Savannah, 8'-10' H		9.68	2.479		294	106		400	495
0600	Juniper, andorra, 18"-24"		80	.300		58.50	12.85		71.35	85
0620	Wiltoni, 15"-18"		80	.300		28.50	12.85		41.35	52
0640	Skyrocket, 4-1/2'-5'	B-17	55	.582		111	27	11.65	149.65	178
0660	Blue pfitzer, 2'-2-1/2'	B-1	44	.545		41	23.50		64.50	83
0680	Ketleerie, 2-1/2'-3'		50	.480		58.50	20.50		79	97.50
0700	Pine, black, 2-1/2'-3'		50	.480		68	20.50		88.50	108
0720	Mugo, 18"-24"		60	.400		64.50	17.10		81.60	98.50
0740	White, 4'-5'	B-17	75	.427		55.50	19.85	8.55	83.90	102
0800	Spruce, blue, 18"-24"	B-1	60	.400		71.50	17.10		88.60	107
0840	Norway, 4'-5'	B-17	75	.427		93.50	19.85	8.55	121.90	144
0900	Yew, denisforma, 12"-15"	B-1	60	.400		38.50	17.10		55.60	70
1000	Capitata, 18"-24"		30	.800		34.50	34		68.50	93
1100	Hicksi, 2'-2-1/2'		30	.800		102	34		136	167

32 93 33.20 Shrubs

	Crew	Daily Output	Labor-Hours	Unit	Material	2020 Bare Costs Labor	Equipment	Total	Total Incl O&P
0010	**SHRUBS**									
0011	Broadleaf Evergreen, planted in prepared beds									
0100	Andromeda, 15"-18", cont	B-1	96	.250	Ea.	33.50	10.70		44.20	54
0200	Azalea, 15"-18", cont		96	.250		34	10.70		44.70	54.50
0300	Barberry, 9"-12", cont		130	.185		18.60	7.90		26.50	33
0400	Boxwood, 15"-18", B&B		96	.250		48	10.70		58.70	69.50
0500	Euonymus, emerald gaiety, 12"-15", cont		115	.209		28	8.95		36.95	45.50
0600	Holly, 15"-18", B&B		96	.250		41.50	10.70		52.20	62.50
0900	Mount laurel, 18"-24", B&B		80	.300		76.50	12.85		89.35	105
1000	Paxistema, 9"-12" H		130	.185		23	7.90		30.90	38
1100	Rhododendron, 18"-24", cont		48	.500		39.50	21.50		61	78
1200	Rosemary, 1 gal. cont		600	.040		19.80	1.71		21.51	24.50
2000	Deciduous, planted in prepared beds, amelanchier, 2'-3', B&B		57	.421		132	18		150	174
2100	Azalea, 15"-18", B&B		96	.250		30.50	10.70		41.20	50.50
2300	Bayberry, 2'-3', B&B		57	.421		35	18		53	67.50
2600	Cotoneaster, 15"-18", B&B		80	.300		26.50	12.85		39.35	50
2800	Dogwood, 3'-4', B&B	B-17	40	.800		36	37	16.05	89.05	116
2900	Euonymus, alatus compacta, 15"-18", cont	B-1	80	.300		28	12.85		40.85	51.50
3200	Forsythia, 2'-3', cont	"	60	.400		17.65	17.10		34.75	47
3300	Hibiscus, 3'-4', B&B	B-17	75	.427		45	19.85	8.55	73.40	90.50
3400	Honeysuckle, 3'-4', B&B	B-1	60	.400		29	17.10		46.10	59.50
3500	Hydrangea, 2'-3', B&B	"	57	.421		33	18		51	65
3600	Lilac, 3'-4', B&B	B-17	40	.800		22	37	16.05	75.05	101
3900	Privet, bare root, 18"-24"	B-1	80	.300		17.50	12.85		30.35	40
4100	Quince, 2'-3', B&B	"	57	.421		30.50	18		48.50	62.50
4200	Russian olive, 3'-4', B&B	B-17	75	.427		26	19.85	8.55	54.40	70
4400	Spirea, 3'-4', B&B	B-1	70	.343		21.50	14.65		36.15	47
4500	Viburnum, 3'-4', B&B	B-17	40	.800		27	37	16.05	80.05	107

571

For customer support on your Commercial Renovation Costs with RSMeans data, call 800.448.8182.

32 93 Plants

32 93 43 – Trees

32 93 43.20 Trees

			Daily	Labor-			2020 Bare Costs			Total
	32 93 43.20 Trees	Crew	Output	Hours	Unit	Material	Labor	Equipment	Total	Incl O&P
0010	**TREES**									
0011	Deciduous, in prep. beds, balled & burlapped (B&B)									
0100	Ash, 2" caliper	G B-17	8	4	Ea.	213	186	80	479	620
0200	Beech, 5'-6'	G	50	.640		240	30	12.85	282.85	325
0300	Birch, 6'-8', 3 stems	G	20	1.600		175	74.50	32	281.50	345
0500	Crabapple, 6'-8'	G	20	1.600		160	74.50	32	266.50	330
0600	Dogwood, 4'-5'	G	40	.800		140	37	16.05	193.05	231
0700	Eastern redbud, 4'-5'	G	40	.800		151	37	16.05	204.05	243
0800	Elm, 8'-10'	G	20	1.600		350	74.50	32	456.50	540
0900	Ginkgo, 6'-7'	G	24	1.333		161	62	27	250	305
1000	Hawthorn, 8'-10', 1" caliper	G	20	1.600		157	74.50	32	263.50	325
1100	Honeylocust, 10'-12', 1-1/2" caliper	G	10	3.200		227	149	64	440	555
1300	Larch, 8'	G	32	1		135	46.50	20	201.50	245
1400	Linden, 8'-10', 1" caliper	G	20	1.600		169	74.50	32	275.50	340
1500	Magnolia, 4'-5'	G	20	1.600		118	74.50	32	224.50	284
1600	Maple, red, 8'-10', 1-1/2" caliper	G	10	3.200		200	149	64	413	530
1700	Mountain ash, 8'-10', 1" caliper	G	16	2		196	93	40	329	410
1800	Oak, 2-1/2"-3" caliper	G	6	5.333		355	248	107	710	905
2100	Planetree, 9'-11', 1-1/4" caliper	G	10	3.200		278	149	64	491	615
2200	Plum, 6'-8', 1" caliper	G	20	1.600		82	74.50	32	188.50	244
2300	Poplar, 9'-11', 1-1/4" caliper	G	10	3.200		102	149	64	315	420
2500	Sumac, 2'-3'	G	75	.427		45.50	19.85	8.55	73.90	91
2700	Tulip, 5'-6'	G	40	.800		48.50	37	16.05	101.55	130
2800	Willow, 6'-8', 1" caliper	G	20	1.600		103	74.50	32	209.50	267
9000	Minimum labor/equipment charge	1 Clab	4	2	Job		84		84	135

32 94 Planting Accessories

32 94 13 – Landscape Edging

32 94 13.20 Edging

			Daily	Labor-			2020 Bare Costs			Total
		Crew	Output	Hours	Unit	Material	Labor	Equipment	Total	Incl O&P
0010	**EDGING**									
0050	Aluminum alloy, including stakes, 1/8" x 4", mill finish	B-1	390	.062	L.F.	2.15	2.63		4.78	6.60
0051	Black paint		390	.062		2.49	2.63		5.12	6.95
0052	Black anodized		390	.062		2.88	2.63		5.51	7.40
0100	Brick, set horizontally, 1-1/2 bricks per L.F.	D-1	370	.043		1.70	2.03		3.73	5.15
0150	Set vertically, 3 bricks per L.F.	"	135	.119		3.80	5.55		9.35	13.20
0200	Corrugated aluminum, roll, 4" wide	1 Carp	650	.012		2.25	.65		2.90	3.53
0250	6" wide	"	550	.015		2.81	.77		3.58	4.33
0600	Railroad ties, 6" x 8"	2 Carp	170	.094		2.48	5		7.48	10.75
0650	7" x 9"		136	.118		2.75	6.25		9	13.05
0750	Redwood 2" x 4"		330	.048		2.31	2.58		4.89	6.65
0800	Steel edge strips, incl. stakes, 1/4" x 5"	B-1	390	.062		4.58	2.63		7.21	9.25
0850	3/16" x 4"	"	390	.062		3.62	2.63		6.25	8.20
9000	Minimum labor/equipment charge	1 Carp	4	2	Job		106		106	170

32 94 50 – Tree Guying

32 94 50.10 Tree Guying Systems

			Daily	Labor-			2020 Bare Costs			Total
		Crew	Output	Hours	Unit	Material	Labor	Equipment	Total	Incl O&P
0010	**TREE GUYING SYSTEMS**									
0015	Tree guying including stakes, guy wire and wrap									
0100	Less than 3" caliper, 2 stakes	2 Clab	35	.457	Ea.	14.15	19.25		33.40	46.50
0200	3" to 4" caliper, 3 stakes	"	21	.762	"	20	32		52	73.50
1000	Including arrowhead anchor, cable, turnbuckles and wrap									

32 94 Planting Accessories

32 94 50 - Tree Guying

32 94 50.10 Tree Guying Systems	Crew	Daily Output	Labor-Hours	Unit	Material	2020 Bare Costs Labor	2020 Bare Costs Equipment	Total	Total Incl O&P
1100	Less than 3" caliper, 3" anchors	2 Clab	20	.800	Ea.	23.50	33.50		57	79.50
1200	3" to 6" caliper, 4" anchors		15	1.067		34	45		79	109
1300	6" caliper, 6" anchors		12	1.333		27.50	56		83.50	120
1400	8" caliper, 8" anchors		9	1.778		115	75		190	246

32 96 Transplanting

32 96 23 - Plant and Bulb Transplanting

32 96 23.23 Planting

		Crew	Daily Output	Labor-Hours	Unit	Material	Labor	Equipment	Total	Total Incl O&P
0010	**PLANTING**									
0012	Moving shrubs on site, 12" ball	B-62	28	.857	Ea.	39	6.35		45.35	69.50
0100	24" ball	"	22	1.091	"	50	8.05		58.05	88

32 96 23.43 Moving Trees

		Crew	Daily Output	Labor-Hours	Unit	Material	Labor	Equipment	Total	Total Incl O&P
0010	**MOVING TREES**, On site									
0300	Moving trees on site, 36" ball	B-6	3.75	6.400	Ea.		293	57	350	530
0400	60" ball	"	1	24	"		1,100	214	1,314	1,975

32 96 43 - Tree Transplanting

32 96 43.20 Tree Removal

		Crew	Daily Output	Labor-Hours	Unit	Material	Labor	Equipment	Total	Total Incl O&P
0010	**TREE REMOVAL**									
0100	Dig & lace, shrubs, broadleaf evergreen, 18"-24" high	B-1	55	.436	Ea.		18.65		18.65	30
0200	2'-3'	"	35	.686			29.50		29.50	47
0300	3'-4'	B-6	30	.800			36.50	7.15	43.65	66
0400	4'-5'	"	20	1.200			55	10.70	65.70	99
1000	Deciduous, 12"-15"	B-1	110	.218			9.35		9.35	14.95
1100	18"-24"		65	.369			15.80		15.80	25.50
1200	2'-3'		55	.436			18.65		18.65	30
1300	3'-4'	B-6	50	.480			22	4.28	26.28	39.50
2000	Evergreen, 18"-24"	B-1	55	.436			18.65		18.65	30
2100	2'-0" to 2'-6"		50	.480			20.50		20.50	33
2200	2'-6" to 3'-0"		35	.686			29.50		29.50	47
2300	3'-0" to 3'-6"		20	1.200			51.50		51.50	82
3000	Trees, deciduous, small, 2'-3'		55	.436			18.65		18.65	30
3100	3'-4'	B-6	50	.480			22	4.28	26.28	39.50
3200	4'-5'		35	.686			31.50	6.10	37.60	56.50
3300	5'-6'		30	.800			36.50	7.15	43.65	66
4000	Shade, 5'-6'		50	.480			22	4.28	26.28	39.50
4100	6'-8'		35	.686			31.50	6.10	37.60	56.50
4200	8'-10'		25	.960			44	8.55	52.55	79
4300	2" caliper		12	2			91.50	17.80	109.30	165
5000	Evergreen, 4'-5'		35	.686			31.50	6.10	37.60	56.50
5100	5'-6'		25	.960			44	8.55	52.55	79
5200	6'-7'		19	1.263			58	11.25	69.25	104
5300	7'-8'		15	1.600			73	14.25	87.25	132
5400	8'-10'		11	2.182			100	19.45	119.45	180

For customer support on your Commercial Renovation Costs with RSMeans data, call 800.448.8182.

573

Division Notes

	CREW	DAILY OUTPUT	LABOR-HOURS	UNIT	BARE COSTS				TOTAL INCL O&P
					MAT.	LABOR	EQUIP.	TOTAL	

Estimating Tips

33 10 00 Water Utilities
33 30 00 Sanitary Sewerage Utilities
33 40 00 Storm Drainage Utilities

- Never assume that the water, sewer, and drainage lines will go in at the early stages of the project. Consider the site access needs before dividing the site in half with open trenches, loose pipe, and machinery obstructions. Always inspect the site to establish that the site drawings are complete. Check off all existing utilities on your drawings as you locate them. Be especially careful with underground utilities because appurtenances are sometimes buried during regrading or repaving operations. If you find any discrepancies, mark up the site plan for further research. Differing site conditions can be very costly if discovered later in the project.

- See also Section 33 01 00 for restoration of pipe where removal/replacement may be undesirable. Use of new types of piping materials can reduce the overall project cost. Owners/design engineers should consider the installing contractor as a valuable source of current information on utility products and local conditions that could lead to significant cost savings.

Reference Numbers

Reference numbers are shown at the beginning of some major classifications. These numbers refer to related items in the Reference Section. The reference information may be an estimating procedure, an alternate pricing method, or technical information.

Note: Not all subdivisions listed here necessarily appear. ■

Same Data. Simplified.

Enjoy the convenience and efficiency of accessing your costs anywhere:

- **Skip the multiplier** by setting your location
- **Quickly search,** edit, favorite and share costs
- **Stay on top of price changes** with automatic updates

Discover more at rsmeans.com/online

---

33 01 30.11 Television Inspection of Sewers	Crew	Daily Output	Labor-Hours	Unit	Material	2020 Bare Costs Labor	Equipment	Total	Total Incl O&P
0010	**TELEVISION INSPECTION OF SEWERS**								
0100	Pipe internal cleaning & inspection, cleaning, pressure pipe systems								
0120	Pig method, lengths 1000' to 10,000'								
0140	4" diameter thru 24" diameter, minimum				L.F.				3.60	4.14
0160	Maximum				"				18	21
6000	Sewage/sanitary systems								
6100	Power rodder with header & cutters								
6110	Mobilization charge, minimum				Total				695	800
6120	Mobilization charge, maximum				"				9,125	10,600
6140	Cleaning 4"-12" diameter				L.F.				6	6.60
6190	14"-24" diameter								8	8.80
6240	30" diameter								9.60	10.50
6250	36" diameter								12	13.20
6260	48" diameter								16	17.60
6270	60" diameter								24	26.50
6280	72" diameter								48	53
9000	Inspection, television camera with video								
9060	up to 500 linear feet				Total				715	820

33 01 30.72 Cured-In-Place Pipe Lining

		Crew	Daily Output	Labor-Hours	Unit	Material	Labor	Equipment	Total	Total Incl O&P
0010	**CURED-IN-PLACE PIPE LINING**									
0011	Not incl. bypass or cleaning									
0020	Less than 10,000 L.F., urban, 6" to 10"	C-17E	130	.615	L.F.	9.40	34	.28	43.68	64.50
0050	10" to 12"		125	.640		11.60	35.50	.29	47.39	69.50
0070	12" to 16"		115	.696		11.85	38.50	.31	50.66	74.50
0100	16" to 20"		95	.842		13.95	46.50	.38	60.83	89.50
0200	24" to 36"		90	.889		15.30	49	.40	64.70	95.50
0300	48" to 72"		80	1		24	55.50	.45	79.95	115
0500	Rural, 6" to 10"		180	.444		9.40	24.50	.20	34.10	49.50
0550	10" to 12"		175	.457		11.60	25.50	.21	37.31	53
0570	12" to 16"		160	.500		12.15	27.50	.23	39.88	57.50
0600	16" to 20"		135	.593		14.95	32.50	.27	47.72	68.50
0700	24" to 36"		125	.640		15.35	35.50	.29	51.14	73.50
0800	48" to 72"		100	.800		24	44	.36	68.36	97.50
1000	Greater than 10,000 L.F., urban, 6" to 10"		160	.500		9.40	27.50	.23	37.13	54.50
1050	10" to 12"		155	.516		11.55	28.50	.23	40.28	58.50
1070	12" to 16"		140	.571		11.85	31.50	.26	43.61	64
1100	16" to 20"		120	.667		14.95	37	.30	52.25	75
1200	24" to 36"		115	.696		15.35	38.50	.31	54.16	78
1300	48" to 72"		95	.842		24	46.50	.38	70.88	101
1500	Rural, 6" to 10"		215	.372		9.40	20.50	.17	30.07	43.50
1550	10" to 12"		210	.381		11.60	21	.17	32.77	46.50
1570	12" to 16"		185	.432		11.85	24	.19	36.04	51.50
1600	16" to 20"		150	.533		14.95	29.50	.24	44.69	63.50
1700	24" to 36"		140	.571		15.35	31.50	.26	47.11	67.50
1800	48" to 72"		120	.667		24.50	37	.30	61.80	86
2000	Cured in place pipe, non-pressure, flexible felt resin									
2100	6" diameter				L.F.				25.50	28
2200	8" diameter								26.50	29
2201	8" diameter									
2300	10" diameter								29	32
2400	12" diameter								34.50	38
2500	15" diameter								59	65

33 01 Operation and Maintenance of Utilities

33 01 30 – Operation and Maintenance of Sewer Utilities

33 01 30.72 Cured-In-Place Pipe Lining	Crew	Daily Output	Labor-Hours	Unit	Material	2020 Bare Costs Labor	Equipment	Total	Total Incl O&P
2600	18" diameter				L.F.				82	90
2700	21" diameter							105	115
2800	24" diameter							177	195
2900	30" diameter							195	215
3000	36" diameter							205	225
3100	48" diameter							218	240

33 05 Common Work Results for Utilities

33 05 07 – Trenchless Installation of Utility Piping

33 05 07.23 Utility Boring and Jacking

		Crew	Daily Output	Labor-Hours	Unit	Material	Labor	Equipment	Total	Total Incl O&P
0010	**UTILITY BORING AND JACKING**									
0011	Casing only, 100' minimum,									
0020	not incl. jacking pits or dewatering									
0100	Roadwork, 1/2" thick wall, 24" diameter casing	B-42	20	3.200	L.F.	122	152	52	326	435
0200	36" diameter		16	4		215	190	65	470	615
0300	48" diameter		15	4.267		300	203	69.50	572.50	730
0500	Railroad work, 24" diameter		15	4.267		122	203	69.50	394.50	535
0600	36" diameter		14	4.571		215	217	74.50	506.50	665
0700	48" diameter		12	5.333		300	253	87	640	830
0900	For ledge, add								20%	20%
1000	Small diameter boring, 3", sandy soil	B-82	900	.018		22	.85	.20	23.05	26
1040	Rocky soil	"	500	.032		22	1.52	.36	23.88	27.50

33 05 07.36 Microtunneling

		Crew	Daily Output	Labor-Hours	Unit	Material	Labor	Equipment	Total	Total Incl O&P
0010	**MICROTUNNELING**									
0011	Not including excavation, backfill, shoring,									
0020	or dewatering, average 50'/day, slurry method									
0100	24" to 48" outside diameter, minimum				L.F.				965	965
0110	Adverse conditions, add				"				500	500
1000	Rent microtunneling machine, average monthly lease				Month				97,500	107,000
1010	Operating technician				Day				630	705
1100	Mobilization and demobilization, minimum				Job				41,200	45,900
1110	Maximum				"				445,500	490,500

33 05 61 – Concrete Manholes

33 05 61.10 Storm Drainage Manholes, Frames and Covers

		Crew	Daily Output	Labor-Hours	Unit	Material	Labor	Equipment	Total	Total Incl O&P
0010	**STORM DRAINAGE MANHOLES, FRAMES & COVERS**									
0020	Excludes footing, excavation, backfill (See line items for frame & cover)									
0050	Brick, 4' inside diameter, 4' deep	D-1	1	16	Ea.	615	750		1,365	1,875
0100	6' deep		.70	22.857		870	1,075		1,945	2,675
0150	8' deep		.50	32		1,125	1,500		2,625	3,650
0200	For depths over 8', add		4	4	V.L.F.	95.50	188		283.50	410
1110	Precast, 4' ID, 4' deep	B-22	4.10	7.317	Ea.	880	360	69	1,309	1,625
1120	6' deep		3	10		1,025	495	94.50	1,614.50	2,025
1130	8' deep		2	15		1,200	740	142	2,082	2,650
1140	For depths over 8', add		16	1.875	V.L.F.	134	92.50	17.70	244.20	315

33 05 63 – Concrete Vaults and Chambers

33 05 63.13 Precast Concrete Utility Structures

		Crew	Daily Output	Labor-Hours	Unit	Material	Labor	Equipment	Total	Total Incl O&P
0010	**PRECAST CONCRETE UTILITY STRUCTURES**, 6" thick									
0050	5' x 10' x 6' high, ID	B-13	2	28	Ea.	1,850	1,300	290	3,440	4,425
0350	Hand hole, precast concrete, 1-1/2" thick									
0400	1'-0" x 2'-0" x 1'-9", ID, light duty	B-1	4	6	Ea.	540	257		797	1,000

For customer support on your Commercial Renovation Costs with RSMeans data, call 800.448.8182.

577

33 05 Common Work Results for Utilities

33 05 63 – Concrete Vaults and Chambers

33 05 63.13 Precast Concrete Utility Structures

		Crew	Daily Output	Labor-Hours	Unit	Material	2020 Bare Costs Labor	Equipment	Total	Total Incl O&P
0450	4'-6" x 3'-2" x 2'-0", OD, heavy duty	B-6	3	8	Ea.	1,625	365	71.50	2,061.50	2,425

33 05 97 – Identification and Signage for Utilities

33 05 97.05 Utility Connection

		Crew	Daily Output	Labor-Hours	Unit	Material	2020 Bare Costs Labor	Equipment	Total	Total Incl O&P
0010	**UTILITY CONNECTION**									
0020	Water, sanitary, stormwater, gas, single connection	B-14	1	48	Ea.	3,050	2,125	214	5,389	6,950
0030	Telecommunication	"	3	16	"	400	710	71	1,181	1,650

33 05 97.10 Utility Accessories

		Crew	Daily Output	Labor-Hours	Unit	Material	2020 Bare Costs Labor	Equipment	Total	Total Incl O&P
0010	**UTILITY ACCESSORIES** R312316-40									
0400	Underground tape, detectable, reinforced, alum. foil core, 2"	1 Clab	150	.053	C.L.F.	9	2.25		11.25	13.50
0500	6"		140	.057	"	38	2.41		40.41	45.50
9000	Minimum labor/equipment charge		4	2	Job		84		84	135

33 11 Groundwater Sources

33 11 13 – Potable Water Supply Wells

33 11 13.10 Wells and Accessories

		Crew	Daily Output	Labor-Hours	Unit	Material	2020 Bare Costs Labor	Equipment	Total	Total Incl O&P
0010	**WELLS & ACCESSORIES** R221113-50									
0011	Domestic									
0100	Drilled, 4" to 6" diameter	B-23	120	.333	L.F.		14.15	13.25	27.40	37
0200	8" diameter	"	95.20	.420	"		17.85	16.70	34.55	47
1400	Remove & reset pump, minimum	B-21	4	7	Ea.		340	47.50	387.50	595
1420	Maximum	"	2	14	"		685	94.50	779.50	1,175
1500	Pumps, installed in wells to 100' deep, 4" submersible									
1510	1/2 HP	Q-1	3.22	4.969	Ea.	660	288		948	1,175
1520	3/4 HP		2.66	6.015		795	350		1,145	1,425
1600	1 HP		2.29	6.987		975	405		1,380	1,700
1700	1-1/2 HP	Q-22	1.60	10		2,150	580	294	3,024	3,600
1800	2 HP		1.33	12.030		1,800	700	355	2,855	3,475
1900	3 HP		1.14	14.035		2,425	815	415	3,655	4,375
2000	5 HP		1.14	14.035		2,850	815	415	4,080	4,825
2050	Remove and install motor only, 4 HP		1.14	14.035		1,450	815	415	2,680	3,300
5000	Wells to 180' deep, 4" submersible, 1 HP	B-21	1.10	25.455		2,425	1,250	172	3,847	4,825
5500	2 HP		1.10	25.455		3,875	1,250	172	5,297	6,425
6000	3 HP		1	28		4,500	1,375	189	6,064	7,325
7000	5 HP		.90	31.111		7,100	1,525	210	8,835	10,500
9000	Minimum labor/equipment charge		1.80	15.556	Job		760	105	865	1,325

33 14 Water Utility Transmission and Distribution

33 14 13 – Public Water Utility Distribution Piping

33 14 13.15 Water Supply, Ductile Iron Pipe

		Crew	Daily Output	Labor-Hours	Unit	Material	2020 Bare Costs Labor	Equipment	Total	Total Incl O&P
0010	**WATER SUPPLY, DUCTILE IRON PIPE** R331113-80									
0020	Not including excavation or backfill									
2000	Pipe, class 50 water piping, 18' lengths									
2020	Mechanical joint, 4" diameter	B-21A	200	.200	L.F.	42.50	10.45	2.14	55.09	65.50
3000	Push-on joint, 4" diameter		400	.100		23	5.25	1.07	29.32	35
3020	6" diameter		333.33	.120		23.50	6.25	1.29	31.04	37.50
8000	Piping, fittings, mechanical joint, AWWA C110									
8006	90° bend, 4" diameter	B-20A	16	2	Ea.	176	101		277	350
8020	6" diameter		12.80	2.500		256	126		382	480

33 14 Water Utility Transmission and Distribution

33 14 13 – Public Water Utility Distribution Piping

33 14 13.15 Water Supply, Ductile Iron Pipe

		Crew	Daily Output	Labor-Hours	Unit	Material	2020 Bare Costs Labor	2020 Bare Costs Equipment	Total	Total Incl O&P
8200	Wye or tee, 4" diameter	B-20A	10.67	2.999	Ea.	390	152		542	665
8220	6" diameter		8.53	3.751		585	190		775	940
8398	45° bend, 4" diameter		16	2		206	101		307	385
8400	6" diameter		12.80	2.500		295	126		421	525
8405	8" diameter		10.67	2.999		430	152		582	715
8450	Decreaser, 6" x 4" diameter		14.22	2.250		262	114		376	465
8460	8" x 6" diameter	↓	11.64	2.749	↓	350	139		489	605
8550	Piping, butterfly valves, cast iron									
8560	4" diameter	B-20	6	4	Ea.	455	188		643	800
9600	Steel sleeve with tap, 4" diameter		3	8		505	375		880	1,150
9620	6" diameter	↓	2	12	↓	555	565		1,120	1,525

33 14 17 – Site Water Utility Service Laterals

33 14 17.15 Tapping, Crosses and Sleeves

		Crew	Daily Output	Labor-Hours	Unit	Material	2020 Bare Costs Labor	2020 Bare Costs Equipment	Total	Total Incl O&P
0010	**TAPPING, CROSSES AND SLEEVES**									
4000	Drill and tap pressurized main (labor only)									
4100	6" main, 1" to 2" service	Q-1	3	5.333	Ea.		310		310	480
4150	8" main, 1" to 2" service	"	2.75	5.818	"		335		335	520
4500	Tap and insert gate valve									
4600	8" main, 4" branch	B-21	3.20	8.750	Ea.		425	59	484	745
4650	6" branch		2.70	10.370			505	70	575	880
4800	12" main, 6" branch	↓	2.35	11.915	↓		580	80.50	660.50	1,025

33 31 Sanitary Sewerage Piping

33 31 11 – Public Sanitary Sewerage Gravity Piping

33 31 11.15 Sewage Collection, Concrete Pipe

0010	**SEWAGE COLLECTION, CONCRETE PIPE**
0020	See Section 33 42 11.60 for sewage/drainage collection, concrete pipe

33 31 11.25 Sewage Collection, Polyvinyl Chloride Pipe

		Crew	Daily Output	Labor-Hours	Unit	Material	2020 Bare Costs Labor	2020 Bare Costs Equipment	Total	Total Incl O&P
0010	**SEWAGE COLLECTION, POLYVINYL CHLORIDE PIPE**									
0020	Not including excavation or backfill									
2000	20' lengths, SDR 35, B&S, 4" diameter	B-20	375	.064	L.F.	1.67	3.01		4.68	6.65
2040	6" diameter		350	.069		4.05	3.22		7.27	9.60
2080	13' lengths, SDR 35, B&S, 8" diameter	↓	335	.072		6.70	3.37		10.07	12.75
2120	10" diameter	B-21	330	.085		11.30	4.14	.57	16.01	19.70
4000	Piping, DWV PVC, no exc./bkfill., 10' L, Sch 40, 4" diameter	B-20	375	.064		3.83	3.01		6.84	9.05
4010	6" diameter		350	.069		8.25	3.22		11.47	14.20
4020	8" diameter	↓	335	.072	↓	13.40	3.37		16.77	20

33 34 Onsite Wastewater Disposal

33 34 13 – Septic Tanks

33 34 13.13 Concrete Septic Tanks

		Crew	Daily Output	Labor-Hours	Unit	Material	2020 Bare Costs Labor	2020 Bare Costs Equipment	Total	Total Incl O&P
0010	**CONCRETE SEPTIC TANKS**									
0011	Not including excavation or piping									
0015	Septic tanks, precast, 1,000 gallon	B-21	8	3.500	Ea.	1,050	171	23.50	1,244.50	1,475
0020	1,250 gallon		8	3.500		1,425	171	23.50	1,619.50	1,875
0060	1,500 gallon		7	4		1,625	195	27	1,847	2,150
0100	2,000 gallon		5	5.600		2,500	273	38	2,811	3,225
0140	2,500 gallon	↓	5	5.600	↓	2,375	273	38	2,686	3,075

33 34 Onsite Wastewater Disposal

33 34 13 – Septic Tanks

33 34 13.13 Concrete Septic Tanks

		Crew	Daily Output	Labor-Hours	Unit	Material	2020 Bare Costs Labor	2020 Bare Costs Equipment	Total	Total Incl O&P
0180	4,000 gallon	B-21	4	7	Ea.	5,850	340	47.50	6,237.50	7,050
0220	5,000 gallon, 4 piece	B-13	3	18.667		9,025	860	194	10,079	11,500
0300	15,000 gallon, 4 piece	B-13B	1.70	32.941		23,700	1,525	575	25,800	29,100
0400	25,000 gallon, 4 piece		1.10	50.909		47,900	2,350	890	51,140	57,000
0500	40,000 gallon, 4 piece	↓	.80	70		53,500	3,225	1,225	57,950	65,000
0520	50,000 gallon, 5 piece	B-13C	.60	93.333		61,500	4,300	3,800	69,600	78,500
0640	75,000 gallon, cast in place	C-14C	.25	448		74,500	22,600	108	97,208	118,000
0660	100,000 gallon	"	.15	747		92,500	37,600	179	130,279	161,500
1150	Leaching field chambers, 13' x 3'-7" x 1'-4", standard	B-13	16	3.500		500	161	36.50	697.50	845
1200	Heavy duty, 8' x 4' x 1'-6"		14	4		310	184	41.50	535.50	680
1300	13' x 3'-9" x 1'-6"		12	4.667		1,350	215	48.50	1,613.50	1,875
1350	20' x 4' x 1'-6"	↓	5	11.200		1,225	515	116	1,856	2,300
1420	Leaching pit, precast concrete, 6' diameter, 3' deep	B-21	4.70	5.957		940	290	40	1,270	1,525
1600	Leaching pit, 6'-6" diameter, 6' deep		5	5.600		1,075	273	38	1,386	1,650
1620	8' deep		4	7		1,175	340	47.50	1,562.50	1,900
1700	8' diameter, H-20 load, 6' deep		4	7		1,475	340	47.50	1,862.50	2,200
1720	8' deep		3	9.333		2,700	455	63	3,218	3,775
2000	Velocity reducing pit, precast conc., 6' diameter, 3' deep	↓	4.70	5.957	↓	1,825	290	40	2,155	2,525

33 34 13.33 Polyethylene Septic Tanks

		Crew	Daily Output	Labor-Hours	Unit	Material	2020 Bare Costs Labor	2020 Bare Costs Equipment	Total	Total Incl O&P
0010	**POLYETHYLENE SEPTIC TANKS**									
0015	High density polyethylene, 1,000 gallon	B-21	8	3.500	Ea.	1,475	171	23.50	1,669.50	1,925
0020	1,250 gallon		8	3.500		1,325	171	23.50	1,519.50	1,750
0025	1,500 gallon	↓	7	4	↓	1,375	195	27	1,597	1,875

33 34 16 – Septic Tank Effluent Filters

33 34 16.13 Septic Tank Gravity Effluent Filters

		Crew	Daily Output	Labor-Hours	Unit	Material	2020 Bare Costs Labor	2020 Bare Costs Equipment	Total	Total Incl O&P
0010	**SEPTIC TANK GRAVITY EFFLUENT FILTERS**									
3000	Effluent filter, 4" diameter	1 Skwk	8	1	Ea.	44	55		99	136
3020	6" diameter		7	1.143		42.50	62.50		105	147
3030	8" diameter		7	1.143		252	62.50		314.50	375
3040	8" diameter, very fine		7	1.143		500	62.50		562.50	650
3050	10" diameter, very fine		6	1.333		237	73		310	380
3060	10" diameter		6	1.333		276	73		349	420
3080	12" diameter		6	1.333		655	73		728	835
3090	15" diameter	↓	5	1.600	↓	1,075	88		1,163	1,350

33 34 51 – Drainage Field Systems

33 34 51.10 Drainage Field Excavation and Fill

		Crew	Daily Output	Labor-Hours	Unit	Material	2020 Bare Costs Labor	2020 Bare Costs Equipment	Total	Total Incl O&P
0010	**DRAINAGE FIELD EXCAVATION AND FILL**									
2200	Septic tank & drainage field excavation with 3/4 C.Y. backhoe	B-12F	145	.110	C.Y.		5.60	4.79	10.39	14.10
2400	4' trench for disposal field, 3/4 C.Y. backhoe	"	335	.048	L.F.		2.42	2.07	4.49	6.10
2600	Gravel fill, run of bank	B-6	150	.160	C.Y.	16.90	7.30	1.43	25.63	32
2800	Crushed stone, 3/4"	"	150	.160	"	42	7.30	1.43	50.73	59

33 34 51.13 Utility Septic Tank Tile Drainage Field

		Crew	Daily Output	Labor-Hours	Unit	Material	2020 Bare Costs Labor	2020 Bare Costs Equipment	Total	Total Incl O&P
0010	**UTILITY SEPTIC TANK TILE DRAINAGE FIELD**									
0015	Distribution box, concrete, 5 outlets	2 Clab	20	.800	Ea.	93.50	33.50		127	157
0020	7 outlets		16	1		103	42		145	182
0025	9 outlets		8	2		545	84		629	735
0115	Distribution boxes, HDPE, 5 outlets		20	.800		76.50	33.50		110	138
0117	6 outlets		15	1.067		81	45		126	161
0118	7 outlets		15	1.067		75	45		120	155
0120	8 outlets	↓	10	1.600		79	67.50		146.50	195
0240	Distribution boxes, outlet flow leveler	1 Clab	50	.160	↓	2.53	6.75		9.28	13.60

33 34 Onsite Wastewater Disposal

33 34 51 – Drainage Field Systems

33 34 51.13 Utility Septic Tank Tile Drainage Field

		Crew	Daily Output	Labor-Hours	Unit	Material	2020 Bare Costs Labor	Equipment	Total	Total Incl O&P
0300	Precast concrete, galley, 4' x 4' x 4'	B-21	16	1.750	Ea.	246	85.50	11.80	343.30	420
0350	HDPE infiltration chamber 12" H x 15" W	2 Clab	300	.053	L.F.	7	2.25		9.25	11.30
0351	12" H x 15" W end cap	1 Clab	32	.250	Ea.	18.90	10.55		29.45	38
0355	chamber 12" H x 22" W	2 Clab	300	.053	L.F.	6.65	2.25		8.90	10.95
0356	12" H x 22" W end cap	1 Clab	32	.250	Ea.	16.60	10.55		27.15	35
0360	chamber 13" H x 34" W	2 Clab	300	.053	L.F.	14.30	2.25		16.55	19.35
0361	13" H x 34" W end cap	1 Clab	32	.250	Ea.	50.50	10.55		61.05	73
0365	chamber 16" H x 34" W	2 Clab	300	.053	L.F.	19.25	2.25		21.50	24.50
0366	16" H x 34" W end cap	1 Clab	32	.250	Ea.	16.55	10.55		27.10	35
0370	chamber 8" H x 16" W	2 Clab	300	.053	L.F.	9.90	2.25		12.15	14.50
0371	8" H x 16" W end cap	1 Clab	32	.250	Ea.	11.10	10.55		21.65	29

33 41 Subdrainage

33 41 16 – Subdrainage Piping

33 41 16.10 Piping, Subdrainage, Vitrified Clay

		Crew	Daily Output	Labor-Hours	Unit	Material	2020 Bare Costs Labor	Equipment	Total	Total Incl O&P
0010	**PIPING, SUBDRAINAGE, VITRIFIED CLAY**									
0020	Not including excavation and backfill									
1000	Foundation drain, 6" diameter	B-14	315	.152	L.F.	6.25	6.75	.68	13.68	18.40
1010	8" diameter		290	.166		9.30	7.30	.74	17.34	22.50
1020	12" diameter		275	.175		20	7.70	.78	28.48	35
3000	Perforated, 5' lengths, C700, 4" diameter		400	.120		4.70	5.30	.53	10.53	14.20
3020	6" diameter		315	.152		6.65	6.75	.68	14.08	18.85
3040	8" diameter		290	.166		7.25	7.30	.74	15.29	20.50
3060	12" diameter		275	.175		20.50	7.70	.78	28.98	35.50

33 42 Stormwater Conveyance

33 42 11 – Stormwater Gravity Piping

33 42 11.40 Piping, Storm Drainage, Corrugated Metal

		Crew	Daily Output	Labor-Hours	Unit	Material	2020 Bare Costs Labor	Equipment	Total	Total Incl O&P
0010	**PIPING, STORM DRAINAGE, CORRUGATED METAL** R221113-50									
0020	Not including excavation or backfill									
2000	Corrugated metal pipe, galvanized									
2020	Bituminous coated with paved invert, 20' lengths									
2040	8" diameter, 16 ga.	B-14	330	.145	L.F.	8.35	6.45	.65	15.45	20
2080	12" diameter, 16 ga.	"	210	.229	"	12.40	10.10	1.02	23.52	31

33 42 11.60 Sewage/Drainage Collection, Concrete Pipe

		Crew	Daily Output	Labor-Hours	Unit	Material	2020 Bare Costs Labor	Equipment	Total	Total Incl O&P
0010	**SEWAGE/DRAINAGE COLLECTION, CONCRETE PIPE**									
0020	Not including excavation or backfill									
1000	Non-reinforced pipe, extra strength, B&S or T&G joints									
1010	6" diameter	B-14	265.04	.181	L.F.	7.80	8	.81	16.61	22
1020	8" diameter		224	.214		8.55	9.50	.95	19	25.50
1040	12" diameter		200	.240		10.05	10.60	1.07	21.72	29
2000	Reinforced culvert, class 3, no gaskets									
2010	12" diameter	B-14	150	.320	L.F.	16.15	14.15	1.42	31.72	42
2020	15" diameter	"	150	.320	"	22	14.15	1.42	37.57	48

33 42 Stormwater Conveyance

33 42 33 – Stormwater Curbside Drains and Inlets

33 42 33.13 Catch Basins	Crew	Daily Output	Labor-Hours	Unit	Material	2020 Bare Costs Labor	Equipment	Total	Total Incl O&P
0010 **CATCH BASINS**									
0011 Not including footing & excavation									
1600 Frames & grates, C.I., 24" square, 500 lb.	B-6	7.80	3.077	Ea.	375	141	27.50	543.50	665
1700 26" D shape, 600 lb.	"	7	3.429	"	720	157	30.50	907.50	1,075
3320 Frames and covers, existing, raised for paving, 2", including									
3340 row of brick, concrete collar, up to 12" wide frame	B-6	18	1.333	Ea.	52	61	11.90	124.90	167

33 42 33.15 Remove and Replace Catch Basin Cover

	Crew	Daily Output	Labor-Hours	Unit	Material	2020 Bare Costs Labor	Equipment	Total	Total Incl O&P
0010 **REMOVE AND REPLACE CATCH BASIN COVER**									
0023 Remove catch basin cover	1 Clab	80	.100	Ea.		4.21		4.21	6.75
0033 Replace catch basin cover	"	80	.100	"		4.21		4.21	6.75

33 52 Hydrocarbon Transmission and Distribution

33 52 13 – Liquid Hydrocarbon Piping

33 52 13.16 Gasoline Piping	Crew	Daily Output	Labor-Hours	Unit	Material	2020 Bare Costs Labor	Equipment	Total	Total Incl O&P
0010 **GASOLINE PIPING**									
0020 Primary containment pipe, fiberglass-reinforced									
0030 Plastic pipe 15' & 30' lengths									
0040 2" diameter	Q-6	425	.056	L.F.	7	3.45		10.45	13.05
0050 3" diameter		400	.060		10.65	3.67		14.32	17.40
0060 4" diameter		375	.064		13.85	3.92		17.77	21.50
0100 Fittings									
0110 Elbows, 90° & 45°, bell ends, 2"	Q-6	24	1	Ea.	40.50	61		101.50	140
0120 3" diameter		22	1.091		56	66.50		122.50	165
0130 4" diameter		20	1.200		70	73.50		143.50	192
0200 Tees, bell ends, 2"		21	1.143		57	70		127	171
0210 3" diameter		18	1.333		68	81.50		149.50	201
0220 4" diameter		15	1.600		81.50	98		179.50	241
0230 Flanges bell ends, 2"		24	1		31.50	61		92.50	129
0240 3" diameter		22	1.091		38.50	66.50		105	145
0250 4" diameter		20	1.200		43.50	73.50		117	162
0260 Sleeve couplings, 2"		21	1.143		12.05	70		82.05	121
0270 3" diameter		18	1.333		18.05	81.50		99.55	146
0280 4" diameter		15	1.600		22.50	98		120.50	176
0290 Threaded adapters, 2"		21	1.143		19.25	70		89.25	129
0300 3" diameter		18	1.333		32	81.50		113.50	162
0310 4" diameter		15	1.600		35.50	98		133.50	190
0320 Reducers, 2"		27	.889		27	54.50		81.50	114
0330 3" diameter		22	1.091		27	66.50		93.50	133
0340 4" diameter		20	1.200		36	73.50		109.50	154
1010 Gas station product line for secondary containment (double wall)									
1100 Fiberglass reinforced plastic pipe 25' lengths									
1120 Pipe, plain end, 3" diameter	Q-6	375	.064	L.F.	29.50	3.92		33.42	38.50
1130 4" diameter		350	.069		33.50	4.20		37.70	43.50
1140 5" diameter		325	.074		35.50	4.52		40.02	46.50
1150 6" diameter		300	.080		39.50	4.89		44.39	51
1200 Fittings									
1230 Elbows, 90° & 45°, 3" diameter	Q-6	18	1.333	Ea.	131	81.50		212.50	270
1240 4" diameter		16	1.500		163	92		255	320
1250 5" diameter		14	1.714		170	105		275	350
1260 6" diameter		12	2		211	122		333	420

33 52 Hydrocarbon Transmission and Distribution

33 52 13 – Liquid Hydrocarbon Piping

33 52 13.16 Gasoline Piping

		Crew	Daily Output	Labor-Hours	Unit	Material	2020 Bare Costs Labor	Equipment	Total	Total Incl O&P
1270	Tees, 3" diameter	Q-6	15	1.600	Ea.	157	98		255	325
1280	4" diameter		12	2		184	122		306	390
1290	5" diameter		9	2.667		299	163		462	580
1300	6" diameter		6	4		340	245		585	755
1310	Couplings, 3" diameter		18	1.333		54.50	81.50		136	186
1320	4" diameter		16	1.500		117	92		209	271
1330	5" diameter		14	1.714		218	105		323	400
1340	6" diameter		12	2		320	122		442	540
1350	Cross-over nipples, 3" diameter		18	1.333		10.30	81.50		91.80	137
1360	4" diameter		16	1.500		12.55	92		104.55	156
1370	5" diameter		14	1.714		16.30	105		121.30	180
1380	6" diameter		12	2		17.55	122		139.55	208
1400	Telescoping, reducers, concentric 4" x 3"		18	1.333		44.50	81.50		126	175
1410	5" x 4"		17	1.412		94.50	86.50		181	238
1420	6" x 5"		16	1.500		237	92		329	405

33 52 16 – Gas Hydrocarbon Piping

33 52 16.20 Piping, Gas Service and Distribution, P.E.

		Crew	Daily Output	Labor-Hours	Unit	Material	2020 Bare Costs Labor	Equipment	Total	Total Incl O&P
0010	**PIPING, GAS SERVICE AND DISTRIBUTION, POLYETHYLENE**									
0020	Not including excavation or backfill									
1000	60 psi coils, compression coupling @ 100', 1/2" diameter, SDR 11	B-20A	608	.053	L.F.	.49	2.66		3.15	4.72
1010	1" diameter, SDR 11		544	.059		1.14	2.97		4.11	5.90
1040	1-1/4" diameter, SDR 11		544	.059		1.64	2.97		4.61	6.45
1100	2" diameter, SDR 11		488	.066		2.48	3.31		5.79	7.90
1160	3" diameter, SDR 11		408	.078		6.20	3.96		10.16	13
1500	60 psi 40' joints with coupling, 3" diameter, SDR 11	B-21A	408	.098		7.35	5.15	1.05	13.55	17.30
1540	4" diameter, SDR 11		352	.114		13.35	5.95	1.22	20.52	25.50
1600	6" diameter, SDR 11		328	.122		35.50	6.40	1.31	43.21	50.50
1640	8" diameter, SDR 11		272	.147		54	7.70	1.58	63.28	73.50
9000	Minimum labor/equipment charge	B-20	2	12	Job		565		565	900

33 71 Electrical Utility Transmission and Distribution

33 71 19 – Electrical Underground Ducts and Manholes

33 71 19.17 Electric and Telephone Underground

		Crew	Daily Output	Labor-Hours	Unit	Material	2020 Bare Costs Labor	Equipment	Total	Total Incl O&P
0010	**ELECTRIC AND TELEPHONE UNDERGROUND**									
0011	Not including excavation									
0200	backfill and cast in place concrete									
4200	Underground duct, banks ready for concrete fill, min. of 7.5"									
4400	between conduits, center to center									
4600	2 @ 2" diameter	2 Elec	240	.067	L.F.	2.23	4.09		6.32	8.75
4800	4 @ 2" diameter		120	.133		4.47	8.20		12.67	17.50
5600	4 @ 4" diameter		80	.200		7.45	12.25		19.70	27
6200	Rigid galvanized steel, 2 @ 2" diameter		180	.089		15.75	5.45		21.20	25.50
6400	4 @ 2" diameter		90	.178		31.50	10.90		42.40	51.50
7400	4 @ 4" diameter		34	.471		91.50	29		120.50	145
9990	Minimum labor/equipment charge	1 Elec	3.50	2.286	Job		140		140	216

583

For customer support on your Commercial Renovation Costs with RSMeans data, call 800.448.8182.

Division Notes

		CREW	DAILY OUTPUT	LABOR-HOURS	UNIT	BARE COSTS				TOTAL INCL O&P
						MAT.	LABOR	EQUIP.	TOTAL	

Estimating Tips
34 11 00 Rail Tracks
This subdivision includes items that may involve either repair of existing or construction of new railroad tracks. Additional preparation work, such as the roadbed earthwork, would be found in Division 31. Additional new construction siding and turnouts are found in Subdivision 34 72. Maintenance of railroads is found under 34 01 23 Operation and Maintenance of Railways.

34 40 00 Traffic Signals
This subdivision includes traffic signal systems. Other traffic control devices such as traffic signs are found in Subdivision 10 14 53 Traffic Signage.

34 70 00 Vehicle Barriers
This subdivision includes security vehicle barriers, guide and guard rails, crash barriers, and delineators. The actual maintenance and construction of concrete and asphalt pavement are found in Division 32.

Reference Numbers
Reference numbers are shown at the beginning of some major classifications. These numbers refer to related items in the Reference Section. The reference information may be an estimating procedure, an alternate pricing method, or technical information.

Note: Not all subdivisions listed here necessarily appear. ■

Same Data. Simplified.

Enjoy the convenience and efficiency of accessing your costs anywhere:

- **Skip the multiplier** by setting your location
- **Quickly search,** edit, favorite and share costs
- **Stay on top of price changes** with automatic updates

Discover more at rsmeans.com/online

34 71 Roadway Construction

34 71 13 – Vehicle Barriers

34 71 13.26 Vehicle Guide Rails	Crew	Daily Output	Labor-Hours	Unit	Material	2020 Bare Costs Labor	Equipment	Total	Total Incl O&P
0010 **VEHICLE GUIDE RAILS**									
0012 Corrugated stl., galv. stl. posts, 6'-3" OC	B-80	850	.038	L.F.	27.50	1.75	1.20	30.45	34
0200 End sections, galvanized, flared		50	.640	Ea.	99	30	20.50	149.50	179
0300 Wrap around end		50	.640	"	141	30	20.50	191.50	226
0400 Timber guide rail, 4" x 8" with 6" x 8" wood posts, treated		960	.033	L.F.	13.45	1.55	1.06	16.06	18.40

Estimating Tips

Products such as conveyors, material handling cranes and hoists, and other items specified in this division require trained installers. The general contractor may not have any choice as to who will perform the installation or when it will be performed. Long lead times are often required for these products, making early decisions in purchasing and scheduling necessary. The installation of this type of equipment may require the embedment of mounting hardware during the construction of floors, structural walls, or interior walls/partitions. Electrical connections will require coordination with the electrical contractor.

Reference Numbers

Reference numbers are shown at the beginning of some major classifications. These numbers refer to related items in the Reference Section. The reference information may be an estimating procedure, an alternate pricing method, or technical information.

41 22 Cranes and Hoists

41 22 13 – Cranes

41 22 13.10 Crane Rail		Crew	Daily Output	Labor-Hours	Unit	Material	2020 Bare Costs Labor	Equipment	Total	Total Incl O&P
0010	**CRANE RAIL**									
0020	Box beam bridge, no equipment included	E-4	3400	.009	Lb.	1.44	.55	.04	2.03	2.53
0200	Running track only, 104 lb. per yard		5600	.006	"	.72	.33	.03	1.08	1.36
0210	Running track only, 104 lb. per yard, 20' piece		160	.200	L.F.	25	11.65	.92	37.57	47.50

41 22 13.13 Bridge Cranes

41 22 13.13 Bridge Cranes		Crew	Daily Output	Labor-Hours	Unit	Material	2020 Bare Costs Labor	Equipment	Total	Total Incl O&P
0010	**BRIDGE CRANES**									
0100	1 girder, 20' span, 3 ton	M-3	1	34	Ea.	28,000	2,200	146	30,346	34,300
0125	5 ton		1	34		30,900	2,200	146	33,246	37,400
0150	7.5 ton		1	34		36,400	2,200	146	38,746	43,600
0175	10 ton		.80	42.500		48,900	2,725	182	51,807	58,000
0176	12.5 ton		.80	42.500		58,500	2,725	182	61,407	69,000
0200	15 ton		.80	42.500		63,000	2,725	182	65,907	74,000
0225	30' span, 3 ton		1	34		29,200	2,200	146	31,546	35,600
0250	5 ton		1	34		32,200	2,200	146	34,546	39,000
0275	7.5 ton		1	34		38,300	2,200	146	40,646	45,700
0300	10 ton		.80	42.500		50,500	2,725	182	53,407	60,000
0325	15 ton		.80	42.500		66,000	2,725	182	68,907	77,000
0350	2 girder, 40' span, 3 ton	M-4	.50	72		47,500	4,600	375	52,475	59,500
0375	5 ton		.50	72		49,800	4,600	375	54,775	62,500
0400	7.5 ton		.50	72		54,500	4,600	375	59,475	67,500
0425	10 ton		.40	90		52,000	5,750	470	58,220	67,000
0450	15 ton		.40	90		88,000	5,750	470	94,220	106,500
0475	25 ton		.30	120		103,000	7,650	630	111,280	125,500
0500	50' span, 3 ton		.50	72		51,500	4,600	375	56,475	64,500
0525	5 ton		.50	72		56,500	4,600	375	61,475	70,000
0550	7.5 ton		.50	72		61,000	4,600	375	65,975	74,500
0575	10 ton		.40	90		70,500	5,750	470	76,720	87,000
0600	15 ton		.40	90		92,000	5,750	470	98,220	111,000
0625	25 ton		.30	120		108,000	7,650	630	116,280	131,000

41 22 13.19 Jib Cranes

41 22 13.19 Jib Cranes		Crew	Daily Output	Labor-Hours	Unit	Material	2020 Bare Costs Labor	Equipment	Total	Total Incl O&P
0010	**JIB CRANES**									
0020	Jib crane, wall cantilever, 500 lb. capacity, 8' span	2 Mill	1	16	Ea.	1,325	895		2,220	2,825
0040	12' span		1	16		1,450	895		2,345	2,975
0060	16' span		1	16		1,700	895		2,595	3,225
0080	20' span		1	16		2,025	895		2,920	3,600
0100	1,000 lb. capacity, 8' span		1	16		1,425	895		2,320	2,925
0120	12' span		1	16		1,550	895		2,445	3,075
0130	16' span		1	16		2,150	895		3,045	3,725
0150	20' span		1	16		2,400	895		3,295	4,000

Estimating Tips

- When estimating costs for the installation of electrical power generation equipment, factors to review include access to the job site, access and setting up at the installation site, required connections, uncrating pads, anchors, leveling, final assembly of the components, and temporary protection from physical damage, such as environmental exposure.

- Be aware of the costs of equipment supports, concrete pads, and vibration isolators. Cross-reference them against other trades' specifications. Also, review site and structural drawings for items that must be included in the estimates.

- It is important to include items that are not documented in the plans and specifications but must be priced. These items include, but are not limited to, testing, dust protection, roof penetration, core drilling concrete floors and walls, patching, cleanup, and final adjustments. Add a contingency or allowance for utility company fees for power hookups, if needed.

- The project size and scope of electrical power generation equipment will have a significant impact on cost. The intent of RSMeans cost data is to provide a benchmark cost so that owners, engineers, and electrical contractors will have a comfortable number with which to start a project. Additionally, there are many websites available to use for research and to obtain a vendor's quote to finalize costs.

Reference Numbers

Reference numbers are shown at the beginning of some major classifications. These numbers refer to related items in the Reference Section. The reference information may be an estimating procedure, an alternate pricing method, or technical information.

Same Data. Simplified.

Enjoy the convenience and efficiency of accessing your costs anywhere:

- **Skip the multiplier** by setting your location
- **Quickly search,** edit, favorite and share costs
- **Stay on top of price changes** with automatic updates

Discover more at rsmeans.com/online

Note: Trade Service, in part, has been used as a reference source for some of the material prices used in Division 48.

48 13 Hydroelectric Plant Electrical Power Generation Equipment

48 13 13 – Hydroelectric Power Plant Water Turbines

48 13 13.10 Water Turbines and Components	Crew	Daily Output	Labor-Hours	Unit	Material	2020 Bare Costs Labor	Equipment	Total	Total Incl O&P
0010	**WATER TURBINES & COMPONENTS**								
0500	Water turbine, 100-500 watt, 3P, 48VDC, 54"W x 24"H, 260LBS	2 Elec	2	8	Ea.	10,200	490		10,690	12,000
0600	Water turbine, 1500-2000 watt, 3P, 130-200VAC to DC, 95LBS	2 Elec	3	5.333	Ea.	4,025	325		4,350	4,925

Assemblies Section

Table of Contents

Table of Contents

RSMeans data: Assemblies— How They Work

Assemblies estimating provides a fast and reasonably accurate way to develop construction costs. An assembly is the grouping of individual work items—with appropriate quantities— to provide a cost for a major construction component in a convenient unit of measure.

An assemblies estimate is often used during early stages of design development to compare the cost impact of various design alternatives on total building cost.

Assemblies estimates are also used as an efficient tool to verify construction estimates.

Assemblies estimates do not require a completed design or detailed drawings. Instead, they are based on the general size of the structure and other known parameters of the project. The degree of accuracy of an assemblies estimate is generally within +/- 15%.

Most assemblies consist of three major elements: a graphic, the system components, and the cost data itself. The **Graphic** is a visual representation showing the typical appearance of the assembly

❶ Unique 12-character Identifier

Our assemblies are identified by a **unique 12-character identifier**. The assemblies are numbered using UNIFORMAT II, ASTM Standard E1557. The first 5 characters represent this system to Level 3. The last 7 characters represent further breakdown in order to arrange items in understandable groups of similar tasks. Line numbers are consistent across all of our publications, so a line number in any assemblies data set will always refer to the same item.

❷ Narrative Descriptions

Our assemblies descriptions appear in two formats: narrative and table. **Narrative descriptions** are shown in a hierarchical structure to make them readable. In order to read a complete description, read up through the indents to the top of the section. Include everything that is above and to the left that is not contradicted by information below.

A10 Foundations

A1010   Standard Foundations

This page illustrates and describes a spread footing system including concrete, forms, reinforcing and anchor bolts. Lines within System Components give the unit price and total price on a cost each basis for this system. Prices for spread footing systems are on Line Items A1010 230 1200 thru 2300. Both material quantities and labor costs have been adjusted for the system listed.

Factors: To adjust for job conditions other than normal working situations use Lines A1010 230 2900 thru 4000.

Example: You are to install the system in an existing occupied building. Access to the site and protection to the building are mandatory. Go to Lines A1010 230 3600 and 3800 and apply these percentages to the appropriate MAT. and INST. costs.

System Components	QUANTITY	UNIT	COST EACH MAT.	COST EACH INST.	COST EACH TOTAL
Interior column footing, 3' square 1' thick, 2000 psi concrete, Including forms, reinforcing, and anchor bolts.					
Concrete, 2000 psi	.330	C.Y.	38.61		38.61
Placing concrete	.330	C.Y.		20.48	20.48
Forms, footing, 4 uses	12.000	SFCA	9.84	75	84.84
Reinforcing	11.000	Lb.	6.82	7.48	14.30
Anchor bolts, 3/4" diameter	2.000	Ea.	3.56	10.20	13.76
TOTAL			58.83	113.16	171.99

❶

A1010 230	Spread Footing	COST EACH MAT.	COST EACH INST.	COST EACH TOTAL
1200	3' square x 1' thick, 2000 psi concrete	59	113	172
1300	3000 psi concrete	67	113	180
1400	4000 psi concrete	66.50	113	179.50
1600	For alternate footing systems:			
1700	4' square x 1' thick, 2000 psi concrete	98	161	259
1800	3000 psi concrete	113	161	274
1900	4000 psi concrete	111	161	272
2100	5' square x 1'-3" thick, 2000 psi concrete	183	264	447
2200	3000 psi concrete	212	264	476
2300	4000 psi concrete	210	264	474
2600				
2700				
2900	Cut & patch to match existing construction, add, minimum	2%	3%	
3000	Maximum	5%	9%	
3100	Dust protection, add, minimum	1%	2%	
3200	Maximum	4%	11%	
3300	Equipment usage curtailment, add, minimum	1%	1%	
3400	Maximum	3%	10%	
3500	Material handling & storage limitation, add, minimum	1%	1%	
3600	Maximum	6%	7%	
3700	Protection of existing work, add, minimum	2%	2%	
3800	Maximum	5%	7%	
3900	Shift work requirements, add, minimum		5%	
4000	Maximum		30%	

❷

For supplemental customizable square foot estimating forms, visit: **RSMeans.com/2020books** in question. It is frequently accompanied by additional explanatory technical information describing the class of items. The **System Components** is a listing of the individual tasks that make up the assembly, including the quantity and unit of measure for each item, along with the cost of material and installation. The **Assemblies** data below lists prices for other similar systems with dimensional and/or size variations.

All of our assemblies costs represent the cost for the installing contractor. An allowance for profit has been added to all material, labor, and equipment rental costs. A markup for labor burdens, including workers' compensation, fixed overhead, and business overhead, is included with installation costs.

The information in RSMeans cost data represents a "national average" cost. This data should be modified to the project location using the **City Cost Indexes** or **Location Factors** tables found in the Reference Section.

A10 Foundations

A1010 Standard Foundations

This page illustrates and describes a spread footing system including concrete, forms, reinforcing and anchor bolts. Lines within System Components give the unit price and total price on a cost each basis for this system. Prices for spread footing systems are on Line Items A1010 230 1200 thru 2300. Both material quantities and labor costs have been adjusted for the system listed.

Factors: To adjust for job conditions other than normal working situations use Lines A1010 230 2900 thru 4000.

Example: You are to install the system in an existing occupied building. Access to the site and protection to the building are mandatory. Go to Lines A1010 230 3600 and 3800 and apply these percentages to the appropriate MAT. and INST. costs.

3 Unit of Measure

All RSMeans data: Assemblies include a typical **Unit of Measure** used for estimating that item. For instance, for continuous footings or foundation walls the unit is linear feet (L.F.). For spread footings the unit is each (Ea.). The estimator needs to take special care that the unit in the data matches the unit in the takeoff. Abbreviations and unit conversions can be found in the Reference Section.

4 System Components

System components are listed separately to detail what is included in the development of the total system price.

System Components				COST EACH		
	QUANTITY	UNIT	MAT.	INST.	TOTAL
Interior column footing, 3' square 1' thick, 2000 psi concrete, Including forms, reinforcing, and anchor bolts.					
Concrete, 2000 psi	.330	C.Y.	38.61		38.61
Placing concrete	.330	C.Y.		20.48	20.48
Forms, footing, 4 uses	12.000	SFCA	9.84	75	84.84
Reinforcing	11.000	Lb.	6.82	7.48	14.30
Anchor bolts, 3/4" diameter	2.000	Ea.	3.56	10.20	13.76
TOTAL			58.83	113.16	171.99

A1010 230	Spread Footing	COST EACH		
		MAT.	INST.	TOTAL
1200	3' square x 1' thick, 2000 psi concrete	59	113	172
1300	3000 psi concrete	67	113	180
1400	4000 psi concrete	66.50	113	179.50
1600	For alternate footing systems:			
1700	4' square x 1' thick, 2000 psi concrete	98	161	259
1800	3000 psi concrete	113	161	274
1900	4000 psi concrete	111	161	272
2100	5' square x 1'-3" thick, 2000 psi concrete	183	264	447
2200	3000 psi concrete	212	264	476
2300	4000 psi concrete	210	264	474
2600				
2700				
2900	Cut & patch to match existing construction, add, minimum	2%	3%	
3000	Maximum	5%	9%	
3100	Dust protection, add, minimum	1%	2%	
3200	Maximum	4%	11%	
3300	Equipment usage curtailment, add, minimum	1%	1%	
3400	Maximum	3%	10%	
3500	Material handling & storage limitation, add, minimum	1%	1%	
3600	Maximum	6%	7%	
3700	Protection of existing work, add, minimum	2%	2%	
3800	Maximum	5%	7%	
3900	Shift work requirements, add, minimum		5%	
4000	Maximum		30%	

Sample Estimate

This sample demonstrates the elements of an estimate, including a tally of the RSMeans data lines. Published assemblies costs include all markups for labor burden and profit for the installing contractor. This estimate adds a summary of the markups applied by a general contractor on the installing contractor's work. These figures represent the total cost to the owner. The location factor with RSMeans data is applied at the bottom of the estimate to adjust the cost of the work to a specific location.

Project Name:	**Interior Fit-out, ABC Office**			
Location:	**Anywhere, USA**	**Date: 1/1/2020**		**R+R**
Assembly Number	**Description**	**Qty.**	**Unit**	**Subtotal**
❶ C1010 124 1200	Wood partition, 2 x 4 @ 16" OC w/5/8" FR gypsum board	560	S.F.	$3,130.40
C1020 114 1800	Metal door & frame, flush hollow core, 3'-0" x 7'-0"	2	Ea.	$2,680.00
C3010 230 0080	Painting, brushwork, primer & 2 coats	1,120	S.F.	$1,545.60
C3020 410 0140	Carpet, tufted, nylon, roll goods, 12' wide, 26 oz	240	S.F.	$943.20
C3030 210 6000	Acoustic ceilings, 24" x 48" tile, tee grid suspension	200	S.F.	$1,352.00
D5020 125 0560	Receptacles incl plate, box, conduit, wire, 20 A duplex	8	Ea.	$2,500.00
D5020 125 0720	Light switch incl plate, box, conduit, wire, 20 A single pole	2	Ea.	$609.00
D5020 210 0560	Fluorescent fixtures, recess mounted, 20 per 1000 SF	200	S.F.	$2,390.00
	Assembly Subtotal			**$15,150.20**
	Sales Tax @ ❷	5 %		$ 378.76
	General Requirements @ ❸	7 %		$ 1,060.51
	Subtotal A			**$16,589.47**
	GC Overhead @ ❹	5 %		$ 829.47
	Subtotal B			**$17,418.94**
	GC Profit @ ❺	5 %		$ 870.95
	Subtotal C			**$18,289.89**
	Adjusted by Location Factor ❻	115.5		$ 21,124.82
	Architects Fee @ ❼	8 %		$ 1,689.99
	Contingency @ ❽	15 %		$ 3,168.72
	Project Total Cost			**$ 25,983.53**

This estimate is based on an interactive spreadsheet. You are free to download it and adjust it to your methodology. A copy of this spreadsheet is available at **RSMeans.com/2020books.**

1. Work Performed

The body of the estimate shows the RSMeans data selected, including line numbers, a brief description of each item, its takeoff quantity and unit, and the total installed cost, including the installing contractor's overhead and profit.

2. Sales Tax

If the work is subject to state or local sales taxes, the amount must be added to the estimate. In a conceptual estimate it can be assumed that one half of the total represents material costs. Therefore, apply the sales tax rate to 50% of the assembly subtotal.

3. General Requirements

This item covers project-wide needs provided by the general contractor. These items vary by project but may include temporary facilities and utilities, security, testing, project cleanup, etc. In assemblies estimates a percentage is used—typically between 5% and 15% of project cost.

4. General Contractor Overhead

This entry represents the general contractor's markup on all work to cover project administration costs.

5. General Contractor Profit

This entry represents the GC's profit on all work performed. The value included here can vary widely by project and is influenced by the GC's perception of the project's financial risk and market conditions.

6. Location Factor

RSMeans published data are based on national average costs. If necessary, adjust the total cost of the project using a location factor from the "Location Factor" table or the "City Cost Indexes" table found in the Reference Section. Use location factors if the work is general, covering the work of multiple trades. If the work is by a single trade (e.g., masonry) use the more specific data found in the City Cost Indexes.

To adjust costs by location factors, multiply the base cost by the factor and divide by 100.

7. Architect's Fee

If appropriate, add the design cost to the project estimate. These fees vary based on project complexity and size. Typical design and engineering fees can be found in the Reference Section.

8. Contingency

A factor for contingency may be added to any estimate to represent the cost of unknowns that may occur between the time that the estimate is performed and the time the project is constructed. The amount of the allowance will depend on the stage of design at which the estimate is done, as well as the contractor's assessment of the risk involved.

A1010 Standard Foundations

This page illustrates and describes a strip footing system including concrete, forms, reinforcing, keyway and dowels. Lines within System Components give the unit price and total price per linear foot for this system. Prices for strip footing systems are on Line Items A1010 120 1400 thru 2500. Both material quantities and labor costs have been adjusted for the system listed.

Factors: To adjust for job conditions other than normal working situations use Lines A1010 120 2900 thru 4000.

Example: You are to install this footing, and due to a lack of accessibility, only hand tools can be used. Material handling is also a problem. Go to Lines A1010 120 3400 and 3600 and apply these percentages to the appropriate MAT. and INST. costs.

System Components	QUANTITY	UNIT	COST PER L.F.		
			MAT.	INST.	TOTAL
Strip footing, 2'-0" wide x 1'-0" thick, 2000 psi concrete including forms					
Reinforcing, keyway, and dowels.					
Concrete, 2000 psi	.074	C.Y.	8.66		8.66
Placing concrete	.074	C.Y.		2.11	2.11
Forms, footing, 4 uses	2.000	S.F.	4.92	10.70	15.62
Reinforcing	3.170	Lb.	1.97	2.16	4.13
Keyway, 2" x 4", 4 uses	1.000	L.F.	.28	1.28	1.56
Dowels, #4 bars, 2' long, 24" O.C.	.500	Ea.	.46	1.48	1.94
TOTAL			16.29	17.73	34.02

A1010 120	Strip Footing	COST PER L.F.		
		MAT.	INST.	TOTAL
1400	2'-0" wide x 1' thick, 2000 psi concrete	16.30	17.75	34.05
1500	3000 psi concrete	18.15	17.75	35.90
1600	4000 psi concrete	18	17.75	35.75
1800	For alternate footing systems:			
1900	2'-6" wide x 1' thick, 2000 psi concrete	19	18.85	37.85
2000	3000 psi concrete	21.50	18.85	40.35
2100	4000 psi concrete	21	18.85	39.85
2300	3'-0" wide x 1' thick, 2000 psi concrete	21.50	19.85	41.35
2400	3000 psi concrete	24.50	19.85	44.35
2500	4000 psi concrete	24	19.85	43.85
2700				
2800				
2900	Cut & patch to match existing construction, add, minimum	2%	3%	
3000	Maximum	5%	9%	
3100	Dust protection, add, minimum	1%	2%	
3200	Maximum	4%	11%	
3300	Equipment usage curtailment, add, minimum	1%	1%	
3400	Maximum	3%	10%	
3500	Material handling & storage limitation, add, minimum	1%	1%	
3600	Maximum	6%	7%	
3700	Protection of existing work, add, minimum	2%	2%	
3800	Maximum	5%	7%	
3900	Shift work requirements, add, minimum		5%	
4000	Maximum		30%	

A1010 Standard Foundations

This page illustrates and describes a spread footing system including concrete, forms, reinforcing and anchor bolts. Lines within System Components give the unit price and total price on a cost each basis for this system. Prices for spread footing systems are on Line Items A1010 230 1200 thru 2300. Both material quantities and labor costs have been adjusted for the system listed.

Factors: To adjust for job conditions other than normal working situations use Lines A1010 230 2900 thru 4000.

Example: You are to install the system in an existing occupied building. Access to the site and protection to the building are mandatory. Go to Lines A1010 230 3600 and 3800 and apply these percentages to the appropriate MAT. and INST. costs.

System Components	QUANTITY	UNIT	COST EACH		
			MAT.	INST.	TOTAL
Interior column footing, 3' square 1' thick, 2000 psi concrete,					
Including forms, reinforcing, and anchor bolts.					
Concrete, 2000 psi	.330	C.Y.	38.61		38.61
Placing concrete	.330	C.Y.		20.48	20.48
Forms, footing, 4 uses	12.000	SFCA	9.84	75	84.84
Reinforcing	11.000	Lb.	6.82	7.48	14.30
Anchor bolts, 3/4" diameter	2.000	Ea.	3.56	10.20	13.76
TOTAL			58.83	113.16	171.99

A1010 230	Spread Footing	COST EACH		
		MAT.	INST.	TOTAL
1200	3' square x 1' thick, 2000 psi concrete	59	113	172
1300	3000 psi concrete	67	113	180
1400	4000 psi concrete	66.50	113	179.50
1600	For alternate footing systems:			
1700	4' square x 1' thick, 2000 psi concrete	98	161	259
1800	3000 psi concrete	113	161	274
1900	4000 psi concrete	111	161	272
2100	5' square x 1'-3" thick, 2000 psi concrete	183	264	447
2200	3000 psi concrete	212	264	476
2300	4000 psi concrete	210	264	474
2600				
2700				
2900	Cut & patch to match existing construction, add, minimum	2%	3%	
3000	Maximum	5%	9%	
3100	Dust protection, add, minimum	1%	2%	
3200	Maximum	4%	11%	
3300	Equipment usage curtailment, add, minimum	1%	1%	
3400	Maximum	3%	10%	
3500	Material handling & storage limitation, add, minimum	1%	1%	
3600	Maximum	6%	7%	
3700	Protection of existing work, add, minimum	2%	2%	
3800	Maximum	5%	7%	
3900	Shift work requirements, add, minimum		5%	
4000	Maximum		30%	

A1010 Standard Foundations

General: Footing drains can be placed either inside or outside of foundation walls depending upon the source of water to be intercepted. If the source of subsurface water is principally from grade or a subsurface stream above the bottom of the footing, outside drains should be used. For high water tables, use inside drains or both inside and outside.

The effectiveness of underdrains depends on good waterproofing. This must be carefully installed and protected during construction.

Costs below include the labor and materials for the pipe and 6″ of crushed stone around pipe. Excavation and backfill are not included.

System Components	QUANTITY	UNIT	COST PER L.F.		
			MAT.	INST.	TOTAL
Foundation underdrain, outside, PVC 4″ diameter.					
PVC pipe 4″ diam. S.D.R. 35	1.000	L.F.	1.84	4.81	6.65
Crushed stone 3/4″ to 1/2″	.070	C.Y.	2.14	.92	3.06
TOTAL			3.98	5.73	9.71

A1010 310	Foundation Underdrain	COST PER L.F.		
		MAT.	INST.	TOTAL
1000	Foundation underdrain, outside only, PVC, 4″ diameter	3.98	5.70	9.68
1100	6″ diameter	7.20	6.35	13.55
1400	Perforated HDPE, 6″ diameter	5.05	2.38	7.43
1450	8″ diameter	7.90	2.99	10.89
1500	12″ diameter	12.45	7.40	19.85
1600	Corrugated metal, 16 ga. asphalt coated, 6″ diameter	9.80	5.95	15.75
1650	8″ diameter	13.05	6.30	19.35
1700	10″ diameter	16.05	8.35	24.40
3000	Outside and inside, PVC, 4″ diameter	7.95	11.45	19.40
3100	6″ diameter	14.40	12.70	27.10
3400	Perforated HDPE, 6″ diameter	10.10	4.77	14.87
3450	8″ diameter	15.75	6	21.75
3500	12″ diameter	25	14.80	39.80
3600	Corrugated metal, 16 ga., asphalt coated, 6″ diameter	19.60	11.90	31.50
3650	8″ diameter	26	12.65	38.65
3700	10″ diameter	32	16.65	48.65
4700	Cut & patch to match existing construction, add, minimum	2%	3%	
4800	Maximum	5%	9%	
4900	Dust protection, add, minimum	1%	2%	
5000	Maximum	4%	11%	
5100	Equipment usage curtailment, add, minimum	1%	1%	
5200	Maximum	3%	10%	
5300	Material handling & storage limitation, add, minimum	1%	1%	
5400	Maximum	6%	7%	
5500	Protection of existing work, add, minimum	2%	2%	
5600	Maximum	5%	7%	
5700	Shift work requirements, add, minimum		5%	
5800	Maximum		30%	
5900	Temporary shoring and bracing, add, minimum	2%	5%	
6000	Maximum	5%	12%	

A1030 Slab on Grade

This page illustrates and describes a slab on grade system including slab, bank run gravel, bulkhead forms, placing concrete, welded wire fabric, vapor barrier, steel trowel finish and curing paper. Lines within System Components give the unit price and total price per square foot for this system. Prices for slab on grade systems are on Line Items A1030 110 1400 thru 2600. Both material quantities and labor costs have been adjusted for the system listed.

Factors: To adjust for job conditions other than normal working situations use Lines A1030 110 2900 thru 4000.

Example: You are to install the system at a site where protection of the existing building is required. Go to Line A1030 110 3800 and apply these percentages to the appropriate MAT. and INST. costs.

System Components	QUANTITY	UNIT	COST PER S.F.		
			MAT.	INST.	TOTAL
Ground slab, 4″ thick, 3000 psi concrete, 4″ granular base, vapor barrier					
Welded wire fabric, screed and steel trowel finish.					
Concrete, 4″ thick, 3000 psi concrete	.012	C.Y.	1.70		1.70
Bank run gravel, 4″ deep	.012	C.Y.	.25	.10	.35
Polyethylene vapor barrier, 10 mil.	.011	C.S.F.	.12	.20	.32
Bulkhead forms, expansion material	.100	L.F.	.04	.43	.47
Welded wire fabric, 6 x 6 - #10/10	.011	C.S.F.	.20	.45	.65
Place concrete	.012	C.Y.		.38	.38
Screed & steel trowel finish	1.000	S.F.		1.11	1.11
TOTAL			2.31	2.67	4.98

A1030 110	Interior Slab on Grade	COST PER S.F.		
		MAT.	INST.	TOTAL
1400	4″ thick slab, 3000 PSI concrete, 4″ deep bank run gravel	2.31	2.67	4.98
1500	6″ deep bank run gravel	2.13	2.59	4.72
1600	12″ deep bank run gravel	2.35	2.62	4.97
1700				
1800				
1900				
2000	For alternate slab systems:			
2100	5″ thick slab, 3000 psi concrete, 6″ deep bank run gravel	2.93	2.77	5.70
2200	12″ deep bank run gravel	3.36	2.81	6.17
2300				
2400				
2500	6″ thick slab, 3000 psi concrete, 6″ deep bank run gravel	3.50	2.89	6.39
2600	12″ deep bank run gravel	3.93	2.93	6.86
2700				
2900	Cut & patch to match existing construction, add, minimum	2%	3%	
3000	Maximum	5%	9%	
3100	Dust protection, add, minimum	1%	2%	
3200	Maximum	4%	11%	
3300	Equipment usage curtailment, add, minimum	1%	1%	
3400	Maximum	3%	10%	
3500	Material handling & storage limitation, add, minimum	1%	1%	
3600	Maximum	6%	7%	
3700	Protection of existing work, add, minimum	2%	2%	
3800	Maximum	5%	7%	
3900	Shift work requirements, add, minimum		5%	
4000	Maximum		30%	

A2010 Basement Excavation

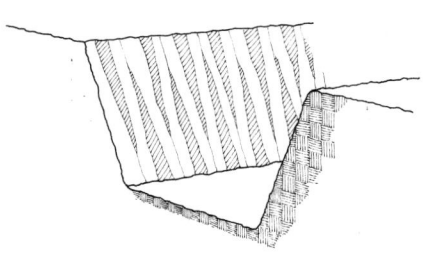

This page illustrates and describes continuous footing and trench excavation systems including a wheel mounted backhoe, operator, equipment rental, fuel, oil, mobilization, no hauling or backfill. Lines within System Components give the unit price and total price per linear foot for this system. Prices for alternate continuous footing and trench excavation systems are on Line Items A2010 130 1200 thru 2700. Both material quantities and labor costs have been adjusted for the system listed.

Factors: To adjust for job conditions other than normal working situations use Lines A2010 130 2900 thru 4000.

Example: You are to install the system and supply dust protection. Go to Line A2010 130 3000 and apply this percentage to the appropriate TOTAL costs.

System Components		QUANTITY	UNIT	COST PER L.F.		
				EQUIP.	LABOR	TOTAL
Continuous footing or trench excav. w/ 3/4 C.Y. wheel mntd backhoe Including operator, equipment rental, fuel, oil and mobilization. Trench Is 4' wide at bottom, 3' deep w/sides sloped 1 to 2 and 200' long. Costs Are based on a production rate of 240 C.Y./day. No hauling/backfill incl.						
	3/4 C.Y. backhoe, wheel mounted	.500	Day	117.50	625	742.50
	Mobilization	1.000	Ea.	1,075	2,025	3,100
	TOTAL			1,192.50	2,650	3,842.50
	COST PER L.F.		L.F.	5.96	13.25	19.21

A2010 130	Excavation, Footings or Trench		COST PER L.F.		
			EQUIP.	LABOR	TOTAL
1200	For alternate trench sizes, 4' bottom with sloped sides:				
1300	2' deep, 50' long		7.20	25.50	32.70
1400	100' long		3.61	12.70	16.31
1500	300' long		3.97	8.85	12.82
1600	4' deep, 50' long		7.20	25.50	32.70
1700	100' long		3.61	12.70	16.31
1800	300' long		1.62	6.45	8.07
1900	6' deep, 50' long		7.20	25.50	32.70
2000	100' long		3.96	14.60	18.56
2100	300' long		2.38	10.50	12.88
2200	8' deep, 50' long		7.20	25.50	32.70
2300	100' long		4.78	18.95	23.73
2400	300' long		3.16	14.65	17.81
2500	10' deep, 50' long		8.40	31.50	39.90
2600	100' long		5.80	24.50	30.30
2700	300' long		4.10	19.65	23.75
2800					
2900	Dust protection, add, minimum			2%	2%
3000	Maximum			11%	11%
3100	Equipment usage curtailment, add, minimum			1%	1%
3200	Maximum			10%	10%
3300	Material handling & storage limitation, add, minimum			1%	1%
3400	Maximum			7%	7%
3500	Protection of existing work, add, minimum			2%	2%
3600	Maximum			7%	7%
3700	Shift work requirements, add, minimum			5%	5%
3800	Maximum			30%	30%
3900	Temporary shoring and bracing, add, minimum			5%	5%
4000	Maximum			12%	12%

A2010 Basement Excavation

This page illustrates and describes foundation excavation systems including a backhoe-loader, operator, equipment rental, fuel, oil, mobilization, hauling material, no backfilling. Lines within System Components give the unit price and total price per cubic yard for this system. Prices for foundation excavation systems are on Line Items A2010 140 1600 thru 2200. Both material quantities and labor costs have been adjusted for the system listed.

Factors: To adjust for job conditions other than normal working situations use Lines A2010 140 2900 thru 4000.

Example: You are to install the system with the use of temporary shoring and bracing. Go to Line A2010 140 3900 and apply these percentages to the appropriate TOTAL costs.

System Components	QUANTITY	UNIT	COST PER C.Y. EQUIP.	COST PER C.Y. LABOR	COST PER C.Y. TOTAL
Foundation excav w/ 3/4 C.Y. backhoe-loader, incl. operator, equip Rental, fuel, oil, and mobilization. Hauling of excavated material is Included. Prices based on one day production of 360 C.Y. in medium soil Without backfilling.					
Backhoe-loader, 3/4 C.Y.	1.000	Day	235	1,250	1,485
Hauling, 12 C.Y. trucks, 1 mile round trip	360.000	L.C.Y.	900	892.80	1,792.80
Mobilization	1.000	Ea.	243	645	888
TOTAL			1,378	2,787.80	4,165.80
COST PER C.Y.		C.Y.	3.82	7.72	11.57

A2010 140	Excavation, Foundation	COST PER C.Y. EQUIP.	COST PER C.Y. LABOR	COST PER C.Y. TOTAL
1600	100 C.Y.	5.60	12.45	18.05
1700	200 C.Y.	4.38	9.20	13.58
1800	300 C.Y.	4	8.30	12.30
1850	360 C.Y.	3.82	7.78	11.60
1900	400 C.Y.	3.76	7.55	11.31
2000	500 C.Y.	3.65	7.25	10.90
2100	600 C.Y.	3.54	6.90	10.44
2200	700 C.Y.	3.49	6.80	10.29
2300				
2400				
2900	Dust protection, add, minimum		2%	2%
3000	Maximum		11%	11%
3100	Equipment usage curtailment, add, minimum		1%	1%
3200	Maximum		10%	10%
3300	Material handling & storage limitation, add, minimum		1%	1%
3400	Maximum		7%	7%
3500	Protection of existing work, add, minimum		2%	2%
3600	Maximum		7%	7%
3700	Shift work requirements, add, minimum		5%	5%
3800	Maximum		30%	30%
3900	Temporary shoring and bracing, add, minimum		5%	5%
4000	Maximum		12%	12%

A2020 Basement Walls

This page illustrates and describes a concrete wall system including concrete, placing concrete, forms, reinforcing, insulation, waterproofing and anchor bolts. Lines within System Components give the unit price and total price per linear foot for this system. Prices for concrete wall systems are on Line Items A2020 120 1400 thru 2600. Both material quantities and labor costs have been adjusted for the system listed.

Factors: To adjust for job conditions other than normal working situations use Lines A2020 120 2900 thru 4000.

Example: You are to install this wall system where delivery of material is difficult. Go to Line A2020 120 3600 and apply these percentages to the appropriate MAT. and INST. costs.

System Components			COST PER L.F.		
	QUANTITY	UNIT	MAT.	INST.	TOTAL
Cast in place concrete foundation wall, 8" thick, 3' high, 2500 psi					
Concrete including forms, reinforcing, waterproofing, and anchor bolts.					
Concrete, 2500 psi, 8" thick, 3' high	.070	C.Y.	8.47		8.47
Forms, wall, 4 uses	6.000	S.F.	7.44	47.10	54.54
Reinforcing	6.000	Lb.	3.72	2.82	6.54
Placing concrete	.070	C.Y.		3.58	3.58
Waterproofing	3.000	S.F.	1.56	3.87	5.43
Rigid insulaton, 1" polystyrene	3.000	S.F.	.96	3	3.96
Anchor bolts, 1/2" diameter, 4' O.C.	.250	Ea.	.50	1.29	1.79
TOTAL			22.65	61.66	84.31

A2020 120	Concrete Wall	COST PER L.F.		
		MAT.	INST.	TOTAL
1400	8" thick, 2500 psi concrete, 3' high	22.50	62	84.50
1500	4' high	31	82	113
1600	6' high	46	123	169
1700	8' high	61	163	224
1800	3500 psi concrete, 4' high	32.50	82	114.50
1900	6' high	48.50	123	171.50
2000	8' high	64.50	163	227.50
2100	12" thick, 2500 psi concrete, 4' high	39.50	87	126.50
2200	6' high	57	128	185
2300	8' high	76	170	246
2400	3500 psi concrete, 4' high	42	87	129
2500	8' high	80.50	170	250.50
2600	10' high	100	212	312
2700				
2900	Cut & patch to match existing construction, add, minimum	2%	3%	
3000	Maximum	5%	9%	
3100	Dust protection, add, minimum	1%	2%	
3200	Maximum	4%	11%	
3300	Equipment usage curtailment, add, minimum	1%	1%	
3400	Maximum	3%	10%	
3500	Material handling & storage limitation, add, minimum	1%	1%	
3600	Maximum	6%	7%	
3700	Protection of existing work, add, minimum	2%	2%	
3800	Maximum	5%	7%	
3900	Shift work requirements, add, minimum		5%	
4000	Maximum		30%	

A2020 Basement Walls

This page illustrates and describes a concrete block wall system including concrete block, masonry reinforcing, parging, waterproofing insulation and anchor bolts. Lines within System Components give the unit price and total price per linear foot for this system. Prices for concrete block wall systems are on Line Items A2020 140 1200 thru 2600. Both material quantities and labor costs have been adjusted for the system listed.

Factors: To adjust for job conditions other than normal working situations use Lines A2020 140 2900 thru 4000.

Example: You are to install the system to match an existing foundation wall. Go to Line A2020 140 3000 and apply these percentages to the appropriate MAT. and INST. costs.

System Components			COST PER L.F.		
	QUANTITY	UNIT	MAT.	INST.	TOTAL
Concrete block, 8″ thick, masonry reinforcing, parged and					
Waterproofed, insulation and anchor bolts, wall 2′-8″ high.					
Concrete block, 8″ x 8″ x 16″	2.670	S.F.	9.48	19.49	28.97
Masonry reinforcing	2.000	L.F.	.58	.44	1.02
Parging	.193	S.Y.	.90	2.97	3.87
Waterproofing	2.670	S.F.	1.39	3.44	4.83
Insulation, 1″ rigid polystyrene	2.670	S.F.	.85	2.67	3.52
Anchor bolts, 1/2″ diameter, 4′ O.C.	.250	Ea.	.50	1.29	1.79
TOTAL			13.70	30.30	44

A2020 140	Concrete Block Wall	COST PER L.F.		
		MAT.	INST.	TOTAL
1200	8″ thick block, 2′-8″ high	13.70	30	43.70
1300	4′ high	20.50	44.50	65
1400	6′ high	30	66.50	96.50
1500	8′ high	40	88.50	128.50
1600	Grouted solid, 4′ high	26.50	60.50	87
1700	6′ high	39	89.50	128.50
1800	8′ high	52	119	171
2100	12″ thick block, 4′ high	29	64	93
2200	6′ high	43	96	139
2300	8′ high	57.50	127	184.50
2400	Grouted solid, 4′ high	39	81	120
2500	6′ high	58	120	178
2600	8′ high	77.50	160	237.50
2700				
2900	Cut & patch to match existing construction, add, minimum	2%	3%	
3000	Maximum	5%	9%	
3100	Dust protection, add, minimum	1%	2%	
3200	Maximum	4%	11%	
3300	Equipment usage curtailment, add, minimum	1%	1%	
3400	Maximum	3%	10%	
3500	Material handling & storage limitation, add, minimum	1%	1%	
3600	Maximum	6%	7%	
3700	Protection of existing work, add, minimum	2%	2%	
3800	Maximum	5%	7%	
3900	Shift work requirements, add, minimum		5%	
4000	Maximum		30%	

B1010 Floor Construction

This page illustrates and describes a reinforced concrete slab system including concrete, placing concrete, formwork, reinforcing, steel trowel finish, V.A. floor tile and acoustical spray ceiling finish. Lines within System Components give the unit price per square foot for this system. Prices for reinforced concrete slab systems are on Line Items B1010 225 1400 thru 1800. Both material quantities and labor costs have been adjusted for the system listed.

Factors: To adjust for job conditions other than normal working situations use Lines B1010 225 2700 thru 4000.

Example: You are to install the system to match an existing floor system. Go to Line B1010 225 2800 and apply these percentages to the appropriate MAT. and INST. costs.

System Components	QUANTITY	UNIT	COST PER S.F. MAT.	COST PER S.F. INST.	COST PER S.F. TOTAL
Flat slab system, reinforced concrete with vinyl tile floor, sprayed Acoustical ceiling finish, not including columns.					
Concrete, 4000 psi, 6" thick	.020	C.Y.	2.80		2.80
Placing concrete	.020	C.Y.		.73	.73
Formwork, 4 use	1.000	S.F.	1.46	7.10	8.56
Edge form	.120	S.F.	.08	1.38	1.46
Reinforcing steel	3.000	Lb.	1.86	2.04	3.90
Steel trowel finish	1.000	S.F.		1.42	1.42
Vinyl floor tile, 1/8" thick, premium colors/patterns	1.000	S.F.	7.60	1.24	8.84
Acoustical spray ceiling finish, solvent based	1.000	S.F.	.31	.19	.50
TOTAL			14.11	14.10	28.21

B1010 225	Floor - Ceiling, Concrete Slab		COST PER S.F. MAT.	COST PER S.F. INST.	COST PER S.F. TOTAL
1400	Concrete, 4000 psi, 6" thick		14.10	14.10	28.20
1500	7" thick		14.65	14.70	29.35
1600	8" thick		15.45	15.45	30.90
1700	9" thick		16.15	16.05	32.20
1800	10" thick		16.95	16.75	33.70
1900					
2000					
2100					
2200					
2300					
2400					
2500					
2700	Cut & patch to match existing construction, add, minimum		2%	3%	
2800	Maximum		5%	9%	
2900	Dust protection, add, minimum		1%	2%	
3000	Maximum		4%	11%	
3100	Equipment usage curtailment, add, minimum		1%	1%	
3200	Maximum		3%	10%	
3300	Material handling & storage limitation, add, minimum		1%	1%	
3400	Maximum		6%	7%	
3500	Protection of existing work, add, minimum		2%	2%	
3600	Maximum		5%	7%	
3700	Shift work requirements, add, minimum			5%	
3800	Maximum			30%	
3900	Temporary shoring and bracing, add, minimum		2%	5%	
4000	Maximum		5%	12%	

B1010 Floor Construction

This page illustrates and describes a hollow core prestressed concrete panel system including hollow core slab, grout, carpet, carpet padding, and sprayed ceiling. Lines within System Components give the unit price and total price per square foot for this system. Prices for hollow core prestressed concrete panel systems are on Line Items B1010 232 1200 thru 1600. Both material and labor costs have been adjusted for the system listed.

Factors: To adjust for job conditions other than normal working situations use Lines B1010 232 2700 thru 3800.

Example: You are to install the system where dust control is a major concern. Go to Line B1010 232 2800 and apply these percentages to the appropriate MAT. and INST. costs.

System Components				QUANTITY	UNIT	COST PER S.F.		
						MAT.	INST.	TOTAL
Precast hollow core plank with carpeted floors, padding And sprayed textured ceiling.								
Hollow core plank, 4" thick with grout topping				1.000	S.F.	9.45	3.83	13.28
Nylon carpet, 26 oz. medium traffic				.110	S.Y.	4.95	.91	5.86
Carpet padding, 20 oz./sq. yd.				.110	S.Y.	.59	.46	1.05
Sprayed texture ceiling				1.000	S.F.	.04	.82	.86
			TOTAL			15.03	6.02	21.05

B1010 232	Floor - Ceiling, Conc. Panel	COST PER S.F.		
		MAT.	INST.	TOTAL
1200	Hollow core concrete plank, with grout, 4" thick	15.05	6	21.05
1300	6" thick	15	5.45	20.45
1400	8" thick	18.65	5.05	23.70
1500	10" thick	16.50	4.74	21.24
1600	12" thick	18.40	4.48	22.88
1700				
1800				
1900				
2000				
2100				
2200				
2300				
2400				
2500				
2700	Dust protection, add, minimum	1%	2%	
2800	Maximum	4%	11%	
2900	Equipment usage curtailment, add, minimum	1%	1%	
3000	Maximum	3%	10%	
3100	Material handling & storage limitation, add, minimum	1%	1%	
3200	Maximum	6%	7%	
3300	Protection of existing work, add, minimum	2%	2%	
3400	Maximum	5%	7%	
3500	Shift work requirements, add, minimum		5%	
3600	Maximum		30%	
3700	Temporary shoring and bracing, add, minimum	2%	5%	
3800	Maximum	5%	12%	

B1010   Floor Construction

This page illustrates and describes a hollow core plank system including precast hollow core plank, concrete topping, V.A. tile, concealed suspension system and ceiling tile. Lines within System Components give the unit price and total price per square foot for this system. Prices for hollow core plank systems are on Line Items B1010 233 1400 thru 1800. Both material quantities and labor costs have been adjusted for the system listed.

Factors: To adjust for job conditions other than normal working situations use Lines B1010 233 2900 thru 4000.

Example: You are to install the system where equipment usage is a problem. Go to Line B1010 233 3200 and apply these percentages to the appropriate MAT. and INST. costs.

System Components	QUANTITY	UNIT	COST PER S.F.		
			MAT.	INST.	TOTAL
Precast hollow core plank with 2″ topping, vinyl composition floor tile **And suspended acoustical tile ceiling.**					
Hollow core concrete plank, 4″ thick	1.000	S.F.	9.45	3.83	13.28
Finishing floors, granolithic topping, 1:1:1-1/2 mix, 2″ thick	1.000	S.F.	1.81	6.44	8.25
Vinyl composition tile, .08″ thick	1.000	S.F.	1.68	1.24	2.92
Concealed Z bar suspension system, 12″ module	1.000	S.F.	.95	1.31	2.26
Ceiling tile, mineral fiber, 3/4″ thick	1.000	S.F.	3.47	2.27	5.74
TOTAL			17.36	15.09	32.45

B1010 233	Floor - Ceiling, Conc. Plank	COST PER S.F.		
		MAT.	INST.	TOTAL
1400	Hollow core concrete plank, 4″ thick	17.35	15.10	32.45
1500	6″ thick	17.30	14.55	31.85
1600	8″ thick	21	14.10	35.10
1700	10″ thick	18.80	13.85	32.65
1800	12″ thick	20.50	13.55	34.05
1900				
2000				
2100				
2200				
2300				
2400				
2500				
2600				
2700				
2900	Dust protection, add, minimum	1%	2%	
3000	Maximum	4%	11%	
3100	Equipment usage curtailment, add, minimum	1%	1%	
3200	Maximum	3%	10%	
3300	Material handling & storage limitation, add, minimum	1%	1%	
3400	Maximum	6%	7%	
3500	Protection of existing work, add, minimum	2%	2%	
3600	Maximum	5%	7%	
3700	Shift work requirements, add, minimum		5%	
3800	Maximum		30%	
3900	Temporary shoring and bracing, add, minimum	2%	5%	
4000	Maximum	5%	12%	

B1010 Floor Construction

This page illustrates and describes a structural steel w/metal decking and concrete system including steel beams, steel decking, shear studs, concrete, placing concrete, edge form, steel trowel finish, curing, wire fabric, fireproofing, beams and decking, tile floor, and suspended ceiling. Lines within System Components give the unit price and total price per square foot for this system. Prices for alternate structural steel w/metal decking and concrete systems are on Line Items B1010 243 1900 thru 2400. Both material quantities and labor costs have been adjusted for the system listed.

Factors: To adjust for job conditions other than normal working situations use Lines B1010 243 2900 thru 4000.

Example: You are to install the system where material handling and storage are a problem. Go to Line B1010 243 3400 and apply these percentages to the appropriate MAT. and INST. costs.

System Components	QUANTITY	UNIT	COST PER S.F.		
			MAT.	INST.	TOTAL
Composite structural beams, 20 ga. 3″ deep steel decking, shear studs, 3000 psi, concrete, placing concrete, edge forms, steel trowel finish, Welded wire fabric fireproofing, tile floor and suspended ceiling.					
Steel deck, 20 gage, 3″ deep, galvanized	1.000	S.F.	4.09	1.07	5.16
Structural steel framing	.004	Ton	12.60	2.95	15.55
Shear studs, 3/4″	.250	Ea.	.14	.63	.77
Concrete, 3000 psi, 5″ thick	.015	C.Y.	2.13		2.13
Place concrete	.015	C.Y.		.55	.55
Edge form	.140	L.F.	.03	.72	.75
Steel trowel finish	1.000	S.F.		1.42	1.42
Welded wire fabric 6 x 6 - #10/10	.011	C.S.F.	.20	.45	.65
Fireproofing, sprayed	1.880	S.F.	1.22	2.38	3.60
Vinyl composition floor tile	1.000	S.F.	1.35	1.24	2.59
Suspended acoustical ceiling	1.000	S.F.	4.43	1.79	6.22
TOTAL			26.19	13.20	39.39

B1010 243	Floor - Ceiling, Struc. Steel	COST PER S.F.		
		MAT.	INST.	TOTAL
1900	Composite deck, galvanized, 3″ deep, 22 ga.	25	13.15	38.15
1950	20 ga.	26	13.20	39.20
2000	18 ga.	25.50	13.25	38.75
2100	16 ga.	28	13.30	41.30
2200	Composite deck, galvanized, 2″ deep, 22 ga.	24.50	12.95	37.45
2300	20 ga.	26	13.05	39.05
2400	16 ga.	27.50	13.15	40.65
2500				
2900	Dust protection, add, minimum	1%	2%	
3000	Maximum	4%	11%	
3100	Equipment usage curtailment, add, minimum	1%	1%	
3200	Maximum	3%	10%	
3300	Material handling & storage limitation, add, minimum	1%	1%	
3400	Maximum	6%	7%	
3500	Protection of existing work, add, minimum	2%	2%	
3600	Maximum	5%	7%	
3700	Shift work requirements, add, minimum		5%	
3800	Maximum		30%	
3900	Temporary shoring and bracing, add, minimum	2%	5%	
4000	Maximum	5%	12%	

This page illustrates and describes an open web joist and steel slab-form system including open web steel joist, slab form, concrete, placing concrete, wire fabric, steel trowel finish, tile floor and plasterboard. Lines within System Components give the unit price and total price per square foot for this system. Prices for open web joists and steel slab-form systems are on Line Items B1010 245 1800 thru 2200. Both material quantities and labor costs have been adjusted for the systems listed.

Factors: To adjust for job conditions other than normal working situations use Lines B1010 245 2900 thru 4000.

Example: You are to install the system in a congested commercial area and most work will be done at night. Go to Line B1010 245 3800 and apply this percentage to the appropriate INST. cost.

System Components	QUANTITY	UNIT	COST PER S.F. MAT.	COST PER S.F. INST.	COST PER S.F. TOTAL
Open web steel joists, 24″ O.C., slab form, 3000 psi concrete, welded					
Wire fabric, steel trowel finish, floor tile, 5/8″ drywall ceiling.					
Open web steel joists, 12″ deep, 5.2#/L.F., 24″ O.C.	2.630	Lb.	3.10	1.26	4.36
Slab form, 28 gage, 9/16″ deep, galvanized	1.000	S.F.	1.85	.80	2.65
Concrete, 3000 psi, 2-1/2″ thick	.008	C.Y.	1.14		1.14
Placing concrete	.008	C.Y.		.29	.29
Welded wire fabric 6 x 6 - #10/10	.011	C.S.F.	.20	.45	.65
Steel trowel finish	1.000	S.F.		1.42	1.42
Vinyl composition floor tile	1.000	S.F.	1.35	1.24	2.59
Ceiling furring, 3/4″ channels, 24″ O.C.	1.000	S.F.	.27	1.55	1.82
Gypsum drywall, 5/8″ thick, finished	1.000	S.F.	.42	1.78	2.20
Paint ceiling	1.000	S.F.	.23	.74	.97
TOTAL			8.56	9.53	18.09

B1010 245	Floor - Ceiling, Steel Joists	MAT.	INST.	TOTAL
1800	Open web joists, 12″ deep, 5.2#/L.F.	8.55	9.50	18.05
1900	16″ deep, 6.6#/L.F.	9.35	9.85	19.20
2000	20″ deep, 8.4# L.F.	10.40	10.30	20.70
2100	24″ deep, 11.5# L.F.	12.25	11.05	23.30
2200	26″ deep, 12.8# L.F.	13	11.30	24.30
2300				
2400				
2500				
2600				
2700				
2900	Dust protection, add, minimum	1%	2%	
3000	Maximum	4%	11%	
3100	Equipment usage curtailment, add, minimum	1%	1%	
3200	Maximum	3%	10%	
3300	Material handling & storage limitation, add, minimum	1%	1%	
3400	Maximum	6%	7%	
3500	Protection of existing work, add, minimum	2%	2%	
3600	Maximum	5%	7%	
3700	Shift work requirements, add, minimum		5%	
3800	Maximum		30%	
3900	Temporary shoring and bracing, add, minimum	2%	5%	
4000	Maximum	5%	12%	

B1010 Floor Construction

Open Web Wood Joists

This page illustrates and describes floor framing systems including sills, joist hangers, joists, bridging, subfloor and insulation. Lines within systems components give the unit price and total price per square foot for this system. Prices for alternate joist types are on line items B1010 262 0100 through 0700. Both material quantities and labor costs have been adjusted for the system listed.

Factors: To adjust for job conditions other than normal working situations use lines B1010 262 2700 through 4000.

Example: You are to install the system where delivery of material is difficult. Go to line B1010 262 3400 and apply these percentages to the appropriate MAT. and INST. costs.

System Components	QUANTITY	UNIT	COST PER S.F. MAT.	INST.	TOTAL
Open web wood joists, sill, butt to girder, 5/8″ subfloor, insul.					
Anchor bolt, hook type, with nut and washer, 1/2″ x 8″ long	.025	Ea.	.04	.13	.17
Joist & beam hanger, 18 ga. galvanized	.075	Ea.	.13	.31	.44
Bridging, steel, compression, for 2″ x 12″ joists	.010	C.Pr.	.02	.03	.05
Framing, open web wood joists, 12″ deep	.001	M.L.F.	4.25	1.55	5.80
Framing, sills, 2″ x 8″	.001	M.B.F.	.82	2.03	2.85
Sub-floor, plywood, CDX, 3/4″ thick	1.000	S.F.	.85	1.01	1.86
Insulation, fiberglass, kraft face, 9″ thick x 23″ wide	1.000	S.F.	.95	.50	1.45
TOTAL			7.06	5.56	12.62

B1010 262	Open Web Wood Joists	COST PER S.F. MAT.	INST.	TOTAL
0100	12″ open web joists, sill, butt to girder, 5/8″ sub-floor	7.05	5.55	12.60
0200	14″ joists	7.10	5.65	12.75
0300	16″ joists	7.45	5.75	13.20
0400	18″ joists	7.65	5.85	13.50
0450	Wood struc." I " joists, 24″ O.C., to 24' span, 50 PSF LL, butt to girder	4.80	5.15	9.95
0500	55 PSF LL	5.30	5.25	10.55
0600	to 30' span, 45 PSF LL	5.75	5.05	10.80
0700	55 PSF LL	8.05	5.15	13.20
2700	Cut & patch to match existing construction, add, minimum	2%	3%	
2800	Maximum	5%	9%	
2900	Dust protection, add, minimum	1%	2%	
3000	Maximum	4%	11%	
3100	Equipment usage curtailment, add, minimum	1%	1%	
3200	Maximum	3%	10%	
3300	Material handling & storage limitation, add, minimum	1%	1%	
3400	Maximum	6%	7%	
3500	Protection of existing work, add, minimum	2%	2%	
3600	Maximum	5%	7%	
3700	Shift work requirements, add, minimum		5%	
3800	Maximum		30%	
3900	Temporary shoring and bracing, add, minimum	2%	5%	
4000	Maximum	5%	12%	

615

B1010 Floor Construction

This page illustrates and describes a wood joist floor system including wood joist, oak floor, sub-floor, bridging, sand and finish floor, furring, plasterboard, taped, finished and painted ceiling. Lines within System Components give the unit price and total price per square foot for this system. Prices for wood joist floor systems are on Line Items B1010 263 1700 thru 2300. Both material quantities and labor costs have been adjusted for the system listed.

Factors: To adjust for job conditions other than normal working situations use Lines B1010 263 2700 thru 4000.

Example: You are to install the system during off peak hours, 6 P.M. to 2 A.M. Go to Line B1010 263 3800 and apply this percentage to the appropriate INST. costs.

System Components	QUANTITY	UNIT	COST PER S.F.		
			MAT.	INST.	TOTAL
Wood joists, 2″ x 8″, 16″ O.C.,oak floor (sanded & finished),1/2″ sub Floor, 1″x 3″ bridging, furring, 5/8″ drywall, taped, finished and painted.					
Wood joists, 2″ x 8″, 16″ O.C.	1.000	L.F.	1.10	1.24	2.34
Subfloor plywood CDX 1/2″	1.000	SF Flr.	.75	.91	1.66
Bridging 1″ x 3″	.150	Pr.	.12	.79	.91
Oak flooring, No. 1 common	1.000	S.F.	4.87	4.01	8.88
Sand and finish floor	1.000	S.F.	.25	1.83	2.08
Wood furring, 1″ x 3″, 16″ O.C.	1.000	L.F.	.48	1.95	2.43
Gypsum drywall, 5/8″ thick	1.000	S.F.	.36	.76	1.12
Tape and finishing	1.000	S.F.	.06	.68	.74
Paint ceiling	1.000	S.F.	.23	.74	.97
TOTAL			8.22	12.91	21.13

B1010 263	Floor-Ceiling, Wood Joists	COST PER S.F.		
		MAT.	INST.	TOTAL
1700	16″ on center, 2″ x 6″ joists	7.85	12.75	20.60
1750	2″ x 8″	8.20	12.90	21.10
1800	2″ x 10″	8.60	13.20	21.80
1900	2″ x 12″	9.05	13.25	22.30
2000	2″ x 14″	9.75	13.45	23.20
2100	12″ on center, 2″ x 10″ joists	8.95	13.55	22.50
2200	2″ x 12″	9.50	13.60	23.10
2300	2″ x 14″	10.40	13.90	24.30
2400				
2500				
2700	Cut & patch to match existing construction, add, minimum	2%	3%	
2800	Maximum	5%	9%	
2900	Dust protection, add, minimum	1%	2%	
3000	Maximum	4%	11%	
3100	Equipment usage curtailment, add, minimum	1%	1%	
3200	Maximum	3%	10%	
3300	Material handling & storage limitation, add, minimum	1%	1%	
3400	Maximum	6%	7%	
3500	Protection of existing work, add, minimum	2%	2%	
3600	Maximum	5%	7%	
3700	Shift work requirements, add, minimum		5%	
3800	Maximum		30%	
3900	Temporary shoring and bracing, add, minimum	2%	5%	
4000	Maximum	5%	12%	

B1020 Roof Construction

This page illustrates and describes a flat roof bar joist system including asphalt and gravel roof, lightweight concrete, metal decking, open web joists, and suspended acoustic ceiling. Lines within System Components give the unit price and total price per square foot for this system. Prices for flat roof bar joist systems are on Line Items B1020 106 1200 thru 1600. Both material quantities and labor costs have been adjusted for the system listed.

Factors: To adjust for job conditions other than normal working situations use Lines B1020 106 2700 thru 4000.

Example: You are to install the system where material handling and storage present some problem. Go to Line B1020 106 3300 and apply these percentages to the appropriate MAT. and INST. costs.

System Components	QUANTITY	UNIT	COST PER S.F. MAT.	COST PER S.F. INST.	COST PER S.F. TOTAL
Three ply asphalt and gravel roof on lightweight concrete, metal decking on Web joist 4' O.C., with suspended acoustic ceiling.					
Open web joist, 10" deep, 4' O.C.	1.250	Lb.	1.48	.60	2.08
Metal decking, 1-1/2" deep, 22 Ga., galvanized	1.000	S.F.	2.95	.72	3.67
Gravel roofing, 3 ply asphalt	.010	C.S.F.	1.39	2.21	3.60
Perlite roof fill, 3" thick	1.000	S.F.	2.42	.60	3.02
Suspended ceiling, 3/4" mineral fiber on "Z" bar suspension	1.000	S.F.	2.68	4.54	7.22
TOTAL			10.92	8.67	19.59

B1020 106	Steel Joist Roof & Ceiling	COST PER S.F. MAT.	COST PER S.F. INST.	COST PER S.F. TOTAL
1200	Open web bar joist, 10" deep, 5#/L.F.	10.90	8.70	19.60
1300	16" deep, 5#/L.F.	11.40	8.85	20.25
1400	18" deep, 6.6#/L.F.	11.80	9	20.80
1500	20" deep, 9.6#/L.F.	12.25	9.20	21.45
1600	24" deep, 11.52#/L.F.	12.85	9.45	22.30
1700				
1800				
1900				
2000				
2100				
2200				
2300				
2400				
2500				
2700	Cut & patch to match existing construction, add, minimum	2%	3%	
2800	Maximum	5%	9%	
2900	Dust protection, add, minimum	1%	2%	
3000	Maximum	4%	11%	
3100	Equipment usage curtailment, add, minimum	1%	1%	
3200	Maximum	3%	10%	
3300	Material handling & storage limitation, add, minimum	1%	1%	
3400	Maximum	6%	7%	
3500	Protection of existing work, add, minimum	2%	2%	
3600	Maximum	5%	7%	
3700	Shift work requirements, add, minimum		5%	
3800	Maximum		30%	
3900	Temporary shoring and bracing, add, minimum	2%	5%	
4000	Maximum	5%	12%	

B1020 Roof Construction

This page illustrates and describes a wood frame roof system including rafters, ceiling joists, sheathing, building paper, asphalt shingles, roof trim, furring, insulation, plaster and paint. Lines within System Components give the unit price and total price per square foot for this system. Prices for wood frame roof systems are on Line Items B1020 202 1800 thru 2700. Both material quantities and labor costs have been adjusted for the system listed.

Factors: To adjust for job conditions other than normal working situations use Lines B1020 202 3300 thru 4000.

Example: You are to install the system while protecting existing work. Go to Line B1020 202 3800 and apply these percentages to the appropriate MAT. and INST. costs.

System Components	QUANTITY	UNIT	COST PER S.F.		
			MAT.	INST.	TOTAL
Wood frame roof system, 4 in 12 pitch, including rafters, sheathing, Shingles, insulation, drywall, thin coat plaster, and painting.					
Rafters, 2" x 6", 16" O.C., 4 in 12 pitch	1.080	L.F.	.81	1.47	2.28
Ceiling joists, 2" x 6", 16 O.C.	1.000	L.F.	.75	1.09	1.84
Sheathing, 1/2" CDX	1.080	S.F.	.81	1.05	1.86
Building paper, 15# felt	.011	C.S.F.	.06	.20	.26
Asphalt, standard strip shingles, inorganic, class A, 30 year	.011	C.S.F.	1.34	1.42	2.76
Roof trim	.100	L.F.	.18	.30	.48
Furring, 1" x 3", 16" O.C.	1.000	L.F.	.48	1.95	2.43
Fiberglass insulation, 6" batts	1.000	S.F.	.69	.50	1.19
Gypsum board, 1/2" thick	1.000	S.F.	.36	.76	1.12
Thin coat plaster	1.000	S.F.	.11	.86	.97
Paint, roller, 2 coats	1.000	S.F.	.23	.74	.97
TOTAL			5.82	10.34	16.16

B1020 202	Wood Frame Roof & Ceiling	COST PER S.F.		
		MAT.	INST.	TOTAL
1800	Rafters 16" O.C., 2" x 6"	5.80	10.35	16.15
1900	2" x 8"	6.20	10.40	16.60
2000	2" x 10"	6.60	11.20	17.80
2100	2" x 12"	7.05	11.45	18.50
2200	Rafters 24" O.C., 2" x 6"	5.40	9.70	15.10
2300	2" x 8"	5.70	9.75	15.45
2400	2" x 10"	6	10.35	16.35
2500	2" x 12"	6.35	10.50	16.85
2600	Roof pitch, 6 in 12, add	3%	10%	
2700	8 in 12, add	5%	12%	
2800				
2900				
3000				
3100				
3300	Cut & patch to match existing construction, add, minimum	2%	3%	
3400	Maximum	5%	9%	
3500	Material handling & storage limitation, add, minimum	1%	1%	
3600	Maximum	6%	7%	
3700	Protection of existing work, add, minimum	2%	2%	
3800	Maximum	5%	7%	
3900	Shift work requirements, add, minimum		5%	
4000	Maximum		30%	

B1020 Roof Construction

5/8" Plywood Sheathing

Truss Anchors

Roof Trusses at 2'-0" o.c.

9" Fiberglass Insulation

1" x 8" Fascia Board

Wood Fabricated Roof Truss System

This page illustrates and describes a sloped roof truss framing system including trusses, truss anchors, sheathing, fascia and insulation. Lines B1020 210 1500 through 1800 are for a flat roof framing system including flat roof trusses, truss anchors, sheathing, fascia and insulation. Lines within system components give the unit price and total price per square foot for this system. Prices for alternate sizes and systems are given in line items B1020 210 0100 through 1800. Both material quantities and labor costs have been adjusted for the system listed.

Factors: To adjust for job conditions other than normal working situations use lines B1020 210 2700 through 4000.

Example: You are to install the system where crane placement and movement will be difficult. Go to line B1020 210 3200 and apply these percentages to the appropriate MAT. and INST. costs.

System Components	QUANTITY	UNIT	COST PER S.F.		
			MAT.	INST.	TOTAL
Roof truss, 2' O.C., 4/12 slope, 12' span, 5/8 sheath., fascia, insul.					
Timber connectors, rafter anchors, galv., 1 1/2" x 5 1/4"	.084	Ea.	.04	.39	.43
Plywood sheathing, CDX, 5/8" thick	1.054	S.F.	.90	1.11	2.01
Truss, 2' O.C., metal plate connectors, 12' span	.042	Ea.	1.68	2.10	3.78
Molding, fascia trim, 1" x 8"	.167	L.F.	.30	.49	.79
Insulation, fiberglass, kraft face, 9" thick x 23" wide	1.000	S.F.	.95	.50	1.45
TOTAL			3.87	4.59	8.46

B1020 210	Roof Truss, Wood	COST PER S.F.		
		MAT.	INST.	TOTAL
0100	Roof truss, 2'O.C., 4/12p, 1'ovhg, 12'span, 5/8"sheath, fascia, insul.	3.87	4.59	8.46
0200	20' span	4.11	3.80	7.91
0300	24' span	3.86	3.54	7.40
0400	26' span	5.35	4.71	10.06
0500	28' span	4.14	3.47	7.61
0600	30' span	4.02	3.43	7.45
0700	32' span	4.02	3.38	7.40
0800	34' span	3.80	3.33	7.13
0890	60' span	5.05	3.37	8.42
0900	8/12 pitch, 20' span	4.74	4.08	8.82
1000	24' span	4.64	3.82	8.46
1100	26' span	4.34	3.73	8.07
1200	28' span	4.62	3.72	8.34
1300	32' span	4.63	3.65	8.28
1400	36' span	4.73	3.59	8.32
1500	Flat roof frame, fab.strl.joists, 2' O.C., 15'-24' span, 50 PSF LL	4.01	3.12	7.13
1600	55 PSF LL	4.51	3.19	7.70
1700	24'-30'span, 45 PSF LL	4.93	2.98	7.91
1800	55 PSF LL	7.25	3.07	10.32
1900				
2700	Cut & patch to match existing construction, add, minimum	2%	3%	
2800	Maximum	5%	9%	
2900	Dust protection, add, minimum	1%	2%	
3000	Maximum	4%	11%	
3100	Equipment usage curtailment, add, minimum	1%	1%	
3200	Maximum	3%	10%	
3300	Material handling & storage limitation, add, minimum	1%	1%	
3400	Maximum	6%	7%	

B10 Superstructure

B1020 Roof Construction

B1020 210	Roof Truss, Wood	COST PER S.F.		
		MAT.	INST.	TOTAL
3500	Protection of existing work, add, minimum	2%	2%	
3600	Maximum	5%	7%	
3700	Shift work requirements, add, minimum		5%	
3800	Maximum		30%	
3900	Temporary shoring and bracing, add, minimum	2%	5%	
4000	Maximum	5%	12%	

B2010 Exterior Walls

This page illustrates and describes a masonry concrete block system including concrete block wall, pointed, reinforcing, waterproofing, gypsum plaster on gypsum lath, and furring. Lines within System Components give the unit price and total price per square foot for this system. Prices for masonry concrete block wall systems are on Line Items B2010 108 1400 thru 2200. Both material quantities and labor costs have been adjusted for the system listed.

Factors: To adjust for job conditions other than normal working situations use Lines B2010 108 2700 thru 4000.

Example: You are to install the system and match existing construction at several locations. Go to Line B2010 108 2800 and apply these percentages to the appropriate MAT. and INST. costs.

System Components	QUANTITY	UNIT	COST PER S.F.		
			MAT.	INST.	TOTAL
Concrete block, 8" thick, reinforced every 2 courses, waterproofing gypsum					
Plaster over gypsum lath on 1" x 3" furring, interior painting & baseboard.					
Concrete block, 8" x 8" x 16", reinforced	1.000	S.F.	5.20	7.45	12.65
Silicone waterproofing, 2 coats	1.000	S.F.	.87	.22	1.09
Bituminous coating, 1/16" thick	1.000	S.F.	.45	1.29	1.74
On concrete	1.000	L.F.	.52	2.62	3.14
Gypsum lath, 3/8" thick	.110	S.Y.	.42	.84	1.26
Gypsum plaster, 2 coats	.110	S.Y.	.47	3.22	3.69
Painting, 2 coats	1.000	S.F.	.15	.58	.73
Baseboard wood, 9/16" x 2-5/8"	.100	L.F.	.14	.28	.42
Paint baseboard, primer + 1 coat enamel	.100	L.F.	.02	.07	.09
TOTAL			8.24	16.57	24.81

B2010 108	Masonry Wall, Concrete Block	COST PER S.F.		
		MAT.	INST.	TOTAL
1400	8" thick block, regular	8.25	16.60	24.85
1500	Fluted 2 sides	9.50	19.50	29
1600	Deep grooved	7.95	19.50	27.45
1700	Slump block	11.85	16.95	28.80
1800	Split rib	10.30	19.50	29.80
1900				
2000	12" thick block, regular	8.15	19.85	28
2100	Slump block	16.85	18.50	35.35
2200	Split rib	11.30	22	33.30
2300				
2700	Cut & patch to match existing construction, add, minimum	2%	3%	
2800	Maximum	5%	9%	
2900	Dust protection, add, minimum	1%	2%	
3000	Maximum	4%	11%	
3100	Equipment usage curtailment, add, minimum	1%	1%	
3200	Maximum	3%	10%	
3300	Material handling & storage limitation, add, minimum	1%	1%	
3400	Maximum	6%	7%	
3500	Protection of existing work, add, minimum	2%	2%	
3600	Maximum	5%	7%	
3700	Shift work requirements, add, minimum		5%	
3800	Maximum		30%	
3900	Temporary shoring and bracing, add, minimum	2%	5%	
4000	Maximum	5%	12%	

B2010 Exterior Walls

This page illustrates and describes a masonry wall, brick-stone system including brick, concrete block, durawall, insulation, plasterboard, taped and finished, furring, baseboard and painting interior. Lines within System Components give the unit price and total price per square foot for this system. Prices for masonry wall, brick-stone systems are on Line Item B2010 129 1400 thru 2500. Both material quantities and labor costs have been adjusted for the system listed.

Factors: To adjust for job conditions other than normal working situations use Lines B2010 129 3100 thru 4200.

Example: You are to install the system without damaging the existing work. Go to Line B2010 129 3900 and apply these percentages to the appropriate MAT. and INST. costs.

System Components	QUANTITY	UNIT	COST PER S.F.		
			MAT.	INST.	TOTAL
Face brick, 4"thick, concrete block back-up, reinforce every second course, 3/4"insulation, furring, 1/2"drywall, taped, finish, and painted, baseboard					
Face brick, 4" brick	1.000	S.F.	5.35	14.10	19.45
Concrete back-up block, reinforced 8" thick	1.000	S.F.	3.23	7.85	11.08
3/4" rigid polystyrene insulation	1.000	S.F.	.68	.85	1.53
Furring, 1" x 3", wood, 16" O.C.	1.000	L.F.	.52	1.38	1.90
Drywall, 1/2" thick	1.000	S.F.	.36	.68	1.04
Taping & finishing	1.000	S.F.	.06	.68	.74
Painting, 2 coats	1.000	S.F.	.23	.74	.97
Baseboard, wood, 9/16" x 2-5/8"	.100	L.F.	.14	.28	.42
Paint baseboard, primer + 1 coat enamel	.100	L.F.	.02	.07	.09
TOTAL			10.59	26.63	37.22

B2010 131	Masonry Wall, Brick - Stone	COST PER S.F.		
		MAT.	INST.	TOTAL
1400	Face brick, standard, red, 4" x 2-2/3" x 8" (6.75/S.F.)	10.60	26.50	37.10
1500	Norman, 4" x 2-2/3" x 12" (4.5 per S.F.)	12.60	22	34.60
1600	Roman, 4" x 2" x 12" (6.0 per S.F.)	13.70	25	38.70
1700	Engineer, 4" x 3-1/5" x 8" (5.63 per S.F.)	9.50	24.50	34
1800	S.C.R., 6" x 2-2/3" x 12" (4.5 per S.F.)	12.20	22.50	34.70
1900	Jumbo, 6" x 4" x 12" (3.0 per S.F.)	11.15	19.65	30.80
2000	Norwegian, 6" x 3-1/5" x 12" (3.75 per S.F.)	11.90	21	32.90
2100				
2200				
2300	Stone, veneer, fieldstone, 6" thick	13	22.50	35.50
2400	Marble, 2" thick	58.50	38.50	97
2500	Limestone, 2" thick	37.50	23.50	61
3100	Cut & patch to match existing construction, add, minimum	2%	3%	
3200	Maximum	5%	9%	
3300	Dust protection, add, minimum	1%	2%	
3400	Maximum	4%	11%	
3500	Equipment usage curtailment, add, minimum	1%	1%	
3600	Maximum	3%	10%	
3700	Material handling & storage limitation, add, minimum	1%	1%	
3800	Maximum	6%	7%	
3900	Protection of existing work, add, minimum	2%	2%	
4000	Maximum	5%	7%	
4100	Shift work requirements, add, minimum		5%	
4200	Maximum		30%	

B2010 Exterior Walls

This page illustrates and describes a parapet wall system including wall, coping, flashing and cant strip. Lines within System Components give the unit price and cost per lineal foot for this system. Prices for parapet wall systems are on Lines B2010 142 1300 thru 2400. Both material quantities and labor costs have been adjusted for the system listed.

Factors: To adjust for job conditions other than normal working situations, use Lines B2010 142 2900 thru 4000.

Example: You are to install the system without damaging the adjacent property. Go to Line B2010 142 3600 and apply these percentages to the appropriate MAT. and INST. costs.

System Components	QUANTITY	UNIT	COST PER L.F. MAT.	COST PER L.F. INST.	COST PER L.F. TOTAL
Concrete block parapet, incl. reinf., coping, flashing & cant strip, 2'high					
8" concrete block	2.000	S.F.	10.40	14.90	25.30
Masonry reinforcing	1.000	L.F.	.29	.22	.51
Coping, precast	1.000	L.F.	30	17.30	47.30
Roof cant	1.000	L.F.	.61	1.99	2.60
16 oz.	1.000	L.F.	7.20	3.03	10.23
Cap flashing	1.000	L.F.	9	5.60	14.60
TOTAL			57.50	43.04	100.54

B2010 142	Parapet Wall	COST PER L.F. MAT.	COST PER L.F. INST.	COST PER L.F. TOTAL
1300	Concrete block, 8" thick	57.50	43	100.50
1400	12" thick	71	56	127
1500	Split rib block, 8" thick	61.50	49	110.50
1600	12" thick	70.50	57.50	128
1700	Brick, common brick, 8" wall	68	74	142
1800	12" wall	78.50	93	171.50
1900	4" brick, 4" backup block	63	70.50	133.50
2000	8" backup block	64	72.50	136.50
2100	Stucco on masonry	58	48	106
2200	On wood frame	22	39.50	61.50
2300	Wood, T 1-11 siding	34.50	47	81.50
2400	Boards, 1" x 6" cedar	42.50	52	94.50
2500				
2600				
2900	Dust protection, add, minimum	1%	2%	
3000	Maximum	4%	11%	
3100	Equipment usage curtailment, add, minimum	1%	1%	
3200	Maximum	3%	10%	
3300	Material handling & storage limitation, add, minimum	1%	1%	
3400	Maximum	6%	7%	
3500	Protection of existing work, add, minimum	2%	2%	
3600	Maximum	5%	7%	
3700	Shift work requirements, add, minimum		5%	
3800	Maximum		30%	
3900	Temporary shoring and bracing, add, minimum	2%	5%	
4000	Maximum	5%	12%	

This page illustrates and describes a masonry cleaning and restoration system, including staging, cleaning, repointing. Lines within System Components give the unit price and cost per square foot for this system. Prices for systems are on Lines B2010 143 1300 thru 2500. Both material quantities and labor costs have been adjusted for the system listed.

Factors: To adjust for conditions other than normal working situations, use Lines B2010 143 2900 thru 4200.

Example: You are to clean a wall and be concerned about dust control. Go to line B2010 143 3100 and apply these percentages to the appropriate MAT. and INST. costs.

System Components	QUANTITY	UNIT	COST PER S.F. MAT.	COST PER S.F. INST.	COST PER S.F. TOTAL
Repoint existing building, brick, common bond, high pressure cleaning, Water only, soft old mortar.					
Scaffolding, steel tubular, bldg ext wall face, labor only, 1 to 5 stories	.010	C.S.F.		2.55	2.55
Cleaning masonry, high pressure wash, water only, avg soil, bio staining	1.100	S.F.		1.89	1.89
Repoint, brick common bond	1.000	S.F.	.63	7	7.63
TOTAL			.63	11.44	12.07

B2010 143	Masonry Restoration - Cleaning	COST PER S.F. MAT.	COST PER S.F. INST.	COST PER S.F. TOTAL
1300	Brick, common bond	.63	11.45	12.08
1400	Flemish bond	.66	11.95	12.61
1500	English bond	.66	12.65	13.31
1700	Stone, 2' x 2' blocks	.83	8.65	9.48
1800	2' x 4' blocks	.62	7.60	8.22
2000	Add to above prices for alternate cleaning systems:			
2100	Chemical brush and wash	.05	.76	.81
2200	High pressure chemical and water	.09	.65	.74
2300	Sandblasting, wet system	.43	1.82	2.25
2400	Dry system	.40	1.06	1.46
2500	Steam cleaning		.50	.50
2600				
2900	Cut & patch to match existing construction, add, minimum	2%	3%	
3000	Maximum	5%	9%	
3100	Dust protection, add, minimum	1%	2%	
3200	Maximum	4%	11%	
3300	Equipment usage curtailment, add, minimum	1%	1%	
3400	Maximum	3%	10%	
3500	Material handling & storage limitation, add, minimum	1%	1%	
3600	Maximum	6%	7%	
3900	Protection of existing work, add, minimum	2%	2%	
4000	Maximum	5%	7%	
4100	Shift work requirements, add, minimum		5%	
4200	Maximum		30%	

B20 Exterior Enclosure

B2010 Exterior Walls

This page illustrates and describes a wood frame exterior wall system including wood studs, sheathing, felt, insulation, plasterboard, taped and finished, baseboard and painted interior. Lines within System Components give the unit price and total price per square foot for this system. Prices for wood frame exterior wall systems are on Line Items B2010 150 1600 thru 2700. Both material quantities and labor costs have been adjusted for the system listed.

Factors: To adjust for job conditions other than normal working situations use Lines B2010 150 3100 thru 4200.

Example: You are to install the system with need for complete temporary bracing. Go to Line B2010 150 4200 and apply these percentages to the appropriate MAT. and INST. costs.

System Components	QUANTITY	UNIT	COST PER S.F. MAT.	COST PER S.F. INST.	COST PER S.F. TOTAL
Wood stud wall, cedar shingle siding, building paper, plywood sheathing,					
Insulation, 5/8" drywall, taped, finished and painted, baseboard.					
2" x 4" wood studs, 16" O.C.	.100	L.F.	.55	1.36	1.91
1/2" CDX sheathing	1.000	S.F.	.75	.97	1.72
18" No. 1 red cedar shingles, 7-1/2" exposure	.008	C.S.F.	1.85	2.44	4.29
15# felt paper	.010	C.S.F.	.06	.18	.24
3-1/2" fiberglass insulation	1.000	S.F.	.52	.50	1.02
5/8" drywall	1.000	S.F.	.36	.68	1.04
Drywall finishing adder	1.000	S.F.	.06	.68	.74
Baseboard trim, stock pine, 9/16" x 3-1/2", painted	.100	L.F.	.14	.28	.42
Paint baseboard, primer + 1 coat enamel	.100	L.F.	.02	.07	.09
Paint, 2 coats, interior	1.000	S.F.	.23	.74	.97
TOTAL			4.54	7.90	12.44

B2010 150	Wood Frame Exterior Wall		COST PER S.F. MAT.	COST PER S.F. INST.	COST PER S.F. TOTAL
1600		Cedar shingle siding, painted	4.54	7.90	12.44
1700		Aluminum siding, horizontal clapboard	5.70	8.10	13.80
1800		Cedar bevel siding, 1/2" x 6", vertical , painted	7.60	7.75	15.35
1900		Redwood siding 1" x 4" to 1" x 6" vertical, T & G	8.20	8.55	16.75
2000		Board and batten	7.95	7.10	15.05
2100		Ship lap siding	8.35	7.50	15.85
2200		Plywood, grooved (T1-11) fir	4.02	7.50	11.52
2300		Redwood	5.55	7.50	13.05
2400		Southern yellow pine	4.27	7.50	11.77
2500		Masonry on stud wall, stucco, wire and plaster	3.61	12	15.61
2600		Stone veneer	10.45	15.45	25.90
2700		Brick veneer, brick $275 per M	8.05	19.55	27.60
3100	Cut & patch to match existing construction, add, minimum		2%	3%	
3200		Maximum	5%	9%	
3300	Dust protection, add, minimum		1%	2%	
3400		Maximum	4%	11%	
3500	Material handling & storage limitation, add, minimum		1%	1%	
3600		Maximum	6%	7%	
3700	Protection of existing work, add, minimum		2%	2%	
3800		Maximum	5%	7%	
3900	Shift work requirements, add, minimum			5%	
4000		Maximum		30%	
4100	Temporary shoring and bracing, add, minimum		2%	5%	
4200		Maximum	5%	12%	

B2010 Exterior Walls

B2010 190	Selective Price Sheet	COST PER S.F.		
		MAT.	INST.	TOTAL
0100	Exterior surface, masonry, concrete block, standard 4" thick	2.27	7.20	9.47
0200	6" thick	2.88	7.75	10.63
0300	8" thick	2.89	8.25	11.14
0400	12" thick	5.10	10.70	15.80
0500	Split rib, 4" thick	5.10	9	14.10
0600	8" thick	7.25	10.35	17.60
0700	Brick running bond, standard size, 6.75/S.F.	5.35	14.10	19.45
0800	Buff, 6.75/S.F.	5.65	14.10	19.75
0900	Stucco, on frame	.93	6.60	7.53
1000	On masonry	.37	5.10	5.47
1100	Metal, aluminum, horizontal, plain	2.99	2.65	5.64
1200	Insulated	3.49	2.65	6.14
1300	Vertical, plain	2.60	2.65	5.25
1400	Insulated	2.71	2.76	5.47
1500	Wood, beveled siding, "A" grade cedar, 1/2" x 6"	4.91	2.31	7.22
1600	1/2" x 8"	9.35	2.06	11.41
1700	Shingles, 16" #1 red, 7-1/2" exposure	2.34	3.30	5.64
1800	18" perfections, 7-1/2" exposure	2.27	2.78	5.05
1900	Handsplit, 10" exposure	3.75	2.72	6.47
2000	White cedar, 7-1/2" exposure	1.52	3.40	4.92
2100	Vertical, board & batten, redwood	5.50	3.10	8.60
2200	White pine	2.81	2.06	4.87
2300	T. & G. boards, redwood, 1" x 4"	7.80	3.10	10.90
2400	1' x 8"	9.15	2.06	11.21
2500				
2600	Interior surface, drywall, taped & finished, standard, 1/2"	.42	1.41	1.83
2700	5/8" thick	.42	1.41	1.83
2800	Fire resistant, 1/2" thick	.47	1.41	1.88
2900	5/8" thick	.47	1.41	1.88
3000	Moisture resistant, 1/2" thick	.51	1.41	1.92
3100	5/8" thick	.54	1.41	1.95
3200	Core board, 1" thick	.98	2.84	3.82
3300	Plaster, gypsum, 2 coats	.48	3.25	3.73
3400	3 coats	.68	3.94	4.62
3500	Perlite or vermiculite, 2 coats	.73	3.71	4.44
3600	3 coats	.82	4.57	5.39
3700	Gypsum lath, standard, 3/8" thick	.43	.85	1.28
3800	1/2" thick	.33	.90	1.23
3900	Fire resistant, 3/8" thick	.31	1.03	1.34
4000	1/2" thick	.33	1.12	1.45
4100	Metal lath, diamond, 2.5 lb.	.49	.85	1.34
4200	Rib, 3.4 lb.	.59	1.03	1.62
4300	Framing metal studs including top and bottom			
4400	Runners, walls 10' high			
4500	24" O.C., non load bearing 20 gauge, 2-1/2" wide	.36	1.14	1.50
4600	3-5/8" wide	.42	1.15	1.57
4700	4" wide	.51	1.15	1.66
4800	6" wide	.62	1.17	1.79
4900	Load bearing 18 gauge, 2-1/2" wide	.65	1.28	1.93
5000	3-5/8" wide	.77	1.30	2.07
5100	4" wide	.85	1.66	2.51
5200	6" wide	1.03	1.35	2.38
5300	16" O.C., non load bearing 20 gauge, 2-1/2" wide	.50	1.75	2.25
5400	3-5/8" wide	.57	1.77	2.34
5500	4" wide	.70	1.79	2.49
5600	6" wide	.84	1.82	2.66
5700	Load bearing 18 gauge, 2-1/2" wide	.88	1.77	2.65
5800	3-5/8" wide	1.04	1.80	2.84

B2010 Exterior Walls

B2010 190	Selective Price Sheet	COST PER S.F.		
		MAT.	INST.	TOTAL
5900	4" wide	1.08	1.84	2.92
6000	6" wide	1.39	1.87	3.26
6100	Framing wood studs incl. double top plate and			
6200	Single bottom plate, walls 10' high			
6300	24" O.C., 2" x 4"	.42	1.09	1.51
6400	2" x 6"	.61	1.19	1.80
6500	16" O.C., 2" x 4"	.55	1.36	1.91
6600	2" x 6"	.79	1.52	2.31
6700	Sheathing, boards, 1" x 6"	1.97	2.10	4.07
6800	1" x 8"	2.26	1.78	4.04
6900	Plywood, 3/8" thick	.58	1.14	1.72
7000	1/2" thick	.75	1.21	1.96
7100	5/8" thick	.85	1.30	2.15
7200	3/4" thick	1.04	1.40	2.44
7300	Wood fiber, 5/8" thick	.77	1.14	1.91
7400	Gypsum weatherproof, 1/2" thick	.58	1.21	1.79
7500	Insulation, fiberglass batts, 3-1/2" thick, R13	.45	.97	1.42
7600	6" thick, R19	.56	1.14	1.70
7700	Poured 4" thick, fiberglass wool, R4/inch	.77	3.41	4.18
7800	Mineral wool, R3/inch	.62	3.41	4.03
7900	Polystyrene, R4/inch	1.72	3.41	5.13
8000	Perlite or vermiculite, R2.7/inch	5.80	3.41	9.21
8100	Rigid, fiberglass, R4.3/inch, 1" thick	.63	.68	1.31
8200	R8.7/inch, 2" thick	1.20	.77	1.97

B2020 Exterior Windows

B2020 108	Vinyl Clad Windows

	MATERIAL	TYPE	GLAZING	SIZE	DETAIL	COST PER UNIT		
						MAT.	INST.	TOTAL
1000	Vinyl clad	casement	insul. glass	1'-4" x 4'-0"		455	157	612
1010				2'-0" x 3'-0"		375	157	532
1020				2'-0" x 4'-0"		430	165	595
1030				2'-0" x 5'-0"		465	174	639
1040				2'-0" x 6'-0"		515	174	689
1050				2'-4" x 4'-0"		465	174	639
1060				2'-6" x 5'-0"		590	169	759
1070				3'-0" x 5'-0"		820	174	994
1080				4'-0" x 3'-0"		910	174	1,084
1090				4'-0" x 4'-0"		775	174	949
1100				4'-8" x 4'-0"		845	174	1,019
1110				4'-8" x 5'-0"		980	203	1,183
1120				4'-8" x 6'-0"		1,075	203	1,278
1130				6'-0" x 4'-0"		980	203	1,183
1140				6'-0" x 5'-0"		1,200	203	1,403
3000		double-hung	insul. glass	2'-0" x 4'-0"		520	151	671
3050				2'-0" x 5'-0"		535	151	686
3100				2'-4" x 4'-0"		545	157	702
3150				2'-4" x 4'-8"		505	157	662
3200				2'-4" x 6'-0"		535	157	692
3250				2'-6" x 4'-0"		485	157	642
3300				2'-8" x 4'-0"		610	203	813
3350				2'-8" x 5'-0"		585	174	759
3375				2'-8" x 6'-0"		570	174	744
3400				3'-0" x 3'-6"		455	157	612
3450				3'-0" x 4'-0"		510	165	675
3500				3'-0" x 4'-8"		570	165	735
3550				3'-0" x 5'-0"		580	174	754
3600				3'-0" x 6'-0"		510	174	684
3700				4'-0" x 5'-0"		700	187	887
3800				4'-0" x 6'-0"		855	187	1,042

For customer support on your Commercial Renovation Costs with RSMeans data, call 800.448.8182.

B2020 Exterior Windows

This page illustrates and describes an aluminum window system including double hung aluminum window, exterior and interior trim, hardware and insulating glass. Lines within System Components give the unit price and total price on a cost each basis for this system. Prices for aluminum window systems are on Line Items B2020 110 1000 thru 2400. Both material quantities and labor costs have been adjusted for the system listed.

Factors: To adjust for job conditions other than normal working situations use Lines B2020 110 3100 thru 4000.

Example: You are to install the system and cut and patch to match existing construction. Go to Line B2020 110 3200 and apply these percentages to the appropriate MAT. and INST. costs.

System Components	QUANTITY	UNIT	COST EACH MAT.	COST EACH INST.	COST EACH TOTAL
Double hung aluminum window, 2'-4" x 2'-6" exterior and interior trim,					
Hardware glazed with insulating glass.					
Double-hung aluminum window, 2'-4" x 2' 6"	1.000	Ea.	271.10	44.02	315.12
Interior Trim	1.000	Set	36.50	52.50	89
Hardware	1.000	Set	6.80	28.50	35.30
TOTAL			314.40	125.02	439.42

B2020 110	Windows - Aluminum	COST EACH MAT.	COST EACH INST.	COST EACH TOTAL
1000	Aluminum, double hung, 2'-4" x 2'-6"	315	125	440
1100	2'-8" x 4'-6"	610	187	797
1200	3'-4" x 5'-6"	945	281	1,226
1300	Casement, 3'-6" x 2'-4"	425	143	568
1400	4'-6" x 2'-4"	545	176	721
1500	5'-6" x 2'-4"	695	239	934
1600	Projected window, 2'-1" x 3'-0"	310	128	438
1700	3'-5" x 3'-0"	485	174	659
1800	4'-0" x 4'-0"	775	263	1,038
1900	Horizontal sliding 3'-0" x 3'-0"	273	149	422
2000	3'-6" x 4'-0"	405	202	607
2100	5'-0" x 6'-0"	860	370	1,230
2200	Picture window, 3'-8" x 3'-1"	286	166	452
2300	4'-4" x 4'-5"	460	241	701
2400	5'-8" x 4'-9"	670	345	1,015
2500				
2600				
2700				
2800				
2900				
3100	Cut & patch to match existing construction, add, minimum	2%	3%	
3200	Maximum	5%	9%	
3300	Dust protection, add, minimum	1%	2%	
3400	Maximum	4%	11%	
3500	Material handling & storage limitation, add, minimum	1%	1%	
3600	Maximum	6%	7%	
3700	Protection of existing work, add, minimum	2%	2%	
3800	Maximum	5%	7%	
3900	Shift work requirements, add, minimum		5%	
4000	Maximum		30%	

B2020 Exterior Windows

This page illustrates and describes a wood window system including wood picture window, exterior and interior trim, hardware and insulating glass. Lines within System Components give the unit price and total price on a cost each basis for this system. Prices for wood window systems are on Line Items B2020 112 1200 thru 2900. Both material quantities and labor costs have been adjusted for the system listed.

Factors: To adjust for job conditions other than normal working situations use Lines B2020 112 3100 thru 4000.

Example: You are to install the above system where dust control is vital. Go to Line B2020 112 3400 and apply these percentages to the appropriate MAT. and INST. costs.

System Components	QUANTITY	UNIT	COST EACH		
			MAT.	INST.	TOTAL
Wood picture window 4'-0" x 4'-6", exterior and interior trim, hardware, Glazed with insulating glass.					
4'-0" x 4'-6" wood picture window with insulating glass	1.000	Ea.	615	124	739
Oak interior trim	1.000	Set	87.50	114	201.50
Hardware	1.000	Set	6.80	28.50	35.30
TOTAL			709.30	266.50	975.80

B2020 112	Windows - Wood	COST EACH		
		MAT.	INST.	TOTAL
1200	Picture window, 4'-0" x 4'-6"	710	267	977
1300	5'-0" x 4'-0"	775	267	1,042
1400	6'-0" x 4'-6"	805	279	1,084
1500	Bow, bay window, 8'-0" x 5'-0", standard	2,700	279	2,979
1600	Deluxe	2,000	279	2,279
1700	Bow, bay window, 12'-0" x 6'-0" standard	3,350	370	3,720
1800	Deluxe	2,800	370	3,170
1900	Double hung, 3'-0" x 4'-0"	375	172	547
2000	4'-0" x 4'-6"	485	228	713
2100	Casement 2'-0" x 3'-0"	360	149	509
2200	2 leaf, 4'-0" x 4'-0"	1,025	248	1,273
2300	3 leaf, 6'-0" x 6'-0"	1,575	400	1,975
2400	Awning, 2'-10" x 1'-10"	405	149	554
2500	3'-6" x 2'-4"	435	172	607
2600	4'-0" x 3'-0"	650	228	878
2700	Horizontal sliding 3'-0" x 2'-0"	390	149	539
2800	4'-0" x 3'-6"	515	172	687
2900	6'-0" x 5'-0"	760	228	988
3100	Cut & patch to match existing construction, add, minimum	2%	3%	
3200	Maximum	5%	9%	
3300	Dust protection, add, minimum	1%	2%	
3400	Maximum	4%	11%	
3500	Material handling & storage limitation, add, minimum	1%	1%	
3600	Maximum	6%	7%	
3700	Protection of existing work, add, minimum	2%	2%	
3800	Maximum	5%	7%	
3900	Shift work requirements, add, minimum		5%	
4000	Maximum		30%	

This page illustrates and describes storm window and door systems based on a cost each price. Prices for storm window and door systems are on Line Items B2020 116 0400 thru 2400. Both material quantities and labor costs have been adjusted for the system listed.

Factors: To adjust for job conditions other than normal working situations use Lines B2020 116 3100 thru 4000 and apply these percentages to the appropriate MAT and INST. costs.

Example: You are to install the system and protect all existing construction. Go to Line B2020 116 3800 and apply these percentages to the appropriate MAT. and INST. costs.

B2020 116	Storm Windows & Doors	COST EACH		
		MAT.	INST.	TOTAL
0350	Storm Window and door systems, single glazed			
0400	Window, custom aluminum anodized, 2'-0" x 3'-5"	107	45.50	152.50
0500	2'-6" x 5'-0"	147	48.50	195.50
0600	4'-0" x 6'-0"	269	54.50	323.50
0700	White painted aluminum, 2'-0" x 3'-5"	141	45.50	186.50
0800	2'-6" x 5'-0"	208	48.50	256.50
0900	4'-0" x 6'-0"	325	54.50	379.50
1000	Average quality aluminum, anodized, 2'-0" x 3'-5"	108	45.50	153.50
1100	2'-6" x 5'-0"	138	48.50	186.50
1200	4'-0" x 6'-0"	146	54.50	200.50
1300	White painted aluminum, 2'-0" x 3'-5"	114	45.50	159.50
1400	2'-6" x 5'-0"	112	48.50	160.50
1500	4'-0" x 6'-0"	132	54.50	186.50
1600	Mill finish, 2'-0" x 3'-5"	95	45.50	140.50
1700	2'-6" x 5'-0"	113	48.50	161.50
1800	4'-0" x 6'-0"	144	54.50	198.50
1900				
2000	Door, aluminum anodized 3'-0" x 6'-8"	211	97.50	308.50
2100	White painted aluminum, 3'-0" x 6'-8"	345	97.50	442.50
2200	Mill finish, 3'-0" x 6'-8"	295	97.50	392.50
2300				
2400	Wood, storm and screen, painted	380	151	531
2500				
2600				
2700				
2800				
3100	Cut & patch to match existing construction, add, minimum	2%	3%	
3200	Maximum	5%	9%	
3300	Dust protection, add, minimum	1%	2%	
3400	Maximum	4%	11%	
3500	Material handling & storage limitation, add, minimum	1%	1%	
3600	Maximum	6%	7%	
3700	Protection of existing work, add, minimum	2%	2%	
3800	Maximum	5%	7%	
3900	Shift work requirements, add, minimum		5%	
4000	Maximum		30%	

B2020 Exterior Windows

This page illustrates and describes a window wall system including aluminum tube framing, caulking, and glass. Lines within System Components give the unit price and total price per square foot for this system. Prices for window wall systems are on Line Items B2020 124 1300 thru 2600. Both material quantities and labor costs have been adjusted for the system listed.

Factors: To adjust for job conditions other than normal working situations use Lines B2020 124 3100 thru 4000.

Example: You are to install the system with need for complete dust protection. Go to Line B2020 124 3400 and apply these percentages to the appropriate MAT. and INST. costs.

System Components	QUANTITY	UNIT	COST PER S.F.		
			MAT.	INST.	TOTAL
Window wall, including aluminum header, sill, mullions, caulking And glass					
Header, mill finish, 1-3/4" x 4-1/2" deep	.167	L.F.	3.84	2.62	6.46
Sill, mill finish, 1-3/4" x 4-1/2" deep	.167	L.F.	3.67	2.56	6.23
Vertical mullion, 1-3/4" x 4-1/2" deep, 6' O.C.	.191	L.F.	9.49	6.39	15.88
Caulking	.381	L.F.	.10	.89	.99
Glass, 1/4" plate	1.000	S.F.	9.20	10.85	20.05
TOTAL			26.30	23.31	49.61

B2020 124	Aluminum Frame, Window Wall	COST PER S.F.		
		MAT.	INST.	TOTAL
1300	Mill finish, 1-3/4" x 4-1/2" deep, 1/4" plate glass	26.50	23.50	50
1400	2" x 4-1/2" deep, insulating glass	24.50	27	51.50
1500	Thermo-break, 2-1/4" x 4-1/2" deep, insulating glass	25.50	27.50	53
1600	Bronze finish, 1-3/4" x 4-1/2" deep, 1/4" plate glass	29.50	25.50	55
1700	2" x 4-1/2" deep, insulating glass	27	29	56
1800	Thermo-break, 2-1/4" x 4-1/2" deep, insulating glass	28	30	58
2000	Black finish, 1-3/4" x 4-1/2" deep, 1/4" plate glass	31	26.50	57.50
2100	2" x 4-1/2" deep, insulating glass	28	30.50	58.50
2200	Thermo-break, 2-1/4" x 4-1/2" deep, insulating glass	29.50	31	60.50
2300				
2400	Stainless steel, 1-3/4" x 4-1/2" deep, 1/4" plate glass	39	32	71
2500	2" x 4-1/2" deep, insulating glass	34	36.50	70.50
2600	Thermo-break, 2-1/4" x 4-1/2" deep, insulating glass	36	37	73
2700				
3100	Cut & patch to match existing construction, add, minimum	2%	3%	
3200	Maximum	5%	9%	
3300	Dust protection, add, minimum	1%	2%	
3400	Maximum	4%	11%	
3500	Material handling, & storage limitation, add, minimum	1%	1%	
3600	Maximum	6%	7%	
3700	Protection of existing work, add, minimum	2%	2%	
3800	Maximum	5%	7%	
3900	Shift work requirements, add, minimum		5%	
4000	Maximum		30%	

B2030 Exterior Doors

This page illustrates and describes a commercial metal door system, including a single aluminum and glass door, narrow stiles, jamb, hardware weatherstripping, panic hardware and closer. Lines within System Components give the unit price and total price on a cost each basis for this system. Prices for commercial metal door systems are on Line Items B2030 125 1200 thru 2500. Both material quantities and labor costs have been adjusted for the system listed.

Factors: To adjust for job conditions other than normal working situations, use Lines B2030 125 3100 thru 4000.

Example: You are to install the system and cut and patch to match existing construction. Go to Line B2030 125 3200 and apply these percentages to the appropriate MAT. and INST. costs.

System Components	QUANTITY	UNIT	COST EACH MAT.	COST EACH INST.	COST EACH TOTAL
Single aluminum and glass door, 3'-0"x7'-0", with narrow stiles, ext. jamb. **Weatherstripping, 1/2" tempered insul. glass, panic hardware, and closer.**					
Aluminum door, 3'-0" x 7'-0" x 1-3/4", narrow stiles	1.000	Ea.	1,175	755	1,930
Tempered insulating glass, 1/2" thick	20.000	S.F.	610	470	1,080
Panic hardware	1.000	Set	925	114	1,039
Surface mount regular arm	1.000	Ea.	176	105	281
TOTAL			2,886	1,444	4,330

B2030 125	Doors, Metal - Commercial	COST EACH MAT.	COST EACH INST.	COST EACH TOTAL
1200	Single aluminum and glass, 3'-0" x 7'-0"	2,875	1,450	4,325
1300	With transom, 3'-0" x 10'-0"	3,425	1,725	5,150
1400	Anodized aluminum and glass, 3'-0" x 7'-0"	3,300	1,700	5,000
1500	With transom, 3'-0" x 10'-0"	3,950	2,000	5,950
1600	Steel, deluxe, hollow metal 3'-0" x 7'-0"	2,050	570	2,620
1700	With transom 3'-0" x 10'-0"	2,525	645	3,170
1800	Fire door, "A" label, 3'-0" x 7'-0"	1,875	570	2,445
1900	Double, aluminum and glass, 6'-0" x 7'-0"	4,125	2,475	6,600
2000	With transom, 6'-0" x 10'-0"	5,075	3,075	8,150
2100	Anodized aluminum and glass 6'-0" x 7'-0"	4,525	2,875	7,400
2200	With transom, 6'-0" x 10'-0"	5,675	3,550	9,225
2300	Steel, deluxe, hollow metal, 6'-0" x 7'-0"	3,500	985	4,485
2400	With transom, 6'-0" x 10'-0"	4,475	1,150	5,625
2500	Fire door, "A" label, 6'-0" x 7'-0"	3,075	975	4,050
2800				
2900				
3100	Cut & patch to match existing construction, add, minimum	2%	3%	
3200	Maximum	5%	9%	
3300	Dust protection, add, minimum	1%	2%	
3400	Maximum	4%	11%	
3500	Material handling & storage limitation, add, minimum	1%	1%	
3600	Maximum	6%	7%	
3700	Protection of existing work, add, minimum	2%	2%	
3800	Maximum	5%	7%	
3900	Shift work requirements, add, minimum		5%	
4000	Maximum		30%	

B2030 Exterior Doors

This page illustrates and describes sliding door systems including a sliding door, frame, interior and exterior trim with exterior staining. Lines within System Components give the unit price and total price on a cost each basis for this system. Prices for sliding door systems are on Line Items B2030 150 1000 thru 2400. Both material quantities and labor costs have been adjusted for the system listed.

Factors: To adjust for job conditions other than normal working situations use Lines B2030 150 2700 thru 4000.

Example: You are to install the system with temporary shoring and bracing. Go to Line B2030 150 3900 and apply these percentages to the appropriate MAT. and INST. costs.

System Components	QUANTITY	UNIT	COST EACH MAT.	COST EACH INST.	COST EACH TOTAL
Sliding wood door, 6'-0" x 6'-8", with wood frame, interior and exterior					
Trim and exterior staining.					
Sliding wood door, standard, 6'-0" x 6'-8", insulated glass	1.000	Ea.	1,375	340	1,715
Interior & exterior trim	1.000	Set	31	54.40	85.40
Stain door	1.000	Ea.	2.79	33	35.79
Stain trim	1.000	Ea.	3.20	14.20	17.40
TOTAL			1,411.99	441.60	1,853.59

B2030 150	Doors, Sliding - Patio	MAT.	INST.	TOTAL
1000	Wood, standard, 6'-0" x 6'-8", insulated glass	1,400	440	1,840
1100	8'-0" x 6'-8"	1,800	565	2,365
1200	12'-0" x 6'-8"	2,975	685	3,660
1300	Vinyl coated, 6'-0" x 6'-8"	1,700	440	2,140
1400	8'-0" x 6'-8"	2,200	565	2,765
1500	12'-0" x 6'-8"	3,650	685	4,335
1700	Aluminum, standard, 6'-0" x 6'-8", insulated glass	1,025	475	1,500
1800	8'-0" x 6'-8"	1,800	605	2,405
1900	12'-0" x 6'-8"	1,900	725	2,625
2000	Anodized, 6'-0" x 6'-8"	1,825	475	2,300
2100	8'-0" x 6'-8"	2,025	605	2,630
2200	12'-0" x 6'-8"	3,500	725	4,225
2300				
2400	Deduct for single glazing	113		113
2500				
2700	Cut & patch to match existing construction, add, minimum	2%	3%	
2800	Maximum	5%	9%	
2900	Dust protection, add, minimum	1%	2%	
3000	Maximum	4%	11%	
3100	Equipment usage curtailment, add, minimum	1%	1%	
3200	Maximum	3%	10%	
3300	Material handling & storage limitation, add, minimum	1%	1%	
3400	Maximum	6%	7%	
3500	Protection of existing, work, add, minimum	2%	2%	
3600	Maximum	5%	7%	
3700	Shift work requirements, add, minimum		5%	
3800	Maximum		30%	
3900	Temporary shoring and bracing, add, minimum	2%	5%	
4000	Maximum	5%	12%	

B2030 Exterior Doors

This page illustrates and describes residential door systems including a door, frame, trim, hardware, weatherstripping, stained and finished. Lines within System Components give the unit price and total price on a cost each basis for this system. Prices for residential door systems are on Line Items B2030 240 1500 thru 2200. Both material quantities and labor costs have been adjusted for the system listed.

Factors: To adjust for job conditions other than normal working situations use Lines B2030 240 3100 thru 4000.

Example: You are to install the system with a material handling and storage limitation. Go to Line B2030 240 3600 and apply these percentages to the appropriate MAT. and INST. costs.

System Components	QUANTITY	UNIT	COST EACH MAT.	COST EACH INST.	COST EACH TOTAL
Single solid wood colonial door, 3' x 6'-8" with wood frame and trim,					
Stained and finished, including hardware and weatherstripping.					
Solid wood colonial door, fir 3' x 6'-8" x 1-3/4", hinges	1.000	Ea.	670	91	761
Hinges, ball bearing	1.000	Ea.	40.50		40.50
Exterior frame, with trim	1.000	Set	187	61.71	248.71
Interior trim	1.000	Set	26	57	83
Average quality	1.000	Set	55.50	48.50	104
Sill, oak 8" deep	3.500	Ea.	82.25	47.60	129.85
Weatherstripping	1.000	Set	26.50	89.50	116
Stained and finished	1.000	Ea.	5.95	70.50	76.45
TOTAL			1,093.70	465.81	1,559.51

B2030 240	Doors, Residential - Exterior	COST EACH MAT.	COST EACH INST.	COST EACH TOTAL
1400	Single doors, 3' x 6'-8"			
1500	Solid colonial door, fir	1,100	465	1,565
1600	Hollow metal exterior door, plain	970	435	1,405
1700	Solid core wood door, plain	590	445	1,035
1800				
1900	Double doors, 6' x 6'-8"			
2000	Solid colonial double doors, fir	1,875	620	2,495
2100	Hollow metal exterior doors, plain	1,625	580	2,205
2200	Hollow core wood doors, plain	870	600	1,470
2300				
2800				
2900				
3100	Cut & patch to match existing construction, add, minimum	2%	3%	
3200	Maximum	5%	9%	
3300	Dust protection, add, minimum	1%	2%	
3400	Maximum	4%	11%	
3500	Material handling & storage limitation, add, minimum	1%	1%	
3600	Maximum	6%	7%	
3700	Protection of existing work, add, minimum	2%	2%	
3800	Maximum	5%	7%	
3900	Shift work requirements, add, minimum		5%	
4000	Maximum		30%	

635

B2030 Exterior Doors

This page illustrates and describes overhead door systems including an overhead door, track, hardware, trim, and electric door opener. Lines within System Components give the unit price and total price on a cost each basis for this system. Prices for overhead door systems are on Line Items B2030 410 1200 thru 2300. Both material quantities and labor costs have been adjusted for the system listed.

Factors: To adjust for job conditions other than normal working situations use Lines B2030 410 3100 thru 4000.

Example: You are to install the system and match existing construction. Go to Line B2030 410 3200 and apply these percentages to the appropriate MAT. and INST. costs.

System Components	QUANTITY	UNIT	COST EACH		
			MAT.	INST.	TOTAL
Wood overhead door, commercial sectional, including track, door, hardware And trim, electrically operated.					
Commercial, heavy duty wood door, 8' x 8' x 1-3/4" thick	1.000	Ea.	1,375	680	2,055
Frame, 2 x 8, pressure treated	1.000	Set	32.40	86.16	118.56
Wood trim	1.000	Set	33.84	68.16	102
Painting, two coats	1.000	Ea.	24.48	226.08	250.56
Electric trolley operator	1.000	Ea.	1,225	340	1,565
TOTAL			2,690.72	1,400.40	4,091.12

B2030 410	Doors, Overhead	COST EACH		
		MAT.	INST.	TOTAL
1200	Commercial, wood, 1-3/4" thick, 8' x 8'	2,700	1,400	4,100
1300	12' x 12'	4,425	1,975	6,400
1400	14' x 14'	5,600	2,325	7,925
1500	Fiberglass & aluminum 12' x 12'	5,125	2,000	7,125
1600	20' x 20'	10,200	5,100	15,300
1700	Residential, wood, 9' x 7'	1,250	530	1,780
1800	16' x 7'	2,025	790	2,815
1900	Hardboard faced, 9' x 7'	900	530	1,430
2000	16' x 7'	1,550	790	2,340
2100	Fiberglass & aluminum, 9' x 7'	1,150	630	1,780
2200	16' x 7'	2,250	790	3,040
2300	For residential electric opener, add	700	85	785
2400				
2500				
2600				
2700				
2800				
2900				
3100	Cut & patch to match existing construction, add, minumum	2%	3%	
3200	Maximum	5%	9%	
3300	Dust protection, add, minimum	1%	2%	
3400	Maximum	4%	11%	
3500	Material handling & storage limitation, add, minimum	1%	1%	
3600	Maximum	6%	7%	
3700	Protection of existing work, add, minimum	2%	2%	
3800	Maximum	5%	7%	
3900	Shift work requirements, add, minimum		5%	
4000	Maximum		30%	

B2030 Exterior Doors

B2030 810	Selective Price Sheet	COST EACH		
		MAT.	INST.	TOTAL
0100	Door closer, rack and pinion	232	114	346
0200	Backcheck and adjustable power	232	114	346
0300	Regular, hinge face mount, all sizes, regular arm	232	114	346
0400	Hold open arm	109	114	223
0500	Top jamb mount, all sizes, regular arm	232	114	346
0600	Hold open arm	109	114	223
0700	Stop face mount, all sizes, regular arm	232	114	346
0800	Hold open arm	109	114	223
0900	Fusible link, hinge face mount, all sizes, regular arm	292	114	406
1000	Hold open arm	169	114	283
1100	Top jamb mount, all sizes, regular arm	292	114	406
1200	Hold open arm	169	114	283
1300	Stop face mount, all sizes, regular arm	292	114	406
1400	Hold open arm	169	114	283
1500				
1600				
1700	Door stops			
1800				
1900	Holder & bumper, floor or wall	40	21.50	61.50
2000	Wall bumper	14.90	21.50	36.40
2100	Floor bumper	4.33	21.50	25.83
2200	Plunger type, door mounted	35.50	21.50	57
2300	Hinges, full mortise, material only, per pair			
2400	Low frequency, 4-1/2" x 4-1/2", steel base, USP	27.50		27.50
2500	Brass base, US10	64.50		64.50
2600	Stainless steel base, US32	88		88
2700	Average frequency, 4-1/2" x 4-1/2", steel base, USP	52.50		52.50
2800	Brass base, US10	81		81
2900	Stainless steel base, US32	102		102
3000	High frequency, 4-1/2" x 4-1/2", steel base, USP	74		74
3100	Brass base, US10	82		82
3200	Stainless steel base, US32	117		117
3300	Kick plate			
3400				
3500	6" high, for 3'-0" door, aluminum	37	45.50	82.50
3600	Bronze	71	45.50	116.50
3700	Panic device			
3800				
3900	For rim locks, single door, exit	925	114	1,039
4000	Outside key and pull	1,350	136	1,486
4100	Bar and vertical rod, exit only	1,225	136	1,361
4200	Outside key and pull	1,475	170	1,645
4300	Lockset			
4400				
4500	Heavy duty, cylindrical, passage doors	142	57	199
4600	Classroom	640	85	725
4700	Bedroom, bathroom, and inner office doors	175	57	232
4800	Apartment, office, and corridor doors	355	68	423
4900	Standard duty, cylindrical, exit doors	179	68	247
5100	Passage doors	85.50	57	142.50
5200	Public restroom, classroom, & office doors	231	85	316
5300	Deadlock, mortise, heavy duty	213	75.50	288.50
5400	Double cylinder	212	75.50	287.50
5500	Entrance lock, cylinder, deadlocking latch	209	75.50	284.50
5600	Deadbolt	214	85	299
5700	Commercial, mortise, wrought knob, keyed, minimum	219	85	304
5800	Maximum	595	97.50	692.50
5900	Cast knob, keyed, minimum	300	75.50	375.50

B2030 810	Selective Price Sheet	COST EACH		
		MAT.	INST.	TOTAL
6000	Maximum	610	75.50	685.50
6100	Push-pull			
6200				
6300	Aluminum	17.05	57	74.05
6400	Bronze	29.50	57	86.50
6500	Door pull, designer style, minimum	29	57	86
6600	Maximum	274	124	398
6700	Threshold			
6800				
6900	3'-0" long door saddles, aluminum, minimum	12.20	14.20	26.40
7000	Maximum	62	57	119
7100	Bronze, minimum	45	11.35	56.35
7200	Maximum	182	57	239
7300	Rubber, 1/2" thick, 5-1/2" wide	47.50	34	81.50
7400	2-3/4" wide	59.50	34	93.50
7500	Weatherstripping, per set			
7600				
7700	Doors, wood frame, interlocking for 3' x 7' door, zinc	64.50	227	291.50
7800	Bronze	78	227	305
7900	Wood frame, spring type for 3' x 7' door, bronze	26.50	89.50	116
8000	Metal frame, spring type for 3' x 7' door, bronze	57	227	284
8100	For stainless steel, spring type, add	133%		
8200				
8300	Metal frame, extruded sections, 3' x 7' door, aluminum	34	227	261
8400	Bronze	99	227	326

B3010 Roof Coverings

B3010 160	Selective Price Sheet	COST PER S.F.		
		MAT.	INST.	TOTAL
0100	Roofing, built-up, asphalt roll roof, 3 ply organic/mineral surface	.76	1.80	2.56
0200	3 plies glass fiber felt type iv, 1 ply mineral surface	1.33	1.94	3.27
0300	Cold applied, 3 ply		.63	.63
0400	Coal tar pitch, 4 ply tarred felt	2.24	2.31	4.55
0500	Mopped, 3 ply glass fiber	1.84	2.56	4.40
0600	4 ply organic felt	2.24	2.31	4.55
0700	Elastomeric, hypalon, neoprene unreinforced	1.44	4.28	5.72
0800	Polyester reinforced	1.58	5.05	6.63
0900	Neoprene, 5 coats 60 mils	4.80	14.90	19.70
1000	Over 10,000 S.F.	4.80	7.75	12.55
1600				
1700	Asphalt, strip, 210-235#/sq.	.83	1.18	2.01
1800	235-240#/sq.	1.22	1.29	2.51
1900	Class A laminated	1.13	1.44	2.57
2000	Class C laminated	1.13	1.62	2.75
2100	Slate, buckingham, 3/16" thick	6.15	3.70	9.85
2200	Black, 1/4" thick	6.15	3.70	9.85
2300	Wood, shingles, 16" no. 1, 5" exp.	3.50	2.72	6.22
2400	Red cedar, 18" perfections	3.15	2.48	5.63
2500	Shakes, 24", 10" exposure	3.75	2.72	6.47
2600	18", 8-1/2" exposure	3.25	3.40	6.65
2700	Insulation, ceiling batts, fiberglass, 3-1/2" thick, R13	.37	.50	.87
2800	6" thick, R19	.48	.50	.98
2900	9" thick, R30	.95	.59	1.54
3000	12" thick, R38	1.14	.59	1.73
3100	Mineral fiber, 3-1/2" thick, R13	.90	.43	1.33
3200	6" thick, R19	1.42	.43	1.85
3300	Roof deck, fiberboard, 1" thick, R2.78	.88	.83	1.71
3400	Mineral fiber, 2" thick, R5.26	1.43	.83	2.26
3500	Perlite boards, 3/4" thick, R2.08	.55	.83	1.38
3600	2" thick, R5.26	1.32	.92	2.24
3700	Polystyrene extruded, 15 PSI, 1" thick, R5.26	.69	.54	1.23
3800	2" thick, R10	.86	.61	1.47
3900	40 PSI, 1" thick, R5	1.08	.54	1.62
4000	Tapered for drainage	1.09	.57	1.66
4300	Ceiling, plaster, gypsum, 2 coats	.48	3.71	4.19
4400	3 coats	.68	4.34	5.02
4500	Perlite or vermiculite, 2 coats	.73	4.29	5.02
4600	3 coats	.82	5.40	6.22
4700	Gypsum lath, plain 3/8" thick	.43	.85	1.28
4800	1/2" thick	.33	.90	1.23
4900	Firestop, 3/8" thick	.31	1.03	1.34
5000	1/2" thick	.33	1.12	1.45
5100	Metal lath, rib, 2.75 lb.	.43	.97	1.40
5200	3.40 lb.	.59	1.03	1.62
5300	Diamond, 2.50 lb.	.49	.97	1.46
5400	3.40 lb.	.52	1.20	1.72
5500	Drywall, taped and finished standard, 1/2" thick	.42	1.78	2.20
5600	5/8" thick	.42	1.78	2.20
5700	Fire resistant, 1/2" thick	.47	1.78	2.25
5800	5/8" thick	.47	1.78	2.25
5900	Water resist., 1/2" thick	.51	1.78	2.29
6000	5/8" thick	.54	1.78	2.32
6100	Finish, instead of taping			
6200	For thin coat plaster, add	.11	.86	.97
6300	Finish textured spray, add	.04	.82	.86
6500	Tile, stapled glued, mineral fiber plastic coated, 5/8" thick	2.45	2.27	4.72
6600	3/4" thick	3.47	2.27	5.74

639

B3010 Roof Coverings

B3010 160	Selective Price Sheet	COST PER S.F.		
		MAT.	INST.	TOTAL
6700	Wood fiber, 1/2" thick	1.35	1.70	3.05
6800	3/4" thick	1.33	1.70	3.03
6900	Suspended, fiberglass film faced, 5/8" thick	1.47	1.09	2.56
7000	3" thick	4.26	1.51	5.77
7100	Mineral fiber 5/8" thick, standard face	1.16	1.01	2.17
7200	Aluminum faced	4.28	1.14	5.42
7500	Ceiling suspension systems, for tile, "T" bar, class "A", 2' x 4' grid	1.03	.85	1.88
7600	2' x 2' grid	1.28	1.05	2.33
7700	Concealed "Z" bar, 12" module	.95	1.31	2.26
7800				
7900	Plaster/drywall, 3/4" channels, steel furring, 16" O.C.	.41	2.25	2.66
8000	24" O.C.	.27	1.55	1.82
8100	1-1/2" channels, 16" O.C.	.55	2.51	3.06
8200	24" O.C.	.37	1.67	2.04
8300	Ceiling framing, 2" x 4" studs, 16" O.C.	.39	1.13	1.52
8400	24" O.C.	.26	.76	1.02

B3020 Roof Openings

This page illustrates and describes a roof hatch system. Lines within System Components give the unit price and total cost each for this system. Prices for systems are on Lines B3020 230 1400 thru 2400. Both material quantities and labor costs have been adjusted for the system listed.

Factors: To adjust for job conditions other than normal working situations use Lines B3020 230 2900 thru 4000.

Example: You are to install the system and cut and patch to match existing construction. Use line B3020 230 3000 and apply these percentages to the appropriate MAT. and INST. costs.

System Components	QUANTITY	UNIT	COST EACH MAT.	COST EACH INST.	COST EACH TOTAL
Roof hatch, 2'-6" x 3'-0", aluminum, curb and cover included.					
Through steel construction.					
Roof hatch, 2'-6" x 3'-0", aluminum	1.000	Ea.	990	264	1,254
Cutout decking	11.000	L.F.	4.73	27.28	32.01
Frame opening	44.000	L.F.	41.80	407.44	449.24
Flashing, 16 oz. copper	16.000	S.F.	144	89.60	233.60
Cant strip, 4 x 4, mineral fiber	12.000	L.F.	7.32	23.88	31.20
TOTAL			1,187.85	812.20	2,000.05

B3020 230	Roof Hatches, Skylights	COST EACH MAT.	COST EACH INST.	COST EACH TOTAL
1400	Roof hatch, aluminum, curb and cover, 2'-6" x 3'-0"	1,200	815	2,015
1500	2'-6" x 4'-6"	1,525	990	2,515
1600	2'-6" x 8'-0"	2,750	1,425	4,175
1700	Skylight, plexiglass dome, with curb mounting, 30" x 32"	735	770	1,505
1800	30" x 45"	825	960	1,785
1900	40" x 45"	1,025	1,275	2,300
2000	Smoke hatch, unlabeled, 2'-6" x 3'-0"	1,425	880	2,305
2100	2'-6" x 4'-6"	1,850	1,050	2,900
2200	2'-6" x 8'-0"	2,975	1,475	4,450
2300				
2400	For steel ladder, add per vertical linear foot	46.50	38	84.50
2450				
2600				
2700				
2900	Cut & patch to match existing construction, add, minimum	2%	3%	
3000	Maximum	5%	9%	
3100	Dust protection, add, minimum	1%	2%	
3200	Maximum	4%	11%	
3300	Equipment usage curtailment, add, minimum	1%	1%	
3400	Maximum	3%	10%	
3500	Material handling & storage limitation, add, minimum	1%	1%	
3600	Maximum	6%	7%	
3700	Protection of existing work, add, minimum	2%	2%	
3800	Maximum	5%	7%	
3900	Shift work requirements, add, minimum		5%	
4000	Maximum		30%	

B3020 Roof Openings

B3020 240	Selective Price Sheet	COST PER UNIT		
		MAT.	INST.	TOTAL
0100	Downspouts per L.F., aluminum, enameled .024" thick, 2" x 3"	2.39	4.34	6.73
0200	3" x 4"	2.52	5.60	8.12
0300	Round .025" thick, 3" diam.	2.34	4.11	6.45
0400	4" diam.	3.53	5.60	9.13
0500	Copper, round 16 oz. stock, 2" diam.	8.80	4.11	12.91
0600	3" diam.	9.90	4.11	14.01
0700	4" diam.	11.75	5.40	17.15
0800	5" diam.	16.35	6	22.35
0900	Rectangular, 2" x 3"	14.60	4.11	18.71
1000	3" x 4"	15.20	5.40	20.60
1100	Lead coated copper, round, 2" diam.	28	4.11	32.11
1200	3" diam.	27	4.11	31.11
1300	4" diam.	30.50	5.40	35.90
1400	5" diam.	28.50	6	34.50
1500	Rectangular, 2" x 3"	30.50	4.11	34.61
1600	3" x 4"	44.50	5.40	49.90
1700	Steel galvanized, round 28 gauge, 3" diam.	2.53	4.11	6.64
1800	4" diam.	2.41	5.40	7.81
1900	5" diam.	4.25	6	10.25
2000	6" diam.	4.09	7.45	11.54
2100	Rectangular, 2" x 3"	4.38	4.11	8.49
2200	3" x 4"	5.30	5.40	10.70
2300	Elbows, aluminum, round, 3" diam.	3.47	7.80	11.27
2400	4" diam.	7.70	7.80	15.50
2500	Rectangular, 2" x 3"	1.10	7.80	8.90
2600	3" x 4"	3.26	7.80	11.06
2700	Copper, round 16 oz., 2" diam.	12	7.80	19.80
2800	3" diam.	11.35	7.80	19.15
2900	4" diam.	14.20	7.80	22
3100	Rectangular, 2" x 3"	11.25	7.80	19.05
3200	3" x 4"	16.70	7.80	24.50
3300	Drip edge per L.F., aluminum, 5" wide	.65	1.70	2.35
3400	8" wide	1.63	1.70	3.33
3500	28" wide	8.80	6.80	15.60
3600				
3700	Steel galvanized, 5" wide	.68	1.70	2.38
3800	8" wide	.92	1.70	2.62
3900				
4000				
4100				
4200				
4300	Flashing 12" wide per S.F., aluminum, mill finish, .013" thick	1.11	4.46	5.57
4400	.019" thick	1.65	4.46	6.11
4500	.040" thick	2.56	4.46	7.02
4600	.050" thick	3.54	4.46	8
4700	Copper, mill finish, 16 oz.	9	5.60	14.60
4800	20 oz.	11.50	5.90	17.40
4900	24 oz.	16.80	6.15	22.95
5000	32 oz.	23	6.45	29.45
5100	Lead, 2.5 lb./S.F., 12" wide	6.70	4.79	11.49
5200	Over 12" wide	4.46	4.79	9.25
5900	Polyvinyl chloride, black, .010" thick	.30	2.27	2.57
6000	.020" thick	.29	2.27	2.56
6100	.030" thick	.36	2.27	2.63
6200	.056" thick	.97	2.27	3.24
6300	Steel, galvanized, 20 gauge	1.34	4.98	6.32
6400	30 gauge	.68	4.04	4.72
6500	Stainless, 32 gauge, .010" thick	3.86	4.17	8.03

B3020 Roof Openings

B3020 240	Selective Price Sheet	COST PER UNIT		
		MAT.	INST.	TOTAL
6600	28 gauge, .015" thick	5.50	4.17	9.67
6700	26 gauge, .018" thick	5.60	4.17	9.77
6800	24 gauge, .025" thick	6.40	4.17	10.57
6900	Gutters per L.F., aluminum, 5" box, .027" thick	3.25	6.45	9.70
7000	.032" thick	4.17	6.25	10.42
7100	Copper, half round, 4" wide	9.50	6.25	15.75
7200	6" wide	11.95	6.60	18.55
7300	Steel, 26 gauge galvanized, 5" wide	2.93	6.25	9.18
7400	6" wide	2.90	6.25	9.15
7500	Wood, treated hem-fir, 3" x 4"	13.95	6.80	20.75
7600	4" x 5"	27	6.80	33.80
7700	Reglet per L.F., aluminum, .025" thick	2.02	3.03	5.05
7800	Copper, 10 oz.	7.20	3.03	10.23
7900	Steel, galvanized, 24 gauge	1.51	3.03	4.54
8000	Stainless, .020" thick	4.63	3.03	7.66
8100	Counter flashing 12" wide per L.F., aluminum, .032" thick	2.30	5.20	7.50
8200	Aluminum, .025" thick	2.30	5.20	7.50
8300	Steel, galvanized, 24 gauge	1.46	5.20	6.66
8400	Stainless, .020" thick	7.50	5.20	12.70

C1010 Partitions

This page illustrates and describes a concrete block wall system including concrete block, horizontal reinforcing alternate courses, mortar, and tooled joints both sides. Lines within System Components give the unit price and total price per square foot for this system. Prices for concrete block wall systems are on Line Items C1010 106 0800 thru 2100. Both material quantities and labor costs have been adjusted for the system listed.

Factors: To adjust for job conditions other than normal working situations use Lines C1010 106 2900 thru 4000.

Example: You are to install the system and protect all existing work. Go to Line C1010 106 3600 and apply these percentages to the appropriate MAT. and INST. costs.

System Components	QUANTITY	UNIT	COST PER S.F. MAT.	COST PER S.F. INST.	COST PER S.F. TOTAL
Concrete block partition, including horizontal reinforcing every second					
Course, mortar, tooled joints both sides.					
8″ x 16″ concrete block, normal weight, 4″ thick	1.000	S.F.	2.27	7.20	9.47
Horizontal reinforcing every second course	.750	L.F.	.22	.17	.39
TOTAL			2.49	7.37	9.86

C1010 106	Partitions, Concrete Block	MAT.	INST.	TOTAL
0800	8″ x 16″ concrete block, normal weight, 4″ thick	2.49	7.35	9.84
0900	6″ thick	3.10	7.90	11
1000	8″ thick	3.11	8.40	11.51
1100	10″ thick	3.68	8.75	12.43
1200	12″ thick	5.30	10.85	16.15
1300	8″ x 16″ concrete block, lightweight, 4″ thick	2.42	7.20	9.62
1400	6″ thick	3.26	7.70	10.96
1500	8″ thick	4.34	8.20	12.54
1600	10″ thick	4.82	8.50	13.32
1700	12″ thick	4.93	10.55	15.48
1800	8″ x 16″ glazed concrete block, 4″ thick	14	9.15	23.15
1900	8″ thick	17.10	10.15	27.25
2000	Structural facing tile, 6T series, glazed 2 sides, 4″ thick	18.20	16.05	34.25
2100	6″ thick	23	16.90	39.90
2200				
2300				
2400				
2500				
2600				
2900	Cut & patch to match existing construction, add, minimum	2%	3%	
3000	Maximum	5%	9%	
3100	Dust protection, add, minimum	1%	2%	
3200	Maximum	4%	11%	
3300	Material handling & storage limitation, add, minimum	1%	1%	
3400	Maximum	6%	7%	
3500	Protection of existing work, add, minimum	2%	2%	
3600	Maximum	5%	7%	
3700	Shift work requirements, add, minimum		5%	
3800	Maximum		30%	
3900	Temporary shoring and bracing, add, minimum	2%	5%	
4000	Maximum	5%	12%	

C1010 Partitions

This page illustrates and describes a wood stud partition system including wood studs with plates, gypsum plasterboard – taped and finished, insulation, baseboard and painting. Lines within System Components give the unit price and total price per square foot for this system. Prices for wood stud partition systems are on Line Items C1010 132 1300 thru 2700. Both material quantities and labor costs have been adjusted for the system listed.

Factors: To adjust for job conditions other than normal working situations use Lines C1010 132 2900 thru 4000.

Example: You are to install the system where material handling and storage present a serious problem. Go to Line C1010 132 3400 and apply these percentages to the appropriate MAT. and INST. costs.

System Components	QUANTITY	UNIT	COST PER S.F. MAT.	COST PER S.F. INST.	COST PER S.F. TOTAL
Wood stud wall, 2"x4", 16" O.C., dbl. top plate, sngl bot. plate, 5/8" dwl.					
taped, finished and painted on 2 faces, insulation, baseboard, wall 8' high.					
Wood studs, 2" x 4", 16" O.C., 8' high	1.000	S.F.	.59	1.70	2.29
Gypsum drywall, 5/8" thick	2.000	S.F.	.72	1.36	2.08
Taping and finishing	2.000	S.F.	.12	1.36	1.48
Insulation, 3-1/2" fiberglass batts	1.000	S.F.	.52	.50	1.02
Baseboard	.250	L.F.	.35	.71	1.06
Paint baseboard, primer + 2 coats	.250	L.F.	.06	.42	.48
Painting, roller, 2 coats	2.000	S.F.	.46	1.48	1.94
TOTAL			2.82	7.53	10.35

C1010 132	Partitions, Wood Stud	COST PER S.F. MAT.	COST PER S.F. INST.	COST PER S.F. TOTAL
1300	2" x 3" studs, 8' high, 16" O.C.	2.78	7.45	10.23
1400	24" O.C.	2.66	7.10	9.76
1500	10' high, 16" O.C.	2.75	7.10	9.85
1600	24" O.C.	2.63	6.85	9.48
1650	2" x 4" studs, 8' high, 16" O.C.	2.82	7.55	10.37
1700	24" O.C.	2.69	7.20	9.89
1800	10' high, 16" O.C.	2.78	7.20	9.98
1900	24" O.C.	2.65	6.90	9.55
2000	12' high, 16" O.C.	2.73	7.20	9.93
2100	24" O.C.	2.61	6.90	9.51
2200	2" x 6" studs, 8' high, 16" O.C.	3.08	7.70	10.78
2300	24" O.C.	2.89	7.30	10.19
2400	10' high, 16" O.C.	3.02	7.35	10.37
2500	24" O.C.	2.84	7	9.84
2600	12' high, 16" O.C.	2.95	7.40	10.35
2700	24" O.C.	2.77	7.05	9.82
2900	Cut & patch to match existing construction, add, minimum	2%	3%	
3000	Maximum	5%	9%	
3100	Dust protection, add, minimum	1%	2%	
3200	Maximum	4%	11%	
3300	Material handling & storage limitation, add, minimum	1%	1%	
3400	Maximum	6%	7%	
3500	Protection of existing work, add, minimum	2%	2%	
3600	Maximum	5%	7%	
3700	Shift work requirements, add, minimum		5%	
3800	Maximum		30%	
3900	Temporary shoring and bracing, add, minimum	2%	5%	
4000	Maximum	5%	12%	

C10 Interior Construction

C1010 Partitions

This page illustrates and describes a non-load bearing metal stud partition system including metal studs with runners, gypsum plasterboard, taped and finished, insulation, baseboard and painting. Lines within System Components give the unit price and total price per square foot for this system. Prices for non-load bearing metal stud partition systems are on Line Items C1010 134 1300 thru 2300. Both material quantities and labor costs have been adjusted for the system listed.

Factors: To adjust for job conditions other than normal working situations use Lines C1010 134 2900 thru 4000.

Example: You are to install the system and cut and patch to match existing construction. Go to Line C1010 134 3000 and apply these percentages to the appropriate MAT. and INST. costs.

System Components	QUANTITY	UNIT	COST PER S.F.		
			MAT.	INST.	TOTAL
Non-load bearing metal studs, including top & bottom runners, 5/8" drywall, Taped, finished and painted 2 faces, insulation, painted baseboard.					
Metal studs, 25 ga., 3-5/8" wide, 24" O.C.	1.000	S.F.	.33	.92	1.25
Gypsum drywall, 5/8" thick	2.000	S.F.	.72	1.36	2.08
Taping & finishing	2.000	S.F.	.12	1.36	1.48
Insulation, 3-1/2" fiberglass batts	1.000	S.F.	.52	.50	1.02
Baseboard	.200	L.F.	.28	.57	.85
Paint baseboard, primer + 2 coats	.200	L.F.	.05	.33	.38
Painting, roller work, 2 coats	2.000	S.F.	.46	1.48	1.94
TOTAL			2.48	6.52	9

C1010 134	Partitions, Metal Stud, NLB	COST PER S.F.		
		MAT.	INST.	TOTAL
1300	Non-load bearing, 25 ga., 24" O.C., 2-1/2" wide	2.43	6.50	8.93
1350	3-5/8" wide	2.48	6.50	8.98
1400	6" wide	2.59	6.55	9.14
1500	16" O.C., 2-1/2" wide	2.50	6.75	9.25
1600	3-5/8" wide	2.56	6.75	9.31
1700	6" wide	2.70	6.80	9.50
1800	20 ga., 24" O.C., 2-1/2" wide	2.51	6.75	9.26
1900	3-5/8" wide	2.66	6.75	9.41
2000	6" wide	2.77	6.75	9.52
2100	16" O.C., 2-1/2" wide	2.60	7.05	9.65
2200	3-5/8" wide	2.79	7.05	9.84
2300	6" wide	2.93	7.05	9.98
2400				
2500				
2600				
2700				
2900	Cut & patch to match existing construction, add, minimum	2%	3%	
3000	Maximum	5%	9%	
3100	Dust protection, add, minimum	1%	2%	
3200	Maximum	4%	11%	
3300	Material handling & storage limitation, add, minimum	1%	1%	
3400	Maximum	6%	7%	
3500	Protection of existing work, add, minimum	2%	2%	
3600	Maximum	5%	7%	
3700	Shift work requirements, add, minimum		5%	
3800	Maximum		30%	
3900	Temporary shoring and bracing, add, minimum	2%	5%	
4000	Maximum	5%	12%	

C1010 Partitions

This page illustrates and describes a drywall system including gypsum plasterboard, taped and finished, metal studs with runners, insulation, baseboard and painting. Lines within System Components give the unit price and total price per square foot for this system. Prices for drywall systems are on Line Items C1010 136 1200 thru 1900. Both material quantities and labor costs have been adjusted for the system listed.

Factors: To adjust for job conditions other than normal working situations use Lines C1010 136 2900 thru 4000.

Example: You are to install the system and control dust in the work area. Go to Line C1010 136 3100 and apply these percentages to the appropriate MAT. and INST. costs.

System Components	QUANTITY	UNIT	COST PER S.F. MAT.	COST PER S.F. INST.	COST PER S.F. TOTAL
Gypsum drywall, taped, finished and painted 2 faces, galvanized metal studs					
Including top & bottom runners, insulation, painted baseboard, wall 10'high					
Gypsum drywall, 5/8" thick, standard	2.000	S.F.	.72	1.36	2.08
Taping and finishing	2.000	S.F.	.12	1.36	1.48
Metal studs, 20 ga., 3-5/8" wide, 24" O.C.	1.000	S.F.	.42	1.15	1.57
Insulation, 3-1/2" fiberglass batts	1.000	S.F.	.52	.50	1.02
Baseboard	.200	L.F.	.28	.57	.85
Paint baseboard, primer + 2 coats	.200	L.F.	.03	.14	.17
Painting, roller 2 coats	2.000	S.F.	.34	1.42	1.76
TOTAL			2.43	6.50	8.93

C1010 136	Partitions, Drywall	COST PER S.F. MAT.	COST PER S.F. INST.	COST PER S.F. TOTAL
1200	Gypsum drywall, 5/8" thick, standard	2.43	6.50	8.93
1300	Fire resistant	2.46	6.25	8.71
1400	Water resistant	2.58	6.25	8.83
1500	1/2" thick, standard	2.34	6.25	8.59
1600	Fire resistant	2.46	6.25	8.71
1700	Water resistant	2.52	6.25	8.77
1800	3/8" thick, vinyl faced, standard	2.76	7.95	10.71
1900	5/8" thick, vinyl faced, fire resistant	3.72	7.95	11.67
2000				
2100				
2200				
2300				
2400				
2500				
2600				
2700				
2900	Cut & patch to match existing construction, add, minimum	2%	3%	
3000	Maximum	5%	9%	
3100	Dust protection, add, minimum	1%	2%	
3200	Maximum	4%	11%	
3300	Material handling & storage limitation, add, minimum	1%	1%	
3400	Maximum	6%	7%	
3500	Protection of existing work, add, minimum	2%	2%	
3600	Maximum	5%	7%	
3700	Shift work requirements, add, minimum		5%	
3800	Maximum		30%	
3900	Temporary shoring and bracing, add, minimum	2%	5%	
4000	Maximum	5%	12%	

C1010 Partitions

This page illustrates and describes a load bearing metal stud wall system including metal studs, sheetrock–taped and finished, insulation, baseboard and painting. Lines within System Components give the unit price and total price per square foot for this system. Prices for load bearing metal stud wall systems are on Line Items C1010 138 1500 thru 2500. Both material quantities and labor costs have been adjusted for the system listed.

Factors: To adjust for job conditions other than normal working situations use Lines C1010 138 2900 thru 4000.

Example: You are to install the system using temporary shoring and bracing. Go to Line C1010 138 3900 and apply these percentages to the appropriate MAT. and INST. costs.

System Components			COST PER S.F.		
	QUANTITY	UNIT	MAT.	INST.	TOTAL
Load bearing, 18 ga., 3-5/8", galvanized metal studs, 24" O.C.,including Top and bottom runners, 1/2" drywall, taped, finished and painted 2 Faces, 3" insulation, and painted baseboard, wall 10' high.					
Metal studs, 24" OC, 18 ga., 3-5/8" wide, galvanized	1.000	S.F.	.77	1.30	2.07
Gypsum drywall, 1/2" thick	2.000	S.F.	.72	1.36	2.08
Taping and finishing	2.000	S.F.	.12	1.36	1.48
Insulation, 3-1/2" fiberglass batts	1.000	S.F.	.52	.50	1.02
Baseboard	.200	L.F.	.28	.57	.85
Paint baseboard, primer + 2 coats	.200	L.F.	.05	.33	.38
Painting, roller 2 coats	2.000	S.F.	.46	1.48	1.94
TOTAL			2.92	6.90	9.82

C1010 138	Partitions, Metal Stud, LB	COST PER S.F.		
		MAT.	INST.	TOTAL
1500	Load bearing, 18 ga., 24" O.C., 2-1/2" wide	2.80	6.90	9.70
1550	3-5/8" wide	2.92	6.90	9.82
1600	6" wide	3.18	6.95	10.13
1700	16" O.C. 2-1/2" wide	2.96	7.20	10.16
1800	3-5/8" wide	3.11	7.25	10.36
1900	6" wide	3.44	7.30	10.74
2000	16 ga., 24" O.C., 2-1/2" wide	2.95	7.05	10
2100	3-5/8" wide	3.09	7.10	10.19
2200	6" wide	3.39	7.15	10.54
2300	16" O.C., 2-1/2" wide	3.14	7.40	10.54
2400	3-5/8" wide	3.32	7.45	10.77
2500	6" wide	3.70	7.55	11.25
2600				
2700				
2900	Cut & patch to match existing construction, add, minimum	2%	3%	
3000	Maximum	5%	9%	
3100	Dust protection, add, minumum	1%	2%	
3200	Maximum	4%	11%	
3300	Material handling & storage limitation, add, minimum	1%	1%	
3400	Maximum	6%	7%	
3500	Protection of existing work, add, minimum	2%	2%	
3600	Maximum	5%	7%	
3700	Shift work requirements, add, minimum		5%	
3800	Maximum		30%	
3900	Temporary shoring and bracing, add, minimum	2%	5%	
4000	Maximum	5%	12%	

C1010 Partitions

This page illustrates and describes a plaster and lath system including gypsum plaster, gypsum lath, wood studs with plates, insulation, baseboard and painting. Lines within System Components give the unit price and price per square foot for this system. Prices for plaster and lath systems are on Line Items C1010 148 1400 thru 2500. Both material quantities and labor costs have been adjusted for the system listed.

Factors: To adjust for job conditions other than normal working situations use Lines C1010 148 2900 thru 4000.

Example: You are to install the system during evening hours only. Go to Line C1010 148 3800 and apply this percentage to the appropriate INST. costs.

System Components	QUANTITY	UNIT	COST PER S.F.		
			MAT.	INST.	TOTAL
Gypsum plaster, 2 coats, over 3/8" lath, 2 faces, 2" x 4" wood					
Stud partition, 24"O.C. incl. double top plate, single bottom plate, 3-1/2"					
Insulation, baseboard and painting.					
Gypsum plaster, 2 coats	.220	S.Y.	.94	6.43	7.37
Lath, gypsum, 3/8" thick	.220	S.Y.	.85	1.68	2.53
Wood studs, 2" x 4", 24" O.C.	1.000	S.F.	.42	1.09	1.51
Insulation, 3-1/2" fiberglass batts	1.000	S.F.	.52	.50	1.02
Baseboard, 9/16" x 3-1/2"	.200	L.F.	.28	.57	.85
Paint baseboard, primer + 2 coats	.200	L.F.	.03	.14	.17
Paint, 2 coats	2.000	S.F.	.46	1.48	1.94
TOTAL			3.50	11.89	15.39

C1010 148	Partitions, Plaster & Lath	COST PER S.F.		
		MAT.	INST.	TOTAL
1400	Gypsum plaster, 2 coats	3.50	11.85	15.35
1500	3 coats	3.90	13.30	17.20
1600	Perlite plaster, 2 coats	4	12.80	16.80
1700	3 coats	4.18	14.55	18.73
1800				
1900				
2000	For alternate lath systems:			
2100	Gypsum lath, 1/2" thick	3.80	12.90	16.70
2200	Foil back, 3/8" thick	3.89	13.05	16.94
2300	1/2" thick	3.93	13.15	17.08
2400	Metal lath, 2.5 Lb. diamond	4.13	12.80	16.93
2500	3.4 Lb. diamond	4.19	13.05	17.24
2600				
2700				
2900	Cut & patch to match existing construction, add, minimum	2%	3%	
3000	Maximum	5%	9%	
3100	Dust protection, add, minimum	1%	2%	
3200	Maximum	4%	11%	
3300	Material handling & storage limitation, add, minimum	1%	1%	
3400	Maximum	6%	7%	
3500	Protection of existing work, add, minimum	2%	2%	
3600	Maximum	5%	7%	
3700	Shift work requirements, add, minimum		5%	
3800	Maximum		30%	
3900	Temporary shoring and bracing, add, minimum	2%	5%	
4000	Maximum	5%	12%	

C10 Interior Construction

C1010 Partitions

C1010 170	Selective Price Sheet	COST PER S.F.		
		MAT.	INST.	TOTAL
0100	Studs			
0200				
0300	24" O.C. metal, 10' high wall, including			
0400	Top and bottom runners			
0500	Non load bearing, galvanized 25 Ga., 1-5/8" wide	.21	.90	1.11
0600	2-1/2" wide	.28	.91	1.19
0700	3-5/8" wide	.33	.92	1.25
0800	4" wide	.37	.92	1.29
0900	6" wide	.44	.94	1.38
1000				
1100	Galvanized 20 Ga., 2-1/2" wide	.36	1.14	1.50
1200	3-5/8" wide	.42	1.15	1.57
1300	4" wide	.51	1.15	1.66
1400	6" wide	.62	1.17	1.79
1500	Load bearing, painted 18 Ga., 2-1/2" wide	.65	1.28	1.93
1600	3-5/8" wide	.77	1.30	2.07
1700	4" wide	.68	1.33	2.01
1800	6" wide	1.03	1.35	2.38
1900	Galvanized 18 Ga., 2-1/2" wide	.65	1.28	1.93
2000	3-5/8" wide	.77	1.30	2.07
2100	4" wide	.68	1.33	2.01
2200	6" wide	1.03	1.35	2.38
2300	Galvanized 16 Ga., 2-1/2" wide	.80	1.45	2.25
2400	3-5/8" wide	.94	1.48	2.42
2500	4" wide	.99	1.52	2.51
2600	6" wide	1.24	1.55	2.79
2700				
2800				
2900	24" O.C. wood, 10' high wall, including			
3000	Double top plate and shoe			
3100	2" x 4"	.49	1.29	1.78
3200	2" x 6"	.70	1.37	2.07
3300				
3400				
3500	Furring, 24" O.C. 10' high wall, metal, 3/4" channels	.27	1.55	1.82
3600	1-1/2" channels	.37	1.67	2.04
3700	Wood, on wood, 1" x 2" strips	.16	.62	.78
3800	1" x 3" strips	.24	.62	.86
3900	On masonry, 1" x 2" strips	.18	.69	.87
4000	1" x 3" strips	.26	.69	.95
4100	On concrete, 1" x 2" strips	.18	1.31	1.49
4200	1" x 3" strips	.26	1.31	1.57
4300	Studs			
4400				
4500	16" O.C., metal, 10' high wall, including			
4600	Top and bottom runners			
4700	Non load bearing, galvanized 25 Ga., 1-5/8" wide	.29	1.38	1.67
4800	2-1/2" wide	.38	1.39	1.77
4900	3-5/8" wide	.45	1.42	1.87
5000	4" wide	.50	1.43	1.93
5100	6" wide	.60	1.45	2.05
5200				
5300	Galvanized 20 Ga., 2-1/2" wide	.50	1.75	2.25
5400	3-5/8" wide	.57	1.77	2.34
5500	4" wide	.70	1.79	2.49
5600	6" wide	.84	1.82	2.66
5700	Load bearing, painted 18 Ga., 2-1/2" wide	.88	1.77	2.65
5800	3-5/8" wide	1.04	1.80	2.84

For customer support on your Commercial Renovation Costs with RSMeans data, call 800.448.8182.

C1010 Partitions

C1010 170	Selective Price Sheet	COST PER S.F.		
		MAT.	INST.	TOTAL
5900	4″ wide	1.08	1.84	2.92
6000	6″ wide	1.39	1.87	3.26
6100	Galvanized 18 Ga., 2-1/2″ wide	.88	1.77	2.65
6200	3-5/8″ wide	1.04	1.80	2.84
6300	4″ wide	1.08	1.84	2.92
6400	6″ wide	1.39	1.87	3.26
6500	Galvanized 16 Ga., 2-1/2″ wide	1.10	2	3.10
6600	3-5/8″ wide	1.29	2.05	3.34
6700	4″ wide	1.36	2.10	3.46
6800	6″ wide	1.70	2.15	3.85
6900				
7000				
7100	16″ O.C., wood, 10′ high wall, including			
7200	Double top plate and shoe			
7300	2″ x 4″	.59	1.48	2.07
7400	2″ x 6″	.70	1.33	2.03
7500				
7600				
7700	Furring, 16″ O.C. 10′ high wall, metal, 3/4″ channels	.41	2.25	2.66
7800	1-1/2″ channels	.55	2.51	3.06
7900	Wood, on wood, 1″ x 2″ strips	.24	.93	1.17
8000	1″ x 3″ strips	.36	.93	1.29
8100	On masonry, 1″ x 2″ strips	.26	1.04	1.30
8200	1″ x 3″ strips	.39	1.04	1.43
8300	On concrete, 1″ x 2″ strips	.26	1.97	2.23
8400	1″ x 3″ strips	.39	1.97	2.36

C1010 180	Selective Price Sheet, Drywall & Plaster	COST PER S.F.		
		MAT.	INST.	TOTAL
0100	Lath, gypsum perforated			
0200				
0300	Lath, gypsum, perforated, regular, 3/8" thick	.43	.85	1.28
0400	1/2" thick	.33	1.03	1.36
0500	Fire resistant, 3/8" thick	.31	1.03	1.34
0600	1/2" thick	.33	1.12	1.45
0700	Foil back, 3/8" thick	.37	.97	1.34
0800	1/2" thick	.40	1.03	1.43
0900				
1000	Metal lath			
1100	Metal lath, diamond, painted, 2.5 lb.	.49	.85	1.34
1200	3.4 lb.	.52	.97	1.49
1300	Rib painted, 2.75 lb	.43	.97	1.40
1400	3.40 lb	.59	1.03	1.62
1500				
1600				
1700	Plaster, gypsum, 2 coats	.48	3.25	3.73
1800	3 coats	.68	3.94	4.62
1900	Perlite/vermiculite, 2 coats	.73	3.71	4.44
2000	3 coats	.82	4.57	5.39
2100	Bondcrete, 1 coat	.52	1.70	2.22
2200				
2500				
2600				
2700	Drywall, standard, 3/8" thick, no finish included	.39	.68	1.07
2800	1/2" thick, no finish included	.36	.68	1.04
2900	Taped and finished	.42	1.41	1.83
3000	5/8" thick, no finish included	.36	.68	1.04
3100	Taped and finished	.42	1.41	1.83
3200	Fire resistant, 1/2" thick, no finish included	.42	.68	1.10
3300	Taped and finished	.47	1.41	1.88
3400	5/8" thick, no finish included	.42	.68	1.10
3500	Taped and finished	.47	1.41	1.88
3600	Water resistant, 1/2" thick, no finish included	.45	.68	1.13
3700	Taped and finished	.51	1.41	1.92
3800	5/8" thick, no finish included	.48	.68	1.16
3900	Taped and finished	.54	1.41	1.95
4000	Finish, instead of taping			
4100	For thin coat plaster, add	.11	.86	.97
4200	Finish, textured spray, add	.04	.82	.86

C1010 Partitions

This page illustrates and describes a movable office partition system including demountable office partitions of various styles and sizes based on a cost per square foot basis. Prices for movable office partition systems are on Line Items C1010 210 0300 thru 3600.

C1010 210	Partitions, Movable Office	COST PER S.F.		
		MAT.	INST.	TOTAL
0250	Office partition, demountable, no deduction for door opening, add for doors			
0300	Air wall, cork finish, semi acoustic, 1-5/8" thick, minimum	43.50	4.19	47.69
0400	Maximum	54.50	7.15	61.65
0500	Acoustic, 2" thick, minimum	46.50	4.47	50.97
0600	Maximum	65	6.05	71.05
0700				
0800				
0900	In-plant modular office system, w/prehung steel door			
1000	3" thick honeycomb core panels			
1100	12' x 12', 2 wall	10.95	.57	11.52
1200	4 wall	23.50	1.71	25.21
1300				
1400				
1500	Gypsum, demountable, 3" to 3-3/4" thick x 9' high, vinyl clad	7.45	3.16	10.61
1600	Fabric clad	18.55	3.44	21.99
1700	1.75 system, vinyl clad hardboard, paper honeycomb core panel			
1800	1-3/4" to 2-1/2" thick x 9' high	12.45	3.16	15.61
1900	Unitized gypsum panel, 2" to 2-1/2" thick x 9' high, vinyl clad	16.10	3.16	19.26
2000	Fabric clad	26.50	3.44	29.94
2100				
2200				
2300	Unitized mineral fiber panel system, 2-1/4" thick x 9' high			
2400	Vinyl clad mineral fiber	16	3.16	19.16
2500	Fabric clad mineral fiber	24	3.44	27.44
2600				
2700	Movable steel walls, modular system			
2800	Unitized panels, 48" wide x 9' high			
2900	Baked enamel, pre-finished	18.10	2.50	20.60
3000	Fabric clad	26	2.72	28.72
3100	For acoustical partitions, add, unsealed	2.62		2.62
3200	Sealed	12.20		12.20
3300				
3400				
3500	Note: For door prices, see Div. 08			
3600	For door hardware prices, see Div. 08			
3700				
3800				
3900				
4000				

C10 Interior Construction

C1020 Interior Doors

This page illustrates and describes flush interior door systems including hollow core door, jamb, header and trim with hardware. Lines within System Components give the unit price and total price on a cost each basis for this system. Prices for flush interior door systems are on Line items C1020 106 1000 thru 2400. Both material quantities and labor costs have been adjusted for the system listed.

Factors: To adjust for job conditions other than normal working situations use Lines C1020 106 2700 thru 3800.

Example: You are to install the system in an area where dust protection is vital. Go to Line C1020 106 3000 and apply these percentages to the appropriate MAT. and INST. costs.

System Components			COST EACH		
	QUANTITY	UNIT	MAT.	INST.	TOTAL
Single hollow core door, include jamb, header, trim and hardware, painted.					
Hollow core Lauan, 1-3/8" thick, 2'-0" x 6'-8", painted	1.000	Ea.	69.50	75.50	145
Paint door and frame, 1 coat	1.000	Ea.	2.79	33	35.79
Wood jamb, 4-9/16" deep	1.000	Set	95.20	58.08	153.28
Trim, casing	1.000	Set	49.60	87.04	136.64
Hardware, hinges	1.000	Set	27.50		27.50
Hardware, lockset	1.000	Set	45.50	42.50	88
TOTAL			290.09	296.12	586.21

C1020 106	Doors, Interior Flush, Wood	COST EACH		
		MAT.	INST.	TOTAL
1000	Lauan (Mahogany) hollow core, 1-3/8" x 2'-0" x 6'-8"	290	296	586
1100	2'-6" x 6'-8"	305	300	605
1200	2'-8" x 6'-8"	310	305	615
1300	3'-0" x 6'-8"	330	315	645
1500	Birch, hollow core, 1-3/8" x 2'-0" x 6'-8"	315	296	611
1600	2'-6" x 6'-8"	320	300	620
1700	2'-8" x 6'-8"	330	305	635
1800	3'-0" x 6'-8"	335	315	650
1900	Solid core, pre-hung, 1-3/8" x 2'-6" x 6'-8"	570	305	875
2000	2'-8" x 6'-8"	460	310	770
2100	3'-0" x 6'-8"	580	320	900
2200				
2400	For metal frame instead of wood, add	50%	20%	
2600				
2700	Cut & patch to match existing construction, add, minimum	2%	3%	
2800	Maximum	5%	9%	
2900	Dust protection, add, minimum	1%	2%	
3000	Maximum	4%	11%	
3100	Equipment usage curtailment, add, minimum	1%	1%	
3200	Maximum	3%	10%	
3300	Material handling & storage limitation, add, minimum	1%	1%	
3400	Maximum	6%	7%	
3500	Protection of existing work, add, minimum	2%	2%	
3600	Maximum	5%	7%	
3700	Shift work requirements, add, minimum		5%	
3800	Maximum		30%	
3900				

C1020 Interior Doors

This page illustrates and describes interior, solid and louvered door systems including a pine panel door, wood jambs, header, and trim with hardware. Lines within System Components give the unit price and total price on a cost each basis for this system. Prices for interior, solid and louvered systems are on Line Items C1020 108 1100 thru 2400. Both material quantities and labor costs have been adjusted for the system listed.

Factors: To adjust for job conditions other than normal working situations use Lines C1020 108 2900 thru 4000.

Example: You are to install the system during night hours only. Go to Line C1020 108 4000 and apply these percentages to the appropriate INST. costs.

System Components	QUANTITY	UNIT	COST EACH MAT.	COST EACH INST.	COST EACH TOTAL
Single interior door, including jamb, header, trim and hardware, painted.					
Solid pine panel door, 1-3/8″ thick, 2′-0″ x 6′-8″	1.000	Ea.	276	75.50	351.50
Paint door and frame, 1 coat	1.000	Ea.	2.79	33	35.79
Wooden jamb, 4-5/8″ deep	1.000	Set	95.20	58.08	153.28
Trim, casing	1.000	Set	49.60	87.04	136.64
Hardware, hinges	1.000	Set	27.50		27.50
Hardware, lockset	1.000	Set	45.50	42.50	88
TOTAL			496.59	296.12	792.71

C1020 108	Doors, Interior Solid & Louvered	COST EACH MAT.	COST EACH INST.	COST EACH TOTAL
1000				
1100	Solid pine, painted raised panel, 1-3/8″ x 2′-0″ x 6′-8″	495	296	791
1200	2′-6″ x 6′-8″	515	300	815
1300	2′-8″ x 6′-8″	525	305	830
1400	3′-0″ x 6′-8″	555	315	870
1500				
1600	Louvered pine, painted 1′-6″ x 6′-8″	395	292	687
1700	2′-0″ x 6′-8″	420	300	720
1800	2′-6″ x 6′-8″	450	305	755
1900	3′-0″ x 6′-8″	490	315	805
2200	For prehung door, deduct	5%	30%	
2300				
2400	For metal frame instead of wood, add	50%	20%	
2500				
2900	Cut & patch to match existing construction, add, minimum	2%	3%	
3000	Maximum	5%	9%	
3100	Dust protection, add, minimum	1%	2%	
3200	Maximum	4%	11%	
3300	Equipment usage curtailment, add, minimum	1%	1%	
3400	Maximum	3%	10%	
3500	Material handling & storage limitation, add, minimum	1%	1%	
3600	Maximum	6%	7%	
3700	Protection of existing work, add, minimum	2%	2%	
3800	Maximum	5%	7%	
3900	Shift work requirements, add, minimum		5%	
4000	Maximum		30%	

C1020 Interior Doors

This page illustrates and describes interior metal door systems including a metal door, metal frame and hardware. Lines within System Components give the unit price and total price on a cost each for this system. Prices for interior metal door systems are on Line Items C1020 110 1000 thru 2000. Both material quantities and labor costs have been adjusted for the system listed.

Factors: To adjust for job conditions other than normal working situations use Lines C1020 110 2900 thru 4000.

Example: You are to install the system while protecting existing construction. Go to Line C1020 110 3700 and apply these percentages to the appropriate MAT. and INST. costs.

System Components	QUANTITY	UNIT	COST EACH		
			MAT.	INST.	TOTAL
Single metal door, including frame and hardware.					
Hollow metal door, 1-3/8" thick, 2'-6" x 6'-8", painted	1.000	Ea.	330	91	421
Metal frame, 6-1/4" deep	1.000	Set	204	85	289
Paint door and frame, 1 coat	1.000	Ea.	2.79	33	35.79
Hardware, hinges	1.000	Set	85.50		85.50
Hardware, passage lockset	1.000	Set	45.50	42.50	88
TOTAL			667.79	251.50	919.29

C1020 110	Doors, Interior Flush, Metal	COST EACH		
		MAT.	INST.	TOTAL
1000	Hollow metal doors, 1-3/8" thick, 2'-6" x 6'-8"	670	252	922
1100	1-3/8" thick, 2'-8" x 6'-8"	675	252	927
1200	3'-0" x 7'-0"	665	264	929
1300				
1400	Interior fire door, 1-3/8" thick, 2'-6" x 6'-8"			
1500	2'-8" x 6'-8"			
1600	3'-0" x 7'-0"	665	264	929
1700				
1800	Add to fire doors:			
1900	Baked enamel finish	30%	15%	
2000	Galvanizing	20%		
2200				
2300				
2400				
2900	Cut & patch to match existing construction, add, minimum	2%	3%	
3000	Maximum	5%	9%	
3100	Dust protection, add, minimum	1%	2%	
3200	Maximum	4%	11%	
3300	Equipment usage curtailment, add, minimum	1%	1%	
3400	Maximum	3%	10%	
3500	Material handling & storage limitation, add, minimum	1%	1%	
3600	Maximum	6%	7%	
3700	Protection of existing work, add, minimum	2%	2%	
3800	Maximum	5%	7%	
3900	Shift work requirements, add, minimum		5%	
4000	Maximum		30%	

C1020 Interior Doors

This page illustrates and describes an interior closet door system including an interior closet door, painted, with trim and hardware. Prices for interior closet door systems are on Line Items C1020 112 0500 thru 2200. Both material quantities and labor costs have been adjusted for the system listed.

Factors: To adjust for job conditions other than normal working situations use Lines C1020 112 2900 thru 4000.

Example: You are to install the system and match the existing construction. Go to Line C1020 112 2900 and apply these percentages to the appropriate MAT. and INST. costs.

C1020 112	Doors, Closet	COST PER SET		
		MAT.	INST.	TOTAL
0350	Interior closet door painted, including frame, trim and hardware, prehung.			
0400	Bi-fold doors			
0500	Pine paneled, 3'-0" x 6'-8"	445	246	691
0600	6'-0" x 6'-8"	645	330	975
0700	Birch, hollow core, 3'-0" x 6'-8"	212	267	479
0800	6'-0" x 6'-8"	310	355	665
0900	Lauan, hollow core, 3'-0" x 6'-8"	195	246	441
1000	6'-0" x 6'-8"	282	330	612
1100	Louvered pine, 3'-0" x 6'-8"	425	246	671
1200	6'-0" x 6'-8"	495	330	825
1300				
1400	Sliding, bi-passing closet doors			
1500	Pine paneled, 4'-0" x 6'-8"	670	261	931
1600	6'-0" x 6'-8"	1,125	330	1,455
1700	Birch, hollow core, 4'-0" x 6'-8"	410	261	671
1800	6'-0" x 6'-8"	525	330	855
1900	Lauan, hollow core, 4'-0" x 6'-8"	310	261	571
2000	6'-0" x 6'-8"	325	330	655
2100	Louvered pine, 4'-0" x 6'-8"	690	261	951
2200	6'-0" x 6'-8"	975	330	1,305
2300				
2400				
2500				
2600				
2700				
2800				
2900	Cut & patch to match existing construction, add, minimum	2%	3%	
3000	Maximum	5%	9%	
3100	Dust protection, add, minimum	1%	2%	
3200	Maximum	4%	11%	
3300	Equipment usage curtailment, add, minimum	1%	1%	
3400	Maximum	3%	10%	
3500	Material handling & storage limitation, add, minimum	1%	1%	
3600	Maximum	6%	7%	
3700	Protection of existing work, add, minimum	2%	2%	
3800	Maximum	5%	7%	
3900	Shift work requirements, add, minimum		5%	
4000	Maximum		30%	

C2010   Stair Construction

This page illustrates and describes a stair system based on a cost per flight price. Prices for various stair systems are on Line Items C2010 130 0700 thru 3200. Both material quantities and labor costs have been adjusted for the system listed.

Factors: To adjust for job conditions other than normal working situations use Lines C2010 130 3500 thru 4200.

Example: You are to install the system during evenings only. Go to Line C2010 130 4200 and apply this percentage to the appropriate MAT. and INST. costs.

System Components	QUANTITY	UNIT	COST PER FLIGHT		
			MAT.	INST.	TOTAL
Below are various stair systems based on cost per flight of stairs, no side Walls. Stairs are 4'-0" wide, railings are included unless otherwise noted.					

C2010 130	Stairs	COST PER FLIGHT		
		MAT.	INST.	TOTAL
0700	Concrete, cast in place, no nosings, no railings, 12 risers	335	2,300	2,635
0800	24 risers	670	4,575	5,245
0900	Add for 1 intermediate landing	181	635	816
1000	Concrete, cast in place, with nosings, no railings, 12 risers	1,325	2,550	3,875
1100	24 risers	2,650	5,075	7,725
1200	Add for 1 intermediate landing	263	655	918
1300	Steel, grating tread, safety nosing, 12 risers	5,675	1,200	6,875
1400	24 risers	11,400	2,425	13,825
1500	Add for intermediate landing	1,400	320	1,720
1600	Steel, metal pan tread for concrete in-fill, 12 risers	8,700	1,200	9,900
1700	24 risers	17,400	2,425	19,825
1800	Add for intermediate landing	1,400	320	1,720
1900	Spiral, industrial, 4' - 6" diameter, 12 risers	9,175	860	10,035
2000	24 risers	18,400	1,700	20,100
2100	Wood, box stairs, oak treads, 12 risers	2,525	770	3,295
2200	24 risers	5,050	1,550	6,600
2300	Add for 1 intermediate landing	219	174	393
2400	Wood, basement stairs, no risers, 12 steps	840	310	1,150
2500	24 steps	1,675	625	2,300
2600	Add for 1 intermediate landing	46	33	79
2700	Wood, open, rough sawn cedar, 12 steps	2,900	625	3,525
2800	24 steps	5,775	1,250	7,025
2900	Add for 1 intermediate landing	26.50	34.50	61
3000	Wood, residential, oak treads, 12 risers	2,725	3,400	6,125
3100	24 risers	5,450	6,825	12,275
3200	Add for 1 intermediate landing	196	158	354
3500	Dust protection, add, minimum	1%	2%	
3600	Maximum	4%	11%	
3700	Material handling & storage limitation, add, minimum	1%	1%	
3800	Maximum	6%	7%	
3900	Protection of existing work, add, minimum	2%	2%	
4000	Maximum	5%	7%	
4100	Shift work requirements, add, minimum		5%	
4200	Maximum		30%	

C3010 Wall Finishes

C3010 210	Selective Price Sheet	COST PER S.F.		
		MAT.	INST.	TOTAL
0100	Painting, on plaster or drywall, brushwork, primer and 1 ct.	.17	.83	1
0200	Primer and 2 ct.	.24	1.32	1.56
0300	Rollerwork, primer and 1 ct.	.17	.71	.88
0400	Primer and 2 ct.	.24	1.13	1.37
0500	Woodwork incl. puttying, brushwork, primer and 1 ct.	.15	1.25	1.40
0600	Primer and 2 ct.	.23	1.66	1.89
0700	Wood trim to 6" wide, enamel, primer and 1 ct.	.16	.71	.87
0800	Primer and 2 ct.	.25	.90	1.15
0900	Cabinets and casework, enamel, primer and 1 ct.	.16	1.41	1.57
1000	Primer and 2 ct.	.25	1.74	1.99
1100	On masonry or concrete, latex, brushwork, primer and 1 ct.	.31	1.18	1.49
1200	Primer and 2 ct.	.41	1.68	2.09
1300	For block filler, add	.20	1.45	1.65
1400				
1500	Varnish, wood trim, sealer 1 ct., sanding, puttying, quality work	.14	2.54	2.68
1600	Medium work	.11	1.97	2.08
1700	Without sanding	.17	.30	.47
1800				
1900	Wall coverings, wall paper, at low price per double roll, avg workmanship	.69	.88	1.57
2000	At medium price per double roll, average workmanship	1.41	1.06	2.47
2100	At high price per double roll, quality workmanship	2.42	1.30	3.72
2200				
2300	Grass cloths with lining paper	1.43	1.42	2.85
2400	Premium texture/color	3.62	1.62	5.24
2500	Vinyl, fabric backed, light weight	1.60	.88	2.48
2600	Medium weight	1.17	1.18	2.35
2700	Heavy weight	1.75	1.30	3.05
2800				
2900	Cork tiles, 12" x 12", 3/16" thick	4.69	2.36	7.05
3000	5/16" thick	3.74	2.41	6.15
3100	Granular surface, 12" x 36", 1/2" thick	1.52	1.47	2.99
3200	1" thick	1.89	1.53	3.42
3300	Aluminum foil	1.16	2.06	3.22
3400				
3500	Tile, ceramic, adhesive set, 4-1/4" x 4-1/4"	3.41	5.85	9.26
3600	6" x 6"	4.49	6.35	10.84
3700	Decorated, 4-1/4" x 4-1/4", minimum	3.87	4.10	7.97
3800	Color group 4	60	6.15	66.15
3900	For epoxy grout, add	.46	1.38	1.84
4000	Pregrouted sheets	6.30	4.61	10.91
4100	Glass mosaics, 3/4" tile on 12" sheets, minimum	23	15.15	38.15
4200	Color group 8	46.50	17.30	63.80
4300	Metal, tile pattern, 4' x 4' sheet, 24 ga., nailed			
4400	Stainless steel	31	2.66	33.66
4500	Aluminized steel	22.50	2.66	25.16
4600				
4700	Brick, interior veneer, 4" face brick, running bond, minimum	5.25	14.40	19.65
4800	Maximum	5.25	14.40	19.65
4900	Simulated, urethane pieces, set in mastic	9.90	4.54	14.44
5000	Fiberglass panels	10.90	3.41	14.31
5100	Wall coating, on drywall, thin coat, plain	.11	.86	.97
5200	Stipple	.11	.86	.97
5300	Textured spray	.04	.82	.86
5400				
5500	Paneling not incl. furring or trim, hardboard, tempered, 1/8" thick	.55	2.72	3.27
5600	1/4" thick	.64	2.72	3.36
5700	Plastic faced, 1/8" thick	.69	2.72	3.41
5800	1/4" thick	.96	2.72	3.68

C30 Interior Finishes

C3010 Wall Finishes

C3010 210	Selective Price Sheet	COST PER S.F.		
		MAT.	INST.	TOTAL
5900	Woodgrained, 1/4" thick, minimum	.75	2.72	3.47
6000	Maximum	1.57	3.21	4.78
6100	Plywood, 4' x 8' shts. 1/4" thick, prefin., birch faced, min.	1.86	2.72	4.58
6200	Maximum	1.19	3.89	5.08
6300	Walnut, minimum	2.90	2.72	5.62
6400	Maximum	2.62	3.41	6.03
6500	Mahogany, African	2.67	3.41	6.08
6600	Philippine	.63	2.72	3.35
6700	Chestnut	6	3.63	9.63
6800	Pecan	2.83	3.41	6.24
6900	Rosewood	3.39	4.26	7.65
7000	Teak	3.59	3.41	7
7100	Aromatic cedar, plywood	2.49	3.41	5.90
7200	Particle board	1.28	3.41	4.69
7300	Wood board, 3/4" thick, knotty pine	2.28	4.54	6.82
7400	Rough sawn cedar	3.83	4.54	8.37
7500	Redwood, clear	5.55	4.54	10.09
7600	Aromatic cedar	2.51	4.95	7.46

C3020 430	Selective Price Sheet	COST PER S.F.		
		MAT.	INST.	TOTAL
0100	Flooring, carpet, acrylic, 26 oz. light traffic	3	.92	3.92
0200	35 oz. heavy traffic	5.85	.92	6.77
0300	Nylon anti-static, 15 oz. light traffic	2.61	.92	3.53
0400	22 oz. medium traffic	4.55	.92	5.47
0500	26 oz. heavy traffic	6.55	.98	7.53
0600	28 oz. heavy traffic	7.55	.98	8.53
0700	Tile, foamed back, needle punch	4.61	1.09	5.70
0800	Tufted loop	3.65	1.09	4.74
0900	Wool, 36 oz. medium traffic	12.45	.98	13.43
1000	42 oz. heavy traffic	13.65	.98	14.63
1100	Composition, epoxy, with colored chips, minimum	3.54	5.05	8.59
1200	Maximum	4.97	6.95	11.92
1300	Trowelled, minimum	4	6.10	10.10
1400	Maximum	6.75	7.10	13.85
1500	Terrazzo, 1/4" thick, chemical resistant, minimum	7.25	7.10	14.35
1600	Maximum	11.45	9.45	20.90
1700	Resilient, asphalt tile, 1/8" thick	1.61	1.55	3.16
1800	Conductive flooring, rubber, 1/8" thick	8.10	1.97	10.07
1900	Cork tile 1/8" thick, standard finish	5.75	1.97	7.72
2000	Urethane finish	5.60	1.97	7.57
2100	PVC sheet goods for gyms, 1/4" thick	8	7.75	15.75
2200	3/8" thick	10.95	10.35	21.30
2300	Vinyl composition 12" x 12" tile, plain, 1/16" thick	1.35	1.24	2.59
2400	1/8" thick	1.93	1.24	3.17
2500	Vinyl tile, 12" x 12" x 1/8" thick, minimum	7.60	1.24	8.84
2600	Premium colors/pattern	7.95	1.24	9.19
2700	Vinyl sheet goods, backed, .080" thick, plain pattern/colors	4.57	2.70	7.27
2900	Slate, random rectangular, 1/4" thick	25	8.10	33.10
3000	1/2" thick	23	11.55	34.55
3100	Natural cleft, irregular, 3/4" thick	10.70	13.20	23.90
3200	For sand rubbed finish, add	10.65		10.65
3300	Terrazzo, cast in place, bonded 1-3/4" thick, gray cement	4.22	18.90	23.12
3400	White cement	4.98	18.90	23.88
3500	Not bonded 3" thick, gray cement	5.20	20.50	25.70
3600	White cement	6.10	20.50	26.60
3700	Precast, 12" x 12" x 1" thick	28	20	48
3800	1-1/4" thick	29	20	49
3900	16" x 16" x 1-1/4" thick	31	24.50	55.50
4000	1-1/2" thick	28.50	27	55.50
4100	Marble travertine, standard, 12" x 12" x 3/4" thick	10.45	18.45	28.90
4200				
4300	Tile, ceramic, natural clay, thin set	5	6.05	11.05
4400	Porcelain, thin set	7	5.85	12.85
4500	Specialty, decorator finish	14.10	6.05	20.15
4600				
4700	Quarry, red, mud set, 4" x 4" x 1/2" thick	9.40	9.20	18.60
4800	6" x 6" x 1/2" thick	9.60	7.90	17.50
4900	Brown, imported, 6" x 6" x 3/4" thick	7.90	9.20	17.10
5000	8" x 8" x 1" thick	10.60	10.05	20.65
5100	Slate, Vermont, thin set, 6" x 6" x 1/4" thick	8.60	6.15	14.75
5200				
5300	Wood, maple strip, 25/32" x 2-1/4", finished, select	5.75	4.31	10.06
5400	2nd and better	5.80	4.31	10.11
5500	Oak, 25/32" x 2-1/4" finished, clear	4.15	4.31	8.46
5600	No. 1 common	5.05	4.31	9.36
5700	Parquet, standard, 5/16", finished, minimum	6.20	4.56	10.76
5800	Maximum	11.75	7.10	18.85
5900	Custom, finished, plain pattern	19.55	6.80	26.35

C3020 Floor Finishes

C3020 430	Selective Price Sheet	COST PER S.F.		
		MAT.	INST.	TOTAL
6000	Intricate pattern	27	13.60	40.60
6100	Prefinished, oak, 2-1/4" wide	7.25	4.01	11.26
6200	Ranch plank	7.60	4.70	12.30
6300	Sleepers on concrete, treated, 24" O.C., 1" x 2"	.26	.45	.71
6400	1" x 3"	.47	.58	1.05
6500	2" x 4"	.59	.80	1.39
6600	2" x 6"	.82	.92	1.74
6700	Refinish old floors, soft wood	.25	1.35	1.60
6800	Hard wood	.25	4.15	4.40
6900	Subflooring, plywood, CDX, 1/2" thick	.75	.91	1.66
7000	5/8" thick	.85	1.01	1.86
7100	3/4" thick	1.04	1.09	2.13
7200				
7300	1" x 10" boards, S4S, laid regular	2.74	1.24	3.98
7400	Laid diagonal	2.74	1.51	4.25
7500	1" x 8" boards, S4S, laid regular	2.26	1.36	3.62
7600	Laid diagonal	2.26	1.60	3.86
7700	Underlayment, plywood, underlayment grade, 3/8" thick	1.16	.91	2.07
7800	1/2" thick	1.30	.94	2.24
7900	5/8" thick	1.69	.97	2.66
8000	3/4" thick	1.64	1.05	2.69
8100	Particle board, 3/8" thick	.43	.91	1.34
8200	1/2" thick	.45	.94	1.39
8300	5/8" thick	.67	.97	1.64
8400	3/4" thick	.81	1.05	1.86

C3030 Ceiling Finishes

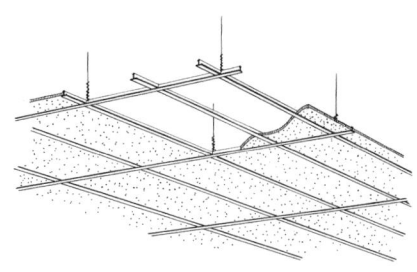

This page illustrates suspended acoustical board systems including acoustic ceiling board, hangers, and T bar suspension. Lines within System Components give the unit price and total price per square foot for this system. Prices for suspended acoustical board systems are on Line Items C3030 210 1000 thru 2400. Both material quantities and labor costs have been adjusted for the system listed.

Factors: To adjust for job conditions other than normal working situations use Lines C3030 210 2900 thru 4000.

Example: You are to install the system and protect existing construction. Go to Line C3030 210 3800 and apply these percentages to the appropriate MAT. and INST. costs.

System Components			COST PER S.F.		
	QUANTITY	UNIT	MAT.	INST.	TOTAL
Suspended acoustical ceiling board installed on exposed grid system.					
Fiberglass boards, film faced, 2' x 4', 5/8" thick	1.000	S.F.	1.47	1.09	2.56
Hangers, #12 wire	1.000	S.F.	.21	.66	.87
T bar suspension system, 2' x 4' grid	1.000	S.F.	1.03	.85	1.88
TOTAL			2.71	2.60	5.31

C3030 210	Ceiling, Suspended Acoustical	COST PER S.F.		
		MAT.	INST.	TOTAL
1000	2' x 4' grid, fiberglass board, film faced, 5/8" thick	2.71	2.60	5.31
1100	Mineral fiber board, aluminum faced	5.90	2.56	8.46
1200	Standard faced	2.40	2.52	4.92
1300	Plastic faced	4.50	3.21	7.71
1400	Fiberglass, film faced, 3" thick, R11	5.50	3.02	8.52
1500	Grass cloth faced, 3/4" thick	5.15	2.87	8.02
1600	1" thick	5.05	2.91	7.96
1700	1-1/2" thick, nubby face	4.32	2.94	7.26
2200				
2300				
2400	Add for 2' x 2' grid system	.26	.21	.47
2500				
2600				
2700				
2900	Cut & patch to match existing construction, add, minimum	2%	3%	
3000	Maximum	5%	9%	
3100	Dust protection, add, minimum	1%	2%	
3200	Maximum	4%	11%	
3300	Equipment usage curtailment, add, minimum	1%	1%	
3400	Maximum	3%	10%	
3500	Material handling & storage limitation, add, minimum	1%	1%	
3600	Maximum	6%	7%	
3700	Protection of existing work, add, minimum	2%	2%	
3800	Maximum	5%	7%	
3900	Shift work requirements, add, minimum		5%	
4000	Maximum		30%	

C3030 Ceiling Finishes

This page illustrates and describes suspended gypsum board systems including gypsum board, metal furring, taping, finished and painted. Lines within System Components give the unit price and total price per square foot for this system. Prices for suspended gypsum board systems are on Line Items C3030 220 1400 thru 1700. Both material quantities and labor costs have been adjusted for the system listed.

Factors: To adjust for job conditions other than normal working situations use Lines C3030 220 2900 thru 4000.

Example: You are to install the system and control dust in the working area. Go to Line C3030 220 3200 and apply these percentages to the appropriate MAT. and INST. costs.

System Components	QUANTITY	UNIT	COST PER S.F.		
			MAT.	INST.	TOTAL
Suspended ceiling gypsum board, 4' x 8' x 5/8" thick,					
On metal furring, taped, finished and painted.					
Gypsum drywall, 4' x 8', 5/8" thick, screwed	1.000	S.F.	.36	.76	1.12
Main runners, 1-1/2" C.R.C., 4' O.C.	.500	S.F.	.19	.84	1.03
25 ga., channels, 2' O.C.	1.000	S.F.	.27	1.55	1.82
Taped and finished	1.000	S.F.	.06	.68	.74
Paint, 2 coats, roller work	1.000	S.F.	.23	.74	.97
TOTAL			1.11	4.57	5.68

C3030 220	Ceilings, Suspended Gypsum Board	COST PER S.F.		
		MAT.	INST.	TOTAL
1400	4' x 8' x 5/8", gypsum drywall, 2 coats paint	1.11	4.57	5.68
1500	Thin coat plaster, 2 coats paint	.93	4.01	4.94
1600	Spray-on sand finish, no paint	.86	3.97	4.83
1700	12" x 12" x 3/4" acoustical wood fiber tile	2.15	4.85	7
1800				
1900				
2000				
2100				
2200				
2300				
2400				
2500				
2600				
2700				
2900	Cut & patch to match existing construction, add, minimum	2%	3%	
3000	Maximum	5%	9%	
3100	Dust protection, add, minimum	1%	2%	
3200	Maximum	4%	11%	
3300	Equipment usage curtailment, add, minimum	1%	1%	
3400	Maximum	3%	10%	
3500	Material handling & storage limitation, add, minimum	1%	1%	
3600	Maximum	6%	7%	
3700	Protection of existing work, add, minimum	2%	2%	
3800	Maximum	5%	7%	
3900	Shift work requirements, add, minimum		5%	
4000	Maximum		30%	

C3030 Ceiling Finishes

This page illustrates and describes suspended plaster and lath systems including gypsum plaster, lath, furring and runners with ceiling painted. Lines within System Components give the unit price and total price per square foot for this system. Prices for suspended plaster and lath systems are on Line Items C3030 230 1200 thru 2300. Both material quantities and labor costs have been adjusted for the system listed.

Factors: To adjust for job conditions other than normal working situations use Lines C3030 230 2900 thru 4000.

Example: You are to install the system to match existing construction. Go to Line C3030 230 2900 and apply these percentages to the appropriate MAT. and INST. costs.

System Components			QUANTITY	UNIT	COST PER S.F.		
					MAT.	INST.	TOTAL
Gypsum plaster, 3 coats, on 3.4# rib lath, on 3/4" C.R.C. furring on 1-1/2" main runners, ceiling painted.							
Gypsum plaster, 3 coats			.110	S.Y.	.67	4.31	4.98
3.4# rib lath			.110	S.Y.	.58	1.02	1.60
Main runners, 1-1/2" C.R.C. 24" O.C.			1.000	S.F.	.37	1.67	2.04
Furring 3/4" C.R.C. 16" OC			1.000	S.F.	.41	2.25	2.66
Painting, 2 coats, roller work			1.000	S.F.	.23	.74	.97
		TOTAL			2.26	9.99	12.25

C3030 230	Ceiling, Suspended Plaster	COST PER S.F.		
		MAT.	INST.	TOTAL
1200	Gypsum plaster, 3 coats, on 3.4# rib lath	2.26	10	12.26
1300	On 2.5# diamond lath	2.17	9.95	12.12
1400	On 3/8" gypsum lath	2.10	9.95	12.05
1500	2 coats, on 3.4# rib lath	2.01	10	12.01
1600	On 2.5# diamond lath	1.97	9.30	11.27
1700	On 3/8" gypsum lath	1.90	9.30	11.20
1800	Perlite plaster, 3 coats, on 3.4# rib lath	2.40	11.05	13.45
1900	On 2.5# diamond lath	2.31	11	13.31
2000	On 3/8" gypsum lath	2.24	11	13.24
2100	2 coats, on 3.4# rib lath	2.31	9.95	12.26
2200	On 2.5# diamond lath	2.22	9.90	12.12
2300	On 3/8" gypsum lath	2.15	9.90	12.05
2400				
2500				
2600				
2700				
2900	Cut & patch to match existing construction, add, minimum	2%	3%	
3000	Maximum	5%	9%	
3100	Dust protection, add, minimum	1%	2%	
3200	Maximum	4%	11%	
3300	Equipment usage curtailment, add, minimum	1%	1%	
3400	Maximum	3%	10%	
3500	Material handling & storage limitation, add, minimum	1%	1%	
3600	Maximum	6%	7%	
3700	Protection of existing work, add, minimum	2%	2%	
3800	Maximum	5%	7%	
3900	Shift work requirements, add, minimum		5%	
4000	Maximum		30%	

667

C3030 Ceiling Finishes

C3030 250	Selective Price Sheet	COST PER S.F.		
		MAT.	INST.	TOTAL
0100	Ceiling, plaster, gypsum, 2 coats	.48	3.71	4.19
0200	3 coats	.68	4.34	5.02
0300	Perlite or vermiculite, 2 coats	.73	4.29	5.02
0400	3 coats	.82	5.40	6.22
0500	Gypsum lath, plain, 3/8" thick	.43	.85	1.28
0600	1/2" thick	.33	.90	1.23
0700	Firestop, 3/8" thick	.31	1.03	1.34
0800	1/2" thick	.33	1.12	1.45
0900	Metal lath, rib, 2.75 lb.	.43	.97	1.40
1000	3.40 lb.	.59	1.03	1.62
1100	Diamond, 2.50 lb.	.49	.97	1.46
1200	3.40 lb.	.52	1.20	1.72
1500	Drywall, standard, 1/2" thick, no finish included	.36	.76	1.12
1600	Taped and finished	.42	1.78	2.20
1700	5/8" thick, no finish included	.36	.76	1.12
1800	Taped and finished	.42	1.78	2.20
1900	Fire resistant, 1/2" thick, no finish included	.42	.76	1.18
2000	Taped and finished	.47	1.78	2.25
2100	5/8" thick, no finish included	.42	.76	1.18
2200	Taped and finished	.47	1.78	2.25
2300	Water resistant, 1/2" thick, no finish included	.45	.76	1.21
2400	Taped and finished	.51	1.78	2.29
2500	5/8" thick, no finish included	.48	.76	1.24
2600	Taped and finished	.54	1.78	2.32
2700	Finish, instead of taping			
2800	For thin coat plaster, add	.11	.86	.97
2900	Finish, textured spray, add	.04	.82	.86
3000				
3100				
3200				
3300	Tile, stapled or glued, mineral fiber plastic coated, 5/8" thick	2.45	2.27	4.72
3400	3/4" thick	3.47	2.27	5.74
3500	Wood fiber, 1/2" thick	1.35	1.70	3.05
3600	3/4" thick	1.33	1.70	3.03
3700	Suspended, fiberglass, film faced, 5/8" thick	1.47	1.09	2.56
3800	3" thick	4.26	1.51	5.77
3900	Mineral fiber, 5/8" thick, standard	1.16	1.01	2.17
4000	Aluminum	4.28	1.14	5.42
4100	Wood fiber, reveal edge, 1" thick			
4200	3" thick			
4300	Framing, metal furring, 3/4" channels, 12" O.C.	.46	3.11	3.57
4400	16" O.C.	.41	2.25	2.66
4500	24" O.C.	.27	1.55	1.82
4600				
4700	1-1/2" channels, 12" O.C.	.62	3.43	4.05
4800	16" O.C.	.55	2.51	3.06
4900	24" O.C.	.37	1.67	2.04
5000				
5100				
5200				
5300	Ceiling suspension systems, for tile,			
5400	Concealed "Z" bar, 12" module	.95	1.31	2.26
5500	Class A, "T" bar 2'-0" x 4'-0" grid	1.03	.85	1.88
5600	2'-0" x 2'-0" grid	1.28	1.05	2.33
5700	Carrier channels for lighting fixtures, add	.24	1.48	1.72
5800				
5900				
6000				

C3030 Ceiling Finishes

C3030 250	Selective Price Sheet	COST PER S.F.		
		MAT.	INST.	TOTAL
6100	Wood, furring 1" x 3", on wood, 12" O.C.	.48	1.95	2.43
6200	16" O.C..	.36	1.46	1.82
6300	24" O.C.	.24	.98	1.22
6400				
6500	On concrete, 12" O.C.	.52	3.24	3.76
6600	16" O.C.	.39	2.43	2.82
6700	24" O.C.	.26	1.62	1.88
6800				
6900	Joists, 2" x 4", 12" O.C.	.67	1.41	2.08
7000	16" O.C.	.56	1.18	1.74
7100	24" O.C.	.46	.95	1.41
7200	32" O.C.	.35	.73	1.08
7300	2" x 6", 12" O.C.	.98	1.42	2.40
7400	16" O.C.	.81	1.18	1.99
7500	24" O.C.	.65	.94	1.59
7600	32" O.C.	.48	.70	1.18

Same Data. Simplified.

Enjoy the convenience and efficiency of accessing your costs anywhere:

- **Skip the multiplier** by setting your location
- **Quickly search,** edit, favorite and share costs
- **Stay on top of price changes** with automatic updates

Discover more at rsmeans.com/online

This page illustrates and describes oil hydraulic elevator systems. Prices for various oil hydraulic elevator systems are on Line Items D1010 130 0400 thru 2000. Both material quantities and labor costs have been adjusted for the system listed.

Factors: To adjust for job conditions other than normal working situations use Lines D1010 130 3100 thru 4000.

Example: You are to install the system with limited equipment usage. Go to Line D1010 130 3400 and apply this percentage to the appropriate TOTAL costs.

D1010 130	Oil Hydraulic Elevators	COST EACH		
		MAT.	INST.	TOTAL
0310	Oil-hydraulic elevator systems			
0320	Including piston and piston shaft			
0330				
0400	1500 Lb. passenger, 2 floors	44,800	21,000	65,800
0500	3 floors	55,500	32,500	88,000
0600	4 floors	80,500	45,700	126,200
0700	5 floors	90,500	57,000	147,500
0800	6 floors	100,500	68,000	168,500
0900				
1000	2500 Lb. passenger, 2 floors	48,300	21,000	69,300
1100	3 floors	59,000	32,500	91,500
1200	4 floors	79,000	45,700	124,700
1300	5 floors	89,500	57,000	146,500
1400	6 floors	104,000	68,000	172,000
1500				
1600	4000 Lb. passenger, 2 floors	55,000	21,000	76,000
1700	3 floors	67,000	32,500	99,500
1800	4 floors	90,500	45,700	136,200
1900	5 floors	101,000	57,000	158,000
2000	6 floors	110,500	68,000	178,500
2100				
2200				
2300				
2400				
2500				
2600				
2700				
2800				
2900				
3000				
3100	Dust protection, add, minimum	1%	2%	
3200	Maximum	4%	11%	
3300	Equipment usage curtailment, add, minimum	1%	1%	
3400	Maximum	3%	10%	
3500	Material handling & storage limitation, add, minimum	1%	1%	
3600	Maximum	6%	7%	
3700	Protection of existing work, add, minimum	2%	2%	
3800	Maximum	5%	7%	
3900	Shift work requirements, add, minimum		5%	
4000	Maximum		30%	

D2010 Plumbing Fixtures

This page illustrates and describes a women's public restroom system including a water closet, lavatory, accessories, and service piping. Lines within System Components give the unit price and total price on a cost each basis for this system. Prices for women's public restroom systems are on Line Items D2010 956 1300 thru 1700. Both material quantities and labor costs have been adjusted for the system listed.

Factors: To adjust for job conditions other than normal working situations use Lines D2010 956 2900 thru 4000.

Example: You are to install the system and protect surrounding area from dust. Go to Line D2010 956 3100 and apply these percentages to the MAT. and INST. costs.

System Components	QUANTITY	UNIT	COST EACH MAT.	COST EACH INST.	COST EACH TOTAL
Public women's restroom incl. water closet, lavatory, accessories and Necessary service piping to install this system in one wall.					
Water closet, wall mounted, one piece	1.000	Ea.	1,200	247	1,447
Rough-in waste and vent for water closet	1.000	Set	1,450	560	2,010
Lavatory, 20″ x 18″ P.E. cast iron with accessories	1.000	Ea.	271	179	450
Rough-in waste and vent for lavatory	1.000	Set	520	865	1,385
Toilet partition, painted metal between walls, floor mounted	1.000	Ea.	425	227	652
For handicap unit, add	1.000	Ea.	360		360
Grab bar, 36″ long	2.000	Ea.	76	68	144
Mirror, 18″ x 24″, with stainless steel shelf	1.000	Ea.	203	34	237
Napkin/tampon dispenser, recessed	1.000	Ea.	615	45.50	660.50
Soap Dispenser, chrome, surface mounted, liquid	1.000	Ea.	64	34	98
Toilet tissue dispenser, surface mounted, stainless steel	1.000	Ea.	17.60	22.50	40.10
Towel dispenser, surface mounted, stainless steel	1.000	Ea.	46	42.50	88.50
TOTAL			5,247.60	2,324.50	7,572.10

D2010 956	Plumbing - Public Restroom	COST EACH MAT.	COST EACH INST.	COST EACH TOTAL
1200				
1300	Public women's restroom, one water closet, one lavatory	5,250	2,325	7,575
1400	Two water closets, two lavatories	9,900	4,425	14,325
1500				
1600	For each additional water closet over 2, add	3,350	1,025	4,375
1700	For each additional lavatory over 2, add	1,050	1,100	2,150
1800				
1900				
2400	NOTE: PLUMBING APPROXIMATIONS			
2500	WATER CONTROL: water meter, backflow preventer,			
2600	Shock absorbers, vacuum breakers, mixer....10 to 15% of fixtures			
2700	PIPE AND FITTINGS: .... 30 to 60% of fixtures			
2800				
2900	Cut & patch to match existing construction, add, minimum	2%	3%	
3000	Maximum	5%	9%	
3100	Dust protection, add, minimum	1%	2%	
3200	Maximum	4%	11%	
3300	Equipment usage curtailment, add, minimum	1%	1%	
3400	Maximum	3%	10%	
3500	Material handling & storage limitation, add, minimum	1%	1%	

D20 Plumbing

D2010 Plumbing Fixtures

D2010 956	Plumbing - Public Restroom	COST EACH		
		MAT.	INST.	TOTAL
3600	Maximum	6%	7%	
3700	Protection of existing work, add, minimum	2%	2%	
3800	Maximum	5%	7%	
3900	Shift work requirements, add, minimum		5%	
4000	Maximum		30%	

D2010 Plumbing Fixtures

This page illustrates and describes a men's public restroom system including a water closet, urinal, lavatory, accessories and service piping. Lines within System Components give the unit price and total price on a cost each basis for this system. Prices for men's public restroom systems are on Line Items D2010 957 1700 thru 2200. Both material quantities and labor costs have been adjusted for the system listed.

Factors: To adjust for job conditions other than normal working situations use Lines D2010 957 2900 thru 4000.

Example: You are to install the system and match existing construction. Go to Line D2010 957 3000 and apply these percentages to the appropriate MAT. and INST. costs.

System Components	QUANTITY	UNIT	COST EACH		
			MAT.	INST.	TOTAL
Public men's restroom incl. water closet, urinal, lavatory, accessories, And necessary service piping to install this system in one wall.					
Water closet, wall mounted, one piece	1.000	Ea.	1,200	247	1,447
Rough-in waste & vent for water closet	1.000	Set	1,450	560	2,010
Urinal, wall hung	1.000	Ea.	345	480	825
Rough-in waste & vent for urinal	1.000	Set	835	505	1,340
Lavatory, 20" x 18", P.E. cast iron with accessories	1.000	Ea.	271	179	450
Rough-in waste & vent for lavatory	1.000	Set	520	865	1,385
Partition, painted mtl., between walls, floor mntd.	1.000	Ea.	425	227	652
For handicap unit, add	1.000	Ea.	360		360
Grab bars	2.000	Ea.	76	68	144
Urinal screen, powder coated steel, wall mounted	1.000	Ea.	193	136	329
Mirror, 18" x 24" with stainless steel shelf	1.000	Ea.	203	34	237
Soap dispenser, surface mounted, liquid	1.000	Ea.	64	34	98
Toilet tissue dispenser, surface mounted, stainless steel	1.000	Ea.	17.60	22.50	40.10
Towel dispenser, surface mounted, stainless steel	1.000	Ea.	46	42.50	88.50
TOTAL			6,005.60	3,400	9,405.60

D2010 957	Plumbing - Public Restroom	COST EACH		
		MAT.	INST.	TOTAL
1600				
1700	Public men's restroom, one water closet, one urinal, one lavatory	6,000	3,400	9,400
1800	Two water closets, two urinals, two lavatories	12,000	6,625	18,625
1900				
2000	For each additional water closet over 2, add	3,350	1,025	4,375
2100	For each additional urinal over 2, add	1,375	1,125	2,500
2200	For each additional lavatory over 2, add	1,050	1,100	2,150
2300				
2400	NOTE: PLUMBING APPROXIMATIONS			
2500	WATER CONTROL: water meter, backflow preventer,			
2600	Shock absorbers, vacuum breakers, mixer....10 to 15% of fixtures			
2700	PIPE AND FITTINGS: .... 30 to 60% of fixtures			
2800				
2900	Cut & patch to match existing construction, add, minimum	2%	3%	
3000	Maximum	5%	9%	
3100	Dust protection, add, minimum	1%	2%	
3200	Maximum	4%	11%	
3300	Equipment usage curtailment, add, minimum	1%	1%	

D2010 Plumbing Fixtures

D2010 957	Plumbing - Public Restroom	COST EACH		
		MAT.	INST.	TOTAL
3400	Maximum	3%	10%	
3500	Material handling, & storage limitation, add, minimum	1%	1%	
3600	Maximum	6%	7%	
3700	Protection of existing work, add, minimum	2%	2%	
3800	Maximum	5%	7%	
3900	Shift work requirements, add, minimum		5%	
4000	Maximum		30%	

D2010 Plumbing Fixtures

This page illustrates and describes a two fixture lavatory system including a water closet, lavatory, accessories and all service piping. Lines within System Components give the unit price and total price on a cost each basis for this system. Prices for two fixture lavatory systems are on Line Items D2010 958 1800 and 1900. Both material quantities and labor costs have been adjusted for the system listed.

Factors: To adjust for job conditions other than normal working situations use Lines D2010 958 2900 thru 4000.

Example: You are to install the system while controlling dust in the work area. Go to Line D2010 958 3200 and apply these percentages to the appropriate MAT. and INST. costs.

System Components	QUANTITY	UNIT	COST EACH MAT.	COST EACH INST.	COST EACH TOTAL
Two fixture bathroom incl. water closet, lavatory, accessories and Necessary service piping to install this system in 2 walls.					
Water closet, floor mounted, 2 piece, close coupled	1.000	Ea.	241	271	512
Rough in waste & vent for water closet	1.000	Set	375	470	845
Lavatory, 20" x 18", P.E. cast iron with accessories	1.000	Ea.	271	179	450
Rough in waste & vent for lavatory	1.000	Set	520	865	1,385
Additional service piping					
1/2" copper pipe with sweat solder joints	10.000	L.F.	40.50	98.50	139
2" black schedule 40 steel pipe with threaded couplings	12.000	L.F.	91.20	270	361.20
4" cast iron soil pipe with lead and oakum joints	7.000	L.F.	189	182	371
Accessories					
Toilet tissue dispenser, chrome, single roll	1.000	Ea.	17.60	22.50	40.10
18" long stainless steel towel bar	1.000	Ea.	44	29.50	73.50
Medicine cabinet with mirror, 20" x 16", unlighted	1.000	Ea.	108	48.50	156.50
TOTAL			1,897.30	2,436	4,333.30

D2010 958	Plumbing - Two Fixture Bathroom	COST EACH MAT.	COST EACH INST.	COST EACH TOTAL
1800	Water closet, lavavatory & accessories, in 2 walls with all service piping	1,900	2,425	4,325
1900	Above system installed in 1 wall	1,775	2,125	3,900
2000				
2400	NOTE: PLUMBING APPROXIMATIONS			
2500	WATER CONTROL: water meter, backflow preventer,			
2600	Shock absorbers, vacuum breakers, mixer....10 to 15% of fixtures			
2700	PIPE AND FITTINGS: .... 30 to 60% of fixtures			
2800				
2900	Cut & patch to match existing construction, add, minimum	2%	3%	
3000	Maximum	5%	9%	
3100	Dust protection, add, minimum	1%	2%	
3200	Maximum	4%	11%	
3300	Equipment usage curtailment, add, minimum	1%	1%	
3400	Maximum	3%	10%	
3500	Material handling & storage limitation, add, minimum	1%	1%	
3600	Maximum	6%	7%	
3700	Protection of existing work, add, minimum	2%	2%	
3800	Maximum	5%	7%	
3900	Shift work requirements, add, minimum		5%	
4000	Maximum		30%	

D2010 Plumbing Fixtures

This page illustrates and describes a three fixture bathroom system including a water closet, tub, lavatory, accessories and service piping. Lines within System Components give the unit price and total price on a cost each basis for this system. Prices for a three fixture bathroom system is on Line Item D2010 959 1700. Both material quantities and labor costs have been adjusted for the system listed.

Factors: To adjust for job conditions other than normal working situations use Lines D2010 959 2900 thru 4000.

Example: You are to install the system and protect all existing work. Go to Line D2010 959 3800 and apply these percentages to the appropriate MAT. and INST. costs.

System Components			COST EACH		
	QUANTITY	UNIT	MAT.	INST.	TOTAL
Three fixture bathroom incl. water closet, bathtub, lavatory, accessories, And necessary service piping to install this system in 1 wall.					
Water closet, tank vit china flr mtd close cpld, 2 pc. w/seat supply & stop	1.000	Ea.	241	271	512
For rough in, supply, waste and vent	.930	Set	348.75	437.10	785.85
Bath tub, P.E., cast iron, 5' long, with accessories	1.000	Ea.	1,400	325	1,725
Rough-in, supply, waste and vent, for all above tubs, add	.950	Set	465.50	660.25	1,125.75
Lavatory, 20" x 18" P.E. cast iron with accessories	1.000	Ea.	271	179	450
Rough-in, supply, waste and vent for above lavatories	1.000	Set	520	865	1,385
Accessories					
Toilet tissue dispenser, surface mounted, SS, single roll	1.000	Ea.	17.60	22.50	40.10
Towel bar, stainless steel, 18" long	2.000	Ea.	88	59	147
Medicine cabinets with mirror, sst. frame, 16" x 22", unlighted	1.000	Ea.	108	48.50	156.50
TOTAL			3,459.85	2,867.35	6,327.20

D2010 959	Plumbing - Three Fixture Bathroom	COST EACH		
		MAT.	INST.	TOTAL
1600				
1700	Water closet, lavatory, bathtub, in 1 wall with all service piping	3,450	2,875	6,325
2400	NOTE: PLUMBING APPROXIMATIONS			
2500	WATER CONTROL: water meter, backflow preventer,			
2600	Shock absorbers, vacuum breakers, mixer....10 to 15% of fixtures			
2700	PIPE AND FITTINGS: .... 30 to 60% of fixtures			
2800				
2900	Cut & patch to match existing construction, add, minimum	2%	3%	
3000	Maximum	5%	9%	
3100	Dust protection, add, minimum	1%	2%	
3200	Maximum	4%	11%	
3300	Equipment usage curtailment, add, minimum	1%	1%	
3400	Maximum	3%	10%	
3500	Material handling & storage limitation, add, minimum	1%	1%	
3600	Maximum	6%	7%	
3700	Protection of existing work, add, minimum	2%	2%	
3800	Maximum	5%	7%	
3900	Shift work requirements, add, minimum		5%	
4000	Maximum		30%	

D2010 Plumbing Fixtures

This page illustrates and describes a three fixture bathroom system including a water closet, tub, lavatory, accessories, and service piping. Lines within System Components give the unit price and total price on a cost each basis for this system. Prices for a three fixture bathroom system are on Line Item D2010 960 1900 thru 2000. Both material quantities and labor costs have been adjusted for the system listed.

Factors: To adjust for job conditions other than normal working situations use Lines D2010 960 2900 thru 4000.

Example: You are to install the system and protect the surrounding area from dust. Go to Line D2010 960 3100 and apply these percentages to the appropriate MAT. and INST. costs.

System Components	QUANTITY	UNIT	COST EACH MAT.	COST EACH INST.	COST EACH TOTAL
Three fixture bathroom incl. water closet, bathtub, lavatory, accessories, And necessary service piping to install this system in 2 walls.					
Water closet, floor mounted, 2 piece, close coupled	1.000	Ea.	241	271	512
Rough-in waste & vent for water closet	1.000	Set	375	470	845
Bathtub, P.E. cast iron, 5' long with accessories	1.000	Ea.	1,400	325	1,725
Rough-in waste & vent for bathtub	1.000	Set	490	695	1,185
Lavatory, 20" x 18" P.E. cast iron with accessories	1.000	Ea.	271	179	450
Rough-in waste & vent for lavatory	1.000	Set	520	865	1,385
Additional service piping					
1-1/4" copper DWV type tubing, sweat solder joints	6.000	L.F.	82.20	79.80	162
2" black schedule 40 steel pipe with threaded couplings	12.000	L.F.	91.20	270	361.20
Accessories					
Toilet tissue dispenser, chrome, single roll	1.000	Ea.	17.60	22.50	40.10
18" long stainless steel towel bar	2.000	Ea.	88	59	147
Medicine cabinet with mirror, 20" x 16", unlighted	1.000	Ea.	108	48.50	156.50
TOTAL			3,684	3,284.80	6,968.80

D2010 960	Plumbing - Three Fixture Bathroom	COST EACH MAT.	COST EACH INST.	COST EACH TOTAL
1900	Water closet, lavatory, bathtub, in 2 walls with all service piping	3,675	3,275	6,950
2000	Above system with corner contour tub, P.E. cast iron	5,475	3,275	8,750
2300				
2400	NOTE: PLUMBING APPROXIMATIONS			
2500	WATER CONTROL: water meter, backflow preventer,			
2600	Shock absorbers, vacuum breakers, mixer....10 to 15% of fixtures			
2700	PIPE AND FITTINGS: .... 30 to 60% of fixtures			
2800				
2900	Cut & patch to match existing construction, add, minimum	2%	3%	
3000	Maximum	5%	9%	
3100	Dust protection, add, minimum	1%	2%	
3200	Maximum	4%	11%	
3300	Equipment usage curtailment, add, minimum	1%	1%	
3400	Maximum	3%	10%	
3500	Material handling & storage limitation, add, minimum	1%	1%	
3600	Maximum	6%	7%	
3700	Protection of existing work, add, minimum	2%	2%	
3800	Maximum	5%	7%	
3900	Shift work requirements, add, minimum		5%	
4000	Maximum		30%	

D2010 Plumbing Fixtures

This page illustrates and describes a three fixture bathroom system including a water closet, shower, lavatory, accessories, and service piping, Lines within System Components give the unit price and total price on a cost each basis for this system. Prices for three fixture bathroom systems are on Line Items D2010 961 2100 thru 2200. Both material quantities and labor costs have been adjusted for the system listed.

Factors: To adjust for job conditions other than normal working situations use Lines D2010 961 2900 thru 4000.

Example: You are to install the system and protect existing construction. Go to Line D2010 961 3700 and apply these percentages to the appropriate MAT. and INST. costs.

System Components	QUANTITY	UNIT	COST EACH MAT.	COST EACH INST.	COST EACH TOTAL
Three fixture bathroom incl. water closet, shower, lavatory, accessories,					
And necessary service piping to install this system in 2 walls.					
Water closet, floor mounted, 2 piece, close coupled	1.000	Ea.	241	271	512
Rough-in waste & vent for water closet	1.000	Set	375	470	845
32" shower, enameled stall, molded stone receptor	1.000	Ea.	1,275	287	1,562
Rough-in waste & vent for shower	1.000	Set	450	700	1,150
Lavatory, 20" x 18" P.E. cast iron with accessories	1.000	Ea.	271	179	450
Rough-in waste & vent for lavatory	1.000	Set	520	865	1,385
Additional service piping					
1-1/4" Copper DWV type tubing, sweat solder joints	4.000	L.F.	54.80	53.20	108
2" black schedule 40 steel pipe with threaded couplings	6.000	L.F.	45.60	135	180.60
4" cast iron soil with lead and oakum joints	7.000	L.F.	189	182	371
Accessories					
Toilet tissue dispenser, chrome, single roll	1.000	Ea.	17.60	22.50	40.10
18" long stainless steel towel bar	2.000	Ea.	88	59	147
Medicine cabinet with mirror, 20" x 16", unlighted	1.000	Ea.	108	48.50	156.50
TOTAL			3,635	3,272.20	6,907.20

D2010 961	Plumbing - Three Fixture Bathroom	COST EACH MAT.	COST EACH INST.	COST EACH TOTAL
2100	Water closet, 32" shower, lavavatory, in 2 walls with all service piping	3,625	3,275	6,900
2200	Above system with 36" corner angle shower	4,600	3,275	7,875
2300				
2400	NOTE: PLUMBING APPROXIMATIONS			
2500	WATER CONTROL: water meter, backflow preventer,			
2600	Shock absorbers, vacuum breakers, mixer....10 to 15% of fixtures			
2700	PIPE AND FITTINGS: .... 30 to 60% of fixtures			
2800				
2900	Cut & patch to match existing construction, add, minimum	2%	3%	
3000	Maximum	5%	9%	
3100	Dust protection, add, minimum	1%	2%	
3200	Maximum	4%	11%	
3300	Equipment usage curtailment, add, minimum	1%	1%	
3400	Maximum	3%	10%	
3500	Material handling & storage limitation, add, minimum	1%	1%	
3600	Maximum	6%	7%	
3700	Protection of existing work, add, minimum	2%	2%	
3800	Maximum	5%	7%	
3900	Shift work requirements, add, minimum		5%	
4000	Maximum		30%	

D2010 Plumbing Fixtures

This page illustrates and describes a four fixture bathroom system including a water closet, shower, bathtub, lavatory, accessories, and service piping. Lines within System Components give the unit price and total price on a cost each basis for this system. Prices for four fixture bathroom systems are on Line Items D2010 962 1800 thru 1900. Both material quantities and labor costs have been adjusted for the system listed.

Factors: To adjust for job conditions other than normal working situations use Lines D2010 962 2900 thru 4000.

Example: You are to install the system during weekends and evenings. Go to Line D2010 962 4000 and apply these percentages to the appropriate MAT. and INST. costs.

System Components	QUANTITY	UNIT	COST EACH		
			MAT.	INST.	TOTAL
Four fixture bathroom incl. water closet, shower, bathtub, lavatory					
Accessories and necessary service piping to install this system in 2 walls.					
Water closet, floor mounted, 2 piece, close coupled	1.000	Ea.	241	271	512
Rough-in waste & vent for water closet	1.000	Set	375	470	845
32″ shower, enameled steel stall, molded stone receptor	1.000	Ea.	1,275	287	1,562
Rough-in waste & vent for shower	1.000	Set	450	700	1,150
Bathtub, P.E. cast iron, 5′ long with accessories	1.000	Ea.	1,400	325	1,725
Rough-in waste & vent for bathtub	1.000	Set	490	695	1,185
Lavatory, 20″ x 18″ P.E. cast iron with accessories	1.000	Ea.	271	179	450
Rough-in waste & vent for lavatory	1.000	Set	520	865	1,385
Accessories					
Toilet tissue dispenser, chrome, single roll	1.000	Ea.	17.60	22.50	40.10
18″ long stainless steel towel bar	2.000	Ea.	88	59	147
Medicine cabinet with mirror, 20″ x 16″ unlighted	1.000	Ea.	108	48.50	156.50
TOTAL			5,235.60	3,922	9,157.60

D2010 962	Plumbing - Four Fixture Bathroom	COST EACH		
		MAT.	INST.	TOTAL
1800	Water closet, 32″ shower, bathtub, lavavatory, in 2 walls	5,225	3,925	9,150
1900	Above system with 36″ corner angle shower, plumbing in 3 walls	6,350	4,300	10,650
2000				
2300				
2400	NOTE: PLUMBING APPROXIMATIONS			
2500	WATER CONTROL: water meter, backflow preventer,			
2600	Shock absorbers, vacuum breakers, mixer....10 to 15% of fixtures			
2700	PIPE AND FITTINGS: .... 30 to 60% of fixtures			
2800				
2900	Cut & patch to match existing construction, add, minimum	2%	3%	
3000	Maximum	5%	9%	
3100	Dust protection, add, minimum	1%	2%	
3200	Maximum	4%	11%	
3300	Equipment usage curtailment, add, minimum	1%	1%	
3400	Maximum	3%	10%	
3500	Material handling & storage limitation, add, minimum	1%	1%	
3600	Maximum	6%	7%	
3700	Protection of existing work, add, minimum	2%	2%	
3800	Maximum	5%	7%	
3900	Shift work requirements, add, minimum		5%	
4000	Maximum		30%	

681

D2010 Plumbing Fixtures

This page illustrates and describes a five fixture bathroom system including a water closet, shower, bathtub, lavatories, accessories, and service piping. Lines within System Components give the unit price and total price on a cost each basis for this system. Prices for five fixture bathroom systems are on Line Items D2010 963 1800 thru 1900. Both material quantities and labor costs have been adjusted for the system listed.

Factors: To adjust for job conditions other than normal working situations use Lines D2010 963 2900 thru 4000.

Example: You are to install and match any existing construction. Go to Line D2010 963 2900 and apply these percentages to the appropriate MAT. and INST. costs.

System Components	QUANTITY	UNIT	COST EACH		
			MAT.	INST.	TOTAL
Five fixture bathroom incl. water closet, shower, bathtub, 2 lavatories,					
Accessories and necessary service piping to install this system in 1 wall.					
Water closet, floor mounted, 2 piece, close coupled	1.000	Ea.	241	271	512
Rough-in waste & vent for water closet	1.000	Set	375	470	845
36" corner angle shower, enameled steel stall, molded stone receptor	1.000	Ea.	2,250	299	2,549
Rough-in waste & vent for shower	1.000	Set	450	700	1,150
Bathtub, P.E. cast iron, 5' long with accessories	1.000	Ea.	1,400	325	1,725
Rough-in waste & vent for bathtub	1.000	Set	490	695	1,185
Countertop, plastic laminate, with backsplash	2.000	Ea.	202	90	292
Vanity base, 2 doors, 24" wide	2.000	Ea.	840	248	1,088
Lavatories, 20" x 18" cabinet mntd, PECI	2.000	Ea.	680	448	1,128
Rough-in waste & vent for lavatories	1.600	Set	488	1,000	1,488
Accessories					
Toilet tissue dispenser, chrome, single roll	1.000	Ea.	17.60	22.50	40.10
18" long stainless steel towel bars	2.000	Ea.	88	59	147
Medicine cabinet with mirror, 20" x 16", unlighted	2.000	Ea.	216	97	313
TOTAL			7,737.60	4,724.50	12,462.10

D2010 963	Plumbing - Five Fixture Bathroom	COST EACH		
		MAT.	INST.	TOTAL
1800	Water closet, shower, bathtub, 2 lavatories, on 1 wall	7,750	4,725	12,475
1900	Above system installed in 2 walls with all necessary service piping	7,800	4,875	12,675
2000				
2100				
2200				
2300				
2400	NOTE: PLUMBING APPROXIMATIONS			
2500	WATER CONTROL: water meter, backflow preventer,			
2600	Shock absorbers, vacuum breakers, mixer....10 to 15% of fixtures			
2700	PIPE AND FITTINGS: .... 30 to 60% of fixtures			
2800				
2900	Cut & patch to match existing construction, add, minimum	2%	3%	
3000	Maximum	5%	9%	
3100	Dust protection, add, minimum	1%	2%	
3200	Maximum	4%	11%	
3300	Equipment usage curtailment, add, minimum	1%	1%	
3400	Maximum	3%	10%	
3500	Material handling & storage limitation, add, minimum	1%	1%	

D20 Plumbing

D2010 Plumbing Fixtures

D2010 963	Plumbing - Five Fixture Bathroom	COST EACH		
		MAT.	INST.	TOTAL
3600	Maximum	6%	7%	
3700	Protection of existing work, add, minimum	2%	2%	
3800	Maximum	5%	7%	
3900	Shift work requirements, add, minimum		5%	
4000	Maximum		30%	

This page illustrates and describes an oil fired hot water baseboard system including an oil fired boiler, fin tube radiation and all fittings and piping. Lines within System Components give the unit price and total price per square foot for this system. Prices for oil fired hot water baseboard systems are on Line Items D3010 540 1600 thru 2100. Both material quantities and labor costs have been adjusted for the system listed.

Factors: To adjust for job conditions other than normal working situations use Lines D3010 540 2900 thru 4000.

Example: You are to install the system while protecting all existing work. Go to Line D3010 540 3700 and apply these percentages to the appropriate MAT. and INST. costs.

System Components			COST PER S.F.		
	QUANTITY	UNIT	MAT.	INST.	TOTAL
Oil fired hot water baseboard system including boiler, fin tube radiation					
And all necessary fittings and piping.					
Area to 800 S.F.					
Boiler, cast iron, w/oil piping, 97 MBH	1.000	Ea.	2,750	1,781.25	4,531.25
Copper piping	130.000	L.F.	1,716	1,787.50	3,503.50
Fin tube radiation	68.000	L.F.	601.80	1,700	2,301.80
Circulator	1.000	Ea.	785	239	1,024
Oil tank	1.000	Ea.	560	292	852
Expansion tank, ASME	1.000	Ea.	835	104	939
TOTAL			7,247.80	5,903.75	13,151.55
COST PER S.F.		S.F.	9.06	7.38	16.44

D3010 540	Heating - Oil Fired Hot Water	COST PER S.F.		
		MAT.	INST.	TOTAL
1600	Cast iron boiler, area to 800 S.F.	9.06	7.38	16.44
1700	To 1000 S.F.	7.85	6.35	14.20
1800	To 1200 S.F.	6.80	5.75	12.55
1900	To 1600 S.F.	8.55	5.75	14.30
2000	To 2000 S.F.	7.10	5.10	12.20
2100	To 3000 S.F.	5.20	4.29	9.49
2200				
2300				
2700				
2800				
2900	Cut & patch to match existing construction, add, minimum	2%	3%	
3000	Maximum	5%	9%	
3100	Dust protection, add, minimum	1%	2%	
3200	Maximum	4%	11%	
3300	Equipment usage curtailment, add, minimum	1%	1%	
3400	Maximum	3%	10%	
3500	Material handling & storage limitation, add, minimum	1%	1%	
3600	Maximum	6%	7%	
3700	Protection of existing work, add, minimum	2%	2%	
3800	Maximum	5%	7%	
3900	Shift work requirements, add, minimum		5%	
4000	Maximum		30%	

D3010 Energy Supply

This page illustrates and describes a gas fired hot water baseboard system including a gas fired boiler, fin tube radiation, fittings and piping. Lines within System Components give the unit price and total price per square foot for this system. Prices for gas fired hot water baseboard systems are on Line Items D3010 550 1400 thru 1900. Both material quantities and labor costs have been adjusted for the system listed.

Factors: To adjust for job conditions other than normal working situations use Lines D3010 550 2900 thru 4000.

Example: You are to install the system with minimal equipment usage. Go to Line D3010 550 3400 and apply these percentages to the appropriate MAT. and INST. costs.

System Components	QUANTITY	UNIT	COST PER S.F. MAT.	COST PER S.F. INST.	COST PER S.F. TOTAL
Gas fired hot water baseboard system including boiler, fin tube radiation,					
All necessary fittings and piping.					
Area to 800 S.F.					
Cast iron boiler, insulating jacket, gas piping, 80 MBH	1.000	Ea.	2,718.75	2,656.25	5,375
Copper piping	130.000	L.F.	1,716	1,787.50	3,503.50
Fin tube radiation	68.000	L.F.	601.80	1,700	2,301.80
Circulator, flange connection	1.000	Ea.	785	239	1,024
Expansion tank, ASME	1.000	Ea.	835	104	939
TOTAL			6,656.55	6,486.75	13,143.30
COST PER S.F.		S.F.	8.32	8.11	16.43

D3010 550	Heating - Gas Fired Hot Water	COST PER S.F. MAT.	COST PER S.F. INST.	COST PER S.F. TOTAL
1400	Cast iron boiler, area to 800 S.F.	8.31	8.09	16.40
1500	To 1000 S.F.	7	7.10	14.10
1600	To 1200 S.F.	6.10	6.40	12.50
1700	To 1600 S.F.	8	6.20	14.20
1800	To 2000 S.F.	7	5.55	12.55
1900	To 3000 S.F.	5.40	4.88	10.28
2000				
2100				
2900	Cut & patch to match existing construction, add, minimum	2%	3%	
3000	Maximum	5%	9%	
3100	Dust protection, add, minimum	1%	2%	
3200	Maximum	4%	11%	
3300	Equipment usage curtailment, add, minimum	1%	1%	
3400	Maximum	3%	10%	
3500	Material handling & storage limitation, add, minimum	1%	1%	
3600	Maximum	6%	7%	
3700	Protection of existing work, add, minimum	2%	2%	
3800	Maximum	5%	7%	
3900	Shift work requirements, add, minimum		5%	
4000	Maximum		30%	

D3010 Energy Supply

This page illustrates and describes a wood burning stove system including a free standing stove, preformed hearth, masonry chimney, and necessary piping and fittings. Lines within System Components give the unit price and total price on a cost each basis for this system. Prices for wood burning stove systems are on Line Items D3010 991 1400 thru 2000. Both material quantities and labor costs have been adjusted for the system listed.

Factors: To adjust for job conditions other than normal working situations use Lines D3010 991 2900 thru 4000.

Example: You are to install the system and protect existing construction. Go to Line D3010 991 3500 and apply these percentages to the appropriate MAT., and INST. costs.

System Components	QUANTITY	UNIT	COST EACH MAT.	COST EACH INST.	COST EACH TOTAL
Cast iron, free standing, wood burning stove with preformed hearth, masonry Chimney, and all necessary piping and fittings to install in chimney.					
Cast iron wood burning stove, stove pipe	1.000	Ea.	1,650	1,050	2,700
Wall panel or hearth, non-combustible	1.000	Ea.	1,035.50	191.90	1,227.40
16″ x 16″ brick chimney, 8″ x 8″ flue	20.000	V.L.F.	725	1,662.50	2,387.50
Concrete in place, chimney foundations	.500	C.Y.	97.50	139.95	237.45
TOTAL			3,508	3,044.35	6,552.35

D3010 991	Wood Burning Stoves	COST EACH MAT.	COST EACH INST.	COST EACH TOTAL
1400	Cast iron, free standing with preformed hearth	3,500	3,050	6,550
1500				
1600	Installed in existing fireplace	2,200	1,150	3,350
1700				
1800	Installed with insulated metal chimney			
1900	System including ceiling package,			
2000	Metal chimney to 10 L.F.	4,025	1,625	5,650
2100				
2200				
2300				
2400				
2500				
2600				
2700				
2900	Dust protection, add, minimum	1%	2%	
3000	Maximum	4%	11%	
3100	Equipment usage curtailment, add, minimum	1%	1%	
3200	Maximum	3%	10%	
3300	Material handling & storage limitation, add, minimum	1%	1%	
3400	Maximum	6%	7%	
3500	Protection of existing work, add, minimum	2%	2%	
3600	Maximum	5%	7%	
3700	Shift work requirements, add, minimum		5%	
3800	Maximum		30%	
3900	Temporary shoring and bracing, add, minimum	2%	5%	
4000	Maximum	5%	12%	

This page illustrates and describes masonry fireplace systems including a brick fireplace, footing, foundation, hearth, firebox, chimney and flue. Lines within System Components give the unit price and total price on a cost each basis for this system. Prices for masonry fireplace systems are on Line Items D3010 992 1200 thru 1500. Both material quantities and labor costs have been adjusted for the system listed.

Factors: To adjust for job conditions other than normal working situations use Lines D3010 992 2700 thru 4000.

Example: You are to install the system with some temporary shoring and bracing. Go to Line D3010 992 3900 and apply these percentages to the appropriate MAT. and INST. costs.

System Components	QUANTITY	UNIT	COST EACH MAT.	COST EACH INST.	COST EACH TOTAL
Brick masonry fireplace, including footing, foundation, hearth, firebox, Chimney and flue, chimney 12' above firebox.					
Footing, 4' x 7' x 12" thick, 3000 psi concrete	1.040	C.Y.	232.96	333.88	566.84
Foundation, 12" concrete block	180.000	S.F.	1,035	2,178	3,213
Fireplace, brick faced, 6'-0" wide x 5'-0" high	1.000	Ea.	685	3,025	3,710
Hearth	1.000	Ea.	236	605	841
Chimney, 20" x 20", one 12" x 12" flue, 12 V.L.F.	12.000	V.L.F.	654	1,062	1,716
Mantle, wood	7.000	L.F.	67.90	132.30	200.20
TOTAL			2,910.86	7,336.18	10,247.04

D3010 992	Masonry Fireplace	COST EACH MAT.	COST EACH INST.	COST EACH TOTAL
1200	Chimney, 20" x 20", one 12" x 12" flue, 12 V.L.F.	2,900	7,325	10,225
1300	18 V.L.F.	3,250	7,875	11,125
1400	24 V.L.F.	3,575	8,400	11,975
1500	Fieldstone face instead of brick	3,250	7,325	10,575
1600				
1700				
1800				
1900				
2000				
2100				
2200				
2300				
2700	Cut & patch to match existing construction, add, minimum	2%	3%	
2800	Maximum	5%	9%	
2900	Dust protection, add, minimum	1%	2%	
3000	Maximum	4%	11%	
3100	Equipment usage curtailment, add, minimum	1%	1%	
3200	Maximum	3%	10%	
3300	Material handling & storage limitation, add, minimum	1%	1%	
3400	Maximum	6%	7%	
3500	Protection of existing work, add, minimum	2%	2%	
3600	Maximum	5%	7%	
3700	Shift work requirements, add, minimum		5%	
3800	Maximum		30%	
3900	Temporary shoring and bracing, add, minimum	2%	5%	
4000	Maximum	5%	12%	

D3020 Heat Generating Systems

This page illustrates and describes an oil fired forced air system including an oil fired furnace, ductwork, registers and hookups. Lines within System Components give the unit price and total price per square foot for this system. Prices for oil fired forced air systems are on Line Items D3020 122 1600 thru 2800. Both material quantities and labor costs have been adjusted for the system listed.

Factors: To adjust for job conditions other than normal working situations use Lines D3020 122 3100 thru 4200.

Example: You are to install the system during evenings and weekends. Go to Line D3020 122 4200 and apply this percentage to the appropriate INST. cost.

System Components	QUANTITY	UNIT	COST PER S.F. MAT.	COST PER S.F. INST.	COST PER S.F. TOTAL
Oil fired hot air heating system including furnace, ductwork, registers					
And all necessary hookups.					
Area to 800 S.F., heat only					
Furnace, oil, atomizing gun type burner, w/oil piping	1.000	Ea.	4,500	487.50	4,987.50
Oil tank, steel, 275 gallon	1.000	Ea.	560	292	852
Duct, galvanized, steel	312.000	Lb.	205.92	2,901.60	3,107.52
Insulation, blanket type, duct work	270.000	S.F.	1,333.80	4,306.50	5,640.30
Flexible duct, 6" diameter, insulated	100.000	L.F.	397	540	937
Registers, baseboard, gravity, 12" x 6"	8.000	Ea.	100.80	272	372.80
Return, damper, 18" x 36"	1.000	Ea.	188	78	266
TOTAL			7,285.52	8,877.60	16,163.12
COST PER S.F.		S.F.	9.11	11.10	20.21

D3020 122	Heating-Cooling, Oil, Forced Air	COST PER S.F. MAT.	COST PER S.F. INST.	COST PER S.F. TOTAL
1600	Oil fired, area to 800 S.F.	9.09	11.09	20.18
1700	To 1000 S.F.	7.30	8.95	16.25
1800	To 1200 S.F.	6.30	7.75	14.05
1900	To 1600 S.F.	4.99	6.20	11.19
2000	To 2000 S.F.	5.10	8.50	13.60
2100	To 3000 S.F.	3.87	6.80	10.67
2200	For combined heating and cooling systems:			
2300	Oil fired, heating and cooling, area to 800 S.F.	12	12.40	24.40
2400	To 1000 S.F.	9.75	10.05	19.80
2500	To 1200 S.F.	8.30	8.70	17
2600	To 1600 S.F.	6.70	6.95	13.65
2700	To 2000 S.F.	6.45	9.20	15.65
2800	To 3000 S.F.	4.81	7.25	12.06
3100	Cut & patch to match existing construction, add, minimum	2%	3%	
3200	Maximum	5%	9%	
3300	Dust protection, add, minimum	1%	2%	
3400	Maximum	4%	11%	
3500	Equipment usage curtailment, add, minimum	1%	1%	
3600	Maximum	3%	10%	
3700	Material handling & storage limitation, add, minimum	1%	1%	
3800	Maximum	6%	7%	
3900	Protection of existing work, add, minimum	2%	2%	
4000	Maximum	5%	7%	
4100	Shift work requirements, add, minimum		5%	
4200	Maximum		30%	

D3020 Heat Generating Systems

This page illustrates and describes a gas fired forced air system including a gas fired furnace, ductwork, registers and hookups. Lines within System Components give the unit price and total price per square foot for this system. Prices for gas fired forced air systems are on Line Items D3020 124 1400 thru 2600. Both material quantities and labor costs have been adjusted for the system listed.

Factors: To adjust for job conditions other than normal working situations use Lines D3020 124 2900 thru 4000.

Example: You are to install the system with material handling and storage limitations. Go to Line D3020 124 3500 and apply these percentages to the appropriate MAT. and INST. costs.

System Components	QUANTITY	UNIT	COST PER S.F. MAT.	COST PER S.F. INST.	COST PER S.F. TOTAL
Gas fired hot air heating system including furnace, ductwork, registers					
And all necessary hookups.					
Area to 800 S.F., heat only					
Furnace, gas, AGA certified, direct drive, w/gas piping, 44 MBH	1.000	Ea.	962.50	437.50	1,400
Duct, galvanized steel	312.000	Lb.	205.92	2,901.60	3,107.52
Insulation, blanket type, ductwork	270.000	S.F.	1,333.80	4,306.50	5,640.30
Flexible duct, 6" diameter, insulated	100.000	L.F.	397	540	937
Registers, baseboard, gravity, 12" x 6"	8.000	Ea.	100.80	272	372.80
Return, damper, 18" x 36"	1.000	Ea.	188	78	266
TOTAL			3,188.02	8,535.60	11,723.62
COST PER S.F.		S.F.	3.99	10.67	14.66

D3020 124	Heating-Cooling, Gas, Forced Air	COST PER S.F. MAT.	COST PER S.F. INST.	COST PER S.F. TOTAL
1400	Gas fired, area to 800 S.F.	4	10.66	14.66
1500	To 1000 S.F.	3.26	8.65	11.91
1600	To 1200 S.F.	2.90	7.50	10.40
1700	To 1600 S.F.	2.79	6.80	9.59
1800	To 2000 S.F.	3.05	8.35	11.40
1900	To 3000 S.F.	2.39	6.70	9.09
2000	For combined heating and cooling systems:			
2100	Gas fired, heating and cooling, area to 800 S.F.	6.90	12	18.90
2200	To 1000 S.F.	5.65	9.75	15.40
2300	To 1200 S.F.	4.91	8.40	13.31
2400	To 1600 S.F.	4.51	7.55	12.06
2500	To 2000 S.F.	4.41	9.05	13.46
2600	To 3000 S.F.	3.33	7.15	10.48
2900	Cut & patch to match existing construction, add, minimum	2%	3%	
3000	Maximum	5%	9%	
3100	Dust protection, add, minimum	1%	2%	
3200	Maximum	4%	11%	
3300	Equipment usage curtailment, add, minimum	1%	1%	
3400	Maximum	3%	10%	
3500	Material handling & storage limitation, add, minimum	1%	1%	
3600	Maximum	6%	7%	
3700	Protection of existing work, add, minimum	2%	2%	
3800	Maximum	5%	7%	
3900	Shift work requirements, add, minimum		5%	
4000	Maximum		30%	

D3020 Heat Generating Systems

- Expansion Tank
- Supply Piping
- Gas Supply
- Pressure Relief Valve
- Return Piping
- Circulating Pump

Cast Iron Boiler, Hot Water, Gas Fired

This page illustrates and describes boilers including expansion tank, circulating pump and all service piping. Lines within Systems Components give the material and installation price on a cost each basis for the components. Prices for boiler systems are on Line Items D3020 130 1010 thru D3020 138 1040. Material quantities and labor costs have been adjusted for the system listed.

Factors: To adjust for job conditions other than normal working situations use Line D3020 138 2700 thru 4000.

System Components	QUANTITY	UNIT	COST EACH		
			MAT.	INST.	TOTAL
Cast iron boiler, including piping, chimney and accessories					
Boilers, gas fired, std controls, CI, insulated, HW, gross output 100 MBH	1.000	Ea.	2,825	2,275	5,100
Pipe, black steel, Sch 40, threaded, W/coupling & hangers, 10' OC, 3/4" dia	22.500	L.F.	108.90	293.63	402.53
Pipe, black steel, Sch 40, threaded, W/coupling & hangers, 10' OC, 1" dia.	21.500	L.F.	183.83	323.58	507.41
Elbow, 90°, black, straight, 3/4" dia.	9.000	Ea.	40.77	513	553.77
Elbow, 90°, black, straight, 1" dia.	6.000	Ea.	55.20	369	424.20
Tee, black, straight, 3/4" dia.	2.000	Ea.	17.20	177	194.20
Tee, black, straight, 1" dia.	2.000	Ea.	29.50	199	228.50
Tee, black, reducing, 1" dia.	2.000	Ea.	38	199	237
Pipe cap, black, 3/4" dia.	1.000	Ea.	6.15	25	31.15
Union, black with brass seat, 3/4" dia.	2.000	Ea.	42	123	165
Union, black with brass seat, 1" dia.	2.000	Ea.	54	133	187
Pipe nipples, black, 3/4" dia.	5.000	Ea.	12.10	32.63	44.73
Pipe nipples, black, 1" dia.	3.000	Ea.	12.83	22.58	35.41
Valves, bronze, gate, N.R.S., threaded, class 150, 3/4" size	2.000	Ea.	228	80	308
Valves, bronze, gate, N.R.S., threaded, class 150, 1" size	2.000	Ea.	296	84	380
Gas cock, brass, 3/4" size	1.000	Ea.	19.90	36	55.90
Thermometer, stem type, 9" case, 8" stem, 3/4" NPT	2.000	Ea.	412	58	470
Tank, steel, liquid expansion, ASME, painted, 15 gallon capacity	1.000	Ea.	940	86	1,026
Pump, circulating, bronze, flange connection, 3/4" to 1-1/2" size, 1/8 HP	1.000	Ea.	1,375	239	1,614
Vent chimney, all fuel, pressure tight, double wall, SS, 6" dia.	20.000	L.F.	1,330	470	1,800
Vent chimney, elbow, 90° fixed, 6" dia.	2.000	Ea.	930	94	1,024
Vent chimney, Tee, 6" dia.	2.000	Ea.	554	117	671
Vent chimney, ventilated roof thimble, 6" dia.	1.000	Ea.	290	54	344
Vent chimney, flat roof flashing, 6" dia.	1.000	Ea.	122	47	169
Vent chimney, stack cap, 6" diameter	1.000	Ea.	335	30.50	365.50
Insulation, fiberglass pipe covering, 1" wall, 1" IPS	20.000	L.F.	23.80	122	145.80
TOTAL			10,281.18	6,202.92	16,484.10

D3020 130	Boiler, Cast Iron, Hot Water, Gas	COST EACH		
		MAT.	INST.	TOTAL
1000	For alternate boilers:			
1010	Boiler, cast iron, gas, hot water, 100 MBH	10,300	6,200	16,500
1020	200 MBH	12,000	7,125	19,125
1030	320 MBH	10,900	7,725	18,625
1040	440 MBH	15,100	9,975	25,075
1050	544 MBH	24,600	15,800	40,400
1060	765 MBH	29,600	16,900	46,500
1070	1088 MBH	33,100	18,100	51,200

D3020 Heat Generating Systems

D3020 130	Boiler, Cast Iron, Hot Water, Gas	COST EACH		
		MAT.	INST.	TOTAL
1080	1530 MBH	39,600	23,900	63,500
1090	2312 MBH	49,500	28,100	77,600

D3020 134	Boiler, Cast Iron, Steam, Gas	COST EACH		
		MAT.	INST.	TOTAL
1010	Boiler, cast iron, gas, steam, 100 MBH	8,125	6,025	14,150
1020	200 MBH	20,200	11,300	31,500
1030	320 MBH	21,800	12,700	34,500
1040	544 MBH	29,200	20,000	49,200
1050	765 MBH	32,900	20,900	53,800
1060	1275 MBH	39,000	22,900	61,900
1070	2675 MBH	55,500	33,400	88,900

D3020 136	Boiler, Cast Iron, Hot Water, Gas/Oil	COST EACH		
		MAT.	INST.	TOTAL
1010	Boiler, cast iron, gas & oil, hot water, 584 MBH	30,900	17,500	48,400
1020	876 MBH	41,000	18,800	59,800
1030	1168 MBH	49,200	21,100	70,300
1040	1460 MBH	55,500	25,500	81,000
1050	2044 MBH	63,000	26,400	89,400
1060	2628 MBH	80,000	31,900	111,900

D3020 138	Boiler, Cast Iron, Steam, Gas/Oil	COST EACH		
		MAT.	INST.	TOTAL
1010	Boiler, cast iron, gas & oil, steam, 810 MBH	37,500	22,700	60,200
1020	1360 MBH	42,900	24,200	67,100
1030	2040 MBH	50,500	31,100	81,600
1040	2700 MBH	53,500	34,500	88,000
2700	Cut & patch to match existing construction, add, minimum	2%	3%	
2800	Maximum	5%	9%	
2900	Dust protection, add, minimum	1%	2%	
3000	Maximum	4%	11%	
3100	Equipment usage curtailment, add, minimum	1%	1%	
3200	Maximum	3%	10%	
3300	Material handling & storage limitation, add, minimum	1%	1%	
3400	Maximum	6%	7%	
3500	Protection of existing work, add, minimum	2%	2%	
3600	Maximum	5%	7%	
3700	Shift work requirements, add, minimum		5%	
3800	Maximum		30%	
3900	Temporary shoring and bracing, add, minimum	2%	5%	
4000	Maximum	5%	12%	

691

For customer support on your Commercial Renovation Costs with RSMeans data, call 800.448.8182.

D3020 Heat Generating Systems

Flow ↓   ↑ Flow

Gate Valve →

Y Strainer →

Pressure Gauge →

Expansion Joint →

← Combination Valve

← Pressure Gauge

← Expansion Joint

Base Mounted End-Suction Pump

This page illustrates and describes end suction base mounted circulating pumps, including strainer and piping. Lines within system components give the material and installation price on a cost each basis for the components. Prices for circulating pumps are on Line items D3020 330 1010 thru 1050. Material quantities and labor costs have been adjusting for the system listed.

Factors: To adjust for job conditions other than normal working situations uses Lines D3020 330 2700 thru 4000.

System Components	QUANTITY	UNIT	COST EACH		
			MAT.	INST.	TOTAL
Circulator pump, end suction, including pipe,					
Valves, fittings and guages.					
Pump, circulating, CI, base mounted, 2-1/2" size, 3 HP, to 150 GPM	1.000	Ea.	11,400	795	12,195
Pipe, black steel, Sch. 40, on yoke & roll hangers, 10' O.C., 2-1/2" dia	12.000	L.F.	156	395.88	551.88
Elbow, 90°, weld joint, steel, 2-1/2" pipe size	1.000	Ea.	38.50	193.65	232.15
Flange, weld neck, 150 LB, 2-1/2" pipe size	8.000	Ea.	328	774.40	1,102.40
Valve, iron body, gate, 125 lb., N.R.S., flanged, 2-1/2" size	1.000	Ea.	690	287	977
Strainer, Y type, iron body, flanged, 125 lb., 2-1/2" pipe size	1.000	Ea.	177	292	469
Multipurpose valve, CI body, 2" size	1.000	Ea.	630	101	731
Expansion joint, flanged spool, 6" F to F, 2-1/2" dia	2.000	Ea.	610	236	846
T-O-L, weld joint, socket, 1/4" pipe size, nozzle	2.000	Ea.	18.20	135.20	153.40
T-O-L, weld joint, socket, 1/2" pipe size, nozzle	2.000	Ea.	18.20	140.60	158.80
Control gauges, pressure or vacuum, 3-1/2" diameter dial	2.000	Ea.	40	51	91
Pressure/temperature relief plug, 316 SS, 3/4" OD, 7-1/2" insertion	2.000	Ea.	128	51	179
Insulation, fiberglass pipe covering, 1-1/2" wall, 2-1/2" IPS	12.000	L.F.	39.36	89.40	128.76
Water flow	1.000	Ea.	1,400	1,225	2,625
Pump balancing	1.000	Ea.		355	355
TOTAL			15,673.26	5,122.13	20,795.39

D3020 330	Circulating Pump Systems, End Suction	COST EACH		
		MAT.	INST.	TOTAL
1000	For alternate pump sizes:			
1010	Pump, base mtd with motor, end-suction, 2-1/2" size, 3 HP, to 150 GPM	15,700	5,125	20,825
1020	3" size, 5 HP, to 225 GPM	19,100	5,775	24,875
1030	4" size, 7-1/2 HP, to 350 GPM	21,700	6,875	28,575
1040	5" size, 15 HP, to 1000 GPM	32,400	10,200	42,600
1050	6" size, 25 HP, to 1550 GPM	43,500	12,300	55,800
2700	Cut & patch to match existing construction, add, minimum	2%	3%	
2800	Maximum	5%	9%	
2900	Dust protection, add, minimum	1%	2%	
3000	Maximum	4%	11%	
3100	Equipment usage curtailment, add, minimum	1%	1%	
3200	Maximum	3%	10%	
3300	Material handling & storage limitation, add, minimum	1%	1%	
3400	Maximum	6%	7%	
3500	Protection of existing work, add, minimum	2%	2%	
3600	Maximum	5%	7%	
3700	Shift work requirements, add, minimum		5%	
3800	Maximum		30%	

D3020 Heat Generating Systems

D3020 330	Circulating Pump Systems, End Suction	COST EACH		
		MAT.	INST.	TOTAL
3900	Temporary shoring and bracing, add, minimum	2%	5%	
4000	Maximum	5%	12%	

D3040 Distribution Systems

Galvanized Steel Duct

Fiberglass Duct Insulation

Flexible Fiberglass Duct

Horizontal Fan Coil Air Conditioning System

This page illustrates and describes fan coil air conditioners including duct work, duct installation, piping and diffusers. Lines within Systems Components give the material and installation price on a cost each basis for the components. Prices for fan coil A/C unit systems are on Line Items D3040 118 1010 thru D3040 126 1120. Material quantities and labor costs have been adjusted for the system listed.

Factors: To adjust for job conditions other than normal working situations use Lines D3040 126 2700 thru 4000.

System Components	QUANTITY	UNIT	COST EACH		
			MAT.	INST.	TOTAL
Fan coil A/C system, horizontal, with housing, controls, 2 pipe, 1/2 ton					
Fan coil A/C, horizontal housing, filters, chilled water, 1/2 ton cooling	1.000	Ea.	1,575	182	1,757
Pipe, black steel, Sch 40, threaded, W/cplgs & hangers, 10' OC, 3/4" dia.	21.500	L.F.	104.06	280.58	384.64
Elbow, 90°, black, straight, 3/4" dia.	6.000	Ea.	27.18	342	369.18
Tee, black, straight, 3/4" dia.	2.000	Ea.	17.20	177	194.20
Union, black with brass seat, 3/4" dia.	2.000	Ea.	42	123	165
Pipe nipples, 3/4" diam	3.000	Ea.	7.26	19.58	26.84
Valves, bronze, gate, N.R.S., threaded, class 150, 3/4" size	2.000	Ea.	228	80	308
Circuit setter, balance valve, bronze body, threaded, 3/4" pipe size	1.000	Ea.	99	40.50	139.50
Insulation, fiberglass pipe covering, 1" wall, 3/4" IPS	20.000	L.F.	21.60	116	137.60
Ductwork, 12" x 8" fabricated, galvanized steel, 12 LF	55.000	Lb.	36.30	511.50	547.80
Insulation, ductwork, blanket type, fiberglass, 1" thk, 1-1/2 LB density	40.000	S.F.	197.60	638	835.60
Diffusers, aluminum, OB damper, ceiling, perf, 24"x24" panel size, 6"x6"	2.000	Ea.	364	98	462
Ductwork, flexible, fiberglass fabric, insulated, 1"thk, PE jacket, 6" dia	16.000	L.F.	63.52	86.40	149.92
Round volume control damper 6" dia.	2.000	Ea.	73	71	144
Fan coil unit balancing	1.000	Ea.		101	101
Re-heat coil balancing	1.000	Ea.		150	150
Diffuser/register, high, balancing	2.000	Ea.		282	282
TOTAL			2,855.72	3,298.56	6,154.28

D3040 118	Fan Coil A/C Unit, Two Pipe	COST EACH		
		MAT.	INST.	TOTAL
1000	For alternate A/C units:			
1010	Fan coil A/C system, cabinet mounted, controls, 2 pipe, 1/2 Ton	1,175	1,500	2,675
1020	1 Ton	1,675	1,675	3,350
1030	1-1/2 Ton	1,725	1,700	3,425
1040	2 Ton	2,725	2,075	4,800
1050	3 Ton	3,625	2,150	5,775

D3040 120	Fan Coil A/C Unit, Two Pipe, Electric Heat	COST EACH		
		MAT.	INST.	TOTAL
1010	Fan coil A/C system, cabinet mntd, elect. ht, controls, 2 pipe, 1/2 Ton	2,175	1,500	3,675
1020	1 Ton	2,700	1,675	4,375
1030	1-1/2 Ton	3,150	1,700	4,850
1040	2 Ton	4,600	2,100	6,700
1050	3 Ton	7,550	2,150	9,700

D3040 Distribution Systems

D3040 122	Fan Coil A/C Unit, Four Pipe	COST EACH		
		MAT.	INST.	TOTAL
1010	Fan coil A/C system, cabinet mounted, controls, 4 pipe, 1/2 Ton	2,000	2,925	4,925
1020	1 Ton	2,600	3,100	5,700
1030	1-1/2 Ton	2,650	3,150	5,800
1040	2 Ton	3,275	3,250	6,525
1050	3 Ton	4,775	3,500	8,275

D3040 124	Fan Coil A/C, Horizontal, Duct Mount, 2 Pipe	COST EACH		
		MAT.	INST.	TOTAL
1010	Fan coil A/C system, horizontal w/housing, controls, 2 pipe, 1/2 Ton	2,850	3,300	6,150
1020	1 Ton	3,900	5,250	9,150
1030	1-1/2 Ton	4,825	7,050	11,875
1040	2 Ton	6,175	8,750	14,925
1050	3 Ton	7,950	12,100	20,050
1060	3-1/2 Ton	8,275	12,400	20,675
1070	4 Ton	8,700	14,500	23,200
1080	5 Ton	10,500	18,600	29,100
1090	6 Ton	10,700	19,900	30,600
1100	7 Ton	10,900	20,400	31,300
1110	8 Ton	10,900	20,700	31,600
1120	10 Ton	11,200	21,300	32,500

D3040 126	Fan Coil A/C, Horiz., Duct Mount, 2 Pipe, Elec. Ht.	COST EACH		
		MAT.	INST.	TOTAL
1010	Fan coil A/C system, horiz. hsng, elect. ht, ctrls, 2 pipe, 1/2 Ton	3,125	3,300	6,425
1020	1 Ton	4,300	5,250	9,550
1030	1-1/2 Ton	5,250	7,050	12,300
1040	2 Ton	6,850	8,750	15,600
1050	3 Ton	10,700	12,400	23,100
1060	3-1/2 Ton	11,800	12,400	24,200
1070	4 Ton	13,200	14,500	27,700
1080	5 Ton	16,200	18,600	34,800
1090	6 Ton	16,300	19,900	36,200
1100	7 Ton	17,000	20,700	37,700
1110	8 Ton	17,200	21,300	38,500
1120	10 Ton	18,200	20,400	38,600
2700	Cut & patch to match existing construction, add, minimum	2%	3%	
2800	Maximum	5%	9%	
2900	Dust protection, add, minimum	1%	2%	
3000	Maximum	4%	11%	
3100	Equipment usage curtailment, add, minimum	1%	1%	
3200	Maximum	3%	10%	
3300	Material handling & storage limitation, add, minimum	1%	1%	
3400	Maximum	6%	7%	
3500	Protection of existing work, add, minimum	2%	2%	
3600	Maximum	5%	7%	
3700	Shift work requirements, add, minimum		5%	
3800	Maximum		30%	
3900	Temporary shoring and bracing, add, minimum	2%	5%	
4000	Maximum	5%	12%	

D3040   Distribution Systems

Plate Heat Exchanger

Shell and Tube Heat Exchanger

This page illustrates and describes plate and shell and tube type heat exchangers including pressure gauge and exchanger control system. Lines within Systems Components give the material and installation price on a cost each basis for the components. Prices for heat exchanger systems are on Line Items D3040 610 1010 thru D3040 620 1040. Material quantities and labor costs have been adjusted for the system listed.

Factors To adjust for job conditions other then normal working situations use Lines D3040 620 2700 thru 4000.

System Components	QUANTITY	UNIT	COST EACH		
			MAT.	INST.	TOTAL
Shell and tube type heat exchanger					
with related valves and piping.					
Heat exchanger, 4 pass, 3/4" O.D. copper tubes, by steam at 10 psi, 40 GPM	1.000	Ea.	7,025	365	7,390
Pipe, black steel, Sch 40, threaded, w/couplings & hangers, 10' OC, 2" dia.	50.500	L.F.	383.80	1,136.25	1,520.05
Elbow, 90°, straight, 2" dia.	16.000	Ea.	448	1,272	1,720
Reducer, black steel, concentric, 2" dia.	2.000	Ea.	66	137	203
Valves, bronze, gate, rising stem, threaded, class 150, 2" size	6.000	Ea.	1,482	435	1,917
Valves, bronze, globe, class 150, rising stem, threaded, 2" size	2.000	Ea.	1,520	145	1,665
Strainers, Y type, bronze body, screwed, 125 lb, 2" pipe size	2.000	Ea.	418	125	543
Valve, electric motor actuated, brass, 2 way, screwed, 1-1/2" pipe size	1.000	Ea.	1,050	62	1,112
Union, black with brass seat, 2" dia.	8.000	Ea.	524	676	1,200
Tee, black, straight, 2" dia.	9.000	Ea.	400.50	1,170	1,570.50
Tee, black, reducing run and outlet, 2" dia.	4.000	Ea.	226	520	746
Pipe nipple, black, 2" dia.	2.000	Ea.	3.80	11.25	15.05
Thermometers, stem type, 9" case, 8" stem, 3/4" NPT	2.000	Ea.	412	58	470
Gauges, pressure or vacuum, 3-1/2" diameter dial	2.000	Ea.	40	51	91
Insulation, fiberglass pipe covering, 1" wall, 2" IPS	40.000	L.F.	78.40	268	346.40
Heat exchanger control system	1.000	Ea.	3,125	2,575	5,700
Coil balancing	1.000	Ea.		150	150
TOTAL			17,202.50	9,156.50	26,359

D3040 610	Heat Exchanger, Plate Type	COST EACH		
		MAT.	INST.	TOTAL
1000	For alternate heat exchangers:			
1010	Plate heat exchanger, 400 GPM	66,500	18,700	85,200
1020	800 GPM	107,000	23,800	130,800
1030	1200 GPM	160,500	32,200	192,700
1040	1800 GPM	216,000	40,100	256,100

D3040 620	Heat Exchanger, Shell & Tube	COST EACH		
		MAT.	INST.	TOTAL
1010	Shell & tube heat exchanger, 40 GPM	17,200	9,150	26,350
1020	96 GPM	28,700	16,500	45,200
1030	240 GPM	54,000	22,300	76,300
1040	600 GPM	108,500	31,300	139,800
2700	Cut & patch to match existing construction, add, minimum	2%	3%	
2800	Maximum	5%	9%	
2900	Dust protection, add, minimum	1%	2%	
3000	Maximum	4%	11%	

D3040 Distribution Systems

D3040 620	Heat Exchanger, Shell & Tube	COST EACH		
		MAT.	INST.	TOTAL
3100	Equipment usage curtailment, add, minimum	1%	1%	
3200	Maximum	3%	10%	
3300	Material handling & storage limitation, add, minimum	1%	1%	
3400	Maximum	6%	7%	
3500	Protection of existing work, add, minimum	2%	2%	
3600	Maximum	5%	7%	
3700	Shift work requirements, add, minimum		5%	
3800	Maximum		30%	
3900	Temporary shoring and bracing, add, minimum	2%	5%	
4000	Maximum	5%	12%	

D3050 Terminal & Package Units

Gas Unit Heater

Weather Cap Roof Vent →

Roof Flashing

Vent Pipe

90° Elbow

Tee

Tee Cap

Gas Cock →

Hydronic Unit Heater

This page illustrates and describes unit heaters including piping and vents. Lines within Systems Components give the material and installation price on a cost each basis for this system. Prices for unit heater systems are on Line items D3050 120 1010 thru D3050 140 1040. Material quantities and labor costs have been adjusted for the system listed.

Factors: To adjust for job conditions other than normal working situations use Lines D3050 140 2700 thru 4000.

System Components	QUANTITY	UNIT	COST EACH		
			MAT.	INST.	TOTAL
Unit heaters complete with piping, thermostats and chimney as required					
Space heater, propeller fan, 20 MBH output	1.000	Ea.	1,900	172	2,072
Pipe, black steel, Sch 40, threaded, w/coupling & hangers, 10' OC, 3/4" dia	21.000	L.F.	101.64	274.05	375.69
Elbow, 90°, black steel, straight, 1/2" dia.	3.000	Ea.	13.20	159	172.20
Elbow, 90°, black steel, straight, 3/4" dia.	3.000	Ea.	13.59	171	184.59
Tee, black steel, reducing, 3/4" dia.	1.000	Ea.	14.45	88.50	102.95
Union, black with brass seat, 3/4" dia.	1.000	Ea.	21	61.50	82.50
Pipe nipples, black, 1/2" dia	2.000	Ea.	8.86	25.30	34.16
Pipe nipples, black, 3/4" dia	2.000	Ea.	4.84	13.05	17.89
Pipe cap, black, 3/4" dia.	1.000	Ea.	6.15	25	31.15
Gas cock, brass, 3/4" size	1.000	Ea.	19.90	36	55.90
Thermostat, 1 set back, electric, timed	1.000	Ea.	85.50	101	186.50
Wiring, thermostat hook-up,25' of #18-3	1.000	Ea.	7.45	31.50	38.95
Vent chimney, prefab metal, U.L. listed, gas, double wall, galv. st, 4" dia	12.000	L.F.	123.60	246	369.60
Vent chimney, gas, double wall, galv. steel, elbow 90°, 4" dia.	2.000	Ea.	59	83	142
Vent chimney, gas, double wall, galv. steel, Tee, 4" dia.	1.000	Ea.	41.50	54	95.50
Vent chimney, gas, double wall, galv. steel, T cap, 4" dia.	1.000	Ea.	2.83	33.50	36.33
Vent chimney, gas, double wall, galv. steel, roof flashing, 4" dia.	1.000	Ea.	10.05	41.50	51.55
Vent chimney, gas, double wall, galv. steel, top, 4" dia.	1.000	Ea.	20.50	32	52.50
TOTAL			2,454.06	1,647.90	4,101.96

D3050 120	Unit Heaters, Gas	COST EACH		
		MAT.	INST.	TOTAL
1000	For alternate unit heaters:			
1010	Space heater, suspended, gas fired, propeller fan, 20 MBH	2,450	1,650	4,100
1020	60 MBH	2,625	1,725	4,350
1030	100 MBH	3,000	1,825	4,825
1040	160 MBH	3,625	2,000	5,625
1050	200 MBH	4,525	2,225	6,750
1060	280 MBH	5,275	2,375	7,650
1070	320 MBH	6,950	2,675	9,625

D3050 130	Unit Heaters, Hydronic	COST EACH		
		MAT.	INST.	TOTAL
1010	Space heater, suspended, horiz. mount, HW, prop. fan, 20 MBH	2,475	1,775	4,250
1020	60 MBH	3,400	2,000	5,400
1030	100 MBH	3,875	2,175	6,050
1040	150 MBH	4,300	2,400	6,700

D3050 Terminal & Package Units

D3050 130	Unit Heaters, Hydronic	COST EACH		
		MAT.	INST.	TOTAL
1050	200 MBH	5,250	2,700	7,950
1060	300 MBH	5,950	3,175	9,125

D3050 140	Cabinet Unit Heaters, Hydronic	COST EACH		
		MAT.	INST.	TOTAL
1010	Unit heater, cabinet type, horizontal blower, hot water, 20 MBH	3,100	1,725	4,825
1020	60 MBH	4,575	1,900	6,475
1030	100 MBH	4,850	2,100	6,950
1040	120 MBH	4,850	2,175	7,025
2700	Cut & patch to match existing construction, add, minimum	2%	3%	
2800	Maximum	5%	9%	
2900	Dust protection, add, minimum	1%	2%	
3000	Maximum	4%	11%	
3100	Equipment usage curtailment, add, minimum	1%	1%	
3200	Maximum	3%	10%	
3300	Material handling & storage limitation, add, minimum	1%	1%	
3400	Maximum	6%	7%	
3500	Protection of existing work, add, minimum	2%	2%	
3600	Maximum	5%	7%	
3700	Shift work requirements, add, minimum		5%	
3800	Maximum		30%	
3900	Temporary shoring and bracing, add, minimum	2%	5%	
4000	Maximum	5%	12%	

Rooftop Air Conditioner Unit

Gas Cock

Roof Curb

Gas Piping

This page illustrates and describes packaged air conditioners including rooftop DX unit, pipes and fittings. Lines within Systems Components give the material and installation price on a cost each basis for the components. Prices for A/C systems are on Line Items D3050 175 1010 thru D3050 180 1060. Material quantities and labor costs have been adjusted for the system listed.

Factors: To adjust for job conditions other than normal working situations use Lines D3050 175 2700 thru 4000.

System Components	QUANTITY	UNIT	COST EACH		
			MAT.	INST.	TOTAL
Packaged rooftop DX unit, gas, with related piping and valves.					
Roof top A/C, curb, economizer, sgl zone, elec cool, gas ht, 5 ton, 112 MBH	1.000	Ea.	4,925	2,600	7,525
Pipe, black steel, Sch 40, threaded, w/cplgs & hangers, 10' OC, 1" dia	21.000	L.F.	179.55	316.05	495.60
Elbow, 90°, black, straight, 3/4" dia.	3.000	Ea.	13.59	171	184.59
Elbow, 90°, black, straight, 1" dia.	3.000	Ea.	27.60	184.50	212.10
Tee, black, reducing, 1" dia.	1.000	Ea.	19	99.50	118.50
Union, black with brass seat, 1" dia.	1.000	Ea.	27	66.50	93.50
Pipe nipple, black, 3/4" dia	2.000	Ea.	9.68	26.10	35.78
Pipe nipple, black, 1" dia	2.000	Ea.	8.55	15.05	23.60
Cap, black, 1" dia.	1.000	Ea.	7.45	26.50	33.95
Gas cock, brass, 1" size	1.000	Ea.	37.50	42	79.50
Control system, pneumatic, Rooftop A/C unit	1.000	Ea.	3,225	2,850	6,075
Rooftop unit heat/cool balancing	1.000	Ea.		545	545
TOTAL			8,479.92	6,942.20	15,422.12

D3050 175	Rooftop Air Conditioner, Const. Volume	COST EACH		
		MAT.	INST.	TOTAL
1000	For alternate A/C systems:			
1010	A/C, Rooftop, DX cool, gas heat, curb, economizer, fltrs, 5 Ton	8,475	6,950	15,425
1020	7-1/2 Ton	9,850	7,225	17,075
1030	12-1/2 Ton	15,400	7,950	23,350
1040	18 Ton	19,100	8,700	27,800
1050	25 Ton	42,000	9,900	51,900
1060	40 Ton	52,500	13,200	65,700

D3050 180	Rooftop Air Conditioner, Variable Air Volume	COST EACH		
		MAT.	INST.	TOTAL
1010	A/C, Rooftop, DX cool, gas heat, curb, ecmizr, fltrs, VAV, 12-1/2 Ton	18,400	8,475	26,875
1020	18 Ton	22,900	9,350	32,250
1030	25 Ton	46,800	10,700	57,500
1040	40 Ton	60,000	14,500	74,500
1050	60 Ton	85,500	19,800	105,300
1060	80 Ton	109,000	24,800	133,800
2700	Cut & patch to match existing construction, add, minimum	2%	3%	
2800	Maximum	5%	9%	
2900	Dust protection, add, minimum	1%	2%	
3000	Maximum	4%	11%	
3100	Equipment usage curtailment, add, minimum	1%	1%	
3200	Maximum	3%	10%	

D3050 Terminal & Package Units

D3050 180	Rooftop Air Conditioner, Variable Air Volume	COST EACH		
		MAT.	INST.	TOTAL
3300	Material handling & storage limitation, add, minimum	1%	1%	
3400	Maximum	6%	7%	
3500	Protection of existing work, add, minimum	2%	2%	
3600	Maximum	5%	7%	
3700	Shift work requirements, add, minimum		5%	
3800	Maximum		30%	
3900	Temporary shoring and bracing, add, minimum	2%	5%	
4000	Maximum	5%	12%	

D3050 Terminal & Package Units

This page illustrates and describes packaged electrical air conditioning units including multizone control and related piping. Lines within Systems Components give the material and installation price on a cost each basis for the components. Prices for packaged electrical A/C unit systems are on Line Items D3050 210 1010 thru D3050 225 1060. Material quantities and labor costs have been adjusted for the system listed.

Factors: To adjust for job conditions other than normal working situations use Lines D3050 225 2700 thru 4000.

Condenser Supply Water Piping   Condenser Return Water Piping

Self–Contained Air Conditioner Unit, Water Cooled

System Components	QUANTITY	UNIT	COST EACH MAT.	COST EACH INST.	COST EACH TOTAL
Packaged water cooled electric air conditioning unit					
with related piping and valves.					
Self-contained, water cooled, elect. heat, not inc tower, 5 ton, const vol	1.000	Ea.	9,525	1,900	11,425
Pipe, black steel, Sch 40, threaded, w/cplg & hangers, 10' OC, 1-1/4" dia.	21.500	L.F.	193.50	346.15	539.65
Elbow, 90°, black, straight, 1-1/4" dia.	6.000	Ea.	91.50	390	481.50
Tee, black, straight, 1-1/4" dia.	2.000	Ea.	48	204	252
Tee, black, reducing, 1-1/4" dia.	2.000	Ea.	82	204	286
Thermometers, stem type, 9" case, 8" stem, 3/4" NPT	2.000	Ea.	412	58	470
Union, black with brass seat, 1-1/4" dia.	2.000	Ea.	91	137	228
Pipe nipple, black, 1-1/4" dia	3.000	Ea.	13.50	24.15	37.65
Valves, bronze, gate, N.R.S., threaded, class 150, 1-1/4" size	2.000	Ea.	302	106	408
Circuit setter, bal valve, bronze body, threaded, 1-1/4" pipe size	1.000	Ea.	189	54	243
Insulation, fiberglass pipe covering, 1" wall, 1-1/4" IPS	20.000	L.F.	25.60	128	153.60
Control system, pneumatic, A/C Unit with heat	1.000	Ea.	3,225	2,850	6,075
Re-heat coil balancing	1.000	Ea.		150	150
Rooftop unit heat/cool balancing	1.000	Ea.		545	545
TOTAL			14,198.10	7,096.30	21,294.40

D3050 210	AC Unit, Package, Elec. Ht., Water Cooled	COST EACH MAT.	COST EACH INST.	COST EACH TOTAL
1000	For alternate A/C systems:			
1010	A/C, Self contained, single pkg., water cooled, elect. heat, 5 Ton	14,200	7,100	21,300
1020	10 Ton	21,200	8,750	29,950
1030	20 Ton	28,500	11,200	39,700
1040	30 Ton	50,500	12,200	62,700
1050	40 Ton	63,500	13,500	77,000
1060	50 Ton	85,000	16,200	101,200

D3050 215	AC Unit, Package, Elec. Ht., Water Cooled, VAV	COST EACH MAT.	COST EACH INST.	COST EACH TOTAL
1010	A/C, Self contained, single pkg., water cooled, elect. ht, VAV, 10 Ton	23,900	8,750	32,650
1020	20 Ton	33,200	11,200	44,400
1030	30 Ton	57,000	12,200	69,200
1040	40 Ton	71,500	13,500	85,000
1050	50 Ton	96,000	16,200	112,200
1060	60 Ton	110,500	19,600	130,100

D30 HVAC

D3050 Terminal & Package Units

D3050 220	AC Unit, Package, Hot Water Coil, Water Cooled	COST EACH		
		MAT.	INST.	TOTAL
1010	A/C, Self contn'd, single pkg., water cool, H/W ht, const. vol, 5 Ton	12,700	9,825	22,525
1020	10 Ton	17,500	11,700	29,200
1030	20 Ton	42,500	14,800	57,300
1040	30 Ton	55,500	18,700	74,200
1050	40 Ton	65,500	22,700	88,200
1060	50 Ton	74,500	28,100	102,600

D3050 225	AC Unit, Package, HW Coil, Water Cooled, VAV	COST EACH		
		MAT.	INST.	TOTAL
1010	A/C, Self contn'd, single pkg., water cool, H/W ht, VAV, 10 Ton	20,300	11,000	31,300
1020	20 Ton	28,900	14,400	43,300
1030	30 Ton	50,000	18,000	68,000
1040	40 Ton	64,500	19,600	84,100
1050	50 Ton	85,500	23,200	108,700
1060	60 Ton	102,000	25,200	127,200
2700	Cut & patch to match existing construction, add, minimum	2%	3%	
2800	Maximum	5%	9%	
2900	Dust protection, add, minimum	1%	2%	
3000	Maximum	4%	11%	
3100	Equipment usage curtailment, add, minimum	1%	1%	
3200	Maximum	3%	10%	
3300	Material handling & storage limitation, add, minimum	1%	1%	
3400	Maximum	6%	7%	
3500	Protection of existing work, add, minimum	2%	2%	
3600	Maximum	5%	7%	
3700	Shift work requirements, add, minimum		5%	
3800	Maximum		30%	
3900	Temporary shoring and bracing, add, minimum	2%	5%	
4000	Maximum	5%	12%	

D3050 Terminal & Package Units

Thru-Wall Heat Pump   Thru-Wall Air Conditioner

This page illustrates and describes thru-wall air conditioning and heat pump units. Lines within Systems Components give the material and installation price on a cost each basis for the components. Prices for thru-wall unit systems are on Line Items D3050 255 1010 thru D3050 260 1050. Material quantities and labor costs have been adjusted for the system listed.

Factors: To adjust for job conditions other than normal working conditions use Lines D3050 255 2700 thru 4000.

System Components		QUANTITY	UNIT	COST EACH		
				MAT.	INST.	TOTAL
Electric thru-wall air conditioning unit.						
A/C unit, thru-wall, electric heat., cabinet, louver, 1/2 ton		1.000	Ea.	800	243	1,043
	TOTAL			800	243	1,043

D3050 255	Thru-Wall A/C Unit	COST EACH		
		MAT.	INST.	TOTAL
1000	For alternate A/C units:			
1010	A/C Unit, thru-the-wall, supplemental elect. heat, cabinet, louver, 1/2 Ton	800	243	1,043
1020	3/4 Ton	1,075	292	1,367
1030	1 Ton	1,700	365	2,065
1040	1-1/2 Ton	2,975	610	3,585
1050	2 Ton	3,000	770	3,770

D3050 260	Thru-Wall Heat Pump	COST EACH		
		MAT.	INST.	TOTAL
1010	Heat pump, thru-the-wall, cabinet, louver, 1/2 Ton	3,900	182	4,082
1020	3/4 Ton	4,575	243	4,818
1030	1 Ton	3,775	365	4,140
1040	Supplemental. elect. heat, 1-1/2 Ton	3,725	940	4,665
1050	2 Ton	3,675	975	4,650
2700	Cut & patch to match existing construction, add, minimum	2%	3%	
2800	Maximum	5%	9%	
2900	Dust protection, add, minimum	1%	2%	
3000	Maximum	4%	11%	
3100	Equipment usage curtailment, add, minimum	1%	1%	
3200	Maximum	3%	10%	
3300	Material handling & storage limitation, add, minimum	1%	1%	
3400	Maximum	6%	7%	
3500	Protection of existing work, add, minimum	2%	2%	
3600	Maximum	5%	7%	
3700	Shift work requirements, add, minimum		5%	
3800	Maximum		30%	
3900	Temporary shoring and bracing, add, minimum	2%	5%	
4000	Maximum	5%	12%	

D4010 Sprinklers

This page illustrates and describes a dry type fire sprinkler system. Lines within System Components give the unit cost of a system for a 2,000 square foot building. Lines D4010 320 1700 thru 2400 give the square foot costs for alternate systems. Both material quantities and labor costs have been adjusted for the system listed.

Factors: To adjust for conditions other than normal working conditions, use Lines D4010 320 2900 thru 4000.

System Components	QUANTITY	UNIT	COST EACH		
			MAT.	INST.	TOTAL
Pre-action fire sprinkler system, ordinary hazard, open area to 2000 S.F. On one floor.					
Air compressor	1.000	Ea.	1,325	605	1,930
4" OS & Y valve	1.000	Ea.	755	480	1,235
Pipe riser 4" diameter	10.000	L.F.	265	421.70	686.70
Sprinkler head supply piping, 1"	163.000	L.F.	641.25	1,128.75	1,770
Sprinkler head supply piping, 1-1/4"	163.000	L.F.	495	885.50	1,380.50
Sprinkler head supply piping, 2-1/2"	163.000	L.F.	476.90	1,083	1,559.90
Pipe fittings, 1"	8.000	Ea.	118	796	914
Pipe fittings, 1-1/4"	8.000	Ea.	192	816	1,008
Pipe fittings, 2-1/2"	4.000	Ea.	436	636	1,072
Pipe fittings, 4"	2.000	Ea.	780	720	1,500
Detectors	2.000	Ea.	1,760	98	1,858
Sprinkler heads	16.000	Ea.	197.60	784	981.60
Fire department connection	1.000	Ea.	850	282	1,132
TOTAL			8,291.75	8,735.95	17,027.70

D4010 320	Fire Sprinkler Systems, Dry	COST PER S.F.		
		MAT.	INST.	TOTAL
1700	Ordinary hazard, one floor, area to 2000 S.F./floor	4.16	4.37	8.53
1800	For each additional floor, add per floor	1.81	3.64	5.45
1900	Area to 3200 S.F./ floor	3.27	4.35	7.62
2000	For each additional floor, add per floor	1.80	3.89	5.69
2100	Area to 5000 S.F./ floor	3.63	4.41	8.04
2200	For each additional floor, add per floor	2.34	4.09	6.43
2300	Area to 8000 S.F./ floor	2.80	3.72	6.52
2400	For each additional floor, add per floor	1.88	3.51	5.39
2500				
2600				
2700				
2800				
2900	Cut & patch to match existing construction, add, minimum	2%	3%	
3000	Maximum	5%	9%	
3100	Dust protection, add, minimum	1%	2%	
3200	Maximum	4%	11%	
3300	Equipment usage curtailment, add, minimum	1%	1%	
3400	Maximum	3%	10%	

D40 Fire Protection

D4010 Sprinklers

D4010 320	Fire Sprinkler Systems, Dry	COST PER S.F.		
		MAT.	INST.	TOTAL
3500	Material handling & storage limitation, add, minimum	1%	1%	
3600	Maximum	6%	7%	
3700	Protection of existing work, add, minimum	2%	2%	
3800	Maximum	5%	7%	
3900	Shift work requirements, add, minimum		5%	
4000	Maximum		30%	

This page illustrates and describes a wet type fire sprinkler system. Lines within System Components give the unit cost of a system for a 2,000 square foot building. Lines D4010 420 1900 thru 2600 give the square foot costs for alternate systems. Both material quantities and labor costs have been adjusted for the system listed.

Factors: To adjust for conditions other than normal working conditions, use Lines D4010 420 2900 thru 4000.

System Components	QUANTITY	UNIT	COST EACH MAT.	COST EACH INST.	COST EACH TOTAL
Wet pipe fire sprinkler system, ordinary hazard, open area to 2000 S.F.					
On one floor.					
4" OS & Y valve	1.000	Ea.	755	480	1,235
Wet pipe alarm valve	1.000	Ea.	2,550	705	3,255
Water motor alarm	1.000	Ea.	530	196	726
3" check valve	1.000	Ea.	460	470	930
Pipe riser, 4" diameter	10.000	L.F.	265	421.70	686.70
Water gauges and trim	1.000	Set	3,625	1,400	5,025
Electric fire horn	1.000	Ea.	59.50	113	172.50
Sprinkler head supply piping, 1" diameter	168.000	L.F.	641.25	1,128.75	1,770
Sprinkler head supply piping, 1-1/2" diameter	168.000	L.F.	528	987.25	1,515.25
Sprinkler head supply piping, 2-1/2" diameter	168.000	L.F.	476.90	1,083	1,559.90
Pipe fittings, 1"	8.000	Ea.	118	796	914
Pipe fittings, 1-1/2"	8.000	Ea.	236	880	1,116
Pipe fittings, 2-1/2"	4.000	Ea.	436	636	1,072
Pipe fittings, 4"	2.000	Ea.	780	720	1,500
Sprinkler heads	16.000	Ea.	197.60	784	981.60
Fire department connection	1.000	Ea.	850	282	1,132
TOTAL			12,508.25	11,082.70	23,590.95

D4010 420	Fire Sprinkler Systems, Wet	COST PER S.F. MAT.	COST PER S.F. INST.	COST PER S.F. TOTAL
1900	Ordinary hazard, one floor, area to 2000 S.F./floor	9.25	5.85	15.10
2000	For each additional floor, add per floor	1.84	3.72	5.56
2100	Area to 3200 S.F./floor	6.45	5.25	11.70
2200	For each additional floor, add per floor	1.80	3.89	5.69
2300	Area to 5000 S.F./ floor	5.30	4.95	10.25
2400	For each additional floor, add per floor	2.34	4.09	6.43
2500	Area to 8000 S.F./ floor	3.74	4.05	7.79
2600	For each additional floor, add per floor	1.99	3.73	5.72
2700				
2900	Cut & patch to match existing construction, add minimum	2%	3%	
3000	Maximum	5%	9%	
3100	Dust protection, add, minimum	1%	2%	
3200	Maximum	4%	11%	
3300	Equipment usage curtailment, add, minimum	1%	1%	
3400	Maximum	3%	10%	
3500	Material handling & storage limitation, add, minimum	1%	1%	
3600	Maximum	6%	7%	

D5010 Electrical Service/Distribution

This page illustrates and describes commercial service systems including a meter socket, service head and cable, entrance switch, steel conduit, copper wire, panel board, ground rod, wire, and conduit. Lines within System Components give the unit price and total price on a cost each basis for this system. Prices for commercial service systems are on Line Items D5010 210 1500 thru 1700. Both material quantities and labor costs have been adjusted for the system listed.

Factors: To adjust for job conditions other than normal working situations use Lines D5010 210 2900 thru 4000.

Example: You are to install the system with maximum equipment usage curtailment. Go to Line D5010 210 3400 and apply these percentages to the appropriate MAT. and INST. costs.

System Components	QUANTITY	UNIT	COST EACH		
			MAT.	INST.	TOTAL
Commercial electric service including service breakers, metering 120/208 Volt, 3 phase, 4 wire, feeder, and panel board.					
100 Amp Service					
Meter socket	1.000	Ea.	58	236	294
Service head	.200	C.L.F.	31.60	151	182.60
Service entrance switch	1.000	Ea.	192	400	592
Rigid steel conduit	20.000	L.F.	86.80	232	318.80
600 volt copper wire #3	1.000	C.L.F.	129	151	280
Panel board, 24 circuits, 20 Amp breakers	1.000	Ea.	1,400	1,250	2,650
Ground rod	1.000	Ea.	25	137	162
TOTAL			1,922.40	2,557	4,479.40

D5010 210	Commercial Service - 3 Phase	COST EACH		
		MAT.	INST.	TOTAL
1500	120/208 Volt, 3 phase, 4 wire service, 60 Amp	1,475	2,000	3,475
1550	100 Amp	1,925	2,550	4,475
1600	200 Amp	3,875	4,525	8,400
1700	400 Amp	6,925	5,650	12,575
1800				
1900				
2000				
2100				
2200				
2300				
2400				
2500				
2900	Cut & patch to match existing construction, add, minimum	2%	3%	
3000	Maximum	5%	9%	
3100	Dust protection, add, minimum	1%	2%	
3200	Maximum	4%	11%	
3300	Equipment usage curtailment, add, minimum	1%	1%	
3400	Maximum	3%	10%	
3500	Material handling & storage limitation, add, minimum	1%	1%	
3600	Maximum	6%	7%	
3700	Protection of existing work, add, minimum	2%	2%	
3800	Maximum	5%	7%	
3900	Shift work requirements, add, minimum		5%	
4000	Maximum		30%	

D5010 Electrical Service/Distribution

This page illustrates and describes a residential, single phase system including a weather cap, service entrance cable, meter socket, entrance switch, ground rod, ground cable, EMT, and panelboard. Lines within System Components give the unit price and total price on a cost each basis for this system. Prices for a residential, single phase system are also given. Both material quantities and labor costs have been adjusted for the system listed.

Factors: To adjust for job conditions other than normal working situations use Lines D5010 220 2900 thru 4000.

Example: You are to install the system with a minimum equipment usage curtailment. Go to Line D5010 220 3300 and apply these percentages to the appropriate MAT. and INST. costs.

System Components	QUANTITY	UNIT	COST EACH		
			MAT.	INST.	TOTAL
100 Amp Service, single phase					
Weathercap	1.000	Ea.	8.65	63	71.65
Service entrance cable	.200	C.L.F.	31.60	151	182.60
Meter socket	1.000	Ea.	58	236	294
Entrance disconnect switch	1.000	Ea.	192	400	592
Ground rod, with clamp	1.000	Ea.	27	157	184
Ground cable	.100	C.L.F.	16	47	63
Panelboard, 12 circuit	1.000	Ea.	140	630	770
TOTAL			473.25	1,684	2,157.25
200 Amp Service , single phase					
Weathercap	1.000	Ea.	19.60	94.50	114.10
Service entrance cable	.200	C.L.F.	64	215	279
Meter socket	1.000	Ea.	121	400	521
Entrance disconnect switch	1.000	Ea.	410	580	990
Ground rod, with clamp	1.000	Ea.	58.50	172	230.50
Ground cable	.100	C.L.F.	43	26.10	69.10
3/4" EMT	10.000	L.F.	11.80	58	69.80
Panelboard, 24 circuit	1.000	Ea.	293	1,050	1,343
TOTAL			1,020.90	2,595.60	3,616.50

D5010 220	Residential Service - Single Phase	COST EACH		
		MAT.	INST.	TOTAL
2800				
2810	100 Amp Service, single phase	475	1,675	2,150
2820	200 Amp Service, single phase	1,025	2,600	3,625
2900	Cut & patch to match existing construction, add, minimum	2%	3%	
3000	Maximum	5%	9%	
3100	Dust protection, add, minimum	1%	2%	
3200	Maximum	4%	11%	
3300	Equipment usage curtailment, add, minimum	1%	1%	
3400	Maximum	3%	10%	
3500	Material handling & storage limitation, add, minimum	1%	1%	
3600	Maximum	6%	7%	
3700	Protection of existing work, add, minimum	2%	2%	
3800	Maximum	5%	7%	
3900	Shift work requirements, add, minimum		5%	
4000	Maximum		30%	

D5020 Lighting and Branch Wiring

D5020 180	Selective Price Sheet	COST EACH		
		MAT.	INST.	TOTAL
0100	Using non-metallic sheathed, cable, air conditioning receptacle	21.50	75.50	97
0200	Disposal wiring	22	84	106
0300	Dryer circuit	32.50	137	169.50
0400	Duplex receptacle	21.50	58	79.50
0500	Fire alarm or smoke detector	91.50	75.50	167
0600	Furnace circuit & switch	29	126	155
0700	Ground fault receptacle	62.50	94.50	157
0800	Heater circuit	24.50	94.50	119
0900	Lighting wiring	28	47	75
1000	Range circuit	78	189	267
1100	Switches single pole	13.55	47	60.55
1200	3-way	16.85	63	79.85
1300	Water heater circuit	22.50	151	173.50
1400	Weatherproof receptacle	148	126	274
1500	Using BX cable, air conditioning receptacle	32	91	123
1600	Disposal wiring	32	101	133
1700	Dryer circuit	41	164	205
1800	Duplex receptacle	32	70	102
1900	Fire alarm or smoke detector	91.50	75.50	167
2000	Furnace circuit & switch	41	151	192
2100	Ground fault receptacle	71.50	114	185.50
2200	Heater circuit	30.50	114	144.50
2300	Lighting wiring	33	57	90
2400	Range circuit	118	229	347
2500	Switches, single pole	24	57	81
2600	3-way	23	75.50	98.50
2700	Water heater circuit	36.50	180	216.50
2800	Weatherproof receptacle	154	151	305
2900	Using EMT conduit, air conditioning receptacle	46	113	159
3000	Disposal wiring	48	126	174
3100	Dryer circuit	52.50	204	256.50
3200	Duplex receptacle	46	87	133
3300	Fire alarm or smoke detector	110	113	223
3400	Furnace circuit & switch	52	189	241
3500	Ground fault receptacle	87.50	140	227.50
3600	Heater circuit	42.50	140	182.50
3700	Lighting wiring	40	70.50	110.50
3800	Range circuit	94.50	280	374.50
3900	Switches, single pole	38.50	70.50	109
4000	3-way	34	94.50	128.50
4100	Water heater circuit	43	222	265
4200	Weatherproof receptacle	169	189	358
4300	Using aluminum conduit, air conditioning receptacle	44.50	151	195.50
4400	Disposal wiring	46.50	168	214.50
4500	Dryer circuit	58.50	270	328.50
4600	Duplex receptacle	43	116	159
4700	Fire alarm or smoke detector	114	151	265
4800	Furnace circuit & switch	54	252	306
4900	Ground fault receptacle	92.50	189	281.50
5000	Heater circuit	41.50	189	230.50
5100	Lighting wiring	46.50	94.50	141
5200	Range circuit	107	380	487
5300	Switches, single pole	44.50	94.50	139
5400	3-way	40	126	166
5500	Water heater circuit	48	300	348
5600	Weatherproof receptacle	181	252	433
5700	Using galvanized steel conduit	50.50	161	211.50
5800	Disposal wiring	49	180	229

D5020 Lighting and Branch Wiring

D5020 180	Selective Price Sheet	COST EACH		
		MAT.	INST.	TOTAL
5900	Dryer circuit	114	291	405
6000	Duplex receptacle	50.50	124	174.50
6100	Fire alarm or smoke detector	140	161	301
6200	Furnace circuit & switch	79.50	270	349.50
6300	Ground fault receptacle	102	199	301
6400	Heater circuit	71	199	270
6500	Lighting wiring	73	101	174
6600	Range circuit	116	400	516
6700	Switches, single pole	53.50	101	154.50
6800	3-way	93.50	130	223.50
6900	Water heater circuit	57.50	315	372.50
7000	Weatherproof receptacle	185	270	455
7100				
7200				
7300				
7400				
7500				
7600				
7700				
7800				
7900				
8000				
8100				
8200				
8300				
8400				

D5020 Lighting and Branch Wiring

This page illustrates and describes fluorescent lighting systems including a fixture, lamp, outlet box and wiring. Lines within System Components give the unit price and total price on a cost each basis for this system. Prices for fluorescent lighting systems are on Line Items D5020 248 1200 thru 1500. Both material quantities and labor costs have been adjusted for the system listed.

Factors: To adjust for job conditions other than normal working situations use Lines D5020 248 2900 thru 4000.

Example: You are to install the system during evening hours. Go to Line D5020 248 3900 and apply this percentage to the appropriate INST. cost.

System Components	QUANTITY	UNIT	COST EACH		
			MAT.	INST.	TOTAL
Fluorescent lighting, including fixture, lamp, outlet box and wiring.					
Recessed lighting fixture, on suspended system	1.000	Ea.	71	161	232
Outlet box	1.000	Ea.	7.55	42	49.55
#12 wire	.660	C.L.F.	7.29	45.21	52.50
Conduit, EMT, 1/2" conduit	20.000	L.F.	16.60	88.80	105.40
TOTAL			102.44	337.01	439.45

D5020 248	Lighting, Fluorescent	COST EACH		
		MAT.	INST.	TOTAL
1200	Recessed lighting fixture, on suspended system	102	335	437
1300	Surface mounted, 2' x 4', acrylic prismatic diffuser	126	355	481
1400	Strip fixture, 8' long, two 8' lamps	109	335	444
1500	Pendant mounted, industrial, 8' long, with reflectors	142	385	527
1600				
1700				
1800				
1900				
2000				
2100				
2200				
2300				
2900	Cut & patch to match existing construction, add, minimum	2%	3%	
3000	Maximum	5%	9%	
3100	Dust protection, add, minimum	1%	2%	
3200	Maximum	4%	11%	
3300	Equipment usage curtailment, add, minimum	1%	1%	
3400	Maximum	3%	10%	
3500	Material handling & storage limitation, add, minimum	1%	1%	
3600	Maximum	6%	7%	
3700	Protection of existing work, add, minimum	2%	2%	
3800	Maximum	5%	7%	
3900	Shift work requirements, add, minimum		5%	
4000	Maximum		30%	

D5020 Lighting and Branch Wiring

This page illustrates and describes incandescent lighting systems including a fixture, lamp, outlet box, conduit and wiring. Lines within System Components give the unit price and total price on a cost each basis for this system. Prices for an alternate incandescent lighting system are also given. Both material quantities and labor costs have been adjusted for the system listed.

Factors: To adjust for conditions other than normal working situations use Lines D5020 250 2900 thru 4000.

Example: You are to install the system and cut and match existing construction. Go to Line D5020 250 3000 and apply this percentage to the appropriate INST. costs.

System Components		QUANTITY	UNIT	COST EACH MAT.	COST EACH INST.	COST EACH TOTAL
Incandescent light fixture, including lamp, outlet box, conduit and wiring.						
Recessed wide reflector with flat glass lens		1.000	Ea.	107	113	220
Outlet box		1.000	Ea.	7.55	42	49.55
Armored cable, 3 wire		.200	C.L.F.	14.40	69	83.40
	TOTAL			128.95	224	352.95
Recessed flood light fixture, including lamp, outlet box, conduit & wire						
Recessed, R-40 flood lamp with refractor		1.000	Ea.	114	94.50	208.50
150 watt R-40 flood lamp		.010	Ea.	7.65	5.80	13.45
Outlet box		1.000	Ea.	7.55	42	49.55
Outlet box cover		1.000	Ea.	.85	11.80	12.65
Romex, 12-2 with ground		.200	C.L.F.	5.80	69	74.80
Conduit, 1/2" EMT		20.000	L.F.	16.60	88.80	105.40
	TOTAL			152.45	311.90	464.35

D5020 250	Lighting, Incandescent	COST EACH MAT.	COST EACH INST.	COST EACH TOTAL
2810	Incandescent light fixture, including lamp, outlet box, conduit and wiring.	129	224	353
2820	Recessed flood light fixture, including lamp, outlet box, conduit & wire	152	310	462
2900	Cut & patch to match existing construction, add, minimum	2%	3%	
3000	Maximum	5%	9%	
3100	Dust protection, add, minimum	1%	2%	
3200	Maximum	4%	11%	
3300	Equipment usage curtailment, add, minimum	1%	1%	
3400	Maximum	3%	10%	
3500	Material handling & storage limitation, add, minimum	1%	1%	
3600	Maximum	6%	7%	
3700	Protection of existing work, add, minimum	2%	2%	
3800	Maximum	5%	7%	
3900	Shift work requirements, add, minimum		5%	
4000	Maximum		30%	

D50 Electrical

D5020 Lighting and Branch Wiring

This page illustrates and describes high intensity lighting systems including a lamp, EMT conduit, EMT "T" fitting with cover, and wire. Lines within System Components give the unit price and total price on a cost each basis for this system. Prices for high intensity lighting systems are on Line Items D5020 252 1100 thru 1800. Both material quantities and labor costs have been adjusted for the system listed.

Factors: To adjust for job conditions other than normal working situations use Lines D5020 252 2900 thru 4000.

Example: You are to install the system and protect existing construction. Go to Line D5020 252 3700 and apply these percentages to the appropriate MAT. and INST. costs.

System Components			COST EACH		
	QUANTITY	UNIT	MAT.	INST.	TOTAL
High intensity lighting system consisting of 400 watt metal halide fixture					
And lamp with 1/2" EMT conduit and fittings using #12 wire, high bay.					
Single unit, 400 watt	1.000	Ea.	465	330	795
Electric metallic tubing (EMT), 1/2" diameter	30.000	Ea.	24.90	133.20	158.10
1/2" EMT "T" fitting with cover	1.000	Ea.	6.15	47	53.15
Wire, 600 volt, type THWN-THHN, copper, solid, #12	.600	Ea.	6.63	41.10	47.73
TOTAL			502.68	551.30	1,053.98

D5020 252	Lighting, High Intensity	COST EACH		
		MAT.	INST.	TOTAL
1100	High bay: 400 watt, metal halide fixture and lamp	505	550	1,055
1200	High pressure sodium and lamp	470	550	1,020
1400	1000 watt, metal halide	710	600	1,310
1500	High pressure sodium	640	600	1,240
1700	Low bay: 250 watt, metal halide	440	455	895
1800	150 watt, high pressure sodium	410	455	865
1900				
2000				
2100				
2200				
2300				
2400				
2500				
2600				
2700				
2900	Cut & patch to match existing construction, add, minimum	2%	3%	
3000	Maximum	5%	9%	
3100	Dust protection, add, minimum	1%	2%	
3200	Maximum	4%	11%	
3300	Equipment usage curtailment, add, minimum	1%	1%	
3400	Maximum	3%	10%	
3500	Material handling & storage limitation, add, minimum	1%	1%	
3600	Maximum	6%	7%	
3700	Protection of existing work, add, minimum	2%	2%	
3800	Maximum	5%	7%	
3900	Shift work requirements, add, minimum		5%	
4000	Maximum		30%	

D5090 Other Electrical Systems

This page illustrates and describes baseboard heat systems including a thermostat, outlet box, breaker, and feed. Lines within System Components give the unit price and total price on a cost each basis for this system. Prices for baseboard heat systems are on Line Items D5090 430 1500 thru 1800. Both material quantities and labor costs have been adjusted for the system listed.

Factors: To adjust for job conditions other than normal working situations use Lines D5090 430 2900 thru 4000.

Example: You are to install the system during evenings and weekends only. Go to Line D5090 430 4000 and apply this percentage to the appropriate INST. cost.

System Components	QUANTITY	UNIT	COST EACH		
			MAT.	INST.	TOTAL
Baseboard heat including thermostat, outlet box, breaker and feed.					
Electric baseboard heater, 4' long	1.000	Ea.	45	113	158
Thermostat, integral	1.000	Ea.	32	47	79
Romex, 12-3 with ground	.400	C.L.F.	18	152	170
Panel board breaker, 15-50 amp	1.000	Ea.	62	94.50	156.50
Outlet box	1.000	Ea.	7.55	42	49.55
TOTAL			164.55	448.50	613.05

D5090 530	Heat, Baseboard	COST EACH		
		MAT.	INST.	TOTAL
1400	Heat, electric baseboard			
1500	2' long	151	430	581
1550	4' long	165	450	615
1600	6' long	183	485	668
1700	8' long	197	525	722
1800	10' long	298	565	863
1900				
2000				
2100				
2200				
2300				
2400				
2500				
2600				
2700				
2900	Cut & patch to match existing construction, add, minimum	2%	3%	
3000	Maximum	5%	9%	
3100	Dust protection, add, minimum	1%	2%	
3200	Maximum	4%	11%	
3300	Equipment usage curtailment, add, minimum	1%	1%	
3400	Maximum	3%	10%	
3500	Material handling & storage limitation, add, minimum	1%	1%	
3600	Maximum	6%	7%	
3700	Protection of existing work, add, minimum	2%	2%	
3800	Maximum	5%	7%	
3900	Shift work requirements, add, minimum		5%	
4000	Maximum		30%	

715

E1090   Other Equipment

This page illustrates and describes kitchen systems including top and bottom cabinets, custom laminated plastic top, single bowl sink, and appliances. Lines within System Components give the unit price and total price on a cost each basis for this system. Prices for kitchen systems are on Line Items E1090 310 1400 thru 1600. Both material quantities and labor costs have been adjusted for the system listed.

Factors: To adjust for job conditions other than normal working situations use Lines E1090 310 2900 thru 4000.

Example: You are to install the system and protect the work area from dust. Go to Line E1090 310 3200 and apply these percentages to the appropriate MAT. and INST. costs.

System Components	QUANTITY	UNIT	COST EACH MAT.	COST EACH INST.	COST EACH TOTAL
Kitchen cabinets including wall and base cabinets, custom laminated					
Plastic top,sink & appliances,no plumbing or electrical rough-in included.					
Prefinished wood cabinets, average quality, wall and base	20.000	L.F.	4,720	910	5,630
Custom laminated plastic counter top	20.000	L.F.	390	450	840
Stainless steel sink, 22" x 25"	1.000	Ea.	760	256	1,016
Faucet, top mount	1.000	Ea.	92.50	79.50	172
Dishwasher, built-in	1.000	Ea.	340	390	730
Compactor, built-in	1.000	Ea.	775	136	911
Range hood, vented, 30" wide	1.000	Ea.	118	290	408
TOTAL			7,195.50	2,511.50	9,707

E1090 310	Kitchens	COST EACH MAT.	COST EACH INST.	COST EACH TOTAL
1400	Prefinished wood cabinets, average quality	7,200	2,500	9,700
1500	Prefinished wood cabinets, high quality	13,200	2,875	16,075
1600	Custom cabinets, built in place, high quality	17,900	3,325	21,225
1700				
1800				
1900				
2000				
2100				
2200	NOTE: No plumbing or electric rough-ins are included in the above			
2300	Prices, for plumbing see Division 22, for electric see Division 26.			
2400				
2500				
2600				
2700				
2900	Cut & patch to match existing construction, add, minimum	2%	3%	
3000	Maximum	5%	9%	
3100	Dust protection, add, minimum	1%	2%	
3200	Maximum	4%	11%	
3300	Equipment usage curtailment, add, minimum	1%	1%	
3400	Maximum	3%	10%	
3500	Material handling & storage limitation, add, minimum	1%	1%	
3600	Maximum	6%	7%	
3700	Protection of existing work, add, minimum	2%	2%	
3800	Maximum	5%	7%	
3900	Shift work requirements, add, minimum		5%	
4000	Maximum		30%	

E10 Equipment

E1090 Other Equipment

E1090 315	Selective Price Sheet	COST EACH		
		MAT.	INST.	TOTAL
0100	Cabinets standard wood, base, one drawer one door, 12" wide	355	55	410
0200	15" wide	425	57	482
0300	18" wide	330	58.50	388.50
0400	21" wide	345	60	405
0500	24" wide	480	61	541
0600				
0700	Two drawers two doors, 27" wide	500	62	562
0800	30" wide	570	63.50	633.50
0900	33" wide	600	65	665
1000	36" wide	630	67	697
1100	42" wide	655	69	724
1200	48" wide	765	72	837
1300	Drawer base (4 drawers), 12" wide	365	55	420
1400	15" wide	370	57	427
1500	18" wide	405	58.50	463.50
1600	24" wide	460	61	521
1700	Sink or range base, 30" wide	505	63.50	568.50
1800	33" wide	535	65	600
1900	36" wide	545	67	612
2000	42" wide	595	69	664
2100	Corner base, 36" wide	870	75.50	945.50
2200	Lazy Susan with revolving door	1,125	82.50	1,207.50
2300	Cabinets standard wood, wall two doors, 12" high, 30" wide	305	55	360
2500	15" high, 30" high	315	57	372
2600	36" wide	390	60	450
2700	24" high, 30" wide	460	58.50	518.50
2800	36" wide	535	60	595
2900	30" high, 30" wide	535	70.50	605.50
3000	36" wide	510	72.50	582.50
3100	42" wide	585	73.50	658.50
3200	48" wide	685	74	759
3300	One door, 30" high, 12" wide	310	62	372
3400	15" wide	320	63.50	383.50
3500	18" wide	350	65	415
3600	24" wide	415	67	482
3700	Corner, 30" high, 24" wide	480	75.50	555.50
3800	36" wide	525	82.50	607.50
3900	Broom, 84" high, 24" deep, 18" wide	865	136	1,001
4000	Oven, 84" high, 24" deep, 27" wide	1,175	170	1,345
4100	Valance board, 4' long	92	13.75	105.75
4200	6" long	138	20.50	158.50
4300	Counter tops, laminated plastic, stock 25" wide w/backsplash, min.	19.50	22.50	42
4400	Maximum	46.50	27.50	74
4500	Custom, 7/8" thick, no splash	39	22.50	61.50
4600	Cove splash	51.50	22.50	74
4700	1-1/4" thick, no splash	43.50	24.50	68
4800	Square splash	48.50	24.50	73
4900	Post formed	13.80	22.50	36.30
5000				
5100	Maple laminated 1-1/2" thick, no splash	101	24.50	125.50
5200	Square splash	120	24.50	144.50
5300				
5400				
5500	Appliances, range, free standing, minimum	510	108	618
5600	Maximum	2,425	270	2,695
5700	Built-in, minimum	955	126	1,081
5800	Maximum	2,200	680	2,880
5900	Counter top range 4 burner, minimum	370	126	496

E1090 Other Equipment

E1090 315	Selective Price Sheet	COST EACH		
		MAT.	INST.	TOTAL
6000	Maximum	2,225	252	2,477
6100	Compactor, built-in, minimum	775	136	911
6200	Maximum	1,275	227	1,502
6300	Dishwasher, built-in, minimum	340	390	730
6400	Maximum	490	775	1,265
6500	Garbage disposer, sink-pipe, minimum	122	155	277
6600	Maximum	237	155	392
6700	Range hood, 30″ wide, 2 speed, minimum	118	290	408
6800	Maximum	1,100	485	1,585
6900	Refrigerator, no frost, 12 cu. ft.	500	108	608
7000	20 cu. ft.	620	180	800
7100	Plumb. not incl. rough-ins, sinks porc. C.I., single bowl, 21″ x 24″	340	256	596
7200	21″ x 30″	945	256	1,201
7300	Double bowl, 20″ x 32″	430	299	729
7400				
7500	Stainless steel, single bowl, 19″ x 18″	685	256	941
7600	22″ x 25″	760	256	1,016
7700				
7800				
7900				
8000				
8100				
8200				
8300				
8400				

G1030   Site Earthwork

This page illustrates and describes utility trench excavation with backfill systems including a trench, backfill, excavated material, utility pipe or ductwork not included. Lines within System Components give the unit price and total price on a linear foot basis for this system. Prices for utility trench excavation with backfill systems are on Line Items G1030 810 1100 thru 2500. Both material

quantities and labor costs have been adjusted for the system listed.

Factors: To adjust for job conditions other than normal working situations use Lines G1030 810 3100 thru 4000.

Example: You are to install the system and protect existing construction. Go to Line G1030 810 3600 and apply these percentages to the appropriate TOTAL cost.

System Components			COST PER L.F.		
	QUANTITY	UNIT	MAT.	INST.	TOTAL
Cont. utility trench excav. with 3/8 C.Y. wheel mtd. backhoe, Backfilling with excav. mat. and comp. in 12″ lifts. Trench is 2′ wide by 4′ dp. Cost of utility piping or ductwork is not incl. Cost based on an Excavation production rate of 150 C.Y. daily no hauling included.					
Machine excavate trench, 2′ wide by 4′ deep	.296	C.Y.	.46	2.47	2.93
Dozer backfill with excavated material	.296	C.Y.	.31	.68	.99
Compact in 12″ lifts, vibrating plate	.296	C.Y.	.30	2.71	3.01
TOTAL			1.07	5.86	6.93

G1030 810	Excavation, Utility Trench	COST PER L.F.		
		MAT.	INST.	TOTAL
1100	2′ wide x 2′ deep	.53	2.93	3.46
1200	3′ deep	.80	4.39	5.20
1250	4′ deep	1.07	5.85	6.90
1300	With sloping sides, 2′ wide x 5′ deep	3.02	16.50	19.50
1400	6′ deep	4.02	22	26
1500	7′ deep	5.15	28.50	33.50
1600	8′ deep	6.45	35	41.50
1700	9′ deep	7.85	43	51
1800	10′ deep	9.40	51.50	61
2000	For hauling excavated material up to 2 miles & backfilling w/ gravel			
2100	Gravel, 2′ wide by 2′ deep, add	3.40	3.98	7.38
2200	4′ deep, add	6.80	7.95	14.75
2300	6′ deep, add	25.50	30	55.50
2400	8′ deep, add	41	48	89
2500	10′ deep, add	59.50	70	129.50
2600				
2700				
2800	For shallow, hand excavated trenches, no backfill, to 6′ dp, light soil		67.50	67.50
2900	Heavy soil		135	135
3000				
3100	Equipment usage curtailment, add, minimum	1%	1%	
3200	Maximum	3%	10%	
3300	Material handling & storage limitation, add, minimum	1%	1%	
3400	Maximum	6%	7%	
3500	Protection of existing work, add, minimum	2%	2%	
3600	Maximum	5%	7%	
3700	Shift work requirements, add, minimum		5%	
3800	Maximum		30%	
3900	Temporary shoring and bracing, add, minimum	2%	5%	
4000	Maximum	5%	12%	

G2010 Roadways

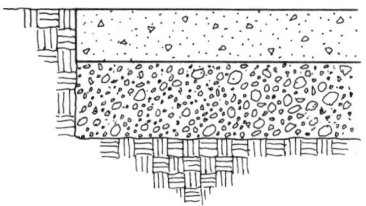

This page illustrates and describes driveway systems including concrete slab, gravel base, broom finish, compaction, and joists. Lines within System Components give the unit price and total price on a cost each basis for this system. Prices for driveway systems are on Line Items G2010 240 1200 thru 2700. Both material quantities and labor costs have been adjusted for the system listed.

Factors: To adjust for job conditions other than normal working situations use Lines G2010 240 2900 thru 4000.

Example: You are to install the system and match existing construction. Go to Line G2010 240 3000 and apply these percentages to the appropriate MAT. and INST. costs.

System Components	QUANTITY	UNIT	COST EACH MAT.	COST EACH INST.	COST EACH TOTAL
Complete driveway, 10' x 30', 4" concrete slab, 3000 psi, on 6" crushed Stone base, concrete broom finished and cured with 10' x 10' joints.					
Grade and compact subgrade	33.000	S.Y.		85.14	85.14
Place and compact 6" crushed stone base	33.000	S.Y.	196.35	52.47	248.82
Place and remove edgeforms (4 use)	80.000	L.F.	28.80	344.80	373.60
Concrete for slab, 3000 psi	4.000	C.Y.	568		568
Place concrete, 4" thick slab, direct chute	4.000	C.Y.		124.16	124.16
Screed, float and broom finish slab	4.000	C.Y.		291	291
Spray on membrane curing compound	3.000	S.F.	41.10	34.05	75.15
Saw cut 3" deep joints	60.000	L.F.	9	147	156
TOTAL			843.25	1,078.62	1,921.87

G2010 240	Driveways	COST EACH MAT.	COST EACH INST.	COST EACH TOTAL
1200	Concrete: 10' x 30' with 4" concrete on 6" crushed stone	845	1,075	1,920
1300	With 6" concrete	1,125	1,150	2,275
1400	20' x 30' with 4" concrete	1,675	2,125	3,800
1500	With 6" concrete	2,250	2,250	4,500
1600	10' wide, for each additional 10' length over 30', 4" thick, add	269	355	624
1700	6" thick, add	375	375	750
1800	20' wide, for each additional 10' length over 30', 4" thick, add	535	695	1,230
1900	6" thick, add	715	730	1,445
2000	Asphalt: 10' x 30' with 2" binder, 1" topping on 6" cr. st., sealed	660	1,175	1,835
2100	3" binder, 1" topping	795	1,175	1,970
2200	20' x 30' with 2" binder, 1" topping	1,325	1,450	2,775
2300	3" binding, 1" topping	1,600	1,475	3,075
2400	10' wide for each add'l. 10' length over 30', 2" b + 1" t, add	220	94.50	314.50
2500	3" binder, 1" topping	265	99	364
2600	20' wide for each add'l. 10' length over 30', 2" b + 1" t, add	440	190	630
2700	3" binder, 1" topping	530	199	729
2800				
2900	Cut & patch to match existing construction, add, minimum	2%	3%	
3000	Maximum	5%	9%	
3100	Dust protection, add, minimum	1%	2%	
3200	Maximum	4%	11%	
3300	Equipment usage curtailment, add, minimum	1%	1%	
3400	Maximum	3%	10%	
3500	Material handling & storage limitation, add, minimum	1%	1%	
3600	Maximum	6%	7%	
3700	Protection of existing work, add, minimum	2%	2%	
3800	Maximum	5%	7%	
3900	Shift work requirements, add, minimum		5%	
4000	Maximum		30%	

G20 Site Improvements

G2020 Parking Lots

This page illustrates and describes asphalt parking lot systems including asphalt binder, topping, crushed stone base, painted parking stripes and concrete parking blocks. Lines within System Components give the unit price and total price per square yard for this system. Prices for asphalt parking lot systems are on Line Items G2020 230 1400 thru 2500. Both material quantities and labor costs have been adjusted for the system listed.

Factors: To adjust for job conditions other than normal working situations use Lines G2020 230 2900 thru 4000.

Example: You are to install the system and match existing construction. Go to Line G2020 230 2900 and apply these percentages to the appropriate MAT. and INST. costs.

System Components	QUANTITY	UNIT	COST PER S.Y.		
			MAT.	INST.	TOTAL
Parking lot consisting of 2″ asphalt binder and 1″ topping on 6″					
Crushed stone base with painted parking stripes and concrete parking blocks.					
Fine grade and compact subgrade	1.000	S.Y.		2.58	2.58
6″ crushed stone base, stone	.320	Ton	1.90	.51	2.41
2″ asphalt binder	1.000	S.Y.	8.10	1.49	9.59
1″ asphalt topping	1.000	S.Y.	4.03	.99	5.02
Paint parking stripes	.500	L.F.	.06	.12	.18
6″ x 10″ x 6′ precast concrete parking blocks	.020	Ea.	1.39	.45	1.84
Mobilization of equipment	.005	Ea.		4.45	4.45
TOTAL			15.48	10.59	26.07

G2020 230	Parking Lots, Asphalt	COST PER S.Y.		
		MAT.	INST.	TOTAL
1400	2″ binder plus 1″ topping on 6″ crushed stone	15.50	10.60	26.10
1500	9″ crushed stone	22.50	11.80	34.30
1600	12″ crushed stone	25.50	11.95	37.45
1700	On bank run gravel, 6″ deep	17.55	11	28.55
1800	9″ deep	19.55	11.30	30.85
1900	12″ deep	21.50	11.45	32.95
2000	3″ binder plus 1″ topping on 6″ crushed stone	23.50	12.10	35.60
2100	9″ deep crushed stone	26.50	12.25	38.75
2200	12″ deep crushed stone	29.50	12.40	41.90
2300	On bank run gravel, 6″ deep	21.50	11.45	32.95
2400	9″ deep	23.50	11.70	35.20
2500	12″ deep	25.50	11.85	37.35
2600				
2700				
2900	Cut & patch to match existing construction, add, minimum	2%	3%	
3000	Maximum	5%	9%	
3100	Dust protection, add, minimum	1%	2%	
3200	Maximum	4%	11%	
3300	Equipment usage curtailment, add, minimum	1%	1%	
3400	Maximum	3%	10%	
3500	Material handling & storage limitation, add, minimum	1%	1%	
3600	Maximum	6%	7%	
3700	Protection of existing work, add, minimum	2%	2%	
3800	Maximum	5%	7%	
3900	Shift work requirements, add, minimum		5%	
4000	Maximum		30%	

G2020 Parking Lots

This page illustrates and describes concrete parking lot systems including a concrete slab, unreinforced, compacted gravel base, broom finished, cured, joints, painted parking stripes, and parking blocks. Lines within System Components give the unit price and total price per square yard for this system. Prices for concrete parking lot systems are on Line Items G2020 240 1500 thru 2500. All materials have been adjusted according to the system listed.

Factors: To adjust for job conditions other than normal working situations use Lines G2020 240 2900 thru 4000.

Example: You are to install the system and protect existing construction. Go to Line G2020 240 3800 and apply these percentages to the appropriate MAT. and INST. costs.

System Components	QUANTITY	UNIT	COST PER S.Y. MAT.	COST PER S.Y. INST.	COST PER S.Y. TOTAL
Parking lot, 4" slab, 3000 psi concrete on 6" crushed stone base					
With 12' x 12' joints, painted parking strips, conc. parking blocks.					
Grade and compact subgrade	1.000	S.Y.		2.58	2.58
6" crushed stone base, stone	.320	Ton	1.90	.51	2.41
Place and remove edge forms, 4 use	.250	L.F.	.09	1.08	1.17
Concrete for slab, 3000 psi	1.000	S.F.	15.62		15.62
Place concrete, 4" thick slab, direct chute	1.000	S.F.		3.42	3.42
Spray on membrane curing compound	1.000	S.F.	1.23	.72	1.95
Saw cut 3" deep joints	2.250	L.F.	.34	5.51	5.85
Paint parking stripes	.500	L.F.	.06	.12	.18
6" x 10" x 6' precast concrete parking blocks	.020	Ea.	1.39	.45	1.84
TOTAL			20.63	14.39	35.02

G2020 240	Parking Lots, Concrete	COST PER S.Y. MAT.	COST PER S.Y. INST.	COST PER S.Y. TOTAL
1500	4" concrete slab, 3000 psi on 3" crushed stone	21.50	15.40	36.90
1550	6" crushed stone	20.50	14.35	34.85
1600	9" crushed stone	27.50	15.65	43.15
1700	On bank run gravel, 6" deep	22.50	14.85	37.35
1800	9" deep	24.50	15.05	39.55
1900	On compacted subgrade only	18.75	13.85	32.60
2000	6" concrete slab, 3000 psi on 3" crushed stone	29	17.05	46.05
2100	6" crushed stone	32	17.15	49.15
2200	9" crushed stone	35	17.25	52.25
2300	On bank run gravel, 6" deep	30	16.45	46.45
2400	9" deep	32	16.70	48.70
2500	On compacted subgrade only	32	17.15	49.15
2600				
2700				
2900	Cut & patch to match existing construction, add, minimum	2%	3%	
3000	Maximum	5%	9%	
3100	Dust protection, add, minimum	1%	2%	
3200	Maximum	4%	11%	
3300	Equipment usage curtailment, add, minimum	1%	1%	
3400	Maximum	3%	10%	
3500	Material handling & storage limitation, add, minimum	1%	1%	
3600	Maximum	6%	7%	
3700	Protection of existing work, add, minimum	2%	2%	
3800	Maximum	5%	7%	
3900	Shift work requirements, add, minimum		5%	
4000	Maximum		30%	

725

G2030 Pedestrian Paving

This page illustrates and describes sidewalk systems including concrete, welded wire and broom finish. Lines within System Components give unit price and total price per square foot for this system. Prices for sidewalk systems are on Line Items G2030 105 1700 thru 1900. Both material quantities and labor costs have been adjusted for the system listed.

Factors: To adjust for job conditions other than normal working situations use Lines G2030 105 2900 thru 4000.

Example: You are to install the system and match existing construction. Go to Line G2030 105 2900 and apply these percentages to the appropriate MAT. and INST. costs.

System Components	QUANTITY	UNIT	COST PER S.F.		
			MAT.	INST.	TOTAL
4″ thick concrete sidewalk with welded wire fabric					
3000 psi air entrained concrete, broom finish.					
Gravel fill, 4″ deep	.012	C.Y.	.25	.10	.35
Compact fill	.012	C.Y.		.03	.03
Hand grade	1.000	S.F.		2.61	2.61
Edge form	.250	L.F.	.09	1.08	1.17
Welded wire fabric	.011	S.F.	.20	.45	.65
Concrete, 3000 psi air entrained	.012	C.Y.	1.70		1.70
Place concrete	.012	C.Y.		.38	.38
Broom finish	1.000	S.F.		.97	.97
TOTAL			2.24	5.62	7.86

G2030 105	Sidewalks	COST PER S.F.		
		MAT.	INST.	TOTAL
1700	Asphalt (bituminous), 2″ thick	1.12	3.30	4.42
1750	4″ thick	2.24	5.65	7.89
1800	Brick, on sand, bed, 4.5 brick per S.F.	3.36	12.10	15.46
1900	Flagstone, slate, 1″ thick, rectangular	17.25	13.20	30.45
2000				
2100				
2200				
2300				
2400				
2500				
2600				
2700				
2900	Cut & patch to match existing construction, add, minimum	2%	3%	
3000	Maximum	5%	9%	
3100	Dust protection, add, minimum	1%	2%	
3200	Maximum	4%	11%	
3300	Equipment usage curtailment, add, minimum	1%	1%	
3400	Maximum	3%	10%	
3500	Material handling & storage limitation, add, minimum	1%	1%	
3600	Maximum	6%	7%	
3700	Protection of existing work, add, minimum	2%	2%	
3800	Maximum	5%	7%	
3900	Shift work requirements, add, minimum		5%	
4000	Maximum		30%	

G2050 Landscaping

This page describes landscaping—lawn establishment systems including loam, lime, fertilizer, side and top mulching. Lines within system components give the unit price and total price per square yard for this system. Prices for landscaping—lawn establishment systems are on Line Items G2050 420 1000 thru 1200. Both material quantities and labor costs have been adjusted for the system listed.

Factors: To adjust for job conditions other than normal working situations use Lines G2050 420 2900 thru 4000.

Example: You are to install the system and provide dust protection. Go to Line G2050 420 3200 and apply these percentages to the appropriate MAT. and INST. costs.

System Components			COST PER S.Y.		
	QUANTITY	UNIT	MAT.	INST.	TOTAL
Establishing lawns with loam, lime, fertilizer, seed and top mulching					
On rough graded areas.					
Furnish and place loam 4″ deep	.110	C.Y.	4.02	1.11	5.13
Fine grade, lime, fertilize and seed	1.000	S.Y.	.54	3.63	4.17
Hay mulch, 1 bale/M.S.F.	1.000	S.Y.	.34	.66	1
Rolling with hand roller	1.000	S.Y.		1.16	1.16
TOTAL			4.90	6.56	11.46

G2050 420	Landscaping - Lawn Establishment	COST PER S.Y.		
		MAT.	INST.	TOTAL
1000	Establishing lawns with loam, lime, fertilizer, seed and hay mulch	4.90	6.60	11.50
1100	Above system with jute mesh in place of hay mulch	5.70	6.95	12.65
1200	Above system with sod in place of seed	6.40	2.39	8.79
1300				
1400				
1500				
1600				
1700				
1800				
1900				
2000				
2100				
2200				
2300				
2400				
2500				
2600				
2700				
2900	Cut & patch to match existing construction, add, minimum	2%	3%	
3000	Maximum	5%	9%	
3100	Dust protection, add, minimum	1%	2%	
3200	Maximum	4%	11%	
3300	Equipment usage curtailment, add, minimum	1%	1%	
3400	Maximum	3%	10%	
3500	Material handling & storage limitation, add, minimum	1%	1%	
3600	Maximum	6%	7%	
3700	Protection of existing work, add, minimum	2%	2%	
3800	Maximum	5%	7%	
3900	Shift work requirements, add, minimum		5%	
4000	Maximum		30%	

G2050 Landscaping

G2050 510	Selective Price Sheet	COST PER UNIT		
		MAT.	INST.	TOTAL
0100	Shrubs and trees, Evergreen, in prepared beds			
0200				
0300	Arborvitae pyramidal, 4'-5'	115	103	218
0400	Globe, 12"-15"	26	17.15	43.15
0500	Cedar, blue, 8'-10'	266	171	437
0600	Hemlock, Canadian, 2-1/2'-3'	36.50	45.50	82
0700	Juniper, andora, 18"-24"	64.50	20.50	85
0800	Wiltoni, 15"-18"	31.50	20.50	52
0900	Skyrocket, 4-1/2'-5'	122	56	178
1000	Blue pfitzer, 2'-2-1/2'	45.50	37.50	83
1100	Ketleerie, 2-1/2'-3'	64.50	33	97.50
1200	Pine, black, 2-1/2'-3'	74.50	33	107.50
1300	Mugo, 18"-24"	71	27.50	98.50
1400	White, 4'-5'	61	41	102
1500	Spruce, blue, 18"-24"	79	27.50	106.50
1600	Norway, 4'-5'	103	41	144
1700	Yew, denisforma, 12"-15"	42.50	27.50	70
1800	Capitata, 18"-24"	38	55	93
1900	Hicksi, 2'-2-1/2'	112	55	167
2000				
2100	Trees, Deciduous, in prepared beds			
2200				
2300	Beech, 5'-6'	264	61.50	325.50
2400	Dogwood, 4'-5'	154	76.50	230.50
2500	Elm, 8'-10'	385	154	539
2600	Magnolia, 4'-5'	130	154	284
2700	Maple, red, 8'-10', 1-1/2" caliper	220	310	530
2800	Oak, 2-1/2"-3" caliper	390	515	905
2900	Willow, 6'-8', 1" caliper	113	154	267
3000				
3100	Shrubs, Broadleaf Evergreen, in prepared beds			
3200				
3300	Andromeda, 15"-18", container	37	17.15	54.15
3400	Azalea, 15"-18", container	37.50	17.15	54.65
3500	Barberry, 9"-12", container	20.50	12.65	33.15
3600	Boxwood, 15"-18", B & B	52.50	17.15	69.65
3700	Euonymus, emerald gaiety, 12"-15", container	31	14.30	45.30
3800	Holly, 15"-18", B & B	45.50	17.15	62.65
3900	Mount laurel, 18"-24", B & B	84	20.50	104.50
4000	Privet, 18"-24", B & B	25.50	12.65	38.15
4100	Rhododendron, 18"-24", container	43.50	34.50	78
4200	Rosemary, 1 gal. container	22	2.74	24.74
4300	Deciduous, amalanchier, 2'-3', B & B	145	29	174
4400	Azalea, 15"-18", B & B	33.50	17.15	50.65
4500	Bayberry, 2'-3', B & B	38.50	29	67.50
4600	Cotoneaster, 15"-18", B & B	29.50	20.50	50
4700	Dogwood, 3'-4', B & B	39.50	76.50	116
4800	Euonymus, alatus compacta, 15"-18", container	31	20.50	51.50
4900	Forsythia, 2'-3', container	19.40	27.50	46.90
5000	Honeysuckle, 3'-4', B & B	32	27.50	59.50
5100	Hydrangea, 2'-3', B & B	36	29	65
5200	Lilac, 3'-4', B & B	24	76.50	100.50
5300	Quince, 2'-3', B & B	33.50	29	62.50
5400				
5500	Ground Cover			
5600				
5700	Plants, Pachysandra, prepared beds, per hundred	97	110	207
5800	Vinca minor or English ivy, per hundred	90	137	227

G2050 Landscaping

G2050 510	Selective Price Sheet	COST PER UNIT		
		MAT.	INST.	TOTAL
5900	Plant bed prep. 18" dp., mach., per square foot	2.85	.49	3.34
6000	By hand, per square foot	2.85	3.48	6.33
6100	Stone chips, Georgia marble, 50# bags, per bag	5.25	3.16	8.41
6200	Onyx gemstone, per bag	18.55	6.35	24.90
6300	Quartz, per bag	18.85	6.35	25.20
6400	Pea gravel, truck load lots, per cubic yard	13	58.50	71.50
6500	Mulch, polyethylene mulch, per square yard	.56	.54	1.10
6600	Wood chips, 2" deep, per square yard	1.85	2.45	4.30
6700	Peat moss, 1" deep, per square yard	5.65	.60	6.25
6800	Erosion control per square yard, Jute mesh stapled	1.13	1.05	2.18
6900	Plastic netting, stapled	.30	.66	.96
7000	Polypropylene mesh, stapled	1.87	.66	2.53
7100	Tobacco netting, stapled	.34	.66	1
7200	Lawns per square yard, seeding incl. fine grade, limestone, fert. and seed	.54	3.63	4.17
7300	Sodding, incl. fine grade, level ground	2.21	1.23	3.44
7500	Edging, per linear foot, Redwood, untreated, 1" x 4"	1.29	2.72	4.01
7600	2" x 4"	2.54	4.13	6.67
7700	Stl. edge strips, 1/4" x 5" inc. stakes	4.77	4.13	8.90
7800	3/16" x 4"	4.66	4.13	8.79
7900	Brick edging, set on edge	3.73	9	12.73
8000	Set flat	1.87	3.28	5.15
8100				
8200				
8300				
8400				

G3020 Sanitary Sewer

This page illustrates and describes drainage and utilities – septic system including a tank, distribution box, excavation, piping, crushed stone and backfill. Lines within System Components give the unit price and total price on a cost each basis for this system. Prices for drainage and utilities – septic systems are on Line Items G3020 610 1600 thru 2000. Both material quantities and labor costs have been adjusted for the system listed.

Factors: To adjust for job conditions other than normal working situations use Lines G3020 610 3100 thru 4000.

Example: You are to install the system with a material handling limitation. Go to Line G3020 610 3500 and apply these percentages to the appropriate MAT. and INST. costs.

System Components	QUANTITY	UNIT	COST EACH MAT.	COST EACH INST.	COST EACH TOTAL
Septic system including tank, distribution box, excavation, piping, crushed Stone and backfill for a 1000 S.F. leaching field.					
Precast concrete septic tank, 1000 gal. capacity	1.000	Ea.	1,175	298	1,473
Concrete distribution box	1.000	Ea.	114	67.50	181.50
Sch. 40 sewer pipe, PVC, 4″ diameter	170.000	L.F.	312.80	817.70	1,130.50
Sch. 40 sewer pipe fittings, PVC, 4″ diameter	8.000	Ea.	164	700	864
Excavation for septic tank	18.000	C.Y.		253.80	253.80
Excavation for disposal field	101.000	C.Y.		616.10	616.10
Crushed stone backfill	76.000	C.Y.	1,413.60	1,000.92	2,414.52
Backfill with excavated material	26.000	C.Y.		57.20	57.20
Asphalt felt paper, #15	6.000	Sq.	34.80	110.40	145.20
TOTAL			3,214.20	3,921.62	7,135.82

G3020 610	Septic Systems	COST EACH MAT.	COST EACH INST.	COST EACH TOTAL
1600	1000 gal. tank with 1000 S.F. field	3,225	3,925	7,150
1700	1000 gal. tank with 2000 S.F. field	4,450	5,750	10,200
1800	With leaching pits	3,475	2,375	5,850
1900	2000 gal. tank with 2000 S.F. field	6,025	5,950	11,975
2000	With leaching pits	5,150	3,050	8,200
2100				
2200				
2300				
2400				
2500				
2600				
2700				
2800				
2900				
3100	Dust protection, add, minimum	1%	2%	
3200	Maximum	4%	11%	
3300	Equipment usage curtailment, add, minimum	1%	1%	
3400	Maximum	3%	10%	
3500	Material handling & storage limitation, add, minimum	1%	1%	
3600	Maximum	6%	7%	
3700	Protection of existing work, add, minimum	2%	2%	
3800	Maximum	5%	7%	
3900	Shift work requirements, add, minimum		5%	
4000	Maximum		30%	

G3060 Fuel Distribution

This page illustrates and describes fiberglass tanks including hold down slab, reinforcing steel and peastone gravel. Lines within Systems Components give the material and installation price on a cost each basis for the components. Prices for fiberglass tank systems are on Line Items G3060 305 1010 thru G3060 310 1100. Material quantities and labor costs have been adjusted for the system listed.

Factors: To adjust for job conditions other than normal working situations use Lines G3060 310 2700 thru 4000.

Fiberglass Underground Fuel Storage Tank, Single Wall

System Components	QUANTITY	UNIT	COST EACH		
			MAT.	INST.	TOTAL
Double wall underground fiberglass storage tanks including, excavation, hold down slab, backfill, manway extension and related piping.					
Excavating trench, no sheeting or dewatering, 6'-10 ' D, 3/4 CY hyd backhoe	18.000	Ea.		163.62	163.62
Corrosion resistance wrap & coat, small diam. pipe, 1" diam., add	120.000	Ea.	133.20		133.20
Fiberglass-reinforced elbows 90 & 45 degree, 2 inch	7.000	Ea.	315	661.50	976.50
Fiberglass-reinforced threaded adapters, 2 inch	2.000	Ea.	42	216	258
Remote tank gauging system, 30', 5" pointer travel	1.000	Ea.	6,325	405	6,730
Corrosion resistance wrap & coat, small diam. pipe. 4 " diam. add	30.000	Ea.	78.60		78.60
Pea gravel	9.000	Ea.	1,134	526.50	1,660.50
Reinforcing in hold down pad, #3 to #7	120.000	Ea.	74.40	81.60	156
Tank leak detector system, 8 channel, external monitoring	1.000	Ea.	1,350		1,350
Tubbing copper, type L, 3/4" diam.	120.000	Ea.	624	1,260	1,884
Elbow 90 degree ,wrought cu x cu 3/4 " diam.	4.000	Ea.	11.80	168	179.80
Pipe, black steel welded, sch 40, 3" diam.	30.000	Ea.	549	1,086.60	1,635.60
Elbow 90 degree, steel welded butt joint, 3 " diam.	3.000	Ea.	99	665.25	764.25
Fiberglass-reinforced plastic pipe, 2 inch	156.000	Ea.	1,201.20	834.60	2,035.80
Fiberglass-reinforced plastic pipe, 3 inch	36.000	Ea.	1,170	217.80	1,387.80
Fiberglass-reinforced elbows 90 & 45 degree, 3 inch	3.000	Ea.	432	378	810
Tank leak detection system, probes, well monitoring, liquid phase detection	2.000	Ea.	1,600		1,600
Double wall fiberglass annular space	1.000	Ea.	470		470
Flange, weld neck, 150 lb, 3 " pipe size	1.000	Ea.	45	110.35	155.35
Vent protector/breather, 2" diam.	1.000	Ea.	59	25.50	84.50
Corrosion resistance wrap & coat, small diam. pipe, 2" diam., add	36.000	Ea.	50.76		50.76
Concrete hold down pad, 6' x 5' x 8" thick	30.000	Ea.	112.50	54	166.50
600 gallon capacity, underground storage tank, double wall fiberglass	1.000	Ea.	10,000	605	10,605
Tank, for hold-downs 500-4,000 gal., add	1.000	Ea.	555	193	748
Foot valve, single poppet, 3/4" diam.	1.000	Ea.	124	45	169
Fuel fill box, locking inner cover, 3" diam.	1.000	Ea.	131	162	293
Fuel oil specialties, valve, ball chk, globe type fusible, 3/4" diam.	2.000	Ea.	174	81	255
TOTAL			26,860.46	7,940.32	34,800.78

G3060 310	Fiberglass Fuel Tank, Double Wall	COST EACH		
		MAT.	INST.	TOTAL
1010	Storage tank, fuel, underground, double wall fiberglass, 600 Gal.	26,900	7,950	34,850
1020	2500 Gal.	41,000	10,900	51,900
1030	4000 Gal.	49,000	13,500	62,500
1040	6000 Gal.	58,000	15,200	73,200
1050	8000 Gal.	72,000	18,400	90,400
1060	10,000 Gal.	82,500	21,200	103,700

731

G30 Site Mechanical Utilities

G3060 Fuel Distribution

G3060 310	Fiberglass Fuel Tank, Double Wall	COST EACH		
		MAT.	INST.	TOTAL
1070	15,000 Gal.	118,000	24,300	142,300
1080	20,000 Gal.	142,500	29,400	171,900
1090	25,000 Gal.	163,500	34,200	197,700
1100	30,000 Gal.	187,500	36,200	223,700
2700	Cut & patch to match existing construction, add, minimum	2%	3%	
2800	Maximum	5%	9%	
2900	Dust protection, add, minimum	1%	2%	
3000	Maximum	4%	11%	
3100	Equipment usage curtailment, add, minimum	1%	1%	
3200	Maximum	3%	10%	
3300	Material handling & storage limitation, add, minimum	1%	1%	
3400	Maximum	6%	7%	
3500	Protection of existing work, add, minimum	2%	2%	
3600	Maximum	5%	7%	
3700	Shift work requirements, add, minimum		5%	
3800	Maximum		30%	
3900	Temporary shoring and bracing, add, minimum	2%	5%	
4000	Maximum	5%	12%	

G3060 Fuel Distribution

Steel Underground Fuel Storage Tank, Double Wall

System Components	QUANTITY	UNIT	COST EACH		
			MAT.	INST.	TOTAL
Double wall underground steel storage tank including, excavation, hold down slab, backfill, manway extension and related piping.					
Fiberglass-reinforced plastic pipe, 2 inch	156.000	Ea.	1,201.20	834.60	2,035.80
Fiberglass-reinforced plastic pipe, 3 inch	36.000	Ea.	1,170	217.80	1,387.80
Fiberglass-reinforced elbows 90 & 45 degree, 3 inch	3.000	Ea.	432	378	810
Tank leak detection system, probes, well monitoring, liquid phase detection	2.000	Ea.	1,600		1,600
Flange, weld neck, 150 lb, 3" pipe size	1.000	Ea.	45	110.35	155.35
Vent protector/breather, 2" diam.	1.000	Ea.	59	25.50	84.50
Concrete hold down pad, 6' x 5' x 8" thick	30.000	Ea.	112.50	54	166.50
Tanks, for hold-downs, 500-2,000 gal, add	1.000	Ea.	249	193	442
Double wall steel tank annular space	1.000	Ea.	310		310
Foot valve, single poppet, 3/4" diam.	1.000	Ea.	124	45	169
Fuel fill box, locking inner cover, 3" diam.	1.000	Ea.	131	162	293
Fuel oil specialties, valve, ball chk, globe type fusible, 3/4" diam.	2.000	Ea.	174	81	255
Pea gravel	9.000	Ea.	1,134	526.50	1,660.50
Reinforcing in hold down pad, #3 to #7	120.000	Ea.	74.40	81.60	156
Tank, for manways, add	1.000	Ea.	1,175		1,175
Tank leak detector system, 8 channel, external monitoring	1.000	Ea.	1,350		1,350
Tubbing copper, type L, 3/4" diam.	120.000	Ea.	624	1,260	1,884
Elbow 90 degree, wrought cu x cu 3/4" diam.	8.000	Ea.	23.60	336	359.60
Pipe, black steel welded, sch 40, 3" diam.	30.000	Ea.	549	1,086.60	1,635.60
Elbow 90 degree, steel welded butt joint, 3" diam.	3.000	Ea.	99	665.25	764.25
Excavating trench, no sheeting or dewatering, 6'-10' D, 3/4 CY hyd backhoe	18.000	Ea.		163.62	163.62
Fiberglass-reinforced elbows 90 & 45 degree, 2 inch	7.000	Ea.	315	661.50	976.50
Fiberglass-reinforced threaded adapters, 2 inch	2.000	Ea.	42	216	258
500 gallon capacity, underground storage tank, double wall steel	1.000	Ea.	3,650	610	4,260
Remote tank gauging system, 30', 5" pointer travel	1.000	Ea.	6,325	405	6,730
TOTAL			20,968.70	8,113.32	29,082.02

G3060 320	Steel Fuel Tank, Double Wall	COST EACH		
		MAT.	INST.	TOTAL
1010	Storage tank, fuel, underground, double wall steel, 500 Gal.	21,000	8,100	29,100
1020	2000 Gal.	26,800	10,600	37,400
1030	4000 Gal.	37,000	13,700	50,700
1040	6000 Gal.	42,200	15,100	57,300
1050	8000 Gal.	48,300	18,700	67,000
1060	10,000 Gal.	55,500	21,400	76,900

733

For customer support on your Commercial Renovation Costs with RSMeans data, call 800.448.8182.

G3060 Fuel Distribution

G3060 320	Steel Fuel Tank, Double Wall	COST EACH		
		MAT.	INST.	TOTAL
1070	15,000 Gal.	67,500	24,900	92,400
1080	20,000 Gal.	93,000	29,800	122,800
1090	25,000 Gal.	103,000	34,200	137,200
1100	30,000 Gal.	106,000	36,400	142,400
1110	40,000 Gal.	203,500	45,800	249,300
1120	50,000 Gal.	246,500	55,000	301,500

G3060 325	Steel Tank, Above Ground, Single Wall	COST EACH		
		MAT.	INST.	TOTAL
1010	Storage tank, fuel, above ground, single wall steel, 550 Gal.	15,400	5,750	21,150
1020	2000 Gal.	25,500	5,875	31,375
1030	5000 Gal.	36,100	9,025	45,125
1040	10,000 Gal.	57,500	9,600	67,100
1050	15,000 Gal.	75,500	9,875	85,375
1060	20,000 Gal.	94,000	10,200	104,200
1070	25,000 Gal.	105,500	10,400	115,900
1080	30,000 Gal.	127,000	10,900	137,900

G3060 330	Steel Tank, Above Ground, Double Wall	COST EACH		
		MAT.	INST.	TOTAL
1010	Storage tank, fuel, above ground, double wall steel, 500 Gal.	14,500	5,800	20,300
1020	2000 Gal.	19,600	5,950	25,550
1030	4000 Gal.	28,500	6,050	34,550
1040	6000 Gal.	32,600	9,325	41,925
1050	8000 Gal.	36,300	9,600	45,900
1060	10,000 Gal.	48,300	9,775	58,075
1070	15,000 Gal.	60,000	10,100	70,100
1080	20,000 Gal.	68,000	10,400	78,400
1090	25,000 Gal.	81,000	10,700	91,700
1100	30,000 Gal.	88,500	11,200	99,700
2700	Cut & patch to match existing construction, add, minimum	2%	3%	
2800	Maximum	5%	9%	
2900	Dust protection, add, minimum	1%	2%	
3000	Maximum	4%	11%	
3100	Equipment usage curtailment, add, minimum	1%	1%	
3200	Maximum	3%	10%	
3300	Material handling & storage limitation, add, minimum	1%	1%	
3400	Maximum	6%	7%	
3500	Protection of existing work, add, minimum	2%	2%	
3600	Maximum	5%	7%	
3700	Shift work requirements, add, minimum		5%	
3800	Maximum		30%	
3900	Temporary shoring and bracing, add, minimum	2%	5%	
4000	Maximum	5%	12%	

G4020 Site Lighting

This page illustrates and describes light poles for parking or walkway area lighting. Included are aluminum or steel light poles, single or multiple fixture bracket arms, excavation, concrete footing, backfill and compaction. Lines within system components give the unit price and total price on a cost per each basis. Prices for systems are shown on lines G4020 210 0200 through 1440. Both material quantities and labor costs have been adjusted for the system listed.

Factors: To adjust for job conditions other than normal working situations use lines G4020 210 2700 through 3600.

Example: You are to install the system where you must be careful not to damage existing walks or parking areas. Go to line G4020 210 330 and apply these percentages to the appropriate MAT. and INST. costs.

System Components	QUANTITY	UNIT	COST EACH		
			MAT.	INST.	TOTAL
Light poles, aluminum, 20' high, 1 arm bracket					
Aluminum light pole, 20', no concrete base	1.000	Ea.	1,200	721.50	1,921.50
Bracket arm for aluminum light pole	1.000	Ea.	155	94.50	249.50
Excavation by hand, pits to 6' deep, heavy soil or clay	2.368	C.Y.		319.68	319.68
Footing, concrete incl forms, reinforcing, spread, under 1 C.Y.	.465	C.Y.	104.16	149.28	253.44
Backfill by hand	1.903	C.Y.		93.25	93.25
Compaction vibrating plate	1.903	C.Y.		15.26	15.26
TOTAL			1,459.16	1,393.47	2,852.63

G4020 210	Light Pole (Installed)	COST EACH		
		MAT.	INST.	TOTAL
0200	Light pole, aluminum, 20' high, 1 arm bracket	1,450	1,400	2,850
0240	2 arm brackets	1,625	1,400	3,025
0280	3 arm brackets	1,775	1,450	3,225
0320	4 arm brackets	1,925	1,450	3,375
0360	30' high, 1 arm bracket	2,550	1,750	4,300
0400	2 arm brackets	2,725	1,750	4,475
0440	3 arm brackets	2,875	1,800	4,675
0480	4 arm brackets	3,025	1,825	4,850
0680	40' high, 1 arm bracket	3,325	2,350	5,675
0720	2 arm brackets	3,500	2,350	5,850
0760	3 arm brackets	3,650	2,400	6,050
0800	4 arm brackets	3,800	2,425	6,225
0840	Steel, 20' high, 1 arm bracket	1,750	1,475	3,225
0880	2 arm brackets	1,850	1,475	3,325
0920	3 arm brackets	1,750	1,525	3,275
0960	4 arm brackets	1,900	1,525	3,425
1000	30' high, 1 arm bracket	1,775	1,875	3,650
1040	2 arm brackets	1,875	1,875	3,750
1080	3 arm brackets	1,775	1,925	3,700
1120	4 arm brackets	1,925	1,925	3,850
1320	40' high, 1 arm bracket	2,275	2,525	4,800
1360	2 arm brackets	2,375	2,525	4,900
1400	3 arm brackets	2,275	2,575	4,850
1440	4 arm brackets	2,425	2,575	5,000
2700	Cut & patch to match existing construction, add, minimum	2%	3%	
2800	Maximum	5%	9%	
2900	Equipment usage curtailment, add, minimum	1%	1%	
3000	Maximum	3%	10%	

735

G40 Site Electrical Utilities

G4020 Site Lighting

G4020 210	Light Pole (Installed)	COST EACH		
		MAT.	INST.	TOTAL
3100	Material handling & storage limitation, add, minimum	1%	1%	
3200	Maximum	6%	7%	
3300	Protection of existing work, add, minimum	2%	2%	
3400	Maximum	5%	7%	
3500	Shift work requirements, add, minimum		5%	
3600	Maximum		30%	

Reference Section

All the reference information is in one section, making it easy to find what you need to know ... and easy to use the data set on a daily basis. This section is visually identified by a vertical black bar on the page edges.

In this Reference Section, we've included Equipment Rental Costs, a listing of rental and operating costs; Crew Listings, a full listing of all crews and equipment, and their costs; Historical Cost Indexes for cost comparisons over time; City Cost Indexes and Location Factors for adjusting costs to the region you are in; Reference Tables, where you will find explanations, estimating information and procedures, or technical data; Change Orders, information on pricing changes to contract documents; and an explanation of all the Abbreviations in the data set.

Table of Contents

Reference Tables (cont.)

Estimating Tips

- This section contains the average costs to rent and operate hundreds of pieces of construction equipment. This is useful information when one is estimating the time and material requirements of any particular operation in order to establish a unit or total cost. Bare equipment costs shown on a unit cost line include, not only rental, but also operating costs for equipment under normal use.

Rental Costs

- Equipment rental rates are obtained from the following industry sources throughout North America: contractors, suppliers, dealers, manufacturers, and distributors.
- Rental rates vary throughout the country, with larger cities generally having lower rates. Lease plans for new equipment are available for periods in excess of six months, with a percentage of payments applying toward purchase.
- Monthly rental rates vary from 2% to 5% of the purchase price of the equipment depending on the anticipated life of the equipment and its wearing parts.
- Weekly rental rates are about 1/3 of the monthly rates, and daily rental rates are about 1/3 of the weekly rate.
- Rental rates can also be treated as reimbursement costs for contractor-owned equipment. Owned equipment costs include depreciation, loan payments, interest, taxes, insurance, storage, and major repairs.

Operating Costs

- The operating costs include parts and labor for routine servicing, such as the repair and replacement of pumps, filters, and worn lines. Normal operating expendables, such as fuel, lubricants, tires, and electricity (where applicable), are also included.
- Extraordinary operating expendables with highly variable wear patterns, such as diamond bits and blades, are excluded. These costs can be found as material costs in the Unit Price section.
- The hourly operating costs listed do not include the operator's wages.

Equipment Cost/Day

- Any power equipment required by a crew is shown in the Crew Listings with a daily cost.
- This daily cost of equipment needed by a crew includes both the rental cost and the operating cost and is based on dividing the weekly rental rate by 5 (the number of working days in the week), then adding the hourly operating cost multiplied by 8 (the number of hours in a day). This "Equipment Cost/Day" is shown in the far right column of the Equipment Rental section.
- If equipment is needed for only one or two days, it is best to develop your own cost by including components for daily rent and hourly operating costs. This is important when the listed Crew for a task does not contain the equipment needed, such as a crane for lifting mechanical heating/cooling equipment up onto a roof.

- If the quantity of work is less than the crew's Daily Output shown for a Unit Price line item that includes a bare unit equipment cost, the recommendation is to estimate one day's rental cost and operating cost for equipment shown in the Crew Listing for that line item.
- Please note, in some cases the equipment description in the crew is followed by a time period in parenthesis. For example: (daily) or (monthly). In these cases the equipment cost/day is calculated by adding the rental cost per time period to the hourly operating cost multiplied by 8.

Mobilization, Demobilization Costs

- The cost to move construction equipment from an equipment yard or rental company to the job site and back again is not included in equipment rental costs listed in the Reference Section. It is also not included in the bare equipment cost of any Unit Price line item or in any equipment costs shown in the Crew Listings.
- Mobilization (to the site) and demobilization (from the site) costs can be found in the Unit Price section.
- If a piece of equipment is already at the job site, it is not appropriate to utilize mobilization or demobilization costs again in an estimate. ■

Same Data. Simplified.

Enjoy the convenience and efficiency of accessing your costs anywhere:

- **Skip the multiplier** by setting your location
- **Quickly search,** edit, favorite and share costs
- **Stay on top of price changes** with automatic updates

Discover more at rsmeans.com/online

		01 54 33 | Equipment Rental		UNIT	HOURLY OPER. COST	RENT PER DAY	RENT PER WEEK	RENT PER MONTH	EQUIPMENT COST/DAY	
10	0010	**CONCRETE EQUIPMENT RENTAL** without operators	R015433 -10							**10**
	0200	Bucket, concrete lightweight, 1/2 C.Y.		Ea.	.87	38.50	115	345	30	
	0300	1 C.Y.			.98	63.50	190	570	45.80	
	0400	1-1/2 C.Y.			1.23	60	180	540	45.85	
	0500	2 C.Y.			1.34	73.50	220	660	54.70	
	0580	8 C.Y.			6.47	93.50	280	840	107.75	
	0600	Cart, concrete, self-propelled, operator walking, 10 C.F.			2.85	175	525	1,575	127.85	
	0700	Operator riding, 18 C.F.			4.81	192	575	1,725	153.45	
	0800	Conveyer for concrete, portable, gas, 16" wide, 26' long			10.61	160	480	1,450	180.85	
	0900	46' long			10.98	175	525	1,575	192.85	
	1000	56' long			11.15	192	575	1,725	204.15	
	1100	Core drill, electric, 2-1/2 H.P., 1" to 8" bit diameter			1.56	83.50	250	750	62.50	
	1150	11 H.P., 8" to 18" cores			5.38	119	356.32	1,075	114.30	
	1200	Finisher, concrete floor, gas, riding trowel, 96" wide			9.64	153	459.60	1,375	169	
	1300	Gas, walk-behind, 3 blade, 36" trowel			2.03	96	287.50	865	73.75	
	1400	4 blade, 48" trowel			3.06	104	312.50	940	87	
	1500	Float, hand-operated (Bull float), 48" wide			.08	12.35	37	111	8.05	
	1570	Curb builder, 14 H.P., gas, single screw			14.00	253	760	2,275	263.95	
	1590	Double screw			15.00	253	760	2,275	272	
	1600	Floor grinder, concrete and terrazzo, electric, 22" path			3.03	134	401.75	1,200	104.60	
	1700	Edger, concrete, electric, 7" path			1.18	57.50	172.50	520	43.95	
	1750	Vacuum pick-up system for floor grinders, wet/dry			1.61	102	305.71	915	74.05	
	1800	Mixer, powered, mortar and concrete, gas, 6 C.F., 18 H.P.			7.39	97	291	875	117.35	
	1900	10 C.F., 25 H.P.			8.97	114	342.50	1,025	140.30	
	2000	16 C.F.			9.33	144	432.50	1,300	161.15	
	2100	Concrete, stationary, tilt drum, 2 C.Y.			7.21	80	240	720	105.70	
	2120	Pump, concrete, truck mounted, 4" line, 80' boom			29.79	287	860	2,575	410.35	
	2140	5" line, 110' boom			37.34	287	860	2,575	470.75	
	2160	Mud jack, 50 C.F. per hr.			6.43	228	685	2,050	188.45	
	2180	225 C.F. per hr.			8.52	293	880	2,650	244.15	
	2190	Shotcrete pump rig, 12 C.Y./hr.			13.93	223	670	2,000	245.40	
	2200	35 C.Y./hr.			15.75	287	860	2,575	298	
	2600	Saw, concrete, manual, gas, 18 H.P.			5.52	112	337	1,000	111.55	
	2650	Self-propelled, gas, 30 H.P.			7.87	81	242.71	730	111.50	
	2675	V-groove crack chaser, manual, gas, 6 H.P.			1.64	100	300	900	73.10	
	2700	Vibrators, concrete, electric, 60 cycle, 2 H.P.			.47	73	218.50	655	47.45	
	2800	3 H.P.			.56	73.50	221	665	48.70	
	2900	Gas engine, 5 H.P.			1.54	16.85	50.61	152	22.45	
	3000	8 H.P.			2.08	17.05	51.12	153	26.85	
	3050	Vibrating screed, gas engine, 8 H.P.			2.80	88	263.50	790	75.10	
	3120	Concrete transit mixer, 6 x 4, 250 H.P., 8 C.Y., rear discharge			50.57	70	210	630	446.55	
	3200	Front discharge			58.71	135	405	1,225	550.65	
	3300	6 x 6, 285 H.P., 12 C.Y., rear discharge			57.97	150	450	1,350	553.80	
	3400	Front discharge			60.41	170	510	1,525	585.25	
20	0010	**EARTHWORK EQUIPMENT RENTAL** without operators	R015433 -10							**20**
	0040	Aggregate spreader, push type, 8' to 12' wide		Ea.	2.59	75	225	675	65.75	
	0045	Tailgate type, 8' wide			2.54	63.50	190	570	58.30	
	0055	Earth auger, truck mounted, for fence & sign posts, utility poles			13.81	150	450	1,350	200.50	
	0060	For borings and monitoring wells			42.52	83.50	250	750	390.20	
	0070	Portable, trailer mounted			2.29	100	300	900	78.35	
	0075	Truck mounted, for caissons, water wells			85.14	150	450	1,350	771.10	
	0080	Horizontal boring machine, 12" to 36" diameter, 45 H.P.			22.70	104	312	935	244	
	0090	12" to 48" diameter, 65 H.P.			31.16	108	325	975	314.25	
	0095	Auger, for fence posts, gas engine, hand held			.45	84	251.50	755	53.85	
	0100	Excavator, diesel hydraulic, crawler mounted, 1/2 C.Y. cap.			21.66	465	1,394.28	4,175	452.10	
	0120	5/8 C.Y. capacity			28.95	610	1,833.22	5,500	598.25	
	0140	3/4 C.Y. capacity			32.56	725	2,168.88	6,500	694.25	
	0150	1 C.Y. capacity			41.09	780	2,333	7,000	795.30	

01 54 33	Equipment Rental	UNIT	HOURLY OPER. COST	RENT PER DAY	RENT PER WEEK	RENT PER MONTH	EQUIPMENT COST/DAY	
20	0200	1-1/2 C.Y. capacity	Ea.	48.44	500	1,500	4,500	687.55	20
	0300	2 C.Y. capacity		56.41	835	2,500	7,500	951.30	
	0320	2-1/2 C.Y. capacity		82.39	1,350	4,027.92	12,100	1,465	
	0325	3-1/2 C.Y. capacity		119.76	2,000	6,000	18,000	2,158	
	0330	4-1/2 C.Y. capacity		151.17	3,675	11,000	33,000	3,409	
	0335	6 C.Y. capacity		191.81	3,225	9,650	29,000	3,464	
	0340	7 C.Y. capacity		174.67	3,400	10,200	30,600	3,437	
	0342	Excavator attachments, bucket thumbs		3.39	258	774.60	2,325	182.05	
	0345	Grapples		3.13	222	666.16	2,000	158.30	
	0346	Hydraulic hammer for boom mounting, 4000 ft lb.		13.44	890	2,670	8,000	641.50	
	0347	5000 ft lb.		15.90	950	2,850	8,550	697.25	
	0348	8000 ft lb.		23.47	1,200	3,600	10,800	907.75	
	0349	12,000 ft lb.		25.64	1,100	3,333	10,000	871.75	
	0350	Gradall type, truck mounted, 3 ton @ 15' radius, 5/8 C.Y.		43.31	835	2,500	7,500	846.50	
	0370	1 C.Y. capacity		59.22	835	2,500	7,500	973.80	
	0400	Backhoe-loader, 40 to 45 H.P., 5/8 C.Y. capacity		11.86	244	732.50	2,200	241.40	
	0450	45 H.P. to 60 H.P., 3/4 C.Y. capacity		17.97	117	350	1,050	213.75	
	0460	80 H.P., 1-1/4 C.Y. capacity		20.30	117	350	1,050	232.40	
	0470	112 H.P., 1-1/2 C.Y. capacity		32.89	610	1,833.22	5,500	629.75	
	0482	Backhoe-loader attachment, compactor, 20,000 lb.		6.42	155	464.76	1,400	144.30	
	0485	Hydraulic hammer, 750 ft lb.		3.67	107	320.17	960	93.40	
	0486	Hydraulic hammer, 1200 ft lb.		6.53	205	614.52	1,850	175.15	
	0500	Brush chipper, gas engine, 6" cutter head, 35 H.P.		9.14	247	740	2,225	221.10	
	0550	Diesel engine, 12" cutter head, 130 H.P.		23.60	340	1,020	3,050	392.80	
	0600	15" cutter head, 165 H.P.		26.51	415	1,239.36	3,725	459.90	
	0750	Bucket, clamshell, general purpose, 3/8 C.Y.		1.40	91.50	275	825	66.15	
	0800	1/2 C.Y.		1.51	91.50	275	825	67.10	
	0850	3/4 C.Y.		1.64	91.50	275	825	68.10	
	0900	1 C.Y.		1.70	91.50	275	825	68.55	
	0950	1-1/2 C.Y.		2.78	91.50	275	825	77.25	
	1000	2 C.Y.		2.91	91.50	275	825	78.30	
	1010	Bucket, dragline, medium duty, 1/2 C.Y.		.82	91.50	275	825	61.55	
	1020	3/4 C.Y.		.78	91.50	275	825	61.25	
	1030	1 C.Y.		.80	91.50	275	825	61.35	
	1040	1-1/2 C.Y.		1.26	91.50	275	825	65.05	
	1050	2 C.Y.		1.29	91.50	275	825	65.30	
	1070	3 C.Y.		2.07	91.50	275	825	71.60	
	1200	Compactor, manually guided 2-drum vibratory smooth roller, 7.5 H.P.		7.20	181	542.50	1,625	166.10	
	1250	Rammer/tamper, gas, 8"		2.20	48	144.59	435	46.50	
	1260	15"		2.62	55	165.25	495	54	
	1300	Vibratory plate, gas, 18" plate, 3000 lb. blow		2.12	24.50	72.81	218	31.55	
	1350	21" plate, 5000 lb. blow		2.61	241	722.50	2,175	165.35	
	1370	Curb builder/extruder, 14 H.P., gas, single screw		13.99	253	760	2,275	263.95	
	1390	Double screw		14.99	253	760	2,275	271.95	
	1500	Disc harrow attachment, for tractor		.47	82.50	246.84	740	53.15	
	1810	Feller buncher, shearing & accumulating trees, 100 H.P.		39.08	460	1,380	4,150	588.60	
	1860	Grader, self-propelled, 25,000 lb.		33.25	1,100	3,333	10,000	932.60	
	1910	30,000 lb.		32.76	1,325	4,000	12,000	1,062	
	1920	40,000 lb.		51.73	1,550	4,667	14,000	1,347	
	1930	55,000 lb.		66.73	1,775	5,333	16,000	1,600	
	1950	Hammer, pavement breaker, self-propelled, diesel, 1000 to 1250 lb.		28.31	600	1,800	5,400	586.50	
	2000	1300 to 1500 lb.		42.67	1,000	3,020.94	9,075	945.55	
	2050	Pile driving hammer, steam or air, 4150 ft lb. @ 225 bpm		12.11	500	1,500	4,500	396.90	
	2100	8750 ft lb. @ 145 bpm		14.30	700	2,100	6,300	534.45	
	2150	15,000 ft lb. @ 60 bpm		14.63	835	2,500	7,500	617.05	
	2200	24,450 ft lb. @ 111 bpm		15.64	965	2,900	8,700	705.15	
	2250	Leads, 60' high for pile driving hammers up to 20,000 ft lb.		3.66	300	900	2,700	209.25	
	2300	90' high for hammers over 20,000 ft lb.		5.43	540	1,620	4,850	367.45	

01 54 33 | Equipment Rental

		UNIT	HOURLY OPER. COST	RENT PER DAY	RENT PER WEEK	RENT PER MONTH	EQUIPMENT COST/DAY	
20	2350	Diesel type hammer, 22,400 ft lb.	Ea.	17.76	490	1,471.74	4,425	436.45	**20**
	2400	41,300 ft lb.		25.61	620	1,859.04	5,575	576.65	
	2450	141,000 ft lb.		41.20	980	2,943.48	8,825	918.35	
	2500	Vib. elec. hammer/extractor, 200 kW diesel generator, 34 H.P.		41.25	715	2,143.06	6,425	758.60	
	2550	80 H.P.		72.81	1,025	3,098.40	9,300	1,202	
	2600	150 H.P.		134.74	2,000	5,964.42	17,900	2,271	
	2700	Hydro Excavator w/EXT boom 12 C.Y., 1200 gallons		37.70	1,600	4,800	14,400	1,262	
	2800	Log chipper, up to 22″ diameter, 600 H.P.		46.13	305	915	2,750	552	
	2850	Logger, for skidding & stacking logs, 150 H.P.		43.40	930	2,785	8,350	904.20	
	2860	Mulcher, diesel powered, trailer mounted		17.99	305	915	2,750	326.90	
	2900	Rake, spring tooth, with tractor		14.67	370	1,110.26	3,325	339.45	
	3000	Roller, vibratory, tandem, smooth drum, 20 H.P.		7.78	320	967	2,900	255.65	
	3050	35 H.P.		10.10	260	779.76	2,350	236.75	
	3100	Towed type vibratory compactor, smooth drum, 50 H.P.		25.20	520	1,566	4,700	514.75	
	3150	Sheepsfoot, 50 H.P.		25.56	385	1,161.90	3,475	436.90	
	3170	Landfill compactor, 220 H.P.		69.80	1,650	4,985	15,000	1,555	
	3200	Pneumatic tire roller, 80 H.P.		12.88	405	1,213.54	3,650	345.75	
	3250	120 H.P.		19.33	665	1,988.14	5,975	552.25	
	3300	Sheepsfoot vibratory roller, 240 H.P.		62.02	1,425	4,260.30	12,800	1,348	
	3320	340 H.P.		83.57	2,175	6,500	19,500	1,969	
	3350	Smooth drum vibratory roller, 75 H.P.		23.27	655	1,962.32	5,875	578.60	
	3400	125 H.P.		27.53	740	2,220.52	6,650	664.35	
	3410	Rotary mower, brush, 60″, with tractor		18.73	360	1,084.44	3,250	366.75	
	3420	Rototiller, walk-behind, gas, 5 H.P.		2.13	60	180	540	53.05	
	3422	8 H.P.		2.80	132	395	1,175	101.40	
	3440	Scrapers, towed type, 7 C.Y. capacity		6.42	127	382.14	1,150	127.80	
	3450	10 C.Y. capacity		7.18	170	511.24	1,525	159.70	
	3500	15 C.Y. capacity		7.38	196	588.70	1,775	176.75	
	3525	Self-propelled, single engine, 14 C.Y. capacity		132.89	2,225	6,660	20,000	2,395	
	3550	Dual engine, 21 C.Y. capacity		140.95	2,500	7,500	22,500	2,628	
	3600	31 C.Y. capacity		187.28	3,625	10,844.40	32,500	3,667	
	3640	44 C.Y. capacity		231.98	4,650	13,942.80	41,800	4,644	
	3650	Elevating type, single engine, 11 C.Y. capacity		61.68	1,075	3,200	9,600	1,133	
	3700	22 C.Y. capacity		114.28	1,625	4,850	14,600	1,884	
	3710	Screening plant, 110 H.P. w/5′ x 10′ screen		21.07	645	1,933	5,800	555.15	
	3720	5′ x 16′ screen		26.60	1,325	4,000	12,000	1,013	
	3850	Shovel, crawler-mounted, front-loading, 7 C.Y. capacity		218.00	3,925	11,773.92	35,300	4,099	
	3855	12 C.Y. capacity		335.89	5,450	16,318.24	49,000	5,951	
	3860	Shovel/backhoe bucket, 1/2 C.Y.		2.68	73	218.95	655	65.25	
	3870	3/4 C.Y.		2.66	82	245.81	735	70.40	
	3880	1 C.Y.		2.75	91	272.66	820	76.50	
	3890	1-1/2 C.Y.		2.94	107	320.17	960	87.60	
	3910	3 C.Y.		3.43	145	433.78	1,300	114.15	
	3950	Stump chipper, 18″ deep, 30 H.P.		6.88	232	697	2,100	194.45	
	4110	Dozer, crawler, torque converter, diesel 80 H.P.		25.18	335	1,000	3,000	401.40	
	4150	105 H.P.		34.23	600	1,800	5,400	633.85	
	4200	140 H.P.		41.15	720	2,166	6,500	762.40	
	4260	200 H.P.		62.97	1,675	5,000	15,000	1,504	
	4310	300 H.P.		80.49	1,875	5,600	16,800	1,764	
	4360	410 H.P.		106.42	3,200	9,630	28,900	2,777	
	4370	500 H.P.		132.98	3,900	11,670	35,000	3,398	
	4380	700 H.P.		229.47	5,475	16,421.52	49,300	5,120	
	4400	Loader, crawler, torque conv., diesel, 1-1/2 C.Y., 80 H.P.		29.45	550	1,651	4,950	565.80	
	4450	1-1/2 to 1-3/4 C.Y., 95 H.P.		30.18	695	2,091.42	6,275	659.75	
	4510	1-3/4 to 2-1/4 C.Y., 130 H.P.		47.61	965	2,900	8,700	960.90	
	4530	2-1/2 to 3-1/4 C.Y., 190 H.P.		57.61	1,175	3,540	10,600	1,169	
	4560	3-1/2 to 5 C.Y., 275 H.P.		71.19	1,525	4,595.96	13,800	1,489	
	4610	Front end loader, 4WD, articulated frame, diesel, 1 to 1-1/4 C.Y., 70 H.P.		16.58	282	846.90	2,550	302	

01 54 33 | Equipment Rental

			UNIT	HOURLY OPER. COST	RENT PER DAY	RENT PER WEEK	RENT PER MONTH	EQUIPMENT COST/DAY	
20	4620	1-1/2 to 1-3/4 C.Y., 95 H.P.	Ea.	19.94	440	1,320	3,950	423.50	20
	4650	1-3/4 to 2 C.Y., 130 H.P.		21.00	395	1,187.72	3,575	405.55	
	4710	2-1/2 to 3-1/2 C.Y., 145 H.P.		29.44	780	2,333	7,000	702.10	
	4730	3 to 4-1/2 C.Y., 185 H.P.		31.99	890	2,667	8,000	789.35	
	4760	5-1/4 to 5-3/4 C.Y., 270 H.P.		53.03	890	2,666	8,000	957.50	
	4810	7 to 9 C.Y., 475 H.P.		90.90	2,550	7,667	23,000	2,261	
	4870	9 to 11 C.Y., 620 H.P.		131.52	2,700	8,107.48	24,300	2,674	
	4880	Skid-steer loader, wheeled, 10 C.F., 30 H.P. gas		9.54	169	506.07	1,525	177.55	
	4890	1 C.Y., 78 H.P., diesel		18.38	420	1,265.18	3,800	400.10	
	4892	Skid-steer attachment, auger		.74	145	433.50	1,300	92.65	
	4893	Backhoe		.74	122	366.64	1,100	79.25	
	4894	Broom		.70	140	420.25	1,250	89.70	
	4895	Forks		.15	32	96	288	20.45	
	4896	Grapple		.72	88.50	265.25	795	58.80	
	4897	Concrete hammer		1.05	183	550	1,650	118.40	
	4898	Tree spade		.60	103	309.84	930	66.75	
	4899	Trencher		.65	102	305	915	66.20	
	4900	Trencher, chain, boom type, gas, operator walking, 12 H.P.		4.16	206	618.25	1,850	156.95	
	4910	Operator riding, 40 H.P.		16.64	450	1,343.75	4,025	401.85	
	5000	Wheel type, diesel, 4' deep, 12" wide		68.50	965	2,891.84	8,675	1,126	
	5100	6' deep, 20" wide		87.32	1,050	3,127.50	9,375	1,324	
	5150	Chain type, diesel, 5' deep, 8" wide		16.25	360	1,084.44	3,250	346.90	
	5200	Diesel, 8' deep, 16" wide		89.39	1,925	5,783.68	17,400	1,872	
	5202	Rock trencher, wheel type, 6" wide x 18" deep		46.98	90	270	810	429.85	
	5206	Chain type, 18" wide x 7' deep		104.38	283	850	2,550	1,005	
	5210	Tree spade, self-propelled		13.65	230	690	2,075	247.20	
	5250	Truck, dump, 2-axle, 12 ton, 8 C.Y. payload, 220 H.P.		23.88	395	1,185	3,550	428.05	
	5300	Three axle dump, 16 ton, 12 C.Y. payload, 400 H.P.		44.50	360	1,084.44	3,250	572.90	
	5310	Four axle dump, 25 ton, 18 C.Y. payload, 450 H.P.		49.85	525	1,575.02	4,725	713.80	
	5350	Dump trailer only, rear dump, 16-1/2 C.Y.		5.73	151	454.43	1,375	136.70	
	5400	20 C.Y.		6.18	170	511.24	1,525	151.70	
	5450	Flatbed, single axle, 1-1/2 ton rating		19.00	73.50	221.02	665	196.15	
	5500	3 ton rating		23.05	1,050	3,180	9,550	820.40	
	5550	Off highway rear dump, 25 ton capacity		62.67	1,475	4,389.40	13,200	1,379	
	5600	35 ton capacity		66.90	665	2,000	6,000	935.20	
	5610	50 ton capacity		83.87	1,825	5,499.66	16,500	1,771	
	5620	65 ton capacity		89.56	2,000	5,990.24	18,000	1,915	
	5630	100 ton capacity		121.24	2,950	8,830.44	26,500	2,736	
	6000	Vibratory plow, 25 H.P., walking		6.77	300	900	2,700	234.20	
40	0010	**GENERAL EQUIPMENT RENTAL** without operators							40
	0020	Aerial lift, scissor type, to 20' high, 1200 lb. capacity, electric	Ea.	3.48	129	385.75	1,150	105	
	0030	To 30' high, 1200 lb. capacity		3.77	203	607.67	1,825	151.70	
	0040	Over 30' high, 1500 lb. capacity		5.13	243	727.50	2,175	186.60	
	0070	Articulating boom, to 45' high, 500 lb. capacity, diesel		9.92	250	750	2,250	229.35	
	0075	To 60' high, 500 lb. capacity		13.66	300	900	2,700	289.25	
	0080	To 80' high, 500 lb. capacity		16.05	900	2,702.25	8,100	668.85	
	0085	To 125' high, 500 lb. capacity		18.34	1,525	4,603.50	13,800	1,067	
	0100	Telescoping boom to 40' high, 500 lb. capacity, diesel		11.24	315	945	2,825	278.90	
	0105	To 45' high, 500 lb. capacity		12.51	320	965	2,900	293.05	
	0110	To 60' high, 500 lb. capacity		16.36	300	900	2,700	310.90	
	0115	To 80' high, 500 lb. capacity		21.27	355	1,067	3,200	383.55	
	0120	To 100' high, 500 lb. capacity		28.71	865	2,587.75	7,775	747.25	
	0125	To 120' high, 500 lb. capacity		29.16	1,450	4,348.50	13,000	1,103	
	0195	Air compressor, portable, 6.5 CFM, electric		.90	44.50	133	400	33.80	
	0196	Gasoline		.65	56	167.50	505	38.70	
	0200	Towed type, gas engine, 60 CFM		9.43	129	387.50	1,175	152.95	
	0300	160 CFM		10.47	198	595	1,775	202.80	

Note: R015433-10 (row 0010), R015433-15 (row 0070)

For customer support on your Commercial Renovation Costs with RSMeans data, call 800.448.8182.

743

		UNIT	HOURLY OPER. COST	RENT PER DAY	RENT PER WEEK	RENT PER MONTH	EQUIPMENT COST/DAY	
40	0400	Diesel engine, rotary screw, 250 CFM	Ea.	12.08	175	524	1,575	201.50	**40**
	0500	365 CFM		16.00	310	937	2,800	315.40	
	0550	450 CFM		19.95	277	832.25	2,500	326.05	
	0600	600 CFM		34.10	248	743.62	2,225	421.50	
	0700	750 CFM		34.62	435	1,306.50	3,925	538.25	
	0930	Air tools, breaker, pavement, 60 lb.		.57	81.50	245	735	53.50	
	0940	80 lb.		.56	81	242.50	730	53	
	0950	Drills, hand (jackhammer), 65 lb.		.67	68	203.50	610	46.05	
	0960	Track or wagon, swing boom, 4" drifter		54.66	1,025	3,104	9,300	1,058	
	0970	5" drifter		63.30	1,025	3,104	9,300	1,127	
	0975	Track mounted quarry drill, 6" diameter drill		101.91	1,900	5,665	17,000	1,948	
	0980	Dust control per drill		1.04	25	75.50	227	23.40	
	0990	Hammer, chipping, 12 lb.		.60	46	138	415	32.40	
	1000	Hose, air with couplings, 50' long, 3/4" diameter		.07	12	36	108	7.75	
	1100	1" diameter		.08	12.35	37	111	8	
	1200	1-1/2" diameter		.22	37.50	112.50	340	24.25	
	1300	2" diameter		.24	45	135	405	28.90	
	1400	2-1/2" diameter		.36	57.50	172.50	520	37.35	
	1410	3" diameter		.42	58.50	175	525	38.35	
	1450	Drill, steel, 7/8" x 2'		.08	12.90	38.73	116	8.40	
	1460	7/8" x 6'		.12	19.60	58.87	177	12.70	
	1520	Moil points		.03	7	21	63	4.40	
	1525	Pneumatic nailer w/accessories		.48	39.50	118	355	27.40	
	1530	Sheeting driver for 60 lb. breaker		.04	7.75	23.24	69.50	5	
	1540	For 90 lb. breaker		.13	10.50	31.50	94.50	7.35	
	1550	Spade, 25 lb.		.50	7.40	22.21	66.50	8.45	
	1560	Tamper, single, 35 lb.		.59	48.50	145.75	435	33.85	
	1570	Triple, 140 lb.		.89	61.50	184.87	555	44.05	
	1580	Wrenches, impact, air powered, up to 3/4" bolt		.43	49.50	148.25	445	33.05	
	1590	Up to 1-1/4" bolt		.58	79.50	238.50	715	52.30	
	1600	Barricades, barrels, reflectorized, 1 to 99 barrels		.03	4	12	36	2.65	
	1610	100 to 200 barrels		.02	4.41	13.22	39.50	2.85	
	1620	Barrels with flashers, 1 to 99 barrels		.03	6.40	19.16	57.50	4.10	
	1630	100 to 200 barrels		.03	5.10	15.34	46	3.30	
	1640	Barrels with steady burn type C lights		.05	8.45	25.30	76	5.45	
	1650	Illuminated board, trailer mounted, with generator		3.28	139	418.28	1,250	109.85	
	1670	Portable barricade, stock, with flashers, 1 to 6 units		.03	6.35	19.11	57.50	4.10	
	1680	25 to 50 units		.03	5.95	17.82	53.50	3.85	
	1685	Butt fusion machine, wheeled, 1.5 HP electric, 2" - 8" diameter pipe		2.63	225	675	2,025	156.05	
	1690	Tracked, 20 HP diesel, 4"-12" diameter pipe		11.23	560	1,685	5,050	426.85	
	1695	83 HP diesel, 8" - 24" diameter pipe		51.32	1,100	3,325	9,975	1,076	
	1700	Carts, brick, gas engine, 1000 lb. capacity		2.94	65	195	585	62.55	
	1800	1500 lb., 7-1/2' lift		2.92	69.50	208	625	64.95	
	1822	Dehumidifier, medium, 6 lb./hr., 150 CFM		1.19	76.50	229.28	690	55.35	
	1824	Large, 18 lb./hr., 600 CFM		2.19	585	1,750	5,250	367.55	
	1830	Distributor, asphalt, trailer mounted, 2000 gal., 38 H.P. diesel		10.99	355	1,058.62	3,175	299.65	
	1840	3000 gal., 38 H.P. diesel		12.87	380	1,136.08	3,400	330.15	
	1850	Drill, rotary hammer, electric		1.11	71.50	214	640	51.70	
	1860	Carbide bit, 1-1/2" diameter, add to electric rotary hammer		.03	41.50	125	375	25.25	
	1865	Rotary, crawler, 250 H.P.		135.77	2,300	6,868.12	20,600	2,460	
	1870	Emulsion sprayer, 65 gal., 5 H.P. gas engine		2.77	107	320.17	960	86.15	
	1880	200 gal., 5 H.P. engine		7.22	179	537.06	1,600	165.20	
	1900	Floor auto-scrubbing machine, walk-behind, 28" path		5.62	222	667	2,000	178.40	
	1930	Floodlight, mercury vapor, or quartz, on tripod, 1000 watt		.46	36.50	110	330	25.65	
	1940	2000 watt		.59	28	84.69	254	21.65	
	1950	Floodlights, trailer mounted with generator, 1 - 300 watt light		3.54	78.50	235.48	705	75.45	
	1960	2 - 1000 watt lights		4.49	87.50	262.33	785	88.35	
	2000	4 - 300 watt lights		4.24	100	299.51	900	93.85	

			UNIT	HOURLY OPER. COST	RENT PER DAY	RENT PER WEEK	RENT PER MONTH	EQUIPMENT COST/DAY	
40		**01 54 33	Equipment Rental**							**40**
	2005	Foam spray rig, incl. box trailer, compressor, generator, proportioner	Ea.	25.46	535	1,600.84	4,800	523.85	
	2015	Forklift, pneumatic tire, rough terr, straight mast, 5000 lb, 12' lift, gas		18.59	219	655.83	1,975	279.90	
	2025	8000 lb, 12' lift		22.68	360	1,084.44	3,250	398.35	
	2030	5000 lb, 12' lift, diesel		15.41	244	733.29	2,200	269.90	
	2035	8000 lb, 12' lift, diesel		16.70	277	831.40	2,500	299.90	
	2045	All terrain, telescoping boom, diesel, 5000 lb, 10' reach, 19' lift		17.20	233	700	2,100	277.60	
	2055	6600 lb, 29' reach, 42' lift		21.04	233	700	2,100	308.30	
	2065	10,000 lb, 31' reach, 45' lift		23.03	315	950	2,850	374.20	
	2070	Cushion tire, smooth floor, gas, 5000 lb capacity		8.23	247	741.50	2,225	214.10	
	2075	8000 lb capacity		11.33	275	826.25	2,475	255.90	
	2085	Diesel, 5000 lb capacity		7.73	210	629.25	1,900	187.65	
	2090	12,000 lb capacity		12.01	400	1,194.50	3,575	335	
	2095	20,000 lb capacity		17.20	660	1,980	5,950	533.60	
	2100	Generator, electric, gas engine, 1.5 kW to 3 kW		2.57	46.50	140	420	48.55	
	2200	5 kW		3.21	91	272.50	820	80.15	
	2300	10 kW		5.91	108	322.50	970	111.80	
	2400	25 kW		7.38	405	1,210	3,625	301.10	
	2500	Diesel engine, 20 kW		9.18	229	687.50	2,075	210.95	
	2600	50 kW		15.90	370	1,110	3,325	349.20	
	2700	100 kW		28.51	445	1,340.50	4,025	496.20	
	2800	250 kW		54.19	750	2,249.33	6,750	883.40	
	2850	Hammer, hydraulic, for mounting on boom, to 500 ft lb.		2.89	93.50	279.89	840	79.10	
	2860	1000 ft lb.		4.59	139	418.28	1,250	120.40	
	2900	Heaters, space, oil or electric, 50 MBH		1.46	46.50	140	420	39.65	
	3000	100 MBH		2.71	46.50	140	420	49.70	
	3100	300 MBH		7.90	135	405	1,225	144.20	
	3150	500 MBH		13.12	200	600	1,800	224.95	
	3200	Hose, water, suction with coupling, 20' long, 2" diameter		.02	5.65	17	51	3.55	
	3210	3" diameter		.03	14.15	42.50	128	8.75	
	3220	4" diameter		.03	28	84	252	17.05	
	3230	6" diameter		.11	40.50	121.50	365	25.20	
	3240	8" diameter		.27	53.50	160	480	34.15	
	3250	Discharge hose with coupling, 50' long, 2" diameter		.01	6.50	19.50	58.50	4	
	3260	3" diameter		.01	7.35	22	66	4.50	
	3270	4" diameter		.02	21	62.50	188	12.65	
	3280	6" diameter		.06	29	87	261	17.90	
	3290	8" diameter		.24	37.50	112.50	340	24.40	
	3295	Insulation blower		.83	117	350	1,050	76.65	
	3300	Ladders, extension type, 16' to 36' long		.18	41.50	125	375	26.45	
	3400	40' to 60' long		.64	120	360.50	1,075	77.20	
	3405	Lance for cutting concrete		2.20	65	195	585	56.60	
	3407	Lawn mower, rotary, 22", 5 H.P.		1.05	38.50	115	345	31.40	
	3408	48" self-propelled		2.89	138	415	1,250	106.10	
	3410	Level, electronic, automatic, with tripod and leveling rod		1.05	37.50	112	335	30.80	
	3430	Laser type, for pipe and sewer line and grade		2.17	117	350	1,050	87.35	
	3440	Rotating beam for interior control		.90	64	192.50	580	45.70	
	3460	Builder's optical transit, with tripod and rod		.10	37.50	112	335	23.20	
	3500	Light towers, towable, with diesel generator, 2000 watt		4.25	101	303.64	910	94.75	
	3600	4000 watt		4.50	165	495	1,475	135	
	3700	Mixer, powered, plaster and mortar, 6 C.F., 7 H.P.		2.05	83.50	250	750	66.40	
	3800	10 C.F., 9 H.P.		2.24	124	372.50	1,125	92.40	
	3850	Nailer, pneumatic		.48	33.50	100.18	300	23.85	
	3900	Paint sprayers complete, 8 CFM		.85	61.50	184.87	555	43.75	
	4000	17 CFM		1.60	110	330.50	990	78.85	
	4020	Pavers, bituminous, rubber tires, 8' wide, 50 H.P., diesel		31.93	570	1,704.12	5,100	596.25	
	4030	10' wide, 150 H.P.		95.62	1,950	5,835.32	17,500	1,932	
	4050	Crawler, 8' wide, 100 H.P., diesel		87.59	2,050	6,170.98	18,500	1,935	
	4060	10' wide, 150 H.P.		103.97	2,350	7,048.86	21,100	2,241	

		UNIT	HOURLY OPER. COST	RENT PER DAY	RENT PER WEEK	RENT PER MONTH	EQUIPMENT COST/DAY	
40	4070	Concrete paver, 12' to 24' wide, 250 H.P.	Ea.	87.62	1,675	5,060.72	15,200	1,713	**40**
	4080	Placer-spreader-trimmer, 24' wide, 300 H.P.		117.51	2,550	7,668.54	23,000	2,474	
	4100	Pump, centrifugal gas pump, 1-1/2" diam., 65 GPM		3.92	54	162.15	485	63.80	
	4200	2" diameter, 130 GPM		4.98	44	132.50	400	66.35	
	4300	3" diameter, 250 GPM		5.12	56	167.50	505	74.45	
	4400	6" diameter, 1500 GPM		22.24	91.50	275	825	232.90	
	4500	Submersible electric pump, 1-1/4" diameter, 55 GPM		.40	36	107.50	325	24.70	
	4600	1-1/2" diameter, 83 GPM		.44	43.50	130	390	29.55	
	4700	2" diameter, 120 GPM		1.64	61.50	185	555	50.15	
	4800	3" diameter, 300 GPM		3.03	109	327.50	985	89.75	
	4900	4" diameter, 560 GPM		14.75	61.50	185	555	155	
	5000	6" diameter, 1590 GPM		22.08	65	195	585	215.60	
	5100	Diaphragm pump, gas, single, 1-1/2" diameter		1.13	38.50	115	345	32	
	5200	2" diameter		3.98	91.50	275	825	86.80	
	5300	3" diameter		4.05	95	285	855	89.40	
	5400	Double, 4" diameter		6.03	95	285	855	105.25	
	5450	Pressure washer 5 GPM, 3000 psi		3.87	110	330	990	96.95	
	5460	7 GPM, 3000 psi		4.94	85	255	765	90.50	
	5470	High pressure water jet 10 ksi		39.55	720	2,160	6,475	748.40	
	5480	40 ksi		27.88	980	2,940	8,825	811.05	
	5500	Trash pump, self-priming, gas, 2" diameter		3.82	108	325	975	95.55	
	5600	Diesel, 4" diameter		6.68	162	485	1,450	150.40	
	5650	Diesel, 6" diameter		16.85	162	485	1,450	231.80	
	5655	Grout Pump		18.70	281	841.73	2,525	317.90	
	5700	Salamanders, L.P. gas fired, 100,000 BTU		2.88	57.50	172.50	520	57.55	
	5705	50,000 BTU		1.66	23.50	71	213	27.50	
	5720	Sandblaster, portable, open top, 3 C.F. capacity		.60	132	395	1,175	83.80	
	5730	6 C.F. capacity		1.00	132	395	1,175	87	
	5740	Accessories for above		.14	24	71.26	214	15.35	
	5750	Sander, floor		.77	73.50	220	660	50.15	
	5760	Edger		.52	35	105	315	25.15	
	5800	Saw, chain, gas engine, 18" long		1.75	63.50	190	570	52	
	5900	Hydraulic powered, 36" long		.78	58.50	175	525	41.25	
	5950	60" long		.78	65	195	585	45.25	
	6000	Masonry, table mounted, 14" diameter, 5 H.P.		1.32	76.50	230	690	56.55	
	6050	Portable cut-off, 8 H.P.		1.81	77.50	232.50	700	61	
	6100	Circular, hand held, electric, 7-1/4" diameter		.23	13.85	41.50	125	10.10	
	6200	12" diameter		.24	41	122.50	370	26.40	
	6250	Wall saw, w/hydraulic power, 10 H.P.		3.29	98.50	296	890	85.50	
	6275	Shot blaster, walk-behind, 20" wide		4.73	281	841.73	2,525	206.20	
	6280	Sidewalk broom, walk-behind		2.24	82.50	247.87	745	67.50	
	6300	Steam cleaner, 100 gallons per hour		3.34	82.50	247.87	745	76.30	
	6310	200 gallons per hour		4.33	100	299.51	900	94.55	
	6340	Tar Kettle/Pot, 400 gallons		16.48	127	380	1,150	207.85	
	6350	Torch, cutting, acetylene-oxygen, 150' hose, excludes gases		.45	15.30	45.96	138	12.80	
	6360	Hourly operating cost includes tips and gas		20.92	7	20.98	63	171.55	
	6410	Toilet, portable chemical		.13	23	69.20	208	14.90	
	6420	Recycle flush type		.16	28.50	85.72	257	18.45	
	6430	Toilet, fresh water flush, garden hose,		.19	34	102.25	305	22	
	6440	Hoisted, non-flush, for high rise		.16	28	83.66	251	18	
	6465	Tractor, farm with attachment		17.37	390	1,172.50	3,525	373.40	
	6480	Trailers, platform, flush deck, 2 axle, 3 ton capacity		1.69	94.50	284	850	70.30	
	6500	25 ton capacity		6.23	143	428.61	1,275	135.55	
	6600	40 ton capacity		8.04	203	609.35	1,825	186.20	
	6700	3 axle, 50 ton capacity		8.72	225	676.48	2,025	205.05	
	6800	75 ton capacity		11.08	300	898.54	2,700	268.30	
	6810	Trailer mounted cable reel for high voltage line work		5.89	28.50	85	255	64.10	
	6820	Trailer mounted cable tensioning rig		11.67	28.50	85	255	110.40	

01 54 33 | Equipment Rental

		Description	UNIT	HOURLY OPER. COST	RENT PER DAY	RENT PER WEEK	RENT PER MONTH	EQUIPMENT COST/DAY	
40	6830	Cable pulling rig	Ea.	73.77	28.50	85	255	607.15	**40**
	6850	Portable cable/wire puller, 8000 lb max pulling capacity		3.70	120	360	1,075	101.60	
	6900	Water tank trailer, engine driven discharge, 5000 gallons		7.16	158	475.09	1,425	152.25	
	6925	10,000 gallons		9.75	215	645.50	1,925	207.10	
	6950	Water truck, off highway, 6000 gallons		71.75	835	2,504.54	7,525	1,075	
	7010	Tram car for high voltage line work, powered, 2 conductor		6.88	28.50	85	255	72.05	
	7020	Transit (builder's level) with tripod		.10	17.55	52.67	158	11.30	
	7030	Trench box, 3000 lb., 6' x 8'		.56	96.50	290.22	870	62.50	
	7040	7200 lb., 6' x 20'		.72	187	560	1,675	117.75	
	7050	8000 lb., 8' x 16'		1.08	186	557.71	1,675	120.15	
	7060	9500 lb., 8' x 20'		1.20	232	697.14	2,100	149.05	
	7065	11,000 lb., 8' x 24'		1.26	219	655.83	1,975	141.25	
	7070	12,000 lb., 10' x 20'		1.49	263	790.09	2,375	169.95	
	7100	Truck, pickup, 3/4 ton, 2 wheel drive		9.24	61.50	184.87	555	110.85	
	7200	4 wheel drive		9.48	167	500	1,500	175.85	
	7250	Crew carrier, 9 passenger		12.66	108	325	975	166.25	
	7290	Flat bed truck, 20,000 lb. GVW		15.26	133	397.63	1,200	201.60	
	7300	Tractor, 4 x 2, 220 H.P.		22.25	215	645.50	1,925	307.10	
	7410	330 H.P.		32.33	294	883.04	2,650	435.25	
	7500	6 x 4, 380 H.P.		36.09	340	1,022.47	3,075	493.25	
	7600	450 H.P.		44.23	415	1,239.36	3,725	601.75	
	7610	Tractor, with A frame, boom and winch, 225 H.P.		24.74	293	877.88	2,625	373.50	
	7620	Vacuum truck, hazardous material, 2500 gallons		12.79	310	929.52	2,800	288.25	
	7625	5,000 gallons		13.02	440	1,316.82	3,950	367.55	
	7650	Vacuum, HEPA, 16 gallon, wet/dry		.85	122	365	1,100	79.80	
	7655	55 gallon, wet/dry		.78	25.50	76.50	230	21.50	
	7660	Water tank, portable		.73	160	480.25	1,450	101.90	
	7690	Sewer/catch basin vacuum, 14 C.Y., 1500 gallons		17.31	665	1,988.14	5,975	536.15	
	7700	Welder, electric, 200 amp		3.81	33.50	100	300	50.50	
	7800	300 amp		5.55	103	310	930	106.40	
	7900	Gas engine, 200 amp		8.95	58.50	175	525	106.55	
	8000	300 amp		10.13	110	330	990	147	
	8100	Wheelbarrow, any size		.06	11.15	33.50	101	7.20	
	8200	Wrecking ball, 4000 lb.		2.50	60	180	540	56	
50	0010	**HIGHWAY EQUIPMENT RENTAL** without operators							**50**
	0050	Asphalt batch plant, portable drum mixer, 100 ton/hr. R015433 -10	Ea.	88.41	1,550	4,621.78	13,900	1,632	
	0060	200 ton/hr.		101.99	1,650	4,931.62	14,800	1,802	
	0070	300 ton/hr.		119.86	1,925	5,783.68	17,400	2,116	
	0100	Backhoe attachment, long stick, up to 185 H.P., 10.5' long		.37	25.50	76.43	229	18.25	
	0140	Up to 250 H.P., 12' long		.41	28.50	85.72	257	20.45	
	0180	Over 250 H.P., 15' long		.56	39	116.71	350	27.85	
	0200	Special dipper arm, up to 100 H.P., 32' long		1.16	79.50	238.58	715	56.95	
	0240	Over 100 H.P., 33' long		1.44	100	299.51	900	71.45	
	0280	Catch basin/sewer cleaning truck, 3 ton, 9 C.Y., 1000 gal.		35.39	420	1,265.18	3,800	536.15	
	0300	Concrete batch plant, portable, electric, 200 C.Y./hr.		24.18	560	1,678.30	5,025	529.15	
	0520	Grader/dozer attachment, ripper/scarifier, rear mounted, up to 135 H.P.		3.15	63.50	190.04	570	63.20	
	0540	Up to 180 H.P.		4.13	95.50	287.12	860	90.50	
	0580	Up to 250 H.P.		5.85	153	459.60	1,375	138.75	
	0700	Pvmt. removal bucket, for hyd. excavator, up to 90 H.P.		2.16	58	174.54	525	52.20	
	0740	Up to 200 H.P.		2.31	74.50	223.08	670	63.05	
	0780	Over 200 H.P.		2.52	91	273.69	820	74.90	
	0900	Aggregate spreader, self-propelled, 187 H.P.		50.60	740	2,220.52	6,650	848.90	
	1000	Chemical spreader, 3 C.Y.		3.17	96.50	290	870	83.35	
	1900	Hammermill, traveling, 250 H.P.		67.35	515	1,550	4,650	848.80	
	2000	Horizontal borer, 3" diameter, 13 H.P. gas driven		5.42	232	695	2,075	182.35	
	2150	Horizontal directional drill, 20,000 lb. thrust, 78 H.P. diesel		27.58	530	1,590	4,775	538.65	
	2160	30,000 lb. thrust, 115 H.P.		33.90	615	1,850	5,550	641.20	
	2170	50,000 lb. thrust, 170 H.P.		48.60	710	2,135	6,400	815.80	

01 54 33 | Equipment Rental

		UNIT	HOURLY OPER. COST	RENT PER DAY	RENT PER WEEK	RENT PER MONTH	EQUIPMENT COST/DAY	
50	2190	Mud trailer for HDD, 1500 gallons, 175 H.P., gas	Ea.	25.50	175	525	1,575	309	**50**
	2200	Hydromulcher, diesel, 3000 gallon, for truck mounting		17.43	227	680	2,050	275.45	
	2300	Gas, 600 gallon		7.49	95	285	855	116.95	
	2400	Joint & crack cleaner, walk behind, 25 H.P.		3.16	45.50	136	410	52.45	
	2500	Filler, trailer mounted, 400 gallons, 20 H.P.		8.35	147	440	1,325	154.75	
	3000	Paint striper, self-propelled, 40 gallon, 22 H.P.		6.76	122	365	1,100	127.10	
	3100	120 gallon, 120 H.P.		19.23	380	1,140	3,425	381.80	
	3200	Post drivers, 6" I-Beam frame, for truck mounting		12.41	320	960	2,875	291.30	
	3400	Road sweeper, self-propelled, 8' wide, 90 H.P.		35.91	715	2,143.06	6,425	715.85	
	3450	Road sweeper, vacuum assisted, 4 C.Y., 220 gallons		58.28	670	2,013.96	6,050	869	
	4000	Road mixer, self-propelled, 130 H.P.		46.23	825	2,478.72	7,425	865.60	
	4100	310 H.P.		75.01	2,150	6,480.82	19,400	1,896	
	4220	Cold mix paver, incl. pug mill and bitumen tank, 165 H.P.		94.97	2,325	6,945.58	20,800	2,149	
	4240	Pavement brush, towed		3.43	100	299.51	900	87.30	
	4250	Paver, asphalt, wheel or crawler, 130 H.P., diesel		94.23	2,275	6,816.48	20,400	2,117	
	4300	Paver, road widener, gas, 1' to 6', 67 H.P.		46.66	975	2,917.66	8,750	956.80	
	4400	Diesel, 2' to 14', 88 H.P.		56.38	1,150	3,459.88	10,400	1,143	
	4600	Slipform pavers, curb and gutter, 2 track, 75 H.P.		57.83	1,250	3,769.72	11,300	1,217	
	4700	4 track, 165 H.P.		35.69	845	2,530.36	7,600	791.55	
	4800	Median barrier, 215 H.P.		58.43	1,350	4,027.92	12,100	1,273	
	4901	Trailer, low bed, 75 ton capacity		10.71	282	846.90	2,550	255.05	
	5000	Road planer, walk behind, 10" cutting width, 10 H.P.		2.45	243	730	2,200	165.65	
	5100	Self-propelled, 12" cutting width, 64 H.P.		8.26	190	570	1,700	180.05	
	5120	Traffic line remover, metal ball blaster, truck mounted, 115 H.P.		46.56	905	2,720	8,150	916.50	
	5140	Grinder, truck mounted, 115 H.P.		50.89	905	2,720	8,150	951.15	
	5160	Walk-behind, 11 H.P.		3.56	142	425	1,275	113.45	
	5200	Pavement profiler, 4' to 6' wide, 450 H.P.		216.58	1,275	3,800	11,400	2,493	
	5300	8' to 10' wide, 750 H.P.		331.58	1,325	3,975	11,900	3,448	
	5400	Roadway plate, steel, 1" x 8' x 20'		.09	61	182.50	550	37.20	
	5600	Stabilizer, self-propelled, 150 H.P.		41.14	1,025	3,100	9,300	949.10	
	5700	310 H.P.		76.18	1,300	3,900	11,700	1,389	
	5800	Striper, truck mounted, 120 gallon paint, 460 H.P.		48.74	340	1,015	3,050	592.95	
	5900	Thermal paint heating kettle, 115 gallons		7.71	61.50	185	555	98.65	
	6000	Tar kettle, 330 gallon, trailer mounted		12.27	96.50	290	870	156.20	
	7000	Tunnel locomotive, diesel, 8 to 12 ton		29.76	620	1,859.04	5,575	609.85	
	7005	Electric, 10 ton		29.25	705	2,117.24	6,350	657.40	
	7010	Muck cars, 1/2 C.Y. capacity		2.30	26.50	80.04	240	34.40	
	7020	1 C.Y. capacity		2.51	35	104.31	315	40.95	
	7030	2 C.Y. capacity		2.66	39	116.71	350	44.60	
	7040	Side dump, 2 C.Y. capacity		2.87	48	144.59	435	51.90	
	7050	3 C.Y. capacity		3.85	53	159.05	475	62.65	
	7060	5 C.Y. capacity		5.62	68.50	205.53	615	86.10	
	7100	Ventilating blower for tunnel, 7-1/2 H.P.		2.14	52.50	158.02	475	48.70	
	7110	10 H.P.		2.42	55	165.25	495	52.40	
	7120	20 H.P.		3.54	71.50	214.82	645	71.30	
	7140	40 H.P.		6.14	94.50	284.02	850	105.90	
	7160	60 H.P.		8.69	102	304.68	915	130.45	
	7175	75 H.P.		10.37	158	475.09	1,425	177.95	
	7180	200 H.P.		20.78	310	934.68	2,800	353.20	
	7800	Windrow loader, elevating		53.94	1,650	4,975	14,900	1,427	
60	0010	**LIFTING AND HOISTING EQUIPMENT RENTAL** without operators							**60**
		R015433 -10							
	0150	Crane, flatbed mounted, 3 ton capacity	Ea.	14.41	201	604.19	1,825	236.10	
	0200	Crane, climbing, 106' jib, 6000 lb. capacity, 410 fpm		39.72	2,600	7,800	23,400	1,878	
		R312316 -45							
	0300	101' jib, 10,250 lb. capacity, 270 fpm		46.43	2,275	6,800	20,400	1,731	
	0500	Tower, static, 130' high, 106' jib, 6200 lb. capacity at 400 fpm		45.16	2,250	6,715	20,100	1,704	
	0520	Mini crawler spider crane, up to 24" wide, 1990 lb. lifting capacity		12.50	550	1,652.48	4,950	430.50	
	0525	Up to 30" wide, 6450 lb. lifting capacity		14.52	655	1,962.32	5,875	508.65	
	0530	Up to 52" wide, 6680 lb. lifting capacity		23.10	800	2,401.26	7,200	665.05	

01 54 33 | Equipment Rental

		UNIT	HOURLY OPER. COST	RENT PER DAY	RENT PER WEEK	RENT PER MONTH	EQUIPMENT COST/DAY	
60	0535	Up to 55" wide, 8920 lb. lifting capacity	Ea.	25.79	885	2,659.46	7,975	738.25	60
	0540	Up to 66" wide, 13,350 lb. lifting capacity		34.92	1,375	4,131.20	12,400	1,106	
	0600	Crawler mounted, lattice boom, 1/2 C.Y., 15 tons at 12' radius		36.96	830	2,483	7,450	792.30	
	0700	3/4 C.Y., 20 tons at 12' radius		50.42	930	2,790	8,375	961.35	
	0800	1 C.Y., 25 tons at 12' radius		67.42	985	2,950	8,850	1,129	
	0900	1-1/2 C.Y., 40 tons at 12' radius		66.31	1,125	3,375	10,100	1,206	
	1000	2 C.Y., 50 tons at 12' radius		88.77	1,325	4,000	12,000	1,510	
	1100	3 C.Y., 75 tons at 12' radius		75.26	2,325	7,000	21,000	2,002	
	1200	100 ton capacity, 60' boom		85.91	2,675	8,000	24,000	2,287	
	1300	165 ton capacity, 60' boom		106.10	3,000	9,000	27,000	2,649	
	1400	200 ton capacity, 70' boom		138.21	3,825	11,500	34,500	3,406	
	1500	350 ton capacity, 80' boom		182.20	4,175	12,500	37,500	3,958	
	1600	Truck mounted, lattice boom, 6 x 4, 20 tons at 10' radius		39.76	1,950	5,850	17,600	1,488	
	1700	25 tons at 10' radius		42.73	2,325	7,000	21,000	1,742	
	1800	8 x 4, 30 tons at 10' radius		45.54	2,500	7,500	22,500	1,864	
	1900	40 tons at 12' radius		48.55	2,725	8,200	24,600	2,028	
	2000	60 tons at 15' radius		53.69	1,650	4,950	14,900	1,419	
	2050	82 tons at 15' radius		59.43	1,775	5,350	16,100	1,545	
	2100	90 tons at 15' radius		66.39	1,950	5,825	17,500	1,696	
	2200	115 tons at 15' radius		74.90	2,175	6,525	19,600	1,904	
	2300	150 tons at 18' radius		81.09	2,700	8,100	24,300	2,269	
	2350	165 tons at 18' radius		87.03	2,425	7,275	21,800	2,151	
	2400	Truck mounted, hydraulic, 12 ton capacity		29.50	390	1,175	3,525	471	
	2500	25 ton capacity		36.36	485	1,450	4,350	580.85	
	2550	33 ton capacity		50.67	900	2,700	8,100	945.35	
	2560	40 ton capacity		49.47	900	2,700	8,100	935.80	
	2600	55 ton capacity		53.78	915	2,750	8,250	980.20	
	2700	80 ton capacity		75.71	1,475	4,400	13,200	1,486	
	2720	100 ton capacity		74.96	1,550	4,675	14,000	1,535	
	2740	120 ton capacity		102.81	1,825	5,500	16,500	1,922	
	2760	150 ton capacity		109.92	2,050	6,125	18,400	2,104	
	2800	Self-propelled, 4 x 4, with telescoping boom, 5 ton		15.14	430	1,285	3,850	378.10	
	2900	12-1/2 ton capacity		21.42	430	1,285	3,850	428.30	
	3000	15 ton capacity		34.42	450	1,350	4,050	545.35	
	3050	20 ton capacity		24.02	650	1,950	5,850	582.20	
	3100	25 ton capacity		36.69	1,425	4,250	12,800	1,144	
	3150	40 ton capacity		44.90	660	1,975	5,925	754.20	
	3200	Derricks, guy, 20 ton capacity, 60' boom, 75' mast		22.74	1,425	4,250	12,800	1,032	
	3300	100' boom, 115' mast		36.04	2,000	6,000	18,000	1,488	
	3400	Stiffleg, 20 ton capacity, 70' boom, 37' mast		25.41	615	1,850	5,550	573.25	
	3500	100' boom, 47' mast		39.32	665	2,000	6,000	714.55	
	3550	Helicopter, small, lift to 1250 lb. maximum, w/pilot		99.14	2,150	6,435	19,300	2,080	
	3600	Hoists, chain type, overhead, manual, 3/4 ton		.14	10.25	30.70	92	7.30	
	3900	10 ton		.79	6.20	18.59	56	10	
	4000	Hoist and tower, 5000 lb. cap., portable electric, 40' high		5.12	142	426	1,275	126.20	
	4100	For each added 10' section, add		.12	31.50	95	285	19.95	
	4200	Hoist and single tubular tower, 5000 lb. electric, 100' high		6.96	105	315	945	118.65	
	4300	For each added 6'-6" section, add		.21	38.50	115	345	24.65	
	4400	Hoist and double tubular tower, 5000 lb., 100' high		7.57	105	315	945	123.60	
	4500	For each added 6'-6" section, add		.23	41.50	125	375	26.80	
	4550	Hoist and tower, mast type, 6000 lb., 100' high		8.24	94.50	284	850	122.70	
	4570	For each added 10' section, add		.13	31.50	95	285	20.05	
	4600	Hoist and tower, personnel, electric, 2000 lb., 100' @ 125 fpm		17.50	25	75	225	155	
	4700	3000 lb., 100' @ 200 fpm		20.02	25	75	225	175.15	
	4800	3000 lb., 150' @ 300 fpm		22.22	25	75	225	192.75	
	4900	4000 lb., 100' @ 300 fpm		22.98	25	75	225	198.85	
	5000	6000 lb., 100' @ 275 fpm	▼	24.70	25	75	225	212.60	
	5100	For added heights up to 500', add	L.F.	.01	3.33	10	30	2.10	

01 54 33 | Equipment Rental

		UNIT	HOURLY OPER. COST	RENT PER DAY	RENT PER WEEK	RENT PER MONTH	EQUIPMENT COST/DAY	
60	5200	Jacks, hydraulic, 20 ton	Ea.	.05	19.65	59	177	12.20	60
	5500	100 ton		.40	26	78.50	236	18.90	
	6100	Jacks, hydraulic, climbing w/50' jackrods, control console, 30 ton cap.		2.17	31	93	279	35.90	
	6150	For each added 10' jackrod section, add		.05	5	15	45	3.40	
	6300	50 ton capacity		3.48	33.50	100	300	47.85	
	6350	For each added 10' jackrod section, add		.06	5	15	45	3.50	
	6500	125 ton capacity		9.10	51.50	155	465	103.85	
	6550	For each added 10' jackrod section, add		.61	5	15	45	7.90	
	6600	Cable jack, 10 ton capacity with 200' cable		1.82	35.50	107	320	35.95	
	6650	For each added 50' of cable, add		.22	15	45	135	10.75	
70	0010	**WELLPOINT EQUIPMENT RENTAL** without operators							70
	0020	Based on 2 months rental							
	0100	Combination jetting & wellpoint pump, 60 H.P. diesel	Ea.	15.67	298	895	2,675	304.35	
	0200	High pressure gas jet pump, 200 H.P., 300 psi	"	33.83	275	825	2,475	435.65	
	0300	Discharge pipe, 8" diameter	L.F.	.01	1.40	4.20	12.60	.90	
	0350	12" diameter		.01	2.07	6.20	18.60	1.35	
	0400	Header pipe, flows up to 150 GPM, 4" diameter		.01	.73	2.20	6.60	.50	
	0500	400 GPM, 6" diameter		.01	1.07	3.20	9.60	.70	
	0600	800 GPM, 8" diameter		.01	1.40	4.20	12.60	.95	
	0700	1500 GPM, 10" diameter		.01	1.73	5.20	15.60	1.15	
	0800	2500 GPM, 12" diameter		.03	2.07	6.20	18.60	1.45	
	0900	4500 GPM, 16" diameter		.03	2.40	7.20	21.50	1.70	
	0950	For quick coupling aluminum and plastic pipe, add		.03	9.35	28	84	5.85	
	1100	Wellpoint, 25' long, with fittings & riser pipe, 1-1/2" or 2" diameter	Ea.	.07	132	395	1,175	79.55	
	1200	Wellpoint pump, diesel powered, 4" suction, 20 H.P.		7.00	150	450	1,350	146	
	1300	6" suction, 30 H.P.		9.39	167	500	1,500	175.15	
	1400	8" suction, 40 H.P.		12.73	250	750	2,250	251.80	
	1500	10" suction, 75 H.P.		18.77	265	795	2,375	309.20	
	1600	12" suction, 100 H.P.		27.24	298	895	2,675	396.90	
	1700	12" suction, 175 H.P.		38.98	315	950	2,850	501.80	
80	0010	**MARINE EQUIPMENT RENTAL** without operators							80
	0200	Barge, 400 Ton, 30' wide x 90' long	Ea.	17.63	1,200	3,588.98	10,800	858.85	
	0240	800 Ton, 45' wide x 90' long		22.14	1,475	4,415.22	13,200	1,060	
	2000	Tugboat, diesel, 100 H.P.		29.57	238	712.63	2,150	379.10	
	2040	250 H.P.		57.41	430	1,291	3,875	717.50	
	2080	380 H.P.		124.99	1,300	3,873	11,600	1,774	
	3000	Small work boat, gas, 16-foot, 50 H.P.		11.35	48	143.56	430	119.50	
	4000	Large, diesel, 48-foot, 200 H.P.		74.68	1,375	4,105.38	12,300	1,418	

Note: R015433-10 reference appears at rows 0010 (Wellpoint) and 0010 (Marine).

Crews - Renovation

| Crew No. | Bare Costs | | Incl. Subs O&P | | Cost Per Labor-Hour | |

Crew A-1

Crew A-1	Hr.	Daily	Hr.	Daily	Bare Costs	Incl. O&P
1 Building Laborer	$42.10	$336.80	$67.45	$539.60	$42.10	$67.45
1 Concrete Saw, Gas Manual		111.55		122.71	13.94	15.34
8 L.H., Daily Totals		$448.35		$662.30	$56.04	$82.79

Crew A-1A

Crew A-1A	Hr.	Daily	Hr.	Daily	Bare Costs	Incl. O&P
1 Skilled Worker	$54.85	$438.80	$87.40	$699.20	$54.85	$87.40
1 Shot Blaster, 20"		206.20		226.82	25.77	28.35
8 L.H., Daily Totals		$645.00		$926.02	$80.63	$115.75

Crew A-1B

Crew A-1B	Hr.	Daily	Hr.	Daily	Bare Costs	Incl. O&P
1 Building Laborer	$42.10	$336.80	$67.45	$539.60	$42.10	$67.45
1 Concrete Saw		111.50		122.65	13.94	15.33
8 L.H., Daily Totals		$448.30		$662.25	$56.04	$82.78

Crew A-1C

Crew A-1C	Hr.	Daily	Hr.	Daily	Bare Costs	Incl. O&P
1 Building Laborer	$42.10	$336.80	$67.45	$539.60	$42.10	$67.45
1 Chain Saw, Gas, 18"		52.00		57.20	6.50	7.15
8 L.H., Daily Totals		$388.80		$596.80	$48.60	$74.60

Crew A-1D

Crew A-1D	Hr.	Daily	Hr.	Daily	Bare Costs	Incl. O&P
1 Building Laborer	$42.10	$336.80	$67.45	$539.60	$42.10	$67.45
1 Vibrating Plate, Gas, 18"		31.55		34.70	3.94	4.34
8 L.H., Daily Totals		$368.35		$574.30	$46.04	$71.79

Crew A-1E

Crew A-1E	Hr.	Daily	Hr.	Daily	Bare Costs	Incl. O&P
1 Building Laborer	$42.10	$336.80	$67.45	$539.60	$42.10	$67.45
1 Vibrating Plate, Gas, 21"		165.35		181.88	20.67	22.74
8 L.H., Daily Totals		$502.15		$721.49	$62.77	$90.19

Crew A-1F

Crew A-1F	Hr.	Daily	Hr.	Daily	Bare Costs	Incl. O&P
1 Building Laborer	$42.10	$336.80	$67.45	$539.60	$42.10	$67.45
1 Rammer/Tamper, Gas, 8"		46.50		51.15	5.81	6.39
8 L.H., Daily Totals		$383.30		$590.75	$47.91	$73.84

Crew A-1G

Crew A-1G	Hr.	Daily	Hr.	Daily	Bare Costs	Incl. O&P
1 Building Laborer	$42.10	$336.80	$67.45	$539.60	$42.10	$67.45
1 Rammer/Tamper, Gas, 15"		54.00		59.40	6.75	7.42
8 L.H., Daily Totals		$390.80		$599.00	$48.85	$74.88

Crew A-1H

Crew A-1H	Hr.	Daily	Hr.	Daily	Bare Costs	Incl. O&P
1 Building Laborer	$42.10	$336.80	$67.45	$539.60	$42.10	$67.45
1 Exterior Steam Cleaner		76.30		83.93	9.54	10.49
8 L.H., Daily Totals		$413.10		$623.53	$51.64	$77.94

Crew A-1J

Crew A-1J	Hr.	Daily	Hr.	Daily	Bare Costs	Incl. O&P
1 Building Laborer	$42.10	$336.80	$67.45	$539.60	$42.10	$67.45
1 Cultivator, Walk-Behind, 5 H.P.		53.05		58.35	6.63	7.29
8 L.H., Daily Totals		$389.85		$597.96	$48.73	$74.74

Crew A-1K

Crew A-1K	Hr.	Daily	Hr.	Daily	Bare Costs	Incl. O&P
1 Building Laborer	$42.10	$336.80	$67.45	$539.60	$42.10	$67.45
1 Cultivator, Walk-Behind, 8 H.P.		101.40		111.54	12.68	13.94
8 L.H., Daily Totals		$438.20		$651.14	$54.77	$81.39

Crew A-1M

Crew A-1M	Hr.	Daily	Hr.	Daily	Bare Costs	Incl. O&P
1 Building Laborer	$42.10	$336.80	$67.45	$539.60	$42.10	$67.45
1 Snow Blower, Walk-Behind		67.50		74.25	8.44	9.28
8 L.H., Daily Totals		$404.30		$613.85	$50.54	$76.73

Crew A-2

Crew A-2	Hr.	Daily	Hr.	Daily	Bare Costs	Incl. O&P
2 Laborers	$42.10	$673.60	$67.45	$1079.20	$43.82	$70.15
1 Truck Driver (light)	47.25	378.00	75.55	604.40		
1 Flatbed Truck, Gas, 1.5 Ton		196.15		215.76	8.17	8.99
24 L.H., Daily Totals		$1247.75		$1899.37	$51.99	$79.14

Crew A-2A

Crew A-2A	Hr.	Daily	Hr.	Daily	Bare Costs	Incl. O&P
2 Laborers	$42.10	$673.60	$67.45	$1079.20	$43.82	$70.15
1 Truck Driver (light)	47.25	378.00	75.55	604.40		
1 Flatbed Truck, Gas, 1.5 Ton		196.15		215.76		
1 Concrete Saw		111.50		122.65	12.82	14.10
24 L.H., Daily Totals		$1359.25		$2022.02	$56.64	$84.25

Crew A-2B

Crew A-2B	Hr.	Daily	Hr.	Daily	Bare Costs	Incl. O&P
1 Truck Driver (light)	$47.25	$378.00	$75.55	$604.40	$47.25	$75.55
1 Flatbed Truck, Gas, 1.5 Ton		196.15		215.76	24.52	26.97
8 L.H., Daily Totals		$574.15		$820.16	$71.77	$102.52

Crew A-3A

Crew A-3A	Hr.	Daily	Hr.	Daily	Bare Costs	Incl. O&P
1 Equip. Oper. (light)	$53.00	$424.00	$82.95	$663.60	$53.00	$82.95
1 Pickup Truck, 4x4, 3/4 Ton		175.85		193.44	21.98	24.18
8 L.H., Daily Totals		$599.85		$857.03	$74.98	$107.13

Crew A-3B

Crew A-3B	Hr.	Daily	Hr.	Daily	Bare Costs	Incl. O&P
1 Equip. Oper. (medium)	$56.75	$454.00	$88.80	$710.40	$52.85	$83.53
1 Truck Driver (heavy)	48.95	391.60	78.25	626.00		
1 Dump Truck, 12 C.Y., 400 H.P.		572.90		630.19		
1 F.E. Loader, W.M., 2.5 C.Y.		702.10		772.31	79.69	87.66
16 L.H., Daily Totals		$2120.60		$2738.90	$132.54	$171.18

Crew A-3C

Crew A-3C	Hr.	Daily	Hr.	Daily	Bare Costs	Incl. O&P
1 Equip. Oper. (light)	$53.00	$424.00	$82.95	$663.60	$53.00	$82.95
1 Loader, Skid Steer, 78 H.P.		400.10		440.11	50.01	55.01
8 L.H., Daily Totals		$824.10		$1103.71	$103.01	$137.96

Crew A-3D

Crew A-3D	Hr.	Daily	Hr.	Daily	Bare Costs	Incl. O&P
1 Truck Driver (light)	$47.25	$378.00	$75.55	$604.40	$47.25	$75.55
1 Pickup Truck, 4x4, 3/4 Ton		175.85		193.44		
1 Flatbed Trailer, 25 Ton		135.55		149.10	38.92	42.82
8 L.H., Daily Totals		$689.40		$946.94	$86.17	$118.37

Crew A-3E

Crew A-3E	Hr.	Daily	Hr.	Daily	Bare Costs	Incl. O&P
1 Equip. Oper. (crane)	$59.20	$473.60	$92.65	$741.20	$54.08	$85.45
1 Truck Driver (heavy)	48.95	391.60	78.25	626.00		
1 Pickup Truck, 4x4, 3/4 Ton		175.85		193.44	10.99	12.09
16 L.H., Daily Totals		$1041.05		$1560.64	$65.07	$97.54

Crew A-3F

Crew A-3F	Hr.	Daily	Hr.	Daily	Bare Costs	Incl. O&P
1 Equip. Oper. (crane)	$59.20	$473.60	$92.65	$741.20	$54.08	$85.45
1 Truck Driver (heavy)	48.95	391.60	78.25	626.00		
1 Pickup Truck, 4x4, 3/4 Ton		175.85		193.44		
1 Truck Tractor, 6x4, 380 H.P.		493.25		542.58		
1 Lowbed Trailer, 75 Ton		255.05		280.56	57.76	63.54
16 L.H., Daily Totals		$1789.35		$2383.76	$111.83	$148.99

For customer support on your Commercial Renovation Costs with RSMeans data, call 800.448.8182.

751

Crew No.	Bare Costs Hr.	Daily	Incl. Subs O&P Hr.	Daily	Cost Per Labor-Hour Bare Costs	Incl. O&P
Crew A-3G	Hr.	Daily	Hr.	Daily	Bare Costs	Incl. O&P
1 Equip. Oper. (crane)	$59.20	$473.60	$92.65	$741.20	$54.08	$85.45
1 Truck Driver (heavy)	48.95	391.60	78.25	626.00		
1 Pickup Truck, 4x4, 3/4 Ton		175.85		193.44		
1 Truck Tractor, 6x4, 450 H.P.		601.75		661.92		
1 Lowbed Trailer, 75 Ton		255.05		280.56	64.54	70.99
16 L.H., Daily Totals		$1897.85		$2503.11	$118.62	$156.44

Crew A-3H	Hr.	Daily	Hr.	Daily	Bare Costs	Incl. O&P
1 Equip. Oper. (crane)	$59.20	$473.60	$92.65	$741.20	$59.20	$92.65
1 Hyd. Crane, 12 Ton (Daily)		724.85		797.34	90.61	99.67
8 L.H., Daily Totals		$1198.45		$1538.54	$149.81	$192.32

Crew A-3I	Hr.	Daily	Hr.	Daily	Bare Costs	Incl. O&P
1 Equip. Oper. (crane)	$59.20	$473.60	$92.65	$741.20	$59.20	$92.65
1 Hyd. Crane, 25 Ton (Daily)		801.40		881.54	100.18	110.19
8 L.H., Daily Totals		$1275.00		$1622.74	$159.38	$202.84

Crew A-3J	Hr.	Daily	Hr.	Daily	Bare Costs	Incl. O&P
1 Equip. Oper. (crane)	$59.20	$473.60	$92.65	$741.20	$59.20	$92.65
1 Hyd. Crane, 40 Ton (Daily)		1272.00		1399.20	159.00	174.90
8 L.H., Daily Totals		$1745.60		$2140.40	$218.20	$267.55

Crew A-3K	Hr.	Daily	Hr.	Daily	Bare Costs	Incl. O&P
1 Equip. Oper. (crane)	$59.20	$473.60	$92.65	$741.20	$54.88	$85.88
1 Equip. Oper. (oiler)	50.55	404.40	79.10	632.80		
1 Hyd. Crane, 55 Ton (Daily)		1362.00		1498.20		
1 P/U Truck, 3/4 Ton (Daily)		142.15		156.37	94.01	103.41
16 L.H., Daily Totals		$2382.15		$3028.57	$148.88	$189.29

Crew A-3L	Hr.	Daily	Hr.	Daily	Bare Costs	Incl. O&P
1 Equip. Oper. (crane)	$59.20	$473.60	$92.65	$741.20	$54.88	$85.88
1 Equip. Oper. (oiler)	50.55	404.40	79.10	632.80		
1 Hyd. Crane, 80 Ton (Daily)		2101.00		2311.10		
1 P/U Truck, 3/4 Ton (Daily)		142.15		156.37	140.20	154.22
16 L.H., Daily Totals		$3121.15		$3841.47	$195.07	$240.09

Crew A-3M	Hr.	Daily	Hr.	Daily	Bare Costs	Incl. O&P
1 Equip. Oper. (crane)	$59.20	$473.60	$92.65	$741.20	$54.88	$85.88
1 Equip. Oper. (oiler)	50.55	404.40	79.10	632.80		
1 Hyd. Crane, 100 Ton (Daily)		2227.00		2449.70		
1 P/U Truck, 3/4 Ton (Daily)		142.15		156.37	148.07	162.88
16 L.H., Daily Totals		$3247.15		$3980.07	$202.95	$248.75

Crew A-3N	Hr.	Daily	Hr.	Daily	Bare Costs	Incl. O&P
1 Equip. Oper. (crane)	$59.20	$473.60	$92.65	$741.20	$59.20	$92.65
1 Tower Crane (monthly)		1693.00		1862.30	211.63	232.79
8 L.H., Daily Totals		$2166.60		$2603.50	$270.82	$325.44

Crew A-3P	Hr.	Daily	Hr.	Daily	Bare Costs	Incl. O&P
1 Equip. Oper. (light)	$53.00	$424.00	$82.95	$663.60	$53.00	$82.95
1 A.T. Forklift, 31' reach, 45' lift		374.20		411.62	46.77	51.45
8 L.H., Daily Totals		$798.20		$1075.22	$99.78	$134.40

Crew A-3Q	Hr.	Daily	Hr.	Daily	Bare Costs	Incl. O&P
1 Equip. Oper. (light)	$53.00	$424.00	$82.95	$663.60	$53.00	$82.95
1 Pickup Truck, 4x4, 3/4 Ton		175.85		193.44		
1 Flatbed Trailer, 3 Ton		70.30		77.33	30.77	33.85
8 L.H., Daily Totals		$670.15		$934.37	$83.77	$116.80

Crew A-3R	Hr.	Daily	Hr.	Daily	Bare Costs	Incl. O&P
1 Equip. Oper. (light)	$53.00	$424.00	$82.95	$663.60	$53.00	$82.95
1 Forklift, Smooth Floor, 8,000 Lb.		255.90		281.49	31.99	35.19
8 L.H., Daily Totals		$679.90		$945.09	$84.99	$118.14

Crew A-4	Hr.	Daily	Hr.	Daily	Bare Costs	Incl. O&P
2 Carpenters	$53.15	$850.40	$85.15	$1362.40	$50.23	$80.27
1 Painter, Ordinary	44.40	355.20	70.50	564.00		
24 L.H., Daily Totals		$1205.60		$1926.40	$50.23	$80.27

Crew A-5	Hr.	Daily	Hr.	Daily	Bare Costs	Incl. O&P
2 Laborers	$42.10	$673.60	$67.45	$1079.20	$42.67	$68.35
.25 Truck Driver (light)	47.25	94.50	75.55	151.10		
.25 Flatbed Truck, Gas, 1.5 Ton		49.04		53.94	2.72	3.00
18 L.H., Daily Totals		$817.14		$1284.24	$45.40	$71.35

Crew A-6	Hr.	Daily	Hr.	Daily	Bare Costs	Incl. O&P
1 Instrument Man	$54.85	$438.80	$87.40	$699.20	$52.92	$84.45
1 Rodman/Chainman	51.00	408.00	81.50	652.00		
1 Level, Electronic		30.80		33.88	1.93	2.12
16 L.H., Daily Totals		$877.60		$1385.08	$54.85	$86.57

Crew A-7	Hr.	Daily	Hr.	Daily	Bare Costs	Incl. O&P
1 Chief of Party	$66.05	$528.40	$102.30	$818.40	$57.30	$90.40
1 Instrument Man	54.85	438.80	87.40	699.20		
1 Rodman/Chainman	51.00	408.00	81.50	652.00		
1 Level, Electronic		30.80		33.88	1.28	1.41
24 L.H., Daily Totals		$1406.00		$2203.48	$58.58	$91.81

Crew A-8	Hr.	Daily	Hr.	Daily	Bare Costs	Incl. O&P
1 Chief of Party	$66.05	$528.40	$102.30	$818.40	$55.73	$88.17
1 Instrument Man	54.85	438.80	87.40	699.20		
2 Rodmen/Chainmen	51.00	816.00	81.50	1304.00		
1 Level, Electronic		30.80		33.88	.96	1.06
32 L.H., Daily Totals		$1814.00		$2855.48	$56.69	$89.23

Crew A-9	Hr.	Daily	Hr.	Daily	Bare Costs	Incl. O&P
1 Asbestos Foreman	$59.15	$473.20	$93.75	$750.00	$58.71	$93.09
7 Asbestos Workers	58.65	3284.40	93.00	5208.00		
64 L.H., Daily Totals		$3757.60		$5958.00	$58.71	$93.09

Crew A-10A	Hr.	Daily	Hr.	Daily	Bare Costs	Incl. O&P
1 Asbestos Foreman	$59.15	$473.20	$93.75	$750.00	$58.82	$93.25
2 Asbestos Workers	58.65	938.40	93.00	1488.00		
24 L.H., Daily Totals		$1411.60		$2238.00	$58.82	$93.25

Crew A-10B	Hr.	Daily	Hr.	Daily	Bare Costs	Incl. O&P
1 Asbestos Foreman	$59.15	$473.20	$93.75	$750.00	$58.77	$93.19
3 Asbestos Workers	58.65	1407.60	93.00	2232.00		
32 L.H., Daily Totals		$1880.80		$2982.00	$58.77	$93.19

Crew A-10C	Hr.	Daily	Hr.	Daily	Bare Costs	Incl. O&P
3 Asbestos Workers	$58.65	$1407.60	$93.00	$2232.00	$58.65	$93.00
1 Flatbed Truck, Gas, 1.5 Ton		196.15		215.76	8.17	8.99
24 L.H., Daily Totals		$1603.75		$2447.76	$66.82	$101.99

For customer support on your Commercial Renovation Costs with RSMeans data, call 800.448.8182.

Crew No.	Bare Costs Hr.	Bare Costs Daily	Incl. Subs O&P Hr.	Incl. Subs O&P Daily	Cost Per Labor-Hour Bare Costs	Cost Per Labor-Hour Incl. O&P
Crew A-10D	Hr.	Daily	Hr.	Daily	Bare Costs	Incl. O&P
2 Asbestos Workers	$58.65	$938.40	$93.00	$1488.00	$56.76	$89.44
1 Equip. Oper. (crane)	59.20	473.60	92.65	741.20		
1 Equip. Oper. (oiler)	50.55	404.40	79.10	632.80		
1 Hydraulic Crane, 33 Ton		945.35		1039.89	29.54	32.50
32 L.H., Daily Totals		$2761.75		$3901.89	$86.30	$121.93
Crew A-11	Hr.	Daily	Hr.	Daily	Bare Costs	Incl. O&P
1 Asbestos Foreman	$59.15	$473.20	$93.75	$750.00	$58.71	$93.09
7 Asbestos Workers	58.65	3284.40	93.00	5208.00		
2 Chip Hammers, 12 Lb., Elec.		64.80		71.28	1.01	1.11
64 L.H., Daily Totals		$3822.40		$6029.28	$59.73	$94.21
Crew A-12	Hr.	Daily	Hr.	Daily	Bare Costs	Incl. O&P
1 Asbestos Foreman	$59.15	$473.20	$93.75	$750.00	$58.71	$93.09
7 Asbestos Workers	58.65	3284.40	93.00	5208.00		
1 Trk-Mtd Vac, 14 CY, 1500 Gal.		536.15		589.76		
1 Flatbed Truck, 20,000 GVW		201.60		221.76	11.53	12.68
64 L.H., Daily Totals		$4495.35		$6769.52	$70.24	$105.77
Crew A-13	Hr.	Daily	Hr.	Daily	Bare Costs	Incl. O&P
1 Equip. Oper. (light)	$53.00	$424.00	$82.95	$663.60	$53.00	$82.95
1 Trk-Mtd Vac, 14 CY, 1500 Gal.		536.15		589.76		
1 Flatbed Truck, 20,000 GVW		201.60		221.76	92.22	101.44
8 L.H., Daily Totals		$1161.75		$1475.13	$145.22	$184.39
Crew B-1	Hr.	Daily	Hr.	Daily	Bare Costs	Incl. O&P
1 Labor Foreman (outside)	$44.10	$352.80	$70.65	$565.20	$42.77	$68.52
2 Laborers	42.10	673.60	67.45	1079.20		
24 L.H., Daily Totals		$1026.40		$1644.40	$42.77	$68.52
Crew B-1A	Hr.	Daily	Hr.	Daily	Bare Costs	Incl. O&P
1 Labor Foreman (outside)	$44.10	$352.80	$70.65	$565.20	$42.77	$68.52
2 Laborers	42.10	673.60	67.45	1079.20		
2 Cutting Torches		25.60		28.16		
2 Sets of Gases		343.10		377.41	15.36	16.90
24 L.H., Daily Totals		$1395.10		$2049.97	$58.13	$85.42
Crew B-1B	Hr.	Daily	Hr.	Daily	Bare Costs	Incl. O&P
1 Labor Foreman (outside)	$44.10	$352.80	$70.65	$565.20	$46.88	$74.55
2 Laborers	42.10	673.60	67.45	1079.20		
1 Equip. Oper. (crane)	59.20	473.60	92.65	741.20		
2 Cutting Torches		25.60		28.16		
2 Sets of Gases		343.10		377.41		
1 Hyd. Crane, 12 Ton		471.00		518.10	26.24	28.86
32 L.H., Daily Totals		$2339.70		$3309.27	$73.12	$103.41
Crew B-1C	Hr.	Daily	Hr.	Daily	Bare Costs	Incl. O&P
1 Labor Foreman (outside)	$44.10	$352.80	$70.65	$565.20	$42.77	$68.52
2 Laborers	42.10	673.60	67.45	1079.20		
1 Telescoping Boom Lift, to 60'		310.90		341.99	12.95	14.25
24 L.H., Daily Totals		$1337.30		$1986.39	$55.72	$82.77
Crew B-1D	Hr.	Daily	Hr.	Daily	Bare Costs	Incl. O&P
2 Laborers	$42.10	$673.60	$67.45	$1079.20	$42.10	$67.45
1 Small Work Boat, Gas, 50 H.P.		119.50		131.45		
1 Pressure Washer, 7 GPM		90.50		99.55	13.13	14.44
16 L.H., Daily Totals		$883.60		$1310.20	$55.23	$81.89

Crew No.	Bare Costs Hr.	Bare Costs Daily	Incl. Subs O&P Hr.	Incl. Subs O&P Daily	Cost Per Labor-Hour Bare Costs	Cost Per Labor-Hour Incl. O&P
Crew B-1E	Hr.	Daily	Hr.	Daily	Bare Costs	Incl. O&P
1 Labor Foreman (outside)	$44.10	$352.80	$70.65	$565.20	$42.60	$68.25
3 Laborers	42.10	1010.40	67.45	1618.80		
1 Work Boat, Diesel, 200 H.P.		1418.00		1559.80		
2 Pressure Washers, 7 GPM		181.00		199.10	49.97	54.97
32 L.H., Daily Totals		$2962.20		$3942.90	$92.57	$123.22
Crew B-1F	Hr.	Daily	Hr.	Daily	Bare Costs	Incl. O&P
2 Skilled Workers	$54.85	$877.60	$87.40	$1398.40	$50.60	$80.75
1 Laborer	42.10	336.80	67.45	539.60		
1 Small Work Boat, Gas, 50 H.P.		119.50		131.45		
1 Pressure Washer, 7 GPM		90.50		99.55	8.75	9.63
24 L.H., Daily Totals		$1424.40		$2169.00	$59.35	$90.38
Crew B-1G	Hr.	Daily	Hr.	Daily	Bare Costs	Incl. O&P
2 Laborers	$42.10	$673.60	$67.45	$1079.20	$42.10	$67.45
1 Small Work Boat, Gas, 50 H.P.		119.50		131.45	7.47	8.22
16 L.H., Daily Totals		$793.10		$1210.65	$49.57	$75.67
Crew B-1H	Hr.	Daily	Hr.	Daily	Bare Costs	Incl. O&P
2 Skilled Workers	$54.85	$877.60	$87.40	$1398.40	$50.60	$80.75
1 Laborer	42.10	336.80	67.45	539.60		
1 Small Work Boat, Gas, 50 H.P.		119.50		131.45	4.98	5.48
24 L.H., Daily Totals		$1333.90		$2069.45	$55.58	$86.23
Crew B-1J	Hr.	Daily	Hr.	Daily	Bare Costs	Incl. O&P
1 Labor Foreman (inside)	$42.60	$340.80	$68.25	$546.00	$42.35	$67.85
1 Laborer	42.10	336.80	67.45	539.60		
16 L.H., Daily Totals		$677.60		$1085.60	$42.35	$67.85
Crew B-1K	Hr.	Daily	Hr.	Daily	Bare Costs	Incl. O&P
1 Carpenter Foreman (inside)	$53.65	$429.20	$85.95	$687.60	$53.40	$85.55
1 Carpenter	53.15	425.20	85.15	681.20		
16 L.H., Daily Totals		$854.40		$1368.80	$53.40	$85.55
Crew B-2	Hr.	Daily	Hr.	Daily	Bare Costs	Incl. O&P
1 Labor Foreman (outside)	$44.10	$352.80	$70.65	$565.20	$42.50	$68.09
4 Laborers	42.10	1347.20	67.45	2158.40		
40 L.H., Daily Totals		$1700.00		$2723.60	$42.50	$68.09
Crew B-2A	Hr.	Daily	Hr.	Daily	Bare Costs	Incl. O&P
1 Labor Foreman (outside)	$44.10	$352.80	$70.65	$565.20	$42.77	$68.52
2 Laborers	42.10	673.60	67.45	1079.20		
1 Telescoping Boom Lift, to 60'		310.90		341.99	12.95	14.25
24 L.H., Daily Totals		$1337.30		$1986.39	$55.72	$82.77
Crew B-3	Hr.	Daily	Hr.	Daily	Bare Costs	Incl. O&P
1 Labor Foreman (outside)	$44.10	$352.80	$70.65	$565.20	$47.16	$75.14
2 Laborers	42.10	673.60	67.45	1079.20		
1 Equip. Oper. (medium)	56.75	454.00	88.80	710.40		
2 Truck Drivers (heavy)	48.95	783.20	78.25	1252.00		
1 Crawler Loader, 3 C.Y.		1169.00		1285.90		
2 Dump Trucks, 12 C.Y., 400 H.P.		1145.80		1260.38	48.23	53.05
48 L.H., Daily Totals		$4578.40		$6153.08	$95.38	$128.19
Crew B-3A	Hr.	Daily	Hr.	Daily	Bare Costs	Incl. O&P
4 Laborers	$42.10	$1347.20	$67.45	$2158.40	$45.03	$71.72
1 Equip. Oper. (medium)	56.75	454.00	88.80	710.40		
1 Hyd. Excavator, 1.5 C.Y.		687.55		756.30	17.19	18.91
40 L.H., Daily Totals		$2488.75		$3625.11	$62.22	$90.63

Crew No.	Bare Costs		Incl. Subs O&P		Cost Per Labor-Hour	
Crew B-3B	Hr.	Daily	Hr.	Daily	Bare Costs	Incl. O&P
2 Laborers	$42.10	$673.60	$67.45	$1079.20	$47.48	$75.49
1 Equip. Oper. (medium)	56.75	454.00	88.80	710.40		
1 Truck Driver (heavy)	48.95	391.60	78.25	626.00		
1 Backhoe Loader, 80 H.P.		232.40		255.64		
1 Dump Truck, 12 C.Y., 400 H.P.		572.90		630.19	25.17	27.68
32 L.H., Daily Totals		$2324.50		$3301.43	$72.64	$103.17
Crew B-3C	Hr.	Daily	Hr.	Daily	Bare Costs	Incl. O&P
3 Laborers	$42.10	$1010.40	$67.45	$1618.80	$45.76	$72.79
1 Equip. Oper. (medium)	56.75	454.00	88.80	710.40		
1 Crawler Loader, 4 C.Y.		1489.00		1637.90	46.53	51.18
32 L.H., Daily Totals		$2953.40		$3967.10	$92.29	$123.97
Crew B-4	Hr.	Daily	Hr.	Daily	Bare Costs	Incl. O&P
1 Labor Foreman (outside)	$44.10	$352.80	$70.65	$565.20	$43.58	$69.78
4 Laborers	42.10	1347.20	67.45	2158.40		
1 Truck Driver (heavy)	48.95	391.60	78.25	626.00		
1 Truck Tractor, 220 H.P.		307.10		337.81		
1 Flatbed Trailer, 40 Ton		186.20		204.82	10.28	11.30
48 L.H., Daily Totals		$2584.90		$3892.23	$53.85	$81.09
Crew B-5	Hr.	Daily	Hr.	Daily	Bare Costs	Incl. O&P
1 Labor Foreman (outside)	$44.10	$352.80	$70.65	$565.20	$46.57	$74.01
4 Laborers	42.10	1347.20	67.45	2158.40		
2 Equip. Oper. (medium)	56.75	908.00	88.80	1420.80		
1 Air Compressor, 250 cfm		201.50		221.65		
2 Breakers, Pavement, 60 lb.		107.00		117.70		
2 -50' Air Hoses, 1.5"		48.50		53.35		
1 Crawler Loader, 3 C.Y.		1169.00		1285.90	27.25	29.98
56 L.H., Daily Totals		$4134.00		$5823.00	$73.82	$103.98
Crew B-5D	Hr.	Daily	Hr.	Daily	Bare Costs	Incl. O&P
1 Labor Foreman (outside)	$44.10	$352.80	$70.65	$565.20	$46.87	$74.54
4 Laborers	42.10	1347.20	67.45	2158.40		
2 Equip. Oper. (medium)	56.75	908.00	88.80	1420.80		
1 Truck Driver (heavy)	48.95	391.60	78.25	626.00		
1 Air Compressor, 250 cfm		201.50		221.65		
2 Breakers, Pavement, 60 lb.		107.00		117.70		
2 -50' Air Hoses, 1.5"		48.50		53.35		
1 Crawler Loader, 3 C.Y.		1169.00		1285.90		
1 Dump Truck, 12 C.Y., 400 H.P.		572.90		630.19	32.80	36.07
64 L.H., Daily Totals		$5098.50		$7079.19	$79.66	$110.61
Crew B-5E	Hr.	Daily	Hr.	Daily	Bare Costs	Incl. O&P
1 Labor Foreman (outside)	$44.10	$352.80	$70.65	$565.20	$46.87	$74.54
4 Laborers	42.10	1347.20	67.45	2158.40		
2 Equip. Oper. (medium)	56.75	908.00	88.80	1420.80		
1 Truck Driver (heavy)	48.95	391.60	78.25	626.00		
1 Water Tank Trailer, 5000 Gal.		152.25		167.47		
1 High Pressure Water Jet, 40 KSI		811.05		892.16		
2 -50' Air Hoses, 1.5"		48.50		53.35		
1 Crawler Loader, 3 C.Y.		1169.00		1285.90		
1 Dump Truck, 12 C.Y., 400 H.P.		572.90		630.19	43.03	47.33
64 L.H., Daily Totals		$5753.21		$7799.38	$89.90	$121.87
Crew B-6	Hr.	Daily	Hr.	Daily	Bare Costs	Incl. O&P
2 Laborers	$42.10	$673.60	$67.45	$1079.20	$45.73	$72.62
1 Equip. Oper. (light)	53.00	424.00	82.95	663.60		
1 Backhoe Loader, 48 H.P.		213.75		235.13	8.91	9.80
24 L.H., Daily Totals		$1311.35		$1977.93	$54.64	$82.41

Crew No.	Bare Costs		Incl. Subs O&P		Cost Per Labor-Hour	
Crew B-6A	Hr.	Daily	Hr.	Daily	Bare Costs	Incl. O&P
.5 Labor Foreman (outside)	$44.10	$176.40	$70.65	$282.60	$48.36	$76.63
1 Laborer	42.10	336.80	67.45	539.60		
1 Equip. Oper. (medium)	56.75	454.00	88.80	710.40		
1 Vacuum Truck, 5000 Gal.		367.55		404.31	18.38	20.22
20 L.H., Daily Totals		$1334.75		$1936.91	$66.74	$96.85
Crew B-6B	Hr.	Daily	Hr.	Daily	Bare Costs	Incl. O&P
2 Labor Foremen (outside)	$44.10	$705.60	$70.65	$1130.40	$42.77	$68.52
4 Laborers	42.10	1347.20	67.45	2158.40		
1 S.P. Crane, 4x4, 5 Ton		378.10		415.91		
1 Flatbed Truck, Gas, 1.5 Ton		196.15		215.76		
1 Butt Fusion Mach., 4"-12" diam.		426.85		469.54	20.86	22.94
48 L.H., Daily Totals		$3053.90		$4390.01	$63.62	$91.46
Crew B-6C	Hr.	Daily	Hr.	Daily	Bare Costs	Incl. O&P
2 Labor Foremen (outside)	$44.10	$705.60	$70.65	$1130.40	$42.77	$68.52
4 Laborers	42.10	1347.20	67.45	2158.40		
1 S.P. Crane, 4x4, 12 Ton		428.30		471.13		
1 Flatbed Truck, Gas, 3 Ton		820.40		902.44		
1 Butt Fusion Mach., 8"-24" diam.		1076.00		1183.60	48.43	53.27
48 L.H., Daily Totals		$4377.50		$5845.97	$91.20	$121.79
Crew B-6D	Hr.	Daily	Hr.	Daily	Bare Costs	Incl. O&P
0.5 Labor Foreman (outside)	$44.10	$176.40	$70.65	$282.60	$48.36	$76.63
1 Laborer	42.10	336.80	67.45	539.60		
1 Equip. Oper. (medium)	56.75	454.00	88.80	710.40		
1 Hydro Excavator, 12 C.Y.		1262.00		1388.20	63.10	69.41
20 L.H., Daily Totals		$2228.80		$2920.36	$111.46	$146.04
Crew B-7	Hr.	Daily	Hr.	Daily	Bare Costs	Incl. O&P
1 Labor Foreman (outside)	$44.10	$352.80	$70.65	$565.20	$44.88	$71.54
4 Laborers	42.10	1347.20	67.45	2158.40		
1 Equip. Oper. (medium)	56.75	454.00	88.80	710.40		
1 Brush Chipper, 12", 130 H.P.		392.80		432.08		
1 Crawler Loader, 3 C.Y.		1169.00		1285.90		
2 Chain Saws, Gas, 36" Long		82.50		90.75	34.26	37.68
48 L.H., Daily Totals		$3798.30		$5242.73	$79.13	$109.22
Crew B-7A	Hr.	Daily	Hr.	Daily	Bare Costs	Incl. O&P
2 Laborers	$42.10	$673.60	$67.45	$1079.20	$45.73	$72.62
1 Equip. Oper. (light)	53.00	424.00	82.95	663.60		
1 Rake w/Tractor		339.45		373.39		
2 Chain Saws, Gas, 18"		104.00		114.40	18.48	20.32
24 L.H., Daily Totals		$1541.05		$2230.59	$64.21	$92.94
Crew B-7B	Hr.	Daily	Hr.	Daily	Bare Costs	Incl. O&P
1 Labor Foreman (outside)	$44.10	$352.80	$70.65	$565.20	$45.46	$72.50
4 Laborers	42.10	1347.20	67.45	2158.40		
1 Equip. Oper. (medium)	56.75	454.00	88.80	710.40		
1 Truck Driver (heavy)	48.95	391.60	78.25	626.00		
1 Brush Chipper, 12", 130 H.P.		392.80		432.08		
1 Crawler Loader, 3 C.Y.		1169.00		1285.90		
2 Chain Saws, Gas, 36" Long		82.50		90.75		
1 Dump Truck, 8 C.Y., 220 H.P.		428.05		470.86	37.01	40.71
56 L.H., Daily Totals		$4617.95		$6339.59	$82.46	$113.21

Crews - Renovation

Crew No.	Bare Costs		Incl. Subs O&P		Cost Per Labor-Hour	

Crew B-7C	Hr.	Daily	Hr.	Daily	Bare Costs	Incl. O&P
1 Labor Foreman (outside)	$44.10	$352.80	$70.65	$565.20	$45.46	$72.50
4 Laborers	42.10	1347.20	67.45	2158.40		
1 Equip. Oper. (medium)	56.75	454.00	88.80	710.40		
1 Truck Driver (heavy)	48.95	391.60	78.25	626.00		
1 Brush Chipper, 12", 130 H.P.		392.80		432.08		
1 Crawler Loader, 3 C.Y.		1169.00		1285.90		
2 Chain Saws, Gas, 36" Long		82.50		90.75		
1 Dump Truck, 12 C.Y., 400 H.P.		572.90		630.19	39.59	43.55
56 L.H., Daily Totals		$4762.80		$6498.92	$85.05	$116.05

Crew B-8	Hr.	Daily	Hr.	Daily	Bare Costs	Incl. O&P
1 Labor Foreman (outside)	$44.10	$352.80	$70.65	$565.20	$48.78	$77.34
2 Laborers	42.10	673.60	67.45	1079.20		
2 Equip. Oper. (medium)	56.75	908.00	88.80	1420.80		
1 Equip. Oper. (oiler)	50.55	404.40	79.10	632.80		
2 Truck Drivers (heavy)	48.95	783.20	78.25	1252.00		
1 Hyd. Crane, 25 Ton		580.85		638.93		
1 Crawler Loader, 3 C.Y.		1169.00		1285.90		
2 Dump Trucks, 12 C.Y., 400 H.P.		1145.80		1260.38	45.24	49.77
64 L.H., Daily Totals		$6017.65		$8135.22	$94.03	$127.11

Crew B-9	Hr.	Daily	Hr.	Daily	Bare Costs	Incl. O&P
1 Labor Foreman (outside)	$44.10	$352.80	$70.65	$565.20	$42.50	$68.09
4 Laborers	42.10	1347.20	67.45	2158.40		
1 Air Compressor, 250 cfm		201.50		221.65		
2 Breakers, Pavement, 60 lb.		107.00		117.70		
2 -50' Air Hoses, 1.5"		48.50		53.35	8.93	9.82
40 L.H., Daily Totals		$2057.00		$3116.30	$51.42	$77.91

Crew B-9A	Hr.	Daily	Hr.	Daily	Bare Costs	Incl. O&P
2 Laborers	$42.10	$673.60	$67.45	$1079.20	$44.38	$71.05
1 Truck Driver (heavy)	48.95	391.60	78.25	626.00		
1 Water Tank Trailer, 5000 Gal.		152.25		167.47		
1 Truck Tractor, 220 H.P.		307.10		337.81		
2 -50' Discharge Hoses, 3"		9.00		9.90	19.51	21.47
24 L.H., Daily Totals		$1533.55		$2220.39	$63.90	$92.52

Crew B-9B	Hr.	Daily	Hr.	Daily	Bare Costs	Incl. O&P
2 Laborers	$42.10	$673.60	$67.45	$1079.20	$44.38	$71.05
1 Truck Driver (heavy)	48.95	391.60	78.25	626.00		
2 -50' Discharge Hoses, 3"		9.00		9.90		
1 Water Tank Trailer, 5000 Gal.		152.25		167.47		
1 Truck Tractor, 220 H.P.		307.10		337.81		
1 Pressure Washer		96.95		106.65	23.55	25.91
24 L.H., Daily Totals		$1630.50		$2327.03	$67.94	$96.96

Crew B-9D	Hr.	Daily	Hr.	Daily	Bare Costs	Incl. O&P
1 Labor Foreman (outside)	$44.10	$352.80	$70.65	$565.20	$42.50	$68.09
4 Common Laborers	42.10	1347.20	67.45	2158.40		
1 Air Compressor, 250 cfm		201.50		221.65		
2 -50' Air Hoses, 1.5"		48.50		53.35		
2 Air Powered Tampers		67.70		74.47	7.94	8.74
40 L.H., Daily Totals		$2017.70		$3073.07	$50.44	$76.83

Crew B-9E	Hr.	Daily	Hr.	Daily	Bare Costs	Incl. O&P
1 Cement Finisher	$49.95	$399.60	$78.45	$627.60	$46.02	$72.95
1 Laborer	42.10	336.80	67.45	539.60		
1 Chip. Hammers, 12 Lb., Elec.		32.40		35.64	2.02	2.23
16 L.H., Daily Totals		$768.80		$1202.84	$48.05	$75.18

Crew B-10	Hr.	Daily	Hr.	Daily	Bare Costs	Incl. O&P
1 Equip. Oper. (medium)	$56.75	$454.00	$88.80	$710.40	$51.87	$81.68
.5 Laborer	42.10	168.40	67.45	269.80		
12 L.H., Daily Totals		$622.40		$980.20	$51.87	$81.68

Crew B-10A	Hr.	Daily	Hr.	Daily	Bare Costs	Incl. O&P
1 Equip. Oper. (medium)	$56.75	$454.00	$88.80	$710.40	$51.87	$81.68
.5 Laborer	42.10	168.40	67.45	269.80		
1 Roller, 2-Drum, W.B., 7.5 H.P.		166.10		182.71	13.84	15.23
12 L.H., Daily Totals		$788.50		$1162.91	$65.71	$96.91

Crew B-10B	Hr.	Daily	Hr.	Daily	Bare Costs	Incl. O&P
1 Equip. Oper. (medium)	$56.75	$454.00	$88.80	$710.40	$51.87	$81.68
.5 Laborer	42.10	168.40	67.45	269.80		
1 Dozer, 200 H.P.		1504.00		1654.40	125.33	137.87
12 L.H., Daily Totals		$2126.40		$2634.60	$177.20	$219.55

Crew B-10C	Hr.	Daily	Hr.	Daily	Bare Costs	Incl. O&P
1 Equip. Oper. (medium)	$56.75	$454.00	$88.80	$710.40	$51.87	$81.68
.5 Laborer	42.10	168.40	67.45	269.80		
1 Dozer, 200 H.P.		1504.00		1654.40		
1 Vibratory Roller, Towed, 23 Ton		514.75		566.23	168.23	185.05
12 L.H., Daily Totals		$2641.15		$3200.82	$220.10	$266.74

Crew B-10D	Hr.	Daily	Hr.	Daily	Bare Costs	Incl. O&P
1 Equip. Oper. (medium)	$56.75	$454.00	$88.80	$710.40	$51.87	$81.68
.5 Laborer	42.10	168.40	67.45	269.80		
1 Dozer, 200 H.P.		1504.00		1654.40		
1 Sheepsft. Roller, Towed		436.90		480.59	161.74	177.92
12 L.H., Daily Totals		$2563.30		$3115.19	$213.61	$259.60

Crew B-10E	Hr.	Daily	Hr.	Daily	Bare Costs	Incl. O&P
1 Equip. Oper. (medium)	$56.75	$454.00	$88.80	$710.40	$51.87	$81.68
.5 Laborer	42.10	168.40	67.45	269.80		
1 Tandem Roller, 5 Ton		255.65		281.21	21.30	23.43
12 L.H., Daily Totals		$878.05		$1261.42	$73.17	$105.12

Crew B-10F	Hr.	Daily	Hr.	Daily	Bare Costs	Incl. O&P
1 Equip. Oper. (medium)	$56.75	$454.00	$88.80	$710.40	$51.87	$81.68
.5 Laborer	42.10	168.40	67.45	269.80		
1 Tandem Roller, 10 Ton		236.75		260.43	19.73	21.70
12 L.H., Daily Totals		$859.15		$1240.63	$71.60	$103.39

Crew B-10G	Hr.	Daily	Hr.	Daily	Bare Costs	Incl. O&P
1 Equip. Oper. (medium)	$56.75	$454.00	$88.80	$710.40	$51.87	$81.68
.5 Laborer	42.10	168.40	67.45	269.80		
1 Sheepsfoot Roller, 240 H.P.		1348.00		1482.80	112.33	123.57
12 L.H., Daily Totals		$1970.40		$2463.00	$164.20	$205.25

Crew B-10H	Hr.	Daily	Hr.	Daily	Bare Costs	Incl. O&P
1 Equip. Oper. (medium)	$56.75	$454.00	$88.80	$710.40	$51.87	$81.68
.5 Laborer	42.10	168.40	67.45	269.80		
1 Diaphragm Water Pump, 2"		86.80		95.48		
1 -20' Suction Hose, 2"		3.55		3.90		
2 -50' Discharge Hoses, 2"		8.00		8.80	8.20	9.02
12 L.H., Daily Totals		$720.75		$1088.39	$60.06	$90.70

For customer support on your Commercial Renovation Costs with RSMeans data, call 800.448.8182.

755

Crew B-10I	Hr.	Daily	Hr.	Daily	Bare Costs	Incl. O&P
1 Equip. Oper. (medium)	$56.75	$454.00	$88.80	$710.40	$51.87	$81.68
.5 Laborer	42.10	168.40	67.45	269.80		
1 Diaphragm Water Pump, 4"		105.25		115.78		
1 -20' Suction Hose, 4"		17.05		18.75		
2 -50' Discharge Hoses, 4"		25.30		27.83	12.30	13.53
12 L.H., Daily Totals		$770.00		$1142.56	$64.17	$95.21

Crew B-10J	Hr.	Daily	Hr.	Daily	Bare Costs	Incl. O&P
1 Equip. Oper. (medium)	$56.75	$454.00	$88.80	$710.40	$51.87	$81.68
.5 Laborer	42.10	168.40	67.45	269.80		
1 Centrifugal Water Pump, 3"		74.45		81.89		
1 -20' Suction Hose, 3"		8.75		9.63		
2 -50' Discharge Hoses, 3"		9.00		9.90	7.68	8.45
12 L.H., Daily Totals		$714.60		$1081.62	$59.55	$90.14

Crew B-10K	Hr.	Daily	Hr.	Daily	Bare Costs	Incl. O&P
1 Equip. Oper. (medium)	$56.75	$454.00	$88.80	$710.40	$51.87	$81.68
.5 Laborer	42.10	168.40	67.45	269.80		
1 Centr. Water Pump, 6"		232.90		256.19		
1 -20' Suction Hose, 6"		25.20		27.72		
2 -50' Discharge Hoses, 6"		35.80		39.38	24.49	26.94
12 L.H., Daily Totals		$916.30		$1303.49	$76.36	$108.62

Crew B-10L	Hr.	Daily	Hr.	Daily	Bare Costs	Incl. O&P
1 Equip. Oper. (medium)	$56.75	$454.00	$88.80	$710.40	$51.87	$81.68
.5 Laborer	42.10	168.40	67.45	269.80		
1 Dozer, 80 H.P.		401.40		441.54	33.45	36.80
12 L.H., Daily Totals		$1023.80		$1421.74	$85.32	$118.48

Crew B-10M	Hr.	Daily	Hr.	Daily	Bare Costs	Incl. O&P
1 Equip. Oper. (medium)	$56.75	$454.00	$88.80	$710.40	$51.87	$81.68
.5 Laborer	42.10	168.40	67.45	269.80		
1 Dozer, 300 H.P.		1764.00		1940.40	147.00	161.70
12 L.H., Daily Totals		$2386.40		$2920.60	$198.87	$243.38

Crew B-10N	Hr.	Daily	Hr.	Daily	Bare Costs	Incl. O&P
1 Equip. Oper. (medium)	$56.75	$454.00	$88.80	$710.40	$51.87	$81.68
.5 Laborer	42.10	168.40	67.45	269.80		
1 F.E. Loader, T.M., 1.5 C.Y.		565.80		622.38	47.15	51.87
12 L.H., Daily Totals		$1188.20		$1602.58	$99.02	$133.55

Crew B-10O	Hr.	Daily	Hr.	Daily	Bare Costs	Incl. O&P
1 Equip. Oper. (medium)	$56.75	$454.00	$88.80	$710.40	$51.87	$81.68
.5 Laborer	42.10	168.40	67.45	269.80		
1 F.E. Loader, T.M., 2.25 C.Y.		960.90		1056.99	80.08	88.08
12 L.H., Daily Totals		$1583.30		$2037.19	$131.94	$169.77

Crew B-10P	Hr.	Daily	Hr.	Daily	Bare Costs	Incl. O&P
1 Equip. Oper. (medium)	$56.75	$454.00	$88.80	$710.40	$51.87	$81.68
.5 Laborer	42.10	168.40	67.45	269.80		
1 Crawler Loader, 3 C.Y.		1169.00		1285.90	97.42	107.16
12 L.H., Daily Totals		$1791.40		$2266.10	$149.28	$188.84

Crew B-10Q	Hr.	Daily	Hr.	Daily	Bare Costs	Incl. O&P
1 Equip. Oper. (medium)	$56.75	$454.00	$88.80	$710.40	$51.87	$81.68
.5 Laborer	42.10	168.40	67.45	269.80		
1 Crawler Loader, 4 C.Y.		1489.00		1637.90	124.08	136.49
12 L.H., Daily Totals		$2111.40		$2618.10	$175.95	$218.18

Crew B-10R	Hr.	Daily	Hr.	Daily	Bare Costs	Incl. O&P
1 Equip. Oper. (medium)	$56.75	$454.00	$88.80	$710.40	$51.87	$81.68
.5 Laborer	42.10	168.40	67.45	269.80		
1 F.E. Loader, W.M., 1 C.Y.		302.00		332.20	25.17	27.68
12 L.H., Daily Totals		$924.40		$1312.40	$77.03	$109.37

Crew B-10S	Hr.	Daily	Hr.	Daily	Bare Costs	Incl. O&P
1 Equip. Oper. (medium)	$56.75	$454.00	$88.80	$710.40	$51.87	$81.68
.5 Laborer	42.10	168.40	67.45	269.80		
1 F.E. Loader, W.M., 1.5 C.Y.		423.50		465.85	35.29	38.82
12 L.H., Daily Totals		$1045.90		$1446.05	$87.16	$120.50

Crew B-10T	Hr.	Daily	Hr.	Daily	Bare Costs	Incl. O&P
1 Equip. Oper. (medium)	$56.75	$454.00	$88.80	$710.40	$51.87	$81.68
.5 Laborer	42.10	168.40	67.45	269.80		
1 F.E. Loader, W.M., 2.5 C.Y.		702.10		772.31	58.51	64.36
12 L.H., Daily Totals		$1324.50		$1752.51	$110.38	$146.04

Crew B-10U	Hr.	Daily	Hr.	Daily	Bare Costs	Incl. O&P
1 Equip. Oper. (medium)	$56.75	$454.00	$88.80	$710.40	$51.87	$81.68
.5 Laborer	42.10	168.40	67.45	269.80		
1 F.E. Loader, W.M., 5.5 C.Y.		957.50		1053.25	79.79	87.77
12 L.H., Daily Totals		$1579.90		$2033.45	$131.66	$169.45

Crew B-10V	Hr.	Daily	Hr.	Daily	Bare Costs	Incl. O&P
1 Equip. Oper. (medium)	$56.75	$454.00	$88.80	$710.40	$51.87	$81.68
.5 Laborer	42.10	168.40	67.45	269.80		
1 Dozer, 700 H.P.		5120.00		5632.00	426.67	469.33
12 L.H., Daily Totals		$5742.40		$6612.20	$478.53	$551.02

Crew B-10W	Hr.	Daily	Hr.	Daily	Bare Costs	Incl. O&P
1 Equip. Oper. (medium)	$56.75	$454.00	$88.80	$710.40	$51.87	$81.68
.5 Laborer	42.10	168.40	67.45	269.80		
1 Dozer, 105 H.P.		633.85		697.24	52.82	58.10
12 L.H., Daily Totals		$1256.25		$1677.43	$104.69	$139.79

Crew B-10X	Hr.	Daily	Hr.	Daily	Bare Costs	Incl. O&P
1 Equip. Oper. (medium)	$56.75	$454.00	$88.80	$710.40	$51.87	$81.68
.5 Laborer	42.10	168.40	67.45	269.80		
1 Dozer, 410 H.P.		2777.00		3054.70	231.42	254.56
12 L.H., Daily Totals		$3399.40		$4034.90	$283.28	$336.24

Crew B-10Y	Hr.	Daily	Hr.	Daily	Bare Costs	Incl. O&P
1 Equip. Oper. (medium)	$56.75	$454.00	$88.80	$710.40	$51.87	$81.68
.5 Laborer	42.10	168.40	67.45	269.80		
1 Vibr. Roller, Towed, 12 Ton		578.60		636.46	48.22	53.04
12 L.H., Daily Totals		$1201.00		$1616.66	$100.08	$134.72

Crew B-11A	Hr.	Daily	Hr.	Daily	Bare Costs	Incl. O&P
1 Equipment Oper. (med.)	$56.75	$454.00	$88.80	$710.40	$49.42	$78.13
1 Laborer	42.10	336.80	67.45	539.60		
1 Dozer, 200 H.P.		1504.00		1654.40	94.00	103.40
16 L.H., Daily Totals		$2294.80		$2904.40	$143.43	$181.53

Crew B-11B	Hr.	Daily	Hr.	Daily	Bare Costs	Incl. O&P
1 Equipment Oper. (light)	$53.00	$424.00	$82.95	$663.60	$47.55	$75.20
1 Laborer	42.10	336.80	67.45	539.60		
1 Air Powered Tamper		33.85		37.23		
1 Air Compressor, 365 cfm		315.40		346.94		
2 -50' Air Hoses, 1.5"		48.50		53.35	24.86	27.35
16 L.H., Daily Totals		$1158.55		$1640.72	$72.41	$102.55

For customer support on your Commercial Renovation Costs with RSMeans data, call 800.448.8182.

Crews - Renovation

Crew B-11C	Hr.	Daily	Hr.	Daily	Bare Costs	Incl. O&P
1 Equipment Oper. (med.)	$56.75	$454.00	$88.80	$710.40	$49.42	$78.13
1 Laborer	42.10	336.80	67.45	539.60		
1 Backhoe Loader, 48 H.P.		213.75		235.13	13.36	14.70
16 L.H., Daily Totals		$1004.55		$1485.13	$62.78	$92.82

Crew B-11K	Hr.	Daily	Hr.	Daily	Bare Costs	Incl. O&P
1 Equipment Oper. (med.)	$56.75	$454.00	$88.80	$710.40	$49.42	$78.13
1 Laborer	42.10	336.80	67.45	539.60		
1 Trencher, Chain Type, 8' D		1872.00		2059.20	117.00	128.70
16 L.H., Daily Totals		$2662.80		$3309.20	$166.43	$206.82

Crew B-11L	Hr.	Daily	Hr.	Daily	Bare Costs	Incl. O&P
1 Equipment Oper. (med.)	$56.75	$454.00	$88.80	$710.40	$49.42	$78.13
1 Laborer	42.10	336.80	67.45	539.60		
1 Grader, 30,000 Lbs.		1062.00		1168.20	66.38	73.01
16 L.H., Daily Totals		$1852.80		$2418.20	$115.80	$151.14

Crew B-11M	Hr.	Daily	Hr.	Daily	Bare Costs	Incl. O&P
1 Equipment Oper. (med.)	$56.75	$454.00	$88.80	$710.40	$49.42	$78.13
1 Laborer	42.10	336.80	67.45	539.60		
1 Backhoe Loader, 80 H.P.		232.40		255.64	14.53	15.98
16 L.H., Daily Totals		$1023.20		$1505.64	$63.95	$94.10

Crew B-11S	Hr.	Daily	Hr.	Daily	Bare Costs	Incl. O&P
1 Equipment Operator (med.)	$56.75	$454.00	$88.80	$710.40	$51.87	$81.68
.5 Laborer	42.10	168.40	67.45	269.80		
1 Dozer, 300 H.P.		1764.00		1940.40		
1 Ripper, Beam & 1 Shank		90.50		99.55	154.54	170.00
12 L.H., Daily Totals		$2476.90		$3020.15	$206.41	$251.68

Crew B-11T	Hr.	Daily	Hr.	Daily	Bare Costs	Incl. O&P
1 Equipment Operator (med.)	$56.75	$454.00	$88.80	$710.40	$51.87	$81.68
.5 Laborer	42.10	168.40	67.45	269.80		
1 Dozer, 410 H.P.		2777.00		3054.70		
1 Ripper, Beam & 2 Shanks		138.75		152.63	242.98	267.28
12 L.H., Daily Totals		$3538.15		$4187.52	$294.85	$348.96

Crew B-11W	Hr.	Daily	Hr.	Daily	Bare Costs	Incl. O&P
1 Equipment Operator (med.)	$56.75	$454.00	$88.80	$710.40	$49.03	$78.23
1 Common Laborer	42.10	336.80	67.45	539.60		
10 Truck Drivers (heavy)	48.95	3916.00	78.25	6260.00		
1 Dozer, 200 H.P.		1504.00		1654.40		
1 Vibratory Roller, Towed, 23 Ton		514.75		566.23		
10 Dump Trucks, 8 C.Y., 220 H.P.		4280.50		4708.55	65.62	72.18
96 L.H., Daily Totals		$11006.05		$14439.17	$114.65	$150.41

Crew B-11Y	Hr.	Daily	Hr.	Daily	Bare Costs	Incl. O&P
1 Labor Foreman (outside)	$44.10	$352.80	$70.65	$565.20	$47.21	$74.92
5 Common Laborers	42.10	1684.00	67.45	2698.00		
3 Equipment Operators (med.)	56.75	1362.00	88.80	2131.20		
1 Dozer, 80 H.P.		401.40		441.54		
2 Rollers, 2-Drum, W.B., 7.5 H.P.		332.20		365.42		
4 Vibrating Plates, Gas, 21"		661.40		727.54	19.38	21.31
72 L.H., Daily Totals		$4793.80		$6928.90	$66.58	$96.23

Crew B-12A	Hr.	Daily	Hr.	Daily	Bare Costs	Incl. O&P
1 Equip. Oper. (crane)	$59.20	$473.60	$92.65	$741.20	$50.65	$80.05
1 Laborer	42.10	336.80	67.45	539.60		
1 Hyd. Excavator, 1 C.Y.		795.30		874.83	49.71	54.68
16 L.H., Daily Totals		$1605.70		$2155.63	$100.36	$134.73

Crew B-12B	Hr.	Daily	Hr.	Daily	Bare Costs	Incl. O&P
1 Equip. Oper. (crane)	$59.20	$473.60	$92.65	$741.20	$50.65	$80.05
1 Laborer	42.10	336.80	67.45	539.60		
1 Hyd. Excavator, 1.5 C.Y.		687.55		756.30	42.97	47.27
16 L.H., Daily Totals		$1497.95		$2037.11	$93.62	$127.32

Crew B-12C	Hr.	Daily	Hr.	Daily	Bare Costs	Incl. O&P
1 Equip. Oper. (crane)	$59.20	$473.60	$92.65	$741.20	$50.65	$80.05
1 Laborer	42.10	336.80	67.45	539.60		
1 Hyd. Excavator, 2 C.Y.		951.30		1046.43	59.46	65.40
16 L.H., Daily Totals		$1761.70		$2327.23	$110.11	$145.45

Crew B-12D	Hr.	Daily	Hr.	Daily	Bare Costs	Incl. O&P
1 Equip. Oper. (crane)	$59.20	$473.60	$92.65	$741.20	$50.65	$80.05
1 Laborer	42.10	336.80	67.45	539.60		
1 Hyd. Excavator, 3.5 C.Y.		2158.00		2373.80	134.88	148.36
16 L.H., Daily Totals		$2968.40		$3654.60	$185.53	$228.41

Crew B-12E	Hr.	Daily	Hr.	Daily	Bare Costs	Incl. O&P
1 Equip. Oper. (crane)	$59.20	$473.60	$92.65	$741.20	$50.65	$80.05
1 Laborer	42.10	336.80	67.45	539.60		
1 Hyd. Excavator, .5 C.Y.		452.10		497.31	28.26	31.08
16 L.H., Daily Totals		$1262.50		$1778.11	$78.91	$111.13

Crew B-12F	Hr.	Daily	Hr.	Daily	Bare Costs	Incl. O&P
1 Equip. Oper. (crane)	$59.20	$473.60	$92.65	$741.20	$50.65	$80.05
1 Laborer	42.10	336.80	67.45	539.60		
1 Hyd. Excavator, .75 C.Y.		694.25		763.67	43.39	47.73
16 L.H., Daily Totals		$1504.65		$2044.47	$94.04	$127.78

Crew B-12G	Hr.	Daily	Hr.	Daily	Bare Costs	Incl. O&P
1 Equip. Oper. (crane)	$59.20	$473.60	$92.65	$741.20	$50.65	$80.05
1 Laborer	42.10	336.80	67.45	539.60		
1 Crawler Crane, 15 Ton		792.30		871.53		
1 Clamshell Bucket, .5 C.Y.		67.10		73.81	53.71	59.08
16 L.H., Daily Totals		$1669.80		$2226.14	$104.36	$139.13

Crew B-12H	Hr.	Daily	Hr.	Daily	Bare Costs	Incl. O&P
1 Equip. Oper. (crane)	$59.20	$473.60	$92.65	$741.20	$50.65	$80.05
1 Laborer	42.10	336.80	67.45	539.60		
1 Crawler Crane, 25 Ton		1129.00		1241.90		
1 Clamshell Bucket, 1 C.Y.		68.55		75.41	74.85	82.33
16 L.H., Daily Totals		$2007.95		$2598.11	$125.50	$162.38

Crew B-12I	Hr.	Daily	Hr.	Daily	Bare Costs	Incl. O&P
1 Equip. Oper. (crane)	$59.20	$473.60	$92.65	$741.20	$50.65	$80.05
1 Laborer	42.10	336.80	67.45	539.60		
1 Crawler Crane, 20 Ton		961.35		1057.48		
1 Dragline Bucket, .75 C.Y.		61.25		67.38	63.91	70.30
16 L.H., Daily Totals		$1833.00		$2405.66	$114.56	$150.35

Crew B-12J	Hr.	Daily	Hr.	Daily	Bare Costs	Incl. O&P
1 Equip. Oper. (crane)	$59.20	$473.60	$92.65	$741.20	$50.65	$80.05
1 Laborer	42.10	336.80	67.45	539.60		
1 Gradall, 5/8 C.Y.		846.50		931.15	52.91	58.20
16 L.H., Daily Totals		$1656.90		$2211.95	$103.56	$138.25

For customer support on your Commercial Renovation Costs with RSMeans data, call 800.448.8182.

Crew No.	Bare Costs Hr.	Daily	Incl. Subs O&P Hr.	Daily	Cost Per Labor-Hour Bare Costs	Incl. O&P
Crew B-12K	Hr.	Daily	Hr.	Daily	Bare Costs	Incl. O&P
1 Equip. Oper. (crane)	$59.20	$473.60	$92.65	$741.20	$50.65	$80.05
1 Laborer	42.10	336.80	67.45	539.60		
1 Gradall, 3 Ton, 1 C.Y.		973.80		1071.18	60.86	66.95
16 L.H., Daily Totals		$1784.20		$2351.98	$111.51	$147.00
Crew B-12L	Hr.	Daily	Hr.	Daily	Bare Costs	Incl. O&P
1 Equip. Oper. (crane)	$59.20	$473.60	$92.65	$741.20	$50.65	$80.05
1 Laborer	42.10	336.80	67.45	539.60		
1 Crawler Crane, 15 Ton		792.30		871.53		
1 F.E. Attachment, .5 C.Y.		65.25		71.78	53.60	58.96
16 L.H., Daily Totals		$1667.95		$2224.11	$104.25	$139.01
Crew B-12M	Hr.	Daily	Hr.	Daily	Bare Costs	Incl. O&P
1 Equip. Oper. (crane)	$59.20	$473.60	$92.65	$741.20	$50.65	$80.05
1 Laborer	42.10	336.80	67.45	539.60		
1 Crawler Crane, 20 Ton		961.35		1057.48		
1 F.E. Attachment, .75 C.Y.		70.40		77.44	64.48	70.93
16 L.H., Daily Totals		$1842.15		$2415.72	$115.13	$150.98
Crew B-12N	Hr.	Daily	Hr.	Daily	Bare Costs	Incl. O&P
1 Equip. Oper. (crane)	$59.20	$473.60	$92.65	$741.20	$50.65	$80.05
1 Laborer	42.10	336.80	67.45	539.60		
1 Crawler Crane, 25 Ton		1129.00		1241.90		
1 F.E. Attachment, 1 C.Y.		76.50		84.15	75.34	82.88
16 L.H., Daily Totals		$2015.90		$2606.85	$125.99	$162.93
Crew B-12O	Hr.	Daily	Hr.	Daily	Bare Costs	Incl. O&P
1 Equip. Oper. (crane)	$59.20	$473.60	$92.65	$741.20	$50.65	$80.05
1 Laborer	42.10	336.80	67.45	539.60		
1 Crawler Crane, 40 Ton		1206.00		1326.60		
1 F.E. Attachment, 1.5 C.Y.		87.60		96.36	80.85	88.94
16 L.H., Daily Totals		$2104.00		$2703.76	$131.50	$168.99
Crew B-12P	Hr.	Daily	Hr.	Daily	Bare Costs	Incl. O&P
1 Equip. Oper. (crane)	$59.20	$473.60	$92.65	$741.20	$50.65	$80.05
1 Laborer	42.10	336.80	67.45	539.60		
1 Crawler Crane, 40 Ton		1206.00		1326.60		
1 Dragline Bucket, 1.5 C.Y.		65.05		71.56	79.44	87.38
16 L.H., Daily Totals		$2081.45		$2678.95	$130.09	$167.43
Crew B-12Q	Hr.	Daily	Hr.	Daily	Bare Costs	Incl. O&P
1 Equip. Oper. (crane)	$59.20	$473.60	$92.65	$741.20	$50.65	$80.05
1 Laborer	42.10	336.80	67.45	539.60		
1 Hyd. Excavator, 5/8 C.Y.		598.25		658.08	37.39	41.13
16 L.H., Daily Totals		$1408.65		$1938.88	$88.04	$121.18
Crew B-12S	Hr.	Daily	Hr.	Daily	Bare Costs	Incl. O&P
1 Equip. Oper. (crane)	$59.20	$473.60	$92.65	$741.20	$50.65	$80.05
1 Laborer	42.10	336.80	67.45	539.60		
1 Hyd. Excavator, 2.5 C.Y.		1465.00		1611.50	91.56	100.72
16 L.H., Daily Totals		$2275.40		$2892.30	$142.21	$180.77
Crew B-12T	Hr.	Daily	Hr.	Daily	Bare Costs	Incl. O&P
1 Equip. Oper. (crane)	$59.20	$473.60	$92.65	$741.20	$50.65	$80.05
1 Laborer	42.10	336.80	67.45	539.60		
1 Crawler Crane, 75 Ton		2002.00		2202.20		
1 F.E. Attachment, 3 C.Y.		114.15		125.57	132.26	145.49
16 L.H., Daily Totals		$2926.55		$3608.57	$182.91	$225.54

Crew No.	Bare Costs Hr.	Daily	Incl. Subs O&P Hr.	Daily	Cost Per Labor-Hour Bare Costs	Incl. O&P
Crew B-12V	Hr.	Daily	Hr.	Daily	Bare Costs	Incl. O&P
1 Equip. Oper. (crane)	$59.20	$473.60	$92.65	$741.20	$50.65	$80.05
1 Laborer	42.10	336.80	67.45	539.60		
1 Crawler Crane, 75 Ton		2002.00		2202.20		
1 Dragline Bucket, 3 C.Y.		71.60		78.76	129.60	142.56
16 L.H., Daily Totals		$2884.00		$3561.76	$180.25	$222.61
Crew B-12Y	Hr.	Daily	Hr.	Daily	Bare Costs	Incl. O&P
1 Equip. Oper. (crane)	$59.20	$473.60	$92.65	$741.20	$47.80	$75.85
2 Laborers	42.10	673.60	67.45	1079.20		
1 Hyd. Excavator, 3.5 C.Y.		2158.00		2373.80	89.92	98.91
24 L.H., Daily Totals		$3305.20		$4194.20	$137.72	$174.76
Crew B-12Z	Hr.	Daily	Hr.	Daily	Bare Costs	Incl. O&P
1 Equip. Oper. (crane)	$59.20	$473.60	$92.65	$741.20	$47.80	$75.85
2 Laborers	42.10	673.60	67.45	1079.20		
1 Hyd. Excavator, 2.5 C.Y.		1465.00		1611.50	61.04	67.15
24 L.H., Daily Totals		$2612.20		$3431.90	$108.84	$143.00
Crew B-13	Hr.	Daily	Hr.	Daily	Bare Costs	Incl. O&P
1 Labor Foreman (outside)	$44.10	$352.80	$70.65	$565.20	$46.04	$73.17
4 Laborers	42.10	1347.20	67.45	2158.40		
1 Equip. Oper. (crane)	59.20	473.60	92.65	741.20		
1 Equip. Oper. (oiler)	50.55	404.40	79.10	632.80		
1 Hyd. Crane, 25 Ton		580.85		638.93	10.37	11.41
56 L.H., Daily Totals		$3158.85		$4736.53	$56.41	$84.58
Crew B-13A	Hr.	Daily	Hr.	Daily	Bare Costs	Incl. O&P
1 Labor Foreman (outside)	$44.10	$352.80	$70.65	$565.20	$48.53	$77.09
2 Laborers	42.10	673.60	67.45	1079.20		
2 Equipment Operators (med.)	56.75	908.00	88.80	1420.80		
2 Truck Drivers (heavy)	48.95	783.20	78.25	1252.00		
1 Crawler Crane, 75 Ton		2002.00		2202.20		
1 Crawler Loader, 4 C.Y.		1489.00		1637.90		
2 Dump Trucks, 8 C.Y., 220 H.P.		856.10		941.71	77.63	85.39
56 L.H., Daily Totals		$7064.70		$9099.01	$126.16	$162.48
Crew B-13B	Hr.	Daily	Hr.	Daily	Bare Costs	Incl. O&P
1 Labor Foreman (outside)	$44.10	$352.80	$70.65	$565.20	$46.04	$73.17
4 Laborers	42.10	1347.20	67.45	2158.40		
1 Equip. Oper. (crane)	59.20	473.60	92.65	741.20		
1 Equip. Oper. (oiler)	50.55	404.40	79.10	632.80		
1 Hyd. Crane, 55 Ton		980.20		1078.22	17.50	19.25
56 L.H., Daily Totals		$3558.20		$5175.82	$63.54	$92.43
Crew B-13C	Hr.	Daily	Hr.	Daily	Bare Costs	Incl. O&P
1 Labor Foreman (outside)	$44.10	$352.80	$70.65	$565.20	$46.04	$73.17
4 Laborers	42.10	1347.20	67.45	2158.40		
1 Equip. Oper. (crane)	59.20	473.60	92.65	741.20		
1 Equip. Oper. (oiler)	50.55	404.40	79.10	632.80		
1 Crawler Crane, 100 Ton		2287.00		2515.70	40.84	44.92
56 L.H., Daily Totals		$4865.00		$6613.30	$86.88	$118.09
Crew B-13D	Hr.	Daily	Hr.	Daily	Bare Costs	Incl. O&P
1 Laborer	$42.10	$336.80	$67.45	$539.60	$50.65	$80.05
1 Equip. Oper. (crane)	59.20	473.60	92.65	741.20		
1 Hyd. Excavator, 1 C.Y.		795.30		874.83		
1 Trench Box		117.75		129.53	57.07	62.77
16 L.H., Daily Totals		$1723.45		$2285.16	$107.72	$142.82

Crew No.	Bare Costs Hr.	Daily	Incl. Subs O&P Hr.	Daily	Cost Per Labor-Hour Bare Costs	Incl. O&P
Crew B-13E	Hr.	Daily	Hr.	Daily	Bare Costs	Incl. O&P
1 Laborer	$42.10	$336.80	$67.45	$539.60	$50.65	$80.05
1 Equip. Oper. (crane)	59.20	473.60	92.65	741.20		
1 Hyd. Excavator, 1.5 C.Y.		687.55		756.30		
1 Trench Box		117.75		129.53	50.33	55.36
16 L.H., Daily Totals		$1615.70		$2166.63	$100.98	$135.41
Crew B-13F	Hr.	Daily	Hr.	Daily	Bare Costs	Incl. O&P
1 Laborer	$42.10	$336.80	$67.45	$539.60	$50.65	$80.05
1 Equip. Oper. (crane)	59.20	473.60	92.65	741.20		
1 Hyd. Excavator, 3.5 C.Y.		2158.00		2373.80		
1 Trench Box		117.75		129.53	142.23	156.46
16 L.H., Daily Totals		$3086.15		$3784.13	$192.88	$236.51
Crew B-13G	Hr.	Daily	Hr.	Daily	Bare Costs	Incl. O&P
1 Laborer	$42.10	$336.80	$67.45	$539.60	$50.65	$80.05
1 Equip. Oper. (crane)	59.20	473.60	92.65	741.20		
1 Hyd. Excavator, .75 C.Y.		694.25		763.67		
1 Trench Box		117.75		129.53	50.75	55.83
16 L.H., Daily Totals		$1622.40		$2174.00	$101.40	$135.88
Crew B-13H	Hr.	Daily	Hr.	Daily	Bare Costs	Incl. O&P
1 Laborer	$42.10	$336.80	$67.45	$539.60	$50.65	$80.05
1 Equip. Oper. (crane)	59.20	473.60	92.65	741.20		
1 Gradall, 5/8 C.Y.		846.50		931.15		
1 Trench Box		117.75		129.53	60.27	66.29
16 L.H., Daily Totals		$1774.65		$2341.47	$110.92	$146.34
Crew B-13I	Hr.	Daily	Hr.	Daily	Bare Costs	Incl. O&P
1 Laborer	$42.10	$336.80	$67.45	$539.60	$50.65	$80.05
1 Equip. Oper. (crane)	59.20	473.60	92.65	741.20		
1 Gradall, 3 Ton, 1 C.Y.		973.80		1071.18		
1 Trench Box		117.75		129.53	68.22	75.04
16 L.H., Daily Totals		$1901.95		$2481.51	$118.87	$155.09
Crew B-13J	Hr.	Daily	Hr.	Daily	Bare Costs	Incl. O&P
1 Laborer	$42.10	$336.80	$67.45	$539.60	$50.65	$80.05
1 Equip. Oper. (crane)	59.20	473.60	92.65	741.20		
1 Hyd. Excavator, 2.5 C.Y.		1465.00		1611.50		
1 Trench Box		117.75		129.53	98.92	108.81
16 L.H., Daily Totals		$2393.15		$3021.82	$149.57	$188.86
Crew B-13K	Hr.	Daily	Hr.	Daily	Bare Costs	Incl. O&P
2 Equip. Opers. (crane)	$59.20	$947.20	$92.65	$1482.40	$59.20	$92.65
1 Hyd. Excavator, .75 C.Y.		694.25		763.67		
1 Hyd. Hammer, 4000 ft-lb		641.50		705.65		
1 Hyd. Excavator, .75 C.Y.		694.25		763.67	126.88	139.56
16 L.H., Daily Totals		$2977.20		$3715.40	$186.07	$232.21
Crew B-13L	Hr.	Daily	Hr.	Daily	Bare Costs	Incl. O&P
2 Equip. Opers. (crane)	$59.20	$947.20	$92.65	$1482.40	$59.20	$92.65
1 Hyd. Excavator, 1.5 C.Y.		687.55		756.30		
1 Hyd. Hammer, 5000 ft-lb		697.25		766.98		
1 Hyd. Excavator, .75 C.Y.		694.25		763.67	129.94	142.93
16 L.H., Daily Totals		$3026.25		$3769.36	$189.14	$235.58

Crew No.	Bare Costs Hr.	Daily	Incl. Subs O&P Hr.	Daily	Cost Per Labor-Hour Bare Costs	Incl. O&P
Crew B-13M	Hr.	Daily	Hr.	Daily	Bare Costs	Incl. O&P
2 Equip. Opers. (crane)	$59.20	$947.20	$92.65	$1482.40	$59.20	$92.65
1 Hyd. Excavator, 2.5 C.Y.		1465.00		1611.50		
1 Hyd. Hammer, 8000 ft-lb		907.75		998.52		
1 Hyd. Excavator, 1.5 C.Y.		687.55		756.30	191.27	210.40
16 L.H., Daily Totals		$4007.50		$4848.73	$250.47	$303.05
Crew B-13N	Hr.	Daily	Hr.	Daily	Bare Costs	Incl. O&P
2 Equip. Opers. (crane)	$59.20	$947.20	$92.65	$1482.40	$59.20	$92.65
1 Hyd. Excavator, 3.5 C.Y.		2158.00		2373.80		
1 Hyd. Hammer, 12,000 ft-lb		871.75		958.92		
1 Hyd. Excavator, 1.5 C.Y.		687.55		756.30	232.33	255.56
16 L.H., Daily Totals		$4664.50		$5571.43	$291.53	$348.21
Crew B-14	Hr.	Daily	Hr.	Daily	Bare Costs	Incl. O&P
1 Labor Foreman (outside)	$44.10	$352.80	$70.65	$565.20	$44.25	$70.57
4 Laborers	42.10	1347.20	67.45	2158.40		
1 Equip. Oper. (light)	53.00	424.00	82.95	663.60		
1 Backhoe Loader, 48 H.P.		213.75		235.13	4.45	4.90
48 L.H., Daily Totals		$2337.75		$3622.32	$48.70	$75.47
Crew B-14A	Hr.	Daily	Hr.	Daily	Bare Costs	Incl. O&P
1 Equip. Oper. (crane)	$59.20	$473.60	$92.65	$741.20	$53.50	$84.25
.5 Laborer	42.10	168.40	67.45	269.80		
1 Hyd. Excavator, 4.5 C.Y.		3409.00		3749.90	284.08	312.49
12 L.H., Daily Totals		$4051.00		$4760.90	$337.58	$396.74
Crew B-14B	Hr.	Daily	Hr.	Daily	Bare Costs	Incl. O&P
1 Equip. Oper. (crane)	$59.20	$473.60	$92.65	$741.20	$53.50	$84.25
.5 Laborer	42.10	168.40	67.45	269.80		
1 Hyd. Excavator, 6 C.Y.		3464.00		3810.40	288.67	317.53
12 L.H., Daily Totals		$4106.00		$4821.40	$342.17	$401.78
Crew B-14C	Hr.	Daily	Hr.	Daily	Bare Costs	Incl. O&P
1 Equip. Oper. (crane)	$59.20	$473.60	$92.65	$741.20	$53.50	$84.25
.5 Laborer	42.10	168.40	67.45	269.80		
1 Hyd. Excavator, 7 C.Y.		3437.00		3780.70	286.42	315.06
12 L.H., Daily Totals		$4079.00		$4791.70	$339.92	$399.31
Crew B-14F	Hr.	Daily	Hr.	Daily	Bare Costs	Incl. O&P
1 Equip. Oper. (crane)	$59.20	$473.60	$92.65	$741.20	$53.50	$84.25
.5 Laborer	42.10	168.40	67.45	269.80		
1 Hyd. Shovel, 7 C.Y.		4099.00		4508.90	341.58	375.74
12 L.H., Daily Totals		$4741.00		$5519.90	$395.08	$459.99
Crew B-14G	Hr.	Daily	Hr.	Daily	Bare Costs	Incl. O&P
1 Equip. Oper. (crane)	$59.20	$473.60	$92.65	$741.20	$53.50	$84.25
.5 Laborer	42.10	168.40	67.45	269.80		
1 Hyd. Shovel, 12 C.Y.		5951.00		6546.10	495.92	545.51
12 L.H., Daily Totals		$6593.00		$7557.10	$549.42	$629.76
Crew B-14J	Hr.	Daily	Hr.	Daily	Bare Costs	Incl. O&P
1 Equip. Oper. (medium)	$56.75	$454.00	$88.80	$710.40	$51.87	$81.68
.5 Laborer	42.10	168.40	67.45	269.80		
1 F.E. Loader, 8 C.Y.		2261.00		2487.10	188.42	207.26
12 L.H., Daily Totals		$2883.40		$3467.30	$240.28	$288.94

For customer support on your Commercial Renovation Costs with RSMeans data, call 800.448.8182.

759

Crew No.	Bare Costs		Incl. Subs O&P		Cost Per Labor-Hour	

Crew B-14K

Crew B-14K	Hr.	Daily	Hr.	Daily	Bare Costs	Incl. O&P
1 Equip. Oper. (medium)	$56.75	$454.00	$88.80	$710.40	$51.87	$81.68
.5 Laborer	42.10	168.40	67.45	269.80		
1 F.E. Loader, 10 C.Y.		2674.00		2941.40	222.83	245.12
12 L.H., Daily Totals		$3296.40		$3921.60	$274.70	$326.80

Crew B-15	Hr.	Daily	Hr.	Daily	Bare Costs	Incl. O&P
1 Equipment Oper. (med.)	$56.75	$454.00	$88.80	$710.40	$50.20	$79.72
.5 Laborer	42.10	168.40	67.45	269.80		
2 Truck Drivers (heavy)	48.95	783.20	78.25	1252.00		
2 Dump Trucks, 12 C.Y., 400 H.P.		1145.80		1260.38		
1 Dozer, 200 H.P.		1504.00		1654.40	94.64	104.10
28 L.H., Daily Totals		$4055.40		$5146.98	$144.84	$183.82

Crew B-16	Hr.	Daily	Hr.	Daily	Bare Costs	Incl. O&P
1 Labor Foreman (outside)	$44.10	$352.80	$70.65	$565.20	$44.31	$70.95
2 Laborers	42.10	673.60	67.45	1079.20		
1 Truck Driver (heavy)	48.95	391.60	78.25	626.00		
1 Dump Truck, 12 C.Y., 400 H.P.		572.90		630.19	17.90	19.69
32 L.H., Daily Totals		$1990.90		$2900.59	$62.22	$90.64

Crew B-17	Hr.	Daily	Hr.	Daily	Bare Costs	Incl. O&P
2 Laborers	$42.10	$673.60	$67.45	$1079.20	$46.54	$74.03
1 Equip. Oper. (light)	53.00	424.00	82.95	663.60		
1 Truck Driver (heavy)	48.95	391.60	78.25	626.00		
1 Backhoe Loader, 48 H.P.		213.75		235.13		
1 Dump Truck, 8 C.Y., 220 H.P.		428.05		470.86	20.06	22.06
32 L.H., Daily Totals		$2131.00		$3074.78	$66.59	$96.09

Crew B-17A	Hr.	Daily	Hr.	Daily	Bare Costs	Incl. O&P
2 Labor Foremen (outside)	$44.10	$705.60	$70.65	$1130.40	$45.25	$72.39
6 Laborers	42.10	2020.80	67.45	3237.60		
1 Skilled Worker Foreman (out)	56.85	454.80	90.55	724.40		
1 Skilled Worker	54.85	438.80	87.40	699.20		
80 L.H., Daily Totals		$3620.00		$5791.60	$45.25	$72.39

Crew B-17B	Hr.	Daily	Hr.	Daily	Bare Costs	Incl. O&P
2 Laborers	$42.10	$673.60	$67.45	$1079.20	$46.54	$74.03
1 Equip. Oper. (light)	53.00	424.00	82.95	663.60		
1 Truck Driver (heavy)	48.95	391.60	78.25	626.00		
1 Backhoe Loader, 48 H.P.		213.75		235.13		
1 Dump Truck, 12 C.Y., 400 H.P.		572.90		630.19	24.58	27.04
32 L.H., Daily Totals		$2275.85		$3234.11	$71.12	$101.07

Crew B-18	Hr.	Daily	Hr.	Daily	Bare Costs	Incl. O&P
1 Labor Foreman (outside)	$44.10	$352.80	$70.65	$565.20	$42.77	$68.52
2 Laborers	42.10	673.60	67.45	1079.20		
1 Vibrating Plate, Gas, 21"		165.35		181.88	6.89	7.58
24 L.H., Daily Totals		$1191.75		$1826.29	$49.66	$76.10

Crew B-19	Hr.	Daily	Hr.	Daily	Bare Costs	Incl. O&P
1 Pile Driver Foreman (outside)	$56.20	$449.60	$90.15	$721.20	$55.24	$87.77
4 Pile Drivers	54.20	1734.40	86.90	2780.80		
2 Equip. Oper. (crane)	59.20	947.20	92.65	1482.40		
1 Equip. Oper. (oiler)	50.55	404.40	79.10	632.80		
1 Crawler Crane, 40 Ton		1206.00		1326.60		
1 Lead, 90' High		367.45		404.19		
1 Hammer, Diesel, 22k ft-lb		436.45		480.10	31.40	34.55
64 L.H., Daily Totals		$5545.50		$7828.09	$86.65	$122.31

Crew B-19A	Hr.	Daily	Hr.	Daily	Bare Costs	Incl. O&P
1 Pile Driver Foreman (outside)	$56.20	$449.60	$90.15	$721.20	$55.24	$87.77
4 Pile Drivers	54.20	1734.40	86.90	2780.80		
2 Equip. Oper. (crane)	59.20	947.20	92.65	1482.40		
1 Equip. Oper. (oiler)	50.55	404.40	79.10	632.80		
1 Crawler Crane, 75 Ton		2002.00		2202.20		
1 Lead, 90' High		367.45		404.19		
1 Hammer, Diesel, 41k ft-lb		576.65		634.32	46.03	50.64
64 L.H., Daily Totals		$6481.70		$8857.91	$101.28	$138.40

Crew B-19B	Hr.	Daily	Hr.	Daily	Bare Costs	Incl. O&P
1 Pile Driver Foreman (outside)	$56.20	$449.60	$90.15	$721.20	$55.24	$87.77
4 Pile Drivers	54.20	1734.40	86.90	2780.80		
2 Equip. Oper. (crane)	59.20	947.20	92.65	1482.40		
1 Equip. Oper. (oiler)	50.55	404.40	79.10	632.80		
1 Crawler Crane, 40 Ton		1206.00		1326.60		
1 Lead, 90' High		367.45		404.19		
1 Hammer, Diesel, 22k ft-lb		436.45		480.10		
1 Barge, 400 Ton		858.85		944.74	44.82	49.31
64 L.H., Daily Totals		$6404.35		$8772.83	$100.07	$137.08

Crew B-19C	Hr.	Daily	Hr.	Daily	Bare Costs	Incl. O&P
1 Pile Driver Foreman (outside)	$56.20	$449.60	$90.15	$721.20	$55.24	$87.77
4 Pile Drivers	54.20	1734.40	86.90	2780.80		
2 Equip. Oper. (crane)	59.20	947.20	92.65	1482.40		
1 Equip. Oper. (oiler)	50.55	404.40	79.10	632.80		
1 Crawler Crane, 75 Ton		2002.00		2202.20		
1 Lead, 90' High		367.45		404.19		
1 Hammer, Diesel, 41k ft-lb		576.65		634.32		
1 Barge, 400 Ton		858.85		944.74	59.45	65.40
64 L.H., Daily Totals		$7340.55		$9802.65	$114.70	$153.17

Crew B-20	Hr.	Daily	Hr.	Daily	Bare Costs	Incl. O&P
1 Labor Foreman (outside)	$44.10	$352.80	$70.65	$565.20	$47.02	$75.17
1 Skilled Worker	54.85	438.80	87.40	699.20		
1 Laborer	42.10	336.80	67.45	539.60		
24 L.H., Daily Totals		$1128.40		$1804.00	$47.02	$75.17

Crew B-20A	Hr.	Daily	Hr.	Daily	Bare Costs	Incl. O&P
1 Labor Foreman (outside)	$44.10	$352.80	$70.65	$565.20	$50.55	$79.36
1 Laborer	42.10	336.80	67.45	539.60		
1 Plumber	64.45	515.60	99.65	797.20		
1 Plumber Apprentice	51.55	412.40	79.70	637.60		
32 L.H., Daily Totals		$1617.60		$2539.60	$50.55	$79.36

Crew B-21	Hr.	Daily	Hr.	Daily	Bare Costs	Incl. O&P
1 Labor Foreman (outside)	$44.10	$352.80	$70.65	$565.20	$48.76	$77.66
1 Skilled Worker	54.85	438.80	87.40	699.20		
1 Laborer	42.10	336.80	67.45	539.60		
.5 Equip. Oper. (crane)	59.20	236.80	92.65	370.60		
.5 S.P. Crane, 4x4, 5 Ton		189.05		207.96	6.75	7.43
28 L.H., Daily Totals		$1554.25		$2382.55	$55.51	$85.09

Crew B-21A	Hr.	Daily	Hr.	Daily	Bare Costs	Incl. O&P
1 Labor Foreman (outside)	$44.10	$352.80	$70.65	$565.20	$52.28	$82.02
1 Laborer	42.10	336.80	67.45	539.60		
1 Plumber	64.45	515.60	99.65	797.20		
1 Plumber Apprentice	51.55	412.40	79.70	637.60		
1 Equip. Oper. (crane)	59.20	473.60	92.65	741.20		
1 S.P. Crane, 4x4, 12 Ton		428.30		471.13	10.71	11.78
40 L.H., Daily Totals		$2519.50		$3751.93	$62.99	$93.80

Crew No.	Bare Costs		Incl. Subs O&P		Cost Per Labor-Hour	
	Hr.	Daily	Hr.	Daily	Bare Costs	Incl. O&P
Crew B-21B	Hr.	Daily	Hr.	Daily	Bare Costs	Incl. O&P
1 Labor Foreman (outside)	$44.10	$352.80	$70.65	$565.20	$45.92	$73.13
3 Laborers	42.10	1010.40	67.45	1618.80		
1 Equip. Oper. (crane)	59.20	473.60	92.65	741.20		
1 Hyd. Crane, 12 Ton		471.00		518.10	11.78	12.95
40 L.H., Daily Totals		$2307.80		$3443.30	$57.70	$86.08
Crew B-21C	Hr.	Daily	Hr.	Daily	Bare Costs	Incl. O&P
1 Labor Foreman (outside)	$44.10	$352.80	$70.65	$565.20	$46.04	$73.17
4 Laborers	42.10	1347.20	67.45	2158.40		
1 Equip. Oper. (crane)	59.20	473.60	92.65	741.20		
1 Equip. Oper. (oiler)	50.55	404.40	79.10	632.80		
2 Cutting Torches		25.60		28.16		
2 Sets of Gases		343.10		377.41		
1 Lattice Boom Crane, 90 Ton		1696.00		1865.60	36.87	40.56
56 L.H., Daily Totals		$4642.70		$6368.77	$82.91	$113.73
Crew B-22	Hr.	Daily	Hr.	Daily	Bare Costs	Incl. O&P
1 Labor Foreman (outside)	$44.10	$352.80	$70.65	$565.20	$49.45	$78.66
1 Skilled Worker	54.85	438.80	87.40	699.20		
1 Laborer	42.10	336.80	67.45	539.60		
.75 Equip. Oper. (crane)	59.20	355.20	92.65	555.90		
.75 S.P. Crane, 4x4, 5 Ton		283.57		311.93	9.45	10.40
30 L.H., Daily Totals		$1767.18		$2671.83	$58.91	$89.06
Crew B-22A	Hr.	Daily	Hr.	Daily	Bare Costs	Incl. O&P
1 Labor Foreman (outside)	$44.10	$352.80	$70.65	$565.20	$48.47	$77.12
1 Skilled Worker	54.85	438.80	87.40	699.20		
2 Laborers	42.10	673.60	67.45	1079.20		
1 Equipment Operator, Crane	59.20	473.60	92.65	741.20		
1 S.P. Crane, 4x4, 5 Ton		378.10		415.91		
1 Butt Fusion Mach., 4"-12" diam.		426.85		469.54	20.12	22.14
40 L.H., Daily Totals		$2743.75		$3970.24	$68.59	$99.26
Crew B-22B	Hr.	Daily	Hr.	Daily	Bare Costs	Incl. O&P
1 Labor Foreman (outside)	$44.10	$352.80	$70.65	$565.20	$48.47	$77.12
1 Skilled Worker	54.85	438.80	87.40	699.20		
2 Laborers	42.10	673.60	67.45	1079.20		
1 Equip. Oper. (crane)	59.20	473.60	92.65	741.20		
1 S.P. Crane, 4x4, 5 Ton		378.10		415.91		
1 Butt Fusion Mach., 8"-24" diam.		1076.00		1183.60	36.35	39.99
40 L.H., Daily Totals		$3392.90		$4684.31	$84.82	$117.11
Crew B-22C	Hr.	Daily	Hr.	Daily	Bare Costs	Incl. O&P
1 Skilled Worker	$54.85	$438.80	$87.40	$699.20	$48.48	$77.42
1 Laborer	42.10	336.80	67.45	539.60		
1 Butt Fusion Mach., 2"-8" diam.		156.05		171.66	9.75	10.73
16 L.H., Daily Totals		$931.65		$1410.45	$58.23	$88.15
Crew B-23	Hr.	Daily	Hr.	Daily	Bare Costs	Incl. O&P
1 Labor Foreman (outside)	$44.10	$352.80	$70.65	$565.20	$42.50	$68.09
4 Laborers	42.10	1347.20	67.45	2158.40		
1 Drill Rig, Truck-Mounted		771.10		848.21		
1 Flatbed Truck, Gas, 3 Ton		820.40		902.44	39.79	43.77
40 L.H., Daily Totals		$3291.50		$4474.25	$82.29	$111.86

Crew No.	Bare Costs		Incl. Subs O&P		Cost Per Labor-Hour	
	Hr.	Daily	Hr.	Daily	Bare Costs	Incl. O&P
Crew B-23A	Hr.	Daily	Hr.	Daily	Bare Costs	Incl. O&P
1 Labor Foreman (outside)	$44.10	$352.80	$70.65	$565.20	$47.65	$75.63
1 Laborer	42.10	336.80	67.45	539.60		
1 Equip. Oper. (medium)	56.75	454.00	88.80	710.40		
1 Drill Rig, Truck-Mounted		771.10		848.21		
1 Pickup Truck, 3/4 Ton		110.85		121.94	36.75	40.42
24 L.H., Daily Totals		$2025.55		$2785.34	$84.40	$116.06
Crew B-23B	Hr.	Daily	Hr.	Daily	Bare Costs	Incl. O&P
1 Labor Foreman (outside)	$44.10	$352.80	$70.65	$565.20	$47.65	$75.63
1 Laborer	42.10	336.80	67.45	539.60		
1 Equip. Oper. (medium)	56.75	454.00	88.80	710.40		
1 Drill Rig, Truck-Mounted		771.10		848.21		
1 Pickup Truck, 3/4 Ton		110.85		121.94		
1 Centr. Water Pump, 6"		232.90		256.19	46.45	51.10
24 L.H., Daily Totals		$2258.45		$3041.53	$94.10	$126.73
Crew B-24	Hr.	Daily	Hr.	Daily	Bare Costs	Incl. O&P
1 Cement Finisher	$49.95	$399.60	$78.45	$627.60	$48.40	$77.02
1 Laborer	42.10	336.80	67.45	539.60		
1 Carpenter	53.15	425.20	85.15	681.20		
24 L.H., Daily Totals		$1161.60		$1848.40	$48.40	$77.02
Crew B-25	Hr.	Daily	Hr.	Daily	Bare Costs	Incl. O&P
1 Labor Foreman (outside)	$44.10	$352.80	$70.65	$565.20	$46.28	$73.56
7 Laborers	42.10	2357.60	67.45	3777.20		
3 Equip. Oper. (medium)	56.75	1362.00	88.80	2131.20		
1 Asphalt Paver, 130 H.P.		2117.00		2328.70		
1 Tandem Roller, 10 Ton		236.75		260.43		
1 Roller, Pneum. Whl., 12 Ton		345.75		380.32	30.68	33.74
88 L.H., Daily Totals		$6771.90		$9443.05	$76.95	$107.31
Crew B-25B	Hr.	Daily	Hr.	Daily	Bare Costs	Incl. O&P
1 Labor Foreman (outside)	$44.10	$352.80	$70.65	$565.20	$47.15	$74.83
7 Laborers	42.10	2357.60	67.45	3777.20		
4 Equip. Oper. (medium)	56.75	1816.00	88.80	2841.60		
1 Asphalt Paver, 130 H.P.		2117.00		2328.70		
2 Tandem Rollers, 10 Ton		473.50		520.85		
1 Roller, Pneum. Whl., 12 Ton		345.75		380.32	30.59	33.64
96 L.H., Daily Totals		$7462.65		$10413.88	$77.74	$108.48
Crew B-25C	Hr.	Daily	Hr.	Daily	Bare Costs	Incl. O&P
1 Labor Foreman (outside)	$44.10	$352.80	$70.65	$565.20	$47.32	$75.10
3 Laborers	42.10	1010.40	67.45	1618.80		
2 Equip. Oper. (medium)	56.75	908.00	88.80	1420.80		
1 Asphalt Paver, 130 H.P.		2117.00		2328.70		
1 Tandem Roller, 10 Ton		236.75		260.43	49.04	53.94
48 L.H., Daily Totals		$4624.95		$6193.93	$96.35	$129.04
Crew B-25D	Hr.	Daily	Hr.	Daily	Bare Costs	Incl. O&P
1 Labor Foreman (outside)	$44.10	$352.80	$70.65	$565.20	$47.54	$75.44
3 Laborers	42.10	1010.40	67.45	1618.80		
2.125 Equip. Oper. (medium)	56.75	964.75	88.80	1509.60		
.125 Truck Driver (heavy)	48.95	48.95	78.25	78.25		
.125 Truck Tractor, 6x4, 380 H.P.		61.66		67.82		
.125 Dist. Tanker, 3000 Gallon		41.27		45.40		
1 Asphalt Paver, 130 H.P.		2117.00		2328.70		
1 Tandem Roller, 10 Ton		236.75		260.43	49.13	54.05
50 L.H., Daily Totals		$4833.58		$6474.19	$96.67	$129.48

Crew No.	Bare Costs		Incl. Subs O&P		Cost Per Labor-Hour	

Crew B-25E	Hr.	Daily	Hr.	Daily	Bare Costs	Incl. O&P
1 Labor Foreman (outside)	$44.10	$352.80	$70.65	$565.20	$47.74	$75.75
3 Laborers	42.10	1010.40	67.45	1618.80		
2.250 Equip. Oper. (medium)	56.75	1021.50	88.80	1598.40		
.25 Truck Driver (heavy)	48.95	97.90	78.25	156.50		
.25 Truck Tractor, 6x4, 380 H.P.		123.31		135.64		
.25 Dist. Tanker, 3000 Gallon		82.54		90.79		
1 Asphalt Paver, 130 H.P.		2117.00		2328.70		
1 Tandem Roller, 10 Ton		236.75		260.43	49.22	54.15
52 L.H., Daily Totals		$5042.20		$6754.46	$96.97	$129.89

Crew B-26	Hr.	Daily	Hr.	Daily	Bare Costs	Incl. O&P
1 Labor Foreman (outside)	$44.10	$352.80	$70.65	$565.20	$46.96	$74.57
6 Laborers	42.10	2020.80	67.45	3237.60		
2 Equip. Oper. (medium)	56.75	908.00	88.80	1420.80		
1 Rodman (reinf.)	56.40	451.20	88.85	710.80		
1 Cement Finisher	49.95	399.60	78.45	627.60		
1 Grader, 30,000 Lbs.		1062.00		1168.20		
1 Paving Mach. & Equip.		2474.00		2721.40	40.18	44.20
88 L.H., Daily Totals		$7668.40		$10451.60	$87.14	$118.77

Crew B-26A	Hr.	Daily	Hr.	Daily	Bare Costs	Incl. O&P
1 Labor Foreman (outside)	$44.10	$352.80	$70.65	$565.20	$46.96	$74.57
6 Laborers	42.10	2020.80	67.45	3237.60		
2 Equip. Oper. (medium)	56.75	908.00	88.80	1420.80		
1 Rodman (reinf.)	56.40	451.20	88.85	710.80		
1 Cement Finisher	49.95	399.60	78.45	627.60		
1 Grader, 30,000 Lbs.		1062.00		1168.20		
1 Paving Mach. & Equip.		2474.00		2721.40		
1 Concrete Saw		111.50		122.65	41.45	45.59
88 L.H., Daily Totals		$7779.90		$10574.25	$88.41	$120.16

Crew B-26B	Hr.	Daily	Hr.	Daily	Bare Costs	Incl. O&P
1 Labor Foreman (outside)	$44.10	$352.80	$70.65	$565.20	$47.77	$75.75
6 Laborers	42.10	2020.80	67.45	3237.60		
3 Equip. Oper. (medium)	56.75	1362.00	88.80	2131.20		
1 Rodman (reinf.)	56.40	451.20	88.85	710.80		
1 Cement Finisher	49.95	399.60	78.45	627.60		
1 Grader, 30,000 Lbs.		1062.00		1168.20		
1 Paving Mach. & Equip.		2474.00		2721.40		
1 Concrete Pump, 110' Boom		470.75		517.83	41.74	45.91
96 L.H., Daily Totals		$8593.15		$11679.83	$89.51	$121.66

Crew B-26C	Hr.	Daily	Hr.	Daily	Bare Costs	Incl. O&P
1 Labor Foreman (outside)	$44.10	$352.80	$70.65	$565.20	$45.98	$73.14
6 Laborers	42.10	2020.80	67.45	3237.60		
1 Equip. Oper. (medium)	56.75	454.00	88.80	710.40		
1 Rodman (reinf.)	56.40	451.20	88.85	710.80		
1 Cement Finisher	49.95	399.60	78.45	627.60		
1 Paving Mach. & Equip.		2474.00		2721.40		
1 Concrete Saw		111.50		122.65	32.32	35.55
80 L.H., Daily Totals		$6263.90		$8695.65	$78.30	$108.70

Crew B-27	Hr.	Daily	Hr.	Daily	Bare Costs	Incl. O&P
1 Labor Foreman (outside)	$44.10	$352.80	$70.65	$565.20	$42.60	$68.25
3 Laborers	42.10	1010.40	67.45	1618.80		
1 Berm Machine		271.95		299.14	8.50	9.35
32 L.H., Daily Totals		$1635.15		$2483.15	$51.10	$77.60

Crew B-28	Hr.	Daily	Hr.	Daily	Bare Costs	Incl. O&P
2 Carpenters	$53.15	$850.40	$85.15	$1362.40	$49.47	$79.25
1 Laborer	42.10	336.80	67.45	539.60		
24 L.H., Daily Totals		$1187.20		$1902.00	$49.47	$79.25

Crew B-29	Hr.	Daily	Hr.	Daily	Bare Costs	Incl. O&P
1 Labor Foreman (outside)	$44.10	$352.80	$70.65	$565.20	$46.04	$73.17
4 Laborers	42.10	1347.20	67.45	2158.40		
1 Equip. Oper. (crane)	59.20	473.60	92.65	741.20		
1 Equip. Oper. (oiler)	50.55	404.40	79.10	632.80		
1 Gradall, 5/8 C.Y.		846.50		931.15	15.12	16.63
56 L.H., Daily Totals		$3424.50		$5028.75	$61.15	$89.80

Crew B-30	Hr.	Daily	Hr.	Daily	Bare Costs	Incl. O&P
1 Equip. Oper. (medium)	$56.75	$454.00	$88.80	$710.40	$51.55	$81.77
2 Truck Drivers (heavy)	48.95	783.20	78.25	1252.00		
1 Hyd. Excavator, 1.5 C.Y.		687.55		756.30		
2 Dump Trucks, 12 C.Y., 400 H.P.		1145.80		1260.38	76.39	84.03
24 L.H., Daily Totals		$3070.55		$3979.09	$127.94	$165.80

Crew B-31	Hr.	Daily	Hr.	Daily	Bare Costs	Incl. O&P
1 Labor Foreman (outside)	$44.10	$352.80	$70.65	$565.20	$44.71	$71.63
3 Laborers	42.10	1010.40	67.45	1618.80		
1 Carpenter	53.15	425.20	85.15	681.20		
1 Air Compressor, 250 cfm		201.50		221.65		
1 Sheeting Driver		7.35		8.09		
2 -50' Air Hoses, 1.5"		48.50		53.35	6.43	7.08
40 L.H., Daily Totals		$2045.75		$3148.28	$51.14	$78.71

Crew B-32	Hr.	Daily	Hr.	Daily	Bare Costs	Incl. O&P
1 Laborer	$42.10	$336.80	$67.45	$539.60	$53.09	$83.46
3 Equip. Oper. (medium)	56.75	1362.00	88.80	2131.20		
1 Grader, 30,000 Lbs.		1062.00		1168.20		
1 Tandem Roller, 10 Ton		236.75		260.43		
1 Dozer, 200 H.P.		1504.00		1654.40	87.59	96.34
32 L.H., Daily Totals		$4501.55		$5753.82	$140.67	$179.81

Crew B-32A	Hr.	Daily	Hr.	Daily	Bare Costs	Incl. O&P
1 Laborer	$42.10	$336.80	$67.45	$539.60	$51.87	$81.68
2 Equip. Oper. (medium)	56.75	908.00	88.80	1420.80		
1 Grader, 30,000 Lbs.		1062.00		1168.20		
1 Roller, Vibratory, 25 Ton		664.35		730.78	71.93	79.12
24 L.H., Daily Totals		$2971.15		$3859.39	$123.80	$160.81

Crew B-32B	Hr.	Daily	Hr.	Daily	Bare Costs	Incl. O&P
1 Laborer	$42.10	$336.80	$67.45	$539.60	$51.87	$81.68
2 Equip. Oper. (medium)	56.75	908.00	88.80	1420.80		
1 Dozer, 200 H.P.		1504.00		1654.40		
1 Roller, Vibratory, 25 Ton		664.35		730.78	90.35	99.38
24 L.H., Daily Totals		$3413.15		$4345.59	$142.21	$181.07

Crew B-32C	Hr.	Daily	Hr.	Daily	Bare Costs	Incl. O&P
1 Labor Foreman (outside)	$44.10	$352.80	$70.65	$565.20	$49.76	$78.66
2 Laborers	42.10	673.60	67.45	1079.20		
3 Equip. Oper. (medium)	56.75	1362.00	88.80	2131.20		
1 Grader, 30,000 Lbs.		1062.00		1168.20		
1 Tandem Roller, 10 Ton		236.75		260.43		
1 Dozer, 200 H.P.		1504.00		1654.40	58.39	64.23
48 L.H., Daily Totals		$5191.15		$6858.63	$108.15	$142.89

For customer support on your Commercial Renovation Costs with RSMeans data, call 800.448.8182.

Crew No.	Bare Costs		Incl. Subs O&P		Cost Per Labor-Hour	
	Hr.	Daily	Hr.	Daily	Bare Costs	Incl. O&P
Crew B-33A						
1 Equip. Oper. (medium)	$56.75	$454.00	$88.80	$710.40	$52.56	$82.70
.5 Laborer	42.10	168.40	67.45	269.80		
.25 Equip. Oper. (medium)	56.75	113.50	88.80	177.60		
1 Scraper, Towed, 7 C.Y.		127.80		140.58		
1.25 Dozers, 300 H.P.		2205.00		2425.50	166.63	183.29
14 L.H., Daily Totals		$3068.70		$3723.88	$219.19	$265.99
Crew B-33B						
1 Equip. Oper. (medium)	$56.75	$454.00	$88.80	$710.40	$52.56	$82.70
.5 Laborer	42.10	168.40	67.45	269.80		
.25 Equip. Oper. (medium)	56.75	113.50	88.80	177.60		
1 Scraper, Towed, 10 C.Y.		159.70		175.67		
1.25 Dozers, 300 H.P.		2205.00		2425.50	168.91	185.80
14 L.H., Daily Totals		$3100.60		$3758.97	$221.47	$268.50
Crew B-33C						
1 Equip. Oper. (medium)	$56.75	$454.00	$88.80	$710.40	$52.56	$82.70
.5 Laborer	42.10	168.40	67.45	269.80		
.25 Equip. Oper. (medium)	56.75	113.50	88.80	177.60		
1 Scraper, Towed, 15 C.Y.		176.75		194.43		
1.25 Dozers, 300 H.P.		2205.00		2425.50	170.13	187.14
14 L.H., Daily Totals		$3117.65		$3777.72	$222.69	$269.84
Crew B-33D						
1 Equip. Oper. (medium)	$56.75	$454.00	$88.80	$710.40	$52.56	$82.70
.5 Laborer	42.10	168.40	67.45	269.80		
.25 Equip. Oper. (medium)	56.75	113.50	88.80	177.60		
1 S.P. Scraper, 14 C.Y.		2395.00		2634.50		
.25 Dozer, 300 H.P.		441.00		485.10	202.57	222.83
14 L.H., Daily Totals		$3571.90		$4277.40	$255.14	$305.53
Crew B-33E						
1 Equip. Oper. (medium)	$56.75	$454.00	$88.80	$710.40	$52.56	$82.70
.5 Laborer	42.10	168.40	67.45	269.80		
.25 Equip. Oper. (medium)	56.75	113.50	88.80	177.60		
1 S.P. Scraper, 21 C.Y.		2628.00		2890.80		
.25 Dozer, 300 H.P.		441.00		485.10	219.21	241.14
14 L.H., Daily Totals		$3804.90		$4533.70	$271.78	$323.84
Crew B-33F						
1 Equip. Oper. (medium)	$56.75	$454.00	$88.80	$710.40	$52.56	$82.70
.5 Laborer	42.10	168.40	67.45	269.80		
.25 Equip. Oper. (medium)	56.75	113.50	88.80	177.60		
1 Elev. Scraper, 11 C.Y.		1133.00		1246.30		
.25 Dozer, 300 H.P.		441.00		485.10	112.43	123.67
14 L.H., Daily Totals		$2309.90		$2889.20	$164.99	$206.37
Crew B-33G						
1 Equip. Oper. (medium)	$56.75	$454.00	$88.80	$710.40	$52.56	$82.70
.5 Laborer	42.10	168.40	67.45	269.80		
.25 Equip. Oper. (medium)	56.75	113.50	88.80	177.60		
1 Elev. Scraper, 22 C.Y.		1884.00		2072.40		
.25 Dozer, 300 H.P.		441.00		485.10	166.07	182.68
14 L.H., Daily Totals		$3060.90		$3715.30	$218.64	$265.38

Crew No.	Bare Costs		Incl. Subs O&P		Cost Per Labor-Hour	
	Hr.	Daily	Hr.	Daily	Bare Costs	Incl. O&P
Crew B-33K						
1 Equipment Operator (med.)	$56.75	$454.00	$88.80	$710.40	$52.56	$82.70
.25 Equipment Operator (med.)	56.75	113.50	88.80	177.60		
.5 Laborer	42.10	168.40	67.45	269.80		
1 S.P. Scraper, 31 C.Y.		3667.00		4033.70		
.25 Dozer, 410 H.P.		694.25		763.67	311.52	342.67
14 L.H., Daily Totals		$5097.15		$5955.18	$364.08	$425.37
Crew B-34A						
1 Truck Driver (heavy)	$48.95	$391.60	$78.25	$626.00	$48.95	$78.25
1 Dump Truck, 8 C.Y., 220 H.P.		428.05		470.86	53.51	58.86
8 L.H., Daily Totals		$819.65		$1096.86	$102.46	$137.11
Crew B-34B						
1 Truck Driver (heavy)	$48.95	$391.60	$78.25	$626.00	$48.95	$78.25
1 Dump Truck, 12 C.Y., 400 H.P.		572.90		630.19	71.61	78.77
8 L.H., Daily Totals		$964.50		$1256.19	$120.56	$157.02
Crew B-34C						
1 Truck Driver (heavy)	$48.95	$391.60	$78.25	$626.00	$48.95	$78.25
1 Truck Tractor, 6x4, 380 H.P.		493.25		542.58		
1 Dump Trailer, 16.5 C.Y.		136.70		150.37	78.74	86.62
8 L.H., Daily Totals		$1021.55		$1318.94	$127.69	$164.87
Crew B-34D						
1 Truck Driver (heavy)	$48.95	$391.60	$78.25	$626.00	$48.95	$78.25
1 Truck Tractor, 6x4, 380 H.P.		493.25		542.58		
1 Dump Trailer, 20 C.Y.		151.70		166.87	80.62	88.68
8 L.H., Daily Totals		$1036.55		$1335.44	$129.57	$166.93
Crew B-34E						
1 Truck Driver (heavy)	$48.95	$391.60	$78.25	$626.00	$48.95	$78.25
1 Dump Truck, Off Hwy., 25 Ton		1379.00		1516.90	172.38	189.61
8 L.H., Daily Totals		$1770.60		$2142.90	$221.32	$267.86
Crew B-34F						
1 Truck Driver (heavy)	$48.95	$391.60	$78.25	$626.00	$48.95	$78.25
1 Dump Truck, Off Hwy., 35 Ton		935.20		1028.72	116.90	128.59
8 L.H., Daily Totals		$1326.80		$1654.72	$165.85	$206.84
Crew B-34G						
1 Truck Driver (heavy)	$48.95	$391.60	$78.25	$626.00	$48.95	$78.25
1 Dump Truck, Off Hwy., 50 Ton		1771.00		1948.10	221.38	243.51
8 L.H., Daily Totals		$2162.60		$2574.10	$270.32	$321.76
Crew B-34H						
1 Truck Driver (heavy)	$48.95	$391.60	$78.25	$626.00	$48.95	$78.25
1 Dump Truck, Off Hwy., 65 Ton		1915.00		2106.50	239.38	263.31
8 L.H., Daily Totals		$2306.60		$2732.50	$288.32	$341.56
Crew B-34I						
1 Truck Driver (heavy)	$48.95	$391.60	$78.25	$626.00	$48.95	$78.25
1 Dump Truck, 18 C.Y., 450 H.P.		713.80		785.18	89.22	98.15
8 L.H., Daily Totals		$1105.40		$1411.18	$138.18	$176.40
Crew B-34J						
1 Truck Driver (heavy)	$48.95	$391.60	$78.25	$626.00	$48.95	$78.25
1 Dump Truck, Off Hwy., 100 Ton		2736.00		3009.60	342.00	376.20
8 L.H., Daily Totals		$3127.60		$3635.60	$390.95	$454.45

Crew B-34K	Hr.	Daily	Hr.	Daily	Bare Costs	Incl. O&P
1 Truck Driver (heavy)	$48.95	$391.60	$78.25	$626.00	$48.95	$78.25
1 Truck Tractor, 6x4, 450 H.P.		601.75		661.92		
1 Lowbed Trailer, 75 Ton		255.05		280.56	107.10	117.81
8 L.H., Daily Totals		$1248.40		$1568.48	$156.05	$196.06

Crew B-34L	Hr.	Daily	Hr.	Daily	Bare Costs	Incl. O&P
1 Equip. Oper. (light)	$53.00	$424.00	$82.95	$663.60	$53.00	$82.95
1 Flatbed Truck, Gas, 1.5 Ton		196.15		215.76	24.52	26.97
8 L.H., Daily Totals		$620.15		$879.37	$77.52	$109.92

Crew B-34M	Hr.	Daily	Hr.	Daily	Bare Costs	Incl. O&P
1 Equip. Oper. (light)	$53.00	$424.00	$82.95	$663.60	$53.00	$82.95
1 Flatbed Truck, Gas, 3 Ton		820.40		902.44	102.55	112.81
8 L.H., Daily Totals		$1244.40		$1566.04	$155.55	$195.76

Crew B-34N	Hr.	Daily	Hr.	Daily	Bare Costs	Incl. O&P
1 Truck Driver (heavy)	$48.95	$391.60	$78.25	$626.00	$52.85	$83.53
1 Equip. Oper. (medium)	56.75	454.00	88.80	710.40		
1 Truck Tractor, 6x4, 380 H.P.		493.25		542.58		
1 Flatbed Trailer, 40 Ton		186.20		204.82	42.47	46.71
16 L.H., Daily Totals		$1525.05		$2083.80	$95.32	$130.24

Crew B-34P	Hr.	Daily	Hr.	Daily	Bare Costs	Incl. O&P
1 Pipe Fitter	$65.55	$524.40	$101.35	$810.80	$56.52	$88.57
1 Truck Driver (light)	47.25	378.00	75.55	604.40		
1 Equip. Oper. (medium)	56.75	454.00	88.80	710.40		
1 Flatbed Truck, Gas, 3 Ton		820.40		902.44		
1 Backhoe Loader, 48 H.P.		213.75		235.13	43.09	47.40
24 L.H., Daily Totals		$2390.55		$3263.17	$99.61	$135.97

Crew B-34Q	Hr.	Daily	Hr.	Daily	Bare Costs	Incl. O&P
1 Pipe Fitter	$65.55	$524.40	$101.35	$810.80	$57.33	$89.85
1 Truck Driver (light)	47.25	378.00	75.55	604.40		
1 Equip. Oper. (crane)	59.20	473.60	92.65	741.20		
1 Flatbed Trailer, 25 Ton		135.55		149.10		
1 Dump Truck, 8 C.Y., 220 H.P.		428.05		470.86		
1 Hyd. Crane, 25 Ton		580.85		638.93	47.69	52.45
24 L.H., Daily Totals		$2520.45		$3415.30	$105.02	$142.30

Crew B-34R	Hr.	Daily	Hr.	Daily	Bare Costs	Incl. O&P
1 Pipe Fitter	$65.55	$524.40	$101.35	$810.80	$57.33	$89.85
1 Truck Driver (light)	47.25	378.00	75.55	604.40		
1 Equip. Oper. (crane)	59.20	473.60	92.65	741.20		
1 Flatbed Trailer, 25 Ton		135.55		149.10		
1 Dump Truck, 8 C.Y., 220 H.P.		428.05		470.86		
1 Hyd. Crane, 25 Ton		580.85		638.93		
1 Hyd. Excavator, 1 C.Y.		795.30		874.83	80.82	88.91
24 L.H., Daily Totals		$3315.75		$4290.13	$138.16	$178.76

Crew B-34S	Hr.	Daily	Hr.	Daily	Bare Costs	Incl. O&P
2 Pipe Fitters	$65.55	$1048.80	$101.35	$1621.60	$59.81	$93.40
1 Truck Driver (heavy)	48.95	391.60	78.25	626.00		
1 Equip. Oper. (crane)	59.20	473.60	92.65	741.20		
1 Flatbed Trailer, 40 Ton		186.20		204.82		
1 Truck Tractor, 6x4, 380 H.P.		493.25		542.58		
1 Hyd. Crane, 80 Ton		1486.00		1634.60		
1 Hyd. Excavator, 2 C.Y.		951.30		1046.43	97.40	107.14
32 L.H., Daily Totals		$5030.75		$6417.23	$157.21	$200.54

Crew B-34T	Hr.	Daily	Hr.	Daily	Bare Costs	Incl. O&P
2 Pipe Fitters	$65.55	$1048.80	$101.35	$1621.60	$59.81	$93.40
1 Truck Driver (heavy)	48.95	391.60	78.25	626.00		
1 Equip. Oper. (crane)	59.20	473.60	92.65	741.20		
1 Flatbed Trailer, 40 Ton		186.20		204.82		
1 Truck Tractor, 6x4, 380 H.P.		493.25		542.58		
1 Hyd. Crane, 80 Ton		1486.00		1634.60	67.67	74.44
32 L.H., Daily Totals		$4079.45		$5370.80	$127.48	$167.84

Crew B-34U	Hr.	Daily	Hr.	Daily	Bare Costs	Incl. O&P
1 Truck Driver (heavy)	$48.95	$391.60	$78.25	$626.00	$50.98	$80.60
1 Equip. Oper. (light)	53.00	424.00	82.95	663.60		
1 Truck Tractor, 220 H.P.		307.10		337.81		
1 Flatbed Trailer, 25 Ton		135.55		149.10	27.67	30.43
16 L.H., Daily Totals		$1258.25		$1776.52	$78.64	$111.03

Crew B-34V	Hr.	Daily	Hr.	Daily	Bare Costs	Incl. O&P
1 Truck Driver (heavy)	$48.95	$391.60	$78.25	$626.00	$53.72	$84.62
1 Equip. Oper. (crane)	59.20	473.60	92.65	741.20		
1 Equip. Oper. (light)	53.00	424.00	82.95	663.60		
1 Truck Tractor, 6x4, 450 H.P.		601.75		661.92		
1 Equipment Trailer, 50 Ton		205.05		225.56		
1 Pickup Truck, 4x4, 3/4 Ton		175.85		193.44	40.94	45.04
24 L.H., Daily Totals		$2271.85		$3111.72	$94.66	$129.65

Crew B-34W	Hr.	Daily	Hr.	Daily	Bare Costs	Incl. O&P
5 Truck Drivers (heavy)	$48.95	$1958.00	$78.25	$3130.00	$51.60	$81.84
2 Equip. Opers. (crane)	59.20	947.20	92.65	1482.40		
1 Equip. Oper. (mechanic)	59.15	473.20	92.55	740.40		
1 Laborer	42.10	336.80	67.45	539.60		
4 Truck Tractors, 6x4, 380 H.P.		1973.00		2170.30		
2 Equipment Trailers, 50 Ton		410.10		451.11		
2 Flatbed Trailers, 40 Ton		372.40		409.64		
1 Pickup Truck, 4x4, 3/4 Ton		175.85		193.44		
1 S.P. Crane, 4x4, 20 Ton		582.20		640.42	48.80	53.68
72 L.H., Daily Totals		$7228.75		$9757.31	$100.40	$135.52

Crew B-35	Hr.	Daily	Hr.	Daily	Bare Costs	Incl. O&P
1 Labor Foreman (outside)	$44.10	$352.80	$70.65	$565.20	$52.54	$82.82
1 Skilled Worker	54.85	438.80	87.40	699.20		
1 Welder (plumber)	64.45	515.60	99.65	797.20		
1 Laborer	42.10	336.80	67.45	539.60		
1 Equip. Oper. (crane)	59.20	473.60	92.65	741.20		
1 Equip. Oper. (oiler)	50.55	404.40	79.10	632.80		
1 Welder, Electric, 300 amp		106.40		117.04		
1 Hyd. Excavator, .75 C.Y.		694.25		763.67	16.68	18.35
48 L.H., Daily Totals		$3322.65		$4855.92	$69.22	$101.16

Crew B-35A	Hr.	Daily	Hr.	Daily	Bare Costs	Incl. O&P
1 Labor Foreman (outside)	$44.10	$352.80	$70.65	$565.20	$51.05	$80.62
2 Laborers	42.10	673.60	67.45	1079.20		
1 Skilled Worker	54.85	438.80	87.40	699.20		
1 Welder (plumber)	64.45	515.60	99.65	797.20		
1 Equip. Oper. (crane)	59.20	473.60	92.65	741.20		
1 Equip. Oper. (oiler)	50.55	404.40	79.10	632.80		
1 Welder, Gas Engine, 300 amp		147.00		161.70		
1 Crawler Crane, 75 Ton		2002.00		2202.20	38.38	42.21
56 L.H., Daily Totals		$5007.80		$6878.70	$89.42	$122.83

Crew No.	Bare Costs Hr.	Daily	Incl. Subs O&P Hr.	Daily	Bare Costs	Incl. O&P
Crew B-36	**Hr.**	**Daily**	**Hr.**	**Daily**	**Bare Costs**	**Incl. O&P**
1 Labor Foreman (outside)	$44.10	$352.80	$70.65	$565.20	$48.36	$76.63
2 Laborers	42.10	673.60	67.45	1079.20		
2 Equip. Oper. (medium)	56.75	908.00	88.80	1420.80		
1 Dozer, 200 H.P.		1504.00		1654.40		
1 Aggregate Spreader		65.75		72.33		
1 Tandem Roller, 10 Ton		236.75		260.43	45.16	49.68
40 L.H., Daily Totals		$3740.90		$5052.35	$93.52	$126.31
Crew B-36A	**Hr.**	**Daily**	**Hr.**	**Daily**	**Bare Costs**	**Incl. O&P**
1 Labor Foreman (outside)	$44.10	$352.80	$70.65	$565.20	$50.76	$80.11
2 Laborers	42.10	673.60	67.45	1079.20		
4 Equip. Oper. (medium)	56.75	1816.00	88.80	2841.60		
1 Dozer, 200 H.P.		1504.00		1654.40		
1 Aggregate Spreader		65.75		72.33		
1 Tandem Roller, 10 Ton		236.75		260.43		
1 Roller, Pneum. Whl., 12 Ton		345.75		380.32	38.43	42.28
56 L.H., Daily Totals		$4994.65		$6853.48	$89.19	$122.38
Crew B-36B	**Hr.**	**Daily**	**Hr.**	**Daily**	**Bare Costs**	**Incl. O&P**
1 Labor Foreman (outside)	$44.10	$352.80	$70.65	$565.20	$50.53	$79.88
2 Laborers	42.10	673.60	67.45	1079.20		
4 Equip. Oper. (medium)	56.75	1816.00	88.80	2841.60		
1 Truck Driver (heavy)	48.95	391.60	78.25	626.00		
1 Grader, 30,000 Lbs.		1062.00		1168.20		
1 F.E. Loader, Crl, 1.5 C.Y.		659.75		725.73		
1 Dozer, 300 H.P.		1764.00		1940.40		
1 Roller, Vibratory, 25 Ton		664.35		730.78		
1 Truck Tractor, 6x4, 450 H.P.		601.75		661.92		
1 Water Tank Trailer, 5000 Gal.		152.25		167.47	76.63	84.29
64 L.H., Daily Totals		$8138.10		$10506.51	$127.16	$164.16
Crew B-36C	**Hr.**	**Daily**	**Hr.**	**Daily**	**Bare Costs**	**Incl. O&P**
1 Labor Foreman (outside)	$44.10	$352.80	$70.65	$565.20	$52.66	$83.06
3 Equip. Oper. (medium)	56.75	1362.00	88.80	2131.20		
1 Truck Driver (heavy)	48.95	391.60	78.25	626.00		
1 Grader, 30,000 Lbs.		1062.00		1168.20		
1 Dozer, 300 H.P.		1764.00		1940.40		
1 Roller, Vibratory, 25 Ton		664.35		730.78		
1 Truck Tractor, 6x4, 450 H.P.		601.75		661.92		
1 Water Tank Trailer, 5000 Gal.		152.25		167.47	106.11	116.72
40 L.H., Daily Totals		$6350.75		$7991.19	$158.77	$199.78
Crew B-37	**Hr.**	**Daily**	**Hr.**	**Daily**	**Bare Costs**	**Incl. O&P**
1 Labor Foreman (outside)	$44.10	$352.80	$70.65	$565.20	$44.25	$70.57
4 Laborers	42.10	1347.20	67.45	2158.40		
1 Equip. Oper. (light)	53.00	424.00	82.95	663.60		
1 Tandem Roller, 5 Ton		255.65		281.21	5.33	5.86
48 L.H., Daily Totals		$2379.65		$3668.42	$49.58	$76.43
Crew B-37A	**Hr.**	**Daily**	**Hr.**	**Daily**	**Bare Costs**	**Incl. O&P**
2 Laborers	$42.10	$673.60	$67.45	$1079.20	$43.82	$70.15
1 Truck Driver (light)	47.25	378.00	75.55	604.40		
1 Flatbed Truck, Gas, 1.5 Ton		196.15		215.76		
1 Tar Kettle, T.M.		156.20		171.82	14.68	16.15
24 L.H., Daily Totals		$1403.95		$2071.18	$58.50	$86.30

Crew No.	Bare Costs Hr.	Daily	Incl. Subs O&P Hr.	Daily	Bare Costs	Incl. O&P
Crew B-37B	**Hr.**	**Daily**	**Hr.**	**Daily**	**Bare Costs**	**Incl. O&P**
3 Laborers	$42.10	$1010.40	$67.45	$1618.80	$43.39	$69.47
1 Truck Driver (light)	47.25	378.00	75.55	604.40		
1 Flatbed Truck, Gas, 1.5 Ton		196.15		215.76		
1 Tar Kettle, T.M.		156.20		171.82	11.01	12.11
32 L.H., Daily Totals		$1740.75		$2610.78	$54.40	$81.59
Crew B-37C	**Hr.**	**Daily**	**Hr.**	**Daily**	**Bare Costs**	**Incl. O&P**
2 Laborers	$42.10	$673.60	$67.45	$1079.20	$44.67	$71.50
2 Truck Drivers (light)	47.25	756.00	75.55	1208.80		
2 Flatbed Trucks, Gas, 1.5 Ton		392.30		431.53		
1 Tar Kettle, T.M.		156.20		171.82	17.14	18.85
32 L.H., Daily Totals		$1978.10		$2891.35	$61.82	$90.35
Crew B-37D	**Hr.**	**Daily**	**Hr.**	**Daily**	**Bare Costs**	**Incl. O&P**
1 Laborer	$42.10	$336.80	$67.45	$539.60	$44.67	$71.50
1 Truck Driver (light)	47.25	378.00	75.55	604.40		
1 Pickup Truck, 3/4 Ton		110.85		121.94	6.93	7.62
16 L.H., Daily Totals		$825.65		$1265.93	$51.60	$79.12
Crew B-37E	**Hr.**	**Daily**	**Hr.**	**Daily**	**Bare Costs**	**Incl. O&P**
3 Laborers	$42.10	$1010.40	$67.45	$1618.80	$47.22	$75.03
1 Equip. Oper. (light)	53.00	424.00	82.95	663.60		
1 Equip. Oper. (medium)	56.75	454.00	88.80	710.40		
2 Truck Drivers (light)	47.25	756.00	75.55	1208.80		
4 Barrels w/ Flasher		16.40		18.04		
1 Concrete Saw		111.50		122.65		
1 Rotary Hammer Drill		51.70		56.87		
1 Hammer Drill Bit		25.25		27.77		
1 Loader, Skid Steer, 30 H.P.		177.55		195.31		
1 Conc. Hammer Attach.		118.40		130.24		
1 Vibrating Plate, Gas, 18"		31.55		34.70		
2 Flatbed Trucks, Gas, 1.5 Ton		392.30		431.53	16.51	18.16
56 L.H., Daily Totals		$3569.05		$5218.72	$63.73	$93.19
Crew B-37F	**Hr.**	**Daily**	**Hr.**	**Daily**	**Bare Costs**	**Incl. O&P**
3 Laborers	$42.10	$1010.40	$67.45	$1618.80	$43.39	$69.47
1 Truck Driver (light)	47.25	378.00	75.55	604.40		
4 Barrels w/ Flasher		16.40		18.04		
1 Concrete Mixer, 10 C.F.		140.30		154.33		
1 Air Compressor, 60 cfm		152.95		168.25		
1 -50' Air Hose, 3/4"		7.75		8.53		
1 Spade (Chipper)		8.45		9.29		
1 Flatbed Truck, Gas, 1.5 Ton		196.15		215.76	16.31	17.94
32 L.H., Daily Totals		$1910.40		$2797.40	$59.70	$87.42
Crew B-37G	**Hr.**	**Daily**	**Hr.**	**Daily**	**Bare Costs**	**Incl. O&P**
1 Labor Foreman (outside)	$44.10	$352.80	$70.65	$565.20	$44.25	$70.57
4 Laborers	42.10	1347.20	67.45	2158.40		
1 Equip. Oper. (light)	53.00	424.00	82.95	663.60		
1 Berm Machine		271.95		299.14		
1 Tandem Roller, 5 Ton		255.65		281.21	10.99	12.09
48 L.H., Daily Totals		$2651.60		$3967.56	$55.24	$82.66
Crew B-37H	**Hr.**	**Daily**	**Hr.**	**Daily**	**Bare Costs**	**Incl. O&P**
1 Labor Foreman (outside)	$44.10	$352.80	$70.65	$565.20	$44.25	$70.57
4 Laborers	42.10	1347.20	67.45	2158.40		
1 Equip. Oper. (light)	53.00	424.00	82.95	663.60		
1 Tandem Roller, 5 Ton		255.65		281.21		
1 Flatbed Truck, Gas, 1.5 Ton		196.15		215.76		
1 Tar Kettle, T.M.		156.20		171.82	12.67	13.93
48 L.H., Daily Totals		$2732.00		$4056.00	$56.92	$84.50

Crew No.	Bare Costs		Incl. Subs O&P		Cost Per Labor-Hour	

Crew B-37I

	Hr.	Daily	Hr.	Daily	Bare Costs	Incl. O&P
3 Laborers	$42.10	$1010.40	$67.45	$1618.80	$47.22	$75.03
1 Equip. Oper. (light)	53.00	424.00	82.95	663.60		
1 Equip. Oper. (medium)	56.75	454.00	88.80	710.40		
2 Truck Drivers (light)	47.25	756.00	75.55	1208.80		
4 Barrels w/ Flasher		16.40		18.04		
1 Concrete Saw		111.50		122.65		
1 Rotary Hammer Drill		51.70		56.87		
1 Hammer Drill Bit		25.25		27.77		
1 Air Compressor, 60 cfm		152.95		168.25		
1 -50' Air Hose, 3/4"		7.75		8.53		
1 Spade (Chipper)		8.45		9.29		
1 Loader, Skid Steer, 30 H.P.		177.55		195.31		
1 Conc. Hammer Attach.		118.40		130.24		
1 Concrete Mixer, 10 C.F.		140.30		154.33		
1 Vibrating Plate, Gas, 18"		31.55		34.70		
2 Flatbed Trucks, Gas, 1.5 Ton		392.30		431.53	22.04	24.24
56 L.H., Daily Totals		$3878.50		$5559.11	$69.26	$99.27

Crew B-37J

	Hr.	Daily	Hr.	Daily	Bare Costs	Incl. O&P
1 Labor Foreman (outside)	$44.10	$352.80	$70.65	$565.20	$44.25	$70.57
4 Laborers	42.10	1347.20	67.45	2158.40		
1 Equip. Oper. (light)	53.00	424.00	82.95	663.60		
1 Air Compressor, 60 cfm		152.95		168.25		
1 -50' Air Hose, 3/4"		7.75		8.53		
2 Concrete Mixers, 10 C.F.		280.60		308.66		
2 Flatbed Trucks, Gas, 1.5 Ton		392.30		431.53		
1 Shot Blaster, 20"		206.20		226.82	21.66	23.83
48 L.H., Daily Totals		$3163.80		$4530.98	$65.91	$94.40

Crew B-37K

	Hr.	Daily	Hr.	Daily	Bare Costs	Incl. O&P
1 Labor Foreman (outside)	$44.10	$352.80	$70.65	$565.20	$44.25	$70.57
4 Laborers	42.10	1347.20	67.45	2158.40		
1 Equip. Oper. (light)	53.00	424.00	82.95	663.60		
1 Air Compressor, 60 cfm		152.95		168.25		
1 -50' Air Hose, 3/4"		7.75		8.53		
2 Flatbed Trucks, Gas, 1.5 Ton		392.30		431.53		
1 Shot Blaster, 20"		206.20		226.82	15.82	17.40
48 L.H., Daily Totals		$2883.20		$4222.32	$60.07	$87.97

Crew B-38

	Hr.	Daily	Hr.	Daily	Bare Costs	Incl. O&P
1 Labor Foreman (outside)	$44.10	$352.80	$70.65	$565.20	$47.61	$75.46
2 Laborers	42.10	673.60	67.45	1079.20		
1 Equip. Oper. (light)	53.00	424.00	82.95	663.60		
1 Equip. Oper. (medium)	56.75	454.00	88.80	710.40		
1 Backhoe Loader, 48 H.P.		213.75		235.13		
1 Hyd. Hammer (1200 lb.)		175.15		192.66		
1 F.E. Loader, W.M., 4 C.Y.		789.35		868.28		
1 Pvmt. Rem. Bucket		63.05		69.36	31.03	34.14
40 L.H., Daily Totals		$3145.70		$4383.83	$78.64	$109.60

Crew B-39

	Hr.	Daily	Hr.	Daily	Bare Costs	Incl. O&P
1 Labor Foreman (outside)	$44.10	$352.80	$70.65	$565.20	$44.25	$70.57
4 Laborers	42.10	1347.20	67.45	2158.40		
1 Equip. Oper. (light)	53.00	424.00	82.95	663.60		
1 Air Compressor, 250 cfm		201.50		221.65		
2 Breakers, Pavement, 60 lb.		107.00		117.70		
2 -50' Air Hoses, 1.5"		48.50		53.35	7.44	8.18
48 L.H., Daily Totals		$2481.00		$3779.90	$51.69	$78.75

Crew B-40

	Hr.	Daily	Hr.	Daily	Bare Costs	Incl. O&P
1 Pile Driver Foreman (outside)	$56.20	$449.60	$90.15	$721.20	$55.24	$87.77
4 Pile Drivers	54.20	1734.40	86.90	2780.80		
2 Equip. Oper. (crane)	59.20	947.20	92.65	1482.40		
1 Equip. Oper. (oiler)	50.55	404.40	79.10	632.80		
1 Crawler Crane, 40 Ton		1206.00		1326.60		
1 Vibratory Hammer & Gen.		2271.00		2498.10	54.33	59.76
64 L.H., Daily Totals		$7012.60		$9441.90	$109.57	$147.53

Crew B-40B

	Hr.	Daily	Hr.	Daily	Bare Costs	Incl. O&P
1 Labor Foreman (outside)	$44.10	$352.80	$70.65	$565.20	$46.69	$74.13
3 Laborers	42.10	1010.40	67.45	1618.80		
1 Equip. Oper. (crane)	59.20	473.60	92.65	741.20		
1 Equip. Oper. (oiler)	50.55	404.40	79.10	632.80		
1 Lattice Boom Crane, 40 Ton		2028.00		2230.80	42.25	46.48
48 L.H., Daily Totals		$4269.20		$5788.80	$88.94	$120.60

Crew B-41

	Hr.	Daily	Hr.	Daily	Bare Costs	Incl. O&P
1 Labor Foreman (outside)	$44.10	$352.80	$70.65	$565.20	$43.63	$69.71
4 Laborers	42.10	1347.20	67.45	2158.40		
.25 Equip. Oper. (crane)	59.20	118.40	92.65	185.30		
.25 Equip. Oper. (oiler)	50.55	101.10	79.10	158.20		
.25 Crawler Crane, 40 Ton		301.50		331.65	6.85	7.54
44 L.H., Daily Totals		$2221.00		$3398.75	$50.48	$77.24

Crew B-42

	Hr.	Daily	Hr.	Daily	Bare Costs	Incl. O&P
1 Labor Foreman (outside)	$44.10	$352.80	$70.65	$565.20	$47.46	$75.79
4 Laborers	42.10	1347.20	67.45	2158.40		
1 Equip. Oper. (crane)	59.20	473.60	92.65	741.20		
1 Equip. Oper. (oiler)	50.55	404.40	79.10	632.80		
1 Welder	57.45	459.60	94.15	753.20		
1 Hyd. Crane, 25 Ton		580.85		638.93		
1 Welder, Gas Engine, 300 amp		147.00		161.70		
1 Horz. Boring Csg. Mch.		314.25		345.68	16.28	17.91
64 L.H., Daily Totals		$4079.70		$5997.11	$63.75	$93.70

Crew B-43

	Hr.	Daily	Hr.	Daily	Bare Costs	Incl. O&P
1 Labor Foreman (outside)	$44.10	$352.80	$70.65	$565.20	$46.69	$74.13
3 Laborers	42.10	1010.40	67.45	1618.80		
1 Equip. Oper. (crane)	59.20	473.60	92.65	741.20		
1 Equip. Oper. (oiler)	50.55	404.40	79.10	632.80		
1 Drill Rig, Truck-Mounted		771.10		848.21	16.06	17.67
48 L.H., Daily Totals		$3012.30		$4406.21	$62.76	$91.80

Crew B-44

	Hr.	Daily	Hr.	Daily	Bare Costs	Incl. O&P
1 Pile Driver Foreman (outside)	$56.20	$449.60	$90.15	$721.20	$54.19	$86.31
4 Pile Drivers	54.20	1734.40	86.90	2780.80		
2 Equip. Oper. (crane)	59.20	947.20	92.65	1482.40		
1 Laborer	42.10	336.80	67.45	539.60		
1 Crawler Crane, 40 Ton		1206.00		1326.60		
1 Lead, 60' High		209.25		230.18		
1 Hammer, Diesel, 15K ft.-lbs.		617.05		678.76	31.75	34.93
64 L.H., Daily Totals		$5500.30		$7759.53	$85.94	$121.24

Crew B-45

	Hr.	Daily	Hr.	Daily	Bare Costs	Incl. O&P
1 Equip. Oper. (medium)	$56.75	$454.00	$88.80	$710.40	$52.85	$83.53
1 Truck Driver (heavy)	48.95	391.60	78.25	626.00		
1 Dist. Tanker, 3000 Gallon		330.15		363.17		
1 Truck Tractor, 6x4, 380 H.P.		493.25		542.58	51.46	56.61
16 L.H., Daily Totals		$1669.00		$2242.14	$104.31	$140.13

Crew No.	Bare Costs Hr.	Daily	Incl. Subs O&P Hr.	Daily	Cost Per Labor-Hour Bare Costs	Incl. O&P
Crew B-46	Hr.	Daily	Hr.	Daily	Bare Costs	Incl. O&P
1 Pile Driver Foreman (outside)	$56.20	$449.60	$90.15	$721.20	$48.48	$77.72
2 Pile Drivers	54.20	867.20	86.90	1390.40		
3 Laborers	42.10	1010.40	67.45	1618.80		
1 Chain Saw, Gas, 36" Long		41.25		45.38	.86	.95
48 L.H., Daily Totals		$2368.45		$3775.78	$49.34	$78.66
Crew B-47	Hr.	Daily	Hr.	Daily	Bare Costs	Incl. O&P
1 Blast Foreman (outside)	$44.10	$352.80	$70.65	$565.20	$46.40	$73.68
1 Driller	42.10	336.80	67.45	539.60		
1 Equip. Oper. (light)	53.00	424.00	82.95	663.60		
1 Air Track Drill, 4"		1058.00		1163.80		
1 Air Compressor, 600 cfm		421.50		463.65		
2 -50' Air Hoses, 3"		76.70		84.37	64.84	71.33
24 L.H., Daily Totals		$2669.80		$3480.22	$111.24	$145.01
Crew B-47A	Hr.	Daily	Hr.	Daily	Bare Costs	Incl. O&P
1 Drilling Foreman (outside)	$44.10	$352.80	$70.65	$565.20	$51.28	$80.80
1 Equip. Oper. (heavy)	59.20	473.60	92.65	741.20		
1 Equip. Oper. (oiler)	50.55	404.40	79.10	632.80		
1 Air Track Drill, 5"		1127.00		1239.70	46.96	51.65
24 L.H., Daily Totals		$2357.80		$3178.90	$98.24	$132.45
Crew B-47C	Hr.	Daily	Hr.	Daily	Bare Costs	Incl. O&P
1 Laborer	$42.10	$336.80	$67.45	$539.60	$47.55	$75.20
1 Equip. Oper. (light)	53.00	424.00	82.95	663.60		
1 Air Compressor, 750 cfm		538.25		592.08		
2 -50' Air Hoses, 3"		76.70		84.37		
1 Air Track Drill, 4"		1058.00		1163.80	104.56	115.02
16 L.H., Daily Totals		$2433.75		$3043.45	$152.11	$190.22
Crew B-47E	Hr.	Daily	Hr.	Daily	Bare Costs	Incl. O&P
1 Labor Foreman (outside)	$44.10	$352.80	$70.65	$565.20	$42.60	$68.25
3 Laborers	42.10	1010.40	67.45	1618.80		
1 Flatbed Truck, Gas, 3 Ton		820.40		902.44	25.64	28.20
32 L.H., Daily Totals		$2183.60		$3086.44	$68.24	$96.45
Crew B-47G	Hr.	Daily	Hr.	Daily	Bare Costs	Incl. O&P
1 Labor Foreman (outside)	$44.10	$352.80	$70.65	$565.20	$45.33	$72.13
2 Laborers	42.10	673.60	67.45	1079.20		
1 Equip. Oper. (light)	53.00	424.00	82.95	663.60		
1 Air Track Drill, 4"		1058.00		1163.80		
1 Air Compressor, 600 cfm		421.50		463.65		
2 -50' Air Hoses, 3"		76.70		84.37		
1 Gunite Pump Rig		317.90		349.69	58.57	64.42
32 L.H., Daily Totals		$3324.50		$4369.51	$103.89	$136.55
Crew B-47H	Hr.	Daily	Hr.	Daily	Bare Costs	Incl. O&P
1 Skilled Worker Foreman (out)	$56.85	$454.80	$90.55	$724.40	$55.35	$88.19
3 Skilled Workers	54.85	1316.40	87.40	2097.60		
1 Flatbed Truck, Gas, 3 Ton		820.40		902.44	25.64	28.20
32 L.H., Daily Totals		$2591.60		$3724.44	$80.99	$116.39

Crew No.	Bare Costs Hr.	Daily	Incl. Subs O&P Hr.	Daily	Cost Per Labor-Hour Bare Costs	Incl. O&P
Crew B-48	Hr.	Daily	Hr.	Daily	Bare Costs	Incl. O&P
1 Labor Foreman (outside)	$44.10	$352.80	$70.65	$565.20	$47.59	$75.39
3 Laborers	42.10	1010.40	67.45	1618.80		
1 Equip. Oper. (crane)	59.20	473.60	92.65	741.20		
1 Equip. Oper. (oiler)	50.55	404.40	79.10	632.80		
1 Equip. Oper. (light)	53.00	424.00	82.95	663.60		
1 Centr. Water Pump, 6"		232.90		256.19		
1 -20' Suction Hose, 6"		25.20		27.72		
1 -50' Discharge Hose, 6"		17.90		19.69		
1 Drill Rig, Truck-Mounted		771.10		848.21	18.70	20.57
56 L.H., Daily Totals		$3712.30		$5373.41	$66.29	$95.95
Crew B-49	Hr.	Daily	Hr.	Daily	Bare Costs	Incl. O&P
1 Labor Foreman (outside)	$44.10	$352.80	$70.65	$565.20	$50.12	$79.39
3 Laborers	42.10	1010.40	67.45	1618.80		
2 Equip. Oper. (crane)	59.20	947.20	92.65	1482.40		
2 Equip. Oper. (oilers)	50.55	808.80	79.10	1265.60		
1 Equip. Oper. (light)	53.00	424.00	82.95	663.60		
2 Pile Drivers	54.20	867.20	86.90	1390.40		
1 Hyd. Crane, 25 Ton		580.85		638.93		
1 Centr. Water Pump, 6"		232.90		256.19		
1 -20' Suction Hose, 6"		25.20		27.72		
1 -50' Discharge Hose, 6"		17.90		19.69		
1 Drill Rig, Truck-Mounted		771.10		848.21	18.50	20.35
88 L.H., Daily Totals		$6038.35		$8776.75	$68.62	$99.74
Crew B-50	Hr.	Daily	Hr.	Daily	Bare Costs	Incl. O&P
2 Pile Driver Foremen (outside)	$56.20	$899.20	$90.15	$1442.40	$52.35	$83.46
6 Pile Drivers	54.20	2601.60	86.90	4171.20		
2 Equip. Oper. (crane)	59.20	947.20	92.65	1482.40		
1 Equip. Oper. (oiler)	50.55	404.40	79.10	632.80		
3 Laborers	42.10	1010.40	67.45	1618.80		
1 Crawler Crane, 40 Ton		1206.00		1326.60		
1 Lead, 60' High		209.25		230.18		
1 Hammer, Diesel, 15K ft.-lbs.		617.05		678.76		
1 Air Compressor, 600 cfm		421.50		463.65		
2 -50' Air Hoses, 3"		76.70		84.37		
1 Chain Saw, Gas, 36" Long		41.25		45.38	22.96	25.26
112 L.H., Daily Totals		$8434.55		$12176.53	$75.31	$108.72
Crew B-51	Hr.	Daily	Hr.	Daily	Bare Costs	Incl. O&P
1 Labor Foreman (outside)	$44.10	$352.80	$70.65	$565.20	$43.29	$69.33
4 Laborers	42.10	1347.20	67.45	2158.40		
1 Truck Driver (light)	47.25	378.00	75.55	604.40		
1 Flatbed Truck, Gas, 1.5 Ton		196.15		215.76	4.09	4.50
48 L.H., Daily Totals		$2274.15		$3543.76	$47.38	$73.83
Crew B-52	Hr.	Daily	Hr.	Daily	Bare Costs	Incl. O&P
1 Carpenter Foreman (outside)	$55.15	$441.20	$88.40	$707.20	$48.73	$77.60
1 Carpenter	53.15	425.20	85.15	681.20		
3 Laborers	42.10	1010.40	67.45	1618.80		
1 Cement Finisher	49.95	399.60	78.45	627.60		
.5 Rodman (reinf.)	56.40	225.60	88.85	355.40		
.5 Equip. Oper. (medium)	56.75	227.00	88.80	355.20		
.5 Crawler Loader, 3 C.Y.		584.50		642.95	10.44	11.48
56 L.H., Daily Totals		$3313.50		$4988.35	$59.17	$89.08
Crew B-53	Hr.	Daily	Hr.	Daily	Bare Costs	Incl. O&P
1 Equip. Oper. (light)	$53.00	$424.00	$82.95	$663.60	$53.00	$82.95
1 Trencher, Chain, 12 H.P.		156.95		172.65	19.62	21.58
8 L.H., Daily Totals		$580.95		$836.25	$72.62	$104.53

Crew No.	Bare Costs		Incl. Subs O&P		Cost Per Labor-Hour	
Crew B-54	Hr.	Daily	Hr.	Daily	Bare Costs	Incl. O&P
1 Equip. Oper. (light)	$53.00	$424.00	$82.95	$663.60	$53.00	$82.95
1 Trencher, Chain, 40 H.P.		401.85		442.04	50.23	55.25
8 L.H., Daily Totals		$825.85		$1105.64	$103.23	$138.20

Crew No.	Bare Costs		Incl. Subs O&P		Cost Per Labor-Hour	
Crew B-54A	Hr.	Daily	Hr.	Daily	Bare Costs	Incl. O&P
.17 Labor Foreman (outside)	$44.10	$59.98	$70.65	$96.08	$54.91	$86.16
1 Equipment Operator (med.)	56.75	454.00	88.80	710.40		
1 Wheel Trencher, 67 H.P.		1126.00		1238.60	120.30	132.33
9.36 L.H., Daily Totals		$1639.98		$2045.08	$175.21	$218.49

Crew No.	Bare Costs		Incl. Subs O&P		Cost Per Labor-Hour	
Crew B-54B	Hr.	Daily	Hr.	Daily	Bare Costs	Incl. O&P
.25 Labor Foreman (outside)	$44.10	$88.20	$70.65	$141.30	$54.22	$85.17
1 Equipment Operator (med.)	56.75	454.00	88.80	710.40		
1 Wheel Trencher, 150 H.P.		1324.00		1456.40	132.40	145.64
10 L.H., Daily Totals		$1866.20		$2308.10	$186.62	$230.81

Crew No.	Bare Costs		Incl. Subs O&P		Cost Per Labor-Hour	
Crew B-54D	Hr.	Daily	Hr.	Daily	Bare Costs	Incl. O&P
1 Laborer	$42.10	$336.80	$67.45	$539.60	$49.42	$78.13
1 Equipment Operator (med.)	56.75	454.00	88.80	710.40		
1 Rock Trencher, 6" Width		429.85		472.83	26.87	29.55
16 L.H., Daily Totals		$1220.65		$1722.84	$76.29	$107.68

Crew No.	Bare Costs		Incl. Subs O&P		Cost Per Labor-Hour	
Crew B-54E	Hr.	Daily	Hr.	Daily	Bare Costs	Incl. O&P
1 Laborer	$42.10	$336.80	$67.45	$539.60	$49.42	$78.13
1 Equipment Operator (med.)	56.75	454.00	88.80	710.40		
1 Rock Trencher, 18" Width		1005.00		1105.50	62.81	69.09
16 L.H., Daily Totals		$1795.80		$2355.50	$112.24	$147.22

Crew No.	Bare Costs		Incl. Subs O&P		Cost Per Labor-Hour	
Crew B-55	Hr.	Daily	Hr.	Daily	Bare Costs	Incl. O&P
2 Laborers	$42.10	$673.60	$67.45	$1079.20	$43.82	$70.15
1 Truck Driver (light)	47.25	378.00	75.55	604.40		
1 Truck-Mounted Earth Auger		390.20		429.22		
1 Flatbed Truck, Gas, 3 Ton		820.40		902.44	50.44	55.49
24 L.H., Daily Totals		$2262.20		$3015.26	$94.26	$125.64

Crew No.	Bare Costs		Incl. Subs O&P		Cost Per Labor-Hour	
Crew B-56	Hr.	Daily	Hr.	Daily	Bare Costs	Incl. O&P
1 Laborer	$42.10	$336.80	$67.45	$539.60	$47.55	$75.20
1 Equip. Oper. (light)	53.00	424.00	82.95	663.60		
1 Air Track Drill, 4"		1058.00		1163.80		
1 Air Compressor, 600 cfm		421.50		463.65		
1 -50' Air Hose, 3"		38.35		42.19	94.87	104.35
16 L.H., Daily Totals		$2278.65		$2872.84	$142.42	$179.55

Crew No.	Bare Costs		Incl. Subs O&P		Cost Per Labor-Hour	
Crew B-57	Hr.	Daily	Hr.	Daily	Bare Costs	Incl. O&P
1 Labor Foreman (outside)	$44.10	$352.80	$70.65	$565.20	$48.51	$76.71
2 Laborers	42.10	673.60	67.45	1079.20		
1 Equip. Oper. (crane)	59.20	473.60	92.65	741.20		
1 Equip. Oper. (light)	53.00	424.00	82.95	663.60		
1 Equip. Oper. (oiler)	50.55	404.40	79.10	632.80		
1 Crawler Crane, 25 Ton		1129.00		1241.90		
1 Clamshell Bucket, 1 C.Y.		68.55		75.41		
1 Centr. Water Pump, 6"		232.90		256.19		
1 -20' Suction Hose, 6"		25.20		27.72		
20 -50' Discharge Hoses, 6"		358.00		393.80	37.78	41.56
48 L.H., Daily Totals		$4142.05		$5677.02	$86.29	$118.27

Crew No.	Bare Costs		Incl. Subs O&P		Cost Per Labor-Hour	
Crew B-58	Hr.	Daily	Hr.	Daily	Bare Costs	Incl. O&P
2 Laborers	$42.10	$673.60	$67.45	$1079.20	$45.73	$72.62
1 Equip. Oper. (light)	53.00	424.00	82.95	663.60		
1 Backhoe Loader, 48 H.P.		213.75		235.13		
1 Small Helicopter, w/ Pilot		2080.00		2288.00	95.57	105.13
24 L.H., Daily Totals		$3391.35		$4265.93	$141.31	$177.75

Crew No.	Bare Costs		Incl. Subs O&P		Cost Per Labor-Hour	
Crew B-59	Hr.	Daily	Hr.	Daily	Bare Costs	Incl. O&P
1 Truck Driver (heavy)	$48.95	$391.60	$78.25	$626.00	$48.95	$78.25
1 Truck Tractor, 220 H.P.		307.10		337.81		
1 Water Tank Trailer, 5000 Gal.		152.25		167.47	57.42	63.16
8 L.H., Daily Totals		$850.95		$1131.29	$106.37	$141.41

Crew No.	Bare Costs		Incl. Subs O&P		Cost Per Labor-Hour	
Crew B-59A	Hr.	Daily	Hr.	Daily	Bare Costs	Incl. O&P
2 Laborers	$42.10	$673.60	$67.45	$1079.20	$44.38	$71.05
1 Truck Driver (heavy)	48.95	391.60	78.25	626.00		
1 Water Tank Trailer, 5000 Gal.		152.25		167.47		
1 Truck Tractor, 220 H.P.		307.10		337.81	19.14	21.05
24 L.H., Daily Totals		$1524.55		$2210.49	$63.52	$92.10

Crew No.	Bare Costs		Incl. Subs O&P		Cost Per Labor-Hour	
Crew B-60	Hr.	Daily	Hr.	Daily	Bare Costs	Incl. O&P
1 Labor Foreman (outside)	$44.10	$352.80	$70.65	$565.20	$49.15	$77.60
2 Laborers	42.10	673.60	67.45	1079.20		
1 Equip. Oper. (crane)	59.20	473.60	92.65	741.20		
2 Equip. Oper. (light)	53.00	848.00	82.95	1327.20		
1 Equip. Oper. (oiler)	50.55	404.40	79.10	632.80		
1 Crawler Crane, 40 Ton		1206.00		1326.60		
1 Lead, 60' High		209.25		230.18		
1 Hammer, Diesel, 15K ft.-lbs.		617.05		678.76		
1 Backhoe Loader, 48 H.P.		213.75		235.13	40.11	44.12
56 L.H., Daily Totals		$4998.45		$6816.26	$89.26	$121.72

Crew No.	Bare Costs		Incl. Subs O&P		Cost Per Labor-Hour	
Crew B-61	Hr.	Daily	Hr.	Daily	Bare Costs	Incl. O&P
1 Labor Foreman (outside)	$44.10	$352.80	$70.65	$565.20	$44.68	$71.19
3 Laborers	42.10	1010.40	67.45	1618.80		
1 Equip. Oper. (light)	53.00	424.00	82.95	663.60		
1 Cement Mixer, 2 C.Y.		105.70		116.27		
1 Air Compressor, 160 cfm		202.80		223.08	7.71	8.48
40 L.H., Daily Totals		$2095.70		$3186.95	$52.39	$79.67

Crew No.	Bare Costs		Incl. Subs O&P		Cost Per Labor-Hour	
Crew B-62	Hr.	Daily	Hr.	Daily	Bare Costs	Incl. O&P
2 Laborers	$42.10	$673.60	$67.45	$1079.20	$45.73	$72.62
1 Equip. Oper. (light)	53.00	424.00	82.95	663.60		
1 Loader, Skid Steer, 30 H.P.		177.55		195.31	7.40	8.14
24 L.H., Daily Totals		$1275.15		$1938.11	$53.13	$80.75

Crew No.	Bare Costs		Incl. Subs O&P		Cost Per Labor-Hour	
Crew B-62A	Hr.	Daily	Hr.	Daily	Bare Costs	Incl. O&P
2 Laborers	$42.10	$673.60	$67.45	$1079.20	$45.73	$72.62
1 Equip. Oper. (light)	53.00	424.00	82.95	663.60		
1 Loader, Skid Steer, 30 H.P.		177.55		195.31		
1 Trencher Attachment		66.20		72.82	10.16	11.17
24 L.H., Daily Totals		$1341.35		$2010.93	$55.89	$83.79

Crew No.	Bare Costs		Incl. Subs O&P		Cost Per Labor-Hour	
Crew B-63	Hr.	Daily	Hr.	Daily	Bare Costs	Incl. O&P
4 Laborers	$42.10	$1347.20	$67.45	$2158.40	$44.28	$70.55
1 Equip. Oper. (light)	53.00	424.00	82.95	663.60		
1 Loader, Skid Steer, 30 H.P.		177.55		195.31	4.44	4.88
40 L.H., Daily Totals		$1948.75		$3017.30	$48.72	$75.43

Crew No.	Bare Costs		Incl. Subs O&P		Cost Per Labor-Hour	
Crew B-63B	Hr.	Daily	Hr.	Daily	Bare Costs	Incl. O&P
1 Labor Foreman (inside)	$42.60	$340.80	$68.25	$546.00	$44.95	$71.53
2 Laborers	42.10	673.60	67.45	1079.20		
1 Equip. Oper. (light)	53.00	424.00	82.95	663.60		
1 Loader, Skid Steer, 78 H.P.		400.10		440.11	12.50	13.75
32 L.H., Daily Totals		$1838.50		$2728.91	$57.45	$85.28
Crew B-64	Hr.	Daily	Hr.	Daily	Bare Costs	Incl. O&P
1 Laborer	$42.10	$336.80	$67.45	$539.60	$44.67	$71.50
1 Truck Driver (light)	47.25	378.00	75.55	604.40		
1 Power Mulcher (small)		221.10		243.21		
1 Flatbed Truck, Gas, 1.5 Ton		196.15		215.76	26.08	28.69
16 L.H., Daily Totals		$1132.05		$1602.97	$70.75	$100.19
Crew B-65	Hr.	Daily	Hr.	Daily	Bare Costs	Incl. O&P
1 Laborer	$42.10	$336.80	$67.45	$539.60	$44.67	$71.50
1 Truck Driver (light)	47.25	378.00	75.55	604.40		
1 Power Mulcher (Large)		326.90		359.59		
1 Flatbed Truck, Gas, 1.5 Ton		196.15		215.76	32.69	35.96
16 L.H., Daily Totals		$1237.85		$1719.36	$77.37	$107.46
Crew B-66	Hr.	Daily	Hr.	Daily	Bare Costs	Incl. O&P
1 Equip. Oper. (light)	$53.00	$424.00	$82.95	$663.60	$53.00	$82.95
1 Loader-Backhoe, 40 H.P.		241.40		265.54	30.18	33.19
8 L.H., Daily Totals		$665.40		$929.14	$83.17	$116.14
Crew B-67	Hr.	Daily	Hr.	Daily	Bare Costs	Incl. O&P
1 Millwright	$55.90	$447.20	$86.60	$692.80	$54.45	$84.78
1 Equip. Oper. (light)	53.00	424.00	82.95	663.60		
1 R.T. Forklift, 5,000 Lb., diesel		269.90		296.89	16.87	18.56
16 L.H., Daily Totals		$1141.10		$1653.29	$71.32	$103.33
Crew B-67B	Hr.	Daily	Hr.	Daily	Bare Costs	Incl. O&P
1 Millwright Foreman (inside)	$56.40	$451.20	$87.35	$698.80	$56.15	$86.97
1 Millwright	55.90	447.20	86.60	692.80		
16 L.H., Daily Totals		$898.40		$1391.60	$56.15	$86.97
Crew B-68	Hr.	Daily	Hr.	Daily	Bare Costs	Incl. O&P
2 Millwrights	$55.90	$894.40	$86.60	$1385.60	$54.93	$85.38
1 Equip. Oper. (light)	53.00	424.00	82.95	663.60		
1 R.T. Forklift, 5,000 Lb., diesel		269.90		296.89	11.25	12.37
24 L.H., Daily Totals		$1588.30		$2346.09	$66.18	$97.75
Crew B-68A	Hr.	Daily	Hr.	Daily	Bare Costs	Incl. O&P
1 Millwright Foreman (inside)	$56.40	$451.20	$87.35	$698.80	$56.07	$86.85
2 Millwrights	55.90	894.40	86.60	1385.60		
1 Forklift, Smooth Floor, 8,000 Lb.		255.90		281.49	10.66	11.73
24 L.H., Daily Totals		$1601.50		$2365.89	$66.73	$98.58
Crew B-68B	Hr.	Daily	Hr.	Daily	Bare Costs	Incl. O&P
1 Millwright Foreman (inside)	$56.40	$451.20	$87.35	$698.80	$59.97	$92.68
2 Millwrights	55.90	894.40	86.60	1385.60		
2 Electricians	61.35	981.60	94.45	1511.20		
2 Plumbers	64.45	1031.20	99.65	1594.40		
1 R.T. Forklift, 5,000 Lb., gas		279.90		307.89	5.00	5.50
56 L.H., Daily Totals		$3638.30		$5497.89	$64.97	$98.18

Crew No.	Bare Costs		Incl. Subs O&P		Cost Per Labor-Hour	
Crew B-68C	Hr.	Daily	Hr.	Daily	Bare Costs	Incl. O&P
1 Millwright Foreman (inside)	$56.40	$451.20	$87.35	$698.80	$59.52	$92.01
1 Millwright	55.90	447.20	86.60	692.80		
1 Electrician	61.35	490.80	94.45	755.60		
1 Plumber	64.45	515.60	99.65	797.20		
1 R.T. Forklift, 5,000 Lb., gas		279.90		307.89	8.75	9.62
32 L.H., Daily Totals		$2184.70		$3252.29	$68.27	$101.63
Crew B-68D	Hr.	Daily	Hr.	Daily	Bare Costs	Incl. O&P
1 Labor Foreman (inside)	$42.60	$340.80	$68.25	$546.00	$45.90	$72.88
1 Laborer	42.10	336.80	67.45	539.60		
1 Equip. Oper. (light)	53.00	424.00	82.95	663.60		
1 R.T. Forklift, 5,000 Lb., gas		279.90		307.89	11.66	12.83
24 L.H., Daily Totals		$1381.50		$2057.09	$57.56	$85.71
Crew B-68E	Hr.	Daily	Hr.	Daily	Bare Costs	Incl. O&P
1 Struc. Steel Foreman (inside)	$58.15	$465.20	$95.30	$762.40	$57.71	$94.56
3 Struc. Steel Workers	57.65	1383.60	94.45	2266.80		
1 Welder	57.45	459.60	94.15	753.20		
1 Forklift, Smooth Floor, 8,000 Lb.		255.90		281.49	6.40	7.04
40 L.H., Daily Totals		$2564.30		$4063.89	$64.11	$101.60
Crew B-68F	Hr.	Daily	Hr.	Daily	Bare Costs	Incl. O&P
1 Skilled Worker Foreman (out)	$56.85	$454.80	$90.55	$724.40	$55.52	$88.45
2 Skilled Workers	54.85	877.60	87.40	1398.40		
1 R.T. Forklift, 5,000 Lb., gas		279.90		307.89	11.66	12.83
24 L.H., Daily Totals		$1612.30		$2430.69	$67.18	$101.28
Crew B-68G	Hr.	Daily	Hr.	Daily	Bare Costs	Incl. O&P
2 Structural Steel Workers	$57.65	$922.40	$94.45	$1511.20	$57.65	$94.45
1 R.T. Forklift, 5,000 Lb., gas		279.90		307.89	17.49	19.24
16 L.H., Daily Totals		$1202.30		$1819.09	$75.14	$113.69
Crew B-69	Hr.	Daily	Hr.	Daily	Bare Costs	Incl. O&P
1 Labor Foreman (outside)	$44.10	$352.80	$70.65	$565.20	$46.69	$74.13
3 Laborers	42.10	1010.40	67.45	1618.80		
1 Equip. Oper. (crane)	59.20	473.60	92.65	741.20		
1 Equip. Oper. (oiler)	50.55	404.40	79.10	632.80		
1 Hyd. Crane, 80 Ton		1486.00		1634.60	30.96	34.05
48 L.H., Daily Totals		$3727.20		$5192.60	$77.65	$108.18
Crew B-69A	Hr.	Daily	Hr.	Daily	Bare Costs	Incl. O&P
1 Labor Foreman (outside)	$44.10	$352.80	$70.65	$565.20	$46.18	$73.38
3 Laborers	42.10	1010.40	67.45	1618.80		
1 Equip. Oper. (medium)	56.75	454.00	88.80	710.40		
1 Concrete Finisher	49.95	399.60	78.45	627.60		
1 Curb/Gutter Paver, 2-Track		1217.00		1338.70	25.35	27.89
48 L.H., Daily Totals		$3433.80		$4860.70	$71.54	$101.26
Crew B-69B	Hr.	Daily	Hr.	Daily	Bare Costs	Incl. O&P
1 Labor Foreman (outside)	$44.10	$352.80	$70.65	$565.20	$46.18	$73.38
3 Laborers	42.10	1010.40	67.45	1618.80		
1 Equip. Oper. (medium)	56.75	454.00	88.80	710.40		
1 Cement Finisher	49.95	399.60	78.45	627.60		
1 Curb/Gutter Paver, 4-Track		791.55		870.71	16.49	18.14
48 L.H., Daily Totals		$3008.35		$4392.70	$62.67	$91.51

Crew No.	Bare Costs		Incl. Subs O&P		Cost Per Labor-Hour	
	Hr.	Daily	Hr.	Daily	Bare Costs	Incl. O&P
Crew B-70						
1 Labor Foreman (outside)	$44.10	$352.80	$70.65	$565.20	$48.66	$77.06
3 Laborers	42.10	1010.40	67.45	1618.80		
3 Equip. Oper. (medium)	56.75	1362.00	88.80	2131.20		
1 Grader, 30,000 Lbs.		1062.00		1168.20		
1 Ripper, Beam & 1 Shank		90.50		99.55		
1 Road Sweeper, S.P., 8' wide		715.85		787.43		
1 F.E. Loader, W.M., 1.5 C.Y.		423.50		465.85	40.93	45.02
56 L.H., Daily Totals		$5017.05		$6836.23	$89.59	$122.08
Crew B-71					Bare Costs	Incl. O&P
1 Labor Foreman (outside)	$44.10	$352.80	$70.65	$565.20	$48.66	$77.06
3 Laborers	42.10	1010.40	67.45	1618.80		
3 Equip. Oper. (medium)	56.75	1362.00	88.80	2131.20		
1 Pvmt. Profiler, 750 H.P.		3448.00		3792.80		
1 Road Sweeper, S.P., 8' wide		715.85		787.43		
1 F.E. Loader, W.M., 1.5 C.Y.		423.50		465.85	81.92	90.11
56 L.H., Daily Totals		$7312.55		$9361.28	$130.58	$167.17
Crew B-72					Bare Costs	Incl. O&P
1 Labor Foreman (outside)	$44.10	$352.80	$70.65	$565.20	$49.67	$78.53
3 Laborers	42.10	1010.40	67.45	1618.80		
4 Equip. Oper. (medium)	56.75	1816.00	88.80	2841.60		
1 Pvmt. Profiler, 750 H.P.		3448.00		3792.80		
1 Hammermill, 250 H.P.		848.80		933.68		
1 Windrow Loader		1427.00		1569.70		
1 Mix Paver, 165 H.P.		2149.00		2363.90		
1 Roller, Pneum. Whl., 12 Ton		345.75		380.32	128.41	141.26
64 L.H., Daily Totals		$11397.75		$14066.00	$178.09	$219.78
Crew B-73					Bare Costs	Incl. O&P
1 Labor Foreman (outside)	$44.10	$352.80	$70.65	$565.20	$51.51	$81.19
2 Laborers	42.10	673.60	67.45	1079.20		
5 Equip. Oper. (medium)	56.75	2270.00	88.80	3552.00		
1 Road Mixer, 310 H.P.		1896.00		2085.60		
1 Tandem Roller, 10 Ton		236.75		260.43		
1 Hammermill, 250 H.P.		848.80		933.68		
1 Grader, 30,000 Lbs.		1062.00		1168.20		
.5 F.E. Loader, W.M., 1.5 C.Y.		211.75		232.93		
.5 Truck Tractor, 220 H.P.		153.55		168.91		
.5 Water Tank Trailer, 5000 Gal.		76.13		83.74	70.08	77.09
64 L.H., Daily Totals		$7781.38		$10129.87	$121.58	$158.28
Crew B-74					Bare Costs	Incl. O&P
1 Labor Foreman (outside)	$44.10	$352.80	$70.65	$565.20	$51.39	$81.22
1 Laborer	42.10	336.80	67.45	539.60		
4 Equip. Oper. (medium)	56.75	1816.00	88.80	2841.60		
2 Truck Drivers (heavy)	48.95	783.20	78.25	1252.00		
1 Grader, 30,000 Lbs.		1062.00		1168.20		
1 Ripper, Beam & 1 Shank		90.50		99.55		
2 Stabilizers, 310 H.P.		2778.00		3055.80		
1 Flatbed Truck, Gas, 3 Ton		820.40		902.44		
1 Chem. Spreader, Towed		83.35		91.69		
1 Roller, Vibratory, 25 Ton		664.35		730.78		
1 Water Tank Trailer, 5000 Gal.		152.25		167.47		
1 Truck Tractor, 220 H.P.		307.10		337.81	93.09	102.40
64 L.H., Daily Totals		$9246.75		$11752.15	$144.48	$183.63

Crew No.	Bare Costs		Incl. Subs O&P		Cost Per Labor-Hour	
	Hr.	Daily	Hr.	Daily	Bare Costs	Incl. O&P
Crew B-75						
1 Labor Foreman (outside)	$44.10	$352.80	$70.65	$565.20	$51.74	$81.65
1 Laborer	42.10	336.80	67.45	539.60		
4 Equip. Oper. (medium)	56.75	1816.00	88.80	2841.60		
1 Truck Driver (heavy)	48.95	391.60	78.25	626.00		
1 Grader, 30,000 Lbs.		1062.00		1168.20		
1 Ripper, Beam & 1 Shank		90.50		99.55		
2 Stabilizers, 310 H.P.		2778.00		3055.80		
1 Dist. Tanker, 3000 Gallon		330.15		363.17		
1 Truck Tractor, 6x4, 380 H.P.		493.25		542.58		
1 Roller, Vibratory, 25 Ton		664.35		730.78	96.75	106.43
56 L.H., Daily Totals		$8315.45		$10532.48	$148.49	$188.08
Crew B-76					Bare Costs	Incl. O&P
1 Dock Builder Foreman (outside)	$56.20	$449.60	$90.15	$721.20	$55.13	$87.67
5 Dock Builders	54.20	2168.00	86.90	3476.00		
2 Equip. Oper. (crane)	59.20	947.20	92.65	1482.40		
1 Equip. Oper. (oiler)	50.55	404.40	79.10	632.80		
1 Crawler Crane, 50 Ton		1510.00		1661.00		
1 Barge, 400 Ton		858.85		944.74		
1 Hammer, Diesel, 15K ft.-lbs.		617.05		678.76		
1 Lead, 60' High		209.25		230.18		
1 Air Compressor, 600 cfm		421.50		463.65		
2 -50' Air Hoses, 3"		76.70		84.37	51.30	56.43
72 L.H., Daily Totals		$7662.55		$10375.08	$106.42	$144.10
Crew B-76A					Bare Costs	Incl. O&P
1 Labor Foreman (outside)	$44.10	$352.80	$70.65	$565.20	$45.54	$72.46
5 Laborers	42.10	1684.00	67.45	2698.00		
1 Equip. Oper. (crane)	59.20	473.60	92.65	741.20		
1 Equip. Oper. (oiler)	50.55	404.40	79.10	632.80		
1 Crawler Crane, 50 Ton		1510.00		1661.00		
1 Barge, 400 Ton		858.85		944.74	37.01	40.71
64 L.H., Daily Totals		$5283.65		$7242.94	$82.56	$113.17
Crew B-77					Bare Costs	Incl. O&P
1 Labor Foreman (outside)	$44.10	$352.80	$70.65	$565.20	$43.53	$69.71
3 Laborers	42.10	1010.40	67.45	1618.80		
1 Truck Driver (light)	47.25	378.00	75.55	604.40		
1 Crack Cleaner, 25 H.P.		52.45		57.70		
1 Crack Filler, Trailer Mtd.		154.75		170.22		
1 Flatbed Truck, Gas, 3 Ton		820.40		902.44	25.69	28.26
40 L.H., Daily Totals		$2768.80		$3918.76	$69.22	$97.97
Crew B-78					Bare Costs	Incl. O&P
1 Labor Foreman (outside)	$44.10	$352.80	$70.65	$565.20	$43.29	$69.33
4 Laborers	42.10	1347.20	67.45	2158.40		
1 Truck Driver (light)	47.25	378.00	75.55	604.40		
1 Paint Striper, S.P., 40 Gallon		127.10		139.81		
1 Flatbed Truck, Gas, 3 Ton		820.40		902.44		
1 Pickup Truck, 3/4 Ton		110.85		121.94	22.05	24.25
48 L.H., Daily Totals		$3136.35		$4492.19	$65.34	$93.59
Crew B-78A					Bare Costs	Incl. O&P
1 Equip. Oper. (light)	$53.00	$424.00	$82.95	$663.60	$53.00	$82.95
1 Line Rem. (Metal Balls) 115 H.P.		916.50		1008.15	114.56	126.02
8 L.H., Daily Totals		$1340.50		$1671.75	$167.56	$208.97

Crew No.	Bare Costs		Incl. Subs O&P		Cost Per Labor-Hour	
Crew B-78B	Hr.	Daily	Hr.	Daily	Bare Costs	Incl. O&P
2 Laborers	$42.10	$673.60	$67.45	$1079.20	$43.31	$69.17
.25 Equip. Oper. (light)	53.00	106.00	82.95	165.90		
1 Pickup Truck, 3/4 Ton		110.85		121.94		
1 Line Rem.,11 H.P.,Walk Behind		113.45		124.80		
.25 Road Sweeper, S.P., 8' wide		178.96		196.86	22.40	24.64
18 L.H., Daily Totals		$1182.86		$1688.69	$65.71	$93.82
Crew B-78C	Hr.	Daily	Hr.	Daily	Bare Costs	Incl. O&P
1 Labor Foreman (outside)	$44.10	$352.80	$70.65	$565.20	$43.29	$69.33
4 Laborers	42.10	1347.20	67.45	2158.40		
1 Truck Driver (light)	47.25	378.00	75.55	604.40		
1 Paint Striper, T.M., 120 Gal.		592.95		652.25		
1 Flatbed Truck, Gas, 3 Ton		820.40		902.44		
1 Pickup Truck, 3/4 Ton		110.85		121.94	31.75	34.93
48 L.H., Daily Totals		$3602.20		$5004.62	$75.05	$104.26
Crew B-78D	Hr.	Daily	Hr.	Daily	Bare Costs	Incl. O&P
2 Labor Foremen (outside)	$44.10	$705.60	$70.65	$1130.40	$43.02	$68.90
7 Laborers	42.10	2357.60	67.45	3777.20		
1 Truck Driver (light)	47.25	378.00	75.55	604.40		
1 Paint Striper, T.M., 120 Gal.		592.95		652.25		
1 Flatbed Truck, Gas, 3 Ton		820.40		902.44		
3 Pickup Trucks, 3/4 Ton		332.55		365.81		
1 Air Compressor, 60 cfm		152.95		168.25		
1 -50' Air Hose, 3/4"		7.75		8.53		
1 Breaker, Pavement, 60 lb.		53.50		58.85	24.50	26.95
80 L.H., Daily Totals		$5401.30		$7668.11	$67.52	$95.85
Crew B-78E	Hr.	Daily	Hr.	Daily	Bare Costs	Incl. O&P
2 Labor Foremen (outside)	$44.10	$705.60	$70.65	$1130.40	$42.86	$68.66
9 Laborers	42.10	3031.20	67.45	4856.40		
1 Truck Driver (light)	47.25	378.00	75.55	604.40		
1 Paint Striper, T.M., 120 Gal.		592.95		652.25		
1 Flatbed Truck, Gas, 3 Ton		820.40		902.44		
4 Pickup Trucks, 3/4 Ton		443.40		487.74		
2 Air Compressors, 60 cfm		305.90		336.49		
2 -50' Air Hoses, 3/4"		15.50		17.05		
2 Breakers, Pavement, 60 lb.		107.00		117.70	23.80	26.18
96 L.H., Daily Totals		$6399.95		$9104.86	$66.67	$94.84
Crew B-78F	Hr.	Daily	Hr.	Daily	Bare Costs	Incl. O&P
2 Labor Foremen (outside)	$44.10	$705.60	$70.65	$1130.40	$42.75	$68.49
11 Laborers	42.10	3704.80	67.45	5935.60		
1 Truck Driver (light)	47.25	378.00	75.55	604.40		
1 Paint Striper, T.M., 120 Gal.		592.95		652.25		
1 Flatbed Truck, Gas, 3 Ton		820.40		902.44		
7 Pickup Trucks, 3/4 Ton		775.95		853.54		
3 Air Compressors, 60 cfm		458.85		504.74		
3 -50' Air Hoses, 3/4"		23.25		25.57		
3 Breakers, Pavement, 60 lb.		160.50		176.55	25.28	27.81
112 L.H., Daily Totals		$7620.30		$10785.49	$68.04	$96.30
Crew B-79	Hr.	Daily	Hr.	Daily	Bare Costs	Incl. O&P
1 Labor Foreman (outside)	$44.10	$352.80	$70.65	$565.20	$43.53	$69.71
3 Laborers	42.10	1010.40	67.45	1618.80		
1 Truck Driver (light)	47.25	378.00	75.55	604.40		
1 Paint Striper, T.M., 120 Gal.		592.95		652.25		
1 Heating Kettle, 115 Gallon		98.65		108.52		
1 Flatbed Truck, Gas, 3 Ton		820.40		902.44		
2 Pickup Trucks, 3/4 Ton		221.70		243.87	43.34	47.68
40 L.H., Daily Totals		$3474.90		$4695.47	$86.87	$117.39

Crew No.	Bare Costs		Incl. Subs O&P		Cost Per Labor-Hour	
Crew B-79A	Hr.	Daily	Hr.	Daily	Bare Costs	Incl. O&P
1.5 Equip. Oper. (light)	$53.00	$636.00	$82.95	$995.40	$53.00	$82.95
.5 Line Remov. (Grinder) 115 H.P.		475.57		523.13		
1 Line Rem. (Metal Balls) 115 H.P.		916.50		1008.15	116.01	127.61
12 L.H., Daily Totals		$2028.08		$2526.68	$169.01	$210.56
Crew B-79B	Hr.	Daily	Hr.	Daily	Bare Costs	Incl. O&P
1 Laborer	$42.10	$336.80	$67.45	$539.60	$42.10	$67.45
1 Set of Gases		171.55		188.71	21.44	23.59
8 L.H., Daily Totals		$508.35		$728.30	$63.54	$91.04
Crew B-79C	Hr.	Daily	Hr.	Daily	Bare Costs	Incl. O&P
1 Labor Foreman (outside)	$44.10	$352.80	$70.65	$565.20	$43.12	$69.06
5 Laborers	42.10	1684.00	67.45	2698.00		
1 Truck Driver (light)	47.25	378.00	75.55	604.40		
1 Paint Striper, T.M., 120 Gal.		592.95		652.25		
1 Heating Kettle, 115 Gallon		98.65		108.52		
1 Flatbed Truck, Gas, 3 Ton		820.40		902.44		
3 Pickup Trucks, 3/4 Ton		332.55		365.81		
1 Air Compressor, 60 cfm		152.95		168.25		
1 -50' Air Hose, 3/4"		7.75		8.53		
1 Breaker, Pavement, 60 lb.		53.50		58.85	36.76	40.44
56 L.H., Daily Totals		$4473.55		$6132.23	$79.88	$109.50
Crew B-79D	Hr.	Daily	Hr.	Daily	Bare Costs	Incl. O&P
2 Labor Foremen (outside)	$44.10	$705.60	$70.65	$1130.40	$43.24	$69.26
5 Laborers	42.10	1684.00	67.45	2698.00		
1 Truck Driver (light)	47.25	378.00	75.55	604.40		
1 Paint Striper, T.M., 120 Gal.		592.95		652.25		
1 Heating Kettle, 115 Gallon		98.65		108.52		
1 Flatbed Truck, Gas, 3 Ton		820.40		902.44		
4 Pickup Trucks, 3/4 Ton		443.40		487.74		
1 Air Compressor, 60 cfm		152.95		168.25		
1 -50' Air Hose, 3/4"		7.75		8.53		
1 Breaker, Pavement, 60 lb.		53.50		58.85	33.90	37.29
64 L.H., Daily Totals		$4937.20		$6819.36	$77.14	$106.55
Crew B-79E	Hr.	Daily	Hr.	Daily	Bare Costs	Incl. O&P
2 Labor Foremen (outside)	$44.10	$705.60	$70.65	$1130.40	$43.02	$68.90
7 Laborers	42.10	2357.60	67.45	3777.20		
1 Truck Driver (light)	47.25	378.00	75.55	604.40		
1 Paint Striper, T.M., 120 Gal.		592.95		652.25		
1 Heating Kettle, 115 Gallon		98.65		108.52		
1 Flatbed Truck, Gas, 3 Ton		820.40		902.44		
5 Pickup Trucks, 3/4 Ton		554.25		609.67		
2 Air Compressors, 60 cfm		305.90		336.49		
2 -50' Air Hoses, 3/4"		15.50		17.05		
2 Breakers, Pavement, 60 lb.		107.00		117.70	31.18	34.30
80 L.H., Daily Totals		$5935.85		$8256.11	$74.20	$103.20
Crew B-80	Hr.	Daily	Hr.	Daily	Bare Costs	Incl. O&P
1 Labor Foreman (outside)	$44.10	$352.80	$70.65	$565.20	$46.61	$74.15
1 Laborer	42.10	336.80	67.45	539.60		
1 Truck Driver (light)	47.25	378.00	75.55	604.40		
1 Equip. Oper. (light)	53.00	424.00	82.95	663.60		
1 Flatbed Truck, Gas, 3 Ton		820.40		902.44		
1 Earth Auger, Truck-Mtd.		200.50		220.55	31.90	35.09
32 L.H., Daily Totals		$2512.50		$3495.79	$78.52	$109.24

Crew B-80A

Crew No.	Bare Costs Hr.	Daily	Incl. Subs O&P Hr.	Daily	Cost Per Labor-Hour Bare Costs	Incl. O&P
3 Laborers	$42.10	$1010.40	$67.45	$1618.80	$42.10	$67.45
1 Flatbed Truck, Gas, 3 Ton		820.40		902.44	34.18	37.60
24 L.H., Daily Totals		$1830.80		$2521.24	$76.28	$105.05

Crew B-80B

Crew No.	Bare Costs Hr.	Daily	Incl. Subs O&P Hr.	Daily	Cost Per Labor-Hour Bare Costs	Incl. O&P
3 Laborers	$42.10	$1010.40	$67.45	$1618.80	$44.83	$71.33
1 Equip. Oper. (light)	53.00	424.00	82.95	663.60		
1 Crane, Flatbed Mounted, 3 Ton		236.10		259.71	7.38	8.12
32 L.H., Daily Totals		$1670.50		$2542.11	$52.20	$79.44

Crew B-80C

Crew No.	Bare Costs Hr.	Daily	Incl. Subs O&P Hr.	Daily	Cost Per Labor-Hour Bare Costs	Incl. O&P
2 Laborers	$42.10	$673.60	$67.45	$1079.20	$43.82	$70.15
1 Truck Driver (light)	47.25	378.00	75.55	604.40		
1 Flatbed Truck, Gas, 1.5 Ton		196.15		215.76		
1 Manual Fence Post Auger, Gas		53.85		59.23	10.42	11.46
24 L.H., Daily Totals		$1301.60		$1958.60	$54.23	$81.61

Crew B-81

Crew No.	Bare Costs Hr.	Daily	Incl. Subs O&P Hr.	Daily	Cost Per Labor-Hour Bare Costs	Incl. O&P
1 Laborer	$42.10	$336.80	$67.45	$539.60	$49.27	$78.17
1 Equip. Oper. (medium)	56.75	454.00	88.80	710.40		
1 Truck Driver (heavy)	48.95	391.60	78.25	626.00		
1 Hydromulcher, T.M., 3000 Gal.		275.45		303.00		
1 Truck Tractor, 220 H.P.		307.10		337.81	24.27	26.70
24 L.H., Daily Totals		$1764.95		$2516.80	$73.54	$104.87

Crew B-81A

Crew No.	Bare Costs Hr.	Daily	Incl. Subs O&P Hr.	Daily	Cost Per Labor-Hour Bare Costs	Incl. O&P
1 Laborer	$42.10	$336.80	$67.45	$539.60	$44.67	$71.50
1 Truck Driver (light)	47.25	378.00	75.55	604.40		
1 Hydromulcher, T.M., 600 Gal.		116.95		128.65		
1 Flatbed Truck, Gas, 3 Ton		820.40		902.44	58.58	64.44
16 L.H., Daily Totals		$1652.15		$2175.09	$103.26	$135.94

Crew B-82

Crew No.	Bare Costs Hr.	Daily	Incl. Subs O&P Hr.	Daily	Cost Per Labor-Hour Bare Costs	Incl. O&P
1 Laborer	$42.10	$336.80	$67.45	$539.60	$47.55	$75.20
1 Equip. Oper. (light)	53.00	424.00	82.95	663.60		
1 Horiz. Borer, 6 H.P.		182.35		200.59	11.40	12.54
16 L.H., Daily Totals		$943.15		$1403.79	$58.95	$87.74

Crew B-82A

Crew No.	Bare Costs Hr.	Daily	Incl. Subs O&P Hr.	Daily	Cost Per Labor-Hour Bare Costs	Incl. O&P
2 Laborers	$42.10	$673.60	$67.45	$1079.20	$47.55	$75.20
2 Equip. Opers. (light)	53.00	848.00	82.95	1327.20		
2 Dump Truck, 8 C.Y., 220 H.P.		856.10		941.71		
1 Flatbed Trailer, 25 Ton		135.55		149.10		
1 Horiz. Dir. Drill, 20k lb. Thrust		538.65		592.51		
1 Mud Trailer for HDD, 1500 Gal.		309.00		339.90		
1 Pickup Truck, 4x4, 3/4 Ton		175.85		193.44		
1 Flatbed Trailer, 3 Ton		70.30		77.33		
1 Loader, Skid Steer, 78 H.P.		400.10		440.11	77.67	85.44
32 L.H., Daily Totals		$4007.15		$5140.51	$125.22	$160.64

Crew B-82B

Crew No.	Bare Costs Hr.	Daily	Incl. Subs O&P Hr.	Daily	Cost Per Labor-Hour Bare Costs	Incl. O&P
2 Laborers	$42.10	$673.60	$67.45	$1079.20	$47.55	$75.20
2 Equip. Opers. (light)	53.00	848.00	82.95	1327.20		
2 Dump Truck, 8 C.Y., 220 H.P.		856.10		941.71		
1 Flatbed Trailer, 25 Ton		135.55		149.10		
1 Horiz. Dir. Drill, 30k lb. Thrust		641.20		705.32		
1 Mud Trailer for HDD, 1500 Gal.		309.00		339.90		
1 Pickup Truck, 4x4, 3/4 Ton		175.85		193.44		
1 Flatbed Trailer, 3 Ton		70.30		77.33		
1 Loader, Skid Steer, 78 H.P.		400.10		440.11	80.88	88.97
32 L.H., Daily Totals		$4109.70		$5253.31	$128.43	$164.17

Crew B-82C

Crew No.	Bare Costs Hr.	Daily	Incl. Subs O&P Hr.	Daily	Cost Per Labor-Hour Bare Costs	Incl. O&P
2 Laborers	$42.10	$673.60	$67.45	$1079.20	$47.55	$75.20
2 Equip. Opers. (light)	53.00	848.00	82.95	1327.20		
2 Dump Truck, 8 C.Y., 220 H.P.		856.10		941.71		
1 Flatbed Trailer, 25 Ton		135.55		149.10		
1 Horiz. Dir. Drill, 50k lb. Thrust		815.80		897.38		
1 Mud Trailer for HDD, 1500 Gal.		309.00		339.90		
1 Pickup Truck, 4x4, 3/4 Ton		175.85		193.44		
1 Flatbed Trailer, 3 Ton		70.30		77.33		
1 Loader, Skid Steer, 78 H.P.		400.10		440.11	86.33	94.97
32 L.H., Daily Totals		$4284.30		$5445.37	$133.88	$170.17

Crew B-82D

Crew No.	Bare Costs Hr.	Daily	Incl. Subs O&P Hr.	Daily	Cost Per Labor-Hour Bare Costs	Incl. O&P
1 Equip. Oper. (light)	$53.00	$424.00	$82.95	$663.60	$53.00	$82.95
1 Mud Trailer for HDD, 1500 Gal.		309.00		339.90	38.63	42.49
8 L.H., Daily Totals		$733.00		$1003.50	$91.63	$125.44

Crew B-83

Crew No.	Bare Costs Hr.	Daily	Incl. Subs O&P Hr.	Daily	Cost Per Labor-Hour Bare Costs	Incl. O&P
1 Tugboat Captain	$56.75	$454.00	$88.80	$710.40	$49.42	$78.13
1 Tugboat Hand	42.10	336.80	67.45	539.60		
1 Tugboat, 250 H.P.		717.50		789.25	44.84	49.33
16 L.H., Daily Totals		$1508.30		$2039.25	$94.27	$127.45

Crew B-84

Crew No.	Bare Costs Hr.	Daily	Incl. Subs O&P Hr.	Daily	Cost Per Labor-Hour Bare Costs	Incl. O&P
1 Equip. Oper. (medium)	$56.75	$454.00	$88.80	$710.40	$56.75	$88.80
1 Rotary Mower/Tractor		366.75		403.43	45.84	50.43
8 L.H., Daily Totals		$820.75		$1113.83	$102.59	$139.23

Crew B-85

Crew No.	Bare Costs Hr.	Daily	Incl. Subs O&P Hr.	Daily	Cost Per Labor-Hour Bare Costs	Incl. O&P
3 Laborers	$42.10	$1010.40	$67.45	$1618.80	$46.40	$73.88
1 Equip. Oper. (medium)	56.75	454.00	88.80	710.40		
1 Truck Driver (heavy)	48.95	391.60	78.25	626.00		
1 Telescoping Boom Lift, to 80'		383.55		421.90		
1 Brush Chipper, 12", 130 H.P.		392.80		432.08		
1 Pruning Saw, Rotary		26.40		29.04	20.07	22.08
40 L.H., Daily Totals		$2658.75		$3838.22	$66.47	$95.96

Crew B-86

Crew No.	Bare Costs Hr.	Daily	Incl. Subs O&P Hr.	Daily	Cost Per Labor-Hour Bare Costs	Incl. O&P
1 Equip. Oper. (medium)	$56.75	$454.00	$88.80	$710.40	$56.75	$88.80
1 Stump Chipper, S.P.		194.45		213.90	24.31	26.74
8 L.H., Daily Totals		$648.45		$924.29	$81.06	$115.54

Crew B-86A

Crew No.	Bare Costs Hr.	Daily	Incl. Subs O&P Hr.	Daily	Cost Per Labor-Hour Bare Costs	Incl. O&P
1 Equip. Oper. (medium)	$56.75	$454.00	$88.80	$710.40	$56.75	$88.80
1 Grader, 30,000 Lbs.		1062.00		1168.20	132.75	146.03
8 L.H., Daily Totals		$1516.00		$1878.60	$189.50	$234.82

Crew B-86B

Crew No.	Bare Costs Hr.	Daily	Incl. Subs O&P Hr.	Daily	Cost Per Labor-Hour Bare Costs	Incl. O&P
1 Equip. Oper. (medium)	$56.75	$454.00	$88.80	$710.40	$56.75	$88.80
1 Dozer, 200 H.P.		1504.00		1654.40	188.00	206.80
8 L.H., Daily Totals		$1958.00		$2364.80	$244.75	$295.60

Crew B-87

Crew No.	Bare Costs Hr.	Daily	Incl. Subs O&P Hr.	Daily	Cost Per Labor-Hour Bare Costs	Incl. O&P
1 Laborer	$42.10	$336.80	$67.45	$539.60	$53.82	$84.53
4 Equip. Oper. (medium)	56.75	1816.00	88.80	2841.60		
2 Feller Bunchers, 100 H.P.		1177.20		1294.92		
1 Log Chipper, 22" Tree		552.00		607.20		
1 Dozer, 105 H.P.		633.85		697.24		
1 Chain Saw, Gas, 36" Long		41.25		45.38	60.11	66.12
40 L.H., Daily Totals		$4557.10		$6025.93	$113.93	$150.65

For customer support on your Commercial Renovation Costs with RSMeans data, call 800.448.8182.

Crews - Renovation

Crew B-88	Hr.	Daily	Hr.	Daily	Bare Costs	Incl. O&P
1 Laborer	$42.10	$336.80	$67.45	$539.60	$54.66	$85.75
6 Equip. Oper. (medium)	56.75	2724.00	88.80	4262.40		
2 Feller Bunchers, 100 H.P.		1177.20		1294.92		
1 Log Chipper, 22" Tree		552.00		607.20		
2 Log Skidders, 50 H.P.		1808.40		1989.24		
1 Dozer, 105 H.P.		633.85		697.24		
1 Chain Saw, Gas, 36" Long		41.25		45.38	75.23	82.75
56 L.H., Daily Totals		$7273.50		$9435.97	$129.88	$168.50

Crew B-89	Hr.	Daily	Hr.	Daily	Bare Costs	Incl. O&P
1 Equip. Oper. (light)	$53.00	$424.00	$82.95	$663.60	$50.13	$79.25
1 Truck Driver (light)	47.25	378.00	75.55	604.40		
1 Flatbed Truck, Gas, 3 Ton		820.40		902.44		
1 Concrete Saw		111.50		122.65		
1 Water Tank, 65 Gal.		101.90		112.09	64.61	71.07
16 L.H., Daily Totals		$1835.80		$2405.18	$114.74	$150.32

Crew B-89A	Hr.	Daily	Hr.	Daily	Bare Costs	Incl. O&P
1 Skilled Worker	$54.85	$438.80	$87.40	$699.20	$48.48	$77.42
1 Laborer	42.10	336.80	67.45	539.60		
1 Core Drill (Large)		114.30		125.73	7.14	7.86
16 L.H., Daily Totals		$889.90		$1364.53	$55.62	$85.28

Crew B-89B	Hr.	Daily	Hr.	Daily	Bare Costs	Incl. O&P
1 Equip. Oper. (light)	$53.00	$424.00	$82.95	$663.60	$50.13	$79.25
1 Truck Driver (light)	47.25	378.00	75.55	604.40		
1 Wall Saw, Hydraulic, 10 H.P.		85.50		94.05		
1 Generator, Diesel, 100 kW		496.20		545.82		
1 Water Tank, 65 Gal.		101.90		112.09		
1 Flatbed Truck, Gas, 3 Ton		820.40		902.44	94.00	103.40
16 L.H., Daily Totals		$2306.00		$2922.40	$144.13	$182.65

Crew B-89C	Hr.	Daily	Hr.	Daily	Bare Costs	Incl. O&P
1 Cement Finisher	$49.95	$399.60	$78.45	$627.60	$49.95	$78.45
1 Masonry cut-off saw, gas		61.00		67.10	7.63	8.39
8 L.H., Daily Totals		$460.60		$694.70	$57.58	$86.84

Crew B-90	Hr.	Daily	Hr.	Daily	Bare Costs	Incl. O&P
1 Labor Foreman (outside)	$44.10	$352.80	$70.65	$565.20	$46.79	$74.42
3 Laborers	42.10	1010.40	67.45	1618.80		
2 Equip. Oper. (light)	53.00	848.00	82.95	1327.20		
2 Truck Drivers (heavy)	48.95	783.20	78.25	1252.00		
1 Road Mixer, 310 H.P.		1896.00		2085.60		
1 Dist. Truck, 2000 Gal.		299.65		329.62	34.31	37.74
64 L.H., Daily Totals		$5190.05		$7178.42	$81.09	$112.16

Crew B-90A	Hr.	Daily	Hr.	Daily	Bare Costs	Incl. O&P
1 Labor Foreman (outside)	$44.10	$352.80	$70.65	$565.20	$50.76	$80.11
2 Laborers	42.10	673.60	67.45	1079.20		
4 Equip. Oper. (medium)	56.75	1816.00	88.80	2841.60		
2 Graders, 30,000 Lbs.		2124.00		2336.40		
1 Tandem Roller, 10 Ton		236.75		260.43		
1 Roller, Pneum. Whl., 12 Ton		345.75		380.32	48.33	53.16
56 L.H., Daily Totals		$5548.90		$7463.15	$99.09	$133.27

Crew B-90B	Hr.	Daily	Hr.	Daily	Bare Costs	Incl. O&P
1 Labor Foreman (outside)	$44.10	$352.80	$70.65	$565.20	$49.76	$78.66
2 Laborers	42.10	673.60	67.45	1079.20		
3 Equip. Oper. (medium)	56.75	1362.00	88.80	2131.20		
1 Roller, Pneum. Whl., 12 Ton		345.75		380.32		
1 Road Mixer, 310 H.P.		1896.00		2085.60	46.70	51.37
48 L.H., Daily Totals		$4630.15		$6241.52	$96.46	$130.03

Crew B-90C	Hr.	Daily	Hr.	Daily	Bare Costs	Incl. O&P
1 Labor Foreman (outside)	$44.10	$352.80	$70.65	$565.20	$48.15	$76.51
4 Laborers	42.10	1347.20	67.45	2158.40		
3 Equip. Oper. (medium)	56.75	1362.00	88.80	2131.20		
3 Truck Drivers (heavy)	48.95	1174.80	78.25	1878.00		
3 Road Mixers, 310 H.P.		5688.00		6256.80	64.64	71.10
88 L.H., Daily Totals		$9924.80		$12989.60	$112.78	$147.61

Crew B-90D	Hr.	Daily	Hr.	Daily	Bare Costs	Incl. O&P
1 Labor Foreman (outside)	$44.10	$352.80	$70.65	$565.20	$47.22	$75.12
6 Laborers	42.10	2020.80	67.45	3237.60		
3 Equip. Oper. (medium)	56.75	1362.00	88.80	2131.20		
3 Truck Drivers (heavy)	48.95	1174.80	78.25	1878.00		
3 Road Mixers, 310 H.P.		5688.00		6256.80	54.69	60.16
104 L.H., Daily Totals		$10598.40		$14068.80	$101.91	$135.28

Crew B-90E	Hr.	Daily	Hr.	Daily	Bare Costs	Incl. O&P
1 Labor Foreman (outside)	$44.10	$352.80	$70.65	$565.20	$47.97	$76.12
4 Laborers	42.10	1347.20	67.45	2158.40		
3 Equip. Oper. (medium)	56.75	1362.00	88.80	2131.20		
1 Truck Driver (heavy)	48.95	391.60	78.25	626.00		
1 Road Mixer, 310 H.P.		1896.00		2085.60	26.33	28.97
72 L.H., Daily Totals		$5349.60		$7566.40	$74.30	$105.09

Crew B-91	Hr.	Daily	Hr.	Daily	Bare Costs	Incl. O&P
1 Labor Foreman (outside)	$44.10	$352.80	$70.65	$565.20	$50.53	$79.88
2 Laborers	42.10	673.60	67.45	1079.20		
4 Equip. Oper. (medium)	56.75	1816.00	88.80	2841.60		
1 Truck Driver (heavy)	48.95	391.60	78.25	626.00		
1 Dist. Tanker, 3000 Gallon		330.15		363.17		
1 Truck Tractor, 6x4, 380 H.P.		493.25		542.58		
1 Aggreg. Spreader, S.P.		848.90		933.79		
1 Roller, Pneum. Whl., 12 Ton		345.75		380.32		
1 Tandem Roller, 10 Ton		236.75		260.43	35.23	38.75
64 L.H., Daily Totals		$5488.80		$7592.28	$85.76	$118.63

Crew B-91B	Hr.	Daily	Hr.	Daily	Bare Costs	Incl. O&P
1 Laborer	$42.10	$336.80	$67.45	$539.60	$49.42	$78.13
1 Equipment Oper. (med.)	56.75	454.00	88.80	710.40		
1 Road Sweeper, Vac. Assist.		869.00		955.90	54.31	59.74
16 L.H., Daily Totals		$1659.80		$2205.90	$103.74	$137.87

Crew B-91C	Hr.	Daily	Hr.	Daily	Bare Costs	Incl. O&P
1 Laborer	$42.10	$336.80	$67.45	$539.60	$44.67	$71.50
1 Truck Driver (light)	47.25	378.00	75.55	604.40		
1 Catch Basin Cleaning Truck		536.15		589.76	33.51	36.86
16 L.H., Daily Totals		$1250.95		$1733.77	$78.18	$108.36

For customer support on your Commercial Renovation Costs with RSMeans data, call 800.448.8182.

773

Crew No.	Bare Costs		Incl. Subs O&P		Cost Per Labor-Hour	

Crew B-91D	Hr.	Daily	Hr.	Daily	Bare Costs	Incl. O&P
1 Labor Foreman (outside)	$44.10	$352.80	$70.65	$565.20	$48.94	$77.57
5 Laborers	42.10	1684.00	67.45	2698.00		
5 Equip. Oper. (medium)	56.75	2270.00	88.80	3552.00		
2 Truck Drivers (heavy)	48.95	783.20	78.25	1252.00		
1 Aggreg. Spreader, S.P.		848.90		933.79		
2 Truck Tractors, 6x4, 380 H.P.		986.50		1085.15		
2 Dist. Tankers, 3000 Gallon		660.30		726.33		
2 Pavement Brushes, Towed		174.60		192.06		
2 Rollers Pneum. Whl., 12 Ton		691.50		760.65	32.33	35.56
104 L.H., Daily Totals		$8451.80		$11765.18	$81.27	$113.13

Crew B-92	Hr.	Daily	Hr.	Daily	Bare Costs	Incl. O&P
1 Labor Foreman (outside)	$44.10	$352.80	$70.65	$565.20	$42.60	$68.25
3 Laborers	42.10	1010.40	67.45	1618.80		
1 Crack Cleaner, 25 H.P.		52.45		57.70		
1 Air Compressor, 60 cfm		152.95		168.25		
1 Tar Kettle, T.M.		156.20		171.82		
1 Flatbed Truck, Gas, 3 Ton		820.40		902.44	36.94	40.63
32 L.H., Daily Totals		$2545.20		$3484.20	$79.54	$108.88

Crew B-93	Hr.	Daily	Hr.	Daily	Bare Costs	Incl. O&P
1 Equip. Oper. (medium)	$56.75	$454.00	$88.80	$710.40	$56.75	$88.80
1 Feller Buncher, 100 H.P.		588.60		647.46	73.58	80.93
8 L.H., Daily Totals		$1042.60		$1357.86	$130.32	$169.73

Crew B-94A	Hr.	Daily	Hr.	Daily	Bare Costs	Incl. O&P
1 Laborer	$42.10	$336.80	$67.45	$539.60	$42.10	$67.45
1 Diaphragm Water Pump, 2"		86.80		95.48		
1 -20' Suction Hose, 2"		3.55		3.90		
2 -50' Discharge Hoses, 2"		8.00		8.80	12.29	13.52
8 L.H., Daily Totals		$435.15		$647.78	$54.39	$80.97

Crew B-94B	Hr.	Daily	Hr.	Daily	Bare Costs	Incl. O&P
1 Laborer	$42.10	$336.80	$67.45	$539.60	$42.10	$67.45
1 Diaphragm Water Pump, 4"		105.25		115.78		
1 -20' Suction Hose, 4"		17.05		18.75		
2 -50' Discharge Hoses, 4"		25.30		27.83	18.45	20.30
8 L.H., Daily Totals		$484.40		$701.96	$60.55	$87.75

Crew B-94C	Hr.	Daily	Hr.	Daily	Bare Costs	Incl. O&P
1 Laborer	$42.10	$336.80	$67.45	$539.60	$42.10	$67.45
1 Centrifugal Water Pump, 3"		74.45		81.89		
1 -20' Suction Hose, 3"		8.75		9.63		
2 -50' Discharge Hoses, 3"		9.00		9.90	11.53	12.68
8 L.H., Daily Totals		$429.00		$641.02	$53.63	$80.13

Crew B-94D	Hr.	Daily	Hr.	Daily	Bare Costs	Incl. O&P
1 Laborer	$42.10	$336.80	$67.45	$539.60	$42.10	$67.45
1 Centr. Water Pump, 6"		232.90		256.19		
1 -20' Suction Hose, 6"		25.20		27.72		
2 -50' Discharge Hoses, 6"		35.80		39.38	36.74	40.41
8 L.H., Daily Totals		$630.70		$862.89	$78.84	$107.86

Crew C-1	Hr.	Daily	Hr.	Daily	Bare Costs	Incl. O&P
3 Carpenters	$53.15	$1275.60	$85.15	$2043.60	$50.39	$80.72
1 Laborer	42.10	336.80	67.45	539.60		
32 L.H., Daily Totals		$1612.40		$2583.20	$50.39	$80.72

Crew C-2	Hr.	Daily	Hr.	Daily	Bare Costs	Incl. O&P
1 Carpenter Foreman (outside)	$55.15	$441.20	$88.40	$707.20	$51.64	$82.74
4 Carpenters	53.15	1700.80	85.15	2724.80		
1 Laborer	42.10	336.80	67.45	539.60		
48 L.H., Daily Totals		$2478.80		$3971.60	$51.64	$82.74

Crew C-2A	Hr.	Daily	Hr.	Daily	Bare Costs	Incl. O&P
1 Carpenter Foreman (outside)	$55.15	$441.20	$88.40	$707.20	$51.11	$81.63
3 Carpenters	53.15	1275.60	85.15	2043.60		
1 Cement Finisher	49.95	399.60	78.45	627.60		
1 Laborer	42.10	336.80	67.45	539.60		
48 L.H., Daily Totals		$2453.20		$3918.00	$51.11	$81.63

Crew C-3	Hr.	Daily	Hr.	Daily	Bare Costs	Incl. O&P
1 Rodman Foreman (outside)	$58.40	$467.20	$92.00	$736.00	$52.65	$83.16
4 Rodmen (reinf.)	56.40	1804.80	88.85	2843.20		
1 Equip. Oper. (light)	53.00	424.00	82.95	663.60		
2 Laborers	42.10	673.60	67.45	1079.20		
3 Stressing Equipment		56.70		62.37		
.5 Grouting Equipment		122.08		134.28	2.79	3.07
64 L.H., Daily Totals		$3548.38		$5518.65	$55.44	$86.23

Crew C-4	Hr.	Daily	Hr.	Daily	Bare Costs	Incl. O&P
1 Rodman Foreman (outside)	$58.40	$467.20	$92.00	$736.00	$56.90	$89.64
3 Rodmen (reinf.)	56.40	1353.60	88.85	2132.40		
3 Stressing Equipment		56.70		62.37	1.77	1.95
32 L.H., Daily Totals		$1877.50		$2930.77	$58.67	$91.59

Crew C-4A	Hr.	Daily	Hr.	Daily	Bare Costs	Incl. O&P
2 Rodmen (reinf.)	$56.40	$902.40	$88.85	$1421.60	$56.40	$88.85
4 Stressing Equipment		75.60		83.16	4.72	5.20
16 L.H., Daily Totals		$978.00		$1504.76	$61.13	$94.05

Crew C-5	Hr.	Daily	Hr.	Daily	Bare Costs	Incl. O&P
1 Rodman Foreman (outside)	$58.40	$467.20	$92.00	$736.00	$56.25	$88.45
4 Rodmen (reinf.)	56.40	1804.80	88.85	2843.20		
1 Equip. Oper. (crane)	59.20	473.60	92.65	741.20		
1 Equip. Oper. (oiler)	50.55	404.40	79.10	632.80		
1 Hyd. Crane, 25 Ton		580.85		638.93	10.37	11.41
56 L.H., Daily Totals		$3730.85		$5592.14	$66.62	$99.86

Crew C-6	Hr.	Daily	Hr.	Daily	Bare Costs	Incl. O&P
1 Labor Foreman (outside)	$44.10	$352.80	$70.65	$565.20	$43.74	$69.82
4 Laborers	42.10	1347.20	67.45	2158.40		
1 Cement Finisher	49.95	399.60	78.45	627.60		
2 Gas Engine Vibrators		53.70		59.07	1.12	1.23
48 L.H., Daily Totals		$2153.30		$3410.27	$44.86	$71.05

Crew C-6A	Hr.	Daily	Hr.	Daily	Bare Costs	Incl. O&P
2 Cement Finishers	$49.95	$799.20	$78.45	$1255.20	$49.95	$78.45
1 Concrete Vibrator, Elec, 2 HP		47.45		52.20	2.97	3.26
16 L.H., Daily Totals		$846.65		$1307.40	$52.92	$81.71

Crew No.	Bare Costs		Incl. Subs O&P		Cost Per Labor-Hour	

Crew C-7	Hr.	Daily	Hr.	Daily	Bare Costs	Incl. O&P
1 Labor Foreman (outside)	$44.10	$352.80	$70.65	$565.20	$45.76	$72.69
5 Laborers	42.10	1684.00	67.45	2698.00		
1 Cement Finisher	49.95	399.60	78.45	627.60		
1 Equip. Oper. (medium)	56.75	454.00	88.80	710.40		
1 Equip. Oper. (oiler)	50.55	404.40	79.10	632.80		
2 Gas Engine Vibrators		53.70		59.07		
1 Concrete Bucket, 1 C.Y.		45.80		50.38		
1 Hyd. Crane, 55 Ton		980.20		1078.22	15.00	16.50
72 L.H., Daily Totals		$4374.50		$6421.67	$60.76	$89.19

Crew C-8	Hr.	Daily	Hr.	Daily	Bare Costs	Incl. O&P
1 Labor Foreman (outside)	$44.10	$352.80	$70.65	$565.20	$46.72	$74.10
3 Laborers	42.10	1010.40	67.45	1618.80		
2 Cement Finishers	49.95	799.20	78.45	1255.20		
1 Equip. Oper. (medium)	56.75	454.00	88.80	710.40		
1 Concrete Pump (Small)		410.35		451.38	7.33	8.06
56 L.H., Daily Totals		$3026.75		$4600.98	$54.05	$82.16

Crew C-8A	Hr.	Daily	Hr.	Daily	Bare Costs	Incl. O&P
1 Labor Foreman (outside)	$44.10	$352.80	$70.65	$565.20	$45.05	$71.65
3 Laborers	42.10	1010.40	67.45	1618.80		
2 Cement Finishers	49.95	799.20	78.45	1255.20		
48 L.H., Daily Totals		$2162.40		$3439.20	$45.05	$71.65

Crew C-8B	Hr.	Daily	Hr.	Daily	Bare Costs	Incl. O&P
1 Labor Foreman (outside)	$44.10	$352.80	$70.65	$565.20	$45.43	$72.36
3 Laborers	42.10	1010.40	67.45	1618.80		
1 Equip. Oper. (medium)	56.75	454.00	88.80	710.40		
1 Vibrating Power Screed		75.10		82.61		
1 Roller, Vibratory, 25 Ton		664.35		730.78		
1 Dozer, 200 H.P.		1504.00		1654.40	56.09	61.69
40 L.H., Daily Totals		$4060.65		$5362.19	$101.52	$134.05

Crew C-8C	Hr.	Daily	Hr.	Daily	Bare Costs	Incl. O&P
1 Labor Foreman (outside)	$44.10	$352.80	$70.65	$565.20	$46.18	$73.38
3 Laborers	42.10	1010.40	67.45	1618.80		
1 Cement Finisher	49.95	399.60	78.45	627.60		
1 Equip. Oper. (medium)	56.75	454.00	88.80	710.40		
1 Shotcrete Rig, 12 C.Y./hr		245.40		269.94		
1 Air Compressor, 160 cfm		202.80		223.08		
4 -50' Air Hoses, 1"		32.00		35.20		
4 -50' Air Hoses, 2"		115.60		127.16	12.41	13.65
48 L.H., Daily Totals		$2812.60		$4177.38	$58.60	$87.03

Crew C-8D	Hr.	Daily	Hr.	Daily	Bare Costs	Incl. O&P
1 Labor Foreman (outside)	$44.10	$352.80	$70.65	$565.20	$47.29	$74.88
1 Laborer	42.10	336.80	67.45	539.60		
1 Cement Finisher	49.95	399.60	78.45	627.60		
1 Equipment Oper. (light)	53.00	424.00	82.95	663.60		
1 Air Compressor, 250 cfm		201.50		221.65		
2 -50' Air Hoses, 1"		16.00		17.60	6.80	7.48
32 L.H., Daily Totals		$1730.70		$2635.25	$54.08	$82.35

Crew C-8E	Hr.	Daily	Hr.	Daily	Bare Costs	Incl. O&P
1 Labor Foreman (outside)	$44.10	$352.80	$70.65	$565.20	$45.56	$72.40
3 Laborers	42.10	1010.40	67.45	1618.80		
1 Cement Finisher	49.95	399.60	78.45	627.60		
1 Equipment Oper. (light)	53.00	424.00	82.95	663.60		
1 Shotcrete Rig, 35 C.Y./hr		298.00		327.80		
1 Air Compressor, 250 cfm		201.50		221.65		
4 -50' Air Hoses, 1"		32.00		35.20		
4 -50' Air Hoses, 2"		115.60		127.16	13.48	14.83
48 L.H., Daily Totals		$2833.90		$4187.01	$59.04	$87.23

Crew C-9	Hr.	Daily	Hr.	Daily	Bare Costs	Incl. O&P
1 Cement Finisher	$49.95	$399.60	$78.45	$627.60	$46.79	$74.08
2 Laborers	42.10	673.60	67.45	1079.20		
1 Equipment Oper. (light)	53.00	424.00	82.95	663.60		
1 Grout Pump, 50 C.F./hr.		188.45		207.29		
1 Air Compressor, 160 cfm		202.80		223.08		
2 -50' Air Hoses, 1"		16.00		17.60		
2 -50' Air Hoses, 2"		57.80		63.58	14.53	15.99
32 L.H., Daily Totals		$1962.25		$2881.95	$61.32	$90.06

Crew C-10	Hr.	Daily	Hr.	Daily	Bare Costs	Incl. O&P
1 Laborer	$42.10	$336.80	$67.45	$539.60	$47.33	$74.78
2 Cement Finishers	49.95	799.20	78.45	1255.20		
24 L.H., Daily Totals		$1136.00		$1794.80	$47.33	$74.78

Crew C-10B	Hr.	Daily	Hr.	Daily	Bare Costs	Incl. O&P
3 Laborers	$42.10	$1010.40	$67.45	$1618.80	$45.24	$71.85
2 Cement Finishers	49.95	799.20	78.45	1255.20		
1 Concrete Mixer, 10 C.F.		140.30		154.33		
2 Trowels, 48" Walk-Behind		174.00		191.40	7.86	8.64
40 L.H., Daily Totals		$2123.90		$3219.73	$53.10	$80.49

Crew C-10C	Hr.	Daily	Hr.	Daily	Bare Costs	Incl. O&P
1 Laborer	$42.10	$336.80	$67.45	$539.60	$47.33	$74.78
2 Cement Finishers	49.95	799.20	78.45	1255.20		
1 Trowel, 48" Walk-Behind		87.00		95.70	3.63	3.99
24 L.H., Daily Totals		$1223.00		$1890.50	$50.96	$78.77

Crew C-10D	Hr.	Daily	Hr.	Daily	Bare Costs	Incl. O&P
1 Laborer	$42.10	$336.80	$67.45	$539.60	$47.33	$74.78
2 Cement Finishers	49.95	799.20	78.45	1255.20		
1 Vibrating Power Screed		75.10		82.61		
1 Trowel, 48" Walk-Behind		87.00		95.70	6.75	7.43
24 L.H., Daily Totals		$1298.10		$1973.11	$54.09	$82.21

Crew C-10E	Hr.	Daily	Hr.	Daily	Bare Costs	Incl. O&P
1 Laborer	$42.10	$336.80	$67.45	$539.60	$47.33	$74.78
2 Cement Finishers	49.95	799.20	78.45	1255.20		
1 Vibrating Power Screed		75.10		82.61		
1 Cement Trowel, 96" Ride-On		169.00		185.90	10.17	11.19
24 L.H., Daily Totals		$1380.10		$2063.31	$57.50	$85.97

Crew C-10F	Hr.	Daily	Hr.	Daily	Bare Costs	Incl. O&P
1 Laborer	$42.10	$336.80	$67.45	$539.60	$47.33	$74.78
2 Cement Finishers	49.95	799.20	78.45	1255.20		
1 Telescoping Boom Lift, to 60'		310.90		341.99	12.95	14.25
24 L.H., Daily Totals		$1446.90		$2136.79	$60.29	$89.03

Crews - Renovation

Crew No.	Bare Costs Hr.	Bare Costs Daily	Incl. Subs O&P Hr.	Incl. Subs O&P Daily	Cost Per Labor-Hour Bare Costs	Cost Per Labor-Hour Incl. O&P
Crew C-11						
1 Struc. Steel Foreman (outside)	$59.65	$477.20	$97.75	$782.00	$57.26	$92.91
6 Struc. Steel Workers	57.65	2767.20	94.45	4533.60		
1 Equip. Oper. (crane)	59.20	473.60	92.65	741.20		
1 Equip. Oper. (oiler)	50.55	404.40	79.10	632.80		
1 Lattice Boom Crane, 150 Ton		2269.00		2495.90	31.51	34.67
72 L.H., Daily Totals		$6391.40		$9185.50	$88.77	$127.58
Crew C-12						
1 Carpenter Foreman (outside)	$55.15	$441.20	$88.40	$707.20	$52.65	$83.99
3 Carpenters	53.15	1275.60	85.15	2043.60		
1 Laborer	42.10	336.80	67.45	539.60		
1 Equip. Oper. (crane)	59.20	473.60	92.65	741.20		
1 Hyd. Crane, 12 Ton		471.00		518.10	9.81	10.79
48 L.H., Daily Totals		$2998.20		$4549.70	$62.46	$94.79
Crew C-13						
1 Struc. Steel Worker	$57.65	$461.20	$94.45	$755.60	$56.08	$91.25
1 Welder	57.45	459.60	94.15	753.20		
1 Carpenter	53.15	425.20	85.15	681.20		
1 Welder, Gas Engine, 300 amp		147.00		161.70	6.13	6.74
24 L.H., Daily Totals		$1493.00		$2351.70	$62.21	$97.99
Crew C-14						
1 Carpenter Foreman (outside)	$55.15	$441.20	$88.40	$707.20	$51.36	$81.56
5 Carpenters	53.15	2126.00	85.15	3406.00		
4 Laborers	42.10	1347.20	67.45	2158.40		
4 Rodmen (reinf.)	56.40	1804.80	88.85	2843.20		
2 Cement Finishers	49.95	799.20	78.45	1255.20		
1 Equip. Oper. (crane)	59.20	473.60	92.65	741.20		
1 Equip. Oper. (oiler)	50.55	404.40	79.10	632.80		
1 Hyd. Crane, 80 Ton		1486.00		1634.60	10.32	11.35
144 L.H., Daily Totals		$8882.40		$13378.60	$61.68	$92.91
Crew C-14A						
1 Carpenter Foreman (outside)	$55.15	$441.20	$88.40	$707.20	$52.88	$84.33
16 Carpenters	53.15	6803.20	85.15	10899.20		
4 Rodmen (reinf.)	56.40	1804.80	88.85	2843.20		
2 Laborers	42.10	673.60	67.45	1079.20		
1 Cement Finisher	49.95	399.60	78.45	627.60		
1 Equip. Oper. (medium)	56.75	454.00	88.80	710.40		
1 Gas Engine Vibrator		26.85		29.54		
1 Concrete Pump (Small)		410.35		451.38	2.19	2.40
200 L.H., Daily Totals		$11013.60		$17347.72	$55.07	$86.74
Crew C-14B						
1 Carpenter Foreman (outside)	$55.15	$441.20	$88.40	$707.20	$52.77	$84.11
16 Carpenters	53.15	6803.20	85.15	10899.20		
4 Rodmen (reinf.)	56.40	1804.80	88.85	2843.20		
2 Laborers	42.10	673.60	67.45	1079.20		
2 Cement Finishers	49.95	799.20	78.45	1255.20		
1 Equip. Oper. (medium)	56.75	454.00	88.80	710.40		
1 Gas Engine Vibrator		26.85		29.54		
1 Concrete Pump (Small)		410.35		451.38	2.10	2.31
208 L.H., Daily Totals		$11413.20		$17975.32	$54.87	$86.42
Crew C-14C						
1 Carpenter Foreman (outside)	$55.15	$441.20	$88.40	$707.20	$50.37	$80.38
6 Carpenters	53.15	2551.20	85.15	4087.20		
2 Rodmen (reinf.)	56.40	902.40	88.85	1421.60		
4 Laborers	42.10	1347.20	67.45	2158.40		
1 Cement Finisher	49.95	399.60	78.45	627.60		
1 Gas Engine Vibrator		26.85		29.54	.24	.26
112 L.H., Daily Totals		$5668.45		$9031.53	$50.61	$80.64
Crew C-14D						
1 Carpenter Foreman (outside)	$55.15	$441.20	$88.40	$707.20	$52.62	$84.04
18 Carpenters	53.15	7653.60	85.15	12261.60		
2 Rodmen (reinf.)	56.40	902.40	88.85	1421.60		
2 Laborers	42.10	673.60	67.45	1079.20		
1 Cement Finisher	49.95	399.60	78.45	627.60		
1 Equip. Oper. (medium)	56.75	454.00	88.80	710.40		
1 Gas Engine Vibrator		26.85		29.54		
1 Concrete Pump (Small)		410.35		451.38	2.19	2.40
200 L.H., Daily Totals		$10961.60		$17288.52	$54.81	$86.44
Crew C-14E						
1 Carpenter Foreman (outside)	$55.15	$441.20	$88.40	$707.20	$51.21	$81.35
2 Carpenters	53.15	850.40	85.15	1362.40		
4 Rodmen (reinf.)	56.40	1804.80	88.85	2843.20		
3 Laborers	42.10	1010.40	67.45	1618.80		
1 Cement Finisher	49.95	399.60	78.45	627.60		
1 Gas Engine Vibrator		26.85		29.54	.31	.34
88 L.H., Daily Totals		$4533.25		$7188.73	$51.51	$81.69
Crew C-14F						
1 Labor Foreman (outside)	$44.10	$352.80	$70.65	$565.20	$47.56	$75.14
2 Laborers	42.10	673.60	67.45	1079.20		
6 Cement Finishers	49.95	2397.60	78.45	3765.60		
1 Gas Engine Vibrator		26.85		29.54	.37	.41
72 L.H., Daily Totals		$3450.85		$5439.53	$47.93	$75.55
Crew C-14G						
1 Labor Foreman (outside)	$44.10	$352.80	$70.65	$565.20	$46.87	$74.19
2 Laborers	42.10	673.60	67.45	1079.20		
4 Cement Finishers	49.95	1598.40	78.45	2510.40		
1 Gas Engine Vibrator		26.85		29.54	.48	.53
56 L.H., Daily Totals		$2651.65		$4184.34	$47.35	$74.72
Crew C-14H						
1 Carpenter Foreman (outside)	$55.15	$441.20	$88.40	$707.20	$51.65	$82.24
2 Carpenters	53.15	850.40	85.15	1362.40		
1 Rodman (reinf.)	56.40	451.20	88.85	710.80		
1 Laborer	42.10	336.80	67.45	539.60		
1 Cement Finisher	49.95	399.60	78.45	627.60		
1 Gas Engine Vibrator		26.85		29.54	.56	.62
48 L.H., Daily Totals		$2506.05		$3977.14	$52.21	$82.86
Crew C-14L						
1 Carpenter Foreman (outside)	$55.15	$441.20	$88.40	$707.20	$49.37	$78.96
6 Carpenters	53.15	2551.20	85.15	4087.20		
4 Laborers	42.10	1347.20	67.45	2158.40		
1 Cement Finisher	49.95	399.60	78.45	627.60		
1 Gas Engine Vibrator		26.85		29.54	.28	.31
96 L.H., Daily Totals		$4766.05		$7609.94	$49.65	$79.27

For customer support on your Commercial Renovation Costs with RSMeans data, call 800.448.8182.

Crew C-14M

	Bare Costs Hr.	Bare Costs Daily	Incl. Subs O&P Hr.	Incl. Subs O&P Daily	Cost Per Labor-Hour Bare Costs	Cost Per Labor-Hour Incl. O&P
1 Carpenter Foreman (outside)	$55.15	$441.20	$88.40	$707.20	$51.09	$81.21
2 Carpenters	53.15	850.40	85.15	1362.40		
1 Rodman (reinf.)	56.40	451.20	88.85	710.80		
2 Laborers	42.10	673.60	67.45	1079.20		
1 Cement Finisher	49.95	399.60	78.45	627.60		
1 Equip. Oper. (medium)	56.75	454.00	88.80	710.40		
1 Gas Engine Vibrator		26.85		29.54		
1 Concrete Pump (Small)		410.35		451.38	6.83	7.51
64 L.H., Daily Totals		$3707.20		$5678.52	$57.93	$88.73

Crew C-15

	Bare Costs Hr.	Bare Costs Daily	Incl. Subs O&P Hr.	Incl. Subs O&P Daily	Cost Per Labor-Hour Bare Costs	Cost Per Labor-Hour Incl. O&P
1 Carpenter Foreman (outside)	$55.15	$441.20	$88.40	$707.20	$49.34	$78.53
2 Carpenters	53.15	850.40	85.15	1362.40		
3 Laborers	42.10	1010.40	67.45	1618.80		
2 Cement Finishers	49.95	799.20	78.45	1255.20		
1 Rodman (reinf.)	56.40	451.20	88.85	710.80		
72 L.H., Daily Totals		$3552.40		$5654.40	$49.34	$78.53

Crew C-16

	Bare Costs Hr.	Bare Costs Daily	Incl. Subs O&P Hr.	Incl. Subs O&P Daily	Cost Per Labor-Hour Bare Costs	Cost Per Labor-Hour Incl. O&P
1 Labor Foreman (outside)	$44.10	$352.80	$70.65	$565.20	$46.72	$74.10
3 Laborers	42.10	1010.40	67.45	1618.80		
2 Cement Finishers	49.95	799.20	78.45	1255.20		
1 Equip. Oper. (medium)	56.75	454.00	88.80	710.40		
1 Gunite Pump Rig		317.90		349.69		
2 -50' Air Hoses, 3/4"		15.50		17.05		
2 -50' Air Hoses, 2"		57.80		63.58	6.99	7.68
56 L.H., Daily Totals		$3007.60		$4579.92	$53.71	$81.78

Crew C-16A

	Bare Costs Hr.	Bare Costs Daily	Incl. Subs O&P Hr.	Incl. Subs O&P Daily	Cost Per Labor-Hour Bare Costs	Cost Per Labor-Hour Incl. O&P
1 Laborer	$42.10	$336.80	$67.45	$539.60	$49.69	$78.29
2 Cement Finishers	49.95	799.20	78.45	1255.20		
1 Equip. Oper. (medium)	56.75	454.00	88.80	710.40		
1 Gunite Pump Rig		317.90		349.69		
2 -50' Air Hoses, 3/4"		15.50		17.05		
2 -50' Air Hoses, 2"		57.80		63.58		
1 Telescoping Boom Lift, to 60'		310.90		341.99	21.94	24.13
32 L.H., Daily Totals		$2292.10		$3277.51	$71.63	$102.42

Crew C-17

	Bare Costs Hr.	Bare Costs Daily	Incl. Subs O&P Hr.	Incl. Subs O&P Daily	Cost Per Labor-Hour Bare Costs	Cost Per Labor-Hour Incl. O&P
2 Skilled Worker Foremen (out)	$56.85	$909.60	$90.55	$1448.80	$55.25	$88.03
8 Skilled Workers	54.85	3510.40	87.40	5593.60		
80 L.H., Daily Totals		$4420.00		$7042.40	$55.25	$88.03

Crew C-17A

	Bare Costs Hr.	Bare Costs Daily	Incl. Subs O&P Hr.	Incl. Subs O&P Daily	Cost Per Labor-Hour Bare Costs	Cost Per Labor-Hour Incl. O&P
2 Skilled Worker Foremen (out)	$56.85	$909.60	$90.55	$1448.80	$55.30	$88.09
8 Skilled Workers	54.85	3510.40	87.40	5593.60		
.125 Equip. Oper. (crane)	59.20	59.20	92.65	92.65		
.125 Hyd. Crane, 80 Ton		185.75		204.32	2.29	2.52
81 L.H., Daily Totals		$4664.95		$7339.38	$57.59	$90.61

Crew C-17B

	Bare Costs Hr.	Bare Costs Daily	Incl. Subs O&P Hr.	Incl. Subs O&P Daily	Cost Per Labor-Hour Bare Costs	Cost Per Labor-Hour Incl. O&P
2 Skilled Worker Foremen (out)	$56.85	$909.60	$90.55	$1448.80	$55.35	$88.14
8 Skilled Workers	54.85	3510.40	87.40	5593.60		
.25 Equip. Oper. (crane)	59.20	118.40	92.65	185.30		
.25 Hyd. Crane, 80 Ton		371.50		408.65		
.25 Trowel, 48" Walk-Behind		21.75		23.93	4.80	5.28
82 L.H., Daily Totals		$4931.65		$7660.27	$60.14	$93.42

Crew C-17C

	Bare Costs Hr.	Bare Costs Daily	Incl. Subs O&P Hr.	Incl. Subs O&P Daily	Cost Per Labor-Hour Bare Costs	Cost Per Labor-Hour Incl. O&P
2 Skilled Worker Foremen (out)	$56.85	$909.60	$90.55	$1448.80	$55.39	$88.20
8 Skilled Workers	54.85	3510.40	87.40	5593.60		
.375 Equip. Oper. (crane)	59.20	177.60	92.65	277.95		
.375 Hyd. Crane, 80 Ton		557.25		612.98	6.71	7.39
83 L.H., Daily Totals		$5154.85		$7933.32	$62.11	$95.58

Crew C-17D

	Bare Costs Hr.	Bare Costs Daily	Incl. Subs O&P Hr.	Incl. Subs O&P Daily	Cost Per Labor-Hour Bare Costs	Cost Per Labor-Hour Incl. O&P
2 Skilled Worker Foremen (out)	$56.85	$909.60	$90.55	$1448.80	$55.44	$88.25
8 Skilled Workers	54.85	3510.40	87.40	5593.60		
.5 Equip. Oper. (crane)	59.20	236.80	92.65	370.60		
.5 Hyd. Crane, 80 Ton		743.00		817.30	8.85	9.73
84 L.H., Daily Totals		$5399.80		$8230.30	$64.28	$97.98

Crew C-17E

	Bare Costs Hr.	Bare Costs Daily	Incl. Subs O&P Hr.	Incl. Subs O&P Daily	Cost Per Labor-Hour Bare Costs	Cost Per Labor-Hour Incl. O&P
2 Skilled Worker Foremen (out)	$56.85	$909.60	$90.55	$1448.80	$55.25	$88.03
8 Skilled Workers	54.85	3510.40	87.40	5593.60		
1 Hyd. Jack with Rods		35.90		39.49	.45	.49
80 L.H., Daily Totals		$4455.90		$7081.89	$55.70	$88.52

Crew C-18

	Bare Costs Hr.	Bare Costs Daily	Incl. Subs O&P Hr.	Incl. Subs O&P Daily	Cost Per Labor-Hour Bare Costs	Cost Per Labor-Hour Incl. O&P
.125 Labor Foreman (outside)	$44.10	$44.10	$70.65	$70.65	$42.32	$67.81
1 Laborer	42.10	336.80	67.45	539.60		
1 Concrete Cart, 10 C.F.		127.85		140.63	14.21	15.63
9 L.H., Daily Totals		$508.75		$750.88	$56.53	$83.43

Crew C-19

	Bare Costs Hr.	Bare Costs Daily	Incl. Subs O&P Hr.	Incl. Subs O&P Daily	Cost Per Labor-Hour Bare Costs	Cost Per Labor-Hour Incl. O&P
.125 Labor Foreman (outside)	$44.10	$44.10	$70.65	$70.65	$42.32	$67.81
1 Laborer	42.10	336.80	67.45	539.60		
1 Concrete Cart, 18 C.F.		153.45		168.79	17.05	18.75
9 L.H., Daily Totals		$534.35		$779.04	$59.37	$86.56

Crew C-20

	Bare Costs Hr.	Bare Costs Daily	Incl. Subs O&P Hr.	Incl. Subs O&P Daily	Cost Per Labor-Hour Bare Costs	Cost Per Labor-Hour Incl. O&P
1 Labor Foreman (outside)	$44.10	$352.80	$70.65	$565.20	$45.16	$71.89
5 Laborers	42.10	1684.00	67.45	2698.00		
1 Cement Finisher	49.95	399.60	78.45	627.60		
1 Equip. Oper. (medium)	56.75	454.00	88.80	710.40		
2 Gas Engine Vibrators		53.70		59.07		
1 Concrete Pump (Small)		410.35		451.38	7.25	7.98
64 L.H., Daily Totals		$3354.45		$5111.65	$52.41	$79.87

Crew C-21

	Bare Costs Hr.	Bare Costs Daily	Incl. Subs O&P Hr.	Incl. Subs O&P Daily	Cost Per Labor-Hour Bare Costs	Cost Per Labor-Hour Incl. O&P
1 Labor Foreman (outside)	$44.10	$352.80	$70.65	$565.20	$45.16	$71.89
5 Laborers	42.10	1684.00	67.45	2698.00		
1 Cement Finisher	49.95	399.60	78.45	627.60		
1 Equip. Oper. (medium)	56.75	454.00	88.80	710.40		
2 Gas Engine Vibrators		53.70		59.07		
1 Concrete Conveyer		204.15		224.57	4.03	4.43
64 L.H., Daily Totals		$3148.25		$4884.84	$49.19	$76.33

Crew C-22

	Bare Costs Hr.	Bare Costs Daily	Incl. Subs O&P Hr.	Incl. Subs O&P Daily	Cost Per Labor-Hour Bare Costs	Cost Per Labor-Hour Incl. O&P
1 Rodman Foreman (outside)	$58.40	$467.20	$92.00	$736.00	$56.71	$89.31
4 Rodmen (reinf.)	56.40	1804.80	88.85	2843.20		
.125 Equip. Oper. (crane)	59.20	59.20	92.65	92.65		
.125 Equip. Oper. (oiler)	50.55	50.55	79.10	79.10		
.125 Hyd. Crane, 25 Ton		72.61		79.87	1.73	1.90
42 L.H., Daily Totals		$2454.36		$3830.82	$58.44	$91.21

Crew No.	Bare Costs		Incl. Subs O&P		Cost Per Labor-Hour	
Crew C-23	Hr.	Daily	Hr.	Daily	Bare Costs	Incl. O&P
2 Skilled Worker Foremen (out)	$56.85	$909.60	$90.55	$1448.80	$55.26	$87.72
6 Skilled Workers	54.85	2632.80	87.40	4195.20		
1 Equip. Oper. (crane)	59.20	473.60	92.65	741.20		
1 Equip. Oper. (oiler)	50.55	404.40	79.10	632.80		
1 Lattice Boom Crane, 90 Ton		1696.00		1865.60	21.20	23.32
80 L.H., Daily Totals		$6116.40		$8883.60	$76.45	$111.05
Crew C-24	Hr.	Daily	Hr.	Daily	Bare Costs	Incl. O&P
2 Skilled Worker Foremen (out)	$56.85	$909.60	$90.55	$1448.80	$55.26	$87.72
6 Skilled Workers	54.85	2632.80	87.40	4195.20		
1 Equip. Oper. (crane)	59.20	473.60	92.65	741.20		
1 Equip. Oper. (oiler)	50.55	404.40	79.10	632.80		
1 Lattice Boom Crane, 150 Ton		2269.00		2495.90	28.36	31.20
80 L.H., Daily Totals		$6689.40		$9513.90	$83.62	$118.92
Crew C-25	Hr.	Daily	Hr.	Daily	Bare Costs	Incl. O&P
2 Rodmen (reinf.)	$56.40	$902.40	$88.85	$1421.60	$45.50	$74.70
2 Rodmen Helpers	34.60	553.60	60.55	968.80		
32 L.H., Daily Totals		$1456.00		$2390.40	$45.50	$74.70
Crew C-27	Hr.	Daily	Hr.	Daily	Bare Costs	Incl. O&P
2 Cement Finishers	$49.95	$799.20	$78.45	$1255.20	$49.95	$78.45
1 Concrete Saw		111.50		122.65	6.97	7.67
16 L.H., Daily Totals		$910.70		$1377.85	$56.92	$86.12
Crew C-28	Hr.	Daily	Hr.	Daily	Bare Costs	Incl. O&P
1 Cement Finisher	$49.95	$399.60	$78.45	$627.60	$49.95	$78.45
1 Portable Air Compressor, Gas		38.70		42.57	4.84	5.32
8 L.H., Daily Totals		$438.30		$670.17	$54.79	$83.77
Crew C-29	Hr.	Daily	Hr.	Daily	Bare Costs	Incl. O&P
1 Laborer	$42.10	$336.80	$67.45	$539.60	$42.10	$67.45
1 Pressure Washer		96.95		106.65	12.12	13.33
8 L.H., Daily Totals		$433.75		$646.25	$54.22	$80.78
Crew C-30	Hr.	Daily	Hr.	Daily	Bare Costs	Incl. O&P
1 Laborer	$42.10	$336.80	$67.45	$539.60	$42.10	$67.45
1 Concrete Mixer, 10 C.F.		140.30		154.33	17.54	19.29
8 L.H., Daily Totals		$477.10		$693.93	$59.64	$86.74
Crew C-31	Hr.	Daily	Hr.	Daily	Bare Costs	Incl. O&P
1 Cement Finisher	$49.95	$399.60	$78.45	$627.60	$49.95	$78.45
1 Grout Pump		317.90		349.69	39.74	43.71
8 L.H., Daily Totals		$717.50		$977.29	$89.69	$122.16
Crew C-32	Hr.	Daily	Hr.	Daily	Bare Costs	Incl. O&P
1 Cement Finisher	$49.95	$399.60	$78.45	$627.60	$46.02	$72.95
1 Laborer	42.10	336.80	67.45	539.60		
1 Crack Chaser Saw, Gas, 6 H.P.		73.10		80.41		
1 Vacuum Pick-Up System		74.05		81.45	9.20	10.12
16 L.H., Daily Totals		$883.55		$1329.07	$55.22	$83.07
Crew D-1	Hr.	Daily	Hr.	Daily	Bare Costs	Incl. O&P
1 Bricklayer	$52.05	$416.40	$84.10	$672.80	$46.90	$75.78
1 Bricklayer Helper	41.75	334.00	67.45	539.60		
16 L.H., Daily Totals		$750.40		$1212.40	$46.90	$75.78

Crew No.	Bare Costs		Incl. Subs O&P		Cost Per Labor-Hour	
Crew D-2	Hr.	Daily	Hr.	Daily	Bare Costs	Incl. O&P
3 Bricklayers	$52.05	$1249.20	$84.10	$2018.40	$48.40	$78.14
2 Bricklayer Helpers	41.75	668.00	67.45	1079.20		
.5 Carpenter	53.15	212.60	85.15	340.60		
44 L.H., Daily Totals		$2129.80		$3438.20	$48.40	$78.14
Crew D-3	Hr.	Daily	Hr.	Daily	Bare Costs	Incl. O&P
3 Bricklayers	$52.05	$1249.20	$84.10	$2018.40	$48.18	$77.81
2 Bricklayer Helpers	41.75	668.00	67.45	1079.20		
.25 Carpenter	53.15	106.30	85.15	170.30		
42 L.H., Daily Totals		$2023.50		$3267.90	$48.18	$77.81
Crew D-4	Hr.	Daily	Hr.	Daily	Bare Costs	Incl. O&P
1 Bricklayer	$52.05	$416.40	$84.10	$672.80	$47.14	$75.49
2 Bricklayer Helpers	41.75	668.00	67.45	1079.20		
1 Equip. Oper. (light)	53.00	424.00	82.95	663.60		
1 Grout Pump, 50 C.F./hr.		188.45		207.29	5.89	6.48
32 L.H., Daily Totals		$1696.85		$2622.90	$53.03	$81.97
Crew D-5	Hr.	Daily	Hr.	Daily	Bare Costs	Incl. O&P
1 Bricklayer	52.05	416.40	84.10	672.80	52.05	84.10
8 L.H., Daily Totals		$416.40		$672.80	$52.05	$84.10
Crew D-6	Hr.	Daily	Hr.	Daily	Bare Costs	Incl. O&P
3 Bricklayers	$52.05	$1249.20	$84.10	$2018.40	$47.15	$76.15
3 Bricklayer Helpers	41.75	1002.00	67.45	1618.80		
.25 Carpenter	53.15	106.30	85.15	170.30		
50 L.H., Daily Totals		$2357.50		$3807.50	$47.15	$76.15
Crew D-7	Hr.	Daily	Hr.	Daily	Bare Costs	Incl. O&P
1 Tile Layer	$49.50	$396.00	$77.55	$620.40	$44.15	$69.17
1 Tile Layer Helper	38.80	310.40	60.80	486.40		
16 L.H., Daily Totals		$706.40		$1106.80	$44.15	$69.17
Crew D-8	Hr.	Daily	Hr.	Daily	Bare Costs	Incl. O&P
3 Bricklayers	$52.05	$1249.20	$84.10	$2018.40	$47.93	$77.44
2 Bricklayer Helpers	41.75	668.00	67.45	1079.20		
40 L.H., Daily Totals		$1917.20		$3097.60	$47.93	$77.44
Crew D-9	Hr.	Daily	Hr.	Daily	Bare Costs	Incl. O&P
3 Bricklayers	$52.05	$1249.20	$84.10	$2018.40	$46.90	$75.78
3 Bricklayer Helpers	41.75	1002.00	67.45	1618.80		
48 L.H., Daily Totals		$2251.20		$3637.20	$46.90	$75.78
Crew D-10	Hr.	Daily	Hr.	Daily	Bare Costs	Incl. O&P
1 Bricklayer Foreman (outside)	$54.05	$432.40	$87.35	$698.80	$51.76	$82.89
1 Bricklayer	52.05	416.40	84.10	672.80		
1 Bricklayer Helper	41.75	334.00	67.45	539.60		
1 Equip. Oper. (crane)	59.20	473.60	92.65	741.20		
1 S.P. Crane, 4x4, 12 Ton		428.30		471.13	13.38	14.72
32 L.H., Daily Totals		$2084.70		$3123.53	$65.15	$97.61
Crew D-11	Hr.	Daily	Hr.	Daily	Bare Costs	Incl. O&P
1 Bricklayer Foreman (outside)	$54.05	$432.40	$87.35	$698.80	$49.28	$79.63
1 Bricklayer	52.05	416.40	84.10	672.80		
1 Bricklayer Helper	41.75	334.00	67.45	539.60		
24 L.H., Daily Totals		$1182.80		$1911.20	$49.28	$79.63

Crew No.	Bare Costs		Incl. Subs O&P		Cost Per Labor-Hour	
Crew D-12	Hr.	Daily	Hr.	Daily	Bare Costs	Incl. O&P
1 Bricklayer Foreman (outside)	$54.05	$432.40	$87.35	$698.80	$47.40	$76.59
1 Bricklayer	52.05	416.40	84.10	672.80		
2 Bricklayer Helpers	41.75	668.00	67.45	1079.20		
32 L.H., Daily Totals		$1516.80		$2450.80	$47.40	$76.59
Crew D-13	Hr.	Daily	Hr.	Daily	Bare Costs	Incl. O&P
1 Bricklayer Foreman (outside)	$54.05	$432.40	$87.35	$698.80	$50.33	$80.69
1 Bricklayer	52.05	416.40	84.10	672.80		
2 Bricklayer Helpers	41.75	668.00	67.45	1079.20		
1 Carpenter	53.15	425.20	85.15	681.20		
1 Equip. Oper. (crane)	59.20	473.60	92.65	741.20		
1 S.P. Crane, 4x4, 12 Ton		428.30		471.13	8.92	9.82
48 L.H., Daily Totals		$2843.90		$4344.33	$59.25	$90.51
Crew D-14	Hr.	Daily	Hr.	Daily	Bare Costs	Incl. O&P
3 Bricklayers	$52.05	$1249.20	$84.10	$2018.40	$49.48	$79.94
1 Bricklayer Helper	41.75	334.00	67.45	539.60		
32 L.H., Daily Totals		$1583.20		$2558.00	$49.48	$79.94
Crew E-1	Hr.	Daily	Hr.	Daily	Bare Costs	Incl. O&P
1 Welder Foreman (outside)	$59.45	$475.60	$97.40	$779.20	$56.63	$91.50
1 Welder	57.45	459.60	94.15	753.20		
1 Equip. Oper. (light)	53.00	424.00	82.95	663.60		
1 Welder, Gas Engine, 300 amp		147.00		161.70	6.13	6.74
24 L.H., Daily Totals		$1506.20		$2357.70	$62.76	$98.24
Crew E-2	Hr.	Daily	Hr.	Daily	Bare Costs	Incl. O&P
1 Struc. Steel Foreman (outside)	$59.65	$477.20	$97.75	$782.00	$57.14	$92.47
4 Struc. Steel Workers	57.65	1844.80	94.45	3022.40		
1 Equip. Oper. (crane)	59.20	473.60	92.65	741.20		
1 Equip. Oper. (oiler)	50.55	404.40	79.10	632.80		
1 Lattice Boom Crane, 90 Ton		1696.00		1865.60	30.29	33.31
56 L.H., Daily Totals		$4896.00		$7044.00	$87.43	$125.79
Crew E-3	Hr.	Daily	Hr.	Daily	Bare Costs	Incl. O&P
1 Struc. Steel Foreman (outside)	$59.65	$477.20	$97.75	$782.00	$58.25	$95.45
1 Struc. Steel Worker	57.65	461.20	94.45	755.60		
1 Welder	57.45	459.60	94.15	753.20		
1 Welder, Gas Engine, 300 amp		147.00		161.70	6.13	6.74
24 L.H., Daily Totals		$1545.00		$2452.50	$64.38	$102.19
Crew E-3A	Hr.	Daily	Hr.	Daily	Bare Costs	Incl. O&P
1 Struc. Steel Foreman (outside)	$59.65	$477.20	$97.75	$782.00	$58.25	$95.45
1 Struc. Steel Worker	57.65	461.20	94.45	755.60		
1 Welder	57.45	459.60	94.15	753.20		
1 Welder, Gas Engine, 300 amp		147.00		161.70		
1 Telescoping Boom Lift, to 40'		278.90		306.79	17.75	19.52
24 L.H., Daily Totals		$1823.90		$2759.29	$76.00	$114.97
Crew E-4	Hr.	Daily	Hr.	Daily	Bare Costs	Incl. O&P
1 Struc. Steel Foreman (outside)	$59.65	$477.20	$97.75	$782.00	$58.15	$95.28
3 Struc. Steel Workers	57.65	1383.60	94.45	2266.80		
1 Welder, Gas Engine, 300 amp		147.00		161.70	4.59	5.05
32 L.H., Daily Totals		$2007.80		$3210.50	$62.74	$100.33

Crew No.	Bare Costs		Incl. Subs O&P		Cost Per Labor-Hour	
Crew E-5	Hr.	Daily	Hr.	Daily	Bare Costs	Incl. O&P
2 Struc. Steel Foremen (outside)	$59.65	$954.40	$97.75	$1564.00	$57.48	$93.36
5 Struc. Steel Workers	57.65	2306.00	94.45	3778.00		
1 Equip. Oper. (crane)	59.20	473.60	92.65	741.20		
1 Welder	57.45	459.60	94.15	753.20		
1 Equip. Oper. (oiler)	50.55	404.40	79.10	632.80		
1 Lattice Boom Crane, 90 Ton		1696.00		1865.60		
1 Welder, Gas Engine, 300 amp		147.00		161.70	23.04	25.34
80 L.H., Daily Totals		$6441.00		$9496.50	$80.51	$118.71
Crew E-6	Hr.	Daily	Hr.	Daily	Bare Costs	Incl. O&P
3 Struc. Steel Foremen (outside)	$59.65	$1431.60	$97.75	$2346.00	$57.38	$93.26
9 Struc. Steel Workers	57.65	4150.80	94.45	6800.40		
1 Equip. Oper. (crane)	59.20	473.60	92.65	741.20		
1 Welder	57.45	459.60	94.15	753.20		
1 Equip. Oper. (oiler)	50.55	404.40	79.10	632.80		
1 Equip. Oper. (light)	53.00	424.00	82.95	663.60		
1 Lattice Boom Crane, 90 Ton		1696.00		1865.60		
1 Welder, Gas Engine, 300 amp		147.00		161.70		
1 Air Compressor, 160 cfm		202.80		223.08		
2 Impact Wrenches		104.60		115.06	16.80	18.48
128 L.H., Daily Totals		$9494.40		$14302.64	$74.17	$111.74
Crew E-7	Hr.	Daily	Hr.	Daily	Bare Costs	Incl. O&P
1 Struc. Steel Foreman (outside)	$59.65	$477.20	$97.75	$782.00	$57.44	$93.30
4 Struc. Steel Workers	57.65	1844.80	94.45	3022.40		
1 Equip. Oper. (crane)	59.20	473.60	92.65	741.20		
1 Equip. Oper. (oiler)	50.55	404.40	79.10	632.80		
1 Welder Foreman (outside)	59.45	475.60	97.40	779.20		
2 Welders	57.45	919.20	94.15	1506.40		
1 Lattice Boom Crane, 90 Ton		1696.00		1865.60		
2 Welder, Gas Engine, 300 amp		294.00		323.40	24.88	27.36
80 L.H., Daily Totals		$6584.80		$9653.00	$82.31	$120.66
Crew E-8	Hr.	Daily	Hr.	Daily	Bare Costs	Incl. O&P
1 Struc. Steel Foreman (outside)	$59.65	$477.20	$97.75	$782.00	$57.10	$92.63
4 Struc. Steel Workers	57.65	1844.80	94.45	3022.40		
1 Welder Foreman (outside)	59.45	475.60	97.40	779.20		
4 Welders	57.45	1838.40	94.15	3012.80		
1 Equip. Oper. (crane)	59.20	473.60	92.65	741.20		
1 Equip. Oper. (oiler)	50.55	404.40	79.10	632.80		
1 Equip. Oper. (light)	53.00	424.00	82.95	663.60		
1 Lattice Boom Crane, 90 Ton		1696.00		1865.60		
4 Welder, Gas Engine, 300 amp		588.00		646.80	21.96	24.16
104 L.H., Daily Totals		$8222.00		$12146.40	$79.06	$116.79
Crew E-9	Hr.	Daily	Hr.	Daily	Bare Costs	Incl. O&P
2 Struc. Steel Foremen (outside)	$59.65	$954.40	$97.75	$1564.00	$57.31	$93.16
5 Struc. Steel Workers	57.65	2306.00	94.45	3778.00		
1 Welder Foreman (outside)	59.45	475.60	97.40	779.20		
5 Welders	57.45	2298.00	94.15	3766.00		
1 Equip. Oper. (crane)	59.20	473.60	92.65	741.20		
1 Equip. Oper. (oiler)	50.55	404.40	79.10	632.80		
1 Equip. Oper. (light)	53.00	424.00	82.95	663.60		
1 Lattice Boom Crane, 90 Ton		1696.00		1865.60		
5 Welder, Gas Engine, 300 amp		735.00		808.50	18.99	20.89
128 L.H., Daily Totals		$9767.00		$14598.90	$76.30	$114.05

For customer support on your Commercial Renovation Costs with RSMeans data, call 800.448.8182.

779

Crew No.	Bare Costs		Incl. Subs O&P		Cost Per Labor-Hour	

Crew E-10	Hr.	Daily	Hr.	Daily	Bare Costs	Incl. O&P
1 Welder Foreman (outside)	$59.45	$475.60	$97.40	$779.20	$58.45	$95.78
1 Welder	57.45	459.60	94.15	753.20		
1 Welder, Gas Engine, 300 amp		147.00		161.70		
1 Flatbed Truck, Gas, 3 Ton		820.40		902.44	60.46	66.51
16 L.H., Daily Totals		$1902.60		$2596.54	$118.91	$162.28

Crew E-11	Hr.	Daily	Hr.	Daily	Bare Costs	Incl. O&P
2 Painters, Struc. Steel	$45.85	$733.60	$79.00	$1264.00	$46.70	$77.10
1 Building Laborer	42.10	336.80	67.45	539.60		
1 Equip. Oper. (light)	53.00	424.00	82.95	663.60		
1 Air Compressor, 250 cfm		201.50		221.65		
1 Sandblaster, Portable, 3 C.F.		83.80		92.18		
1 Set Sand Blasting Accessories		15.35		16.89	9.40	10.33
32 L.H., Daily Totals		$1795.05		$2797.92	$56.10	$87.43

Crew E-11A	Hr.	Daily	Hr.	Daily	Bare Costs	Incl. O&P
2 Painters, Struc. Steel	$45.85	$733.60	$79.00	$1264.00	$46.70	$77.10
1 Building Laborer	42.10	336.80	67.45	539.60		
1 Equip. Oper. (light)	53.00	424.00	82.95	663.60		
1 Air Compressor, 250 cfm		201.50		221.65		
1 Sandblaster, Portable, 3 C.F.		83.80		92.18		
1 Set Sand Blasting Accessories		15.35		16.89		
1 Telescoping Boom Lift, to 60'		310.90		341.99	19.11	21.02
32 L.H., Daily Totals		$2105.95		$3139.91	$65.81	$98.12

Crew E-11B	Hr.	Daily	Hr.	Daily	Bare Costs	Incl. O&P
2 Painters, Struc. Steel	$45.85	$733.60	$79.00	$1264.00	$44.60	$75.15
1 Building Laborer	42.10	336.80	67.45	539.60		
2 Paint Sprayer, 8 C.F.M.		87.50		96.25		
1 Telescoping Boom Lift, to 60'		310.90		341.99	16.60	18.26
24 L.H., Daily Totals		$1468.80		$2241.84	$61.20	$93.41

Crew E-12	Hr.	Daily	Hr.	Daily	Bare Costs	Incl. O&P
1 Welder Foreman (outside)	$59.45	$475.60	$97.40	$779.20	$56.23	$90.17
1 Equip. Oper. (light)	53.00	424.00	82.95	663.60		
1 Welder, Gas Engine, 300 amp		147.00		161.70	9.19	10.11
16 L.H., Daily Totals		$1046.60		$1604.50	$65.41	$100.28

Crew E-13	Hr.	Daily	Hr.	Daily	Bare Costs	Incl. O&P
1 Welder Foreman (outside)	$59.45	$475.60	$97.40	$779.20	$57.30	$92.58
.5 Equip. Oper. (light)	53.00	212.00	82.95	331.80		
1 Welder, Gas Engine, 300 amp		147.00		161.70	12.25	13.48
12 L.H., Daily Totals		$834.60		$1272.70	$69.55	$106.06

Crew E-14	Hr.	Daily	Hr.	Daily	Bare Costs	Incl. O&P
1 Welder Foreman (outside)	$59.45	$475.60	$97.40	$779.20	$59.45	$97.40
1 Welder, Gas Engine, 300 amp		147.00		161.70	18.38	20.21
8 L.H., Daily Totals		$622.60		$940.90	$77.83	$117.61

Crew E-16	Hr.	Daily	Hr.	Daily	Bare Costs	Incl. O&P
1 Welder Foreman (outside)	$59.45	$475.60	$97.40	$779.20	$58.45	$95.78
1 Welder	57.45	459.60	94.15	753.20		
1 Welder, Gas Engine, 300 amp		147.00		161.70	9.19	10.11
16 L.H., Daily Totals		$1082.20		$1694.10	$67.64	$105.88

Crew E-17	Hr.	Daily	Hr.	Daily	Bare Costs	Incl. O&P
1 Struc. Steel Foreman (outside)	$59.65	$477.20	$97.75	$782.00	$58.65	$96.10
1 Structural Steel Worker	57.65	461.20	94.45	755.60		
16 L.H., Daily Totals		$938.40		$1537.60	$58.65	$96.10

Crew E-18	Hr.	Daily	Hr.	Daily	Bare Costs	Incl. O&P
1 Struc. Steel Foreman (outside)	$59.65	$477.20	$97.75	$782.00	$57.87	$93.98
3 Structural Steel Workers	57.65	1383.60	94.45	2266.80		
1 Equipment Operator (med.)	56.75	454.00	88.80	710.40		
1 Lattice Boom Crane, 20 Ton		1488.00		1636.80	37.20	40.92
40 L.H., Daily Totals		$3802.80		$5396.00	$95.07	$134.90

Crew E-19	Hr.	Daily	Hr.	Daily	Bare Costs	Incl. O&P
1 Struc. Steel Foreman (outside)	$59.65	$477.20	$97.75	$782.00	$56.77	$91.72
1 Structural Steel Worker	57.65	461.20	94.45	755.60		
1 Equip. Oper. (light)	53.00	424.00	82.95	663.60		
1 Lattice Boom Crane, 20 Ton		1488.00		1636.80	62.00	68.20
24 L.H., Daily Totals		$2850.40		$3838.00	$118.77	$159.92

Crew E-20	Hr.	Daily	Hr.	Daily	Bare Costs	Incl. O&P
1 Struc. Steel Foreman (outside)	$59.65	$477.20	$97.75	$782.00	$57.21	$92.72
5 Structural Steel Workers	57.65	2306.00	94.45	3778.00		
1 Equip. Oper. (crane)	59.20	473.60	92.65	741.20		
1 Equip. Oper. (oiler)	50.55	404.40	79.10	632.80		
1 Lattice Boom Crane, 40 Ton		2028.00		2230.80	31.69	34.86
64 L.H., Daily Totals		$5689.20		$8164.80	$88.89	$127.58

Crew E-22	Hr.	Daily	Hr.	Daily	Bare Costs	Incl. O&P
1 Skilled Worker Foreman (out)	$56.85	$454.80	$90.55	$724.40	$55.52	$88.45
2 Skilled Workers	54.85	877.60	87.40	1398.40		
24 L.H., Daily Totals		$1332.40		$2122.80	$55.52	$88.45

Crew E-24	Hr.	Daily	Hr.	Daily	Bare Costs	Incl. O&P
3 Structural Steel Workers	$57.65	$1383.60	$94.45	$2266.80	$57.42	$93.04
1 Equipment Operator (med.)	56.75	454.00	88.80	710.40		
1 Hyd. Crane, 25 Ton		580.85		638.93	18.15	19.97
32 L.H., Daily Totals		$2418.45		$3616.14	$75.58	$113.00

Crew E-25	Hr.	Daily	Hr.	Daily	Bare Costs	Incl. O&P
1 Welder Foreman (outside)	$59.45	$475.60	$97.40	$779.20	$59.45	$97.40
1 Cutting Torch		12.80		14.08	1.60	1.76
8 L.H., Daily Totals		$488.40		$793.28	$61.05	$99.16

Crew E-26	Hr.	Daily	Hr.	Daily	Bare Costs	Incl. O&P
1 Struc. Steel Foreman (outside)	$59.65	$477.20	$97.75	$782.00	$58.91	$95.68
1 Struc. Steel Worker	57.65	461.20	94.45	755.60		
1 Welder	57.45	459.60	94.15	753.20		
.25 Electrician	61.35	122.70	94.45	188.90		
.25 Plumber	64.45	128.90	99.65	199.30		
1 Welder, Gas Engine, 300 amp		147.00		161.70	5.25	5.78
28 L.H., Daily Totals		$1796.60		$2840.70	$64.16	$101.45

Crew E-27	Hr.	Daily	Hr.	Daily	Bare Costs	Incl. O&P
1 Struc. Steel Foreman (outside)	$59.65	$477.20	$97.75	$782.00	$57.21	$92.72
5 Struc. Steel Workers	57.65	2306.00	94.45	3778.00		
1 Equip. Oper. (crane)	59.20	473.60	92.65	741.20		
1 Equip. Oper. (oiler)	50.55	404.40	79.10	632.80		
1 Hyd. Crane, 12 Ton		471.00		518.10		
1 Hyd. Crane, 80 Ton		1486.00		1634.60	30.58	33.64
64 L.H., Daily Totals		$5618.20		$8086.70	$87.78	$126.35

Crew F-3	Hr.	Daily	Hr.	Daily	Bare Costs	Incl. O&P
4 Carpenters	$53.15	$1700.80	$85.15	$2724.80	$54.36	$86.65
1 Equip. Oper. (crane)	59.20	473.60	92.65	741.20		
1 Hyd. Crane, 12 Ton		471.00		518.10	11.78	12.95
40 L.H., Daily Totals		$2645.40		$3984.10	$66.14	$99.60

Crew No.	Bare Costs		Incl. Subs O&P		Cost Per Labor-Hour	
Crew F-4	Hr.	Daily	Hr.	Daily	Bare Costs	Incl. O&P
4 Carpenters	$53.15	$1700.80	$85.15	$2724.80	$53.73	$85.39
1 Equip. Oper. (crane)	59.20	473.60	92.65	741.20		
1 Equip. Oper. (oiler)	50.55	404.40	79.10	632.80		
1 Hyd. Crane, 55 Ton		980.20		1078.22	20.42	22.46
48 L.H., Daily Totals		$3559.00		$5177.02	$74.15	$107.85
Crew F-5	Hr.	Daily	Hr.	Daily	Bare Costs	Incl. O&P
1 Carpenter Foreman (outside)	$55.15	$441.20	$88.40	$707.20	$53.65	$85.96
3 Carpenters	53.15	1275.60	85.15	2043.60		
32 L.H., Daily Totals		$1716.80		$2750.80	$53.65	$85.96
Crew F-6	Hr.	Daily	Hr.	Daily	Bare Costs	Incl. O&P
2 Carpenters	$53.15	$850.40	$85.15	$1362.40	$49.94	$79.57
2 Building Laborers	42.10	673.60	67.45	1079.20		
1 Equip. Oper. (crane)	59.20	473.60	92.65	741.20		
1 Hyd. Crane, 12 Ton		471.00		518.10	11.78	12.95
40 L.H., Daily Totals		$2468.60		$3700.90	$61.72	$92.52
Crew F-7	Hr.	Daily	Hr.	Daily	Bare Costs	Incl. O&P
2 Carpenters	$53.15	$850.40	$85.15	$1362.40	$47.63	$76.30
2 Building Laborers	42.10	673.60	67.45	1079.20		
32 L.H., Daily Totals		$1524.00		$2441.60	$47.63	$76.30
Crew G-1	Hr.	Daily	Hr.	Daily	Bare Costs	Incl. O&P
1 Roofer Foreman (outside)	$48.20	$385.60	$84.35	$674.80	$43.17	$75.55
4 Roofers Composition	46.20	1478.40	80.85	2587.20		
2 Roofer Helpers	34.60	553.60	60.55	968.80		
1 Application Equipment		192.85		212.13		
1 Tar Kettle/Pot		207.85		228.63		
1 Crew Truck		166.25		182.88	10.12	11.14
56 L.H., Daily Totals		$2984.55		$4854.44	$53.30	$86.69
Crew G-2	Hr.	Daily	Hr.	Daily	Bare Costs	Incl. O&P
1 Plasterer	$48.60	$388.80	$77.40	$619.20	$44.37	$70.78
1 Plasterer Helper	42.40	339.20	67.50	540.00		
1 Building Laborer	42.10	336.80	67.45	539.60		
1 Grout Pump, 50 C.F./hr.		188.45		207.29	7.85	8.64
24 L.H., Daily Totals		$1253.25		$1906.10	$52.22	$79.42
Crew G-2A	Hr.	Daily	Hr.	Daily	Bare Costs	Incl. O&P
1 Roofer Composition	$46.20	$369.60	$80.85	$646.80	$40.97	$69.62
1 Roofer Helper	34.60	276.80	60.55	484.40		
1 Building Laborer	42.10	336.80	67.45	539.60		
1 Foam Spray Rig, Trailer-Mtd.		523.85		576.24		
1 Pickup Truck, 3/4 Ton		110.85		121.94	26.45	29.09
24 L.H., Daily Totals		$1617.90		$2368.97	$67.41	$98.71
Crew G-3	Hr.	Daily	Hr.	Daily	Bare Costs	Incl. O&P
2 Sheet Metal Workers	$62.30	$996.80	$97.60	$1561.60	$52.20	$82.53
2 Building Laborers	42.10	673.60	67.45	1079.20		
32 L.H., Daily Totals		$1670.40		$2640.80	$52.20	$82.53
Crew G-4	Hr.	Daily	Hr.	Daily	Bare Costs	Incl. O&P
1 Labor Foreman (outside)	$44.10	$352.80	$70.65	$565.20	$42.77	$68.52
2 Building Laborers	42.10	673.60	67.45	1079.20		
1 Flatbed Truck, Gas, 1.5 Ton		196.15		215.76		
1 Air Compressor, 160 cfm		202.80		223.08	16.62	18.29
24 L.H., Daily Totals		$1425.35		$2083.24	$59.39	$86.80
Crew G-5	Hr.	Daily	Hr.	Daily	Bare Costs	Incl. O&P
1 Roofer Foreman (outside)	$48.20	$385.60	$84.35	$674.80	$41.96	$73.43
2 Roofers Composition	46.20	739.20	80.85	1293.60		
2 Roofer Helpers	34.60	553.60	60.55	968.80		
1 Application Equipment		192.85		212.13	4.82	5.30
40 L.H., Daily Totals		$1871.25		$3149.34	$46.78	$78.73
Crew G-6A	Hr.	Daily	Hr.	Daily	Bare Costs	Incl. O&P
2 Roofers Composition	$46.20	$739.20	$80.85	$1293.60	$46.20	$80.85
1 Small Compressor, Electric		33.80		37.18		
2 Pneumatic Nailers		54.80		60.28	5.54	6.09
16 L.H., Daily Totals		$827.80		$1391.06	$51.74	$86.94
Crew G-7	Hr.	Daily	Hr.	Daily	Bare Costs	Incl. O&P
1 Carpenter	$53.15	$425.20	$85.15	$681.20	$53.15	$85.15
1 Small Compressor, Electric		33.80		37.18		
1 Pneumatic Nailer		27.40		30.14	7.65	8.41
8 L.H., Daily Totals		$486.40		$748.52	$60.80	$93.56
Crew H-1	Hr.	Daily	Hr.	Daily	Bare Costs	Incl. O&P
2 Glaziers	$51.00	$816.00	$81.50	$1304.00	$54.33	$87.97
2 Struc. Steel Workers	57.65	922.40	94.45	1511.20		
32 L.H., Daily Totals		$1738.40		$2815.20	$54.33	$87.97
Crew H-2	Hr.	Daily	Hr.	Daily	Bare Costs	Incl. O&P
2 Glaziers	$51.00	$816.00	$81.50	$1304.00	$48.03	$76.82
1 Building Laborer	42.10	336.80	67.45	539.60		
24 L.H., Daily Totals		$1152.80		$1843.60	$48.03	$76.82
Crew H-3	Hr.	Daily	Hr.	Daily	Bare Costs	Incl. O&P
1 Glazier	$51.00	$408.00	$81.50	$652.00	$45.48	$73.22
1 Helper	39.95	319.60	64.95	519.60		
16 L.H., Daily Totals		$727.60		$1171.60	$45.48	$73.22
Crew H-4	Hr.	Daily	Hr.	Daily	Bare Costs	Incl. O&P
1 Carpenter	$53.15	$425.20	$85.15	$681.20	$49.51	$78.93
1 Carpenter Helper	39.95	319.60	64.95	519.60		
.5 Electrician	61.35	245.40	94.45	377.80		
20 L.H., Daily Totals		$990.20		$1578.60	$49.51	$78.93
Crew J-1	Hr.	Daily	Hr.	Daily	Bare Costs	Incl. O&P
3 Plasterers	$48.60	$1166.40	$77.40	$1857.60	$46.12	$73.44
2 Plasterer Helpers	42.40	678.40	67.50	1080.00		
1 Mixing Machine, 6 C.F.		117.35		129.09	2.93	3.23
40 L.H., Daily Totals		$1962.15		$3066.68	$49.05	$76.67
Crew J-2	Hr.	Daily	Hr.	Daily	Bare Costs	Incl. O&P
3 Plasterers	$48.60	$1166.40	$77.40	$1857.60	$47.11	$74.78
2 Plasterer Helpers	42.40	678.40	67.50	1080.00		
1 Lather	52.05	416.40	81.50	652.00		
1 Mixing Machine, 6 C.F.		117.35		129.09	2.44	2.69
48 L.H., Daily Totals		$2378.55		$3718.68	$49.55	$77.47
Crew J-3	Hr.	Daily	Hr.	Daily	Bare Costs	Incl. O&P
1 Terrazzo Worker	$49.40	$395.20	$77.40	$619.20	$44.95	$70.42
1 Terrazzo Helper	40.50	324.00	63.45	507.60		
1 Floor Grinder, 22" Path		104.60		115.06		
1 Terrazzo Mixer		161.15		177.26	16.61	18.27
16 L.H., Daily Totals		$984.95		$1419.13	$61.56	$88.70

For customer support on your Commercial Renovation Costs with RSMeans data, call 800.448.8182.

781

Crew No.	Bare Costs Hr.	Bare Costs Daily	Incl. Subs O&P Hr.	Incl. Subs O&P Daily	Cost Per Labor-Hour Bare Costs	Cost Per Labor-Hour Incl. O&P
Crew J-4	Hr.	Daily	Hr.	Daily	Bare Costs	Incl. O&P
2 Cement Finishers	$49.95	$799.20	$78.45	$1255.20	$47.33	$74.78
1 Laborer	42.10	336.80	67.45	539.60		
1 Floor Grinder, 22" Path		104.60		115.06		
1 Floor Edger, 7" Path		43.95		48.34		
1 Vacuum Pick-Up System		74.05		81.45	9.28	10.20
24 L.H., Daily Totals		$1358.60		$2039.66	$56.61	$84.99
Crew J-4A	Hr.	Daily	Hr.	Daily	Bare Costs	Incl. O&P
2 Cement Finishers	$49.95	$799.20	$78.45	$1255.20	$46.02	$72.95
2 Laborers	42.10	673.60	67.45	1079.20		
1 Floor Grinder, 22" Path		104.60		115.06		
1 Floor Edger, 7" Path		43.95		48.34		
1 Vacuum Pick-Up System		74.05		81.45		
1 Floor Auto Scrubber		178.40		196.24	12.53	13.78
32 L.H., Daily Totals		$1873.80		$2775.50	$58.56	$86.73
Crew J-4B	Hr.	Daily	Hr.	Daily	Bare Costs	Incl. O&P
1 Laborer	$42.10	$336.80	$67.45	$539.60	$42.10	$67.45
1 Floor Auto Scrubber		178.40		196.24	22.30	24.53
8 L.H., Daily Totals		$515.20		$735.84	$64.40	$91.98
Crew J-6	Hr.	Daily	Hr.	Daily	Bare Costs	Incl. O&P
2 Painters	$44.40	$710.40	$70.50	$1128.00	$45.98	$72.85
1 Building Laborer	42.10	336.80	67.45	539.60		
1 Equip. Oper. (light)	53.00	424.00	82.95	663.60		
1 Air Compressor, 250 cfm		201.50		221.65		
1 Sandblaster, Portable, 3 C.F.		83.80		92.18		
1 Set Sand Blasting Accessories		15.35		16.89	9.40	10.33
32 L.H., Daily Totals		$1771.85		$2661.92	$55.37	$83.18
Crew J-7	Hr.	Daily	Hr.	Daily	Bare Costs	Incl. O&P
2 Painters	$44.40	$710.40	$70.50	$1128.00	$44.40	$70.50
1 Floor Belt Sander		50.15		55.16		
1 Floor Sanding Edger		25.15		27.66	4.71	5.18
16 L.H., Daily Totals		$785.70		$1210.83	$49.11	$75.68
Crew K-1	Hr.	Daily	Hr.	Daily	Bare Costs	Incl. O&P
1 Carpenter	$53.15	$425.20	$85.15	$681.20	$50.20	$80.35
1 Truck Driver (light)	47.25	378.00	75.55	604.40		
1 Flatbed Truck, Gas, 3 Ton		820.40		902.44	51.27	56.40
16 L.H., Daily Totals		$1623.60		$2188.04	$101.47	$136.75
Crew K-2	Hr.	Daily	Hr.	Daily	Bare Costs	Incl. O&P
1 Struc. Steel Foreman (outside)	$59.65	$477.20	$97.75	$782.00	$54.85	$89.25
1 Struc. Steel Worker	57.65	461.20	94.45	755.60		
1 Truck Driver (light)	47.25	378.00	75.55	604.40		
1 Flatbed Truck, Gas, 3 Ton		820.40		902.44	34.18	37.60
24 L.H., Daily Totals		$2136.80		$3044.44	$89.03	$126.85
Crew L-1	Hr.	Daily	Hr.	Daily	Bare Costs	Incl. O&P
1 Electrician	$61.35	$490.80	$94.45	$755.60	$62.90	$97.05
1 Plumber	64.45	515.60	99.65	797.20		
16 L.H., Daily Totals		$1006.40		$1552.80	$62.90	$97.05
Crew L-2	Hr.	Daily	Hr.	Daily	Bare Costs	Incl. O&P
1 Carpenter	$53.15	$425.20	$85.15	$681.20	$46.55	$75.05
1 Carpenter Helper	39.95	319.60	64.95	519.60		
16 L.H., Daily Totals		$744.80		$1200.80	$46.55	$75.05

Crew No.	Bare Costs Hr.	Bare Costs Daily	Incl. Subs O&P Hr.	Incl. Subs O&P Daily	Cost Per Labor-Hour Bare Costs	Cost Per Labor-Hour Incl. O&P
Crew L-3	Hr.	Daily	Hr.	Daily	Bare Costs	Incl. O&P
1 Carpenter	$53.15	$425.20	$85.15	$681.20	$57.49	$90.59
.5 Electrician	61.35	245.40	94.45	377.80		
.5 Sheet Metal Worker	62.30	249.20	97.60	390.40		
16 L.H., Daily Totals		$919.80		$1449.40	$57.49	$90.59
Crew L-3A	Hr.	Daily	Hr.	Daily	Bare Costs	Incl. O&P
1 Carpenter Foreman (outside)	$55.15	$441.20	$88.40	$707.20	$57.53	$91.47
.5 Sheet Metal Worker	62.30	249.20	97.60	390.40		
12 L.H., Daily Totals		$690.40		$1097.60	$57.53	$91.47
Crew L-4	Hr.	Daily	Hr.	Daily	Bare Costs	Incl. O&P
2 Skilled Workers	$54.85	$877.60	$87.40	$1398.40	$49.88	$79.92
1 Helper	39.95	319.60	64.95	519.60		
24 L.H., Daily Totals		$1197.20		$1918.00	$49.88	$79.92
Crew L-5	Hr.	Daily	Hr.	Daily	Bare Costs	Incl. O&P
1 Struc. Steel Foreman (outside)	$59.65	$477.20	$97.75	$782.00	$58.16	$94.66
5 Struc. Steel Workers	57.65	2306.00	94.45	3778.00		
1 Equip. Oper. (crane)	59.20	473.60	92.65	741.20		
1 Hyd. Crane, 25 Ton		580.85		638.93	10.37	11.41
56 L.H., Daily Totals		$3837.65		$5940.14	$68.53	$106.07
Crew L-5A	Hr.	Daily	Hr.	Daily	Bare Costs	Incl. O&P
1 Struc. Steel Foreman (outside)	$59.65	$477.20	$97.75	$782.00	$58.54	$94.83
2 Structural Steel Workers	57.65	922.40	94.45	1511.20		
1 Equip. Oper. (crane)	59.20	473.60	92.65	741.20		
1 S.P. Crane, 4x4, 25 Ton		1144.00		1258.40	35.75	39.33
32 L.H., Daily Totals		$3017.20		$4292.80	$94.29	$134.15
Crew L-5B	Hr.	Daily	Hr.	Daily	Bare Costs	Incl. O&P
1 Struc. Steel Foreman (outside)	$59.65	$477.20	$97.75	$782.00	$59.83	$94.44
2 Structural Steel Workers	57.65	922.40	94.45	1511.20		
2 Electricians	61.35	981.60	94.45	1511.20		
2 Steamfitters/Pipefitters	65.55	1048.80	101.35	1621.60		
1 Equip. Oper. (crane)	59.20	473.60	92.65	741.20		
1 Equip. Oper. (oiler)	50.55	404.40	79.10	632.80		
1 Hyd. Crane, 80 Ton		1486.00		1634.60	20.64	22.70
72 L.H., Daily Totals		$5794.00		$8434.60	$80.47	$117.15
Crew L-6	Hr.	Daily	Hr.	Daily	Bare Costs	Incl. O&P
1 Plumber	$64.45	$515.60	$99.65	$797.20	$63.42	$97.92
.5 Electrician	61.35	245.40	94.45	377.80		
12 L.H., Daily Totals		$761.00		$1175.00	$63.42	$97.92
Crew L-7	Hr.	Daily	Hr.	Daily	Bare Costs	Incl. O&P
2 Carpenters	$53.15	$850.40	$85.15	$1362.40	$51.16	$81.42
1 Building Laborer	42.10	336.80	67.45	539.60		
.5 Electrician	61.35	245.40	94.45	377.80		
28 L.H., Daily Totals		$1432.60		$2279.80	$51.16	$81.42
Crew L-8	Hr.	Daily	Hr.	Daily	Bare Costs	Incl. O&P
2 Carpenters	$53.15	$850.40	$85.15	$1362.40	$55.41	$88.05
.5 Plumber	64.45	257.80	99.65	398.60		
20 L.H., Daily Totals		$1108.20		$1761.00	$55.41	$88.05

For customer support on your Commercial Renovation Costs with RSMeans data, call 800.448.8182.

Crew L-9

Crew No.	Bare Costs Hr.	Daily	Incl. Subs O&P Hr.	Daily	Cost Per Labor-Hour Bare Costs	Incl. O&P
1 Labor Foreman (inside)	$42.60	$340.80	$68.25	$546.00	$47.81	$76.63
2 Building Laborers	42.10	673.60	67.45	1079.20		
1 Struc. Steel Worker	57.65	461.20	94.45	755.60		
.5 Electrician	61.35	245.40	94.45	377.80		
36 L.H., Daily Totals		$1721.00		$2758.60	$47.81	$76.63

Crew L-10

Crew No.	Bare Costs Hr.	Daily	Incl. Subs O&P Hr.	Daily	Cost Per Labor-Hour Bare Costs	Incl. O&P
1 Struc. Steel Foreman (outside)	$59.65	$477.20	$97.75	$782.00	$58.83	$94.95
1 Structural Steel Worker	57.65	461.20	94.45	755.60		
1 Equip. Oper. (crane)	59.20	473.60	92.65	741.20		
1 Hyd. Crane, 12 Ton		471.00		518.10	19.63	21.59
24 L.H., Daily Totals		$1883.00		$2796.90	$78.46	$116.54

Crew L-11

Crew No.	Bare Costs Hr.	Daily	Incl. Subs O&P Hr.	Daily	Cost Per Labor-Hour Bare Costs	Incl. O&P
2 Wreckers	$42.10	$673.60	$69.05	$1104.80	$49.10	$78.42
1 Equip. Oper. (crane)	59.20	473.60	92.65	741.20		
1 Equip. Oper. (light)	53.00	424.00	82.95	663.60		
1 Hyd. Excavator, 2.5 C.Y.		1465.00		1611.50		
1 Loader, Skid Steer, 78 H.P.		400.10		440.11	58.28	64.11
32 L.H., Daily Totals		$3436.30		$4561.21	$107.38	$142.54

Crew M-1

Crew No.	Bare Costs Hr.	Daily	Incl. Subs O&P Hr.	Daily	Cost Per Labor-Hour Bare Costs	Incl. O&P
3 Elevator Constructors	$85.55	$2053.20	$131.25	$3150.00	$81.28	$124.69
1 Elevator Apprentice	68.45	547.60	105.00	840.00		
5 Hand Tools		50.00		55.00	1.56	1.72
32 L.H., Daily Totals		$2650.80		$4045.00	$82.84	$126.41

Crew M-3

Crew No.	Bare Costs Hr.	Daily	Incl. Subs O&P Hr.	Daily	Cost Per Labor-Hour Bare Costs	Incl. O&P
1 Electrician Foreman (outside)	$63.35	$506.80	$97.50	$780.00	$64.39	$99.62
1 Common Laborer	42.10	336.80	67.45	539.60		
.25 Equipment Operator (med.)	56.75	113.50	88.80	177.60		
1 Elevator Constructor	85.55	684.40	131.25	1050.00		
1 Elevator Apprentice	68.45	547.60	105.00	840.00		
.25 S.P. Crane, 4x4, 20 Ton		145.55		160.10	4.28	4.71
34 L.H., Daily Totals		$2334.65		$3547.30	$68.67	$104.33

Crew M-4

Crew No.	Bare Costs Hr.	Daily	Incl. Subs O&P Hr.	Daily	Cost Per Labor-Hour Bare Costs	Incl. O&P
1 Electrician Foreman (outside)	$63.35	$506.80	$97.50	$780.00	$63.75	$98.70
1 Common Laborer	42.10	336.80	67.45	539.60		
.25 Equipment Operator, Crane	59.20	118.40	92.65	185.30		
.25 Equip. Oper. (oiler)	50.55	101.10	79.10	158.20		
1 Elevator Constructor	85.55	684.40	131.25	1050.00		
1 Elevator Apprentice	68.45	547.60	105.00	840.00		
.25 S.P. Crane, 4x4, 40 Ton		188.55		207.41	5.24	5.76
36 L.H., Daily Totals		$2483.65		$3760.51	$68.99	$104.46

Crew Q-1

Crew No.	Bare Costs Hr.	Daily	Incl. Subs O&P Hr.	Daily	Cost Per Labor-Hour Bare Costs	Incl. O&P
1 Plumber	$64.45	$515.60	$99.65	$797.20	$58.00	$89.67
1 Plumber Apprentice	51.55	412.40	79.70	637.60		
16 L.H., Daily Totals		$928.00		$1434.80	$58.00	$89.67

Crew Q-1A

Crew No.	Bare Costs Hr.	Daily	Incl. Subs O&P Hr.	Daily	Cost Per Labor-Hour Bare Costs	Incl. O&P
.25 Plumber Foreman (outside)	$66.45	$132.90	$102.75	$205.50	$64.85	$100.27
1 Plumber	64.45	515.60	99.65	797.20		
10 L.H., Daily Totals		$648.50		$1002.70	$64.85	$100.27

Crew Q-1C

Crew No.	Bare Costs Hr.	Daily	Incl. Subs O&P Hr.	Daily	Cost Per Labor-Hour Bare Costs	Incl. O&P
1 Plumber	$64.45	$515.60	$99.65	$797.20	$57.58	$89.38
1 Plumber Apprentice	51.55	412.40	79.70	637.60		
1 Equip. Oper. (medium)	56.75	454.00	88.80	710.40		
1 Trencher, Chain Type, 8' D		1872.00		2059.20	78.00	85.80
24 L.H., Daily Totals		$3254.00		$4204.40	$135.58	$175.18

Crew Q-2

Crew No.	Bare Costs Hr.	Daily	Incl. Subs O&P Hr.	Daily	Cost Per Labor-Hour Bare Costs	Incl. O&P
2 Plumbers	$64.45	$1031.20	$99.65	$1594.40	$60.15	$93.00
1 Plumber Apprentice	51.55	412.40	79.70	637.60		
24 L.H., Daily Totals		$1443.60		$2232.00	$60.15	$93.00

Crew Q-3

Crew No.	Bare Costs Hr.	Daily	Incl. Subs O&P Hr.	Daily	Cost Per Labor-Hour Bare Costs	Incl. O&P
1 Plumber Foreman (inside)	$64.95	$519.60	$100.45	$803.60	$61.35	$94.86
2 Plumbers	64.45	1031.20	99.65	1594.40		
1 Plumber Apprentice	51.55	412.40	79.70	637.60		
32 L.H., Daily Totals		$1963.20		$3035.60	$61.35	$94.86

Crew Q-4

Crew No.	Bare Costs Hr.	Daily	Incl. Subs O&P Hr.	Daily	Cost Per Labor-Hour Bare Costs	Incl. O&P
1 Plumber Foreman (inside)	$64.95	$519.60	$100.45	$803.60	$61.35	$94.86
1 Plumber	64.45	515.60	99.65	797.20		
1 Welder (plumber)	64.45	515.60	99.65	797.20		
1 Plumber Apprentice	51.55	412.40	79.70	637.60		
1 Welder, Electric, 300 amp		106.40		117.04	3.33	3.66
32 L.H., Daily Totals		$2069.60		$3152.64	$64.67	$98.52

Crew Q-5

Crew No.	Bare Costs Hr.	Daily	Incl. Subs O&P Hr.	Daily	Cost Per Labor-Hour Bare Costs	Incl. O&P
1 Steamfitter	$65.55	$524.40	$101.35	$810.80	$59.00	$91.22
1 Steamfitter Apprentice	52.45	419.60	81.10	648.80		
16 L.H., Daily Totals		$944.00		$1459.60	$59.00	$91.22

Crew Q-6

Crew No.	Bare Costs Hr.	Daily	Incl. Subs O&P Hr.	Daily	Cost Per Labor-Hour Bare Costs	Incl. O&P
2 Steamfitters	$65.55	$1048.80	$101.35	$1621.60	$61.18	$94.60
1 Steamfitter Apprentice	52.45	419.60	81.10	648.80		
24 L.H., Daily Totals		$1468.40		$2270.40	$61.18	$94.60

Crew Q-7

Crew No.	Bare Costs Hr.	Daily	Incl. Subs O&P Hr.	Daily	Cost Per Labor-Hour Bare Costs	Incl. O&P
1 Steamfitter Foreman (inside)	$66.05	$528.40	$102.15	$817.20	$62.40	$96.49
2 Steamfitters	65.55	1048.80	101.35	1621.60		
1 Steamfitter Apprentice	52.45	419.60	81.10	648.80		
32 L.H., Daily Totals		$1996.80		$3087.60	$62.40	$96.49

Crew Q-8

Crew No.	Bare Costs Hr.	Daily	Incl. Subs O&P Hr.	Daily	Cost Per Labor-Hour Bare Costs	Incl. O&P
1 Steamfitter Foreman (inside)	$66.05	$528.40	$102.15	$817.20	$62.40	$96.49
1 Steamfitter	65.55	524.40	101.35	810.80		
1 Welder (steamfitter)	65.55	524.40	101.35	810.80		
1 Steamfitter Apprentice	52.45	419.60	81.10	648.80		
1 Welder, Electric, 300 amp		106.40		117.04	3.33	3.66
32 L.H., Daily Totals		$2103.20		$3204.64	$65.72	$100.15

Crew Q-9

Crew No.	Bare Costs Hr.	Daily	Incl. Subs O&P Hr.	Daily	Cost Per Labor-Hour Bare Costs	Incl. O&P
1 Sheet Metal Worker	$62.30	$498.40	$97.60	$780.80	$56.08	$87.85
1 Sheet Metal Apprentice	49.85	398.80	78.10	624.80		
16 L.H., Daily Totals		$897.20		$1405.60	$56.08	$87.85

Crew Q-10

Crew No.	Bare Costs Hr.	Daily	Incl. Subs O&P Hr.	Daily	Cost Per Labor-Hour Bare Costs	Incl. O&P
2 Sheet Metal Workers	$62.30	$996.80	$97.60	$1561.60	$58.15	$91.10
1 Sheet Metal Apprentice	49.85	398.80	78.10	624.80		
24 L.H., Daily Totals		$1395.60		$2186.40	$58.15	$91.10

Crew No.	Bare Costs		Incl. Subs O&P		Cost Per Labor-Hour	
Crew Q-11	Hr.	Daily	Hr.	Daily	Bare Costs	Incl. O&P
1 Sheet Metal Foreman (inside)	$62.80	$502.40	$98.40	$787.20	$59.31	$92.92
2 Sheet Metal Workers	62.30	996.80	97.60	1561.60		
1 Sheet Metal Apprentice	49.85	398.80	78.10	624.80		
32 L.H., Daily Totals		$1898.00		$2973.60	$59.31	$92.92
Crew Q-12	Hr.	Daily	Hr.	Daily	Bare Costs	Incl. O&P
1 Sprinkler Installer	$63.25	$506.00	$97.95	$783.60	$56.92	$88.17
1 Sprinkler Apprentice	50.60	404.80	78.40	627.20		
16 L.H., Daily Totals		$910.80		$1410.80	$56.92	$88.17
Crew Q-13	Hr.	Daily	Hr.	Daily	Bare Costs	Incl. O&P
1 Sprinkler Foreman (inside)	$63.75	$510.00	$98.75	$790.00	$60.21	$93.26
2 Sprinkler Installers	63.25	1012.00	97.95	1567.20		
1 Sprinkler Apprentice	50.60	404.80	78.40	627.20		
32 L.H., Daily Totals		$1926.80		$2984.40	$60.21	$93.26
Crew Q-14	Hr.	Daily	Hr.	Daily	Bare Costs	Incl. O&P
1 Asbestos Worker	$58.65	$469.20	$93.00	$744.00	$52.77	$83.67
1 Asbestos Apprentice	46.90	375.20	74.35	594.80		
16 L.H., Daily Totals		$844.40		$1338.80	$52.77	$83.67
Crew Q-15	Hr.	Daily	Hr.	Daily	Bare Costs	Incl. O&P
1 Plumber	$64.45	$515.60	$99.65	$797.20	$58.00	$89.67
1 Plumber Apprentice	51.55	412.40	79.70	637.60		
1 Welder, Electric, 300 amp		106.40		117.04	6.65	7.32
16 L.H., Daily Totals		$1034.40		$1551.84	$64.65	$96.99
Crew Q-16	Hr.	Daily	Hr.	Daily	Bare Costs	Incl. O&P
2 Plumbers	$64.45	$1031.20	$99.65	$1594.40	$60.15	$93.00
1 Plumber Apprentice	51.55	412.40	79.70	637.60		
1 Welder, Electric, 300 amp		106.40		117.04	4.43	4.88
24 L.H., Daily Totals		$1550.00		$2349.04	$64.58	$97.88
Crew Q-17	Hr.	Daily	Hr.	Daily	Bare Costs	Incl. O&P
1 Steamfitter	$65.55	$524.40	$101.35	$810.80	$59.00	$91.22
1 Steamfitter Apprentice	52.45	419.60	81.10	648.80		
1 Welder, Electric, 300 amp		106.40		117.04	6.65	7.32
16 L.H., Daily Totals		$1050.40		$1576.64	$65.65	$98.54
Crew Q-17A	Hr.	Daily	Hr.	Daily	Bare Costs	Incl. O&P
1 Steamfitter	$65.55	$524.40	$101.35	$810.80	$59.07	$91.70
1 Steamfitter Apprentice	52.45	419.60	81.10	648.80		
1 Equip. Oper. (crane)	59.20	473.60	92.65	741.20		
1 Hyd. Crane, 12 Ton		471.00		518.10		
1 Welder, Electric, 300 amp		106.40		117.04	24.06	26.46
24 L.H., Daily Totals		$1995.00		$2835.94	$83.13	$118.16
Crew Q-18	Hr.	Daily	Hr.	Daily	Bare Costs	Incl. O&P
2 Steamfitters	$65.55	$1048.80	$101.35	$1621.60	$61.18	$94.60
1 Steamfitter Apprentice	52.45	419.60	81.10	648.80		
1 Welder, Electric, 300 amp		106.40		117.04	4.43	4.88
24 L.H., Daily Totals		$1574.80		$2387.44	$65.62	$99.48
Crew Q-19	Hr.	Daily	Hr.	Daily	Bare Costs	Incl. O&P
1 Steamfitter	$65.55	$524.40	$101.35	$810.80	$59.78	$92.30
1 Steamfitter Apprentice	52.45	419.60	81.10	648.80		
1 Electrician	61.35	490.80	94.45	755.60		
24 L.H., Daily Totals		$1434.80		$2215.20	$59.78	$92.30

Crew No.	Bare Costs		Incl. Subs O&P		Cost Per Labor-Hour	
Crew Q-20	Hr.	Daily	Hr.	Daily	Bare Costs	Incl. O&P
1 Sheet Metal Worker	$62.30	$498.40	$97.60	$780.80	$57.13	$89.17
1 Sheet Metal Apprentice	49.85	398.80	78.10	624.80		
.5 Electrician	61.35	245.40	94.45	377.80		
20 L.H., Daily Totals		$1142.60		$1783.40	$57.13	$89.17
Crew Q-21	Hr.	Daily	Hr.	Daily	Bare Costs	Incl. O&P
2 Steamfitters	$65.55	$1048.80	$101.35	$1621.60	$61.23	$94.56
1 Steamfitter Apprentice	52.45	419.60	81.10	648.80		
1 Electrician	61.35	490.80	94.45	755.60		
32 L.H., Daily Totals		$1959.20		$3026.00	$61.23	$94.56
Crew Q-22	Hr.	Daily	Hr.	Daily	Bare Costs	Incl. O&P
1 Plumber	$64.45	$515.60	$99.65	$797.20	$58.00	$89.67
1 Plumber Apprentice	51.55	412.40	79.70	637.60		
1 Hyd. Crane, 12 Ton		471.00		518.10	29.44	32.38
16 L.H., Daily Totals		$1399.00		$1952.90	$87.44	$122.06
Crew Q-22A	Hr.	Daily	Hr.	Daily	Bare Costs	Incl. O&P
1 Plumber	$64.45	$515.60	$99.65	$797.20	$54.33	$84.86
1 Plumber Apprentice	51.55	412.40	79.70	637.60		
1 Laborer	42.10	336.80	67.45	539.60		
1 Equip. Oper. (crane)	59.20	473.60	92.65	741.20		
1 Hyd. Crane, 12 Ton		471.00		518.10	14.72	16.19
32 L.H., Daily Totals		$2209.40		$3233.70	$69.04	$101.05
Crew Q-23	Hr.	Daily	Hr.	Daily	Bare Costs	Incl. O&P
1 Plumber Foreman (outside)	$66.45	$531.60	$102.75	$822.00	$62.55	$97.07
1 Plumber	64.45	515.60	99.65	797.20		
1 Equip. Oper. (medium)	56.75	454.00	88.80	710.40		
1 Lattice Boom Crane, 20 Ton		1488.00		1636.80	62.00	68.20
24 L.H., Daily Totals		$2989.20		$3966.40	$124.55	$165.27
Crew R-1	Hr.	Daily	Hr.	Daily	Bare Costs	Incl. O&P
1 Electrician Foreman	$61.85	$494.80	$95.20	$761.60	$57.35	$88.28
3 Electricians	61.35	1472.40	94.45	2266.80		
2 Electrician Apprentices	49.10	785.60	75.55	1208.80		
48 L.H., Daily Totals		$2752.80		$4237.20	$57.35	$88.28
Crew R-1A	Hr.	Daily	Hr.	Daily	Bare Costs	Incl. O&P
1 Electrician	$61.35	$490.80	$94.45	$755.60	$55.23	$85.00
1 Electrician Apprentice	49.10	392.80	75.55	604.40		
16 L.H., Daily Totals		$883.60		$1360.00	$55.23	$85.00
Crew R-1B	Hr.	Daily	Hr.	Daily	Bare Costs	Incl. O&P
1 Electrician	$61.35	$490.80	$94.45	$755.60	$53.18	$81.85
2 Electrician Apprentices	49.10	785.60	75.55	1208.80		
24 L.H., Daily Totals		$1276.40		$1964.40	$53.18	$81.85
Crew R-1C	Hr.	Daily	Hr.	Daily	Bare Costs	Incl. O&P
2 Electricians	$61.35	$981.60	$94.45	$1511.20	$55.23	$85.00
2 Electrician Apprentices	49.10	785.60	75.55	1208.80		
1 Portable cable puller, 8000 lb.		101.60		111.76	3.17	3.49
32 L.H., Daily Totals		$1868.80		$2831.76	$58.40	$88.49
Crew R-1D	Hr.	Daily	Hr.	Daily	Bare Costs	Incl. O&P
1 Electrician	$61.35	$490.80	$94.45	$755.60	$55.23	$85.00
1 Electrician Apprentice	49.10	392.80	75.55	604.40		
1 Aerial lift		105.00		115.50	6.56	7.22
16 L.H., Daily Totals		$988.60		$1475.50	$61.79	$92.22

Crew No.	Bare Costs		Incl. Subs O&P		Cost Per Labor-Hour	

Crew R-2	Hr.	Daily	Hr.	Daily	Bare Costs	Incl. O&P
1 Electrician Foreman	$61.85	$494.80	$95.20	$761.60	$57.61	$88.90
3 Electricians	61.35	1472.40	94.45	2266.80		
2 Electrician Apprentices	49.10	785.60	75.55	1208.80		
1 Equip. Oper. (crane)	59.20	473.60	92.65	741.20		
1 S.P. Crane, 4x4, 5 Ton		378.10		415.91	6.75	7.43
56 L.H., Daily Totals		$3604.50		$5394.31	$64.37	$96.33

Crew R-3	Hr.	Daily	Hr.	Daily	Bare Costs	Incl. O&P
1 Electrician Foreman	$61.85	$494.80	$95.20	$761.60	$61.12	$94.39
1 Electrician	61.35	490.80	94.45	755.60		
.5 Equip. Oper. (crane)	59.20	236.80	92.65	370.60		
.5 S.P. Crane, 4x4, 5 Ton		189.05		207.96	9.45	10.40
20 L.H., Daily Totals		$1411.45		$2095.76	$70.57	$104.79

Crew R-4	Hr.	Daily	Hr.	Daily	Bare Costs	Incl. O&P
1 Struc. Steel Foreman (outside)	$59.65	$477.20	$97.75	$782.00	$58.79	$95.11
3 Struc. Steel Workers	57.65	1383.60	94.45	2266.80		
1 Electrician	61.35	490.80	94.45	755.60		
1 Welder, Gas Engine, 300 amp		147.00		161.70	3.67	4.04
40 L.H., Daily Totals		$2498.60		$3966.10	$62.47	$99.15

Crew R-5	Hr.	Daily	Hr.	Daily	Bare Costs	Incl. O&P
1 Electrician Foreman	$61.85	$494.80	$95.20	$761.60	$53.61	$83.79
4 Electrician Linemen	61.35	1963.20	94.45	3022.40		
2 Electrician Operators	61.35	981.60	94.45	1511.20		
4 Electrician Groundmen	39.95	1278.40	64.95	2078.40		
1 Crew Truck		166.25		182.88		
1 Flatbed Truck, 20,000 GVW		201.60		221.76		
1 Pickup Truck, 3/4 Ton		110.85		121.94		
.2 Hyd. Crane, 55 Ton		196.04		215.64		
.2 Hyd. Crane, 12 Ton		94.20		103.62		
.2 Earth Auger, Truck-Mtd.		40.10		44.11		
1 Tractor w/Winch		373.50		410.85	13.44	14.78
88 L.H., Daily Totals		$5900.54		$8674.39	$67.05	$98.57

Crew R-6	Hr.	Daily	Hr.	Daily	Bare Costs	Incl. O&P
1 Electrician Foreman	$61.85	$494.80	$95.20	$761.60	$53.61	$83.79
4 Electrician Linemen	61.35	1963.20	94.45	3022.40		
2 Electrician Operators	61.35	981.60	94.45	1511.20		
4 Electrician Groundmen	39.95	1278.40	64.95	2078.40		
1 Crew Truck		166.25		182.88		
1 Flatbed Truck, 20,000 GVW		201.60		221.76		
1 Pickup Truck, 3/4 Ton		110.85		121.94		
.2 Hyd. Crane, 55 Ton		196.04		215.64		
.2 Hyd. Crane, 12 Ton		94.20		103.62		
.2 Earth Auger, Truck-Mtd.		40.10		44.11		
1 Tractor w/Winch		373.50		410.85		
3 Cable Trailers		192.30		211.53		
.5 Tensioning Rig		55.20		60.72		
.5 Cable Pulling Rig		303.57		333.93	19.70	21.67
88 L.H., Daily Totals		$6451.61		$9280.58	$73.31	$105.46

Crew R-7	Hr.	Daily	Hr.	Daily	Bare Costs	Incl. O&P
1 Electrician Foreman	$61.85	$494.80	$95.20	$761.60	$43.60	$69.99
5 Electrician Groundmen	39.95	1598.00	64.95	2598.00		
1 Crew Truck		166.25		182.88	3.46	3.81
48 L.H., Daily Totals		$2259.05		$3542.47	$47.06	$73.80

Crew R-8	Hr.	Daily	Hr.	Daily	Bare Costs	Incl. O&P
1 Electrician Foreman	$61.85	$494.80	$95.20	$761.60	$54.30	$84.74
3 Electrician Linemen	61.35	1472.40	94.45	2266.80		
2 Electrician Groundmen	39.95	639.20	64.95	1039.20		
1 Pickup Truck, 3/4 Ton		110.85		121.94		
1 Crew Truck		166.25		182.88	5.77	6.35
48 L.H., Daily Totals		$2883.50		$4372.41	$60.07	$91.09

Crew R-9	Hr.	Daily	Hr.	Daily	Bare Costs	Incl. O&P
1 Electrician Foreman	$61.85	$494.80	$95.20	$761.60	$50.71	$79.79
1 Electrician Lineman	61.35	490.80	94.45	755.60		
2 Electrician Operators	61.35	981.60	94.45	1511.20		
4 Electrician Groundmen	39.95	1278.40	64.95	2078.40		
1 Pickup Truck, 3/4 Ton		110.85		121.94		
1 Crew Truck		166.25		182.88	4.33	4.76
64 L.H., Daily Totals		$3522.70		$5411.61	$55.04	$84.56

Crew R-10	Hr.	Daily	Hr.	Daily	Bare Costs	Incl. O&P
1 Electrician Foreman	$61.85	$494.80	$95.20	$761.60	$57.87	$89.66
4 Electrician Linemen	61.35	1963.20	94.45	3022.40		
1 Electrician Groundman	39.95	319.60	64.95	519.60		
1 Crew Truck		166.25		182.88		
3 Tram Cars		216.15		237.76	7.97	8.76
48 L.H., Daily Totals		$3160.00		$4724.24	$65.83	$98.42

Crew R-11	Hr.	Daily	Hr.	Daily	Bare Costs	Incl. O&P
1 Electrician Foreman	$61.85	$494.80	$95.20	$761.60	$58.36	$90.44
4 Electricians	61.35	1963.20	94.45	3022.40		
1 Equip. Oper. (crane)	59.20	473.60	92.65	741.20		
1 Common Laborer	42.10	336.80	67.45	539.60		
1 Crew Truck		166.25		182.88		
1 Hyd. Crane, 12 Ton		471.00		518.10	11.38	12.52
56 L.H., Daily Totals		$3905.65		$5765.77	$69.74	$102.96

Crew R-12	Hr.	Daily	Hr.	Daily	Bare Costs	Incl. O&P
1 Carpenter Foreman (inside)	$53.65	$429.20	$85.95	$687.60	$49.91	$79.96
4 Carpenters	53.15	1700.80	85.15	2724.80		
4 Common Laborers	42.10	1347.20	67.45	2158.40		
1 Equip. Oper. (medium)	56.75	454.00	88.80	710.40		
1 Steel Worker	57.65	461.20	94.45	755.60		
1 Dozer, 200 H.P.		1504.00		1654.40		
1 Pickup Truck, 3/4 Ton		110.85		121.94	18.35	20.19
88 L.H., Daily Totals		$6007.25		$8813.14	$68.26	$100.15

Crew R-13	Hr.	Daily	Hr.	Daily	Bare Costs	Incl. O&P
1 Electrician Foreman	$61.85	$494.80	$95.20	$761.60	$59.29	$91.58
3 Electricians	61.35	1472.40	94.45	2266.80		
.25 Equip. Oper. (crane)	59.20	118.40	92.65	185.30		
1 Equipment Oiler	50.55	404.40	79.10	632.80		
.25 Hydraulic Crane, 33 Ton		236.34		259.97	5.63	6.19
42 L.H., Daily Totals		$2726.34		$4106.47	$64.91	$97.77

Crew R-15	Hr.	Daily	Hr.	Daily	Bare Costs	Incl. O&P
1 Electrician Foreman	$61.85	$494.80	$95.20	$761.60	$60.04	$92.66
4 Electricians	61.35	1963.20	94.45	3022.40		
1 Equipment Oper. (light)	53.00	424.00	82.95	663.60		
1 Telescoping Boom Lift, to 40'		278.90		306.79	5.81	6.39
48 L.H., Daily Totals		$3160.90		$4754.39	$65.85	$99.05

For customer support on your Commercial Renovation Costs with RSMeans data, call 800.448.8182.

785

Crew No.	Bare Costs		Incl. Subs O & P		Cost Per Labor-Hour	

Crew R-18	Hr.	Daily	Hr.	Daily	Bare Costs	Incl. O&P
.25 Electrician Foreman	$61.85	$123.70	$95.20	$190.40	$53.85	$82.88
1 Electrician	61.35	490.80	94.45	755.60		
2 Electrician Apprentices	49.10	785.60	75.55	1208.80		
26 L.H., Daily Totals		$1400.10		$2154.80	$53.85	$82.88

Crew R-19	Hr.	Daily	Hr.	Daily	Bare Costs	Incl. O&P
.5 Electrician Foreman	$61.85	$247.40	$95.20	$380.80	$61.45	$94.60
2 Electricians	61.35	981.60	94.45	1511.20		
20 L.H., Daily Totals		$1229.00		$1892.00	$61.45	$94.60

Crew R-21	Hr.	Daily	Hr.	Daily	Bare Costs	Incl. O&P
1 Electrician Foreman	$61.85	$494.80	$95.20	$761.60	$61.36	$94.50
3 Electricians	61.35	1472.40	94.45	2266.80		
.1 Equip. Oper. (medium)	56.75	45.40	88.80	71.04		
.1 S.P. Crane, 4x4, 25 Ton		114.40		125.84	3.49	3.84
32.8 L.H., Daily Totals		$2127.00		$3225.28	$64.85	$98.33

Crew R-22	Hr.	Daily	Hr.	Daily	Bare Costs	Incl. O&P
.66 Electrician Foreman	$61.85	$326.57	$95.20	$502.66	$56.16	$86.44
2 Electricians	61.35	981.60	94.45	1511.20		
2 Electrician Apprentices	49.10	785.60	75.55	1208.80		
37.28 L.H., Daily Totals		$2093.77		$3222.66	$56.16	$86.44

Crew R-30	Hr.	Daily	Hr.	Daily	Bare Costs	Incl. O&P
.25 Electrician Foreman (outside)	$63.35	$126.70	$97.50	$195.00	$49.66	$78.07
1 Electrician	61.35	490.80	94.45	755.60		
2 Laborers (Semi-Skilled)	42.10	673.60	67.45	1079.20		
26 L.H., Daily Totals		$1291.10		$2029.80	$49.66	$78.07

Crew R-31	Hr.	Daily	Hr.	Daily	Bare Costs	Incl. O&P
1 Electrician	$61.35	$490.80	$94.45	$755.60	$61.35	$94.45
1 Core Drill, Electric, 2.5 H.P.		62.50		68.75	7.81	8.59
8 L.H., Daily Totals		$553.30		$824.35	$69.16	$103.04

For customer support on your Commercial Renovation Costs with RSMeans data, call 800.448.8182.

Historical Cost Indexes

The table below lists both the RSMeans® historical cost index based on Jan. 1, 1993 = 100 as well as the computed value of an index based on Jan. 1, 2020 costs. Since the Jan. 1, 2020 figure is estimated, space is left to write in the actual index figures as they become available through the quarterly *RSMeans Construction Cost Indexes*.

To compute the actual index based on Jan. 1, 2020 = 100, divide the historical cost index for a particular year by the actual Jan. 1, 2020 construction cost index. Space has been left to advance the index figures as the year progresses.

Year	Historical Cost Index Jan. 1, 1993 = 100		Current Index Based on Jan. 1, 2020 = 100		Year	Historical Cost Index Jan. 1, 1993 = 100	Current Index Based on Jan. 1, 2020 = 100		Year	Historical Cost Index Jan. 1, 1993 = 100	Current Index Based on Jan. 1, 2020 = 100	
	Est.	Actual	Est.	Actual		Actual	Est.	Actual		Actual	Est.	Actual
Oct 2020*					July 2005	151.6	63.4		July 1987	87.7	36.7	
July 2020*					2004	143.7	60.1		1986	84.2	35.2	
Apr 2020*					2003	132.0	55.2		1985	82.6	34.6	
Jan 2020*	239.1		100.0	100.0	2002	128.7	53.8		1984	82.0	34.3	
July 2019		232.2	97.1		2001	125.1	52.3		1983	80.2	33.5	
2018		222.9	93.2		2000	120.9	50.6		1982	76.1	31.8	
2017		213.6	89.3		1999	117.6	49.2		1981	70.0	29.3	
2016		207.3	86.7		1998	115.1	48.1		1980	62.9	26.3	
2015		206.2	86.2		1997	112.8	47.2		1979	57.8	24.2	
2014		204.9	85.7		1996	110.2	46.1		1978	53.5	22.4	
2013		201.2	84.1		1995	107.6	45.0		1977	49.5	20.7	
2012		194.6	81.4		1994	104.4	43.7		1976	46.9	19.6	
2011		191.2	80.0		1993	101.7	42.5		1975	44.8	18.7	
2010		183.5	76.7		1992	99.4	41.6		1974	41.4	17.3	
2009		180.1	75.3		1991	96.8	40.5		1973	37.7	15.8	
2008		180.4	75.4		1990	94.3	39.4		1972	34.8	14.6	
2007		169.4	70.8		1989	92.1	38.5		1971	32.1	13.4	
2006		162.0	67.8		1988	89.9	37.6		1970	28.7	12.0	

Adjustments to Costs

The "Historical Cost Index" can be used to convert national average building costs at a particular time to the approximate building costs for some other time.

Example:

Estimate and compare construction costs for different years in the same city.

To estimate the national average construction cost of a building in 1970, knowing that it cost $900,000 in 2020:

INDEX in 1970 = 28.7

INDEX in 2020 = 239.1

Note: The city cost indexes for Canada can be used to convert U.S. national averages to local costs in Canadian dollars.

Example:

To estimate and compare the cost of a building in Toronto, ON in 2020 with the known cost of $600,000 (US$) in New York, NY in 2020:

INDEX Toronto = 115.6

INDEX New York = 137.1

$$\frac{\text{INDEX Toronto}}{\text{INDEX New York}} \times \text{Cost New York} = \text{Cost Toronto}$$

$$\frac{115.6}{137.1} \times \$600,000 = .843 \times \$600,000 = \$505,908$$

The construction cost of the building in Toronto is $505,908 (CN$).

Time Adjustment Using the Historical Cost Indexes:

$$\frac{\text{Index for Year A}}{\text{Index for Year B}} \times \text{Cost in Year B} = \text{Cost in Year A}$$

$$\frac{\text{INDEX 1970}}{\text{INDEX 2020}} \times \text{Cost 2020} = \text{Cost 1970}$$

$$\frac{28.7}{239.1} \times \$900,000 = .120 \times \$900,000 = \$108,000$$

The construction cost of the building in 1970 was $108,000.

*Historical Cost Index updates and other resources are provided on the following website:
http://info.thegordiangroup.com/RSMeans.html

How to Use the City Cost Indexes

What you should know before you begin

RSMeans City Cost Indexes (CCI) are an extremely useful tool for when you want to compare costs from city to city and region to region.

This publication contains average construction cost indexes for 731 U.S. and Canadian cities covering over 930 three-digit zip code locations, as listed directly under each city.

Keep in mind that a City Cost Index number is a percentage ratio of a specific city's cost to the national average cost of the same item at a stated time period.

In other words, these index figures represent relative construction factors (or, if you prefer, multipliers) for material and installation costs, as well as the weighted average for Total In Place costs for each CSI MasterFormat division. Installation costs include both labor and equipment rental costs. When estimating equipment rental rates only for a specific location, use 01 54 33 EQUIPMENT RENTAL COSTS in the Reference Section.

The 30 City Average Index is the average of 30 major U.S. cities and serves as a national average.

Index figures for both material and installation are based on the 30 major city average of 100 and represent the cost relationship as of July 1, 2019. The index for each division is computed from representative material and labor quantities for that division. The weighted average for each city is a weighted total of the components listed above it. It does not include relative productivity between trades or cities.

As changes occur in local material prices, labor rates, and equipment rental rates (including fuel costs), the impact of these changes should be accurately measured by the change in the City Cost Index for each particular city (as compared to the 30 city average).

Therefore, if you know (or have estimated) building costs in one city today, you can easily convert those costs to expected building costs in another city.

In addition, by using the Historical Cost Index, you can easily convert national average building costs at a particular time to the approximate building costs for some other time. The City Cost Indexes can then be applied to calculate the costs for a particular city.

Quick calculations

Location Adjustment Using the City Cost Indexes:

$$\frac{\text{Index for City A}}{\text{Index for City B}} \times \text{Cost in City B} = \text{Cost in City A}$$

Time Adjustment for the National Average Using the Historical Cost Index:

$$\frac{\text{Index for Year A}}{\text{Index for Year B}} \times \text{Cost in Year B} = \text{Cost in Year A}$$

Adjustment from the National Average:

$$\frac{\text{Index for City A}}{100} \times \text{National Average Cost} = \text{Cost in City A}$$

Since each of the other RSMeans data sets contains many different items, any *one* item multiplied by the particular city index may give incorrect results. However, the larger the number of items compiled, the closer the results should be to actual costs for that particular city.

The City Cost Indexes for Canadian cities are calculated using Canadian material and equipment prices and labor rates in Canadian dollars. Therefore, indexes for Canadian cities can be used to convert U.S. national average prices to local costs in Canadian dollars.

How to use this section

1. Compare costs from city to city.

In using the RSMeans Indexes, remember that an index number is not a fixed number but a ratio: It's a percentage ratio of a building component's cost at any stated time to the national average cost of that same component at the same time period. Put in the form of an equation:

$$\frac{\text{Specific City Cost}}{\text{National Average Cost}} \times 100 = \text{City Index Number}$$

Therefore, when making cost comparisons between cities, do not subtract one city's index number from the index number of another city and read the result as a percentage difference. Instead, divide one city's index number by that of the other city. The resulting number may then be used as a multiplier to calculate cost differences from city to city.

The formula used to find cost differences between cities for the purpose of comparison is as follows:

$$\frac{\text{City A Index}}{\text{City B Index}} \times \text{City B Cost (Known)} = \text{City A Cost (Unknown)}$$

In addition, you can use RSMeans CCI to calculate and compare costs division by division between cities using the same basic formula. (Just be sure that you're comparing similar divisions.)

2. Compare a specific city's construction costs with the national average.

When you're studying construction location feasibility, it's advisable to compare a prospective project's cost index with an index of the national average cost.

For example, divide the weighted average index of construction costs of a specific city by that of the 30 City Average, which = 100.

$$\frac{\text{City Index}}{100} = \text{\% of National Average}$$

As a result, you get a ratio that indicates the relative cost of construction in that city in comparison with the national average.

3. Convert U.S. national average to actual costs in Canadian City.

$$\frac{\text{Index for Canadian City}}{100} \times \text{National Average Cost} = \text{Cost in Canadian City in \$ CAN}$$

4. Adjust construction cost data based on a national average.

When you use a source of construction cost data which is based on a national average (such as RSMeans cost data), it is necessary to adjust those costs to a specific location.

$$\frac{\text{City Index}}{100} \quad \times \quad \frac{\text{Cost Based on}}{\text{National Average Costs}} \quad = \quad \frac{\text{City Cost}}{\text{(Unknown)}}$$

5. When applying the City Cost Indexes to demolition projects, use the appropriate division installation index. For example, for removal of existing doors and windows, use the Division 8 (Openings) index.

What you might like to know about how we developed the Indexes

The information presented in the CCI is organized according to the Construction Specifications Institute (CSI) MasterFormat 2018 classification system.

To create a reliable index, RSMeans researched the building type most often constructed in the United States and Canada. Because it was concluded that no one type of building completely represented the building construction industry, nine different types of buildings were combined to create a composite model.

The exact material, labor, and equipment quantities are based on detailed analyses of these nine building types, and then each quantity is weighted in proportion to expected usage. These various material items, labor hours, and equipment rental rates are thus combined to form a composite building representing as closely as possible the actual usage of materials, labor, and equipment in the North American building construction industry.

The following structures were chosen to make up that composite model:

1. Factory, 1 story
2. Office, 2–4 stories
3. Store, Retail
4. Town Hall, 2–3 stories
5. High School, 2–3 stories
6. Hospital, 4–8 stories
7. Garage, Parking
8. Apartment, 1–3 stories
9. Hotel/Motel, 2–3 stories

For the purposes of ensuring the timeliness of the data, the components of the index for the composite model have been streamlined. They currently consist of:

- specific quantities of 66 commonly used construction materials;
- specific labor-hours for 21 building construction trades; and
- specific days of equipment rental for 6 types of construction equipment (normally used to install the 66 material items by the 21 trades.) Fuel costs and routine maintenance costs are included in the equipment cost.

Material and equipment price quotations are gathered quarterly from cities in the United States and Canada. These prices and the latest negotiated labor wage rates for 21 different building trades are used to compile the quarterly update of the City Cost Index.

The 30 major U.S. cities used to calculate the national average are:

Atlanta, GA	Memphis, TN
Baltimore, MD	Milwaukee, WI
Boston, MA	Minneapolis, MN
Buffalo, NY	Nashville, TN
Chicago, IL	New Orleans, LA
Cincinnati, OH	New York, NY
Cleveland, OH	Philadelphia, PA
Columbus, OH	Phoenix, AZ
Dallas, TX	Pittsburgh, PA
Denver, CO	St. Louis, MO
Detroit, MI	San Antonio, TX
Houston, TX	San Diego, CA
Indianapolis, IN	San Francisco, CA
Kansas City, MO	Seattle, WA
Los Angeles, CA	Washington, DC

What the CCI does not indicate

The weighted average for each city is a total of the divisional components weighted to reflect typical usage. It does not include the productivity variations between trades or cities.

In addition, the CCI does not take into consideration factors such as the following:

- managerial efficiency
- competitive conditions
- automation
- restrictive union practices
- unique local requirements
- regional variations due to specific building codes

City Cost Indexes

		UNITED STATES			ALABAMA														
	DIVISION	30 CITY AVERAGE			ANNISTON 362			BIRMINGHAM 350 - 352			BUTLER 369			DECATUR 356			DOTHAN 363		
		MAT.	INST.	TOTAL	MAT.	INST.	TOTAL	MAT.	INST.	TOTAL	MAT.	INST.	TOTAL	MAT.	INST.	TOTAL	MAT.	INST.	TOTAL
015433	CONTRACTOR EQUIPMENT		100.0	100.0		101.9	101.9		104.7	104.7		99.5	99.5		101.9	101.9		99.5	99.5
0241, 31 - 34	SITE & INFRASTRUCTURE, DEMOLITION	100.0	100.0	100.0	90.9	89.5	89.9	91.2	94.5	93.5	105.4	84.4	90.9	84.5	88.1	87.0	102.9	84.8	90.4
0310	Concrete Forming & Accessories	100.0	100.0	100.0	85.6	68.7	71.2	90.6	69.7	72.8	82.3	68.8	70.8	90.9	62.8	66.9	90.3	69.1	72.2
0320	Concrete Reinforcing	100.0	100.0	100.0	88.8	71.2	80.3	94.2	71.3	83.1	93.9	70.5	82.6	88.1	67.7	78.2	93.9	71.3	83.0
0330	Cast-in-Place Concrete	100.0	100.0	100.0	82.3	67.7	76.9	103.8	69.3	90.9	80.3	67.8	75.6	101.9	67.2	89.0	80.3	67.7	75.6
03	CONCRETE	100.0	100.0	100.0	89.2	70.4	81.0	92.2	71.4	83.1	90.3	70.4	81.6	92.0	67.0	81.0	89.5	70.7	81.2
04	MASONRY	100.0	100.0	100.0	89.2	63.1	73.2	87.1	64.3	73.1	93.7	63.1	74.9	85.2	62.4	71.2	95.1	62.9	75.3
05	METALS	100.0	100.0	100.0	101.8	94.8	99.7	100.5	94.2	98.7	100.7	94.8	98.9	102.9	93.1	100.0	100.7	95.4	99.1
06	WOOD, PLASTICS & COMPOSITES	100.0	100.0	100.0	82.0	69.9	75.3	91.0	70.0	79.4	77.1	69.9	73.1	95.7	62.1	77.2	88.1	69.9	78.1
07	THERMAL & MOISTURE PROTECTION	100.0	100.0	100.0	96.0	63.0	81.6	94.0	67.6	82.5	96.1	65.9	82.9	93.1	64.5	80.7	96.0	65.7	82.8
08	OPENINGS	100.0	100.0	100.0	94.7	69.2	88.7	102.6	69.8	94.9	94.7	69.5	88.8	106.9	64.6	97.0	94.7	69.7	88.9
0920	Plaster & Gypsum Board	100.0	100.0	100.0	81.8	69.5	73.5	88.9	69.5	75.9	79.3	69.5	72.7	92.7	61.6	71.7	88.7	69.5	75.8
0950, 0980	Ceilings & Acoustic Treatment	100.0	100.0	100.0	77.1	69.5	72.0	84.8	69.5	74.5	77.1	69.5	72.0	83.2	61.6	68.6	77.1	69.5	72.0
0960	Flooring	100.0	100.0	100.0	82.0	69.1	78.3	99.6	69.1	91.0	86.5	69.1	81.6	89.9	69.1	84.1	91.6	69.1	85.3
0970, 0990	Wall Finishes & Painting/Coating	100.0	100.0	100.0	89.1	66.7	75.9	87.0	66.7	75.0	89.1	45.9	63.5	79.4	61.0	68.5	89.1	80.4	84.0
09	FINISHES	100.0	100.0	100.0	80.4	68.8	74.1	91.2	69.2	79.3	83.6	66.6	74.4	85.3	63.5	73.5	86.1	70.3	77.6
COVERS	DIVS. 10 - 14, 25, 28, 41, 43, 44, 46	100.0	100.0	100.0	100.0	74.0	94.2	100.0	83.6	96.3	100.0	74.0	94.2	100.0	72.9	94.0	100.0	74.0	94.2
21, 22, 23	FIRE SUPPRESSION, PLUMBING & HVAC	100.0	100.0	100.0	101.0	51.8	81.7	100.0	66.3	86.8	98.1	67.0	85.9	100.0	64.5	86.1	98.1	65.0	85.1
26, 27, 3370	ELECTRICAL, COMMUNICATIONS & UTIL.	100.0	100.0	100.0	97.7	59.9	79.1	98.4	59.7	79.3	99.8	61.7	81.0	94.5	63.8	79.4	98.6	76.1	87.5
MF2018	WEIGHTED AVERAGE	100.0	100.0	100.0	95.9	67.7	84.0	97.4	71.8	86.6	96.1	70.6	85.3	97.1	69.2	85.3	96.2	72.7	86.3

		ALABAMA																	
	DIVISION	EVERGREEN 364			GADSDEN 359			HUNTSVILLE 357 - 358			JASPER 355			MOBILE 365 - 366			MONTGOMERY 360 - 361		
		MAT.	INST.	TOTAL	MAT.	INST.	TOTAL	MAT.	INST.	TOTAL	MAT.	INST.	TOTAL	MAT.	INST.	TOTAL	MAT.	INST.	TOTAL
015433	CONTRACTOR EQUIPMENT		99.5	99.5		101.9	101.9		101.9	101.9		101.9	101.9		99.5	99.5		102.0	102.0
0241, 31 - 34	SITE & INFRASTRUCTURE, DEMOLITION	105.9	84.4	91.0	89.9	89.5	89.6	84.2	89.4	87.8	89.9	89.5	89.6	97.7	85.4	89.2	95.6	89.9	91.7
0310	Concrete Forming & Accessories	79.4	68.6	70.2	83.9	69.1	71.3	90.9	65.9	69.6	88.5	64.7	68.3	89.4	68.4	71.5	92.8	69.4	72.8
0320	Concrete Reinforcing	94.0	70.4	82.6	93.5	71.3	82.8	88.1	76.3	82.4	88.1	71.3	80.0	91.5	70.5	81.3	99.7	71.4	86.0
0330	Cast-in-Place Concrete	80.3	67.6	75.6	101.9	67.8	89.2	99.2	67.8	87.5	113.0	67.8	96.1	84.2	66.9	77.7	81.5	68.7	76.7
03	CONCRETE	90.7	70.2	81.7	96.5	70.7	85.1	90.8	70.1	81.7	100.1	68.7	86.3	85.5	69.9	78.7	84.1	71.1	78.4
04	MASONRY	93.8	62.9	74.8	83.8	63.1	71.1	86.3	63.3	72.2	81.4	63.1	70.2	92.3	61.4	73.3	89.4	63.2	73.3
05	METALS	100.7	94.6	98.9	100.7	95.2	99.1	102.9	96.7	101.0	100.7	95.1	99.0	102.8	94.9	100.5	101.7	94.3	99.5
06	WOOD, PLASTICS & COMPOSITES	73.8	69.9	71.6	86.7	69.9	77.4	95.7	65.3	79.0	93.0	64.1	77.1	86.7	69.9	77.4	90.3	70.0	79.1
07	THERMAL & MOISTURE PROTECTION	96.0	65.4	82.7	93.3	66.3	81.6	93.0	65.9	81.2	93.3	64.4	80.7	95.6	65.4	82.4	94.2	66.9	82.3
08	OPENINGS	94.7	69.3	88.8	103.4	69.7	95.6	106.6	68.5	97.7	103.4	66.6	94.8	97.3	69.5	90.8	96.2	69.8	90.0
0920	Plaster & Gypsum Board	78.3	69.5	72.4	85.1	69.5	74.6	92.7	64.8	73.9	89.5	63.6	72.1	85.6	69.5	74.8	85.9	69.5	74.9
0950, 0980	Ceilings & Acoustic Treatment	77.1	69.5	72.0	79.1	69.5	72.7	85.0	64.8	71.4	79.1	63.6	68.6	83.0	69.5	73.9	84.6	69.5	74.4
0960	Flooring	84.5	69.1	80.1	86.4	69.1	81.5	89.9	69.1	84.1	88.4	69.1	82.9	91.1	69.1	84.9	91.7	69.1	85.3
0970, 0990	Wall Finishes & Painting/Coating	89.1	45.9	63.5	79.4	57.4	66.4	79.4	63.7	70.1	79.4	66.7	71.9	92.3	45.9	64.8	91.1	66.7	76.7
09	FINISHES	82.9	66.5	74.1	82.9	67.8	74.7	85.7	65.9	75.0	84.0	65.4	73.9	86.2	66.2	75.4	88.3	68.9	77.8
COVERS	DIVS. 10 - 14, 25, 28, 41, 43, 44, 46	100.0	73.9	94.2	100.0	82.9	96.2	100.0	82.5	96.1	100.0	73.4	94.1	100.0	82.5	96.1	100.0	83.3	96.3
21, 22, 23	FIRE SUPPRESSION, PLUMBING & HVAC	98.1	59.8	83.0	102.3	62.3	86.6	100.0	66.9	87.0	102.3	65.7	87.9	100.0	61.4	84.8	100.0	64.3	86.0
26, 27, 3370	ELECTRICAL, COMMUNICATIONS & UTIL.	97.1	56.8	77.2	94.5	59.7	77.4	95.4	65.9	80.8	94.2	59.9	77.3	100.4	56.8	78.9	101.1	76.1	88.8
MF2018	WEIGHTED AVERAGE	95.8	68.3	84.2	97.3	70.2	85.9	97.1	71.7	86.4	97.8	69.8	86.0	96.7	68.7	84.9	96.3	73.1	86.5

		ALABAMA									ALASKA								
	DIVISION	PHENIX CITY 368			SELMA 367			TUSCALOOSA 354			ANCHORAGE 995 - 996			FAIRBANKS 997			JUNEAU 998		
		MAT.	INST.	TOTAL	MAT.	INST.	TOTAL	MAT.	INST.	TOTAL	MAT.	INST.	TOTAL	MAT.	INST.	TOTAL	MAT.	INST.	TOTAL
015433	CONTRACTOR EQUIPMENT		99.5	99.5		99.5	99.5		101.9	101.9		110.5	110.5		112.9	112.9		110.5	110.5
0241, 31 - 34	SITE & INFRASTRUCTURE, DEMOLITION	109.8	85.5	93.0	102.7	85.5	90.8	84.8	89.5	88.0	118.4	121.7	120.7	117.8	124.5	122.5	133.6	121.7	125.4
0310	Concrete Forming & Accessories	85.6	67.6	70.2	83.5	69.1	71.2	90.8	69.2	72.4	120.5	116.2	116.8	124.7	115.0	116.4	125.7	116.2	117.6
0320	Concrete Reinforcing	93.9	66.9	80.9	93.9	71.3	83.0	88.1	71.3	80.0	148.4	120.7	135.0	147.2	120.7	134.4	138.9	120.7	130.1
0330	Cast-in-Place Concrete	80.3	67.9	75.7	80.3	67.8	75.6	103.4	67.9	90.2	105.1	117.5	109.7	109.7	115.6	111.9	116.1	117.5	116.6
03	CONCRETE	93.6	69.3	82.9	89.0	70.7	81.0	92.7	70.8	83.1	107.5	116.7	111.6	101.4	115.5	107.6	114.1	116.7	115.2
04	MASONRY	93.7	63.1	74.9	97.4	63.1	76.3	85.4	63.1	71.7	171.5	119.4	139.5	181.4	117.8	142.3	162.5	119.4	136.0
05	METALS	100.6	93.7	98.6	100.6	95.2	99.0	102.1	95.3	100.1	127.3	104.4	120.5	122.6	105.4	117.5	113.1	104.4	110.6
06	WOOD, PLASTICS & COMPOSITES	81.7	67.7	74.0	78.9	69.9	73.9	95.7	69.9	81.5	112.3	113.5	112.9	126.9	112.7	119.1	121.9	113.5	117.3
07	THERMAL & MOISTURE PROTECTION	96.5	66.1	83.2	95.9	65.9	82.8	93.1	66.3	81.4	176.9	116.9	150.8	186.2	115.8	155.6	188.4	116.9	157.3
08	OPENINGS	94.7	67.4	88.3	94.7	69.7	88.8	106.6	69.7	98.0	130.8	115.8	127.3	132.8	114.7	128.6	131.0	115.8	127.5
0920	Plaster & Gypsum Board	83.0	67.3	72.4	81.1	69.5	73.3	92.7	69.5	77.1	144.1	113.8	123.7	177.5	112.9	134.0	158.2	113.8	128.3
0950, 0980	Ceilings & Acoustic Treatment	77.1	67.3	70.5	77.1	69.5	72.0	85.0	69.5	74.6	131.3	113.8	119.5	123.9	112.9	116.5	136.1	113.8	121.0
0960	Flooring	88.4	69.1	83.0	87.0	69.1	81.9	89.9	69.1	84.1	118.4	119.4	118.6	116.7	119.4	117.4	125.0	119.4	123.5
0970, 0990	Wall Finishes & Painting/Coating	89.1	66.7	75.9	89.1	66.7	75.9	79.4	66.7	71.9	113.1	119.1	116.6	111.6	116.9	114.7	108.7	119.1	114.8
09	FINISHES	85.2	68.9	76.4	83.7	68.8	75.7	85.6	68.8	76.6	127.0	117.1	121.7	128.6	116.0	121.8	128.9	117.1	122.5
COVERS	DIVS. 10 - 14, 25, 28, 41, 43, 44, 46	100.0	82.7	96.1	100.0	70.4	93.4	100.0	82.9	96.2	100.0	114.1	103.2	100.0	113.6	103.0	100.0	114.1	103.2
21, 22, 23	FIRE SUPPRESSION, PLUMBING & HVAC	98.1	64.2	84.8	98.1	65.1	85.1	100.0	67.1	87.1	100.5	107.1	103.1	100.2	107.9	103.2	100.5	107.1	103.1
26, 27, 3370	ELECTRICAL, COMMUNICATIONS & UTIL.	99.2	65.4	82.6	98.3	76.1	87.3	95.0	59.7	77.6	117.2	109.4	113.3	124.2	109.4	116.9	110.2	109.4	109.8
MF2018	WEIGHTED AVERAGE	96.8	70.8	85.8	95.9	72.5	86.0	97.1	71.4	86.3	118.5	113.1	116.2	118.8	113.0	116.3	116.8	113.1	115.2

City Cost Indexes

ALASKA / ARIZONA

DIVISION		ALASKA KETCHIKAN 999 MAT.	INST.	TOTAL	ARIZONA CHAMBERS 865 MAT.	INST.	TOTAL	FLAGSTAFF 860 MAT.	INST.	TOTAL	GLOBE 855 MAT.	INST.	TOTAL	KINGMAN 864 MAT.	INST.	TOTAL	MESA/TEMPE 852 MAT.	INST.	TOTAL
015433	CONTRACTOR EQUIPMENT		112.9	112.9		88.2	88.2		88.2	88.2		89.3	89.3		88.2	88.2		89.3	89.3
0241, 31 - 34	SITE & INFRASTRUCTURE, DEMOLITION	169.1	124.6	138.4	71.5	90.1	84.4	91.2	90.1	90.4	109.0	91.0	96.5	71.4	90.1	84.3	97.8	91.0	93.1
0310	Concrete Forming & Accessories	116.3	116.2	116.2	99.2	70.3	74.6	104.6	67.4	72.8	93.5	70.4	73.8	97.4	66.1	70.7	96.5	70.5	74.3
0320	Concrete Reinforcing	110.6	120.7	115.5	105.5	76.9	91.7	105.4	76.9	91.6	110.5	76.9	94.2	105.7	76.9	91.7	111.2	76.9	94.6
0330	Cast-in-Place Concrete	221.8	117.3	182.9	89.8	67.3	81.4	89.8	67.4	81.5	83.3	67.0	77.2	89.5	67.3	81.2	84.0	67.0	77.7
03	CONCRETE	170.9	116.7	147.1	93.4	70.3	83.3	114.1	69.0	94.3	101.9	70.3	88.0	93.1	68.4	82.3	93.1	70.4	83.1
04	MASONRY	188.6	119.4	146.1	96.0	59.4	73.5	96.1	59.4	73.6	101.7	59.3	75.7	96.0	59.4	73.5	101.9	59.3	75.8
05	METALS	122.8	105.4	117.7	103.0	71.6	93.8	103.5	71.7	94.2	105.8	72.5	96.0	103.7	71.6	94.3	106.2	72.6	96.3
06	WOOD, PLASTICS & COMPOSITES	117.1	113.5	115.1	101.8	73.0	85.9	108.2	68.8	86.5	92.7	73.1	81.9	96.4	67.3	80.4	96.7	73.1	83.7
07	THERMAL & MOISTURE PROTECTION	191.6	116.4	158.9	99.9	69.7	86.8	101.8	69.3	87.7	99.1	68.3	85.7	99.9	69.1	86.5	99.0	68.3	85.6
08	OPENINGS	128.6	115.8	125.6	105.8	69.3	97.3	105.9	70.1	97.6	94.9	69.4	88.9	106.0	67.9	97.1	94.9	69.4	89.0
0920	Plaster & Gypsum Board	161.0	113.8	129.2	92.8	72.5	79.1	96.5	68.2	77.5	88.2	72.5	77.6	84.2	66.6	72.4	91.7	72.5	78.7
0950, 0980	Ceilings & Acoustic Treatment	118.4	113.8	115.3	106.8	72.5	83.6	107.6	68.2	81.1	95.5	72.5	80.0	107.6	66.6	80.0	95.5	72.5	80.0
0960	Flooring	116.4	119.4	117.2	89.7	64.5	82.6	92.0	64.5	84.3	98.6	64.5	89.0	88.4	64.5	81.6	100.4	64.5	90.3
0970, 0990	Wall Finishes & Painting/Coating	111.6	119.1	116.0	87.9	58.6	70.5	87.9	58.6	70.5	90.6	58.6	71.6	87.9	58.6	70.5	90.6	58.6	71.6
09	FINISHES	129.2	117.1	122.7	91.3	68.1	78.8	94.5	65.7	78.9	96.6	68.2	81.2	89.9	64.8	76.3	96.4	68.2	81.2
COVERS	DIVS. 10 - 14, 25, 28, 41, 43, 44, 46	100.0	114.2	103.2	100.0	81.8	95.9	100.0	81.3	95.8	100.0	82.1	96.0	100.0	81.2	95.8	100.0	82.1	96.0
21, 22, 23	FIRE SUPPRESSION, PLUMBING & HVAC	98.1	107.2	101.7	97.7	76.8	89.5	100.3	77.0	91.1	96.6	76.9	88.9	97.7	76.8	89.5	100.1	77.0	91.1
26, 27, 3370	ELECTRICAL, COMMUNICATIONS & UTIL.	124.2	109.4	116.9	102.3	66.5	84.7	101.3	60.7	81.3	97.1	62.9	80.2	102.3	62.9	82.9	94.2	62.9	78.7
MF2018	WEIGHTED AVERAGE	128.4	113.4	122.1	98.3	71.8	87.1	102.4	70.5	88.9	99.5	71.4	87.6	98.3	70.4	87.2	98.7	71.5	87.2

ARIZONA / ARKANSAS

DIVISION		PHOENIX 850,853 MAT.	INST.	TOTAL	PRESCOTT 863 MAT.	INST.	TOTAL	SHOW LOW 859 MAT.	INST.	TOTAL	TUCSON 856 - 857 MAT.	INST.	TOTAL	BATESVILLE 725 MAT.	INST.	TOTAL	CAMDEN 717 MAT.	INST.	TOTAL
015433	CONTRACTOR EQUIPMENT		92.5	92.5		88.2	88.2		89.3	89.3		89.3	89.3		89.7	89.7		89.7	89.7
0241, 31 - 34	SITE & INFRASTRUCTURE, DEMOLITION	98.3	94.3	95.6	78.4	90.1	86.5	111.4	91.0	97.3	92.9	91.0	91.6	73.0	85.7	81.8	77.2	85.8	83.1
0310	Concrete Forming & Accessories	101.2	70.8	75.3	100.8	70.3	74.8	100.6	70.5	74.9	97.1	67.5	71.8	82.5	61.4	64.5	78.6	61.9	64.4
0320	Concrete Reinforcing	109.1	76.9	93.6	105.4	76.9	91.6	111.2	76.9	94.6	92.3	76.9	84.9	84.4	69.4	77.1	91.0	69.4	80.5
0330	Cast-in-Place Concrete	83.6	67.8	77.7	89.8	67.3	81.4	83.3	67.0	77.3	86.5	67.0	79.2	72.8	74.6	73.5	76.4	74.6	75.7
03	CONCRETE	95.5	70.8	84.6	99.0	70.3	86.4	104.2	70.3	89.3	91.6	69.0	81.7	73.2	68.1	70.9	75.7	68.3	72.4
04	MASONRY	94.8	62.1	74.7	96.1	59.4	73.6	101.8	59.3	75.7	88.9	59.3	70.7	94.1	60.8	73.6	104.7	60.8	77.7
05	METALS	107.8	73.7	97.8	103.5	71.6	94.2	105.6	72.5	95.9	107.0	72.6	96.9	94.7	76.0	89.2	101.1	76.1	93.8
06	WOOD, PLASTICS & COMPOSITES	102.1	72.0	85.5	103.3	73.0	86.6	101.2	73.1	85.7	96.9	69.0	81.6	91.1	63.5	75.9	89.2	64.0	75.4
07	THERMAL & MOISTURE PROTECTION	100.3	69.6	86.9	100.5	69.7	87.1	99.3	68.3	85.8	99.9	67.9	86.0	102.7	61.9	84.9	97.5	61.9	82.0
08	OPENINGS	100.7	71.4	93.9	105.9	69.3	97.4	94.2	71.1	88.8	91.4	70.2	86.5	100.0	61.0	90.9	103.1	61.3	93.4
0920	Plaster & Gypsum Board	99.4	71.2	80.4	93.3	72.5	79.3	93.5	72.5	79.4	96.3	68.2	77.4	73.7	62.9	66.5	84.7	63.5	70.4
0950, 0980	Ceilings & Acoustic Treatment	106.5	71.2	82.7	106.0	72.5	83.4	95.5	72.5	80.0	96.4	68.2	77.4	88.1	62.9	71.1	86.3	63.5	70.9
0960	Flooring	102.4	64.5	91.7	90.6	64.5	83.2	102.2	64.5	91.6	91.8	64.8	84.3	81.8	70.9	78.7	86.1	70.9	81.9
0970, 0990	Wall Finishes & Painting/Coating	95.3	57.2	72.8	87.9	58.6	70.5	90.6	58.6	71.6	91.3	58.6	71.9	92.2	54.5	69.9	88.8	54.5	68.5
09	FINISHES	101.0	68.1	83.2	92.0	68.1	79.1	98.6	68.2	82.2	94.5	65.9	79.0	76.3	62.6	68.9	79.8	62.9	70.7
COVERS	DIVS. 10 - 14, 25, 28, 41, 43, 44, 46	100.0	83.2	96.2	100.0	81.8	95.9	100.0	82.1	96.0	100.0	81.7	95.9	100.0	66.4	92.5	100.0	66.5	92.5
21, 22, 23	FIRE SUPPRESSION, PLUMBING & HVAC	100.0	78.5	91.6	100.3	76.8	91.0	96.6	76.9	88.9	100.1	74.8	90.2	96.5	51.2	78.7	96.4	57.5	81.2
26, 27, 3370	ELECTRICAL, COMMUNICATIONS & UTIL.	99.8	60.7	80.5	101.0	62.9	82.2	94.5	62.9	78.9	96.2	60.6	78.7	95.5	60.5	78.3	95.8	58.1	77.2
MF2018	WEIGHTED AVERAGE	100.6	72.3	88.6	99.9	71.3	87.8	99.7	71.5	87.8	97.7	70.2	86.1	91.6	63.6	79.8	94.0	64.7	81.6

ARKANSAS

DIVISION		FAYETTEVILLE 727 MAT.	INST.	TOTAL	FORT SMITH 729 MAT.	INST.	TOTAL	HARRISON 726 MAT.	INST.	TOTAL	HOT SPRINGS 719 MAT.	INST.	TOTAL	JONESBORO 724 MAT.	INST.	TOTAL	LITTLE ROCK 720 - 722 MAT.	INST.	TOTAL
015433	CONTRACTOR EQUIPMENT		89.7	89.7		89.7	89.7		89.7	89.7		89.7	89.7		112.7	112.7		92.4	92.4
0241, 31 - 34	SITE & INFRASTRUCTURE, DEMOLITION	72.5	85.7	81.6	77.7	85.5	83.1	77.7	85.7	83.3	80.2	85.7	84.0	96.7	102.4	100.7	85.3	90.3	88.8
0310	Concrete Forming & Accessories	78.4	61.8	64.3	95.7	61.6	66.7	86.5	61.8	65.4	76.3	61.8	64.0	85.7	61.9	65.4	93.1	62.4	66.9
0320	Concrete Reinforcing	84.4	62.1	73.6	85.4	66.3	76.2	84.0	69.4	76.9	89.3	69.4	79.7	81.5	65.1	73.6	90.4	69.5	80.3
0330	Cast-in-Place Concrete	72.9	74.6	73.5	83.3	75.5	80.4	80.8	74.5	78.5	78.1	74.6	76.8	79.4	75.8	78.0	80.7	76.9	79.3
03	CONCRETE	72.9	67.0	70.3	80.1	68.0	74.8	79.7	68.2	74.6	78.9	68.2	74.2	77.3	69.0	73.7	81.0	69.3	75.9
04	MASONRY	85.0	60.8	70.1	91.6	60.8	72.7	94.3	60.8	73.7	78.6	60.8	67.7	86.5	60.8	70.7	87.7	62.1	72.0
05	METALS	94.7	73.6	88.5	96.9	74.9	90.4	95.8	76.0	90.0	101.1	76.1	93.7	91.3	89.4	90.8	97.2	76.1	91.0
06	WOOD, PLASTICS & COMPOSITES	87.7	64.0	74.7	107.1	64.0	83.4	96.8	64.0	78.7	86.7	64.0	74.2	95.2	64.4	78.3	98.4	64.2	79.5
07	THERMAL & MOISTURE PROTECTION	103.5	61.9	85.4	103.8	61.7	85.5	103.0	61.9	85.1	97.8	61.9	82.2	108.4	61.9	88.2	98.5	63.1	83.1
08	OPENINGS	100.0	59.6	90.6	102.0	60.1	92.3	100.8	61.3	91.6	103.1	61.3	93.3	105.6	60.5	95.1	95.6	60.7	87.5
0920	Plaster & Gypsum Board	73.0	63.5	66.6	79.7	63.5	68.8	78.8	63.5	68.5	83.4	63.5	70.0	87.3	63.5	71.2	92.0	63.5	72.8
0950, 0980	Ceilings & Acoustic Treatment	88.1	63.5	71.5	89.8	63.5	72.0	89.8	63.5	72.0	86.3	63.5	70.9	92.1	63.5	72.8	87.3	63.5	71.2
0960	Flooring	78.9	70.9	76.7	88.3	70.9	83.4	84.1	70.9	80.4	85.1	70.9	81.1	59.3	70.9	62.6	89.2	81.0	86.9
0970, 0990	Wall Finishes & Painting/Coating	92.2	54.5	69.9	92.2	52.1	68.5	92.2	54.5	69.9	88.8	54.5	68.5	81.7	54.5	65.6	92.8	52.4	68.9
09	FINISHES	75.5	62.9	68.7	79.6	62.6	70.4	78.3	62.9	70.0	79.6	62.9	70.6	74.6	63.2	68.4	85.3	64.9	74.3
COVERS	DIVS. 10 - 14, 25, 28, 41, 43, 44, 46	100.0	66.1	92.4	100.0	77.7	95.0	100.0	68.5	93.0	100.0	66.2	92.5	100.0	67.0	92.6	100.0	78.2	95.1
21, 22, 23	FIRE SUPPRESSION, PLUMBING & HVAC	96.5	60.0	82.2	100.0	48.0	79.6	96.5	49.1	77.8	96.4	51.5	78.8	100.4	51.2	81.1	99.9	49.1	79.9
26, 27, 3370	ELECTRICAL, COMMUNICATIONS & UTIL.	90.0	53.9	72.2	93.1	57.0	75.3	94.2	57.5	76.1	97.7	65.2	81.7	99.0	62.3	81.0	100.1	66.7	83.7
MF2018	WEIGHTED AVERAGE	90.5	64.1	79.3	94.1	62.6	80.8	92.9	62.8	80.2	93.5	64.4	81.2	93.8	66.6	82.3	94.5	65.4	82.2

For customer support on your Commercial Renovation Costs with RSMeans data, call 800.448.8182.

791

City Cost Indexes

DIVISION		ARKANSAS												CALIFORNIA					
		PINE BLUFF			RUSSELLVILLE			TEXARKANA			WEST MEMPHIS			ALHAMBRA			ANAHEIM		
		716			728			718			723			917 - 918			928		
		MAT.	INST.	TOTAL	MAT.	INST.	TOTAL	MAT.	INST.	TOTAL	MAT.	INST.	TOTAL	MAT.	INST.	TOTAL	MAT.	INST.	TOTAL
015433	CONTRACTOR EQUIPMENT		89.7	89.7		89.7	89.7		90.9	90.9		112.7	112.7		95.4	95.4		99.2	99.2
0241, 31 - 34	SITE & INFRASTRUCTURE, DEMOLITION	82.3	85.6	84.6	74.2	85.7	82.2	92.9	87.8	89.3	103.1	102.8	102.9	100.8	105.3	103.9	97.7	105.5	103.1
0310	Concrete Forming & Accessories	76.1	61.9	64.0	83.1	61.7	64.8	81.4	61.5	64.5	90.6	62.2	66.4	114.4	139.5	135.8	102.9	139.7	134.3
0320	Concrete Reinforcing	90.9	69.4	80.5	84.9	69.4	77.4	90.5	69.3	80.2	81.5	63.3	72.7	105.0	133.6	118.9	95.2	133.5	113.8
0330	Cast-in-Place Concrete	78.1	74.6	76.8	76.3	74.5	75.6	85.2	74.4	81.2	83.2	75.9	80.5	83.3	127.8	99.9	86.7	131.0	103.2
03	CONCRETE	79.7	68.3	74.7	76.0	68.1	72.6	77.8	68.1	73.5	83.8	68.8	77.2	94.2	133.1	111.3	96.2	134.3	112.9
04	MASONRY	110.7	60.8	80.0	90.9	60.8	72.4	92.6	60.8	73.0	75.3	60.8	66.4	106.9	139.4	126.9	78.2	138.0	115.0
05	METALS	101.9	76.1	94.3	94.7	75.9	89.1	93.8	75.8	88.5	90.4	89.1	90.0	79.9	113.2	89.7	107.7	114.2	109.6
06	WOOD, PLASTICS & COMPOSITES	86.4	64.0	74.1	92.3	64.0	76.8	93.4	64.0	77.2	100.7	64.4	80.7	99.3	138.0	120.6	97.5	138.3	119.9
07	THERMAL & MOISTURE PROTECTION	97.9	62.0	82.3	103.7	61.9	85.5	98.5	61.9	82.6	108.9	61.9	88.4	104.2	131.8	116.2	108.9	134.9	120.2
08	OPENINGS	104.3	60.8	94.1	100.0	61.3	91.0	109.2	61.3	98.0	103.2	60.1	93.1	87.8	137.4	99.4	103.9	137.6	111.8
0920	Plaster & Gypsum Board	83.1	63.5	69.9	73.7	63.5	66.8	85.7	63.5	70.7	89.2	63.5	71.9	94.5	139.3	124.6	112.7	139.3	130.6
0950, 0980	Ceilings & Acoustic Treatment	86.3	63.5	70.9	88.1	63.5	71.5	89.6	63.5	72.0	90.2	63.5	72.2	111.0	139.3	130.1	109.3	139.3	129.5
0960	Flooring	84.9	70.9	80.9	81.3	70.9	78.4	86.9	43.4	74.7	61.6	70.9	64.2	104.8	121.1	109.4	99.2	121.1	105.3
0970, 0990	Wall Finishes & Painting/Coating	88.8	52.1	67.1	92.2	54.5	69.9	88.8	54.5	68.5	81.7	54.5	65.6	103.9	120.3	113.6	92.6	120.3	109.0
09	FINISHES	79.6	62.6	70.4	76.4	62.9	69.1	81.7	57.2	68.4	75.9	63.2	69.0	101.8	134.1	119.3	99.4	134.2	118.3
COVERS	DIVS. 10 - 14, 25, 28, 41, 43, 44, 46	100.0	77.7	95.0	100.0	66.5	92.5	100.0	66.1	92.5	100.0	67.0	92.6	100.0	119.6	104.4	100.0	120.2	104.5
21, 22, 23	FIRE SUPPRESSION, PLUMBING & HVAC	100.0	51.4	80.9	96.5	50.9	78.6	100.0	54.5	82.1	96.8	64.6	84.2	96.5	130.9	110.0	99.9	130.9	112.1
26, 27, 3370	ELECTRICAL, COMMUNICATIONS & UTIL.	95.9	57.5	77.0	93.1	57.5	75.6	97.6	57.5	77.9	100.5	65.3	83.1	121.0	129.2	125.0	91.5	113.3	102.2
MF2018	WEIGHTED AVERAGE	96.0	63.6	82.3	91.7	63.1	79.6	94.8	63.3	81.5	93.3	69.8	83.3	96.7	128.6	110.2	99.3	126.6	110.9

DIVISION		CALIFORNIA																	
		BAKERSFIELD			BERKELEY			EUREKA			FRESNO			INGLEWOOD			LONG BEACH		
		932 - 933			947			955			936 - 938			903 - 905			906 - 908		
		MAT.	INST.	TOTAL	MAT.	INST.	TOTAL	MAT.	INST.	TOTAL	MAT.	INST.	TOTAL	MAT.	INST.	TOTAL	MAT.	INST.	TOTAL
015433	CONTRACTOR EQUIPMENT		98.7	98.7		100.2	100.2		97.1	97.1		97.3	97.3		96.9	96.9		96.9	96.9
0241, 31 - 34	SITE & INFRASTRUCTURE, DEMOLITION	96.1	106.8	103.5	109.8	106.0	107.2	109.0	102.7	104.7	98.9	103.2	101.8	87.7	102.0	97.6	94.7	102.0	99.7
0310	Concrete Forming & Accessories	105.8	139.2	134.3	113.0	168.1	160.0	111.2	154.1	147.8	102.9	153.3	145.8	107.3	139.9	135.1	102.0	139.9	134.3
0320	Concrete Reinforcing	99.8	133.4	116.0	90.7	135.2	112.2	103.7	136.6	119.6	83.3	134.2	107.9	99.8	133.7	116.2	99.0	133.7	115.8
0330	Cast-in-Place Concrete	88.1	130.2	103.8	110.4	133.5	119.0	94.1	129.6	107.3	95.5	129.2	108.0	81.5	130.2	99.6	92.7	130.2	106.7
03	CONCRETE	91.7	133.7	110.2	105.1	148.2	124.0	106.8	140.8	121.7	95.5	139.9	115.0	90.9	134.2	109.9	100.5	134.2	115.3
04	MASONRY	92.1	137.5	120.0	122.7	154.9	142.5	103.3	153.4	134.1	96.9	145.0	126.5	73.1	139.5	113.9	81.7	139.5	117.2
05	METALS	102.5	112.5	105.4	109.9	116.5	111.9	107.4	116.1	109.9	102.8	115.2	106.4	88.0	114.8	95.9	87.9	114.8	95.8
06	WOOD, PLASTICS & COMPOSITES	96.9	138.3	119.7	104.3	173.4	142.3	111.2	157.3	136.6	101.2	157.3	132.1	99.0	138.4	120.7	92.5	138.4	117.8
07	THERMAL & MOISTURE PROTECTION	108.5	124.3	115.4	112.2	155.7	131.2	112.6	151.2	129.4	98.9	132.9	113.7	107.2	133.4	118.6	107.5	133.4	118.8
08	OPENINGS	92.6	134.3	102.3	91.0	161.7	107.5	103.2	143.2	112.5	94.7	145.1	106.5	87.6	137.6	99.2	87.5	137.6	99.2
0920	Plaster & Gypsum Board	94.4	139.3	124.6	113.4	175.0	154.8	117.4	158.8	145.2	90.9	158.8	136.6	100.3	139.3	126.5	96.2	139.3	125.2
0950, 0980	Ceilings & Acoustic Treatment	100.7	139.3	126.8	101.2	175.0	151.0	113.3	158.8	144.0	98.1	158.8	139.0	114.0	139.3	131.1	114.0	139.3	131.1
0960	Flooring	102.2	121.1	107.5	124.0	143.3	129.4	103.2	143.3	114.4	100.4	121.5	106.3	114.7	121.1	116.5	111.5	121.1	114.2
0970, 0990	Wall Finishes & Painting/Coating	92.5	108.6	102.0	111.6	171.3	146.9	94.1	138.6	120.4	101.3	120.7	112.8	115.1	120.3	118.2	115.1	120.3	118.2
09	FINISHES	96.4	133.0	116.2	110.1	165.4	140.0	104.6	152.0	130.3	94.6	146.3	122.6	108.2	134.3	122.4	107.2	134.3	121.9
COVERS	DIVS. 10 - 14, 25, 28, 41, 43, 44, 46	100.0	117.7	103.9	100.0	134.0	107.6	100.0	130.8	106.9	100.0	130.8	106.9	100.0	120.5	104.6	100.0	120.5	104.6
21, 22, 23	FIRE SUPPRESSION, PLUMBING & HVAC	100.1	128.8	111.4	96.7	169.6	125.3	96.4	130.6	109.9	100.1	130.1	111.9	96.1	131.0	109.8	96.1	131.0	109.8
26, 27, 3370	ELECTRICAL, COMMUNICATIONS & UTIL.	107.6	108.1	107.8	99.9	161.8	130.4	98.4	124.9	111.5	97.0	107.4	102.2	98.5	129.2	113.6	98.3	129.2	113.5
MF2018	WEIGHTED AVERAGE	98.9	124.6	109.8	103.0	151.9	123.7	102.5	134.1	115.9	98.5	129.4	111.5	93.8	128.8	108.6	95.4	128.8	109.5

DIVISION		CALIFORNIA																	
		LOS ANGELES			MARYSVILLE			MODESTO			MOJAVE			OAKLAND			OXNARD		
		900 - 902			959			953			935			946			930		
		MAT.	INST.	TOTAL	MAT.	INST.	TOTAL	MAT.	INST.	TOTAL	MAT.	INST.	TOTAL	MAT.	INST.	TOTAL	MAT.	INST.	TOTAL
015433	CONTRACTOR EQUIPMENT		103.3	103.3		97.1	97.1		97.1	97.1		97.3	97.3		100.2	100.2		96.3	96.3
0241, 31 - 34	SITE & INFRASTRUCTURE, DEMOLITION	93.7	109.2	104.4	105.6	102.6	103.5	100.6	102.6	102.0	91.9	103.4	99.8	115.2	106.0	108.8	99.2	101.6	100.8
0310	Concrete Forming & Accessories	105.4	140.1	135.0	101.6	153.6	145.9	98.1	153.7	145.5	113.6	139.3	135.5	102.3	168.2	158.4	105.5	139.8	134.7
0320	Concrete Reinforcing	100.7	133.7	116.6	103.7	134.1	118.4	107.4	134.1	120.3	100.4	133.4	116.4	92.8	136.9	114.1	98.6	133.5	115.5
0330	Cast-in-Place Concrete	84.1	131.4	101.7	105.2	129.2	114.1	94.2	129.3	107.2	82.6	130.1	100.2	104.8	133.5	115.4	96.1	130.4	108.9
03	CONCRETE	98.3	134.7	114.3	107.1	140.0	121.6	98.2	140.1	116.6	87.4	133.8	107.8	106.3	148.5	124.8	95.5	134.1	112.5
04	MASONRY	88.6	139.5	119.9	104.2	142.9	128.0	101.5	142.9	126.9	94.9	137.4	121.0	131.5	154.9	145.9	97.9	136.8	121.8
05	METALS	94.3	115.0	100.4	106.9	114.7	109.2	103.6	114.8	106.9	100.0	113.4	104.0	105.2	117.2	108.7	98.0	113.9	102.7
06	WOOD, PLASTICS & COMPOSITES	105.6	138.6	123.7	98.3	157.3	130.8	93.9	157.3	128.8	100.8	138.4	121.5	92.8	173.4	137.2	95.6	138.4	119.1
07	THERMAL & MOISTURE PROTECTION	103.4	139.0	119.3	112.0	138.9	123.7	111.6	139.5	123.5	105.6	123.1	113.5	110.8	155.7	130.7	108.6	133.3	119.3
08	OPENINGS	98.1	136.9	107.2	102.5	145.6	112.6	101.3	147.4	112.1	89.8	134.3	100.2	91.1	162.1	107.7	92.4	137.6	102.9
0920	Plaster & Gypsum Board	95.6	139.3	125.0	110.0	158.8	142.8	112.7	158.8	143.7	100.5	139.3	126.6	107.6	175.0	152.9	94.5	139.3	124.7
0950, 0980	Ceilings & Acoustic Treatment	121.4	139.3	133.5	112.5	158.8	143.7	109.3	158.8	142.6	99.4	139.3	126.3	103.8	175.0	151.8	100.8	139.3	126.8
0960	Flooring	112.5	122.2	115.3	98.8	119.5	104.6	99.2	133.0	108.7	101.3	121.1	106.8	117.0	143.3	124.4	93.8	121.1	101.5
0970, 0990	Wall Finishes & Painting/Coating	111.6	120.3	116.7	94.1	134.1	117.8	94.1	138.6	120.4	87.2	109.6	100.5	111.6	171.3	146.9	87.2	114.8	103.6
09	FINISHES	109.4	134.7	123.1	101.7	147.4	126.4	101.0	150.2	127.7	95.0	133.2	115.6	108.1	165.4	139.1	92.5	133.7	114.8
COVERS	DIVS. 10 - 14, 25, 28, 41, 43, 44, 46	100.0	121.0	104.7	100.0	130.8	106.9	100.0	130.8	106.9	100.0	117.9	104.0	100.0	134.0	107.6	100.0	120.4	104.5
21, 22, 23	FIRE SUPPRESSION, PLUMBING & HVAC	99.9	131.0	112.1	96.4	130.0	109.6	99.9	131.7	112.4	96.5	128.8	109.2	100.2	169.6	127.5	100.0	131.0	112.2
26, 27, 3370	ELECTRICAL, COMMUNICATIONS & UTIL.	96.9	131.0	113.7	95.1	120.7	107.7	97.5	109.9	103.6	95.5	108.1	101.7	99.1	161.8	130.0	101.4	117.9	109.5
MF2018	WEIGHTED AVERAGE	98.5	129.8	111.7	101.7	131.3	114.2	100.6	130.6	113.3	95.4	124.4	107.7	103.4	152.0	123.9	98.0	126.7	110.1

792

CALIFORNIA

DIVISION		PALM SPRINGS 922			PALO ALTO 943			PASADENA 910 - 912			REDDING 960			RICHMOND 948			RIVERSIDE 925		
		MAT.	INST.	TOTAL	MAT.	INST.	TOTAL	MAT.	INST.	TOTAL	MAT.	INST.	TOTAL	MAT.	INST.	TOTAL	MAT.	INST.	TOTAL
015433	CONTRACTOR EQUIPMENT		98.1	98.1		100.2	100.2		95.4	95.4		97.1	97.1		100.2	100.2		98.1	98.1
0241, 31 - 34	SITE & INFRASTRUCTURE, DEMOLITION	89.3	103.6	99.2	106.1	106.0	106.0	97.5	105.3	102.9	122.2	102.6	108.7	114.3	106.0	108.5	96.2	103.6	101.3
0310	Concrete Forming & Accessories	99.8	136.1	130.7	100.5	168.2	158.2	103.4	139.6	134.2	105.0	153.6	146.5	115.5	167.8	160.1	103.5	139.7	134.4
0320	Concrete Reinforcing	109.0	133.5	120.9	90.7	136.7	113.0	106.0	133.6	119.3	137.8	134.1	136.0	90.7	136.7	112.9	105.9	133.5	119.3
0330	Cast-in-Place Concrete	82.7	131.0	100.6	93.4	133.5	108.4	79.1	127.8	97.2	107.2	129.3	115.4	107.4	133.3	117.1	89.9	131.0	105.2
03	CONCRETE	90.5	132.7	109.0	95.3	148.5	118.7	89.9	133.1	108.9	114.0	140.1	125.4	108.0	148.2	125.7	96.4	134.3	113.1
04	MASONRY	76.5	137.7	114.1	105.1	151.5	133.6	93.3	139.4	121.6	131.9	142.9	138.7	122.5	151.5	140.3	77.5	137.7	114.5
05	METALS	108.3	114.1	110.0	102.7	116.9	106.9	80.0	113.2	89.7	103.9	114.7	107.1	102.8	116.6	106.8	107.7	114.2	109.6
06	WOOD, PLASTICS & COMPOSITES	92.1	133.5	114.9	90.4	173.4	136.1	85.4	138.0	114.4	107.8	157.3	135.0	107.5	173.4	143.8	97.5	138.3	119.9
07	THERMAL & MOISTURE PROTECTION	108.1	134.3	119.5	110.0	155.0	129.6	103.8	131.8	116.0	128.2	138.9	132.9	110.6	154.8	129.8	109.0	134.8	120.2
08	OPENINGS	99.9	134.9	108.1	91.1	160.2	107.2	87.8	137.4	99.4	115.9	145.6	122.8	91.1	160.2	107.2	102.6	137.6	110.8
0920	Plaster & Gypsum Board	107.2	134.4	125.5	106.0	175.0	152.4	88.8	139.3	122.8	112.6	158.8	143.7	114.5	175.0	155.2	111.9	139.3	130.4
0950, 0980	Ceilings & Acoustic Treatment	106.0	134.4	125.1	102.0	175.0	151.2	111.0	139.3	130.1	148.8	158.8	155.5	102.0	175.0	151.2	114.2	139.3	131.1
0960	Flooring	101.5	121.1	107.0	115.6	143.3	123.4	98.5	121.1	104.9	87.8	128.6	99.3	126.5	143.3	131.3	102.9	121.1	108.0
0970, 0990	Wall Finishes & Painting/Coating	91.1	120.3	108.4	111.6	171.3	146.9	103.9	120.3	113.6	105.2	134.1	122.3	111.6	171.3	146.9	91.1	120.3	108.4
09	FINISHES	98.0	131.4	116.1	106.5	165.4	138.4	99.0	134.1	118.0	107.2	149.0	129.8	111.6	165.4	140.7	101.0	134.2	119.0
COVERS	DIVS. 10 - 14, 25, 28, 41, 43, 44, 46	100.0	119.6	104.4	100.0	134.0	107.6	100.0	119.6	104.4	100.0	130.8	106.9	100.0	133.8	107.5	100.0	119.3	104.3
21, 22, 23	FIRE SUPPRESSION, PLUMBING & HVAC	96.4	130.9	109.9	96.7	170.6	125.7	96.5	130.9	110.0	100.3	130.1	112.0	96.7	164.0	123.1	99.9	130.9	112.1
26, 27, 3370	ELECTRICAL, COMMUNICATIONS & UTIL.	94.6	110.8	102.6	99.0	176.6	137.2	117.5	129.2	123.2	101.3	120.8	110.9	99.6	133.2	116.2	91.3	114.4	102.7
MF2018	WEIGHTED AVERAGE	97.4	125.3	109.2	99.2	153.8	122.3	94.8	128.6	109.1	107.6	131.5	117.7	102.4	146.2	120.9	99.3	126.6	110.8

CALIFORNIA

DIVISION		SACRAMENTO 942, 956 - 958			SALINAS 939			SAN BERNARDINO 923 - 924			SAN DIEGO 919 - 921			SAN FRANCISCO 940 - 941			SAN JOSE 951		
		MAT.	INST.	TOTAL	MAT.	INST.	TOTAL	MAT.	INST.	TOTAL	MAT.	INST.	TOTAL	MAT.	INST.	TOTAL	MAT.	INST.	TOTAL
015433	CONTRACTOR EQUIPMENT		99.0	99.0		97.3	97.3		98.1	98.1		102.2	102.2		109.8	109.8		98.7	98.7
0241, 31 - 34	SITE & INFRASTRUCTURE, DEMOLITION	95.6	111.3	106.4	112.3	103.2	106.0	76.1	103.6	95.1	105.7	107.8	107.2	116.4	113.1	114.1	131.6	97.9	108.3
0310	Concrete Forming & Accessories	100.5	156.0	147.8	108.9	156.6	149.6	107.1	139.7	134.9	103.8	129.1	125.4	106.1	168.3	159.1	103.9	167.9	158.5
0320	Concrete Reinforcing	86.0	134.4	109.4	99.1	134.6	116.3	105.9	133.5	119.3	102.6	133.4	117.5	105.6	132.1	118.4	94.7	135.0	114.2
0330	Cast-in-Place Concrete	88.6	130.2	104.1	95.0	129.6	107.8	62.1	131.0	87.8	90.5	124.5	103.2	116.9	133.3	123.0	109.4	132.4	118.0
03	CONCRETE	97.9	141.1	116.9	105.8	141.6	121.5	71.1	134.3	98.9	100.2	127.3	112.1	117.8	148.1	131.1	103.9	148.1	123.3
04	MASONRY	106.1	145.9	130.6	95.1	148.4	127.9	83.6	137.7	116.8	86.2	133.4	115.2	140.9	154.6	149.3	131.5	151.5	143.8
05	METALS	100.8	109.8	103.4	102.7	115.8	106.5	107.7	114.2	109.6	94.8	114.1	100.5	110.7	122.4	114.2	101.8	121.6	107.6
06	WOOD, PLASTICS & COMPOSITES	87.1	160.2	127.3	100.6	160.2	133.4	101.2	138.3	121.6	95.0	126.1	112.1	95.1	173.4	138.2	106.0	173.2	143.0
07	THERMAL & MOISTURE PROTECTION	122.2	142.5	131.0	106.2	145.4	123.3	107.1	134.8	119.2	108.2	121.5	114.0	116.5	155.7	133.5	108.0	155.1	128.5
08	OPENINGS	103.9	149.0	114.4	93.5	154.4	107.7	100.0	137.6	108.7	99.8	127.8	106.3	99.5	159.0	113.3	92.6	161.5	108.7
0920	Plaster & Gypsum Board	103.4	161.5	142.5	95.4	161.8	140.1	113.3	139.3	130.8	94.4	126.7	116.1	105.8	175.0	152.4	108.3	175.0	153.2
0950, 0980	Ceilings & Acoustic Treatment	102.0	161.5	142.1	99.4	161.8	141.5	109.3	139.3	129.5	121.1	126.7	124.8	111.7	175.0	154.4	109.3	175.0	153.6
0960	Flooring	115.8	128.6	119.4	95.9	143.3	109.2	105.0	121.1	109.5	108.9	121.1	112.3	117.5	143.3	124.7	92.4	143.3	106.7
0970, 0990	Wall Finishes & Painting/Coating	107.9	134.1	123.4	88.1	171.3	137.3	91.1	120.3	108.4	101.5	118.0	111.2	109.6	174.8	148.2	94.4	171.3	139.9
09	FINISHES	105.3	150.7	129.9	94.6	157.3	128.5	99.4	134.2	118.3	106.6	126.1	117.2	109.9	165.8	140.1	99.9	165.2	135.3
COVERS	DIVS. 10 - 14, 25, 28, 41, 43, 44, 46	100.0	131.7	107.1	100.0	131.2	107.0	100.0	117.7	103.9	100.0	115.7	103.5	100.0	128.4	106.3	100.0	133.4	107.5
21, 22, 23	FIRE SUPPRESSION, PLUMBING & HVAC	100.1	131.4	112.4	96.5	137.6	112.7	96.4	130.9	110.0	100.0	129.4	111.5	100.1	182.8	132.6	99.9	170.5	127.7
26, 27, 3370	ELECTRICAL, COMMUNICATIONS & UTIL.	94.8	120.8	107.6	96.5	132.1	114.0	94.6	122.6	103.5	103.8	104.3	104.0	99.2	182.8	140.4	100.7	176.6	138.1
MF2018	WEIGHTED AVERAGE	100.0	133.0	114.4	99.1	137.2	115.2	95.0	126.3	108.2	99.7	121.7	109.0	107.4	158.4	129.0	102.6	153.5	124.2

CALIFORNIA

DIVISION		SAN LUIS OBISPO 934			SAN MATEO 944			SAN RAFAEL 949			SANTA ANA 926 - 927			SANTA BARBARA 931			SANTA CRUZ 950		
		MAT.	INST.	TOTAL	MAT.	INST.	TOTAL	MAT.	INST.	TOTAL	MAT.	INST.	TOTAL	MAT.	INST.	TOTAL	MAT.	INST.	TOTAL
015433	CONTRACTOR EQUIPMENT		97.3	97.3		100.2	100.2		99.4	99.4		98.1	98.1		97.3	97.3		98.7	98.7
0241, 31 - 34	SITE & INFRASTRUCTURE, DEMOLITION	104.4	103.4	103.7	112.1	106.0	107.9	107.6	111.2	110.1	87.8	103.6	98.7	99.1	103.4	102.1	131.3	97.7	108.1
0310	Concrete Forming & Accessories	115.3	139.7	136.1	106.3	168.2	159.0	111.5	167.8	159.5	107.4	139.7	134.9	106.1	139.7	134.8	103.9	156.9	149.0
0320	Concrete Reinforcing	100.4	133.4	116.4	90.7	136.9	113.0	91.3	135.1	112.5	109.6	133.5	121.2	98.6	133.4	115.4	117.1	134.6	125.6
0330	Cast-in-Place Concrete	102.1	130.2	112.6	104.0	133.5	115.0	121.1	132.3	125.2	79.3	131.0	98.6	95.8	130.2	108.6	108.7	131.4	117.1
03	CONCRETE	103.3	134.0	116.8	104.6	148.5	123.9	128.2	147.4	136.6	88.0	134.3	108.3	95.4	134.0	112.4	106.4	142.6	122.3
04	MASONRY	96.5	135.9	120.7	122.2	154.6	142.1	99.1	154.6	133.2	73.4	138.0	113.1	95.2	135.9	120.2	135.3	148.6	143.4
05	METALS	100.7	113.8	104.6	102.6	117.1	106.9	103.7	112.4	106.3	107.8	114.2	109.7	98.5	113.8	103.0	109.3	120.2	112.5
06	WOOD, PLASTICS & COMPOSITES	103.0	138.4	122.4	97.9	173.4	139.5	95.0	173.2	138.0	103.0	138.3	122.4	95.6	138.4	119.1	106.0	160.4	135.9
07	THERMAL & MOISTURE PROTECTION	106.4	132.6	117.8	110.4	156.4	130.4	115.0	155.4	132.5	108.4	134.9	119.9	105.8	132.6	117.5	107.6	147.9	125.1
08	OPENINGS	91.7	134.3	101.6	91.1	162.1	107.6	101.5	159.7	115.1	99.2	137.6	108.2	93.2	137.6	103.6	93.9	154.4	108.0
0920	Plaster & Gypsum Board	101.1	139.3	126.8	111.7	175.0	154.3	113.2	175.0	154.8	114.6	139.3	131.2	94.5	139.3	124.7	116.8	161.8	147.1
0950, 0980	Ceilings & Acoustic Treatment	99.4	139.3	126.3	102.0	175.0	151.2	109.3	175.0	153.6	109.3	139.3	129.5	100.8	139.3	126.8	110.9	161.8	145.2
0960	Flooring	102.0	121.1	107.4	119.5	143.3	126.2	132.6	143.3	135.6	105.6	121.1	110.0	95.0	121.1	102.4	96.6	143.3	109.7
0970, 0990	Wall Finishes & Painting/Coating	87.2	114.8	103.6	111.6	171.3	146.9	107.8	165.5	141.9	91.1	118.0	107.0	87.2	114.8	103.6	94.5	171.3	139.9
09	FINISHES	96.3	133.7	116.5	108.9	165.4	139.5	112.0	164.6	140.5	100.9	134.0	118.8	93.0	133.7	115.0	102.6	157.4	132.3
COVERS	DIVS. 10 - 14, 25, 28, 41, 43, 44, 46	100.0	126.5	105.9	100.0	134.0	107.6	100.0	133.3	107.4	100.0	120.2	104.5	100.0	119.5	104.3	100.0	131.5	107.0
21, 22, 23	FIRE SUPPRESSION, PLUMBING & HVAC	96.5	131.0	110.0	96.7	165.7	123.8	96.7	182.8	130.5	96.4	130.9	110.0	100.0	131.0	112.2	99.9	137.9	114.8
26, 27, 3370	ELECTRICAL, COMMUNICATIONS & UTIL.	95.5	110.8	103.1	99.0	168.0	133.0	96.2	124.9	110.3	94.6	111.9	103.1	94.5	113.7	103.9	99.8	132.1	115.7
MF2018	WEIGHTED AVERAGE	98.2	125.8	109.9	101.5	152.0	122.9	104.5	149.1	123.4	97.0	126.3	109.4	97.2	126.1	109.4	104.6	137.5	118.5

CALIFORNIA / COLORADO

DIVISION		SANTA ROSA 954 MAT.	INST.	TOTAL	STOCKTON 952 MAT.	INST.	TOTAL	SUSANVILLE 961 MAT.	INST.	TOTAL	VALLEJO 945 MAT.	INST.	TOTAL	VAN NUYS 913 - 916 MAT.	INST.	TOTAL	ALAMOSA 811 MAT.	INST.	TOTAL
015433	CONTRACTOR EQUIPMENT		97.6	97.6		97.1	97.1		97.1	97.1		99.4	99.4		95.4	95.4		90.2	90.2
0241, 31 - 34	SITE & INFRASTRUCTURE, DEMOLITION	101.4	102.6	102.2	100.3	102.6	101.9	129.5	102.6	110.9	96.3	111.1	106.5	115.7	105.3	108.5	141.3	84.4	102.0
0310	Concrete Forming & Accessories	100.7	167.1	157.3	101.9	155.5	147.6	106.2	153.9	146.8	102.3	166.8	157.3	109.8	139.5	135.2	103.8	66.0	71.5
0320	Concrete Reinforcing	104.6	135.1	119.3	107.4	134.1	120.3	137.8	134.1	136.0	92.5	135.0	113.0	106.0	133.6	119.3	114.0	67.9	91.7
0330	Cast-in-Place Concrete	103.3	130.8	113.5	91.7	129.2	105.7	97.5	129.3	109.3	96.4	131.4	109.5	83.4	127.8	99.9	100.1	73.1	90.1
03	CONCRETE	107.1	146.9	124.6	97.3	140.9	116.5	117.0	140.2	127.2	102.4	146.6	121.8	104.4	133.1	117.0	112.5	69.4	93.6
04	MASONRY	101.8	153.5	133.6	101.4	142.9	126.9	130.4	142.9	138.1	77.4	153.5	124.1	106.9	139.4	126.9	130.1	59.5	86.7
05	METALS	108.1	117.7	110.9	103.9	114.7	107.0	102.9	114.7	106.4	103.7	111.7	106.0	79.2	113.2	89.2	105.2	78.5	97.3
06	WOOD, PLASTICS & COMPOSITES	93.6	173.0	137.3	99.5	159.9	132.8	109.6	157.6	136.0	84.8	173.2	133.4	94.1	138.0	118.3	98.5	67.4	81.4
07	THERMAL & MOISTURE PROTECTION	108.9	154.1	128.6	112.1	138.8	123.7	130.2	138.9	134.0	113.1	154.1	130.9	105.0	131.8	116.6	111.9	67.2	92.5
08	OPENINGS	100.8	161.4	114.9	101.3	147.0	112.0	116.8	145.8	123.6	103.2	159.7	116.3	87.7	137.4	99.3	96.3	68.5	89.8
0920	Plaster & Gypsum Board	109.2	175.0	153.5	112.7	161.5	145.5	113.1	159.1	144.1	107.5	175.0	152.9	92.8	139.3	124.1	80.8	66.5	71.2
0950, 0980	Ceilings & Acoustic Treatment	109.3	175.0	153.6	117.5	161.5	147.1	139.9	159.1	152.9	111.1	175.0	154.2	107.8	139.3	129.1	102.7	66.5	78.3
0960	Flooring	102.1	136.5	111.7	99.2	133.0	108.7	88.3	128.6	99.6	126.0	143.3	130.9	101.9	121.1	107.3	107.8	70.1	97.2
0970, 0990	Wall Finishes & Painting/Coating	91.1	165.5	135.2	94.1	134.1	117.8	105.2	134.1	122.3	108.7	165.5	142.3	103.9	120.3	113.6	101.7	76.9	87.0
09	FINISHES	100.0	163.0	134.1	102.6	151.3	129.0	106.7	149.2	129.7	108.4	164.4	138.7	101.2	134.1	119.0	99.8	67.5	82.3
COVERS	DIVS. 10 - 14, 25, 28, 41, 43, 44, 46	100.0	132.4	107.2	100.0	128.0	106.2	100.0	130.9	106.9	100.0	132.9	107.3	100.0	119.6	104.4	100.0	83.0	96.2
21, 22, 23	FIRE SUPPRESSION, PLUMBING & HVAC	96.4	182.2	130.1	99.9	131.7	112.4	96.8	130.1	109.9	100.2	147.2	118.7	96.5	130.9	110.0	96.5	72.4	87.0
26, 27, 3370	ELECTRICAL, COMMUNICATIONS & UTIL.	94.9	124.9	109.7	97.5	114.3	105.8	101.7	120.6	111.1	92.4	127.5	109.7	117.5	129.2	123.2	96.3	63.2	80.0
MF2018	WEIGHTED AVERAGE	101.2	148.4	121.2	100.7	131.3	113.7	107.3	131.6	117.5	100.2	141.6	117.7	97.8	128.6	110.8	103.6	70.3	89.5

COLORADO

DIVISION		BOULDER 803 MAT.	INST.	TOTAL	COLORADO SPRINGS 808 - 809 MAT.	INST.	TOTAL	DENVER 800 - 802 MAT.	INST.	TOTAL	DURANGO 813 MAT.	INST.	TOTAL	FORT COLLINS 805 MAT.	INST.	TOTAL	FORT MORGAN 807 MAT.	INST.	TOTAL
015433	CONTRACTOR EQUIPMENT		92.2	92.2		90.2	90.2		99.6	99.6		90.2	90.2		92.2	92.2		92.2	92.2
0241, 31 - 34	SITE & INFRASTRUCTURE, DEMOLITION	97.1	91.4	93.2	99.2	86.7	90.6	104.1	102.5	103.0	134.3	84.4	99.9	109.6	91.2	96.9	99.3	90.9	93.5
0310	Concrete Forming & Accessories	107.0	77.4	81.8	97.3	63.5	68.5	106.1	66.4	72.2	109.9	65.7	72.2	104.5	65.8	71.5	107.6	65.7	71.9
0320	Concrete Reinforcing	108.7	68.1	89.0	107.9	67.9	88.6	107.9	70.1	89.6	114.0	67.9	91.7	108.8	68.0	89.1	108.9	68.0	89.2
0330	Cast-in-Place Concrete	113.8	73.9	98.9	116.8	73.4	100.7	121.8	73.6	103.9	115.1	73.1	99.5	128.5	72.5	107.6	103.7	72.5	97.1
03	CONCRETE	105.2	74.9	91.9	108.7	68.4	91.0	111.4	70.2	93.3	114.2	69.3	94.5	116.3	69.2	95.6	103.7	69.1	88.5
04	MASONRY	101.6	65.4	79.4	103.7	62.3	78.3	104.9	64.1	79.9	117.5	60.6	82.5	118.9	63.1	84.6	116.7	63.1	83.8
05	METALS	94.4	77.9	89.6	97.5	77.7	91.7	99.9	79.0	93.8	105.2	78.4	97.3	95.7	77.7	90.4	94.2	77.8	89.4
06	WOOD, PLASTICS & COMPOSITES	107.1	81.4	93.0	96.6	63.6	78.5	108.4	67.1	85.7	108.4	67.4	85.8	104.6	67.0	83.9	107.1	67.0	85.0
07	THERMAL & MOISTURE PROTECTION	108.7	73.2	93.2	109.4	69.7	92.1	107.1	71.7	91.7	111.9	67.5	92.6	109.0	70.6	92.3	108.6	69.6	91.6
08	OPENINGS	95.4	76.2	90.9	99.5	66.4	91.8	102.5	68.8	94.6	103.1	68.5	95.0	95.3	68.3	89.0	95.3	68.2	89.0
0920	Plaster & Gypsum Board	116.3	81.3	92.8	100.3	62.9	75.2	111.7	66.6	81.4	95.3	66.5	75.9	110.3	66.6	80.9	116.3	66.5	82.8
0950, 0980	Ceilings & Acoustic Treatment	91.9	81.3	84.8	100.0	62.9	75.0	104.4	66.6	78.9	102.7	66.5	78.3	91.9	66.6	74.8	91.9	66.5	74.8
0960	Flooring	111.2	76.7	101.5	102.2	69.0	92.9	108.4	76.7	99.5	113.1	69.0	100.7	107.3	76.7	98.7	111.7	76.7	101.8
0970, 0990	Wall Finishes & Painting/Coating	97.1	76.9	85.1	96.8	76.9	85.0	103.4	76.9	87.7	101.7	76.9	87.0	97.1	76.9	85.1	97.1	76.9	85.1
09	FINISHES	100.6	77.7	88.2	97.4	65.3	80.0	102.5	68.9	84.3	102.4	67.4	83.5	99.2	68.6	82.6	100.7	68.5	83.3
COVERS	DIVS. 10 - 14, 25, 28, 41, 43, 44, 46	100.0	84.5	96.5	100.0	82.2	96.0	100.0	86.3	96.9	100.0	82.9	96.2	100.0	82.1	96.0	100.0	82.1	96.0
21, 22, 23	FIRE SUPPRESSION, PLUMBING & HVAC	96.5	74.8	87.9	100.1	72.2	89.1	99.9	73.8	89.7	96.5	58.7	81.6	100.0	71.7	88.9	96.5	73.5	87.4
26, 27, 3370	ELECTRICAL, COMMUNICATIONS & UTIL.	99.6	80.6	90.3	103.2	73.1	88.4	105.0	80.4	92.9	95.8	50.3	73.4	99.6	80.6	90.2	100.0	78.1	89.2
MF2018	WEIGHTED AVERAGE	98.7	77.0	89.6	101.2	71.5	88.6	103.0	75.4	91.3	103.9	65.6	87.8	102.2	73.5	90.1	99.3	73.5	88.4

COLORADO

DIVISION		GLENWOOD SPRINGS 816 MAT.	INST.	TOTAL	GOLDEN 804 MAT.	INST.	TOTAL	GRAND JUNCTION 815 MAT.	INST.	TOTAL	GREELEY 806 MAT.	INST.	TOTAL	MONTROSE 814 MAT.	INST.	TOTAL	PUEBLO 810 MAT.	INST.	TOTAL
015433	CONTRACTOR EQUIPMENT		93.3	93.3		92.2	92.2		93.3	93.3		92.2	92.2		91.7	91.7		90.2	90.2
0241, 31 - 34	SITE & INFRASTRUCTURE, DEMOLITION	150.9	91.9	110.2	110.4	91.4	97.3	133.7	92.1	105.0	95.9	91.2	92.7	144.0	87.9	105.3	125.2	84.5	97.1
0310	Concrete Forming & Accessories	100.6	65.5	70.7	99.7	66.7	71.5	108.7	66.6	72.8	102.4	65.8	71.2	100.1	65.5	70.6	106.1	63.7	70.0
0320	Concrete Reinforcing	112.8	68.0	91.2	108.9	68.0	89.2	113.2	67.8	91.2	108.7	68.0	89.0	112.7	67.8	91.0	109.0	67.9	89.1
0330	Cast-in-Place Concrete	100.1	72.1	89.6	111.7	73.9	97.6	110.8	73.5	96.9	107.3	72.5	94.3	100.1	72.6	89.8	99.4	73.7	89.8
03	CONCRETE	118.0	68.9	96.4	114.1	70.0	94.7	110.7	69.9	92.8	100.3	69.1	86.6	108.5	69.0	91.1	100.9	68.7	86.7
04	MASONRY	103.1	63.0	78.5	119.7	65.4	86.3	137.8	61.9	91.2	111.9	62.7	81.7	110.0	59.6	79.0	99.5	60.3	75.4
05	METALS	104.8	78.2	97.0	94.3	77.8	89.5	106.1	78.2	98.2	95.7	77.7	90.4	103.9	78.1	96.3	108.3	78.7	99.6
06	WOOD, PLASTICS & COMPOSITES	93.6	67.1	79.0	98.9	67.0	81.3	106.1	67.2	84.7	101.7	67.0	82.6	94.6	67.2	79.6	101.2	63.9	80.7
07	THERMAL & MOISTURE PROTECTION	111.8	68.2	92.8	109.8	71.6	93.2	110.9	69.3	92.8	108.2	70.4	91.8	112.0	67.2	92.5	110.4	67.8	91.8
08	OPENINGS	102.1	68.3	94.2	95.4	68.2	89.0	102.8	68.4	94.7	95.3	68.3	89.0	103.2	68.4	95.1	98.0	66.6	90.7
0920	Plaster & Gypsum Board	121.2	66.5	84.4	107.5	66.5	79.9	135.4	66.6	89.1	108.7	66.6	80.4	79.9	66.5	70.9	84.7	62.9	70.1
0950, 0980	Ceilings & Acoustic Treatment	101.9	66.5	78.0	91.9	66.5	74.8	101.9	66.6	78.1	91.9	66.6	74.8	102.7	66.5	78.3	110.0	62.9	78.3
0960	Flooring	106.9	76.7	98.4	104.9	76.7	97.0	112.4	70.1	100.5	106.2	73.1	96.9	110.4	73.1	100.0	109.1	73.1	99.0
0970, 0990	Wall Finishes & Painting/Coating	101.7	76.9	87.0	97.1	76.9	85.1	101.7	76.9	87.0	97.1	76.9	85.1	101.7	76.9	87.0	101.7	76.9	87.0
09	FINISHES	105.2	68.7	85.4	98.9	69.1	82.8	106.8	67.9	85.8	97.9	67.9	81.6	100.5	68.0	82.9	100.2	66.3	81.9
COVERS	DIVS. 10 - 14, 25, 28, 41, 43, 44, 46	100.0	82.3	96.1	100.0	82.9	96.2	100.0	83.1	96.2	100.0	82.1	96.0	100.0	82.6	96.1	100.0	82.9	96.2
21, 22, 23	FIRE SUPPRESSION, PLUMBING & HVAC	96.5	58.7	81.6	96.5	74.8	87.9	100.0	74.8	90.1	100.0	71.7	88.9	96.5	58.7	81.6	100.0	72.2	89.1
26, 27, 3370	ELECTRICAL, COMMUNICATIONS & UTIL.	93.4	52.3	73.1	100.0	80.6	90.4	95.5	54.6	75.3	99.6	80.6	90.2	95.5	50.3	73.2	96.4	63.3	80.0
MF2018	WEIGHTED AVERAGE	103.9	66.8	88.2	100.9	74.7	89.8	105.7	70.5	90.8	99.3	73.4	88.4	102.6	65.8	87.1	101.8	70.0	88.3

For customer support on your Commercial Renovation Costs with RSMeans data, call 800.448.8182.

Table 1

DIVISION		COLORADO SALIDA 812			CONNECTICUT BRIDGEPORT 066			BRISTOL 060			HARTFORD 061			MERIDEN 064			NEW BRITAIN 060		
		MAT.	INST.	TOTAL	MAT.	INST.	TOTAL	MAT.	INST.	TOTAL	MAT.	INST.	TOTAL	MAT.	INST.	TOTAL	MAT.	INST.	TOTAL
015433	CONTRACTOR EQUIPMENT		91.7	91.7		95.7	95.7		95.7	95.7		99.0	99.0		96.1	96.1		95.7	95.7
0241, 31 - 34	SITE & INFRASTRUCTURE, DEMOLITION	134.0	88.2	102.4	106.1	97.8	100.3	105.1	97.7	100.0	101.6	102.4	102.2	102.8	98.5	99.8	105.3	97.7	100.1
0310	Concrete Forming & Accessories	108.6	65.9	72.2	103.4	116.7	114.7	103.4	116.6	114.6	103.5	116.8	114.8	103.1	116.6	114.6	103.9	116.6	114.7
0320	Concrete Reinforcing	112.4	67.9	90.9	116.3	144.5	130.0	116.3	144.5	130.0	111.6	144.6	127.5	116.3	144.5	130.0	116.3	144.5	130.0
0330	Cast-in-Place Concrete	114.7	72.8	99.1	108.5	127.5	115.6	101.6	127.5	111.2	106.5	128.5	114.7	97.8	127.5	108.8	103.3	127.5	112.3
03	CONCRETE	109.1	69.3	91.6	104.1	124.5	113.0	100.9	124.4	111.2	101.1	124.8	111.5	99.1	124.4	110.2	101.7	124.4	111.7
04	MASONRY	138.6	60.6	90.7	109.4	130.9	122.6	100.8	130.9	119.3	100.5	131.0	119.2	100.4	130.9	119.2	102.5	130.9	120.0
05	METALS	103.6	78.6	96.2	98.1	116.6	103.5	98.1	116.5	103.5	103.1	116.0	106.9	95.4	116.5	101.6	94.5	116.5	100.9
06	WOOD, PLASTICS & COMPOSITES	102.4	67.2	83.0	105.0	114.2	110.1	105.0	114.2	110.1	96.2	114.4	106.2	105.0	114.2	110.1	105.0	114.2	110.1
07	THERMAL & MOISTURE PROTECTION	110.8	67.5	92.0	99.7	123.5	110.0	99.8	120.6	108.9	104.4	121.1	111.7	99.8	120.6	108.9	99.8	120.7	108.9
08	OPENINGS	96.3	68.4	89.8	96.4	120.6	102.0	96.4	120.6	102.0	98.1	120.7	103.4	98.5	120.6	103.7	96.4	120.6	102.0
0920	Plaster & Gypsum Board	80.2	66.5	71.0	115.7	114.3	114.8	115.7	114.3	114.8	102.5	114.3	110.5	117.3	114.3	115.3	115.7	114.3	114.8
0950, 0980	Ceilings & Acoustic Treatment	102.7	66.5	78.3	102.0	114.3	110.3	102.0	114.3	110.3	98.8	114.3	109.3	107.5	114.3	112.1	102.0	114.3	110.3
0960	Flooring	115.9	69.9	103.0	92.6	127.0	102.3	92.6	124.5	101.6	96.9	127.0	105.4	92.6	124.5	101.6	92.6	124.5	101.6
0970, 0990	Wall Finishes & Painting/Coating	101.7	76.9	87.0	92.0	125.5	111.8	92.0	129.1	113.9	98.1	129.1	116.4	92.0	129.1	113.9	92.0	129.1	113.9
09	FINISHES	101.1	67.4	82.8	94.7	118.9	107.8	94.7	118.8	107.8	96.2	119.4	108.8	96.0	118.8	108.4	94.7	118.8	107.8
COVERS	DIVS. 10 - 14, 25, 28, 41, 43, 44, 46	100.0	82.7	96.1	100.0	113.9	103.1	100.0	113.9	103.1	100.0	114.4	103.2	100.0	114.0	103.1	100.0	113.9	103.1
21, 22, 23	FIRE SUPPRESSION, PLUMBING & HVAC	96.5	73.6	87.5	100.1	118.6	107.4	100.1	118.6	107.4	100.0	118.6	107.3	100.1	118.6	105.2	100.1	118.6	107.4
26, 27, 3370	ELECTRICAL, COMMUNICATIONS & UTIL.	95.7	63.2	79.7	93.0	105.0	98.9	93.0	103.7	98.3	92.7	109.4	100.9	92.9	105.8	99.3	93.0	103.7	98.3
MF2018	WEIGHTED AVERAGE	103.1	70.9	89.5	99.3	117.0	106.8	98.5	116.7	106.2	99.6	118.0	107.3	97.2	117.0	105.6	98.1	116.7	106.0

Table 2

DIVISION		CONNECTICUT NEW HAVEN 065			NEW LONDON 063			NORWALK 068			STAMFORD 069			WATERBURY 067			WILLIMANTIC 062		
		MAT.	INST.	TOTAL	MAT.	INST.	TOTAL	MAT.	INST.	TOTAL	MAT.	INST.	TOTAL	MAT.	INST.	TOTAL	MAT.	INST.	TOTAL
015433	CONTRACTOR EQUIPMENT		96.1	96.1		96.1	96.1		95.7	95.7		95.7	95.7		95.7	95.7		95.7	95.7
0241, 31 - 34	SITE & INFRASTRUCTURE, DEMOLITION	105.3	98.4	100.5	97.2	98.1	97.8	105.8	97.6	100.2	106.5	97.7	100.4	105.8	97.8	100.2	105.8	97.7	100.2
0310	Concrete Forming & Accessories	103.2	116.5	114.6	103.1	116.5	114.5	103.4	116.6	114.7	103.4	116.9	114.9	103.4	116.6	114.7	103.4	116.4	114.5
0320	Concrete Reinforcing	116.3	144.5	130.0	91.2	144.5	117.0	116.3	144.5	130.0	116.3	144.5	130.0	116.3	144.5	130.0	116.3	144.5	130.0
0330	Cast-in-Place Concrete	105.1	125.5	112.7	89.5	125.5	102.9	106.7	126.7	114.2	108.5	126.8	115.3	108.5	127.5	115.6	101.3	125.5	110.3
03	CONCRETE	116.3	123.7	119.6	89.2	123.7	104.4	103.3	124.1	112.4	104.1	124.3	113.0	104.1	124.4	113.0	100.8	123.6	110.8
04	MASONRY	101.1	130.9	119.4	99.3	130.9	118.7	100.6	130.1	118.7	101.4	130.1	119.0	101.4	130.9	119.5	100.6	130.9	119.3
05	METALS	94.7	116.4	101.1	94.4	116.4	100.9	98.1	116.5	103.5	98.1	116.8	103.6	98.1	116.5	103.5	97.9	116.3	103.3
06	WOOD, PLASTICS & COMPOSITES	105.0	114.2	110.1	105.0	114.2	110.1	105.0	114.2	110.1	105.0	114.2	110.1	105.0	114.2	110.1	105.0	114.2	110.1
07	THERMAL & MOISTURE PROTECTION	99.9	120.8	109.0	99.7	120.3	108.7	99.9	123.2	110.0	99.8	123.2	110.0	99.8	121.1	109.1	100.0	120.3	108.8
08	OPENINGS	96.4	120.6	102.0	98.7	120.6	103.8	96.4	120.6	102.0	96.4	120.6	102.0	96.4	120.6	102.0	98.8	120.6	103.9
0920	Plaster & Gypsum Board	115.7	114.3	114.8	115.7	114.3	114.8	115.7	114.3	114.8	115.7	114.3	114.8	115.7	114.3	114.8	115.7	114.3	114.8
0950, 0980	Ceilings & Acoustic Treatment	102.0	114.3	110.3	100.2	114.3	109.7	102.0	114.3	110.3	102.0	114.3	110.3	102.0	114.3	110.3	100.2	114.3	109.7
0960	Flooring	92.6	127.0	102.3	92.6	127.0	102.3	92.6	124.5	101.6	92.6	127.0	102.3	92.6	127.0	102.3	92.6	121.1	100.6
0970, 0990	Wall Finishes & Painting/Coating	92.0	125.5	111.8	92.0	129.1	113.9	92.0	125.5	111.8	92.0	125.5	111.8	92.0	129.1	113.9	92.0	129.1	113.9
09	FINISHES	94.7	118.9	107.8	93.9	119.3	107.6	94.7	118.5	107.6	94.8	118.9	107.8	94.6	119.3	107.9	94.5	118.3	107.3
COVERS	DIVS. 10 - 14, 25, 28, 41, 43, 44, 46	100.0	114.0	103.1	100.0	114.0	103.1	100.0	113.9	103.1	100.0	114.1	103.2	100.0	114.0	103.1	100.0	114.0	103.1
21, 22, 23	FIRE SUPPRESSION, PLUMBING & HVAC	100.1	118.6	107.4	96.6	118.6	105.2	100.1	118.6	107.4	100.1	118.6	107.4	100.1	118.6	107.4	100.1	118.4	107.3
26, 27, 3370	ELECTRICAL, COMMUNICATIONS & UTIL.	92.9	106.0	99.4	90.5	107.9	99.1	93.0	104.8	98.8	93.0	150.1	121.1	92.6	107.1	99.7	93.0	108.8	100.8
MF2018	WEIGHTED AVERAGE	99.9	117.0	107.1	95.2	117.2	104.5	98.8	116.7	106.4	99.0	123.2	109.2	98.9	117.2	106.6	98.7	117.1	106.5

Table 3

DIVISION		D.C. WASHINGTON 200 - 205			DELAWARE DOVER 199			NEWARK 197			WILMINGTON 198			FLORIDA DAYTONA BEACH 321			FORT LAUDERDALE 333		
		MAT.	INST.	TOTAL	MAT.	INST.	TOTAL	MAT.	INST.	TOTAL	MAT.	INST.	TOTAL	MAT.	INST.	TOTAL	MAT.	INST.	TOTAL
015433	CONTRACTOR EQUIPMENT		105.3	105.3		118.2	118.2		119.0	119.0		118.4	118.4		99.5	99.5		92.8	92.8
0241, 31 - 34	SITE & INFRASTRUCTURE, DEMOLITION	103.0	97.8	99.4	106.0	107.5	107.0	104.5	108.7	107.4	102.2	107.8	106.1	117.6	85.0	95.1	95.0	73.2	80.0
0310	Concrete Forming & Accessories	97.7	73.4	77.0	98.0	99.4	99.2	97.0	99.2	98.9	95.9	99.4	98.9	97.1	61.6	66.8	93.4	59.4	64.4
0320	Concrete Reinforcing	103.5	86.8	95.4	101.4	115.1	108.0	97.2	115.1	105.9	101.0	115.1	107.8	96.0	61.0	79.1	94.6	59.9	77.8
0330	Cast-in-Place Concrete	103.6	79.0	94.5	108.1	107.7	107.9	91.2	106.7	97.0	102.6	107.7	104.5	91.8	66.0	82.2	96.3	62.3	83.6
03	CONCRETE	106.0	78.6	94.0	100.4	106.0	102.9	93.4	105.7	98.8	97.7	106.0	101.4	90.5	64.8	79.2	93.4	62.3	79.7
04	MASONRY	97.6	87.8	91.6	99.0	98.5	98.7	101.0	98.5	99.5	94.4	98.5	96.9	86.4	60.9	70.7	87.6	54.4	67.2
05	METALS	103.9	95.6	101.5	102.9	121.9	108.5	104.7	123.1	110.1	103.1	121.9	108.6	101.9	88.9	98.1	96.9	88.2	94.4
06	WOOD, PLASTICS & COMPOSITES	95.4	72.2	82.6	92.3	97.6	95.2	91.0	97.4	94.5	85.7	97.6	92.2	93.2	60.1	75.0	78.3	60.6	68.6
07	THERMAL & MOISTURE PROTECTION	104.3	87.1	96.8	105.2	110.4	107.4	109.2	109.8	109.4	105.1	110.4	107.4	101.4	65.4	85.7	106.3	60.4	86.3
08	OPENINGS	100.7	74.6	94.6	91.2	107.4	95.0	91.2	107.3	94.9	88.6	107.4	93.0	92.9	59.4	85.1	94.2	59.5	86.1
0920	Plaster & Gypsum Board	104.3	71.4	82.2	96.0	97.4	96.9	96.7	97.4	97.2	97.3	97.4	97.4	92.7	59.5	70.4	107.2	60.0	75.5
0950, 0980	Ceilings & Acoustic Treatment	110.9	71.4	84.3	100.2	97.4	98.3	96.0	97.4	96.9	90.8	97.4	95.2	78.3	59.5	65.6	84.8	60.0	68.1
0960	Flooring	95.4	75.5	89.8	96.9	105.4	99.3	91.3	105.4	95.3	96.0	105.4	98.7	99.5	62.7	89.2	97.1	62.7	87.5
0970, 0990	Wall Finishes & Painting/Coating	103.7	73.1	85.6	92.6	119.1	108.3	88.2	119.1	106.4	95.3	119.1	109.4	103.8	61.4	78.7	96.5	58.3	73.9
09	FINISHES	98.5	72.8	84.6	95.9	101.6	99.0	89.7	101.5	96.1	94.2	101.6	98.2	91.5	61.1	75.1	91.3	59.4	74.1
COVERS	DIVS. 10 - 14, 25, 28, 41, 43, 44, 46	100.0	94.5	98.8	100.0	103.7	100.8	100.0	103.4	100.8	100.0	103.7	100.8	100.0	81.4	95.9	100.0	80.7	95.7
21, 22, 23	FIRE SUPPRESSION, PLUMBING & HVAC	100.0	88.4	95.4	100.0	119.4	107.6	100.3	119.3	107.7	100.1	119.4	107.7	99.9	75.6	90.4	100.0	66.7	86.9
26, 27, 3370	ELECTRICAL, COMMUNICATIONS & UTIL.	97.7	100.0	98.8	95.1	109.7	102.3	96.9	109.7	103.2	95.1	109.7	102.3	96.7	58.9	78.1	94.7	67.4	81.3
MF2018	WEIGHTED AVERAGE	101.1	87.5	95.4	99.0	109.6	103.5	98.3	109.7	103.1	97.9	109.7	102.9	97.2	69.5	85.5	96.1	66.4	83.6

FLORIDA

DIVISION		FORT MYERS 339, 341			GAINESVILLE 326, 344			JACKSONVILLE 320, 322			LAKELAND 338			MELBOURNE 329			MIAMI 330 - 332, 340		
		MAT.	INST.	TOTAL	MAT.	INST.	TOTAL	MAT.	INST.	TOTAL	MAT.	INST.	TOTAL	MAT.	INST.	TOTAL	MAT.	INST.	TOTAL
015433	CONTRACTOR EQUIPMENT		99.5	99.5		99.5	99.5		99.5	99.5		99.5	99.5		99.5	99.5		94.9	94.9
0241, 31 - 34	SITE & INFRASTRUCTURE, DEMOLITION	106.7	85.2	91.9	127.1	84.8	97.9	117.6	85.2	95.2	108.7	85.2	92.5	125.6	85.1	97.7	96.5	78.9	84.3
0310	Concrete Forming & Accessories	89.6	63.8	67.6	92.5	57.1	62.4	96.9	60.2	65.6	86.4	63.8	67.1	93.7	61.7	66.4	99.3	60.2	66.0
0320	Concrete Reinforcing	95.7	76.1	86.2	101.8	61.5	82.3	96.0	61.3	79.2	98.0	76.4	87.6	97.1	65.6	81.9	101.6	60.0	81.5
0330	Cast-in-Place Concrete	100.4	66.0	87.6	105.4	65.7	90.6	92.7	65.8	82.7	102.7	66.1	89.1	110.7	66.0	94.1	93.1	63.9	82.2
03	CONCRETE	94.0	68.3	82.7	101.7	62.7	84.6	90.9	64.2	79.2	95.8	68.5	83.8	102.1	65.6	86.1	91.7	63.2	79.2
04	MASONRY	82.5	58.8	67.9	99.3	60.8	75.7	86.1	58.7	69.3	96.8	61.0	74.8	84.2	60.9	69.9	88.0	55.5	68.0
05	METALS	99.0	93.6	97.4	100.7	88.4	97.1	100.4	88.6	96.9	98.9	94.7	97.7	110.6	90.6	104.7	97.1	87.6	94.3
06	WOOD, PLASTICS & COMPOSITES	75.4	63.2	68.6	87.2	54.6	69.3	93.2	58.6	74.2	71.0	63.2	66.7	88.9	60.1	73.0	92.9	60.7	75.2
07	THERMAL & MOISTURE PROTECTION	106.1	63.2	87.4	101.8	62.5	84.7	101.7	62.4	84.6	106.0	63.8	87.6	101.9	64.1	85.5	106.1	61.4	86.6
08	OPENINGS	95.5	64.6	88.3	92.5	56.5	84.1	92.9	58.8	84.9	95.5	64.7	88.3	92.2	60.5	84.8	96.6	59.5	88.0
0920	Plaster & Gypsum Board	103.3	62.7	76.0	89.6	53.8	65.6	92.7	58.0	69.3	100.2	62.7	75.0	89.6	59.5	69.4	96.2	60.0	71.9
0950, 0980	Ceilings & Acoustic Treatment	79.7	62.7	68.2	72.9	53.8	60.1	78.3	58.0	64.6	79.7	62.7	68.2	77.5	59.5	65.4	85.6	60.0	68.3
0960	Flooring	94.1	77.2	89.4	97.0	62.7	87.4	99.5	62.7	89.2	92.2	62.7	83.9	97.2	62.7	87.5	98.6	62.7	88.5
0970, 0990	Wall Finishes & Painting/Coating	101.4	63.2	78.8	103.8	63.2	79.8	103.8	63.2	79.8	101.4	63.2	78.8	103.8	80.9	90.2	100.2	58.3	75.4
09	FINISHES	91.1	66.1	77.6	90.4	58.1	72.9	91.6	60.4	74.7	90.2	63.1	75.6	91.0	63.2	76.0	90.9	59.8	74.1
COVERS	DIVS. 10 - 14, 25, 28, 41, 43, 44, 46	100.0	80.0	95.5	100.0	80.8	95.7	100.0	78.0	95.1	100.0	80.0	95.5	100.0	81.4	95.9	100.0	81.5	95.9
21, 22, 23	FIRE SUPPRESSION, PLUMBING & HVAC	98.1	59.3	82.9	98.6	63.1	84.7	99.9	63.1	85.5	98.1	59.9	83.1	99.9	74.2	89.8	100.0	63.8	85.8
26, 27, 3370	ELECTRICAL, COMMUNICATIONS & UTIL.	96.6	60.0	78.6	97.0	58.6	78.0	96.4	61.9	79.4	95.0	58.5	77.0	97.7	62.7	80.5	98.4	77.9	88.3
MF2018	WEIGHTED AVERAGE	96.5	67.7	84.3	98.8	65.8	84.8	96.9	66.6	84.1	97.1	67.6	84.6	100.1	70.3	87.5	96.7	68.0	84.6

FLORIDA

DIVISION		ORLANDO 327 - 328, 347			PANAMA CITY 324			PENSACOLA 325			SARASOTA 342			ST. PETERSBURG 337			TALLAHASSEE 323		
		MAT.	INST.	TOTAL	MAT.	INST.	TOTAL	MAT.	INST.	TOTAL	MAT.	INST.	TOTAL	MAT.	INST.	TOTAL	MAT.	INST.	TOTAL
015433	CONTRACTOR EQUIPMENT		102.0	102.0		99.5	99.5		99.5	99.5		99.5	99.5		99.5	99.5		102.0	102.0
0241, 31 - 34	SITE & INFRASTRUCTURE, DEMOLITION	115.8	89.6	97.7	131.4	85.2	99.5	131.3	85.0	99.3	119.0	85.2	95.7	110.4	85.0	92.9	110.2	89.8	96.1
0310	Concrete Forming & Accessories	101.8	61.5	67.4	96.2	66.2	70.6	94.2	61.6	66.4	93.7	63.7	68.1	92.8	60.8	65.5	99.0	62.2	67.6
0320	Concrete Reinforcing	104.9	65.6	85.9	100.3	67.4	84.4	102.8	67.4	85.7	96.6	76.2	86.7	98.0	76.0	87.4	98.4	61.5	80.5
0330	Cast-in-Place Concrete	111.6	66.7	94.9	97.4	65.9	85.7	120.4	65.3	99.9	105.1	66.1	90.6	103.8	65.8	89.7	92.3	66.7	82.8
03	CONCRETE	100.8	65.7	85.4	100.0	67.9	85.9	109.7	65.6	90.3	95.3	68.4	83.5	97.3	66.9	83.9	90.0	65.3	79.2
04	MASONRY	94.8	60.9	74.0	90.6	60.3	72.0	109.1	59.9	78.9	88.7	61.0	71.6	132.4	58.7	87.1	85.2	60.3	69.9
05	METALS	96.7	89.1	94.5	101.5	90.9	98.4	102.6	90.8	99.1	102.5	93.9	100.0	99.8	93.5	98.0	101.7	87.7	97.6
06	WOOD, PLASTICS & COMPOSITES	91.7	60.2	74.4	92.0	66.6	78.0	90.3	61.1	74.2	92.5	63.2	76.4	79.5	59.3	68.4	95.6	61.2	76.7
07	THERMAL & MOISTURE PROTECTION	109.2	66.0	90.4	102.0	64.1	85.5	101.9	63.2	85.1	100.1	63.8	84.3	106.2	62.2	87.1	96.7	64.0	82.4
08	OPENINGS	97.8	60.6	89.1	91.0	64.5	84.9	91.0	61.4	84.1	97.8	64.2	90.0	94.3	62.5	86.9	97.9	60.2	89.1
0920	Plaster & Gypsum Board	93.3	59.5	70.6	91.9	66.1	74.6	99.6	60.5	73.3	97.2	62.7	74.0	105.6	58.7	74.0	100.0	60.5	73.4
0950, 0980	Ceilings & Acoustic Treatment	89.7	59.5	69.3	77.5	66.1	69.8	77.5	60.5	66.0	83.9	62.7	69.6	81.6	58.7	66.1	86.2	60.5	68.9
0960	Flooring	95.7	62.7	86.4	99.1	72.5	91.6	95.1	62.7	86.0	103.5	54.9	89.9	96.1	60.8	86.2	97.6	61.7	87.5
0970, 0990	Wall Finishes & Painting/Coating	95.6	60.3	74.7	103.8	63.2	79.8	103.8	63.2	79.8	99.0	63.2	77.8	101.4	63.2	78.8	101.6	63.2	78.9
09	FINISHES	93.2	61.1	75.8	92.6	66.8	78.6	92.1	61.6	75.6	96.0	61.8	77.5	92.6	60.5	75.2	94.4	61.8	76.7
COVERS	DIVS. 10 - 14, 25, 28, 41, 43, 44, 46	100.0	81.7	95.9	100.0	80.6	95.7	100.0	79.7	95.5	100.0	80.0	95.5	100.0	79.6	95.4	100.0	81.5	95.9
21, 22, 23	FIRE SUPPRESSION, PLUMBING & HVAC	100.0	56.4	82.9	99.9	63.2	85.5	99.9	62.7	85.3	99.9	58.5	83.6	100.0	59.9	84.2	100.0	65.5	86.5
26, 27, 3370	ELECTRICAL, COMMUNICATIONS & UTIL.	98.4	63.1	81.0	95.5	58.6	77.3	99.0	50.7	75.2	96.9	58.5	78.0	95.0	60.6	78.0	103.7	58.6	81.4
MF2018	WEIGHTED AVERAGE	99.0	66.6	85.3	98.6	68.3	85.8	101.4	65.8	86.2	98.8	67.0	85.4	99.7	66.8	85.8	98.2	67.7	85.3

		FLORIDA						GEORGIA												
DIVISION		TAMPA 335 - 336, 346			WEST PALM BEACH 334, 349			ALBANY 317, 398			ATHENS 306			ATLANTA 300 - 303, 399			AUGUSTA 308 - 309		
		MAT.	INST.	TOTAL	MAT.	INST.	TOTAL	MAT.	INST.	TOTAL	MAT.	INST.	TOTAL	MAT.	INST.	TOTAL	MAT.	INST.	TOTAL
015433	CONTRACTOR EQUIPMENT		99.5	99.5		92.8	92.8		93.7	93.7		92.6	92.6		96.6	96.6		92.6	92.6
0241, 31 - 34	SITE & INFRASTRUCTURE, DEMOLITION	110.9	87.0	94.4	91.6	73.2	78.9	107.3	76.7	86.1	102.8	90.8	94.5	99.7	96.0	97.1	95.9	91.9	93.2
0310	Concrete Forming & Accessories	95.4	64.1	68.7	96.7	59.1	64.7	90.6	66.5	70.1	90.0	43.6	50.4	94.3	72.3	75.5	90.9	73.0	75.6
0320	Concrete Reinforcing	94.6	76.5	85.8	97.3	57.5	78.1	96.5	71.3	84.3	94.5	63.5	79.5	93.9	71.3	83.0	94.9	70.7	83.2
0330	Cast-in-Place Concrete	101.5	66.2	88.3	91.5	62.2	80.6	94.6	69.1	85.1	109.7	69.5	94.7	113.1	71.1	97.5	103.6	70.0	91.1
03	CONCRETE	95.9	68.7	83.9	90.1	61.8	77.7	86.9	70.1	79.5	99.6	57.4	81.1	101.8	72.4	88.9	92.5	72.2	83.6
04	MASONRY	88.3	61.0	71.5	87.1	52.2	65.7	92.5	68.7	77.9	75.0	77.0	76.2	87.8	68.9	76.2	88.1	68.8	76.2
05	METALS	99.0	86.7	95.4	95.9	87.0	93.3	106.2	97.4	103.6	95.3	79.1	90.6	96.2	83.5	92.5	95.0	82.6	91.4
06	WOOD, PLASTICS & COMPOSITES	83.3	63.2	72.2	83.1	60.6	70.7	80.1	66.4	72.6	92.5	35.8	61.3	97.8	74.0	84.7	93.8	75.3	83.6
07	THERMAL & MOISTURE PROTECTION	106.5	63.8	87.9	106.0	60.6	86.2	100.6	68.0	86.4	95.4	68.3	83.6	96.8	73.8	86.8	95.1	71.4	84.8
08	OPENINGS	95.5	64.7	88.3	93.8	59.0	85.7	85.7	68.5	81.7	91.3	50.2	81.7	99.4	73.5	93.3	91.4	74.0	87.3
0920	Plaster & Gypsum Board	108.2	62.7	77.6	111.9	60.0	77.0	101.1	66.0	77.5	93.0	34.4	53.6	95.0	73.6	80.6	94.0	75.0	81.2
0950, 0980	Ceilings & Acoustic Treatment	84.8	62.7	69.9	79.7	60.0	66.4	77.5	66.0	69.8	96.6	34.4	54.7	89.7	73.6	78.8	97.5	75.0	82.4
0960	Flooring	97.1	62.7	87.5	98.9	54.9	86.6	99.4	70.8	91.4	96.2	81.9	92.2	99.1	70.8	91.1	96.4	70.8	89.2
0970, 0990	Wall Finishes & Painting/Coating	101.4	63.2	78.8	96.5	58.3	73.9	88.7	94.0	91.8	92.4	94.0	93.3	96.1	97.3	96.8	92.4	90.2	91.1
09	FINISHES	93.9	63.1	77.3	91.2	57.8	73.2	91.7	69.7	79.8	95.0	53.9	72.7	95.0	74.5	83.9	94.7	74.5	83.7
COVERS	DIVS. 10 - 14, 25, 28, 41, 43, 44, 46	100.0	80.0	95.6	100.0	80.7	95.7	100.0	85.9	96.8	100.0	82.7	96.1	100.0	87.2	97.2	100.0	87.0	97.1
21, 22, 23	FIRE SUPPRESSION, PLUMBING & HVAC	100.0	59.9	84.3	98.1	58.5	82.6	100.0	68.9	87.8	96.6	66.3	84.7	100.0	69.4	88.0	100.1	69.0	87.9
26, 27, 3370	ELECTRICAL, COMMUNICATIONS & UTIL.	94.7	63.1	79.1	95.7	67.4	81.8	95.5	61.9	79.0	97.8	63.9	81.1	97.6	71.5	84.7	98.3	67.3	83.1
MF2018	WEIGHTED AVERAGE	97.6	68.4	85.3	95.1	64.1	82.0	96.5	71.9	86.1	95.6	66.9	83.5	98.2	75.0	88.4	96.0	73.9	86.6

City Cost Indexes

GEORGIA

DIVISION		COLUMBUS 318 - 319			DALTON 307			GAINESVILLE 305			MACON 310 - 312			SAVANNAH 313 - 314			STATESBORO 304		
		MAT.	INST.	TOTAL	MAT.	INST.	TOTAL	MAT.	INST.	TOTAL	MAT.	INST.	TOTAL	MAT.	INST.	TOTAL	MAT.	INST.	TOTAL
015433	CONTRACTOR EQUIPMENT		93.7	93.7		107.6	107.6		92.6	92.6		103.2	103.2		96.7	96.7		95.5	95.5
0241, 31 - 34	SITE & INFRASTRUCTURE, DEMOLITION	107.2	76.7	86.2	102.6	95.8	97.9	102.5	90.7	94.3	108.6	91.0	96.4	107.0	82.0	89.7	103.7	78.0	85.9
0310	Concrete Forming & Accessories	90.5	66.6	70.1	83.2	64.5	67.3	93.3	40.7	48.5	90.3	66.6	70.1	93.4	72.0	75.2	78.5	52.2	56.1
0320	Concrete Reinforcing	96.6	71.3	84.4	94.1	60.9	78.0	94.4	63.4	79.4	97.8	71.3	85.0	104.1	70.7	88.0	93.7	71.2	82.8
0330	Cast-in-Place Concrete	94.2	68.8	84.8	106.6	68.0	92.2	115.3	68.7	98.0	93.0	68.3	83.8	98.3	69.5	87.6	109.5	68.6	94.3
03	CONCRETE	86.8	70.0	79.4	98.9	67.0	84.9	101.4	55.8	81.4	86.3	69.8	79.1	88.5	72.5	81.5	99.3	63.4	83.5
04	MASONRY	92.5	68.7	77.9	75.8	75.5	75.6	83.2	76.2	78.9	104.8	68.9	82.7	87.4	68.8	75.9	77.8	77.0	77.3
05	METALS	105.9	97.6	103.4	96.4	94.0	95.7	94.6	78.5	89.9	100.9	97.6	100.0	102.1	96.0	100.3	100.0	98.9	99.7
06	WOOD, PLASTICS & COMPOSITES	80.1	66.4	72.6	75.9	65.2	70.0	96.1	33.2	61.5	85.9	66.5	75.2	87.4	73.9	80.0	69.7	47.2	57.3
07	THERMAL & MOISTURE PROTECTION	100.5	68.9	86.8	97.3	70.1	85.5	95.4	67.7	83.3	99.0	70.6	86.6	98.5	70.4	86.3	95.9	67.1	83.4
08	OPENINGS	85.7	68.3	81.9	92.8	65.8	86.5	91.3	48.8	81.4	85.5	69.3	81.7	95.3	73.2	90.2	93.7	58.5	85.5
0920	Plaster & Gypsum Board	101.1	66.0	77.5	81.9	64.7	70.3	94.9	31.8	52.4	104.7	66.0	78.7	103.3	73.6	83.3	83.6	46.2	58.4
0950, 0980	Ceilings & Acoustic Treatment	77.5	66.0	69.8	110.2	64.7	79.5	96.6	31.8	52.9	72.8	66.0	68.3	87.3	73.6	78.0	105.8	46.2	65.6
0960	Flooring	99.4	70.8	91.4	97.0	81.9	92.8	97.8	81.9	93.3	77.9	70.8	75.9	95.6	70.8	88.6	114.6	81.9	105.4
0970, 0990	Wall Finishes & Painting/Coating	88.7	87.4	87.9	82.8	70.1	75.3	92.4	94.0	93.3	90.7	94.0	92.6	88.0	86.1	86.9	90.8	70.1	78.5
09	FINISHES	91.6	68.9	79.3	104.5	68.2	84.9	95.5	52.1	72.0	80.5	69.7	74.7	92.9	73.2	82.3	107.9	57.9	80.9
COVERS	DIVS. 10 - 14, 25, 28, 41, 43, 44, 46	100.0	85.9	96.9	100.0	85.3	96.7	100.0	35.5	85.6	100.0	86.0	96.9	100.0	87.0	97.1	100.0	83.8	96.4
21, 22, 23	FIRE SUPPRESSION, PLUMBING & HVAC	100.1	66.8	87.0	96.7	60.8	82.6	96.6	65.6	84.4	100.1	68.2	87.6	100.1	64.8	86.3	97.2	67.3	85.5
26, 27, 3370	ELECTRICAL, COMMUNICATIONS & UTIL.	95.7	67.7	81.9	106.9	62.2	84.9	97.8	70.8	84.5	94.2	62.7	78.6	99.1	63.7	81.7	98.3	63.4	81.1
MF2018	WEIGHTED AVERAGE	96.5	72.2	86.2	97.5	71.4	86.5	96.2	65.6	83.2	95.2	73.1	85.8	97.3	72.7	86.9	97.8	69.7	85.9

GEORGIA / HAWAII / IDAHO

DIVISION		VALDOSTA 316 (GEORGIA)			WAYCROSS 315 (GEORGIA)			HILO 967 (HAWAII)			HONOLULU 968 (HAWAII)			STATES & POSS., GUAM 969 (HAWAII)			BOISE 836 - 837 (IDAHO)		
		MAT.	INST.	TOTAL	MAT.	INST.	TOTAL	MAT.	INST.	TOTAL	MAT.	INST.	TOTAL	MAT.	INST.	TOTAL	MAT.	INST.	TOTAL
015433	CONTRACTOR EQUIPMENT		93.7	93.7		93.7	93.7		97.8	97.8		99.3	99.3		160.5	160.5		94.6	94.6
0241, 31 - 34	SITE & INFRASTRUCTURE, DEMOLITION	116.9	76.7	89.1	113.7	77.0	88.3	142.5	103.1	115.3	150.0	106.6	120.0	186.3	98.5	125.7	86.8	92.0	90.4
0310	Concrete Forming & Accessories	81.7	41.6	47.5	83.5	64.0	66.8	108.2	123.8	121.5	120.2	123.7	123.2	110.5	51.5	60.2	100.2	81.1	83.9
0320	Concrete Reinforcing	98.7	59.5	79.8	98.7	59.8	79.9	143.4	127.0	135.5	166.8	127.0	147.6	253.7	27.4	144.2	107.6	79.5	94.0
0330	Cast-in-Place Concrete	92.6	68.7	83.7	104.7	68.6	91.2	180.3	123.3	159.1	145.0	123.5	137.0	156.3	95.7	133.7	90.6	96.6	92.8
03	CONCRETE	91.2	56.6	76.0	94.5	66.5	82.2	144.7	123.2	135.3	137.9	123.1	131.4	147.9	63.9	111.0	97.6	86.4	92.7
04	MASONRY	97.8	77.1	85.1	98.7	77.0	85.4	148.6	122.1	132.3	135.5	122.2	127.3	206.7	34.8	101.0	126.3	86.2	101.7
05	METALS	105.3	93.0	101.7	104.4	88.9	99.8	110.8	108.1	110.0	124.1	107.1	119.1	143.1	76.0	123.4	109.6	82.7	101.7
06	WOOD, PLASTICS & COMPOSITES	68.7	33.0	49.0	70.2	63.4	66.5	111.4	124.3	118.5	132.8	124.2	128.1	123.4	53.9	85.1	93.5	79.9	86.0
07	THERMAL & MOISTURE PROTECTION	100.8	66.6	86.0	100.6	69.1	86.9	128.7	119.3	124.6	146.3	120.0	134.9	150.7	58.2	110.4	101.4	86.6	94.9
08	OPENINGS	82.5	47.6	74.4	82.8	61.8	77.9	114.1	123.2	116.2	128.1	123.2	126.9	118.4	43.8	101.0	96.9	75.9	92.0
0920	Plaster & Gypsum Board	93.8	31.6	52.0	93.8	62.9	73.0	115.2	124.8	121.7	160.8	124.8	136.6	236.1	42.5	105.8	92.3	79.4	83.6
0950, 0980	Ceilings & Acoustic Treatment	75.9	31.6	46.0	74.0	62.9	66.5	136.6	124.8	128.6	144.8	124.8	131.3	257.2	42.5	112.4	102.8	79.4	87.0
0960	Flooring	93.0	83.5	90.3	94.3	81.9	90.8	103.8	139.3	113.8	121.3	139.3	126.4	121.2	39.5	98.2	92.9	90.7	92.3
0970, 0990	Wall Finishes & Painting/Coating	88.7	94.0	91.8	88.7	70.1	77.7	99.2	143.8	125.6	110.1	143.8	130.1	105.2	31.5	61.6	92.6	39.5	61.2
09	FINISHES	89.4	52.5	69.4	89.0	67.5	77.4	110.8	129.1	120.7	125.5	129.1	127.4	185.1	48.0	110.9	92.7	78.2	84.9
COVERS	DIVS. 10 - 14, 25, 28, 41, 43, 44, 46	100.0	82.2	96.0	100.0	85.5	96.8	100.0	113.7	103.1	100.0	113.6	103.0	100.0	66.0	92.4	100.0	88.6	97.4
21, 22, 23	FIRE SUPPRESSION, PLUMBING & HVAC	100.1	68.5	87.7	97.9	63.6	84.4	100.3	112.4	105.0	100.4	112.3	105.1	102.6	33.5	75.5	100.1	74.2	89.9
26, 27, 3370	ELECTRICAL, COMMUNICATIONS & UTIL.	94.1	57.8	76.2	97.7	63.4	80.8	108.1	124.6	116.2	109.8	124.5	117.1	158.1	35.9	97.9	97.2	68.2	82.9
MF2018	WEIGHTED AVERAGE	96.7	66.2	83.8	96.8	70.0	85.5	114.8	118.4	116.3	119.1	118.5	118.9	137.1	51.4	100.9	101.0	79.9	92.1

IDAHO / ILLINOIS

DIVISION		COEUR D'ALENE 838 (IDAHO)			IDAHO FALLS 834 (IDAHO)			LEWISTON 835 (IDAHO)			POCATELLO 832 (IDAHO)			TWIN FALLS 833 (IDAHO)			BLOOMINGTON 617 (ILLINOIS)		
		MAT.	INST.	TOTAL	MAT.	INST.	TOTAL	MAT.	INST.	TOTAL	MAT.	INST.	TOTAL	MAT.	INST.	TOTAL	MAT.	INST.	TOTAL
015433	CONTRACTOR EQUIPMENT		89.7	89.7		94.6	94.6		89.7	89.7		94.6	94.6		94.6	94.6		101.9	101.9
0241, 31 - 34	SITE & INFRASTRUCTURE, DEMOLITION	85.5	86.5	86.2	85.1	92.1	89.9	92.2	87.2	88.7	88.3	92.0	90.9	95.4	91.8	92.9	94.9	96.0	95.7
0310	Concrete Forming & Accessories	108.7	81.4	85.4	94.1	78.3	80.6	113.9	81.1	85.9	100.4	80.5	83.4	101.6	76.5	80.2	84.0	116.0	111.2
0320	Concrete Reinforcing	115.7	96.2	106.3	109.6	77.9	94.3	115.7	96.2	106.3	108.0	79.4	94.2	110.0	77.8	94.4	93.0	102.4	97.5
0330	Cast-in-Place Concrete	98.0	85.2	93.2	86.3	83.4	85.2	101.8	82.0	94.5	93.1	96.4	94.3	95.6	82.3	90.6	98.0	113.7	103.9
03	CONCRETE	103.9	85.3	95.7	89.6	80.2	85.5	107.5	84.1	97.2	96.7	86.0	92.0	104.6	79.0	93.4	91.9	113.6	101.4
04	MASONRY	128.0	87.3	103.0	121.5	87.2	100.4	128.5	84.7	101.6	124.0	86.2	100.8	126.7	82.8	99.7	110.2	121.7	117.3
05	METALS	103.0	88.1	98.6	118.1	81.6	107.4	102.4	88.2	98.3	118.2	82.0	107.6	118.2	81.4	107.4	95.2	123.1	103.4
06	WOOD, PLASTICS & COMPOSITES	96.7	79.8	87.4	87.2	76.3	81.2	102.6	79.8	90.1	93.5	79.9	86.0	94.6	76.3	84.5	81.7	113.0	99.0
07	THERMAL & MOISTURE PROTECTION	158.9	84.3	126.4	100.4	75.3	89.5	159.2	83.1	126.1	101.0	75.0	89.7	101.8	80.5	92.5	95.4	112.1	102.6
08	OPENINGS	113.8	75.9	105.0	99.8	71.3	93.2	106.7	79.6	100.4	97.6	68.8	90.9	100.6	62.4	91.7	91.3	118.7	97.7
0920	Plaster & Gypsum Board	167.0	79.4	108.1	80.3	75.7	77.2	168.9	79.4	108.7	81.9	79.4	80.2	84.0	75.7	78.4	89.1	113.5	105.5
0950, 0980	Ceilings & Acoustic Treatment	136.3	79.4	98.0	104.3	75.7	85.0	136.3	79.4	98.0	110.0	79.4	89.4	106.8	75.7	85.8	83.9	113.5	103.9
0960	Flooring	130.9	77.0	115.8	92.7	77.0	88.3	134.3	77.0	118.2	96.2	77.0	90.8	97.5	77.0	91.7	86.0	120.8	95.8
0970, 0990	Wall Finishes & Painting/Coating	111.2	73.8	89.1	92.6	39.5	61.2	111.2	71.3	87.6	92.5	39.5	61.1	92.6	39.5	61.2	87.6	134.8	115.5
09	FINISHES	158.3	79.5	115.7	90.9	74.0	81.8	159.7	78.9	116.0	93.7	75.9	84.1	94.5	72.9	82.8	86.1	119.0	103.9
COVERS	DIVS. 10 - 14, 25, 28, 41, 43, 44, 46	100.0	93.0	98.4	100.0	87.4	97.2	100.0	92.6	98.3	100.0	88.5	97.4	100.0	85.9	96.9	100.0	104.5	101.0
21, 22, 23	FIRE SUPPRESSION, PLUMBING & HVAC	99.4	85.5	94.0	101.1	81.8	93.5	99.9	85.1	94.6	100.0	73.6	89.6	100.0	72.2	89.1	96.5	105.4	100.0
26, 27, 3370	ELECTRICAL, COMMUNICATIONS & UTIL.	88.7	81.5	85.2	88.3	70.9	79.7	86.9	78.2	82.6	94.4	66.0	80.4	89.6	70.9	80.4	95.9	90.3	93.1
MF2018	WEIGHTED AVERAGE	108.1	84.4	98.1	100.5	79.9	91.8	108.2	83.5	97.8	102.0	78.4	92.0	103.2	76.8	92.1	95.0	109.5	101.1

For customer support on your Commercial Renovation Costs with RSMeans data, call 800.448.8182.

797

City Cost Indexes

ILLINOIS

DIVISION		CARBONDALE 629			CENTRALIA 628			CHAMPAIGN 618 - 619			CHICAGO 606 - 608			DECATUR 625			EAST ST. LOUIS 620 - 622		
		MAT.	INST.	TOTAL	MAT.	INST.	TOTAL	MAT.	INST.	TOTAL	MAT.	INST.	TOTAL	MAT.	INST.	TOTAL	MAT.	INST.	TOTAL
015433	CONTRACTOR EQUIPMENT		110.3	110.3		110.3	110.3		102.8	102.8		100.3	100.3		102.8	102.8		110.3	110.3
0241, 31 - 34	SITE & INFRASTRUCTURE, DEMOLITION	100.9	96.8	98.1	101.3	98.6	99.5	104.0	97.2	99.3	105.4	104.4	104.7	94.8	97.0	96.3	103.5	97.7	99.5
0310	Concrete Forming & Accessories	90.0	108.9	106.1	91.6	114.2	110.9	90.3	116.7	112.8	97.4	159.7	150.5	92.1	117.9	114.1	87.5	114.2	110.2
0320	Concrete Reinforcing	82.2	103.4	92.5	82.2	104.0	92.8	93.0	101.6	97.2	103.8	153.3	127.7	80.6	99.8	89.9	82.2	104.0	92.7
0330	Cast-in-Place Concrete	91.5	102.7	95.7	92.0	119.6	102.3	113.7	110.8	112.6	126.0	155.6	137.0	99.9	114.6	105.4	93.6	118.4	103.0
03	CONCRETE	79.9	107.3	91.9	80.4	115.8	96.0	104.1	112.7	107.9	108.0	156.5	129.3	90.3	114.3	100.8	81.4	115.4	96.3
04	MASONRY	75.0	111.5	97.4	75.0	122.9	104.5	133.1	125.4	128.4	105.4	163.4	141.0	70.9	122.8	102.8	75.3	123.0	104.6
05	METALS	100.9	130.4	109.6	100.9	133.5	110.5	95.2	120.3	102.6	95.9	146.8	110.8	104.9	120.7	109.5	102.1	133.5	111.3
06	WOOD, PLASTICS & COMPOSITES	87.7	105.8	97.7	90.2	111.2	101.7	88.5	114.0	102.5	100.0	158.4	132.1	88.7	116.8	104.2	84.8	111.2	99.3
07	THERMAL & MOISTURE PROTECTION	90.6	101.3	95.3	90.7	111.5	99.8	96.2	115.6	104.6	95.2	150.7	119.4	95.9	113.3	103.5	90.7	110.7	99.4
08	OPENINGS	87.3	118.1	94.1	87.3	119.2	94.8	91.8	116.9	97.7	100.9	169.3	116.9	98.2	118.2	102.9	87.4	118.4	94.6
0920	Plaster & Gypsum Board	93.7	106.2	102.1	95.0	111.6	106.2	91.0	114.5	106.8	101.3	160.0	140.8	96.5	117.4	110.6	92.2	111.6	105.3
0950, 0980	Ceilings & Acoustic Treatment	80.6	106.2	97.8	80.6	111.6	101.5	83.9	114.5	104.5	100.4	160.0	140.6	87.1	117.4	107.5	80.6	111.6	101.5
0960	Flooring	113.7	116.6	114.5	114.6	116.6	115.1	89.2	116.6	96.9	91.3	163.8	111.7	101.8	117.1	106.1	112.6	116.6	113.7
0970, 0990	Wall Finishes & Painting/Coating	103.1	105.0	104.2	103.1	111.6	108.2	87.6	114.9	103.7	96.7	170.3	140.2	95.5	117.8	108.7	103.1	111.6	108.2
09	FINISHES	92.1	110.7	102.1	92.5	114.4	104.4	88.1	116.9	103.7	95.8	162.4	131.8	92.2	118.3	106.3	91.6	114.4	103.9
COVERS	DIVS. 10 - 14, 25, 28, 41, 43, 44, 46	100.0	106.9	101.5	100.0	105.8	101.3	100.0	109.2	102.0	100.0	129.3	106.5	100.0	108.6	101.9	100.0	108.3	101.9
21, 22, 23	FIRE SUPPRESSION, PLUMBING & HVAC	96.3	106.8	100.4	96.3	97.8	96.9	96.5	107.0	100.6	99.9	137.4	114.6	99.9	98.8	99.5	99.9	99.2	99.6
26, 27, 3370	ELECTRICAL, COMMUNICATIONS & UTIL.	91.9	107.8	99.8	93.1	106.3	99.6	99.0	95.4	97.2	97.9	136.2	116.8	95.3	103.7	99.5	92.7	99.7	96.2
MF2018	WEIGHTED AVERAGE	92.5	109.6	99.7	92.8	111.2	100.5	98.5	110.5	103.6	100.1	145.6	119.3	96.6	110.1	102.3	93.8	110.4	100.9

ILLINOIS

DIVISION		EFFINGHAM 624			GALESBURG 614			JOLIET 604			KANKAKEE 609			LA SALLE 613			NORTH SUBURBAN 600 - 603		
		MAT.	INST.	TOTAL	MAT.	INST.	TOTAL	MAT.	INST.	TOTAL	MAT.	INST.	TOTAL	MAT.	INST.	TOTAL	MAT.	INST.	TOTAL
015433	CONTRACTOR EQUIPMENT		102.8	102.8		101.9	101.9		93.7	93.7		93.7	93.7		101.9	101.9		93.7	93.7
0241, 31 - 34	SITE & INFRASTRUCTURE, DEMOLITION	99.3	95.9	97.0	97.3	95.8	96.3	101.2	97.2	98.5	94.7	97.1	96.4	96.7	96.8	96.8	100.5	97.9	98.7
0310	Concrete Forming & Accessories	96.5	114.6	111.9	90.2	117.0	113.0	96.8	156.7	147.8	90.5	145.4	137.3	103.7	125.3	122.1	96.1	158.5	149.3
0320	Concrete Reinforcing	83.3	93.8	88.4	92.5	105.6	98.8	103.8	136.6	119.7	104.6	135.1	119.3	92.7	132.9	112.1	103.8	149.1	125.7
0330	Cast-in-Place Concrete	99.6	109.9	103.4	100.9	109.2	104.0	114.6	143.9	125.5	106.8	135.8	117.6	100.8	121.8	108.6	114.6	152.9	128.9
03	CONCRETE	91.0	109.9	99.3	94.8	113.0	102.8	101.0	147.9	121.6	95.0	139.7	114.7	95.6	126.1	109.0	101.0	154.1	124.3
04	MASONRY	79.1	115.3	101.3	110.4	122.4	117.8	104.5	156.7	136.6	100.6	147.0	129.1	110.4	129.7	122.3	101.2	158.8	136.6
05	METALS	101.9	114.9	105.7	95.2	123.6	103.5	94.1	134.7	106.0	94.1	133.4	105.6	95.3	142.0	109.0	95.1	141.6	108.8
06	WOOD, PLASTICS & COMPOSITES	91.0	114.0	103.6	88.3	115.4	103.2	96.6	157.7	130.2	90.0	144.7	120.1	103.8	123.2	114.5	95.4	158.2	130.0
07	THERMAL & MOISTURE PROTECTION	95.4	109.1	101.4	95.6	110.0	101.9	99.5	144.2	119.0	98.6	139.5	116.4	95.8	122.5	107.4	99.9	147.6	120.7
08	OPENINGS	93.3	114.5	98.2	91.3	115.9	97.0	99.3	161.7	113.9	92.4	156.2	107.3	91.3	133.3	101.1	99.4	167.9	115.4
0920	Plaster & Gypsum Board	96.3	114.5	108.5	91.0	115.9	107.8	93.7	159.5	138.0	92.5	146.1	128.6	98.3	124.0	115.6	97.2	160.0	139.5
0950, 0980	Ceilings & Acoustic Treatment	80.6	114.5	103.4	83.9	115.9	105.5	100.9	159.5	140.4	100.9	146.1	131.4	83.9	124.0	110.9	100.9	160.0	140.8
0960	Flooring	103.0	116.6	106.8	89.1	120.8	98.0	89.4	149.8	106.4	86.3	150.7	104.4	95.7	124.8	103.9	89.9	159.5	109.4
0970, 0990	Wall Finishes & Painting/Coating	95.5	109.3	103.7	87.6	98.5	94.0	89.0	135.7	116.6	89.0	135.7	116.6	87.6	133.0	114.5	91.0	163.9	134.1
09	FINISHES	91.3	115.1	104.2	87.4	116.3	103.1	91.6	157.4	127.2	90.0	145.8	120.2	90.4	124.7	109.0	92.2	160.1	129.0
COVERS	DIVS. 10 - 14, 25, 28, 41, 43, 44, 46	100.0	104.5	101.0	100.0	104.4	101.0	100.0	122.6	105.0	100.0	120.6	104.6	100.0	105.6	101.2	100.0	124.1	105.4
21, 22, 23	FIRE SUPPRESSION, PLUMBING & HVAC	96.4	104.5	99.6	96.5	105.7	100.1	100.0	130.0	111.8	96.5	129.6	109.5	96.5	128.4	109.0	99.9	136.2	114.1
26, 27, 3370	ELECTRICAL, COMMUNICATIONS & UTIL.	93.5	107.8	100.5	96.7	86.5	91.7	96.9	133.3	114.8	92.4	137.0	114.4	94.0	137.0	115.2	96.7	135.5	115.9
MF2018	WEIGHTED AVERAGE	95.1	109.1	101.0	95.7	108.5	101.1	98.3	138.8	115.4	95.0	134.9	111.9	95.9	127.0	109.0	98.3	142.9	117.1

ILLINOIS

DIVISION		PEORIA 615 - 616			QUINCY 623			ROCK ISLAND 612			ROCKFORD 610 - 611			SOUTH SUBURBAN 605			SPRINGFIELD 626 - 627		
		MAT.	INST.	TOTAL	MAT.	INST.	TOTAL	MAT.	INST.	TOTAL	MAT.	INST.	TOTAL	MAT.	INST.	TOTAL	MAT.	INST.	TOTAL
015433	CONTRACTOR EQUIPMENT		101.9	101.9		102.8	102.8		101.9	101.9		101.9	101.9		93.7	93.7		105.1	105.1
0241, 31 - 34	SITE & INFRASTRUCTURE, DEMOLITION	97.9	96.0	96.6	98.1	96.4	96.9	95.5	94.8	95.0	97.4	96.8	97.0	100.5	97.8	98.6	99.8	100.5	100.3
0310	Concrete Forming & Accessories	93.0	115.9	112.5	94.3	113.6	110.8	91.6	99.1	98.0	97.2	127.7	123.2	96.1	158.5	149.3	93.2	115.6	112.3
0320	Concrete Reinforcing	90.1	104.7	97.2	83.0	86.2	84.5	92.5	99.0	95.6	85.3	134.1	108.9	103.8	149.0	125.7	83.2	99.8	91.2
0330	Cast-in-Place Concrete	97.9	115.6	104.5	99.8	106.6	102.3	98.7	98.4	98.6	100.2	126.6	110.0	114.6	152.8	128.8	94.9	109.2	100.2
03	CONCRETE	92.0	114.6	102.0	90.7	107.1	97.9	92.7	100.0	95.9	92.7	129.0	108.7	101.0	154.0	124.3	88.5	111.2	98.5
04	MASONRY	109.7	121.5	116.9	98.9	113.2	107.7	110.2	100.2	104.1	85.6	142.3	120.4	101.2	158.7	136.6	80.7	122.2	106.2
05	METALS	97.9	124.2	105.6	101.9	112.6	105.1	95.2	118.5	102.1	97.9	141.5	110.7	95.1	141.4	108.7	102.5	119.1	107.3
06	WOOD, PLASTICS & COMPOSITES	96.4	113.2	105.6	88.6	114.0	102.6	90.0	97.5	94.1	96.4	123.6	111.3	95.4	158.2	130.0	90.0	114.1	103.3
07	THERMAL & MOISTURE PROTECTION	96.3	118.4	103.9	95.3	106.8	100.3	94.9	96.8	95.5	98.8	129.0	111.9	99.9	147.6	120.7	98.3	114.4	105.3
08	OPENINGS	96.6	119.5	102.0	94.0	112.1	98.2	91.3	103.3	94.1	96.6	136.5	105.9	99.4	167.9	115.4	99.2	114.6	103.2
0920	Plaster & Gypsum Board	95.5	113.7	107.7	95.0	114.5	108.1	91.0	97.6	95.5	95.5	124.4	114.9	97.2	160.0	139.5	99.9	114.5	109.7
0950, 0980	Ceilings & Acoustic Treatment	89.6	113.7	105.8	80.6	114.5	103.4	83.9	97.6	93.2	89.6	124.4	113.1	100.9	160.0	140.8	91.3	114.5	107.0
0960	Flooring	92.6	119.5	100.2	101.8	113.6	105.1	90.2	94.6	91.4	92.6	124.9	101.7	89.9	159.5	109.4	104.8	117.1	108.3
0970, 0990	Wall Finishes & Painting/Coating	87.6	134.8	115.5	95.5	111.5	105.0	87.6	95.5	92.3	87.6	142.7	120.2	91.0	163.9	134.1	98.0	111.5	106.0
09	FINISHES	90.2	118.7	105.6	90.7	114.2	103.4	87.7	97.6	93.0	90.2	128.5	110.9	92.2	160.1	129.0	98.0	115.9	107.7
COVERS	DIVS. 10 - 14, 25, 28, 41, 43, 44, 46	100.0	104.4	101.0	100.0	106.9	101.5	100.0	97.5	99.4	100.0	114.2	103.2	100.0	124.1	105.4	100.0	108.3	101.9
21, 22, 23	FIRE SUPPRESSION, PLUMBING & HVAC	100.0	100.2	100.1	96.4	102.9	98.9	96.5	97.5	96.9	100.1	116.1	106.4	99.9	136.2	114.1	99.9	103.7	101.4
26, 27, 3370	ELECTRICAL, COMMUNICATIONS & UTIL.	97.7	91.5	94.6	91.4	82.2	86.9	89.3	93.1	91.2	98.0	125.6	111.6	96.7	135.5	115.9	98.1	88.7	93.5
MF2018	WEIGHTED AVERAGE	97.6	108.8	102.3	95.7	104.2	99.3	94.6	99.5	96.7	96.6	125.5	108.8	98.3	142.8	117.1	97.5	108.3	102.1

City Cost Indexes

INDIANA

DIVISION		ANDERSON 460 MAT.	INST.	TOTAL	BLOOMINGTON 474 MAT.	INST.	TOTAL	COLUMBUS 472 MAT.	INST.	TOTAL	EVANSVILLE 476-477 MAT.	INST.	TOTAL	FORT WAYNE 467-468 MAT.	INST.	TOTAL	GARY 463-464 MAT.	INST.	TOTAL
015433	CONTRACTOR EQUIPMENT		94.5	94.5		81.5	81.5		81.5	81.5		110.3	110.3		94.5	94.5		94.5	94.5
0241, 31 - 34	SITE & INFRASTRUCTURE, DEMOLITION	98.7	89.0	92.0	86.5	87.7	87.3	83.2	87.4	86.1	91.9	115.1	107.9	99.8	88.8	92.2	99.3	92.8	94.8
0310	Concrete Forming & Accessories	94.9	77.4	80.0	100.8	82.1	84.9	95.0	80.2	82.4	94.3	78.6	80.9	92.9	73.7	76.5	95.0	111.1	108.7
0320	Concrete Reinforcing	104.9	83.0	94.3	90.4	86.9	88.7	90.8	86.9	88.9	99.0	81.0	90.3	104.9	77.9	91.9	104.9	115.8	110.2
0330	Cast-in-Place Concrete	106.3	74.9	94.6	101.3	75.7	91.7	100.8	76.5	91.8	96.8	83.2	91.7	113.0	75.2	98.9	111.1	112.2	111.5
03	CONCRETE	95.7	78.1	88.0	99.2	80.3	90.9	98.5	79.6	90.2	99.5	80.8	91.3	98.6	75.7	88.6	97.9	112.1	104.1
04	MASONRY	87.4	74.0	79.2	88.9	74.4	80.0	88.8	74.3	79.9	84.3	77.3	80.0	90.7	71.4	78.8	88.8	110.3	102.0
05	METALS	97.7	88.2	94.9	99.1	75.7	92.2	99.1	74.7	91.9	91.9	83.2	89.3	97.7	86.3	94.3	97.7	107.1	100.5
06	WOOD, PLASTICS & COMPOSITES	93.8	77.6	84.9	109.5	82.9	94.9	104.0	80.5	91.1	90.8	77.7	83.6	93.7	73.8	82.7	91.3	109.1	101.1
07	THERMAL & MOISTURE PROTECTION	109.2	75.0	94.4	95.8	77.9	88.0	95.2	78.7	88.0	100.0	81.7	92.1	108.9	76.8	94.9	107.7	106.2	107.0
08	OPENINGS	93.2	75.8	89.2	97.5	80.0	93.4	93.9	78.6	90.3	91.8	76.0	88.1	93.2	71.2	88.1	93.2	114.6	98.2
0920	Plaster & Gypsum Board	106.4	77.3	86.8	98.6	83.2	88.2	95.4	80.7	85.5	94.2	76.7	82.4	105.8	73.4	84.0	99.5	109.7	106.4
0950, 0980	Ceilings & Acoustic Treatment	89.8	77.3	81.4	77.9	83.2	81.5	77.9	80.7	79.8	81.6	76.7	78.3	89.8	73.4	78.7	89.8	109.7	103.2
0960	Flooring	93.5	75.0	88.3	99.6	83.2	95.0	94.4	83.2	91.2	94.1	71.4	87.7	93.5	71.2	87.2	93.5	110.5	98.3
0970, 0990	Wall Finishes & Painting/Coating	92.8	65.8	76.9	84.4	81.1	82.5	84.4	81.1	82.5	90.3	84.4	86.8	92.8	70.2	79.5	92.8	120.7	109.3
09	FINISHES	91.3	75.7	82.9	90.6	82.4	86.2	88.6	81.0	84.5	89.3	77.7	83.0	91.1	72.9	81.3	90.3	111.8	102.0
COVERS	DIVS. 10 - 14, 25, 28, 41, 43, 44, 46	100.0	87.4	97.2	100.0	88.3	97.4	100.0	88.0	97.3	100.0	91.2	98.0	100.0	87.2	97.1	100.0	107.9	101.8
21, 22, 23	FIRE SUPPRESSION, PLUMBING & HVAC	100.0	77.0	91.0	99.7	79.6	91.8	96.2	79.1	89.5	99.9	78.6	91.5	100.0	72.2	89.1	100.0	106.2	102.4
26, 27, 3370	ELECTRICAL, COMMUNICATIONS & UTIL.	87.7	83.2	85.5	99.9	86.6	93.3	99.1	86.7	93.0	95.7	81.4	88.7	88.4	74.9	81.7	99.5	110.1	104.7
MF2018	WEIGHTED AVERAGE	96.0	79.8	89.1	97.8	81.1	90.7	96.1	80.5	89.5	95.5	82.7	90.1	96.6	76.3	88.0	97.5	108.1	102.0

INDIANA

DIVISION		INDIANAPOLIS 461-462 MAT.	INST.	TOTAL	KOKOMO 469 MAT.	INST.	TOTAL	LAFAYETTE 479 MAT.	INST.	TOTAL	LAWRENCEBURG 470 MAT.	INST.	TOTAL	MUNCIE 473 MAT.	INST.	TOTAL	NEW ALBANY 471 MAT.	INST.	TOTAL
015433	CONTRACTOR EQUIPMENT		86.1	86.1		94.5	94.5		81.5	81.5		100.6	100.6		92.8	92.8		90.6	90.6
0241, 31 - 34	SITE & INFRASTRUCTURE, DEMOLITION	99.8	91.3	93.9	94.9	89.0	90.8	83.9	87.4	86.3	81.2	102.2	95.7	86.6	87.9	87.5	78.5	89.8	86.3
0310	Concrete Forming & Accessories	100.4	85.2	87.4	98.0	78.2	81.1	92.5	78.6	80.7	91.5	78.4	80.3	92.3	76.9	79.2	90.1	76.9	78.9
0320	Concrete Reinforcing	106.1	87.2	97.0	94.9	87.2	91.1	90.4	82.9	86.8	89.7	79.8	84.9	100.0	82.9	91.7	91.0	82.6	86.9
0330	Cast-in-Place Concrete	101.2	84.8	95.1	105.2	81.8	96.5	101.4	77.6	92.5	94.8	74.9	87.4	106.4	73.9	94.3	97.9	73.1	88.7
03	CONCRETE	99.0	84.8	92.8	92.5	81.6	87.7	98.7	78.7	89.9	91.8	78.0	85.7	97.4	77.6	88.7	97.3	76.8	88.3
04	MASONRY	89.8	78.7	83.0	87.1	77.2	81.0	94.0	74.0	81.7	73.8	74.2	74.1	90.5	74.1	80.4	80.2	71.1	74.6
05	METALS	94.7	76.0	89.2	94.2	89.8	92.9	97.4	73.7	90.5	93.9	86.2	91.7	100.8	88.0	97.0	96.0	82.2	91.9
06	WOOD, PLASTICS & COMPOSITES	100.1	86.0	92.4	96.8	77.4	86.1	101.2	79.2	89.1	89.1	78.3	83.2	102.6	77.1	88.6	91.2	77.6	83.7
07	THERMAL & MOISTURE PROTECTION	98.7	82.0	91.4	108.1	77.5	94.8	95.2	77.2	87.4	100.8	76.9	90.4	98.2	76.4	88.7	87.6	72.3	80.9
08	OPENINGS	103.9	81.9	98.8	88.4	76.9	85.7	92.4	76.7	88.8	93.6	75.0	89.3	91.0	75.6	87.4	91.4	76.3	87.9
0920	Plaster & Gypsum Board	96.8	85.8	89.4	111.0	77.1	88.2	93.3	79.4	83.9	71.7	78.4	76.2	94.2	77.3	82.8	92.0	77.5	82.2
0950, 0980	Ceilings & Acoustic Treatment	92.7	85.8	88.0	90.6	77.1	81.5	73.9	79.4	77.6	84.0	78.4	80.2	77.9	77.3	77.5	81.6	77.5	78.8
0960	Flooring	97.2	83.2	93.2	97.3	88.4	94.8	93.3	79.2	89.3	68.9	83.2	72.9	93.7	75.0	88.5	91.6	54.0	81.1
0970, 0990	Wall Finishes & Painting/Coating	96.5	81.1	87.4	92.8	68.8	78.6	84.4	81.4	82.6	85.2	73.0	78.0	84.4	65.8	73.4	90.3	66.3	76.1
09	FINISHES	94.9	84.6	89.4	93.1	79.1	85.5	87.2	79.2	82.8	78.8	78.9	78.8	87.9	75.4	81.1	88.5	71.4	79.2
COVERS	DIVS. 10 - 14, 25, 28, 41, 43, 44, 46	100.0	93.7	98.6	100.0	88.6	97.5	100.0	86.6	97.0	100.0	88.0	97.3	100.0	86.3	97.0	100.0	87.7	97.3
21, 22, 23	FIRE SUPPRESSION, PLUMBING & HVAC	100.1	79.9	92.2	96.5	79.3	89.7	96.2	76.2	88.3	97.0	75.9	88.7	99.7	76.9	90.7	96.4	78.6	89.4
26, 27, 3370	ELECTRICAL, COMMUNICATIONS & UTIL.	101.9	86.7	94.4	92.0	78.3	85.3	98.6	78.6	88.7	93.6	74.1	84.0	91.4	74.0	82.9	94.3	74.6	84.6
MF2018	WEIGHTED AVERAGE	98.7	83.2	92.1	94.2	81.1	88.6	95.8	78.1	88.3	92.5	79.6	87.0	96.2	78.2	88.6	93.9	77.3	86.9

INDIANA / IOWA

DIVISION		SOUTH BEND 465-466 MAT.	INST.	TOTAL	TERRE HAUTE 478 MAT.	INST.	TOTAL	WASHINGTON 475 MAT.	INST.	TOTAL	BURLINGTON 526 MAT.	INST.	TOTAL	CARROLL 514 MAT.	INST.	TOTAL	CEDAR RAPIDS 522-524 MAT.	INST.	TOTAL
015433	CONTRACTOR EQUIPMENT		108.2	108.2		110.3	110.3		110.3	110.3		98.8	98.8		98.8	98.8		95.7	95.7
0241, 31 - 34	SITE & INFRASTRUCTURE, DEMOLITION	97.7	94.2	95.3	93.4	115.4	108.6	93.2	115.7	108.7	98.0	91.8	93.7	87.1	92.6	90.9	99.9	91.4	94.0
0310	Concrete Forming & Accessories	95.1	76.0	78.9	95.0	77.1	79.7	96.0	80.5	82.8	96.5	94.2	94.5	84.1	83.9	83.9	102.5	83.9	86.6
0320	Concrete Reinforcing	104.1	83.6	94.2	99.0	83.0	91.3	91.6	82.6	87.2	94.7	97.8	96.2	95.4	85.8	90.8	95.3	80.2	88.0
0330	Cast-in-Place Concrete	108.2	78.4	97.1	93.7	78.3	88.0	102.0	85.0	95.6	107.3	54.8	87.8	107.3	82.2	97.9	107.6	83.4	98.6
03	CONCRETE	97.7	79.7	89.8	102.5	78.8	92.1	108.3	82.6	97.0	95.8	81.5	89.6	94.7	84.3	90.1	95.9	83.8	90.6
04	MASONRY	91.2	74.0	80.6	92.0	73.3	80.5	84.5	79.6	81.5	99.5	71.9	82.5	101.1	72.4	83.5	105.0	80.0	89.6
05	METALS	101.5	104.2	101.7	92.6	84.4	90.2	87.1	84.7	86.4	87.9	99.1	91.2	87.9	95.4	90.1	90.3	93.1	91.2
06	WOOD, PLASTICS & COMPOSITES	96.6	75.2	84.8	93.0	76.7	84.0	93.2	79.2	85.5	91.6	98.1	95.2	77.8	88.4	83.6	98.9	83.6	90.6
07	THERMAL & MOISTURE PROTECTION	103.2	78.7	92.5	100.1	79.1	91.0	100.0	83.5	92.8	104.4	77.1	92.5	104.7	78.8	93.4	105.5	80.1	94.4
08	OPENINGS	93.1	75.5	89.0	92.3	75.3	88.3	89.3	77.1	86.4	94.5	95.8	94.8	98.8	84.2	95.4	99.3	81.2	95.1
0920	Plaster & Gypsum Board	97.7	74.7	82.2	94.2	75.7	81.8	94.0	78.3	83.4	105.0	98.3	100.5	100.9	88.3	92.4	110.3	83.6	92.3
0950, 0980	Ceilings & Acoustic Treatment	94.2	74.7	81.1	81.6	75.7	77.6	76.7	78.3	77.8	99.0	98.3	98.5	99.0	88.3	91.8	101.4	83.6	89.4
0960	Flooring	90.7	86.2	89.5	94.1	75.7	88.9	95.1	83.2	91.7	93.7	69.1	86.8	88.0	79.5	85.6	107.9	84.3	101.3
0970, 0990	Wall Finishes & Painting/Coating	94.3	82.5	87.3	90.3	79.8	84.1	90.3	84.4	86.8	92.8	84.6	87.9	92.8	84.6	87.9	94.7	70.8	80.6
09	FINISHES	92.5	78.3	84.8	89.3	76.9	82.6	88.9	81.3	84.7	93.9	89.7	91.6	90.1	83.6	86.6	99.3	82.7	90.3
COVERS	DIVS. 10 - 14, 25, 28, 41, 43, 44, 46	100.0	88.7	97.5	100.0	88.7	97.5	100.0	91.8	98.2	100.0	93.4	98.5	100.0	89.1	97.6	100.0	93.4	98.5
21, 22, 23	FIRE SUPPRESSION, PLUMBING & HVAC	99.9	75.0	90.1	99.9	76.5	90.7	96.4	81.0	90.4	96.5	83.8	91.5	96.5	77.8	89.2	100.0	81.2	92.6
26, 27, 3370	ELECTRICAL, COMMUNICATIONS & UTIL.	101.3	83.9	92.7	94.1	83.5	88.8	94.5	83.6	89.2	101.1	72.2	86.9	101.8	76.9	89.5	98.8	80.1	89.5
MF2018	WEIGHTED AVERAGE	98.4	81.8	91.4	96.2	81.7	90.1	94.6	84.8	90.4	95.7	84.3	90.9	95.5	82.4	89.9	98.0	83.7	92.0

For customer support on your Commercial Renovation Costs with RSMeans data, call 800.448.8182.

799

City Cost Indexes

IOWA

DIVISION		COUNCIL BLUFFS 515			CRESTON 508			DAVENPORT 527 - 528			DECORAH 521			DES MOINES 500 - 503,509			DUBUQUE 520		
		MAT.	INST.	TOTAL	MAT.	INST.	TOTAL	MAT.	INST.	TOTAL	MAT.	INST.	TOTAL	MAT.	INST.	TOTAL	MAT.	INST.	TOTAL
015433	CONTRACTOR EQUIPMENT		95.0	95.0		98.8	98.8		98.8	98.8		98.8	98.8		102.4	102.4		94.5	94.5
0241, 31 - 34	SITE & INFRASTRUCTURE, DEMOLITION	104.5	88.5	93.4	92.9	93.6	93.4	98.4	94.6	95.8	96.6	91.6	93.2	98.3	99.3	99.0	97.7	88.7	91.5
0310	Concrete Forming & Accessories	83.5	73.6	75.0	79.2	86.7	85.6	102.0	95.9	96.8	94.0	71.8	75.1	95.6	89.5	90.4	84.9	81.5	82.0
0320	Concrete Reinforcing	97.3	79.8	88.8	96.3	85.9	91.3	95.3	99.0	97.1	94.7	85.2	90.1	101.0	101.6	101.3	94.0	79.9	87.2
0330	Cast-in-Place Concrete	111.8	79.1	99.6	114.6	85.7	103.8	103.6	95.2	100.5	104.4	76.0	93.8	97.6	92.1	95.6	105.3	83.0	97.0
03	CONCRETE	98.1	77.6	89.1	97.7	86.8	92.9	94.0	96.7	95.2	93.8	76.6	86.3	91.0	92.8	91.8	92.7	82.6	88.3
04	MASONRY	106.5	76.2	87.8	99.7	79.9	87.5	102.0	92.2	96.0	120.3	69.2	88.9	87.7	89.0	88.5	105.9	68.8	83.1
05	METALS	95.2	92.8	94.5	88.4	96.0	90.6	90.3	105.5	94.8	88.0	94.2	89.9	94.1	96.8	94.9	88.9	92.4	89.9
06	WOOD, PLASTICS & COMPOSITES	76.5	72.1	74.1	69.7	88.4	80.0	98.9	95.3	96.9	88.5	70.9	78.8	87.6	88.5	88.1	78.3	82.2	80.4
07	THERMAL & MOISTURE PROTECTION	104.8	74.8	91.8	106.2	82.2	95.8	104.9	91.6	99.1	104.6	69.9	89.5	98.5	87.6	93.8	105.1	76.5	92.6
08	OPENINGS	98.4	76.8	93.3	108.6	85.9	103.3	99.3	96.7	98.7	97.4	77.9	92.9	100.8	89.6	98.2	98.4	82.4	94.6
0920	Plaster & Gypsum Board	100.9	71.8	81.3	95.8	88.3	90.8	110.3	95.4	100.2	103.7	70.3	81.2	92.8	88.3	89.8	100.9	82.1	88.3
0950, 0980	Ceilings & Acoustic Treatment	99.0	71.8	80.6	90.2	88.3	88.9	101.4	95.4	97.3	99.0	70.3	79.6	93.4	88.3	90.0	99.0	82.1	87.6
0960	Flooring	87.0	84.3	86.2	81.3	69.1	77.8	96.1	89.4	94.2	93.3	69.1	86.5	95.2	94.5	95.0	99.1	69.1	90.7
0970, 0990	Wall Finishes & Painting/Coating	89.6	59.8	72.0	83.6	84.6	84.2	92.8	90.5	91.4	92.8	84.6	87.9	91.6	84.6	87.5	93.9	78.4	84.8
09	FINISHES	91.0	73.7	81.7	85.1	83.4	84.2	95.7	94.1	94.8	93.5	72.4	82.1	93.6	90.0	91.6	94.8	78.9	86.2
COVERS	DIVS. 10 - 14, 25, 28, 41, 43, 44, 46	100.0	91.4	98.1	100.0	91.6	98.1	100.0	96.9	99.3	100.0	88.6	97.5	100.0	95.6	99.0	100.0	92.3	98.3
21, 22, 23	FIRE SUPPRESSION, PLUMBING & HVAC	100.0	72.8	89.4	96.3	78.9	89.8	100.0	93.9	97.6	96.5	74.5	87.9	99.8	86.8	94.7	100.0	75.6	90.4
26, 27, 3370	ELECTRICAL, COMMUNICATIONS & UTIL.	103.9	82.7	93.4	93.5	76.9	85.3	96.9	87.6	92.3	98.8	47.2	73.4	105.1	84.3	94.8	102.5	76.9	89.9
MF2018	WEIGHTED AVERAGE	98.9	79.2	90.6	95.7	84.2	90.8	97.1	94.5	95.9	96.4	73.7	86.8	97.2	90.2	94.3	97.2	79.9	89.9

IOWA

DIVISION		FORT DODGE 505			MASON CITY 504			OTTUMWA 525			SHENANDOAH 516			SIBLEY 512			SIOUX CITY 510 - 511		
		MAT.	INST.	TOTAL	MAT.	INST.	TOTAL	MAT.	INST.	TOTAL	MAT.	INST.	TOTAL	MAT.	INST.	TOTAL	MAT.	INST.	TOTAL
015433	CONTRACTOR EQUIPMENT		98.8	98.8		98.8	98.8		94.5	94.5		95.0	95.0		98.8	98.8		98.8	98.8
0241, 31 - 34	SITE & INFRASTRUCTURE, DEMOLITION	101.4	90.5	93.9	101.5	91.5	94.6	98.0	86.7	90.2	102.6	88.6	92.9	107.7	91.6	96.6	109.6	93.3	98.3
0310	Concrete Forming & Accessories	79.8	78.0	78.3	83.6	71.4	73.2	91.9	87.1	87.8	85.1	77.2	78.3	85.7	37.8	44.9	102.5	75.5	79.4
0320	Concrete Reinforcing	96.3	85.2	90.9	96.2	85.2	90.9	94.7	98.0	96.3	97.3	85.8	91.7	97.3	85.1	91.4	95.3	100.8	98.0
0330	Cast-in-Place Concrete	107.6	41.9	83.1	107.6	71.2	94.0	108.0	63.9	91.6	108.1	82.6	98.6	105.9	55.5	87.2	106.6	89.3	100.2
03	CONCRETE	93.1	67.6	81.9	93.3	74.8	86.3	95.4	81.5	89.3	95.6	81.5	89.4	94.7	54.1	76.8	95.3	85.6	91.0
04	MASONRY	98.6	51.9	69.9	111.6	68.0	84.8	102.3	55.2	73.3	106.0	74.5	86.6	124.6	52.2	80.1	99.2	70.0	81.2
05	METALS	88.5	94.0	90.1	88.5	94.2	90.2	87.9	99.7	91.3	94.2	95.5	94.6	88.1	93.3	89.7	90.3	101.3	93.6
06	WOOD, PLASTICS & COMPOSITES	70.1	88.4	80.2	74.0	70.9	72.3	85.7	97.9	92.4	78.3	78.5	78.4	79.1	34.2	54.4	98.9	74.4	85.4
07	THERMAL & MOISTURE PROTECTION	105.5	65.2	88.0	105.0	70.5	90.0	105.3	70.3	90.1	104.0	73.3	90.7	104.3	55.2	82.9	104.9	73.4	91.2
08	OPENINGS	102.5	75.9	96.3	94.7	77.9	90.8	98.8	92.8	97.4	90.2	76.4	87.0	95.6	46.0	84.0	99.3	81.6	95.1
0920	Plaster & Gypsum Board	95.8	88.3	90.8	95.8	70.3	78.6	101.5	98.3	99.3	100.9	78.3	85.7	100.9	32.5	54.9	110.3	73.8	85.8
0950, 0980	Ceilings & Acoustic Treatment	90.2	88.3	89.0	90.2	70.3	76.8	99.0	98.3	98.5	99.0	78.3	85.0	99.0	32.5	54.2	101.4	73.8	82.8
0960	Flooring	82.6	69.1	78.8	84.6	69.1	80.4	102.2	69.1	92.9	87.7	73.9	83.8	88.9	69.1	83.3	96.1	73.3	89.7
0970, 0990	Wall Finishes & Painting/Coating	83.6	81.3	82.2	83.6	84.6	84.2	93.9	84.6	88.4	89.6	84.6	86.6	92.8	83.6	87.3	92.8	69.1	78.8
09	FINISHES	87.0	78.0	82.1	87.6	72.1	79.2	95.9	84.8	89.9	91.1	77.1	83.5	93.3	46.4	67.9	97.2	73.8	84.5
COVERS	DIVS. 10 - 14, 25, 28, 41, 43, 44, 46	100.0	84.8	96.6	100.0	88.2	97.4	100.0	87.3	97.2	100.0	88.2	97.4	100.0	78.9	95.3	100.0	91.4	98.1
21, 22, 23	FIRE SUPPRESSION, PLUMBING & HVAC	96.3	71.4	86.6	96.3	77.7	89.0	96.5	73.1	87.3	96.5	83.9	91.6	96.5	71.5	86.7	100.0	81.2	92.6
26, 27, 3370	ELECTRICAL, COMMUNICATIONS & UTIL.	99.8	69.4	84.8	98.9	47.2	73.4	100.9	71.0	86.2	98.8	78.4	88.7	98.8	47.2	73.4	98.8	74.0	86.6
MF2018	WEIGHTED AVERAGE	95.5	73.7	86.3	95.3	74.0	86.3	96.3	78.7	88.9	96.1	81.9	90.1	96.8	62.4	82.3	97.7	81.5	90.9

IOWA / KANSAS

DIVISION		IOWA SPENCER 513			IOWA WATERLOO 506 - 507			KANSAS BELLEVILLE 669			KANSAS COLBY 677			KANSAS DODGE CITY 678			KANSAS EMPORIA 668		
		MAT.	INST.	TOTAL	MAT.	INST.	TOTAL	MAT.	INST.	TOTAL	MAT.	INST.	TOTAL	MAT.	INST.	TOTAL	MAT.	INST.	TOTAL
015433	CONTRACTOR EQUIPMENT		98.8	98.8		98.8	98.8		103.7	103.7		103.7	103.7		103.7	103.7		101.9	101.9
0241, 31 - 34	SITE & INFRASTRUCTURE, DEMOLITION	107.8	90.4	95.7	106.9	92.7	97.1	111.3	92.0	98.0	106.7	92.0	97.0	109.4	91.7	97.2	103.1	89.7	93.8
0310	Concrete Forming & Accessories	91.8	37.5	45.5	94.4	68.7	72.5	94.8	54.3	60.3	97.5	61.2	66.6	91.1	61.1	65.5	86.2	70.0	72.4
0320	Concrete Reinforcing	97.3	85.1	91.4	96.9	80.3	88.9	98.9	102.6	100.7	97.1	102.6	99.8	94.7	102.3	98.4	97.5	102.8	100.1
0330	Cast-in-Place Concrete	105.9	65.5	90.9	115.2	84.0	103.6	121.0	83.5	107.1	119.5	87.2	107.5	121.6	86.9	108.7	117.0	87.3	106.0
03	CONCRETE	95.0	57.4	78.5	99.0	77.2	89.4	110.5	74.3	94.6	108.3	78.7	95.3	109.7	78.5	96.0	103.1	82.8	94.2
04	MASONRY	124.6	52.2	80.1	99.4	75.1	84.5	86.4	58.5	69.3	100.5	64.6	78.4	110.7	59.6	79.3	91.8	66.0	76.0
05	METALS	88.1	93.0	89.5	90.7	93.0	91.4	95.6	99.2	96.7	92.8	100.0	94.9	94.2	98.6	95.5	95.3	100.2	96.8
06	WOOD, PLASTICS & COMPOSITES	85.5	34.2	57.3	87.2	64.4	74.6	93.3	51.2	70.1	96.9	57.4	75.2	88.9	57.4	71.5	84.3	68.8	75.8
07	THERMAL & MOISTURE PROTECTION	105.3	55.8	83.8	105.3	77.2	93.0	91.4	62.1	78.7	97.8	65.4	83.7	97.8	66.6	84.2	89.5	75.4	83.3
08	OPENINGS	106.7	46.0	92.5	95.1	72.9	90.0	94.1	62.5	86.7	97.9	65.9	90.4	97.8	65.9	90.4	92.1	74.4	88.0
0920	Plaster & Gypsum Board	101.5	32.5	55.1	103.8	63.6	76.7	88.7	50.0	62.6	96.8	56.3	69.6	90.5	56.3	67.5	85.5	68.1	73.8
0950, 0980	Ceilings & Acoustic Treatment	99.0	32.5	54.2	91.8	63.6	72.8	76.5	50.0	58.6	77.3	56.3	63.2	77.3	56.3	63.2	76.5	68.1	70.8
0960	Flooring	91.6	69.1	85.3	89.6	78.9	86.6	92.9	68.4	86.0	87.6	68.4	82.2	83.8	68.4	79.4	88.3	68.4	82.7
0970, 0990	Wall Finishes & Painting/Coating	92.8	57.2	71.7	83.6	81.3	82.2	89.7	57.9	70.9	96.1	57.9	73.5	96.1	57.9	73.5	89.7	57.9	70.9
09	FINISHES	94.2	42.3	66.1	90.7	70.9	80.0	86.1	55.7	69.6	84.7	60.8	71.8	82.8	60.8	70.9	83.3	67.6	74.8
COVERS	DIVS. 10 - 14, 25, 28, 41, 43, 44, 46	100.0	78.9	95.3	100.0	91.0	98.0	100.0	83.3	96.3	100.0	86.1	96.9	100.0	86.0	96.9	100.0	85.9	96.9
21, 22, 23	FIRE SUPPRESSION, PLUMBING & HVAC	96.5	71.5	86.7	99.9	80.1	92.1	96.4	70.3	86.1	96.5	71.1	86.5	100.0	71.1	88.7	96.4	77.6	87.4
26, 27, 3370	ELECTRICAL, COMMUNICATIONS & UTIL.	100.4	47.2	74.2	95.4	61.6	78.7	104.1	62.7	83.7	95.5	67.9	81.9	92.9	67.9	80.6	101.4	66.9	84.4
MF2018	WEIGHTED AVERAGE	98.2	62.3	83.0	96.8	77.4	88.6	97.9	70.7	86.4	97.3	74.0	87.4	98.6	73.3	87.9	96.1	76.5	87.8

KANSAS

DIVISION		FORT SCOTT 667 MAT.	INST.	TOTAL	HAYS 676 MAT.	INST.	TOTAL	HUTCHINSON 675 MAT.	INST.	TOTAL	INDEPENDENCE 673 MAT.	INST.	TOTAL	KANSAS CITY 660-662 MAT.	INST.	TOTAL	LIBERAL 679 MAT.	INST.	TOTAL
015433	CONTRACTOR EQUIPMENT		102.8	102.8		103.7	103.7		103.7	103.7		103.7	103.7		100.4	100.4		103.7	103.7
0241, 31 - 34	SITE & INFRASTRUCTURE, DEMOLITION	99.9	89.7	92.8	111.4	92.2	98.2	90.6	92.7	92.0	110.2	92.7	98.1	94.3	90.2	91.5	111.3	92.3	98.2
0310	Concrete Forming & Accessories	102.8	82.9	85.9	95.2	58.9	64.3	86.0	56.6	61.0	105.8	67.8	73.4	99.4	98.4	98.6	91.5	58.8	63.7
0320	Concrete Reinforcing	96.9	100.8	98.8	94.7	102.6	98.5	94.7	102.6	98.5	94.1	100.8	97.3	94.0	105.3	99.4	96.1	102.3	99.1
0330	Cast-in-Place Concrete	108.5	82.4	98.8	94.2	83.8	90.3	87.3	86.9	87.2	122.1	87.1	109.1	92.9	97.4	94.6	94.2	83.5	90.3
03	CONCRETE	98.4	86.6	93.2	99.8	76.5	89.6	83.0	76.5	80.2	110.8	81.3	97.8	90.7	99.7	94.6	101.7	76.3	90.5
04	MASONRY	92.7	56.1	70.2	109.5	58.6	78.2	100.2	64.5	78.3	97.8	64.5	77.3	93.5	98.1	96.3	108.2	53.6	74.6
05	METALS	95.3	99.0	96.4	92.3	99.8	94.5	92.1	99.0	94.1	92.0	98.8	94.0	103.1	106.2	104.0	92.6	98.7	94.4
06	WOOD, PLASTICS & COMPOSITES	103.8	90.5	96.5	93.9	57.4	73.8	84.0	51.4	66.0	107.3	65.9	84.5	99.6	98.5	99.0	89.6	57.4	71.9
07	THERMAL & MOISTURE PROTECTION	90.5	72.8	82.8	98.1	62.8	82.8	96.6	64.3	82.5	97.8	74.9	87.9	90.3	99.3	94.2	98.3	60.9	82.0
08	OPENINGS	92.1	85.9	90.6	97.8	65.9	90.4	97.2	62.6	89.6	95.9	70.2	89.9	93.2	97.4	94.2	97.9	65.9	90.4
0920	Plaster & Gypsum Board	91.2	90.4	90.7	94.3	56.3	68.7	89.2	50.2	63.0	105.0	65.1	78.1	84.8	98.6	94.1	91.4	56.3	67.8
0950, 0980	Ceilings & Acoustic Treatment	76.5	90.4	85.9	77.3	56.3	63.2	77.3	50.2	59.0	77.3	65.1	69.0	76.5	98.6	91.4	77.3	56.3	63.2
0960	Flooring	102.6	66.9	92.5	86.4	68.4	81.3	80.9	68.4	77.3	92.0	66.9	84.9	82.4	95.3	86.0	84.1	68.4	79.6
0970, 0990	Wall Finishes & Painting/Coating	91.3	77.5	83.1	96.1	57.9	73.5	96.1	57.9	73.5	96.1	57.9	73.5	97.1	102.7	100.4	96.1	57.9	73.5
09	FINISHES	88.5	80.7	84.3	84.5	59.3	70.9	80.3	57.3	67.9	87.3	66.1	75.8	83.6	98.3	91.5	83.7	59.3	70.5
COVERS	DIVS. 10 - 14, 25, 28, 41, 43, 44, 46	100.0	86.8	97.1	100.0	84.0	96.4	100.0	85.4	96.7	100.0	87.0	97.1	100.0	94.5	98.8	100.0	84.0	96.4
21, 22, 23	FIRE SUPPRESSION, PLUMBING & HVAC	96.4	67.4	85.0	96.5	67.4	85.1	96.5	71.1	86.5	96.5	70.4	86.3	99.9	99.1	99.6	96.5	69.3	85.8
26, 27, 3370	ELECTRICAL, COMMUNICATIONS & UTIL.	100.6	67.9	84.5	94.6	67.9	81.4	90.5	61.0	75.9	92.3	72.8	82.7	106.2	98.1	102.2	92.9	67.9	80.6
MF2018	WEIGHTED AVERAGE	96.0	77.1	88.0	96.5	72.0	86.1	92.5	71.9	83.8	97.2	76.0	88.2	97.3	98.6	97.8	96.5	71.7	86.0

KANSAS / KENTUCKY

DIVISION		SALINA 674 MAT.	INST.	TOTAL	TOPEKA 664-666 MAT.	INST.	TOTAL	WICHITA 670-672 MAT.	INST.	TOTAL	ASHLAND 411-412 MAT.	INST.	TOTAL	BOWLING GREEN 421-422 MAT.	INST.	TOTAL	CAMPTON 413-414 MAT.	INST.	TOTAL
015433	CONTRACTOR EQUIPMENT		103.7	103.7		105.0	105.0		106.7	106.7		97.4	97.4		90.6	90.6		96.8	96.8
0241, 31 - 34	SITE & INFRASTRUCTURE, DEMOLITION	99.8	92.2	94.6	98.4	94.4	95.6	96.8	98.5	97.9	114.0	78.8	89.7	78.7	89.4	86.1	87.5	90.7	89.7
0310	Concrete Forming & Accessories	87.8	61.4	65.3	96.7	66.1	70.7	92.9	57.0	62.3	87.2	89.6	89.3	86.5	78.8	79.9	89.3	80.9	82.2
0320	Concrete Reinforcing	94.1	100.5	97.2	93.5	103.8	98.5	92.4	100.4	96.3	92.5	94.2	93.3	89.7	76.8	83.5	90.6	93.0	91.7
0330	Cast-in-Place Concrete	105.6	83.8	97.5	97.6	86.4	93.4	98.7	79.5	91.5	89.1	94.6	91.1	88.4	69.7	81.4	98.6	68.1	87.2
03	CONCRETE	96.8	77.3	88.2	92.3	80.8	87.3	92.4	73.7	84.2	93.1	93.3	93.2	91.6	75.7	84.6	95.0	78.8	87.9
04	MASONRY	125.4	54.9	82.1	87.7	64.3	73.3	97.5	54.1	70.8	90.7	92.0	91.5	92.7	69.2	78.3	89.7	54.7	68.2
05	METALS	94.0	98.9	95.5	99.5	100.0	99.6	96.2	97.5	96.6	95.1	107.7	98.8	96.7	82.8	92.6	96.0	89.8	94.2
06	WOOD, PLASTICS & COMPOSITES	85.4	60.7	71.8	99.9	65.2	80.8	98.7	54.4	74.3	73.4	87.7	81.3	85.7	80.6	82.9	84.2	88.3	86.5
07	THERMAL & MOISTURE PROTECTION	97.2	62.1	82.0	94.0	74.7	85.6	96.2	60.8	80.8	91.2	88.3	90.0	87.6	77.9	83.4	100.2	67.8	86.1
08	OPENINGS	97.8	67.3	90.7	103.5	74.1	96.7	101.3	63.8	92.5	90.6	89.0	90.2	91.4	77.1	88.1	92.6	87.4	91.4
0920	Plaster & Gypsum Board	89.2	59.8	69.4	99.5	64.2	75.7	95.7	53.1	67.0	59.3	87.6	78.4	87.6	80.6	82.9	87.6	87.6	87.6
0950, 0980	Ceilings & Acoustic Treatment	77.3	59.8	65.5	87.2	64.2	71.7	88.0	53.1	64.5	77.0	87.6	84.2	81.6	80.6	80.9	81.6	87.6	85.7
0960	Flooring	82.2	68.4	78.3	98.2	68.4	89.8	94.3	68.4	87.0	73.8	81.3	75.9	89.5	61.9	81.8	91.6	64.3	83.9
0970, 0990	Wall Finishes & Painting/Coating	96.1	57.9	73.5	92.8	67.7	78.0	94.9	57.9	73.0	91.6	90.4	90.9	90.3	67.6	76.9	90.3	53.9	68.8
09	FINISHES	81.5	61.3	70.6	93.9	65.7	78.6	90.3	57.7	72.7	76.1	88.1	82.6	87.1	74.6	80.3	87.9	75.9	81.4
COVERS	DIVS. 10 - 14, 25, 28, 41, 43, 44, 46	100.0	85.3	96.7	100.0	80.3	95.6	100.0	85.1	96.7	100.0	88.2	97.4	100.0	84.4	96.5	100.0	47.8	88.4
21, 22, 23	FIRE SUPPRESSION, PLUMBING & HVAC	100.0	70.5	88.5	100.0	73.2	89.5	99.8	71.0	88.5	96.2	85.0	91.8	99.9	77.0	90.9	96.4	76.4	88.6
26, 27, 3370	ELECTRICAL, COMMUNICATIONS & UTIL.	92.7	73.8	83.4	105.1	71.9	88.7	95.9	73.8	85.0	91.9	88.3	90.1	94.6	74.7	84.8	92.1	88.2	90.2
MF2018	WEIGHTED AVERAGE	97.2	73.5	87.2	98.6	76.5	89.3	97.0	72.6	86.7	93.2	89.7	91.7	94.6	77.2	87.3	94.4	78.0	87.5

KENTUCKY

DIVISION		CORBIN 407-409 MAT.	INST.	TOTAL	COVINGTON 410 MAT.	INST.	TOTAL	ELIZABETHTOWN 427 MAT.	INST.	TOTAL	FRANKFORT 406 MAT.	INST.	TOTAL	HAZARD 417-418 MAT.	INST.	TOTAL	HENDERSON 424 MAT.	INST.	TOTAL
015433	CONTRACTOR EQUIPMENT		96.8	96.8		100.6	100.6		90.6	90.6		100.1	100.1		96.8	96.8		110.3	110.3
0241, 31 - 34	SITE & INFRASTRUCTURE, DEMOLITION	91.6	91.2	91.3	82.7	101.9	95.9	72.9	89.3	84.3	90.0	96.8	94.7	85.3	91.8	89.7	81.7	114.6	104.5
0310	Concrete Forming & Accessories	85.0	76.0	77.3	84.9	70.8	72.8	81.3	72.1	73.5	102.0	78.5	82.0	85.8	81.8	82.4	92.3	77.1	79.3
0320	Concrete Reinforcing	89.6	92.3	90.9	89.3	78.6	84.1	90.2	83.7	87.0	94.0	84.0	89.2	91.0	92.5	91.7	89.8	84.4	87.2
0330	Cast-in-Place Concrete	91.0	72.5	84.1	94.3	79.6	88.9	79.9	67.6	75.3	91.1	77.0	85.8	94.8	69.6	85.4	78.1	84.5	80.5
03	CONCRETE	84.2	78.0	81.5	93.4	76.1	85.8	83.5	73.1	79.0	87.1	79.3	83.6	91.8	79.7	86.5	88.7	81.3	85.5
04	MASONRY	81.9	60.0	68.4	104.0	72.0	84.3	77.4	61.8	67.7	78.6	74.2	75.9	88.6	56.8	69.0	96.0	78.7	85.4
05	METALS	90.9	89.3	90.5	93.9	88.1	92.2	95.9	85.3	92.7	93.1	86.4	91.1	96.0	89.8	94.2	86.8	87.0	86.9
06	WOOD, PLASTICS & COMPOSITES	72.1	77.6	75.1	81.8	68.3	74.3	80.6	73.9	76.9	101.5	77.7	88.4	80.9	88.3	84.9	88.4	75.8	81.5
07	THERMAL & MOISTURE PROTECTION	103.8	69.5	88.9	101.1	72.1	88.5	87.0	68.5	79.0	101.8	75.7	90.5	100.1	69.5	86.8	99.3	82.8	92.1
08	OPENINGS	87.3	70.5	83.4	94.5	71.8	89.2	91.4	73.0	87.1	98.9	78.4	94.1	93.0	87.2	91.6	89.6	78.5	87.0
0920	Plaster & Gypsum Board	88.4	76.6	80.5	68.2	68.0	68.1	86.3	73.7	77.8	91.6	76.6	81.5	86.3	87.6	87.2	90.9	74.8	80.0
0950, 0980	Ceilings & Acoustic Treatment	74.7	76.6	76.0	84.0	68.0	73.2	81.6	73.7	76.2	90.5	76.6	81.1	81.6	87.6	85.7	76.7	74.8	75.4
0960	Flooring	88.4	64.3	81.6	66.3	83.5	71.1	86.9	73.8	83.2	98.9	78.6	93.2	89.8	64.3	82.6	93.2	76.0	88.4
0970, 0990	Wall Finishes & Painting/Coating	87.2	62.0	72.3	85.2	71.1	76.9	90.3	68.1	77.2	94.2	91.7	92.8	90.3	53.9	68.8	90.3	86.7	88.2
09	FINISHES	83.0	72.5	77.3	77.8	72.7	75.0	85.8	71.6	78.1	93.2	80.2	86.2	87.0	76.4	81.3	87.1	77.9	82.1
COVERS	DIVS. 10 - 14, 25, 28, 41, 43, 44, 46	100.0	89.8	97.7	100.0	85.9	96.8	100.0	70.0	93.3	100.0	58.5	90.8	100.0	48.5	88.5	100.0	55.8	90.1
21, 22, 23	FIRE SUPPRESSION, PLUMBING & HVAC	96.5	74.2	87.7	97.0	76.0	88.8	96.6	77.9	89.3	99.9	81.0	92.5	96.4	77.5	89.0	96.6	79.3	89.8
26, 27, 3370	ELECTRICAL, COMMUNICATIONS & UTIL.	90.0	88.2	89.1	95.6	72.4	84.2	91.9	77.9	85.0	100.6	77.9	89.4	92.1	88.2	90.2	94.0	77.9	86.1
MF2018	WEIGHTED AVERAGE	90.8	78.0	85.4	94.3	77.8	87.3	91.3	75.6	84.7	95.4	80.3	89.0	93.8	78.7	87.4	92.2	81.9	87.8

For customer support on your Commercial Renovation Costs with RSMeans data, call 800.448.8182.

801

KENTUCKY

DIVISION		LEXINGTON 403 - 405			LOUISVILLE 400 - 402			OWENSBORO 423			PADUCAH 420			PIKEVILLE 415 - 416			SOMERSET 425 - 426		
		MAT.	INST.	TOTAL	MAT.	INST.	TOTAL	MAT.	INST.	TOTAL	MAT.	INST.	TOTAL	MAT.	INST.	TOTAL	MAT.	INST.	TOTAL
015433	CONTRACTOR EQUIPMENT		96.8	96.8		93.7	93.7		110.3	110.3		110.3	110.3		97.4	97.4		96.8	96.8
0241, 31 - 34	SITE & INFRASTRUCTURE, DEMOLITION	93.8	92.8	93.1	87.8	94.4	92.3	91.8	115.3	108.0	84.4	114.8	105.4	125.2	78.0	92.6	77.9	91.2	87.1
0310	Concrete Forming & Accessories	96.5	73.7	77.1	93.5	79.1	81.2	90.8	76.8	78.9	88.8	81.2	82.3	96.0	85.6	87.2	86.8	76.6	78.1
0320	Concrete Reinforcing	98.4	84.0	91.4	94.8	84.2	89.6	89.8	77.3	83.8	90.4	82.5	86.6	93.0	94.1	93.5	90.2	92.5	91.3
0330	Cast-in-Place Concrete	93.1	84.4	89.9	89.3	70.7	82.4	91.1	83.9	88.4	83.2	81.1	82.4	97.9	89.4	94.8	78.1	87.9	81.7
03	CONCRETE	86.9	79.7	83.8	85.8	77.4	82.1	100.7	79.8	91.5	93.3	81.7	88.2	106.8	89.7	99.3	78.6	83.7	80.8
04	MASONRY	80.6	71.1	74.8	78.7	72.0	74.6	88.8	78.6	82.5	91.5	79.2	83.9	88.5	83.2	85.3	83.3	63.2	71.0
05	METALS	93.3	86.7	91.3	95.5	85.5	92.5	88.3	84.7	87.3	85.3	87.5	86.0	95.0	107.5	98.7	95.9	90.0	94.2
06	WOOD, PLASTICS & COMPOSITES	87.3	71.6	78.6	88.8	80.8	84.4	86.3	75.8	80.5	84.0	81.4	82.6	83.3	87.7	85.7	81.5	77.6	79.3
07	THERMAL & MOISTURE PROTECTION	104.1	73.9	90.9	101.0	75.7	90.0	100.1	76.6	89.9	99.4	82.6	92.1	92.1	80.5	87.0	99.3	71.5	87.2
08	OPENINGS	87.5	74.6	84.5	88.8	76.8	86.0	89.6	76.7	86.6	89.0	81.1	87.1	91.2	85.7	89.9	92.1	77.1	88.6
0920	Plaster & Gypsum Board	97.7	70.4	79.4	93.7	80.6	84.9	89.3	74.8	79.5	88.6	80.6	83.2	63.1	87.6	79.6	86.3	76.6	79.8
0950, 0980	Ceilings & Acoustic Treatment	77.9	70.4	72.9	82.8	80.6	81.3	76.7	74.8	75.4	76.7	80.6	79.3	77.0	87.6	84.2	81.6	76.6	78.2
0960	Flooring	93.4	64.3	85.2	95.0	62.9	86.0	92.6	61.9	84.0	91.5	76.0	87.2	78.1	64.3	74.2	90.1	64.3	82.8
0970, 0990	Wall Finishes & Painting/Coating	87.2	79.7	82.8	93.9	68.1	78.7	90.3	86.7	88.2	90.3	73.1	80.2	91.6	91.7	91.7	90.3	68.1	77.2
09	FINISHES	86.4	72.0	78.6	90.8	74.9	82.2	87.3	74.7	80.5	86.5	79.4	82.7	78.7	82.6	80.8	86.4	73.3	79.3
COVERS	DIVS. 10 - 14, 25, 28, 41, 43, 44, 46	100.0	92.1	98.2	100.0	90.5	97.9	100.0	93.0	98.4	100.0	91.1	98.0	100.0	48.2	88.4	100.0	89.2	97.6
21, 22, 23	FIRE SUPPRESSION, PLUMBING & HVAC	100.0	76.4	90.8	99.9	80.0	92.1	99.9	77.1	91.0	96.6	78.2	89.4	96.2	82.3	90.7	96.6	74.0	87.7
26, 27, 3370	ELECTRICAL, COMMUNICATIONS & UTIL.	92.5	74.7	83.7	96.4	77.9	87.3	94.1	73.7	84.0	96.3	77.3	86.9	94.8	88.3	91.6	92.5	88.2	90.4
MF2018	WEIGHTED AVERAGE	93.1	78.1	86.7	93.9	79.6	87.8	94.7	81.0	88.9	92.4	83.2	88.6	95.8	85.4	91.4	91.6	79.5	86.5

LOUISIANA

DIVISION		ALEXANDRIA 713 - 714			BATON ROUGE 707 - 708			HAMMOND 704			LAFAYETTE 705			LAKE CHARLES 706			MONROE 712		
		MAT.	INST.	TOTAL	MAT.	INST.	TOTAL	MAT.	INST.	TOTAL	MAT.	INST.	TOTAL	MAT.	INST.	TOTAL	MAT.	INST.	TOTAL
015433	CONTRACTOR EQUIPMENT		90.9	90.9		91.9	91.9		89.3	89.3		89.3	89.3		88.8	88.8		90.9	90.9
0241, 31 - 34	SITE & INFRASTRUCTURE, DEMOLITION	98.7	87.7	91.1	102.1	91.2	94.6	101.2	85.7	90.5	102.5	87.9	92.4	103.2	87.1	92.0	98.7	87.6	91.1
0310	Concrete Forming & Accessories	77.8	60.0	62.6	99.0	71.8	75.8	79.0	54.9	58.5	95.6	67.8	71.9	96.4	67.8	72.0	77.4	59.4	62.1
0320	Concrete Reinforcing	92.2	54.3	73.9	93.3	54.3	74.5	92.1	54.3	73.8	93.4	54.3	74.5	93.4	54.3	74.5	91.2	54.3	73.3
0330	Cast-in-Place Concrete	89.0	60.8	80.5	88.8	73.1	82.9	87.6	62.6	78.3	87.1	66.1	79.3	91.7	66.1	82.2	89.0	64.1	79.7
03	CONCRETE	82.8	62.0	73.6	85.5	69.8	78.6	85.8	58.3	73.7	86.8	65.4	77.4	89.0	65.4	78.6	82.6	61.0	73.1
04	MASONRY	109.3	64.3	81.6	91.7	66.7	76.3	95.3	60.0	73.6	95.3	66.6	77.7	94.7	66.6	77.5	103.9	63.1	78.8
05	METALS	92.6	70.9	86.2	96.9	74.0	90.1	88.4	69.3	82.8	87.7	69.6	82.4	87.7	69.6	82.4	92.6	70.8	86.2
06	WOOD, PLASTICS & COMPOSITES	89.2	58.8	72.5	103.0	73.2	86.6	83.0	54.3	67.2	103.7	68.0	84.1	101.9	68.0	83.3	88.6	58.8	72.2
07	THERMAL & MOISTURE PROTECTION	99.0	65.6	84.5	97.7	69.2	85.3	96.7	62.9	82.0	97.3	67.4	84.3	97.1	67.2	84.1	99.0	64.8	84.1
08	OPENINGS	110.8	58.6	98.6	97.9	70.4	91.5	96.2	54.9	86.6	99.8	62.2	91.0	99.8	62.2	91.0	110.7	57.1	98.2
0920	Plaster & Gypsum Board	83.6	58.1	66.4	101.0	72.7	82.0	100.0	53.4	68.7	108.6	67.6	81.0	108.6	67.6	81.0	83.3	58.1	66.3
0950, 0980	Ceilings & Acoustic Treatment	87.1	58.1	67.6	92.9	72.7	79.3	96.9	53.4	67.6	94.4	67.6	76.3	95.3	67.6	76.6	87.1	58.1	67.6
0960	Flooring	85.1	68.2	80.4	93.0	68.2	86.0	89.9	68.2	83.8	98.2	68.2	89.8	98.2	68.2	89.8	84.7	68.2	80.1
0970, 0990	Wall Finishes & Painting/Coating	88.8	59.7	71.6	94.7	59.7	74.0	97.7	59.9	75.3	97.7	59.7	75.2	97.7	59.7	75.2	88.8	59.7	71.6
09	FINISHES	80.9	61.2	70.2	92.2	70.3	80.3	90.4	57.5	72.6	93.5	67.2	79.3	93.7	67.2	79.4	80.7	60.9	70.0
COVERS	DIVS. 10 - 14, 25, 28, 41, 43, 44, 46	100.0	79.1	95.3	100.0	84.2	96.5	100.0	80.7	95.7	100.0	83.3	96.3	100.0	83.3	96.3	100.0	78.9	95.3
21, 22, 23	FIRE SUPPRESSION, PLUMBING & HVAC	100.0	63.2	85.6	99.9	64.1	85.9	96.7	61.0	82.7	100.2	64.2	86.0	100.2	64.5	86.2	100.0	61.9	85.0
26, 27, 3370	ELECTRICAL, COMMUNICATIONS & UTIL.	94.3	61.4	78.1	101.7	58.0	80.2	97.0	69.6	83.5	98.1	63.8	81.2	97.7	65.9	82.0	95.9	57.3	76.9
MF2018	WEIGHTED AVERAGE	95.9	65.6	83.1	96.6	69.3	85.1	93.7	64.4	81.3	95.4	67.9	83.8	95.6	68.2	84.1	95.8	64.4	82.5

LOUISIANA / MAINE

DIVISION		NEW ORLEANS 700 - 701			SHREVEPORT 710 - 711			THIBODAUX 703			AUGUSTA 043			BANGOR 044			BATH 045		
		MAT.	INST.	TOTAL	MAT.	INST.	TOTAL	MAT.	INST.	TOTAL	MAT.	INST.	TOTAL	MAT.	INST.	TOTAL	MAT.	INST.	TOTAL
015433	CONTRACTOR EQUIPMENT		88.6	88.6		93.9	93.9		89.3	89.3		98.4	98.4		95.7	95.7		95.7	95.7
0241, 31 - 34	SITE & INFRASTRUCTURE, DEMOLITION	104.3	93.9	97.1	101.1	92.3	95.0	103.6	87.6	92.6	88.2	98.0	95.0	90.6	95.5	94.0	88.3	94.2	92.4
0310	Concrete Forming & Accessories	95.7	68.9	72.8	92.9	60.3	65.1	90.2	64.5	68.2	98.6	78.6	81.5	93.0	76.5	79.0	89.1	78.4	80.0
0320	Concrete Reinforcing	92.9	54.4	74.3	92.5	54.3	74.0	92.1	54.3	73.8	102.6	81.1	92.2	94.1	81.1	87.8	93.1	81.1	87.3
0330	Cast-in-Place Concrete	90.7	69.7	82.9	92.0	65.2	82.0	94.1	64.3	83.0	89.8	115.3	99.3	70.9	114.1	87.0	70.9	114.3	87.1
03	CONCRETE	94.4	66.6	82.2	86.2	61.7	75.4	90.6	63.2	78.5	95.5	92.3	94.1	88.5	91.0	89.6	88.6	91.9	90.1
04	MASONRY	99.8	62.6	77.0	94.5	63.4	75.4	119.9	62.8	84.8	101.0	94.1	96.8	115.8	93.7	102.2	122.5	94.1	105.0
05	METALS	99.1	62.6	88.4	96.7	70.2	88.9	88.4	69.5	82.8	105.4	92.3	101.6	93.7	93.0	93.5	92.1	92.9	92.3
06	WOOD, PLASTICS & COMPOSITES	100.3	71.1	84.2	98.0	59.7	76.9	90.5	65.5	76.8	93.5	75.6	83.6	90.7	73.0	80.9	85.1	75.4	79.8
07	THERMAL & MOISTURE PROTECTION	95.0	68.0	83.3	97.4	65.6	83.6	96.9	65.2	83.1	111.0	100.6	106.5	109.0	100.0	105.1	108.9	100.1	105.1
08	OPENINGS	98.8	64.2	90.7	104.7	56.6	93.5	100.7	56.9	90.5	102.7	78.5	97.1	96.4	77.1	91.9	96.4	78.4	92.2
0920	Plaster & Gypsum Board	96.5	70.5	79.0	91.5	58.8	69.5	101.6	65.0	77.0	109.7	74.5	86.0	116.8	71.9	86.6	112.0	74.5	86.8
0950, 0980	Ceilings & Acoustic Treatment	97.5	70.5	79.3	92.8	58.8	69.9	96.9	65.0	75.4	107.1	74.5	85.1	89.6	71.9	77.7	88.6	74.5	79.1
0960	Flooring	103.6	68.2	93.7	91.2	68.2	84.8	95.8	68.2	88.0	89.9	110.6	95.7	82.5	110.6	90.4	80.9	110.6	89.2
0970, 0990	Wall Finishes & Painting/Coating	103.6	60.6	78.2	87.7	59.7	71.1	98.9	59.9	75.8	96.0	88.9	91.8	90.6	88.9	89.6	90.6	88.9	89.6
09	FINISHES	96.6	68.2	81.2	87.7	61.5	73.5	92.6	64.8	77.6	95.8	84.7	89.8	89.7	83.0	86.1	88.3	84.6	86.3
COVERS	DIVS. 10 - 14, 25, 28, 41, 43, 44, 46	100.0	86.5	97.0	100.0	82.6	96.1	100.0	81.8	95.9	100.0	98.0	99.6	100.0	99.6	99.9	100.0	100.9	100.2
21, 22, 23	FIRE SUPPRESSION, PLUMBING & HVAC	100.1	62.5	85.4	99.9	62.8	85.3	96.7	62.5	83.2	100.0	74.3	89.9	100.3	74.1	90.0	96.7	74.3	87.9
26, 27, 3370	ELECTRICAL, COMMUNICATIONS & UTIL.	102.1	70.6	86.6	101.8	65.5	83.9	95.7	69.6	82.8	101.6	76.8	89.3	100.0	69.4	84.9	98.2	76.8	87.6
MF2018	WEIGHTED AVERAGE	99.0	68.5	86.1	97.1	66.3	84.1	96.0	67.0	83.8	100.4	85.6	94.2	97.1	83.9	91.5	95.9	85.4	91.4

DIVISION		MAINE																
		HOULTON			KITTERY			LEWISTON			MACHIAS			PORTLAND			ROCKLAND		
		047			039			042			046			040 - 041			048		
		MAT.	INST.	TOTAL	MAT.	INST.	TOTAL	MAT.	INST.	TOTAL	MAT.	INST.	TOTAL	MAT.	INST.	TOTAL	MAT.	INST.	TOTAL
015433	CONTRACTOR EQUIPMENT		95.7	95.7		95.7	95.7		95.7	95.7		95.7	95.7		99.2	99.2		95.7	95.7
0241, 31 - 34	SITE & INFRASTRUCTURE, DEMOLITION	90.2	94.2	93.0	79.5	94.3	89.7	88.2	95.5	93.3	89.5	94.2	92.8	87.4	100.6	96.5	86.0	94.2	91.7
0310	Concrete Forming & Accessories	96.7	78.4	81.1	88.8	78.7	80.2	98.2	76.6	79.8	93.8	78.4	80.7	99.8	76.8	80.2	94.9	78.4	80.9
0320	Concrete Reinforcing	94.1	81.1	87.8	89.7	81.1	85.5	115.6	81.1	98.9	94.1	81.1	87.8	106.5	81.1	94.2	94.1	81.1	87.8
0330	Cast-in-Place Concrete	70.9	113.2	86.7	70.7	114.4	87.0	72.4	114.1	87.9	70.9	114.3	87.1	85.8	115.2	96.8	72.4	114.3	88.0
03	CONCRETE	89.5	91.5	90.4	82.1	92.1	86.5	88.1	91.0	89.4	89.0	91.9	90.3	91.5	91.4	91.5	86.0	91.9	88.6
04	MASONRY	97.9	94.1	95.6	106.5	94.1	98.9	98.5	93.7	95.5	97.9	94.1	95.6	104.7	93.7	98.0	92.5	94.1	93.5
05	METALS	92.3	92.8	92.4	86.8	93.1	88.6	97.1	93.0	95.9	92.3	92.8	92.5	103.0	92.5	99.9	92.2	92.9	92.4
06	WOOD, PLASTICS & COMPOSITES	94.5	75.4	84.0	88.2	75.4	81.2	96.3	73.0	83.5	91.5	75.4	82.7	95.8	73.1	83.3	92.4	75.4	83.1
07	THERMAL & MOISTURE PROTECTION	109.1	100.1	105.1	107.8	100.1	104.4	108.7	100.0	104.9	109.0	100.1	105.1	113.6	100.6	107.9	108.6	100.1	104.9
08	OPENINGS	96.5	78.4	92.3	96.1	82.0	92.8	99.3	77.1	94.1	96.5	78.4	92.3	96.8	77.2	92.2	96.5	78.4	92.2
0920	Plaster & Gypsum Board	119.0	74.5	89.0	106.1	74.5	84.8	121.8	71.9	88.2	117.4	74.5	88.5	108.0	71.9	83.7	117.4	74.5	88.5
0950, 0980	Ceilings & Acoustic Treatment	88.6	74.5	79.1	101.8	74.5	83.4	99.3	71.9	80.8	88.6	74.5	79.1	107.4	71.9	83.5	88.6	74.5	79.1
0960	Flooring	83.7	110.6	91.2	86.6	110.6	93.4	85.2	110.6	92.3	82.9	110.6	90.7	89.6	110.6	95.5	83.3	110.6	91.0
0970, 0990	Wall Finishes & Painting/Coating	90.6	88.9	89.6	81.8	102.1	93.8	90.6	88.9	89.6	90.6	88.9	89.6	94.6	88.9	91.2	90.6	88.9	89.6
09	FINISHES	90.2	84.6	87.2	91.6	86.0	88.6	92.6	83.0	87.4	89.7	84.6	86.9	96.2	83.2	89.1	89.4	84.6	86.8
COVERS	DIVS. 10 - 14, 25, 28, 41, 43, 44, 46	100.0	97.6	99.5	100.0	100.9	100.2	100.0	99.7	99.9	100.0	97.6	99.5	100.0	100.1	100.0	100.0	100.9	100.2
21, 22, 23	FIRE SUPPRESSION, PLUMBING & HVAC	96.7	74.3	87.9	96.7	80.4	90.3	100.3	74.1	90.0	96.7	74.3	87.9	100.0	74.1	89.9	96.7	74.3	87.9
26, 27, 3370	ELECTRICAL, COMMUNICATIONS & UTIL.	101.9	76.8	89.5	90.3	76.8	83.6	101.9	76.7	89.5	101.9	76.8	89.5	105.0	76.7	91.0	101.8	76.8	89.5
MF2018	WEIGHTED AVERAGE	95.6	85.2	91.2	92.6	87.0	90.2	97.5	84.9	92.2	95.4	85.3	91.1	99.6	85.4	93.6	94.6	85.4	90.7

DIVISION		MAINE			MARYLAND														
		WATERVILLE			ANNAPOLIS			BALTIMORE			COLLEGE PARK			CUMBERLAND			EASTON		
		049			214			210 - 212			207 - 208			215			216		
		MAT.	INST.	TOTAL	MAT.	INST.	TOTAL	MAT.	INST.	TOTAL	MAT.	INST.	TOTAL	MAT.	INST.	TOTAL	MAT.	INST.	TOTAL
015433	CONTRACTOR EQUIPMENT		95.7	95.7		104.5	104.5		104.1	104.1		107.8	107.8		102.4	102.4		102.4	102.4
0241, 31 - 34	SITE & INFRASTRUCTURE, DEMOLITION	90.0	94.2	92.9	100.0	93.5	95.5	100.1	97.7	98.4	100.1	93.6	95.6	92.0	90.3	90.8	98.8	87.7	91.1
0310	Concrete Forming & Accessories	88.6	78.4	79.9	101.0	78.1	81.5	98.4	76.2	79.5	85.1	76.2	77.5	92.3	81.9	83.5	90.4	73.6	76.1
0320	Concrete Reinforcing	94.1	81.1	87.8	108.6	94.9	102.0	101.2	87.3	94.5	103.5	94.0	98.9	93.7	87.4	90.6	92.9	87.1	90.1
0330	Cast-in-Place Concrete	70.9	114.3	87.1	114.1	79.4	101.2	113.8	77.1	100.1	109.4	76.4	97.2	98.0	82.9	92.4	108.8	64.8	92.4
03	CONCRETE	90.2	91.9	90.9	101.4	82.7	93.2	108.2	79.4	95.5	105.2	80.9	94.5	91.4	84.5	88.4	99.7	74.2	88.5
04	MASONRY	109.8	94.1	100.2	94.8	76.6	83.6	104.7	73.1	85.2	111.1	74.6	88.7	99.3	85.4	90.8	113.5	58.6	79.7
05	METALS	92.3	92.8	92.4	105.5	105.2	105.4	103.8	96.6	101.7	90.2	108.3	95.5	100.7	104.2	101.8	101.0	99.7	100.6
06	WOOD, PLASTICS & COMPOSITES	84.6	75.4	79.5	97.1	77.2	86.1	102.5	76.7	88.3	77.3	75.7	76.4	86.7	81.0	83.6	84.5	80.8	82.5
07	THERMAL & MOISTURE PROTECTION	109.1	100.1	105.2	103.5	83.4	94.8	102.7	80.7	93.2	105.3	81.7	95.0	100.9	81.9	92.6	101.0	73.3	89.0
08	OPENINGS	96.5	78.4	92.3	102.3	83.0	97.8	100.5	80.7	95.9	92.7	82.4	90.3	97.6	84.6	94.6	96.0	83.0	92.9
0920	Plaster & Gypsum Board	112.0	74.5	86.8	94.4	76.8	82.5	100.3	76.2	84.0	95.3	75.2	81.8	99.9	80.8	87.0	99.9	80.6	86.9
0950, 0980	Ceilings & Acoustic Treatment	88.6	74.5	79.1	89.1	76.8	80.8	100.5	76.2	84.1	115.1	75.2	88.2	99.0	80.8	86.7	99.0	80.6	86.6
0960	Flooring	80.6	110.6	89.0	90.7	79.5	87.6	92.8	75.7	88.0	87.9	78.4	85.3	85.8	93.1	87.9	85.0	75.7	82.4
0970, 0990	Wall Finishes & Painting/Coating	90.6	88.9	89.6	99.2	72.7	83.5	101.3	75.3	85.9	104.8	72.5	85.7	98.2	84.7	90.2	98.2	72.5	83.0
09	FINISHES	88.4	84.6	86.4	88.5	77.3	82.4	97.0	75.7	85.5	95.6	75.6	84.8	93.3	84.2	88.4	93.5	74.4	83.2
COVERS	DIVS. 10 - 14, 25, 28, 41, 43, 44, 46	100.0	97.5	99.4	100.0	90.3	97.8	100.0	86.8	97.1	100.0	86.7	97.0	100.0	91.3	98.1	100.0	82.9	96.2
21, 22, 23	FIRE SUPPRESSION, PLUMBING & HVAC	96.7	74.3	87.9	100.0	87.0	94.9	100.1	82.0	93.0	96.5	87.4	93.0	96.4	72.0	86.8	96.4	72.8	87.1
26, 27, 3370	ELECTRICAL, COMMUNICATIONS & UTIL.	101.9	76.8	89.5	101.6	88.9	95.3	99.6	86.9	93.3	97.3	101.9	99.5	98.2	80.8	89.6	97.7	62.3	80.3
MF2018	WEIGHTED AVERAGE	96.0	85.2	91.4	100.4	86.3	94.4	101.8	83.2	93.9	97.3	87.7	93.2	96.9	83.8	91.4	98.6	74.8	88.5

DIVISION		MARYLAND														MASSACHUSETTS		
		ELKTON			HAGERSTOWN			SALISBURY			SILVER SPRING			WALDORF			BOSTON		
		219			217			218			209			206			020 - 022, 024		
		MAT.	INST.	TOTAL	MAT.	INST.	TOTAL	MAT.	INST.	TOTAL	MAT.	INST.	TOTAL	MAT.	INST.	TOTAL	MAT.	INST.	TOTAL
015433	CONTRACTOR EQUIPMENT		102.4	102.4		102.4	102.4		102.4	102.4		99.7	99.7		99.7	99.7		103.0	103.0
0241, 31 - 34	SITE & INFRASTRUCTURE, DEMOLITION	86.3	89.0	88.1	90.6	90.4	90.4	98.8	87.6	91.1	88.9	85.7	86.7	95.1	85.7	88.6	92.7	102.1	99.2
0310	Concrete Forming & Accessories	96.2	92.3	92.9	91.5	78.8	80.7	104.5	50.2	58.2	93.0	75.5	78.1	99.7	75.4	79.0	105.1	139.7	134.6
0320	Concrete Reinforcing	92.9	117.3	104.7	93.7	87.4	90.6	92.9	65.2	79.5	102.3	93.9	98.3	103.0	93.9	98.6	117.7	156.9	136.7
0330	Cast-in-Place Concrete	88.1	72.2	82.2	93.3	82.9	89.5	108.8	62.9	91.7	112.1	76.8	98.9	125.6	76.7	107.4	99.2	141.8	115.1
03	CONCRETE	84.0	90.6	86.9	87.7	83.1	85.7	100.5	59.2	82.4	102.8	80.5	93.0	113.3	80.5	98.9	105.4	142.7	121.8
04	MASONRY	98.6	67.0	79.2	105.1	85.4	93.0	112.9	55.2	77.4	110.6	75.0	88.7	94.7	75.0	82.6	112.0	146.3	133.1
05	METALS	101.0	114.0	104.8	100.9	104.3	101.9	101.0	90.9	98.1	94.8	104.3	97.6	94.8	104.3	97.5	102.8	136.1	112.6
06	WOOD, PLASTICS & COMPOSITES	91.9	101.3	97.0	85.8	76.7	80.8	102.6	51.4	74.4	84.0	75.0	79.0	90.8	75.0	82.1	103.8	140.2	123.8
07	THERMAL & MOISTURE PROTECTION	100.5	79.4	91.3	101.2	83.0	93.3	101.3	68.5	87.0	108.4	86.5	98.8	109.0	86.5	99.2	107.2	137.1	120.2
08	OPENINGS	96.0	102.1	97.4	95.9	81.2	92.5	96.2	60.8	87.9	84.9	82.1	84.2	85.5	81.5	84.5	99.9	147.3	111.0
0920	Plaster & Gypsum Board	102.8	101.7	102.0	99.9	76.4	84.1	109.4	50.4	69.7	101.9	75.2	83.9	104.8	75.2	84.8	109.0	141.0	130.6
0950, 0980	Ceilings & Acoustic Treatment	99.0	101.7	100.8	101.7	76.4	84.6	99.0	50.4	66.2	123.2	75.2	90.8	123.2	75.2	90.8	100.5	141.0	127.8
0960	Flooring	87.2	75.7	84.0	85.4	93.1	87.6	90.7	75.7	86.5	93.8	78.4	89.5	97.2	78.4	92.0	90.5	163.5	111.0
0970, 0990	Wall Finishes & Painting/Coating	98.2	72.5	83.0	98.2	72.5	83.0	98.2	72.5	83.0	112.7	72.5	88.9	112.7	72.5	88.9	94.3	160.4	133.4
09	FINISHES	93.7	88.5	90.9	93.5	80.4	86.4	96.4	56.2	74.6	95.8	75.0	84.5	97.5	75.1	85.4	98.7	147.0	124.8
COVERS	DIVS. 10 - 14, 25, 28, 41, 43, 44, 46	100.0	56.7	90.4	100.0	90.8	98.0	100.0	76.5	94.8	100.0	82.5	96.2	100.0	83.1	96.2	100.0	119.1	104.3
21, 22, 23	FIRE SUPPRESSION, PLUMBING & HVAC	96.4	78.7	89.5	99.9	83.1	93.3	96.4	70.7	86.3	96.5	87.9	93.1	96.5	87.8	93.1	100.1	127.4	110.8
26, 27, 3370	ELECTRICAL, COMMUNICATIONS & UTIL.	99.4	86.9	93.2	98.0	80.8	89.5	96.6	60.1	78.6	94.6	101.9	98.2	92.2	101.9	97.0	102.6	129.5	115.9
MF2018	WEIGHTED AVERAGE	95.9	86.2	91.8	97.4	85.3	92.3	99.0	66.9	85.4	96.5	86.7	92.4	97.3	86.6	92.8	101.9	134.0	115.5

For customer support on your Commercial Renovation Costs with RSMeans data, call 800.448.8182.

803

City Cost Indexes

MASSACHUSETTS

DIVISION		BROCKTON 023			BUZZARDS BAY 025			FALL RIVER 027			FITCHBURG 014			FRAMINGHAM 017			GREENFIELD 013		
		MAT.	INST.	TOTAL	MAT.	INST.	TOTAL	MAT.	INST.	TOTAL	MAT.	INST.	TOTAL	MAT.	INST.	TOTAL	MAT.	INST.	TOTAL
015433	CONTRACTOR EQUIPMENT		98.1	98.1		98.1	98.1		98.9	98.9		95.7	95.7		97.4	97.4		95.7	95.7
0241, 31 - 34	SITE & INFRASTRUCTURE, DEMOLITION	91.1	98.0	95.8	81.2	98.1	92.9	90.3	98.0	95.6	82.7	97.8	93.1	79.5	97.8	92.1	86.4	97.3	93.9
0310	Concrete Forming & Accessories	100.7	121.1	118.1	98.5	120.6	117.4	100.7	120.8	117.8	94.0	118.5	114.9	101.4	121.3	118.4	92.4	117.7	114.0
0320	Concrete Reinforcing	109.7	142.7	125.7	88.0	122.3	104.6	109.7	122.3	115.8	87.9	142.5	114.3	87.9	142.8	114.4	91.5	126.5	108.4
0330	Cast-in-Place Concrete	88.1	136.6	106.2	73.2	136.4	96.7	85.2	136.9	104.4	78.4	136.5	100.0	78.4	136.7	100.1	80.5	122.2	96.0
03	CONCRETE	95.1	129.8	110.3	80.7	126.0	100.6	93.8	126.2	108.0	79.5	128.3	101.0	82.3	130.0	103.2	83.0	120.3	99.4
04	MASONRY	105.4	135.1	123.6	97.9	135.1	120.8	105.9	135.1	123.8	98.2	131.9	118.9	104.7	136.1	124.0	102.7	119.1	112.8
05	METALS	98.8	125.6	106.7	93.6	117.0	100.5	98.8	117.3	104.3	95.3	122.1	103.2	95.3	125.8	104.3	97.8	113.8	102.5
06	WOOD, PLASTICS & COMPOSITES	98.1	119.7	109.9	94.7	119.7	108.4	98.1	119.9	110.1	94.5	116.5	106.6	101.0	119.4	111.2	92.5	119.7	107.4
07	THERMAL & MOISTURE PROTECTION	103.2	128.4	114.2	102.0	125.7	112.3	103.0	124.9	112.5	103.2	122.1	111.4	103.3	128.3	114.2	103.3	109.9	106.2
08	OPENINGS	97.4	128.4	104.6	93.4	118.9	99.3	97.4	118.3	102.2	99.2	126.7	105.6	99.8	128.3	105.8	94.9	120.3	104.3
0920	Plaster & Gypsum Board	92.3	119.9	110.9	87.3	119.9	109.3	92.3	119.9	110.9	111.3	116.7	114.9	114.4	119.9	118.1	112.6	119.9	117.5
0950, 0980	Ceilings & Acoustic Treatment	104.3	119.9	114.9	86.2	119.9	108.9	104.3	119.9	114.9	93.9	116.7	109.3	93.9	119.9	111.5	103.7	119.9	114.6
0960	Flooring	85.0	162.0	106.6	82.6	162.0	104.9	83.8	162.0	105.8	86.7	162.0	107.8	88.2	162.0	108.9	85.9	135.0	99.7
0970, 0990	Wall Finishes & Painting/Coating	85.0	139.2	117.1	85.0	139.2	117.1	85.0	139.2	117.1	86.2	139.2	117.6	87.1	139.2	117.9	86.2	109.3	99.9
09	FINISHES	90.8	130.8	112.5	85.2	130.8	109.9	90.5	131.0	112.4	89.5	128.9	110.9	90.2	130.6	112.1	91.7	120.9	107.5
COVERS	DIVS. 10 - 14, 25, 28, 41, 43, 44, 46	100.0	109.7	102.2	100.0	109.7	102.2	100.0	110.2	102.3	100.0	104.2	100.9	100.0	109.3	102.1	100.0	102.9	100.7
21, 22, 23	FIRE SUPPRESSION, PLUMBING & HVAC	100.4	103.9	101.7	96.8	103.5	99.4	100.4	103.5	101.6	97.0	104.5	99.9	97.0	121.1	106.4	97.0	98.5	97.6
26, 27, 3370	ELECTRICAL, COMMUNICATIONS & UTIL.	101.3	97.7	99.5	98.5	100.1	99.3	101.2	97.7	99.4	100.6	104.1	102.3	97.4	124.4	110.7	100.5	96.1	98.4
MF2018	WEIGHTED AVERAGE	98.5	116.7	106.2	93.2	115.2	102.5	98.3	114.9	105.9	94.6	116.2	103.7	94.0	124.2	106.8	95.9	109.0	101.4

MASSACHUSETTS

DIVISION		HYANNIS 026			LAWRENCE 019			LOWELL 018			NEW BEDFORD 027			PITTSFIELD 012			SPRINGFIELD 010 - 011		
		MAT.	INST.	TOTAL	MAT.	INST.	TOTAL	MAT.	INST.	TOTAL	MAT.	INST.	TOTAL	MAT.	INST.	TOTAL	MAT.	INST.	TOTAL
015433	CONTRACTOR EQUIPMENT		98.1	98.1		98.1	98.1		95.7	95.7		98.9	98.9		95.7	95.7		95.7	95.7
0241, 31 - 34	SITE & INFRASTRUCTURE, DEMOLITION	87.4	98.1	94.8	91.6	98.0	96.0	90.6	98.1	95.8	88.7	98.0	95.1	91.6	96.7	95.2	91.1	97.1	95.2
0310	Concrete Forming & Accessories	92.7	120.6	116.5	102.5	121.6	118.8	99.0	122.6	119.1	100.7	120.8	117.9	99.0	102.9	102.3	99.3	117.7	115.0
0320	Concrete Reinforcing	88.0	122.3	104.6	108.8	148.3	127.9	109.7	148.1	128.2	109.7	122.4	115.8	90.8	114.1	102.1	109.7	126.5	117.8
0330	Cast-in-Place Concrete	80.4	136.4	101.3	90.7	136.8	107.9	82.4	138.5	103.3	75.1	136.9	98.1	90.0	113.4	98.7	85.8	122.2	99.4
03	CONCRETE	86.7	126.0	104.0	96.2	131.1	111.5	88.0	131.9	107.3	89.1	126.3	105.4	89.0	108.3	97.5	89.5	122.0	103.0
04	MASONRY	104.3	135.1	123.2	110.1	135.9	125.9	97.3	134.9	120.4	104.2	135.7	123.6	98.0	111.5	106.3	97.6	119.1	110.8
05	METALS	95.1	117.0	101.5	98.2	128.4	107.1	98.2	125.0	106.1	98.8	117.4	104.3	98.0	108.2	101.0	100.9	113.7	104.6
06	WOOD, PLASTICS & COMPOSITES	88.0	119.7	105.4	101.5	119.7	111.5	100.8	119.7	111.2	98.1	119.9	110.1	100.8	102.5	101.8	100.8	119.7	111.2
07	THERMAL & MOISTURE PROTECTION	102.5	125.7	112.6	104.0	128.6	114.7	103.7	128.0	114.3	102.9	125.1	112.6	103.8	104.3	104.0	103.7	109.9	106.4
08	OPENINGS	93.9	118.9	99.8	93.4	129.9	101.9	100.4	129.9	107.3	97.4	123.0	103.3	100.4	107.6	102.1	100.4	120.3	105.1
0920	Plaster & Gypsum Board	83.2	119.9	107.9	117.3	119.9	119.1	117.3	119.9	119.1	92.3	119.9	110.9	117.3	102.3	107.2	117.3	119.9	119.1
0950, 0980	Ceilings & Acoustic Treatment	96.2	119.9	112.2	105.5	119.9	115.2	105.5	119.9	115.2	104.3	119.9	114.9	105.5	102.3	103.4	105.5	119.9	115.2
0960	Flooring	80.0	162.0	103.0	88.8	162.0	109.4	88.8	162.0	109.4	83.8	162.0	105.8	89.1	129.8	100.6	88.2	135.0	101.3
0970, 0990	Wall Finishes & Painting/Coating	85.0	139.2	117.1	86.3	139.2	117.6	86.2	139.2	117.6	85.0	139.2	117.1	86.2	109.3	99.9	87.0	109.3	100.2
09	FINISHES	86.3	130.8	110.4	93.8	130.8	113.8	93.7	131.6	114.2	90.4	131.0	112.3	93.8	108.6	101.8	93.6	120.9	108.4
COVERS	DIVS. 10 - 14, 25, 28, 41, 43, 44, 46	100.0	109.7	102.2	100.0	109.8	102.2	100.0	110.8	102.4	100.0	110.2	102.3	100.0	99.5	99.9	100.0	102.9	100.7
21, 22, 23	FIRE SUPPRESSION, PLUMBING & HVAC	100.4	104.2	101.9	100.1	119.7	107.8	100.1	121.2	108.4	100.4	103.4	101.5	100.1	94.2	97.8	100.1	98.5	99.5
26, 27, 3370	ELECTRICAL, COMMUNICATIONS & UTIL.	98.9	97.7	98.3	99.7	122.9	111.1	100.1	119.3	109.6	102.1	97.7	99.9	100.1	96.1	98.2	100.1	93.0	96.6
MF2018	WEIGHTED AVERAGE	95.7	115.0	103.8	98.4	124.2	109.3	97.5	123.8	108.6	97.7	115.2	105.1	97.7	102.6	99.7	98.2	108.5	102.5

MASSACHUSETTS / MICHIGAN

DIVISION		WORCESTER 015 - 016			ANN ARBOR 481			BATTLE CREEK 490			BAY CITY 487			DEARBORN 481			DETROIT 482		
		MAT.	INST.	TOTAL	MAT.	INST.	TOTAL	MAT.	INST.	TOTAL	MAT.	INST.	TOTAL	MAT.	INST.	TOTAL	MAT.	INST.	TOTAL
015433	CONTRACTOR EQUIPMENT		95.7	95.7		109.1	109.1		96.1	96.1		109.1	109.1		109.1	109.1		97.8	97.8
0241, 31 - 34	SITE & INFRASTRUCTURE, DEMOLITION	91.0	97.9	95.7	81.0	92.2	88.7	93.6	81.7	85.3	72.5	91.7	85.8	80.8	92.3	88.7	98.2	101.2	100.2
0310	Concrete Forming & Accessories	99.6	118.5	115.7	96.3	104.1	103.0	96.0	77.3	80.0	96.4	81.5	83.7	96.2	104.7	103.5	100.4	105.3	104.6
0320	Concrete Reinforcing	109.7	155.5	131.9	101.2	103.5	102.3	98.5	80.9	90.0	101.2	102.7	101.9	101.2	103.6	102.3	101.4	104.8	103.0
0330	Cast-in-Place Concrete	85.3	136.5	104.4	88.3	96.4	91.3	86.1	91.0	87.9	84.5	83.5	84.1	86.4	97.2	90.4	104.5	101.3	103.3
03	CONCRETE	89.3	130.6	107.5	89.8	102.2	95.2	85.7	82.6	84.3	88.0	87.3	87.7	88.9	102.8	95.0	103.1	103.2	103.2
04	MASONRY	97.2	131.9	118.5	99.6	96.4	97.6	99.1	78.4	86.4	99.2	77.7	86.0	99.5	98.2	98.7	102.6	99.8	100.9
05	METALS	100.9	127.4	108.7	102.7	115.7	106.5	103.7	83.7	97.9	103.3	114.0	106.4	102.8	115.8	106.6	104.3	94.3	101.3
06	WOOD, PLASTICS & COMPOSITES	101.2	116.5	109.5	89.2	106.6	98.8	88.4	75.6	81.3	89.2	81.9	85.2	89.2	106.6	98.8	97.1	106.7	102.4
07	THERMAL & MOISTURE PROTECTION	103.8	122.1	111.8	105.3	98.2	102.2	96.1	79.0	88.7	102.7	81.0	93.2	103.6	100.7	102.3	102.6	102.9	102.7
08	OPENINGS	100.4	130.2	107.4	94.2	100.8	95.7	86.5	74.1	83.6	94.2	83.5	91.7	94.2	101.1	95.8	97.2	101.9	98.3
0920	Plaster & Gypsum Board	117.3	116.7	116.9	107.4	106.5	106.8	91.7	71.7	78.2	107.4	81.1	89.7	107.4	106.5	106.8	104.1	106.5	105.7
0950, 0980	Ceilings & Acoustic Treatment	105.5	116.7	113.0	84.8	106.5	99.4	78.7	71.7	74.0	85.7	81.1	82.6	84.8	106.5	99.4	96.9	106.5	103.4
0960	Flooring	88.8	154.9	107.4	91.2	107.2	95.7	91.5	70.4	85.6	91.2	77.0	87.2	90.6	101.6	93.6	95.7	106.7	98.8
0970, 0990	Wall Finishes & Painting/Coating	86.2	139.2	117.6	82.7	95.5	90.3	82.3	76.3	78.7	82.7	79.1	80.6	82.7	93.8	89.3	93.1	100.1	97.3
09	FINISHES	93.7	127.5	112.0	90.8	104.3	98.1	84.7	75.2	79.6	90.6	79.9	84.8	90.6	103.5	97.6	97.7	105.4	101.9
COVERS	DIVS. 10 - 14, 25, 28, 41, 43, 44, 46	100.0	104.2	100.9	100.0	93.7	98.6	100.0	93.7	98.6	100.0	88.0	97.3	100.0	94.1	98.7	100.0	102.4	100.5
21, 22, 23	FIRE SUPPRESSION, PLUMBING & HVAC	100.1	104.1	101.7	100.1	91.3	96.6	100.1	82.6	93.2	100.1	78.6	91.6	100.0	100.0	100.0	99.9	102.7	101.0
26, 27, 3370	ELECTRICAL, COMMUNICATIONS & UTIL.	100.1	104.1	102.1	97.3	103.0	100.1	96.6	77.7	87.3	96.4	82.5	89.5	97.3	96.1	96.7	99.9	103.5	101.6
MF2018	WEIGHTED AVERAGE	98.1	117.0	106.1	97.1	99.8	98.2	95.5	80.4	89.1	96.6	85.3	91.8	96.9	100.9	98.6	100.7	102.0	101.3

City Cost Indexes

MICHIGAN

DIVISION		FLINT 484-485 MAT.	INST.	TOTAL	GAYLORD 497 MAT.	INST.	TOTAL	GRAND RAPIDS 493,495 MAT.	INST.	TOTAL	IRON MOUNTAIN 498-499 MAT.	INST.	TOTAL	JACKSON 492 MAT.	INST.	TOTAL	KALAMAZOO 491 MAT.	INST.	TOTAL
015433	CONTRACTOR EQUIPMENT		109.1	109.1		103.6	103.6		98.3	98.3		91.1	91.1		103.6	103.6		96.1	96.1
0241, 31 - 34	SITE & INFRASTRUCTURE, DEMOLITION	70.2	91.8	85.1	87.4	79.6	82.0	92.9	85.9	88.1	96.4	88.5	90.9	111.0	81.2	90.4	93.9	81.6	85.4
0310	Concrete Forming & Accessories	99.1	81.8	84.4	95.0	73.9	77.0	95.6	75.7	78.6	86.6	78.6	79.8	91.6	82.3	83.7	96.0	77.1	79.9
0320	Concrete Reinforcing	101.2	103.1	102.1	91.9	93.0	92.4	103.4	80.7	92.4	91.6	85.0	88.4	89.3	103.1	96.0	98.5	76.2	87.8
0330	Cast-in-Place Concrete	88.9	85.8	87.8	85.8	78.4	83.1	90.4	89.0	89.9	101.7	67.6	89.0	85.7	89.8	87.2	87.8	91.0	89.0
03	CONCRETE	90.2	88.4	89.4	82.6	80.5	81.6	89.7	81.1	85.9	90.9	76.6	84.6	77.8	89.8	83.1	88.9	81.7	85.7
04	MASONRY	99.7	85.9	91.2	109.7	72.7	87.0	92.0	74.1	81.0	95.4	79.7	85.7	89.2	86.9	87.8	97.6	78.4	85.8
05	METALS	102.8	114.6	106.3	105.5	110.8	107.0	100.8	83.2	95.6	104.8	91.1	100.8	105.7	112.7	107.8	103.7	81.9	97.3
06	WOOD, PLASTICS & COMPOSITES	92.6	80.7	86.1	81.9	73.4	77.2	92.2	73.9	82.2	78.2	78.9	78.6	80.8	79.8	80.2	88.4	75.6	81.3
07	THERMAL & MOISTURE PROTECTION	103.0	84.5	95.0	94.7	74.7	86.0	98.0	71.8	86.6	98.4	74.8	88.1	94.0	88.3	91.5	96.1	79.0	88.7
08	OPENINGS	94.2	82.6	91.5	86.0	78.7	84.3	101.8	73.8	95.2	92.5	69.8	87.2	85.2	85.5	85.3	86.5	73.5	83.4
0920	Plaster & Gypsum Board	109.0	79.8	89.4	91.3	71.8	78.2	99.8	70.2	79.8	51.3	79.0	69.9	90.1	78.3	82.1	91.7	71.7	78.2
0950, 0980	Ceilings & Acoustic Treatment	84.8	79.8	81.5	77.1	71.8	73.5	91.1	70.2	77.0	76.0	79.0	78.0	77.1	78.3	77.9	78.7	71.7	74.0
0960	Flooring	91.2	86.0	89.7	84.9	85.6	85.1	95.7	76.1	90.2	102.0	91.0	98.9	83.4	78.7	82.1	91.5	70.4	85.6
0970, 0990	Wall Finishes & Painting/Coating	82.7	78.3	80.1	78.7	81.6	80.4	89.3	76.9	82.0	94.8	69.6	79.9	78.7	93.8	87.7	82.3	76.3	78.7
09	FINISHES	90.1	81.6	85.5	84.1	76.2	79.8	91.4	75.2	82.6	84.6	80.0	82.1	85.2	81.9	83.4	84.7	75.2	79.6
COVERS	DIVS. 10 - 14, 25, 28, 41, 43, 44, 46	100.0	88.8	97.5	100.0	77.9	95.1	100.0	93.0	98.4	100.0	85.9	96.8	100.0	91.6	98.1	100.0	93.7	98.6
21, 22, 23	FIRE SUPPRESSION, PLUMBING & HVAC	100.0	83.3	93.4	96.8	79.0	89.8	100.0	79.6	92.0	96.7	83.9	91.7	96.8	85.7	92.4	100.1	78.2	91.5
26, 27, 3370	ELECTRICAL, COMMUNICATIONS & UTIL.	97.3	89.1	93.2	94.5	76.3	85.5	102.2	82.9	92.7	100.8	80.7	90.9	98.3	103.0	100.6	96.4	74.1	85.4
MF2018	WEIGHTED AVERAGE	96.8	88.6	93.3	94.5	80.6	88.6	97.9	79.9	90.3	96.5	81.7	90.2	94.0	90.7	92.6	95.8	78.6	88.6

MICHIGAN / MINNESOTA

DIVISION		LANSING 488-489 MAT.	INST.	TOTAL	MUSKEGON 494 MAT.	INST.	TOTAL	ROYAL OAK 480,483 MAT.	INST.	TOTAL	SAGINAW 486 MAT.	INST.	TOTAL	TRAVERSE CITY 496 MAT.	INST.	TOTAL	BEMIDJI 566 MAT.	INST.	TOTAL
015433	CONTRACTOR EQUIPMENT		111.0	111.0		96.1	96.1		89.8	89.8		109.1	109.1		91.1	91.1		97.7	97.7
0241, 31 - 34	SITE & INFRASTRUCTURE, DEMOLITION	90.7	96.5	94.7	91.5	81.7	84.8	84.8	92.6	90.2	73.5	91.7	86.1	81.8	87.5	85.7	95.1	95.1	95.1
0310	Concrete Forming & Accessories	93.8	74.9	77.7	96.3	79.4	81.9	92.3	105.2	103.2	96.3	79.7	82.1	86.6	72.4	74.5	88.0	82.8	83.6
0320	Concrete Reinforcing	104.8	102.9	103.8	99.3	80.7	90.3	91.7	104.8	98.0	101.2	102.7	101.9	93.1	80.2	86.9	97.4	101.8	99.5
0330	Cast-in-Place Concrete	97.9	85.1	93.1	85.8	90.5	87.5	77.3	97.5	84.8	87.3	83.4	85.8	79.3	76.7	78.3	101.8	104.5	102.8
03	CONCRETE	91.4	84.9	88.5	84.2	83.3	83.8	77.1	101.6	87.9	89.3	86.5	88.1	74.8	76.1	75.4	91.0	95.0	92.8
04	MASONRY	94.5	84.6	88.4	96.3	76.2	83.9	93.2	99.5	97.1	101.2	77.7	86.7	93.5	74.8	82.0	98.9	102.4	101.0
05	METALS	101.6	112.7	104.9	101.5	83.1	96.1	106.3	92.2	102.2	102.8	113.8	106.0	104.8	88.9	100.1	91.2	117.9	99.0
06	WOOD, PLASTICS & COMPOSITES	87.0	71.2	78.3	85.4	78.0	81.3	85.0	106.6	96.9	85.9	79.7	82.5	78.2	71.9	74.8	68.8	77.0	73.3
07	THERMAL & MOISTURE PROTECTION	102.0	83.8	94.1	95.0	73.5	85.7	101.2	100.9	101.1	103.9	80.9	93.9	97.3	74.2	87.3	106.3	90.7	99.5
08	OPENINGS	102.7	77.2	96.8	85.8	76.5	83.6	94.0	101.3	95.7	92.4	82.2	90.0	92.5	65.4	86.2	99.7	101.3	100.1
0920	Plaster & Gypsum Board	100.1	70.2	79.9	71.8	74.2	73.4	104.8	106.5	106.0	107.4	78.8	88.2	51.3	71.8	65.1	104.3	76.8	85.8
0950, 0980	Ceilings & Acoustic Treatment	83.9	70.2	74.6	78.7	74.2	75.6	84.1	106.5	99.2	84.8	78.8	80.8	76.0	71.8	73.2	128.8	76.8	93.7
0960	Flooring	97.2	78.7	92.0	90.2	72.3	85.1	88.2	106.7	93.4	91.2	77.0	87.2	102.0	85.6	97.4	90.7	91.4	90.9
0970, 0990	Wall Finishes & Painting/Coating	91.1	76.9	82.7	80.5	79.7	80.0	84.2	93.8	89.9	82.7	79.1	80.6	94.8	39.4	62.1	89.6	105.8	99.2
09	FINISHES	90.2	74.6	81.7	81.1	77.1	78.9	89.7	104.3	97.6	90.5	78.6	84.0	83.5	71.0	76.7	98.2	86.0	91.6
COVERS	DIVS. 10 - 14, 25, 28, 41, 43, 44, 46	100.0	88.8	97.5	100.0	94.5	98.8	100.0	100.0	100.0	100.0	87.7	97.3	100.0	84.4	96.5	100.0	95.6	99.0
21, 22, 23	FIRE SUPPRESSION, PLUMBING & HVAC	99.9	83.7	93.5	99.9	83.4	93.4	96.7	101.4	98.5	100.0	78.3	91.5	96.7	78.9	89.7	96.7	84.8	92.0
26, 27, 3370	ELECTRICAL, COMMUNICATIONS & UTIL.	99.1	87.6	93.4	96.8	73.0	85.1	99.4	103.1	101.3	95.4	84.5	90.0	96.2	76.2	86.4	105.0	100.7	102.8
MF2018	WEIGHTED AVERAGE	98.0	86.8	93.3	94.3	80.0	88.3	95.1	100.3	97.3	96.5	85.1	91.7	93.4	77.7	86.8	96.7	95.4	96.1

MINNESOTA

DIVISION		BRAINERD 564 MAT.	INST.	TOTAL	DETROIT LAKES 565 MAT.	INST.	TOTAL	DULUTH 556-558 MAT.	INST.	TOTAL	MANKATO 560 MAT.	INST.	TOTAL	MINNEAPOLIS 553-555 MAT.	INST.	TOTAL	ROCHESTER 559 MAT.	INST.	TOTAL
015433	CONTRACTOR EQUIPMENT		100.2	100.2		97.7	97.7		102.9	102.9		100.2	100.2		108.4	108.4		100.7	100.7
0241, 31 - 34	SITE & INFRASTRUCTURE, DEMOLITION	96.0	99.9	98.7	93.3	95.4	94.7	98.2	101.7	100.6	92.6	99.4	97.3	95.0	108.8	104.5	96.0	97.6	97.1
0310	Concrete Forming & Accessories	88.7	83.6	84.4	84.9	82.8	83.1	98.4	94.6	95.2	96.9	96.6	96.6	99.3	114.0	111.8	99.2	94.8	95.4
0320	Concrete Reinforcing	96.2	102.2	99.1	97.4	101.8	99.6	102.6	102.0	102.3	96.0	109.2	102.4	96.1	109.4	102.5	98.9	108.8	103.7
0330	Cast-in-Place Concrete	110.7	109.1	110.1	98.9	107.5	102.1	94.3	97.7	95.6	101.9	99.5	101.0	95.6	115.3	102.9	94.8	95.3	95.0
03	CONCRETE	94.8	97.1	95.8	88.6	96.1	91.9	95.3	98.1	96.5	90.5	100.8	95.0	97.6	114.3	104.9	91.4	98.6	94.6
04	MASONRY	122.5	116.3	118.7	122.6	109.8	114.8	95.3	105.7	101.7	111.3	110.8	111.0	114.5	119.5	117.6	102.9	103.5	103.3
05	METALS	92.2	118.3	99.9	91.1	117.5	98.9	100.2	119.1	105.8	92.1	121.6	100.7	98.7	124.5	106.3	99.3	123.1	106.3
06	WOOD, PLASTICS & COMPOSITES	84.9	74.2	79.1	65.8	74.4	70.6	93.9	91.9	92.8	94.5	95.4	95.0	96.1	110.8	104.2	95.2	92.9	93.9
07	THERMAL & MOISTURE PROTECTION	104.7	102.4	103.7	106.1	90.1	103.2	97.9	102.2	99.8	105.1	93.4	100.1	101.2	115.9	107.6	105.3	90.0	98.6
08	OPENINGS	86.4	99.8	89.5	99.6	99.9	99.7	103.3	104.6	103.6	90.9	112.2	95.9	99.3	121.5	104.5	98.6	112.5	101.8
0920	Plaster & Gypsum Board	91.5	74.1	79.8	103.7	74.1	83.8	94.5	92.1	92.9	95.6	95.9	95.8	101.1	111.2	107.9	103.2	93.2	96.5
0950, 0980	Ceilings & Acoustic Treatment	58.8	74.1	69.1	128.8	74.1	91.9	89.2	92.1	91.1	58.8	95.9	83.8	103.1	111.2	108.6	93.9	93.2	93.4
0960	Flooring	89.6	85.6	88.4	89.5	85.6	88.4	93.2	123.9	101.8	91.4	81.8	88.7	100.6	114.6	104.5	92.3	81.8	89.3
0970, 0990	Wall Finishes & Painting/Coating	84.0	105.8	96.9	89.6	83.4	85.9	86.0	104.4	96.9	95.0	103.7	100.2	101.3	123.8	114.6	85.3	98.2	92.9
09	FINISHES	82.1	85.2	83.8	97.6	82.5	89.4	89.4	100.2	95.3	83.7	94.9	89.7	98.9	115.0	107.6	90.0	93.1	91.7
COVERS	DIVS. 10 - 14, 25, 28, 41, 43, 44, 46	100.0	97.3	99.4	100.0	91.1	99.4	100.0	94.6	98.8	100.0	97.5	99.4	100.0	105.8	101.3	100.0	101.8	100.4
21, 22, 23	FIRE SUPPRESSION, PLUMBING & HVAC	96.0	88.6	93.1	96.7	87.5	93.1	99.8	94.5	97.7	96.0	84.6	91.5	100.0	110.6	104.2	100.0	91.6	96.7
26, 27, 3370	ELECTRICAL, COMMUNICATIONS & UTIL.	102.7	101.7	102.2	104.8	71.1	88.2	100.3	101.7	101.0	109.1	93.1	101.2	107.2	110.3	108.7	103.0	93.1	98.1
MF2018	WEIGHTED AVERAGE	95.6	98.6	96.8	97.4	92.4	95.3	98.6	101.3	99.7	95.7	98.5	96.9	100.7	114.1	106.3	98.3	98.7	98.5

City Cost Indexes

MINNESOTA / MISSISSIPPI

DIVISION		SAINT PAUL 550-551 MAT.	INST.	TOTAL	ST. CLOUD 563 MAT.	INST.	TOTAL	THIEF RIVER FALLS 567 MAT.	INST.	TOTAL	WILLMAR 562 MAT.	INST.	TOTAL	WINDOM 561 MAT.	INST.	TOTAL	BILOXI 395 MAT.	INST.	TOTAL
015433	CONTRACTOR EQUIPMENT		102.9	102.9		100.2	100.2		97.7	97.7		100.2	100.2		100.2	100.2		100.1	100.1
0241, 31 - 34	SITE & INFRASTRUCTURE, DEMOLITION	96.5	102.5	100.7	91.4	100.8	97.9	94.0	94.9	94.7	90.6	99.7	96.9	84.5	98.9	94.4	105.2	85.9	91.9
0310	Concrete Forming & Accessories	98.6	117.0	114.3	86.0	113.5	109.5	88.8	81.9	82.9	85.8	86.8	86.7	90.1	82.2	83.4	92.8	64.1	68.4
0320	Concrete Reinforcing	105.4	109.9	107.6	96.2	109.4	102.6	97.8	101.6	99.6	95.8	109.2	102.3	95.8	108.1	101.8	93.4	48.8	71.8
0330	Cast-in-Place Concrete	99.9	116.3	106.0	97.7	115.0	104.1	100.9	81.3	93.6	99.2	79.5	91.9	85.8	83.3	84.8	114.2	66.3	96.4
03	CONCRETE	95.3	116.1	104.4	86.5	114.0	98.5	89.8	86.5	88.3	86.3	89.5	87.7	77.6	88.5	82.4	94.4	64.0	81.0
04	MASONRY	105.3	127.4	118.9	107.3	121.8	116.2	98.9	102.3	101.0	112.1	114.5	113.6	122.0	87.9	101.0	90.4	62.8	73.4
05	METALS	100.1	125.0	107.4	92.9	123.1	101.8	91.3	116.7	98.7	92.0	121.5	100.7	91.9	119.5	100.0	90.3	84.5	88.6
06	WOOD, PLASTICS & COMPOSITES	95.0	114.2	105.6	82.1	110.3	97.6	70.0	77.0	73.8	81.8	79.9	80.8	86.3	79.9	82.8	93.1	64.5	77.3
07	THERMAL & MOISTURE PROTECTION	102.0	120.5	110.1	104.9	111.7	107.9	107.1	88.1	98.8	104.7	101.0	103.1	104.6	83.0	95.2	100.0	63.1	83.9
08	OPENINGS	98.2	124.3	104.3	91.4	122.1	98.5	99.7	101.3	100.1	88.4	102.8	91.7	91.9	102.8	94.4	96.3	55.0	86.7
0920	Plaster & Gypsum Board	94.5	115.0	108.3	91.5	111.2	104.7	104.3	76.8	85.8	91.5	79.9	83.7	91.5	79.9	83.7	103.8	64.0	77.0
0950, 0980	Ceilings & Acoustic Treatment	91.6	115.0	107.4	58.8	111.2	94.1	128.8	76.8	93.7	58.8	79.9	73.0	58.8	79.9	73.0	86.4	64.0	71.3
0960	Flooring	92.4	122.8	101.0	86.4	120.9	96.1	90.4	85.6	89.0	87.8	85.6	87.2	90.3	85.6	89.0	94.0	65.0	85.8
0970, 0990	Wall Finishes & Painting/Coating	92.1	129.1	114.0	95.0	123.8	112.0	89.6	83.4	85.9	89.6	83.4	85.9	89.6	103.7	98.0	86.0	46.9	62.8
09	FINISHES	91.1	119.1	106.3	81.5	115.9	100.1	98.1	82.6	89.7	81.6	85.3	83.6	81.8	84.8	83.4	89.4	62.8	75.0
COVERS	DIVS. 10 - 14, 25, 28, 41, 43, 44, 46	100.0	108.7	101.9	100.0	103.2	100.7	100.0	95.5	99.0	100.0	97.3	99.4	100.0	93.7	98.6	100.0	71.9	93.7
21, 22, 23	FIRE SUPPRESSION, PLUMBING & HVAC	99.9	116.7	106.5	99.5	115.0	103.8	96.7	84.4	91.9	96.0	101.6	98.2	96.0	81.0	90.1	100.0	55.3	82.4
26, 27, 3370	ELECTRICAL, COMMUNICATIONS & UTIL.	102.9	115.1	108.9	102.7	115.1	108.8	102.1	71.0	86.8	102.7	83.7	93.4	109.1	93.1	101.2	101.0	53.7	77.7
MF2018	WEIGHTED AVERAGE	99.0	117.6	106.9	95.1	114.2	103.2	96.3	89.3	93.3	94.0	98.0	95.7	94.3	91.2	93.0	96.2	63.9	82.6

MISSISSIPPI

DIVISION		CLARKSDALE 386 MAT.	INST.	TOTAL	COLUMBUS 397 MAT.	INST.	TOTAL	GREENVILLE 387 MAT.	INST.	TOTAL	GREENWOOD 389 MAT.	INST.	TOTAL	JACKSON 390 - 392 MAT.	INST.	TOTAL	LAUREL 394 MAT.	INST.	TOTAL
015433	CONTRACTOR EQUIPMENT		100.1	100.1		100.1	100.1		100.1	100.1		100.1	100.1		102.6	102.6		100.1	100.1
0241, 31 - 34	SITE & INFRASTRUCTURE, DEMOLITION	101.4	84.4	89.7	103.9	85.2	91.0	107.5	86.1	92.7	104.5	84.1	90.4	101.1	90.6	93.9	109.6	84.5	92.2
0310	Concrete Forming & Accessories	83.8	43.9	49.8	81.9	46.3	51.6	80.7	62.7	65.4	92.5	44.3	51.4	92.0	64.6	68.6	82.0	59.5	62.8
0320	Concrete Reinforcing	104.2	66.5	85.9	99.9	66.6	83.8	104.7	66.7	86.3	104.2	66.5	86.0	104.0	51.7	78.7	100.6	32.4	67.6
0330	Cast-in-Place Concrete	104.1	58.8	87.2	116.5	61.1	95.9	107.2	65.7	91.7	112.0	58.9	92.2	99.6	66.5	87.3	113.9	60.1	93.9
03	CONCRETE	93.8	55.1	76.8	95.8	57.1	78.8	99.3	66.2	84.8	100.1	55.4	80.4	89.2	64.7	78.4	97.9	56.9	79.9
04	MASONRY	89.1	50.8	65.6	115.4	53.6	77.4	132.2	61.4	88.7	89.8	50.7	65.8	95.7	61.5	74.7	111.5	49.4	73.4
05	METALS	92.5	87.9	91.1	87.4	90.4	88.3	93.6	91.2	92.8	92.5	87.8	91.1	96.9	84.6	93.3	87.5	76.0	84.1
06	WOOD, PLASTICS & COMPOSITES	79.4	44.4	60.1	78.8	45.2	60.3	76.2	62.8	68.8	91.8	44.4	65.7	96.3	65.5	79.4	79.9	64.5	71.4
07	THERMAL & MOISTURE PROTECTION	97.7	52.0	77.8	100.0	55.4	80.6	98.2	62.1	82.5	98.1	54.2	79.0	98.3	62.9	82.9	100.2	56.5	81.2
08	OPENINGS	95.1	48.2	84.1	95.9	48.6	84.9	94.8	58.1	86.2	95.1	48.2	84.1	99.5	56.0	89.4	93.2	51.8	83.5
0920	Plaster & Gypsum Board	89.9	43.3	58.6	94.3	44.2	60.6	89.6	62.3	71.2	100.6	43.3	62.1	90.7	64.9	73.3	94.3	64.0	73.9
0950, 0980	Ceilings & Acoustic Treatment	82.3	43.3	56.0	81.5	44.2	56.3	85.0	62.3	69.7	82.3	43.3	56.0	89.7	64.9	73.0	81.5	64.0	69.7
0960	Flooring	97.8	45.0	83.0	87.9	65.0	81.4	96.2	65.0	87.4	103.5	65.0	92.7	92.1	65.0	84.5	86.6	65.0	80.5
0970, 0990	Wall Finishes & Painting/Coating	94.6	46.9	66.4	86.0	46.9	62.8	94.6	56.7	72.1	94.6	46.9	66.4	89.9	56.7	70.3	86.0	46.9	62.8
09	FINISHES	90.2	43.9	65.1	85.4	49.1	65.8	90.8	62.5	75.5	93.7	48.0	68.9	89.9	64.1	75.9	85.6	60.5	72.0
COVERS	DIVS. 10 - 14, 25, 28, 41, 43, 44, 46	100.0	48.0	88.4	100.0	49.0	88.6	100.0	71.3	93.6	100.0	48.0	88.4	100.0	71.9	93.7	100.0	34.2	85.3
21, 22, 23	FIRE SUPPRESSION, PLUMBING & HVAC	98.4	51.3	79.9	98.0	53.1	80.4	100.0	57.4	83.3	98.4	51.7	80.1	100.0	59.1	83.9	98.1	47.0	78.0
26, 27, 3370	ELECTRICAL, COMMUNICATIONS & UTIL.	96.8	41.6	69.6	98.6	55.1	77.2	96.8	55.9	76.6	96.8	39.0	68.3	102.7	55.9	79.6	100.1	57.9	79.3
MF2018	WEIGHTED AVERAGE	95.3	55.2	78.4	95.9	59.1	80.4	98.7	65.5	84.7	96.6	55.5	79.2	97.3	65.6	83.9	96.1	57.8	79.9

MISSISSIPPI / MISSOURI

DIVISION		MCCOMB 396 MAT.	INST.	TOTAL	MERIDIAN 393 MAT.	INST.	TOTAL	TUPELO 388 MAT.	INST.	TOTAL	BOWLING GREEN 633 MAT.	INST.	TOTAL	CAPE GIRARDEAU 637 MAT.	INST.	TOTAL	CHILLICOTHE 646 MAT.	INST.	TOTAL
015433	CONTRACTOR EQUIPMENT		100.1	100.1		100.1	100.1		100.1	100.1		107.1	107.1		107.1	107.1		101.5	101.5
0241, 31 - 34	SITE & INFRASTRUCTURE, DEMOLITION	96.3	84.3	88.0	100.7	86.1	90.7	98.9	84.3	88.8	87.6	90.0	89.3	89.3	89.9	89.7	101.2	88.4	92.3
0310	Concrete Forming & Accessories	81.9	45.5	50.9	79.4	62.8	65.2	81.2	45.9	51.1	96.0	94.0	94.3	88.9	82.0	83.0	86.2	95.7	94.3
0320	Concrete Reinforcing	101.2	34.3	68.8	99.9	51.6	76.6	101.9	66.5	84.8	93.4	96.5	94.9	94.6	79.6	87.3	100.4	102.6	101.5
0330	Cast-in-Place Concrete	101.1	58.4	85.2	108.2	66.0	92.5	104.1	67.9	90.6	90.7	95.5	92.5	89.8	86.6	88.6	91.7	86.0	89.6
03	CONCRETE	85.9	50.2	70.2	90.1	63.8	78.6	93.4	59.2	78.4	91.2	96.4	93.5	90.4	85.0	88.0	95.0	94.3	94.7
04	MASONRY	116.7	50.3	75.9	90.0	62.1	72.9	121.7	53.3	79.6	109.6	97.9	102.4	105.9	80.5	90.3	99.2	92.2	94.9
05	METALS	87.6	75.6	84.1	88.5	85.6	87.7	92.4	89.7	91.1	91.8	117.7	99.4	92.9	109.4	97.8	85.1	110.6	92.5
06	WOOD, PLASTICS & COMPOSITES	78.8	46.8	61.2	76.2	62.8	68.8	76.7	45.2	59.4	96.7	94.4	95.4	89.2	80.0	84.1	94.0	97.0	95.7
07	THERMAL & MOISTURE PROTECTION	99.4	54.2	79.8	99.6	62.4	83.4	97.6	53.9	78.6	96.4	99.2	97.6	95.9	85.3	91.3	92.7	92.7	92.7
08	OPENINGS	96.0	41.3	83.2	95.6	54.8	86.1	95.0	51.4	84.9	98.7	98.2	98.5	98.6	76.1	93.4	86.0	96.2	88.3
0920	Plaster & Gypsum Board	94.3	45.8	61.7	94.3	62.3	72.8	89.6	44.2	59.0	101.7	94.6	96.9	101.1	79.8	86.8	98.2	96.9	97.3
0950, 0980	Ceilings & Acoustic Treatment	81.5	45.8	57.4	82.4	62.3	68.8	82.3	44.2	56.6	88.9	94.6	92.8	88.9	79.8	82.8	87.4	96.9	93.8
0960	Flooring	87.9	65.0	81.4	86.5	65.0	80.5	96.5	65.0	87.6	95.7	96.9	96.0	92.4	84.5	90.2	94.9	99.7	96.2
0970, 0990	Wall Finishes & Painting/Coating	86.0	46.9	62.8	86.0	46.9	62.8	94.6	45.2	65.4	95.4	106.3	101.8	95.4	67.1	78.6	94.3	103.2	99.6
09	FINISHES	84.8	49.2	65.5	84.8	61.6	72.3	89.7	48.9	67.6	98.2	95.7	96.8	97.0	79.8	87.7	97.1	97.6	97.4
COVERS	DIVS. 10 - 14, 25, 28, 41, 43, 44, 46	100.0	50.9	88.4	100.0	71.5	93.6	100.0	49.0	88.6	100.0	81.4	95.8	100.0	94.3	98.7	100.0	81.3	95.8
21, 22, 23	FIRE SUPPRESSION, PLUMBING & HVAC	98.0	50.7	79.5	100.0	57.6	83.3	98.5	52.7	80.5	96.5	99.5	97.7	100.0	98.7	99.5	96.6	99.9	97.9
26, 27, 3370	ELECTRICAL, COMMUNICATIONS & UTIL.	97.3	56.0	76.9	100.1	55.6	78.2	96.5	55.0	76.1	99.2	76.8	88.2	99.2	98.5	98.9	94.9	75.8	85.5
MF2018	WEIGHTED AVERAGE	94.4	55.8	78.0	94.6	64.5	81.9	96.6	59.0	80.7	96.4	95.5	96.0	97.0	91.2	94.5	93.7	93.8	93.7

For customer support on your Commercial Renovation Costs with RSMeans data, call 800.448.8182.

City Cost Indexes

MISSOURI

DIVISION		COLUMBIA 652			FLAT RIVER 636			HANNIBAL 634			HARRISONVILLE 647			JEFFERSON CITY 650 - 651			JOPLIN 648		
		MAT.	INST.	TOTAL	MAT.	INST.	TOTAL	MAT.	INST.	TOTAL	MAT.	INST.	TOTAL	MAT.	INST.	TOTAL	MAT.	INST.	TOTAL
015433	CONTRACTOR EQUIPMENT		110.3	110.3		107.1	107.1		107.1	107.1		101.5	101.5		113.1	113.1		104.8	104.8
0241, 31 - 34	SITE & INFRASTRUCTURE, DEMOLITION	94.2	93.8	93.9	90.2	89.7	89.8	85.4	89.8	88.4	92.9	89.5	90.6	94.4	98.7	97.4	102.0	92.6	95.5
0310	Concrete Forming & Accessories	84.4	81.0	81.5	102.4	88.3	90.4	94.3	82.5	84.3	83.4	99.0	96.7	96.3	78.6	81.2	97.4	74.2	77.7
0320	Concrete Reinforcing	85.6	93.3	89.3	94.6	102.9	98.6	92.9	96.5	94.6	100.0	110.9	105.3	92.4	93.3	92.9	103.7	84.3	94.3
0330	Cast-in-Place Concrete	88.6	84.8	87.2	93.7	90.6	92.5	85.9	93.5	88.7	93.9	99.7	96.0	93.9	84.1	90.2	99.3	75.5	90.4
03	CONCRETE	80.8	86.3	83.2	94.1	93.3	93.8	87.6	90.5	88.9	91.1	102.0	95.9	87.5	84.8	86.3	93.9	77.7	86.7
04	MASONRY	136.9	86.4	105.9	106.5	76.4	88.0	101.6	94.6	97.3	94.0	99.4	97.3	96.0	86.5	90.1	92.4	80.2	84.9
05	METALS	96.3	115.4	101.9	91.7	119.2	99.8	91.8	117.2	99.3	85.4	115.1	94.1	95.8	114.0	101.1	87.9	98.8	91.1
06	WOOD, PLASTICS & COMPOSITES	80.3	78.6	79.3	105.2	89.6	96.6	94.9	81.1	87.3	90.4	98.7	95.0	97.1	75.6	85.2	106.1	73.5	88.2
07	THERMAL & MOISTURE PROTECTION	90.6	86.1	88.7	96.6	91.0	94.2	96.2	93.4	95.0	91.8	100.8	95.7	97.5	86.1	92.5	91.9	81.2	87.3
08	OPENINGS	94.4	81.7	91.4	98.6	97.4	98.4	98.7	83.7	95.2	85.7	103.2	89.8	94.5	80.0	91.1	86.9	76.3	84.5
0920	Plaster & Gypsum Board	84.7	78.2	80.3	107.7	89.7	95.6	101.4	81.0	87.7	94.1	98.6	97.1	96.2	74.9	81.9	105.4	72.7	83.4
0950, 0980	Ceilings & Acoustic Treatment	88.5	78.2	81.5	88.9	89.7	89.4	88.9	81.0	83.6	87.4	98.6	95.1	93.3	74.9	80.9	88.3	72.7	77.8
0960	Flooring	90.2	98.6	92.5	98.9	84.5	94.9	95.0	96.9	95.5	90.3	100.8	93.2	97.6	71.5	90.3	120.1	72.2	106.6
0970, 0990	Wall Finishes & Painting/Coating	93.1	81.9	86.4	95.4	71.7	81.4	95.4	94.1	94.6	98.6	107.8	104.0	91.2	81.9	85.7	93.9	75.9	83.2
09	FINISHES	84.0	83.5	83.7	100.0	85.1	92.0	97.7	85.3	91.0	94.8	100.2	97.7	91.5	77.1	83.7	103.8	74.4	87.9
COVERS	DIVS. 10 - 14, 25, 28, 41, 43, 44, 46	100.0	93.6	98.6	100.0	94.0	98.7	100.0	78.8	95.3	100.0	82.7	96.2	100.0	93.6	98.6	100.0	79.5	95.4
21, 22, 23	FIRE SUPPRESSION, PLUMBING & HVAC	100.0	96.9	98.8	96.5	96.8	96.6	96.5	97.7	97.0	96.6	100.8	98.2	100.0	96.9	98.8	100.1	71.1	88.7
26, 27, 3370	ELECTRICAL, COMMUNICATIONS & UTIL.	97.8	81.4	89.7	103.8	98.5	101.2	98.0	76.8	87.5	101.5	101.2	101.4	103.2	81.4	92.4	92.9	66.1	79.7
MF2018	WEIGHTED AVERAGE	96.1	90.7	93.8	97.3	94.2	96.0	95.3	91.6	93.2	93.2	100.8	96.4	96.4	89.9	93.7	95.1	77.7	87.7

MISSOURI

DIVISION		KANSAS CITY 640 - 641			KIRKSVILLE 635			POPLAR BLUFF 639			ROLLA 654 - 655			SEDALIA 653			SIKESTON 638		
		MAT.	INST.	TOTAL	MAT.	INST.	TOTAL	MAT.	INST.	TOTAL	MAT.	INST.	TOTAL	MAT.	INST.	TOTAL	MAT.	INST.	TOTAL
015433	CONTRACTOR EQUIPMENT		104.4	104.4		97.4	97.4		99.6	99.6		110.3	110.3		100.4	100.4		99.6	99.6
0241, 31 - 34	SITE & INFRASTRUCTURE, DEMOLITION	94.6	98.6	97.3	88.8	84.9	86.1	76.1	88.5	84.7	92.9	94.1	93.8	92.1	89.3	90.2	79.6	89.1	86.2
0310	Concrete Forming & Accessories	97.2	103.5	102.6	86.9	78.4	79.6	87.4	79.3	80.5	91.6	94.0	93.6	89.6	78.2	79.9	88.2	79.4	80.7
0320	Concrete Reinforcing	98.5	111.1	104.6	93.8	83.3	88.7	96.7	76.0	86.7	86.0	93.4	89.6	84.8	110.3	97.1	96.0	76.0	86.4
0330	Cast-in-Place Concrete	97.3	102.6	99.3	93.6	82.2	89.4	72.0	83.5	76.3	90.6	94.7	92.1	94.7	81.3	89.7	77.0	83.5	79.4
03	CONCRETE	93.8	104.9	98.7	106.9	81.8	95.9	81.5	81.4	81.5	82.5	95.6	88.2	95.7	86.1	91.5	85.2	81.5	83.6
04	MASONRY	99.6	103.3	101.9	113.2	84.8	95.8	104.5	74.1	85.8	110.8	85.3	95.2	117.1	81.6	95.3	104.2	74.1	85.7
05	METALS	94.9	112.8	100.2	91.5	101.0	94.3	92.1	97.8	93.7	95.7	115.9	101.6	94.5	112.8	99.9	92.4	97.9	94.0
06	WOOD, PLASTICS & COMPOSITES	100.1	103.8	102.2	82.2	76.5	79.1	81.3	80.0	80.6	88.0	95.6	92.2	81.6	76.7	78.9	82.9	80.0	81.3
07	THERMAL & MOISTURE PROTECTION	91.8	105.2	97.7	102.2	90.8	97.3	100.5	83.0	92.9	90.9	92.9	91.8	96.1	87.8	92.5	100.7	82.1	92.6
08	OPENINGS	96.5	106.0	98.7	104.1	78.8	98.2	105.0	75.2	98.1	94.4	91.1	93.6	99.4	87.0	96.5	105.0	75.2	98.1
0920	Plaster & Gypsum Board	100.9	104.0	103.0	96.3	76.2	82.8	96.6	79.8	85.3	86.9	95.6	92.8	80.5	76.2	77.6	98.5	79.8	85.9
0950, 0980	Ceilings & Acoustic Treatment	90.5	104.0	99.6	87.3	76.2	79.8	88.9	79.8	82.8	88.5	95.6	93.3	88.5	76.2	80.2	88.9	79.8	82.8
0960	Flooring	95.9	101.8	97.6	73.8	96.4	80.2	87.3	84.5	86.5	93.8	96.4	94.5	72.5	72.9	72.6	87.8	84.5	86.9
0970, 0990	Wall Finishes & Painting/Coating	97.2	107.8	103.5	91.3	78.3	83.6	90.6	67.1	76.7	93.1	90.2	91.4	93.1	103.2	99.1	90.6	67.1	76.7
09	FINISHES	98.1	103.6	101.1	97.4	81.1	88.6	96.9	78.4	86.9	85.4	94.0	90.1	82.9	79.1	80.8	97.5	78.7	87.3
COVERS	DIVS. 10 - 14, 25, 28, 41, 43, 44, 46	100.0	98.4	99.7	100.0	78.3	95.2	100.0	92.2	98.3	100.0	96.7	99.3	100.0	87.9	97.3	100.0	92.2	98.3
21, 22, 23	FIRE SUPPRESSION, PLUMBING & HVAC	100.0	103.4	101.4	96.6	97.1	96.8	96.6	94.9	95.9	96.5	98.6	97.3	96.4	94.9	95.8	96.6	95.0	95.9
26, 27, 3370	ELECTRICAL, COMMUNICATIONS & UTIL.	103.3	101.2	102.2	98.2	76.8	87.6	98.4	98.4	98.4	96.4	79.4	88.0	97.2	101.2	99.2	97.5	98.4	97.9
MF2018	WEIGHTED AVERAGE	97.9	103.8	100.4	98.9	86.5	93.7	95.1	87.7	92.0	94.1	94.3	94.2	96.4	91.5	94.3	95.7	87.8	92.3

		MISSOURI									MONTANA								
DIVISION		SPRINGFIELD 656 - 658			ST. JOSEPH 644 - 645			ST. LOUIS 630 - 631			BILLINGS 590 - 591			BUTTE 597			GREAT FALLS 594		
		MAT.	INST.	TOTAL	MAT.	INST.	TOTAL	MAT.	INST.	TOTAL	MAT.	INST.	TOTAL	MAT.	INST.	TOTAL	MAT.	INST.	TOTAL
015433	CONTRACTOR EQUIPMENT		102.8	102.8		101.5	101.5		108.6	108.6		97.9	97.9		97.7	97.7		97.7	97.7
0241, 31 - 34	SITE & INFRASTRUCTURE, DEMOLITION	94.4	91.4	92.4	96.7	87.4	90.3	94.4	97.8	96.7	91.8	93.5	93.0	97.3	93.3	94.5	101.0	93.3	95.7
0310	Concrete Forming & Accessories	97.7	75.4	78.7	96.3	89.1	90.2	99.5	102.5	102.1	99.3	68.3	72.9	86.5	68.2	70.9	99.2	68.2	72.8
0320	Concrete Reinforcing	82.5	92.9	87.5	97.3	110.4	103.6	86.5	105.2	95.5	98.0	81.8	90.2	106.3	81.7	94.4	98.0	81.8	90.2
0330	Cast-in-Place Concrete	96.3	73.7	87.9	92.2	95.5	93.4	101.3	102.5	101.7	118.7	72.7	101.6	130.6	72.5	109.0	138.1	72.5	113.7
03	CONCRETE	92.1	79.0	86.3	90.3	96.0	92.8	97.1	104.0	100.1	98.3	73.1	87.2	101.9	73.0	89.2	106.9	73.0	92.0
04	MASONRY	89.1	79.9	83.5	95.3	88.7	91.2	91.1	107.1	100.9	125.8	83.8	100.0	120.9	82.7	97.4	125.4	83.8	99.8
05	METALS	100.7	101.7	101.0	91.5	113.8	98.0	97.4	119.0	103.8	107.0	89.6	101.9	101.2	89.4	97.7	104.4	89.5	100.0
06	WOOD, PLASTICS & COMPOSITES	89.4	75.0	81.5	106.1	89.6	96.5	97.6	101.0	99.5	90.1	64.6	76.1	76.8	64.6	70.1	91.4	64.6	76.6
07	THERMAL & MOISTURE PROTECTION	94.6	75.5	86.3	92.2	88.7	90.7	94.3	105.5	99.2	108.3	73.7	93.2	107.8	73.3	92.8	108.5	73.7	93.4
08	OPENINGS	101.7	84.3	97.6	89.9	96.7	91.5	100.3	105.8	101.6	98.6	65.9	91.0	96.9	65.9	89.7	99.8	65.9	91.9
0920	Plaster & Gypsum Board	87.8	74.4	78.8	106.9	88.3	94.4	106.5	101.5	103.1	118.3	64.0	81.8	118.8	64.0	81.9	128.7	64.0	85.2
0950, 0980	Ceilings & Acoustic Treatment	88.5	74.4	79.0	94.8	88.3	90.4	90.5	101.5	97.9	96.6	64.0	74.6	104.1	64.0	77.0	105.7	64.0	77.6
0960	Flooring	92.8	71.5	86.8	99.4	99.0	99.3	98.0	99.1	98.3	88.1	80.8	86.0	85.4	90.9	87.0	91.9	80.8	88.8
0970, 0990	Wall Finishes & Painting/Coating	87.6	99.4	94.6	94.3	107.8	102.3	96.4	106.3	102.3	93.8	91.0	92.2	92.3	67.8	77.8	92.3	91.0	91.5
09	FINISHES	87.6	77.6	82.2	100.3	92.8	96.3	100.5	101.9	101.3	92.5	72.3	81.6	93.0	71.8	81.5	96.8	72.3	83.5
COVERS	DIVS. 10 - 14, 25, 28, 41, 43, 44, 46	100.0	90.5	97.9	100.0	88.3	98.7	100.0	102.8	100.6	100.0	91.6	98.1	100.0	91.6	98.1	100.0	91.6	98.1
21, 22, 23	FIRE SUPPRESSION, PLUMBING & HVAC	100.0	69.4	88.0	100.1	87.5	95.2	100.0	105.5	102.2	100.0	76.4	90.7	100.0	71.2	88.7	100.0	71.2	88.7
26, 27, 3370	ELECTRICAL, COMMUNICATIONS & UTIL.	101.3	68.8	85.3	101.5	75.8	88.9	101.8	98.5	100.2	102.3	75.2	88.9	109.2	71.3	90.5	101.7	70.8	86.5
MF2018	WEIGHTED AVERAGE	97.6	79.0	89.7	96.1	90.9	93.9	98.8	104.5	101.2	101.6	78.4	91.8	101.5	76.5	90.9	102.9	76.6	91.8

MONTANA

DIVISION		HAVRE 595			HELENA 596			KALISPELL 599			MILES CITY 593			MISSOULA 598			WOLF POINT 592		
		MAT.	INST.	TOTAL	MAT.	INST.	TOTAL	MAT.	INST.	TOTAL	MAT.	INST.	TOTAL	MAT.	INST.	TOTAL	MAT.	INST.	TOTAL
015433	CONTRACTOR EQUIPMENT		97.7	97.7		99.9	99.9		97.7	97.7		97.7	97.7		97.7	97.7		97.7	97.7
0241, 31 - 34	SITE & INFRASTRUCTURE, DEMOLITION	104.2	93.1	96.6	91.1	96.8	95.0	87.5	93.3	91.5	93.5	93.2	93.3	80.5	93.3	89.3	110.2	93.2	98.4
0310	Concrete Forming & Accessories	79.6	67.2	69.0	100.2	68.3	73.0	89.7	68.3	71.4	97.7	67.3	71.8	89.7	68.4	71.5	90.4	67.3	70.7
0320	Concrete Reinforcing	107.1	78.9	93.5	113.0	81.8	97.9	109.1	85.8	97.8	106.7	78.9	93.3	108.1	85.8	97.3	108.2	78.2	93.7
0330	Cast-in-Place Concrete	140.7	71.2	114.9	102.1	73.3	91.4	113.4	72.5	98.2	124.2	71.3	104.5	96.2	72.6	87.4	139.1	70.2	113.5
03	CONCRETE	109.8	71.6	93.0	94.7	73.3	85.3	91.7	73.7	83.8	98.8	71.7	86.9	80.5	73.8	77.6	113.2	71.2	94.8
04	MASONRY	121.9	80.6	96.5	114.2	82.7	94.8	119.6	82.7	96.9	127.6	80.6	98.7	144.5	82.7	106.5	128.9	80.6	99.2
05	METALS	97.3	88.2	94.6	103.2	88.5	98.9	97.2	90.8	95.3	96.5	88.4	94.1	97.7	91.0	95.7	96.6	88.1	94.1
06	WOOD, PLASTICS & COMPOSITES	68.1	64.6	66.2	93.8	64.7	77.8	80.2	64.6	71.6	88.4	64.6	75.3	80.2	64.6	71.6	79.5	64.6	71.3
07	THERMAL & MOISTURE PROTECTION	108.3	66.2	90.0	103.1	73.9	90.4	107.4	73.6	92.7	107.8	67.7	90.3	106.9	75.6	93.3	108.9	67.5	90.9
08	OPENINGS	97.4	65.2	89.9	96.3	65.9	89.2	97.4	66.8	90.3	96.9	65.2	89.5	96.9	66.8	89.9	96.9	65.1	89.5
0920	Plaster & Gypsum Board	114.1	64.0	80.4	113.5	64.0	80.2	118.8	64.0	81.9	127.8	64.0	84.9	118.8	64.0	81.9	121.8	64.0	82.9
0950, 0980	Ceilings & Acoustic Treatment	104.1	64.0	77.0	107.9	64.0	78.3	104.1	64.0	77.0	101.6	64.0	76.3	104.1	64.0	77.0	101.6	64.0	76.3
0960	Flooring	83.0	90.9	85.3	95.9	90.9	94.5	87.1	90.9	88.2	91.8	90.9	91.6	87.1	90.9	88.2	88.4	90.9	89.1
0970, 0990	Wall Finishes & Painting/Coating	92.3	67.8	77.8	97.6	67.8	79.9	92.3	91.0	91.5	92.3	67.8	77.8	92.3	91.0	91.5	92.3	67.8	77.8
09	FINISHES	92.3	71.3	80.9	100.1	71.8	84.8	92.9	74.3	82.8	95.4	71.3	82.3	92.3	74.3	82.6	95.0	71.3	82.1
COVERS	DIVS. 10 - 14, 25, 28, 41, 43, 44, 46	100.0	88.8	97.5	100.0	91.9	98.2	100.0	89.5	97.7	100.0	88.9	97.5	100.0	92.4	98.3	100.0	88.9	97.5
21, 22, 23	FIRE SUPPRESSION, PLUMBING & HVAC	96.5	68.9	85.6	100.0	71.2	88.7	96.5	69.4	85.8	96.5	75.2	88.1	100.0	71.2	88.7	96.5	75.2	88.1
26, 27, 3370	ELECTRICAL, COMMUNICATIONS & UTIL.	101.7	69.2	85.7	109.0	70.8	90.2	106.0	68.1	87.3	101.7	74.4	88.3	107.0	70.2	88.9	101.7	74.4	88.3
MF2018	WEIGHTED AVERAGE	100.4	74.8	89.6	100.9	76.7	90.7	98.1	76.2	88.9	99.2	77.0	89.8	98.6	77.1	89.5	101.4	76.9	91.1

NEBRASKA

DIVISION		ALLIANCE 693			COLUMBUS 686			GRAND ISLAND 688			HASTINGS 689			LINCOLN 683 - 685			MCCOOK 690		
		MAT.	INST.	TOTAL	MAT.	INST.	TOTAL	MAT.	INST.	TOTAL	MAT.	INST.	TOTAL	MAT.	INST.	TOTAL	MAT.	INST.	TOTAL
015433	CONTRACTOR EQUIPMENT		95.3	95.3		101.9	101.9		101.9	101.9		101.9	101.9		105.0	105.0		101.9	101.9
0241, 31 - 34	SITE & INFRASTRUCTURE, DEMOLITION	98.2	96.2	96.8	101.8	90.6	94.1	106.8	90.7	95.7	105.4	90.6	95.2	94.9	95.7	95.4	101.5	90.6	94.0
0310	Concrete Forming & Accessories	87.0	55.3	60.0	96.5	75.0	78.2	96.1	68.8	72.9	99.3	72.5	76.5	94.4	75.5	78.3	92.3	55.9	61.2
0320	Concrete Reinforcing	107.4	87.4	97.7	100.4	85.9	93.4	99.8	76.1	88.4	99.8	76.1	88.4	97.6	76.4	87.3	100.1	76.3	88.6
0330	Cast-in-Place Concrete	108.1	81.8	98.3	107.3	81.8	97.8	113.6	77.7	100.2	113.6	74.0	98.1	87.2	83.1	85.6	116.9	73.9	100.9
03	CONCRETE	114.3	70.9	95.2	100.2	80.4	91.5	104.9	74.5	91.5	105.1	74.9	91.8	87.5	79.4	83.9	103.7	67.4	87.7
04	MASONRY	106.0	77.4	88.4	111.1	77.4	90.4	104.3	73.8	85.6	112.8	74.0	88.9	93.6	78.2	84.1	101.9	77.4	86.8
05	METALS	99.1	85.2	95.0	91.4	98.0	93.3	93.2	93.7	93.3	94.0	93.7	93.9	95.0	93.3	94.5	93.9	93.8	93.9
06	WOOD, PLASTICS & COMPOSITES	82.6	48.7	64.0	94.3	74.9	83.6	93.4	66.3	78.5	97.2	72.1	83.4	99.0	75.0	85.8	90.7	49.7	68.1
07	THERMAL & MOISTURE PROTECTION	100.7	75.8	89.9	101.6	80.4	92.4	101.8	78.3	91.5	101.8	77.6	91.3	100.1	81.1	91.8	96.0	76.1	87.3
08	OPENINGS	90.8	58.1	83.2	91.3	73.7	87.2	91.3	66.8	85.6	91.3	69.8	86.3	105.4	68.8	96.9	91.2	55.8	83.0
0920	Plaster & Gypsum Board	79.8	47.5	58.1	93.4	74.3	80.6	92.5	65.5	74.3	94.4	71.4	78.9	102.7	74.3	83.6	91.0	48.4	62.4
0950, 0980	Ceilings & Acoustic Treatment	88.6	47.5	60.9	83.4	74.3	77.3	83.4	65.5	71.3	83.4	71.4	75.3	98.3	74.3	82.1	83.9	48.4	60.0
0960	Flooring	91.4	87.3	90.2	84.0	87.3	84.9	83.7	78.9	82.4	85.0	81.3	84.0	96.3	86.6	93.6	90.0	87.3	89.2
0970, 0990	Wall Finishes & Painting/Coating	155.8	51.7	94.2	74.9	60.1	66.1	74.9	62.5	67.6	74.9	60.1	66.1	94.4	79.3	85.5	87.6	45.0	62.4
09	FINISHES	90.8	59.2	73.7	84.9	75.5	79.8	84.9	69.3	76.5	85.6	72.7	78.6	95.6	77.9	86.0	88.4	59.1	72.5
COVERS	DIVS. 10 - 14, 25, 28, 41, 43, 44, 46	100.0	83.9	96.4	100.0	83.5	96.3	100.0	89.7	97.7	100.0	83.1	96.2	100.0	91.0	98.0	100.0	84.1	96.5
21, 22, 23	FIRE SUPPRESSION, PLUMBING & HVAC	96.7	75.0	88.2	96.6	75.3	88.2	100.1	79.5	92.0	96.6	74.6	87.9	100.0	79.6	92.0	96.5	75.2	88.1
26, 27, 3370	ELECTRICAL, COMMUNICATIONS & UTIL.	93.6	65.6	79.8	97.3	81.4	89.5	95.9	65.7	81.0	95.3	79.4	87.5	109.2	65.7	87.8	95.6	65.7	80.9
MF2018	WEIGHTED AVERAGE	98.7	73.3	87.9	96.0	80.7	89.6	97.4	76.8	88.7	97.1	78.2	89.1	98.3	79.7	90.5	96.3	73.1	86.5

NEBRASKA / NEVADA

DIVISION		NORFOLK 687			NORTH PLATTE 691			OMAHA 680 - 681			VALENTINE 692			CARSON CITY 897			ELKO 898		
		MAT.	INST.	TOTAL	MAT.	INST.	TOTAL	MAT.	INST.	TOTAL	MAT.	INST.	TOTAL	MAT.	INST.	TOTAL	MAT.	INST.	TOTAL
015433	CONTRACTOR EQUIPMENT		91.7	91.7		101.9	101.9		94.2	94.2		95.0	95.0		97.9	97.9		94.6	94.6
0241, 31 - 34	SITE & INFRASTRUCTURE, DEMOLITION	84.1	89.7	88.0	102.9	90.4	94.3	89.1	93.9	92.4	86.3	95.2	92.4	83.3	97.4	93.0	67.8	91.3	84.0
0310	Concrete Forming & Accessories	83.2	74.1	75.5	94.8	75.2	78.1	93.1	76.7	79.1	83.5	53.1	57.6	106.1	76.3	80.7	111.4	96.3	98.6
0320	Concrete Reinforcing	100.5	66.5	84.1	99.5	76.2	88.3	100.8	76.5	89.0	100.1	66.3	83.7	115.8	116.1	116.0	124.5	114.5	119.7
0330	Cast-in-Place Concrete	107.9	71.3	94.3	116.9	60.6	96.0	91.4	78.6	86.7	103.1	54.0	84.8	98.0	82.2	92.1	95.0	72.4	86.6
03	CONCRETE	98.8	72.4	87.2	103.8	71.4	89.6	89.8	77.8	84.5	101.2	56.9	81.7	101.0	85.3	94.1	98.5	90.8	95.1
04	MASONRY	117.1	77.3	92.7	90.4	76.9	82.1	93.0	80.0	85.0	101.4	76.8	86.3	118.2	68.8	87.8	125.0	67.7	89.8
05	METALS	94.8	80.4	90.6	93.1	93.2	93.2	95.0	83.9	91.7	105.1	79.4	97.5	111.4	94.1	106.3	115.0	93.8	108.7
06	WOOD, PLASTICS & COMPOSITES	78.4	74.4	76.2	92.7	76.9	84.0	93.4	76.6	84.2	77.5	47.4	61.0	93.6	74.4	83.0	104.7	102.3	103.4
07	THERMAL & MOISTURE PROTECTION	101.2	78.3	91.2	95.9	78.4	88.3	96.7	80.8	89.8	96.5	73.9	86.7	118.1	79.1	101.2	114.2	73.6	96.5
08	OPENINGS	92.7	68.7	87.1	90.6	72.1	86.3	99.8	76.8	94.4	92.7	53.4	83.6	101.1	77.6	95.6	102.5	92.6	100.2
0920	Plaster & Gypsum Board	93.2	74.3	80.5	91.0	76.4	81.2	102.9	76.4	85.1	92.5	46.6	61.6	100.8	73.7	82.6	106.9	102.5	103.9
0950, 0980	Ceilings & Acoustic Treatment	97.6	74.3	81.9	83.9	76.4	78.8	101.0	76.4	84.4	100.4	46.6	64.1	96.0	73.7	80.9	94.5	102.5	99.9
0960	Flooring	104.1	87.3	99.4	91.1	78.9	87.7	100.6	87.3	96.9	116.8	87.3	108.5	99.5	67.1	90.4	103.0	67.1	92.9
0970, 0990	Wall Finishes & Painting/Coating	132.5	60.1	89.7	87.6	58.9	70.6	104.2	61.0	78.6	155.3	61.0	99.5	95.8	79.3	86.1	94.2	79.3	85.4
09	FINISHES	101.8	75.2	87.4	88.7	74.8	81.2	100.4	76.9	87.7	109.1	59.3	82.1	96.5	74.0	84.3	94.0	90.4	92.0
COVERS	DIVS. 10 - 14, 25, 28, 41, 43, 44, 46	100.0	85.1	96.7	100.0	61.1	91.3	100.0	90.2	97.8	100.0	56.1	90.2	100.0	100.2	100.0	100.0	91.0	98.0
21, 22, 23	FIRE SUPPRESSION, PLUMBING & HVAC	96.3	74.3	87.7	100.0	73.7	89.6	100.0	80.3	92.3	96.1	73.5	87.2	100.0	77.0	91.0	98.3	76.9	89.9
26, 27, 3370	ELECTRICAL, COMMUNICATIONS & UTIL.	96.2	81.4	88.9	93.9	65.7	80.0	103.9	83.7	94.0	91.4	65.7	78.7	102.2	88.3	95.3	98.8	88.3	93.6
MF2018	WEIGHTED AVERAGE	97.4	77.5	89.0	96.3	75.6	87.6	97.6	81.5	90.8	98.6	69.2	86.2	102.9	82.5	94.3	102.3	85.2	95.0

City Cost Indexes

DIVISION		NEVADA									NEW HAMPSHIRE								
		ELY 893			LAS VEGAS 889-891			RENO 894-895			CHARLESTON 036			CLAREMONT 037			CONCORD 032-033		
		MAT.	INST.	TOTAL	MAT.	INST.	TOTAL	MAT.	INST.	TOTAL	MAT.	INST.	TOTAL	MAT.	INST.	TOTAL	MAT.	INST.	TOTAL
015433	CONTRACTOR EQUIPMENT		94.6	94.6		94.6	94.6		94.6	94.6		95.7	95.7		95.7	95.7		98.4	98.4
0241, 31 - 34	SITE & INFRASTRUCTURE, DEMOLITION	73.4	92.5	86.6	76.6	95.5	89.7	73.4	92.6	86.7	80.6	96.1	91.3	74.6	96.1	89.5	88.7	102.4	97.3
0310	Concrete Forming & Accessories	104.3	101.6	102.0	105.2	105.6	105.5	100.6	76.3	79.9	86.6	83.1	83.6	92.1	83.2	84.5	97.2	94.8	95.1
0320	Concrete Reinforcing	123.1	114.7	119.0	113.5	125.9	119.5	116.4	124.2	120.1	89.7	88.6	89.1	89.7	88.6	89.1	100.9	88.7	95.0
0330	Cast-in-Place Concrete	102.0	95.2	99.4	98.7	108.2	102.2	107.9	81.1	97.9	84.7	116.1	96.4	77.8	116.1	92.1	100.9	117.8	107.2
03	CONCRETE	107.5	101.2	104.7	102.4	109.5	105.5	106.7	86.3	97.8	90.4	95.8	92.7	82.8	95.8	88.5	96.8	101.6	98.9
04	MASONRY	130.5	74.8	96.3	117.4	106.2	110.5	124.2	68.7	90.1	89.8	99.5	95.8	89.9	99.5	95.8	97.8	101.4	100.0
05	METALS	115.0	96.5	109.5	124.1	103.9	118.2	116.7	97.4	111.1	94.2	92.1	93.5	94.2	92.1	93.6	101.3	91.8	98.5
06	WOOD, PLASTICS & COMPOSITES	94.8	104.4	100.1	92.4	102.3	97.8	88.4	74.3	80.6	86.6	79.4	82.6	92.4	79.4	85.3	93.3	93.6	93.5
07	THERMAL & MOISTURE PROTECTION	114.7	92.2	104.9	129.4	101.9	117.4	114.2	78.3	98.6	107.3	106.5	107.0	107.1	106.5	106.8	111.3	109.5	110.5
08	OPENINGS	102.4	93.8	100.4	101.4	109.6	103.3	100.2	79.2	95.3	96.1	80.6	92.5	97.2	80.6	93.3	97.0	88.5	95.1
0920	Plaster & Gypsum Board	101.1	104.6	103.5	93.9	102.5	99.7	89.0	73.7	78.7	106.1	78.5	87.5	107.0	78.5	87.9	108.7	93.1	98.2
0950, 0980	Ceilings & Acoustic Treatment	94.5	104.6	101.3	101.0	102.5	102.0	97.8	73.7	81.5	101.8	78.5	86.1	101.8	78.5	86.1	106.1	93.1	97.4
0960	Flooring	100.5	67.1	91.1	92.0	104.1	95.4	97.3	67.1	88.8	85.1	113.1	92.9	87.6	113.1	94.7	96.4	113.1	101.1
0970, 0990	Wall Finishes & Painting/Coating	94.2	112.0	104.7	96.8	119.4	110.2	94.2	79.3	85.4	81.8	90.0	86.6	81.8	90.0	86.6	95.9	90.0	92.4
09	FINISHES	92.9	97.0	95.1	90.8	106.4	99.3	90.8	73.9	81.7	90.2	88.5	89.3	90.6	88.5	89.5	95.9	97.4	96.7
COVERS	DIVS. 10 - 14, 25, 28, 41, 43, 44, 46	100.0	70.7	93.5	100.0	105.1	101.1	100.0	99.8	100.0	100.0	88.5	97.4	100.0	88.5	97.4	100.0	109.5	102.1
21, 22, 23	FIRE SUPPRESSION, PLUMBING & HVAC	98.3	96.2	97.4	100.1	103.0	101.3	100.0	77.1	91.0	96.7	82.2	91.0	96.7	82.2	91.0	100.0	87.2	95.0
26, 27, 3370	ELECTRICAL, COMMUNICATIONS & UTIL.	99.0	93.0	96.0	103.5	107.8	105.6	99.4	88.3	93.9	91.4	75.2	83.4	91.4	75.2	83.4	90.2	75.2	82.8
MF2018	WEIGHTED AVERAGE	103.7	93.2	99.2	105.0	105.1	105.0	103.6	82.6	94.7	94.1	88.4	91.7	93.1	88.4	91.1	98.0	93.0	96.0

DIVISION		NEW HAMPSHIRE															NEW JERSEY		
		KEENE 034			LITTLETON 035			MANCHESTER 031			NASHUA 030			PORTSMOUTH 038			ATLANTIC CITY 082, 084		
		MAT.	INST.	TOTAL	MAT.	INST.	TOTAL	MAT.	INST.	TOTAL	MAT.	INST.	TOTAL	MAT.	INST.	TOTAL	MAT.	INST.	TOTAL
015433	CONTRACTOR EQUIPMENT		95.7	95.7		95.7	95.7		99.0	99.0		95.7	95.7		95.7	95.7		93.3	93.3
0241, 31 - 34	SITE & INFRASTRUCTURE, DEMOLITION	88.0	96.2	93.6	74.7	95.2	88.8	87.8	101.3	97.1	89.5	96.3	94.2	83.3	97.0	92.7	88.2	99.7	96.1
0310	Concrete Forming & Accessories	90.7	83.4	84.5	102.4	77.3	81.0	97.5	95.1	95.5	98.5	94.9	95.4	87.9	94.1	93.2	114.8	142.0	138.0
0320	Concrete Reinforcing	89.7	88.6	89.2	90.4	88.6	89.5	108.5	88.7	98.9	112.0	88.7	100.7	89.7	88.7	89.2	81.1	138.0	108.6
0330	Cast-in-Place Concrete	85.1	116.2	96.7	76.3	107.5	87.9	99.2	118.3	106.3	80.5	117.4	94.3	76.3	117.2	91.5	78.5	137.1	100.3
03	CONCRETE	89.9	95.9	92.6	82.2	90.1	85.7	97.0	101.9	99.2	90.0	101.6	95.1	82.4	101.2	90.7	87.7	138.1	109.8
04	MASONRY	92.5	99.5	96.8	100.9	84.1	90.6	94.4	101.4	98.7	94.4	101.4	98.7	90.2	99.8	96.1	104.1	140.0	126.1
05	METALS	94.8	92.5	94.2	94.9	92.2	94.1	102.0	92.2	99.1	100.5	92.7	98.2	96.4	94.3	95.8	98.8	115.0	103.6
06	WOOD, PLASTICS & COMPOSITES	90.9	79.4	84.6	101.9	79.4	89.5	93.6	93.7	93.7	100.2	93.6	96.6	87.8	93.6	91.0	122.8	143.7	134.3
07	THERMAL & MOISTURE PROTECTION	107.9	106.4	107.2	107.2	99.5	103.8	112.5	109.5	111.2	108.3	108.9	108.6	107.8	108.1	107.9	102.1	134.0	116.0
08	OPENINGS	94.8	84.1	92.3	98.1	80.6	94.0	95.7	93.3	95.2	99.0	91.3	97.2	99.6	81.9	95.5	96.4	139.2	106.4
0920	Plaster & Gypsum Board	106.4	78.5	87.6	120.6	78.5	92.3	99.7	93.1	95.3	115.6	93.1	100.5	106.1	93.1	97.4	114.9	144.6	134.9
0950, 0980	Ceilings & Acoustic Treatment	101.8	78.5	86.1	101.8	78.5	86.1	101.2	93.1	95.8	114.1	93.1	100.0	102.7	93.1	96.3	96.4	144.6	128.9
0960	Flooring	87.2	113.1	94.6	81.9	113.1	102.2	93.8	113.1	99.2	91.0	113.1	97.2	85.2	113.1	93.0	97.5	162.5	115.8
0970, 0990	Wall Finishes & Painting/Coating	81.8	103.4	94.6	81.8	90.0	86.6	95.3	119.0	109.3	81.8	102.2	93.9	81.8	102.2	93.9	81.9	141.8	117.4
09	FINISHES	92.1	89.9	90.9	95.5	84.6	89.6	94.8	100.6	97.9	96.8	98.7	97.8	91.1	98.2	95.0	93.5	147.2	122.6
COVERS	DIVS. 10 - 14, 25, 28, 41, 43, 44, 46	100.0	94.7	98.8	100.0	95.6	99.0	100.0	109.7	102.2	100.0	109.4	102.1	100.0	108.7	101.9	100.0	114.9	103.3
21, 22, 23	FIRE SUPPRESSION, PLUMBING & HVAC	96.7	82.3	91.0	96.7	74.1	87.8	100.1	87.3	95.1	100.2	87.3	95.1	100.2	86.3	94.7	99.7	133.1	112.8
26, 27, 3370	ELECTRICAL, COMMUNICATIONS & UTIL.	91.4	75.2	83.4	92.2	49.1	71.0	93.9	79.3	86.7	93.2	79.3	86.4	91.7	75.3	83.6	94.2	143.4	118.5
MF2018	WEIGHTED AVERAGE	94.5	89.0	92.2	94.4	80.2	88.4	98.2	94.5	96.6	97.5	93.7	95.9	94.9	92.4	93.8	96.8	133.2	112.2

DIVISION		NEW JERSEY																	
		CAMDEN 081			DOVER 078			ELIZABETH 072			HACKENSACK 076			JERSEY CITY 073			LONG BRANCH 077		
		MAT.	INST.	TOTAL	MAT.	INST.	TOTAL	MAT.	INST.	TOTAL	MAT.	INST.	TOTAL	MAT.	INST.	TOTAL	MAT.	INST.	TOTAL
015433	CONTRACTOR EQUIPMENT		93.3	93.3		95.7	95.7		95.7	95.7		95.7	95.7		93.3	93.3		92.9	92.9
0241, 31 - 34	SITE & INFRASTRUCTURE, DEMOLITION	89.3	99.3	96.2	103.8	101.4	102.2	108.9	101.4	103.7	104.8	101.4	102.5	95.2	101.4	99.5	100.0	100.5	100.4
0310	Concrete Forming & Accessories	105.0	137.3	132.5	96.4	148.3	140.6	108.0	148.5	142.5	96.4	148.3	140.7	100.6	148.4	141.3	101.1	139.0	133.4
0320	Concrete Reinforcing	106.5	128.5	117.1	75.3	156.5	114.6	75.3	156.5	114.6	75.3	156.5	114.6	97.7	156.5	126.2	75.3	156.1	114.4
0330	Cast-in-Place Concrete	76.1	133.8	97.6	81.3	131.5	100.0	69.8	140.3	96.1	79.4	140.3	102.1	63.4	131.6	88.8	70.5	133.8	94.0
03	CONCRETE	88.2	133.2	108.0	86.1	142.3	110.8	83.3	145.5	110.6	84.5	145.4	111.2	81.0	142.2	107.9	85.1	138.6	108.6
04	MASONRY	94.2	135.0	119.3	91.2	145.6	124.7	107.2	145.6	130.8	95.3	145.6	126.2	85.3	145.6	122.4	99.6	135.5	121.6
05	METALS	104.7	111.0	106.5	96.5	125.1	104.9	98.0	125.3	106.0	96.5	125.1	104.9	102.4	122.2	108.2	96.6	121.0	103.7
06	WOOD, PLASTICS & COMPOSITES	110.1	137.9	125.4	94.9	148.3	124.3	109.7	148.3	131.0	94.9	148.3	124.3	95.8	148.3	124.7	96.9	139.3	120.2
07	THERMAL & MOISTURE PROTECTION	102.0	133.0	115.5	103.7	142.8	120.7	104.0	144.1	121.4	103.5	136.7	117.9	103.2	142.8	120.4	103.4	129.0	114.5
08	OPENINGS	98.4	132.0	106.3	102.0	145.1	112.1	100.4	145.1	110.8	99.8	145.1	110.4	98.5	145.1	109.4	94.6	138.5	104.8
0920	Plaster & Gypsum Board	110.9	138.6	129.6	111.5	149.4	137.0	119.0	149.4	139.5	111.5	149.4	137.0	114.6	149.4	138.0	113.7	140.2	131.5
0950, 0980	Ceilings & Acoustic Treatment	106.3	138.6	128.1	92.3	149.4	130.8	94.1	149.4	131.4	92.3	149.4	130.8	102.2	149.4	134.0	92.3	140.2	124.6
0960	Flooring	93.7	162.5	113.0	83.7	188.8	113.2	88.7	188.8	116.8	83.7	188.8	113.2	84.7	188.8	114.0	84.9	176.9	110.8
0970, 0990	Wall Finishes & Painting/Coating	81.9	141.8	117.4	83.8	148.6	122.1	83.8	148.6	122.1	83.8	148.6	122.1	83.9	148.6	122.2	83.9	141.8	118.2
09	FINISHES	93.7	143.7	120.8	90.0	156.2	125.8	93.4	156.2	127.4	89.9	156.2	125.8	92.4	156.6	127.1	91.0	147.2	121.4
COVERS	DIVS. 10 - 14, 25, 28, 41, 43, 44, 46	100.0	114.3	103.2	100.0	131.7	107.1	100.0	131.7	107.1	100.0	131.7	107.1	100.0	131.7	107.1	100.0	114.5	103.2
21, 22, 23	FIRE SUPPRESSION, PLUMBING & HVAC	100.0	126.3	110.3	99.7	135.7	113.8	100.0	137.1	114.6	99.7	135.7	113.8	100.0	135.7	114.0	99.7	131.1	112.1
26, 27, 3370	ELECTRICAL, COMMUNICATIONS & UTIL.	98.4	132.1	115.0	96.0	138.5	116.9	96.6	138.5	117.2	96.0	142.3	118.8	100.5	142.3	121.1	95.7	129.9	112.6
MF2018	WEIGHTED AVERAGE	98.0	127.8	110.6	96.4	137.5	113.7	97.5	138.3	114.7	96.1	138.3	114.0	96.5	137.8	114.0	95.8	130.9	110.7

For customer support on your Commercial Renovation Costs with RSMeans data, call 800.448.8182.

809

NEW JERSEY

DIVISION		NEW BRUNSWICK 088 - 089			NEWARK 070 - 071			PATERSON 074 - 075			POINT PLEASANT 087			SUMMIT 079			TRENTON 085 - 086		
		MAT.	INST.	TOTAL	MAT.	INST.	TOTAL	MAT.	INST.	TOTAL	MAT.	INST.	TOTAL	MAT.	INST.	TOTAL	MAT.	INST.	TOTAL
015433	CONTRACTOR EQUIPMENT		92.9	92.9		98.3	98.3		95.7	95.7		92.9	92.9		95.7	95.7		97.0	97.0
0241, 31 - 34	SITE & INFRASTRUCTURE, DEMOLITION	101.1	100.9	100.9	110.0	106.2	107.4	106.6	101.4	103.0	102.6	100.5	101.2	106.3	101.4	102.9	88.6	105.4	100.2
0310	Concrete Forming & Accessories	108.5	148.0	142.2	100.7	148.7	141.6	98.3	148.3	140.9	102.4	139.1	133.6	99.0	148.5	141.2	103.5	138.8	133.6
0320	Concrete Reinforcing	82.1	156.1	117.9	96.2	156.6	125.4	97.7	156.5	126.2	82.1	156.1	117.9	75.3	156.5	114.6	107.7	116.2	111.8
0330	Cast-in-Place Concrete	97.0	136.9	111.8	89.8	140.9	108.8	80.8	140.2	102.9	97.0	135.8	111.5	67.5	140.3	94.6	95.6	134.4	110.0
03	CONCRETE	104.2	143.8	121.6	93.1	145.7	116.2	88.8	145.3	113.7	103.9	139.4	119.5	80.6	145.5	109.1	97.4	132.1	112.6
04	MASONRY	102.5	140.6	125.9	94.9	145.7	126.1	91.7	145.6	124.9	91.4	135.5	118.5	93.8	145.6	125.6	95.4	135.5	120.0
05	METALS	98.9	121.4	105.5	104.2	124.7	110.3	97.5	125.1	105.6	98.9	121.0	105.4	96.5	125.3	104.9	104.3	108.8	105.6
06	WOOD, PLASTICS & COMPOSITES	115.7	148.3	133.6	97.8	148.4	125.6	97.5	148.3	125.5	107.3	139.3	124.9	98.7	148.3	126.0	101.6	139.4	122.4
07	THERMAL & MOISTURE PROTECTION	102.5	138.2	118.0	105.2	144.7	122.4	103.8	136.7	118.1	102.5	131.3	115.0	104.1	144.1	121.5	102.9	135.2	117.0
08	OPENINGS	91.6	145.1	104.1	100.5	145.1	110.9	105.0	145.1	114.3	93.4	141.0	104.5	106.1	145.1	115.2	96.9	130.0	104.6
0920	Plaster & Gypsum Board	113.0	149.4	137.5	106.2	149.4	135.3	114.6	149.4	138.0	107.9	140.2	129.6	113.7	149.4	137.7	101.6	140.2	127.5
0950, 0980	Ceilings & Acoustic Treatment	96.4	149.4	132.2	103.0	149.4	134.3	102.2	149.4	134.0	96.4	140.2	125.9	92.3	149.4	130.8	101.2	140.2	127.5
0960	Flooring	95.1	188.8	121.4	96.7	188.8	122.6	84.7	188.8	114.0	92.5	162.5	112.2	84.9	188.8	114.1	100.1	176.9	121.7
0970, 0990	Wall Finishes & Painting/Coating	81.9	148.6	121.4	94.1	148.6	126.3	83.8	148.6	122.1	81.9	141.8	117.4	83.8	148.6	122.1	94.9	141.8	122.7
09	FINISHES	93.8	156.2	127.6	97.2	156.7	129.4	92.5	156.2	127.0	92.3	144.7	120.7	91.0	156.2	126.3	97.0	147.3	124.2
COVERS	DIVS. 10 - 14, 25, 28, 41, 43, 44, 46	100.0	131.5	107.0	100.0	131.8	107.1	100.0	131.7	107.1	100.0	111.6	102.6	100.0	131.7	107.1	100.0	114.7	103.3
21, 22, 23	FIRE SUPPRESSION, PLUMBING & HVAC	99.7	133.8	113.1	100.0	139.2	115.4	100.0	135.5	114.0	99.7	131.1	112.0	99.8	137.1	114.4	100.1	130.8	112.1
26, 27, 3370	ELECTRICAL, COMMUNICATIONS & UTIL.	94.8	133.7	114.0	105.0	142.3	123.4	100.5	138.5	119.2	94.2	132.9	111.8	96.6	138.5	117.2	102.6	128.7	115.4
MF2018	WEIGHTED AVERAGE	98.7	135.6	114.3	100.3	139.7	117.0	98.0	137.7	114.8	98.1	130.8	111.9	96.4	138.3	114.1	99.7	128.9	112.0

DIVISION		NEW JERSEY VINELAND 080, 083			NEW MEXICO ALBUQUERQUE 870 - 872			CARRIZOZO 883			CLOVIS 881			FARMINGTON 874			GALLUP 873		
		MAT.	INST.	TOTAL	MAT.	INST.	TOTAL	MAT.	INST.	TOTAL	MAT.	INST.	TOTAL	MAT.	INST.	TOTAL	MAT.	INST.	TOTAL
015433	CONTRACTOR EQUIPMENT		93.3	93.3		107.5	107.5		107.5	107.5		107.5	107.5		107.5	107.5		107.5	107.5
0241, 31 - 34	SITE & INFRASTRUCTURE, DEMOLITION	92.5	99.4	97.3	91.1	97.5	95.5	109.7	97.5	101.3	96.6	97.5	97.2	97.3	97.5	97.5	107.2	97.5	100.5
0310	Concrete Forming & Accessories	99.2	137.6	131.9	101.0	66.2	71.3	98.6	66.2	71.0	98.6	66.1	70.9	101.1	66.2	71.4	101.1	66.2	71.4
0320	Concrete Reinforcing	81.1	131.2	105.3	99.9	71.5	86.1	116.0	71.5	94.5	117.3	71.4	95.1	109.2	71.5	90.9	104.5	71.5	88.5
0330	Cast-in-Place Concrete	84.6	133.9	103.0	90.5	70.1	82.9	94.7	70.1	85.6	94.7	70.1	85.5	91.3	70.1	83.4	86.0	70.1	80.1
03	CONCRETE	92.2	133.8	110.5	92.9	69.8	82.7	116.7	69.8	96.0	104.3	69.7	89.1	96.4	69.8	84.7	103.3	69.8	88.6
04	MASONRY	92.6	135.5	118.9	106.0	60.5	78.1	108.5	60.5	79.0	108.5	60.5	79.0	113.8	60.5	81.1	101.6	60.5	76.4
05	METALS	98.8	112.6	102.8	108.1	90.6	102.9	106.6	90.6	101.9	106.3	90.5	101.6	105.7	90.6	101.3	104.8	90.6	100.7
06	WOOD, PLASTICS & COMPOSITES	103.7	137.9	122.5	102.1	67.2	82.9	93.5	67.2	79.0	93.5	67.2	79.0	102.2	67.2	82.9	102.2	67.2	82.9
07	THERMAL & MOISTURE PROTECTION	101.9	132.2	115.1	101.0	72.8	88.7	106.3	72.8	91.7	105.0	72.8	91.0	101.3	72.8	88.9	102.5	72.8	89.6
08	OPENINGS	92.9	134.9	102.7	98.5	67.0	91.1	96.5	67.0	89.6	96.6	67.0	89.7	100.8	67.0	92.9	100.8	67.0	93.0
0920	Plaster & Gypsum Board	106.4	138.6	128.1	113.0	66.1	81.5	80.2	66.1	70.7	80.2	66.1	70.7	99.4	66.1	77.0	99.4	66.1	77.0
0950, 0980	Ceilings & Acoustic Treatment	96.4	138.6	124.9	99.8	66.1	77.1	102.7	66.1	78.1	102.7	66.1	78.1	98.6	66.1	76.7	98.6	66.1	76.7
0960	Flooring	91.7	162.5	111.6	88.9	66.8	82.7	97.2	66.8	88.7	97.2	66.8	88.7	90.4	66.8	83.8	90.4	66.8	83.8
0970, 0990	Wall Finishes & Painting/Coating	81.9	141.8	117.4	97.8	52.6	71.1	92.6	52.6	68.9	92.6	52.6	68.9	92.1	52.6	68.7	92.1	52.6	68.7
09	FINISHES	91.1	143.9	119.6	92.3	64.7	77.4	94.2	64.7	78.3	92.9	64.7	77.7	90.7	64.7	76.6	92.1	64.7	77.3
COVERS	DIVS. 10 - 14, 25, 28, 41, 43, 44, 46	100.0	114.5	103.2	100.0	85.1	96.7	100.0	85.1	96.7	100.0	85.1	96.7	100.0	85.1	96.7	100.0	85.1	96.7
21, 22, 23	FIRE SUPPRESSION, PLUMBING & HVAC	99.7	126.6	110.3	100.3	69.0	88.0	97.9	69.0	86.5	97.9	68.6	86.4	100.2	69.0	88.0	98.1	69.0	86.6
26, 27, 3370	ELECTRICAL, COMMUNICATIONS & UTIL.	94.2	143.3	118.5	87.9	69.5	78.8	90.1	69.5	80.0	87.9	69.5	78.8	85.9	69.5	77.8	85.3	69.5	77.5
MF2018	WEIGHTED AVERAGE	96.3	129.8	110.5	98.6	72.5	87.6	101.6	72.5	89.3	99.3	72.4	87.9	99.0	72.5	87.8	99.0	72.5	87.8

NEW MEXICO

DIVISION		LAS CRUCES 880			LAS VEGAS 877			ROSWELL 882			SANTA FE 875			SOCORRO 878			TRUTH/CONSEQUENCES 879		
		MAT.	INST.	TOTAL	MAT.	INST.	TOTAL	MAT.	INST.	TOTAL	MAT.	INST.	TOTAL	MAT.	INST.	TOTAL	MAT.	INST.	TOTAL
015433	CONTRACTOR EQUIPMENT		83.5	83.5		107.5	107.5		107.5	107.5		110.4	110.4		107.5	107.5		83.5	83.5
0241, 31 - 34	SITE & INFRASTRUCTURE, DEMOLITION	97.8	77.5	83.8	96.2	97.5	97.1	99.0	97.5	98.0	99.1	102.8	101.6	92.8	97.5	96.1	112.5	77.5	88.3
0310	Concrete Forming & Accessories	95.4	65.2	69.7	101.1	66.2	71.4	98.6	66.2	71.0	100.0	66.3	71.2	101.1	66.2	71.4	98.8	65.1	70.1
0320	Concrete Reinforcing	113.1	71.3	92.9	106.3	71.5	89.5	117.3	71.5	95.1	99.3	71.5	85.9	108.4	71.5	90.5	101.7	71.3	87.0
0330	Cast-in-Place Concrete	89.5	63.1	79.7	88.8	70.1	81.9	94.7	70.1	85.6	94.3	70.9	85.6	87.0	70.1	80.7	95.4	63.1	83.4
03	CONCRETE	82.1	66.5	75.3	94.0	69.8	83.4	105.3	69.8	89.7	92.9	70.0	82.9	92.9	69.8	82.8	86.1	66.4	77.4
04	MASONRY	104.0	60.2	77.1	101.9	60.5	76.5	119.5	60.5	83.3	95.0	60.6	73.9	101.8	60.5	76.4	98.7	60.2	75.0
05	METALS	105.0	83.0	98.5	104.5	90.6	100.4	107.5	90.6	102.6	101.6	89.6	98.1	104.8	90.6	100.7	104.4	83.0	98.1
06	WOOD, PLASTICS & COMPOSITES	82.9	66.2	73.7	102.2	67.2	82.9	93.5	67.2	79.0	99.0	67.2	81.5	102.2	67.2	82.9	93.5	66.2	78.5
07	THERMAL & MOISTURE PROTECTION	92.2	68.2	81.7	100.8	72.8	88.6	105.2	72.8	91.1	103.5	73.7	90.5	100.8	72.8	88.6	90.1	68.2	80.5
08	OPENINGS	92.0	66.4	86.1	97.3	67.0	90.3	96.4	67.0	89.6	98.8	67.0	91.4	97.2	67.0	90.1	90.6	66.4	85.0
0920	Plaster & Gypsum Board	78.5	66.1	70.2	99.4	66.1	77.0	80.2	66.1	70.7	114.1	66.1	81.8	99.4	66.1	77.0	100.6	66.1	77.4
0950, 0980	Ceilings & Acoustic Treatment	87.3	66.1	73.0	98.6	66.1	76.7	102.7	66.1	78.1	95.2	66.1	75.6	98.6	66.1	76.7	86.2	66.1	72.7
0960	Flooring	128.1	66.8	110.9	90.4	66.8	83.8	97.2	66.8	88.7	100.0	66.8	91.3	90.4	66.8	83.8	119.2	66.8	104.5
0970, 0990	Wall Finishes & Painting/Coating	81.8	52.6	64.5	92.1	52.6	68.7	92.6	52.6	68.9	99.8	52.6	71.9	92.1	52.6	68.7	84.5	52.6	65.6
09	FINISHES	103.0	64.0	81.9	90.5	64.7	76.6	93.0	64.7	77.7	98.2	64.8	80.1	90.4	64.7	76.5	102.4	64.0	81.6
COVERS	DIVS. 10 - 14, 25, 28, 41, 43, 44, 46	100.0	82.7	96.1	100.0	85.1	96.7	100.0	85.1	96.7	100.0	85.2	96.7	100.0	85.1	96.7	100.0	82.6	96.1
21, 22, 23	FIRE SUPPRESSION, PLUMBING & HVAC	100.4	68.7	88.0	98.1	69.0	86.6	100.0	69.0	87.8	100.3	69.0	88.0	98.1	69.0	86.6	98.0	68.7	86.5
26, 27, 3370	ELECTRICAL, COMMUNICATIONS & UTIL.	90.1	83.8	87.0	87.4	69.5	78.6	89.2	69.5	79.5	100.3	69.5	85.2	85.7	69.5	77.7	89.5	69.5	79.6
MF2018	WEIGHTED AVERAGE	96.9	71.4	86.1	97.2	72.5	86.8	100.8	72.5	88.9	99.0	72.9	88.0	96.8	72.5	86.6	96.6	69.4	85.1

		NEW MEXICO			NEW YORK															
		TUCUMCARI			ALBANY			BINGHAMTON			BRONX			BROOKLYN			BUFFALO		
	DIVISION	884			120 - 122			137 - 139			104			112			140 - 142		
		MAT.	INST.	TOTAL	MAT.	INST.	TOTAL	MAT.	INST.	TOTAL	MAT.	INST.	TOTAL	MAT.	INST.	TOTAL	MAT.	INST.	TOTAL
015433	CONTRACTOR EQUIPMENT		107.5	107.5		115.6	115.6		117.6	117.6		104.4	104.4		109.4	109.4		100.5	100.5
0241, 31 - 34	SITE & INFRASTRUCTURE, DEMOLITION	96.2	97.5	97.1	80.0	105.1	97.4	94.3	89.3	90.8	97.3	109.7	105.9	118.1	120.6	119.8	97.3	102.3	100.7
0310	Concrete Forming & Accessories	98.6	66.1	70.9	97.3	107.3	105.8	99.4	93.4	94.3	95.7	189.1	175.3	105.1	189.1	176.7	101.3	117.1	114.7
0320	Concrete Reinforcing	115.0	71.4	93.9	101.3	115.2	108.0	95.8	106.3	100.9	94.8	182.0	136.9	97.2	242.5	167.5	98.9	116.4	107.3
0330	Cast-in-Place Concrete	94.7	70.1	85.5	81.0	116.4	94.2	106.3	106.9	106.5	85.6	172.0	117.8	108.3	170.5	131.5	109.1	123.5	114.4
03	CONCRETE	103.6	69.7	88.7	87.7	112.8	98.7	94.3	102.6	97.9	88.0	181.1	128.9	106.4	189.9	143.1	104.9	118.7	111.0
04	MASONRY	119.9	60.5	83.4	88.0	117.9	106.4	104.3	104.9	104.7	88.8	188.9	150.3	114.8	188.8	160.3	112.6	122.2	118.5
05	METALS	106.3	90.5	101.6	102.2	125.0	108.9	95.0	133.6	106.3	86.4	174.6	112.3	103.1	175.0	124.2	97.0	107.5	100.1
06	WOOD, PLASTICS & COMPOSITES	93.5	67.2	79.0	94.6	104.0	99.8	101.7	90.2	95.4	97.2	188.6	147.5	103.7	188.3	150.3	98.6	115.8	108.1
07	THERMAL & MOISTURE PROTECTION	105.0	72.8	91.0	105.4	110.4	107.6	109.2	94.8	102.9	102.5	168.8	131.3	109.3	168.2	134.9	102.5	111.7	106.5
08	OPENINGS	96.4	67.0	89.5	96.4	103.8	98.1	90.4	93.7	91.2	92.3	197.7	116.9	87.8	197.5	113.4	99.8	111.4	102.5
0920	Plaster & Gypsum Board	80.2	66.1	70.7	97.8	103.9	101.9	109.2	89.7	96.1	102.8	191.0	162.1	103.3	191.0	162.3	105.8	116.1	112.7
0950, 0980	Ceilings & Acoustic Treatment	102.7	66.1	78.1	97.1	103.9	101.7	96.6	89.7	92.0	89.1	191.0	157.8	91.8	191.0	158.7	104.2	116.1	112.2
0960	Flooring	97.2	66.8	88.7	90.9	110.3	96.4	103.1	102.4	102.9	99.9	182.5	123.1	110.2	182.5	130.5	96.3	116.6	102.0
0970, 0990	Wall Finishes & Painting/Coating	92.6	52.6	68.9	95.9	104.4	100.9	89.3	106.3	99.4	106.2	167.4	142.4	116.0	167.4	146.4	96.9	117.8	109.3
09	FINISHES	92.8	64.7	77.6	90.6	107.1	99.5	93.6	95.9	94.8	98.0	186.2	145.7	107.1	186.0	149.8	102.0	117.6	110.5
COVERS	DIVS. 10 - 14, 25, 28, 41, 43, 44, 46	100.0	85.1	96.7	100.0	103.8	100.9	100.0	96.6	99.2	100.0	141.5	109.3	100.0	140.8	109.1	100.0	106.5	101.4
21, 22, 23	FIRE SUPPRESSION, PLUMBING & HVAC	97.9	68.6	86.4	100.1	110.3	104.1	100.6	97.0	99.2	100.3	177.9	130.7	99.8	177.9	130.5	100.0	102.4	100.9
26, 27, 3370	ELECTRICAL, COMMUNICATIONS & UTIL.	90.1	69.5	80.0	100.0	108.4	104.1	98.7	101.0	99.8	90.6	184.5	136.9	98.5	184.5	140.9	99.5	105.3	102.4
MF2018	WEIGHTED AVERAGE	99.9	72.4	88.3	96.8	111.2	102.8	97.3	101.5	99.1	93.8	175.3	128.3	101.9	177.3	133.8	100.8	110.3	104.8

		NEW YORK																	
		ELMIRA			FAR ROCKAWAY			FLUSHING			GLENS FALLS			HICKSVILLE			JAMAICA		
	DIVISION	148 - 149			116			113			128			115, 117, 118			114		
		MAT.	INST.	TOTAL	MAT.	INST.	TOTAL	MAT.	INST.	TOTAL	MAT.	INST.	TOTAL	MAT.	INST.	TOTAL	MAT.	INST.	TOTAL
015433	CONTRACTOR EQUIPMENT		120.0	120.0		109.4	109.4		109.4	109.4		112.6	112.6		109.4	109.4		109.4	109.4
0241, 31 - 34	SITE & INFRASTRUCTURE, DEMOLITION	97.4	89.6	92.0	121.2	120.6	120.8	121.3	120.6	120.8	71.4	99.6	90.9	111.1	119.1	116.6	115.5	120.6	119.0
0310	Concrete Forming & Accessories	85.0	96.3	94.6	92.2	189.1	174.8	95.9	189.1	175.3	81.5	99.9	97.2	88.8	157.2	147.1	95.9	189.1	175.3
0320	Concrete Reinforcing	98.6	105.3	101.8	97.2	242.5	167.5	98.9	242.5	168.4	96.9	112.7	104.5	97.2	174.4	134.6	97.2	242.5	167.5
0330	Cast-in-Place Concrete	98.8	106.4	101.6	117.2	170.5	137.1	117.2	170.5	137.1	78.1	111.6	90.6	99.5	161.5	122.6	108.3	170.5	131.5
03	CONCRETE	90.9	104.2	96.7	112.8	189.9	146.7	113.2	189.9	146.9	81.2	107.4	92.8	98.2	161.8	126.2	105.8	189.9	142.8
04	MASONRY	101.4	106.7	104.6	118.6	188.8	161.8	112.8	188.8	159.5	93.4	112.6	105.2	109.0	174.4	149.2	117.2	188.8	161.2
05	METALS	95.5	135.8	107.3	103.2	175.0	124.3	103.2	175.0	124.3	95.5	124.5	104.0	104.7	172.4	124.6	103.2	175.0	124.3
06	WOOD, PLASTICS & COMPOSITES	84.1	94.4	89.8	87.1	188.3	142.8	91.7	188.3	144.9	82.7	96.1	90.1	83.7	154.0	122.4	91.7	188.3	144.9
07	THERMAL & MOISTURE PROTECTION	108.9	95.2	102.9	109.2	168.3	134.9	109.3	168.3	134.9	98.6	107.1	102.3	108.9	157.8	130.2	109.1	168.3	134.8
08	OPENINGS	96.7	95.7	96.4	86.6	197.5	112.5	86.6	197.5	112.5	90.0	99.1	92.1	87.0	178.6	108.3	86.6	197.5	112.5
0920	Plaster & Gypsum Board	99.5	94.2	96.0	91.7	191.0	158.5	94.2	191.0	159.3	91.0	96.0	94.4	91.4	155.6	134.6	94.2	191.0	159.3
0950, 0980	Ceilings & Acoustic Treatment	106.0	94.2	98.1	81.2	191.0	155.2	81.2	191.0	155.2	86.4	96.0	92.9	80.3	155.6	131.1	81.2	191.0	155.2
0960	Flooring	87.6	102.4	91.8	105.0	182.5	126.8	106.5	182.5	127.9	82.3	107.7	89.5	104.0	179.7	125.2	106.5	182.5	127.9
0970, 0990	Wall Finishes & Painting/Coating	97.2	93.3	94.9	116.0	167.4	146.4	116.0	167.4	146.4	89.3	104.4	98.2	116.0	167.4	146.4	116.0	167.4	146.4
09	FINISHES	94.2	96.9	95.7	102.4	186.0	147.6	103.2	186.0	148.0	84.0	101.1	93.2	101.0	161.6	133.8	102.7	186.0	147.8
COVERS	DIVS. 10 - 14, 25, 28, 41, 43, 44, 46	100.0	97.6	99.5	100.0	140.8	109.1	100.0	140.8	109.1	100.0	96.9	99.3	100.0	132.1	107.1	100.0	140.8	109.1
21, 22, 23	FIRE SUPPRESSION, PLUMBING & HVAC	96.6	95.5	96.2	96.2	177.9	128.3	96.2	177.9	128.3	96.7	109.3	101.6	99.8	161.0	123.8	96.2	177.9	128.3
26, 27, 3370	ELECTRICAL, COMMUNICATIONS & UTIL.	96.4	104.1	100.2	105.3	184.5	144.4	105.3	184.5	144.4	94.3	105.0	99.6	98.0	141.0	119.2	97.0	184.5	140.1
MF2018	WEIGHTED AVERAGE	96.3	102.5	98.9	102.2	177.3	133.9	102.1	177.3	133.9	92.0	107.3	98.5	99.9	157.1	124.1	100.3	177.3	132.8

		NEW YORK																	
		JAMESTOWN			KINGSTON			LONG ISLAND CITY			MONTICELLO			MOUNT VERNON			NEW ROCHELLE		
	DIVISION	147			124			111			127			105			108		
		MAT.	INST.	TOTAL	MAT.	INST.	TOTAL	MAT.	INST.	TOTAL	MAT.	INST.	TOTAL	MAT.	INST.	TOTAL	MAT.	INST.	TOTAL
015433	CONTRACTOR EQUIPMENT		90.6	90.6		109.4	109.4		109.4	109.4		109.4	109.4		104.4	104.4		104.4	104.4
0241, 31 - 34	SITE & INFRASTRUCTURE, DEMOLITION	98.8	89.9	92.6	139.6	115.9	123.2	119.2	120.6	120.2	134.9	115.8	121.7	102.7	105.4	104.5	102.2	105.3	104.3
0310	Concrete Forming & Accessories	85.1	90.1	89.3	82.7	130.7	123.7	100.0	189.1	176.0	89.3	130.8	124.7	86.7	138.8	131.1	100.5	135.3	130.1
0320	Concrete Reinforcing	98.9	111.2	104.8	97.3	160.4	127.8	97.2	242.5	167.5	96.5	160.4	127.4	93.5	180.8	135.7	93.6	180.6	135.7
0330	Cast-in-Place Concrete	102.4	105.2	103.4	105.5	146.3	120.7	111.8	170.5	133.7	98.7	146.3	116.5	95.6	150.0	115.8	95.6	149.7	115.7
03	CONCRETE	93.9	99.0	96.2	102.4	140.8	119.3	109.0	189.9	144.5	97.3	140.8	116.4	97.2	150.4	120.6	96.4	148.7	119.4
04	MASONRY	110.2	104.6	106.8	108.1	155.4	137.2	111.7	188.8	159.1	100.8	155.4	134.3	94.6	156.4	132.6	94.6	156.4	132.6
05	METALS	93.0	101.9	95.6	104.0	135.9	113.4	103.1	175.0	124.2	104.0	136.0	113.4	86.2	169.8	110.7	86.4	169.0	110.7
06	WOOD, PLASTICS & COMPOSITES	82.9	86.2	84.7	83.9	124.0	106.0	97.8	188.3	147.6	90.4	124.0	108.9	87.6	132.4	112.3	104.5	128.7	117.8
07	THERMAL & MOISTURE PROTECTION	108.5	95.4	102.8	121.5	145.0	131.7	109.2	168.3	134.9	121.1	145.0	131.5	103.4	147.3	122.5	103.5	144.7	121.4
08	OPENINGS	96.5	93.0	95.7	92.1	142.5	103.9	86.6	197.5	112.5	88.0	142.4	100.7	92.3	168.4	110.1	92.4	166.3	109.6
0920	Plaster & Gypsum Board	89.2	85.8	86.9	91.0	124.9	113.8	99.3	191.0	161.0	91.7	124.8	114.0	98.3	133.2	121.8	111.0	129.3	123.3
0950, 0980	Ceilings & Acoustic Treatment	102.7	85.8	91.3	77.5	124.9	109.4	81.2	191.0	155.2	77.5	124.8	109.4	87.5	133.2	118.3	87.5	129.3	115.7
0960	Flooring	90.3	102.4	93.7	98.6	160.5	116.0	108.2	182.5	129.1	101.0	160.5	117.8	91.2	179.7	116.1	99.2	164.0	117.4
0970, 0990	Wall Finishes & Painting/Coating	98.8	100.6	99.9	116.6	130.9	125.0	116.0	167.4	146.4	116.6	123.4	120.6	104.6	167.4	141.8	104.6	167.4	141.8
09	FINISHES	93.2	92.5	92.9	97.9	135.4	118.2	104.1	186.0	148.4	98.3	134.6	117.9	95.0	148.0	123.7	98.9	142.6	122.5
COVERS	DIVS. 10 - 14, 25, 28, 41, 43, 44, 46	100.0	96.6	99.3	100.0	121.2	104.7	100.0	140.8	109.1	100.0	121.3	104.7	100.0	129.2	106.5	100.0	113.0	102.9
21, 22, 23	FIRE SUPPRESSION, PLUMBING & HVAC	96.5	93.6	95.4	96.6	134.6	111.5	99.8	177.9	130.5	96.6	138.9	113.2	96.8	145.9	116.1	96.8	145.6	116.0
26, 27, 3370	ELECTRICAL, COMMUNICATIONS & UTIL.	95.4	94.9	95.1	95.7	116.3	105.9	97.5	184.5	140.4	95.7	116.3	105.9	88.8	170.6	129.1	88.8	141.7	114.9
MF2018	WEIGHTED AVERAGE	96.4	96.0	96.3	100.6	133.8	114.6	101.6	177.3	133.6	99.1	134.6	114.1	94.1	150.6	118.0	94.4	144.8	115.7

For customer support on your Commercial Renovation Costs with RSMeans data, call 800.448.8182.

811

City Cost Indexes

DIVISION		NEW YORK																	
		NEW YORK 100 - 102			NIAGARA FALLS 143			PLATTSBURGH 129			POUGHKEEPSIE 125 - 126			QUEENS 110			RIVERHEAD 119		
		MAT.	INST.	TOTAL	MAT.	INST.	TOTAL	MAT.	INST.	TOTAL	MAT.	INST.	TOTAL	MAT.	INST.	TOTAL	MAT.	INST.	TOTAL
015433	CONTRACTOR EQUIPMENT		104.6	104.6		90.6	90.6		94.5	94.5		109.4	109.4		109.4	109.4		109.4	109.4
0241, 31 - 34	SITE & INFRASTRUCTURE, DEMOLITION	106.6	111.9	110.3	100.9	90.8	94.0	109.3	96.7	100.6	135.9	114.8	121.3	114.3	120.6	118.7	112.1	118.6	116.6
0310	Concrete Forming & Accessories	104.3	188.9	176.4	85.0	113.1	108.9	86.8	91.4	90.7	82.7	167.6	155.1	89.0	189.1	174.3	93.2	156.2	146.9
0320	Concrete Reinforcing	100.3	178.3	138.0	97.5	112.0	104.5	101.4	112.0	106.5	97.3	160.0	127.6	98.9	242.5	168.4	99.1	215.6	155.5
0330	Cast-in-Place Concrete	98.9	172.2	126.2	106.0	124.1	112.7	95.5	100.6	97.4	102.2	135.4	114.5	102.9	170.5	128.1	101.2	159.7	123.0
03	CONCRETE	101.6	180.3	136.2	96.2	116.1	105.0	94.3	98.0	95.9	99.7	153.6	123.4	101.5	189.9	140.3	98.7	166.5	128.5
04	MASONRY	102.1	188.8	155.4	117.5	125.1	122.2	88.3	97.4	93.9	100.6	139.5	124.5	106.0	188.8	156.9	114.3	173.3	150.6
05	METALS	97.5	171.0	119.1	95.6	102.4	97.6	99.8	99.5	99.7	104.0	135.6	113.3	103.1	175.0	124.2	105.1	160.6	121.4
06	WOOD, PLASTICS & COMPOSITES	102.0	188.6	149.7	82.8	108.2	96.8	88.8	88.2	88.5	83.9	179.3	136.4	83.8	188.3	141.3	88.7	154.0	124.6
07	THERMAL & MOISTURE PROTECTION	105.2	170.0	133.4	108.6	110.0	109.2	116.0	97.2	107.8	121.4	144.1	131.3	108.8	168.3	134.7	110.0	154.8	129.5
08	OPENINGS	96.0	197.5	119.7	96.5	104.7	98.4	97.4	94.7	96.8	92.1	170.5	110.4	86.6	197.5	112.5	87.0	172.1	106.8
0920	Plaster & Gypsum Board	108.6	191.0	164.0	89.2	108.4	102.1	108.2	87.4	94.2	91.0	181.7	152.0	91.4	191.0	158.4	92.5	155.6	135.0
0950, 0980	Ceilings & Acoustic Treatment	104.5	191.0	162.8	102.7	108.4	106.6	107.3	87.4	93.9	77.5	181.7	147.7	81.2	191.0	155.2	81.1	155.6	131.4
0960	Flooring	101.3	182.5	124.1	90.3	112.1	96.4	104.2	105.1	104.4	98.6	158.6	115.5	104.0	182.5	126.0	105.0	166.9	122.4
0970, 0990	Wall Finishes & Painting/Coating	102.7	167.4	141.0	98.8	111.5	106.3	111.2	95.1	101.7	116.6	123.4	120.6	116.0	167.4	146.4	116.0	167.4	146.4
09	FINISHES	102.9	186.2	148.0	93.3	112.9	103.9	95.0	93.5	94.2	97.7	164.2	133.7	101.3	186.0	147.1	101.5	158.6	132.4
COVERS	DIVS. 10 - 14, 25, 28, 41, 43, 44, 46	100.0	141.5	109.3	100.0	102.7	100.6	100.0	92.7	98.4	100.0	116.5	103.7	100.0	140.8	109.1	100.0	120.1	104.5
21, 22, 23	FIRE SUPPRESSION, PLUMBING & HVAC	100.2	178.3	130.9	96.5	105.1	99.9	96.6	100.3	98.1	96.6	120.3	105.9	99.8	177.9	130.5	100.0	155.9	121.9
26, 27, 3370	ELECTRICAL, COMMUNICATIONS & UTIL.	96.8	184.5	140.0	94.1	98.9	96.4	92.1	91.7	91.9	95.7	121.2	108.3	98.0	184.5	140.6	99.5	134.9	116.9
MF2018	WEIGHTED AVERAGE	99.7	175.2	131.6	97.4	107.5	101.6	97.0	96.6	96.8	99.8	136.8	115.4	100.0	177.3	132.6	100.6	153.5	122.9

DIVISION		NEW YORK																	
		ROCHESTER 144 - 146			SCHENECTADY 123			STATEN ISLAND 103			SUFFERN 109			SYRACUSE 130 - 132			UTICA 133 - 135		
		MAT.	INST.	TOTAL	MAT.	INST.	TOTAL	MAT.	INST.	TOTAL	MAT.	INST.	TOTAL	MAT.	INST.	TOTAL	MAT.	INST.	TOTAL
015433	CONTRACTOR EQUIPMENT		117.6	117.6		112.6	112.6		104.4	104.4		104.4	104.4		112.6	112.6		112.6	112.6
0241, 31 - 34	SITE & INFRASTRUCTURE, DEMOLITION	88.8	105.1	100.0	81.4	100.1	94.3	107.0	109.7	108.9	99.4	103.7	102.4	93.0	98.5	96.8	71.9	98.1	90.0
0310	Concrete Forming & Accessories	105.8	100.3	101.2	96.4	107.1	105.5	86.1	189.3	174.1	94.4	140.8	133.9	98.5	89.1	90.5	99.5	86.3	88.2
0320	Concrete Reinforcing	101.4	105.0	103.1	95.9	115.1	105.2	94.8	216.0	153.4	93.6	152.0	121.9	96.8	106.1	101.3	96.8	100.3	98.5
0330	Cast-in-Place Concrete	97.8	103.3	99.9	94.4	115.4	102.2	95.6	172.1	124.1	92.7	138.9	109.9	98.7	102.3	100.0	90.3	101.0	94.3
03	CONCRETE	99.5	103.3	101.2	94.9	112.4	102.6	99.1	186.4	137.5	93.8	141.3	114.7	97.6	98.1	97.8	95.8	95.4	95.6
04	MASONRY	93.6	103.4	99.7	90.7	117.9	107.4	100.9	188.9	154.9	94.5	142.2	123.8	97.0	101.0	99.5	88.4	99.4	95.2
05	METALS	104.6	119.3	108.9	99.7	126.1	107.4	84.6	174.8	111.1	84.6	131.9	98.5	98.6	119.8	104.9	96.7	117.5	102.9
06	WOOD, PLASTICS & COMPOSITES	108.4	100.2	103.9	100.1	103.8	102.1	86.2	188.6	142.6	96.9	142.5	122.0	98.2	86.0	91.5	98.2	82.7	89.7
07	THERMAL & MOISTURE PROTECTION	116.4	99.8	109.2	100.1	109.8	104.3	102.9	168.8	131.5	103.4	140.9	119.7	104.0	94.1	99.7	92.7	94.0	93.3
08	OPENINGS	100.7	99.3	100.3	95.3	103.7	97.3	92.3	197.7	116.9	92.4	150.7	106.0	92.4	89.5	91.8	95.2	86.3	93.1
0920	Plaster & Gypsum Board	103.8	100.1	101.3	100.3	103.9	102.7	98.4	191.0	160.7	102.1	143.5	130.0	96.8	85.6	89.3	96.8	82.2	87.0
0950, 0980	Ceilings & Acoustic Treatment	100.4	100.1	100.2	92.8	103.9	100.3	89.1	191.0	157.8	87.5	143.5	125.2	96.6	85.6	89.2	96.6	82.2	86.9
0960	Flooring	90.4	105.0	94.5	89.6	101.3	95.4	95.1	182.5	119.6	95.0	171.6	116.5	92.0	92.3	92.1	90.0	92.4	90.7
0970, 0990	Wall Finishes & Painting/Coating	97.7	98.4	98.1	89.3	104.4	98.2	106.2	167.4	142.4	104.6	126.9	117.8	91.9	98.7	95.9	85.6	98.7	93.4
09	FINISHES	95.7	101.4	98.8	89.2	107.0	98.9	96.7	186.2	145.2	96.3	145.8	123.1	92.4	90.0	91.1	90.8	87.9	89.2
COVERS	DIVS. 10 - 14, 25, 28, 41, 43, 44, 46	100.0	98.9	99.7	100.0	103.5	100.8	100.0	141.5	109.3	100.0	114.7	103.3	100.0	95.1	98.9	100.0	89.8	97.7
21, 22, 23	FIRE SUPPRESSION, PLUMBING & HVAC	99.9	88.1	95.3	100.2	106.2	102.6	100.3	177.9	130.8	96.8	123.7	107.4	100.3	94.4	98.0	100.3	92.5	97.2
26, 27, 3370	ELECTRICAL, COMMUNICATIONS & UTIL.	99.5	89.9	94.8	98.7	108.4	103.5	90.6	184.5	136.9	94.8	114.6	104.6	98.6	101.0	99.8	96.6	101.0	98.8
MF2018	WEIGHTED AVERAGE	100.3	99.2	99.8	97.0	109.9	102.4	95.6	176.1	129.6	94.1	130.2	109.4	97.8	98.3	98.0	96.0	96.6	96.2

DIVISION		NEW YORK									NORTH CAROLINA								
		WATERTOWN 136			WHITE PLAINS 106			YONKERS 107			ASHEVILLE 287 - 288			CHARLOTTE 281 - 282			DURHAM 277		
		MAT.	INST.	TOTAL	MAT.	INST.	TOTAL	MAT.	INST.	TOTAL	MAT.	INST.	TOTAL	MAT.	INST.	TOTAL	MAT.	INST.	TOTAL
015433	CONTRACTOR EQUIPMENT		112.6	112.6		104.4	104.4		104.4	104.4		99.0	99.0		100.4	100.4		104.1	104.1
0241, 31 - 34	SITE & INFRASTRUCTURE, DEMOLITION	79.5	98.6	92.7	97.1	105.4	102.8	104.5	105.4	105.1	96.4	77.8	83.5	98.6	81.9	87.1	98.5	86.3	90.1
0310	Concrete Forming & Accessories	85.3	93.0	91.9	99.2	149.3	141.9	99.4	149.1	141.8	90.5	61.4	65.7	96.6	61.5	66.6	96.2	61.4	66.5
0320	Concrete Reinforcing	97.5	106.1	101.6	93.6	180.8	135.8	97.3	180.8	137.7	95.4	66.0	81.2	99.6	66.8	83.7	104.4	66.8	86.2
0330	Cast-in-Place Concrete	105.1	104.7	105.0	84.9	150.0	109.1	95.0	150.0	115.4	105.9	71.5	93.1	108.0	71.6	94.4	106.3	71.2	93.2
03	CONCRETE	108.8	100.7	105.3	87.6	155.2	117.3	96.6	155.1	122.3	91.1	67.4	80.7	91.2	67.5	80.8	95.2	67.4	83.0
04	MASONRY	89.7	104.8	99.0	93.9	156.4	132.3	97.1	156.4	133.6	84.9	64.4	72.3	88.6	64.4	73.8	84.3	64.4	72.0
05	METALS	96.8	119.8	104.6	86.1	169.8	110.7	94.1	169.9	116.4	101.2	89.9	97.9	102.1	88.9	98.2	117.9	90.2	109.7
06	WOOD, PLASTICS & COMPOSITES	80.3	89.9	85.6	102.4	146.7	126.8	102.3	146.3	126.5	87.6	59.0	71.9	89.9	59.0	72.9	90.6	59.0	73.2
07	THERMAL & MOISTURE PROTECTION	93.0	97.2	94.8	103.2	148.7	123.0	103.5	149.3	123.4	99.1	63.8	83.8	93.7	64.5	81.0	102.8	63.8	85.8
08	OPENINGS	95.2	93.8	94.8	92.4	176.2	112.0	96.0	176.5	114.8	89.0	59.8	82.2	99.0	60.0	89.9	95.6	60.0	87.3
0920	Plaster & Gypsum Board	87.6	89.6	89.0	105.3	147.8	133.9	108.7	147.4	134.7	102.2	57.8	72.3	98.1	57.8	71.0	89.9	57.8	68.3
0950, 0980	Ceilings & Acoustic Treatment	96.6	89.6	91.9	87.5	147.8	128.2	103.7	147.4	133.2	79.4	57.8	64.9	83.5	57.8	66.2	82.8	57.8	66.0
0960	Flooring	83.1	92.4	85.7	97.5	179.7	120.6	96.9	182.5	121.0	94.9	66.8	87.0	95.3	66.8	87.3	102.0	66.8	92.1
0970, 0990	Wall Finishes & Painting/Coating	85.6	93.0	90.0	104.6	167.4	141.8	104.6	167.4	141.8	102.5	57.1	75.7	95.9	57.1	72.9	104.4	57.1	76.4
09	FINISHES	88.2	92.4	90.5	97.1	156.4	129.2	97.9	156.7	131.1	88.8	61.6	74.1	88.3	61.6	73.9	90.4	61.6	74.8
COVERS	DIVS. 10 - 14, 25, 28, 41, 43, 44, 46	100.0	96.4	99.2	100.0	127.5	106.1	100.0	130.7	106.8	100.0	82.5	96.1	100.0	82.5	96.1	100.0	82.5	96.1
21, 22, 23	FIRE SUPPRESSION, PLUMBING & HVAC	100.3	87.8	95.4	100.4	145.9	118.3	100.4	146.0	118.3	100.4	61.3	85.0	99.9	63.1	85.5	100.5	61.3	85.1
26, 27, 3370	ELECTRICAL, COMMUNICATIONS & UTIL.	98.6	90.2	94.4	88.8	170.6	129.1	94.9	170.6	132.2	100.8	57.8	79.6	100.0	60.3	80.4	96.5	57.4	77.2
MF2018	WEIGHTED AVERAGE	97.7	96.8	97.4	93.8	152.8	118.8	97.9	153.0	121.2	96.4	66.6	83.8	97.4	67.6	84.8	100.1	67.2	86.2

NORTH CAROLINA

DIVISION		ELIZABETH CITY 279			FAYETTEVILLE 283			GASTONIA 280			GREENSBORO 270, 272 - 274			HICKORY 286			KINSTON 285		
		MAT.	INST.	TOTAL	MAT.	INST.	TOTAL	MAT.	INST.	TOTAL	MAT.	INST.	TOTAL	MAT.	INST.	TOTAL	MAT.	INST.	TOTAL
015433	CONTRACTOR EQUIPMENT		108.0	108.0		104.1	104.1		99.0	99.0		104.1	104.1		104.1	104.1		104.1	104.1
0241, 31 - 34	SITE & INFRASTRUCTURE, DEMOLITION	102.7	87.9	92.4	95.6	86.1	89.0	96.1	78.1	83.7	98.3	86.4	90.1	95.1	85.1	88.2	93.9	84.9	87.7
0310	Concrete Forming & Accessories	82.6	63.6	66.4	90.1	60.1	64.5	96.4	61.4	66.6	96.0	61.3	66.5	87.1	61.2	65.0	83.6	59.9	63.4
0320	Concrete Reinforcing	102.3	69.8	86.6	99.1	66.8	83.4	95.8	66.8	81.7	103.2	66.8	85.6	95.4	66.7	81.5	94.9	66.7	81.3
0330	Cast-in-Place Concrete	106.5	71.7	93.5	111.1	69.3	95.5	103.6	71.2	91.5	105.5	71.2	92.7	105.9	71.2	93.0	102.2	69.2	90.0
03	CONCRETE	95.4	69.1	83.8	93.1	66.2	81.3	89.8	67.5	80.0	94.7	67.4	82.7	90.9	67.3	80.5	88.1	66.1	78.4
04	MASONRY	95.4	61.0	74.2	88.3	61.0	71.5	89.4	64.4	74.0	81.8	64.4	71.1	74.6	64.4	68.3	81.2	61.0	68.8
05	METALS	103.5	92.3	100.2	122.2	90.2	112.8	101.8	90.2	98.4	110.0	90.2	104.1	101.3	90.0	98.0	100.0	90.0	97.1
06	WOOD, PLASTICS & COMPOSITES	75.1	63.4	68.7	86.5	59.0	71.4	95.4	59.0	75.4	90.2	59.0	73.1	82.4	59.0	69.6	79.3	59.0	68.2
07	THERMAL & MOISTURE PROTECTION	102.0	62.5	84.8	98.6	62.3	82.8	99.3	63.8	83.9	102.6	63.8	85.7	99.5	63.8	84.0	99.3	62.3	83.2
08	OPENINGS	92.9	63.1	86.0	89.1	60.0	82.3	92.3	60.0	84.8	95.6	60.0	87.3	89.0	60.0	82.3	89.1	60.0	82.3
0920	Plaster & Gypsum Board	84.4	61.7	69.1	105.8	57.8	73.5	107.9	57.8	74.2	91.1	57.8	68.7	102.2	57.8	72.3	100.2	57.8	72.3
0950, 0980	Ceilings & Acoustic Treatment	82.8	61.7	68.5	81.9	57.8	65.7	83.5	57.8	66.2	82.8	57.8	66.0	79.4	57.8	64.9	83.5	57.8	66.2
0960	Flooring	94.1	66.8	86.4	95.1	66.8	87.1	98.0	66.8	89.2	102.0	66.8	92.1	94.8	66.8	86.9	92.2	66.8	85.0
0970, 0990	Wall Finishes & Painting/Coating	104.4	57.1	76.4	102.5	57.1	75.7	102.5	57.1	75.7	104.4	57.1	76.4	102.5	57.1	75.7	102.5	57.1	75.7
09	FINISHES	87.7	63.4	74.6	89.7	60.8	74.0	91.1	61.6	75.2	90.6	61.6	74.9	88.9	61.6	74.1	88.7	60.8	73.6
COVERS	DIVS. 10 - 14, 25, 28, 41, 43, 44, 46	100.0	85.4	96.8	100.0	81.4	95.8	100.0	82.5	96.1	100.0	79.8	95.5	100.0	82.5	96.1	100.0	81.3	95.8
21, 22, 23	FIRE SUPPRESSION, PLUMBING & HVAC	96.9	58.6	81.9	100.2	59.5	84.2	100.4	60.1	84.6	100.4	61.3	85.0	96.9	60.2	82.5	96.9	58.3	81.7
26, 27, 3370	ELECTRICAL, COMMUNICATIONS & UTIL.	96.2	64.6	80.6	100.6	57.4	79.3	100.3	60.4	80.6	95.7	57.8	77.0	98.6	60.4	79.8	98.4	55.8	77.4
MF2018	WEIGHTED AVERAGE	96.9	68.3	84.8	100.2	66.1	85.8	97.0	66.8	84.2	98.5	67.2	85.3	94.7	67.3	83.1	94.4	65.5	82.2

NORTH CAROLINA / NORTH DAKOTA

DIVISION		MURPHY 289			RALEIGH 275 - 276			ROCKY MOUNT 278			WILMINGTON 284			WINSTON-SALEM 271			BISMARCK 585		
		MAT.	INST.	TOTAL	MAT.	INST.	TOTAL	MAT.	INST.	TOTAL	MAT.	INST.	TOTAL	MAT.	INST.	TOTAL	MAT.	INST.	TOTAL
015433	CONTRACTOR EQUIPMENT		99.0	99.0		106.5	106.5		104.1	104.1		99.0	99.0		104.1	104.1		99.9	99.9
0241, 31 - 34	SITE & INFRASTRUCTURE, DEMOLITION	97.4	76.3	82.8	98.8	90.8	93.3	100.7	86.3	90.8	97.4	77.6	83.7	98.7	86.4	90.2	99.8	97.8	98.4
0310	Concrete Forming & Accessories	97.0	59.8	65.3	96.9	61.0	66.3	88.5	60.9	64.9	91.8	60.1	64.8	97.9	61.4	66.8	110.6	76.6	81.6
0320	Concrete Reinforcing	94.9	64.3	80.1	105.4	66.8	86.7	102.3	66.8	85.1	96.1	66.8	81.9	103.2	66.8	85.6	92.6	98.3	95.3
0330	Cast-in-Place Concrete	109.6	69.2	94.6	109.3	71.3	95.1	104.2	70.5	91.6	105.5	69.3	92.0	108.1	71.2	94.4	106.3	86.6	98.9
03	CONCRETE	93.9	65.6	81.4	93.9	67.2	82.2	95.9	66.9	83.2	91.1	66.2	80.1	96.0	67.4	83.5	93.3	84.4	89.4
04	MASONRY	77.3	61.0	67.3	80.5	63.1	69.8	74.4	63.1	67.4	75.1	61.0	66.4	82.0	64.4	71.2	105.3	83.4	91.8
05	METALS	99.0	89.1	96.1	102.2	89.1	98.3	102.6	90.2	99.0	100.7	90.2	97.6	107.0	90.2	102.1	95.6	94.2	95.2
06	WOOD, PLASTICS & COMPOSITES	96.1	58.9	75.6	90.9	59.2	73.4	82.0	59.0	69.4	89.2	59.0	72.6	90.2	59.0	73.1	106.6	72.7	88.0
07	THERMAL & MOISTURE PROTECTION	99.3	62.3	83.2	97.2	63.8	82.7	102.5	63.2	85.4	99.1	62.3	83.1	102.6	63.8	85.7	109.3	86.0	99.2
08	OPENINGS	89.0	59.4	82.1	97.6	60.1	88.8	92.2	60.0	84.7	89.1	60.0	82.3	95.6	60.0	87.3	103.7	81.9	98.6
0920	Plaster & Gypsum Board	106.9	57.7	73.8	86.5	57.8	67.2	85.8	57.8	67.0	103.7	57.8	72.8	91.1	57.8	68.7	102.4	72.2	82.1
0950, 0980	Ceilings & Acoustic Treatment	79.4	57.7	64.8	83.5	57.8	66.2	80.4	57.8	65.2	81.9	57.8	65.7	82.8	57.8	66.0	108.5	72.2	84.1
0960	Flooring	98.3	66.8	89.4	96.8	66.8	88.4	97.6	66.8	89.6	96.8	66.8	89.6	102.0	66.8	92.1	86.9	53.7	77.6
0970, 0990	Wall Finishes & Painting/Coating	102.5	57.1	75.7	97.8	57.1	73.7	104.4	57.1	76.4	102.5	57.1	75.7	104.4	57.1	76.4	90.9	59.8	72.5
09	FINISHES	90.6	60.7	74.4	90.0	61.4	74.5	88.6	61.3	73.8	89.6	60.8	74.0	90.6	61.6	74.9	96.1	71.1	82.5
COVERS	DIVS. 10 - 14, 25, 28, 41, 43, 44, 46	100.0	81.3	95.8	100.0	82.3	96.1	100.0	82.1	96.0	100.0	81.4	95.8	100.0	82.5	96.1	100.0	94.8	98.8
21, 22, 23	FIRE SUPPRESSION, PLUMBING & HVAC	96.9	58.3	81.7	100.0	60.6	84.5	96.9	59.5	82.2	100.4	59.5	84.3	100.4	61.3	85.0	99.9	77.0	90.9
26, 27, 3370	ELECTRICAL, COMMUNICATIONS & UTIL.	101.7	55.8	79.1	99.1	56.3	78.0	98.0	57.4	78.0	101.1	55.8	78.8	95.7	57.8	77.0	98.7	73.4	86.3
MF2018	WEIGHTED AVERAGE	95.5	64.7	82.5	97.4	67.0	84.5	96.0	66.6	83.6	96.0	65.2	83.0	98.2	67.3	85.2	98.9	81.6	91.6

NORTH DAKOTA

DIVISION		DEVILS LAKE 583			DICKINSON 586			FARGO 580 - 581			GRAND FORKS 582			JAMESTOWN 584			MINOT 587		
		MAT.	INST.	TOTAL	MAT.	INST.	TOTAL	MAT.	INST.	TOTAL	MAT.	INST.	TOTAL	MAT.	INST.	TOTAL	MAT.	INST.	TOTAL
015433	CONTRACTOR EQUIPMENT		97.7	97.7		97.7	97.7		99.9	99.9		97.7	97.7		97.7	97.7		97.7	97.7
0241, 31 - 34	SITE & INFRASTRUCTURE, DEMOLITION	107.5	93.7	98.0	115.7	93.7	100.5	101.3	97.6	98.8	111.5	93.7	99.2	106.5	93.7	97.7	108.9	94.1	98.7
0310	Concrete Forming & Accessories	107.0	71.6	76.8	95.5	71.5	75.1	99.4	72.1	76.1	99.7	71.6	75.7	97.1	71.6	75.4	95.1	71.8	75.2
0320	Concrete Reinforcing	94.8	98.0	96.4	95.7	98.2	96.9	95.7	98.7	97.2	93.4	98.0	95.6	95.4	98.7	97.0	96.6	98.1	97.4
0330	Cast-in-Place Concrete	126.2	81.6	109.6	114.1	81.5	102.0	102.5	85.6	96.2	114.1	81.5	102.0	124.6	81.6	108.6	114.1	81.8	102.1
03	CONCRETE	104.4	80.4	93.8	103.5	80.4	93.3	98.3	82.1	91.2	100.9	80.3	91.8	102.9	80.5	93.1	99.4	80.6	91.1
04	MASONRY	118.0	79.2	94.2	120.1	81.4	96.4	102.7	90.0	94.9	112.1	79.2	91.9	130.9	89.9	105.7	110.1	80.7	92.1
05	METALS	94.7	94.1	94.5	94.6	94.0	94.4	100.2	94.8	98.6	94.6	93.8	94.4	94.6	94.9	94.7	94.9	94.8	94.9
06	WOOD, PLASTICS & COMPOSITES	99.0	67.5	81.7	85.6	67.5	75.7	95.9	67.6	80.4	90.4	67.5	77.8	87.7	67.5	76.6	95.3	67.5	75.5
07	THERMAL & MOISTURE PROTECTION	107.1	82.3	96.3	107.7	83.7	97.2	104.1	87.4	96.8	107.3	83.0	96.7	106.9	85.7	97.7	107.0	83.9	97.0
08	OPENINGS	100.9	79.0	95.8	100.9	79.0	95.8	101.0	79.1	95.9	99.5	79.0	94.7	100.9	79.0	95.8	99.7	79.0	94.8
0920	Plaster & Gypsum Board	118.2	67.0	83.8	109.1	67.0	80.8	101.1	67.0	78.2	110.3	67.0	81.2	110.0	67.0	81.1	109.1	67.0	80.8
0950, 0980	Ceilings & Acoustic Treatment	105.6	67.0	79.6	105.6	67.0	79.6	95.4	67.0	76.3	105.6	67.0	79.6	105.6	67.0	79.6	105.6	67.0	79.6
0960	Flooring	93.8	53.7	82.5	87.4	53.7	77.9	101.2	53.7	87.9	89.2	53.7	79.2	88.0	53.7	78.4	87.1	53.7	77.7
0970, 0990	Wall Finishes & Painting/Coating	86.6	56.4	68.7	86.6	56.4	68.7	93.1	67.4	77.9	86.6	66.0	74.4	86.6	56.4	68.7	86.6	57.5	69.4
09	FINISHES	95.2	66.8	79.8	93.0	66.8	78.8	96.2	68.1	81.0	93.2	67.8	79.5	92.3	66.8	78.5	92.1	66.9	78.5
COVERS	DIVS. 10 - 14, 25, 28, 41, 43, 44, 46	100.0	87.2	97.2	100.0	87.2	97.2	100.0	93.6	98.6	100.0	87.2	97.2	100.0	87.2	97.0	100.0	93.3	98.5
21, 22, 23	FIRE SUPPRESSION, PLUMBING & HVAC	96.6	78.3	89.4	96.6	73.3	87.4	100.0	74.1	89.8	100.1	72.3	89.2	96.6	72.3	87.0	100.1	73.1	89.1
26, 27, 3370	ELECTRICAL, COMMUNICATIONS & UTIL.	94.3	68.3	81.5	101.9	67.9	85.1	98.5	69.4	84.2	97.3	68.3	83.0	94.3	68.3	81.5	100.1	72.6	86.6
MF2018	WEIGHTED AVERAGE	99.2	78.8	90.6	99.9	77.9	90.6	99.7	80.3	91.5	99.4	77.6	90.2	99.2	78.8	90.6	99.3	78.6	90.5

City Cost Indexes

OHIO (and North Dakota)

DIVISION		WILLISTON 588 MAT.	INST.	TOTAL	AKRON 442-443 MAT.	INST.	TOTAL	ATHENS 457 MAT.	INST.	TOTAL	CANTON 446-447 MAT.	INST.	TOTAL	CHILLICOTHE 456 MAT.	INST.	TOTAL	CINCINNATI 451-452 MAT.	INST.	TOTAL
015433	CONTRACTOR EQUIPMENT		97.7	97.7		88.7	88.7		84.9	84.9		88.7	88.7		95.5	95.5		95.9	95.9
0241, 31 - 34	SITE & INFRASTRUCTURE, DEMOLITION	109.2	91.5	97.0	96.2	94.1	94.8	107.3	85.3	92.1	96.3	93.9	94.7	93.8	94.9	94.6	90.0	98.2	95.7
0310	Concrete Forming & Accessories	101.7	71.3	75.8	102.8	82.8	85.8	95.3	78.9	81.4	102.8	74.3	78.5	98.0	81.5	84.0	101.5	80.2	83.3
0320	Concrete Reinforcing	97.5	98.1	97.8	94.7	91.0	92.9	86.9	89.2	88.0	94.7	75.0	85.2	84.0	88.8	86.3	89.1	77.8	83.6
0330	Cast-in-Place Concrete	114.1	81.4	101.9	102.6	89.0	97.6	111.2	95.8	105.5	103.6	87.2	97.5	100.9	92.4	97.8	96.4	77.5	89.3
03	CONCRETE	100.8	80.2	91.7	99.8	85.9	93.7	101.0	86.3	94.6	100.3	78.7	90.8	95.6	86.9	91.8	94.7	79.0	87.8
04	MASONRY	105.1	80.7	90.1	90.9	89.4	90.0	73.6	97.2	88.1	91.5	80.7	84.9	80.2	90.5	86.5	83.0	79.8	81.0
05	METALS	94.8	93.6	94.4	97.5	80.2	92.4	97.1	80.5	92.3	97.5	73.7	90.5	89.4	89.8	89.5	91.5	81.7	88.6
06	WOOD, PLASTICS & COMPOSITES	92.1	67.5	78.6	105.9	81.1	92.2	87.4	74.0	80.0	106.2	71.8	87.3	100.3	78.0	88.0	102.3	80.1	90.1
07	THERMAL & MOISTURE PROTECTION	107.3	83.5	96.9	104.9	91.1	98.9	102.9	91.5	98.0	106.0	87.2	97.8	104.7	88.4	97.6	102.7	80.9	93.2
08	OPENINGS	101.0	79.0	95.8	107.5	82.7	101.7	96.2	75.8	91.5	101.3	69.8	93.9	88.1	78.0	85.7	98.3	76.4	93.2
0920	Plaster & Gypsum Board	110.3	67.0	81.2	98.3	80.5	86.3	93.1	73.2	79.7	99.2	71.0	80.2	96.6	77.8	83.9	96.1	79.9	85.2
0950, 0980	Ceilings & Acoustic Treatment	105.6	67.0	79.6	90.6	80.5	83.8	105.9	73.2	83.8	90.6	71.0	77.4	98.7	77.8	84.6	93.6	79.9	84.3
0960	Flooring	90.1	53.7	79.9	91.7	83.3	89.3	120.7	75.3	108.0	91.8	74.1	86.8	97.5	75.3	91.3	98.7	78.5	93.0
0970, 0990	Wall Finishes & Painting/Coating	86.6	56.4	68.7	98.3	90.9	93.9	100.1	89.1	93.6	98.3	73.9	83.9	97.4	89.1	92.5	97.0	71.7	82.0
09	FINISHES	93.4	66.8	79.0	95.0	83.6	88.8	99.4	78.7	88.2	95.2	73.4	83.4	96.5	80.4	87.8	95.9	79.1	86.8
COVERS	DIVS. 10 - 14, 25, 28, 41, 43, 44, 46	100.0	87.2	97.1	100.0	90.3	97.8	100.0	87.9	97.3	100.0	88.4	97.4	100.0	86.8	97.1	100.0	88.4	97.4
21, 22, 23	FIRE SUPPRESSION, PLUMBING & HVAC	96.6	73.2	87.4	100.0	87.8	95.2	96.5	81.2	90.5	100.0	78.7	91.6	97.0	92.4	95.2	99.9	76.9	90.9
26, 27, 3370	ELECTRICAL, COMMUNICATIONS & UTIL.	97.7	68.7	83.4	98.6	81.8	90.4	97.9	89.7	93.9	97.9	85.0	91.5	97.2	89.7	93.5	96.3	74.2	85.4
MF2018	WEIGHTED AVERAGE	98.4	77.7	89.6	99.4	86.0	93.8	97.1	84.8	91.9	98.9	79.9	90.9	94.3	88.5	91.8	96.1	80.0	89.3

OHIO

DIVISION		CLEVELAND 441 MAT.	INST.	TOTAL	COLUMBUS 430-432 MAT.	INST.	TOTAL	DAYTON 453-454 MAT.	INST.	TOTAL	HAMILTON 450 MAT.	INST.	TOTAL	LIMA 458 MAT.	INST.	TOTAL	LORAIN 440 MAT.	INST.	TOTAL
015433	CONTRACTOR EQUIPMENT		92.0	92.0		92.3	92.3		88.9	88.9		95.5	95.5		88.3	88.3		88.7	88.7
0241, 31 - 34	SITE & INFRASTRUCTURE, DEMOLITION	94.9	96.2	95.8	101.8	92.3	95.3	90.0	94.2	92.9	90.0	94.5	93.1	101.1	84.9	89.9	95.5	94.3	94.7
0310	Concrete Forming & Accessories	102.3	90.2	92.0	98.7	78.1	81.1	99.9	77.2	80.6	99.9	77.4	80.7	95.3	76.3	79.1	102.8	74.6	78.7
0320	Concrete Reinforcing	95.2	91.4	93.4	99.9	78.7	89.7	89.1	79.1	84.2	89.1	77.2	83.3	86.9	79.3	83.2	94.7	91.3	93.1
0330	Cast-in-Place Concrete	99.5	97.2	98.6	102.1	80.9	94.2	86.5	80.6	84.3	92.6	80.9	88.3	102.1	89.5	97.4	97.7	91.4	95.4
03	CONCRETE	100.1	92.4	96.7	98.6	79.2	90.0	87.0	78.6	83.3	89.9	79.1	85.1	93.9	81.5	88.5	97.5	83.1	91.2
04	MASONRY	96.9	97.9	97.5	91.4	85.2	87.6	79.1	76.4	77.4	79.5	81.2	80.5	98.8	78.1	86.1	87.8	93.1	91.0
05	METALS	99.1	83.3	94.5	98.1	79.3	92.5	90.9	76.8	86.7	90.9	85.6	89.4	97.2	80.1	92.2	98.1	81.6	93.3
06	WOOD, PLASTICS & COMPOSITES	99.8	88.5	93.6	98.5	77.7	87.0	104.2	77.1	89.3	103.0	77.1	88.7	87.3	75.1	80.6	105.9	70.1	86.2
07	THERMAL & MOISTURE PROTECTION	102.0	96.9	99.8	93.6	83.8	89.3	108.8	79.5	96.1	104.8	79.9	94.0	102.4	83.7	94.3	105.9	91.0	99.4
08	OPENINGS	100.3	86.4	97.0	98.3	74.0	92.7	95.0	74.6	90.3	92.8	74.8	88.6	96.3	73.7	91.0	101.3	76.6	95.5
0920	Plaster & Gypsum Board	97.9	88.1	91.3	94.1	77.0	82.6	98.1	76.9	83.9	98.1	76.9	83.9	93.1	74.3	80.5	98.3	69.2	78.7
0950, 0980	Ceilings & Acoustic Treatment	86.3	88.1	87.5	93.5	77.0	82.4	99.6	76.9	84.3	98.7	76.9	84.0	105.9	74.3	84.6	90.6	69.2	76.2
0960	Flooring	93.1	89.8	92.2	95.1	75.3	89.5	101.2	72.4	93.1	98.5	78.5	92.9	119.8	77.2	107.8	91.8	89.8	91.2
0970, 0990	Wall Finishes & Painting/Coating	101.2	90.9	95.1	99.5	79.6	87.7	97.4	70.7	81.6	97.4	71.3	82.0	100.2	76.5	86.2	98.3	90.9	93.9
09	FINISHES	95.3	90.1	92.5	95.0	77.5	85.5	97.7	75.4	85.6	96.7	76.9	86.0	98.7	76.0	86.4	95.0	78.0	85.8
COVERS	DIVS. 10 - 14, 25, 28, 41, 43, 44, 46	100.0	96.4	99.2	100.0	88.2	97.4	100.0	85.1	96.7	100.0	85.3	96.7	100.0	84.4	96.5	100.0	90.9	98.0
21, 22, 23	FIRE SUPPRESSION, PLUMBING & HVAC	100.0	89.9	96.0	100.0	84.1	93.7	100.7	81.2	93.0	100.5	74.9	90.5	96.5	91.7	94.6	100.0	88.4	95.4
26, 27, 3370	ELECTRICAL, COMMUNICATIONS & UTIL.	98.3	92.7	95.5	99.9	80.2	90.2	94.9	75.7	85.4	95.2	75.5	85.5	98.2	75.7	87.1	98.0	78.2	88.2
MF2018	WEIGHTED AVERAGE	99.1	91.6	95.9	98.4	82.0	91.4	94.8	79.2	88.2	94.8	79.5	88.3	97.2	81.7	90.7	98.4	84.7	92.6

OHIO

DIVISION		MANSFIELD 448-449 MAT.	INST.	TOTAL	MARION 433 MAT.	INST.	TOTAL	SPRINGFIELD 455 MAT.	INST.	TOTAL	STEUBENVILLE 439 MAT.	INST.	TOTAL	TOLEDO 434-436 MAT.	INST.	TOTAL	YOUNGSTOWN 444-445 MAT.	INST.	TOTAL
015433	CONTRACTOR EQUIPMENT		88.7	88.7		88.7	88.7		88.9	88.9		92.4	92.4		92.4	92.4		88.7	88.7
0241, 31 - 34	SITE & INFRASTRUCTURE, DEMOLITION	91.7	94.0	93.3	95.7	90.6	92.2	90.3	94.2	93.0	141.5	98.2	111.6	99.8	90.9	93.6	96.1	93.9	94.6
0310	Concrete Forming & Accessories	92.1	73.5	76.3	96.6	79.4	81.9	99.9	77.1	80.5	97.6	79.0	81.7	100.3	84.9	87.1	102.8	76.6	80.5
0320	Concrete Reinforcing	86.2	79.0	82.7	92.1	79.0	85.8	89.1	79.1	84.2	89.7	94.9	92.2	99.9	82.9	91.7	94.7	85.2	90.1
0330	Cast-in-Place Concrete	95.1	87.8	92.3	88.5	88.1	88.3	88.8	80.3	85.6	96.1	89.3	93.6	96.9	90.2	94.4	101.7	87.2	96.3
03	CONCRETE	92.1	79.3	86.4	85.4	82.2	84.0	88.1	78.5	83.9	89.4	84.9	87.4	93.2	86.4	90.2	99.4	81.5	91.5
04	MASONRY	89.9	88.9	89.3	92.7	89.9	90.9	79.3	75.9	77.2	80.7	91.1	87.1	98.9	91.8	94.5	91.1	86.6	88.4
05	METALS	98.4	76.0	91.8	97.1	78.9	91.7	90.9	76.8	86.7	93.6	83.8	90.6	97.8	85.3	94.2	97.5	77.9	91.8
06	WOOD, PLASTICS & COMPOSITES	92.8	70.1	80.3	92.8	77.5	84.4	105.6	77.1	89.9	88.1	76.4	81.6	96.9	83.9	89.8	105.9	74.6	88.7
07	THERMAL & MOISTURE PROTECTION	104.1	88.4	97.3	90.5	88.8	89.8	108.7	79.2	95.9	102.0	87.4	95.7	91.0	91.2	91.1	106.1	88.2	98.3
08	OPENINGS	102.3	71.0	95.0	91.2	75.1	87.5	93.2	74.6	88.9	91.3	78.4	88.3	93.8	81.3	90.9	101.3	75.2	95.2
0920	Plaster & Gypsum Board	90.9	69.2	76.3	95.3	77.0	83.0	98.1	76.9	83.9	93.5	75.4	81.3	97.2	83.6	88.0	98.3	73.9	81.9
0950, 0980	Ceilings & Acoustic Treatment	91.4	69.2	76.4	99.5	77.0	84.3	99.6	76.9	84.3	96.7	75.4	82.3	99.5	83.6	88.8	90.6	73.9	79.3
0960	Flooring	86.6	92.7	88.3	94.4	92.7	93.9	101.2	72.4	93.1	122.5	92.7	114.1	94.8	94.4	94.7	91.8	90.0	91.3
0970, 0990	Wall Finishes & Painting/Coating	98.3	77.8	86.2	104.0	77.8	88.5	97.4	70.7	81.6	117.8	87.1	99.6	104.0	87.0	94.0	98.3	80.0	87.5
09	FINISHES	92.4	76.6	83.8	96.3	81.2	88.1	97.7	75.3	85.6	114.5	81.6	96.7	97.0	86.6	91.4	95.1	78.8	86.3
COVERS	DIVS. 10 - 14, 25, 28, 41, 43, 44, 46	100.0	88.3	97.4	100.0	85.9	96.8	100.0	84.9	96.6	100.0	86.2	96.9	100.0	91.3	98.1	100.0	88.5	97.4
21, 22, 23	FIRE SUPPRESSION, PLUMBING & HVAC	96.5	87.5	92.9	96.5	93.1	95.1	100.7	80.9	92.9	96.9	91.3	94.7	100.0	93.6	97.5	100.0	83.9	93.7
26, 27, 3370	ELECTRICAL, COMMUNICATIONS & UTIL.	95.7	90.8	93.3	94.0	90.8	92.4	94.9	80.2	87.7	88.7	106.4	97.4	100.0	103.4	101.7	98.0	74.3	86.3
MF2018	WEIGHTED AVERAGE	96.4	84.2	91.3	94.3	86.7	91.1	94.8	79.7	88.4	96.1	90.0	93.5	97.6	91.2	94.9	98.7	81.9	91.6

DIVISION		OHIO ZANESVILLE 437-438			OKLAHOMA ARDMORE 734			CLINTON 736			DURANT 747			ENID 737			GUYMON 739		
		MAT.	INST.	TOTAL	MAT.	INST.	TOTAL	MAT.	INST.	TOTAL	MAT.	INST.	TOTAL	MAT.	INST.	TOTAL	MAT.	INST.	TOTAL
015433	CONTRACTOR EQUIPMENT		88.7	88.7		80.7	80.7		79.9	79.9		79.9	79.9		79.9	79.9		79.9	79.9
0241, 31 - 34	SITE & INFRASTRUCTURE, DEMOLITION	98.9	90.5	93.1	96.9	92.6	94.0	98.1	91.2	93.3	93.4	88.6	90.1	100.0	91.2	93.9	102.4	90.5	94.2
0310	Concrete Forming & Accessories	93.7	78.3	80.6	88.0	55.4	60.2	86.8	55.4	60.1	82.9	55.0	59.1	89.9	55.6	60.7	92.6	55.1	60.7
0320	Concrete Reinforcing	91.6	92.6	92.1	79.5	67.7	73.8	80.0	67.7	74.0	88.6	63.6	76.5	79.5	67.7	73.8	80.0	63.3	71.9
0330	Cast-in-Place Concrete	93.2	86.5	90.7	96.6	70.5	86.9	93.4	70.5	84.9	88.6	70.3	81.8	93.4	70.6	84.9	93.4	70.0	84.7
03	CONCRETE	89.2	83.5	86.7	86.7	62.8	76.2	86.2	62.8	75.9	84.1	61.8	74.3	86.6	62.9	76.2	89.4	61.7	77.2
04	MASONRY	90.7	86.6	88.2	95.2	57.0	71.8	119.3	57.0	81.0	88.6	62.1	72.3	101.3	57.0	74.1	97.5	55.4	71.7
05	METALS	98.5	84.2	94.3	97.3	62.2	87.0	97.4	62.2	87.0	92.7	60.6	83.2	98.8	62.4	88.1	97.8	59.3	86.5
06	WOOD, PLASTICS & COMPOSITES	88.2	77.5	82.3	97.4	54.4	73.4	96.5	54.4	73.4	87.8	54.4	69.5	99.7	54.4	74.8	102.9	54.4	76.2
07	THERMAL & MOISTURE PROTECTION	90.6	84.5	88.0	100.7	64.8	85.1	100.8	64.8	85.1	96.4	64.6	82.6	100.9	64.8	85.2	101.3	62.2	84.3
08	OPENINGS	91.2	79.0	88.4	103.7	55.0	92.3	103.7	55.0	92.3	96.0	54.0	86.2	104.9	55.9	93.5	103.8	54.0	92.2
0920	Plaster & Gypsum Board	91.2	77.0	81.7	90.5	53.7	65.8	90.2	53.7	65.7	78.5	53.7	61.8	91.1	53.7	66.0	91.1	53.7	66.0
0950, 0980	Ceilings & Acoustic Treatment	99.5	77.0	84.3	89.1	53.7	65.3	89.1	53.7	65.3	83.0	53.7	63.3	89.1	53.7	65.3	89.1	53.7	65.3
0960	Flooring	92.6	75.3	87.7	82.7	57.3	75.6	81.7	57.3	74.9	89.7	51.7	79.0	83.3	72.9	80.4	84.8	57.3	77.1
0970, 0990	Wall Finishes & Painting/Coating	104.0	89.1	95.2	86.4	44.8	61.8	86.4	44.8	61.8	92.8	44.8	64.4	86.4	44.8	61.8	86.4	42.2	60.3
09	FINISHES	95.4	78.2	86.1	82.7	53.5	66.9	82.5	53.5	66.8	83.2	52.4	66.5	83.2	56.7	68.8	84.0	54.5	68.0
COVERS	DIVS. 10 - 14, 25, 28, 41, 43, 44, 46	100.0	83.3	96.3	100.0	79.4	95.4	100.0	79.4	95.4	100.0	79.3	95.4	100.0	79.4	95.4	100.0	79.4	95.4
21, 22, 23	FIRE SUPPRESSION, PLUMBING & HVAC	96.5	89.8	93.9	96.5	67.5	85.1	96.5	67.5	85.1	96.6	67.4	85.2	100.0	67.5	87.2	96.5	63.6	83.6
26, 27, 3370	ELECTRICAL, COMMUNICATIONS & UTIL.	94.1	89.7	92.0	95.0	71.0	83.2	95.9	71.0	83.7	97.3	69.2	83.4	95.9	71.0	83.7	97.6	65.6	81.8
MF2018	WEIGHTED AVERAGE	94.9	85.7	91.0	95.2	65.6	82.7	96.4	65.5	83.3	93.1	65.1	81.3	97.0	66.0	83.9	96.3	63.3	82.4

DIVISION		OKLAHOMA LAWTON 735			MCALESTER 745			MIAMI 743			MUSKOGEE 744			OKLAHOMA CITY 730 - 731			PONCA CITY 746		
		MAT.	INST.	TOTAL	MAT.	INST.	TOTAL	MAT.	INST.	TOTAL	MAT.	INST.	TOTAL	MAT.	INST.	TOTAL	MAT.	INST.	TOTAL
015433	CONTRACTOR EQUIPMENT		80.7	80.7		79.9	79.9		90.9	90.9		90.9	90.9		85.6	85.6		79.9	79.9
0241, 31 - 34	SITE & INFRASTRUCTURE, DEMOLITION	96.2	92.7	93.7	86.8	90.5	89.4	88.1	88.1	88.1	88.7	87.9	88.1	94.2	98.0	96.8	93.9	90.9	91.8
0310	Concrete Forming & Accessories	92.5	55.6	61.1	81.1	41.4	47.3	93.2	55.2	60.8	97.3	54.8	61.1	92.8	63.8	68.0	89.0	55.4	60.3
0320	Concrete Reinforcing	79.7	67.7	73.9	88.3	63.3	76.2	86.9	67.7	77.6	87.7	62.8	75.7	87.9	67.8	78.2	87.7	67.7	78.0
0330	Cast-in-Place Concrete	90.3	70.6	83.0	77.7	70.4	74.8	81.4	71.4	77.7	82.4	71.1	78.2	90.3	73.4	84.0	91.0	70.5	83.4
03	CONCRETE	83.0	62.9	74.2	75.3	55.4	66.6	79.7	64.0	72.8	81.2	62.8	73.1	86.1	67.6	78.0	86.1	62.7	75.8
04	MASONRY	97.4	57.0	72.6	106.1	56.9	75.9	91.3	57.1	70.3	108.1	49.2	71.9	99.8	57.4	73.8	84.1	57.0	67.5
05	METALS	102.5	62.3	90.7	92.6	59.4	82.9	92.6	77.0	88.0	94.1	73.9	88.2	95.7	63.7	86.3	92.6	62.1	83.6
06	WOOD, PLASTICS & COMPOSITES	102.0	54.4	75.8	85.5	36.2	58.4	100.0	54.6	75.0	104.3	54.6	76.9	92.5	65.1	77.4	95.6	54.4	72.9
07	THERMAL & MOISTURE PROTECTION	100.7	64.8	85.1	96.0	61.2	80.9	96.5	64.1	82.4	96.8	61.1	81.2	92.1	66.9	81.1	96.6	65.0	82.9
08	OPENINGS	106.6	55.9	94.8	96.0	43.9	83.9	96.0	55.0	86.5	97.2	54.0	87.1	100.0	61.8	91.1	96.0	55.0	86.4
0920	Plaster & Gypsum Board	92.8	53.7	66.5	77.6	35.0	48.9	84.2	53.7	63.7	86.4	53.7	64.4	94.7	64.6	74.4	82.9	53.7	63.3
0950, 0980	Ceilings & Acoustic Treatment	96.4	53.7	67.6	83.0	35.0	50.6	83.0	53.7	63.3	92.7	53.7	66.4	87.4	64.6	72.0	83.0	53.7	63.3
0960	Flooring	85.0	72.9	81.6	88.6	57.3	79.8	95.7	51.7	83.3	98.2	33.2	79.9	86.4	72.9	82.6	92.8	57.3	82.8
0970, 0990	Wall Finishes & Painting/Coating	86.4	44.8	61.8	92.8	42.2	62.9	92.8	42.2	62.9	92.8	42.2	62.9	89.5	44.8	63.1	92.8	44.8	64.4
09	FINISHES	84.9	56.7	69.6	82.2	42.4	60.7	85.1	52.2	67.3	88.1	48.8	66.9	86.3	63.1	73.7	84.9	54.4	68.4
COVERS	DIVS. 10 - 14, 25, 28, 41, 43, 44, 46	100.0	79.4	95.4	100.0	77.4	95.0	100.0	76.3	94.7	100.0	76.3	94.7	100.0	80.9	95.8	100.0	79.4	95.4
21, 22, 23	FIRE SUPPRESSION, PLUMBING & HVAC	100.0	67.5	87.2	96.6	63.5	83.6	96.6	63.6	83.7	100.1	61.3	84.9	99.9	67.8	87.3	96.6	63.6	83.7
26, 27, 3370	ELECTRICAL, COMMUNICATIONS & UTIL.	97.6	69.2	83.6	95.8	67.0	81.6	97.1	67.0	82.3	95.4	67.2	81.5	103.4	71.1	87.5	95.4	65.6	80.7
MF2018	WEIGHTED AVERAGE	97.3	65.8	84.0	92.4	60.5	78.9	92.7	65.1	81.1	95.0	62.9	81.4	96.4	68.7	84.7	93.1	64.0	80.8

DIVISION		OKLAHOMA POTEAU 749			SHAWNEE 748			TULSA 740 - 741			WOODWARD 738			OREGON BEND 977			EUGENE 974		
		MAT.	INST.	TOTAL	MAT.	INST.	TOTAL	MAT.	INST.	TOTAL	MAT.	INST.	TOTAL	MAT.	INST.	TOTAL	MAT.	INST.	TOTAL
015433	CONTRACTOR EQUIPMENT		89.7	89.7		79.9	79.9		90.9	90.9		79.9	79.9		97.3	97.3		97.3	97.3
0241, 31 - 34	SITE & INFRASTRUCTURE, DEMOLITION	74.8	83.8	81.0	96.9	90.9	92.7	94.8	87.2	89.5	98.5	91.2	93.5	105.9	98.5	100.8	96.6	98.5	97.9
0310	Concrete Forming & Accessories	87.0	54.9	59.7	82.8	55.3	59.4	97.4	56.0	62.1	86.9	43.7	50.1	107.5	96.6	98.2	104.2	96.6	97.7
0320	Concrete Reinforcing	88.7	67.7	78.5	87.7	65.4	76.9	88.0	67.7	78.2	79.5	67.7	73.8	91.0	112.8	101.5	94.9	112.8	103.5
0330	Cast-in-Place Concrete	81.4	71.3	77.7	93.9	70.5	85.2	89.7	73.2	83.6	93.4	70.5	84.9	117.6	100.6	111.3	113.9	100.6	108.9
03	CONCRETE	81.9	63.8	73.9	87.7	62.3	76.6	86.2	65.0	76.9	86.5	57.5	73.7	110.5	100.3	106.0	101.3	100.3	100.9
04	MASONRY	91.6	57.1	70.4	107.5	57.0	76.5	92.2	57.1	70.6	90.8	57.0	70.0	105.1	100.6	102.3	102.1	100.6	101.1
05	METALS	92.6	76.7	87.9	92.5	61.3	83.3	97.0	77.1	91.2	97.4	62.1	87.0	108.0	96.0	104.5	108.7	95.9	105.0
06	WOOD, PLASTICS & COMPOSITES	92.4	54.6	71.6	87.7	54.4	69.4	103.5	55.6	77.2	96.6	38.7	64.8	99.6	96.1	97.7	95.4	96.1	95.8
07	THERMAL & MOISTURE PROTECTION	96.6	64.1	82.5	96.6	63.9	82.4	96.8	64.3	82.7	100.9	63.3	84.5	118.3	100.2	110.4	117.4	102.7	111.0
08	OPENINGS	96.0	55.0	86.5	96.0	54.4	86.3	98.9	56.2	88.9	103.7	46.3	90.3	96.9	100.0	97.6	97.2	100.0	97.8
0920	Plaster & Gypsum Board	81.3	53.7	62.8	78.5	53.7	61.8	86.4	54.8	65.1	90.2	37.6	54.8	122.7	95.9	104.6	120.8	95.9	104.0
0950, 0980	Ceilings & Acoustic Treatment	83.0	53.7	63.3	83.0	53.7	63.3	92.7	54.8	67.1	89.1	37.6	54.4	85.7	95.9	92.6	86.6	95.9	92.9
0960	Flooring	92.2	51.7	80.8	89.7	57.3	80.6	97.0	61.6	87.1	81.7	54.6	74.1	105.8	105.7	105.7	104.2	105.7	104.6
0970, 0990	Wall Finishes & Painting/Coating	92.8	44.8	64.4	92.8	42.2	62.9	92.8	51.3	68.2	86.4	44.8	61.8	98.6	77.3	86.0	98.6	69.6	81.4
09	FINISHES	82.9	52.5	66.4	83.5	53.2	67.1	88.0	55.8	70.5	82.6	43.7	61.5	100.1	94.9	97.8	98.6	95.0	96.7
COVERS	DIVS. 10 - 14, 25, 28, 41, 43, 44, 46	100.0	79.6	95.5	100.0	79.4	95.4	100.0	76.4	94.7	100.0	77.7	95.0	100.0	102.0	100.4	100.0	102.0	100.4
21, 22, 23	FIRE SUPPRESSION, PLUMBING & HVAC	96.6	63.6	83.7	96.6	67.5	85.2	100.1	63.7	85.8	96.5	67.5	85.1	100.0	105.2	100.0	100.1	98.6	99.5
26, 27, 3370	ELECTRICAL, COMMUNICATIONS & UTIL.	95.5	67.0	81.5	97.4	71.0	84.4	97.3	67.0	82.4	97.4	71.0	84.4	102.0	99.8	100.9	100.7	99.8	100.2
MF2018	WEIGHTED AVERAGE	92.3	64.9	80.7	94.5	65.2	82.2	95.9	65.8	83.2	95.3	62.9	81.6	102.6	100.2	101.6	101.7	98.7	100.5

For customer support on your Commercial Renovation Costs with RSMeans data, call 800.448.8182.

815

		OREGON																
DIVISION		KLAMATH FALLS			MEDFORD			PENDLETON			PORTLAND			SALEM			VALE		
		976			975			978			970 - 972			973			979		
		MAT.	INST.	TOTAL	MAT.	INST.	TOTAL	MAT.	INST.	TOTAL	MAT.	INST.	TOTAL	MAT.	INST.	TOTAL	MAT.	INST.	TOTAL
015433	CONTRACTOR EQUIPMENT		97.3	97.3		97.3	97.3		94.9	94.9		97.3	97.3		99.9	99.9		94.9	94.9
0241, 31 - 34	SITE & INFRASTRUCTURE, DEMOLITION	109.8	98.4	102.0	104.0	98.4	100.2	103.0	91.9	95.4	99.1	98.5	98.7	92.6	102.1	99.2	90.6	91.8	91.5
0310	Concrete Forming & Accessories	100.7	96.3	97.0	99.7	96.3	96.8	101.2	96.6	97.3	105.3	96.8	98.1	105.8	96.8	98.1	107.5	95.3	97.1
0320	Concrete Reinforcing	91.0	112.7	101.5	92.6	112.7	102.3	90.3	112.8	101.2	95.6	112.8	103.9	102.7	112.8	107.6	88.1	112.5	99.9
0330	Cast-in-Place Concrete	117.6	97.0	110.0	117.6	100.4	111.2	118.4	97.5	110.7	117.0	100.6	110.9	107.6	101.7	105.4	93.2	97.8	94.9
03	CONCRETE	113.5	98.9	107.1	107.8	100.1	104.4	93.7	99.3	96.2	103.0	100.4	101.8	98.7	100.7	99.6	77.8	98.8	87.0
04	MASONRY	119.1	100.6	107.7	99.0	100.6	100.0	109.9	100.6	104.2	104.0	100.6	101.9	108.2	100.6	103.5	107.7	100.6	103.4
05	METALS	108.0	95.6	104.4	108.3	95.6	104.6	116.3	96.4	110.4	110.1	96.1	106.0	117.1	95.4	110.8	116.2	95.1	110.0
06	WOOD, PLASTICS & COMPOSITES	90.4	96.1	93.5	89.3	96.1	93.0	92.3	96.2	94.5	96.4	96.1	96.3	90.3	96.3	93.6	100.9	96.2	98.3
07	THERMAL & MOISTURE PROTECTION	118.5	96.8	109.1	118.1	96.1	108.5	110.7	95.4	104.1	117.3	100.2	109.9	114.1	100.9	108.3	110.1	91.4	102.0
08	OPENINGS	96.9	100.0	97.6	99.6	100.0	99.7	93.2	100.0	94.8	95.1	100.0	96.3	102.6	100.0	102.0	93.2	89.0	92.2
0920	Plaster & Gypsum Board	117.1	95.9	102.8	116.5	95.9	102.6	102.4	95.9	98.0	120.3	95.9	103.9	115.4	95.9	102.3	109.0	95.9	100.2
0950, 0980	Ceilings & Acoustic Treatment	93.0	95.9	95.0	98.5	95.9	96.7	61.9	95.9	84.8	88.5	95.9	93.5	96.0	95.9	95.9	61.9	95.9	84.8
0960	Flooring	102.8	105.7	103.6	102.2	105.7	103.2	70.8	105.7	80.6	101.9	105.7	102.9	107.4	105.7	106.9	72.9	105.7	82.1
0970, 0990	Wall Finishes & Painting/Coating	98.6	65.2	78.8	98.6	65.2	78.8	88.3	77.3	81.8	98.4	77.3	85.9	97.3	75.2	84.2	88.3	77.3	81.8
09	FINISHES	100.3	94.6	97.2	100.4	94.6	97.3	70.4	96.0	84.2	98.2	95.9	97.0	99.7	95.8	97.6	71.0	96.0	84.5
COVERS	DIVS. 10 - 14, 25, 28, 41, 43, 44, 46	100.0	101.9	100.4	100.0	101.9	100.4	100.0	96.5	99.2	100.0	102.1	100.5	100.0	102.4	100.5	100.0	102.2	100.5
21, 22, 23	FIRE SUPPRESSION, PLUMBING & HVAC	96.6	100.2	98.0	100.1	100.5	100.2	100.1	87.1	95.0	100.1	111.5	104.6	100.1	106.9	102.8	98.7	71.1	87.9
26, 27, 3370	ELECTRICAL, COMMUNICATIONS & UTIL.	100.7	82.6	91.8	104.1	82.6	93.5	92.8	96.8	94.8	100.9	108.7	104.7	109.3	99.8	104.6	92.8	68.1	80.6
MF2018	WEIGHTED AVERAGE	103.5	97.3	100.9	103.3	97.4	100.8	98.6	100.3	99.3	102.1	102.8	102.4	104.4	100.8	102.9	96.2	87.0	92.3

		PENNSYLVANIA																
DIVISION		ALLENTOWN			ALTOONA			BEDFORD			BRADFORD			BUTLER			CHAMBERSBURG		
		181			166			155			167			160			172		
		MAT.	INST.	TOTAL	MAT.	INST.	TOTAL	MAT.	INST.	TOTAL	MAT.	INST.	TOTAL	MAT.	INST.	TOTAL	MAT.	INST.	TOTAL
015433	CONTRACTOR EQUIPMENT		112.6	112.6		112.6	112.6		110.7	110.7		112.6	112.6		112.6	112.6		111.8	111.8
0241, 31 - 34	SITE & INFRASTRUCTURE, DEMOLITION	91.7	97.2	95.5	95.1	97.1	96.5	103.6	94.8	97.6	90.5	96.0	94.3	86.3	97.8	94.2	86.3	95.3	92.5
0310	Concrete Forming & Accessories	97.9	108.2	106.7	83.7	82.6	82.8	82.0	81.2	81.3	85.8	96.5	94.9	85.1	94.1	92.8	88.0	76.8	78.5
0320	Concrete Reinforcing	96.8	113.3	104.8	93.8	105.7	99.6	93.0	105.7	99.1	95.8	105.8	100.6	94.4	119.3	106.5	94.5	112.2	103.1
0330	Cast-in-Place Concrete	89.4	99.4	93.3	99.7	86.9	94.9	109.1	85.9	100.5	95.2	90.6	93.5	88.0	96.2	91.0	93.0	93.8	93.3
03	CONCRETE	92.3	107.0	98.8	87.5	89.5	88.4	102.2	88.5	96.2	94.2	96.0	95.5	79.5	100.3	88.7	97.1	90.5	94.2
04	MASONRY	92.1	93.1	92.7	95.7	83.4	88.1	107.1	80.2	90.6	93.1	82.0	86.3	97.6	93.0	94.8	93.8	79.1	84.8
05	METALS	98.9	120.5	105.3	92.9	114.8	99.3	102.2	113.4	105.5	96.7	113.2	101.6	92.6	121.7	101.2	98.3	118.8	104.3
06	WOOD, PLASTICS & COMPOSITES	97.6	111.5	105.3	76.7	81.8	79.5	79.4	81.7	80.7	82.7	101.2	92.9	78.1	94.1	86.9	84.3	75.6	79.5
07	THERMAL & MOISTURE PROTECTION	104.0	107.8	105.7	102.7	90.0	97.2	98.3	87.4	93.6	103.8	88.5	97.2	102.2	94.4	98.8	96.1	83.1	90.4
08	OPENINGS	92.4	107.3	95.9	86.3	85.6	86.1	93.4	85.6	91.6	92.4	95.3	93.1	86.2	99.3	89.3	89.2	81.4	87.4
0920	Plaster & Gypsum Board	94.9	111.8	106.3	86.3	81.2	82.9	97.5	81.2	86.6	87.1	101.2	96.6	86.3	93.9	91.4	108.9	74.9	86.0
0950, 0980	Ceilings & Acoustic Treatment	88.5	111.8	104.2	90.9	81.2	84.4	101.1	81.2	87.7	91.0	101.2	97.9	91.8	93.9	93.2	96.0	74.9	81.8
0960	Flooring	92.0	95.1	92.8	85.3	98.9	89.2	95.0	103.7	97.5	86.2	103.7	91.1	86.3	105.3	91.6	90.5	78.0	87.0
0970, 0990	Wall Finishes & Painting/Coating	91.9	101.8	97.8	87.2	106.2	98.4	97.6	105.7	102.4	91.9	105.7	100.0	87.2	106.2	98.4	92.0	100.7	97.1
09	FINISHES	90.6	105.8	98.8	88.2	87.5	87.8	100.6	87.4	93.5	88.4	99.4	94.3	88.1	97.2	93.0	90.8	78.8	84.3
COVERS	DIVS. 10 - 14, 25, 28, 41, 43, 44, 46	100.0	99.6	99.9	100.0	94.5	98.8	100.0	93.2	98.5	100.0	96.1	99.1	100.0	97.3	99.4	100.0	92.7	98.4
21, 22, 23	FIRE SUPPRESSION, PLUMBING & HVAC	100.3	114.5	105.9	99.8	83.6	93.4	96.6	84.4	91.8	96.8	90.6	94.4	96.3	95.4	96.0	96.7	89.7	94.0
26, 27, 3370	ELECTRICAL, COMMUNICATIONS & UTIL.	98.0	95.8	96.9	88.7	107.7	98.0	94.2	107.7	100.9	92.1	107.7	99.8	89.2	107.7	98.3	92.4	85.8	89.2
MF2018	WEIGHTED AVERAGE	96.7	105.8	100.6	93.3	92.8	93.1	98.8	92.1	96.0	94.9	97.1	95.8	91.3	100.6	95.2	95.1	89.4	92.7

		PENNSYLVANIA																
DIVISION		DOYLESTOWN			DUBOIS			ERIE			GREENSBURG			HARRISBURG			HAZLETON		
		189			158			164 - 165			156			170 - 171			182		
		MAT.	INST.	TOTAL	MAT.	INST.	TOTAL	MAT.	INST.	TOTAL	MAT.	INST.	TOTAL	MAT.	INST.	TOTAL	MAT.	INST.	TOTAL
015433	CONTRACTOR EQUIPMENT		91.5	91.5		110.7	110.7		112.6	112.6		110.7	110.7		114.7	114.7		112.6	112.6
0241, 31 - 34	SITE & INFRASTRUCTURE, DEMOLITION	104.8	85.9	91.7	108.4	95.2	99.3	92.1	97.4	95.7	99.6	96.7	97.6	86.6	100.3	96.1	85.0	97.1	93.3
0310	Concrete Forming & Accessories	82.9	126.8	120.3	81.6	83.7	83.4	97.2	85.8	87.5	87.8	88.6	88.5	100.5	85.7	87.9	80.6	88.8	87.6
0320	Concrete Reinforcing	93.6	150.9	121.3	92.3	119.4	105.4	95.8	107.3	101.4	92.3	119.0	105.2	103.4	111.0	107.1	94.0	113.1	103.2
0330	Cast-in-Place Concrete	84.5	129.7	101.3	105.2	94.1	101.1	98.0	88.7	94.5	101.3	95.7	99.2	91.7	96.9	93.6	84.5	95.6	88.6
03	CONCRETE	88.0	131.2	107.0	104.4	94.7	100.1	86.4	91.9	88.8	97.2	97.5	97.4	91.9	95.3	93.4	85.0	96.7	90.2
04	MASONRY	95.4	134.9	119.7	107.8	93.4	98.9	85.0	86.8	86.1	117.5	89.1	100.0	88.8	83.9	85.8	104.7	89.5	95.4
05	METALS	96.5	128.4	104.8	102.2	118.8	107.0	93.1	115.6	99.7	102.1	119.7	107.3	105.1	117.4	108.7	98.7	119.7	104.8
06	WOOD, PLASTICS & COMPOSITES	78.5	125.8	104.5	78.3	81.7	80.2	94.0	84.4	88.7	85.3	87.1	86.3	101.8	85.3	92.7	77.1	87.0	82.6
07	THERMAL & MOISTURE PROTECTION	101.1	131.7	114.4	98.7	92.9	96.1	103.2	88.6	96.8	98.2	92.2	95.6	98.9	100.4	99.6	103.3	100.7	102.2
08	OPENINGS	94.8	133.2	103.7	93.4	88.7	92.3	86.4	88.0	86.8	93.4	95.5	93.9	100.9	86.5	97.5	93.0	93.3	93.0
0920	Plaster & Gypsum Board	85.4	126.5	113.0	96.3	81.2	86.2	94.9	83.9	87.5	98.3	86.8	90.5	113.8	84.7	94.2	85.8	86.6	86.4
0950, 0980	Ceilings & Acoustic Treatment	87.7	126.5	113.8	101.1	81.2	87.7	88.5	83.9	85.4	100.3	86.8	91.2	104.0	84.7	91.0	89.4	86.6	87.5
0960	Flooring	76.2	139.3	93.9	94.8	103.7	97.3	92.2	98.9	94.1	98.4	78.0	92.7	94.7	90.4	93.5	83.5	89.3	85.1
0970, 0990	Wall Finishes & Painting/Coating	91.5	143.7	122.4	97.6	106.2	102.7	98.2	92.3	94.7	97.6	106.2	102.7	95.8	84.8	89.3	91.9	104.1	99.1
09	FINISHES	81.9	130.5	108.2	100.9	88.8	94.4	91.6	88.4	89.9	101.3	88.1	94.1	96.0	85.8	90.5	86.6	90.2	88.6
COVERS	DIVS. 10 - 14, 25, 28, 41, 43, 44, 46	100.0	113.0	102.9	100.0	95.2	98.9	100.0	96.6	99.2	100.0	96.4	99.2	100.0	95.4	99.0	100.0	95.5	99.0
21, 22, 23	FIRE SUPPRESSION, PLUMBING & HVAC	96.3	132.6	110.6	96.6	87.5	93.0	99.8	93.6	97.4	96.6	87.3	92.9	100.1	91.7	96.8	96.8	95.5	96.3
26, 27, 3370	ELECTRICAL, COMMUNICATIONS & UTIL.	91.6	132.4	111.7	94.8	107.7	101.1	90.3	93.1	91.7	94.8	107.7	101.2	99.7	85.8	92.9	92.9	90.0	91.5
MF2018	WEIGHTED AVERAGE	94.0	127.1	108.0	99.3	96.0	97.9	93.2	93.9	93.5	98.7	96.3	97.7	98.7	93.0	96.3	94.4	95.9	95.0

PENNSYLVANIA

DIVISION		INDIANA 157			JOHNSTOWN 159			KITTANNING 162			LANCASTER 175 - 176			LEHIGH VALLEY 180			MONTROSE 188		
		MAT.	INST.	TOTAL	MAT.	INST.	TOTAL	MAT.	INST.	TOTAL	MAT.	INST.	TOTAL	MAT.	INST.	TOTAL	MAT.	INST.	TOTAL
015433	CONTRACTOR EQUIPMENT		110.7	110.7		110.7	110.7		112.6	112.6		111.8	111.8		112.6	112.6		112.6	112.6
0241, 31 - 34	SITE & INFRASTRUCTURE, DEMOLITION	97.8	95.8	96.4	104.1	96.0	98.5	88.8	97.4	94.7	78.4	95.6	90.3	88.9	97.5	94.8	87.5	97.2	94.2
0310	Concrete Forming & Accessories	82.6	92.9	91.4	81.6	82.5	82.4	85.1	89.7	89.0	90.0	86.4	86.9	91.7	110.3	107.6	81.5	89.3	88.2
0320	Concrete Reinforcing	91.7	119.4	105.1	93.0	118.9	105.5	94.4	119.2	106.4	94.1	111.0	102.3	94.0	108.0	100.8	98.4	116.2	107.0
0330	Cast-in-Place Concrete	99.3	94.4	97.5	110.1	86.7	101.4	91.5	94.7	92.7	79.1	96.7	85.7	91.5	99.2	94.3	89.7	93.0	90.9
03	CONCRETE	94.7	99.1	96.6	103.2	91.6	98.1	82.1	97.7	88.9	85.2	95.6	89.8	91.4	106.9	98.2	90.1	96.6	93.0
04	MASONRY	103.8	93.5	97.5	104.7	83.7	91.8	100.1	88.8	93.1	99.6	86.7	91.7	92.1	96.5	94.8	92.0	91.5	91.7
05	METALS	102.3	119.9	107.4	102.2	117.9	106.8	92.7	120.6	100.9	98.3	118.7	104.3	98.6	119.0	104.6	96.8	120.4	103.7
06	WOOD, PLASTICS & COMPOSITES	80.1	94.0	87.7	78.3	81.7	80.2	78.1	89.3	84.3	87.4	85.1	86.2	89.1	112.6	102.0	77.8	87.0	82.9
07	THERMAL & MOISTURE PROTECTION	98.1	94.2	96.4	98.4	88.5	94.1	102.3	92.4	98.0	95.4	101.0	97.8	103.7	111.9	107.3	103.3	91.0	97.9
08	OPENINGS	93.4	95.4	93.9	93.4	88.7	92.3	86.3	96.7	88.7	89.2	86.4	88.5	92.9	106.6	96.1	89.8	91.9	90.3
0920	Plaster & Gypsum Board	97.8	93.9	95.2	96.1	81.2	86.1	86.3	89.0	88.1	111.8	84.7	93.6	88.0	112.6	104.7	86.2	86.6	86.5
0950, 0980	Ceilings & Acoustic Treatment	101.1	93.9	96.2	100.3	81.2	87.5	91.8	89.0	89.9	96.0	84.7	88.4	89.4	112.9	105.2	91.0	86.6	88.1
0960	Flooring	95.6	103.7	97.8	94.8	78.0	90.1	86.3	103.7	91.2	91.5	95.7	92.7	88.9	94.1	90.4	84.1	103.7	89.6
0970, 0990	Wall Finishes & Painting/Coating	97.6	106.2	102.7	97.6	106.2	102.7	87.2	106.2	98.4	92.0	86.6	88.8	91.9	101.0	97.3	91.9	104.1	99.1
09	FINISHES	100.4	96.1	98.1	100.3	83.9	91.4	88.3	93.4	91.0	90.9	87.3	88.9	88.8	106.5	98.4	87.4	92.7	90.3
COVERS	DIVS. 10 - 14, 25, 28, 41, 43, 44, 46	100.0	96.5	99.2	100.0	94.4	98.8	100.0	96.2	99.1	100.0	95.7	99.0	100.0	102.5	100.6	100.0	96.2	99.2
21, 22, 23	FIRE SUPPRESSION, PLUMBING & HVAC	96.6	85.4	92.2	96.6	78.9	89.6	96.3	90.7	94.1	96.7	92.7	95.1	96.8	117.6	104.9	96.8	94.8	96.0
26, 27, 3370	ELECTRICAL, COMMUNICATIONS & UTIL.	94.8	107.7	101.2	94.8	107.7	101.1	88.7	107.7	98.1	93.7	93.3	93.5	92.9	129.4	110.9	92.1	95.2	93.6
MF2018	WEIGHTED AVERAGE	97.6	97.7	97.7	98.9	91.9	95.9	91.8	97.9	94.4	93.8	94.5	94.1	94.9	111.6	102.0	93.8	96.8	95.1

PENNSYLVANIA

DIVISION		NEW CASTLE 161			NORRISTOWN 194			OIL CITY 163			PHILADELPHIA 190 - 191			PITTSBURGH 150 - 152			POTTSVILLE 179		
		MAT.	INST.	TOTAL	MAT.	INST.	TOTAL	MAT.	INST.	TOTAL	MAT.	INST.	TOTAL	MAT.	INST.	TOTAL	MAT.	INST.	TOTAL
015433	CONTRACTOR EQUIPMENT		112.6	112.6		97.5	97.5		112.6	112.6		99.8	99.8		99.7	99.7		111.8	111.8
0241, 31 - 34	SITE & INFRASTRUCTURE, DEMOLITION	86.7	97.7	94.3	97.3	95.1	95.8	85.2	96.3	92.9	99.8	101.1	100.7	103.7	96.1	98.5	81.3	95.8	91.3
0310	Concrete Forming & Accessories	85.1	93.6	92.3	83.7	125.3	119.1	85.1	92.7	91.6	99.8	142.4	136.2	97.5	97.5	97.5	82.0	87.7	86.8
0320	Concrete Reinforcing	93.3	100.2	96.7	95.9	150.8	122.5	94.4	94.9	94.7	100.4	143.1	121.0	93.5	122.9	107.7	93.4	114.1	103.4
0330	Cast-in-Place Concrete	88.8	94.3	90.8	86.1	127.4	101.5	86.3	94.4	89.3	89.8	135.6	106.8	108.6	101.2	105.8	84.3	96.4	88.8
03	CONCRETE	79.9	96.2	87.0	88.0	129.7	106.3	78.3	95.0	85.6	99.4	139.1	116.8	103.8	103.1	103.5	88.7	96.7	92.2
04	MASONRY	97.3	92.6	94.4	108.3	131.2	122.4	96.5	88.7	91.7	100.2	139.7	124.5	100.4	101.4	101.0	93.3	87.6	89.8
05	METALS	92.7	114.3	99.0	100.3	124.0	107.3	92.7	113.2	98.7	103.8	125.3	110.1	103.7	107.4	104.8	98.6	120.4	105.0
06	WOOD, PLASTICS & COMPOSITES	78.1	94.1	86.9	76.2	125.7	103.4	78.1	94.1	86.9	99.1	143.5	123.6	99.6	96.7	98.0	77.2	85.1	81.6
07	THERMAL & MOISTURE PROTECTION	102.2	92.0	97.8	108.7	130.1	118.0	102.1	90.7	97.1	103.4	139.7	119.2	98.6	99.3	98.9	95.6	99.6	97.3
08	OPENINGS	86.3	91.4	87.5	86.5	133.1	97.4	86.3	94.9	88.3	97.6	144.9	108.6	96.8	102.4	98.1	89.2	93.0	90.1
0920	Plaster & Gypsum Board	86.3	93.9	91.4	84.5	126.5	112.7	86.3	93.9	91.4	103.0	144.6	131.0	97.5	96.4	96.7	106.4	84.7	91.8
0950, 0980	Ceilings & Acoustic Treatment	91.8	93.9	93.2	91.1	126.5	114.9	91.8	93.9	93.2	104.2	144.6	131.5	95.2	96.4	96.0	96.0	84.7	88.4
0960	Flooring	86.3	105.3	91.6	86.7	139.3	101.5	86.3	103.7	91.2	97.8	156.2	114.2	103.1	107.3	104.3	87.5	103.7	92.1
0970, 0990	Wall Finishes & Painting/Coating	87.2	106.2	98.4	89.4	143.7	121.5	87.2	106.2	98.4	97.2	152.6	130.0	100.7	111.3	107.0	92.0	104.1	99.1
09	FINISHES	88.2	96.9	92.9	85.3	129.5	109.2	88.0	96.1	92.4	98.3	146.5	124.4	102.4	100.3	101.2	89.2	91.3	90.3
COVERS	DIVS. 10 - 14, 25, 28, 41, 43, 44, 46	100.0	97.3	99.4	100.0	109.1	102.0	100.0	96.7	99.3	100.0	119.2	104.3	100.0	103.0	100.7	100.0	97.5	99.4
21, 22, 23	FIRE SUPPRESSION, PLUMBING & HVAC	96.3	96.8	96.5	96.6	130.6	110.0	96.3	94.2	95.5	100.1	141.2	116.3	100.0	99.8	99.9	96.7	95.8	96.3
26, 27, 3370	ELECTRICAL, COMMUNICATIONS & UTIL.	89.2	98.2	93.7	91.9	142.0	116.6	90.9	107.7	99.2	98.8	159.0	128.4	97.1	111.3	104.1	92.0	91.3	91.6
MF2018	WEIGHTED AVERAGE	91.4	97.8	94.1	94.7	128.0	108.8	91.3	97.8	94.0	100.2	138.7	116.5	100.7	102.7	101.5	93.7	96.1	94.7

PENNSYLVANIA

DIVISION		READING 195 - 196			SCRANTON 184 - 185			STATE COLLEGE 168			STROUDSBURG 183			SUNBURY 178			UNIONTOWN 154		
		MAT.	INST.	TOTAL	MAT.	INST.	TOTAL	MAT.	INST.	TOTAL	MAT.	INST.	TOTAL	MAT.	INST.	TOTAL	MAT.	INST.	TOTAL
015433	CONTRACTOR EQUIPMENT		119.0	119.0		112.6	112.6		111.8	111.8		112.6	112.6		112.6	112.6		110.7	110.7
0241, 31 - 34	SITE & INFRASTRUCTURE, DEMOLITION	101.8	107.7	105.9	92.2	97.2	95.7	82.7	95.7	91.7	86.6	97.3	94.0	92.6	96.7	95.4	98.3	96.5	97.1
0310	Concrete Forming & Accessories	98.8	86.2	88.0	98.0	86.6	88.3	84.3	82.8	83.0	86.5	90.5	89.9	93.1	84.8	86.0	76.5	93.9	91.4
0320	Concrete Reinforcing	97.2	148.6	122.1	96.8	116.3	106.2	95.1	105.8	100.3	97.1	116.4	106.4	96.0	112.3	103.9	92.3	119.5	105.5
0330	Cast-in-Place Concrete	77.2	97.0	84.6	93.4	93.2	93.3	90.0	87.0	88.9	88.0	94.5	90.4	92.2	94.9	93.2	99.3	95.6	97.9
03	CONCRETE	87.1	102.1	93.7	94.1	95.4	94.7	94.6	89.7	92.5	88.8	97.7	92.7	92.0	94.5	93.1	94.4	100.0	96.9
04	MASONRY	97.5	91.5	93.8	92.5	93.8	93.3	97.8	83.9	89.3	90.3	96.3	94.0	93.6	81.0	85.8	119.3	95.4	104.6
05	METALS	100.6	133.0	110.1	100.9	120.5	106.6	96.6	115.1	102.0	98.7	120.9	105.2	98.3	119.4	104.5	102.0	120.2	107.3
06	WOOD, PLASTICS & COMPOSITES	93.7	82.4	87.5	97.6	83.3	89.8	84.5	81.8	83.0	83.5	87.0	85.5	85.4	85.1	85.2	72.8	94.0	84.4
07	THERMAL & MOISTURE PROTECTION	109.0	103.3	106.6	103.9	91.3	98.4	102.9	97.6	100.6	103.5	91.3	98.2	96.8	92.7	95.0	98.0	95.1	96.7
08	OPENINGS	90.9	99.4	92.9	92.4	89.8	91.8	89.6	85.6	88.7	93.0	94.5	93.3	89.3	89.9	89.4	93.3	99.3	94.7
0920	Plaster & Gypsum Board	94.1	81.9	85.9	96.8	82.8	87.4	87.8	81.2	83.4	86.7	86.6	86.6	105.9	84.7	91.6	94.2	93.9	94.0
0950, 0980	Ceilings & Acoustic Treatment	83.7	81.9	82.5	96.6	82.8	87.3	88.5	81.2	83.6	87.7	86.6	87.0	93.6	84.7	87.6	100.3	93.9	96.0
0960	Flooring	91.3	97.9	93.2	92.0	92.9	92.2	89.4	97.9	91.8	86.6	94.1	88.8	88.3	103.7	92.6	92.0	103.7	95.3
0970, 0990	Wall Finishes & Painting/Coating	88.2	101.8	96.3	91.9	108.3	101.6	91.9	106.2	100.4	91.9	104.1	99.1	92.0	104.1	99.1	97.6	110.0	104.9
09	FINISHES	86.8	88.4	87.7	92.4	89.3	90.7	88.0	87.3	87.6	87.5	91.8	89.8	90.2	89.6	89.9	98.7	97.0	97.8
COVERS	DIVS. 10 - 14, 25, 28, 41, 43, 44, 46	100.0	95.8	99.1	100.0	94.6	98.8	100.0	94.6	98.8	100.0	97.2	99.4	100.0	93.1	98.5	100.0	97.2	99.4
21, 22, 23	FIRE SUPPRESSION, PLUMBING & HVAC	100.3	106.8	102.8	100.3	96.0	98.6	96.8	91.4	94.7	96.8	97.9	97.2	96.7	87.4	93.0	96.6	91.7	94.6
26, 27, 3370	ELECTRICAL, COMMUNICATIONS & UTIL.	97.8	91.3	94.6	98.0	95.2	96.6	91.3	107.7	99.4	92.9	141.8	117.0	92.4	91.9	92.1	91.8	107.7	99.7
MF2018	WEIGHTED AVERAGE	96.5	101.5	98.6	97.4	96.5	97.1	94.5	94.6	94.5	94.3	104.7	98.7	94.6	92.7	93.8	97.8	99.8	98.6

City Cost Indexes

PENNSYLVANIA

DIVISION		WASHINGTON 153 MAT.	INST.	TOTAL	WELLSBORO 169 MAT.	INST.	TOTAL	WESTCHESTER 193 MAT.	INST.	TOTAL	WILKES-BARRE 186-187 MAT.	INST.	TOTAL	WILLIAMSPORT 177 MAT.	INST.	TOTAL	YORK 173-174 MAT.	INST.	TOTAL
015433	CONTRACTOR EQUIPMENT		110.7	110.7		112.6	112.6		97.5	97.5		112.6	112.6		112.6	112.6		111.8	111.8
0241, 31 - 34	SITE & INFRASTRUCTURE, DEMOLITION	98.3	96.8	97.3	94.0	96.6	95.8	103.4	96.1	98.4	84.5	97.2	93.3	83.7	96.8	92.8	82.4	95.6	91.5
0310	Concrete Forming & Accessories	82.8	94.0	92.4	85.2	84.2	84.3	90.0	126.6	121.2	89.1	85.3	85.8	90.1	85.6	86.3	85.1	86.3	86.1
0320	Concrete Reinforcing	92.3	119.5	105.5	95.1	116.1	105.3	95.0	150.8	122.0	95.8	116.3	105.7	95.3	110.9	102.8	96.0	111.0	103.3
0330	Cast-in-Place Concrete	99.3	95.9	98.0	94.4	88.2	92.1	95.4	129.5	108.1	84.5	93.0	87.6	78.0	90.0	82.5	84.7	96.7	89.2
03	CONCRETE	94.8	100.1	97.2	96.9	92.5	95.0	95.8	131.0	111.3	85.8	94.7	89.7	80.2	92.9	85.8	89.8	95.6	92.4
04	MASONRY	103.6	95.5	98.6	98.7	81.4	88.1	102.8	134.9	122.5	105.0	93.5	97.9	85.9	86.1	86.0	94.5	86.7	89.7
05	METALS	101.9	120.6	107.4	96.6	119.7	103.4	100.3	124.0	107.3	96.8	120.4	103.7	98.3	118.1	104.1	100.0	118.6	105.5
06	WOOD, PLASTICS & COMPOSITES	80.2	94.0	87.8	82.1	84.4	83.4	82.8	125.7	106.4	85.9	81.7	83.6	81.9	85.1	83.7	80.7	85.1	83.1
07	THERMAL & MOISTURE PROTECTION	98.1	95.1	96.8	104.1	86.0	96.2	109.1	127.0	116.9	103.3	90.9	97.9	96.1	88.3	92.7	95.6	101.0	98.0
08	OPENINGS	93.3	99.3	94.7	92.4	90.0	91.8	86.5	133.1	97.4	89.8	88.9	89.6	89.3	89.6	89.4	89.2	86.4	88.5
0920	Plaster & Gypsum Board	97.7	93.9	95.1	86.5	83.9	84.8	85.2	126.5	112.9	87.4	81.2	83.2	106.4	84.7	91.8	106.7	84.7	91.9
0950, 0980	Ceilings & Acoustic Treatment	100.3	93.9	96.0	88.5	83.9	85.4	91.1	126.5	114.9	91.0	81.2	84.4	96.0	84.7	88.4	95.1	84.7	88.1
0960	Flooring	95.7	103.7	97.9	85.8	103.7	90.9	89.6	139.3	103.6	87.7	92.9	89.1	87.2	86.8	87.1	88.9	95.7	90.8
0970, 0990	Wall Finishes & Painting/Coating	97.6	110.0	104.9	91.9	104.1	99.1	89.4	143.7	121.5	91.9	101.8	97.8	92.0	101.8	97.8	92.0	84.8	87.7
09	FINISHES	100.3	97.0	98.5	88.1	89.6	88.9	86.7	130.5	110.4	88.4	87.6	88.0	89.7	86.7	88.0	89.5	87.1	88.2
COVERS	DIVS. 10 - 14, 25, 28, 41, 43, 44, 46	100.0	97.2	99.4	100.0	96.2	99.2	100.0	115.3	103.4	100.0	94.3	98.7	100.0	95.5	99.0	100.0	95.6	99.0
21, 22, 23	FIRE SUPPRESSION, PLUMBING & HVAC	96.6	93.2	95.2	96.8	87.4	93.1	96.6	132.6	110.7	96.8	96.6	96.7	96.7	92.2	94.9	100.2	92.7	97.3
26, 27, 3370	ELECTRICAL, COMMUNICATIONS & UTIL.	94.2	107.7	100.9	92.1	95.2	93.6	91.7	132.4	111.8	92.9	90.0	91.5	92.8	83.4	88.2	93.7	82.9	88.4
MF2018	WEIGHTED AVERAGE	97.5	100.2	98.6	95.5	92.9	94.4	95.8	127.9	109.3	94.0	95.5	94.7	92.5	92.3	92.4	95.2	93.0	94.3

PUERTO RICO / RHODE ISLAND / SOUTH CAROLINA

DIVISION		SAN JUAN 009 MAT.	INST.	TOTAL	NEWPORT 028 MAT.	INST.	TOTAL	PROVIDENCE 029 MAT.	INST.	TOTAL	AIKEN 298 MAT.	INST.	TOTAL	BEAUFORT 299 MAT.	INST.	TOTAL	CHARLESTON 294 MAT.	INST.	TOTAL
015433	CONTRACTOR EQUIPMENT		87.3	87.3		97.9	97.9		100.1	100.1		103.7	103.7		103.7	103.7		103.7	103.7
0241, 31 - 34	SITE & INFRASTRUCTURE, DEMOLITION	81.4	87.9	85.9	86.3	98.0	94.4	89.4	102.2	98.3	128.3	85.2	98.5	123.4	84.7	96.7	108.2	85.2	92.3
0310	Concrete Forming & Accessories	105.9	21.6	34.0	100.7	124.7	121.1	99.9	124.7	121.0	94.0	64.6	68.9	93.1	38.9	46.9	92.3	64.9	69.0
0320	Concrete Reinforcing	114.9	18.7	68.4	109.7	127.1	118.1	103.7	127.1	115.0	94.3	60.6	78.0	93.4	66.7	80.5	93.2	67.3	80.7
0330	Cast-in-Place Concrete	70.9	32.7	56.6	71.7	118.5	89.1	87.5	119.0	99.2	92.9	66.9	83.2	92.9	66.8	83.2	109.1	67.2	93.5
03	CONCRETE	89.8	26.1	61.8	87.5	122.4	102.9	91.8	122.5	105.3	101.0	66.4	85.8	98.4	55.7	79.7	94.9	67.8	83.0
04	MASONRY	110.0	24.2	57.3	99.2	123.1	113.8	103.4	123.1	115.5	79.9	64.8	70.6	94.9	64.8	76.4	96.4	67.5	78.6
05	METALS	116.8	40.6	94.4	98.8	116.7	104.1	103.2	115.9	106.9	102.6	89.1	98.7	102.7	90.7	99.2	104.6	92.1	100.9
06	WOOD, PLASTICS & COMPOSITES	145.4	20.5	76.6	98.0	124.0	112.3	98.4	124.0	112.5	89.7	66.8	77.1	88.4	32.5	57.6	87.2	66.8	76.0
07	THERMAL & MOISTURE PROTECTION	134.1	27.7	87.8	102.6	123.0	111.5	108.3	123.5	114.9	98.1	64.2	83.4	97.8	56.2	79.7	96.8	66.1	83.5
08	OPENINGS	96.8	19.1	78.7	97.4	124.4	103.7	97.4	124.4	104.9	94.0	62.5	86.7	94.1	44.9	82.6	97.6	65.1	90.0
0920	Plaster & Gypsum Board	76.4	25.2	48.0	91.4	124.3	113.6	104.9	124.3	118.0	83.4	65.8	71.6	87.0	30.5	49.0	88.1	65.8	73.1
0950, 0980	Ceilings & Acoustic Treatment	158.8	19.3	64.7	90.3	124.3	113.2	104.1	124.3	117.7	78.8	65.8	70.1	82.1	30.5	47.3	82.1	65.8	71.1
0960	Flooring	117.8	29.7	93.1	83.8	127.3	96.0	86.4	127.3	97.9	97.5	89.8	95.3	99.0	75.8	92.5	98.6	81.9	93.9
0970, 0990	Wall Finishes & Painting/Coating	90.2	24.6	51.4	85.0	118.9	105.0	90.8	118.9	107.4	95.2	66.0	77.9	95.2	56.9	72.5	95.2	70.0	80.3
09	FINISHES	106.0	23.2	61.2	87.7	125.1	107.9	93.7	125.1	110.7	89.0	69.7	78.5	90.1	46.0	66.2	88.2	68.9	77.8
COVERS	DIVS. 10 - 14, 25, 28, 41, 43, 44, 46	100.0	26.6	83.6	100.0	110.2	102.3	100.0	110.3	102.3	100.0	70.3	93.4	100.0	80.1	95.6	100.0	70.3	93.4
21, 22, 23	FIRE SUPPRESSION, PLUMBING & HVAC	88.5	18.8	61.1	100.4	114.8	106.0	100.2	114.8	105.9	96.9	55.2	80.5	96.9	55.6	80.7	100.4	58.4	83.9
26, 27, 3370	ELECTRICAL, COMMUNICATIONS & UTIL.	84.4	25.0	55.2	102.1	95.5	98.9	102.3	95.5	99.0	96.2	61.9	79.3	99.7	30.9	65.8	98.1	58.9	78.8
MF2018	WEIGHTED AVERAGE	98.5	29.8	69.4	97.0	114.7	104.5	99.3	115.0	105.9	97.6	67.1	84.7	98.3	57.5	81.1	98.7	68.2	85.8

SOUTH CAROLINA / SOUTH DAKOTA

DIVISION		COLUMBIA 290-292 MAT.	INST.	TOTAL	FLORENCE 295 MAT.	INST.	TOTAL	GREENVILLE 296 MAT.	INST.	TOTAL	ROCK HILL 297 MAT.	INST.	TOTAL	SPARTANBURG 293 MAT.	INST.	TOTAL	ABERDEEN 574 MAT.	INST.	TOTAL
015433	CONTRACTOR EQUIPMENT		105.6	105.6		103.7	103.7		103.7	103.7		103.7	103.7		103.7	103.7		97.7	97.7
0241, 31 - 34	SITE & INFRASTRUCTURE, DEMOLITION	105.9	89.5	94.6	117.6	84.9	95.1	112.6	85.4	93.8	110.0	84.3	92.2	112.4	85.4	93.8	99.2	93.4	95.2
0310	Concrete Forming & Accessories	93.2	64.9	69.1	81.1	64.8	67.2	91.8	64.9	68.8	90.1	64.0	67.9	94.4	64.9	69.2	98.9	75.7	79.1
0320	Concrete Reinforcing	98.3	67.3	83.3	92.8	67.2	80.4	92.8	64.4	79.1	93.5	65.8	80.1	92.8	67.3	80.4	101.0	71.4	86.7
0330	Cast-in-Place Concrete	110.3	67.5	94.4	92.9	67.0	83.2	92.9	67.2	83.3	92.8	66.4	83.0	92.9	67.2	83.3	109.5	78.8	98.1
03	CONCRETE	93.2	67.8	82.1	93.3	67.6	82.0	92.0	67.1	81.2	90.1	66.8	79.9	92.2	67.8	81.5	98.7	76.7	89.1
04	MASONRY	89.9	67.5	76.1	79.9	67.4	72.2	77.9	67.5	71.5	101.6	64.0	78.5	79.9	67.5	72.3	109.4	73.3	87.2
05	METALS	101.6	91.0	98.5	103.5	91.4	99.9	103.5	91.7	100.0	102.7	90.5	99.1	103.5	92.1	100.1	93.8	82.1	90.4
06	WOOD, PLASTICS & COMPOSITES	88.6	66.9	76.7	74.8	66.8	70.4	86.8	66.8	75.8	85.4	66.8	75.2	90.6	66.8	77.5	97.6	74.2	84.7
07	THERMAL & MOISTURE PROTECTION	93.3	66.3	81.5	97.2	66.1	83.7	97.1	66.1	83.6	96.9	59.2	80.5	97.1	66.1	83.6	103.1	77.2	91.8
08	OPENINGS	99.7	65.1	91.7	94.1	65.1	87.3	94.0	64.8	87.2	94.1	63.6	87.0	94.1	65.1	87.3	97.2	62.2	89.0
0920	Plaster & Gypsum Board	88.7	65.8	73.3	78.2	65.8	69.9	82.1	65.8	71.2	81.8	65.8	71.1	84.7	65.8	72.0	110.1	73.9	85.7
0950, 0980	Ceilings & Acoustic Treatment	86.2	65.8	72.5	79.6	65.8	70.3	78.8	65.8	70.1	78.8	65.8	70.1	78.8	65.8	70.1	98.8	73.9	82.0
0960	Flooring	90.7	80.4	87.8	89.9	80.4	87.2	96.3	80.4	91.8	95.2	75.8	89.7	97.6	80.4	92.8	91.1	43.8	77.8
0970, 0990	Wall Finishes & Painting/Coating	93.9	70.0	79.2	95.2	70.0	80.3	95.2	70.0	80.3	95.2	66.0	77.9	95.2	70.0	80.3	89.8	33.4	56.4
09	FINISHES	86.5	68.7	76.8	85.0	68.7	76.2	86.9	68.7	77.0	86.3	67.1	75.9	87.6	68.7	77.4	92.4	65.9	78.1
COVERS	DIVS. 10 - 14, 25, 28, 41, 43, 44, 46	100.0	70.5	93.4	100.0	70.3	93.4	100.0	70.3	93.4	100.0	70.0	93.3	100.0	70.3	93.4	100.0	84.8	96.6
21, 22, 23	FIRE SUPPRESSION, PLUMBING & HVAC	100.0	58.3	83.6	100.4	58.2	83.9	100.4	58.2	83.9	96.9	54.2	80.4	100.4	58.2	83.9	100.0	54.2	82.0
26, 27, 3370	ELECTRICAL, COMMUNICATIONS & UTIL.	98.5	62.8	80.9	96.1	62.8	79.7	98.2	59.8	79.3	98.2	59.9	79.3	98.2	59.8	79.3	101.1	65.3	83.4
MF2018	WEIGHTED AVERAGE	97.6	68.9	85.5	96.9	68.6	84.9	97.0	68.2	84.8	96.7	66.4	83.9	97.2	68.3	85.0	98.6	70.1	86.5

SOUTH DAKOTA

DIVISION		MITCHELL 573			MOBRIDGE 576			PIERRE 575			RAPID CITY 577			SIOUX FALLS 570 - 571			WATERTOWN 572		
		MAT.	INST.	TOTAL	MAT.	INST.	TOTAL	MAT.	INST.	TOTAL	MAT.	INST.	TOTAL	MAT.	INST.	TOTAL	MAT.	INST.	TOTAL
015433	CONTRACTOR EQUIPMENT		97.7	97.7		97.7	97.7		99.9	99.9		97.7	97.7		100.9	100.9		97.7	97.7
0241, 31 - 34	SITE & INFRASTRUCTURE, DEMOLITION	95.8	93.2	94.0	95.7	93.2	94.0	97.2	96.3	96.6	97.5	92.8	94.3	92.5	98.7	96.8	95.6	93.2	94.0
0310	Concrete Forming & Accessories	97.9	44.2	52.1	88.2	44.6	51.0	99.7	45.6	53.6	106.9	56.6	64.0	103.3	78.1	81.8	84.6	74.5	76.0
0320	Concrete Reinforcing	100.4	70.5	86.0	103.0	70.5	87.3	97.8	100.2	99.0	94.4	100.4	97.3	106.6	100.6	103.7	97.6	70.5	84.5
0330	Cast-in-Place Concrete	106.4	52.3	86.3	106.4	77.2	95.6	103.1	77.8	93.7	105.6	77.2	95.0	88.3	78.8	84.7	106.4	77.2	95.5
03	CONCRETE	96.5	53.1	77.4	96.2	62.0	81.2	91.8	67.4	81.1	95.9	72.3	85.5	88.6	82.7	86.0	95.3	75.5	86.6
04	MASONRY	97.8	75.5	84.1	106.5	72.8	85.8	106.5	72.6	85.7	106.0	74.0	86.3	91.1	75.6	81.5	132.0	73.7	96.2
05	METALS	92.9	81.6	89.6	92.9	82.0	89.7	96.3	89.5	94.3	95.5	91.2	94.2	95.3	92.6	94.5	92.9	81.9	89.6
06	WOOD, PLASTICS & COMPOSITES	96.6	33.7	62.0	84.4	33.4	56.4	104.2	34.0	65.6	102.2	48.0	72.3	101.6	76.5	87.8	80.5	74.2	77.0
07	THERMAL & MOISTURE PROTECTION	102.8	72.9	89.8	102.9	75.1	90.8	104.1	72.7	90.4	103.6	78.5	92.7	103.2	83.3	94.6	102.6	76.7	91.3
08	OPENINGS	96.3	39.6	83.1	98.9	39.0	84.9	99.8	58.4	90.2	101.0	66.1	92.9	104.2	81.8	99.0	96.3	61.5	88.2
0920	Plaster & Gypsum Board	109.0	32.2	57.3	102.2	32.0	54.9	108.1	32.5	57.2	109.8	46.9	67.5	103.6	76.1	85.1	100.5	73.9	82.6
0950, 0980	Ceilings & Acoustic Treatment	95.5	32.2	52.8	98.8	32.0	53.7	98.6	32.5	54.0	100.4	46.9	64.3	97.7	76.1	83.1	95.5	73.9	80.9
0960	Flooring	90.7	43.8	77.6	86.5	46.4	75.3	98.2	31.2	79.4	90.5	72.7	85.5	93.9	75.0	88.6	85.2	43.8	73.6
0970, 0990	Wall Finishes & Painting/Coating	89.8	36.6	58.3	89.8	37.5	58.8	97.1	95.6	96.2	89.8	95.6	93.2	100.3	95.6	97.5	89.8	33.4	56.4
09	FINISHES	91.4	41.1	64.2	89.9	41.6	63.7	98.4	45.2	69.6	92.5	61.8	75.9	96.4	79.1	87.1	88.6	64.7	75.7
COVERS	DIVS. 10 - 14, 25, 28, 41, 43, 44, 46	100.0	82.0	96.0	100.0	81.9	96.0	100.0	82.3	96.1	100.0	83.5	96.3	100.0	87.0	97.1	100.0	85.8	96.8
21, 22, 23	FIRE SUPPRESSION, PLUMBING & HVAC	96.5	49.0	77.8	96.5	69.0	85.7	100.0	76.8	90.9	100.0	76.8	90.9	99.9	70.2	88.2	96.5	52.1	79.0
26, 27, 3370	ELECTRICAL, COMMUNICATIONS & UTIL.	99.4	63.4	81.7	101.1	39.8	70.9	103.9	47.7	76.2	97.6	47.7	73.0	102.6	63.4	83.3	98.6	63.4	81.2
MF2018	WEIGHTED AVERAGE	96.3	60.9	81.3	96.9	62.9	82.5	99.0	68.5	86.1	99.0	72.0	87.2	97.7	78.4	89.5	97.3	69.0	85.4

TENNESSEE

DIVISION		CHATTANOOGA 373 - 374			COLUMBIA 384			COOKEVILLE 385			JACKSON 383			JOHNSON CITY 376			KNOXVILLE 377 - 379		
		MAT.	INST.	TOTAL	MAT.	INST.	TOTAL	MAT.	INST.	TOTAL	MAT.	INST.	TOTAL	MAT.	INST.	TOTAL	MAT.	INST.	TOTAL
015433	CONTRACTOR EQUIPMENT		104.7	104.7		99.5	99.5		99.5	99.5		105.5	105.5		98.8	98.8		98.8	98.8
0241, 31 - 34	SITE & INFRASTRUCTURE, DEMOLITION	106.6	93.8	97.8	91.5	84.6	86.7	97.3	81.6	86.5	100.9	94.0	96.2	113.6	81.1	91.2	92.6	83.8	86.5
0310	Concrete Forming & Accessories	95.8	57.3	63.0	80.9	62.4	65.1	81.1	31.9	39.2	87.5	41.0	47.8	82.7	57.5	61.2	94.3	61.7	66.5
0320	Concrete Reinforcing	101.6	65.1	83.9	91.4	65.0	78.6	91.4	64.6	78.4	91.3	67.2	79.7	102.2	61.9	82.7	101.6	61.8	82.3
0330	Cast-in-Place Concrete	97.5	64.1	85.1	92.8	63.8	82.0	105.2	56.7	87.2	102.7	66.0	89.0	78.4	58.0	70.8	91.6	64.0	81.3
03	CONCRETE	93.1	62.8	79.8	92.4	65.1	80.4	102.4	48.7	78.8	93.2	56.5	77.0	102.7	60.2	84.0	90.7	64.2	79.0
04	MASONRY	98.6	56.3	72.6	111.4	55.5	77.0	106.7	39.4	65.3	111.7	44.4	70.4	111.0	43.2	69.3	76.3	50.7	60.6
05	METALS	93.0	88.1	91.6	93.5	88.9	92.1	93.5	87.6	91.8	96.0	89.5	94.1	90.3	86.4	89.2	93.4	86.5	91.4
06	WOOD, PLASTICS & COMPOSITES	106.5	56.3	78.9	67.8	63.4	65.4	68.0	29.5	46.8	82.9	39.7	59.1	78.7	62.6	69.8	93.1	62.6	76.4
07	THERMAL & MOISTURE PROTECTION	99.2	61.7	82.9	95.6	62.9	81.4	96.1	52.6	77.2	98.0	55.5	79.5	95.3	55.6	78.0	92.8	60.8	78.9
08	OPENINGS	102.0	57.5	91.6	91.0	53.3	82.2	91.1	35.5	78.1	97.7	44.3	85.2	98.3	59.5	89.3	95.7	55.2	86.2
0920	Plaster & Gypsum Board	80.0	55.6	63.6	87.4	62.9	70.9	87.4	28.0	47.4	90.0	38.5	55.3	101.0	62.1	74.8	109.1	62.1	77.5
0950, 0980	Ceilings & Acoustic Treatment	97.9	55.6	69.4	75.8	62.9	67.1	75.8	28.0	43.6	83.9	38.5	53.3	93.5	62.1	72.3	94.3	62.1	72.6
0960	Flooring	98.3	55.2	86.2	81.7	53.9	73.9	81.7	50.0	72.8	81.0	54.2	73.5	93.3	43.9	79.4	98.4	49.7	84.7
0970, 0990	Wall Finishes & Painting/Coating	98.0	59.5	75.2	85.0	56.5	68.1	85.0	56.5	68.1	87.0	56.5	68.9	94.9	58.7	73.5	94.9	57.6	72.8
09	FINISHES	94.4	56.6	73.9	86.2	60.1	72.1	86.8	36.2	59.4	86.1	43.4	63.0	98.5	55.3	75.1	91.4	58.9	73.8
COVERS	DIVS. 10 - 14, 25, 28, 41, 43, 44, 46	100.0	67.8	92.8	100.0	68.0	92.9	100.0	35.9	85.7	100.0	62.0	91.5	100.0	75.8	94.6	100.0	79.3	95.4
21, 22, 23	FIRE SUPPRESSION, PLUMBING & HVAC	100.1	59.3	84.1	97.8	73.6	88.3	97.8	67.1	85.8	100.1	61.4	84.9	100.0	55.5	82.5	100.0	60.5	84.5
26, 27, 3370	ELECTRICAL, COMMUNICATIONS & UTIL.	100.8	82.9	92.0	94.0	50.6	72.6	95.5	61.3	78.6	100.5	65.7	83.3	91.5	40.8	66.5	97.0	54.9	76.2
MF2018	WEIGHTED AVERAGE	98.0	68.1	85.3	94.8	66.5	82.8	96.2	56.9	79.6	97.6	61.4	82.3	98.1	58.6	81.4	94.8	63.7	81.6

TENNESSEE / TEXAS

DIVISION		MCKENZIE 382			MEMPHIS 375, 380 - 381			NASHVILLE 370 - 372			ABILENE 795 - 796			AMARILLO 790 - 791			AUSTIN 786 - 787		
		MAT.	INST.	TOTAL	MAT.	INST.	TOTAL	MAT.	INST.	TOTAL	MAT.	INST.	TOTAL	MAT.	INST.	TOTAL	MAT.	INST.	TOTAL
015433	CONTRACTOR EQUIPMENT		99.5	99.5		102.0	102.0		105.5	105.5		90.9	90.9		93.9	93.9		92.5	92.5
0241, 31 - 34	SITE & INFRASTRUCTURE, DEMOLITION	96.9	81.8	86.5	91.1	95.4	94.0	102.8	98.1	99.6	91.9	87.4	88.8	91.3	91.3	91.3	95.5	90.1	91.7
0310	Concrete Forming & Accessories	88.4	34.1	42.1	94.6	64.7	69.1	95.9	68.1	72.2	94.0	59.2	64.3	95.6	51.7	58.2	95.3	53.0	59.3
0320	Concrete Reinforcing	91.5	65.5	78.9	110.9	66.0	89.2	107.1	66.8	87.6	88.5	51.7	70.7	95.3	49.6	73.2	86.8	47.2	67.6
0330	Cast-in-Place Concrete	102.9	57.9	86.2	96.5	76.7	89.1	88.2	69.0	81.1	84.6	64.9	77.3	83.2	65.7	76.7	94.0	65.9	83.5
03	CONCRETE	101.0	50.2	78.7	96.5	70.2	84.9	92.9	69.4	82.6	84.5	60.7	74.1	87.1	57.2	74.0	89.4	57.4	75.4
04	MASONRY	110.5	45.1	70.3	92.6	56.3	70.3	87.7	55.9	68.2	99.9	60.6	75.8	97.6	60.2	74.6	91.7	60.7	72.7
05	METALS	93.5	88.0	91.9	87.2	82.5	85.8	91.6	85.1	96.7	107.1	70.4	96.3	101.9	68.7	92.1	100.7	67.4	90.9
06	WOOD, PLASTICS & COMPOSITES	76.6	31.7	51.9	98.9	65.4	80.4	100.1	70.1	83.6	99.2	60.8	78.1	102.6	51.0	74.2	92.9	52.6	70.7
07	THERMAL & MOISTURE PROTECTION	96.1	53.6	77.6	93.0	66.4	81.4	96.9	65.0	83.0	98.3	63.4	83.1	100.1	62.3	83.6	95.1	63.8	81.5
08	OPENINGS	91.1	37.2	78.5	99.5	61.3	90.6	101.0	67.5	93.2	101.8	56.7	91.2	103.2	50.8	91.0	101.1	49.6	89.1
0920	Plaster & Gypsum Board	90.9	30.3	50.1	90.9	64.5	73.1	95.9	69.3	78.0	88.2	60.1	69.3	100.7	50.0	66.6	88.9	51.6	63.8
0950, 0980	Ceilings & Acoustic Treatment	75.8	30.3	45.1	90.8	64.5	73.0	94.1	69.3	77.4	99.2	60.1	72.9	100.7	50.0	66.5	93.3	51.6	65.2
0960	Flooring	84.5	53.9	75.9	97.8	55.2	85.9	97.1	57.5	86.0	93.2	67.6	86.0	93.7	63.8	85.3	94.1	63.8	85.6
0970, 0990	Wall Finishes & Painting/Coating	85.0	44.2	60.8	93.2	59.3	73.1	100.7	68.6	81.7	92.8	51.7	68.5	91.0	51.7	67.7	95.7	44.2	65.2
09	FINISHES	88.0	37.5	60.7	95.9	61.9	77.5	97.8	66.1	80.6	86.1	60.1	72.0	92.6	53.5	71.4	91.1	53.6	70.8
COVERS	DIVS. 10 - 14, 25, 28, 41, 43, 44, 46	100.0	24.0	83.1	100.0	80.5	95.7	100.0	81.1	95.8	100.0	78.9	95.3	100.0	72.7	93.9	100.0	76.5	94.8
21, 22, 23	FIRE SUPPRESSION, PLUMBING & HVAC	97.8	58.6	82.4	100.1	70.3	88.4	99.9	75.1	90.2	100.1	52.1	81.3	99.9	51.1	80.7	99.9	58.7	83.8
26, 27, 3370	ELECTRICAL, COMMUNICATIONS & UTIL.	95.3	55.4	75.6	102.9	63.6	83.5	93.5	63.2	78.6	97.9	54.0	76.3	100.9	58.9	80.2	97.5	57.2	77.6
MF2018	WEIGHTED AVERAGE	96.3	55.1	78.9	96.6	69.8	85.3	98.0	71.8	86.9	97.8	61.4	82.4	98.2	60.0	82.1	97.2	61.4	82.1

TEXAS

DIVISION		BEAUMONT 776 - 777			BROWNWOOD 768			BRYAN 778			CHILDRESS 792			CORPUS CHRISTI 783 - 784			DALLAS 752 - 753		
		MAT.	INST.	TOTAL	MAT.	INST.	TOTAL	MAT.	INST.	TOTAL	MAT.	INST.	TOTAL	MAT.	INST.	TOTAL	MAT.	INST.	TOTAL
015433	CONTRACTOR EQUIPMENT		95.3	95.3		90.9	90.9		95.3	95.3		90.9	90.9		100.6	100.6		106.9	106.9
0241, 31 - 34	SITE & INFRASTRUCTURE, DEMOLITION	90.6	91.8	91.4	101.3	87.3	91.7	81.5	91.9	88.7	101.8	85.8	90.8	144.2	82.8	101.8	106.0	97.6	100.2
0310	Concrete Forming & Accessories	102.3	54.6	61.7	97.6	58.7	64.4	82.1	58.2	61.7	92.6	58.6	63.6	98.4	52.2	59.0	95.0	62.3	67.1
0320	Concrete Reinforcing	85.7	62.0	74.2	81.9	51.1	67.0	87.9	51.2	70.2	88.7	51.0	70.5	79.8	50.9	65.8	91.4	52.3	72.5
0330	Cast-in-Place Concrete	90.7	65.2	81.2	96.8	58.6	82.6	72.6	59.6	67.8	86.7	58.6	76.3	114.1	67.1	96.6	92.6	71.4	84.7
03	CONCRETE	91.0	60.7	77.7	90.3	58.2	76.2	75.8	58.5	68.2	93.4	58.2	77.9	95.2	59.1	79.4	93.3	65.3	81.0
04	MASONRY	102.9	61.7	77.6	122.8	60.6	84.6	143.3	60.6	92.5	103.5	60.2	76.9	81.4	60.7	68.7	99.5	60.2	75.3
05	METALS	99.4	76.2	92.6	102.6	69.8	92.9	99.5	72.1	91.4	104.4	69.6	94.2	97.3	84.4	93.5	101.7	82.6	96.1
06	WOOD, PLASTICS & COMPOSITES	113.6	54.1	80.9	105.8	60.8	81.0	78.6	59.3	68.0	98.6	60.8	77.8	115.6	51.5	80.3	100.8	63.6	80.3
07	THERMAL & MOISTURE PROTECTION	92.8	64.3	80.4	92.0	61.9	78.9	85.2	63.8	75.9	99.0	60.4	82.2	97.7	62.2	82.2	87.3	67.2	78.6
08	OPENINGS	95.0	55.0	85.6	100.4	56.6	90.2	95.7	56.0	86.4	96.1	56.6	86.9	101.3	50.0	89.4	96.5	59.5	87.9
0920	Plaster & Gypsum Board	101.0	53.2	68.9	82.6	60.1	67.5	88.2	58.7	68.3	87.6	60.1	69.1	96.5	50.5	65.5	92.7	62.5	72.4
0950, 0980	Ceilings & Acoustic Treatment	105.3	53.2	70.2	86.6	60.1	68.7	96.7	58.7	71.1	96.8	60.1	72.1	98.8	50.5	66.2	102.4	62.5	75.5
0960	Flooring	116.9	74.6	105.0	75.7	54.6	69.8	86.7	67.0	81.1	91.6	54.6	81.2	107.4	67.7	96.3	94.2	69.4	87.2
0970, 0990	Wall Finishes & Painting/Coating	94.9	53.9	70.6	92.8	51.7	68.4	92.2	57.4	71.6	92.8	51.7	68.5	110.1	44.2	71.1	96.1	54.8	71.6
09	FINISHES	98.0	57.8	76.2	77.6	57.4	66.7	85.4	59.4	71.3	86.2	57.4	70.6	101.9	53.8	75.9	97.9	62.7	78.8
COVERS	DIVS. 10 - 14, 25, 28, 41, 43, 44, 46	100.0	77.8	95.0	100.0	73.4	94.1	100.0	77.0	94.9	100.0	73.4	94.1	100.0	77.8	95.0	100.0	82.5	96.1
21, 22, 23	FIRE SUPPRESSION, PLUMBING & HVAC	100.0	63.0	85.5	96.5	49.5	78.0	96.5	62.4	83.1	96.6	51.7	79.0	100.0	48.7	79.9	99.9	62.3	85.1
26, 27, 3370	ELECTRICAL, COMMUNICATIONS & UTIL.	100.1	63.4	82.0	98.6	46.1	72.7	97.8	63.4	80.8	97.9	55.8	77.1	94.0	63.4	78.9	95.2	63.2	79.4
MF2018	WEIGHTED AVERAGE	97.9	65.5	84.2	97.4	58.7	81.1	95.2	64.9	82.4	97.6	60.3	81.8	98.9	61.4	83.1	98.2	68.0	85.4

TEXAS

DIVISION		DEL RIO 788			DENTON 762			EASTLAND 764			EL PASO 798 - 799, 885			FORT WORTH 760 - 761			GALVESTON 775		
		MAT.	INST.	TOTAL	MAT.	INST.	TOTAL	MAT.	INST.	TOTAL	MAT.	INST.	TOTAL	MAT.	INST.	TOTAL	MAT.	INST.	TOTAL
015433	CONTRACTOR EQUIPMENT		89.9	89.9		100.2	100.2		90.9	90.9		93.9	93.9		93.9	93.9		107.9	107.9
0241, 31 - 34	SITE & INFRASTRUCTURE, DEMOLITION	122.2	85.5	96.9	101.4	79.7	86.4	104.0	85.5	91.2	89.7	92.7	91.8	97.4	92.7	94.2	108.4	90.6	96.1
0310	Concrete Forming & Accessories	95.1	51.0	57.5	105.0	58.8	65.7	98.4	58.6	64.4	95.6	56.7	62.5	98.2	59.5	65.2	90.9	55.8	61.0
0320	Concrete Reinforcing	80.4	47.6	64.5	83.2	51.1	67.7	82.1	51.1	67.1	94.2	51.1	73.4	88.1	50.9	70.1	87.5	59.7	74.0
0330	Cast-in-Place Concrete	122.9	58.4	98.9	74.9	59.7	69.2	102.3	58.5	86.0	73.9	68.0	71.7	89.3	65.9	80.6	96.4	60.6	83.1
03	CONCRETE	115.8	54.0	88.6	70.6	59.7	65.8	94.8	58.1	78.7	82.7	60.5	72.9	86.1	61.0	75.1	92.1	60.3	78.1
04	MASONRY	95.4	60.5	74.0	131.0	59.3	86.9	92.7	60.6	73.0	91.6	62.7	73.8	92.5	59.3	72.1	99.0	60.7	75.5
05	METALS	97.2	67.2	88.4	102.1	84.9	97.1	102.4	69.6	92.7	102.1	69.4	92.5	104.3	69.5	94.1	101.2	91.4	98.3
06	WOOD, PLASTICS & COMPOSITES	96.1	50.4	71.0	117.8	60.9	86.5	113.0	60.8	84.3	90.2	57.1	72.0	99.6	60.9	78.3	95.3	55.9	73.6
07	THERMAL & MOISTURE PROTECTION	94.7	62.0	80.5	90.0	62.4	78.0	92.4	61.9	79.1	91.9	64.4	80.0	89.7	64.4	78.7	84.6	64.4	75.8
08	OPENINGS	96.6	47.8	85.2	118.7	56.2	104.2	70.6	56.6	67.4	96.3	52.8	86.2	101.2	56.6	90.8	99.7	56.0	89.5
0920	Plaster & Gypsum Board	93.3	49.5	63.8	87.3	60.1	69.0	82.6	60.1	67.5	97.3	56.2	69.6	89.8	60.1	69.8	94.8	55.0	68.0
0950, 0980	Ceilings & Acoustic Treatment	95.3	49.5	64.4	90.7	60.1	70.1	86.6	60.1	68.7	94.2	56.2	68.6	92.7	60.1	70.7	100.8	55.0	69.9
0960	Flooring	91.9	54.6	81.4	71.9	54.6	67.1	95.9	54.6	84.3	97.3	69.7	89.5	94.5	64.5	86.0	102.4	67.0	92.5
0970, 0990	Wall Finishes & Painting/Coating	97.8	42.6	65.1	103.3	48.9	71.1	94.2	51.7	69.0	98.6	49.6	69.6	101.2	51.8	72.0	104.6	53.7	74.5
09	FINISHES	94.5	50.3	70.6	76.3	57.2	66.0	84.1	57.4	69.6	92.3	58.1	73.8	89.6	59.5	73.3	95.4	56.9	74.6
COVERS	DIVS. 10 - 14, 25, 28, 41, 43, 44, 46	100.0	73.6	94.1	100.0	79.2	95.4	100.0	71.7	93.7	100.0	76.6	94.8	100.0	79.2	95.4	100.0	78.3	95.2
21, 22, 23	FIRE SUPPRESSION, PLUMBING & HVAC	96.5	59.4	81.9	96.5	55.8	80.5	96.5	49.5	78.0	99.8	65.3	86.3	99.9	55.4	82.4	96.5	62.5	83.2
26, 27, 3370	ELECTRICAL, COMMUNICATIONS & UTIL.	95.9	59.8	78.1	101.0	58.7	80.1	98.5	58.6	78.8	97.5	50.5	74.3	100.2	58.6	79.7	99.6	63.5	81.8
MF2018	WEIGHTED AVERAGE	99.7	60.4	83.1	97.3	62.9	82.7	94.2	60.3	79.8	95.9	63.6	82.3	97.6	62.9	82.9	97.7	66.5	84.5

TEXAS

DIVISION		GIDDINGS 789			GREENVILLE 754			HOUSTON 770 - 772			HUNTSVILLE 773			LAREDO 780			LONGVIEW 756		
		MAT.	INST.	TOTAL	MAT.	INST.	TOTAL	MAT.	INST.	TOTAL	MAT.	INST.	TOTAL	MAT.	INST.	TOTAL	MAT.	INST.	TOTAL
015433	CONTRACTOR EQUIPMENT		89.9	89.9		100.4	100.4		103.3	103.3		95.3	95.3		89.9	89.9		91.8	91.8
0241, 31 - 34	SITE & INFRASTRUCTURE, DEMOLITION	106.8	85.8	92.3	96.4	82.4	86.7	107.7	95.5	99.3	96.8	91.6	93.2	100.2	86.2	90.5	93.7	88.6	90.2
0310	Concrete Forming & Accessories	92.6	51.1	57.3	86.9	57.1	61.5	96.0	58.9	64.4	89.2	54.1	59.2	95.1	52.1	58.5	82.9	58.8	62.3
0320	Concrete Reinforcing	80.9	48.4	65.2	91.9	51.1	72.2	87.3	52.2	70.3	88.2	51.1	70.3	80.4	50.9	66.1	90.9	50.7	71.5
0330	Cast-in-Place Concrete	104.1	58.7	87.2	92.1	59.6	80.0	89.6	68.9	81.9	99.9	59.5	84.9	87.8	65.0	79.3	107.1	58.5	89.0
03	CONCRETE	92.6	54.3	75.8	86.7	58.7	74.4	91.4	62.5	78.7	98.7	56.6	80.2	87.3	57.4	74.2	102.7	58.1	83.1
04	MASONRY	102.2	60.6	76.7	158.6	59.3	97.5	98.0	63.1	76.6	141.3	60.6	91.7	88.9	60.6	71.5	154.2	59.3	95.9
05	METALS	96.7	68.2	88.4	99.2	83.2	94.5	103.9	77.3	96.1	99.4	71.8	91.3	99.6	69.4	90.7	91.8	68.4	84.9
06	WOOD, PLASTICS & COMPOSITES	95.3	50.4	70.6	87.2	58.4	71.4	100.6	59.2	77.8	88.1	54.1	69.4	96.1	51.4	71.5	81.2	60.8	70.0
07	THERMAL & MOISTURE PROTECTION	95.0	62.7	81.0	87.3	61.5	76.1	86.3	68.5	78.5	86.5	63.2	76.4	93.8	63.0	80.4	89.1	60.9	76.8
08	OPENINGS	95.8	48.7	84.8	92.2	55.3	83.6	106.6	56.4	94.9	95.7	52.5	85.6	96.7	49.8	85.8	83.4	56.2	77.0
0920	Plaster & Gypsum Board	92.4	49.5	63.5	82.8	57.5	65.8	94.0	58.0	69.8	92.3	53.2	66.0	94.1	50.5	64.7	81.7	60.1	67.2
0950, 0980	Ceilings & Acoustic Treatment	95.3	49.5	64.4	98.1	57.5	70.7	97.9	58.0	71.0	96.7	53.2	67.4	98.6	50.5	66.1	93.2	60.1	70.9
0960	Flooring	91.7	54.6	81.2	87.7	54.6	78.4	103.5	71.2	94.4	91.3	54.6	81.0	91.8	63.8	83.9	93.7	54.6	82.7
0970, 0990	Wall Finishes & Painting/Coating	97.8	44.2	66.1	90.4	51.7	67.5	102.1	58.6	76.3	92.2	57.4	71.6	97.8	44.2	66.1	98.1	48.9	62.4
09	FINISHES	93.1	50.5	70.0	92.3	56.1	72.7	102.7	60.8	80.1	88.3	54.1	69.8	93.4	52.9	71.5	98.1	57.2	76.0
COVERS	DIVS. 10 - 14, 25, 28, 41, 43, 44, 46	100.0	70.3	93.4	100.0	76.5	94.8	100.0	83.1	96.2	100.0	69.8	93.3	100.0	75.7	94.6	100.0	79.0	95.3
21, 22, 23	FIRE SUPPRESSION, PLUMBING & HVAC	96.5	61.4	82.7	96.5	58.0	81.3	100.0	65.3	86.3	96.5	62.4	83.1	100.0	58.9	83.8	96.4	57.5	81.1
26, 27, 3370	ELECTRICAL, COMMUNICATIONS & UTIL.	93.1	55.6	74.6	92.1	58.7	75.6	101.8	67.8	85.1	97.8	59.1	78.7	96.0	57.1	76.8	92.6	50.9	72.0
MF2018	WEIGHTED AVERAGE	96.1	60.4	81.0	97.4	62.9	82.8	100.4	68.2	86.8	98.7	62.8	83.6	96.4	61.1	81.5	97.5	61.0	82.1

TEXAS

DIVISION		LUBBOCK 793-794			LUFKIN 759			MCALLEN 785			MCKINNEY 750			MIDLAND 797			ODESSA 797		
		MAT.	INST.	TOTAL	MAT.	INST.	TOTAL	MAT.	INST.	TOTAL	MAT.	INST.	TOTAL	MAT.	INST.	TOTAL	MAT.	INST.	TOTAL
015433	CONTRACTOR EQUIPMENT		102.9	102.9		91.8	91.8		100.7	100.7		100.4	100.4		102.9	102.9		90.9	90.9
0241, 31-34	SITE & INFRASTRUCTURE, DEMOLITION	115.0	86.7	95.5	88.6	90.0	89.5	149.0	82.9	103.3	92.8	82.4	85.6	117.7	85.8	95.6	92.1	88.1	89.3
0310	Concrete Forming & Accessories	92.9	53.0	58.9	85.9	54.1	58.8	99.1	51.2	58.3	86.0	57.1	61.4	96.4	59.1	64.6	93.9	59.2	64.3
0320	Concrete Reinforcing	89.7	51.8	71.4	92.6	66.7	80.1	80.0	50.9	65.9	91.9	51.1	72.2	90.7	51.7	71.8	88.5	51.6	70.7
0330	Cast-in-Place Concrete	84.7	68.2	78.6	95.8	58.3	81.8	124.0	59.7	100.1	86.3	59.6	76.4	90.0	65.9	81.1	84.6	64.9	77.3
03	CONCRETE	83.1	60.1	73.0	95.1	58.6	79.0	102.8	56.2	82.3	82.1	58.8	71.9	87.1	62.1	76.1	84.5	60.7	74.1
04	MASONRY	99.2	61.0	75.7	117.3	60.5	82.4	95.1	60.7	74.0	170.8	59.3	102.3	116.2	60.2	81.8	99.9	60.2	75.5
05	METALS	110.9	85.8	103.5	99.4	73.3	91.7	97.1	84.2	93.3	99.2	83.3	94.5	109.2	85.4	102.2	106.4	70.4	95.8
06	WOOD, PLASTICS & COMPOSITES	98.6	52.1	73.0	88.6	54.2	69.6	114.3	50.5	79.2	86.1	58.4	70.9	103.0	60.9	79.8	99.2	60.8	78.1
07	THERMAL & MOISTURE PROTECTION	88.2	63.5	77.5	88.8	60.0	76.3	97.6	62.9	82.5	87.1	61.5	76.0	88.5	63.3	77.5	98.3	62.7	82.8
08	OPENINGS	108.6	52.4	95.5	63.3	56.6	61.7	100.4	48.6	88.3	92.2	55.3	83.6	110.0	56.7	97.6	101.8	56.7	91.2
0920	Plaster & Gypsum Board	88.4	51.1	63.3	80.6	53.2	62.2	97.5	49.5	65.2	82.8	57.5	65.8	90.0	60.1	69.9	88.2	60.1	69.3
0950, 0980	Ceilings & Acoustic Treatment	100.0	51.1	67.0	88.3	53.2	64.7	98.6	49.5	65.5	98.1	57.5	70.7	97.6	60.1	72.3	99.2	60.1	72.9
0960	Flooring	87.5	69.0	82.3	126.2	54.6	106.1	106.9	72.8	97.3	87.2	54.6	78.1	88.6	63.8	81.7	93.2	63.8	84.9
0970, 0990	Wall Finishes & Painting/Coating	103.4	53.7	74.0	82.1	51.7	64.1	110.1	42.6	70.2	90.4	51.7	67.5	103.4	51.7	72.8	92.8	51.7	68.5
09	FINISHES	88.5	55.5	70.6	106.5	53.9	78.0	102.4	54.1	76.3	91.9	56.1	72.5	89.0	59.4	72.9	86.1	59.3	71.6
COVERS	DIVS. 10-14, 25, 28, 41, 43, 44, 46	100.0	78.3	95.2	100.0	70.1	93.3	100.0	75.8	94.6	100.0	79.1	95.3	100.0	73.8	94.1	100.0	73.5	94.1
21, 22, 23	FIRE SUPPRESSION, PLUMBING & HVAC	99.7	53.1	81.4	96.4	58.1	81.4	96.5	48.7	77.7	96.5	60.2	82.2	96.1	47.4	77.0	100.1	52.5	81.4
26, 27, 3370	ELECTRICAL, COMMUNICATIONS & UTIL.	96.7	60.1	78.6	93.9	61.4	77.8	93.8	31.6	63.1	92.2	58.7	75.6	96.7	58.9	78.1	98.0	58.9	78.7
MF2018	WEIGHTED AVERAGE	99.2	62.9	83.9	94.7	62.6	81.1	99.7	56.5	81.4	97.2	63.5	82.9	99.7	62.2	83.8	97.7	61.8	82.6

TEXAS

DIVISION		PALESTINE 758			SAN ANGELO 769			SAN ANTONIO 781-782			TEMPLE 765			TEXARKANA 755			TYLER 757		
		MAT.	INST.	TOTAL	MAT.	INST.	TOTAL	MAT.	INST.	TOTAL	MAT.	INST.	TOTAL	MAT.	INST.	TOTAL	MAT.	INST.	TOTAL
015433	CONTRACTOR EQUIPMENT		91.8	91.8		90.9	90.9		94.1	94.1		90.9	90.9		91.8	91.8		91.8	91.8
0241, 31-34	SITE & INFRASTRUCTURE, DEMOLITION	94.2	88.8	90.5	97.8	87.7	90.8	101.5	93.7	96.1	86.7	87.6	87.3	83.4	91.1	88.7	92.9	88.7	90.0
0310	Concrete Forming & Accessories	77.5	58.8	61.5	97.9	51.4	58.8	94.7	52.6	58.8	101.2	51.2	58.5	93.0	59.4	64.3	87.6	58.9	63.1
0320	Concrete Reinforcing	90.2	51.0	71.3	81.8	51.6	67.2	89.6	47.7	69.3	82.0	50.8	66.9	90.1	50.8	71.1	90.9	50.8	71.5
0330	Cast-in-Place Concrete	87.5	58.5	76.7	91.4	64.9	81.5	85.7	68.3	79.2	74.9	58.6	68.8	88.2	64.9	79.5	105.2	58.6	87.8
03	CONCRETE	97.1	58.1	80.0	86.1	57.2	73.4	88.0	58.0	74.8	73.4	54.7	65.2	88.0	60.6	75.9	102.3	58.2	82.9
04	MASONRY	111.9	59.2	79.5	119.4	60.6	83.3	89.7	60.8	71.9	130.1	60.6	87.4	175.3	59.2	103.9	164.4	59.2	99.7
05	METALS	99.1	68.5	90.1	102.7	70.3	93.2	102.0	65.4	91.3	102.4	69.2	92.7	91.7	69.0	85.0	99.0	68.5	90.1
06	WOOD, PLASTICS & COMPOSITES	79.3	60.8	69.1	106.1	50.4	75.5	98.1	51.7	72.5	115.7	50.4	79.8	93.7	60.8	75.6	90.3	60.8	74.1
07	THERMAL & MOISTURE PROTECTION	89.4	61.8	77.4	91.8	62.6	79.1	86.8	65.6	77.5	91.3	61.7	78.4	88.6	63.0	77.5	89.2	61.8	77.3
08	OPENINGS	63.3	56.6	61.7	100.5	51.0	88.9	102.1	49.5	89.8	67.3	50.8	63.4	83.3	56.5	77.1	63.2	56.5	61.7
0920	Plaster & Gypsum Board	79.0	60.1	66.3	82.6	49.5	60.3	93.3	50.5	64.5	82.6	49.5	60.3	86.8	60.1	68.8	80.6	60.1	66.8
0950, 0980	Ceilings & Acoustic Treatment	88.3	60.1	69.3	86.6	49.5	61.6	95.2	50.5	65.1	86.6	49.5	61.6	93.2	60.1	70.9	88.3	60.1	69.3
0960	Flooring	116.4	54.6	99.0	75.8	54.6	69.9	101.2	67.7	91.8	97.5	54.6	85.4	102.7	63.8	91.7	128.7	54.6	107.8
0970, 0990	Wall Finishes & Painting/Coating	82.1	51.7	64.1	92.8	51.7	68.4	99.4	44.2	66.7	94.2	44.2	64.6	82.1	51.7	64.1	82.1	51.7	64.1
09	FINISHES	104.0	57.5	78.8	77.4	51.3	63.3	104.2	54.0	77.0	83.3	50.5	65.5	100.9	59.3	78.4	107.5	57.5	80.4
COVERS	DIVS. 10-14, 25, 28, 41, 43, 44, 46	100.0	73.5	94.1	100.0	76.0	94.7	100.0	78.0	95.1	100.0	72.3	93.8	100.0	79.1	95.3	100.0	79.0	95.3
21, 22, 23	FIRE SUPPRESSION, PLUMBING & HVAC	96.4	56.7	80.8	96.5	52.2	79.1	100.0	59.8	84.2	96.5	53.2	79.5	96.4	55.5	80.3	96.4	58.2	81.4
26, 27, 3370	ELECTRICAL, COMMUNICATIONS & UTIL.	90.1	48.4	69.5	102.8	54.6	79.1	97.5	59.8	78.9	99.8	52.0	76.2	93.7	54.5	74.4	92.6	53.7	73.4
MF2018	WEIGHTED AVERAGE	94.2	60.4	79.9	97.1	59.4	81.2	98.3	62.3	83.1	92.5	58.5	78.2	96.7	62.0	82.1	97.8	61.6	82.5

TEXAS / UTAH

DIVISION		VICTORIA 779			WACO 766-767			WAXAHACHIE 751			WHARTON 774			WICHITA FALLS 763			LOGAN 843		
		MAT.	INST.	TOTAL	MAT.	INST.	TOTAL	MAT.	INST.	TOTAL	MAT.	INST.	TOTAL	MAT.	INST.	TOTAL	MAT.	INST.	TOTAL
015433	CONTRACTOR EQUIPMENT		106.6	106.6		90.9	90.9		100.4	100.4		107.9	107.9		90.9	90.9		93.9	93.9
0241, 31-34	SITE & INFRASTRUCTURE, DEMOLITION	113.4	88.0	95.8	95.1	88.0	90.2	94.7	82.5	86.3	118.7	90.0	98.9	95.8	88.1	90.5	99.8	89.6	92.7
0310	Concrete Forming & Accessories	90.6	51.3	57.1	99.8	59.2	65.2	86.0	59.0	63.0	85.7	53.1	57.9	99.8	59.2	65.2	103.5	67.2	72.6
0320	Concrete Reinforcing	84.0	50.9	68.0	81.6	47.2	65.0	91.9	51.1	72.2	87.4	50.9	69.7	81.6	51.7	67.1	108.0	84.4	96.6
0330	Cast-in-Place Concrete	108.3	60.5	90.5	81.0	64.9	75.0	91.1	59.8	79.5	111.3	59.5	92.0	86.5	65.0	78.5	86.4	74.3	81.9
03	CONCRETE	99.7	56.7	80.8	79.6	60.0	71.0	85.8	59.7	74.3	103.8	57.1	83.3	82.2	60.8	72.8	105.4	73.2	91.2
04	MASONRY	116.9	60.7	82.4	90.9	60.6	72.3	159.2	59.3	97.8	103.0	60.6	75.9	91.4	60.1	72.2	109.5	60.5	79.4
05	METALS	99.4	87.8	96.0	104.8	68.4	94.1	99.2	83.6	94.6	101.2	87.4	97.1	104.7	70.4	94.6	119.5	82.5	101.5
06	WOOD, PLASTICS & COMPOSITES	98.4	50.5	72.1	113.9	60.8	84.7	86.1	61.0	72.3	88.4	52.7	68.8	113.9	60.8	84.7	85.9	67.1	75.5
07	THERMAL & MOISTURE PROTECTION	88.0	61.1	76.3	92.2	63.7	79.8	87.2	61.6	76.1	85.0	63.7	75.7	92.2	62.9	79.5	103.0	68.8	88.1
08	OPENINGS	99.5	51.1	88.2	78.4	55.6	73.1	92.2	56.7	83.9	99.7	52.3	88.6	78.4	56.7	73.4	92.2	66.1	86.1
0920	Plaster & Gypsum Board	90.9	49.5	63.0	82.7	60.1	67.5	83.2	60.1	67.7	90.4	51.7	64.4	82.7	60.1	67.5	80.6	66.2	70.9
0950, 0980	Ceilings & Acoustic Treatment	101.6	49.5	66.5	87.4	60.1	69.0	99.7	60.1	73.0	100.8	51.7	67.7	87.4	60.1	69.0	104.3	66.2	78.7
0960	Flooring	100.7	54.6	87.7	96.8	63.8	87.5	87.2	54.6	78.1	99.5	66.6	90.2	97.4	71.9	90.2	97.2	60.5	86.9
0970, 0990	Wall Finishes & Painting/Coating	104.8	57.4	76.8	94.2	51.7	69.0	90.4	51.7	67.5	104.6	55.7	75.7	96.1	51.7	69.8	92.6	60.7	73.7
09	FINISHES	92.4	52.0	70.6	83.7	59.3	70.5	92.5	57.6	73.6	94.7	55.2	73.3	84.1	60.9	71.5	93.9	64.9	78.2
COVERS	DIVS. 10-14, 25, 28, 41, 43, 44, 46	100.0	69.7	93.2	100.0	78.9	95.3	100.0	79.4	95.4	100.0	70.0	93.3	100.0	73.5	94.1	100.0	84.4	96.5
21, 22, 23	FIRE SUPPRESSION, PLUMBING & HVAC	96.5	62.3	83.1	100.0	58.7	83.8	96.5	58.0	81.4	96.5	61.9	82.9	100.0	52.1	81.2	100.0	69.2	87.9
26, 27, 3370	ELECTRICAL, COMMUNICATIONS & UTIL.	104.9	54.0	79.8	103.0	54.1	78.9	92.1	58.7	75.6	103.9	59.2	81.9	105.0	56.0	80.8	95.2	71.3	83.4
MF2018	WEIGHTED AVERAGE	99.7	62.9	84.2	94.4	62.4	80.9	97.2	63.5	83.0	99.8	64.3	84.8	95.0	61.6	80.9	100.9	71.7	88.5

For customer support on your Commercial Renovation Costs with RSMeans data, call 800.448.8182.

821

UTAH / VERMONT

DIVISION		OGDEN 842, 844 MAT.	INST.	TOTAL	PRICE 845 MAT.	INST.	TOTAL	PROVO 846 - 847 MAT.	INST.	TOTAL	SALT LAKE CITY 840 - 841 MAT.	INST.	TOTAL	BELLOWS FALLS 051 MAT.	INST.	TOTAL	BENNINGTON 052 MAT.	INST.	TOTAL
015433	CONTRACTOR EQUIPMENT		93.9	93.9		92.9	92.9		92.9	92.9		93.8	93.8		95.7	95.7		95.7	95.7
0241, 31 - 34	SITE & INFRASTRUCTURE, DEMOLITION	87.3	89.6	88.9	97.1	87.4	90.4	95.9	88.0	90.5	87.0	89.5	88.7	86.8	96.1	93.3	86.2	96.1	93.1
0310	Concrete Forming & Accessories	103.5	67.2	72.6	106.0	64.2	70.4	105.3	67.2	72.9	105.8	67.2	72.9	96.5	101.7	100.9	94.0	101.6	100.4
0320	Concrete Reinforcing	107.6	84.4	96.4	115.8	84.0	100.4	116.8	84.4	101.1	110.0	84.4	97.6	85.1	86.3	85.7	85.1	86.3	85.6
0330	Cast-in-Place Concrete	87.7	74.3	82.7	86.5	71.6	80.9	86.5	74.3	81.9	95.8	74.3	87.8	89.6	115.9	99.4	89.6	115.8	99.4
03	CONCRETE	94.4	73.2	85.1	106.6	70.8	90.8	104.9	73.2	91.0	114.2	73.2	96.2	88.7	103.7	95.3	88.5	103.6	95.2
04	MASONRY	104.0	60.5	77.3	115.1	65.1	84.4	115.3	60.5	81.6	118.1	60.5	82.7	96.3	86.2	90.1	105.1	86.2	93.5
05	METALS	110.0	82.5	101.9	106.5	82.5	99.5	107.5	82.5	100.1	113.7	82.5	104.5	95.2	91.1	94.0	95.1	90.9	93.9
06	WOOD, PLASTICS & COMPOSITES	85.9	67.1	75.5	89.2	64.0	75.3	87.5	67.1	76.3	87.8	67.1	76.4	102.8	106.8	105.0	100.1	106.8	103.8
07	THERMAL & MOISTURE PROTECTION	101.8	68.8	87.4	104.8	68.3	88.9	104.9	68.8	89.2	101.6	68.8	92.1	100.9	89.1	95.8	100.9	89.1	95.8
08	OPENINGS	92.2	66.1	86.1	96.1	76.3	91.5	96.1	66.1	89.1	94.1	66.1	87.5	100.1	97.5	99.5	100.1	97.5	99.5
0920	Plaster & Gypsum Board	80.6	66.2	70.9	83.4	63.1	69.8	81.2	66.2	71.1	90.9	66.2	74.3	108.9	106.7	107.4	107.9	106.7	107.1
0950, 0980	Ceilings & Acoustic Treatment	104.3	66.2	78.7	104.3	63.1	76.5	104.3	66.2	78.7	97.1	66.2	76.3	97.2	106.7	103.6	97.2	106.7	103.6
0960	Flooring	95.2	60.5	85.4	98.4	56.1	86.5	98.2	60.5	87.6	99.0	60.5	88.1	92.0	98.9	94.0	91.2	98.9	93.4
0970, 0990	Wall Finishes & Painting/Coating	92.6	60.7	73.7	92.6	60.7	73.7	92.6	60.7	73.7	95.7	60.7	75.0	86.4	101.5	95.3	86.4	101.5	95.3
09	FINISHES	92.0	64.9	77.3	95.0	61.7	77.0	94.5	64.9	78.5	93.7	64.9	78.1	90.6	101.8	96.7	90.3	101.8	96.5
COVERS	DIVS. 10 - 14, 25, 28, 41, 43, 44, 46	100.0	84.4	96.5	100.0	81.6	95.9	100.0	84.4	96.5	100.0	84.4	96.5	100.0	91.8	98.2	100.0	91.7	98.2
21, 22, 23	FIRE SUPPRESSION, PLUMBING & HVAC	100.0	69.2	87.7	98.1	63.2	84.4	100.0	69.2	87.9	100.1	69.2	88.0	96.7	87.6	93.1	96.7	87.6	93.1
26, 27, 3370	ELECTRICAL, COMMUNICATIONS & UTIL.	95.5	71.3	83.6	100.2	68.8	84.7	95.7	68.6	82.3	98.1	71.3	84.9	108.3	79.5	94.1	108.3	54.5	81.8
MF2018	WEIGHTED AVERAGE	98.8	71.7	87.4	101.4	70.0	88.1	101.2	71.2	88.5	103.5	71.7	90.1	96.7	92.1	94.8	97.0	88.6	93.4

VERMONT

DIVISION		BRATTLEBORO 053 MAT.	INST.	TOTAL	BURLINGTON 054 MAT.	INST.	TOTAL	GUILDHALL 059 MAT.	INST.	TOTAL	MONTPELIER 056 MAT.	INST.	TOTAL	RUTLAND 057 MAT.	INST.	TOTAL	ST. JOHNSBURY 058 MAT.	INST.	TOTAL
015433	CONTRACTOR EQUIPMENT		95.7	95.7		99.0	99.0		95.7	95.7		99.0	99.0		95.7	95.7		95.7	95.7
0241, 31 - 34	SITE & INFRASTRUCTURE, DEMOLITION	87.8	96.1	93.5	92.6	101.0	98.4	85.8	91.9	90.0	89.7	100.8	97.3	90.4	96.1	94.3	85.9	95.2	92.3
0310	Concrete Forming & Accessories	96.7	101.7	101.0	99.4	80.2	83.0	94.6	95.1	95.1	97.3	101.3	100.7	97.0	80.0	82.5	93.5	95.6	95.2
0320	Concrete Reinforcing	84.1	86.3	85.2	107.3	86.2	97.1	85.8	86.1	86.0	101.6	86.2	94.2	106.4	86.1	96.6	84.1	86.2	85.1
0330	Cast-in-Place Concrete	92.5	115.9	101.2	109.6	116.7	112.2	86.9	107.2	94.4	104.5	116.7	109.0	87.8	115.6	98.2	86.9	107.3	94.5
03	CONCRETE	90.7	103.7	96.4	101.7	94.2	98.4	86.2	97.7	91.3	98.5	103.7	100.8	91.4	93.8	92.5	85.9	97.9	91.2
04	MASONRY	104.6	86.2	93.3	103.9	86.4	93.1	104.8	71.5	84.4	101.4	86.4	92.2	87.2	86.3	86.7	131.1	71.5	94.5
05	METALS	95.1	91.1	93.9	102.8	89.8	99.0	95.2	90.0	93.7	101.2	89.9	97.9	101.1	90.3	97.9	95.2	90.4	93.8
06	WOOD, PLASTICS & COMPOSITES	103.1	106.8	105.1	99.2	78.5	87.8	99.0	106.8	103.3	96.0	106.9	102.0	103.2	78.3	89.5	94.1	106.8	101.0
07	THERMAL & MOISTURE PROTECTION	101.0	89.1	95.8	107.4	86.6	98.3	100.7	82.4	92.8	109.4	89.7	100.8	101.1	86.1	94.6	100.6	82.4	92.7
08	OPENINGS	100.1	97.8	99.6	102.6	78.4	96.9	100.1	94.0	98.7	103.1	94.1	101.0	103.2	78.3	97.4	100.1	94.0	98.7
0920	Plaster & Gypsum Board	108.9	106.7	107.4	109.9	77.4	88.1	114.5	106.7	109.2	110.4	106.7	107.9	109.4	77.4	87.9	115.8	106.7	109.6
0950, 0980	Ceilings & Acoustic Treatment	97.2	106.7	103.6	103.9	77.4	86.0	97.2	106.7	103.6	104.9	106.7	106.1	103.2	77.4	85.8	97.2	106.7	103.6
0960	Flooring	92.2	98.9	94.1	101.8	98.9	101.0	95.0	98.9	96.1	98.5	98.9	98.6	92.0	98.9	94.1	98.3	98.9	98.5
0970, 0990	Wall Finishes & Painting/Coating	86.4	101.5	95.3	98.0	88.3	92.3	86.4	88.3	87.5	97.8	88.3	92.2	86.4	88.3	87.5	86.4	88.3	87.5
09	FINISHES	90.8	101.8	96.8	99.1	83.8	90.8	92.2	96.7	94.6	97.9	100.6	99.4	92.0	83.7	87.5	93.3	96.7	95.2
COVERS	DIVS. 10 - 14, 25, 28, 41, 43, 44, 46	100.0	91.8	98.2	100.0	89.1	97.6	100.0	86.7	97.0	100.0	92.2	98.3	100.0	88.6	97.5	100.0	86.8	97.0
21, 22, 23	FIRE SUPPRESSION, PLUMBING & HVAC	96.7	87.6	93.1	100.0	68.3	87.6	96.7	60.5	82.5	96.5	68.3	85.4	100.2	68.2	87.7	96.7	60.5	82.5
26, 27, 3370	ELECTRICAL, COMMUNICATIONS & UTIL.	108.3	79.5	94.1	107.6	53.6	81.0	108.3	54.5	81.8	108.1	53.6	81.3	108.4	53.6	81.4	108.3	54.5	81.8
MF2018	WEIGHTED AVERAGE	97.4	92.1	95.2	101.9	79.8	92.5	96.8	79.0	89.3	100.2	84.4	93.5	98.9	79.4	90.7	98.1	79.3	90.2

VERMONT / VIRGINIA

DIVISION		WHITE RIVER JCT. 050 MAT.	INST.	TOTAL	ALEXANDRIA 223 MAT.	INST.	TOTAL	ARLINGTON 222 MAT.	INST.	TOTAL	BRISTOL 242 MAT.	INST.	TOTAL	CHARLOTTESVILLE 229 MAT.	INST.	TOTAL	CULPEPER 227 MAT.	INST.	TOTAL
015433	CONTRACTOR EQUIPMENT		95.7	95.7		105.3	105.3		104.1	104.1		104.1	104.1		108.1	108.1		104.1	104.1
0241, 31 - 34	SITE & INFRASTRUCTURE, DEMOLITION	90.3	95.2	93.7	113.7	89.0	96.7	123.9	87.0	98.5	108.2	86.0	92.8	112.7	88.3	95.9	111.4	86.5	94.2
0310	Concrete Forming & Accessories	91.4	96.1	95.4	91.5	71.9	74.8	90.5	71.6	74.4	86.6	64.1	67.5	85.1	61.7	65.2	82.4	71.0	72.7
0320	Concrete Reinforcing	85.1	86.2	85.6	86.6	85.0	85.9	97.8	85.1	91.6	97.7	83.7	91.0	97.1	73.0	85.4	97.8	84.9	91.6
0330	Cast-in-Place Concrete	92.5	108.1	98.3	108.1	77.7	96.8	105.2	77.2	94.8	104.8	44.5	82.3	109.0	75.5	96.5	107.7	74.4	95.3
03	CONCRETE	92.6	98.4	95.2	99.9	77.7	90.1	104.5	77.4	92.6	100.8	62.4	83.9	101.3	70.2	87.7	98.4	76.0	88.6
04	MASONRY	117.2	72.9	90.0	89.8	73.7	79.9	102.3	72.6	84.0	92.3	45.4	63.5	116.9	55.3	79.0	105.1	70.8	84.1
05	METALS	95.2	90.4	93.8	106.0	100.8	104.5	104.6	101.8	103.8	103.4	97.8	101.8	103.7	95.8	101.4	103.7	99.6	102.5
06	WOOD, PLASTICS & COMPOSITES	97.1	106.8	102.4	91.0	70.1	79.5	87.5	70.1	78.0	80.2	69.8	74.5	78.7	59.8	68.3	77.4	70.1	73.4
07	THERMAL & MOISTURE PROTECTION	101.1	83.0	93.2	102.5	81.1	93.2	104.6	80.5	94.1	103.9	58.8	84.3	103.5	69.7	88.8	103.7	78.9	92.9
08	OPENINGS	100.1	94.0	98.7	95.4	74.3	90.5	93.7	74.3	89.2	96.4	67.9	89.7	94.8	65.7	88.0	95.1	73.8	90.1
0920	Plaster & Gypsum Board	106.6	106.7	106.7	98.0	69.2	78.6	94.8	69.2	77.6	91.1	68.9	76.2	91.1	57.9	68.8	91.3	69.2	76.4
0950, 0980	Ceilings & Acoustic Treatment	97.2	106.7	103.6	92.4	69.2	76.8	90.8	69.2	76.2	90.0	68.9	75.8	90.0	57.9	68.4	90.8	69.2	76.2
0960	Flooring	90.2	98.9	92.6	97.1	77.4	91.5	95.6	75.7	90.0	92.3	55.7	82.0	90.8	55.7	80.9	90.8	74.4	86.2
0970, 0990	Wall Finishes & Painting/Coating	86.4	88.3	87.5	115.8	76.0	92.3	115.8	73.1	90.6	102.1	55.3	74.4	102.1	58.6	76.3	115.8	73.1	90.6
09	FINISHES	90.1	97.1	93.9	93.8	72.6	82.3	93.8	71.6	81.8	90.6	62.3	75.1	90.1	59.4	73.5	90.8	71.2	80.2
COVERS	DIVS. 10 - 14, 25, 28, 41, 43, 44, 46	100.0	87.2	97.1	100.0	86.6	97.0	100.0	84.4	96.5	100.0	75.1	94.5	100.0	83.2	96.2	100.0	83.6	96.3
21, 22, 23	FIRE SUPPRESSION, PLUMBING & HVAC	96.7	61.2	82.8	100.4	86.2	94.8	100.4	85.6	94.6	96.8	46.0	76.9	96.8	69.1	86.0	96.8	85.3	92.3
26, 27, 3370	ELECTRICAL, COMMUNICATIONS & UTIL.	108.3	54.3	81.7	96.6	100.2	98.4	94.4	100.2	97.2	96.2	36.3	66.7	96.2	70.4	83.5	98.5	96.9	97.7
MF2018	WEIGHTED AVERAGE	98.2	79.8	90.4	99.6	84.8	93.3	100.4	84.2	93.6	98.2	59.3	81.8	99.4	71.0	87.4	98.7	83.0	92.1

VIRGINIA

DIVISION		FAIRFAX 220 - 221			FARMVILLE 239			FREDERICKSBURG 224 - 225			GRUNDY 246			HARRISONBURG 228			LYNCHBURG 245		
		MAT.	INST.	TOTAL	MAT.	INST.	TOTAL	MAT.	INST.	TOTAL	MAT.	INST.	TOTAL	MAT.	INST.	TOTAL	MAT.	INST.	TOTAL
015433	CONTRACTOR EQUIPMENT		104.1	104.1		108.1	108.1		104.1	104.1		104.1	104.1		104.1	104.1		104.1	104.1
0241, 31 - 34	SITE & INFRASTRUCTURE, DEMOLITION	122.4	86.7	97.8	108.5	87.3	93.9	110.9	86.4	93.9	106.0	85.3	91.7	119.3	85.0	95.6	106.8	86.7	92.9
0310	Concrete Forming & Accessories	85.1	63.8	67.0	98.9	60.3	66.0	85.1	65.4	68.3	89.6	38.2	45.8	81.4	56.6	60.2	86.6	70.5	72.9
0320	Concrete Reinforcing	97.8	90.3	94.2	92.9	67.9	80.8	98.5	80.1	89.6	96.4	44.1	71.1	97.8	57.3	78.2	97.1	84.5	91.0
0330	Cast-in-Place Concrete	105.2	78.5	95.2	106.8	83.9	98.3	106.8	74.2	94.7	104.8	50.9	84.7	105.2	52.6	85.6	104.8	74.3	93.4
03	CONCRETE	104.2	75.2	91.5	101.8	71.5	88.5	98.1	72.7	86.9	99.3	45.8	75.8	102.0	57.1	82.3	99.2	75.8	88.9
04	MASONRY	102.2	73.6	84.6	101.8	50.0	69.9	104.0	66.7	81.1	94.9	55.4	70.6	101.4	57.9	74.6	109.3	63.9	81.4
05	METALS	103.8	102.5	103.4	101.3	90.8	98.2	103.8	98.0	102.1	103.4	75.2	95.1	103.7	87.2	98.9	103.6	99.8	102.5
06	WOOD, PLASTICS & COMPOSITES	80.2	59.2	68.6	93.0	59.8	74.7	80.2	64.6	71.6	83.1	30.6	54.2	76.3	59.2	66.9	80.2	71.5	75.4
07	THERMAL & MOISTURE PROTECTION	104.4	79.9	93.7	105.2	69.4	89.6	103.7	76.3	91.8	103.9	57.3	83.6	104.1	69.6	89.1	103.7	70.4	89.2
08	OPENINGS	93.7	69.6	88.0	94.3	54.9	85.1	94.8	68.5	88.7	96.4	30.1	80.9	95.0	57.0	86.2	95.0	68.8	88.9
0920	Plaster & Gypsum Board	91.3	57.9	68.9	103.1	57.9	72.7	91.3	63.6	72.7	91.1	28.5	49.0	91.1	57.9	68.8	91.1	70.7	77.4
0950, 0980	Ceilings & Acoustic Treatment	90.8	57.9	68.6	86.5	57.9	67.2	90.8	63.6	72.4	90.0	28.5	48.6	90.0	57.9	68.4	90.0	70.7	77.0
0960	Flooring	92.5	76.7	88.1	95.6	50.5	82.9	92.5	72.4	86.9	93.5	28.2	75.2	90.6	74.4	86.1	92.3	65.4	84.7
0970, 0990	Wall Finishes & Painting/Coating	115.8	73.6	90.8	99.6	57.2	74.5	115.8	57.0	81.0	102.1	31.2	60.2	115.8	48.3	75.9	102.1	55.3	74.4
09	FINISHES	92.4	65.7	77.9	90.6	57.7	72.8	91.3	65.0	77.1	90.7	33.9	60.0	91.3	58.9	73.7	90.4	67.8	78.2
COVERS	DIVS. 10 - 14, 25, 28, 41, 43, 44, 46	100.0	84.2	96.5	100.0	77.1	94.9	100.0	78.3	95.2	100.0	73.6	94.1	100.0	62.8	91.7	100.0	78.8	95.3
21, 22, 23	FIRE SUPPRESSION, PLUMBING & HVAC	96.8	88.1	93.4	96.9	52.0	79.3	96.8	79.9	90.2	96.8	65.7	84.6	96.8	65.4	84.5	96.8	69.5	86.1
26, 27, 3370	ELECTRICAL, COMMUNICATIONS & UTIL.	97.3	101.1	99.1	91.2	68.3	79.9	94.5	91.8	93.2	96.2	68.3	82.5	96.4	86.4	91.5	97.2	67.8	82.7
MF2018	WEIGHTED AVERAGE	99.5	83.6	92.8	97.9	65.3	84.1	98.3	78.7	90.0	98.1	58.8	81.5	99.0	68.8	86.3	98.7	73.8	88.2

VIRGINIA

DIVISION		NEWPORT NEWS 236			NORFOLK 233 - 235			PETERSBURG 238			PORTSMOUTH 237			PULASKI 243			RICHMOND 230 - 232		
		MAT.	INST.	TOTAL	MAT.	INST.	TOTAL	MAT.	INST.	TOTAL	MAT.	INST.	TOTAL	MAT.	INST.	TOTAL	MAT.	INST.	TOTAL
015433	CONTRACTOR EQUIPMENT		108.2	108.2		110.2	110.2		108.1	108.1		108.0	108.0		104.1	104.1		109.6	109.6
0241, 31 - 34	SITE & INFRASTRUCTURE, DEMOLITION	107.6	88.4	94.3	106.9	93.8	97.9	111.1	88.4	95.4	106.2	88.2	93.8	105.4	85.7	91.8	103.4	92.8	96.1
0310	Concrete Forming & Accessories	98.1	61.1	66.6	103.6	61.2	67.5	91.8	61.8	66.2	87.8	61.3	65.2	89.6	40.4	47.7	94.2	61.9	66.7
0320	Concrete Reinforcing	92.8	66.2	79.9	101.2	66.2	84.3	92.4	73.6	83.3	92.4	66.2	79.7	96.4	83.8	90.3	102.1	73.6	88.3
0330	Cast-in-Place Concrete	103.8	64.5	89.2	111.0	76.1	98.0	110.2	76.6	97.7	102.8	65.6	89.0	104.8	82.4	96.5	95.3	76.6	88.4
03	CONCRETE	98.9	65.0	84.0	102.1	69.0	87.6	104.5	70.8	89.7	97.7	65.5	83.6	99.3	64.7	84.1	94.5	70.7	84.0
04	MASONRY	97.3	55.0	71.3	97.1	55.0	71.3	110.2	64.6	82.2	103.0	55.6	73.9	87.9	54.3	67.3	93.3	64.7	75.7
05	METALS	103.8	93.1	100.6	105.3	91.8	101.3	101.4	96.7	100.0	102.7	93.1	99.9	103.5	96.0	101.3	105.9	95.3	102.8
06	WOOD, PLASTICS & COMPOSITES	91.8	59.8	74.2	99.3	59.7	77.5	83.2	59.5	70.2	79.6	59.8	68.7	83.1	32.0	55.0	93.7	59.7	75.0
07	THERMAL & MOISTURE PROTECTION	105.1	67.7	88.9	104.4	70.0	89.5	105.1	72.3	90.9	105.1	68.0	89.0	103.9	66.6	87.6	101.0	72.9	88.8
08	OPENINGS	94.6	60.1	86.6	95.2	60.1	87.0	94.0	65.6	87.4	94.7	56.0	85.7	96.4	42.9	83.9	101.7	65.7	93.3
0920	Plaster & Gypsum Board	104.3	57.9	73.1	98.5	57.9	71.2	97.8	57.7	70.8	98.6	57.9	71.2	91.1	30.0	50.0	99.5	57.9	71.5
0950, 0980	Ceilings & Acoustic Treatment	91.4	57.9	68.8	88.8	57.9	68.0	88.2	57.7	67.6	91.4	57.9	68.8	90.0	30.0	49.6	89.5	57.9	68.2
0960	Flooring	95.6	51.1	83.1	96.2	51.1	83.5	91.7	71.7	86.1	88.9	51.7	78.4	93.5	55.7	82.9	94.0	71.7	87.8
0970, 0990	Wall Finishes & Painting/Coating	99.6	59.7	76.0	101.2	59.7	76.6	99.6	58.6	75.3	99.6	59.7	76.0	102.1	48.3	70.3	95.4	58.6	73.6
09	FINISHES	91.5	58.4	73.6	90.5	58.3	73.1	89.3	62.5	74.8	88.8	58.6	72.5	90.7	40.9	63.8	91.9	62.5	76.0
COVERS	DIVS. 10 - 14, 25, 28, 41, 43, 44, 46	100.0	83.4	96.3	100.0	83.2	96.3	100.0	83.1	96.2	100.0	83.6	96.3	100.0	73.2	94.0	100.0	83.0	96.2
21, 22, 23	FIRE SUPPRESSION, PLUMBING & HVAC	100.4	66.8	87.2	100.1	66.8	87.0	96.9	69.1	86.0	100.4	67.1	87.4	96.8	66.0	84.7	100.0	69.1	87.9
26, 27, 3370	ELECTRICAL, COMMUNICATIONS & UTIL.	93.6	64.2	79.1	98.0	60.7	79.6	93.8	70.5	82.3	92.1	60.7	76.7	96.2	83.4	89.9	96.8	70.4	83.8
MF2018	WEIGHTED AVERAGE	98.8	68.3	85.9	99.9	68.7	86.7	98.7	72.6	87.7	98.3	67.8	85.4	97.8	67.1	84.8	99.2	72.8	88.1

VIRGINIA / WASHINGTON

DIVISION		ROANOKE 240 - 241			STAUNTON 244			WINCHESTER 226			CLARKSTON 994			EVERETT 982			OLYMPIA 985		
		MAT.	INST.	TOTAL	MAT.	INST.	TOTAL	MAT.	INST.	TOTAL	MAT.	INST.	TOTAL	MAT.	INST.	TOTAL	MAT.	INST.	TOTAL
015433	CONTRACTOR EQUIPMENT		104.1	104.1		108.1	108.1		104.1	104.1		90.2	90.2		99.3	99.3		101.6	101.6
0241, 31 - 34	SITE & INFRASTRUCTURE, DEMOLITION	106.0	86.8	92.7	109.3	87.3	94.1	118.1	86.4	96.2	98.1	86.1	89.8	92.1	105.6	101.4	92.8	108.2	103.5
0310	Concrete Forming & Accessories	95.7	71.2	74.8	89.3	61.4	65.5	83.6	66.4	68.9	107.0	64.2	70.5	111.6	104.6	105.7	101.6	104.3	103.9
0320	Concrete Reinforcing	97.5	84.6	91.2	97.1	83.8	90.7	97.1	80.2	88.9	108.4	95.9	102.3	109.2	114.5	111.8	119.5	114.4	117.0
0330	Cast-in-Place Concrete	119.0	86.2	106.8	109.0	85.5	100.2	105.2	63.9	89.8	82.3	82.9	82.5	102.6	109.4	105.1	95.3	109.8	100.7
03	CONCRETE	103.5	80.2	93.3	100.7	75.2	89.5	101.5	69.5	87.4	87.6	76.6	82.8	95.1	107.4	100.5	93.7	107.2	99.7
04	MASONRY	96.3	65.9	77.6	105.0	55.3	74.5	98.7	69.8	80.9	97.8	91.7	94.0	121.0	101.8	109.2	115.3	101.7	107.0
05	METALS	105.8	100.2	104.2	103.6	96.0	101.4	103.8	97.9	102.1	92.8	87.3	91.2	116.8	98.4	111.4	117.8	96.1	111.4
06	WOOD, PLASTICS & COMPOSITES	91.9	71.5	80.7	83.1	59.8	70.3	78.7	64.8	71.1	104.4	58.5	79.2	107.5	104.7	105.9	97.5	104.8	101.5
07	THERMAL & MOISTURE PROTECTION	103.6	76.0	91.6	103.4	67.5	87.8	104.2	75.9	91.9	159.7	82.9	126.3	112.4	105.9	109.6	108.3	103.0	106.0
08	OPENINGS	95.4	68.8	89.2	95.0	58.5	86.5	96.5	68.3	89.9	116.4	65.0	104.4	106.0	107.3	106.3	109.6	104.8	108.5
0920	Plaster & Gypsum Board	98.0	70.7	79.6	91.1	57.9	68.8	91.3	63.8	72.8	154.4	57.3	89.1	112.7	104.9	107.5	106.5	104.9	105.4
0950, 0980	Ceilings & Acoustic Treatment	92.4	70.7	77.8	90.0	57.9	68.4	90.8	63.8	72.6	103.8	57.3	72.4	101.2	104.9	103.7	98.7	104.9	102.9
0960	Flooring	97.1	67.3	88.7	93.1	33.1	76.2	92.0	74.4	87.1	85.4	77.5	83.2	109.9	96.0	106.0	97.6	96.0	97.1
0970, 0990	Wall Finishes & Painting/Coating	102.1	55.3	74.4	102.1	30.7	59.8	115.8	77.6	93.2	84.4	72.1	77.2	97.6	95.2	96.2	99.6	93.1	95.8
09	FINISHES	92.7	68.5	79.6	90.6	52.6	70.0	91.8	68.3	79.1	108.2	65.8	85.2	105.4	101.6	103.4	94.9	101.5	98.5
COVERS	DIVS. 10 - 14, 25, 28, 41, 43, 44, 46	100.0	79.2	95.4	100.0	78.2	95.1	100.0	82.7	96.1	100.0	89.3	97.6	100.0	103.7	100.8	100.0	99.9	100.0
21, 22, 23	FIRE SUPPRESSION, PLUMBING & HVAC	100.4	67.9	87.6	96.8	59.3	82.1	96.8	81.4	90.8	96.7	80.6	90.4	100.0	105.2	102.0	99.9	100.6	100.2
26, 27, 3370	ELECTRICAL, COMMUNICATIONS & UTIL.	96.2	60.8	78.8	95.2	68.4	82.0	94.9	91.8	93.4	89.4	95.6	92.5	107.0	104.1	105.6	105.4	97.7	101.6
MF2018	WEIGHTED AVERAGE	100.1	73.6	88.9	98.6	67.9	85.6	98.9	79.5	90.7	99.1	81.9	91.8	105.1	104.0	104.6	104.0	101.8	103.1

For customer support on your Commercial Renovation Costs with RSMeans data, call 800.448.8182.

823

WASHINGTON

DIVISION		RICHLAND 993			SEATTLE 980 - 981, 987			SPOKANE 990 - 992			TACOMA 983 - 984			VANCOUVER 986			WENATCHEE 988		
		MAT.	INST.	TOTAL	MAT.	INST.	TOTAL	MAT.	INST.	TOTAL	MAT.	INST.	TOTAL	MAT.	INST.	TOTAL	MAT.	INST.	TOTAL
015433	CONTRACTOR EQUIPMENT		90.2	90.2		102.4	102.4		90.2	90.2		99.3	99.3		95.8	95.8		99.3	99.3
0241, 31 - 34	SITE & INFRASTRUCTURE, DEMOLITION	100.3	86.3	90.6	97.8	108.9	105.5	99.7	86.3	90.4	95.2	105.1	102.1	105.2	93.1	96.8	104.0	102.8	103.2
0310	Concrete Forming & Accessories	107.1	80.0	84.0	107.4	109.5	109.2	112.0	79.5	84.3	102.3	104.1	103.9	102.7	95.7	96.7	104.0	76.5	80.5
0320	Concrete Reinforcing	103.9	95.4	99.8	109.2	117.3	113.1	104.6	95.3	100.1	107.8	114.3	110.9	108.8	114.8	111.7	108.8	95.9	102.5
0330	Cast-in-Place Concrete	82.5	84.4	83.2	105.9	113.9	108.9	85.7	84.2	85.2	105.5	108.8	106.7	117.8	99.9	111.2	107.7	90.8	101.4
03	CONCRETE	87.2	84.2	85.9	102.8	111.8	106.7	89.2	83.9	86.8	97.0	106.9	101.4	107.2	100.2	104.1	105.3	84.9	96.3
04	MASONRY	98.8	83.7	89.5	123.5	107.2	113.5	99.4	83.7	89.7	117.6	100.2	106.9	118.8	105.5	110.6	120.9	94.1	104.4
05	METALS	93.2	87.1	91.4	116.1	101.9	111.9	95.4	86.6	92.8	118.7	96.5	112.2	116.0	99.3	111.1	116.0	87.3	107.6
06	WOOD, PLASTICS & COMPOSITES	104.6	77.9	89.9	103.4	108.7	106.3	113.5	77.9	93.9	97.1	104.7	101.2	89.5	94.9	92.5	98.7	73.3	84.8
07	THERMAL & MOISTURE PROTECTION	161.1	84.4	127.7	111.6	110.8	111.3	157.4	84.7	125.8	112.1	102.1	107.8	111.8	101.6	107.4	111.6	88.3	101.5
08	OPENINGS	114.4	74.7	105.2	106.1	110.7	107.2	115.0	74.7	105.6	106.7	104.7	106.2	102.8	101.5	102.5	106.2	73.0	98.5
0920	Plaster & Gypsum Board	154.4	77.2	102.5	111.2	109.1	109.8	144.2	77.2	99.1	110.6	104.9	106.8	109.0	95.2	99.7	113.9	72.7	86.2
0950, 0980	Ceilings & Acoustic Treatment	109.2	77.2	87.6	106.6	109.1	108.3	105.6	77.2	86.4	104.5	104.9	104.8	102.4	95.2	97.5	97.8	72.7	80.9
0960	Flooring	85.7	77.5	83.4	109.5	103.4	107.8	84.8	77.5	82.7	102.2	96.0	100.5	107.4	107.1	107.3	105.3	77.5	97.5
0970, 0990	Wall Finishes & Painting/Coating	84.4	75.1	78.9	107.1	95.8	100.4	84.6	76.4	79.8	97.6	93.1	95.0	100.3	79.9	88.2	97.6	70.9	81.8
09	FINISHES	109.5	78.1	92.5	108.9	106.4	107.6	107.3	78.2	91.6	103.6	101.4	102.4	101.3	95.6	98.3	104.7	74.9	88.5
COVERS	DIVS. 10 - 14, 25, 28, 41, 43, 44, 46	100.0	92.2	98.3	100.0	105.2	101.2	100.0	92.2	98.3	100.0	99.6	99.9	100.0	101.5	100.3	100.0	91.2	98.0
21, 22, 23	FIRE SUPPRESSION, PLUMBING & HVAC	100.3	108.6	103.6	99.9	123.0	109.0	100.3	83.7	93.8	100.0	100.6	100.2	100.0	110.7	104.2	96.5	92.0	94.7
26, 27, 3370	ELECTRICAL, COMMUNICATIONS & UTIL.	87.2	92.8	90.0	106.2	118.9	112.5	85.4	77.1	81.3	106.8	97.7	102.3	113.1	107.3	110.2	107.7	94.5	101.2
MF2018	WEIGHTED AVERAGE	99.8	90.0	95.6	106.3	112.5	108.9	100.1	82.4	92.6	105.4	101.3	103.7	106.6	102.8	105.0	105.6	88.7	98.5

DIVISION		WASHINGTON YAKIMA 989			WEST VIRGINIA BECKLEY 258 - 259			BLUEFIELD 247 - 248			BUCKHANNON 262			CHARLESTON 250 - 253			CLARKSBURG 263 - 264		
		MAT.	INST.	TOTAL	MAT.	INST.	TOTAL	MAT.	INST.	TOTAL	MAT.	INST.	TOTAL	MAT.	INST.	TOTAL	MAT.	INST.	TOTAL
015433	CONTRACTOR EQUIPMENT		99.3	99.3		104.1	104.1		104.1	104.1		104.1	104.1		106.8	106.8		104.1	104.1
0241, 31 - 34	SITE & INFRASTRUCTURE, DEMOLITION	97.6	103.8	101.9	100.8	87.4	91.6	99.9	87.4	91.3	106.0	87.4	93.1	100.4	93.0	95.3	106.8	87.4	93.4
0310	Concrete Forming & Accessories	102.9	99.2	99.8	85.0	87.1	86.8	86.6	87.1	87.0	86.1	83.7	84.0	96.0	88.6	89.7	83.6	83.8	83.7
0320	Concrete Reinforcing	108.3	95.4	102.1	98.0	87.7	93.1	96.0	82.5	89.5	96.7	82.4	89.8	106.1	87.8	97.2	96.7	95.3	96.0
0330	Cast-in-Place Concrete	112.7	83.0	101.7	99.2	92.5	96.7	102.5	92.5	98.8	102.1	92.0	98.4	95.9	96.6	96.2	111.9	87.6	102.9
03	CONCRETE	101.8	92.4	97.7	91.8	90.2	91.1	95.1	89.3	92.6	98.2	87.5	93.5	90.7	92.2	91.4	102.2	88.2	96.1
04	MASONRY	110.3	81.0	92.3	91.0	89.0	89.8	89.5	89.0	89.2	100.6	85.8	91.5	88.4	90.7	89.8	104.1	85.8	92.9
05	METALS	116.7	87.3	108.0	98.2	103.3	99.7	103.6	101.5	103.0	103.8	101.6	103.2	96.8	102.5	98.5	103.8	106.0	104.5
06	WOOD, PLASTICS & COMPOSITES	97.4	104.7	101.4	80.2	87.4	84.2	82.1	87.4	85.0	81.4	83.0	82.3	89.3	88.2	88.7	78.0	83.0	80.8
07	THERMAL & MOISTURE PROTECTION	112.3	86.1	100.9	105.0	86.5	96.9	103.6	86.5	96.1	104.0	86.6	96.4	101.2	88.1	95.5	103.8	86.3	96.2
08	OPENINGS	106.2	100.5	104.9	94.5	83.0	91.8	96.9	81.8	93.4	96.9	79.4	92.8	94.9	83.4	92.2	96.9	82.1	93.4
0920	Plaster & Gypsum Board	110.4	104.9	106.7	93.5	87.0	89.1	90.2	87.0	88.0	90.8	82.5	85.2	98.7	87.6	91.2	88.6	82.5	84.5
0950, 0980	Ceilings & Acoustic Treatment	99.6	104.9	103.2	78.1	87.0	84.1	87.5	87.0	87.2	90.0	82.5	84.9	89.6	87.6	88.2	90.0	82.5	84.9
0960	Flooring	103.2	77.5	96.0	91.0	98.9	93.2	90.1	98.9	92.6	89.8	95.2	91.3	97.1	98.9	97.6	88.8	95.2	90.6
0970, 0990	Wall Finishes & Painting/Coating	97.6	76.4	85.1	89.3	92.1	91.0	102.1	89.8	94.8	102.1	88.0	93.7	88.6	91.9	90.5	102.1	88.0	93.7
09	FINISHES	103.1	93.3	97.8	86.3	90.2	88.4	88.7	89.9	89.4	89.7	86.1	87.8	92.8	91.0	91.9	89.0	86.1	87.4
COVERS	DIVS. 10 - 14, 25, 28, 41, 43, 44, 46	100.0	96.1	99.1	100.0	90.3	97.8	100.0	90.3	97.8	100.0	90.6	97.9	100.0	92.5	98.3	100.0	90.6	97.9
21, 22, 23	FIRE SUPPRESSION, PLUMBING & HVAC	100.0	107.4	102.9	97.1	80.3	90.5	96.8	84.0	91.8	96.8	89.7	94.0	100.1	81.4	92.8	96.8	89.8	94.1
26, 27, 3370	ELECTRICAL, COMMUNICATIONS & UTIL.	109.9	92.9	101.5	92.9	84.1	88.6	95.2	84.1	89.8	96.6	89.3	93.0	98.0	87.1	92.6	96.6	89.3	93.0
MF2018	WEIGHTED AVERAGE	105.6	95.4	101.3	95.2	87.7	92.0	97.0	88.1	93.2	98.3	88.8	94.3	96.5	89.4	93.5	98.9	89.5	94.9

WEST VIRGINIA

DIVISION		GASSAWAY 266			HUNTINGTON 255 - 257			LEWISBURG 249			MARTINSBURG 254			MORGANTOWN 265			PARKERSBURG 261		
		MAT.	INST.	TOTAL	MAT.	INST.	TOTAL	MAT.	INST.	TOTAL	MAT.	INST.	TOTAL	MAT.	INST.	TOTAL	MAT.	INST.	TOTAL
015433	CONTRACTOR EQUIPMENT		104.1	104.1		104.1	104.1		104.1	104.1		104.1	104.1		104.1	104.1		104.1	104.1
0241, 31 - 34	SITE & INFRASTRUCTURE, DEMOLITION	103.5	87.4	92.4	105.4	88.5	93.7	115.5	87.4	96.1	104.7	88.0	93.2	100.9	88.2	92.1	109.3	88.4	94.9
0310	Concrete Forming & Accessories	85.5	86.9	86.7	96.0	89.3	90.3	83.9	86.6	86.2	85.0	79.5	80.3	83.9	83.8	83.9	88.0	86.0	86.3
0320	Concrete Reinforcing	96.7	91.6	94.2	99.5	93.1	96.4	96.7	82.4	89.8	98.0	95.1	96.6	96.7	95.3	96.0	96.0	82.3	89.4
0330	Cast-in-Place Concrete	106.9	94.0	102.1	108.2	94.0	102.9	102.5	92.3	98.7	104.0	88.3	98.1	102.1	92.0	98.4	104.4	89.0	98.7
03	CONCRETE	98.6	91.3	95.4	96.8	92.6	95.0	105.1	89.0	98.0	95.3	86.5	91.4	94.9	89.8	92.7	100.3	87.5	94.7
04	MASONRY	105.0	87.9	94.5	91.7	93.4	92.7	94.3	89.0	91.1	93.1	83.9	87.5	122.6	85.8	100.0	80.1	85.2	83.2
05	METALS	103.8	104.8	104.1	100.6	105.7	102.1	103.7	101.0	102.9	98.6	104.4	100.3	103.9	106.0	104.5	104.5	101.4	103.6
06	WOOD, PLASTICS & COMPOSITES	80.4	87.4	84.3	90.8	88.7	89.6	78.3	87.4	83.3	80.2	78.6	79.3	78.3	83.0	80.9	81.5	85.6	83.8
07	THERMAL & MOISTURE PROTECTION	103.6	87.9	96.8	105.2	88.4	97.9	104.6	86.4	96.7	105.3	80.8	94.6	103.7	86.6	96.2	103.9	86.7	96.4
08	OPENINGS	95.2	83.9	92.6	93.7	84.8	91.7	96.9	81.8	93.4	96.2	73.8	91.0	98.0	82.1	94.3	95.8	80.4	92.2
0920	Plaster & Gypsum Board	89.9	87.0	87.9	99.5	88.3	91.9	88.6	87.0	87.5	93.5	77.9	83.0	88.6	82.5	84.5	91.1	85.1	87.1
0950, 0980	Ceilings & Acoustic Treatment	90.0	87.0	88.0	78.1	88.3	85.0	90.0	87.0	88.0	78.1	77.9	78.0	90.0	82.5	84.9	90.0	85.1	86.7
0960	Flooring	89.6	98.9	92.2	98.4	103.3	99.7	88.9	98.9	91.7	91.0	98.9	93.2	88.9	95.2	90.7	92.6	91.9	92.4
0970, 0990	Wall Finishes & Painting/Coating	102.1	91.9	96.0	89.3	89.8	89.6	102.1	64.5	88.5	99.3	83.7	86.0	102.1	88.0	93.7	102.1	83.7	91.2
09	FINISHES	89.2	89.9	89.6	89.3	92.0	90.8	90.1	87.2	88.5	86.6	83.0	84.6	88.6	86.1	87.3	90.6	86.8	88.6
COVERS	DIVS. 10 - 14, 25, 28, 41, 43, 44, 46	100.0	91.1	98.0	100.0	92.3	98.3	100.0	65.4	92.3	100.0	81.4	95.8	100.0	83.5	96.3	100.0	91.3	98.1
21, 22, 23	FIRE SUPPRESSION, PLUMBING & HVAC	96.8	83.4	91.6	100.6	90.6	96.7	96.8	80.2	90.3	97.1	80.9	90.8	96.8	89.8	94.1	100.3	88.3	95.6
26, 27, 3370	ELECTRICAL, COMMUNICATIONS & UTIL.	96.6	87.1	91.9	96.6	89.6	93.1	92.8	84.1	88.5	98.6	76.9	87.9	96.7	89.3	93.1	96.7	88.3	92.5
MF2018	WEIGHTED AVERAGE	98.3	89.0	94.3	97.8	92.2	95.4	98.7	86.1	93.4	96.7	84.1	91.4	98.8	89.6	94.9	98.6	88.6	94.4

For customer support on your Commercial Renovation Costs with RSMeans data, call 800.448.8182.

WEST VIRGINIA / WISCONSIN

DIVISION		PETERSBURG 268 MAT.	INST.	TOTAL	ROMNEY 267 MAT.	INST.	TOTAL	WHEELING 260 MAT.	INST.	TOTAL	BELOIT 535 MAT.	INST.	TOTAL	EAU CLAIRE 547 MAT.	INST.	TOTAL	GREEN BAY 541-543 MAT.	INST.	TOTAL
015433	CONTRACTOR EQUIPMENT		104.1	104.1		104.1	104.1		104.1	104.1		99.0	99.0		99.9	99.9		97.7	97.7
0241, 31 - 34	SITE & INFRASTRUCTURE, DEMOLITION	100.2	88.3	92.0	103.0	88.3	92.9	109.9	88.2	94.9	97.1	101.7	100.3	96.0	99.0	98.1	99.5	95.3	96.6
0310	Concrete Forming & Accessories	87.0	86.4	86.5	83.3	86.8	86.3	89.7	83.8	84.7	99.8	96.2	96.7	98.5	94.7	95.3	108.2	97.5	99.1
0320	Concrete Reinforcing	96.0	91.4	93.8	96.7	95.4	96.1	95.5	95.3	95.4	95.0	137.1	115.4	94.2	114.2	103.9	92.3	109.4	100.5
0330	Cast-in-Place Concrete	102.1	91.8	98.3	106.9	83.5	98.2	104.4	92.0	99.8	105.8	100.2	103.7	99.9	94.8	98.0	103.3	99.9	102.1
03	CONCRETE	95.0	90.2	92.9	98.4	88.2	93.9	100.3	89.8	95.7	96.7	104.7	100.2	93.1	98.4	95.4	96.2	100.6	98.2
04	MASONRY	96.1	85.8	89.8	94.7	87.9	90.5	103.4	85.7	92.5	99.3	100.3	99.9	91.4	96.1	94.3	122.7	97.8	107.4
05	METALS	103.9	104.1	104.0	104.0	105.9	104.5	104.7	106.1	105.1	98.6	112.0	102.5	93.8	105.8	97.3	96.4	104.2	98.7
06	WOOD, PLASTICS & COMPOSITES	82.3	87.4	85.1	77.5	87.4	82.9	83.1	83.0	83.1	96.9	94.1	95.4	100.6	94.2	97.1	107.8	97.2	101.9
07	THERMAL & MOISTURE PROTECTION	103.7	82.5	94.5	103.8	83.8	95.1	104.1	86.7	96.6	105.4	93.4	100.2	104.0	94.6	99.9	106.1	97.4	102.3
08	OPENINGS	98.1	83.6	94.7	98.0	84.5	94.8	96.6	82.1	93.2	98.1	108.3	100.5	103.1	99.3	102.2	99.1	101.2	99.6
0920	Plaster & Gypsum Board	90.8	87.0	88.2	88.3	87.0	87.4	91.1	82.5	85.3	91.6	94.5	93.5	105.5	94.5	98.1	103.3	97.5	99.4
0950, 0980	Ceilings & Acoustic Treatment	90.0	87.0	88.0	90.0	87.0	88.0	90.0	82.5	84.9	84.4	94.5	91.2	90.9	94.5	93.3	83.6	97.5	93.0
0960	Flooring	90.6	95.2	91.9	88.7	98.9	91.6	93.5	95.2	94.0	94.8	118.5	101.4	80.5	106.5	87.8	96.0	114.8	101.3
0970, 0990	Wall Finishes & Painting/Coating	102.1	88.0	93.7	102.1	88.0	93.7	102.1	88.0	93.7	95.0	103.6	100.1	83.3	76.4	79.2	92.0	80.0	84.9
09	FINISHES	89.4	88.7	89.0	88.7	89.5	89.1	90.9	86.1	88.3	92.3	100.7	96.9	87.3	94.9	91.4	91.4	99.1	95.6
COVERS	DIVS. 10 - 14, 25, 28, 41, 43, 44, 46	100.0	55.0	90.2	100.0	91.0	98.0	100.0	84.4	96.5	100.0	98.6	99.7	100.0	94.1	98.7	100.0	95.7	99.0
21, 22, 23	FIRE SUPPRESSION, PLUMBING & HVAC	96.8	83.7	91.7	96.8	83.6	91.6	100.4	89.8	96.2	99.8	95.5	98.1	100.0	88.2	95.4	100.2	83.4	93.6
26, 27, 3370	ELECTRICAL, COMMUNICATIONS & UTIL.	99.7	76.9	88.4	99.0	76.9	88.1	94.1	89.3	91.7	100.6	84.3	92.5	104.3	84.4	94.5	99.0	81.2	90.2
MF2018	WEIGHTED AVERAGE	98.0	85.9	92.9	98.3	87.2	93.6	99.6	89.6	95.4	98.6	98.9	98.7	97.5	94.1	96.1	99.4	93.4	96.9

WISCONSIN

DIVISION		KENOSHA 531 MAT.	INST.	TOTAL	LA CROSSE 546 MAT.	INST.	TOTAL	LANCASTER 538 MAT.	INST.	TOTAL	MADISON 537 MAT.	INST.	TOTAL	MILWAUKEE 530, 532 MAT.	INST.	TOTAL	NEW RICHMOND 540 MAT.	INST.	TOTAL
015433	CONTRACTOR EQUIPMENT		97.1	97.1		99.9	99.9		99.0	99.0		101.4	101.4		90.2	90.2		100.2	100.2
0241, 31 - 34	SITE & INFRASTRUCTURE, DEMOLITION	103.0	98.4	99.9	89.8	99.0	96.1	96.0	101.1	99.5	94.7	105.3	102.0	93.6	96.0	95.2	93.3	99.6	97.7
0310	Concrete Forming & Accessories	108.7	106.1	106.5	85.2	94.5	93.2	99.1	95.2	95.7	103.8	95.8	97.0	102.9	113.9	112.3	93.6	92.5	92.6
0320	Concrete Reinforcing	94.8	110.5	102.4	93.9	106.2	99.8	96.3	106.2	101.1	97.8	106.4	101.9	99.9	115.1	107.2	91.3	114.1	102.3
0330	Cast-in-Place Concrete	115.2	102.7	110.5	89.8	96.5	92.3	105.2	99.7	103.2	101.4	100.3	101.0	93.9	112.1	100.7	104.0	102.7	103.5
03	CONCRETE	101.5	105.5	103.3	84.6	97.5	90.3	96.4	98.8	97.4	96.3	99.2	97.6	94.5	112.5	102.4	90.8	100.1	94.9
04	MASONRY	97.1	108.0	103.8	90.5	96.1	93.9	99.4	100.1	99.9	97.4	99.1	98.5	102.3	117.2	111.4	117.9	98.4	105.9
05	METALS	99.6	103.3	100.7	93.7	102.8	96.4	96.0	100.3	97.3	102.8	100.4	102.1	97.5	97.5	97.5	94.1	105.4	97.4
06	WOOD, PLASTICS & COMPOSITES	103.9	105.7	104.9	85.2	94.2	90.2	96.1	94.1	95.0	99.4	94.3	96.6	101.2	112.3	107.3	89.5	90.2	89.9
07	THERMAL & MOISTURE PROTECTION	105.5	104.0	104.9	103.4	93.5	99.1	105.2	93.7	100.2	106.3	100.5	103.7	104.8	112.7	108.2	104.9	100.2	102.9
08	OPENINGS	92.5	105.4	95.5	103.1	92.0	100.5	94.0	91.9	93.5	101.6	98.9	100.9	103.0	111.5	105.0	88.6	90.3	89.0
0920	Plaster & Gypsum Board	80.2	104.6	97.8	100.9	94.5	96.3	90.9	94.5	93.3	101.7	94.5	96.8	96.4	112.6	107.3	91.7	90.5	90.9
0950, 0980	Ceilings & Acoustic Treatment	84.4	106.4	99.2	90.0	94.5	93.0	81.1	94.5	90.1	86.5	94.5	91.9	93.1	112.6	106.3	57.1	90.5	79.6
0960	Flooring	112.8	114.8	113.4	74.7	111.6	85.1	94.4	111.0	99.0	94.3	111.0	99.0	101.7	118.6	106.4	89.7	114.5	96.7
0970, 0990	Wall Finishes & Painting/Coating	106.4	118.7	113.7	83.3	81.7	82.4	95.0	97.7	96.6	94.9	103.6	100.0	104.5	124.3	116.3	95.0	81.7	87.1
09	FINISHES	97.1	109.4	103.7	84.4	96.5	90.9	91.4	98.4	95.2	93.7	99.3	96.7	99.5	116.1	108.5	82.5	95.2	89.4
COVERS	DIVS. 10 - 14, 25, 28, 41, 43, 44, 46	100.0	97.3	99.4	100.0	92.8	98.4	100.0	84.2	96.5	100.0	96.8	99.3	100.0	105.2	101.2	100.0	95.1	98.9
21, 22, 23	FIRE SUPPRESSION, PLUMBING & HVAC	100.0	95.9	98.4	100.0	88.1	95.3	96.3	89.2	93.5	99.8	95.3	98.0	99.8	107.0	102.6	96.0	89.0	93.2
26, 27, 3370	ELECTRICAL, COMMUNICATIONS & UTIL.	100.9	97.1	99.0	104.6	84.4	94.7	100.3	84.9	92.7	101.5	94.0	97.8	100.2	102.4	101.3	102.4	84.9	93.7
MF2018	WEIGHTED AVERAGE	99.4	102.0	100.5	95.9	93.5	94.9	96.7	94.3	95.7	99.7	98.1	99.0	99.2	107.9	102.9	95.4	94.6	95.0

WISCONSIN

DIVISION		OSHKOSH 549 MAT.	INST.	TOTAL	PORTAGE 539 MAT.	INST.	TOTAL	RACINE 534 MAT.	INST.	TOTAL	RHINELANDER 545 MAT.	INST.	TOTAL	SUPERIOR 548 MAT.	INST.	TOTAL	WAUSAU 544 MAT.	INST.	TOTAL
015433	CONTRACTOR EQUIPMENT		97.7	97.7		99.0	99.0		99.0	99.0		97.7	97.7		100.2	100.2		97.7	97.7
0241, 31 - 34	SITE & INFRASTRUCTURE, DEMOLITION	91.0	95.3	94.0	86.9	100.9	96.6	96.9	102.1	100.5	103.2	95.3	97.7	90.1	99.3	96.5	86.9	95.8	93.0
0310	Concrete Forming & Accessories	90.7	94.7	94.1	91.3	95.7	95.1	100.1	106.1	105.2	88.3	94.6	93.7	91.6	88.6	89.0	90.1	97.4	96.3
0320	Concrete Reinforcing	92.4	109.3	100.6	96.4	106.4	101.2	95.0	110.5	102.5	92.6	106.4	99.3	91.3	110.2	100.4	92.6	106.2	99.2
0330	Cast-in-Place Concrete	95.8	98.0	96.6	90.3	98.3	93.3	103.8	102.5	103.3	108.6	97.9	104.6	98.0	98.3	98.1	89.3	92.4	90.5
03	CONCRETE	86.2	98.6	91.7	84.2	98.7	90.6	95.8	105.4	100.0	98.0	98.1	98.1	85.5	96.2	90.2	81.4	97.4	88.4
04	MASONRY	104.7	97.8	100.4	98.2	100.2	99.4	99.3	108.0	104.6	121.3	97.1	106.8	95.1	104.6	100.9	104.2	97.6	100.1
05	METALS	94.3	103.5	97.0	96.7	104.3	98.9	100.2	103.3	101.1	94.2	102.5	96.6	95.1	104.6	97.9	94.0	102.9	96.6
06	WOOD, PLASTICS & COMPOSITES	86.7	94.2	90.9	85.8	94.1	90.4	97.2	105.7	101.9	84.2	94.2	89.7	87.9	85.8	86.7	86.2	97.2	92.2
07	THERMAL & MOISTURE PROTECTION	105.0	83.3	95.6	104.5	100.1	102.6	105.5	102.7	104.3	106.0	83.1	96.0	104.5	95.9	100.8	104.8	95.1	100.6
08	OPENINGS	95.6	96.2	95.7	94.1	99.7	95.4	98.1	105.4	99.8	95.6	91.9	94.7	88.0	91.9	88.9	95.8	97.1	96.1
0920	Plaster & Gypsum Board	90.7	94.5	93.2	83.8	94.5	91.0	91.6	106.4	101.5	90.7	94.5	93.2	92.1	86.0	88.0	90.7	97.5	95.3
0950, 0980	Ceilings & Acoustic Treatment	83.6	94.5	90.9	83.6	94.5	90.9	84.4	106.4	99.2	83.6	94.5	90.9	58.8	86.0	77.1	83.6	97.5	93.0
0960	Flooring	87.8	114.8	95.4	90.5	114.5	97.3	94.8	114.8	100.4	87.2	114.5	94.9	90.8	123.1	99.9	87.7	114.5	95.2
0970, 0990	Wall Finishes & Painting/Coating	89.6	104.3	98.3	95.0	97.7	96.6	95.0	115.9	107.4	89.6	80.0	83.9	84.0	105.8	96.9	89.6	83.2	85.8
09	FINISHES	86.5	100.0	93.8	89.2	99.4	94.7	92.3	109.1	101.4	87.4	97.3	92.7	82.1	96.4	89.8	86.2	99.4	93.3
COVERS	DIVS. 10 - 14, 25, 28, 41, 43, 44, 46	100.0	85.6	96.8	100.0	85.6	96.8	100.0	97.3	99.4	100.0	85.9	96.8	100.0	93.9	98.6	100.0	95.7	99.0
21, 22, 23	FIRE SUPPRESSION, PLUMBING & HVAC	96.7	83.1	91.3	96.3	95.8	96.1	99.8	96.0	98.3	96.7	88.6	93.5	96.0	91.1	94.1	96.7	88.8	93.6
26, 27, 3370	ELECTRICAL, COMMUNICATIONS & UTIL.	103.0	79.1	91.2	104.0	94.0	99.1	100.4	98.0	99.2	102.3	78.1	90.4	107.2	98.5	102.9	104.1	78.1	91.3
MF2018	WEIGHTED AVERAGE	95.4	91.9	93.9	95.2	98.0	96.4	98.7	102.3	100.2	97.9	92.2	95.5	95.1	96.5	95.7	94.7	93.4	94.2

For customer support on your Commercial Renovation Costs with RSMeans data, call 800.448.8182.

825

WYOMING

DIVISION		CASPER 826			CHEYENNE 820			NEWCASTLE 827			RAWLINS 823			RIVERTON 825			ROCK SPRINGS 829 - 831		
		MAT.	INST.	TOTAL	MAT.	INST.	TOTAL	MAT.	INST.	TOTAL	MAT.	INST.	TOTAL	MAT.	INST.	TOTAL	MAT.	INST.	TOTAL
015433	CONTRACTOR EQUIPMENT		97.3	97.3		94.6	94.6		94.6	94.6		94.6	94.6		94.6	94.6		94.6	94.6
0241, 31 - 34	SITE & INFRASTRUCTURE, DEMOLITION	97.8	94.7	95.7	91.6	90.1	90.6	83.6	89.3	87.6	97.5	89.3	91.9	91.1	89.3	89.8	87.4	89.5	88.9
0310	Concrete Forming & Accessories	101.3	53.6	60.6	102.5	64.6	70.2	93.7	70.5	74.0	97.6	70.8	74.7	92.7	60.3	65.0	99.7	64.7	69.9
0320	Concrete Reinforcing	111.1	83.4	97.7	106.4	83.4	95.3	114.5	83.2	99.3	114.2	83.2	99.2	115.2	83.2	99.7	115.2	81.6	99.0
0330	Cast-in-Place Concrete	103.8	77.8	94.1	97.9	77.4	90.3	99.0	72.8	89.2	99.0	72.9	89.3	99.0	72.8	89.2	99.0	75.0	90.0
03	CONCRETE	103.3	67.9	87.7	100.9	72.8	88.6	101.1	73.8	89.1	116.0	74.0	97.5	110.1	69.2	92.1	101.6	71.7	88.4
04	MASONRY	102.0	64.6	79.0	105.3	65.2	80.7	102.2	59.3	75.8	102.2	59.3	75.8	102.2	59.3	75.8	164.5	61.8	101.4
05	METALS	102.8	80.1	96.1	105.3	80.8	98.1	101.4	79.9	95.1	101.5	80.1	95.2	101.6	79.9	95.2	102.3	79.5	95.6
06	WOOD, PLASTICS & COMPOSITES	95.2	47.8	69.1	91.8	62.6	75.7	82.4	72.7	77.1	86.2	72.7	78.8	81.4	58.7	68.9	91.3	62.6	75.5
07	THERMAL & MOISTURE PROTECTION	112.7	65.9	92.3	107.2	68.0	90.2	108.8	63.9	89.3	110.3	74.9	94.9	109.7	67.7	91.4	108.9	69.5	91.8
08	OPENINGS	109.3	59.4	97.7	107.2	67.7	98.0	111.4	72.3	102.3	111.1	72.3	102.0	111.3	64.6	100.4	111.8	66.5	101.2
0920	Plaster & Gypsum Board	101.7	46.2	64.4	90.6	61.7	71.2	87.6	72.0	77.1	87.9	72.0	77.2	87.6	57.7	67.5	100.2	61.7	74.3
0950, 0980	Ceilings & Acoustic Treatment	105.3	46.2	65.5	98.5	61.7	73.7	100.3	72.0	81.2	100.3	72.0	81.2	100.3	57.7	71.6	100.3	61.7	74.3
0960	Flooring	102.2	73.8	94.2	100.3	73.8	92.8	94.2	44.5	80.2	97.0	44.5	82.3	93.6	60.3	84.2	99.3	59.9	88.3
0970, 0990	Wall Finishes & Painting/Coating	95.7	58.7	73.8	97.9	58.7	74.7	94.6	60.5	74.4	94.6	60.5	74.4	94.6	60.5	74.4	94.6	60.5	74.4
09	FINISHES	99.6	56.3	76.2	96.3	65.1	79.4	91.4	64.6	76.9	93.6	64.6	77.9	92.0	59.5	74.4	100.6	62.7	77.3
COVERS	DIVS. 10 - 14, 25, 28, 41, 43, 44, 46	100.0	92.5	98.3	100.0	87.4	97.2	100.0	91.8	98.2	100.0	91.8	98.2	100.0	77.3	94.9	100.0	86.9	97.1
21, 22, 23	FIRE SUPPRESSION, PLUMBING & HVAC	100.0	70.5	88.4	100.1	73.3	89.6	98.2	71.8	87.8	98.2	71.9	87.8	98.2	71.8	87.8	100.0	73.9	89.8
26, 27, 3370	ELECTRICAL, COMMUNICATIONS & UTIL.	94.1	60.3	77.5	95.2	69.0	82.3	94.0	62.3	78.4	94.0	62.3	78.4	94.0	59.1	76.8	92.5	66.0	79.4
MF2018	WEIGHTED AVERAGE	101.5	68.9	87.7	101.1	72.7	89.1	99.6	71.1	87.5	102.0	71.5	89.1	101.0	68.6	87.3	103.4	71.4	89.9

WYOMING / CANADA

DIVISION		SHERIDAN 828			WHEATLAND 822			WORLAND 824			YELLOWSTONE NAT'L PA 821			BARRIE, ONTARIO			BATHURST, NEW BRUNSWICK		
		MAT.	INST.	TOTAL	MAT.	INST.	TOTAL	MAT.	INST.	TOTAL	MAT.	INST.	TOTAL	MAT.	INST.	TOTAL	MAT.	INST.	TOTAL
015433	CONTRACTOR EQUIPMENT		94.6	94.6		94.6	94.6		94.6	94.6		94.6	94.6		100.1	100.1		99.7	99.7
0241, 31 - 34	SITE & INFRASTRUCTURE, DEMOLITION	91.5	90.1	90.5	88.1	89.3	88.9	85.5	89.3	88.1	85.6	89.3	88.1	117.0	94.7	101.6	105.1	90.6	95.1
0310	Concrete Forming & Accessories	100.5	61.6	67.4	95.4	48.4	55.3	95.5	61.1	66.1	95.5	61.1	66.2	125.6	82.3	88.7	106.1	57.8	64.9
0320	Concrete Reinforcing	115.2	83.2	99.7	114.5	82.6	99.1	115.2	83.2	99.7	117.2	82.3	100.3	175.2	86.9	132.5	140.5	57.5	100.4
0330	Cast-in-Place Concrete	102.3	77.4	93.1	103.3	72.7	91.9	99.0	72.7	89.2	99.0	72.8	89.2	157.7	82.1	129.6	114.0	56.2	92.5
03	CONCRETE	109.9	71.4	93.0	106.3	63.7	87.6	101.3	69.5	87.3	101.6	69.4	87.4	138.9	83.5	114.5	110.3	58.3	87.4
04	MASONRY	102.5	62.0	77.6	102.6	51.2	71.0	102.2	59.3	75.8	102.2	59.3	75.8	165.6	89.3	118.7	159.8	57.4	96.9
05	METALS	105.1	80.5	97.8	101.4	79.1	94.8	101.6	79.8	95.2	102.2	79.1	95.4	111.1	91.5	105.3	114.1	73.6	102.2
06	WOOD, PLASTICS & COMPOSITES	93.3	58.7	74.3	84.2	43.0	61.5	84.3	60.0	70.9	84.3	60.0	70.9	116.5	80.8	96.9	97.5	57.9	75.7
07	THERMAL & MOISTURE PROTECTION	110.1	66.7	91.2	109.1	58.1	86.9	108.9	64.8	89.7	108.3	64.8	89.4	115.3	86.1	102.6	111.8	58.4	88.6
08	OPENINGS	112.0	64.6	101.0	110.0	55.9	97.4	111.6	65.3	100.8	104.5	64.8	95.2	90.5	80.9	88.3	84.0	51.2	76.3
0920	Plaster & Gypsum Board	111.2	57.7	75.2	87.6	41.5	56.6	87.6	58.9	68.3	87.8	58.9	68.4	153.5	80.3	104.3	124.2	56.7	78.8
0950, 0980	Ceilings & Acoustic Treatment	103.9	57.7	72.7	100.3	41.5	60.6	100.3	58.9	72.4	101.1	58.9	72.7	94.3	80.3	84.9	116.8	56.7	76.3
0960	Flooring	97.9	60.3	87.4	95.6	43.8	81.0	95.6	44.5	81.3	95.6	44.5	81.3	119.3	87.3	110.3	99.6	41.5	83.3
0970, 0990	Wall Finishes & Painting/Coating	96.8	58.7	74.2	94.6	58.4	73.2	94.6	60.5	74.4	94.6	60.5	74.4	105.6	83.8	92.7	111.3	47.9	73.8
09	FINISHES	99.2	60.4	78.2	92.1	46.7	67.5	91.8	57.1	73.0	92.0	57.2	73.2	111.3	83.5	96.2	107.7	53.8	78.5
COVERS	DIVS. 10 - 14, 25, 28, 41, 43, 44, 46	100.0	93.4	98.5	100.0	81.4	95.9	100.0	76.1	94.7	100.0	76.2	94.7	139.2	65.0	122.7	131.1	57.9	114.8
21, 22, 23	FIRE SUPPRESSION, PLUMBING & HVAC	98.2	73.3	88.4	98.2	71.8	87.8	98.2	71.8	87.8	98.2	71.8	87.8	103.7	93.5	99.7	103.8	64.8	88.5
26, 27, 3370	ELECTRICAL, COMMUNICATIONS & UTIL.	96.1	59.1	77.9	94.0	76.6	85.4	94.0	76.6	85.4	93.1	86.1	89.7	117.0	82.9	100.2	113.5	55.9	85.1
MF2018	WEIGHTED AVERAGE	102.5	70.2	88.8	100.3	67.1	86.2	99.7	70.7	87.5	99.0	71.9	87.6	116.5	87.1	104.1	110.7	62.3	90.2

CANADA

DIVISION		BRANDON, MANITOBA			BRANTFORD, ONTARIO			BRIDGEWATER, NOVA SCOTIA			CALGARY, ALBERTA			CAP-DE-LA-MADELEINE, QUEBEC			CHARLESBOURG, QUEBEC		
		MAT.	INST.	TOTAL	MAT.	INST.	TOTAL	MAT.	INST.	TOTAL	MAT.	INST.	TOTAL	MAT.	INST.	TOTAL	MAT.	INST.	TOTAL
015433	CONTRACTOR EQUIPMENT		101.5	101.5		99.7	99.7		99.4	99.4		127.1	127.1		100.2	100.2		100.2	100.2
0241, 31 - 34	SITE & INFRASTRUCTURE, DEMOLITION	126.4	92.9	103.3	116.9	94.8	101.6	101.1	92.3	95.0	125.5	118.4	120.6	96.9	93.7	94.7	96.9	93.7	94.7
0310	Concrete Forming & Accessories	145.4	66.1	77.8	127.2	89.1	94.7	99.7	67.9	72.6	125.0	94.9	99.3	132.9	79.2	87.1	132.9	79.2	87.1
0320	Concrete Reinforcing	170.2	54.5	114.3	163.7	85.6	125.9	139.3	47.6	95.0	136.4	82.0	110.1	139.3	72.4	107.0	139.3	72.4	107.0
0330	Cast-in-Place Concrete	109.4	70.2	94.8	130.7	101.8	119.9	134.3	66.9	109.2	136.4	104.3	124.4	105.5	87.5	98.8	105.5	87.5	98.8
03	CONCRETE	118.1	66.6	95.5	121.8	93.2	109.3	120.1	65.1	95.9	126.3	96.8	113.3	107.5	81.5	96.1	107.5	81.5	96.1
04	MASONRY	212.8	60.7	119.3	166.1	93.4	121.4	161.8	65.6	102.6	212.4	89.8	137.1	162.4	76.3	109.5	162.4	76.3	109.5
05	METALS	127.2	78.2	112.8	112.3	92.0	106.3	111.6	76.0	101.1	131.0	102.4	122.6	110.7	85.0	103.1	110.7	85.0	103.1
06	WOOD, PLASTICS & COMPOSITES	150.5	66.8	104.4	120.3	87.7	102.4	89.6	67.4	77.4	102.6	94.4	98.1	131.6	79.0	102.7	131.6	79.0	102.7
07	THERMAL & MOISTURE PROTECTION	128.4	68.7	102.4	121.6	91.4	108.5	116.1	67.6	95.0	132.0	98.0	117.2	114.8	84.1	101.5	114.8	84.1	101.5
08	OPENINGS	100.3	59.9	90.8	87.7	86.6	87.5	82.4	61.3	77.5	82.3	83.7	82.6	89.1	72.3	85.2	89.1	72.3	85.2
0920	Plaster & Gypsum Board	115.4	65.6	81.9	115.8	87.4	96.7	124.3	66.5	85.4	126.4	93.3	104.1	147.6	78.4	101.0	147.6	78.4	101.0
0950, 0980	Ceilings & Acoustic Treatment	123.8	65.6	84.6	103.5	87.4	92.7	103.5	66.5	78.6	149.2	93.3	111.5	103.5	78.4	86.6	103.5	78.4	86.6
0960	Flooring	131.2	61.9	111.7	113.6	87.3	106.2	95.7	58.9	85.4	118.1	84.8	108.7	113.6	86.0	105.8	113.6	86.0	105.8
0970, 0990	Wall Finishes & Painting/Coating	117.9	54.1	80.1	110.3	92.0	99.5	110.3	59.3	80.1	113.7	106.4	109.4	110.3	83.6	94.5	110.3	83.6	94.5
09	FINISHES	122.6	61.4	89.5	108.7	89.3	98.2	103.8	65.5	83.1	124.3	94.8	108.0	112.2	81.2	95.4	112.2	81.2	95.4
COVERS	DIVS. 10 - 14, 25, 28, 41, 43, 44, 46	131.1	60.1	115.3	131.1	66.9	116.8	131.1	60.8	115.5	131.1	92.7	122.6	131.1	75.6	118.8	131.1	75.6	118.8
21, 22, 23	FIRE SUPPRESSION, PLUMBING & HVAC	104.0	78.9	94.1	103.8	96.3	100.9	103.8	79.5	94.3	105.1	90.2	99.2	104.2	84.2	96.4	104.2	84.2	96.4
26, 27, 3370	ELECTRICAL, COMMUNICATIONS & UTIL.	113.9	63.1	88.8	111.5	82.5	97.2	115.8	58.7	87.6	107.1	93.9	100.6	110.2	65.2	88.0	110.2	65.2	88.0
MF2018	WEIGHTED AVERAGE	120.6	70.5	99.4	113.1	90.7	103.6	111.3	70.3	94.0	119.7	95.6	109.5	110.6	80.0	97.7	110.6	80.0	97.7

CANADA

DIVISION		CHARLOTTETOWN, PRINCE EDWARD ISLAND			CHICOUTIMI, QUEBEC			CORNER BROOK, NEWFOUNDLAND			CORNWALL, ONTARIO			DALHOUSIE, NEW BRUNSWICK			DARTMOUTH, NOVA SCOTIA		
		MAT.	INST.	TOTAL	MAT.	INST.	TOTAL	MAT.	INST.	TOTAL	MAT.	INST.	TOTAL	MAT.	INST.	TOTAL	MAT.	INST.	TOTAL
015433	CONTRACTOR EQUIPMENT		116.8	116.8		100.3	100.3		100.0	100.0		99.7	99.7		99.7	99.7		98.9	98.9
0241, 31 - 34	SITE & INFRASTRUCTURE, DEMOLITION	134.6	101.5	111.7	102.0	93.6	96.2	130.4	90.5	102.8	115.1	94.3	100.7	101.0	90.6	93.8	117.9	92.0	100.0
0310	Concrete Forming & Accessories	123.7	53.1	63.5	134.9	89.8	96.4	121.9	75.6	82.4	125.0	82.4	88.7	105.9	58.0	65.1	111.8	67.8	74.3
0320	Concrete Reinforcing	156.7	46.6	103.5	106.0	94.5	100.4	156.0	48.8	104.2	163.7	85.3	125.8	142.5	57.6	101.4	163.2	47.6	107.3
0330	Cast-in-Place Concrete	152.1	57.6	116.9	107.4	94.7	102.7	127.3	63.6	103.6	117.5	92.8	108.3	110.5	56.3	90.3	122.9	66.8	102.0
03	CONCRETE	136.8	55.4	101.0	101.6	92.4	97.6	149.1	67.5	113.3	115.7	87.0	103.1	113.1	58.4	89.0	131.9	65.0	102.5
04	MASONRY	187.9	54.9	106.2	161.9	89.8	117.6	208.6	74.3	126.1	165.0	85.6	116.2	163.3	57.4	98.3	222.6	65.6	126.1
05	METALS	136.4	79.6	119.8	113.9	91.9	107.4	127.7	75.2	112.2	112.1	90.7	105.8	105.6	73.7	96.2	128.1	75.7	112.7
06	WOOD, PLASTICS & COMPOSITES	104.3	52.5	75.8	131.8	90.3	108.9	128.1	81.4	102.4	118.5	81.6	98.2	95.8	57.9	74.9	115.9	67.4	89.2
07	THERMAL & MOISTURE PROTECTION	137.8	57.4	102.8	112.3	96.4	105.4	133.0	65.9	103.8	121.4	86.2	106.1	121.0	58.4	93.8	130.1	67.6	102.9
08	OPENINGS	85.7	45.8	76.4	87.8	76.3	85.2	106.3	66.6	97.0	89.1	80.6	87.1	85.5	51.2	77.5	91.3	61.3	84.3
0920	Plaster & Gypsum Board	126.3	50.5	75.3	143.9	89.9	107.6	151.2	80.9	103.9	174.9	81.1	111.8	129.5	56.7	80.5	146.6	66.5	92.7
0950, 0980	Ceilings & Acoustic Treatment	128.9	50.5	76.1	116.0	89.9	98.4	124.6	80.9	95.1	107.6	81.1	89.8	106.6	56.7	72.9	131.1	66.5	87.5
0960	Flooring	115.1	55.3	98.3	115.7	86.0	107.3	112.8	49.5	95.0	113.6	86.0	105.8	101.9	63.4	91.1	107.6	58.9	93.9
0970, 0990	Wall Finishes & Painting/Coating	117.3	39.9	71.5	111.3	104.1	107.1	117.8	56.5	81.5	110.3	85.7	95.7	113.8	47.9	74.8	117.8	59.3	83.2
09	FINISHES	119.8	52.3	83.3	114.6	91.6	102.1	122.6	70.2	94.2	117.3	83.6	99.0	108.4	58.2	81.2	120.3	65.5	90.7
COVERS	DIVS. 10 - 14, 25, 28, 41, 43, 44, 46	131.1	58.4	114.9	131.1	81.5	120.1	131.1	60.6	115.4	131.1	64.6	116.3	131.1	57.8	114.8	131.1	60.8	115.5
21, 22, 23	FIRE SUPPRESSION, PLUMBING & HVAC	104.1	59.0	86.4	103.8	81.0	94.9	104.0	65.9	89.0	104.2	94.3	100.3	103.9	64.8	88.5	104.0	79.5	94.4
26, 27, 3370	ELECTRICAL, COMMUNICATIONS & UTIL.	110.0	47.5	79.2	109.9	84.1	97.2	110.9	51.5	81.6	111.6	83.4	97.7	113.9	52.8	83.8	115.0	58.7	87.2
MF2018	WEIGHTED AVERAGE	121.3	60.2	95.5	110.4	87.6	100.7	124.7	68.4	100.9	113.1	87.3	102.2	110.2	62.5	90.0	121.5	70.2	99.8

CANADA

DIVISION		EDMONTON, ALBERTA			FORT MCMURRAY, ALBERTA			FREDERICTON, NEW BRUNSWICK			GATINEAU, QUEBEC			GRANBY, QUEBEC			HALIFAX, NOVA SCOTIA		
		MAT.	INST.	TOTAL	MAT.	INST.	TOTAL	MAT.	INST.	TOTAL	MAT.	INST.	TOTAL	MAT.	INST.	TOTAL	MAT.	INST.	TOTAL
015433	CONTRACTOR EQUIPMENT		129.1	129.1		101.9	101.9		113.9	113.9		100.2	100.2		100.2	100.2		115.6	115.6
0241, 31 - 34	SITE & INFRASTRUCTURE, DEMOLITION	125.1	121.8	122.8	121.7	95.4	103.5	116.0	100.8	105.5	96.7	93.7	94.6	97.2	93.7	94.8	102.8	103.6	103.4
0310	Concrete Forming & Accessories	128.3	94.9	99.8	125.0	87.5	93.1	126.9	58.6	68.7	132.9	79.1	87.0	132.9	79.0	87.0	124.1	87.7	93.0
0320	Concrete Reinforcing	136.0	82.0	109.9	151.5	81.9	117.8	141.0	57.8	100.8	147.4	72.4	111.1	147.4	72.4	111.1	156.3	77.9	118.3
0330	Cast-in-Place Concrete	145.8	104.3	130.4	173.5	98.2	145.5	112.7	57.5	92.1	104.0	87.5	97.9	107.5	87.4	100.0	99.4	85.1	94.0
03	CONCRETE	130.8	96.8	115.9	139.8	90.6	118.2	116.9	59.7	91.8	107.9	81.4	96.2	109.5	81.4	97.1	111.9	86.1	100.6
04	MASONRY	212.7	89.8	137.2	206.1	86.3	132.5	189.1	58.8	109.0	162.3	76.3	109.5	162.6	76.3	109.6	185.4	86.4	124.6
05	METALS	134.6	102.4	125.1	139.8	90.9	125.5	136.6	83.3	121.0	110.7	84.8	103.1	110.9	84.8	103.2	136.3	99.2	125.4
06	WOOD, PLASTICS & COMPOSITES	101.6	94.4	97.6	114.5	86.7	99.2	106.5	58.2	79.9	131.6	79.0	102.7	131.6	79.0	102.7	103.7	87.8	95.0
07	THERMAL & MOISTURE PROTECTION	143.5	98.0	123.7	129.8	92.6	113.6	137.4	59.4	103.5	114.8	84.1	101.5	114.8	82.6	100.8	137.4	88.7	116.2
08	OPENINGS	80.9	83.7	81.5	89.1	79.5	86.9	88.0	50.3	79.2	89.1	67.9	84.2	89.1	67.9	84.2	88.9	79.2	86.7
0920	Plaster & Gypsum Board	129.8	93.3	105.2	116.6	86.1	96.1	130.3	56.7	80.8	114.4	78.4	90.2	114.4	78.4	90.2	117.8	87.1	97.1
0950, 0980	Ceilings & Acoustic Treatment	144.5	93.3	110.0	112.5	86.1	94.7	127.2	56.7	82.9	103.5	78.4	86.6	103.5	78.4	86.6	139.8	87.1	104.2
0960	Flooring	121.8	84.8	111.4	113.6	84.8	105.5	115.6	66.2	101.7	113.6	86.0	105.8	113.6	86.0	105.8	109.5	81.5	101.6
0970, 0990	Wall Finishes & Painting/Coating	115.6	107.4	110.8	110.4	88.3	97.3	118.8	61.1	84.6	110.3	83.6	94.5	110.3	83.6	94.5	118.4	90.9	102.1
09	FINISHES	123.6	94.9	108.0	111.8	87.4	98.6	125.1	60.4	90.1	107.8	81.2	93.4	107.8	81.2	93.4	117.7	87.5	101.3
COVERS	DIVS. 10 - 14, 25, 28, 41, 43, 44, 46	131.1	93.6	122.8	131.1	90.1	122.0	131.1	58.5	115.0	131.1	75.6	118.8	131.1	75.6	118.8	131.1	68.9	117.3
21, 22, 23	FIRE SUPPRESSION, PLUMBING & HVAC	104.5	90.4	99.2	104.2	93.2	99.9	104.2	73.7	92.2	104.2	84.2	96.3	103.8	84.2	96.1	104.9	82.3	96.0
26, 27, 3370	ELECTRICAL, COMMUNICATIONS & UTIL.	115.0	93.9	104.6	105.4	77.5	91.7	112.2	69.7	91.2	110.2	65.2	88.0	110.9	65.2	88.4	113.3	90.6	102.1
MF2018	WEIGHTED AVERAGE	121.8	95.9	110.9	121.9	88.4	107.7	119.3	69.0	98.0	110.3	79.8	97.4	110.5	79.7	97.5	117.9	88.0	105.2

CANADA

DIVISION		HAMILTON, ONTARIO			HULL, QUEBEC			JOLIETTE, QUEBEC			KAMLOOPS, BRITISH COLUMBIA			KINGSTON, ONTARIO			KITCHENER, ONTARIO		
		MAT.	INST.	TOTAL	MAT.	INST.	TOTAL	MAT.	INST.	TOTAL	MAT.	INST.	TOTAL	MAT.	INST.	TOTAL	MAT.	INST.	TOTAL
015433	CONTRACTOR EQUIPMENT		115.5	115.5		100.2	100.2		100.2	100.2		103.3	103.3		101.9	101.9		101.6	101.6
0241, 31 - 34	SITE & INFRASTRUCTURE, DEMOLITION	105.8	107.0	106.6	96.7	93.7	94.6	97.3	93.7	94.8	120.1	97.1	104.3	115.0	98.0	103.3	94.2	99.8	98.0
0310	Concrete Forming & Accessories	130.0	94.7	99.9	132.9	79.1	87.0	132.9	79.2	87.1	124.3	82.3	88.5	125.1	82.5	88.7	117.3	88.0	92.3
0320	Concrete Reinforcing	136.9	103.3	120.7	147.4	72.4	111.1	139.3	72.4	107.0	109.3	75.8	93.1	163.7	85.3	125.8	96.4	103.1	99.6
0330	Cast-in-Place Concrete	107.6	102.1	105.6	104.0	87.5	97.9	108.4	87.5	100.6	93.8	91.8	93.1	117.5	92.8	108.3	111.7	97.2	106.3
03	CONCRETE	110.2	99.3	105.4	107.9	81.4	96.2	108.8	81.5	96.8	117.2	85.0	103.0	117.5	87.0	104.1	97.1	94.1	95.8
04	MASONRY	173.2	100.5	128.5	162.3	76.3	109.5	162.7	76.3	109.6	169.4	84.0	116.9	171.9	85.7	118.9	142.2	98.6	115.4
05	METALS	134.7	106.4	126.4	110.9	84.8	103.2	110.9	85.0	103.3	112.8	87.1	105.3	113.5	90.7	106.8	124.1	98.0	116.5
06	WOOD, PLASTICS & COMPOSITES	111.4	94.0	101.8	131.6	79.0	102.7	131.6	79.0	102.7	102.2	80.7	90.3	118.5	81.7	98.3	110.3	85.9	96.9
07	THERMAL & MOISTURE PROTECTION	134.0	101.5	119.8	114.8	84.1	101.5	114.8	84.1	101.5	131.2	81.7	109.7	121.4	87.3	106.6	116.4	98.4	108.5
08	OPENINGS	85.3	93.2	87.1	89.1	67.9	84.2	89.1	72.3	85.2	86.1	78.5	84.3	89.1	80.0	87.1	80.0	87.1	81.7
0920	Plaster & Gypsum Board	131.0	93.1	105.5	114.4	78.4	90.2	147.6	78.4	101.0	101.1	79.8	86.8	177.6	81.2	112.8	107.2	85.6	92.6
0950, 0980	Ceilings & Acoustic Treatment	139.9	93.1	108.4	103.5	78.4	86.6	103.5	78.4	86.6	103.5	79.8	87.5	119.0	81.2	93.5	107.9	85.6	92.9
0960	Flooring	115.3	95.0	109.6	113.6	86.0	105.8	113.6	86.0	105.8	112.5	49.2	94.7	113.6	86.0	105.8	101.3	95.1	99.6
0970, 0990	Wall Finishes & Painting/Coating	113.0	104.1	107.8	110.3	83.6	94.5	110.3	83.6	94.5	110.3	76.4	90.3	110.3	79.2	91.9	104.4	93.9	98.2
09	FINISHES	121.0	95.6	107.3	107.8	81.2	93.4	112.2	81.2	95.4	108.3	75.8	90.7	119.9	82.9	99.9	102.1	89.4	95.2
COVERS	DIVS. 10 - 14, 25, 28, 41, 43, 44, 46	131.1	92.5	122.5	131.1	75.6	118.8	131.1	75.6	118.8	131.1	84.4	120.7	131.1	64.6	116.3	131.1	89.6	121.9
21, 22, 23	FIRE SUPPRESSION, PLUMBING & HVAC	105.2	92.1	100.0	103.8	84.2	96.1	103.8	84.2	96.1	103.8	87.5	97.4	104.2	94.4	100.4	104.2	91.3	99.1
26, 27, 3370	ELECTRICAL, COMMUNICATIONS & UTIL.	108.5	103.6	106.1	112.3	65.2	89.1	110.9	65.2	88.4	114.3	74.2	94.5	111.6	82.1	97.1	109.6	99.8	104.8
MF2018	WEIGHTED AVERAGE	116.3	98.8	108.9	110.5	79.8	97.5	110.8	80.0	97.8	113.0	83.5	100.5	114.1	87.4	102.8	108.6	94.6	102.7

DIVISION		LAVAL, QUEBEC			LETHBRIDGE, ALBERTA			LLOYDMINSTER, ALBERTA			LONDON, ONTARIO			MEDICINE HAT, ALBERTA			MONCTON, NEW BRUNSWICK		
		MAT.	INST.	TOTAL	MAT.	INST.	TOTAL	MAT.	INST.	TOTAL	MAT.	INST.	TOTAL	MAT.	INST.	TOTAL	MAT.	INST.	TOTAL
015433	CONTRACTOR EQUIPMENT		100.2	100.2		101.9	101.9		101.9	101.9		116.7	116.7		101.9	101.9		99.7	99.7
0241, 31 - 34	SITE & INFRASTRUCTURE, DEMOLITION	97.2	93.9	94.9	114.5	96.0	101.7	114.5	95.4	101.3	103.5	106.8	105.8	113.2	95.5	101.0	104.5	92.5	96.2
0310	Concrete Forming & Accessories	133.2	81.1	88.8	126.4	87.6	93.3	124.6	78.3	85.1	129.1	89.1	95.0	126.4	78.2	85.3	106.1	66.1	72.0
0320	Concrete Reinforcing	147.4	74.4	112.1	151.5	81.9	117.8	151.5	81.8	117.8	126.7	102.1	114.8	151.5	81.8	117.8	140.5	69.8	106.3
0330	Cast-in-Place Concrete	107.5	89.4	100.7	130.1	98.2	118.2	120.7	94.6	111.0	120.2	100.8	113.0	120.7	94.6	111.0	109.8	68.0	94.3
03	CONCRETE	109.5	83.4	98.0	120.0	90.6	107.1	115.5	85.2	102.2	114.6	96.2	106.5	115.7	85.2	102.3	108.4	68.5	90.9
04	MASONRY	162.6	78.4	110.8	180.1	86.3	122.4	161.6	80.1	111.6	183.8	99.5	132.0	161.6	80.1	111.6	159.4	74.6	107.3
05	METALS	110.7	85.9	103.5	134.1	91.0	121.5	112.8	90.8	106.4	134.7	106.8	126.5	113.0	90.8	106.5	114.1	86.8	106.1
06	WOOD, PLASTICS & COMPOSITES	131.7	81.1	103.9	118.0	86.7	100.8	114.5	77.4	94.1	114.8	86.8	99.4	118.0	77.4	95.7	97.5	65.0	79.6
07	THERMAL & MOISTURE PROTECTION	115.4	86.0	102.6	127.1	92.6	112.1	123.5	88.2	108.1	129.9	99.0	116.5	130.5	88.2	112.1	116.6	72.4	97.4
08	OPENINGS	89.1	69.8	84.6	89.1	79.5	86.9	89.1	74.4	85.7	81.6	87.9	83.1	89.1	74.4	85.7	84.0	61.9	78.8
0920	Plaster & Gypsum Board	114.7	80.5	91.7	106.4	86.1	92.7	102.1	76.6	84.9	134.4	85.8	101.7	104.3	76.6	85.6	124.2	64.1	83.7
0950, 0980	Ceilings & Acoustic Treatment	103.5	80.5	88.0	112.5	86.1	94.7	103.5	76.6	85.3	138.5	85.8	102.9	103.5	76.6	85.3	116.8	64.1	81.2
0960	Flooring	113.6	88.4	106.5	113.6	84.8	105.5	113.6	84.8	105.5	109.9	95.1	105.7	113.6	84.8	105.5	99.6	65.6	90.1
0970, 0990	Wall Finishes & Painting/Coating	110.3	85.9	95.8	110.2	96.4	102.1	110.4	75.1	89.5	113.5	100.5	105.8	110.2	75.1	89.4	111.3	83.6	94.9
09	FINISHES	107.8	83.3	94.5	109.6	88.3	98.0	107.4	78.9	92.0	121.5	91.0	105.0	107.5	78.9	92.0	107.7	67.5	86.0
COVERS	DIVS. 10 - 14, 25, 28, 41, 43, 44, 46	131.1	77.5	119.2	131.1	89.2	121.8	131.1	87.1	121.3	131.1	91.6	122.3	131.1	86.1	121.1	131.1	61.0	115.5
21, 22, 23	FIRE SUPPRESSION, PLUMBING & HVAC	104.3	86.5	97.3	104.1	90.1	98.6	104.2	90.0	98.6	105.1	89.2	98.8	103.8	86.8	97.2	103.8	73.7	92.0
26, 27, 3370	ELECTRICAL, COMMUNICATIONS & UTIL.	110.9	67.0	89.3	107.1	77.5	92.5	104.6	77.5	91.2	103.5	102.0	102.8	104.6	77.5	91.2	117.3	88.5	103.1
MF2018	WEIGHTED AVERAGE	110.7	81.6	98.4	117.0	87.9	104.7	111.5	84.7	100.2	116.3	96.5	108.0	111.7	84.0	100.0	110.9	76.0	96.2

DIVISION		MONTREAL, QUEBEC			MOOSE JAW, SASKATCHEWAN			NEW GLASGOW, NOVA SCOTIA			NEWCASTLE, NEW BRUNSWICK			NORTH BAY, ONTARIO			OSHAWA, ONTARIO		
		MAT.	INST.	TOTAL	MAT.	INST.	TOTAL	MAT.	INST.	TOTAL	MAT.	INST.	TOTAL	MAT.	INST.	TOTAL	MAT.	INST.	TOTAL
015433	CONTRACTOR EQUIPMENT		117.7	117.7		98.3	98.3		98.9	98.9		99.7	99.7		99.3	99.3		101.6	101.6
0241, 31 - 34	SITE & INFRASTRUCTURE, DEMOLITION	113.0	105.5	107.8	114.4	90.0	97.5	111.9	92.0	98.1	105.1	90.6	95.1	126.8	93.6	103.9	104.6	99.6	101.2
0310	Concrete Forming & Accessories	130.9	90.4	96.4	109.6	54.6	62.7	111.8	67.8	74.3	106.1	58.0	65.1	146.2	80.0	89.8	122.8	91.3	96.0
0320	Concrete Reinforcing	126.8	94.6	111.3	106.9	61.4	84.9	156.0	47.6	103.6	140.5	57.6	100.4	184.7	84.8	136.4	152.7	103.7	129.0
0330	Cast-in-Place Concrete	124.4	97.4	114.3	117.1	63.7	97.2	122.9	66.8	102.0	114.0	56.3	92.6	118.4	80.1	104.1	129.2	106.4	120.7
03	CONCRETE	116.7	94.3	106.8	101.4	60.0	83.2	131.0	65.0	102.0	110.3	58.4	87.5	132.7	81.4	110.2	118.3	98.8	109.8
04	MASONRY	178.8	89.8	124.1	160.3	56.4	96.4	208.2	65.6	120.5	159.8	57.4	96.9	214.4	81.9	133.0	145.1	101.9	118.6
05	METALS	144.3	102.2	131.9	109.6	75.0	99.4	125.6	75.7	111.0	114.1	73.8	102.3	126.5	90.2	115.8	114.2	98.6	109.6
06	WOOD, PLASTICS & COMPOSITES	111.3	90.8	100.0	99.4	53.2	74.0	115.9	67.4	89.2	97.5	57.9	75.7	153.5	80.2	113.2	117.6	89.4	102.1
07	THERMAL & MOISTURE PROTECTION	125.9	97.2	113.4	114.1	60.1	90.6	130.1	67.6	102.9	116.6	58.4	91.3	136.4	82.7	113.0	117.3	104.1	111.6
08	OPENINGS	87.4	78.7	85.4	85.3	51.4	77.4	91.3	61.3	84.3	84.0	51.2	76.3	98.6	78.2	93.8	85.0	90.2	86.2
0920	Plaster & Gypsum Board	126.2	89.9	101.8	98.2	51.9	67.1	145.0	66.5	92.2	124.2	56.7	78.8	140.6	79.8	99.7	110.7	89.2	96.2
0950, 0980	Ceilings & Acoustic Treatment	146.1	89.9	108.2	103.5	51.9	68.7	123.8	66.5	85.2	116.8	56.7	76.3	123.8	79.8	94.1	103.9	89.2	94.0
0960	Flooring	112.4	90.5	106.2	103.6	53.7	89.6	107.6	58.9	93.9	99.6	63.4	89.5	131.2	86.0	118.5	104.2	97.6	102.4
0970, 0990	Wall Finishes & Painting/Coating	115.0	104.1	108.5	110.3	61.0	81.1	117.8	59.3	83.2	111.3	47.9	73.8	117.8	85.0	98.4	104.4	108.1	106.6
09	FINISHES	121.6	92.8	106.0	103.7	54.8	77.2	118.7	65.5	89.9	107.7	58.2	80.9	125.9	81.9	102.1	103.2	93.7	98.1
COVERS	DIVS. 10 - 14, 25, 28, 41, 43, 44, 46	131.1	82.8	120.4	131.1	57.5	114.7	131.1	60.8	115.5	131.1	57.9	114.8	131.1	63.4	116.0	131.1	90.4	122.0
21, 22, 23	FIRE SUPPRESSION, PLUMBING & HVAC	105.1	81.2	95.7	104.2	70.3	90.9	104.0	79.5	94.4	103.8	64.8	88.5	104.0	92.4	99.4	104.2	91.9	99.3
26, 27, 3370	ELECTRICAL, COMMUNICATIONS & UTIL.	107.9	84.1	96.2	113.4	56.3	85.3	111.3	58.7	85.4	112.9	55.9	84.8	112.6	83.3	98.1	110.6	103.8	107.3
MF2018	WEIGHTED AVERAGE	119.1	90.1	106.9	109.0	63.9	89.9	119.6	70.2	98.7	110.7	62.9	90.5	122.5	85.2	106.8	110.8	97.2	105.0

DIVISION		OTTAWA, ONTARIO			OWEN SOUND, ONTARIO			PETERBOROUGH, ONTARIO			PORTAGE LA PRAIRIE, MANITOBA			PRINCE ALBERT, SASKATCHEWAN			PRINCE GEORGE, BRITISH COLUMBIA		
		MAT.	INST.	TOTAL	MAT.	INST.	TOTAL	MAT.	INST.	TOTAL	MAT.	INST.	TOTAL	MAT.	INST.	TOTAL	MAT.	INST.	TOTAL
015433	CONTRACTOR EQUIPMENT		117.8	117.8		100.1	100.1		99.7	99.7		101.9	101.9		98.3	98.3		103.3	103.3
0241, 31 - 34	SITE & INFRASTRUCTURE, DEMOLITION	107.0	107.2	107.1	117.0	94.6	101.5	116.9	94.2	101.3	115.4	93.2	100.1	110.2	90.1	96.3	123.4	97.1	105.3
0310	Concrete Forming & Accessories	124.8	88.4	93.7	125.6	78.7	85.6	127.2	81.0	87.8	126.6	65.7	74.7	109.6	54.4	62.5	115.1	77.5	83.0
0320	Concrete Reinforcing	137.0	102.0	120.1	175.2	86.9	132.5	163.7	85.3	125.8	151.5	54.5	104.6	111.7	61.4	87.3	109.3	75.8	93.1
0330	Cast-in-Place Concrete	122.3	103.0	115.1	157.7	76.6	127.5	130.7	81.8	112.5	120.7	69.8	101.8	106.2	63.6	90.3	117.6	91.8	108.0
03	CONCRETE	116.6	96.7	107.9	138.9	79.9	113.0	121.8	82.5	104.6	109.5	66.2	90.5	96.9	59.9	80.7	127.5	82.8	107.9
04	MASONRY	173.2	99.4	127.8	165.6	87.1	117.3	166.1	88.0	118.1	164.8	59.7	100.2	159.5	56.4	96.1	171.4	84.0	117.7
05	METALS	135.5	108.9	127.7	111.1	91.3	105.3	112.3	90.8	106.0	113.0	78.4	102.8	109.7	74.8	99.4	112.8	87.2	105.3
06	WOOD, PLASTICS & COMPOSITES	108.4	85.7	95.9	116.5	77.3	94.9	120.3	79.0	97.6	118.0	66.8	89.8	99.4	53.2	74.0	102.2	74.1	86.7
07	THERMAL & MOISTURE PROTECTION	137.3	99.7	121.0	115.3	83.5	101.5	121.6	88.0	107.0	114.6	68.2	94.4	114.0	59.0	90.1	124.5	81.0	105.6
08	OPENINGS	90.3	87.5	89.7	90.5	77.6	87.5	87.7	79.9	85.9	89.1	59.9	82.3	84.3	51.4	76.6	86.1	74.9	83.5
0920	Plaster & Gypsum Board	129.1	84.6	99.1	153.5	76.7	101.8	118.5	78.5	90.7	104.0	65.6	78.2	98.2	51.9	67.1	101.1	73.0	82.2
0950, 0980	Ceilings & Acoustic Treatment	142.3	84.6	103.4	94.3	76.7	82.4	103.5	78.5	86.7	103.5	65.6	78.0	103.5	51.9	68.7	103.5	73.0	83.0
0960	Flooring	108.3	90.6	103.4	119.3	87.3	110.3	113.6	86.0	105.8	113.6	61.9	99.1	103.6	53.7	89.6	109.0	67.4	97.3
0970, 0990	Wall Finishes & Painting/Coating	113.5	95.9	103.1	105.6	83.8	92.7	110.3	87.1	96.6	110.4	54.1	77.1	110.3	52.0	75.8	110.3	76.4	90.3
09	FINISHES	122.3	89.0	104.3	111.3	80.8	94.8	108.7	82.6	94.5	107.4	64.1	84.0	103.7	53.8	76.7	107.2	75.0	89.8
COVERS	DIVS. 10 - 14, 25, 28, 41, 43, 44, 46	131.1	89.5	121.9	139.2	63.9	122.4	131.1	64.8	116.3	131.1	59.8	115.2	131.1	57.5	114.7	131.1	83.7	120.6
21, 22, 23	FIRE SUPPRESSION, PLUMBING & HVAC	105.1	90.7	99.4	103.7	92.4	99.3	103.8	95.8	100.7	103.8	78.4	93.8	104.2	63.4	88.2	103.8	87.5	97.4
26, 27, 3370	ELECTRICAL, COMMUNICATIONS & UTIL.	104.4	101.9	103.2	118.5	82.1	100.6	111.5	82.9	97.4	113.0	54.4	84.2	113.4	56.3	85.3	111.5	74.2	93.1
MF2018	WEIGHTED AVERAGE	117.5	96.8	108.7	116.7	85.4	103.4	113.1	87.1	102.1	111.6	69.1	93.6	108.2	62.2	88.8	114.0	82.8	100.8

For customer support on your Commercial Renovation Costs with RSMeans data, call 800.448.8182.

CANADA

DIVISION		QUEBEC CITY, QUEBEC			RED DEER, ALBERTA			REGINA, SASKATCHEWAN			RIMOUSKI, QUEBEC			ROUYN-NORANDA, QUEBEC			SAINT HYACINTHE, QUEBEC		
		MAT.	INST.	TOTAL	MAT.	INST.	TOTAL	MAT.	INST.	TOTAL	MAT.	INST.	TOTAL	MAT.	INST.	TOTAL	MAT.	INST.	TOTAL
015433	CONTRACTOR EQUIPMENT		117.4	117.4		101.9	101.9		128.3	128.3		100.2	100.2		100.2	100.2		100.2	100.2
0241, 31 - 34	SITE & INFRASTRUCTURE, DEMOLITION	114.1	104.7	107.6	113.2	95.5	101.0	126.4	119.6	121.7	97.0	93.6	94.6	96.7	93.7	94.6	97.2	93.7	94.8
0310	Concrete Forming & Accessories	127.4	90.5	95.9	139.7	78.2	87.3	131.0	91.5	97.4	132.9	89.8	96.2	132.9	79.1	87.0	132.9	79.1	87.0
0320	Concrete Reinforcing	124.2	94.6	109.9	151.5	81.8	117.8	123.8	88.8	106.9	105.1	94.5	100.0	147.4	72.4	111.1	147.4	72.4	111.1
0330	Cast-in-Place Concrete	131.6	97.0	118.7	120.7	94.6	111.0	153.8	97.7	132.9	109.4	94.7	103.9	104.0	87.5	97.9	107.5	87.5	100.0
03	CONCRETE	119.4	94.3	108.4	116.5	85.2	102.7	131.3	94.0	114.9	105.0	92.4	99.5	107.9	81.4	96.2	109.5	81.4	97.1
04	MASONRY	173.6	89.8	122.1	161.6	80.1	111.6	205.2	86.7	132.4	162.1	89.8	117.6	162.3	76.3	109.5	162.6	76.3	109.6
05	METALS	138.0	103.3	127.8	113.0	90.8	106.5	139.1	101.4	128.0	110.4	91.8	104.9	110.9	84.8	103.2	110.9	84.8	103.2
06	WOOD, PLASTICS & COMPOSITES	114.2	90.7	101.3	118.0	77.4	95.7	108.0	92.2	99.3	131.6	90.3	108.9	131.6	79.0	102.7	131.6	79.0	102.7
07	THERMAL & MOISTURE PROTECTION	125.1	97.2	112.9	141.4	88.2	118.3	143.8	86.6	118.9	114.8	96.4	106.8	114.8	84.1	101.5	115.2	84.1	101.7
08	OPENINGS	88.4	86.0	87.8	89.1	74.4	85.7	87.9	79.8	86.1	88.7	76.3	85.8	89.1	67.9	84.2	89.1	67.9	84.2
0920	Plaster & Gypsum Board	128.5	89.9	102.6	104.3	76.6	85.6	140.8	91.2	107.4	147.4	89.9	108.7	114.2	78.4	90.1	114.2	78.4	90.1
0950, 0980	Ceilings & Acoustic Treatment	131.7	89.9	103.5	103.5	76.6	85.3	154.3	91.2	111.7	102.7	89.9	94.1	102.7	78.4	86.3	102.7	78.4	86.3
0960	Flooring	108.9	90.5	103.8	116.0	84.8	107.2	123.6	95.1	115.6	114.8	86.0	106.7	113.6	86.0	105.8	113.6	86.0	105.8
0970, 0990	Wall Finishes & Painting/Coating	120.8	104.1	111.0	110.2	75.1	89.4	118.0	89.0	100.9	113.4	104.1	107.9	110.3	83.6	94.5	110.3	83.6	94.5
09	FINISHES	117.4	92.7	104.0	108.2	78.9	92.4	130.9	92.9	110.3	112.6	91.6	101.2	107.6	81.2	93.3	107.6	81.2	93.3
COVERS	DIVS. 10 - 14, 25, 28, 41, 43, 44, 46	131.1	82.5	120.3	131.1	86.1	121.1	131.1	70.9	117.7	131.1	81.5	120.1	131.1	75.6	118.8	131.1	75.6	118.8
21, 22, 23	FIRE SUPPRESSION, PLUMBING & HVAC	105.1	81.2	95.7	103.8	86.8	97.2	104.8	86.9	97.8	103.8	81.0	94.9	103.8	84.2	96.1	100.5	84.2	94.1
26, 27, 3370	ELECTRICAL, COMMUNICATIONS & UTIL.	114.7	84.1	99.6	104.6	77.5	91.2	110.8	91.5	101.3	110.9	84.1	97.7	110.9	65.2	88.4	111.5	65.2	88.7
MF2018	WEIGHTED AVERAGE	118.7	90.4	106.7	112.1	84.0	100.2	123.1	92.4	110.2	110.2	87.6	100.6	110.3	79.8	97.4	109.8	79.8	97.1

CANADA

DIVISION		SAINT JOHN, NEW BRUNSWICK			SARNIA, ONTARIO			SASKATOON, SASKATCHEWAN			SAULT STE MARIE, ONTARIO			SHERBROOKE, QUEBEC			SOREL, QUEBEC		
		MAT.	INST.	TOTAL	MAT.	INST.	TOTAL	MAT.	INST.	TOTAL	MAT.	INST.	TOTAL	MAT.	INST.	TOTAL	MAT.	INST.	TOTAL
015433	CONTRACTOR EQUIPMENT		99.7	99.7		99.7	99.7		98.4	98.4		99.7	99.7		100.2	100.2		100.2	100.2
0241, 31 - 34	SITE & INFRASTRUCTURE, DEMOLITION	105.2	92.5	96.4	115.5	94.3	100.9	112.7	92.8	98.9	105.6	93.9	97.5	97.2	93.7	94.8	97.3	93.7	94.8
0310	Concrete Forming & Accessories	126.2	63.4	72.7	125.9	87.6	93.3	109.7	90.7	93.5	115.1	86.2	90.5	132.9	79.1	87.0	132.9	79.2	87.1
0320	Concrete Reinforcing	140.5	69.8	106.3	116.5	86.7	102.1	114.2	88.7	101.9	105.1	85.4	95.6	147.4	72.4	111.1	139.3	72.4	107.0
0330	Cast-in-Place Concrete	112.4	68.0	95.9	120.6	94.2	110.8	113.5	93.1	105.9	108.3	80.5	98.0	107.5	87.5	100.0	108.4	87.5	100.6
03	CONCRETE	110.9	67.3	91.7	111.1	90.0	101.8	101.3	91.2	96.9	97.4	84.7	91.8	109.5	81.4	97.1	108.8	81.5	96.8
04	MASONRY	180.8	74.4	115.4	177.6	90.5	124.1	172.1	86.6	119.5	162.9	90.1	118.2	162.6	76.3	109.6	162.7	76.3	109.6
05	METALS	114.0	86.8	106.0	112.3	91.2	106.1	106.5	88.6	101.2	111.5	95.2	106.7	110.7	84.8	103.1	110.9	85.0	103.3
06	WOOD, PLASTICS & COMPOSITES	120.7	61.4	88.0	119.4	86.7	101.4	96.0	91.3	93.4	106.5	87.9	96.3	131.6	79.0	102.7	131.6	79.0	102.7
07	THERMAL & MOISTURE PROTECTION	116.9	72.2	97.5	121.7	91.5	108.6	117.0	84.7	103.0	120.4	87.3	106.0	114.8	84.1	101.5	114.8	84.1	101.5
08	OPENINGS	83.9	58.9	78.0	90.5	83.5	88.8	85.3	79.4	83.9	82.4	86.1	83.2	89.1	67.9	84.2	89.1	72.3	85.2
0920	Plaster & Gypsum Board	137.9	60.3	85.7	143.2	86.4	105.0	115.6	91.2	99.1	107.1	87.6	94.0	114.2	78.4	90.1	147.4	78.4	101.0
0950, 0980	Ceilings & Acoustic Treatment	121.7	60.3	80.3	109.2	86.4	93.9	124.5	91.2	102.0	103.5	87.6	92.8	102.7	78.4	86.3	102.7	78.4	86.3
0960	Flooring	111.4	65.6	98.5	113.6	94.1	108.1	106.5	95.1	103.3	106.5	92.1	102.4	113.6	86.0	105.8	113.6	86.0	105.8
0970, 0990	Wall Finishes & Painting/Coating	111.3	83.6	94.9	110.3	98.5	103.3	113.8	89.0	99.2	110.3	91.3	99.1	110.3	83.6	94.5	110.3	83.6	94.5
09	FINISHES	114.0	65.4	87.7	113.5	90.2	100.9	112.1	92.3	101.4	104.7	88.2	95.8	107.6	81.2	93.3	112.0	81.2	95.3
COVERS	DIVS. 10 - 14, 25, 28, 41, 43, 44, 46	131.1	61.0	115.5	131.1	66.1	116.6	131.1	69.3	117.4	131.1	87.9	121.5	131.1	75.6	118.8	131.1	75.6	118.8
21, 22, 23	FIRE SUPPRESSION, PLUMBING & HVAC	103.8	75.1	92.6	103.8	101.5	102.9	104.4	86.7	97.5	103.8	90.5	98.6	104.2	84.2	96.3	103.8	84.2	96.1
26, 27, 3370	ELECTRICAL, COMMUNICATIONS & UTIL.	119.9	88.5	104.4	114.7	85.0	100.0	114.6	91.4	103.2	113.2	83.3	98.5	110.9	65.2	88.4	110.9	65.2	88.4
MF2018	WEIGHTED AVERAGE	113.2	75.7	97.3	113.2	91.3	103.9	109.9	88.5	100.9	108.6	88.7	100.2	110.6	79.8	97.6	110.8	80.0	97.8

CANADA

DIVISION		ST. CATHARINES, ONTARIO			ST JEROME, QUEBEC			ST. JOHN'S, NEWFOUNDLAND			SUDBURY, ONTARIO			SUMMERSIDE, PRINCE EDWARD ISLAND			SYDNEY, NOVA SCOTIA		
		MAT.	INST.	TOTAL	MAT.	INST.	TOTAL	MAT.	INST.	TOTAL	MAT.	INST.	TOTAL	MAT.	INST.	TOTAL	MAT.	INST.	TOTAL
015433	CONTRACTOR EQUIPMENT		99.5	99.5		100.2	100.2		123.4	123.4		99.5	99.5		98.8	98.8		98.9	98.9
0241, 31 - 34	SITE & INFRASTRUCTURE, DEMOLITION	94.2	96.1	95.6	96.7	93.7	94.6	115.4	111.8	112.9	94.4	95.7	95.3	121.8	89.4	99.5	108.0	92.0	96.9
0310	Concrete Forming & Accessories	115.3	93.8	97.0	132.9	79.1	87.0	125.3	85.2	91.1	111.0	88.9	92.1	112.0	52.5	61.3	111.8	67.8	74.3
0320	Concrete Reinforcing	97.2	103.2	100.1	147.4	72.4	111.1	172.1	82.4	128.7	98.0	101.4	99.7	154.0	46.5	102.0	156.0	47.6	103.6
0330	Cast-in-Place Concrete	106.7	98.7	103.7	104.0	87.5	97.9	137.1	98.5	122.8	107.6	95.3	103.0	116.7	54.7	93.6	94.8	66.8	84.4
03	CONCRETE	94.8	97.2	95.8	107.9	81.4	96.2	132.7	90.3	114.1	95.0	93.5	94.3	139.1	53.4	101.5	118.1	65.0	94.8
04	MASONRY	141.8	100.8	116.6	162.3	76.3	109.5	200.1	89.9	132.4	141.9	94.9	113.0	207.4	54.8	113.7	205.8	65.6	119.6
05	METALS	114.2	97.9	109.4	110.9	84.8	103.2	137.9	99.1	126.5	113.6	97.1	108.8	125.6	69.2	109.1	125.6	75.7	111.0
06	WOOD, PLASTICS & COMPOSITES	107.8	93.2	99.8	131.6	79.0	102.7	106.9	83.4	94.0	103.5	87.9	94.9	116.4	51.9	80.9	115.9	67.4	89.2
07	THERMAL & MOISTURE PROTECTION	116.4	102.0	110.1	114.8	84.1	101.5	140.2	94.7	120.4	115.7	96.3	107.3	129.4	57.3	98.0	130.1	67.6	102.9
08	OPENINGS	79.5	91.4	82.3	89.1	67.9	84.2	85.4	74.3	82.8	80.2	86.7	81.7	102.5	45.4	89.2	91.3	61.3	84.3
0920	Plaster & Gypsum Board	100.8	93.1	95.6	114.2	78.4	90.1	150.0	82.1	104.3	100.0	87.6	91.7	145.9	50.5	81.8	145.0	66.5	92.2
0950, 0980	Ceilings & Acoustic Treatment	103.9	93.1	96.6	102.7	78.4	86.3	143.7	82.1	102.2	98.2	87.6	91.0	123.8	50.5	74.4	123.8	66.5	85.1
0960	Flooring	100.2	91.9	97.8	113.6	86.0	105.8	116.8	52.2	98.7	98.3	92.1	96.6	107.6	55.3	92.9	107.6	58.9	93.9
0970, 0990	Wall Finishes & Painting/Coating	104.4	104.1	104.2	110.3	83.6	94.5	120.8	98.6	107.7	104.4	95.3	99.0	117.8	39.9	71.7	117.8	59.3	83.2
09	FINISHES	100.1	94.6	97.1	107.6	81.2	93.3	129.2	80.4	102.8	98.3	89.9	93.7	119.8	51.9	83.1	118.7	65.5	89.9
COVERS	DIVS. 10 - 14, 25, 28, 41, 43, 44, 46	131.1	69.7	117.4	131.1	75.6	118.8	131.1	68.6	117.2	131.1	89.6	121.9	131.1	57.0	114.6	131.1	60.8	115.5
21, 22, 23	FIRE SUPPRESSION, PLUMBING & HVAC	104.2	91.1	99.0	103.8	84.2	96.1	104.8	83.0	96.2	103.6	89.4	98.1	104.0	58.9	86.2	104.0	79.5	94.4
26, 27, 3370	ELECTRICAL, COMMUNICATIONS & UTIL.	111.3	101.8	106.6	111.6	65.2	88.7	115.6	81.4	98.7	109.5	102.1	105.9	110.2	47.4	79.2	111.3	58.7	85.4
MF2018	WEIGHTED AVERAGE	106.6	95.6	101.9	110.4	79.8	97.4	122.6	87.4	107.7	106.1	93.6	100.8	121.9	57.9	94.9	117.8	70.2	97.7

For customer support on your Commercial Renovation Costs with RSMeans data, call 800.448.8182.

829

City Cost Indexes

DIVISION		THUNDER BAY, ONTARIO			TIMMINS, ONTARIO			TORONTO, ONTARIO			TROIS RIVIERES, QUEBEC			TRURO, NOVA SCOTIA			VANCOUVER, BRITISH COLUMBIA		
		MAT.	INST.	TOTAL	MAT.	INST.	TOTAL	MAT.	INST.	TOTAL	MAT.	INST.	TOTAL	MAT.	INST.	TOTAL	MAT.	INST.	TOTAL
015433	CONTRACTOR EQUIPMENT		99.5	99.5		99.7	99.7		117.5	117.5		99.7	99.7		99.4	99.4		135.3	135.3
0241, 31 - 34	SITE & INFRASTRUCTURE, DEMOLITION	98.8	96.0	96.9	116.9	93.9	101.0	108.4	107.7	107.9	108.2	93.4	97.9	101.3	92.3	95.1	117.0	122.8	121.0
0310	Concrete Forming & Accessories	122.8	92.5	97.0	127.2	80.0	87.0	128.7	102.5	106.3	153.7	79.1	90.1	99.7	67.9	72.6	131.0	91.6	97.4
0320	Concrete Reinforcing	86.9	102.6	94.5	163.7	84.8	125.6	136.9	103.9	120.9	156.0	72.4	115.6	139.3	47.6	95.0	132.5	95.2	114.5
0330	Cast-in-Place Concrete	117.5	98.0	110.2	130.7	80.2	111.9	108.4	113.7	110.4	98.2	87.4	94.2	135.7	66.9	110.1	123.5	90.6	111.3
03	CONCRETE	101.9	96.2	99.4	121.8	81.5	104.1	110.5	107.1	109.0	120.3	81.4	103.2	120.8	65.1	96.3	119.7	93.0	108.0
04	MASONRY	142.4	100.5	116.7	166.1	81.9	114.4	173.3	108.7	133.6	209.8	76.3	127.8	161.9	65.6	102.7	165.3	85.6	116.3
05	METALS	114.2	97.0	109.1	112.3	90.4	105.9	135.3	109.4	127.7	124.9	84.7	113.1	111.6	76.0	101.1	133.5	109.8	126.6
06	WOOD, PLASTICS & COMPOSITES	117.6	91.3	103.1	120.3	80.3	98.3	114.8	100.5	107.0	168.2	79.0	119.1	89.6	67.4	77.4	110.3	91.0	99.7
07	THERMAL & MOISTURE PROTECTION	116.6	99.3	109.1	121.6	82.7	104.7	135.6	109.4	124.2	128.6	84.1	109.2	116.1	67.6	95.0	141.7	88.6	118.6
08	OPENINGS	78.9	89.8	81.4	87.7	78.2	85.5	83.9	98.9	87.4	100.3	72.3	93.7	82.4	61.3	77.5	84.5	88.7	85.5
0920	Plaster & Gypsum Board	128.0	91.2	103.2	115.8	79.8	91.6	129.1	99.9	109.5	174.3	78.4	109.8	124.3	66.5	85.4	127.2	89.7	102.0
0950, 0980	Ceilings & Acoustic Treatment	98.2	91.2	93.4	103.5	79.8	87.5	139.9	99.9	112.9	122.2	78.4	92.6	103.5	66.5	78.6	151.0	89.7	109.7
0960	Flooring	104.2	98.5	102.6	113.6	86.0	105.8	109.1	100.8	106.8	131.2	86.0	118.5	95.7	58.9	85.4	124.7	88.5	114.6
0970, 0990	Wall Finishes & Painting/Coating	104.4	96.2	99.5	110.3	85.0	95.3	109.8	108.1	108.8	117.8	83.6	97.6	110.3	59.3	80.1	113.7	100.3	105.8
09	FINISHES	104.1	93.9	98.6	108.7	81.9	94.2	117.8	102.8	109.7	129.3	81.1	103.2	103.8	65.5	83.1	128.0	92.1	108.6
COVERS	DIVS. 10 - 14, 25, 28, 41, 43, 44, 46	131.1	69.7	117.4	131.1	63.4	116.0	131.1	95.4	123.2	131.1	75.6	118.8	131.1	60.8	115.5	131.1	90.8	122.1
21, 22, 23	FIRE SUPPRESSION, PLUMBING & HVAC	104.2	91.2	99.1	103.8	92.4	99.3	105.1	99.6	103.0	104.0	84.2	96.2	103.8	79.5	94.3	105.2	87.4	98.2
26, 27, 3370	ELECTRICAL, COMMUNICATIONS & UTIL.	109.6	101.2	105.4	113.2	83.3	98.5	106.8	103.8	105.3	111.6	65.2	88.8	110.5	58.7	85.0	108.7	78.6	93.9
MF2018	WEIGHTED AVERAGE	107.8	95.0	102.4	113.3	85.3	101.4	116.0	104.2	111.0	120.2	79.9	103.2	110.8	70.3	93.7	117.9	92.4	107.2

CANADA

DIVISION		VICTORIA, BRITISH COLUMBIA			WHITEHORSE, YUKON			WINDSOR, ONTARIO			WINNIPEG, MANITOBA			YARMOUTH, NOVA SCOTIA			YELLOWKNIFE, NWT		
		MAT.	INST.	TOTAL	MAT.	INST.	TOTAL	MAT.	INST.	TOTAL	MAT.	INST.	TOTAL	MAT.	INST.	TOTAL	MAT.	INST.	TOTAL
015433	CONTRACTOR EQUIPMENT		106.3	106.3		132.8	132.8		99.5	99.5		125.5	125.5		98.9	98.9		130.7	130.7
0241, 31 - 34	SITE & INFRASTRUCTURE, DEMOLITION	124.3	101.3	108.4	135.7	119.6	124.6	90.8	96.0	94.4	113.7	115.3	114.8	111.7	92.0	98.1	146.3	123.6	130.6
0310	Concrete Forming & Accessories	114.9	88.9	92.7	135.0	56.7	68.2	122.8	90.6	95.3	130.4	65.4	75.0	111.8	67.8	74.3	137.3	75.8	84.8
0320	Concrete Reinforcing	111.7	95.0	103.6	167.7	63.3	117.2	95.2	101.9	98.5	127.3	61.0	95.3	156.0	47.6	103.6	138.6	65.8	103.4
0330	Cast-in-Place Concrete	117.6	87.0	106.2	152.4	71.2	122.2	109.3	99.7	105.7	143.6	72.0	116.9	121.6	66.8	101.2	174.9	87.6	142.4
03	CONCRETE	129.4	89.5	111.9	144.0	64.8	109.1	96.2	95.9	96.0	127.0	68.4	101.3	130.4	65.0	101.7	150.7	79.5	119.4
04	MASONRY	175.0	85.5	120.0	246.5	58.1	130.7	141.9	100.1	116.3	194.6	65.3	115.2	208.1	65.6	120.5	234.9	68.9	132.9
05	METALS	107.7	92.2	103.2	145.1	90.5	129.1	114.2	97.3	109.2	141.6	88.0	125.8	125.6	75.7	111.0	140.5	92.7	126.4
06	WOOD, PLASTICS & COMPOSITES	100.0	88.3	93.6	121.3	55.3	85.0	117.6	89.0	101.9	107.7	66.0	84.8	115.9	67.4	89.2	130.1	77.1	100.9
07	THERMAL & MOISTURE PROTECTION	126.9	86.3	109.2	145.8	63.3	109.9	116.4	99.1	108.9	136.1	70.3	107.5	130.1	67.6	102.9	143.7	78.8	115.5
08	OPENINGS	86.8	82.9	85.9	99.5	53.0	88.6	78.7	88.9	81.1	87.5	59.7	81.0	91.3	61.3	84.3	93.9	65.5	87.3
0920	Plaster & Gypsum Board	109.1	87.7	94.7	178.3	53.0	94.0	111.4	88.8	96.2	133.0	64.2	86.7	145.0	66.5	92.2	192.2	75.5	113.7
0950, 0980	Ceilings & Acoustic Treatment	105.8	87.7	93.6	169.3	53.0	90.9	98.2	88.8	91.8	145.3	64.2	90.6	123.8	66.5	85.2	169.0	75.5	105.9
0960	Flooring	112.0	67.4	99.5	122.4	55.2	103.5	104.2	95.9	101.9	115.9	68.5	102.6	107.6	58.9	93.9	121.9	82.9	111.0
0970, 0990	Wall Finishes & Painting/Coating	113.8	100.3	105.8	120.7	53.1	80.7	104.4	97.1	100.1	113.9	53.5	78.1	117.8	59.3	83.2	123.3	75.9	95.2
09	FINISHES	110.7	86.8	97.8	143.7	55.8	96.1	101.6	92.1	96.5	126.9	65.0	93.4	118.7	65.5	89.9	147.5	76.6	109.1
COVERS	DIVS. 10 - 14, 25, 28, 41, 43, 44, 46	131.1	65.8	116.6	131.1	60.6	115.4	131.1	69.2	117.3	131.1	63.8	116.1	131.1	60.8	115.5	131.1	63.9	116.1
21, 22, 23	FIRE SUPPRESSION, PLUMBING & HVAC	103.9	88.7	97.9	104.3	71.6	91.5	104.2	90.7	98.9	105.2	64.6	89.2	104.0	79.5	94.4	104.8	88.5	98.4
26, 27, 3370	ELECTRICAL, COMMUNICATIONS & UTIL.	112.8	82.1	97.7	130.5	57.1	94.3	114.5	100.4	107.6	113.1	63.4	88.6	111.3	58.7	85.4	125.8	77.4	102.0
MF2018	WEIGHTED AVERAGE	114.1	87.6	102.9	132.2	69.2	105.6	107.1	94.4	101.8	122.0	71.2	100.5	119.5	70.2	98.7	131.4	83.2	111.0

For customer support on your Commercial Renovation Costs with RSMeans data, call 800.448.8182.

Costs shown in RSMeans cost data publications are based on national averages for materials and installation. To adjust these costs to a specific location, simply multiply the base cost by the factor and divide by 100 for that city. The data is arranged alphabetically by state and postal zip code numbers. For a city not listed, use the factor for a nearby city with similar economic characteristics.

STATE/ZIP	CITY	MAT.	INST.	TOTAL
ALABAMA				
350-352	Birmingham	97.4	71.8	86.6
354	Tuscaloosa	97.1	71.4	86.3
355	Jasper	97.8	69.8	86.0
356	Decatur	97.1	69.2	85.3
357-358	Huntsville	97.1	71.7	86.4
359	Gadsden	97.3	70.2	85.9
360-361	Montgomery	96.3	73.1	86.5
362	Anniston	95.9	67.7	84.0
363	Dothan	96.2	72.7	86.3
364	Evergreen	95.8	68.3	84.2
365-366	Mobile	96.7	68.7	84.9
367	Selma	95.9	72.5	86.0
368	Phenix City	96.8	70.8	85.8
369	Butler	96.1	70.6	85.3
ALASKA				
995-996	Anchorage	118.5	113.1	116.2
997	Fairbanks	118.8	113.0	116.3
998	Juneau	116.8	113.1	115.2
999	Ketchikan	128.4	113.4	122.1
ARIZONA				
850,853	Phoenix	100.6	72.3	88.6
851,852	Mesa/Tempe	98.7	71.5	87.2
855	Globe	99.5	71.4	87.6
856-857	Tucson	97.7	70.2	86.1
859	Show Low	99.7	71.5	87.8
860	Flagstaff	102.4	70.5	88.9
863	Prescott	99.9	71.3	87.8
864	Kingman	98.3	70.4	86.5
865	Chambers	98.3	71.8	87.1
ARKANSAS				
716	Pine Bluff	96.0	63.6	82.3
717	Camden	94.0	64.7	81.6
718	Texarkana	94.8	63.3	81.5
719	Hot Springs	93.5	64.4	81.2
720-722	Little Rock	94.5	65.4	82.2
723	West Memphis	93.3	69.8	83.3
724	Jonesboro	93.8	66.6	82.3
725	Batesville	91.6	63.6	79.8
726	Harrison	92.9	62.8	80.2
727	Fayetteville	90.5	64.1	79.3
728	Russellville	91.7	63.1	79.6
729	Fort Smith	94.1	62.6	80.8
CALIFORNIA				
900-902	Los Angeles	98.5	129.8	111.7
903-905	Inglewood	93.8	128.8	108.6
906-908	Long Beach	95.4	128.8	109.5
910-912	Pasadena	94.8	128.6	109.1
913-916	Van Nuys	97.8	128.6	110.8
917-918	Alhambra	96.7	128.6	110.2
919-921	San Diego	99.7	121.7	109.0
922	Palm Springs	97.4	125.3	109.2
923-924	San Bernardino	95.0	126.3	108.2
925	Riverside	99.3	126.6	110.8
926-927	Santa Ana	97.0	126.3	109.4
928	Anaheim	99.3	126.6	110.9
930	Oxnard	98.0	126.7	110.1
931	Santa Barbara	97.2	126.1	109.4
932-933	Bakersfield	98.9	124.6	109.8
934	San Luis Obispo	98.2	125.8	109.9
935	Mojave	95.4	124.4	107.7
936-938	Fresno	98.5	129.4	111.5
939	Salinas	99.1	137.2	115.2
940-941	San Francisco	107.4	158.4	129.0
942,956-958	Sacramento	100.8	133.0	114.4
943	Palo Alto	99.2	153.8	122.3
944	San Mateo	101.5	152.0	122.9
945	Vallejo	100.2	141.6	117.7
946	Oakland	103.4	152.0	123.9
947	Berkeley	103.0	151.9	123.7
948	Richmond	102.4	146.2	120.9
949	San Rafael	104.5	149.1	123.4
950	Santa Cruz	104.6	137.5	118.5

STATE/ZIP	CITY	MAT.	INST.	TOTAL
CALIFORNIA (CONT'D)				
951	San Jose	102.6	153.5	124.2
952	Stockton	100.7	131.3	113.7
953	Modesto	100.6	130.6	113.3
954	Santa Rosa	101.2	148.4	121.2
955	Eureka	102.5	134.1	115.9
959	Marysville	101.7	131.3	114.2
960	Redding	107.6	131.5	117.7
961	Susanville	107.3	131.6	117.5
COLORADO				
800-802	Denver	103.0	75.4	91.3
803	Boulder	98.7	77.0	89.6
804	Golden	100.9	74.7	89.8
805	Fort Collins	102.2	73.5	90.1
806	Greeley	99.3	73.4	88.4
807	Fort Morgan	99.3	73.5	88.4
808-809	Colorado Springs	101.2	71.5	88.6
810	Pueblo	101.8	70.0	88.3
811	Alamosa	103.6	70.3	89.5
812	Salida	103.1	70.9	89.5
813	Durango	103.9	65.6	87.8
814	Montrose	102.6	65.8	87.1
815	Grand Junction	105.7	70.5	90.8
816	Glenwood Springs	103.9	66.8	88.2
CONNECTICUT				
060	New Britain	98.1	116.7	106.0
061	Hartford	99.6	118.0	107.3
062	Willimantic	98.7	117.1	106.5
063	New London	95.2	117.2	104.5
064	Meriden	97.2	117.0	105.6
065	New Haven	99.9	117.0	107.1
066	Bridgeport	99.3	117.0	106.8
067	Waterbury	98.9	117.2	106.6
068	Norwalk	98.8	116.7	106.4
069	Stamford	99.0	123.2	109.2
D.C.				
200-205	Washington	101.1	87.5	95.4
DELAWARE				
197	Newark	98.3	109.7	103.1
198	Wilmington	97.9	109.7	102.9
199	Dover	99.0	109.6	103.5
FLORIDA				
320,322	Jacksonville	96.9	66.6	84.1
321	Daytona Beach	97.2	69.5	85.5
323	Tallahassee	98.2	67.7	85.3
324	Panama City	98.6	68.3	85.8
325	Pensacola	101.2	65.8	86.2
326,344	Gainesville	98.8	65.8	84.8
327-328,347	Orlando	99.0	66.6	85.3
329	Melbourne	100.1	70.3	87.5
330-332,340	Miami	96.7	68.0	84.6
333	Fort Lauderdale	96.1	66.4	83.6
334,349	West Palm Beach	95.1	64.1	82.0
335-336,346	Tampa	97.6	68.4	85.3
337	St. Petersburg	99.7	66.8	85.8
338	Lakeland	97.1	67.6	84.6
339,341	Fort Myers	96.5	67.7	84.3
342	Sarasota	98.8	67.0	85.4
GEORGIA				
300-303,399	Atlanta	98.2	75.0	88.4
304	Statesboro	97.8	69.7	85.9
305	Gainesville	96.2	65.6	83.2
306	Athens	95.6	66.9	83.5
307	Dalton	97.5	71.4	86.5
308-309	Augusta	96.0	73.9	86.6
310-312	Macon	95.2	73.1	85.8
313-314	Savannah	97.3	72.7	86.9
315	Waycross	96.8	70.0	85.5
316	Valdosta	96.7	66.2	83.8
317,398	Albany	96.5	71.9	86.1
318-319	Columbus	96.5	72.2	86.2

For customer support on your Commercial Renovation Costs with RSMeans data, call 800.448.8182.

831

Location Factors - Commercial

STATE/ZIP	CITY	MAT.	INST.	TOTAL
HAWAII				
967	Hilo	114.8	118.4	116.3
968	Honolulu	119.1	118.5	118.9
STATES & POSS.				
969	Guam	137.1	51.4	100.9
IDAHO				
832	Pocatello	102.0	78.4	92.0
833	Twin Falls	103.2	76.8	92.1
834	Idaho Falls	100.5	79.9	91.8
835	Lewiston	108.2	83.5	97.8
836-837	Boise	101.0	79.9	92.1
838	Coeur d'Alene	108.1	84.4	98.1
ILLINOIS				
600-603	North Suburban	98.3	142.9	117.1
604	Joliet	98.3	138.8	115.4
605	South Suburban	98.3	142.8	117.1
606-608	Chicago	100.1	145.6	119.3
609	Kankakee	95.0	134.9	111.9
610-611	Rockford	96.6	125.5	108.8
612	Rock Island	94.6	99.5	96.7
613	La Salle	95.9	127.0	109.0
614	Galesburg	95.7	108.5	101.1
615-616	Peoria	97.6	108.8	102.3
617	Bloomington	95.0	109.5	101.1
618-619	Champaign	98.5	110.5	103.6
620-622	East St. Louis	93.8	110.4	100.9
623	Quincy	95.7	104.2	99.3
624	Effingham	95.1	109.1	101.0
625	Decatur	96.6	110.1	102.3
626-627	Springfield	97.5	108.3	102.1
628	Centralia	92.8	111.2	100.5
629	Carbondale	92.5	109.6	99.7
INDIANA				
460	Anderson	96.0	79.8	89.1
461-462	Indianapolis	98.7	83.2	92.1
463-464	Gary	97.5	108.1	102.0
465-466	South Bend	98.4	81.8	91.4
467-468	Fort Wayne	96.6	76.3	88.0
469	Kokomo	94.2	81.1	88.6
470	Lawrenceburg	92.5	79.6	87.0
471	New Albany	93.9	77.3	86.9
472	Columbus	96.1	80.5	89.5
473	Muncie	96.2	78.2	88.6
474	Bloomington	97.8	81.1	90.7
475	Washington	94.6	84.8	90.4
476-477	Evansville	95.5	82.7	90.1
478	Terre Haute	96.2	81.7	90.1
479	Lafayette	95.8	78.1	88.3
IOWA				
500-503,509	Des Moines	97.2	90.2	94.3
504	Mason City	95.3	74.0	86.3
505	Fort Dodge	95.5	73.7	86.3
506-507	Waterloo	96.8	77.4	88.6
508	Creston	95.7	84.2	90.8
510-511	Sioux City	97.7	81.5	90.9
512	Sibley	96.8	62.4	82.3
513	Spencer	98.2	62.3	83.0
514	Carroll	95.5	82.4	89.9
515	Council Bluffs	98.9	79.2	90.6
516	Shenandoah	96.1	81.9	90.1
520	Dubuque	97.2	79.9	89.9
521	Decorah	96.4	73.7	86.8
522-524	Cedar Rapids	98.0	83.7	92.0
525	Ottumwa	96.3	78.7	88.9
526	Burlington	95.7	84.3	90.9
527-528	Davenport	97.1	94.5	96.0
KANSAS				
660-662	Kansas City	97.3	98.6	97.8
664-666	Topeka	98.6	76.5	89.3
667	Fort Scott	96.0	77.1	88.0
668	Emporia	96.1	76.5	87.8
669	Belleville	97.9	70.7	86.4
670-672	Wichita	97.0	72.6	86.7
673	Independence	97.2	76.0	88.2
674	Salina	97.2	73.5	87.2
675	Hutchinson	92.5	71.9	83.8
676	Hays	96.5	72.0	86.1
677	Colby	97.3	74.0	87.4

STATE/ZIP	CITY	MAT.	INST.	TOTAL
KANSAS (CONT'D)				
678	Dodge City	98.6	73.3	87.9
679	Liberal	96.5	71.7	86.0
KENTUCKY				
400-402	Louisville	93.9	79.6	87.8
403-405	Lexington	93.1	78.1	86.7
406	Frankfort	95.4	80.3	89.0
407-409	Corbin	90.8	78.0	85.4
410	Covington	94.3	77.8	87.3
411-412	Ashland	93.2	89.7	91.7
413-414	Campton	94.4	78.0	87.5
415-416	Pikeville	95.8	85.4	91.4
417-418	Hazard	93.8	78.7	87.4
420	Paducah	92.4	83.2	88.6
421-422	Bowling Green	94.6	77.2	87.3
423	Owensboro	94.7	81.0	88.9
424	Henderson	92.2	81.9	87.8
425-426	Somerset	91.6	79.5	86.5
427	Elizabethtown	91.3	75.6	84.7
LOUISIANA				
700-701	New Orleans	99.0	68.5	86.1
703	Thibodaux	96.0	67.0	83.8
704	Hammond	93.7	64.4	81.3
705	Lafayette	95.4	67.9	83.8
706	Lake Charles	95.6	68.2	84.1
707-708	Baton Rouge	96.6	69.3	85.1
710-711	Shreveport	97.1	66.3	84.1
712	Monroe	95.8	64.4	82.5
713-714	Alexandria	95.9	65.6	83.1
MAINE				
039	Kittery	92.6	87.0	90.2
040-041	Portland	99.6	85.4	93.6
042	Lewiston	97.5	84.9	92.2
043	Augusta	100.4	85.6	94.2
044	Bangor	97.1	83.9	91.5
045	Bath	95.9	85.4	91.4
046	Machias	95.4	85.3	91.1
047	Houlton	95.6	85.2	91.2
048	Rockland	94.6	85.4	90.7
049	Waterville	96.0	85.2	91.4
MARYLAND				
206	Waldorf	97.3	86.6	92.8
207-208	College Park	97.3	87.7	93.2
209	Silver Spring	96.5	86.7	92.4
210-212	Baltimore	101.8	83.2	93.9
214	Annapolis	100.4	86.3	94.4
215	Cumberland	96.9	83.8	91.4
216	Easton	98.6	74.8	88.5
217	Hagerstown	97.4	85.3	92.3
218	Salisbury	99.0	66.9	85.4
219	Elkton	95.9	86.2	91.8
MASSACHUSETTS				
010-011	Springfield	98.2	108.5	102.5
012	Pittsfield	97.7	102.6	99.7
013	Greenfield	95.9	109.0	101.4
014	Fitchburg	94.6	116.2	103.7
015-016	Worcester	98.1	117.0	106.1
017	Framingham	94.0	124.2	106.8
018	Lowell	97.5	123.8	108.6
019	Lawrence	98.4	124.2	109.3
020-022, 024	Boston	101.9	134.0	115.5
023	Brockton	98.5	116.7	106.2
025	Buzzards Bay	93.2	115.2	102.5
026	Hyannis	95.7	115.0	103.8
027	New Bedford	97.7	115.2	105.1
MICHIGAN				
480,483	Royal Oak	95.1	100.3	97.3
481	Ann Arbor	97.1	99.8	98.2
482	Detroit	100.7	102.0	101.3
484-485	Flint	96.8	88.6	93.3
486	Saginaw	96.5	85.1	91.7
487	Bay City	96.6	85.3	91.8
488-489	Lansing	98.0	86.8	93.3
490	Battle Creek	95.5	80.4	89.1
491	Kalamazoo	95.8	78.6	88.6
492	Jackson	94.0	90.7	92.6
493,495	Grand Rapids	97.9	79.9	90.3
494	Muskegon	94.3	80.0	88.3

For customer support on your Commercial Renovation Costs with RSMeans data, call 800.448.8182.

STATE/ZIP	CITY	MAT.	INST.	TOTAL
MICHIGAN (CONT'D)				
496	Traverse City	93.4	77.7	86.8
497	Gaylord	94.5	80.6	88.6
498-499	Iron Mountain	96.5	81.7	90.2
MINNESOTA				
550-551	Saint Paul	99.0	117.6	106.9
553-555	Minneapolis	100.7	114.1	106.3
556-558	Duluth	98.6	101.3	99.7
559	Rochester	98.3	98.7	98.5
560	Mankato	95.7	98.5	96.9
561	Windom	94.3	91.2	93.0
562	Willmar	94.0	98.0	95.7
563	St. Cloud	95.1	114.2	103.2
564	Brainerd	95.6	98.6	96.8
565	Detroit Lakes	97.4	92.4	95.3
566	Bemidji	96.7	95.4	96.1
567	Thief River Falls	96.3	89.3	93.3
MISSISSIPPI				
386	Clarksdale	95.3	55.2	78.4
387	Greenville	98.7	65.5	84.7
388	Tupelo	96.6	59.0	80.7
389	Greenwood	96.6	55.5	79.2
390-392	Jackson	97.3	65.6	83.9
393	Meridian	94.6	64.5	81.9
394	Laurel	96.1	57.8	79.9
395	Biloxi	96.2	63.9	82.6
396	McComb	94.4	55.8	78.0
397	Columbus	95.9	59.1	80.4
MISSOURI				
630-631	St. Louis	98.8	104.5	101.2
633	Bowling Green	96.4	95.5	96.0
634	Hannibal	95.3	91.6	93.7
635	Kirksville	98.9	86.5	93.7
636	Flat River	97.3	94.2	96.0
637	Cape Girardeau	97.0	91.2	94.5
638	Sikeston	95.7	87.8	92.3
639	Poplar Bluff	95.1	87.7	92.0
640-641	Kansas City	97.9	103.8	100.4
644-645	St. Joseph	96.1	90.9	93.9
646	Chillicothe	93.7	93.8	93.7
647	Harrisonville	93.2	100.8	96.4
648	Joplin	95.1	77.7	87.7
650-651	Jefferson City	96.4	89.9	93.7
652	Columbia	96.1	90.7	93.8
653	Sedalia	96.4	91.5	94.3
654-655	Rolla	94.1	94.3	94.2
656-658	Springfield	97.6	79.0	89.7
MONTANA				
590-591	Billings	101.6	78.4	91.8
592	Wolf Point	101.4	76.9	91.1
593	Miles City	99.2	77.0	89.8
594	Great Falls	102.9	76.6	91.8
595	Havre	100.4	74.8	89.6
596	Helena	100.9	76.7	90.7
597	Butte	101.5	76.5	90.9
598	Missoula	98.6	77.1	89.5
599	Kalispell	98.1	76.2	88.9
NEBRASKA				
680-681	Omaha	97.6	81.5	90.8
683-685	Lincoln	98.3	79.7	90.5
686	Columbus	96.0	80.7	89.6
687	Norfolk	97.4	77.5	89.0
688	Grand Island	97.4	76.8	88.7
689	Hastings	97.1	78.2	89.1
690	McCook	96.3	73.1	86.5
691	North Platte	96.3	75.6	87.6
692	Valentine	98.6	69.2	86.2
693	Alliance	98.7	73.3	87.9
NEVADA				
889-891	Las Vegas	105.0	105.1	105.0
893	Ely	103.7	93.2	99.2
894-895	Reno	103.6	82.6	94.7
897	Carson City	102.9	82.5	94.3
898	Elko	102.3	85.2	95.0
NEW HAMPSHIRE				
030	Nashua	97.5	93.7	95.9
031	Manchester	98.2	94.5	96.6

STATE/ZIP	CITY	MAT.	INST.	TOTAL
NEW HAMPSHIRE (CONT'D)				
032-033	Concord	98.0	93.2	96.0
034	Keene	94.5	89.0	92.2
035	Littleton	94.4	80.2	88.4
036	Charleston	94.1	88.4	91.7
037	Claremont	93.1	88.4	91.1
038	Portsmouth	94.9	92.4	93.8
NEW JERSEY				
070-071	Newark	100.3	139.7	117.0
072	Elizabeth	97.5	138.5	114.7
073	Jersey City	96.5	137.8	114.0
074-075	Paterson	98.0	137.7	114.8
076	Hackensack	96.1	138.3	114.0
077	Long Branch	95.8	130.9	110.7
078	Dover	96.4	137.5	113.7
079	Summit	96.4	138.3	114.1
080,083	Vineland	96.3	129.8	110.5
081	Camden	98.0	127.8	110.6
082,084	Atlantic City	96.8	133.2	112.2
085-086	Trenton	99.7	128.9	112.0
087	Point Pleasant	98.1	130.8	111.9
088-089	New Brunswick	98.7	135.6	114.3
NEW MEXICO				
870-872	Albuquerque	98.6	72.5	87.6
873	Gallup	99.0	72.5	87.8
874	Farmington	99.0	72.5	87.8
875	Santa Fe	99.0	72.9	88.0
877	Las Vegas	97.2	72.5	86.8
878	Socorro	96.8	72.5	86.6
879	Truth/Consequences	96.6	69.4	85.1
880	Las Cruces	96.9	71.4	86.1
881	Clovis	99.3	72.4	87.9
882	Roswell	100.8	72.5	88.9
883	Carrizozo	101.6	72.5	89.3
884	Tucumcari	99.9	72.4	88.3
NEW YORK				
100-102	New York	99.7	175.2	131.6
103	Staten Island	95.6	176.1	129.6
104	Bronx	93.8	175.3	128.3
105	Mount Vernon	94.1	150.6	118.0
106	White Plains	93.8	152.8	118.8
107	Yonkers	97.9	153.0	121.2
108	New Rochelle	94.4	144.8	115.7
109	Suffern	94.1	130.2	109.4
110	Queens	100.0	177.3	132.6
111	Long Island City	101.6	177.3	133.6
112	Brooklyn	101.9	177.3	133.8
113	Flushing	102.1	177.3	133.9
114	Jamaica	100.3	177.3	132.8
115,117,118	Hicksville	99.9	157.1	124.1
116	Far Rockaway	102.2	177.3	133.9
119	Riverhead	100.6	153.5	122.9
120-122	Albany	96.8	111.2	102.8
123	Schenectady	97.0	109.9	102.4
124	Kingston	100.6	133.8	114.6
125-126	Poughkeepsie	99.8	136.8	115.4
127	Monticello	99.1	134.6	114.1
128	Glens Falls	92.0	107.3	98.5
129	Plattsburgh	97.0	96.6	96.8
130-132	Syracuse	97.8	98.3	98.0
133-135	Utica	96.0	96.6	96.2
136	Watertown	97.7	96.8	97.4
137-139	Binghamton	97.3	101.5	99.1
140-142	Buffalo	100.8	110.3	104.8
143	Niagara Falls	97.4	107.5	101.6
144-146	Rochester	100.3	99.2	99.8
147	Jamestown	96.4	96.0	96.3
148-149	Elmira	96.3	102.5	98.9
NORTH CAROLINA				
270,272-274	Greensboro	98.5	67.2	85.3
271	Winston-Salem	98.2	67.3	85.2
275-276	Raleigh	97.4	67.0	84.5
277	Durham	100.1	67.2	86.2
278	Rocky Mount	96.0	66.6	83.6
279	Elizabeth City	96.9	68.3	84.8
280	Gastonia	97.0	66.8	84.2
281-282	Charlotte	97.4	67.6	84.8
283	Fayetteville	100.2	66.1	85.8
284	Wilmington	96.0	65.2	83.0
285	Kinston	94.4	65.5	82.2

STATE/ZIP	CITY	MAT.	INST.	TOTAL
NORTH CAROLINA (CONT'D)				
286	Hickory	94.7	67.3	83.1
287-288	Asheville	96.4	66.6	83.8
289	Murphy	95.5	64.7	82.5
NORTH DAKOTA				
580-581	Fargo	99.7	80.3	91.5
582	Grand Forks	99.4	77.6	90.2
583	Devils Lake	99.2	78.8	90.6
584	Jamestown	99.2	78.8	90.6
585	Bismarck	98.9	81.6	91.6
586	Dickinson	99.9	77.9	90.6
587	Minot	99.3	78.6	90.5
588	Williston	98.4	77.7	89.6
OHIO				
430-432	Columbus	98.4	82.0	91.4
433	Marion	94.3	86.7	91.1
434-436	Toledo	97.6	91.2	94.9
437-438	Zanesville	94.9	85.7	91.0
439	Steubenville	96.1	90.0	93.5
440	Lorain	98.4	84.7	92.6
441	Cleveland	99.1	91.6	95.9
442-443	Akron	99.4	86.0	93.8
444-445	Youngstown	98.7	81.9	91.6
446-447	Canton	98.9	79.9	90.9
448-449	Mansfield	96.4	84.2	91.3
450	Hamilton	94.8	79.5	88.3
451-452	Cincinnati	96.1	80.0	89.3
453-454	Dayton	94.8	79.2	88.2
455	Springfield	94.8	79.7	88.4
456	Chillicothe	94.3	88.5	91.8
457	Athens	97.1	84.8	91.9
458	Lima	97.2	81.7	90.7
OKLAHOMA				
730-731	Oklahoma City	96.4	68.7	84.7
734	Ardmore	95.2	65.6	82.7
735	Lawton	97.3	65.8	84.0
736	Clinton	96.4	65.5	83.3
737	Enid	97.0	66.0	83.9
738	Woodward	95.3	62.9	81.6
739	Guymon	96.3	63.3	82.4
740-741	Tulsa	95.9	65.8	83.2
743	Miami	92.7	65.1	81.1
744	Muskogee	95.0	62.9	81.4
745	McAlester	92.4	60.5	78.9
746	Ponca City	93.1	64.0	80.8
747	Durant	93.1	65.1	81.3
748	Shawnee	94.5	65.2	82.2
749	Poteau	92.3	64.9	80.7
OREGON				
970-972	Portland	102.1	102.8	102.4
973	Salem	104.4	100.8	102.9
974	Eugene	101.7	98.7	100.5
975	Medford	103.3	97.4	100.8
976	Klamath Falls	103.5	97.3	100.9
977	Bend	102.6	100.2	101.6
978	Pendleton	98.6	100.3	99.3
979	Vale	96.2	87.0	92.3
PENNSYLVANIA				
150-152	Pittsburgh	100.7	102.7	101.5
153	Washington	97.5	100.2	98.6
154	Uniontown	97.8	99.8	98.6
155	Bedford	98.8	92.1	96.0
156	Greensburg	98.7	96.3	97.7
157	Indiana	97.6	97.7	97.7
158	Dubois	99.3	96.0	97.9
159	Johnstown	98.9	91.9	95.9
160	Butler	91.3	100.6	95.2
161	New Castle	91.4	97.8	94.1
162	Kittanning	91.8	97.9	94.4
163	Oil City	91.3	97.8	94.0
164-165	Erie	93.2	93.9	93.5
166	Altoona	93.3	92.8	93.1
167	Bradford	94.9	97.1	95.8
168	State College	94.5	94.6	94.5
169	Wellsboro	95.5	92.9	94.4
170-171	Harrisburg	98.7	93.0	96.3
172	Chambersburg	95.1	89.4	92.7
173-174	York	95.2	93.0	94.3
175-176	Lancaster	93.8	94.5	94.1

STATE/ZIP	CITY	MAT.	INST.	TOTAL
PENNSYLVANIA (CONT'D)				
177	Williamsport	92.5	92.3	92.4
178	Sunbury	94.6	92.7	93.8
179	Pottsville	93.7	96.1	94.7
180	Lehigh Valley	94.9	111.6	102.0
181	Allentown	96.7	105.8	100.6
182	Hazleton	94.4	95.9	95.0
183	Stroudsburg	94.3	104.7	98.7
184-185	Scranton	97.4	96.5	97.1
186-187	Wilkes-Barre	94.0	95.5	94.7
188	Montrose	93.8	96.8	95.1
189	Doylestown	94.0	127.1	108.0
190-191	Philadelphia	100.2	138.7	116.5
193	Westchester	95.8	127.9	109.3
194	Norristown	94.7	128.0	108.8
195-196	Reading	96.5	101.5	98.6
PUERTO RICO				
009	San Juan	98.5	29.8	69.4
RHODE ISLAND				
028	Newport	97.0	114.7	104.5
029	Providence	99.3	115.0	105.9
SOUTH CAROLINA				
290-292	Columbia	97.6	68.9	85.5
293	Spartanburg	97.2	68.3	85.0
294	Charleston	98.7	68.2	85.8
295	Florence	96.9	68.6	84.9
296	Greenville	97.0	68.2	84.8
297	Rock Hill	96.7	66.4	83.9
298	Aiken	97.6	67.1	84.7
299	Beaufort	98.3	57.5	81.1
SOUTH DAKOTA				
570-571	Sioux Falls	97.7	78.4	89.5
572	Watertown	97.3	69.0	85.4
573	Mitchell	96.3	60.9	81.3
574	Aberdeen	98.6	70.1	86.5
575	Pierre	99.0	68.5	86.1
576	Mobridge	96.9	62.9	82.5
577	Rapid City	98.3	72.0	87.2
TENNESSEE				
370-372	Nashville	98.0	71.8	86.9
373-374	Chattanooga	98.0	68.1	85.3
375,380-381	Memphis	96.6	69.8	85.3
376	Johnson City	98.1	58.6	81.4
377-379	Knoxville	94.8	63.7	81.6
382	McKenzie	96.3	55.1	78.9
383	Jackson	97.6	61.4	82.3
384	Columbia	94.8	66.5	82.8
385	Cookeville	96.2	56.9	79.6
TEXAS				
750	McKinney	97.2	63.5	82.9
751	Waxahachie	97.2	63.5	83.0
752-753	Dallas	98.2	68.0	85.4
754	Greenville	97.4	62.9	82.8
755	Texarkana	96.7	62.0	82.1
756	Longview	97.5	61.0	82.1
757	Tyler	97.8	61.6	82.5
758	Palestine	94.2	60.4	79.9
759	Lufkin	94.7	62.6	81.1
760-761	Fort Worth	97.6	62.9	82.9
762	Denton	97.3	62.9	82.7
763	Wichita Falls	95.0	61.6	80.9
764	Eastland	94.2	60.3	79.8
765	Temple	92.5	58.5	78.2
766-767	Waco	94.4	62.4	80.9
768	Brownwood	97.4	58.7	81.1
769	San Angelo	97.1	59.4	81.2
770-772	Houston	100.4	68.2	86.8
773	Huntsville	98.7	62.8	83.6
774	Wharton	99.8	64.3	84.8
775	Galveston	97.7	66.5	84.5
776-777	Beaumont	97.9	65.5	84.2
778	Bryan	95.2	64.9	82.4
779	Victoria	99.7	62.9	84.2
780	Laredo	96.4	61.1	81.5
781-782	San Antonio	98.3	62.3	83.1
783-784	Corpus Christi	98.9	61.4	83.1
785	McAllen	99.7	56.5	81.4
786-787	Austin	97.2	61.4	82.1

STATE/ZIP	CITY	MAT.	INST.	TOTAL
TEXAS (CONT'D)				
788	Del Rio	99.7	60.4	83.1
789	Giddings	96.1	60.4	81.0
790-791	Amarillo	98.2	60.0	82.1
792	Childress	97.6	60.3	81.8
793-794	Lubbock	99.2	62.9	83.9
795-796	Abilene	97.8	61.4	82.4
797	Midland	99.7	62.2	83.8
798-799,885	El Paso	95.9	63.6	82.3
UTAH				
840-841	Salt Lake City	103.5	71.7	90.1
842,844	Ogden	98.8	71.7	87.4
843	Logan	100.9	71.7	88.5
845	Price	101.4	70.0	88.1
846-847	Provo	101.2	71.2	88.5
VERMONT				
050	White River Jct.	98.2	79.8	90.4
051	Bellows Falls	96.7	92.1	94.8
052	Bennington	97.0	88.6	93.4
053	Brattleboro	97.4	92.1	95.2
054	Burlington	101.9	79.8	92.5
056	Montpelier	100.2	84.4	93.5
057	Rutland	98.9	79.4	90.7
058	St. Johnsbury	98.1	79.3	90.2
059	Guildhall	96.8	79.0	89.3
VIRGINIA				
220-221	Fairfax	99.5	83.6	92.8
222	Arlington	100.4	84.2	93.6
223	Alexandria	99.6	84.8	93.3
224-225	Fredericksburg	98.3	78.7	90.0
226	Winchester	98.9	79.5	90.7
227	Culpeper	98.7	83.0	92.1
228	Harrisonburg	99.0	68.8	86.3
229	Charlottesville	99.4	71.0	87.4
230-232	Richmond	99.2	72.8	88.1
233-235	Norfolk	99.9	68.7	86.7
236	Newport News	98.8	68.3	85.9
237	Portsmouth	98.3	67.8	85.4
238	Petersburg	98.7	72.6	87.7
239	Farmville	97.9	65.3	84.1
240-241	Roanoke	100.1	73.6	88.9
242	Bristol	98.2	59.3	81.8
243	Pulaski	97.8	67.1	84.8
244	Staunton	98.6	67.9	85.6
245	Lynchburg	98.7	73.8	88.2
246	Grundy	98.1	58.8	81.5
WASHINGTON				
980-981,987	Seattle	106.3	112.5	108.9
982	Everett	105.1	104.0	104.6
983-984	Tacoma	105.4	101.3	103.7
985	Olympia	104.0	101.8	103.1
986	Vancouver	106.6	102.8	105.0
988	Wenatchee	105.8	88.7	98.5
989	Yakima	105.6	95.4	101.3
990-992	Spokane	100.1	82.4	92.6
993	Richland	99.8	90.0	95.6
994	Clarkston	99.1	81.9	91.8
WEST VIRGINIA				
247-248	Bluefield	97.0	88.1	93.2
249	Lewisburg	98.7	86.1	93.4
250-253	Charleston	96.5	89.4	93.5
254	Martinsburg	96.7	84.1	91.4
255-257	Huntington	97.8	92.2	95.4
258-259	Beckley	95.2	87.7	92.0
260	Wheeling	99.6	89.6	95.4
261	Parkersburg	98.6	88.6	94.4
262	Buckhannon	98.3	88.8	94.3
263-264	Clarksburg	98.9	89.5	94.9
265	Morgantown	98.8	89.6	94.9
266	Gassaway	98.3	89.0	94.3
267	Romney	98.3	87.2	93.6
268	Petersburg	98.0	85.9	92.9
WISCONSIN				
530,532	Milwaukee	99.2	107.9	102.9
531	Kenosha	99.4	102.0	100.5
534	Racine	98.7	102.3	100.2
535	Beloit	98.6	98.9	98.7
537	Madison	99.7	98.1	99.0

STATE/ZIP	CITY	MAT.	INST.	TOTAL
WISCONSIN (CONT'D)				
538	Lancaster	96.7	94.3	95.7
539	Portage	95.2	98.0	96.4
540	New Richmond	95.4	94.6	95.0
541-543	Green Bay	99.4	93.4	96.9
544	Wausau	94.7	93.4	94.2
545	Rhinelander	97.9	92.2	95.5
546	La Crosse	95.9	93.5	94.9
547	Eau Claire	97.5	94.1	96.1
548	Superior	95.1	96.5	95.7
549	Oshkosh	95.4	91.9	93.9
WYOMING				
820	Cheyenne	101.1	72.7	89.1
821	Yellowstone Nat'l Park	99.0	71.9	87.6
822	Wheatland	100.3	67.1	86.2
823	Rawlins	102.0	71.5	89.1
824	Worland	99.7	70.7	87.5
825	Riverton	101.0	68.6	87.3
826	Casper	101.5	68.9	87.7
827	Newcastle	99.6	71.1	87.5
828	Sheridan	102.5	70.2	88.8
829-831	Rock Springs	103.4	71.4	89.9
CANADIAN FACTORS (reflect Canadian currency)				
ALBERTA				
	Calgary	119.7	95.6	109.5
	Edmonton	121.8	95.9	110.9
	Fort McMurray	121.9	88.4	107.7
	Lethbridge	117.0	87.9	104.7
	Lloydminster	111.5	84.7	100.2
	Medicine Hat	111.7	84.0	100.0
	Red Deer	112.1	84.0	100.2
BRITISH COLUMBIA				
	Kamloops	113.0	83.5	100.5
	Prince George	114.0	82.8	100.8
	Vancouver	117.9	92.4	107.2
	Victoria	114.1	87.6	102.9
MANITOBA				
	Brandon	120.6	70.5	99.4
	Portage la Prairie	111.6	69.1	93.6
	Winnipeg	122.0	71.2	100.5
NEW BRUNSWICK				
	Bathurst	110.7	62.3	90.2
	Dalhousie	110.2	62.5	90.0
	Fredericton	119.3	69.0	98.0
	Moncton	110.9	76.0	96.2
	Newcastle	110.7	62.9	90.5
	Saint John	113.2	75.7	97.3
NEWFOUNDLAND				
	Corner Brook	124.7	68.4	100.9
	St. John's	122.6	87.4	107.7
NORTHWEST TERRITORIES				
	Yellowknife	131.4	83.2	111.0
NOVA SCOTIA				
	Bridgewater	111.3	70.3	94.0
	Dartmouth	121.5	70.2	99.8
	Halifax	117.9	88.0	105.2
	New Glasgow	119.6	70.2	98.7
	Sydney	117.8	70.2	97.7
	Truro	110.8	70.3	93.7
	Yarmouth	119.5	70.2	98.7
ONTARIO				
	Barrie	116.5	87.1	104.1
	Brantford	113.1	90.7	103.6
	Cornwall	113.1	87.3	102.2
	Hamilton	116.3	98.8	108.9
	Kingston	114.1	87.4	102.8
	Kitchener	108.6	94.6	102.7
	London	116.3	96.5	108.0
	North Bay	122.5	85.2	106.8
	Oshawa	110.8	97.2	105.0
	Ottawa	117.5	96.8	108.7
	Owen Sound	116.7	85.4	103.4
	Peterborough	113.1	87.1	102.1
	Sarnia	113.2	91.3	103.9

STATE/ZIP	CITY	MAT.	INST.	TOTAL
ONTARIO (CONT'D)				
	Sault Ste. Marie	108.6	88.7	100.2
	St. Catharines	106.6	95.6	101.9
	Sudbury	106.1	93.6	100.8
	Thunder Bay	107.8	95.0	102.4
	Timmins	113.3	85.3	101.4
	Toronto	116.0	104.2	111.0
	Windsor	107.1	94.4	101.8
PRINCE EDWARD ISLAND				
	Charlottetown	121.3	60.2	95.5
	Summerside	121.9	57.9	94.9
QUEBEC				
	Cap-de-la-Madeleine	110.6	80.0	97.7
	Charlesbourg	110.6	80.0	97.7
	Chicoutimi	110.4	87.6	100.7
	Gatineau	110.3	79.8	97.4
	Granby	110.5	79.7	97.5
	Hull	110.5	79.8	97.5
	Joliette	110.8	80.0	97.8
	Laval	110.7	81.6	98.4
	Montreal	119.1	90.1	106.9
	Quebec City	118.7	90.4	106.7
	Rimouski	110.2	87.6	100.6
	Rouyn-Noranda	110.3	79.8	97.4
	Saint-Hyacinthe	109.8	79.8	97.1
	Sherbrooke	110.6	79.8	97.6
	Sorel	110.8	80.0	97.8
	Saint-Jerome	110.4	79.8	97.4
	Trois-Rivieres	120.2	79.9	103.2
SASKATCHEWAN				
	Moose Jaw	109.0	63.9	89.9
	Prince Albert	108.2	62.2	88.8
	Regina	123.1	92.4	110.2
	Saskatoon	109.9	88.5	100.9
YUKON				
	Whitehorse	132.2	69.2	105.6

For customer support on your Commercial Renovation Costs with RSMeans data, call 800.448.8182.

R011105-05   Tips for Accurate Estimating

1. Use pre-printed or columnar forms for orderly sequence of dimensions and locations and for recording telephone quotations.

2. Use only the front side of each paper or form except for certain pre-printed summary forms.

3. Be consistent in listing dimensions: For example, length x width x height. This helps in rechecking to ensure that, the total length of partitions is appropriate for the building area.

4. Use printed (rather than measured) dimensions where given.

5. Add up multiple printed dimensions for a single entry where possible.

6. Measure all other dimensions carefully.

7. Use each set of dimensions to calculate multiple related quantities.

8. Convert foot and inch measurements to decimal feet when listing. Memorize decimal equivalents to .01 parts of a foot (1/8″ equals approximately .01′).

9. Do not "round off" quantities until the final summary.

10. Mark drawings with different colors as items are taken off.

11. Keep similar items together, different items separate.

12. Identify location and drawing numbers to aid in future checking for completeness.

13. Measure or list everything on the drawings or mentioned in the specifications.

14. It may be necessary to list items not called for to make the job complete.

15. Be alert for: Notes on plans such as N.T.S. (not to scale); changes in scale throughout the drawings; reduced size drawings; discrepancies between the specifications and the drawings.

16. Develop a consistent pattern of performing an estimate. For example:
    a. Start the quantity takeoff at the lower floor and move to the next higher floor.
    b. Proceed from the main section of the building to the wings.
    c. Proceed from south to north or vice versa, clockwise or counterclockwise.
    d. Take off floor plan quantities first, elevations next, then detail drawings.

17. List all gross dimensions that can be either used again for different quantities, or used as a rough check of other quantities for verification (exterior perimeter, gross floor area, individual floor areas, etc.).

18. Utilize design symmetry or repetition (repetitive floors, repetitive wings, symmetrical design around a center line, similar room layouts, etc.). Note: Extreme caution is needed here so as not to omit or duplicate an area.

19. Do not convert units until the final total is obtained. For instance, when estimating concrete work, keep all units to the nearest cubic foot, then summarize and convert to cubic yards.

20. When figuring alternatives, it is best to total all items involved in the basic system, then total all items involved in the alternates. Therefore you work with positive numbers in all cases. When adds and deducts are used, it is often confusing whether to add or subtract a portion of an item; especially on a complicated or involved alternate.

R011105-50   Metric Conversion Factors

Description: This table is primarily for converting customary U.S. units in the left hand column to Sl metric units in the right hand column. In addition, conversion factors for some commonly encountered Canadian and non-Sl metric units are included.

	If You Know		Multiply By		To Find
Length	Inches	x	25.4[a]	=	Millimeters
	Feet	x	0.3048[a]	=	Meters
	Yards	x	0.9144[a]	=	Meters
	Miles (statute)	x	1.609	=	Kilometers
Area	Square inches	x	645.2	=	Square millimeters
	Square feet	x	0.0929	=	Square meters
	Square yards	x	0.8361	=	Square meters
Volume (Capacity)	Cubic inches	x	16,387	=	Cubic millimeters
	Cubic feet	x	0.02832	=	Cubic meters
	Cubic yards	x	0.7646	=	Cubic meters
	Gallons (U.S. liquids)[b]	x	0.003785	=	Cubic meters[c]
	Gallons (Canadian liquid)[b]	x	0.004546	=	Cubic meters[c]
	Ounces (U.S. liquid)[b]	x	29.57	=	Milliliters[c, d]
	Quarts (U.S. liquid)[b]	x	0.9464	=	Liters[c, d]
	Gallons (U.S. liquid)[b]	x	3.785	=	Liters[c, d]
Force	Kilograms force[d]	x	9.807	=	Newtons
	Pounds force	x	4.448	=	Newtons
	Pounds force	x	0.4536	=	Kilograms force[d]
	Kips	x	4448	=	Newtons
	Kips	x	453.6	=	Kilograms force[d]
Pressure, Stress, Strength (Force per unit area)	Kilograms force per square centimeter[d]	x	0.09807	=	Megapascals
	Pounds force per square inch (psi)	x	0.006895	=	Megapascals
	Kips per square inch	x	6.895	=	Megapascals
	Pounds force per square inch (psi)	x	0.07031	=	Kilograms force per square centimeter[d]
	Pounds force per square foot	x	47.88	=	Pascals
	Pounds force per square foot	x	4.882	=	Kilograms force per square meter[d]
Flow	Cubic feet per minute	x	0.4719	=	Liters per second
	Gallons per minute	x	0.0631	=	Liters per second
	Gallons per hour	x	1.05	=	Milliliters per second
Bending Moment Or Torque	Inch-pounds force	x	0.01152	=	Meter-kilograms force[d]
	Inch-pounds force	x	0.1130	=	Newton-meters
	Foot-pounds force	x	0.1383	=	Meter-kilograms force[d]
	Foot-pounds force	x	1.356	=	Newton-meters
	Meter-kilograms force[d]	x	9.807	=	Newton-meters
Mass	Ounces (avoirdupois)	x	28.35	=	Grams
	Pounds (avoirdupois)	x	0.4536	=	Kilograms
	Tons (metric)	x	1000	=	Kilograms
	Tons, short (2000 pounds)	x	907.2	=	Kilograms
	Tons, short (2000 pounds)	x	0.9072	=	Megagrams[e]
Mass per Unit Volume	Pounds mass per cubic foot	x	16.02	=	Kilograms per cubic meter
	Pounds mass per cubic yard	x	0.5933	=	Kilograms per cubic meter
	Pounds mass per gallon (U.S. liquid)[b]	x	119.8	=	Kilograms per cubic meter
	Pounds mass per gallon (Canadian liquid)[b]	x	99.78	=	Kilograms per cubic meter
Temperature	Degrees Fahrenheit	(F-32)/1.8		=	Degrees Celsius
	Degrees Fahrenheit	(F+459.67)/1.8		=	Degrees Kelvin
	Degrees Celsius	C+273.15		=	Degrees Kelvin

[a]The factor given is exact
[b]One U.S. gallon = 0.8327 Canadian gallon
[c]1 liter = 1000 milliliters = 1000 cubic centimeters
  1 cubic decimeter = 0.001 cubic meter

[d]Metric but not Sl unit
[e]Called "tonne" in England and
  "metric ton" in other metric countries

R011105-60 Weights and Measures

Measures of Length
1 Mile = 1760 Yards = 5280 Feet
1 Yard = 3 Feet = 36 inches
1 Foot = 12 Inches
1 Mil = 0.001 Inch
1 Fathom = 2 Yards = 6 Feet
1 Rod = 5.5 Yards = 16.5 Feet
1 Hand = 4 Inches
1 Span = 9 Inches
1 Micro-inch = One Millionth Inch or 0.000001 Inch
1 Micron = One Millionth Meter + 0.00003937 Inch

Surveyor's Measure
1 Mile = 8 Furlongs = 80 Chains
1 Furlong = 10 Chains = 220 Yards
1 Chain = 4 Rods = 22 Yards = 66 Feet = 100 Links
1 Link = 7.92 Inches

Square Measure
1 Square Mile = 640 Acres = 6400 Square Chains
1 Acre = 10 Square Chains = 4840 Square Yards =
  43,560 Sq. Ft.
1 Square Chain = 16 Square Rods = 484 Square Yards =
  4356 Sq. Ft.
1 Square Rod = 30.25 Square Yards = 272.25 Square Feet = 625 Square
  Lines
1 Square Yard = 9 Square Feet
1 Square Foot = 144 Square Inches
An Acre equals a Square 208.7 Feet per Side

Cubic Measure
1 Cubic Yard = 27 Cubic Feet
1 Cubic Foot = 1728 Cubic Inches
1 Cord of Wood = 4 x 4 x 8 Feet = 128 Cubic Feet
1 Perch of Masonry = 16½ x 1½ x 1 Foot = 24.75 Cubic Feet

Avoirdupois or Commercial Weight
1 Gross or Long Ton = 2240 Pounds
1 Net or Short Ton = 2000 Pounds
1 Pound = 16 Ounces = 7000 Grains
1 Ounce = 16 Drachms = 437.5 Grains
1 Stone = 14 Pounds

Power
1 British Thermal Unit per Hour = 0.2931 Watts
1 Ton (Refrigeration) = 3.517 Kilowatts
1 Horsepower (Boiler) = 9.81 Kilowatts
1 Horsepower (550 ft-lb/s) = 0.746 Kilowatts

Shipping Measure
For Measuring Internal Capacity of a Vessel:
  1 Register Ton = 100 Cubic Feet

For Measurement of Cargo:
  Approximately 40 Cubic Feet of Merchandise is considered a Shipping
  Ton, unless that bulk would weigh more than 2000 Pounds, in which case
  Freight Charge may be based upon weight.

40 Cubic Feet = 32.143 U.S. Bushels = 31.16 Imp. Bushels

Liquid Measure
1 Imperial Gallon = 1.2009 U.S. Gallon = 277.42 Cu. In.
1 Cubic Foot = 7.48 U.S. Gallons

R011110-10 Architectural Fees

Tabulated below are typical percentage fees by project size, for good professional architectural service. Fees may vary from those listed depending upon degree of design difficulty and economic conditions in any particular area.

Rates can be interpolated horizontally and vertically. Various portions of the same project requiring different rates should be adjusted proportionately. For alterations, add 50% to the fee for the first $500,000 of project cost and add 25% to the fee for project cost over $500,000.

Architectural fees tabulated below include Structural, Mechanical and Electrical Engineering Fees. They do not include the fees for special consultants such as kitchen planning, security, acoustical, interior design, etc.

Civil Engineering fees are included in the Architectural fee for project sites requiring minimal design such as city sites. However, separate Civil Engineering fees must be added when utility connections require design, drainage calculations are needed, stepped foundations are required, or provisions are required to protect adjacent wetlands.

Building Types	Total Project Size in Thousands of Dollars						
	100	250	500	1,000	5,000	10,000	50,000
Factories, garages, warehouses, repetitive housing	9.0%	8.0%	7.0%	6.2%	5.3%	4.9%	4.5%
Apartments, banks, schools, libraries, offices, municipal buildings	12.2	12.3	9.2	8.0	7.0	6.6	6.2
Churches, hospitals, homes, laboratories, museums, research	15.0	13.6	12.7	11.9	9.5	8.8	8.0
Memorials, monumental work, decorative furnishings	—	16.0	14.5	13.1	10.0	9.0	8.3

For customer support on your Commercial Renovation Costs with RSMeans data, call 800.448.8182.

839

R011110-30   Engineering Fees

Typical **Structural Engineering Fees** based on type of construction and total project size. These fees are included in Architectural Fees.

Type of Construction	Total Project Size (in thousands of dollars)			
	$500	$500-$1,000	$1,000-$5,000	Over $5000
Industrial buildings, factories & warehouses	Technical payroll times 2.0 to 2.5	1.60%	1.25%	1.00%
Hotels, apartments, offices, dormitories, hospitals, public buildings, food stores		2.00%	1.70%	1.20%
Museums, banks, churches and cathedrals		2.00%	1.75%	1.25%
Thin shells, prestressed concrete, earthquake resistive		2.00%	1.75%	1.50%
Parking ramps, auditoriums, stadiums, convention halls, hangars & boiler houses		2.50%	2.00%	1.75%
Special buildings, major alterations, underpinning & future expansion		Add to above 0.5%	Add to above 0.5%	Add to above 0.5%

For complex reinforced concrete or unusually complicated structures, add 20% to 50%.

Typical **Mechanical and Electrical Engineering Fees** are based on the size of the subcontract. The fee structure for both is shown below. These fees are included in Architectural Fees.

Type of Construction	Subcontract Size							
	$25,000	$50,000	$100,000	$225,000	$350,000	$500,000	$750,000	$1,000,000
Simple structures	6.4%	5.7%	4.8%	4.5%	4.4%	4.3%	4.2%	4.1%
Intermediate structures	8.0	7.3	6.5	5.6	5.1	5.0	4.9	4.8
Complex structures	10.1	9.0	9.0	8.0	7.5	7.5	7.0	7.0

For renovations, add 15% to 25% to applicable fee.

R012153-10   Repair and Remodeling

Cost figures are based on new construction utilizing the most cost-effective combination of labor, equipment and material with the work scheduled in proper sequence to allow the various trades to accomplish their work in an efficient manner.

The costs for repair and remodeling work must be modified due to the following factors that may be present in any given repair and remodeling project.

1. Equipment usage curtailment due to the physical limitations of the project, with only hand-operated equipment being used.

2. Increased requirement for shoring and bracing to hold up the building while structural changes are being made and to allow for temporary storage of construction materials on above-grade floors.

3. Material handling becomes more costly due to having to move within the confines of an enclosed building. For multi-story construction, low capacity elevators and stairwells may be the only access to the upper floors.

4. Large amount of cutting and patching and attempting to match the existing construction is required. It is often more economical to remove entire walls rather than create many new door and window openings. This sort of trade-off has to be carefully analyzed.

5. Cost of protection of completed work is increased since the usual sequence of construction usually cannot be accomplished.

6. Economies of scale usually associated with new construction may not be present. If small quantities of components must be custom fabricated due to job requirements, unit costs will naturally increase. Also, if only small work areas are available at a given time, job scheduling between trades becomes difficult and subcontractor quotations may reflect the excessive start-up and shut-down phases of the job.

7. Work may have to be done on other than normal shifts and may have to be done around an existing production facility which has to stay in production during the course of the repair and remodeling.

8. Dust and noise protection of adjoining non-construction areas can involve substantial special protection and alter usual construction methods.

9. Job may be delayed due to unexpected conditions discovered during demolition or removal. These delays ultimately increase construction costs.

10. Piping and ductwork runs may not be as simple as for new construction. Wiring may have to be snaked through walls and floors.

11. Matching "existing construction" may be impossible because materials may no longer be manufactured. Substitutions may be expensive.

12. Weather protection of existing structure requires additional temporary structures to protect building at openings.

13. On small projects, because of local conditions, it may be necessary to pay a tradesman for a minimum of four hours for a task that is completed in one hour.

All of the above areas can contribute to increased costs for a repair and remodeling project. Each of the above factors should be considered in the planning, bidding and construction stage in order to minimize the increased costs associated with repair and remodeling jobs.

R012153-60   Security Factors

Contractors entering, working in, and exiting secure facilities often lose productive time during a normal workday. The recommended allowances in this section are intended to provide for the loss of productivity by increasing labor costs. Note that different costs are associated with searches upon entry only and searches upon entry and exit. Time spent in a queue is unpredictable and not part of these allowances. Contractors should plan ahead for this situation.

Security checkpoints are designed to reflect the level of security required to gain access or egress. An extreme example is when contractors, along with any materials, tools, equipment, and vehicles, must be physically searched and have all materials, tools, equipment, and vehicles inventoried and documented prior to both entry and exit.

Physical searches without going through the documentation process represent the next level and take up less time.

Electronic searches—passing through a detector or x-ray machine with no documentation of materials, tools, equipment, and vehicles—take less time than physical searches.

Visual searches of materials, tools, equipment, and vehicles represent the next level of security.

Finally, access by means of an ID card or displayed sticker takes the least amount of time.

Another consideration is if the searches described above are performed each and every day, or if they are performed only on the first day with access granted by ID card or displayed sticker for the remainder of the project. The figures for this situation have been calculated to represent the initial check-in as described and subsequent entry by ID card or displayed sticker for up to 20 days on site. For the situation described above, where the time period is beyond 20 days, the impact on labor cost is negligible.

There are situations where tradespeople must be accompanied by an escort and observed during the work day. The loss of freedom of movement will slow down productivity for the tradesperson. Costs for the observer have not been included. Those costs are normally born by the owner.

R012157-20   Construction Time Requirements

The table below lists the construction durations for various building types along with their respective project sizes and project values. Design time runs 25% to 40% of construction time.

Building Type	Size S.F.	Project Value	Construction Duration
Industrial/Warehouse	100,000	$8,000,000	14 months
	500,000	$32,000,000	19 months
	1,000,000	$75,000,000	21 months
Offices/Retail	50,000	$7,000,000	15 months
	250,000	$28,000,000	23 months
	500,000	$58,000,000	34 months
Institutional/Hospitals/Laboratory	200,000	$45,000,000	31 months
	500,000	$110,000,000	52 months
	750,000	$160,000,000	55 months
	1,000,000	$210,000,000	60 months

R012909-80   Sales Tax by State

State sales tax on materials is tabulated below (5 states have no sales tax). Many states allow local jurisdictions, such as a county or city, to levy additional sales tax.

Some projects may be sales tax exempt, particularly those constructed with public funds.

State	Tax (%)	State	Tax (%)	State	Tax (%)	State	Tax (%)
Alabama	4	Illinois	6.25	Montana	0	Rhode Island	7
Alaska	0	Indiana	7	Nebraska	5.5	South Carolina	6
Arizona	5.6	Iowa	6	Nevada	6.85	South Dakota	4.5
Arkansas	6.5	Kansas	6.5	New Hampshire	0	Tennessee	7
California	7.25	Kentucky	6	New Jersey	6.625	Texas	6.25
Colorado	2.9	Louisiana	4.45	New Mexico	5.125	Utah	4.85
Connecticut	6.35	Maine	5.5	New York	4	Vermont	6
Delaware	0	Maryland	6	North Carolina	4.75	Virginia	4.3
District of Columbia	6	Massachusetts	6.25	North Dakota	5	Washington	6.5
Florida	6	Michigan	6	Ohio	5.75	West Virginia	6
Georgia	4	Minnesota	6.875	Oklahoma	4.5	Wisconsin	5
Hawaii	4	Mississippi	7	Oregon	0	Wyoming	4
Idaho	6	Missouri	4.225	Pennsylvania	6	Average	5.06 %

Sales Tax by Province (Canada)

GST - a value-added tax, which the government imposes on most goods and services provided in or imported into Canada. PST - a retail sales tax, which five of the provinces impose on the prices of most goods and some services. QST - a value-added tax, similar to the federal GST, which Quebec imposes. HST - Three provinces have combined their retail sales taxes with the federal GST into one harmonized tax.

Province	PST (%)	QST(%)	GST(%)	HST (%)
Alberta	0	0	5	0
British Columbia	7	0	5	0
Manitoba	7	0	5	0
New Brunswick	0	0	0	15
Newfoundland	0	0	0	15
Northwest Territories	0	0	5	0
Nova Scotia	0	0	0	15
Ontario	0	0	0	13
Prince Edward Island	0	0	0	15
Quebec	9.975	0	5	14.975
Saskatchewan	6	0	5	0
Yukon	0	0	5	0

General Requirements   R0129 Payment Procedures

R012909-85   Unemployment Taxes and Social Security Taxes

State unemployment tax rates vary not only from state to state, but also with the experience rating of the contractor. The federal unemployment tax rate is 6.0% of the first $7,000 of wages. This is reduced by a credit of up to 5.4% for timely payment to the state. The minimum federal unemployment tax is 0.6% after all credits.

Social security (FICA) for 2020 is estimated at time of publication to be 7.65% of wages up to $137,100.

R012909-86   Unemployment Tax by State

Information is from the U.S. Department of Labor, state unemployment tax rates.

State	Tax (%)	State	Tax (%)	State	Tax (%)	State	Tax (%)
Alabama	6.80	Illinois	6.93	Montana	6.30	Rhode Island	9.49
Alaska	5.4	Indiana	7.4	Nebraska	5.4	South Carolina	5.46
Arizona	12.76	Iowa	7.5	Nevada	5.4	South Dakota	9.35
Arkansas	14.3	Kansas	7.6	New Hampshire	7.5	Tennessee	10.0
California	6.2	Kentucky	9.3	New Jersey	5.8	Texas	6.5
Colorado	8.15	Louisiana	6.2	New Mexico	5.4	Utah	7.1
Connecticut	6.8	Maine	5.46	New York	9.1	Vermont	7.7
Delaware	8.20	Maryland	7.50	North Carolina	5.76	Virginia	6.21
District of Columbia	7	Massachusetts	12.65	North Dakota	10.74	Washington	7.73
Florida	5.4	Michigan	10.3	Ohio	9	West Virginia	8.5
Georgia	5.4	Minnesota	9.1	Oklahoma	5.5	Wisconsin	12.0
Hawaii	5.6	Mississippi	5.6	Oregon	5.4	Wyoming	8.8
Idaho	5.4	Missouri	8.37	Pennsylvania	11.03	Median	7.40%

General Requirements   R0131 Project Management & Coordination

R013113-40   Builder's Risk Insurance

Builder's risk insurance is insurance on a building during construction. Premiums are paid by the owner or the contractor. Blasting, collapse and underground insurance would raise total insurance costs.

For customer support on your Commercial Renovation Costs with RSMeans data, call 800.448.8182.

843

R013113-50   General Contractor's Overhead

There are two distinct types of overhead on a construction project: Project overhead and main office overhead. Project overhead includes those costs at a construction site not directly associated with the installation of construction materials. Examples of project overhead costs include the following:

1. Superintendent
2. Construction office and storage trailers
3. Temporary sanitary facilities
4. Temporary utilities
5. Security fencing
6. Photographs
7. Cleanup
8. Performance and payment bonds

The above project overhead items are also referred to as general requirements and therefore are estimated in Division 1. Division 1 is the first division listed in the CSI MasterFormat but it is usually the last division estimated. The sum of the costs in Divisions 1 through 49 is referred to as the sum of the direct costs.

All construction projects also include indirect costs. The primary components of indirect costs are the contractor's main office overhead and profit. The amount of the main office overhead expense varies depending on the following:

1. Owner's compensation
2. Project managers' and estimators' wages
3. Clerical support wages
4. Office rent and utilities
5. Corporate legal and accounting costs
6. Advertising
7. Automobile expenses
8. Association dues
9. Travel and entertainment expenses

These costs are usually calculated as a percentage of annual sales volume. This percentage can range from 35% for a small contractor doing less than $500,000 to 5% for a large contractor with sales in excess of $100 million.

R013113-55   Installing Contractor's Overhead

Installing contractors (subcontractors) also incur costs for general requirements and main office overhead.

Included within the total incl. overhead and profit costs is a percent mark-up for overhead that includes:

1. Compensation and benefits for office staff and project managers
2. Office rent, utilities, business equipment, and maintenance
3. Corporate legal and accounting costs

4. Advertising
5. Vehicle expenses (for office staff and project managers)
6. Association dues
7. Travel, entertainment
8. Insurance
9. Small tools and equipment

For customer support on your Commercial Renovation Costs with RSMeans data, call 800.448.8182.

R013113-60   Workers' Compensation Insurance Rates by Trade

The table below tabulates the national averages for workers' compensation insurance rates by trade and type of building. The average "Insurance Rate" is multiplied by the "% of Building Cost" for each trade. This produces the "Workers' Compensation" cost by % of total labor cost, to be added for each trade by building type to determine the weighted average workers' compensation rate for the building types analyzed.

Trade	Insurance Rate (% Labor Cost) Range		Insurance Rate (% Labor Cost) Average	% of Building Cost Office Bldgs.	% of Building Cost Schools & Apts.	% of Building Cost Mfg.	Workers' Compensation Office Bldgs.	Workers' Compensation Schools & Apts.	Workers' Compensation Mfg.
Excavation, Grading, etc.	2.3 % to	16.9%	9.6%	4.8%	4.9%	4.5%	0.46%	0.47%	0.43%
Piles & Foundations	3.3 to	28.0	15.7	7.1	5.2	8.7	1.11	0.82	1.37
Concrete	3.1 to	25.8	14.4	5.0	14.8	3.7	0.72	2.13	0.53
Masonry	3.3 to	52.1	27.7	6.9	7.5	1.9	1.91	2.08	0.53
Structural Steel	4.4 to	31.8	18.1	10.7	3.9	17.6	1.94	0.71	3.19
Miscellaneous & Ornamental Metals	2.9 to	21.4	12.2	2.8	4.0	3.6	0.34	0.49	0.44
Carpentry & Millwork	3.4 to	29.1	16.2	3.7	4.0	0.5	0.60	0.65	0.08
Metal or Composition Siding	4.8 to	124.6	64.7	2.3	0.3	4.3	1.49	0.19	2.78
Roofing	4.8 to	110.4	57.6	2.3	2.6	3.1	1.32	1.50	1.79
Doors & Hardware	3.1 to	29.1	16.1	0.9	1.4	0.4	0.14	0.23	0.06
Sash & Glazing	3.9 to	21.6	12.8	3.5	4.0	1.0	0.45	0.51	0.13
Lath & Plaster	2.7 to	30.1	16.4	3.3	6.9	0.8	0.54	1.13	0.13
Tile, Marble & Floors	2.0 to	19.0	10.5	2.6	3.0	0.5	0.27	0.32	0.05
Acoustical Ceilings	1.7 to	29.7	15.7	2.4	0.2	0.3	0.38	0.03	0.05
Painting	3.3 to	37.4	20.3	1.5	1.6	1.6	0.30	0.32	0.32
Interior Partitions	3.4 to	29.1	16.2	3.9	4.3	4.4	0.63	0.70	0.71
Miscellaneous Items	1.9 to	97.7	10.3	5.2	3.7	9.7	0.54	0.38	1.00
Elevators	1.3 to	9.0	5.1	2.1	1.1	2.2	0.11	0.06	0.11
Sprinklers	1.8 to	14.6	8.2	0.5	—	2.0	0.04	—	0.16
Plumbing	1.4 to	13.3	7.4	4.9	7.2	5.2	0.36	0.53	0.38
Heat., Vent., Air Conditioning	2.8 to	15.8	9.3	13.5	11.0	12.9	1.26	1.02	1.20
Electrical	1.7 to	10.7	6.2	10.1	8.4	11.1	0.63	0.52	0.69
Total	1.3 % to	124.6%	—	100.0%	100.0%	100.0%	15.54%	14.79%	16.13%
						Overall Weighted Average	15.49%		

Workers' Compensation Insurance Rates by States

The table below lists the weighted average Workers' Compensation base rate for each state with a factor comparing this with the national average of 9.9%.

State	Weighted Average	Factor	State	Weighted Average	Factor	State	Weighted Average	Factor
Alabama	12.4%	126	Kentucky	10.5%	108	North Dakota	6.4%	65
Alaska	8.6	88	Louisiana	17.5	178	Ohio	5.5	56
Arizona	7.7	79	Maine	8.0	81	Oklahoma	7.4	76
Arkansas	5.1	52	Maryland	9.7	99	Oregon	7.5	76
California	19.6	200	Massachusetts	8.8	90	Pennsylvania	17.2	175
Colorado	6.2	63	Michigan	6.3	64	Rhode Island	10.0	102
Connecticut	12.4	126	Minnesota	13.4	137	South Carolina	16.2	165
Delaware	11.8	120	Mississippi	9.2	94	South Dakota	7.6	77
District of Columbia	7.7	78	Missouri	12.0	122	Tennessee	6.0	61
Florida	8.4	86	Montana	7.1	73	Texas	4.9	50
Georgia	30.0	307	Nebraska	11.7	120	Utah	5.7	58
Hawaii	8.4	86	Nevada	7.6	77	Vermont	8.9	91
Idaho	7.8	79	New Hampshire	8.8	90	Virginia	6.7	69
Illinois	17.5	179	New Jersey	14.2	145	Washington	7.3	75
Indiana	3.1	32	New Mexico	13.0	132	West Virginia	4.2	42
Iowa	9.9	101	New York	15.6	159	Wisconsin	11.2	115
Kansas	5.6	57	North Carolina	11.6	119	Wyoming	4.8	49
			Weighted Average for U.S. is	9.9% of payroll = 100%				

The weighted average skilled worker rate for 35 trades is 9.8%. For bidding purposes, apply the full value of Workers' Compensation directly to total labor costs, or if labor is 38%, materials 42% and overhead and profit 20% of total cost, carry 38/80 x 9.8% = 4.66% of cost (before overhead and profit)

into overhead. Rates vary not only from state to state but also with the experience rating of the contractor.

Rates are the most current available at the time of publication.

General Requirements   R0131 Project Management & Coordination

R013113-80   Performance Bond

This table shows the cost of a Performance Bond for a construction job scheduled to be completed in 12 months. Add 1% of the premium cost per month for jobs requiring more than 12 months to complete. The rates are "standard" rates offered to contractors that the bonding company considers financially sound and capable of doing the work. Preferred rates are offered by some bonding companies based upon financial strength of the contractor. Actual rates vary from contractor to contractor and from bonding company to bonding company. Contractors should prequalify through a bonding agency before submitting a bid on a contract that requires a bond.

Contract Amount	Building Construction Class B Projects			Highways & Bridges					
				Class A New Construction			Class A-1 Highway Resurfacing		
First $ 100,000 bid	$25.00 per M			$15.00 per M			$9.40 per M		
Next 400,000 bid	$ 2,500	plus $15.00	per M	$ 1,500	plus $10.00	per M	$ 940	plus $7.20	per M
Next 2,000,000 bid	8,500	plus 10.00	per M	5,500	plus 7.00	per M	3,820	plus 5.00	per M
Next 2,500,000 bid	28,500	plus 7.50	per M	19,500	plus 5.50	per M	15,820	plus 4.50	per M
Next 2,500,000 bid	47,250	plus 7.00	per M	33,250	plus 5.00	per M	28,320	plus 4.50	per M
Over 7,500,000 bid	64,750	plus 6.00	per M	45,750	plus 4.50	per M	39,570	plus 4.00	per M

General Requirements   R0154 Construction Aids

R015423-10   Steel Tubular Scaffolding

On new construction, tubular scaffolding is efficient up to 60' high or five stories. Above this it is usually better to use a hung scaffolding if construction permits. Swing scaffolding operations may interfere with tenants. In this case, the tubular is more practical at all heights.

In repairing or cleaning the front of an existing building the cost of tubular scaffolding per S.F. of building front increases as the height increases above the first tier. The first tier cost is relatively high due to leveling and alignment.

The minimum efficient crew for erecting and dismantling is three workers. They can set up and remove 18 frame sections per day up to 5 stories high. For 6 to 12 stories high, a crew of four is most efficient. Use two or more on top and two on the bottom for handing up or hoisting. They can also set up and remove 18 frame sections per day. At 7' horizontal spacing, this will run about 800 S.F. per day of erecting and dismantling. Time for placing and removing planks must be added to the above. A crew of three can place and remove 72 planks per day up to 5 stories. For over 5 stories, a crew of four can place and remove 80 planks per day.

The table below shows the number of pieces required to erect tubular steel scaffolding for 1000 S.F. of building frontage. This area is made up of a scaffolding system that is 12 frames (11 bays) long by 2 frames high.

For jobs under twenty-five frames, add 50% to rental cost. Rental rates will be lower for jobs over three months duration. Large quantities for long periods can reduce rental rates by 20%.

Description of Component	Number of Pieces for 1000 S.F. of Building Front	Unit
5' Wide Standard Frame, 6'-4" High	24	Ea.
Leveling Jack & Plate	24	
Cross Brace	44	
Side Arm Bracket, 21"	12	
Guardrail Post	12	
Guardrail, 7' section	22	
Stairway Section	2	
Stairway Starter Bar	1	
Stairway Inside Handrail	2	
Stairway Outside Handrail	2	
Walk-Thru Frame Guardrail	2	

Scaffolding is often used as falsework over 15' high during construction of cast-in-place concrete beams and slabs. Two foot wide scaffolding is generally used for heavy beam construction. The span between frames depends upon the load to be carried with a maximum span of 5'.

Heavy duty shoring frames with a capacity of 10,000#/leg can be spaced up to 10' O.C. depending upon form support design and loading.

Scaffolding used as horizontal shoring requires less than half the material required with conventional shoring.

On new construction, erection is done by carpenters.

Rolling towers supporting horizontal shores can reduce labor and speed the job. For maintenance work, catwalks with spans up to 70' can be supported by the rolling towers.

R015423-20 Pump Staging

Pump staging is generally not available for rent. The table below shows the number of pieces required to erect pump staging for 2400 S.F. of building frontage. This area is made up of a pump jack system that is 3 poles (2 bays) wide by 2 poles high.

Item	Number of Pieces for 2400 S.F. of Building Front	Unit
Aluminum pole section, 24' long	6	Ea.
Aluminum splice joint, 6' long	3	
Aluminum foldable brace	3	
Aluminum pump jack	3	
Aluminum support for workbench/back safety rail	3	
Aluminum scaffold plank/workbench, 14" wide x 24' long	4	
Safety net, 22' long	2	
Aluminum plank end safety rail	2	↓

The cost in place for this 2400 S.F. will depend on how many uses are realized during the life of the equipment.

For customer support on your Commercial Renovation Costs with RSMeans data, call 800.448.8182.

847

R015433-10 Contractor Equipment

Rental Rates shown elsewhere in the data set pertain to late model high quality machines in excellent working condition, rented from equipment dealers. Rental rates from contractors may be substantially lower than the rental rates from equipment dealers depending upon economic conditions; for older, less productive machines, reduce rates by a maximum of 15%. Any overtime must be added to the base rates. For shift work, rates are lower. Usual rule of thumb is 150% of one shift rate for two shifts; 200% for three shifts.

For periods of less than one week, operated equipment is usually more economical to rent than renting bare equipment and hiring an operator.

Costs to move equipment to a job site (mobilization) or from a job site (demobilization) are not included in rental rates, nor in any Equipment costs on any Unit Price line items or crew listings. These costs can be found elsewhere. If a piece of equipment is already at a job site, it is not appropriate to utilize mob/demob costs in an estimate again.

Rental rates vary throughout the country with larger cities generally having lower rates. Lease plans for new equipment are available for periods in excess of six months with a percentage of payments applying toward purchase.

Rental rates can also be treated as reimbursement costs for contractor-owned equipment. Owned equipment costs include depreciation, loan payments, interest, taxes, insurance, storage, and major repairs.

Monthly rental rates vary from 2% to 5% of the cost of the equipment depending on the anticipated life of the equipment and its wearing parts. Weekly rates are about 1/3 the monthly rates and daily rental rates are about 1/3 the weekly rates.

The hourly operating costs for each piece of equipment include costs to the user such as fuel, oil, lubrication, normal expendables for the equipment, and a percentage of the mechanic's wages chargeable to maintenance. The hourly operating costs listed do not include the operator's wages.

The daily cost for equipment used in the standard crews is figured by dividing the weekly rate by five, then adding eight times the hourly operating cost to give the total daily equipment cost, not including the operator. This figure is in the right hand column of the Equipment listings under Equipment Cost/Day.

Pile Driving rates shown for the pile hammer and extractor do not include leads, cranes, boilers or compressors. Vibratory pile driving requires an added field specialist during set-up and pile driving operation for the electric model. The hydraulic model requires a field specialist for set-up only. Up to 125 reuses of sheet piling are possible using vibratory drivers. For normal conditions, crane capacity for hammer type and size is as follows.

Crane Capacity	Hammer Type and Size		
	Air or Steam	Diesel	Vibratory
25 ton	to 8,750 ft.-lb.		70 H.P.
40 ton	15,000 ft.-lb.	to 32,000 ft.-lb.	170 H.P.
60 ton	25,000 ft.-lb.		300 H.P.
100 ton		112,000 ft.-lb.	

Cranes should be specified for the job by size, building and site characteristics, availability, performance characteristics, and duration of time required.

Backhoes & Shovels rent for about the same as equivalent size cranes but maintenance and operating expenses are higher. The crane operator's rate must be adjusted for high boom heights. Average adjustments: for 150' boom add 2% per hour; over 185', add 4% per hour; over 210', add 6% per hour; over 250', add 8% per hour and over 295', add 12% per hour.

Tower Cranes of the climbing or static type have jibs from 50' to 200' and capacities at maximum reach range from 4,000 to 14,000 pounds. Lifting capacities increase up to maximum load as the hook radius decreases.

Typical rental rates, based on purchase price, are about 2% to 3% per month. Erection and dismantling run between 500 and 2000 labor hours. Climbing operation takes 10 labor hours per 20' climb. Crane dead time is about 5 hours per 40' climb. If crane is bolted to side of the building add cost of ties and extra mast sections. Climbing cranes have from 80' to 180' of mast while static cranes have 80' to 800' of mast.

Truck Cranes can be converted to tower cranes by using tower attachments. Mast heights over 400' have been used.

A single 100' high material **Hoist and Tower** can be erected and dismantled in about 400 labor hours; a double 100' high hoist and tower in about 600 labor hours. Erection times for additional heights are 3 and 4 labor hours

per vertical foot respectively up to 150', and 4 to 5 labor hours per vertical foot over 150' high. A 40' high portable Buck hoist takes about 160 labor hours to erect and dismantle. Additional heights take 2 labor hours per vertical foot to 80' and 3 labor hours per vertical foot for the next 100'. Most material hoists do not meet local code requirements for carrying personnel.

A 150' high **Personnel Hoist** requires about 500 to 800 labor hours to erect and dismantle. Budget erection time at 5 labor hours per vertical foot for all trades. Local code requirements or labor scarcity requiring overtime can add up to 50% to any of the above erection costs.

Earthmoving Equipment: The selection of earthmoving equipment depends upon the type and quantity of material, moisture content, haul distance, haul road, time available, and equipment available. Short haul cut and fill operations may require dozers only, while another operation may require excavators, a fleet of trucks, and spreading and compaction equipment. Stockpiled material and granular material are easily excavated with front end loaders. Scrapers are most economically used with hauls between 300' and 1-1/2 miles if adequate haul roads can be maintained. Shovels are often used for blasted rock and any material where a vertical face of 8' or more can be excavated. Special conditions may dictate the use of draglines, clamshells, or backhoes. Spreading and compaction equipment must be matched to the soil characteristics, the compaction required and the rate the fill is being supplied.

R015436-50   Mobilization

Costs to move rented construction equipment to a job site from an equipment dealer's or contractor's yard (mobilization) or off the job site (demobilization) are not included in the rental or operating rates, nor in the equipment cost on a unit price line or in a crew listing. These costs can be found consolidated in the Mobilization section of the data and elsewhere in particular site work sections. If a piece of equipment is already on the job site, it is not appropriate to include mob/demob costs in a new estimate that requires use of that equipment. The following table identifies approximate sizes of rented construction equipment that would be hauled on a towed trailer. Because this listing is not all-encompassing, the user can infer as to what size trailer might be required for a piece of equipment not listed.

3-ton Trailer	20-ton Trailer	40-ton Trailer	50-ton Trailer
20 H.P. Excavator	110 H.P. Excavator	200 H.P. Excavator	270 H.P. Excavator
50 H.P. Skid Steer	165 H.P. Dozer	300 H.P. Dozer	Small Crawler Crane
35 H.P. Roller	150 H.P. Roller	400 H.P. Scraper	500 H.P. Scraper
40 H.P. Trencher	Backhoe	450 H.P. Art. Dump Truck	500 H.P. Art. Dump Truck

R024119-10   Demolition Defined

Whole Building Demolition - Demolition of the whole building with no concern for any particular building element, component, or material type being demolished. This type of demolition is accomplished with large pieces of construction equipment that break up the structure, load it into trucks and haul it to a disposal site, but disposal or dump fees are not included. Demolition of below-grade foundation elements, such as footings, foundation walls, grade beams, slabs on grade, etc., is not included. Certain mechanical equipment containing flammable liquids or ozone-depleting refrigerants, electric lighting elements, communication equipment components, and other building elements may contain hazardous waste, and must be removed, either selectively or carefully, as hazardous waste before the building can be demolished.

Foundation Demolition - Demolition of below-grade foundation footings, foundation walls, grade beams, and slabs on grade. This type of demolition is accomplished by hand or pneumatic hand tools, and does not include saw cutting, or handling, loading, hauling, or disposal of the debris.

Gutting - Removal of building interior finishes and electrical/mechanical systems down to the load-bearing and sub-floor elements of the rough building frame, with no concern for any particular building element, component, or material type being demolished. This type of demolition is accomplished by hand or pneumatic hand tools, and includes loading into trucks, but not hauling, disposal or dump fees, scaffolding, or shoring. Certain mechanical equipment containing flammable liquids or ozone-depleting refrigerants, electric lighting elements, communication equipment components, and other building elements may contain hazardous waste, and must be removed, either selectively or carefully, as hazardous waste, before the building is gutted.

Selective Demolition - Demolition of a selected building element, component, or finish, with some concern for surrounding or adjacent elements, components, or finishes (see the first Subdivision (s) at the beginning of appropriate Divisions). This type of demolition is accomplished by hand or pneumatic hand tools, and does not include handling, loading, storing, hauling, or disposal of the debris, scaffolding, or shoring. "Gutting" methods may be used in order to save time, but damage that is caused to surrounding or adjacent elements, components, or finishes may have to be repaired at a later time.

Careful Removal - Removal of a piece of service equipment, building element or component, or material type, with great concern for both the removed item and surrounding or adjacent elements, components or finishes. The purpose of careful removal may be to protect the removed item for later re-use, preserve a higher salvage value of the removed item, or replace an item while taking care to protect surrounding or adjacent elements, components, connections, or finishes from cosmetic and/or structural damage. An approximation of the time required to perform this type of removal is 1/3 to 1/2 the time it would take to install a new item of like kind (see Reference Number R220105-10). This type of removal is accomplished by hand or pneumatic hand tools, and does not include loading, hauling, or storing the removed item, scaffolding, shoring, or lifting equipment.

Cutout Demolition - Demolition of a small quantity of floor, wall, roof, or other assembly, with concern for the appearance and structural integrity of the surrounding materials. This type of demolition is accomplished by hand or pneumatic hand tools, and does not include saw cutting, handling, loading, hauling, or disposal of debris, scaffolding, or shoring.

Rubbish Handling - Work activities that involve handling, loading or hauling of debris. Generally, the cost of rubbish handling must be added to the cost of all types of demolition, with the exception of whole building demolition.

Minor Site Demolition - Demolition of site elements outside the footprint of a building. This type of demolition is accomplished by hand or pneumatic hand tools, or with larger pieces of construction equipment, and may include loading a removed item onto a truck (check the Crew for equipment used). It does not include saw cutting, hauling or disposal of debris, and, sometimes, handling or loading.

R024119-20   Dumpsters

Dumpster rental costs on construction sites are presented in two ways.

The cost per week rental includes the delivery of the dumpster; its pulling or emptying once per week, and its final removal. The assumption is made that the dumpster contractor could choose to empty a dumpster by simply bringing in an empty unit and removing the full one. These costs also include the disposal of the materials in the dumpster.

The Alternate Pricing can be used when actual planned conditions are not approximated by the weekly numbers. For example, these lines can be used when a dumpster is needed for 4 weeks and will need to be emptied 2 or 3 times per week. Conversely the Alternate Pricing lines can be used when a dumpster will be rented for several weeks or months but needs to be emptied only a few times over this period.

R024119-30   Rubbish Handling Chutes

To correctly estimate the cost of rubbish handling chute systems, the individual components must be priced separately. First choose the size of the system; a 30-inch diameter chute is quite common, but the sizes range from 18 to 36 inches in diameter. The 30-inch chute comes in a standard weight and two thinner weights. The thinner weight chutes are sometimes chosen for cost savings, but they are more easily damaged.

There are several types of major chute pieces that make up the chute system. The first component to consider is the top chute section (top intake hopper) where the material is dropped into the chute at the highest point. After determining the top chute, the intermediate chute pieces called the regular chute sections are priced. Next, the number of chute control door sections (intermediate intake hoppers) must be determined. In the more complex systems, a chute control door section is provided at each floor level. The last major component to consider is bolt down frames; these are usually provided at every other floor level.

There are a number of accessories to consider for safe operation and control. There are covers for the top chute and the chute door sections. The top chute can have a trough that allows for better loading of the chute. For the safest operation, a chute warning light system can be added that will warn the other chute intake locations not to load while another is being used. There are dust control devices that spray a water mist to keep down the dust as the debris is loaded into a Dumpster. There are special breakaway cords that are used to prevent damage to the chute if the dumpster is removed without disconnecting from the chute. There are chute liners that can be installed to protect the chute structure from physical damage from rough abrasive materials. Warning signs can be posted at each floor level that is provided with a chute control door section.

In summary, a complete rubbish handling chute system will include one top section, several intermediate regular sections, several intermediate control door (intake hopper) sections and bolt down frames at every other floor level starting with the top floor. If so desired, the system can also include covers and a light warning system for a safer operation. The bottom of the chute should always be above the Dumpster and should be tied off with a breakaway cord to the Dumpster.

R026510-20 Underground Storage Tank Removal

Underground storage tank removal can be divided into two categories: non-leaking and leaking. Prior to removing an underground storage tank, tests should be made, with the proper authorities present, to determine whether a tank has been leaking or the surrounding soil has been contaminated.

To safely remove liquid underground storage tanks:
1. Excavate to the top of the tank.
2. Disconnect all piping.
3. Open all tank vents and access ports.
4. Remove all liquids and/or sludge.
5. Purge the tank with an inert gas.
6. Provide access to the inside of the tank and clean out the interior using proper personal protective equipment (PPE).
7. Excavate soil surrounding the tank using proper PPE for on-site personnel.
8. Pull and properly dispose of the tank.
9. Clean up the site of all contaminated material.
10. Install new tanks or close the excavation.

Existing Conditions R0282 Asbestos Remediation

R028213-20 Asbestos Removal Process

Asbestos removal is accomplished by a specialty contractor who understands the federal and state regulations regarding the handling and disposal of the material. The process of asbestos removal is divided into many individual steps. An accurate estimate can be calculated only after all the steps have been priced.

The steps are generally as follows:
1. Obtain an asbestos abatement plan from an industrial hygienist.
2. Monitor the air quality in and around the removal area and along the path of travel between the removal area and transport area. This establishes the background contamination.
3. Construct a two part decontamination chamber at entrance to removal area.
4. Install a HEPA filter to create a negative pressure in the removal area.
5. Install wall, floor and ceiling protection as required by the plan, usually 2 layers of fireproof 6 mil polyethylene.
6. Industrial hygienist visually inspects work area to verify compliance with plan.
7. Provide temporary supports for conduit and piping affected by the removal process.
8. Proceed with asbestos removal and bagging process. Monitor air quality as described in Step #2. Discontinue operations when contaminate levels exceed applicable standards.
9. Document the legal disposal of materials in accordance with EPA standards.
10. Thoroughly clean removal area including all ledges, crevices and surfaces.
11. Post abatement inspection by industrial hygienist to verify plan compliance.
12. Provide a certificate from a licensed industrial hygienist attesting that contaminate levels are within acceptable standards before returning area to regular use.

For customer support on your Commercial Renovation Costs with RSMeans data, call 800.448.8182.

851

R028319-60   Lead Paint Remediation Methods

Lead paint remediation can be accomplished by the following methods.

1. Abrasive blast
2. Chemical stripping
3. Power tool cleaning with vacuum collection system
4. Encapsulation
5. Remove and replace
6. Enclosure

Each of these methods has strengths and weaknesses depending on the specific circumstances of the project. The following is an overview of each method.

1. **Abrasive blasting** is usually accomplished with sand or recyclable metallic blast. Before work can begin, the area must be contained to ensure the blast material with lead does not escape to the atmosphere. The use of vacuum blast greatly reduces the containment requirements. Lead abatement equipment that may be associated with this work includes a negative air machine. In addition, it is necessary to have an industrial hygienist monitor the project on a continual basis. When the work is complete, the spent blast sand with lead must be disposed of as a hazardous material. If metallic shot was used, the lead is separated from the shot and disposed of as hazardous material. Worker protection includes disposable clothing and respiratory protection.

2. **Chemical stripping** requires strong chemicals to be applied to the surface to remove the lead paint. Before the work can begin, the area under/adjacent to the work area must be covered to catch the chemical and removed lead. After the chemical is applied to the painted surface it is usually covered with paper. The chemical is left in place for the specified period, then the paper with lead paint is pulled or scraped off. The process may require several chemical applications. The paper with chemicals and lead paint adhered to it, plus the containment and loose scrapings collected by a HEPA (High Efficiency Particulate Air Filter) vac, must be disposed of as hazardous material. The chemical stripping process usually requires a neutralizing agent and several wash downs after the paint is removed. Worker protection includes a neoprene or other compatible protective clothing and respiratory

protection with face shield. An industrial hygienist is required intermittently during the process.

3. **Power tool cleaning** is accomplished using shrouded needle blasting guns. The shrouding with different end configurations is held up against the surface to be cleaned. The area is blasted with hardened needles and the shroud captures the lead with a HEPA vac and deposits it in a holding tank. An industrial hygienist monitors the project. Protective clothing and a respirator are required until air samples prove otherwise. When the work is complete the lead must be disposed of as hazardous material.

4. **Encapsulation** is a method that leaves the well bonded lead paint in place after the peeling paint has been removed. Before the work can begin, the area under/adjacent to the work must be covered to catch the scrapings. The scraped surface is then washed with a detergent and rinsed. The prepared surface is covered with approximately 10 mils of paint. A reinforcing fabric can also be embedded in the paint covering. The scraped paint and containment must be disposed of as hazardous material. Workers must wear protective clothing and respirators.

5. **Removing and replacing** are effective ways to remove lead paint from windows, gypsum walls and concrete masonry surfaces. The painted materials are removed and new materials are installed. Workers should wear a respirator and tyvek suit. The demolished materials must be disposed of as hazardous waste if it fails the TCLP (Toxicity Characteristic Leachate Process) test.

6. **Enclosure** is the process that permanently seals lead painted materials in place. This process has many applications such as covering lead painted drywall with new drywall, covering exterior construction with tyvek paper then residing, or covering lead painted structural members with aluminum or plastic. The seams on all enclosing materials must be securely sealed. An industrial hygienist monitors the project, and protective clothing and a respirator are required until air samples prove otherwise.

All the processes require clearance monitoring and wipe testing as required by the hygienist.

R031113-10 Wall Form Materials

Aluminum Forms

Approximate weight is 3 lbs. per S.F.C.A. Standard widths are available from 4″ to 36″ with 36″ most common. Standard lengths of 2′, 4′, 6′ to 8′ are available. Forms are lightweight and fewer ties are needed with the wider widths. The form face is either smooth or textured.

Metal Framed Plywood Forms

Manufacturers claim over 75 reuses of plywood and over 300 reuses of steel frames. Many specials such as corners, fillers, pilasters, etc. are available. Monthly rental is generally about 15% of purchase price for first month and 9% per month thereafter with 90% of rental applied to purchase for the first month and decreasing percentages thereafter. Aluminum framed forms cost 25% to 30% more than steel framed.

After the first month, extra days may be prorated from the monthly charge. Rental rates do not include ties, accessories, cleaning, loss of hardware or freight in and out. Approximate weight is 5 lbs. per S.F. for steel; 3 lbs. per S.F. for aluminum.

Forms can be rented with option to buy.

Plywood Forms, Job Fabricated

There are two types of plywood used for concrete forms.

1. Exterior plyform which is completely waterproof. This is face oiled to facilitate stripping. Ten reuses can be expected with this type with 25 reuses possible.
2. An overlaid type consists of a resin fiber fused to exterior plyform. No oiling is required except to facilitate cleaning. This is available in both high density (HDO) and medium density overlaid (MDO). Using HDO, 50 reuses can be expected with 200 possible.

Plyform is available in 5/8″ and 3/4″ thickness. High density overlaid is available in 3/8″, 1/2″, 5/8″ and 3/4″ thickness.

5/8″ thick is sufficient for most building forms, while 3/4″ is best on heavy construction.

Plywood Forms, Modular, Prefabricated

There are many plywood forming systems without frames. Most of these are manufactured from 1-1/8″ (HDO) plywood and have some hardware attached. These are used principally for foundation walls 8′ or less high. With care and maintenance, 100 reuses can be attained with decreasing quality of surface finish.

Steel Forms

Approximate weight is 6-1/2 lbs. per S.F.C.A. including accessories. Standard widths are available from 2″ to 24″, with 24″ most common. Standard lengths are from 2′ to 8′, with 4′ the most common. Forms are easily ganged into modular units.

Forms are usually leased for 15% of the purchase price per month prorated daily over 30 days.

Rental may be applied to sale price, and usually rental forms are bought. With careful handling and cleaning 200 to 400 reuses are possible.

Straight wall gang forms up to 12′ x 20′ or 8′ x 30′ can be fabricated. These crane handled forms usually lease for approx. 9% per month.

Individual job analysis is available from the manufacturer at no charge.

R031113-40 Forms for Reinforced Concrete

Design Economy

Avoid many sizes in proportioning beams and columns.

From story to story avoid changing column dimensions. Gain strength by adding steel or using a richer mix. If a change in size of column is necessary, vary one dimension only to minimize form alterations. Keep beams and columns the same width.

From floor to floor in a multi-story building vary beam depth, not width, as that will leave the slab panel form unchanged. It is cheaper to vary the strength of a beam from floor to floor by means of a steel area than by 2″ changes in either width or depth.

Cost Factors

Material includes the cost of lumber, cost of rent for metal pans or forms if used, nails, form ties, form oil, bolts and accessories.

Labor includes the cost of carpenters to make up, erect, remove and repair, plus common labor to clean and move. Having carpenters remove forms minimizes repairs.

Improper alignment and condition of forms will increase finishing cost. When forms are heavily oiled, concrete surfaces must be neutralized before finishing. Special curing compounds will cause spillages to spall off in first frost. Gang forming methods will reduce costs on large projects.

Materials Used

Boards are seldom used unless their architectural finish is required. Generally, steel, fiberglass and plywood are used for contact surfaces. Labor on plywood is 10% less than with boards. The plywood is backed up with

2 x 4's at 12″ to 32″ O.C. Walers are generally 2 - 2 x 4's. Column forms are held together with steel yokes or bands. Shoring is with adjustable shoring or scaffolding for high ceilings.

Reuse

Floor and column forms can be reused four or possibly five times without excessive repair. Remember to allow for 10% waste on each reuse.

When modular sized wall forms are made, up to twenty uses can be expected with exterior plyform.

When forms are reused, the cost to erect, strip, clean and move will not be affected. 10% replacement of lumber should be included and about one hour of carpenter time for repairs on each reuse per 100 S.F.

The reuse cost for certain accessory items normally rented on a monthly basis will be lower than the cost for the first use.

After the fifth use, new material required plus time needed for repair prevent the form cost from dropping further; it may go up. Much depends on care in stripping, the number of special bays, changes in beam or column sizes and other factors.

Costs for multiple use of formwork may be developed as follows:

2 Uses	3 Uses	4 Uses
$\dfrac{\text{(1st Use + Reuse)}}{2} = \text{avg. cost/2 uses}$	$\dfrac{\text{(1st Use + 2 Reuses)}}{3} = \text{avg. cost/3 uses}$	$\dfrac{\text{(1st use + 3 Reuses)}}{4} = \text{avg. cost/4 uses}$

For customer support on your Commercial Renovation Costs with RSMeans data, call 800.448.8182.

853

R031113-60 Formwork Labor-Hours

Item	Unit	Hours Required Fabricate	Erect & Strip	Clean & Move	Total Hours 1 Use	Multiple Use 2 Use	3 Use	4 Use
Beam and Girder, interior beams, 12" wide	100 S.F.	6.4	8.3	1.3	16.0	13.3	12.4	12.0
Hung from steel beams		5.8	7.7	1.3	14.8	12.4	11.6	11.2
Beam sides only, 36" high		5.8	7.2	1.3	14.3	11.9	11.1	10.7
Beam bottoms only, 24" wide		6.6	13.0	1.3	20.9	18.1	17.2	16.7
Box out for openings		9.9	10.0	1.1	21.0	16.6	15.1	14.3
Buttress forms, to 8' high		6.0	6.5	1.2	13.7	11.2	10.4	10.0
Centering, steel, 3/4" rib lath			1.0		1.0			
3/8" rib lath or slab form	↓		0.9		0.9			
Chamfer strip or keyway	100 L.F.		1.5		1.5	1.5	1.5	1.5
Columns, fiber tube 8" diameter			20.6		20.6			
12"			21.3		21.3			
16"			22.9		22.9			
20"			23.7		23.7			
24"			24.6		24.6			
30"	↓		25.6		25.6			
Columns, round steel, 12" diameter			22.0		22.0	22.0	22.0	22.0
16"			25.6		25.6	25.6	25.6	25.6
20"			30.5		30.5	30.5	30.5	30.5
24"	↓		37.7		37.7	37.7	37.7	37.7
Columns, plywood 8" x 8"	100 S.F.	7.0	11.0	1.2	19.2	16.2	15.2	14.7
12" x 12"		6.0	10.5	1.2	17.7	15.2	14.4	14.0
16" x 16"		5.9	10.0	1.2	17.1	14.7	13.8	13.4
24" x 24"		5.8	9.8	1.2	16.8	14.4	13.6	13.2
Columns, steel framed plywood 8" x 8"			10.0	1.0	11.0	11.0	11.0	11.0
12" x 12"			9.3	1.0	10.3	10.3	10.3	10.3
16" x 16"			8.5	1.0	9.5	9.5	9.5	9.5
24" x 24"			7.8	1.0	8.8	8.8	8.8	8.8
Drop head forms, plywood		9.0	12.5	1.5	23.0	19.0	17.7	17.0
Coping forms		8.5	15.0	1.5	25.0	21.3	20.0	19.4
Culvert, box			14.5	4.3	18.8	18.8	18.8	18.8
Curb forms, 6" to 12" high, on grade		5.0	8.5	1.2	14.7	12.7	12.1	11.7
On elevated slabs	↓	6.0	10.8	1.2	18.0	15.5	14.7	14.3
Edge forms to 6" high, on grade	100 L.F.	2.0	3.5	0.6	6.1	5.6	5.4	5.3
7" to 12" high	100 S.F.	2.5	5.0	1.0	8.5	7.8	7.5	7.4
Equipment foundations		10.0	18.0	2.0	30.0	25.5	24.0	23.3
Flat slabs, including drops		3.5	6.0	1.2	10.7	9.5	9.0	8.8
Hung from steel		3.0	5.5	1.2	9.7	8.7	8.4	8.2
Closed deck for domes		3.0	5.8	1.2	10.0	9.0	8.7	8.5
Open deck for pans		2.2	5.3	1.0	8.5	7.9	7.7	7.6
Footings, continuous, 12" high		3.5	3.5	1.5	8.5	7.3	6.8	6.6
Spread, 12" high		4.7	4.2	1.6	10.5	8.7	8.0	7.7
Pile caps, square or rectangular		4.5	5.0	1.5	11.0	9.3	8.7	8.4
Grade beams, 24" deep		2.5	5.3	1.2	9.0	8.3	8.0	7.9
Lintel or Sill forms		8.0	17.0	2.0	27.0	23.5	22.3	21.8
Spandrel beams, 12" wide		9.0	11.2	1.3	21.5	17.5	16.2	15.5
Stairs			25.0	4.0	29.0	29.0	29.0	29.0
Trench forms in floor		4.5	14.0	1.5	20.0	18.3	17.7	17.4
Walls, Plywood, at grade, to 8' high		5.0	6.5	1.5	13.0	11.0	9.7	9.5
8' to 16'		7.5	8.0	1.5	17.0	13.8	12.7	12.1
16' to 20'		9.0	10.0	1.5	20.5	16.5	15.2	14.5
Foundation walls, to 8' high		4.5	6.5	1.0	12.0	10.3	9.7	9.4
8' to 16' high		5.5	7.5	1.0	14.0	11.8	11.0	10.6
Retaining wall to 12' high, battered		6.0	8.5	1.5	16.0	13.5	12.7	12.3
Radial walls to 12' high, smooth		8.0	9.5	2.0	19.5	16.0	14.8	14.3
2' chords		7.0	8.0	1.5	16.5	13.5	12.5	12.0
Prefabricated modular, to 8' high		—	4.3	1.0	5.3	5.3	5.3	5.3
Steel, to 8' high		—	6.8	1.2	8.0	8.0	8.0	8.0
8' to 16' high		—	9.1	1.5	10.6	10.3	10.2	10.2
Steel framed plywood to 8' high		—	6.8	1.2	8.0	7.5	7.3	7.2
8' to 16' high	↓	—	9.3	1.2	10.5	9.5	9.2	9.0

R032110-10   Reinforcing Steel Weights and Measures

Bar Designation No.**	Nominal Weight Lb./Ft.	U.S. Customary Units			SI Units			
		Nominal Dimensions*			Nominal Dimensions*			
		Diameter in.	Cross Sectional Area, in.^{2}	Perimeter in.	Nominal Weight kg/m	Diameter mm	Cross Sectional Area, cm^2	Perimeter mm
3	.376	.375	.11	1.178	.560	9.52	.71	29.9
4	.668	.500	.20	1.571	.994	12.70	1.29	39.9
5	1.043	.625	.31	1.963	1.552	15.88	2.00	49.9
6	1.502	.750	.44	2.356	2.235	19.05	2.84	59.8
7	2.044	.875	.60	2.749	3.042	22.22	3.87	69.8
8	2.670	1.000	.79	3.142	3.973	25.40	5.10	79.8
9	3.400	1.128	1.00	3.544	5.059	28.65	6.45	90.0
10	4.303	1.270	1.27	3.990	6.403	32.26	8.19	101.4
11	5.313	1.410	1.56	4.430	7.906	35.81	10.06	112.5
14	7.650	1.693	2.25	5.320	11.384	43.00	14.52	135.1
18	13.600	2.257	4.00	7.090	20.238	57.33	25.81	180.1

* The nominal dimensions of a deformed bar are equivalent to those of a plain round bar having the same weight per foot as the deformed bar.

** Bar numbers are based on the number of eighths of an inch included in the nominal diameter of the bars.

R032110-80   Shop-Fabricated Reinforcing Steel

The material prices for reinforcing, shown in the unit cost sections of the data set, are for 50 tons or more of shop-fabricated reinforcing steel and include:

1. Mill base price of reinforcing steel
2. Mill grade/size/length extras
3. Mill delivery to the fabrication shop
4. Shop storage and handling
5. Shop drafting/detailing
6. Shop shearing and bending
7. Shop listing
8. Shop delivery to the job site

Both material and installation costs can be considerably higher for small jobs consisting primarily of smaller bars, while material costs may be slightly lower for larger jobs.

R032205-30   Common Stock Styles of Welded Wire Fabric

This table provides some of the basic specifications, sizes, and weights of welded wire fabric used for reinforcing concrete.

	New Designation	Old Designation		Steel Area per Foot				Approximate Weight per 100 S.F.	
	Spacing — Cross Sectional Area (in.) — (Sq. in. 100)	Spacing — Wire Gauge (in.) — (AS & W)		Longitudinal		Transverse			
				in.	cm	in.	cm	lbs	kg
Rolls	6 x 6 — W1.4 x W1.4	6 x 6 — 10 x 10		.028	.071	.028	.071	21	9.53
	6 x 6 — W2.0 x W2.0	6 x 6 — 8 x 8	1	.040	.102	.040	.102	29	13.15
	6 x 6 — W2.9 x W2.9	6 x 6 — 6 x 6		.058	.147	.058	.147	42	19.05
	6 x 6 — W4.0 x W4.0	6 x 6 — 4 x 4		.080	.203	.080	.203	58	26.91
	4 x 4 — W1.4 x W1.4	4 x 4 — 10 x 10		.042	.107	.042	.107	31	14.06
	4 x 4 — W2.0 x W2.0	4 x 4 — 8 x 8	1	.060	.152	.060	.152	43	19.50
	4 x 4 — W2.9 x W2.9	4 x 4 — 6 x 6		.087	.227	.087	.227	62	28.12
	4 x 4 — W4.0 x W4.0	4 x 4 — 4 x 4		.120	.305	.120	.305	85	38.56
Sheets	6 x 6 — W2.9 x W2.9	6 x 6 — 6 x 6		.058	.147	.058	.147	42	19.05
	6 x 6 — W4.0 x W4.0	6 x 6 — 4 x 4		.080	.203	.080	.203	58	26.31
	6 x 6 — W5.5 x W5.5	6 x 6 — 2 x 2	2	.110	.279	.110	.279	80	36.29
	4 x 4 — W1.4 x W1.4	4 x 4 — 4 x 4		.120	.305	.120	.305	85	38.56

NOTES: 1. Exact W—number size for 8 gauge is W2.1
2. Exact W—number size for 2 gauge is W5.4

The above table was compiled with the following excerpts from the WRI Manual of Standard Practices, 7th Edition, Copyright 2006. Reproduced with permission of the Wire Reinforcement Institute, Inc.:

1. Chapter 3, page 7, Table 1 Common Styles of Metric Wire Reinforcement (WWR) With Equivalent US Customary Units
2. Chapter 6, page 19, Table 5 Customary Units
3. Chapter 6, Page 23, Table 7 Customary Units (in.) Welded Plain Wire Reinforcement
4. Chapter 6, Page 25 Table 8 Wire Size Comparison
5. Chapter 9, Page 30, Table 9 Weight of Longitudinal Wires Weight (Mass) Estimating Tables
6. Chapter 9, Page 31, Table 9M Weight of Longitudinal Wires Weight (Mass) Estimating Tables
7. Chapter 9, Page 32, Table 10 Weight of Transverse Wires Based on 62″ lengths of transverse wire (60″ width plus 1″ overhand each side)
8. Chapter 9, Page 33, Table 10M Weight of Transverse Wires

R033105-10   Proportionate Quantities

The tables below show both quantities per S.F. of floor areas as well as form and reinforcing quantities per C.Y. Unusual structural requirements would increase the ratios below. High strength reinforcing would reduce the steel weights. Figures are for 3000 psi concrete and 60,000 psi reinforcing unless specified otherwise.

Type of Construction	Live Load	Span	Per S.F. of Floor Area				Per C.Y. of Concrete		
			Concrete	Forms	Reinf.	Pans	Forms	Reinf.	Pans
Flat Plate	50 psf	15 Ft.	.46 C.F.	1.06 S.F.	1.71 lb.		62 S.F.	101 lb.	
		20	.63	1.02	2.40		44	104	
		25	.79	1.02	3.03		35	104	
	100	15	.46	1.04	2.14		61	126	
		20	.71	1.02	2.72		39	104	
		25	.83	1.01	3.47		33	113	
Flat Plate (waffle construction) 20" domes	50	20	.43	1.00	2.10	.84 S.F.	63	135	53 S.F.
		25	.52	1.00	2.90	.89	52	150	46
		30	.64	1.00	3.70	.87	42	155	37
	100	20	.51	1.00	2.30	.84	53	125	45
		25	.64	1.00	3.20	.83	42	135	35
		30	.76	1.00	4.40	.81	36	160	29
Waffle Construction 30" domes	50	25	.69	1.06	1.83	.68	42	72	40
		30	.74	1.06	2.39	.69	39	87	39
		35	.86	1.05	2.71	.69	33	85	39
		40	.78	1.00	4.80	.68	35	165	40
Flat Slab (two way with drop panels)	50	20	.62	1.03	2.34		45	102	
		25	.77	1.03	2.99		36	105	
		30	.95	1.03	4.09		29	116	
	100	20	.64	1.03	2.83		43	119	
		25	.79	1.03	3.88		35	133	
		30	.96	1.03	4.66		29	131	
	200	20	.73	1.03	3.03		38	112	
		25	.86	1.03	4.23		32	133	
		30	1.06	1.03	5.30		26	135	
One Way Joists 20" Pans	50	15	.36	1.04	1.40	.93	78	105	70
		20	.42	1.05	1.80	.94	67	120	60
		25	.47	1.05	2.60	.94	60	150	54
	100	15	.38	1.07	1.90	.93	77	140	66
		20	.44	1.08	2.40	.94	67	150	58
		25	.52	1.07	3.50	.94	55	185	49
One Way Joists 8" x 16" filler blocks	50	15	.34	1.06	1.80	.81 Ea.	84	145	64 Ea.
		20	.40	1.08	2.20	.82	73	145	55
		25	.46	1.07	3.20	.83	63	190	49
	100	15	.39	1.07	1.90	.81	74	130	56
		20	.46	1.09	2.80	.82	64	160	48
		25	.53	1.10	3.60	.83	56	190	42
One Way Beam & Slab	50	15	.42	1.30	1.73		84	111	
		20	.51	1.28	2.61		68	138	
		25	.64	1.25	2.78		53	117	
	100	15	.42	1.30	1.90		84	122	
		20	.54	1.35	2.69		68	154	
		25	.69	1.37	3.93		54	145	
	200	15	.44	1.31	2.24		80	137	
		20	.58	1.40	3.30		65	163	
		25	.69	1.42	4.89		53	183	
Two Way Beam & Slab	100	15	.47	1.20	2.26		69	130	
		20	.63	1.29	3.06		55	131	
		25	.83	1.33	3.79		43	123	
	200	15	.49	1.25	2.70		41	149	
		20	.66	1.32	4.04		54	165	
		25	.88	1.32	6.08		41	187	

R033105-10 Proportionate Quantities (cont.)

4000 psi Concrete and 60,000 psi Reinforcing—Form and Reinforcing Quantities per C.Y.

Item	Size	Forms	Reinforcing	Minimum	Maximum
	10″ x 10″	130 S.F.C.A.	#5 to #11	220 lbs.	875 lbs.
	12″ x 12″	108	#6 to #14	200	955
	14″ x 14″	92	#7 to #14	190	900
	16″ x 16″	81	#6 to #14	187	1082
	18″ x 18″	72	#6 to #14	170	906
	20″ x 20″	65	#7 to #18	150	1080
Columns	22″ x 22″	59	#8 to #18	153	902
(square tied)	24″ x 24″	54	#8 to #18	164	884
	26″ x 26″	50	#9 to #18	169	994
	28″ x 28″	46	#9 to #18	147	864
	30″ x 30″	43	#10 to #18	146	983
	32″ x 32″	40	#10 to #18	175	866
	34″ x 34″	38	#10 to #18	157	772
	36″ x 36″	36	#10 to #18	175	852
	38″ x 38″	34	#10 to #18	158	765
	40″ x 40″	32	#10 to #18	143	692

Item	Size	Form	Spiral	Reinforcing	Minimum	Maximum
	12″ diameter	34.5 L.F.	190 lbs.	#4 to #11	165 lbs.	1505 lb.
		34.5	190	#14 & #18	—	1100
	14″	25	170	#4 to #11	150	970
		25	170	#14 & #18	800	1000
	16″	19	160	#4 to #11	160	950
		19	160	#14 & #18	605	1080
	18″	15	150	#4 to #11	160	915
		15	150	#14 & #18	480	1075
	20″	12	130	#4 to #11	155	865
		12	130	#14 & #18	385	1020
	22″	10	125	#4 to #11	165	775
		10	125	#14 & #18	320	995
	24″	9	120	#4 to #11	195	800
		9	120	#14 & #18	290	1150
Columns	26″	7.3	100	#4 to #11	200	729
(spirally reinforced)		7.3	100	#14 & #18	235	1035
	28″	6.3	95	#4 to #11	175	700
		6.3	95	#14 & #18	200	1075
	30″	5.5	90	#4 to #11	180	670
		5.5	90	#14 & #18	175	1015
	32″	4.8	85	#4 to #11	185	615
		4.8	85	#14 & #18	155	955
	34″	4.3	80	#4 to #11	180	600
		4.3	80	#14 & #18	170	855
	36″	3.8	75	#4 to #11	165	570
		3.8	75	#14 & #18	155	865
	40″	3.0	70	#4 to #11	165	500
		3.0	70	#14 & #18	145	765

R033105-10   Proportionate Quantities (cont.)

		3000 psi Concrete and 60,000 psi Reinforcing—Form and Reinforcing Quantities per C.Y.				
Item	**Type**	**Loading**	**Height**	**C.Y./L.F.**	**Forms/C.Y.**	**Reinf./C.Y.**
Retaining Walls	Cantilever	Level Backfill	4 Ft.	0.2 C.Y.	49 S.F.	35 lbs.
			8	0.5	42	45
			12	0.8	35	70
			16	1.1	32	85
			20	1.6	28	105
		Highway Surcharge	4	0.3	41	35
			8	0.5	36	55
			12	0.8	33	90
			16	1.2	30	120
			20	1.7	27	155
		Railroad Surcharge	4	0.4	28	45
			8	0.8	25	65
			12	1.3	22	90
			16	1.9	20	100
			20	2.6	18	120
	Gravity, with Vertical Face	Level Backfill	4	0.4	37	None
			7	0.6	27	
			10	1.2	20	
		Sloping Backfill	4	0.3	31	
			7	0.8	21	
			10	1.6	15	↓

		Live Load in Kips per Linear Foot							
	Span	**Under 1 Kip**		**2 to 3 Kips**		**4 to 5 Kips**		**6 to 7 Kips**	
		Forms	**Reinf.**	**Forms**	**Reinf.**	**Forms**	**Reinf.**	**Forms**	**Reinf.**
Beams	10 Ft.	—	—	90 S.F.	170 #	85 S.F.	175 #	75 S.F.	185 #
	16	130 S.F.	165 #	85	180	75	180	65	225
	20	110	170	75	185	62	200	51	200
	26	90	170	65	215	62	215	—	—
	30	85	175	60	200	—	—	—	—

Item	**Size**	**Type**	**Forms per C.Y.**	**Reinforcing per C.Y.**
Spread Footings	Under 1 C.Y.	1,000 psf soil	24 S.F.	44 lbs.
		5,000	24	42
		10,000	24	52
	1 C.Y. to 5 C.Y.	1,000	14	49
		5,000	14	50
		10,000	14	50
	Over 5 C.Y.	1,000	9	54
		5,000	9	52
		10,000	9	56
Pile Caps (30 Ton Concrete Piles)	Under 5 C.Y.	shallow caps	20	65
		medium	20	50
		deep	20	40
	5 C.Y. to 10 C.Y.	shallow	14	55
		medium	15	45
		deep	15	40
	10 C.Y. to 20 C.Y.	shallow	11	60
		medium	11	45
		deep	12	35
	Over 20 C.Y.	shallow	9	60
		medium	9	45
		deep	10	40

R033105-10 Proportionate Quantities (cont.)

3000 psi Concrete and 60,000 psi Reinforcing — Form and Reinforcing Quantities per C.Y.						
Item	Size	Pile Spacing	50 T Pile	100 T Pile	50 T Pile	100 T Pile
Pile Caps (Steel H Piles)	Under 5 C.Y.	24" O.C.	24 S.F.	24 S.F.	75 lbs.	90 lbs.
		30"	25	25	80	100
		36"	24	24	80	110
	5 C.Y. to 10 C.Y.	24"	15	15	80	110
		30"	15	15	85	110
		36"	15	15	75	90
	Over 10 C.Y.	24"	13	13	85	90
		30"	11	11	85	95
		36"	10	10	85	90

		8" Thick		10" Thick		12" Thick		15" Thick	
	Height	Forms	Reinf.	Forms	Reinf.	Forms	Reinf.	Forms	Reinf.
Basement Walls	7 Ft.	81 S.F.	44 lbs.	65 S.F.	45 lbs.	54 S.F.	44 lbs.	41 S.F.	43 lbs.
	8		44		45		44		43
	9		46		45		44		43
	10		57		45		44		43
	12		83		50		52		43
	14		116		65		64		51
	16				86		90		65
	18						106		70

R033105-20 Materials for One C.Y. of Concrete

This is an approximate method of figuring quantities of cement, sand and coarse aggregate for a field mix with waste allowance included.

With crushed gravel as coarse aggregate, to determine barrels of cement required, divide 10 by total mix; that is, for 1:2:4 mix, 10 divided by 7 = 1-3/7 barrels.

If the coarse aggregate is crushed stone, use 10-1/2 instead of 10 as given for gravel.

To determine tons of sand required, multiply barrels of cement by parts of sand and then by 0.2; that is, for the 1:2:4 mix, as above, 1-3/7 x 2 x .2 = .57 tons.

Tons of crushed gravel are in the same ratio to tons of sand as parts in the mix, or 4/2 x .57 = 1.14 tons.

1 bag cement = 94#	1 C.Y. sand or crushed gravel = 2700#	1 C.Y. crushed stone = 2575#
4 bags = 1 barrel	1 ton sand or crushed gravel = 20 C.F.	1 ton crushed stone = 21 C.F.

Average carload of cement is 692 bags; of sand or gravel is 56 tons.

Do not stack stored cement over 10 bags high.

R033105-70 Placing Ready-Mixed Concrete

For ground pours allow for 5% waste when figuring quantities.

Prices in the front of the data set assume normal deliveries. If deliveries are made before 8 A.M. or after 5 P.M. or on Saturday afternoons add 30%. Negotiated discounts for large volumes are not included in prices in front of the data set.

For the lower floors without truck access, concrete may be wheeled in rubber-tired buggies, conveyer handled, crane handled or pumped. Pumping is economical if there is top steel. Conveyers are more efficient for thick slabs.

At higher floors the rubber-tired buggies may be hoisted by a hoisting tower and wheeled to the location. Placement by a conveyer is limited to three

floors and is best for high-volume pours. Pumped concrete is best when the building has no crane access. Concrete may be pumped directly as high as thirty-six stories using special pumping techniques. Normal maximum height is about fifteen stories.

The best pumping aggregate is screened and graded bank gravel rather than crushed stone.

Pumping downward is more difficult than pumping upward. The horizontal distance from pump to pour may increase preparation time prior to pour. Placing by cranes, either mobile, climbing or tower types, continues as the most efficient method for high-rise concrete buildings.

R033105-85   Lift Slabs

The cost advantage of the lift slab method is due to placing all concrete, reinforcing steel, inserts and electrical conduit at ground level and in reduction of formwork. Minimum economical project size is about 30,000 S.F. Slabs may be tilted for parking garage ramps.

It is now used in all types of buildings and has gone up to 22 stories high in apartment buildings. Current trend is to use post-tensioned flat plate slabs with spans from 22' to 35'. Cylindrical void forms are used when deep slabs are required. One pound of prestressing steel is about equal to seven pounds of conventional reinforcing.

To be considered cured for stressing and lifting, a slab must have attained 75% of design strength. Seven days are usually sufficient with four to five days possible if high early strength cement is used. Slabs can be stacked using two coats of a non-bonding agent to insure that slabs do not stick to each other. Lifting is done by companies specializing in this work. Lift rate is 5' to 15' per hour with an average of 10' per hour. Total areas up to 33,000 S.F. have been lifted at one time. 24 to 36 jacking columns are common. Most economical bay sizes are 24' to 28' with four to fourteen stories most efficient. Continuous design reduces reinforcing steel cost. Use of post-tensioned slabs allows larger bay sizes.

R033543-10   Polished Concrete Floors

A polished concrete floor has a glossy mirror-like appearance and is created by grinding the concrete floor with finer and finer diamond grits, similar to sanding wood, until the desired level of reflective clarity and sheen are achieved. The technical term for this type of polished concrete is bonded abrasive polished concrete. The basic piece of equipment used in the polishing process is a walk-behind planetary grinder for working large floor areas. This grinder drives diamond-impregnated abrasive discs, which progress from coarse- to fine-grit discs.

The process begins with the use of very coarse diamond segments or discs bonded in a metallic matrix. These segments are coarse enough to allow the removal of pits, blemishes, stains, and light coatings from the floor surface in preparation for final smoothing. The condition of the original concrete surface will dictate the grit coarseness of the initial grinding step which will generally end up being a three- to four-step process using ever finer grits. The purpose of this initial grinding step is to remove surface coatings and blemishes and to cut down into the cream for very fine aggregate exposure, or deeper into the fine aggregate layer just below the cream layer, or even deeper into the coarse aggregate layer. These initial grinding steps will progress up to the 100/120 grit. If wet grinding is done, a waste slurry is produced that must be removed between grit changes and disposed of properly. If dry grinding is done, a high performance vacuum will pick up the dust during grinding and collect it in bags which must be disposed of properly.

The process continues with honing the floor in a series of steps that progresses from 100-grit to 400-grit diamond abrasive discs embedded in a plastic or resin matrix. At some point during, or just prior to, the honing step, one or two coats of stain or dye can be sprayed onto the surface to give color to the concrete, and two coats of densifier/hardener must be applied to the floor surface and allowed to dry. This sprayed-on densifier/hardener will penetrate about 1/8" into the concrete to make the surface harder, denser and more abrasion-resistant.

The process ends with polishing the floor surface in a series of steps that progresses from resin-impregnated 800-grit (medium polish) to 1500-grit (high polish) to 3000-grit (very high polish), depending on the desired level of reflective clarity and sheen.

The Concrete Polishing Association of America (CPAA) has defined the flooring options available when processing concrete to a desired finish. The first category is aggregate exposure, the grinding of a concrete surface with bonded abrasives, in as many abrasive grits necessary, to achieve one of the following classes:

©Concrete Polishing Association of America "Glossary."

A. Cream – very little surface cut depth; little aggregate exposure

B. Fine aggregate (salt and pepper) – surface cut depth of 1/16"; fine aggregate exposure with little or no medium aggregate exposure at random locations

C. Medium aggregate – surface cut depth of 1/8"; medium aggregate exposure with little or no large aggregate exposure at random locations

D. Large aggregate – surface cut depth of 1/4"; large aggregate exposure with little or no fine aggregate exposure at random locations

The second CPAA defined category is reflective clarity and sheen, the polishing of a concrete surface with the minimum number of bonded abrasives as indicated to achieve one of the following levels:

1. Ground – flat appearance with none to very slight diffused reflection; none to very low reflective sheen; using a minimum total of 4 grit levels up to 100-grit

2. Honed – matte appearance with or without slight diffused reflection; low to medium reflective sheen; using a minimum total of 5 grit levels up to 400-grit

3. Semi-polished – objects being reflected are not quite sharp and crisp but can be easily identified; medium to high reflective sheen; using a minimum total of 6 grit levels up to 800-grit

4. Highly-polished – objects being reflected are sharp and crisp as would be seen in a mirror-like reflection; high to highest reflective sheen; using a minimum total of up to 8 grit levels up to 1500-grit or 3000-grit

The CPAA defines reflective clarity as the degree of sharpness and crispness of the reflection of overhead objects when viewed 5' above and perpendicular to the floor surface. Reflective sheen is the degree of gloss reflected from a surface when viewed at least 20' from and at an angle to the floor surface. These terms are relatively subjective. The final outcome depends on the internal makeup and surface condition of the original concrete floor, the experience of the floor polishing crew, and the expectations of the owner. Before the grinding, honing, and polishing work commences on the main floor area, it might be beneficial to do a mock-up panel in the same floor but in an out of the way place to demonstrate the sequence of steps with increasingly fine abrasive grits and to demonstrate the final reflective clarity and reflective sheen. This mock-up panel will be within the area of, and part of, the final work.

R034105-30 Prestressed Precast Concrete Structural Units

Type	Location	Depth	Span in Ft.		Live Load Lb. per S.F.
Double Tee 8' to 10'	Floor	28" to 34"	60 to 80		50 to 80
	Roof	12" to 24"	30 to 50		40
	Wall	Width 8'	Up to 55' high		Wind
Multiple Tee 8'	Roof	8" to 12"	15 to 40		40
	Floor	8" to 12"	15 to 30		100
Plank or	Roof or Floor		Roof	Floor	
		4"	13	12	40 for Roof
		6"	22	18	
		8"	26	25	
		10"	33	29	100 for Floor
		12"	42	32	
Single Tee 8' to 10'	Roof	28" 32" 36" 48"	40 80 100 120		40
AASHO Girder	Bridges	Type 4 5 6	100 110 125		Highway
Box Beam 4'	Bridges	15" 27" 33"	40 to 100		Highway

The majority of precast projects today utilize double tees rather than single tees because of speed and ease of installation. As a result casting beds at manufacturing plants are normally formed for double tees. Single tee projects will therefore require an initial set up charge to be spread over the individual single tee costs.

For floors, a 2″ to 3″ topping is field cast over the shapes. For roofs, insulating concrete or rigid insulation is placed over the shapes.

Member lengths up to 40′ are standard haul, 40′ to 60′ require special permits and lengths over 60′ must be escorted. Excessive width and/or length can add up to 100% on hauling costs.

Large heavy members may require two cranes for lifting which would increase erection costs by about 45%. An eight man crew can install 12 to 20 double tees, or 45 to 70 quad tees or planks per day.

Grouting of connections must also be included.

Several system buildings utilizing precast members are available. Heights can go up to 22 stories for apartment buildings. The optimum design ratio is 3 S.F. of surface to 1 S.F. of floor area.

R035216-10 Lightweight Concrete

Lightweight aggregate concrete is usually purchased ready mixed, but it can also be field mixed.

Vermiculite or Perlite comes in bags of 4 C.F. under various trade names. The weight is about 8 lbs. per C.F. For insulating roof fill use 1:6 mix. For a structural deck use 1:4 mix over gypsum boards, steeltex, steel centering, etc., supported by closely spaced joists or bulb trees. For structural slabs use 1:3:2 vermiculite sand concrete over steeltex, metal lath, steel centering, etc., on joists spaced 2'-0" O.C. for maximum L.L. of 80 P.S.F. Use same mix for slab base fill over steel flooring or regular reinforced concrete slab when tile, terrazzo or other finish is to be laid over.

For slabs on grade use 1:3:2 mix when tile, etc., finish is to be laid over. If radiant heating units are installed use a 1:6 mix for a base. After coils are in place, cover with a regular granolithic finish (mix 1:3:2) to a minimum depth of 1-1/2" over top of units.

Reinforce all slabs with 6 x 6 or 10 x 10 welded wire mesh.

Masonry | R0421 Clay Unit Masonry

R042110-20 Common and Face Brick

Common building brick manufactured according to ASTM C62 and facing brick manufactured according to ASTM C216 are the two standard bricks available for general building use.

Building brick is made in three grades: SW, where high resistance to damage caused by cyclic freezing is required; MW, where moderate resistance to cyclic freezing is needed; and NW, where little resistance to cyclic freezing is needed. Facing brick is made in only the two grades SW and MW. Additionally, facing brick is available in three types: FBS, for general use; FBX, for general use where a higher degree of precision and lower permissible variation in size than FBS are needed; and FBA, for general use to produce characteristic architectural effects resulting from non-uniformity in size and texture of the units.

In figuring the material cost of brickwork, an allowance of 25% mortar waste and 3% brick breakage was included. If bricks are delivered palletized with 280 to 300 per pallet, or packaged, allow only 1-1/2% for breakage. Packaged or palletized delivery is practical when a job is big enough to have a crane or other equipment available to handle a package of brick. This is so on all industrial work but not always true on small commercial buildings.

The use of buff and gray face is increasing, and there is a continuing trend to the Norman, Roman, Jumbo and SCR brick.

Common red clay brick for backup is not used that often. Concrete block is the most usual backup material with occasional use of sand lime or cement brick. Building brick is commonly used in solid walls for strength and as a fire stop.

Brick panels built on the ground and then crane erected to the upper floors have proven to be economical. This allows the work to be done under cover and without scaffolding.

R042110-50 Brick, Block & Mortar Quantities

Running Bond							For Other Bonds Standard Size Add to S.F. Quantities in Table to Left		
Number of Brick per S.F. of Wall - Single Wythe with 3/8" Joints					C.F. of Mortar per M Bricks, Waste Included				
Type Brick	Nominal Size (incl. mortar) L H W	Modular Coursing	Number of Brick per S.F.		3/8" Joint	1/2" Joint	Bond Type	Description	Factor
Standard	8 x 2-2/3 x 4	3C=8"	6.75		8.1	10.3	Common	full header every fifth course	+20%
Economy	8 x 4 x 4	1C=4"	4.50		9.1	11.6		full header every sixth course	+16.7%
Engineer	8 x 3-1/5 x 4	5C=16"	5.63		8.5	10.8	English	full header every second course	+50%
Fire	9 x 2-1/2 x 4-1/2	2C=5"	6.40		550 # Fireclay	—	Flemish	alternate headers every course	+33.3%
Jumbo	12 x 4 x 6 or 8	1C=4"	3.00		22.5	29.2		every sixth course	+5.6%
Norman	12 x 2-2/3 x 4	3C=8"	4.50		11.2	14.3	Header = W x H exposed		+100%
Norwegian	12 x 3-1/5 x 4	5C=16"	3.75		11.7	14.9	Rowlock = H x W exposed		+100%
Roman	12 x 2 x 4	2C=4"	6.00		10.7	13.7	Rowlock stretcher = L x W exposed		+33.3%
SCR	12 x 2-2/3 x 6	3C=8"	4.50		21.8	28.0	Soldier = H x L exposed		—
Utility	12 x 4 x 4	1C=4"	3.00		12.3	15.7	Sailor = W x L exposed		-33.3%

Concrete Blocks Nominal Size		Approximate Weight per S.F.		Blocks per 100 S.F.	Mortar per M block, waste included	
		Standard	Lightweight		Partitions	Back up
2"	x 8" x 16"	20 PSF	15 PSF	113	27 C.F.	36 C.F.
4"		30	20		41	51
6"		42	30		56	66
8"		55	38		72	82
10"		70	47		87	97
12"		85	55		102	112

Brick & Mortar Quantities
©Brick Industry Association. 2009 Feb. Technical Notes on Brick Construction 10:
  Dimensioning and Estimating Brick Masonry. Reston (VA): BIA. Table 1
  Modular Brick Sizes and Table 4 Quantity Estimates for Brick Masonry.

R050521-20   Welded Structural Steel

Usual weight reductions with welded design run 10% to 20% compared with bolted or riveted connections. This amounts to about the same total cost compared with bolted structures since field welding is more expensive than bolts. For normal spans of 18′ to 24′ figure 6 to 7 connections per ton.

Trusses — For welded trusses add 4% to weight of main members for connections. Up to 15% less steel can be expected in a welded truss compared to one that is shop bolted. Cost of erection is the same whether shop bolted or welded.

General — Typical electrodes for structural steel welding are E6010, E6011, E60T and E70T. Typical buildings vary between 2# to 8# of weld rod per

ton of steel. Buildings utilizing continuous design require about three times as much welding as conventional welded structures. In estimating field erection by welding, it is best to use the average linear feet of weld per ton to arrive at the welding cost per ton. The type, size and position of the weld will have a direct bearing on the cost per linear foot. A typical field welder will deposit 1.8# to 2# of weld rod per hour manually. Using semiautomatic methods can increase production by as much as 50% to 75%.

R051223-10   Structural Steel

The bare material prices for structural steel, shown in the unit cost sections of the data set, are for 100 tons of shop-fabricated structural steel and include:

1. Mill base price of structural steel
2. Mill scrap/grade/size/length extras
3. Mill delivery to a metals service center (warehouse)
4. Service center storage and handling
5. Service center delivery to a fabrication shop
6. Shop storage and handling
7. Shop drafting/detailing
8. Shop fabrication

9. Shop coat of primer paint
10. Shop listing
11. Shop delivery to the job site

In unit cost sections of the data set that contain items for field fabrication of steel components, the bare material cost of steel includes:

1. Mill base price of structural steel
2. Mill scrap/grade/size/length extras
3. Mill delivery to a metals service center (warehouse)
4. Service center storage and handling
5. Service center delivery to the job site

R051223-15   Structural Steel Estimating for Repair and Remodeling Projects

The correct approach to estimating structural steel is dependent upon the amount of steel required for the particular project. If the project is a sizable addition to a building, the data can be used directly from the data set. This is not the case however if the project requires a small amount of steel, for instance to reinforce existing roof or floor structural systems. To better understand this, please refer to the unit price line for a W16x31 beam with bolted connections. Assume your project requires the reinforcement of the structural members of the roof system of a 3 story building required for the installation a new piece of HVAC equipment. The project will need 4 pieces of W16x31 that are 30′ long. After pricing this 120 L.F. job using the unit prices for a W16x31, an analysis will reveal that the price is wholly inadequate.

The first problem is apparent if you examine the amount of steel that can be installed per day. The unit price line indicates 900 linear feet per day can be installed by a 5 man crew with an 90 ton crane. This productivity is correct for new construction but certainly is not for repair and remodeling work. Installation of new structural steel considers that each member is installed from the foundation to the roof of the structure in a planned and systematic manner with a crane having unrestricted access to all parts of the project. Additionally each connection is planned and detailed with full field access for fit-up and final bolting. The erection is planned and progresses such that interferences and conflicts with other structural members are minimized, if not completely eliminated. All of these assumptions are clearly not the case with a repair and remodeling job, and a significant decrease in the stated productivity will be observed.

A crane will certainly be needed to lift the members into the general area of the project but in most cases will not be able to place the beams into

their final position. An opening in the existing roof may not be large enough to permit the beams to pass through and it may be necessary to bring them into the building through existing windows or doors. Moving the beams to the actual area where they will be installed may involve hand labor and the use of dollies. Finally, hoists and/or jacks may be needed for final positioning.

The connection of new members to existing can often be accomplished by field bolting with accurate field measurements and good planning but in many cases access to both sides of existing members is not possible and field welding becomes the only alternative. In addition to the cost of the actual welding, protection of existing finishes, systems and structure and fire protection must be considered.

New beams can never be installed tight to the existing decks or floors which they must support and the use of shims and tack welding becomes necessary. Additionally, further planning and cost are involved in assuring that existing loads are minimized during the installation and shimming process.

It is apparent that installation of structural steel as part of a repair and remodeling project involves more than simply installing the members and estimating the cost in the same manner as new construction. The best procedure for estimating the total cost are adequate planning and coordination of each process and activity that will be needed. Unit costs for the materials, labor and equipment can then be attached to each needed activity and a final, complete price can be determined.

R053100-10   Decking Descriptions

General - All Deck Products

A steel deck is made by cold forming structural grade sheet steel into a repeating pattern of parallel ribs. The strength and stiffness of the panels are the result of the ribs and the material properties of the steel. Deck lengths can be varied to suit job conditions, but because of shipping considerations, are usually less than 40 feet. Standard deck width varies with the product used but full sheets are usually 12", 18", 24", 30", or 36". The deck is typically furnished in a standard width with the ends cut square. Any cutting for width, such as at openings or for angular fit, is done at the job site.

The deck is typically attached to the building frame with arc puddle welds, self-drilling screws, or powder or pneumatically driven pins. Sheet to sheet fastening is done with screws, button punching (crimping), or welds.

Composite Floor Deck

After installation and adequate fastening, a floor deck serves several purposes. It (a) acts as a working platform, (b) stabilizes the frame, (c) serves as a concrete form for the slab, and (d) reinforces the slab to carry the design loads applied during the life of the building. Composite decks are distinguished by the presence of shear connector devices as part of the deck. These devices are designed to mechanically lock the concrete and deck together so that the concrete and the deck work together to carry subsequent floor loads. These shear connector devices can be rolled-in embossments, lugs, holes, or wires welded to the panels. The deck profile can also be used to interlock concrete and steel.

Composite deck finishes are either galvanized (zinc coated) or phosphatized/painted. Galvanized deck has a zinc coating on both the top and bottom surfaces. The phosphatized/painted deck has a bare (phosphatized) top surface that will come into contact with the concrete. This bare top surface can be expected to develop rust before the concrete is placed. The bottom side of the deck has a primer coat of paint.

A composite floor deck is normally installed so the panel ends do not overlap on the supporting beams. Shear lugs or panel profile shapes often prevent a tight metal to metal fit if the panel ends overlap; the air gap caused by overlapping will prevent proper fusion with the structural steel supports when the panel end laps are shear stud welded.

Adequate end bearing of the deck must be obtained as shown on the drawings. If bearing is actually less in the field than shown on the drawings, further investigation is required.

Roof Deck

A roof deck is not designed to act compositely with other materials. A roof deck acts alone in transferring horizontal and vertical loads into the building frame. Roof deck rib openings are usually narrower than floor deck rib openings. This provides adequate support of the rigid thermal insulation board.

A roof deck is typically installed to endlap approximately 2" over supports. However, it can be butted (or lapped more than 2") to solve field fit problems. Since designers frequently use the installed deck system as part of the horizontal bracing system (the deck as a diaphragm), any fastening substitution or change should be approved by the designer. Continuous perimeter support of the deck is necessary to limit edge deflection in the finished roof and may be required for diaphragm shear transfer.

Standard roof deck finishes are galvanized or primer painted. The standard factory applied paint for roof decks is a primer paint and is not intended to weather for extended periods of time. Field painting or touching up of abrasions and deterioration of the primer coat or other protective finishes is the responsibility of the contractor.

Cellular Deck

A cellular deck is made by attaching a bottom steel sheet to a roof deck or composite floor deck panel. A cellular deck can be used in the same manner as a floor deck. Electrical, telephone, and data wires are easily run through the chase created between the deck panel and the bottom sheet.

When used as part of the electrical distribution system, the cellular deck must be installed so that the ribs line up and create a smooth cell transition at abutting ends. The joint that occurs at butting cell ends must be taped or otherwise sealed to prevent wet concrete from seeping into the cell. Cell interiors must be free of welding burrs, or other sharp intrusions, to prevent damage to wires.

When used as a roof deck, the bottom flat plate is usually left exposed to view. Care must be maintained during erection to keep good alignment and prevent damage.

A cellular deck is sometimes used with the flat plate on the top side to provide a flat working surface. Installation of the deck for this purpose requires special methods for attachment to the frame because the flat plate, now on the top, can prevent direct access to the deck material that is bearing on the structural steel. It may be advisable to treat the flat top surface to prevent slipping.

A cellular deck is always furnished galvanized or painted over galvanized.

Form Deck

A form deck can be any floor or roof deck product used as a concrete form. Connections to the frame are by the same methods used to anchor floor and roof decks. Welding washers are recommended when welding a deck that is less than 20 gauge thickness.

A form deck is furnished galvanized, prime painted, or uncoated. Galvanized deck must be used for those roof deck systems where a form deck is used to carry a lightweight insulating concrete fill.

Wood, Plastics & Comp. | R0611 Wood Framing

R061110-30   Lumber Product Material Prices

The price of forest products fluctuates widely from location to location and from season to season depending upon economic conditions. The bare material prices in the unit cost sections of the data set show the National Average material prices in effect Jan. 1 of this data year. It must be noted that lumber prices in general may change significantly during the year.

Availability of certain items depends upon geographic location and must be checked prior to firm-price bidding.

Wood, Plastics & Comp. | R0616 Sheathing

R061636-20   Plywood

There are two types of plywood used in construction: interior, which is moisture-resistant but not waterproofed, and exterior, which is waterproofed.

The grade of the exterior surface of the plywood sheets is designated by the first letter: A, for smooth surface with patches allowed; B, for solid surface with patches and plugs allowed; C, which may be surface plugged or may have knot holes up to 1″ wide; and D, which is used only for interior type plywood and may have knot holes up to 2-1/2″ wide. "Structural Grade" is specifically designed for engineered applications such as box beams. All CC & DD grades have roof and floor spans marked on them.

Underlayment-grade plywood runs from 1/4″ to 1-1/4″ thick. Thicknesses 5/8″ and over have optional tongue and groove joints which eliminate the need for blocking the edges. Underlayment 19/32″ and over may be referred to as Sturd-i-Floor.

The price of plywood can fluctuate widely due to geographic and economic conditions.

Typical uses for various plywood grades are as follows:

AA-AD Interior — cupboards, shelving, paneling, furniture

BB Plyform — concrete form plywood

CDX — wall and roof sheathing

Structural — box beams, girders, stressed skin panels

AA-AC Exterior — fences, signs, siding, soffits, etc.

Underlayment — base for resilient floor coverings

Overlaid HDO — high density for concrete forms & highway signs

Overlaid MDO — medium density for painting, siding, soffits & signs

303 Siding — exterior siding, textured, striated, embossed, etc.

Thermal & Moist. Protec. | R0751 Built-Up Bituminous Roofing

R075113-20   Built-Up Roofing

Asphalt is available in kegs of 100 lbs. each; coal tar pitch in 560 lb. kegs. Prepared roofing felts are available in a wide range of sizes, weights and characteristics. However, the most commonly used are #15 (432 S.F. per roll, 13 lbs. per square) and #30 (216 S.F. per roll, 27 lbs. per square).

Inter-ply bitumen varies from 24 lbs. per sq. (asphalt) to 30 lbs. per sq. (coal tar) per ply, MF4@ 25%. Flood coat bitumen also varies from 60 lbs. per sq. (asphalt) to 75 lbs. per sq. (coal tar), MF4@ 25%. Expendable equipment (mops, brooms, screeds, etc.) runs about 16% of the bitumen cost. For new, inexperienced crews this factor may be much higher.

A rigid insulation board is typically applied in two layers. The first is mechanically attached to nailable decks or spot or solid mopped to non-nailable decks; the second layer is then spot or solid mopped to the first layer. Membrane application follows the insulation, except in protected membrane roofs, where the membrane goes down first and the insulation on top, followed with ballast (stone or concrete pavers). Insulation and related labor costs are NOT included in prices for built-up roofing.

R075213-30   Modified Bitumen Roofing

The cost of modified bitumen roofing is highly dependent on the type of installation that is planned. Installation is based on the type of modifier used in the bitumen. The two most popular modifiers are atactic polypropylene (APP) and styrene butadiene styrene (SBS). The modifiers are added to heated bitumen during the manufacturing process to change its characteristics. A polyethylene, polyester or fiberglass reinforcing sheet is then sandwiched between layers of this bitumen. When completed, the result is a pre-assembled, built-up roof that has increased elasticity and weatherability. Some manufacturers include a surfacing material such as ceramic or mineral granules, metal particles or sand.

The preferred method of adhering SBS-modified bitumen roofing to the substrate is with hot-mopped asphalt (much the same as built-up roofing). This installation method requires a tar kettle/pot to heat the asphalt, as well as the labor, tools and equipment necessary to distribute and spread the hot asphalt.

The alternative method for applying APP and SBS modified bitumen is as follows. A skilled installer uses a torch to melt a small pool of bitumen off the membrane. This pool must form across the entire roll for proper adhesion. The installer must unroll the roofing at a pace slow enough to melt the bitumen, but fast enough to prevent damage to the rest of the membrane.

Modified bitumen roofing provides the advantages of both built-up and single-ply roofing. Labor costs are reduced over those of built-up roofing because only a single ply is necessary. The elasticity of single-ply roofing is attained with the reinforcing sheet and polymer modifiers. Modifieds have some self-healing characteristics and because of their multi-layer construction, they offer the reliability and safety of built-up roofing.

R078413-30   Firestopping

Firestopping is the sealing of structural, mechanical, electrical and other penetrations through fire-rated assemblies. The basic components of firestop systems are safing insulation and firestop sealant on both sides of wall penetrations and the top side of floor penetrations.

Pipe penetrations are assumed to be through concrete, grout, or joint compound and can be sleeved or unsleeved. Costs for the penetrations and sleeves are not included. An annular space of 1" is assumed. Escutcheons are not included.

A metallic pipe is assumed to be copper, aluminum, cast iron or similar metallic material. An insulated metallic pipe is assumed to be covered with a thermal insulating jacket of varying thickness and materials.

A non-metallic pipe is assumed to be PVC, CPVC, FR Polypropylene or similar plastic piping material. Intumescent firestop sealants or wrap strips are included. Collars on both sides of wall penetrations and a sheet metal plate on the underside of floor penetrations are included.

Ductwork is assumed to be sheet metal, stainless steel or similar metallic material. Duct penetrations are assumed to be through concrete, grout or joint compound. Costs for penetrations and sleeves are not included. An annular space of 1/2" is assumed.

Multi-trade openings include costs for sheet metal forms, firestop mortar, wrap strips, collars and sealants as necessary.

Structural penetrations joints are assumed to be 1/2" or less. CMU walls are assumed to be within 1-1/2" of the metal deck. Drywall walls are assumed to be tight to the underside of metal decking.

Metal panel, glass or curtain wall systems include a spandrel area of 5' filled with mineral wool foil-faced insulation. Fasteners and stiffeners are included.

R081313-20 Steel Door Selection Guide

Standard steel doors are classified into four levels, as recommended by the Steel Door Institute in the chart below. Each of the four levels offers a range of construction models and designs to meet architectural requirements for preference and appearance, including full flush, seamless, and stile & rail. Recommended minimum gauge requirements are also included.

For complete standard steel door construction specifications and available sizes, refer to the Steel Door Institute Technical Data Series, ANSI A250.8-98 (SDI-100), and ANSI A250.4-94 Test Procedure and Acceptance Criteria for Physical Endurance of Steel Door and Hardware Reinforcements.

Level		Model	Construction	For Full Flush or Seamless		
				Min. Gauge	Thickness (in)	Thickness (mm)
I	Standard Duty	1	Full Flush	20	0.032	0.8
		2	Seamless			
II	Heavy Duty	1	Full Flush	18	0.042	1.0
		2	Seamless			
III	Extra Heavy Duty	1	Full Flush	16	0.053	1.3
		2	Seamless			
		3	*Stile & Rail			
IV	Maximum Duty	1	Full Flush	14	0.067	1.6
		2	Seamless			

*Stiles & rails are 16 gauge; flush panels, when specified, are 18 gauge

R085123-10 Steel Sash

An ironworker crew will erect 25 S.F. or 1.3 sash unit per hour, whichever is less.

A mechanic will point 30 L.F. per hour.

A painter will paint 90 S.F. per coat per hour.

A Glazier production depends on light size.

Allow 1 lb. special steel sash putty per 16″ x 20″ light.

R085313-20 Replacement Windows

Replacement windows are typically measured per United Inch.

United Inches are calculated by rounding the width and height of the window opening up to the nearest inch, then adding the two figures.

The labor cost for replacement windows includes removal of sash, existing sash balance or weights, parting bead where necessary and installation of new window.

Debris hauling and dump fees are not included.

R087110-10   Hardware Finishes

This table describes hardware finishes used throughout the industry. It also shows the base metal and the respective symbols in the three predominate systems of identification. Many of these are used in pricing descriptions in Division Eight.

US″	BMHA*	CDN	Base	Description
US P	600	CP	Steel	Primed for Painting
US 1B	601	C1B	Steel	Bright Black Japanned
US 2C	602	C2C	Steel	Zinc Plated
US 2G	603	C2G	Steel	Zinc Plated
US 3	605	C3	Brass	Bright Brass, Clear Coated
US 4	606	C4	Brass	Satin Brass, Clear Coated
US 5	609	C5	Brass	Satin Brass, Blackened, Satin Relieved, Clear Coated
US 7	610	C7	Brass	Satin Brass, Blackened, Bright Relieved, Clear Coated
US 9	611	C9	Bronze	Bright Bronze, Clear Coated
US 10	612	C10	Bronze	Satin Bronze, Clear Coated
US 10A	641	C10A	Steel	Antiqued Bronze, Oiled and Lacquered
US 10B	613	C10B	Bronze	Antiqued Bronze, Oiled
US 11	616	C11	Bronze	Satin Bronze, Blackened, Satin Relieved, Clear Coated
US 14	618	C14	Brass/Bronze	Bright Nickel Plated, Clear Coated
US 15	619	C15	Brass/Bronze	Satin Nickel, Clear Coated
US 15A	620	C15A	Brass/Bronze	Satin Nickel Plated, Blackened, Satin Relieved, Clear Coated
US 17A	621	C17A	Brass/Bronze	Nickel Plated, Blackened, Relieved, Clear Coated
US 19	622	C19	Brass/Bronze	Flat Black Coated
US 20	623	C20	Brass/Bronze	Statuary Bronze, Light
US 20A	624	C20A	Brass/Bronze	Statuary Bronze, Dark
US 26	625	C26	Brass/Bronze	Bright Chromium
US 26D	626	C26D	Brass/Bronze	Satin Chromium
US 20	627	C27	Aluminum	Satin Aluminum Clear
US 28	628	C28	Aluminum	Anodized Dull Aluminum
US 32	629	C32	Stainless Steel	Bright Stainless Steel
US 32D	630	C32D	Stainless Steel	Stainless Steel
US 3	632	C3	Steel	Bright Brass Plated, Clear Coated
US 4	633	C4	Steel	Satin Brass, Clear Coated
US 7	636	C7	Steel	Satin Brass Plated, Blackened, Bright Relieved, Clear Coated
US 9	637	C9	Steel	Bright Bronze Plated, Clear Coated
US 5	638	C5	Steel	Satin Brass Plated, Blackened, Bright Relieved, Clear Coated
US 10	639	C10	Steel	Satin Bronze Plated, Clear Coated
US 10B	640	C10B	Steel	Antique Bronze, Oiled
US 10A	641	C10A	Steel	Antiqued Bronze, Oiled and Lacquered
US 11	643	C11	Steel	Satin Bronze Plated, Blackened, Bright Relieved, Clear Coated
US 14	645	C14	Steel	Bright Nickel Plated, Clear Coated
US 15	646	C15	Steel	Satin Nickel Plated, Clear Coated
US 15A	647	C15A	Steel	Nickel Plated, Blackened, Bright Relieved, Clear Coated
US 17A	648	C17A	Steel	Nickel Plated, Blackened, Relieved, Clear Coated
US 20	649	C20	Steel	Statuary Bronze, Light
US 20A	650	C20A	Steel	Statuary Bronze, Dark
US 26	651	C26	Steel	Bright Chromium Plated
US 26D	652	C26D	Steel	Satin Chromium Plated

* - BMHA Builders Hardware Manufacturing Association
″ - US Equivalent
^ - Canadian Equivalent
Japanning is imitating Asian lacquer work

R088110-10 Glazing Productivity

Some glass sizes are estimated by the "united inch" (height + width). The table below shows the number of lights glazed in an eight-hour period by the crew size indicated, for glass up to 1/4" thick. Square or nearly square lights are more economical on a S.F. basis. Long slender lights will have a high S.F. installation cost. For insulated glass reduce production by 33%. For 1/2" float glass reduce production by 50%. Production time for glazing with two glaziers per day averages: 1/4" float glass 120 S.F.; 1/2" float glass 55 S.F.; 1/2" insulated glass 95 S.F.; 3/4" insulated glass 75 S.F.

Glazing Method	United Inches per Light							
	40″	60″	80″	100″	135″	165″	200″	240″
Number of Men in Crew	1	1	1	1	2	3	3	4
Industrial sash, putty	60	45	24	15	18	—	—	—
With stops, putty bed	50	36	21	12	16	8	4	3
Wood stops, rubber	40	27	15	9	11	6	3	2
Metal stops, rubber	30	24	14	9	9	6	3	2
Structural glass	10	7	4	3	—	—	—	—
Corrugated glass	12	9	7	4	4	4	3	—
Storefronts	16	15	13	11	7	6	4	4
Skylights, putty glass	60	36	21	12	16	—	—	—
Thiokol set	15	15	11	9	9	6	3	2
Vinyl set, snap on	18	18	13	12	12	7	5	4
Maximum area per light	2.8 S.F.	6.3 S.F.	11.1 S.F.	17.4 S.F.	31.6 S.F.	47 S.F.	69 S.F.	100 S.F.

R092000-50 Lath, Plaster and Gypsum Board

Gypsum board lath is available in 3/8″ thick x 16″ wide x 4′ long sheets as a base material for multi-layer plaster applications. It is also available as a base for either multi-layer or veneer plaster applications in 1/2″ and 5/8″ thick–4′ wide x 8′, 10′ or 12′ long sheets. Fasteners are screws or blued ring shank nails for wood framing and screws for metal framing.

Metal lath is available in diamond mesh patterns with flat or self-furring profiles. Paper backing is available for applications where excessive plaster waste needs to be avoided. A slotted mesh ribbed lath should be used in areas where the span between structural supports is greater than normal. Most metal lath comes in 27″ x 96″ sheets. Diamond mesh weighs 1.75, 2.5 or 3.4 pounds per square yard, slotted mesh lath weighs 2.75 or 3.4 pounds per square yard. Metal lath can be nailed, screwed or tied in place.

Many **accessories** are available. Corner beads, flat reinforcing strips, casing beads, control and expansion joints, furring brackets and channels are some examples. Note that accessories are not included in plaster or stucco line items.

Plaster is defined as a material or combination of materials that when mixed with a suitable amount of water, forms a plastic mass or paste. When applied to a surface, the paste adheres to it and subsequently hardens, preserving in a rigid state the form or texture imposed during the period of elasticity.

Gypsum plaster is made from ground calcined gypsum. It is mixed with aggregates and water for use as a base coat plaster.

Vermiculite plaster is a fire-retardant plaster covering used on steel beams, concrete slabs and other heavy construction materials. Vermiculite is a group name for certain clay minerals, hydrous silicates or aluminum, magnesium and iron that have been expanded by heat.

Perlite plaster is a plaster using perlite as an aggregate instead of sand. Perlite is a volcanic glass that has been expanded by heat.

Gauging plaster is a mix of gypsum plaster and lime putty that when applied produces a quick drying finish coat.

Veneer plaster is a one or two component gypsum plaster used as a thin finish coat over special gypsum board.

Keenes cement is a white cementitious material manufactured from gypsum that has been burned at a high temperature and ground to a fine powder. Alum is added to accelerate the set. The resulting plaster is hard and strong and accepts and maintains a high polish, hence it is used as a finishing plaster.

Stucco is a Portland cement based plaster used primarily as an exterior finish.

Plaster is used on both interior and exterior surfaces. Generally it is applied in multiple-coat systems. A three-coat system uses the terms scratch, brown and finish to identify each coat. A two-coat system uses base and finish to describe each coat. Each type of plaster and application system has attributes that are chosen by the designer to best fit the intended use.

Gypsum Plaster Quantities for 100 S.Y.	2 Coat, 5/8″ Thick		3 Coat, 3/4″ Thick		
	Base	Finish	Scratch	Brown	Finish
	1:3 Mix	2:1 Mix	1:2 Mix	1:3 Mix	2:1 Mix
Gypsum plaster	1,300 lb.		1,350 lb.	650 lb.	
Sand	1.75 C.Y.		1.85 C.Y.	1.35 C.Y.	
Finish hydrated lime		340 lb.			340 lb.
Gauging plaster		170 lb.			170 lb.

Vermiculite or Perlite Plaster Quantities for 100 S.Y.	2 Coat, 5/8″ Thick		3 Coat, 3/4″ Thick		
	Base	Finish	Scratch	Brown	Finish
Gypsum plaster	1,250 lb.		1,450 lb.	800 lb.	
Vermiculite or perlite	7.8 bags		8.0 bags	3.3 bags	
Finish hydrated lime		340 lb.			340 lb.
Gauging plaster		170 lb.			170 lb.

Stucco–Three-Coat System Quantities for 100 S.Y.	On Wood Frame	On Masonry
Portland cement	29 bags	21 bags
Sand	2.6 C.Y.	2.0 C.Y.
Hydrated lime	180 lb.	120 lb.

For customer support on your Commercial Renovation Costs with RSMeans data, call 800.448.8182.

871

R092910-10 Levels of Gypsum Board Finish

In the past, contract documents often used phrases such as "industry standard" and "workmanlike finish" to specify the expected quality of gypsum board wall and ceiling installations. The vagueness of these descriptions led to unacceptable work and disputes.

In order to resolve this problem, four major trade associations concerned with the manufacture, erection, finish and decoration of gypsum board wall and ceiling systems developed an industry-wide *Recommended Levels of Gypsum Board Finish*.

The finish of gypsum board walls and ceilings for specific final decoration is dependent on a number of factors. A primary consideration is the location of the surface and the degree of decorative treatment desired. Painted and unpainted surfaces in warehouses and other areas where appearance is normally not critical may simply require the taping of wallboard joints and 'spotting' of fastener heads. Blemish-free, smooth, monolithic surfaces often intended for painted and decorated walls and ceilings in habitated structures, ranging from single-family dwellings through monumental buildings, require additional finishing prior to the application of the final decoration.

Other factors to be considered in determining the level of finish of the gypsum board surface are (1) the type of angle of surface illumination (both natural and artificial lighting), and (2) the paint and method of application or the type and finish of wallcovering specified as the final decoration. Critical lighting conditions, gloss paints, and thin wall coverings require a higher level of gypsum board finish than heavily textured surfaces which are subsequently painted or surfaces which are to be decorated with heavy grade wall coverings.

The following descriptions were developed by the Association of the Wall and Ceiling Industries-International (AWCI), Ceiling & Interior Systems Construction Association (CISCA), Gypsum Association (GA), and Painting and Decorating Contractors of America (PDCA) as a guide.

Level 0: Used in temporary construction or wherever the final decoration has not been determined. Unfinished. No taping, finishing or corner beads are required. Also could be used where non-predecorated panels will be used in demountable-type partitions that are to be painted as a final finish.

Level 1: Frequently used in plenum areas above ceilings, in attics, in areas where the assembly would generally be concealed, or in building service corridors and other areas not normally open to public view. Some degree of sound and smoke control is provided; in some geographic areas, this level is referred to as "fire-taping," although this level of finish does not typically meet fire-resistant assembly requirements. Where a fire resistance rating is required for the gypsum board assembly, details of construction should be in accordance with reports of fire tests of assemblies that have met the requirements of the fire rating acceptable.

All joints and interior angles shall have tape embedded in joint compound. Accessories are optional at specifier discretion in corridors and other areas with pedestrian traffic. Tape and fastener heads need not be covered with joint compound. Surface shall be free of excess joint compound. Tool marks and ridges are acceptable.

Level 2: It may be specified for standard gypsum board surfaces in garages, warehouse storage, or other similar areas where surface appearance is not of primary importance.

All joints and interior angles shall have tape embedded in joint compound and shall be immediately wiped with a joint knife or trowel, leaving a thin coating of joint compound over all joints and interior angles. Fastener heads and accessories shall be covered with a coat of joint compound. Surface shall be free of excess joint compound. Tool marks and ridges are acceptable.

Level 3: Typically used in areas receiving heavy texture (spray or hand applied) finishes before final painting, or where commercial-grade (heavy duty) wall coverings are to be applied as the final decoration. This level of finish should not be used where smooth painted surfaces or where lighter weight wall coverings are specified. The prepared surface shall be coated with a drywall primer prior to the application of final finishes.

All joints and interior angles shall have tape embedded in joint compound and shall be immediately wiped with a joint knife or trowel, leaving a thin coating of joint compound over all joints and interior angles. One additional coat of joint compound shall be applied over all joints and interior angles. Fastener heads and accessories shall be covered with two separate coats of joint compound. All joint compounds shall be smooth and free of tool marks and ridges. The prepared surface shall be covered with a drywall primer prior to the application of the final decoration.

Level 4: This level should be used where residential grade (light duty) wall coverings, flat paints, or light textures are to be applied. The prepared surface shall be coated with a drywall primer prior to the application of final finishes. Release agents for wall coverings are specifically formulated to minimize damage if coverings are subsequently removed.

The weight, texture, and sheen level of the wall covering material selected should be taken into consideration when specifying wall coverings over this level of drywall treatment. Joints and fasteners must be sufficiently concealed if the wall covering material is lightweight, contains limited pattern, has a glossy finish, or has any combination of these features. In critical lighting areas, flat paints applied over light textures tend to reduce joint photographing. Gloss, semi-gloss, and enamel paints are not recommended over this level of finish.

All joints and interior angles shall have tape embedded in joint compound and shall be immediately wiped with a joint knife or trowel, leaving a thin coating of joint compound over all joints and interior angles. In addition, two separate coats of joint compound shall be applied over all flat joints and one separate coat of joint compound applied over interior angles. Fastener heads and accessories shall be covered with three separate coats of joint compound. All joint compounds shall be smooth and free of tool marks and ridges. The prepared surface shall be covered with a drywall primer like Sheetrock first coat prior to the application of the final decoration.

Level 5: The highest quality finish is the most effective method to provide a uniform surface and minimize the possibility of joint photographing and of fasteners showing through the final decoration. This level of finish is required where gloss, semi-gloss, or enamel is specified; when flat joints are specified over an untextured surface; or where critical lighting conditions occur. The prepared surface shall be coated with a drywall primer prior to the application of the final decoration.

All joints and interior angles shall have tape embedded in joint compound and be immediately wiped with a joint knife or trowel, leaving a thin coating of joint compound over all joints and interior angles. Two separate coats of joint compound shall be applied over all flat joints and one separate coat of joint compound applied over interior angles. Fastener heads and accessories shall be covered with three separate coats of joint compound.

A thin skim coat of joint compound shall be trowel applied to the entire surface. Excess compound is immediately troweled off, leaving a film or skim coating of compound completely covering the paper. As an alternative to a skim coat, a material manufactured especially for this purpose may be applied such as Sheetrock Tuff-Hide primer surfacer. The surface must be smooth and free of tool marks and ridges. The prepared surface shall be covered with a drywall primer prior to the application of the final decoration.

R142000-10   Freight Elevators

Capacities run from 2,000 lbs. to over 100,000 lbs. with 3,000 lbs. to 10,000 lbs. most common. Travel speeds are generally lower and the control is less intricate than on passenger elevators. Costs in the unit price sections are for hydraulic and geared elevators.

R142000-20   Elevator Selective Costs   See R142000-40 for cost development.

A. Base Unit	Passenger		Freight		Hospital	
	Hydraulic	**Electric**	**Hydraulic**	**Electric**	**Hydraulic**	**Electric**
Capacity	1,500 lb.	2,000 lb.	2,000 lb.	4,000 lb.	4,000 lb.	4,000 lb.
Speed	100 F.P.M.	200 F.P.M.	100 F.P.M.	200 F.P.M.	100 F.P.M.	200 F.P.M.
#Stops/Travel Ft.	2/12	4/40	2/20	4/40	2/20	4/40
Push Button Oper.	Yes	Yes	Yes	Yes	Yes	Yes
Telephone Box & Wire	"	"	"	"	"	"
Emergency Lighting	"	"	No	No	"	"
Cab	Plastic Lam. Walls	Plastic Lam. Walls	Painted Steel	Painted Steel	Plastic Lam. Walls	Plastic Lam. Walls
Cove Lighting	Yes	Yes	No	No	Yes	Yes
Floor	V.C.T.	V.C.T.	Wood w/Safety Treads	Wood w/Safety Treads	V.C.T.	V.C.T.
Doors, & Speedside Slide	Yes	Yes	Yes	Yes	Yes	Yes
Gates, Manual	No	No	No	No	No	No
Signals, Lighted Buttons	Car and Hall	Car and Hall	Car and Hall	Car and Hall	Car and Hall	Car and Hall
O.H. Geared Machine	N.A.	Yes	N.A.	Yes	N.A.	Yes
Variable Voltage Contr.	"	"	N.A.	"	"	"
Emergency Alarm	Yes	"	Yes	"	Yes	"
Class "A" Loading	N.A.	N.A.	"	"	N.A.	N.A.

R142000-30   Passenger Elevators

Electric elevators are used generally but hydraulic elevators can be used for lifts up to 70' and where large capacities are required. Hydraulic speeds are limited to 200 F.P.M. but cars are self leveling at the stops. On low rises, hydraulic installation runs about 15% less than standard electric types but on higher rises this installation cost advantage is reduced. Maintenance of hydraulic elevators is about the same as the electric type but the underground portion is not included in the maintenance contract.

In electric elevators there are several control systems available, the choice of which will be based upon elevator use, size, speed and cost criteria. The two types of drives are geared for low speeds and gearless for 450 F.P.M. and over.

The tables on the preceding pages illustrate typical installed costs of the various types of elevators available.

R142000-40   Elevator Cost Development

To price a new car or truck from the factory, you must start with the manufacturer's basic model, then add or exchange optional equipment and features. The same is true for pricing elevators.

Requirement: One-passenger elevator, five-story hydraulic, 2,500 lb. capacity, 12' floor to floor, speed 150 F.P.M., emergency power switching and maintenance contract.

Example:

Description	Adjustment
A. Base Elevator: Hydraulic Passenger, 1500 lb. Capacity, 100 fpm, 2 Stops, Standard Finish	1 Ea.
B. Capacity Adjustment (2,500 lb.)	1 Ea.
C. Excess Travel Adjustment: 48' Total Travel (4 x 12') minus 12' Base Unit Travel =	36 V.L.F.
D. Stops Adjustment: 5 Total Stops minus 2 Stops (Base Unit) =	3 Stops
E. Speed Adjustment (150 F.P.M.)	1 Ea.
F. Options:	
1. Intercom Service	1 Ea.
2. Emergency Power Switching, Automatic	1 Ea.
3. Stainless Steel Entrance Doors	5 Ea.
4. Maintenance Contract (12 Months)	1 Ea.
5. Position Indicator for main floor level (none indicated in Base Unit)	1 Ea.

R220102-20   Labor Adjustment Factors

Labor Adjustment Factors are provided for Divisions 21, 22, and 23 to assist the mechanical estimator account for the various complexities and special conditions of any particular project. While a single percentage has been entered on each line of Division 22 01 02.20, it should be understood that these are just suggested midpoints of ranges of values commonly used by mechanical estimators. They may be increased or decreased depending on the severity of the special conditions.

The group for "existing occupied buildings" has been the subject of requests for explanation. Actually there are two stages to this group: buildings that are existing and "finished" but unoccupied, and those that also are occupied. Buildings that are "finished" may result in higher labor costs due to the

workers having to be more careful not to damage finished walls, ceilings, floors, etc. and may necessitate special protective coverings and barriers. Also corridor bends and doorways may not accommodate long pieces of pipe or larger pieces of equipment. Work above an already hung ceiling can be very time consuming. The addition of occupants may force the work to be done on premium time (nights and/or weekends), eliminate the possible use of some preferred tools such as pneumatic drivers, powder charged drivers, etc. The estimator should evaluate the access to the work area and just how the work is going to be accomplished to arrive at an increase in labor costs over "normal" new construction productivity.

R220105-10   Demolition (Selective vs. Removal for Replacement)

Demolition can be divided into two basic categories.

One type of demolition involves the removal of material with no concern for its replacement. The labor-hours to estimate this work are found under "Selective Demolition" in the Fire Protection, Plumbing and HVAC Divisions. It is selective in that individual items or all the material installed as a system or trade grouping such as plumbing or heating systems are removed. This may be accomplished by the easiest way possible, such as sawing, torch cutting, or sledge hammering as well as simple unbolting.

The second type of demolition is the removal of some items for repair or replacement. This removal may involve careful draining, opening of unions,

disconnecting and tagging electrical connections, capping pipes/ducts to prevent entry of debris or leakage of the material contained as well as transporting the item away from its in-place location to a truck/dumpster. An approximation of the time required to accomplish this type of demolition is to use half of the time indicated as necessary to install a new unit. For example: installation of a new pump might be listed as requiring 6 labor-hours so if we had to estimate the removal of the old pump we would allow an additional 3 hours for a total of 9 hours. That is, the complete replacement of a defective pump with a new pump would be estimated to take 9 labor-hours.

R221113-50   Pipe Material Considerations

1. Malleable fittings should be used for gas service.
2. Malleable fittings are used where there are stresses/strains due to expansion and vibration.
3. Cast fittings may be broken as an aid to disassembling heating lines frozen by long use, temperature and minerals.
4. A cast iron pipe is extensively used for underground and submerged service.
5. Type M (light wall) copper tubing is available in hard temper only and is used for nonpressure and less severe applications than K and L.

6. Type L (medium wall) copper tubing, available hard or soft for interior service.
7. Type K (heavy wall) copper tubing, available in hard or soft temper for use where conditions are severe. For underground and interior service.
8. Hard drawn tubing requires fewer hangers or supports but should not be bent. Silver brazed fittings are recommended, but soft solder is normally used.
9. Type DMV (very light wall) copper tubing designed for drainage, waste and vent plus other non-critical pressure services.

Domestic/Imported Pipe and Fittings Costs

The prices shown in this publication for steel/cast iron pipe and steel, cast iron, and malleable iron fittings are based on domestic production sold at the normal trade discounts. The above listed items of foreign manufacture may be available at prices 1/3 to 1/2 of those shown. Some imported items after minor machining or finishing operations are being sold as domestic to further complicate the system.

Caution: Most pipe prices in this data set also include a coupling and pipe hangers which for the larger sizes can add significantly to the per foot cost and should be taken into account when comparing "book cost" with the quoted supplier's cost.

R260105-30   Electrical Demolition (Removal for Replacement)

The purpose of this reference number is to provide a guide to users for electrical "removal for replacement" by applying the rule of thumb: 1/3 of new installation time (typical range from 20% to 50%) for removal. Remember to use reasonable judgment when applying the suggested percentage factor. For example:

Contractors have been requested to remove an existing fluorescent lighting fixture and replace with a new fixture utilizing energy saver lamps and electronic ballast:

In order to fully understand the extent of the project, contractors should visit the job site and estimate the time to perform the renovation work in accordance with applicable national, state and local regulation and codes.

The contractor may need to add extra labor hours to his estimate if he discovers unknown concealed conditions such as: contaminated asbestos ceiling, broken acoustical ceiling tile and need to repair, patch and touch-up paint all the damaged or disturbed areas, tasks normally assigned to general contractors. In addition, the owner could request that the contractors salvage the materials removed and turn over the materials to the owner or dispose of the materials to a reclamation station. The normal removal item is 0.5 labor hour for a lighting fixture and 1.5 labor-hours for new installation time. Revise the estimate times from 2 labor-hours work up to a minimum 4 labor-hours work for just fluorescent lighting fixture.

For removal of large concentrations of lighting fixtures in the same area, apply an "economy of scale" to reduce estimating labor hours.

R260519-90   Wire

Wire quantities are taken off by either measuring each cable run or by extending the conduit and raceway quantities times the number of conductors in the raceway. Ten percent should be added for waste and tie-ins. Keep in mind that the unit of measure of wire is C.L.F. not L.F. as in raceways so the formula would read:

$$\frac{(\text{L.F. Raceway} \times \text{No. of Conductors}) \times 1.10}{100} = \text{C.L.F.}$$

Price per C.L.F. of wire includes:
1. Setting up wire coils or spools on racks
2. Attaching wire to pull in means
3. Measuring and cutting wire
4. Pulling wire into a raceway
5. Identifying and tagging

Price does not include:
1. Connections to breakers, panelboards, or equipment
2. Splices

Job Conditions: Productivity is based on new construction to a height of 15' using rolling staging in an unobstructed area. Material staging is assumed to be within 100' of work being performed.

Economy of Scale: If more than three wires at a time are being pulled, deduct the following percentages from the labor of that grouping:

4-5 wires	25%
6-10 wires	30%
11-15 wires	35%
over 15	40%

If a wire pull is less than 100' in length and is interrupted several times by boxes, lighting outlets, etc., it may be necessary to add the following lengths to each wire being pulled:

Junction box to junction box	2 L.F.
Lighting panel to junction box	6 L.F.
Distribution panel to sub panel	8 L.F.
Switchboard to distribution panel	12 L.F.
Switchboard to motor control center	20 L.F.
Switchboard to cable tray	40 L.F.

Measure of Drops and Riser: It is important when taking off wire quantities to include the wire for drops to electrical equipment. If heights of electrical equipment are not clearly stated, use the following guide:

	Bottom A.F.F.	Top A.F.F.	Inside Cabinet
Safety switch to 100A	5'	6'	2'
Safety switch 400 to 600A	4'	6'	3'
100A panel 12 to 30 circuit	4'	6'	3'
42 circuit panel	3'	6'	4'
Switch box	3'	3'6"	1'
Switchgear	0'	8'	8'
Motor control centers	0'	8'	8'
Transformers - wall mount	4'	8'	2'
Transformers - floor mount	0'	12'	4'

R260533-22 Conductors in Conduit

The table below lists the maximum number of conductors for various sized conduit using THW, TW or THWN insulations.

Copper Wire Size	Conduit Size																											
	1/2"			3/4"			1"			1-1/4"			1-1/2"			2"			2-1/2"			3"		3-1/2"		4"	
	TW	THW	THWN	TW	THW	THWN	TW	THW	THWN	TW	THW	THWN	TW	THW	THWN	TW	THW	THWN	TW	THW	THWN	THW	THWN	THW	THWN	THW	THWN
#14	9	6	13	15	10	24	25	16	39	44	29	69	60	40	94	99	65	154	142	93		143		192			
#12	7	4	10	12	8	18	19	13	29	35	24	51	47	32	70	78	53	114	111	76	164	117		157			
#10	5	4	6	9	6	11	15	11	18	26	19	32	36	26	44	60	43	73	85	61	104	95	160	127		163	
#8	2	1	3	4	3	5	7	5	9	12	10	16	17	13	22	28	22	36	40	32	51	49	79	66	106	85	136
#6		1	1		2	4		4	6		7	11		10	15		16	26		23	37	36	57	48	76	62	98
#4		1	1		1	2		3	4		5	7		7	9		12	16		17	22	27	35	36	47	47	60
#3		1	1		1	1		2	3		4	6		6	8		10	13		15	19	23	29	31	39	40	51
#2		1	1		1	1		2	3		4	5		5	7		9	11		13	16	20	25	27	33	34	43
#1					1	1		1	1		3	3		4	5		6	8		9	12	14	18	19	25	25	32
1/0					1	1		1	1		2	3		3	4		5	7		8	10	12	15	16	21	21	27
2/0					1	1		1	1		1	2		3	3		5	6		7	8	10	13	14	17	18	22
3/0					1	1		1	1		1	1		2	3		4	5		6	7	9	11	12	14	15	18
4/0						1		1	1		1	1		1	2		3	4		5	6	7	9	10	12	13	15
250 kcmil								1	1		1	1		1	1		2	3		4	4	6	7	8	10	10	12
300								1	1		1	1		1	1		2	3		3	4	5	6	7	8	9	11
350									1		1	1		1	1		1	2		3	3	4	5	6	7	8	9
400											1	1		1	1		1	1		2	3	4	5	5	6	7	8
500											1	1		1	1		1	1		1	2	3	4	4	5	6	7
600												1		1	1		1	1		1	1	3	3	4	4	5	5
700														1	1		1	1		1	1	2	3	3	4	4	5
750														1	1		1	1		1	1	2	2	3	3	4	4

Reprinted with permission from NFPA 70-2014, *National Electrical Code®*, Copyright © 2013, National Fire Protection Association, Quincy, MA. This reprinted material is not the complete and official position of the NFPA on the referenced subject, which is represented solely by the standard in its entirety.

R312316-40 Excavating

The selection of equipment used for structural excavation and bulk excavation or for grading is determined by the following factors.
1. Quantity of material
2. Type of material
3. Depth or height of cut
4. Length of haul
5. Condition of haul road
6. Accessibility of site
7. Moisture content and dewatering requirements
8. Availability of excavating and hauling equipment

Some additional costs must be allowed for hand trimming the sides and bottom of concrete pours and other excavation below the general excavation.

Number of B.C.Y. per truck = 1.5 C.Y. bucket x 8 passes = 12 loose C.Y.

$$= 12 \text{ x } \frac{100}{118} = 10.2 \text{ B.C.Y. per truck}$$

Truck Haul Cycle:

Load truck, 8 passes	=	4 minutes
Haul distance, 1 mile	=	9 minutes
Dump time	=	2 minutes
Return, 1 mile	=	7 minutes
Spot under machine	=	1 minute
		23 minute cycle

Add the mobilization and demobilization costs to the total excavation costs. When equipment is rented for more than three days, there is often no mobilization charge by the equipment dealer. On larger jobs outside of urban areas, scrapers can move earth economically provided a dump site or fill area and adequate haul roads are available. Excavation within sheeting bracing or cofferdam bracing is usually done with a clamshell and production

When planning excavation and fill, the following should also be considered.
1. Swell factor
2. Compaction factor
3. Moisture content
4. Density requirements

A typical example for scheduling and estimating the cost of excavation of a 15' deep basement on a dry site when the material must be hauled off the site is outlined below.

Assumptions:
1. Swell factor, 18%
2. No mobilization or demobilization
3. Allowance included for idle time and moving on job
4. No dewatering, sheeting, or bracing
5. No truck spotter or hand trimming

Fleet Haul Production per day in B.C.Y.

$$4 \text{ trucks x } \frac{50 \text{ min. hour}}{23 \text{ min. haul cycle}} \text{ x 8 hrs. x 10.2 B.C.Y.}$$

$$= 4 \text{ x } 2.2 \text{ x } 8 \text{ x } 10.2 = 718 \text{ B.C.Y./day}$$

is low, since the clamshell may have to be guided by hand between the bracing. When excavating or filling an area enclosed with a wellpoint system, add 10% to 15% to the cost to allow for restricted access. When estimating earth excavation quantities for structures, allow work space outside the building footprint for construction of the foundation and a slope of 1:1 unless sheeting is used.

For customer support on your Commercial Renovation Costs with RSMeans data, call 800.448.8182.

877

R312316-45 Excavating Equipment

The table below lists theoretical hourly production in C.Y./hr. bank measure for some typical excavation equipment. Figures assume 50 minute hours, 83% job efficiency, 100% operator efficiency, 90° swing and properly sized hauling units, which must be modified for adverse digging and loading conditions. Actual production costs in the front of the data set average about 50% of the theoretical values listed here.

Equipment	Soil Type	B.C.Y. Weight	% Swell	1 C.Y.	1-1/2 C.Y.	2 C.Y.	2-1/2 C.Y.	3 C.Y.	3-1/2 C.Y.	4 C.Y.
Hydraulic Excavator	Moist loam, sandy clay	3400 lb.	40%	165	195	200	275	330	385	440
"Backhoe"	Sand and gravel	3100	18	140	170	225	240	285	330	380
15' Deep Cut	Common earth	2800	30	150	180	230	250	300	350	400
	Clay, hard, dense	3000	33	120	140	190	200	240	260	320
Power Shovel Optimum Cut (Ft.)	Moist loam, sandy clay	3400	40	170 (6.0)	245 (7.0)	295 (7.8)	335 (8.4)	385 (8.8)	435 (9.1)	475 (9.4)
	Sand and gravel	3100	18	165 (6.0)	225 (7.0)	275 (7.8)	325 (8.4)	375 (8.8)	420 (9.1)	460 (9.4)
	Common earth	2800	30	145 (7.8)	200 (9.2)	250 (10.2)	295 (11.2)	335 (12.1)	375 (13.0)	425 (13.8)
	Clay, hard, dense	3000	33	120 (9.0)	175 (10.7)	220 (12.2)	255 (13.3)	300 (14.2)	335 (15.1)	375 (16.0)
Drag Line Optimum Cut (Ft.)	Moist loam, sandy clay	3400	40	130 (6.6)	180 (7.4)	220 (8.0)	250 (8.5)	290 (9.0)	325 (9.5)	385 (10.0)
	Sand and gravel	3100	18	130 (6.6)	175 (7.4)	210 (8.0)	245 (8.5)	280 (9.0)	315 (9.5)	375 (10.0)
	Common earth	2800	30	110 (8.0)	160 (9.0)	190 (9.9)	220 (10.5)	250 (11.0)	280 (11.5)	310 (12.0)
	Clay, hard, dense	3000	33	90 (9.3)	130 (10.7)	160 (11.8)	190 (12.3)	225 (12.8)	250 (13.3)	280 (12.0)

				Wheel Loaders				Track Loaders		
				3 C.Y.	4 C.Y.	6 C.Y.	8 C.Y.	2-1/4 C.Y.	3 C.Y.	4 C.Y.
Loading Tractors	Moist loam, sandy clay	3400	40	260	340	510	690	135	180	250
	Sand and gravel	3100	18	245	320	480	650	130	170	235
	Common earth	2800	30	230	300	460	620	120	155	220
	Clay, hard, dense	3000	33	200	270	415	560	110	145	200
	Rock, well-blasted	4000	50	180	245	380	520	100	130	180

R312323-30 Compacting Backfill

Compaction of fill in embankments, around structures, in trenches, and under slabs is important to control settlement. Factors affecting compaction are:
1. Soil gradation
2. Moisture content
3. Equipment used
4. Depth of fill per lift
5. Density required

Production Rate:

$$\frac{1.75' \text{ plate width} \times 50 \text{ F.P.M.} \times 50 \text{ min./hr.} \times .67' \text{ lift}}{27 \text{ C.F. per C.Y.}} = 108.5 \text{ C.Y./hr.}$$

Production Rate for 4 Passes:

$$\frac{108.5 \text{ C.Y.}}{4 \text{ passes}} = 27.125 \text{ C.Y./hr.} \times 8 \text{ hrs.} = 217 \text{ C.Y./day}$$

Example:

Compact granular fill around a building foundation using a 21" wide x 24" vibratory plate in 8" lifts. Operator moves at 50 F.P.M. working a 50 minute hour to develop 95% Modified Proctor Density with 4 passes.

R314116-40   Wood Sheet Piling

Wood sheet piling may be used for depths to 20' where there is no ground water. If moderate ground water is encountered Tongue & Groove sheeting will help to keep it out. When considerable ground water is present, steel sheeting must be used.

For estimating purposes on trench excavation, sizes are as follows:

Depth	Sheeting	Wales	Braces	B.F. per S.F.
To 8'	3 x 12's	6 x 8's, 2 line	6 x 8's, @ 10'	4.0 @ 8'
8' x 12'	3 x 12's	10 x 10's, 2 line	10 x 10's, @ 9'	5.0 average
12' to 20'	3 x 12's	12 x 12's, 3 line	12 x 12's, @ 8'	7.0 average

Sheeting to be toed in at least 2' depending upon soil conditions. A five person crew with an air compressor and sheeting driver can drive and brace 440 SF/day at 8' deep, 360 SF/day at 12' deep, and 320 SF/day at 16' deep.

For normal soils, piling can be pulled in 1/3 the time to install. Pulling difficulty increases with the time in the ground. Production can be increased by high pressure jetting.

R316326-60   Caissons

The three principal types of caissons are:

(1) Belled Caissons, which except for shallow depths and poor soil conditions, are generally recommended. They provide more bearing than shaft area. Because of its conical shape, no horizontal reinforcement of the bell is required.

(2) Straight Shaft Caissons are used where relatively light loads are to be supported by caissons that rest on high value bearing strata. While the shaft is larger in diameter than for belled types this is more than offset by the savings in time and labor.

(3) Keyed Caissons are used when extremely heavy loads are to be carried. A keyed or socketed caisson transfers its load into rock by a combination of end-bearing and shear reinforcing of the shaft. The most economical shaft often consists of a steel casing, a steel wide flange core and concrete. Allowable compressive stresses of .225 f'c for concrete, 16,000 psi for the wide flange core, and 9,000 psi for the steel casing are commonly used. The usual range of shaft diameter is 18" to 84". The number of sizes specified for any one project should be limited due to the problems of casing and auger storage. When hand work is to be performed, shaft diameters should not be less than 32". When inspection of borings is required a minimum shaft diameter of 30" is recommended. Concrete caissons are intended to be poured against earth excavation so permanent forms, which add to cost, should not be used if the excavation is clean and the earth is sufficiently impervious to prevent excessive loss of concrete.

Soil Conditions for Belling		
Good	Requires Handwork	Not Recommended
Clay	Hard Shale	Silt
Sandy Clay	Limestone	Sand
Silty Clay	Sandstone	Gravel
Clayey Silt	Weathered Mica	Igneous Rock
Hard-pan		
Soft Shale		
Decomposed Rock		

R329219-50   Seeding

The type of grass is determined by light, shade and moisture content of soil plus intended use. Fertilizer should be disked 4" before seeding. For steep slopes disk five tons of mulch and lay two tons of hay or straw on surface per acre after seeding. Surface mulch can be staked, lightly disked or tar emulsion sprayed. Material for mulch can be wood chips, peat moss, partially rotted hay or straw, wood fibers and sprayed emulsions. Hemp seed blankets with fertilizer are also available. For spring seeding, watering is necessary. Late fall seeding may have to be reseeded in the spring. Hydraulic seeding, power mulching, and aerial seeding can be used on large areas.

R331113-80   Piping Designations

There are several systems currently in use to describe pipe and fittings. The following paragraphs will help to identify and clarify classifications of piping systems used for water distribution.

Piping may be classified by schedule. Piping schedules include 5S, 10S, 10, 20, 30, Standard, 40, 60, Extra Strong, 80, 100, 120, 140, 160 and Double Extra Strong. These schedules are dependent upon the pipe wall thickness. The wall thickness of a particular schedule may vary with pipe size.

Ductile iron pipe for water distribution is classified by Pressure Classes such as Class 150, 200, 250, 300 and 350. These classes are actually the rated water working pressure of the pipe in pounds per square inch (psi). The pipe in these pressure classes is designed to withstand the rated water working pressure plus a surge allowance of 100 psi.

The American Water Works Association (AWWA) provides standards for various types of **plastic pipe.** C-900 is the specification for polyvinyl chloride (PVC) piping used for water distribution in sizes ranging from 4″ through 12″. C-901 is the specification for polyethylene (PE) pressure pipe, tubing and fittings used for water distribution in sizes ranging from 1/2″ through 3″. C-905 is the specification for PVC piping sizes 14″ and greater.

PVC pressure-rated pipe is identified using the standard dimensional ratio (SDR) method. This method is defined by the American Society for Testing and Materials (ASTM) Standard D 2241. This pipe is available in SDR numbers 64, 41, 32.5, 26, 21, 17, and 13.5. A pipe with an SDR of 64 will have the thinnest wall while a pipe with an SDR of 13.5 will have the thickest wall. When the pressure rating (PR) of a pipe is given in psi, it is based on a line supplying water at 73 degrees F.

The National Sanitation Foundation (NSF) seal of approval is applied to products that can be used with potable water. These products have been tested to ANSI/NSF Standard 14.

Valves and strainers are classified by American National Standards Institute (ANSI) Classes. These Classes are 125, 150, 200, 250, 300, 400, 600, 900, 1500 and 2500. Within each class there is an operating pressure range dependent upon temperature. Design parameters should be compared to the appropriate material dependent, pressure-temperature rating chart for accurate valve selection.

Change Orders

Change Order Considerations

A change order is a written document usually prepared by the design professional and signed by the owner, the architect/engineer, and the contractor. A change order states the agreement of the parties to: an addition, deletion, or revision in the work; an adjustment in the contract sum, if any; or an adjustment in the contract time, if any. Change orders, or "extras", in the construction process occur after execution of the construction contract and impact architects/engineers, contractors, and owners.

Change orders that are properly recognized and managed can ensure orderly, professional, and profitable progress for everyone involved in the project. There are many causes for change orders and change order requests. In all cases, change orders or change order requests should be addressed promptly and in a precise and prescribed manner. The following paragraphs include information regarding change order pricing and procedures.

The Causes of Change Orders

Reasons for issuing change orders include:

- Unforeseen field conditions that require a change in the work
- Correction of design discrepancies, errors, or omissions in the contract documents
- Owner-requested changes, either by design criteria, scope of work, or project objectives
- Completion date changes for reasons unrelated to the construction process
- Changes in building code interpretations, or other public authority requirements that require a change in the work
- Changes in availability of existing or new materials and products

Procedures

Properly written contract documents must include the correct change order procedures for all parties—owners, design professionals, and contractors—to follow in order to avoid costly delays and litigation.

Being "in the right" is not always a sufficient or acceptable defense. The contract provisions requiring notification and documentation must be adhered to within a defined or reasonable time frame.

The appropriate method of handling change orders is by a written proposal and acceptance by all parties involved. Prior to starting work on a project, all parties should identify their authorized agents who may sign and accept change orders, as well as any limits placed on their authority.

Time may be a critical factor when the need for a change arises. For such cases, the contractor might be directed to proceed on a "time and materials" basis, rather than wait for all paperwork to be processed—a delay that could impede progress. In this situation, the contractor must still follow the prescribed change order procedures including, but not limited to, notification and documentation.

Lack of documentation can be very costly, especially if legal judgments are to be made, and if certain field personnel are no longer available. For time and material change orders, the contractor should keep accurate daily records of all labor and material allocated to the change.

Owners or awarding authorities who do considerable and continual building construction (such as the federal government) realize the inevitability of change orders for numerous reasons, both predictable and unpredictable. As a result, the federal government, the American Institute of Architects (AIA), the Engineers Joint Contract Documents Committee (EJCDC), and other contractor, legal, and technical organizations have developed standards and procedures to be followed by all parties to achieve contract continuance and timely completion, while being financially fair to all concerned.

Pricing Change Orders

When pricing change orders, regardless of their cause, the most significant factor is when the change occurs. The need for a change may be perceived in the field or requested by the architect/engineer *before* any of the actual installation has begun, or may evolve or appear *during* construction when the item of work in question is partially installed. In the latter cases, the original sequence of construction is disrupted, along with all contiguous and supporting systems. Change orders cause the greatest impact when they occur *after* the installation has been completed and must be uncovered, or even replaced. Post-completion changes may be caused by necessary design changes, product failure, or changes in the owner's requirements that are not discovered until the building or the systems begin to function.

Specified procedures of notification and record keeping must be adhered to and enforced regardless of the stage of construction: *before, during,* or *after* installation. Some bidding documents anticipate change orders by requiring that unit prices including overhead and profit percentages—for additional as well as deductible changes—be listed. Generally these unit prices do not fully take into account the ripple effect, or impact on other trades, and should be used for general guidance only.

When pricing change orders, it is important to classify the time frame in which the change occurs. There are two basic time frames for change orders: *pre-installation change orders,* which occur before the start of construction, and *post-installation change orders,* which involve reworking after the original installation. Change orders that occur between these stages may be priced according to the extent of work completed using a combination of techniques developed for pricing *pre-* and *post-installation* changes.

Factors To Consider When Pricing Change Orders

As an estimator begins to prepare a change order, the following questions should be reviewed to determine their impact on the final price.

General

- *Is the change order work* pre-installation *or* post-installation?

  Change order work costs vary according to how much of the installation has been completed. Once workers have the project scoped in their minds, even though they have not started, it can be difficult to refocus. Consequently they may spend more than the normal amount of time understanding the change. Also, modifications to work in place, such as trimming or refitting, usually take more time than was initially estimated. The greater the amount of work in place, the more reluctant workers are to change it. Psychologically they may resent the change and as a result the rework takes longer than normal. Post-installation change order estimates must include demolition of existing work as required to accomplish the change. If the work is performed at a later time, additional obstacles, such as building finishes, may be present which must be protected. Regardless of whether the change occurs pre-installation or post-installation, attempt to isolate the identifiable factors and price them separately. For example, add shipping costs that may be required pre-installation or any demolition required post-installation. Then analyze the potential impact on productivity of psychological and/or learning curve factors and adjust the output rates accordingly. One approach is to break down the typical workday into segments and quantify the impact on each segment.

Change Order Installation Efficiency

The labor-hours expressed (for new construction) are based on average installation time, using an efficiency level. For change order situations, adjustments to this efficiency level should reflect the daily labor-hour allocation for that particular occurrence.

- *Will the change substantially delay the original completion date?*

A significant change in the project may cause the original completion date to be extended. The extended schedule may subject the contractor to new wage rates dictated by relevant labor contracts. Project supervision and other project overhead must also be extended beyond the original completion date. The schedule extension may also put installation into a new weather season. For example, underground piping scheduled for October installation was delayed until January. As a result, frost penetrated the trench area, thereby changing the degree of difficulty of the task. Changes and delays may have a ripple effect throughout the project. This effect must be analyzed and negotiated with the owner.

- *What is the net effect of a deduct change order?*

In most cases, change orders resulting in a deduction or credit reflect only bare costs. The contractor may retain the overhead and profit based on the original bid.

Materials

- *Will you have to pay more or less for the new material, required by the change order, than you paid for the original purchase?*

The same material prices or discounts will usually apply to materials purchased for change orders as new construction. In some

instances, however, the contractor may forfeit the advantages of competitive pricing for change orders. Consider the following example:

A contractor purchased over $20,000 worth of fan coil units for an installation and obtained the maximum discount. Some time later it was determined the project required an additional matching unit. The contractor has to purchase this unit from the original supplier to ensure a match. The supplier at this time may not discount the unit because of the small quantity, and he is no longer in a competitive situation. The impact of quantity on purchase can add between 0% and 25% to material prices and/or subcontractor quotes.

- *If materials have been ordered or delivered to the job site, will they be subject to a cancellation charge or restocking fee?*

Check with the supplier to determine if ordered materials are subject to a cancellation charge. Delivered materials not used as a result of a change order may be subject to a restocking fee if returned to the supplier. Common restocking charges run between 20% and 40%. Also, delivery charges to return the goods to the supplier must be added.

Labor

- *How efficient is the existing crew at the actual installation?*

Is the same crew that performed the initial work going to do the change order? Possibly the change consists of the installation of a unit identical to one already installed; therefore, the change should take less time. Be sure to consider this potential productivity increase and modify the productivity rates accordingly.

- *If the crew size is increased, what impact will that have on supervision requirements?*

Under most bargaining agreements or management practices, there is a point at which a working foreman is replaced by a nonworking foreman. This replacement increases project overhead by adding a nonproductive worker. If additional workers are added to accelerate the project or to perform changes while maintaining the schedule, be sure to add additional supervision time if warranted. Calculate the

hours involved and the additional cost directly if possible.

- *What are the other impacts of increased crew size?*

The larger the crew, the greater the potential for productivity to decrease. Some of the factors that cause this productivity loss are: overcrowding (producing restrictive conditions in the working space) and possibly a shortage of any special tools and equipment required. Such factors affect not only the crew working on the elements directly involved in the change order, but other crews whose movements may also be hampered. As the crew increases, check its basic composition for changes by the addition or deletion of apprentices or nonworking foreman, and quantify the potential effects of equipment shortages or other logistical factors.

- *As new crews, unfamiliar with the project, are brought onto the site, how long will it take them to become oriented to the project requirements?*

The orientation time for a new crew to become 100% effective varies with the site and type of project. Orientation is easiest at a new construction site and most difficult at existing, very restrictive renovation sites. The type of work also affects orientation time. When all elements of the work are exposed, such as concrete or masonry work, orientation is decreased. When the work is concealed or less visible, such as existing electrical systems, orientation takes longer. Usually orientation can be accomplished in one day or less. Costs for added orientation should be itemized and added to the total estimated cost.

- *How much actual production can be gained by working overtime?*

Short term overtime can be used effectively to accomplish more work in a day. However, as overtime is scheduled to run beyond several weeks, studies have shown marked decreases in output. The following chart shows the effect of long term overtime on worker efficiency. If the anticipated change requires extended overtime to keep the job on schedule, these factors can be used as a guide to predict the impact on time and cost. Add project overhead, particularly supervision, that may also be incurred.

For customer support on your Commercial Renovation Costs with RSMeans data, call 800.448.8182.

Days per Week	Hours per Day	Production Efficiency					Payroll Cost Factors	
		1st Week	2nd Week	3rd Week	4th Week	Average 4 Weeks	@ 1-1/2 Times	@ 2 Times
5	8	100%	100%	100%	100%	100%	100%	100%
	9	100	100	95	90	96.25	105.6	111.1
	10	100	95	90	85	92.50	110.0	120.0
	11	95	90	75	65	81.25	113.6	127.3
	12	90	85	70	60	76.25	116.7	133.3
6	8	100	100	95	90	96.25	108.3	116.7
	9	100	95	90	85	92.50	113.0	125.9
	10	95	90	85	80	87.50	116.7	133.3
	11	95	85	70	65	78.75	119.7	139.4
	12	90	80	65	60	73.75	122.2	144.4
7	8	100	95	85	75	88.75	114.3	128.6
	9	95	90	80	70	83.75	118.3	136.5
	10	90	85	75	65	78.75	121.4	142.9
	11	85	80	65	60	72.50	124.0	148.1
	12	85	75	60	55	68.75	126.2	152.4

Effects of Overtime

Caution: Under many labor agreements, Sundays and holidays are paid at a higher premium than the normal overtime rate.

The use of long-term overtime is counterproductive on almost any construction job; that is, the longer the period of overtime, the lower the actual production rate. Numerous studies have been conducted, and while they have resulted in slightly different numbers, all reach the same conclusion. The figure above tabulates the effects of overtime work on efficiency.

As illustrated, there can be a difference between the *actual* payroll cost per hour and the *effective* cost per hour for overtime work. This is due to the reduced production efficiency with the increase in weekly hours beyond 40. This difference between actual and effective cost results from overtime work over a prolonged period. Short-term overtime work does not result in as great a reduction in efficiency and, in such cases, effective cost may not vary significantly from the actual payroll cost. As the total hours per week are increased on a regular basis, more time is lost due to fatigue, lowered morale, and an increased accident rate.

As an example, assume a project where workers are working 6 days a week, 10 hours per day. From the figure above (based on productivity studies), the average effective productive hours over a 4-week period are:

$$0.875 \times 60 = 52.5$$

Depending upon the locale and day of week, overtime hours may be paid at time and a half or double time. For time and a half, the overall (average) *actual* payroll cost (including regular and overtime hours) is determined as follows:

$$\frac{40 \text{ reg. hrs.} + (20 \text{ overtime hrs.} \times 1.5)}{60 \text{ hrs.}} = 1.167$$

Based on 60 hours, the payroll cost per hour will be 116.7% of the normal rate at 40 hours per week. However, because the effective production (efficiency) for 60 hours is reduced to the equivalent of 52.5 hours, the effective cost of overtime is calculated as follows:

For time and a half:

$$\frac{40 \text{ reg. hrs.} + (20 \text{ overtime hrs.} \times 1.5)}{52.5 \text{ hrs.}} = 1.33$$

The installed cost will be 133% of the normal rate (for labor).

Thus, when figuring overtime, the actual cost per unit of work will be higher than the apparent overtime payroll dollar increase, due to the reduced productivity of the longer work week. These efficiency calculations are true only for those cost factors determined by hours worked. Costs that are applied weekly or monthly, such as equipment rentals, will not be similarly affected.

Equipment

- *What equipment is required to complete the change order?*

Change orders may require extending the rental period of equipment already on the job site, or the addition of special equipment brought in to accomplish the change work. In either case, the additional rental charges and operator labor charges must be added.

Summary

The preceding considerations and others you deem appropriate should be analyzed and applied to a change order estimate. The impact of each should be quantified and listed on the estimate to form an audit trail.

Change orders that are properly identified, documented, and managed help to ensure the orderly, professional, and profitable progress of the work. They also minimize potential claims or disputes at the end of the project.

Back by customer demand!

You asked and we listened. For customer convenience and estimating ease, we have made the 2020 Project Costs available for download at **RSMeans.com/2020books**. You will also find sample estimates, an RSMeans data overview video, and a book registration form to receive quarterly data updates throughout 2020.

Estimating Tips

- The cost figures available in the download were derived from hundreds of projects contained in the RSMeans database of completed construction projects. They include the contractor's overhead and profit. The figures have been adjusted to January of the current year.

- These projects were located throughout the U.S. and reflect a tremendous variation in square foot (S.F.) costs. This is due to differences, not only in labor and material costs, but also in individual owners' requirements. For instance, a bank in a large city would have different features than one in a rural area. This is true of all the different types of buildings analyzed. Therefore, caution should be exercised when using these Project Costs. For example, for courthouses, costs in the database are local courthouse costs and will not apply to the larger, more elaborate federal courthouses.

- None of the figures "go with" any others. All individual cost items were computed and tabulated separately. Thus, the sum of the median figures for plumbing, HVAC, and electrical will not normally total up to the total mechanical and electrical costs arrived at by separate analysis and tabulation of the projects.

- Each building was analyzed as to total and component costs and percentages. The figures were arranged in ascending order with the results tabulated as shown. The 1/4 column shows that 25% of the projects had lower costs and 75% had higher. The 3/4 column shows that 75% of the projects had lower costs and 25% had higher. The median column shows that 50% of the projects had lower costs and 50% had higher.

- Project Costs are useful in the conceptual stage when no details are available. As soon as details become available in the project design, the square foot approach should be discontinued and the project should be priced as to its particular components. When more precision is required, or for estimating the replacement cost of specific buildings, the current edition of *Square Foot Costs with RSMeans data* should be used.

- In using the figures in this section, it is recommended that the median column be used for preliminary figures if no additional information is available. The median figures, when multiplied by the total city construction cost index figures (see City Cost Indexes) and then multiplied by the project size modifier at the end of this section, should present a fairly accurate base figure, which would then have to be adjusted in view of the estimator's experience, local economic conditions, code requirements, and the owner's particular requirements. There is no need to factor in the percentage figures, as these should remain constant from city to city.

- The editors of this data would greatly appreciate receiving cost figures on one or more of your recent projects, which would then be included in the averages for next year. All cost figures received will be kept confidential, except that they will be averaged with other similar projects to arrive at square foot cost figures for next year.

See the website above for details and the discount available for submitting one or more of your projects.

Same Data. Simplified.

Enjoy the convenience and efficiency of accessing your costs anywhere:

- **Skip the multiplier** by setting your location
- **Quickly search,** edit, favorite and share costs
- **Stay on top of price changes** with automatic updates

Discover more at rsmeans.com/online

50 17 00 | Project Costs

			UNIT	UNIT COSTS			% OF TOTAL			
				1/4	MEDIAN	3/4	1/4	MEDIAN	3/4	
01	0000	**Auto Sales with Repair**	S.F.							01
	0100	Architectural		106	119	128	58%	64%	67%	
	0200	Plumbing		8.90	9.30	12.40	4.84%	5.20%	6.80%	
	0300	Mechanical		11.90	15.95	17.60	6.40%	8.70%	10.15%	
	0400	Electrical		18.30	22.50	28.50	9.05%	11.70%	15.90%	
	0500	Total Project Costs		178	186	191				
02	0000	**Banking Institutions**	S.F.							02
	0100	Architectural		160	197	239	59%	65%	69%	
	0200	Plumbing		6.45	9	12.50	2.12%	3.39%	4.19%	
	0300	Mechanical		12.80	17.70	21	4.41%	5.10%	10.75%	
	0400	Electrical		31.50	38	58	10.45%	13.05%	15.90%	
	0500	Total Project Costs		266	299	370				
03	0000	**Court House**	S.F.							03
	0100	Architectural		84.50	166	166	54.50%	58.50%	58.50%	
	0200	Plumbing		3.19	3.19	3.19	2.07%	2.07%	2.07%	
	0300	Mechanical		19.95	19.95	19.95	12.95%	12.95%	12.95%	
	0400	Electrical		25.50	25.50	25.50	16.60%	16.60%	16.60%	
	0500	Total Project Costs		154	284	284				
04	0000	**Data Centers**	S.F.							04
	0100	Architectural		191	191	191	68%	68%	68%	
	0200	Plumbing		10.45	10.45	10.45	3.71%	3.71%	3.71%	
	0300	Mechanical		26.50	26.50	26.50	9.45%	9.45%	9.45%	
	0400	Electrical		25.50	25.50	25.50	9%	9%	9%	
	0500	Total Project Costs		281	281	281				
05	0000	**Detention Centers**	S.F.							05
	0100	Architectural		177	187	198	52%	53%	60.50%	
	0200	Plumbing		18.65	22.50	27.50	5.15%	7.10%	7.25%	
	0300	Mechanical		23.50	34	40.50	7.55%	9.50%	13.80%	
	0400	Electrical		39	46	60	10.90%	14.85%	17.95%	
	0500	Total Project Costs		299	315	370				
06	0000	**Fire Stations**	S.F.							06
	0100	Architectural		100	129	187	49%	54.50%	61.50%	
	0200	Plumbing		10.75	13.95	16.95	4.67%	5.60%	6.30%	
	0300	Mechanical		14.95	21	29.50	6.10%	8.45%	10.20%	
	0400	Electrical		23	29.50	35.50	10.75%	12.75%	14.95%	
	0500	Total Project Costs		210	238	320				
07	0000	**Gymnasium**	S.F.							07
	0100	Architectural		88.50	116	116	57%	64.50%	64.50%	
	0200	Plumbing		2.17	7.10	7.10	1.58%	3.48%	3.48%	
	0300	Mechanical		3.32	30	30	2.42%	14.65%	14.65%	
	0400	Electrical		10.90	21	21	7.95%	10.35%	10.35%	
	0500	Total Project Costs		137	204	204				
08	0000	**Hospitals**	S.F.							08
	0100	Architectural		107	176	191	43%	47.50%	48%	
	0200	Plumbing		7.85	15.05	32.50	6%	7.45%	7.65%	
	0300	Mechanical		52	58.50	76.50	14.20%	17.95%	23.50%	
	0400	Electrical		23.50	47.50	61.50	10.95%	13.75%	16.85%	
	0500	Total Project Costs		250	375	405				
09	0000	**Industrial Buildings**	S.F.							09
	0100	Architectural		45	71.50	232	46%	54%	56.50%	
	0200	Plumbing		1.74	6.55	13.25	2%	3.06%	6.30%	
	0300	Mechanical		4.81	9.15	43.50	4.77%	5.55%	14.80%	
	0400	Electrical		7.35	8.40	70	7.85%	13.55%	16.20%	
	0500	Total Project Costs		80	104	430				
10	0000	**Medical Clinics & Offices**	S.F.							10
	0100	Architectural		89.50	121	161	48.50%	55.50%	62.50%	
	0200	Plumbing		8.90	13.15	21	4.47%	6.60%	8.65%	
	0300	Mechanical		13.95	22.50	44	7.80%	10.95%	16.10%	
	0400	Electrical		20	27	37.50	9.70%	11.65%	14.05%	
	0500	Total Project Costs		166	217	292				

For customer support on your Commercial Renovation Costs with RSMeans data, call 800.448.8182.

		50 17 00 | **Project Costs**	UNIT	UNIT COSTS			% OF TOTAL		
				1/4	MEDIAN	3/4	1/4	MEDIAN	3/4
11	0000	**Mixed Use**	S.F.						
	0100	Architectural		91.50	129	211	48.50%	57%	62.50%
	0200	Plumbing		6.15	9.35	11.80	3.23%	3.44%	4.18%
	0300	Mechanical		16.40	24.50	47.50	6.10%	13.75%	18.85%
	0400	Electrical		16	26.50	41.50	8.30%	11.40%	14%
	0500	Total Project Costs		189	216	345			
12	0000	**Multi-Family Housing**	S.F.						
	0100	Architectural		76	114	170	56.50%	62%	66.50%
	0200	Plumbing		6.55	11.55	15.05	5.30%	6.85%	8%
	0300	Mechanical		7	9.45	27	4.92%	6.90%	10.40%
	0400	Electrical		9.90	15.45	19.85	6.20%	7.90%	10.25%
	0500	Total Project Costs		111	208	271			
13	0000	**Nursing Home & Assisted Living**	S.F.						
	0100	Architectural		71.50	94	118	51.50%	55.50%	63.50%
	0200	Plumbing		7.75	11.60	12.80	6.25%	7.40%	8.80%
	0300	Mechanical		6.35	9.35	18.30	4.04%	6.70%	9.55%
	0400	Electrical		10.50	16.55	23	7%	10.75%	13.10%
	0500	Total Project Costs		122	160	192			
14	0000	**Office Buildings**	S.F.						
	0100	Architectural		94.50	129	177	56%	61%	69%
	0200	Plumbing		5.10	8	15.05	2.65%	3.56%	5.85%
	0300	Mechanical		11	17	26	5.60%	8.20%	11.10%
	0400	Electrical		12.75	21.50	34	7.75%	10%	12.70%
	0500	Total Project Costs		161	200	282			
15	0000	**Parking Garage**	S.F.						
	0100	Architectural		31.50	38.50	40.50	70%	79%	88%
	0200	Plumbing		1.04	1.09	2.04	2.05%	2.70%	2.83%
	0300	Mechanical		.81	1.24	4.72	2.11%	3.62%	3.81%
	0400	Electrical		2.78	3.05	6.35	5.30%	6.35%	7.95%
	0500	Total Project Costs		38.50	47	51			
16	0000	**Parking Garage/Mixed Use**	S.F.						
	0100	Architectural		103	112	114	61%	62%	65.50%
	0200	Plumbing		3.30	4.32	6.60	2.47%	2.72%	3.66%
	0300	Mechanical		14.10	15.85	23	7.80%	13.10%	13.60%
	0400	Electrical		14.75	21	22	8.20%	12.65%	18.15%
	0500	Total Project Costs		168	175	181			
17	0000	**Police Stations**	S.F.						
	0100	Architectural		116	130	164	49%	56.50%	61%
	0200	Plumbing		15.30	18.40	18.50	5.05%	5.55%	9.05%
	0300	Mechanical		34.50	48.50	50.50	13%	14.55%	16.55%
	0400	Electrical		26	28.50	30.50	9.15%	12.10%	14%
	0500	Total Project Costs		217	267	305			
18	0000	**Police/Fire**	S.F.						
	0100	Architectural		113	113	345	55.50%	66%	68%
	0200	Plumbing		9.05	9.35	34.50	5.45%	5.50%	5.55%
	0300	Mechanical		13.85	22	79	8.35%	12.70%	12.80%
	0400	Electrical		15.75	20	90.50	9.50%	11.75%	14.55%
	0500	Total Project Costs		166	171	625			
19	0000	**Public Assembly Buildings**	S.F.						
	0100	Architectural		115	159	238	57.50%	61.50%	66%
	0200	Plumbing		6.10	8.90	13.40	2.60%	3.36%	4.79%
	0300	Mechanical		12.85	23	35.50	6.55%	8.95%	12.45%
	0400	Electrical		19	26	41	8.60%	10.75%	13%
	0500	Total Project Costs		185	253	375			
20	0000	**Recreational**	S.F.						
	0100	Architectural		106	168	236	53.50%	60%	66%
	0200	Plumbing		7.80	14.75	21.50	3.08%	5%	6.85%
	0300	Mechanical		13.15	20	31.50	5.15%	6.95%	11.70%
	0400	Electrical		15.20	26.50	39.50	7.35%	8.95%	10.75%
	0500	Total Project Costs		168	281	445			

For customer support on your Commercial Renovation Costs with RSMeans data, call 800.448.8182.

887

		50 17 00 | Project Costs	UNIT	UNIT COSTS			% OF TOTAL			
				1/4	MEDIAN	3/4	1/4	MEDIAN	3/4	
21	0000	**Restaurants**	S.F.							21
	0100	Architectural		126	193	248	59%	60%	63.50%	
	0200	Plumbing		13.80	31.50	40	7.35%	7.75%	8.95%	
	0300	Mechanical		14.85	17.55	37	6.50%	8.15%	11.15%	
	0400	Electrical		14.80	24	48.50	7.10%	10.30%	11.60%	
	0500	Total Project Costs		208	305	415				
22	0000	**Retail**	S.F.							22
	0100	Architectural		56	86	127	60%	62%	64.50%	
	0200	Plumbing		7	9.25	12.10	5.05%	6.70%	9%	
	0300	Mechanical		6.55	9.25	17.05	5.70%	6.20%	10.20%	
	0400	Electrical		10.45	18.95	31.50	8.05%	11.25%	12.45%	
	0500	Total Project Costs		85.50	114	186				
23	0000	**Schools**	S.F.							23
	0100	Architectural		97	125	164	52.50%	56%	61%	
	0200	Plumbing		7.75	10.60	16.05	3.85%	4.82%	7.25%	
	0300	Mechanical		18.90	26.50	38.50	9.50%	12.35%	15.10%	
	0400	Electrical		17.85	25	32.50	9.45%	11.45%	13.30%	
	0500	Total Project Costs		169	227	300				
24	0000	**University, College & Private School Classroom & Admin Buildings**	S.F.							24
	0100	Architectural		124	153	192	50.50%	55%	59.50%	
	0200	Plumbing		7.05	10.95	15.45	2.74%	4.30%	6.35%	
	0300	Mechanical		26.50	38.50	46	10.10%	12.15%	14.70%	
	0400	Electrical		19.95	28	34	7.65%	9.50%	11.55%	
	0500	Total Project Costs		205	284	375				
25	0000	**University, College & Private School Dormitories**	S.F.							25
	0100	Architectural		81	142	151	54.50%	65%	68.50%	
	0200	Plumbing		10.70	15.10	22.50	6.45%	7.30%	9.15%	
	0300	Mechanical		4.79	20.50	32.50	4.13%	9%	12.05%	
	0400	Electrical		5.70	19.75	30	4.75%	7.35%	12.30%	
	0500	Total Project Costs		119	227	268				
26	0000	**University, College & Private School Science, Eng. & Lab Buildings**	S.F.							26
	0100	Architectural		143	164	193	50.50%	56.50%	58%	
	0200	Plumbing		9.60	15.15	26.50	3.29%	3.95%	8.40%	
	0300	Mechanical		43.50	68.50	70	12.75%	19.40%	23.50%	
	0400	Electrical		29	33.50	38.50	9%	11.55%	13.15%	
	0500	Total Project Costs		290	315	365				
27	0000	**University, College & Private School Student Union Buildings**	S.F.							27
	0100	Architectural		110	289	289	54.50%	54.50%	59.50%	
	0200	Plumbing		16.65	16.65	25	3.13%	4.27%	11.45%	
	0300	Mechanical		31.50	51	51	9.60%	9.60%	14.55%	
	0400	Electrical		27.50	48	48	9.05%	12.80%	13.15%	
	0500	Total Project Costs		217	530	530				
28	0000	**Warehouses**	S.F.							28
	0100	Architectural		47	73.50	174	61.50%	67.50%	72%	
	0200	Plumbing		2.45	5.25	10.10	2.82%	3.72%	5%	
	0300	Mechanical		2.91	16.55	26	4.56%	8.20%	10.70%	
	0400	Electrical		5.30	19.85	33	7.50%	10.10%	18.30%	
	0500	Total Project Costs		70.50	125	243				

Square Foot Project Size Modifier

One factor that affects the S.F. cost of a particular building is the size. In general, for buildings built to the same specifications in the same locality, the larger building will have the lower S.F. cost. This is due mainly to the decreasing contribution of the exterior walls plus the economy of scale usually achievable in larger buildings. The Area Conversion Scale shown below will give a factor to convert costs for the typical size building to an adjusted cost for the particular project.

The Square Foot Base Size lists the median costs, most typical project size in our accumulated data, and the range in size of the projects.

The Size Factor for your project is determined by dividing your project area in S.F. by the typical project size for the particular Building Type. With this factor, enter the Area Conversion Scale at the appropriate Size Factor and determine the appropriate Cost Multiplier for your building size.

Example: Determine the cost per S.F. for a 152,600 S.F. Multi-family housing.

$$\frac{\text{Proposed building area} = 152,600 \text{ S.F.}}{\text{Typical size from below} = 49,900 \text{ S.F.}} = 2.00$$

Enter Area Conversion Scale at 2.0, intersect curve, read horizontally the appropriate cost multiplier of .94. Size adjusted cost becomes .94 x $208.00 = $195.52 based on national average costs.

Note: For Size Factors less than .50, the Cost Multiplier is 1.1
      For Size Factors greater than 3.5, the Cost Multiplier is .90

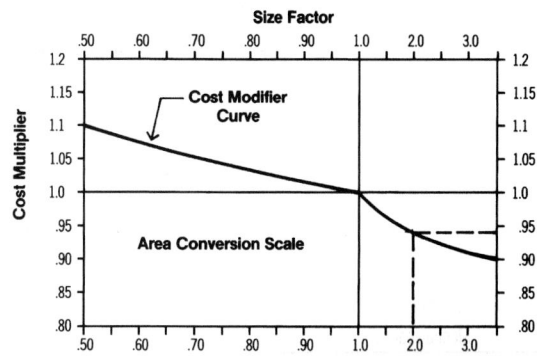

System	Median Cost (Total Project Costs)	Typical Size Gross S.F. (Median of Projects)	Typical Range (Low – High) (Projects)
Auto Sales with Repair	$186.00	24,900	4,700 – 29,300
Banking Institutions	299.00	9,300	3,300 – 38,100
Court House	284.00	47,600	24,700 – 70,500
Data Centers	281.00	14,400	14,400 – 14,400
Detention Centers	315.00	37,800	12,300 – 183,300
Fire Stations	238.00	13,100	6,300 – 49,600
Gymnasium	204.00	52,400	22,800 – 82,000
Hospitals	375.00	87,100	22,400 – 410,300
Industrial Buildings	104.00	22,100	5,100 – 200,600
Medical Clinics & Offices	217.00	20,900	2,300 – 327,000
Mixed Use	216.00	28,500	7,200 – 188,900
Multi-Family Housing	208.00	49,900	2,500 – 1,161,500
Nursing Home & Assisted Living	160.00	38,200	1,500 – 242,600
Office Buildings	200.00	20,500	1,100 – 930,000
Parking Garage	47.00	151,800	99,900 – 287,000
Parking Garage/Mixed Use	175.00	254,200	5,300 – 318,000
Police Stations	267.00	28,500	15,400 – 88,600
Police/Fire	171.00	44,300	8,600 – 50,300
Public Assembly Buildings	253.00	21,000	2,200 – 235,300
Recreational	281.00	28,800	1,000 – 223,800
Restaurants	305.00	6,000	4,000 – 42,000
Retail	114.00	28,700	5,200 – 84,300
Schools	227.00	70,600	1,300 – 410,800
University, College & Private School Classroom & Admin Buildings	284.00	48,300	9,400 – 196,200
University, College & Private School Dormitories	227.00	28,900	1,500 – 126,900
University, College & Private School Science, Eng. & Lab Buildings	315.00	39,800	5,300 – 117,600
University, College & Private School Student Union Buildings	530.00	48,700	42,100 – 50,000
Warehouses	125.00	10,400	600 – 303,800

A	Area Square Feet; Ampere	Brk., brk	Brick	Csc	Cosecant
AAFES	Army and Air Force Exchange Service	brkt	Bracket	C.S.F.	Hundred Square Feet
		Brs.	Brass	CSI	Construction Specifications Institute
ABS	Acrylonitrile Butadiene Stryrene; Asbestos Bonded Steel	Brz.	Bronze		
		Bsn.	Basin	CT	Current Transformer
A.C., AC	Alternating Current;	Btr.	Better	CTS	Copper Tube Size
	Air-Conditioning;	BTU	British Thermal Unit	Cu	Copper, Cubic
	Asbestos Cement;	BTUH	BTU per Hour	Cu. Ft.	Cubic Foot
	Plywood Grade A & C	Bu.	Bushels	cw	Continuous Wave
ACI	American Concrete Institute	BUR	Built-up Roofing	C.W.	Cool White; Cold Water
ACR	Air Conditioning Refrigeration	BX	Interlocked Armored Cable	Cwt.	100 Pounds
ADA	Americans with Disabilities Act	°C	Degree Centigrade	C.W.X.	Cool White Deluxe
AD	Plywood, Grade A & D	c	Conductivity, Copper Sweat	C.Y.	Cubic Yard (27 cubic feet)
Addit.	Additional	C	Hundred; Centigrade	C.Y./Hr.	Cubic Yard per Hour
Adh.	Adhesive	C/C	Center to Center, Cedar on Cedar	Cyl.	Cylinder
Adj.	Adjustable	C-C	Center to Center	d	Penny (nail size)
af	Audio-frequency	Cab	Cabinet	D	Deep; Depth; Discharge
AFFF	Aqueous Film Forming Foam	Cair.	Air Tool Laborer	Dis., Disch.	Discharge
AFUE	Annual Fuel Utilization Efficiency	Cal.	Caliper	Db	Decibel
AGA	American Gas Association	Calc	Calculated	Dbl.	Double
Agg.	Aggregate	Cap.	Capacity	DC	Direct Current
A.H., Ah	Ampere Hours	Carp.	Carpenter	DDC	Direct Digital Control
A hr.	Ampere-hour	C.B.	Circuit Breaker	Demob.	Demobilization
A.H.U., AHU	Air Handling Unit	C.C.A.	Chromate Copper Arsenate	d.f.t.	Dry Film Thickness
A.I.A.	American Institute of Architects	C.C.F.	Hundred Cubic Feet	d.f.u.	Drainage Fixture Units
AIC	Ampere Interrupting Capacity	cd	Candela	D.H.	Double Hung
Allow.	Allowance	cd/sf	Candela per Square Foot	DHW	Domestic Hot Water
alt., alt	Alternate	CD	Grade of Plywood Face & Back	DI	Ductile Iron
Alum.	Aluminum	CDX	Plywood, Grade C & D, exterior glue	Diag.	Diagonal
a.m.	Ante Meridiem			Diam., Dia	Diameter
Amp.	Ampere	Cefi.	Cement Finisher	Distrib.	Distribution
Anod.	Anodized	Cem.	Cement	Div.	Division
ANSI	American National Standards Institute	CF	Hundred Feet	Dk.	Deck
		C.F.	Cubic Feet	D.L.	Dead Load; Diesel
APA	American Plywood Association	CFM	Cubic Feet per Minute	DLH	Deep Long Span Bar Joist
Approx.	Approximate	CFRP	Carbon Fiber Reinforced Plastic	dlx	Deluxe
Apt.	Apartment	c.g.	Center of Gravity	Do.	Ditto
Asb.	Asbestos	CHW	Chilled Water; Commercial Hot Water	DOP	Dioctyl Phthalate Penetration Test (Air Filters)
A.S.B.C.	American Standard Building Code				
Asbe.	Asbestos Worker	C.I., CI	Cast Iron	Dp., dp	Depth
ASCE	American Society of Civil Engineers	C.I.P., CIP	Cast in Place	D.P.S.T.	Double Pole, Single Throw
A.S.H.R.A.E.	American Society of Heating, Refrig. & AC Engineers	Circ.	Circuit	Dr.	Drive
		C.L.	Carload Lot	DR	Dimension Ratio
ASME	American Society of Mechanical Engineers	CL	Chain Link	Drink.	Drinking
		Clab.	Common Laborer	D.S.	Double Strength
ASTM	American Society for Testing and Materials	Clam	Common Maintenance Laborer	D.S.A.	Double Strength A Grade
		C.L.F.	Hundred Linear Feet	D.S.B.	Double Strength B Grade
Attchmt.	Attachment	CLF	Current Limiting Fuse	Dty.	Duty
Avg., Ave.	Average	CLP	Cross Linked Polyethylene	DWV	Drain Waste Vent
AWG	American Wire Gauge	cm	Centimeter	DX	Deluxe White, Direct Expansion
AWWA	American Water Works Assoc.	CMP	Corr. Metal Pipe	dyn	Dyne
Bbl.	Barrel	CMU	Concrete Masonry Unit	e	Eccentricity
B&B, BB	Grade B and Better; Balled & Burlapped	CN	Change Notice	E	Equipment Only; East; Emissivity
		Col.	Column	Ea.	Each
B&S	Bell and Spigot	CO₂	Carbon Dioxide	EB	Encased Burial
B.&W.	Black and White	Comb.	Combination	Econ.	Economy
b.c.c.	Body-centered Cubic	comm.	Commercial, Communication	E.C.Y	Embankment Cubic Yards
B.C.Y.	Bank Cubic Yards	Compr.	Compressor	EDP	Electronic Data Processing
BE	Bevel End	Conc.	Concrete	EIFS	Exterior Insulation Finish System
B.F.	Board Feet	Cont., cont	Continuous; Continued, Container	E.D.R.	Equiv. Direct Radiation
Bg. cem.	Bag of Cement	Corkbd.	Cork Board	Eq.	Equation
BHP	Boiler Horsepower; Brake Horsepower	Corr.	Corrugated	EL	Elevation
		Cos	Cosine	Elec.	Electrician; Electrical
B.I.	Black Iron	Cot	Cotangent	Elev.	Elevator; Elevating
bidir.	bidirectional	Cov.	Cover	EMT	Electrical Metallic Conduit; Thin Wall Conduit
Bit., Bitum.	Bituminous	C/P	Cedar on Paneling		
Bit., Conc.	Bituminous Concrete	CPA	Control Point Adjustment	Eng.	Engine, Engineered
Bk.	Backed	Cplg.	Coupling	EPDM	Ethylene Propylene Diene Monomer
Bkrs.	Breakers	CPM	Critical Path Method		
Bldg., bldg	Building	CPVC	Chlorinated Polyvinyl Chloride	EPS	Expanded Polystyrene
Blk.	Block	C.Pr.	Hundred Pair	Eqhv.	Equip. Oper., Heavy
Bm.	Beam	CRC	Cold Rolled Channel	Eqlt.	Equip. Oper., Light
Boil.	Boilermaker	Creos.	Creosote	Eqmd.	Equip. Oper., Medium
bpm	Blows per Minute	Crpt.	Carpet & Linoleum Layer	Eqmm.	Equip. Oper., Master Mechanic
BR	Bedroom	CRT	Cathode-ray Tube	Eqol.	Equip. Oper., Oilers
Brg., brng.	Bearing	CS	Carbon Steel, Constant	Equip.	Equipment
Brhe.	Bricklayer Helper		Shear Bar Joist	ERW	Electric Resistance Welded
Bric.	Bricklayer				

E.S.	Energy Saver	H	High Henry	Lath.	Lather
Est.	Estimated	HC	High Capacity	Lav.	Lavatory
esu	Electrostatic Units	H.D., HD	Heavy Duty; High Density	lb.; #	Pound
E.W.	Each Way	H.D.O.	High Density Overlaid	L.B., LB	Load Bearing; L Conduit Body
EWT	Entering Water Temperature	HDPE	High Density Polyethylene Plastic	L. & E.	Labor & Equipment
Excav.	Excavation	Hdr.	Header	lb./hr.	Pounds per Hour
excl	Excluding	Hdwe.	Hardware	lb./L.F.	Pounds per Linear Foot
Exp., exp	Expansion, Exposure	H.I.D., HID	High Intensity Discharge	lbf/sq.in.	Pound-force per Square Inch
Ext., ext	Exterior; Extension	Help.	Helper Average	L.C.L.	Less than Carload Lot
Extru.	Extrusion	HEPA	High Efficiency Particulate Air	L.C.Y.	Loose Cubic Yard
f.	Fiber Stress		Filter	Ld.	Load
F	Fahrenheit; Female; Fill	Hg	Mercury	LE	Lead Equivalent
Fab., fab	Fabricated; Fabric	HIC	High Interrupting Capacity	LED	Light Emitting Diode
FBGS	Fiberglass	HM	Hollow Metal	L.F.	Linear Foot
F.C.	Footcandles	HMWPE	High Molecular Weight	L.F. Hdr	Linear Feet of Header
f.c.c.	Face-centered Cubic		Polyethylene	L.F. Nose	Linear Foot of Stair Nosing
f'c.	Compressive Stress in Concrete;	HO	High Output	L.F. Rsr	Linear Foot of Stair Riser
	Extreme Compressive Stress	Horiz.	Horizontal	Lg.	Long; Length; Large
F.E.	Front End	H.P., HP	Horsepower; High Pressure	L & H	Light and Heat
FEP	Fluorinated Ethylene Propylene	H.P.F.	High Power Factor	LH	Long Span Bar Joist
	(Teflon)	Hr.	Hour	L.H.	Labor Hours
F.G.	Flat Grain	Hrs./Day	Hours per Day	L.L., LL	Live Load
F.H.A.	Federal Housing Administration	HSC	High Short Circuit	L.L.D.	Lamp Lumen Depreciation
Fig.	Figure	Ht.	Height	lm	Lumen
Fin.	Finished	Htg.	Heating	lm/sf	Lumen per Square Foot
FIPS	Female Iron Pipe Size	Htrs.	Heaters	lm/W	Lumen per Watt
Fixt.	Fixture	HVAC	Heating, Ventilation & Air-	LOA	Length Over All
FJP	Finger jointed and primed		Conditioning	log	Logarithm
Fl. Oz.	Fluid Ounces	Hvy.	Heavy	L-O-L	Lateralolet
Flr.	Floor	HW	Hot Water	long.	Longitude
Flrs.	Floors	Hyd.; Hydr.	Hydraulic	L.P., LP	Liquefied Petroleum; Low Pressure
FM	Frequency Modulation;	Hz	Hertz (cycles)	L.P.F.	Low Power Factor
	Factory Mutual	I.	Moment of Inertia	LR	Long Radius
Fmg.	Framing	IBC	International Building Code	L.S.	Lump Sum
FM/UL	Factory Mutual/Underwriters Labs	I.C.	Interrupting Capacity	Lt.	Light
Fdn.	Foundation	ID	Inside Diameter	Lt. Ga.	Light Gauge
FNPT	Female National Pipe Thread	I.D.	Inside Dimension; Identification	L.T.L.	Less than Truckload Lot
Fori.	Foreman, Inside	I.F.	Inside Frosted	Lt. Wt.	Lightweight
Foro.	Foreman, Outside	I.M.C.	Intermediate Metal Conduit	L.V.	Low Voltage
Fount.	Fountain	In.	Inch	M	Thousand; Material; Male;
fpm	Feet per Minute	Incan.	Incandescent		Light Wall Copper Tubing
FPT	Female Pipe Thread	Incl.	Included; Including	M²CA	Meters Squared Contact Area
Fr	Frame	Int.	Interior	m/hr.; M.H.	Man-hour
F.R.	Fire Rating	Inst.	Installation	mA	Milliampere
FRK	Foil Reinforced Kraft	Insul., insul	Insulation/Insulated	Mach.	Machine
FSK	Foil/Scrim/Kraft	I.P.	Iron Pipe	Mag. Str.	Magnetic Starter
FRP	Fiberglass Reinforced Plastic	I.P.S., IPS	Iron Pipe Size	Maint.	Maintenance
FS	Forged Steel	IPT	Iron Pipe Threaded	Marb.	Marble Setter
FSC	Cast Body; Cast Switch Box	I.W.	Indirect Waste	Mat; Mat'l.	Material
Ft., ft	Foot; Feet	J	Joule	Max.	Maximum
Ftng.	Fitting	J.I.C.	Joint Industrial Council	MBF	Thousand Board Feet
Ftg.	Footing	K	Thousand; Thousand Pounds;	MBH	Thousand BTU's per hr.
Ft lb.	Foot Pound		Heavy Wall Copper Tubing, Kelvin	MC	Metal Clad Cable
Furn.	Furniture	K.A.H.	Thousand Amp. Hours	MCC	Motor Control Center
FVNR	Full Voltage Non-Reversing	kcmil	Thousand Circular Mils	M.C.F.	Thousand Cubic Feet
FVR	Full Voltage Reversing	KD	Knock Down	MCFM	Thousand Cubic Feet per Minute
FXM	Female by Male	K.D.A.T.	Kiln Dried After Treatment	M.C.M.	Thousand Circular Mils
Fy.	Minimum Yield Stress of Steel	kg	Kilogram	MCP	Motor Circuit Protector
g	Gram	kG	Kilogauss	MD	Medium Duty
G	Gauss	kgf	Kilogram Force	MDF	Medium-density fibreboard
Ga.	Gauge	kHz	Kilohertz	M.D.O.	Medium Density Overlaid
Gal., gal.	Gallon	Kip	1000 Pounds	Med.	Medium
Galv., galv	Galvanized	KJ	Kilojoule	MF	Thousand Feet
GC/MS	Gas Chromatograph/Mass	K.L.	Effective Length Factor	M.F.B.M.	Thousand Feet Board Measure
	Spectrometer	K.L.F.	Kips per Linear Foot	Mfg.	Manufacturing
Gen.	General	Km	Kilometer	Mfrs.	Manufacturers
GFI	Ground Fault Interrupter	KO	Knock Out	mg	Milligram
GFRC	Glass Fiber Reinforced Concrete	K.S.F.	Kips per Square Foot	MGD	Million Gallons per Day
Glaz.	Glazier	K.S.I.	Kips per Square Inch	MGPH	Million Gallons per Hour
GPD	Gallons per Day	kV	Kilovolt	MH, M.H.	Manhole; Metal Halide; Man-Hour
gpf	Gallon per Flush	kVA	Kilovolt Ampere	MHz	Megahertz
GPH	Gallons per Hour	kVAR	Kilovar (Reactance)	Mi.	Mile
gpm, GPM	Gallons per Minute	KW	Kilowatt	MI	Malleable Iron; Mineral Insulated
GR	Grade	KWh	Kilowatt-hour	MIPS	Male Iron Pipe Size
Gran.	Granular	L	Labor Only; Length; Long;	mj	Mechanical Joint
Grnd.	Ground		Medium Wall Copper Tubing	m	Meter
GVW	Gross Vehicle Weight	Lab.	Labor	mm	Millimeter
GWB	Gypsum Wall Board	lat	Latitude	Mill.	Millwright
				Min., min.	Minimum, Minute

Misc.	Miscellaneous	PCM	Phase Contrast Microscopy	SBS	Styrene Butadiere Styrene
ml	Milliliter, Mainline	PDCA	Painting and Decorating	SC	Screw Cover
M.L.F.	Thousand Linear Feet		Contractors of America	SCFM	Standard Cubic Feet per Minute
Mo.	Month	P.E., PE	Professional Engineer;	Scaf.	Scaffold
Mobil.	Mobilization		Porcelain Enamel;	Sch., Sched.	Schedule
Mog.	Mogul Base		Polyethylene; Plain End	S.C.R.	Modular Brick
MPH	Miles per Hour	P.E.C.I.	Porcelain Enamel on Cast Iron	S.D.	Sound Deadening
MPT	Male Pipe Thread	Perf.	Perforated	SDR	Standard Dimension Ratio
MRGWB	Moisture Resistant Gypsum	PEX	Cross Linked Polyethylene	S.E.	Surfaced Edge
	Wallboard	Ph.	Phase	Sel.	Select
MRT	Mile Round Trip	P.I.	Pressure Injected	SER, SEU	Service Entrance Cable
ms	Millisecond	Pile.	Pile Driver	S.F.	Square Foot
M.S.F.	Thousand Square Feet	Pkg.	Package	S.F.C.A.	Square Foot Contact Area
Mstz.	Mosaic & Terrazzo Worker	Pl.	Plate	S.F. Flr.	Square Foot of Floor
M.S.Y.	Thousand Square Yards	Plah.	Plasterer Helper	S.F.G.	Square Foot of Ground
Mtd., mtd., mtd	Mounted	Plas.	Plasterer	S.F. Hor.	Square Foot Horizontal
Mthe.	Mosaic & Terrazzo Helper	plf	Pounds Per Linear Foot	SFR	Square Feet of Radiation
Mtng.	Mounting	Pluh.	Plumber Helper	S.F. Shlf.	Square Foot of Shelf
Mult.	Multi; Multiply	Plum.	Plumber	S4S	Surface 4 Sides
MUTCD	Manual on Uniform Traffic Control	Ply.	Plywood	Shee.	Sheet Metal Worker
	Devices	p.m.	Post Meridiem	Sin.	Sine
M.V.A.	Million Volt Amperes	Pntd.	Painted	Skwk.	Skilled Worker
M.V.A.R.	Million Volt Amperes Reactance	Pord.	Painter, Ordinary	SL	Saran Lined
MV	Megavolt	pp	Pages	S.L.	Slimline
MW	Megawatt	PP, PPL	Polypropylene	Sldr.	Solder
MXM	Male by Male	P.P.M.	Parts per Million	SLH	Super Long Span Bar Joist
MYD	Thousand Yards	Pr.	Pair	S.N.	Solid Neutral
N	Natural; North	P.E.S.B.	Pre-engineered Steel Building	SO	Stranded with oil resistant inside
nA	Nanoampere	Prefab.	Prefabricated		insulation
NA	Not Available; Not Applicable	Prefin.	Prefinished	S-O-L	Socketolet
N.B.C.	National Building Code	Prop.	Propelled	sp	Standpipe
NC	Normally Closed	PSF, psf	Pounds per Square Foot	S.P.	Static Pressure; Single Pole; Self-
NEMA	National Electrical Manufacturers	PSI, psi	Pounds per Square Inch		Propelled
	Assoc.	PSIG	Pounds per Square Inch Gauge	Spri.	Sprinkler Installer
NEHB	Bolted Circuit Breaker to 600V.	PSP	Plastic Sewer Pipe	spwg	Static Pressure Water Gauge
NFPA	National Fire Protection Association	Pspr.	Painter, Spray	S.P.D.T.	Single Pole, Double Throw
NLB	Non-Load-Bearing	Psst.	Painter, Structural Steel	SPF	Spruce Pine Fir; Sprayed
NM	Non-Metallic Cable	P.T.	Potential Transformer		Polyurethane Foam
nm	Nanometer	P. & T.	Pressure & Temperature	S.P.S.T.	Single Pole, Single Throw
No.	Number	Ptd.	Painted	SPT	Standard Pipe Thread
NO	Normally Open	Ptns.	Partitions	Sq.	Square; 100 Square Feet
N.O.C.	Not Otherwise Classified	Pu	Ultimate Load	Sq. Hd.	Square Head
Nose.	Nosing	PVC	Polyvinyl Chloride	Sq. In.	Square Inch
NPT	National Pipe Thread	Pvmt.	Pavement	S.S.	Single Strength; Stainless Steel
NQOD	Combination Plug-on/Bolt on	PRV	Pressure Relief Valve	S.S.B.	Single Strength B Grade
	Circuit Breaker to 240V.	Pwr.	Power	sst, ss	Stainless Steel
N.R.C., NRC	Noise Reduction Coefficient/	Q	Quantity Heat Flow	Sswk.	Structural Steel Worker
	Nuclear Regulator Commission	Qt.	Quart	Sswl.	Structural Steel Welder
N.R.S.	Non Rising Stem	Quan., Qty.	Quantity	St.; Stl.	Steel
ns	Nanosecond	Q.C.	Quick Coupling	STC	Sound Transmission Coefficient
NTP	Notice to Proceed	r	Radius of Gyration	Std.	Standard
nW	Nanowatt	R	Resistance	Stg.	Staging
OB	Opposing Blade	R.C.P.	Reinforced Concrete Pipe	STK	Select Tight Knot
OC	On Center	Rect.	Rectangle	STP	Standard Temperature & Pressure
OD	Outside Diameter	recpt.	Receptacle	Stpi.	Steamfitter, Pipefitter
O.D.	Outside Dimension	Reg.	Regular	Str.	Strength; Starter; Straight
ODS	Overhead Distribution System	Reinf.	Reinforced	Strd.	Stranded
O.G.	Ogee	Req'd.	Required	Struct.	Structural
O.H.	Overhead	Res.	Resistant	Sty.	Story
O&P	Overhead and Profit	Resi.	Residential	Subj.	Subject
Oper.	Operator	RF	Radio Frequency	Subs.	Subcontractors
Opng.	Opening	RFID	Radio-frequency Identification	Surf.	Surface
Orna.	Ornamental	Rgh.	Rough	Sw.	Switch
OSB	Oriented Strand Board	RGS	Rigid Galvanized Steel	Swbd.	Switchboard
OS&Y	Outside Screw and Yoke	RHW	Rubber, Heat & Water Resistant;	S.Y.	Square Yard
OSHA	Occupational Safety and Health		Residential Hot Water	Syn.	Synthetic
	Act	rms	Root Mean Square	S.Y.P.	Southern Yellow Pine
Ovhd.	Overhead	Rnd.	Round	Sys.	System
OWG	Oil, Water or Gas	Rodm.	Rodman	t.	Thickness
Oz.	Ounce	Rofc.	Roofer, Composition	T	Temperature; Ton
P.	Pole; Applied Load; Projection	Rofp.	Roofer, Precast	Tan	Tangent
p.	Page	Rohe.	Roofer Helpers (Composition)	T.C.	Terra Cotta
Pape.	Paperhanger	Rots.	Roofer, Tile & Slate	T & C	Threaded and Coupled
P.A.P.R.	Powered Air Purifying Respirator	R.O.W.	Right of Way	T.D.	Temperature Difference
PAR	Parabolic Reflector	RPM	Revolutions per Minute	TDD	Telecommunications Device for
P.B., PB	Push Button	R.S.	Rapid Start		the Deaf
Pc., Pcs.	Piece, Pieces	Rsr	Riser	T.E.M.	Transmission Electron Microscopy
P.C.	Portland Cement; Power Connector	RT	Round Trip	temp	Temperature, Tempered, Temporary
P.C.F.	Pounds per Cubic Foot	S.	Suction; Single Entrance; South	TFFN	Nylon Jacketed Wire

Abbreviations

TFE	Tetrafluoroethylene (Teflon)	U.L., UL	Underwriters Laboratory	w/	With
T. & G.	Tongue & Groove;	Uld.	Unloading	W.C., WC	Water Column; Water Closet
	Tar & Gravel	Unfin.	Unfinished	W.F.	Wide Flange
Th., Thk.	Thick	UPS	Uninterruptible Power Supply	W.G.	Water Gauge
Thn.	Thin	URD	Underground Residential	Wldg.	Welding
Thrded	Threaded		Distribution	W. Mile	Wire Mile
Tilf.	Tile Layer, Floor	US	United States	W-O-L	Weldolet
Tilh.	Tile Layer, Helper	USGBC	U.S. Green Building Council	W.R.	Water Resistant
THHN	Nylon Jacketed Wire	USP	United States Primed	Wrck.	Wrecker
THW.	Insulated Strand Wire	UTMCD	Uniform Traffic Manual For Control	WSFU	Water Supply Fixture Unit
THWN	Nylon Jacketed Wire		Devices	W.S.P.	Water, Steam, Petroleum
T.L., TL	Truckload	UTP	Unshielded Twisted Pair	WT., Wt.	Weight
T.M.	Track Mounted	V	Volt	WWF	Welded Wire Fabric
Tot.	Total	VA	Volt Amperes	XFER	Transfer
T-O-L	Threadolet	VAT	Vinyl Asbestos Tile	XFMR	Transformer
tmpd	Tempered	V.C.T.	Vinyl Composition Tile	XHD	Extra Heavy Duty
TPO	Thermoplastic Polyolefin	VAV	Variable Air Volume	XHHW	Cross-Linked Polyethylene Wire
T.S.	Trigger Start	VC	Veneer Core	XLPE	Insulation
Tr.	Trade	VDC	Volts Direct Current	XLP	Cross-linked Polyethylene
Transf.	Transformer	Vent.	Ventilation	Xport	Transport
Trhv.	Truck Driver, Heavy	Vert.	Vertical	Y	Wye
Trlr	Trailer	V.F.	Vinyl Faced	yd	Yard
Trlt.	Truck Driver, Light	V.G.	Vertical Grain	yr	Year
TTY	Teletypewriter	VHF	Very High Frequency	Δ	Delta
TV	Television	VHO	Very High Output	%	Percent
T.W.	Thermoplastic Water Resistant	Vib.	Vibrating	~	Approximately
	Wire	VLF	Vertical Linear Foot	Ø	Phase; diameter
UCI	Uniform Construction Index	VOC	Volatile Organic Compound	@	At
UF	Underground Feeder	Vol.	Volume	#	Pound; Number
UGND	Underground Feeder	VRP	Vinyl Reinforced Polyester	<	Less Than
UHF	Ultra High Frequency	W	Wire; Watt; Wide; West	>	Greater Than
U.I.	United Inch			Z	Zone

For customer support on your Commercial Renovation Costs with RSMeans data, call 800.448.8182.

893

Index

For customer support on your Commercial Renovation Costs with RSMeans data, call 800.448.8182.

Index

For customer support on your Commercial Renovation Costs with RSMeans data, call 800.448.8182.

Index

Index

For customer support on your Commercial Renovation Costs with RSMeans data, call 800.448.8182.

Index

For customer support on your Commercial Renovation Costs with RSMeans data, call 800.448.8182.

Index

For customer support on your Commercial Renovation Costs with RSMeans data, call 800.448.8182.

Index

Index

For customer support on your Commercial Renovation Costs with RSMeans data, call 800.448.8182.

For customer support on your Commercial Renovation Costs with RSMeans data, call 800.448.8182.

Index

Division Notes

		CREW	DAILY OUTPUT	LABOR-HOURS	UNIT	BARE COSTS				TOTAL INCL O&P
						MAT.	LABOR	EQUIP.	TOTAL	

Division Notes

		CREW	DAILY OUTPUT	LABOR-HOURS	UNIT	BARE COSTS				TOTAL INCL O&P
						MAT.	LABOR	EQUIP.	TOTAL	

Division Notes

	CREW	DAILY OUTPUT	LABOR-HOURS	UNIT	BARE COSTS				TOTAL INCL O&P
					MAT.	LABOR	EQUIP.	TOTAL	

Division Notes